# For All Practical Purposes

D0217560

**PROJECT DIRECTOR**

Solomon Garfunkel,
*Consortium for Mathematics and its Applications*

**CONTRIBUTING AUTHORS**

PART I Management Science
Joseph Malkevitch, York College, CUNY

PART II Statistics: The Science of Data
Lawrence M. Lesser, The University of Texas at El Paso
David S. Moore, Purdue University

PART III Voting and Social Choice
Alan D. Taylor, Union College
Bruce P. Conrad, Temple University
Steven J. Brams, New York University

PART IV Fairness and Game Theory
Alan D. Taylor, Union College
Bruce P. Conrad, Temple University
Steven J. Brams, New York University

PART V The Digital Revolution
Joseph A. Gallian, University of Minnesota Duluth

PART VI On Size and Growth
Paul J. Campbell, Beloit College

PART VII Your Money and Resources
Paul J. Campbell, Beloit College

# For All Practical Purposes

## Mathematical Literacy In Today's World

EIGHTH EDITION

W. H. FREEMAN AND COMPANY
New York

*Senior Publisher:* Craig Bleyer
*Publisher:* Ruth Baruth
*Senior Acquisitions Editor:* Terri Ward
*Executive Marketing Manager:* Jennifer Somerville
*Freelance Development Editor:* Lisa Collette
*Associate Editor:* Laura Capuano
*Senior Media Editor:* Roland Cheyney
*Assistant Editor:* Brian Tedesco
*Editorial Assistant:* Katrina Wilhelm
*Photo Editor and Researcher:* Christine Buese, Rae Grant
*Cover Designer:* Blake Logan
*Project Editor:* Vivien Weiss
*Director of Production:* Ellen Cash
*Printing and Binding:* Quebecor

**COMAP Production Department:**
*Production Manager:* George Ward
*Copy Editor:* Joyce Barnes
*Text Design and Composition:* Daiva Chauhan

Library of Congress Control Number: 2008936372
©2009 by COMAP, Inc.
All rights reserved.

ISBN-13: 978-1-4292-0900-7
ISBN-10: 1-4292-0900-3

**Chapter Openers** 1: Angelo Cavalli/age footstock; 2: Reuters/Corbis; 3: Jerzy Dabrowski/dpa/Corbis; 4: Mark E. Gibson/Corbis; 5: Studio DL/Corbis; 6: Robert Michael/Corbis; 7: Charles Gupton/Corbis; 8: D. Hurst/Alamy; 9: William Manning/www.williammanning.com/Corbis; 10: Getty Images; 11: Getty Images; 12: Paul J. Richards/AFP/Getty Images; 13: Corbis Super RF/Alamy; 14: Achim Scheidemann/Corbis; 15: Jerry Cooke/Corbis; 16: Steve Krongard/STONE/Getty Images; 17: NASA/Corby Waste; 18: Juniors Bildarchiv/Alamy; 19: Erik Von Weber/The Image Bank/Getty Images; 20: Michos Tzovaras/Art Resource, NY; 21: Premium Stock/Corbis; 22: Yellow Dog Productions/The Image Bank/Getty Images; 23: James L. Stanfield/Getty Images

Printed in the United States of America
First printing

W. H. Freeman and Company
41 Madison Avenue
New York, NY 10010
Houndmills, Basingstoke RG21 6XS, England

www.whfreeman.com

# BRIEF CONTENTS

# CONTENTS

Scott Andrews/Science Faction/Getty Images

Photo by Rich Pilling/ MLB Photos via Getty Images

CHANGE
WE CAN BELIEVE IN
CHANGE
WE CAN BELIEVE IN

Mark Hirsch/Getty Images

Brooks Kraft/Corbis

Getty Images

Craig Tuttle/Corbis

Dan Lamont/Corbis

# PREFACE

## To the Student

For *All Practical Purposes*, Eighth Edition, continues our effort to bring the excitement of contemporary mathematical thinking to the nonspecialist. In science and industry, mathematical models are the main tools for analyzing and solving problems that arise. In this book, our goal is to convey the power of mathematics by showing you the great variety of problems that can be modeled and solved by quantitative means. An extensive supplements package designed to make study time supremely effective complements the eighth edition. Highlights of the supplements package include the *Student Study Guide* and *Student Solutions Manual.* Between the text and the available resources, *For All Practical Purposes* offers you the tools to succeed in the course and apply your new knowledge to daily life experiences.

There are many ways to talk about why mathematics and its applications matter. You will hear expressions such as "mathematical literacy" or "quantitative literacy." They mean, essentially, that math is important. It is important because knowing it can make your life easier. In other words, it can help to explain how your world works. We created this course and this book because we know that not everyone looks at mathematics in this way.

In school, you spent a great deal of time learning the tools of mathematics—how to manipulate symbols, how to solve equations. In this course, you will spend time learning the uses of mathematics and the power of mathematics to help us to understand so many different parts of our everyday lives and the world itself. We hope that this exploration will give you a broader sense of what our subject is about and why we wanted you to take a math course every year you were in school. It's "for all practical purposes," because, in a sense, you've learned to hammer nails and saw wood. Now we're going to build houses.

Enjoy!

## To the Instructor

Because *For All Practical Purposes* stresses the connections between contemporary mathematics and modern society, our text must be flexible enough to accommodate new ideas in mathematics and their new applications to our daily lives. We maintain this flexibility in the eighth edition.

Our primary goal for this edition was to further improve the ease of use for instructors and students alike. An extensive supplements package is available, including the new MathPortal, available packaged with the text or sold separately. This innovative online resource brings together the complete text and its media in one easy-to-use learning space. From the eBook and Assignment Center to the full array of Resources, including practice quizzes, exercise solutions, interactive applets, flashcards, video clips, and much more, it's new, it's innovative, it's a must have!

D5MBJ3d6

# NEW TO THE EIGHTH EDITION

## Improved Pedagogical Structure

The enhanced pedagogy makes it easier to navigate the text. More examples are called out for students, more key terms are in definition boxes, and more key theorems, procedures, and rules have been identified.

## New Examples

Each chapter offers two to three new examples, all based on real-world scenarios of particular interest to students, such as:

- Does Running Lead to Winning in Football? (Chapter 6, Example 9).
- Simple Interest on a Student Loan (Chapter 21, Example 1).
- Chaos in Manhattan (Chapter 23, Example 8).

Examples provide new topics for class discussion and new ways of relating to essential concepts.

## New Exercises

- **All exercise sets** have been updated and refreshed.
- **End-of-chapter exercises** now include the correlating section number, making it easier to assign homework and create tests.
- The **skills check exercises** have been updated and now include multiple choice and fill-in-the-blank questions, providing a greater variation in assessment formats.

## New MathPortal

Available packaged with *For All Practical Purposes*, Eighth Edition, or for purchase online. **MathPortal** brings together *For All Practical Purposes*, Eighth Edition, and its media in one affordable, easy-to-use learning space that offers a range of assessment and course management features. It is organized into three main areas:

- **The FAPP interactive eBook**
- **Resources**
- **Assignment Center**

## Content Update Highlights

### PART I Management Science

- **Enhanced** presentation of the material related to the Four Color Theorem in Chapter 3.

### PART II Statistics

- **New** Technology Spotlights were added to aid in calculating standard deviation, the 5-number summary, correlation, line of best fit, combinatorics, and factorials (covering graphing, scientific, and nonscientific calculators).
- **Revised:** Key formulas are now provided in both computational and conceptual versions (Standard Deviation in Chapter 5, and Correlation in Chapter 6).

- **Expanded coverage** of sample space (e.g., tree diagrams), combinatorics (e.g., counting when order does not matter), probability rules (e.g., multiplication rule for independent events) and descriptive statistics (e.g., mode change).
- Notation used is **more explicitly** explained and is aligned with already-familiar notation, e.g., line of fit uses notation from algebra class: $y = mx + b$.

### PART III Voting and Social Choice

- Arrow's Impossibility Theorem to Organ Transplant Policy example has been **expanded**.
- **New** discussion of the National Popular Vote law.
- **Expansion of coverage** on the discussion of positioning in presidential primaries.

### PART IV Fairness and Game Theory

- **New section** on Fair Division and Organ Transplant Policies.
- **New** discussion of the work of the winners of the 2005 and 2007 Nobel Prize in Economics.

### PART V The Digital Revolution

- **New** discussion of 13-digit ISBN number and new postal Bar Code.
- **New** spotlights on Bar Coding DNA and on Morse Code.
- **New** sub-section on Message Routing.
- **New** error detection method.
- **Updated** material on Web Searches.

### PART VI On Size and Growth

- **New** Spotlight: Fitness Test.
- **New** Discussions of rotation symmetry and of vertex type.
- **New** Example: A Group of Non-Numbers.

### PART VII Your Money and Resources

- **Updated** treatment of Simple Interest on a Student Loan.
- **Updated** formulae for compound interest, saving and payment.
- **New** sub-section on Real Growth Under Inflation.
- **Updated** real world data.
- **New** Spotlight on The Mortgage Crisis.
- **Expanded** treatment of Chaos.

## Focus on Accuracy

For this edition we once again implemented a detailed accuracy checking plan to sustain the quality of the exercises and improve the solutions. To this end, we are very grateful to **John Samons of Florida Community College at Jacksonville.** He tirelessly worked with the authors to ensure accuracy in this edition of the text. John once again collaborated with the supplements author, **Heidi Howard of Florida Community College at Jacksonville**, to ensure both accuracy and consistency between the text and supplements package.

We are also grateful to **Scott Inch of Bloomsburg University**, **Rosalie Abraham of Florida Community College at Jacksonville**, and **Paul Lorczak** for their participation in a detailed line edit review of the eighth edition.

## NEW CUSTOM OPTIONS

In addition to the extensive topics covered in the text, more traditional chapters (including **Problem Solving**, **Sets**, **Logic**, **Geometry**, **Counting and Probability**, **Numeration Systems and Personal Finance**) are available with *FAPP* through custom publishing. For more information, please contact your W. H. Freeman representative or go to www.whfreeman.com/fapp8e. Restrictions apply.

## MEDIA AND SUPPLEMENTS

The media and supplements package for the eighth edition has been updated to reflect changes in the book. Both instructors and students will benefit from the innovative materials available to them.

### Student Resources

**NEW! MathPortal**

**http://courses.bfwpub.com/fapp8e** (Access code required. Available packaged with *For All Practical Purposes,* Eighth Edition, or for purchase online.)

**MathPortal** brings together the text *For All Practical Purposes,* Eighth Edition, and its media in one affordable, easy-to-use learning space that offers a range of assessment and course management features. It is organized into three main areas:

- **The FAPP interactive eBook:** integrates a complete and customizable online version of the text with all of its media resources. Students can quickly search the text, and can personalize the eBook with highlighting, bookmarking, and note-taking features. Instructors can add, hide, and reorder content, integrate their own material, and highlight key text.
- **Resources:** organizes all student and instructor resources in one easily searchable location. Resources include self quizzes, interactive applets, flashcards, and games as well as video clips, news feeds, projects, and PowerPoint® sets.
- **Assignment Center:** organizes assignments and guides instructors through an easy-to-create assignment process. Exercises come from the Test Bank, Web Quizzes, and the text, and include many algorithmic problems.

**NEW! *For All Practical Purposes*, Eighth Edition, eBook**

The complete eBook is also available stand-alone, outside of MathPortal, at approximately one-half the cost of the printed textbook.

**NEW! Online Study Center**

**www.whfreeman.com/osc/fapp8e** (Access code or online purchase required.)

This premium Web-based study alternative helps make study time supremely efficient. Students take a pre-chapter Self-Test that generates a Personalized Study

Plan linking to the online resources relevant to the questions they missed. Instructors have access to an easy-to-manage gradebook and all media resources to help them track student progress and prepare lectures or course Web pages.

**Student Study Guide, ISBN: 1-4292-2650-1**
Heidi Howard, Florida Community College at Jacksonville
Offers study tips and tools to help students gain a better understanding of course material.

**Student Solutions Manual, ISBN: 1-4292-2646-3**
Heidi Howard, Florida Community College at Jacksonville
Contains full, worked solutions to the odd-numbered problems in the text.

**Book Companion Site: www.whfreeman.com/fapp8e**
The complimentary site provides students with access to study tools and instructors a range of assessment, presentation, and course management resources.

### Instructor Resources

**Instructor's Manual with Full Solutions, ISBN: 1-4292-2649-8**
Heidi Howard, Florida Community College at Jacksonville
Includes teaching support for each chapter *and* full solutions for all problems in the text.

**Teaching Guide for First-Time Instructors, ISBN: 1-4292-2645-5**
Heidi Howard, Florida Community College at Jacksonville
This guide for new instructors, adjuncts, and teaching assistants will help make planning your course and teaching with *FAPP* easier and more effective. Ideas set forth in this guide also offer fresh perspective and ideas to experienced instructors.

**Enhanced Instructor's Resource CD-ROM (IRCD), ISBN: 1-4292-2654-4**
Created to help instructors develop lecture presentations, course Web sites, and other resources, this CD-ROM allows instructors to **search** and **export** all the resources contained below by key term or chapter:

- All text images.
- Applets, movies, flashcards, spreadsheet projects, self-quizzes available on the Web site.
- Instructor's Manual with Full Solutions.

## Assessment

### Test Bank

**CD-ROM (Windows and Macintosh): 1-4292-2653-6**
**Printed: 1-4292-2651-X**
John Emert, Ball State University
The *Test Bank* offers 75 multiple-choice and fill-in-the-blank questions and 35 short-answer questions per chapter. The easy-to-use CD includes Windows and Macintosh versions on a single disc, in a format that lets you add, edit, and re-sequence questions to suit your needs.

### Course Management

**WebCT and Blackboard**
All the book's Web and testing materials are compatible with WebCT and Blackboard. We offer the electronic content as a service to adopters; please contact your local sales representative.

# ACKNOWLEDGEMENTS

*For All Practical Purposes* continues to evolve in great part due to our many friends and colleagues who have offered suggestions, comments, and corrections. We are grateful to them all.

Rosalie Abraham, *Florida Community College at Jacksonville*
Alison Ahlgren, *University of Illinois at Urbana-Champaign*
Scott Balcomb, *St. Joseph's University*
Nancy Balle, *Ball State University*
Richard Bedient, *Hamilton College*
Rebecca Bergs, *Ball State University*
Terence R. Blows, *Northern Arizona University*
Raouf N. Boules, *Towson University*
Kristina K. Bowers, *Florida State University*
Terry Boyd, *University of Indianapolis*
Linda Braddy, *East Central University*
Barry Brunson, *Western Kentucky University*
Paul Buckelew, *Oklahoma City Community College*
Annette M. Burden, *Youngstown State University*
Shana Calaway, *Shoreline Community College*
Tim Carroll, *Eastern Michigan University*
G. Andy Chang, *Youngstown State University*
Yi Cheng, *Indiana University South Bend*
Leo Chouinard, *University of Nebraska–Lincoln*
Karen Clark, *Tacoma Community College*
Valerie Morgan-Crick, *Tacoma Community College*
Greg Crow, *Point Loma Nazarene University*
Sloan Despeaux, *Western Carolina University*
Rob Donnelly, *Murray State University*
Daniel Dreibelbis, *University of North Florida*
Gina Poore Dunn, *Lander University*
Nancy Eaton, *University of Rhode Island*
Kristy J. Eisenhart, *Western Michigan University*
John W. Emert, *Ball State University*
Sandra Fillebrown, *Saint Joseph's University*
Joseph Fox, *Salem State College*
W. Bart Frye, *Ball State University*
Martha Gady, *Whitworth College*
Monica Pierri-Galvao, *Gannon University*
Steve Gendler, *Clarion University*
Marty Getz, *University of Alaska, Fairbanks*
Carol E. Gibbons, *Salve Regina University*
T. R. Hamlett, *East Central University*
Geoffrey Hagopian, *College of the Desert*
Mohammad Halim, *Ball State University*
Frederick Hoffman, *Florida Atlantic University*
Michael Hull, *Northern Arizona University*
Scott Inch, *Bloomsburg University of Pennsylvania*
Peter Johnson, *Auburn University*
W. T. Kiley, *George Mason University*
Julie Killingbeck, *Ball State University*
Nancy Kitt, *Ball State University*
Samuel Kohn, *Thomas Edison State College*
Kathy Lewis, *State University of New York, Oswego*
Monica Liddle, *State University of New York, Delhi*
Jay Malmstrom, *Oklahoma City Community College*
Barbara Margoulius, *Cleveland State University*
Vania Mascioni, *Ball State University*
Mary T. McMahon, *North Central College*
Christopher McCord, *University of Cincinnati*
Ricardo Moena, *University of Cincinnati*
Steve Morics, *University of Redlands*
Dean Morrow, *Washington and Jefferson College*
Anne Marie Mosher, *St. Louis Community College*
Ellen Mulqueeny, *Cleveland State University*
Mika Munakata, *Montclair State University*
Chris Oehrlein, *Oklahoma City Community College*
Steven Ohs, *Western Michigan University*
Patricia Parkison, *Ball State University*
Deb Pearson, *Ball State University*
Andrew B. Perry, *Springfield College*
Marilyn Reba, *Clemson University*
Leo Robinson, *Ball State University*
Chris Rodger, *Auburn University*
Jennifer Marie Rodin, *University of South Carolina Aiken*
Robin Ruffato, *Ball State University*
Daniel Russow, *Arizona Western University*
Steven Schecter, *North Carolina State*
Brian Siebenaler, *Ball State University*
Debora J. Simonson, *University of North Florida*
Samuel Bruce Smith, *St. Joseph's University*
Patricia Stanley, *Ball State University*
James D. Stoops, *Ball State University*
William R. Stout, *Salve Regina University*
Tamas Szabo, *Weber State University*
Robert Terrell, *Cornell University*
Helen Thorwarth, *Northern Kentucky University*
Aaron K. Trautwein, *Carthage College*
David Urion, *Winona State University*
Bonnie Wachhaus, *Messiah College*
W. D. Wallis, *Southern Illinois University*
Kim Ward, *Eastern Connecticut State University*
John Weglarz, *Kirkwood Community College*
Gideon Weinstein, *Montclair State University*
Cheryl Whitelaw, *Southern Utah University*
Liz Whittern, *Ball State University*
Scott Wilde, *Baylor University*
Meredith Wort, *East Central University*
Mingqing Xiao, *Southern Illinois University*
Christian Yankov, *Eastern Connecticut State University*
Janet Yi, *Ball State University*
Laurie Margaret Zack, *High Point University*
John Zerger, *Catawba College*
Cathleen M. Zucco-Teveloff, *Rowan University*

We owe our appreciation to the people at W. H. Freeman and Company who participated in the development and production of this edition. We wish especially to thank the editorial staff for their tireless efforts and support. Among them are Craig Bleyer, Senior Publisher; Ruth Baruth, Publisher; Terri Ward, Senior Acquisitions Editor; Laura Capuano, Associate Editor; Vivien Weiss, Project Editor; Blake Logan, Cover Designer; Christine Buese, Photo Editor; Roland Cheney, Senior Media Editor; and Ellen Cash, Director of Production. We would also like to extend our appreciation to outside Development Editor, Lisa Collette.

The efforts of the COMAP staff must also be recognized. We thank our production staff, George Ward, Joyce Barnes, and Daiva Chauhan.

To everyone who helped make our purposes practical, we offer our appreciation for an exciting and exhilarating time.

***Solomon Garfunkel, COMAP***

# Management Science

# PART I

On November 7, 2007 the space shuttle Discovery returned home from a highly successful, if not exactly trouble free 15-day mission. The National Aeronautics and Space Administration (NASA) had reason to be proud of the crew's accomplishments. This shuttle flight was supporting one of mankind's most ambitious ventures on the last frontier of human exploration–the construction of a space station in Earth orbit. This space station is not only a laboratory for new technologies that are being pioneered for ventures into space: It also tests the ability of human beings to live and work in space for extended periods of time. The mission made the history books when a team of women astronauts served as commanders of Discovery and the Space Station for the first time. The cool-headed leaders of this mission, supported by a team of experts back on Earth, scheduled a successful (but not initially planned) space walk to deal with a tear in one of the two "solar wings" designed for the space station.

This flexibility to handle the unexpected illustrated a hidden mathematical story behind this mission. Missions of this complexity require planning, scheduling, resource allocation, and cost minimization on a vast scale. To get a feel for the size of this project, some estimate its cost at about $100 billion. The branch of mathematics concerned with helping governments, businesses, and individuals operate as efficiently as possible is known as **management science** or **operations research (OR)**. The tools of this subject (graph theory, linear algebra, probability theory, and so forth) have evolved over more than a 100 years and often build on simple but clever mathematical ideas. However, it was only about 50 years ago that OR was identified as a distinct branch of knowledge and the systematic study of the subject began.

The chapters that follow chart a wide variety of ways that mathematical ideas make our lives more enjoyable and satisfying. At the same time these ideas challenge us with "simple puzzles" that are just fun to think about. ■

CHAPTER
1



# Urban Services

The underlying theme of management science, also called **operations research,** is finding the best method for solving some problem–what mathematicians call the **optimal solution**. In some cases, the goal may be to finish a job as quickly as possible. In other situations, the objective might be to maximize profit or minimize cost. In this chapter, our goal is to save time in traversing a street network while checking parking meters, delivering mail, removing snow, collecting bottles for recycling, or inspecting for potholes.

Let's begin by assisting the parking department of a city government. Most cities and many small towns have parking meters that must be regularly checked for parking violations or emptied of coins. We will use an imaginary town to show how management science techniques can help to make parking control more efficient.

## 1.1 Euler Circuits

The street map in Figure 1.1 is typical of many towns across the United States, with streets, residential blocks, and a town park. Our job, or that of the commissioner of parking, is to find the most efficient route for the parking-control officer, who travels on foot, to check the meters in an area. Efficient routes save money. Our map shows only a small area, allowing us to start with an easy problem. But the problem occurs on a larger scale in all cities and towns and for larger areas. The bigger the region involved, the greater the potential for cost savings.

The commissioner has two goals in mind: (1) The parking-control officer must cover all the sidewalks that have parking meters without retracing any more steps than are necessary; and (2) the route should end at the same point at which it began, perhaps where the officer's patrol car is parked. To be specific, suppose there are only two blocks that have parking meters, the two lightly shaded blocks that are side by side toward the top of Figure 1.1. Suppose further that the parking-control officer must start and end at the upper left corner of the left-hand

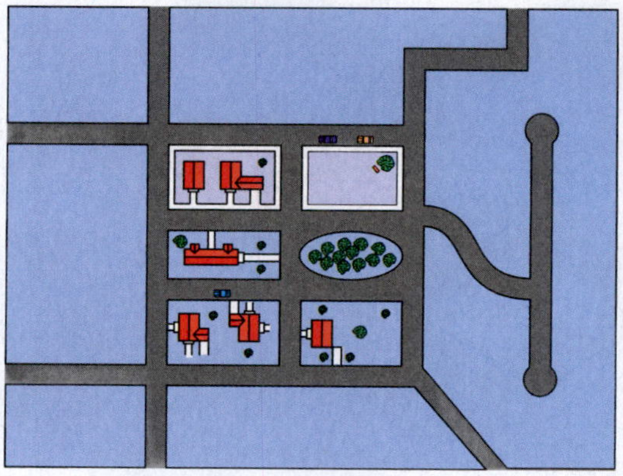

**FIGURE 1.1** A street map for part of a town.

block. You might enjoy working out some routes by trial and error and evaluating their good and bad features. We are going to leave this problem for the moment and establish some concepts that will give us a better method than trial and error to deal with this problem.

### Describing a Graph

DEFINITION

A **graph** is a finite set of dots and connecting links. The dots are called **vertices** (a single dot is called a **vertex**), and the links are called **edges**. Each edge must connect two different vertices. A **path** is a connected sequence of edges showing a route on the graph that starts at a vertex and ends at a vertex; a path is usually described by naming in turn the vertices visited in traversing it. A path that starts and ends at the same vertex is called a **circuit**. A graph can represent our city map, a communications network, a system of air routes, or electrical power lines.

## EXAMPLE 1 ■ Parts of a Graph

We can use the graph in Figure 1.2 to help explain these technical terms. The graph shown has 5 vertices and 8 edges. The vertices represent cities, and the edges represent nonstop airline routes between them. We see that there is a nonstop flight between Berlin and Rome, but no such flight between New York and Berlin. There are several paths that describe how a person might travel with this airline from New York to Berlin. The path that seems most direct is New York, London, Berlin, but New York, Miami, Rome, Berlin is also a path. We can describe these two paths as *NLB* and *NMRB*. Another path would be New York, Miami, Rome, London, Berlin, which can be written as *NMRLB*. An example of a circuit is Miami, Rome, London, Miami. It is a circuit because the path starts and ends at the same vertex. This circuit can best be described in symbols by *MRLM*. Another example of a circuit in this graph would be *LRBL*, which is the circuit involving the cities London, Rome, Berlin, and back to London. In this chapter, we are especially interested in circuits, just as we are in real life. Most of us end our day in the same location where we start it–at home!

Notice that the edges *MB* (which could also be denoted *BM*) and *RL* shown in Figure 1.2 meet at a point that has no label. Furthermore, this point does not have a dark dot. This is because this point does not represent a vertex of our graph; it does not represent a city. It arises as an "accidental" consequence of the way this diagram has been drawn. We could join *M* and *B* with a curved line segment so that

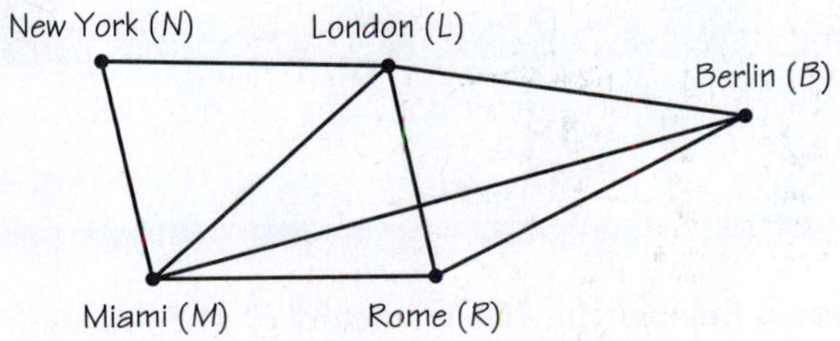


**FIGURE 1.2** The edges of the graph show nonstop routes that an airline might offer.

the edges *LR* and *MB* do not cross, or redraw the diagram so as to avoid a crossing in this case (but not in all graphs we might wish to draw). We will be working often in situations where graphs can be drawn without accidental crossings and we will try to avoid such crossings when it is convenient to do so.

Returning to the case of parking control in Figure 1.1, we can use a graph to represent the whole territory to be patrolled: Think of each street intersection as a vertex and each sidewalk that contains meters as an edge, as in Figure 1.3. Notice in Figure 1.3b that the width of the street separating the blocks is not explicitly represented; it has been shrunk to nothing. In effect, we are simplifying our problem by ignoring any distance traveled in crossing streets.

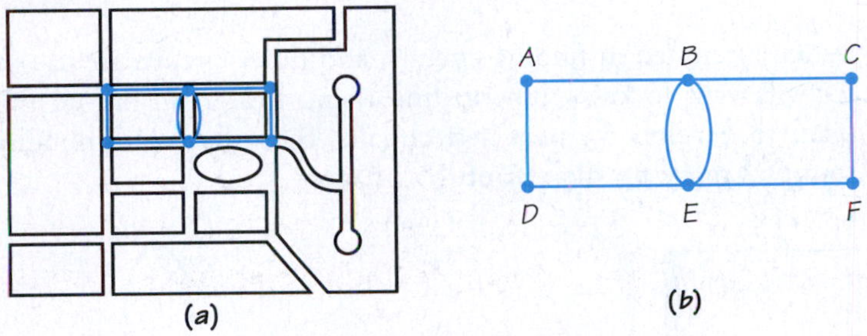


**FIGURE 1.3** (a) A graph superimposed upon a street map. The edges show which sidewalks have parking meters. (b) The same graph enlarged.

The numbered sequence of edges in Figure 1.4a shows one circuit that covers all the meters (note that it is a circuit because its path returns to its starting point). However, one edge is traversed three times. Figure 1.4b shows another solution that is better because its circuit covers every edge (sidewalk) exactly once. In Figure 1.4b, no edge is covered more than once, or *deadheaded* (a term borrowed from shipping, which means making a return trip without a load).

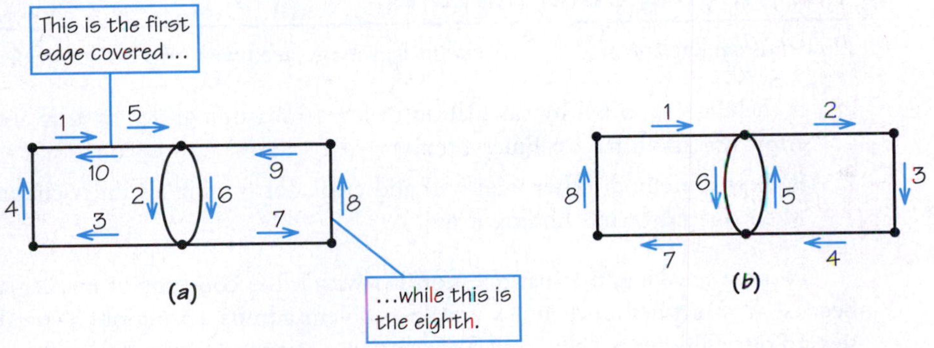


**FIGURE 1.4** (a) A circuit and (b) an Euler circuit.

**Euler Circuit** DEFINITION

A circuit that covers each edge of a graph once but not more than once is called an **Euler circuit**.

Figure 1.4b shows an Euler circuit. These circuits get their name from the great eighteenth-century mathematician Leonhard Euler (pronounced *oy' lur*), who first studied them (see Spotlight 1.1). Euler was the founder of the theory of graphs, or graph theory. One of his first discoveries was that some graphs have no Euler circuits at all. For example, in the graph in Figure 1.5b, it would be impossible to start at one point, return to that starting vertex and cover all the edges without retracing some steps: If we try to start a circuit at the leftmost vertex, we discover that once we have left the vertex, we have "used up" the only edge meeting it. We have no way to return to our starting point except to reuse that edge. But this is not allowed in an Euler circuit. If we try to start a circuit at one of the other two vertices, we likewise can't complete it to form an Euler circuit.

As mentioned in Spotlight 1.2, realistic problems of this type will involve larger neighborhoods that might require the use of a computer. In addition, there may be other complications that might take us beyond the simple mathematics we want to stick to.

Because we are interested in finding circuits, and Euler circuits are the most efficient ones, we will want to know how to find them. If a graph has no Euler circuit, we will want to develop the next best circuits, those having minimum deadheading. These topics make up the rest of this chapter.

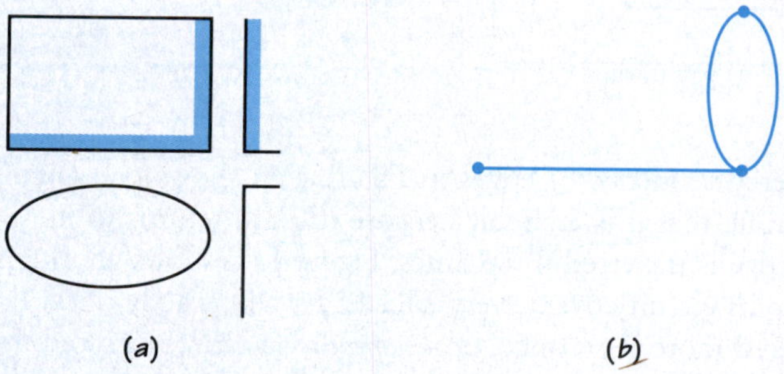


FIGURE 1.5 (a) The three shaded sidewalks cannot be covered by an Euler circuit. (b) The graph of the shaded sidewalks in part (a).

## 1.2 Finding Euler Circuits

Now that we know what an Euler circuit is, we are faced with two obvious questions:

1. Is there a way to tell by calculation or logical reasoning, not by trial and error, if a graph has an Euler circuit?
2. Is there a method, other than trial and error, for finding an Euler circuit when one exists and finding it quickly?

Loosely speaking, the first question lies within the concerns of mathematicians because it asks whether or not a certain problem admits a solution. Typically, the second question lies in the domain of computer science because it concerns finding the actual answer to a complex version of a problem in a short enough time to be useful.

## SPOTLIGHT 1.1 Leonhard Euler

Leonhard Euler (1707–1783) was one of those rare individuals who was remarkable in many ways. He was extremely prolific, publishing over 500 works in his lifetime. But he wasn't devoted just to mathematics; he was a people person, too. He was extremely fond of children and had thirteen of his own, of whom only five survived childhood. It is said that he often wrote difficult mathematical works with a child or two in his lap.

Human interest stories about Euler have been handed down through two centuries. He was a prodigy at doing complex mathematical calculations under less than ideal conditions, and he continued to do them even after he became totally blind later in life. His blindness diminished neither the quantity nor the quality of his output. Throughout his life, he was able to mentally calculate in a short time what would have taken ordinary mathematicians hours of pencil-and-paper work. A contemporary claimed that Euler could calculate effortlessly, "just as men breathe, as eagles sustain themselves in the air." His collected works are not yet fully published.

**Leonhard Euler**

*(Portrait by Emanuel Handmann, Bildnis des Mathematikers, 1753, Oeffentliche Kunstsammlung Basel, Kunstmuseum.)*

Euler invented the idea of a graph in 1736 when he solved a problem in "recreational mathematics." He showed that it was impossible to stroll a route visiting the seven bridges of the German town of Königsberg exactly once. Ironically, in 1752 he discovered that three-dimensional polyhedra obey the remarkable formula $V - E + F = 2$ (that is, number of vertices − number of edges + number of faces = 2) but failed to give a proof because he did not analyze the situation using graph theory methods.

Euler investigated these questions in 1735 by using the concepts of **valence** and **connectedness.**

### Valence — DEFINITION

The **valence** of a vertex in a graph is the number of edges meeting at the vertex.

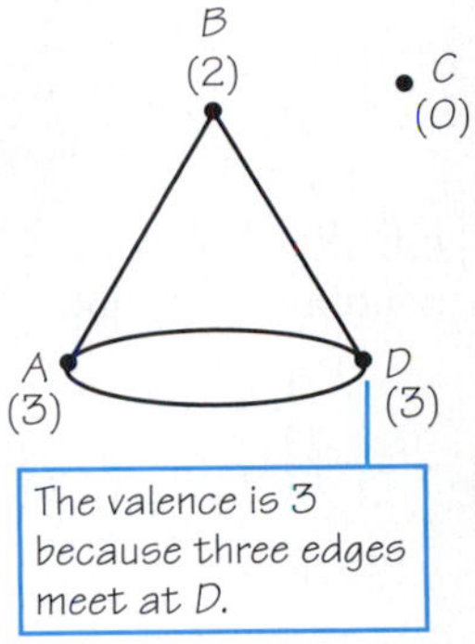
B
(2)
C
(0)
A
(3)
D
(3)
The valence is 3 because three edges meet at D.

**FIGURE 1.6**
Valences of vertices.

Figure 1.6 illustrates the concept of valence, with vertices $A$ and $D$ having valence 3, vertex $B$ having valence 2, and vertex $C$ having valence 0. Isolated vertices such as vertex $C$ are an annoyance in Euler circuit theory. Because they don't occur in typical applications, we henceforth assume that our graphs have no vertices of valence 0.

Figure 1.3b has four vertices of valence 2, namely, $A$, $C$, $F$, and $D$. This graph also has two vertices, $B$ and $E$, of valence 4. Notice that each vertex has a valence that is an even number. We'll soon see that this is very significant.

### Connected Graph — DEFINITION

A graph is said to be **connected** if for every pair of its vertices there is at least one path connecting the two vertices.

## SPOTLIGHT 1.2 The Human Aspect of Problem Solving

Thomas Magnanti, professor of operations research and management, heads the Department of Management Science at MIT's Sloan School of Management. Here are some of his observations:

Thomas Magnanti
*(Courtesy of Thomas Magnanti.)*

Typically, a management science approach has several different ingredients. One is just structuring the problem—understanding that the problem is an Euler circuit problem or a related management science problem. After that, one has to develop the solution methods.

But one should also recognize that you don't just push a button and get the answer. In using these underlying mathematical tools, we never want to lose sight of our common sense, of understanding, intuition, and judgment. The computer provides certain kinds of insights. It deals with some of the combinatorial complexities of these problems very nicely. But a model such as an Euler circuit can never capture the full essence of a decision-making problem.

Typically, when we solve the mathematical problem, we see that it doesn't quite correspond to the real problem we want to solve. So we make modifications in the underlying model. It is an interactive approach, using the best of what computers and mathematics have to offer and the best of what we, as human beings, with our own decision-making capabilities, have to offer.

Given a graph, if we can find even one pair of vertices not connected by a path, then we say that the graph is not connected. For example, the graph in Figure 1.7 is not connected because we are unable to join $A$ to $D$ with a path of edges. However, the graph does consist of two "pieces" or connected components, one containing the vertices $A$, $B$, $F$, and $G$, the other containing $C$, $D$, and $E$. A connected graph will contain a single connected component. Notice that the parking-control graph of Figure 1.3b is connected.

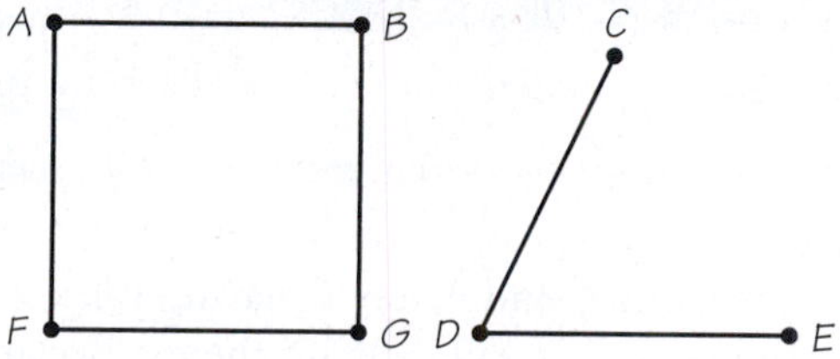


**FIGURE 1.7** A nonconnected graph.

We can now state Euler's theorem, his simple answer to the problem of detecting when a graph $G$ has an Euler circuit.

### Euler Circuit Theorem — THEOREM

1. If $G$ is connected and has all valences even, then $G$ has an Euler circuit.
2. Conversely, if $G$ has an Euler circuit, then $G$ must be connected and all its valences must be even numbers.

Because the parking-control graph of Figure 1.3b conforms to the connectedness and even-valence conditions, Euler's theorem tells us that it has an Euler circuit. We already have found an Euler circuit for the graph shown in Figure 1.4b by trial and error. For a very large graph, however, trial and error may take a long time. It is usually quicker to check whether the graph is connected and even-valent than to find out if it has an Euler circuit.

Once we know there is an Euler circuit in a certain graph, how do we find it? Many people find that, after a little practice, they can find Euler circuits by trial and error, and they don't need detailed instructions on how to proceed. At this point you should see if you can develop this skill by trying to find Euler circuits in Figure 1.8a, Figure 1.9a, and Figure 1.10. In doing your experiments, draw your graph in ink and the circuit in pencil so you can erase if necessary.

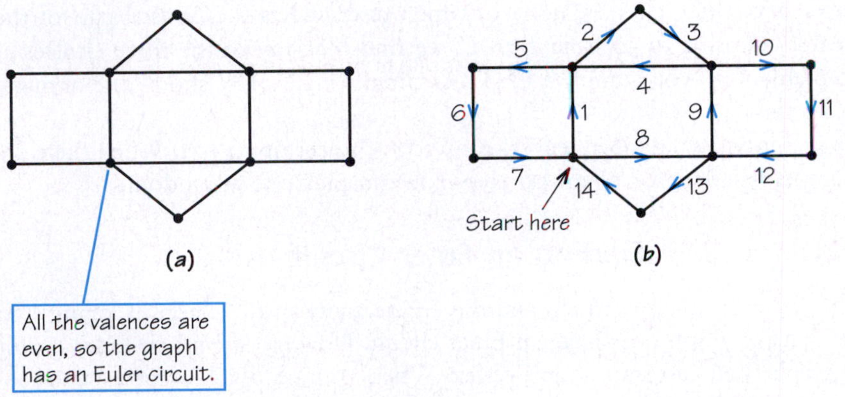


**FIGURE 1.8** (a) A graph having (b) an Euler circuit.

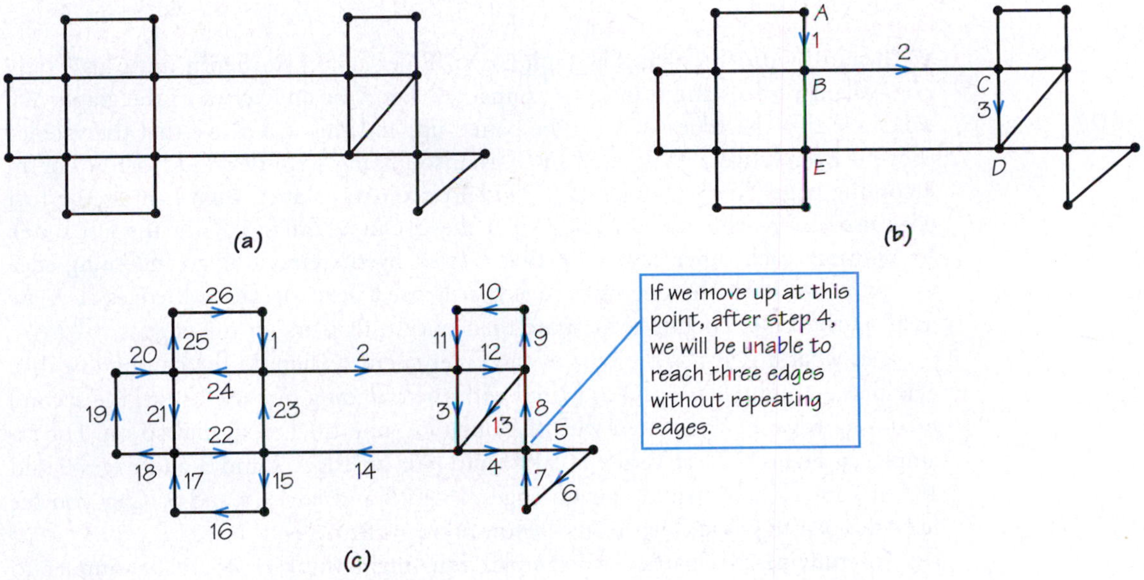


**FIGURE 1.9** (a) A graph that has an Euler circuit. (b) A critical junction in finding an Euler circuit in this graph, starting from vertex *A*. (c) A description of a full Euler circuit for this graph.

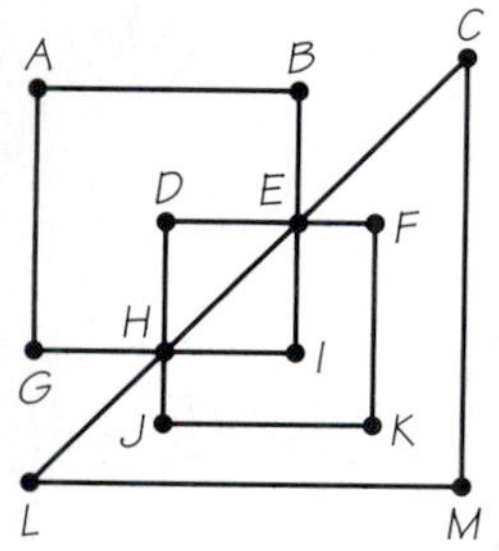


**FIGURE 1.10** A graph with an Euler circuit.

If you would like more guidance on how to find an Euler circuit without trial and error, here is a method that works: Never use an edge that is the only link between two parts of the graph that still need to be covered. Figure 1.9b illustrates this. Here we have started the circuit at *A* and gotten to *D* via *B* and *C*, and we want to know what to do next. Going to *E* would be a bad idea because the uncovered part of the graph would then be disconnected into left and right portions. You will never be able to get from the left part back to the right part because you have just used the last remaining link between these parts. Therefore, you should stay on the right side and finish that before using the edge from *D* to *E*. This kind of thinking needs to be applied every time you need to choose a new edge.

Let's see how this works, starting at the beginning at *A*. From vertex *A* there are two possible edges, and neither of them disconnects the unused portion of the graph. Thus, we could have gone either to the left or down. Having gone down to *B*, we now have three choices, none of which disconnects the unused part of the graph. After choosing to go from *B* to *C*, we find that any of the three choices at *C* is acceptable. Can you complete the Euler circuit? Figure 1.9c shows one of many ways to do this.

The method just described leaves many edge choices up to you. When there are many acceptable edges for your next step, you can pick one at random.

## EXAMPLE 2 ■ Finding an Euler Circuit

Check the valences of the vertices and the connectivity of the graph in Figure 1.8a to verify that the graph does have an Euler circuit. Now try to find an Euler circuit for that graph. You can start at any vertex. When you are done, compare your solution with the Euler circuit given in Figure 1.8b. If your path covers each edge exactly once and returns to its original vertex (is a circuit), then it is an Euler circuit, even if it is not the same as the one we give. ■

## Proving Euler's Theorem

We'll start by proving that if a graph has an Euler circuit *R*, then it must have only even-valent vertices and it must be connected. Let *X* be any vertex of the graph. We will show that the edges at *X* can be paired up, and this will prove that the valence is even. Every edge at *X* is used by *R* as an outgoing edge (leaving from *X*) or an incoming edge (arriving at *X*). If the Euler circuit starts at *X*, then pair up the first edge used by *R* with the last one (when the circuit returns to *X* for the last time). In addition, each other edge at *X* that is used by the circuit as an incoming edge will be paired with the outgoing edge that is used next. Because all edges at *X* are used by the Euler circuit, none more than once, this pairs up the edges.

But what if *X* is not the start of the Euler circuit? Then do the pairing like this: The first incoming edge at *X* is paired with the outgoing one used next, the second incoming edge at *X* is paired with the outgoing one used next, and so on. For example, in Figure 1.11 at vertex *B*, we would pair up edges 2 and 3 and edges 9 and 10. At vertex *C*, we would pair up edges 4 and 5 and edges 8 and 9. Can you see how the pairings would work at *D*? How about vertex *A*?

In studying this particular example, you might think it would be simpler to count the edges at a vertex to see that the valence is even. True, but our pairing method works for a graph about which we know nothing except that it has an Euler circuit.

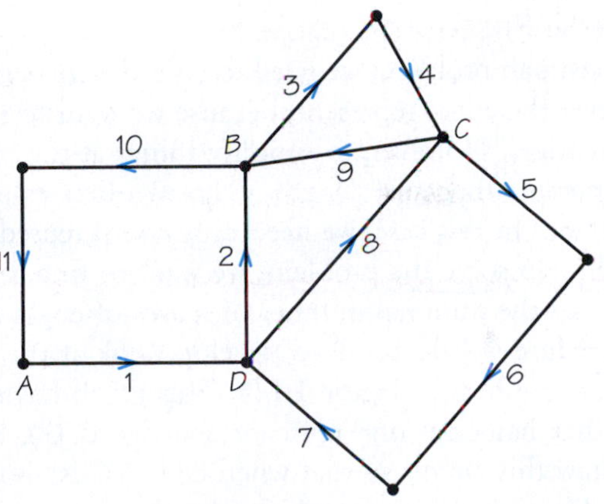


FIGURE 1.11 An Euler circuit starting and ending at *A*.

To see that a graph with an Euler circuit is connected, note that by following the Euler circuit around we can get from any edge to any other edge (it covers them all) using a portion of the Euler circuit. Because every vertex is on an edge (there are no vertices of valence 0), we can get from any vertex to any other using a portion of the Euler circuit.

So far, this is not a complete proof of Euler's theorem. To complete the proof we would need to prove that if a graph has all vertices even-valent and is connected, then an Euler circuit can be found for it.

## 1.3 Beyond Euler Circuits

Now let's see what Euler's theorem tells us about the three-block neighborhood with parking meters, represented by dots in Figure 1.12a. Figure 1.12b shows the corresponding graph. (Because we use edges to represent only sidewalks along which the officer must walk, the sidewalk with no meters is not represented by any edge in the graph.) This graph has vertices with odd valences (at vertices *C* and *G*), so Euler's theorem tells us that there is no Euler circuit for this graph.

Because we must reuse some edges in this graph to cover all edges in a circuit, for efficiency we need to keep the total length of reused edges to a minimum. This type of problem, in which we want to minimize the length of a circuit by carefully choosing which edges to retrace, is often called the **Chinese postman problem** (like parking-control routes, mail routes need to be efficient). The problem was first studied by the Chinese mathematician Meigu Guan in 1962–hence the name. The remainder of this chapter is dedicated to solving the Chinese postman problem and discussing applications beyond parking control.

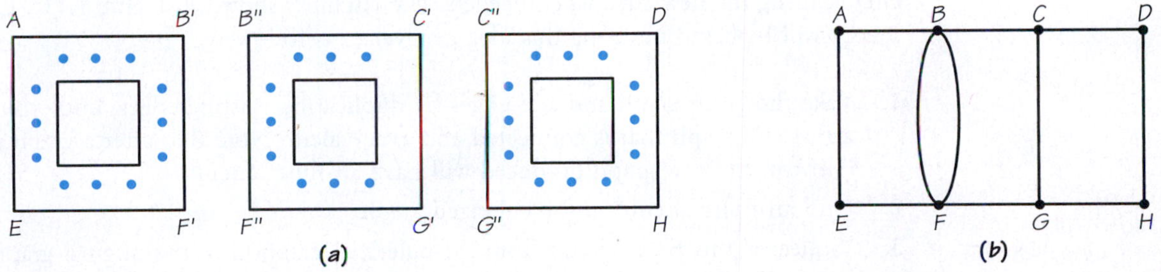


FIGURE 1.12 (a) A street network and (b) its graphic representation. Locations such as *B'* and *B''*, *C'* and *C''*, *F'* and *F''*, and *G'* and *G''* are merged to form the vertices *B*, *C*, *F*, and *G*. The dots shown represent parking meters.

## Solving the Chinese Postman Problem

In a realistic Chinese postman problem, we need to consider the lengths of the sidewalks, streets, or whatever the edges represent, because we want to minimize the total length of the reused edges. However, to simplify things at the start, we can suppose that all edges represent the same length. (This is often called the *simplified* Chinese postman problem.) In this case, we need only count reused edges and need not add up their lengths. To solve the problem, we want to find a circuit that covers each edge and that has the minimal number of reuses of edges already covered.

To follow the procedure we are going to develop, look at the graph in Figure 1.13a, which is the same graph as in Figure 1.12b. This graph has no Euler circuit, but there is a circuit that has only one reuse of an edge (*CG*), namely, *ABCDHGCGFBFEA*. Let's draw this circuit so that when edge *CG* is about to be reused, we install a new, extra, blue edge in the graph for the circuit to use. By duplicating edge *CG*, we can avoid reusing the edge. To duplicate an edge, we must add an edge that joins the two vertices that are already joined by the edge we want to duplicate. (It makes no sense to join vertices that are not already connected by an edge, because such edges would not represent sidewalk sections with meters; see Figure 1.15.)

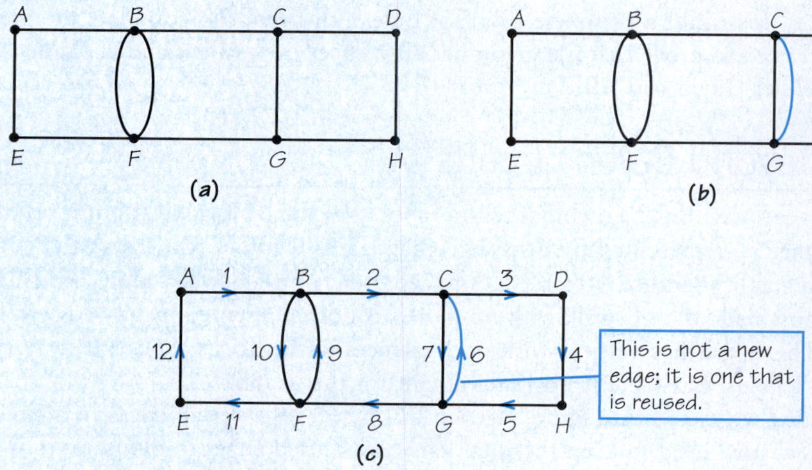


**FIGURE 1.13** Making a circuit by reusing an edge.

We have now created the graph of Figure 1.13b. In the graphs we draw, the edges that are added will be shown in color to distinguish them from the original edges, which are shown in black. (You may want to use a similar scheme to help you remember which edges are the originals and which are duplicated in the graphs you draw.) In the graph of Figure 1.13b, the original circuit can be traced as an Euler circuit, using the new edge when needed. The circuit is shown in Figure 1.13c. Our theory will be based on using this idea in reverse, as follows:

1. Take the given graph and add edges by duplicating existing edges, until you arrive at a graph that is connected and even-valent. Note that after a graph is *eulerized,* the new graph produced will have an Euler circuit.
2. Find an Euler circuit on the eulerized graph.
3. "Squeeze" this Euler circuit from the eulerized graph onto the original graph by reusing an edge of the original graph each time the circuit on the eulerized graph uses an added edge.

**Eulerizing a Graph** DEFINITION

Adding edges that duplicate existing edges to a connected graph to make all valences even is called **eulerizing** the graph.

## EXAMPLE 3 ■ Eulerizing a Graph

Suppose we want to eulerize the graph of Figure 1.14a. When we eulerize a graph, we first locate the vertices with odd valence. The graph in Figure 1.14a has two, *B* and *C*. Next, we add one end of an edge at each such vertex, matching up the new edge with an existing edge in the original graph. Figure 1.14b shows one way to eulerize the graph. Note that *B* and *C* have even valence in the second graph. After eulerization, each vertex has even valence. To see an Euler circuit on the eulerized graph in Figure 1.14c, simply follow the edges in numerical order and in the direction of the arrows, beginning and ending at vertex *A*. The final step, shown in Figure 1.14d, is to "squeeze" our Euler circuit onto the original graph. There are two reuses of previously covered edges. Notice that each reuse of an edge corresponds to an added edge.

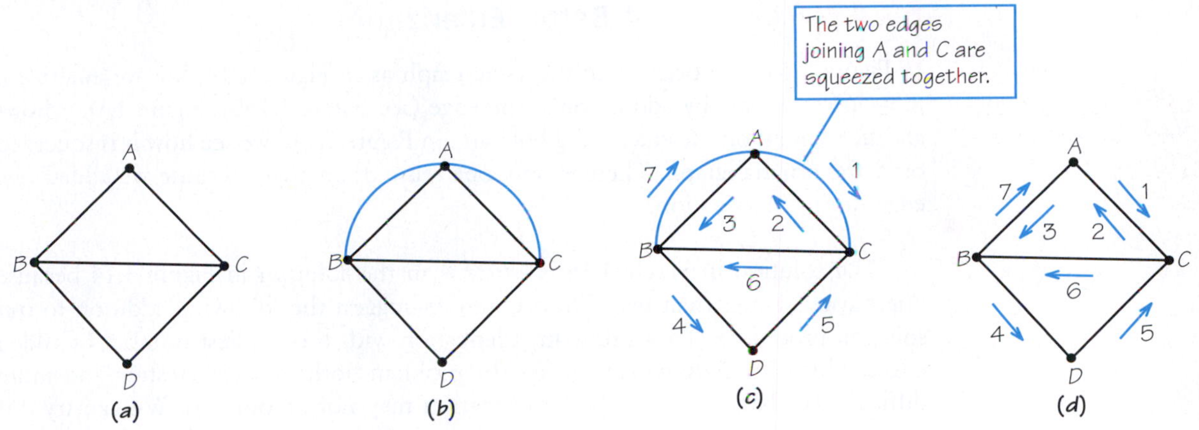


**FIGURE 1.14** Eulerizing a graph.

In the previous example, we noticed that we could count how many reuses we needed by counting added edges. This is generally true in this type of problem: *If you add the new edges correctly, the number of reuses of edges equals the number of edges added during eulerization.*

Adding new edges correctly means adding only edges that are duplicates of existing edges. Doing this makes the rule, just stated in italics, always true, and so it is easy to count the needed reuses.

To see why we add only duplicate edges, examine Figure 1.15a. We need to alter the valences of vertices *X* and *Y* by adding edges so that they become even-valent. Adding one long edge from *X* to *Y* (Figure 1.15b) might seem like an attractive idea, but adding this edge is equivalent to asking a snowplow, say, to get from *X* to *Y* without moving along existing streets. At times it is necessary to traverse sections of the graph that have been previously traversed. This is the significance of the duplicated edges. Here the structure of the graph forces us to repeat some edges.

We cannot get away with fewer than three repeats—the three edges *XU*, *UV*, and *VY* (Figure 1.15c). The duplicated edges are shown in color.

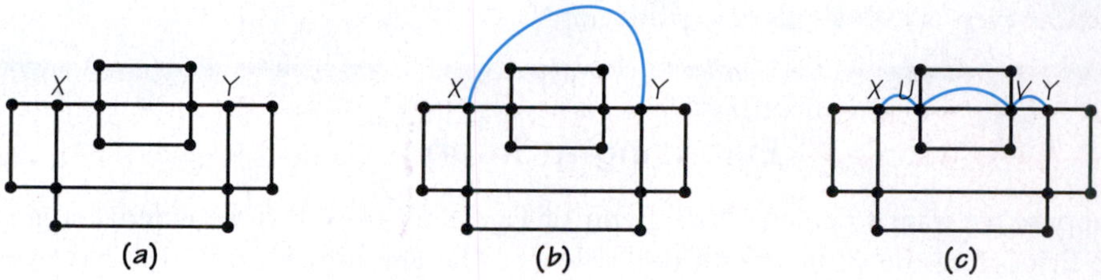


**FIGURE 1.15** Eulerizing when the vertices are more than one edge apart.

Now that we have learned to eulerize, the next step is to try to get a best eulerization we can—one with the fewest added edges. It turns out that there are many ways to eulerize a graph. It is even possible that the smallest number of added edges can be achieved with two different eulerizations. This is the reason we use the phrase "a best eulerization" rather than "the best eulerization." Remember, we want a best eulerization because this enables us to find the circuit for the original graph that has the minimum number of reuses of edges.

## EXAMPLE 4 ■ A Better Eulerization

In Figure 1.16a, we begin with the same graph as in Figure 1.14, but we eulerize it in a different way—by adding only one edge (see Figure 1.16b). Figure 1.16c shows an Euler circuit on the eulerized graph, and in Figure 1.14d we see how it is squeezed onto the original graph. There is only one reuse of an edge, because we added one edge during eulerization.

The solution in Figure 1.16 is better than the solution in Figure 1.14 because one reuse is better than two. These examples suggest the following addition to our solution procedure: Try to find an eulerization with the smallest number of added edges. This extra requirement makes the problem both more interesting and more difficult. For large graphs, a best eulerization may not be obvious. We can try out a few and pick the best among the ones we find, but there may be an even better one that our haphazard search does not turn up.

A systematic procedure for finding a best eulerization does exist, but the process is complicated. There is an especially easy technique for eulerizing the following special category of networks often found in our neighborhoods.

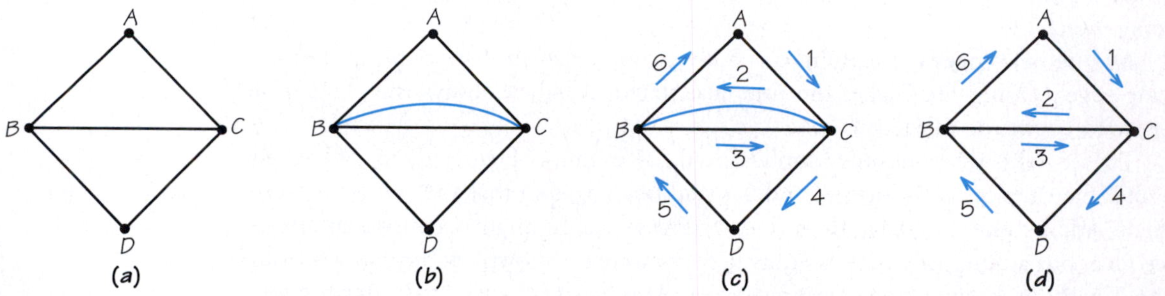


**FIGURE 1.16** A better eulerization of Figure 1.14.

**Rectangular Network** DEFINITION

If a street network is composed of a series of rectangular blocks that form a large rectangle a certain number of blocks high and a certain number of blocks wide, the network is called **rectangular**.

Examples of rectangular street networks (a 3-by-3, a 3-by-4, and a 4-by-4) are shown in Figure 1.17. The graph on the right in each pair shows a best eulerization for the rectangular street network on the left. There appear to be three different eulerization patterns, depending upon whether the rectangle height and width in the original graph are odd or even numbers. In Figure 1.17a, both lengths are 3, both odd; in Figure 1.17b, one length is odd (3) and one is even (4); in Figure 1.17c, both lengths are 4, an even number.

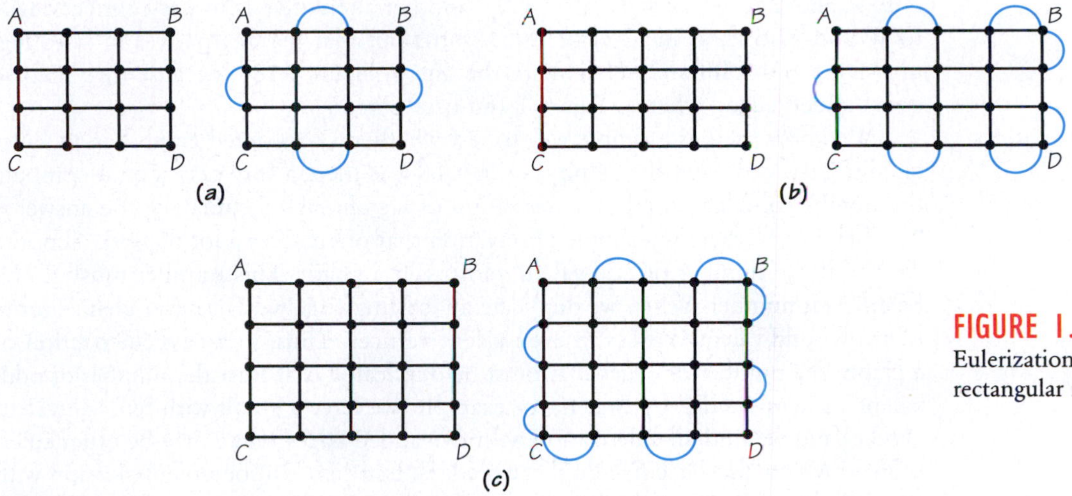


FIGURE 1.17 Eulerizations of three rectangular networks.

Although the patterns appear different, one technique can be used to create all of them. This technique can be thought of as involving an "edge walker" who walks around the outer boundary of the large rectangle in some direction, say, clockwise. He starts at any corner, say, the upper left corner. As he goes around, he adds edges by the following rules. When he comes to an odd-valent vertex, he links it to the next vertex with an added edge. This next vertex now becomes either even or odd. If it became even, he skips it and continues around, looking for an odd vertex. If it became odd (this could happen only at a corner of the big rectangle), the edge walker links it to the next vertex and then checks this vertex to see whether it is even or odd. Each of the three parts of Figure 1.17 has been eulerized by this method.

In a street network that is not rectangular, the eulerization process is started by locating all the vertices with odd valence and then pairing these vertices with each other and finding the length of the shortest path between each pair. We look for the shortest paths because each edge on the connecting paths will be duplicated. The idea is to choose the pairings cleverly so that the sum of the lengths of those paths is the smallest it can be. With a little practice, most people can find a best or nearly best eulerization using only this idea together with trial and error and some ingenuity.

## Finding Good Eulerizations

Suppose we want a perfect procedure for eulerizing a graph. What theoretical ideas and methods could we use to build such a tool?

One building block we could use is a method for finding the shortest path between two given vertices of a graph. For example, let us focus on vertices *X* and *Y* in Figure 1.18a. Both have odd valence. We can connect them with a pattern of duplicate edges, as in Figure 1.18b. The cost of this is the length of the path we duplicated from *X* to *Y*. A shorter path from *X* to *Y*, such as the one shown in Figure 1.18c, would be better. Fortunately, the *shortest-path problem* has been well studied, and we have many good procedures for solving it exactly, even in large, complex graphs.

But there is more to eulerizing the graph in Figure 1.18a than dealing with *X* and *Y*. Notice that we have odd valences at *Z* and *W*. Should we connect *X* and *Y* with a path, and then connect *Z* and *W*, as in Figure 1.18d? Or should we connect *X* to *Z* and *Y* to *W*, as in Figure 1.18e? Another alternative is to use connections *X* to *W* and *Y* to *Z*, as in Figure 1.18f. It turns out that the alternatives in both Figures 1.18e and 1.18f are preferable to the one in Figure 1.18d, because they involve seven added edges, whereas Figure 1.18d uses nine.

We know there is a simple way to test whether a connected graph has an Euler circuit: Check to see if the graph is even-valent. Is there a very easy way to compute the number of edges in a best eulerization of a graph? Unfortunately, the answer is "no." However, there is a simple observation that often saves a lot of work. Suppose we count the number of odd-valent vertices in a graph. This number must always be an even number. When we duplicate an existing edge we can never change more than two odd-valent vertices to even-valent vertices. Thus, in a best eulerization of a graph, the number of edges that must be duplicated is at least the number of odd-valent vertices divided by two. If, for example, we have a graph with ten odd-valent vertices and we find an eulerization with five added edges, there may be other eulerizations which also have five duplicated edges, but there can be no eulerization with fewer than five duplicated edges.

Remember that when an unweighted graph is eulerized in an optimal way, then the total cost of traversing each edge at least once can be found by adding the total number of edges in the graph to the number of edges that are reused (duplicated). Small problems involving eulerization can be carried out by trial-and-error methods. Unfortunately, although there is an algorithm that can be applied to find the best eulerization for large problems, the details of this algorithm are quite complex. However, the procedure works quickly not only for graphs without weights but also for graphs with weights on the edges.

# 1.4 Urban Graph Traversal Problems

Euler circuits and eulerizing have many more practical applications than just checking parking meters. Almost anytime services must be delivered along streets or roads, our theory can make the job more efficient. Examples include collecting garbage, salting icy roads, plowing snow, inspecting railroad tracks for flaws, and reading electric meters (see Spotlight 1.3).

Each of these problems has its own special requirements that may call for modifications in the theory. For example, in the case of garbage collection, the edges of our graph will represent streets, not sidewalks. If some of the streets are one-way, we need to put arrows on the corresponding edges, resulting in a directed graph, or

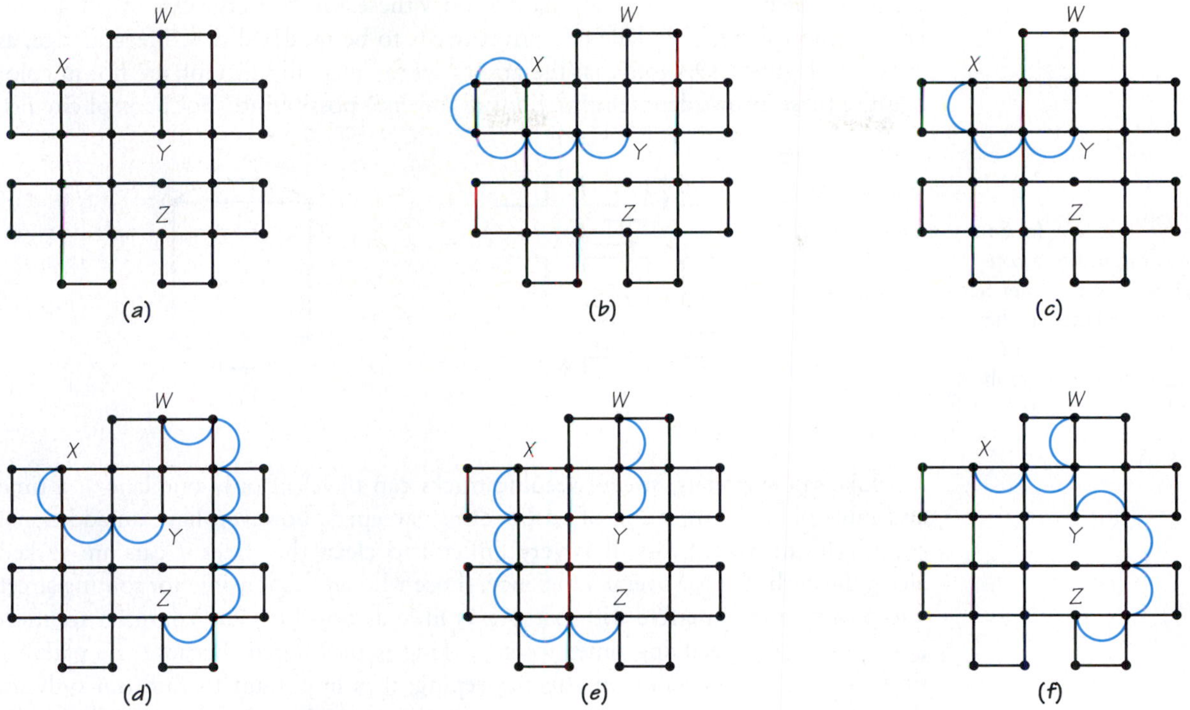


**FIGURE 1.18** Choosing among eulerizations.

## SPOTLIGHT 1.3 Israel Electric Company Reduces Meter-Reading Task

The Beersheba branch of Israel's major electric company wanted to make the job of meter reading more efficient. When the branch managers decided to minimize the number of people required to read the electric meters in the houses of one particular neighborhood, they set a precedent by applying management science. Formerly, each person's route had been worked out by trial and error and intuition, with no help from mathematics. The whole job required 24 people, each doing a part of the neighborhood in a five-hour shift.

At first, it looks as though one would find a more efficient way of doing the work the same way as in the Chinese postman problem, but there are two important differences. First, the neighborhood was big enough to negate any possibility of having only one route assigned to one person. Instead, it was necessary to find a number of routes that, taken together, covered all the edges (sidewalks). Second, a meter reader who was done with a route was allowed to return home directly. Thus, there was no reason for the individual routes to return to their starting points; therefore, routes could be paths instead of circuits.

The Beersheba researchers found solutions to these problems by modifying the basic ideas we have described in this chapter. They managed to cover the neighborhood with 15 five-hour routes, a 40% reduction of the original 24 five-hour routes. Altogether, these routes involve a total of 4338 minutes of walking time, of which 41 minutes (less than 1%) is deadheading.

**digraph**. The circuits we seek will have to obey these arrows. In the case of salt spreaders and snowplows, each lane of a street needs to be modeled as a directed edge, as shown in Figure 1.19. Note that the arrows on the map and digraph are not in color because these arrows denote restrictions in traversal possibilities, not parts of circuits.

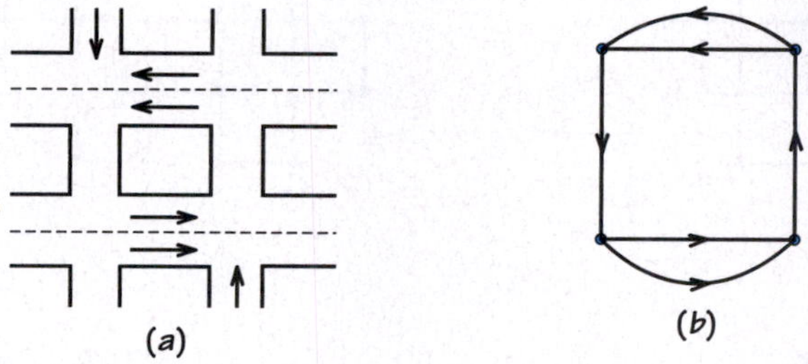


**FIGURE 1.19** (a) Salt-spreading route, where each west–west street has two traffic lanes in the same direction, and (b) an appropriate digraph model.

Like salt spreaders, street-sweeping trucks can travel in only one lane at a time and must obey the direction of traffic. Street sweepers, however, have an additional complication: parked cars. It is very difficult to clean the street if cars are parked along the curb. Yet for overall efficiency, those who are responsible for routing street sweepers want to interfere with parking as little as possible. The common solution is to post signs specifying times when parking is prohibited. Because the parking-time factor is a constraint on street sweeping, it is important to find not only an Euler circuit, or a circuit with very few duplications, but also a circuit that visits streets when they are free of cars. Once again, mathematicians have developed techniques to handle this constraint.

Finally, because towns and cities of any size will have more than one street sweeper, parking officer, or garbage truck, a single best route may not suffice. Instead, it becomes necessary to divide the territory into multiple routes. The general goal is to find optimal solutions while taking into account traffic direction, number of lanes, parking-time restrictions, and divided routes (see Figure 1.20).

Management science makes all this possible. For example, a pilot study done in the 1970s in New York City showed that applying these techniques to street sweepers in just one district could save about $30,000 per year. With 57 sanitation districts in New York, this would amount to a savings of more than $1.5 million in a single year. This translates to about $5 million in 2008 dollars. In addition, the same principles could be extended to garbage collection, parking control, and other services carried out on street networks.

This plan was not adopted when first proposed. Because city services take place in a political context, several other factors come into play. For example, union leaders try to protect the jobs of city workers, bureaucrats might try to keep their departmental budgets high, and elected politicians rarely want to be accused of cutting the jobs of their constituents. Thus, political obstacles can overrule management science. As mentioned in Spotlight 1.2, such human factors often arise when applying management science. Perhaps a more acceptable street-sweeping plan would have been devised for New York City if more attention had been paid to the human factors earlier.

Despite the complications of real-world problems, management science principles provide ways to understand these problems by using graphs as models. We can reason about the graphs and then return to the real-world problem with a workable solution. The results we get can have a lasting effect on the efficiency and economic well-being of any organization or community.

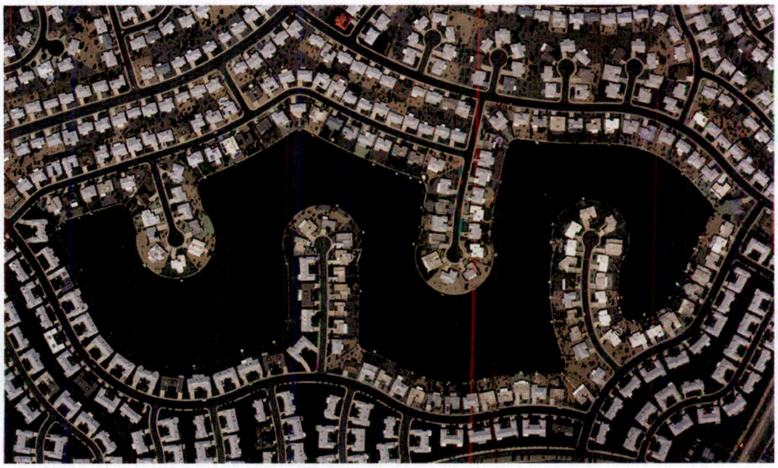

(a)

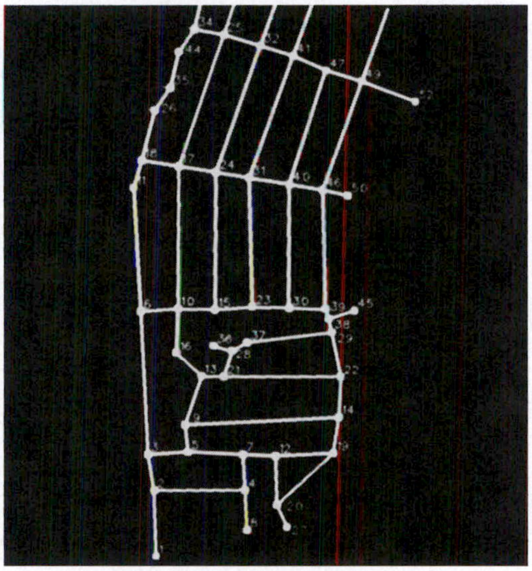

(b)

**FIGURE 1.20**
(a) Residential neighborhoods, whether they be in cities or the suburbs, require many services such as mail delivery, garbage collection, street sweeping, meter reading, or sewage systems. The mathematical techniques of operations research make it possible to provide these services as cheaply as possible. When optimal solutions to providing such services can be found, everyone is a winner. (*Brand X Pictures/ Picturequest.*)
(b) Computers can be used to extract the essential information needed to solve routing problems from photographs.

## REVIEW VOCABULARY

**Chinese postman problem** The problem of finding a circuit on a graph that covers every edge of the graph at least once and that has the shortest possible length. (p. 11)

**Circuit** A path that starts and ends at the same vertex. (p. 4)

**Connected graph** A graph is connected if it is possible to reach any vertex from any specified starting vertex by traversing edges. (p. 7)

**Digraph** A graph in which each edge has an arrow indicating the direction of the edge. Such directed edges are appropriate when the relationship is "one-sided" rather than symmetric (for instance, one-way streets as opposed to regular streets). (p. 18)

**Edge** A link joining two vertices in a graph. (p. 4)

**Euler circuit** A circuit that traverses each edge of a graph exactly once. (p. 6)

**Eulerizing** Adding new edges, which duplicate existing edges, to a connected graph so as to make a graph that possesses an Euler circuit. (p. 13)

**Graph** A mathematical structure in which points (called vertices) are used to represent things of interest and in

which links (called edges) are used to connect vertices, denoting that the connected vertices have a certain relationship. (p. 4)
**Management science** A discipline in which mathematical methods are applied to management problems in pursuit of optimal solutions that cannot readily be obtained by common sense. (p. 1)
**Operations research (OR)** Another name for management science. (p. 1)
**Optimal solution** When a problem has various solutions that can be ranked in preference order (perhaps according to some numerical measure of "goodness"), the optimal solution is the best-ranking solution. (p. 3)
**Path** A connected sequence of edges in a graph. (p. 4)
**Valence** (of a vertex) The number of edges touching that vertex. (p. 7)
**Vertex** A point in a graph where one or more edges end. (p. 4)

## ✓ SKILLS CHECK

**1.** What is the valence of vertex *A* in the graph below?

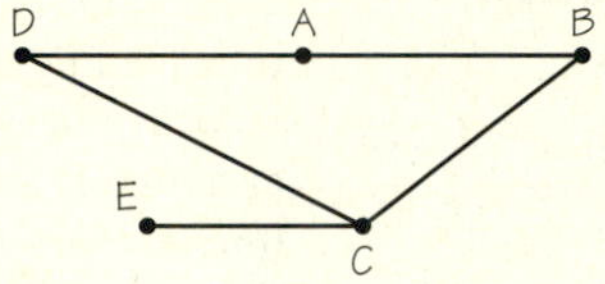


**(a)** 2
**(b)** 1
**(c)** 3

**2.** The number of vertices in the graph below is _____ , while the number of edges in this graph is __________ .

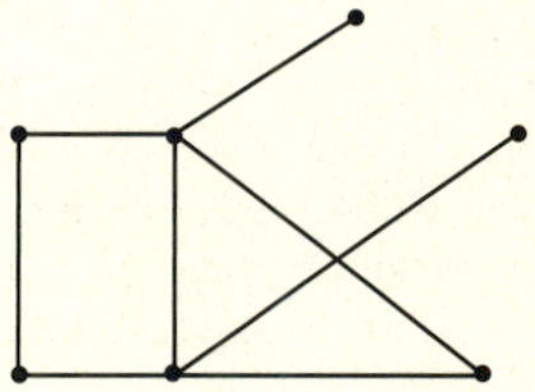

**3.** The valences of the vertices in the accompanying graph listed in non-increasing order are

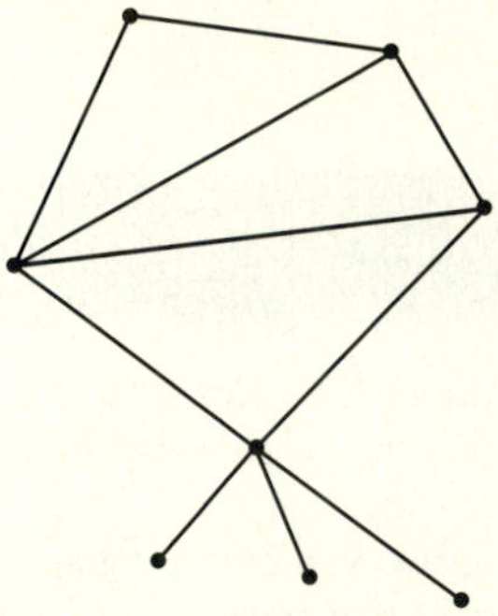

**(a)** 5, 4, 3, 3, 2, 1, 1, 1.
**(b)** 1, 3, 4, 4, 5, 5.
**(c)** 5, 5, 4, 3, 3, 1

**4.** The graph shown below is not connected because it consists of ________ parts.

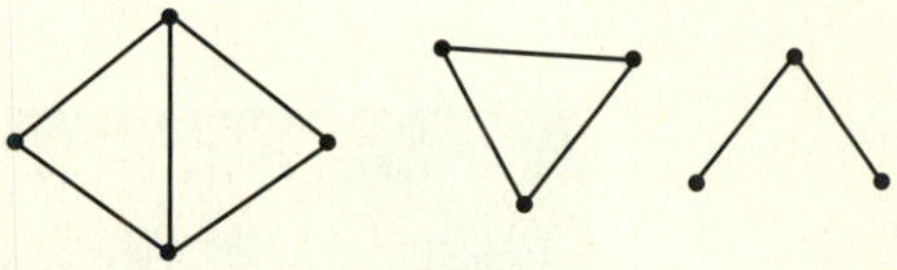

**5.** The graph below has

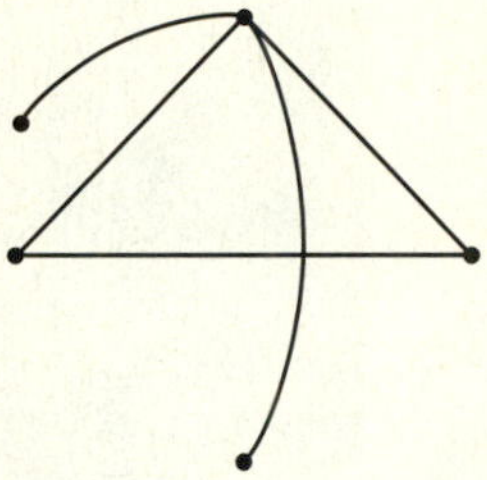

**(a)** four vertices and six edges.
**(b)** four vertices and four edges.
**(c)** five vertices and five edges.

**6.** The alphabetically ordered list of even-valent vertices of the graph below is ______ , ______ .

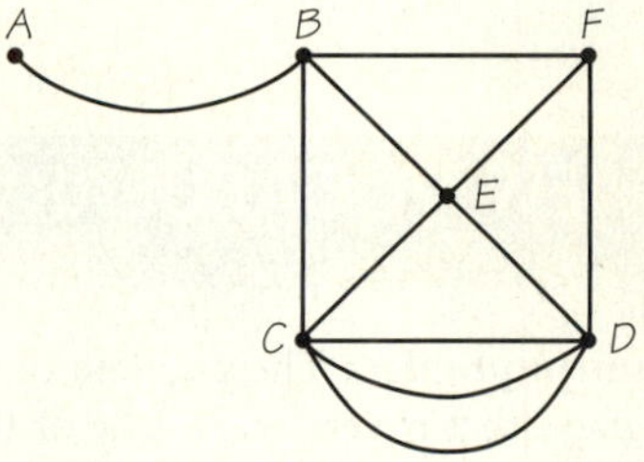


**7.** Which of the following statements is true about a *path?*

**(a)** A path always forms a circuit.
**(b)** A path is always connected.
**(c)** A path can visit any vertex only once.

**8.** If a graph consists of four vertices and every pair of vertices is connected by a single edge, the number of edges in the graph is exactly ________ .

**9.** It is not possible for a graph to have five vertices of valence 3 and six vertices of valence 4 because

**(a)** there are no graphs with exactly 11 vertices.
**(b)** a graph cannot have an even number of 4-valent vertices.
**(c)** a graph cannot have an odd number of odd-valent vertices.

**10.** If a graph is connected and has seven vertices, the graph must have at least __________ edges.

**11.** For which of the situations below is it most desirable to find an Euler circuit or an efficient eulerization of the graph?

**(a)** Sweeping the sidewalks of a small town
**(b)** Planning a new highway
**(c)** Planning a parade route in Muncie, Indiana

**12.** The minimum number of edges which must be duplicated to create a best possible eulerization of the following graph is _______ .

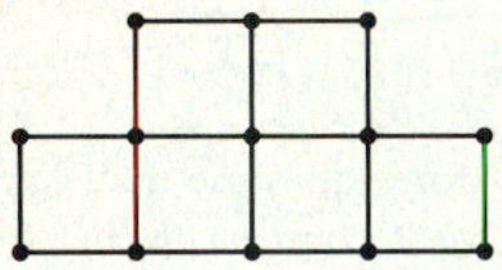

**13.** Consider the path represented by the sequence of numbered edges on the graph below. Which statement is correct?

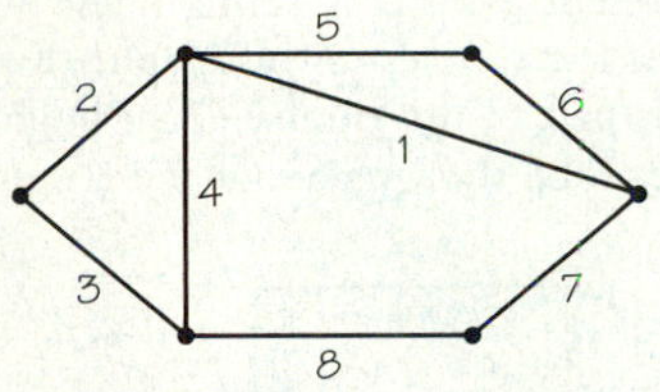


**(a)** The sequence of numbered edges forms an Euler circuit.
**(b)** The sequence of numbered edges traverses each edge exactly once but is not an Euler circuit.
**(c)** The sequence of numbered edges forms a circuit but not an Euler circuit.

**14.** For the graph below, the minimum *total* number of edges which constitutes a tour of the graph, starting and ending at the same vertex, and which visits each edge at least once, is _______ .

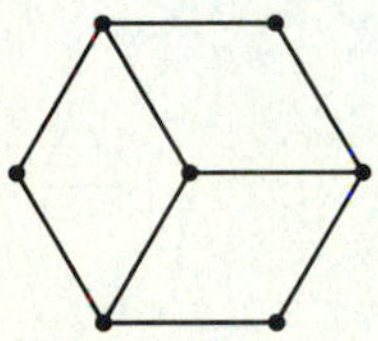

**15.** Suppose each vertex of a graph represents a baseball team and each edge represents a game played by two baseball teams. If the resulting graph is not connected, which of the following statements must be true?

**(a)** At least one pair of teams never played a game.
**(b)** At least one team played every other team.
**(c)** The teams play in distinct leagues.

**16.** If a graph has six vertices of odd valence, the absolute minimum number of edges that must be added (duplicated) to eulerize the graph is ________ .

**17.** Suppose the edges of a graph represent streets that must be plowed after a snowstorm. To eulerize the graph, four edges must be added. The real-world interpretation of this is that

**(a)** four streets will not be plowed.
**(b)** four streets will be traversed twice.
**(c)** four new streets would be built.

**18.** For each of the following situations, decide whether a graph or a digraph seems a more reasonable model:

**(a)** A system of hiking trails: ____________________ .
**(b)** An electrical wiring plan for a home: __________ .
**(c)** A bus route map: ____________________ .

**19.** Suppose a civic club offers several craft courses, and each club member can choose to participate in up to two different courses. Let each vertex of a graph represent one of these courses and each edge represent a club member who wants to take the two courses represented by the vertices at its endpoints. What can be said about the vertices in the resulting graph whose valence is zero?

**(a)** There are no vertices whose valence is zero.
**(b)** These vertices represent courses that can occur at the same time without displeasing any club member.
**(c)** These vertices represent the least popular courses.

**20.** If the valences of the vertices of a graph $G$ are: 5, 4, 4, 4, 3, 2, 2, and 2, the number of vertices of $G$ is ______ and the number of edges of $G$ is _______ .

# CHAPTER 1 EXERCISES

■ Challenge ◆ Discussion

## 1.1 Euler Circuits

## 1.2 Finding Euler Circuits

**1.** In the graph below, the vertices represent houses and two vertices are joined by an edge if it is possible to drive between the two houses in under 10 minutes.

**(a)** How many vertices does the graph have?
**(b)** How many edges does the graph have?
**(c)** What are the valences of the vertices in this graph?
**(d)** Based on the information given by the graph below, for which houses, if any, is it possible to drive to all the other houses in less than 20 minutes?
**(e)** Based on the graph below, from house *B* which houses require a trip of longer than 20 minutes?

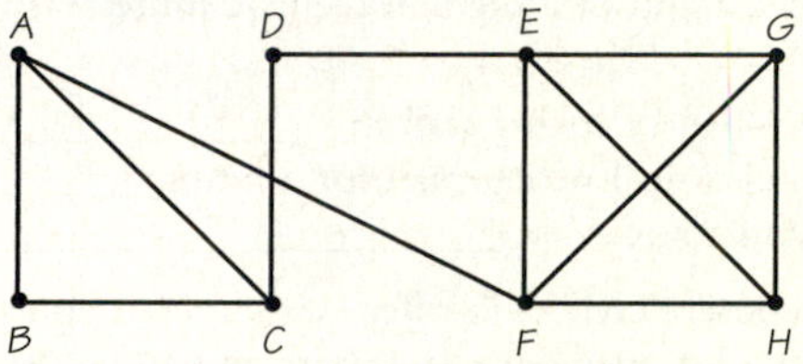


**2. (a)** Redraw the graph in Figure 1.2 to obtain a graph which has the same information where the edges only meet other edges at vertices.
**(b)** List all the routes that start on the U.S. side of the Atlantic Ocean and cross the ocean once and immediately.

**3. (a)** Is the figure below a graph? Explain your answer.

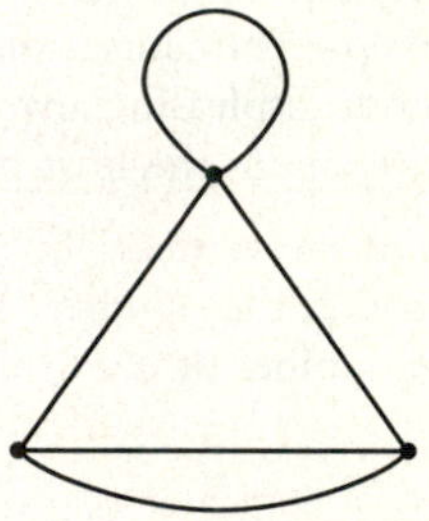

**(b)** The graph below has edges that "cross" at points that are not vertices of the graph. Which edges are these?

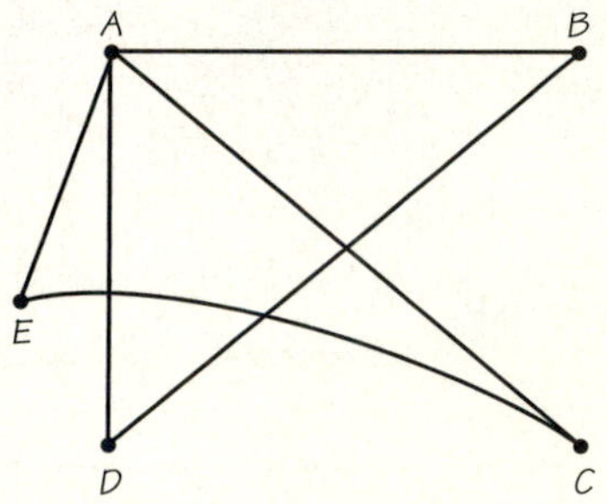


**(c)** How many vertices and edges are there in the prededing graph?

**4.** The graph below shows the stores and roads connecting them in a small shopping mall.

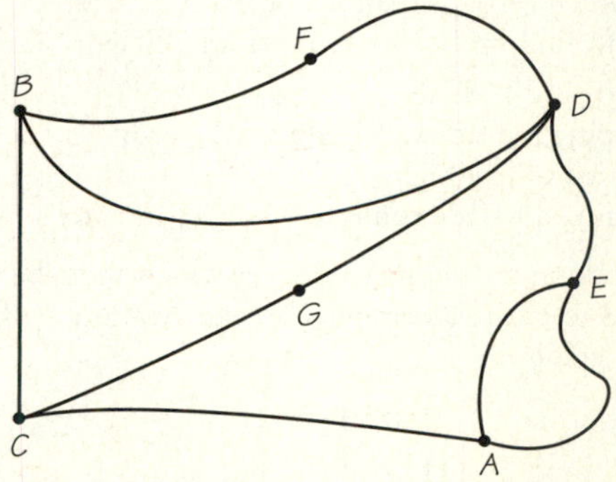


**(a)** How many stores does the mall have?
**(b)** How many roads connect up the stores in the mall?
**(c)** Write down a path from *C* to *F*.
**(d)** Write down a path from *E* to *B*.

**5.** In the graph below, the vertices represent cities and the edges represent roads connecting them. What are the valences of the vertices in this graph? (Keep in mind that *E* is part of the graph.) What might the valence of city *E* be showing about the geography?

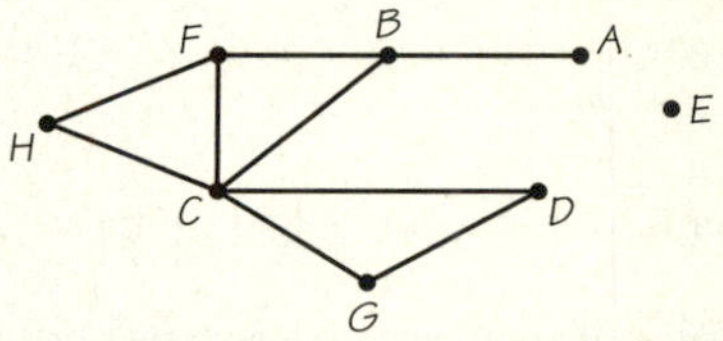


**6.** In the two graphs below, the vertices represent cities and the edges represent roads connecting them. In which graphs could a person located in city *A* choose any other city and then find a sequence of roads to get from *A* to that other city?

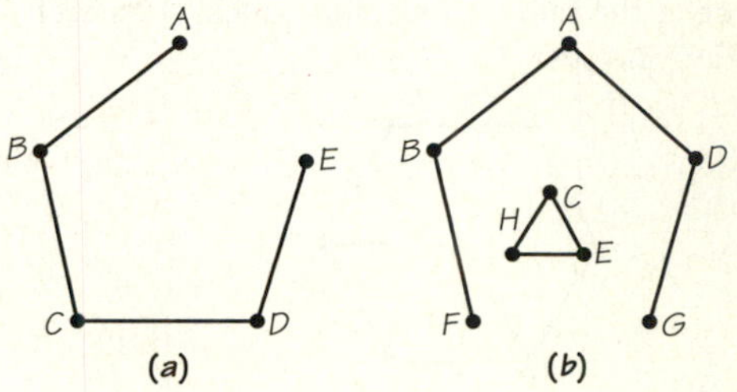

**7.** Refer to the figure in Exercise 4.

**(a)** Write down a circuit that includes the vertices *C* and *D* but does not start or end at either of these vertices.
**(b)** If two paths are considered different if they use different edges, write down:

**(i)** two different paths from *B* to *D*.
**(ii)** three different paths from *C* to *F*.
**(iii)** a circuit that has four edges.

**8.** Jack and Jill are located in Miami and want to fly to Berlin (see Figure 1.2).

**(a)** Find three paths for them to carry out this trip.
**(b)** What is the largest number of paths that can be used to carry out this trip that do not repeat a vertex (city)?
**(c)** Explain why it is reasonable not to want to repeat a vertex in this situation.

**9. (a)** How many vertices and edges does the graph in Figure 1.6 have?
**(b)** How many vertices and edges does the graph in Figure 1.7 have?
**(c)** How many vertices and edges does the graph in Figure 1.8a have?

**10. (a)** Add up the numbers you get for the valences of the vertices in Figure 1.6.
**(b)** Add up the numbers you get for the valences of the vertices in Figure 1.7.
**(c)** Add up the numbers you get for the valences of the vertices in Figure 1.8a.
**(d)** Describe the pattern you see in the answers you got for parts (a) through (c).
**(e)** Show that the pattern describes a fact that is true for any graph. (*Hint:* How many endpoints does an edge have?)

**11.** In the graph in Figure 1.8a, find the smallest possible number of edges you could remove that would disconnect the graph.

**12.** In the graphs in Figure 1.17, find the smallest possible number of edges you could remove that would disconnect the graph.

**13.** Draw a graph with eight vertices that is connected where

**(a)** each vertex has valence 3.
**(b)** each vertex has valence 4.
**(c)** Do all graphs with eight vertices having valence 2 have the same number of edges?

**14.** Is it possible that a street network gives rise to a disconnected graph? If so, draw such a network of blocks and streets and parking meters (in the style of Figure 1.12a). Then draw the disconnected graph it gives rise to.

**15. (a)** Draw a connected graph with six vertices, all of whose vertices have valence 2.
**(b)** Draw a disconnected graph with 6 vertices, all of whose vertices have valence 2.

**16. (a)** Draw a graph where every vertex has valence of at least 3 but where removing a single edge disconnects the graph.
**(b)** In what urban settings might a road network be represented by a graph that has an edge whose removal would disconnect the graph?

**17. (a)** Find a graph where the valences of the seven vertices of the graph are 1, 2, 2, 3, 3, 3, 4.
**(b)** Find another graph with the same valences as above that is "different" from the one you found for part (a).

**18.** For some services provided along streets, it may matter whether the roads are one-way or two-way. Give some examples where the street directions do and do not matter for our graph model analysis.

◆ **19.** A postal worker is supposed to deliver mail on all streets represented by edges in the graph below by traversing each edge exactly once. The first day the worker traverses the numbered edges in the order shown in (a), but the supervisor is not satisfied–why? The second day the worker follows the path indicated in (b), and the worker is unhappy–why? Is the original job description realistic? Why?

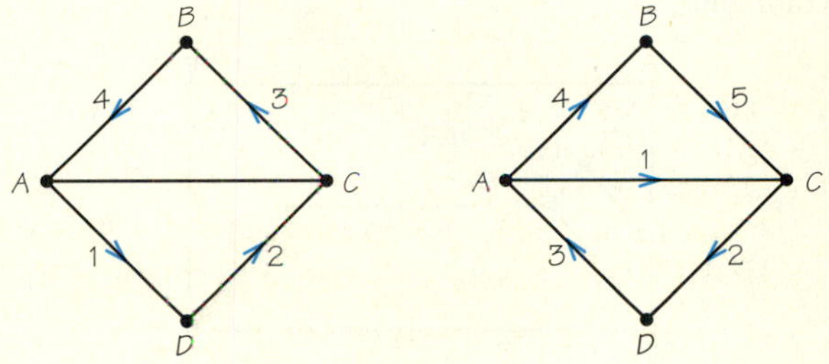


**20.** For the street network in Exercise 19, draw the graph that would be useful for routing a snowplow. Assume that all streets are two-way, one lane in each direction, and that you need to pass down each lane separately.

**21.** Find an efficient route for the snowplow to follow in the graph you drew in Exercise 20.

**22. (a)** Give examples of services that could be performed by a vehicle that moved in the direction of traffic down either lane of a two-way street.
**(b)** Give examples of services that would probably require a vehicle to travel down each of the lanes of a two-way street (in the direction of traffic for that lane) to perform the service.

**23.** For the street network shown below, draw the graph that would be useful for finding an efficient route for checking parking meters. (*Hint:* Notice that not every sidewalk has a meter; see Figure 1.12.)

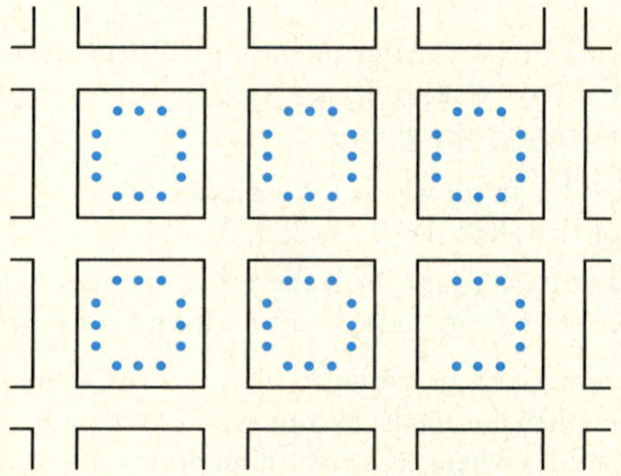

**24.** **(a)** For the street network in Exercise 23, draw a graph that would be useful for routing a garbage truck. Assume that all streets are two-way and that passing once down a street suffices to collect from both sides.

**(b)** Do the same problem on the assumption that one pass down the street suffices to collect from only one side.

**25.** **(a)** In the graph below, find the largest number of paths from $A$ to $F$ that do not have any edges in common.

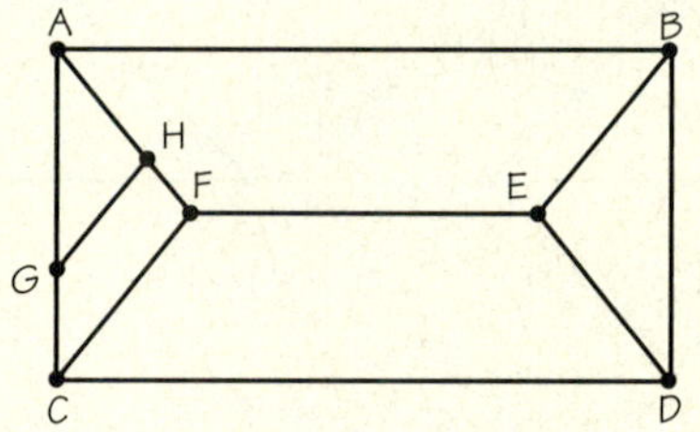


**(b)** Verify that the largest number of paths with no edges in common between any pair of vertices in this graph is the same.

**(c)** Why might one want to be able to design graphs such that one can move between two vertices of the graph using paths that have no edges in common?

**26.** Examine the paths represented by the numbered sequences of edges in both parts of the figure below. Determine whether each path is a circuit. If it is a circuit, determine if it is an Euler circuit.

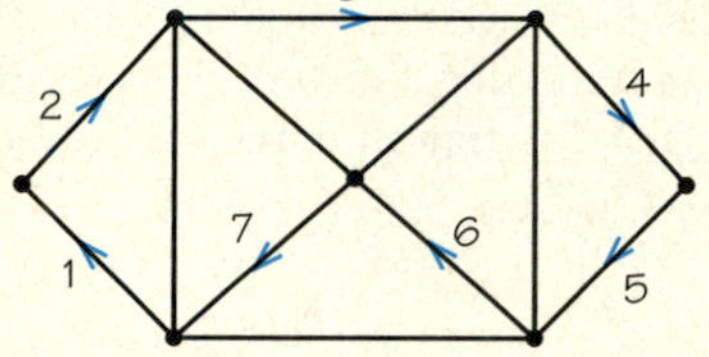


(*a*)

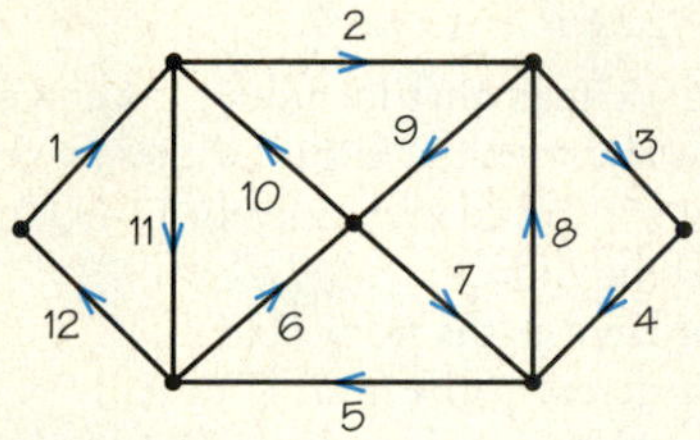


(*b*)

**27.** In Figure 1.13c, suppose we started an Euler circuit using this sequence of edges: 6, 7, 8, 9 (ignore existing arrows on the edges). What does our guideline for finding Euler circuits tell you *not* to do next?

**28.** In Figure 1.8b, suppose we started an Euler circuit using this sequence of edges: 14, 13, 8, 1, 4 (ignore existing arrows on the edges). What does our guideline for finding Euler circuits tell you *not* to do next?

**29.** Find an Euler circuit on the graph of Figure 1.15c (including the blue edges).

**30.** Find Euler circuits in the right-hand graphs in Figures 1.17a and 1.17b.

**31.** In the following graph, we see a territory for a parking-control officer that has no Euler circuit. How many sidewalks (edges) need to be omitted in order to enable us to find an Euler circuit? What effect would this have in the associated real-world situation?

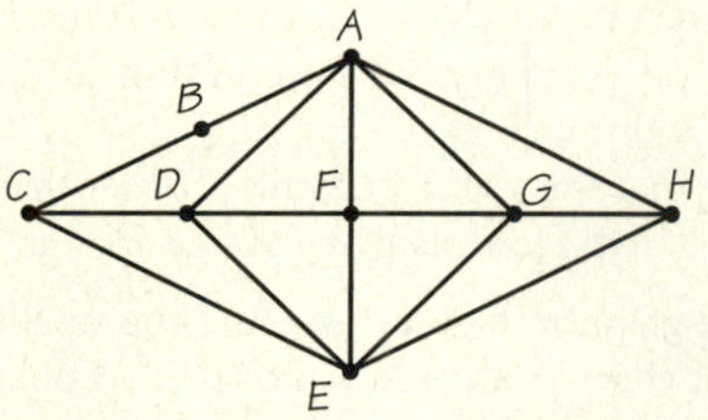


**32.** An Euler circuit visits a four-valent vertex $X$, such as the one in the accompanying graph, by using the edges $AX$ and $XB$ consecutively, and then using $CX$ and $XD$ consecutively. When this happens, we say that the Euler circuit cuts through at $X$.

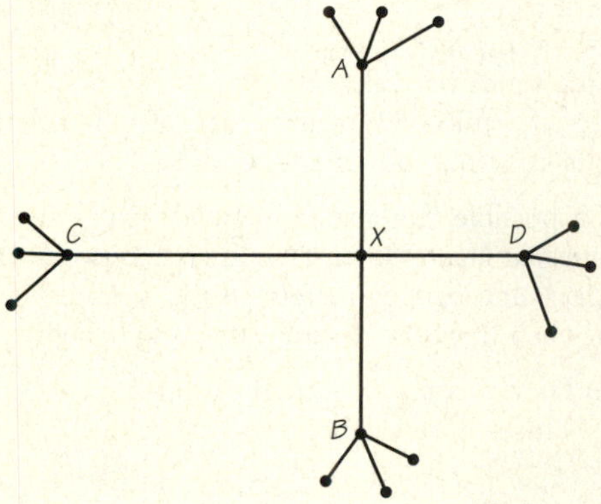

Suppose $G$ is a four-valent graph such as that in the diagram below. Is it possible to find an Euler circuit of this graph that never cuts through any vertex? Explain why it might be desirable to find an Euler circuit of this special kind in an applied situation.

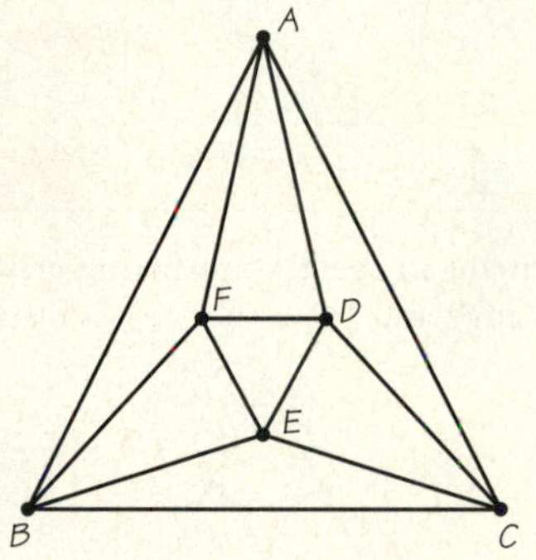


## 1.3 Beyond Euler Circuits

## 1.4 Urban Graph Traversal Problems

**33.** Find an Euler circuit on the eulerized graph (b) of the following figure. Use it to find a circuit on the original graph (a) that covers all edges and reuses edges only five times. Can fewer than five reused edges be achieved?

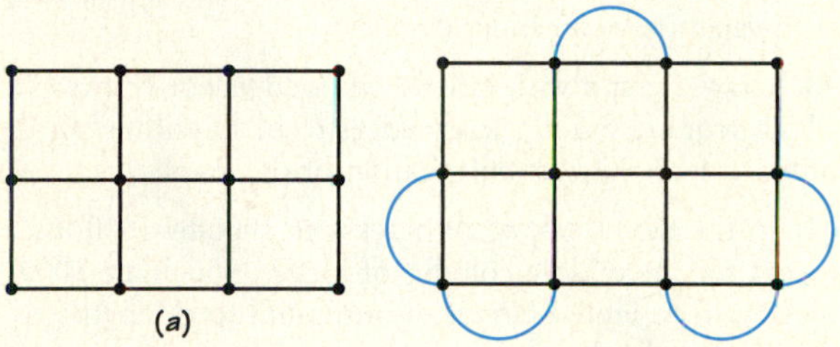

(*a*) (*b*)

**34.** Squeeze the circuit shown in graph (a) below onto graph (b). Show your answers by writing numbered arrows on the edges and by listing a sequence of vertices (for example, $ABEB \ldots A$).

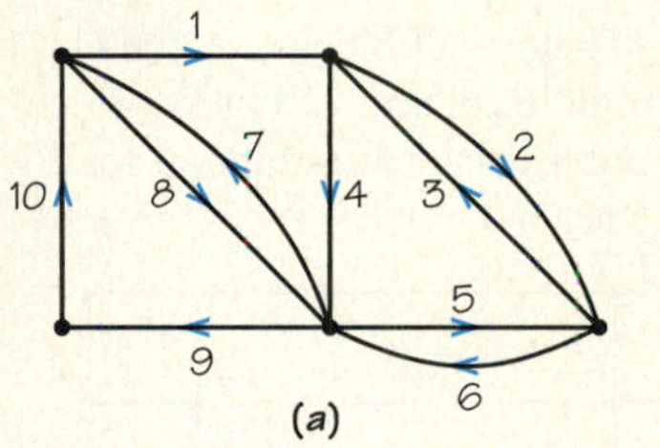


(*a*)

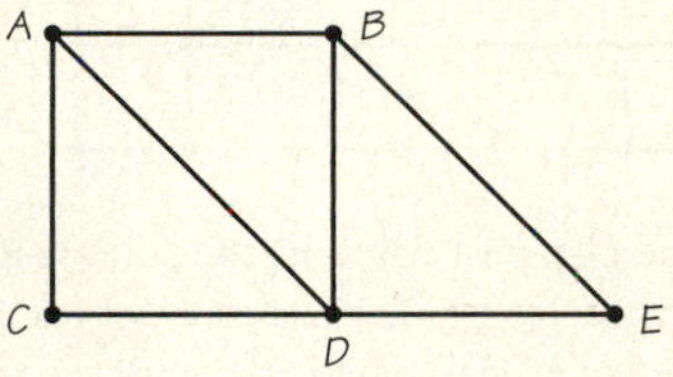


(*b*)

Then squeeze the circuit shown in graph (c) onto graph (d). Show your answers by writing numbered arrows on the edges and by listing a sequence of vertices.

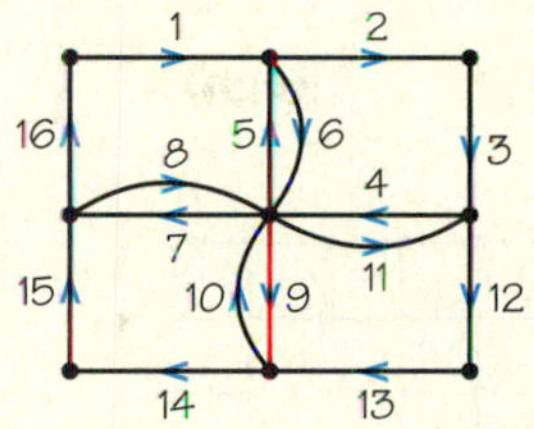


(*c*)

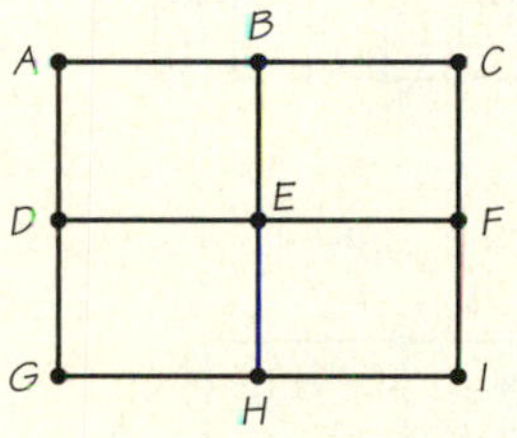


(*d*)

**35.** A college campus has a central square with sides arranged as shown by the edges in the graph below. Show how all these sidewalks can be traversed at least once in a tour that starts and ends at the same vertex.

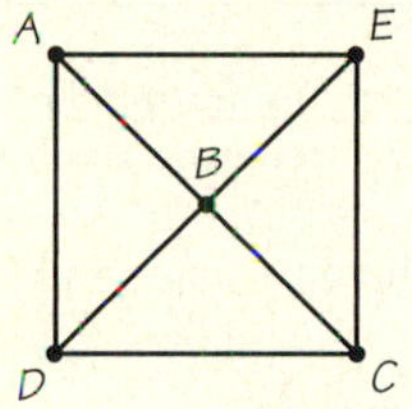


**36.** In the graph below, add one or more edges to produce a graph that has an Euler circuit.

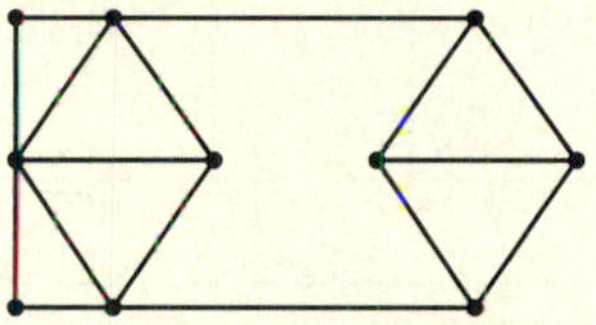

**37.** Eulerize these rectangular street networks using the same patterns that would be used by the edge walker described in the text.

**(a)** A $5 \times 5$ rectangle
**(b)** A $4 \times 5$ rectangle
**(c)** A $6 \times 6$ rectangle
**(d)** Can you find an eulerization with nine added edges for a 2-by-7-block rectangular street network? Can you do better than nine added edges?

**38.** Find good eulerizations for the following graphs, using as few duplicated edges as you can. See "Finding Good Eulerizations" for hints.

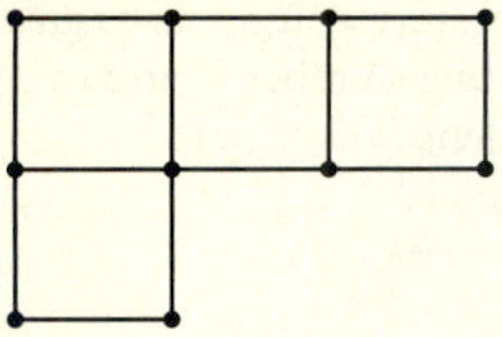

(*a*)

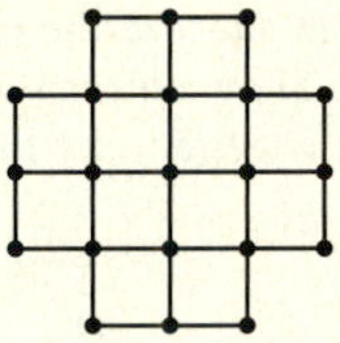

(*b*)

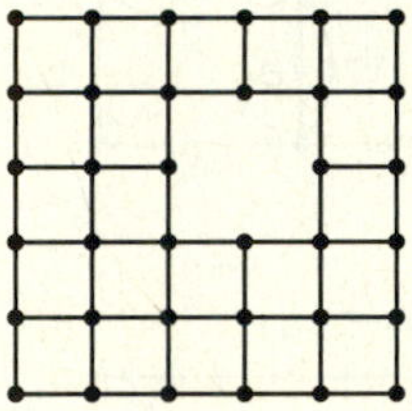

(*c*)

**39.** For the following graph:

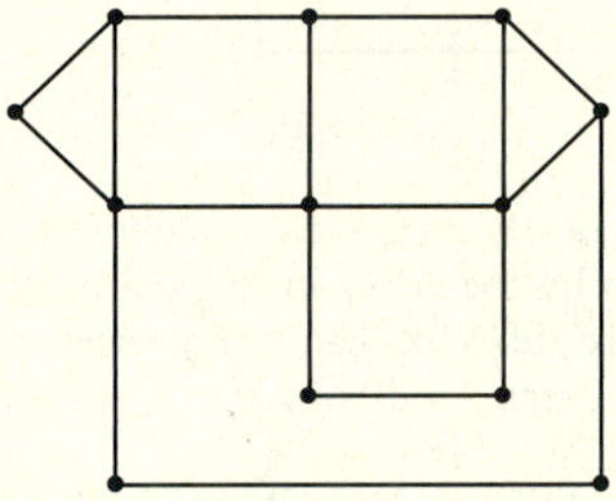

**(a)** Determine the minimum number of edges that have to be removed for the resulting graph to have all even-valent vertices.

**(b)** Does the graph you obtain in part (a) have an Euler circuit?

For the graph below:

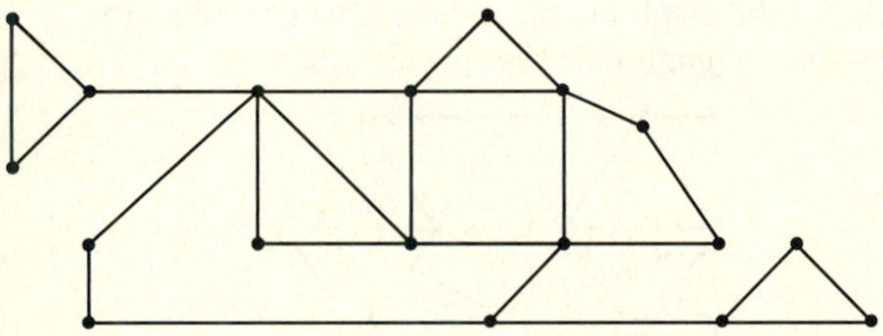

**(c)** Determine the minimum number of edges that have to be removed for the resulting graph to have all even-valent vertices.

**(d)** Does the graph you obtain in part (c) have an Euler circuit?

**40.** The following figure shows a river, some islands, and bridges connecting the islands and riverbanks. A charity is sponsoring a race in which entrants have to start at *A*, go over each bridge at least once, and end at *A*. Draw a graph that would be useful for finding a route that requires the least recrossing of bridges. Show what that route would be. (*Historical note:* This situation resembles the one that inspired Leonhard Euler's 1736 "recreational mathematics" problem that resulted in the first work in graph theory.)

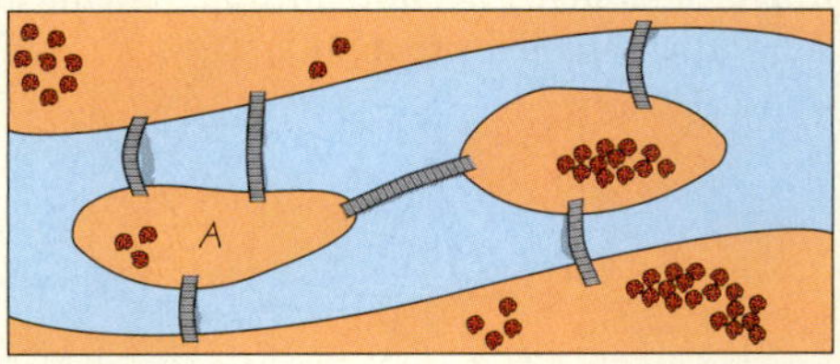


**41.** Find a circuit in the accompanying graph that covers every edge and has as few reuses as possible.

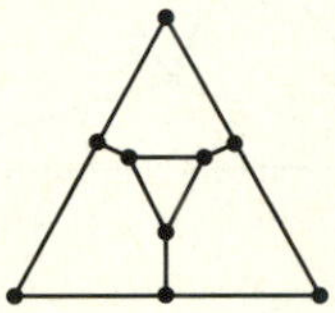

**42.** **(a)** Discuss the difference between the problem of:

**(i)** Adding the minimum number of edges to a graph to make all its vertices even-valent, and

**(ii)** Finding the best eulerization of a connected graph.

**(b)** In (i) must the graph that results from adding a minimum number of edges to make all the vertices even-valent have an Euler circuit?

**43.** Draw a graph with exactly two odd-valent vertices which requires exactly seven edges to be duplicated in order to find the best eulerization of the graph.

**44.** In the figure below, all blocks are 1000-by-1000 feet, except for the middle column of blocks, which are 1000-by-4000 feet. Find a circuit of minimum total length that covers all edges.

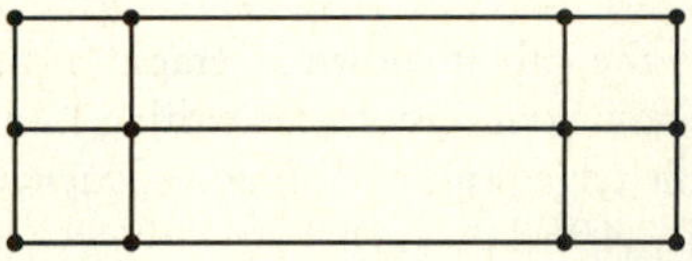

**45.** In the figure below, all blocks are 1000-by-1000 feet, except for the middle column of blocks, which are 1000-by-4000 feet. Find a circuit of minimum total length that covers all edges.

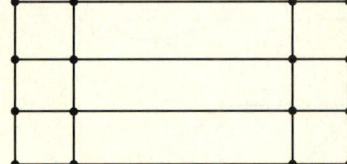

**46.** **(a)** Find the cheapest route in the following graph, where one starts at vertex *A*, finishes at vertex *A*, and traverses each edge at least once. The cost of a route is computed by summing the numbers along the edges that one uses.

**(b)** How many edges are repeated in the minimal-cost route?

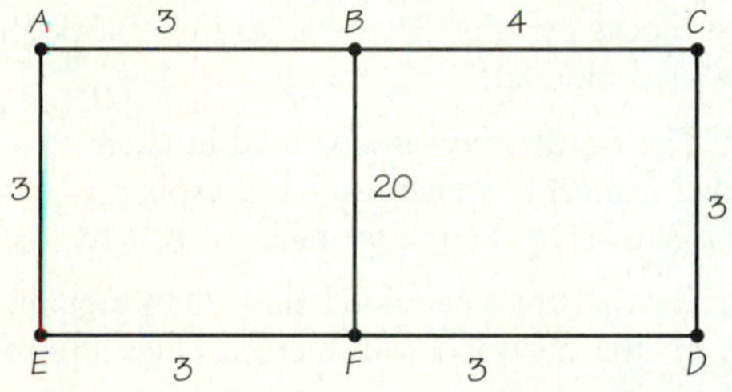


**(c)** Discuss the implications of this example for the relation between finding good eulerizations of graphs and the problem of finding cheap routes that start and end at the same vertex and traverse each edge at least once.

**(d)** The physical edge with cost 20 in the diagram is not physically longer than other edges with lower costs attached to them. Explain why in an urban setting it might make sense to assign two stretches of street of similar length very different "costs" for traversing them.

**(e)** What are some different meanings that "weights" (for example, traffic volume) potentially assigned to edges in a graph might have in an urban setting?

**47.** Which graphs (see figures below) have Euler circuits? In the ones that do, find the Euler circuits by numbering the edges in the order the Euler circuit uses them. For the ones that don't, explain why no Euler circuit is possible.

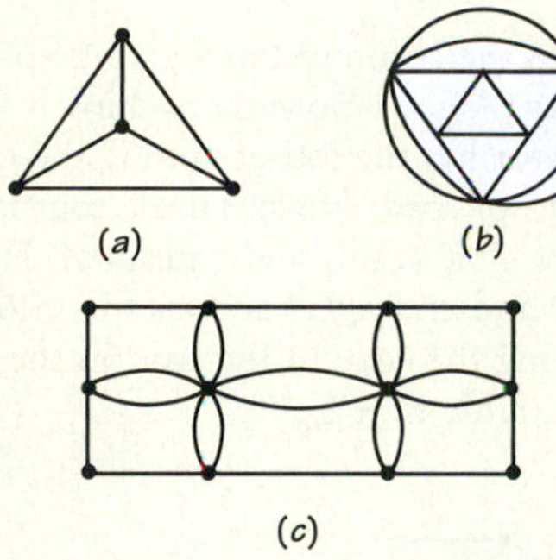


**48.** Eulerize the graph below by using four new edges. Find an Euler circuit in the eulerized graph and use that circuit to find a circuit of the original graph that covers all edges but reuses edges only four times. How many different ways can the four edges be chosen?

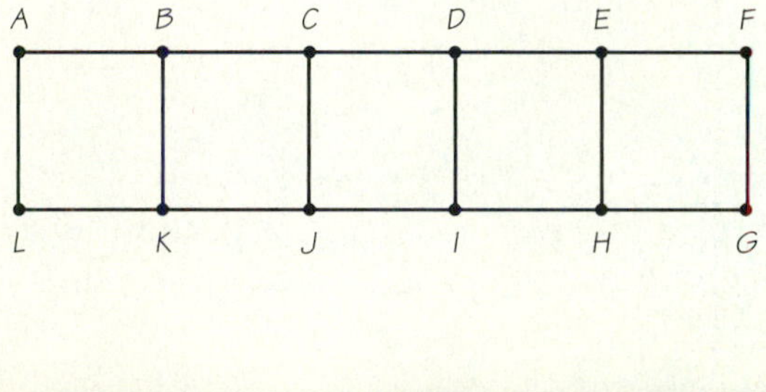


■ **49.** A graph *G* represents a street network to be traveled by a postal worker who must traverse every street twice, once for each side of the street. In graph *G*, the edges represent sidewalks. Does such a graph always have an Euler circuit? Explain your answer.

**50.** In the graph below, find a circuit that covers every edge and has as few reuses as possible.

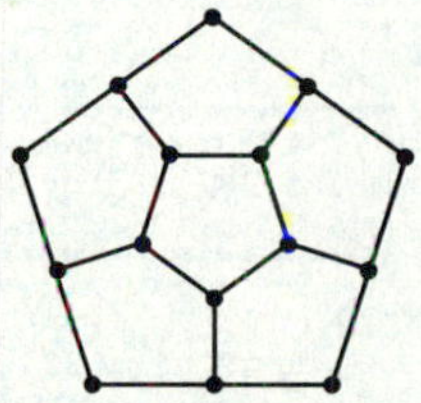

**51. (a)** Find the best eulerizations you can for the two graphs below.

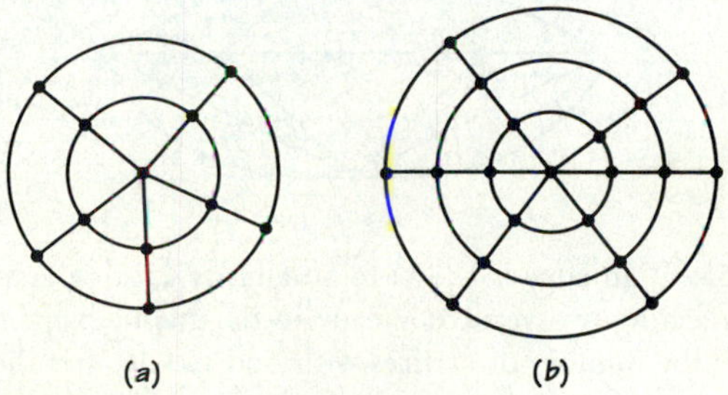


**(b)** Graph (a) can be thought of as having five rays and two circles, and graph (b) as having six rays and three circles. Draw a graph with four rays and four circles and find the best eulerization you can for this graph.

**(c)** Find a "formula" involving *r* and *s* for the smallest number of edges needed to eulerize a graph of this type having *r* rays and *s* circles.

■ **52.** Suppose that for a certain connected graph it is possible to disconnect it by removing one edge. Explain why such a graph (before the edge is removed) must have at least one vertex of odd valence. (*Hint:* Show that it cannot have an Euler circuit.)

**53.** Can you draw a graph with six vertices where the valence of each vertex is 5?

**54.** Each of the following graphs represent the sidewalks to be cleaned in a fancy garden (one pass over a sidewalk will clean it). Can the cleaning be done using an Euler circuit? If so, show the circuit in each graph by numbering the edges in the order the Euler circuit uses them. If not, explain why no Euler circuit is possible.

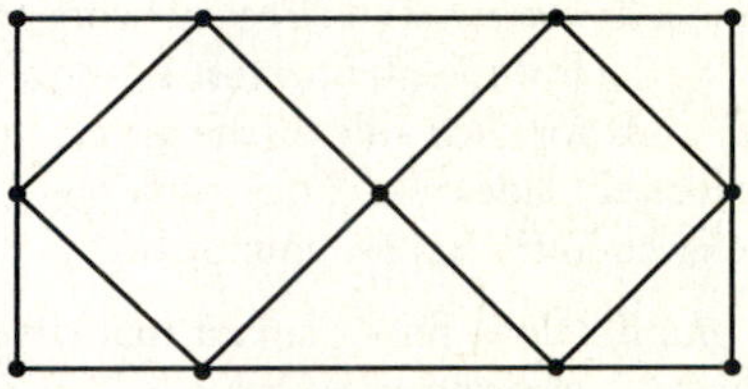

(a)

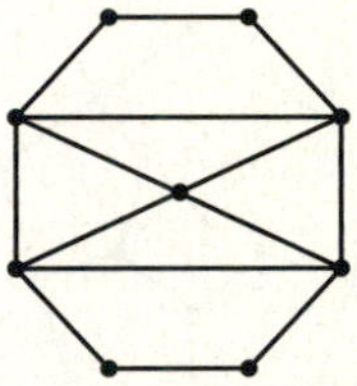

(b)

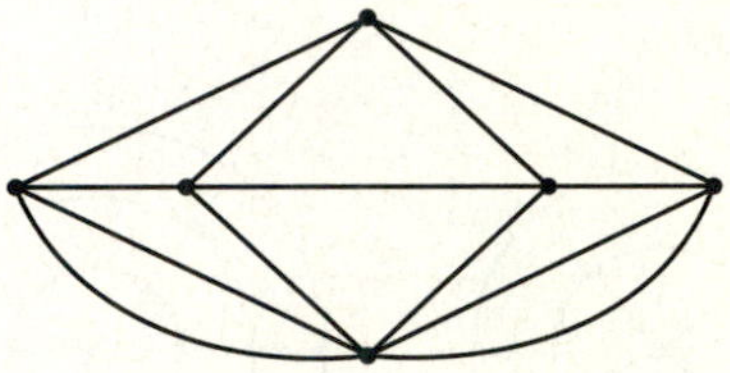

(c)

■ **55.** If an edge is added to an already existing graph, connecting two vertices already in the graph, explain why the number of vertices with odd valence has the same parity before and after. (This means if it was even before, it is even after, while if it was odd before, it remains odd.)

■ **56.** Any graph can be built in the following fashion: Put down dots for the vertices, then add edges connecting the dots as needed. When you have put down the dots, and before any edges have been added, is the number of vertices with odd valence an even number or an odd number? What is the number of vertices with odd valence when all the edges have been added (see Exercise 55)?

**57.** Draw the graph for the parking-control territory shown in the figure below. Label each vertex with its valence and determine if the graph is connected.

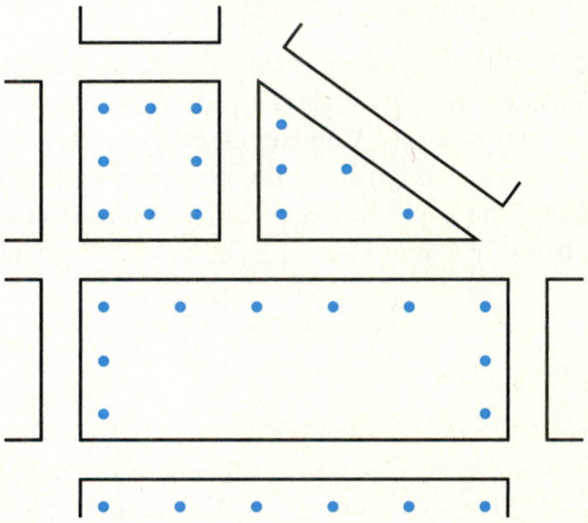

**58.** If a rectangular street network is $r$ blocks by $s$ blocks, find a formula for the minimum number of edges that must be added to eulerize a graph representing the network in terms of $r$ and $s$. (*Hint:* Treat the case $r = 1$ separately. Test your formula with the cases 6 blocks by 5 blocks, 6 blocks by 6 blocks, and 5 blocks by 3 blocks.)

◆ **59.** The word *valence* is also used in chemistry. Find out what it means in chemistry and explain how this usage is similar to the use we make of it here.

■ **60.** For the street network below, draw a graph that represents the sidewalks with meters. Then find the minimum-length circuit that covers all sidewalks with meters. If you drew the graph as we recommended, you would find that the shortest circuit has length 18 (it reuses every edge).

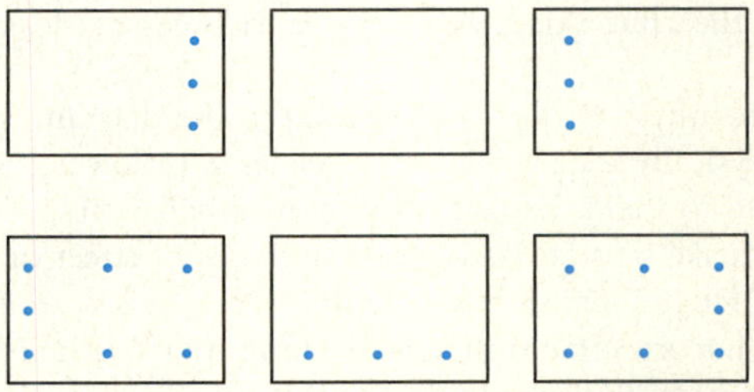

But the meter checker comes to you and says: "I don't know anything about your theories, but I have found a way to cover the sidewalks with meters using a circuit of length 10. My trick is that I don't rule out walking on sidewalks with no meters." Explain what he means and discuss whether his strategy can be used in other problems.

**61.** Each edge of the accompanying graph represents a two-lane highway. A grass-mowing machine is located at *A*, and its operator has the job of cutting the grass along each of the edges of road shown. Find a tour for the mowing machine that begins and ends at *A*. Find such a tour that begins and ends at *A* and, as the mowing is done, moves along the edge of the road in the same direction as the traffic is going.

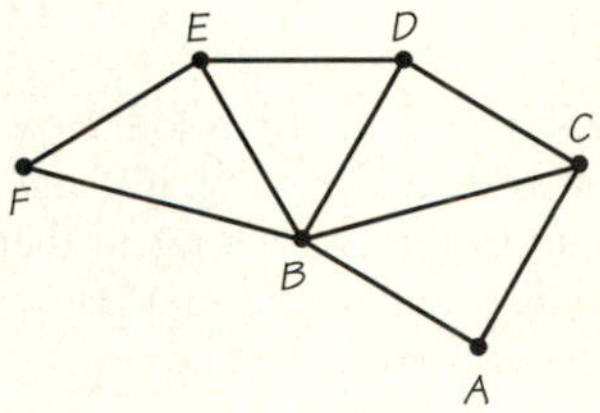

## APPLET EXERCISES

To do these exercises, go to www.whfreeman.com/fapp8e.

### Eulerizing a Graph

We learned that if a graph has exactly two vertices with odd valences, then an Euler circuit does not exist–but an Euler path does. It is also possible to produce an Euler circuit through the process of eulerization, by duplicating certain edges of the graph. But how many duplications are necessary to obtain an Euler circuit? Investigate this problem and more general related topics using the *Eulerizing a Graph* applet.

### Euler Circuits

We know that if all the vertices have even valence, then an Euler circuit exists. Try your hand at finding such circuits in the *Euler Circuit* applet.

## WRITING PROJECTS

**1.** Write a memo to your local department of parking control (or police department) in which you suggest that management science techniques like the ones in this chapter be used to plan routes. Assume that the person to whom you are writing is not extensively trained in mathematics but is willing to read through some technical material, provided you make it seem worth the trouble.

**2.** Do the same as in Writing Project 1, but to the department in charge of spreading salt on roads after snowstorms.

**3.** If you were making a recommendation to the mayor of New York City concerning proposed new street-sweeping routes, designed using the theory of this chapter, would you recommend that the changes be adopted or not? Write a memo that outlines the pros and cons as fairly as you can, and then conclude with your recommendation.

## SUGGESTED READINGS

BELTRAMI, EDWARD J., *Models for Public Systems Analysis,* Academic Press, New York, 1977. This book gives a good overview of the way that operations research has provided and continues to provide new tools for solving societal problems. Among the ideas discussed are police patrol tactics, organization of emergency services, and scheduling. Some of the mathematics used is advanced.

MALKEVITCH, JOSEPH, and WALTER MEYER, *Graphs, Models and Finite Mathematics,* Prentice-Hall, Englewood Cliffs, NJ, 1974. This introductory book includes much of the same material as presented here but provides more details of the proofs and uses somewhat different algorithms for solving the problems involved.

The following books treat many of the topics discussed here as well as shortest-path problems and matching problems, and they formulate some problems in more realistic terms:

ROBERTS, FRED S., AND BARRY TESMAN, *Applied Combinatorics,* Second Edition, Pearson Prentice Hall, Upper Saddle River, NJ, 2004.

TUCKER, ALAN. *Applied Combinatorics,* Third Edition, Wiley, New York, 1995.

## SUGGESTED WEB SITES

**www.hsor.org/what_is_or.cfm** This site discusses the history of operations research (OR) and some of the areas where OR is being applied. Be sure to follow the "Networks Routing" link to see applications of the Chinese postman problem.

**www.geom.uiuc.edu/~doty/applications** This Web page provides some examples of how to apply Euler circuits.

**www-gap.dcs.st-and.ac.uk/~history/Mathematicians/Euler.html** This essay discusses the numerous contributions that Euler made to mathematics, and provides biographical information about him.

**www.ams.org/featurecolumn/archive/urban-geom.html** This Web page includes an introduction to how graph theory has provided tools for urban operations research.

# CHAPTER 2



# Business Efficiency

In the previous chapter, we saw that there was an easy way of telling whether a connected graph has a circuit that traverses each of the edges of a graph exactly once–for example, a route for a snowplow that covers the streets of a section of a town. However, the situation changes radically if we make a seemingly small change in the problem: When is it possible to find a route along distinct edges of a graph that visits each *vertex* once and only once in a simple circuit? Perhaps there has been a hurricane and it is important to check whether or not the storm sewers at every corner in town are clogged.

This problem is called the *Hamiltonian circuit problem,* and, like the Euler circuit problem, it is a graph theory problem. The Hamiltonian circuit problem has many applications. Suppose inspections or deliveries need to be made at each vertex (rather than along each edge) of a graph. An "efficient" tour of the graph would be a route that started and ended at the same vertex and passed through all the vertices without reuse, or repetition; that is, the route would be a **Hamiltonian circuit.** Such routes would be useful for inspecting traffic signals or for delivering mail to drop-off boxes, which hold heavy loads of mail so that urban postal carriers do not have to carry them long distances, or delivering Meals on Wheels to the elderly.

**Hamiltonian Circuit** DEFINITION

A tour, like the ones marked by wiggly edges in Figure 2.1, that starts at a vertex of a graph and visits each vertex once and only once, returning to where it started, is called a **Hamiltonian circuit**.

For example, the wiggly line in Figure 2.1a shows a circuit we can take to tour that graph, visiting each vertex once and only once. This tour can be written *ABDGIHFECA.* Note that another way of writing the same circuit would be *EFHIGDBACE.* A different circuit visiting each vertex once and only once would be *CDBIGFEHAC* (Figure 2.1b). Do not be confused because *C* is written twice when we write down this list of vertices. We can think of the circuit as starting at any of its vertices, but we do start and end at the same vertex.

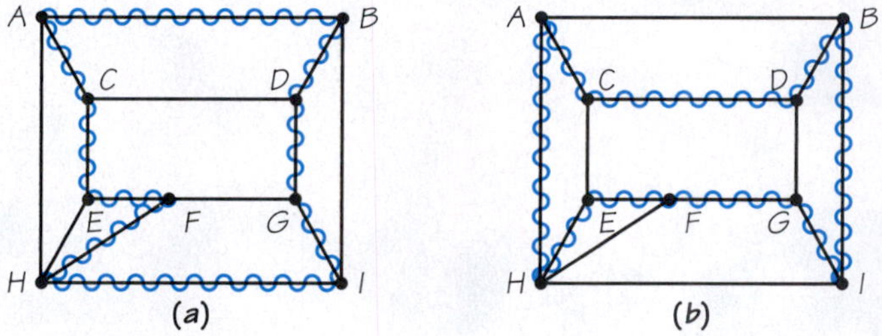


**FIGURE 2.1** Wiggly edges illustrate Hamiltonian circuits.

## 2.1 Hamiltonian Circuits

The concept is named for the Irish mathematician William Rowan Hamilton (1805–1865), who was one of the first to study it. We now know that the concept was discovered somewhat earlier by Thomas Kirkman (1806–1895), a British minister with a penchant for mathematics.

The concepts of Euler and Hamiltonian circuits are similar in that both forbid reuse. An Euler circuit forbids the reuse of edges, while a Hamiltonian circuit forbids the reuse of vertices. However, it is far more difficult to determine which connected graphs possess a Hamiltonian circuit than to determine which connected graphs have Euler circuits. As we saw in Chapter 1, looking at the valences of vertices tells us whether a connected graph has an Euler circuit, but we have no such simple method for telling whether or not a graph has a Hamiltonian circuit.

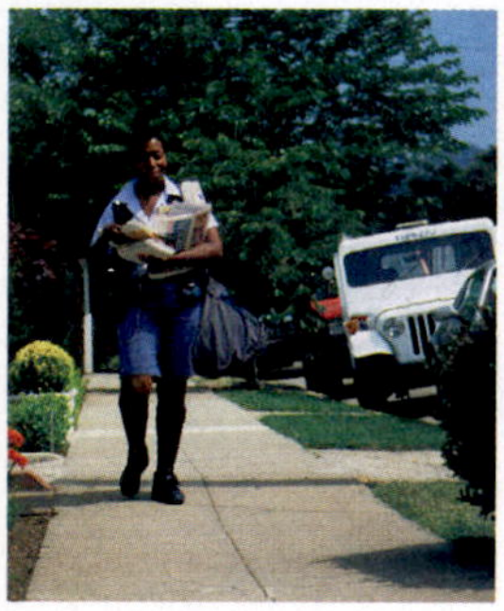

It may seem that delivering mail in a cheap and timely manner should not be that hard. However, finding the optimal way to deliver mail over a variety of environments, rural, suburban, and urban, is very complex. How should a large geographic area be divided into smaller sections? Should each mail carrier use a truck as a "depot" to resupply small amounts of mail for delivery or should there be deposit boxes on street corners? Mathematics can be used to find answers to such questions. (*Lawrence Migdale/Photo Researchers, Inc.*)

Some special classes of graphs are known to have Hamiltonian circuits, and some special classes of graphs are known to lack them. For example, here is a method for constructing an infinite family of graphs where each graph in the family cannot have a Hamiltonian circuit. Construct a vertical column of $m$ vertices and a parallel column of $n$ vertices, where $m$ is bigger than $n$, as shown in Figure 2.2a. The figure illustrates a typical case where $m = 4$ and $n = 2$. Now join each vertex on the left in the figure to every vertex on the right. As $m$ and $n$ vary, we get a family of different graphs.

No graph obtained in this manner can have a Hamiltonian circuit. If a Hamiltonian circuit existed, it would have to alternately include vertices on the left and right of the figure. This is not possible because the number of vertices on the left and right are not the same. It is unlikely that a method will ever be found to easily determine whether or not a graph has a Hamiltonian circuit. If Hamiltonian circuits were easy to find in any graph at all, many applied problems could be solved in a less costly way.

In many urban operations research situations "grid graphs" such as the one in Figure 2.2b are of interest. If we wanted an efficient route (circuit) to inspect traffic surveillance cameras located at urban street intersections, we would need to find a Hamiltonian circuit for the graph in Figure 2.2b. Note that in going from one vertex to another we move from a vertex of one color to a vertex of the other color. Since colors would alternate in a Hamiltonian circuit, it follows that the number of vertices of each color would have to be the same if there is a Hamiltonian circuit in this graph. Since the number of vertices of the two colors is *not* the same, there is no Hamiltonian circuit and, hence, no fully efficient route for inspecting the traffic control cameras.

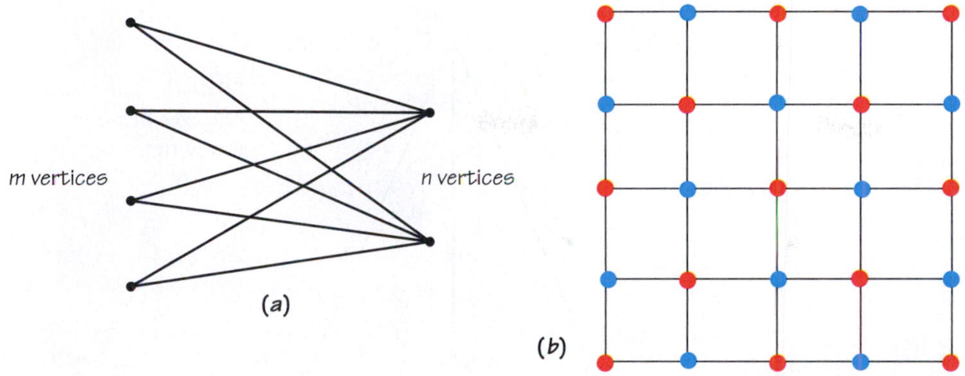


**FIGURE 2.2** (a) An example of one graph from a family of graphs that has no Hamiltonian circuit. The number of vertices $m$ on the left is chosen to be greater than the number of vertices $n$ on the right. The case $m = 4$ and $n = 2$ is shown. (b) A graph used to model a portion of a city. Since the graph reflects the block structure of the city, it is known as a "grid graph."

The Hamiltonian circuit problem itself has many applications. This is not unusual in mathematics. Often mathematics used to solve a particular real-world problem leads to new mathematics that suggests applications to other real-world situations. One class of problems to which we can apply Hamiltonian circuits is vacation planning.

## EXAMPLE 1 ■ Vacation Planning

Let's imagine that you are a college student studying in Chicago. During spring break you and a group of friends have decided to take a car trip to visit other friends in Minneapolis, Cleveland, and St. Louis. There are many choices as to the order of visiting the cities and returning to Chicago, but you want to design a route that minimizes the distance you have to travel. Presumably, you also want a route that cuts costs, and you know that minimizing distance will minimize the cost of gasoline for the trip. Similar problems with different complications would arise for bus, railroad, or airplane trips.

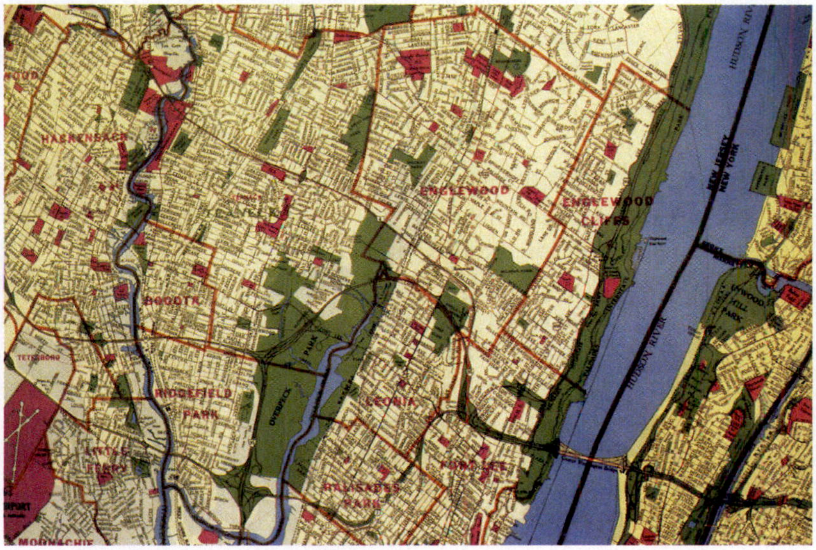


Express mail and parcel post delivery companies need to make complicated patterns of deliveries and pick-ups. To do this they need to know driving distances between the various geographical locations involved. Using this information, together with driving times, they can use mathematics to cut costs and to make the pick-ups and deliveries on time. (© *Rhoda Sidney/PhotoEdit.*)

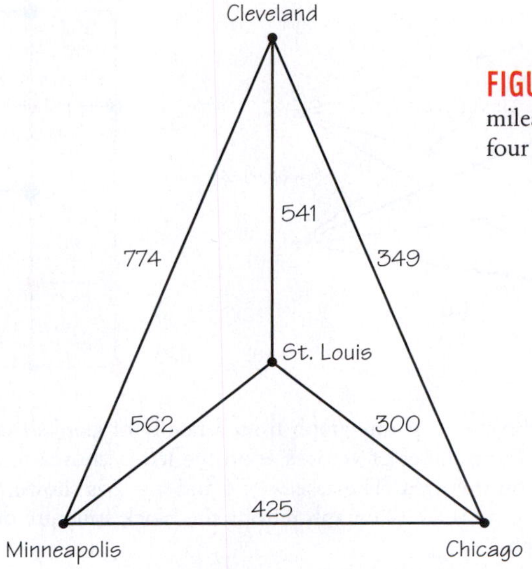


**FIGURE 2.3** Road mileages between four cities.

Imagine now that the local automobile club has provided you with the intercity driving distances between Chicago, Minneapolis, Cleveland, and St. Louis. We can construct a graph model with this information, representing each city by a vertex and the legs of the journey between the cities by edges joining the vertices. To complete the model, we add a number called a **weight** to each graph edge, as in Figure 2.3. In this example, the weights represent the distances between the cities, each of which corresponds to one of the endpoints of the edges in the graph. (In other examples the weight might represent a cost, time, satisfaction rating, or profit.) We want to find a minimum-cost tour that starts and ends in Chicago and visits each other city only once. Using our earlier terminology, what we wish to find is a **minimum-cost Hamiltonian circuit**–a Hamiltonian circuit with the lowest possible sum of the weights of its edges.

## Finding a Minimum-Cost Hamiltonian Circuit — PROCEDURE

How can we determine which Hamiltonian circuit has minimum cost? There is a conceptually easy **algorithm**, or mechanical step-by-step process, for solving this problem:

1. Generate all possible Hamiltonian tours (starting from Chicago).
2. Add up the distances on the edges of each tour.
3. Choose a tour with total distance being a minimum, that is, as small as possible.

Steps 2 and 3 of the algorithm are straightforward. Thus, we need worry only about Step 1, generating all the possible Hamiltonian circuits in a systematic way. To find the Hamiltonian tours, we will use the **method of trees**, as follows. Starting from Chicago, we can choose any of the three cities to visit after leaving Chicago. The first stage of the enumeration tree is shown in Figure 2.4. If Minneapolis is chosen as the first city to visit, then there are two possible cities to visit next, Cleveland and St. Louis. The possible branchings of the **tree** at this stage are shown in Figure 2.5. In this second stage, however, for each choice of first city to visit, there are two choices from this city to the second city to visit. This would lead to the diagram in Figure 2.6.

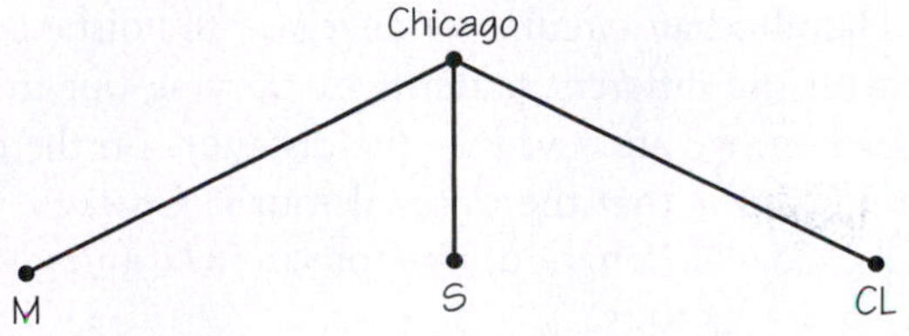


**FIGURE 2.4** First stage in finding vacation-planning routes.

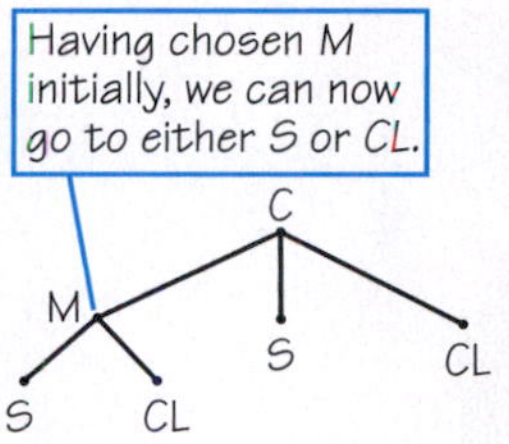


**FIGURE 2.5** Part of the second stage in finding vacation-planning routes.

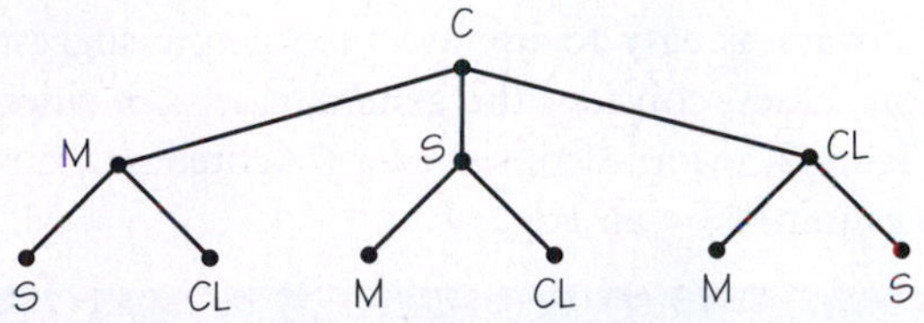


**FIGURE 2.6** Complete second stage in finding vacation-planning routes.

Having chosen the order of the first two cities to visit, and knowing that no revisits (reuses) can occur in a Hamiltonian circuit, there is only one choice left for the next city. From this city we return to Chicago. The complete tree diagram showing the third and fourth stages for these routes is given in Figure 2.7. Notice, however, that because we can traverse a circular tour in either of two directions, the paths shown in the tree diagram of Figure 2.7 do not correspond to different Hamiltonian circuits. For example, the leftmost path (C–M–S–CL–C) and the rightmost path (C–CL–S–M–C) represent the same Hamiltonian circuit. Thus, among what appear to be six different paths in the tree diagram, only three in fact correspond to different Hamiltonian circuits. These three distinct Hamiltonian circuits are shown in Figure 2.8.

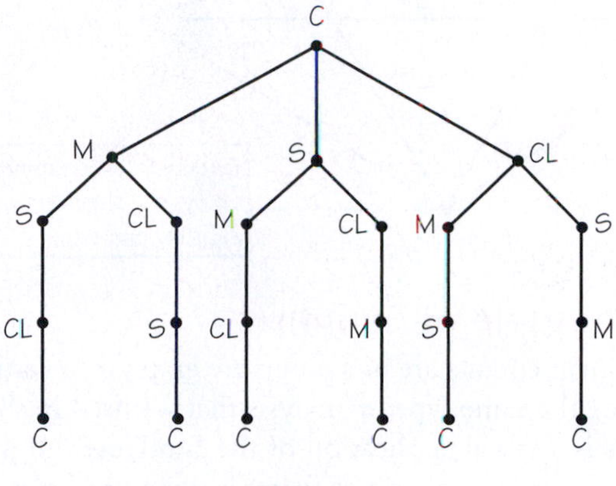


**FIGURE 2.7** Completed enumeration of routes using the method of trees for the vacation-planning problem.

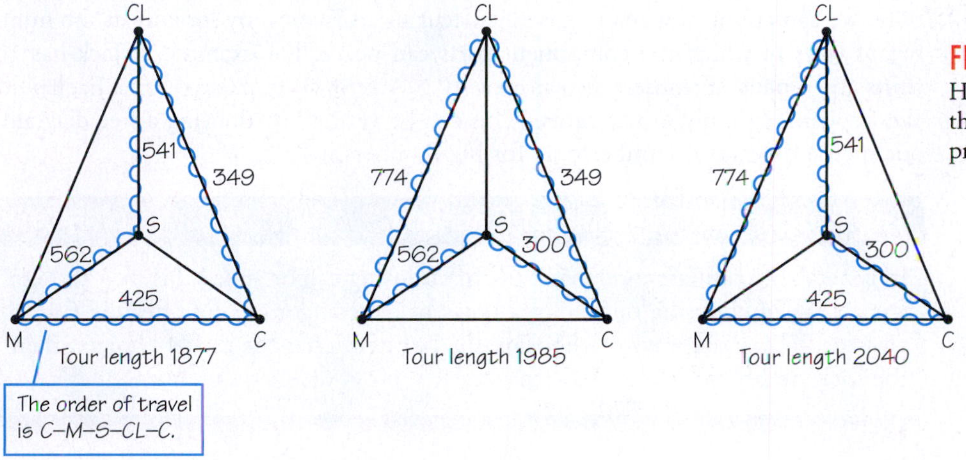


**FIGURE 2.8** The three Hamiltonian circuits for the vacation-planning problem of Figure 2.3.

Note that in generating the Hamiltonian circuits we disregard the distances involved. We are concerned only with the different patterns of carrying out the visits. To find the optimal route, however, we must add up the distances on the edges to get each tour's length. Figure 2.8 shows that the optimal tour is Chicago, Minneapolis, St. Louis, Cleveland, Chicago. The length of this tour is 1877 miles, which saves 163 miles over the longest choice of tour.

The method of trees is not always as easy to use as our example suggests. Instead of doing our analysis for four cities, consider the general case of $n$ cities. The graph model similar to that in Figure 2.3 would consist of a weighted graph with $n$ vertices, with every pair of vertices joined by an edge.

### Complete Graph

DEFINITION

A graph is called **complete** if there is exactly one edge between each pair of vertices in the graph.

A complete graph with five vertices is illustrated in Figure 2.9. The graph in Figure 2.3 is a weighted complete graph with four vertices.

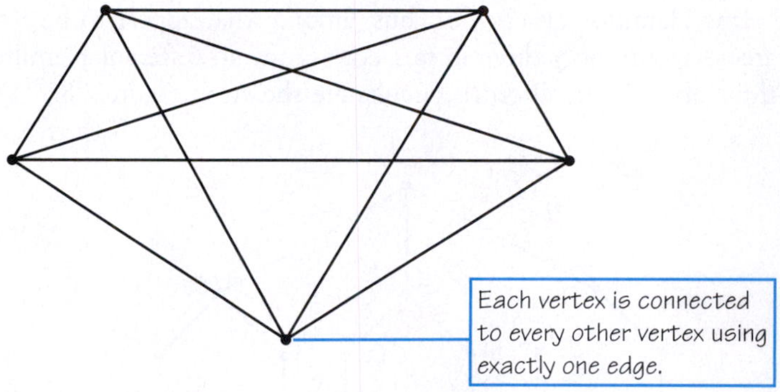

Each vertex is connected to every other vertex using exactly one edge.


**FIGURE 2.9** A complete graph with five vertices. Every pair of vertices is joined by an edge.

## Fundamental Principle of Counting

How many Hamiltonian circuits are in a complete graph of $n$ vertices? We can solve this problem by using the same type of analysis that we used in the method of trees. The method of trees is a visual application of the **fundamental principle of counting**, a procedure for counting outcomes in multistage processes. Using this procedure, we can count how many patterns occur in a situation by looking at the number of ways in which the component parts can occur. For example, if Jack has 10 shirts and 4 pairs of trousers, he can wear $10 \times 4 = 40$ shirt–pants outfits. Each shirt can be worn with any of the pants. (This can be verified by drawing a tree diagram, but such a diagram is cumbersome for big numbers.)

### The Fundamental Principle of Counting

DEFINITION

In general, the **fundamental principle of counting** can be stated this way. If there are $a$ ways of choosing one thing, $b$ ways of choosing a second after the first is chosen, . . . , and $z$ ways of choosing the last item after the earlier choices, then the total number of choice patterns is $a \times b \times c \times \cdots \times z$.

## EXAMPLE 2 ■ Counting

Here are some examples of how to use the fundamental principle of counting:

1. In a restaurant there are 4 kinds of soup, 12 entrees, 6 desserts, and 3 drinks. How many different four-course meals can a patron choose? The four choices can be made in 4, 12, 6, and 3 ways, respectively. Hence, applying the fundamental principle of counting, there are $4 \times 12 \times 6 \times 3 = 864$ possible meals.
2. In a state lottery a contestant gets to pick a four-digit number that does not contain a zero followed by an uppercase or lowercase letter. How many such sequences of digits and a letter are there? Each of the four digits can be chosen in 9 ways (that is, 1, 2, . . . , 9), and the letter can be chosen in 52 ways (that is, A, B, . . . , Z plus a, b, . . . , z). Hence, there are $9 \times 9 \times 9 \times 9 \times 52 = 341{,}172$ possible patterns.
3. A corporation is planning a musical logo consisting of four different ordered notes from the scale C, D, E, F, G, A, and B. How many logos are there to choose from? The first note can be chosen in 7 ways, but because reuse is not allowed, the next note can be chosen in only 6 ways. The remaining two notes can be chosen in 5 and 4 ways, respectively. Using the fundamental principle of counting, $7 \times 6 \times 5 \times 4 = 840$ musical logos are possible. If reuse of notes is allowed, $7 \times 7 \times 7 \times 7 = 2401$ logos are possible.

Let's now return to the problem of enumerating Hamiltonian circuits for the complete graph with $n$ vertices. The city visited first after the home city can be chosen in $n - 1$ ways, the next city in $n - 2$ ways, and so on, until only one choice remains. Using the fundamental principle of counting, there are $(n - 1)! = (n - 1)(n - 2) \times \cdots \times 3 \times 2 \times 1$ routes. The exclamation mark in $(n - 1)!$ is read "factorial" and is shorthand notation for the product $(n - 1)(n - 2) \times \cdots \times 3 \times 2 \times 1$. For example, $5! = 5 \times 4 \times 3 \times 2 \times 1 = 120$.

As we saw in Figure 2.7, pairs of routes correspond to the same Hamiltonian circuit because one route can be obtained from the other by traversing the cities in reverse order. Thus, although there are $(n - 1)!$ possible routes, there are only half as many, or $(n - 1)!/2$, different Hamiltonian circuits. Now, if we have only a few cities to visit, $(n - 1)!/2$ Hamiltonian circuits can be listed and examined in a reasonable amount of time. Analysis of a six-city problem would require generation of $(6 - 1)!/2 = 5!/2 = 120/2 = 60$ tours. But for, say, 25 cities, $24!/2$ is approximately $3 \times 10^{23}$. Even if these tours could be generated at the rate of 1 million a second, it would take 10 billion years to generate them all. Because it would take so long to solve large vacation-planning problems using this method, it is sometimes referred to as a **brute force method** (that is, trying all the possibilities). Computer scientists and engineers have made it possible to market faster and faster computers. However, governments and businesses need to solve larger scale problems; say, for example, finding a Hamiltonian circuit in a graph with 10,000 vertices. If the methods one knows for solving such problems are not much better than brute force, then it's unlikely that even these faster computers can solve large versions of such problems. Mathematicians and computer scientists are actively seeking procedures that will significantly improve our ability to solve large versions of important problems.

## 2.2 Traveling Salesman Problem

If the only benefit were saving money and time in vacation planning, the difficulty of finding a minimum-cost Hamiltonian circuit in a complete graph with $n$ vertices for large values of $n$ would not be of great concern. However, the problem we are discussing is one of the most common in *operations research,* the branch of mathematics concerned with getting governments and businesses to operate more efficiently. This problem is usually called the **traveling salesman problem (TSP)** because of its early formulation: Determine the trip of minimum cost that a salesperson can make to visit the cities in a sales territory, starting and ending the trip in the same city.

Many situations require solving a TSP:

1. A lobster fisherman has set out traps at various locations and wishes to pick up his catch.
2. The telephone company wishes to pick up the coins from its pay telephone booths. (To avoid the high cost of picking up these coins, phone companies in many countries have adopted a system that uses prepurchased phone cards to operate phones. This means that there are no coins to collect.) Due to the increased use of cell phones, fewer pay phones are available.
3. The electric (or gas) company needs to design a route for its meter readers.
4. A minibus must pick up six day campers, deliver them to camp, and return them home later in the day.
5. In drilling holes in a series of plates, the drill press operator (perhaps a robot) must drill the holes in a predetermined order.
6. Physical records generated at automated teller machine (ATM) locations–as backup in case of failure of the electronic systems–must be picked up periodically.
7. A limousine service with a van located at an airport must pick up five customers and deliver them to the airport in time to catch their flights.

Perhaps surprisingly, TSP problems are also solved regularly in the design of computer chips. The components must be located so that the machines involved in the assembly can insert them on the chips as efficiently as possible. Because many chips are manufactured, even a small improvement in the time needed to make a chip can save a lot of money.

The meaning of *cost* can vary from one formulation of a TSP to another. We can measure cost as distance, airplane ticket prices, time, or any other factor that is to be optimized. In many situations, the TSP arises as a subproblem of a more complicated problem. For example, a supermarket chain may have a very large number of stores to be served from a single large warehouse. If there are fewer trucks than stores, the stores must be grouped into clusters so that one truck can serve each cluster. If we then solve the TSP for every truck, we can minimize total costs for the supermarket chain. Similar vehicle-routing problems–for dial-a-ride services that transport senior citizens to activity centers, for example, or that deliver children to their schools or camps–often involve solving the TSP as a subproblem.

## 2.3 Helping Traveling Salesmen

Because the traveling salesman problem arises so often in situations where the associated complete graphs would be very large, we must find a faster method than the brute force method we have described. We need to look at our original problem in Figure 2.3 and try to find an alternative algorithm for solving it. Recall that our goal is to find the minimum-cost Hamiltonian circuit.

**Nearest-Neighbor Algorithm** PROCEDURE

Starting from the home city, first visit the nearest city, then visit the nearest city that has not already been visited. We return to the start city when no other choice is available. This approach is called the **nearest-neighbor algorithm**.

### EXAMPLE 3 ■ Applying the Nearest-Neighbor Algorithm

Applying this algorithm to the TSP in Figure 2.3 quickly leads to the tour of Chicago, St. Louis, Cleveland, Minneapolis, and Chicago, with a length of 2040 miles. Here is how this tour is determined. Because we are starting in Chicago, there is a choice of going to a city that is 425, 300, or 349 miles away. Because the smallest of these numbers is 300, we next visit St. Louis, which is the nearest neighbor of Chicago not already visited. At St. Louis, we have a choice of visiting next cities that are 541 or 562 miles away. Hence, Cleveland, which is nearer (541 miles), is visited. To complete the tour, we visit Minneapolis and return to Chicago, thereby adding 774 and 425 miles to the length of the tour. ■

The nearest-neighbor algorithm is an example of a **greedy algorithm**, because at each stage a best (greedy) choice, based on an appropriate criterion, is made. Unfortunately, this is not the optimal tour, which we saw was C–M–S–CL–C, for a total length of 1877 miles. Making the best choice at each stage may not yield the best "global" solution. However, even for a large TSP, one can always find a nearest-neighbor route quickly.

### EXAMPLE 4 ■ Applying the Nearest-Neighbor Algorithm Revisited

Figure 2.10 again illustrates the ease of applying the nearest-neighbor algorithm, this time to a weighted complete graph with five vertices. Starting at vertex *A*, we get the tour *ADECBA* (cost 2800) (Figure 2.10a). Note that the nearest-neighbor algorithm starting at vertex *B* yields the tour *BCADEB* (cost 3050) (Figure 2.10b). ■

This example illustrates that a nearest-neighbor tour can be computed for each vertex of the complete graph being considered and that different nearest-neighbor tours can be obtained starting at different vertices. Thus, even though we may seek a tour starting at a particular vertex–say, *A* in Figure 2.10–because a Hamiltonian circuit can be thought of as starting at any of its vertices, we can just as easily apply the nearest-neighbor procedure starting at vertex *B* (rather than at *A*). The Hamiltonian circuit we get can still be thought of as beginning at vertex *A* rather than *B*. Even for complete graphs with a large number of vertices, it would still be faster to

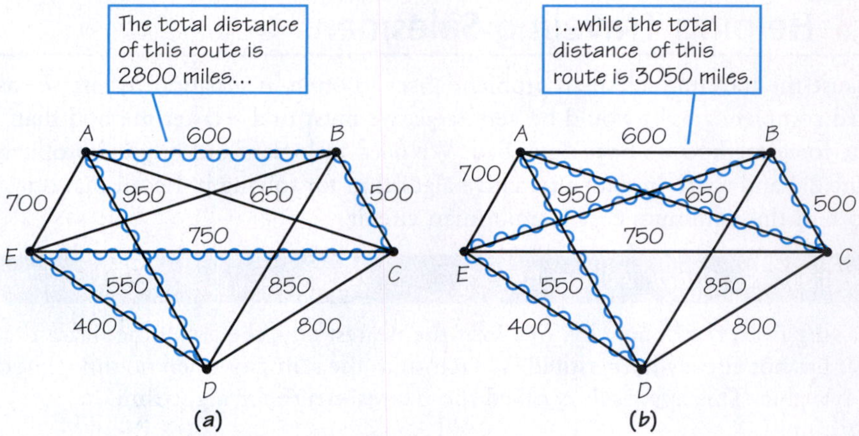


**FIGURE 2.10** (a) A weighted complete graph with five vertices that illustrates the use of the nearest-neighbor algorithm (starting at *A*). (b) TSP tour generated by the nearest-neighbor algorithm (starting at *B*).

apply the nearest-neighbor algorithm for each vertex and pick the cheapest of the tours generated (though such a tour might not be optimal) than to apply the brute force method.

**Sorted-Edges Algorithm** PROCEDURE

Start by sorting or arranging the edges of the complete graph in order of increasing cost (or, equivalently, arranging the intercity distances in order of increasing distance). Then at each stage select an edge that has not been previously chosen of least cost that (1) never requires that three used edges meet at a vertex (because a Hamiltonian circuit uses up exactly two edges at each vertex) and that (2) never closes up a circular tour that doesn't include all the vertices. This algorithm is called the **sorted-edges algorithm**.

## EXAMPLE 5 ■ Applying the Sorted-Edges Algorithm

Applying the sorted-edges algorithm to the TSP in Figure 2.3 works as follows: First, the six weights on the edges listed in increasing order would be 300, 349, 425, 541, 562, and 774. Because the cheapest edge in this sorted list is 300, this is the first edge we select for the tour we are building. Next we add the edge with weight 349 to the tour. The next-cheapest edge would be 425, but using this edge together with those already selected would result in having three edges at a vertex (Figure 2.11a), which is not consistent with having a Hamiltonian circuit. Hence, we do not use this edge. The next-cheapest edge, 541, used together with the edges already selected, would create a circuit (see Figure 2.11b) that does not include all the vertices. Thus, this edge, too, would be skipped over. However, we are able to add the edges 562 and 774 without either creating a circuit shorter than one including all the vertices or having three edges at a vertex. Hence, the tour we arrive at is Chicago, St. Louis, Minneapolis, Cleveland, and Chicago. Again, this solution is not optimal because its length is 1985. Note that this algorithm, like the nearest-neighbor, is greedy. ■

## EXAMPLE 6 ■ Applying the Sorted-Edges Algorithm Revisited

Although the edges selected by applying the sorted-edges method to the example in Figure 2.3 are connected to each other at every stage, this does not always happen.

For example, if we apply the sorted-edges algorithm to the graph in Figure 2.10a, we build up the tour first with edge *ED* (400) and then edge *BC* (500), which do not touch. The edges that are then selected are *AD*, *AB*, and *EC*, giving the circuit *EDABCE*, which is the same as the nearest-neighbor circuit starting at vertex *A*. ■

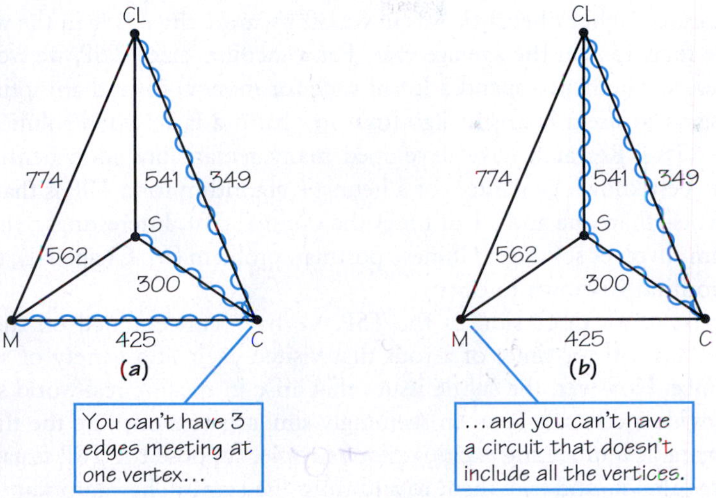


**FIGURE 2.11** (a) When three shortest edges are added in order of increasing distance, three edges at a vertex are selected, which is not allowed as part of a Hamiltonian circuit. (b) When the edges of distances 300, 349, and 541 are selected, a circuit that does not include all vertices results.

Many "quick-and-dirty" methods for solving the TSP have been suggested; while some methods give an optimal solution in some cases, none of these methods guarantees an optimal solution. Surprisingly, most experts believe that no efficient method that guarantees an optimal solution for the TSP will ever be found (see Spotlight 2.1).

## SPOTLIGHT 2.1 NP-Complete Problems

Steven Cook, a computer scientist at the University of Toronto, showed in 1971 that certain computational problems are equivalently difficult. This class of problems, now referred to as **NP-complete problems**, has the following characteristic: If a "fast" algorithm for solving one of these problems could be found, then a fast method would exist for all these problems.

In this context, "fast" means that as the size $n$ of the problem grows (the number of cities gives the problem size in the traveling salesman problem), the amount of time needed to solve the problem grows no more rapidly than a polynomial function in $n$. (A polynomial function has the form $a_k n^k + a_{k-1} n^{k-1} + \cdots + a_1 n + a_0$.) On the other hand, if it could be shown that any problem in the class of NP-complete problems required an amount of time that grows faster than any polynomial (an exponential function, such as $3^n$, is an example of a function that grows faster than any polynomial) as the problem size increased, then all problems in the NP-complete class would share this characteristic. The TSP, along with a wide variety of other practical problems, is known to be NP-complete. It is widely believed that large versions of these problems cannot be solved quickly. Furthermore, the security of some recent cryptographical systems relies on the hope that large NP-complete problems are actually as time consuming to solve as they appear to be. The Clay Foundation is offering a $1 million prize for determining whether NP-complete problems are truly computationally hard. The prize is still unclaimed! Researchers are also exploring whether the development of new approaches to computer design, such as quantum computing, will offer faster ways to solve very difficult problems.

Recently, mathematical researchers have adopted a somewhat different strategy for dealing with TSP problems. If finding a fast algorithm to generate optimal solutions for large problems is unlikely, perhaps we can show that the quick-and-dirty methods, usually called **heuristic algorithms**, come close enough to giving optimal solutions to be important for practical use. For example, suppose we could prove that the nearest-neighbor heuristic was never off by more than 25% in the worst case or by more than 15% in the average case. For a medium-sized TSP, we would then have to choose whether to spend a lot of time (or money) to find an optimal solution or instead to use a heuristic algorithm to obtain a fairly good solution. Investigators at AT&T Research have developed many remarkably good heuristic algorithms. The best-known guarantee for a heuristic algorithm for a TSP is that it yields a cost no worse than one and a half times the optimal cost. Interestingly, this heuristic algorithm involves solving a Chinese postman problem (see Chapter 1), for which a "fast" algorithm is known to exist.

Throughout our discussion of the TSP, we have concentrated on the goal of minimizing the cost (or time) of a tour that visited each of a variety of sites once and only once. However, the subtle issues that arise in specific real-world situations (or that provide a contrast between seemingly similar situations) are the things that make mathematical modeling exciting. For example, suppose the TSP situation is to pick up day campers and take them to and from the camp. The camp wants to minimize the total length of time that the bus needs to pick up the campers. The parents of the campers, however, may want to minimize the time their children spend on the bus. For some problems, the tour that minimizes the mean (average) time that a child spends on the bus may not be the same tour that minimizes the total time of the tour. (Specifically, if the bus first picks up the child who lives the farthest from the camp, and then picks up the other children, this may yield a relatively short time on the bus for the kids but a relatively long time for the tour itself.) Mathematicians return to examine these subtleties between problems at a later time, after the basic structure of the main problem itself is well understood. It is in this way that mathematics continues to grow, explore new ideas, and find new applications.

## 2.4 Minimum-Cost Spanning Trees

The traveling salesman problem is but one of many graph theory optimization problems that have grown out of real-world problems in both government and industry. Here is another.

Some videophone (pictaphone) and video conferencing requires the creation of specialized networks for which the techniques of finding a minimum-cost spanning tree are required. (*Digital Vision/Punchstock.*)

### EXAMPLE 7 ■ Pictaphone Service

Imagine that Pictaphone service (telephone service that provides a video image of the callers) will be set up on an experimental basis among five cities. The graph in Figure 2.12 shows the possible links that might be included in the Pictaphone network, with each edge showing the cost in millions of dollars to create that particular link. To send a Pictaphone message between two cities, a direct communication link is not necessary because it is possible to send a message indirectly via another city. Thus, in Figure 2.12, sending a message from $A$ to $C$ could be achieved by sending the message from $A$ to $B$, from $B$ to $E$, and from $E$ to $C$, provided the links $AB$, $BE$, and $EC$ are part of the network. We assume that the cost of relaying a message, compared with the cost of the direct communication link, is so small that we can neglect this amount. The problem that concerns us, therefore, is to provide service between any pair of cities in a way that minimizes the total cost of the links.

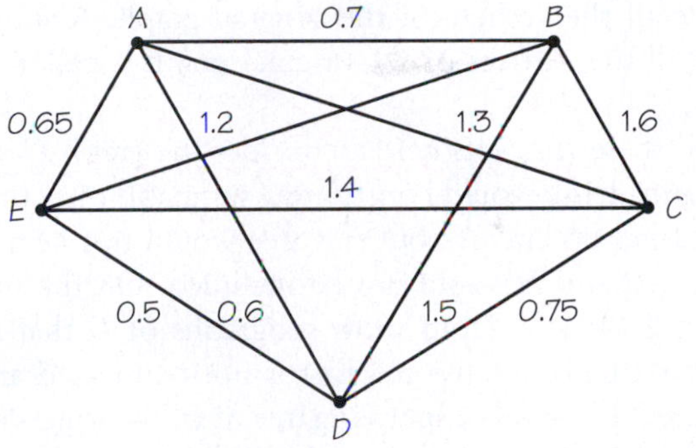


**FIGURE 2.12** Costs (in millions of dollars) of installing Pictaphone service among five cities.

Our first guess at a solution is to put in the cheapest possible links between cities first, until all cities could send messages to any other city. Such an approach is analogous to the sorted-edges method that was used to study the traveling salesman problem. In our example, if the cheapest links are added until all cities are joined, we obtain the connections shown in Figure 2.13a.

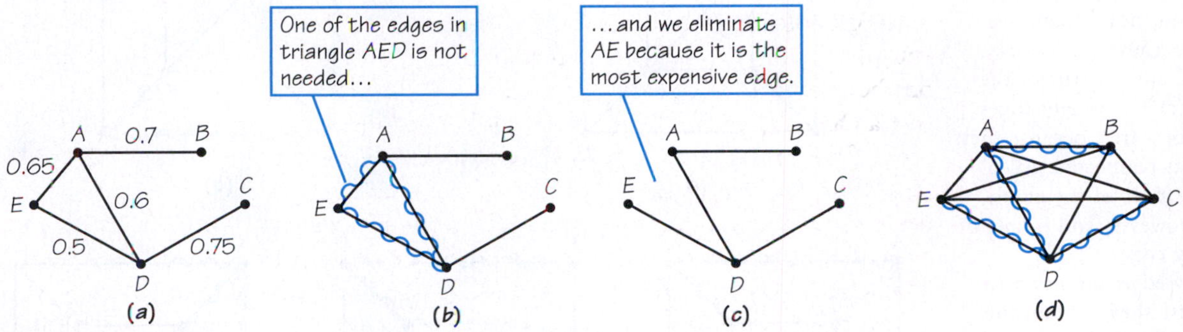


**FIGURE 2.13** (a) Cities are linked in order of increasing cost until all cities are connected. (b) Circuit in part (a) highlighted. (c) Most expensive link in circuit in part (a) deleted. (d) Highlighted edges show, as a subgraph of the original graph, those links connecting the cities with minimum cost, obtained using Kruskal's algorithm.

The links were added in the order *ED, AD, AE, AB, DC*. However, because this graph contains the circuit *ADEA* (wiggly edges in Figure 2.13b), it has redundant edges: We can still send messages between any pair of cities using relays after omitting the most expensive edge in the circuit–*AE*. After deleting an edge of a circuit, a message can still be relayed among the cities of the circuit by sending signals the long way around. After *AE* is deleted, messages from *A* to *E* can be sent via *D* (Figure 2.13c). These ideas constitute a procedure developed by Joseph Kruskal (AT&T Research) in 1956.

## Kruskal's Algorithm

PROCEDURE

**Kruskal's algorithm**: Add links in order of cheapest cost so that no circuits form and so that every vertex belongs to some link added (Figure 2.13d).

In Kruskal's procedure, as in the sorted-edges method for the TSP, the edges that are added need not be connected to each other until the end. A subgraph formed in this way will be a **tree;** that is, it will consist of one piece and contain no circuits. It will also include all the vertices of the original graph. A subgraph that is a tree and that contains all the vertices of the original graph is called a **spanning tree** of the original graph.

To understand these concepts better, consider the graph $G$ in Figure 2.14a. The wiggly edges in Figure 2.14b would constitute a subgraph of $G$ that is a tree (because it is connected and has no circuit), but this tree would not be a spanning tree of $G$ because the vertices $D$ and $E$ would not be included. On the other hand, the wiggly edges in Figure 2.14c and 2.14d show subgraphs of $G$ that include all the vertices of $G$ but are not trees because the first is not connected and the second contains a circuit. Figure 2.14e shows a spanning tree of $G$; the wiggly edges are connected and contain no circuit, and every vertex of the original graph is an endpoint of some wiggly edge.

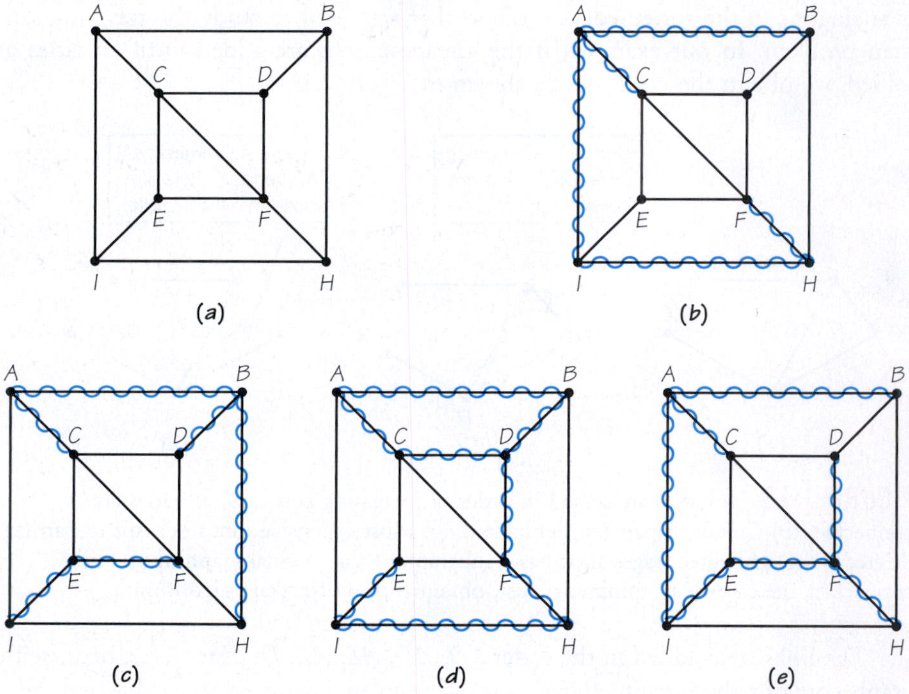


**FIGURE 2.14** (a) A graph to help illustrate the concept of a spanning tree. (b) The wiggly edges are a tree, but not a spanning tree, because vertices $D$ and $E$ are not part of the tree. (c) The wiggly edges are not a tree, because they are not connected. All of the vertices of the graph are, however, end points of wiggly edges. (d) The wiggly edges are not a tree, because they contain the edges of the circuit *BDCAB*. All the vertices of the graph are, however, endpoints of wiggly edges. (e) The wiggly edges form a tree and include all of the vertices of the graph as endpoints of wiggly edges. Thus, the wiggly edges are a spanning tree.

Finding a **minimum-cost spanning tree**–that is, a spanning tree whose edge weights sum to a minimum value–solves the Pictaphone problem. Note that having a different goal in the Pictaphone problem led to a different mathematical question from that of finding a Chinese postman tour or TSP tour. This application required that we find a minimum-cost spanning tree. In Figure 2.15a we have a graph model showing the costs of putting in roads to connect new houses in a suburban land-development project. Applying Kruskal's algorithm–adding the edges in the order of increasing cost, but avoiding the creation of a circuit–yields a minimum-cost spanning tree, indicated by the wiggly edges in Figure 2.15b. This tree is the cheapest one that makes it possible to drive between any pair of homes, though the driving distance between some of the homes will be relatively large, because only roads corresponding to wiggly edges will be built.

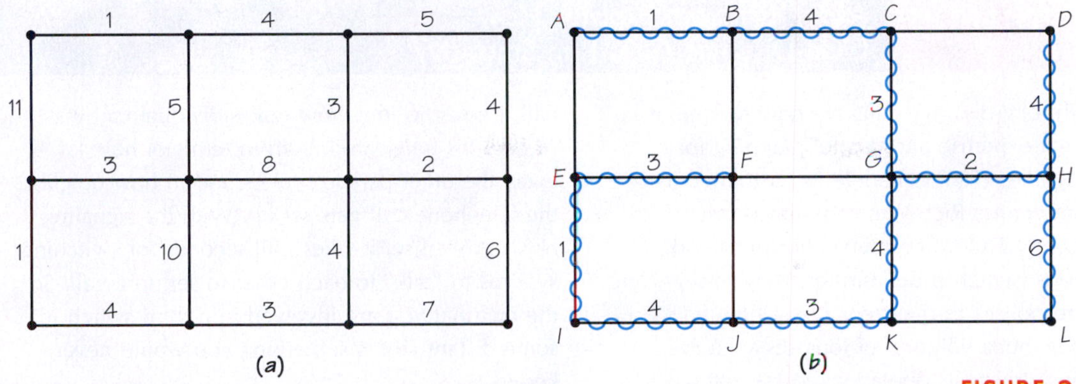


**FIGURE 2.15** (a) A graph showing costs for construction of roads between houses. (b) Wiggly edges show a minimum-cost spanning tree for the graph in part (a).

Remember that the weights on the edges of the graph in Figure 2.15a represent the costs of building roads, not the driving distances between the houses. Note that Figure 2.15a is not a complete graph, one in which all possible edges are included. Edges that correspond to roads that would be economically prohibitive to build have not been shown in the graph model. Also, in Figure 2.15b, the two edges of weight 5 (shown in Figure 2.15a) do not become part of the minimum-cost spanning tree, because they would create circuits with edges already chosen.

Although Kruskal's algorithm worked in our example, how do we know that the spanning tree found by this algorithm will always achieve the minimum possible cost? While this sounds very plausible, our experience with the TSP should suggest caution. Remember that for the TSP, the sorted-edges algorithm did not necessarily give an optimal solution even though it is a greedy algorithm like Kruskal's. On what basis should we have more faith in Kruskal's algorithm?

Kruskal proposed his algorithm as a way to solve a pure mathematics problem put forward by Czechoslovakian mathematician Otakar Borůvka. In mathematics it is surprising but not uncommon to find that ideas used to solve problems with no apparent application often turn out to have many real-world uses. Kruskal's solution to the problem of finding a minimum-cost spanning tree in a graph with weights is a good example of this phenomenon.

Kruskal showed that the greedy algorithm described does yield the minimum answer, and his work led to applications of these and related ideas in designing minimum-cost computer networks, phone connections, sewer systems, and road and railway systems. For additional discussion of operations research in the communications industry, see Spotlight 2.2. To explore how one can reconstruct full information from partial information using the tree concept, see Spotlight 2.3.

In our discussion of routing problems in graphs, we have not touched on one of the most obvious ones: finding the path between two specified, distinct vertices while keeping the sum of the weights of the edges in the path as small as possible. (Here there is no need to cover all vertices or to cover all edges.) We have seen that the weights on the edges have many possible interpretations, including time, distance, and cost. The following are some of the many possible applications:

1. Design routes to be used by an ambulance, police car, or fire engine to get to an emergency as quickly as possible.
2. Design delivery routes that minimize gasoline use.
3. Design routes to bring soldiers to the front as quickly as possible.
4. Design a route for a truck carrying nuclear waste.

### SPOTLIGHT 2.2 AT&T Manager Explains How Long-Distance Calls Run Smoothly

Although long-distance calls are now routine, it takes great expertise and careful planning for a company like AT&T to handle its vast amounts of telephone traffic. Rich Wetmore was district manager of AT&T's Communications Network Operations Center in Bedminster, New Jersey. Here are his responses to questions about how AT&T handles its huge volumes of long-distance traffic and how it tracks its operations to keep things running smoothly.

**How do you make sure that a customer doesn't run into a delayed signal when attempting a long-distance call?**

We monitor the performance of our AT&T network by displaying data collected from all over the country on a special wallboard. The wallboard is configured to tell us if a customer's call is not going through because the network doesn't have enough capacity to handle it.

That's when we step in and take control to correct the problem. The typical control we use is to reroute the call. Instead of sending the customer's call directly to its destination, we'll route it via a third city—to someplace else in the country that has the capacity to complete the call.

**It would seem that routing via another city would take longer. Is the customer aware of this process?**

Routing a call via a third city is entirely transparent [imperceptible] to the customer. I'm an expert about the network, and even when I make a phone call, I have no idea how that individual call was routed. It's transparent both in terms of how far away the other person sounds and in how quickly the telephone call gets set up. With the signaling network we use, it takes milliseconds for switching systems to "talk" to each other to set up a call. So the fact that you are involved in a third switch in some distant city is something you would never know.

**You want to be sure to keep costs down while supplying enough service to customers. So how do you balance company benefits with customer benefit?**

In terms of making the network efficient, we want to do two things. First, we want our customers to be happy with our service and for all their calls to go through, which means we must build enough capacity in the network to allow that to happen. Second, we want to be efficient for stockholders and not spend more money than we need to for the network to be at the optimum size.

There are basically two costs in terms of building the network. There is the cost of switching systems and the cost of the circuits that connect the switching systems. Basically, you can use operations-research techniques and mathematics to determine cost trade-offs. It may make sense to build direct routing between two switching systems and use a lot of circuits, or maybe to involve three switching systems, with fewer circuits between the main two, and so on.

Many people find it increasingly convenient to use the Web or software installed in their cars to get driving directions and driving time estimates to a place they wish to visit. The software that provides this information relies on algorithms that compute the shortest-path route in an appropriate weighted graph, which involves distances or times.

The need to find shortest paths seems natural. Next we investigate a situation in which finding a *longest* path is the right tool.

## 2.5 Critical-Path Analysis

Mathematics can confirm the obvious in certain situations while showing that our intuition is wrong in other circumstances. Our next group of applications will illustrate this point.

## SPOTLIGHT 2.3 Common Ancestors?

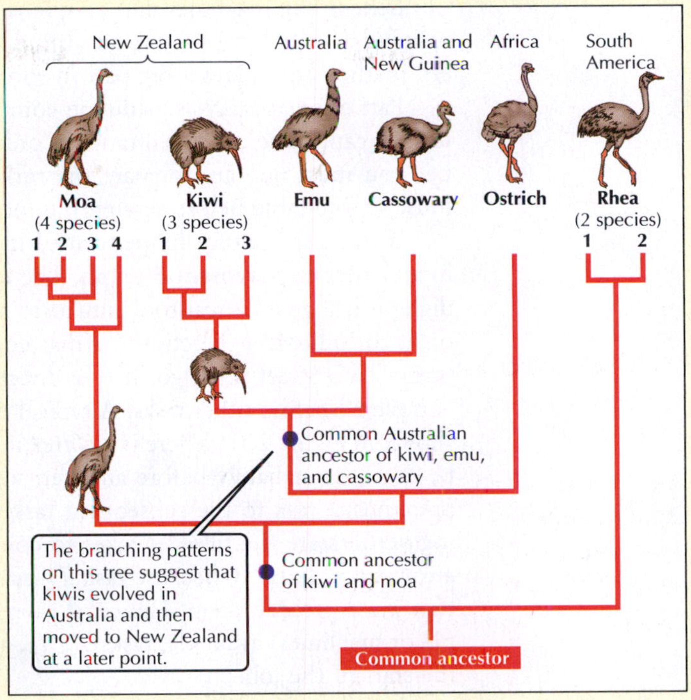


In the study of ancient manuscripts, different manuscripts of the same book are available, even though the original manuscript upon which they are based has been lost. Examples of this include Euclid's *Elements* and Chaucer's *Canterbury Tales*. What interests scholars is reconstructing the relationships between the manuscripts and the common ancestors of the manuscripts, even when some of the ancestors are now missing.

Similarly, perceptual psychologists may be interested in which colors people perceive as being closely related and comparing these perceptions with those of people who are color-blind. Linguists are interested in the connections between languages that seem very different today, but have some words that are similar. Finally, in studying different species, biologists are interested in determining which species are more closely related to each other, including species known only in fossil form, and constructing a "tree" of life that shows which species were ancestors of others.

Reconstructions of this kind are made possible by using graph theory, specifically using the graph theory concept of a tree. The value of the graph theory in these and many other situations lies in using the distance between pairs of vertices in the tree as a way of reflecting the closeness of the relationships that pairs of manuscripts, pairs of colors, or pairs of species have. The distance between two vertices in a tree is the sum of the weights along the one path that joins the two vertices. If there are no weights on the edges, the distance is the number of edges in the path. In some reconstruction problems, a special vertex of the tree called the *root* is singled out. This root plays the role of the original common ancestors, and distances to the root are of critical interest.

In the case of species, trees of family relatedness were traditionally constructed based on similarities of bones and physical appearance. With the discovery of molecular biology, many new avenues have been opened up. We can now draw trees of relatedness based on an organism's genetic material, DNA, or the proteins that the DNA codes for. The traditional trees based on physical traits often show different species as being more closely related than trees based on newer molecular biological approaches. These differences focus scholars on how to resolve the discrepancies and thereby reach a deeper understanding of the unity of life.

A characteristic of American life is its fast pace. People are interested in getting things done quickly and efficiently. This means that when you take your car in to be repaired before going to work, you want to know for sure that the repairs will be done when you pick the car up. You want the trains and the bus that take you to your doctor's appointment to run on time. When you arrive at the doctor's office, you want a technician to be free to take a blood sample and a throat culture. You want your outpatient appointment for an X-ray at the local hospital to occur on schedule. You want the X-ray to be interpreted quickly and the results reported back to your internist.

Scheduling machines and people is a big part of modern life. It is involved in running a school, a hospital, an airline, or in landing a person on Mars, and modern mathematics plays a big part in solving scheduling problems.

Part of what makes scheduling complicated is that the tasks that make up a job usually cannot be done in a random order. For example, to make Thanksgiving dinner you must buy and prepare the turkey before putting it in the oven, and you must set the table before serving the food.

If the tasks cannot be performed in a random order, we can specify the order in an **order-requirement digraph**. The term *digraph* is short for "directed graph." A digraph is a geometrical tool similar to a graph except that each edge has an arrow on it to indicate a direction for that edge. Digraphs can be used to illustrate that traffic on a street must go in one direction or that certain tasks in a job must be completed before other tasks. A typical example of an order-requirement digraph is shown in Figure 2.16. There is a vertex in this digraph for each task. If one task must be done immediately before another, we draw a directed edge, or arrow, from the prerequisite task to the subsequent task. The numbers within the circles representing vertices are the times it takes to complete the tasks. In Figure 2.16 there is no arrow from $T_1$ to $T_5$ because task $T_2$ intervenes. Also, $T_1$, $T_7$, and $T_8$ have no tasks that must precede them. Hence, if there are at least three processors (such as people or machines) available, tasks $T_1$, $T_7$, and $T_8$ can be worked on simultaneously at the start of the job.

Let's investigate a typical scheduling problem faced by a business.

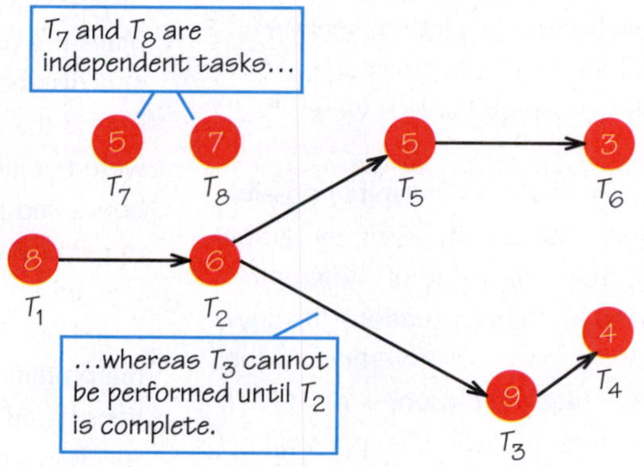


**FIGURE 2.16** A typical order-requirement digraph.

## EXAMPLE 8 ■ Turning a Plane Around

Consider an airplane that carries both freight and passengers. The plane must have its passengers and freight unloaded and new passengers and cargo loaded before it can take off again. Also, the cabin must be cleaned before departure can occur. Thus, the job of "turning the plane around" requires completion of five tasks:

| | | |
|---|---|---|
| Task *A* | Unload passengers | 13 minutes |
| Task *B* | Unload cargo | 25 minutes |
| Task *C* | Clean cabin | 15 minutes |
| Task *D* | Load new cargo | 22 minutes |
| Task *E* | Load new passengers | 27 minutes |

Turning a plane around, which involves such tasks as refueling, unloading, and then reloading cargo and passengers, entails very careful scheduling to avoid time slippage. (*David Butow/Corbis Saba.*)

The order-requirement digraph for the problem of turning an airplane around is shown in Figure 2.17. The presence or absence of an edge in the order-requirement digraph depends on the analysis made as part of the modeling process for the problem. It seems natural that we need an arrow between task $A$ and task $C$, because the passengers have to be unloaded before the cabin is cleaned. Other arrows may not seem natural–say, perhaps the arrow from task $B$ (unload the cargo) to task $E$ (load new passengers). This arrow may be due to government rules or union requirements.

What matters is that the mathematics of solving the problem does not depend on the reason that the order-requirement digraph looks the way that it does. The person solving the problem constructs the order-requirement digraph and then the mathematical techniques we will develop can be applied, regardless of whether or not another business faced with a similar problem might model the problem in a different way. Because we want to find the earliest completion time, it might seem that finding the shortest path in the digraph (path $BD$ with time length $25 + 22 = 47$) would solve the problem. But this approach shows the danger of ignoring the relationship between the mathematical model (the digraph) and the original problem.

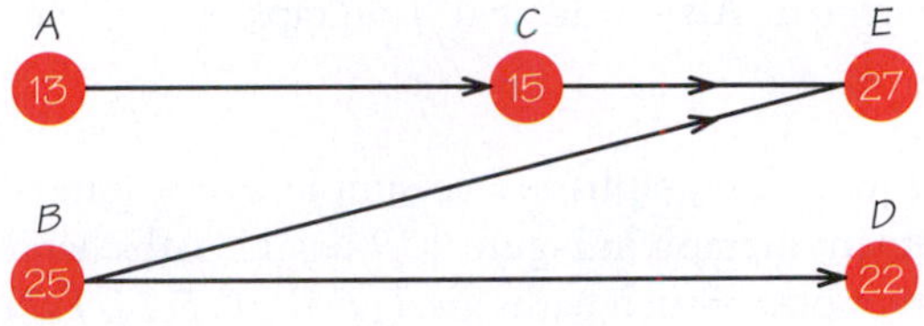


**FIGURE 2.17** An order-requirement digraph for turning an airplane around after landing.

The time required to complete all the tasks, $A$ through $E$, must be at least as long as the time necessary to do the tasks on any particular path. Consider the path $BD$, which has length $25 + 22 = 47$. Recall that here *length* of a path refers to the sum of the times of the tasks that lie along the path. Because task $B$ must be done before task $D$ can begin, the two tasks $B$ and $D$ cannot be completed before time 47. Hence, even if work on other tasks (such as $A$, $C$, and $E$) proceeds during this period, all the tasks cannot be finished before the tasks on path $BD$ are finished. The same statement is true for every other path in the order-requirement digraph. Thus, the earliest completion time actually corresponds to the length of the longest path. In the airplane example, this earliest completion time is 55 ($= 13 + 15 + 27$) minutes, corresponding to the path $ACE$. We call $ACE$ the **critical path** because the times of the tasks on this path determine the earliest completion time.

## Critical Path

DEFINITION

A **critical path** in an order-requirement digraph is a longest path. The length is measured in terms of summing the task times of the tasks making up the path.

Note that if none of these tasks could go on simultaneously, the time to complete all the tasks would be $13 + 25 + 15 + 22 + 27 = 102$ minutes. However, even though some tasks may be performed simultaneously, the fact that the length of the critical path is 55 means that completion of the tasks in less than 55 minutes is not possible. Only by speeding up the times to complete the critical-path tasks themselves can a completion time less than 55 minutes be achieved.

Suppose it were desirable to speed the turnaround of the plane to less than 55 minutes. One way to do this might be to build a second jetway to help unload passengers more quickly. For example, we could unload passengers (task *A*) in 7 minutes instead of 13. However, reducing task *A* to 7 minutes does not reduce the completion time by 6 minutes, because in the new digraph (Figure 2.18) *ACE* is no longer the critical (longest) path. The longest path is now *BE,* which has a length of 52 minutes. Thus, shortening task *A* by 6 minutes results in only a 3-minute saving in completion time. This may mean that building a new jetway is uneconomical.

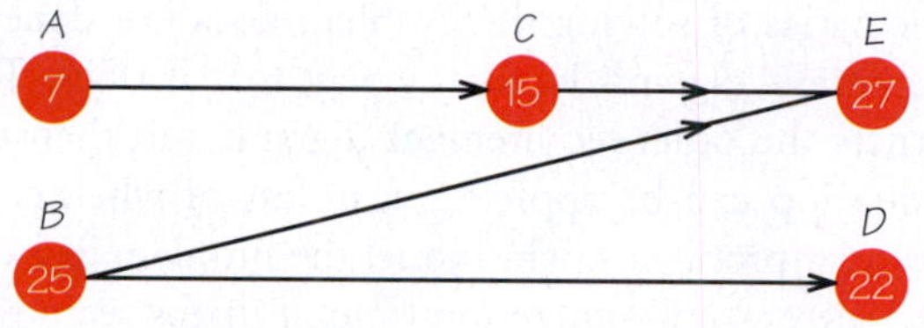


**FIGURE 2.18** An order-requirement digraph for turning an airplane around in reduced time due to construction of a new jetway.

Note also that shortening the time to complete tasks that are not on the original critical path *ACE* will not shorten the completion time at all. Speeding tasks on the critical path will shorten completion time of the job only up to the point where a new critical path is created. Also note that a digraph may have more than one longest path.

Not all order-requirement digraphs are as simple as the one shown in Figure 2.17. The order-requirement digraph in Figure 2.19 has 12 paths, which can be found by exhaustive search. Examples of such paths are $T_1T_2T_3$, $T_1T_5T_9$, $T_4T_5T_9$, and $T_7T_5T_3$. (Although we have not discussed them here, fast algorithms for finding longest and shortest paths in graphs are known.) The critical path is $T_7T_8T_6$ (length 21), and the earliest completion time for all nine tasks is time 21, though the actual completion time may be later than time 21 depending on the resources available to carry out the tasks. Completing the tasks by time 21 depends on having sufficient resources available so that some of the tasks can be worked on simultaneously.

These examples are typical of many scheduling problems that occur in practice (see Spotlight 2.4). Perhaps the most dramatic use of critical-path analysis is in the construction trades. No major new building project is now carried out without a critical-path analysis first being performed to ensure that the proper personnel and materials are available at the right times in order to have the project finished as quickly as possible. Many such problems are too large and complicated to be solved without the aid of computers.

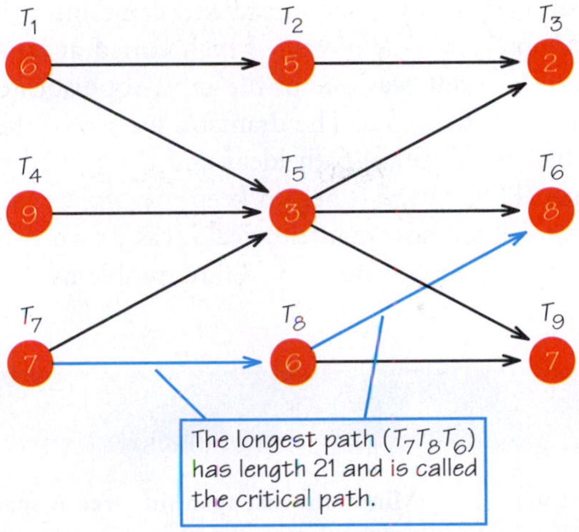


**FIGURE 2.19** An order-requirement digraph with 12 paths, to examine how to find the length of the longest path.

## SPOTLIGHT 2.4 Every Moment Counts in Rigorous Airline Scheduling

When people think of airline scheduling, the first thing that comes to mind is how quickly a particular plane can safely reach its destination. But using ground time efficiently is just as important to an airline's timetable as the time spent in flight. Bill Rodenhizer was the manager of control operations for an airline that provided shuttle service between Boston and New York. He is considered to be an expert on airplane turnaround time, the process by which an airplane is prepared for almost immediate takeoff once it has landed. He tells us how this well-orchestrated effort works:

> Scheduling, to the airline, is just about the whole ball game. Everything is scheduled right to the minute. The whole fleet operates on a strict schedule. Each of the departments responsible for turning around an aircraft has an allotted period of time in which to perform its function. Manpower is geared to the amount of ground time scheduled for that aircraft. This would be adjusted during off-weather or bad-weather days or during heavy air-traffic delays.
>
> Most of our aircraft in Boston are scheduled for a 42- to 65-minute ground time. Boston is the end of the line, so it is a "terminating and originating station." In plain talk, that means almost every aircraft that comes in must be fully unloaded, refueled, serviced, and dispatched within roughly an hour's time.
>
> This is how the process works: In the larger aircraft, it takes passengers roughly 20 minutes to load and 20 minutes to unload. During this period, we will have completely cleaned the aircraft and unloaded the cargo, and the caterers will have taken care of the food. The ramp service may take 20 to 30 minutes to unload the baggage, mail, and cargo from underneath the plane, and it will take the same amount of time to load it up again. We double-crew those aircraft with heavier weights so that the workload will fit the time it takes passengers to load and unload upstairs.
>
> While this has been going on, the fueler has fueled the aircraft. As to repairs, most major maintenance is done during the midnight shift, when [most of our] several hundred aircraft are inactive. We all work under a very strict time frame.

New security requirements in the wake of the World Trade Center attack (9/11/2001) have increased the difficulty of adhering to timetables in operating shuttle services between East Coast cities such as New York and Boston. This makes it even more important to use analytical tools in keeping operations on schedule.

The critical-path method was popularized and came into wider use as a consequence of the *Apollo* project. This project, which aimed at landing a man on the moon within 10 years of 1960, was one of the most sophisticated projects in planning and scheduling ever attempted. The dramatic success of the project can be attributed partly to the use of critical-path ideas and the related program evaluation and review technique (PERT), which helped keep the project on schedule.

In Chapter 3, we will see how mathematical ideas drawn from outside of graph theory can be used to gain insight into scheduling problems.

## REVIEW VOCABULARY

**Algorithm** A step-by-step description of how to solve a problem. (p. 34)

**Brute force method** The method that solves the traveling salesman problem (TSP) by enumerating all the Hamiltonian circuits and then selecting the one with minimum cost. (p. 37)

**Complete graph** A graph in which every pair of vertices is joined by an edge. (p. 36)

**Critical path** The longest path in an order-requirement digraph. The length of this path gives the earliest completion time for all the tasks making up the job consisting of the tasks in the digraph. (p. 49)

**Fundamental principle of counting** A method for counting outcomes of multistage processes. (p. 36)

**Greedy algorithm** An approach for solving an optimization problem, where at each stage of the algorithm the best (or cheapest) action is taken. Unfortunately, greedy algorithms do not always lead to optimal solutions. (p. 39)

**Hamiltonian circuit** A circuit using distinct edges of a graph that starts and ends at a particular vertex of the graph and visits each vertex once and only once. A Hamiltonian circuit can start at any one of its vertices. (p. 31)

**Heuristic algorithm** A method of solving an optimization problem that is "fast" but does not guarantee an optimal answer to the problem. (p. 42)

**Kruskal's algorithm** An algorithm developed by Joseph Kruskal (AT&T Research) that solves the minimum-cost spanning-tree problem by selecting edges in order of increasing cost, but in such a way that no edge forms a circuit with edges chosen earlier. It can be proved that this algorithm always produces an optimal solution. (p. 43)

**Method of trees** A visual method of carrying out the fundamental principle of counting. (p. 34)

**Minimum-cost Hamiltonian circuit** A Hamiltonian circuit in a graph with weights on the edges, for which the sum of the weights of the edges of the Hamiltonian circuit is as small as possible. (p. 34)

**Minimum-cost spanning tree** A spanning tree of a weighted connected graph having minimum cost. The cost of a tree is the sum of the weights on the edges of the tree. (p. 44)

**Nearest-neighbor algorithm** An algorithm for attempting to solve the TSP that begins at a "home" vertex and visits next that vertex not already visited that can be reached most cheaply. When all other vertices have been visited, the tour returns to home. This method may not give an optimal answer. (p. 39)

**NP-complete problems** A collection of problems, which includes the TSP, that appear to be very hard to solve quickly for an optimal solution. (p. 41)

**Order-requirement digraph** A directed graph that shows which tasks precede other tasks among the collection of tasks making up a job. (p. 48)

**Sorted-edges algorithm** An algorithm for attempting to solve the TSP where the edges added to the circuit being built up are selected in order of increasing cost, but no edge is chosen that would prevent a Hamiltonian circuit from forming. These edges must all be connected at the end, but not necessarily at earlier stages. The tour obtained may not have the lowest possible cost. (p. 40)

**Spanning tree** A subgraph of a connected graph that is a tree and includes all the vertices of the original graph. (p. 44)

**Traveling salesman problem (TSP)** The problem of finding a minimum-cost Hamiltonian circuit in a complete graph where each edge has been assigned a cost (or weight). (p. 38)

**Tree** A connected graph with no circuits. (p. 34)

**Weight** A number assigned to an edge of a graph that can be thought of as a cost, distance, or time associated with that edge. (p. 34)

## ✓ SKILLS CHECK

**1.** Which of the following describes a Hamiltonian circuit for the graph below?

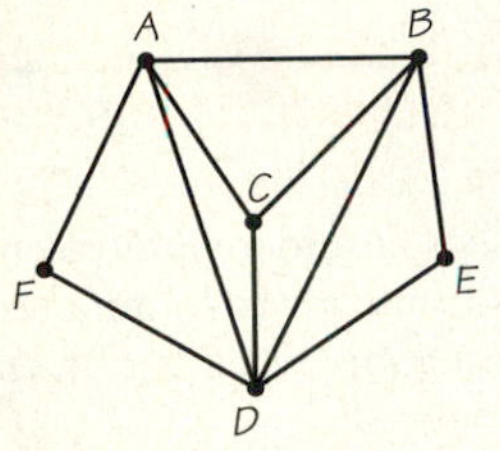


(a) *ABCDFA*
(b) *AFDCBE*
(c) *ACBEDFA*
(d) *ACEBDFA*

**2.** The cost of the nearest-neighbor tour (Hamiltonian circuit) that starts at vertex *A* for the graph below is ______________________________ .

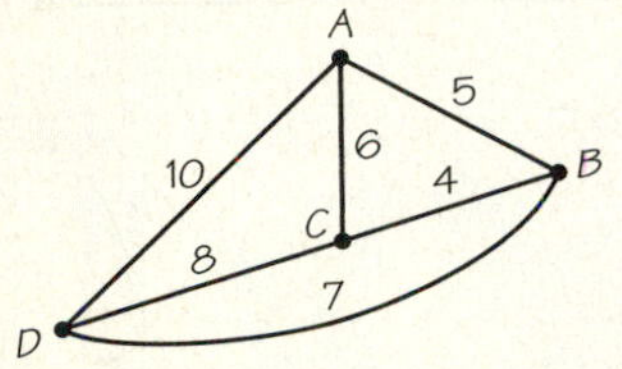


**3.** Suppose that after a hurricane, a van is dispatched to pick up five nurses at their homes and bring them to work at the local hospital. Which of these techniques is most likely to be useful in solving this problem?

(a) Finding an Euler circuit in a graph
(b) Solving a TSP (traveling salesman problem)
(c) Finding a minimum-cost spanning tree in a graph

**4.** The cost of the sorted-edges tour (Hamiltonian circuit) for the graph below is __________________ .

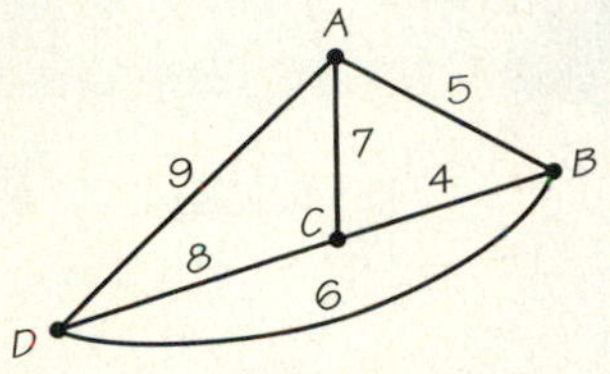


**5.** The graph shown below has

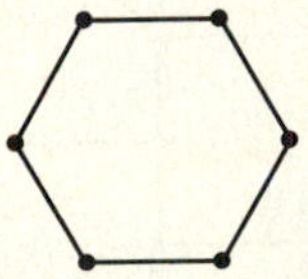

(a) no Hamiltonian circuit and no Euler circuit.
(b) an Euler circuit and a Hamiltonian circuit.
(c) no Hamiltonian circuit, but it has an Euler circuit.

**6.** The cost of the nearest-neighbor traveling salesman tour that starts at *B* for the following graph is _______ .

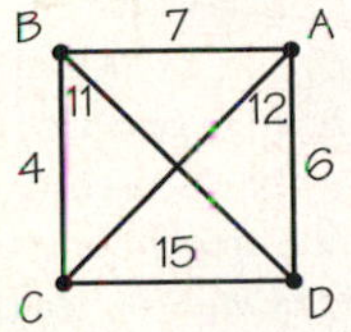


**7.** When the sorted-edges method and nearest-neighbor method are applied to a complete graph on seven vertices with nonnegative weights,

(a) both methods always give the same optimal answer.
(b) both methods always give the same answer but that answer may not be optimal.
(c) neither method may give an optimal answer.

**8.** If a graph has *E* edges and *V* vertices as well as a Hamiltonian circuit, then the number of edges in the Hamiltonian circuit is ____________________ .

**9.** Paul has packed four ties, three shirts, and two pairs of pants for a trip. How many different outfits can he create if he never wears a tie?

(a) Fewer than 10
(b) Between 10 and 25
(c) More than 25

**10.** The number of different lunches that Jules can design by selecting one of three meats, one of three salads, and one of six vegetables is exactly ________________ .

**11.** An ice-cream shop offers 3 types of cones, 20 flavors, and 4 different toppings (crushed peanuts, crushed almonds, chocolate bits, or corn flakes). If a customer is allergic to nuts, how many different choices can she choose from?

(a) 240
(b) 120
(c) 25

**12.** If a three-character password system must begin with a lowercase letter of the English alphabet followed by two decimal digits that may be repeated, the number of different possible passwords is __________________ .

**13.** Assuming a graph with *E* edges and *V* vertices has a minimum-cost spanning tree *T*, which of the following statements *must* be true?

(a) The tree *T* has exactly *V* edges.
(b) The tree *T* includes every minimum-cost edge.
(c) The graph is connected.

**14.** When arranged in increasing order, the weights of the edges in the following graph that are not part of the minimum-cost spanning tree selected when Kruskal's algorithm is applied are ________ , ________ , ________ .

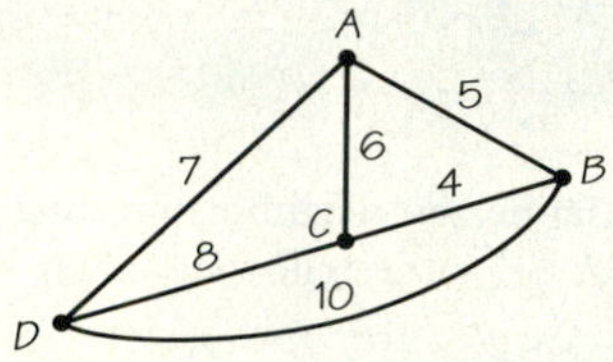


**15.** Assume that every edge of a graph $G$ has a different cost. If Kruskal's algorithm is used to find the minimum-cost spanning tree $T$ for graph $G$, which of the following statements *must* be true?

**(a)** Any other spanning tree for graph $G$ will have more edges than $T$.

**(b)** Any other spanning tree for graph $G$ will have a greater cost than $T$.

**(c)** The edge of graph $G$ having greatest weight is included in $T$.

**16.** The smallest positive integer valued weight that $x$ can have in the graph below so that it could not be selected by Kruskal's algorithm as an edge of a minimum-cost spanning tree is ____________________ .

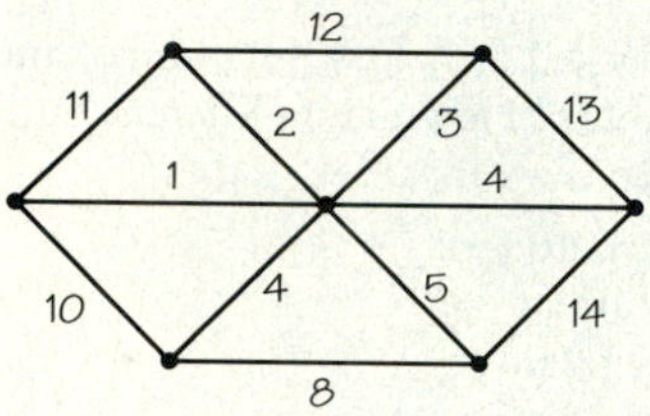


**17.** If a graph contains a circuit, which of the following statements is true?

**(a)** The graph cannot be a tree.

**(b)** The graph must have the same number of vertices as edges.

**(c)** The graph is not connected.

**18.** The earliest completion time (in minutes) for a job with the following order-requirement digraph is ______ .

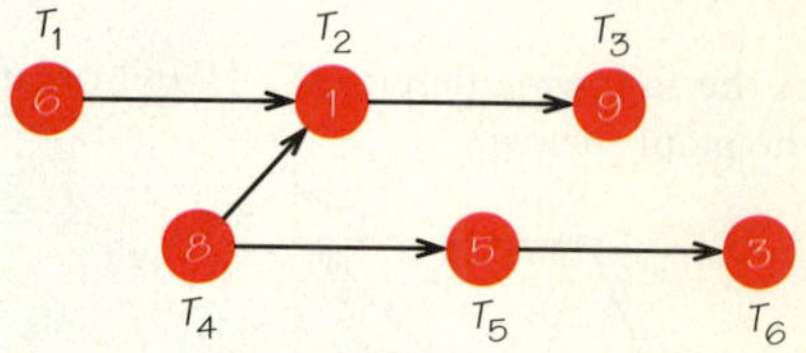


**19.** Assume a job has an order-requirement digraph with five tasks whose critical path is 25 minutes in length. Based on this information, what can be said about the tasks?

**(a)** Each task takes exactly 5 minutes.

**(b)** Some task takes 25 minutes.

**(c)** The five tasks in total take at least 25 minutes.

**20.** The length of the critical path in the order-requirement digraph below is ____________________ minutes.

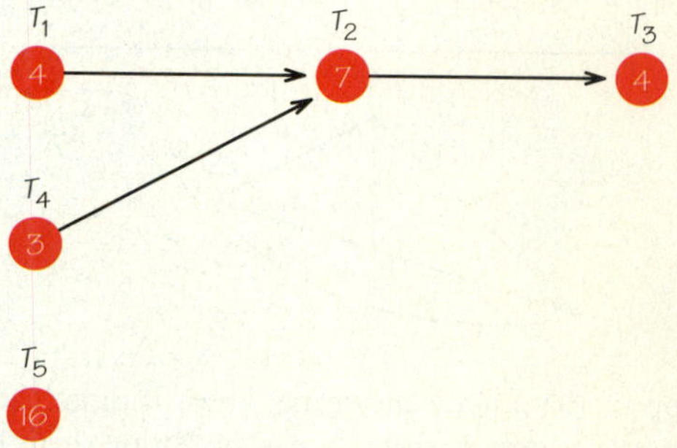


# CHAPTER 2 EXERCISES

■ Challenge ◆ Discussion

## 2.1 Hamiltonian Circuits

**1.** For the accompanying graphs (a) through (c), write a Hamiltonian circuit starting at $X_5$.

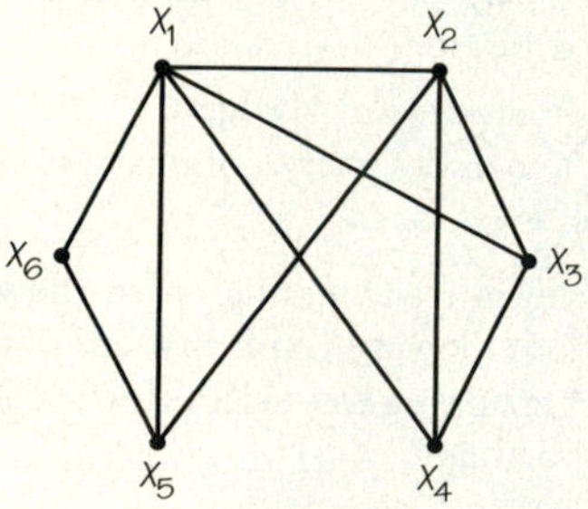


*(a)*

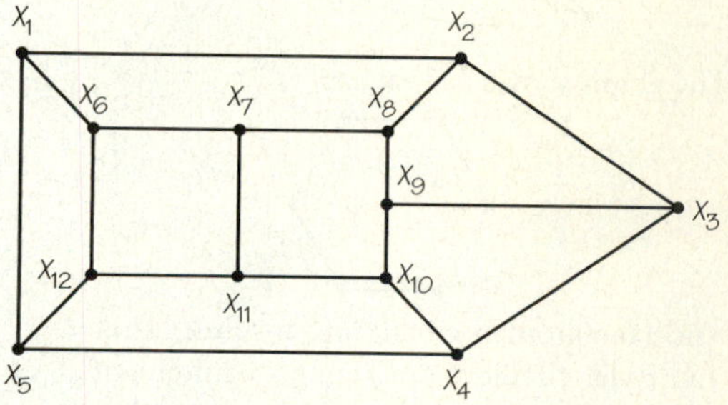


*(b)*

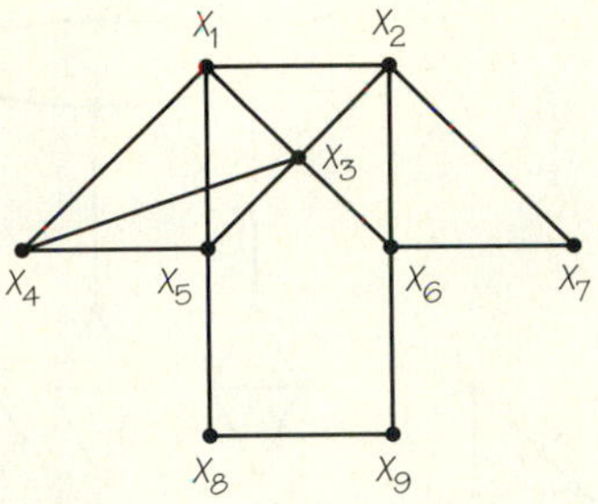


(c)

2. For the accompanying graphs (a) through (c) write down a Hamiltonian circuit starting at $X_5$.

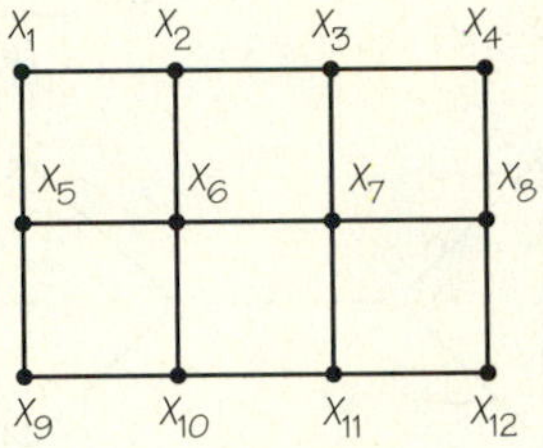


(a)

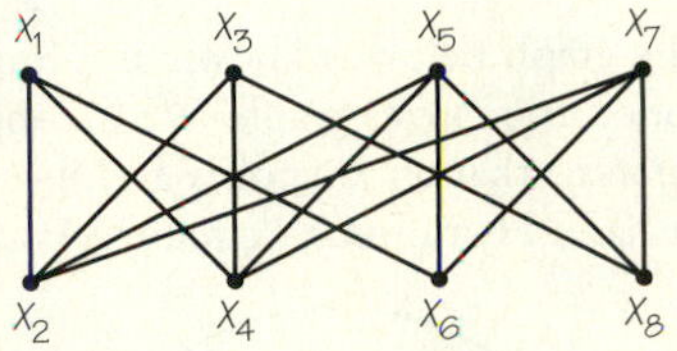


(b)

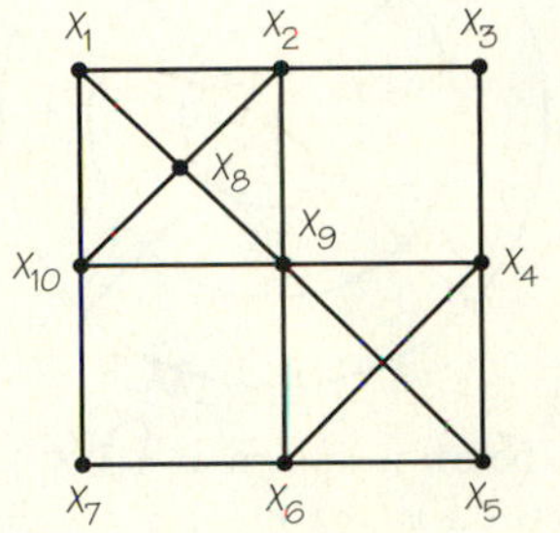


(c)

3. If the edge $X_2X_3$ is erased from each of the graphs in Exercise 1, does the resulting graph still have a Hamiltonian circuit?

4. **(a)** If the vertex $X_6$ and the edges attached to $X_6$ are removed from the graphs in Exercise 1, do the new graphs that result still have Hamiltonian circuits?
**(b)** If you think of the graphs in Exercise 1 as communications networks, what interpretation might be given to the "removal" of a vertex and the edges attached as described in part (a)?

5. **(a)** If the edge $X_6X_7$ is removed (erased) from each of the graphs in Exercise 2, do the new graphs that result still have Hamiltonian circuits?
**(b)** If you think of the graphs in Exercise 2 as communications networks, what interpretation might be given to the "removal" of an edge as described in part (a)?

6. **(a)** Give examples of real-world situations that can be modeled using a graph and for which finding a Hamiltonian circuit in the graph would be of interest.
**(b)** For each of the examples you mention in part (a), can you adapt the question about the real-world situation involved so that finding an Eulerian circuit in the same graph would be of interest?

7. Suppose two Hamiltonian circuits are considered different if the collections of edges that they use are different. How many other Hamiltonian circuits can you find in the graph in Figure 2.1 that are different from the two discussed?

8. For each of the following graphs, add wiggly edges to indicate a Hamiltonian circuit.

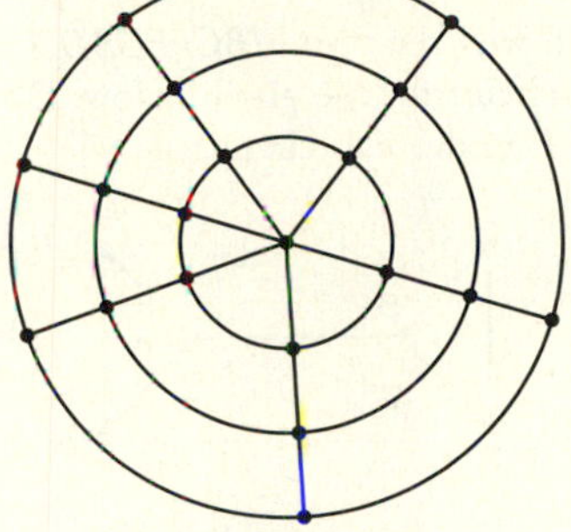

(a)

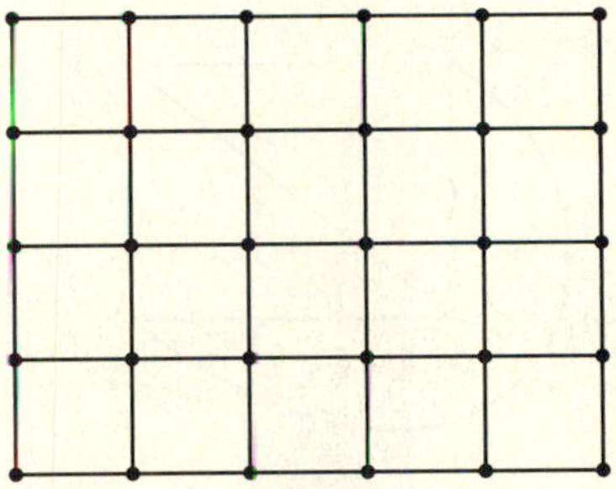

(b)

9. **(a)** Neither of the following graphs has a Hamiltonian circuit. Is it possible to add a single new edge to these graphs to obtain a new graph that has a Hamiltonian circuit?

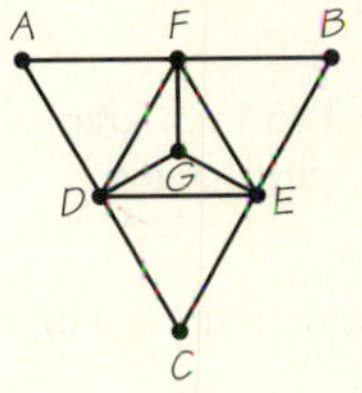


(a)

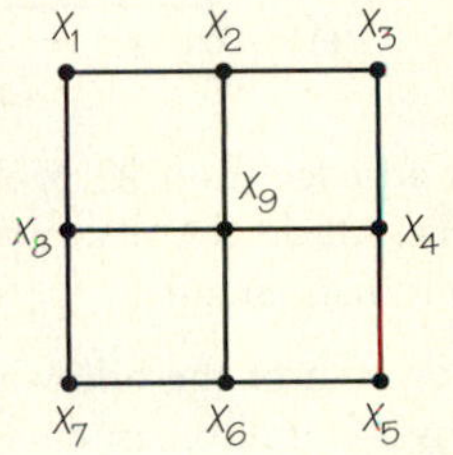


(b)

**(b)** Find an example of a graph that has no Hamiltonian circuit and will still have no Hamiltonian circuit no matter what single edge is added to it.
**(c)** Show that it is possible to add 4 additional edges to the graph diagram in part (b) above so that the resulting new graph will still have no Hamiltonian circuit.

**10.** Explain why the graph below has no Hamiltonian circuit.

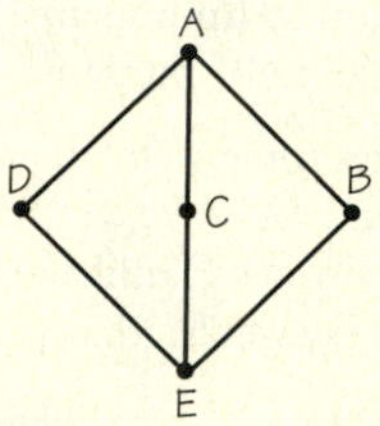


**11.** Use the graph shown in Exercise 10 to help you construct a connected graph for which every vertex has valence 3 and that does not have a Hamiltonian circuit.

■ **12.** Explain why the tour *ABCFECBDA* is not a Hamiltonian circuit for the graph below. Does this graph have a Hamiltonian circuit?

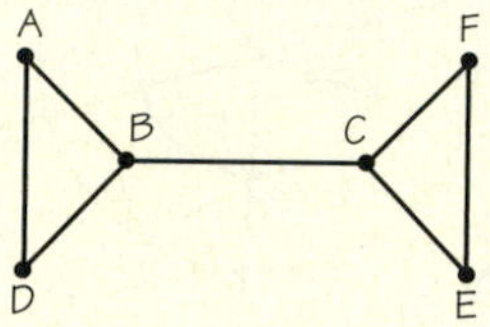


■ **13.** Do the following graphs have Hamiltonian circuits? If not, can you demonstrate why not?

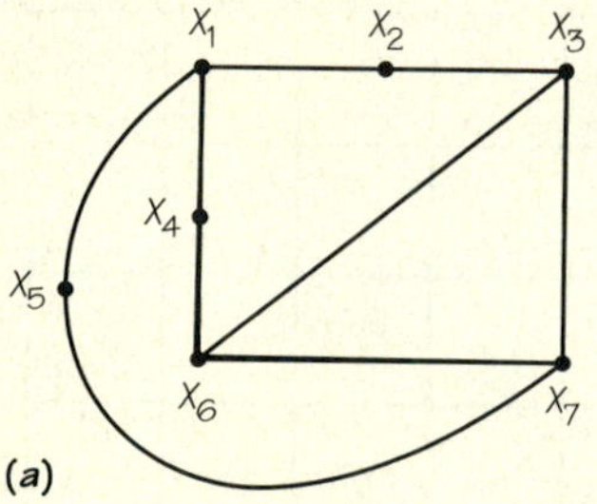


(*a*)

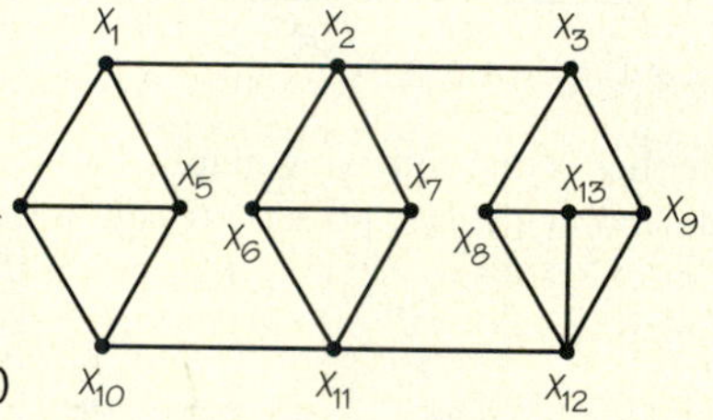


(*b*)

**14.** If an edge from $X_2$ to $X_5$ is added to each graph in Exercise 13, do the new graphs that result have a Hamiltonian circuit?

**15.** For each of the following graphs, determine whether there is a Hamiltonian circuit.

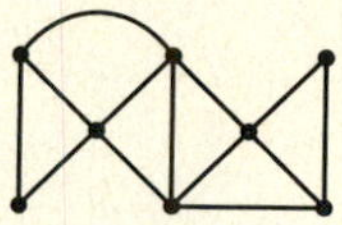

(*a*)

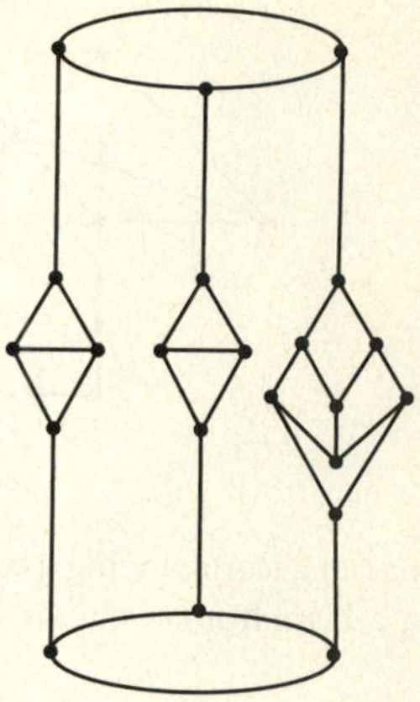

(*b*)

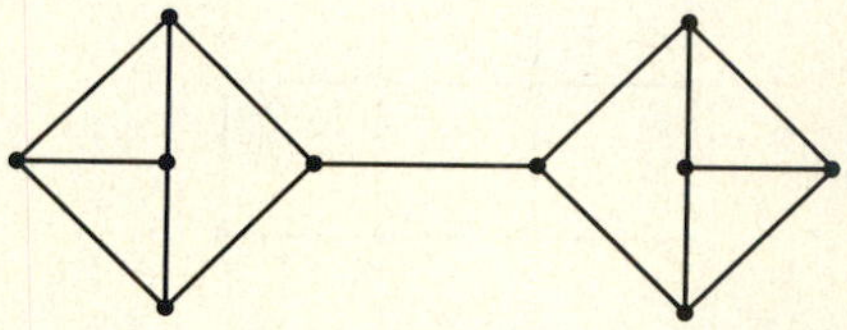

(*c*)

■ **16.** **(a)** The graph below is known as a four spokes and three concentric circles graph. What conditions on $m$ and $n$ guarantee that an $m$ spokes and $n$ concentric circles graph has a Hamiltonian circuit? (Assume $m \geq 2$, $n \geq 1$.)

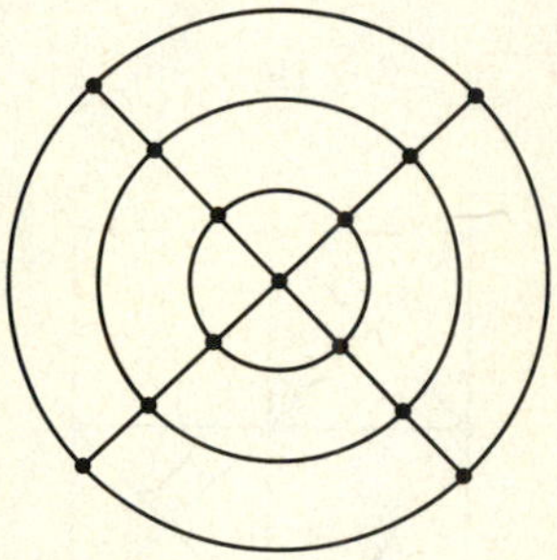

**(b)** The graph below is known as a $3 \times 4$ grid graph. What conditions on $m$ and $n$ guarantee that an $m \times n$ grid graph has a Hamiltonian circuit?

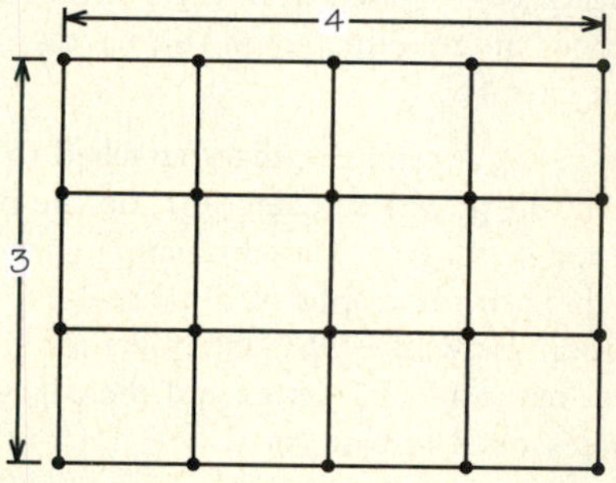


Can you think of a real-world situation in which finding a Hamiltonian circuit in an $m \times n$ grid graph would represent a solution to the problem? If an $m \times n$ grid graph has no Hamiltonian circuit, can you find a tour that repeats a minimum number of vertices and starts and ends at the same vertex?

**17.** A Hamiltonian path in a graph is a tour of the vertices of the graph that visits each vertex once and only once and starts and ends at different vertices.

**(a)** For each of the graphs shown in Exercise 13, does the graph have a Hamiltonian path?
**(b)** Does each of these graphs have a Hamiltonian path that starts at $X_1$ and ends at $X_2$?
**(c)** Describe three real-world situations where finding a Hamiltonian path in a graph would be required.

**18.** Using the terminology of Exercise 17, draw a graph that has

**(a)** a Hamiltonian path but no Hamiltonian circuit.
**(b)** an Euler circuit but no Hamiltonian path.
**(c)** a Hamiltonian path but no Euler circuit.

**19.** To practice your understanding of the concepts of Euler circuits and Hamiltonian circuits, determine for the following graphs (a) through (d) whether there is an Euler circuit and/or a Hamiltonian circuit. If so, write it down.

(*a*)

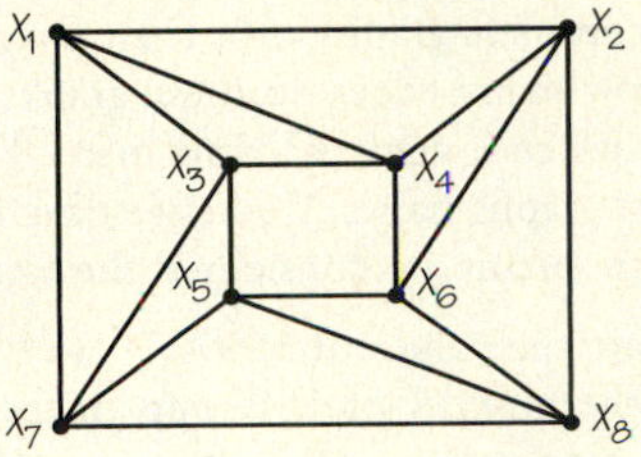


(*b*)

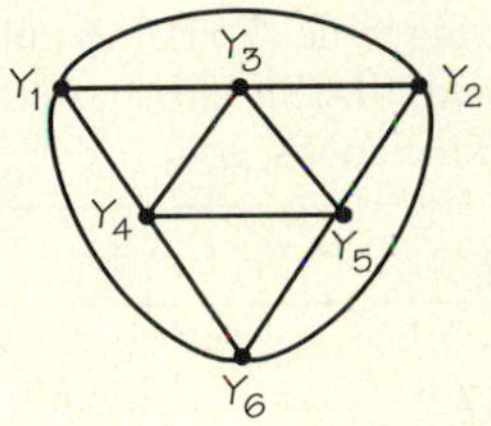


(*c*)

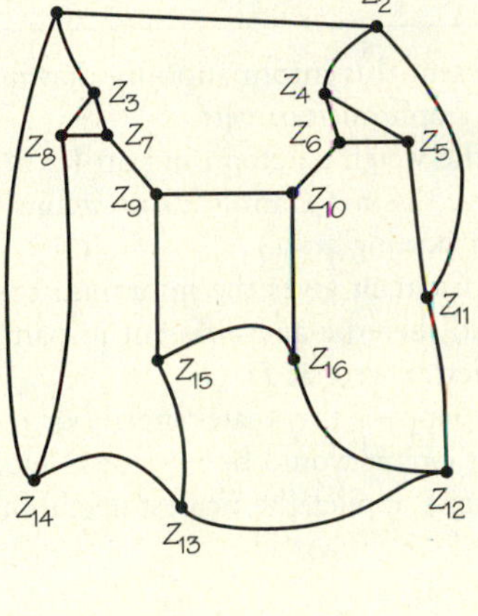


(*d*)

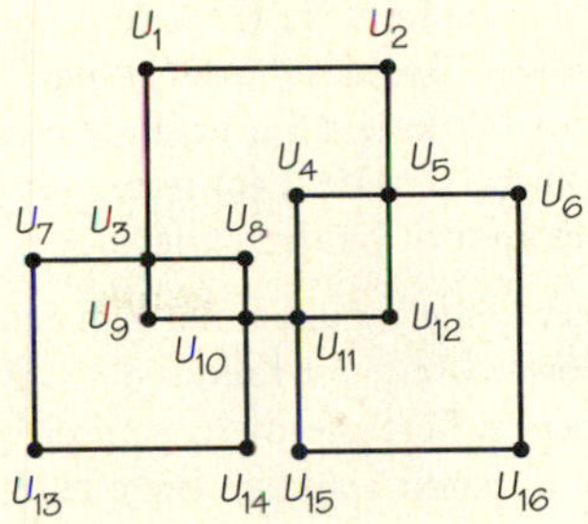


■ **20.** **(a)** The $n$-dimensional cube is obtained from two copies of an $(n-1)$-dimensional cube by joining corresponding vertices. (The process is illustrated for the 3-cube and the 4-cube in the following figure.) Can you show that every $n$-cube has a Hamiltonian circuit? [*Hint:* Show that if you know how to find a Hamiltonian circuit on an $(n-1)$-cube, then you can use two copies of this to build a Hamiltonian circuit on an $n$-cube.]

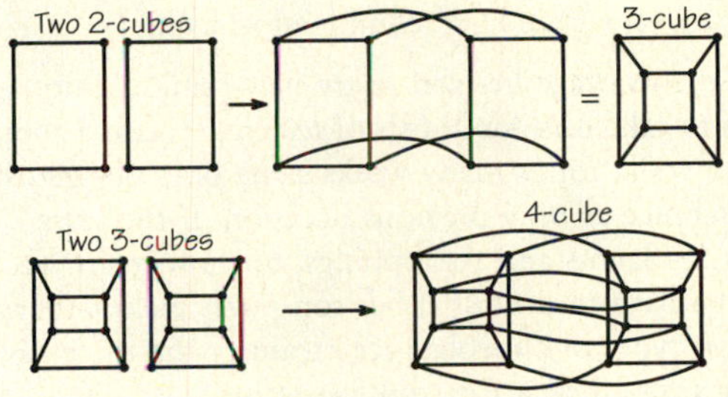


**(b)** Find formulas for the number of vertices and the number of edges of an $n$-cube.

**21.** If an edge is added from the vertex with subscript 4 to the vertex with subscript 5 in each graph in Exercise 19, which of the resulting graphs will have Hamiltonian circuits and which will have Euler circuits?

**22.** Find a family of graphs none of which have Hamiltonian circuits but for which adding a single edge to the first graph in the family creates a Hamiltonian circuit, adding two edges to the second graph in the family creates a Hamiltonian circuit, and so forth.

**23.** A Hamiltonian path in a graph is a tour of the vertices that visits each vertex once and only once and that starts and ends at different vertices.

**(a)** Draw an example of a graph that has no Hamiltonian path and where all the vertices are 3-valent.
**(b)** Draw a graph that has no Hamiltonian path but that does have an Euler circuit.
**(c)** By analogy with the Hamiltonian path, develop a definition of "Euler path."

**24.** **(a)** When going outside on a cold winter day, Jill can choose from three winter coats, five wool scarves, four pairs of boots, and three ski hats. How many outfits might her friends see her in?
**(b)** If Jill always insists on wearing her green wool scarf, how many outfits might her friends see her in?

**25.** The notes C, D, E, F, G, A, and B are to be used to form an ordered five-note musical logo. In how many ways can this be done if **(a)** no note can be repeated; **(b)** notes can be repeated; **(c)** notes can be repeated but all the notes cannot be the same?

**26.** A lottery game requires that a person select an upper- or lowercase letter followed by five different two-digit numbers (where the digits cannot both be zero). How many different ways are there to fill out a lottery ticket?

**27.** **(a)** In designing a security system for its accounts, a bank asks each customer to choose a five-digit number, all the digits to be distinct and nonzero. How many choices can a customer make?
**(b)** A suitcase with a liquid-crystal display allows one to unlock it with a specific combination of three capital letters that are not necessarily different. How many choices would a thief have to go through to be sure that all the possibilities had been tried? How does this compare to a "standard" combination lock?

**28.** To encourage her son to try new things, a mother offers to take him for a dish of ice cream with a topping once a week, for as many weeks as he does not get the same choice as on a previous occasion. If the store offers 12 flavors and six toppings, for how many weeks will she have to do this if her son never picks either of the two types of chocolate ice cream or the three types of nut topping that the store carries?

**29.** A large corporation has found that it has "outgrown" its current code system for routing interoffice mail. The current system places a code of three ordered, distinct nonzero digits on the mail. The new proposal calls for the use of two ordered capital letters. Does the new system have more code numbers than the old system? If so, how many more locations will the new system enable the company to encode over the current system?

**30.** Repeat Exercise 25a, except that exactly one of the notes in the musical logo must be a sharp and the note chosen to be sharped cannot appear elsewhere (for example, BCD#AG, where D# denotes D sharp).

◆ **31.** **(a)** In New York State, one type of license plate has three letters followed by three numbers. Suppose the digits from 0, 1, . . . , 9 can be used, except that all three digits cannot be zero, and that any letter from A to Z (repeats allowed) can be used. How many plates are possible?
**(b)** Investigate what schemes for license plates are used in your state and determine how many different plates are possible.

**32.** A restaurant offers 5 soups, 10 entrees, and 8 desserts. How many different choices for a meal can a customer make if one selection is made from each category? If 3 of the desserts are pies and the customer will never order pie, how many meals can the customer choose?

**33.** In the last several years, heavily populated regions that previously had only one area code have been divided into service areas with more than one area code. What is the largest number of different phone numbers that can be served using one area code? If an area code cannot begin with a zero, how many different area codes are possible?

**34.** **(a)** A credit-card company makes it easier for customers to memorize their PIN (personal identification number) by using a four-digit PIN that consists of three different digits selected from 0, 1, 2, . . . , 9 where one of the digits must be a zero, another is a nonzero digit that is repeated, and another is a digit different from these two. How many different PINs of this kind are there?
**(b)** How many PINs are possible if there are no restrictions on repeats of the 10 possible digits that can be used?

## 2.2 Traveling Salesman Problem

## 2.3 Helping Traveling Salesmen

**35.** Draw complete graphs with four, five, and six vertices. How many edges do these graphs have? Can you generalize to $n$ vertices? How many TSP tours would these graphs have? (Tours yielding the same Hamiltonian circuit are considered the same.)

**36.** Calculate the values of 5!, 6!, 7!, 8!, 9!, and 10!. Then find the number of TSP tours in the complete graph with nine vertices.

**37.** The following table shows the mileage between four cities: Springfield, Ill. ($S$); Urbana, Ill. ($U$); Effingham, Ill. ($E$); and Indianapolis, Ind. ($I$).

| | $E$ | $I$ | $S$ | $U$ |
|---|---|---|---|---|
| $E$ | – | 147 | 92 | 79 |
| $I$ | 147 | – | 190 | 119 |
| $S$ | 92 | 190 | – | 88 |
| $U$ | 79 | 119 | 88 | – |

**(a)** Represent this information by drawing a weighted complete graph on four vertices.
**(b)** Use the weighted graph in part (a) to find the cost of the three distinct Hamiltonian circuits in the graph. (List them starting at $U$.)
**(c)** Which circuit gives the minimum cost?
**(d)** Would there be any different in parts (b) and (c) if the start vertex were at $I$?
**(e)** If one applies the nearest-neighbor method starting at $U$, what circuit would be obtained? Does the answer change if one applies the nearest-neighbor algorithm starting at $S$? At $E$? At $I$?

**(f)** If one applies the sorted-edges method, what circuit would be obtained? Does one get the optimal answer?

**38.** After a party at her house, Francine (*F*) has agreed to drive Mary (*M*), Rachel (*R*), and Constance (*C*) home. If the times (in minutes) to drive between her friends' homes are shown below, what route gets Francine back home the quickest?

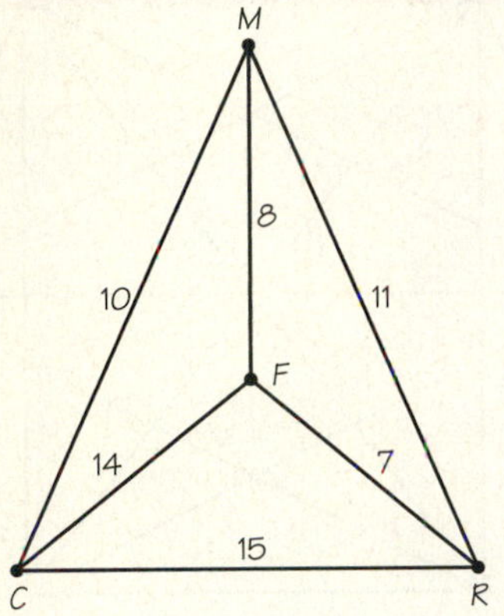


**39.** In Exercise 38, what route would Francine have to follow to get home as quickly as possible, assuming she promised to drive Mary home first?

**40.** In Exercise 38, Francine is planning to deliver her friends home and then spend the night at Rachel's house. What would her fastest route be?

**41.** Starting from the location where she moors her boat (*M*), a fisherwoman wishes to visit three areas–*A*, *B*, and *C*–where she has set fishing nets. If the times (in minutes) between the locales are given in the figure below, what route to visit the three sites and return to the mooring place would be optimal?

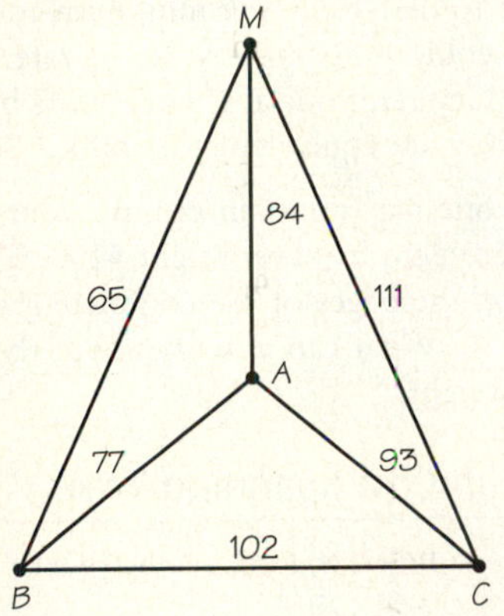


**42. (a)** For the two complete graphs that follow, find the costs of the nearest-neighbor tour starting at *B* and of the tour generated by the sorted-edges algorithm.

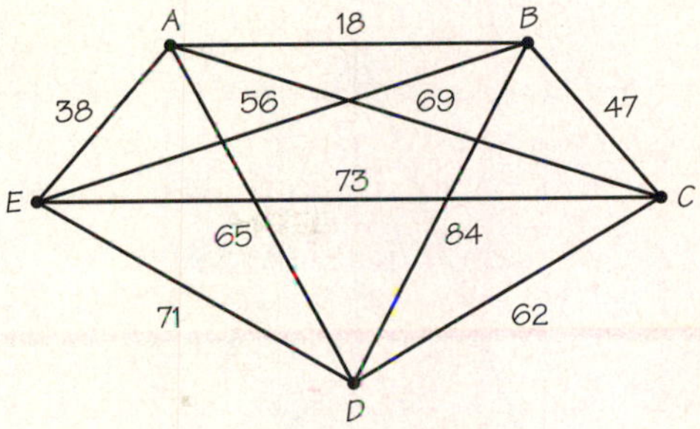

(*a*)

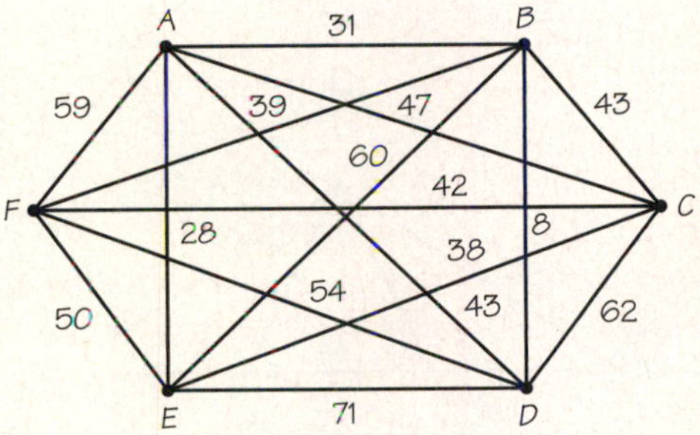

(*b*)

■ **(b)** How many Hamiltonian circuits would have to be examined to find a shortest route for part (a) by the brute force method?

**(c)** Invent an algorithm different from the sorted-edges and nearest-neighbor algorithms that is easy to apply for finding TSP solutions.

**43.** An airport limo must take its five passengers from the airport to different downtown hotels. Is this a traveling salesman problem, a Chinese postman problem, or an Euler circuit problem?

**44.** For each of the following graphs with weights, apply the nearest-neighbor method (starting at vertex *A*) and the sorted-edges method to find (it is hoped) a cheap tour.

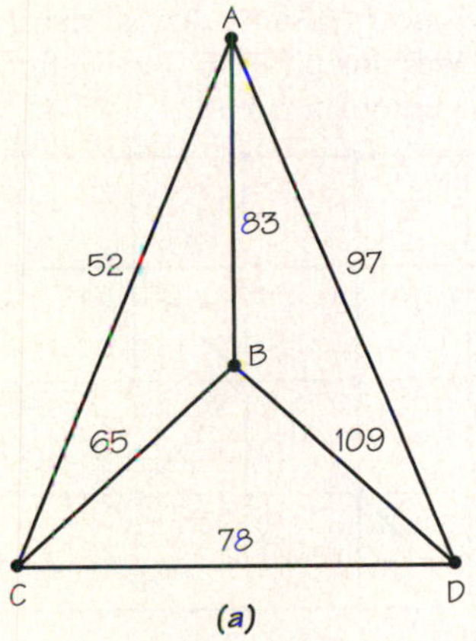

(*a*)

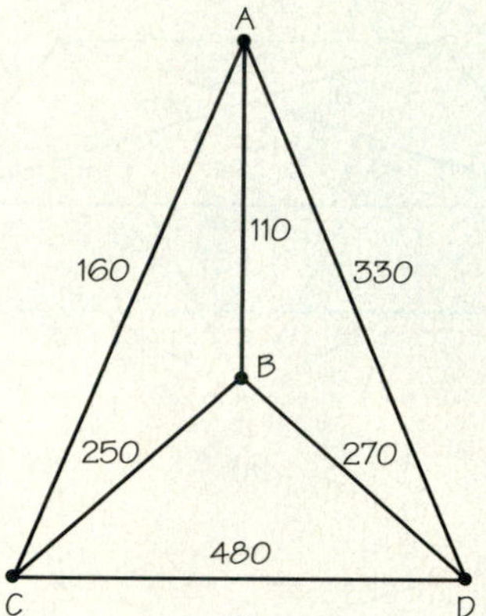


(b)

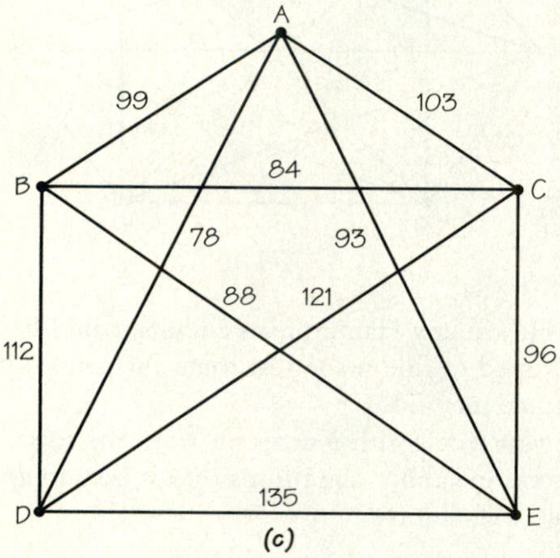


(c)

**45.** The following figure represents a town where there is a sewer located at each corner (where two or more streets meet). After every thunderstorm, the department of public works wishes to have a truck start at its headquarters (at vertex $H$) and make an inspection of sewer drains to be sure that leaves are not clogging them. Can a route start and end at $H$ that visits each corner exactly once? (Assume that all the streets are two-way streets.) Does this problem involve finding an Euler circuit or a Hamiltonian circuit?

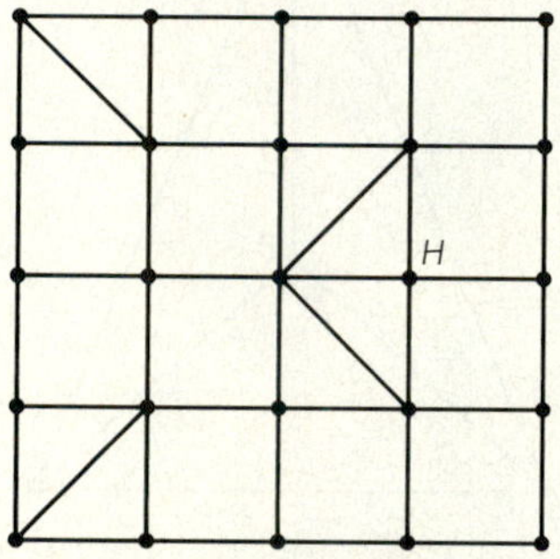


Assume that at equally spaced intervals along the blocks in this graph there are storm sewers that must be inspected after each thunderstorm to see if they are clogged. Is this a Hamiltonian circuit problem, an Euler circuit problem, or a Chinese postman problem? Find an optimal tour to do this inspection.

**46.** **(a)** Solve the six-city TSP shown in the diagram using the nearest-neighbor algorithm starting at vertex $A$ and starting at vertex $B$.
**(b)** Apply the sorted-edges method.

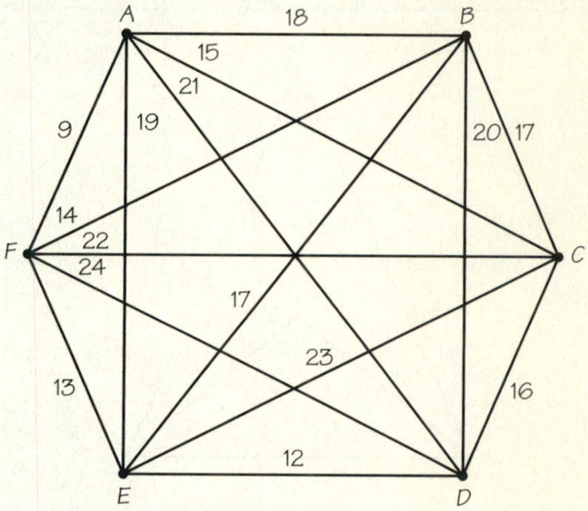


■ **47.** Construct an example of a complete graph of five vertices, with distinct weights on the edges for which the nearest-neighbor algorithm starting at a particular vertex and the sorted-edges algorithm yield different solutions for the traveling salesman problem. Can you find a five-vertex complete graph with weights on the edges in which the optimal solution, the nearest-neighbor solution, and the sorted-edges algorithm solution are all different?

■ **48.** If the brute force method of solving a 20-city TSP is employed, use a calculator to determine how many Hamiltonian circuits must be examined. How long would it take to determine the minimum-cost tour if the cost of tours could be computed at the rate of 1 billion per second? (Convert your answer to years by seeing how many years are equivalent to a billion seconds!)

**49.** Suppose one has found an optimal tour for a given 10-city TSP problem to have weight 4200. Now suppose the weights on the edges of the complete graph are increased by 50. What can you say about the optimal tour and its weight?

## 2.4 Minimum-Cost Spanning Trees

**50.** For each graph below, explain why it is or is not a tree.

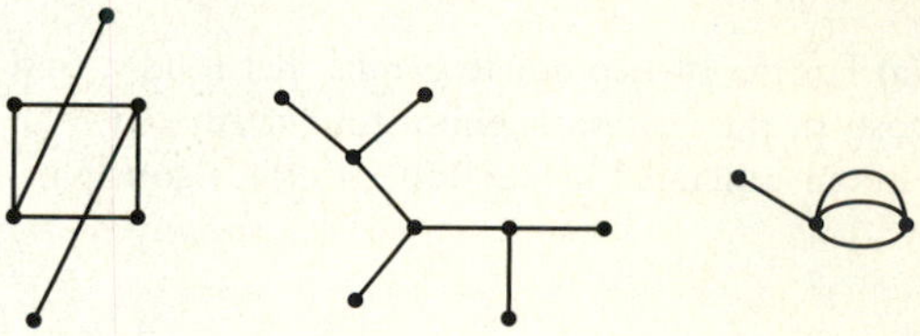

(a) (b) (c)

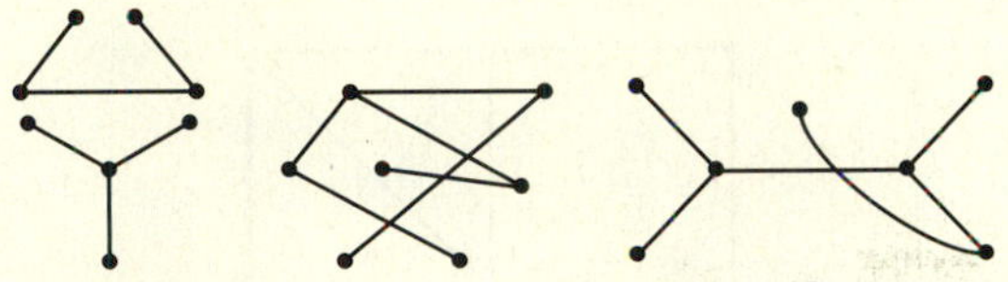

(d) (e) (f)

**51.** For each of the diagrams below, explain why the wiggly edges are not

**(a)** a spanning tree.

**(b)** a Hamiltonian circuit.

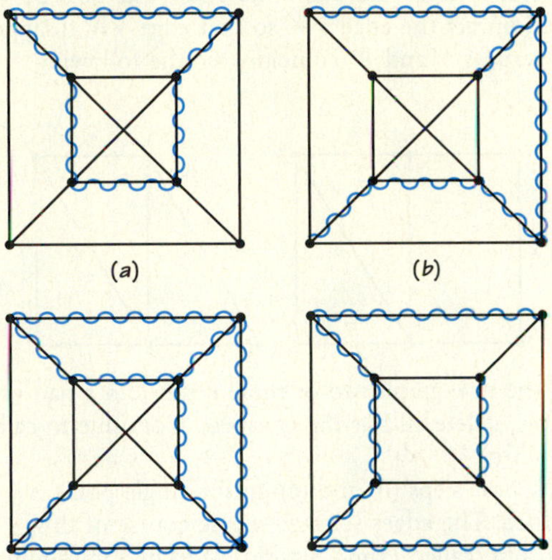

(a) (b) (c) (d)

**52.** Find all the spanning trees in the graphs below.

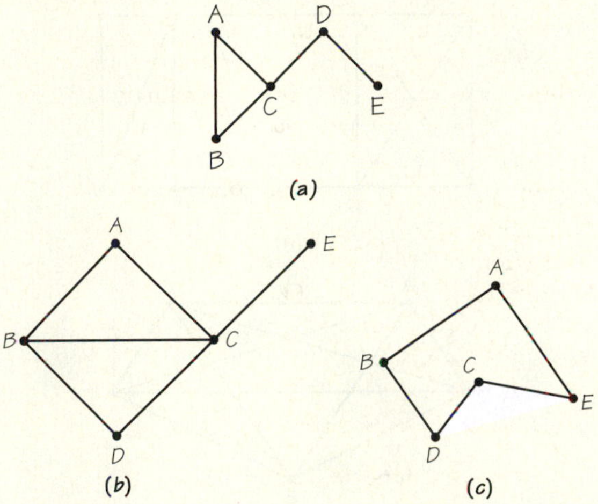


(a) (b) (c)

**53.** Use Kruskal's algorithm to find a minimum-cost spanning tree for the following graphs (a), (b), (c), and (d). In each case, what is the cost associated with the tree?

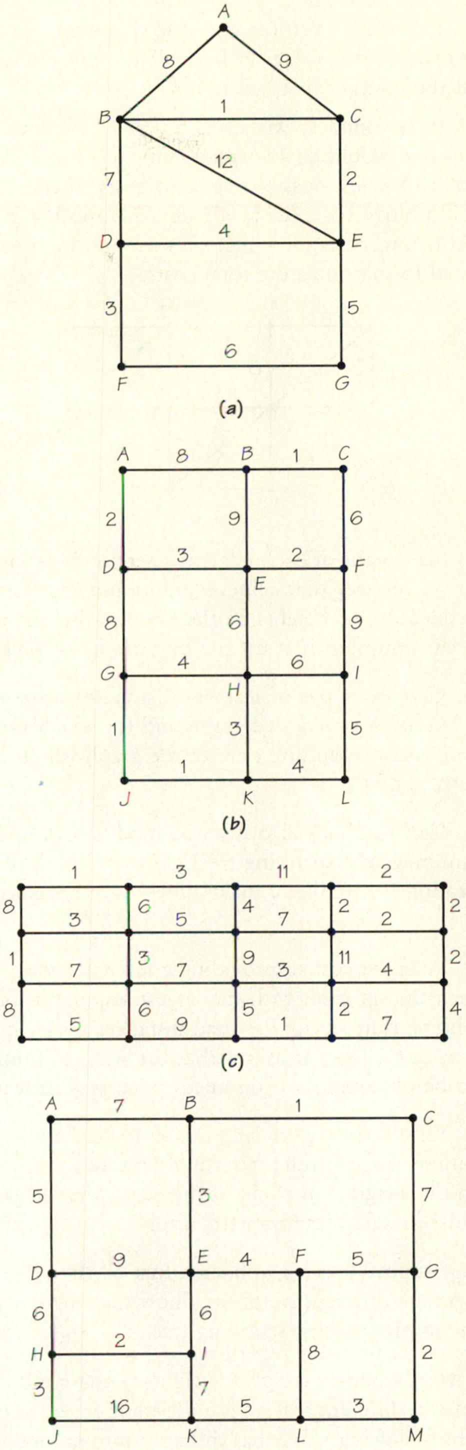


(a) (b) (c) (d)

**54.** A connected graph $G$ has 16 vertices. How many edges does a spanning tree of $G$ have? How many vertices does a spanning tree of $G$ have? What can one say about the number of edges $G$ has?

**55.** A connected graph $H$ has a spanning tree with 26 edges. How many vertices does the spanning tree have? How many vertices does $H$ have? What can one say about the number of edges $H$ has?

**56.** A large company wishes to install a pneumatic tube system that would enable small items to be sent between any of 10 locales, possibly by using relay. If the nonprohibitive costs (in \$100) are shown in the graph model below, between which sites should the tube be installed to minimize the total cost?

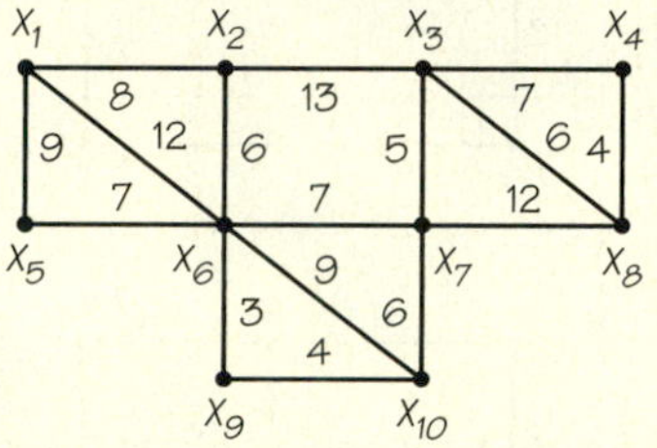


**57.** If the weight of each edge in Exercise 56 is increased by 3, will the tree that achieves minimum cost for the new collection of weights be the same as the one that achieves minimum cost for the original set of weights?

◆ **58.** Give examples of real-world situations that can be modeled using a weighted graph and for which finding a minimum-cost spanning tree for the graph would be of interest.

■ **59.** Can Kruskal's algorithm be modified to find a maximum-weight spanning tree? Can you think of an application for finding a maximum-weight spanning tree?

◆ **60.** Find the cost of providing a relay network between the six cities with the largest populations in your home state, using the road distances between the cities as costs. Does it follow that the same solution would be obtained if air distances were used instead?

■ **61.** Would there ever be a reason to find a minimum-cost spanning tree for a weighted graph in which the weights on some of the edges were negative? Would Kruskal's algorithm still apply?

■ **62.** Suppose $G$ is a graph such that all the weights on its edges are different numbers. Show that there is a unique minimum-cost spanning tree.

**63.** Two spanning trees of a (weighted) graph are considered different if they use different edges. Show that the following graph has different minimum-cost spanning trees, though all these different trees have the same cost.

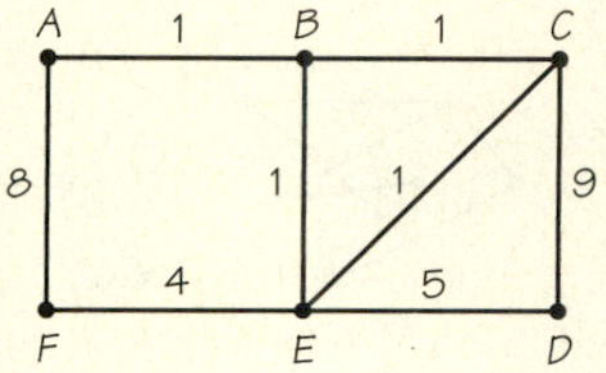


**64.** Let $G$ be a graph with weights assigned to each edge. Consider the following algorithm:

**(a)** Pick any vertex $V$ of $G$.

**(b)** Select an edge $E$ with a vertex at $V$ that has a minimum weight. Let the other endpoints of $E$ be $W$.

**(c)** Contract the edge $VW$ so that edge $VW$ disappears and vertices $V$ and $W$ coincide (see the following figures).

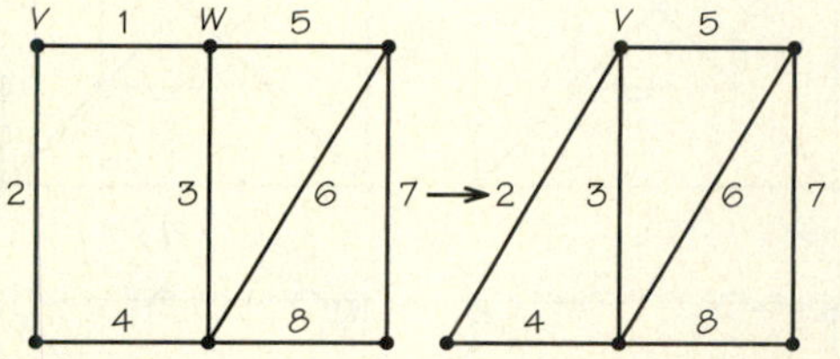


If in the new graph two or more edges join a pair of vertices, delete all but the cheapest. Continue to call the new vertex $V$.

**(d)** Repeat steps (b) and (c) until a single point is obtained. The edges selected in the course of this algorithm (called Prim's algorithm) form a minimum-cost spanning tree. Apply this algorithm to the following graphs.

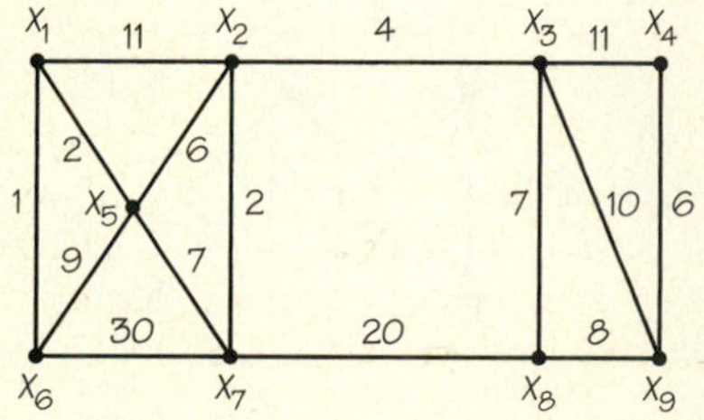


(*a*)

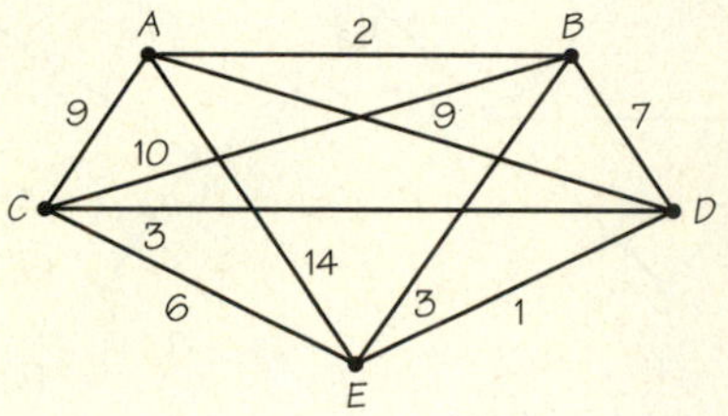


(*b*)

**65.** Determine whether each of the following statements is true or false for a minimum-cost spanning tree $T$ for a weighted connected graph $G$:

**(a)** $T$ contains a cheapest edge in the graph.

**(b)** $T$ cannot contain a most expensive edge in the graph.

**(c)** $T$ contains one fewer edge than there are vertices in $G$.
**(d)** There is some vertex in $T$ to which all others are joined by edges.
**(e)** There is some vertex in $T$ that has valence 3.

■ **66.** In the following graphs, the number in the circle for each vertex is the cost of installing equipment at the vertex if relaying must be done at the vertex, while the number on an edge indicates the cost of providing service between the endpoints of the edge.

In each case, find the minimum cost (allowing relays) for sending messages between any pair of vertices, taking vertex relay costs into account.

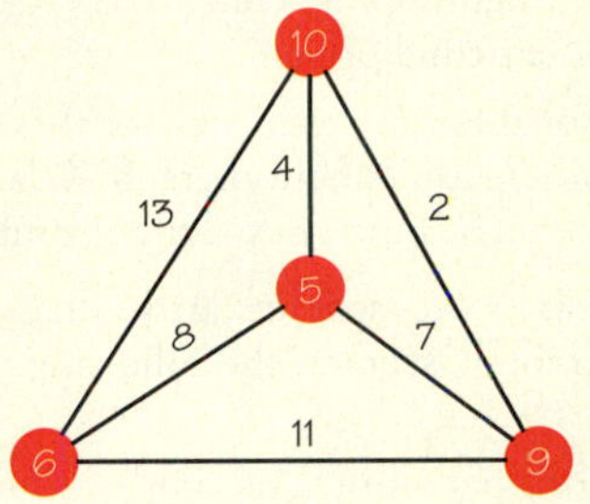


**(a)**

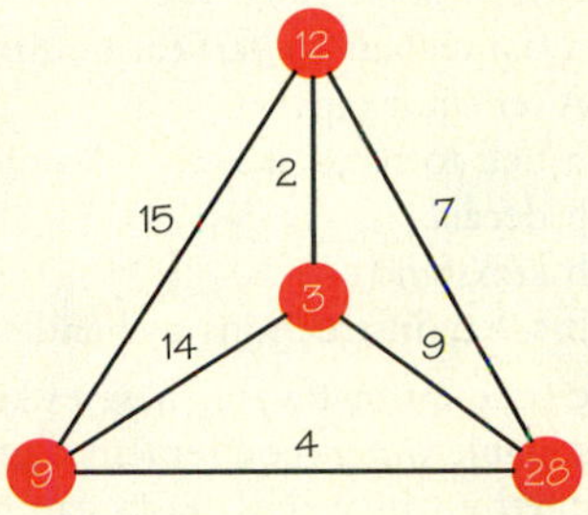


**(b)**

Would your answer be different if vertex relay costs were neglected? (*Warning:* Kruskal's algorithm cannot be used to answer the first question. This problem illustrates the value of having an algorithm over relying on "brute force.")

**67. (a)** Show that for each edge of graph $J$ below there is a spanning tree of $J$ that avoids that edge.
**(b)** For each spanning tree that you found in graph $J$, count the number of vertices and edges. Do you notice any pattern?
**(c)** For graph $H$ below and each edge in the graph, is there a spanning tree that does not include that edge of $H$?

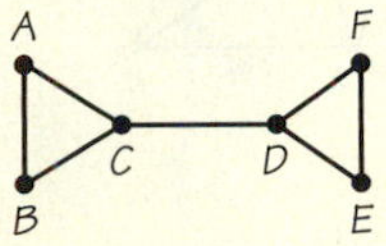


*Graph H*

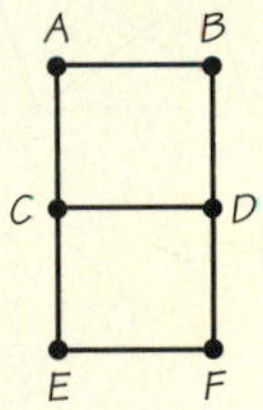


*Graph J*

**68. (a)** The table shown gives the "closeness" or distance values between four objects. Construct a four-vertex tree with weights on its edges such that the distances between pairs of vertices of the tree (as measured by the sum of the weights on the path in the tree between these vertices) give rise to this table.

| | A | B | C | D |
|---|---|---|---|---|
| A | 0 | 3 | 10 | 14 |
| B | 3 | 0 | 7 | 11 |
| C | 10 | 7 | 0 | 4 |
| D | 14 | 11 | 4 | 0 |

**(b)** Produce several real-world contexts that might give rise to the situation described here.

**69.** The figure below represents four objects using a tree with weights on the edges. Construct a table with four rows and four columns recording how "close" pairs of vertices in the tree are to each other. To find how close a pair of objects is, add together the weights along the path that joins these two objects.

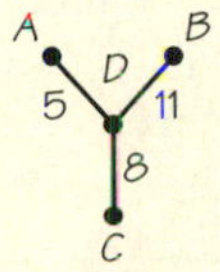


## 2.5 Critical-Path Analysis

**70.** Find the earliest completion time and critical paths for the order-requirement digraphs below.

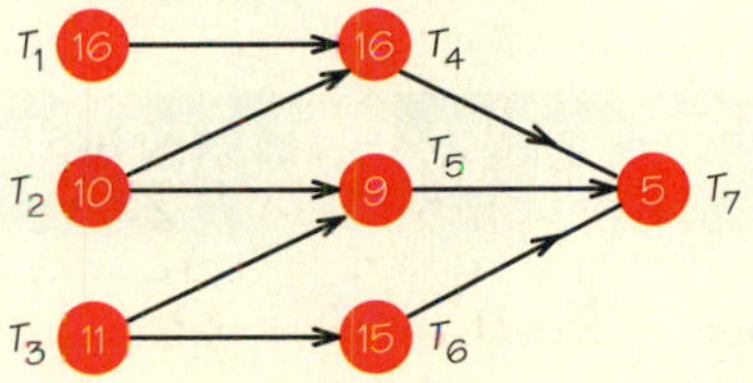


*(a)*

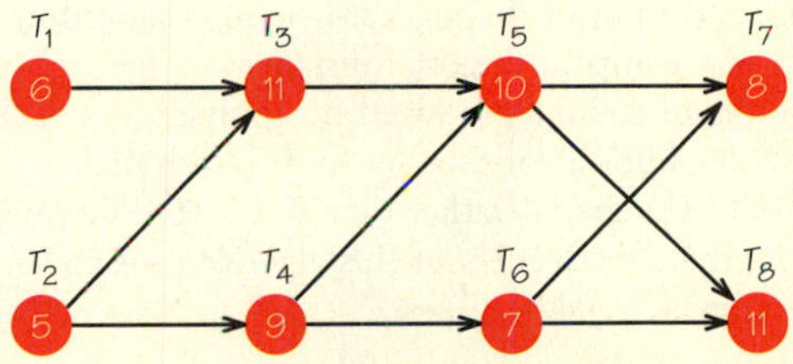


*(b)*

**71.** Find the earliest completion time and critical paths for the following order-requirement digraphs.

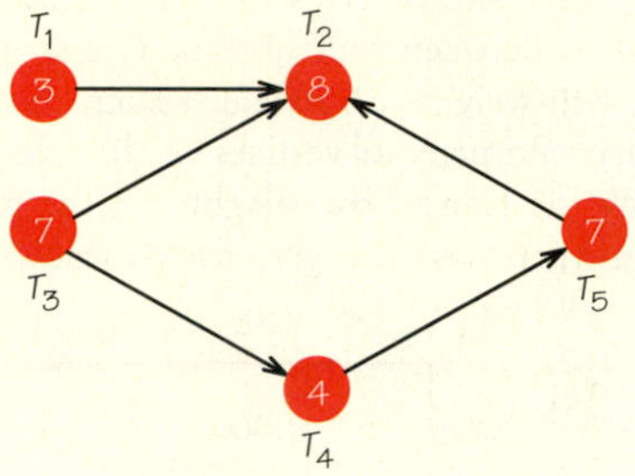

(*a*)

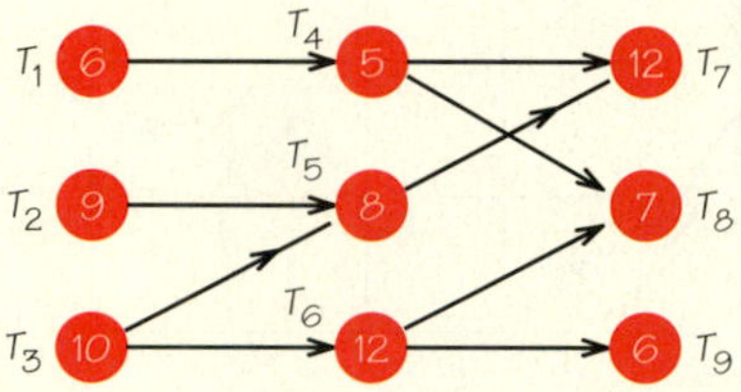

(*b*)

**72.** Construct an example of an order-requirement digraph with two different critical paths.

**73.** In the order-requirement digraph below, determine which tasks, if shortened, would reduce the earliest completion time and which would not. Then find the earliest completion time if task $T_5$ is reduced to time length 7. What is the new critical path?

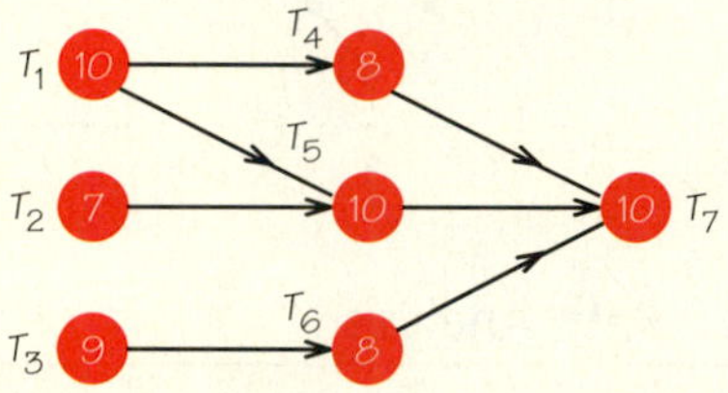


**74.** For the order-requirement digraph in Exercise 73, find the critical path and the task(s) in the critical path whose time, when reduced the least, creates a new critical path.

**75.** To build a new addition on a house, the following tasks must be completed:

**(a)** Lay foundation.
**(b)** Erect sidewalls.
**(c)** Erect roof.
**(d)** Install plumbing.
**(e)** Install electric wiring.
**(f)** Lay tile flooring.
**(g)** Obtain building permits.
**(h)** Put in door that connects new room to existing house.
**(i)** Install track lighting on ceiling.
**(j)** Install wall air-conditioner.

Construct reasonable time estimates for these tasks and a reasonable order-requirement digraph. What is the fastest time in which these tasks can be completed?

**76.** At a large toy store, scooters arrive unassembled in boxes. To assemble a scooter, the following tasks must be performed:

TASK 1. Remove parts from the box.
TASK 2. Attach wheels to the footboard.
TASK 3. Attach vertical housing.
TASK 4. Attach handlebars to vertical housing.
TASK 5. Put on reflector tape.
TASK 6. Attach bell to handlebars.
TASK 7. Attach decals.
TASK 8. Attach kickstand.
TASK 9. Attach safety instructions to handlebars.

Give reasonable time estimates for these tasks and construct a reasonable order-requirement digraph. What is the earliest time by which these tasks can be completed?

**77.** Construct an order-requirement digraph with six tasks that has three critical paths of length 26.

## APPLET EXERCISES

**To do these exercises, go to www.whfreeman.com/fapp8e.**

**1. TSP: Nearest-Neighbor Algorithm.** There is an extended version of the nearest-neighbor algorithm, in which you compare the total distances of the Hamiltonian circuits produced by applying the ordinary nearest-neighbor algorithm starting at each of the vertices of the graph (rather than just a specific one). Explore the effectiveness of this algorithm using the *TSP: Nearest-Neighbor* applet.

**2. TSP: Sorted-Edges Algorithm.** Go to the *TSP: Sorted Edges* applet, where you can apply the sorted-edges algorithm to see if it solves the traveling salesman problem for the following graphs (and others):

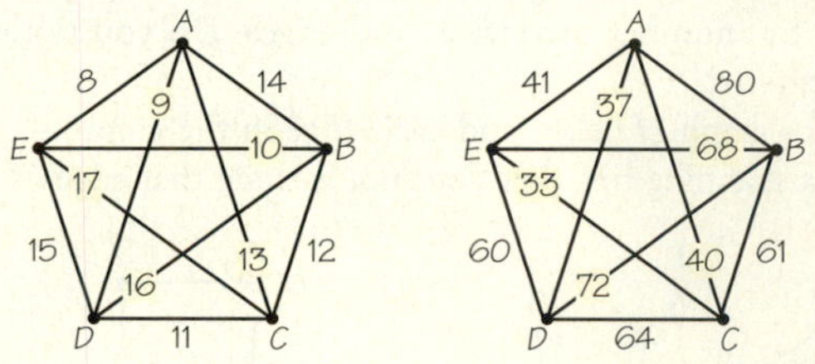

(*a*) (*b*)

**3. Kruskal's Algorithm.** Go to the *Kruskal's Algorithm* applet, where you can apply Kruskal's algorithm to find the minimum-cost spanning trees in the following graphs (and others):

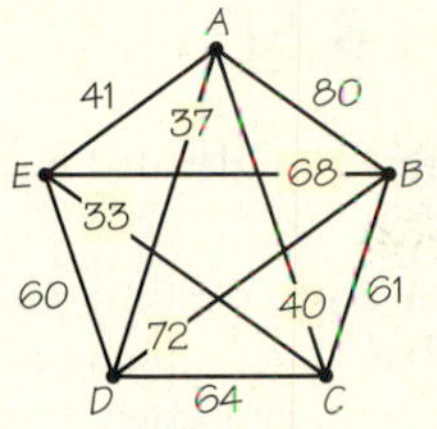

(*a*)

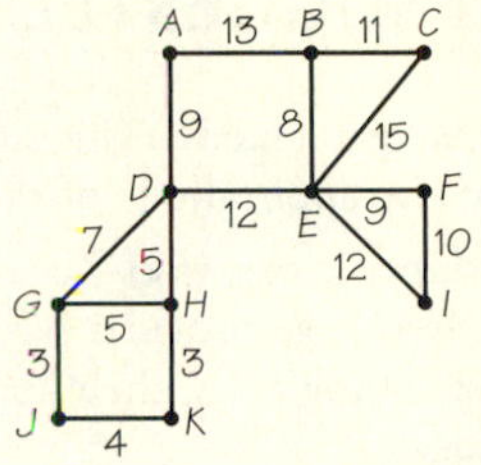

(*b*)

## WRITING PROJECTS

**1.** Write an essay about a variety of situations in which you are personally involved for which a solution of the TSP is (perhaps implicitly) required. Explain under what circumstances it might be valuable to carry out a formal mathematical solution to such TSPs rather than use an ad hoc solution.

**2.** Construct an example, of the kind suggested on page 42, that shows that in a situation where three day campers must be picked up and brought to camp, it may make a difference if the optimization criterion is minimizing distance traveled by the camp bus versus minimizing average time that the children spend on the bus.

**3.** Determine the six largest cities in the state in which you live. By consulting a road atlas (or by some other means) construct the graph that represents the road distances between your hometown and these six other cities. Now apply (a) the nearest-neighbor method, (b) the sorted-edges method, and (c) the nearest neighbor from each city, and pick the minimum tour method to solve the associated TSP. Do you have reason to believe that the answers you get might include an optimum solution among them?

## SUGGESTED READINGS

BODIN, LAWRENCE. Twenty years of routing and scheduling, *Operations Research,* 38 (1990): 571–579. A survey of real-world situations where routing and scheduling were used, written by a pioneer in this area.

DOLAN, ALAN, and JOAN ALDUS. *Networks and Algorithms: An Introductory Approach,* Wiley, Chichester, England, 1993. An excellent introduction to graph theory algorithms.

GUSFIELD, DAN. *Algorithms on Strings, Trees, and Sequences,* Cambridge University Press, New York, 1997. Details applications of graph theory in pattern recognition and reconstruction problems.

JONES, NEIL C., and PAVEL A. PEVZNER, *An Introduction to Bioinformatics Algorithms,* MIT Press, Cambridge, Mass., 2004. This book has material on how graph theory ideas, particularly those related to Hamiltonian circuits, are being used in molecular genetics and computational biology.

LAWLER, EUGENE, J. LENSTRA, RINNOY KAN, and D. SHMOYS, eds. *The Traveling Salesman Problem,* Prentice-Hall, Englewood Cliffs, N.J., 1985. Includes survey and technical articles on all aspects of the TSP.

LUCAS, WILLIAM, FRED ROBERTS, and ROBERT THRALL, eds. *Discrete and Systems Models,* vol. 3: *Modules in Applied Mathematics,* Springer-Verlag, New York, 1983. Chapter 6, "A Model for Municipal Street Sweeping Operations," by A. Tucker and L. Bodin, describes street-sweeping and related models in detail. Other chapters detail many recent applications of mathematics.

ROBERTS, FRED S., and BARRY TESMAN, *Applied Combinatorics,* 2nd ed., Pearson Prentice Hall, Upper Saddle River, N.J., 2004. The material on network-optimization problems is excellent.

ROBERTS, FRED. *Graph Theory and Its Applications to Problems of Society,* Society for Industrial and Applied Mathematics, Philadelphia, 1978. A very readable account of how graph theory is finding a wide variety of applications.

## SUGGESTED WEB SITES

**www.tsp.gatech.edu** This site provides a detailed history and many applications of the TSP.

**en.wikipedia.org/wiki/Minimum_cost_spanning_tree** This Web page provides basic ideas about minimum-cost spanning trees, their applications, and extensions of this idea.

**www-gap.dcs.st-and.ac.uk/~history/Mathematicians/Hamilton.html** This site provides biographical information about William Rowan Hamilton, for whom Hamiltonian circuits are named.

**www.ams.org/featurecolumn/archive/tsp.html**
**www.ams.org/featurecolumn/archive/trees.html** These sites provide some history and information about applications of the Traveling Salesman Problem and of minimum-cost spanning trees.

CHAPTER 3



# Planning and Scheduling

In a society as complex as ours, everyday problems such as providing services efficiently and on time require accurate planning of both people and machines. Take the example of a hospital in a major city. Around-the-clock scheduling of nurses and doctors must be provided to guarantee that people with particular expertise are available during each shift. The operating rooms must be scheduled in a manner flexible enough to deal with emergencies. Equipment used for X-ray, CT, or MRI scans must be scheduled for maximum efficiency.

Although many scheduling problems are often solved on an ad hoc basis, we can also use mathematical ideas to gain insight into the complications that arise in scheduling. The ideas we develop in this chapter have practical value in a relatively narrow range of applications, but they shed light on many characteristics of more realistic, and hence more complex, scheduling problems.

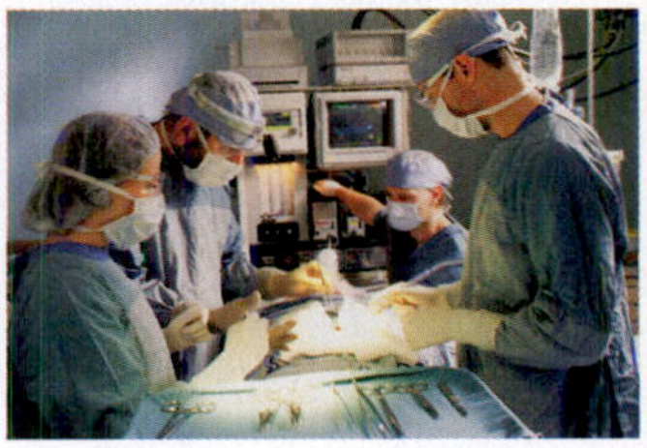

Hospitals are increasingly making use of mathematical techniques applied to scheduling problems. To make efficient use of one or more operating rooms requires the complicated assembly of a team of doctors, nurses, equipment, and support staff. Mathematical techniques for scheduling have made it possible to carry out more operations in less time. (*Eyewire/Punchstock.*)

## 3.1 Scheduling Tasks

Assume that a certain number of identical **processors** (machines, humans, or robots) work on a series of tasks that make up a job. Associated with each task is a specified amount of time required to complete the task. For simplicity, we assume that any of the processors can work on any of the tasks. Our problem, known as the **machine-scheduling problem**, is to decide how the tasks should be scheduled so that the completion time for the tasks collectively is as early as possible.

Even with these simplifying assumptions, complications in scheduling will arise. Some tasks may be more important than others and perhaps should be scheduled first. When "ties" occur, they must be resolved by special rules. As an example, suppose we are scheduling patients to be seen in a hospital emergency room staffed by one doctor. If two patients arrive simultaneously, one with a bleeding foot, the other with a bleeding arm, which patient should be examined first? Suppose the doctor treats the arm patient first, and while treatment is going on, a person in cardiac arrest arrives. Scheduling rules must establish appropriate priorities for cases such as these.

Another common complication arises with jobs consisting of several tasks that cannot be done in an arbitrary order. For example, if the job of putting up a new house is treated as a scheduling problem, the task of laying the foundation must precede the task of putting up the walls, which in turn must be completed before work on the roof can begin. The plumbing system can be scheduled for installation later.

## Assumptions and Goals

To simplify our analysis, we need to make clear and explicit assumptions:

1. If a processor starts work on a task, the work on that task will continue without interruption until the task is completed.
2. No processor stays voluntarily idle. In other words, if there is a processor free and a task available to be worked on, then that processor will immediately begin work on that task.
3. The requirements for ordering the tasks are given by an order-requirement digraph. (A typical example is shown in Figure 3.1, with task times highlighted within each vertex. The ordering of the tasks imposed by the order-requirement digraph often represents constraints of physical reality. For example, you cannot fly a plane until it has taken fuel on board.)
4. The tasks are arranged in a **priority list** that is independent of the order requirements. (The priority list is a ranking of the tasks according to some criterion of "importance.")

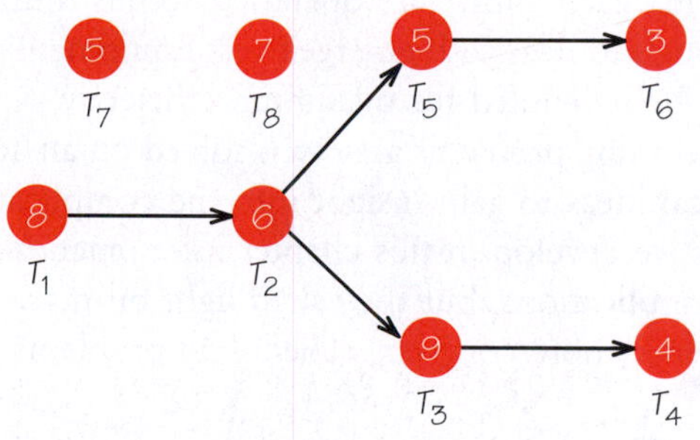


**FIGURE 3.1** A typical order-requirement digraph. Tasks with no edges entering or leaving the vertices representing them ($T_7$, $T_8$) can be more flexibly scheduled than the other tasks.

## EXAMPLE 1 ■ Home Construction

Let's see how these assumptions might work for an example involving a home construction project. In this case, the processors are human workers with identical skills. Assumption 1 means that once a worker begins a task, the work on this task is finished without interruption. Assumption 2 means that no worker will stay idle if there is some task for which the predecessors are finished. Assumption 3 requires that the ordering of the tasks be summarized in an order-requirement digraph. This digraph would code facts such as that the site must be cleared before the task of laying the foundation is begun. Assumption 4 requires that the tasks be ranked in a list from some perspective, perhaps a subjective view.

The task with highest priority rank is listed first in the list, followed left to right by the other tasks in priority rank. The priority list might be based on the size of the payments made to the construction company when a task is completed, even though these payments have no relation to the way the tasks must be done, as indicated in the order-requirement digraph. Alternatively, the priority list might reflect an attempt to find an algorithm to schedule the tasks needed to complete the whole job more quickly.

When considering a scheduling problem, there are various goals we might want to achieve. Among these are:

**Goal 1.** Minimizing the completion time of the job.

**Goal 2.** Minimizing the total time that processors are idle.

**Goal 3.** Finding the minimum number of processors necessary to finish the job by a specified time.

In the context of the construction example, goal 1 would complete the home as quickly as possible. Goal 2 would ensure that workers, who are perhaps paid by the hour, were not paid for doing nothing. One way of accomplishing this would be to hire one fewer worker even if it means the house takes longer to finish. Goal 3 might be reasonable if the family wants the house done before the first day of school, even if they have to pay a lot more workers to get the house done by this time.

For now we will concentrate on goal 1, finishing all the tasks at the earliest possible time. Note, however, that optimizing with respect to one goal may not optimize with respect to another. Our discussion here goes beyond what was discussed in Chapter 2 (see section 2.4) by dealing with how to assign tasks in a job to the processors that do the work. To build a new skyscraper involves designing a schedule for who will do what work when.

## List-Processing Algorithm

The scheduling problem we have described sounds more complicated than the traveling salesman problem (TSP). Indeed, like the TSP, it is known to be NP-complete. This means that it is unlikely that anyone will ever find a computationally fast algorithm that can find an optimal solution. Thus, we will be content to seek a solution method that is computationally fast and gives only approximately optimal answers.

### List-Processing Algorithm: Part I and Ready Task — PROCEDURE

The algorithm we use to schedule tasks is the **list-processing algorithm**. In describing it, we will call a task **ready** at a particular time if all its predecessors as indicated in the order-requirement digraph have been completed at that time. In Figure 3.1 at time 0, the ready tasks are $T_1$, $T_7$, and $T_8$, while $T_2$ cannot be ready until 8 time units after $T_1$ is started. The algorithm works as follows: At a given time, assign to the lowest-numbered free processor the first task on the priority list that is *ready* at that time and that hasn't already been assigned to a processor.

In applying this algorithm, we will need to develop skill at coordinating the use of the information in the order-requirement digraph and the priority list. It will be helpful to cross out the tasks in the priority list as they are assigned to a processor to keep track of which tasks remain to be scheduled.

## EXAMPLE 2 ■ Applying the List-Processing Algorithm

Let's apply the list-processing algorithm to one possible priority list—$T_8$, $T_7$, $T_6$, . . . , $T_1$—using two processors and the order-requirement digraph in Figure 3.1. The result is the schedule shown in Figure 3.2, where idle processor time (time during which a processor is not at work on a task) is indicated by white. How does the list-processing algorithm generate this schedule?

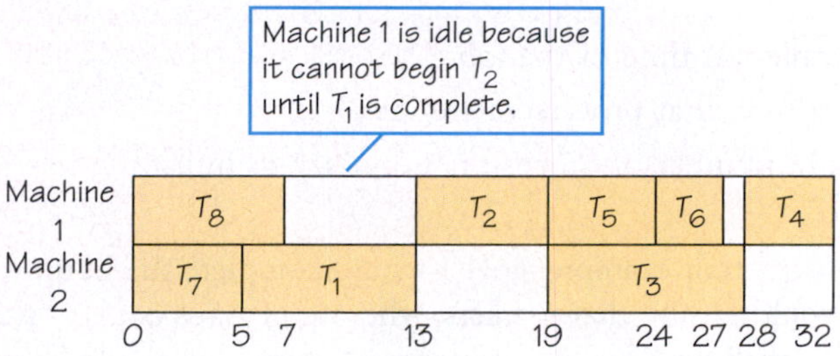


**FIGURE 3.2** The schedule produced by applying the list-processing algorithm to the order-requirement digraph in Figure 3.1 using the list $T_8$, $T_7$, . . . , $T_1$.

$T_8$ (task 8) is first on the priority list and ready at time 0 since it has no predecessors. It is assigned to the lowest-numbered free processor, processor 1. Task 7, next on the priority list, is also ready at time 0 and thus is assigned to processor 2. The first processor to become free is processor 2 at time 5. Recall that by assumption 1, once a processor starts work on a task, its work cannot be interrupted until the task is complete. Task 6, the next unassigned task on the list, is not ready at time 5, as can be seen by consulting Figure 3.1. The reason task 6 is not ready at time 5 is that task 5 has not been completed by time 5. In fact, at time 5, the only ready task on the list is $T_1$, so that task is assigned to processor 2. At time 7, processor 1 becomes free, but no task becomes ready until time 13.

Thus, processor 1 stays idle from time 7 to time 13. At this time, because $T_2$ is the first ready task on the list not already scheduled, it is assigned to processor 1. Processor 2, however, stays idle because no other ready task is available at this time. The remainder of the scheduling shown in Figure 3.2 is completed in this manner. ■

We can summarize this procedure as follows:

**List-Processing Algorithm: Part II** PROCEDURE

As the priority list is scanned from left to right to assign a task to a processor at a particular time, we pass over tasks that are not ready to find ones that are ready. If no task can be assigned in this manner, we keep one or more processors idle until such time that, reading the priority list from the left, there is a ready task not already assigned. After a task is assigned to a processor, we resume scanning the priority list, starting over at the far left, for unassigned tasks.

## When Is a Schedule Optimal?

The schedule in Figure 3.2 has a lot of idle time, so it may not be optimal. Indeed, if we apply the list-processing algorithm for two processors to another possible priority list $T_1$, . . . , $T_8$, using the digraph in Figure 3.1, the resulting schedule is that shown in Figure 3.3.

Here are the details of how we arrived at this schedule. Remember that we must coordinate the list $T_1$, $T_2$, . . . , $T_8$ with the information in the order-requirement digraph shown in Figure 3.3a. At time 0, task $T_1$ is ready, so this task is assigned to processor 1. However, at time 0, tasks $T_2, T_3, \ldots, T_6$ are not ready because their predecessors are not done. For example, $T_2$ is not ready at time 0 because $T_1$, which precedes it, is not done at time 0. The first ready task on the list, reading from left to right, that is not already assigned is $T_7$, so task $T_7$ gets assigned to processor 2. Both processors are now busy until time 5, at which point processor 2 becomes available to work on another task (Figure 3.3b).

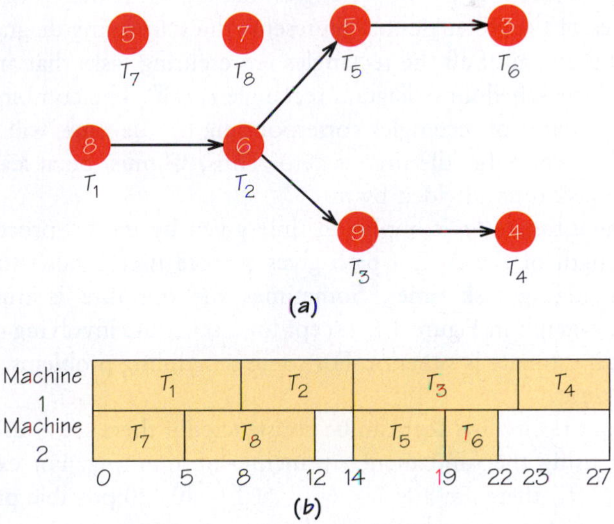


**FIGURE 3.3** (a) A typical order-requirement digraph (repeat of Figure 3.1). (b) The schedule produced by applying the list-processing algorithm to the order-requirement digraph in Figure 3.3a using the list $T_1, T_2, \ldots, T_8$.

Tasks $T_1$ and $T_7$ have been assigned. Reading from left to right along the list, the first task not already assigned whose predecessors are done by time 5 is $T_8$, so this task is started at time 5 on processor 2; processor 2 will continue to work on this task until time 12, because the task time for this task is 7 time units. At time 8, processor 1 becomes free, and reading the list from left to right we find that $T_2$ is ready (because $T_1$ has just been completed). Thus, $T_2$ is assigned to processor 1, which will stay busy on this task until time 14. At time 12, processor 2 becomes free, but the tasks that have not already been assigned from the list, $T_3$, $T_4$, $T_5$, $T_6$ are not ready, because they depend on $T_2$ being completed before these tasks can start. Thus, processor 2 must stay idle until time 14. At this time, $T_3$ and $T_5$ become ready. Since both processors 1 and 2 are idle at time 14, the lower numbered of the two, processor 1, gets to start on $T_3$ because it is the first ready task left to be assigned on the list scanned from left to right. Task $T_5$ gets assigned to processor 2 at time 14. The remaining tasks are assigned in a similar manner.

The schedule shown in Figure 3.3b is optimal because the path $T_1$, $T_2$, $T_3$, $T_4$, with length 27, is the critical path in the order-requirement digraph. As we saw in Chapter 2, the earliest completion time for the job made up of all the tasks is the length of the longest path in the order-requirement digraph.

There is another way of relating optimal completion time to the completion time that is yielded by the list-processing algorithm. Suppose that we add all the task times given in the order-requirement digraph and divide by the number of processors. The completion time using the list-processing algorithm must be at least as large as this number. For example, the task times for the order-requirement digraph in Figure 3.3a sum to 47. Thus, if these tasks are scheduled on two processors, the completion time is at least $\frac{47}{2} = 23.5$ (in fact, 24, because the list-processing algorithm applied to integer task times must yield an integer solution), while for three processors the completion time is at least $\frac{47}{3}$ (in fact, 16).

Why is it helpful to take the total time to do all the tasks in a job and divide this number by the number of processors? Think of each task that must be scheduled as a rectangle that is 1 unit high and $t$ units wide, where $t$ is the time allotted for the task. Think of the scheduling diagram with $m$ processors as a rectangle that is $m$ units high and whose width, $W$, is the completion time for the tasks. The scheduling diagram is to be filled up by the rectangles that represent the tasks. How small can $W$ be? The area of the rectangle that represents the scheduling diagram must be at least as large as the sum of all the rectangles representing tasks that are "packed" into it. The area of the scheduling diagram rectangle is $mW$. The combined areas of all the tasks, plus the area of rectangles corresponding to idle time, will equal $mW$. Width $W$ is smallest when the idle time is zero. Thus, $W$ must be at least as big as the sum of all the task times divided by $m$.

Sometimes the estimate for completion time given by the list-processing algorithm from the length of the critical path gives a more useful value than the approach based on adding task times. Sometimes the opposite is true. For the order-requirement digraph in Figure 3.1, except for a schedule involving one processor, the critical-path estimate is superior. For some scheduling problems, both these estimates may be poor.

The number of priority lists that can be constructed if there are $n$ tasks is $n!$ and can be computed using the fundamental principle of counting. For example, for eight tasks, $T_1, \ldots, T_8$, there are $8 \times 7 \times 6 \cdots \times 1 = 40{,}320$ possible priority lists. For different choices of the priority list, the list-processing algorithm may schedule the tasks, subject to the constraints of the order-requirement digraph, in different ways. More specifically, two different lists may yield different completion times or the same completion time, but the order in which the tasks are carried out will be different. It is also possible that two different lists produce identical ordering of the assignments of the tasks to processors and completion times. A little later we will see a method that can be used to select a list that, if we are lucky, will give a schedule with a relatively good completion time. In fact, no method is known, except for very specialized cases, of how to choose a list that can be guaranteed to produce an optimal schedule when the list algorithm is applied to it.

## Strange Happenings

The list-processing algorithm involves four factors that affect the final schedule. The answer we get depends on the following:

1. The times to carry out the tasks
2. Number of processors
3. Order-requirement digraph
4. Ordering of the tasks on the priority list

To see the interplay of these four factors, consider another scheduling problem, this time asociated with the order-requirement digraph shown in Figure 3.4 (the highlighted numbers are task time lengths). The schedule generated by the list-processing algorithm applied to the list $T_1, T_2, \ldots, T_9$, using three processors, is given in Figure 3.5.

*Treating the list $T_1, \ldots, T_9$ as fixed,* how might we make the completion time earlier? Our alternatives are to pursue one or more of these strategies:

1. Reduce task times.
2. Use more processors.
3. "Loosen" the constraints by having fewer directed edges in the order-requirement digraph.

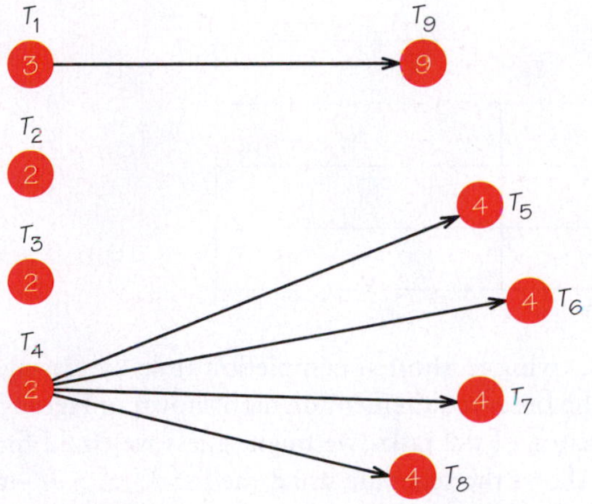


**FIGURE 3.4** An order-requirement digraph designed to help illustrate some paradoxical behavior produced by the list-processing algorithm.

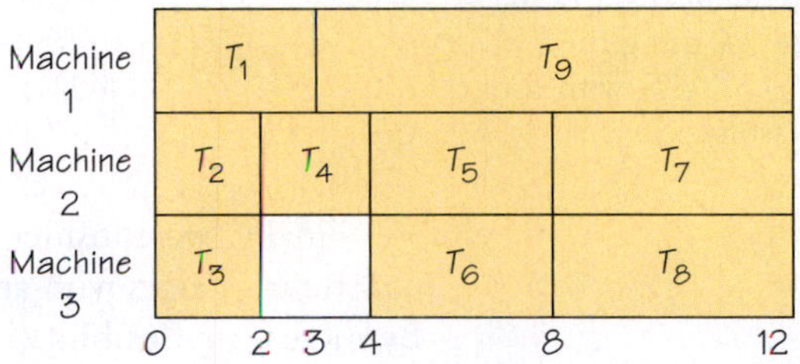


**FIGURE 3.5** The schedule produced by applying the list-processing algorithm to the order-requirement digraph in Figure 3.4 using the list $T_1, T_2, \ldots, T_9$ with three processors.

Let's consider each alternative in turn, changing one feature of the original problem at a time, and see what happens to the resulting schedule. If we use strategy 1 and reduce the time of each task by one unit, we would expect the completion time to go down. Figure 3.6 shows the new order-requirement digraph, and Figure 3.7 shows the schedule produced for this problem, using the list-processing algorithm with three processors applied to the list $T_1, \ldots, T_9$.

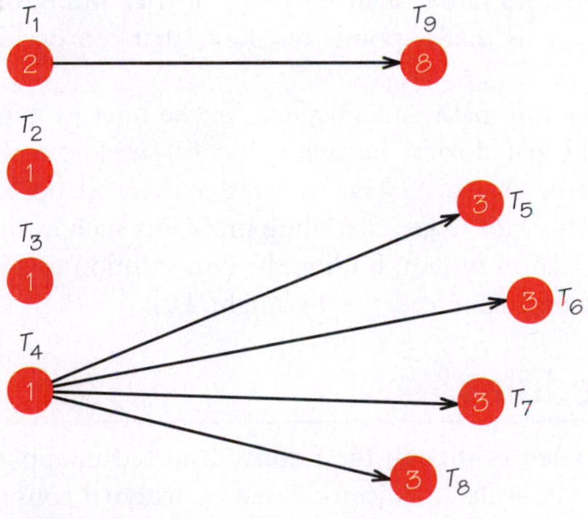


**FIGURE 3.6** The order-requirement digraph obtained from the one in Figure 3.4 by reducing by one unit each of the task times shown there.

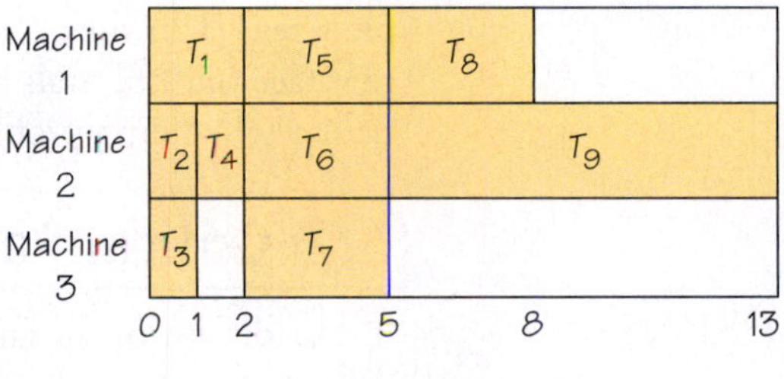


**FIGURE 3.7** The schedule produced by applying the list-processing algorithm to the order-requirement digraph in Figure 3.6 using the list $T_1, T_2, \ldots, T_9$ with three processors.

The completion time is now 13–longer than the completion time of 12 for the case (Figure 3.5) with longer task times. Here is something unexpected! Let's explore further and see what happens.

Next we consider strategy 2, increasing the number of machines. Surely this should speed matters up. When we apply the list-processing algorithm to the original graph in Figure 3.4, using the list $T_1, \ldots, T_9$ and four machines, we get the schedule shown in Figure 3.8. The completion time is now 15–an even later completion time than for the previous alteration.

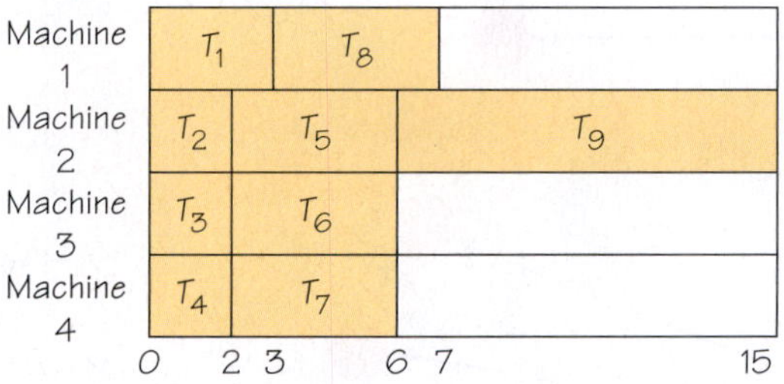


**FIGURE 3.8** The schedule produced by applying the list-processing algorithm to the order-requirement digraph in Figure 3.4 using the list $T_1, T_2, \ldots, T_9$ with four processors.

Finally, we consider strategy 3, trying to shorten completion time by erasing all constraints (edges with arrows) in the order-requirement digraph shown in Figure 3.4. By increasing flexibility of the ordering of the tasks, we might guess we could finish our tasks more quickly. Figure 3.9 shows the schedule using the list $T_1, \ldots, T_9$–now it takes 16 units! This is the worst of our three strategies to reduce completion time.

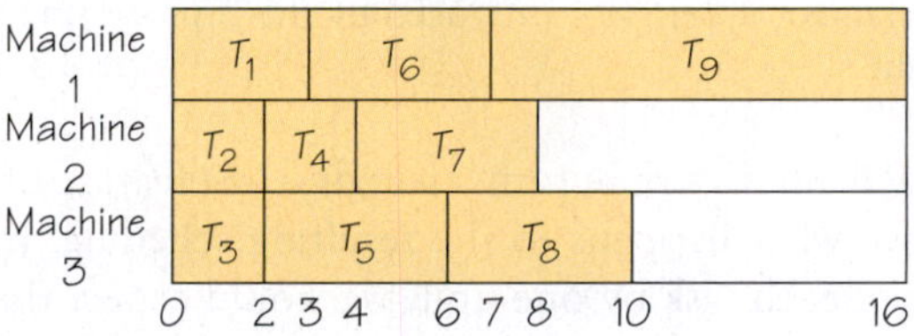


**FIGURE 3.9** The schedule produced by applying the list-processing algorithm to the order-requirement digraph in Figure 3.4, modified by erasing all its directed edges, using the list $T_1, T_2, \ldots, T_9$ with three processors.

The failures we have seen here are surprising at first glance, but they are typical of what can happen when a situation is too complex to analyze with naïve intuition. The value of using mathematics rather than intuition or trial and error to study scheduling and other problems is that it points out flaws that can occur in unguarded intuitive reasoning.

It is tempting to believe that we can make an adjustment in the rules for scheduling that we adopted to avoid the paradoxical behavior that has just been illustrated. Unfortunately, operations research experts have shown that there are no "simple fixes." This means that, in practice, for large scheduling problems such as those that face our hospitals and transportation system, finding the best solution to a particular scheduling problem cannot be guaranteed (see Spotlight 3.1).

## 3.2 Critical-Path Schedules

In our discussion so far, we have acted as though the priority list used in applying the list-processing algorithm was given to us in advance based on external considerations. Let's now consider the question of whether there is a systematic method of *choosing* a priority list that yields optimal or nearly optimal schedules. We will show how to construct a specific priority list based on this principle, to which the list-processing algorithm can then be applied.

Recall from our discussion of critical-path analysis in Chapter 2 that no matter how a schedule is constructed, the finish time cannot be earlier than the length of the longest path in the order-requirement digraph. This suggests that we should try

## SPOTLIGHT 3.1 Management Science and Disaster Recovery

The city of New York depends on a public transportation system of subways and roads to bring hundreds of thousands of people who live in the four outer boroughs (Queens, Brooklyn, the Bronx, and Staten Island) into Manhattan to work and "play." New York City also has a communication system of telephones, radio and television stations, and computer networks. These systems speed information between New York's citizens and people outside the city and around the world. The area in southern Manhattan, in the vicinity of the World Trade Center (WTC), was a center for banking, insurance, financial markets, and domestic and international commerce. The attack on the World Trade Center on September 11, 2001, disrupted these networks and markets but did not destroy them, partly because the principles of operations research and management science were used in the design and development of these systems over a long period of time.

The diagram below shows a very simple subway (train) system between an eastern and a western terminus.

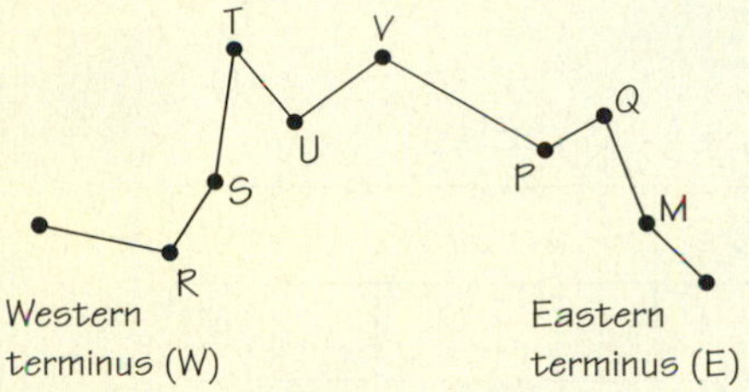


There are two tracks, each dedicated for use by westbound or eastbound trains to run between the two termini. The only place where trains can be turned around is at these termini. Simple graph theory tells us that in such a system, if a vertex is "destroyed" or out of service, or an edge is "destroyed" or out of service, the system totally breaks down. However, the simple provision that trains can be turned around at *U*, even though this is usually only one stop on the way from *W* to *E*, gives much greater flexibility to the system if there is a water main break, or a gas leak, etc. Thanks to simple principles of this kind and creating routes that use independent lines with many transfer points, New Yorkers were able to use the subway system in a flexible way after the World Trade Center disaster. In the days right after the WTC collapsed, trains were not allowed past the geographic area near the WTC for fear that the tunnels' structural foundation had been weakened and that subway vibrations could cause the collapse of damaged buildings. After it was ascertained that running the subways was safe both for partially damaged buildings and for the subways themselves, routes were altered several times to give rescue workers and people returning to their daily routines maximum support. One line's tunnels did collapse, and several stations had to be closed for extended periods, but due to the redundancy and flexibility of the design of the system, a remarkable amount of service was quickly restored.

*(AP/World Wide Photo.)*

Good planning and wise application of the principles of management science make it possible to minimize the effects of natural and manmade disasters.

to schedule first those tasks that occur early in long paths, because they might be a bottleneck for the other tasks. This idea leads to **critical-path scheduling**.

## EXAMPLE 3 ■ Scheduling Two Processors

To illustrate this method, consider the order-requirement digraph in Figure 3.10a. Suppose we wish to schedule these tasks on two processors. Initially, there are two critical paths of length 64: $T_1$, $T_2$, $T_3$ and $T_1$, $T_4$, $T_3$. Thus, we place $T_1$ first on the priority list. With $T_1$ "gone," there is a new critical path of length 60 ($T_5$, $T_6$, $T_4$, $T_3$) that starts with $T_5$, so $T_5$ is placed second on the priority list. At this stage, with $T_1$ and $T_5$ removed, we have the residual order-requirement digraph shown in Figure 3.10b. In this diagram there are paths of length 50 ($T_2$, $T_3$), 56 ($T_6$, $T_4$, $T_3$), 36 ($T_6$, $T_4$, $T_7$), and 24 ($T_8$, $T_4$, $T_{10}$). Because $T_6$ heads the path that is currently longest in length, it is placed third in the priority list. Once $T_6$ is removed from Figure 3.10b, there is a tie for which is the longest path remaining, because both $T_2$, $T_3$ and $T_4$, $T_3$ are paths of length 50.

When there is a tie between two longest paths, we place next on the priority list in the lowest-numbered task heading a longest path. In the example shown here, this means that $T_2$ is placed next on the priority list, to be followed by $T_4$. Continuing in this fashion, we obtain the priority list $T_1$, $T_5$, $T_6$, $T_2$, $T_4$, $T_3$, $T_8$, $T_9$, $T_7$, $T_{10}$. Note that the order of $T_7$ and $T_{10}$ was decided using the rule for breaking ties. The list-processing algorithm is now applied using this priority list and the order-requirement digraph in Figure 3.10a. We obtain the schedule in Figure 3.11.

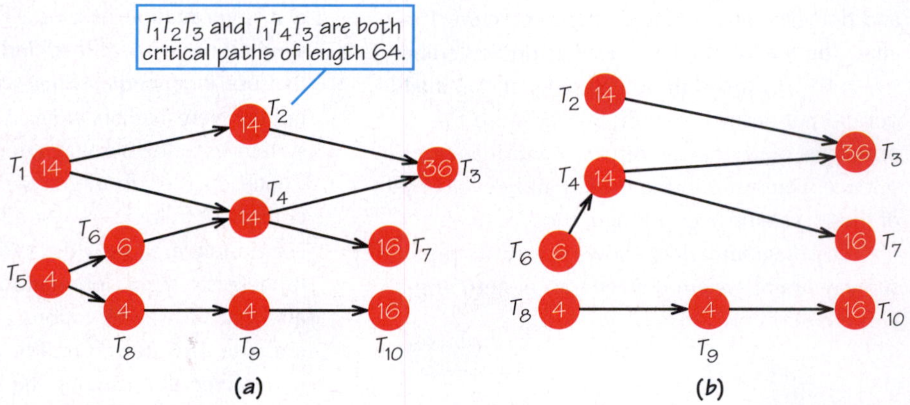


**FIGURE 3.10** (a) An order-requirement digraph used to illustrate the critical-path scheduling method. (b) Residual order-requirement digraph after tasks $T_1$ and $T_5$ have been removed.

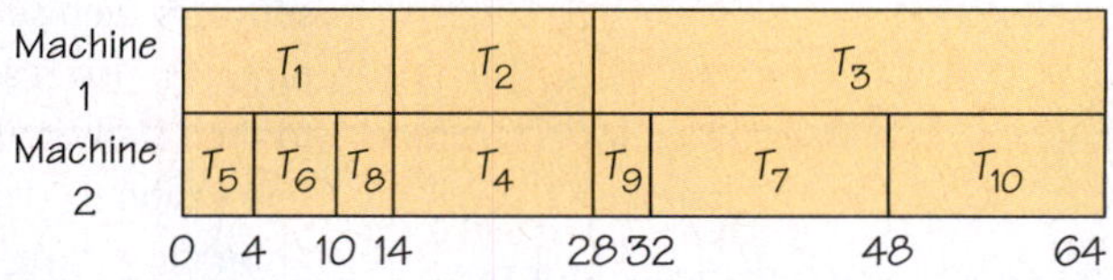


**FIGURE 3.11** The optimal schedule produced by applying the critical-path scheduling method to the order-requirement digraph in Figure 3.10. The list used was $T_1$, $T_5$, $T_6$, $T_2$, $T_4$, $T_3$, $T_8$, $T_9$, $T_7$, $T_{10}$.

## Critical-Path Scheduling — PROCEDURE

The **critical-path scheduling** algorithm applies the list-processing algorithm using the priority list $L$ obtained as follows:

1. Find a task that heads a critical (longest) path in the order-requirement digraph. If there is a tie, choose the task with the lower number.
2. Place the task found in step 1 next on the list $L$. (The first time through the process this task will head the list.)
3. Remove the task found in step 1 and the edges attached to it from the current order-requirement digraph, obtaining a new (modified) order-requirement digraph.
4. If there are no vertices left in the new order-requirement digraph, the procedure is complete; if there are vertices left, go to step 1.

This procedure will terminate when all the tasks in the original order-requirement digraph have been placed on the list $L$.

The preceding example shows that critical-path scheduling can sometimes yield optimal solutions. Unfortunately, this algorithm does not always perform well. For example, the critical-path method employing four processors applied to the order-requirement digraph shown in Figure 3.12 yields the list $T_1$, $T_8$, $T_9$, $T_{10}$, $T_{11}$, $T_5$, $T_6$, $T_7$, $T_{12}$, $T_2$, $T_3$, $T_4$ and then the schedule in Figure 13.3. (Note that $T_5$, $T_6$, $T_7$ are thought of as heading paths of length 10.) In fact, there can be no worse schedule than this one. An optimal schedule is shown in Figure 3.14.

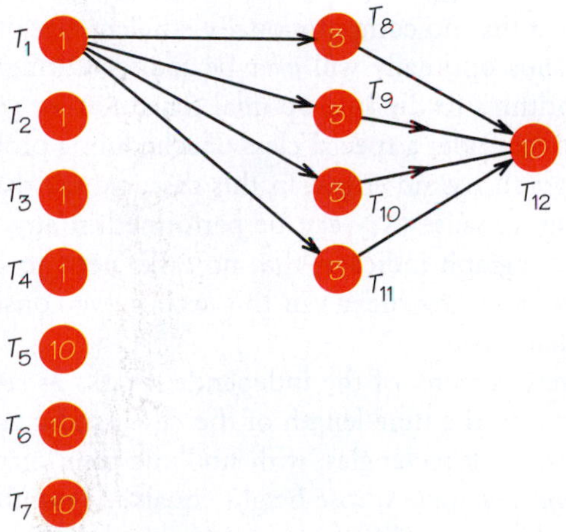


**FIGURE 3.12** An order-requirement digraph used to illustrate how poorly the critical-path scheduling method can sometimes behave.

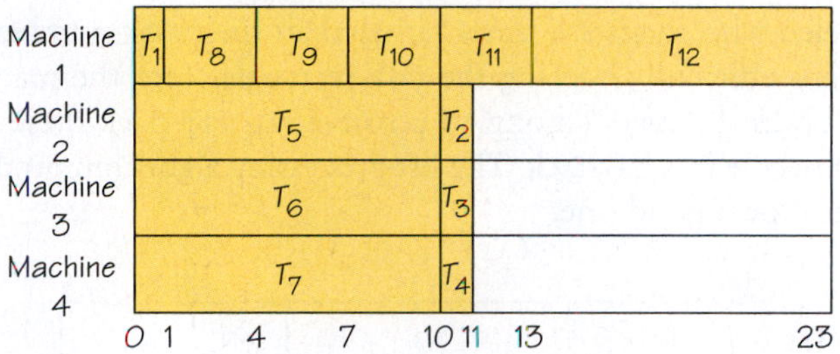


**FIGURE 3.13** The schedule produced by applying the critical-path scheduling method to the order-requirement digraph in Figure 3.12 using four processors. The list used was $T_1$, $T_8$, $T_9$, $T_{10}$, $T_{11}$, $T_5$, $T_6$, $T_7$, $T_{12}$, $T_2$, $T_3$, $T_4$.

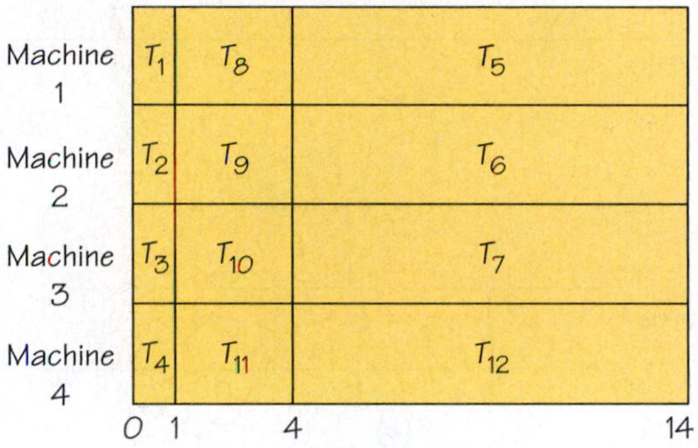


**FIGURE 3.14** An optimal schedule for the order-requirement digraph in Figure 3.12 using four processors.

Many of the results we have examined so far are negative because we are dealing with a general class of problems that defy our using computationally efficient algorithms to find an optimal schedule. But we can close on a more positive note. Consider an arbitrary order-requirement digraph, but assume all the tasks take equal time. It turns out that we can always construct an optimal schedule using two processors in this situation. Ironically, we can choose among many algorithms to produce these optimal schedules. The algorithms are easy to understand (though not easy to prove optimal) and have all been discovered since 1969! Many people think that mathematics is a subject that is no longer alive, and that all its ideas and methods were discovered hundreds of years ago—but as we have just seen, this is not true. In fact, more new mathematics has been discovered and published in the last 30 years than during any previous 30-year period.

## 3.3 Independent Tasks

Mathematicians suspect that no computationally efficient algorithm for solving general scheduling problems optimally will ever be found. Owing to our limited success in designing algorithms for finding optimal schedules for general order-requirement digraphs, we will consider a special class of scheduling problems for which the order-requirement digraph has no edges. In this case, we say that the tasks are *independent* of one another, because they can be performed in any order. (No edges in the order-requirement digraph indicates that no tasks need to precede others; that is, the tasks can be done in any order.) In this section we consider the problem of scheduling **independent tasks.**

Geometrically, we can think of the independent tasks as rectangles of height 1 whose lengths are equal to the time length of the task. Finding an optimal schedule amounts to packing the task rectangles, with no "idle time" gaps between adjacent rectangles, into a longer rectangle whose height equals the number of machines. For example, Figure 3.15 shows two different ways to schedule tasks of length 10, 4, 5, 9, 7, 7 on two machines. (For convenience, the rectangles in the case of independent tasks are labeled with their task times rather than their task numbers.) Scheduling basically means efficiently packing the task rectangles into the machine rectangle. Finding the optimal answer among all possible ways to pack these rectangles is like looking for a needle in a haystack. The list-processing algorithm produces a packing, but it may not be a good one.

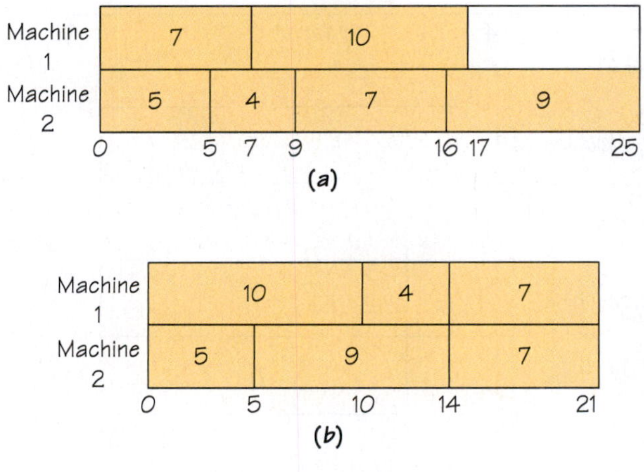


**FIGURE 3.15** (a) A nonoptimal way to schedule independent tasks of time lengths 10, 4, 5, 9, 7, 7 using two processors. (b) An optimal way to schedule independent tasks of time lengths 10, 4, 5, 9, 7, 7 using two processors.

There are two approaches we can consider. To study **average-case analysis,** we might ask: *Is the average (mean) of the completion times arrived at by using the list-processing algorithm with all the possible different lists close to the optimal possible completion time?* To study **worst-case analysis,** we might ask: *How far from optimal is a schedule obtained using the list-processing algorithm with one particular priority list?* What is being contrasted with these two points of view is that an algorithm may work well most of the time (give an answer close to optimal) even though there may be a few cases in which it performs very badly. Average-case analysis is amenable to mathematical solution but requires methods of great sophistication.

## Decreasing-Time Lists

Is there some way of choosing a priority list for independent tasks that consistently yields relatively good schedules? The surprising answer is yes! The idea is that when long tasks appear toward the end of the list, they often seem to "stick out" on the right end, as in Figure 3.15a. This suggests that before one tries to schedule a collection of tasks, the tasks should be placed in a list where the longest tasks are listed first.

**Decreasing-Time-List Algorithm** PROCEDURE

The list-processing algorithm applied to a list of task times arranged in order of nonincreasing size is called the **decreasing-time-list algorithm**.

If we apply it to the set of tasks listed previously (10, 4, 5, 9, 7, 7), we obtain the times 10, 9, 7, 7, 5, 4 and the schedule (packing) shown in Figure 3.16. This packing is again optimal, but it is different from the optimal scheduling in Figure 3.15b. It is worth noting that the decreasing-time list and the list obtained by the critical-path method discussed earlier will coincide in the case of independent tasks. The decreasing-time list can also be constructed for the case in which the tasks are not independent. For general order-requirement digraphs, the decreasing-time list does not produce particularly good schedules.

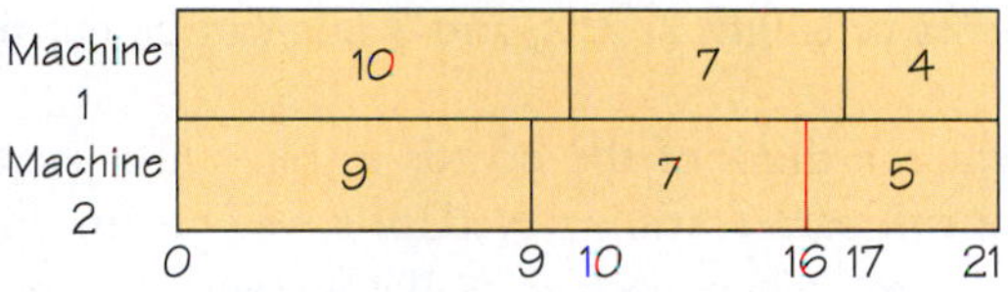


**FIGURE 3.16** The optimal schedule resulting from applying the decreasing-time-list algorithm to a collection of independent tasks. The list used, written in terms of task times only, is 10, 9, 7, 7, 5, 4.

It is important to remember that the decreasing-time-list algorithm does not *guarantee* optimal solutions. This can be seen by scheduling the tasks with times 11, 10, 9, 6, 4 (Figure 3.17). The schedule has a completion time of 21. However, the rearranged list 9, 4, 6, 11, 10 yields the schedule in Figure 3.18, which finishes at time 20. This solution is obviously optimal because the machines finish at the same time and there is no idle time. Note that when tasks are independent, if there are $m$ machines available, the completion time cannot be less than the sum of the task times divided by $m$.

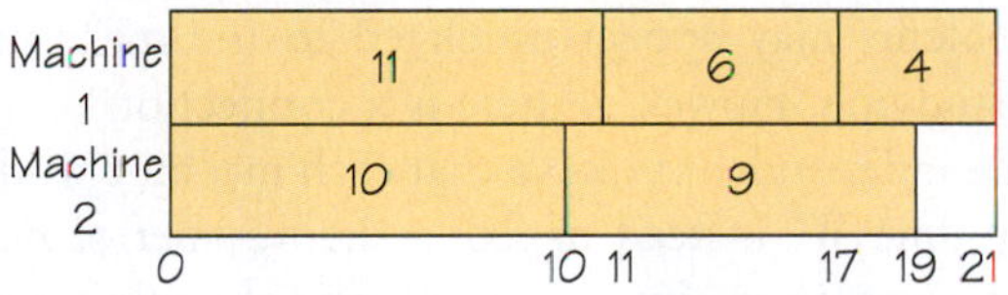


**FIGURE 3.17** The nonoptimal schedule resulting from applying the decreasing-time-list algorithm to a collection of independent tasks. The list used, written in terms of task times only, is 11, 10, 9, 6, 4.

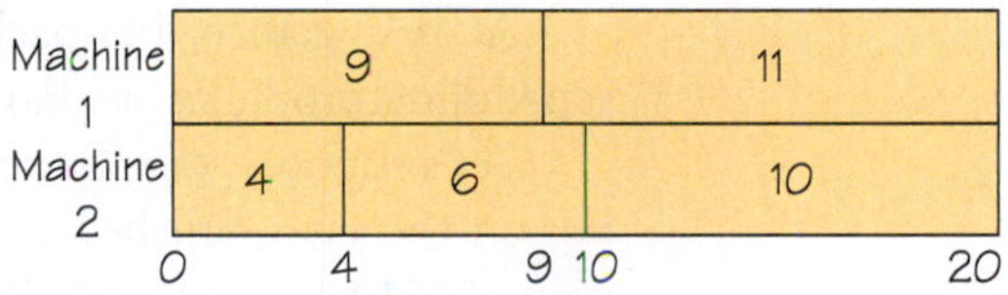


**FIGURE 3.18** The optimal schedule resulting from applying the list-processing algorithm to a collection of independent tasks. The list used, written in terms of task times only, is 9, 4, 6, 11, 10.

## EXAMPLE 4 ■ Photocopy Shop and Data Entry Problems

A modern copy shop provides a wide array of services ranging from copying a few sheets for a "drop in" customer, to printing elaborate reports for small businesses, to publishing monographs and advertising flyers. Using mathematical scheduling techniques can save time and cost by ensuring the many tasks are completed most efficiently. (*Christopher Robbins/ Digital Vision/Getty Images.*)

Imagine a photocopy shop with three photocopiers. Photocopying tasks that must be completed overnight are accepted until 5 P.M. The tasks are to be done in any manner that minimizes the finish time for all the work. Because this problem involves scheduling machines for independent tasks, the decreasing-time-list algorithm would be a good heuristic to apply.

For another example, consider a data entry pool at a large corporation or college, where individual entry tasks can be assigned to any data entry specialist. In this setting, however, the assumption that the data entry workers are identical in skill is less likely to be true. Hence, the tasks might have different times with different processors. This phenomenon, which occurs in real-world scheduling problems, violates one of the assumptions of our mathematical model.

## 3.4 Bin Packing

Suppose you plan to build a wall system for your books, CDs, DVDs, and stereo set. It requires 24 wooden shelves of various lengths: 6, 6, 5, 5, 5, 4, 4, 4, 4, 2, 2, 2, 2, 3, 3, 7, 7, 5, 5, 8, 8, 4, 4, and 5 feet. The lumberyard, however, sells wood only in boards of length 9 feet. If each board costs $8, what is the minimum cost to buy sufficient wood for this wall system?

Because all shelves required for the wall system are shorter than the boards sold at the lumberyard, the largest number of boards needed is 24, the precise number of shelves needed for the wall system. Buying 24 boards would, of course, be a waste of wood and money because several of the shelves you need could be cut from one board. For example, pieces of length 2, 2, 2, and 3 feet can be cut from one 9-foot board.

To be more efficient, we think of the boards as bins of capacity $W$ (9 feet in this case) into which we will pack (without overlap) $n$ weights (in this case, lengths) whose values are $w_1, \ldots, w_n$, where each $w_i \leq W$. We wish to find the minimum number of bins into which the weights can be packed. In this formulation, the problem is known as the **bin-packing problem**.

**Bin-Packing Problem** DEFINITION

The **bin-packing problem** involves finding the minimum number of bins of weight capacity $W$ into which weights $w_1, w_2, \ldots, w_n$ (each less than or equal to $W$) can be packed without exceeding the capacity of the bins.

At first glance, bin-packing problems may appear unrelated to the machine-scheduling problems we have been studying. However, there is a connection.

Let's suppose we want to schedule independent tasks so that each machine working on the tasks finishes its work by time $W$. Instead of fixing the number of machines and trying to find the earliest completion time, we must find the minimum number of machines that will guarantee completion by the fixed completion time ($W$). Despite this similarity between the machine-scheduling problem and the bin-packing problem, the discussion that follows will use the traditional terminology of bin packing.

By now, it should come as no surprise to learn that no one knows a fast algorithm that always picks the optimal (smallest) number of bins (boards). In fact, the bin-packing problem belongs to the class of NP-complete problems (see Spotlight 2.1), which means that most experts think it unlikely that any fast optimal algorithm will ever be found. Relatively good algorithms for problems that come up in actual applications are known.

## Bin-Packing Heuristics

We will think of the items to be packed, in any particular order, as constituting a list. In what follows we will use the list of 24 shelf lengths given for the wall system. We will consider various **heuristic algorithms**, namely, methods that can be carried out quickly but cannot be guaranteed to produce optimal results. Probably the easiest approach is simply to put the weights into the first bin until the next weight won't fit, and then start a new bin. (Once you open a new bin, don't use leftover space in an earlier, partially filled bin.) Continue in the same way until as many bins as necessary are used.

The resulting solution is shown in Figure 3.19. This algorithm, called **next-fit (NF)**, has the advantage of not requiring knowledge of all the weights in advance. Only the remaining space in the bin currently being packed must be remembered. The disadvantage of this heuristic is that a bin packed early on may have had room for small items that come later in the list.

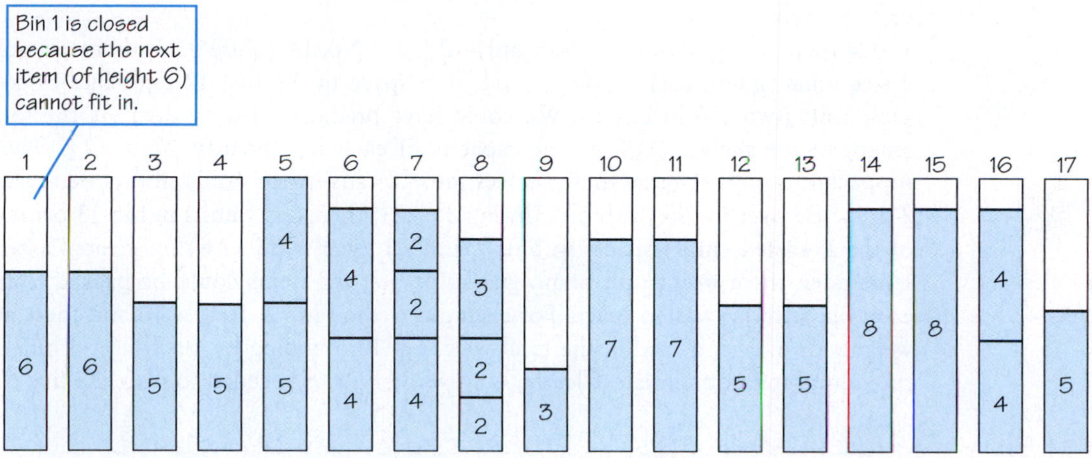


**FIGURE 3.19** The list 6, 6, 5, 5, 5, 4, 4, 4, 4, 2, 2, 2, 2, 3, 3, 7, 7, 5, 5, 8, 8, 4, 4, 5 packed in bins using next fit.

Our wish to avoid permanently closing a bin too early suggests a different heuristic–**first-fit (FF)**: Put the next weight into the first bin already opened that has room for this weight. If no such bin exists, start a new bin. Note that a computer program to carry out first fit would have to keep track of how much room was left in all the previously opened bins. For the 24 wall-system shelves, the first-fit algorithm would generate a solution that uses only 14 bins (see Figure 3.20) instead of the 17 bins generated by the next-fit algorithm.

If we are keeping track of how much room remains in each partially filled bin, we can put the next item to be packed into the bin that currently has the most room available. This heuristic will be called **worst-fit (WF)**. The name *worst fit* refers to the fact that an item is packed into a bin with the most room available, that is, into which it fits "worst," rather than into a bin that will leave little room left over after it is placed in that bin ("best fit"). The solution generated by this approach looks the same as that shown in Figure 3.20. Although this heuristic also leads to 14 bins, the items are packed in a different order. For example, the first item of size 2, the tenth item in the list, is put into bin 6 in worst fit, but into bin 1 in first fit.

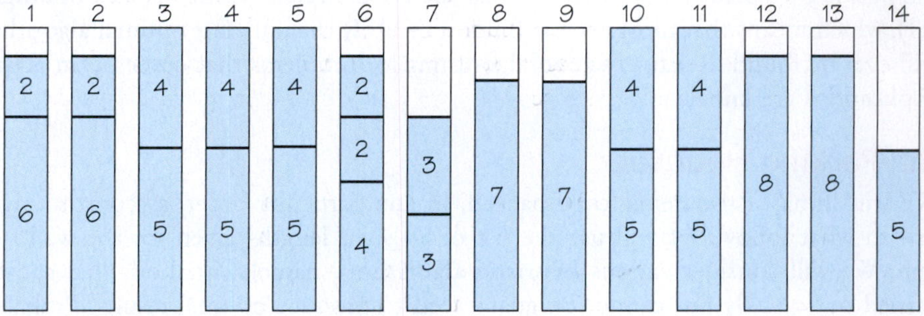


**FIGURE 3.20** The list 6, 6, 5, 5, 5, 4, 4, 4, 4, 2, 2, 2, 2, 3, 3, 7, 7, 5, 5, 8, 8, 4, 4, 5 packed in bins using first fit. Worst fit would yield a packing that would look identical.

## Decreasing-Time Heuristics

One difficulty with all three of these heuristics is that large weights that appear late in the list can't be packed efficiently. Therefore, we should first sort the items to be packed in order of decreasing size, assuming that all items are known in advance. We can then pack large items first and the smaller items into leftover spaces. This approach yields three new heuristics: **next-fit decreasing (NFD), first-fit decreasing (FFD)**, and **worst-fit decreasing (WFD)**. Here is the original list sorted by decreasing size: 8, 8, 7, 7, 6, 6, 5, 5, 5, 5, 5, 5, 4, 4, 4, 4, 4, 4, 3, 3, 2, 2, 2, 2. Packing using first-fit-decreasing order yields the solution in Figure 3.21. This solution uses only 13 bins.

Is there any packing that uses only 12 bins? No. In Figure 3.21, there are only 2 free units (1 unit each in bins 1 and 2) of space in the first 12 bins, but 4 occupied units (two 2's) in bin 13. We could have predicted this by dividing the total length of the shelves (110) by the capacity of each bin (board): $\frac{110}{9} = 12\frac{2}{9}$. Thus, no packing could squeeze these shelves into 12 bins–there would always be at least 2 units left over for the 13th bin. (In Figure 3.21, there are 4 units in bin 13 because of the 2 wasted empty spaces in bins 1 and 2.) Even if this division created a zero remainder, there would still be no guarantee that the items could be packed to fill each bin without wasted space. For example, if the bin capacity is 10 and there are weights of 6, 6, 6, 6, and 6, the total weight is 30; dividing by 10, we get 3 bins as the minimum requirement. Clearly, however, 5 bins are needed to pack the five 6's.

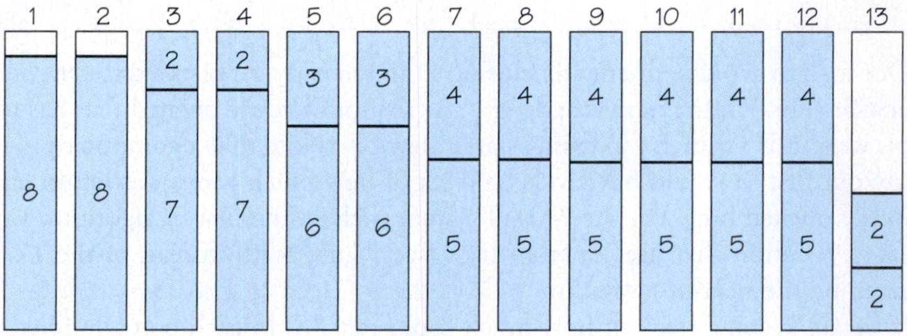


**FIGURE 3.21** The bin packing resulting from applying first-fit decreasing to the wall-system numbers. The list involved, which uses the original list sorted in decreasing order, is 8, 8, 7, 7, 6, 6, 5, 5, 5, 5, 5, 5, 4, 4, 4, 4, 4, 4, 3, 3, 2, 2, 2, 2.

None of the six heuristic methods shown will necessarily find the optimal number of bins for an arbitrary problem. How can we decide which heuristic to use? One approach is to see how far from the optimal solution each method might stray.

Various formulas have been discovered to calculate the maximum discrepancy between what a bin-packing algorithm actually produces and the best possible result. For example, in situations where a large number of bins are to be packed, FF can be off by as much as 70%, but FFD is never off by more than 22%. Of course, FFD doesn't give an answer as quickly as FF, because extra time for sorting a large collection of weights may be considerable. Also, FFD requires knowing the whole list of weights in advance, whereas FF does not. It is important to emphasize that a 22% margin of error is a worst-case figure. In many cases, FFD will perform much better. Results obtained by computer simulation indicate excellent average-case performance for this algorithm.

When solving real-world problems, we always have to look at the relationship between mathematics and the real world. Thus, first-fit decreasing usually results in fewer bins than next fit, but next fit can be used even when all the weights are not known in advance. Next fit also requires much less computer storage than first fit, because once a bin is packed, it need never be looked at again.

Fine-tuning of the conditions of the actual problem often results in better practical solutions and in interesting new mathematics as well. See Spotlight 3.2 for a discussion of some of the tools mathematicians use to verify and even extend mathematical truths by raising new mathematical problems.

## SPOTLIGHT 3.2 Using Mathematical Tools

The tools of a carpenter include the saw, T square, level, and hammer. A mathematician also requires tools of the trade. Some of these tools are the proof techniques that enable verification of mathematical truths. Another set of tools consists of strategies to sharpen or extend the mathematical truths already known. For example, suppose that if *A* and *B* hold, then *C* is true. What happens if only *A* holds? Will *C* still be true? Similarly, if only *B* holds, will *C* still be true?

This type of thinking is of value because such questions will result either in more general cases where *C* holds or in examples showing that *B* alone and/or *A* alone can't imply *C*. For example, we saw that if a graph *G* is connected (hypothesis *A*) and even-valent (hypothesis *B*), then *G* has a circuit which uses each edge only once (conclusion *C*). If either hypothesis is omitted, the conclusion fails to hold. The figures illustrate this point. On the left is an even-valent but nonconnected graph; on the right, a connected graph with two odd-valent vertices. Neither graph has an Euler circuit.

Here is another way that a mathematician might approach extending mathematical knowledge. If *A* and *B* imply *C*, will *A* and *B* imply both *C* and *D*, where *D* extends the conclusion of *C*? For example, not only can we prove that a connected, even-valent (hypotheses *A* and *B*) graph has an Euler circuit, but we can also show that the first edge of the Euler circuit can be chosen arbitrarily (conclusions *C* and *D*).

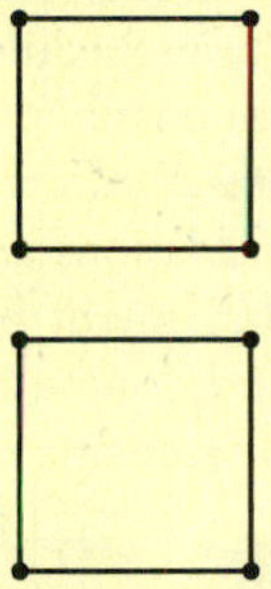

No Euler circuit
Nonconnected
Even-valent

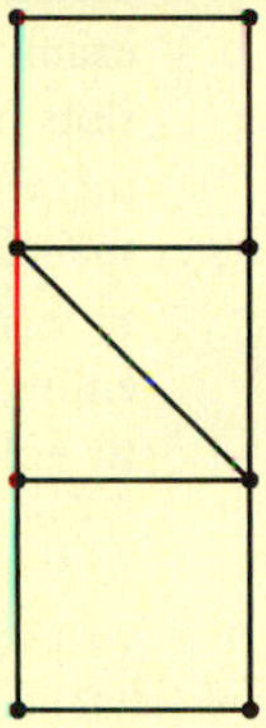

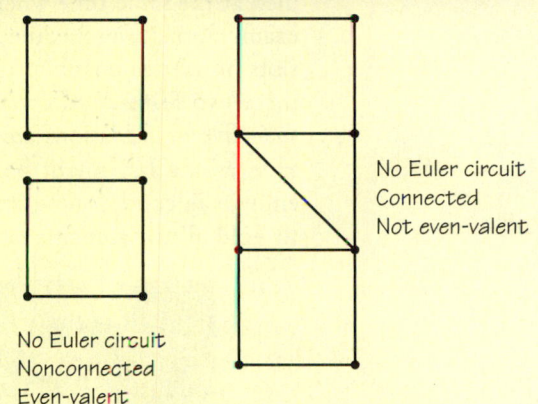

No Euler circuit
Connected
Not even-valent

It turns out that being able to specify the first two edges of the Euler circuit may not always be possible. Mathematicians are trained to vary the hypotheses and conclusions of results they prove, in an attempt to clarify and sharpen the range of applicability of the results.

We have seen that machine scheduling and bin packing are probably computationally difficult to solve because they are NP-complete. A mathematician could then try to find the simplest version of a bin-packing problem that would still be NP-complete: What if the items to be packed can have only eight weights? What if the weights are only 1 and 2? Asking questions like these is part of the mathematician's craft. Such questions help to extend the domain of mathematics and hence the applications of mathematics.

## 3.5 Resolving Conflict via Coloring

In attempting to understand situations that involve scheduling, one might desire to achieve a wide variety of goals. For example, in certain types of scheduling problems, as we have seen here, one is interested in optimization issues. What is the earliest completion time for getting a collection of tasks done on two identical processors? However, in other situations a different goal may arise. For example, in sports, consider a league of baseball teams. Each team has to play some games during the day, some at night, some at home, and some away from home. In the interests of *equity*, it may be desirable for each team to play the same number of day games and night games both at home and away against each of the other teams in the league. If, for example, team $A$ plays 8 games away against team $B$ and 2 games at home against $B$, then if $A$ wins both home games but loses 7 out of 8 away games, it may appear that $B$ had an advantage due to the way its games against $A$ were scheduled.

Another goal of scheduling, other than optimization and equity, may be to prevent conflicts from occurring. We can use our knowledge of graph theory to solve some interesting scheduling problems where the goal is "conflict resolution." For example, at most colleges, every semester and summer session final examinations must be scheduled. From the point of view of students and faculty both, it would be desirable to schedule these examinations so that (1) no two examinations are scheduled at the same time when a student is enrolled in both of the courses and (2) the examinations are scheduled in as "compact" a way as possible, that is, in as few time slots or days as possible. The administration of the college may share the desire for these two features and want still another property for the scheduling: (3) no more than five examinations are scheduled for any time slot. The reason for a condition such as the last might be that during the summer only five rooms with reliable enough air conditioning are available (or there might be only five rooms large enough to hold all the students taking the common final for multiple-section courses).

Graph theory can be used to resolve scheduling conflicts that occur in trying to provide students access to limited database or computer resources. (*Bananastock/Picturequest.*)

### EXAMPLE 5 ■ Scheduling Examinations

Small State is offering eight courses during its summer session. The table shows with an X which pairs of courses have one or more students in common. Only two air-conditioned lecture halls are available for use at any one time. To design an efficient way to schedule the final examinations, we can represent the information in this table by using a graph, as shown in Figure 3.22a. In the graph, courses are represented by vertices and two courses are joined by an edge if there is any student enrolled in both courses.

| | F | M | H | P | E | I | S | C |
|---|---|---|---|---|---|---|---|---|
| French (F) | | X | | X | X | X | | X |
| Mathematics (M) | X | | | | X | X | | |
| History (H) | | | | | | X | X | X |
| Philosophy (P) | X | | | | | | | X |
| English (E) | X | X | | | | X | | |
| Italian (I) | X | X | X | | X | | X | |
| Spanish (S) | | | X | | | X | | |
| Chemistry (C) | X | | X | X | | | | |

We are faced with the following graph theory problem: Can we assign labels to the vertices of the graph in such a way that vertices that are joined by an edge get different labels? We think of the labels as the time slots the courses are assigned for final examinations. Traditionally, in graph theory such labels are referred to as *colors.* In this language we seek to color the vertices of the graph so that vertices that are joined by an edge get different colors. Such a coloring is called a **vertex coloring**.

### Vertex Coloring — DEFINITION

The **vertex coloring** problem for a graph requires assigning each vertex of the graph a color (label) such that two vertices joined by an edge are assigned different colors.

Figure 3.22b shows one way to color the vertices of the graph so that each vertex gets a different color. Note that numbers are being used to represent the different colors. This solution is not very valuable, however, because it means that each course must be given its own time slot.

To minimize the number of time slots used, we assign colors so that no two vertices that are joined by an edge get the same color. Thus, vertices *F, M, I,* and *E* must get four different colors. These four colors can then be used to color the remaining vertices, ensuring that no two connected vertices have the same color.

The coloring in Figure 3.22c is a major improvement over the one in Figure 3.22b. It uses only four colors. In fact, this is the smallest number of colors that can be used. To see this, notice that the vertices *F, M, I, E* in Figure 3.22a are all joined by edges to each other. Thus, in any coloring of this graph they would require four different colors. The improved coloring in Figure 3.22c was found by trial and error.

### Chromatic Number — DEFINITION

The **chromatic number** is the minimum number of colors needed to label the vertices of a graph so that no two vertices of the graph joined by an edge get the same color.

The examination graph we have been studying has chromatic number 4; hence, we can schedule the eight examinations in four time slots without a conflict. Notice, however, that the coloring in Figure 3.22c schedules three different courses for the time slot corresponding to color 2. This means that not enough rooms with air conditioning will be available. Is there a way to recolor the graph with four colors so that each of the four colors is used only twice? Figure 3.22d shows that the answer is yes.

Thus, we are able to schedule the eight final examinations in four time slots, using only two air-conditioned rooms, and no student will have a conflict under this schedule!

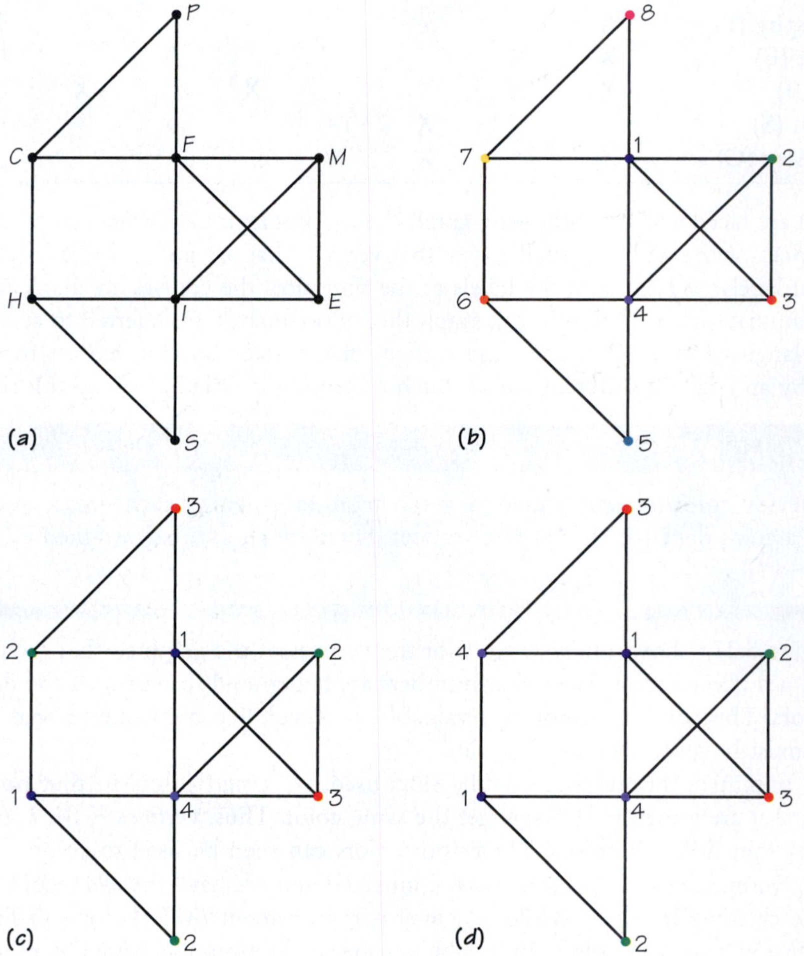


**FIGURE 3.22** (a) A graph used to represent conflict information about courses. When two courses have a common student, an edge is drawn between the vertices that represent these courses. (b) A coloring of the scheduling graph with 8 colors, representing 8 time slots. Using this coloring would lead to a schedule where 8 time slots are used to schedule the examinations. This number is far from optimal. (c) A coloring of the scheduling graph with 4 colors. This translates into a way of scheduling the examinations during 4 time slots, and it is not possible to design a schedule with fewer time slots. However, this schedule calls for the use of three different rooms, because three examinations are scheduled during time slot 2. (d) A coloring of the scheduling graph with 4 colors. This means that the examinations can be scheduled in 4 time slots. However, because each color appears only twice, all the examinations can be scheduled in two air-conditioned rooms.

Realistic problems in scheduling government committees, high school and university final examinations, and job interviews (see Spotlight 3.4) are usually so large that graph coloring algorithms have to be incorporated into elaborate software packages to solve them.

Mathematicians have examined many kinds of coloring problems. Many developments about coloring graphs have been an outgrowth of work on the Four Color Problem (see Spotlight 3.3). One can study problems that involve the coloring of the edges of a graph rather than its vertices. Using techniques that have emerged from the study of coloring problems, problems involving such diverse contexts as scheduling government committees, using runways at airports efficiently, assigning frequencies for use by mobile pagers and cell phones, and designing timetables for public transportation have been solved—all these benefits from a problem that at first glance looks as if it belongs to recreational mathematics!

## SPOTLIGHT 3.3 Four Color Problem

Many people perceive mathematics as complex because it often uses strange notations and algebraic symbols. Thus, it may come as a surprise that a problem that is relatively easy to state and understand without complex symbolism eluded solution for about 100 years. When it was finally solved, it set off a "firestorm" that it had not truly been solved. More importantly, many of the ideas that have been developed in the theory of graphs were expanded or developed in the course of trying to prove this "guess."

When a graph can be drawn on a flat piece of paper so that edges meet only at vertices, we can talk about not only the vertices and edges of the graph, but also about its regions or *faces*. Such graphs are known as *plane graphs*. Two examples of plane graphs are shown in the diagram below.

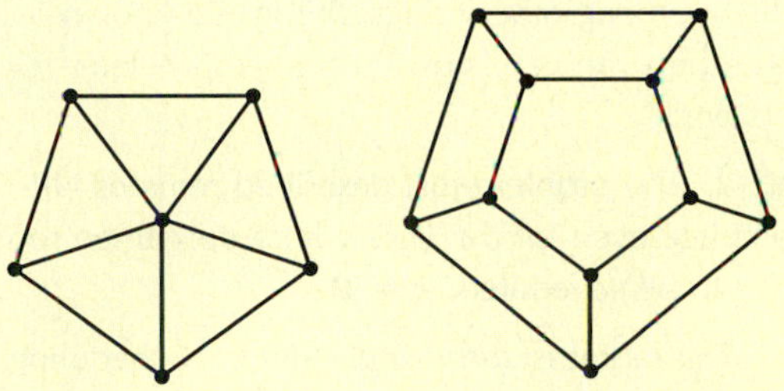

(*a*) (*b*)

Graph (a) has 6 regions (the area "outside" of the graph is counted as one of the 6 regions), 5 of which have 3 sides and 1 of which has 5 sides, while graph (b) has 7 regions, 2 of which have 5 sides and 5 of which have 4 sides. To count the number of sides of a region, imagine you are a small ant and are following the edges around the region, starting at some vertex *w*. You count edges until you get back to *w*. Note that for each of these graphs there is one *unbounded* (goes off to "infinity") region, in addition to the other regions. When you color the regions of a plane map do not forget to assign a color to the unbounded region.

If you think of the regions of a plane graph as being distinct countries on a page that is to appear in an atlas, it would be nice if countries that share a border got different colors so that they can be distinguished. Countries that meet at a vertex, but do not share an edge representing a common border, can be colored with the same color. It is convenient to use the term map for the regions created by the drawing of a plane graph.

The following provocative question was raised in a letter (1852) from Augustus De Morgan to William Rowan Hamilton that was based on a problem posed to De Morgan by his student Fredrick Guthrie, who heard the question from his brother Francis:

Can the regions of any (plane) map always be colored with four or fewer colors?

A clever proposal as to how to prove the "Four Color Conjecture" was proposed by Alfred Kempe. Kempe's "proof" had a subtle error, which defied detection for many years, showing that proofs in mathematics really depend on the community of mathematicians to guarantee their accuracy. The British mathematician Percy Heawood discovered the error Kempe made. Heawood adapted Kempe's proof to show correctly that any map can be colored with five or fewer colors. Approximately 100 years elapsed before a proof that the Four Color Conjecture was true was found. This occurred in 1976, but there was a curious loose end: The proof found by Wolfgang Haken and Kenneth Appel required that a computer verify a large collection of "calculations," which were too numerous to be done by hand. This proof troubled some philosophers and mathematicians, but has been widely accepted by the mathematics community. In 1995, Neil Robertson, Daniel Sanders, Paul Seymour, and Robin Thomas found another proof. This proof, while simpler and shorter than the earlier Haken–Appel proof, also required computer calculations too numerous to be checked by "hand." Though it is possible that some new approach to the Four Color Conjecture will avoid the use of computers, this is not widely thought to be likely. However, human ingenuity sometimes surprises us!

## SPOTLIGHT 3.4 Scheduling Job Interviews

A group of companies is coming to campus for job interviews. Different companies may want different numbers of time slots to hold their interviews. In each time slot one student can be interviewed. In the example below, all the companies have requested contiguous time slots for the interviews, but this need not be the case. Due to the fact that classes are going on at the same time, five departmental conference rooms have been made available to the companies to conduct their interviews.

The interviews will follow the school's regular hourly periods, which start at 9 A.M. and end at 4 P.M. (Companies will be scheduled for continuous interviews during lunch-hour times. Interviews cannot be scheduled beyond the end of the period that starts at 4 P.M. and ends at 5 P.M.)

| Company | | Time Slot Requested |
|---|---|---|
| A | (Apricot Computers) | 7 |
| B | (Big Green) | 1 |
| C | (Challenge Insurance) | 4, 5 |
| D | (Daisy Printers) | 7, 8 |
| E | (Earnest Engine) | 4, 5, 6 |
| F | (Flexible Systems) | 2, 3 |
| G | (Gutter Leaders) | 1, 2 |
| H | (Halley's Combs) | 6, 7 |
| I | (Indelible Ink Corporation) | 7, 8 |
| J | (Jay's Produce) | 4, 5 |
| K | (Kelly's Detective Agency) | 2, 3 |
| L | (Large Clothes) | 4, 5, 6 |
| M | (Metropolitan TV) | 1, 2 |
| N | (Nationwide Bank) | 4, 5, 6, 7 |

Look at the list of time blocks that the companies requested (where 1 = 9–10 A.M., . . . , 8 = 4–5 P.M.). Is it possible to accommodate all the companies that wish to do interviewing in the five rooms available while meeting their desired schedule times?

Problems of this kind seem simple enough, and you should try your hand at solving this particular one, for which a schedule does exist! However, this situation is not simple at all. The following facts are known about problems of this kind.

**Fact 1.** Suppose there are $i$ interviews, $p$ time periods, and $r$ rooms where interviews can be scheduled. Each interviewer has specified periods during which he or she wishes to conduct interviews. Is it possible to design a schedule that meets the desired specifications? It turns out that this problem is NP-complete (see Spotlight 2.1), that is, it belongs to a large group of problems for which, among other things, the fastest known algorithms run very slowly on large-problem versions.

**Fact 2.** The problem just described remains NP-complete even for the case where only three rooms have to be scheduled ($p = 3$).

The moral is *surprisingly simple:* Scheduling problems are very hard to solve.

However, the situation is not always as hopeless as it might seem. If you look at the list of time requests for the corporations, you will note again that, not surprisingly, each company has requested a contiguous block of times. It turns out that when this condition holds, it is possible to determine whether there is a feasible schedule using an algorithm that works relatively quickly.

## REVIEW VOCABULARY

**Average-case analysis** The study of the list-processing algorithm (more generally, any algorithm) from the point of view of how well it performs in all the types of problems it may be used for and seeing on average how well it does. *See also* worst-case analysis. (p. 78)

**Bin-packing problem** The problem of determining the minimum number of containers of capacity $W$ into which objects of size $w_1, \ldots, w_n$ ($w_i \leq W$) can be packed. (p. 80)

**Chromatic number** The chromatic number of a graph $G$ is the minimum number of colors (labels) needed in any vertex coloring of $G$. (p. 85)
**Critical-path scheduling** A heuristic algorithm for solving scheduling problems where the list-processing algorithm is applied to the priority list obtained by listing next in the priority list a task that heads a longest path in the order-requirement digraph. This task is then deleted from the order-requirement digraph, and the next task placed in the priority list is obtained by repeating the process. (p. 75)
**Decreasing-time-list algorithm** The heuristic algorithm that applies the list-processing algorithm to the priority list obtained by listing the tasks in decreasing order of their time length. (p. 79)
**First-fit (FF)** A heuristic algorithm for bin packing in which the next weight to be packed is placed in the lowest-numbered bin already opened into which it will fit. If it fits in no open bin, a new bin is opened. (p. 81)
**First-fit decreasing (FFD)** A heuristic algorithm for bin packing where the first-fit algorithm is applied to the list of weights sorted so that they appear in decreasing order. (p. 82)
**Heuristic algorithm** An algorithm that is fast to carry out but that doesn't necessarily give an optimal solution to an optimization problem. (p. 81)
**Independent tasks** Tasks are independent when there are no edges in the order-requirement digraph. These are tasks that can be performed in any order. (p. 78)
**List-processing algorithm** A heuristic algorithm for assigning tasks to processors: Assign the first ready task on the priority list that has not already been assigned to the lowest-numbered processor that is not working on a task. (p. 69)
**Machine scheduling problem** The problem of assigning tasks to processors so as to complete the tasks by the earliest time possible. (p. 67)
**Next-fit (NF)** A heuristic algorithm for bin packing in which a new bin is opened if the weight to be packed next will not fit in the bin that is currently being filled; the current bin is then closed. (p. 81)
**Next-fit decreasing (NFD)** A heuristic algorithm for bin packing where the next-fit algorithm is applied to the list of weights sorted so that they appear in decreasing order. (p. 82)
**Priority list** An ordering of the collection of tasks to be scheduled for the purpose of attaining a particular scheduling goal. One such goal is minimizing completion time when the list-processing algorithm is applied. (p. 68)
**Processor** A person, machine, robot, operating room, or runway with time that must be scheduled. (p. 67)
**Ready task** A task is called ready at a particular time if its predecessors, as given by the order-requirement digraph, have been completed by that time. (p. 69)
**Vertex coloring** A vertex coloring of a graph $G$ is an assignment of labels, which can be thought of as "colors," to the vertices of $G$ so that vertices joined by an edge get different labels (colors). (p. 85)
**Worst-case analysis** The study of the list-processing algorithm (more generally, any algorithm) from the point of view of how well it performs on the hardest problems it may be used on. *See also* average-case analysis. (p. 78)
**Worst-fit (WF)** A heuristic algorithm for bin packing in which the next weight to be packed is placed into the open bin with the largest amount of room remaining. If the weight fits in no open bin, a new bin is opened. (p. 81)
**Worst-fit decreasing (WFD)** A heuristic algorithm for bin packing where the worst-fit algorithm is applied to the list of weights sorted so that they appear in decreasing order. (p. 82)

## SKILLS CHECK

**1.** What is the minimum time required to complete 8 independent tasks with a total task time of 64 minutes on 4 machines?

**(a)** Less than 8 minutes
**(b)** Between 8 and 10 minutes
**(c)** More than 12 minutes

**2.** Given the order-requirement digraph below (time in minutes) and the priority list $T_1$, $T_2$, $T_3$, $T_4$, $T_5$, $T_6$, apply the list-processing algorithm to construct a schedule using two processors. **The completion time of the resulting schedule is ____________________.**

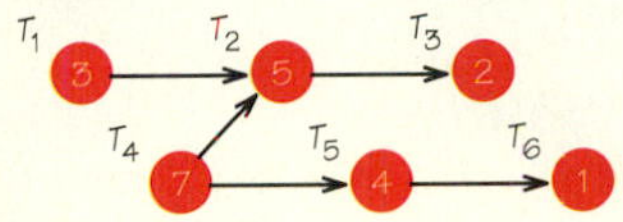


**3.** The following digraph cannot be an order-requirement digraph because

**(a)** no vertex has three edges that enter that particular vertex.
**(b)** it has a directed circuit.
**(c)** all the tasks require the same time to complete.

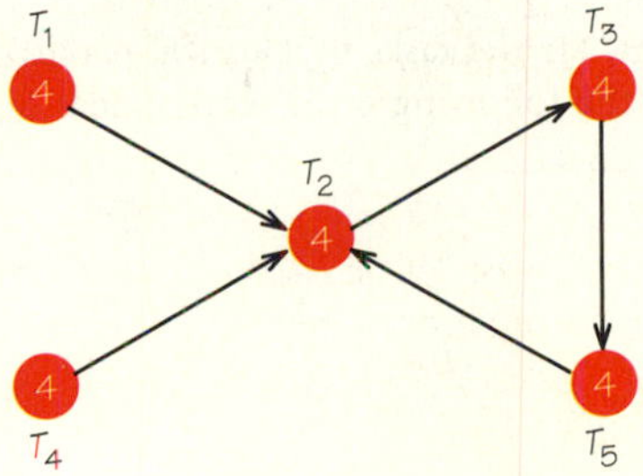

**4.** Suppose that a crew can complete in a minimum amount of time the job whose order-requirement digraph is shown below. If task $T_2$ is shortened from 5 minutes to 2 minutes, then the maximum amount by which the completion time for the entire job can be shortened is ______________________________ .

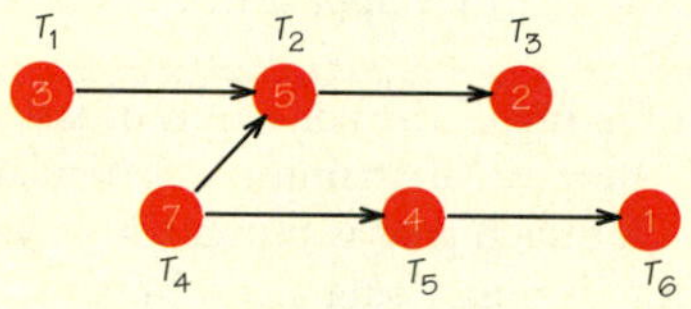


**5.** Suppose that independent tasks require a total of 30 minutes, while only one task takes as long as 10 minutes. If these tasks are scheduled on two machines,

**(a)** the tasks might take longer than 16 minutes to complete.
**(b)** the tasks can never take longer than 15 minutes to complete.
**(c)** the tasks can always be completed within 16 minutes.

**6.** The subscripts for the tasks that make up a critical path for the order-requirement digraph below are: ____ , ____ , ____ .

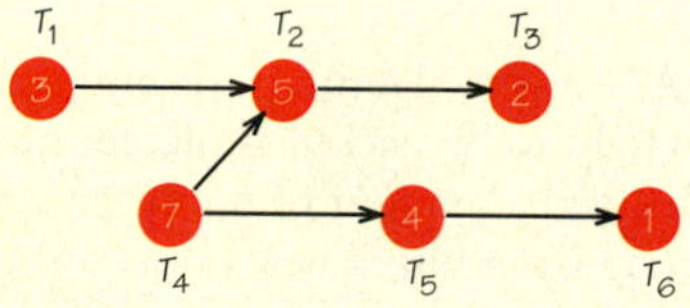


**7.** Which statement is true for the following digraph?

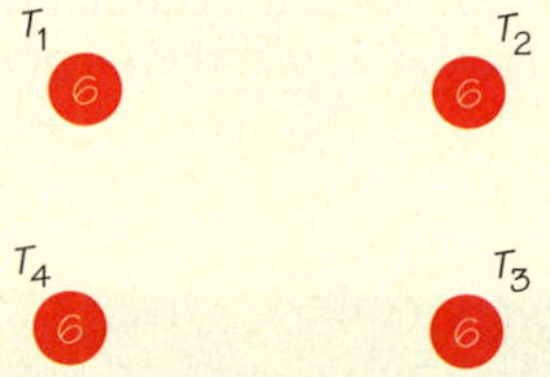


**(a)** This digraph cannot be the order-requirement digraph for a scheduling problem because the digraph has no (directed) edges.
**(b)** This digraph cannot be the order-requirement digraph for a scheduling problem because it is not allowed for all the tasks to have the same time length.
**(c)** This digraph can be the order-requirement digraph for a scheduling problem.

**8.** The subscripts for the tasks in a critical path list associated with the following order-requirement digraph are: _____ , _____ , _____ .

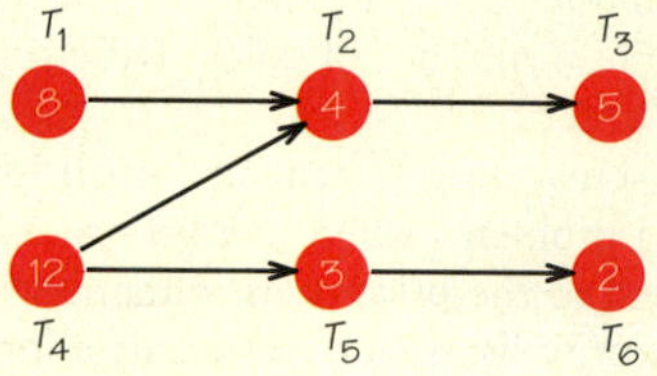


**9.** Assume an order-requirement digraph has a critical path with length 20 minutes. Based on this information, when the tasks are scheduled on two machines, how much time will be required?

**(a)** Exactly 10 minutes
**(b)** Exactly 20 minutes
**(c)** At least 20 minutes

**10.** The list-processing algorithm is used to schedule independent tasks lasting 6, 7, 4, 3, and 6 minutes on three machines, using these times as given for a list. The completion time for all the tasks will be ___________ .

**11.** Assume a job consists of six independent tasks ranging in time from 2 to 10 minutes and totaling 27 minutes. Efficiently scheduled on three machines, how much time will the job require?

**(a)** Exactly 9 minutes
**(b)** Exactly 10 minutes
**(c)** More than 10 minutes

**12.** When the decreasing-time-list algorithm is used to schedule independent tasks lasting 6 minutes, 7 minutes, 4 minutes, 3 minutes, and 6 minutes, on two machines, a schedule results where the tasks are completed after ________ minutes.

**13.** A radio announcer has 10 songs of various lengths to schedule into several segments. The announcer must identify the station at least once every 15 minutes, so the segments cannot be longer than 15 minutes. This job can be solved using the

**(a)** list-processing algorithm for independent tasks.
**(b)** critical-path scheduling algorithm.
**(c)** first-fit algorithm for bin packing.

**14.** Use the first-fit (FF) bin-packing algorithm to pack the following weights into bins that can hold no more than 10 lb: 6 lb, 7 lb, 4 lb, 3 lb, 6 lb. The number of bins required is ______________________________ .

**15.** Use the worst-fit-decreasing (WFD) bin-packing algorithm to pack the following weights into bins that can hold no more than 10 lb: 6 lb, 7 lb, 4 lb, 3 lb, 6 lb. How many bins are holding a full 10 lb?

**(a)** 0 bins
**(b)** 1 bin
**(c)** 2 bins

**16.** The first-fit decreasing (FFD) bin-packing algorithm is applied to the weight list 1, 2, 3, 4, 5, 5, 6, 8 for packing into bins of capacity 10. The item of weight 2 is packed into the bin numbered ______ when the packed bins are numbered from left to right.

**17.** A vertex coloring seeks to color the vertices of a graph to ensure which of the following traits?

**(a)** Every color is used.
**(b)** Every edge connects vertices of the same color.
**(c)** Vertices of the same color are never connected by an edge.

**18.** Assume the 8 corners of a cube represent vertices of a graph and the 12 edges of a cube represent the cube's edges. The chromatic number of this graph is ______________________ .

**19.** The minimum number of colors needed to color the vertices of the accompanying graph is

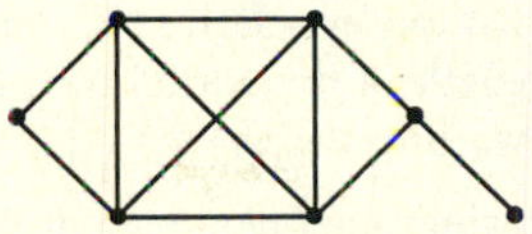

**(a)** 4.
**(b)** 2.
**(c)** 3.

**20.** A graph that has a circuit of length 3 can always be vertex colored with no fewer than _____ colors.

# CHAPTER 3 EXERCISES

■ Challenge ◆ Discussion

## 3.1 Scheduling Tasks

## 3.2 Critical-Path Schedules

**1.** List as many scheduling situations as you can for these environments:

**(a)** Your school
**(b)** Hospital
**(c)** Train station
**(d)** Police station
**(e)** Bookstore
**(f)** Internet café
**(g)** Firehouse
**(h)** Television studio

**2.** Compare and contrast the scheduling problems which arise at a

**(a)** Fast food restaurant.
**(b)** Standard sit-down restaurant.

**3.** You and your two housemates are planning to have a party this Friday night at your apartment. Eight guests are expected and there are plans to serve a small homemade dinner. List the tasks involved in carrying out such a party, and the types of processors to be used to carry out the tasks. Can any of the tasks be done simultaneously?

◆ **4.** Jane is planning a getaway weekend at a ski resort. She plans to leave work in Manhattan at 1 P.M. and must make her way to a local airport for a 5 P.M. shuttle plane to Boston. She then hopes to get a bus to the nearby resort. Discuss the tasks that Jane must complete to be at the resort by 10 P.M. What are the different types of processors involved in getting these tasks done? Can any of these tasks be done simultaneously?

**5.** Use the list-processing algorithm to schedule the tasks in the following order-requirement digraph on

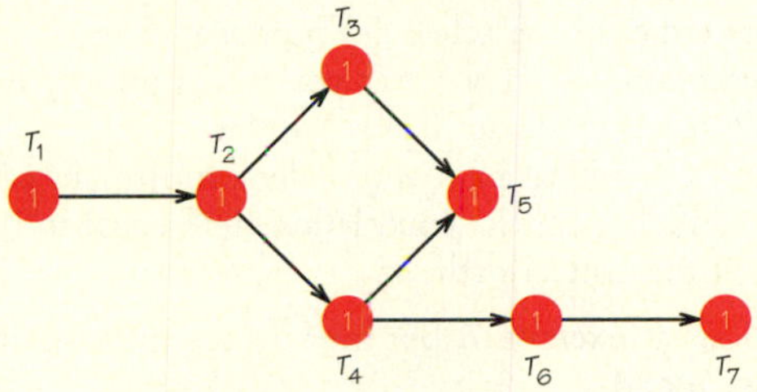


**(a)** two processors using the list $T_1, \ldots, T_7$.
**(b)** two processors using the list $T_1, T_2, T_3, T_4, T_6, T_5, T_7$.
**(c)** Is either of the schedules that you obtain optimal?
**(d)** Will adding a third processor enable the tasks to be finished earlier?
**(e)** Which tasks in this order-requirement digraph can be shortened and not affect the completion time of all the tasks?

**6.** Consider the following order-requirement digraph:

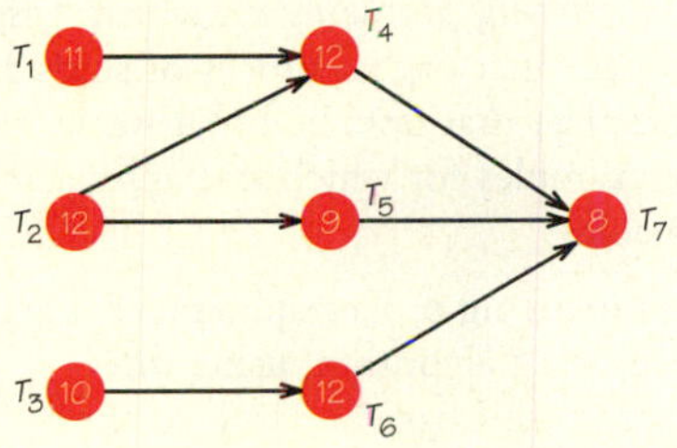


**(a)** Find the length of the critical path.
**(b)** Schedule these seven tasks on two processors using the list algorithm and the lists:
**(i)** $T_1, T_2, T_3, T_4, T_5, T_6, T_7$
**(ii)** $T_2, T_1, T_3, T_6, T_5, T_4, T_7$

**(c)** Does either list lead to a completion time that equals the length of the critical path?

**(d)** Show that no list can ever lead to a completion time equal to the length of the critical path (providing the schedule uses two processors).

**7.** **(a)** Use the following order-requirement digraph to schedule the 6 tasks $T_1$, $T_2$, $T_3$, $T_4$, $T_5$, $T_6$ on two processors with the priority lists:

**(i)** $T_1$, $T_2$, $T_3$, $T_4$, $T_5$, $T_6$

**(ii)** $T_1$, $T_6$, $T_3$, $T_5$, $T_4$, $T_2$

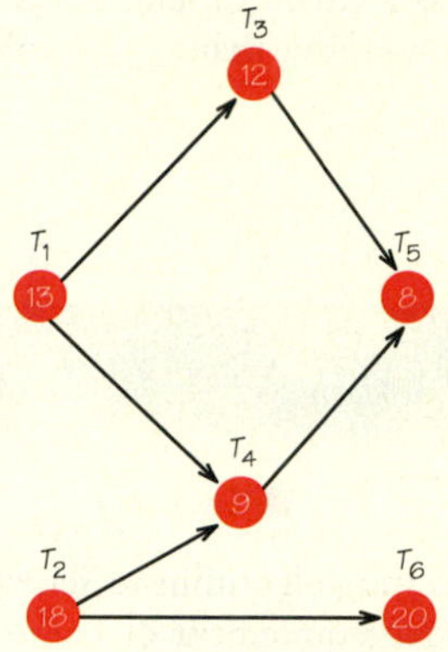


**(b)** Are either of the schedules produced from these lists optimal? If not, can you find a priority list that will result in an optimal schedule?

**(c)** Find the critical path and its length. Explain why no schedule has earliest completion time equal to the length of the critical path.

**8.** **(a)** Repeat Exercise 7, but interchange the task times of tasks $T_2$ and $T_6$.

**(b)** How does the completion time for an optimum schedule for this situation compare with the optimum schedule for Exercise 7?

**9.** **(a)** If one adds a new directed edge to an order-requirement digraph *D*, can the critical path in the the new order-requirement digraph *D′* have longer length?

**(b)** If one adds a new directed edge to an order-requirement digraph *D*, can the critical path in the the new order-requirement digraph *D′* have shorter length?

◆ **10.** Discuss scheduling problems for which it is not reasonable to assume that once a processor starts a task, it will always complete that task, before it works on any other task. Give examples for which this approach would be reasonable.

**11.** For the accompanying order-requirement digraph, apply the list-processing algorithm, using three processors for lists (a) through (c). How do the completion times obtained compare with the length of the critical path?

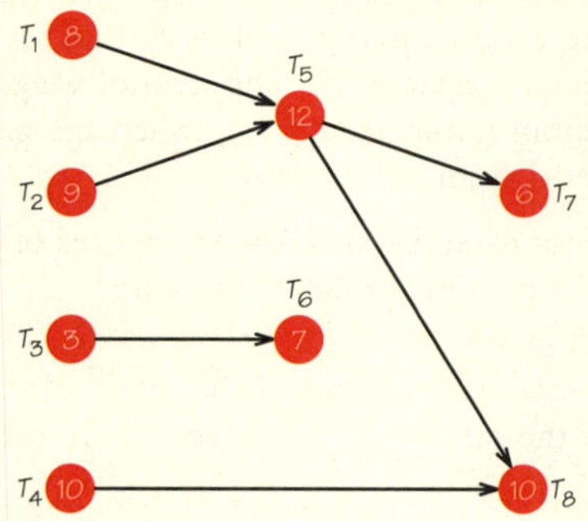


**(a)** $T_1$, $T_2$, $T_3$, $T_4$, $T_5$, $T_6$, $T_7$, $T_8$

**(b)** $T_1$, $T_3$, $T_5$, $T_7$, $T_2$, $T_4$, $T_6$, $T_8$

**(c)** $T_8$, $T_6$, $T_4$, $T_2$, $T_1$, $T_3$, $T_5$, $T_7$

**12.** **(a)** Can you find an order-requirement digraph with four tasks for which every priority list used to schedule the tasks on two machines assigns task $T_4$ to machine 1 at time 0?

**(b)** Can you choose the order-requirement digraph in part (a) so that machine 2 stays idle for all lists from time 0 to time 3?

**13.** Can you give examples of scheduling problems for which it seems reasonable to assume that all the task times are the same?

**14.** Use the list-processing algorithm to schedule the tasks in the following order-requirement digraph on

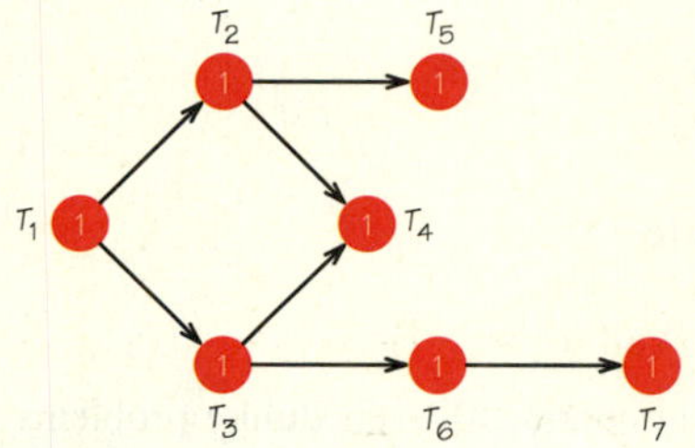


**(a)** two processors using the list $T_1, \ldots, T_7$.

**(b)** two processors using the list $T_1$, $T_2$, $T_3$, $T_4$, $T_6$, $T_5$, $T_7$.

**(c)** Is either of the schedules that you obtain optimal?

**15.** Can you find a list that gives rise to the optimal schedule shown in Figure 3.14 for the order-requirement digraph in Figure 3.12?

**16.** Consider the following order-requirement digraph:

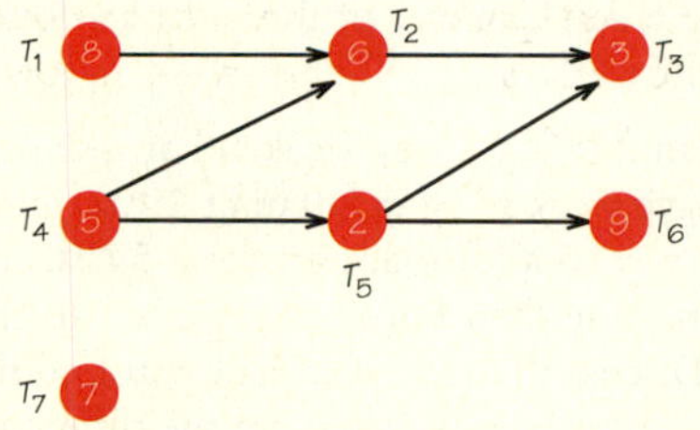


**(a)** Find the critical path(s).

**(b)** Schedule these tasks on one processor using the critical-path scheduling method.
**(c)** Schedule these tasks on one processor using the priority list obtained by listing the tasks in order of decreasing time.
**(d)** Does either of these schedules have idle time? How do their completion times compare?
**(e)** If two different schedules have the same completion time, what criteria can be used to say one schedule is superior to the other?
**(f)** Schedule these tasks on two processors using the order-requirement digraph shown and the priority list from part (b).
**(g)** Does the schedule produced in part (f) finish in half the time that the schedule in part (b) did, which might be expected, since the number of processors has doubled?
**(h)** Schedule the tasks on (i) one processor and (ii) two processors (using the decreasing-time list), assuming that each task time has been reduced by one. Do the changes in completion time agree with your expectations?

**17.** **(a)** Can all the processors being used to schedule tasks be simultaneously idle at a time before the completion time of a collection of tasks scheduled using the list-processing algorithm?
**(b)** Explain why the list-processing algorithm cannot give rise to the schedule below, regardless of what priority list was used to schedule the tasks on the two processors.

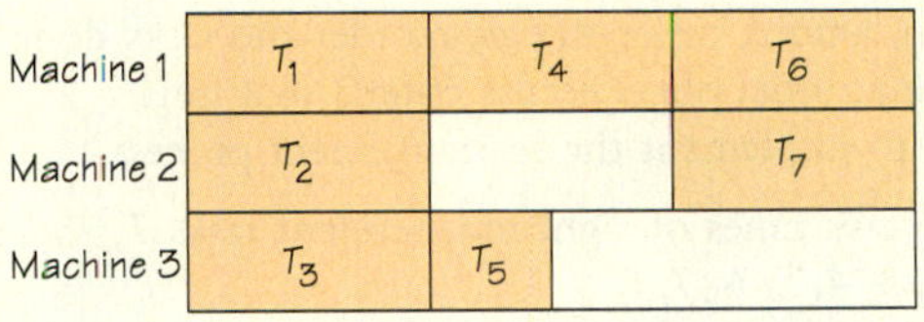


**(c)** Construct an order-requirement digraph and a priority list that will give rise to the following schedule on two processors.

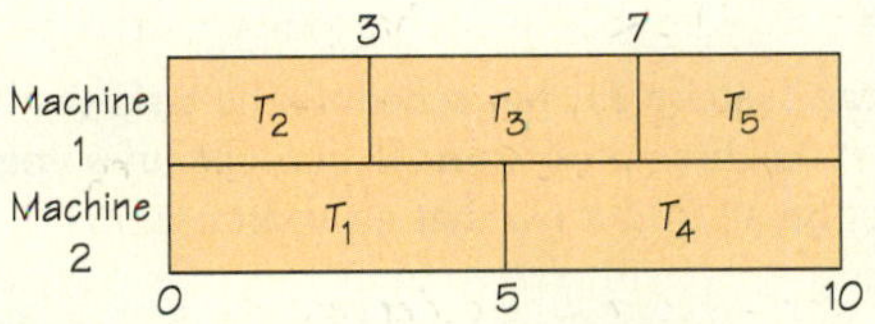


**18.** To prepare a meal quickly involves carrying out the tasks shown (time lengths in minutes) in the following order-requirement digraph:

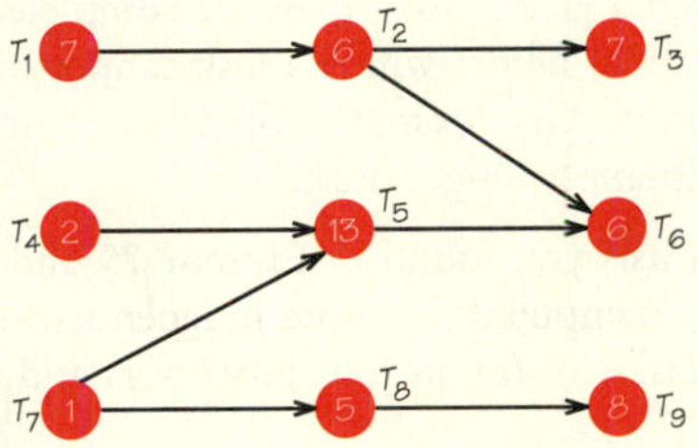


**(a)** If Mike prepares the meal alone, how long will it take?
**(b)** If Mike can talk Mary into helping him prepare the meal, how long will it take them if the tasks are scheduled using the list $T_5$, $T_9$, $T_1$, $T_3$, $T_2$, $T_6$, $T_8$, $T_4$, $T_7$ and the list-processing algorithm?
**(c)** If Mike can talk Mary and Jack into helping him prepare the meal, how long will it take if the tasks are scheduled using the same list as in part (b)?
**(d)** What would be a reasonable set of criteria for choosing a priority list in this situation?

**19.** **(a)** Making use of the order-requirement digraph below, determine at time 0 which tasks are ready.

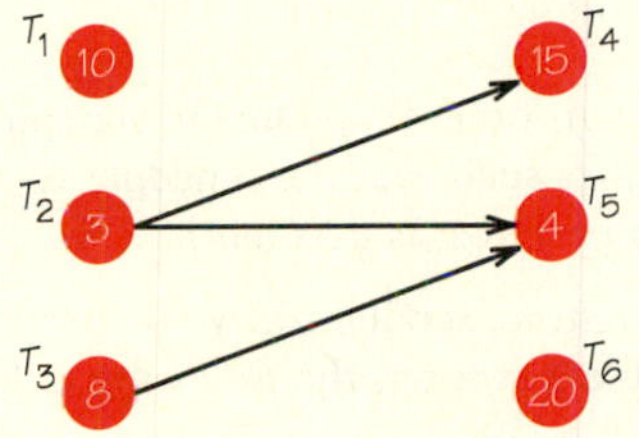


**(b)** What is special about tasks $T_1$ and $T_6$?
**(c)** What is the critical path, and what is its length?
**(d)** Schedule the tasks on three processors with the priority list $T_1, \ldots, T_6$.
**(e)** Is the schedule found in part (d) optimal?
**(f)** Schedule the tasks on three processors using the priority list $T_6, \ldots, T_1$.
**(g)** Is the schedule found in part (f) optimal?
**(h)** Can you find a priority list that yields an optimal schedule?

**20.** **(a)** In Exercise 19, what priority list would be used if you applied the critical-path scheduling method?
**(b)** Use this priority list to schedule the tasks on three processors. Is this schedule optimal?
**(c)** How does this schedule compare with the schedules that you found using the lists in Exercise 19?

**21.** Consider the order-requirement digraph below. Suppose one plans to schedule these tasks on two identical processors.

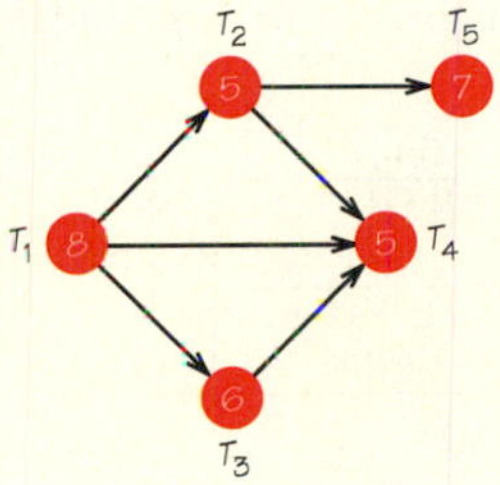


**(a)** How many different priority lists can be used to schedule the tasks?
**(b)** Can all these priority lists lead to different schedules? If not, why not?
**(c)** Can an optimal schedule have no idle time? Can you give two different reasons why an optimal schedule must have some idle time?

**(d)** Is there any list that produces a schedule where the second processor has no idle time?

**22.** **(a)** In Exercise 21, how many different lists are there that do not list $T_1$ first?
**(b)** Would it make any sense not to list $T_1$ first in a list?
**(c)** Construct a list and schedule the tasks on two processors.
**(d)** Can you find another list that leads to a different completion time than the schedule you found for part (c)?
**(e)** Find a list that leads to an optimal schedule.

**23.** Can you find an order-requirement digraph with five tasks for which every possible list yields exactly the same schedule?

**24.** Can you find an order-requirement digraph involving three tasks such that the schedule corresponding to every list is different?

**25.** At a large toy store, scooters arrive unassembled in boxes. To assemble a scooter, the following tasks must be performed:

Task 1. Remove parts from the box.
Task 2. Attach wheels to the footboard.
Task 3. Attach vertical housing.
Task 4. Attach handlebars to vertical housing.
Task 5. Put on reflector tape.
Task 6. Attach bell to handlebars.
Task 7. Attach decals.
Task 8. Attach kickstand.
Task 9. Attach safety instructions to handlebars.

**(a)** Give reasonable time estimates for these tasks and construct a reasonable order-requirement digraph. What is the earliest time by which these tasks can be completed?
**(b)** Schedule this job on two processors (humans) using the decreasing-time-list algorithm.

**26.** If two schedules for the same number of processors have the same completion time, can one schedule have more idle time than the other?

**27.** Could the schedule below be obtained by applying the list-scheduling algorithm to some order-requirement digraph?

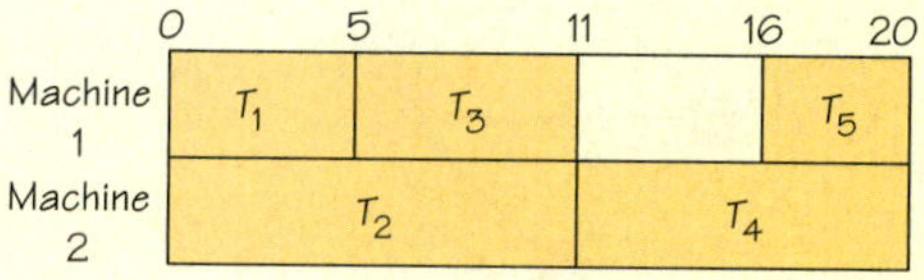


## 3.3 Independent Tasks

**28.** Could the following schedule be obtained by applying the list-scheduling algorithm to some order-requirement digraph?

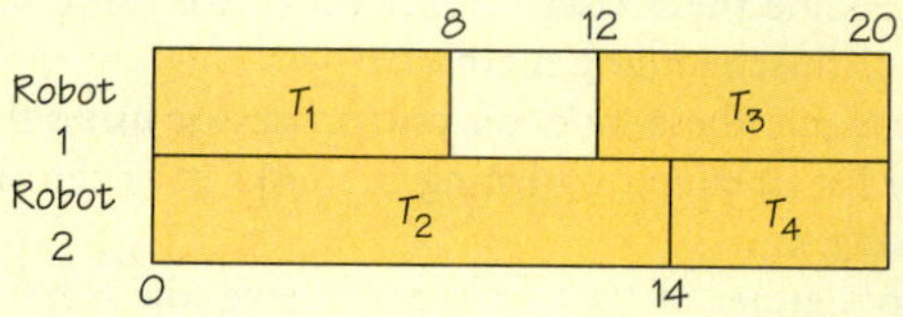


**29.** For the following schedules, can you produce a list so that the list-processing algorithm produces the schedule shown when the tasks are independent? What are the task times for each task?

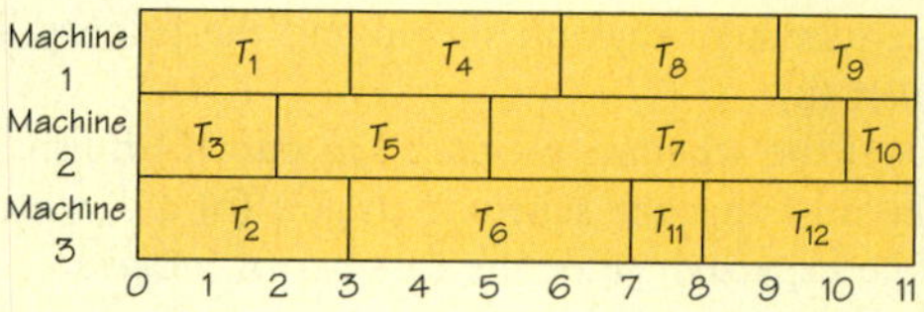

(a)

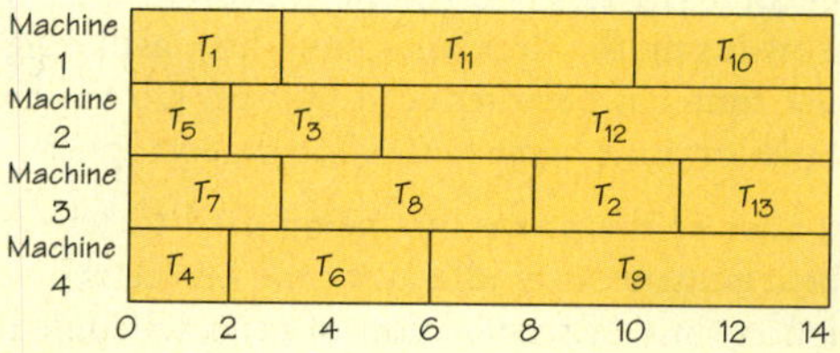

(b)

◆ **30.** Once an optimal schedule has been found for independent tasks (see diagrams in Exercise 29), usually the scheduling of the tasks can be rearranged and the same optimal time achieved.
One can, among other things, reorder the tasks done by a particular processor. Discuss criteria that might be used to implement the rearrangement process.

**31.** The task times of eight independent tasks $T_1$ to $T_8$ are 1, 2, 3, 4, 5, 6, 7, 8.

**(a)** Schedule the tasks on two processors using the lists (i) $T_1, T_2, \ldots, T_8$ and (ii) $T_8, T_7, \ldots, T_1$.
**(b)** Is either of the schedules you get in part (a) optimal? If not, find a list that gives an optimal schedule.

**32.** Repeat Exercise 31, but schedule the tasks (with the same lists) on three processors. If the schedules you get are not optimal, find a list that gives an optimal schedule.

◆ **33.** Discuss different criteria that might be used to construct a priority list for a scheduling problem.

◆ **34.** Some scheduling projects have due dates for tasks (times by which a given task should be completed) and release dates (times before which a task cannot have work begun on it). Give examples of circumstances where these situations might arise.

**35.** Using the lists you found in Exercise 29 and the task times you computed for those independent tasks, schedule the tasks for (a) on four processors and the tasks for (b) on five processors. Can you see why for any schedule you may produce for (a) on four processors and (b) on five processors there must be some idle time for one or more processors?

**36.** Given the following order-requirement digraph:

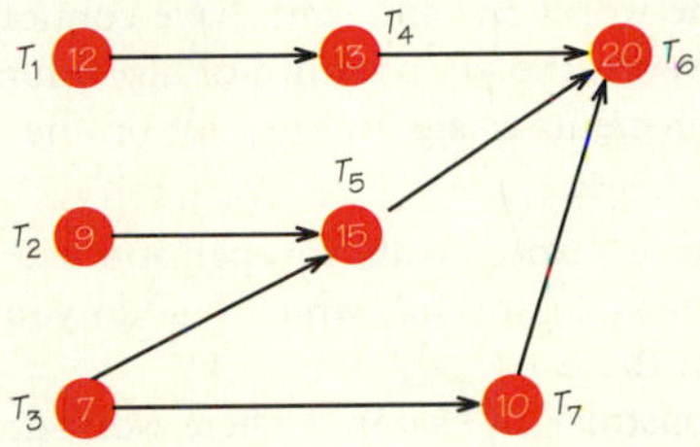


**(a)** Use the list-processing algorithm to schedule these seven tasks on two processors using these lists:

**(i)** $T_1, T_3, T_7, T_2, T_4, T_5, T_6$

**(ii)** $T_1, T_3, T_2, T_4, T_5, T_6, T_7$

**(iii)** The list obtained by listing the tasks in order of decreasing time

**(b)** Try to determine if any of the resulting schedules are optimal.

**(c)** Schedule the tasks using the critical-path scheduling method. Try to determine if this schedule is optimal.

**37.** Repeat the questions in Exercise 36 using the order-requirement digraph obtained by erasing all the (directed) edges shown there. How do the schedules you get compare with the ones you originally got?

**38. (a)** Find the completion time for independent tasks of length 8, 11, 17, 14, 16, 9, 2, 1, 18, 5, 3, 7, 6, 2, 1 on two processors, using the list-processing algorithm.

**(b)** Find the completion time for the tasks in part (a) on two processors, using the decreasing-time-list algorithm.

**(c)** Does either algorithm give rise to an optimal schedule?

**(d)** Repeat for tasks of lengths 19, 19, 20, 20, 1, 1, 2, 2, 3, 3, 5, 5, 11, 11, 17, 18, 18, 17, 2, 16, 16, 2.

**39.** Repeat parts (a)–(c) of Exercise 38 for independent tasks of lengths 19, 19, 20, 20, 1, 1, 2, 2, 3, 3, 5, 5, 11, 11, 17, 17, 18, 18, 17, 2, 16, 16, 2.

**40.** Suppose that independent tasks require a total of 36 minutes, while only one of the tasks takes as long as 12 minutes. If these tasks are scheduled on two machines, show by an example that the earliest completion time may be as long as 22 minutes.

**41.** A photocopy shop must schedule independent batches of documents to be copied. The times for the different sets of documents are (in minutes): 12, 23, 32, 13, 24, 45, 23, 23, 14, 21, 34, 53, 18, 63, 47, 25, 74, 23, 43, 43, 16, 16, 76.

**(a)** Construct a schedule using the list-processing algorithm on three machines.

**(b)** Construct a schedule using the list-processing algorithm on four machines.

**(c)** Repeat parts (a) and (b), but use the decreasing-time-list algorithm.

**(d)** Suppose union regulations require that an 8-minute rest period be allowed for any photocopy task over 45 minutes. Use the decreasing-time-list algorithm, with the preceding times modified to take into account the union requirement, to schedule the tasks on three human-operated machines.

**42.** Find a list that produces the following optimal schedule when the list-processing algorithm is applied to this list. (Assume the tasks are independent.)

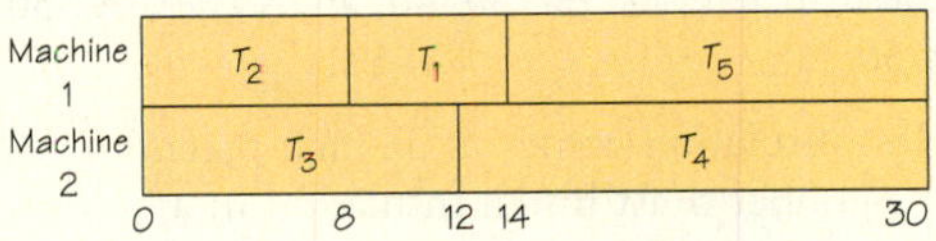


What completion time and schedule are obtained when the decreasing-time-list algorithm is applied to this list?

**43.** Can you think of situations other than those mentioned in the text where scheduling independent tasks on processors occurs?

**44.** Can you think of real-world scheduling situations in which all the tasks have the same time and are independent? Find an algorithm for solving this problem optimally. (If there are $n$ independent tasks of time length $k$, when will all the tasks be finished?)

**45.** Show that when tasks to be scheduled are independent, the critical-path method and the decreasing-time-list method are identical.

## 3.4 Bin Packing

**46.** Two wooden wall systems are to be made of pieces of wood with lengths shown in the accompanying diagram. If wood is sold in 10-foot planks and can be cut with no waste, what number of boards would be purchased if one uses the first-fit-decreasing, next-fit-decreasing, and worst-fit-decreasing heuristics, respectively?

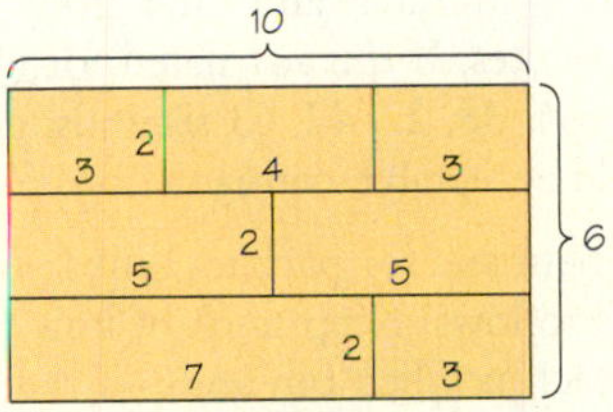


In solving this problem, does it make a difference if the 10-foot horizontal shelves and 6-foot vertical boards employ single-length pieces as compared with using pieces of boards that add up to 10- and 6-foot lengths?

**47.** It takes 4 seconds to photocopy one page. Manuscripts of 10, 8, 15, 24, 22, 24, 20, 14, 19, 12, 16, 30, 15, and 16 pages are to be photocopied. How many photocopy machines would be required, using the first-fit-decreasing algorithm, to guarantee that all manuscripts are photocopied in 2 minutes or less? Would the solution differ if worst-fit decreasing were used?

**48.** A radio station's policy allows advertising breaks of no longer than 2 minutes, 15 seconds. Using first-fit and first-fit-decreasing algorithms, determine the minimum number of breaks into which the following ads will fit (lengths given in seconds): 80, 90, 130, 50, 60, 20, 90, 30, 30, 40. Can you find the optimal solution? Do the same for these ad lengths: 60, 50, 40, 40, 60, 90, 90, 50, 20, 30, 30, 50.

**49.** Fiberglass insulation comes in 36-inch precut sections. A plumber must install insulation in a basement on piping that is interrupted often by joints. The distances between the joints on the stretches of pipe that must be insulated are 12, 15, 16, 12, 9, 11, 15, 17, 12, 14, 17, 18, 19, 21, 31, 7, 21, 9, 23, 24, 15, 16, 12, 9, 8, 27, 22, 18 inches. How many precut sections would he have to use to provide the insulation if he bases his decision on

**(a)** next-fit?
**(b)** next-fit decreasing?
**(c)** worst-fit?
**(d)** worst-fit decreasing?

**50.** The files that a company has for its employees dealing with utilities occupy 100, 120, 60, 90, 110, 45, 30, 70, 60, 50, 40, 25, 65, 25, 55, 35, 45, 60, 75, 30, 120, 100, 60, 90, 85 sectors. If, after operating systems are installed, a disk can store up to 480 sectors, determine the number of disks needed to store the utilities if each of these heuristics is used to pack the disk with files:

**(a)** next-fit
**(b)** next-fit decreasing
**(c)** first-fit
**(d)** first-fit decreasing

**51.** Advertisements for the TV show *Q* are permitted to last up to a total of 8 minutes, and each group of ads can last up to 2 minutes. If the ads slated for *Q* last 63, 32, 11, 19, 24, 87, 64, 36, 27, 42, 63 seconds, determine if FF and FFD yield acceptable configurations for the ads.

**52.** Consider the heuristic for packing bins known as *best-fit* described as follows: Keep track of how much room remains in each unfilled bin and put the next item to be packed into that bin that would leave the least room left over after the item is put into the bin. (For example, suppose that bin 4 had 6 units left, bin 7 had 5 units left, and bin 9 had 8 units left. If the next item in the list had size 5, then first-fit would place this item in bin 4, worst-fit would place the item in bin 9, while best-fit would place the item in bin 7.) If there is a tie, place the item into the bin with the lowest number. Apply this heuristic to the list 8, 7, 1, 9, 2, 5, 7, 3, 6, 4, where the bins have capacity 10.

◆ **53.** We have described two algorithms for bin packing called worst-fit and best-fit (see page 81 and Exercise 52). The words *best* and *worst* have connotations in English. However, the performance of algorithms depends on their merits as algorithms, not on the names we give them.

**(a)** On the basis of experiments you perform with the best-fit and worst-fit algorithms, which one do you think is the "better" of the two?
**(b)** Can you construct an example where worst-fit uses fewer bins than best-fit?

**54.** The best-fit heuristic (see Exercise 52) also has a "decreasing" version, where the list is first sorted in decreasing order. Using bins of capacity 10, apply the best-fit heuristic and its decreasing version to the following list: 6, 9, 5, 8, 3, 2, 1, 9, 2, 7, 2, 5, 4, 3, 7, 6, 2, 8, 3, 7, 1, 6, 4, 2, 5, 3, 7, 2, 5, 2, 3, 6, 2, 7, 1, 3, 5, 4, 2, 6.

■ **55.** One pianist's recording of the complete Mozart piano sonatas takes the following times (given in minutes and seconds): 13:46, 6:15, 3:29, 5:37, 7:52, 2:55, 5:00, 4:28, 4:21, 7:39, 7:55, 6:42, 4:23, 3:52, 4:21, 4:20, 5:46, 6:29, 5:34, 6:23, 6:39, 7:19, 5:54, 6:54, 2:58, 5:22, 1:42, 5:00, 1:29, 5:47, 7:30, 8:19, 4:44, 4:57, 4:09, 14:31, 3:55, 4:04, 4:01, 6:06, 6:50, 5:27, 4:28, 5:40, 2:52, 5:16, 5:34, 3:10, 7:22, 4:40, 3:08, 6:32, 4:47, 6:59, 5:38, 7:57, 3:38. If the maximum time that can be recorded on a compact disc is 70:30, can all the music be performed on four compact discs? Can all the music be performed on five compact discs?

■ **56.** In the wall-system example in the text, first-fit and worst-fit required equal numbers of bins (see Figure 3.20). Can you find an example where first-fit and worst-fit yield different numbers of bins? Can you find an example where first-fit, worst-fit, and next-fit yield answers with different numbers of bins?

◆ **57.** A common suggestion for heuristics for the bin-packing problem with bins of capacity $W$ involves finding weights that sum to exactly $W$. Discuss the pros and cons of a heuristic of this type.

■ **58.** A recording company wishes to record all the Beethoven string quartets (16 quartets, each consisting of several consecutive parts called movements) on LPs. It wishes to complete the project on as few records as possible. Recording can be done on two sides as long as the movements are consecutive. Is this an example of a bin-packing problem? (Defend your answer.) If the project were to record the quartets on (standard) tape cassettes or compact discs, would your answer be different?

**59.** Give examples where it would be realistic to keep bins open as more items "arrive" to be packed, rather than to close a bin permanently based on some criterion.

**60.** Give examples where it would be unrealistic to keep bins open as more items "arrive" to be packed, rather than to close a bin permanently based on some criterion.

**61.** A data entry group must handle 30 (independent) tasks that will take the following amounts of time (in minutes) to type: 25, 18, 13, 19, 30, 32, 12, 36, 25, 17, 18, 26, 12, 15, 31, 18, 15, 18, 16, 19, 30, 12, 16, 15, 24, 16, 27, 18, 9, 14. Using these times as a priority list:

**(a)** Use the list-processing algorithm to find the completion time for scheduling tasks with four secretaries. Also, solve with five secretaries.
**(b)** Repeat the scheduling using the decreasing-time-list algorithm.
**(c)** Can you show that any of the schedules that you get are optimal?

If one needs to finish the typing in one hour:
**(d)** Use the FFD heuristic to find how many typists would be needed.
**(e)** Repeat for the NFD and WFD heuristics.
**(f)** Can you show that any of the solutions you get are optimal?

**62.** Find the minimum number of bins necessary to pack items of size 8, 5, 3, 4, 3, 7, 8, 8, 6, 5, 3, 2, 1, 2, 1, 2, 1, 3, 5, 2, 4, 2, 6, 5, 3, 4, 2, 6, 7, 7, 8, 6, 5, 4, 6, 1, 4, 7, 5, 1, 2, 4 in bins of capacity (a) through (d) using the first-fit and first-fit-decreasing algorithms. Can you determine if any of the packings you get are optimal?

**(a)** 9
**(b)** 10
**(c)** 11
**(d)** 12

■ **63.** Two-dimensional bin packing refers to the problem of packing rectangles of various sizes into a minimum number of $m \times n$ rectangles, with the sides of the packed rectangles parallel to those of the containing rectangle.

**(a)** Suggest some possible real-world applications of this problem.
**(b)** Devise a heuristic algorithm for this problem.
**(c)** Give an argument to show that the problem is at least as hard to solve as the usual bin-packing problem.
**(d)** If you have $1 \times m$ rectangles with total area $W$ to be packed into a single rectangle of area $p \times q = W$, can the packing always be accomplished?

◆ **64.** In what situations would packing bins of different capacities be the appropriate model for real-world situations? Suggest some possible algorithms for this type of problem.

■ **65.** Find an example of weights that, when packed into bins using first-fit, use fewer bins than the number of bins used when the first-fit algorithm is applied with the first weight on the list removed.

◆ **66.** Formulate "paradoxical" situations for bin packing that are analogous to those we found for scheduling processors.

## 3.5 Resolving Conflict via Coloring

**67.** For each of the graphs below:

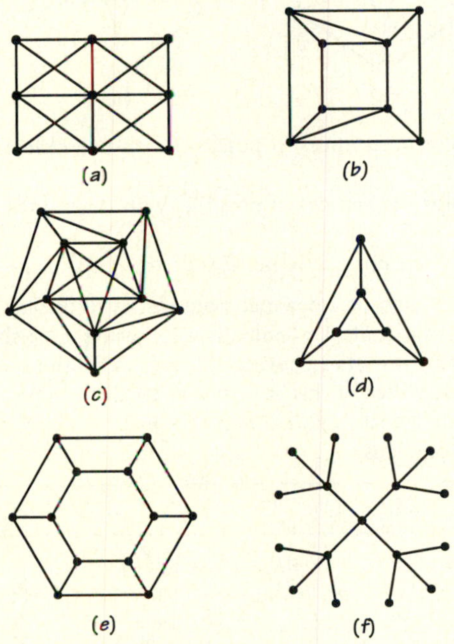


**(a)** Color the vertices (if possible) with three different colors.
**(b)** Color the vertices (if possible) with four different colors.
**(c)** Find the chromatic number of the graph.

**68.** For each of the following graphs:

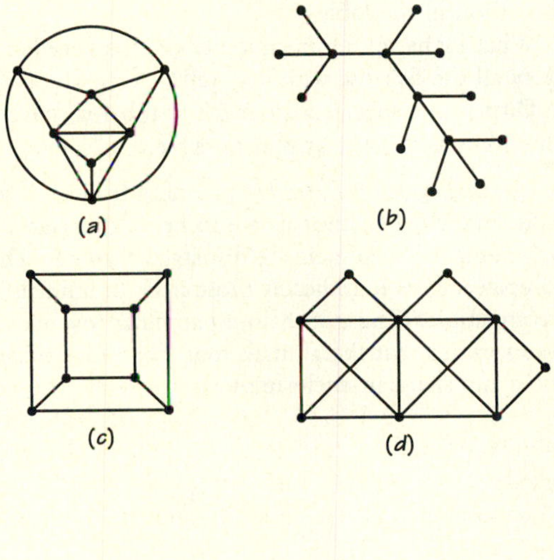

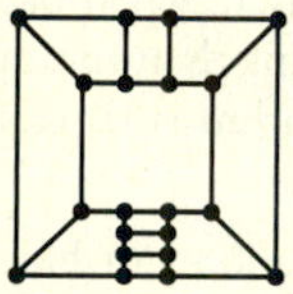

(e)

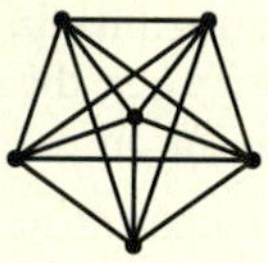

(f)

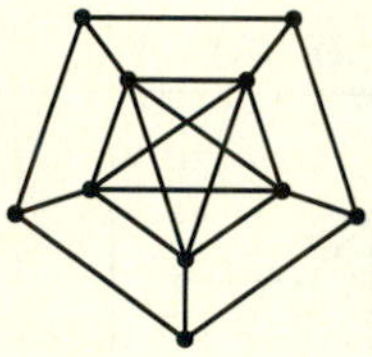

(g)

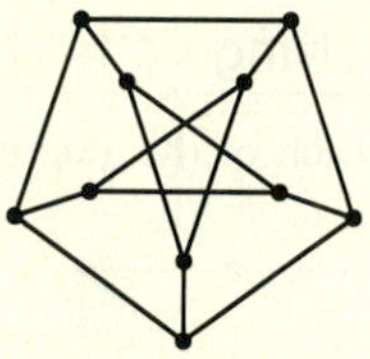

(h)

**(a)** Color the vertices (if possible) with two different colors.
**(b)** Color the vertices (if possible) with three different colors.
**(c)** Find the chromatic number of the graph.

**69.** The owner of a new pet store wishes to display tropical fish in display tanks. The following table shows the incompatibilities between the species, in the sense that an X indicates that it is unwise to allow those species in the row and column that meet at the X to be in the same tank.

| | A | B | C | D | E | F | G | H | I |
|---|---|---|---|---|---|---|---|---|---|
| A | | | | | | X | X | | X |
| B | | | X | | | | | X | |
| C | | X | | | X | | | X | |
| D | | | | | X | X | | X | |
| E | | | X | X | | | X | | |
| F | X | | | X | | | X | | X |
| G | X | | | | X | X | | X | X |
| H | | X | X | X | | | X | | |
| I | X | | | | | X | X | | |

**(a)** Draw an appropriate graph to represent the information in the table.
**(b)** What is the minimum number of tanks needed to display all the fish she wishes to sell?
**(c)** Display the species so that the number of species in each tank is as nearly equal as possible.

**70.** The managers of a zoo are planning to open a small satellite branch. The animals are to be in enclosures in which compatible animals are displayed together. The accompanying table indicates those pairs of animals that are compatible. (Thus, an X in a particular row and column means that the animals that label this row and column *can* share an enclosure.)

| | A | B | C | D | E | F | G | H | I | J |
|---|---|---|---|---|---|---|---|---|---|---|
| A | X | X | | X | X | X | X | | | |
| B | X | X | | | X | X | X | | X | X |
| C | | | X | | X | X | X | | | |
| D | X | | | X | X | X | X | | X | X |
| E | X | X | X | X | X | | | X | X | |
| F | X | X | X | X | | X | X | X | X | |
| G | X | X | X | X | | X | X | X | | |
| H | | | | | X | X | X | X | | |
| I | | X | | X | X | X | | | X | |
| J | | X | | X | | | | | | X |

**(a)** Draw an appropriate graph to represent the information in the table.
**(b)** What is the minimum number of enclosures needed to avoid housing incompatible animals in the same enclosure?
**(c)** Is it possible to enclose the animals in such a way that each enclosure contains the same number of animals?
**(d)** Why might that be desirable? Why might this approach to grouping the animals not be ideal?

**71.** The nine standing committees of a state legislature are designing a schedule for when the committees can meet. The matrix shown in the following table has an X in a position where the committees corresponding to the row and column have a common member and, hence, should not be scheduled to meet at the same hour. The committees involved are Agriculture (A), Commerce (C), Consumer Affairs (CA), Education (E), Forests (F), Health (H), Justice ( J), Labor (L), and Rules (R).

| | A | C | CA | E | F | H | J | L | R |
|---|---|---|---|---|---|---|---|---|---|
| A | | X | X | | | X | | | |
| C | X | | X | X | X | | | | |
| CA | X | X | | | | | X | | X |
| E | | X | | | X | X | | | |
| F | | X | | X | | X | X | | |
| H | X | | | X | X | | | X | |
| J | | | X | | X | | | X | X |
| L | | | | | | X | X | | X |
| R | | | X | | | | X | X | |

**(a)** Draw a graph that will be of value in determining the minimum number of time slots the committees can meet in without any legislator having to be in two places at one time.
**(b)** What is the minimum number of time slots in which the committees can be scheduled without a conflict?
**(c)** How many different rooms are needed at any time that a committee is scheduled to meet? (Why might this issue matter?)

**72.** Determine the minimum number of colors, and how often each color is used, in a vertex coloring of the graphs below.

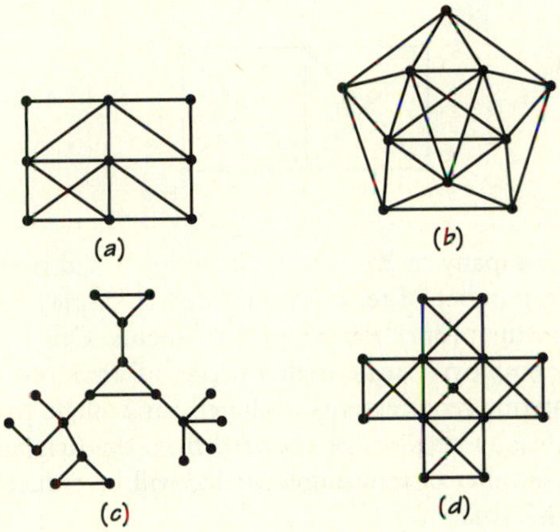


**73.** The faculty–student governing council at All State College has nine standing committees (such as Curriculum, Academic Standards, Campus Life) that are designed A, B, C, D, . . . , I for convenience. The following table shows which committees have no member in common.

| | A | B | C | D | E | F | G | H | I |
|---|---|---|---|---|---|---|---|---|---|
| A | | X | | X | | X | X | | X |
| B | X | | | | X | X | | X | X |
| C | | | | X | | X | X | X | X |
| D | X | | X | | | X | | X | |
| E | | X | | | | | X | X | X |
| F | X | X | X | X | | | | | |
| G | X | | X | | X | | | X | |
| H | | X | X | X | X | | X | | X |
| I | X | X | X | | X | | | X | |

**(a)** Draw an appropriate graph to represent the information in the table.
**(b)** What is the minimum number of time slots in which all the committee meetings can be scheduled?
**(c)** How many rooms are needed during each time slot to accommodate the committees that are scheduled to meet in that time slot?

**74.** When two towns are within 145 miles of each other, the frequency used by a certain type of emergency response system for the towns requires that they be on different frequencies to avoid possible interference with each other. The following table shows the mileage distances between six towns.

| | E | F | G | I | S | T |
|---|---|---|---|---|---|---|
| **Evansville (E)** | | 290 | 277 | 168 | 303 | 133 |
| **Ft. Wayne (F)** | 290 | | 132 | 83 | 79 | 201 |
| **Gary (G)** | 277 | 132 | | 153 | 58 | 164 |
| **Indianapolis (I)** | 168 | 83 | 153 | | 140 | 71 |
| **South Bend (S)** | 303 | 79 | 50 | 140 | | 196 |
| **Terre Haute (T)** | 113 | 201 | 164 | 71 | 196 | |

**(a)** What would be the minimum number of frequencies that are needed for each town to have its emergency broadcasts not conflict with those of any other town using this system?
**(b)** How many different towns would be assigned to each frequency used?

**75.** Show that the vertices of any tree can be colored with two colors.

**76.** Can you find a family of graphs $H_n$ ($n \geq 1$) that require $n$ colors to color their vertices?

**77.** The edge-coloring number of a graph $G$ is the minimum number of colors needed to color the edges of $G$ so that edges that share a common vertex get different colors. Determine the edge-coloring number for each of the graphs in Exercise 67. Can you make a conjecture about the value of the minimum number of colors needed to color the edges of any graph?

**78.** Can you think of any applications that require determining the minimum number of colors needed to color the edges of a graph?

**79.** When a graph has been drawn on a piece of paper so that edges meet only at vertices, the graph divides the paper up into regions called *faces*. The faces include one called the "infinite" face, which surrounds the whole graph. The face-coloring number of a graph $G$ (which can be drawn in this special way) is the minimum number of colors needed to color the faces of $G$ so that two faces that share an edge receive different colors. (Note that if two faces meet only at a vertex, they can be colored the same color.)

**(a)** Determine the minimum number of colors needed to color the faces of the following graphs. In each case, remember to color the infinite face, which is labeled $I$ (for "infinite").

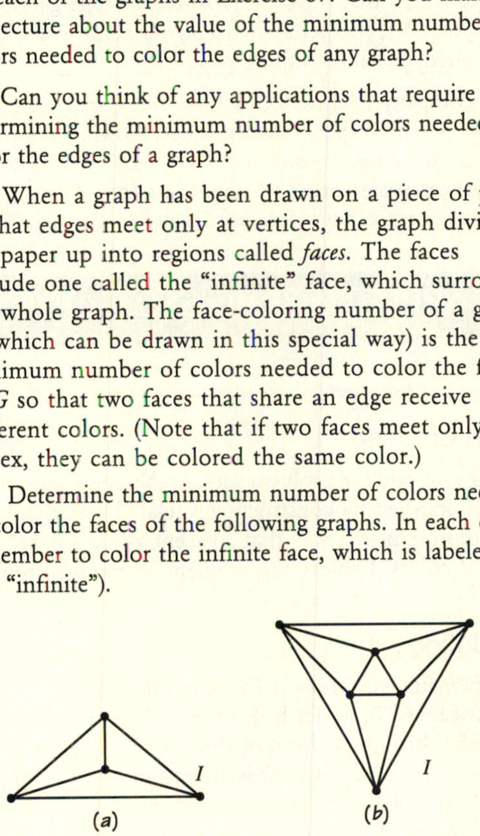

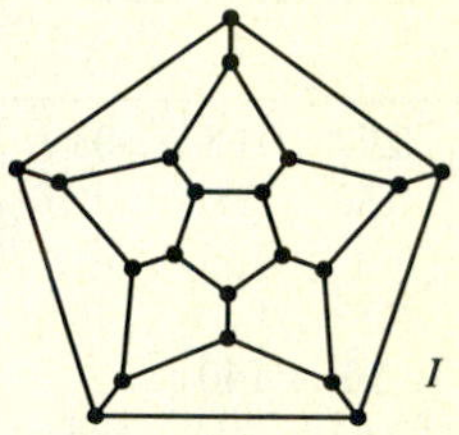


(c)

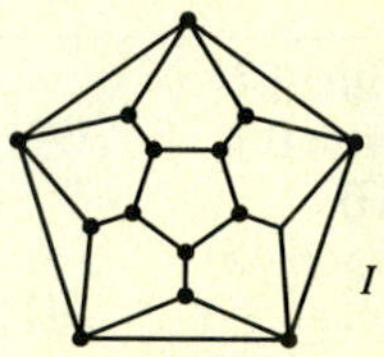


(d)

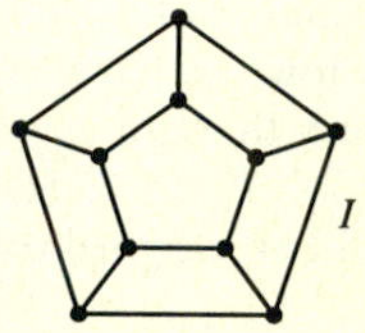


(e)

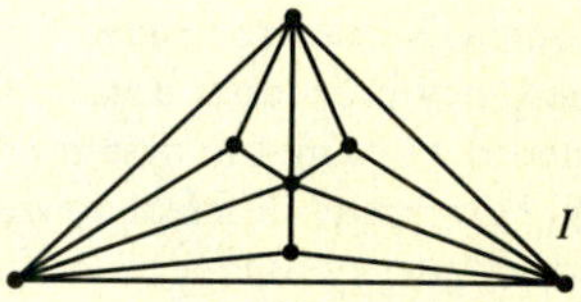


(f)

**(b)** Can you think of an application of the problem of coloring the faces of a graph with a minimum number of colors?

**80.** For each of the graphs in Exercise 68 where the graph shown has edges that meet only at vertices, verify that the Four Color Theorem holds by showing that the regions (faces) of the graph can be colored with four or fewer colors so that regions that share an edge get different colors. (Remember to assign a color to the unbounded, so-called infinite region.)

**81.** A company sells herbs, each of which requires a certain level of proper watering. The following graph is constructed by having one vertex for each type of herb. The vertices representing two herbs are joined by an edge if they must have different levels of watering. What is the minimum number of terrariums that the herbs can be displayed in so that herbs in the same terrarium can be watered at the same level?

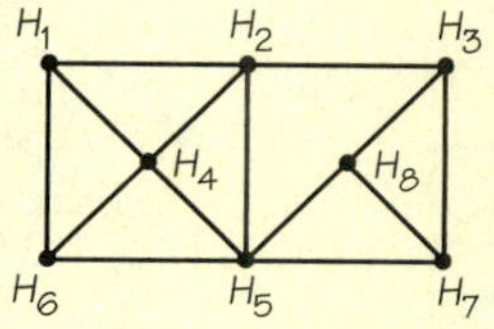


**82.** The company in Exercise 81 is disappointed by the minimum number of terrariums needed to display the herbs with the proper watering requirements. One company employee suggests that if the information about watering requirements is altered for a single pair of herbs (e.g., a single edge is erased from the diagram), then the number of terrariums needed will be reduced by 1. Is this true?

**83.** Each vertex in the graph below represents a child who attends a day care center. An edge between two children indicates these children tend to cause problems when they are in the same play group. What is the minimum number of play groups that will ensure that no conflicts arise? Can conflict-free play groups with the same number of children in each group be formed?

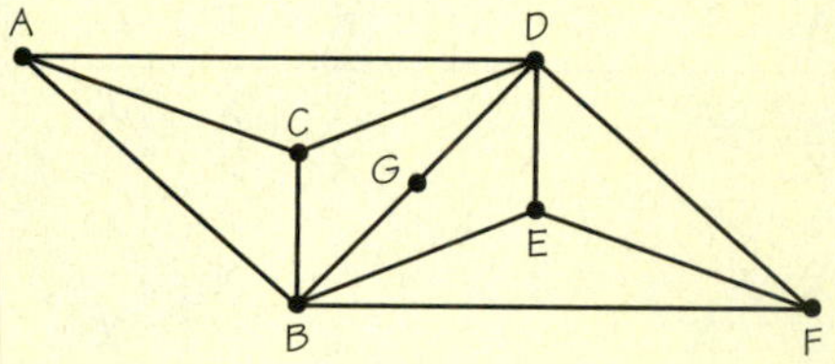


## APPLET EXERCISES

**To do these exercises, go to www.whfreeman.com/fapp8e.**

### Graph Coloring

Solving a scheduling problem such as the one below can be accomplished by constructing a related graph and then coloring it in a way that adjacent vertices have different colors. Explore the problem of graph coloring in the *Graph Coloring* applet.

### Scheduling

A mathematics department has seven faculty committees–A, B, C, D, E, F and G. Because there is overlap in the composition of the committees, the chairman of the department is attempting to work out a schedule that will avoid conflicts among the committees. The following chart indicates the overlapping committee structure:

| | A | B | C | D | E | F | G |
|---|---|---|---|---|---|---|---|
| A | | X | | X | | X | |
| B | X | | X | | | X | |
| C | | X | | | X | | X |
| D | X | | | | | | X |
| E | | | X | | | X | X |
| F | X | X | | | X | | |
| G | | | X | X | X | | |

Help the chairman arrange a schedule without conflicts in the *Scheduling* applet.

## WRITING PROJECTS

**1.** Scheduling is important for hospitals, schools, transportation systems, police services, and fire services. Pick one of these areas and write about the different scheduling situations that come up, types of processors, and extent to which the assumptions of the list-processing model hold for the area you pick.

**2.** Compare and contrast the basic scheduling problem we investigated with the scheduling version of the bin-packing problem.

**3.** One of the oversimplifications made in our discussion of scheduling was that there were no "due dates" involved for the tasks making up a job. Develop an algorithm for solving a scheduling problem under the assumption that each task has a due date as well as a time length. You will probably want to decide on a penalty amount that will occur when a due date is exceeded.

**4.** Consider the problem of scheduling tasks on a single machine. Design different algorithms for achieving different goals. You will probably wish to assume that each task has a due date such that if the task is not finished by this date, some penalty payment must be made.

**5.** Discuss the role of graph colorings for scheduling committee meetings so as to avoid conflicts. Research whether or not these ideas are used in the legislature of your home state.

**6.** In choosing a location (vertex) for trains to turn around in the graph shown in Spotlight 3.1, explain why it seems to be a much better choice to use $V$ as a place to allow the turn arounds, rather than at $M$ or at $R$.

## SUGGESTED READINGS


BRUCKER, P. *Scheduling Algorithms,* 4th ed., Springer-Verlag, Heidelberg, Germany, 2004. A detailed mathematical look at scheduling.

GRAHAM, RONALD. Combinatorial scheduling theory, in Lynn Steen (ed.), *Mathematics Today,* Springer-Verlag, New York, 1978, pp. 183–211. This essay on scheduling is one of many excellent accounts of recent developments in mathematics in this book.

GRAHAM, RONALD. The combinatorial mathematics of scheduling. *Scientific American,* March 1978, pp. 124–132. A very readable introduction to scheduling and bin packing.

JENSEN, T. R., and BJARNE TOFT. *Graph Coloring Problems,* Wiley, New York, 1995. A detailed summary of what is known about coloring problems and many questions that await answering.

LAWLER, E., et al. Sequencing and scheduling algorithms and complexity, in S. C. Graves et al. (eds.), *Handbooks in OR and MS,* vol. 4, Elsevier, New York, 1993, pp. 445–522. A recent survey of results about scheduling.

LEUNG, JOSEPH Y-T., *Handbook of Scheduling,* Chapman & Hall/CRC, Boca Raton, Florida, 2004. This book has an encyclopedic treatment of scheduling algorithms and the great variety of situations where mathematical analysis has assisted schedulers, ranging from sports to hospitals.

PARKER, R. GARY, *Deterministic Scheduling Theory,* Chapman & Hall, London, 1995. A wide-ranging look at scheduling methods and their applications.


## SUGGESTED WEB SITES

**www.ctl.ua.edu/math103/scheduling/schedmnu.htm** This site provides an overview of scheduling as discussed in this chapter.

**www.ams.org/featurecolumn/archive/machines1.html, www.ams.org/featurecolumn/archive/packings1.html, www.ams.org/featurecolumn/archive/bins1.html** These Web pages describe mathematical aspects of machine scheduling and bin packing and give a discussion of the relationship between these two mathematical problems.

**www.ie.bilkent.edu.tr/~ie672/docs/resources.html** This Web page contains links to many aspects of scheduling theory, including research on the frontier.

# CHAPTER 4



# Linear Programming

A manager's job often calls for making very complicated decisions. One set of decisions involves planning what products the business is to make and determining what resources are needed. In the modern business world, diversification of products provides a company with stability in a climate of changing tastes and needs. So it is not surprising that companies would produce many products, some of which share resource needs. For example, any bakery uses many resources—like butter, sugar, eggs, and flour—to make its products such as cookies, cakes, pies, and breads. Similarly, car manufacturers use many kinds of metals in the different models of cars they make, and manufacturers of gasoline use different kinds of crude oils to make their product.

Resources can include more than just raw materials. A labor force with appropriate skills, farmland, time, and machinery are also resources. Typically, resources are limited: A farmer owns only so much land; there are only so many hours in a day; in a year of drought the wheat crop is very small. Resource availability is also limited by location and competition.

Because resources are limited, management faces important questions: How should the available resources be shared among the possible products? One goal of management is to maximize profit. How can that determine how much of each product should be produced? There are usually so many alternative product mixes that it is impossible to evaluate them all individually. Despite this complexity, millions of dollars may ride on management's decision.

Many business and government agencies must deal with supply-and-demand problems. The general idea is that goods or services can be provided by different providers to individuals or businesses who need these goods or services. There are varying costs to the suppliers to provide different recipients with these goods or services. The goal is to find how to meet the demands for the supplies as cheaply as possible. For example, what is the cheapest way for a company with several oil refineries to provide oil distributors, in many different geographical locations, with the oil they need?

In this chapter, we learn about **linear programming**, a management science technique that helps a business allocate the resources it has on hand to make a particular mix of products that will maximize profit.

**Linear Programming** DEFINITION

**Linear programming** is a tool for maximizing or minimizing a quantity, typically a profit or a cost, subject to constraints.

The technique is so powerful that linear programming is said to account for over 50% and perhaps as much as 90% of all computing time used for management decisions in business.

Linear programming is an example of "new" mathematics. It came into being, along with many other management science techniques, during and shortly after World War II, in the 1940s. It is quite young as intellectual ideas go. Yet, during its short history, linear programming has changed the way businesses and governments make decisions, from "seat-of-the-pants" methods based on guesswork and intuition to using an algorithm based on available data and guaranteed to produce an optimal decision.

Linear programming is but one operations research tool belonging to a family of tools known as mathematical programming. Another such tool is integer programming. The difference between linear programming and integer programming is that for linear programming, the quantities being studied can take on values such as $\pi = 3.14159\ldots$ or $7\frac{1}{8}$; in integer programming, the values are confined to whole numbers such as 8, 50, or 1,102,362. Whole numbers are conceptually easier than the broader group consisting of all numbers that can be represented by decimals (1.32, 1.455555 . . . ), yet integer-programming problems have proved much harder to solve.

The assembly of an automobile requires many complicated steps and processes. Using linear-programming techniques enables the robots and humans to carry out their tasks faster and more accurately than would be possible without the use of mathematics. This makes American cars more competitive and of a higher quality than otherwise would be the case. (*Digital Vision/Getty Images.*)

In the discussions that follow, we often describe "relaxed" versions of integer-programming problems as linear-programming problems. For example, it would make no sense to produce 3.24 dolls to sell. So, strictly speaking, we must find an optimum whole number of dolls to produce. If we are "lucky," the linear-programming problem associated with an integer-programming problem has an integer solution. In this case, we have also found the correct answer to the integer-programming problem. Some other examples that fall into this category are discussed below.

Linear programming has saved businesses and governments billions of dollars. Of all the management science techniques presented in this book, linear programming is far and away the most frequently used. It can be applied in a variety of situations, in addition to the one we study in this chapter. Some of the problems studied in Chapters 1, 2, and 3–for example, the TSP and scheduling problems–can be viewed as linear-programming problems. Linear programming is an excellent example of a mathematical technique useful for solving many different kinds of problems that at first do not seem to be similar problems at all. It has been suggested that without linear programming, management science would not exist.

### SPOTLIGHT 4.1 Case Studies in Linear Programming

Linear programming is not limited to mixture problems. Here are two case studies that do not involve mixture problems, yet where applying linear-programming techniques produced impressive savings:

- The Exxon Corporation spends several million dollars per day running refineries in the United States. Because running a refinery takes a lot of energy, energy-saving measures can have a large effect. Managers at Exxon's Baton Rouge plant had over 600 energy-saving projects under consideration. They couldn't implement them all because some conflicted with others, and there were so many ways of making a selection from the 600 that it was impossible to evaluate all selections individually.

  Exxon used linear programming to select an optimal configuration of about 200 projects, resulting in millions of dollars in savings.

- Edwards Lifesciences uses heart valves from pigs to produce artificial heart valves for human beings. Pig heart valves come in different sizes. Shipments of pig heart valves often contain too many of some sizes and too few of others. However, each supplier tends to ship roughly the same imbalance of valve sizes on every order, so the company can expect consistently different imbalances from the different suppliers. Thus, if they order shipments from all the suppliers, the imbalances could cancel each other out in a fairly predictable way. The amount of cancellation will depend on the sizes of the individual shipments. Unfortunately, there are too many combinations of shipment sizes to consider all combinations individually.

  Edwards Lifesciences used linear programming to figure out which combination of shipment sizes would give the best cancellation effect. This reduced the company's annual cost by $1.5 million.

## 4.1 Mixture Problems: Combining Resources to Maximize Profit

In this chapter, we study how to use linear programming to solve a special kind of problem–a **mixture problem**. Realistic versions of such problems would be much more involved. Our discussion is designed to give you the flavor of what is actually done. Realistic examples of what follows are commonly used in the manufacture of different kinds of breads from the grain flours available, and in the making of different kinds of sausages from meats such as beef and pork.

**Mixture Problem** DEFINITION

In a **mixture problem**, limited resources are combined into products so that the profit from selling those products is a maximum.

Mixture problems are widespread because nearly every product in our economy is created by combining resources. A typical example would be how different kinds of aviation fuel are manufactured using different kinds of crude oil.

Let's analyze small versions of the kinds of problems that might confront a toy or a beverage manufacturer. Both manufacturers can sell many different products on which each company can make a profit. There could be dozens of possible products and many resources. A manufacturer must periodically look at the quantities and prices of resources and then determine which products should be produced in which quantities in order to gain the greatest, or optimum, profit. This is an enormous task that usually requires a computer to solve.

What does it mean to find a solution to a linear-programming mixture problem? A solution to a mixture problem is a production policy that tells us how many units of each product to make. An **optimal production policy** has two properties:

1. It is possible; that is, it does not violate any of the limitations under which the manufacturer operates, such as availability of resources.
2. The optimal production policy gives the maximum profit.

## Common Features of Mixture Problems

Although our first mixture problem (Example 1) has only two products and one resource, it does contain the essential features that are common to *all* mixture problems:

- *Resources.* Definite resources are available in limited, known quantities for the time period in question. The resource in Example 1 is containers of plastic.
- *Products.* Definite products can be made by combining, or mixing, the resources. In Example 1, the products are skateboards and dolls.
- *Recipes.* A recipe for each product specifies how many units of each resource are needed to make one unit of that product. Each skateboard in Example 1 uses five units of plastic, and each doll uses two units.
- *Profits.* Each product earns a known profit per unit. (We assume that every unit produced can be sold. More complicated mathematical models, which we will not discuss here, are needed if we want to consider the possibility of items being produced but not sold.)
- *Objective.* The objective in a mixture problem is to find how much of each product to make so as to maximize the profit without exceeding any of the resource limitations.

The examples we show are not designed to be realistic. Rather, our goal is to demonstrate how ideas whose roots are in basic algebra and geometry can solve, when scaled up to realistic versions, problems that save Americans much time and money and make our government and American businesses more efficient.

## EXAMPLE 1 ■ Making Skateboards and Dolls

(*Patrik Giardino/Corbis.*)

A toy manufacturer can manufacture only skateboards, only dolls, or some mixture of skateboards and dolls. Skateboards require five units of plastic and can be sold for a profit of \$1, while dolls require two units of plastic and can be sold for a \$0.55 profit. If 60 units of plastic are available, what numbers of skateboards and/or dolls should be manufactured for the company to maximize its profit?

Attacking this and other mixture problems requires the carrying out of a series of steps that get at the essence of the problem.

As a first step, we need to take the "verbal" information that we have been given and display it in a form that makes it easier to convert into the mathematics necessary to solve the problem. This is done by making a **mixture chart** for the information we are given (see Figure 4.1).

In the rows of this chart we display the products we want to make, and in the column of the chart we display the resources and the profit margin information that is available. In this case we have two products, so we have two rows. We have one resource, which accounts for one column.

| PRODUCTS | RESOURCE(S)<br>Containers of Plastic<br>60 | PROFIT |
|---|---|---|
| **Skateboards** (x units) | 5 | \$1.00 |
| **Dolls** (y units) | 2 | \$0.55 |

**FIGURE 4.1** Mixture chart for Example 1.

The other column is reserved for profit information. Since this information is used in a somewhat different way from the information about the resources, we will separate the resource column(s) from the profit column by a double bar.

Since we do not know the number of skateboards the company should make, we will use a letter $x$ to represent the unknown number of skateboard units that the company might manufacture. Similarly, $y$ will represent the number of dolls that the company might manufacture. We enter these letters as part of the labels of the rows of our table.

We can now enter the numbers about resources in the columns based on the information we have been given. In this case, there is one resource, plastic. Thus, for the five units of plastic needed for a skateboard we record a 5 in the Skateboards row and the Containers of Plastic column. Similarly, we enter a 2 in the second row and first column because dolls need two units of plastic for each doll. Since we have 60 units of plastic available, we display this fact by placing the number 60 at the top of this column. We complete the table with the information about profit. We enter \$1 in the Skateboards row and Profit column and \$0.55 in the Dolls row and Profit column.

## EXAMPLE 2 ■ Making a Mixture Chart

Make a mixture chart to display this situation: A clothing manufacturer has 60 yards of cloth available to make shirts and decorated vests. Each shirt requires 3 yards of cloth and provides a profit of $5. Each vest requires 2 yards of cloth and provides a profit of $3.

**SOLUTION:** See the mixture chart in Figure 4.2.

| PRODUCTS | RESOURCE(S) Yards of Cloth 60 | PROFIT |
|---|---|---|
| Shirts ($x$ units) | 3 | $5 |
| Vests ($y$ units) | 2 | $3 |

**FIGURE 4.2** Mixture chart for the clothing manufacturer.

## Translating Mixture Charts into Mathematical Form

Consider again the mixture chart in Figure 4.1. What can we say about the numbers of skateboards and dolls that might be manufactured? Clearly we cannot make negative numbers of skateboards or dolls. Since we are using the letter $x$ to represent the number of skateboards we plan to make, we can write down the algebraic expression that $x \geq 0$. Here we are using the standard symbol $\geq$ for "greater than or equal to."

Algebraic expressions that involve the symbol $\geq$ or its companion symbols $\leq$ (less than or equal to), $>$ (greater than), and $<$ (less than) are known as *inequalities.* We can also write down an inequality for $y$ dolls we plan to make, based on the fact that we cannot make a negative number of dolls. Thus, we must have $y \geq 0$. We will use the phrase **minimum constraints** for these two inequalities, $x \geq 0$ and $y \geq 0$, which say simply that one cannot manufacture negative numbers of objects.

However, we also have only a limited number of units of plastic available. How can we represent this information? Consulting the mixture table, we see that we need five units of plastic for every skateboard we make. Thus, we will need $5x$ (5 times $x$) units of plastic for the skateboards we make. Similarly, we will need $2y$ (2 times $y$) units of plastic for the dolls we make. Hence we will need $5x + 2y$ units of plastic for the mixture of skateboards and dolls we make. We added the $5x$ and $2y$, because we need to find the total plastic used when we make a mixture of skateboards and dolls.

Reading from the table, we see that we are limited in having only 60 units of plastic. So we can express the **resource constraint** imposed by the limited number of units of plastic by writing that $5x + 2y \leq 60$. Here we use the symbol for less than or equal to, $\leq$, to express the fact that we cannot use more than the amount of plastic we have available.

Notice that all of the numbers in this inequality can be obtained from a column of the mixture chart. One of the reasons we construct a mixture chart is that it helps us speed up the conversion into inequalities of the information about the problem we wish to solve.

In addition to the resource inequalities (of which realistic problems will often have hundreds), there is one additional algebraic expression, this time an equality, that the mixture table allows us to create. Using the mixture table, we can compute the profit that will be produced when we manufacture different mixtures. For each skateboard, we make a profit of $1, so if $x$ skateboards are made, the profit is $1x$ (1 times $x$). For each doll made, the profit is $0.55. So if $y$ dolls are made, the profit is $0.55y$ (0.55 times $y$). Denoting by $P$ the total profit from making $x$ skateboards and $y$ dolls, we get the equation

$$P = 1x + 0.55y$$

Note that unlike the situation for the resources where we got an inequality, here we get an expression for what the profit will be as we vary the numbers of skateboards and dolls manufactured. Our goal is to find which values of $x$ and $y$ (skateboards and dolls) make this profit as large as possible.

## EXAMPLE 3 ■ Revisiting Our Clothing Manufacturer

We can also translate the information in the mixture chart shown in Figure 4.2 into inequalities and an equation for expressing the profit in terms of how many shirts and vests are produced. Using the first column of the mixture chart and the fact that for cloth there are only 60 units available, we can write

$$3x + 2y \leq 60$$

And using the last column, we get the following expression for the profit $P$:

$$P = 5x + 3y$$

Now that we have the information from the original problems represented in mathematical terms, we will return our attention to finding a solution to the problems.

Finding the best (largest profit) mixture of skateboards and/or dolls to make can be carried out in two phases.

1. Determine those mixtures of skateboards and/or dolls that can be manufactured subject to the limited resources that are available. This step involves finding the **feasible set** for the mixture problem.

**Feasible Set or Feasible Region** DEFINITION

The **feasible set**, also called the **feasible region**, for a linear-programming problem is the collection of all physically possible solution choices that can be made.

We can use a geometric diagram such as the one in Figure 4.3 to help us understand the feasible set of options that the manufacturer of skateboards and dolls has available. The geometric diagram we draw will have as many "dimensions" as there are products being manufactured. We have two products represented by the variables $x$ and $y$ so we use a two-dimensional picture. Even diagrams involving three variables are hard to draw and visualize. Though these diagrams helped with developing algorithms for solving mixture problems, they are of little practical use for realistic problems.

2. Determine how to pick out, from the feasible set, the mixture (or mixtures) that gives rise to the largest profit.

### Representing the Feasible Region with a Picture

After we have constructed inequalities using a mixture chart or have the inequalities that must be obeyed for a more general linear-programming problem, we can draw a helpful picture to visualize the choices to be made in solving the problem. This picture will show in a convenient way the different choices that are available in solving the linear-programming problem at hand. To get the picture that will help us, we need to draw graphs of the inequalities associated with the linear-programming problem.

To draw the graph of an inequality, let's first review how to draw the graph of the equation of a straight line. Remember that two points can be used to uniquely determine a straight line. Let's use the equation associated with the less-than-or-equal-to inequality

$$5x + 2y \leq 60$$

namely,

$$5x + 2y = 60$$

There are two points that are easy to find on this line. When $x = 0$, this gives rise to one point on the line, and when $y = 0$, we can find another point. (See Figure 4.3.) When $x = 0$, if we substitute this value in the equation $5x + 2y = 60$, we get $5(0) + 2y = 60$. Solving this equation, we discover that $y = 30$. Similarly, if we substitute $y = 0$ in the equation $5x + 2y = 60$, we get $5x + 2(0) = 60$, from which we conclude that $x = 12$. We now have two points (0, 30) and (12, 0) that lie on the line $5x + 2y = 60$.

We are using the usual convention that when we write a pair such as (3, 10), we are describing a point that has $x = 3$ and $y = 10$. We always list the $x$-value first in such a pair and the $y$-value second. Furthermore, when a point $(x, y)$ is represented in a diagram, larger values of $x$ are shown farther to the right (east) and larger value of $y$ are shown farther up (north).

Using the two points we find on the line $5x + 2y = 60$, we can draw the graph shown in Figure 4.3a, where we have also displayed the point (3, 10). How do we know that (3, 10) is not on the line? We can see this by replacing $x$ by 3 and $y$ by 10 in the equation $5x + 2y = 60$, getting $5(3) + 2(10) = 15 + 20 = 35$. To have been on the line, we would have to have had the value 60. Furthermore, we know that the point (3, 10) is below the line $5x + 2y = 60$ because when $x = 3$, and we replace $x$ by this value in the equation $5x + 2y = 60$, we get $5(3) + 2y = 60$, which means that $y = 45/2$. Since 45/2 is greater than 10, the $y$-value for the point (3, 10), we conclude that (3, 10) is below the line.

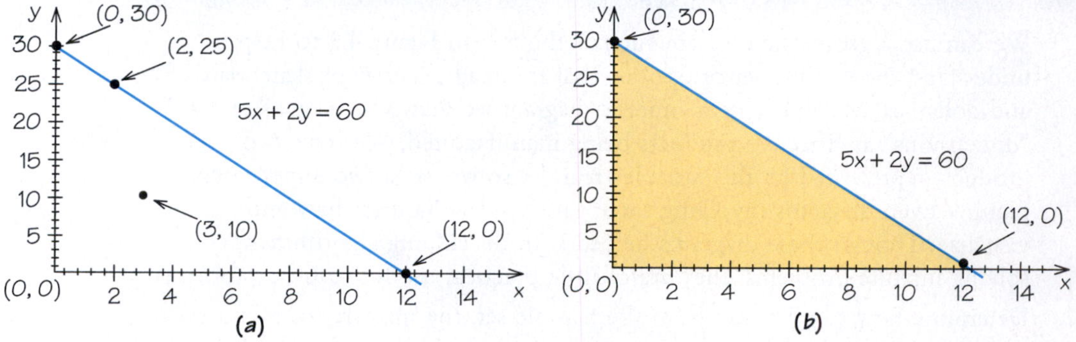


**FIGURE 4.3** The feasible region for Example 3. (a) Graph of $5x + 2y = 60$. (b) Shading of the half-plane $5x + 2y < 60$, and where $x \geq 0$, $y \geq 0$.

Now that we know what the graph of the equation $5x + 2y = 60$ looks like, we can think through where points $(x, y)$ that satisfy $5x + 2y < 60$ are located. The points that are either on the line $5x + 2y = 60$ or satisfy $5x + 2y < 60$ will satisfy $5x + 2y \leq 60$.

Any line, for example, $5x + 2y = 60$, divides the $xy$-plane into three parts: those points on the line, and the points in one of two half-planes. In one of these half-planes we have the points for which $5x + 2y < 60$, and in the other we have the points for which $5x + 2y > 60$. How can we tell which of the two half-planes is above the line $5x + 2y = 60$ and which is below?

The key is the use of a test point $(x, y)$ that is not on the line and whose half-planes we wish to distinguish. We saw above that (3, 10) is not on the line $5x + 2y = 60$ and is below the line. This enables us to see that the half-plane for which $5x + 2y < 60$ consists of the points below the line $5x + 2y = 60$.

To complete the drawing of the points that are feasible for the skateboard and dolls manufacturing problem, we also have to know which points satisfy the constraints that state that the number of skateboards produced $x$ cannot be negative ($x \geq 0$) and the number of dolls produced $y$ cannot be negative ($y \geq 0$). Each of these inequalities corresponds to a half-plane, and we can again test which of the half-planes associated with the line $x = 0$ is determined by $x \geq 0$.

This can be done using the point (3, 10) as a test point again. Since $x = 3$ is greater than 0, $x \geq 0$ determines the half-plane to the right of the line $x = 0$ (the $y$-axis). Similarly, using the point (3, 10) we see that since $y = 10$ is greater than 0, $y \geq 0$ determines the half-plane above the line $y = 0$ (the $x$-axis).

Putting this information together leads us to the conclusion that the collection of points $(x, y)$ that meets the three inequalities involved ($x \geq 0, y \geq 0, 5x + 2y \leq 60$) corresponds to the shaded region in Figure 4.3b.

Note that since the minimality conditions are always present in the kind of linear-programming problems we are dealing with, the points that are feasible for these problems are always in the upper right region (quadrant) that is created by the $x$-axis and $y$-axis. Next, we draw the feasible region for the clothing manufacturing problem.

## EXAMPLE 4 ■ Drawing a Feasible Region

In the earlier clothing manufacturer example, we developed a resource constraint of $3x + 2y \leq 60$. Draw the feasible region corresponding to that resource constraint, using the reality minimums of $x \geq 0$ and $y \geq 0$.

**SOLUTION:** First we find the two points where the line, $3x + 2y = 60$, crosses the axes. When $x = 0$, we get $3(0) + 2y = 60$, giving $y = \frac{60}{2} = 30$, yielding the point (0, 30). For $y = 0$, we get $3x + 2(0) = 60$, or $x = \frac{60}{3} = 20$, so we have the point (20, 0). We draw the line connecting those points. Testing the point (0, 0), we find that the down side of the line we have drawn corresponds to $3x + 2y < 60$. The feasible region is shown in Figure 4.4.

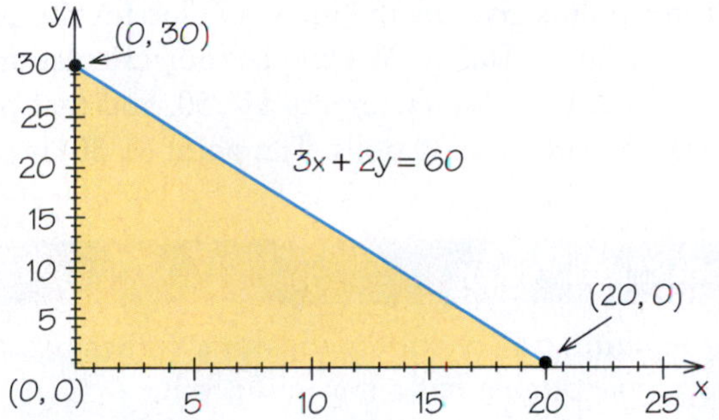


**FIGURE 4.4** Feasible region for the clothing manufacturer.

# 4.2 Finding the Optimal Production Policy

Our next step is that we still must find the *optimal production policy,* a point within the feasible region that gives a maximum profit. There are a lot of points in that region. If you consider points with only whole numbers as values for $x$ or $y$, there are many points, but, in fact, either $x$ or $y$ or both of them could be some fractional number. There are so many points in this feasible region that to consider the profit at each one of them would require us to calculate profits from now until we grow very old, and still the calculations would not be done. Here is where the genius of the linear-programming technique comes in, with the **corner point principle**, which we define in terms of our mixture problems.

### Corner Point Principle

DEFINITION

The **corner point principle** states that in a linear-programming problem, the maximum value for the profit formula always corresponds to a corner point of the feasible region.

The corner point principle is probably the most important insight into the theory of linear programming. Later in this chapter, we will explain why this principle works. The geometric nature of this principle explains the value of creating a geometric model from the data in a mixture chart.

The corner point principle gives us the following method to solve a linear-programming problem:

1. Determine the corner points of the feasible region.
2. Evaluate the profit at each corner point of the feasible region.
3. Choose the corner point with the highest profit as the production policy.

**TABLE 4.1** Calculation of the Profit Formula for Skateboards and Dolls

| Corner Point | Value of the Profit Formula: $\$1.00x + \$0.55y$ |
|---|---|
| (0, 0) | \$1.00(0) + \$0.55(0) = \$0.00 + \$0.00 = \$0.00 |
| (0, 30) | \$1.00(0) + \$0.55(30) = \$0.00 + \$16.50 = \$16.50 |
| (12, 0) | \$1.00(12) + \$0.55(0) = \$12.00 + \$0.00 = \$12.00 |

Let's look at the feasible region we drew in Figure 4.3. It is a triangle having three corners, namely, (0, 0), (0, 30), and (12, 0). Now all we need to do is find out which of these three points gives us the highest value for the profit formula, which in this problem is $\$1.00x + \$0.55y$. We display our calculations in Table 4.1. The maximum profit for the toy manufacturer is \$16.50, and that happens if the manufacturer makes 0 skateboards and 30 dolls. The point (0, 30) is called the *optimal production policy*.

### Optimal Production Policy

DEFINITION

An **optimal production policy** corresponds to a corner point of the feasible region where the profit formula has a maximum value.

## EXAMPLE 5 ▪ Finding the Optimal Production Policy

Our analysis of the clothing manufacturer problem resulted in a feasible region with three corner points (0, 0), (0, 30), and (20, 0). Which of these maximizes the profit formula, \$5*x* + \$3*y*, and what does that corner represent in terms of how many shirts and vests to manufacture?

**SOLUTION:** The evaluation of the profit formula at the corner points is shown in Table 4.2. The maximum profit of \$100 occurs at the corner point (20, 0), which represents making 20 shirts and no vests.

**TABLE 4.2 Evaluating the Profit Formula in the Clothing Example**

| Corner Point | Value of the Profit Formula: $5x + $3y | | | | | | | | |
|---|---|---|---|---|---|---|---|---|---|
| (0, 0) | \$5(0) | + | \$3(0) | = | \$0 | + | \$0 | = | \$0 |
| (0, 30) | \$5(0) | + | \$3(30) | = | \$0 | + | \$90 | = | \$90 |
| (20, 0) | \$5(20) | + | \$3(0) | = | \$100 | + | \$0 | = | \$100 |

## General Shape of Feasible Regions

The shape of a feasible region for a linear-programming mixture problem has some important characteristics, without which the corner point principle would not work:

1. The feasible region is a polygon in the first quadrant, where both $x \geq 0$ and $y \geq 0$. This is because the minimum constraints require that both $x$ and $y$ be nonnegative.
2. The region is a polygon that has neither dents (as in Figure 4.5a) nor holes (as in Figure 4.5b). Figure 4.5c is a typical example. Such polygons are called *convex*.

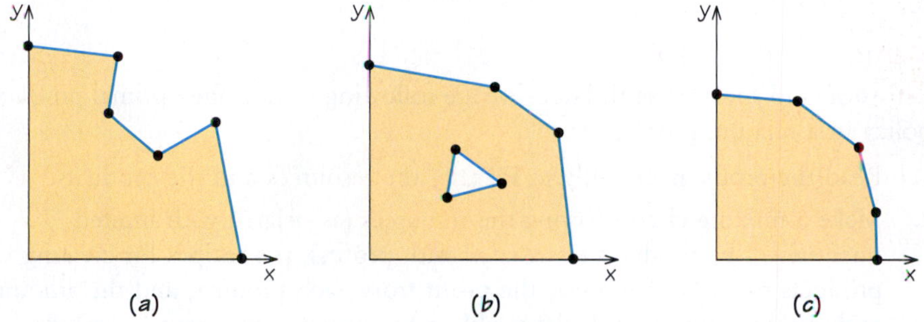


**FIGURE 4.5** A feasible region may not have (a) dents or (b) holes. Graph (c) shows a typical feasible region.

## The Role of the Profit Formula: Skateboards and Dolls

In practice, there are often different amounts of resources available in different time periods. The selling price for the products can also change. For example, if competition forces us to cut our selling price, the profit per unit can decrease. To maximize profit, it is usually necessary for a manufacturer to redo the mixture problem calculations whenever any of the numbers change.

Suppose that business conditions change and now the profits per skateboard and doll are, respectively, \$1.05 and \$0.40. Let us keep everything else about the skateboards and dolls problem the same. The change in profits would give us a new profit formula of \$1.05*x* + \$0.40*y*. When we evaluate the new profit formula at the corner points, we get the results shown in Table 4.3. This time the optimal production policy, the point that gives the maximum value for the profit formula, is the point (12, 0). To get the maximum profit of \$12.60, the toy manufacturer should now make 12 skateboards and 0 dolls.

**TABLE 4.3 A Different Profit Formula: Skateboards and Dolls**

| Corner Point | Value of the Profit Formula: $1.05*x* + $0.40*y* |
|---|---|
| (0, 0) | $1.05(0) + $0.40(0) = $0.00 + $0.00 = $0.00 |
| (0, 30) | $1.05(0) + $0.40(30) = $0.00 + $12.00 = $12.00 |
| (12, 0) | $1.05(12) + $0.40(0) = $12.60 + $0.00 = $12.60 |

We see from this example that the shape of the feasible region, and thus the corner points we test, are determined by the constraint inequalities. The profit formula is used to choose an optimal point from among the corner points, so it is not surprising that different profit formulas might give us different optimal production policies.

We started the exploration of skateboard and doll production with the idea that a toy manufacturer has a product line with either one to two products. But both linear-programming solutions we have found tell the manufacturer that to maximize profit, make just one product. This is probably not an acceptable result for the manufacturer, who might want to produce both products for business reasons other than profit, such as establishing brand loyalty. And it certainly would be very difficult for the manufacturer to be ready to switch back and forth between producing either skateboards or dolls every time the profit formula changed. Linear programming is a flexible enough technique that it can accommodate the desire for there to be both products in the optimal production policy. This is done by specifying that there be nonzero minimum quantities for each period.

## Summary of the Pictorial Method Using a Feasible Region

Let's stop and summarize the steps we are following to find the optimal production policy in a mixture problem:

1. Read the problem carefully to identify the resources and the products.
2. Make a mixture chart showing the resources (associated with limited quantities), the products (associated with profits), the recipes for creating the products from the resources, the profit from each product, and the amount of each resource on hand. If the problem has nonzero minimums, include a column for those as well.
3. Assign an unknown quantity, $x$ or $y$, to each product. Use the mixture chart to write down the resource constraints, the minimum constraints, and the profit formula.
4. Graph the line corresponding to each resource constraint and determine which side of the line is in the feasible region. If there are nonzero minimum constraints, graph lines for them also, and determine which side of each is in the feasible region. Sketch the feasible region by finding the common points in the half-planes from all the resource constraints plus the minimum constraints. (This process is called finding the "intersection" of the half-planes.)
5. Find the coordinates of all the corner points of the feasible region. Some of these may have been calculated so that you can graph the individual lines. Proceed in order around the boundary of the feasible region. Be sure that every point you consider is part of the feasible region.

6. Evaluate the profit formula for each of the corner points. The production policy that maximizes profit is the one that gives the biggest value to the profit formula.

## Two Products and Two Resources: Skateboards and Dolls

We return to the toy manufacturer, now to consider two limited resources instead of one. The second limited resource will be time, the number of person-minutes available to prepare the products. Suppose that there are 360 person-minutes of labor available and that making one skateboard requires 15 person-minutes and making one doll requires 18 person-minutes. We will continue to use the original figures regarding containers of plastic, the first of our two profit formulas, and to keep the problem relatively simple, we use the zero minimum constraints: $x \geq 0$ and $y \geq 0$. We need a new mixture chart. In general, we will include a column for minimums in a mixture chart only if there are any nonzero minimum constraints. In Figure 4.6 we have the mixture chart for this problem. Using the mixture chart, we can write the two resource constraints:

$$5x + 2y \leq 60 \qquad \text{for containers of plastic}$$

and

$$15x + 18y \leq 360 \qquad \text{for person-minutes}$$

We can also write the profit formula: $\$1.00x + \$0.55y$.

| PRODUCTS | RESOURCE(S): Containers of Plastic 60 | RESOURCE(S): Person-minutes 360 | PROFIT |
|---|---|---|---|
| Skateboards (x units) | 5 | 15 | $1.00 |
| Dolls (y units) | 2 | 18 | $0.55 |

**FIGURE 4.6** Mixture chart for Skateboards and Dolls (two resources).

The half-plane corresponding to the plastic resource is shown in Figure 4.7a. We now need to graph the half-plane corresponding to the time constraint. We find where the line $15x + 18y = 360$ intersects the two axes by substituting first $x = 0$ and then $y = 0$ into that equation.

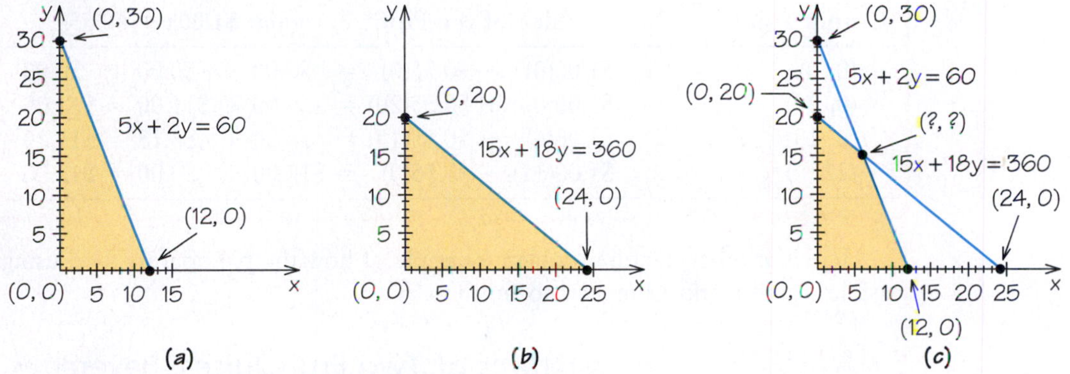


**FIGURE 4.7** Feasible region for Skateboards and Dolls (two resources). (a) Half-plane for the plastic resource constraint. (b) Half-plane for the time resource constraint. (c) Intersection for the two half-planes.

The line corresponding to the time constraint contains the two points (0, 20) and (24, 0). When we substitute the point (0, 0) into the inequality $15x + 18y < 360$, we get $15(0) + 18(0) < 360$, or $0 < 360$, which is true, so (0, 0) is on the side of the line that we shade. Putting all this together, we get the half-plane in Figure 4.7b as the correct half-plane for the time resource constraint.

We are not permitted to exceed the supply of even a single resource. Therefore, the feasible region must be made up of points that are shaded twice–both in the half-plane for the plastic resource constraint, shown in Figure 4.7a, and in the half-plane for the time resource constraint in Figure 4.7b. The procedure with several half-plane constraints is that we build our feasible region by finding the intersection, or overlap, of the individual half-planes in the problem. In Figure 4.7c we show the result of intersecting the half-planes from the two resource constraints. Because this problem has minimums that are zeroes, the shaded region in Figure 4.7c is in fact the feasible region for the problem.

The next step that we need to carry out to use the pictorial method for solving this problem is to find the corner points of the feasible region. This is done by using the algebra necessary to solve two equations in two unknowns. This leads to the points (0, 0), (0, 20), (6, 15), and (12, 0). Three of these points have a zero value for one or more of the unknowns, so the calculations are easy.

To find the coordinates of the point (6, 15), it is necessary to solve for the point that satisfies both of the equations $5x + 2y = 60$ and $15x + 18y = 360$. In order to solve these two equations simultaneously, you must multiply one or both of the equations by a number to create equivalent equations, so that when added together, one of the variables will cancel out. One way to do this is to multiply the first of these equations by $-3$, obtaining the equation $-15x - 6y = -180$. When this is added to the second equation, the $x$ term "drops out" and we can solve $12y = 180$, to get $y = 15$. Now it is an easy matter to substitute this value into either of the original equations to get the $x$ value of 6.

We are ready to finish the problem. In Table 4.4 we have evaluated the profit formula at the four corner points of the feasible regions. The optimal production policy for the toy manufacturer would be to make 6 skateboards and 15 dolls, for a maximum profit of \$14.25.

**TABLE 4.4 The Profit at the Four Corner Points**

| Corner Point | Value of the Profit Formula: $1.00x + $0.55y |
|---|---|
| (0, 0) | \$1.00(0) + \$0.55(0) = \$0.00 + \$0.00 = \$0.00 |
| (0, 20) | \$1.00(0) + \$0.55(20) = \$0.00 + \$11.00 = \$11.00 |
| (6, 15) | \$1.00(6) + \$0.55(15) = \$6.00 + \$8.25 = \$14.25 |
| (12, 0) | \$1.00(12) + \$0.55(0) = \$12.00 + \$0.00 = \$12.00 |

Here is another mixture problem example of how the pictorial method using a feasible region works from start to finish.

## EXAMPLE 6 ■ Mixtures of Two Fruit Juices: Beverages

A juice manufacturer produces and sells two fruit beverages: 1 gallon of cranapple is made from 3 quarts of cranberry juice and 1 quart of apple juice; and 1 gallon of appleberry is made from 2 quarts of apple juice and 2 quarts of cranberry juice. The manufacturer makes a profit of 3 cents on a gallon of cranapple and 4 cents on a gallon of appleberry. Today, there are 200 quarts of cranberry juice and 100 quarts of apple juice available. How many gallons of cranapple and how many gallons of appleberry should be produced to obtain the highest profit without exceeding available supplies? We use zeroes as "reality minimums." The mixture chart for this problem is shown in Figure 4.8.

| PRODUCTS | RESOURCE(S): Cranberry 200 quarts | RESOURCE(S): Apple 100 quarts | PROFIT |
|---|---|---|---|
| Cranapple (x gallons) | 3 quarts | 1 quart | 3 cents/gallon |
| Appleberry (y gallons) | 2 quarts | 2 quarts | 4 cents/gallon |

**FIGURE 4.8** A mixture chart for Example 6.

For each resource, we develop a resource constraint reflecting the fact that the manufacturer cannot use more of that resource than is available. The number of quarts of cranberry juice needed for $x$ gallons of cranapple is $3x$. Similarly, $2y$ quarts of cranberry are needed for making $y$ gallons of appleberry. So if the manufacturer makes $x$ gallons of cranapple and $y$ gallons of appleberry, then $3x + 2y$ quarts of cranberry juice will be used. Because there are only 200 quarts of cranberry available, we get the cranberry resource constraint $3x + 2y \leq 200$. Note that the numbers 3, 2, and 200 are all in the Cranberry column. We get another resource constraint from the column for the apple juice resource: $1x + 2y \leq 100$. We also have these minimum constraints: $x \geq 0$ and $y \geq 0$.

Finally, we have the profit formula. Because $3x$ is the profit from making $x$ units of cranapple and $4y$ is the profit from making $y$ units of appleberry, we get the profit formula $3x + 4y$.

We summarize our analysis of the juice mixture problem. Maximize the profit formula, $3x + 4y$, given these constraints:

| | |
|---|---|
| cranberry: | $3x + 2y \leq 200$ |
| apple: | $1x + 2y \leq 100$ |
| minimums: | $x \geq 0$ and $y \geq 0$ |

Remember, in a mixture problem, our job is to find a production policy $(x, y)$, that makes all the constraints true and maximizes the profit.

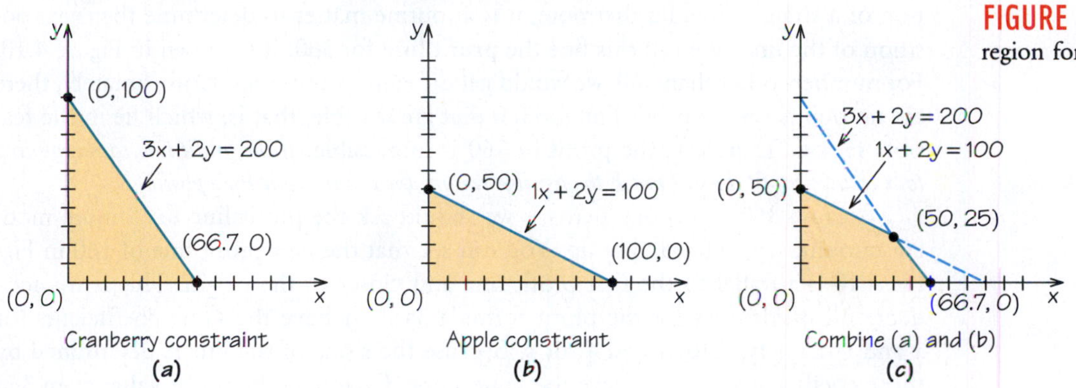


**FIGURE 4.9** Feasible region for Example 6.

Figure 4.9a shows the result of graphing the constraint associated with the cranberry resource, while Figure 4.9b shows the result of graphing the constraint associated with the apple resource, taking into account that the amounts of these resources used cannot be negative. When these two diagrams are superimposed, we get the diagram in Figure 4.9c. Now, to carry out the pictorial method, we need to find the profits associated with the four corner points shown. This is done in Table 4.5.

When we evaluate the profit formula at the four corner points, we see that the optimal production policy is to make 50 gallons of cranapple and 25 gallons of appleberry for a profit of 250 cents. ■

**TABLE 4.5 Finding the Optimal Production Policy for Beverages**

| Corner Point | Value of the Profit Formula: $3x + 4y$ cents |
|---|---|
| (0, 0) | 3(0) + 4(0) = 0 cents |
| (0, 50) | 3(0) + 4(50) = 200 cents |
| (50, 25) | 3(50) + 4(25) = 250 cents |
| (66.7, 0) | 3(66.7) + 4(0) = 200 cents (rounded) |

## 4.3 Why the Corner Point Principle Works

In finding solutions to our mixture problems, we have been using the corner point principle, which says that the highest profit value on a polygonal feasible region is always at a corner point. A feasible region has infinitely many points, making it impossible to compute the profit for each point. The corner point principle gives us a finite set of points, making the calculation possible.

You can visualize a mathematical proof of the corner point principle by imagining that each point of the plane is a tiny light bulb that is capable of lighting up. For the juice mixture example, whose feasible region is shown in Figure 4.9c, imagine what would happen if we ask this question: Will all points with profit = 360 please light up? What geometric figure do these lit-up points form?

In algebraic terms, we can restate the profit question in this way: Will all points $(x, y)$ with $3x + 4y = 360$ please light up? As it happens, this version of the profit question is one mathematicians learned to answer hundreds of years before linear programming was born.

The points that light up make a straight line because $3x + 4y = 360$ is the equation of a straight line. Furthermore, it is a routine matter to determine the exact position of the line. We call this line the **profit line** for 360; it is shown in Figure 4.10. For numbers other than 360, we would get different profit lines. Unfortunately, there are no points on the profit line for 360 that are feasible, that is, which lie in the feasible region. Therefore, the profit of 360 is impossible. *If the profit line corresponding to a certain profit doesn't touch the feasible region, then that profit isn't possible.*

Because 360 is too big, perhaps we should ask the profit line for a more modest amount, say, 160, to light up. You can see that the new profit line of 160 in Figure 4.10 is parallel to the first profit line and closer to the origin. This is no accident: All profit lines for the profit formula $3x + 4y$ have the same coefficients for $x$ and $y$–namely, 3 for $x$ and 4 for $y$. Because the slope of the line is determined by those coefficients, they all have the same slope. Changing the profit value from 360 to 160 has the effect of changing where the line intersects the $y$-axis, but it does not affect the slope. These different profit lines are parallel to each other.

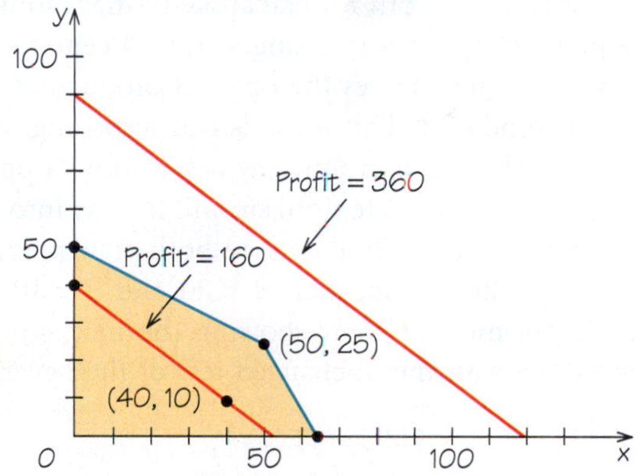


**FIGURE 4.10** The profit line for 360 lies outside the feasible region, whereas the profit line for 160 passes through the region.

The most important feature of the profit line for 160 is that it has points in common with the interior of the feasible region. For example, (40, 10) is on that profit line because 3(40) + 4(10) = 160; in addition, (40, 10) is a feasible point. This means that it is possible to make 40 gallons of cranapple and 10 gallons of appleberry and that if we do so, we will have a profit of 160.

Can we do better than a 160 profit? As we slowly increase our desired profit from 160 toward 360, the location of the profit line that lights up shifts smoothly upward away from the origin. As long as the line continues to cross the feasible region, we are happy to see it move away from the origin, because the more it moves, the higher the profit represented by the line. We would like to stop the movement of the line at the last possible instant, while the line still has one or more points in common with the feasible region. It should be obvious that this will occur when the line is just touching the feasible region either at a corner point (Figure 4.11a) or along a line segment joining two corners (Figure 4.11b). That point or line segment corresponds to the production policy or policies with the maximum achievable profit. This is just what the corner point principle says: The maximum profit always occurs at a corner or along an edge of the feasible region.

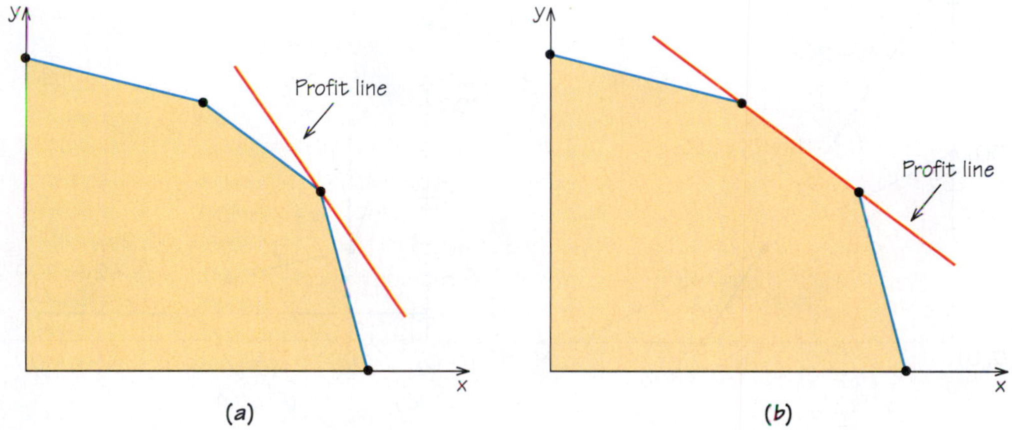


**FIGURE 4.11** The highest profit will occur when the profit is just touching the feasible region, either (a) at the corner point or (b) along a line segment, which will include corner points.

## EXAMPLE 7 ■ Adding Nonzero Minimums: Beverages

Suppose that in Example 6 the profit for cranapple changes from 3 cents per gallon to 2 cents and the profit of appleberry changes from 4 cents per gallon to 5 cents. You can verify that this change moves the optimal production policy to the point (0, 50)–no cranapple is produced. This result is not surprising: Appleberry is giving a higher profit and the policy is to produce as much of it as possible. But suppose the manufacturer wants to incorporate nonzero minimums into the linear-programming specifications so that there will always be both cranapple, $x$, and appleberry, $y$, produced. Specifically, they decide that $x \geq 20$ and $y \geq 10$ are desirable minimums. Figure 4.12 is the mixture chart showing the new profit formula and the nonzero minimums along with the unchanged rest of the beverage problem.

**FIGURE 4.12** Mixture chart for Example 7.

| | | RESOURCE(S) | | | |
|---|---|---|---|---|---|
| | | Cranberry Juice 200 quarts | Apple Juice 100 quarts | MINIMUMS | PROFIT |
| PRODUCTS | **Cranapple** (x gallons) | 3 | 1 | 20 | 2 cents |
| | **Appleberry** (y gallons) | 2 | 2 | 10 | 5 cents |

The feasible region for Beverages, Example 6, is shown in Figure 4.13a. The feasible region for Beverages, Example 7, is shown in Figure 4.13b. You can verify that, starting at the lower left corner of the new feasible region and moving clockwise around its boundary, we have corner points (20, 10), (20, 40), (50, 25), and (60, 10). (One of those points was also a corner point of the old feasible region. Can you explain why?) Table 4.6 shows the evaluation of the profit formula at these corner points. For this modified problem the optimal production policy is to produce 20 gallons of cranapple and 40 of appleberry for a maximum profit of 240 cents.

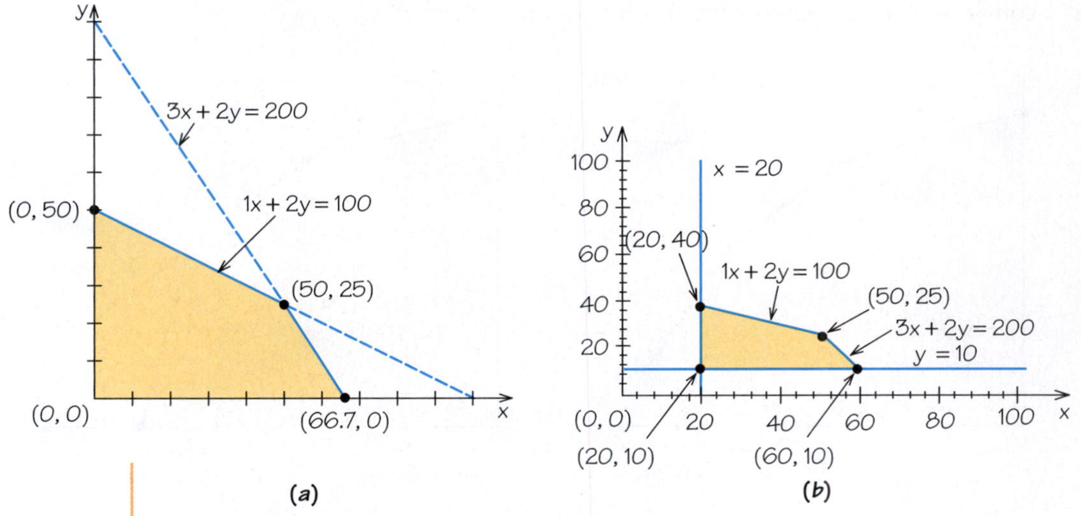


**FIGURE 4.13** Feasible region for Examples 6 and 7. (a) Zero minimums. (b) Nonzero minimums.

**TABLE 4.6 Profit Evaluation for Beverages**

| Corner Point | Value of the Profit Formula: $2x + 5y$ |
|---|---|
| (20, 10) | $2(20) + 5(10) = 40 + 50 = 90$ cents |
| (20, 40) | $2(20) + 5(40) = 40 + 200 = 240$ cents |
| (50, 25) | $2(50) + 5(25) = 100 + 125 = 225$ cents |
| (60, 10) | $2(60) + 5(10) = 120 + 50 = 170$ cents |

One final note about this solution concerns the resources. The point (20, 40) is on the resource constraint line for the apple juice resource, so it represents using up all the available apple juice. We can see this by substituting into the apple juice resource constraint: $1(20) + 2(40) = 100$ is true. However, (20, 40) is *below* the line for the cranberry juice resource, indicating that there will be *slack*, or leftover, amounts of cranberry juice. Specifically, substituting (20, 40) into the cranberry juice constraint gives $3x + 2y = 3(20) + 2(40) = 60 + 80 = 140$, which is 60 quarts less than the 200 quarts available. The slack is 60 quarts of cranberry juice. Dealing with slack can be an important consideration for manufacturers. Can you see why?

## 4.4 Linear Programming: Life Is Complicated

Every algorithm for solving a linear-programming problem has the following three characteristics, which hold true regardless of the number of products or the number of resources in the problem:

1. The algorithm can distinguish between "good" production policies–those in the feasible set that satisfy all the constraints–and those that violate some constraint(s) and are thus not feasible. There are usually many good points, each of which corresponds to some production policy; for example, "Make $x$ units of product 1 and $y$ units of product 2."
2. The algorithm makes use of some geometric principles–one such principle is the corner point principle–to select a special subset of the feasible set.
3. The algorithm evaluates the profit formula at points in the special subset to find which corner point actually gives the maximum profit.

The various algorithms for linear programming differ in how they process the feasible set and in how quickly the algorithm finds the production policy–corner point–that gives the optimal profit.

In practical linear-programming problems, the feasible region will not be as simple as the ones we have examined here. There are two ways the feasible region can be more complex:

1. Sometimes, as in Figure 4.14, we have a great many corners. The more corners there are, the more calculations we need to determine the coordinates of all of them and the profit at each one. The number of corners literally can exceed the number of grains of sand on the earth. Even with the fastest computer, computing the profit of every corner is impossible.

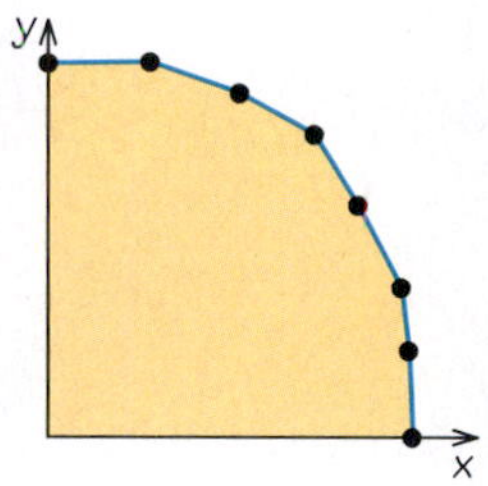


**FIGURE 4.14** A feasible region with many corners.

2. It is not possible to visualize the feasible region as a part of two-dimensional space when there are more than two products. Each product is represented by an unknown, and each unknown is represented by a dimension of space. If we have 50 products, we would need 50 dimensions and couldn't visualize the feasible region.

Another type of complication can occur even in simple two-dimensional regions: Corner points can have fractional coordinates, not the integer ones we see in the specially constructed problems in this text. Making 3.75 skateboards and 5.45 dolls is not possible. Integer programming, a special type of linear programming, is used when it is not possible to use fractional answers.

## The Simplex Method

Several methods are used for the typically large linear-programming problems solved in practice. The oldest method is the **simplex method**, which is still the most commonly used. Devised by the American mathematician George Dantzig (see Spotlight 4.2), this ingenious mathematical invention makes it possible to find the best corner point by evaluating only a tiny fraction of all the corners. With the use of the simplex method, a problem that might be impossible to solve if each corner point had to be checked can be solved in a few minutes or even a few seconds on a typical business computer.

FIGURE 4.15 The simplex method can be compared to an ant crawling along the edges of a polyhedron, looking for the "target"–the optimal corner point.

The operation of the simplex method may be likened to the behavior of an ant crawling on the edges of a polyhedron (a solid with flat sides) looking for an optimal corner point–one that gives the highest profit (Figure 4.15). The ant cannot see where the optimal corner is. As a result, if it were to wander along the edges randomly, it might take a long time to reach that corner. The ant will do much better if it has a temperature clue to let it know it is getting warmer (closer to the optimal corner) or colder (farther from the optimal corner).

Think of the simplex method as a way of calculating these temperature hints. We begin at any corner. All neighboring corners are evaluated to see which ones are warmer and which are colder. A new corner is chosen from among the warmer ones, and the evaluation of neighbors is repeated–this time checking neighbors of the new corner. The process ends when we arrive at a corner all of whose neighbors are colder than it is.

Part of what the simplex method has going for it is that it works faster in practice than its worst-case behavior would lead us to believe. Although mathematicians have devised artificial cases for which the simplex method bogs down in unacceptable amounts of arithmetic, the examples arising from real applications are never like that. This may be the world's most impressive counterexample to Murphy's law, which says that if something can go wrong, it will.

Although the simplex method usually avoids visiting every corner, it may require visiting many intermediate ones as it moves from the starting corner to the optimal one. The simplex method has to search along edges on the boundary of the polyhedron. If it happens that there are a great many small edges lying between the starting corner and the optimal one, the simplex method must operate like a slow-moving bus that stops on every block.

Many computer programs are available that will use the simplex method to produce an optimal production policy if we just supply the computer with the constraint inequalities and profit formula. Simplex method programs can be found in a variety of places, among which are spreadsheets, packages of mathematics programs designed for business applications or finite mathematics courses, and large "all-purpose" mathematics packages. A graphical solution is possible only for problems limited to two products; these special exercises involve more than two products.

## SPOTLIGHT 4.2 Father of Linear Programming Recalls Its Origins

George Dantzig, who died in 2005, spent most of his career as a professor of operations research and computer science at Stanford University. He is credited with inventing the linear-programming technique called the simplex method. Since its invention in the 1940s, the simplex method has provided solutions to linear-programming problems that have saved both industry and the military time and money. Here Dantzig talks about the background of his famous technique:

**George Dantzig**

**George Dantzig (left), sometimes referred to as the "father" of linear programming, shown with Leonid Khaciyan (right) who developed an important new approach to solving linear programming problems.** (*Kees Roos.*)

> Initially, all the work we did had to do with military planning. During World War II, we were planning on a very extensive scale. The civilian population and the military were all performing scheduling and planning tasks, perhaps on a larger scale than at any time in history. And this was the case up until about 1950. From 1950 on, the whole emphasis shifted from military planning to practical planning for the civilian population, and industry picked it up.
>
> The first areas of industry to use linear programming were the petroleum refineries. They used it for blending gasoline. Nowadays, all of the refineries in the world (except for one) use linear programming methods. They are one of the biggest users of it, and it's been picked up by every other industry you can think of—the forestry industry, the steel industry—you could fill up a book with all the different places it's used.
>
> The question of why linear programming wasn't invented before World War II is an interesting one. In the postwar period, various technologies just evolved that had never been there before. Computers were one example. These technologies were talked about before. You can go back in history and you'll find papers on them, but these were isolated cases that never went anywhere. . . .
>
> The problems we solve nowadays have thousands of equations, sometimes a million variables. One of the things that still amazes me is to see a program run on the computer—and to see the answer come out. If we think of the number of combinations of different solutions that we're trying to choose the best of, it's akin to the stars in the heavens. Yet we solve them in a matter of moments. This, to me, is staggering. Not that we can solve them—but that we can solve them so rapidly and efficiently.
>
> The simplex method has been used now for roughly 70 years. There has been steady work going on trying to use different versions of the simplex method, nonlinear methods, and interior methods. It has been recognized that certain classes of problems can be solved much more rapidly by special algorithms than by using the simplex method. If I were to say what my field of specialty is, it is in looking at these different methods and seeing which are more promising than others. There's a lot of promise in this—there's always something new to be looked at.

## An Alternative to the Simplex Method

In 1984, Narendra Karmarkar (see Figure 4.16), a mathematician working at Bell Laboratories, devised an alternative method for linear programming that finds the optimal corner point in fewer steps than the simplex algorithm by making use of search routes through the interior of the feasible region. The applications of Karmarkar's algorithm are important to a lot of industries, including telephone communications and the airlines (see Spotlight 4.3). Routing millions of long-distance calls, for example, means deciding how to use the resources of long-distance landlines, repeater amplifiers, and satellite terminals to best advantage. The problem is similar to the juice company's need to find the best use of its stocks of juice to create the most profitable mix of products.

**FIGURE 4.16** Narendra Karmarkar, a researcher at AT&T Bell Laboratories, invented a powerful new linear-programming algorithm that solves many complex linear-programming problems faster and more efficiently than any previous method. (*Courtesy of AT&T Labs.*)

Many airlines use software based on Karmarkar's algorithm to reduce fuel costs and deal with delays caused by storms.

In the 1980s, scientists at Bell Labs applied Karmarkar's algorithm to a problem of unprecedented complexity: deciding how to economically build telephone links between cities so that calls can get from any city to any other, possibly being relayed through intermediate cities. Figure 4.17 shows one such linking. The number of possible linkings is unimaginably large, so picking the most economical one is difficult. For any given linking, there is also the problem of deciding how to economically route calls through the network to reach their destinations.

Now, similar approaches are being used to route email packets and phone calls over the Internet.

**FIGURE 4.17** A map of the United States showing one conceivable network of major communication lines connecting major cities. Routing millions of calls over this immense network requires sophisticated linear-programming techniques and high-speed computers. (*Courtesy of AT&T Labs.*)

## SPOTLIGHT 4.3 Finding Fast Algorithms Means Better Airline Service

Linear-programming techniques have a direct impact on the efficiency and profitability of major airlines. Thomas Cook, once director of operations research at American Airlines, made these comments concerning why optimal solutions are essential to the airline business:

> Finding an optimal solution means finding the best solution. Let's say you are trying to minimize a cost function of some kind. For example, we may want to minimize the excess costs related to scheduling crews, hotels, and other costs that are not associated with flight time. So we try to minimize that excess cost, subject to a lot of constraints, such as the amount of time a pilot can fly, how much rest time is needed, and so forth.
>
> An optimal solution, then, is either a minimum-cost solution or a maximizing solution. For example, we might want to maximize the profit associated with assigning aircrafts to the schedule; so we assign large aircraft to high-need segments and small aircraft to low-load segments.
>
> The simplex method, which was developed some 50 years ago by George Dantzig, has been very useful at American Airlines and, indeed, at a lot of large businesses. The difference between his method and Narendra Karmarkar's is speed. Finding fast solutions to linear-programming problems is also essential. With an algorithm like Karmarkar's, which is 50 to 100 times faster than the simplex method, we could do a lot of things that we couldn't do otherwise. For example, some applications could be real-time applications, as opposed to batch applications. So instead of running a job overnight and getting an answer the next morning, we could actually key in the data or access the database, generate the matrix, and come up with a solution that could be implemented a few minutes after keying in the data.
>
> A good example of this kind of application is what we call a major weather disruption. If we get a major weather disruption at one of the hubs, such as Dallas or Chicago, then a lot of flights may get canceled, which means we have a lot of crews and airplanes in the wrong places. What we need is a way to put that whole operation back together again so that the crews and airplanes are in the right places. That way, we minimize the cost of the disruption as well as passenger inconvenience.

## 4.5 A Transportation Problem: Delivering Perishables

A supermarket chain gets bread deliveries from a bakery chain that does its baking in different places. Each supermarket store needs a certain number of loaves each day, and the supplier bakes in total enough breads to exactly meet the demands. Figure 4.18 shows the cost to ship a loaf from a particular baking location to the store involved. How many breads should be shipped from each locale to each of the stores to stay within the demands and to minimize the cost?

Similarly, after a long holiday weekend, a car rental company will have extra cars in some cities and too few cars in other cities. It is faced with the problem of reshuffling the cars at minimal cost so that each city has the right number of cars. Problems such as these go under the general name of **transportation problems** and they form a special class of linear-programming problems that can be solved by a specialized method.

**Transportation Problem** — DEFINITION

A group of suppliers must meet the needs of users of these supplies. There is a cost for shipping from a particular supplier to a particular user (demander). The **transportation problem** involves minimizing the total shipping cost of meeting the required demands from the supplies available.

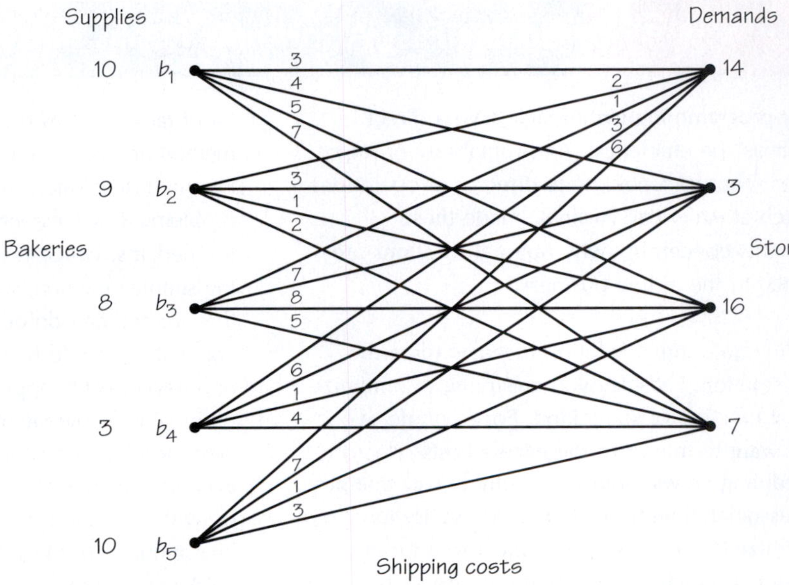


**FIGURE 4.18** A graph theory representation of a supply-and-demand transportation problem that involves shipping breads from bakeries to stores.

## EXAMPLE 8 ■ Delivering Bread

Imagine we have three bakeries and three stores, though the ideas we develop will also solve problems where the number of stores and bakeries need not be the same. The three stores require 3 dozen, 7 dozen, and 1 dozen loaves of bread, respectively, while the three bakeries can supply 8 dozen, 1 dozen, and 2 dozen loaves, respectively. The information given so far can be displayed in Figure 4.19, where the "suppliers" are represented by the rows of the table (labeled with Roman numerals) and the "demanders" are represented by the columns (labeled with Hindu-Arabic numerals).

The numbers of breads available and the numbers being required are shown on the right side and bottom of the table and will be referred to as **rim conditions**. Each entry of the table shown in Figure 4.19 is known as a cell. It is convenient to have a name for each of these cells. For example, the cell in the third row and second column will be denoted (III, 2). The first number always corresponds to a row, the second to a column. Thus cell (I, 2) refers to bakery I and store 2.

Stores

| Bakeries | 1 | 2 | 3 | Breads baked (supply available) |
|---|---|---|---|---|
| I | 8 | 9 | 3 | 8 |
| II | 15 | 1 | 12 | 1 |
| III | 1 | 3 | 5 | 2 |
| Breads needed (demands) | 3 | 7 | 1 | 11 |

Equal totals

**FIGURE 4.19** A representation of a specific problem involving meeting the demands of three stores for breads from the supplies available at three bakeries. Shipping costs between bakeries and stores are also shown.

In deciding which bakeries should ship to which stores, it seems natural to take into account the costs of shipping a dozen breads from a particular bakery to a particular store. If bakery I is farther from store 2 than is bakery II, it seems reasonable that the shipping cost for I will be higher than for II when shipping to that particular store.

However, the costs of shipping may also involve time considerations. (The distance to a store may be shorter, but it may be that this route is a very slow one.) Also, it may take extra time for a truck coming from I to park when making the next delivery.

The numbers we use in our diagrams are "aggregate" costs. The nice thing about what we are doing is that the solution method works independently of the way the costs are computed or arrived at. These costs (see Figure 4.19) are shown in the upper-right-hand corner of a cell. Thus the 9 shown in the cell (I, 2) means that it costs nine units to ship a bread from bakery I to store 2. Our goal will be to supply the stores with the breads they require from the supplies available at the bakeries so that the total cost of providing the breads to the stores is as small as possible (a minimum).

The tools for solving transportation problems like these were developed during World War II in conjunction with getting supplies from different ports in the United States to different ports in Europe (mostly the United Kingdom) in as efficient a manner as possible. (The U.S. ports were like the bakeries, and the British ports were like the stores that needed the breads.)

We can think of finding a solution to a problem like this as a special kind of linear-programming problem, because we can express the objective of minimizing the cost using a linear relationship. The constraints that express that the rim conditions are met can also be expressed using linear equations. However, it turns out there are algorithms that make it possible to solve problems of this kind that are rather larger than general linear-programming problems that can be solved by hand. These algorithms are intuitively appealing.

We can divide the problem of finding a solution to a transportation problem into two phases, as we did for general linear-programming problems. First, find a solution that is feasible (that is, a solution that does not violate any of the constraints of the problem). Second, if the current solution is not optimal, we move to a better one. Thus, we will first find a solution that meets the constraints and then try to find an improved solution. If there is no better solution than the one we have, under suitable circumstances we show that there is never a better one. Thus, the solution that we have found is an optimal solution. We will work our way through a simple example that is typical of what is required in general transportation problems.

Let's turn to the table shown in Figure 4.20, where certain numbers have been inserted with circles around them.

**Tableau** DEFINITION

A table showing costs and rim conditions for a transportation problem is known as a **tableau**.

When we see a circled number such as the 6 in row I, column 2, this means that we plan to ship six breads to store 2 from bakery I. Similarly, the circled number 1 in row III and column 3 means we plan to ship one bread from bakery III to store 3. The cells that have no circled numbers are thought of as having zero entries; no breads are being shipped between these stores and these bakeries. Note, for example, that the row sum of the circled numbers in the first row is 8. This means that all the breads available at bakery I are being shipped to some store.

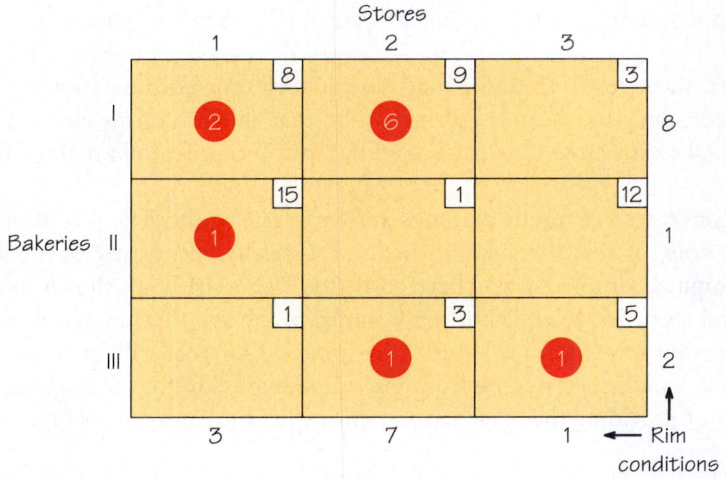


**FIGURE 4.20** A possible solution to meeting the needs of three stores for bread from supplies available at three bakeries. The circled numbers show the amounts shipped from the bakeries to the stores.

Similarly, the fact that circled entries in column 2 add up to 7 means that all the breads needed by store 2 are being supplied to it. You can verify that all the row sums and column sums add to exactly the numbers that we want to ship from each bakery to each store. Note that 11 breads have been shipped by the bakeries and received by the stores. When this happens, the circled numbers are said to be a *feasible* solution to the problem.

How much will it cost to ship these amounts of breads (see Figure 4.20) to the stores? The number can be computed by multiplying the circled numbers by the cost shown in the associated cell. For example, to ship two breads from bakery I to store 1 costs $2(8) = 16$ because the cost associated with the cell in which the 2 appears is 8. The cost of shipping six breads from bakery I to store 2 is $6(9) = 54$. To get the total cost of this "shipment plan," we sum all the shipped amounts by the associated costs to get

$$2(8) + 6(9) + 1(15) + 1(3) + 1(5) = 16 + 54 + 15 + 3 + 5 = 93$$

However, at this point we do not know if there is a cheaper way to ship the breads to the stores. Notice that the number of cells with circled numbers is exactly equal to the number of rows $m$ plus the number of columns $n$ minus 1. This is the general pattern with transportation problems. Cells that are used for shipping are circled. On occasion, we ship a zero amount because the procedure works only when $m + n - 1$ cells are circled.

If we look at the pattern of circled numbers in the tableau in Figure 4.21, we see that there is a difficulty even though 11 breads are involved (the sum of all the circled numbers).

The numbers in the first row add to 6, which means that there will be breads left over at bakery I that have not been shipped. In row 2 the sum of the circled numbers is 2, but this means that something is wrong. How can bakery II, which has a supply of only one bread, ship two breads? Furthermore, column 3 sums to 3, which means that three breads have been shipped to store 1 despite the fact that it only requested one bread! These facts add up to the realization that this assignment of numbers to the cells violates the rules we are requiring. This proposed shipment plan also violates our rule that we are not allowed to circle more than five cells.

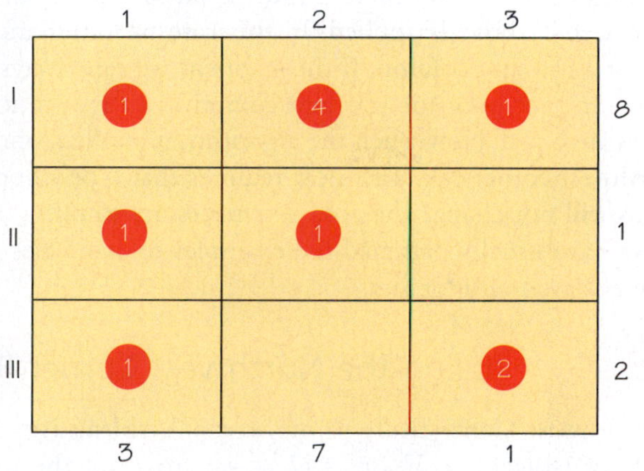


**FIGURE 4.21** The circled numbers are not a possible solution to this transportation problem because the rim conditions are not satisfied.

How can we find a solution that meets the constraints of the problem (the rim conditions)? We will show two ways to do this. The first is "fast and dirty" but typically does not find a very good solution with which to start. The second (developed in the exercises) usually gives a better "initial" solution but is a little harder to carry out.

This pair of approaches displays a common tension in problem solving: the ease of getting started but requiring more work later on, or more work at the start, which often proves to be a good investment of extra effort because less work is needed to find an optimal solution. If we know in advance the method being chosen to solve a problem, we can often find an example where this particular method does poorly. Mathematicians work hard to find methods that work well on the kinds of problems that come up in genuine applications.

## Northwest Corner Rule

The easier approach involves what is called the **Northwest Corner Rule (NCR)**. This rule is simple because it is based on the geometry of the table that is involved and does not even look at the costs associated with the cells in the table, which in the long run cannot be a good idea, because these costs come into play when trying to get an optimal solution.

How does the Northwest Corner Rule work? The algorithm carries out the following procedure until exactly one cell remains in the "altered tableau."

### Northwest Corner Rule — PROCEDURE

1. Locate that cell of the current tableau that is as far to the top and to the left as possible (that is, in the northwest corner). Ship via this cell the smaller of the two rim values (call the value $s$) associated with the row and column of this cell. (Indicate that this cell is being used by putting a circle around the entry in the tableau.)
2. Cross out the row or column that had rim value $s$ and reduce the other rim value for this cell by $s$.
3. When a single cell remains, there will be a tie for the rim conditions of both the row and column involved, and this amount is entered into the cell and circled.

Note that it is possible (when there is more than one cell at the start) for there to be a tie when step 1 above is applied. In this case we simultaneously fulfill the rim conditions for a row and column. If this happens we can always choose to cross out, say, the column (not both the row and column) and reduce to 0 the rim condition for the row involved. Now when the algorithm is applied, one has a rim value of 0 for the northwest corner cell. This now requires that 0 be shipped via that cell. Even though this will not change the cost, it is necessary to put a 0 in this cell and circle it. Here we have usually designed the examples to avoid ties so as to make it easier to get the essential ideas across.

## EXAMPLE 9 ■ Using the Northwest Corner Rule

Applying the Northwest Corner Rule to our original tableau (see Figure 4.19), we get the sequence of tableaux in Figure 4.22 as we cross out the rows or columns, where for clarity the costs associated with the cells are suppressed. The last diagram in the sequence shows the results on the original tableau, with the cost restored. Note that, at the steps in between, the costs played no role. It is a good idea to check that the circled numbers in each row and column really add up to the rim value for that row and column and that exactly $m + n - 1$ cells are filled.

We can now compute the cost of the associated solution that we have found (feasible solution), which obeys the rim conditions. As we did previously, we add up the cost multiplied by the amount shipped for each cell with a circled entry. We get the following calculation:

$$3(8) + 5(9) + 1(1) + 1(3) + 1(5) = 78$$

This shows a cost that is smaller than the solution we found earlier. That solution involved a cost of 93. But is this solution the cheapest one? Since finding this feasible solution did not make use of the costs on the cells, it suggests that it is not very likely.

## Improving the Feasible Solution

The next phase of the transportation problem algorithm attempts to answer the question of how to tell if the feasible solution found by using the Northwest Corner Rule is the best. If this solution is not the best, we should be able to find a way to improve it.

Suppose we decided to ship an additional bread from bakery II to store 3. Now, this would violate the fact that we had shipped exactly the right numbers of breads before this new additional shipment, so we have shipped one bread too many from bakery II. We can compensate for this by reducing from 1 to 0 the bread shipped from bakery II to store 2. But this now means that store 2 has not gotten all the breads it needs. We can take care of this by shipping one more bread from bakery III to store 2, but again we now have one extra bread shipped from bakery III. We can compensate for this by reducing the number of breads shipped via cell (III, 3)–that is, from bakery III to store 3. This step will ensure that the rim conditions will hold for the circled numbers. This is because we have located a circuit–(II, 3), (II, 2), (III, 2), (III, 3), (II, 3)–where, if we increase and decrease the breads alternately going around that circuit, we maintain the rim conditions (see Figure 4.23).

Check for yourself that the tableau on the right in Figure 4.23 with the circled entries meets the rim conditions. To the left we show the circuit of cells with plus and minus signs (+ and −) where we have increased the amounts in the cells with +

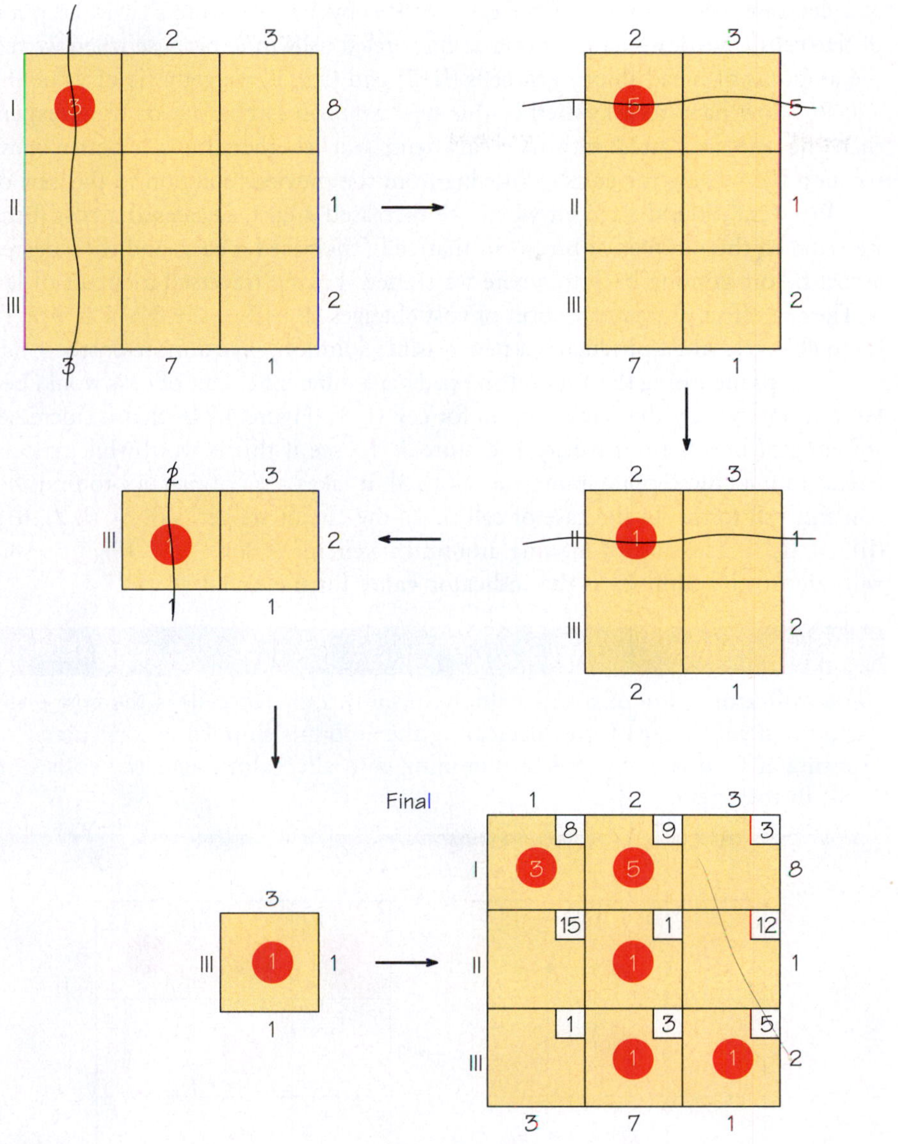


**FIGURE 4.22** The construction of an initial solution to a transportation problem using the Northwest Corner Rule.

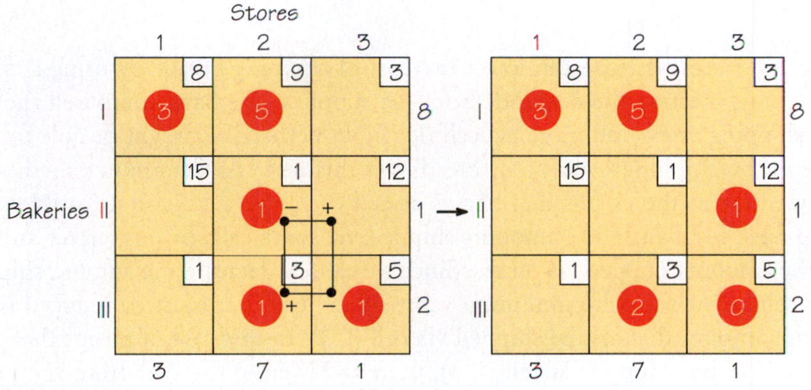


**FIGURE 4.23** An illustration of how to take a current solution to a transportation problem and try to get an improved cheaper solution that still meets the rim conditions.

and decreased the amounts in the cells with − by 1 unit. (Note that to keep a total of five cells circled, we have set one of the circled cells to 0, because when we reduce the amounts of bread shipped in cells (II, 2) and (III, 3), we get a tie of value 0.)

We now have to ask whether this new solution is cheaper or more expensive than the one we started with. We can figure out whether this is a better or worse solution by tracking the costs of moving from the previous solution to the new one.

We went around a circuit where we increased a cost, decreased a cost (because we reduced the number of breads in that cell), increased a cost, and then decreased a cost before coming back to where we started, having traversed a circuit of length 4. The net effect of this collection of cost changes is $+12 - 1 + 3 - 5 = +9$. Thus, these changes, while producing a new feasible solution, give a more costly solution!

Perhaps increasing the amount shipped via a different circuit of cells would be better. Suppose we try the same process for cell (I, 3) (Figure 4.24)–that is, increase the shipping of breads from bakery I to store 3. To see if this is worthwhile, check the circuit formed by shipping more via cell (I, 3). It takes a bit of practice to find the circuit this cell forms. In the case of cell (I, 3), the circuit we get is (I, 3), (I, 2), (III, 2), (III, 3), (I, 3). The cost of moving around this circuit is $+3 - 9 + 3 - 5 = -8$. We will refer to this number as the **indicator value** for this cell.

## Indicator Value

DEFINITION

The **indicator value of a cell** *C* (not currently a circled cell) is the cost change associated with increasing or decreasing the amounts shipped in a circuit of cells starting at *C*. It is computed by summing with alternating signs the costs of the cells in the circuit.

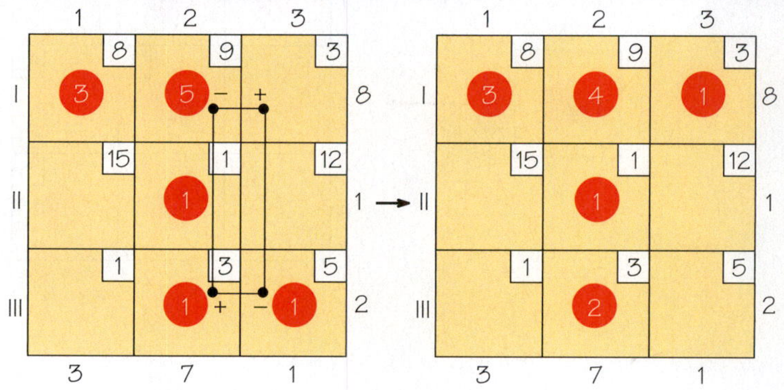


**FIGURE 4.24** Using a cell with a negative indicator value, we can find a cheaper way of meeting the demands from the supplies.

The −8 means that we can lower the cost of shipping breads by using a different pattern of meeting the demands from the supplies. We have computed the saving for shipping one bread more via cell (I, 3), but perhaps we might be able to save even more by shipping even more breads via this cell. To determine whether we could, we look at the circuit that begins at cell (I, 3). To maintain a feasible solution, we have to increase the amounts shipped via some cells of this circuit and decrease the amounts shipped via others. Since we cannot decrease the amount shipped via any cell below zero, the minimum value of any cell that must be reduced is the maximum amount that can be shipped via cell (I, 3). In this case, it means that only one bread can be shipped via cell (I, 3), thereby lowering the cost from the previous solution by 8.

When we looked to improve the solution shown in Figure 4.22, we have now seen that by shipping via cell (I, 3) we can get a better solution. However, there might be several cells in the solution shown in Figure 4.22 that would lead to improvement. Which one should we choose? The answer is that we should adopt a greedy point of view. If there are several cells with a negative indicator value, pick the one that is "most negative" to improve the solution.

Given a current feasible solution (one that satisfies the rim condition), we check each cell that does not have a circled number for improvement if we ship via that cell. If a cell leads to a positive indicator value with the circuit associated with it, no improvement is possible. If a cell has a negative indicator value associated with the circuit for that cell, we can get an improvement. We select as the cell to increase that cell with the largest negative indicator value. We now have a new feasible solution that is cheaper than the one we started with and can repeat our procedure just described starting from this new feasible solution.

It turns out that there was no better cell than (I, 3) (using this greedy approach) to get an improved solution. We will take the current best solution and see if we can improve it further. It turns out that for the current tableau (Figure 4.24), all the cells have a positive indicator except for cell (III, 1):

$$\text{Indicator for cell (III, 1):} \quad +1 - 3 + 9 - 8 = -1$$

Since the minimum of the circled numbers in the cell with a negative label is 2 in cell (III, 2), we can increase by 2 the amount shipped in cell (III, 1) and get a new solution as shown in Figure 4.25.

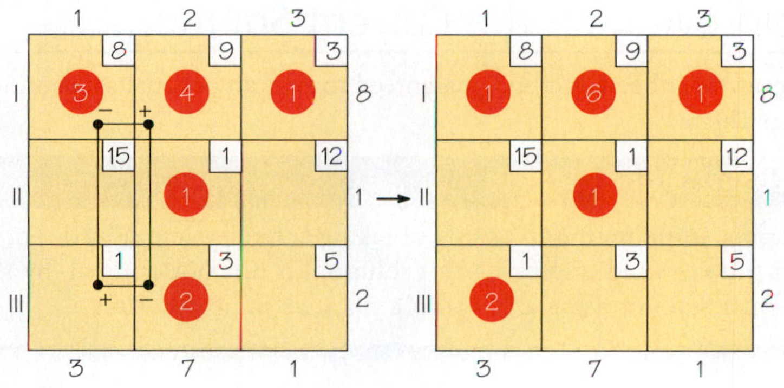

**FIGURE 4.25** We can find an even cheaper way of meeting the demands from the supplies available using a cell with a negative indicator value.

Now, for this tableau, all the empty cells have positive indicator values.

$$\begin{aligned} &\text{Indicator (II, 1):} && +15 - 1 + 9 - 8 = 15 \\ &\text{Indicator (II, 3):} && +12 - 3 + 9 - 1 = 17 \\ &\text{Indicator (III, 2):} && +3 - 9 + 8 - 1 = 1 \\ &\text{Indicator (III, 3):} && +5 - 3 + 8 - 1 = 9 \end{aligned}$$

This means that the current solution is optimal. The cost of this solution is

$$1(8) + 6(9) + 1(3) + 1(1) + 2(1) = 8 + 54 + 3 + 1 + 2 = 68$$

It turns out that if all the cells associated with a feasible solution have positive indicator values, then the solution one has reached is optimal. (Cells with zero indicator value show that there are other solutions that achieve the same optimal value.)

### How to Recognize an Optimal Solution — THEOREM

We are given a transportation problem with $m$ suppliers and $n$ demanders where the amount of the supplies equals the amount of demands. A collection of $m + n - 1$ circled cells is optimal (that is, the circled cells determine a minimum cost solution) if the indicator value associated with the empty cells is positive. If some indicator cells are positive and some are zero, there are multiple solutions for an optimal value.

This theorem is the analog of the result for linear programming that states that if a corner point is feasible, and if no neighbor of the corner point has a better value of the objective function, then the corner point we are at is already an optimal one. Note that there may be other optimal solutions that use a different number of cells than $m + n - 1$, but we can never do any better in terms of the cheapness of a solution than what we have described above.

For those interested in the exciting fact that one piece of mathematics is often useful for other mathematics, we see an example of that here. The reason an empty cell gives rise to a unique circuit with which we can try to improve the current solution of a transportation problem results from the fact that when an edge not in a tree is added to a tree, it creates a unique circuit (see Chapter 3). Since we have $m$ rows and $n$ columns, a tree associated with a graph on $m + n$ vertices has $m + n - 1$ edges, exactly the number of cells we need to fill in a transportation problem!

## 4.6 Improving on the Current Solution

We have now described a method guaranteed to find an optimal solution to a transportation problem.

### The Stepping Stone Method — DEFINITION

The **stepping stone method** consists of taking some feasible solution of a transportation problem and improving this solution, if it is not optimal, by shipping an additional amount using a cell with a negative indicator value.

### EXAMPLE 10 ■ Applying the Stepping Stone Method

We will work out another small example to illustrate the technique of applying the Northwest Corner Rule to get an initial solution, and then improving this solution if it is not optimal. Again, we do so by computing the indicator values of the cells and improving the current solution by shipping using a cell with a negative indicator value.

We start with an initial tableau where there are two mines that can supply ore to three companies that extract ore. There are 10 units of ore being mined and the extractors need 10 units to run at full capacity. The initial tableau for the problem is displayed in Figure 4.26.

Using the Northwest Corner Rule we find an initial feasible solution as shown in Figure 4.27. When applying the Northwest Corner Rule we eliminate a row or column as follows: first column 1, then column 2, then row I, and we are now left with a single cell. The cost of the feasible solution shown is

$$2(7) + 4(1) + 1(3) + 3(12) = 14 + 4 + 3 + 36 = 57$$

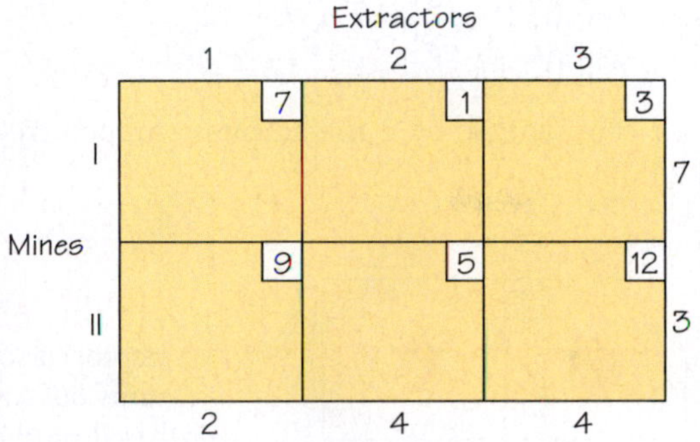


**FIGURE 4.26** A transportation problem where two mines supply ore to three companies that extract metal from the ore. The shipping costs are indicated.

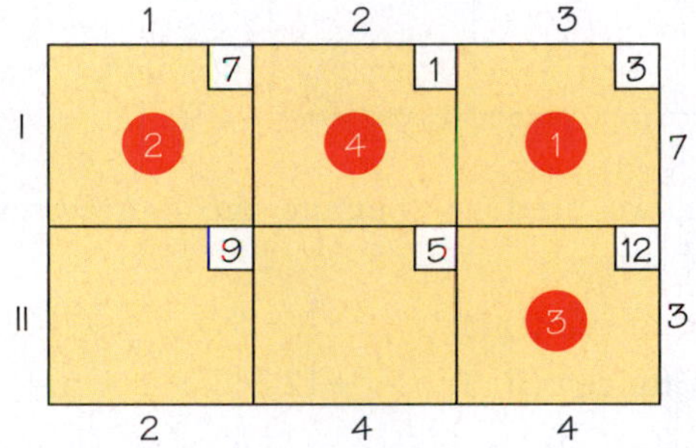


**FIGURE 4.27** The Northwest Corner Rule has been used to find a possible way to meet the demands from the supplies for the tableau in Figure 4.26.

The two empty cells we have are (II, 1) and (II, 2). We compute the indicator value for each of these cells:

$$\text{Indicator for cell (II, 1):} \quad +9 - 12 + 3 - 7 = -7$$

$$\text{Indicator for cell (II, 2):} \quad +5 - 12 + 3 - 1 = -5$$

Since cell (II, 1) has a more negative indicator value, we can reduce the cost more by using that cell. Increasing by 2 (since this is the minimum of circled numbers with negative signs in the computation of the indicator) the amount of metal shipped via cell (II, 1) and cell (I, 3) and reducing by 2 the amount in cells (I, 1) and (II, 3), we obtain the new tableau in Figure 4.28. This has cost

$$4(1) + 3(3) + 2(9) + 1(12) = 4 + 9 + 18 + 12 = 43$$

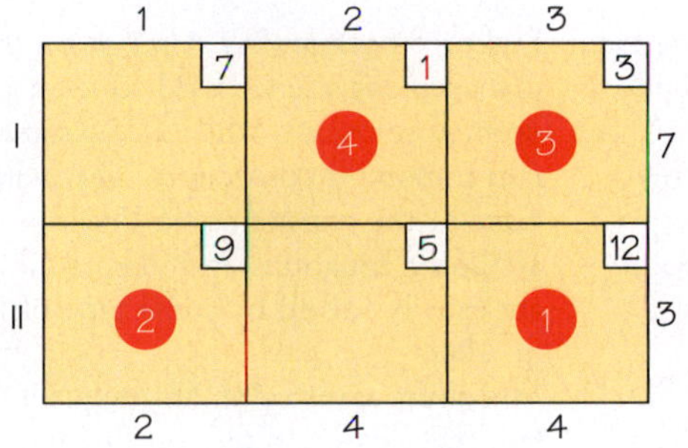


**FIGURE 4.28** An improved solution based on the negative indicator value for cell (II, 1) in Figure 4.27.

Note that as a partial check on our work, if we multiply the indicator (−7) by 2, this is −14 and 57 − 43 = 14, so we reduced the cost of our first solution by 14, as expected.

We now repeat this procedure for this new tableau. We must compute the indicator value of cells (I, 1) and (II, 2).

Indicator for cell (I, 1): $+7 - 9 + 12 - 3 = +7$

Indicator for cell (II, 2): $+5 - 12 + 3 - 1 = -5$

Thus, it turns out that we can increase by 1 the amount shipped by (II, 2), and get the tableau in Figure 4.29.

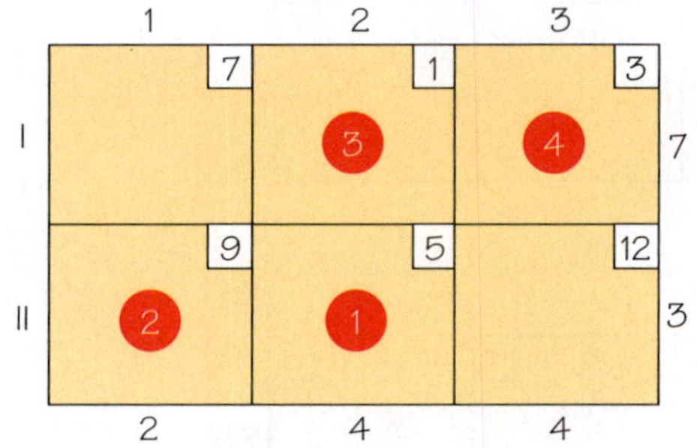


**FIGURE 4.29** An improved solution, which turns out to be optimal, based on the negative indicator value for cell (II, 2) in Figure 4.28.

From this tableau, we need to compute the indicator values for the cells (I, 1) and (II, 3). We obtain

Indicator for cell (I, 1): $+7 - 9 + 5 - 1 = +2$

Indicator for cell (II, 3): $+12 - 3 + 1 - 5 = +5$

Not surprisingly, the cell (II, 2) has a positive indicator value because in the previous tableau, that cell was the one that, when we shipped less via it, enabled us to reduce the cost. The fact that both of these indicator values are positive means that the current shipping schedule is an optimal one; that is, using a shipping schedule that ships via only four cells, we cannot find any other solution with the same value. ■

Transportation problems arise in a very large range of situations including shipping milk from dairies to supermarkets, vegetables to health food stores, and vitamins to your local drug store. The next time you sit down to breakfast, think about how many mathematics problems were solved for you to have a healthy breakfast!

## REVIEW VOCABULARY

**Corner point principle** The principle states that there is a corner point of the feasible region that yields the optimal solution. (p. 112)

**Feasible points** A possible solution (but not necessarily the best) to a linear-programming problem. With just two products, we can think of a feasible point as a point on the plane. (p. 136)

**Feasible region** The set of all **feasible points**, that is, possible solutions to a linear-programming problem. For problems with just two products, the feasible region is a part of the plane. Also called **feasible set.** (p. 109)

**Indicator value of a cell** The change in cost due to shipping an increased or decreased amount, using the cells in a transportation tableau that form a circuit consisting of circled cells together with a selected cell that is not circled. When an indicator value is negative, a cheaper solution can be found by shipping using this cell. (p. 132)

**Linear programming** A set of organized methods of management science used to solve problems of finding optimal solutions, while at the same time respecting certain important constraints. The mathematical formulations of the constraints in linear-programming problems are linear equations and inequalities. Mixture problems are usually solved by some type of linear programming. (p. 104)

**Minimum constraint** An inequality in a mixture problem that gives a minimum quantity of a product. Negative quantities can never be produced. (p. 108)

**Mixture chart** A table displaying the relevant data in a linear-programming mixture problem. The table has a row for each product and a column for each resource, for any nonzero minimums, and for the profit. (p. 107)

**Mixture problem** A problem in which a variety of resources available in limited quantities can be combined in different ways to make different products. It is usually desired to find the way of combining the resources that produces the most profit. (p. 105)

**Northwest Corner Rule (NCR)** A method for finding an initial but rarely optimal solution to a transportation problem starting from a tableau with rim conditions. The amounts to be shipped between the suppliers and demanders are indicated by circling numbers in the cells in the tableau. The number of cells circled after applying the method will equal the number of rows plus the number of columns minus 1. The method depends on locating at each stage the "northwest corner" of the original tableau or a part of it. (p. 129)

**Optimal production policy** A corner point of the feasible region where the profit formula has a maximum value. (p. 106)

**Profit line** In a two-dimensional, two-product, linear-programming problem, the set of all feasible points that yield the same profit. (p. 118)

**Resource constraint** An inequality in a mixture problem that reflects the fact that no more of a resource can be used than what is available. (p. 108)

**Rim conditions** The supplies available (listed in a column at the right of a transportation tableau) and demands required (listed in a row at the bottom of a transportation tableau) in a transportation problem. The supplies available are usually taken to exactly meet the demands required. (p. 126)

**Simplex method** One of a number of algorithms for solving linear-programming problems. (p. 122)

**Stepping stone method** A method for solving a transportation problem that improves the current solution, when it is not optimal, by increasing the amount shipped using a cell with a negative indicator value. (p. 134)

**Tableau** A table for a transportation problem indicating the supplies available and demands required, as well as the cost of shipping from a supplier to a demander. The amounts to be shipped from different suppliers to different users are indicated by circled cells in the tableau. The number of such circled cells is always the number of rows plus the number of columns diminished by 1 for the tableau. (p. 127)

**Transportation problem** A special type of linear-programming problem where we have sources of supplies and users of, or demand for, these supplies. There is a cost to ship an item from a supplier to a demander. The goal is to minimize the total shipping cost to meet the demands from the supplies. (p. 125)

## ✓ SKILLS CHECK

**1.** Where do the lines $6x + 2y = 26$ and $2x + 3y = 18$ intersect?

**(a)** At the point (3, 4)
**(b)** At the point (6, 2)
**(c)** At the point (3, 2)

**2.** The lines $x + 3y = 12$ and $y = 2$ intersect at the point with $x$-coordinate _____ and $y$-coordinate _____ .

**3.** Which of these points lie in the region $4x + 3y \geq 24$, $x \geq 0, y \geq 0$?

**(a)** Points (5, 2) and (3, 4)
**(b)** Points (2, 5) and (3, 4)
**(c)** Points (5, 2) and (2, 5)

**4.** Producing a bench ($x$) requires 2 boards, and producing a table ($y$) requires 5 boards. There are 25 boards available. The resource constraint associated with this situation is _____ $x$ + _____ $y \leq 25$.

**5.** A tart requires 3 oz of fruit and 2 oz of dough; a pie requires 13 oz of fruit and 7 oz of dough. There are 140 oz of fruit and 90 oz of dough available. Each tart earns 6 cents profit; each pie earns 25 cents profit. What are the resource inequalities of this situation?

**(a)** $3x + 2y \leq 140$
$13x + 7y \leq 90$
$x \geq 0, y \geq 0$

**(b)** $3x + 13y \leq 140$
$2x + 7y \leq 90$
$x \geq 0, y \geq 0$

**(c)** $3x + 2y \leq 6$
$13x + 7y \leq 25$
$x \geq 0, y \geq 0$

**6.** A tart requires 3 oz of fruit and 2 oz of dough; a pie requires 13 oz of fruit and 7 oz of dough. There are 140 oz of fruit and 90 oz of dough available. Each tart earns 6 cents profit; each pie earns 25 cents profit. The profit formula for this situation, if $x$ represents the numbers of pies produced and $y$ represents the number of tarts produced, is given by $P$ (in cents) = _____ $x$ + _____ $y$.

**7.** Graph the feasible region identified by the following inequalities:

$2x + 4y \leq 20$
$4x + 2y \leq 16$
$x \geq 0, y \geq 0$

Which of these points is *not* in the feasible region of the graph drawn?

**(a)** (2, 4)
**(b)** (1, 1)
**(c)** (10, 0)

**8.** Suppose the feasible region has four corners, at points (0, 0), (4, 0), (0, 3), and (3, 2). If the profit formula is \$3$x$ − \$2$y$, the maximum value for the profit is ________________________________.

**9.** Suppose the feasible region has four corners, at points (0, 0), (4, 0), (0, 3), and (3, 2). For which of these profit formulas is the profit maximized by producing a mix of products?

**(a)** \$2$x$ − \$2$y$
**(b)** \$$x$ + \$2$y$
**(c)** \$2$x$ − \$$y$

**10.** The corner point method cannot be applied to find the optimal answer for the value of the profit $P = 3x + 7y$, where the feasible region is shown in the diagram, because the feasible region is not ________________.

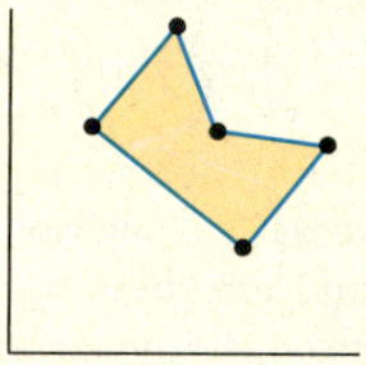

**11.** The shaded region in the accompanying diagram is an example of a region

**(a)** whose area is not bounded.
**(b)** that is not convex.
**(c)** that is not bounded by straight-line segments.

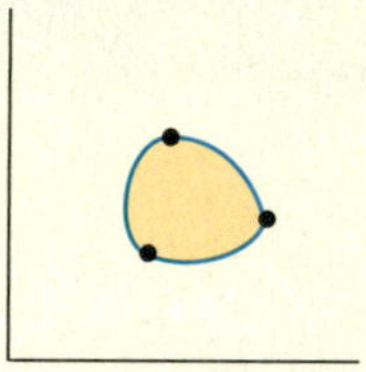

**12.** Suppose the feasible region has five corners, at points (1, 1), (2, 1), (3, 2), (2, 4), (1, 5). If the profit formula is \$5$x$ − \$3$y$, the corner point which maximizes the profit has $x$-coordinate ____ and $y$-coordinate ____ .

**13.** Suppose the feasible region has five corners, at points (1, 1), (2, 1), (3, 2), (2, 4), (1, 5). Which of these points is *not* in the feasible region?

**(a)** (1, 3)
**(b)** (2, 2)
**(c)** (0, 0)

**14.** Consider the feasible region identified by the inequalities $x \geq 0, y \geq 0, 3x + y \leq 10, x + 2y \leq 6$. The corner point of this region, which is not (0, 0), that has $x$-coordinate 0 has $y$-coordinate ________________.

**15.** How does the line representing the maximum feasible profit intersect the feasible region?

**(a)** No points of intersection
**(b)** Only one point of intersection
**(c)** At least one point of intersection, and sometimes more than one point of intersection

**16.** Consider the feasible region for a linear programming problem involving the inequalities $x \geq 0$, $y \geq 0$, $3x + y \leq 10$, $x + 2y \leq 5$. The corner point for this feasible region that has no zero coordinates has $x$-coordinate ______ and $y$-coordinate ______ .

**17.** When the Northwest Corner Rule is applied to the accompanying transportation problem tableau, the cells that remain empty are

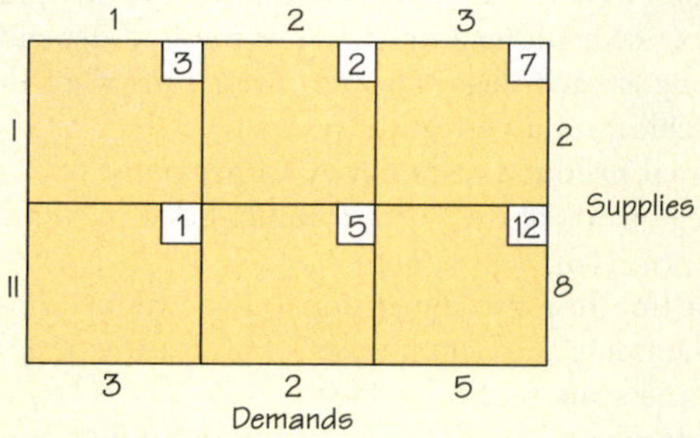


**(a)** cell (II, 2) and cell (I, 2).
**(b)** cell (I, 2) and cell (II, 3).
**(c)** cell (I, 2) and cell (I, 3).

**18.** The circled cells in the accompanying tableau give a solution that satisfies the rim conditions. The cost associated with this solution is ________________.

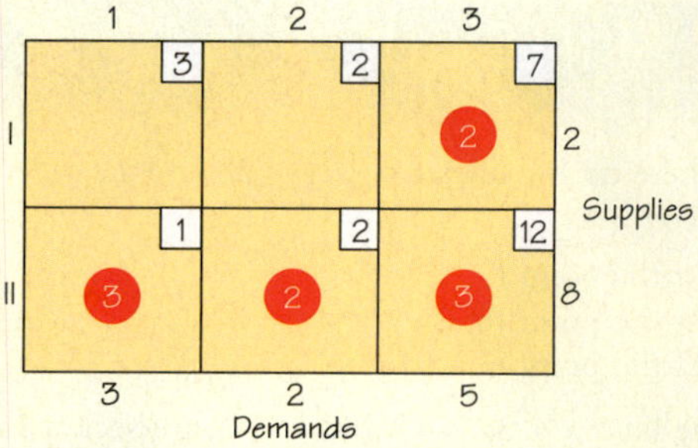


**19.** The circled cells in the accompanying tableau satisfy the rim conditions. When the indicator value for cell (I, 2) is computed,

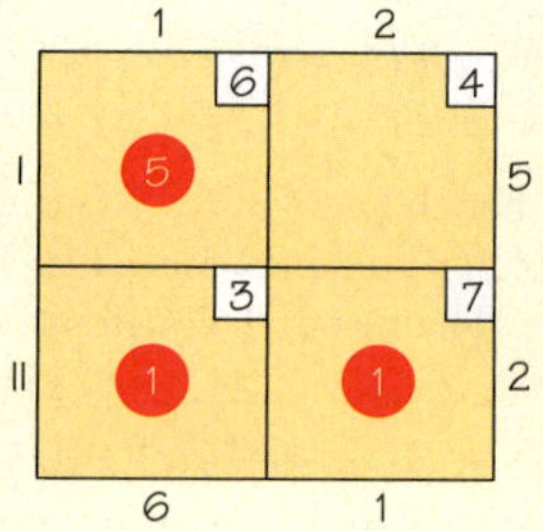


**(a)** the result being positive, the current solution is optimal.
**(b)** the result being positive, this tableau has no minimal cost solution.
**(c)** the result being negative means this tableau does not give a minimal cost solution.

**20.** The indicator value associated with cell (I, 2) of the accompanying tableau is ______________ .

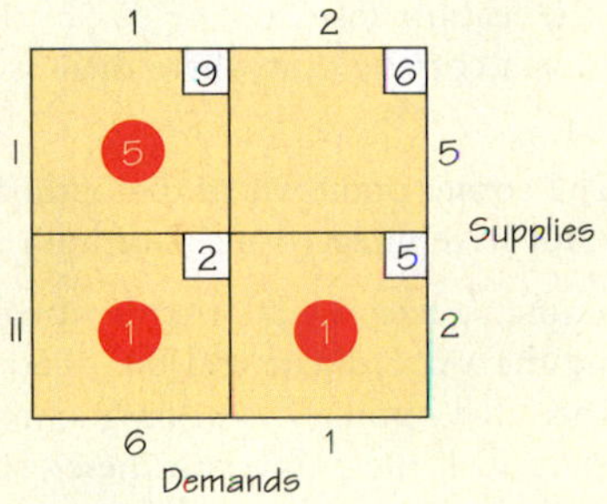


# CHAPTER 4 EXERCISES

■ Challenge ◆ Discussion

## 4.1 Mixture Problems: Combining Resources to Maximize Profit

**1.** Using intercepts, the points where the lines cross the axes, graph each line.

(a) $2x + 3y = 12$
(b) $3x + 5y = 30$
(c) $4x + 3y = 24$

**2.** Using intercepts, the points where the lines cross the axes, graph each line.

(a) $7x + 4y = 42$
(b) $x = -3$
(c) $y = 6$

**3.** Graph both lines on the same axes. Put a dot where the lines intersect. Use algebra to find the $x$- and $y$-coordinates of the point of intersection.

(a) $4x + 3y = 18$ and $x = 0$
(b) $5x + 3y = 45$ and $y = -5$
(c) $5x + 3y = 45$ and $x = 3$

**4.** Graph both lines on the same axes. Put a dot where the lines intersect. Use algebra to find the $x$- and $y$-coordinates of the point of intersection.

(a) $x = 3$ and $y = -4$
(b) $3x + 5y = 45$ and $x = -5$
(c) $5x + 3y = 45$ and $x = -3$

**5.** Graph both lines on the same axes. Put a dot where the lines intersect. Use algebra to find the $x$- and $y$-coordinates of the point of intersection.

(a) $x + y = 10$ and $x + 2y = 14$
(b) $y - 2x = 0$ and $x = 2$

**6.** Graph the line and half-plane corresponding to the inequality, a typical constraint from a mixture problem.

(a) $x \geq 7$ (c) $5x + 3y \leq 15$
(b) $y \geq 4$ (d) $4x + 5y \leq 30$

**7.** Graph the line and half-plane corresponding to the inequality, a typical constraint from a mixture problem.

(a) $x \geq 3$ (c) $3x + 2y \leq 18$
(b) $y \geq 8$ (d) $7x + 2y \leq 42$

In Exercises 8–10, for each description, write one or more appropriate resource-constraint inequalities. The unknown to use for each product is given in parentheses.

**8. (a)** One bridesmaid's bouquet ($x$) requires 2 roses, and one corsage ($y$) requires 4 roses. There are 28 roses available.
**(b)** Maintaining a large tree ($x$) takes 2 hours of pruning time and 30 minutes of shredder time; maintaining a small tree ($y$) takes 30 minutes of pruning time and 15 minutes of shredder time. There are 40 hours of pruning time and 2 hours of shredder time available.

**9. (a)** Manufacturing one package of hot dogs ($x$) requires 6 oz of beef, and manufacturing one package of bologna ($y$) requires 4 oz of beef. There are 300 oz of beef available.
**(b)** It takes 30 ft of 12-in. board to make one bookcase ($x$); it takes 72 ft of 12-in. board to make one table ($y$). There are 420 ft of 12-in. board available.

**10.** Manufacturing one salami ($x$) requires 12 oz of beef and 4 oz of pork. Manufacturing one bologna ($y$) requires 10 oz of beef and 3 oz of pork. There are 40 lb of beef and 480 oz of pork available.

In Exercises 11–16, graph the feasible region, label each line segment bounding it with the appropriate equation, and give the coordinates of every corner point.

**11.** $x \geq 0; y \geq 0; 2x + y \leq 10$

**12.** $x \geq 0; y \geq 0; x + 2y \leq 12$

**13.** $x \geq 0; y \geq 0; 2x + 5y \leq 60$

**14.** $x \geq 10; y \geq 0; 3x + 5y \leq 120$

**15.** $x \geq 0; y \geq 4; x + y \leq 20$

**16.** $x \geq 2; y \geq 6; 3x + 2y \leq 30$

In Exercises 17–18, determine whether the points (2, 4) and/or (10, 6) are points of the given feasible regions of:

**17.** Exercises 11, 13, and 15.

**18.** Exercises 12, 14, and 16.

**19.** In the toy problem, $x$ represents the number of skateboards and $y$ the number of dolls. Using the version of that problem whose feasible region is presented in Figure 4.3b, with the profit formula $\$2.30x + \$3.70y$, write a sentence giving the maximum profit and describing the production policy that gives that profit.

**20.** In the toy problem, $x$ represents the number of skateboards and $y$ the number of dolls. Using the version of that problem whose feasible region is presented in Figure 4.3b, with the profit formula $\$5.50x + \$1.80y$, write a sentence giving the maximum profit and describing the production policy that gives that profit.

**21.** Graph both lines on the same axes. Put a dot where the lines intersect. Use algebra to find the $x$- and $y$-coordinates of the point of intersection.

**(a)** $5x + 4y = 22$ and $5x + 10y = 40$
**(b)** $x + y = 7$ and $3x + 4y = 24$

In Exercises 22–25, graph the feasible region, label each line segment bounding it with the appropriate equation, and give the coordinates of every corner point.

**22.** $x \geq 0; y \geq 0; 3x + y \leq 9; x + y \leq 7$

**23.** $x \geq 0; y \geq 0; 2x + y \leq 4; 4x + 4y \leq 12$

**24.** $x \geq 0; y \geq 2; 5x + y \leq 14; x + 2y \leq 10$

**25.** $x \geq 4; y \geq 0; 5x + 4y \leq 60; x + y \leq 13$

**26.** Determine whether the points (4, 2) and/or (1, 3) are points of the given feasible regions of Exercises 23 and 25.

## 4.2 Finding the Optimal Production Policy

## 4.3 Why the Corner Point Principle Works

## 4.4 Linear Programming: Life Is Complicated

**27.** Find the maximum value of $P$ where $P = 3x + 2y$ subject to the constraints $x \geq 3, y \geq 2, x + y \leq 10, 2x + 3y \leq 24$.

**28.** Find the maximum value of $P$ where $P = 3x - 2y$ subject to the constraints $x \geq 2, y \geq 3, 3x + y \leq 18, 6x + 4y \leq 48$.

**29.** Find the maximum value of $P$ where $P = 5x + 2y$ subject to the constraints $x \geq 2, y \geq 4, x + y \leq 10$.

**30.** Given profit $P = 21x + 11y$ subject to the constraints $x \geq 0, y \geq 0, 7x + 4y \leq 13$:

**(a)** Graph the feasible region.
**(b)** Determine a corner point where there is an optimal solution.

(*Warning:* The corner point where the optimal solution occurs may not have integer values for both $x$ and $y$.)

**31. (a)** Referring to Exercise 30, use the usual rounding rule to round the $x$-coordinate and the $y$-coordinate of the point where the optimal linear-programming solution occurs. Call the point with these coordinates $Q$.
**(b)** Determine if $Q$'s coordinates define a feasible point by checking them against the constraints.
**(c)** Evaluate the profit value $P$ at point $Q$. How does the profit value compare with the point where the optimal value occurred in Exercise 30?
**(d)** Let $R$ be the point with coordinates (0, 3). Is $R$ in the feasible region? Evaluate $P$ at point $R$ and compare the result with the answer at $Q$ and where the optimum linear-programming value occurred.
**(e)** Explain the significance of the situation here for solving maximization problems where $P = ax + by$ ($a$ and $b$ are known in advance) is subject to linear constraints but where the variables must be nonnegative integers rather than arbitrary nonnegative decimal numbers.

Exercises 32–43 each have several steps leading to a complete solution to a mixture problem. Practice a specific step of the solution algorithm by working out just that step for several problems. The steps are:

**(a)** Make a mixture chart for the problem.
**(b)** Using the mixture chart, write the profit formula and the resource- and minimum-constraint inequalities.
**(c)** Draw the feasible region for those constraints and find the coordinates of the corner points.
**(d)** Evaluate the profit information at the corner points to determine the production policy that best answers the question.
**(e)** (Requires technology) Compare your answer with the one you get from running the same problem on a simplex algorithm computer program.

**32.** A clothing manufacturer has 600 yd of cloth available to make shirts and decorated vests. Each shirt requires 3 yd of material and provides a profit of \$5. Each vest requires 2 yd of material and provides a profit of \$2. The manufacturer wants to guarantee that under all circumstances there are minimums of 100 shirts and 30 vests produced. How many of each garment should be made to maximize profit? If there are no minimum quantities, how, if at all, does the optimal production policy change?

**33.** A car maintenance shop must decide how many oil changes and how many tune-ups can be scheduled in a typical week. The oil change takes 20 min, and the tune-up requires 100 min. The maintenance shop makes a profit of $15 on an oil change and $65 on a tune-up. What mix of services should the shop schedule if the typical week has available 8000 min for these two types of services? How, if at all, do the maximum profit and optimal production policy change if the shop is required to schedule at least 50 oil changes and 20 tune-ups?

**34.** A clerk in a bookstore has 90 min at the end of each workday to process orders received by mail or on voice mail. The store has found that a typical mail order brings in a profit of $30 and a typical voice-mail order brings in a profit of $40. Each mail order takes 10 min to process and each voice-mail order takes 15 min. How many of each type of order should the clerk process? How, if at all, do the maximum profit and optimal processing policy change if the clerk must process at least three mail orders and two voice-mail orders?

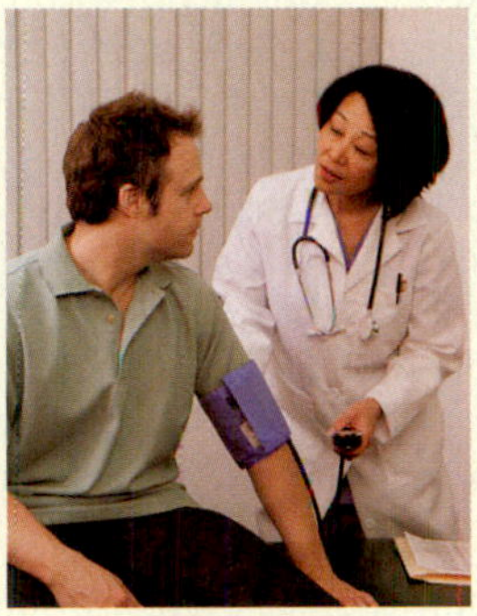
*(Hill Street Studios/Stock This Way/Corbis.)*

**35.** In a certain medical office, a routine office visit requires 5 min of doctors' time and a comprehensive office visit requires 25 min of doctors' time. In a typical week, there are 1800 min of doctors' time available. If the medical office clears $30 from a routine visit and $50 from a comprehensive visit, how many of each should be scheduled per week? How, if at all, do the maximum profit and optimal production policy change if the office is required to schedule at least 20 routine visits and 30 comprehensive ones?

**36.** A bakery makes 600 specialty breads–multigrain or herb–each week. Standing orders from restaurants are for 100 multigrain breads and 200 herb breads. The profit on each multigrain bread is $8 and on herb bread, $10. How many breads of each type should the bakery make in order to maximize profit? How, if at all, do the maximum profit and optimal production policy change if the bakery has no standing orders?

**37.** A student has decided that passing a mathematics course will, in the long run, be twice as valuable as passing any other kind of course. The student estimates that to pass a typical math course will require 12 hr a week to study and do homework. The student estimates that any other course will require only 8 hr a week. The student has available 48 hr for study per week. How many of each kind of course should the student take? (*Hint:* The profit could be viewed as 2 "value points" for passing a math course and 1 "value point" for passing any other course.) How, if at all, do the maximum value and optimal course mix change if the student decides to take at least two math courses and two other courses?

Exercises 38–43 require finding the point of intersection of two lines, each corresponding to a resource constraint.

**38.** The firm WebsAreUs creates and maintains Web sites for client companies. There are two types of Web sites: "Hot" sites change their layout frequently but keep their content for long times; "cool" sites keep their layout for a while but frequently change their content. To maintain a hot site requires 1.5 hr of layout time and 1 hr for content changes. To maintain a cool site requires 1 hr of layout time and 2 hr for content changes. Every day, WebsAreUs has available 12 hr for layout changes and 16 hr for content changes. Net profit is $50 for a set of changes on a hot site and $250 for a set of changes on a cool site. In order to maximize profit, how many of each type of site should WebsAreUs maintain daily? How, if at all, do the maximum profit and optimal policy change if the company must maintain at least two hot and three cool sites daily?

**39.** A paper recycling company uses scrap cloth and scrap paper to make two different grades of recycled paper. A single batch of grade A recycled paper is made from 25 lb of scrap cloth and 10 lb of scrap paper, whereas one batch of grade B recycle paper is made from 10 lb of scrap cloth and 20 lb of scrap paper. The company has 100 lb of scrap cloth and 120 lb of scrap paper on hand. A batch of grade A paper brings a profit of $500, whereas a batch of grade B paper brings a profit of $250. What amounts of each grade should be made? How, if at all, do the maximum profit and optimal production policy change if the company is required to produce at least one batch of each type?

**40.** Jerry Wolfe has a 100-acre farm that he is dividing into one-acre plots, on each of which he builds a house. He then sells the house and land. It costs him $20,000 to build a modest house and $40,000 to build a deluxe house. He has $2,600,000 to cover these costs. The profits are $25,000 for a modest house and $60,000 for a deluxe house. How many of each type of house should he build to maximize profit? How, if at all, do the maximum profit and optimal production policy change if Wolfe is required to build at least 20 of each type of house?

**41.** The maximum production of a soft-drink bottling company is 5000 cartons per day. The company produces regular and diet drinks, and must make at least 600 cartons of regular and 1000 cartons of diet per day. Production costs are $1.00 per carton of regular and $1.20 per carton of diet. The daily operating budget is $5400. How many cartons of each type of drink should be produced if the profit is $0.10 per regular and $0.11 per diet? How, if at all, do the maximum profit and optimal bottling policy change if the company has no minimum required production?

**42.** Wild Things raises pheasants and partridges to restock the woodlands and has room to raise 100 birds during the season. The cost of raising one bird is \$20 per pheasant and \$30 per partridge. The Wildlife Foundation pays Wild Things for the birds; the latter clears a profit of \$14 per pheasant and \$16 per partridge. Wild Things has \$2400 available to cover costs. How many of each type of bird should they raise? How, if at all, do the maximum profit and optimal restocking policy change if Wild Things is required to raise at least 20 pheasants and 10 partridges?

**43.** Lights Afire makes desk lamps and floor lamps, on which the profits are \$2.65 and \$4.67, respectively. The company has 1200 hr of labor and \$4200 for materials each week. A desk lamp takes 0.8 hr of labor and \$4 for materials; a floor lamp takes 1.0 hr of labor and \$3 for materials. What production policy maximizes profit? How, if at all, do the maximum profit and optimal production policy change if Lights Afire wants to produce at least 150 desk lamps and 200 floor lamps per week?

In Exercises 44–47, there are more than two products in the problem. Although you cannot solve these problems using the two-dimensional graphical method, you can follow these steps:

**(a)** Make a mixture chart for each problem.
**(b)** Using the mixture chart, write the resource- and minimum-constraint inequalities. Also write the profit formula.
**(c)** (Requires software) If you have a simplex method program available, run the program to obtain the optimal production policy.

**44.** A toy company makes three types of toys, each of which must be processed by three machines: a shaper, a smoother, and a painter. Each Toy A requires 1 hr in the shaper, 2 hr in the smoother, and 1 hr in the painter, and brings in a \$4 profit. Each Toy B requires 2 hr in the shaper, 1 hr in the smoother, and 3 hr in the painter, and brings in a \$5 profit. Each Toy C requires 3 hr in the shaper, 2 hr in the smoother, and 1 hr in the painter, and brings in a \$9 profit. The shaper can work at most 50 hr per week, the smoother 40 hr, and the painter 60 hr. What production policy would maximize the toy company's profit?

**45.** A rustic furniture company handcrafts chairs, tables, and beds. It has three workers, Chris, Sue, and Juan. Chris can work only 80 hr per month, but Sue and Juan can each put in 200 hr. Each of these artisans has special skills. To make a chair takes 1 hr of Chris's time, 3 from Sue, and 2 from Juan. A table needs 3 hr from Chris, 5 from Sue, and 4 from Juan. A bed requires 5 hr from Chris, 4 from Sue, and 8 from Juan. Even artisans are concerned about maximizing their profit, so what product mix should they stick with if they get \$100 profit per chair, \$250 per table, and \$350 per bed?

(Atlantide Phototravel/Corbis.)

**46.** A candy manufacturer has 1000 lb of chocolate, 200 lb of nuts, and 100 lb of fruit in stock. The Special Mix requires 3 lb of chocolate, 1 lb each of nuts and fruit, and it brings in \$10. The Regular Mix requires 4 lb of chocolate, 0.5 lb of nuts, and no fruit, and brings in \$6. The Purist Mix requires 5 lb of chocolate, no nuts or fruit, and brings in \$4. How many boxes of each type should be produced to maximize profit?

**47.** A gourmet coffee distributor has on hand 17,600 lb of African coffee, 21,120 oz of Brazilian coffee, and 12,320 oz of Colombian coffee. It sells four blends–Excellent, Southern, World, and Special–on which it makes these per-pound profits, respectively: \$1.80, \$1.40, \$1.20, and \$1.00. One pound of Excellent is 16 oz of Colombian; it is not a blend at all. One pound of Southern consists of 12 oz of Brazilian and 4 oz of Colombian. One pound of World requires 6 oz of African, 8 of Brazilian, and 2 of Colombian. One pound of Special is made up of 10 oz of African and 6 oz of Brazilian. What product mix should the gourmet coffee distributor prepare in order to maximize profit?

In Exercises 48 and 49, use the fact that the corner point approach can also solve minimization problems to minimize the given expression for cost $C$.

**48.** Minimize $C$ given by $C = 7x + 8y$ over the feasible region for Exercise 27.

**49.** Minimize $C$ given by $C = 5x + 11y$ over the feasible region for Exercise 28.

**50.** Show by example that a feasible region that has the nonnegativity constraints $x \geq 0$, $y \geq 0$, and $x + y \leq 0.5$ can have no feasible points with integer coordinates other than (0, 0).

**51.** Courtesy Calls makes telephone calls for businesses and charities. A profit of \$0.50 is made for each business call and \$0.40 for each charity call. It takes 4 min (on average) to make a business call and 6 min (on average) to make a charity call. If there are 240 min of calling time to be distributed each day, how should that time be spent so that Courtesy Calls makes a maximum profit? What changes, if any, occur in the maximum profit and optimal production policy if every day they must make at least 12 business and 10 charity calls?

**52.** A refinery mixes high-octane and low-octane fuels to produce regular and premium gasolines. The profits per gallon on the two gasolines are \$0.30 and \$0.40, respectively. One gallon of premium gasoline is produced by mixing 0.5 gal of each of the fuels. One gallon of regular gasoline is produced by mixing 0.25 gal of high octane with 0.75 gal of low octane. If there are 500 gal of high octane and 600 gal of low octane available, how many gallons of each gasoline should the refinery make? How, if at all, do the maximum profit and optimal production policy change if the refinery is required to produce at least 100 gal of each gasoline?

**53.** A toy manufacturer makes bikes, for a profit of \$12, and wagons, for a profit of \$10. To produce a bike requires 2 hr of machine time and 4 hr painting time. To produce a wagon requires 3 hr machine time and 2 hr painting time. There are 12 hr of machine time and 16 hr of painting time available per day. How many of each toy should be produced to maximize profit? How, if at all, do the maximum profit and optimal production policy change if the manufacturer must daily produce at least two bikes and two wagons?

## 4.5 A Transportation Problem: Delivering Perishables

## 4.6 Improving on the Current Solution

**54.** Apply the Northwest Corner Rule, thereby finding a feasible solution that obeys the rim conditions, to the following transportation problem tableaux.

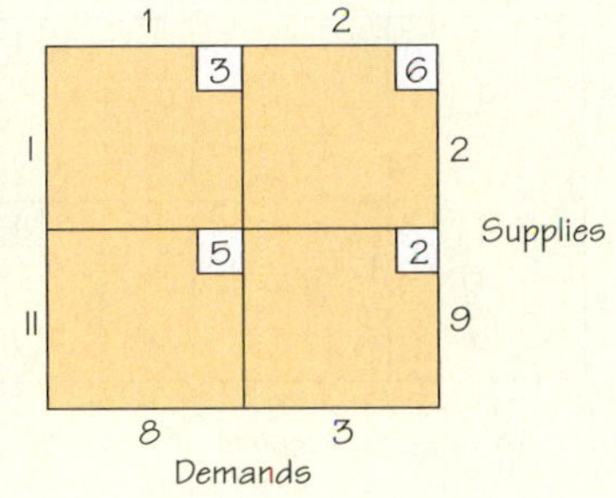

*(a)*

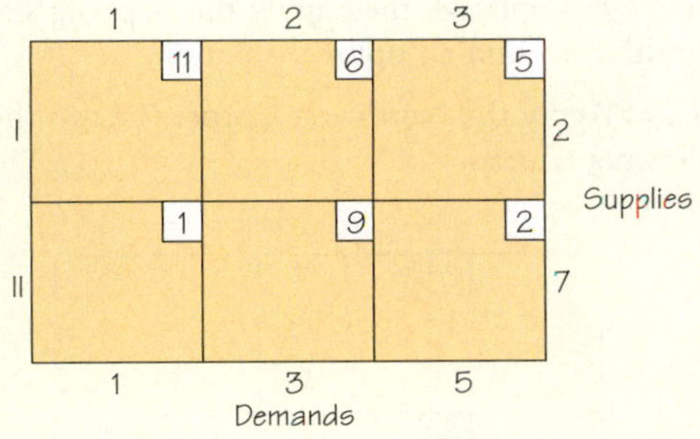

*(b)*

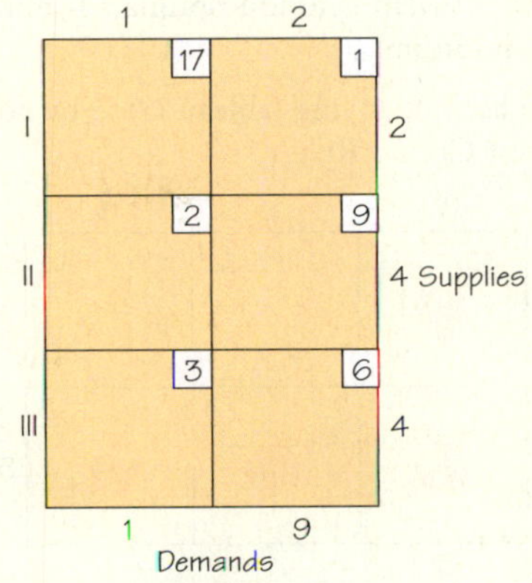

*(c)*

**(d)** For each tableau, give a possible real-world setting for the problem.
**(e)** For each tableau, find the cost of shipping using the cells that were circled when you used the Northwest Corner Rule.

**55. (a)** Apply the Northwest Corner Rule, thereby finding a feasible solution that obeys the rim conditions, to the accompanying tableau, which arose from meeting the demands of fruit stands for peaches from supplies available from local orchards.

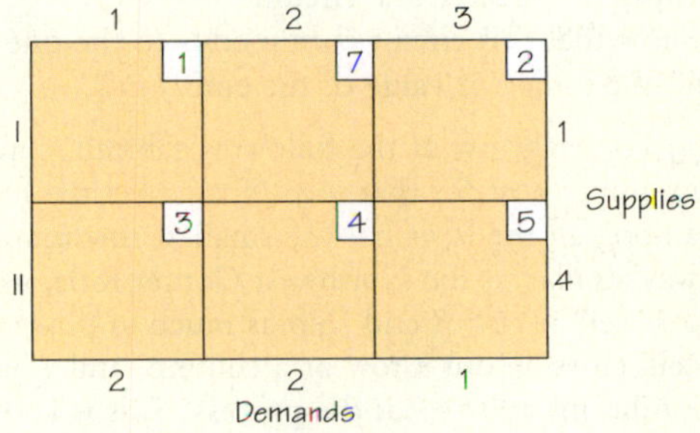


**(b)** Determine the cost associated with the solution you found.
**(c)** Compute the indicator value for each noncircled cell.

**56.** The accompanying tableau represents the shipping costs and supply-and-demand constraints for supplies of purified water to be shipped to companies that resell the water to office buildings.

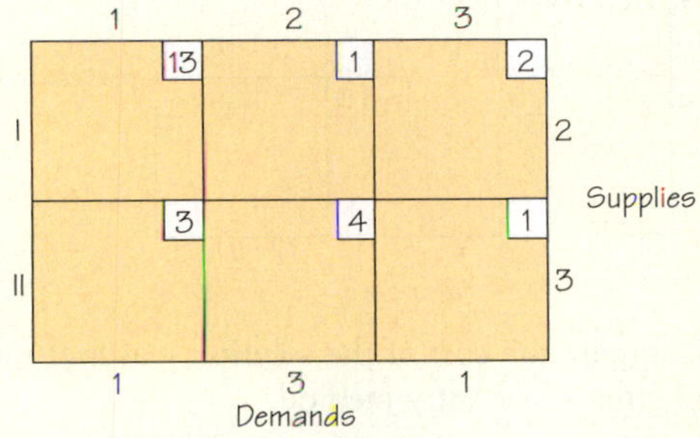


**(a)** Find the Northwest Corner Rule initial solution.
**(b)** Determine the indicator value for each noncircled cell.

**(c)** Is the current solution optimal? If not, find a cheaper solution.

**57.** The accompanying tableau arose by applying the Northwest Corner Rule.

| | 1 | 2 | 3 | Supplies |
|---|---|---|---|---|
| I | cost 12, circled 2 | cost 1 | cost 2 | 2 |
| II | cost 1, circled 1 | cost 4, circled 1 | cost 3, circled 1 | 3 |
| Demands | 3 | 1 | 1 | |

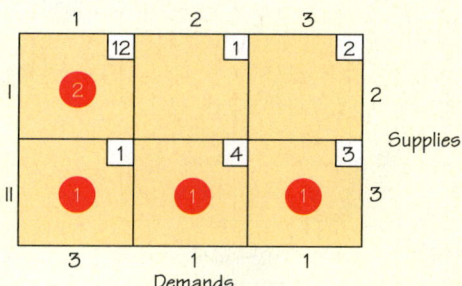

The accompanying graph was constructed so that there is one edge for each circled vertex in the tableau above.

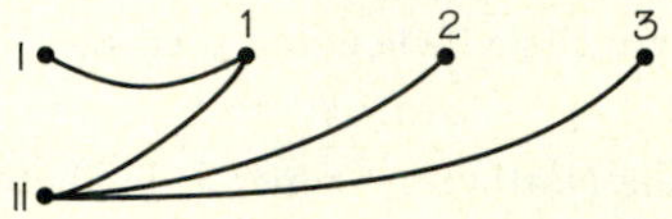


**(a)** Verify that the graph is a tree.
**(b)** Show that for each empty cell in the tableau, adding to the graph the unique edge that corresponds to the empty cell creates one circuit.
**(c)** Show that this circuit corresponds to the one used to find the indicator value of the empty cell.

**58. (a)** For each row of the following tableau, compute the minimum cost for that row. Now select the row *R* that among all the rows has the smallest row minimum. In a way similar to the Northwest Corner Rule, use the cheapest cell in row *R* and ship as much as possible via that cell, crossing out a row or a column, and adjust the rim conditions and repeat the process. This is known as the ***minimum row entry method.*** Use the minimum row entry method to find an initial solution to the following transportation problem, which shows the costs of returning rental cars from cities that have more cars than necessary to cities that have too few cars.

| | 1 | 2 | 3 | Extra cars |
|---|---|---|---|---|
| I | 13 | 1 | 3 | 3 |
| II | 2 | 5 | 4 | 5 |
| Need cars | 4 | 2 | 2 | |

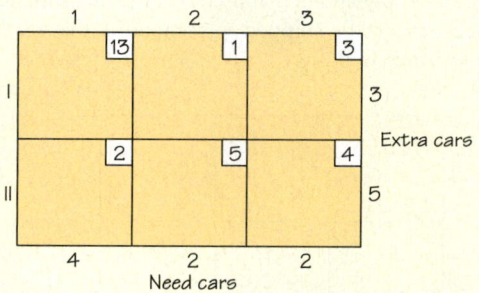

**(b)** Compute the cost of the solution you find using the minimum row entry method.
**(c)** Compare the cost found in part (b) with the cost of the initial solution obtained using the Northwest Corner Rule.

**59. (a)** For each of the following tableaux, find an initial solution using the Northwest Corner Rule.

| | 1 | 2 | 3 | Supplies |
|---|---|---|---|---|
| I | 1 | 2 | 3 | 3 |
| II | 4 | 5 | 6 | 5 |
| Demands | 1 | 1 | 6 | |

(i)

| | 1 | 2 | 3 | Supplies |
|---|---|---|---|---|
| I | 6 | 5 | 4 | 3 |
| II | 3 | 2 | 1 | 5 |
| Demands | 1 | 1 | 6 | |

(ii)

| | 1 | 2 | 3 | Supplies |
|---|---|---|---|---|
| I | 6 | 5 | 1 | 5 |
| II | 2 | 3 | 4 | 4 |
| Demands | 2 | 1 | 6 | |

(iii)

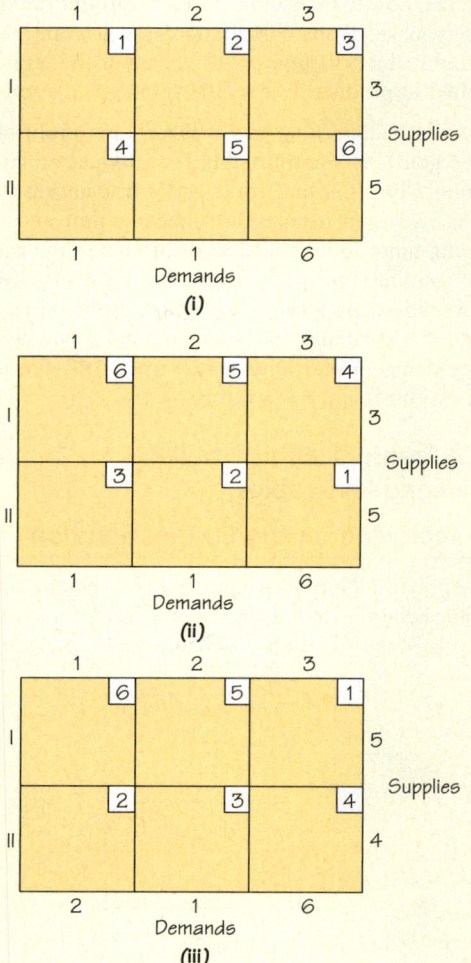

**(b)** If the solution you find using the Northwest Corner Rule is not optimal, then apply the stepping stone algorithm to find an optimal solution.

**60. (a)** Apply the Northwest Corner Rule to the following tableau.

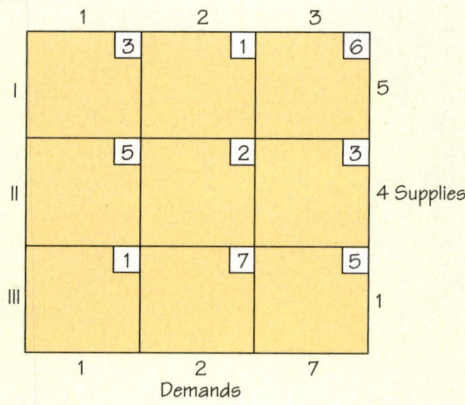

| | 1 | 2 | 3 | Supplies |
|---|---|---|---|---|
| I | 3 | 1 | 6 | 5 |
| II | 5 | 2 | 3 | 4 |
| III | 1 | 7 | 5 | 1 |
| Demands | 1 | 2 | 7 | |

**(b)** Determine the cost associated with the solution you found.
**(c)** Compute the indicator value for each noncircled cell.
**(d)** Does the Northwest Corner Rule give rise to an optimal solution?

**61.** For each tableau at right, a solution for the associated transportation problem has been proposed using the circled cells that obey the rim conditions.
**(a)** Determine the cost associated with the indicated feasible solution.
**(b)** Is the solution shown optimal?
**(c)** If it is not an optimal solution, apply the stepping stone method to obtain an optimal solution.

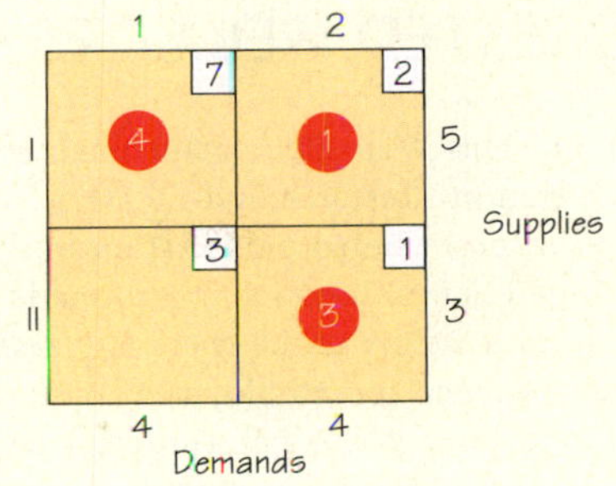

(i)

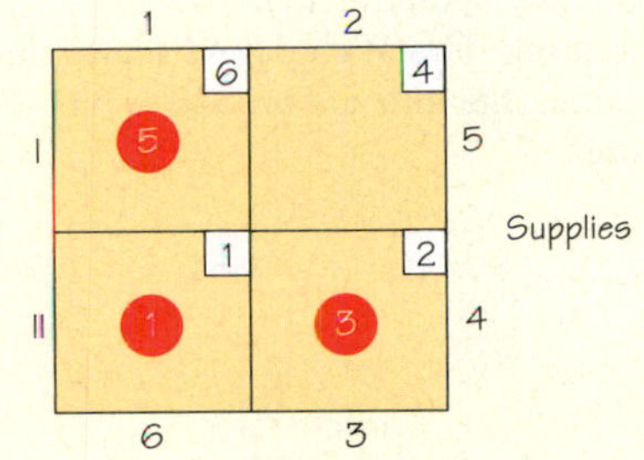

(ii)

# WRITING PROJECTS

**1.** Interview a local businessperson who is in charge of deciding the product mix for a business. Must this business take into consideration situations other than minimum and resource constraints? If so, what are these considerations? Find out what methods the person uses to make production policy decisions. Is linear programming used? Are other methods used? If so, what are they? Write a report of your findings, and add some of your own conclusions about the usefulness of linear programming for this business.

**2.** In economics, it is often useful to distinguish between a firm that has a monopoly (for example, is the only supplier of a product) and firms that supply only a small share of the market. How would the presence of a monopoly affect the relation between production and price? Would the presence of a monopoly tend to ensure the fixed-profit assumption of linear programming, or would it make it more likely that the interplay of supply and demand would have to be considered in order to have a truly realistic model? Write an essay addressing these issues.

# SUGGESTED READINGS

ANDERSON, DAVID R., DENNIS J. SWEENEY, and THOMAS A. WILLIAMS. *An Introduction to Management Science: Quantitative Approaches to Decision Making,* West, St. Paul, Minn., 1985. A business management text with seven chapters on linear programming.

DOLAN, ALAN, and JOAN ALDUS, *Networks and Algorithms: An Introductory Approach,* Wiley, New York, 1993. A graph theoretical approach to network optimization problems, including the transportation problem.

GASS, SAUL I. *Decision Making, Models, and Algorithms,* Krieger, Melbourne, Fla., 1991. This book demonstrates how to use linear programming and related ideas to solve a variety of industrial and governmental problems.

HARDWICK, I., *Decision and Discrete Mathematics,* Albion Publishing, Chichester, England, 1996. A survey of situations that can be modeled using graphs in the area of operations research. It treats both the simplex method for solving linear-programming problems and the transportation problem.

***Note:*** Simplex software can be found in *Maple* (keyword is *simplex*), *Mathematica* (keyword is *Linear Programming*), in both *Lotus 1-2-3* and *MSExcel* via *Solver,* and in other software packages, especially those intended for quantitative mathematics courses focusing on business applications.

## SUGGESTED WEB SITES

**www.informs.org** This Web site is maintained by the Institute for Operations Research and the Management Sciences, the main professional organization in these fields in the United States. It contains information on (and/or links to) news items about operations research and management science and employment opportunities and summer internships; it also has a student newsletter. Much of the material is written in a nontechnical style.

**www.hsor.org/what_is_or.cfm?name=linear_programming** This Web page discusses how linear programming fits into the broader subject of operations research.

**www-gap.dcs.st-and.ac.uk/~history/Mathematicians/Dantzig_George.html** This site contains biographical information about George Dantzig, who, by developing the simplex method, greatly expanded the use and applicability of linear programming.

**www-unix.mcs.anl.gov/otc/Guide/faq/linear-programming-faq.html** This site is the "frequently asked questions" section of an online newsgroup for people interested in linear programming.

**en.wikipedia.org/wiki/Linear_programming** This Web page outlines the theory of linear programming.

# Statistics: The Science of Data

# PART II

What CDs are big sellers this week? When you buy a CD, the checkout scanner probably reports your choice to a company that tallies sales and reports the winners. Are there genetic differences between two related types of cancer? To find out, biologists use "microarrays" to report the activity of thousands of genes at once. Checkout scanners and microarrays produce immense amounts of *data*, numerical facts. So do opinion polls, medical studies, and even the sports pages. *Statistics* is the science of collecting, organizing, and interpreting data.

Chapters 5 and 6 concern *data analysis*, the art of seeing what data say. We learn from data by making graphs and doing calculations, guided by principles that help us decide what graphs to make, what to look for in our graphs, and what calculations are helpful based on what we see. Sometimes we want to know more: An opinion poll or a medical study looks at only some people, but we want conclusions that apply to all voters or all patients. This is called *statistical inference*, because we infer conclusions about a large group from data on a small part of the group. Chapter 7 discusses inference from beginning to end: from how to produce data when we have inference in mind to how to say just how much confidence we can have in our conclusions. Confidence, uncertainty, risk, chance–the mathematics that describes all these ideas is *probability theory*, the topic of Chapter 8. Probability is the mathematics behind statistical inference, but that's just a small part of its usefulness. ■

CHAPTER 5



# Exploring Data: Distributions

A flood of data is a prominent feature of modern society. Data are essential for making decisions in almost every area of life and work. Like other great floods, the flood of numbers threatens to overwhelm us. We must control the flood by careful organization and interpretation. A corporate database, for example, contains an immense volume of data–on employees, sales, inventories, customer accounts, equipment, taxes, and other topics. These data are useful only if we can organize them and present them so that their meaning is clear. The penalties for ignoring data can be severe–several banks have suffered billion-dollar losses from unauthorized trades in financial markets by their employees, trades that were hidden in a mass of data that the banks' management did not examine carefully.

This chapter and the next show you how to use graphs and numerical summaries to work with data. Always remember that although knowing which tools to use and how to use them is important, interpreting your work is even more important. Data come from the real world, and in the end, your goal is to use data to learn something about the real world.

Any set of data contains information about some group of **individuals**. The information is organized in **variables**.

**Individuals** — DEFINITION

**Individuals** are the objects described by a set of data. Individuals may be people, but they may also be animals or things.

**Variable** — DEFINITION

A **variable** is any characteristic of an individual. A variable can take different values for different individuals.

**FIGURE 5.1** Part of a data set as displayed by the Excel spreadsheet program.

| | A | B | C | D | E | F |
|---|---|---|---|---|---|---|
| 1 | SEX | HAND | HEIGHT | STUDY | COINS | |
| 2 | F | L | 65 | 200 | 50 | |
| 3 | M | L | 72 | 30 | 35 | |
| 4 | M | R | 62 | 95 | 35 | |
| 5 | F | L | 64 | 120 | 0 | |
| 6 | M | R | 63 | 220 | 0 | |
| 7 | F | R | 58 | 60 | 76 | |
| 8 | F | R | 67 | 150 | 215 | |
| 9 | | | | | | |

Sheet1 / Sheet2 / Sheet3

## EXAMPLE 1 ▪ Data from a Student Questionnaire

Figure 5.1 is a small part of a data set that describes the students in a large statistics class. The data come from anonymous responses to a class questionnaire. Each row records data on one *individual,* that is, one student. Each column contains the values of one *variable* for all the individuals. There are five variables. Sex (female or male) and handedness (left-handed or right-handed) are variables that are usually described as categorical or qualitative because they categorize individuals by traits and do not take numerical values. You have probably already had much experience with the usual ways to summarize categorical data (for example, proportions, pie charts, and bar graphs), so we will not take time to repeat that here.

The remaining three variables are described as measurement or quantitative variables because they do take numerical values. They are: height (inches), time spent studying (in minutes per weeknight), and "How many cents in coins (not bills) are you carrying?" Our main focus will be on variables involving numerical data.

Most data tables follow this format–each row is an individual, and each column is a variable. This data set appears in a *spreadsheet* program that has rows and columns ready for your use. Spreadsheets are commonly used to enter and transmit data, and spreadsheet programs also have functions for basic statistics.

Knowing the context of the data–that these are student responses to a class questionnaire–helps us make sense of them. For example, one student claimed to study 30,000 minutes on a typical night. We know that this is impossible!

Statistical tools and ideas help us examine data in order to describe their main features. This examination is called **exploratory data analysis**. Like an explorer crossing unknown lands, we want first to describe simply what we see. In this chapter and the next, we use both numbers and graphs to explore data. Here are two principles that provide the tactics for exploratory analysis of data.

### Exploring Data — PROCEDURE

1. Begin by examining each variable by itself. Then move on to study the relationships among the variables.
2. Begin with a graph or graphs. Then add numerical summaries of specific aspects of the data.

These principles also organize the material in Chapters 5 and 6. In this chapter, we look at data on a single variable. Chapter 6 moves on to relations among several variables. In each chapter, we first display data in graphs, then add numerical summaries.

## 5.1 Displaying Distributions: Histograms

Data analysis begins with graphical displays of the values of a single variable. For example, you may want to compare the study times claimed by female and male students. Because individual study times vary so much, we are interested in the **distribution** of study time for female and male students.

**Distribution of a Variable** DEFINITION

The **distribution** of a variable tells us what values the variable takes and how often it takes these values.

Numerical variables often take many values. A graph of the distribution is clearer if nearby values are grouped together. The most common graph of the distribution of one numerical variable is a **histogram**.

**Histogram** DEFINITION

A **histogram** is a graph of the distribution of outcomes (often divided into classes) for a single numerical variable. The height of each bar is the number of observations in the class of outcomes covered by the base of the bar. All classes should have the same width and each observation must fall into exactly one class.

### EXAMPLE 2 ■ Population Distribution

Every 10 years, the Census Bureau (www.census.gov) tries to contact every household in the United States. One of the most striking findings of the 2000 Census was the growth of the Hispanic population of the United States. Table 5.1 presents the percent of adult (age 18 and over) residents in each of the 50 states who identified themselves in the 2000 Census as "Spanish/Hispanic/Latino." The *individuals* in this data set are the 50 states. The *variable* is the percent of Hispanics in a state's population. To make a histogram of the distribution of this variable, proceed as follows:

#### Making a Histogram

**Step 1. Choose the classes.** Divide the range of the data into some reasonable number of classes of equal width. The data in Table 5.1 range from 0.7 to 42.1, so here's one way to choose classes:

$$0.0 \leq \text{percent Hispanic} < 5.0$$
$$5.0 \leq \text{percent Hispanic} < 10.0$$
$$\vdots$$
$$40.0 \leq \text{percent Hispanic} < 45.0$$

Be sure to specify the classes precisely so that each individual falls into exactly one class. A state with 4.9% Hispanic residents would fall into the first class, but a state with 5.0% falls into the second.

**TABLE 5.1 Percent of Adult Population of Hispanic Origin, by State (2000 Census)**

| State | Percent | State | Percent | State | Percent |
|---|---|---|---|---|---|
| Alabama | 1.5 | Louisiana | 2.4 | Ohio | 1.9 |
| Alaska | 4.1 | Maine | 0.7 | Oklahoma | 5.2 |
| Arizona | 25.3 | Maryland | 4.3 | Oregon | 8.0 |
| Arkansas | 2.8 | Massachusetts | 6.8 | Pennsylvania | 3.2 |
| California | 32.4 | Michigan | 3.3 | Rhode Island | 8.7 |
| Colorado | 17.1 | Minnesota | 2.9 | South Carolina | 2.4 |
| Connecticut | 9.4 | Mississippi | 1.3 | South Dakota | 1.4 |
| Delaware | 4.8 | Missouri | 2.1 | Tennessee | 2.0 |
| Florida | 16.8 | Montana | 2.0 | Texas | 32.0 |
| Georgia | 5.3 | Nebraska | 5.5 | Utah | 9.0 |
| Hawaii | 7.2 | Nevada | 19.7 | Vermont | 0.9 |
| Idaho | 7.9 | New Hampshire | 1.7 | Virginia | 4.7 |
| Illinois | 10.7 | New Jersey | 13.3 | Washington | 7.2 |
| Indiana | 3.5 | New Mexico | 42.1 | West Virginia | 0.7 |
| Iowa | 2.8 | New York | 15.1 | Wisconsin | 3.6 |
| Kansas | 7.0 | North Carolina | 4.7 | Wyoming | 6.4 |
| Kentucky | 1.5 | North Dakota | 1.2 | | |

**Step 2. Count the individuals in each class.** Here are the counts (sometimes called frequencies, since they tell how frequently values fall in a class):

| Class | Count | Class | Count | Class | Count |
|---|---|---|---|---|---|
| 0.0 to 4.9 | 27 | 15.0 to 19.9 | 4 | 30.0 to 34.9 | 2 |
| 5.0 to 9.9 | 13 | 20.0 to 24.9 | 0 | 35.0 to 39.9 | 0 |
| 10.0 to 14.9 | 2 | 25.0 to 29.9 | 1 | 40.0 to 44.9 | 1 |

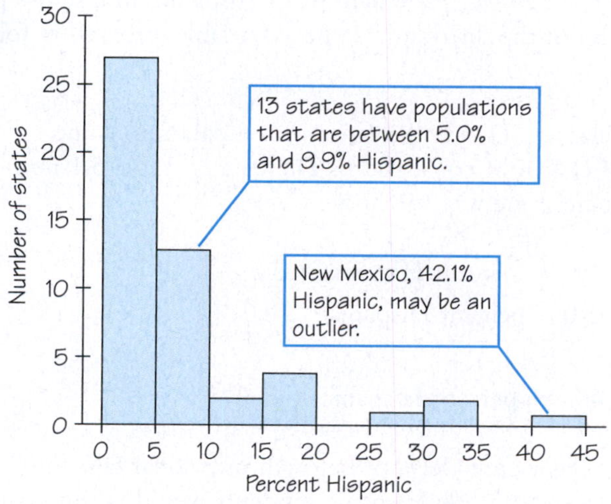


**FIGURE 5.2** Histogram of the percent of Hispanics among the adult residents of the states.

**Step 3. Draw the histogram.** First mark the scale for the variable whose distribution you are displaying on the horizontal axis. That's the percent of a state's population who are Hispanic. The scale runs from 0 to 45 because that is the span of the classes we chose. The vertical axis contains the scale of counts. Each bar represents a class. The base of the bar covers the class, and the bar height is the class count. There is no horizontal space between the bars unless a class is empty, so that its bar has height zero. Figure 5.2 is our histogram.

The bars of a histogram should cover the entire range of values of a variable. When the possible values of a variable have gaps between them, extend the bases of the bars to meet halfway between two adjacent possible values. For example, in a histogram of the ages in years of university faculty, the bars representing 25 to 29 years and 30 to 34 years would meet at 29.5.

Our eyes respond to the *area* of the bars in a histogram. Because the classes are all the same width, area is determined by height and all classes are fairly represented. There is no one right choice of the classes in a histogram. Too few classes will give a "skyscraper" graph, with all values in a few classes with tall bars. Too many will produce a "pancake" graph, with most classes having one or no observations. Neither choice will give a good picture of the shape of the distribution. You must use your judgment in choosing classes to display the shape. Statistics software will choose the classes for you. The computer's choice is usually a good one, but you can change it if you want.

## 5.2 Interpreting Histograms

Making a statistical graph is not an end in itself. The purpose of the graph is to help us understand the data. After you make a graph, always ask, "What do I see?" Once you have displayed a distribution, you can see its important features as follows.

**Outlier** DEFINITION

In any graph of data, look for the overall pattern and for striking deviations from that pattern. You can describe the overall pattern of a distribution by its shape, center, and spread. An important kind of deviation is an **outlier**, an individual value that falls outside the overall pattern.

We will soon learn how to describe center and spread numerically. For now, you can describe the center of a distribution by its middle value–with roughly half the observations taking smaller values and half taking larger values. You can give a rough description of the spread of a distribution by giving the *smallest and largest values.*

### EXAMPLE 3 ■ Describing a Distribution

Look again at the histogram in Figure 5.2. **Shape:** The distribution has a *single peak*, which represents states in which less than 5% of adults are Hispanic. The distribution is *skewed to the right.* Most states have no more than 10% Hispanics, but some

(*Corbis/Punchstock.*)

states have much higher percentages, so that the graph trails off to the right. **Center:** Table 5.1 shows that about half the states have less than 4.7% Hispanics among their adult residents and half have more. So the midpoint of the distribution is close to 4.7%. **Spread:** The spread is from about 0% to 42%, but only four states fall above 20%. **Outliers:** Arizona, California, New Mexico, and Texas stand out. Whether these are outliers or just part of the long right tail of the distribution is a matter of judgment. There is no universal rule for calling an observation an outlier. Once you have spotted possible outliers, look for an explanation. Some outliers are due to mistakes, such as typing 4.2 as 42. Other outliers point to the special nature of some observations. These four states are heavily Hispanic by history and location. ■

When you describe a distribution, concentrate on the main features. Look for major peaks, not for minor ups and downs in the bars of the histogram. Look for clear outliers, not just for the smallest and largest observations. Look for rough **symmetry** or clear **skewness**.

Some variables have distributions with predictable shapes. For example, incomes, house prices and other money amounts usually have right-skewed distributions because there are always CEOs and celebrities well to the right of the rest of us! A simple test designed to measure basic achievement may yield a left-skewed distribution because most students will cluster together with high scores, but there are usually still a few people who perform low (due to lack of attendance or effort) and give the distribution a tail stretching out to the left.

### Skewed Distribution — DEFINITION

A distribution is **skewed to the right** if the longer tail of the histogram is on the right side. (Because positive numbers are on the right side of a number line, such a distribution is also referred to as positively-skewed.) For example, see Figure 5.2 or Figure 5.7.

Similarly, a distribution is **skewed to the left** (or negatively-skewed) if the longer tail is on the left side.

Other distributions, however, are *symmetric* and may have little or no skewness. For example, the distribution of heights (or handspans) in an adult population may look like two hills of equal size (visualize a two-humped camel) if males cluster around one value and females cluster around another. A more common and more important symmetric shape is the bell-shaped histogram (Figure 5.3) yielded by many standardized tests as well as by many biological measurements (such as, height, length of thigh bone, and so on) on specimens from the same species and sex.

### Symmetric Distribution — DEFINITION

A distribution is **symmetric** if the right and left sides of the histogram are approximately mirror images of each other. (That is, if you "folded" the histogram in half, the left half would fall roughly onto the right half.) For example, see Figure 5.3.

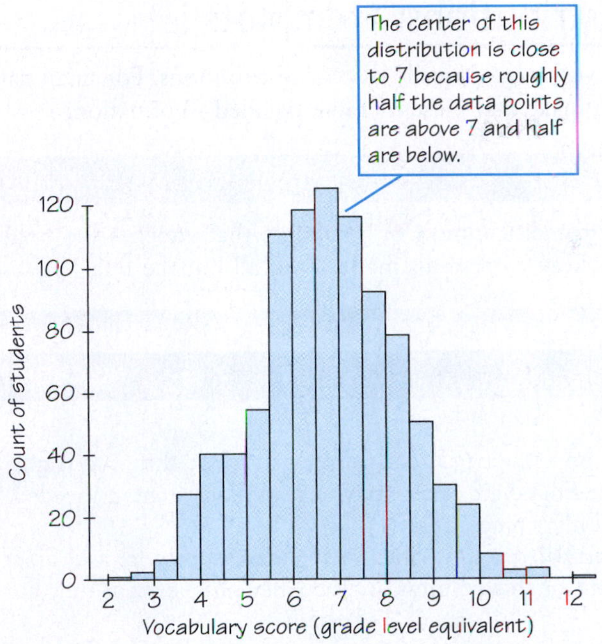


**FIGURE 5.3** Histogram of the Iowa Test of Basic Skills vocabulary scores for 947 seventh-grade students.

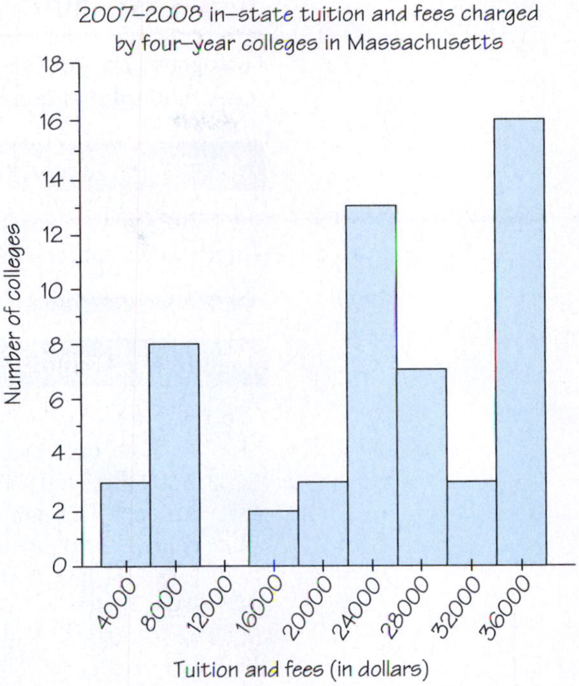


**FIGURE 5.4** Histogram of the tuition and fees charged by four-year colleges in Massachusetts.

## EXAMPLE 4 ■ Iowa Test Scores

Figure 5.3 displays the scores of all 947 seventh-grade students in the public schools of Gary, Indiana, on the vocabulary part of the Iowa Test of Basic Skills. The distribution is *single-peaked* and *symmetric.* In mathematics, the two sides of symmetric patterns are exact mirror images, but real-life data are almost never exactly symmetric. We are content to describe Figure 5.3 as symmetric. The center (half above, half below) is close to 7. This is a seventh-grade reading level. The scores range from 2.0 (second-grade level) to 12.1 (twelfth-grade level). ■

## EXAMPLE 5 ■ College Tuition

Jeanna plans to attend college in her home state of Massachusetts. She looks up the tuition and fees for the 2007–2008 academic year for all 55 four-year colleges in Massachusetts (omitting art schools and other special colleges). Figure 5.4 is a histogram of the data. For example, the tallest bar tells us there are 16 colleges charging between $34,000 and $38,000. As is often the case, we can't call this irregular distribution either symmetric or skewed. It does show two separate *clusters* of colleges, 11 with tuition less than $10,000 and the remaining 44 costing more than $16,000. Clusters suggest that two types of individuals are mixed in the data set. In fact, the histogram distinguishes the 11 state colleges in Massachusetts from the 44 private colleges, which charge much more. ■

## 5.3 Displaying Distributions: Stemplots

Histograms are not the only way to graphically display distributions. For small data sets, a **stemplot** is quicker to make and presents more detailed information.

**Stemplot** DEFINITION

A **stemplot** is a display of the distribution of a variable that attaches the final digits of the observations as leaves on stems made up of all but the final digit.

**Making a Stemplot** PROCEDURE

To make a **stemplot**:

1. Separate each observation into a *stem* consisting of all but the final (rightmost) digit and a *leaf*, the final digit. Stems may have as many digits as needed, but each leaf contains only a single digit.
2. Write the stems in a vertical column with the smallest at the top, and draw a vertical line at the right of this column. Include all stems, even if they are not used.
3. Write each leaf in the row to the right of its stem, in increasing order out from the stem.

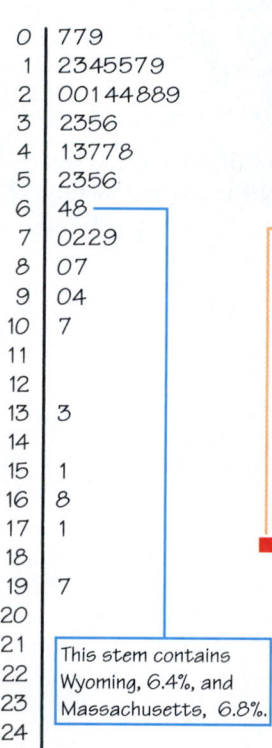

| | |
|---|---|
| 0 | 779 |
| 1 | 2345579 |
| 2 | 00144889 |
| 3 | 2356 |
| 4 | 13778 |
| 5 | 2356 |
| 6 | 48 |
| 7 | 0229 |
| 8 | 07 |
| 9 | 04 |
| 10 | 7 |
| 11 | |
| 12 | |
| 13 | 3 |
| 14 | |
| 15 | 1 |
| 16 | 8 |
| 17 | 1 |
| 18 | |
| 19 | 7 |
| 20 | |
| 21 | |
| 22 | |
| 23 | |
| 24 | |
| 25 | 3 |

This stem contains Wyoming, 6.4%, and Massachusetts, 6.8%.

High 32.0 32.4 42.1

**FIGURE 5.5** Stemplot of the percent of Hispanics among the adult residents of the states.

### EXAMPLE 6 ■ Making a Stemplot

For the "percent Hispanic" percents in Table 5.1, take the whole-number part of the percent as the stem and the final digit (tenths) as the leaf. The Massachusetts entry, 6.8%, has stem 6 and leaf 8. Wyoming, at 6.4%, places leaf 4 on the same stem. These are the only observations on this stem. We then arrange the leaves in order, as 48, so that 6|48 is one row in the stemplot. Figure 5.5 is the complete stemplot for the data in Table 5.1. To save space, we left out California, Texas, and New Mexico, which have stems 32 and 42. These observations are listed as "High" below the stemplot. ■

If we rotate Figure 5.5 a quarter-turn counterclockwise, the stemplot would look like a histogram (of a distribution skewed to the right). Comparing the stemplot in Figure 5.5 with the histogram in Figure 5.2 reveals the strengths and weaknesses of stemplots. The stemplot, unlike the histogram, preserves the actual value of each observation. But you can choose the classes in a histogram, whereas the classes (the stems) of a stemplot are forced on you. Whether the large number of classes in Figure 5.5 is an improvement over Figure 5.2 is a matter of taste. Stemplots do not work well for large data sets like the 947 Iowa Test scores in Figure 5.3, because each stem must hold a large number of leaves.

When the observed values have many digits, it is often best to *round* the numbers to just a few digits before making a stemplot. For example, a stemplot of data like

3.468 2.567 2.981 1.095 . . .

would have very many stems and no leaves or just one leaf on most stems. You can round these data to

3.5 2.6 3.0 1.1 . . .

before making a stemplot.

Graphical summaries are good for analyzing the shape of distribution of values. To answer precise questions about features of a dataset, such as its center, however, it helps to have numerical summaries as well. We explore this next.

## 5.4 Describing Center: Mean and Median

What kind of gas mileage do you get with the new cars in the Environmental Protection Agency's "midsized cars" category? Table 5.2 gives the city and highway gas mileage for a representative sample of Model Year 2008 midsized cars.

**TABLE 5.2 Fuel Economy (Miles per Gallon) for Model Year 2008 Vehicles**

| Model | City Mileage | Highway Mileage |
|---|---|---|
| Acura RL | 16 | 24 |
| BMW 550i | 15 | 23 |
| Chevrolet Malibu | 22 | 30 |
| Dodge Avenger | 21 | 30 |
| Hyundai Elantra | 24 | 33 |
| Lexus ES 350 | 19 | 27 |
| Mercury Milan | 20 | 29 |
| Mitsubishi Galant | 20 | 27 |
| Nissan Sentra | 21 | 29 |
| Nissan Versa | 27 | 33 |
| Pontiac Grand Prix | 18 | 28 |
| Toyota Camry | 21 | 31 |
| Toyota Prius | 48 | 45 |

We start with graphs. Figure 5.6 is a *dotplot* of the city mileages of the 13 cars in the sample of midsized cars. As is often the case when there only a few observations, the shape of the distribution is irregular. The most striking feature is a high outlier on the right end of the dotplot. Upon closer examination, to see if the outlier value may be a typographical error, we see that the Toyota Prius is the only hybrid gas-electric car in the sample.

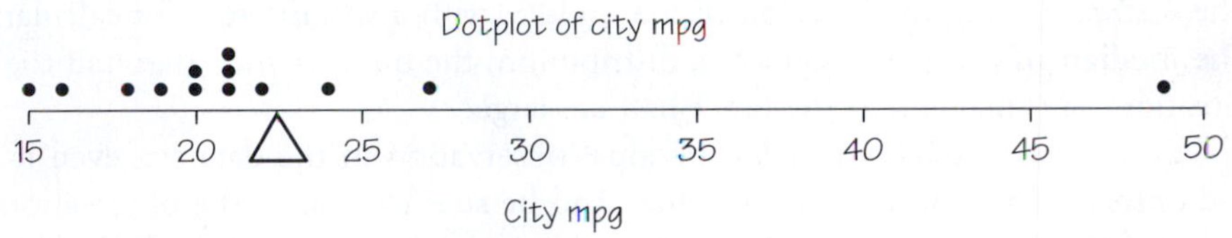


**FIGURE 5.6** Dotplot of the city gas mileages of the sample of midsized cars. The Toyota Prius is an outlier.

Numerical summaries make the comparison we want more specific. Numerical description of a distribution begins with a measure of its center. The two most common measures of center are the **mean** and the **median**. Basically, the mean is the arithmetic "average value" and the median is the "middle value." We need to explore the precise procedures for calculating these measures and observe how they behave differently.

## Finding the Mean $\bar{x}$ — PROCEDURE

To find the **mean** of a set of observations, add their values and divide by the number of observations. If the $n$ observations are $x_1, x_2, \ldots, x_n$, their mean is

$$\bar{x} = \frac{x_1 + x_2 + \cdots + x_n}{n}$$

The bar over the $x$ indicates the mean of all the $x$-values. Pronounce the mean $\bar{x}$ as "$x$-bar." This notation is very common. When writers who are discussing data use $\bar{x}$ or $\bar{y}$, they are talking about a mean.

One way to visualize the value of the mean of a dataset is to imagine where the fulcrum would have to be placed for its dotplot to "balance." This metaphor tells us the mean is always between the largest and smallest values, and by visual inspection we can further estimate that the balance point appears to be somewhere between 20 and 25. Let's see what exact value the formula yields.

## EXAMPLE 7 ■ Calculating the Mean

The mean city mileage for the 13 midsized cars in Table 5.2 is

$$\begin{aligned}\bar{x} &= \frac{x_1 + x_2 + \cdots + x_n}{n}\\ &= \frac{16 + 15 + 22 + 21 + 24 + 19 + 20 + 20 + 21 + 27 + 18 + 21 + 48}{13}\\ &= \frac{292}{13} = 22.5 \text{ miles per gallon (mpg)}\end{aligned}$$

We said that the Toyota Prius may not belong with the other cars. If we exclude the Prius, the mean city mileage drops to $\frac{244}{12} = 20.3$ mpg. The single outlier adds more than 2 mpg to the mean city mileage. This illustrates an important weakness of the mean as a measure of center: *The mean is sensitive to the influence of a few extreme observations.* These may be outliers, but a skewed distribution that has no outliers will also pull the mean toward its long tail.

We have used the middle of a distribution as an informal measure of center. The *median* is the formal version of the middle, with a specific rule for calculation. The **median** $M$ is the midpoint of a distribution, the number such that half the observations are smaller and the other half are larger.

Be sure to write down each individual observation in the data set, even if several observations repeat the same value. And be sure to arrange the observations in order of size before locating the median. Note that the recipe $(n + 1)/2$ gives the *position* of the median in the ordered list of observations, *not* the median itself.

## Finding the Median $M$ — PROCEDURE

To find the median of a distribution:

1. Arrange all observations in increasing order (from smallest to largest).
2. If the number of observations is *odd*, the median $M$ is the center observation in the ordered list.
3. If the number of observations is *even*, the median $M$ is the average of the two center observations in the ordered list.

## EXAMPLE 8 ■ Calculating the Median

Since we're exploring the gas mileage cars get on the road, you might notice the connection that just as a median divides a road into two halves (with opposite directions of travel), a median divides a dataset into two halves! To find the median city mileage for 2008 midsized cars, arrange the data in increasing order:

15 16 18 19 20 20 **21** 21 21 22 24 27 48

The median is the bold **21**, which you can find by eye–there are 6 observations to the left and 6 to the right. Or visualize (or construct) a long paper strip divided into 13 equal-sized squares, where the sorted values are written into the squares. When the strip is folded in half, the fold line falls on the median!

What happens if we drop the Toyota Prius? The remaining 12 cars have city mileages:

15 16 18 19 20 **20** **21** 21 21 22 24 27

Because the number of observations $n = 12$ is even, there is no single center observation. There is a center *pair* of observations (20 and 21) that has 5 observations to its left and 5 to its right. The median $M$ is the mean of the center pair, which is $(20 + 21)/2 = 20.5$.

You see that the median resists the influence of extreme observations better than the mean does. A very high value like the Toyota Prius is simply one observation to the right of center and removing it hardly changed the median at all. In fact, removing an extreme outlier can leave the median completely unchanged while significantly changing the mean. The *Mean and Median* applet (at www.whfreeman.com/fapp8e) is an excellent way to compare the resistance of $M$ and $\bar{x}$. See Applet Exercises 1 and 2.

(*Dan Forer/Beateworks/Corbis.*)

The median and mean are the most common measures of the center of a distribution. The mean and median of a symmetric distribution are close together. If the distribution is exactly symmetric, the mean and median are exactly the same. In a skewed distribution, the mean is generally farther out in the long tail than is the median (see Figure 5.7). For example, the distribution of house prices is skewed to the right. There are many moderately priced houses and a few very expensive mansions. The mansions pull the mean up but do not affect the median. The mean price of existing single-family houses in the U.S. sold in September 2007 was \$257,800, but the median price for these same houses was only \$211,700.

Another common numerical summary of a distribution is the **mode**–the most frequently occurring value. Like the mean and median, it is a measure of location

## SPOTLIGHT 5.1 Think!

Computer software will do all of your statistical graphs and calculations for you. But you can think and the computer can't. Here are some examples of the importance of thinking as well as calculating.

**What's not on the plot matters** Abraham Wald (1902–1950), like many statisticians, worked on war problems during World War II. Wald invented some statistical methods that were military secrets until the war ended. Here is one of his simpler contributions. Asked where extra armor should be added to airplanes, Wald studied the location of enemy bullet holes in planes returning from combat. He plotted the locations on an outline of the plane. As data accumulated, most of the outline filled up. Put the armor in the few spots with no bullet holes, said Wald. That's where bullets hit the planes that didn't make it back.

**What are your units?** Not paying attention to units of measurement can get you into trouble. In 1999, the *Mars Climate Orbiter* spacecraft burned up in the Martian atmosphere. It was supposed to be 93 miles (150 kilometers) above the planet, but was in fact only 35 miles (57 kilometers) up. It seems that Lockheed Martin, which built the *Orbiter*, specified important measurements in English units (for example, pounds and miles). The National Aeronautics and Space Administration team who directed the spacecraft thought the numbers were in metric system units (for example, kilograms and kilometers). There went $125 million!

**What are we counting?** A news report says that 22% of American high school students smoke. That sounds serious. But what counts as "smoking"? It turns out that this is the percent who smoked at least once in the past month. If we say that a smoker is someone who smoked on at least 20 of the past 30 days, only 9.7% of high school students smoke. The number you get depends on how you define your variable—and you can make a problem sound more or less serious by changing the definition. Interest groups do this all the time, so pay attention.

for a distribution. However, the mode is not necessarily a good measure of center. Consider a skewed histogram with its maximum height at one end and a long tail on the other end, or a symmetric histogram with a tall peak at each end and a valley in between!

### The Mode — DEFINITION

The **mode** is the most frequently occurring value in a set of numerical observations.

The mode of the city mileage numbers in Table 5.2 would be 21, because it occurs 3 times and no other value occurs more than twice. Some datasets, however, may have two or more values that "tie" for the honor of being a mode, such as the values 27, 29, 30, and 33 in the Table 5.2 highway mileage numbers. We can also identify the mode(s) of a histogram. For example, Figure 5.4 does not give the raw data of individual values, but we can say that the modal class is $34,000–$38,000.

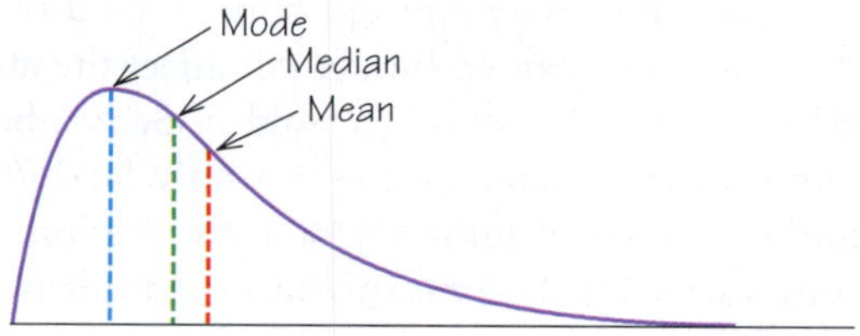


**FIGURE 5.7** This smooth curve describes a right-skewed distribution with one mode. The mean is pulled toward the long tail more than the median is.

## 5.5 Describing Spread: The Quartiles

The mean and median provide two different measures of the center of a distribution. But a measure of center alone can be misleading. Two neighborhoods with median house price $193,000 are very different if one has both mansions and modest homes and the other has little variation among houses. We are interested in the spread or variability of house prices as well as their centers. *The simplest useful numerical description of a distribution consists of both a measure of center and a measure of spread.*

The simplest way to measure spread is with the **range**, which is the difference between the smallest and largest observations. For example, the percents of Hispanics in the states are as low as 0.7% (Maine and West Virginia) and as high as 42.1% (New Mexico), so the range would be 42.1% − 0.7% = 41.4%. Also, the range of the city mileage numbers in Table 5.2 is 48 − 15 = 33 mpg. The range tells us the full span of the data, but it may be greatly affected by one or more outliers. Without the Toyota Prius, the preceding answer becomes 27 − 15 = 12 mpg.

### The Range — DEFINITION

The **range** is a measure of spread of a set of observations. It is obtained by subtracting the smallest observation from the largest observation.

We can improve our description of spread by also looking at the spread of the middle half of the data. The first and third *quartiles* separate out the middle half. At the end of the first quarter of a football game, one quarter of the game is complete. Similarly, the first quartile of a distribution or dataset is the point that exceeds one-quarter (or 25%) of the values. The third quartile is the point that exceeds three quarters (or 75%) of the values. (The second quartile exceeds two quarters (or 50%) of the values, and so is equivalent to the median.) To make the idea of quartiles more exact, we need a rule to find them:

### Calculating the Quartiles $Q_1$ and $Q_3$ — PROCEDURE

To calculate the **quartiles**:

1. Use the median to split the data set into two halves–an upper half and a lower half.
2. The **first quartile** $Q_1$ is the median of the lower half.
3. The **third quartile** $Q_3$ is the median of the upper half.

## EXAMPLE 9 ■ Calculating Quartiles

The city mileages of the 12 gasoline-powered midsized cars, after sorting:

15 16 18 19 20 20 21 21 21 22 24 27

The first quartile is the median of the 6 observations in the lower half, so $Q_1 = 18.5$. Similarly, the third quartile is the median of the upper half: $Q_3 = 21.5$.

For an example with an odd number of observations, try the city mileages of all 13 midsized cars in Table 5.2. Below are the mileages in increasing order with **bold** used to denote the median which will be excluded to form two equal-sized groups:

15 16 18 19 20 20 **21** 21 21 22 24 27 48

Ignoring the bold **21**, we find the quartiles by finding the median of each half of the dataset: $Q_1 = 18.5$ and $Q_3 = 23$.

Some software packages or calculators use a slightly different rule to find the quartiles, so computer results may be a bit different from your own work. Don't worry about this. The differences will always be too small to be important.

## 5.6 The Five-Number Summary and Boxplots

We started by using the smallest and largest observations to indicate the spread of a distribution. These single observations tell us little about the distribution as a whole, but they give information about the tails of the distribution that is missing if we know only $Q_1$, $M$, and $Q_3$. To get a quick summary of both center and spread, combine all five numbers.

### The Five-Number Summary

DEFINITION

The **five-number summary** of a distribution consists of the smallest observation, the first quartile, the median, the third quartile, and the largest observation, written in order from smallest to largest. In symbols, the five-number summary is

$$\text{Minimum} \quad Q_1 \quad M \quad Q_3 \quad \text{Maximum}$$

These five numbers offer a reasonably complete description of center and spread. For the 12 gasoline-powered midsized cars, you can verify that the five-number summary for city gas mileage is

15 18.5 20.5 21.5 27

and is

23 27 29 30.5 33

for highway gas mileage.

The five-number summary breaks the dataset into four equal groups with equal numbers of observations. A **boxplot** (sometimes called "box and whisker plot") can visually represent the spread of the data across these groups from the five-number summary. Figure 5.8 shows boxplots for both city and highway gas mileages for midsized cars.

### Boxplot

DEFINITION

A **boxplot** is a graph of the five-number summary.

- A central box spans the quartiles $Q_1$ and $Q_3$.
- A line somewhere in the middle of the box marks the median $M$ of the data set.
- Lines extend from the box out to the smallest and largest observations.

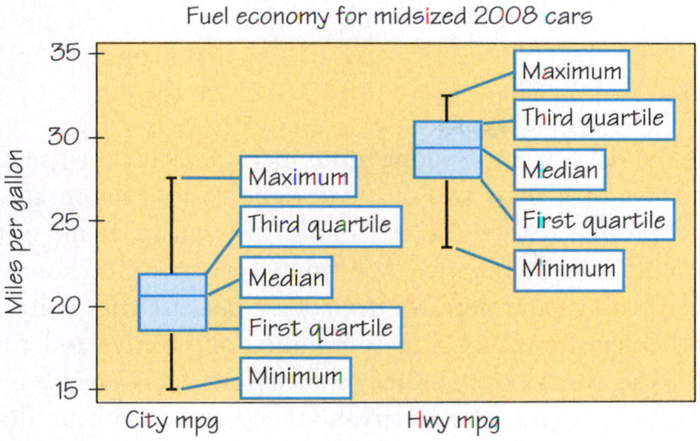


**FIGURE 5.8** Boxplots of the highway and city gas mileages for cars classified as midsized by the Environmental Protection Agency. These boxplots are drawn vertically, but it is equally correct to draw them horizontally.

Because boxplots show less detail than histograms or stemplots, they are best used for side-by-side comparison of more than one distribution, as in Figure 5.8. When you look at a boxplot, first locate the median, which marks the center of the distribution. Then look at the spread. The quartiles show the spread of the middle half of the data, and the extremes (the smallest and largest observations) show the spread of the entire data set. So for non-hybrid cars, is there really much of a difference in gas mileages between city and highway? From the boxplots, we see at once that highway mileages are noticeably higher than city mileages: The maximum city mileage reaches only the first quartile of highway mileages! We also see that the spread of highway mileages is roughly the same as the spread of city mileages. Boxplots can also be an indicator of a distribution's skewness, so we have gotten a lot of mileage from this vehicle for exploratory data analysis.

Be aware that some calculators and software packages offer an alternative option for boxplots in which the lines go to the furthest values within 1.5 box lengths of the quartiles, but do not automatically go out to the minimum and maximum values. The advantage of this is that any values beyond these can be individually marked as outliers.

## 5.7 Describing Spread: The Standard Deviation

Although the five-number summary is the most generally useful numerical description of a distribution, it is not the most common. That distinction belongs to the combination of the mean with the **standard deviation**. The mean, like the median, is a measure of center. The standard deviation, like the quartiles and extremes in the five-number summary, measures spread. The standard deviation and its close relative, the *variance*, measure spread by looking at how far the observations are from their mean.

(*AP Photo/Robert E. Klein.*)

### EXAMPLE 10 ■ Understanding the Standard Deviation

Starting October 10, 2007, the English rock band Radiohead allowed people to choose how much they wanted to pay to download a digital copy of the group's new album *In Rainbows*. According to a survey reported by Associated Press, about 60% of Americans worldwide who downloaded the album chose to pay nothing, but the remaining 40% voluntarily paid an average payment of $8.05. A sample (sorted in increasing order) of the dollar amounts Americans in the "paying group" paid is:

3 4 5 7 10 12 15

Figure 5.9 displays the data as points along a number line, with their mean marked by an asterisk (*). The arrows mark two of the deviations from the mean. These deviations show how spread out the data are around their mean. Some of the deviations are positive and some are negative. We won't get a useful measure of spread by totaling up these positive and negative deviations, because they will always sum to zero! Squaring the deviations makes these numbers all positive and a reasonable measure of spread is the average of the squared deviations. This average is called the *variance*. The variance is large if the observations are widely spread around their mean. It is small if the observations are all close to the mean.

But the variance does not have meaningful units. With the Radiohead sales data measured in dollars, the variance of the purchase prices has units of "squared dollars." Taking the square root of the variance yields the standard deviation, which gets us back to dollars. The standard deviation has other uses as well, as discussed in Section 5.9.

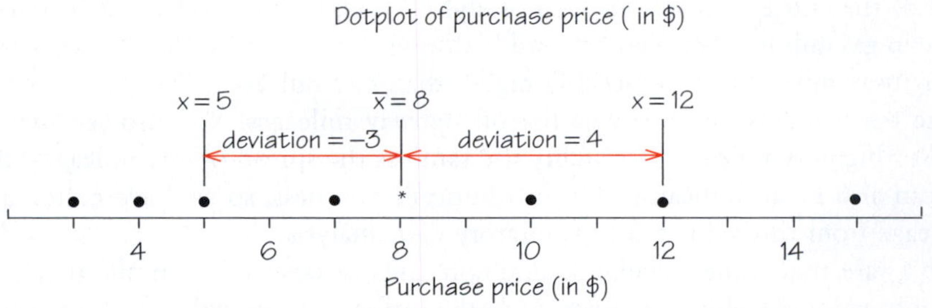


**FIGURE 5.9** The variance and standard deviation measure spread by looking at the deviations of observations from their mean.

## The Standard Deviation s — DEFINITION

The **standard deviation** is a kind of "standard" or average amount that observed data values deviate from their mean. More precisely, it is the square root of the mean of the squared deviations, except that the mean involves dividing by $n-1$ instead of $n$. (It turns out $n-1$ makes the formula more accurate, but the justification is beyond the scope of this book.) In symbols, the standard deviation $s$ of $n$ observations $x_1, x_2, \cdots, x_n$ is

$$s = \sqrt{\frac{(x_1 - \bar{x})^2 + (x_2 - \bar{x})^2 + \cdots + (x_n - \bar{x})^2}{n - 1}}$$

In practice (especially with large data sets), you will use your calculator or software to obtain the standard deviation with a single command. However, going through all the steps of an example first will give you understanding about how the standard deviation works.

## EXAMPLE 11 ■ Calculating the Standard Deviation

To find the standard deviation of the 7 purchase prices, first find the mean:

$$\overline{x} = \frac{3 + 4 + 5 + 7 + 10 + 12 + 15}{7} = \frac{56}{7} = 8 \text{ dollars}$$

The deviations shown in Figure 5.9 are the starting point for calculating the standard deviation.

| Observations $x_i$ | Deviations (of observation from mean) $x_i - \overline{x}$ | Squared Deviations $(x_i - \overline{x})^2$ |
|---|---|---|
| 3 | $3 - 8 = -5$ | $(-5)^2 = 25$ |
| 4 | $4 - 8 = -4$ | $(-4)^2 = 16$ |
| 5 | $5 - 8 = -3$ | $(-3)^2 = 9$ |
| 7 | $7 - 8 = -1$ | $(-1)^2 = 1$ |
| 10 | $10 - 8 = 2$ | $2^2 = 4$ |
| 12 | $12 - 8 = 4$ | $4^2 = 16$ |
| 15 | $15 - 8 = 7$ | $7^2 = 49$ |
| | | sum = 120 |

**FIGURE 5.10** This table shows a step-by-step approach to form the building blocks for the calculation of standard deviation, whether done by hand, calculator, or spreadsheet.

The variance is the sum of the squared deviations divided by 1 less than the number of observations, so it would be $\frac{120}{7-1} = 20$. The standard deviation is the square root of the variance, and so we obtain $s = \sqrt{20} = 4.47$ dollars. This value can be considered large for this context and range, which suggests that people vary quite a bit in what they consider a fair price for music when they have the power to choose.

If the 7 observations still had a mean of \$8, but were spread out further, their deviations from 8 would be larger and the standard deviation would then be even larger. To explore this relationship, redo the calculation after decreasing \$5 to \$2 and increasing \$15 to \$18. The mean remains the same, but the resulting standard deviation will be larger since the numbers are more spread out.

## SPOTLIGHT 5.2 Calculating Standard Deviation

While the formula in the Definition box for standard deviation has conceptual clarity and a straightforward implementation (as in Figure 5.10), it can be tedious to apply to large datasets. Even with the most basic calculator, you'll get the same answer faster using this more computationally-oriented formula:

$$\sqrt{\frac{(x_1^2 + x_2^2 + \cdots + x_n^2) - n(\overline{x})^2}{n-1}}.$$

If you have a *scientific calculator*, put it into a "STAT MODE" if required, clear out any old data, then enter your data one number at a time (after each number, press your calculator's data-entry button—it may say DATA or have a symbol such as [Σ+] or [M+]). Once the data is entered, you can find the standard deviation by hitting the key labeled something like [σn−1] or [σxn−1] or [s].

If you have a *graphing calculator* in the TI-83/84+ family (and you already used (STAT)→ EDIT to enter one variable of quantitative data in a column, say, L1), then hit this sequence of buttons: [2ND] (CATALOG) (it's above the "0" key)→stdDev (ENTER) [2ND] (L1) (it's above the "1" key) (ENTER). If you use the alternative command sequence (STAT)→CALC→1-Var Stats [2ND] (L1) (ENTER), you will obtain not only the standard deviation $s_x$ but also other descriptive statistics, including the five-number summary. Keystrokes for other specific models can be found online, for example, at http://www.geocities.com/calculatorhelp/.

More important than the details of hand calculation are the properties that determine the usefulness of the standard deviation:

- $s$ measures spread about the mean $\bar{x}$. Use $s$ to describe the spread of a distribution only when you use $\bar{x}$ to describe the center.
- $s = 0$ only when there is *no spread*. This happens only when all observations have the same value. Otherwise $s > 0$. As the observations become more spread out about their mean, $s$ gets larger.
- $s$ has the same units of measurement as the original observations. For example, if you measure metabolic rates in calories, both the mean $\bar{x}$ and the standard deviation $s$ are also in calories.
- The use of squared deviations makes $s$ even more sensitive than $\bar{x}$ to a few extreme observations. For example, dropping the Toyota Prius from our list of midsized cars cuts the standard deviation of city mileages more than half, from 8.3 mpg with the Prius to 3.3 mpg without it. Distributions with outliers and strongly skewed distributions have large standard deviations. The number $s$ does not give much helpful information about such distributions.

We now have a choice between two descriptions of the center and spread of a distribution: the five-number summary, or $\bar{x}$ and $s$. Because $\bar{x}$ and $s$ are sensitive to extreme observations, they can be misleading when a distribution is strongly skewed or has outliers. In fact, because the two sides of a skewed distribution have different spreads, no single number such as $s$ describes the spread well. The five-number summary, with its two quartiles and two extremes, does a better job.

**Choosing a Summary** RULE

The five-number summary is usually better than the mean and standard deviation for describing a skewed distribution or a distribution with outliers. Use $\bar{x}$ and $s$ only for reasonably symmetric distributions that are free of outliers.

Although the standard deviation is widely used, it is not a natural or convenient measure of the spread of a distribution. The real reason for the popularity of the standard deviation is that it is the natural measure of spread for **normal distributions**, an important class of distributions that we will meet next.

Do remember that a graph gives the best overall picture of a distribution. Numerical measures of center and spread report specific facts about a distribution, but they do not describe its entire shape. Numerical summaries do not disclose the presence of clusters, for example. *Always start with a graph of your data.*

## 5.8 Normal Distributions

We now have a kit of graphical and numerical tools for describing distributions. What is more, we have a clear strategy for exploring data on a single numerical variable:

1. Always plot your data: make a graph, usually a histogram, dotplot, or a stemplot.

2. Look for the overall pattern (shape, center, spread) and for striking deviations such as outliers.
3. Calculate a numerical summary to give some description of center and spread.

Here is one more step to add to this strategy:

4. Sometimes the overall pattern of a large number of observations is so regular that we can describe it by a smooth curve.

Figure 5.3 is a histogram of the Iowa Test vocabulary scores of 947 seventh-grade students. Like most histograms from national standardized tests, the histogram is symmetric, is single-peaked, and has a distinctive bell shape. In Figure 5.11, we draw a smooth curve through the tops of the histogram bars to describe the shape. The curve is an idealized description of the distribution. It gives a compact picture of the overall pattern of the data but ignores minor irregularities as well as any outliers. The curve in Figure 5.11 is a *normal curve.* A distribution whose shape is described by a normal curve is a *normal distribution.*

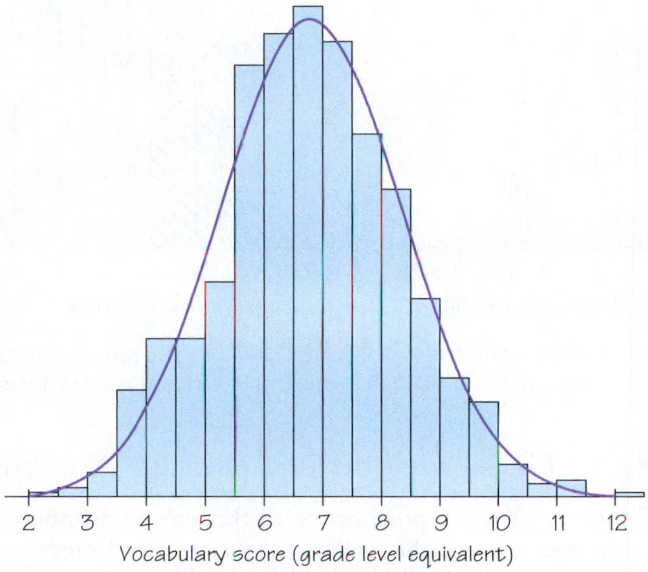


**FIGURE 5.11** Histogram of the vocabulary scores of 947 seventh-grade students in Gary, Indiana. The smooth curve shows the overall shape of the distribution.

**Normal Distribution** DEFINITION

The *distribution* of a variable tells us what values the variable takes and how often it takes these values. A **normal distribution** is described by a *normal* curve. The area under the curve above any interval of values tells us what proportion of all values of the variable lie in that interval. The total area under the curve is exactly 1.

## EXAMPLE 12 ■ From Histogram to Normal Curve

You can think of a normal curve as a smoothed-out histogram when there is symmetry and one mode. Our eyes respond to the *areas* of the bars in a histogram. The bar areas represent proportions of the observations. Figure 5.12a is a copy of Figure 5.11

with the leftmost bars shaded. The area of the shaded bars in Figure 5.12a represents the students with vocabulary scores 6.0 or lower. There are 287 such students, who make up the proportion 287/947 = 0.303 of all Gary seventh graders.

Now look at the curve drawn through the bars. In Figure 5.12b, the area under the curve to the left of 6.0 is shaded. We know that the areas of histogram bars represent proportions of all the observations, but we don't worry about the actual total area. Note that all the bars together represent 100% of the students and so we treat the total area under the normal curve as 1 = 100%. Now areas under the curve actually *are* proportions of the observations. This curve is a normal curve. The shaded area under the normal curve in Figure 5.12b is the proportion of students with score 6.0 or lower. This area turns out to be 0.293, only 0.010 away from the histogram result. You see that areas under the normal curve give quite good approximations of areas given by the histogram. ■

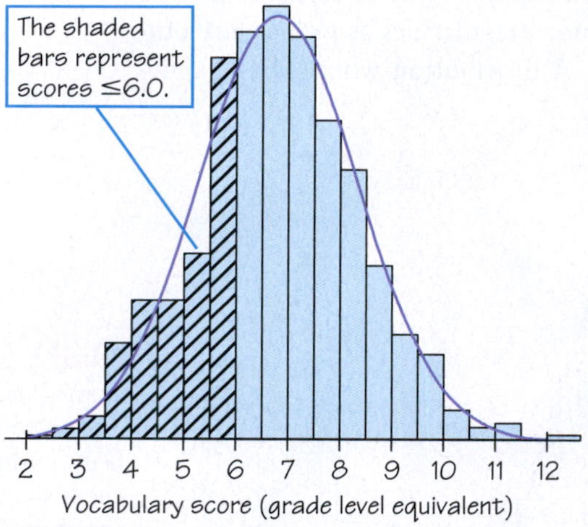


**FIGURE 5.12a** The proportion of scores less than or equal to 6.0 from the histogram is 0.303.

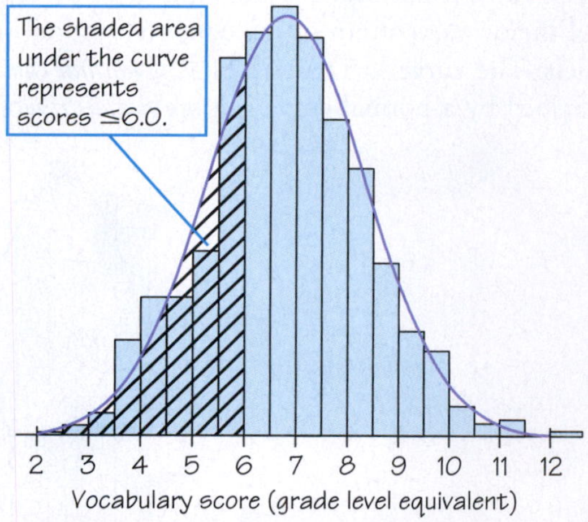


**FIGURE 5.12b** The proportion of scores less than or equal to 6.0 from the normal curve is 0.293.

## EXAMPLE 13 ■ Heights of American Women

The normal curve is a good approximation of the real-life distribution for a variety of biological measures (height, weight, heart rate, blood pressure, and so on), when examined for a particular species and gender. Figure 5.13 shows the heights of American women between the ages 18–24. The proportion of young women who are between 60 inches (5 feet) and 65 inches tall is given by the area under the curve between 60 and 65. This area is about 0.54, so approximately 54% of these women are between 60 and 65 inches tall. ■

Normal distributions play a large role in statistics, but they are rather special and not at all "normal" in the everyday sense of being typical or natural. Normal curves can be specified exactly by an equation, but we will be content with pictures. Figure 5.14 shows two normal curves. All normal curves have the same overall *shape.* They are symmetric and bell-shaped, with tails that fall off rapidly from a central peak. The *center* of the normal curve is the center of the distribution in several senses. It is the mean of the distribution. It is also the median since half the observations (half the area under the curve) lie on each side of the center.

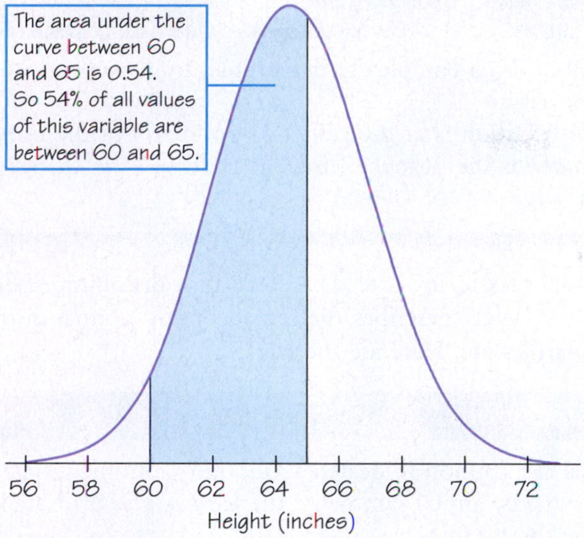
The area under the curve between 60 and 65 is 0.54. So 54% of all values of this variable are between 60 and 65.
56
58
60
62
64
66
68
70
72
Height (inches)

**FIGURE 5.13** Areas under a normal curve describe a normal distribution. This normal curve describes the distribution of heights of American women.

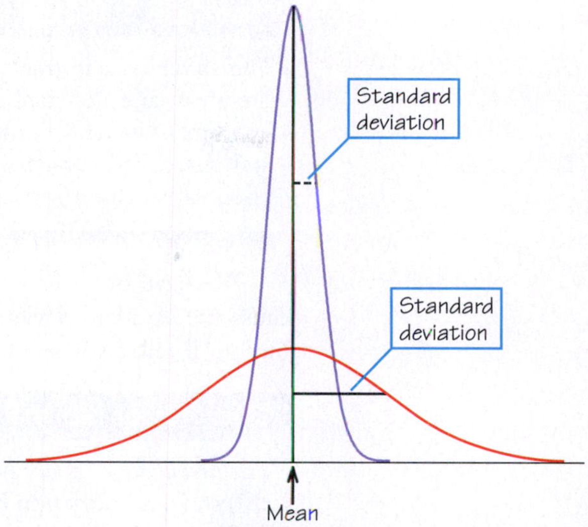
Standard deviation
Standard deviation
Mean

**FIGURE 5.14** Two normal curves with the same mean but different standard deviations. The standard deviation for each curve is the distance from the center (the mean) to the change-of-curvature point on one side of the center.

What about the *spread* of a normal curve? Normal curves have the special property that their spread is completely determined by a single number, the standard deviation. We have learned how to calculate the standard deviation from a set of observations. For normal distributions, the standard deviation, like the mean, can be found directly from the curve. Here's how. Imagine that you are skiing down a mountain that has the shape of a normal curve. At first, you descend at an ever-steeper angle as you go out from the peak:

Fortunately, before you find yourself going straight down, the slope now begins to grow flatter rather than steeper as you continue downhill:

*The points at which this change of curvature takes place are located one standard deviation from the mean on either side.* You can feel the change as you run your finger along a normal curve, and so find the standard deviation. Try it with the two normal curves in Figure 5.14. Normal curves with the same standard deviation have exactly the same shape. Changing the mean just moves the center of the curve to a new location. Changing the standard deviation changes the spread of the curve, as Figure 5.14 shows.

## Mean and Standard Deviation of Normal Distributions — DEFINITION

The shape of a normal distribution is completely determined by two numbers, the mean and the standard deviation.

The *mean* of a normal distribution is at the center of symmetry of the normal curve. The *standard deviation* is the distance from the center to the change-of-curvature points on either side.

We have often used the quartiles to indicate the spread of a distribution. Because the standard deviation completely describes the spread of any normal distribution, it tells us where the quartiles are. Here are the facts.

## Quartiles of Normal Distributions — DEFINITION

The *first quartile* of any normal distribution is located about 0.67 (a bit more than 2/3) standard deviation below the mean; by symmetry, the *third quartile* is located 0.67 standard deviation above the mean.

## EXAMPLE 14 ■ Heights of American Women

The distribution of heights of American women (aged 18–24) is approximately normal with mean 64.5 inches and standard deviation 2.5 inches. Figure 5.15 shows this normal curve. The quartiles are 0.67 standard deviation, or

$$(0.67)(2.5 \text{ inches}) = 1.7 \text{ inches}$$

away from the mean. The first quartile is 64.5 − 1.7, or 62.8 inches. The third quartile is 64.5 + 1.7, or 66.2 inches. The middle 50% of women's heights lie approximately between 62.8 inches and 66.2 inches. These numbers are exact for the normal distribution with mean 64.5 inches and standard deviation 2.5 inches, but only approximately true for the actual heights of the women because real-life distributions of biological measurements such as heights are only approximately normal. ■

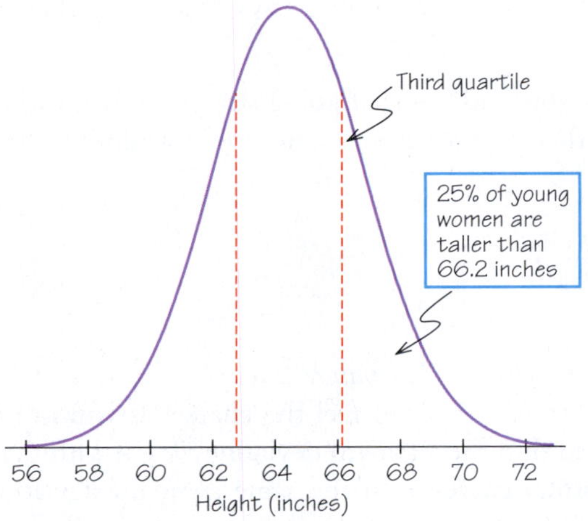


**FIGURE 5.15** The quartiles of a normal distribution are located 0.67 standard deviation on either side of the mean. For this normal curve, the mean is 64.5 inches and the standard deviation is 2.5 inches.

## SPOTLIGHT 5.3 Density Estimation

Smooth curves that describe the overall pattern of distributions of data are called *density curves.* Normal curves are one type of density curve. There are many other types used for different purposes. However, you don't have to call for a specific type such as the normal curves. Clever software for "density estimation" will calculate a density curve to describe any set of observations you give it.

The figure shows a strongly skewed distribution, the survival times of 72 guinea pigs in a medical experiment. Two graphs of the distribution are overlaid: a histogram and a density curve produced by software from the data. The histogram and density curve agree on the overall shape and on the "bumps" in the long right tail. The density curve shows a higher single peak as a main feature of the distribution. The histogram divides the observations near the peak between two bars, thus reducing the height of the peak. Because density estimators don't depend on dividing the data into classes, as histograms do, many statisticians prefer them when they need a picture of a distribution.

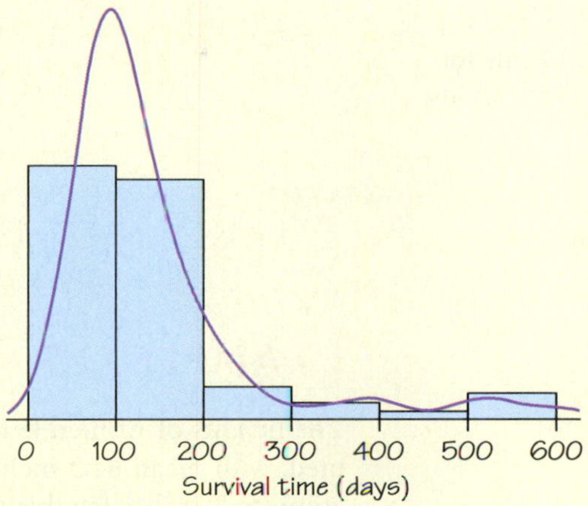


Density estimation software fits this smooth curve to data on the survival time of 72 guinea pigs.

Why are the normal distributions important in statistics? Here are two reasons. First, normal distributions are good models or approximations for some distributions of *real data*. Distributions that are often close to normal include scores on tests taken by many people (such as SAT exams and many psychological tests), repeated careful measurements of the same quantity, and characteristics of biological populations (such as heights of young women and yields of corn). Second, normal distributions are good approximations to the results of many kinds of *chance outcomes*, such as tossing a coin many times. We will return to normal curves in Chapter 8 when we study probability, the mathematics of chance. However, many sets of data do not follow a normal distribution. Most income distributions, for example, are skewed to the right and so are not normal.

## 5.9 The 68–95–99.7 Rule

Because any particular normal distribution is completely determined by its mean and standard deviation, it is not surprising that all normal distributions are the same in terms of what proportion of observations are any given number of standard deviations from the mean. Here is an important rule based on this fact.

### The 68–95–99.7 Rule for Normal Distributions — RULE

According to the **68–95–99.7 rule**, in any normal distribution:

- about 68% of the observations fall within 1 standard deviation of the mean.
- about 95% of the observations fall within 2 standard deviations of the mean.
- about 99.7% of the observations fall within 3 standard deviations of the mean.

Figure 5.16 illustrates the 68–95–99.7 rule. By remembering these three numbers, you can think about normal distributions without making detailed calculations. For more detailed information, you can use tables or software that give areas under normal curves, but the 68–95–99.7 rule is adequate for our purposes.

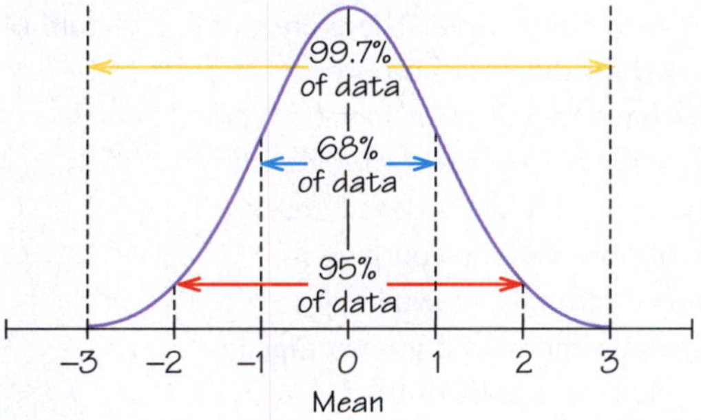


**FIGURE 5.16** The 68–95–99.7 rule for normal distributions.

## EXAMPLE 15 ■ Heights of American Women

The heights of women between the ages of 18 and 24 are roughly normally distributed, with mean 64.5 inches and standard deviation 2.5 inches. Two standard deviations is 5 inches for this distribution. The 95 part of the 68–95–99.7 rule says that the middle 95% of young women are between 64.5 − 5 and 64.5 + 5 inches tall, that is, between 59.5 inches and 69.5 inches. This fact is exactly true for an exactly normal distribution. It is approximately true for the heights of young women because the distribution of heights is approximately normal.

The other 5% of American women have heights outside the range from 59.5 to 69.5 inches. Because the normal distributions are symmetric, half of these women are on the tall side and half on the short side. So the tallest 2.5% of young women are taller than 69.5 inches.

■

## EXAMPLE 16 ■ SAT Reasoning Test Scores

The distribution of scores on tests such as the SAT college entrance examination is close to normal. Scores on each of the three sections (math, critical reading, writing) of the SAT are adjusted so that the mean score is about $\mu = 500$ and the standard deviation is about $\sigma = 100$. This information allows us to answer many questions about SAT scores.

▸ *How high must a student score to fall in the top 25%?*

The third quartile is (0.67)(100) = 67 points above the mean. So scores above 567 are in the top 25%.

▸ *What percent of scores fall between 200 and 800?*

Scores of 200 and 800 are 3 standard deviations on either side of the mean. The 99.7 part of the 68–95–99.7 rule says that 99.7% of all scores lie in this range. (In fact, 200 and 800 are the lowest and highest scores that are reported on the SAT. The few scores higher than 800 are reported as 800.)

▸ *What percent of scores are above 700?*

A score of 700 is 2 standard deviations above the mean. By the 95 part of the 68–95–99.7 rule, 95% of all scores fall between 300 and 700 and 5% fall below 300 or above 700. Because normal curves are symmetric, half of this 5% are above 700. So a score above 700 places a student in the top 2.5% of test-takers.

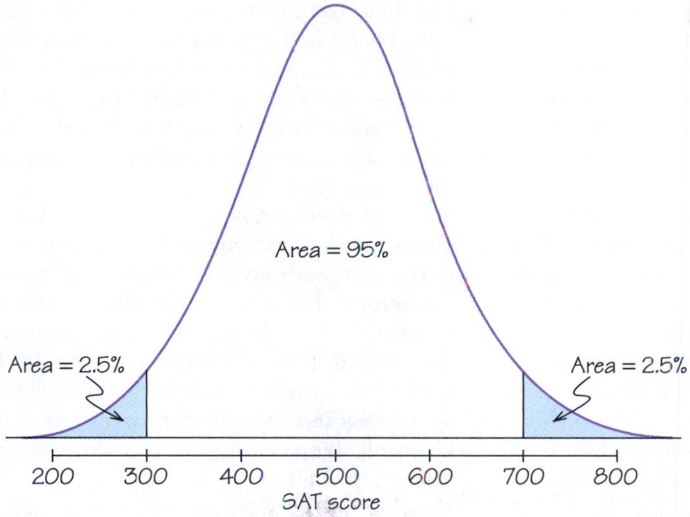


**FIGURE 5.17** Using the 68–95–99.7 rule to find the percent of SAT scores that are above 700. This normal curve has mean 500 and standard deviation 100.

Sketching a normal curve with the points 1, 2, and 3 standard deviations from the mean marked can help you use the 68–95–99.7 rule. Figure 5.17 shows the distribution of SAT scores with the areas needed to find the percent of scores above 700. Note that the tails of Figure 5.17, like any bell curve, technically stretch out forever in both directions (even as the amount of far-away area becomes vanishingly small). This is another reminder that the bell-curve is a very good, but not perfect, model of reality. We know that real-life SAT subtest scores are scaled so that they do not go beyond 200 or 800.

The 68–95–99.7 rule allows you to find selected areas under a normal curve—areas for outcomes bounded by 1, 2, or 3 standard deviations away from the mean. You can use software, a graphing calculator, or the *Normal Curve* applet to find any area under a normal curve. See Applet Exercises 3 to 5 for use of the applet.

## REVIEW VOCABULARY

**Boxplot** A graph of the five-number summary. A box spans the quartiles, with an interior line marking the median. Lines extend out from this box to the extreme high and low observations. (p. 162)

**Distribution** The pattern of outcomes of a variable. The distribution describes what values the variable takes and how often each value occurs. (p. 151)

**Exploratory data analysis** The practice of using graphs and numbers to examine data for overall patterns and special features, without necessarily seeking answers to specific questions. (p. 150)

**Five-number summary** A summary of a distribution that gives the smallest observation, first quartile, median, third quartile, and largest observation, in that order. (p. 162)

**Histogram** A graph of the distribution of outcomes (often divided into classes) for a single numerical variable. The height of each bar is the number of observations in the class of outcomes covered by the base of the bar. All classes should have the same width and each observation must fall into exactly one class. (p. 151)

**Individuals** The people, animals, or things described by a data set. (p. 149)

**Mean** The ordinary arithmetic average of a set of observations. To find the mean, add all the observations and divide the sum by the number of observations summed. (p. 158)

**Median** The middle of a set of ordered observations. Half the observations fall below the median and half fall above. (p. 158)
**Mode** The most frequently occurring value in a set of numerical observations. (p. 159)
**Normal distributions** A family of distributions that describe how often a variable takes its values by areas under a curve. The normal curves are symmetric and bell-shaped. A specific normal curve is completely described by giving its mean and its standard deviation. (p. 167)
**Outlier** A data point that falls clearly outside the overall pattern of a set of data. (p. 153)
**Quartiles** The first quartile ($Q_1$) of a distribution is the point with one quarter of the observations falling below it; the third quartile ($Q_3$) is the point with three quarters below it. $Q_1$ is the median of the lower half of the observations; $Q_3$ is the median of the upper half. (p. 161)
**Range** Measure of spread obtained by subtracting the smallest observation from the largest observation. (p. 161)
**68–95–99.7 rule** In any normal distribution, 68% of the observations lie within 1 standard deviation on either side of the mean, 95% lie within 2 standard deviations of the mean, and 99.7% lie within 3 standard deviations of the mean. (p. 171)
**Skewed distribution** A distribution in which observations on one side of the median extend notably farther from the median than do observations on the other side. In a right-skewed distribution, the larger observations extend farther to the right of the median than the smaller observations extend to the left. (p. 154)
**Standard deviation** A measure of the spread of a distribution about its mean as center. It is the square root of the average squared deviation of the observations from their mean. (p. 163)
**Standard deviation of a normal curve** The standard deviation of a normal curve is the distance from the mean to the change-of-curvature points on either side. (p. 170)
**Stemplot** A display of the distribution of a variable that attaches the final digits of the observations as leaves on stems made up of all but the final digit. (p. 156)
**Symmetric distribution** A distribution with a histogram or stemplot in which the part to the left of the median is roughly a mirror image of the part to the right of the median. (p. 154)
**Variable** Any characteristic of an individual. (p. 149)

## ✓ SKILLS CHECK

**1.** Here are the first lines of a professor's data set at the end of a mathematics course:

| Name | Major | Points | Grade |
|---|---|---|---|
| ADVANI, SURA | COMM | 397 | B |
| BARTON, DAVID | HIST | 323 | C |
| BOAZ, JUDAH | BIOL | 446 | A |
| CHIU, SUN | PSYC | 405 | B |
| DAVIS, LAUREN | PSYC | 461 | A |

The individuals in these data are

**(a)** the students.
**(b)** the total points.
**(c)** the course grades.

*Figure 5.4 is a histogram of the tuition and fee charges for the 2007–2008 academic year for 55 four-year colleges in Massachusetts. Exercises 2 and 3 are based on this histogram.*

**2.** The number of colleges with tuition and fee charges covered by the leftmost bar in the histogram is ______ .

**3.** The leftmost bar in the histogram covers tuition and fee charges ranging from about

**(a)** $2000 to $6000.
**(b)** $3000 to $5000.
**(c)** $4000 to $8000.

**4.** The distribution in Figure 5.2 is best described as _______-skewed.

**5.** You look at real estate ads for houses in Sarasota, Florida. There are many houses ranging from $200,000 to $400,000 in price. The few houses on the water, however, have prices up to $15 million. The distribution of house prices will be

**(a)** skewed to the left.
**(b)** roughly symmetric.
**(c)** skewed to the right.

**6.** Here are the systolic blood pressures of 10 randomly chosen adults:

| | | | | |
|---|---|---|---|---|
| 147 | 141 | 120 | 124 | 127 |
| 132 | 98 | 112 | 120 | 128 |

In a stemplot of these scores, the largest stem is _____ .

**7.** For Figure 5.5, interpret the meaning of 10|7.

**(a)** 10 states have 7% Hispanic population.
**(b)** 7 states have 10% Hispanic population.
**(c)** 1 state has 10.7% Hispanic population.

**8.** The mean blood pressure of the 10 adults in Exercise 6 is _________ .

**9.** The median of the blood pressures in Exercise 6 is

**(a)** 127.
**(b)** 125.5.
**(c)** 124.9.

**10.** If a single-peaked distribution is skewed to the right, the median is generally to the __________ of the mean.

**11.** The mode of the 10 blood pressures in Exercise 6 is
(a) 147.
(b) 120.
(c) 2.

**12.** Between the first quartile and the third quartile lie _________ percent of the observations in a distribution.

**13.** Which of these is *not* in a five-number summary?
(a) Median
(b) Minimum
(c) Mean

**14.** The five-number summary of the 10 blood pressures in Exercise 6 is _____. (Remember to list the 5 numbers in increasing order).

**15.** The standard deviation of the 10 blood pressures in Exercise 6 (use your calculator) is
(a) 13.23.
(b) 13.95.
(c) 194.6.

**16.** You have data on the weights (measured in grams) of 5 crackers. The correct units for the standard deviation of these weights are: ___________________ .

**17.** What are all the values that a standard deviation $s$ can possibly take?
(a) $0 \leq s$
(b) $0 \leq s \leq 1$
(c) $-1 \leq s \leq 1$

**18.** To completely specify the shape of a normal distribution, you must give its mean and its _________ .

**19.** The scale of scores on an IQ test is approximately normal with mean 100 and standard deviation 15. The organization MENSA, which calls itself "the high IQ society," requires an IQ score of 130 or higher for membership. What percent of adults would qualify for membership?
(a) 95%
(b) 5%
(c) 2.5%

**20.** The length of human pregnancies from conception to birth varies according to a distribution that is approximately normal with mean 266 days and standard deviation 16 days. We can expect that about _________ percent of all completed pregnancies are between 234 and 298 days.

# CHAPTER 5 EXERCISES

■ Challenge ◆ Discussion

| Make and Model | Vehicle Type | Transmission Type | Number of Cylinders | City mpg | Highway mpg |
|---|---|---|---|---|---|
| Mazda MX-5 | Two-seater | Manual | 4 | 22 | 27 |
| Toyota Yaris | Subcompact | Automatic | 4 | 29 | 35 |
| Bentley Azure | Compact | Automatic | 12 | 9 | 15 |
| Audi S4 Avant | Small Station Wagon | Manual | 8 | 13 | 20 |

Some exercises require use of a calculator (or software or Internet applet) that will find mean and standard deviation from keyed-in data.

**1.** Above is a small part of a data set that describes the fuel economy (in miles per gallon) of year 2008 model motor vehicles:
(a) What are the individuals in this data set?
(b) For each individual, what variables are given? For which of these variables would a histogram be helpful? (That is, which variables do not yield categorical data)?

## 5.1 Displaying Distributions: Histograms

## 5.2 Interpreting Histograms

◆ **2.** Figure 5.18 is a histogram of the lengths of words used in Shakespeare's plays. Because there are so many words in the plays, the vertical axis of the graph is the percent that are of each length, rather than the count. What is the overall shape of this distribution? What does this shape say about word lengths in Shakespeare? Do you expect other authors to have word-length distributions of the same general shape? Why?

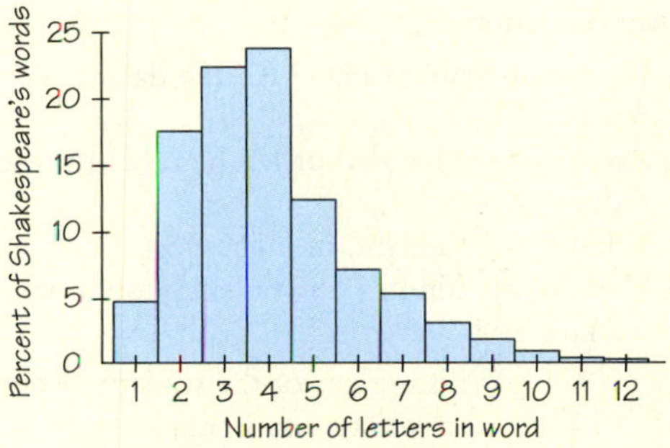


**FIGURE 5.18** Histogram of the lengths of words used in Shakespeare's plays, for Exercise 2.

## TABLE 5.3 Carbon Dioxide Emissions, Metric Tons per Person

| Country | $CO_2$ | Country | $CO_2$ | Country | $CO_2$ | Country | $CO_2$ |
|---|---|---|---|---|---|---|---|
| Algeria | 2.3 | Germany | 10.0 | Mexico | 3.7 | South Africa | 8.1 |
| Argentina | 3.9 | Ghana | 0.2 | Morocco | 1.0 | Spain | 6.8 |
| Australia | 17.0 | India | 0.9 | Myanmar | 0.2 | Sudan | 0.2 |
| Bangladesh | 0.2 | Indonesia | 1.2 | Nepal | 0.1 | Tanzania | 0.1 |
| Brazil | 1.8 | Iran | 3.8 | Nigeria | 0.3 | Thailand | 2.5 |
| Canada | 16.0 | Iraq | 3.6 | Pakistan | 0.7 | Turkey | 2.8 |
| China | 2.5 | Italy | 7.3 | Peru | 0.8 | Ukraine | 7.6 |
| Colombia | 1.4 | Japan | 9.1 | Philippines | 0.9 | United Kingdom | 9.0 |
| Congo | 0.0 | Kenya | 0.3 | Poland | 8.0 | United States | 19.9 |
| Egypt | 1.7 | Korea, North | 9.7 | Romania | 3.9 | Uzbekistan | 4.8 |
| Ethiopia | 0.0 | Korea, South | 8.8 | Russia | 10.2 | Venezuela | 5.1 |
| France | 6.1 | Malaysia | 4.6 | Saudi Arabia | 11.0 | Vietnam | 0.5 |

◆ **3.** Suppose that you and your friends emptied your pockets of coins and recorded the year marked on each coin. Would you expect the histogram for the distribution of dates to be skewed to the left or right? Explain your answer and make a sketch of this histogram.

**4.** Make a histogram of the city gas mileages of the midsized cars in Table 5.2 on page 157. Use classes with width 5 mpg. Do you prefer the histogram or the representation in Figure 5.6 of the same data? Why?

**5.** Burning fuels in power plants or motor vehicles emits carbon dioxide ($CO_2$), which contributes to global warming. Table 5.3 displays $CO_2$ emissions per person from countries with population at least 20 million.

**(a)** Why do you think we choose to measure emissions per person rather than total $CO_2$ emissions for each country?
**(b)** Display the data of Table 5.3 in a histogram. Describe the shape, center, and spread of the distribution. Which countries appear to be outliers?

■ **6.** A survey of a large college class asked the following questions:

1. Are you female or male? (In the data, male = 0, female = 1.)
2. Are you right-handed or left-handed? (In the data, right = 0, left = 1.)
3. What is your height, in inches?
4. How many minutes do you study on a typical weeknight?

Figure 5.19 shows histograms of the student responses, in scrambled order and without scale markings. Which histogram goes with each variable? Explain your reasoning.

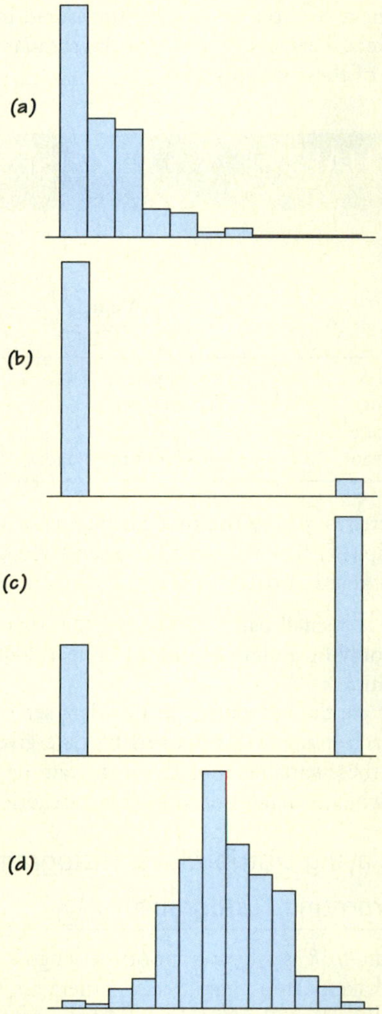


**FIGURE 5.19** Match each histogram with its variable, for Exercise 6.

## 5.3 Displaying Distributions: Stemplots

**7.** The population of the United States is aging, though less rapidly than in other developed countries. Figure 5.20 is a stemplot of the percents of residents aged 65 and over in the 50 states, according to the 2000 census. The stems are whole percents and the leaves are tenths of a percent.

| | |
|---|---|
| 5 | 7 |
| 6 | |
| 7 | |
| 8 | 5 |
| 9 | 679 |
| 10 | 6 |
| 11 | 02233677 |
| 12 | 0011113445789 |
| 13 | 00012233345568 |
| 14 | 034579 |
| 15 | 36 |
| 16 | |
| 17 | 6 |

FIGURE 5.20 Stemplot of the percentages of residents aged 65 and over in the 50 states, for Exercise 7.

**(a)** There are two outliers: Alaska has the lowest percent of older residents, and Florida has the highest. What are the percents for these two states?
**(b)** Ignoring Alaska and Florida, describe the shape, center, and spread of this distribution.

**8.** People with diabetes must monitor and control their blood glucose level. The goal is to maintain "fasting plasma glucose" between about 90 and 130 milligrams per deciliter (mg/dl). Here are the fasting plasma glucose levels for 18 diabetics enrolled in a diabetes control class, five months after the end of the class:

| | | | | | |
|---|---|---|---|---|---|
| 78 | 103 | 141 | 148 | 172 | 255 |
| 95 | 112 | 145 | 153 | 172 | 271 |
| 96 | 134 | 147 | 158 | 200 | 359 |

Round these values to the nearest 10 and then drop the zero. For example, 141 rounds to 14 and 158 rounds to 16. Make a stemplot of the rounded data. Describe the main features of the distribution. Are there outliers? How well is the group as a whole achieving the goal for controlling glucose levels?

**9.** The Survey of Study Habits and Attitudes (SSHA) is a psychological test that evaluates college students' motivation, study habits, and attitudes toward school. A private college gives the SSHA to 18 of its incoming first-year women students. Their scores are (sorted):

| | | | | | |
|---|---|---|---|---|---|
| 101 | 115 | 129 | 140 | 154 | 165 |
| 103 | 126 | 137 | 148 | 154 | 178 |
| 109 | 126 | 137 | 152 | 165 | 200 |

Make a stemplot of these data. The overall shape of the distribution is irregular, as often happens when only a few observations are available. Are there any outliers? About where is the center of the distribution (the score with half the scores above it and half below)? What is the spread of the scores (ignoring any outliers)?

**10.** In 1798 the English scientist Henry Cavendish measured the density of the earth in a careful experiment with a torsion balance. In sorted order here are his 29 measurements of the same quantity (the density of the earth relative to that of water) made with the same instrument. [S. M. Stigler, Do robust estimators work with real data? *Annals of Statistics*, 5 (1977): 1055–1078.]

| | | | | | |
|---|---|---|---|---|---|
| 4.88 | 5.29 | 5.36 | 5.47 | 5.58 | 5.68 |
| 5.07 | 5.29 | 5.39 | 5.50 | 5.61 | 5.75 |
| 5.10 | 5.30 | 5.42 | 5.53 | 5.62 | 5.79 |
| 5.26 | 5.34 | 5.44 | 5.55 | 5.63 | 5.85 |
| 5.27 | 5.34 | 5.46 | 5.57 | 5.65 | |

Make a stemplot of the data. Describe the distribution: Is it approximately symmetric or distinctly skewed? Are there gaps or outliers?

**11.** Here is a stemplot for percentage of live births to unmarried mothers for each state in the United States in 2006. (*Source*: Centers for Disease Control Web site)

| | |
|---|---|
| 1 | 9 |
| 2 | 4 |
| 2 | 89 |
| 3 | 222223344444 |
| 3 | 5556677778888999 |
| 4 | 00011111224 |
| 4 | 5669 |
| 5 | 13 |

**(a)** Explain how and why there are repeated stems.
**(b)** Describe the shape of the distribution.

## 5.4 Describing Center: Mean and Median

■ **12.** In Malay, the expression for the *mean* is *sama rata*, which roughly translates as "same level." To understand this cultural and conceptual connection, take some poker chips (or other equal-sized, stackable objects) and make stacks with 3, 7 and 8 chips. Redistribute chips among the stacks until they are at the same level and explain how this relates to the mean. (*Optional extension*: How might such redistribution enter into discussions of social justice?)

**13.** Refer to the data and the stemplot from Exercise 9:

| | | | | | |
|---|---|---|---|---|---|
| 101 | 115 | 129 | 140 | 154 | 165 |
| 103 | 126 | 137 | 148 | 154 | 178 |
| 109 | 126 | 137 | 152 | 165 | 200 |

**(a)** Find the mean score from the formula for the mean. That is, add the 18 scores, record the sum, and divide by 18.
**(b)** Your stemplot of the scores suggests that the score 200 is an outlier. Use your calculator to find the mean for the 17 observations that remain when you drop the outlier. How does the outlier change the mean?

**14.** The Major League Baseball career and single-season home run records are held by Barry Bonds of the San Francisco Giants. Here are Bonds's home run totals from 1986 (his first year) through 2007:

| | | | | | | | |
|---|---|---|---|---|---|---|---|
| 16 | 25 | 24 | 19 | 33 | 25 | 34 | 46 |
| 37 | 33 | 42 | 40 | 37 | 34 | 49 | 73 |
| 46 | 45 | 45 | 5 | 26 | 28 | | |

(Lucy Nicholson/Reuters/Corbis.)

**(a)** Make a stemplot of the data. Are there any outliers?
**(b)** Find his career mean and median number of home runs. How do these change when you drop 73? What general fact about the mean and median does your result illustrate?

◆ **15.** The distribution of income in the United States is skewed to the right. According to a Census Bureau report, the mean and median incomes of American households were \$48,201 and \$66,570 in 2006. Which of these numbers is the mean and which is the median? Explain your reasoning.

◆ **16.** Which team is #1? In addition to polls of coaches and journalists, rankings from six computer programs, which have various ways to value factors such as the quality of the opponent played, determine the Bowl Championship Series (BCS) Standings in major college football.

**(a)** At the end of the 2007 regular season, Hawaii (the only undefeated team) received these computer rankings: 12th, 8th, 14th, 10th, 8th, 13th. The BCS formula throws out the high and low of the six computer rankings and uses the mean of the remaining four ranks. Find this mean.
**(b)** Why do you think the high and low values are excluded from the mean? Is your reason connected to why the median is sometimes preferred to the mean?

**17.** Make up an example of a small set of data for which the mean lies in the top 25% of the observations.

■ **18.** According to the 1.5 × *IQR* rule explained in Exercise 29, which countries in Table 5.3 are suspected outliers? Based on your histogram (Exercise 5), do you agree with the rule's suggestions about which countries are and are not outliers?

## 5.5 Describing Spread: The Quartiles

## 5.6 The Five-Number Summary and Boxplots

**19.** The stemplot in Figure 5.20 (p. 177) displays the distribution of the percents of residents aged 65 and over in the 50 states. Stemplots help you find the five-number summary because they arrange the observations in increasing order. Give the five-number summary of this distribution.

**20.** In chronological order, here are the percents of the popular vote won by each successful candidate in the last 15 presidential elections, starting with 1948:

| | | | | |
|---|---|---|---|---|
| 49.6 | 55.1 | 57.4 | 49.7 | 61.1 |
| 43.4 | 60.7 | 50.1 | 50.7 | 58.8 |
| 53.9 | 43.2 | 49.2 | 47.9 | 51.2 |

**(a)** Make a stemplot of the winners' percents.
**(b)** What is the median percent of the vote won by the successful candidate in presidential elections?
**(c)** Call an election a landslide if the winner's percent falls at or above the third quartile. Find the third quartile. Which elections were landslides?
**(d)** Find the range.

**21.** Figure 5.4 is a histogram of the tuition and fees charged by the 55 four-year colleges in the state of Massachusetts. Here are those charges (in dollars), arranged in increasing order. [Data for 2007–2008, from the College Board Web site, www.collegeboard.com.]

| | | | | |
|---|---|---|---|---|
| 5799 | 5864 | 5992 | 6034 | 6124 |
| 6168 | 6210 | 8595 | 8732 | 8840 |
| 9924 | 16080 | 17750 | 20000 | 21330 |
| 21850 | 22073 | 22500 | 22950 | 23600 |
| 23755 | 24075 | 24250 | 24617 | 25748 |
| 25755 | 25850 | 25942 | 25990 | 26080 |
| 26250 | 27485 | 27497 | 28302 | 28440 |
| 29810 | 31899 | 32865 | 32896 | 34186 |
| 34830 | 34986 | 34994 | 34998 | 35142 |
| 35418 | 35670 | 35674 | 35702 | 35940 |
| 36232 | 36550 | 36645 | 36690 | 36700 |

Find the five-number summary and make a boxplot. What distinctive feature of the histogram do these summaries miss? Remember that numerical summaries are not a substitute for looking at the data.

**22.** Find the five-number summary of Cavendish's measurements of the density of the earth in Exercise 10. How is the symmetry of the distribution reflected in the five-number summary?

**23.** Table 5.3 gives carbon dioxide ($CO_2$) emissions per person for countries with population at least 20 million. The distribution is strongly skewed to the right. The United States and several other countries appear to be high outliers. Give the five-number summary. Explain why this summary suggests that the distribution is right-skewed.

**24.** Find the five-number summary of the data from Exercise 8.

◆ **25.** Figure 5.21 shows boxplots of the incomes of a large sample of people who have a high school diploma but no further education and another large group of people with a bachelor's degree but no higher degree. The data come from a Census Bureau survey, so that they represent all people aged 25 to 64 in the United States. Because there are a few extremely high incomes, the boxplot leaves out the highest 5% in each group. Based on the plot, compare the distributions of income for these two levels of education. Comment on both center and spread.

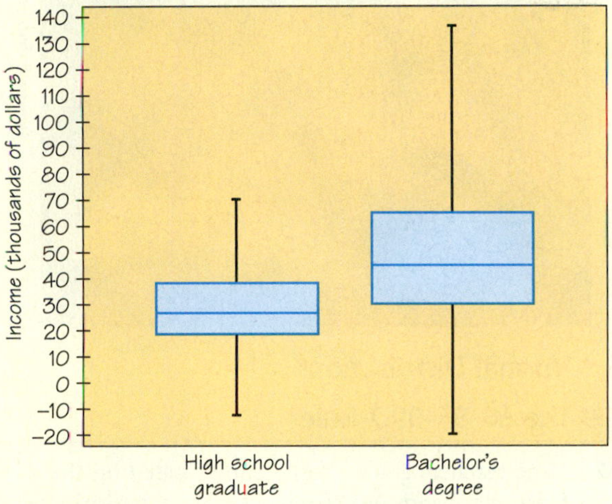


FIGURE 5.21 Boxplots comparing the incomes (in thousands of dollars) of people aged 25 to 64 years who worked full time for two levels of education. Because the highest incomes in any large group are very high indeed, the plot omits the top 5% of incomes in each group.

**26.** The data for Figure 5.21 include the incomes of 14,959 people whose highest level of education is a bachelor's degree.

**(a)** What is the position of the median in the ordered list of incomes (1 to 14,959)? From the boxplot, about what is the median income of people with a bachelor's degree?
**(b)** What is the position of the first and third quartiles in the ordered list of incomes for these people? About what are the numerical values of $Q_1$ and $Q_3$?

■ **27.** How much oil the wells in a given field will ultimately produce is key information in deciding whether to drill more wells. Here are the estimated total amounts of oil recovered from 64 wells in the Devonian Richmond Dolomite area of the Michigan basin, in thousands of barrels. [J. Marcus Jobe and Hutch Jobe, A statistical approach for additional infill development, *Energy Exploration and Exploitation*, 18 (2000): 89–103.]

| | | | | |
|---|---|---|---|---|
| 2.0 | 18.5 | 34.6 | 47.6 | 69.5 |
| 2.5 | 20.1 | 34.6 | 49.4 | 69.8 |
| 3.0 | 21.3 | 35.1 | 50.4 | 79.5 |
| 7.1 | 21.7 | 36.6 | 51.9 | 81.1 |
| 10.1 | 24.9 | 37.0 | 53.2 | 82.2 |
| 10.3 | 26.9 | 37.7 | 54.2 | 92.2 |
| 12.0 | 28.3 | 37.9 | 56.4 | 97.7 |
| 12.1 | 29.1 | 38.6 | 57.4 | 103.1 |
| 12.9 | 30.5 | 42.7 | 58.8 | 118.2 |
| 14.7 | 31.4 | 43.4 | 61.4 | 156.5 |
| 14.8 | 32.5 | 44.5 | 63.1 | 196.0 |
| 17.6 | 32.9 | 44.9 | 64.9 | 204.9 |
| 18.0 | 33.7 | 46.4 | 65.6 | |

**(a)** Make a histogram and describe its main features.
**(b)** Find the mean and median of the amounts recovered. Explain how the relationship between the mean and the median reflects the shape of the distribution.
**(c)** Give the five-number summary and explain briefly how it reflects the shape of the distribution.

■ **28.** Look at the histogram of lengths of words in Shakespeare's plays, Figure 5.18. The heights of the bars tell us what percent of words have each length. (Analysis of writing tendencies can help determine authorship of a newly-discovered manuscript.) The median length is the middle, the length with half of all words shorter and half longer. What is the median length of words used by Shakespeare? Similarly, what are the quartiles? Give the five-number summary for Shakespeare's word lengths.

■ **29.** A common criterion for identifying an outlier in a set of data is if an observation falls more than 1.5 × *IQR* above the third quartile or below the first quartile. (IQR stands for the interquartile range, which is the difference between the quartiles: $Q_3 - Q_1$.)

So which states are suspected outliers in the distribution of percent of Hispanics among adult residents, Table 5.1?

## 5.7 Describing Spread: The Standard Deviation

**30.** Do you think the standard deviation of the tuition and fees of the public colleges in Massachusetts is likely to be bigger or smaller than the standard deviation for the private colleges? Why?

**31.** Many standard statistical methods are intended for use with distributions that are symmetric and have no outliers. These methods start with the mean and standard deviation, $\bar{x}$ and $s$. An example of scientific data for which standard methods should work well are Cavendish's measurements of the density of the earth in Exercise 10.

**(a)** Summarize this data set by giving $\bar{x}$ and $s$.
**(b)** Find the median. Is the median quite close to the mean, as we expect it to be for symmetric distributions?

**32.** The level of various substances in the blood influences our health. Here are measurements of the level of phosphate in the blood of a patient, in milligrams of phosphate per deciliter of blood, made on six consecutive visits to a clinic.

5.6 5.2 4.6 4.9 5.7 6.4

**(a)** Find the mean.
**(b)** Find the standard deviation.

**33.** The mean $\bar{x}$ and standard deviation $s$ measure center and spread but are not a complete description of a distribution. Data sets with different shapes can have the same mean and standard deviation. To demonstrate this fact, use your calculator to find $\bar{x}$ and $s$ for these two small data sets. Then make a stemplot of each and comment on the shape of each distribution.

| | | | | |
|---|---|---|---|---|
| Data A: | 9.14 | 8.14 | 8.74 | 8.77 |
| | 9.26 | 8.10 | 6.13 | 3.10 |
| | 9.13 | 7.26 | 4.74 | |
| Data B: | 7.46 | 6.77 | 12.74 | 7.11 |
| | 7.81 | 8.84 | 6.08 | 5.39 |
| | 8.15 | 6.42 | 5.73 | |

**34.** Your data consist of observations on the age of several subjects (measured in years) and the reaction times of these subjects (measured in seconds). In what units are each of the following descriptive statistics measured?

**(a)** The mean age of the subjects
**(b)** The standard deviation of the subjects' reaction times
**(c)** The variance of the subjects' reaction times
**(d)** The median age of the subjects

■ **35.** This is a standard deviation contest! You must choose four numbers from the whole numbers 0 to 10, with repeats allowed.

**(a)** Choose four numbers that have the smallest possible standard deviation.
**(b)** Choose four numbers that have the largest possible standard deviation.
**(c)** Is more than one choice possible in part (a)? Explain.
**(d)** Is more than one choice possible in part (b)? Explain.

■ **36.** "Conservationists have despaired over destruction of tropical rainforest by logging, clearing, and burning." These words begin a report on a statistical study of the effects of logging in Borneo. [C. H. Cannon, D. R. Peart, and M. Leighton, Tree species diversity in commercially logged Bornean rainforest, *Science,* 281 (1998): 1366–1367.] Researchers compared forest plots that had never been logged (Group 1) with similar plots nearby that had been logged 1 year earlier (Group 2) and 8 years earlier (Group 3). All plots were 0.1 hectare in area. Here are the counts of trees for plots in each group, courtesy of Charles Cannon:

| | | | | | | |
|---|---|---|---|---|---|---|
| Group 1: | 27 | 22 | 29 | 21 | 19 | 33 |
| | 16 | 20 | 24 | 27 | 28 | 19 |
| Group 2: | 12 | 12 | 15 | 9 | 20 | 18 |
| | 17 | 14 | 14 | 2 | 17 | 19 |
| Group 3: | 18 | 4 | 22 | 15 | 18 | |
| | 19 | 22 | 12 | 12 | | |

Give a complete comparison of the three distributions, using both graphs and numerical summaries. To what extent has logging affected the count of trees? The researchers used an analysis based on $\bar{x}$ and $s$. Explain why this is reasonably well justified.

(*Edward Parker/Alamy.*)

## 5.8 Normal Distributions

## 5.9 The 68–95–99.7 Rule

**37.** Some teachers graded "on a curve" based on the belief that classroom test scores are normally distributed. One way of doing this is to assign a "C" to all scores within 1 standard deviation of the mean. Then, the teacher would assign a "B" to all scores between 1 and 2 standard deviations above the mean, an "A" to all scores more than 2 standard deviations above the mean, and use symmetry to define the regions for "D" and "F" on the left side of the normal curve. If 200 students take an exam, determine the number of students who would receive a B.

**38.** The length of human pregnancies from conception to birth varies according to a distribution that is approximately normal, with mean 266 days and standard deviation 16 days. Draw a normal curve for this distribution on which the mean and standard deviation are correctly located. (*Hint:* First draw the curve, then mark the axis.)

**39.** Figure 5.22 shows a smooth curve used to describe a distribution that is not symmetric. The mean and median do not coincide. Which of the points marked is the mean of the distribution, and which is the median? Explain your answer.

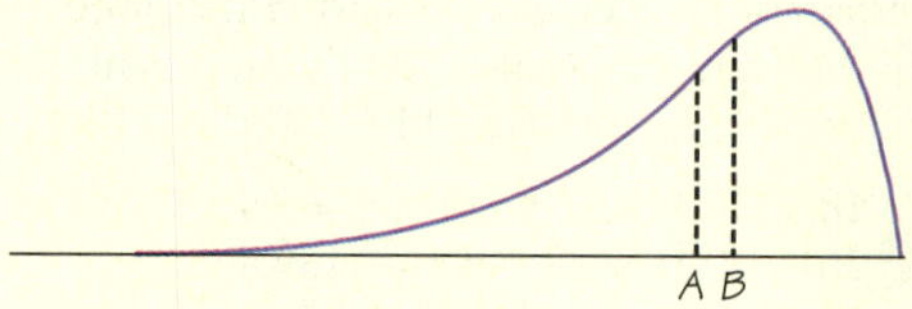


**FIGURE 5.22** A curve describing a skewed distribution, for Exercise 39.

**40.** Sketch a smooth curve that describes a distribution that is symmetric but has two peaks (that is, two strong clusters of observations).

**41.** Bigger animals tend to carry their young longer before birth. The length of horse pregnancies from conception to birth varies according to a roughly normal distribution, with mean 336 days and standard deviation 3 days. Use the 68–95–99.7 rule to answer the following questions.

**(a)** Almost all (99.7%) horse pregnancies fall in what range of lengths?
**(b)** What percent of horse pregnancies are longer than 339 days?

**42.** Scores on the three-section SAT Reasoning college entrance test for the class of 2007 were roughly normal, with mean 1511 and standard deviation 194.

**(a)** What was the range of the middle 68% of SAT scores?
**(b)** How high must a student score to be in the top 2.5% of SAT scores?

**43.** What are the quartiles of SAT Reasoning scores, according to the distribution in Exercise 42?

**44.** The Wechsler Adult Intelligence Scale (WAIS) is the most common "IQ test." The scale of scores is set separately for each age group and is approximately normal, with mean 100 and standard deviation 15. People with WAIS scores below 70 are considered mentally retarded for purposes of applying for Social Security disability benefits. By this criterion, what percent of adults are retarded?

**45.** The yearly rate of return on the Standard & Poor's 500 (an index of 500 large-cap corporations) is approximately normal. From January 1956 through September 2007, the S&P 500 had a mean yearly return of 10.51%, with a standard deviation of about 15.51%. Take this normal distribution to be the distribution of yearly returns over a long period.

**(a)** In what range do the middle 95% of all yearly returns lie?
**(b)** Stocks can go down as well as up. What are the worst 2.5% of annual returns?

**46.** What is the range of the middle 50% of annual returns on stocks, according to the distribution given in the previous exercise? (*Hint:* What two numbers mark off the middle 50% of any distribution?)

**47.** The concentration of the active ingredient in capsules of a prescription painkiller varies according to a normal distribution with $\mu = 10\%$ and $\sigma = 0.2\%$.

**(a)** What is the median concentration? Explain your answer.
**(b)** What range of concentrations covers the middle 95% of all the capsules?
**(c)** What range covers the middle half of all capsules?

**48.** Answer the following questions for the painkiller in Exercise 47.

**(a)** What percent of all capsules have a concentration of active ingredient higher than 10.4%?
**(b)** What percent have a concentration higher than 10.6%?

**49.** One reason that normal distributions are important is that they describe how the results of an opinion poll would vary if the poll were repeated many times. About 40% of adult Americans say they are afraid to go out at night because of crime. Take many randomly chosen samples of 1050 people. The proportions of people in these samples who stay home for fear of crime will follow the normal distribution with mean 0.4 and standard deviation 0.015. Use this fact and the 68–95–99.7 rule to answer these questions.

**(a)** In many samples, what percent of samples give results above 0.4? Above 0.43?
**(b)** In a large number of samples, what range contains the central 95% of proportions of people who stay home because of crime?

■ **50.** You can compare observations from different normal distributions if you measure in standard deviations away from the mean. Scores expressed in standard deviation units are called *standard scores* (or *z-scores*).

**(a)** Scores on the ACT college entrance exam in a recent year were roughly normal, with mean 21.2 and standard deviation 4.8. Jermaine scores 27 on the ACT. Express his score in standard deviation units by calculating

$$\text{standard score} = \frac{\text{score} - \text{mean}}{\text{standard deviation}}$$

**(b)** Scores on the SAT Reasoning college entrance exam in the same year were roughly normal, with mean 1511 and standard deviation 194. Tonya scores 1718 on the SAT. What is her standard score?
**(c)** Assuming that the ACT and the SAT measure the same thing, did Jermaine or Tonya have the higher score?

## Chapter Review

Different varieties of the tropical flower *Heliconia* are fertilized by different species of hummingbirds. Over time, the lengths of the flowers and the form of the hummingbirds' beaks have evolved to match each other. Here are data on the lengths in millimeters of two varieties of these flowers on the island of Dominica.

***H. caribaea*** **Red**

| | | | | |
|---|---|---|---|---|
| 37.40 | 38.07 | 38.87 | 40.66 | 41.93 |
| 37.78 | 38.10 | 39.16 | 41.47 | 42.01 |
| 37.87 | 38.20 | 39.63 | 41.69 | 42.18 |
| 37.97 | 38.23 | 39.78 | 41.90 | 43.09 |
| 38.01 | 38.79 | 40.57 | | |

*H. caribaea* Yellow

| | | | | |
|---|---|---|---|---|
| 34.57 | 35.45 | 36.03 | 36.66 | 37.02 |
| 34.63 | 35.68 | 36.11 | 36.78 | 37.10 |
| 35.17 | 36.03 | 36.52 | 36.82 | 38.13 |

SOURCE: Thanks to Ethan J. Temeles of Amherst College for providing the data. His work is described in Ethan J. Temeles and W. John Kress, Adaptation in a plant-hummingbird association, *Science,* 300 (2003): 630–633.

Exercises 51 to 55 use these data.

**51.** Make stemplots of the lengths of each of the two varieties (red and yellow). Briefly describe the overall shape of the two distributions.

**52.** Find the five-number summaries of the two distributions of flower lengths. Make side-by-side boxplots to give a quick picture that compares the two distributions.

**53.** The biologists who collected the flower length data compared the two *Heliconia* varieties using statistical methods based on the mean and standard deviation. Find $\bar{x}$ and $s$ for each variety. Based on your stemplots in Exercise 51, which distribution is more suitable for use of $\bar{x}$ and $s$ as summaries? Why?

**54.** Your stemplot in Exercise 51 suggests that the distribution of lengths of yellow *Heliconia* flowers is roughly normal. Suppose that the distribution is exactly normal. Use the mean and standard deviation you found in Exercise 53 as the $\mu$ and $\sigma$ of the distribution.

**(a)** What range of lengths covers the middle 50% of yellow flowers?

**(b)** What range of lengths covers the middle 95% of yellow flowers?

■ **55.** Continue to work with the normal distribution of lengths of yellow flowers from the previous exercise. The shortest red flower was 37.4 millimeters long. Using the 68–95–99.7 rule and the location of the quartiles in normal distributions, what can you say about what percent of yellow flowers that are longer than 37.4 millimeters?

**56.** By hand, find the standard deviation of these five numbers: 0, 1, 3, 4, 12. Use the approach in the standard deviation definition box on page 164 and Figure 5.9.

**57.** If every number in a data set is increased by 10, which of these will increase: range, standard deviation, mode, mean, median?

## APPLET EXERCISES

**To do these exercises, go to www.whfreeman.com/fapp8e.**

**1.** The *Mean and Median* applet allows you to place observations on a line and see their mean and median visually. Place two observations on the line, by clicking below it. Why does only one arrow appear?

**2.** In the *Mean and Median* applet, place three observations on the line by clicking below it, two close together near the center of the line and one somewhat to the right of these two. Pull the single rightmost observation out to the right. (Place the cursor on the point, hold down a mouse button, and drag the point.) How does the mean behave? How does the median behave? Explain briefly why each measure acts as it does.

**3.** In Example 16 we used the fact that SAT scores are close to normal and are adjusted so that the mean is close to 500 and the standard deviation is close to 100. (Actual scores in a particular year have slightly different mean and standard deviation.) Use the *Normal Curve* applet with $\mu = 500$ and $\sigma = 100$ to answer these questions:

**(a)** What proportion of SAT scores are above 640?

**(b)** What proportion of SAT scores are between 420 and 640? (If you drag one flag across the other, the applet shows the area between the flags.)

**4.** Because Web browsers have limited resolution, the *Normal Curve* applet can't always get exactly the values you want. Use the applet to come close to exact answers to these questions:

**(a)** How high must an SAT score be to fall in the top 10% of all scores?

**(b)** How high must an SAT score be to fall in the top 1% of all scores?

**5.** The 68–95–99.7 rule for normal distributions is a useful approximation. You can use the *Normal Curve* applet to see how accurate the rule is. Drag one flag across the other so that the applet shows the area under the curve between the two flags.

**(a)** Place the flags 1 standard deviation on either side of the mean. What is the area between these two values? What does the 68–95–99.7 rule say this area is?

**(b)** Repeat for locations 2 and 3 standard deviations on either side of the mean. Again compare the 68–95–99.7 rule with the area given by the applet.

# WRITING PROJECTS

**1.** Many social issues involve data and interpreting data. For example, income inequality (roughly speaking, the gap in income between people toward the top of the income scale and people toward the bottom) has increased in the past few decades. A good place to find data is on the Web site of the Census Bureau, www.census.gov. Click on "Income" and look for the latest report on income in the United States. Select a few facts from this detailed collection of income data to describe the extent of income inequality. Write a few paragraphs based on these facts.

**2.** Let's produce some data and describe them in order to gain insight into chance behavior. The mathematics of chance is the topic of Chapter 8, but for now we will concentrate on data rather than math. You need two things: a standard six-sided die (raid your Monopoly game) and a thumbtack with a rounded back (like a satellite dish). Toss the thumbtack 100 times (to speed things up, you could do 10 tosses of 10 tacks each) and record each outcome (pointing straight up or angled down). Also, toss the die 180 times and record each outcome (1, 2, 3, 4, 5, or 6). Use graphs and numbers to describe each set of results. Is the die roughly balanced, so that all six outcomes come up about equally? What about the thumbtack: Is point up or point down much more common?

# SUGGESTED READINGS

CLEVELAND, WILLIAM S. *The Elements of Graphing Data,* rev. ed., Hobart Press, Summit, N.J., 1994. A careful study of the most effective elementary ways to present data graphically, with much sound advice on improving simple graphs.

LESSER, LAWRENCE M. Critical values and transforming data: Teaching statistics with social justice, *Journal of Statistics Education* (2007): www.amstat.org/publications/jse/v15n1/lesser.html. Article filled with resources for finding social justice data to expand upon Writing Project 1.

MOORE, DAVID S. *The Basic Practice of Statistics,* 3rd ed., W. H. Freeman, New York, 2004. This text is a natural next step for more detail on all the material in Part II at about the same mathematical level. The first three chapters provide a more extensive treatment of the material of Chapter 5.

ROSSMAN, ALLAN J., and BETH L. CHANCE. *Workshop Statistics: Discovery with Data,* 3rd ed., Springer, New York, 2008. A different approach to basic data analysis, using hands-on activities. There are several versions, keyed to graphing calculators and to several different software packages.

# SUGGESTED WEB SITES

The Web site of the U.S. Census Bureau, www.census.gov, is a prime source of information on many topics. The latest estimates for the populations of the United States and the world are on the home page, updated regularly. Data appear under many headings, not only as numbers but as maps. Try clicking on "American Fact Finder," then on "Maps and Geography," then on "Reference Maps." Enter your zip code, then zoom in on your street. The 1000 tables in the *Statistical Abstract of the United States* are also available. If you need data for a report, this is the place to start. Canadians can find similar help at the Web site of Statistics Canada: www.statcan.ca.

Interested in data about schools, colleges, and students? The National Center for Education Statistics, nces.ed.gov, is the place to look. Go to the "What's New" section. There are useful statistics applets at www.shodor.org/interactivate/activities/.

# CHAPTER 6



# Exploring Data: Relationships

A medical study finds that short women are more likely to have heart attacks than women of average height, while tall women have fewer heart attacks. An insurance group reports that heavier cars have fewer accident deaths per 100,000 vehicles registered than do lighter cars. These and many other statistical studies look at the *relationship between two variables.* To understand such a relationship, we must often examine other variables as well. To conclude that shorter women have higher risk from heart attacks, for example, the researchers had to eliminate the effect of other variables such as weight, diet and exercise habits. Our topic in this chapter is relationships between variables.

To study the relationship between two variables, we measure both variables on the *same individuals.* If we measure both the height and the weight of each of a large group of people, we know which height goes with each weight. These data allow us to study the connection between height and weight. A list of heights and a separate list of weights, two sets of single-variable data, do not show the connection between the two variables.

Height and weight are connected: Taller people also tend to be heavier. Neither height nor weight explains or causes the other. They go together in describing bigger or smaller people. Smoking and life expectancy are also connected, and in this case we think that smoking does explain or influence life expectancy: People who smoke more cigarettes per day tend not to live as long as those who smoke fewer. So we call smoking an **explanatory variable** and life expectancy a **response variable**.

**Response Variable** — DEFINITION

A **response variable** measures an outcome or result of a study.

**Explanatory Variable** DEFINITION

An **explanatory variable** is a variable that we think explains or causes changes in the response variables.

## 6.1 Displaying Relationships: Scatterplot

The most useful graph for displaying the relationship (whether it fits a trend perfectly or not) between two numerical variables is a **scatterplot**.

**Scatterplot** DEFINITION

A **scatterplot** shows the relationship between two numerical variables measured on the same individuals. The values of one variable appear on the horizontal axis, and the values of the other variable appear on the vertical axis. Each individual in the data appears as the point in the plot fixed by the values of both variables for that individual.

Always plot the explanatory variable, if there is one, on the horizontal axis (the $x$-axis) of a scatterplot. As a reminder, we usually call the explanatory variable $x$ and the response variable $y$. If the variables don't naturally fall into "explanatory" and "response," either variable can go on the horizontal axis.

### EXAMPLE 1 ■ Beer and Blood Alcohol

How well does the number of beers a student drinks predict his or her blood alcohol content? In a study at The Ohio State University, 16 student volunteers drank a randomly assigned number of cans of beer. Thirty minutes later, a police officer measured their blood alcohol content (BAC) in grams of alcohol per deciliter of blood. In all states of the U.S., the legal BAC limit is 0.08. Here are the data:

| **Student** | **1** | **2** | **3** | **4** | **5** | **6** | **7** | **8** |
|---|---|---|---|---|---|---|---|---|
| Beers | 5 | 2 | 9 | 8 | 3 | 7 | 3 | 5 |
| BAC | 0.10 | 0.03 | 0.19 | 0.12 | 0.04 | 0.095 | 0.07 | 0.06 |
| **Student** | **9** | **10** | **11** | **12** | **13** | **14** | **15** | **16** |
| Beers | 3 | 5 | 4 | 6 | 5 | 7 | 1 | 4 |
| BAC | 0.02 | 0.05 | 0.07 | 0.10 | 0.085 | 0.09 | 0.01 | 0.05 |

The students were equally divided between men and women and differed in weight and usual drinking habits. Because of this variation, many students don't believe that number of drinks predicts blood alcohol well. What do the data say?

Figure 6.1 is a scatterplot of these data. Because we think that number of beers helps explain BAC, it is the explanatory variable. We plot number of beers on the horizontal axis. One student drank 2 beers and had BAC 0.03. This student's point on the scatterplot is (2, 0.03), above $x = 2$ and to the right of $y = 0.03$. We have marked this point in Figure 6.1. ■

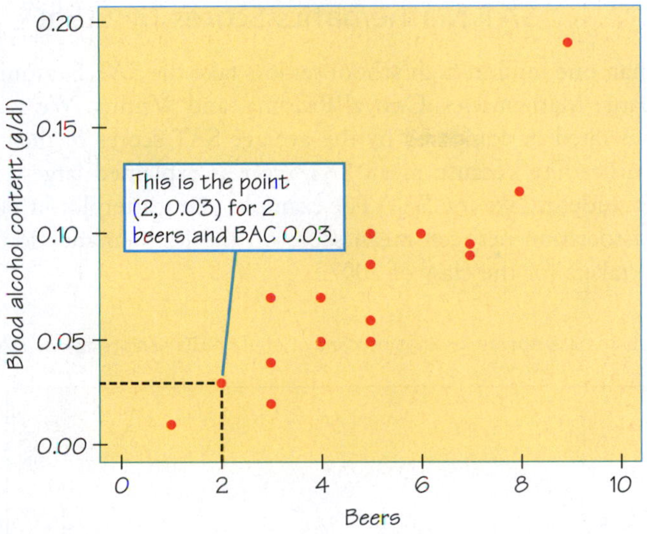


**FIGURE 6.1** Scatterplot of blood alcohol content (response variable) against the number of beers a student drinks (explanatory variable).

To interpret a scatterplot, apply the usual strategies of data analysis.

### Examining a Scatterplot — PROCEDURE

In any graph of data, look for the *overall pattern* and for striking *deviations* from that pattern.

You can describe the overall pattern of a scatterplot by the *form*, *direction*, and *strength* of the relationship.

An important kind of deviation is an **outlier**, an individual value that falls outside the overall pattern of the relationship.

The *form* of the relationship in Figure 6.1 is roughly a straight-line pattern. If you look ahead a bit, Figure 6.3 draws a line through the plot to describe the overall pattern. The *direction* of the relationship is clear: As number of beers increases, BAC also increases. We call this a **positive association** between the two variables.

### Positive Association — DEFINITION

Two variables are **positively associated** if an increase in one variable tends to accompany an *increase* in the other variable.

### Negative Association — DEFINITION

Two variables are **negatively associated** if an increase in one variable tends to accompany a *decrease* in the other variable.

The *strength* of a relationship describes how closely the points in a scatterplot follow a simple form such as a straight line. Figure 6.1 shows only a small amount of scatter about the straight line, so the relationship is moderately strong. We will soon learn a numerical measure of the strength of a straight-line relationship.

## EXAMPLE 2 ■ SAT Mathematics Scores by State

(*Cleve Bryant/PhotoEdit Inc.*)

Each year, more than one million high school seniors take the SAT Reasoning Test, which has three parts: Mathematics, Critical Reading, and Writing. We sometimes see individual states rated or compared by the average SAT scores of their seniors. However, this is misleading because mean SAT score is explained largely by what percent of a state's students take the SAT. For example, the scatterplot in Figure 6.2 shows a negative association between mean score on the Mathematics section and the percent of test takers for the class of 2007.

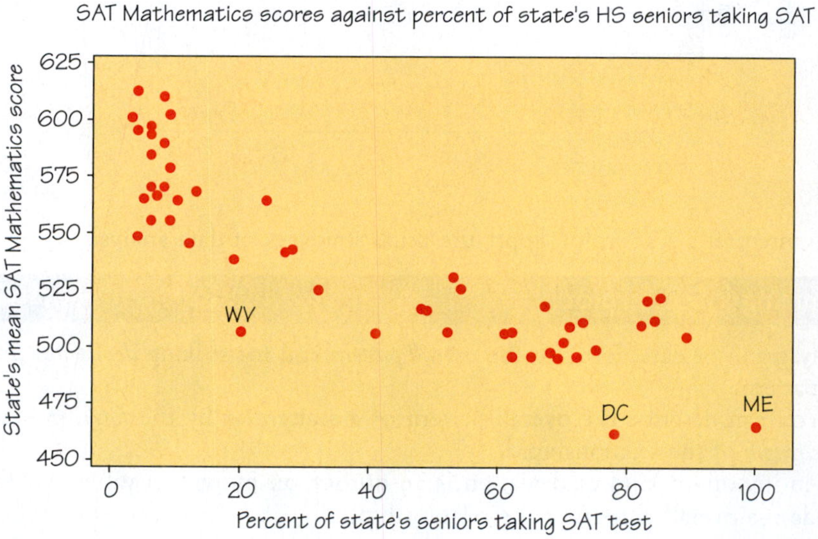


**FIGURE 6.2** Scatterplot of the mean SAT Mathematics scores (response variable) against the percent of high school seniors who take the SAT (explanatory variable).

The *form* of Figure 6.2 is a bit irregular, but there are two distinct *clusters* of states. In one cluster, more than half of high school seniors take the SAT, and the mean scores are low. Fewer than 40% of seniors in states in the other cluster take the SAT—fewer than 20% in most of these states—and these states have higher mean scores. Clusters in a graph suggest that the data describe several distinct kinds of individuals. The two clusters in Figure 6.2 do in fact describe two distinct sets of states. There are two common college entrance examinations, the SAT and the ACT. Each state tends to prefer one or the other. In ACT states (the left cluster in Figure 6.2) most students who take the SAT are applying to selective out-of-state colleges. This select group performs well. In SAT states (the right cluster), many seniors take the SAT, and this broader group has a lower mean score.

The relationship in Figure 6.2 also has a clear *direction:* States in which a higher percent of students take the SAT tend to have lower mean scores. This is true both between the clusters and within each cluster. That is, there is a **negative association** between the two variables.

There are no clear *outliers* in Figure 6.2, but each cluster does include a state whose mean SAT Mathematics score is lower than we would expect from the percent of its students who take the SAT. In the cluster of ACT states, this occurs with West Virginia (WV). In the cluster of SAT states, this occurs with the District of Columbia (DC)—a city, not a state—and Maine (ME).

## 6.2 Making Predictions: Regression Line

If a scatterplot shows a straight-line relationship, we would like to summarize this overall pattern by drawing a line on the scatterplot. A **regression line** summarizes the relationship between two variables, but only in a specific setting: when one of the variables helps explain or predict the other. That is, regression describes a relationship between an explanatory variable and a response variable.

**Regression Line** DEFINITION

A **regression line** is a straight line that describes how a response variable $y$ changes as an explanatory variable $x$ changes. We often use a regression line to *predict* the value of $y$ for a given value of $x$.

### EXAMPLE 3 ■ Predicting Blood Alcohol

Figure 6.1 shows a straight-line relationship between how many beers a student drinks and his or her blood alcohol content (BAC) 30 minutes later. Figure 6.3 repeats this scatterplot, and adds a regression line that we can use to predict BAC for a student based on the number of beers consumed.

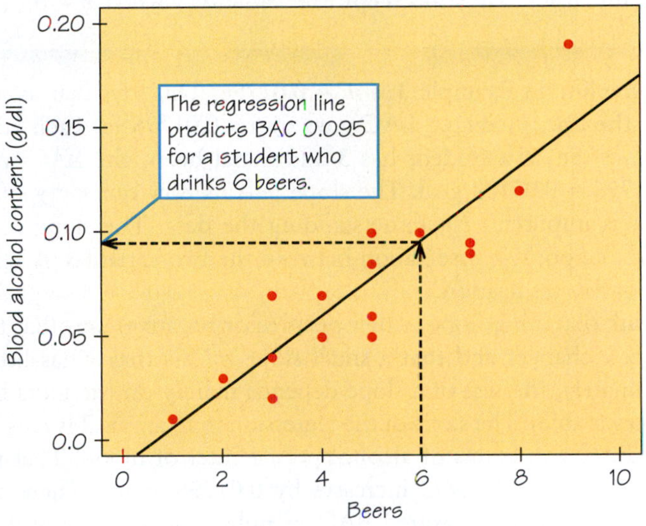


**FIGURE 6.3** Regression line for predicting blood alcohol content from the number of beers a student drinks.

Figure 6.3 shows the prediction in graphical form for a student who drinks 6 beers. Start at $x = 6$, go up to the line, then head left to the $y$-axis. We hit the $y$-axis at BAC 0.095. This is the BAC that corresponds to 6 beers, according to the regression line. (Recall that the legal limit for driving is 0.08.) The line represents only the overall pattern of the data, so the BAC of a randomly chosen student after 6 beers will probably not be exactly 0.095. But because the points for the 16 students in the Ohio State study are not far from the line, we expect the prediction to be reasonably accurate.

It is easier to use the *equation of the line* for prediction. Applying formulas that will be given in Section 6.4, the equation of the line in Figure 6.3 is

$$\text{predicted BAC} = -0.0127 + 0.01796 \times \text{beers}$$

For a student who drinks 6 beers, we have

$$\text{predicted BAC} = -0.0127 + (0.01796)(6) = 0.095$$

You can plot a line from its equation by substituting two values of $x$, such as $x = 2$ and $x = 8$. Find the corresponding values of $y$, plot the two points, and draw the line through them.

Statistical software and many calculators will give you the equation of a regression line from keyed-in data. You should know how to use a regression line even if you don't look into the details needed to calculate the line from data. First, recall some basic facts about the (slope and intercept) coefficients in the equation of a line.

## Equation of a Regression Line

DEFINITION

Suppose that $y$ is a response variable (plotted on the vertical axis) and $x$ is an explanatory variable (plotted on the horizontal axis). If we call $\hat{y}$ the predicted value of $y$, then the resulting regression line for predicting $y$ from $x$ has an equation of the form[1]:

$$\hat{y} = mx + b$$

In this equation, $m$ is the **slope**, the amount by which $y$ changes when $x$ increases by 1 unit. The number $b$ is the **$y$-intercept**, the value of $y$ when $x = 0$.

The slope of the line in Example 3 is $m = 0.01796$. This says that as we move to the right along the line, predicted BAC goes up by 0.01796 for each additional beer a student drinks. So, if a student has 3 additional beers, the BAC would increase by $3 \times 0.01796 = 0.05388$ g/dl. The slope tells us how quickly $y$ changes as we change $x$, so it is important for understanding the data. The slope is positive ($m > 0$) when there is a positive association between the two variables. It is negative when there is a negative association.

You might think that a big slope (either positive or negative) says that there are big changes in $y$ as $x$ changes and that a small slope means that $x$ has little influence on $y$. Unfortunately, the size of a slope depends mainly on the units in which we measure the two variables. The slope of the regression in Example 3 is $m = 0.01796$ when we measure BAC $y$ in grams of alcohol per deciliter of blood. That is, when beers consumed increases by 1, BAC increases by 0.01796 grams. There are 1000 milligrams in a gram, so if we measured BAC in milligrams of alcohol, the slope would be 1000 times as large, $m = 17.96$. When beers consumed increases by 1, BAC increases by 17.96 milligrams. *You can't say how important a relationship is by looking at how big the slope is.*

The intercept of the regression line in Example 3 is $b = -0.0127$. This is the predicted value of $y$ when $x = 0$. Although we need the value of the intercept to draw the line, it is statistically meaningful only when $x$ can actually take values close to zero. Even then, you should think of the intercept as describing the line rather than taking it seriously as a prediction. If a student drinks no beers, his or her blood alcohol should be exactly zero. The intercept of the regression line in Example 3 is close to zero, but it is not exactly zero.

[1] The letters $m$ and $b$ are from the slope-intercept form from algebra class, but be aware that some books and technologies use different letters, such as $b$ and $a$. To be safe, check that the letter used for slope corresponds to the number multiplied by the explanatory variable.

## 6.3 Correlation

A scatterplot displays the form, direction, and strength of the relationship ("co-relation") between two numerical variables. Straight-line relations are particularly important because a straight line is a simple pattern that is quite common. We say a straight-line association is strong if the points lie close to a line, and weak if they are widely scattered about a line. This is vague. We need to follow our strategy for data analysis by using a numerical measure along with the graph. **Correlation** is the measure we use. Correlation is usually written as ***r***, thanks to nineteenth-century statistician Sir Francis Galton, who was studying related ideas of ***r***egression and ***r***eversion.

### Correlation — DEFINITION

The **correlation** measures the direction and strength of the straight-line relationship between two numerical variables.

A correlation ***r*** is always a number between $-1$ and 1, inclusive. It has the same sign as the slope of a regression line: $r > 0$ for positive association and $r < 0$ for negative association.

Perfect correlation $r = 1$ or $r = -1$ occurs only when all points lie exactly on a straight line. The correlation moves away from 1 or $-1$ as the straight-line relationship gets weaker. Correlation $r = 0$ indicates no straight-line relationship.

### SPOTLIGHT 6.1 Scatterplot Smoothers: Crash Test Dummies

Our eyes are good at seeing the overall pattern of a scatterplot. Sometimes the pattern has a simple form, such as a straight line, that we can draw on the plot to summarize the pattern. Clever software for "smoothing" a scatterplot can pick out much more complex overall patterns.

Crash a motorcycle into a wall. The rider, fortunately, is a dummy with an instrument to measure acceleration (change of velocity) mounted in its head. The figure is a scatterplot of the acceleration of the dummy's head against time. Acceleration is measured in *g*'s, or multiples of the acceleration due to gravity at the earth's surface. The motorcycle approaches the wall at a constant speed (acceleration near 0). As it hits, the dummy's head snaps forward and decelerates violently (negative acceleration reaching more than 100 *g*'s), then snaps back again (up to 75 *g*'s) and wobbles a bit before coming to rest. We see that the plot has a clear, but complicated, overall pattern. A scatterplot smoother picks out this pattern and draws a curve on the plot to display it.

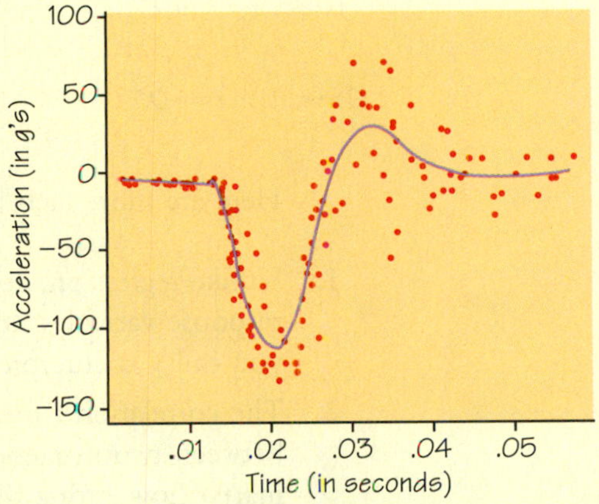


Smoothing a scatterplot. Software draws the smooth curve to describe the overall pattern of the relationship.

### EXAMPLE 4 ■ Scatterplots and Correlation

The scatterplots in Figure 6.4 illustrate how values of $r$ closer to 1 or $-1$ correspond to stronger straight-line relationships. To make the meaning of $r$ clearer, the standard deviations of both variables in these plots are equal and the horizontal and vertical scales are the same. In general, it is not so easy to guess the value of $r$ from the appearance of a scatterplot. Changing the plotting scales in a scatterplot can alter the appearance of the graph, but it does not change the correlation.

We said that Figure 6.1 shows a moderately strong positive straight-line relationship between how many beers a student drinks and his or her blood alcohol content. The correlation between these variables is $r = 0.894$. Figure 6.2, despite the clusters, also shows a quite strong straight-line relationship between the percent of a state's high school seniors who take the SAT exam and their mean SAT score. The association is negative: Higher percents taking the SAT go with lower mean scores. The correlation is $r = -0.877$.

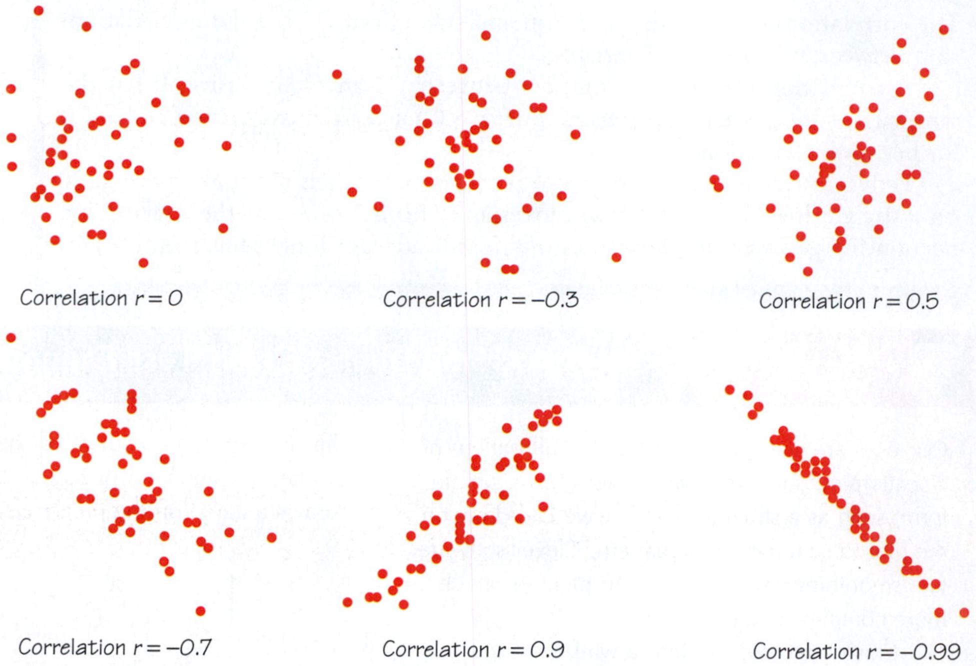


**FIGURE 6.4** How the correlation $r$ measures the direction and strength of straight-line association.

Here are more facts about the correlation $r$:

1. Unlike regression, correlation makes no distinction between explanatory and response variables. It makes no difference which variable you call $x$ and which you call $y$ in interpreting a correlation.
2. The correlation $r$ measures the strength of only a straight-line relationship between two numerical variables. It does not describe curved relationships, no matter how strong they are. Correlation makes no sense for non-numerical variables such as ethnicity and occupation.
3. The correlation $r$ does not change when we change the units of measurement of $x$, $y$, or both. Measuring height in inches rather than centimeters and weight in pounds rather than kilograms does not change the correlation between height and weight. The correlation $r$ itself has no unit of measurement; it is just a number. So correlation is easier to interpret than the slope of a regression line, which changes when the units of $x$ and $y$ change.

4. Like the mean and standard deviation, the correlation is strongly affected by a few outlying observations. Use $r$ with caution when outliers appear in the scatterplot.

In practice, you will use a calculator or software to find the correlation from keyed-in data. That's fortunate, because using the formula for correlation is quite a bit of work. Nonetheless, all the properties of $r$ come from the formula that defines it.

### Formula for Correlation

PROCEDURE

Suppose that we have data on variables $x$ and $y$ for $n$ individuals. The means and standard deviations of the two variables are $\bar{x}$ and $s_x$ for the $x$-values, and $\bar{y}$ and $s_y$ for the $y$-values. The correlation $r$ between $x$ and $y$ is

$$r = \frac{1}{n-1}\left[\left(\frac{x_1-\bar{x}}{s_x}\right)\left(\frac{y_1-\bar{y}}{s_y}\right) + \left(\frac{x_2-\bar{x}}{s_x}\right)\left(\frac{y_2-\bar{y}}{s_y}\right) + \cdots + \left(\frac{x_n-\bar{x}}{s_x}\right)\left(\frac{y_n-\bar{y}}{s_y}\right)\right]$$

This formula starts by *standardizing* each observation value (as you did in Exercise 50 in Chapter 5). That is, subtract the mean for that variable from the observation and then divide by the standard deviation. Standardizing turns each original data value into "number of standard deviations above the mean." A value of, say, −2 indicates 2 standard deviations below the mean. This removes the original units and explains why $r$ has no units and doesn't change when we change the units of $x$ or $y$. The formula says that the correlation is an average of the products of the standardized $x$ and $y$ values for $n$ individuals. Exercise 26 asks you to calculate a correlation step by step from the formula to solidify your understanding.

## SPOTLIGHT 6.2 Calculating Correlation

While the preceding formula for correlation has conceptual clarity, it can be tedious to apply to large datasets. Even with the most basic calculator, you'll get the same answer faster using this more computationally-oriented formula:

$$\frac{(x_1y_1 + x_2y_2 + \cdots + x_ny_n) - n\bar{x}\bar{y}}{(n-1)s_xs_y}.$$

If you have a scientific calculator, select the calculator mode to be able to do two-variable regression statistics, clear out any old data, and then enter your new $x$ and $y$ values. Once the data is entered, you can find the correlation (or regression line slope and $y$-intercept, for that matter) by hitting the appropriate key (probably with some kind of SHIFT or 2nd button). On the Internet, you can find several Web sites such as www.geocities.com/calculatorhelp that can help with keystrokes for specific models.

If you have a graphing calculator in the TI-83/84+ family (and you have the two variables entered in two columns using (STAT)→EDIT), then hit this sequence of buttons: (STAT)→TESTS→LinRegTTest (ENTER). Enter what data columns are your independent(Xlist) and dependent(Ylist) variables (note that the labels L1 through L6 are above the keys (1) through (6), respectively). Then select Calculate, hit (ENTER), and scroll to the end of the output to see the correlation $r$. (Note that the output also contains the slope and $y$-intercept coefficients for the least-squares regression line.)

## 6.4 Least-Squares Regression

In Example 3, we used the straight line given by the equation

$$\text{predicted BAC} = -0.0127 + 0.01796 \times \text{beers}$$

to predict blood alcohol content from the number of beers consumed. Where did this equation come from? We will now see that the equation is the result of saying what we mean by the *best* line for predicting BAC from beers consumed. Once we say exactly what we mean by the best line, finding that line becomes a mathematical problem. The line in Example 3 is the solution to this problem for the beer and BAC data.

Different people might draw different lines by eye on a scatterplot. This is especially true when the points are widely scattered. We need a way to draw a regression line that doesn't depend on our guess as to where the line should go. No line will pass exactly through all the points, but we want one that is as close as possible. We will use the line to predict $y$ from $x$, so we want a line that is as close as possible to the points in the *vertical* direction. That's because the prediction errors we make are errors in $y$, which is the vertical direction in the scatterplot.

The table in Example 1 shows that student #12 drank 6 beers and was observed to have a BAC of 0.10. However, the regression line equation in Example 3 showed that the predicted BAC for a student who drinks 6 beers is 0.095. These values are close, but are not the same. Indeed, from Figure 6.3, we can see the observed data point (6, 0.10) lies a bit off the line, and the vertical deviation of this gap is the prediction error.

$$\text{prediction error} = \text{observed BAC} - \text{predicted BAC} = 0.10 - .095 = .005$$

### SPOTLIGHT 6.3 Regression in Action

No other statistical method is used as much as regression. Here are some more applications:

**Did the vote counters cheat?** Republican Bruce Marks was ahead of Democrat William Stinson when the voting machines were tallied in their Pennsylvania election. But Stinson was ahead after absentee ballots were counted by the Democrats who controlled the election board. A court fight followed. The court called in a statistician, who used regression with data from past elections to predict the counts of absentee ballots based on the results from the voting machines. Marks's lead of 564 votes from the machines predicted that he would get 133 more absentee votes than Stinson. In fact, Stinson got 1025 more absentee votes than Marks. This looks suspicious.

**Is regression garbage?** No—but garbage can be the setting for regression. The Census Bureau once asked if weighing a neighborhood's garbage would help count its people. So 63 households had their garbage sorted and weighed. It turned out that pounds of plastic in the trash gave the best garbage prediction of the number of people in a neighborhood. Alas, the prediction wasn't good enough to help the Census Bureau.

**Can college success be predicted?** Colleges with more applicants than spaces want to admit students who are most likely to succeed. To predict this, admissions officers consider variables such as high school GPA, scores on standardized tests (ACT or SAT), number of advanced (AP) classes taken, and so on. Multiple regression extends regression to allow more than one explanatory variable to contribute to explaining a response variable. By tracking first-year college GPA or graduation rates of admitted students, colleges can continue to assess and fine-tune their equation.

When the observed response lies above the line (as we saw when the number of beers is 6), the error is positive. And when the response lies below the line (as it does when the number of beers is 8), the error is negative. The most common way to make the collection of prediction errors for the entire dataset as small as possible is **least-squares regression.** The line in Figure 6.3 is the least-squares regression line.

### Least-Squares Regression Line

DEFINITION

The **least-squares regression line** is the line that makes the sum of the *squares* of the vertical distances of the data points from the line the *least* value possible.

The least-squares idea says what we mean by the best-fitting line. How can we find this line from data? Starting with $n$ observations on variables $x$ and $y$, finding the line that makes the sum of the squares of the vertical errors as small as possible is a mathematical problem. Here is the solution to this problem.

### Finding the Least-Squares Regression Line

PROCEDURE

We have data on an explanatory variable $x$ and a response variable $y$ for $n$ individuals. From the data, calculate the means $\bar{x}$ and $\bar{y}$ and the standard deviations $s_x$ and $s_y$ of the two variables, as well as their correlation $r$ (recall Spotlights 5.2 and 6.2). If we call $\hat{y}$ the predicted value of $y$, then the **least-squares regression line** is the line

$$\hat{y} = mx + b$$

with **slope** $m$ given by

$$m = r\frac{s_y}{s_x}$$

and **$y$-intercept** $b$ given by

$$b = \bar{y} - m\bar{x}$$

This equation gives insight into the behavior of least-squares regression by showing that it is related to the means and standard deviations of the $x$ and $y$ observations and to the correlation between $x$ and $y$. For example, it is clear that the slope $m$ always has the same sign as the correlation $r$. In practice, you don't need to calculate the means, standard deviations, and correlation first. Statistical software or your calculator will give the slope $m$, intercept $b$, and equation of the least-squares line from keyed-in values of the variables $x$ and $y$. (Recall the footnote on notation in Section 6.2.) Notice that if you confuse whether $y$ is your explanatory or your response variable, you will get a different slope value!

## EXAMPLE 5

### Least-Squares Regression of BAC on Number of Beers

Go back to the data in Example 1. Use your calculator to verify that the mean and standard deviation of $x$, number of beers consumed, are

$$\bar{x} = 4.8125 \quad \text{and} \quad s_x = 2.1975$$

The mean and standard deviation of $y$, blood alcohol content, are

$$\bar{y} = 0.07375 \quad \text{and} \quad s_y = 0.04414$$

The correlation between number of beers and BAC is $r = 0.8943$. The least-squares regression line of BAC $y$ on number of beers $x$ has slope

$$m = r\frac{s_y}{s_x} = 0.8943 \times \frac{0.04414}{2.1975}$$
$$= 0.01796$$

and intercept

$$b = \bar{y} - m\bar{x} = 0.07375 - (0.01796)(4.8125)$$
$$= -0.0127$$

The equation of the least-squares line is therefore

$$\hat{y} = -0.0127 + 0.01796x$$

just as we claimed earlier.

When doing calculations like this by hand, you may need to carry extra decimal places in the preliminary calculations to get accurate values of the slope and intercept. Using software or a calculator with a regression function eliminates this worry. ■

You now see that correlation and least-squares regression are closely connected. The expression $m = rs_y/s_x$ for the slope says that along the regression line, a change of one standard deviation in $x$ corresponds to a change of $r$ standard deviations in $y$. When the variables are perfectly correlated ($r = 1$ or $r = -1$), the change in the predicted response is the same (in standard deviation units) as the change in $x$. Otherwise, because $-1 \le r \le 1$, the change in the predicted $y$ is less than the change in $x$. As the correlation grows less strong, the prediction moves less in response to changes in $x$.

## 6.5 Interpreting Correlation and Regression

Correlation and regression are among the most-used statistical methods. Here are a few cautions to keep in mind when you use these methods or see others use them.

*Both the correlation r and the least-squares regression line can be strongly influenced by a few outlying points.* Always make a scatterplot before doing any calculations. Here is an artificial example that illustrates what can happen.

### EXAMPLE 6 ■ Beware the Outlier!

Figure 6.5 shows a scatterplot of data that have a strong positive straight-line relationship. In fact, the correlation is $r = 0.987$, close to the $r = 1$ of a perfect straight line. The line on the plot is the least-squares regression line for predicting $y$ from $x$. One point is an extreme outlier in both the $x$ and $y$ directions. Let's examine the influence of this outlier.

First, suppose we drop the outlier. The correlation for the 5 remaining points (the cluster at the lower left) is $r = 0.523$. The outlier extends the straight-line pattern and greatly increases the correlation.

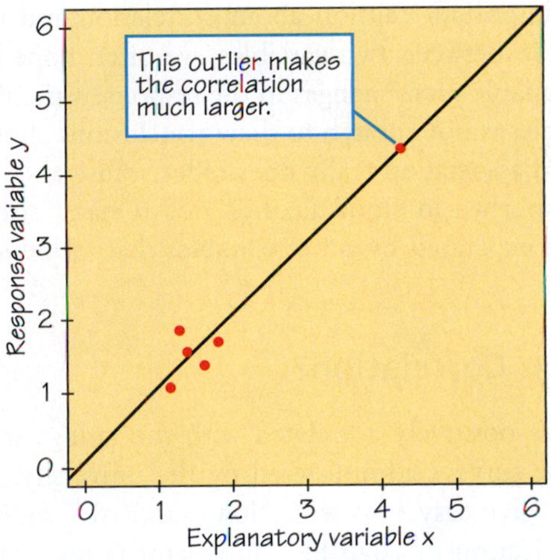


FIGURE 6.5 The outlier increases the correlation and fixes the location of the least-squares line.

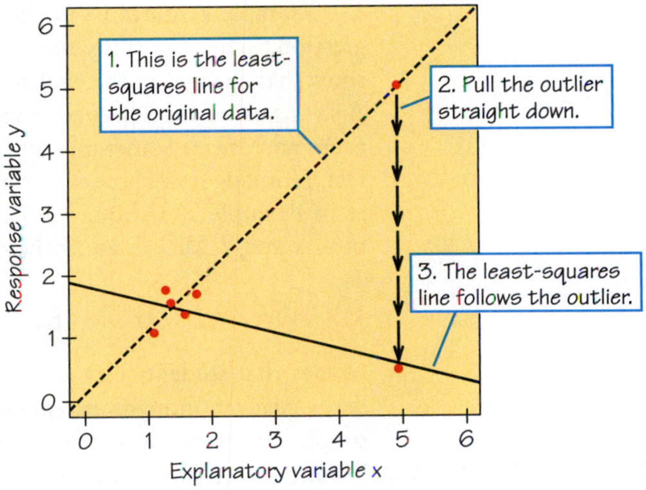


FIGURE 6.6 Moving the outlier unduly changes the correlation and moves the least-squares line.

Next, grab the outlier and pull it straight down, as in Figure 6.6. The least-squares line chases the outlier down, pivoting until it has a negative slope. This is the least-squares idea at work: The line stays close to all 6 points. However, its location is determined almost entirely by the one outlier. Of course, the correlation is now also negative, $r = -0.796$. Never trust a correlation or a regression line if you have not plotted the data. ■

Even if the correlation is strong and there are no outliers in the data we used to find our regression line, we also must not be quick to extrapolate and make predictions well beyond the data collected: Just because the data fits a particular linear trend over a window, there is no guarantee that that trend will continue into the future. For example, the rate of growth of a newborn may fit a line with a steep slope for the first several months, but then the slope (while still positive) starts to decrease.

A good way to see how outlying points can influence the correlation and the regression line is to use the *Correlation and Regression* applet. Applet Exercise 1 asks you to animate Example 6 above, watching $r$ change and the regression line move as you pull the outlier down.

Correlation and regression *describe* relationships. *Interpreting* relationships requires more thought. *Often the relationship between two variables is strongly influenced by other variables.* You should always think about the possible effect of other variables before you draw conclusions based on correlation or regression.

## EXAMPLE 7 ■ Money Helps SAT?

The College Board, which administers the SAT Reasoning Test, offers information on its Web site about college-bound seniors who take the test. This information shows a strong positive association between test score and a test taker's family's income. But there's no direct mechanism—wealthy families are not sending secret bribes to the College Board! It may simply be that children of wealthy parents are more likely to have advantages such as: well-educated role models, high expectations, access to extra tutoring or test preparation, and schools with experienced, qualified teachers and smaller class sizes. ■

Example 7 brings us to the most important caution about correlation and regression. When we study the relationship between two variables, we often hope to show that changes in the explanatory variable *cause* changes in the response variable. A strong association between two variables is not enough to draw conclusions about cause and effect. Sometimes an observed association really does reflect cause and effect. Drinking more beer does cause an increase in blood alcohol. But in many cases, as in Example 7, a strong association is explained by other variables that influence both $x$ and $y$. Here is another example.

## EXAMPLE 8 ■ Evaluation Correlation?

Grades that students earn in courses are positively correlated with the ratings students give on anonymous end-of-course surveys administered by the university. A simple interpretation is that instructors give easy tests with "low standards" which in turn causes students to express appreciation through high instructor ratings. But perhaps there is a third variable that drives the other two variables: A professor who is a skillful teacher and motivator may be more likely both to be rated well and to inspire high performance. Or perhaps a course that includes group projects (rather than only in-class, timed tests) as a significant component of the grade naturally results in higher levels of both performance and satisfaction. Or perhaps courses that have higher grade distributions are more likely to be upper level courses for majors in that subject, and such students would be more favorably inclined towards the course. ■

## EXAMPLE 9 ■ Does Running Lead to Winning in Football?

(*Getty Images.*)

A football broadcaster discussed how often a team wins when it runs the ball at least 30 times in a game. For the most recent NFL regular season, the correlation between wins and number of running plays was indeed strongly positive ($r = 0.643$). Could this mean that running leads to winning–that all any team has to do to improve is to run the ball more? No. To take it to the extreme, if a team chose to run on every single play, the other team would simply adjust its entire defense to focus on and stop the run. Basically, once teams get a good lead in a game (regardless of their mix of running, passing and special team plays), they tend to start running the ball more often as a way to minimize the risk of losing the ball (pass plays are riskier) and to use up clock time faster (an incomplete pass stops the clock). And when teams get far behind late in the game, they begin passing more often as a last chance to get back into the game before time runs out. ■

Correlations such as that in Example 9 are sometimes called "nonsense correlations." The correlation is real, but it is nonsense to conclude that increasing the number of running plays will cause an increase in the number of wins that season. So correlations require thoughtful interpretation, not just computation!

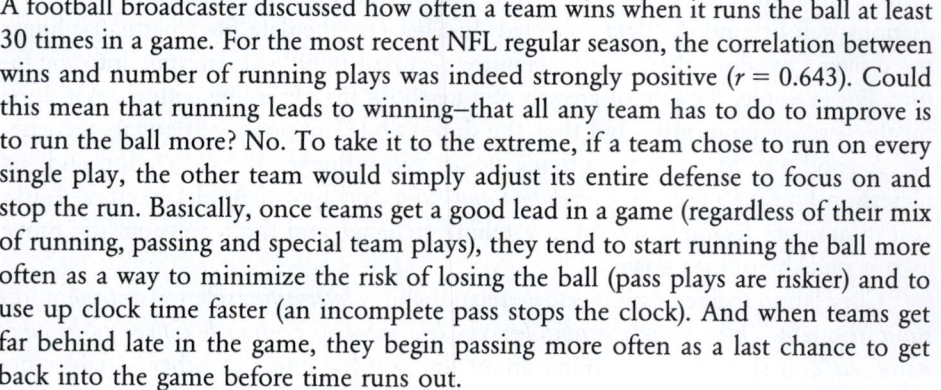

**Association Does Not Imply Causation** RULE

An association between an explanatory variable $x$ and a response variable $y$, even if it is very strong, is not by itself good evidence that changes in $x$ actually cause changes in $y$.

Here is a final example in which we use a scatterplot, correlation, and a regression line to understand data.

## EXAMPLE 10

### What Does Growth Hormone Do in Adults?

In most species, adults stop growing, but still release growth hormone from the pituitary gland to regulate metabolism. Physiologists subjected groups of adult rats to various conditions that activated muscle tissue that was either fast-twitch (like sprinters use) or slow-twitch (like distance runners use). They then measured levels of a bioassayable form of growth hormone (BGH) in the blood and in pituitary tissue. Units are 100s of nanograms per milliliter of blood and micrograms per milligram of tissue, respectively.

Here are the data, courtesy of neurobiologist Kristin Gosselink:

| | | | | | | | | | | |
|---|---|---|---|---|---|---|---|---|---|---|
| blood | 15.8 | 20.0 | 26.7 | 25.0 | 23.0 | 23.8 | 24.7 | 16.3 | 0.8 | 0.8 |
| tissue | 38.0 | 36.7 | 27.8 | 28.3 | 34.9 | 34.1 | 33.2 | 32.7 | 38.1 | 39.1 |
| blood | 0.6 | 10.8 | 37.6 | 41.3 | 39 | 57.5 | 84.8 | 82.8 | 28.8 | 16.5 |
| tissue | 43.9 | 42.8 | 19.3 | 13.7 | 11.2 | 14.2 | 9.7 | 9.5 | 31.7 | 32.8 |

SOURCE: G.E. McCall et al., Muscle afferent-pituitary axis: a novel pathway for modulating the secretion of a pituitary growth factor, *Exercise and Sport Science Reviews* 29 (2001): 164-169.

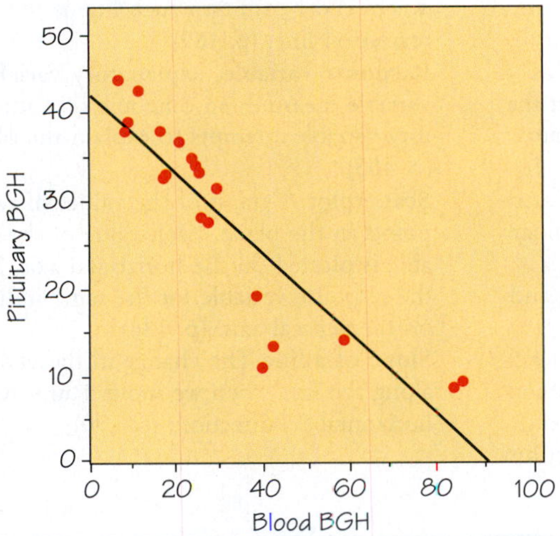


**FIGURE 6.7** This scatterplot of BGH level in pituitary tissue versus BGH level in the blood shows a strong negative association.

Figure 6.7 is a scatterplot of these data. The plot shows a strong negative straight-line association with correlation $r = -0.90$. When there is a higher BGH level in the blood, we can assume that means BGH must have been recently secreted by the pituitary gland so that less BGH is now remaining in pituitary tissue. The least-squares regression line is

$$\hat{y} = 41.081 - 0.43343x, \text{ or}$$

$$\text{predicted pituitary BGH} = 41.081 + ((-0.43343) \times \text{blood BGH}).$$

The slope $m = -0.43343$ is negative, which is consistent with how blood and pituitary tissue levels of BGH move in opposite directions. The y-intercept $b = 41.08$ is the estimated amount of BGH the pituitary gland has if it does not release any into the blood.

Furthermore, the two uppermost points of the scatterplot represent groups of rats whose slow-twitch muscles were activated, the five points on the lower right of the scatterplot involved activation of fast-twitch muscles, and all but one of the remaining points represent groups that were untreated. These data come from an *experiment* that assigned rats randomly to treatment (or no treatment) conditions to make us reasonably confident that slow-twitch muscle activation *causes* a decrease in BGH and that fast-twitch muscle activation *causes* an increase. We will discuss experiments in detail in Chapter 7.

## REVIEW VOCABULARY

**Correlation** A measure of the direction and strength of the straight-line relationship between two numerical variables. Correlations take values between 0 (no straight-line relationship) and $\pm1$ (perfect straight-line relationship). (p. 191)

**Intercept of a line** The vertical ($y$) coordinate of the point on the line above 0 on the horizontal ($x$) axis. (p. 190)

**Least-squares regression line** A line drawn on a scatterplot that makes the sum of the squares of the vertical distances of the data points from the line as small as possible. The regression line can be used to predict the response variable $y$ for a given value of the explanatory variable $x$. (p. 195)

**Negative association** Two variables are negatively associated if an increase in one variable tends to accompany a *decrease* in the other variable. The scatterplot has a northwest-to-southeast pattern, and the correlation and regression slope are both negative. (p. 187)

**Outlier** An outlier in a scatterplot is a point that lies outside the overall pattern of the other points. Outliers sometimes strongly influence the value of the correlation and the position of the least-squares regression line. (p. 187)

**Positive association** Two variables are positively associated if an increase in one variable tends to accompany a *increase* in the other variable. The scatterplot has a southwest-to-northeast pattern, and the correlation and regression slope are both positive. (p. 187)

**Regression line** Any line that describes how a response variable $y$ changes as we change an explanatory variable $x$. The most common such line is the least-squares regression line. (p. 189)

**Response variable, explanatory variable** A response variable measures an outcome of a study. An explanatory variable attempts to explain the observed outcomes. (p. 185)

**Scatterplot** A graph of the values of two variables as points in the plane. Each value of the explanatory variable is plotted on the horizontal axis, and the value of the response variable for the same individual is plotted on the vertical axis. (p. 186)

**Slope of a line** The change in the vertical ($y$) direction along the line when we move 1 unit to the right in the horizontal ($x$) direction. (p. 190)

## SKILLS CHECK

**1.** You have data for many families on the parents' income and the years of education their eldest child completes. When you make a scatterplot, the explanatory variable on the $x$-axis

**(a)** is parents' income.
**(b)** is years of education.
**(c)** doesn't matter.

**2.** You expect to see a ____________ association between parents' income and the years of education their eldest child completes.

**3.** Figure 6.8 is a scatterplot of reading test scores against IQ test scores for 15 fifth-grade children. There is one low outlier in the plot. The IQ and reading scores for this child are

**(a)** IQ = 10, Reading = 124.
**(b)** IQ = 124, Reading = 72.
**(c)** IQ = 124, Reading = 10.

**4.** The line in Figure 6.8 is a regression line for predicting reading score from IQ score. If another child in this class has IQ score 125, then _____ is the multiple

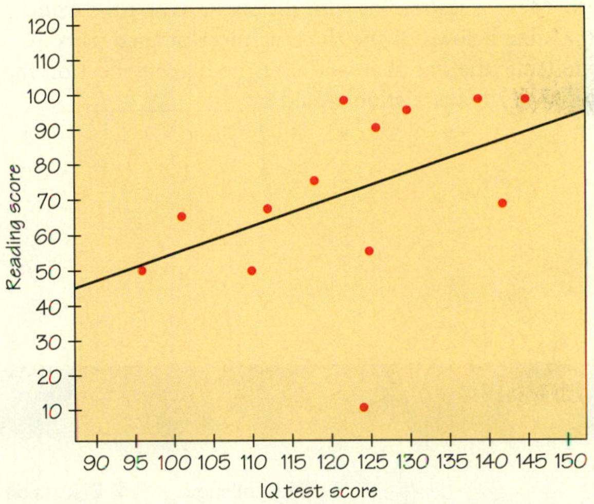


FIGURE 6.8 Scatterplot of the reading test scores of fifth-grade children (response variable) against the children's IQ scores (explanatory variable).

of 10 to which that predicted reading score would be closest.

**5.** The slope of the line in Figure 6.8 is closest to

**(a)** −1.
**(b)** 0.
**(c)** 1.

**6.** The points on a scatterplot lie close to the line whose equation is $y = 2 - 5x$. The slope of this line is ______ .

**7.** Starting with a fresh bar of soap, you weigh the bar each day after you take a shower. Then you find the regression line for predicting weight from number of days elapsed. The slope of this line will be

**(a)** positive.
**(b)** negative.
**(c)** can't tell without seeing the data.

**8.** Fred keeps his savings in his mattress. He began with \$500 from his mother and adds \$100 each year. In the form $y = mx + b$, the equation for his total savings $y$ after $x$ years would be $y =$ ____________________ .

**9.** The amount of water discharged by the Mississippi River has changed over time in roughly a straight-line pattern. A regression line for predicting water discharged (in cubic kilometers) from year is

$$\text{predicted discharge} = -7792 + (4.226 \times \text{year})$$

How much (on the average) does the volume of water increase with each passing year?

**(a)** −7792 cubic kilometers
**(b)** 4.226 cubic kilometers
**(c)** 7792 cubic kilometers

**10.** According to the regression line in the previous exercise, the predicted Mississippi River discharge in the year 2010 is _______ cubic kilometers.

**11.** You have data on the body weight $x$ and brain weight $y$ for many species of mammals. Body weight is given in kilograms and brain weight is given in grams. There are 1000 grams in a kilogram. The slope of the regression line for predicting $y$ from $x$ is $m = 1.4$. If brain weight were given in kilograms, the slope would

**(a)** still be 1.4.
**(b)** change to 0.0014.
**(c)** change to 1400.

**12.** The correlation between brain weight and body weight in Exercise 11 is $r = 0.86$. If brain weight had been measured in kilograms rather than grams, the correlation would have a value of ______________ .

**13.** Given the following set of five ordered pairs, the correlation $r$ equals

| | | | | | |
|---|---|---|---|---|---|
| $x$ | 0 | 1 | 2 | 3 | 4 |
| $y$ | 2 | 3 | 5 | 6 | 14 |

**(a)** 0.3
**(b)** 0.6
**(c)** 0.9

**14.** Suppose $y = 2x + 3$, where $x$ and $y$ are measured in meters. If $x$ is reexpressed in centimeters instead, the equation becomes $y =$ ____________________ .

**15.** The points on a scatterplot lie very close to the line whose equation is $y = 5 - 3x$. The correlation between $x$ and $y$ is close to

**(a)** −3.
**(b)** −1.
**(c)** 1.

**16.** High coffee prices give farmers in Indonesia an incentive to cut forest in order to plant more coffee. Here are data on coffee price $x$ (cents per pound) and percent $y$ of deforestation in a national park for five years:

| | | | | | |
|---|---|---|---|---|---|
| $x$ | 29 | 40 | 54 | 55 | 72 |
| $y$ | 0.49 | 1.59 | 1.69 | 1.82 | 2.98 |

Using a calculator, we can determine that the correlation between $x$ and $y$ has a value of ________ , to the nearer hundredths place.

**17.** Using the Table in Example 1 and the prediction equation in Example 3, the prediction error for Student #4 is what?

**(a)** −.01
**(b)** .01
**(c)** .13

**18.** Look again at the coffee data in Exercise 16. Using your calculator, you can find that the equation (in $y = mx + b$ form, with $m$ and $b$ to the nearest hundredths place) of the least-squares regression line for predicting $y$ from $x$ is: $\hat{y}$ = ______________________________ .

**19.** There is a strong positive correlation between the number of firefighters at a fire and the amount of damage the fire does. The reason for this is that

**(a)** more firefighters cause more damage at the fire scene.
**(b)** bigger fires require more firefighters and also do more damage.
**(c)** more damage requires more firefighters to clean it up.

**20.** Make a scatterplot with the six ordered pairs from the table below. Of the three leftmost ordered pairs in the table, the one that will have the biggest effect on the value of the correlation would be ______________ .

| | | | | | | |
|---|---|---|---|---|---|---|
| $x$ | 0 | 8 | 2 | 0 | 1 | 1 |
| $y$ | 0 | 3 | 4 | 1 | 1 | 0 |

# CHAPTER 6 EXERCISES

■ Challenge ◆ Discussion

Some exercises require use of a calculator (or software or Internet applet) that will find correlation and the slope and intercept of the least-squares regression line from keyed-in data.

**1.** In each of the following situations, is it more reasonable simply to explore the relationship between the two variables or to view one of the variables as an explanatory variable and the other as a response variable? In the latter case, which is the explanatory variable?

**(a)** The amount of time spent studying for a statistics exam and the grade on the exam.
**(b)** The weight in kilograms and height in centimeters of a person.
**(c)** Inches of rain in the growing season and the yield of corn in bushels per acre.
**(d)** A student's scores on the SAT math exam and the SAT verbal exam.

## 6.1 Displaying Relationships: Scatterplot

**2.** Figure 6.9 shows the calories and salt content (milligrams of sodium) in 17 brands of meat hot dogs. Describe the overall pattern (form, direction, and strength) of these data. In what way is the point marked A unusual?

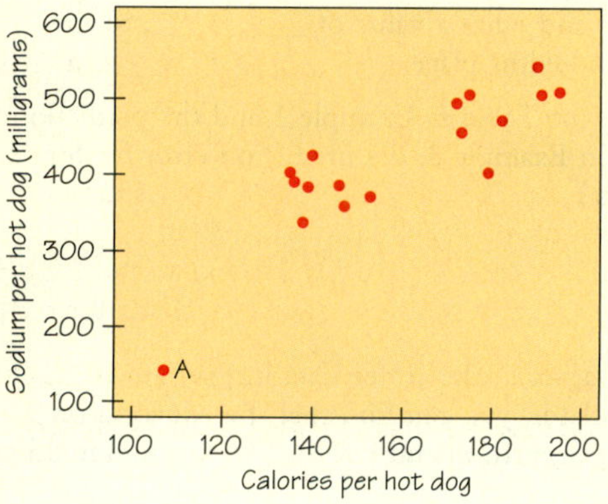


FIGURE 6.9 Scatterplot of sodium content versus calories in 17 brands of meat hot dogs, for Exercise 2.

◆ **3.** Figure 6.10 is a scatterplot of data from the World Bank. The individuals are all the world's nations for which data are available. The explanatory variable is a measure of how rich a country is, the gross domestic product (GDP) per person. GDP is the total value of the goods and services produced in a country, converted into dollars. The response variable is life expectancy at birth. Three African nations are outliers, with lower life expectancy than usual for their GDP. A full study would ask what special circumstances explain these outliers.

**(a)** Describe the direction and form of the relationship. Aside from the outliers, it is moderately strong.
**(b)** Explain why the direction and form of this relationship make sense.

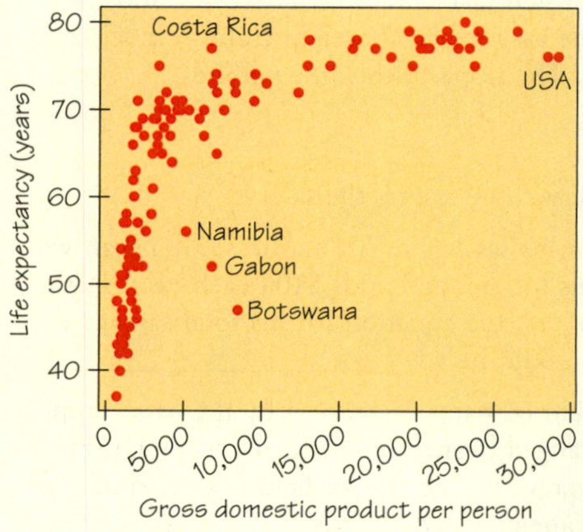


FIGURE 6.10 Scatterplot of the life expectancy of people in many nations against each nation's gross domestic product per person, for Exercise 3.

**4.** Global warming may be due to increased concentrations of greenhouse gases such as carbon dioxide ($CO_2$). Here are data from the National Oceanic and Atmospheric Administration Web site (www.noaa.gov), where $CO_2$ is measured in parts per million by volume:

| | | | | | |
|---|---|---|---|---|---|
| $CO_2$ | 316.91 | 325.68 | 338.70 | 354.16 | 369.41 |
| year | 1960 | 1970 | 1980 | 1990 | 2000 |

**(a)** Which is the explanatory variable?
**(b)** Make a scatterplot. Is the association between these variables positive or negative? Explain why you expect the relationship to have this direction.
**(c)** Describe the form and strength of the relationship.

**5.** Table 5.2 gives the city and highway gas mileages for 13 midsized cars. Omit the hybrid car (Toyota Prius) and make a scatterplot, taking city mileage as the explanatory variable. Describe in words the form, direction, and strength of the relationship between highway mileage and city mileage.

**6.** How fast do icicles grow? Here are data for one set of conditions: no wind, temperature −11°C, and water flowing over the icicle at 12 milligrams per second.

| | | | | | |
|---|---|---|---|---|---|
| Time (minutes) | 10 | 20 | 30 | 40 | 50 |
| Length (centimeters) | 0.6 | 1.8 | 2.9 | 4.0 | 5.0 |
| Time (minutes) | 60 | 70 | 80 | 90 | 100 |
| Length (centimeters) | 6.1 | 7.9 | 10.1 | 10.9 | 12.7 |
| Time (minutes) | 110 | 120 | 130 | 140 | 150 |
| Length (centimeters) | 14.4 | 16.6 | 18.1 | 19.9 | 21.0 |

SOURCE: N. Maeno et al., Growth rates of icicles, *Journal of Glaciology*, 40 (1994): 319–326.

Which is the explanatory variable? Make a scatterplot. Describe in words the direction, form, and strength of the relationship.

◆ **7.** How does the fuel consumption of a car change as its speed increases? Here are data for a British Ford Escort. Fuel consumption is measured in liters of gasoline used per 100 kilometers traveled.

| | | | | | |
|---|---|---|---|---|---|
| Speed | 10 | 20 | 30 | 40 | 50 |
| Fuel | 21.00 | 13.00 | 10.00 | 8.00 | 7.00 |
| Speed | 60 | 70 | 80 | 90 | 100 |
| Fuel | 5.90 | 6.30 | 6.95 | 7.57 | 8.27 |
| Speed | 110 | 120 | 130 | 140 | 150 |
| Fuel | 9.03 | 9.87 | 10.79 | 11.77 | 12.83 |

SOURCE: T. N. Lam, Estimating fuel consumption from engine size, *Journal of Transportation Engineering*, 111 (1985): 339–357.

**(a)** Which is the explanatory variable?
**(b)** Make a scatterplot. Describe the form of the relationship. Explain why the form of the relationship makes sense.
**(c)** How would you describe the direction of this relationship?
**(d)** Is the relationship reasonably strong or quite weak? Explain your answer.

◆ **8.** Give an example of two variables from everyday life that have a positive association. Give an example of two variables that have a negative association.

## 6.2 Making Predictions: Regression Line

**9.** Figure 6.9 shows the calories and salt content (milligrams of sodium) in 17 brands of meat hot dogs. If we ignore the outlying point marked A, a regression line for predicting sodium from calories passes close to these two observations:

calories = 139, sodium = 386 mg
calories = 191, sodium = 506 mg

Use this fact to estimate the slope of this regression line. (*Hint:* Remember that the slope of a line is the "rise" (vertical change) divided by the "run" (horizontal change) for any two points on the line.)

**10.** Exercise 4 gives data on carbon dioxide concentration (in parts per million) over time. A regression line for predicting carbon dioxide from time is

predicted carbon dioxide concentration = 314.276 + 1.3348 × (years elapsed since 1960)

**(a)** What is the slope of this line? What does the slope say about how carbon dioxide is changing over time?
**(b)** Predict the carbon dioxide concentration for 2006. In fact, the observed value was 381.85. How accurate is your prediction?

**11.** Researchers studying acid rain measured the acidity of precipitation in a Colorado wilderness area for 150 consecutive weeks. Acidity is measured by pH. Lower pH values show higher acidity. The acid rain researchers observed a straight-line pattern over time. They reported that the regression line

predicted pH = 5.43 − (0.0053 × weeks)

fit the data well. [W. M. Lewis and M. C. Grant, Acid precipitation in the western United States, *Science*, 207 (1980): 176–177.]

**(a)** Draw a graph of this line. Explain what the line says about how pH was changing over time.
**(b)** According to the regression line, what was the pH at the beginning of the study (weeks = 1)? At the end (weeks = 150)?
**(c)** What is the slope of the regression line? Explain what this slope says about the rate of change in pH.

**12.** A University of Massachusetts Amherst study published in the May 2007 *Journal of Marriage and Family* found that married women do about one less hour of housework a week for every \$7500 they earn as full-time workers outside the home, regardless of the husband's income.

**(a)** What would be the numerical value of the slope coefficient in the regression model that predicts women's housework from their income? What does the sign of the slope (positive or negative) tell us about the relationship between these variables?
**(b)** Suppose Lynette's salary is \$30,000 greater than Gabrielle's. What would you predict to be the difference in hours of housework they each do?

**13.** If heterosexual women always married men who were two years older than they are, what would be the slope of the regression line for predicting husband's age from wife's age? (*Hint:* Draw a scatterplot for several ages.)

**14.** Suppose that the slope of the regression line of weight on height for a group of young men is $m = 1.1$ when we measure height $x$ in centimeters and weight $y$ in kilograms. That is, when height increases by 1 centimeter, weight increases by 1.1 kilograms. There are 1000 grams in a kilogram. If we measured weight in grams, what would be the slope?

## 6.3 Correlation

**15.** Find the correlation between the city and highway gas mileages for the 12 non-hybrid midsized cars in Table 5.2. (That is, omit the Toyota Prius.) Explain why the value of $r$ matches the scatterplot that you made in Exercise 5.

**16.** Exercise 4 gives data on carbon dioxide concentration (in parts per million) over time.

**(a)** Use a calculator to find the correlation $r$. Explain from looking at the scatterplot why this value of $r$ is reasonable.
**(b)** Suppose that the concentration had been recorded in parts per billion instead of parts per million. For example, the value 354.16 would become 354160. How would the value of $r$ change?

**17.** Find the correlation between city and highway mileage for all 13 midsized cars in Table 5.2, including the Toyota Prius. Compare your $r$ with the value you found in Exercise 15. Explain why adding the Prius changes $r$ in this direction.

**18.** Find the correlation between time and icicle length for the data in Exercise 6. Explain why the value of $r$ matches the scatterplot that you made in Exercise 6.

**19.** Exercise 7 gives data on gas used versus speed for a small car. Make a scatterplot, if you did not do so in Exercise 7. Calculate the correlation. Explain why $r$ is close to 0 despite a strong relationship between speed and gas use.

**20.** The length of the icicle in Exercise 6 is measured in centimeters. There are 2.54 centimeters in an inch. If length were measured in inches, how would the correlation you found in Exercise 18 change?

**21.** If heterosexual women always dated men who are three years older than they are, what would be the correlation between the ages of the man and the woman? (*Hint:* Draw a scatterplot for several ages.)

**22.** We want to find the correlation:

**(a)** between the heights of fathers and the heights of their adult sons.
**(b)** between the heights of husbands and the heights of their wives.
**(c)** between the heights of women at age 4 and their heights at age 18.

The answers (in scrambled order) are $r = 0.2$, $r = 0.5$, and $r = 0.8$. Match the answers to the variable pairings and explain your choice.

**23.** For each of the following pairs of variables, would you expect a substantial negative correlation, a substantial positive correlation, or a small correlation?

**(a)** The age of used cars and their prices.
**(b)** The weight of new cars and their gas mileages in miles per gallon.
**(c)** The heights and the weights of adult men.
**(d)** The heights and the IQ scores of adult men.

◆ **24.** Each of the following statements contains a mistake. Explain what is wrong in each case.

**(a)** "There is a high correlation ($r = 0.89$) between the hair color of American workers and their income."
**(b)** "We found a high correlation ($r = 1.09$) between students' ratings of faculty teaching and ratings made by other faculty members."
**(c)** "The correlation between age and income was found to be $r = 0.53$ years."

■ **25.** Mutual-fund reports often give correlations to describe how the prices of different investments are related. You look at the correlations between three Fidelity funds and the Standard & Poor's 500 Stock Index, which describes stocks of large U.S. companies. The three funds are Dividend Growth (stocks of large U.S. companies), Small Cap Stock (stocks of small U.S. companies), and Emerging Markets (stocks in developing countries). For a recent year, the three correlations are $r = 0.35$, $r = 0.81$, and $r = 0.98$.

**(a)** Which correlation goes with each fund? Explain your answer.
**(b)** The correlations of the three funds with the index are all positive. Does this tell you that stocks went up that year? Explain your answer.

■ **26.** *Archaeopteryx* is an extinct beast having feathers like a bird but teeth and a long bony tail like a reptile. Only six fossil specimens are known. If the specimens belong to the same species and differ in size because some are younger than others, there should be a straight-line relationship between the lengths of a pair of bones from all individuals. An outlier from this relationship would suggest a different species. Here are data on the lengths in centimeters of the femur (a leg bone) and the humerus (a bone in the upper arm) for the five specimens that preserve both bones.

| | | | | | |
|---|---|---|---|---|---|
| Femur length $x$ | 38 | 56 | 59 | 64 | 74 |
| Humerus length $y$ | 41 | 63 | 70 | 72 | 84 |

SOURCE: M. A. Houck et al., Allometric scaling in the earliest fossil bird, *Archaeopteryx lithographica*, *Science*, 247 (1990): 195–198.

**(a)** Make a scatterplot. Do you think that all five specimens come from the same species?
**(b)** Find the correlation $r$ step by step. *Step 1*: Find the mean $\bar{x}$ and standard deviation $s_x$ of the five femur lengths. Find the mean $\bar{y}$ and the standard deviation $s_y$ of the five humerus lengths. (Use your calculator.) *Step 2*: Find the standardized values $(x - \bar{x})/s_x$ of each of the five femur lengths, and do the same for the five humerus lengths. *Step 3*: Substitute your numbers from Steps 1 and 2 into the formula for $r$.
**(c)** Now use one of the faster methods in Spotlight 6.2 to find $r$ and check that you get the same result as in part (b).

## 6.4 Least-Squares Regression

**27.** In Exercise 5 you made a scatterplot of city and highway gas mileage for the 12 non-hybrid midsized cars (omitting the Prius) in Table 5.2.

**(a)** What is the least-squares regression line for predicting highway mileage from city mileage?

(David R. Frazier Photolibrary, Photo Researchers.)

**(b)** If a midsized car gets 17 mpg in the city, predict its highway mileage.
**(c)** Based on the scatterplot you made in Exercise 5, do you expect the prediction in part (b) to be quite accurate? Why?

**28.** In Exercise 6 you made a scatterplot of the length of an icicle and the number of minutes water has been flowing over the icicle.

**(a)** What is the equation of the least-squares regression line for predicting icicle length from time?
**(b)** Use your regression line to predict the length of the icicle after 75 minutes.

**29.** Redo your scatterplot of highway mileage against city mileage from Exercise 5. Add your regression line from Exercise 27 to the plot. Be sure to show how you were able to plot the line starting with its equation. Finally, use the "up-and-across" method illustrated in Figure 6.3 to show the predicted highway mileage of a car that gets 18 mpg in the city.

**30.** Redo your scatterplot of icicle length against time from Exercise 6. Add your regression line from Exercise 28 to the plot. Be sure to show how you were able to plot the line starting with its equation. Finally, use the "up-and-across" method illustrated in Figure 6.3 to show the predicted length of the icicle after 75 minutes.

**31.** Exercise 7 gives data on gas used versus speed for a small car. Make a scatterplot, if you did not do so in Exercise 7. The least-squares regression line for these data is

$$\text{predicted fuel} = 11.058 - 0.0147 \times \text{speed}$$

Draw this line on your scatterplot. What are the predicted and observed fuel consumption values for speeds of 10, 70, and 150 kilometers per hour (km/h)? *You can fit a regression line to any set of two-variable data. The line is of little use if the plot does not show a straight-line pattern.*

**32.** The length of the icicle in Exercise 6 is measured in centimeters. There are 2.54 centimeters in an inch. If length were measured in inches, how would the slope of the regression line you found in Exercise 28 change?

**33.** The mean height of American women in their early twenties is about 64.5 inches and the standard deviation is about 2.5 inches. The mean height of men the same age is about 68.5 inches, with standard deviation about 2.7 inches. If the correlation between the heights of husbands and wives is about $r = 0.5$, what is the equation of the regression line of the husband's height on the wife's height in young couples? Predict the height of the husband of a woman who is 67 inches tall.

**34.** This data, from the National Oceanic and Atmospheric Administration Web site (www.noaa.gov), is the mean annual number of named Atlantic storms (hurricanes, tropical storms, tropical depressions), during five-year windows ending with the year shown in the table.

| | | | | | | |
|---|---|---|---|---|---|---|
| 2007 | 2002 | 1997 | 1992 | 1987 | 1982 | 1977 |
| 16.2 | 13.6 | 11 | 10.4 | 8.2 | 10 | 8.8 |
| 1972 | 1967 | 1962 | 1957 | 1952 | 1947 | 1942 |
| 11.2 | 9.2 | 8.8 | 10.6 | 10.4 | 9.4 | 7.4 |

**(a)** What is the slope of the least-squares regression line of named storms on year? What is the intercept?
**(b)** Use the regression line to predict the number of named storms for the five-year window 2008–2012.

■ 35. Use the equation for the least-squares regression line to show that this line always passes through the point $(\bar{x}, \bar{y})$. That is, set $x = \bar{x}$ and show that the line predicts that $y = \bar{y}$.

■ **36.** Exercise 6 gives data on the growth of an icicle.

**(a)** Find the mean and standard deviation of the times and icicle lengths. Find the correlation between the two variables. Use these five numbers to find the equation of the regression line for predicting length from time. Verify that your result agrees with that in Exercise 28.
**(b)** Use the same five numbers to find the equation of the regression line for predicting from an icicle's length the time it has been growing. Use your line to predict the time that an icicle 15 centimeters long has been growing. *There is just one correlation between two variables, but there are two different least-squares lines, depending on which you choose as the response variable.*

■ **37.** Fidelity Investments, like other large mutual fund companies, offers many "sector funds" that concentrate their investments in narrow segments of the stock market. These funds often rise or fall by much more than the market as a whole. Here are the percent returns for 23 Fidelity "Select Portfolios" funds for the years 2002 (when stocks fell) and 2003 (when stocks went up).

| 2002 return | 2003 return | 2002 return | 2003 return | 2002 return | 2003 return |
|---|---|---|---|---|---|
| −17.1 | 23.9 | −0.7 | 36.9 | −37.8 | 59.4 |
| −6.7 | 14.1 | −5.6 | 27.5 | −11.5 | 22.9 |
| −21.1 | 41.8 | −26.9 | 26.1 | −0.7 | 36.9 |
| −12.8 | 43.9 | −42.0 | 62.7 | 64.3 | 32.1 |
| −18.9 | 31.1 | −47.8 | 68.1 | −9.6 | 28.7 |
| −7.7 | 32.3 | −50.5 | 71.9 | −11.7 | 29.5 |
| −17.2 | 36.5 | −49.5 | 57.0 | −2.3 | 19.1 |
| −11.4 | 30.6 | −23.4 | 35.0 | | |

Do a careful statistical analysis of these data using both graphs and whatever numerical measures you think are appropriate. Make a side-by-side comparison of the distributions of returns in 2002 and 2003 and also describe the relationship between the returns of the same funds in these two years. What are your most important findings? (The outlier is Fidelity Gold Fund.)

## 6.5 Interpreting Correlation and Regression

**38.** Make a scatterplot of the following data:

| | | | | | | |
|---|---|---|---|---|---|---|
| $x$ | 1 | 2 | 3 | 4 | 10 | 10 |
| $y$ | 1 | 3 | 3 | 5 | 1 | 11 |

Use your calculator to show that the correlation is about 0.5. What feature of the data is responsible for reducing the correlation to this value despite a strong straight-line association between $x$ and $y$ in most of the observations?

**39.** Table 6.1 on the next page has four data sets prepared by statistician Frank Anscombe to show dangers of calculating without first plotting the data.

**(a)** Without making scatterplots, find the correlation and the least-squares regression line for all four data sets. What do you notice? Use the regression line to predict $y$ for $x = 10$.
**(b)** Make a scatterplot for each of the data sets and add the regression line to each plot.
**(c)** In which of the four cases would you be willing to use the regression line to describe the dependence of $y$ on $x$? Explain your answer in each case.

◆ **40.** Children who watch many hours of television get lower grades in school on the average than those who watch less TV. Explain clearly why this fact does not show that watching TV *causes* poor grades. In particular, suggest some other characteristics of households where children watch lots of TV that may contribute to poor grades.

◆ **41.** People who use artificial sweeteners in place of sugar tend to be heavier than people who use sugar. Does this mean that artificial sweeteners cause weight gain? Give a more plausible explanation for this association.

◆ **42.** "Based on an examination of twenty-two companies that announced large layoffs during 1994, Downs found a strong (.31) correlation between the size of the layoffs and the compensation of the CEOs." [K. Phillips, *Wealth and Democracy*, Broadway Books, New York, 2002, p. 151.] Discuss why this correlation is probably explained by a third variable, the size of the company as measured by its number of employees.

■ **43.** "The positive correlation between health and income per capita is one of the best-known relations in international development. This correlation is commonly thought to reflect a causal link running from income to health. . . . Recently, however, another intriguing possibility has emerged: that the health-income correlation is partly explained by a causal link running the other way–from health to income." [D. E. Bloom and D. Canning, The health and wealth of nations, *Science,* 287 (2000): 1207–1208.] Explain how higher income in a nation can cause better health. Then explain how better health can cause higher national income. There is no simple way to determine the direction of the link.

■ **44.** The effect of an outside variable can be surprising when individuals are divided into groups. In recent years, the mean SAT score of all high school seniors has increased. But the mean SAT score has decreased for students at each level of high school grades (A, B, C, and so on). Explain how grade inflation in high school

**TABLE 6.1** Four Data Sets for Exploring Correlation and Regression

**Data Set A**

| | | | | | | | | | | | |
|---|---|---|---|---|---|---|---|---|---|---|---|
| $x$ | 10 | 8 | 13 | 9 | 11 | 14 | 6 | 4 | 12 | 7 | 5 |
| $y$ | 8.04 | 6.95 | 7.58 | 8.81 | 8.33 | 9.96 | 7.24 | 4.26 | 10.84 | 4.82 | 5.68 |

**Data Set B**

| | | | | | | | | | | | |
|---|---|---|---|---|---|---|---|---|---|---|---|
| $x$ | 10 | 8 | 13 | 9 | 11 | 14 | 6 | 4 | 12 | 7 | 5 |
| $y$ | 9.14 | 8.14 | 8.74 | 8.77 | 9.26 | 8.10 | 6.13 | 3.10 | 9.13 | 7.26 | 4.74 |

**Data Set C**

| | | | | | | | | | | | |
|---|---|---|---|---|---|---|---|---|---|---|---|
| $x$ | 10 | 8 | 13 | 9 | 11 | 14 | 6 | 4 | 12 | 7 | 5 |
| $y$ | 7.46 | 6.77 | 12.74 | 7.11 | 7.81 | 8.84 | 6.08 | 5.39 | 8.15 | 6.42 | 5.73 |

**Data Set D**

| | | | | | | | | | | | |
|---|---|---|---|---|---|---|---|---|---|---|---|
| $x$ | 8 | 8 | 8 | 8 | 8 | 8 | 8 | 8 | 8 | 8 | 19 |
| $y$ | 6.58 | 5.76 | 7.71 | 8.84 | 8.47 | 7.04 | 5.25 | 5.56 | 7.91 | 6.89 | 12.50 |

SOURCE: Frank J. Anscombe, Graphs in statistical analysis, *The American Statistician*, 27 (1973): 17–21.

can account for this pattern. *A relationship that holds for each group within a population need not hold for the population as a whole. In fact, the relationship can even change direction.*

## Chapter Review

**45.** Recent major recalls of toys with lead paint refocused people on the dangers of lead exposure. Below is data from research exploring the association with student achievement for blood lead levels below the "danger threshold" of 10 micrograms/deciliter set by the Centers for Disease Control [M.L. Miranda et al., The relationship between early childhood blood lead levels and performance on end-of-grade tests, *Environmental Health Perspectives*, 115 (2007): 1242-1247].

| Blood lead level | 1 | 2 | 3 | 4 | 5 |
|---|---|---|---|---|---|
| Mean 4th grade reading score | 255.9 | 253.8 | 252.6 | 251.0 | 250.4 |
| Blood lead level | 6 | 7 | 8 | 9 | |
| Mean 4th grade reading score | 249.5 | 248.5 | 247.8 | 249.3 | |

What are the explanatory and response variables? Do you expect a positive or negative association between these variables? Why? Does the scatterplot support this?

**46.** A study of reading ability in schoolchildren chose 60 fifth-grade children at random from a school. The researchers had the children's scores on an IQ test and on a test of reading ability. Figure 6.11 plots reading test score (response) against IQ score (explanatory).

**(a)** Explain why we should expect a positive association between IQ and reading score for children in the same grade. Does the scatterplot show a positive association?
**(b)** A group of four points appear to be outliers. In what way do these children's IQ and reading scores deviate from the overall pattern?
**(c)** Ignoring the outliers, is the form of the association between IQ and reading scores roughly a straight line? Is it very strong? Explain your answers.

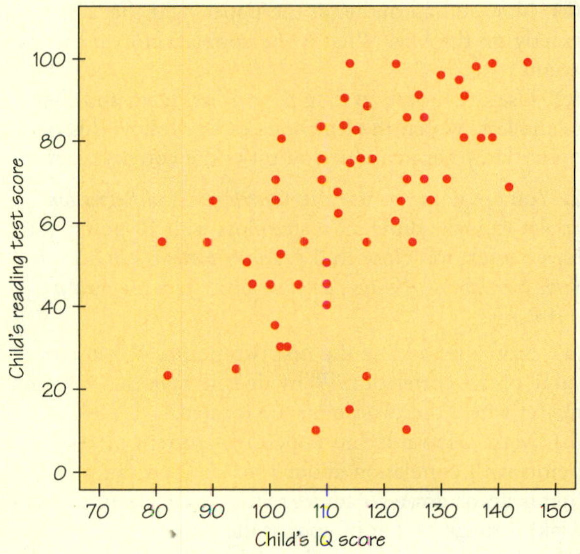


FIGURE 6.11 IQ and reading test scores for 60 fifth-grade children, for Exercise 46.

**47.** A student wonders if tall women tend to date taller people than do short women. She measures herself, her sister, and the women in the adjoining dorm rooms. Then she measures the next person each woman dates and obtains these data (in inches):

| | | | | | | |
|---|---|---|---|---|---|---|
| Heights of women ($x$) | 66 | 64 | 63 | 65 | 70 | 65 |
| Heights of their dates ($y$) | 72 | 68 | 70 | 68 | 71 | 64 |

**(a)** Based on a scatterplot (with the women's heights as the explanatory variable), do you expect the correlation to be positive or negative? Near $\pm 1$ or not?
**(b)** Find the correlation $r$ between the heights of the women and their dates.

**48.** In Exercise 47, you found the correlation $r$ between the heights in inches of several college women and the heights in inches of the next person each woman dates.

**(a)** How would $r$ change if all the dates were 2 inches shorter than the heights given in the table?
**(b)** How would $r$ change if heights were measured in centimeters rather than inches? (There are 2.54 centimeters in an inch.)

**49.** The equation of the least-squares regression line for predicting dates' heights from women's heights for the data in Exercise 47 is

$$\text{predicted height of date} = 41.08 + 0.42 \times \text{woman's height}$$

**(a)** What is the slope of this line? Explain in simple language what the numerical value of the slope tells us about the heights of the people these women date.
**(b)** Use the regression line to predict the height of the next person dated by a woman who is 67 inches tall.

**50.** In *Stats* #49, Schuyler Huck (2008) presents a data set of 100 ordered pairs in which 25 of them are (17, 1), 25 are (18, 2), 25 are (19, 3) and 25 are (20, 4).

**(a)** Without doing much formal calculation, find the value of $r$ and the slope of the least-squares regression line.
**(b)** Now, suppose someone adds the 101st point to the data set: the ordered pair (1, 20). Predict the new value of $r$ and the slope of the regression line, then do a calculation to see how close your answer is.

## APPLET EXERCISES

**To do these exercises, go to www.whfreeman.com/fapp8e.**

**1.** In the *Correlation and Regression* applet, imitate Figure 6.5. That is, click to locate five points at the lower left of the scatterplot, then click "Show least-squares line."

**(a)** What is the correlation $r$ for these five points? If necessary, move points with the mouse to get a value near $r = 0.5$, as in Example 6.
**(b)** Now add an outlier at the upper right that lies exactly on the line. What is the correlation $r$ for the six points?
**(c)** Use the mouse to drag the outlier down and then to the left. Watch the least-squares line follow this one point. How negative can you make the correlation $r$?

**2.** You are going to use the *Correlation and Regression* applet to make different scatterplots with 10 points that have correlation close to 0.7. *Many patterns can have the same correlation. Always plot your data before you trust a correlation.*

**(a)** Stop after adding the first two points. What is the value of the correlation? Why does it have this value no matter where the two points are located?
**(b)** Make a lower-left to upper-right pattern of 10 points with correlation about $r = 0.7$. (You can drag points up or down to adjust $r$ after you have 10 points.) Make a rough sketch of your scatterplot.
**(c)** Make another scatterplot with nine points in a vertical stack at the left of the plot. Add one point far to the right and move it until the correlation is close to 0.7. Make a rough sketch of your scatterplot.
**(d)** Make yet another scatterplot with 10 points in a curved pattern that starts at the lower left, rises to the right, then falls again at the far right. Adjust the points up or down until you have a quite smooth curve with correlation close to 0.7. Make a rough sketch of this scatterplot also.

**3.** It isn't easy to guess the position of the least-squares line by eye. Use the *Correlation and Regression* applet to compare a line you draw with the least-squares line. Click on the scatterplot to create a group of 15 to 20 points from lower left to upper right with a clear, positive straight-line pattern (with correlation around 0.7). Click the "Draw line" button and use the mouse to draw a line through the middle of the cloud of points from lower left to upper right. Note the "thermometer" that appears above the plot. The red portion is the sum of the squared vertical distances from the points in the plot to the least-squares line. The green portion is the "extra" sum of squares for your line–it shows by how much your line misses the smallest possible sum of squares.

**(a)** You drew a line by eye through the middle of the pattern. Yet the right-hand part of the bar is probably almost entirely green. What does that tell you?
**(b)** Now click the "Show least-squares line" box. Is the slope of the least-squares line smaller (the new line is less steep) or larger (line is steeper) than that of your line? If you repeat this exercise several times, you will consistently get the same result. *The least-squares line*

*minimizes the vertical distances of the points from the line. It is not the line through the "middle" of the cloud of points.* This is one reason why it is hard to draw a good regression line by eye.

## WRITING PROJECTS

**1.** Choose two variables that you think have a roughly straight-line relationship. Gather data on these variables and do a statistical analysis: Make a scatterplot, find the correlation, find the regression line (use a statistical calculator or software), and draw the line on your plot. Then write a short report on your work. Some examples of suitable pairs of variables are:

**(a)** The height and arm span of a group of people
**(b)** The height and walking stride length of a group of people
**(c)** The price per ounce and bottle size in ounces for several brands of shampoo and several bottle sizes for each brand

**2.** Can Regression Help Protect Voting Rights? This example is adapted from the author's work as a statistician for the Texas Legislative Council. To comply with the Voting Rights Act, a state cannot redraw its districts in a way that dilutes the voting strength of a protected group. Because we cannot know how individuals voted, we cannot directly measure if minority and majority persons tend to prefer different candidates. While there are technical details and assumptions we cannot fully discuss here, you can begin to understand how this might be estimated by exploring the following data set for 9 equal-sized districts:

| | | | |
|---|---|---|---|
| Y (% of voters that voted for Gomez) | 14 | 7 | 19 |
| X (% of voters who are Hispanic) | 12 | 18 | 24 |
| District # | 1 | 2 | 3 |
| Y (% of voters that voted for Gomez) | 27 | 37 | 36 |
| X (% of voters who are Hispanic) | 36 | 42 | 53 |
| District # | 4 | 5 | 6 |
| Y (% of voters that voted for Gomez) | 53 | 48 | 65 |
| X (% of voters who are Hispanic) | 68 | 79 | 86 |
| District # | 7 | 8 | 9 |

Produce a scatterplot, correlation value, and regression equation. Describe the relationship between concentration of Hispanic population and proportion of votes that went to the Hispanic candidate Gomez. Give a practical interpretation of the value of the slope coefficient. Give a practical interpretation of the value of Y that would be predicted when X = 0 and when X = 100.

## SUGGESTED READINGS

CLEVELAND, WILLIAM S. *The Elements of Graphing Data*, rev. ed., Hobart Press, Summit, N.J., 1994. A careful study of the most effective elementary ways to present data graphically, with much sound advice on improving graphs such as scatterplots.

LESSER, LAWRENCE M. The 'Ys' and 'why nots' of line of best fit, *Teaching Statistics*, 21(2) (1999): 54-55.

MOORE, DAVID S. *The Basic Practice of Statistics*, 4th ed., Freeman, New York, 2006. Chapters 4 and 5 of BPS give more extensive treatment of FAPP Chapter 6 material, at about the same mathematical level.

## SUGGESTED WEB SITES

The Web sites suggested in Chapter 5 as sources for data provide data to investigate relationships as well. How is the number of medical doctors per 100,000 people in the states related to how rich the state is? To infant mortality in the state? To the cost of medical care? You can study these and many other relationships using data from Web sites such as www.census.gov or www.fedstats.gov.

The *Journal of Statistics Education*, www.amstat.org/publications/jse/, is an international electronic journal on teaching statistics, and many articles include interesting data and examples. Look, for example, at "Exploring Relationships in Body Dimensions" by G. Heinz and L. J. Peterson, in Volume 11, Number 2. (Click on "Archive" to see a list of past issues.) Here you will find information about measuring body dimensions, actual data from 247 men and 260 women, and some examples of both distributions and relationships.

CAUSEweb, www.causeweb.org, is a searchable digital library of resources on a wide range of statistics topics operated by the Consortium for the Advancement of Undergraduate Statistics Education.

A useful applet is illuminations.nctm.org/ActivityDetail.aspx?ID=146.

CHAPTER 7



# Data for Decisions

Numerical data are raw material for the growth of knowledge. The discipline of statistics provides tools that help us gain knowledge rather than confusion from data. These tools include methods for producing and interpreting data.

Chapters 5 and 6 explored the art of *data analysis.* They showed how to use graphs and numbers to uncover the nature of a set of data. We can distinguish two different attitudes toward data. We may ask, "What do we see in the data?" Or we may ask, "What answer do the data give to this specific question?" For example, we may want data to answer the question "What car colors are most popular among young adults?" or "Do recent college graduates with higher college debt tend to also have higher incomes?" Data that answer such questions are a product of intelligent effort, just like hybrid tomatoes and DVD players. In this chapter, we first address the issue of *how to produce data* that can be trusted to answer specific questions. You will learn to recognize good and bad methods of producing data. Understanding how to produce trustworthy data is the first–and the most important–step toward the ability to judge whether conclusions based on data are reliable or not.

Statistical techniques for producing data open the door to a further advance in data analysis, *formal statistical inference,* which answers specific questions with a known degree of confidence. **Statistical inference** uses the descriptive tools we have already met, in combination with new kinds of reasoning. We will look briefly at some of this reasoning in the final part of this chapter.

## 7.1 Sampling

- A political scientist wants to know what percent of college-age adults consider themselves to be conservatives.
- An automaker hires a market research firm to learn what percent of adults aged 18 to 35 recall seeing television advertisements for a new sport utility vehicle.

- Government economists inquire about average household income.

In all these cases, we want to gather information about a large group of individuals. Time, cost, and inconvenience preclude contacting every individual. So we gather information about only part of the group in order to draw conclusions about the whole. Also, when an observation is destructive, it is necessary to use only a sample. For example, testing a shipment of fuses to see if they are defective would wipe out the whole shipment if every single fuse were tested. And if your doctor's appointment includes a blood test, you want only *some* of your blood removed!

**Population** DEFINITION

The **population** in a statistical study is the entire group of individuals about which we want information.

**Sample** DEFINITION

A **sample** is a part of the population from which we actually collect information used to draw conclusions about the whole. Sampling refers to the process of choosing a sample from the population.

We often draw conclusions about a whole on the basis of a sample. Everyone has sipped a spoonful of soup and judged the entire bowl on the basis of that taste. But a bowl of soup is homogeneous, so that the taste of a single spoonful represents the whole. On the other hand, a spoonful of salad dressing may be misleading since its elements may separate if the bottle has not been shaken recently. Choosing a representative sample from a large and varied population is not so easy. The first step in a proper *sample survey* is to say carefully just what population we want to describe. The second step is to say exactly what we want to measure. These preliminary steps can be complicated, as the following example illustrates.

## EXAMPLE 1

### How Can a Survey Measure Unemployment?

The monthly unemployment rate comes from the government's Current Population Survey (CPS; www.census.gov/cps/), a sample of about 60,000 households each month conducted by the Census Bureau (see Figure 7.1). To measure unemployment, we must first specify the population we want to describe. Which age groups will we include? Will we include illegal aliens or people in prisons? The CPS defines its population as all U.S. residents (whether citizens or not) 16 years of age and over who are civilians and are not in an institution such as a prison. The civilian unemployment rate announced in the news refers to this specific population.

The second question is harder: What does it mean to be "unemployed"? Someone who is not looking for work–for example, a full-time student–should not be called unemployed just because she is not working for pay. If you are chosen for the CPS sample, the interviewer first asks whether you are available to work and whether you actually looked for work in the past four weeks. If not, you are neither employed nor unemployed–you are not in the labor force.

If you are in the labor force, the interviewer goes on to ask about employment. Any work for pay or in your own business the week of the survey counts you as employed. So does at least 15 hours of unpaid work in a family business. You are also employed if you have a job but didn't work because of vacation, being on strike, or for some other good reason. An unemployment rate of 4.7% means that 4.7% of the sample was unemployed, using the exact CPS definitions of both "labor force" and "unemployed."

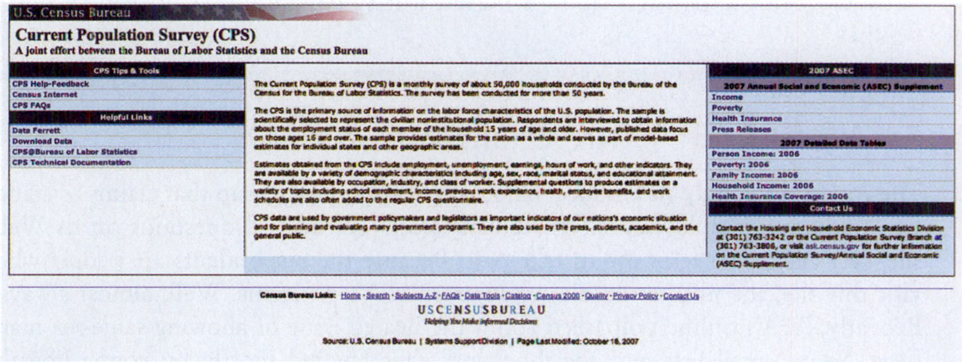


**FIGURE 7.1** The Web page of the Current Population Survey.

## 7.2 Bad Sampling Methods

How can we choose a sample that is truly representative of the population? The easiest–but not the best–way to select a sample is to choose individuals close at hand. If we are interested in finding out how many people have jobs, for example, we might go to a shopping mall and ask people passing by if they are employed.

**Convenience Sample** DEFINITION

A **convenience sample** is a sample of individuals who are selected because they are members of a population who are the most convenient to reach, such as people passing by in the street. Usually, such a sample cannot be trusted to be representative of the population.

### EXAMPLE 2

### The Inconvenient Truth About Convenience Samples

Taking a sample of shoppers at a mall seems like a fast and inexpensive way of finding out Americans' opinions. But people at malls tend to be more prosperous than typical Americans. They are also more likely to be teenagers or retired. Also, when we decide which people to question, we may tend (even unconsciously) to avoid poorly dressed or tough-looking individuals. In short, our shopping mall interviews will result in a sample that is not representative of the entire population because we underrepresent those people we may avoid.

Closer to home, your professor may try to "sample" the understanding the class has about a topic by calling on the next two students who raise their hands or by simply asking the nearest two students on the front row. If students who sit near the front and/or raise their hands have higher levels of preparation, interest and engagement, the professor may overestimate how well the class understands the material.

In both cases, the inaccuracies obtained cannot simply be explained as a sample's "bad luck," but are likely to happen every time with the same pattern because unscientific sampling methods have **bias**. In this context, bias refers matter-of-factly to the built-in systematic error of the procedure itself and not to the kind of political or personal bias an individual human being may have.

**Bias** DEFINITION

The design of a statistical study is **biased** if it systematically favors certain outcomes.

## EXAMPLE 3 ■ Are Online Polls in Line?

The American Family Association (AFA) is a conservative group that claims to stand for "traditional family values." It regularly posts online poll questions on its Web site–just click on a response to take part. Because the respondents are people who visit this site, the poll results always support AFA's positions. Well, almost always. Recently, AFA's online poll asked about the heated issue of allowing same-sex marriage. Soon, email lists and social-network sites favored mostly by young liberals pointed to the AFA poll. Almost 850,000 people responded, and 60% of them favored legalization of same-sex marriage. AFA claimed that homosexual rights groups had skewed its poll. ■

Online polls are now everywhere–some sites will even provide help in conducting your own online poll. As the AFA poll illustrates, you can't trust the results. People who take the trouble to respond to an open invitation are not representative of the entire adult population. That's true of regular visitors to AFA's site, of the activists who made a special effort to vote in the marriage poll, and of the people who bother to respond to write-in, call-in, or online polls in general. Polls like these are examples of **voluntary response sampling**.

**Voluntary Response Sample** DEFINITION

A **voluntary response sample** consists of people who choose themselves by responding to a general appeal. Voluntary response samples are biased because people with strong opinions are most likely to respond.

# 7.3 Simple Random Samples

In a voluntary response sample, people choose whether to respond. In a convenience sample, the interviewer makes the choice. In both cases, personal choice produces bias. The statistician's remedy is to allow impersonal chance to choose the sample. A sample chosen by chance allows neither favoritism by the sampler nor self-selection by respondents. Choosing a sample by chance avoids bias by giving all individuals an equal chance to be chosen. Any individual has the same chance to be in the sample, whether rich or poor, young or old, black or white, and so on.

The simplest way to use chance to select a sample is to place names (the population) in a hat and draw out a handful (the sample). This is the idea of **simple random sampling**.

**Simple Random Sample** DEFINITION

A **simple random sample (SRS)** of size *n* consists of *n* individuals from the population chosen in such a way that every set of *n* individuals has an equal chance to be the sample actually selected.

Picturing drawing names from a hat helps us understand what an SRS is. The same picture helps us see that an SRS is a better method of choosing samples than convenience or voluntary response sampling because it doesn't favor any part of the population. But writing names on slips of paper and drawing them from a hat is slow and inconvenient. That's especially true if, as in the Current Population Survey, we must draw a sample of size 60,000. We can speed up the process by using a **table of random digits**. In practice, samplers use computers to do the work, but we can do it by hand for small samples.

**Table of Random Digits** DEFINITION

A **table of random digits** is a list of the digits 0, 1, 2, 3, 4, 5, 6, 7, 8, 9 with these two properties:

1. Each entry in the table is equally likely to be any of the 10 digits 0 through 9.
2. The entries are independent of one another. That is, knowledge of one part of the table gives no information about any other part.

Table 7.1 is a table of random digits. The digits in the table appear in groups of five to make the table easier to read, and the rows are numbered so we can refer to them, but the groups and row numbers are just for convenience. The entire table is one long string of randomly chosen digits. There are two steps in using the random-digit table to choose a simple random sample:

**Step 1. Label** Give each member of the population a numerical label of the *same length*. Up to 100 items can be labeled with two digits: 01, 02, . . . , 99, 00. Up to 1000 items can be labeled with three digits, and so on.

**Step 2. Table** To choose a simple random sample, read from Table 7.1 successive groups of digits of the length you used as labels. Your sample contains the individuals whose labels you find in the table. This gives all individuals the same chance because all labels of the same length have the same chance of being found in the table. For example, any pair of digits in the table is equally likely to be any of the 100 possible labels 01, 02, . . . , 99, 00. Ignore any group of digits that was not used as a label or that duplicates a label already in the sample.

## EXAMPLE 4 ■ Sampling Songs

Professor Lesser has all 27 songs from *The Beatles One* CD stored on a digital media player and wants to play 4 randomly chosen songs to accompany his morning commute.

**Step 1. Label** Give each song a numerical label. Because two digits are needed to label the 27 songs, all labels will have two digits. In the table below, we have listed the 27 songs with labels from 01 to 27. Always specify how you label the members of the population. If the player had 500 songs, we would label them 001, 002, …, 499, 500.

(*Steve Prezant/Corbis.*)

01 Love Me Do
02 From Me to You
03 She Loves You
04 Ticket to Ride
05 Can't Buy Me Love
06 A Hard Day's Night
07 I Feel Fine
08 Eight Days a Week
09 I Want to Hold Your Hand
10 Help!
11 Yesterday
12 Daytripper
13 We Can Work it Out
14 Paperback Writer
15 Yellow Submarine
16 Eleanor Rigby
17 Penny Lane
18 The Ballad of John and Yoko
19 Hello, Goodbye
20 Lady Madonna
21 Hey Jude
22 Get Back
23 All You Need is Love
24 Something
25 Come Together
26 Let it Be
27 The Long and Winding Road

**Step 2. Table** Go to Table 7.1 and pick any row (we picked line 125). Now read across that row, left to right, two digits at a time, until you have chosen 4 songs. Remember to skip any two-digit groups that repeat numbers already chosen (like 18) and also skip any groups representing numbers beyond the size of our population (27). Here's how you view line 125 with the 4 selected songs in **bold**.

96 74 61 **21** 49 37 82 37 **18** 68 18 44 **23** 51 **19** 62 10 33 92 44

So our media player will play "Hey Jude," "The Ballad of John and Yoko," "All You Need is Love," and "Hello, Goodbye." Instead of using Table 7.1, you can select an SRS using a spreadsheet (for example, the Excel command RANDBETWEEN), statistical software, the *Simple Random Sample* applet (see Applet Exercise 1), or most types of calculators. For example, you can use: (MATH) → PRB → randInt(1, 27, 4) → (ENTER) on the TI-84 calculator. And, of course, many digital media players have a "shuffle" option!

Online polls and mall interviews produce samples. We can't trust results from these samples because they are chosen in ways that invite bias. We have more confidence in results from an SRS because it uses impersonal chance to avoid bias. The first question to ask about any sample is whether it was chosen at random. Opinion polls and other sample surveys carried out by people who know what they are doing use random sampling. Most national sample surveys use sampling schemes more complex than an SRS. They may, for example, dial the last four digits of a telephone number at random separately within each exchange (the area code and first three digits). The national sample is pieced together from many smaller samples. The big idea remains the deliberate use of *chance* to choose the sample. Because simple random sampling is the essential principle behind all random sampling and because it is also the main building block for more complex samples, we will focus our study on simple random sampling.

*(John Kelly/The Image Bank/ Getty Images.)*

## EXAMPLE 5 ▪ Exercising Judgment

A Gallup poll released in 2008 shows that Americans' self-reported rates of physical exercise have changed little since 2001. When asked how often each week they participated in "vigorous sports or physical activities for at least 20 minutes that cause large increases in breathing or heart rate," 45% of Americans answered "not at all." Can we trust that 45%? Ask first how Gallup selected its sample. Later in the press release we learn that the results are based on telephone interviews with a randomly selected national sample of 1014 adults, aged 18 and older, conducted November 11–14, 2007.

It is a good start toward gaining our confidence in the poll to know the intended population, the sample size, the tight window of time (so that there is minimal influence from changes in current events), and—most importantly—random selection. In the next section, we address a few other important considerations.

## TABLE 7.1 Random Digits

| | | | | | | | | |
|---|---|---|---|---|---|---|---|---|
| 101 | 19223 | 95034 | 05756 | 28713 | 96409 | 12531 | 42544 | 82853 |
| 102 | 73676 | 47150 | 99400 | 01927 | 27754 | 42648 | 82425 | 36290 |
| 103 | 45467 | 71709 | 77558 | 00095 | 32863 | 29485 | 82226 | 90056 |
| 104 | 52711 | 38889 | 93074 | 60227 | 40011 | 85848 | 48767 | 52573 |
| 105 | 95592 | 94007 | 69971 | 91481 | 60779 | 53791 | 17297 | 59335 |
| 106 | 68417 | 35013 | 15529 | 72765 | 85089 | 57067 | 50211 | 47487 |
| 107 | 82739 | 57890 | 20807 | 47511 | 81676 | 55300 | 94383 | 14893 |
| 108 | 60940 | 72024 | 17868 | 24943 | 61790 | 90656 | 87964 | 18883 |
| 109 | 36009 | 19365 | 15412 | 39638 | 85453 | 46816 | 83485 | 41979 |
| 110 | 38448 | 48789 | 18338 | 24697 | 39364 | 42006 | 76688 | 08708 |
| 111 | 81486 | 69487 | 60513 | 09297 | 00412 | 71238 | 27649 | 39950 |
| 112 | 59636 | 88804 | 04634 | 71197 | 19352 | 73089 | 84898 | 45785 |
| 113 | 62568 | 70206 | 40325 | 03699 | 71080 | 22553 | 11486 | 11776 |
| 114 | 45149 | 32992 | 75730 | 66280 | 03819 | 56202 | 02938 | 70915 |
| 115 | 61041 | 77684 | 94322 | 24709 | 73698 | 14526 | 31893 | 32592 |
| 116 | 14459 | 26056 | 31424 | 80371 | 65103 | 62253 | 50490 | 61181 |
| 117 | 38167 | 98532 | 62183 | 70632 | 23417 | 26185 | 41448 | 75532 |
| 118 | 73190 | 32533 | 04470 | 29669 | 84407 | 90785 | 65956 | 86382 |
| 119 | 95857 | 07118 | 87664 | 92099 | 58806 | 66979 | 98624 | 84826 |
| 120 | 35476 | 55972 | 39421 | 65850 | 04266 | 35435 | 43742 | 11937 |
| 121 | 71487 | 09984 | 29077 | 14863 | 61683 | 47052 | 62224 | 51025 |
| 122 | 13873 | 81598 | 95052 | 90908 | 73592 | 75186 | 87136 | 95761 |
| 123 | 54580 | 81507 | 27102 | 56027 | 55892 | 33063 | 41842 | 81868 |
| 124 | 71035 | 09001 | 43367 | 49497 | 72719 | 96758 | 27611 | 91596 |
| 125 | 96746 | 12149 | 37823 | 71868 | 18442 | 35119 | 62103 | 39244 |

## SPOTLIGHT 7.1 Is It Really Random?

Are the random digits in Table 7.1 really random? Not a chance. They were produced by a computer program. A computer program implements an algorithm that does exactly what you tell it to do. Give the program the same input and it will produce exactly the same "random" digits. You can get quite respectable random digits by calculating $\pi = 3.14159265358979 \ldots$ to more and more decimal places. Go to http://oldweb.cecm.sfu.ca/pi/pi.html and you will see these digits stream by. You get the same digits on every visit, of course. Clever people have devised algorithms that produce output that *looks* like random digits. These are called "pseudo-random numbers," and that's what Table 7.1 contains. Pseudo-random numbers work fine for statistical randomizing, but they have hidden nonrandom patterns that can mess up more refined uses. *(continued on page 218)*

## Is It Really Random? *(continued)*

For purists, the RAND Corporation long ago published a book titled *One Million Random Digits*. The book lists 1 million digits that were produced by a very elaborate physical randomization and really are random. An employee of RAND once commented that this is not the most boring book that RAND has ever published.

Cryptologists and computer scientists would like an endless supply of really random digits. Really random digits must come from nature, not from a computer program. Radioactive decay is really random, and so is the "thermal noise" in an amplifier, which you can hear as a soft whoosh if you turn up the volume with no music playing. Alas, extracting random digits from these really random sources requires various human devices, and these often impose subtle patterns. As of now, in fact, pseudo-random numbers from the best algorithms actually look more random than numbers refined from the randomness in nature by some human apparatus. "Easy to say, hard to do" applies to making random digits as well as to many other human aspirations.

## 7.4 Cautions About Sample Surveys

Random sampling eliminates bias in the choice of a sample from a list of the population. Sample surveys of large human populations, however, require more than a good sampling design.

To begin, we need an accurate and complete list of the population. Because such a list is rarely available, most samples suffer from some degree of **undercoverage**. A sample survey of households, for example, will miss not only homeless people but also prison inmates and students in dormitories. An opinion poll conducted by telephone will miss the 6% of American households without residential phones. Also, about 3% of the population (often younger adults) have cell phones only, and most random digit dialing does not select cell phones. The results of national sample surveys therefore have some bias if the people not covered–who most often are young or poor people–differ from the rest of the population.

### Undercoverage — DEFINITION

**Undercoverage** occurs when some groups in the population are left out of the process of choosing the sample.

A more serious source of bias in most sample surveys is **nonresponse**, which occurs when a selected individual cannot be contacted or refuses to cooperate. Nonresponse to sample surveys often reaches 50% or more, even with careful planning and several callbacks. Because nonresponse is higher in urban areas, most sample surveys substitute other people in the same area to avoid favoring rural areas in the final sample. If the people contacted differ from those who are rarely at home or who refuse to answer questions, some bias remains.

### Nonresponse — DEFINITION

**Nonresponse** occurs when an individual chosen for the sample can't be contacted or refuses to participate.

## EXAMPLE 6 ▪ How Bad Is Nonresponse?

The Current Population Survey (CPS) has the lowest nonresponse rate of any poll we know: Only about 4% of the households in the CPS sample refuse to take part and another 3% or 4% can't be contacted. People are more likely to respond to a government survey such as the CPS, and the CPS contacts its sample in person before doing later interviews by phone.

What about polls done by the media and by market research and opinion polling firms? We don't know their rates of nonresponse because they won't say. That nondisclosure is a bad sign. The Pew Research Center imitated a careful telephone survey and published the results: Out of 2879 households called, 1658 were never at home, refused, or would not finish the interview. That's a nonresponse rate of 58%. ▪

When people do respond, we can't rely on them to always tell the truth. People know that they should take the trouble to vote, for example, so many who didn't vote in the last election will tell a pollster that they did.

## EXAMPLE 7 ▪ Encouraging Honesty

The Centers for Disease Control and Prevention previously used face-to-face interviews to ask Americans about their sexual activity and illegal drug use. In a new version of the survey, released in 2007, data were gathered using computer-assisted self-interviews in which each participant was alone in a room, heard questions through a headset, and touched a computer screen with responses.

Another tool for encouraging honesty with sensitive topics is "randomized response," invented by sociologist S. L. Warner in 1965. By introducing randomness in a structured way, researchers use their knowledge of probability distributions to get reasonably accurate information about the overall group, while allowing each potentially embarrassing answer to be "camouflaged." ▪

Finally, the *wording of questions* strongly influences the answers given to a sample survey. Confusing or leading questions can introduce strong bias, and even minor changes in wording or order can change a survey's outcome. Here are some examples.

## EXAMPLE 8 ▪ Watch That Wording

How do Americans feel about government help for the poor? Only 13% think we are spending too much on "assistance to the poor," but 44% think we are spending too much on "welfare." How do the Scots feel about the movement to become independent from England? Well, 51% would vote for "independence for Scotland," but only 34% support "an independent Scotland separate from the United Kingdom." It seems that "assistance to the poor" and "independence" are nice, hopeful words. "Welfare" and "separate" are negative words. Other topics that have produced survey results that vary greatly with wording include abortion, gay rights, and affirmative action. ▪

The statistical design of sample surveys is a science, but this science is only part of the art of sampling. Because of nonresponse, false responses, and the difficulty of posing clear and neutral questions, you should analyze critically before fully trusting reports about complicated issues based on surveys of large human populations.

Insist on knowing the exact questions asked, the rate of nonresponse, and the date and method of the survey before you trust a poll result.

## 7.5 Experiments

Sample surveys gather information on part of the population in order to draw conclusions about the whole. When the goal is to describe a population, statistical sampling is the right tool to use.

Suppose, however, that we want to study the response to a stimulus, to see how one variable affects another when we change existing conditions. For example:

- Will a new mathematics curriculum improve the scores of sixth-graders on a standard test of mathematics achievement?
- Will taking small amounts of aspirin daily reduce the risk of suffering a heart attack?
- Does a mother's smoking during pregnancy reduce the IQ of her child?

Studies that simply *observe and describe* are ineffective tools for answering these questions. **Experiments** give us clearer answers.

**Experiment** DEFINITION

An **experiment** deliberately imposes a *treatment* on individuals in order to observe their responses. The purpose of an experiment is to study whether the treatment *causes* a change in the response.

Experiments are the preferred method for examining the effect of one variable on another. By imposing the specific treatment of interest and controlling other influences, we can pin down cause and effect. A sample survey may show that two variables are related, but it cannot demonstrate that one causes the other. Statistics has something to say about how to arrange experiments, just as it suggests methods for sampling.

### EXAMPLE 9 ■ An Uncontrolled Experiment

A college regularly offers a review course to prepare candidates for the Graduate Management Admission Test (GMAT) required by most graduate business schools. This year, it offered only an online version of the course. The average GMAT score of students in the online course was 10% higher than the long-time average for those who took the classroom review course. Is the online course more effective?

This experiment has a very simple design. A group of subjects (the students) were exposed to a treatment (the online course), and the outcome (GMAT scores) was observed. Here is the design:

$$\text{Online course} \longrightarrow \text{Observe GMAT scores}$$

or, in general form

$$\text{Treatment} \longrightarrow \text{Observe response}$$

Most laboratory experiments use a design like that in the example: Apply a treatment and measure the response. In the controlled environment of the laboratory, simple designs often work well. But field experiments and experiments with human subjects are exposed to more variable conditions and deal with more variable subjects. It isn't possible to control outside factors that can influence the outcome. With greater variability comes a greater need for statistical design.

A closer look at the GMAT review course showed that the students in the online review course were quite different from the students who in past years took the classroom course. In particular, they were older and more likely to be employed. An online course appeals to these mature people, but we can't compare their performance with that of the undergraduates who previously dominated the course. The online course might even be less effective than the classroom version. The effect of online versus in-class instruction is hopelessly mixed up with influences lurking in the background. Figure 7.2 shows the mixed-up influences in picture form. We say that student age and background is **confounded** with whatever effect the change to online instruction may have. In everyday usage, someone who is confounded is confused or mixed up. In statistics, confounded variables have their effects mixed together so that it's hard to tell what effect is due to each variable separately.

## Confounding

DEFINITION

Variables, whether intentionally part of a study or not, are said to be **confounded** when their effects on the outcome cannot be distinguished from each other.

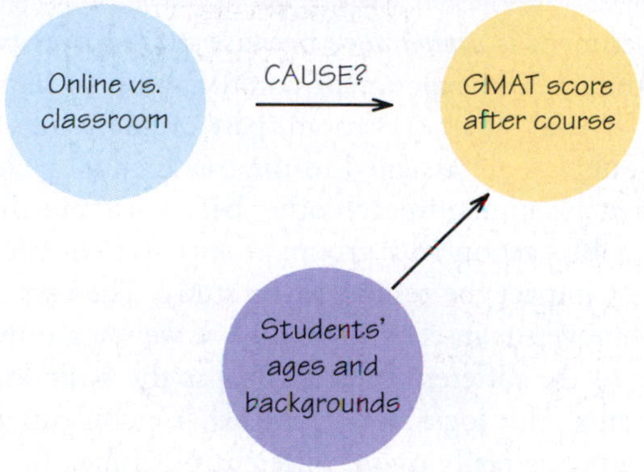

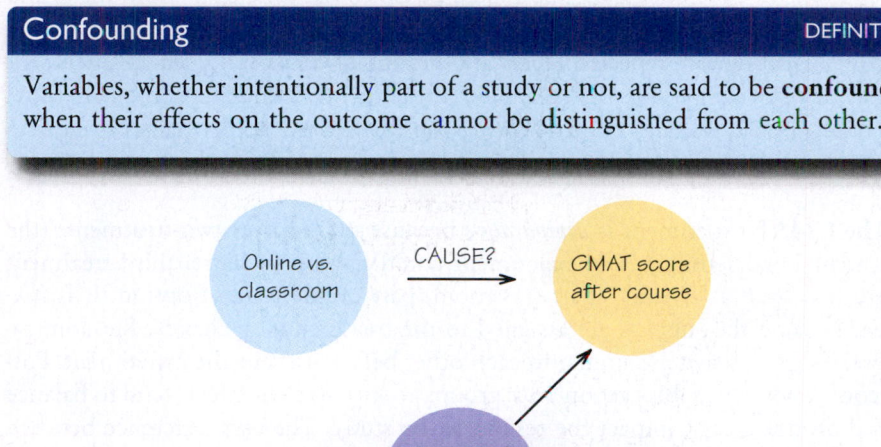

**FIGURE 7.2** Confounding. We can't distinguish the effects of the treatment from those of other influences.

The remedy for confounding is to do a *comparative experiment* in which some students are taught in the classroom and other similar students take the course online. The first group is called a **control group**. Most well-designed experiments compare two or more treatments. Of course, comparison alone isn't enough to produce results we can trust. If the treatments are given to groups that differ markedly when the experiment begins, bias will result. For example, if we allow students to elect online or classroom instruction, older employed students are likely to sign up for the online course. Personal choice will bias our results in the same way that volunteers bias the results of call-in opinion polls. The solution to the problem of bias is the same for experiments and for samples: Use impersonal chance to select the groups.

## EXAMPLE 10

### A Randomized Comparative Experiment

The college decides to compare the progress of 25 on-campus students taught in the classroom with that of 25 students taught the same material online. Select which students will be taught online by taking a simple random sample of size 25 from the 50 available students. The remaining 25 students form the control group. They will receive classroom instruction. The result is a **randomized comparative experiment** with two groups. Figure 7.3 outlines the design in graphical form.

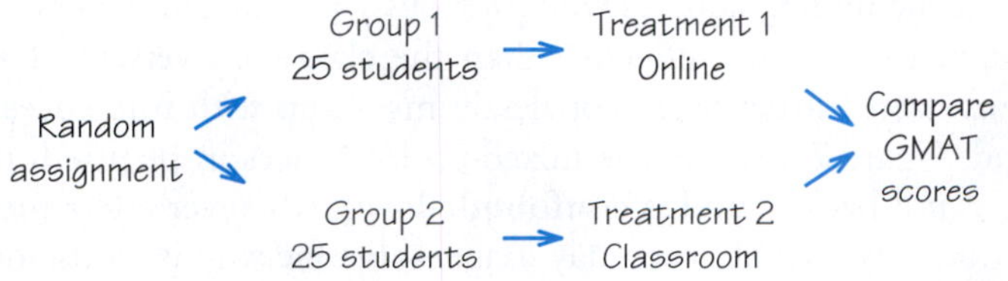


**FIGURE 7.3** The design of a randomized comparative experiment to compare online and classroom instruction.

The selection procedure is exactly the same as it is for sampling: label and table. Step 1: *Label* the 50 students 01 to 50. Step 2: Go to the *table* of random digits and read successive two-digit groups. The first 25 labels encountered select the online group. As usual, ignore repeated labels and groups of digits not used as labels. For example, if you begin at line 125 in Table 7.1, the first five students chosen are those labeled 21, 49, 37, 18, and 44. The *Simple Random Sample* applet makes it particularly easy to choose treatment groups at random.

The GMAT experiment is *comparative* because it compares two treatments (the two instructional settings). The experiment could have even had a third treatment if there had been a "hybrid" (part classroom, part online) course option. It is *randomized* because the subjects are assigned to the treatments by chance. Randomization creates groups that are similar to each other before we start the experiment. Possible confounding variables act on both groups at once so their effects tend to balance out and do not greatly impact the results of the study. The *only* difference between the groups is the online versus in-class setting. So if we see a difference in performance, it must be due to the different setting. That is the basic logic of randomized comparative experiments. This logic shows why experiments can give good evidence that the different treatments really *caused* different outcomes.

There is a fine point: The performance of the two groups will differ even if the treatments are identical, just because the individuals assigned at random to the groups differ. It is only differences *larger than would plausibly occur just by chance* that show the effects of the treatments. The laws of chance allow statisticians to say how big an effect is **statistically significant**. You have intuition for this concept because you would not think it was unusual if the first two children in an extended family were girls, but you would if the first ten were!

**Statistical Significance** DEFINITION

An observed effect so large that it would rarely occur by chance is called **statistically significant**. ("Rarely" usually means $< 5\%$ of the time.)

### EXAMPLE 11 ■ Cervical Cancer Screening

On October 18, 2007, *The New England Journal of Medicine* reported on a randomized comparative experiment in which 10,154 Canadian women (aged 30–69) were given two different cervical cancer screening tests (the standard Pap test and a test for the DNA of the HPV virus) in a randomly assigned sequence. Although it might have seemed "fair" to randomly assign each woman to one treatment or the other, it was deemed better to give each woman the benefit of both medical tests for ethical reasons. Because of the randomization in the sequence, however, the first test was able to be analyzed as if it had been done alone. The DNA-based test proved to be much more powerful than the Pap test in detecting cancer.

## 7.6 Experiments Versus Observational Studies

Randomized comparative experiments are common tools of industrial and academic research. They are also widely used in medical research. For example, federal regulations require that the safety and effectiveness of new drugs be demonstrated by randomized comparative experiments. Let's look at a medical experiment.

### EXAMPLE 12 ■ St. John's Wort for Depression?

Although prescription drugs must pass the test of randomized comparative experiments before being sold, herbs and other "natural remedies" are exempt. Because these treatments are so popular, some are now being studied more carefully. Fans of natural remedies often use extracts of the herb St. John's wort to treat depression. Is the herb safe? Does it work? The *Journal of the American Medical Association* reported a "randomized, double-blind, placebo-controlled clinical trial" in which 200 patients with major depression were assigned at random to take either herb extract or a dummy pill that looked and tasted the same. Results: The herb is safe, but "[i]n this study, St. John's wort was not effective for treatment of major depression."

If you read accounts of medical studies, you will often meet language like "randomized, double-blind, placebo-controlled clinical trial." A clinical trial is a medical experiment with actual patients as subjects. "Randomized" and "controlled" tell us that this was a randomized comparative experiment (that's good). A "placebo" is a fake treatment, the dummy pill in this study. Here we meet a new idea, the importance of the **placebo effect**, a special kind of confounding. The placebo effect is the tendency of patients to respond favorably to any treatment, even a placebo. If depressed patients given St. John's wort are compared with patients who receive no treatment, the first group gets the benefit of both the herb and the placebo effect. Any beneficial effect that St. John's wort may have is confounded with the placebo effect. To prevent confounding, it is important that some treatment be given to all subjects in any medical experiment.

The depression study was a **double-blind experiment**: Neither the subjects nor the experimenters who worked with them knew which treatment any subject received. Subjects might react differently if they knew they were getting "only a placebo." Knowing that a particular subject was getting "only a placebo" could also influence the health workers who interviewed and examined the subjects. So both subjects and workers were kept "blind." Only the study's statistician knew which treatment each subject received.

The difference between the St. John's wort and placebo groups was *not statistically significant*—that is, it was no larger than would be expected when we divide 200 depressed patients at random into two groups and do nothing else. Larger numbers of subjects would give more precise results. It's unlikely that there is exactly *no* difference between St. John's wort and a placebo. If the clinical trial had used 2000 patients rather than 200, it might have picked up a small effect (in either direction). The researchers thought that 200 patients was enough to pick up any effect large enough to be medically important.

The logic of experimentation, the statistical design of experiments, and the laws that govern chance behavior combine to give compelling evidence of cause and effect. Only experimentation can produce the most convincing evidence of causation.

## EXAMPLE 13 ■ Smoking and Health

By way of contrast, consider the statistical evidence linking cigarette smoking to lung cancer. We can't ethically assign groups of people to smoke or not, so a direct experiment isn't possible. The most careful studies have selected samples of smokers and nonsmokers, then followed them for many years, eventually recording the cause of death. These are called **prospective (observational) studies** because they follow the subjects forward in time. (A **retrospective study** looks backward in time.) Prospective studies are comparative, but they are not experiments because the subjects themselves choose whether or not to smoke. A large prospective study of British doctors found that the death rate from lung cancer among cigarette smokers was 20 times that among nonsmokers. Another study of American men aged 40 to 79 found that the lung cancer death rate was 11 times higher among smokers than among nonsmokers. These and many other **observational studies** show a strong connection between smoking and lung cancer.

### Observational Study — DEFINITION

An **observational study** does not try to manipulate the environment (such as by assigning treatments to people), but simply observes the measurements of variables of interest that result from people's free choices. This kind of study is generally done when a treatment is unethical (for example, smoking while pregnant) and/or impossible (such as ethnicity) to assign to a person.

The connection between smoking and lung cancer is statistically significant. That is, it is far stronger than would occur by chance. We can be confident that something other than chance links smoking to cancer. But observation of samples cannot tell us *what* factors other than chance are at work. Perhaps there is something in the genetic makeup of some people that predisposes them both to nicotine addiction and to lung cancer. In that case, we would observe a strong link even if smoking itself had no effect on the lungs.

The statistical evidence that points to cigarette smoking as a cause of lung cancer is about as strong as nonexperimental evidence can be. First, the connection has been observed in many studies in many countries. This eliminates factors peculiar to one group of people or to one specific study design. Second, there is a *dose-response relationship*: People who smoke more are more likely to get lung cancer than those who smoke less, and quitting cigarettes reduces the cancer risk. Third, specific ways in which smoking could cause cancer have been identified–cigarette smoke contains tars that have been shown by experiment to cause tumors in animals. Finally, no plausible alternative explanation is available. For example, the genetic hypothesis cannot explain the increase in lung cancer among women that occurred as more and more women became smokers. Lung cancer, which has long been the leading cause of cancer deaths in men, has now passed breast cancer as the most fatal cancer for women. This evidence is convincing, but it is not quite as strong as the conclusive statistical evidence we get from randomized comparative experiments.

Despite their attractions, experiments can have weaknesses. The most common weakness is a *lack of realism* that makes it hard to say exactly how far the results of an experiment apply beyond controlled or contrived laboratory settings.

### EXAMPLE 14 ■ Is the Experiment Realistic?

Clinical trials give medical treatments to actual patients with the condition that the treatments are supposed to help. Many experiments are less realistic. A psychologist studying the effects of stress on teamwork observes teams of students carrying out tasks in a psychology laboratory under different conditions. The students know it's "just an experiment" and that the stress will only last an hour. Do the conclusions of such experiments apply to real-life stress? An engineer uses a small pilot production process in a laboratory to find the choices of pressure and temperature that maximize yield from a complex chemical reaction. Do the results apply to a full-scale manufacturing plant?

These are not statistical questions. The psychologist and the engineer must use their understanding of psychology and engineering to judge how far their results apply. The statistical design enables us to trust the results for the students and the pilot process but not to generalize the conclusions to other settings.

## 7.7 Inference: From Sample to Population

A market research firm interviews a random sample of 2500 adults. Result: 66% find shopping for clothes frustrating and time consuming. This applies to the 2500 people in the sample. What is the truth about the 230 million American adults who make up the population? Because the sample was chosen at random, it's reasonable to think that these 2500 people represent the entire population fairly well. So the market researchers turn the *fact* that 66% of the *sample* find shopping frustrating into an *estimate* that about 66% of *all adults* feel this way. That's a fundamental operation in statistics: Use a fact about a sample to estimate the truth about the whole population. We call this *statistical inference.*

**Statistical Inference** DEFINITION

**Statistical inference** refers to methods for drawing conclusions about an entire population on the basis of data from a sample. A **confidence interval** is one type of inference method.

If the selected individuals were chosen at random, we think that they fairly represent the population and inference makes sense. If we have data from only a convenience sample or a voluntary response sample, the data do not represent the population and we can't use them for inference. *Statistical inference works only if the data come from a random sample or randomized comparative experiment.* That's why this chapter starts with producing reliable data before moving on to inference from the data to a larger population.

To think about inference, we must keep straight whether a number describes a sample or a population (recall Section 7.1). Here is the vocabulary we use.

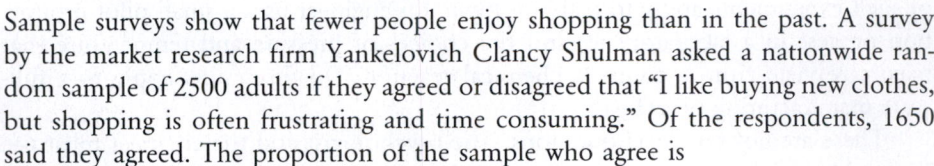

**Parameter** DEFINITION

A **parameter** is a fixed (usually unknown) number that describes a population.

**Statistic** DEFINITION

A **statistic** is a number that describes a sample. The value of a statistic is known when we have taken a sample, but it can change from sample to sample. We often use a statistic to estimate an unknown parameter.

## EXAMPLE 15 ■ Do You Find Shopping Frustrating?

(*Carol Kohen/Getty Images.*)

Sample surveys show that fewer people enjoy shopping than in the past. A survey by the market research firm Yankelovich Clancy Shulman asked a nationwide random sample of 2500 adults if they agreed or disagreed that "I like buying new clothes, but shopping is often frustrating and time consuming." Of the respondents, 1650 said they agreed. The proportion of the sample who agree is

$$\hat{p} = \frac{1650}{2500} = 0.66 = 66\%$$

The symbol $\hat{p}$ is read "p-hat." The ^ symbol here tells us a quantity has been estimated, just as the use of $\hat{y}$ in Chapter 6 told us a value was estimated by using a regression line model. The number $\hat{p} = 0.66$ is a *statistic.* The corresponding *parameter* is the proportion (call it $p$) of all adult U.S. residents who would have said "Agree" if asked the same question. We don't know the value of the parameter $p$, so we use the statistic $\hat{p}$ to estimate it.

If Yankelovich took a second random sample of 2500 adults, the new sample would have different people in it. It is almost certain that there would not be exactly 1650 positive responses. That is, the value of the statistic $\hat{p}$ will vary from sample to sample. If the variation when we take repeat samples from the same population is too great, we can't trust the results of any one sample. We are saved by the second great advantage of random samples. The first advantage is that choosing at random eliminates favoritism. That is, random sampling avoids bias. The second advantage is that if we take lots of random samples of the same size from the same population, the variation from sample to sample will follow a predictable pattern.

*All of statistical inference is based on one idea: to see how trustworthy a procedure is, ask what would happen if we repeated it many times.* So we must ask, "What would happen if we took many samples?" Here's how to answer that question:

- Take a large number of random samples from the same population.
- Calculate the sample proportion $\hat{p}$ for each sample.

- Make a histogram of the values of $\hat{p}$.
- Examine the distribution displayed in the histogram for shape, center, and spread, as well as outliers or other deviations.

In practice it is too expensive to take many samples from a large population such as all adult U.S. residents. But we can use a computer to imitate drawing many samples at random from a population that we specify. This is called *simulation*. Here's what happens when we do this.

## EXAMPLE 16 ■ What Happens in Many Samples?

Figure 7.4 illustrates a result of choosing many samples and finding the sample proportion $\hat{p}$ for each one. The histogram shows the distribution of the values of $\hat{p}$ from 1000 separate SRSs of size 100 drawn from a population that we suppose has a parameter value $p = 0.6$.

Of course, Yankelovich interviewed 2500 people, not just 100. Figure 7.5 is parallel to Figure 7.4. It shows a result from choosing 1000 SRSs, each of size 2500, from a population in which the true proportion is $p = 0.6$. The 1000 values of $\hat{p}$ from these samples form the histogram. Figures 7.4 and 7.5 are drawn on the same scale. Comparing them shows what happens when we increase the size of our samples from 100 to 2500. These histograms display the **sampling distribution** of the statistic $\hat{p}$ for two sample sizes. For intuition, consider a sample size of only 2. There would be only 3 possible $\hat{p}$ values from a sample of size 2: 0, 0.5, or 1. For a sample of size 100, there would be 101 possible $\hat{p}$ values: 0, 0.01, 0.02, ..., 1.00. Of course not all values are equally likely–the ones near 0.6 are most common.

### Sampling Distribution — DEFINITION

The **sampling distribution** of a statistic is the distribution of values taken on by the statistic in all possible samples of the same size from the same population.

Strictly speaking, the sampling distribution is the ideal pattern that would emerge if we looked at all possible samples of the same size from our population. A distribution obtained from a fixed number of trials, like the 1000 trials in these figures, is only an approximation to the sampling distribution. Probability theory, the mathematics of chance behavior, can sometimes describe sampling distributions exactly. Chapter 8 will introduce you to basic probability theory. The interpretation of a sampling distribution is the same, however, whether we obtain it by simulation or by the mathematics of probability.

We can use the tools of data analysis from Chapter 5 to describe any distribution. Let's apply those tools to Figures 7.4 and 7.5.

- **Shape:** The histograms look normal. The normal curves drawn through the histograms describe the overall shape quite well.
- **Center:** In both cases, the values of the sample proportion $\hat{p}$ vary from sample to sample, but the values are centered at 0.6. Recall that we are assuming $p = 0.6$ is the true population parameter. Some samples have a $\hat{p}$ less than 0.6 and some greater, but there is no tendency to be always low or always high. That is, $\hat{p}$ has *no bias* as an estimator of $p$. This is true for both

large and small samples. (Want the details? The mean of the 1000 values of $\hat{p}$ is 0.598 for samples of size 100 and 0.6002 for samples of size 2500. The median value of $\hat{p}$ is exactly 0.6 for samples of both sizes.)

- **Spread:** The values of $\hat{p}$ from samples of size 2500 are much less spread out than the values from samples of size 100. In fact, the standard deviations are 0.051 for Figure 7.4 and 0.0097, or about 0.01, for Figure 7.5.

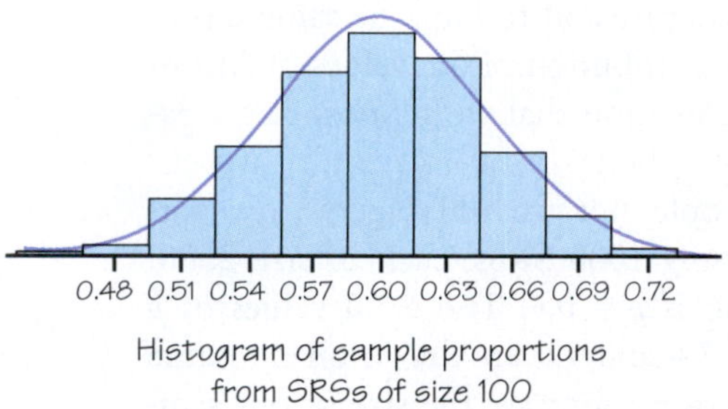


**FIGURE 7.4** Draw 1000 SRSs of size 100 from a population with proportion $p = 0.60$ of successes. The histogram shows the distribution of the 1000 sample proportions $\hat{p}$.

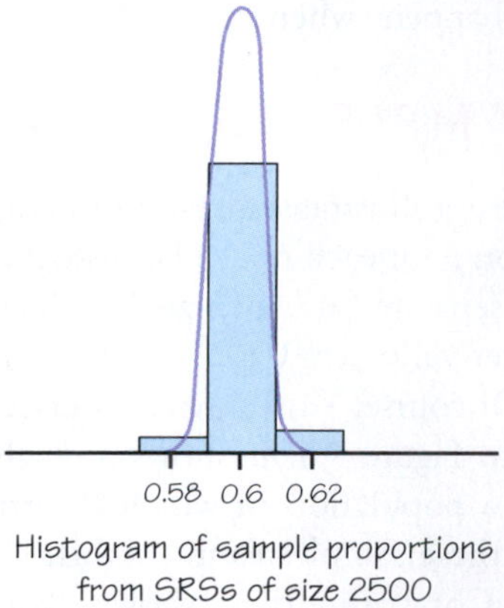

0.58 0.6 0.62
Histogram of sample proportions from SRSs of size 2500


**FIGURE 7.5** Draw 1000 SRSs of size 2500 from the same population as in Figure 7.4. The histogram shows the distribution of the 1000 sample proportions $\hat{p}$, using the same scale as Figure 7.6. The statistic from the larger sample is less variable.

Although these results describe just two sets of simulations, they reflect facts that are true whenever we use random sampling. We now turn to probability theory to learn the mathematical facts that lie behind the simulations. We'll use the word "success" for whatever we are counting, such as "Agree" responses in the shopping survey. Note that "success" does not necessarily have the positive (or negative) association it does in real life, but is simply a convenient way to identify an outcome.

### Sampling Distribution of a Sample Proportion — THEOREM

Choose an SRS of size $n$ from a large population that contains population proportion $p$ of successes. Let $\hat{p}$ be the **sample proportion** of successes,

$$\hat{p} = \frac{\text{count of successes in the sample}}{n}$$

Then:

- **Shape:** For large ($n \geq 30$) sample sizes, the sampling distribution of $\hat{p}$ is *approximately normal.*
- **Center:** The *mean* of the sampling distribution of $\hat{p}$ is $p$.
- **Spread:** The *standard deviation* of the sampling distribution of $\hat{p}$ is

$$\sqrt{\frac{p(1-p)}{n}}$$

(This will be confirmed in Section 8.6.)

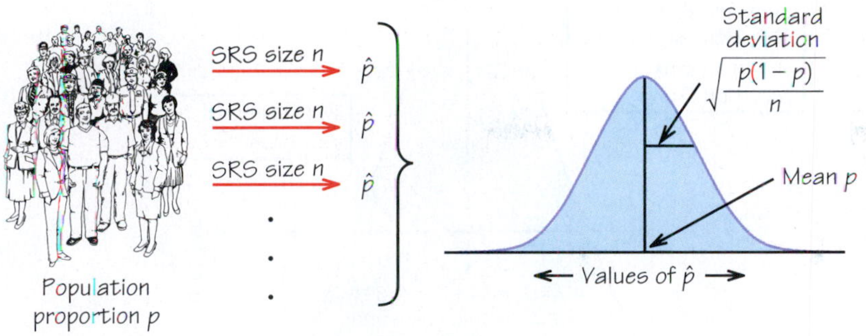


**FIGURE 7.6** Repeat many times the process of selecting an SRS of size $n$ from a population of which the proportion $p$ are successes. The values of the sample proportion of successes $\hat{p}$ have this normal sampling distribution.

Figure 7.6 summarizes these facts in a form that reminds us that a sampling distribution describes the results of lots of samples from the same population.

## EXAMPLE 17 ■ Do You Find Shopping Frustrating?

Suppose that 60% of all adults find shopping for clothes frustrating and time consuming. The population proportion is $p = 0.6$. Take a simple random sample of 2500 adults. That's exactly the setting of the second simulation in Example 16. Now we can apply mathematics to learn how the sample proportion $\hat{p}$ would behave if we took many samples. The distribution of $\hat{p}$ in many samples

- is close to normal;
- has mean 0.6;
- has standard deviation

$$\sqrt{\frac{p(1-p)}{n}} = \sqrt{\frac{(0.6)(0.4)}{2500}} = 0.0098$$

The mean 0.6 and standard deviation 0.0098 from the mathematics are very close to the mean 0.6002 and standard deviation 0.0097 we observed in our simulation. If the simulation used more than 1000 trials, the results would be yet closer to the mathematical truth.

# 7.8 Confidence Intervals

The sampling distribution shows why we can trust the results of a large random sample: Almost all such samples give results that are close to the truth about the population.

## EXAMPLE 18 ■ The 68–95–99.7 Rule Again

In Example 17, the population parameter, the proportion of adults who find shopping frustrating, is $p = 0.6$. If we take SRSs of size 2500, the sample proportions $\hat{p}$ follow the normal distribution with mean 0.6 and standard deviation about 0.01. The 95 part of the 68–95–99.7 rule from Section 5.9 says that 95% of all samples give a $\hat{p}$ within 2 standard deviations of the truth about the population. So in this example, 95% of all samples have $\hat{p}$ within $2 \times 0.01$ of 0.6, that is, between 0.58 and 0.62. Figure 7.7 illustrates this use of the 68–95–99.7 rule.

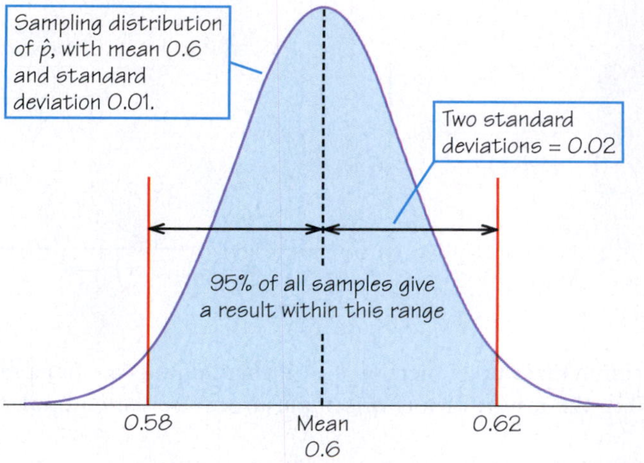
Sampling distribution of $\hat{p}$, with mean 0.6 and standard deviation 0.01.
Two standard deviations = 0.02
95% of all samples give a result within this range
0.58
Mean 0.6
0.62

**FIGURE 7.7** The sampling distribution of $\hat{p}$ for Example 17. By the 68–95–99.7 rule, 95% of all samples have a sample proportion $\hat{p}$ within $\pm 0.02$ of the true population proportion $p = 0.6$.

We can repeat this reasoning for any value of the parameter $p$ and the sample size $n$. It is always true that 95% of all samples give a sample proportion $\hat{p}$ within 2 standard deviations of the population proportion $p$. That is, 95% of all samples catch $p$ in the interval extending 2 standard deviations on either side of $\hat{p}$. That's the interval

$$\hat{p} \pm 2\sqrt{\frac{p(1-p)}{n}}$$

This formula tells us how close the unknown parameter $p$ lies to the observed statistic $\hat{p}$ in 95% of all samples. There is one catch: We can't calculate the interval from the data because the standard deviation involves the population proportion $p$, and in practice we don't know $p$. In Examples 17 and 18, we applied the formula for $p = 0.6$, but this may not be the true $p$ for the actual population of all American adults.

What to do? The standard deviation of the statistic $\hat{p}$ does depend on the parameter $p$, but it doesn't change a lot when $p$ changes. Go back to Example 17 and redo the calculation for other values of $p$. Here's the result:

| Value of $p$ | 0.4 | 0.5 | 0.6 | 0.7 | 0.8 |
|---|---|---|---|---|---|
| Standard deviation | 0.0098 | 0.01 | 0.0098 | 0.0092 | 0.008 |

The standard deviations are all 0.01 when rounded to two places. You see that if we guess a value of $p$ reasonably close to the true value, the standard deviation found from the guessed value will be about right. We know that when we take a large random sample, the statistic $\hat{p}$ is almost always close to the parameter $p$. So we will use $\hat{p}$ as the guessed value of the unknown $p$. Now we have an interval that we can calculate from the sample data. We call it a *confidence interval.*

## Confidence Interval

DEFINITION

A **95% confidence interval** is an interval obtained from the sample data by a method in which 95% of all samples will produce an interval containing the true population parameter.

Choose an SRS of size $n$ from a large population that contains an unknown proportion $p$ of successes. A 95% confidence interval for $p$ is approximately

$$\hat{p} \pm 2\sqrt{\frac{\hat{p}(1-\hat{p})}{n}}$$

The ± sign is read "plus or minus," so 0.5 ± 0.2 yields two numbers, 0.3 and 0.7, which can be written as an interval: (0.3, 0.7).

This formula is only approximately correct but is quite accurate when the sample size $n$ is large ($\geq 030$). Here $\hat{p}$ is the proportion of successes in the sample and $2\sqrt{\hat{p}(1-\hat{p})/n}$ is the **margin of error**.

**Margin of Error** DEFINITION

The **margin of error** is the number to the right of the ± sign in a 95% confidence interval and is equal to half of the width of the full interval. It equals about 2 standard deviations of the sampling distribution of the estimated parameter. If you conducted a very large number of polls, about 95% of the time the difference between a particular poll's result and the true value of the population parameter would be within the margin of error.

This interval is only approximately correct for two reasons. The sampling distribution of the sample proportion $\hat{p}$ isn't exactly normal. And we don't get the standard deviation of $\hat{p}$ exactly right because we used $\hat{p}$ in place of the unknown $p$. Both of these difficulties go away as the sample size $n$ gets larger. Our method works well enough for many practical uses. More important, it shows how we get a confidence interval from the sampling distribution of a statistic. That's the reasoning behind any confidence interval.

## EXAMPLE 19 ■ Risky Behavior in the Age of AIDS

How common is behavior that puts heterosexuals at risk for AIDS? In 1990–1991, the National AIDS Behavioral Survey interviewed a random sample of 2673 adult heterosexuals. Of these people, 170 had had more than one sexual partner in the past year. The sample proportion who admit to multiple partners is

$$\hat{p} = \frac{170}{2673} = 0.0636$$

A 95% confidence interval for the proportion $p$ of all adult heterosexuals with multiple partners is therefore

$$\begin{aligned}\hat{p} \pm 2\sqrt{\frac{\hat{p}(1-\hat{p})}{n}} &= 0.0636 \pm 2\sqrt{\frac{(0.0636)(0.9364)}{2673}} \\ &= 0.0636 \pm 0.0094, \text{ or } 6.36\% \pm 0.94\% \\ &= 0.0542 \text{ to } 0.0730, \text{ or } 5.42\% \text{ to } 7.30\%\end{aligned}$$

A report of these calculations might say, "The study found that 6.36% of heterosexuals had more than one sexual partner. The margin of error for this result is 0.94%."

We got the interval in Example 19 by using a formula that catches the true unknown population proportion in 95% of all samples. The shorthand for this is: We are **95% confident** that the true proportion of heterosexuals with multiple partners lies between 5.42% and 7.30%. The margin of error refers to the spread needed to capture the true $p$ in 95% of all samples. The truth lies outside the interval (5.42%, 7.30%) in 5% of all samples.

Figure 7.8 lays out the meaning of "95% confidence." The vertical line is the true value of the population proportion $p$. The normal curve at the top of the figure is the sampling distribution of the sample statistic $\hat{p}$, which is centered at the true $p$. The 95% confidence intervals from 25 SRSs appear below, one after the other. The central dots are the values of $\hat{p}$, the centers of the intervals. The arrows on either side span the confidence interval. In the long run, 95% of the intervals will cover the true $p$ and 5% will miss. Of the 25 intervals in Figure 7.8, 24 hit and 1 misses. (Remember that the sampling distribution describes what happens in a very large number of samples–we don't expect exactly 95% of 25 intervals to capture the true parameter.) The *Confidence Interval* applet animates Figure 7.8. You can use the applet to watch confidence intervals from one sample after another capture or fail to capture the true parameter.

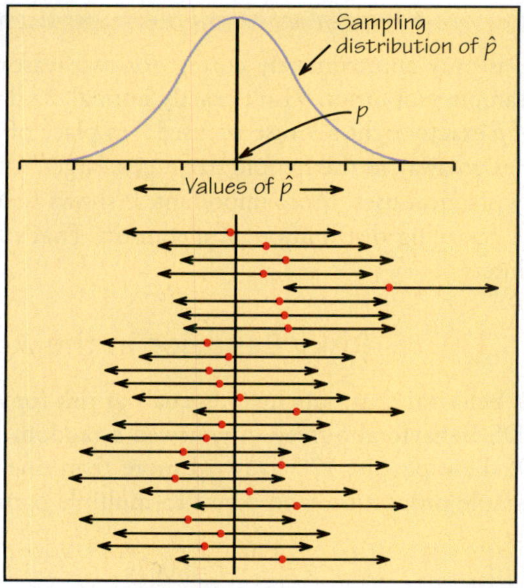


**FIGURE 7.8** Twenty-five samples from the same population give these 95% confidence intervals. In the long run, 95% of all such intervals cover the true population proportion, marked by the vertical line.

The length of a confidence interval depends on the size $n$ of the sample. Larger samples give shorter intervals because of the $\sqrt{n}$ in the denominator of the margin of error. But the interval does *not* depend on the size of the population. This is true as long as the population is much larger than the sample. The confidence interval in Example 19 works for a sample of 2673 from a city with 100,000 adults as well as for a sample of 2673 from a nation of 220 million. What matters is how many people we interview, not what percent of the population we contact.

The length of a confidence interval also depends on how confident we want to be that the interval does capture the true parameter value. It is common to use 95% confidence, but you can ask for higher or lower confidence if you want. Our 95% confidence interval was based on the middle 95% of a normal distribution. A 99% confidence interval requires the middle 99% of the distribution and so is wider (has a larger margin of error).

## EXAMPLE 20 ■ Understanding the News

Here's what the TV news announcer says: "A new Gallup poll on American exercise habits finds that 45% of adults are not engaging in vigorous sports or physical activities. The margin of error for the poll was 3 percentage points." Plus or minus 3% starting at 45% is 42% to 48%. People with minimal statistics knowledge may think that the truth about the entire population must be in that interval, but now we know better!

This is the full background Gallup actually gives: "For results based on this sample, one can say with 95% confidence that the maximum error attributable to sampling and other random effects is 3 percentage points. In addition to sampling error, question wording and practical difficulties in conducting surveys can introduce error or bias into the findings of public opinion polls." That is, Gallup tells us that the margin of error only works for 95% of all its samples. "95% confidence" is shorthand for that. The news report left out the "95% confidence." In fact, *almost all margins of error in the news are for 95% confidence.* If you don't see the confidence level in a scientific poll, it's usually safe to assume 95%.

Gallup's mention of "question wording and practical difficulties" takes us back to our cautions about sample surveys. *The margin of error does not address nonresponse and other practical difficulties.* The margin of error in a confidence interval comes from the sampling distribution of the statistic. The sampling distribution describes the variation of the statistic due to chance in repeated random samples. This random variation is the *only* source of error covered by the margin of error. Real-life samples also suffer from undercoverage and nonresponse. Errors from these practical difficulties are usually more serious and harder to quantify than random sampling error. The actual error in sample surveys may be much larger than the announced margin of error. Worse, we can't say how much larger. Statistical conclusions are approximations to a complicated truth, not mathematical results that are simply true.

## EXAMPLE 21 ■ Measuring Risky Behavior

What about the National AIDS Behavioral Surveys? The interviews were carried out by telephone. This is acceptable for surveys of the general population, because about 94% of American households have telephones. However, some groups at high risk for AIDS, such as intravenous drug users, often don't live in settled households and are underrepresented in the sample. About 30% of the people reached refused to cooperate. A nonresponse rate of 30% is not unusual in large sample surveys, but it may cause some bias if those who refuse differ systematically from those who cooperate. The survey used statistical methods that adjust for unequal response rates in different groups. Finally, some respondents may not have told the truth when asked about their sexual behavior. The survey team tried hard to make respondents feel comfortable. For example, Hispanic women were interviewed by Hispanic women, and Spanish speakers were interviewed by Spanish speakers with the same regional accent (Cuban, Mexican, or Puerto Rican). Nonetheless, the survey report says that some bias is probably present:

> *It is more likely that the present figures are underestimates; some respondents may underreport their numbers of sexual partners and intravenous drug use because of embarrassment and fear of reprisal, or they may forget or not know details of their own*

*or of their partner's HIV risk and their antibody testing history.* [Joseph H. Catania et al., Prevalence of AIDS-related risk factors and condom use in the United States, *Science*, 258 (1992): 1104.]

Reading the report of a large study like the National AIDS Behavioral Surveys reminds us that statistics in practice involves much more than formulas for confidence intervals.

## SPOTLIGHT 7.2 Truth in Polling

Responsible polling organizations tell the public something about both the precision and limitations of their poll results. For example, here is the statement from the Harris Poll Web site that accompanied results of a poll (of about 1000 randomly selected people):

> "In theory, with a probability sample of this size, one can say with 95 percent certainty that the results have a statistical precision of plus or minus 3 percentage points of what they would be if the entire adult population had been polled with complete accuracy. Unfortunately, there are several other possible sources of error in all polls or surveys that are probably more serious than theoretical calculations of sampling error. They include refusals to be interviewed (nonresponse), question wording and question order, interviewer bias, weighting by demographic control data and screening (e.g., for likely voters). It is difficult or impossible to quantify the errors that may result from these factors."

Your college student newspaper may not have the resources to conduct polls using random sampling, but it is refreshing when the polls it publishes from voluntary response samples are accompanied by a disclaimer such as this one from The University of Texas at El Paso's *The Prospector*:

> "This poll is not scientific and reflects the opinions of only those Internet users who have chosen to participate. The results cannot be assumed to represent the opinions of Internet users in general, nor the public as a whole."

Because of this limitation, *The Prospector* simply reports the breakdown of responses given but without any margin of error, since sampling error cannot be quantified from a (voluntary response) sample that is not probability-based.

## REVIEW VOCABULARY

**Bias** A systematic error that tends to cause the observations to deviate in the same direction from the truth about the population whenever a sample or experiment is repeated. (p. 214)

**Confidence interval** An interval of values used to estimate a population parameter with a specific level of confidence. A **95% confidence interval** is an interval computed from a sample by a method that surrounds the unknown parameter 95% of the time, so when we calculate the interval for a single sample, we are 95% confident that the interval contains the unknown parameter. (pp. 225, 230)

**Confounding** Two variables are confounded when their effects on the outcome of a study cannot be distinguished from each other. (p. 221)

**Control group** A group of experimental subjects that is given a standard treatment or no treatment (such as a placebo). (p. 221)

**Convenience sample** A sample that consists of the individuals who are most easily available, such as people passing by in the street. A convenience sample is usually biased. (p. 213)

**Double-blind experiment** An experiment in which neither the experimental subjects nor the persons who interact with them know which treatment each subject received. (p. 223)

**Experiment** A study in which treatments are applied to people, animals, or things in order to observe the effect of the treatments. (p. 220)

**Margin of error** The number to the right of the ± sign in a 95% confidence interval and is equal to half of the width of the full interval. It equals about 2 standard deviations of the sampling distribution of the estimated parameter. If you conducted a very large number of polls, about 95% of the time the difference between a particular poll's result and the true value of the population parameter would be within the margin of error. (p. 231)

**Nonresponse** Some individuals chosen for a sample cannot be contacted or refuse to participate. (p. 218)

**Observational study** A study (such as a sample survey) that observes individuals and measures variables of interest but does not attempt to influence the responses. (p. 224)

**Parameter** A number that describes the population. In statistical inference, the goal is often to estimate an unknown parameter or make a decision about its value. (p. 226)

**Placebo effect** The effect of a dummy treatment (such as an inert pill in a medical experiment) on the response of subjects. (p. 223)

**Population** The entire group of people or things about which we want information. (p. 212)

**Prospective study** An observational study that follows two or more groups of subjects forward in time. (p. 224)

**Randomized comparative experiment** An experiment to compare two or more treatments in which people, animals, or things are assigned to treatments by chance. (p. 222)

**Retrospective study** An observational study that uses interviews or records to collect information about past behaviors of subjects in two or more groups. (p. 224)

**Sample** A part of the population that is actually observed and used to draw conclusions, or inferences, about the entire population. (p. 212)

**Sample proportion** The proportion $\hat{p}$ of the members of a sample having some characteristic (such as agreeing with an opinion poll question). The sample proportion from a simple random sample is used to estimate the corresponding proportion $p$ in the population from which the sample was drawn. (p. 228)

**Sampling distribution** The distribution of values taken by a statistic when all possible random samples of the same size are drawn from the same population. The sampling distributions of sample proportions are approximately normal. (p. 227)

**Simple random sample (SRS)** A sample chosen by chance, so that every possible sample of the same size has an equal chance to be the one selected. (p. 214)

**Statistic** A number that describes a sample. A statistic can be calculated from the sample data alone; it does not involve any unknown parameters of the population. (p. 226)

**Statistical inference** Methods for drawing conclusions about an entire population on the basis of data from a sample. Confidence intervals are one type of inference method. (p. 211)

**Statistical significance** An observed effect is statistically significant if it is so large that it is unlikely to occur just by chance in the absence of a real effect in the population from which the data were drawn. (p. 222)

**Table of random digits** A table whose entries are the digits 0, 1, 2, 3, 4, 5, 6, 7, 8, 9 in a completely random order. That is, each entry is equally likely to be any of the 10 digits and no entry gives information about any other entry. (p. 215)

**Undercoverage** The process of choosing a sample may systematically leave out some groups in the population, such as households without telephones. (p. 218)

**Voluntary response sample** A sample of people who choose themselves by responding to a general invitation to give their opinions. Such a sample is usually strongly biased. (p. 214)

## ✓ SKILLS CHECK

**1.** An opinion poll contacts 1021 adults and asks them, "Which political party do you think has better ideas for leading the country in the twenty-first century?" In all, 723 of the 1021 say, "The Democrats." The sample in this setting is

**(a)** all 220 million adults in the United States.
**(b)** the 1021 people interviewed.
**(c)** the 723 people who chose the Democrats.

**2.** A committee on community relations in a college town plans to survey local businesses about the importance of students as customers. From the 10,000 businesses listed in the telephone book, the committee chooses 150 businesses at random. Of these, 72 return the questionnaire mailed by the committee. The nonresponse rate is ____________________ percent.

**3.** The sample in the setting of the previous exercise is

**(a)** all 10,000 businesses in the college town.
**(b)** the 150 businesses chosen.
**(c)** the 72 businesses that returned the questionnaire.

**4.** A call-in poll asks who people are planning to vote for in the next Presidential election. People who think major change is needed are likely to be represented in this poll ________ than they should be, if the goal is to get results that are representative of all voters.

**5.** On January 2, 2008, the American Idol Web site (www.americanidol.com) had an online poll that asked who you think would win among six former contestants. To become part of the sample, you simply clicked on a response. Of the 941,434 responses to this poll, 55% went to Clay Aiken. We can conclude that:

**(a)** most Americans prefer Clay Aiken out of those former contestants.
**(b)** the sample is too small a fraction of the millions of people who watched the TV show to draw any conclusion.
**(c)** the poll uses voluntary response, so the results tell us little about the population of all adults.

**6.** You are using the table of random digits to choose a simple random sample of 6 students from a class of 30 students. You label the students 01 to 30 in alphabetical order. Go to line 113 of Table 7.1. Of the labels corresponding to the six students selected for your sample, the label that is largest is ________________ .

**7.** You must choose an SRS of 10 of the 420 retail outlets in New York that sell your company's products. How would you label this population in order to use Table 7.1?

**(a)** 001, 002, 003, ..., 419, 420
**(b)** 000, 001, 002, ..., 419, 420
**(c)** 1, 2, 3, ...,419, 420

**8.** From an alphabetical list of the 7200 salaried employees of a corporation, you label the employees 0001 to 7200. Using line 111 of Table 7.1, choose an SRS of 5 of the 7200 salaried employees of a corporation. Of the five employees selected for your sample, the label that is the largest is ____________ .

**9.** A sample of households in a community is selected at random from the telephone directory. In this community, 4% of households have no telephone and another 35% have unlisted telephone numbers. The sample will certainly suffer from

**(a)** nonresponse.
**(b)** undercoverage.
**(c)** false responses.

**10.** To learn about the population of a county containing 100,000 people, a sample of 3000 households was selected to be interviewed. For 1200 of the 3000 households researchers attempted to contact, there was no one home willing to participate. For this survey, the nonresponse rate was _____________ percent.

**11.** A clinical trial compares an antidepression medicine with a placebo for relief of chronic headaches. There are 36 headache patients available to serve as subjects. To choose 18 patients to receive the medicine, you would

**(a)** assign labels 01 to 36 and use Table 7.1 to choose 18.
**(b)** assign labels 01 to 18 because only 18 need be chosen.
**(c)** assign the first 18 who signed up to get the medicine.

**12.** A study of cell phones and the risk of brain cancer looked at a group of 519 people who have brain cancer. The investigators matched each cancer patient with a person of the same sex, age, and race who did not have brain cancer, then asked about use of cell phones. This kind of study is known as ______________________ .

**13.** Studies that follow subjects forward in time are called

**(a)** retrospective.
**(b)** prospective.
**(c)** double-blind.

**14.** A treatment consisting of a "dummy pill" that looks like (but isn't) real medicine is known as a _________ .

**15.** A study of religious practices among college students interviewed a sample of 125 students; 105 of the students said that they prayed at least once in a while. The sample proportion who said they pray is what?

**(a)** 107
**(b)** 84
**(c)** 0.84

**16.** An opinion poll asks a simple random sample of 1000 adults how they view the state of the economy. Suppose that 35% of all adults would say "good" if they were asked. In repeated samples, the sample proportion $\hat{p}$ who say "good" would follow a normal distribution with mean having a value of ____________________ .

**17.** The standard deviation of the distribution of the sample proportion in the previous exercise is about

**(a)** 0.00023.
**(b)** 0.015.
**(c)** 0.03.

**18.** To the nearer half of a percentage point, the margin of error is ______ when we use the result of Exercise 15 to estimate what percent of all college students pray.

**19.** The sample survey in Exercise 15 actually called 150 students, but 23 of the students refused to say whether they pray. This nonresponse could cause the survey result to be in error. The error due to nonresponse

**(a)** is in addition to the margin of error found in Exercise 18.
**(b)** is included in the margin of error found in Exercise 18.
**(c)** can be ignored because it isn't random.

**20.** A survey of folk music fans yields this 95% confidence interval estimate of the proportion of fans who love the music of David Wilcox: 0.74 to 0.86. To the nearer percentage point, the margin of error for this survey is ______________________________ .

# CHAPTER 7 EXERCISES

■ Challenge ◆ Discussion

## 7.1 Sampling

1. A Gallup poll asked, "In general, are you satisfied or dissatisfied with the way things are going in your personal life at this time?" 84% of Americans answered "satisfied." Interestingly, only 27% of Americans in the same survey said they were satisfied with the way things are going in the United States at this time. Gallup's report said, "Results are based on telephone interviews with 1027 national adults, aged 18 and older, conducted Dec. 6–9, 2007."

**(a)** What is the population for this sample survey?
**(b)** What is the sample size?

2. The American Community Survey (ACS) will replace the census "long form" starting with the 2010 census. The main part of the ACS contacts households at 250,000 addresses by mail each month, with follow-up by phone and in person if there is no response. Each household answers questions about its housing, economic, and social status. What is the population for the ACS?

## 7.2 Bad Sampling Methods

◆ **3.** You see a woman student standing in front of the student center, now and then stopping other students to ask them questions. She says that she is collecting student opinions for a class assignment. Explain why this sampling method is almost certainly biased.

◆ **4.** A member of Congress is interested in whether her constituents favor a proposed gun-control bill. Her staff reports that letters on the bill have been received from 361 constituents and that 323 of these oppose the bill. What is the population of interest? What is the sample? Is this sample likely to represent the population well? Explain your answer.

◆ **5.** Highway planners made a main street in a college town one-way. Local businesses were against the change. The local newspaper invited readers to call a telephone number to record their comments. The next day, the paper reported:

> *Readers overwhelmingly prefer two-way traffic flow to one-way streets. By nearly a 7-1 margin, callers to the newspaper's Express Yourself opinion line on Wednesday complained about the one-way streets that have been in place since May. Of the 98 comments received, all but 14 said no to one-way.*

**(a)** What population do you think the newspaper wants information about?
**(b)** Is the proportion of this population who favor one-way streets almost certainly larger or smaller than the proportion 14/98 in the sample? Why?

◆ **6.** Your college wants to gather student opinion about a proposed student fee increase. It isn't practical to contact all students.

**(a)** Give an example of a way to choose a sample of students that is poor practice because it depends on voluntary response.
**(b)** Give another example of a bad way to choose a sample that doesn't use voluntary response.

## 7.3 Simple Random Samples

7. You have just been blessed with triplets (all girls). You decide to select their names using an SRS of three names from the following list of the most popular names given to American girls born in this decade. To do this, use Table 7.1, starting at line 117.

| | | | |
|---|---|---|---|
| 1) Emily | 2) Madison | 3) Hannah | 4) Emma |
| 5) Ashley | 6) Alexis | 7) Samantha | 8) Sarah |
| 9) Abigail | 10) Olivia | 11) Elizabeth | 12) Alyssa |
| 13) Jessica | 14) Grace | 15) Lauren | 16) Taylor |
| 17) Kayla | 18) Brianna | 19) Isabella | 20) Anna |

8. **(a)** Would pulling out and lining up several dollar bills to use the 8-digit serial numbers be a reasonable substitute for Table 7.1? Explain.
**(b)** How about using the telephone numbers on a page of the phone book? Explain.

9. There are approximately 371 active telephone area codes covering Canada, the United States, and some Caribbean areas. (More are created regularly.) You want to choose an SRS of 25 of these area codes for a study of available telephone numbers.

**(a)** How would you label the area codes in order to use Table 7.1?
**(b)** Use Table 7.1, starting at line 125, to choose the first 3 members of this sample.

10. Each March, the Current Population Survey is expanded to gather a wider variety of information. On the Bureau of Labor Statistics Web site, you can find data from this survey on 14,959 people aged 25 to 64 whose highest level of education is a bachelor's degree. Think of these people as a population.

**(a)** In order to select an SRS of these people, how would you assign labels?
**(b)** Use Table 7.1, starting at line 107, to choose the first three members of the SRS.

◆ **11.** In using Table 7.1 repeatedly to choose samples, you should not always begin at the same place, such as line 101. Why not?

■ **12.** Which of the following statements are true of a table of random digits and which are false? Explain your answers.

**(a)** There are exactly four 0's in each row of 40 digits.
**(b)** Each pair of digits has chance 1/100 of being 00.
**(c)** The digits 0000 can never appear as a group because this pattern is not random.

■ **13.** The last stage of the Current Population Survey uses a *systematic sample.* An example will illustrate the idea of a systematic sample. Suppose that we must choose 4 rooms out of the 100 rooms in a dormitory. Because $100/4 = 25$, we can think of the list of 100 rooms as 4 lists of 25 rooms each. Choose 1 of the first 25 rooms at random, using Table 7.1. The sample will contain this room and the rooms 25, 50, and 75 places down the list from it. If 13 is chosen, for example, then the systematic random sample consists of the rooms numbered 13, 38, 63, and 88.

**(a)** Use Table 7.1 to choose a systematic random sample of 5 rooms from a list of 200. Enter the table at line 120.
**(b)** Your sample gives every room the same chance to be chosen. Explain why. Yet this systematic sample is not a simple random sample. Explain why.

■ **14.** At a party there are 30 students over age 21 and 20 students under age 21. You choose at random 3 of those over 21 and separately choose at random 2 of those under 21 to interview about attitudes toward alcohol. You have given every student at the party the same chance to be interviewed: What is that chance? Why is your sample not an SRS?

## 7.4 Cautions About Sample Surveys

◆ **15.** An opinion poll calls 1334 randomly chosen residential telephone numbers, then the interviewer asks to speak with an adult member of the household to ask, "How many movies have you watched in a movie theater in the past 12 months?"

**(a)** What population do you think the poll has in mind?
**(b)** In all, 931 people respond. What is the rate (percent) of nonresponse?
**(c)** Many responses to this question are likely to be inaccurate. Why?

◆ **16.** Randomized Response: Suppose 30 students in a class participate in a survey in which they each flip a coin and do not tell the result. If the result was "heads," the student is supposed to say "yes." If the result was "tails," the student is supposed to give an honest answer to the question "Have you ever used a fake ID?" Suppose the results in the class are 18 "yes" answers and 12 "no" answers.

**(a)** If students follow the procedure correctly, is it true that all students who answered "no" have not used a fake ID?
**(b)** If students follow the procedure correctly, is it true that all students who have not used a fake ID answered "no"?
**(c)** On average, about half of the students who have not used a fake ID flipped "tails," so what is your best estimate of the true number of students who have not used a fake ID?
**(d)** Based on the answer to part (c), what is your estimate of the true number and proportion of students who have used a fake ID?
**(e)** Do we have any way to know which of the 18 "yes" answers are truthful?

◆ **17.** The wording of questions can strongly influence the results of a sample survey. You are writing an opinion poll question about a proposed amendment to the Constitution. You can ask if people are in favor of "changing the Constitution" or "adding to the Constitution" by approving the amendment. One of these choices of wording will likely produce a much higher percent in favor. Which one? Why?

## 7.5 Experiments

◆ **18.** As reported in College Teaching in 2006, R. L. Garner randomly assigned 117 undergraduates to "review lecture videos" on statistics research methods; the videos either did or did not have short bits of humor inserted. Students who viewed the humor-added version of the video gave significantly higher ratings in their opinion of the lesson, how well the lesson communicated information, and quality of the instructor. Even more importantly, that same group of students also recalled and retained significantly more information on the topic. What are the explanatory and response variables? Why is this an experiment? Why were students not initially told that the true purpose of the study was to assess the use of humor? Why do you think the study was done using a fixed video format rather than through live teaching?

◆ **19.** Could the magnetic fields from power lines cause leukemia in children? Investigators spent five years and \$5 million comparing 638 children who had leukemia and 620 who did not. They went into the homes and actually measured the magnetic fields in the children's bedrooms, in other rooms, and at the front door. They recorded facts about nearby power lines for the family home and also for the mother's residence when she was pregnant. Result: No evidence of more than a chance connection between magnetic fields and childhood leukemia. Explain carefully why this study is *not* an experiment and what kind of study it is.

◆ **20.** A typical hour of prime-time television shows three to five violent acts. Linking family interviews and police records shows a clear association between time spent watching TV as a child and later aggressive behavior.

**(a)** Explain why this is an observational study rather than an experiment.
**(b)** Suggest several variables describing a child's home life that may be confounded with how much TV he or she watches. Explain why confounding makes it difficult to conclude that more TV *causes* more aggressive behavior.

◆ **21.** The Nurses' Health Study has interviewed a sample of more than 100,000 female registered nurses every two years since 1976. Beginning in 1980, the study asked questions about diet, including alcohol consumption. The researchers concluded that "light-to-moderate drinkers had a significantly lower risk of death" than either nondrinkers or heavy drinkers.

**(a)** Is the Nurses' Health Study an observational study or an experiment? Why?
**(b)** What does "significant" mean in a statistical report?
**(c)** Suggest some confounding variables that might explain why moderate drinkers have lower death rates than nondrinkers. (The study adjusted for these variables.)

**22.** You can use your computer to make telephone calls over the Internet. How would cost affect the behavior of users of this service? You will offer the service to all 200 rooms in a college dormitory. Some rooms will pay a low flat rate. Others will pay higher rates at peak periods and very low rates off-peak. You are interested in the amount and time of use and in the effect on the congestion of the network. Outline the design of an experiment to study the effect of rate structure.

**23.** Will classroom programs explaining the health advantages of drinking water rather than sugary sodas reduce obesity among children aged 7 to 11 years? Because children are already in school classrooms, we must randomize classes rather than individual children. An experiment assigned 15 classes to receive the program and another 14 to form a control group. After 12 months, obesity had increased in the control group and remained steady in the treatment group. Outline the design of the experiment, label the available classes, and use Table 7.1, beginning at line 103, to carry out the random assignment.

◆ **24.** A college allows students to choose either classroom or self-paced instruction in a basic mathematics course. The college wants to compare the effectiveness of self-paced and regular instruction. Someone proposes giving the same final exam to all students in both versions of the course and comparing the average score of those who took the self-paced option with the average score of students in regular sections.

**(a)** Explain why confounding makes the results of that study worthless.
**(b)** Given 30 students who are willing to use either regular or self-paced instruction, outline an experimental design to compare the two methods of instruction. Then use Table 7.1, starting at line 108, to carry out the randomization.

**25.** Will people spend less on health care if their health insurance requires them to pay some part of the cost themselves? An experiment on this issue asked if the percent of medical costs that are paid by health insurance has an effect either on the amount of medical care that people use or on their health. The treatments were four insurance plans. Each plan paid all medical costs above a ceiling. Below the ceiling, the plans paid 100%, 75%, 50%, or 0% of costs incurred. Outline the design of a randomized comparative experiment suitable for this study.

**26.** Track down a print or online copy of the Bible. The opening chapter of the book of Daniel (especially verses 12–16) appears to have the first clinical trial in recorded history. Outline the design of the experiment. Discuss how you know whether it is an uncontrolled experiment, a comparative experiment, or a randomized comparative experiment.

**27.** Stores advertise price reductions to attract customers. What type of price cut is most attractive? Market researchers prepared ads for athletic shoes announcing different levels of discounts (20%, 40%, or 60%). The student subjects who read the ads were also given "inside information" about the fraction of shoes on sale (50% or 100%). Each subject then rated the attractiveness of the sale on a scale of 1 to 7.

**(a)** Each treatment in this experiment combines values of two explanatory variables, discount level and fraction on sale. List the treatments. How many treatments are there?
**(b)** Outline a randomized comparative experiment using 60 student subjects. Use Table 7.1 at line 123 to choose the subjects for the first treatment.

**28.** Healthcare providers are giving more attention to relieving the pain of cancer patients. An article in the journal *Cancer* surveyed a number of studies and concluded that controlled-release (CR) morphine tablets, which release the painkiller gradually over time, are more effective than giving standard morphine when the patient needs it. The "methods" section of the article begins: "Only those published studies that were controlled (i.e., randomized, double-blind, and comparative), repeated-dose studies with CR morphine tablets in cancer pain patients were considered for this review." Explain the terms in parentheses to someone who knows nothing about medical trials.

**29.** Eye cataracts are responsible for over 40% of blindness around the world. Can drinking tea regularly slow the growth of cataracts? We can't experiment on people, so we use rats as subjects. Researchers injected 14 young rats with a substance that causes cataracts. Half the rats also received tea extract; the other half got a placebo. The response variable was the growth of cataracts over the next six weeks. Yes, the tea extract did slow cataract growth.

**(a)** Outline the design of this experiment.
**(b)** Use Table 7.1, starting at line 108, to assign rats to treatments.

■ **30.** The rats in the previous exercise were labeled 01 to 14 in order to use the table of random digits. Unknown to the researchers, the 5 rats labeled 01 to 05 have a genetic defect that favors cataracts. If we simply put rats 01 to 07 in the tea group, the experiment would be biased against tea. We can observe how random selection works to reduce bias by keeping track of how many of these 5 rats get assigned to the tea group. Carry out the random assignment of 7 rats to the tea group 20 times, keeping track of how many of rats 01 to 05 are in the tea group each time. Make a histogram of the count of rats 01 to 05 assigned to tea. What is the average number in your 20 tries?

## 7.6 Experiments Versus Observational Studies

◆ **31.** People who eat lots of fruits and vegetables have lower rates of colon cancer than those who eat little of these foods. Fruits and vegetables are rich in antioxidants such as vitamins A, C, and E. Will taking antioxidants help prevent colon cancer? A clinical trial studied this question with 864 people who were at risk for colon cancer. The subjects were divided into four groups: daily beta-carotene, daily vitamins C and E, all three vitamins every day, and daily placebo. After four years, the researchers were surprised to find no significant difference in colon cancer among the groups.

**(a)** Outline the design of the experiment. Use your judgment in choosing the group sizes.
**(b)** Assign labels to the 864 subjects and use Table 7.1, starting at line 118, to choose the first five subjects for the beta-carotene group.
**(c)** The study was double-blind. What does this mean?
**(d)** What does "no significant difference" mean in describing the outcome of the study?
**(e)** Suggest some characteristics of the kind of people who eat lots of fruits and vegetables that might explain lower rates of colon cancer. The experiment suggests that these variables, rather than the antioxidants, may be responsible for the observed benefits of fruits and vegetables.

◆ **32.** The financial aid office of a university asks a sample of students about their employment and earnings. The report says that "for academic year earnings, a statistically significant difference was found between the sexes, with men earning more on the average. No significant difference was found between the earnings of black and white students." Explain both of these conclusions, for the effects of sex and of race on average earnings, in language understandable to someone who knows no statistics.

◆ **33.** Do those high center brake lights, required on all cars sold in the United States since 1986, really reduce rear-end collisions? Randomized comparative experiments with fleets of rental and business cars, done before the lights were required, showed that the third brake light reduced rear-end collisions by as much as 50%. Alas, requiring the third light in all cars led to only a 5% drop. Explain why the experiment did not realistically imitate conditions after the lights were required.

◆ **34.** A psychologist studies how much people disclose about themselves to other people met at a party. He arranges for student subjects to be introduced to new people. The subjects are both female and male and both black and white. The results show that "there were no significant race effects, but self-disclosure was significantly higher among females than among males." Explain what this means in language understandable to someone who knows no statistics. Do not use the word *significance* in your answer.

◆ **35.** In the July 15, 2007 issue of *Cancer*, a study reported on 533,715 women at least 40 years old who were diagnosed with invasive breast cancer and reported to the National Cancer Data Base. The study found strong evidence that patients without health insurance were more likely to have a more advanced stage (i.e., III or IV) of cancer. Is this an experiment or observational study and how do you know?

## 7.7 Inference: From Sample to Population

**36.** An opinion poll uses random digit dialing equipment to dial 2000 randomly chosen residential telephone numbers. Of these, <u>631</u> are unlisted numbers. This isn't surprising, because <u>35%</u> of all residential numbers are unlisted. For each underlined number, state whether it is a parameter or a statistic.

**37.** The Tennessee STAR experiment randomly assigned children to regular or small classes during their first four years of school. When these children reached high school, <u>40.2%</u> of blacks from small classes took the ACT or SAT college entrance exams. Only <u>31.7%</u> of blacks from regular classes took one of these exams. For each underlined number, state whether it is a parameter or a statistic.

**38.** The College Alcohol Study interviewed an SRS of 14,941 college students about their drinking habits. Suppose that half of all college students "drink to get drunk" at least once in a while. That is, $p = 0.5$.

**(a)** What are the mean and standard deviation of the proportion $\hat{p}$ of the sample who drink to get drunk?
**(b)** In what range of values do the proportions $\hat{p}$ from 95% of all samples fall?
**(c)** In what range of values do the proportions $\hat{p}$ from 99.7% of all samples fall?

**39.** Harley-Davidson motorcycles make up 14% of all the motorcycles registered in the United States. You plan to interview an SRS of 500 motorcycle owners.

*(Peter Turnley/Corbis.)*

**(a)** What is the approximate distribution of the proportion of your sample who own Harleys?
**(b)** In 95% of all samples like this one, the proportion of the sample who own Harleys will fall between ____________ and ____________. What are the missing numbers?

**40.** Exercise 38 asks what values the sample proportion $\hat{p}$ is likely to take when the population proportion is $p = 0.5$ and the sample size is $n = 14{,}941$. What range covers the middle 95% of values of $\hat{p}$ when $p = 0.5$ and $n = 1000$? When $n = 4000$? When $n = 16{,}000$? What general fact about the behavior of $\hat{p}$ do your results illustrate?

■ **41.** You can use a table of random digits to *simulate* sampling from a population. Suppose that 60% of the population bought a lottery ticket in the last 12 months. We will simulate the behavior of random samples of size 40 from this population.

**(a)** Let each digit in the table stand for one person in this population. Digits 0 to 5 stand for people who bought a lottery ticket, and 6 to 9 stand for people who did not. Why does looking at one digit from Table 7.1 simulate drawing one person at random from a population with 60% "yes"?
**(b)** Each row in Table 7.1 contains 40 digits. So the first 10 rows represent the results of 10 samples. How many digits between 0 and 5 does the top row contain? What is the percent of "yes" responses in this sample? How many of your 10 samples overestimated the population truth 60%? How many underestimated it? You could program a computer to continue this process, say 1000 times, to produce a pattern like that in Figure 7.4.

## 7.8 Confidence Intervals

**42.** In a random sample of students who took the SAT Reasoning college entrance examination twice, it was found that 427 of the respondents had paid for coaching courses and that the remaining 2733 had not. Give a 95% confidence interval for the proportion of coaching among students who retake the SAT.

**43.** A Gallup poll asked each of 1785 randomly selected adults whether she happened to attend a house of worship in the previous seven days. Of the respondents, 750 said "yes." Give a 95% confidence interval for the proportion of all adults who claim that they attended a house of worship during the week preceding the poll. (The proportion who actually attended may be lower–some people say "yes" if they often attend, even if they didn't attend that particular week.)

**44.** *The New York Times* and CBS News conducted a nationwide survey of 1048 randomly selected 13- to 17-year-olds. Of these teenagers, 692 had a television in their room.

**(a)** Give a 95% confidence interval for the proportion of all teens who have a TV set in their room.
**(b)** The news article says, "In theory, in 19 cases out of 20, the survey results will differ by no more than three percentage points in either direction from what would have been obtained by seeking out all American teenagers." Explain how your results agree with this statement.

◆ **45.** A telephone survey of 880 randomly selected drivers asked, "Recalling the last 10 traffic lights you drove through, how many of them were red when you entered the intersections?" Of the 880 respondents, 171 admitted that at least one light had been red.

**(a)** Give a 95% confidence interval for the proportion of all drivers who ran one or more of the last 10 red lights they met.
**(b)** A practical problem with this survey is that people may not give truthful answers. What is the likely direction of the bias: Do you think more or fewer than 171 of the 880 respondents really ran a red light? Why?

◆ **46.** The Harris poll asked a sample of 1009 adults which causes of death they thought would become more common in the future. Topping the list was gun violence: 70% of the sample thought deaths from guns would increase.

**(a)** How many of the 1009 people interviewed thought deaths from gun violence would increase?
**(b)** Harris says that the margin of error for this poll is plus or minus 3 percentage points. Explain to someone

who knows no statistics what "margin of error plus or minus 3 percentage points" means.

**(c)** Give a 95% confidence interval for this survey. Does your margin of error agree with the 3 percentage points announced by Harris?

◆ **47.** Consider the margin of error formula $2\sqrt{\hat{p}(1-\hat{p})/n}$.

**(a)** For a fixed value of $n$, what value of $\hat{p}$ causes this formula to be the largest value it can be?

**(b)** Using the answer to part (a), what is a simplified (and slightly more conservative) formula for the margin of error?

◆ **48.** A news article reports that in a recent Gallup poll, 78% of the sample of 1108 adults said they believe there is a heaven. Only 60% said they believe there is a hell. The news article ends, "The poll's margin of sampling error was plus or minus four percentage points." Can we be certain that between 56% and 64% of all adults believe there is a hell? Explain your answer.

◆ **49.** A survey of Internet users found that males outnumbered females by nearly 2 to 1. This was a surprise, because earlier surveys had put the ratio of men to women closer to 9 to 1. Later in the article we find that surveys were sent to 13,000 organizations and that 1468 of these responded. The survey report claims that "the margin of error is 2.8 percent, with 95 percent confidence."

**(a)** What was the *response rate* for this survey? (The response rate is the percent of the planned sample that responded.)

**(b)** Do you think that the small margin of error is a good measure of the accuracy of the survey's results? Explain your answer.

**50.** A recent Gallup poll found that 68% of adult Americans favor teaching creationism along with evolution in public schools. The Gallup press release says:

> *For results based on samples of this size, one can say with 95 percent confidence that the maximum error attributable to sampling and other random effects is plus or minus 3 percentage points.*

Give one example of a source of error in the poll result that is *not* included in this margin of error.

■ **51.** The Internal Revenue Service plans to examine an SRS of individual income tax returns from each state that were filed electronically. One variable of interest is the proportion of returns that were filed by a tax practitioner rather than by an individual taxpayer. The total number of e-filed tax returns in a state varies from 4.9 million in California to 97,000 in Vermont.

**(a)** Will the margin of error for estimating the proportion change from state to state if an SRS of 1000 e-filed returns is selected in each state? Explain your answer.

**(b)** Will the margin of error change from state to state if an SRS of 1% of all e-filed returns is selected in each state? Explain your answer.

■ **52.** Exercise 46 describes a Harris poll that interviewed 1009 people. Suppose you want a margin of error half as large as the one you found in that exercise. How many people must you plan to interview?

■ **53.** Though opinion polls usually make 95% confidence statements, some sample surveys use other confidence levels. The monthly unemployment rate, for example, is based on the Current Population Survey of about 60,000 households. The margin of error in the unemployment rate is announced as about ±0.15% with 90% confidence. Is the margin of error for 90% confidence larger or smaller than the margin of error for 95% confidence? Why? (*Hint:* Look at Figure 7.7 again.)

## Chapter Review

**54.** The proportion of one's body that is fat is a key indicator of fitness. The many ways to estimate this have different margins of error (given in percentage points):

| method | calipers pinch | bioelectrical impedance | body mass index calculator | hydrostatic weighing (dunk test) |
|---|---|---|---|---|
| margin of error | ±3 | ±4 | ±10 | ±1 |

**(a)** Which of these tests is the least accurate?

**(b)** If the pinch test says you have 21% body fat, what is the 95% confidence interval for this estimate?

**55.** Many medical trials randomly assign patients to either an active treatment or a placebo. These trials are always double-blind. Sometimes the patients can tell whether or not they are getting the active treatment. This defeats the purpose of blinding. Reports of medical research usually ignore this problem. Investigators looked at a random sample of 97 articles reporting on placebo-controlled randomized trials in the top five general medical journals. Only 7 of the 97 discussed the success of blinding. Give a 95% confidence interval for the proportion of all such articles that discuss the success of blinding. [Dean Fergusson et al., Turning a blind eye: The success of blinding reported in a random sample of randomised, placebo-controlled trials, *British Medical Journal*, 328 (2004): 432–436.]

**56.** Tomeka wants to ask a sample of students at her college, "Do you think that Social Security will still be paying benefits when you retire?" She obtains the college email addresses of the 2654 students.

**(a)** How would you label the addresses in order to choose a simple random sample of 100 students?

**(b)** Use Table 7.1, starting at line 103, to choose the first three labels in the sample.

**(c)** Tomeka sends her question by email to the 100 addresses in her sample. Although she has chosen an

SRS, a serious practical difficulty may make it hard to draw clear conclusions from her sample. What practical difficulty do you expect Tomeka to encounter?

■ **57.** Suppose that exactly 10% of all articles in major medical journals that describe placebo-controlled randomized trials discuss the success of blinding. That is, the proportion of "successes" in the population is $p = 0.1$. What is the approximate probability that fewer than 7% of an SRS of 97 articles from this population discuss the success of blinding?

**58.** Ability to grow in shade may help pines found in the dry forests of Arizona resist drought. How well do these pines grow in shade? Investigators planted pine seedlings in a greenhouse in either full light or light reduced to 5% of normal by shade cloth. At the end of the study, they dried the young trees and weighed them.

**(a)** Explain why this study is an experiment.
**(b)** What are the individuals, the treatments, and the response variable in this experiment?
**(c)** You have 200 pine seedlings available. Outline the design you would use for this experiment.

## APPLET EXERCISES

**To do these exercises, go to www.whfreeman.com/fapp8e.**

**1.** Use the *Simple Random Sample* applet to choose the sample of songs in Example 4. Assign labels 01 to 27 by entering 27 in the "Population 1 to" box and clicking "Reset." Then enter 4 in the "Select a sample of size" box and click "Sample." Which songs from the list in Example 4 make up your sample? Click "Reset" and choose another sample. Which songs did you choose this time? You see that random sampling gives different samples each time–what matters is that all songs have the same chance to be chosen.

**2.** The *Simple Random Sample* applet is handy when the population or sample is large. (The applet will handle population sizes up to 500.) Skills Check 5 asks you to choose an SRS of 10 out of 440 retail outlets. Use the applet to do this and report your result.

**3.** You can use the *Simple Random Sample* applet to choose treatment groups at random for a randomized comparative experiment. Exercise 29 asks you to choose the subjects to get the first treatment in an experiment that compares two treatments.

**(a)** Use the applet to choose an SRS of 7 out of 14 to receive the first treatment. Which subjects make up this group?
**(b)** The applet allows you to randomly assign subjects to more than two groups. Suppose you had a total of 36 rats and you wanted to assign a different treatment to each of four 9-rat groups. After you choose the first group, the "Population hopper" contains the 27 subjects that were not chosen, in scrambled order. Click "Sample" again to choose 9 of these remaining subjects to receive the second treatment. Do this once more to choose the third group. The 9 subjects that remain in the "Population hopper" form the fourth group. Which of the 36 subjects will receive each of the four treatments?

**4.** You can use the *Probability* applet to speed up and improve Exercise 41. You have a population in which 60% of the individuals approve of legal gambling. You want to take many samples from this population to observe how the sample proportion that approves of gambling varies from sample to sample. Set the "Probability of heads" in the applet to 0.6 and the number of tosses to 40. This simulates an SRS of size 40 from a large population. Each head in the sample is a person who approves of legal gambling, and each tail is a person who disapproves. By alternating between "Toss" and "Reset" you can take many samples quickly.

**(a)** Take 50 samples, recording the proportion that approves of gambling in each sample. (The applet gives this proportion at the top left of its display.) Make a histogram of the 50 sample proportions.
**(b)** Another population contains only 20% who approve of legal gambling. Take 50 samples of size 40 from this population, record the number in each sample that approves, and make a histogram of the 50 sample proportions. How do the centers of your two histograms reflect the differing truths about the two populations?

**5.** The idea of an 80% confidence interval is that the interval captures the true parameter value in 80% of all samples. That's not high enough confidence for practical use, but 80% hits and 20% misses make it easy to see how a confidence interval behaves in repeated samples from the same population. Go to the *Confidence Interval* applet.

**(a)** Set the confidence level to 80%. Click "Sample" to choose an SRS and calculate the confidence interval. Do this 10 times to simulate 10 SRSs with their 10 confidence intervals. How many of the 10 intervals captured the true mean $\mu$? How many missed?
**(b)** You see that we can't predict whether the next sample will hit or miss. The confidence level, however, tells us what percent will hit in the long run. Reset the applet and click "Sample 50" to get the confidence intervals from 50 SRSs. How many hit? Keep clicking "Sample 50" and record the percent of hits among 100, 200, 300, 400, and 500 SRSs. Even 500 samples is not truly "the long run," but we expect the percent of hits in 500 samples to be fairly close to the confidence level, 80%.

# WRITING PROJECTS

**1.** Go to the Web site of the Gallup Organization (www.gallup.com) and click on "Gallup Poll." You should be able to find a press release you can access and read without charge. Newspapers publish short articles based on press releases. Write a news article about two paragraphs long based on the press release.

**2.** Recall how Example 8 shows how wording can affect survey results. You can explore this by doing an experiment disguised as a survey! Choose a topic, then design two questions with a key difference in wording. Use randomization to choose which version of the question you give each person. Don't reveal the design of the experiment to participants until after they have provided their answers. After you have roughly 20 or more responses to each version, compare and interpret your results. For an example of such an experiment, see John Rubin's article "Weighing Anchors" in the June 1990 issue of *Omni*.

**3.** Do one of the following:

**(a)** How would you design a double-blind experiment in which participants test which of two brands of tissue are preferred? Conduct this experiment and write up the results. How do the results compare with any claims made in advertisements for the products?

**(b)** Choose an issue of current interest to students at your school. Prepare a short questionnaire (no more than five questions) to determine opinions on this issue. Choose a sample of about 25 students, administer your questionnaire, and write a brief description of your findings. Also write a short discussion of your experiences in designing and carrying out the survey. Although 25 students are too few for you to be statistically confident of your results, this project centers on the practical work of a survey. You must first identify a population; if it is not possible to reach a wider student population, use students enrolled in this course. Did the subjects find your questions clear? Did you write the questions so that it was easy to tabulate the responses? At the end, did you wish you had asked different questions?

# SUGGESTED READINGS

ANDERSON-COOK, C. M. and SUNDAR DORAI-RAJ. An active learning in-class demonstration of good experimental design, *Journal of Statistics Education*, 9(1) (2001): http://www.amstat.org/publications/jse/v9n1/anderson-cook.html. A nice example of issues that arise when designing a randomized experiment. This article includes an applet students can use to experience the experiment.

BOCK, DAVID E., PAUL F. VELLEMAN and RICHARD D. DE VEAUX. *Stats*, Addison Wesley, Boston, 2007. Another text at the mathematical level just above ours. Like this chapter, it uses estimating a population proportion to introduce confidence intervals (Chapter 19).

LESSER, LAWRENCE M. and ERIK NORDENHAUG. Ethical statistics and statistical ethics: making an interdisciplinary module, *Journal of Statistical Education*, 12(3) (2004): http://www.amstat.org/publications/jse/v12n3/lesser.html. This article's discussion includes ethical issues associated with surveys, experiments and observational studies.

MOORE, DAVID S. *The Basic Practice of Statistics*, 3rd ed., Freeman, New York, 2004. Chapters 7 and 8 discuss samples and experiments. Chapter 13 presents the reasoning of confidence intervals, and Chapter 18 presents confidence intervals for a population proportion.

# SUGGESTED WEB SITES

The National Council on Public Polls, www.ncpp.org, has a statement on "20 Questions a Journalist Should Ask About a Poll" that makes interesting reading. The explanations expand our cautions about sample surveys in practice. You can find similar information on the Web site of the American Association for Public Opinion Research, www.aapor.org. Click the "For Journalists" box and take a look at the "Guide to Best Practices" and "Survey Practices that AAPOR Condemns" listings. Also, read the American Statistical Association publication "What is a Survey (2nd ed)" at www.whatisasurvey.info.

The single most important sample survey in the United States is probably the government's monthly

Current Population Survey (CPS), carried out by the Census Bureau on behalf of the Bureau of Labor Statistics (BLS). The CPS Web site, www.bls.census.gov/cps/, contains an abundant wealth of information. You might click on "Publications," then on "Chronological List" and look at the latest monthly statement of the BLS Commissioner to Congress about employment and unemployment.

# CHAPTER 8



# Probability:
## The Mathematics of Chance

Have you ever wondered how gambling, which is a recreation or an addiction for individuals, can be a business for the casino? A business requires predictable revenue from the service it offers, even when the service is a game of chance. Individual gamblers may win or lose. They can never say whether a day at the casino will turn a profit or a loss. But the casino isn't gambling. Casinos are consistently profitable, and state governments make money both from running lotteries and from selling licenses for other forms of gambling.

It is a remarkable fact that the aggregate result of many thousands of chance outcomes can be known with near certainty. The casino need not load the dice, mark the cards, or alter the roulette wheel. It knows that in the long run each dollar bet will yield its five cents or so of revenue. It is therefore good business to concentrate on free floor shows or inexpensive bus fares to increase the flow of dollars bet. The flow of profit will follow.

Gambling houses are not alone in profiting from the fact that a chance outcome many times repeated is firmly predictable. For example, although a life insurance company does not know *which* of its policyholders will die next year, it can predict quite accurately *how many* will die. It sets its premiums according to this knowledge, just as the casino sets its jackpots. Statisticians also rely on the regular behavior of chance: A 95% confidence interval works 95% of the time because, in the long run, chance behavior is predictable.

**Random** — DEFINITION

A phenomenon or trial is said to be **random** if individual outcomes are uncertain but the long-term pattern of many individual outcomes is predictable.

To a statistician, "random" does not mean "haphazard." Randomness is a kind of order, an order that emerges only in the long run, over many repetitions. Many phenomena, both natural and of human design, are random. The hair colors of children, the spread of epidemics, and the decay of radioactive substances are examples of natural randomness. Indeed, quantum mechanics asserts that at the subatomic level the natural world is inherently random.

Games of chance are examples of randomness deliberately produced by human effort. Casino dice are carefully machined, and their drilled holes are filled with material equal in density to the plastic body. This guarantees that the side with six spots has the same weight as the opposite side, which has only one spot. Thus, each side is equally likely to land upward. All the odds and payoffs of dice games rest on this carefully planned randomness. Random sampling and randomized comparative experiments are also examples of planned randomness, although they use tables of random digits rather than dice and cards. The reasoning of statistical inference rests on asking, "How often would this method give a correct answer if I used it very many times?" Probability theory, the mathematical description of randomness, is the basis for gambling, insurance, much of modern science, and statistical inference. **Probability** is the topic of this chapter.

## 8.1 Probability Models and Rules

Toss a coin, or choose a simple random sample (SRS). The result can't be predicted in advance, because the result will vary when you toss the coin or choose the sample repeatedly. But there is nonetheless a regular pattern in the results, a pattern that emerges clearly only after many repetitions. This remarkable fact is the basis for the idea of probability.

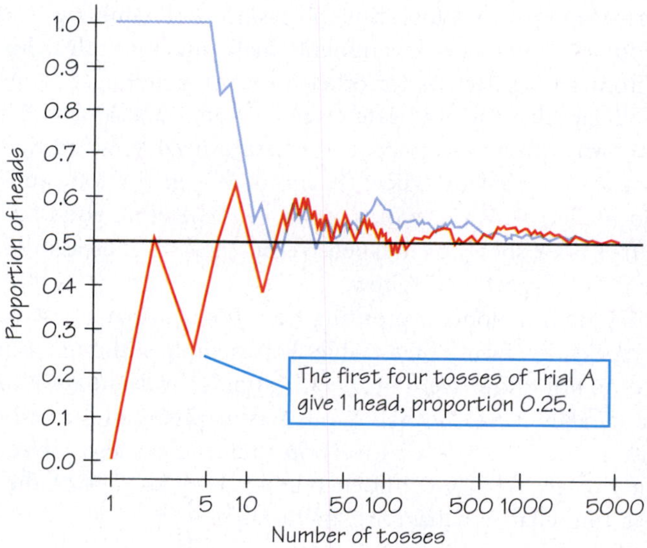


**FIGURE 8.1** The proportion of tosses of a coin that give a head varies as we make more tosses. Eventually, however, the proportion approaches 0.5, the probability of a head. This figure shows the results of two trials of 5000 tosses each. The horizontal scale is transformed using logarithms to show both short-term and long-term behavior.

### EXAMPLE 1 ■ Tossing a Coin

When you toss a coin, there are only two possible outcomes, heads or tails. Figure 8.1 shows the results of tossing a coin 5000 times twice. For each number of tosses from 1 to 5000, we have plotted the proportion of those tosses that gave a head.

Trial $A$ (red line) begins tail, head, tail, tail. You can see that the proportion of heads for Trial $A$ starts at 0 on the first toss, rises to 0.5 when the second toss gives a head, then falls to 0.33 and 0.25 as we get two more tails. Trial $B$ (blue line), on the other hand, starts with five straight heads, so the proportion of heads is 1 until the sixth toss.

The proportion of tosses that produce heads is quite variable at first. Trial $A$ starts low and Trial $B$ starts high. As we make more and more tosses, however, the proportions of heads for both trials get close to 0.5 and stay there. If we made yet a third trial at tossing the coin a great many times, the proportion of heads would again settle down to 0.5 in the long run. We say that 0.5 is the *probability* of a head. The probability 0.5 appears as a horizontal line on the graph.

## Probability — DEFINITION

The **probability** of any outcome of a random phenomenon is the proportion of times the outcome would occur in a very long series of repetitions. We will soon see a concrete expression of this in the procedure box for "Equally Likely Outcomes."

The *Probability* applet (see Applet Exercise 1) animates Figure 8.1. It allows you to choose the probability of a head and simulate any number of tosses of a coin with that probability. Try it. You will see that the proportion of heads gradually settles down close to the probability. Equally important, you will also see that the proportion in a small or moderate number of tosses can be far from the probability. *Probability describes only what happens in the long run.* Random phenomena are irregular and unpredictable in the short run.

We might suspect that a coin has probability 0.5 of coming up heads just because the coin has two sides. As Exercise 1 illustrates, such suspicions are not always correct. The idea of probability is *empirical.* That is, it is based on observation rather than theorizing. Probability describes what happens in very many trials, and we must actually observe many trials to pin down a probability.

Gamblers have known for centuries that the fall of coins, cards, and dice displays clear patterns in the long run. In fact, a question about a gambling game launched probability as a formal branch of mathematics. The idea of probability rests on the observed fact that the average result of many thousands of chance outcomes can be known with near certainty. But a definition of probability as "long-run proportion" is vague. Who can say what "the long run" is? We can always toss the coin another 1000 times. Instead, we give a mathematical description of *how probabilities behave,* based on our understanding of long-run proportions. To see how to proceed, think first about a very simple random phenomenon, tossing a coin once. When we toss a coin, we cannot know the outcome in advance. What do we know? We are willing to say that the outcome will be either heads or tails. We believe that each of these outcomes has probability 1/2. This description of coin tossing has two parts:

- A list of possible outcomes
- A probability for each outcome

This description is the basis for all probability models. Here is the vocabulary we use.

**Sample Space** DEFINITION

The **sample space** $S$ of a random phenomenon is the set of all possible outcomes that cannot be broken down further into simpler components.

**Event** DEFINITION

An **event** is any outcome or any set of outcomes of a random phenomenon. That is, an event is a subset of the sample space.

**Probability Model** DEFINITION

A **probability model** is a mathematical description of a random phenomenon consisting of two parts: a sample space $S$ and a way of assigning probabilities to events.

The sample space $S$ can be very simple or very complex. When we toss a coin once, there are only two outcomes, heads and tails. So the sample space is $S = \{H, T\}$. If we draw a random sample of 1000 U.S. residents age 18 and over, as opinion polls often do, the sample space contains all possible choices of 1000 of the more than 230 million adults in the country. This $S$ is extremely large: $1.3 \times 10^{5794}$. Each member of $S$ is a possible opinion poll sample, which explains the term **sample space**.

## EXAMPLE 2 ■ Tossing Two Coins

Probabilities can be hard to determine without detailing or diagramming the sample space. For example, E. P. Northrop notes that even the great eighteenth-century French mathematician Jean le Rond d'Alembert tripped on the question: "In two coin tosses, what is the probability that heads will appear at least once?" Because the number of heads could be 0, 1 or 2, d'Alembert reasoned (incorrectly) that each of those possibilities would have an equal probability of 1/3, and so he reached the (wrong) answer of 2/3. What went wrong? Well, {0, 1, 2} could not be the fully-detailed sample space because "1 head" can happen in more than one way. For example, if you flip a dime and a penny once each, you could display the sample space with a *table*:

Another way is with a *tree diagram*, in which all possible left-to-right pathways through the branches generate outcomes.

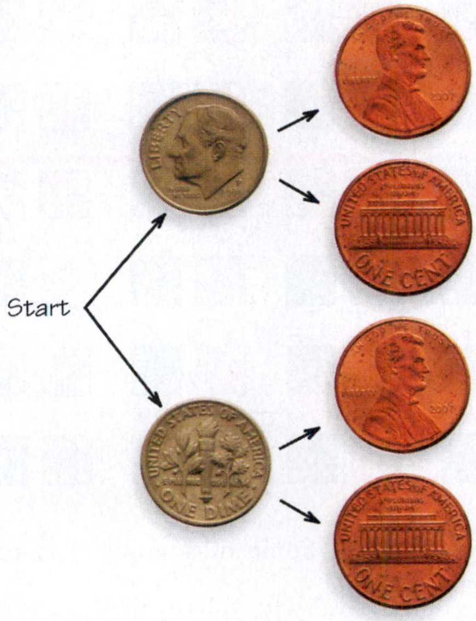


Either way, we can see that the sample space has 4, not 3, equally likely outcomes: {HH, HT, TH, TT}. With the table or tree diagram in front of us, you may already see that the correct probability of at least 1 head is not 2/3, but 3/4.

## EXAMPLE 3

### Pair-a-Dice: Outcomes for Rolling Two Dice

Rolling two dice is a common way to lose money in casinos. There are 36 possible outcomes when we roll two dice and record the up faces in order (first die, second die). Figure 8.2 displays these outcomes. They make up the sample space $S$.

If the dice are carefully made, experience shows that each of the 36 outcomes in Figure 8.2 comes up equally often. So a reasonable probability model assigns probability 1/36 to each outcome.

In craps and most other games, all that matters is the *sum* of the spots on the up faces. Let's change the random outcomes we are interested in: Roll two dice and count the spots on the up faces. Now there are only 11 possible outcomes, from a sum of 2 (for rolling a double 1) to a sum of 12 (for rolling a double 6). The sample space is now

$$S = \{2, 3, 4, 5, 6, 7, 8, 9, 10, 11, 12\}$$

Comparing this $S$ with Figure 8.2 reminds us that we can change $S$ by changing the detailed description of the random phenomenon we are describing. The outcomes in this new sample space are *not* equally likely, because there are six ways to roll a 7 and only one way to roll a 12. The probability aspect of this example is developed further in Example 4.

*(George Diebold/Stone/Getty Images.)*

There are many ways to assign probabilities, so it is convenient to start with some general rules that any assignment of probabilities to outcomes must obey. These facts follow from the idea of probability as "the long-run proportion of

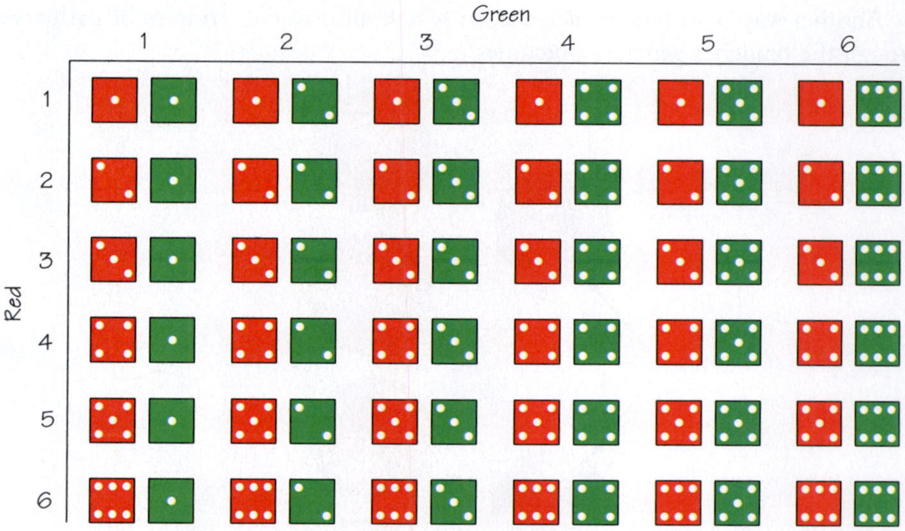


**FIGURE 8.2** The 36 possible outcomes for rolling two dice, for Example 3.

repetitions on which an event occurs." Some rules apply only to special kinds of events, which we define here:

**Complement of an Event** DEFINITION

The **complement of an event** $A$ is the event that $A$ does *not* occur, written as $A^C$.

**Disjoint Events** DEFINITION

Two events are **disjoint events** if they have no outcomes in common. Disjoint events are also called *mutually exclusive events*.

**Independent Events** DEFINITION

Two events are **independent events** if the occurrence of one event has no effect on the probability of the occurrence of the other event.

1. **Any probability is a number between 0 and 1 inclusive**. Any proportion is a number between 0 and 1 inclusive, so any probability is also a number between 0 and 1 inclusive. An event with probability 0 never occurs, and an event with probability 1 always occurs, and an event with probability 0.5 occurs in half the trials in the long run.
2. **All possible outcomes together must have probability 1**. Because some outcome must occur on every trial, the sum of the probabilities for all possible (simplest) outcomes must be exactly 1.
3. **The probability that an event does not occur is 1 minus the probability that the event does occur**. If an event occurs in (say) 70% of all trials, it fails to occur in the other 30%. The probability that an event occurs and the probability that it does not occur always add to 100%, or 1 (see Figure 8.3).

4. **If two events are *independent*, then the probability that one event <u>and</u> the other both occur is the product of their individual probabilities.** Consider event $A$ is "red die is a 1 or 2" and event $B$ is "green die is 6." The red die and green die logically have no influence over each other's outcomes, but we can also look at Figure 8.2 and see that the chance of being in the top two rows does not affect and is not affected by the chance of being in the sixth column. And so Rule 4 for independent events applies and the probability that $A$ and $B$ both happen is the product $(1/3)(1/6) = 1/18$. Note that we can also see from Figure 8.2 that the *intersection* or "overlap" of events $A$ and $B$ happens in 2 of the 36 outcomes and $2/36 = 1/18$. Also, since $A$ and $B$ overlap, they are *not* disjoint, even though the everyday use of the word "independent" might (incorrectly) suggest that kind of separateness.
5. **The probability that one event <u>or</u> the other occurs is the sum of their individual probabilities minus the probability of their intersection.** This general addition rule makes sense if we look at Rule 5 in Figure 8.3. Simply adding the probabilities of the two events would overshoot the answer because we would be incorrectly "double-counting" the overlap. The way to adjust for this is to subtract the overlap so that it is counted only once. Note that the mathematical "or" is inclusive, which means that the event "$A$ or $B$" happens as long as at least one of the two events happens. In the set theory, it is the *union* of $A$ and $B$, which includes $A$'s and $B$'s "separate property" as well as their "community property." Consider event $A$ is "red die is a perfect square," which has probability of 2/6. Consider event $B$ is "red die is an odd number" (that is, 1, 3, or 5), which has probability of 3/6. The intersection of events $A$ and $B$ corresponds to rolling a "1," which has a probability of 1/6. So the probability that $A$ or $B$ occurs is $2/6 + 3/6 - 1/6 = 4/6 = 2/3$. Notice that if events $A$ and $B$ had been disjoint, there would be no intersection to worry about double counting and this rule would simply turn into this next one:
6. **If two events are *disjoint*, the probability that one <u>or</u> the other occurs is the sum of their individual probabilities.** If one event occurs in 40% of all trials, a different event occurs in 25% of all trials, and the two can never occur together, then one or the other occurs on 65% of all trials because 40% + 25% = 65%.

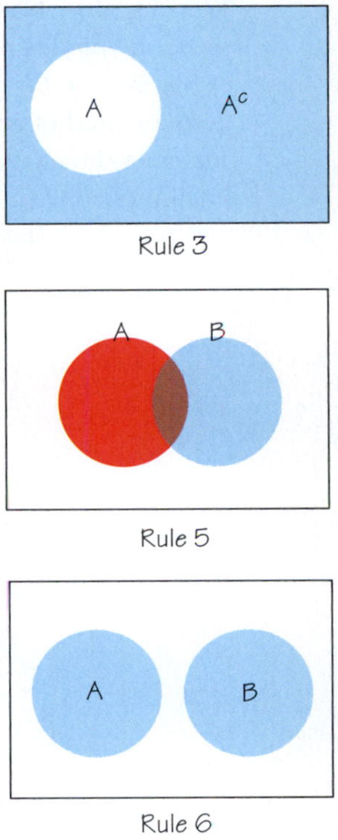


**FIGURE 8.3** Each rectangle represents the whole sample space in these illustrations of Rules 3, 5 and 6.

We can use mathematical notation to state Rules 1 to 6 more concisely. We use capital letters near the beginning of the alphabet to denote events. If $A$ is any event, we write its probability as $P(A)$. Here are our probability facts in formal language. As you apply these rules, remember that they are just another form of intuitively true facts about long-run proportions.

**Probability Rules** RULE

**Rule 1.** The probability $P(A)$ of any event $A$ satisfies $0 \leq P(A) \leq 1$.
**Rule 2.** If $S$ is the sample space in a probability model, then $P(S) = 1$.
**Rule 3.** The **complement rule**: $P(A^C) = 1 - P(A)$
**Rule 4.** The **multiplication rule** for *independent* events: $P(A \text{ and } B) = P(A) \times P(B)$
**Rule 5.** The *general* **addition rule**: $P(A \text{ or } B) = P(A) + P(B) - P(A \text{ and } B)$
**Rule 6.** The *addition rule* for *disjoint* events: $P(A \text{ or } B) = P(A) + P(B)$

## SPOTLIGHT 8.1 Probability and Psychology

Our judgment of probability can be affected by psychological factors. Our desire to become instantly rich may lead us to overestimate the tiny probability of winning the lottery. Our feeling that we are "in control" when we are driving may make us underestimate the probability of an accident. (This may be why some people prefer driving to flying even though flying has a lower probability of death per mile traveled.)

The probability of winning (a share of) the twelve-state Mega Millions jackpot is 1 in 175,711,536. This is like guessing a particular sheet of typing paper from a stack twice the height of Mt. Everest. Or guessing a particular second from a period of about 5.5 years. Without concrete analogies, it is hard to grasp the meaning of very small probabilities and some players may greatly overestimate their chances of winning even if they buy lots of tickets. For example, suppose someone buys 20 $1 Mega Millions tickets every week for 50 years. She would have spent over $50,000 and yet her probability of winning at least one jackpot in that whole time would still be only 1 in 3368. For comparison, the probability of dying in a car accident during a lifetime of driving is about 50 times greater than this!

Andrew Gelman reports that most people say they would not switch to a situation in which they had a small probability *p* of dying and a large probability 1-*p* of gaining $1000. And yet, people will not necessarily spend that much for air bags for their cars. Becoming more aware of our inconsistencies and biases can help us make better use of probability when deciding what risks to take.

## EXAMPLE 4 ■ Probabilities for Rolling Two Dice

Figure 8.2 displays the 36 possible outcomes of rolling two dice. For casino dice, it is reasonable to assign the same probability to each of the 36 outcomes in Figure 8.2. Because all 36 outcomes together must have probability 1 (Rule 2), each outcome must have probability 1/36.

What is the probability of rolling a sum of 5? Because the event "roll a sum of 5" contains the four outcomes displayed in Figure 8.2, the addition rule for disjoint events (Rule 6) says that its probability is

$P$(roll a sum of 5) = $P$(  ) + $P$( 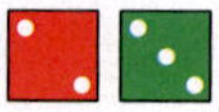 ) + $P$(  ) + $P$(  )

$$= \frac{1}{36} + \frac{1}{36} + \frac{1}{36} + \frac{1}{36}$$

$$= \frac{4}{36} = 0.111$$

Continue using Figure 8.2 in this way to get the full probability model (sample space and assignment of probabilities) for rolling two dice and summing the spots on the up faces. Here it is:

| Outcome | 2 | 3 | 4 | 5 | 6 | 7 | 8 | 9 | 10 | 11 | 12 |
|---|---|---|---|---|---|---|---|---|---|---|---|
| Probability | $\frac{1}{36}$ | $\frac{2}{36}$ | $\frac{3}{36}$ | $\frac{4}{36}$ | $\frac{5}{36}$ | $\frac{6}{36}$ | $\frac{5}{36}$ | $\frac{4}{36}$ | $\frac{3}{36}$ | $\frac{2}{36}$ | $\frac{1}{36}$ |

This model assigns probabilities to individual outcomes. Note that Rule 2 is satisfied because all the probabilities add up to 1. To find the probability of an event, just add the probabilities of the outcomes that make up the event. For example:

$$P(\text{outcome is odd}) = P(3) + P(5) + P(7) + P(9) + P(11)$$
$$= \frac{2}{36} + \frac{4}{36} + \frac{6}{36} + \frac{4}{36} + \frac{2}{36}$$
$$= \frac{18}{36} = \frac{1}{2}$$

What is the probability of rolling any sum other than a 5? The "long way" to find this would be

$$P(2) + P(3) + P(4) + P(6) + P(7) + P(8) + P(9) + P(10) + P(11) + P(12).$$

A much better way would be to use the complement rule (Rule 3):

$$P(\text{roll sum that is } \textit{not } 5) = 1 - P(\text{roll sum of } 5)$$
$$= 1 - \frac{4}{36} = \frac{32}{36} = 0.889$$

Another good time to use the complement rule would be to find the probability of getting a sum greater than 3. Compare the calculation of $P(\text{sum} > 3)$ with $1 - P(\text{sum} \leq 3)$.

For an example of Rule 5, let event $A$ be "sum is odd" and event $B$ be "sum is a multiple of 3." We previously calculated $P(A) = 1/2$. You can verify that $P(B) = 1/3$ and $P(A \text{ and } B)$ is 1/6. And so, $P(A \text{ or } B) = 1/2 + 1/3 - 1/6 = 2/3$.

When the outcomes for a probability model are numbers, we can use a histogram to display the assignment of probabilities to the outcomes. Figure 8.4 is a **probability histogram** of the probability model in Example 4. The height of each bar shows the probability of the outcome at its base. Because the heights are probabilities, they add to 1. Think of Figure 8.4 as an idealized picture of the results of very many rolls of a die. As an idealized picture, it is perfectly symmetric.

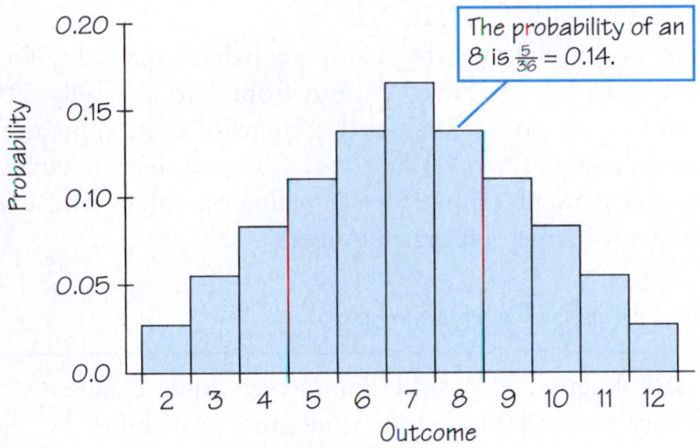


**FIGURE 8.4** Probability histogram showing the probability model for rolling two balanced dice and counting the spots on the up faces.

Example 4 illustrates one way to assign probabilities to events: Assign a probability to every individual outcome, then add these probabilities to find the probability of any event. This idea works well when there are only a finite (fixed and limited) number of outcomes.

## 8.2 Discrete Probability Models

We will work with two kinds of probability models. The first kind is illustrated by Example 4 and is called a **discrete probability model**. (The second kind is in Section 8.4.)

**Discrete Probability Model** DEFINITION

A probability model is called discrete if its sample space has a countable number of outcomes. To assign probabilities in a discrete model, list the probability of all the individual outcomes. By Rules 1 and 2, these probabilities must be numbers between 0 and 1 inclusive and must have sum 1.
The probability of any event is the sum of the probabilities of the outcomes making up the event.

## EXAMPLE 5 ▪ Benford's Law

Faked numbers in tax returns, invoices, or expense account claims often display patterns that aren't present in legitimate records. Some patterns, like too many round numbers, are obvious and easily avoided by a clever crook. Others are more subtle. It is a striking fact that the first (leftmost) digits of numbers in legitimate records often follow a model known as Benford's law. Here it is (note that a first digit can't be 0):

| First digit | 1 | 2 | 3 | 4 | 5 | 6 | 7 | 8 | 9 |
|---|---|---|---|---|---|---|---|---|---|
| Probability | 0.301 | 0.176 | 0.125 | 0.097 | 0.079 | 0.067 | 0.058 | 0.051 | 0.046 |

Check that the probabilities of the outcomes sum exactly to 1. This is therefore a legitimate discrete probability model. Investigators can detect fraud by comparing the first digits in records such as invoices paid by a business with these probabilities. For example, consider the events $A$ = "first digit is 1" and $B$ = "first digit is 2." Applying Rule 6 to the table of probabilities yields $P(A \text{ or } B) = 0.301 + 0.176$, which is 0.477 (almost 50%). Crooks trying to "make up" the numbers probably would not make up numbers starting with 1 or 2 this often.

Let us use some intuition about why first digits behave this way. Note that the increase from 1 to 2 is an increase of 100%, but from 2 to 3 is only 50%, from 3 to 4 is only 33%, and so on. So data values that increase at an approximately constant percentage (which a lot of financial data does, for example) will naturally "spend more time" (within any particular power of 10) taking on values whose left digit is 1, and successively less for larger left-digit numbers.

## 8.3 Equally Likely Outcomes

A simple random sample gives all possible samples an equal chance to be chosen. Rolling two casino dice gives all 36 outcomes the same probability. When randomness is the product of human design, it is often the case that the outcomes in the sample space are all equally likely. Rules 1 and 2 force the assignment of probabilities in this case.

**Finding Probabilities of Equally Likely Outcomes** PROCEDURE

If a random phenomenon has equally likely outcomes, then the probability of event $A$ is

$$P(A) = \frac{\text{count of outcomes in event } A}{\text{count of outcomes in sample space } S}$$

As an aside, a less common way of expressing likelihood that you may encounter in some gambling contexts is *odds*. The *odds* of an event $A$ happening can be expressed as:

$$\frac{\text{count of outcomes in which } A \text{ happens}}{\text{count of outcomes where } A \text{ does not happen.}}$$

The *odds* against an event $A$ happening can be expressed as:

(count of outcomes $A$ does not happen) / (count of outcomes $A$ happens).

For example, let event $A$ be "a die is rolled and lands on a 4 or 5." Since there are twice as many ways $A$ does not happen as there are that $A$ happens, the odds of event $A$ happening are 1:2 and the odds against event $A$ are 2:1. Notice that these numbers are different from the respective probability values $P(A) = 1/3$ and $P(A^C) = 2/3$, but we can express odds in terms of probabilities as follows:

$$\text{odds of } A \text{ happening} = \frac{P(A)}{P(A^C)}, \text{ and odds against } A \text{ happening} = \frac{P(A^C)}{P(A)}.$$

We have included this aside so you will recognize what odds mean in the rare occasions you encounter them, but be aware that odds values do *not* follow the six rules of probability.

## EXAMPLE 6 ■ Are First Digits Equally Likely?

You might think that first (leftmost) digits are distributed "at random" among the digits 1 to 9. Under such a "discrete uniform distribution," the 9 possible outcomes would then be equally likely. The sample space is $S = \{1, 2, 3, 4, 5, 6, 7, 8, 9\}$, and the probability model is:

| First digit | 1 | 2 | 3 | 4 | 5 | 6 | 7 | 8 | 9 |
|---|---|---|---|---|---|---|---|---|---|
| Probability | $\frac{1}{9}$ | $\frac{1}{9}$ | $\frac{1}{9}$ | $\frac{1}{9}$ | $\frac{1}{9}$ | $\frac{1}{9}$ | $\frac{1}{9}$ | $\frac{1}{9}$ | $\frac{1}{9}$ |

The probability of the event that a randomly chosen first digit is a 1 or 2 is

$$\begin{aligned} P(1 \text{ or } 2) &= P(1) + P(2) \\ &= \frac{1}{9} + \frac{1}{9} = \frac{2}{9} = 0.222 \end{aligned}$$

This answer of 0.222 is less than half of what we found for $P(1 \text{ or } 2)$ using the Benford's Law probability model in Example 5–a huge difference that illustrates one way an auditor could easily detect data that was faked–the crook would have too few 1's and 2's. Figure 8.5 displays probability histograms that compare the probability model for random digits with the model given by Benford's law. ■

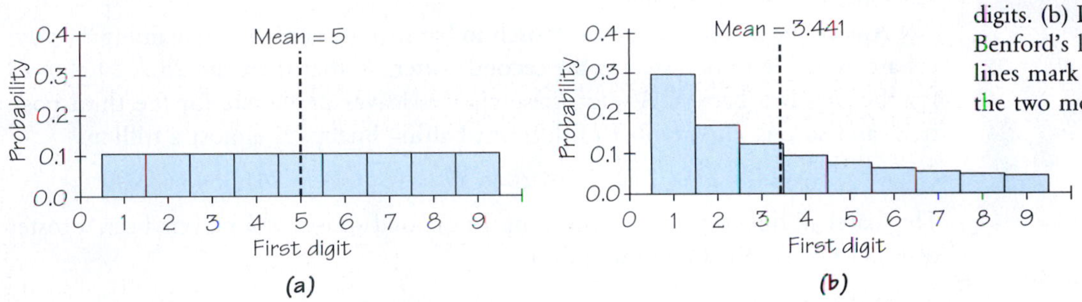


**FIGURE 8.5** Probability histograms of two models for first digits in numerical records. (a) Equally likely digits. (b) Digits follow Benford's law. The vertical lines mark the means of the two models.

When outcomes are equally likely, we find probabilities by counting outcomes. The study of counting methods is called **combinatorics**, and this is mentioned in the Season 1 episode "Noisy Edge" (2005) of the television crime drama *NUMB3RS*.

**Combinatorics** DEFINITION

**Combinatorics** is the study of methods for counting.

One example of a counting method is the **fundamental principle of counting** (from Chapter 2): If there are $a$ ways of choosing one thing, $b$ ways of choosing a second after the first is chosen, . . . , and $z$ ways of choosing the last item after the earlier choices, then the total number of choice sequences is $a \times b \times \ldots \times z$.

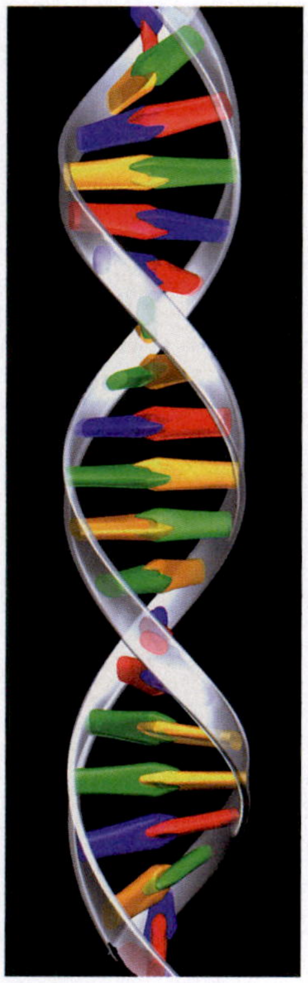

(*Pasieka/Photo Researchers, Inc.*)

## EXAMPLE 7 ■ DNA Sequences

A strand of DNA (deoxyribonucleic acid) is a long sequence of the nucleotides adenine, cytosine, guanine and thymine (abbreviated A, C, G, T). One helical turn of a DNA strand would contain a sequence of 10 of these acids, such as ACTGCCATGT. How many possible sequences of this length are there?

There are 4 letters that can occur in each position in the 10-letter sequence. Any of the 4 letters can be in the first position. Regardless of what is in the first position, any of the 4 letters can be in the second position, and so on. The order of the letters matters, so a sequence that begins AC will be a different sequence than one that begins CA.

The number of different 10-letter sequences is over one million:

$$4 \times 4 \times 4 \times 4 \times 4 \times 4 \times 4 \times 4 \times 4 \times 4 = 4^{10} = 1{,}048{,}576$$

As big as that number is, consider that it would take a DNA sequence about 3 billion letters long to contain your entire genetic "blueprint"! Knowing the number and frequency of DNA sequences has proven important in criminal justice. When skin or bodily fluids from a crime scene are "DNA fingerprinted," the specific DNA sequences in the recovered material are extremely unlikely to be found in any suspect other than the perpetrator. ■

## EXAMPLE 8 ■ Baseball Lineups

(*REUTERS/ Ray Stubblebine/Landov.*)

A Major League Baseball team has 25 players on the active roster who are eligible to play in a game. At the start of the game, the manager gives the officiating crew a list of the team's 9 hitters who will begin the game and in what order they will bat. Like Example 7, order matters here. Unlike Example 7, listing the same item more than once is not allowed.

Any of the 25 players can be chosen to bat first, but only the remaining 24 players are available to be listed as the second batter, so that there are $25 \times 24$ choices for the first two batters. Any of these choices leaves 23 batters for the third position, and so on. The number of different batting lineups is almost a trillion:

$$25 \times 24 \times 23 \times 22 \times 21 \times 20 \times 19 \times 18 \times 17 = 741{,}354{,}768{,}000$$

This baseball lineup scenario—choosing an ordered subset of $k$ players from a roster of $n$ players—is called a **permutation**. ■

## Permutation — DEFINITION

A **permutation** is an ordered arrangement of $k$ items that are chosen without replacement from a collection of $n$ items. It can be notated as $P(n, k)$, ${}_nP_k$ or $P_k^n$ and has the formula

$${}_nP_k = n \times (n-1) \times \cdots \times (n-k+1), \text{ which is Rule B.}$$

Examples 7 and 8 both involve counting the number of arrangements of distinct items. They can each be viewed as specific applications of the fundamental principle of counting, and it is easier to think your way through the counting than to memorize a recipe. Nevertheless, because these two situations occur so often, they deserve to be given their own formal recognition as Rules A and B, respectively:

## Counting Arrangements of Distinct Items — RULE

**Rule A.** Suppose we have a collection of $n$ distinct items. We want to arrange $k$ of these items in order, and the same item can appear several times in the arrangement. The number of possible arrangements is

$$\underbrace{n \times n \times \cdots \times n}_{n \text{ is multiplied by itself } k \text{ times}} = n^k.$$

**Rule B.** (**Permutations**) Suppose we have a collection of $n$ distinct items. We want to arrange $k$ of these items in order, and any item can appear no more than once in the arrangement. The number of possible arrangements is

$$n \times (n-1) \times \cdots \times (n-k+1).$$

## EXAMPLE 9 ▪ Four-Letter Words

Suppose you have 4 cards that are labeled T, S, O, and P. How many four-letter sequences can be created? Since there are only 4 cards, the only way to make a four-letter sequence is to use each letter exactly once, so there are no repeats. So this is a permutation by Counting Rule B, with $n$ and $k$ both equal to 4. To think through the problem, proceed like this: Any of the 4 letters can be chosen first; then any of the 3 that remain can be chosen second; and so on. The number of permutations is therefore $4 \times 3 \times 2 \times 1 = 24$.

It turns out that 6 of these 24 four-letter sequences are actually words in the English language (see if you can find them all), so the probability that a permutation chosen at random will actually be a word would be $6/24 = 1/4$. ■

Example 9 shows us that the permutation of all $n$ elements of a collection yields the product of the first $n$ positive integers. This expression is special enough to have its own name–**factorial**–and is also used in Chapter 11.

## Factorial — DEFINITION

For a positive integer $n$, "$n$ **factorial**" is notated $n!$ and equals the product of the first $n$ positive integers:

$$n \times (n-1) \times (n-2) \times \cdots \times 3 \times 2 \times 1.$$

By convention, we define 0! to equal 1, not 0, which can be interpreted as saying there is one way to arrange zero items.

Factorial notation allows us to write a long string of multiplied factors very compactly. For example, the expression for permutations in Rule B can now be rewritten as $\frac{n!}{(n-k)!}$.

(You can verify this is equivalent by "canceling" the factors common to the numerator and denominator. These common factors are the positive integers from 1 to $n - k$.)

## EXAMPLE 10 ▪ Winning the Lottery?

In a typical state or multi-state lottery game, you win (at least a share of) the jackpot as long as the collection of numbers you pick is the same collection that the Lottery selects. Repetition is not allowed: The same number can't be picked twice in the same drawing. Unlike permutations, order does not matter here. It doesn't matter what order the numbered ping pong balls come out of the mixing chamber–all that matters is what numbers are selected to be in that drawing's group of winners.

So while we can't use the permutation approach of Example 8 here, we can use a modification of it. The number of ordered sets will be much larger than the number of unordered sets since the lottery drawing {2, 14, 15, 21, 30, 33} is the same set of balls as {15, 2, 30, 14, 33, 21}, for example. But from the technique of Example 9, we can see that there would be 6! ways to arrange any particular set of 6 distinct balls. So the number of collections of lottery balls will simply be the number of permutations divided by $k$! In a lottery where a jackpot requires choosing the right set of 6 numbers out of a collection of 46 numbers, there are $\frac{(46)(45)(44)(43)(42)(41)}{(6)(5)(4)(3)(2)(1)} =$ 9,366,819 possible sets of numbers and so the probability of your ticket winning (at least a share of) the jackpot is $\frac{1}{9{,}366{,}819}$. The scenario of choosing an unordered subset of $k$ balls from a collection of $n$ different balls is called a **combination**.

### Combination — DEFINITION

A **combination** is an unordered arrangement of $k$ items that are chosen without replacement from a collection of $n$ items. It is notated as $C(n, k)$, ${}_nC_k$, $C^n_k$, $\binom{n}{k}$, or "$n$ choose $k$".

${}_nC_k = [n \times (n-1) \times \cdots \times (n-k+1)] / k!$ or $\frac{n!}{k!(n-k)!}$, which is Rule D.

If it's hard to remember the difference between combinations (Rule D) and permutations (Rule B), consider this: If you order a "combination platter" at a diner, you're asking for a certain set of foods to be on your plate, but you don't care what order they're in. Also, you can use this memory aid: "Permutations Presume Positions; Combinations Concern Collections." For completeness, we also provide a formula (Rule C) for unordered collections in which repetition *is* allowed, but we cannot give a simple explanation in the space we have, and we will not emphasize it.

### Counting Unordered Collections of Distinct Items — RULE

**Rule C.** Suppose we have a collection of $n$ distinct items. We want to select $k$ of those items with no regard to order, and any item can appear more than once in the collection. The number of possible collections is $\frac{(n+k-1)!}{k!(n-1)!}$.

**Rule D. (Combinations)** Suppose we have a collection of $n$ distinct items. We want to select $k$ of these items with no regard to order, and any item can appear no more than once in the collection. The number of possible selections is $\frac{n!}{k!(n-k)!}$.

This table summarizes all four ways we have seen of choosing items from a collection of distinct items:

| Choosing $k$ items from $n$ distinct items | | |
|---|---|---|
| | Repetition is allowed | Repetition is *not* allowed |
| Order does matter | Rule A: $\underbrace{n \times n \times \cdots \times n}_{n \text{ is multiplied by itself } k \text{ times}} = n^k$ | Rule B (*permutation*): $\frac{n!}{(n-k)!} = n \times (n-1) \times \cdots \times (n-k+1)$ |
| Order does *not* matter | Rule C: $\frac{(n+k-1)!}{k!(n-1)!}$ | Rule D (*combination*): $\frac{n!}{k!(n-k)!}$ |

## SPOTLIGHT 8.2 Combinatorics Calculations

*Factorials* can be tedious to compute for large values of $n$, but a *scientific calculator* should have a key labeled $n!$ or $x!$ and the TI-84+ *graphing calculator* can find 13! with this sequence:
13 (MATH) → PRB → ! (ENTER)(ENTER).

For *permutations*, some (but not all) *scientific calculators* have a command sequence, and Web sites such as http://www.geocities.com/calculatorhelp can offer keystrokes for specific models. If you have only a basic calculator without a factorial key, the expression $n \times (n-1) \times \cdots \times (n-k+1)$ will involve fewer multiplications than $\frac{n!}{(n-k)!}$ because it has already incorporated all the cancellations between numerator and denominator. The TI-84+ *graphing calculator* can find the number of permutations of 3 objects chosen from 8 objects by using this sequence:
8 (MATH) → PRB → nPr(ENTER) 3 (ENTER).
For *combinations*, use nCr instead of nPr.

## 8.4 Continuous Probability Models

When we use the table of random digits to select a digit between 0 and 9, the discrete probability model assigns probability 1/10 to each of the 10 possible outcomes. Suppose that we want to choose a number at random between 0 and 1, allowing *any* number between 0 and 1 as the outcome. Software random-number generators will do this. You can visualize such a random number by thinking of a spinner (Figure 8.6) that turns freely on its axis and slowly comes to a stop. The pointer can come to rest anywhere on a circle that is marked from 0 to 1. Also, your graphing calculator or spreadsheet software may be able to do this with a "rand" command. The sample space is now an entire interval of numbers:

$$S = \{\text{all numbers } x \text{ such that } x \text{ is between 0 and 1}\}$$

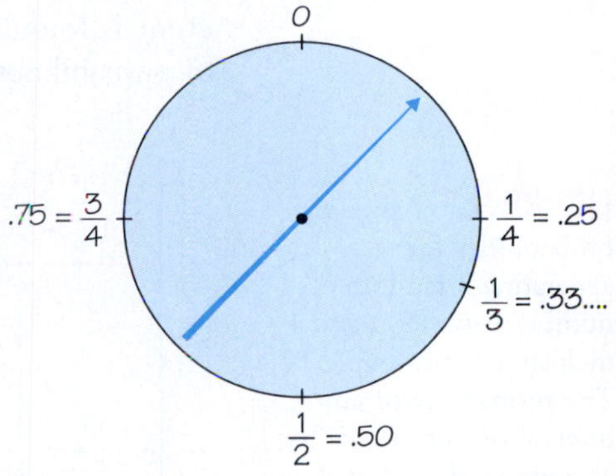


**FIGURE 8.6** This spinner chooses a number between 0 and 1 at random. That is, it is equally likely to stop at any point on the circle.

How can we assign probabilities to events such as $\{0.3 \le x \le 0.7\}$? As in the case of selecting a random digit, we would like all possible outcomes to be equally likely. But we cannot assign probabilities to each individual value of $x$ and then sum, because there are *infinitely* many possible values. Instead we use a second way of assigning probabilities directly to events–as *areas under a curve*. By Probability Rule 2, the curve must have total area 1 underneath it, corresponding to total probability 1. We call such curves **density curves**.

### Density Curve — DEFINITION

A **density curve** is a curve that

- is always on or above the horizontal axis and
- has area exactly 1 underneath it.

A **continuous probability model** assigns probabilities as areas under a density curve. The area under the curve and above any range of values is the probability of an outcome in that range.

## EXAMPLE 11 ■ A Continuous Uniform Model

The random-number generator will spread its output uniformly across the entire interval from 0 to 1 if we allow it to generate many numbers. The results of many trials are represented by the density curve of a *uniform probability model.* This density curve appears in red in Figure 8.7. It has height 1 over the interval from 0 to 1, and height 0 everywhere else. The area under the density curve is 1, the area of a square with base 1 and height 1. The probability of any event is the area under the density curve and above the event in question.

As Figure 8.7a illustrates, the probability that the random-number generator produces a number $X$ between 0.3 and 0.7 inclusive is

$$P(0.3 \leq X \leq 0.7) = 0.4$$

because the rectangular area under the density curve and above the interval from 0.3 to 0.7 is 0.4. The area of a rectangle is the product of height and length and the height of this density curve is 1, so the probability of any interval of outcomes will just be the length of the interval: $0.7 - 0.3 = 0.4$.

Also, we can apply Probability Rule 3 to non-overlapping intervals such as:

$$\begin{aligned} P(X < 0.5 \text{ or } X > 0.8) &= P(X < 0.5) + P(X > 0.8) \\ &= 0.5 \qquad\qquad + 0.2 \qquad\quad = 0.7 \end{aligned}$$

The last event consists of two nonoverlapping intervals, so the total area above the event is found by adding two areas, as illustrated by Figure 8.7b. This assignment of probabilities obeys all of our rules for probability.

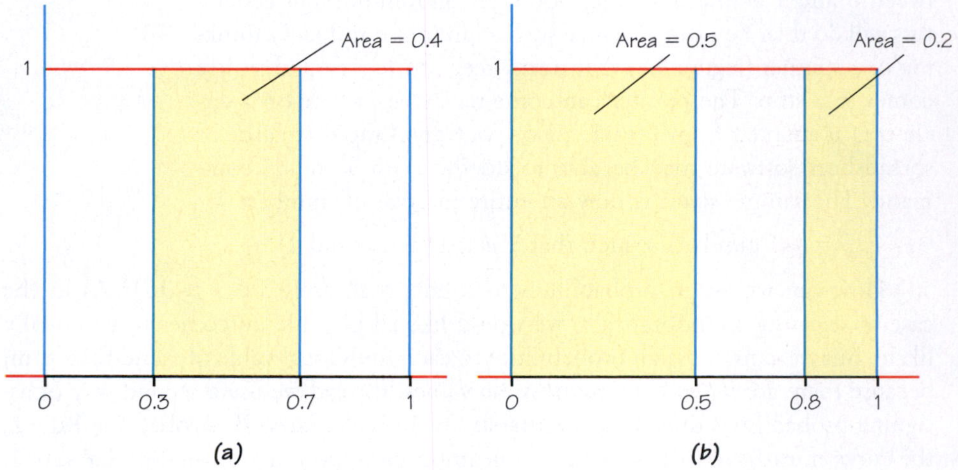


**FIGURE 8.7** Assigning probabilities for generating a random number between 0 and 1 inclusive, for Example 11. The probability of any interval of numbers is the area above the interval and under the density curve. (a) The probability of an outcome between 0.3 and 0.7. (b) The probability of an outcome less than 0.5 or greater than 0.8.

The probability model for a continuous random variable assigns probabilities to *intervals* of outcomes rather than to individual point outcomes. In fact, *all continuous probability models assign probability 0 to every individual outcome.* Only intervals of values have positive probability. To see that this is true, consider a specific outcome such as $P(X = 0.6)$ in Example 11. In this example, the probability of any interval is the same as its length. The point 0.6 has no length, so its probability is 0.

The density curves that are most familiar to us are the normal curves. Because any density curve describes an assignment of probabilities, *normal distributions are continuous probability models.* Recall that a normal curve has total area of 1 underneath. Let's redo Example 17 from Chapter 7, now using the language of probability.

## EXAMPLE 12

### Areas Under a Normal Curve Are Probabilities

Suppose that 60% of adults find shopping for clothes time consuming and frustrating. All adults form a population, with population proportion $p = 0.6$. Interview an SRS of 2500 people from this population and find the proportion $\hat{p}$ of the sample who say that shopping is frustrating. We know that if we take many such samples, the statistic $\hat{p}$ will vary from sample to sample according to a normal distribution with

$$\text{mean} = p = 0.6$$

$$\text{standard deviation} = \sqrt{\frac{p(1-p)}{n}}$$

$$= \sqrt{\frac{(0.6)(0.4)}{2500}} = 0.01 \quad \text{(approximately)}$$

The 68–95–99.7 rule now gives *probabilities* for the value of $\hat{p}$ from a single SRS. The probability is 0.95 that $\hat{p}$ lies between 0.58 and 0.62. Figure 8.8 shows this probability as an area under the normal density curve.

All that is new is the language of probability. "Probability is 0.95" is shorthand for "95% of the time in a very large number of samples."

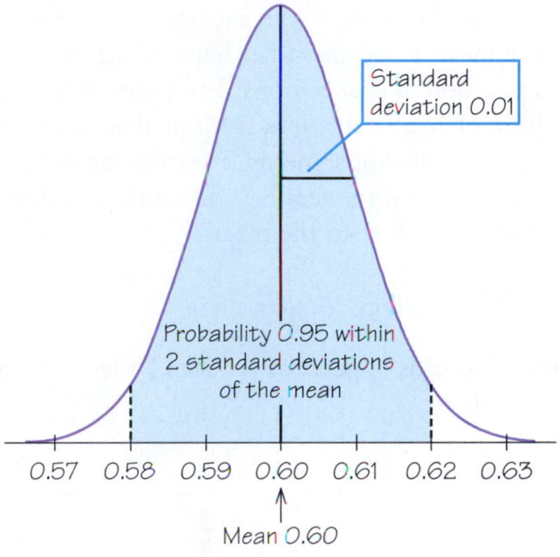


**FIGURE 8.8** Probability as area under a normal curve, for Example 12. The 68–95–99.7 rule gives some probabilities for normal probability models.

# 8.5 The Mean and Standard Deviation of a Probability Model

Suppose you are offered this choice of bets, each costing the same: Bet *A* pays \$10 if you win and you have probability 1/2 of winning, while bet *B* pays \$10,000 and offers probability 1/10 of winning. You would very likely choose *B* even though *A* offers a better chance to win, because *B* pays much more if you win. It would be foolish to decide which bet to make just on the basis of the probability of winning. How much you can win is also important. When a random phenomenon has numerical outcomes, we are concerned with their amounts as well as with their probabilities.

What will be the average payoff of our two bets in many plays? Recall that the probabilities are the long-run proportions of plays in which each outcome occurs. Bet *A* produces \$10 half the time in the long run and nothing half the time. So the average payoff should be

$$\left(\$10 \times \frac{1}{2}\right) + \left(\$0 \times \frac{1}{2}\right) = \$5$$

Bet *B*, on the other hand, pays out \$10,000 on 1/10 of all bets in the long run. So bet *B*'s average payoff is

$$\left(\$10{,}000 \times \frac{1}{10}\right) + \left(\$0 \times \frac{9}{10}\right) = \$1000$$

If you can place many bets, you should certainly choose *B*. Here is a general definition of the kind of "average outcome" we used to compare the two bets.

**Mean of a Discrete Probability Model** DEFINITION

Suppose that the possible outcomes $x_1, x_2, \ldots, x_k$ in a sample space $S$ are numbers and that $p_j$ is the probability of outcome $x_j$. The **mean $\mu$ of a discrete probability model** is

$$\mu = x_1p_1 + x_2p_2 + \cdots + x_kp_k.$$

In Chapter 5, we met the mean $\bar{x}$, the average of $n$ observations that we actually have in hand. The mean $\mu$, on the other hand, describes the probability model rather than any one collection of observations. The Greek letter mu ($\mu$) is pronounced "myoo." You can think of $\mu$ as a theoretical mean that gives the average outcome we expect in the long run. You will sometimes see the mean of a probability model called the *expected value.* This isn't a very helpful name, because we don't necessarily expect the outcome to be close to the mean.

## EXAMPLE 13 ■ First Digits

If first digits in a set of records appear "at random," the probability model for the first digit is as in Example 6:

| First digit | 1 | 2 | 3 | 4 | 5 | 6 | 7 | 8 | 9 |
|---|---|---|---|---|---|---|---|---|---|
| Probability | $\frac{1}{9}$ | $\frac{1}{9}$ | $\frac{1}{9}$ | $\frac{1}{9}$ | $\frac{1}{9}$ | $\frac{1}{9}$ | $\frac{1}{9}$ | $\frac{1}{9}$ | $\frac{1}{9}$ |

The mean of this model is

$$\mu = 1\frac{1}{9} + 2\frac{1}{9} + 3\frac{1}{9} + 4\frac{1}{9} + 5\frac{1}{9} + 6\frac{1}{9} + 7\frac{1}{9} + 8\frac{1}{9} + 9\frac{1}{9}$$
$$= 45 \times \frac{1}{9} = 5$$

If, on the other hand, the records obey Benford's law, the distribution of the first digit is

| First digit | 1 | 2 | 3 | 4 | 5 | 6 | 7 | 8 | 9 |
|---|---|---|---|---|---|---|---|---|---|
| Probability | 0.301 | 0.176 | 0.125 | 0.097 | 0.079 | 0.067 | 0.058 | 0.051 | 0.046 |

The mean is

$$\mu = (1)(0.301) + (2)(0.176) + (3)(0.125) + (4)(0.097) + (5)(0.079) + (6)(0.067) + (7)(0.058) + (8)(0.051) + (9)(0.046)$$
$$= 3.441$$

The means reflect the greater probability of smaller first digits under Benford's law. We have marked the means on the probability histograms in Figure 8.4. Because the histogram for random digits is symmetric, the mean lies at the center of symmetry. We can't locate the mean of the right-skewed Benford's law model by eye; calculation is needed.

What about continuous probability models? Think of the area under a density curve as being cut out of solid homogenous material. The mean $\mu$ is the point at which the shape would balance. Figure 8.9 illustrates this interpretation of the mean. The mean lies at the center of symmetric density curves such as the uniform density in Figure 8.7 and the normal curve in Figure 8.8. Exact calculation of the mean of a distribution with a skewed density curve requires advanced mathematics. The idea that the mean is the balance point of the probabilities applies to discrete models as well (see Section 5.4), but in the discrete case we have a formula that gives us this point.

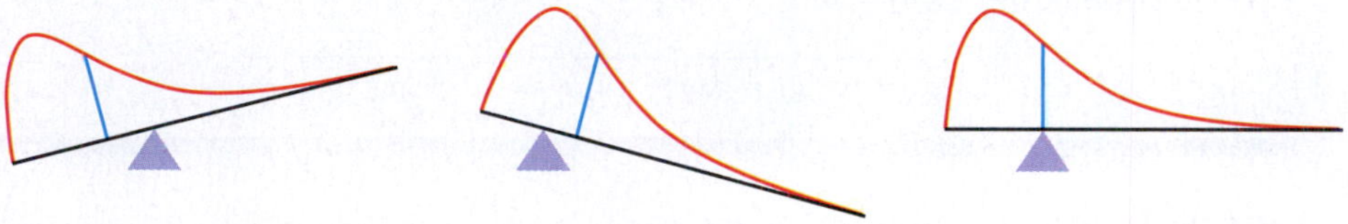

**FIGURE 8.9** The mean of a continuous probability model is the point at which the density curve would balance.

The mean $\mu$ is an average outcome in two senses. The definition for discrete models says that it is the average of the possible outcomes not weighted equally but weighted by their probabilities. More likely outcomes get more weight in the average. An important fact of probability, the **law of large numbers**, says that $\mu$ is the average outcome in another sense as well.

### Law of Large Numbers — THEOREM

Observe any random phenomenon having numerical outcomes with finite mean $\mu$. According to the **law of large numbers**, as the random phenomenon is repeated a large number of times,

- the proportion of trials on which each outcome occurs gets closer and closer to the probability of that outcome, and
- the mean $\bar{x}$ of the observed values gets closer and closer to $\mu$.

These facts can be stated more precisely and then proved mathematically. The law of large numbers brings the idea of probability to a natural completion. We first observed that some phenomena are random in the sense of showing long-run regularity. Then we used the idea of long-run proportions to motivate the basic laws of probability. Those laws are mathematical idealizations that can be used without interpreting probability as proportion in many trials. Now the law of large numbers tells us that in many trials the proportion of trials on which an outcome occurs will always approach its probability.

The law of large numbers also explains why gambling can be a business. The winnings (or losses) of a gambler on a few plays are uncertain–that's why gambling is exciting. It is only *in the long run* that the mean outcome is predictable. The house plays many tens of thousands of times. So the house, unlike individual gamblers, can count on the long-run regularity described by the law of large numbers. The average winnings of the house on tens of thousands of plays will be very close to the mean of the distribution of winnings. Needless to say, gambling games have mean outcomes that guarantee the house a profit.

We know that the simplest description of a distribution of data requires both a measure of center and a measure of spread. The same is true for probability models. The *mean* is the average value for both a set of data and a discrete probability model. All the observations are weighted equally in finding the mean $\bar{x}$ for data, but the values are weighted by their probabilities in finding the mean $\mu$ of a probability model. The measure of spread that goes with the mean is the **standard deviation.** For data, the standard deviation $s$ is the square root of the average squared deviation of the observations from their mean. We apply exactly the same idea to probability models, using probabilities as weights in the average. Here is the definition.

### Standard Deviation of a Discrete Probability Model — DEFINITION

Suppose that the possible outcomes $x_1, x_2, \ldots, x_k$ in a sample space $S$ are numbers, and that $p_j$ is the probability of outcome $x_j$. The **standard deviation $\sigma$ of a discrete probability model** with mean $\mu$ is

$$\sigma = \sqrt{(x_1 - \mu)^2 p_1 + (x_2 - \mu)^2 p_2 + \cdots + (x_k - \mu)^2 p_k}$$

## EXAMPLE 14 ■ First Digits

If the first digits in a set of records obey Benford's law, the discrete probability model is

| First digit | 1 | 2 | 3 | 4 | 5 | 6 | 7 | 8 | 9 |
|---|---|---|---|---|---|---|---|---|---|
| Probability | 0.301 | 0.176 | 0.125 | 0.097 | 0.079 | 0.067 | 0.058 | 0.051 | 0.046 |

We saw in Example 13 that the mean is $\mu = 3.441$. To find the standard deviation,

$$\begin{aligned}\sigma &= \sqrt{(x_1 - \mu)^2 p_1 + (x_2 - \mu)^2 p_2 + \cdots + (x_k - \mu)^2 p_k}\\ &= \sqrt{(1 - 3.441)^2(0.301) + (2 - 3.441)^2(0.176) + \cdots + (9 - 3.441)^2(0.046)}\\ &= \sqrt{1.7935 + 0.3655 + \cdots + 1.4215}\\ &= \sqrt{6.061} = 2.46\end{aligned}$$

You can follow the same pattern to find the standard deviation of the equally likely model and show that the Benford's law model is less spread out than the equally likely model.

Finding the standard deviation of a continuous probability model usually requires advanced mathematics (calculus). Chapter 5 told us the answer in one important case: The standard deviation of a normal curve is the distance from the center (the mean) to the change-of-curvature points on either side.

## SPOTLIGHT 8.3 Birthday Coincidences

If 366 people are gathered, you can see why there's a 100% chance at least two people share the same birthday (ignoring leap days). Now, if only 23 people are gathered, what do you think is the probability of any birthday matches? Guess before reading further.

Now imagine these 23 people enter a room one at a time, adding their birthday to a list in the order they enter. Using $n = 365$ and $k = 23$, Rule A gives us the total number of lists of 23 birthdays, and Rule B gives us how many of those lists have birthdays that are all different. Using the rule for Equally Likely Outcomes (each day of the year is equally likely to be a randomly chosen person's birthday), we conclude that the probability of all birthdays being different is the result from Rule B divided by the result from Rule A:

$$\frac{{}_{365}P_{23}}{365^{23}} = \frac{365 \times 364 \times \ldots \times 343}{365^{23}}$$

Alternatively, we could assume independence of birthdays and use Probability Rule 4. The second person that walks in has a 364/365 chance of not matching person #1. The third person that walks in has a 363/365 chance of not matching persons #1 or #2, and so on. Verify that you get the same product by multiplying this string of fractions:

$$\frac{364}{365} \times \frac{363}{365} \times \ldots \times \frac{343}{365}$$

Either way, our final step to find the probability of getting at least one match is to subtract that answer from 1 (using Probability Rule 3), and we obtain the surprisingly high value of 51%! Maybe it is not so surprising if we consider that the combinations formula tells us that there are 253 ways to choose 2 people from 23 to ask each other if they have the same birthday. Because we underestimate the huge number of potential opportunities for "coincidences," we are surprised that they happen as often as they do. As Jessica Utts points out, if something has a 1 in a million chance of happening to any person on a given day, this rare event will happen to roughly 300 people in the United States each day!

*(Michael Rosenfeld/Photographer's Choice/Getty Images.)*

## 8.6 The Central Limit Theorem

The key to finding a confidence interval that estimates a population proportion (Chapter 7) was the fact that the sampling distribution of a population proportion is close to normal when the sample is large. This fact is an application of one of the most important results of probability theory, the **Central Limit Theorem**. This theorem says that the distribution of any random phenomenon tends to be normal if we average it over a large number of independent repetitions. The Central Limit Theorem allows us to analyze and predict the results of chance phenomena when we average over many observations.

The word "limit" in Central Limit Theorem reflects that the normal curve is the limit or target shape to which the sampling distribution gets closer and closer as the sample size increases. The theorem also tells us the mean of the sampling distribution, and the mean is a measure of "central" tendency.

### Central Limit Theorem — THEOREM

Draw an SRS of size $n$ from any large population with mean $\mu$ and finite standard deviation $\sigma$. Then

- The mean of the sampling distribution of $\bar{x}$ is $\mu$.
- The standard deviation of the sampling distribution of $\bar{x}$ is $\sigma/\sqrt{n}$.
- The **Central Limit Theorem** says that the sampling distribution of $\bar{x}$ is approximately normal when the sample size $n$ is large ($n > 30$).

The first two parts of this statement can be proved from the definitions of the mean and the standard deviation. They are true for any sample size $n$. The Central Limit Theorem is a much deeper result. Pay attention to the fact that the standard deviation of a mean decreases as the number of observations $n$ increases. Together with the Central Limit Theorem, this makes exact two general statements that help us understand a wide variety of random phenomena:

**Averages are less variable than individual observations.**

**Averages are more normal than individual observations.**

The *Central Limit Theorem* applet allows you to watch the Central Limit Theorem in action: It starts with a distribution that is strongly skewed, not at all normal. As you increase the size of the sample, the distribution of the mean $\bar{x}$ gets closer and closer to the normal shape.

Consider dice. Rolls of a single die would have a uniformly flat probability histogram, with each of the six possible values having the probability 1/6. Now consider the mean of rolling a pair of dice. The probability model for the mean of two dice simply divides by 2 the outcome sum in Example 4. (So, the probability that the mean of two dice equals 4.5 must be the same as the probability that their sum equals 9.) And the histogram in Figure 8.4 is certainly less variable and closer to looking "normal" than is the flat histogram for rolling a single die.

## EXAMPLE 15 ■ Heights of Young Women

The distribution of heights of young adult women is approximately normal, with mean 64.5 inches and standard deviation 2.5 inches. This normal distribution describes the population of young women. It is also the probability model for choosing one woman at random from this population and measuring her height. For example, the 68–95–99.7 rule says that the probability is 0.95 that a randomly chosen woman is between 59.5 and 69.5 inches tall.

Now choose an SRS of 25 young women at random and take the mean $\bar{x}$ of their heights. The mean $\bar{x}$ varies in repeated samples–the pattern of variation is the sampling distribution of $\bar{x}$. The sampling distribution has the same center $\mu = 64.5$ inches as the population of young women. In statistical terms, the sample mean $\bar{x}$ has *no bias* as an estimator of the population mean $\mu$. If we take many samples, $\bar{x}$ will sometimes be smaller than $\mu$ and sometimes larger, but it has no systematic tendency to be too small or too large.

The standard deviation of the sampling distribution of $\bar{x}$ is

$$\frac{\sigma}{\sqrt{n}} = \frac{2.5}{\sqrt{25}} = \frac{2.5}{5} = 0.5 \text{ inch}$$

The standard deviation $\sigma$ describes the variation when we measure many individual women. The standard deviation $\sigma/\sqrt{n}$ of the distribution of $\bar{x}$ describes the variation in the average heights of samples of women when we take many samples. The average height is less variable than individual heights.

Figure 8.10 compares the two distributions: Both are normal and both have the same mean, but the average height of 25 randomly chosen women is much less spread out. For example, the 68–95–99.7 rule says that 95% of all averages $\bar{x}$ lie between 63.5 and 65.5 inches because 2 standard deviations of $\bar{x}$ make 1 inch. This 2-inch span is just one-fifth as wide as the 10-inch span that catches the middle 95% of heights for individual women.

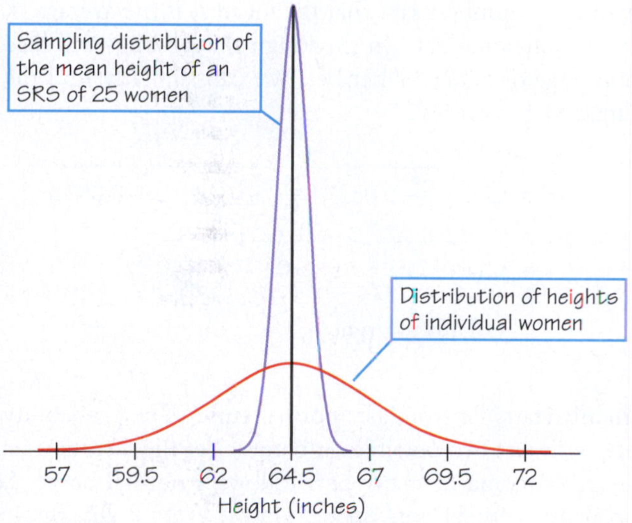


**FIGURE 8.10** The sampling distribution of the average height of an SRS of 25 women has the same center as the distribution of individual heights but is much less spread out.

The Central Limit Theorem says that in large samples the sample mean $\bar{x}$ is approximately normal. In Figure 8.10, we show a normal curve for $\bar{x}$ even though sample size 25 is not very large. Is that acceptable? How large a sample is needed for the Central Limit Theorem to work depends on how far from a normal curve the model we start with is. The closer to normality we start, the quicker the distribution of the sample mean becomes normal. In fact, if individual observations follow a normal curve, the sampling distribution of $\bar{x}$ is exactly normal for any sample size. So Figure 8.10 is accurate. The Central Limit Theorem is a striking result because as $n$ gets large it works for *any* model we may start with, no matter how far from normal. Here is an example that starts very far from normal.

**FIGURE 8.11** A gambler may win or lose at roulette, but in the long run the casino always wins. (*Ingram Publishing/PictureQuest.*)

## EXAMPLE 16 ■ Red or Black in Roulette

An American roulette wheel has 38 slots, of which 18 are black, 18 are red, and 2 are green. The dealer spins the wheel and whirls a small ball in the opposite direction within the wheel. Gamblers bet on where the ball will come to rest (see Figure 8.11). One of the simplest wagers chooses red (or black). A bet of $1 on red pays off an additional $1 if the ball lands in a red slot. Otherwise, the player loses his $1. The two green slots always belong to the house.

Lou bets on red. He wins if the ball stops in one of the 18 red slots. He loses if it lands in one of the 20 slots that are black or green. Because casino roulette wheels are carefully balanced so that all slots are equally likely, the probability model is

| Net Outcome for Gambler | | |
|---|---|---|
| | Win $1 | Lose $1 |
| Probability | 18/38 = .474 | 20/38 = .526 |

The mean outcome of a single $1 bet on red is

$$\mu = (\$1)\left(\frac{18}{38}\right) + (-\$1)\left(\frac{20}{38}\right)$$
$$= -\$\frac{2}{38} = -\ \$0.053 \text{(a loss of 5.3 cents)}$$

The law of large numbers says that the mean $\mu$ is the average outcome of a very large number of individual bets. In the long run, gamblers will lose (and the casino will win) an average of 5.3 cents per bet. We can similarly find the standard deviation for a single $1 bet on red:

$$\sigma = \sqrt{(1 - (-0.053))^2 \frac{18}{38} + (-1 - (-0.053))^2 \frac{20}{38}}$$
$$= \sqrt{(1.053)^2 \frac{18}{38} + (-0.947)^2 \frac{20}{38}}$$
$$= \sqrt{0.9972} = 0.9986$$

Lou certainly starts far from any normal curve. The probability model for each bet is discrete, with just two possible outcomes. Yet the Central Limit Theorem says that the average outcome of many bets follows a normal curve. Lou is a habitual gambler who places fifty $1 bets on red almost every night. Because we know the probability model for a bet on red, we can simulate Lou's experience over many nights at the roulette wheel. The histogram in Figure 8.12 shows Lou's average winnings for 1000 nights. As the Central Limit Theorem says, the distribution looks normal.

## EXAMPLE 17 ■ Lou Gets Entertainment

The normal curve in Figure 8.12 comes from the Central Limit Theorem and the values of the mean $\mu$ and standard deviation $\sigma$ in Example 16. It has

$$\text{mean} = \mu = -0.053$$
$$\text{standard deviation} = \frac{\sigma}{\sqrt{n}} = \frac{0.9986}{\sqrt{50}} = 0.141$$

Apply the 99.7 part of the 68–95–99.7 rule: Almost all average nightly winnings will fall within 3 standard deviations of the mean, that is, between

$$-0.053 - (3)(0.141) = -0.476$$

and

$$-0.053 + (3)(0.141) = 0.370$$

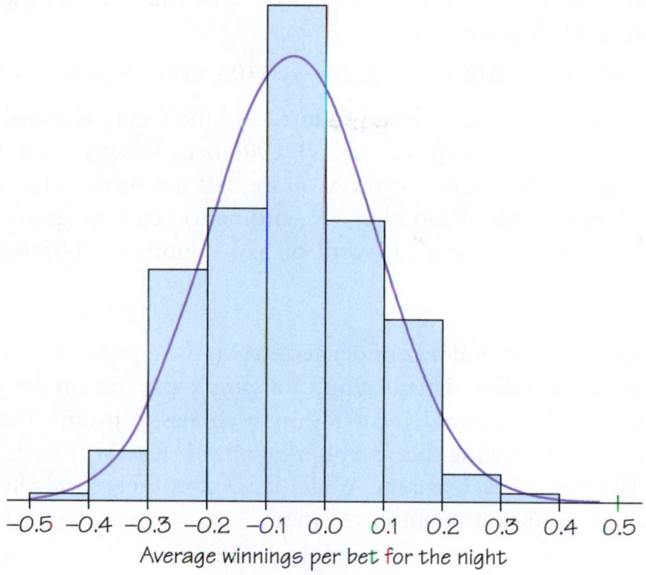


**FIGURE 8.12** A gambler's winnings in a night of 50 bets on red or black in roulette vary from night to night. Here is the distribution for 1000 nights. It is approximately normal.

Lou's total winnings after 50 bets of $1 each will then almost surely fall between

$$(50)(-0.476) = -23.80$$

and

$$(50)(0.370) = 18.50$$

Lou may win as much as $18.50 or lose as much as $23.80. Some find gambling exciting because the outcome, even after an evening of bets, is uncertain. It is possible to walk away a winner. It's all a matter of luck. ■

The casino, however, is in a different position. It doesn't want excitement, just a steady income.

## EXAMPLE 18 ■ The Casino Gets Rich

The casino bets with all its customers–perhaps 100,000 individual bets on black or red in a week. The Central Limit Theorem guarantees that the distribution of average customer winnings on 100,000 bets is very close to normal. The mean is still the mean outcome for one bet, −0.053, a loss of 5.3 cents per dollar bet. The standard deviation is much smaller when we average over 100,000 bets. It is

$$\frac{\sigma}{\sqrt{n}} = \frac{0.9986}{\sqrt{100{,}000}} = 0.003$$

Here is what the spread in the average result looks like after 100,000 bets:

$$\begin{aligned} \text{Spread} &= \text{mean} \pm 3 \text{ standard deviations} \\ &= -0.053 \pm (3)(0.003) \\ &= -0.053 \pm 0.009 \\ &= -0.062 \text{ to } -0.044 \end{aligned}$$

Because the casino covers so many bets, the standard deviation of the average winnings per bet becomes very small. And because the mean is negative, almost all outcomes will be negative. The gamblers' losses and the casino's winnings are almost certain to average between 4.4 and 6.2 cents for every dollar bet.

The gamblers who collectively place those 100,000 bets will lose money. The probable range of their losses is

$$(100{,}000)(-0.062) = -6200 \quad \text{to} \quad (100{,}000)(-0.044) = -4400$$

The gamblers are almost certain to lose–and the casino is almost certain to take in–between \$4400 and \$6200 on those 100,000 bets. What's more, the range of average outcomes continues to narrow as more bets are made. That is how a casino can make a business out of gambling. According to *Forbes* magazine, the third richest American (with an estimated worth of \$28 billion) in 2007 was casino mogul Sheldon Adelson.

In Chapter 7, we based a confidence interval for a population proportion $p$ on the fact that the **sampling distribution** of a sample proportion $\hat{p}$ is close to normal for large samples. The Central Limit Theorem applies to means. How can we apply it to proportions? By seeing that *a proportion is really a mean.* This is our final example of the Central Limit Theorem. While it is more theoretical than our other examples, it gives us an important foundation.

## EXAMPLE 19

### The Sampling Distribution of a Proportion

If we can express the sample proportion of successes as a sample mean, we can apply tools we have learned to derive the formula (in Section 7.7) for the standard deviation of the sample proportion.

Consider an SRS of size $n$ from a population that contains proportion $p$ of "having a particular trait." For each of the $n$ individuals, we can define a simple numerical variable $x_i$ to equal 1 for a success and 0 for a failure. For example, if the third individual has the trait of interest, then $x_3 = 1$. So the sum of all $n$ of the $x_i$ values is the total number of "successes" (that is, people that had the trait of interest). So the proportion $\hat{p}$ of successes is given by

$$\hat{p} = \frac{\textit{number of successes}}{n} = \frac{x_1 + x_2 + \cdots + x_n}{n} = \bar{x}$$

So $\hat{p}$ is really a mean, and so its sampling distribution (by the Central Limit Theorem) is close to normal when the sample size $n$ is large ($n > 30$).

Because $\hat{p}$ is the mean of the $x_i$, we can find the mean and standard deviation of $\hat{p}$ from the mean and standard deviation of one observation $x_i$. Each observation has probability $p$ of being a success, so the probability model for one observation is

| Outcome | Success, $x_i = 1$ | Failure, $x_i = 0$ |
|---|---|---|
| Probability | $p$ | $1 - p$ |

Using the tools of Section 8.5, the mean of $x_i$ is therefore

$$\mu = (1)(p) + (0)(1 - p) = p$$

In the same way, after a bit more algebra, the tools of Section 8.5 show that the standard deviation of one observation $x_i$ is

$$\sigma = \sqrt{(1 - p)^2 p + (0 - p)^2 (1 - p)} = \sqrt{p(1 - p)}$$

From the Central Limit Theorem (Section 8.6), the standard deviation of the mean of $n$ observations is $\frac{\sigma}{\sqrt{n}}$, so we simply substitute in our expression for σ and obtain:

$$\frac{\sigma}{\sqrt{n}} = \frac{\sqrt{p(1-p)}}{\sqrt{n}} = \sqrt{\frac{p(1-p)}{n}}.$$

This last expression is precisely the fact we used in Section 7.7.

Examples 15 to 19 illustrate the importance of the Central Limit Theorem and the reason for the importance of normal distributions. We can often replace tricky calculations about a probability model by simpler calculations for a normal distribution, courtesy of the Central Limit Theorem.

## REVIEW VOCABULARY

**Addition rule** The probability that one event or the other occurs is the sum of their individual probabilities minus the probability of any overlap they have. (p. 253)
**Central Limit Theorem** The average of many independent random outcomes is approximately normally distributed. When we average $n$ independent repetitions of the same random phenomenon, the resulting distribution of outcomes has mean equal to the mean outcome of a single trial and standard deviation proportional to $1/\sqrt{n}$. (p. 267)
**Combination** An unordered collection of $k$ items chosen (without allowing repetition) from a set of $n$ distinct items. (p. 260)
**Combinatorics** The branch of mathematics that counts arrangements of objects. (p. 258)
**Complement of an event** The complement of an event $A$ is the event "$A$ does not occur," which is denoted $A^C$. (p. 252)
**Complement rule** $P(A^C) = 1 - P(A)$. (p. 253)
**Continuous probability model** A probability model that assigns probabilities to events as areas under a density curve. (p. 262)
**Density curve** A curve that is always on or above the horizontal axis and has area exactly 1 underneath it. A density curve describes a continuous probability model. (p. 261)
**Discrete probability model** A probability model that assigns probabilities to each of a finite number of possible outcomes. (p. 255)
**Disjoint events** Events that have no outcomes in common. (Also called *mutually exclusive events.*) (p. 252)
**Event** A collection of possible outcomes of a random phenomenon. A subset of the sample space. (p. 250)
**Factorial** The product of the first $n$ positive integers, denoted "$n$!" (p. 259)
**Fundamental principle of counting** A multiplicative method for counting outcomes of multistage processes. (p. 258)
**Independent events** Events that do not affect each other's probability of occurring. (p. 252)
**Law of large numbers** As a random phenomenon is repeated many times, the mean $\bar{x}$ of the observed outcomes approaches the mean $\mu$ of the probability model. (p. 265)
**Mean of a discrete probability model** The average outcome of a random phenomenon with numerical values. When possible values $x_1, x_2, \ldots, x_k$ have probabilities $p_1, p_2, \cdots, p_k$, the mean is the average of the outcomes weighted by their probabilities, $\mu = x_1p_1 + x_2p_2 + \ldots + x_kp_k$. (Also called *expected value.*) (p. 264)
**Multiplication rule** $P(A \text{ and } B) = P(A) \times P(B)$, when $A$ and $B$ are independent events. (p. 253)
**Permutation** An ordered arrangement of $k$ items chosen (without allowing repetition) from a set of $n$ distinct items. (p. 258)
**Probability** A number between 0 and 1 that gives the long-run proportion of repetitions of a random phenomenon on which an event will occur. (p. 248)
**Probability histogram** A histogram that displays a discrete probability model when the outcomes are numerical. The height of each bar is the probability of the event at the base of the bar. (p. 255)
**Probability model** A sample space $S$ together with an assignment of probabilities to events. The two main types of probability models are *discrete* and *continuous.* (p. 250)
**Random** A phenomenon or trial is random if it is uncertain what the next outcome will be but each outcome nonetheless tends to occur in a fixed proportion of a very long sequence of repetitions. These long-run proportions are the probabilities of the outcomes. (p. 247)
**Sample space** A list of all possible (simplest) outcomes of a random phenomenon. (p. 250)
**Sampling distribution** The distribution of values taken by a statistic when many random samples are drawn under the same circumstances. A sampling distribution

consists of an assignment of probabilities to the possible values of a statistic. (p. 272)

**Standard deviation of a discrete probability model** A measure of the variability of a probability model. When the possible values $x_1, x_2, \dots, x_k$ have probabilities $p_1, p_2, \dots, p_k$, the standard deviation is the square root of the average (weighted by probabilities) of the squared deviations from the mean:

$$\sigma = \sqrt{(x_1 - \mu)^2 p_1 + (x_2 - \mu)^2 p_2 + \cdots + (x_k - \mu)^2 p_k}\,.$$

(p. 266)

## ✓ SKILLS CHECK

**1.** You read in a book on poker that the probability of being dealt three of a kind in a five-card poker hand is 1/50. What does this mean?

**(a)** If you deal thousands of poker hands, the fraction of them that contain three of a kind will be very close to 1/50.
**(b)** If you deal 50 poker hands, exactly one of them will contain three of a kind.
**(c)** If you deal 10,000 poker hands, exactly 200 of them will contain three of a kind.

**2.** If two coins are flipped and then a die is rolled, the sample space would have ________ different outcomes.

Exercises 3 to 5 use this probability model for the blood type of a randomly chosen person in the United States:

| Blood type | O | A | B | AB |
|---|---|---|---|---|
| Probability | 0.45 | 0.40 | 0.11 | ? |

**3.** The probability that a randomly chosen American has type AB blood is

**(a)** 0.044.
**(b)** 0.04.
**(c)** 0.4.

**4.** Maria has type A blood. She can safely receive blood transfusions from people with blood types O and A. The probability that a randomly chosen American can donate blood to Maria is ____________________ .

**5.** What is the probability that a randomly chosen American does not have type O blood?

**(a)** 0.55
**(b)** 0.45
**(c)** 0.04

**6.** Figure 8.2 shows the 36 possible outcomes for rolling two dice. These outcomes are equally likely. A "soft 4" is a roll of 1 on one die and 3 on the other. The probability of rolling a soft 4 is ______________ .

**7.** In a table of random digits such as Table 7.1, each digit is equally likely to be any of 0, 1, 2, 3, 4, 5, 6, 7, 8, or 9. What is the probability that a digit in the table is a 0?

**(a)** 1/9
**(b)** 1/10
**(c)** 9/10

**8.** In a table of random digits such as Table 7.1, each digit is equally likely to be any of 0, 1, 2, 3, 4, 5, 6, 7, 8, or 9. The probability that a digit in the table is 7 or greater is ______________________________ .

**9.** Toward the end of a game of Scrabble, you hold the letters J, U, D, A, and H. In how many orders can you arrange these 5 letters?

**(a)** 5
**(b)** (5)(4)(3)(2)(1) = 120
**(c)** (5)(5)(5)(5)(5) = 3125

**10.** Toward the end of a game of Scrabble, you hold the letters D, O, G, and Q. You can choose 3 of these 4 letters and arrange them in order in ______________ different ways.

**11.** A 52-card deck contains 13 cards from each of the four suits: clubs ♣, diamonds ♦, hearts ♥, and spades ♠. You deal 4 cards without replacement from a well-shuffled deck, so that you are equally likely to deal any 4 cards. What is the probability that all 4 cards are clubs?

**(a)** 1/4, because 1/4 of the cards are clubs.
**(b)** (13)(12)(11)(10)/(52)(51)(50)(49) = 0.0026
**(c)** (13)(12)(11)(10)/(52)(52)(52)(52) = 0.0023

**12.** You deal 4 cards as in the previous exercise. The probability that you deal no clubs is ______________ .

**13.** Figure 5.3 (page 155) shows that the normal distribution with mean $\mu = 6.8$ and standard deviation $\sigma = 1.6$ is a good description of the Iowa Test vocabulary scores of seventh-grade students in Gary, Indiana. The probability that a randomly chosen student has a score higher than 8.4 is

**(a)** 0.68.
**(b)** 0.32.
**(c)** 0.16.

**14.** Figure 8.7 shows the density curve of a continuous probability model for choosing a number at random between 0 and 1 inclusive. The probability that the number chosen is less than or equal to 0.4 is ______________ .

**15.** Annual returns on the more than 5000 common stocks available to investors vary a lot. In a recent year, the mean return was 8.3% and the standard deviation of returns was 28.5%. The law of large numbers says:

**(a)** you can get an average return higher than the mean 8.3% by investing in a large number of stocks.

**(b)** as you invest in more and more stocks chosen at random, your average return on these stocks gets ever closer to 8.3%.
**(c)** if you invest in a large number of stocks chosen at random, your average return will have approximately a normal distribution.

**16.** Suppose you are trying to decide between buying many shares of a promising individual stock and spending that same amount of money on a mutual fund consisting of a variety of different stocks. Choosing the mutual fund would result in an investment that is _____ variable than the individual stock.

**17.** Figure 8.7 shows the density curve of a continuous probability model for choosing a number at random between 0 and 1 inclusive. The mean of this model is

**(a)** 0.5 because the curve is symmetric.
**(b)** 1 because there is area 1 under the curve.
**(c)** can't tell–this requires advanced mathematics.

**18.** Scores on the SAT Reasoning college entrance test in a recent year were roughly normal, with mean 1511 and standard deviation 194. You take an SRS of 100 students and average their SAT scores. If you do this many times, the mean of the average scores you get from all those samples would be ________________ .

**19.** The number of hours a light bulb burns before failing varies from bulb to bulb. The distribution of burnout times is strongly skewed to the right. The Central Limit Theorem says that

**(a)** as we look at more and more bulbs, their average burnout time gets ever closer to the mean $\mu$ for all bulbs of this type.
**(b)** the average burnout time of a large number of bulbs has a distribution of the same shape (strongly skewed) as the distribution for individual bulbs.
**(c)** the average burnout time of a large number of bulbs has a distribution that is close to normal.

**20.** Referring to Question #18, the standard deviation of the average scores you get from all those samples would be ________________________________ .

# CHAPTER 8 EXERCISES

■ Challenge ◆ Discussion

## 8.1 Probability Models and Rules

**1.** Estimating probabilities empirically:

**(a)** Hold a penny upright on its edge under your forefinger on a hard surface, then snap it with your other forefinger so that it spins for some time before falling. Based on 30 spins, estimate the probability of heads.
**(b)** Toss a thumbtack (with a gently curved back) on a hard surface 100 times. (To speed it up, toss 10 at a time.) How many times did it land with the point up? What is the approximate probability of landing point up?

**2.** Some situations refer not to probabilities, but to odds. The odds against an event $E$ are equal to $P(E^C)/P(E)$. If there are 3:2 odds against a particular horse winning a race, what is the probability that the horse wins?

**3.** The table of random digits (Table 7.1) was produced by a random mechanism that gives each digit probability 0.1 of being a 0. What proportion of the first five lines in the table are 0's? This proportion is an estimate of the true probability, which in this case is known to be 0.1.

**4.** Probability is a measure of how likely an event is to occur. Match one of the probabilities that follow with each statement about an event. (The probability is usually a much more exact measure of likelihood than is the verbal statement.)

0, 0.01, 0.3, 0.6, 0.99, 1

**(a)** This event is impossible. It can never occur.
**(b)** This event is certain. It will occur on every trial of the random phenomenon.
**(c)** This event is very unlikely, but it will occur once in a while in a long sequence of trials.
**(d)** This event will occur more often than not.

In each of Exercises 5 to 7, describe a reasonable sample space $S$ for the random phenomena mentioned. In some cases, you must use judgment to choose a reasonable $S$.

**5.** Toss a coin 10 times.

**(a)** Count the number of heads observed.
**(b)** Calculate the percent of heads among the outcomes.
**(c)** Record whether or not at least five heads occurred.

**6.** A randomly chosen subject arrives for a study of exercise and fitness.

**(a)** The subject is either female or male.
**(b)** After 10 minutes on an exercise bicycle, you ask the subject to rate his or her effort on the Rate of Perceived Exertion (RPE) scale. RPE ranges in whole-number steps from 6 (no exertion at all) to 20 (maximal exertion).
**(c)** You also measure the subject's maximum heart rate (beats per minute).

**7.** A basketball player shoots four free throws.

**(a)** You record the sequence of hits and misses.
**(b)** You record the number of shots she makes.

**8.** The Punnett square is a diagram biologists use to determine the probability of offspring having certain genetic makeup. Suppose "B" represents the gene for brown eyes and "b" represents the gene for blue eyes. In genetics, capital letters refer to dominant traits, so a person receiving both "B" and "b" generally has brown eyes. This diagram shows the possibilities for the child of two Bb parents. Each parent gives the child one of its 2 genes with equal probability. What is the probability that this child will receive the genetic makeup for brown eyes? Discuss how this relates to Example 2.

| | Mother gives B | Mother gives b |
|---|---|---|
| Father gives B | BB | Bb |
| Father gives b | Bb | bb |

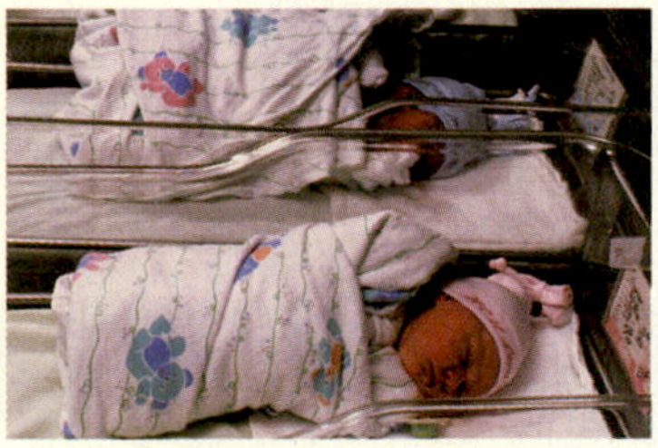
*(Lori Adamski Peek/STONE/Getty Images.)*

**9.** Many email messages are "spam." Choose a spam email message at random. Here is the probability model for the topic of a randomly chosen spam email message:

| Topic | Adult | Financial | Health |
|---|---|---|---|
| Probability | 0.145 | 0.162 | 0.073 |
| Topic | Leisure | Products | Scams |
| Probability | 0.078 | 0.210 | 0.142 |

**(a)** What is the probability that a spam email does not concern one of these topics?
**(b)** Corinne is particularly annoyed by spam offering "adult" content (that is, pornography) and scams. What is the probability that a randomly chosen spam email falls into one of these categories?

**10.** Choose a young adult (age 25 to 34 years) at random. The probability is 0.12 that the person chosen did not complete high school, 0.31 that the person has a high school diploma but no further education, and 0.29 that the person has at least a bachelor's degree.

**(a)** What must be the probability that a randomly chosen young adult has some education beyond high school but does not have a bachelor's degree?
**(b)** What is the probability that a randomly chosen young adult has at least a high school education?

■ **11.** What is the probability that Laurie rolls doubles (both dice match) each of her first three rolls in the game of Monopoly? (This matters because rolling three consecutive doubles sends you right to jail!)

## 8.2 Discrete Probability Models

**12.** Choose a new car or light truck at random and note its color. Here are the probabilities of the most popular colors:

| Color | Silver | White | Black |
|---|---|---|---|
| Probability | 0.201 | 0.184 | 0.116 |
| Color | Gray | Dark blue | Light brown |
| Probability | 0.115 | 0.088 | 0.085 |

**(a)** What is the probability that the car you choose has any color other than the six listed?
**(b)** What is the probability that a randomly chosen car is either silver or white?

**13.** North Carolina State University posts the grade distributions for its courses online. Students in Statistics 101 in a recent semester earned 21% A's, 43% B's, 30% C's, 5% D's, and 1% F's. Here is the probability model for the grade of a randomly chosen Statistics 101 student.

| Grade | 0 (= F) | 1 (= D) | 2 (= C) | 3 (= B) | 4 (= A) |
|---|---|---|---|---|---|
| Probability | 0.01 | 0.05 | 0.30 | 0.43 | 0.21 |

**(a)** Make a probability histogram for this model.
**(b)** What is the probability that the student got a grade of B or better?

**14.** How do rented housing units differ from units occupied by their owners? Here are probability models for the number of rooms for owner-occupied units and renter-occupied units, according to the Census Bureau:

| # of Rooms | 1 | 2 | 3 | 4 | 5 |
|---|---|---|---|---|---|
| Owned | 0.000 | 0.001 | 0.014 | 0.099 | 0.238 |
| Rented | 0.011 | 0.027 | 0.229 | 0.348 | 0.224 |
| # of Rooms | 6 | 7 | 8 | 9 | 10 |
| Owned | 0.266 | 0.178 | 0.107 | 0.050 | 0.047 |
| Rented | 0.105 | 0.035 | 0.012 | 0.004 | 0.005 |

Make probability histograms of these two models, using the same scale. What are the most important differences between the models for owner-occupied and rented housing units?

**15.** In each of the following situations, state whether or not the given assignment of probabilities to individual outcomes is legitimate, that is, satisfies the rules of probability. If not, give specific reasons for your answer.

**(a)** Choose a college student at random and record gender and enrollment status: $P(\text{female full-time}) = 0.56$, $P(\text{female part-time}) = 0.24$, $P(\text{male full-time}) = 0.44$, $P(\text{male part-time}) = 0.17$.

**(b)** Choose a college student at random from a class and record the season of her birth: $P$(spring) = 0.39, $P$(summer) = 0.28, $P$(fall) = 0, $P$(winter) = 0.33.

**16.** What is the probability that a housing unit has five or more rooms? Use the models in Exercise 14 to answer this question for both owner-occupied and rented units.

■ **17.** Balanced six-sided dice with altered labels can produce interesting distributions of outcomes. Construct the probability model (sample space and assignment of probabilities for each sum) for rolling two "weird dice" from a *Math Horizons* article by Joseph Gallian. Instead of using the regular values {1,2,3,4,5,6}, one die has the labels 1,2,2,3,3,4 and the other die has the labels 1,3,4,5,6,8. How does this model compare to the model for regular dice?

■ **18.** Role-playing games like Dungeons & Dragons use many different types of dice. Suppose that a balanced four-sided die has faces marked 1, 2, 3, 4. The intelligence of a character is determined by rolling this die twice and adding 1 to the sum of the spots at the bottom of the die, since a triangular pyramidal die has no up "side," just a point! Give a probability model for the character's intelligence. (Start with a display like Figure 8.2 for the outcomes of the two rolls of the die. These outcomes are equally likely.) What is the probability that the character has intelligence 7 or higher?

## 8.3 Equally Likely Outcomes

**19.** A party host gives a door prize to one guest chosen at random. There are 42 men and 48 women at the party. What is the probability that the prize goes to a woman?

**20.** Abby, Boaz, Carmen, Dani, and Eduardo work in a firm's public relations office. Their employer must choose two of them to attend a conference in Paris. To avoid unfairness, the choice will be made by drawing two names from a hat. (This is an SRS of size 2.)

**(a)** Write down all possible choices of two of the five names. This is the sample space.
**(b)** The random drawing makes all choices equally likely. What is the probability of each choice?
**(c)** What is the probability that Abby is chosen?
**(d)** What is the probability that neither of the two men (Boaz and Eduardo) is chosen?

**21.** You toss a balanced coin 10 times and write down the resulting sequence of heads and tails, such as HTTTHHTHHH.

**(a)** How many possible outcomes are there for the 10 tosses?
**(b)** What is the probability that your 10-toss sequence is either all heads or all tails?

**22.** In the Texas Hold 'Em style of poker, play begins with each player being dealt two cards face down. From a standard 52-card deck, how many possible 2-card hands could be dealt to you?

**23.** A computer assigns three-character log-in IDs that may contain the digits 0 to 9 as well as the letters *a* to *z*, with repeats allowed.

**(a)** What is the probability that your ID contains no *x*?
**(b)** What is the probability that your ID contains no digits?

**24.** Consider a typical "combination lock" on a locker or briefcase.

**(a)** If you ask for the three numbers in the combination needed to open the lock, and they are given to you in numerical order as 3–5–8, why is this not enough information to open the lock?
**(b)** What would be the probability that you could open the lock with one try?
**(c)** Is such a lock accurately named or is it really a "permutation lock"?

**25.** You may have heard that a monkey hitting keys at random on a typewriter keyboard for an infinite amount of time could eventually type a particular chosen text, such as the complete works of Shakespeare. Let's focus on a monkey who just types the letters *a*, *p*, and *s* in random order.

**(a)** How many possible three-letter "words" can the monkey type using only these letters?
**(b)** Which of these are words in an English dictionary?
**(c)** What is the probability that the word the monkey typed is in an English dictionary?

■ **26.** Mozart composed a 16-bar Viennese minuet ("Musical Dice Game") in which bars #1–7 each have 11 choices, bar #8 has 2, bars #9–15 each have 11 and bar #16 has 1. How many possible versions of this minuet are there?

■ **27.** In poker, a royal flush is a 5-card hand containing (in any order) an Ace, King, Queen, Jack and 10 all of the same suit.

**(a)** How many royal flush hands are possible?
**(b)** What is the number of 5-card hands possible from a 52-card deck?
**(c)** What is the probability that 5 cards drawn at random from a 52-card deck yield a royal flush?

■ **28.** Biblical Permutations: The King James Version Old Testament has its 39 books canonized in a different order than the Hebrew Bible does. What mathematical expression would yield the number of possible orders of these 39 books? Is this number larger than you expected?

## 8.4 Continuous Probability Models

**29.** Generate two random numbers between 0 and 1 and take their sum. The sum can take any value between 0 and 2. The density curve is the shaded triangle shown in Figure 8.13.

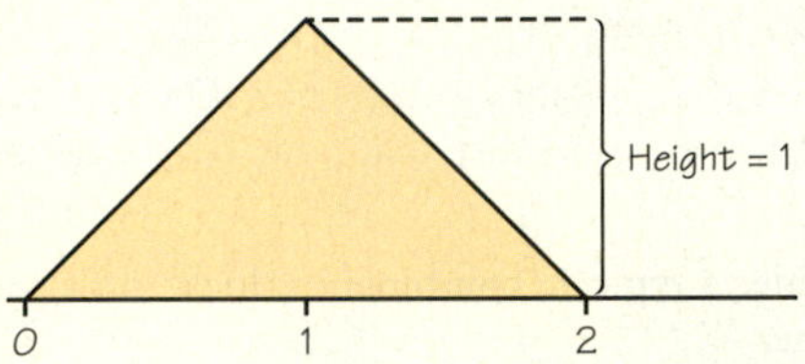


FIGURE 8.13 The density curve for the sum of two random numbers, for Exercise 29.

**(a)** Verify by geometry that the area under this curve is 1.
**(b)** What is the probability that the sum is less than 1? (Sketch the density curve, shade the area that represents the probability, then find that area. Do this for part (c) also.)
**(c)** What is the probability that the sum is less than 0.5?

**30.** Many random-number generators allow users to specify the range of the random numbers to be produced. Suppose that you specify that the range is to be all numbers between 0 and 2. The density curve for the outcome has constant height between 0 and 2, and height 0 elsewhere.

**(a)** What is the height of the density curve between 0 and 2? Draw a graph of the density curve.
**(b)** Use your graph from part (a) and the fact that probability is area under the curve to find the probability of getting an outcome less than or equal to 1.
**(c)** Find the probability of an outcome between 0.5 and 1.3.

**31.** On the TV show *The Price is Right*, there is a pricing game called "Range Game" in which a prize has a price somewhere on a scale with a range of $600. The contestant must try to guess where to position a window spanning $150 of the range so that the price falls within the window. If the actual price is equally likely to be anywhere, what is the probability the contestant will be successful with a random guess?

## 8.5 The Mean and Standard Deviation of a Probability Model

**32.** Linda is a sales associate at a large auto dealership. She expects to earn $400 for each vehicle she sells. Linda motivates herself by using probability estimates of her sales. For a sunny Saturday in April, she estimates her sales as follows:

| Vehicles sold | 0 | 1 | 2 | 3 |
|---|---|---|---|---|
| Probability | 0.3 | 0.4 | 0.2 | 0.1 |

Give the probability model for Linda's earnings. What are her mean earnings?

**33.** Exercise 13 gives a probability model for the grade of a randomly chosen student in Statistics 101 at North Carolina State University, using the 4-point scale. What is the mean grade in this course? What is the standard deviation of the grades?

**34.** In Exercise 18, you gave a probability model for the intelligence of a character in a role-playing game. What is the mean intelligence for these characters?

**35.** Exercise 14 gives probability models for the number of rooms in owner-occupied and rented housing units. Find the mean number of rooms for each type of housing. Make probability histograms for the two models and mark the mean on each histogram. You see that the means describe an important difference between the two models: Owner-occupied units tend to have more rooms.

**36.** Typographical and spelling errors can be either "nonword errors" or "word errors." A nonword error is not a real word, as when "the" is typed as "teh." A word error is a real word, but not the right word, as when "lose" is typed as "loose." When undergraduates are asked to write a 250-word essay (without spell-checking), the number of nonword errors has this probability model:

| Errors | 0 | 1 | 2 | 3 | 4 |
|---|---|---|---|---|---|
| Probability | 0.1 | 0.2 | 0.3 | 0.3 | 0.1 |

The number of word errors has this model:

| Errors | 0 | 1 | 2 | 3 |
|---|---|---|---|---|
| Probability | 0.4 | 0.3 | 0.2 | 0.1 |

What are the mean numbers of nonword errors and word errors in an essay? How does the difference between the means describe the difference between the two models?

**37.** Find (and explain how you found) the mean for:

**(a)** the continuous probability model in Exercise 29.
**(b)** the probability model in Exercise 30.

■ **38.** The idea of insurance is that we all face risks that are unlikely but carry high cost. Think of a fire destroying your home. Insurance spreads the risk: We all pay a small amount, and the insurance policy pays a large amount to those few of us whose homes burn down. An insurance company looks at the records for millions of homeowners and sees that the mean loss from fire in a year is $\mu$ = $250 per person. (The great majority of us have no loss, but a few lose their homes. The $250 is the average loss.) The company plans to sell fire insurance for $250 plus enough to cover its costs and profit. Explain clearly why it would be unwise to sell only 12 policies. Then explain why selling thousands of such policies is a safe business.

◆ **39.** Should you buy the extended warranty on a new washing machine? Suppose there are two outcomes–a 85% probability of needing no repairs, and a 15% probability of needing a $200 repair during the warranty period. Based on the mean outcome for this model, what would be a "break-even" price to you for the extended warranty? (The company, of course, will charge more than this in order to make a profit.)

**40.** An American roulette wheel has 38 slots numbered 0, 00, and 1 to 36. The ball is equally likely to come to rest in any of these slots when the wheel is spun. The slot numbers are laid out on a board on which gamblers place their bets. One column of numbers on the board contains multiples of 3–that is, 3, 6, 9, . . . , 36. Joe places a $1 "column bet" that pays out $3 if any of these numbers comes up.

**(a)** What is the probability model for the outcome of one bet, taking into account the $1 cost of a bet?
**(b)** What are the mean and standard deviation for this model?
**(c)** Joe plays roulette every day for years. What does the law of large numbers tell us about his results?

**41.** This table shows the prizes and respective probabilities for a lottery:

| Net prize | $1,000,000 | $1000 | $100 | $4 |
|---|---|---|---|---|
| Probability | 1/10,000,000 | 1/10,000 | 1/1,000 | 3/100 |

On average, how much money from a $1 ticket comes back to you in prizes?

■ **42.** A state lottery "Pick 3" game offers a choice of several bets. You choose a three-digit number and bet $1. The lottery commission announces the winning three-digit number, chosen at random, at the end of each day. The "box" pays $82.33 if the number you chose has the same digits as the winning number, in any order. Otherwise, you lose your dollar. Find the mean winnings for a bet on the box, taking into account that you paid $1 to play. (Assume that you chose a number having three distinct digits.)

■ **43.** Suppose a test is designed in which each question has 5 possible answer choices (ABCDE).

**(a)** If you get +1 point for every correct answer, what would the "penalty" for a wrong answer need to be if you wanted guessing to neither help nor hurt the score on average? (*Hint:* Consider a test with 5 questions on it.)
**(b)** If you are able to eliminate some but not all of the wrong answers, does it help on average to guess among the remaining choices?

■ **44.** Here is a simple way to create a probability model that has specified mean $\mu$ and standard deviation $\sigma$: There are only two outcomes, $\mu - \sigma$ and $\mu + \sigma$, each with probability 0.5. Use the definition of the mean and variance for probability models to show that this model does have mean $\mu$ and standard deviation $\sigma$.

## 8.6 The Central Limit Theorem

**45.** Newly manufactured automobile radiators may have small leaks. Most have no leaks, but some have one, two, or more. The number of leaks in radiators made by one supplier has mean 0.15 and standard deviation 0.4. The distribution of the number of leaks cannot be normal because only whole-number counts are possible. The supplier ships 400 radiators per day to an auto assembly plant. Take $\bar{x}$ to be the mean number of leaks in these 400 radiators. Over several years of daily shipments, what range of values will contain the middle 95% of the many $\bar{x}$'s?

**46.** The scores of eighth-grade students on the National Assessment of Educational Progress (NAEP) year 2007 mathematics test have a distribution that is approximately normal, with mean $\mu = 281$ and standard deviation $\sigma = 35$.

**(a)** Choose one eighth-grader at random. What is the probability that her score is higher than 281? Higher than 316?
**(b)** Now choose an SRS of four eighth-graders. What is the probability that their mean score is higher than 281? Higher than 316?

◆ **47.** Antonio measures the alcohol content of whiskey for his Chemistry 101 lab. He actually measures the mass of 5 milliliters of whiskey–a chemical calculation then finds the percent alcohol from the mass. The standard deviation of students' measurements of mass is $\sigma = 10$ milligrams (mg). Antonio repeats the measurement three times and records the mean $\bar{x}$ of his three measurements.

**(a)** What is the standard deviation of Antonio's mean result?
**(b)** How many times must Antonio repeat the measurement to reduce the standard deviation of $\bar{x}$ to 5 mg? Explain to someone who knows no statistics the advantage of reporting the average of several measurements rather than the result of a single measurement.

**48.** In Exercise 40 you found the mean and standard deviation of the outcome of a column bet in roulette. The Central Limit Theorem says that the average outcome of a large number of bets has a distribution that is close to normal.

**(a)** What is the spread (mean ± 3 standard deviations) of a gambler's average winnings after 100 bets?
**(b)** What is the spread of a gambler's average winnings after 1000 bets?

**49.** Averages of several measurements are less variable than individual measurements. The true mass of the whiskey sample in Exercise 47 is 4.6 grams, or 4600 milligrams (mg). Antonio's measurements have the normal distribution with mean 4600 mg and standard deviation 10 mg. In this case, the mean of his three measurements also has a normal distribution.

**(a)** Sketch on the same graph the two normal curves, for individual measurements and for means of three measurements. Figure 8.10 is an example of this kind of graph.
**(b)** What spread of values covers the middle 95% of Antonio's measurements?
**(c)** What spread of values covers the middle 95% of averages of three measurements?

**50.** Exercise 18 gives the probability model for the intelligence assigned by chance to a character in a role-playing game. You found the mean intelligence of such characters in Exercise 34. Jermaine plays this character often. What range covers (approximately) the middle 68% of average intelligence scores for 100 of Jermaine's games?

**51.** The scores of high school seniors on the ACT college entrance examination in 2003 were roughly normal with mean $\mu = 20.8$ and standard deviation $\sigma = 4.8$.

**(a)** What is the approximate probability that a single student randomly chosen from all those taking the test scores 25.6 or higher?
**(b)** Now take an SRS of nine students who took the test. What are the mean and standard deviation of the sample mean score $\bar{x}$ of these nine students?
**(c)** What is the approximate probability that the mean score $\bar{x}$ of these nine students is 25.6 or higher?

■ **52.** Although cities encourage carpooling to reduce traffic congestion, most vehicles carry only one person. For example, 70% of vehicles on the roads in the Minneapolis–St. Paul metropolitan area are occupied by just the driver. You choose 84 vehicles at random.

**(a)** What are the mean and standard deviation of the proportion of vehicles in your sample that carry only one person?
**(b)** What is the probability that more than 60% of the vehicles in your sample carry only one person? (Use the Central Limit Theorem.)

## Chapter Review

**53.** License plates in Florida have the form A12BCD, that is, a letter followed by two digits followed by three more letters.

**(a)** How many possible different license plates are there?
**(b)** Jerry would like a plate that ends in AAA. How many such plates are there?
**(c)** If license plates are issued at random from all possible plates, what is the probability that Jerry will get a plate that ends in AAA?

**54.** After you tell Jerry the probability you calculated in the previous exercise, he realizes that he's unlikely to get a plate ending in AAA. So he asks you, "What's the probability I will get a plate in which all four letters are from my name?" These letters are J, E, R, and Y.

**(a)** Suppose Jerry insists that the letters appear in order, so that his plate reads JnmERY, where "n" and "m" stand for any number. What is the probability?
**(b)** Suppose Jerry allows his letters to appear in the plate in any order and also allows repeats. What is now the answer to Jerry's question?

**55.** Choose a person age 19 to 25 years at random and ask, "In the past four days, how many days did you do physical exercise or work out?"
Based on a large sample survey, here is a discrete probability model for the answer you will get:

| # of Days | 0 | 1 | 2 | 3 | 4 |
|---|---|---|---|---|---|
| Probability | 0.61 | 0.17 | 0.10 | 0.08 | 0.04 |

**(a)** What is the probability that the person you choose worked out either two or three days in the past four?
**(b)** What is the probability that the person you choose worked out at least one day in the past four?

**56.** What is the mean number of days that randomly chosen 19- to 25-year-olds worked out in the past four days? (Use the information in the previous exercise.) If you interview many people in this age group, what does the law of large numbers say about the average number of days these people work out?

■ **57.** Use the information in Exercise 55 and your result from Exercise 56 to answer these questions.

**(a)** What is the standard deviation of the number of days in the past four that a randomly chosen 19- to 25-year-old has worked out?
**(b)** You interview 100 randomly chosen 19- to 25-year olds. You ask each how many days in the past four they have worked out and you calculate the average number of days. According to the Central Limit Theorem, there is probability 0.95 that your average will fall between what two values?

**58.** In Example 16, we saw that a \$1 bet on red has a mean outcome of $-\$2/38$. It turns out not all American roulette \$1 bets have the same mean outcome. The "5-number bet" {0,00,1,2,3} pays an additional \$6 if one of those 5 numbers comes up–otherwise the player loses his \$1. Find the expected value for this 5-number bet. Is this 5-number bet a better or worse bet than a bet on red?

**59.** Suppose you select 10 people at random. Find the probability of each event below:

# Voting and Social Choice PART III

The application of mathematics to the study of human beings—their behavior, values, interactions, conflicts, and methods of making decisions—is generally considered to be a recent revolution. Yet the study of voting and social choice, which is very much the root of this revolution, goes back several centuries.

We begin in Chapter 9 with the question of how a group of individuals, each with his or her own set of values, selects one outcome from a list of possibilities. While majority rule is a good system for deciding an election with just two candidates, it turns out that there is no perfect way of deciding an election in which there are three or more candidates.

Group decision-making is often a strategic encounter, and citizens need to be aware of the difficulties that can arise when some participants have an incentive to manipulate the outcome. It is this issue that we turn to in Chapter 10.

In Chapter 11 we consider decision-making bodies in which the individual voters or parties do not have equal power. In particular, we look at weighted voting systems in which a voter's power need not be proportional to the number of votes that he or she is entitled to cast.

In Chapter 12 we analyze not only how the Electoral College influences resource-allocation in a campaign but also how polls and the positioning of candidates on a left–right continuum affect the strategies of candidates and the choices of voters. ■

**(a)** At least one match in the day of the week they were born.
**(b)** At least one match in the day of the month they were born (assume 31 days per month).
**(c)** At least one match in the day of the year they were born.

**60.** Combination Connections:

**(a)** Give an intuitive or algebraic argument to explain why ${}_nC_k = {}_nC_{n-k}$.
**(b)** Generate several small values of ${}_nC_k$ and explain how they relate to the numbers in Pascal's Triangle (see for example, Chapter 11).

## APPLET EXERCISES

**To do these exercises, go to www.whfreeman.com/fapp8e.**

**1.** When we toss a coin, experience shows that the probability (long-term proportion) of a head is close to 1/2. Suppose now that we toss the coin repeatedly until we get a head. What is the probability that the first head comes up in an odd number of tosses (1, 3, 5, and so on)? Use the *Probability* applet to estimate this probability. Set the probability of heads to 0.5. Toss coins one at a time until the first head appears. Do this 50 times (click "Reset" after each trial). What is your estimate of the probability that the first head appears on an odd toss?

**2.** The table of random digits (Table 7.1) was produced by a random mechanism that gives each digit probability 0.1 of being a 0.

**(a)** What proportion of the digits in the first row of Table 7.1 are 0's? This proportion is an estimate, based on 40 repetitions, of the true probability, which in this case is known to be 0.1.

**(b)** The *Probability* applet can imitate random digits. Set the probability of heads in the applet to 0.1. Check "Show true probability" to show this value on the graph. A head stands for a 0 in the random digit table and a tail stands for any other digit. Simulate 200 digits (40 at a time–don't click "Reset"). If you kept going forever, presumably you would get 10% heads. What was the percent of heads in your 200 tosses?

**3.** The basketball player Shaquille O'Neal makes about half of his free throws over an entire season. Use the *Probability* applet to simulate 100 free throws shot independently by a player who has probability 0.5 of making each shot. (Toss 40, 40, and 20 without clicking "Reset.")

**(a)** What percent of the 100 shots did he hit?
**(b)** Examine the sequence of hits and misses after each click on "Toss" and keep track of the longest run of shots made and the longest run of shots missed. How long were the longest runs in the 100 shots taken? (Sequences of random outcomes often show longer runs than our intuition expects.)

## WRITING PROJECTS

**1.** Psychologists have shown that our intuitive understanding of chance behavior is rather poor. Amos Tversky (1937–1996) was a leader in the study of how we make decisions in the face of uncertainty. In its obituary of Tversky, the *New York Times* cited the following example:

> *Tversky asked subjects to choose between two public health programs that affect 600 people. One has probability 1/2 of saving all 600 and probability 1/2 that all 600 will die. The other is guaranteed to save exactly 400 of the 600 people. Most people chose the second program. He then offered a different choice. One program has probability 1/2 of saving all 600 and probability 1/2 of losing all 600, while the other will definitely lose exactly 200 lives. Most people chose the first program.*

Discuss this example. What is the difference between the two choices offered? What is the mean number of people saved by the two options in each choice? What do the reactions of most subjects to these choices show about how people make decisions?

**2.** There are about $1 \times 10^{44}$ air molecules in the atmosphere and about $2 \times 10^{22}$ molecules of air in a single breath taken at rest. What is the probability that the breath you took just now contained at least one molecule of air that was exhaled by Pythagoras in his last breath? What probability rules did you use to calculate this? What assumptions did you make and why do you think they were reasonable?

**3.** Double or Nothing: Gambler's Ruin. We have seen that by betting on "red" in roulette, you have a 18/38 chance of winning, and therefore doubling the money you bet. Suppose you have \$5 and you want to bet until either you reach (and stop with) \$10 or you go broke. Is placing individual \$1 bets on red more, less, or equally likely to reach this goal than just placing a single \$5 bet? First, try to give an answer based on intuition, taking into account the casino's advantage.

You could also explore the following formula that gives the probability of going from $h$ dollars to $N$ dollars by making \$1 bets on red in American roulette:

$$\frac{1 - (20/18)^h}{1 - (20/18)^N}$$

Discuss how the strategy for maximizing the chance of reaching a financial target compares to the strategy for maximizing the length of time your money lasts (for entertainment value).

## SUGGESTED READINGS

COMAP. *Principles and Practices of Mathematics*, Springer, New York, 1997. Chapter 4 of this team-authored text presents combinatorics, and Chapter 8 discusses probability. This is a good choice for going beyond *For All Practical Purposes* in these areas.

LESSER, LAWRENCE M. Take a chance by exploring the statistics in lotteries. *Statistics Teacher Network*, 65 (2004): 6–7. This article shows how lotteries can illustrate all major topics of an introductory statistics course, using a graphing calculator. Available online at www.amstat.org/education/stn/pdfs/STN65.pdf.

MOSTELLER, FREDERICK, ROBERT E. K. ROURKE, and GEORGE B. THOMAS. *Probability with Statistical Applications*, Addison-Wesley, Reading, Mass., 1970. A rich treatment of basic probability that requires only high school algebra but is somewhat sophisticated. Although out of print, this book is a classic that nonetheless deserves mention.

## SUGGESTED WEB SITES

"Buffon's needle" is a probability problem first stated in 1777 by Count Buffon: If you drop a needle on a sheet of lined paper, what is the probability that the needle crosses one of the lines? In the simplest case, the length of the needle is the same as the distance between the lines. Some fairly advanced math shows that the answer is $2/\pi$, or about 0.637. A number of Web sites simulate dropping a needle many times in order to estimate this probability. One good simulation is at the Web site of Charles Stanton, www.math.csusb.edu/faculty/stanton/probstat/buffon.html. (If this site has disappeared, type "Buffon's needle" into Google. Be sure to choose a site that actually pictures the needle's position.)

You may be interested in the debate over legalized gambling. For the case against, visit the National Coalition Against Legalized Gambling at www.ncalg.org. For the defense by the casino industry, visit the American Gaming Association at www.americangaming.org.

# CHAPTER 9



# Social Choice:
## The Impossible Dream

The basic question of *social choice,* of how groups can best arrive at decisions, has occupied social philosophers and political scientists for centuries. One primary example of a social-choice problem is the selection of a "good" voting system. Indeed, voting is a subject that lies at the very heart of representative government and participatory democracy.

Social-choice theory attempts to address the problem of finding good procedures that will turn individual preferences for different candidates into a single choice by the whole group. An example of such a choice would be the selection of a *winner* of an election. The goal is to find such procedures that will result in an outcome that "reflects the will of the people."

This search for good voting systems, as we shall see, is plagued by a variety of counterintuitive results and disturbing outcomes. In fact, it turns out that one can prove (mathematically) that no one will ever find a completely satisfactory voting system for three or more candidates.

The elections with which we are most familiar often involve only two candidates, and we will begin our discussion of voting systems with this two-candidate case.

There are, however, real-world situations in which elections must be held to choose a single winner from among three or more candidates, as in the presidential election of 2004 in which George W. Bush, John Kerry, and Ralph Nader were the candidates.

There are several methods that can be used to elect a single candidate from a choice of three or more, and we will investigate some in this chapter. Most of these methods use a ballot in which a voter provides a rank ordering of the candidates (without ties) that indicates the order in which he or she prefers them.

In the 2004 election, President George W. Bush was challenged by Massachusetts senator John Kerry. President Bush received a slim majority of the vote. The election results have been dogged by charges of irregularities in the important swing state of Ohio.
*(Left: Reuters/Corbis; right: John Gress/Reuters/Corbis.)*

**Preference List Ballot** DEFINITION

A ballot consisting of such a rank ordering of candidates (which we often picture as a vertical list with the most preferred candidate on top and the least preferred on the bottom) is called a **preference list ballot** because it is a statement of the preferences of the individual who is voting.

Preference list ballots allow voters to make a much clearer statement of their preferences than do ballots allowing a single vote. Preference list ballots are already used in a wide range of applications, such as rating football teams and scoring track meets.

Although we do not allow ties in a preference list ballot, most voting rules of interest will, in some elections, result in a tie for the win among two or more of the candidates. In the real world, the number of voters is often so large that ties seldom occur. Nevertheless, to avoid excessive annoyances in the theory we develop, and to simplify what we do in this chapter, we make the following assumption throughout:

**The Number of Voters Assumption** RULE

Throughout this chapter, we assume that the number of voters is odd.

## 9.1 Majority Rule and Condorcet's Method

When a choice is being made between two candidates, the first type of voting system to suggest itself is **majority rule**: Each voter indicates a preference for one of the two candidates and the one with the most votes wins. With two candidates, there is no real distinction between a ballot that indicates a voter's choice for one of the two candidates and what we have called a preference list ballot. The point is that we can, for example, identify a choice for $A$ (however indicated) with the list that has $A$ over $B$ and a choice $B$ with the list that has $B$ over $A$.

Majority rule has at least three desirable properties:

1. All voters are treated equally. That is, if any two voters were to exchange (marked) ballots before submitting them, the outcome of the election would be the same.
2. Both candidates are treated equally. That is, if a new election were held and every voter were to reverse his or her vote, then the outcome of the previous election would be reversed as well.
3. It is **monotone**. If a new election were held and a single voter were to change his or her ballot from being a vote for the loser of the previous election to being a vote for the winner of the previous election, and everyone else voted exactly as before, then the outcome of the new election would be the same as the outcome of the previous election.

It is easy to devise voting systems for two candidates in which these fail, but each such voting system quickly reveals its undesirability. For example, condition 1 is not satisfied by a *dictatorship* (in which all ballots except that of the dictator are ignored); condition 2 is not satisfied by *imposed rule* (in which candidate $X$ wins regardless of who votes for whom); and condition 3 is not satisfied by *minority rule* (in which the candidate with the fewest votes wins).

But maybe there are voting systems in the two-candidate case that are superior to majority rule in the sense of satisfying the three properties just listed *and* some other properties that we might also wish to have satisfied. This, however, turns out not to be the case. In 1952, Kenneth May proved the following:

### May's Theorem — THEOREM

Among all two-candidate voting systems that never result in a tie, majority rule is the *only* one that treats all voters equally, treats both candidates equally, and is monotone.

This is an important and elegant result. Thus, mathematical reasoning spares us the trouble of searching for a better voting system for two candidates.

But what if there are three or more candidates? Perhaps we can design a voting system for this situation that, in some way, builds on the success of majority rule in the two-candidate case. In point of fact, there does exist a voting system that arises from precisely this hope, and it is known today as **Condorcet's method**.

Our description of Condorcet's method begins with the observation that if we have a sequence of preference list ballots, then–for each pair of candidates–we can determine who the winner would have been had the election involved only these two in a one-on-one contest using majority rule.

To illustrate this notion of a one-on-one contest, consider the following preference list ballots:

| **Rank** | **Number of Voters (3)** | | |
|---|---|---|---|
| First | $A$ | $B$ | $C$ |
| Second | $B$ | $C$ | $A$ |
| Third | $C$ | $A$ | $B$ |

In this election, candidate $A$ would defeat candidate $B$ in a one-on-one contest (two votes to one), while $B$ would, in turn, defeat $C$ in a one-on-one contest, again by a score of 2 to 1. We'll return to this example in a moment, but we now have at hand all we need to describe Condorcet's voting system for three or more candidates.

## Description of Condorcet's Method

PROCEDURE

With the voting system known as Condorcet's method, a candidate is a winner precisely when he or she would, on the basis of the ballots cast, defeat every other candidate in a one-on-one contest using majority rule.

Historically, the voting system we are calling Condorcet's method dates back at least to Ramon Llull in the thirteenth century (see Spotlight 9.1). It was rediscovered and popularized in the eighteenth century by the Marquis de Condorcet (1743–1794).

### EXAMPLE 1 ■ Condorcet's Method

Suppose we have four candidates (*GB, AG, RN,* and *PB,* with these initials chosen for a soon-to-be-revealed reason) and the following sequence of preference list ballots, where the heading of "6" indicates that 6 of the 15 voters hold the ballot with *GB* over *AG* over *PB* over *RN,* the heading of "5" indicates that 5 of the 15 voters hold the ballot with *AG* over *RN* over *GB* over *PB,* and so on.

| | Number of Voters (15) | | | |
|---|---|---|---|---|
| **Rank** | **6** | **5** | **3** | **1** |
| First | *GB* | *AG* | *RN* | *PB* |
| Second | *AG* | *RN* | *AG* | *GB* |
| Third | *PB* | *GB* | *GB* | *AG* |
| Fourth | *RN* | *PB* | *PB* | *RN* |

We claim that *AG* is the winner in this election if we use Condorcet's method. Let's check the one-on-one scores for each possible pair of opponents:

*AG* versus *GB: AG* is over *GB* on $5 + 3 = 8$ of the ballots, while the reverse is true on $6 + 1 = 7$ of the ballots. Thus, *AG* defeats *GB* by a score of 8 to 7.

*AG* versus *RN: AG* is over *RN* on $6 + 5 + 1 = 12$ of the ballots, while the reverse is true on 3 of the ballots. Thus, *AG* defeats *RN* by a score of 12 to 3.

*AG* versus *PB: AG* is over *PB* on $6 + 5 + 3 = 14$ of the ballots, while the reverse is true on 1 of the ballots. Thus, *AG* defeats *PB* by a score of 14 to 1.

This shows that *AG* is the winner using Condorcet's method. ■

Like majority rule, Condorcet's method satisfies some very desirable properties, as we'll see later in this section. But it also has a tragic flaw, and this flaw is called **Condorcet's voting paradox.**

**Condorcet's Voting Paradox** THEOREM

With three or more candidates, there are elections in which Condorcet's method yields *no* winners. In particular, the following ballots (often called the "Condorcet voting paradox ballots") constitute an election in which Condorcet's method yields no winner.

| Rank | Number of Voters (3) | | |
|---|---|---|---|
| First | $A$ | $B$ | $C$ |
| Second | $B$ | $C$ | $A$ |
| Third | $C$ | $A$ | $B$ |

The Condorcet voting paradox ballots given above are the same ones we used earlier in illustrating the notion of a one-on-one contest. We pointed out then that $A$ defeats $B$ one on one and $B$ defeats $C$ one on one. The additional observation needed is that we also have $C$ defeating $A$ one on one. Thus $A$ cannot be a winner using Condorcet's method (it loses to $C$), $B$ cannot be a winner (it loses to $A$), and $C$ cannot be a winner (it loses to $B$). We will revisit Condorcet's voting paradox in section 9.3.

Notice that, because of our assumption that the number of voters is odd, Condorcet's method yields either no winner or a unique winner (see Exercise 5).

It is tempting at this point to suggest modifying Condorcet's method as we have presented it by declaring all the candidates to be tied for the win if there is no candidate who defeats each of the others one on one. The drawback to this modification is that a number of the upcoming desirable properties possessed by Condorcet's method then evaporate. We'll explore this in the exercises.

## 9.2 Other Voting Systems for Three or More Candidates

With three or more candidates, we find no shortage of additional procedures that suggest themselves and that seem to represent perfectly reasonable ways to choose a winner. Closer inspection, however, reveals shortcomings with all of these. We illustrate this with a consideration of several well-known procedures. Additional procedures (and additional shortcomings) can be found in the exercises.

### Plurality Voting and the Condorcet Winner Criterion

In **plurality voting**, only first-place votes are considered. Thus, while we will consider plurality voting in the context of preference list ballots, a ballot here might just as well be a single vote for a single candidate. The candidate with the most votes wins, even though this may be considerably fewer than one-half the total votes cast. This is perhaps the most common system in use today. It is how the voters in Florida chose George W. Bush over Al Gore, Ralph Nader, and Patrick Buchanan in the presidential election of 2000.

## SPOTLIGHT 9.1 The Historical Record

The following letter was written by Friedrich Pukelsheim of the University of Augsburg, Germany. He is imagining what Ramon Llull (1232–1316) might say if he were alive today.

Dear Editors:

It is my distinct pleasure to respond "from the beyond" to your kind invitation to set the historical record straight. I was born in 1232 on the Island of Mallorca in the Mediterranean Sea, which in your times is known as a popular tourist place. In my days it was a strong political center of that part of the world, with a population that was a mix of Christians, Jews, and Muslims. It was my dream to persuade people of the virtues of Christian belief by relying, not on force, but on reason.

Unfortunately, people did not find it easy to follow my arguments, so I was more than pleased to discover some down-to-earth applications, including an election system. My idea was to oppose every pair of candidates, one on one, and ask the electors whom of the two they would prefer—very much like a medieval jousting tournament. But how to combine the results from all the duels into a winner of the election? I first proposed electing the candidate who won the most duels, then later suggested a system of successive eliminations.

I wrote three papers on the topic, the second of which I "smuggled" into my novel Blanquerna in 1283. More than a century after my death, in 1428, the young German scholar Nicolaus Cusanus (1401–1464) journeyed to Paris to read my works in libraries there. He even copied out the third of my electoral writings, which I had completed on 1 July 1299 in Paris, and his manuscript is the only copy handed down to your days. Reading my papers, Cusanus was inspired to invent his own electoral system. Did he not understand mine, or just find it inadequate? Who knows?

While I had been concerned with electing church officials, Cusanus sought a system to elect the Holy Roman Emperor. In his system, each elector assigns each candidate a rank score, with the lowest candidate getting a score of 1, the second lowest a score of 2, and the best candidate the highest score possible, that is, 10 when there are 10 candidates. The scores are totaled for each candidate and the candidate with the highest score wins. If you are a soccer player or a hockey player, you will have a good sense for one difference between our systems: Whereas I count victories, Cusanus adds up goals. Cusanus applauded himself for having invented an absolutely ingenious and novel electoral system.

Also, I advocated open voting, whereas Cusanus favored a secret ballot. He was concerned that voters might sell their votes, or that the candidates might pressure the voters. Well, that certainly happened all of the time in elections for worldly authorities! But for election to clerical office, I thought it good enough if electors take an oath to vote for the most worthy candidate and submit themselves to the social control that comes with an open election.

Cusanus was famous in his times, as I was in mine, but fame indeed is transitory. Sure enough, my electoral system was reinvented by the Marquis de Condorcet (1743–1794), and Cusanus's system was proposed afresh by the Chevalier de Borda (1733–1799)—neither of whom, I am sure, wasted a thought on the possibility that "their" systems might already be on record. But, as my works had fallen into oblivion as had those of Cusanus, neither Condorcet nor Borda should be blamed for failing to acknowledge our priority.

My first electoral paper—actually the one that is longest and most detailed, written around 1280—was rediscovered only in 2000, filed away in the Vatican Library. How would you feel if your work attracts fresh attention after more than 700 years? Actually, I am utterly pleased that mine has resurfaced at last! The text was excavated by a mathematician interested in voting systems, Friedrich Pukelsheim of the University of Augsburg, Germany. Since the text is handwritten in Latin, handling it became an interdisciplinary project that brought together experts on medieval manuscripts, Church Latin and theology, and even computer scientists. As a result, my electoral writings are

## The Historical Record *(continued)*

now on the Internet (in the original and in translations into English and German) at www.uni-augsburg.de/llull/.

Looking back on my lack of success in preaching peace among Christians, Jews, and Muslims, and all the writing and copying by hand of my works, I hope you can appreciate how highly I value the printed book (such as this one) and, even more, instant communication worldwide over the Internet. May that ease of communication help facilitate the religious peace that I so dearly sought.

Yours truly,
Ramon Llull (1232–1316)
Left Choir Chapel
San Francisco Cathedral
Palma de Mallorca

## EXAMPLE 2

### Plurality Voting and the 2000 Presidential Election

On the evening of December 12, 2000, Al Gore conceded the presidential election of 2000 to George W. Bush, thus bringing to a close one of the most remarkable elections in modern times. The outcome, ultimately decided in the electoral college, came down to which of Bush or Gore would carry Florida. With more than six million votes cast in Florida, the ultimate margin of victory for George W. Bush was only a few hundred votes.

There is little doubt that if the 2000 presidential election had pitted Al Gore solely against any one of the other three candidates, then Gore would have won both the election in Florida and the presidency. The point is that while most of the Buchanan supporters would have voted for Bush, the far more numerous Nader supporters would have gone largely for Gore. In fact, the illustration of Condorcet's method that we gave in Example 1 is a simplified version of this Florida election (with *GB* standing for George Bush, *AG* for Al Gore, *PB* for Patrick Buchanan, and *RN* for Ralph Nader).

Thus, although plurality voting led to Bush's winning the 2000 election in Florida (and hence the presidency), Gore was, in this example, what is called a **Condorcet winner**: He would have won the election if Condorcet's method had been used.

Governor George W. Bush and Vice President Al Gore debate the issues before the 2000 election, possibly the most controversial election in U.S. history. Gore was the Condorcet winner of the election, but Bush eked out a victory that relied on the rules of the Electoral College. Many voters were suddenly put on notice that the U.S. Constitution makes the election of the president indirect–and not a pure expression of the majority's choice. *(Reuters/Corbis.)*

**Condorcet Winner Criterion (CWC)** DEFINITION

A voting system is said to satisfy the **Condorcet winner criterion** (CWC) provided that, for every possible sequence of preference list ballots, either (1) there is no Condorcet winner (as is often the case) or (2) the voting system produces exactly the same winner for this election as does Condorcet's method.

The CWC is certainly a property that one would like to see satisfied. We record plurality voting's failure in this respect with the following.

**The Failure of the CWC with Plurality Voting** THEOREM

The Florida vote in the 2000 presidential election shows that plurality voting fails to satisfy the Condorcet winner criterion.

Perhaps a more fundamental drawback of plurality voting is the extent to which the ballots provide no opportunity for a voter to express any preferences except for naming his or her top choice. No use is made, for example, of the fact that a candidate may be no one's first choice but everyone's close second choice.

Finally, there is yet another shortcoming of plurality voting: There are elections in which it is to a voter's advantage to submit a ballot that misrepresents his or her true preferences.

**Manipulability** DEFINITION

A voting system is subject to **manipulabity** (or is **manipulable**) if there are elections in which it is to a voter's advantage to submit a ballot that misrepresents his or her true preferences.

For example, in the presidential election of 2000, many voters who ranked Ralph Nader or Patrick Buchanan over George W. Bush and Al Gore chose to vote for Bush or Gore rather than to "throw away" their vote on a candidate they felt had no chance. Condorcet's method, it turns out, is not manipulable, and this is one of its most desirable properties. We'll explore this further in the next chapter.

## The Borda Count and Independence of Irrelevant Alternatives

In many elections that use preference list ballots, the goal is to arrive at a final group rank ordering of all the contestants that best expresses the desires of the electorate. The purpose is not only to determine the winner–say, the class valedictorian–but also to arrive at who finished second, third, and so on, as in the case of one's rank in the senior class. In other applications, such as an election to a hall of fame, the first few finishers each receive the award, while the remaining nominees are also-rans.

One common mechanism for achieving this objective is to assign points to each voter's rankings and then to sum these for all voters to obtain the total points for each candidate. If there are 10 candidates, for example, then we could assign 10 points to each first-place vote for a given candidate, 9 points for each second-place vote, 8 for each third-place vote, and so forth. The candidate with the highest total number of points is the winner. Subsequent positions are assigned to those with the next-highest tallies.

### Description of Rank Methods and the Borda Count — PROCEDURE

A *rank method* of voting assigns points in a nonincreasing manner to the ordered candidates on each voter's preference list ballot and then sums these points to arrive at a group's final ranking. The special case in which there are $n$ candidates with each first-place vote worth $n-1$ points, each second-place vote worth $n-2$ points, and so on down to each last-place vote worth zero points is known as the **Borda count.** The actual point totals are referred to as a candidate's **Borda score.**

The Borda count is named after Jean-Charles de Borda (1733–1799). He was a contemporary of Condorcet.

Rank methods other than the Borda count are common. For example, a track meet can be thought of as an "election" in which each event is a "voter" and each of the schools competing is a "candidate." If the order of finish in the 100-meter dash is school $A$, school $B$, school $C$, school $D$, then points are often awarded to each school as follows: 5 points for first place, 3 for second place, 2 for third place, and 1 for fourth place.

Sports polls often use point assignments that qualify as rank methods according to our definition. The following example provides an illustration of this.

## EXAMPLE 3 ■ Rank Methods and a Basketball Poll

In November of 2007, the Associated Press issued the early-season ranking of the top 25 teams in women's college basketball, shown to the right.

An interesting question is whether or not this is a ranking system. If it is, who are the candidates and how many are there? In fact, this can be regarded as a ranking system, but the number of candidates is not 25. That is, although 25 teams appeared on each ballot, at least one ballot included each of the teams listed at the bottom in the category "Others receiving votes."

To regard this as a ranking system, the set of candidates must include the entire set of eligible collegiate women's basketball teams. We must also infer that each ballot lists all teams other than that voter's top 25 *below* that voter's top 25, perhaps, in alphabetical order. The point assignments are then like those in the newspaper clipping, except that we also assign 0 points for a 26th-place vote, 0 points for a 27th-place vote, and so on. This is why our definition states that a rank method "assigns points in a *nonincreasing* manner" instead of "assigns points in a *decreasing* manner."

We can use this poll to illustrate how total points are arrived at with a ranking method. With the top-ranked team, Tennessee, it's quite easy. Each first-place vote is worth 25 and Tennessee received all 50 first-place votes. This accounts for its total of $25 \times 50 = 1250$ points. But the calculation is more interesting for the second-ranked team, Connecticut, and requires some speculation on our part, since we

**WOMEN'S TOP 25**

The top 25 teams in The Associated Press' women's college basketball poll, with first-place votes in parentheses, records through Sunday, total point based on 25 points for a first-place vote through one point for a 25th-place vote and previous ranking:

| | Record | Pts | Pvs |
|---|---|---|---|
| 1. Tennessee (50) | 1-0 | 1250 | 1 |
| 2. Connecticut | 1-0 | 1187 | 2 |
| 3. Maryland | 2-0 | 1139 | 4 |
| 4. LSU | 2-0 | 1072 | 5 |
| 5. Stanford | 2-0 | 1040 | 7 |
| 6. Rutgers | 0-1 | 973 | 3 |
| 7. North Carolina | 2-0 | 969 | 8 |
| 8. Georgia | 2-0 | 881 | 9 |
| 9. Oklahoma | 0-1 | 833 | 6 |
| 10. Duke | 1-0 | 818 | 10 |
| 11. Texas A&M | 1-0 | 753 | 11 |
| 12. California | 1-0 | 631 | 13 |
| 13. Baylor | 3-0 | 576 | 15 |
| 14. G. Washington | 1-0 | 560 | 14 |
| 15. Arizona St. | 0-1 | 554 | 12 |
| 16. Ohio St. | 1-0 | 506 | 16 |
| 17. Michigan St. | 2-0 | 390 | 17 |
| 18. Florida St. | 2-0 | 365 | 19 |
| 19. West Virginia | 1-0 | 345 | 18 |
| 20. Vanderbilt | 2-0 | 254 | 23 |
| 21. Texas | 1-0 | 247 | 22 |
| 22. Louisville | 1-0 | 212 | 21 |
| 23. Notre Dame | 1-0 | 112 | 24 |
| 24. DePaul | 1-0 | 97 | 25 |
| 25. Wisconsin | 1-0 | 77 | – |

**Others receiving votes:** N.C. State 76, Pittsburgh 76, Purdue 68, Penn St. 53, Auburn 46, Wyoming 21, Oklahoma St. 13, Marquette 9, Middle Tennessee 9, Georgia Tech 8, Old Dominion 8, Illinois 7, Xavier 4, Marist 3, Kentucky 2, New Mexico 2, S. Dakota St. 2, Florida 1, Iowa St. 1.

(*AP Photo/Bill Kostroun.*)

don't actually have the ballots to examine. We know that there were 50 ballots (because there were exactly 50 first-place votes all together), and we know Connecticut had no first-place votes. It stands to reason that Connecticut's 1187 points must have come from a vast majority of the second-place votes, together with a few lower rankings. (We know it didn't receive *all* the second place-votes, otherwise its point total would have been $24 \times 50 = 1200$.)

One possibility is that Connecticut received:

0 first-place votes (at 25 points each)
40 second-place votes (at 24 points each)
8 third-place votes (at 23 points each)
1 fourth-place votes (at 22 points each)
1 fifth-place vote (at 21 points)

The total would then be:

$$0 + 960 + 184 + 22 + 21 = 1187$$

There is an easy way to calculate the Borda score of a candidate. You can count the number of occurrences of other candidate names that are below this candidate's name. For example, consider the following ballots:

| Rank | Number of Voters (5) | | | | | Points |
|---|---|---|---|---|---|---|
| First | *A* | *A* | *A* | *B* | *B* | 2 |
| Second | *B* | *B* | *B* | *C* | *C* | 1 |
| Third | *C* | *C* | *C* | *A* | *A* | 0 |

Because there are three candidates, each first-place vote is $n - 1$, or $3 - 1 = 2$, each second-place vote is $n - 2 = 1$, and each third-place vote is $n - 3 = 0$. If we were to calculate the Borda score of candidate *B* algebraically, we would say that *B* has two first-place votes, worth 2 points each (a total of 4 points), and three second-place votes, worth 1 point each (a total of 3 more points). Thus, the Borda score of candidate *B* is $4 + 3 = 7$.

But instead of calculating this Borda score algebraically, we can mentally replace each occurrence of a letter below *B* by a box, □, and simply count the boxes.

| Rank | Number of Voters (5) | | | | |
|---|---|---|---|---|---|
| First | *A* | *A* | *A* | *B* | *B* |
| Second | *B* | *B* | *B* | □ | □ |
| Third | □ | □ | □ | □ | □ |

Notice that there are seven boxes, giving us the correct value of 7 as the Borda score for candidate *B*. Of course, you don't actually have to draw any boxes. We are just emphasizing the fact that, in the counting process, it is "spaces" that we are counting, without regard to which letter occurs in the space. A quick glance at the original ballots (without the boxes) reveals that the Borda score of candidate *A* is 6 and the Borda score of candidate *C* is 2. When calculating Borda scores this way, be sure that each individual ballot is listed separately, as opposed to using a single list to represent the ballots of several voters (as we often do).

The Borda count certainly seems to be a reasonable way to choose a winner from among several candidates (or to arrive at a group ranking of the candidates). It also has its shortcomings, however, one of which is the failure of a property known as **independence of irrelevant alternatives.**

### Independence of Irrevelant Alternatives (IIA) — DEFINITION

A voting system is said to satisfy independence of **irrelevant alternatives (IIA)** if it is impossible for a candidate $B$ to move from nonwinner status to winner status unless at least one voter reverses the order in which he or she had $B$ and the winning candidate ranked.

To describe this property, suppose that an election yields one candidate (call it $A$) as a winner and another candidate (call it $B$) as a nonwinner. Suppose that a new election is now held and that, although some of the voters may have changed their preference list ballots, no one who had previously ranked $A$ and $B$ changed his or her ballot to rank $B$ over $A$ now.

If this new election were to yield $B$ as a winner, the new outcome would seem strange, especially because none of the relative individual preferences for $A$ over $B$ had changed in $B$'s favor. The ballot changes responsible for the new outcome involve candidates *other than* $A$ or $B$. One could argue that these other candidates ought to be irrelevant to the question of whether $A$ is more desirable than $B$ or $B$ is more desirable than $A$.

Condorcet's method satisfies IIA. That is, if we have a sequence of preference list ballots that yield $A$ as a Condorcet winner and $B$ as a nonwinner, then $A$ defeats every other candidate, and $B$ in particular, in a one-on-one contest according to these ballots. If no voter reverses the order in which he or she ranked $A$ and $B$, then $A$ will still defeat $B$ one on one, and thus $B$ remains a nonwinner.

The following illustration shows that the Borda count, unlike Condorcet's method, fails to satisfy independence of irrelevant alternatives. Suppose the initial five ballots are as follows:

| **Rank** | **Number of Voters (5)** | | | | |
|---|---|---|---|---|---|
| First | $A$ | $A$ | $A$ | $C$ | $C$ |
| Second | $B$ | $B$ | $B$ | $B$ | $B$ |
| Third | $C$ | $C$ | $C$ | $A$ | $A$ |

Our counting procedure shows that the Borda scores are as follows:

Borda score of $A$ is 6
Borda score of $B$ is 5
Borda score of $C$ is 4

The winner is $A$ (with 6 points), and $B$ is a nonwinner (with 5 points). But now suppose that the two voters on the right change their ballots by moving $C$ down between $A$ and $B$. The ballots then become

| Rank | Number of Voters (5) | | | | |
|---|---|---|---|---|---|
| First | $A$ | $A$ | $A$ | $B$ | $B$ |
| Second | $B$ | $B$ | $B$ | $C$ | $C$ |
| Third | $C$ | $C$ | $C$ | $A$ | $A$ |

Our counting procedure shows that the Borda scores are as follows:

Borda score of $A$ is 6

Borda score of $B$ is 7

Borda score of $C$ is 2

The Borda count therefore now yields $B$ as the winner (with 7 points). Thus, $B$ has gone from being a nonwinner to being a winner, even though no one changed his or her mind about whether $B$ is preferred to $A$, or vice versa.

The above discussion establishes the following:

**The Failure of IIA with the Borda Count** THEOREM

The Borda count fails to satisfy independence of irrelevant alternatives.

## Sequential Pairwise Voting and the Pareto Condition

In our voting-theoretic context, an **agenda** will be understood to be a listing (in some order) of the candidates. This listing is not to be confused with any of the preference list ballots, and, to avoid confusion, we will present agendas as horizontal lists and continue to present preference list ballots vertically.

**Description of Sequential Pairwise Voting** PROCEDURE

**Sequential pairwise voting** starts with an agenda and pits the first candidate against the second in a one-on-one contest. The winner then moves on to confront the third candidate in the list, one on one. Losers are deleted. This process continues throughout the entire agenda, and the one remaining at the end wins.

For a given sequence of individual preference list ballots, the particular agenda chosen can greatly affect the outcome of the election, as we'll show in the next chapter. Nevertheless, we will see later in this chapter that sequential pairwise voting arises naturally in the legislative process. Notice also that because of our assumption that the number of voters is odd, there is always a unique winner with sequential pairwise voting.

## EXAMPLE 4 ■ Sequential Pairwise Voting

Assume we have four candidates and that the agenda is *A, B, C, D*. Consider the following sequence of three preference list ballots:

| Rank | Number of Voters (3) | | |
|---|---|---|---|
| First | *A* | *C* | *B* |
| Second | *B* | *A* | *D* |
| Third | *D* | *B* | *C* |
| Fourth | *C* | *D* | *A* |

The first one-on-one pits *A* against *B*, and *A* wins by a score of 2 to 1 (meaning that two of the voters–the two on the left–prefer *A* to *B*, and one of the voters prefers *B* to *A*). Thus, *B* is eliminated and *A* moves on to confront *C*. Because *C* wins this one on one (by a score of 2 to 1), *A* is eliminated. Finally, *C* takes on *D*, and *D* wins by a score of 2 to 1. Thus, *D* is the winner.

There is something very troubling about the outcome of the preceding example, especially if you are candidate *B*. *Everyone* prefers *B* to *D*!

### The Failure of the Pareto Condition with Sequential Pairwise Voting — THEOREM

Sequential pairwise voting fails to satisfy what is called the **Pareto condition,** which says that if everyone prefers one candidate (in this case, *B*) to another candidate (*D*), then this latter candidate (*D*) should not be among the winners of the election.

The Pareto condition is named after Vilfredo Pareto (1848–1923), an Italian economist.

## Runoff Systems and Monotonicity

The voting system known as the **Hare system,** which was introduced by Thomas Hare in 1861, is also known by names such as the "single transferable vote system." In 1862, John Stuart Mill described the Hare system as being "among the greatest improvements yet made in the theory and practice of government." Today, the system is used to elect public officials in Australia, Malta, the Republic of Ireland, and Northern Ireland.

### Description of the Hare System — PROCEDURE

The Hare system proceeds to arrive at a winner by repeatedly deleting candidates that are "least preferred" in the sense of being at the top of the fewest ballots. If a single candidate remains after all others have been eliminated, it alone is the winner. If two or more candidates remain and all of these remaining candidates would be eliminated in the next round (because they all have the same number of first-place votes), then these candidates are declared to be tied for the win.

## EXAMPLE 5 ■ The Hare System

Suppose we have the following sequence of preference list ballots, where, as before, the heading of "5" indicates that 5 of the 13 voters hold the ballot with *A* over *B* over *C*, the heading of "4" indicates that 4 of the 13 voters hold the ballot with *C* over *B* over *A*, and so forth.

| | Number of Voters (13) | | | |
|---|---|---|---|---|
| **Rank** | **5** | **4** | **3** | **1** |
| First | *A* | *C* | *B* | *B* |
| Second | *B* | *B* | *C* | *A* |
| Third | *C* | *A* | *A* | *C* |

Candidates *B* and *C* have only 4 first-place votes (while *A* has 5). Thus, *B* and *C* are eliminated in the first round, and *A* wins the election.

In the preceding example, suppose that the voter in the last column moves candidate *A* up on his list. Let's look at the new election. Notice that, even though *A* won the last election, the only change we are making in ballots for the new election is one that is favorable to *A*. The ballots for the new election are as follows:

| | Number of Voters (13) | | | |
|---|---|---|---|---|
| **Rank** | **5** | **4** | **3** | **1** |
| First | *A* | *C* | *B* | *A* |
| Second | *B* | *B* | *C* | *B* |
| Third | *C* | *A* | *A* | *C* |

If we apply the Hare system again, only *B* is eliminated in round one, as it has 3 first-place votes to 4 for *C* and 6 for *A*. Thus, after this round, the ballots are as follows:

| | Number of Voters (13) | | | |
|---|---|---|---|---|
| **Rank** | **5** | **4** | **3** | **1** |
| First | *A* | *C* | *C* | *A* |
| Second | *C* | *A* | *A* | *C* |

We now have *A* on top of 6 lists and *C* on top of 7 lists. Thus, at stage two, *A* (our previous winner!) is eliminated and *C* is the winner of this new election.

Clearly, this is once again quite counterintuitive. Alternative *A* won the original election, the only change in ballots made was one favorable to *A* (and no one else), and then *A* lost the next election.

### Failure of Monotonicity with the Hare System — THEOREM

Example 5 shows that the Hare system does not satisfy **monotonicity**, which, with three or more candidates, says that if a candidate is a winner, and a new election is held in which the only ballot change made is for some voter to move the former winning candidate higher on his or her ballot, then the original winner should remain a winner.

The fact that the Hare system does not satisfy monotonicity is considered by many–and with good reason–to be a glaring defect. A 17-voter example in which only a single candidate is eliminated in the first round can also be used to show that the Hare system does not satisfy monotonicity–see Exercise 28. For an even more glaring version of this defect, one in which alternative *A* goes from winning to losing because voters move *A* from last place on their ballots to first place on their ballots, see Exercise 29.

In spite of these drawbacks, the Hare system is used in important ways today. For example, it is essentially the method that was used to choose Sydney, Australia, as the site of the 2000 Summer Olympics. Beijing would have been the plurality winner, but after the elimination of Istanbul, Berlin, and Manchester (in that order), Sydney defeated Beijing by a vote of 45 to 43. Four years later, Beijing finally did prevail; it was the site of the 2008 Summer Olympics.

There are other runoff systems, some more frequently used than the Hare system. One such example is the following.

### Description of the Plurality Runoff Method — PROCEDURE

**Plurality runoff** is the voting system in which there is a runoff (that is, a new election using the same ballots) between the two candidates receiving the most first-place votes. If there are ties, then the runoff is among either those tied for the most first-place votes, or the lone candidate with the most first-place votes along with those tied for the second-most first-place votes (and plurality voting is used).

Alas, the plurality runoff method also is not monotone. Exercise 25 asks you to verify this by making use of the following ballots:

| | **Number of Voters (13)** | | | | |
|---|---|---|---|---|---|
| **Rank** | **4** | **3** | **3** | **2** | **1** |
| First | *A* | *B* | *C* | *D* | *E* |
| Second | *B* | *A* | *A* | *B* | *D* |
| Third | *C* | *C* | *B* | *C* | *C* |
| Fourth | *D* | *D* | *D* | *A* | *B* |
| Fifth | *E* | *E* | *E* | *E* | *A* |

## EXAMPLE 6 ■ Plurality Runoff

The plurality runoff method is somewhat similar in spirit to the Hare system. In fact, you might wonder if they aren't just two different descriptions of the same voting system. That is, you might ask if the plurality runoff method and the Hare system always yield the same winner.

The answer is no, however, as we now demonstrate. Consider the following sequence of preference list ballots:

| | Number of Voters (13) | | | |
|---|---|---|---|---|
| **Rank** | **4** | **4** | **3** | **2** |
| First | *A* | *B* | *C* | *D* |
| Second | *B* | *A* | *D* | *C* |
| Third | *C* | *C* | *A* | *A* |
| Fourth | *D* | *D* | *B* | *B* |

With the plurality runoff method, *A* and *B* initially tie with 4 first-place votes each, with 3 for *C* and 2 for *D*. In the runoff between *A* and *B*, the ballots are as follows:

| | Number of Voters (13) | | | |
|---|---|---|---|---|
| **Rank** | **4** | **4** | **3** | **2** |
| First | *A* | *B* | *A* | *A* |
| Second | *B* | *A* | *B* | *B* |

With the plurality runoff method, *A* is the winner after defeating *B* in the runoff by a score of 9 to 5.

On the other hand, with the Hare system we find that the only alternative deleted in the first round is *D*, with only 2 first-place votes. With this deletion of *D*, the ballots are as follows:

| | Number of Voters (13) | | | |
|---|---|---|---|---|
| **Rank** | **4** | **4** | **3** | **2** |
| First | *A* | *B* | *C* | *C* |
| Second | *B* | *A* | *A* | *A* |
| Third | *C* | *C* | *B* | *B* |

*A* and *B* now have only 4 first-place votes compared to the 5 first-place votes that *C* has. Hence, *A* and *B* are now deleted, leaving *C* as the winner with the Hare system.

## 9.3 Insurmountable Difficulties: Arrow's Impossibility Theorem

All of the voting systems for three or more candidates that we have discussed turn out to be flawed in one way or another. You may well ask at this point why we don't simply present *one* voting method for the three-candidate case that has all the desirable properties we want to have satisfied. That is, after all, exactly what we did for the two-candidate case (with majority rule filling the bill, and being the only one to do so by May's theorem).

The answer to this question is extremely important. The difficulties in the three-candidate case are not in any way tied to a few particular systems that we present in a text such as this (or that we choose to use in the real world). The fact is, there are difficulties that will be present *regardless* of what voting system is used, and this applies even to voting systems not yet discovered.

Nothing in the remarkable body of work produced by Nobel laureate Kenneth J. Arrow of Stanford University is as well known or widely acclaimed as the result known as **Arrow's impossibility theorem** (see Spotlight 9.2).

### Arrow's Impossibility Theorem — THEOREM

With three or more candidates and any number of voters, there does not exist—and there never will exist—a voting system that always produces a winner, satisfies the Pareto condition and independence of irrelevant alternatives, and is not a dictatorship.

Arrow's theorem isn't obvious, and we won't be saying anything about the proof. But we can state and prove a much weaker result of some interest in its own right. This version is taken from the 1995 text *Mathematics and Politics,* cited in the Suggested Readings, and replaces Arrow's assumption of the Pareto condition and nondictatorship by the Condorcet winner criterion.

### A Weak Version of Arrow's Impossibility Theorem — THEOREM

With three or more candidates and an odd number of voters, there does not exist—and there never will exist—a voting system that satisfies both the Condorcet winner criterion and independence of irrelevant alternatives and that always produces at least one winner in every election.

To see why this is true, we'll handle only the case of exactly three voters. Our plan will be to assume that we have some kind of hypothetical voting system that satisfies both the Condorcet winner criterion and independence of irrelevant alternatives and to show that, when confronted by the Condorcet voting paradox ballots, it produces *no* winner.

The argument really comes in three separate, but extremely similar, pieces—one for each of the three candidates. Piece 1 argues that $A$ can't be among the winners, piece 2 that $B$ can't be among the winners, and piece 3 that $C$ can't be among the winners. We'll do piece 1 and leave the others for you. The sequence of ballots that we are considering is the following, which we have already seen has no Condorcet winner.

| Rank | Number of Voters (3) | | |
|---|---|---|---|
| First | *A* | *B* | *C* |
| Second | *B* | *C* | *A* |
| Third | *C* | *A* | *B* |

Our starting point, however, will be to ask what our hypothetical voting rule must do when confronted by a slightly different sequence of ballots.

| Rank | Number of Voters (3) | | |
|---|---|---|---|
| First | *A* | *C* | *C* |
| Second | *B* | *B* | *A* |
| Third | *C* | *A* | *B* |

Here, *C* is clearly a Condorcet winner, and thus it must be the unique winner of the election contested under our hypothetical voting rule. Therefore, *C* is a winner and *A* is a nonwinner (for *this* sequence of ballots).

However, because our hypothetical voting rule satisfies independence of irrelevant alternatives, we know that *A* will remain a nonwinner as long as no one reverses his or her ordering of *A* and *C*. But to arrive at the voting paradox ballots we can move *B* (the candidate that is irrelevant to *A* and *C*) up one slot in the second voter's list.

Thus, because of IIA, we know that *A* is a nonwinner when our voting rule is confronted by the voting paradox ballots. This is one-third of the argument. As we mentioned before, similar arguments (see Exercise 34) show that *B* and *C* are also nonwinners when our voting rule is confronted by the voting paradox ballots.

We conclude this section with an example that yields a somewhat surprising application of Arrow's impossibility theorem in the context of what are called *social welfare functions*.

## EXAMPLE 7

### Organ Transplant Policies and Arrow's Impossibility Theorem

Finding an equitable procedure for determining a rank ordering of patients in need of an organ transplant is complicated: there are several criteria that should be considered in arriving at such a "priority ranking." Three such criteria are, for example, (1) the length of time that a patient has been waiting, (2) the probability of success as measured by the numbers of antigens that the patient and donor have matched, and (3) the fraction of the population unsuitable as donors for this potential recipient due to the presence of certain antibodies. A further discussion of these issues occurs in Section 13.3.

Each of the three criteria gives us a ranking (with ties) of the patients according to the more appropriate recipient of the next available organ, according to that particular criterion. Although these rankings are often determined by measurements, the use of different scales for different criteria muddies the water sufficiently so that you might want to work simply with the rankings derived from the measurements, as opposed to working directly with the measurements themselves. This is the context in which we will frame the problem.

So what does the search for a procedure to rank order potential recipients of an organ have to do with voting? In a sense, *everything*, if looked at the right way. We can think of each criterion as a "voter" and each potential recipient as an "alternative." The procedure that we seek is what social choice theorists call a *social welfare function*. It differs from a social choice procedure in that the result of an election is not a single winner or a group tied for the win, but a listing of the alternatives–the priority ranking, in our organ-transplant situation.

For a moment, let's return to the particular task of seeking a priority ranking of the potential recipients of an organ based on how they are ranked according to each of several criteria, like the three we mentioned earlier. What "reasonable" properties might we expect any such procedure to satisfy? Consider the following:

1. If one potential recipient *A* is ranked above another potential recipient *B* with respect to every single criterion, then we should expect *A* to be ranked above *B* in the priority ranking.
2. If potential recipient *A* is ranked above potential recipient *B* in the priority ranking, and there are subsequent changes in how potential recipients are ranked with respect to one or more of the criteria, then potential recipient *B* should not be ranked above potential recipient *A* in the priority ranking unless *B* has moved from being below *A* to being above *A* with respect to at least one criterion.
3. No single criterion should dominate, in the sense that one potential recipient *A*'s ranking above another potential recipient *B*'s ranking, with respect to that criterion, guarantees that *A* will be ranked above *B* in the priority ranking.

If we accept these as being required of any "reasonable" procedure, then we have a striking (and highly non-obvious) fact to report: Our task of finding a reasonable procedure is impossible! In fact, this is precisely the statement of Arrow's impossibility theorem in the context of social welfare functions:

There is no social welfare function (for three or more alternatives) that satisfies Pareto (our first condition above), independence of irrelevant alternatives (our second condition above), and non-dictatorship (our third condition above).

## SPOTLIGHT 9.2 Kenneth J. Arrow

For centuries, mathematicians have been in search of a perfect voting system. Finally, in 1951, economist Kenneth Arrow proved that finding an absolutely fair and decisive voting system is impossible. Arrow is the Joan Kenney Professor of Economics, as well as a professor of operations research, at Stanford University. In 1972, he received the Nobel Memorial Prize in Economic Science for his outstanding work in the theory of general economic equilibrium. His numerous other honors include the 1986 von Neumann Theory Prize for his fundamental contributions to the decision sciences. He has served as president of the American Economic Association, the Institute of Management Sciences, and other organizations. Dr. Arrow talks about the process by which he developed his famous impossibility theorem and his ideas on the laws that govern voting systems:

(continued on page 304)

Kenneth Arrow

## Kenneth J. Arrow *(continued)*

My first interest was in the theory of corporations. In a firm with many owners, how do the owners agree when they have different opinions, for example, about the prospects of the company? I was thinking of stockholders. In the course of this, I realized that there was a paradox involved—that majority voting can lead to cycles. I then dropped that discussion because I was frustrated by it.

I happened to be working with The RAND Corporation one summer about a year or two later. They were very interested in applying concepts of rationality, particularly of game theory, to military and diplomatic affairs. That summer, I felt not like an economist but instead like a general social scientist or a mathematically-oriented social scientist. There was tremendous interest in game theory, which was then new.

Someone there asked me, "What does it mean in terms of national interest?" I said, "That's a very simple matter." He then asked me to write a memorandum on the subject. That memorandum led to a sharper formulation of the social-choice question, and I realized that I had been thinking of it earlier in that other context.

Society must choose among a number of alternative policies. These policies may be thought of as quite comprehensive, covering a number of aspects: foreign policy, budgetary policy, or whatever. Each individual member of the society has a preference, or a set of preferences, over these alternatives. I guess you can say one alternative is better than another. These individual preferences have a property I call rationality or consistency, or more specifically, what is technically known as transitivity: If I prefer a to b, and b to c, then I prefer a to c.

Imagine that society has to make these choices among a set. Each individual has a preference ordering, a ranking of these alternatives. But we really want society, in some sense, to give a ranking of these alternatives. You can always produce a ranking, but you would like it to have some properties. One is that, of course, it be responsive in some sense to the individual rankings. Another is that when you finish, you end up with a real ranking, that is, something that satisfies these consistency, or transitivity, properties. And a third condition is that when choosing between a number of alternatives, all I should take into account are the preferences of the individuals among those alternatives. If certain things are possible and some are impossible, I shouldn't ask individuals whether they care about the impossible alternatives, only the possible ones.

It turns out that if you impose the conditions I just stated, there is no method of putting together the individual preferences that satisfies all of them.

The whole idea of the axiomatic method was very much in the air among anybody who studied mathematics, particularly among those who studied the foundations of mathematics. The idea is that if you want to find out something, to find the properties, you say, "What would I like it to be?" [You do this] instead of trying to investigate special cases. I was really accustomed to this approach. Of course, the actual process did involve trial and error.

But I went in with the idea that there was some method of handling this problem. I started out with some examples. I had already discovered that these led to some problems. The next thing that was reasonable was to write down a condition that I could outlaw. I constructed another example, another method that seemed to meet that problem, and something else didn't seem very right about it. Then I had to postulate that we have some other property. I found I was having difficulty satisfying all of these properties that I thought were desirable, and it occurred to me that they couldn't be satisfied.

After having formulated three or four conditions of this kind, I kept on experimenting. Lo and behold, no matter what I did, there was nothing that would satisfy these axioms. So after a few days of this, I began to get the idea that maybe there was another kind of theorem here, namely, that there was no voting method that would satisfy all the conditions that I regarded as rational and reasonable. It was at this point that I set out to prove it. It turned out to be a matter of only a few days' work.

It should be made clear that my impossibility theorem is really a theorem [showing that] the contradictions are possible, not that they are necessary. What I claim is that given any voting procedure, there will be some possible set of preference orders for individuals that will lead to a contradiction of one of these axioms.

But you say, "Well, okay, since we can't get perfection, let's at least try to find a method that works well most of the time." Then when you do have a problem, you don't notice it as much. So my theorem is not a completely destructive or negative feature any more than the second law of thermodynamics means that people don't work on improving the efficiency of engines. We're told you'll never get 100% efficient engines. That's a fact—and a law. It doesn't mean you wouldn't like to go from 40% to 50%.

## 9.4 A Better Approach? Approval Voting

Elections in which there are only two candidates present no problem. Majority rule is, as we have seen, an eminently successful voting system in both theory and practice. If there are three or more candidates, however, the situation changes quite dramatically. While several voting systems suggest themselves (plurality, the Borda count, sequential pairwise voting, and the Hare system), each fails to satisfy one or more desired properties (the Condorcet winner criterion, independence of irrelevant alternatives, the Pareto condition, and monotonicity). Manipulability is an ever-present problem, as we'll see in the next chapter. Moreover, when all is said and done, Arrow's impossibility theorem says that any search for an ideal voting system of the kind we have discussed is doomed to failure.

Where does this leave us? More than intellectual issues are at stake here: More than 550,000 elected officials serve in approximately 80,000 governments in the United States. Whether it is a small academic department voting on the best senior thesis or a democratic country electing a new leader, multicandidate elections will be contested in one way or another. If there is no perfect voting system–and perhaps not even a best voting system (whatever that may mean; that is, best in what way?)–what can we do?

Perhaps the answer is that different situations lend themselves to different voting systems, and what is required is a judicious blend of common sense with an awareness of what the mathematical theory has to say. For example, while both the Hare system and the Borda count are subject to manipulability, it seems easier to manipulate the latter. Thus, people may tend to vote more sincerely, rather than strategically, if the Hare system is used instead of the Borda count. This may be a consideration when choosing a voting system for a faculty governance system, for example.

For national political elections, there are also practical considerations. The kind of ballot we are considering (a preference list ballot) is certainly more complicated than the ballots we now employ, and preference list ballots cannot be used with existing voting machines. There is, however, a voting system that avoids the practical difficulties caused by the type of ballot being used that has much else to commend it. It is called **approval voting**.

### Description of Approval Voting PROCEDURE

Under approval voting, each voter is allowed to give one vote to as many of the candidates as he or she finds acceptable. No limit is set on the number of candidates for whom an individual can vote. Voters show disapproval of other candidates simply by not voting for them. The winner under approval voting is the candidate who receives the largest number of approval votes. This approach is also appropriate in situations where more than one candidate can win, for example, in electing new members to an exclusive society such as the National Academy of Sciences or the Baseball Hall of Fame.

Approval voting was proposed independently by several analysts in the 1970s. Probably the best-known official elected by approval voting today is the secretary general of the United Nations. In the 1980s, several academic and professional societies initiated the use of approval voting. Examples include the Institute of Electrical and Electronics Engineers (IEEE), with about 400,000 members, and the National Academy of Sciences. In Eastern Europe and some former Soviet republics, approval voting has been used in the form wherein one disapproves of (instead of approving of) as many candidates as one wishes.

Is approval voting the perfect voting system? Certainly not. For example, the type of ballot used limits the extent to which voter preferences can be expressed. However, it is certainly a voting system with much potential, and the reader wishing to explore it in more detail can start with Brams and Fishburn's 1983 monograph, listed in the Suggested Readings.

## REVIEW VOCABULARY

**Agenda** An ordering of the candidates to be considered. Often used in sequential pairwise voting. (p. 296)

**Approval voting** A method of electing one or more candidates from a field of several in which each voter submits a ballot that indicates which candidates he or she approves of. Winning is determined by the total number of approvals a candidate obtains. (p. 305)

**Arrow's impossibility theorem** Kenneth J. Arrow's discovery that any voting system can give undesirable outcomes. (p. 301)

**Borda count** A voting system for elections with several candidates in which points are assigned to voters' preferences and these points are summed for each candidate to determine a winner. The actual point totals are referred to as a candidate's **Borda score.** (p. 293)

**Condorcet's method** A voting system for elections with several candidates in which a candidate is a winner precisely when he or she would, on the basis of the ballots cast, defeat every other candidate in a one-on-one contest. (p. 287)

**Condorcet winner** A Condorcet winner in an election is a candidate who, based on the ballots, would have defeated every other candidate in a one-on-one contest. (p. 291)

**Condorcet winner criterion (CWC)** A voting system satisfies the Condorcet winner criterion if, for every election in which there is a Condorcet winner, it wins the election when that voting system is used. (p. 292)

**Condorcet's voting paradox** The observation that there are elections in which Condorcet's method yields no winner. (p. 288)

**Hare system** A voting system for elections with several candidates in which candidates are successively eliminated in an order based on the number of first-place votes. (p. 297)

**Independence of irrelevant alternatives (IIA)** A voting system satisfies independence of irrelevant alternatives if the only way a candidate (call him $A$) can go from losing one election to being among the winners of a new election (with the same set of candidates and voters) is for at least one voter to reverse his or her ranking of $A$ and the previous winner. (p. 295)

**Manipulability** A voting system is subject to manipulability (or is manipulable) if there are elections in which it is to a voter's advantage to submit a ballot that misrepresents his or her true preferences. (p. 292)

**Majority rule** A voting system for elections with two candidates (and an odd number of voters) in which the candidate preferred by more than half the voters is the winner. (p. 286)

**May's theorem** Kenneth May's discovery that, for two alternatives and an odd number of voters, majority rule is the only voting system satisfying three natural properties. (p. 287)

**Monotonicity** A voting system satisfies monotonicity provided that ballot changes favorable to one candidate (and not favorable to any other candidate) can never hurt that candidate. (p. 287)

**Pareto condition** A voting system satisfies the Pareto condition provided that every voter's ranking of one candidate higher than another precludes the possibility of this latter candidate winning. (p. 297)

**Plurality runoff** A voting system for elections with several candidates in which, assuming there are no ties, there is a runoff between the two candidates receiving the most first-place votes. (p. 299)

**Plurality voting** A voting system for elections with several candidates in which the candidate with the most first-place votes wins. (p. 289)

**Preference list ballot** A ballot that ranks the candidates from most preferred to least preferred, with no ties. (p. 286)

**Sequential pairwise voting** A voting system for elections with several candidates in which one starts with an agenda and pits the candidates against each other in one-on-one contests (based on preference list ballots), with losers being eliminated as one moves along the agenda. (p. 296)

## SKILLS CHECK

1. A preference list ballot

(a) indicates only a voter's top choice.
(b) is a rank ordering of the candidates, with no ties.
(c) will often have ties.

2. To say that a voting system treats all voters equally means that ______________________.

3. To say that a voting system for two candidates treats both candidates equally means that

(a) each wins if he or she receives all the votes.
(b) if all voters reverse their ballots, the election outcome changes.
(c) if any two voters exchange ballots, the election outcome is unchanged.

4. A two-candidate voting system is monotone if ______________________.

5. May's theorem says that, with an odd number of voters, among all two-candidate voting systems that never result in a tie, majority rule is the only one that

(a) treats both candidates equally.
(b) treats both candidates equally and all voters equally.
(c) treats both candidates equally and all voters equally and is monotone.

6. The winner with Condorcet's method is the candidate who ______________________.

7. Which of the following does not satisfy exactly two of the conditions in May's theorem?

(a) A dictatorship
(b) Imposed rule
(c) Minority rule
(d) None of the above

8. The flaw in Condorcet's method is that it ______________________.

9. Condorcet's voting paradox refers to the fact that

(a) people vote even though an individual vote virtually never affects the outcome of an election.
(b) the statement "This statement is false" can be neither true nor false.
(c) there are elections in which there is no winner using Condorcet's method.

10. With plurality voting, the winner is the candidate who ______________________.

11. George W. Bush's defeat of Al Gore in the state of Florida in the 2000 presidential election shows that

(a) plurality voting does not satisfy the Condorcet winner criterion.
(b) majority rule is not monotone.
(c) the Borda count does not satisfy independence of irrelevant alternatives.

12. With the Borda count, the election winner is the candidate who ______________________.

13. Instead of assigning points and doing arithmetic, the Borda score of a candidate can be found by

(a) scanning the ballots and counting the number of occurrences of other candidates below that one.
(b) counting the number of first-place votes and multiplying by 4.
(c) counting the number of candidates that it defeats one on one.

14. Independence of irrelevant alternatives says that a nonwinner can never switch to being a winner unless at least one voter changes his or her ballot in a way that ______________________.

15. The Borda count fails to satisfy

(a) monotonicity.
(b) the Pareto condition.
(c) independence of irrelevant alternatives.

16. Sequential pairwise voting is the voting system in which ______________________.

17. Sequential pairwise voting fails to satisfy

(a) monotonicity.
(b) the Pareto condition.
(c) the Condorcet winner criterion.

18. Both the Hare system and the plurality runoff method are defective in that ______________________.

19. Arrow's theorem says that with three or more candidates and any number of voters, there is no voting system that

(a) is not a dictatorship.
(b) satisfies independence of irrelevant alternatives and is not a dictatorship.
(c) satisfies the Pareto condition and independence of irrelevant alternatives, and is not a dictatorship.
(d) always produces a winner, satisfies the Pareto condition and independence of irrelevant alternatives, and is not a dictatorship.

20. The weak version of Arrow's theorem asserts that, with three or more candidates and an odd number of voters, there is no voting system that ______________________.

# CHAPTER 9 EXERCISES

■ Challenge ◆ Discussion

## 9.1 Majority Rule and Condorcet's Method

**1.** In a few sentences, explain why minority rule (the voting procedure for two alternatives that is described on page 287) satisfies conditions (1) and (2) on page 287, but not (3).

**2.** In a few sentences, explain why imposed rule (the voting procedure for two alternatives that is described on page 287) satisfies conditions (1) and (3) on page 287, but not (2).

**3.** In a few sentences, explain why a dictatorship (the voting procedure for two alternatives that is described on page 287) satisfies conditions (2) and (3) on page 287, but not (1).

**4.** Find (or invent) a voting rule for two alternatives that satisfies

**(a)** condition (1) on page 287, but neither (2) nor (3).
**(b)** condition (2) on page 287, but neither (1) nor (3).
**(c)** condition (3) on page 287, but neither (1) nor (2).

**5.** In a sentence or two, explain why it's impossible, with an odd number of voters, to have two distinct candidates win the same election using Condorcet's method.

**6.** Construct a real-world example (perhaps involving yourself and two friends) where the individual preference lists for three alternatives are as in the voting paradox of Condorcet.

**7.** Condorcet's voting paradox shows that with three voters (or three equal-size groups of voters) and the three alternatives *A*, *B*, and *C*, it is possible to have two-thirds prefer *A* to *B*, two-thirds prefer *B* to *C*, and two-thirds prefer *C* to *A*. Find four preference lists that show that with four voters and the four alternatives *A*, *B*, *C*, and *D*, it is possible to have three-fourths prefer *A* to *B*, three-fourths prefer *B* to *C*, three-fourths prefer *C* to *D*, and three-fourths prefer *D* to *A*.

**8.** Generalize the result in Exercise 7 from four alternatives to *n* alternatives: $A_1, \ldots, A_n$.

## 9.2 Other Voting Systems for Three or More Candidates

**9.** Plurality voting is illustrated by the 1980 U.S. Senate race in New York among Alfonse D'Amato (*D*, a conservative), Elizabeth Holtzman (*H*, a liberal), and Jacob Javits (*J*, also a liberal). Reasonable estimates (based largely on exit polls) suggest that voters ranked the candidates according to the following table:

| 22% | 23% | 15% | 29% | 7% | 4% |
|---|---|---|---|---|---|
| *D* | *D* | *H* | *H* | *J* | *J* |
| *H* | *J* | *D* | *J* | *H* | *D* |
| *J* | *H* | *J* | *D* | *D* | *H* |

**(a)** Is there a Condorcet winner?
**(b)** Who won using plurality voting?

**10.** (Everyone wins.) Consider the following set of preference lists:

| | Number of Voters (9) | | | | | | |
|---|---|---|---|---|---|---|---|
| **Rank** | **3** | **1** | **1** | **1** | **1** | **1** | **1** |
| First | *A* | *A* | *B* | *B* | *C* | *C* | *D* |
| Second | *D* | *B* | *C* | *C* | *B* | *D* | *C* |
| Third | *B* | *C* | *D* | *A* | *D* | *B* | *B* |
| Fourth | *C* | *D* | *A* | *D* | *A* | *A* | *A* |

Note that the first list is held by three voters, not just one. Calculate the winner using

**(a)** plurality voting.
**(b)** the Borda count.
**(c)** the Hare system.
**(d)** sequential pairwise voting with the agenda *A*, *B*, *C*, *D*.

**11.** Consider the following set of preference lists:

| | Number of Voters (7) | | | | |
|---|---|---|---|---|---|
| **Rank** | **2** | **2** | **1** | **1** | **1** |
| First | *C* | *D* | *C* | *B* | *A* |
| Second | *A* | *A* | *D* | *D* | *D* |
| Third | *B* | *C* | *A* | *A* | *B* |
| Fourth | *D* | *B* | *B* | *C* | *C* |

Calculate the winner using

**(a)** plurality voting
**(b)** the Borda count.
**(c)** the Hare system.
**(d)** sequential pairwise voting with the agenda *B*, *D*, *C*, *A*.

**12.** Consider the following set of preference lists:

| | Number of Voters (8) | | | | | |
|---|---|---|---|---|---|---|
| **Rank** | **2** | **2** | **1** | **1** | **1** | **1** |
| First | *A* | *E* | *A* | *B* | *C* | *D* |
| Second | *B* | *B* | *D* | *E* | *E* | *E* |
| Third | *C* | *D* | *C* | *C* | *D* | *A* |
| Fourth | *D* | *C* | *B* | *D* | *A* | *B* |
| Fifth | *E* | *A* | *E* | *A* | *B* | *C* |

Calculate the winner using

**(a)** plurality voting.

**(b)** the Borda count.

**(c)** the Hare system.

**(d)** sequential pairwise voting with the agenda *B, D, C, A, E.*

**13.** Consider the following set of preference lists:

| | Number of Voters (5) | | | | |
|---|---|---|---|---|---|
| **Rank** | **1** | **1** | **1** | **1** | **1** |
| First | *A* | *B* | *C* | *D* | *E* |
| Second | *B* | *C* | *B* | *C* | *D* |
| Third | *E* | *A* | *E* | *A* | *C* |
| Fourth | *D* | *D* | *D* | *E* | *A* |
| Fifth | *C* | *E* | *A* | *B* | *B* |

Calculate the winner using

**(a)** plurality voting.

**(b)** the Borda count.

**(c)** the Hare system.

**(d)** sequential pairwise voting with the agenda *A, B, C, D, E.*

**14.** Consider the following set of preference lists:

| | Number of Voters (7) | | | | |
|---|---|---|---|---|---|
| **Rank** | **2** | **2** | **1** | **1** | **1** |
| First | *A* | *B* | *A* | *C* | *D* |
| Second | *D* | *D* | *B* | *B* | *B* |
| Third | *C* | *A* | *D* | *D* | *A* |
| Fourth | *B* | *C* | *C* | *A* | *C* |

Calculate the winner using

**(a)** plurality voting.

**(b)** the Borda count.

**(c)** the Hare system.

**(d)** sequential pairwise voting with the agenda *B, D, C, A.*

**15.** Consider the following set of preference lists:

| | Number of Voters (7) | | | | |
|---|---|---|---|---|---|
| **Rank** | **2** | **2** | **1** | **1** | **1** |
| First | *C* | *E* | *C* | *D* | *A* |
| Second | *E* | *B* | *A* | *E* | *E* |
| Third | *D* | *D* | *D* | *A* | *C* |
| Fourth | *A* | *C* | *E* | *C* | *D* |
| Fifth | *B* | *A* | *B* | *B* | *B* |

Calculate the winner using

**(a)** plurality voting.

**(b)** the Borda count.

**(c)** the Hare system.

**(d)** sequential pairwise voting with the agenda *A, B, C, D, E.*

**16.** Consider the following set of preference lists:

| | Number of Voters (7) | | | | | | |
|---|---|---|---|---|---|---|---|
| **Rank** | **1** | **1** | **1** | **1** | **1** | **1** | **1** |
| First | *C* | *D* | *C* | *B* | *E* | *D* | *C* |
| Second | *A* | *A* | *E* | *D* | *D* | *E* | *A* |
| Third | *E* | *E* | *D* | *A* | *A* | *A* | *E* |
| Fourth | *B* | *C* | *A* | *E* | *C* | *B* | *B* |
| Fifth | *D* | *B* | *B* | *C* | *B* | *C* | *D* |

Calculate the winner using

**(a)** plurality voting.

**(b)** the Borda count.

**(c)** sequential pairwise voting with the agenda *A, B, C, D, E.*

**(d)** the Hare system.

**17.** An interesting variant of the Hare system was proposed by the psychologist Clyde Coombs. It operates exactly as does the Hare system, but instead of deleting alternatives with the fewest first-place votes, it deletes those with the most last-place votes.

**(a)** Use the Coombs procedure to find the winner if the ballots are as in Exercise 16.

**(b)** Show that for two voters and three alternatives, it is possible to have ballots that result in one candidate winning if the Coombs procedure is used and a tie between the other two if the Hare system is used.

◆ **18.** In a few sentences, explain why Condorcet's rule satisfies

**(a)** the Pareto condition.

**(b)** monotonicity.

◆ **19.** In a few sentences, explain why plurality voting satisfies

**(a)** the Pareto condition.
**(b)** monotonicity.

◆ **20.** In a few sentences, explain why the Borda count satisfies

**(a)** the Pareto condition.
**(b)** monotonicity.

◆ **21.** In a few sentences, explain why sequential pairwise voting satisfies

**(a)** the Condorcet winner criterion.
**(b)** monotonicity.

◆ **22.** In a few sentences, explain why the Hare system satisfies the Pareto condition.

◆ **23.** In a few sentences, explain why the plurality runoff method satisfies the Pareto condition.

■ **24.** Use the following ballots to show that the plurality runoff method does not satisfy the Condorcet winner criterion:

| | **Number of Voters (5)** | | |
|---|---|---|---|
| **Rank** | **2** | **2** | **1** |
| First | *A* | *B* | *C* |
| Second | *C* | *C* | *B* |
| Third | *B* | *A* | *A* |

■ **25.** Use the following ballots to show that the plurality runoff method does not satisfy monotonicity:

| | **Number of Voters (13)** | | | | |
|---|---|---|---|---|---|
| **Rank** | **4** | **3** | **3** | **2** | **1** |
| First | *A* | *B* | *C* | *D* | *E* |
| Second | *B* | *A* | *A* | *B* | *D* |
| Third | *C* | *C* | *B* | *C* | *C* |
| Fourth | *D* | *D* | *D* | *A* | *B* |
| Fifth | *E* | *E* | *E* | *E* | *A* |

**26.** Consider the following two elections among candidates *A, B,* and *C*:

| | **Number of Voters (4)** | | | |
|---|---|---|---|---|
| **Rank** | **1** | **1** | **1** | **1** |
| First | *A* | *A* | *B* | *C* |
| Second | *B* | *B* | *C* | *B* |
| Third | *C* | *C* | *A* | *A* |

| | **Number of Voters (4)** | | | |
|---|---|---|---|---|
| **Rank** | **1** | **1** | **1** | **1** |
| First | *A* | *A* | *B* | *B* |
| Second | *B* | *B* | *C* | *C* |
| Third | *C* | *C* | *A* | *A* |

**(a)** Use these two elections to show that plurality voting does not satisfy independence of irrelevant alternatives.
**(b)** Use these two elections to show that the Hare system does not satisfy independence of irrelevant alternative.

■ **27.** Construct ballots for the alternatives *A, B,* and *C* to show that the Borda count does not satisfy the Condorcet winner criterion.

**28.** Show that the nonmonotonicity of the Hare system can also be demonstrated by the following 17-voter, 4-alternative election. (In a number of recent books, this example is used to show the nonmonotonicity of the Hare system. The 13-voter, 3-alternative example given in the text was pointed out to us by Matt Gendron, an undergraduate at Union College.)

| | **Number of Voters (17)** | | | |
|---|---|---|---|---|
| **Rank** | **7** | **5** | **4** | **1** |
| First | *A* | *C* | *B* | *D* |
| Second | *D* | *A* | *C* | *B* |
| Third | *B* | *B* | *D* | *A* |
| Fourth | *C* | *D* | *A* | *C* |

**29.** The following example illustrates how badly the Hare system can fail to satisfy monotonicity. Consider the following sequence of preference lists:

| | **Number of Voters (21)** | | | |
|---|---|---|---|---|
| **Rank** | **7** | **6** | **5** | **3** |
| First | *A* | *B* | *C* | *D* |
| Second | *B* | *A* | *B* | *C* |
| Third | *C* | *C* | *A* | *B* |
| Fourth | *D* | *D* | *D* | *A* |

**(a)** Show that *A* is the unique winner if the Hare system is used.
**(b)** Find the winner using the Hare system in the new election wherein the three voters on the right all move *A* from last place on their preference lists to first place on their preference lists.

◆ **30.** In a few sentences, explain why, with an odd number of voters,

**(a)** sequential pairwise voting always yields a unique winner.
**(b)** we can never have exactly two winners with the Hare system.

◆ **31.** In a few sentences, explain why the plurality runoff method can never elect a candidate ranked last on a majority of ballots, assuming there are no ties for first or second place in the voting.

**32.** Produce ballots showing that plurality voting can, in fact, elect a candidate ranked last on a majority of the ballots.

**33.** Suppose there are three voters and three alternatives: *A*, *B*, and *C*.
**(a)** If each alternative has exactly one first-place vote, what is the election outcome if the Hare procedure is used? What if plurality runoff is used?
**(b)** If an alternative has two or more first-place votes, what is the election outcome if the Hare procedure is used? What if plurality runoff is used?
**(c)** Can the Hare procedure and plurality runoff yield different election outcomes when there are three voters and three alternatives? Explain your answer in one sentence.

## 9.3 Insurmountable Difficulties: Arrow's Impossibilty Theorem

■ **34.** Complete the proof of the version of Arrow's theorem from the text by showing that neither *B* nor *C* can be a winner in the situation described. (Your argument will be almost word for word the same as the proofs in the text.)

## 9.4 A Better Approach? Approval Voting

**35.** Ten board members vote by approval voting on eight candidates for new positions on their board as indicated in the following table. An X indicates an approval vote. For example, Voter 1, in the first column, approves of candidates *A, D, E, F,* and *G,* and disapproves of *B, C,* and *H.*

| | Voters | | | | | | | | | |
|---|---|---|---|---|---|---|---|---|---|---|
| **Candidate** | **1** | **2** | **3** | **4** | **5** | **6** | **7** | **8** | **9** | **10** |
| *A* | X | X | X | | | X | X | X | | X |
| *B* | | X | X | X | X | X | X | X | X | |
| *C* | | | X | | | | | X | | |
| *D* | X | X | X | X | X | | X | X | X | X |
| *E* | X | | X | | X | | X | | X | |
| *F* | X | | X | X | X | X | X | X | | X |
| *G* | X | X | X | X | X | | | X | | |
| *H* | | X | | X | | X | | X | | X |

**(a)** Which candidate is chosen for the board if just one of them is to be elected?
**(b)** Which candidates are chosen if the top four are selected?
**(c)** Which candidates are elected if 80% approval is necessary and at most four are elected?
**(d)** Which candidates are elected if 60% approval is necessary and at most four are elected?

**36.** The 45 members of a school's football team vote on three nominees, *A, B,* and *C,* by approval voting for the award of "most improved player" as indicated in the following table. An X indicates an approval vote.

| | Number of Voters (45) | | | | | | | |
|---|---|---|---|---|---|---|---|---|
| **Nominee** | **7** | **8** | **9** | **9** | **6** | **3** | **1** | **2** |
| *A* | X | | | X | X | | X | |
| *B* | | X | | X | | X | X | |
| *C* | | | X | | X | X | X | |

**(a)** Which nominee is selected for the award?
**(b)** Which nominee gets announced as runner-up for the award?
**(c)** Note that two of the players "abstained," that is, approved of none of the nominees. Note also that one person approved of all three of the nominees. What would be the difference in the outcome if one were to "abstain" or "approve of everyone"?

## WRITING PROJECTS

**1.** In the 2000 presidential election in Florida, the final results were as follows:

| Candidates | Number of Votes | Percentage of Votes |
|---|---|---|
| Bush | 2,911,872 | 49 |
| Gore | 2,910,942 | 49 |
| Nader | 97,419 | 2 |
| Buchanan | 17,472 | 0 |

Making reasonable assumptions about voters' preference schedules, give a one-page discussion of how the election might have turned out under the different voting methods discussed in this chapter.

**2.** Frequently in presidential campaigns, the winner of the first few primaries is given front-runner status that can lead to the nomination of his or her party. Moreover, there are often several candidates running in early primaries such as New Hampshire. In one page, consider a recent election and discuss how the nominating process might have proceeded through the campaign if approval voting had been used to decide primary winners.

## SUGGESTED READINGS

BLACK, DUNCAN. *The Theory of Committees and Elections,* Kluwer, Dordrecht, The Netherlands, 1986. The historical highlights and development of voting methods in the nineteenth and twentieth centuries are traced in this economist's volume.

BRAMS, STEVEN J., and PETER C. FISHBURN. *Approval Voting,* Birkhäuser, Boston, 1983. This volume is a research-level work on development in the recently popular (but rediscovered) method now called approval voting. The first chapter, however, is an elementary exposition of this voting method and its uses.

NURMI, HANNU. *Comparing Voting Systems,* Reidel, Dordrecht, The Netherlands, 1987. This monograph provides an excellent treatment, at a somewhat more technical level, of the topics dealt with in this chapter.

SAARI, DONALD G. *Chaotic Elections! A Mathematician Looks at Voting,* American Mathematical Society, Providence, R.I., 2001. This expository book begins with the 2000 presidential election and discusses a number of paradoxical results in voting.

TAYLOR, ALAN D. *Mathematics and Politics: Strategy, Voting, Power, and Proof,* Springer-Verlag, New York, 1995. Chapters 5 and 10 give an expanded treatment of the topics considered here, with proofs included. This book is also intended for nonmajors.

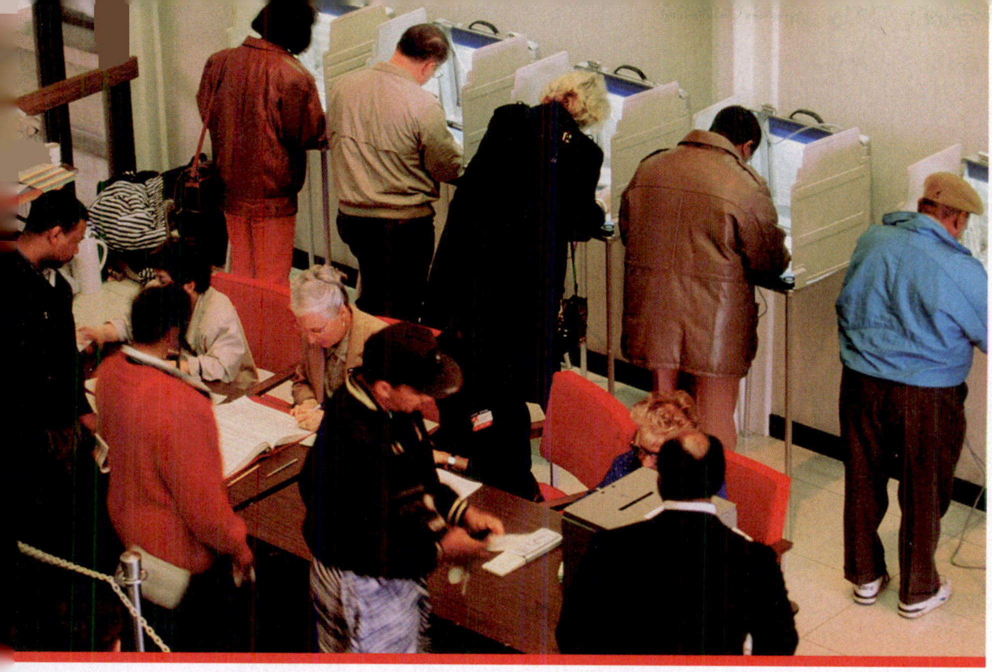

CHAPTER
10



# The Manipulability of Voting Systems

Charles L. Dodgson was a mathematical lecturer at Oxford University. Dodgson, who used the pen name Lewis Carroll, wrote on mathematical topics and even manipulability. But he achieved greater fame for his satirical works. In the *Alice* books, he refers to the mathematical operations as Ambition, Distraction, Uglification, and Derision, and his characters play nonsensical, easily manipulated games. (*Bettmann/Corbis.*)

People know almost by instinct that, sometimes, you can achieve the election result you prefer by submitting a ballot that misrepresents your actual preferences. This type of strategic voting is called **manipulation**, and a ballot that misrepresents a voter's true preferences is referred to as an **insincere** or **disingenuous ballot**.

All three of these terms–manipulation, insincere, disingenuous–are widely used in the social-choice literature, but in daily life we use these terms pejoratively–they aren't exactly warm praise. In point of fact, your choice to manipulate a voting system is typically no more inherently evil than your submission of a sealed bid for a lamp at an auction at a price considerably below its actual worth. "Strategyproof"–a term with considerably less negative content–is sometimes used in place of "nonmanipulable," but "nonmanipulable" is more common so we'll stick with it here.

Historical references to the manipulability of voting systems include a comment by the nineteenth-century mathematician C. L. Dodgson (1832–1898), better known by the pseudonym Lewis Carroll, under which he wrote *Alice's Adventures in Wonderland* (1865). Dogson commented that voters have a tendency to "adopt a principle of voting which makes it more of a game of skill than a true test of the wishes of the electors," and that it would be "better for elections to be decided according to the wishes of the majority than of those who have the most skill at the game."

But the most famous manipulability quote in the history of social choice is Jean Charles de Borda's reply to a colleague who had pointed out to him how easily the Borda count can be manipulated. "My scheme," Borda replied, "is only intended for honest men!"

Let's look at an example to illustrate how the Borda count can be manipulated.

## EXAMPLE 1 ■ Manipulating the Borda Count with Four Candidates and Two Voters

Suppose there are two voters and four candidates, and suppose the true preferences of the voters are reflected in the following ballots:

| Voter 1 | Voter 2 |
|---|---|
| *A* | *B* |
| *B* | *C* |
| *C* | *A* |
| *D* | *D* |

Using the Borda count with point values 3, 2, 1, 0 (or by counting the number of occurrences of other candidates below the one in question, as described in Section 9.2), we see that the Borda scores of the four candidates are as follows:

Borda score of *A* is 4
Borda score of *B* is 5
Borda score of *C* is 3
Borda score of *D* is 0

Thus, Candidate *B* wins this election. Voter 1, however, would have preferred to see Candidate *A*–his top choice, according to his true preferences–win this election rather than Candidate *B*, his second choice.

Assume that Voter 1 had known that Voter 2 planned to submit the ballot that he cast above. Could Voter 1 have secured a victory for Candidate *A* by submitting a disingenuous ballot?

The answer here, as we'll show, turns out to be yes. The intuition is fairly transparent: Voter 1 wants to pretend that *B* is not her second choice, but her last choice. Let's see if this is enough to bring about the desired switch in winner from *B* to *A*. The new ballots and Borda scores are as follows:

| Voter 1 | Voter 2 |
|---|---|
| *A* | *B* |
| *C* | *C* |
| *D* | *A* |
| *B* | *D* |

Borda score of *A* is 4
Borda score of *B* is 3
Borda score of *C* is 4
Borda score of *D* is 1

Close, but not quite what we wanted: Candidates *A* and *C* now tie for the win, and we wanted the winner to be just Candidate *A*. But a moment's inspection reveals that Voter 1 can achieve this if, in addition to plunging Candidate *B* to the bottom of her ballot, she also flip-flops *C* and *D*. That is, the desired ballots (and Borda scores) that yield Candidate *A* as the sole winner are as follows:

| Voter 1 | Voter 2 |
|---|---|
| *A* | *B* |
| *D* | *C* |
| *C* | *A* |
| *B* | *D* |

Borda score of *A* is 4

Borda score of *B* is 3

Borda score of *C* is 3

Borda score of *D* is 2

In presenting an example of a voting system's susceptibility to manipulation, we will typically present two elections–the original one ("Election 1") in which we assume all ballots are sincere, and the one that contains a disingenuous ballot from a voter ("Election 2"). For example, if we collect the pieces of what we just did, this instance of manipulation of the Borda count could be succinctly presented as follows.

**Election 1**

| Rank | Number of Voters (2) | |
|---|---|---|
| First | *A* | *B* |
| Second | *B* | *C* |
| Third | *C* | *A* |
| Fourth | *D* | *D* |

**Election 2**

| Rank | Number of Voters (2) | |
|---|---|---|
| First | *A* | *B* |
| Second | *D* | *C* |
| Third | *C* | *A* |
| Fourth | *B* | *D* |

There are two aspects of manipulation taking place in this example that deserve comment.

First, there is only one voter (the voter on the left, in this example) changing his or her ballot–we call this a **unilateral change** in ballot. An example involving a unilateral change of ballot is sometimes referred to as an instance of "single-voter manipulation" to distinguish it from a situation wherein a group of voters, acting in concert, can change their ballots so that all of them prefer the new winner to the original winner. We'll see examples of group manipulation in Section 10.2.

Second, the original election produced a single winner, as did the new election held after we finished constructing Voter 1's disingenuous ballot. Thus, because we know each voter's sincere preference ranking for the candidates, we also know exactly which of the two election outcomes each voter will prefer. Ties, on the other hand, present a problem. For example, if a voter has sincere preferences that rank *A* over *B* over *C* over *D*, then it's not at all obvious whether this voter will prefer an election outcome that ties *A* and *D* to an election outcome that ties *B* and *C* or vice-versa.

A voting system is **manipulable** if there is at least one scenario in which some voter can achieve a more preferred election outcome by unilaterally changing his or her ballot. The precise definition follows.

**Manipulability** DEFINITION

A voting system is said to be **manipulable** if there exist two sequences of preference list ballots and a voter (call the voter Jane) such that

1. Neither election results in a tie.
2. The only ballot change is by Jane.
3. Jane prefers–assuming that her ballot in the first election represents her true preferences–the outcome of the second election to that of the first election.

In this chapter, as in Chapter 9, we begin with majority rule in the two-candidate case and Condorcet's method in the case of three or more candidates. Condorcet's method again shines, leaving the voting paradox as its only blemish. We then move on to revisit the other voting systems introduced in Chapter 9 that apply to elections with three or more candidates, and we show that each succumbs to some form of manipulation. In fact, there is a striking impossibility result that arises here known as the **Gibbard–Satterthwaite manipulability theorem**. It is related to–indeed, some would say equivalent to–Arrow's impossibility theorem. We conclude with a treatment of a striking first cousin of manipulability known as the **chair's paradox**.

## 10.1 Majority Rule and Condorcet's Method

Throughout this section, we assume that the number of voters is odd. In Section 9.1 we pointed out that with two candidates, majority rule has three very desirable properties: It treats all voters equally, it treats both candidates equally, and it is *monotone*, meaning that a single voter's change in ballot from a vote for the loser to a vote for the winner has no effect on the election outcome. More strikingly, May's theorem told us that among all voting systems in the two-candidate case that never result in a tie, majority rule is the *only* one satisfying these three properties.

But let's consider for a moment what monotonicity is saying in this two-candidate case for voting systems that never yield ties. It says that if you rank $A$ over $B$ on your ballot, and the election winner is $B$, then the election winner will remain $B$ if you switch to a ballot with $B$ over $A$. But there are only two possible choices for a ballot in this two-candidate case: $B$ over $A$ and $A$ over $B$. Monotonicity is thus saying that if you rank $A$ over $B$, then no unilateral change in your ballot can make the outcome $A$. This is simply the assertion that you can't manipulate the voting system!

Thus, in the two-candidate case, nonmanipulability and monotonicity are exactly the same thing. This allows us to restate **May's theorem** from Section 9.1 with "monotonicity" replaced by "nonmanipulability."

### May's Theorem for Manipulability — THEOREM

Among all two-candidate voting systems that never result in a tie, majority rule is the only one that treats all voters equally, treats both candidates equally, and is nonmanipulable.

There are examples of two-candidate voting systems that are manipulable, even though they treat all voters equally and both candidates equally. For example, the voting system that declares the winner to be the alternative with the fewest first-place votes is manipulable, as is the one that declares the winner to be whichever alternative has an odd number of first-place votes, (even if that's fewer than half). Exercises 1 and 2 ask you to provide an example of voter manipulation for each of these systems.

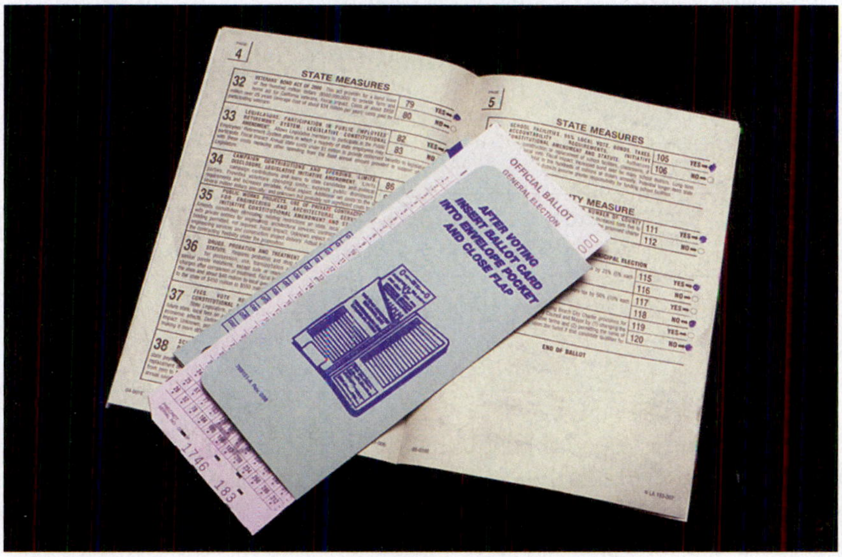


Paper ballots are still used in elections in many states. A lingering controversy from the 2004 elections is over the use of electronic ballots, which do not leave physical evidence and thus may not be possible to recount in disputes about the plurality or majority. (*Jonathan Nourok/PhotoEdit.*)

Turning to the case of three or more candidates, we begin with Condorcet's method, as we did in Chapter 9. Condorcet's method is based on majority rule, and, as we've just seen, majority rule is nonmanipulable. So the following result, as pleasing as it is, comes as no surprise.

### The Nonmanipulability of Condorcet's Method — THEOREM

Condorcet's method is nonmanipulable in the sense that a voter can never unilaterally change an election result from one candidate to another candidate that he or she prefers.

Let's see why Condorcet's method is nonmanipulable, regardless of the number of voters. Suppose that we have an election in which you, as one of the voters, prefer Candidate $A$ to Candidate $B$, but $B$ wins using Condorcet's method. We'll show that any attempt that you might make to manipulate the election so that $A$ becomes the winner is doomed to failure, even if there are more than these two candidates in the election.

Because Candidate $B$ is the winner using Condorcet's method, we know that $B$ defeats every other candidate in a one-on-one contest based on the ballots cast. In particular, $B$ defeats $A$ in a one-on-one contest, even with your original ballot that has $A$ over $B$. This means that more than half of the other voters ranked $B$ over $A$, so, regardless of how you change your ballot, $B$ will *still* defeat $A$ in a one-on-one contest. While this need not ensure that $B$ remains a winner with Condorcet's method, it certainly guarantees that $A$ isn't. Hence, you cannot unilaterally cause $A$ to be a winner using Condorcet's method, and so your attempt at manipulation will have failed.

### EXAMPLE 2 ■ Exploiting the Condorcet Voting Paradox

We had to be careful in stating the above theorem because, as we've seen, there are elections in which there is no winner using Condorcet's method. With three voters and three candidates, it is possible for a voter (the one on the left in this example) to unilaterally change an election from one that yields his or her second choice as the sole winner (Candidate $C$ in the example), to one in which there is no winner at all, as this example shows:

**Election 1**

| Rank | Number of Voters (3) | | |
|---|---|---|---|
| First | $A$ | $B$ | $C$ |
| Second | $C$ | $C$ | $A$ |
| Third | $B$ | $A$ | $B$ |

**Election 2**

| Rank | Number of Voters (3) | | |
|---|---|---|---|
| First | $A$ | $B$ | $C$ |
| Second | $B$ | $C$ | $A$ |
| Third | $C$ | $A$ | $B$ |

A voter's ability to unilaterally bring about this kind of change in an election, however, is not something that falls within the scope of our formal definition of manipulation. Nevertheless, one could argue that there are situations in which you might well prefer having an election with no outcome at all to having an election in which a candidate other than your top choice emerges as the sole winner. ■

We now move on to voting systems with three or more candidates that, unlike Condorcet's method, always produce at least one winner. As you might expect from the results in Chapter 9, these voting systems are not, in terms of nonmanipulability, as perfect as one might hope for.

## 10.2 The Manipulability of Other Voting Systems for Three or More Candidates

### Manipulability and the Borda Count

Example 1 showed how a single voter can manipulate an election in which the Borda count is being used. But Example 1 involved four candidates. Is there a simpler example involving only three candidates?

The answer turns out to be no, provided that we continue to interpret the notion of a "more preferred election outcome" to be a switch from a single winner to another single winner (as opposed to a switch creating or breaking a tie). This negative answer is formalized in the following theorem.

## The Nonmanipulability of the Borda Count with Exactly Three Candidates

THEOREM

With exactly three candidates, the Borda count cannot be manipulated in the sense of a voter unilaterally changing an election outcome from one single winner to another single winner that he or she prefers according to that voter's ballot in the first election, which we take to be sincere preferences.

Let's see why this is true. Suppose the candidates are *A*, *B*, and *C*, and that you prefer *A* to *B*, but *B* is the election winner using the Borda count. We'll show that any attempt you make to manipulate the election by changing your ballot so that *A* emerges as the winner (using the Borda count) is doomed to failure.

Because you prefer *A* to *B*, your sincere ballot can be one of only three possibilities, corresponding to whether *C* is ranked first, second, or third. We'll consider each case in turn.

**Case 1. Your sincere ballot is *A* over *B* over *C*.** No ballot change on your part can increase *A*'s Borda score, and you can only decrease *B*'s Borda score by at most 1. Thus, at best, you can make a unilateral change that results in *A* and *B* having the same Borda score, whereas successful manipulation on your part requires that *A* have a strictly higher Borda score than *B* after your ballot change.

**Case 2. Your sincere ballot is *C* over *A* over *B*.** No ballot change on your part can decrease *B*'s Borda score, and you can only increase *A*'s Borda score by at most 1. Thus, at best, you can make a unilateral change that results in *A* and *B* having the same Borda score, whereas successful manipulation on your part requires that *A* have a strictly higher Borda score than *B* after your ballot change.

**Case 3. Your sincere ballot is *A* over *C* over *B*.** No ballot change on your part can increase *A*'s Borda score or decrease *B*'s Borda score. Thus, after your ballot change *B* will still have a higher Borda score than *A*, so your attempt at manipulation has failed in this case also.

So with three candidates, the Borda count is nonmanipulable. With more than three candidates, the Borda count does not fare as well, regardless of how many voters there are.

## The Manipulability of the Borda Count with Four or More Candidates

THEOREM

With four or more candidates (and two or more voters), the Borda count can be manipulated in the sense that there exists an election in which a voter can unilaterally change the election outcome from one single winner to another single winner that he or she prefers according to that voter's ballot in the first election, which we take to be sincere preferences.

We've already established part of this theorem, as we've seen an example of manipulation of the Borda count in the case of four candidates and two voters. This is really half the battle, as we can readily modify that example to serve in any case in which the number of voters is even as follows.

1. Any candidates in addition to *A*, *B*, *C*, and *D* can be placed below those four on every ballot.
2. The rest of the voters can be paired off with the members of each pair holding ballots that rank the candidates in exactly opposite orders (thus "canceling each other out" in terms of the Borda scores).

The following example illustrates this method of generalizing our earlier instance of manipulation of the Borda count to the case of five candidates and six voters.

## EXAMPLE 3 ■

### Manipulating the Borda Count with Five Candidates and Six Voters

Consider the following two elections:

| Election 1 | | | | | | Election 2 | | | | | |
|---|---|---|---|---|---|---|---|---|---|---|---|
| *A* | *B* | *A* | *E* | *A* | *E* | *A* | *B* | *A* | *E* | *A* | *E* |
| *B* | *C* | *B* | *D* | *B* | *D* | *D* | *C* | *B* | *D* | *B* | *D* |
| *C* | *A* | *C* | *C* | *C* | *C* | *C* | *A* | *C* | *C* | *C* | *C* |
| *D* | *D* | *D* | *B* | *D* | *B* | *B* | *D* | *D* | *B* | *D* | *B* |
| *E* | *E* | *E* | *A* | *E* | *A* | *E* | *E* | *E* | *A* | *E* | *A* |

The ballots of the first two voters (in both elections) are the same as in Example 1 (the manipulation of the Borda count with four candidates and two voters), with the new candidate *E* placed at the bottom of both ballots. The last four voters contribute exactly 8 to the Borda score of each candidate, and so, taken together, they have no effect on who is the winner of the election. This is what we mean by "canceling each other out."

In the first election, as in Example 1, Candidate *B* wins. But if we take these ballots to represent true preferences, the voter on the far left prefers *A* to *B*. Moreover, that voter can achieve this better outcome–Candidate *A*–by submitting the disingenuous ballot that he or she cast in Election 2.

■

To handle the case where the number of voters is odd, we need to start with a four-candidate, three-voter example of manipulation of the Borda count. Exercise 9 provides this. We can then modify this example to work for any odd number of voters by again adding pairs of ballots that cancel each other out exactly as we did before. Exercises 10 and 11 fill in some of the details needed for this part of the argument, and ask you to provide the necessary explanations and calculations.

## Manipulability of Runoff Systems

## EXAMPLE 4 ■ Manipulability of Runoff Systems

Both the plurality runoff rule and the Hare system are manipulable. But rather than give the whole story away, we'll just present the sequences of sincere ballots in each case. Exercises 16 and 17 ask you to figure out how the left-most voter in each case can secure a more preferred outcome by a unilateral change of ballot.

| Election 1 for the Hare System | | | | |
|---|---|---|---|---|
| *A* | *B* | *C* | *C* | *D* |
| *B* | *A* | *B* | *B* | *B* |
| *C* | *C* | *A* | *A* | *C* |
| *D* | *D* | *D* | *D* | *A* |

| Election 1 for the Plurality Runoff Rule | | | | |
|---|---|---|---|---|
| *A* | *A* | *C* | *C* | *B* |
| *B* | *B* | *A* | *A* | *C* |
| *C* | *C* | *B* | *B* | *A* |

## EXAMPLE 5 ■ Manipulating Sequential Pairwise Voting

Sequential pairwise voting can also be manipulated by a single voter, even in the case of three voters and three candidates. For example, consider the following two elections with the agenda *ABC*:

| **Election 1** | | | |
|---|---|---|---|
| **Rank** | **Number of Voters (3)** | | |
| First | *A* | *B* | *C* |
| Second | *B* | *C* | *A* |
| Third | *C* | *A* | *B* |

| **Election 2** | | | |
|---|---|---|---|
| **Rank** | **Number of Voters (3)** | | |
| First | *B* | *B* | *C* |
| Second | *A* | *C* | *A* |
| Third | *C* | *A* | *B* |

In the Election 1, *A* defeats *B* by a score of 2-to-1, so *A* moves on to meet *C*. But *C* defeats *A* by a score of 2-to-1, so *C* is the winner in Election 1. Election 2 is the result of Voter 1 (on the left) submitting a disingenuous ballot in which he has elevated *B* (his actual second choice) to first place on his ballot. It is now clear that *B* first defeats *A* by a score of 2-to-1 and then moves on to defeat *C* by this same score. Hence, *B* is the winner in Election 2. This is an instance of manipulation in which Voter 1 has secured a more preferred outcome by submitting an insincere ballot, because Voter 1 actually prefers *B* to *C* (assuming that his ballot in Election 1 represents his true preferences). This shows that sequential pairwise voting is manipulable. ■

## Sequential Pairwise Voting and Agenda Manipulability

Thus, sequential pairwise voting can also be manipulated by a single voter, even in the case of three voters and three candidates. But there is another aspect of manipulability that arises with this particular voting system that is of even more interest, and this is something called **agenda manipulation**.

### Agenda Manipulation — DEFINITION

**Agenda manipulation** refers to the ability to control who wins an election with sequential pairwise voting by a choice of the agenda.

William H. Riker, in his book *The Art of Political Manipulation*, spoke of the possibility that "those in control of procedures can manipulate the agenda by, for example, restricting alternatives [candidates] or by arranging the order in which they are brought up." The following example provides a striking illustration of this with sequential pairwise voting.

## EXAMPLE 6

### Agenda Manipulation of Sequential Pairwise Voting

Suppose we have four candidates and three voters who we know will be submitting the following preference list ballots:

| Rank | Number of Voters (3) | | |
|---|---|---|---|
| First | *A* | *C* | *B* |
| Second | *B* | *A* | *D* |
| Third | *D* | *B* | *C* |
| Fourth | *C* | *D* | *A* |

Now suppose that we have agenda-setting power in the sense that we get to choose the order in which the one-on-one contests will take place. Remarkably, we can arrange for the winner to be whichever of the four candidates we want!

The intuition behind finding an agenda that will yield a certain candidate as the winner arises from the observation that candidates who appear later in the agenda are favored over candidates who appear early in the agenda. For example, if we want *A* to win, we place *A* last and look for which candidates would, in fact, defeat *A* one on one. Here, only *C* defeats *A*, and so we want to arrange for *C* to be eliminated along the way. But *B* defeats *C* one on one, so if we choose the agenda *BCDA*, we have that *C* is eliminated by *B* in the first round, then *D* is eliminated by *B* in the second round, and finally *B* is eliminated by *A* in the third round, leaving *A* as the winner. Exercise 19 asks you to find the three other agendas that will, in turn, yield *B*, *C*, and *D* as the winner.

## Plurality Voting and Group Manipulability

In the real world, all other voting systems pale in comparison to plurality voting in terms of the significance of the role played by disingenuous voting. "Throwing away your vote"–as some accuse Nader voters in Florida of doing in the 2000 presidential election–represents a choice, conscious or otherwise, to forgo obtaining a more desired outcome through strategic considerations.

Ironically, plurality voting, like Condorcet's method, is nonmanipulable according to the formal definition given on page 346. However, a *group* of voters, acting together, can change an election outcome into something they *all* prefer. We'll record this observation in the following, and then explain why it's true.

**The Group Manipulability of Plurality Voting** THEOREM

Plurality voting cannot be manipulated by a single individual. However, it is **group manipulable** in the sense that there are elections in which a group of voters can change their ballots so that the new winner is preferred to the old winner by everyone in the group, assuming that the original ballots represent the true preferences of each voter in the group.

First of all, let's see why no individual can manipulate plurality voting. Suppose that you prefer *A* to *B*, but *B* is the winner with plurality voting. Then *B* has at least one more first-place vote than $A$. Now, because you prefer $A$ to $B$, we know that $B$ is not on top of your sincere ballot, so no ballot change that you make can subtract from $B$'s number of first-place votes. Moreover, by moving $A$ to the top of your ballot, you only increase $A$'s number of first-place votes by 1. Thus, the best you can do with a unilateral change in ballots is to move $A$ into a tie with $B$.

To see that plurality voting is group manipulable we only have to look at any real-world election in which a third-party candidate acted as the "spoiler." As we've said, Ralph Nader was exactly this in the state of Florida in the 2000 presidential election. Another example occurs in Exercise 20.

At this point, we've seen that several of our familiar voting systems for three or more candidates–the Borda count, runoff systems, sequential pairwise voting–can be manipulated. Can't we do better than this in attempting to improve on Condorcet's method? We turn to this question next.

The Green Party holds its convention. Ralph Nader ran for the presidency as a Green in the 2000 election. By doing so, he brought up many questions of social choice–some would say deliberately. Was Nader a spoiler candidate? Were Nader supporters casting sincere votes for him? Were other voters who liked his positions hedging their bets and voting insincerely if they chose another candidate? (*Mark Leffingwell/AFP/Getty.*)

## 10.3 Impossibility

Condorcet's method, as we've seen, has a number of very desirable properties, including the following four:

1. Elections never result in ties (with an odd number of voters).
2. It satisfies the Pareto condition.
3. It is nonmanipulable.
4. It is not a dictatorship.

Unfortunately, Condorcet's voting paradox on page 289 shows that there are elections in which Condorcet's method produces no winner at all.

Can we find a voting system that satisfies all four of these properties and that, unlike Condorcet's method, always yields a winner? Several possibilities suggest themselves. For example, to avoid ties, we could modify any of our usual methods by agreeing to use a fixed ordering of the candidates to break any ties that occur. Or we could extend Condorcet's method by making the winner be the candidate with the best "win-loss record" in one-on-one contests (a method called *Copeland's rule*).

Alas, any such attempt is doomed. In the early 1970s, Allan Gibbard and Mark Satterthwaite independently proved the following remarkable result.

**The Gibbard-Satterthwaite Theorem** THEOREM

With three or more candidates and any number of voters, there does not exist–and there never will exist–a voting system that always produces a winner, never has ties, satisfies the Pareto condition, is nonmanipulable, and is not a dictatorship.

The Gibbard–Satterthwaite theorem (often called the GS theorem, for short) is a deep result that is related in important ways to Arrow's impossibility theorem. In particular, you shouldn't find it at all obvious, and we won't be saying anything about the proof. But we can state and prove a much weaker result that is of some interest in its own right.

**A Weak Version of the GS Theorem** THEOREM

Any voting system for three candidates that agrees with Condorcet's method whenever there is a Condorcet winner–and that additionally produces a unique winner when confronted by the ballots in the Condorcet voting paradox–is manipulable.

Let's see why this is true. With the Condorcet voting paradox, the winner is either *A* or *B* or *C*. For the moment, we'll assume it is *C* (and leave the other two cases to you–see Exercise 25). Consider the following two elections:

**Election 1**

| Rank | Number of Voters (3) | | |
|---|---|---|---|
| First | *A* | *B* | *C* |
| Second | *B* | *C* | *A* |
| Third | *C* | *A* | *B* |

**Election 2**

| Rank | Number of Voters (3) | | |
|---|---|---|---|
| First | *B* | *B* | *C* |
| Second | *A* | *C* | *A* |
| Third | *C* | *A* | *B* |

In Election 1, the winner is *C* (our assumption in this case) and in Election 2, the winner is *B* (because we are assuming that our voting system agrees with Condorcet's method when there is a Condorcet winner, as *B* is here). Notice that the voter on the left, by a unilateral change in ballot, has improved the election outcome from his or her third choice to being his or her second choice. This is what that voter set out to do and is the desired instance of manipulation.

But the nonintuitive nature of voting and manipulation does not end here. It also turns out that sometimes "more is less" when it comes to "voting power." We illustrate this with the so-called chair's paradox.

## 10.4 The Chair's Paradox

We conclude this chapter with an aspect of manipulability that is so counterintuitive that it is referred to as the chair's paradox. The situation is as follows. Suppose we have three candidates–*A*, *B*, and *C*–and three voters whom we'll call, for simplicity, the chair, you, and me.

Now the chair prefers *A* to *B* to *C*. You prefer *B* to *C* to *A*. I prefer *C* to *A* to *B*. Thus, if we were to cast sincere preference list ballots, they'd be precisely the Condorcet voting paradox ballots. But that's not what we're going to do.

We're going to assume that each of the three of us gets to vote for one of the candidates. Votes are tallied as follows: If any candidate gets at least two of the three votes, he or she wins. But if each gets one vote, then whichever candidate the chair voted for wins. Thus, the chair has what might be called **tie-breaking power**. In particular, the chair clearly has more power than you or I.

The goal now is to analyze the situation and to determine how each of us will vote if we're rational in the sense of being willing to vote strategically (that is, to manipulate the system) if it's in our own best interest. This is really a game-theoretic analysis, and it's useful to borrow a couple of pieces of game-theoretic terminology.

A choice of which candidate to vote for is called a **strategy**. So each of us has three strategies at our disposal: Vote for *A*, vote for *B*, and vote for *C*. It will be useful to have our preferences displayed as if they were ballots. But remember, these are just our preferences, not our ballots.

| Chair | You | Me |
|---|---|---|
| *A* | *B* | *C* |
| *B* | *C* | *A* |
| *C* | *A* | *B* |

Our first observation is that if we're all rational and acting in our own self-interest, none of us will vote for our least-preferred candidate. The point is that voting for either a first or second choice **weakly dominates** the strategies of voting for a third choice in the sense that the former choices always yield outcomes that are either the same as, or better than, the latter.

But now the chair's strategy of voting for *A* weakly dominates his strategy of voting for *B*. That is, if we both vote for *C*, the outcome is *C* regardless of how he votes, but otherwise he does strictly better by voting for *A* rather than *B*. Hence, assuming the chair is rational, we know the chair will, in fact, vote for his top choice, Candidate *A*.

Now, given that we know what the chair will do, the claim is that my strategy of voting for *C* weakly dominates my strategy of voting for *A*. That is, if you vote for *B*, the outcome is *A* regardless of whether I vote for *C* or *A*. On the other hand, if you vote for *C*, then I can secure my best outcome *C* by voting for *C*. Assuming I'm rational, we know that I will, like the chair, vote for my top choice, which is Candidate *C*.

But let's see where these decisions leave you. You know that the chair is voting for *A* and I'm voting for *C*. So if you vote for *B*, then the outcome is *A*–your last choice. However, if you vote for *C* along with me, then the outcome is *C*, your sec-

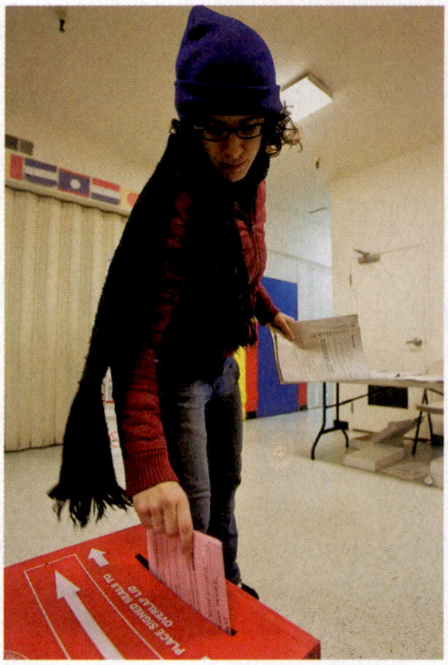

A woman casts her ballot on election day, the most important day in the American civic ritual of political campaigns and elections. Although she is acting as a responsible citizen, she may also contribute to some remarkable and contradictory results: Condorcet's voting paradox and the Gibbard–Satterthwaite theorem warn us that some elections produce strange results! (*David Paul Morris/Getty Images.*)

ond choice. There is no way that you can secure your top choice $B$ as the winner. So if you're rational, then you'll vote for $C$ also, and Candidate $C$ will win the election.

So why is this paradoxical? Well, the chair had the most power, but the eventual winner of the election was his least-preferred candidate! He would have been better off handing over the tie-breaking power to either you or me.

**The Chair's Paradox** THEOREM

With three voters and three candidates, the voter with tie-breaking power can, if all three voters act rationally in their own self-interest, end up with his or her least-preferred candidate as the election winner.

The chair's paradox represents only one of manipulability's first cousins, some of which involve not only the fields of mathematics and political science but psychology as well. I recall a third-grade penmanship contest in which each of us had a writing sample taped to the blackboard, and the teacher, Mrs. Levy, announced that we'd get to vote for the one we thought best, with the proviso that the voter couldn't vote for his or her own paper. She also announced that if two or more were tied, we'd have a runoff among those.

I remember being torn as to which of three particular ones to vote for, all of which I thought were very good and considerably better than the rest, including my own. When the votes were counted, these three were, in fact, tied for the win, with my writing sample alone in fourth, and only one vote out of the tie.

After announcing the results, Mrs. Levy went on to say that the runoff would involve not three of us, but four, as she had decided also to vote, and she was voting for me! I don't remember the final tally, or what Mrs. Levy then said to the class, or what my three classmates, all plenty smart enough to realize what had just happened, later said to me. But I do remember sitting back and smiling—absolutely sure of the outcome—as soon as she had announced her intention to vote for me.

## REVIEW VOCABULARY

**Agenda manipulation** The ability to control who wins an election with sequential pairwise voting by a choice of the agenda–that is, a choice of the order in which the one-on-one contests will be held. (p. 321)

**Chair's paradox** The fact that with three voters and three candidates, the voter with tie-breaking power (the "chair") can, if all three voters act rationally in their own self-interest, end up with her or his least-preferred candidate as the election winner. (p. 316)

**Disingenuous ballot** Any ballot that does not represent a voter's true preferences. Also called an **insincere ballot**. (p. 313)

**Gibbard–Satterthwaite (GS) manipulability theorem** Alan Gibbard and Mark Satterthwaite's independent discovery that every voting system for three or more alternatives and any number of voters that satisfies the Pareto condition, always produces a unique winner, and is not a dictatorship can be manipulated. (p. 316)

**Group manipulability** A voting system is group manipulable if there exists at least one election in which a group of voters can change their ballots (with the ballots of voters not in the group left unchanged) in such a way that they all prefer the winner of the new election to the winner of the old election, assuming that the original ballots represent the true preferences of these voters. (p. 322)

**Manipulation** A voting system is manipulable if there exists at least one election in which a voter can change his or her ballot (with the ballots of all other voters left unchanged) in such a way that he or she prefers the winner of the new election to the winner of the old election, assuming that the original ballots represent the true preferences of the voters. (p. 313)

**May's theorem for manipulability** Kenneth May's discovery that for two candidates and an odd number of voters, majority rule is the only voting system that treats both candidates equally, treats all voters equally, and is nonmanipulable. (p. 317)

**Strategy** In the chair's paradox, a choice of which candidate to vote for is called a strategy. This is a special case of the use of the term in general game-theoretic situations. (p. 325)

**Tie-breaking power** That aspect of the voting rule used in the chair's paradox that says the winner will be whichever candidate the chair votes for if there is a tie (which only happens if each candidate gets exactly one vote). (p. 325)

**Unilateral change** A change (in ballot) by a voter while every other voter keeps her or his ballot exactly as it was. (p. 315)

**Weak-dominance** One strategy (for example, a choice of whom to vote for) weakly dominates another if it yields an outcome that is at least as good, and sometimes better, than the other. (p. 325)

## SKILLS CHECK

**1.** A "unilateral change in ballot" refers to the fact that

**(a)** only one candidate's position is being altered.
**(b)** no communication is taking place.
**(c)** only one voter is changing his or her ballot.

**2.** The quote "My scheme is intended only for honest men!" is from ______________________________ .

**3.** If a voter has sincere preferences of *A* over *B* over *C* over *D*, then

**(a)** she will prefer a tie between *A* and *D* to a tie between *B* and *C*.
**(b)** she will prefer a tie between *B* and *C* to a tie between *A* and *D*.
**(c)** it's not at all clear which tie–*AD* or *BC*–she will prefer.

**4.** A ballot that misrepresents a voter's true preferences is referred to as ______________________________ .

**5.** Suppose that two elections show that a voting system is manipulable. Then

**(a)** neither election results in a tie.
**(b)** the winners are the same in both elections.
**(c)** every voter has changed his or her ballot.

**6.** In the two-candidate case, manipulation is equivalent to ______________________________ .

**7.** Condorcet's method

**(a)** can be manipulated but always produces a winner.
**(b)** is nonmanipulable but sometimes produces no winner.
**(c)** sometimes results in a tie, so manipulability is hard to assess.

**8.** May's theorem for manipulability says that, with an odd number of voters, among all voting systems for two candidates that never result in a tie, majority rule is the only one that is nonmanipulable and ______________ .

**9.** With the Borda count, two ballots "cancel each other out" if

**(a)** they are identical.
**(b)** each is arrived at by turning the other one upside down.
**(c)** other voters also hold these same ballots.

**10.** The Borda count is nonmanipulable in the special case in which ______________________________ .

**11.** A six-voter example of manipulation with the Borda count can be modified to yield a ten-voter example by

**(a)** adding four ballots that are identical to each other.
**(b)** adding four ballots that are identical to Voter 1's ballot.
**(c)** adding two pairs of ballots, with the ballots in each pair canceling each other out.

**12.** With any voting system that satisfies the Pareto condition, an $n$-voter example of manipulation with $k$ candidates can be modified to yield an $n$-voter example with $k + j$ candidates by ______________________________ .

**13.** Of the Hare system and the plurality runoff method,

**(a)** only the Hare system is manipulable.
**(b)** only plurality runoff is manipulable.
**(c)** both are manipulable.

**14.** Sequential pairwise voting is susceptible to a kind of manipulation called ______________________________ .

**15.** Plurality voting

**(a)** cannot be manipulated by a single voter.
**(b)** can be manipulated by a single voter.
**(c)** is subject to agenda manipulation.

**16.** Plurality voting is susceptible to a kind of manipulation called ______________________________ .

**17.** The Gibbard–Satterthwaite theorem says that with three or more candidates and any number of voters, there is no voting system that

**(a)** is not a dictatorship.
**(b)** is nonmanipulable and is not a dictatorship.
**(c)** satisfies the Pareto condition, is nonmanipulable, and is not a dictatorship.
**(d)** always yields a unique winner, satisfies the Pareto condition, is nonmanipulable, and is not a dictatorship.

**18.** The weak version of the Gibbard–Satterthwaite theorem asserts that if we have a voting system that agrees with Condorcet's method whenever there is a Condorcet winner, and that additionally produces a unique winner when confronted by the ballots in the Condorcet voting paradox, then the system is ______________________________ .

**19.** The voters' preferences in the paradox of the chair are

**(a)** precisely the Condorcet voting paradox ballots.
**(b)** all the same.
**(c)** dictated by the chair.

**20.** The chair's paradox is paradoxical because ______________________________ .

# CHAPTER 10 EXERCISES

■ Challenge ◆ Discussion

## 10.1 Majority Rule and Condorcet's Method

**1.** Consider the voting system for two candidates ($A$ and $B$) and three voters in which the candidate with the *fewest* first-place votes wins. Produce two elections that show this voting system is manipulable.

**2.** Consider the voting system for two candidates ($A$ and $B$) and three voters in which the candidate receiving an odd number of first-place votes wins. Produce two elections that show this voting system is manipulable.

**3.** Consider the voting system for two candidates ($A$ and $B$) and three voters in which the candidate receiving an even number of first-place votes wins. Produce two elections that show this voting system is manipulable.

**4.** There are at least two voting systems for two candidates ($A$ and $B$) and three voters that are nonmanipulable and that treat all voters the same (meaning that if two voters were to exchange ballots, then the election outcome would be unchanged).

**(a)** What does May's theorem tell us about such a voting system?
**(b)** In one sentence, give an example of such a voting system (that is, produce the rule that determines which of the two candidates, $A$ or $B$, wins an election).
**(c)** In one sentence, give another example that is different from the example you gave in part (b) in that it produces a different winner for at least one election.

**5.** There are at least three voting systems for two candidates ($A$ and $B$) and three voters that are nonmanipulable and that treat both candidates the same (meaning that if all three voters change their ballots, then the election outcome also changes).

**(a)** What does May's theorem tell us about such a voting system?
**(b)** In one sentence, give an example of such a voting system (that is, produce the rule that determines which of the two candidates wins an election).
**(c)** In one sentence, give two other examples that are different from the example you gave in part (b) in that they produce a different winner for at least one election.

**6.** Alfonse D'Amato (*D*) won the 1980 U.S. Senate race in New York by defeating Elizabeth Holtzman (*H*) and Jacob Javits (*J*). Reasonable estimates (based largely on exit polls) suggest that voters ranked the candidates according to the following table:

| 22% | 23% | 15% | 29% | 7% | 4% |
|---|---|---|---|---|---|
| *D* | *D* | *H* | *H* | *J* | *J* |
| *H* | *J* | *D* | *J* | *H* | *D* |
| *J* | *H* | *J* | *D* | *D* | *H* |

Who would have won if Condorcet's method (instead of plurality voting) had been used?

## 10.2 The Manipulability of Other Voting Systems for Three or More Candidates

**7.** Consider the following election with four candidates and two voters:

| | |
|---|---|
| *B* | *A* |
| *C* | *D* |
| *A* | *C* |
| *D* | *B* |

Show that if the Borda count is being used, the voter on the left can manipulate the outcome (assuming the above ballot represents his true preferences).

**8.** Example 2 showed that the Borda count is manipulable if there are five candidates and six voters. Mimic what was done there in order to construct an example with seven candidates and eight voters.

**9.** Use the following election to illustrate the manipulability of the Borda count with three voters and four candidates:

| | | |
|---|---|---|
| *A* | *B* | *B* |
| *B* | *A* | *A* |
| *C* | *C* | *C* |
| *D* | *D* | *D* |

**10.** Show that the Borda count is manipulable if there are four candidates and five voters. (*Hint:* Start with the ballots in the previous exercise, and then add two ballots that cancel each other out.)

**11.** Building on the idea in the previous exercise, show that the Borda count is manipulable if there are six candidates and nine voters.

**12.** Assume the following ballots give the true preferences of the voters and that the Borda count is being used. Show that at least one of the voters can improve the election outcome from her point of view by a unilateral change in her ballot.

| | | | |
|---|---|---|---|
| *B* | *D* | *C* | *B* |
| *C* | *C* | *A* | *A* |
| *D* | *A* | *B* | *C* |
| *A* | *B* | *D* | *D* |

**13.** There is a modified version of Condorcet's method called the *weak Condorcet rule:* A candidate is among the winners precisely if he would defeat or tie every other candidate in a one-on-one contest. Notice that with an odd number of voters, the weak Condorcet rule is identical to Condorcet's method. Use the following ballots to show that the weak Condorcet rule is manipulable:

| | | | |
|---|---|---|---|
| *A* | *C* | *B* | *D* |
| *B* | *A* | *D* | *C* |
| *C* | *B* | *C* | *A* |
| *D* | *D* | *A* | *B* |

**14.** *Copeland's rule* is a voting system that, like Condorcet's method, looks at one-on-one contests. Copeland's rule, however, takes as the election winner the candidate with the best "win-loss record." Use the following ballots to show that Copeland's rule is manipulable:

| | | | |
|---|---|---|---|
| *A* | *C* | *A* | *D* |
| *B* | *E* | *E* | *B* |
| *C* | *D* | *D* | *E* |
| *D* | *B* | *C* | *C* |
| *E* | *A* | *B* | *A* |

**15.** *Coombs's rule* is the voting system that operates like the Hare system, except that instead of deleting candidates with the *fewest* first-place votes one after another, it deletes candidates with the *most* last-place votes one after another. Use the following ballots to show that Coombs's rule is manipulable:

| | | | | |
|---|---|---|---|---|
| *A* | *B* | *B* | *A* | *A* |
| *B* | *C* | *C* | *C* | *C* |
| *C* | *A* | *A* | *B* | *B* |

**16.** Use the following election to show that the Hare system is manipulable:

| | | | | |
|---|---|---|---|---|
| *A* | *B* | *C* | *C* | *D* |
| *B* | *A* | *B* | *B* | *B* |
| *C* | *C* | *A* | *A* | *C* |
| *D* | *D* | *D* | *D* | *A* |

**17.** Use the following election to show that the plurality runoff rule is manipulable:

| | | | | |
|---|---|---|---|---|
| *A* | *A* | *C* | *C* | *B* |
| *B* | *B* | *A* | *A* | *C* |
| *C* | *C* | *B* | *B* | *A* |

**18.** Use the following election to show that sequential pairwise voting is manipulable. (Assume the agenda is *ABC*.)

| | | |
|---|---|---|
| *A* | *B* | *C* |
| *B* | *C* | *A* |
| *C* | *A* | *B* |

**19.** Given the following ballots:

| | | |
|---|---|---|
| *A* | *C* | *B* |
| *B* | *A* | *D* |
| *D* | *B* | *C* |
| *C* | *D* | *A* |

mimic what was done in Example 6 to find

**(a)** an agenda for which *B* is the winner using sequential pairwise voting.
**(b)** an agenda for which *C* is the winner using sequential pairwise voting.
**(c)** an agenda for which *D* is the winner using sequential pairwise voting.

◆ **20.** Suppose that we have a voting system that satisfies unanimity: If every voter ranks the same candidate first, then that candidate is the unique winner. In a few sentences, explain why it is that, if the system fails to satisfy the Pareto condition, it can be manipulated by some group.

**21.** Use the ballots in Exercise 6 to show that the plurality rule is group manipulable.

**22.** Consider the voting rule in which an alternative is among the winners if it receives at least one first-place vote. In one sentence, explain why this voting system is *not* manipulable.

**23.** Consider the voting rule in which an alternative is among the winners if it has at least two first-place votes.

**(a)** In one sentence, explain why this voting system is *not* manipulable.
**(b)** Explain why the following two elections don't contradict part (a).

**Election 1**

| Rank | Number of Voters (4) | | | |
|---|---|---|---|---|
| First | *B* | *A* | *A* | *C* |
| Second | *C* | *B* | *B* | *B* |
| Third | *A* | *C* | *C* | *A* |

**Election 2**

| Rank | Number of Voters (4) | | | |
|---|---|---|---|---|
| First | *C* | *A* | *A* | *C* |
| Second | *B* | *B* | *B* | *B* |
| Third | *A* | *C* | *C* | *A* |

**(c)** Intuitively, does it seem to you that Voter 1, on the left in part (b), has secured a better outcome by submitting a disingenuous ballot?

**24.** Consider the voting system in which the winner is determined by the total number of first- and second-place votes, with ties broken (when possible) according to the number of first-place votes. Thus, a candidate with no first-place votes and three second-place votes would defeat a candidate with two first-place votes and no second-place votes, but a candidate with two first-place votes and three second-place votes would defeat a candidate with one first-place vote and four second-place votes. Given Election 1 below, find a change in Voter 1's ballot that shows that this voting system is manipulable.

**Election 1**

| Rank | Number of Voters (3) | | |
|---|---|---|---|
| First | *A* | *C* | *E* |
| Second | *B* | *D* | *D* |
| Third | *C* | *A* | *A* |
| Fourth | *D* | *B* | *B* |
| Fifth | *E* | *E* | *C* |

## 10.3 Impossibility

■ **25.** Complete the proof of the weak version of the Gibbard–Satterthwaite theorem by handling the case where

**(a)** the winner with the voting paradox ballots is *A*.
**(b)** the winner with the voting paradox ballots is *B*.

The Gibbard–Satterthwaite theorem says that the following four properties of voting systems can never be simultaneously satisfied:

**(1)** Elections always have unique winners.
**(2)** It satisfies the Pareto condition.
**(3)** It is nonmanipulable.
**(4)** It is not a dictatorship.

**26.** Which of the four properties are satisfied by a dictatorship?

**27.** Which of the four properties are satisfied by an "antidictatorship," where the election winner is whichever candidate Voter 1 ranks *last* on his or her ballot?

**28.** Which of the four properties are satisfied if we use the plurality rule with Voter 1's ballot used to break any ties that occur?

## 10.4 The Chair's Paradox

Consider the ballots from the chair's paradox:

| Chair | You | Me |
|---|---|---|
| *A* | *B* | *C* |
| *B* | *C* | *A* |
| *C* | *A* | *B* |

Assume that we know that the chair will vote for *A*, but that we don't know anything about how I will vote.

◆ **29.** In a sentence or two, explain why your strategy to vote for *B* does not weakly dominate your strategy of voting for *C*.

◆ **30.** In a sentence or two, explain why your strategy to vote for *C* does not weakly dominate your strategy of voting for *B*.

# WRITING PROJECT

In a paragraph or two, explain why Condorcet's method is not group manipulable.

# SUGGESTED READINGS

MOULIN, HERVÉ. *The Strategy of Social Choice*, North Holland, New York, 1983. Manipulability from an economist's point of view.

RIKER, WILLIAM. *The Art of Political Manipulation*, Yale University Press, New Haven and London, 1986. Manipulability from a political scientist's point of view.

TAYLOR, ALAN. *Social Choice and the Mathematics of Manipulation*, Cambridge University Press, Cambridge, U.K., 2005. Manipulability from a mathematician's point of view.

CHAPTER 11



# Weighted Voting Systems

Voting is often used to decide yes or no questions. Legislatures vote on bills, stockholders vote on resolutions presented by the board of directors of a corporation, and juries vote to acquit or convict a defendant. In this chapter, we shall concentrate on situations where there are just two alternatives, such as "yes" or "no." The theorem of Kenneth May quoted in Chapter 9 says that majority rule is the only system with the following properties:

1. All voters are treated equally.
2. Both alternatives are treated equally.
3. If you vote "no," and "yes" wins, then "yes" would still win if you switched your vote to "yes," provided that no other voters switched their votes.
4. A tie cannot occur unless there is an even number of voters.

There are many situations in which one or more of these properties are not valid. For example, in a criminal trial the jury is required to reach a unanimous decision on a motion to convict (or on a motion to acquit); thus, if there is one "no" vote, the motion is not adopted. In this case, the alternatives are not treated equally. Here's another familiar example. Stockholders are allowed one vote per share that they own. If shareholder A owns 10,000 shares and shareholder B owns 100, then this voting system does not treat A and B equally.

Some systems where the voters appear to be unequal in power actually have all of the properties required by May's theorem. Any student of politics will attest that not all legislators are equally powerful (think of the speaker of the U.S. House of Representatives versus a freshman member, or the prime minister versus a backbencher in Parliament). Nevertheless, the voting system actually treats the legislators equally: Each has one vote. Our interest is in the voting system itself and not in the influence that some voters might acquire as a result of experience or accomplishment.

Voting systems that treat participants unequally are often used when the participants are indeed unequal. For example, the Council of Ministers of the European Union accords more power to states such as France, which have large populations, than it does to smaller states, such as Austria. Rather than giving the larger states more representatives, as in the U.S. House of Representatives, the Council of Ministers gives the ministers from the larger states more votes.

We shall find two measures of voting power that apply when voters are not treated equally or alternatives are not treated equally, or both: the **Shapley–Shubik power index**, and the **Banzhaf power index**. The Banzhaf power index is an accurate measure of power when there is no spectrum of opinion. For example, if each voter decides which way to vote by tossing a coin, the Banzhaf power index will indicate each voter's share of power. The Shapley–Shubik index is appropriate in a process where measures are crafted so as to attract enough votes to win.

## 11.1 How Weighted Voting Works

One type of voting system in which the voters or the alternatives may be treated unequally is a **weighted voting system**. Each participant has a specified number of votes, called his or her **weight**. If one voter's voting weight is more than than another's, then the first voter might have more power to influence the outcome, and certainly won't have less. (We will see that voters with different numbers of votes may actually have equal power.) In any voting system, there must be a criterion for deciding whether "yes" or "no" has won. In a weighted voting system, this is done by specifying a number called the **quota**. If the sum of the weights of all the voters who favor a motion is equal to the quota, or exceeds it, then "yes" wins. Otherwise, "no" wins. The quota must be greater than half of the total weight of all the voters, to avoid situations where contradictory motions can pass, and it cannot be greater than the total weight, or no motion would ever pass.

The European Union's Council of Ministers uses weighted voting, but in the United States, it is unusual for a legislative body to use a weighted voting system (see Spotlight 11.4). It cannot be said that there is no weighted voting in the United States, because the Electoral College, which elects the president, functions as a weighted voting system in which the voters are the states. See Spotlight 11.1.

### Notation for Weighted Voting Systems — DEFINITION

To describe a weighted voting system, you must specify the voting **weights** $w_1, w_2, \ldots, w_n$ of the participants, and the **quota**, $q$. The following notation is a shorthand way of making these specifications:

$$[q : w_1, w_2, \ldots, w_n]$$

The weighted voting system [51 : 40, 60] describes a voting system in which there are two voters, with voting weights 40 and 60, and the quota is 51.

### EXAMPLE 1 ■ A Dictator

Suppose there is one voter, $D$, who has all of the power. A motion will pass if and only if $D$ is in favor, and it doesn't matter how the other participants vote. Most weighted voting systems that we will consider do not have a **dictator**, but if there

## SPOTLIGHT 11.1 The Electoral College

In a U.S. presidential election, the voters in each state don't actually cast their votes for the candidates. They vote for electors to represent them in the Electoral College. The number of electors allotted to a state is equal to the size of its congressional delegation, so a state with one congressional district gets 3 electors: one for its representative, and one for each of its two senators. A state with 25 representatives would get 27 electors. The District of Columbia, while not a state, is entitled by the 23rd Amendment to the U.S. Constitution to send 3 electors to the College.

All states except two select their electors in a statewide contest. Thus, all of a state's electors are committed to vote for the presidential candidate favored by a plurality of the voters in the state. For example, the candidate who gets a plurality in California receives all 55 of the state's electoral votes. In Maine and Nebraska, there is a different procedure. Two electors (corresponding to the senators) are chosen statewide, and the electors corresponding to the representatives are chosen by congressional district. Nebraska has three congressional districts. It is possible that one or two of the districts might favor one ticket, while the state as a whole might favor another.

Effectively, the Electoral College functions as a weighted voting systems in which there are 56 participants: the 50 states, the District of Columbia, three Nebraska congressional districts, and two Maine congressional districts. The weights range from 1 for individual congressional districts to 55, and the quota, 270, is a simple majority of the 538 electors.

is one, his or her voting weight must be equal to or more than the quota. The system [51 : 40, 60] has a dictator because the weight-60 voter can pass any motion that she wants.

## EXAMPLE 2 ■ Dummy Voters

A voting system may include some participants–called **dummy** voters–whose votes don't count. For example, the U.S. Congress has a nonvoting delegate who represents the District of Columbia. If a voting system has a dictator, all of the participants except the dictator are dummy voters. In the voting system [8 : 5, 3, 1], the weight-1 voter is a dummy, because a motion will pass only if it has the support of the weight-5 and weight-3 voters, and then the additional 1 vote is not needed. For another example, consider [51 : 26, 26, 26, 22]. The voter with weight 22 is not needed when two of the other voters combine to support a motion; they have enough weight to pass the motion without her. If she joins forces with just one of the other voters, their total weight, 48, is not enough to win. Thus, the weight-22 voter is a dummy.

## EXAMPLE 3 ■ Three More Three-Voter Systems

By adjusting the quota, the distribution of power in a weighted voting system can be altered. We have seen that the weight-1 voter in [8 : 5, 3, 1] is a dummy, but by increasing the quota to 9 we obtain a system in which the power is equally distributed–unanimous support is required to pass a motion in [9 : 5, 3, 1]. The weight-1 voter is also not a dummy in [6 : 5, 3, 1] because he can join the weight-5 voter to pass a motion, even if the weight-3 voter opposes. Finally, consider [51 : 49, 48, 3]. Although it looks as if the weight-3 voter will have relatively little power, and may even be a dummy, in fact she has the same voting power as the other two voters. Any two of the three voters in this system can pass a measure.

### EXAMPLE 4 ▪ Veto Power

A voter whose vote is necessary to pass any motion is said to have **veto power.** For example, in the system [6 : 5, 3, 1], the weight-5 voter has veto power because the other two voters do not have enough combined weight to pass a motion. A dictator always has veto power, and it is possible for more than one voter to have veto power as well. In a criminal trial, each juror has veto power. In the system [8 : 5, 3, 1] the voters with weights 5 and 3 each have veto power. ▪

The voters in the system [6 : 5, 3, 1] are not equally powerful–the weight-5 voter has veto power and the other two don't–and yet none of the voters are dummies. We can't compare power by comparing the voting weights because the weight-3 voter has the same voting power as the weight-1 voter. Together, they can stop the weight-5 voter from passing a motion, and either one can combine with the weight-5 voter to pass a motion. A **power index** gives a way to measure the share of power that each participant in a voting system (weighted or otherwise) has. Spotlight 11.2 is a brief history of power indices.

### SPOTLIGHT 11.2 Power Indices

The first widely accepted numerical index for assessing power in voting systems was the **Shapley–Shubik power index**, developed in 1954 by a mathematician, Lloyd S. Shapley, and an economist, Martin Shubik. A particular voter's power as measured by this index is proportional to the number of different permutations (or orderings) of the voters in which he or she has the potential to cast the pivotal vote—the vote that first turns from losing to winning.

The **Banzhaf power index** was introduced in 1965 by John F. Banzhaf III, a law professor who is also well-known as the founder of the antismoking organization ASH (Action on Smoking and Health). The Banzhaf index is the one most often cited in court rulings, perhaps because Banzhaf brought several cases to court and continues to file *amicus curiae* briefs when courts evaluate weighted voting systems. A voter's Banzhaf index is the number of different possible **voting combinations** in which he or she casts a critical vote—a vote in favor of a motion that is necessary for the motion to pass, or a vote against a motion that is essential for its defeat.

Lloyd S. Shapley

John F. Banzhaf III
*(AP Photos.)*

Martin Shubik
*(Courtesy Yale School of Management Public Affairs.)*

## 11.2 The Shapley–Shubik Power Index

When an election looms, politicians focus on "moderate voters." These are people who could be convinced to favor one side or the other. Moderate voters can make elected officials pay attention while voters who have an extreme commitment to one side or the other are ignored. However, moderate voters achieve their influence as a result of their political position, and we are primarily interested in the power that voters acquire as a result of the system itself. For example, France has 29 votes in the European Union (EU) Council, and Austria has 10. A motion before the Council, in which both countries were moderate voters, would be more likely to be written so as to acquire France's vote rather than Austria's, because France has so many more votes.

In 1954, Lloyd Shapley and Martin Shubik devised a way to gauge the share of decision-making power of each participant in a voting system. A voter's share of power is called his or her *Shapley–Shubik power index*. The index is defined in terms of *permutations*.

**Permutation** DEFINITION

A **permutation** of voters is an ordering of all of the voters in a voting system.

Voters are ordered in accordance with their commitment to an issue, starting with those who are most favorably inclined and ending with those who are most determined to oppose. For example, suppose that the issue is animal rights. Here the spectrum might range from a voter who would outlaw the sale of cow's milk to one who would legalize cockfighting. If an animal rights bill is being drafted, it must be written so as to receive enough votes to meet the quota.

**Pivotal Voter** DEFINITION

The first voter in a permutation who, when joined by those coming before him or her, would have enough voting weight to win is the **pivotal voter** in the permutation. Each permutation has exactly one pivotal voter.

If the issue is taxation instead of animal rights, the spectrum of opinion will probably be completely different. Voters who have moderate positions on animal rights may or may not be at the extremes when the subject is taxes. Each issue being debated corresponds to some permutation–and the pivotal voter on one issue may well not be pivotal on another issue.

A successful tax cut bill must be drafted so as to secure the support of the pivotal voter of the taxation permutation, an animal welfare bill must be drafted so that the the pivotal voter of the animal rights permutation will support it, and so on.

## EXAMPLE 5 ■ The Permutation in the 2004 Election

In 2004 the Electoral College reelected the Bush–Cheney ticket. Spotlight 11.1 explains how the Electoral College operates. Although there are 538 electors in the college, all states except Maine and Nebraska select their electors in a statewide general election, so even if the popular vote in a state was close, all of the electors from that state will vote for the same ticket. There are actually 56 independent votes in the Electoral College.[1]

[1]Strictly speaking, the votes in Maine and Nebraska are not really independent–see Exercise 39 at the end of this chapter.

Each of the 56 voters in the Electoral College is selected by and represents an electorate. Some, such as Nebraska's third district, were heavily in favor of the Bush–Cheney ticket (by more than 3 to 1); others, such as Iowa, New Mexico, and Wisconsin, were almost equally split between the Bush–Cheney and Kerry–Edwards tickets; and still others, such as the District of Columbia, were strongly in the Kerry–Edwards camp (almost 10 to 1). Table 11.1 lists the 56 voters in the Electoral College, ordered by their margin in favor of the Bush–Cheney ticket. The voting weight of each is shown, and a running total of electoral votes gives the total weight of each voter and all who came before it in the table. In listing the states and other voters in this order, we have recorded a permutation of the Electoral College participants. The pivotal voter is the one that brings the running total over the quota (270). If you recall the news about the 2004 election, you will not be surprised to see which voter is pivotal–Ohio.

**TABLE 11.1** The Permutation Resulting from General Election for President of the United States in 2004

| Voter* | Weight | Bush's Margin | Running Total | Voter* | Weight | Bush's Margin | Running Total |
|---|---|---|---|---|---|---|---|
| NE Dist 3 | 1 | 3.158 | 1 | CO | 9 | 1.112 | 222 |
| UT | 5 | 2.695 | 6 | FL | 27 | 1.107 | 249 |
| WY | 3 | 2.370 | 9 | NV | 5 | 1.055 | 254 |
| ID | 4 | 2.257 | 13 | OH | 20 | 1.051 | 274 |
| NE | 2 | 2.023 | 15 | IA | 7 | 1.018 | 281 |
| OK | 7 | 1.904 | 22 | NM | 5 | 1.016 | 286 |
| ND | 3 | 1.771 | 25 | WI | 10 | 0.992 | 296 |
| NE Dist 1 | 1 | 1.752 | 26 | NH | 4 | 0.973 | 300 |
| AK | 3 | 1.721 | 29 | PA | 21 | 0.955 | 321 |
| KS | 6 | 1.705 | 35 | MI | 17 | 0.934 | 338 |
| AL | 9 | 1.697 | 44 | MN | 10 | 0.932 | 348 |
| TX | 34 | 1.599 | 78 | OR | 7 | 0.924 | 355 |
| NE Dist 2 | 1 | 1.574 | 79 | ME Dist 2 | 1 | 0.888 | 356 |
| SD | 3 | 1.558 | 82 | NJ | 15 | 0.879 | 371 |
| MT | 3 | 1.531 | 85 | WA | 11 | 0.864 | 382 |
| IN | 11 | 1.528 | 96 | DE | 3 | 0.858 | 385 |
| MS | 6 | 1.506 | 102 | HI | 4 | 0.839 | 389 |
| KE | 8 | 1.500 | 110 | ME | 2 | 0.836 | 391 |
| SC | 8 | 1.421 | 118 | CA | 55 | 0.818 | 446 |
| GA | 15 | 1.405 | 133 | IL | 21 | 0.815 | 467 |
| LA | 9 | 1.345 | 142 | CT | 7 | 0.810 | 474 |
| TN | 11 | 1.338 | 153 | ME Dist 1 | 1 | 0.783 | 475 |
| WV | 5 | 1.296 | 158 | MD | 10 | 0.778 | 485 |
| NC | 15 | 1.288 | 173 | VT | 3 | 0.701 | 488 |
| AZ | 10 | 1.235 | 183 | NY | 31 | 0.700 | 519 |
| AR | 6 | 1.221 | 189 | RI | 4 | 0.653 | 523 |
| VA | 13 | 1.191 | 202 | MA | 12 | 0.595 | 535 |
| MO | 11 | 1.159 | 213 | DC | 3 | 0.103 | 538 |

*The voters are ordered by decreasing margin for the Bush–Cheney ticket. This margin is the number of popular votes cast for Bush–Cheney divided by the votes cast for Kerry–Edwards.

## SPOTLIGHT 11.3 Battleground States

In United States presidential elections, politicians color the states that are likely to vote Republican in the Electoral College red, and those likely to vote Democratic blue. The states that could go either way are the *battleground states*, and the campaigns put most of their investment in these states. We have colored these states gray. It is the responsibility of a state's legislature to decide how to choose the electors to represent that state. All states, except Maine and Nebraska, award *all* of their electoral votes to the plurality winner in the state's general election. As described in Spotlight 11.1, Maine and Nebraska award a separate electoral vote to each congressional district, with the plurality winner in the state as a whole getting two electoral votes (for the state's two senatorial seats in the Electoral College).

Large blue states have within them some red and some gray congressional districts and large red states have some blue and some gray districts. For example, if the California legislature adopted the system used by Maine and Nebraska, Republican candidates would benefit by getting electoral votes from those red congressional districts.

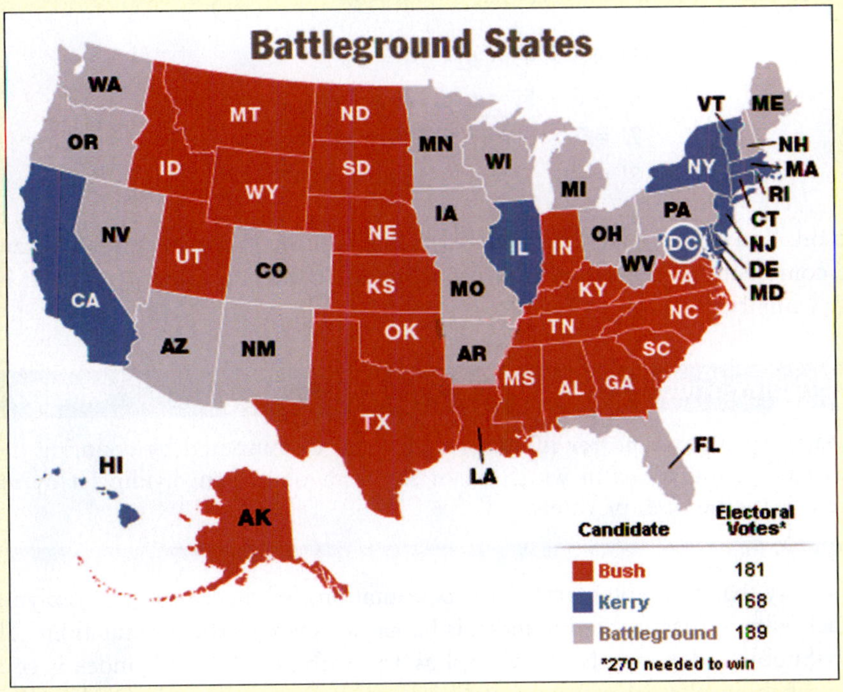


Politicians color the states that are strongly Republican red, and the states that are strongly Democratic blue. We have colored the battleground states—which could go either way—gray. (© *2004 Time Inc. All rights reserved.*)

If there are $n$ voters, the number of permutations is called the **factorial** of $n$ and is denoted $n!$. There is a simple formula for $n!$:

### Factorial Formula — THEOREM

For a positive whole number $n$,

$$n! = n \times (n-1) \times (n-2) \times \cdots \times 2 \times 1$$

and $0! = 1$.

To justify the formula, suppose that we are listing all of the permutations. There are $n$ voters who could be first; when the first voter is selected, there are $n - 1$ remaining voters who could be in second position, then $n - 2$ who could be third, and so on. When it is time to select for the last position, there is one voter left. By the fundamental principle of counting (see Chapter 2), the number of permutations is the product of the numbers of choices that we have had at each stage.

## EXAMPLE 6 ■ Calculating $n!$

Here are the first four, starting with 1!.

$$1! = 1$$
$$2! = 2 \times 1 = 2$$
$$3! = 3 \times 2 \times 1 = 6$$
$$4! = 4 \times 3 \times 2 \times 1 = 24$$

To continue this list, observe that $n! = n \times (n - 1)!$ for $n \geq 1$. Thus

$$5! = 5 \times 4! = 5 \times 24 = 120$$
$$6! = 6 \times 5! = 6 \times 120 = 720$$
$$7! = 7 \times 6! = 7 \times 720 = 5040$$
$$8! = 8 \times 7! = 8 \times 5040 = 40{,}320$$

and so on. You can imagine that $n!$ increases dramatically as $n$ increases–an instance of the combinatorial explosion. You probably don't want to calculate 100!. It is a 158-digit number.

### The Shapley-Shubik Power Index — DEFINITION

The **Shapley–Shubik power index** of each voter is computed by counting the number of permutations in which he or she is pivotal, then dividing it by $n!$, where $n$ is the number of voters.

If we say that each voter "owns" the permutations in which he or she is pivotal, then each voter's Shapley-Shubik index is his or her share of the permutations. The Shapley-Shubik index can also be viewed as a probability. A voter's index is equal to the probability that when a permutation is selected at random, he or she will be the pivotal voter.

## EXAMPLE 7 ■ The Shapley–Shubik Power Index of a Three-Voter System

Let us calculate the Shapley–Shubik power index of the voting system [6: 5, 3, 1]. We will name the participants $A$, $B$, and $C$, and consider their $3! = 6$ permutations. Table 11.2 displays all six permutations. Next to each permutation, the total weights of the first voter, the first two voters, and all three voters are shown in sequence. The first number in the sequence that equals or exceeds the quota (6) is underlined, and the corresponding pivotal voter's symbol is circled. We see that $A$ is pivotal in four permutations, while $B$ and $C$ are each pivotal in one. Hence the Shapley–Shubik index of $A$ is $\frac{4}{6}$, and $B$ and $C$ each have Shapley–Shubik indices of $\frac{1}{6}$.

**TABLE 11.2** Permutations and Pivotal Voters for the Three-Person Committee

| Permutations | | | Weights | | |
|---|---|---|---|---|---|
| *A* | Ⓑ | *C* | 2 | 3 | 4 |
| *A* | Ⓒ | *B* | 2 | 3 | 4 |
| *B* | Ⓐ | *C* | 1 | 3 | 4 |
| *B* | *C* | Ⓐ | 1 | 2 | 4 |
| *C* | Ⓐ | *B* | 1 | 3 | 4 |
| *C* | *B* | Ⓐ | 1 | 2 | 4 |

**TABLE 11.3** Permutations and Pivotal Voters for the Four-Shareholder Corporation

| Permutations | | | | Weights | | | | Pivot |
|---|---|---|---|---|---|---|---|---|
| *A* | Ⓑ | *C* | *D* | 40 | 70 | 90 | 100 | *B* |
| *A* | Ⓑ | *D* | *C* | 40 | 70 | 80 | 100 | *B* |
| *A* | Ⓒ | *B* | *D* | 40 | 60 | 90 | 100 | *C* |
| *A* | Ⓒ | *D* | *B* | 40 | 60 | 70 | 100 | *C* |
| *A* | *D* | Ⓑ | *C* | 40 | 50 | 80 | 100 | *B* |
| *A* | *D* | Ⓒ | *B* | 40 | 50 | 70 | 100 | *C* |
| *B* | Ⓐ | *C* | *D* | 30 | 70 | 90 | 100 | *A* |
| *B* | Ⓐ | *D* | *C* | 30 | 70 | 80 | 100 | *A* |
| *B* | *C* | Ⓐ | *D* | 30 | 50 | 90 | 100 | *A* |
| *B* | *C* | Ⓓ | *A* | 30 | 50 | 60 | 100 | *D* |
| *B* | *D* | Ⓐ | *C* | 30 | 40 | 80 | 100 | *A* |
| *B* | *D* | Ⓒ | *A* | 30 | 40 | 60 | 100 | *C* |
| *C* | Ⓐ | *B* | *D* | 20 | 60 | 90 | 100 | *A* |
| *C* | Ⓐ | *D* | *B* | 20 | 60 | 70 | 100 | *A* |
| *C* | *B* | Ⓐ | *D* | 20 | 50 | 90 | 100 | *A* |
| *C* | *B* | Ⓓ | *A* | 20 | 50 | 60 | 100 | *D* |
| *C* | *D* | Ⓐ | *B* | 20 | 30 | 70 | 100 | *A* |
| *C* | *D* | Ⓑ | *A* | 20 | 50 | 60 | 100 | *B* |
| *D* | *A* | Ⓑ | *C* | 10 | 50 | 80 | 100 | *B* |
| *D* | *A* | Ⓒ | *B* | 10 | 50 | 70 | 100 | *C* |
| *D* | *B* | Ⓐ | *C* | 10 | 40 | 80 | 100 | *A* |
| *D* | *B* | Ⓒ | *A* | 10 | 40 | 60 | 100 | *C* |
| *D* | *C* | Ⓐ | *B* | 10 | 30 | 70 | 100 | *A* |
| *D* | *C* | Ⓑ | *A* | 10 | 30 | 60 | 100 | *B* |

## EXAMPLE 8 ■ The Corporation with Four Shareholders

A corporation has four shareholders, *A*, *B*, *C*, and *D*, with 40, 30, 20, and 10 shares, respectively. The corporation uses the weighted voting system

$$[51 : 40, 30, 20, 10]$$

The 4! = 24 permutations of the shareholders are shown in Table 11.3. In 10 of the permutations, *A* is the pivotal voter; *B* and *C* are each pivotal voters in 6; and *D* is the pivotal voter in 2 permutations. Therefore, the Shapley–Shubik power index for this weighted voting system is

$$\left(\frac{10}{24}, \frac{6}{24}, \frac{6}{24}, \frac{2}{24}\right)$$

## How to Compute the Shapley–Shubik Power Index

It is practical to calculate the Shapley-Shubik power index of a system with up to four voters by making a list of all the voting permutations and identifying the pivotal voter in each, as we have done in the previous two examples. This is the brute force way of determining the Shapley-Shubik power index. With a computer, brute force can be used to determine the Shapley-Shubik power index of somewhat larger systems, but eventually the combinatorial explosion renders the brute force method impossible to execute.

The Shapley-Shubik power index of the Electoral College is shown in Spotlight 11.5. The calculations were performed with a Java applet that is available at www.whfreeman.com/fapp8e. The applet uses an advanced counting method, *generating functions*, that doesn't rely on tallying individual permutations one at a time.

In special cases where all (or almost all) of the voters have the same weight, there are simple methods to determine the Shapley-Shubik power index. They are based on two principles:

- Voters with the same voting weight have the same Shapley–Shubik power index.
- The sum of the Shapley–Shubik power indices of all the voters is 1.

## EXAMPLE 9 ■ A Seven-Person Committee

The chairperson of a committee has 3 votes, and there are six other members, each with 1 vote. The quota for passing a measure is a simple majority, 5 of the 9 votes. In our notation, this voting system is [5 : 3, 1, 1, 1, 1, 1, 1].

Each ordinary member has the same power index. Our strategy is to compute the index of the chair, and then divide the share of power that the chair does *not* have equally among the ordinary members.

There are 7! = 5040 permutations to consider. We will group them by the position occupied by the chairperson. Thus, *CMMMMMM*, in which the chairperson is first, is the first group. Counting from the left, we see that the votes are accumulated in the sequence 3, 4, $\underline{5}$, 6, 7, 8, 9. In these permutations (there are 6! of them), the chairperson is not the pivot; an ordinary member is. In the second group, *MCMMMMM*, the votes accumulate in the sequence 1, 4, $\underline{5}$, 6, 7, 8, 9, and again, the chairperson is not the pivot.

The chairperson is the pivot in the next three groups of permutations, *MMCMMMM*, *MMMCMMM*, and *MMMMCMM*, with vote accumulations 1, 2, $\underline{5}$, 6, 7, 8, 9; 1, 2, 3, $\underline{6}$, 7, 8, 9; and 1, 2, 3, 4, $\underline{7}$, 8, 9, respectively. In the final two groups, *MMMMMCM* and *MMMMMMC*, with vote accumulations 1, 2, 3, 4, $\underline{5}$, 8, 9 and 1, 2, 3, 4, $\underline{5}$, 6, 9, an ordinary member, not the chairperson, is the pivot again.

Each of the 7 groups of permutations is of the same size, 6!, because the 6 ordinary members can appear in any order in each. The chairperson is the pivot in three groups, for a total of $\frac{3}{7}$ of the total number of permutations. His Shapley–Shubik power index is therefore $\frac{3}{7}$. The remaining $\frac{4}{7}$ of the voting power is shared equally by the 6 ordinary members. Therefore each has $\frac{4}{7} \div 6 = \frac{2}{21}$ of the power.

The Shapley–Shubik power index of this weighted voting system is therefore

$$\left(\frac{3}{7}, \frac{2}{21}, \frac{2}{21}, \frac{2}{21}, \frac{2}{21}, \frac{2}{21}, \frac{2}{21}\right)$$

Because $\frac{3}{7} \div \frac{2}{21} = 4\frac{1}{2}$, the Shapley–Shubik model indicates that the chairperson is $4\frac{1}{2}$ times as powerful as an ordinary member, although his voting power is only 3 times as much.

## EXAMPLE 10 ■ A Committee with Two Co-Chairs

A committee has 7 members: two co-chairs who each have 3 votes, and five other members with 1 vote each. The quota is 7, and thus the weighted voting system is [7 : 3, 3, 1, 1, 1, 1, 1]. Suppose $A$ is one of the members with weight-1. We can determine the Shapley–Shubik power index of each member of the committee if we can determine the voting power of $A$. A permutation has $A$ as its pivot if and only if the voters coming before $A$ in the permutation have a combined weight of exactly 6. There are two kinds of permutations that meet this condition:

- first, $C_1C_2AX_1X_2X_3X_4$, where $C_1$ and $C_2$ represent co-chairs, and $X_1, \ldots, X_4$ represent members who are not co-chairs and are not $A$; and
- second, $Y_1Y_2Y_3Y_4AY_5Y_6$, where one of $Y_1, \ldots Y_4$ is a co-chair, one of $Y_5$, $Y_6$ is a co-chair, and the remaining $Y$'s are weight-1 voters.

For the first type of permutation, there are 2 ways to order the co-chairs (if the co-chairs are $P$ and $Q$, then $C_1$ could be $P$ or $Q$ and $C_2$ would be the remaining co-chair), and there are 4! ways of ordering the other 4 members. There are thus $2 \times 4! = 48$ permutations of the first type.

To count the permutations of the second type, let us start with $Y_5$ and $Y_6$. There are 2 ways to choose the co-chair and 4 ways to choose the ordinary member for these positions in the permutation. Once these are chosen, there are 2 ways to put them in order and 4! ways to put the remaining co-chair and three ordinary members in order as $Y_1$, $Y_2$, $Y_3$, and $Y_4$. Thus, the number of permutations of the second type is $2 \times 4 \times 2 \times 4! = 384$.

In all, there are $48 + 384 = 432$ permutations in which $A$ is pivotal. The Shapley–Shubik index of $A$ is therefore $\frac{432}{7!} = \frac{3}{35}$. The other weight-1 voters have the same Shapley–Shubik index, so the combined share of power of the five weight-1 voters is $5 \times \frac{3}{35} = \frac{3}{7}$. The two co-chairs split the remaining $\frac{4}{7}$ of the power, so each has $(1 - \frac{3}{7}) \div 2 = \frac{2}{7}$ of the power. The Shapley–Shubik index of the system is therefore

$$\left(\frac{2}{7}, \frac{2}{7}, \frac{3}{35}, \frac{3}{35}, \frac{3}{35}, \frac{3}{35}, \frac{3}{35}\right)$$

## 11.3 The Banzhaf Power Index

In contrast to the Shapley-Shubik power index, which is based on counting permutations, the Banzhaf power index is based on counting combinations.

**Voting Combination** DEFINITION

A **voting combination** is a list of voters indicating how each voted on an issue.

### EXAMPLE 11 ■ Voting Combinations in the 2004 Presidential Election

The voting combination for the 2004 Electoral College can be determined from Table 11.1. All voters that had Bush's margin greater than 1.000 voted for the Bush–Cheney ticket; those with Bush's margin less than 1.000 voted for the Kerry–Edwards ticket. ■

In any voting combination there may be one or more voters who have the power to change the outcome by switching their votes.

**Critical Voter** DEFINITION

A voter in a given voting combination is a **critical voter** if the outcome would be different if that voter, and no other voter, changed his or her vote.

Although each voting permutation has exactly one pivotal voter, a voting combination may have no critical voters, or it may have many.

### EXAMPLE 12 ■ A Criminal Trial

When the jury is unanimous in favor of a motion to convict (or a motion to acquit), then each juror is a critical voter. On the other hand, if all but one juror is in favor of a motion, then the motion fails, and the lone holdout is a critical voter. Voting combinations in which more than one juror opposes a motion have no critical voters. ■

### EXAMPLE 13 ■ The. U.S. Presidential Elections

Table 11.1 shows that the Bush–Cheney ticket received 286 electoral votes in 2004. The quota for the Electoral College is 270, so the ticket had 16 extra votes. A state that voted for the Bush–Cheney ticket was a critical voter if and only if its voting weight was more than 16: Thus Florida, Ohio, and Texas were critical voters. If one of these states had switched to the Kerry–Edwards ticket, then Kerry–Edwards would have won. States that voted for the Kerry–Edwards ticket were not critical voters, because if they switched to the Bush–Cheney ticket the outcome of the election would not have changed.

In the closer election of 2000, in which the Bush–Cheney ticket won by only two electoral votes, every state that voted for the Bush–Cheney ticket was a critical voter[1]. In the landslide election of 1984 in which the Reagan–Bush ticket beat the

[1]Even Nebraska was critical: See Exercise 22.

Mondale–Ferraro ticket by receiving the largest number of electoral votes in history, 155 votes more than the quota–there were no critical voters at all, because no state has more than 155 electoral votes! Nevertheless, the 1984 voting permutation had a pivot–the state that brought the Reagan–Bush ticket over the 270 vote quota.

### Banzhaf Power Index

DEFINITION

A voter's **Banzhaf power index** is the number of voting combinations in which he or she casts a critical vote.

We have seen that a juror in a criminal trial casts a critical vote in two voting combinations: One in which the jury is unanimously in favor of a motion, and one in which the juror is the lone holdout, voting against a motion that all other jurors support. Thus, each juror has a Banzhaf index of 2.

### SPOTLIGHT 11.4 A Mathematical Quagmire

A county legislature in the United States is usually called a Board of Supervisors. Unlike state legislators, who represent districts that are carefully drawn to be equal in population, supervisors in some counties represent towns within the county. Because the towns differ in population, some countries use weighted voting to compensate for the resulting inequity.

If each supervisor's voting weight is proportional to the population of the town he or she represents, there will be situations in which one or more supervisors on a board are dummy voters, even if no supervisor is dictator. In a 1965 law review article, John F. Banzhaf III showed that three of the six supervisors of Nassau County, New York, were dummies. The article inspired legal action against several elected bodies that employ weighted voting systems.

The first legal challenge to weighted voting was to invalidate the voting system of the Board of Supervisors of Washington County, New York. In its decision, the New York State Court of Appeals provided a way to fix a weighted voting system: Each supervisor's Banzhaf power index, rather than his or her voting weight, should be proportional to the population of the district that he or she represents. The court predicted that its remedy would lead to a "mathematical quagmire."

Five lawsuits, filed over a period of 25 years, challenged weighted voting in the Nassau County Board of Supervisors. These cases proved to be the mathematical quagmire that the appeals court had feared. The courts attempted to force Nassau County to comply with the Washington County decision. Although the county made a sincere attempt to do so, every voting system that it devised faced a new legal challenge. With conflicting expert testimony, the U.S. District Court finally ruled in 1993 that weighted voting was inherently unfair.

Banzhaf's law review article, which initially drew attention to weighted voting in Nassau County, was aptly titled "Weighted Voting Doesn't Work."

Nevertheless, tradition is hard to change. Many boards of supervisors of counties, particularly in the State of New York, still use weighted voting, and legal challenges to the practice, even after the Nassau County decision, have not always been successful.

In more complicated examples, it is helpful to view voting combinations in terms of **coalitions**.

## Winning and Blocking Coalitions

DEFINITION

If a voting combination results in the approval of a motion, the set of voters who support the the motion are said to form a **winning coalition**; the opposing voters are in a **losing coalition**. On the other hand, if the result of the voting combination is to defeat the motion, the voters who oppose are said to form a **blocking coalition**, and the voters in favor are a losing coalition.

If the result of a voting combination is to approve a motion, the critical voters, if there are any, belong to the winning coalition. In a combination that defeats a motion, the critical voters will belong to the blocking coalition.

## EXAMPLE 14 ▪ A Three-Member Committee

Consider a committee of three members, $A$, $B$, and $C$. The chairperson of the committee, $A$, has two votes, while $B$ and $C$ each have one. The quota is three, and this voting system is

$$[3: 2, 1, 1]$$

If the committee votes unanimously in favor of a motion, the winning coalition is $\{A,B,C\}$. Let's identify the critical voters in this coalition. Suppose $A$ switches her vote:

| *A* | *B* | *C* | **Votes** | **Outcome** |
|---|---|---|---|---|
| Yes | Yes | Yes | 4 | Pass |
| ↓ | | | | |
| No | Yes | Yes | 2 | Fail |

By changing her vote, $A$ has changed the outcome. In this coalition, $A$ is a critical voter.

Now let's see what happens if $B$ changes his vote:

| *A* | *B* | *C* | **Votes** | **Outcome** |
|---|---|---|---|---|
| Yes | Yes | Yes | 4 | Pass |
| | ↓ | | | |
| Yes | No | Yes | 3 | Pass |

This time, the outcome doesn't change, so $B$ is not a critical voter in this coalition. Since $C$ has the same power as $B$, he is also not a critical voter in the combination.

Now consider the voting combination in which $A$ and $B$ vote "yes" and $C$ votes "no." Then "yes" wins, with 3 votes, so $\{A,B\}$ is a winning coalition. There are no extra votes, so both $A$ and $B$ are critical voters in this coalition. There is a third voting winning coalition $\{A,C\}$. Again there are no "yes" votes to spare, so both $A$ and $C$ are critical voters.

There are five voting combinations in which "no" wins. The corresponding blocking coalitions are $\{A,B,C\}$, $\{A,B\}$, $\{A,C\}$, $\{A\}$, and $\{B,C\}$. It takes just 2 votes to defeat a motion, so there are no critical voters in $\{A,B,C\}$; if any voter defects, there will still be enough votes to block. In each of the weight-3 blocking coalitions, $\{A,B\}$ and $\{A,C\}$, $A$ is the only critical voter. In the weight-2 blocking coalitions, $\{A\}$ and $\{B,C\}$, all voters are critical.

To determine the Banzhaf index, we count the critical votes in each of the winning or blocking coalitions: $A$ has three critical votes in winning coalitions, and another 3 in blocking coalitions; $B$ and $C$ each have one critical vote in a winning coalition and one critical vote in a blocking coalition. We will say that the Banzhaf index of this system is (6,2,2).

The Banzhaf index provides a comparison of the voting power of the participants in a voting system. Thus, $A$, with a Banzhaf index of 6, is three times as powerful as $B$ or $C$. To determine the way voting power is distributed, we can add the numbers of critical voters for all three voters together to get $6 + 2 + 2 = 10$ critical votes in all. Thus, $A$ has 60% of the voting power, while $B$ and $C$ each have 20%. The Shapley–Shubik model gives $\frac{2}{3}$ of the power to $A$, while $B$ and $C$ each have $\frac{1}{6}$, so the models are in close agreement in this case.

## Counting Combinations

If there are three voters, $A$, $B$, and $C$, and $A$ and $C$ voted "yes" while $B$ voted "no," we might record the voting combination as "Yes,No,Yes." A briefer notation is to visualize voting combinations as **binary numbers**. A whole number $N$ is represented in binary form as a sequence of binary digits, or **bits**, which can be 0 or 1. This sequence expresses the way that $N$ can be expressed as a sum of powers of 2. For example,

$$5 = 2^2 + 2^0$$

can be represented by a binary number where bits 2 and 0 are equal to 1, and the remaining bit 1, is 0. We would say that $5_2 = 101$: This binary number could stand for the voting combination "Yes,No,Yes." The largest number that can be represented with 3 bits is 7, because $7_2 = 111$: Because the smallest number that can be represented in 3 bits is 0 ($0_2 = 000$), there is a total of 8 3-bit binary numbers; thus 8 voting combinations when there are 3 voters.

### Number of Voting Combinations — THEOREM

The number of voting combinations with $n$ voters is $2^n$.

We have seen that the number of voting combinations with 3 voters is equal to the number of 3-bit binary numbers. By the same reasoning, the number of voting combinations with $n$ voters is equal to the number of $n$-bit binary numbers. The largest $n$-bit binary number is the sequence of $n$ ones, which represents $2^n - 1$: Since we start counting with 0, there are $2^n$ $n$-bit binary numbers.

We have interpreted the Shapley-Shubik index of a voter as the probability that he or she will be pivotal in a randomly selected permutation. We can also interpret the Banzhaf index in terms of probability. A voter's Banzhaf index is simply the number of voting combinations in which he or she is a critical voter. Thus, if we divide the voter's Banzhaf index by the number of possible voting combinations, we will get the probability that the voter will cast a critical vote in a randomly selected voting combination.

There is one difference to be aware of: Because each voting permutation has exactly one pivotal voter, the sum of the Shapley-Shubik indices of all voters is 1. However, not all voting combinations have a unique critical voter, so the sum of the probabilities of the voters casting critical votes is usually not equal to 1.

The easiest way to select a voting combination randomly is for each voter to decide how to vote by tossing a coin. Thus, a voter's Banzhaf index is equal to the number of possible voting combinations times the probability that he or she will be a critical voter, provided each voter decides his or her vote by a coin toss.

A 12-member jury has $2^{12} = 4096$ voting combinations. Each juror casts a critical vote in just two combinations; thus, his or her probability of casting a critical vote if the combination is randomly selected is $\frac{2}{4096}$. This is important because we don't want any juror to vote randomly! If he or she does that, we'd like to be sure that there is very little chance that it will make a difference.

## How to Calculate the Banzhaf Power Index

To determine the Banzhaf power index of a voter $A$, we must count all possible winning and blocking coalitions of which $A$ is a member and casts a critical vote. The weight of a winning coalition must be $q$ or more, where $q$ is the quota. A blocking coalition must be large enough to deny the "yes" voters the $q$ votes they need to win. If the total weight of all the voters is $n$, then the weight of the blocking coalition has to be more than $n - q$. Assuming that all weights are integers, this means that the weight of a blocking coalition must be at least $n - q + 1$.

To identify the critical voters in a given winning or blocking coalition, the following principle is useful.

### Extra Votes Principle — THEOREM

A winning coalition with total weight $w$ has $w - q$ **extra votes**. A blocking coalition with total weight $w$ has $w - (n - q + 1)$ extra votes. The critical voters are those whose weight is more than the coalition's extra votes.

We can readily identify the critical voters in any winning or blocking coalition by comparing each voter's weight with the number of extra votes that the coalition has.

### Calculating the Banzhaf Power Index — PROCEDURE

To calculate the Banzhaf power index of a given voting system:

1. Make a list of the winning and blocking coalitions.
2. Use the **extra-votes principle** to identify the critical voters in each coalition.

A voter's Banzhaf power index is then the number of coalitions in which he or she appears as a critical voter.

In the examples that we have discussed so far, each participant has been a critical voter in exactly as many winning coalitions as blocking coalitions. This is not a coincidence, as we will now see.

**Winning-Blocking Duality** THEOREM

The number of winning coalitions in which a given voter is critical is equal to the number of blocking coalitions in which the same voter is critical.

Consider a voter $A$. For every winning coalition $C$ in which $A$ is a critical voter, let $C^*$ be the coalition that would vote against the motion if $A$ switched his vote from"yes" to "no." Because $A$ was a critical voter in $C$, the switch in vote will change the original voting combination, where the result was "yes" to "no." In other words, $C^*$ becomes a blocking coalition, and $A$ is a critical voter in $C^*$: This correspondence,

$$C \longleftrightarrow C^*$$

shows the number of winning coalitions in which $A$ is a critical voter is equal to the number of blocking coalitions in which $A$ is critical. By the winning/blocking duality principle, we can determine a voter's Banzhaf power index by doubling the number of winning coalitions in which he or she is a critical voter–there is no need to count blocking coalitions.

**TABLE 11.4 Winning Coalitions in the Four-Shareholder Corporation**

| | | | Critical Voters | | | |
|---|---|---|---|---|---|---|
| **Coalition** | **Weight** | **Extra Votes** | $A$ | $B$ | $C$ | $D$ |
| $\{A, B, C, D\}$ | 100 | 49 | | | | |
| $\{A, B, C\}$ | 90 | 39 | X | | | |
| $\{A, B, D\}$ | 80 | 29 | X | X | | |
| $\{A, C, D\}$ | 70 | 19 | X | | X | |
| $\{A, B\}$ | 70 | 19 | X | X | | |
| $\{B, C, D\}$ | 60 | 9 | | X | X | X |
| $\{A, C\}$ | 60 | 9 | X | | X | |
| Critical votes | | | 5 | 3 | 3 | 1 |

## EXAMPLE 15 ■ The Corporation with Four Shareholders

The corporation with four shareholders (see Example 8) uses the weighted voting system

$$[51 : 40, 30, 20, 10]$$

Table 11.4 displays a list of all the winning coalitions of shareholders and the number of extra votes that each has. The four columns at the right are marked to indicate the critical voters in each coalition. By doubling the critical votes shown in the table, we arrive at the Banzhaf index of the corporation: $(10, 6, 6, 2)$. In this model, $A$ has

$$\frac{10}{24} \text{ or approximately } 42\%$$

of the voting power, while $B$ and $C$ each have 25% (even though $B$ has more shares than $C$). Shareholder $D$ has the remaining 8% of the voting power, according to the Banzhaf model. In this case, power is distributed exactly as it was by the Shapley–Shubik model.

We have seen that each voting combination for a set of $n$ voters corresponds to a binary number with $n$ bits. Thus, if there are $n$ voters, there will be $2^n$ voting combinations. Obviously there is exactly one voting combination where everyone votes "yes," and one voting combination where everyone votes "no." These correspond to the $n$-bit binary numbers with all bits equal to 1, and all bits equal to 0, respectively.

The number of voting combinations with $n$ voters and exactly $k$ "yes" votes is denoted ${}_nC_k$ (when speaking, ${}_nC_k$ is pronounced "$n$ choose $k$"). Thus, the statement that there is exactly one combination of $n$ voters where everyone votes "yes" would be ${}_nC_n = 1$. Similarly, we have ${}_nC_0 = 1$. There are $n$ combinations with exactly one "yes" vote:

$$100\cdots0,\ 010\cdots0,\ \ldots,\ 000\cdots1,$$

where each combination has one 1 and $n - 1$ zeros. Thus, ${}_nC_1 = n$.

## Duality Formula for Combinations — THEOREM

If each voter in a combination with $k$ "yes" votes and $n - k$ "no" votes were to switch his or her vote to the opposite side, there would be $n - k$ "yes" votes and $k$ "no" votes. Thus, the number of combinations of $n$ voters with $k$ "yes" votes is equal to the number of combinations of $n$ voters with $n - k$ "yes" votes. This proves the following theorem:

$${}_nC_k = {}_nC_{n-k}$$

## Addition Formula for Combinations — THEOREM

Now suppose that there are $n + 1$ voters, one of whom is Zoë. We would like to determine ${}_{n+1}C_k$: The number of combinations with $k$ "yes" votes for a set of $n + 1$ voters. We will divide this set into two parts, depending on how Zoë votes. If she votes "no," there are ${}_nC_k$ voting combinations in which $k$ of the other voters say "yes." If Zoë votes "yes" then a voting combination with $k$ "yes" votes can be assembled by combining Zoë's vote with a combination of the other $n$ voters with $k - 1$ "yes" votes; there are ${}_nC_{k-1}$ of these. Adding these, we obtain a valuable formula:

$${}_{n+1}C_k = {}_nC_k + {}_nC_{k-1}$$

The **addition formula** enables us to calculate the numbers ${}_nC_k$. Starting with ${}_0C_0 = {}_1C_0 = {}_1C_1 = 1$, we obtain ${}_2C_1 = {}_1C_1 + {}_1C_0 = 2$. Continuing, it is convenient to display the results in triangular form:

## Pascal's Triangle

THEOREM

The numbers ${}_nC_k$ can be arranged in the triangle shown below. The number ${}_nC_k$ is located on the $n$th row (rows are numbered downward; the 1 at the summit is the 0th row) and then counting to the $k$th entry from the left (again, the 1 at the left end of the row is the 0th entry).

$$
\begin{array}{c}
1 \\
1 \quad 1 \\
1 \quad 2 \quad 1 \\
1 \quad 3 \quad 3 \quad 1 \\
1 \quad 4 \quad 6 \quad 4 \quad 1 \\
1 \quad 5 \quad 10 \quad 10 \quad 5 \quad 1 \\
1 \quad 6 \quad 15 \quad 20 \quad 15 \quad 6 \quad 1
\end{array}
$$

Each entry in **Pascal's triangle** is determined by adding the two entries above its location on the previous row. For example, ${}_6C_3 = 20$ on the last row in the triangle above is obtained by adding ${}_5C_3 + {}_5C_2 = 10 + 10$ on the previous row. Thus, Pascal's triangle is constructed in accordance with the addition formula. The French mathematician and philosopher, Blaise Pascal (1623–1662) is credited with the discovery of his eponymous triangle.

Pascal's triangle is an intriguing pattern, but it is only useful to calculate ${}_nC_k$ when $n$ is relatively small. The following expression gives a way to calculate ${}_nC_k$ in more general situations:

## Combination Formula

THEOREM

$$ {}_nC_k = \frac{n!}{k!(n-k)!} $$

To use the combination formula, cancel before multiplying.

### EXAMPLE 16 ■ Calculate ${}_{40}C_4$

From the combination formula, ${}_{40}C_4 = \dfrac{40!}{4!\,36!}$. Notice that $40! = 40 \times 39 \times 38 \times 37 \times 36!$. Thus we can cancel 36! and obtain

$$ {}_{40}C_4 = \frac{40 \times 39 \times 38 \times 37}{4 \times 3 \times 2 \times 1} = 91{,}390 $$

■

To verify the combination formula, let ${}_nD_k = \dfrac{n!}{k!(n-k)!}$. It's our job to show that ${}_nC_k = {}_nD_k$. Recalling that $0! = 1$, we have ${}_nD_0 = \frac{n!}{n!0!} = 1$ and ${}_nD_n = \frac{n!}{0!n!} = 1$. Also, we will see that the numbers ${}_nD_k$ obey the addition formula:

$$ {}_{n+1}D_k = {}_nD_k + {}_nD_{k-1} $$

or

$$ \frac{(n+1)!}{k!(n+1-k)!} = \frac{n!}{k!(n-k)!} + \frac{n!}{((k-1)!(n-k+1)!)} $$

To verify this equation, we have to add the two fractions on the right side. Because $k! = k \times (k-1)!$ and $(n-k+1)! = (n-k+1) \times (n-k)!$, the least common denominator is $k!(n-k+1)!$. Therefore

$$\frac{n!}{k!(n-k)!} + \frac{n!}{((k-1)!(n-k+1)!)} =$$

$$\frac{n!(n-k+1)+n!k}{k!(n+1-k)!} = \frac{n!((n-k+1)+k)}{k!(n+1-k)!} = {}_{n+1}D_k$$

It follows that if we arrange the numbers ${}_nD_k$ in a triangle, as we did ${}_nC_k$, we will again get Pascal's triangle, because the left and right edges are filled with 1's, and each interior entry is equal to the sum of the two entries above it. We thus conclude that ${}_nC_k = {}_nD_k$, and hence the combination formula holds. Efficient counting methods make it possible to compute the Banzhaf power index of large weighted voting systems. The method of counting combinations applies to systems in which most of the voters have the same weight, as in the seven-person committee that we considered in Example 9.

## EXAMPLE 17

### The Banzhaf Index of the Seven-Person Committee

The chairperson of this committee has 3 votes. Each of the six other members has 1 vote. The quota is 5, so we are considering the voting system

$$[5 : 3, 1, 1, 1, 1, 1, 1]$$

The chairperson, Alice, is a critical voter in any winning coalition with no more than 2 extra votes. To achieve the quota, Alice's coalition must include at least two weight-1 voters. If there are five or more weight-1 voters in the coalition, then Alice's vote will not be needed: Alice will not be a critical voter. The number of coalitions with two weight-1 voters is ${}_6C_2$, because we are counting the voting combinations of the six weight-1 voters with 2 "yes" votes. Similarly, there are ${}_6C_3$ coalitions consisting of $C$ and three weight-1 voters, and ${}_6C_4$ coalitions with Alice and four weight-1 voters. Referring to Pascal's triangle, displayed on the previous page, there are

$${}_6C_2 + {}_6C_3 + {}_6C_4 = 15 + 20 + 15 = 50$$

winning coalitions in which Alice is critical. Counting an equal number of blocking coalitions, Alice's Banzhaf power index is 100.

When we calculated the Shapley–Shubik power index, we only had to consider the chairperson. The other members' indices could then be determined because the Shapley–Shubik indices of all the members add up to 1. There is no fixed sum of the Banzhaf power indices of all the participants, so we have to calculate the indices of the weight-1 voters separately. These voters do have the same voting power, so we only have to consider one of them, Martin. By the extra votes principle, Martin is a critical voter in a winning coalition only if this coalition has exactly 5 votes. There are two ways to assemble such a winning coalition and include Martin:

- A three-member coalition consisting of Martin, Alice, and one of the other five weight-1 members. There are ${}_5C_1 = 5$ of these coalitions.

- A five-member coalition consisting of Martin and 4 other weight-1 voters. There are ${}_5C_4 = 5$ of these coalitions.

Adding, we find that Martin is a critical voter in 10 winning coalitions. Doubling this to account for the blocking coalitions, the Banzhaf power index of each weight-1 voter is 20. To summarize, the Banzhaf power index of this voting system is

$$(100, 20, 20, 20, 20, 20, 20).$$

The total number of critical votes is $100 + 6 \times 20 = 220$. Thus, according to the Banzhaf model, Alice has $\frac{100}{220}$, or about 45%, of the power in the committee, and each weight-1 voter has $\frac{1}{11}$, or about 9.1%, of the power. This is in pretty close agreement with the Shapley–Shubik model, where we found that Alice had $\frac{3}{7}$, or about 43%, of the power, while each weight-1 voter had $\frac{2}{21}$, or approximately 9.5%, of the power.

In Example 10, we determined the Shapley–Shubik power index of the voting system [7 : 3, 3, 1, 1, 1, 1, 1] (the committee with two co-chairs). In the following example, we will determine the Banzhaf power index of that committee.

## EXAMPLE 18 ■ The Committee with Two Co-Chairs

To determine the voting power of each voter in the system [7 : 3, 3, 1, 1, 1, 1, 1] by the Banzhaf model, let's start with a weight-1 voter, Martin. He will be a critical voter in a winning coalition if and only if the votes of the other members in the coalition add up to exactly 6. There are two ways to achieve this total:

- The two co-chairs, and no other weight-1 voters, could join with Martin. There is exactly one such coalition.
- One of the two co-chairs, and 3 of the other 4 weight-1 voters, could join with Martin. There are ${}_2C_1 \times {}_4C_3 = 8$ such coalitions.

Thus, Martin is a critical voter in 9 winning coalitions; doubling this, we find that his Banzhaf power index is 18. Now we must determine the Banzhaf power index of a weight-3 voter, Alice. She will be a critical voter in a winning coalition in which the other members have a combined total of 4, 5, or 6 votes.

- If she is joined by the other co-chair, the coalition would need 1, 2, or 3 of the 5 weight-1 members. The number of coalitions of this sort is

$${}_5C_1 + {}_5C_2 + {}_5C_3 = 5 + 10 + 10 = 25$$

- If the other co-chair is opposed, she could be joined by 4 or all 5 of the weight-1 members. The number of such coalitions is ${}_5C_4 + {}_5C_5 = 6$.

It follows that $A$ is a critical voter in 31 winning coalitions; her Banzhaf power index is 62. The Banzhaf power index of this committee is $(62, 62, 18, 18, 18, 18, 18)$.

The total number of critical votes in the committee with two co-chairs is 214. Thus, each co-chair has $\frac{62}{214}$, or approximately 29.0%, of the power, and each weight-1 member has $\frac{18}{214}$, or about 8.4%, of the power, by the Banzhaf model. In Example 10 we saw that according to the Shapley–Shubik model, the co-chairs each had about 28.6% of the power and the weight-1 members had about 8.6%. The agreement between the two models is, as in the other examples that we have considered, pretty close.

If we compare the probabilistic interpretations of the two power indices, Examples 10 and 20 do display disagreement between the models. Recall that the Shapley-Shubik index of a voter is equal to the probability that the voter will be pivotal in a randomly selected permutation. As noted, these probabilities are 28.6% and 8.6% for a co-chair and for an ordinary member, respectively. The probability that a voter will be critical in a randomly selected voting combination is equal to the voter's Banzhaf index, divided by the total number of voting combinations ($2^n$ if there are $n$ voters). A co-chair of the committee has a probability of $\frac{62}{2^7} = 48.4\%$ of casting a critical vote, while the probability that an ordinary member will be a critical voter when all voters decide their votes by a coin-toss is $\frac{18}{2^7} = 14.1\%$.

There are situations in which the differences between the models are significant by any measure. In Spotlight 11.5, the Banzhaf and Shapley–Shubik power indices of the United States Electoral College are compared. While the differences may seem small, by the Shapley–Shubik model, California has about 11.0% of the voting power in the college, while by the Banzhaf model, California has 11.4% of the power. The following example presents a situation in which the models give dramatically different results.

## EXAMPLE 19 ■ The Big Shareholder

Alice holds 100,000 shares of stock in a corporation. There is a total of one million shares of stock, and the remaining stock is held by 9000 shareholders, each of whom has 100 shares. A weighted voting system, in which each shareholder's voting weight is equal to the number of shares that he or she owns, is used.

(*vario images GmbH & Co.KG/Alamy.*)

The Shapley–Shubik index of this system is determined by the same strategy that we used in Example 9 (the seven-person committee). This time, the permutations of the stockholders are divided into 9001 groups, depending upon the location of Alice. Each group has the same number of permutations (9000! of them, to be precise), and Alice is pivotal when she appears in the 4002nd through the 5001st position. If she is 4002nd, then there are $4001 \times 100 = 400{,}100$ shares preceding her, and her 100,000 shares bring the total to a bare majority of 500,100 shares. If there are more than 5000 shares ahead of Alice, the 5001st, a small shareholder, would be the pivot. Thus, Alice is pivotal in 1000 of the 9001 groups, and her Shapley–Shubik power index is $\frac{1000}{9001}$, or about 11.1%. The 9000 small shareholders have equal shares of the remaining power: The index of each is

$$\left(1 - \frac{1000}{9001}\right) \div 9000$$

which works out to be 0.0099%.

The Banzhaf index can be approximated by referring to its probabilistic interpretation. Imagine a randomly selected voting combination involving the 9000 small shareholders (there are $2^{9000}$ voting combinations). We could obtain such a combination by having each small shareholder toss a coin.

Alice will be a critical voter in a voting combination if she votes the same way as at least 4001 and not more than 5000 weight-100 voters. Therefore, Alice's probability of casting a critical vote is equal to the probability of getting between 4001 and 5000 heads in 9000 coin tosses. On average there would be 4500 heads, and by the Central Limit Theorem (see Chapter 8) we can see that the standard deviation is $\sqrt{9000} \times \sigma$, where $\sigma$ is the standard deviation of the single-coin toss experiment where heads = 1 and tails = 0. For the single toss $\sigma = \frac{1}{2}$: For the 9000-coin toss the Central Limit Theorem says that the number of heads is normally distributed with mean 4500 and standard deviation $\frac{\sqrt{9000}}{2}$, which is less than 50. Thus, 68% of the voting combinations involving the weight-100 voters will have between 4450 and 4550 "yes" votes, 95% will have between 4400 and 4600 "yes" votes, and 99.7% will have between 4350 and 4650 "yes" votes. A voting combination with fewer than 4001 or more than 5000 "yes" votes would be more than 10 times the standard deviation away from the mean–a very unlikely event. Therefore, in almost 100% of the voting combinations, Alice's vote will be critical.

A weight-100 shareholder, Martin, will be a critical voter in any voting combination when exactly 500,000 shares are voting his way–either Martin is joined by Alice and 4000 weight-100 shareholders, or by 5000 weight-100 shareholders but not Alice. Because the average number of heads in our coin-tossing analogy is 4500, the probability of getting a number of heads so far from this average–and Martin's chance of being a critical voter–is very small. Thus Alice is critical in almost all winning coalitions, and the number of winning coalitions in which Martin is critical is negligible by comparison. In the Banzhaf model Alice has almost 100% of the power in this system.

## 11.4 Comparing Voting Systems

Different weighted voting systems may have identical sets of winning coalitions. A dictatorship is no different if the dictator's weight is exactly equal to the quota or if it is much more. The dictator will have the same Banzhaf power index ($2^n$ if there

## SPOTLIGHT 11.5 The Electoral College: Presidential Elections of 2004 and 2008

The following table displays the Shapley–Shubik (SSPI) and Banzhaf (BPI) power indices of the voters in the Electoral College, as compared with the voter's weight as a percent of 538 (PCT), the total weight of all of the voters. It shows that for the most part, both measures of power agree closely with the actual share of power that a participant in the college has by virtue of its voting weight. There is an exception, though. California, whose voting weight is slightly more than 20% of the quota, has more than its share of power by either measure. The power indices shown were calculated with the Power Index applet that you can find by visiting www.whfreeman.com/fapp8e.

| Voter | Weight | PCT (%) | SSPI (%) | BPI (%) |
|---|---|---|---|---|
| CA | 55 | 10.22 | 11.04 | 11.41 |
| TX | 34 | 6.32 | 6.50 | 6.39 |
| NY | 31 | 5.76 | 5.89 | 5.79 |
| FL | 27 | 5.02 | 5.09 | 5.01 |
| IL, PA | 21 | 3.90 | 3.91 | 3.87 |
| OH | 20 | 3.72 | 3.72 | 3.68 |
| MI | 17 | 3.16 | 3.14 | 3.12 |
| GA, NC, NJ | 15 | 2.79 | 2.76 | 2.74 |
| VA | 13 | 2.42 | 2.38 | 2.37 |
| MA | 12 | 2.23 | 2.20 | 2.19 |
| IN, MO, TN, WA | 11 | 2.04 | 2.01 | 2.01 |
| AZ, MD, MN, WI | 10 | 1.86 | 1.82 | 1.82 |
| AL, CO, LA | 9 | 1.67 | 1.64 | 1.64 |
| KY, SC | 8 | 1.49 | 1.45 | 1.46 |
| CT, IA, OK, OR | 7 | 1.30 | 1.27 | 1.27 |
| AR, KS, MS | 6 | 1.12 | 1.09 | 1.09 |
| NM, NV, UT, WV | 5 | 0.93 | 0.90 | 0.91 |
| HI, ID, NH, RI | 4 | 0.74 | 0.72 | 0.73 |
| AK, DE, DC, MT, ND, SD, VT, WY | 3 | 0.56 | 0.54 | 0.55 |
| ME, NE | 2 | 0.37 | 0.36 | 0.36 |
| Congressional districts (5 in all) | 1 | 0.19 | 0.18 | 0.18 |

are $n$ voters, since the dictator is a critical voter in every voting combination), and the same Shapley–Shubik power index (1, since the dictator is the pivotal voter in every permutation). To compare voting systems–which may be specified with weights or in some other way–we refer to the winning coalitions.

If there are just two voters, $A$ and $B$, the empty coalition, $\{\}$, is surely a losing coalition, and $\{A, B\}$ is a winning coalition. There are only three distinct voting systems in this case: In the first, unanimous consent is required to pass a measure, so the only winning coalition is $\{A, B\}$. In the second, $A$ is a dictator, and $\{A\}$ is also a winning coalition. In the third, $B$ is a dictator, and the winning coalitions are $\{B\}$ and $\{A, B\}$. Although there is an unlimited number of ways to assign weights and a quota to a two-voter system, there are only three ways that the power can be distributed: $A$ as dictator, $B$ as dictator, or consensus rule.

**Equivalent Voting Systems** DEFINITION

Two voting systems are **equivalent** if there is a way for all of the voters of the first system to exchange places with the voters of the second system and preserve all winning coalitions.

The weighted voting systems [50 : 49, 1] and [4 : 3, 3], involving pairs of voters *A*, *B*, and *C*, *D*, respectively, are equivalent because in each system, unanimous support is required to pass a measure. We could have *A* exchange places with *C*, and *B* exchange places with *D*.

Now consider two voting systems [2 : 2, 1] and [5 : 3, 6] involving the same pair of voters, *A* and *B*. In the first, *A* is a dictator, while in the second, *B* dictates. By having *A* and *B* exchange places with each other, we see that the two systems are equivalent. "Equivalent" does not mean "the same." Voter *A* would tell you that the system where he is the dictator is not the same as the system where *B* is the dictator. The systems are equivalent because each has a dictator.

Every two-voter system is equivalent either to a system with a dictator or to one that requires consensus. As the number of voters increases, the number of different types of voting systems increases.

**Minimal Winning Coalitions** DEFINITION

A **minimal winning coalition** is a winning coalition in which each voter is a critical voter.

In a dictatorship, every coalition that includes the dictator is a winning coalition, but the only *minimal* winning coalition is the one that includes the dictator and no other voters.

## EXAMPLE 20 ■

### Minimal Winning Coalitions: A Three-Voter System

The three-member committee from Example 7 uses the voting system [6 : 5, 3, 1]. Let's refer to its members as *A*, *B*, and *C* in order of decreasing weight. There are three winning coalitions. One, {*A*, *B*}, has weight 8, more than the quota, but it is minimal because both voters are critical. Another, {*A*, *C*}, with weight 6, is also minimal. The third winning coalition, {*A*, *B*, *C*}, is not minimal because only *A* is a critical voter.

## EXAMPLE 21 ■ The Four-Shareholder Corporation

Table 11.4 lists the five winning coalitions in the corporation with the voting system [51 : 40, 30, 20, 10]. In each coalition, the critical voters have been identified. The minimal ones are those in which each voter is marked as critical: {*A*, *B*}, {*A*, *C*}, and {*B*, *C*, *D*}. These minimal winning coalitions are displayed in Figure 11.1.

A voting system can be described completely by specifying its minimal winning coalitions. If you want to make up a new voting system, instead of specifying weights and a quota, you could make a list of the minimal winning coalitions. You would have to be careful that your list satisfies the following three requirements:

1. Your list can't be empty. You have to name at least one coalition–otherwise, there would be no way to approve a motion.
2. You can't have one minimal winning coalition that contains another one–otherwise, the larger coalition wouldn't be minimal.
3. Every pair of coalitions in the list has to overlap–otherwise, two opposing motions could pass.

In the four-shareholder corporation (see Figure 11.1), you can see that these requirements are satisfied. Now let's construct some voting systems.

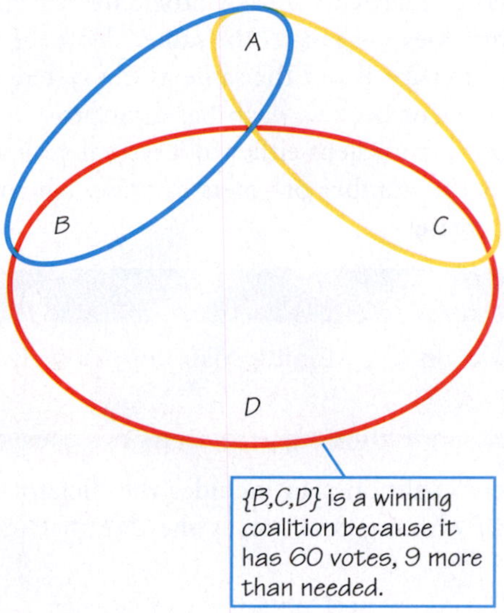
A
B
C
D
{B,C,D} is a winning coalition because it has 60 votes, 9 more than needed.

**FIGURE 11.1** Each oval surrounds a minimal winning coalition for the four-shareholder corporation.

## EXAMPLE 22 ■ Three-Voter Systems

We would like to make a list of all voting systems that have three participants, *A*, *B*, and *C*. To keep the size of the list manageable, we will insist that no two voting systems on the list be equivalent. To start, suppose that {*A*} is a minimal winning coalition. Requirement 3 tells us that every other minimal winning coalition must overlap with {*A*}, but the only way that could happen would be if *A* also belonged to the other coalition. In this case, requirement 2 would be violated. Thus, {*A*} can be the only minimal winning coalition. This is the voting system where *A* is dictator. Systems where *B* or *C* is dictator are not listed because they are equivalent to this one.

Now suppose that there is no dictator. Every minimal winning coalition must contain either two or all three voters. Let's consider the case in which {*A*, *B*, *C*} is a minimal winning coalition. It is the only winning coalition, because any other winning coalition would have to be entirely contained in this coalition, which requirement 2 doesn't allow. In this voting system, a unanimous vote is required to pass a measure. We will call this system *consensus rule*.

Finally, let's suppose that there is a two-voter minimal winning coalition, {*A*, *B*}. If it is the only minimal winning coalition, then a measure will pass if *A* and $B$ both vote "yes" and the vote of $C$ does not matter: In other words, $C$ is a dummy, and $A$ and $B$ make all the arrangements. We will call this system the *clique*. Of course, the clique could be $\{A, C\}$ or $\{B, C\}$, but these systems are equivalent to the one where $\{A, B\}$ is the clique.

There could be two two-voter minimal winning coalitions, say $\{A, B\}$ and $\{A, C\}$. Neither coalition contains the other, and there is an overlap, so all of the requirements are satisfied. In this system, $A$ has veto power. We have encountered this system in Example 5, and we will call it the *chair veto*. There are two other voting systems equivalent to this one, where $B$ or $C$ is chair.

It is possible that all three two-member coalitions are minimal winning coalitions. Because there are only three voters, any two distinct two-member coalitions will overlap, so the requirements are still satisfied. This system is called *majority rule*.

## TABLE 11.5 Voting Systems with Three Participants

| System | Minimal Winning Coalitions | Weights | Banzhaf Index |
|---|---|---|---|
| Dictator | $\{A\}$ | [3: 3, 1, 1] | (8, 0, 0) |
| Clique | $\{A, B\}$ | [4: 2, 2, 1] | (4, 4, 0) |
| Majority | $\{A, B\}$, $\{A, C\}$, $\{B, C\}$ | [2: 1, 1, 1] | (4, 4, 4) |
| Chair veto | $\{A, B\}$, $\{A, C\}$ | [3: 2, 1, 1] | (6, 2, 2) |
| Consensus | $\{A, B, C\}$ | [3: 1, 1, 1] | (2, 2, 2) |

Table 11.5 lists all five of these three-voter systems. Each system can be presented as a weighted voting system, and suitable weights are given in the table. If we want to make a similar list of all types of four-voter systems, we can start by making each three-voter system into a four-voter system. This is done by putting a fourth voter, $D$, into the system, without including him in any of the minimal winning coalitions. This makes $D$ a dummy. You may be interested to know that there are an additional nine 4-voter systems that don't have any dummies. Try to list as many of these systems as you can.

## EXAMPLE 23 ▪ The Scholarship Committee

A university offers scholarships on the basis of either academic excellence or financial need. Each application for a scholarship is reviewed by two professors, who rate the student academically, and two financial aid officers, who rate the applicant's need. If both professors or both financial aid officers recommend the applicant for a scholarship, the Dean of Admissions decides whether or not to award a scholarship. Is it possible to assign weights to the professors, the financial aid officers, and the dean to reflect this decision-making system? The answer is "no."

To see why, let's focus on the minimal winning coalitions. The participants are the two professors, $A$ and $B$; the financial aid officers $E$ and $F$; and the dean $D$. A scholarship will be offered if approved by the professors and the dean, or by the financial aid officers and the dean. Thus, the minimal winning coalitions are

$$\{A, B, D\} \qquad \text{and} \qquad \{D, E, F\}$$

(see Figure 11.2).

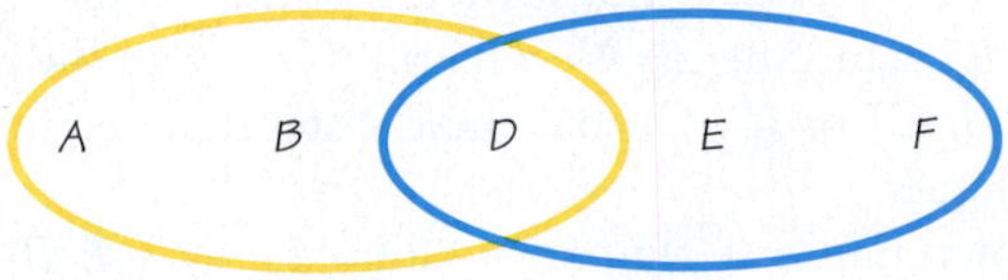


**FIGURE 11.2** The Scholarship Committee: Minimal winning coalitions.

Consider the following two winning coalitions: In the first, all except the financial aid officer $F$ favors an award; while in the second, only the professor $B$ dissents.

$$C_1 = \{A, B, D, E\} \qquad \text{and} \qquad C_2 = \{A, D, E, F\}$$

In $C_1$, we notice that $A$ is a critical voter and $E$ isn't, while in $C_2$ the tables are turned because $E$ is critical while $A$ is not. If this were a weighted voting system, then in any winning coalition, the critical voters would all have greater weight than those who are not critical. Thus $A$ would have to have both more weight than $E$ (because of the situation in $C_1$) and less weight than $E$ (because of $C_2$), which is impossible.

Although the scholarship committee is not equivalent to any weighted voting system, it is possible to determine the Shapley–Shubik and Banzhaf power indices of each participant.

## EXAMPLE 24

### Power Indices of the Scholarship Committee

The dean has veto power. Therefore, she will be the pivot in any permutation where she appears last. If she is second-to-last in a permutation, she will still be the pivot, because either both professors or both financial aid officers must come before her. In the middle position, she will be pivotal if and only if both professors or both aid officers come first. Adding this up, we have $2 \times 4! = 48$ permutations in which the dean is in fourth or fifth position. There are four permutations of the form Prof, Prof, Dean, Aid, Aid because the professors and the aid officers can be in either order, and another four of the form Aid, Aid, Dean, Prof, Prof. The dean is not the pivot when she is first or second, because at least three people have to approve a scholarship. We conclude that the dean is pivotal in $48 + 4 + 4 = 56$ permutations in all. Her Shapley–Shubik power index is therefore $\frac{56}{5!} = \frac{7}{15}$. Each of the other participants is equally powerful, and they share the remaining $\frac{8}{15}$ of the power. Thus each professor and each aid officer has a Shapley–Shubik power index of $\frac{2}{15}$.

To compute the Banzhaf index, let's list all the winning coalitions: There are 7 of them.

$$\{A, B, D\}, \{E, F, D\}, \{B, E, F, D\}, \{A, E, F, D\},$$
$$\{A, B, F, D\}, \{A, B, E, D\}, \text{ and } \{A, B, E, F, D\}$$

The dean, who has veto power, is a critical voter in each winning coalition and in 7 blocking coalitions, so her Banzhaf power index is 14. Professor $A$ is critical in each winning coalition that includes $B$ but not both aid officers: There are 3 of these. He is also critical in 3 blocking coalitions, so his Banzhaf power index is 6. The remaining participants, $B$, $E$, and $F$, have the same power, so the Banzhaf power index of the scholarship committee is

$$(14, 6, 6, 6, 6)$$

## REVIEW VOCABULARY

**Addition formula** ${}_{n+1}C_k = {}_nC_k + {}_nC_{k-1}$ (p. 350)

**Banzhaf power index** A count of all winning or blocking coalitions in which a voter is a critical member. This is a measure of the actual voting power of that voter. (p. 334)

**Bit** A binary digit: 0 or 1. (p. 347)

**Binary number** The expression of a number in base-2 notation. Let $b_n$ denote the $n$th bit of a whole number $N$. Bits are numbered starting with the 0th bit on the right. For example, in the binary number 11001101, $b_0 = 1$, $b_1 = 0$, $b_2 = b_3 = 1$, $b_4 = b_5 = 0$, and $b_6 = b_7 = 1$. The decimal form of this number is

$$2^0 + 2^2 + 2^3 + 2^6 + 2^7 = 205 \text{ (p. 347)}$$

**Blocking coalition** A coalition in opposition to a measure that can prevent the measure from passing. (p. 346)

**${}_nC_k$** The number of voting combinations in a voting system with $n$ voters, in which $k$ voters say "yes" and $n - k$ voters say "no." This number, referred to as "$n$-choose-$k$," is given by the formula

$${}_nC_k = \frac{n!}{k!(n-k)!}. \text{ (p. 346)}$$

**Coalition** The set of participants in a voting system who favor, or who oppose, a given motion. A coalition may be empty (if, for example, the voting body unanimously favors a motion, the opposing coalition is empty), it may contain some but not all voters, or it may consist of all the voters. (p. 345)

**Critical voter** A voter who can change the result of a vote by switching his or her vote. A critical voter in a winning coalition is essential to win; a critical voter in a blocking coalition is essential to block. (p. 344)

**Dictator** A participant in a voting system who can pass any issue even if all other voters oppose it and block any issue even if all other voters approve it. (p. 334)

**Duality formula** ${}_nC_k = {}_nC_{n-k}$ (p. 350)

**Dummy** A participant who has no power in a voting system. A dummy is never a critical voter in any voting combination and is never the pivotal voter in any permutation. (p. 335)

**Equivalent voting systems** Two voting systems are equivalent if there is a way for all the voters of the first system to exchange places with the voters of the second system and preserve all winning coalitions. (p. 357)

**Extra votes** The number of votes on the dominant side of a given voting combination that could be changed without altering the result. (p. 348)

**Extra-votes principle** In any voting combination, the critical voters are those on the dominant side whose voting weights exceed the combination's extra votes. (p. 348)

**Factorial** The number of permutations of $n$ voters (or $n$ distinct objects) is called *n-factorial,* or in symbols, $n!$. Because the empty coalition can be ordered in only one way, $0! = 1$. When $n$ is a positive whole number, $n!$ is equal to the product of all the integers from 1 up to $n$. If $n$ has more than one digit, then $n!$ is a pretty big number: 10! is more than three million, and 1000! has 2568 digits. (p. 339)

**Losing coalition** A coalition that does not have the voting power to get its way. (p. 346)

**Minimal winning coalition** A winning coalition that will become losing if any member defects. Each member is a critical voter. (p. 357)

**Pascal's triangle** A triangular pattern of integers, in which each entry on the left and right edges is 1, and each interior entry is equal to the sum of the two entries above it. The entry that is located $k$ units from the left (starting with $k = 0$), on the row $n$ units below the vertex, is ${}_nC_k$. (p. 351)

**Permutation** A specific ordering from first to last of the elements of a set; for example, an ordering of the participants in a voting system. (p. 337)

**Pivotal voter** The first voter in a permutation who, with his or her predecessors in the permutation, will form a winning coalition. Each permutation has one and only one pivotal voter. (p. 337)

**Power index** A numerical measure of an individual voter's ability to influence a decision, the individual's voting power. (p. 336)

**Quota** The minimum number of votes necessary to pass a measure in a weighted voting system. (p. 334)

**Shapley–Shubik power index** The number of permutations of the voters in which a given voter is pivotal, divided by the number of permutations ($n!$ if there are $n$ participants). (p. 334)

**Veto power** A voter has veto power if no issue can pass without his or her vote. A voter with veto power is a one-person blocking coalition. (p. 336)

**Voting combination** A list of voters indicating the vote of each on an issue. There is a total of $2^n$ combinations in an $n$-element set, and ${}_nC_k$ combinations with $k$ "yes" votes and $n - k$ "no" votes. (p. 336)

**Weight** The number of votes assigned to a voter in a weighted voting system, or the total number of votes of all voters in a coalition. (p. 334)

**Weighted voting system** A voting system in which each participant is assigned a voting weight (different participants may have different voting weights). A quota is specified, and if the sum of the voting weights of the voters supporting a motion is at least equal to that quota, the motion is approved. The notation $[q : w_1, w_2, \ldots w_n]$ is used to denote a system in which there are $n$ voters, with voting weights $w_1, w_2, \ldots w_n$; and the quota is $q$. (p. 334)

**Winning coalition** A set of participants in a voting system who can pass a measure by voting for it. (p. 346)

## SKILLS CHECK

**1.** In the weighted voting system [65 : 60, 30, 10],
**(a)** the weight-60 voter is a dictator.
**(b)** the weight-30 voter has veto power.
**(c)** the weight-10 voter is not a dummy.

**2.** A voting system has 20 voters, and a simple majority is needed to pass a motion. The quota for this system is ______________________________ .

**3.** Two daughters and a son administer a trust fund. Each daughter has six votes, and the son has two votes; the quota for passing a measure is 8.
**(a)** The son is a dummy voter.
**(b)** The son is not a dummy voter but has less power than a daughter.
**(c)** The three siblings have equal voting power.

**4.** Four voters, $A$, $B$, $C$, $D$ use the weighted voting system [6 : 4, 3, 2, 1]. In the permutation $DBCA$, the pivotal voter is ______________________ .

**5.** If the last voter in some permutation is the pivotal voter, then
**(a)** That voter must be the dictator.
**(b)** That voter has veto power.
**(c)** All the other voters must be dummies.

**6.** The Shapley–Shubik index of the weight-3 voter in the voting system [6; 4, 3, 2, 1] is ____________ .

**7.** The number 8! (eight factorial) is
**(a)** More than one million
**(b)** Between 10,000 and one million
**(c)** Less than 10,000

**8.** Six voters can be ordered from first to last in ____ ways.

**9.** In how many ways can six voters respond to a "yes–no" question?
**(a)** 12
**(b)** 36
**(c)** 64

**10.** If a motion passes in the weighted voting system [6 : 4, 3, 2, 1] with only the weight-2 voter dissenting, then the critical voters are ________________ .

**11.** Four voters, $A$, $B$, $C$, $D$ use the weighted voting system [6 : 4, 3, 2, 1]. The Banzhaf index of $B$ is
**(a)** 3
**(b)** 6
**(c)** 14

**12.** In the voting system [19 : 8, 7, 6, 5, 4, 3, 2, 1], there is a total of _____ voting combinations. With the combination 10010101 does the motion pass? _______ .

**13.** In the voting system [6; 4, 1, 1, 1, 1, 1] the Banzhaf index of a weight-1 voter is
**(a)** $2 \times {}_4C_1$
**(b)** $2 \times {}_5C_2$
**(c)** $2 \times {}_6C_3$

**14.** ${}_6C_3 =$ ______________________________ .

**15.** ${}_{12}C_3 =$
**(a)** ${}_{12}C_9$
**(b)** 220
**(c)** Both of the above

**16.** The Banzhaf index of the weight-3 voter in the system [7 : 3, 2, 2, 2] is ________________ .

**17.** If voter $X$ is critical in every winning coalition, then
**(a)** $X$ has veto power.
**(b)** $X$ is a dictator.
**(c)** $X$ will also be a critical voter in every blocking coalition.

**18.** In a system with $n$ voters, $A$ is a dictator. The Banzhaf index of $A$ is ____________________ .

**19.** If a winning coalition is minimal, the number of extra votes is
**(a)** zero.
**(b)** less than the weight of the least powerful member of the coalition.
**(c)** more than the weight of the least powerful member of the opposing coalition.

**20.** List the minimal winning coalitions of the system [7 : 3, 2, 2, 2], with voters named $A$, $B$, $C$, $D$.

## CHAPTER 11 EXERCISES

■ Challenge ◆ Discussion

### 11.1 How Weighted Voting Works

◆ **1.** In the United States Senate, each of the 100 senators has one vote, and the vice president of the United States can vote also if it is necessary to break a tie.
**(a)** A simple majority is needed to pass a bill. What constitutes a winning coalition, and what constitutes a blocking coalition?
**(b)** A three-fifths majority is needed to end a filibuster. What constitutes a blocking coalition in this case?

◆ **2.** Is it possible to have a weighted voting system in which more votes are required to block a measure than to pass a measure?

**3.** Which voters, if any, have veto power in the weighted voting system [9 : 5, 4, 3]? Is any voter a dummy?

**4.** Given a voting system [$q$ : 33, 32, 31, 4] such that exactly one of the voters has veto power,

**(a)** which voter has veto power?
**(b)** find $q$.
**(c)** is any voter a dummy?

**5.** The various weighted voting systems used by the Board of Supervisors of Nassau County, New York turned out to be the mathematical quagmire described in Spotlight 11.4. Before the county's weighted voting was declared unconstitutional by a federal district court in 1993, it was changed several times. The weights in use since 1958 were as follows:

| | | Weights | | | | | |
|---|---|---|---|---|---|---|---|
| **Year** | **Quota** | $H_1$ | $H_2$ | $N$ | $B$ | $G$ | $L$ |
| 1958 | 16 | 9 | 9 | 7 | 3 | 1 | 1 |
| 1964 | 58 | 31 | 31 | 21 | 28 | 2 | 2 |
| 1970 | 63 | 31 | 31 | 21 | 28 | 2 | 2 |
| 1976 | 71 | 35 | 35 | 23 | 32 | 2 | 3 |
| 1982 | 65 | 30 | 28 | 15 | 22 | 6 | 7 |

Here $H_1$ is the presiding supervisor, always from the community of Hempstead; $H_2$ is the second supervisor from Hempstead; and $N$, $B$, $G$, and $L$ are the supervisors from the remaining districts: North Hempstead, Oyster Bay, Glen Cove, and Long Beach.

**(a)** In which years were some supervisors dummy voters?
**(b)** Suppose that the two Hempstead supervisors always vote together. In which years are some of the supervisors dummy voters?

## 11.2 Shapley–Shubik Power Index

**6.** Can a dummy be pivotal in any permutation? Explain why or why not.

**7.** A jury requires a unanimous vote to convict or to acquit. Give a quick way to determine the pivotal voter in any permutation of the jury's members.

**8.** For the weighted voting system [51 : 30, 25, 24, 21]:

**(a)** List all permutations in which the weight-30 voter is pivotal.
**(b)** List all permutations in which the weight-25 voter is pivotal.
**(c)** Calculate the Shapley–Shubik index.

**9.** How would the Shapley–Shubik index in Exercise 8 change if the quota were increased to

**(a)** 52?
**(b)** 55?
**(c)** 58?

**10.** In the voting system [7 : 3, 2, 2, 2, 2, 2]:

**(a)** Describe the set of permutations in which the weight-3 voter is pivotal.
**(b)** How many of these permutations are there?
**(c)** Use the answer that you have given in part (b) to determine the Shapley–Shubik index of the system.

**11.** Refer to the permutation of the 2004 presidential election (see Table 11.1). The Republican Bush–Cheney ticket carried Nevada (NV), 414,939 to 393,372. Which state would be the pivot if, at the last minute, their opponents, Kerry–Edwards, had broadcast an ad that convinced 1000 voters to switch from Bush–Cheney to Kerry–Edwards? Assume that no votes are changed outside of Nevada.

◆ **12.** Show that if a state uses the district system to choose its electors in a two-candidate presidential election, as Maine and Nebraska do, then some electoral permutations are impossible. Give an example of an impossible permutation. How would this affect the calculation of the Shapley–Shubik power index?

## 11.3 Banzhaf Power Index

**13.** **(a)** List the 16 possible combinations of how four voters, $A$, $B$, $C$, and $D$, can vote either "yes" (1) or "no" (0) on an issue.
**(b)** List the 16 subsets of the set $\{A, B, C, D\}$.
**(c)** How do the lists in parts (a) and (b) correspond to each other?
**(d)** In how many of the combinations in part (a) is the vote

**(i)** 4 Y to 0 N?
**(ii)** 3 Y to 1 N?
**(iii)** 2 Y to 2 N?

**14.** Determine the number of extra votes for each winning coalition, and calculate the Banzhaf index for each of the following weighted voting systems.

**(a)** [51: 52, 48]
**(b)** [3: 2, 2, 1]
**(c)** [8: 5, 4, 3]
**(d)** [51: 45, 43, 8, 4]
**(e)** [51: 45, 43, 6, 6]

**15.** Make a table with all winning coalitions in the weighted voting system [51 : 30, 25, 24, 21] in a column, and put the number of extra votes for each coalition in an adjacent column. Use your table to identify the critical voters in each coalition then determine the Banzhaf power index of each participant.

**16.** If the quota for the voting system in Exercise 15 increases, you can quickly modify the table you made by reducing the extra votes–eliminate coalitions when their extra votes become negative. As the number of extra votes decreases, more of a coalition's voters will be critical. Use this method to track changes in the Banzhaf index as the quota increases from the original 51 to 100.

**17.** Calculate the following[1]:

**(a)** ${}_7C_3$
**(b)** ${}_{50}C_{100}$
**(c)** ${}_{15}C_2$
**(d)** ${}_{15}C_{13}$

**18.** Calculate the following:

**(a)** ${}_6C_3$
**(b)** ${}_{100}C_2$
**(c)** ${}_{100}C_{98}$
**(d)** ${}_9C_5$

**19.** Refer to Exercise 5 for the a brief history of weighted voting in the Nassau County Board of Supervisors. Assume that the two Hempstead supervisors always agree, so that the board is in effect a five-voter system. Determine the Banzhaf index of this system in each year. You should be able to do this by hand. If you would like to find the index for the full system each year, you may use the power index calculator on www.whfreeman.com/fapp8e.

**20.** In 1982, a special supermajority of 72 votes was needed to pass measures that ordinarily require a two-thirds majority in the Nassau County Board of Supervisors (see Exercise 5). If the two Hempstead supervisors vote together, what is the Banzhaf index of the resulting five-voter system? Use the power index calculator at www.whfreeman.com/fapp8e to determine the Banzhaf index of the full six-voter system with this quota.

**21.** If each member of a 12-person jury, in which a unanimous decision is required, tosses a coin to determine his or her decision, the probability that a given juror will cast a critical vote is $\frac{1}{2048}$. In some states, civil cases are tried before a 6-person jury, and the quota for a decision is 5 votes. With such a jury, what is the probability that a given juror will cast a critical vote, if each juror uses a coin toss to determine his or her vote?

**22.** Nebraska has 5 electoral votes. Two electors are selected by the candidate who obtained the plurality in the statewide vote, and each of the remaining three electors is selected by the candidate who won a plurality in one of Nebraska's congressional districts. Show that if there are only two candidates on the ballot, then the candidate who won the statewide vote will get at least three electors from Nebraska.

## 11.4 Comparing Voting Systems

**23.** Consider a four-person voting system with voters $A$, $B$, $C$, and $D$. The winning coalitions are

$$\{A, B, C, D\}, \{A, B, C\}, \{A, B, D\}, \{A, C, D\}, \text{ and } \{A, B\}$$

**(a)** List the minimal winning coalitions.
◆ **(b)** Show that $\{A\}$ is a minimal blocking coalition. Are there other minimal blocking coalitions?
**(c)** Determine the Banzhaf power index for this voting system.
■ **(d)** Find an equivalent weighted voting system. *Hint*: If two voters have the same Banzhaf index, give them the same weight.
**(e)** Calculate the Shapley–Shubik index of this system.

**TABLE 11.6** Nassau County Board of Supervisors, 1982

| Supervisor From | Population | Number of Votes | Banzhaf Power Index | |
|---|---|---|---|---|
| Quota | | | 65 | 72 |
| Hempstead (presiding) | 738,517 | 30 | 30 | 26 |
| Hempstead | | 28 | 26 | 22 |
| North Hempstead | 218,624 | 15 | 18 | 18 |
| Oyster Bay | 305,750 | 22 | 22 | 18 |
| Glen Cove | 24,618 | 6 | 2 | 2 |
| Long Beach | 43,073 | 7 | 6 | 6 |
| Totals | 1,320,582 | 108 | 104 | 92 |

[1]**(b)** is not an error–how many ways can you (legitimately) get 100 "yes" votes from 50 voters?

◆ **24.** Must minimal blocking coalitions overlap, as minimal winning coalitions do?

◆ **25.** A five-member committee has the following voting system. The chairperson can pass or block any motion that she supports or opposes, provided that at least one other member is on her side. Show that this voting system is equivalent to the weighted voting system [4: 3, 1, 1, 1, 1].

**26.** Calculate the Banzhaf index for the weighted voting system in Exercise 25.

**27.** Find weighted voting systems that are equivalent to

**(a)** a committee of three faculty members and the dean. To pass a measure, at least two faculty members and the dean must vote "yes."
**(b)** a committee of three faculty members, the dean, and the provost. To pass a measure, two faculty, the dean, and the provost must vote "yes."

◆ **28.** A four-member faculty committee and a three-member administration committee vote separately on each issue. The measure passes if it receives the support of a majority of each of the committees. Show that this system is not equivalent to a weighted voting system.

**29.** Calculate the Banzhaf index of the voting system in Exercise 28. Who is more powerful according to the Banzhaf model, a faculty member or an administrator?

**30.** Determine the Shapley–Shubik index of the system in Exercise 28. Who is more powerful according to the Shapley–Shubik model, a faculty member or an administrator?

◆ **31.** Explain why a voting system in which no voter has veto power must have at least three minimal winning coalitions.

■ **32.** How many *distinct* (nonequivalent) voting systems with four voters can you find? Systems that have dummies don't count. The challenge is to find all nine.

**33.** A corporation has four shareholders and a total of 100 shares. The quota for passing a measure is the votes of shareholders owning 51 or more shares. The number of shares owned are as follows:

| | |
|---|---|
| *A* | 48 shares |
| *B* | 23 shares |
| *C* | 22 shares |
| *D* | 7 shares |

There is also an investor, *E*, who is interested in buying shares but does not own any shares at present. Sales of fractional shares are not permitted.

**(a)** List the winning coalitions and compute the number of extra votes for each. Make a separate list of the losing coalitions, and compute the number of votes that would be needed to make the coalition winning.
**(b)** How many shares can *A* sell to *B* without causing any of the winning coalitions listed in part (a) to lose or any of the losing coalitions in part (a) to win?
**(c)** How many shares can *A* sell to *D* without changing the sets of winning or losing coalitions?
**(d)** How many shares can *A* sell to *E* without changing the winning coalitions? Since *E* is now a dummy, he must remain a dummy after the trade.

**34.** Which of the following voting systems is equivalent to the voting system in use by the corporation in Exercise 28?

**(a)** [3: 1, 1, 1, 1]
**(b)** [3: 2, 1, 1, 1]
**(c)** [5: 3, 1, 1, 1]
**(d)** [5: 3, 2, 1, 1]

**35.** A nine-member committee has a chairperson and eight ordinary members. A motion can pass if and only if it has the support of the chairperson and at least two other members, or if it has the support of all eight ordinary members.

**(a)** Find an equivalent weighted voting system.
**(b)** Determine the Banzhaf power index.
**(c)** Determine the Shapley–Shubik power index.
**(d)** Compare the results of parts (b) and (c): Do the power indices agree on how power is shared in this committee?

**36.** The New York City Board of Estimate consists of the mayor, the comptroller, the city council president, and the presidents of each of the five boroughs. It employed a voting system in which the city officials each had 2 votes and the borough presidents each had 1; the quota to pass a measure was 6. This voting system was declared unconstitutional by the U.S. Supreme Court in 1989 (*Morris* v. *Board of Estimate*).

**(a)** Describe the minimal winning coalitions.
**(b)** Determine the Banzhaf power index.

◆ **37.** Here is a proposed weighted voting system for the New York City Board of Estimate that is based on the populations of the boroughs (see Exercise 36):

[71: 35, 35, 35, 11.3, 7.3, 9.6, 6.0, 1.8]

Find a simpler system of weights that yields an equivalent voting system.

**38.** The United Nations Security Council has 5 permanent members–China, France, Russia, the United Kingdom, and the United States–and 10 other members that serve two-year terms. To resolve a dispute not involving a member of the Security Council, 9 votes, including the votes of each of the permanent members, are required. (Thus, each permanent member has veto power.)

◆ **(a)** Show that this voting system is equivalent to the weighted voting system in which each permanent member has 7 votes, each ordinary member has 1 vote, and the quota is 39.

**(b)** Compute the Banzhaf index for the Security Council.

◆ **(c)** The Security Council originally had 5 permanent members and 6 members who served two-year terms. Each permanent member had veto power, and 6 votes were required to resolve an issue. Devise an equivalent weighted voting system and compute its Banzhaf index. Do you think that the addition of 4 more nonpermanent members involved a significant loss of power by the permanent members?

**39.** Find the minimal winning coalitions of the weighted voting system [7 : 3, 3, 3, 1, 1, 1] and determine the Banzhaf index.

**40.** A new weight-1 voter joins the system of Exercise 39. Again, describe the minimal winning coalitions and determine the Banzhaf power index. Does the presence of this new voter increase or decrease the share of power of each weight-1 voter?

**41.** Compute the Shapley–Shubik power index for the systems in Exercises 39 and 40. How does the addition of the new voter affect the power of the other three weight-1 voters?

◆ **42.** An alumni committee consists of 3 rich alumni and 12 recent graduates. To pass a measure, a majority, including at least 2 of the rich alumni, must approve. Is this equivalent to a weighted voting system? If so, find the weights and a quota; if not, explain why not.

## WRITING PROJECTS

**1.** The most important weighted voting system in the United States is the Electoral College (see Spotlight 11.1). Three alternate methods to elect the president of the United States have been proposed:

- *Direct election.* The Electoral College would be abolished, and the candidate receiving a plurality of the votes would be elected. Most versions of this system include a runoff election or a vote in the House of Representatives in cases where no candidate receives more than 40% of the vote.
- *District system.* This system could be adopted by individual states without amending the Constitution or passing a federal law. It is now in use by two states, Maine and Nebraska. In each congressional district, and in the District of Columbia, the candidate receiving the plurality would select one elector. Furthermore, in each state, including the District of Columbia, the candidate receiving the plurality would receive two electors.
- *Proportional system.* Each state and the District of Columbia would have fractional electoral votes assigned to each candidate in proportion to the number of popular votes received. With this system, if a candidate received 25% of the vote in New Mexico, then that candidate would receive 25%, or 1.25, of New Mexico's 5 electoral votes. Obviously, no actual electors would be involved.

Determine the outcome of a recent election under each of these alternatives. Should the present electoral college, operating under the unit rule, be replaced by one of these systems? Reference: *The Presidential Election Game*, by Steven Brams, which contains useful references to Senate hearings on electoral college reform.

**2.** Write an essay on weighted voting in the Council of Ministers of the European Community. Compute the Banzhaf and Shapley–Shubik indices for the system as it was in 1958. In later years, the number of member nations increased significantly, and you may want to use the power index calculator, available on the Web at www.whfreeman.com/fapp8e. If they differ significantly in their allocation of power, which index represents the true balance of power best?

**3.** California has 10.22% of the votes in Electoral College, but according to Spotlight 11.5 that state has more than 11% of the power in the Electoral College, as measured by either of our power indices. Discuss the appropriateness of each power index as a measure of voting power in the Electoral College. Is the disproportionate power of California in the Electoral College a problem that the United States should address? Assume that California has acquired additional congressional seats as a result of migration. Calculate the Banzhaf index when California has 65, 75, and 100 electors. In each case take the electoral votes that are to be awarded to California are taken from other states. What would happen if all states, except California, adopted the district system for choosing electors? See Writing Project 1 for a discussion of the district system.

**4.** Spotlight 11.6 demonstrates that an individual voter's probability of being a critical voter in a randomly selected voting combination is inversely proportional to the square root of the number of votes cast. If the voter is electing a representative in a weighted voting system (such as the Electoral College or a minister in the European Community), that representative's chance of being a critical voter is proportional to its Banzhaf index. Thus, the individual voter's probability of casting a vote that would change an outcome (of a U.S. Presidential election, for example) is proportional to the

## SPOTLIGHT 11.6 What Are Your Chances of Being a Critical Voter?

An answer to this question was given by Lionel Penrose (1898–1972), a famous geneticist, and father of Sir Roger Penrose, whose tilings you will encounter in Chapter 20. For simplicity, suppose there is an odd number of voters, and let the number be $2n + 1$. Each voter has weight 1 and the quota is a simple majority of $n + 1$ votes. If you belong to a winning coalition and your vote is critical, then the coalition must contain, in addition to you, exactly $n$ other voters. The number of such coalitions is the number of voting combinations of the other $2n$ voters with $n$ "yes" votes; that is,

$$_{2n}C_n = \frac{(2n)!}{(n!)^2}$$

combinations. There is a total of $2^{2n}$ voting combinations involving the other $2n$ voters, so your probability, which we call $P$, of being a critical voter—where your vote *really* makes a difference—is

$$P = \frac{_{2n}C_n}{2^{2n}} = \frac{(2n)!}{2^{2n}(n!)^2}$$

With the number $n$ in the tens of thousands (for a municipal election) up to the tens of millions (for a statewide election in a large state), it seems hopeless to calculate this probability. A formula discovered by a Scottish mathematician, James Stirling (1692–1770) for factorials tames the expression. It gives an approximation of $n!$, which is not very good when $n$ is small, but quite accurate when $n$ is large. If you would like to see his formula, it can be found by searching the Web for "Stirling's Formula." By using this formula, we find that $P$ is approximately $\frac{1}{\sqrt{\pi n}}$. This approximation improves as $n$ increases, as you can see in the following table.

| $n$ | $\frac{_{2n}C_2}{2^{2n}}$ | $\frac{1}{\sqrt{\pi n}}$ |
|---|---|---|
| 1 | 0.5 | 0.56 |
| 5 | 0.246 | 0.252 |
| 25 | 0.1123 | 0.1128 |
| 125 | 0.05041 | 0.05046 |

Thus if two million votes are cast in addition to yours ($n = 1{,}000{,}000$) your probability of casting a critical vote, a situation in which your candidate wins by 1 vote, is $\frac{1}{\sqrt{1{,}000{,}000\pi}} = 0.000564$. This represents a small chance, but your odds are better than they are for winning a big lottery prize!

History has recorded very few elections that were decided by a one-vote margin. Perhaps this is because the mathematical model does not involve politics. In fact, it assumes that voters decide their preferences by tossing coins rather than paying attention to the candidates. The model's purpose is to analyze the voting system itself, and ignore human behavior.

voting weight of the representative, divided by the square root of the number of votes cast in the election of the representative. Using the principle that each individual voter should have the same chance of influencing the outcome, it follows that each representative's Banzhaf index should be proportional to the square root of the state's voting population. This is the Banzhaf square root rule. By this rule, what should the weight of each state be in the Electoral College, or in the Council of Ministers of the European Union? *Suggestion*: Make the voting weights roughly proportional to the square roots of the populations, determine the Banzhaf indices, and then make adjustments as necessary.

## SUGGESTED READINGS

BRAMS, STEVEN J. *Game Theory and Politics,* 2nd Ed., Dover Publications, New York, 2004.

*Iannucci* v. *Board of Supervisors of Washington County* 20 N.Y. 2d 244, 251, 229 N.E. 2d 195, 198, 282 N.Y.S. 2d 502, 507 (1967). This code will help a law librarian find this case for you. It opened a "mathematical quagmire."

FELSENTHAL, DAN S., and MOSHÉ MACHOVER, *The Measurement of Voting Power: Theory and Practice, Problems and Paradoxes,* Edward Elgar, Cheltenham, U.K., 1998. This book has a detailed analysis of the Council of Ministers of the European Community, and a thorough treatment of the power indices mentioned in this chapter.

TAYLOR, ALAN D. *Mathematics and Politics: Strategy, Voting Power, and Proof,* Springer-Verlag, New York, 1995. Chapter 4 covers weighted voting systems and their analysis using the Shapley–Shubik and Banzhaf power indices. It has no mathematical prerequisites, but it does include carefully written logical arguments that must be carefully read.

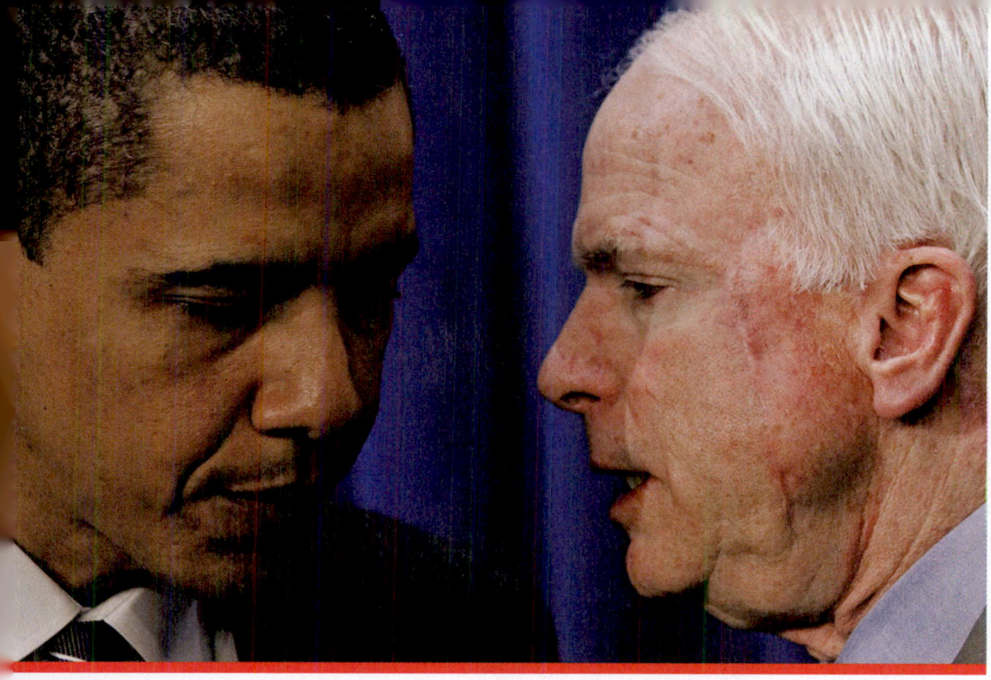

# CHAPTER 12



# Electing the President

Electing the president of the United States has been a tricky business since the founding of the republic. In 1824 none of the four presidential candidates received a majority of votes in the **Electoral College**–the body that elects the president, currently made up of 538 members, which we analyzed in Chapter 11. Consequently, the election went to the House of Representatives, as mandated by the Twelfth Amendment of the U.S. Constitution, wherein John Quincy Adams defeated Andrew Jackson. Although Jackson had won the most popular votes, Henry Clay threw his support to Adams in the Electoral College. Clay was subsequently appointed secretary of state by Adams in what was called the "corrupt bargain."

There were two more instances in the nineteenth century when the popular-vote winner lost the presidential election. In 1876, Rutherford B. Hayes defeated Samuel J. Tilden. In 1888, Benjamin Harrison defeated Grover Cleveland (although Cleveland got his revenge four years later when he beat Harrison). But it was the recurrence of this divided outcome in the 2000 presidential election that provoked the most controversy.

In the 2000 election, Al Gore received 537,000 more popular votes (0.5%) than did George W. Bush, but Bush won the electoral-vote tally by 4 votes. The outcome of this election turned on who would win Florida. Thirty-six days after the election, the Supreme Court, in a 5-4 decision, blocked further vote recounts in disputed Florida counties. By winning in Florida by a razor-thin margin of 537 votes (less than 0.01% of those cast), George W. Bush won the presidency. Although analysts thought there might be another divided outcome in 2004, George W. Bush beat John Kerry both in electoral votes (286 to 252) and in popular votes (by 3.3 million, or 2.8%).

We will not give further details of the 2000 election here but will focus instead on how presidential elections, in general, can be modeled as games, and what insights mathematics can provide about the strategic aspects of campaigning and elections. (In Chapter 15, we provide a more systematic introduction to

(*Getty Images.*)

game theory and discuss a variety of other applications of the theory.) There are several phases in the presidential election process. Because the rules of play change from phase to phase, the optimal strategies candidates pursue in their quest for the presidency change, too.

The first phase begins when Democratic and Republican candidates seek their party's nomination for president by running in state primaries. New Hampshire is the state that kicks off the primary season in January of a presidential election year, although it is preceded by caucuses in Iowa, where people gather in different locations throughout the state to discuss the candidates and then vote for party delegates who represent them. The strategic question in this phase is how to position oneself to gather momentum and thereby beat the competition in a set of contests spread over several months. But the process usually boils down to a heated race in the first few weeks following the New Hampshire primaries.

Because winning in the presidential primaries almost always assures a candidate of his or her party's nomination in its national convention in July or August, the convention is usually just a rubber stamp for the winner in the primary phase. The last exception to this rule occurred in the Democratic Party convention in 1968. After the incumbent president, Lyndon B. Johnson, withdrew from the race following the New Hampshire primary (which he had won!), Vice President Hubert H. Humphrey won his party's presidential nomination without running in the primaries.

There has not been a national party convention that has taken more than one ballot, or round of voting, to decide its presidential nominee since Adlai E. Stevenson defeated Estes Kefauver for the Democratic Party nomination in three ballots in 1952. Conventions are usually staged with consensus-promoting hoopla to unite the different factions of a party, especially if they have been sharply divided in the primaries, in preparation for the final phase of the election.

This occurs with the general election in the fall, which typically involves only two serious contenders–the nominees of the Democratic and Republican parties–but sometimes includes significant other candidates. One example is Ross Perot, who ran with no party affiliation in 1992 and garnered 19% of the popular vote. Although no minor-party candidate has ever won a presidential election, some have affected which of the two major-party nominees did win, as we will see. In the general election, the role of the Electoral College becomes paramount, as the 2000 election vividly demonstrated.

What can mathematics tell us about presidential elections? First, it can clarify what are better and worse campaign strategies in each phase. In addition, it can shed light on the likely effect that different election reforms would have on both campaign strategies and election outcomes. In fact, we will analyze two prominent reform proposals and indicate how they might have changed the outcome in the 2000 election: (1) the use of approval voting and (2) the abolition of the Electoral College and its replacement by direct popular-vote election of the president. We will also briefly discuss a recent third proposal–that states pledge to cast all their electoral votes for the national popular-vote winner. Throughout we will highlight general results that emerge from the mathematical analysis.

We begin by looking at how candidates position themselves in presidential primaries to win their party's nomination. The principal tool of analysis is **spatial models**, which we will describe in the next section and apply to both two-candidate and multicandidate elections.

## 12.1 Spatial Models for Two-Candidate Elections

While two-candidate contests are most common in the general election, sometimes the nomination race in the Democratic Party or Republican Party also comes down to a contest between just two contenders. As a case in point, Gerald Ford faced one major opponent, Ronald Reagan, in the 1976 Republican race. This race was not decided until the Republican national convention in August, when Ford edged out Reagan in a close vote.

To model such elections, we assume that voters respond to the positions that candidates take on issues. This is not to say that other factors, such as personality, ethnicity, religion, and race, have no effect on election outcomes but rather that issues take precedence in a voter's decision.

How can the position of a candidate on issues be represented? We start by assuming that there is a single overriding issue, or set of issues, on which the candidates must take a definite stand, such as the degree of governmental intervention in the economy. We assume that the attitudes of voters on this issue or dimension can be represented along a left–right continuum, ranging from very liberal on the left (much intervention) to very conservative on the right (little intervention).

To derive conclusions about the behavior of voters from their attitudes and the positions candidates take in a campaign, some assumption is necessary about how voters decide for whom to vote. More important than the attitudes of *individual* voters, however, are the *numbers* of voters who have particular attitudes along the left–right continuum.

### Unimodal Distribution

**Voter Distribution** DEFINITION

A **voter distribution** is a curve that gives the number (or percentage) of voters who have attitudes at each point along a horizontal axis.

The greater the vertical height of the curve, the more voters have attitudes at that point. Figure 12.1a shows one such distribution, which is unimodal.

## Mode and Unimodal Distribution

DEFINITION

A distribution that has one peak or highest point, called the **mode,** is **unimodal**.

For simplicity, we picture the distribution as continuous, although in fact, because the number of voters is finite, there cannot be voters at all points along the continuum.

More important than the mode, from the viewpoint of the candidates, is the **median M** of a distribution.

## Median

DEFINITION

The **median *M*** of a voter distribution is the point on the horizontal axis where half the voters have attitudes that lie to the left and half to the right.

The notion of the median of a voter distribution is the same as the notion of a median of a data sample given in Chapter 5.

The Figure 12.1a distribution is *symmetric*–the curve to the left of *M* is a mirror image of the curve to the right. Thus, the same numbers of voters have attitudes that are equal distances to the left and to the right of *M*.

Although the *attitudes* of voters are a fixed quantity in the calculations of the candidates, the *decisions* of voters will depend on the positions that the candidates take. Assuming the candidates know the distribution of voter attitudes, what positions are optimal for them?

Assume candidates *A* (red) and *B* (blue) take the positions along the left–right axis shown in Figure 12.1a, where candidate *A* is to the left of *M* and candidate *B* is to the right. Assume that all voters vote for the candidate whose position is closer to their own, and that all voters vote (we will consider modifications of this assumption later in the exercises). Then *A* will certainly attract all the voters to the left of his position, and *B* all the voters to the right of her position. If both candidates are an equal distance from *M*, as shown in Figure 12.1a, they will split any votes in the middle, with those to the left of *M* going to *A* and those to the right going to *B*.

Can either candidate do better by changing his or her position? If *B*'s position remains fixed to the right of *M*, *A* could move alongside *B*, just to her left, and capture all the votes to *B*'s left, as illustrated in Figure 12.1b. Because *A* would have moved to the right of *M*, he would, by changing his position in this manner, receive a majority of the votes and thereby win the election.

By analogous reasoning, there is no reason for *B* to stick to her original position to the right of *M*. By approaching *A*'s original position to the left of *M*, *B* can capture all the votes to *A*'s right (Figure 12.1c). In other words, both candidates, acting rationally, should approach each other and *M*.

If *A* were to move rightward past *M*, but *B* moved leftward only as far as *M*, their positions would be as shown in Figure 12.1d. Now *B* would receive not only the 50% of the votes to the left of *M* but also some of the votes that lie between *B*'s position at *M* and *A*'s position (now to the right of *M*).

Clearly, *A* loses by crossing *M* from the left. Hence, there is an incentive for both candidates to move toward *M* but not overstep it. In fact, taking a position at *M* maximizes the minimum number of votes a candidate can guarantee for himself or herself.

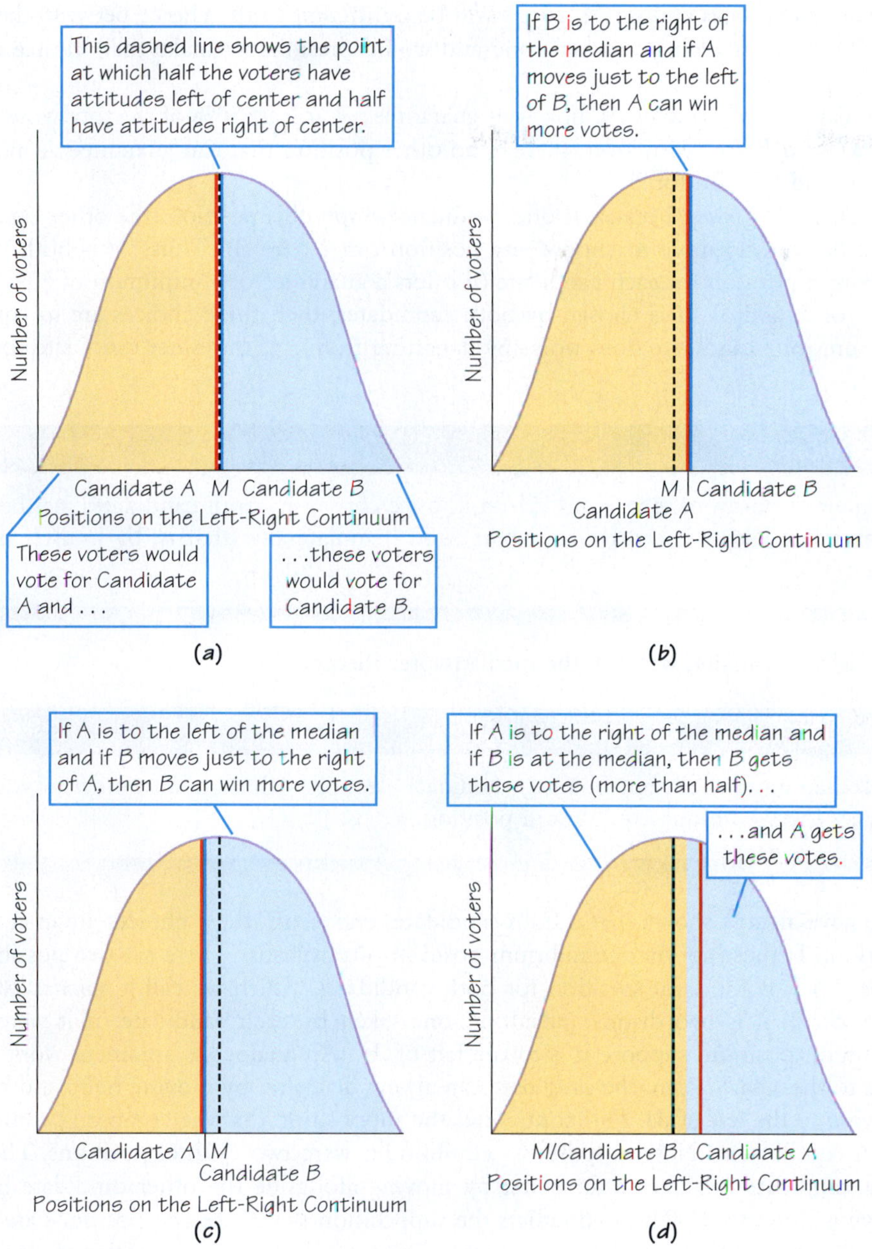


**FIGURE 12.1** Unimodal distribution. The median *M* is the point on the horizontal axis that divides the area under the distribution curve–measuring the number of voters–exactly in half.

### Maximin — DEFINITION

A position is **maximin** for a candidate if there is no other position that can *guarantee* a better outcome–more votes for that candidate–whatever position the other candidate adopts.

If both candidates choose *M*, voters will be indifferent to the choice between them on the basis of their positions alone and would presumably make their choice on other grounds.

Taking a position at *M*, however, guarantees *A* at least 50% of the total vote *no matter what B does*. Moreover, there is no other position that can guarantee *A* more votes, and likewise for *B*.

*M* is also *stable*, because if one candidate adopts this position, the other candidate has no incentive to choose any position other than *M*. Thus, *M* is both the maximin position for each candidate (it offers a guarantee of a minimum of 50% of the votes) and, if *M* is chosen by both candidates, then these choices are in equilibrium (one candidate does worse by departing from it if the other candidate stays at it).

**Equilibrium** DEFINITION

A pair of positions is in **equilibrium** if, once chosen by both candidates, neither candidate has an incentive to depart from it unilaterally (that is, by himself or herself).

More formally, we have the median-voter theorem.

**Median Voter Theorem** THEOREM

**Median-voter theorem**: In a two-candidate election with an odd number of voters, *M* is the unique equilibrium position.

We have already shown that if both candidates choose *M*, these choices are in equilibrium. Is there another equilibrium position or positions? There are two possibilities: (1) It is the same position for both candidates, which we call a *common position*, or (2) it is two distinct positions, one taken by each candidate. If it were a common position, suppose it is to the left of *M*. (An analogous argument works if it is to the right.) Then one candidate can always do better by moving rightward but staying to the left of *M*. This contradicts the supposition that the common position is in equilibrium. Now suppose the equilibrium were two distinct positions. Then one candidate can always do better by moving alongside the other candidate but staying closer to *M*. This contradicts the supposition that these two positions are in equilibrium. Thus, in both cases, one candidate would have an incentive to depart from his or her position–holding the position of the other candidate fixed–so a nonmedian position of one or both candidates cannot be in equilibrium. Therefore, *M* is the only equilibrium position.

## Bimodal Distribution: Median and Mean Different

The median-voter theorem is applicable *whatever* the distribution of the electorate's attitudes. Consider the distribution in Figure 12.2a, which is **bimodal** (two peaks) and is not symmetric. Applying the logic of the previous analysis, *M* is once again the maximin and equilibrium position of two candidates, even though the bulk of voters are concentrated at the two modes.

We next compare the *M* with the mean, which may be quite different:

## Mean of a Voter Distribution

DEFINITION

The **Mean** $\bar{l}$ of a voter distribution is

$$\bar{l} = \frac{1}{n}\sum_{i=1}^{k} n_i l_i$$

where

$k$ = number of different positions $i$ that voters take on the continuum

$n_i$ = number of voters at position $i$

$l_i$ = location of position $i$ on the continuum

$$n = \sum_{i=1}^{k} n_i = n_1 + n_2 + \cdots + n_k = \text{total number of voters}$$

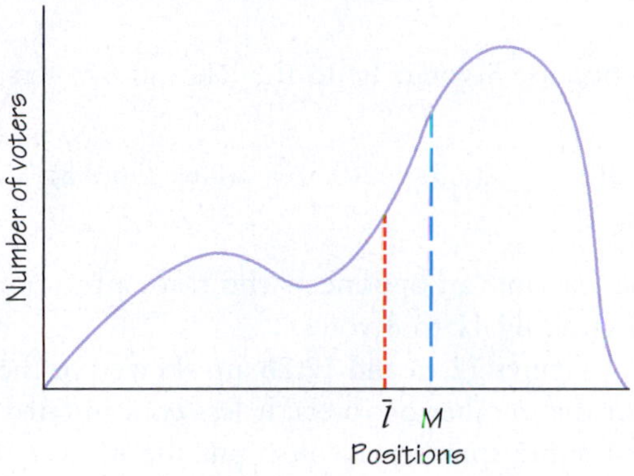


(*a*)

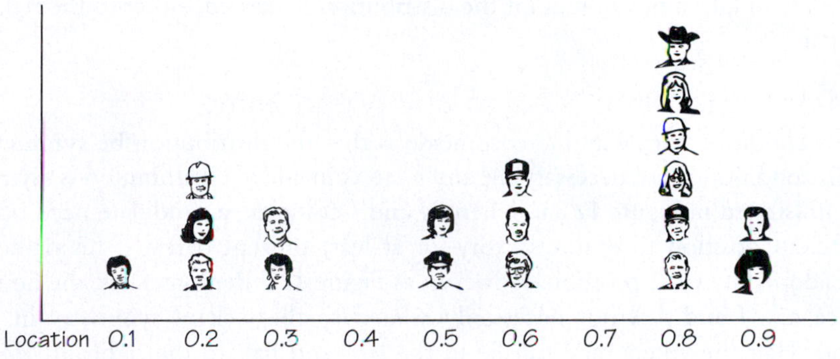


(*b*)

**FIGURE 12.2** Unimodal distribution in which the median and mean are different. The distribution is skewed to the left; the median $M$ is to the right of the mean $\bar{l}$.

The symbol Σ (sigma) is the *summation sign*. It signifies that all subscripted terms to its right (e.g., in the definition of $n$), beginning with the subscript 1 and continuing to the subscript $k$, are summed and a total obtained.

The notion of a mean is the same as that for data samples in Chapter 5, except that here we are calculating a *weighted* average: The location $l_i$ of each position is weighted by the number of voters, $n_i$, at that position. Thus, the mean can be thought of as the position of a typical voter–that is, the expected position of a voter drawn randomly from the set of all voters.

The mean $\bar{l}$ need not coincide with the median $M$. As an illustration of this point, consider the following **discrete distribution** of $n = 19$ voters at $k = 7$ different positions over the interval between 0 and 1, or [0, 1], which is illustrated in Figure 12.2b.

**Discrete Distribution of Voters** DEFINITION

A **discrete distribution of voters** is one in which voters are located at only certain positions–not all points–along the left–right continuum.

| **Position, *i*** | 1 | 2 | 3 | 4 | 5 | 6 | 7 |
|---|---|---|---|---|---|---|---|
| **Location ($l_i$) of position *i*** | 0.1 | 0.2 | 0.3 | 0.5 | 0.6 | 0.8 | 0.9 |
| **Number of voters ($n_i$) at position *i*** | 1 | 3 | 2 | 2 | 3 | 6 | 2 |

Whereas $M$ is 0.6 because 8 voters lie to the left and 8 voters lie to the right,

$$\bar{l} = \left(\frac{1}{19}\right) [1(0.1) + 3(0.2) + 2(0.3) + 2(0.5) + 3(0.6) + 6(0.8) + 2(0.9)] = 0.56$$

Taking a position at 0.56 against an opponent who takes a position at 0.6, a candidate would lose the election by 11 to 8 votes.

The distributions of Figures 12.2a and 12.2b are **skewed** to the left because the area under the curve, or the number of voters, is less concentrated to the left of $M$ than to the right. These more spread-out voters put the mean $\bar{l}$ to the left of the median $M$. The lesson we derived from these figures is that it may *not* be rational for a candidate to take a position at $\bar{l}$ if the distribution is skewed, either to the right or to the left.

## Unimodal Distribution: Median and Mean Same

A sufficient condition for $M$ and $\bar{l}$ to coincide is that the distribution be **symmetric**, but this condition is not necessary: $M$ and $\bar{l}$ can coincide if a distribution is asymmetric, as illustrated in Figure 12.3a. When $M$ and $\bar{l}$ coincide, a candidate need not take a different position to ensure victory–or at least prevent defeat if his or her opponent adopts the same position. However, as Figure 12.3 demonstrates, the noncoincidence of $M$ and $\bar{l}$ is not necessarily related to the lack of symmetry in a distribution: Half the voters may still lie to the left, and half to the right, of $M/\bar{l}$ if the distribution is asymmetric.

What can we say about equilibrium positions if there is an even number of voters? For example, consider the following discrete distribution of $n = 26$ voters at $k = 8$ different positions over the interval [0, 1], which is illustrated in Figure 12.3b.

| **Position, *i*** | 1 | 2 | 3 | 4 | 5 | 6 | 7 | 8 |
|---|---|---|---|---|---|---|---|---|
| **Location ($l_i$) of position *i*** | 0 | 0.2 | 0.3 | 0.4 | 0.5 | 0.7 | 0.8 | 0.9 |
| **Number of voters ($n_i$) at position *i*** | 2 | 3 | 4 | 4 | 2 | 3 | 7 | 1 |

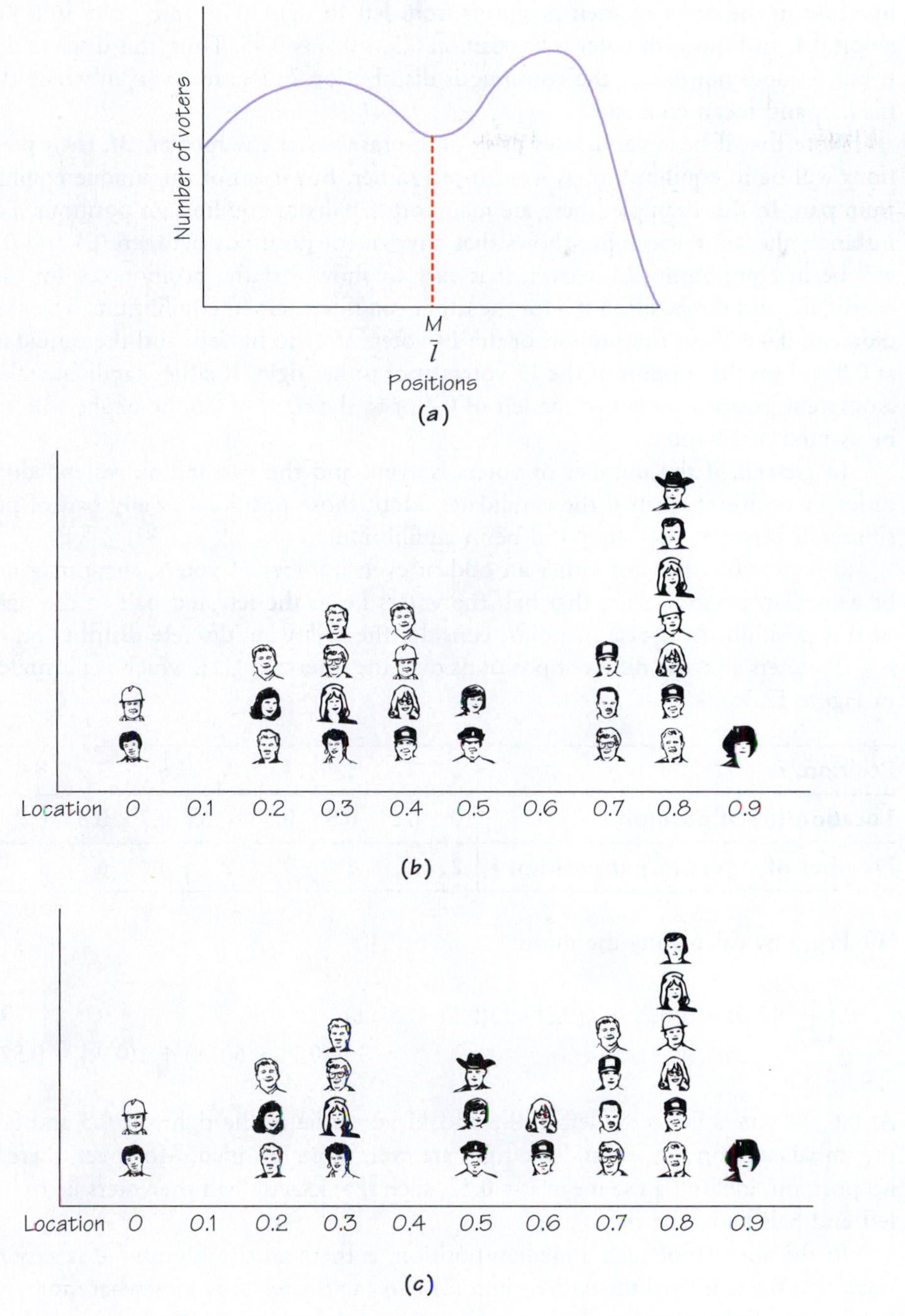


**FIGURE 12.3** Unimodal distribution in which median and mean are the same. It is not necessary for a distribution to be symmetric for the median $M$ and mean $\bar{l}$ to coincide.

We begin by calculating the mean $\bar{l}$:

$$\bar{l} = \left(\frac{1}{26}\right)[2(0) + 3(0.2) + 4(0.3) + 4(0.4) + 2(0.5) + 3(0.7) + 7(0.8) + 1(0.9)] = 0.5$$

The median $M$ is not 0.5. For an even number of voters, as in this example, $M$ is the average of the two middle positions; at this average, 13 voters lie to the left and 13 to the right. The two middle voters are the 13th and 14th voters when they are lined up in the order of their positions from left to right. The 13th voter is at position 0.4, and the 14th voter is at position 0.5, so $M$ is 0.45. Thus, this discrete distribution does not mimic the continuous distribution in Figure 12.3a, in which the median and mean coincide.

Note that if both candidates position themselves at the median $M$, their positions will be in equilibrium, as we showed earlier. But it is not the unique equilibrium pair. In this example, there are many other pairs of equilibrium positions. For instance, the same reasoning shows that any pair of positions between 0.4 and 0.5 will be in equilibrium. Moreover, it is easy to show that the position 0.4 for one candidate, and the position 0.5 for the other candidate, are in equilibrium. The candidate at 0.4 will get the support of the 13 voters at or to his left, and the candidate at 0.5 will get the support of the 13 voters at or to her right. If either candidate takes a different position, either to the left of 0.4 or to the right of 0.5, he or she will not be assured of 13 votes.

In general, if the number of voters is even, and the two middle voters adopt different positions, then if the candidates adopt those positions or any pair of positions in between, then they will be in equilibrium.

It is possible that, for either an odd or even number of voters, there may not be a median position such that half the voters lie to the left and half to the right of this position. As a case in point, consider the following discrete distribution of $n = 25$ voters at $k = 8$ different positions over the interval [0, 1], which is illustrated in Figure 12.3c.

| **Position, $i$** | 1 | 2 | 3 | 4 | 5 | 6 | 7 | 8 |
|---|---|---|---|---|---|---|---|---|
| **Location ($l_i$) of position $i$** | 0 | 0.2 | 0.3 | 0.5 | 0.6 | 0.7 | 0.8 | 0.9 |
| **Number of voters ($n_i$) at position $i$** | 2 | 3 | 4 | 3 | 2 | 4 | 6 | 1 |

We begin by calculating the mean:

$$\bar{l} = \left(\frac{1}{25}\right)[2(0) + 3(0.2) + 4(0.3) + 3(0.5) + 2(0.6) + 4(0.7) + 6(0.8) + 1(0.9)] = 0.52$$

At 0.6, 12 voters lie to the left of 0.6 and 11 voters lie to the right. At 0.5 and 0.7, the imbalances on the left and the right are even more lopsided. Moreover, there is no position, including the mean $\bar{l} = 0.52$, such that exactly half the voters lie to the left and half lie to the right.

In the absence of such a median position, is there an equilibrium? It is easy to show that 0.6 is indeed the equilibrium for two candidates. It is somewhat more difficult to show that if a distribution is discrete and there is no median position, there is still a unique position for both candidates that is in equilibrium. We call this the **extended median** because it extends the median-voter theorem to the discrete case, in which there may be no median position.

**Extended Median** DEFINITION

The **extended median** is the equilibrium position of two candidates in the discrete case when there is no median position.

Given the stability of the median or the extended median in a two-candidate, single-issue election, is it any wonder that candidates who want to win try to avoid extreme positions? As shown in Figures 12.2a and 12.3a, even when the greatest concentration of voters does not lie at $M$ but instead at the mode, a candidate would be foolish to adopt this modal position. For although the right-leaning voters would be very pleased, the candidate's opponent would win the votes of a majority by sidling up to this position but staying just to the left.

Voters on the far left may not be particularly pleased to see both candidates situate themselves at $M$, which is nearer the mode in Figure 12.2a. But in a two-candidate race, they would have nobody else to turn to. Of course, if left-leaning voters felt sufficiently alienated by both candidates, they might decide not to vote at all, which has implications we explore further later.

### EXAMPLE 1 ■ Location of Department Stores

There is a rather different application of the foregoing analysis to business, which in fact was the first substantive area to which spatial modeling was applied. Consider two competitive retail businesses, such as department stores, that consider locating their stores somewhere along the main street that runs through a city. Assume that, because transportation is costly, people will buy at the department store closer to them. Then the analysis says that no matter how the population is distributed along or near the main street, the best location is the median.

(*Andria Patino/Corbis.*)

Thus, if the city's population is symmetrically distributed–that is, not skewed toward one end or the other of the main street–then this location will, of course, be at the center of the main street. Indeed, clusters of similar stores are frequently bunched together near the center of many main streets, although these stores may not be particularly convenient to people who live far from the city's center. Consequently, their location seems not to be in the public interest. Wouldn't it be better to have some of the same kinds of stores near one end of the main street and some near the other, so no people are discriminated against? In an election, by contrast, not every voter can so easily be satisfied if only one candidate is to be elected, so the median seems the most attractive location in this context. ■

## 12.2 Spatial Models for Multicandidate Elections

Primary elections, in which candidates seek the nomination of one of the major parties, tend to attract more than two candidates. In presidential primaries, in particular, many candidates are likely to jump into the fray, especially in the states that go early in the season, if the incumbent president or vice president is not running (as was the case in 2008).

Under what conditions is entry into a multicandidate race attractive? If no positions offer a potential candidate any possibility of success, then it will not be rational for him or her to enter the primary in the first place. Therefore, the rationality of entering a race, and the rationality of the positions he or she might take once there, are really two aspects of the same decision.

Suppose that two candidates have already entered a primary, and they both take positions at $M$. Is there any room for a third candidate?

## EXAMPLE 2

### Entry of a Third Candidate in a Two-Candidate Race

Look at Figure 12.4, where $A$ and $B$ are both at $M$ and therefore split the vote. Now if a third candidate, $C$, enters and takes a position on either side of $M$ (say, to the right), the area under the distribution to $C$'s right may encompass less than $\frac{1}{3}$ of the total area and still enable $C$ to win a plurality of votes.

To show why this is so, consider the portion of the electorate's vote that $A/B$ will receive and the portion that $C$ will receive. If $C$'s area (yellow) is greater than half of $A/B$'s area (blue), $C$ will win more votes than $A$ or $B$, because $C$'s area includes not only the votes to the right of his or her position but also some votes to the left. More precisely, $C$ will attract voters up to the point midway between his or her position on the horizontal axis and that of $A/B$; $A$ and $B$ will split the votes to the left of this midway point. Because $C$ picks up some votes to the left of his or her position, less than $\frac{1}{3}$ of the electorate may lie to the right and still enable $C$ to win a plurality of more than $\frac{1}{3}$ of the total vote.

By similar reasoning, it is possible to show that a fourth candidate, $D$, could take a position to the left of $A/B$ and further chip away at the total of the two centrists. Indeed, $D$ could beat candidate $C$, as well as $A$ and $B$, by moving closer to $A/B$ from the left than $C$ moves from the right.

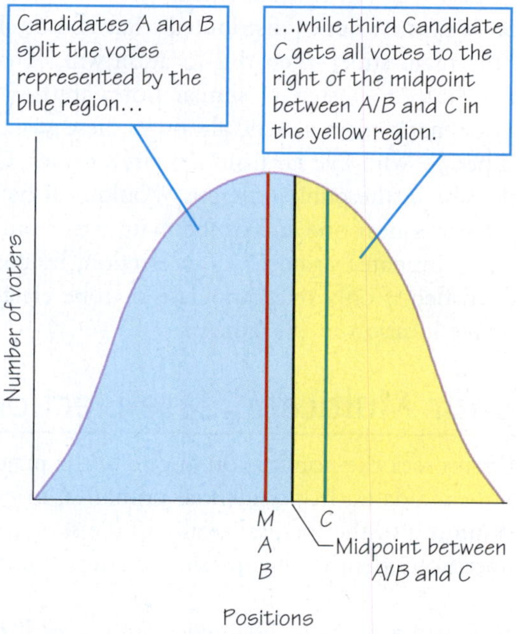


**FIGURE 12.4** Unimodal distribution with three candidates. Candidate $C$ can take a position with less than $\frac{1}{3}$ of the voters to his or her right and still win if candidates $A$ and $B$ at the median $M$ and mean $\bar{l}$ split the remainder of the vote.

Clearly, $M$ has little appeal, and in fact is quite vulnerable, to a third or fourth candidate contemplating a run against two centrists. Indeed, it is not difficult to show that *whatever* positions two candidates adopt—the same or different—at least one of these candidates will be vulnerable to a third candidate.

This is not to say, however, that a third Candidate $C$ will necessarily win against *both* $A$ and $B$. There are both obstacles and opportunities for $C$, which are summarized in Figure 12.5 (the reasoning behind these is explored in Exercises 18 and 19).

## 1/3-Separation Obstacle

DEFINITION

The **1/3-separation obstacle** occurs when there is little room in the middle, enabling *C* to beat *A* or *B* but, in so doing, causing him or her to lose to the other.

## The 1/3-Separation Obstacle and the 2/3-Separation Opportunity

THEOREM

**The 1/3-separation obstacle.** If *A* and *B* are distinct positions that are equidistant from the median of a symmetric distribution and separated from each other by no more than $\frac{1}{3}$ of the area under the curve (so that no more than $\frac{1}{3}$ of the voters lie between *A* and *B*), *C* can take no position that will displace both *A* and *B* and enable *C* to win (see below).

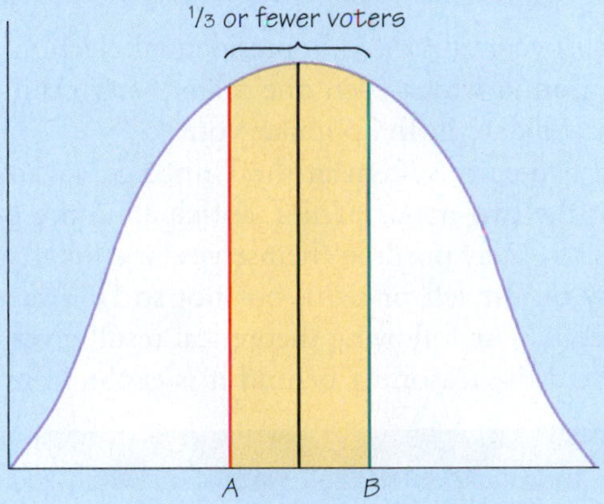


(*a*)

**The 2/3-separation opportunity.** If *A* and *B* are distinct positions that are equidistant from the median of a symmetric unimodal distribution and separated from each other by at least $\frac{2}{3}$ of the area under the curve (so that at least $\frac{2}{3}$ of the voters lies between *A* and *B*), *C* can defeat both *A* and *B* by taking a position at *M* (exactly between them, as shown below).

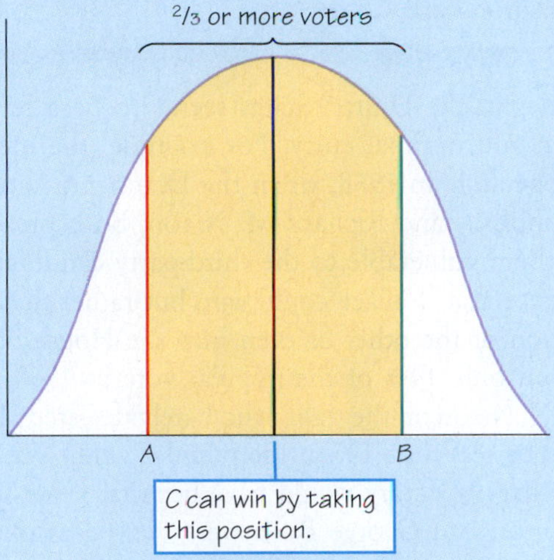


(*b*)

**FIGURE 12.5** The obstacles and opportunities for a third candidate, *C*, to enter a race.

Theodore Roosevelt
*(Hulton Getty/Liaison.)*

This occurred in the 1912 presidential election, when Theodore Roosevelt ran as the Progressive ("Bull Moose") Party candidate after losing the Republican nomination to William Howard Taft. (Roosevelt had previously been president but had lost favor with his party after sitting out one term.) In the general election, Roosevelt received 27% of the popular vote and Taft, 24%. Both candidates were handily defeated by the Democratic candidate, Woodrow Wilson, who got 41% of the popular vote. There was a fourth candidate in this race, socialist Eugene V. Debs, but he received only 6% and was never a serious threat to Wilson on the left. Wilson was also the overwhelming winner in the Electoral College.

### 2/3-Separation Opportunity — DEFINITION

The **2/3-separation opportunity** occurs when there is a wide separation between *A* and *B*, giving enough room in the middle for *C* to win.

This event has never occurred in a U.S. presidential election. In fact, the 1912 election is the only election in which even one major-party candidate has been defeated by a third-party candidate in the popular vote.

The stability of the two-party system in the United States may be partially explained by the fact that the two major parties, anticipating the possible entry of a third-party candidate, deliberately position themselves far enough away from the median to discourage entry on the left or right–but not so far away as to make entry in the middle advantageous. The following theoretical result gives some insight into how this can be done (and the reasoning behind it is explored in Exercise 20).

### Optimal Entry — THEOREM

*The optimal entry of two candidates, anticipating a third entrant.* Assume *A* and *B* are the first candidates to enter an election and anticipate the later entry of *C*. Assume that the distribution of voters is uniform (rectangular) over $[0, 1]$. Then the optimal positions of *A* and *B* are to enter at $\frac{1}{4}$ and $\frac{3}{4}$, whereby $\frac{1}{4}$ of the voters lie to the left of one candidate (say, *A*) and $\frac{1}{4}$ to the right of the other (*B*). Then *C* can do no better than win 25% of the vote: He or she will be indifferent among entering just to the left of *A*, just to the right of *B*, or at any position in between. At none of these positions will *C* win.

Presidential politics in the United States seems to be a reflection of both the median-voter theorem and optimal entry. For example, the median-voter theorem seems to have been operative in 1968, when the Democratic and Republican nominees, Hubert H. Humphrey and Richard M. Nixon, both presented themselves as centrists. This made them vulnerable to the third-party candidate that year, George Wallace–not in the sense that Wallace could win, but rather that he could throw the election in one direction or the other or even into the House of Representatives. In fact, while Wallace won only 14% of the popular vote in 1968, he attracted mostly supporters of Richard N. Nixon on the right, who barely defeated Hubert H. Humphrey that year. (Nixon won by less than 1% in the popular vote.) Without Wallace in the race, polls show that Nixon's victory would have been far more substantial.

In 1992, Bill Clinton and George Bush were viewed as quite far apart on the left–right spectrum. Ross Perot was generally viewed to be between Clinton on the left and Bush on the right, leaving considerable room in the middle that Perot could better exploit than by trying to displace one of the major-party nominees on the left or right. In winning 19% of the popular vote, Perot drew almost equally from each candidate. However, he did not come close to winning, which the optimal entry of the candidates makes difficult. (Clinton won decisively, with 43% of the popular vote to Bush's 38%.)

## 12.3 Narrowing the Field

Up to now, we have looked at the spatial game that candidates play as they vie to position themselves optimally in two-candidate and multicandidate races so as to (1) maximize their vote totals or (2) deter new candidates from entering. We will continue to assume that candidates take positions along a left–right spectrum, but now we consider the game from the point of view of the voters. More specifically, we ask the following question in a multicandidate race: When one candidate drops out, perhaps because of performing below expectations in an early primary, to whom will the dropout's supporters shift their votes?

Beating expectations is the name of the game in the early primaries. Thus, when Senator Edmund S. Muskie from Maine ran for the Democratic Party nomination in 1972, he was expected to do well in the neighboring state of New Hampshire. To "win" in this first primary, he had to exceed these expectations, whereas other candidates, who were not expected to do so well, could afford more mediocre performances. As in a horse race, to which primary elections are often compared, the contenders are handicapped; that is, they must beat expectations about their performances if they are to gain momentum.

Momentum, or what George H. W. Bush (the father) called the "Big Mo," can start a **bandwagon**, or a presumption that a candidate will win.

**Bandwagon Effect** DEFINITION

The **bandwagon effect** induces voters to vote for the presumed winner, independent of his or her merit.

Assume three candidates take positions, from left to right, as follows: *A*–*B*–*C*. Clearly, if *A* or *C* drops out, their supporters mostly likely will switch to *B*, giving the centrist a boost. But what if *B* is the first to drop out? Then it is unclear whether *A*, *C*, or neither will benefit–it depends on the number of *B*'s supporters who prefer *A* next, *C* next, or neither (and hence may not vote at all). In any event, with *B* out of the race, the winner must be one of the candidates on the extremes.

The possibilities become more interesting when there are four candidates arrayed from left to right as follows: *A*–*B*–*C*–*D*. If one of the extremists, *A* or *D*, drops out, then one of the two centrists, *B* or *C*, will benefit. But what if a centrist, say *C*, drops out? Does this benefit one of the extremists, or does the other centrist (*B*) benefit? At first glance, one might think that, with only one centrist remaining, he or she will surely benefit.

This will not be the case, however, if most *C* supporters prefer *D* to *B*, which is certainly possible. Then the extremist *D* will benefit, which will be most upsetting to *A*'s supporters. Conceivably, *A*'s supporters might encourage *A* to withdraw so they can throw their support to *B*, whom they definitely prefer to *D*.

Does this sound implausible? Think back to the 2000 election, in which our four hypothetical candidates are replaced by the following ordering from left to right: Nader–Gore–Bush–Buchanan. (Ralph Nader was the Green Party candidate on the left and Pat Buchanan the Reform Party candidate on the right.) Just before the election, the polls were showing that Buchanan was not much of a threat to Bush, but Nader–who ended up with 2.7% of the popular vote nationwide (Buchanan got only 0.4%)–was definitely a threat to Gore. Despite pleas from some of his supporters, Nader refused to withdraw and, consequently, gave Bush a victory in Florida and a few other close states that won him the presidency.

This 2000 scenario is not the same as the previous four-candidate hypothetical scenario, in which we argued that the extremist on the right, *D*, might win if one of the centrists, *C*, dropped out. In the 2000 scenario, the extremist on the left, Nader, could have dropped out to "save" the centrist closer to him, Gore.

Unfortunately for Gore, Nader not only refused to make this sacrifice but contended afterward that Gore's loss was due to Gore's own poor performance, not Nader's presence in the race. We will return to this issue when we discuss the effects of approval voting and the abolition of the Electoral College as possible remedies to the so-called **spoiler** problem.

**Spoiler** DEFINITION

A **spoiler** is a candidate who cannot win but "spoils" the election for a candidate who otherwise would win.

## 12.4 What Drives Candidates Out?

So far we have considered the possibility that candidates drop out, but not why they do so. Presumably, they do so because of their poor performance in the polls or early primaries, where performance depends in part on expectations of how well they will do. But expectations change over time, and this change in turn affects how voters perceive a race–in particular, who is ahead and who is behind. On this basis, voters choose voting strategies that are likely to benefit their favorites.

Polls make public the standing of candidates in a race, as do the returns from presidential primaries. To be sure, the electorate is different in each primary state, whereas a poll is a sample from the entire electorate. While polls and primaries both provide a glimpse of the "state of the electorate," here we will focus on the change of polls over time. But our results are just as applicable to primaries, whose winnowing-out effects are evident in almost every election in which an incumbent is not running for reelection.

Suppose the election procedure is plurality voting.

**Plurality Voting** PROCEDURE

**Plurality voting** is a voting procedure in which each voter votes for one candidate, and the candidate with the most votes wins.

Assume that voters rank candidates. For example, *A B C* indicates that a voter prefers *A* to *B* to *C*. Before a poll, we assume that each voter votes **sincerely**–for his or her favorite candidate–because, in the absence of poll information, there is no reason to do otherwise.

## EXAMPLE 3 ■ Poll for Three Candidates

Nine voters, who can be divided into three classes, have the following preferences for candidates *A*, *B*, and *C*:

I. 4: *A C B*

II. 3: *B C A*

III. 2: *C A B*

Because the four voters in class I prefer *A* to *C* to *B*, the poll would indicate *A* to have 4 votes (44%), whereas *B* and *C* would have 3 and 2 votes (33% and 22%), respectively, if voters vote sincerely.

After the poll, we make the following assumption:

### Poll Assumption — RULE

Voters adjust, if necessary, their sincere voting strategies to differentiate between the top two candidates revealed in the poll, voting for the one they prefer.

Since class I and class II voters chose one of the top two candidates, only the two class III voters, who voted for *C*, would change their votes. Because they prefer *A* to *B*, it would be in their interest to vote insincerely for *A* (instead of *C*), thereby distinguishing *A* from the other top candidate, *B*, by giving *A* their votes. This would result in *A*'s winning with 6 votes, *B*'s getting 3 votes, and *C*'s getting no votes.

Paradoxically, it is *C* who is the Condorcet winner.

### Condorcet Winner — DEFINITION

A **Condorcet winner** is a candidate who can defeat each of the other candidates in pairwise contests.

*C* is preferred to *A* by class II and III voters (5 to 4) and to *B* by class I and III voters (6 to 3). Hence, if there were a series of pairwise contests (in any order), *C* would win. Yet the poll not only does not make *C* victorious but instead magnifies *A*'s plurality victory (4 votes) by inducing *C*'s supporters, thinking that their candidate is out of the running, to throw their support to *A*, giving *A* a $\frac{2}{3}$-majority victory (6 out of the 9 votes). ■

This example can be generalized to yield the following:

### Condorcet Winner Unsuccessful — THEOREM

Given the poll assumption, a Condorcet winner will always lose if he or she is not one of the top two candidates identified by the poll.

Why this is true is considered in Exercise 28. Here we note that even if *C* were given serious consideration in a tight race, *A* would still win with a plurality of votes.

One might think that $C$'s problem is due solely to the fact that the poll assumption puts him or her out of the running by presuming that his or her supporters "jump ship"–desert $C$ for the apparently more viable candidates, $A$ and $B$. But, surprisingly, even when a Condorcet winner is on top before a poll that distinguishes the top three candidates (instead of the top two), the Condorcet winner may be hurt by the poll after strategic adjustments are made by the voters, as we next illustrate.

### EXAMPLE 4 ▪ Poll for Four Candidates

Add a fourth candidate, $D$, to the three in the preceding example, and assume that there are 12 voters with the following preferences for the four candidates:

I. 3: $A\ C\ B\ D$

II. 3: $B\ C\ A\ D$

III. 4: $C\ A\ B\ D$

IV. 2: $D\ A\ B\ C$

After the poll establishes that $A$, $B$, and $C$ are the top three candidates, the two class IV voters would be motivated to switch to their second choice, $A$. $A$ would thereby increase his or her total from 3 to 5 votes–and win after the poll. Yet $C$ is again the Condorcet winner. In staying the same at 4 votes, $C$ is hurt, relative to $A$, by the poll. In fact, $C$ would lose to $A$ after the poll (because $D$'s supporters would continue to vote for $A$), even though $C$ was the winner before the poll.

Thus, the poll assumption, which induces strategic adjustments that favor the top two candidates, may hurt the Condorcet winner when a larger number of candidates (possibly including the Condorcet winner) are considered to be contenders. However, when one of the top two contenders distinguished by the poll assumption is the Condorcet winner, we have the following result:

**Condorcet Winner Successful** THEOREM

Given the poll assumption, a Condorcet winner will always win if he or she is one of the top two candidates identified by the poll.

Why this is true is considered in Exercise 29. Here we note that a Condorcet winner need not be the winner in the poll but instead can place second and be successful. If a Condorcet winner places second, and the poll has the effect of turning him or her into a majority winner, then it is proper to say that the poll is instrumental in electing this candidate.

We conclude that a poll may either hurt or help a Condorcet winner. If the poll assumption is modified to distinguish more than two candidates, a Condorcet winner may be hurt *even if he or she is among those distinguished by the poll,* as the previous example showed.

## 12.5 Election Reform: Approval Voting

The furor caused by the divided outcome in the 2000 presidential election, in which George W. Bush won the electoral vote and Al Gore won the popular vote, has spurred efforts for reform of the election system. But except for calls for abolition of the Electoral College, whose effects we will analyze in the next section, most of the discussion has centered on making balloting more accurate and reliable and eliminating election irregularities, especially those that discriminate among different classes of voters.

Unfortunately, such reforms ignore a fundamental problem that plagues *multi-candidate elections*–elections with three or more candidates–namely, that the candidate who wins under plurality voting may not be a Condorcet winner. Indeed, there will be no Condorcet winner if each candidate can be beaten by at least one other candidate. In such a situation, who is the rightful winner?

Chapter 9 presents alternatives to plurality voting. Most of these alternatives allow a voter to rank candidates from best to worst; analysts have investigated their ability to elect a Condorcet winner if one exists. Here, however, we will examine in depth a simple election reform, approval voting, that does not require voters to rank candidates.

## Approval Voting

DEFINITION

Under **approval voting**, voters can vote for as many candidates as they like or find acceptable. Each candidate approved of receives one vote, and the candidate with the most approval votes wins the election.

What if approval voting had been used in the 2000 presidential election? Nader supporters, knowing that voting for Nader would be only a protest vote because Nader had no chance of winning, might also have voted for Gore. In fact, because polls show that Gore was the second choice of most Nader voters, Gore almost certainly would have won in Florida if there had been approval voting.

To be sure, Bush would have benefited from the approval votes of Buchanan supporters, but the number of these votes would not have come close to matching the number of votes Gore would have received from Nader supporters. There is therefore little doubt that Gore was the Condorcet winner in this election, because polls show that he could have defeated Bush (with help from Nader voters) as well as each of his less popular opponents in pairwise contests.

Arguments for an election reform like approval voting, however, should not be based on the outcome of only one election. Moreover, even in this one election, we cannot be entirely sure that Gore would have won under approval voting, because the nature of the campaign almost surely would have changed if this reform had been in use.

For example, John McCain, the Republican senator from Arizona who defeated George Bush in the New Hampshire primary but ultimately lost the Republican nomination to him, might have run as an independent candidate if there had been approval voting. As a centrist, he would have been attractive to both Democrats and Republicans and, conceivably, could have won under approval voting. But even if McCain had not run, it is likely that the candidates would have pitched their campaign appeals somewhat differently to try to attract as much approval as possible, especially from Nader and Buchanan supporters.

Although we cannot make precise predictions of the effects of approval voting in a presidential election like that of 2000, we can say what voters will do in certain *types* of situations:

**Voting Only for a Second Choice** THEOREM

In a three-candidate election under approval voting, it is never rational for a voter to vote only for a second choice. If a voter finds a second choice acceptable, he or she should also vote for a first choice.

This is certainly not true under plurality voting. If you were a Nader supporter *and* found Gore also acceptable as a second choice, you should have voted for Gore rather than Nader if (1) you thought Gore could win but Nader could not and (2) electing an acceptable candidate was important to you. (Indeed, some Nader supporters switched to Gore for these reasons.) By comparison, because you lose nothing by voting for *both* Nader and Gore under approval voting–and may gain by doing so (if Nader cannot win, you at least help Gore)–you should never vote for just your second choice (Gore in this example).

Why voting only for a second choice is never rational is considered in Exercise 35. In effect, a voter whose favorite candidate in a three-candidate race seems to be out of the running can have his or her cake and eat it too, by casting a *sincere* vote for a favorite candidate and a *strategic* vote for a second choice (to try to prevent a worst choice from winning). Roughly speaking, **strategic voting** (for instance, by Nader supporters for Gore in the 2000 presidential election) is voting that is not sincere but nevertheless has a strategic purpose–to elect an acceptable candidate if one's first choice is not viable.

Another general result about approval voting that is helpful to know uses the concept of dichotomous preferences:

**Dichotomous Preferences** DEFINITION

A voter has **dichotomous preferences** if he or she divides the set of candidates into two subsets–a preferred subset and a nonpreferred subset–and is indifferent among all candidates in each subset.

In other words, a dichotomous voter sees the world in two colors, white and black, and there is nothing in between. True, most of us see grays, but it is useful to analyze the dichotomous case. In this case, voters have a **dominant strategy**, which is a strategy that is at least as good as, and sometimes better than, any other strategy they might choose. With this definition, we can show a general condition under which Condorcet winners will be elected under approval voting:

**Effect of Dichotomous Preferences** THEOREM

A Condorcet winner will always be elected under approval voting if all voters have dichotomous preferences and choose their dominant strategies.

A dichotomous voter's dominant strategy is to vote for all candidates in his or her preferred subset and no others. This strategy is dominant because the preferred candidates are all assumed to be equally good, so a voter has no reason to distinguish among them. Furthermore, the voter has no reason to vote for a nonpreferred candidate or candidates because they are all equally bad and voting for any one of them could help that candidate win. As an illustration of the effect of dichotomous preferences, consider the following example.

### EXAMPLE 5 ■ Dichotomous Preferences

For each class of voters, the preferred subset of candidates is enclosed in the first set of parentheses, the nonpreferred subset in the second set of parentheses. Thus, the four class I voters prefer $A$ and $B$, between whom they are indifferent, to $C$ and $D$, between whom they are also indifferent:

I. 4: $(A\ B)\ (C\ D)$
II. 3: $(C)\ (A\ B\ D)$
III. 2: $(B\ C\ D)\ (A)$

Assuming that each class of voters chooses its dominant strategy, $B$ wins with 6 votes to 5 votes for $C$ and 4 votes for $A$.

In pairwise contests, notice that $B$ is preferred to $A$ by the two class III voters (class I and II voters are indifferent between these two candidates), so $B$ would defeat $A$ by 2 to 0 votes. (We assume that indifferent voters express no preference.) Because $B$ is preferred to $C$ by the four class I voters, and $C$ is preferred to $B$ by the three class II voters (class III voters are indifferent between these two candidates), $B$ would defeat $C$ by 4 to 3 votes. Thus, $B$, the approval-vote winner, is also the Condorcet winner, which must always be the case when voter preferences are dichotomous and voters vote for all their approved candidates. ■

Insofar as voters in the 2000 presidential election thought equally well (or badly) of Bush and Buchanan on the one hand, and Gore and Nader on the other, they would have preferences like those of class I voters in the previous example. Of course, most voters probably made finer distinctions, which is allowed by "range voting" (see Suggested Web Sites). In this case, there is no guarantee that approval voting will elect a Condorcet winner.

## 12.6 The Electoral College

As we have noted, the Electoral College had a decisive effect in the 2000 presidential election. In winning the popular vote in Florida by the slimmest of margins, George Bush captured all 25 of Florida's electoral votes, which gave him a majority in the Electoral College. This won him the presidency even though he lost the popular vote.

What is the justification for the Electoral College? Its original purpose was to place the selection of a president in the hands of a body that, while its members would be chosen by the people, would be sufficiently removed from them that it could make more deliberative choices. As for its composition, each state gets 2 electoral votes for its two senators (total for all states: 100). In addition, a state receives 1 electoral vote for each of its representatives in the House of Representatives, whose numbers are based on population (see Chapter 11) and range from 1 representative for the seven smallest states to 53 representatives for the largest state, California. The House has a total of 435 representatives. The District of Columbia, like the smallest states, is given 3 electoral votes. Altogether, there are 538 electoral votes, and a candidate needs 270 to win. In 2000, George W. Bush got 271 electoral votes.

Although there is nothing in the U.S. Constitution mandating that the popular-vote winner in a state receive all its electoral votes, this has been the tradition almost from the founding of the republic. Only in Maine and Nebraska can the electoral votes be split among candidates, depending on who wins each of the two congressional districts in Maine and the three congressional districts in Nebraska. Because the statewide winner receives the two senatorial electoral votes, the closest split possible in these two states is 3-1 in Maine and 3-2 in Nebraska. In the actual election, Gore won all of Maine's 4 electoral votes, and Bush won all of Nebraska's 5 electoral votes, so winner-take-all prevailed in all 50 states and the District of Columbia.

Checking ballots in Florida after the 2000 presidential election. (*Reuters/Corbis.*)

Effectively, then, the presidency is decided by 51 players: members of the Electoral College from each of the 50 states and the District of Columbia, who almost always cast their votes as blocs for a single candidate. The voting weights of states, which depend in part on their populations, are related to their voting power (see Chapter 11).

Although the percentage of voting power of a state closely tracks its percentage of electoral votes, this is not the full story. More important is the power of *individual* voters in each state, based on their ability to be pivotal in their states and their states, in turn, to be pivotal in the Electoral College. Amazingly, individual voters in California are about three times as powerful as individual voters in the smallest states, despite the fact that the smallest states (with only one representative) get a 200% (2/1) boost from having 2 "senatorial" electoral votes–besides the 1 electoral vote they are entitled to on the basis of their populations–whereas California receives less than a 4% (2/53) boost.

But there is more to the story than just the size of states. Because California was never considered a close state in the 2000 presidential election (polls indicated that it would almost surely go for Al Gore), it received relatively little attention from both candidates. The real battle was fought in the so-called battleground states, or *toss-up states,* where the outcome was expected to be close (as it certainly was in Florida!). These states received the bulk of the candidates' time, money, and other resources.

Instead of viewing the Electoral College as a 105-million-person game in 2000, in which the voters are the players and their power is a function of the size of the states in which they vote, we view it as a game between the two major-party candidates. We will develop two different models: the first in which the candidates seek to maximize their expected popular vote, and the second in which they seek to maximize their expected electoral vote, in the toss-up states that determine the outcome in a close race.

Common to both models is the assumption that the probability, $p_i$, that a voter in toss-up state $i$ votes for the Democratic candidate is

$$p_i = \frac{d_i}{d_i + r_i}$$

or the proportion of campaign resources that the Democratic candidate ($d_i$), compared with the Republican ($r_i$), spends in state $i$. The probability that a voter in state $i$ votes for the Republican candidate will be the probability $1 - p_i = r_i/(d_i + r_i)$. Thus, we ignore the effects of other candidates in the race and assume that either the Democratic or the Republican candidate will win with certainty: $p_i + (1 - p_i) = 1$. This assumption is plausible in light of the fact that no third-party candidate has ever won the presidency and rarely any states.

## Expected Popular Vote

DEFINITION

The **expected popular vote (*EPV*)** of the Democratic candidate in toss-up states, $EPV_D$, is the number of voters, $n_i$, in toss-up state $i$, multiplied by the probability, $p_i$, that a voter in toss-up state $i$ votes for the Democrat, summed across all toss-up states:

$$EPV_D = \sum_{i=1}^{t} n_i p_i$$

where $t$ is the number of toss-up states. $EPV_R$ can be defined similarly.

$EPV_D$ bears some similarity to the expression for the mean ($\bar{I}$) discussed earlier. Whereas $\bar{I}$ is an average weighted by the proportions, $n_i/n$, $EPV_D$ is an average weighted by the probabilities $p_i$.

The candidates seek strategies for optimally allocating their resources to each toss-up state. Recall that $d_i$ for the Democrat and $r_i$ for the Republican are the resources each candidate allocates to state $i$. For the Democrat, if $d_i$ changes, this affects the value of $p_i$, the probability that a voter votes for him or her, which in turn affects the value of $EPV_D$. Thus, the Democrat seeks a strategy $d_i$ that will make $EPV_D$ as large as possible.

## Proportional Rule

RULE

The strategy of the Democrat that maximizes his or her $EPV_D$ (indicated by the asterisk), given that the Republican also chooses a maximizing strategy, is

$$d_i^* = \left(\frac{n_i}{N}\right) D$$

where $N = \sum_{i=1}^{t} n_i$, the total number of voters in the toss-up states, and $D = \sum_{i=1}^{t} d_i$, the sum of the Democrat's expenditures across all states.

In words, the Democrat should allocate his or her resources in proportion to the size of each state ($n_i/N$) if the Republican behaves similarly by following a strategy of $r_i^* = (n_i/N)R$, where $R = \sum_{i=1}^{t} r_i$.

To show that $d_i^*$ maximizes $EPV_D$ when the Republican chooses $r_i^*$ requires calculus, but we can readily illustrate why departures from $d_i^*$ by the Democrat will cost him or her popular votes.

## EXAMPLE 6

### Departures from a Popular-Vote Maximizing Strategy

Suppose there are three toss-up states with 2, 3, and 4 electoral votes. Assume the candidates accept public financing for the election, and this limits them to spending the same total of $63 million (M). If the Republican follows his or her optimal strategy of spending in the proportion 2:3:4 ($14M:$21M:$28M), but the Democrat, ignoring the smallest state, spends in the proportion of 0:3:4 ($0M:$27M:$36M), the Republican will receive on average

$$EPV_R = 2[14/14] + 3[21/(21 + 27)] + 4[28/(28 + 36)] = 5.06 \text{ votes}$$

or 56% of the 9 votes in the three states.

If the Republican anticipates that the Democrat will spend nothing in the smallest state, the Republican can do even better. By spending only a minuscule amount in the smallest state, the Republican can almost match the Democrat in the other two states and win an average of about 5.5 votes (2 votes from the smallest state and 7/2 = 3.5 votes from the other two states), or 61%.

Now let's assume that the goal of the candidates is not to maximize *EPV* but, instead, their **expected electoral vote (*EEV*)**, which is an entirely different quantity that we define below. To illustrate the difference, a candidate who wants to maximize *EEV* might think of throwing all of his or her resources into the 11 largest states–and ignoring the 39 other states and the District of Columbia if all states are toss-up states–because the 11 largest states have a majority of electoral votes (271). Moreover, the candidate need not win "big" in these states. Winning them by small margins will work just fine, because the candidate will still win *all* their electoral votes and thereby the election.

But this strategy has a problem. An opponent can readily defeat it by spending very small amounts in all the other states, which will defeat the candidate if he or she spends nothing in these states. In addition, by using his or her leftover funds to outspend the candidate in, say, one or two big states, the opponent will end up winning more electoral votes. However, there is a counterstrategy to this strategy, and indeed to every other pure strategy (no randomization–see Chapter 15) one can think of in a winner-take-all system like that of the Electoral College.

To prevent being exploited if an opponent anticipates one's strategy and selects a best counterstrategy against it, each candidate should try to keep secret exactly what he or she intends to do. The best way to keep a secret is to randomize one's choices, using mixed strategies. But mixed strategies are difficult to calculate in a system as complicated as that of the Electoral College, so we make simplifying assumptions. First, however, we need a definition.

To determine $P_i$, we need to count the number of ways that a candidate can win a majority of electoral votes in each state $i$ and then compute the probabilities that each of these ways occurs. These probabilities, in turn, are used to determine $EEV_D$. $EEV_R$ can be defined similarly.

**Expected Electoral Vote** DEFINITION

The **expected electoral vote (*EEV*)** of the Democratic candidate in toss-up states, $EEV_D$, is

$$EEV_D = \sum_{i=1}^{t} v_i P_i$$

where $v_i$ is the number of electoral votes of toss-up state $i$, and $P_i$ is the probability that the Democrat wins *more than* 50% of the popular votes in this state, which would give the Democrat *all* that state's electoral votes, $v_i$. $EEV_R$ can be defined similarly.

## EXAMPLE 7

### Computing the Democrat's Expected Electoral Vote

Consider our earlier example of three states, $A$, $B$, and $C$, with 2, 3, and 4 electoral votes. For simplicity, assume here that the number of electoral votes of each state is equal to the number of voters in that state.

To calculate $P_i$ for each state $i$, we must determine the probabilities that a majority of voters in states $A$, $B$, and $C$ vote Democratic. (We will ignore the possibility of ties in states $A$ and $C$, which have an even number of voters.) To obtain these probabilities, we multiply the probabilities that individual voters in each state, who are assumed to act independently of each other, vote Democratic or Republican (based on the resources the two candidates allocate to each state).

For state $A$ to vote Democratic, for example, both voters in this state must vote Democratic, so $P_A = (p_A)(p_A) = (p_A)^2$. For state $B$ to vote Democratic, either two of the three voters (in 3 possible ways) or all three voters must vote Democratic, so $P_B = 3[(p_B)^2(1 - p_B)] + (p_B)^3$. For state $C$ to vote Democratic, either three of the four voters (in 4 possible ways) or all four voters must vote Democratic, so $P_C = 4[(p_C)^3(1 - p_C)] + (p_C)^4$. Substituting these probabilities into the formula for $EEV_D$, we obtain

$$\begin{aligned} EEV_D &= v_A P_A + v_B P_B + v_C P_C \\ &= 2[(p_A)^2] + 3[3(p_B)^2(1 - p_B) + (p_B)^3] + 4[4(p_C)^3(1 - p_C) + (p_C)^4] \end{aligned}$$

■

We indicated earlier that the strategy of the Democrat that maximizes $EEV_D$, given that the Republican adopts a similar strategy, is mixed, involving randomizing his or her choice of states in which to allocate resources. Because this randomization is difficult to determine, we simplify the task by assuming that $d_i = r_i$ in each toss-up state $i$, and necessarily $D = R$ across all these toss-up states.

This assumption is defensible if the candidates have the same total amount to spend in all the states ($D = R$). If they perceive the value of each toss-up state to be the same, which is reasonable, they will allocate equal resources to each. But how much should these amounts be? It is possible to show the following:

## The 3/2's Rule

RULE

The strategies of the Democratic and the Republican candidates that maximize their *EEVs* are

$$d_i^* = \left(\frac{v_i\sqrt{n_i}}{S}\right)D \qquad r_i^* = \left(\frac{v_i\sqrt{n_i}}{S}\right)R$$

where

$$S = \sum_{i=1}^{t} v_i\sqrt{n_i}$$

In words, the candidates should allocate their resources in proportion to the number of electoral votes of each state ($v_i$) multiplied by the square root of its size ($n_i$). The allocations, $d_i^*$ and $r_i^*$, will be the same if the candidates spend equally in each toss-up state ($d_i = r_i$), as previously assumed.

Because the number of electoral votes in each state $i$ ($v_i$) is approximately proportional to the number of voters in each state (or their populations), $n_i$, the maximizing strategies of the candidates can be approximated by

$$d_i^* = \left(\frac{v_i^{3/2}}{T}\right)D \qquad r_i^* = \left(\frac{v_i^{3/2}}{T}\right)R$$

where

$$T = \sum_{i=1}^{t} v_i^{3/2}$$

Thus, if the candidates allocate the same amount of resources to each of the toss-up states, the 3/2's rule is that they should spend approximately in proportion to the 3/2's power of the electoral votes of these states to maximize *EEV*.

## EXAMPLE 8 ■ Applying the 3/2's Rule

Assume states $A$, $B$, and $C$ have, respectively, 9, 16, and 25 voters, and these are also their numbers of electoral votes. If each of these states is a toss-up state, the 3/2's rule says that the candidates should allocate their resources in the proportions 27:64:125, because the 3/2's powers of their electoral votes are their numbers of voters multiplied by the square root of these numbers. Thus for state $A$, $9^{3/2} = 9\sqrt{9} = (9)(3) = 27$.

To illustrate the difference between the proportional rule (which maximizes *EPV*) and the 3/2's rule (which maximizes *EEV*), assume both candidates can spend 100 units of resources. Then the proportional rule says that states $A$, $B$, and $C$ should get resources in approximately the amounts 18, 32, and 50, whereas the 3/2's rule says these states should get resources in approximately the amounts 13, 30, and 58.

Clearly, the smallest state gets less and the largest state gets more under the 3/2's rule, whereas the middle state stays about the same. In the actual Electoral College, the voters in California, when it is a toss-up state, are about three times as attractive per capita as voters in the smallest states. Following the 3/2's rule, therefore, the candidates should allocate about three times as much per voter to California as to a small toss-up state with only three electoral votes.

In fact, presidential candidates greatly overspend in the largest toss-up states, well out of proportion to their size. This large-state bias is far out of line with the democratic principle of "one person, one vote." For Californians, compared to small-state voters, this principle should read "one person, three votes."

To be sure, it is not the Electoral College itself that creates this bias, but rather, its winner-take-all feature. If this feature were abolished and the electoral votes of a state were split according to the popular votes received by the candidates–insofar as possible–then the large-state bias would disappear.

There is a even better reform to ensure that the electoral-vote winner is also the popular-vote winner. Let each state pass a law that gives all its electoral votes to the *national* popular-vote winner, which becomes effective when states with a majority of electoral votes (that is, 270) pass such a law. Then the popular vote winner would be guaranteed a victory in the Electoral College. As of February 2008, only Maryland and New Jersey had enacted such a law, called the **National Popular Vote law**, which essentially nullifies the winner-take-all effects of the Electoral College.

## EXAMPLE 9

### Departures from an Electoral-Vote Maximizing Strategy

We illustrated earlier how a candidate's departure from the popular-vote maximizing strategy–the proportional rule–lowers that candidate's expected popular vote, given that the candidate's opponent adheres to this rule. Because of the loss candidates suffer when they depart from the proportional rule, it is an equilibrium strategy for both. This is also true of *some* departures from the 3/2's rule: If a candidate's departure from this rule is "small," he or she will lower his or her expected electoral vote, given that the candidate's opponent sticks to that rule.

Suppose, for example, that the Republican follows the 3/2's rule in allocating resources to states $A$, $B$, and $C$ with 9, 16, and 25 voters/electoral votes, respectively. If each candidate has 100 units of resources to spend, we showed in the previous example that this translates into allocating approximately 13, 30, and 58 units to states $A$, $B$, and $C$, respectively.

Now if the Democrat deviates slightly from the 3/2's rule, and allocates 14, 30, and 57 units to these states (more to $A$, less to $C$), he or she increases the chances of winning in $A$ and decreases the chances of winning in $C$. It can be shown that this deviation from the 3/2's rule hurts the Democrat, because even though his or her chances go up more in $A$ than they go down in $C$ (because $A$ is smaller than $C$), $C$ has almost three times as many electoral votes. On balance, this deviation lowers the Democrat's expected electoral vote.

Now suppose the Democrat makes a "large" deviation from the 3/2's rule, ignoring state $B$ entirely and throwing all his or her resources into state $A$ (9 electoral votes) and state $C$ (25 electoral votes). If he or she wins in these two states, this would give the Democrat 34 of the 50 electoral votes, which is more than enough to win.

If the Republican adheres to the 3/2's rule, he or she will put approximately 13% of his or her resources into state $A$ and 58% into state $C$. By following the 3/2's rule in these two states and ignoring state $B$, the Democrat will put approximately 18% into state $A$ and approximately 82% into state $C$. This translates into *each voter* in $A$ and $C$ supporting the Democrat with probability 0.58, which means that the Democrat will almost certainly win in both these states, giving him or her an expected electoral vote of almost 34. The Republican *will* certainly win in state $B$, because the Democrat spends nothing in this state, but this is small consolation if the Republican loses in the two other states and receives an expected electoral vote of somewhat more than 16.

This example illustrates why the 3/2's rule is a local maximum but not a global maximum.

**Local Maximum and Global Maximum** DEFINITION

A **local maximum** is a maximizing strategy from which small deviations are nonoptimal but large deviations may be optimal. A **global maximum** is a maximizing strategy from which *all* deviations (small or large) are nonoptimal. The proportional rule is a global maximum (and equilibrium) for candidates whose goal is to maximize their expected popular vote, whereas the 3/2's rule is only a local maximum for candidates whose goal is to maximize their expected electoral vote.

If the goal of candidates is to maximize their expected electoral vote, there is no *determinate* maximizing strategy–randomizing one's choices is necessary to prevent exploitation by an opponent. How this randomization is done for some simple games is analyzed in Chapter 15.

In the case of the Electoral College, we illustrated how a radical departure from the 3/2's rule by a candidate, who ignores some state or states entirely, may be the best response to an opponent who follows this rule. But then there is a best response to this best response, and so on, so no determinate strategy is invulnerable.

However, insofar as the candidates view the toss-up states in similar terms, the 3/2's rule offers a good rule of thumb as to how much to spend in each, as a function of its size, to maximize their expected electoral vote. But we must remember that it is only a local maximum. Hence, unlike the proportional rule, which is robust against all other popular-vote maximizing strategies, the 3/2's rule is vulnerable to radically different strategies, such as those in which candidates concentrate their efforts on relatively few states.

Usually those who try such strategies, however, take big risks. In 1964, the Republican presidential candidate, Barry Goldwater, said that he would like to "saw off the Eastern seaboard" (Goldwater was from Arizona), but in the end, in his memorable phrase, he went, "shooting where the ducks [voters] are." In doing so, however, he appeared not so much to want to win as to present voters with "a choice, not an echo," by taking relatively extreme (conservative) positions. Is it any wonder, then, that Goldwater lost in a huge landslide to his Democratic opponent, Lyndon Johnson?

## 12.7 Is There a Better Way to Elect a President?

The quest for the presidency is the greatest spectacle in American politics. While there is nothing to match its excitement, especially when the race is close, the quieter gamelike features of a presidential campaign are no less consequential.

We have emphasized these features in this chapter, showing how mathematics can be used to analyze optimal positions in two-candidate and multicandidate races, and how polls and presidential primaries may affect who stays in and who drops out of the race. Sometimes, as we have seen, Condorcet candidates, who can beat every other candidate in pairwise contests, may not survive. And as was dramatically illustrated in the 2000 presidential election, the popular-vote winner may not win in the Electoral College.

Some people think that approval voting would better enable voters to express their preferences, especially in the early presidential primaries, which typically draw many candidates if an incumbent is not running for reelection. However, other election procedures, including those discussed in Chapter 9, possess features that may make them desirable as election reforms.

All these procedures would probably be of more help to centrist candidates, who not only better represent the entire electorate than extremist candidates, but who also are more likely to be a party's strongest contender in the general election. In the past 50 years, the biggest losers in presidential elections, Republican Barry Goldwater in 1964 and Democrat George McGovern in 1972, came from the right and left extremes, respectively, of their parties.

Taking the choice of a president out of the hands of voters and putting it into the hands of members of the Electoral College may no longer be justified. The Electoral College, with its winner-take-all feature, creates a large-state bias, as we showed. In the 2000 presidential election, a few hundred voters in one large toss-up state, Florida, determined the outcome.

If one thinks that the votes of *all* voters, wherever they reside, should count equally, then direct popular-vote election of a president would best accomplish this goal. Allocating electoral votes proportionally in each state would approximate this goal. But a better solution would be for states to enact the National Popular Vote law, which would ensure that the electoral-vote winner is the popular-vote winner if states with a majority of electoral votes passed this law.

If approval voting were used, then it would be approval votes rather than the single votes of each voter that would determine the allocation of electoral votes to the candidates. Because the general election in recent years has drawn major third- and fourth-party candidates, approval voting, or one of the other voting procedures discussed in Chapter 9, seems worthy of consideration if one wants to reduce the role of spoilers.

In summary, mathematics illuminates strategic aspects of campaigning and voting in presidential elections not apparent to the naked eye. It also points the way to possible reforms that may ameliorate some of the problems that affect our current system.

## REVIEW VOCABULARY

**Approval voting** Allows voters to vote for as many candidates as they like or find acceptable. Each candidate approved of receives one vote, and the candidate with the most approval votes wins. (p. 387)

**Bandwagon effect** Voting for a candidate not on the basis of merit but, instead, because of the expectation that he or she will win. (p. 383)

**Condorcet winner** A candidate who can defeat each of the other candidates in pairwise contests. (p. 385)

**Dichotomous preferences** Held by voters who divide the set of candidates into two subsets–a preferred subset and a nonpreferred subset–and are indifferent among all candidates in each subset. (p. 388)

**Discrete distribution of voters** A distribution in which voters are located at only certain positions along the left–right continuum. (p. 376)
**Dominant strategy** A strategy that is at least as good as, and sometimes better than, any other strategy. (p. 388)
**Electoral College** A body of 538 electors that selects a U.S. president. (p. 369)
**Equilibrium position** A position is in equilibrium if no candidate has an incentive to depart from it unilaterally. (p. 374)
**Expected electoral vote (*EEV*)** The number of electoral votes of each toss-up state, multiplied by the probability that the Democratic (or Republican) candidate wins more than 50% of the popular votes in that state, summed across all toss-up states. (p. 392)
**Expected popular vote (*EPV*)** The number of voters in each toss-up state, multiplied by the probability that that voter votes for the Democratic (or Republican) candidate, summed across all toss-up states. (p. 391)
**Extended median** The equilibrium position of two candidates when there is no median. (p. 378)
**Global maximum** A maximizing strategy from which *all* deviations (small or large) are nonoptimal. (p. 396)
**Local maximum** A maximizing strategy from which small deviations are nonoptimal but large deviations may be optimal. (p. 396)
**Maximin position** A position is maximin for a candidate if there is no other position that can guarantee a better outcome–more votes–whatever position another candidate adopts. (p. 373)
**Mean ($\bar{l}$)** A weighted average, wherein the positions of voters are weighted by the fraction of voters at that position. (p. 375)
**Median *M*** The point on the horizontal axis of a voter distribution where half the voters have attitudes that lie to the left and half to the right. (p. 372)
**Median-voter theorem** In a two-candidate election with an odd number of voters, the median is the unique equilibrium position. (p. 374)
**Mode** A peak of a distribution. A distribution is **unimodal** if it has one peak, and **bimodal** if it has two peaks. (p. 372)
**National Popular Vote law** Gives all the electoral votes of a state to the national popular-vote winner if states with a majority of electoral votes enact the law. (p. 395)
**1/3-separation obstacle** An obstacle for the entry of a third candidate created if two previous entrants are sufficiently close together. (p. 381)
**Poll assumption** Voters adjust their sincere voting strategies, if necessary, to differentiate between the top two candidates–as revealed in the poll–by voting for the one they prefer. (p. 385)
**Plurality voting** Allows voters to vote for one candidate, and the candidate with the most votes wins. (p. 384)
**Proportional rule** Presidential candidates allocate their resources to states according to their size. This allocation rule maximizes the expected popular vote of a candidate, given that his or her opponent adheres to it. It is a global maximum. (p. 391)
**Sincere voting** Voting for a favorite candidate, whatever his or her chances are of winning. (p. 384)
**Spatial models** The representation of candidate positions along a left–right continuum in order to determine the equilibrium or optimal positions of the candidates. (p. 371)
**Spoiler** A candidate who cannot win but "spoils" the election for a candidate who otherwise would win. (p. 384)
**Strategic voting** Voting that is not sincere but nevertheless has a strategic purpose–namely, to elect an acceptable candidate if one's first choice is not viable. (p. 388)
**3/2's rule** Presidential candidates allocate their resources to toss-up states according to the 3/2's power of their electoral votes. This allocation rule maximizes the expected electoral vote of a candidate, given that his or her opponent adheres to it. It is a local maximum. (p. 394)
**2/3-separation opportunity** An opportunity for the entry of a third candidate created if two previous entrants are sufficiently far apart. (p. 381)
**Voter distribution** Gives the number (or percentage) of voters who have attitudes at points along the left–right continuum, which can be represented by a curve. The distribution is **symmetric** if the curve to the left of the median is a mirror image of the curve to the right. It is **skewed** to one side if the area under the curve is less concentrated on that side of the median than the other. (p. 371)

## ✓ SKILLS CHECK

**1.** In a two-candidate election, suppose the attitudes of the voters are distributed symmetrically with median *M*. Of the two candidates *A* and *B*, *A* is positioned far to the left of *M* and *B* is positioned just to the right of *M*. Which, if either, candidate will receive more votes?

**(a)** *A* will receive a majority of the votes.
**(b)** *B* will receive a majority of the votes.
**(c)** *A* and *B* will both receive exactly one-half of the votes.

**2.** In a three-candidate election, suppose the attitudes of the voters are distributed symmetrically with median *M*. Of the three candidates *A*, *B*, and *C*, *A* is positioned far to the left of *M*, *B* is positioned just to the right of *M*, and $C$ is positioned at $M$. Candidate _____ will receive the most votes.

**3.** In a two-candidate election, which of the following positions is an optimal position for both candidates $A$ and $B$?

(a) $A$ and $B$ just to the left and right of $M$
(b) $A$ and $B$ far to the left and right of $M$
(c) $A$ and $B$ both at $M$

**4.** In a three-candidate election, if candidates $A$ and $B$ are positioned at $M$, the election-winning position of candidate $C$ is _______________________________ .

**5.** In a three-candidate election, if candidates $A$ and $B$ are positioned just to the left and just to the right of $M$, are there election-winning positions for candidate $C$? What are they?

(a) At $M$
(b) Far to the left or right of $M$
(c) There is no election-winning position for candidate $C$.

**6.** In a four-candidate election, if candidates are aligned in order $A$–$B$–$C$–$D$, candidate _____ benefits if $D$ drops out of the race.

**7.** When is a maximin position not an equilibrium position?

(a) When it is not the median
(b) When it is the median
(c) When it is the mean

**8.** It is desirable that two candidates take median positions, but undesirable that two department stores locate themselves at the center of a main street because _______________________________________ .

**9.** In a three-candidate election, suppose 12 voters can be divided into three classes according to their preferences: Five voters prefer (in order) $A$, $B$, $C$; 4 voters prefer $C$, $A$, $B$; 3 voters prefer $B$, $C$, $A$. To elect one of their top 2 candidates, which group of voters will not vote sincerely (for their first choice)?

(a) The 5 voters
(b) The 4 voters
(c) The 3 voters

**10.** Making the poll assumption, the Condorcet winner will always win when _________________________ .

**11.** In an election with a large number of candidates, approval voting benefits

(a) candidates at the extreme left and right.
(b) candidates at or near $M$.
(c) only candidates precisely at $M$.

**12.** In a four-candidate approval voting election with 12 voters, if 5 voters approve of $A$ and $B$, 4 voters approve of $B$ and $C$, and 3 voters approve of $A$ and $D$, candidate _____ will win the election.

**13.** In Exercise 12, assume that the 5 voters who approve of $A$ and $B$ actually prefer $A$. Would they have an incentive to vote strategically?

(a) Yes
(b) No
(c) Some would and some wouldn't.

**14.** In the election of the president using the Electoral College, voters in smaller toss-up states have less power than voters in large toss-up states because ___________ .

**15.** In the election of the president, if the Democrats believe that the Republicans will not allocate resources in a way that maximizes their $EEV$, then

(a) they can successfully counter by allocating their resources according to the method that maximizes $EEV$.
(b) they can successfully counter by not allocating their resources according to the method that maximizes $EEV$.
(c) they cannot successfully counter.

**16.** Suppose three voting blocs, $A$, $B$, and $C$, control 16, 25, and 36 votes, respectively, and 39 votes are required to win. If each bloc is a toss-up, for every \$1 that is allocated to influence the voters in bloc $A$, about _________ as much should be allocated to influence the voters in bloc $C$.

**17.** In Exercise 16, suppose your opponent spends all his money lobbying blocs $A$ and $B$ and ignores bloc $C$. What would be a good strategy to use in response?

(a) Lobby just $A$
(b) Lobby just $B$
(c) Lobby $A$ and $C$

**18.** Voting for president using the Electoral College would be unbiased if ___________________________ .

**19.** How big is the bias factor in Exercise 18 when comparing small and large states?

(a) Very small
(b) About 2:1
(c) About 3:1

**20.** It is impossible for a candidate for president to win the popular vote and yet lose the Electoral College vote under the National Popular Vote law because _______ .

# CHAPTER 12 EXERCISES

■ Challenge ◆ Discussion

## 12.1 Spatial Models for Two-Candidate Elections

**1.** Why do $M$ and $\bar{l}$ not coincide if a distribution is skewed?

**2.** Show that 0.6 is the equilibrium in Figure 12.3b.

■ **3.** Prove that if a distribution is discrete and there is no median position, there is always an *extended median*. (*Hint:* Show that there is always one position at which a majority of voters lies neither to the left nor to the right, and neither candidate would have an incentive to depart from this position.)

**4.** Assume that the one voter at 0.1 in Figure 12.2b decides not to vote because he is "too far away" from the two candidates who take the median position at 0.6. Would either candidate depart from $M = 0.6$ to try to do better if he or she knew that this voter had decided not to vote–but he or she would vote for the closer of two candidates less than a distance of 0.5 away? What if the candidates knew that the three voters at 0.2 had also decided not to vote–but they would vote for the closer of two candidates less than a distance of 0.4 away?

**5.** If you are a far-left or a far-right voter, are you helping your cause when you announce, like the voters in Exercise 4, that you will not support candidates who are too far away?

**6.** Consider the two most extreme voters at 0 in Figure 12.3b (those who are farthest from the extended median of 0.6). Would their nonvoting change the extended median? How about, as well, the nonvoting of the somewhat less extreme voter at 0.9? Show when, if at all, $M$ or the extended median will change as fewer and fewer extreme voters decide not to vote in this example?

◆ **7.** In Figure 12.2a, $\bar{l}$ is not in equilibrium–one candidate would do better if he or she moved from 0.56 to 0.6. But is 0.6 really a better reflection of the views of the electorate than 0.56?

**8.** Define an outcome to be in equilibrium if, given that one candidate chooses it, the other candidate cannot do better than take the same position. Show that this definition is equivalent to the text's definition of being in equilibrium.

■ **9.** Consider a trimodal distribution (three peaks). When will taking a position at the middle peak be in equilibrium? Is it possible that one of the other peaks can ever be in equilibrium? If so, give a discrete-distribution example.

**10.** Define $A$'s position in a two-candidate race to be *opposition-optimal* if, given that the position of $B$ is fixed, it maximizes $A$'s vote total. Show that $A$'s opposition-optimal position must be adjacent to $B$'s position and closer to $M$, except when $B$ is at the median. (Roughly speaking, being "adjacent" means being a very small distance away.)

◆ **11.** Assume the population along a main street is uniformly distributed over [0, 1], so there are equal numbers of people located at all equally spaced intervals from $M/\bar{l}$. (This makes the distribution rectangular, or "flat.") It has been argued that the "social optimum" for the location of two stores are at the points $\frac{1}{4}$ and $\frac{3}{4}$, because then no person would have to travel more than $\frac{1}{4}$ of the length of the street to buy at one store. Is this desirable if the population is not uniformly distributed?

◆ **12.** What is a social optimum in an election if only one candidate is to be elected? How about five candidates to a city council? Is it better that the city council members' positions all be centered around 0.5, or should they be more spread out?

◆ **13.** Which is better for consumers: (a) to minimize the maximum distance they must travel to a store; or (b) to foster price competition, which would presumably be encouraged if two stores are located at $M = \frac{1}{2}$?

■ **14.** Assume a city comprises three equal-sized districts, each of which elects a candidate to the city council. The mayor is elected by the entire city. Show with an example that the median or extended median for the mayor need not be the median or extended median for any of the three city council districts. Does this explain why mayors and city council members often disagree?

■ **15.** In Exercise 14, must the median or extended median for the mayor be between the leftmost and rightmost medians, or extended medians, of the three districts? How about the mean $\bar{l}$?

## 12.2 Spatial Models for Multicandidate Elections

**16.** Assume that $A$ and $B$ take the *same* nonmedian position. What position should $C$ take to maximize his or her vote total? Is $C$'s position always a winning one?

■ **17.** Assume that $A$ and $B$ take *different* positions, with one possibly being at $M$. What position should $C$ take to maximize his or her vote total? Is $C$'s position always a winning one?

**18.** Is there a 1/3-separation obstacle if the distribution is not symmetric but no more than $\frac{1}{6}$ of the area under the curve separates *A* (on the left) from *M*, and no more than $\frac{1}{6}$ of the area separates *B* (on the right) from *M*? What if these $\frac{1}{6}$-or-less areas on the left and the right are not the same?

**19.** Is there a 2/3-separation opportunity if the distribution is not unimodal but at least $\frac{1}{3}$ of the area under the curve separates *A* (on the left) from *M*, and at least $\frac{1}{3}$ separates *B* (on the right) from *M*? (*Hint:* Start by assuming that the distribution is uniform between *A* and *B*–and hence not unimodal–and that exactly $\frac{2}{3}$ of the voters lie between *A* and *B*. Can *C* always win by taking a position at *M*? If not, is there a distribution that affords *C* this opportunity?)

**20.** Show that *C* cannot win under the conditions for the optimal entry of two candidates, anticipating a third entrant (page 382). (*Hint*: Indicate which candidate will win when *C* enters to the left of *A*, to the right of *B*, or in between.)

**21.** It is known that *A*, *B*, and *C* will enter an election in that order, with *A* announcing his position first, then *B*, and finally *C*. If the distribution is uniform over [0, 1], what position should each candidate take to maximize his or her vote total, anticipating–in the case of *A* and *B*–the entry of future candidates? [*Hint:* Start by assuming that *A* takes a position at $\frac{1}{4}$. Is *B*'s position at $\frac{3}{4}$ optimal, anticipating the entry of *C*? Or can *B* do better at some other position (perhaps by influencing *C*'s choice of a maximizing position)?]

■ **22.** If *A* and *B* are equidistant from the median of a symmetric distribution and separated from each other by exactly $\frac{1}{2}$ of the area under the curve, under what conditions is this separation an obstacle and under what conditions is it an opportunity? (*Hint:* Start by constructing examples of symmetric distributions in which *C* would either win or lose by taking a position at *M*.)

■ **23.** What are the vote-maximizing positions for four candidates to take if it is known that they will enter in the order *A*, *B*, *C*, *D*?

## 12.3 Narrowing the Field

**24.** Assume that the four candidates in the 2000 presidential election can be arrayed from left to right as follows: Nader–Gore–Bush–Buchanan. Suppose a poll reveals Gore at 48%, Bush at 47%, Nader at 3%, and Buchanan at 2%. Would Bush be well advised to offer Buchanan a cabinet position to drop out of the race (as Adams offered Clay the secretary-of-state post after the 1824 election)? What if Bush knew that, after Buchanan dropped out, only half of Buchanan's supporters would switch to him, with most of the remainder not voting, except for a few who would switch to Gore?

**25.** Assuming the same poll results as in Exercise 24, now suppose that Gore offered the same deal to Nader, knowing that only one-third of Nader supporters would switch to him and the rest would not vote. However, suppose Gore also thought that if Nader dropped out, so would Buchanan, and all Buchanan supporters would vote for Bush. Should Gore set off this train of events?

◆ **26.** Is there any evidence that the four presidential candidates in 2000 might have contemplated "deals" of the kind indicated in Exercises 24 and 25? If you cannot find any evidence, do you think this is because the candidates found such ploys unethical or because they thought they might be found out if they tried to engage in them?

◆ **27.** One tactic that was considered by Nader supporters who thought that their votes for Nader might kill Gore's chances in some states was to swap votes: In close states that Gore might lose if Nader supporters stuck with their candidate, these supporters would switch to Gore if Gore supporters in less contested states, where Gore would almost surely win, would switch to Nader. Thereby the popular-vote totals for the two candidates would not change overall, but Gore would be able to win in the close states he might otherwise lose. Is this a sensible way of dealing with problems created by the Electoral College, which puts a premium on winning in large states?

## 12.4 What Drives Candidates Out?

**28.** Show why the Condorcet-winner-unsuccessful result (page 385) is true. Is it true that if the poll assumption was modified to differentiate the top three (rather than the top two) candidates from the rest, and the Condorcet winner was not among the top three, that he or she would still lose?

◆ **29.** Show why the Condorcet-winner-successful result (page 386) is true. Is it proper that the candidate who comes in second in the poll should win after the results of the poll are announced? Why?

◆ **30.** In Example 4 (page 386), after class IV voters switch from *D* to *A*, the vote totals for the top three candidates are *A*–5, *B*–3, and *C*–4. Now assume a second poll is taken, differentiating *A* and *C*, the top two contenders, from *B*. If *B* supporters switch at this point to their second choice, which candidate will win? Do you consider this a desirable outcome?

◆ **31.** Assume there are four classes of voters that rank four candidates as follows:

I. 4: *A D B C*

II. 3: *B D A C*

III. 2: *C D B A*

IV. 1: *D C B A*

Which candidate is the Condorcet winner? Do you find this result strange in the light of what a poll would tell the voters?

**32.** Assume there is a poll that differentiates the top two candidates in Exercise 31. Which candidate will win the election after the poll?

◆ **33.** Assume there is a poll that differentiates the top three candidates in Exercise 31. Which candidate will win the election after the poll? Comment on the different outcomes in this exercise and the previous one.

◆ **34.** Assume there are three classes of voters who rank three candidates as follows:

I. 4: *A B C*

II. 3: *B C A*

III. 2: *C A B*

Show that there is no Condorcet winner. Applying the poll assumption to this example, which candidate will win? Is this fair, given the preferences of class II and III voters, who, together, are a majority?

## 12.5 Election Reform: Approval Voting

■ **35.** Prove the voting-only-for-a-second-choice result (page 388).

**36.** In a three-candidate election, show that your strategy of voting for your top two choices under approval voting is not always better than voting only for your top choice.

**37.** Consider a four-candidate election under approval voting. Is there ever a situation in which a voter would vote for a first and a third choice without also voting for a second choice? [*Hint:* Assume a voter ranks the four candidates *A B C D* and believes that one of two things can happen: The electorate will favor either liberals (say, *A* and *B*) or conservatives (say, *C* and *D*) but never favor each side equally.]

**38.** Is there ever a situation under approval voting in which a voter would vote for a worst choice?

**39.** Is there ever a situation under approval voting in which a voter would *not* vote for a first choice if he or she finds acceptable one or more lower-ranked candidates? (*Note:* This question asks whether the voting-only-for-a-second-choice result can be generalized to more than three candidates.)

■ **40.** Prove the effect-of-dichotomous-preferences result (page 388). (*Hint:* If all voters have dichotomous preferences and vote for all candidates in their preferred subsets, which candidate will get the most approval votes? What does this say about the preferences of voters for the approval-vote winner, compared to their preferences for each of the other candidates?)

**41.** In the following example, class I and II voters have dichotomous preferences, but the class III voter has *trichotomous preferences* (he or she divides the four candidates into three indifference subsets):

I. 2: (*A B*) (*C D*)

II. 2: (*C*) (*A B D*)

III. 1: (*D*) (*C*) (*A B*)

Is it rational for the class III voter to vote only for his or her top choice, *D*? If not, who else should he or she approve of? Which class of voters will be most unhappy if the class III voter does not vote just for *D*? Can voters in this class, by voting strategically, do anything about their situation?

◆ **42.** Assume the class III voter's preferences change to a different trichotomous ordering:

III. 1: (*D*) (*A C*) (*B*)

Suppose, as in Exercise 41, that the class III voter indicates in an initial poll that he or she intends to vote only for *D* but then, in response to the poll, switches to voting for the candidates in his or her second-choice subset as well. If there is a new poll, based on these results, what will be the outcome? What if there is a third poll, fourth poll, and so on? Do you regard this result as desirable? Why?

**43.** In Exercise 31, we saw that under plurality voting the Condorcet winner, *D,* comes in fourth in a poll and, therefore, cannot be helped by subsequent polling, even when the poll distinguishes the top three candidates and voters differentiate among them:

I. 4: *A D B C*

II. 3: *B D A C*

III. 2: *C D B A*

IV. 1: *D C B A*

What are the outcomes under approval voting–both with and without polling–if voters approve of their (i) top-ranked, (ii) two top-ranked, and (iii) three top-ranked candidates initially? [*Note:* In making adjustments to the poll results, assume that voters approve not only of the preferred of their two top-ranked candidates identified by the poll but also of *all* candidates ranked above their preferred candidate. For example, when the poll based on (i) above identifies *A* and *B* as the two top-ranked candidates, with 4 and 3 votes, respectively, the class III and class IV voters after the poll will approve not only of their preferred candidate, *B*, but also of *C* and *D*, because they rank the latter two candidates above *B*.]

◆ **44.** On the basis of your answers to the foregoing problems, do you think approval voting would be beneficial in finding Condorcet winners–either with or without polling–in multicandidate elections?

## 12.6 The Electoral College

**45.** Assume there are three states with 3, 7, and 9 voters, and that they are all toss-up states. If both the Democratic and Republican candidates choose strategies that maximize their expected popular vote (the proportional rule), and they have the same total resources ($D = R$), what is the expected number of votes that each will receive?

**46.** Assume the Republican knows in advance what allocations, $d_i$, to each state $i$ the Democrat will make in Exercise 45. Then the Republican's optimal response can be shown to be

$$r_i = \frac{\sqrt{n_i d_i}}{\sum_{i=1}^{t} \sqrt{n_i d_i}} (R + D) - d_i$$

Suppose that the Democrat ignores the smallest state and makes proportional allocations to the two largest states. (For concreteness, assume both candidates have 100 units of resources.) What is the Republican's optimal response? What if the Democrat makes proportional allocations to all three states?

■ **47.** In Exercise 46, show that if the Democrat makes proportional allocations, and the Republican responds optimally according to the formula given there, this formula simplifies to $r_i = (n_i/N)R$, which does not depend on $d_i$. What does this say about the proportional rule? [*Hint:* If the Republican finds out (say, through a spy) that the Democrat is making proportional allocations, does the Democrat have anything to worry about?]

**48.** In Exercise 47, if the Democrat has only half the resources of the Republican, would you recommend that he or she behave differently from proportional allocations to maximize his or her expected popular vote? Why? If the Republican allocates his or her resources proportionally to the three states, is there any way the Democrat can allocate his or her resources to win a majority of votes in states with more than half the votes?

**49.** Instead of maximizing their expected popular vote, assume the candidates in Exercise 45 want to win in states with more than half the votes. Suppose the candidate who allocates more resources to a state wins that state. Is there any state to which a candidate should not consider allocating resources? Should the states that receive allocations receive equal allocations?

**50.** In Exercise 45, assume you can choose specific voters in each state to whom you can allocate resources. Suppose the candidate who allocates more resources to a voter wins that voter's vote. If your goal is to win the votes of a majority of voters in states that have more than half the votes, which voters would you target, and how much would you spend on each? (*Hint:* First show which states you would target; then show that these states should receive equal allocations, which in turn should be divided equally among a certain set of voters.)

**51.** Assume there are three toss-up states, $A$, $B$, and $C$, with, respectively, 2, 3, and 4 voters, which are also the number of electoral votes of each state. In the text, we gave the formulas for the probabilities, $P_A$, $P_B$, and $P_C$, that the Democrat wins a majority of popular votes in each state and, therefore, wins all the electoral votes of that state. Show that the formula for the probability that the Democrat *wins the election* under the Electoral College, $PWE_D$, is

$$PWE_D = P_A P_B(1 - P_C) + P_A P_C(1 - P_B) + P_B P_C(1 - P_A) + P_A P_B P_C$$

(*Hint:* Winning in any two states is sufficient to win the election.)

■ **52.** Compare the formula for $PWE_D$ with the formula for $EEV_D$ (in the text). Which quantity is it better to maximize? What would be a good resource-allocation strategy for maximizing $PWE_D$?

**53.** Is the square root in the formulas for the $EEV$ maximizing strategies of the Democratic and Republican candidates related to the square-root rule for the Electoral College discussed in Chapter 11?

## WRITING PROJECTS

**1.** Do you think polling is useful in helping voters choose the "best" candidate? Or would it be better, as in some countries, to ban the publication of polls before an election? In one to two pages, discuss these questions in light of the theoretical effects polling has when voters react to polls and possibly change their voting strategies. Is there empirical evidence that voters behave in this way?

**2.** How serious a problem do you think the large-state bias of the Electoral College is? How would you explain the fact that some of the strongest advocates of the Electoral College come from small states? Has the theoretical bias been a reality in the campaign behavior of candidates in recent presidential elections? Discuss in one to two pages.

## SUGGESTED READINGS

BRAMS, STEVEN J. *The Presidential Election Game,* 2nd ed., A K Peters, Wellesley, Mass., 2008. Focuses on the strategic aspects of presidential elections–from primaries to conventions to general elections–and also includes an analysis of the "game" played between President Richard Nixon and the Supreme Court over the release of the Watergate tapes that led to Nixon's resignation in 1974 (Nixon has been the only president to resign the presidency). Approval voting and direct popular-vote election of a president are recommended as election reforms.

BRAMS, STEVEN J., and PETER C. FISHBURN. *Approval Voting,* 2nd ed., Springer, New York, 2007. An in-depth analysis of approval voting, which includes several case studies.

BRAMS, STEVEN J. *Mathematics and Democracy: Designing Better Voting and Fair-Division Procedures*, Princeton University Press, Princeton, N.J., 2008. Shows how mathematics can be used to analyze the properties of different democratic procedures and can help to identify those with the most desirable properties.

HINICH, MELVIN J., and MICHAEL C. MUNGER. *Analytical Politics,* Cambridge University Press, Cambridge, U.K., 1997. Extends spatial modeling to more than one dimension, analyzes probabilistic voting, and introduces game-theoretic solution concepts relevant to the study of elections.

SAARI, DONALD G. *Chaotic Elections! A Mathematician Looks at Voting,* AMS [American Mathematical Society], Providence, R.I., 2001. Argues that elections–in particular, the 2000 presidential election, but others as well–have chaotic features that can be understood through mathematics. The mathematics used is an unusual kind of geometry that will be accessible to those with some mathematical background.

SHEPSLE, KENNETH A., and MARK S. BONACHEK. *Analyzing Politics: Rationality, Behavior, and Institutions,* Norton, New York, 1997. Rational strategies in voting and elections are a major component of this text, but it also includes sections on collective action and political institutions, such as courts and legislatures. Several case studies illustrate the theory.

## SUGGESTED WEB SITES

**www.fec.gov** Federal Election Commission–About Elections and Voting.

**www.ifes.org** International Foundation for Election Systems.

**www.rangevoting.org** The Center for Range Voting.

**wiki.electorama.com/wiki/Election-methods_mailing_list** Election-methods mailing list.

# Fairness and Game Theory PART IV

The central thrust of the first two chapters in Part IV is the fair division of divisible and indivisible objects. Whereas a cake or a parcel of land is divisible, the representatives who are apportioned to the different states are indivisible. Sometimes, however, seemingly indivisible objects, like a car, can be shared, rendering them divisible. By contrast, the game theory chapter focuses on what rational players will choose in different strategic situations, which may be highly unfair to some.

Chapter 13 describes fair-decision schemes in which a group of individuals with different values can be assured of each receiving what he or she views as a fair share when dividing objects like cakes or the goods in an estate.

Chapter 14 discusses the apportionment problem, which is to round a set of fractions to whole numbers while preserving their sum; of course, the sum of the original fractions must be a whole number to start. Apportionment problems occur when resources must be allocated in integer quantities–for instance, when legislators allocated seats in the U.S. House of Representatives to the 50 states.

Chapter 15 introduces the mathematical field called game theory, which describes situations involving two or more decision makers having different goals. Game theory provides a collection of models to assist in the analysis of conflict and cooperation as well as strategies for resolution. Interestingly, you will find that the games covered in this chapter provide us with insights into certain social paradoxes that we routinely encounter in our daily lives. ■

# CHAPTER 13

# Fair Division



When the demands or desires of one party are in conflict with those of another–be it a divorce, a labor–management negotiation, or an international dispute–no one wants to be treated unfairly. And with 1.2 million divorces every year in the United States alone, and crises such as we've seen in the Middle East for decades, it is certainly worth considering how mathematics might help in the search for procedures that can ensure fair and equitable resolutions of such conflicts.

We begin this chapter with one such procedure that was developed in the mid-1990s. The **adjusted winner procedure** allows two parties to settle any dispute involving either issues (as in an international dispute) or objects (as in a divorce or a two-person inheritance) with certain mathematical guarantees of "fairness." Disagreement, it turns out, is both a bad thing and a good thing. On the one hand, disagreement as to how each issue should be resolved typically lies at the heart of a conflict. On the other hand, procedures such as adjusted winner are designed to capitalize on the parties' disagreement as to the importance of each issue, thus allowing each party to end the negotiations thinking it has been met more than halfway.

But adjusted winner is just one of several so-called fair-division procedures that have been developed over the past 65 years. So following our discussion of adjusted winner, we describe a procedure for handling inheritances that was discovered by the Polish mathematician Bronislaw Knaster during World War II. Staying with real-world applications, we next consider the tricky question of finding a priority ranking for potential recipients of an organ that becomes available for transplantation. This is followed by a discussion of an extremely basic fair-division procedure–taking turns–and the question of what the optimal strategy is when taking turns choosing objects.

Bridging the gap between fair-division procedures with obvious real-world potential, such as divorce and inheritance procedures, and procedures that address fundamental mathematical questions of fairness (as do the procedures treated later in this chapter) is the ancient two-person procedure known as divide-and-choose. An application of this procedure to the Law of the Sea Treaty is described.

Divide-and-choose sets the stage for the mathematical investigations of fair division that have gone on for more than half a century. These investigations have often been phrased within the metaphor of "cake cutting." We present three cake-cutting procedures. The first two of these–found by Steinhaus and Banach–Knaster in the 1940s–yield allocations in which each player receives what he or she perceives to be at least his or her fair share of the cake. The last one–found by Selfridge–Conway in 1960–yields allocations in which each player receives what he or she perceives to be a piece at least tied for largest.

## 13.1 The Adjusted Winner Procedure

To illustrate the *adjusted winner procedure*, we will consider an application to the multibillion-dollar world of business mergers. It turns out that one of the most elusive ingredients in the success of a merger is what deal-makers call *social issues*–how power, position, sacrifice, and status are allocated between the merging companies and their executives.

(*Pascal Plessis/AP Photo.*)

As a case in point, let's revisit the 1998 proposed merger between two giant pharmaceutical companies, Glaxo Wellcome and SmithKline Beecham. While most of the details underlying this aborted deal are still unknown to outsiders, the role of social issues is clearly underscored by reports that the companies "saw nearly 19 billion dollars of stock market value vanish in the clash of two corporate egos."

Exactly what kinds of issues might bring on a "clash of two corporate egos"? While not privy to the details of the Glaxo Wellcome–SmithKline Beecham merger attempt, we can speculate as to their nature. For purposes of illustration, let's assume that the following five social issues were paramount:

1. The name that the combined company would use
2. The location of the headquarters of the combined company
3. The question of who would serve as chairman of the combined company
4. The question of who would serve as CEO of the combined company
5. The question of where the necessary layoffs would come from

Each of these five social issues is known to have been a major factor in other recent proposed mergers. For example, when Chrysler merged with Daimler-Benz in 1998, the issue of the choice of a name for the combined company was described as a "standoff" before both sides finally agreed to DaimlerChrysler.

So, let's assume that these were the five social issues confronting Glaxo Wellcome and SmithKline Beecham, and let's see how the adjusted winner procedure would have suggested a resolution. The starting point–and something that is quite difficult when dealing with issues (as in a negotiation) as opposed to objects (as in a divorce)–is to have each side quantify the importance it attaches to getting its own way on each of the issues.

With the adjusted winner procedure, quantification is done by having each side–independently and simultaneously–spread 100 points over the issues in a way that reflects the relative worth of each issue to that party. In our present example, let's assume that the companies allocated their 100 points as shown in Table 13.1. Adjusted winner is now used to decide which side gets its way on which issues, but the procedure requires that a compromise of sorts may have to be reached on one of the issues.

**TABLE 13.1** Applying the Adjusted Winner Procedure to a Merger of Two Companies

| | Point Allocations | |
|---|---|---|
| **Issue** | **Glaxo Wellcome** | **SmithKline Beecham** |
| Name | 5 | 10 |
| Headquarters | 25 | 10 |
| Chairman | 35 | 20 |
| CEO | 15 | 35 |
| Layoffs | 20 | 25 |
| Total | 100 | 100 |

Here's how the procedure works. Suppose we have two parties and a list of either issues to be resolved in one party's favor or the other's (as in our merger example) or objects to be awarded either to one party or to the other (as in a divorce or a two-person inheritance). To have a single word covering both issues and objects, we will speak of "items." The adjusted winner procedure follows these basic steps:

**Basic Steps in the Adjusted Winner Procedure** PROCEDURE

**Step 1.** As described earlier, each party distributes 100 points over the items in a way that reflects their relative worth to that party.

**Step 2.** Each item is initially given to the party that assigned it more points. Each party then assesses how many of his or her own points he or she has received. The party with the fewest points is now given each item on which both parties placed the same number of points.

**Step 3.** Since the point totals are most likely not equal, let $A$ denote the party with the higher point total and $B$ be the other party. Start transferring items from $A$ to $B$, in a certain order, until the point totals are equal. The point at which equality is achieved may involve a fractional transfer of one item.

**Step 4.** The order in which this is done is extremely important and is determined by going through the items in order of increasing **point ratio.** An item's point ratio is the fraction

$$\frac{A\text{'s point value of the item}}{B\text{'s point value of the item}}$$

where $A$ is the party with the higher point total.

Let's demonstrate the adjusted winner procedure by continuing with our analysis of the proposed merger between Glaxo Wellcome and SmithKline Beecham. Why step 4 is so important will be explained later.

1. Assume that Glaxo Wellcome and SmithKline Beecham have given us the point assignments shown in Table 13.1.

2. Because Glaxo Wellcome has placed more points on headquarters (25) and chairman (35), it is initially "given" these issues, while SmithKline Beecham is initially given name (10), CEO (35), and layoffs (25). Notice that SmithKline Beecham now has 10 + 35 + 25 = 70 of its points, whereas Glaxo Wellcome only has 25 + 35 = 60 of its points.
3. We now start transferring issues from SmithKline Beecham to Glaxo Wellcome until the point totals of the two sides are equal. SmithKline Beecham has initially been given three issues (name, CEO, and layoffs), and step 4 will help us decide in what order to start transferring them.
4. Layoffs has point ratio 25/20 = 1.25, name has point ratio 10/5 = 2.00, and CEO has point ratio 35/15 = 2.33. Because layoffs has the lowest point ratio, we start to transfer that item first.

We now see that transferring the entire layoff item (worth 25 to SmithKline Beecham and 20 to Glaxo Wellcome) gives Glaxo Wellcome more points (60 + 20 = 80) than SmithKline Beecham has (70 − 25 = 45).

Thus, the entire layoff item cannot be transferred. Glaxo Wellcome and SmithKline Beecham will need to compromise on the issue of layoffs. But compromise may not mean meeting each other half way. Our goal is to equalize points between the two companies, and a little algebra will tell us exactly the extent to which Glaxo Wellcome and SmithKline Beecham should get their way on the issue of layoffs. Conceptually, it's easier to think of SmithKline Beecham retaining some fraction $x$ of the issue in question and Glaxo Wellcome receiving the complementary fraction $1 - x$ of that same issue.

Because $x$ is the fraction of the issue that SmithKline Beecham retains, the number of points SmithKline Beecham gets from this issue is $x$ times 25. The fraction that Glaxo Wellcome gets is $1 - x$, so the number of points it gets from this issue is $1 - x$ times 20. Thus, if we want a fraction that will make SmithKline Beecham's total points and Glaxo Wellcome's total points equal, then $x$ must satisfy the following equation:

$$10 + 35 + 25x = 25 + 35 + 20(1 - x)$$

We use algebra to solve this equation:

$$\begin{aligned} 45 + 25x &= 60 + 20 - 20x \\ 45 + 25x &= 80 - 20x \\ 45x &= 35 \\ x &= \frac{35}{45} = \frac{7}{9} \end{aligned}$$

Inserting $\frac{7}{9}$ back into the equation, we see that

$$45 + 25\left(\frac{7}{9}\right) = 60 + 20\left(\frac{2}{9}\right)$$

or approximately 64 points for each side. In rough terms, equality of points is achieved when SmithKline Beecham gets about three-fourths ($7/9 \cong 3/4$) of its way on the issue of layoffs and Glaxo Wellcome gets about one-fourth of its way.

Having seen how the adjusted winner procedure works, we must now ask the following question: Exactly what is it about the allocation produced by this scheme that would make someone want to use it? To answer this question, we need three definitions.

**Equitable** DEFINITION

A fair-division procedure, like adjusted winner, is said to be **equitable** if each player believes he or she received the same fractional part of the total value.

**Envy-Free** DEFINITION

A fair-division procedure is said to be **envy-free** if each player has a strategy that can guarantee him or her a share of whatever is being divided that is, in the eyes of that player, at least as large (or at least as desirable) as that received by any other player, no matter what the other players do.

**Pareto-Optimal** DEFINITION

A fair-division procedure is said to be **Pareto-optimal** if it produces an allocation with the property that no other allocation, achieved by any means whatsoever, can make any one player better off without making some other player worse off.

The answer to our earlier question is given by the following theorem (whose proof can be found in *Fair Division* by Brams and Taylor, listed in the Suggested Readings):

**Properties of the Adjusted Winner Allocation** THEOREM

For two parties, the adjusted winner procedure produces an allocation, based on each player's assignment of 100 points over the items to be divided, that has the following properties:

- The allocation is equitable.
- The allocation is envy-free.
- The allocation is Pareto-optimal.

Economists consider Pareto optimality (also named after Vilfredo Pareto) to be an extremely important property, and the order of transfer in step 4 on page 409 of the adjusted winner procedure is so important because it guarantees that the outcome is Pareto-optimal. The fact that the adjusted winner procedure produces an allocation that is efficient in this sense leads us to hope that it can and will play a future role in real-world dispute resolution.

## 13.2 The Knaster Inheritance Procedure

The adjusted winner procedure can be applied in the case of an inheritance if there are only two heirs. For *more than two heirs,* there is quite a different scheme, the **Knaster inheritance procedure**, first proposed by Bronislaw Knaster in 1945. It has a drawback, though, in that is requires the heirs to have a large amount of cash at their disposal.

## EXAMPLE 1 ■ A Four-Person Inheritance

Suppose (for the moment) that there is just one object–a house–and four heirs–Bob, Carol, Ted, and Alice. Knaster's scheme begins with each heir bidding (simultaneously and independently) on the house. Assume, for example, that the bids are

| Bob | Carol | Ted | Alice |
|---|---|---|---|
| $120,000 | $200,000 | $140,000 | $180,000 |

Carol, being the high bidder, is awarded the house. Her fair share, however, is only one-fourth of the $200,000 she thinks the house is worth, and so she places $150,000 (which is three-fourths of the $200,000 she bid) into a temporary "kitty."

Each of the other heirs now withdraws from the kitty his or her fair share, that is, one-fourth of his or her bid:

| | |
|---|---|
| Bob withdraws | $120,000/4 = $30,000 |
| Ted withdraws | $140,000/4 = $35,000 |
| Alice withdraws | $180,000/4 = $45,000 |

Thus, from the $150,000 kitty, a total of $30,000 + $35,000 + $45,000 = $110,000 is withdrawn, and each of the four heirs now feels that he or she has the equivalent of one-fourth of the estate. Moreover, there is a $40,000 surplus ($150,000 kitty − $110,000 withdrawn), which is now divided equally among the four heirs (so each receives an additional $10,000). The final settlement is

| Bob | Carol | Ted | Alice |
|---|---|---|---|
| $40,000 | house − $140,000 | $45,000 | $55,000 |

(*Transtock/Corbis.*)

This illustrates Knaster's procedure for the simple case in which there is only one object. But what if our same four heirs have to divide an estate consisting of, say, a house (as before), a cabin, and a boat? The easiest answer is to handle the estate one object at a time (proceeding for each object as we just did for the house). To illustrate, assume that our four heirs submit the following bids:

| | Bob | Carol | Ted | Alice |
|---|---|---|---|---|
| House | $120,000 | $200,000 | $140,000 | $180,000 |
| Cabin | 60,000 | 40,000 | 90,000 | 50,000 |
| Boat | 30,000 | 24,000 | 20,000 | 20,000 |

We have already settled the house. Let's handle the cabin the same way. Thus, Ted is awarded the cabin based on his high bid of $90,000. His fair share is one-fourth of this, so he places three-fourths of $90,000 (which is $67,500) into the kitty.

Bob withdraws from the kitty $60,000/4 = $15,000. Carol withdraws $40,000/4 = $10,000, and Alice withdraws $50,000/4 = $12,500. Thus, from the $67,500 kitty, a total of $15,000 + $10,000 + $12,500 = $37,500 is withdrawn.

The surplus left in the kitty is thus $30,000, and this is again split equally ($7500 each) among the four heirs. The final settlement on the cabin is

| Bob | Carol | Ted | Alice |
|---|---|---|---|
| $22,500 | $17,500 | cabin − $60,000 | $20,000 |

If we were now to do the same for the boat (we leave the details to you), the corresponding final settlement would be

| Bob | Carol | Ted | Alice |
|---|---|---|---|
| boat − $20,875 | $7625 | $6625 | $6625 |

Putting the three separate analyses (house, cabin, and boat) together, we get a final settlement of

Bob: boat + ($40,000 + $22,500 − $20,875 = $41,625)
Carol: house + (−$140,000 + $17,500 + $7625 = −$114,875)
Ted: cabin + ($45,000 − $60,000 + $6625 = −$8375)
Alice: $55,000 + $20,000 + $6625 = $81,625.

Notice that here, Carol gets the house but must pay $114,875 in cash (and Ted gets the cabin but must put up $8375 in cash). This cash is then disbursed to Bob and Alice. In practice, Carol's having this amount of cash available may be a real problem–the key drawback to Knaster's procedure. Nevertheless, Knaster's procedure shows again that whenever some participants have different evaluations of some objects, there is an allocation in which everyone obtains more than what they would normally consider a fair share.

We summarize Knaster's inheritance procedure as follows.

**Basic Steps in Knaster's Inheritance Procedure with *n* Heirs** PROCEDURE

For each object, the following steps are performed:

**Step 1.** The heirs–independently and simultaneously–submit monetary bids for the object.

**Step 2.** The high bidder is awarded the object, and he or she places all but $1/n$ of his or her bid in a kitty. So, if there are four heirs ($n = 4$), then he or she places all but one-fourth–that is, three-fourths–of his or her bid in a kitty.

**Step 3.** Each of the other heirs withdraws from the kitty $1/n$ of his or her bid.

**Step 4.** The money remaining in the kitty is divided equally among the $n$ heirs.

## 13.3 Fair Division and Organ Transplant Policies

In 1984, the United States Congress passed the National Organ Transplant Act and established a unified transplant network known as the Organ Procurement and Transplantation Network (OPTN). One of the primary goals of the OPTN was to increase the equity in the national system of organ allocation.

Achieving an equitable system of organ allocation is complicated by factors other than demand exceeding supply. For example, should an available organ go to the patient who needs it the most or the one for whom the likelihood of a successful transplant is greatest? Should both of these be taken into consideration, and, if so, how? Questions such as these reveal the extent to which an equitable system of organ allocation is a challenging problem in fair division.

In order to illustrate some of the issues (and paradoxes!) arising in the search for an equitable system for organ allocation, we'll (roughly) follow Peyton Young's synopsis–from his book, listed in Suggested Readings–of the fair division procedure for kidney allocation adopted by the OPTN in the late 1980s.

There were three (main) criteria used in arriving at a final ranking of those needing a kidney, and each potential recipient was awarded points according to a fixed method that we now describe.

- **Criterion 1: Waiting time.** A list of potential recipients was made according to how long they had been waiting for an organ. For each potential recipient, one calculates the fraction of people at or below the spot on the list he or she occupies, and then awards that person a number of points equal to 10 times that fraction. So if there are 5 people on the list, the first (waiting the longest) gets $10 \times 1 = 10$ points, the second gets $10 \times (4/5) = 8$ points, the third gets $10 \times (3/5) = 6$ points, and so on.
- **Criterion 2: Suitability.** The donor and potential recipient each have 6 relevant antigens that are either matched or not matched, with the likelihood of a successful transplant increasing with more matches. Two points are awarded for each match.
- **Criterion 3: Disadvantage.** Each person has antibodies that rule out a certain percentage of the population as being potential donors for that person. For some, only 10% are ruled out, while for others it may be as high as 90%. Those in the latter category are at a serious disadvantage compared to those in the former. Thus, potential recipients are awarded 1 point for each 10% of the population they are "sensitized against."

To illustrate this allocation procedure, let's assume we have 5 potential recipients–A, B, C, D, and E–with the following characteristics:

| Potential recipient | Months waiting | Antigens matched | Percent sensitized |
|---|---|---|---|
| A | 5 | 2 | 10 |
| B | 4.5 | 2 | 20 |
| C | 4 | 0 | 0 |
| D | 2 | 3 | 60 |
| E | 1 | 6 | 90 |

According to the procedure we described, points would be allocated as follows:

| Potential recipient | Months waiting | Antigens matched | Percent sensitized | Total points |
|---|---|---|---|---|
| A | 10 | 4 | 1 | 15 |
| B | 8 | 4 | 2 | 14 |
| C | 6 | 0 | 0 | 6 |
| D | 4 | 6 | 6 | 16 |
| E | 2 | 12 | 9 | 23 |

Thus, if one kidney became available, it would go to E (with 23 points). Presumably, if two became available at the same time, E would get one and D (with 16 points) would get the other.

But now things get interesting. Peyton Young, being well versed in the paradoxes of voting theory, fair division, and apportionment (among other things), observed the following. In the above scenario, what if two kidneys become available, but one is delayed slightly? Presumably, E gets the first one, and then we redo the chart with only A, B, C, and D. This yields the following:

| Potential recipient | Months waiting | Antigens matched | Percent sensitized |
|---|---|---|---|
| A | 5 | 2 | 10 |
| B | 4.5 | 2 | 20 |
| C | 4 | 0 | 0 |
| D | 2 | 3 | 60 |

According to the procedure we described, points would be allocated as follows:

| Potential recipient | Months waiting | Antigens matched | Percent sensitized | Total points |
|---|---|---|---|---|
| A | 10 | 4 | 1 | 15 |
| B | 7.5 | 4 | 2 | 13.5 |
| C | 5 | 0 | 0 | 5 |
| D | 2.5 | 6 | 6 | 14.5 |

Thus, A (not D!) now gets the second kidney, having 15 points to 14.5 for D. This is an example of what is called the "priority paradox." For more on this, we invite the reader to consult Peyton Young's book in the Suggested Readings.

## 13.4 Taking Turns

For many of us, an early lesson in fair division happens in elementary school with the choosing of sides for a spelling bee or when picking teams on the playground. In terms of importance, these pale in comparison with the issue of property settlement in a divorce. Remarkably, however, the same fair-division procedure–*taking turns*–is often used in both.

**Taking turns** is fairly self-explanatory. With two parties (and that's all we'll consider here), one party selects an object, then the other party selects one, then the first party again, and so on. But in this context, there are several interesting questions that suggest themselves:

1. How do we decide who chooses first?
2. Because choosing first is often quite an advantage, shouldn't we compensate the other party in some way, perhaps by giving him or her extra choices at the next turn?
3. Should a player always choose the object he or she most favors from those that remain, or are there strategic considerations that players should take into account?

The answer to question 1 is often "toss a coin," but there are other possibilities–for example, the two parties could "bid" for the right to go first, as in an auction. The answer to question 2 is less clear, but we outline a discussion of the issue it raises in Writing Project 2.

Question 3, on the other hand, is remarkably interesting, and it is this one that we want to pursue. Let's look at an easy example. Suppose that Bob and Carol are getting a divorce, and their four main possessions, ranked from best to worst by each, are as follows:

| | **Bob's Ranking** | **Carol's Ranking** |
|---|---|---|
| Best | Pension | House |
| Second best | House | Investments |
| Third best | Investments | Pension |
| Worst | Vehicles | Vehicles |

If Carol knows nothing of Bob's preferences, then we can assume that she will choose sincerely–selecting at her turn whichever item she most prefers from those not yet chosen. Now, if Bob is also sincere, and if he chooses first, the items will be allocated as follows:

| | |
|---|---|
| First turn: | Bob takes the pension. |
| Second turn: | Carol takes the house. |
| Third turn: | Bob takes the investments. |
| Fourth turn: | Carol is left with the vehicles. |

Hence, Bob gets his first and third favorites (the pension and the investments). However, if Bob opens by choosing the house–and bypassing the pension for the moment–then the allocation will be as follows:

| | |
|---|---|
| First turn: | Bob takes the house. |
| Second turn: | Carol takes the investments. |
| Third turn: | Bob takes the pension. |
| Fourth turn: | Carol is left with the vehicles. |

Thus, by being insincere, Bob does better–getting his first and second favorites (the pension and the house).

In general, then, what is the optimal strategy for rational players to use, assuming that both know the preferences of the other? The answer is something called the **bottom-up strategy**, discovered by the mathematicians D. A. Kohler and R. Chandrasekaran in 1969. We will illustrate it with an example.

Suppose we have five objects–$A, B, C, D, E$–and Bob is choosing first. Suppose that Bob and Carol have the following rankings of the objects (called **preference lists** in what follows):

| Bob | Carol |
|---|---|
| $A$ | $C$ |
| $B$ | $E$ |
| $C$ | $D$ |
| $D$ | $A$ |
| $E$ | $B$ |

It will turn out that Bob should open with $C$ (his third choice) followed by Carol's choice of $D$ (skipping over $E$, for the moment). Bob will then take $A$, Carol will follow with $E$, and finally Bob will get $B$. Bob gets his first, second, and third choices without selecting his first choice first! Where does this strategy come from?

The intuition here is quite easy. Let's make two assumptions about rational players: A rational player will never willingly choose his or her least preferred alternative, and a rational player will avoid wasting a choice on an object that he or she knows will remain available and thus can be chosen later.

With these assumptions as motivation, let's return to the preceding example and think about the mental calculation Bob will go through in deciding what his first choice will be. Bob knows the eventual sequence of choices will fill in all of the following blanks:

| | | | | | |
|---|---|---|---|---|---|
| Bob: | ______ | | ______ | | ______ |
| Carol: | | ______ | | ______ | |

Now, working mentally from right to left, Bob knows that Carol will not choose *B*, because it is the bottom thing on her list. Thus, he will get stuck with *B*, and so he will avoid wasting anything but his last choice on alternative *B*. Thus, Bob can pencil in alternative *B* as his last choice:

| | | | | | |
|---|---|---|---|---|---|
| Bob: | ______ | | ______ | | *B* |
| Carol: | | ______ | | ______ | |

Bob, placing himself momentarily in Carol's shoes, knows she will reason the same way, and thus he pencils Carol in for the bottom alternative, *E*, on his list:

| | | | | | |
|---|---|---|---|---|---|
| Bob: | ______ | | ______ | | *B* |
| Carol: | | ______ | | *E* | |

Mentally now, Bob reasons as if alternatives *B* and *E* never existed (and the choice sequence had been Bob–Carol–Bob) and continues to pencil in alternatives from right to left, with Bob working from bottom to top on Carol's preference list and Carol working from bottom to top on Bob's preference list. This yields the following sequence of choices mentally penciled in by Bob:

| | | | | | |
|---|---|---|---|---|---|
| Bob: | *C* | | *A* | | *B* |
| Carol: | | *D* | | *E* | |

Remember, this is just a mental calculation that Bob went through to decide upon the actual choice–in this case, *C*–with which he will open. Bob has no guarantee that Carol will, in fact, respond with *D*, so the use of this strategy involves some risk on Bob's part.

This bottom-up strategy can also be viewed as a procedure that a mediator could use to specify a division of several objects between two parties. Given the preference lists of both parties, the mediator could construct a list–exactly as we did for Bob and Carol above–and then offer this to the parties as the suggested allocation. In effect, the mediator is simultaneously playing the role of two rational parties who choose to employ optimal strategies.

## 13.5 Divide-and-Choose

There are vast mineral resources under the seabed, all of which, one might argue, should be available to both developed and developing countries. In the absence of some kind of agreement, however, what is to prevent the developed countries from mining all of the most promising tracts before the developing countries have reached a technological level where they can begin their own mining operations? Such an agreement went into effect on November 16, 1994, with 159 signatories (including the United States). It was called the **Convention of the Law of the Sea,** and it protects the interests of developing countries by means of the following fair-division procedure.

Whenever a developed country wants to mine a portion of the seabed, that country must propose a division of the portion into two tracts. An international mining company called the Enterprise, funded by the developed countries but representing the interests of the developing countries through the International Seabed Authority, then chooses one of the two tracts to be reserved for later use by the developing countries.

**Divide-and-Choose** PROCEDURE

With **divide-and-choose**, one party divides the object into two parts in any way that he desires, and the other party chooses whichever part she wants.

As a fair-division procedure, the origins of divide-and-choose go back thousands of years. The Hebrew Bible tells the story of Abram (later to be called Abraham) and Lot, who settled a dispute over land via a proposed division by Abram–"If you go north, I will go south; and if you go south, I will go north" (Gen. 13:8–9)–and a choice (of the plain of Jordan) by Lot. Divide-and-choose resurfaced later in Hesiod's book *Theogony*. The Greek gods Prometheus and Zeus had to divide a portion of meat. Prometheus began by placing the meat into two piles, and Zeus selected one.

Actually, a fair-division procedure consists of both rules and strategies, and all we have described so far are the rules of divide-and-choose. But the natural strategies here are quite obvious: The divider makes the two parts equal in his estimation, and the chooser selects whichever piece she feels is more valuable.

Rules and strategies differ from each other in the following sense: A referee could determine if a rule is being followed, even without knowing the preferences of the players. Strategies represent choices of how players follow the rules, given their individual preferences (and any other knowledge or goals they may have).

The strategies on which we focus in our discussion of fair-division procedures are those that require no knowledge of the preferences of the other players and yet provide some kind of minimal degree of satisfaction even in the face of collusion by the other players. For example, the strategies just given for divide-and-choose guarantee each player a piece that he or she would not wish to trade for that received by the other.

There are, to be sure, other strategic considerations that might be relevant. For example, in divide-and-choose, would you rather be the divider or the chooser? The answer, given our assumptions that nothing is known of the preferences of the others, is to be the chooser. However, if you knew the preferences of your opponent (and how much she may value spite), then you might want to be the divider.

As a final comment on strategic considerations, we need only look to the origins of the well-known expression "the lion's share." It comes from one of Aesop's fables, as reported by Todd Lowry in *Archaeology of Economic Ideas* (1987, p. 130):

> It seems that a lion, a fox, and an ass participated in a joint hunt. On request, the ass divides the kill into three equal shares and invites the others to choose. Enraged, the lion eats the ass, then asks the fox to make the division. The fox piles all the kill into one great heap except for one tiny morsel. Delighted at this division, the lion asks, "Who has taught you, my very excellent fellow, the art of division?" to which the fox replies, "I learnt it from the ass, by witnessing his fate."

## 13.6 Cake-Division Procedures: Proportionality

The modern era of fair division in mathematics began in Poland during World War II (see Spotlight 13.1). At this time, Hugo Steinhaus asked what is, in retrospect, the obvious question: What is the "natural" generalization of divide-and-choose to three or more people? The metaphor that has been used in this context, going back at least to the English political theorist James Harrington (1611–1677), is a cake. We picture different players valuing different parts of the cake differently because of concentrations of certain flavors or depth of frosting.

**Cake-Division Procedure** DEFINITION

A **cake-division procedure** for $n$ players is a procedure that the players can use to allocate a cake among themselves (no outside arbitrators) so that each player has a strategy that will guarantee that player a piece with which he or she is "satisfied," even in the face of collusion by the others.

As we have seen, divide-and-choose is a cake-division procedure for two players, if by "satisfied" we mean either "thinks his piece is of size or value at least one-half" or "does not want to trade what she received for what anyone else received." We define the first notion here; envy-free allocations were defined in Section 13.1.

**Proportional Procedure** DEFINITION

A cake-division procedure (for $n$ players) will be called **proportional** if each player's strategy guarantees that player a piece of size or value at least $1/n$ of the whole in his or her own estimation.

It turns out that for $n = 2$, a procedure is envy-free if and only if it is proportional; that is, for $n = 2$, the two notions of fair division are exactly the same. For $n > 2$, however, all we can say is that an envy-free procedure is automatically proportional. For example, if a three-person allocation is not proportional, then one player (call him Bob) thinks that he received less than one-third. Bob then feels that the other two are sharing more than two-thirds between them, and thus that at least one of the two (call her Carol) must have more than one-third. But then Bob will envy Carol, and so the allocation is not envy-free. Because all nonproportional allocations fail to be envy-free, it follows that if an allocation is envy-free, then it must be proportional.

Many procedures that are proportional, however, fail to be envy-free, as we shall soon show. Thus, proportional procedures are fairly easy to come by, but envy-free procedures are fairly hard to come by.

(*Brand X Pictures/Punchstock.*)

### EXAMPLE 2 ▪ The Steinhaus Proportional Procedure for Three Players (Lone Divider)

Given three players–Bob, Carol, and Ted–we have Bob divide the cake into three pieces, call them $X$, $Y$, and $Z$, each of which he thinks is of size or value exactly one-third. Let's speak of Carol as "approving of a piece" if she thinks it is of size or value at least one-third. Similarly, we will speak of Ted as "approving of a piece" if the same criterion applies. Notice that both Carol and Ted must approve of at least one piece.

If there are distinct pieces–say, $X$ and $Y$–with Carol approving of $X$ and Ted approving of $Y$, then we give the third piece, $Z$, to Bob (and, of course, $X$ to Carol and $Y$ to Ted), and we are done. The problem case is where both Carol and Ted approve of only one piece and it is the *same* piece.

Let's assume that Carol and Ted approve of only one piece, $X$, and hence (of more importance to us) both *disapprove* of piece $Z$. Let $XY$ denote the result of putting piece $X$ and piece $Y$ back together to form a single piece. Notice that both Carol and Ted think that $XY$ is at least two-thirds of the cake because both disapprove of $Z$. Thus, we can give $Z$ to Bob and let Carol and Ted use divide-and-choose on $XY$. Because half of two-thirds is one-third, both Carol and Ted are guaranteed a proportional share (as is Bob, who approved of all three pieces).

The procedure just described, which guarantees proportional shares but is not necessarily envy-free and is sometimes called the **lone-divider method**, was discovered by Hugo Steinhaus around 1944. Unfortunately, it does not extend easily to more than three players. It was left to Steinhaus's students, Stefan Banach and Bronislaw Knaster, to devise a method for more than three players. Picking up where Steinhaus left off (and traveling in quite a different direction), they devised the proportional procedure that today is referred to as the **last-diminisher method**. Like the lone-divider method, it is proportional but not envy-free. We illustrate it for the case of four players (Bob, Carol, Ted, and Alice), and we include both the rules and the strategies that guarantee each player his or her fair share.

## EXAMPLE 3 ■ The Banach–Knaster Proportional Procedure for Four or More Players (Last Diminisher)

Bob cuts from the cake a piece that he thinks is of size one-fourth and hands it to Carol. If Carol thinks the piece handed her is larger than one-fourth, she trims it to size one-fourth in her estimation, places the trimmings back on the cake, and passes the diminished piece to Ted. If Carol thinks the piece handed her is of size at most one-fourth, she passes it unaltered to Ted.

Ted now proceeds exactly as did Carol, trimming the piece to size one-fourth if he thinks it is larger than this and passing it (diminished or unaltered) on to Alice. Alice does the same, but, being the last player, simply holds onto the piece momentarily instead of passing it to anyone.

Notice that everyone now thinks the piece is of size at most one-fourth, and the last person to trim it (or Bob, if no one trimmed it) thinks the piece is of size exactly one-fourth. Thus, the procedure now allocates this piece to the last person who trimmed it (and to Bob if no one trimmed it).

Assume for the moment that it was Ted who trimmed the piece last, so he takes this piece and exits the game. Bob, Carol, and Alice all think that at least three-fourths of the cake is left, so they can start the process over with (say) Bob beginning by cutting a piece from what remains that he thinks is one-fourth of the original cake. Carol and Alice are both given a chance to trim it to size one-fourth in their estimation, and again, the last one to trim it takes that piece and exits the game. The two remaining players both think that at least half the cake is left, so they can use divide-and-choose to divide it between themselves and thus be assured of a piece that is of size at least one-fourth in their estimation.

## SPOTLIGHT 13.1 Sixty Years of Cake Cutting

The modern era of cake cutting began with the investigations of the Polish mathematician Hugo Steinhaus during World War II. His research, and that of dozens of others since, involved dealing with two fundamental difficulties. First, allocation schemes that work in the context of two or three players often do not generalize easily to the context of four or more players. Second, procedures that yield envy-free allocations are considerably harder to obtain than procedures that yield proportional allocations.

The mathematics inspired by these two difficulties constitutes a rather elegant corner of the large and important area of fair division. Steinhaus's investigations in the 1940s led to his observation that there is a rather natural extension of divide-and-choose to the case of three players. This is the "lone-divider procedure" described on page 420. Steinhaus's method was generalized to an arbitrary number of players by Harold W. Kuhn of Princeton University in 1967.

Unable to extend his procedure from three to four players, Steinhaus proposed the problem to some Polish colleagues. Two of them, Stefan Banach and Bronislaw Knaster, solved this problem in the mid-1940s by producing the "last-diminisher procedure" described on page 420.

In addition to the procedures devised by Banach, Knaster, and Kuhn, there are other well-known constructive procedures for obtaining a proportional allocation among four or more players. One of these is by A. M. Fink of Iowa State University and appears in Exercise 29.

Another constructive procedure of note, although different in flavor from the others, is the 1961 recasting by Lester E. Dubins and Edwin H. Spanier of the University of California at Berkeley of the last-diminisher method as a "moving-knife procedure" (illustrated in Exercise 31). The trade-off here involves giving up the "discrete" nature of the last-diminisher method in exchange for the conceptual simplicity of the moving knife.

Although the existence of an envy-free allocation (even for four or more players) was known to Steinhaus in the 1940s, the first constructive procedure for producing an envy-free allocation among three players was not found until around 1960. At that time, John L. Selfridge of Northern Illinois University and, later but independently, John H. Conway of Princeton University found the elegant procedure presented on page 422. Although never published by either, the procedure was quickly and widely disseminated by Richard K. Guy of the University of Calgary and others. Eventually it appeared in several treatments of the problem by different authors.

In 1980, a moving-knife procedure for producing an envy-free allocation among three players was found by Walter R. Stromquist of Daniel Wagner Associates. Then, another procedure, capable of being recast as a moving-knife solution of the three-player case, was found by a law professor at the University of Virginia, Saul X. Levmore, and a former student of his, Elizabeth Early Cook.

In 1992, Steven J. Brams, a political scientist at New York University, and Alan D. Taylor, a mathematician at Union College, succeeded in finding a constructive procedure for producing an envy-free allocation among four or more players. In 1994, Brams, Taylor, and William S. Zwicker (also from Union College) found a moving-knife solution to the four-person envy-free problem. No moving-knife procedure is known that will produce an envy-free allocation among five or more players.

## 13.7 Cake-Division Procedures: The Problem of Envy

Divide-and-choose has a property that neither of the last two procedures possesses: It can ensure that each player receives a piece of cake he or she considers the largest or tied for the largest. In the case of only two players, this means that each player can get what he or she perceives to be at least half the cake, no matter what the other player does. Thus, divide-and-choose is an envy-free procedure.

Steinhaus's $n = 3$ proportional procedure (the lone-divider method) is not envy-free. For example, consider the case where Carol and Ted both find one piece unacceptable (and this piece is given to Bob). Carol and Ted will not envy each other when one divides and the other chooses, but Bob may think that this is not a 50-50 split. Indeed, if Bob divided the cake initially into what he thought was three equal pieces, an unequal split of the remaining two-thirds of the cake by Carol and Ted means that Bob will prefer the larger of these two pieces to the one-third he got. Consequently, Bob will envy the person who got this larger piece.

Nor is the last-diminisher method envy-free. For example, if Bob initially cuts a piece of cake of size one-fourth, and no one else trims it, then Bob receives this piece and exits the game. If Carol is the one to make the next initial cut, she may well cut a piece from the cake that she thinks is of size one-fourth but that Bob thinks is of size considerably more than one-fourth. But Bob is out of the game. Thus, if Ted and Alice think this piece is of size less than one-fourth, then Carol receives it, and so Bob will envy Carol.

Nevertheless, there do exist cake-division procedures that are envy-free. We present one of these in what follows.

### EXAMPLE 4 ▪ The Selfridge–Conway Envy-Free Procedure for Three Players

We start with a cake and three people. The point we wish to arrive at is an envy-free allocation of the entire cake among the three people in a finite number of steps. This task may seem formidable, but quite often in mathematics, an important part of solving a problem involves breaking the problem into identifiable parts. In this case, let's call our starting point $A$ and the final point we wish to reach $C$. Now let's identify an appropriate in-between point $B$ that makes going from $A$ to $C$–via $B$–more manageable. Our in-between point $B$ is the following:

*Point B:* Getting a constructive procedure that gives an envy-free allocation of *part* of the cake.

Can we constructively obtain three pieces of cake, whose union may not be the whole cake, which can be given to the three people so that each thinks he or she received a piece at least tied for largest? This turns out to be quite easy with the solution given by John Selfridge and John Conway. The following process and strategies do the trick:

1. Player 1 cuts the cake into three pieces he considers to be the same size. He hands the three pieces to player 2.
2. Player 2 trims at most one of the three pieces to create at least a two-way tie for largest. Setting the trimmings aside, player 2 hands the three pieces (one of which may have been trimmed) to player 3.

3. Player 3 now chooses, from among the three pieces, one that he considers to be at least tied for largest.
4. Player 2 next chooses, from the two remaining pieces, one that she considers to be at least tied for largest, with the proviso that if she trimmed a piece in step 2, and player 3 did not choose this piece, then she must now choose it.
5. Player 1 receives the remaining piece.

Let's reconsider the five steps of this trimming procedure to assure ourselves that each player experiences no envy. Recall that player 1 cuts the cake into three pieces, and player 2 trims one of these three pieces. Now player 3 chooses, and, as the first to choose, he certainly envies no one. Player 2 created a two-way tie for largest, and at least one of these two pieces is still available after player 3 selects his piece. Hence, player 2 can choose one of the tied pieces she created and will envy no one. Finally, player 1 created a three-way tie for largest and, because of the proviso in step 4, the trimmed piece is not the one left over. Thus, player 1 can choose an untrimmed piece and therefore will envy no one.

So far we have gone from point $A$ to point $B$: Starting with a cake and three players, we have constructively obtained (in finitely many steps) an envy-free allocation of all of the cake, except the part $T$ that player 2 trimmed from one of the pieces. We will now describe how $T$ can be allocated among the three players in such a way that the resulting allocation of the whole cake is envy-free. (This is the rest of the **Selfridge–Conway envy-free procedure**.)

The key observation for the $n = 3$ case is that player 1 will not envy the player who received the trimmed piece, even if that player were to be given all of $T$. Recall that player 1 created a three-way tie and received an untrimmed piece. The union of the trimmed piece and the trimmings yields a piece that player 1 considers to be exactly the same size as the one he received. Thus, assume that it is player 3 who received the trimmed piece (it could as well be player 2). Then player 1 will not envy player 3, no matter how $T$ is allocated.

The next step ensures that neither player 2 nor player 3 will envy another player when it comes time to allocate $T$. Let player 2 cut $T$ into three pieces she considers to be the same size. Let the players choose which of the three pieces they want in the following order: player 3, player 1, player 2.

To see that this yields an envy-free allocation, notice that player 3 envies no one, because he is choosing first. Player 1 does not envy player 2, because he is choosing ahead of her; and player 1 does not envy player 3 because, as pointed out earlier, player 1 will not envy the player who received the trimmed piece. Finally, player 2 envies no one, because she made all three pieces of $T$ the same size.

Hence, for $n = 3$, the Selfridge–Conway procedure will give an envy-free allocation of all the cake except $T$, followed by an allocation of $T$ that gives an envy-free allocation of all the cake.

A naive attempt to generalize to $n = 4$ what we have done for $n = 3$ would proceed as follows: We would begin by having player 1 cut the cake into four pieces he considers to be the same size. Then we would have players 2 and 3 trim some pieces (but how many?) to create ties for the largest. Finally, we would have the players choose from among the pieces–some of which would have been trimmed–in the following order: player 4, player 3, player 2, player 1.

This approach fails because player 1 could be left in a position of envy. To understand how the approach could fail, consider how many pieces player 3 might have to trim to create a sufficient supply of pieces tied for largest so that he is guaranteed to have one available when it is his turn to choose. Player 3 might have to trim one piece to create a two-way tie for largest. Player 2 might need to trim two pieces to create a three-way tie for largest (because if there were only a two-way tie for largest, player 3 might further trim one of these pieces and player 4 might choose the other). This leaves player 1 in a possible position of envy, because we could have a situation where player 2 trims two pieces and player 3 trims a third piece, and player 4 then chooses the only untrimmed piece. If this happens, player 1, by being forced to choose a trimmed piece, will definitely envy player 4.

All is not lost, however, because there are modifications of the Selfridge–Conway procedure that will work for arbitrary $n$. For more on this, see *Fair Division* in the Suggested Readings.

Although we have used the metaphor of cake cutting throughout our discussion of the problem of envy, the idea of successive trimming is nonetheless applicable to problems of fair division other than parceling out the last crumbs of a cake. The main practical problem in applying the trimming procedure is that many fair-division problems involve goods that cannot be divided up at all, much less trimmed in fine amounts. Such goods are said to be *indivisible.*

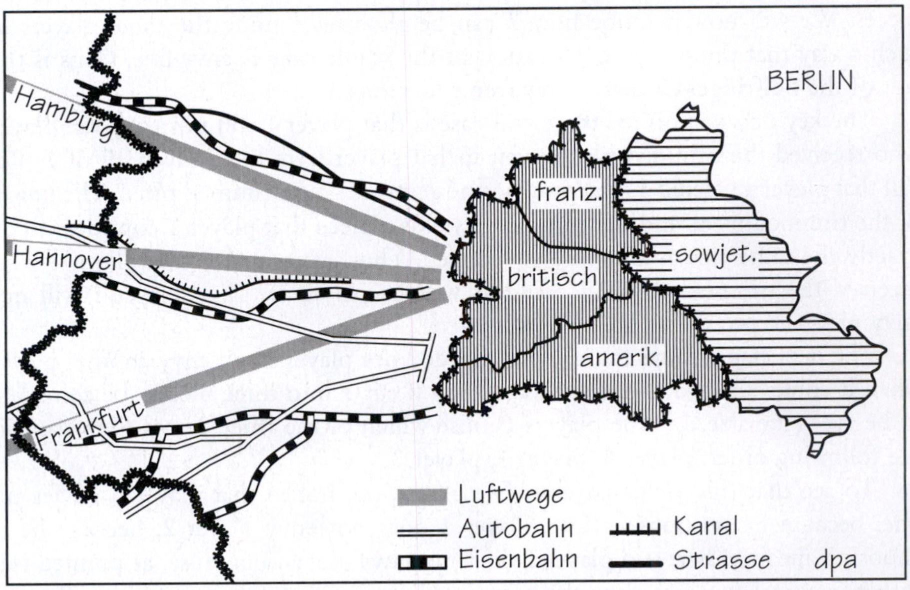


It is interesting to recall that when the Allies agreed in 1944 to partition Germany into sectors after World War II (first stage), they initially did not reach agreement about what to do with Berlin. Subsequently, they decided to partition Berlin itself into sectors (second stage), even though this city fell 110 miles within the Soviet sector. Berlin was simply too valuable a "piece" for the Western Allies (Great Britain, France, and the United States) to cede to the Soviets, which suggests how, after a leftover piece is trimmed off, it can be subsequently divided under the trimming procedure.

Yet, what if a large piece like Berlin is not divisible? In the settlement of an estate, this might be the house, which may be worth half the estate to the claimants. In this situation, there may be no alternative but to sell this big item and use the proceeds to make the remaining estate more liquid or, in our terms, "trimmable."

## REVIEW VOCABULARY

**Adjusted winner procedure** A fair-division procedure introduced by Steven Brams and Alan Taylor in 1993. It works only for two players, and begins by having each player independently spread 100 points over the items to be divided so as to reflect the relative worth of each object to that player. The allocation resulting from this procedure is equitable, envy-free, and Pareto-optimal. It requires no cash from either player, but one of the objects may have to be divided or shared by the two players. (p. 407)

**Bottom-up strategy** A bottom-up strategy is a strategy under an alternating procedure in which sophisticated choices are determined by working backward. (p. 416)

**Cake-division procedure** A fair-division procedure that uses a cake as a metaphor. Such procedures involve finding allocations of a single object that is finely divisible, as opposed to the situation encountered with either the adjusted winner procedure or Knaster's procedure. In a cake-division procedure, each player has a strategy that will guarantee that player a piece with which he or she is "satisfied," even in the face of collusion by the others. (p. 419)

**Convention of the Law of the Sea** An agreement based on divide-and-choose that protects the interests of developing countries in mining operations under the sea. (p. 417)

**Divide-and-choose** A fair-division procedure for dividing an object or several objects between two players. This method produces an allocation that is both proportional and envy-free (the two being equivalent when there are only two players). (p. 418)

**Envy-free** A fair-division procedure is said to be envy-free if each player has a strategy that can guarantee him or her a share of whatever is being divided that is, in the eyes of that player, at least as large (or at least as desirable) as that received by any other player, no matter what the other players do. (p. 411)

**Equitable** A fair-division procedure like adjusted winner is said to be equitable if each player believes he or she received the same fractional part of the total value. (p. 411)

**Knaster inheritance procedure** A fair-division procedure for any number of parties that begins by having each player (independently) assign a dollar value (a "bid") to the item or items to be divided so as to reflect the absolute worth of each object to that player. The allocation resulting from this procedure leaves each party feeling that he or she received a dollar value at least equal to his or her fair share (and often more so). It never requires the dividing or sharing of an object, but it may require that the players have a large amount of cash on hand. (p. 411)

**Last-diminisher method** A cake-division procedure introduced by Stefan Banach and Bronislaw Knaster. It works for any number of players and produces an allocation that is proportional but not, in general, envy-free. (p. 420)

**Lone-divider method** A cake-division procedure introduced by Hugo Steinhaus. It works only for three players and produces an allocation that is proportional but not, in general, envy-free. (p. 420)

**Pareto-optimal** A fair-division procedure is said to be Pareto-optimal if it produces an allocation with the property that no other allocation, achieved by any means whatsoever, can make any one player better off without making some other player worse off. (p. 411)

**Point ratio** The fraction in which the numerator is the number of points one party placed on an object and the denominator is the number of points the other party placed on the object. (p. 409)

**Preference lists** Rankings of the items to be allocated, from best to worst, by each of the participants. (p. 416)

**Proportional** A fair-division procedure is said to be proportional if each of $n$ players has a strategy that can guarantee that player a share of whatever is being divided that he or she considers to be at least $1/n$ of the whole in size or value. (p. 419)

**Selfridge–Conway envy-free procedure** A cake-division procedure introduced independently by John Selfridge and John Conway. It works only for three players but produces an allocation that is envy-free (as well as proportional). (p. 423)

**Taking turns** A fair-division procedure in which two or more parties alternate selecting objects. (p. 415)

## SKILLS CHECK

**1.** The adjusted winner procedure

**(a)** applies only to two-party disputes or disputes that can be recast as two-party disputes.

**(b)** applies to either two-party or three-party disputes.

**(c)** applies to $n$-party disputes for all $n$.

**2.** The starting point with adjusted winner is to have each side–independently and simultaneously–spread 100 points over the issues in such a way that it __________ .

**3.** The "winner" part of the name "adjusted winner" refers to the fact that

**(a)** each party initially "wins" (that is, is given) each issue on which he or she places more points than the other party.
**(b)** objects ultimately go to whichever party bid more.
**(c)** it is impossible for both parties to win with any fair-division scheme.

**4.** The "adjusted" part of the name "adjusted winner" refers to ______________________________ .

**5.** In transferring items from one party to the other with the adjusted winner procedure,

**(a)** the order in which items are transferred is extremely important.
**(b)** only one item will need to be split or shared.
**(c)** the order of transfer is obtained by looking at so-called point ratios.
**(d)** all of the above.

**6.** Chris and Terry must make a fair division of three objects. They assign points to the objects and use the adjusted winner procedure. Chris ends up with ______ .

| Object | Chris | Terry |
|---|---|---|
| Boat | 30 | 20 |
| Land | 50 | 60 |
| Car | 20 | 20 |

**7.** Chris and Terry use the Knaster inheritance procedure to divide a coin collection. Chris bids $1000 and Terry bids $800. What is the outcome?

**(a)** Chris gets the coins and pays Terry $200.
**(b)** Chris gets the coins and pays Terry $450.
**(c)** Chris gets the coins and pays Terry $500.

**8.** Four children bid on two objects. Using the Knaster inheritance procedure, Adam ends up with __________ .

| Object | Adam | Beth | Carl | Dietra |
|---|---|---|---|---|
| House | $80,000 | $75,000 | $90,000 | $60,000 |
| Car | $10,000 | $12,000 | $13,000 | $15,000 |

**9.** With the procedure known as taking turns, the optimal strategy for players is called

**(a)** the sincere strategy.
**(b)** the bottom-up strategy.
**(c)** the top-down strategy.

**10.** With taking turns, we assume that a rational player will ______________________________ .

**11.** Two people use the divide-and-choose procedure to divide a field. Suppose Jeff divides and Karen chooses. Which statement is true?

**(a)** Karen always believes she gets more than her fair share.
**(b)** Karen can guarantee that she always gets at least her fair share.
**(c)** Karen can possibly believe she gets less than her fair share.

**12.** A fair-division procedure is envy-free when each player believes that ______________________________ .

**13.** Using the Steinhaus procedure for three players (lone divider), what happens if there is a single portion that is the only one approved of by both nondividers?

**(a)** One of the other portions is given to the divider.
**(b)** The two nondividers flip a coin to determine who receives the approved portion.
**(c)** All portions are returned to the cake and a different person serves as the new divider.

**14.** Using the Steinhaus procedure for three players (lone divider), if the two nondividers approve different portions, then ______________________________ .

**15.** Using the Banach–Knaster procedure for three or more players (last diminisher), what happens to the first portion after each person has inspected and possibly trimmed it?

**(a)** The portion goes to the last person to approve the portion, whether or not it was trimmed.
**(b)** The portion goes to the last person to trim the portion.
**(c)** The portion goes to the first person to approve and not trim the portion.

**16.** Using the Banach–Knaster procedure for three or more players (last diminisher), the player who receives the first portion ______________________________ .

**17.** Using the Banach–Knaster procedure for three or more players (last diminisher), suppose that Scott initially cuts a piece and passes it among the other people, none of whom trims it. What happens next?

**(a)** Scott gets this piece.
**(b)** The last person who is handed the piece keeps it.
**(c)** The piece is returned to the cake and someone else cuts a piece.

**18.** Using the Banach–Knaster procedure for three or more players (last diminisher), when only two people remain, ______________________________ .

**19.** For the Selfridge–Conway envy-free procedure for three players, which of the following statements is true?

**(a)** Each of the three players has the opportunity to trim the portions if they appear to be unfair.
**(b)** Each player receives a portion that he or she believes to be exactly one-third of the total.
**(c)** The first player may believe that the third player received more than a fair share.

**20.** For the Selfridge–Conway envy-free procedure for three players, the player who will definitely *not* receive the trimmed piece in stage 1 is __________________ .

# CHAPTER 13 EXERCISES

■ Challenge ◆ Discussion

## 13.1 The Adjusted Winner Procedure

**1.** The 1991 divorce of Ivana and Donald Trump was widely covered in the media. The marital assets included a 45-room mansion in Greenwich, Connecticut; the 118-room Mar-a-Lago mansion in Palm Beach, Florida; an apartment in the Trump Plaza; a 50-room Trump Tower triplex; and just over $1 million in cash and jewelry. Assume points are assigned as follows:

| | Point Allocations | |
|---|---|---|
| **Marital asset** | **Donald's Points** | **Ivana's Points** |
| Connecticut estate | 10 | 38 |
| Palm Beach mansion | 40 | 20 |
| Trump Plaza apartment | 10 | 30 |
| Trump Tower triplex | 38 | 10 |
| Cash and jewelry | 2 | 2 |

Use the adjusted winner procedure to determine a fair allocation of the marital assets. (Exercise 1 courtesy of Catherine Duran.)

**2.** Suppose that Calvin and Hobbes discover a sunken pirate ship and must divide their loot. They assign points to the items as follows:

| Object | Calvin's Points | Hobbes's Points |
|---|---|---|
| Cannon | 10 | 5 |
| Anchor | 10 | 20 |
| Unopened chest | 15 | 20 |
| Doubloon | 11 | 14 |
| Figurehead | 20 | 30 |
| Sword | 15 | 6 |
| Cannon ball | 5 | 1 |
| Wooden leg | 2 | 1 |
| Flag | 10 | 2 |
| Crow's nest | 2 | 1 |

Use the adjusted winner procedure to determine a fair allocation of the loot. (Exercise 2 courtesy of Erica DeCarlo.)

**3.** This exercise illustrates how the adjusted winner procedure can be used to resolve disputes as well as to achieve fair allocations. Suppose Mike and Phil are roommates in college, and they encounter serious conflicts during their first week at school. Their resident adviser decides to use the adjusted winner procedure to resolve the dispute. The issues agreed upon, and the (independently assigned) points, turn out to be the following:

| Issue | Mike's Points | Phil's Points |
|---|---|---|
| Stereo level | 4 | 22 |
| Smoking rights | 10 | 20 |
| Room party policy | 50 | 25 |
| Cleanliness | 6 | 3 |
| Alcohol use | 15 | 15 |
| Phone time | 1 | 8 |
| Lights-out time | 10 | 2 |
| Visitor policy | 4 | 5 |

Use the adjusted winner procedure to resolve this dispute. (Exercise 3 courtesy of Erica DeCarlo.)

**4.** Suppose that a labor union and management are trying to resolve a dispute that involves four issues: the base salary of the workers, the annual salary increase that workers can expect, the benefits package the workers will receive, and the amount of vacation time to which each worker will be entitled. Suppose they use adjusted winner to resolve this dispute, with the following point assignments:

| Issue | Labor | Management |
|---|---|---|
| Base salary | 30 | 50 |
| Salary increases | 20 | 40 |
| Benefits | 35 | 5 |
| Vacation time | 15 | 5 |

Use adjusted winner to resolve this dispute.

**5.** Make up an example involving two people and several *objects* for which the adjusted winner procedure can be used, and then use the adjusted winner procedure to determine a fair division.

**6.** Make up an example involving two people and several *issues* for which the adjusted winner procedure can be used, and then use the adjusted winner procedure to determine a fair resolution of the dispute.

■ **7.** Suppose we have three items ($X$, $Y$, and $Z$) and three people (Bob, Carol, and Ted). Assume that each of the people spreads 100 points over the items (as in adjusted winner) to indicate the relative worth of each item to that person:

| Item | Bob | Carol | Ted |
|---|---|---|---|
| $X$ | 40 | 30 | 30 |
| $Y$ | 50 | 40 | 30 |
| $Z$ | 10 | 30 | 40 |

For each of the following allocations, indicate

**(a)** whether or not it is proportional.

**(b)** whether or not it is envy-free.

**(c)** whether or not it is equitable.

**(d)** for the ones that are *not* Pareto-optimal, another allocation that makes one person better off without making anyone else worse off.

Allocation 1: Bob gets *Z*, Carol gets *Y*, and Ted gets *X*. (This is not Pareto-optimal.)

Allocation 2: Bob gets *Y*, Carol gets *Z*, and Ted gets *X*. (This is not Pareto-optimal.)

Allocation 3: Bob gets *X*, *Y*, and *Z*. (This *is* Pareto-optimal; explain why.)

Allocation 4: Bob gets *Y*, Carol gets *X*, and Ted gets *Z*. (This is Pareto-optimal.)

Allocation 5: Bob gets *X*, Carol gets *Y*, and Ted gets *Z*. (This is Pareto-optimal.)

## 13.2 The Knaster Inheritance Procedure

**8.** If John bids $28,225 and Mary bids $32,100 on their aging parents' old classic car, which they no longer drive, how would you reach a fair division?

(Car Culture/Corbis.)

**9.** John and Mary inherit their parents' old house and classic car. John bids $28,225 on the car and $55,900 on the house. Mary bids $32,100 on the car and $59,100 on the house. How should they arrive at a fair division?

**10.** Can you modify your fair-division procedure in Exercise 9 so that both John and Mary receive one of the two objects while still considering the allocation as fair?

**11.** Describe a fair division for three heirs, *A*, *B*, and *C*, who inherit a house in the city, a small farm, and a valuable sculpture, and who submit sealed bids (in dollars) on these objects as follows:

| | *A* | *B* | *C* |
|---|---|---|---|
| House | 145,000 | 149,999 | 165,000 |
| Farm | 135,000 | 130,001 | 128,000 |
| Sculpture | 110,000 | 80,000 | 127,000 |

**12.** Describe a fair division for three children, *E*, *F*, and *G*, who inherit equal shares of their parents' classic car collection and who submit sealed bids (in dollars) on these five cars as follows:

| | *E* | *F* | *G* |
|---|---|---|---|
| Duesenberg | 18,000 | 15,000 | 15,000 |
| Bentley | 18,000 | 24,000 | 20,000 |
| Ferrari | 16,000 | 12,000 | 16,500 |
| Pierce-Arrow | 14,000 | 15,000 | 13,500 |
| Cord | 24,000 | 18,000 | 22,000 |

## 13.3 Fair Division and Organ Transplant Policies

**13.** Construct the table showing how points would be allocated among the following four potential recipients for a kidney transplant, according to the scheme in Section 13.3.

| Potential recipient | Months waiting | Antigens matched | Percent sensitized |
|---|---|---|---|
| A | 9 | 2 | 20 |
| B | 6 | 3 | 0 |
| C | 5 | 4 | 40 |
| D | 2 | 6 | 60 |

**14.** Does the example in Exercise 13 give rise to the same kind of paradox as in Section 13.3? Explain why or why not.

**15.** In the scheme for arriving at a priority ranking for organ transplants, how might one change the way points are assigned for "waiting time" so that the kind of paradox that arose in Section 13.3 could not occur?

## 13.4 Taking Turns

**16.** Suppose that Bob and Carol rank a series of objects, from most preferred to least preferred, as follows:

| Bob | Carol |
|---|---|
| Car | Boat |
| Investments | Investments |
| CD player | Car |
| Boat | Washer–dryer |
| Television | Television |
| Washer–dryer | CD player |

Assume that Bob and Carol use the bottom-up strategy and that Bob gets to choose first. Determine Bob's first choice and the final allocation.

**17.** Repeat Exercise 16 under the assumption that Carol gets to choose first.

**18.** Mark and Fred have inherited a number of items from their parents' estate, with no indication of who gets what. They rank the items from most preferred to least preferred as follows:

| Mark | Fred |
|---|---|
| Truck | Boat |
| Tractor | Tractor |
| Boat | Car |
| Car | Truck |
| Tools | Motorcycle |
| Motorcycle | Tools |

Assume that Mark and Fred use the bottom-up strategy and that Mark gets to choose first. Determine Mark's first choice and the final allocation.

**19.** Repeat Exercise 18 under the assumption that Fred gets to choose first.

**20.** Suppose that Donald and Ivana Trump decide to settle their divorce (described in Exercise 1) by taking turns. Assume they both use the bottom-up strategy, and Donald chooses first. Determine Donald's first choice and the final allocation. (Assume that Donald values the Connecticut estate slightly more than he values the Trump Plaza apartment.)

**21.** Repeat Exercise 20 under the assumption that Ivana gets to choose first.

## 13.5 Divide-and-Choose

◆ **22.** If you and another person are using divide-and-choose to divide something between you, would you rather be the divider or the chooser? (Assume that neither of you knows anything about the preferences of the other.)

◆ **23.** Suppose that Bob is entitled to one-fourth of a cake and Carol is entitled to three-fourths. In a few sentences, explain how divide-and-choose can be used to achieve an allocation in which each party is guaranteed of receiving at least as much as he or she is entitled to.

**24.** Suppose that Bob, Carol, and Ted view a cake as having 18 units of value, with each unit of value represented by a small square (as in the accompanying figure). Suppose, however, that the players value various parts of the cake differently (or that Bob views the cake as being perfectly rectangular, whereas Carol and Ted see it as skewed in opposite ways). We represent this pictorially as follows:

Bob

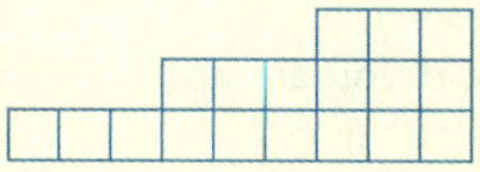

Carol

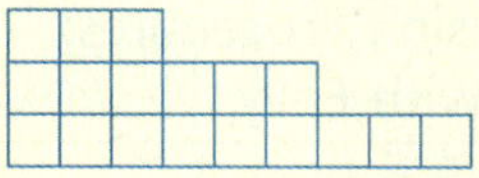

Ted

Assume that all cuts that will be made are vertical.

**(a)** If Bob and Carol use divide-and-choose to divide the cake between them, how large a piece will each receive (assuming they follow the suggested strategies that go with divide-and-choose and that Bob is the divider)?

**(b)** If Carol and Ted use divide-and-choose to divide the cake between them, how large a piece will each receive (assuming they follow the suggested strategies that go with divide-and-choose and that Carol is the divider)?

**25.** Assume that Bob and Carol view the cake as in Exercise 24, but assume also that each knows how the other values the cake, and that neither is spiteful. Suppose they are to divide the cake using the *rules*, but not necessarily the *strategies*, of divide-and-choose.

**(a)** Is Bob better off being the divider or the chooser?

**(b)** Discuss this in relation to Exercise 22.

## 13.6 Cake-Division Procedures: Proportionality

**26.** Suppose that players 1, 2, and 3 view a cake as in Exercise 24. Notice that each player views the cake as having 18 square units of area (or value). Assume that each player regards a piece as acceptable if and only if it is at least $18/3 = 6$ square units of area (his or her "fair share"). Assume also that all cuts made correspond to vertical lines.

*(Angela Wyant/Stone/Getty Images.)*

**(a)** Provide a total of three drawings to show how each player views a division of the cake by player 1 into three pieces he or she considers to be the same size or value. Label the pieces *A*, *B*, and *C*.

**(b)** Identify two of these pieces that player 2 finds acceptable and two that player 3 finds acceptable.

**(c)** Show that a feasible assignment of fair pieces can be achieved by letting the players choose in the following order: player 3, player 2, player 1. Indicate how many square units of value each player thinks he or she received. Is there any other order in which players can choose pieces (in this example) that also results in a feasible assignment?

**27.** Suppose that players 1, 2, and 3 view a cake as follows:

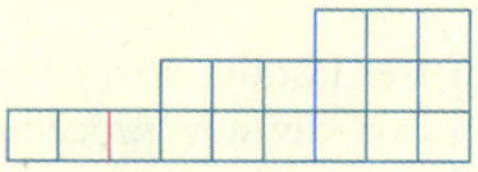

Player 1

Player 2

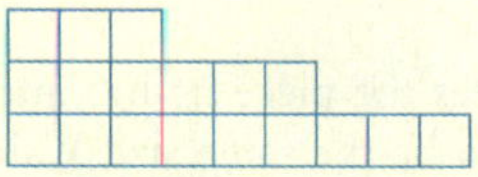

Player 3

**(a)** Provide a total of three drawings to show how each player views a division of the cake by player 1 into three pieces he or she considers to be the same size or value. Label the pieces *A*, *B*, and *C*. (We are still assuming that all cuts correspond to vertical lines, so this will require a cut along a vertical center line of some of the squares.)

**(b)** Show that neither player 2 nor player 3 finds more than one of the three pieces acceptable (with "acceptable" defined as in Exercise 26).

**(c)** Identify a single piece that player 2 and player 3 agree is *not* acceptable. (There are actually two such pieces; for definiteness, find the one on the right.)
**(d)** Assume that players 2 and 3 give the piece from part (c) to player 1. Suppose they reassemble the rest and players 2 and 3 divide it between themselves using divide-and-choose (with a single vertical cut). Determine what size piece each of the three players will think he or she received (1) if player 2 divides and player 3 chooses, and (2) if player 3 divides and player 2 chooses.

**28.** Suppose players 1, 2, and 3 view a cake as in Exercise 27. Illustrate the last-diminisher method (still restricting attention to vertical cuts and, in addition, assuming that the piece potentially being diminished is a piece off the left side of the cake) by following steps (a) through (f) below:

**(a)** Draw a picture showing the third of the cake (6 squares) that player 1 will slice off the cake.
**(b)** Determine whether player 2 will pass or further diminish this piece. If he or she would further diminish it, make a new drawing.
**(c)** Determine whether player 3 will pass or further diminish this piece. If he or she would further diminish it, make a new drawing.
**(d)** Determine who receives the piece cut off the cake and what size or value he or she thinks it is. (Actually, we knew what size the person receiving this first piece would think it was, assuming he or she followed the prescribed strategy. How did we know this?)
**(e)** Finish the last-diminisher method using divide-and-choose on what remains, with the lowest-numbered player who remains doing the dividing.
**(f)** Redo step (e) with the other player doing the dividing.

**29.** The Banach–Knaster last-diminisher method is not the only well-known cake-division procedure that yields a proportional allocation for any number of players. There is also one due to A. M. Fink (sometimes called the *lone-chooser method*). For three players (Bob, Carol, and Ted) it works as follows:

**(i)** Bob and Carol divide the cake into two pieces using divide-and-choose.
**(ii)** Bob now divides the piece he has into three parts that he considers to be the same size. Carol does the same with the piece she has.
**(iii)** Ted now chooses whichever of Bob's three pieces that he (Ted) thinks is largest, and Ted chooses whichever of Carol's three pieces that he thinks is largest.
**(iv)** Bob keeps his remaining two pieces, as does Carol.

**(a)** Explain why Ted thinks he is getting at least one-third of the cake.
**(b)** Explain why Bob and Carol each think they are receiving at least one-third of the cake.
**(c)** Explain why, in general this scheme is not envy-free.

■ **30.** In A. M. Fink's procedure (described in Exercise 29), suppose that a fourth person (Alice) comes along after Bob, Carol, and Ted have already divided the cake among themselves so that each of the three thinks he or she has a piece of size at least one-third. Mimic what was done in the three-person case to obtain an allocation among the four that is proportional. (*Hint:* Begin by having Bob, Carol, and Ted divide the pieces they have into a certain number–how many?–of equal parts.)

**31.** There is a moving-knife version of the Banach–Knaster procedure that is due to Dubins and Spanier. To describe it, we picture the cake as being rectangular, and the procedure beginning with a referee holding a knife along the left edge, as illustrated below.

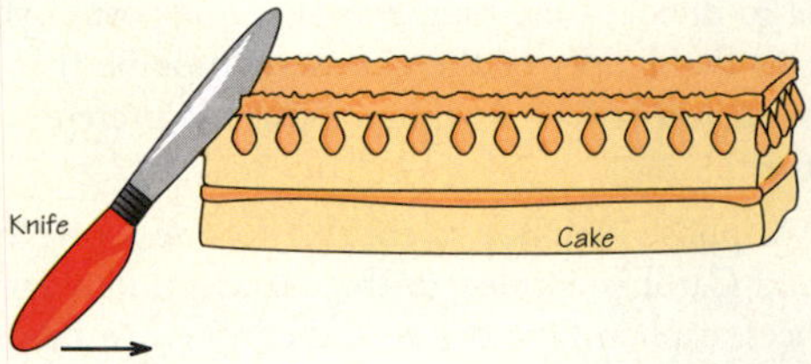


Assume, for the sake of illustration, that there are four players (Bob, Carol, Ted, and Alice). The referee starts moving the knife from left to right over the cake (keeping it parallel to the position in which it started) until one of the players (assume it is Bob) calls "cut." At this time, a cut is made, the piece to the left of the knife is given to Bob, and he exits the game. The knife starts moving again, and the process continues. The strategies are for each player to call "cut" whenever it would yield him or her a piece of size at least one-fourth.

**(a)** Explain why this procedure produces an allocation that is proportional.
**(b)** Explain why the resulting allocation is not, in general, envy-free.
**(c)** Explain why, if you are not the first player to call "cut," there is a strategy different from the one suggested that is never worse for you, and sometimes better.

## 13.7 Cake-Division Procedures: The Problem with Envy

**32.** Suppose players 1, 2, and 3 view the cake as in Exercise 27. Illustrate the envy-free procedure for $n = 3$ (yielding an allocation of part of the cake) by following steps (a) through (c) below. Again, restrict attention to vertical cuts.

**(a)** Provide a total of three drawings to show how each player views a division of the cake by player 1 into three pieces he or she considers to be the same size or value. Label the pieces $A$, $B$, and $C$. (This is the same as Exercise 27a.)
**(b)** Redraw the picture from player 2's view, and illustrate the trimming of piece $A$ that he or she would do. Label the trimmed piece $A'$ and the actual trimmings $T$.

**(c)** Indicate which piece each player would choose (and what he or she thinks its size is) if the players choose in the following order: player 3, player 2, player 1, according to the envy-free procedure.

■ **33.** There is a two-person moving-knife cake-division procedure due to A. K. Austin that leads to each player receiving a piece of cake that he or she considers to be of size exactly one-half. It begins by having one of the two players (Bob) place two knives over the cake, one of which is at the left edge, and the other of which is parallel to the first and placed so that the piece between the knives (*A* in the picture below) is of size exactly one-half in Bob's estimation.

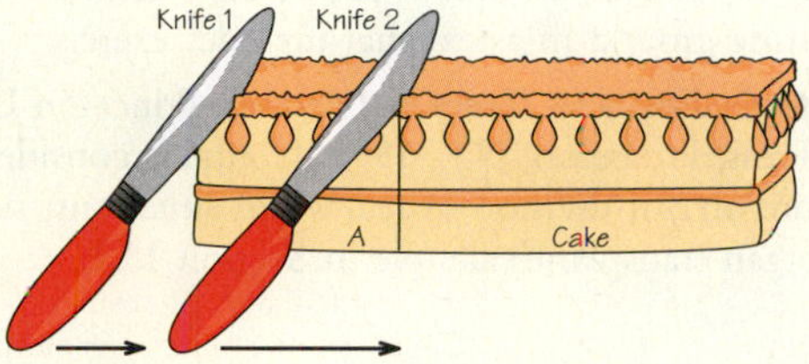


If Carol agrees that this is a 50-50 division, we are done. Otherwise, Bob starts moving both knives to the right—perhaps at different rates—so that the piece between the knives remains of size one-half in his eyes. Carol calls "stop" at the point when she also thinks the piece between the two knives is of size exactly one-half.

**(a)** If the knife on the right were to reach the right-hand edge, where would the knife on the left be?
**(b)** Explain why there definitely is a point where Carol thinks the piece between the two knives is of size exactly one-half. (*Hint:* If Carol thinks the piece is too small at the beginning, what will she think of it at the end?)

## WRITING PROJECTS

**1.** It turns out that there is no way to extend the adjusted winner procedure to three or more players. That is, there are point assignments by three players to three objects so that no allocation satisfies the three desired properties of equability (equal points), envy-freeness, and Pareto optimality. On the other hand, there are separate procedures that will realize any two of the three properties. Thus, trade-offs must be made, and these may depend on the circumstances. In a few paragraphs, discuss the relative importance of the three properties and circumstances that may affect the choice of which two of the three properties one might wish to have satisfied.

**2.** If we use taking turns to divvy up a collection of objects between two people (Bob and Carol), then there is an obvious advantage to going first. Assume that we have decided that Bob will, in fact, choose first (say, by the toss of a coin). Let's think about how Carol might be compensated. First of all, if there are only three objects, then the "choice sequence" Bob–Carol–Carol seems to be the only reasonable one. Do you agree? For four objects, however, there are two choice sequences that suggest themselves: Bob–Carol–Carol–Carol and Bob–Carol–Carol–Bob. Do you think that one of these is obviously more fair than the other? What if there are four identical objects? What if both Bob and Carol value object *A* twice as much as *B*, and *B* twice as much as *C*, and *C* twice as much as *D*? What sequences suggest themselves for five objects? For eight objects? In one page or less, discuss these questions. (For more on this, see *The Win–Win Solution* in the Suggested Readings.)

**3.** One of the most important differences between the three-person and the *n*-person envy-free procedures is that the latter procedure may take more than two stages. And, of course, the more stages there are, the more cuts and trimmings may be necessary. Do you consider this a serious practical problem, or is it mainly a theoretical problem? In one paragraph, explain your reasons.

**4.** One often hears of the importance of "process" versus "product," the latter referring to *what* is achieved and the former referring to *how* it was achieved. In a couple of sentences, comment on the relevance of this to fair division as illustrated by the following rough paraphrasing of an exchange between two old friends, Ralph Kramden (played by Jackie Gleason) and Ed Norton (played by Art Carney) in the 1950s sitcom *The Honeymooners.*

> *Ralph to Ed* (as the two are sitting alone at the dinner table): I can't believe you did that.
>
> *Ed:* Did what, Ralph?
>
> *Ralph:* There were two potatoes there, and you reached right out and took the big one.
>
> *Ed:* What would you have done, Ralph?
>
> *Ralph:* Why, I'd have taken the little one.
>
> *Ed:* You *got* the little one, Ralph.

## SUGGESTED READINGS

BRAMS, S. J., and A. D. TAYLOR. *The Win–Win Solution: Guaranteeing Fair Shares to Everybody,* Norton, New York, 1999. Brams and Taylor do more with adjusted winner, as well as divide-and-choose and taking turns.

BRAMS, S. J., and A. D. TAYLOR. *Fair Division: From Cake-Cutting to Dispute Resolution,* Cambridge University Press, Cambridge, 1996. Brams and Taylor provide a book-length treatment of the kinds of topics introduced in this chapter, as well as divide-and-choose in the political arena, moving-knife procedures for cake cutting, and fairness as it applies to different auction and election procedures.

BRAMS, S. J., and A. D. TAYLOR. An envy-free cake division protocol. *American Mathematical Monthly,* 102 (1995): 9–18. Brams and Taylor describe in detail the finite version of their envy-free procedure for $n = 4$; in addition, they review earlier work on "protocols" (step-by-step procedures) that led up to their constructive solution of the envy-freeness problem for $n > 3$.

ROBERTSON, J., and W. WEBB. *Cake Cutting Algorithms: Be Fair If You Can,* A. K. Peters, Wellesley, Mass., 1998. Robertson and Webb cover a great deal of cake-cutting ground in a text that includes exercises.

YOUNG, P. *Equity in Theory and Practice,* Princeton University Press, Princeton, N.J., 1994. Contains considerably more on fair division in real-world situations, such as the organ transplant example in Section 13.3.

CHAPTER 14



# Apportionment

## 14.1 The Apportionment Problem

Coach is proud of her field hockey team, and has asked for a poster to display its record for the season. The team played 23 games and won 18, lost 4, and tied one. Table 14.1 is a draft of the poster. Coach objects: It looks too complicated. "Just express the percents as whole numbers," she says. We round off the percentages: The winning percentage, 78.26% is rounded down to 78%, the losing percentage, 17.39%, is rounded down to 17%, and the tie percentage, 4.35%, is rounded down to 4%. Because all three percentages were rounded down, the total is only 99%. The coach notices, and changes the winning percentage to 79%. "Now you have 100%!"

**Apportionment Problem** DEFINITION

An **apportionment problem** is to round a set of fractions so that their sum is maintained at its original value. The rounding procedure must not be an arbitrary one, but one that can be applied constantly. Any such rounding procedure is called an **apportionment method**.

When rounding percentages so as to preserve the sum of 100%, we face an apportionment problem. The coach used the apportionment method called "make it look good for the team," but a serious apportionment method should be unbiased, especially when it comes to even *more* critical issues, such as apportioning seats in the U.S. House of Representatives.

The framers of the U.S. Constitution wrote that seats in the House of Representatives "shall be apportioned among the several states within this union according to their respective Numbers. . . ." They may not have realized that the apportionment problem that they set in this clause was of any significance, but it caused trouble right from the start. In 1790, Delaware's population, 55,540, was

Howard Chandler Christy's "Scene at the Signing of the Constitution of the United States." (*Art Resource, NY.*)

**TABLE 14.1 The Field Hockey Team's Season**

| | | Percentage |
|---|---|---|
| Games won | 18 | $\frac{18}{23} \times 100\% = 78.26\%$ |
| Games lost | 4 | $\frac{4}{23} \times 100\% = 17.39\%$ |
| Games tied | 1 | $\frac{1}{23} \times 100\% = 4.35\%$ |
| Games played | 23 | 100.00% |

the least among the 15 states. The total population of the nation at the time was 3,615,920, and the House of Representatives was to have 105 members. Thus, the average congressional district should have a population of 3,615,920 ÷ 105 = 34,437. To obtain its fair share of the House seats, we can divide Delaware's population by the average congressional district population. Ideally, Delaware should have had

$$\frac{55{,}540}{34{,}437} = 1.613 \text{ congressional districts}$$

Table 14.2 displays the apportionment proposed in a bill, written by Alexander Hamilton, that Congress passed and sent to President Washington. Each state's fair share of the 105 congressional districts, or **quota**, was calculated by dividing its population by the average congressional district population of 34,437, as in the case of Delaware, shown above. None of the quotas were whole numbers. The bill gave Delaware two seats in Congress. As you will see in the table, Virginia's quota was 18.310 seats, and Virginia was awarded 18 seats in the House.

**TABLE 14.2 The Congressional Apportionment that George Washington Vetoed**

| State | Population | Quota | Apportionment |
|---|---|---|---|
| Virginia | 630,560 | 18.310 | 18 |
| Massachusetts | 475,327 | 13.803 | 14 |
| Pennsylvania | 432,879 | 12.570 | 13 |
| North Carolina | 353,523 | 10.266 | 10 |
| New York | 331,589 | 9.629 | 10 |
| Maryland | 278,514 | 8.088 | 8 |
| Connecticut | 236,841 | 6.878 | 7 |
| South Carolina | 206,236 | 5.989 | 6 |
| New Jersey | 179,570 | 5.214 | 5 |
| New Hampshire | 141,822 | 4.118 | 4 |
| Vermont | 85,533 | 2.484 | 2 |
| Georgia | 70,835 | 2.057 | 2 |
| Kentucky | 68,705 | 1.995 | 2 |
| Rhode Island | 68,446 | 1.988 | 2 |
| Delaware | 55,540 | 1.613 | 2 |
| Totals | 3,615,920 | 105 | 105 |

Although this apportionment may seem fair enough, President Washington vetoed the bill.[1] Washington came from Virginia, a state that would get less than its quota in the apportionment Congress proposed. It is impossible to determine if he was just biased for his home state–as the field hockey coach was in favor of her team–because, as we will discover, there were substantial reasons for rejecting the bill.

First, let's see how to set up an apportionment problem. Many apportionment problems do not involve the House of Representatives; our terminology refers to *states*, *populations*, and a *house size*. In the problem of rounding percentages so that their sum is 100, the house size is 100. The categories (such as wins, losses, and ties for a field hockey team) correspond to the states, and the numbers in each category (in our field hockey story, 18 wins, 4 losses, and 1 tie) correspond to the populations of the states.

**Standard Divisor** DEFINITION

The quotient of the total population, $p$, divided by the house size is called the **standard divisor**. If $h$ denotes the house size and $s$ is the standard divisor, then

$$s = \frac{p}{h}$$

**Quota** DEFINITION

In an apportionment problem, the **quota** is the exact share that would be allocated *if a whole number were not required.* To obtain a state's quota, divide its population by the standard divisor.

In the following course-scheduling problem, the courses to be taught correspond to the states, the numbers of students enrolled in each course correspond to the populations, and the house size is the total number of sections to be scheduled. Thus, in Example 1, there are three states: geometry, pre-calculus, and calculus, with populations 52, 33, and 15, respectively, and the house size is 5.

## EXAMPLE 1 ■ The High School Mathematics Teacher

A high school has one mathematics teacher who teaches all geometry, precalculus, and calculus classes. She has time to teach a total of five sections. One hundred students are enrolled as follows: 52 for geometry, 33 for pre-calculus, and 15 for calculus. How many sections of each course should be scheduled?

(*LWA-Dann Tardif/Corbis.*)

The number of students enrolled in each course is called the *population*. Thus, the populations of geometry, pre-calculus, and calculus are 52, 33, and 15, respectively. There are five sections for the 100 students, so the average section will have $100 \div 5 = 20$ students. We will call this average section size the *standard divisor*, because each quota can be determined by dividing the corresponding population by this number. Table 14.3 displays these calculations.

As shown in the table, the quotas add up to 5. It is tempting to round each quota to the nearest whole number, as in the right column of the table, but this makes 6 sections in all–too many! The purpose of an apportionment method is to

[1]This apportionment bill was the first bill in U.S. history to be vetoed.

find an equitable way to round a set of numbers such as these quotas without increasing or decreasing the original sum.

**TABLE 14.3 Calculation of the Quotas for High School Mathematics Courses**

| Course | Population | Quota | Rounded |
|---|---|---|---|
| Geometry | 52 | $52 \div 20 = 2.60$ | 3 |
| Pre-calculus | 33 | $33 \div 20 = 1.65$ | 2 |
| Calculus | 15 | $15 \div 20 = 0.75$ | 1 |
| Totals | 100 | 5 | 6 |

### EXAMPLE 2 ■ California's Quota

The Census Bureau recorded the apportionment population[2] of the United States, as of April 1, 2000, to be 281,424,177. There are 435 seats in the House of Representatives; therefore, the standard divisor is

$$s = \frac{281{,}424{,}177}{435} = 646{,}952$$

California's apportionment population was 33,930,798. Its quota is determined by dividing this population by the standard divisor. Thus,

$$\text{California's quota} = \frac{33{,}930{,}798}{646{,}952} = 52.447 \text{ seats}$$

California's apportionment, which is required to be a whole number, was set at 53 seats.

Ideally, each state's apportionment should be close to its quota. It is unrealistic to expect that any state will be apportioned its exact quota because each apportionment is required to be a whole number and the quota is unlikely to be a whole number. In choosing an apportionment method, we must decide what we mean by the phrase "each state's apportionment should be close to its quota."

Apportionment always involves rounding, and there are many ways to round. "Rounding down" means discarding the fractional part of a number $q$ to obtain a whole number that we will denote $\lfloor q \rfloor$. Thus, $\lfloor 7.00001 \rfloor = 7$, $\lfloor 7 \rfloor = 7$, and $\lfloor 6.99999 \rfloor = 6$. "Rounding up" gives the next whole number, $\lceil q \rceil$. Thus, $\lceil 7.00001 \rceil = 8$, but $\lceil 7 \rceil = 7$.

There are numerous different apportionment methods. Each has flaws, and our goal is to understand how to choose a method that is appropriate for a particular apportionment problem.

## 14.2 The Hamilton Method

The congressional apportionment bill that President Washington vetoed was written by Alexander Hamilton. While it may appear that he simply rounded each quota to the nearest whole number, Hamilton was aware that there would be occasions–analogous to the examples of the field hockey team's percentages of wins, losses, and ties not summing to 100%, or the high school teacher receiving an extra class

[2] The apportionment population includes the resident population and the overseas population.

to teach–when the total number of seats apportioned in this way would be either more or less than the statutory house size. Hamilton called his method *largest fractions*.

**Alexander Hamilton's Method and Upper, Lower Quotas** DEFINITIONS

With the **Hamilton method**, each state receives either its **lower quota** $\lfloor q \rfloor$, which is its quota rounded down, or its **upper quota**, $\lceil q \rceil$, obtained by rounding the quota up. The states that receive their upper quotas are those whose quotas have the largest fractional parts.

The apportionment method of largest fractions, also known as the Hamilton method, was named for Alexander Hamilton. (*National Portrait Gallery/Art Resource, NY.*)

Implementing the Hamilton method is a three-step procedure:

1. Calculate each state's quota.
2. Tentatively assign to each state its lower quota of representatives. Each state whose quota is not a whole number loses a fraction of a seat at this stage, so the total number of seats assigned at this point will be less than the house size. This leaves additional seats to be apportioned.
3. Allot the remaining seats, one each, to the states whose quotas have the largest fractional parts, until the house is filled.

It is possible that a tie will occur, with the quotas of two states having identical fractional parts, but in practice, this rarely happens when large populations are involved.

## EXAMPLE 3 ▪ The High School Teacher's Dilemma

Let us use the Hamilton method to determine how many sections of geometry, pre-calculus, and calculus the high school teacher should teach. We have found that the quotas for the three subjects were 2.60, 1.65, and 0.75, respectively (see Table 14.3). The lower quotas are $\lfloor 2.60 \rfloor = 2$, $\lfloor 1.65 \rfloor = 1$, and $\lfloor 0.75 \rfloor = 0$, so we tentatively schedule two sections of geometry and one section of pre-calculus. With three sections apportioned, we have two more to assign to fill the house. These two sections go to the subjects with the largest fractions: Calculus has the largest fraction, 0.75, and gets a section; the second section goes to pre-calculus, whose fraction is 0.65. The final apportionment is as follows: geometry, two sections; pre-calculus, two sections; calculus, one section. ■

## EXAMPLE 4 ▪ The Field Hockey Team

In Table 14.1, we saw that the percentages of wins, losses, and ties for the field hockey team were 78.26%, 17.39%, and 4.35%, respectively. These are the quotas that we must round so that they sum to 100%. To start, we apportion $\lfloor 78.26 \rfloor = 78\%$, $\lfloor 17.39 \rfloor = 17\%$, and $\lfloor 4.35 \rfloor = 4\%$ to the three categories. These lower quotas add up to 99%. The remaining 1% to be apportioned goes to the losses because their fraction, 0.39, is the largest. The final apportionment is 78% wins, 18% losses, and 4% ties. Coach will veto this apportionment. ■

## EXAMPLE 5 ▪ Hamilton's Apportionment

The first congressional apportionment involved 15 states, and the House had 105 seats. According to the 1790 census, the U.S. population was 3,615,920. The standard divisor, $3{,}615{,}920 \div 105 = 34{,}437$, represents the population of the average congressional district.

Table 14.2 displays Alexander Hamilton's proposed apportionment. Each quota shown in the table was calculated by dividing the state's population by this standard divisor. Adding the lower quotas, we find that their sum 97 leaves 8 seats to be apportioned. These go to the 8 states whose quotas had the largest fractional parts.

If you are using a calculator to follow the entries in the table, store the standard divisor in the calculator's memory. Then each quota can be figured by entering the individual state's population and dividing by the divisor recalled from memory. ■

President Washington's veto message stated that the fractions of seats gained by some states in the third step of the Hamilton apportionment were not related to the states' total populations. You may notice that Hamilton could have achieved the same apportionment by simply rounding each quota to the nearest whole number. He was aware that this doesn't always work, and our field hockey and high school examples confirm this. The famous orator and politician, Congressman Daniel Webster, devised a way to adjust the quotas before rounding, and developed an apportionment method that would have given the same result as Hamilton's apportionment–and would have answered Washington's objection. Webster's method is a **divisor method**, and it and two other divisor methods will be described later in this chapter.

Neither President Washington nor Alexander Hamilton were aware of the **Webster method** of apportionment, but even if they had been, the 1791 apportionment bill could have been dismissed on a technicality, because the Constitution requires each congressional district to have a population of at least 30,000. Delaware, with 55,540 inhabitants, was too small to have two districts. Washington was probably aware of this technicality, making it likely that his objection to the Hamilton method was sincere.

The veto prevented the Hamilton method from being used in 1792, but it was adopted by Congress in 1850 and remained in use until 1900. The half-century of experience with the Hamilton method revealed a paradox.

### Paradoxes of the Hamilton Method

A *paradox* is a fact that seems obviously false. The first Hamilton apportionment paradox, called the **Alabama paradox**, was discovered in 1881. As part of the reapportionment procedure mandated by the Constitution, the Census Bureau had supplied Congress with a table of congressional apportionments for a range of different house sizes from 275 to 350, based on the 1880 census. The table revealed a strange phenomenon. Here is a portion of that table, just showing apportionments that changed when the house size increased from 299 to 300:

| | House Size | |
|---|---|---|
| State | 299 | 300 |
| Alabama | 8 | 7 |
| Illinois | 18 | 19 |
| Texas | 9 | 10 |

The paradox was that Alabama's apportionment decreased as a result of an increase in the number of seats in the House of Representatives. To see how this could happen, let's look at the quotas, shown in the following table.

| | House Size | | Increase in quota |
|---|---|---|---|
| State | 299 | 300 | |
| Alabama | 7.646 | 7.671 | 0.025 |
| Illinois | 18.640 | 18.702 | 0.062 |
| Texas | 9.640 | 9.672 | 0.032 |

When the house size increased from 299 to 300, each state's quota increased by a factor of $\frac{300}{299}$. States with larger populations get larger increases, as the table shows. With a 299-seat House of Representatives, Alabama was the last state to get its upper quota. When the house size went to 300, the fractional parts of the quotas of Texas and Illinois advanced past the fractional part of Alabama's quota, so those states were awarded their upper quotas–and that meant Alabama was left with its lower quota. Alabama was left with its lower quota.

### The Alabama Paradox

DEFINITION

The **Alabama paradox** occurs when a state loses a seat as the result of an increase in the house size.

The Alabama paradox validates President Washington's veto message, issued 90 years before its discovery. If the fractional parts of the quotas, which determine the way the last few seats are apportioned, were related to the populations in any sensible way, the paradox could not have occurred.

## EXAMPLE 6

## A Mathematics Department Meets the Alabama Paradox

A mathematics department has 30 teaching assistants to cover recitation sections for College Algebra, Calculus I, Calculus II, Calculus III, and Contemporary Mathematics. The enrollments of these courses are given in Table 14.4. The department will use the Hamilton method to apportion the teaching assistants (TA's) to the five subjects. In this problem, the house size is 30 (the number of TA's) and the population is 750. Therefore, the standard divisor is $750 \div 30 = 25$; this represents the average number of students in each recitation section. Each quota shown in the table was determined by dividing the enrollment of the course by this divisor.

**TABLE 14.4** Apportioning 30 Teaching Assistants

| Course | Enrollment | Quota | Lower Quota | Apportionment |
|---|---|---|---|---|
| College Algebra | 188 | 7.52 | 7 | 7 |
| Calculus I | 142 | 5.68 | 5 ↑ | 6 |
| Calculus II | 138 | 5.52 | 5 | 5 |
| Calculus III | 64 | 2.56 | 2 ↑ | 3 |
| Contemporary Mathematics | 218 | 8.72 | 8 ↑ | 9 |
| Totals | 750 | 30.00 | 27 | 30 |

The lower quotas add up to 27, so the three courses whose quotas have the largest fractional parts, Calculus I and III and Contemporary Mathematics, were given their upper quotas.

After the TA's were given their teaching assignments, the graduate school authorized the department to hire an additional TA. To determine which course should get the new TA, the department had to recalculate the apportionment. With 31 TA's, the standard divisor was $750 \div 31 = 24.19355$. The new quotas, determined by dividing each population by this new divisor, are shown in Table 14.5. Now the lower quotas add up to 28, so again three additional TA's go to the subjects whose quotas have the largest fractions. The Calculus III fraction, which had been larger than the College Algebra fraction when there were just 30 teaching assistants, has been surpassed. The new TA was placed in College Algebra, and one of the Calculus III TA's had to be reassigned to Calculus II.

**TABLE 14.5** Apportioning 31 Teaching Assistants

| Course | Enrollment | Quota | Lower Quota | Apportionment |
|---|---|---|---|---|
| College Algebra | 188 | 7.771 | 7 ↑ | 8 |
| Calculus I | 142 | 5.869 | 5 ↑ | 6 |
| Calculus II | 138 | 5.704 | 5 ↑ | 6 |
| Calculus III | 64 | 2.645 | 2 | 2 |
| Contemporary Mathematics | 218 | 9.011 | 9 | 9 |
| Totals | 750 | 31.000 | 28 | 31 |

The size of the House of Representatives has been fixed by statute at 435 members since Arizona and New Mexico became states on February 14, 1912. Therefore, the Alabama paradox cannot occur. A second paradox, called the **population paradox,** is associated with a fixed house size.

**The Population Paradox** DEFINITION

The **population paradox** occurs when one state's population increases, and its apportionment decreases, while simultaneously another state's population increases proportionally less, or decreases, and its apportionment increases.

## EXAMPLE 7 ▪ Apportioning Seats in Parliament

A country has four political parties. Its parliament has 100 members, and seats are apportioned by the Hamilton method after each election so that the number of seats each party is awarded is as close as possible to being proportional to the number of votes the party receives.

An election is held but the parties are unable to form a government, so a new election is held. Here are the results of the two elections:

| Party | First Election | Repeat Election |
|---|---|---|
| Whigs | 5,525,381 | 5,657,564 |
| Tories | 3,470,152 | 3,507,464 |
| Liberals | 3,864,226 | 3,885,693 |
| Centrists | 201,203 | 201,049 |
| Totals | 13,060,962 | 13,251,770 |

The three major parties, Whigs, Tories, and Liberals, all received more votes in the second election, but the Centrists received fewer. The quotas for each party, shown in the following table, were determined by dividing each party's votes by the the standard divisors

$$13{,}060{,}962 \div 100 = 130{,}609.62$$

for the first election, and 132,517.70 for the second election.

The quotas are given in the following table.

| Party | First Election | Repeat Election |
|---|---|---|
| Whigs | 42.3045 | 42.6929 |
| Tories | 26.5689 | 26.4679 |
| Liberals | 29.5861 | 29.3221 |
| Centrists | 1.5405 | 1.5171 |
| Totals | 100.0000 | 100.0000 |

The lower quotas for the results of the first election were 42, 26, 29, and 1, with a sum of 98–thus the Tories and the Liberals, with the largest fractions, get extra seats. The apportionment after the first election was Whigs, 42; Tories, 27; Liberals, 30; and Centrists, 1.

For the repeat election, the lower quotas were the same, but now the largest fractions belong to the Whigs and the Centrists! Therefore the new apportionment is Whigs, 43; Tories, 26; Liberals, 29; and Centrists, 2.

The Centrists have *gained* a seat, although they received fewer votes in the repeat election, while the Liberals lost a seat even though their vote total increased in the repeat election. This is an instance of the population paradox.

# 14.3 Divisor Methods

## The Jefferson Method

The Constitution requires congressional districts to be drawn so that the population of each is at least 30,000. President Washington could have vetoed the Hamilton apportionment bill because Delaware's population–only 55,540–was too small for the two congressional districts assigned to it. Thomas Jefferson proposed an apportionment method, now called the **Jefferson method,** to replace the Hamilton method.

Thomas Jefferson favored a method of apportionment biased in favor of states with large populations. (*National Portrait Gallery/Art Resource, NY.*)

In any apportionment, the standard divisor, which we will call *s,* obtained by dividing the total population by the house size, represents the average district population. In developing his method, Thomas Jefferson specified the population of the *smallest* district in the nation, which we will now call *d.* Using *d,* rather than *s,* as the divisor, each state receives an **adjusted quota** that is *always* rounded *down* to obtain its apportionment. If *d* is chosen correctly, the apportionments will add up exactly to the statutory house size.

In effect, the Jefferson method apportions to each state the maximum number of congressional districts of population *d* that will be accommodated by the state's population. Any leftover population is divided among these districts.

Once the divisor *d* is known, the Jefferson apportionment is easy to compute. The apportionment for state *X*, with population *V*, is

$$\left\lfloor \frac{V}{d} \right\rfloor$$

The actual divisor that Jefferson used was $d = 33{,}000$. Thus, Virginia's apportionment was

$$\left\lfloor \frac{\text{population of Virginia}}{33{,}000} \right\rfloor = \left\lfloor \frac{630{,}560}{33{,}000} \right\rfloor = \lfloor 19.108 \rfloor = 19$$

Therefore, Virginia received 19 seats, rather than 18, which Hamilton's bill would have allocated. Delaware's apportionment was 1 seat instead of 2, as the following calculation shows.

$$\left\lfloor \frac{\text{population of Delaware}}{33{,}000} \right\rfloor = \left\lfloor \frac{55{,}540}{33{,}000} \right\rfloor = \lfloor 1.683 \rfloor = 1$$

The Jefferson method is one of a class of apportionment methods called *divisor methods.*

**Divisor Methods** DEFINITION

A **divisor method** of apportionment determines each state's apportionment by dividing its population by a common divisor $d$ and rounding the resulting quotient. Divisor methods differ in the rule used to round the quotient.

The divisor must be carefully chosen to achieve the correct house size. In the first congressional apportionment, $d = 30{,}000$ would have resulted in larger apportionments for several states and a house size of 112, while $d = 36{,}000$ would have decreased several apportionments, and the house size would have been 91.

## Critical Divisors

To implement the Jefferson method, we must determine the divisor, $d$. Start by determining the standard divisor, $s$, and the quota for each state, as we did with the Hamilton method. Each state is assigned, as a **tentative apportionment**, its lower quota. Thus, the tentative apportionment of state $X$ is

$$\left\lfloor \frac{\text{Population of } X}{s} \right\rfloor$$

As we noted with the Hamilton method, these tentative apportionments are not enough to fill the house, so the apportionments of some states will have to be increased. To determine which states should receive additional seats, calculate a **critical divisor** for each state.

**Critical Divisor** DEFINITION

A state's **critical divisor** is the number that can be divided into the state's population to produce a number just on the borderline for changing the state's apportionment.

The formula for the critical divisor for a given apportionment method is determined by the rounding method used. The following theorem gives the critical divisor for the Jefferson method.

**Critical Divisor—The Jefferson Method** THEOREM

Let $N$ be the tentative apportionment of a state $X$. The **critical divisor** for $X$ is

$$\frac{\text{Population of } X}{N + 1}$$

## EXAMPLE 8

### Critical Divisors for Virginia and Delaware

Referring to Table 14.2, the lower quota for Virginia was 18 and the lower quota for Delaware was 1. The populations of the two states were 630,560 and 55,540, respectively, so their critical divisors were

$$\frac{630{,}560}{18 + 1} = 33{,}187 \text{ and } \frac{55{,}540}{1 + 1} = 27{,}770\text{, repectively.}$$

The significance of the critical divisor is that if the Jefferson divisor $d$ is equal to the critical divisor of state $X$ (whose tentative apportionment was $N$), then $X$ will contain *exactly* $N + 1$ districts of population $d$. Thus, with $d = 33{,}187$, Virginia's apportionment would be equal to 19. If we took $d = 27{,}770$, then Delaware would get 2 seats. However, the same divisor must be used for all states, so with this lower value of $d$, Virginia would get

$$\left\lfloor \frac{630{,}560}{27{,}770} \right\rfloor = 22 \text{ seats}$$

Once the critical divisors are determined, the state with the largest critical divisor is entitled to another seat, because when that divisor is used, no other state will receive a changed apportionment–a state receives additional seats only when a divisor smaller than or equal to its critical divisor is used. The total apportionment is thus increased by 1 in this step.

If there remain additional seats to be apportioned, we recompute the critical divisor for the state whose tentative apportionment has increased, and repeat the process. When the house is filled, the critical divisor most recently used is the divisor $d$, representing the minimum district population. With the Jefferson method (and any other divisor method), the critical divisors determine the priority of a state for receiving additional seats. *The state with the largest critical divisor gets the next seat, but after receiving that seat, its critical divisor is recomputed, and usually another state will then be first in line.*

## EXAMPLE 9 ▪ The Field Hockey Team

The field hockey team had 18 wins, 4 losses, and 1 tie last season. The Jefferson method can be used to express this record as percentages. The house size is 100%, and the total population is the 23 games played. We've previously determined the quotas (see Table 14.1 and Example 4) to be 78.26% for wins, 17.39% for losses, and 4.35% for ties. The lower quotas are the tentative apportionments: 78, 17, and 4, respectively. The critical divisors are $\frac{18}{78+1} = 0.22785$ for wins, $\frac{4}{17+1} = 0.22222$ for losses, and $\frac{1}{4+1} = 0.20000$ for ties.

The category with the largest critical divisor is "wins"; its apportionment is raised to 79%. The house is now full; we have the apportionment that Coach wanted: 79% wins, 17% losses, and 4% ties.

## EXAMPLE 10 ▪ The High School Mathematics Teacher

The teacher can be assigned five classes. There are 52 students enrolled in geometry, 33 in pre-calculus, and 15 in calculus. Let's use the Jefferson method to determine her teaching assignment. We have previously determined that the lower quotas for the subjects are 2, 1, and 0, respectively (see Table 14.3). Thus, the critical divisors are $\frac{52}{2+1} = 17\frac{1}{3}$ for geometry, $\frac{33}{1+1} = 16\frac{1}{2}$ for pre-calculus, and $\frac{15}{0+1} = 15$ for calculus.

Geometry, with the largest critical divisor, has first priority for a new section, and its tentative apportionment is increased to 3. Its new critical divisor will be $52 \div (3 + 1) = 13$. Now pre-calculus has top priority, and its tentative apportionment is increased to 2. The house is now full, so the final apportionments are 3 sections of geometry and 2 sections of pre-calculus. The minimum section size will be the $16\frac{1}{2}$, because that is the critical divisor for the subject that was the last to receive an increased apportionment: pre-calculus. In practice, the two sections will have 16 and 17 students. Because the enrollment for calculus is less than the minimum section size, there will be no calculus class.

Examples 7, 8, and 9 demonstrate that different methods may yield different apportionments, because the Hamilton method gave different apportionments in each case.

Here is another distinction between the methods. The Hamilton method gives each state either its upper quota or its lower quota. With the Jefferson method, no state can receive less than its lower quota as its apportionment–because the lower quota is the initial tentative apportionment–but a state can be apportioned more than its upper quota.

## EXAMPLE 11 ▪ The 1820 Congressional Apportionment

According to the 1820 census, New York had a population of 1,368,775. The total population of the United States was 8,969,878, and the house size was 213. Therefore, the standard divisor was $8{,}969{,}878 \div 213 = 42{,}112$, and New York's quota was $1{,}368{,}775 \div 42{,}112 = 32.503$. The Hamilton method would have apportioned to New York its upper quota, 33 seats. The divisor for the Jefferson method was $d = 39{,}900$. Thus New York's apportionment was $\lfloor 1{,}368{,}775 \div 39{,}900 \rfloor = 34$ seats.

An apportionment method is said to satisfy the **quota condition** if in every situation each state's apportionment is equal to either its lower quota or its upper quota. It takes only one example like the 1820 apportionment to show that the Jefferson method does not satisfy the quota condition. In fact, if the house had continued to use the Jefferson method, it would have violated the quota condition in every apportionment since 1850.

The Hamilton method satisfies the quota condition. This was obvious to Congress in 1850, so it based its apportionment on the Hamilton method.[3]

The Jefferson method, however, is not troubled by the Alabama and population paradoxes. Consider the Alabama paradox, in which a state loses a seat as a result of an increase in the house size. With any apportionment method, the apportionments of some states must increase when the size of the house increases. The Jefferson method awards seats in order of critical divisors. When the house size increases, the next seat will go to the state with the next largest critical divisor. There is no opportunity for a state to lose a seat.

Congress has never used an apportionment method that satisfies the quota condition and avoids the paradoxes. It would be desirable to use such a method, and in the 1970s, the mathematicians Michel L. Balinski and H. Peyton Young set out to find one. They succeeded in finding a method, which they called the *quota method,* that satisfies the quota condition (as the Hamilton method does) and avoids the Alabama paradox (as the Jefferson method does). However, the population paradox remained. They subsequently proved that the only apportionment methods that are free of the population paradox are the divisor methods. It is known that every divisor method is capable of violating the quota condition, so Balinski and Young have proved an impossibility theorem: *No apportionment method that satisfies the quota*

[3]The origins of the Hamilton method had been forgotten in 1850, and the method was named for Congressman Samuel Vinton, who had rediscovered the method.

*condition is free of paradoxes.* This theorem is like Kenneth Arrow's theorem that there is no completely satisfactory way to decide multicandidate elections based on voter preference schedules (see section 9.4).

The Jefferson method favors the larger states. It is not an accident that in every example that we have considered, the "state" with the largest population fared better with the Jefferson method than it did with the Hamilton method. Virginia got a greater apportionment, and Delaware a smaller apportionment in 1790; the winning percentage for the field hockey team was higher, and the losing percentage lower, when the Jefferson method was used, as compared with the Hamilton method, and there were more sections of geometry, and no sections of calculus when the Jefferson method was substituted for the Hamilton method.

Let's see why the Jefferson method is biased in favor of larger states. In an apportionment problem, let $s$ be the standard divisor and let $d$ be the divisor used in the Jefferson method. The apportionment given to state $X$ is then $\lfloor U \rfloor$, where

$$U = (\text{Population of } X) \div d.$$

We'll call $U$ the state's *adjusted quota.* Comparing $U$ with the quota $q$ for $X$, we will see that the formulas are similar:

$$q = (\text{Population of } X) \div s.$$

In fact, the state's population neatly cancels out of the ratio $U \div q$:

$$U \div q = s \div d.$$

Thus, while $U$ and $q$ have different values for each state, the ratio of $U$ to $q$ is always the same number, $M = s \div d$. Multiplying both sides of the identity $U \div q = M$ by $q$, we obtain $U = M \times q$. Therefore the Jefferson apportionment for state $X$ is

$$\lfloor M \times (\text{quota for } X) \rfloor$$

Consider the congressional apportionment of 1820. The standard divisor was $s = 42{,}112$, and the Jefferson divisor was $d = 39{,}900$. Thus, the quotient, $M$, is $42{,}112 \div 39{,}900 = 1.0554$. Now suppose that a state has a quota of $q$. The state's adjusted quota is $U = (1 + 0.0554) \times q = q + q \times 5.54\%$. In words, this algebraic formula says the adjusted quota is obtained by giving the state a 5.54% raise on its quota. A state with a large quota will get a greater raise than a small state will.

To see how this works with numbers, consider a state $X$ with $q = 18.96$. The adjusted quota is

$$U = 1.0554 \times 18.96 = 20.01$$

The upper quota for $X$ of 19, but $X$ is awarded $\lfloor 20.01 \rfloor = 20$ seats! This violates the quota condition, and in fact, every state whose quota is 18.96 or more will be guaranteed to get at least its upper quota with this value of $M$. If a state has a quota of $2 \times 18.96 = 37.92$, an identical calculation shows that it will receive at least its upper quota plus one seat. On the other hand, consider a small state whose lower quota is 1. To increase its apportionment to 2, its quota must be at least $2 \div M = 1.89502$. Thus, a state with quota 18.96 gets more than its upper quota, and a state with quota 1.89 has to settle for its lower quota.

## The Webster Method

**The Method of Daniel Webster** DEFINITION

The **Webster method** is the divisor method that rounds the quota (adjusted if necessary) to the nearest whole number, rounding up when the fractional part is greater than or equal to $\frac{1}{2}$, and rounding down when the fractional part is less than $\frac{1}{2}$.

Statesman and orator Daniel Webster (1782–1852), who developed a divisor method for apportioning the U.S. House of Representatives. (*National Portrait Gallery/Art Resource, NY.*)

The Webster and Jefferson methods are immune to the Alabama and population paradoxes, but neither satisfies the quota condition. However, the Jefferson method favors the large states, the Webster method is neutral, favoring neither the large nor the small states. Furthermore, the Webster method rarely violates the quota condition by giving a state more than its upper quota, or fewer seats than its lower quota, and would not have done so in any of the 22 congressional apportionments that have occurred so far.

Here is the procedure to calculate an apportionment by the Webster method.

1. Determine the standard divisor, and use it to find the quota for each state.
2. Obtain the tentative apportionments by rounding each quota $q$ to $\lfloor q \rfloor$ if the fractional part of $q$ is less than 0.5; otherwise, round to $\lceil q \rceil$. (This is the standard way of rounding numbers.)
3. Add the rounded quotas. If their sum is equal to the house size, the job is finished. The tentative apportionments calculated in step 2 are the final apportionments.

The procedure for implementing the Webster method of apportionment when the rounded quotas of the states don't add up to the house size is similar to the procedure for the Jefferson method. The quotas must be adjusted before rounding by using a divisor that is either larger than the standard divisor (if the sum of the rounded quotas is greater than the house size), or smaller than the standard divisor (if the sum of the quotas is less than the house size). To determine the correct divisor, one has to calculate a critical divisor for each state. We will denote the critical divisor by $d^+$ if we are looking for a divisor that is smaller than the standard divisor $s$ in order to increase the number of seats apportioned. When a divisor greater than $s$ is needed, to decrease the total apportionment, the critical divisor is denoted $d^-$.

It is never necessary to compute both critical divisors $d^+$ and $d^-$, and when the sum of the rounded quotas is equal to the house size $h$, no critical divisors are needed at all.

**Critical Divisors for the Webster Method** THEOREM

If the tentative apportionments do not fill the house, then the critical divisor for state $X$, with population $V$ and tentative apportionment $N$, is

$$d^{+} = \frac{V}{N + \frac{1}{2}}$$

The state with the largest critical divisor is first in line to receive a seat. If the tentative apportionments overfill the house, then the critical divisor for state $X$ is

$$d^{-} = \frac{V}{N - \frac{1}{2}}$$

The state with the smallest critical divisor is first in line to lose a seat.

When the divisor $d^{+}$ is used, state $X$ will get the adjusted quota

$$\frac{V}{d^{+}} = N + \frac{1}{2}$$

which is the smallest number to be rounded up to $N + 1$. Thus, the tentative apportionment of state $X$ would increase. None of the other states would see any change in their tentative apportionments, because their critical divisors are less than that of $X$. Similarly, if the smallest critical divisor $d^{-}$ belongs to state $Y$ with population $W$, in the resulting apportionment, $\lfloor W \div d^{-} \rceil = N - 1$, because $W \div d^{-} = N - \frac{1}{2}$ is on the borderline for rounding down to $N - 1$. Since $d^{-}$ for $Y$ is the smallest critical divisor, then state $Y$ will be the only state whose tentative apportionment would be reduced by using the divisor $d^{-}$.

In either case, the critical divisor of the state with the changed tentative apportionment must be recomputed. This process is repeated until the tentative apportionments add up to the house size.

## EXAMPLE 12 ■ The Field Hockey Team

As you may remember, the field hockey team has a fine record for the previous season: 18 wins, 4 losses, and 1 tie. We would like to express the record in the form of whole percentages, and Coach has said she is willing to do this with the Webster method. The quotas are 78.26% for wins, 17.39% for losses, and 4.35% for ties; these are rounded to get the tentative percentages 78%, 17%, and 4%, respectively. The total of these tentative percentages is 99%, less than the "house size" of 100%. We therefore compute the critical divisors $d^{+}$:

| Category | Tentative Percentage | $d^{+}$ |
|---|---|---|
| Wins | 78 | $18 \div (78 + \frac{1}{2}) = 0.2293$ |
| Losses | 17 | $4 \div (17 + \frac{1}{2}) = 0.2286$ |
| Ties | 4 | $1 \div (4 + \frac{1}{2}) = 0.2222$ |

The largest critical divisor belongs to the wins, so it receives the extra 1%. The final apportionment is 79% wins, 17% losses, and 4% ties.

## EXAMPLE 13 ■ Apportioning Classes

Let us return to the case of the mathematics teacher who should be assigned to teach a total of five classes in geometry, pre-calculus, and calculus. The enrollments are 52 for geometry, 33 for pre-calculus, and 15 for calculus. With a total of 100 students enrolled, and a house size of 5, the standard divisor is 20. The quotas, determined by dividing the enrollments for the three subjects by the standard divisor, are 2.6, 1.65, and 0.75, respectively. The tentative apportionments are 3, 2, and 1; their total, 6, exceeds the house size. We therefore compute the divisors $d^-$.

| Subject | Tentative Apportionment | $d^-$ |
|---|---|---|
| Geometry | 3 | $52 \div (3 - \frac{1}{2}) = 20.8$ |
| Pre-calculus | 2 | $33 \div (2 - \frac{1}{2}) = 22$ |
| Calculus | 1 | $15 \div (1 - \frac{1}{2}) = 30$ |

The subject with the smallest critical divisor is geometry; it therefore receives the reduced apportionment of 2. The final apportionment is therefore two sections each of geometry and pre-calculus, and one section of calculus.

To see why the Webster method is not biased in favor of large states or small states, let $s$ be the standard divisor, which is equal to the average district population. Let $d$ be the divisor that is used in the Webster method ($d$ is equal to the critical divisor belonging to the last state to gain or lose a seat in the apportionment). Finally, let $M = s \div d$. Each state's adjusted quota $U = M \times q$, where $q$ is the state's quota, as we saw in our discussion of the Jefferson method. When $M > 1$ (this happens when the tentative apportionments are not enough to fill the house), the states all receive an across-the-board increase in their quotas–and just as in a company where the workers receive the same percentage raise–the larger states are favored. When $M < 1$, the reverse is true, because a large number multiplied by $M$ will decrease more than a small number would. Thus, the Webster method favors neither large nor small states when the tentative apportionments exactly fill the house; it favors small states when the tentative apportionments overfill the house, and it favors large states when the tentative apportionments underfill the house. On balance, the Webster method is neutral, because it is equally likely that the house will be overfilled or underfilled by the tentative apportionments.

**The Webster Method Has No Population Bias** THEOREM

Among all divisor methods, the Webster method alone shows no bias with regard to state population.

## 14.4 Which Divisor Method Is Best?

The Jefferson and Webster methods, although both are divisor methods, frequently yield different results. We have just seen that one way to compare apportionment methods is to investigate bias based on population, but disputes about apportionment have usually not focused on this type of bias. Instead, they have cited inequities that could be reduced by another choice of divisor method. It is how inequities are measured that forms the basis for these arguments. For an account of a dispute that occurred in 1991, see Spotlight 14.1.

## SPOTLIGHT 14.1 A Legal Challenge to Apportionment

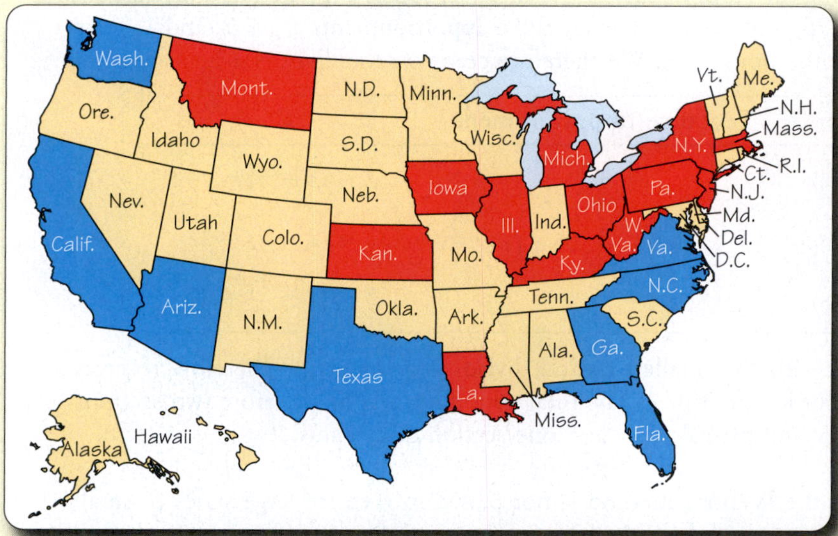


In 1991, the Census Bureau reported the new apportionment for the congressional elections in the years 1992–2000. The states colored red on the above map lost representatives, and those in green gained representatives. New York lost 3 representatives, and Ohio and Pennsylvania lost 2 apiece. Montana, whose apportionment decreased from 2 to 1, sustained the greatest percentage loss, and Montana sued to restore the lost seat. As precedents, Montana referred to the two famous cases, *Baker* v. *Carr* and *Wesberry* v. *Sanders*, in which the U.S. Supreme Court required legislative and congressional district boundaries to be drawn so as to make district populations equal.

Montana argued that the correct apportionment would be the one that met the *Baker* and *Wesberry* criterion of having districts as nearly equal in population as possible, and asked the Court to require the Census Bureau to recompute the apportionments using the Dean method, a divisor method that minimizes differences in district populations. This would have resulted in the transfer of a congressional seat from Washington to Montana.

The Montana case coincided with another federal lawsuit, *Massachusetts* v. *Mosbacher*, which asked the Court to order the apportionment to be calculated by the Webster method. If Massachusetts had won this suit, it would have gained an additional seat, but Montana would not benefit.

In *U.S. Department of Commerce* v. *Montana*, the Supreme Court unanimously rejected Montana's claim. The opinion of the Court, written by Justice John Paul Stevens, pointed out that *intra*state districts, which were the subject of the *Baker* and *Wesberry* cases, could be equalized in population by drawing district boundaries correctly. Because congressional districts can't cross state lines, some inequity is inevitable in congressional apportionment. The opinion conceded that there were alternatives to the Hill–Huntington method but concluded that the choice of apportionment method was best left to Congress.

**Representative Share** DEFINITION

Let $A$ be the apportionment given to a state whose population is $V$. The quotient $A \div V$ is called the **representative share**. It represents the share of a congressional seat given to each citizen of the state.

In an ideal apportionment, every state would have the same representative share. This is impossible, but we can measure how close a given apportionment is to being ideal by using representative share as the standard of comparison. We are given an apportionment done by some method. The method could be Webster's, Jefferson's, Hamilton's, or some method we haven't mentioned yet. We would compute the representative share for each state and identify the state that has the largest share and the state that has the smallest.

The discrepancy between these two values is a measure of how far the apportionment is from being perfectly equitable. It can be shown that among all conceivable apportionments, the one for which this discrepancy is the least is provided by the Webster method.

**Equity and the Webster Method** THEOREM

The Webster method is the most equitable apportionment method when comparisons of equity are based on differences in representative share.

## EXAMPLE 14 ■ Inequity in the 78th Congress

According to the 1940 census, Michigan had a population of 5,256,106 (5.256106 million) and was apportioned 17 seats in the House of Representatives. This apportionment gave Michigan a representative share of

$$\frac{17}{5.256106} = 3.234 \text{ seats per million population}$$

Arkansas–with a population of 1,949,387, or 1.949387 million–received 7 seats; so that state had a representative share of $7 \div 1.949387 = 3.591$ seats per million. Arkansas was favored over Michigan by

$$3.591 - 3.234 = 0.357 \text{ seats per million}$$

If a seat had been transferred from Arkansas to Michigan, then the representative share for Michigan would have been $18 \div 5.256106 = 3.425$ seats per million, while Arkansas would have been left with a representative share of $6 \div 1.949387 = 3.078$ seats per million.

Now Michigan, with the larger representative share, has the advantage, but the discrepancy

$$3.425 - 3.078 = 0.347 \text{ seats per million}$$

is less than it was before the transfer was made. In terms of representative share, it would have been more equitable to have given Arkansas 6 seats and Michigan 18 seats. ■

In an ideal apportionment, each congressional district in the nation would have the same population. This is impossible unless congressional districts are permitted to overlap state lines.

### District Population — DEFINITION

The **district population** of state $X$ is $V \div A$, where $V$ is the state's population, and $A$ is its apportionment. The district population is the average population of a congressional district in the state.

Apportionments can be evaluated by computing differences in district population. The best apportionment by this standard is the one for which the worst difference in district populations between states is minimal.

## EXAMPLE 15 ■ Comparing District Populations

If we consider differences in district population rather than representative share, it was correct to give Michigan 17 seats and Arkansas 7 in the 78th Congress. The district population for Michigan was

$$\frac{\text{population of Michigan}}{\text{Michigan's apportionment}} = \frac{5{,}256{,}106 \text{ people}}{17 \text{ districts}}$$
$$= 309{,}183 \text{ people per district}$$

The district population for Arkansas was $1{,}949{,}387 \div 7 = 278{,}484$. The Arkansas average district population was 30,699 less than Michigan's. If Michigan had 18 seats and Arkansas had 6, Michigan would have the lesser district population, 292,006, while Arkansas's would have increased to 324,898. This adjustment in apportionment would have increased the inequity between the two states, because now Arkansas would be worse off than Michigan by 32,892 in district population.

For state $X$, the representative share is $\frac{A}{V}$ and the district population is $\frac{V}{A}$, where $A$ is the apportionment of $X$ and $V$ is its population. Thus:

$$\text{representative share for state } X = \frac{1}{\text{district population for state } X}$$

It may be surprising that these two ways of evaluating the fairness of an apportionment could disagree, but we have just seen that they can.

A mathematician, Edward V. Huntington, pointed out that if **relative differences** are compared instead of absolute differences, then either district population and representative share would give identical comparisons of apportionments–and he suggested a compromise.

### Absolute and Relative Differences — DEFINITION

Given two positive numbers $A$ and $B$, with $A > B$, the **absolute difference** is $A - B$ and the **relative difference** is the quotient $\frac{(A - B)}{B} \times 100\%$.

For any two states, it turns out that the *relative difference* in district populations is equal to the relative difference in representative share (see Exercise 40). Therefore, an apportionment method that minimizes *relative* difference in representative shares will also minimize the relative difference in district populations.

## EXAMPLE 16 ■ Relative Inequity in the 78th Congress

Recall that Michigan was given 17 seats in the 78th Congress and had a representative share of 3.234 seats per million. Arkansas had 7 seats and a representative share of 3.591 seats per million. The relative difference was

$$\frac{3.591 - 3.234}{3.234} \times 100\% = 11.0\%$$

In terms of district populations, recall that Michigan had a district population of 309,183, and Arkansas had a district population of 278,484. The relative difference in district populations was

$$\frac{309{,}183 - 278{,}484}{278{,}484} \times 100\% = 11.0\%$$

Thus, by either measure, Arkansas was 11.0% better represented in the 78th Congress than Michigan was.

If Michigan had 18 seats and Arkansas had 6, the relative difference in representative shares would be found by subtracting the smaller representative share (Arkansas's) from the larger (Michigan's), and expressing the result as a percentage of the smaller representative share. Thus, the relative difference would have been

$$\frac{3.425 - 3.078}{3.078} \times 100\% = 11.3\%$$

in Michigan's favor. The same relative difference would be found if the district populations were compared. Because the relative inequity was less when Michigan had 17 seats and Arkansas had 7, the 1941 apportionment was preferred from the point of view of Professor Huntington's compromise. ■

To optimize apportionment by the relative difference criterion for equity, Professor Huntington and a statistician from the Bureau of the Census, Joseph Hill, designed a new divisor method. It has been used to apportion seats in the U.S. House of Representatives after each decennial census since 1940.

## The Hill–Huntington Method

Like the Jefferson and Webster methods, the apportionment is calculated by rounding the quotas, after adjusting them if necessary. The only difference between the three divisor methods is in the rounding procedure.

The Hill–Huntington rounding procedure is related to the **geometric mean**.

### Geometric Mean — DEFINITION

The **geometric mean** of two positive numbers $A$ and $B$ is equal to the square root of their product, $\sqrt{A \times B}$.

Consider the rectangle $\mathcal{R}$ displayed in Figure 14.1. The area of $\mathcal{R}$ is the product of the lengths $A$ and $B$, or $A \times B$. The geometric mean of $A$ and $B$ is equal to the length $E$ of the edge of a square $\mathcal{S}$ with the same area as $\mathcal{R}$, because the area of $\mathcal{R}$ is $E^2$, and thus $E^2 = A \times B$. Taking square roots, $E = \sqrt{A \times B}$.

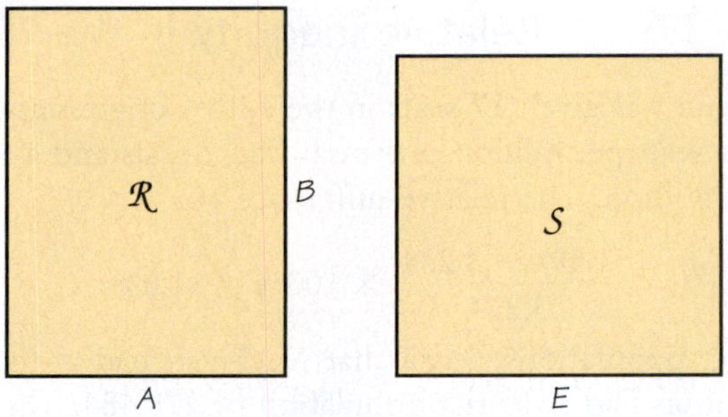


**FIGURE 14.1** The edge of the square is the geometric mean of the edges of the rectangle, because the two figures have the same area.

Given a positive number $q$, let $q^*$ be the geometric mean of $\lfloor q \rfloor$ and $\lceil q \rceil$. This $q^*$ is called the *rounding point* for $q$. The Hill–Huntington rounding of a number $q$ is equal to $\lfloor q \rfloor$ if $q < q^*$, where $q$ is the rounding point, and $\lceil q \rceil$ if $q \geq q^*$.

## EXAMPLE 17 ■ Hill–Huntington Rounding

Suppose that $q = 7.485$. Jefferson and Webster would round 7.485 down to 7. Because $\lfloor 7.485 \rfloor = 7$ and $\lceil 7.485 \rceil = 8$, the rounding point is $q^* = \sqrt{7 \times 8} = 7.48331 \ldots$, which is less than 7.485. Therefore, the Hill–Huntington rounding of 7.485 is $\lceil 7.485 \rceil = 8$.

Hill–Huntington apportionment calculations follow the general plan of the Jefferson and Webster methods. Round each state's quota the Hill–Huntington way to obtain a first tentative apportionment. If the sum of the tentative apportionments is equal to the house size, the job is finished. If not, a list of critical divisors must be constructed, each chosen to be just sufficient to change the corresponding state's apportionment by one seat in the desired direction.

### Critical Divisors for the Hill–Huntington Method — THEOREM

If $N$ is the number of seats apportioned tentatively to state $X$, and the total apportionment is too small, the critical divisor for that state is

$$d^+ = \frac{V}{\sqrt{N(N+1)}}$$

where $V$ is the state's population. The state with the largest critical divisor gains a seat. If the house is still not full, recompute the critical divisor for the state whose apportionment changed, and repeat the process.

If the total apportionment is too large, the critical divisor for state $X$ is

$$d^- = \frac{V}{\sqrt{N(N-1)}}$$

In this case, the state with the smallest critical divisor loses a seat.

A zero apportionment is impossible with the Hill–Huntington method because the rounding point for quotas between 0 and 1 is $\sqrt{0 \times 1} = 0$. Any quota less than 1 will be rounded to 1.

## EXAMPLE 18 ▪ Percent Effort

Faculty members at a certain university must state the percentage of their time spent in several activities. To comply as accurately as possible, Professor Worktorule has requisitioned five stopwatches to keep track of her activities. Here is what she recorded over the course of one week:

| | |
|---|---|
| Instruction | 300 minutes |
| Instructional support | 705 minutes |
| Independent study | 31 minutes |
| Research | 2475 minutes |
| Committee work | 89 minutes |
| Total | 3600 minutes |

The professor is too busy to covert the data into percentages–which the university requires in whole numbers with sum 100%–so we'll do it, using the Hill–Huntington method. As with any percentage apportionment problem, the house size is 100, so the standard divisor–one percentage unit–is 36 minutes. Table 14.6 shows the quotas, the rounding points, and the tentative apportionments, obtained by rounding the quotas up or down, depending on whether the quota exceeds the rounding point or not. The tentative apportionments add up to 101%, so we must determine the critical divisors in order to decide which tentative apportionment to reduce. These are calculated by dividing the minutes of effort in each category by $\sqrt{N(N-1)}$, where $N$ is the tentative apportionment. Research has the smallest critical divisor, $2475 \div \sqrt{69 \times 68} = 36.13$. Therefore the percentage effort in research is reported as 68%.

**TABLE 14.6 Apportioning Professor Worktorule's Effort by the Hill–Huntington Method**

| Effort Category | Effort (min.) | Quota | Rounding Point | Tentative Apportionment | Critical Divisor | Final Apportionment |
|---|---|---|---|---|---|---|
| Instruction | 300 | 8.33% | 8.49 | 8% | 40.09 | 8% |
| Instructional Support | 705 | 19.58% | 19.49 | 20% | 36.17 | 20% |
| Independent Study | 31 | 0.86% | 0 | 1% | ∞ | 1% |
| Research | 2475 | 68.75% | 68.5 | 69% | 36.13 | 68% |
| Committees | 89 | 2.47% | 2.45 | 3% | 36.33 | 3% |
| Totals | 3600 | 100% | – | 101% | – | 100% |

## EXAMPLE 19 ▪ The 435th Seat

When Congress was apportioned as a result of the 1940 census, the last seat in the house was in play. Michigan had 17 seats for its population of 5,256,106. Arkansas, with a population of 1,949,387, had 6 seats. With the Webster method, Michigan's critical divisor was $5{,}256{,}106 \div 17.5 = 300{,}349$, while Arkansas's was $1{,}949{,}387 \div 6.5 = 299{,}906$. Thus the Webster method apportions the 435th seat to Michigan.

By the Hill–Huntington method, the critical divisors for Michigan and Arkansas were $5{,}256{,}106 \div \sqrt{17 \times 18} = 300{,}472$ and $1{,}949{,}387 \div \sqrt{6 \times 7} = 300{,}797$, respectively. Because the Hill–Huntington divisor for Arkansas exceeded Michigan's, the seat went to Arkansas. The politics behind the decision to use the Hill–Huntington method rather than the Webster method is the subject of Spotlight 14.2.

## SPOTLIGHT 14.2 Mathematics and Politics: A Strange Mixture

**Walter F. Willcox**
*(Department of Manuscripts and University Archives, Cornell University Libraries.)*

**Edward V. Huntington**
*(Courtesy of Harvard University Archives.)*

The first American to consider apportionment from a theoretical point of view was Walter Willcox (1861–1964), who strongly advocated the Webster method and had computed the apportionment of 1902. His arguments convinced Congress to use the Webster method again in 1912. In 1911, Joseph Hill, a statistician at the Census Bureau, proposed the Hill–Huntington method—with the strong endorsement of Edward V. Huntington, a mathematics professor at Harvard.

In 1920, the two methods were in competition. There were significant differences in the apportionments determined by the two methods, and the result was Washington gridlock: No apportionment bill passed during the decade, and the 1912 apportionments were retained throughout the 1920s. In preparation for the 1930 census results, the National Academy of Sciences formed a committee to study apportionment. In 1929, the committee endorsed the Hill–Huntington method.

The 1930 census was remarkable in that the apportionments calculated by the Webster method were the same as the Hill–Huntington apportionments. The House was therefore reapportioned, but the method used could be claimed to be either one of the competing methods. The coincidence was almost repeated in the 1940 census, but there was one difference. The Hill–Huntington method gave the last seat to Arkansas, while Webster's method gave it to Michigan (see page 453). At the time, Michigan was a predominantly Republican state, and Arkansas was in the Democratic column. The vote on the apportionment bill split strictly along party lines, with Democrats supporting the Hill–Huntington method and Republicans voting for the Webster method. Because the Democrats had the majority, the Hill–Huntington method became the law.

We have seen that the Jefferson method is biased in favor of populous states and that the Webster method is not biased in regard to state population size. It's natural to ask if the Hill–Huntington method exhibits any bias with respect to state population.

A divisor method will show bias in favor of large states when the quotas are adjusted by using a divisor that is smaller than the standard divisor. If the quotas must be adjusted downward–that is, a divisor larger than the standard divisor is used–small states are favored. Because the rounding point for the Webster method is halfway between whole numbers, it is just as likely for the divisor to be smaller than the standard divisor as it is for it to be larger.

For any positive number $q$, the rounding point used by the Hill–Huntington method is closer to $\lfloor q \rfloor$ than to $\lceil q \rceil$ (see Exercise 36). This means that a random number $q$ is more likely to be above the rounding point and thus rounded up to $\lceil q \rceil$ than it is to be less and thus rounded down to $\lfloor q \rfloor$. The difference between the Webster and Hill–Huntington ways of rounding is not significant for relatively large numbers. For example, the Hill–Huntington rounding point between 50 and 51 is 50.498. Therefore, a number $q$ between 50 and 51 will be rounded up to 51 by Hill–Huntington if it is larger than 50.498. The Webster method would round $q$ to 51 if $q >$ 50.500. The differences are more significant when rounding smaller numbers. Hill–Huntington rounds all numbers between 0 and 1 up to 1; Webster rounds only the numbers in the range 0.500–1 up to 1. When the Hill–Huntington method is used for apportionment, the sum of the tentative apportionments is more likely to exceed the house size than it is to be less, especially if there are many states with small populations. Therefore, the Hill–Huntington method is likely to use a divisor larger than the standard divisor. This favors the less populous states.

In conclusion, the Webster method is the best divisor method for general use. It is the only divisor method that is unbiased regarding population size, and it minimizes differences between representative share. Although the Webster method is capable of violating the quota condition, it is the divisor method least likely to do so.

For apportionment of seats in the U.S. House of Representatives, a slight modification is needed, because no state can receive a zero apportionment. The rounding point for quotas less than 1 is set to zero, rather than 0.5.

There are situations where other apportionment methods could be considered. See Exercise 44 to explore the problem of teaching assignments, and Exercise 34 for apportionment of seats in a parliament.

## REVIEW VOCABULARY

$\lfloor q \rfloor$ The result of rounding a number $q$ down; for example, $\lfloor \pi \rfloor = 3$. (p. 437)

$\lceil q \rceil$ The result of rounding a number $q$ up to the next integer; for example, $\lceil \pi \rceil = 4$. (p. 437)

**Absolute difference** The result of subtracting a smaller number from a larger number. (p. 452)

**Adjusted quota** The result of dividing a state's quota by a divisor other than the standard divisor. The purpose of adjusting the quotas is to correct a failure of the rounded quotas to sum to the house size. (p. 442)

**Alabama paradox** A state loses a representative solely because the size of the House is increased. This paradox is possible with the Hamilton method but not with divisor methods. (p. 438)

**Apportionment method** A systematic way of computing solutions of apportionment problems. (p. 433)

**Apportionment problem** To round a list of fractions to whole numbers in a way that preserves the sum of the original fractions. (p. 433)

**Critical divisor** The number closest to the standard divisor that can be used as a divisor of a state's population to obtain a new tentative apportionment for the state. The following table lists formulas for critical divisors for some divisor methods. In the table, $V$ stands for the state's population, and $N$ is its tentative apportionment. (p. 443)

| | Critical Divisor Causing Tentative Apportionment | |
|---|---|---|
| **Method** | **To Increase** | **To Decrease** |
| Jefferson | $\frac{V}{N+1}$ | Not necessary |
| Webster | $\frac{V}{N+\frac{1}{2}}$ | $\frac{V}{N-\frac{1}{2}}$ |
| Hill–Huntington | $\frac{V}{\sqrt{N(N+1)}}$ | $\frac{V}{\sqrt{N(N-1)}}$ |

**District population** A state's population divided by its apportionment. (p. 452)
**Divisor method** One of many apportionment methods in which the apportionments are determined by dividing the population of each state by a common divisor to obtain adjusted quotas. The apportionments are calculated by rounding the adjusted quotas. Divisor methods differ in the way that the rounding of the quotas is carried out. The methods of Jefferson, Webster, and Hill–Huntington are divisor methods. (p. 443)
**Geometric mean** For positive numbers $A$ and $B$, the geometric mean is defined to be $\sqrt{A \times B}$. (p. 453)
**Hamilton method** An apportionment method that assigns to each state either its lower quota or its upper quota. The states that receive their upper quotas are those whose quotas have the largest fractional parts. (p. 437)
**Hill–Huntington method** A divisor method that minimizes relative differences in both representative shares and district populations. (p. 453)
**Jefferson method** A divisor method based on rounding all fractions down. Thus, if $U$ is the adjusted quota of state $X$, the state's apportionment is $\lfloor U \rfloor$. (p. 442)
**Lower quota** The integer part $\lfloor q \rfloor$ of a state's quota $q$. (p. 437)
**Population paradox** A situation in which the apportionment of one state, $A$, decreases, although its population has increased; while another state, $B$, loses population (or increases population proportionally less than state $A$) and gains a seat. This paradox is possible with all apportionment methods *except* divisor methods. (p. 440)
**Quota** The quotient $V \div s$ of a state's population divided by the standard divisor $s$. The quota is the number of seats a state would receive if fractional seats could be awarded. (p. 435)
**Quota condition** A requirement that an apportionment method should always assign to each state either its lower quota or its upper quota in every situation. The Hamilton method satisfies this condition, but none of the divisor methods do. (p. 445)
**Relative difference** The relative difference between two positive numbers is obtained by subtracting the smaller number from the larger, and expressing the result as a percentage of the smaller number. Thus, the relative difference of 120 and 100 is 20%. (p. 452)
**Representative share** A state's representative share is the state's apportionment divided by its population. It is intended to represent the amount of influence a citizen of that state would have on his or her representative. (p. 451)
**Standard divisor** The ratio $p \div h$ of the total population $p$ to the house size $h$. In a congressional apportionment problem, the standard divisor represents the average district population. (p. 435)
**Tentative apportionment** The result of rounding a state's quota or adjusted quota to obtain a whole number. (p. 443)
**Upper quota** The result of rounding a state's quota *up* to a whole number. A state whose quota is $q$ has an upper quota equal to $\lceil q \rceil$. (p. 437)
**Webster method** A divisor method of apportionment that is based on rounding fractions the usual way. The Webster method minimizes the *absolute* differences of representative share between states. (p. 438)

## ✓ SKILLS CHECK

**1.** A county is divided into 3 districts with the following populations: Southern, 3600; Western, 3100; Northeastern, 1600. There are 6 seats on the county council to be apportioned. What is the quota for the Southern district?

**(a)** 2.6
**(b)** 2.8
**(c)** 3

**2.** Two calculus teachers can teach a total of 8 classes. Enrollments are as follows: Calculus I, 200; Calculus II, 100; Calculus III, 52. In this apportionment problem, the population is ______, the standard divisor is ______, and the quotas are ______ for Calculus I, ______ for Calculus II, and ______ for Calculus III.

**3.** $A$, $B$, and $C$ are arguing about fractions of a cent. On a project, they worked exactly 33, 34, and 35 minutes, respectively, and were paid $100. Use the Hamilton method to see who gets his upper quota (in cents!).

**(a)** $A$
**(b)** $B$
**(c)** $C$

**4.** Round each number in the sum 13.62 + 12.58 + 17.51 + 16.77 + 19.52 = 80 to a whole number.

____ + ____ + ____ + ____ + ____ = 80

**5.** The population paradox occurs when

**(a)** A state's apportionment decreases because the house size increased.
**(b)** A state's apportionment decreases, and its apportionment increases, while another state's apportionment decreases, even though its population has increased.
**(c)** The Jefferson method is used.

**6.** The Alabama paradox occurred when it was noticed that Alabama would lose a seat, in apportionment by the Hamilton method, if the house size was changed from 299 to _______________ .

**7.** When rounding the numbers in the sum $20.45 + 30.30 + 49.25 = 100$ by the Jefferson method, the largest critical divisor belongs to

**(a)** 20.45
**(b)** 30.30
**(c)** 49.25

**8.** Use the Jefferson method to apportion the sum $0.8 + 0.9 + 98.3 = 100$ as a sum of whole numbers.
____ + ____ + ____ = 100

**9.** The Jefferson method frequently

**(a)** gives the smallest state less than its lower quota.
**(b)** gives the largest state more than its upper quota.
**(c)** gives a state a lesser apportionment if the house size increases.

**10.** Use the Webster method to apportion the sum $0.8 + 0.9 + 98.3 = 100$ as a sum of whole numbers.
____ + ____ + ____ = 100

**11.** When rounding the numbers in the sum $20.45 + 30.30 + 49.25 = 100$ by the Webster method, the largest critical divisor belongs to

**(a)** 20.45
**(b)** 30.30
**(c)** 49.25

**12.** States *A* and *B* have populations of 1 million and 2 million, respectively. If their respective apportionments are 2 and 3, then the absolute difference in representative share is ____ per million.

**13.** If the apportionments of the two states in Skills Check 12 were changed to 1 for *A* and 4 for *B* then

**(a)** the absolute difference in representative share would increase.
**(b)** the absolute difference in representative share would decrease.
**(c)** the absolute difference in representative share would be unchanged.

**14.** If the criterion is absolute difference in district population, the most equitable apportionment of 5 seats to states *A* and *B* in Skills Check 12 is ____ for *A*, and ____ for *B*.

**15.** If the initial calculations leading to the Hill–Huntington apportionment result in a sum that is too large, what happens next?

**(a)** A seat is taken from the state with the smallest critical divisor.
**(b)** The largest apportionment is reduced.
**(c)** A different method must be used.

**16.** The ________ method has been used since 1941 to apportion seats in the U.S. House of Representatives.

**17.** Which divisor method never apportions to a state fewer seats than its lower quota?

**(a)** Hill–Huntington
**(b)** Webster
**(c)** Jefferson

**18.** A school principal is apportioning sections of the school's mathematics classes. She wants to set a minimum section size, and to adjust it so that a total of 32 sections are open. She should use the ____ method.

**19.** The Hill–Huntington minimizes relative differences in

**(a)** district population.
**(b)** representative share.
**(c)** both district population and representative share.

**20.** The divisor method that shows the least bias in favor of either large states or small states is the _______ method.

# CHAPTER 14 EXERCISES

■ Challenge ◆ Discussion

## 14.1 The Apportionment Problem

**1.** Jane has decided to track her daily expenses, and finds them to be as listed in the table at right.

Express these as percentages. If rounded to whole numbers, do the percentages add up to 100%?

| | |
|---|---|
| Rent | \$31 |
| Food | 16 |
| Transportation | 7 |
| Gym | 12 |
| Miscellaneous | 5 |

**2.** A mathematics department uses 20 teaching assistants to aid in its four-semester calculus course. The number of teaching assistants assigned to each level of the course depends on enrollment. Here are the fall enrollments:

| | |
|---|---|
| Calculus I | 500 |
| Calculus II | 100 |
| Calculus III | 350 |
| Calculus IV | 175 |
| Total | 1125 |

How many teaching assistants should be assigned to each level of the course?

**3.** Should the mathematics department in Exercise 2 revise the assignments for its TA's? Grades have been posted for the previous semester, and some students need to repeat the previous level of the course. Forty-five students move from Calculus II to Calculus I, 41 students move from Calculus III to Calculus II, and 12 students move from Calculus IV to Calculus III.

◆ **4.** Here is a typical apportionment problem. Round the numbers in the sum to integers:

$$8.37 + 10.33 + 12.38 + 5.47 + 3.45 = 40$$

The rounded numbers must add up to 40. How would you approach this?

## 14.2 The Hamilton Method

**5.** Use the Hamilton method to round each of the following numbers in the sum to a whole number, preserving the total of 10.

$$0.36 + 1.59 + 0.99 + 2.33 + 2.38 + 2.35 = 10$$

**6.** *The 37th pearl.* Three friends have bought a bag guaranteed to contain 36 high-quality pearls for $14,900 at an auction. Abe contributed $5900, Beth's contribution was $7600, and Charles supplied the remaining $1400. After taking the bag to your house, they pour the 36 pearls from the bag onto the kitchen table.

**(a)** How many should each friend get if the Hamilton method is used to apportion the pearls according to the size of the contributions?

**(b)** Charles has noticed the bag isn't empty! Another pearl comes out, and you are asked to recalculate the apportionment.

◆ **(c)** How do you explain the result to Charles?

◆ **7.** A country has three political parties, and apportions seats in its 102 seat parliament by the Hamilton method proportionately to the number of votes each receives. In a recent election, the Pro-UFO party received 254,000 votes, the Anti-UFO party got 153,000 votes, and the Who Cares party polled 103,000 votes. Show that two of the parties are tied.

**8.** A small high school has one mathematics teacher who can teach a total of five sections. The subjects that she teaches, and their enrollments, are as follows: geometry, 52; algebra, 33; calculus, 12. Use the Hamilton method to apportion sections to the subjects.

**9.** Repeat Exercise 8 using the following enrollments: geometry, 77; algebra, 18; calculus, 20.

**10.** Use the Hamilton method to express the summands of the following expression as whole number percentages of the total:

$$2746 + 1725 + 1921 + 100 = 6492.$$

Repeat the calculation for the sum:

$$2814 + 1745 + 1933 + 99 = 6591.$$

Do you see a paradox?

**11.** Abe, Beth, Charles, and David have decided to invest in rare coins. A dealer has offered to sell them a parcel containing 100 identical coins for $10,000. Each person invests all that he or she can afford, but there is not quite enough money, so Charles asks his Aunt Esther to join the group. The coins will be apportioned by the Hamilton method. Here are the amounts contributed:

| | |
|---|---|
| Abe | $3,619 |
| Beth | 1,862 |
| Charles | 2,258 |
| David | 2,010 |
| Esther | 251 |
| Total | $10,000 |

(Alan Carey/Corbis.)

**(a)** How should the coins be apportioned among the five contributors?

**(b)** After the coins are distributed, the dealer mentions that there will be $50 in excise tax! Everyone empties their wallet: Abe finds $16 more, Beth has $2, Charles has $1, and David finds $32. This adds up to $51, so a dollar is returned to Aunt Esther. The apportionment is recalculated and one of the coins changes hands. Who has to give a coin to whom?

◆ **(c)** Explain what happened.?

To see how this situation works out with a different apportionment method, refer to Exercise 21.

**12.** A country has five political parties. Here are the numbers of votes each received in a recent election:

5,576,330; 1,387,342; 3,334,241; 7,512,860; and 310,968. Seats in its parliament are apportioned by the Hamilton method. Calculate the apportionments for house sizes of 82, 83, and 84. Does the Alabama paradox occur?

**13.** Repeat the apportionments in Exercise 12 for house sizes of 89, 90, and 91.

## 14.3 Divisor Methods

◆ **14.** Explain why the tentative Webster apportionment of a state with quota $q$ is $\lfloor q + 0.5 \rfloor$.

**15.** Reapportion the classes in Exercise 9 using the Jefferson method.

**16.** Reapportion the classes in Exercise 8 using the Webster method.

◆ **17.** The three friends who bought the pearls (see Exercise 6) ask you to suggest a different apportionment method to distribute their purchase. Before answering, determine the apportionments given by the Jefferson and Webster methods for the 36- and 37-pearl house sizes. Then make your suggestion.

◆ **18.** The three friends have bought a lot of 36 identical diamonds, at a total cost of \$36,000; Abe's investment was \$15,500, Beth's was \$10,500, and Charles's was \$10,000. They decided to apportion the diamonds using the Webster method, and they can't make it work out. Can you help?

**19.** Round the following to whole percentages using the methods of Hamilton, Jefferson, and Webster:

$$87.85\% + 1.26\% + 1.25\% + 1.24\% + 1.23\% + 1.22\% + 1.21\% + 1.20\% + 1.19\% + 1.18\% + 1.17\% = 100\%$$

Do any of these methods violate the quota condition?

**20.** Round the following percentages to whole numbers, using the methods of Hamilton, Jefferson, and Webster.

$$92.15\% + 1.59\% + 1.58\% + 1.57\% + 1.56\% + 1.55\% = 100\%$$

Do any of these methods violate the quota condition?

**21.** Recalculate the apportionment of the coins in Exercise 11 by the Webster method. Again, after the excise tax is paid, a coin changes hands. Who gives it to whom?

**22.** Reapportion the parliament of the country in Exercise 12 using the Jefferson method for each of the proposed house sizes. Is there a paradox?

**23.** Reapportion the parliament of the country in Exercise 13 using the Webster method, for each of the proposed house sizes. Is there a paradox?

**24.** A country has two political parties, the Liberals and the Tories. The seats in its 99-seat parliament are apportioned to the parties according to the number of votes it receives in the election. If the Liberals receive 49% of the vote, how many seats do the Liberals get with the Hamilton method? With the Webster method? With the Jefferson method?

■ **25.** A country with a parliamentary government has two parties that capture 100% of the vote between them. Each party is awarded seats in proportion to the number of votes received.

**(a)** Explain why the Webster and Hamilton methods will always give the same apportionment in this two-party situation.
**(b)** Explain how to use the result of (a) to show that the Alabama and population paradoxes cannot occur when the Hamilton method is used to apportion seats between two parties or states.
**(c)** Explain why the result of (a) implies that the Webster method satisfies the quota condition when the seats are apportioned between two parties or states.
◆ **(d)** Will the Jefferson and Hill–Huntington methods also yield the same apportionments as the Hamilton method?

## 14.4 Which Divisor Method is Best?

**26.** Determine the relative difference between the numbers 5 and 7.

**27.** Jim is 72 inches tall and Alice is 65 inches tall. What is the relative difference of their heights?

**28.** In the 2001 apportionment of Congress, the average congressional district in North Carolina was given 13 seats, and its average district population was 620,590. Montana, with a population of 905,316, was the most populous state to receive only one district.

**(a)** Which state is the more favored in this apportionment?
**(b)** What are the relative differences in the district sizes?
**(c)** If North Carolina was apportioned 12 seats, and Montana was apportioned 2, determine the absolute and relative differences in district sizes.
**(d)** Does this apportionment minimize absolute differences in district size between these two states?

**29.** The data from Exercise 28 imply that the representative share for North Carolina was 1.6114 representatives per million, and for Montana the representative share was 1.1046 representatives per million.

**(a)** What are the absolute and relative differences in representative share?
**(b)** If North Carolina was apportioned 12 seats, and Montana was apportioned 2, determine the absolute and relative differences in representative share.
**(c)** Does the apportionment minimize absolute differences in representative share between these two states?

**30.** According to the 2000 census, the population of California was 33,930,798 and the population of Utah was 2,236,714. California was apportioned 53 House seats, and Utah received 3 House seats.

**(a)** Determine the average congressional district sizes for these states.

**(b)** Determine the absolute and relative differences in these district sizes.

**(c)** Suppose a seat were transferred from California to Utah, giving Utah 4 seats and California 52. What would now be the absolute and relative differences in district sizes?

**(d)** Does this evidence indicate that the apportionment following the 2000 census fulfilled the criterion of keeping absolute differences in district populations as small as possible? What about relative differences?

**31.** Find the Hill–Huntington rounding points for numbers between 0 and 1; between 1 and 2; between 2 and 3; and between 3 and 4.

**32.** A high school has one math teacher, who can teach five sections. Fifty-six students have enrolled in the algebra class, 28 have signed up for geometry, and 7 students will take calculus. Use the Hill–Huntington method to decide how many sections of each course to schedule.

**33.** One year later, the high school described in Exercise 32 still has just one math teacher who teaches five sections. The enrollments are algebra, 36; geometry, 61; and calculus, 3. Apportion the classes by the Webster and Hill–Huntington methods. Which apportionment do you think the school principal would prefer?

◆ **34.** Seats in parliament are apportioned according to votes received by parties.

**(a)** What would be the drawback of using the Hill–Huntington method?

**(b)** If it is considered undesirable to give small parties much representation in parliament, which divisor method would be preferable–Jefferson or Webster?

■ **35.** Suppose that in 2000, the governor of Utah believed that the population of his state was undercounted. What increase in population would be large enough to entitle Utah to take a seat from California if the apportionment is by the Hill–Huntington method? The data needed for this problem are given in Exercise 30.

◆ **36. (a)** Show that for any positive numbers $A$ and $B$, the geometric mean is less than the arithmetic mean,[4] except when $A = B$; then the two means are equal. (*Hint:* Show that the triangle in Figure 14.2 is a right triangle.)

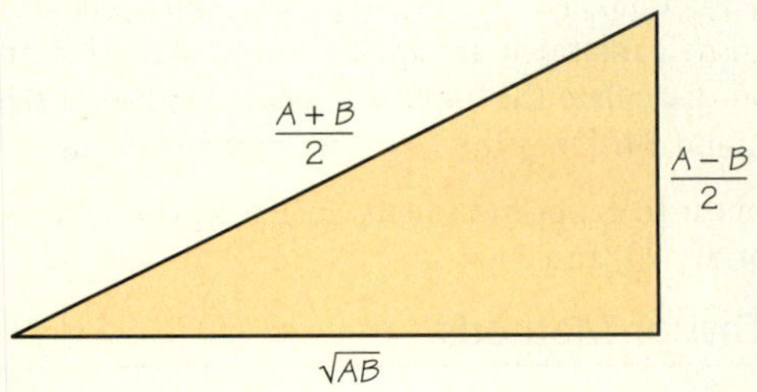


FIGURE 14.2 Is this a right triangle?

**(b)** Compare the Webster and Hill–Huntington roundings: Show that for any positive number $q$, if the two roundings differ, then the Hill–Huntington rounding of $q$ is equal to $\lceil q \rceil$ and the Webster rounding is equal to $\lfloor q \rfloor$.

**(c)** Explain why the fact established in part (a) implies that the Hill–Huntington method is more favorable to small states than the Webster method.

**37.** A city has three districts with populations of 100,000; 600,000; and 700,000. Its council has 20 members, and seats on the council are apportioned according by the Hill–Huntington method to the districts. Show that there is a tie. Would a tie occur with any of the other apportionment methods that we have considered?

**38.** In a 1991 federal lawsuit *Massachusetts* v. *Mosbacher,* Massachusetts claimed that the Hill–Huntington method of apportionment is unconstitutional, because it does not reflect the "one person, one vote" principle as well as the Webster method does. Would Massachusetts have gained a seat from Oklahoma if the Webster method had been used to apportion the House of Representatives in 1991? Use the following populations and Hill–Huntington apportionments:

| State | Population | Apportionment |
|---|---|---|
| Massachusetts | 6,029,051 | 10 |
| Oklahoma | 3,145,585 | 6 |

◆ **39.** The following apportionment method was invented by Congressman William Lowndes of South Carolina in 1822. Lowndes starts, as Hamilton does, by giving each state its lower quota. But where Hamilton apportions the remaining seats to the states whose quotas have the largest fractional parts–in other words, the states for which the *absolute difference* between $q_i$ and $\lfloor q_i \rfloor$ is greatest–Lowndes gives the extra seats to the states where the *relative difference* between $q_i$ and $\lfloor q_i \rfloor$ is greatest, raising as many as necessary to their upper quotas to fill the House.

**(a)** Would this method be more beneficial to states with large populations or small populations, as compared with the Hamilton method?

**(b)** Does the Lowndes method satisfy the quota condition?

[4]The arithmetic mean of $A$ and $B$ is equal to $(A + B)/2$.

**(c)** Would there be any trouble with paradoxes with the Lowndes method?
**(d)** Use the method to apportion the 1790 House of Representatives.

◆ **40.** Let the populations of states $A$ and $B$ be $p_A$ and $p_B$, respectively. The apportionments will be $a_A$ and $a_B$. Assuming that district populations for state $A$ are larger than district populations for state $B$, show that the relative difference in district populations is

$$\left(\frac{a_B p_A}{a_A p_B} - 1\right) \times 100\%$$

Also show that this expression is equal to the relative difference in representative share. Hence the relative difference in district populations is equal to the relative difference in representative shares.

◆ **41.** John Quincy Adams, the sixth president of the United States, proposed that the House of Representatives should be apportioned by a divisor method based on the rounding rule that rounds each fraction up to the next whole number.

**(a)** Is it likely that the initial tentative apportionment will be final?
■ **(b)** Find a formula for a state's critical divisor in terms of the state's tentative apportionment $n_i$ and its population $p_i$.
**(c)** Does the method favor small states or large states?
**(d)** Is it possible for a state to be apportioned zero seats by using this method?

◆ **42.** The U.S. Constitution requires that each state be apportioned at least one seat in the House of Representatives.

**(a)** Show that the Hill–Huntington method is consistent with this requirement. What about the methods of Hamilton, Jefferson, and Webster?
**(b)** Does the Adams method (see Exercise 41) meet the requirement?
■ **(c)** The Dean method (see Writing Project 4) is the divisor method that minimizes absolute differences in district population. Using this information, explain why the Dean method will never give any state zero seats, unless the house size is less than the number of states.

**43.** The Marquis de Condorcet, who proposed a criterion for deciding elections (see Chapter 9), also designed a divisor method for apportionment. His rounding rule was to round down numbers whose fractional parts are less than 0.4 and to round up otherwise.

**(a)** Explain why the Condorcet rounding of a number $q$ is $\lfloor q + 0.6 \rfloor$.
**(b)** Does the method favor large states, small states, or is it neutral?
**(c)** Find a formula for a state's critical divisor in terms of the state's tentative apportionment $n_i$ and its population $p_i$.

◆ **44.** The choice of a divisor method for apportioning classes to subjects according to enrollments, as in the senior high school example, depends on what the school principal considers most important.

**(a)** The principal wants to set a minimum class size. For example, if the minimum class size is 20, and 39 students are signed up for English III, there would be one section, because there are not enough students for two sections with enrollment of at least 20. If there were 40 students, there would be two sections. The minimum class size is adjusted so that as many sections as possible are running. What apportionment method should she use, and what will the minimum class size be?
■ **(b)** The principal prefers to set a maximum class size. For example, if the maximum class size is 33, and 67 students are taking History I, there will be three sections, because there are too many students to fit in two 33-student sections. If there were only 66 taking History I, there would be two sections. The maximum class size is adjusted so that as many sections as possible are running. What apportionment method should she use, and what will the maximum class size be? (*Hint:* This divisor method is not described in the text but is mentioned in one of the previous exercises.)
**(c)** The principal wants to treat students as equitably as possible, so that the differences between students' share of teachers vary as little as possible from course to course. What apportionment method should she use now?
**(d)** The principal wants to minimize relative difference in class size. What divisor method would work best for her?
**(e)** The principal wants to cancel any class that has an enrollment of just one student. Which apportionment methods should she avoid using?

■ **45.** Let $q_1, q_2, \ldots, q_n$ be the quotas for $n$ states in an apportionment problem, and let the apportionments assigned by some apportionment method be denoted $a_1, a_2, \ldots, a_n$. The *absolute deviation* for state $i$ is defined to be $|q_i - a_i|$; it is a measure of the amount by which the state's apportionment differs from its quota. The *maximum absolute deviation* is the largest of these numbers. Explain why the Hamilton method always gives the least possible maximum absolute deviation.

## APPLET EXERCISES

To do these exercises, go to www.whfreeman.com/fapp8e.

A bus company has three lines–*A*, *B*, and *C*–and a total of 48,000 riders. *A* has 21,700 riders daily, *B* has 17,200, and *C* has 9100. The company has 40 buses to allocate to the three lines. Use the applet *Apportionment* to help you find the standard divisor and determine the allocation of the buses according to the methods of apportionment of Hamilton, Jefferson, Webster, and Hill–Huntington.

## WRITING PROJECTS

**1.** Does the Hill–Huntington method best reflect the intentions of the Founding Fathers, as these intentions were set down in the Constitution and in the debate during the 1787 Constitutional Convention? Good sources of information here include all of the publications listed in the Suggested Readings. This writing project requires that you state your answer to the question and make a case for it.

**2.** Suppose that in 2000, Congress had reverted to its nineteenth-century habit of increasing the size of the House of Representatives so that no state would have a decrease in the size of its delegation. How many seats would have been added, and which states would have gotten them? (*Warning:* The apportionments of some states might *increase* as a result of this practice.) As the first step of this project, obtain the populations and apportionments for the 50 states from the Census Bureau Web site (www.census.gov/population/www/censusdata/apportionment.html).

**3.** In 2004, the state of Colorado considered an amendment to its constitution regarding the way electors representing Colorado in the Electoral College would be selected. Until 2000, Colorado's electors were selected by the president/vice president ticket that received a plurality of the votes. The proposed amendment, which would have taken effect in 2004, apportioned electoral votes to each ticket in proportion to the number of popular votes received. The apportionment method specified in the amendment was as follows: Determine each ticket's quota, and round to the nearest whole number. Tickets with quotas less than 0.5 receive no electors. If the number of electors thus apportioned is less than Colorado is entitled to have, give the remaining electors to the ticket that received the most votes. If the number of electors is more than Colorado is entitled to have, take electors from that ticket that, among all who received electors, received the smallest number of popular votes. If more electors need to be removed, take from the ticket that received the next smallest number of popular votes, and so on.

Write an essay exploring the implications of this apportionment method, and compare it to others that Colorado could have chosen. Include a discussion of the consequences that would occur if a third-party candidate received some electoral votes as a result of this procedure.

**4. The Dean method.** The Webster method was proposed in 1832 after New York received an apportionment in excess of its upper quota. Two other apportionment methods were proposed in the same year: the method of John Quincy Adams, which is biased in favor of small states as badly as the Jefferson method is biased in favor of large states, and the Dean method. The latter method, invented by James Dean, a professor of mathematics and astronomy at Dartmouth College, gives the most equitable apportionment when the measure of inequity is absolute difference in district population. Suppose that state *A* has population $p$ and its tentative apportionment is $n$, while state *B* has population $q$ and tentative apportionment $m$. If another seat is to be given to one of these states,

**(a)** Calculate the absolute difference in district populations if *A* gets the seat, and repeat the calculation for the situation when *B* gets the seat.
**(b)** Show that the difference between the two results in (a) is equal to $\frac{p}{n^{\#}} - \frac{q}{m^{\#}}$, where $n^{\#}$ denotes the *harmonic mean* (you may have to look this up) between $n$ and $n + 1$.
**(c)** Explain the mechanics of the Dean method, including why it is a divisor method that rounds a number $r$ down to $\lfloor r \rfloor$ if $r$ is less that the Dean rounding point, and up to $\lceil r \rceil$ otherwise. It's up to you to figure out the Dean rounding point.
**(d)** Is the Dean method biased in favor of large states or small states? Is it possible for any state to get a zero apportionment? Compute the apportionment of the House of Representatives according to the latest decennial census by the Dean method. Is the quota condition satisfied?

## SUGGESTED READINGS

BALINSKI, M. L., and H. P. YOUNG. *Fair Representation: Meeting the Ideal of One Man, One Vote,* Yale University Press, New Haven, Conn., 1982. In the 1970s, Balinski and Young analyzed apportionment methods in depth. Their approach was to postulate the desirable properties of an apportionment method as axioms and to deduce from the axioms which method is best. This book combines an account of the history of apportionment of the U.S. House of Representatives with the results of their research.

ERNST, LAWRENCE R. Apportionment methods for the House of Representatives and the court challenges, *Management Science,* 40 (1994): 1207–1227. Ernst, who wrote briefs for the government in both the *Montana* and the *Massachusetts* cases, reviews the apportionment problem and the arguments in favor of and against each of the divisor methods. The article includes a summary of the arguments used by both sides in the two court cases.

YOUNG, H. PEYTON. *Equity,* Princeton University Press, Princeton, N.J., 1994. Chapter 3 covers apportionment and focuses on which apportionment method is the most equitable.

## SUGGESTED WEB SITES

**www.census.gov/population/www/censusdata/apportionment.html** This site contains a two-page history of apportionment of the Congress.

CHAPTER 15



# Game Theory: The Mathematics of Competition

Conflict has been prevalent throughout human history. It arises whenever two or more individuals, with different values or goals, compete to try to control the course of events. *Game theory* uses mathematical tools to study situations, called games, involving both conflict and cooperation. Its study was greatly stimulated by the publication in 1944 of the monumental *Theory of Games and Economic Behavior* by John von Neumann and Oskar Morgenstern (see Spotlight 15.1).

The *players* in a game, who may be people, organizations, or even countries, choose from a list of options available to them–that is, courses of action they may take–that are called **strategies**. The strategies chosen by the players lead to *outcomes,* which describe the consequences of their choices. We assume that the players have *preferences* for the outcomes: They like some more than others.

Game theory analyzes the **rational choice** of strategies–that is, how players select strategies to obtain preferred outcomes. Among areas to which game theory has been applied are bargaining tactics in labor–management disputes, resource-allocation decisions in political campaigns, military choices in international crises, and the use of threats by animals in habitat acquisition and protection.

Unlike the subject of *individual* decision making, which researchers in psychology, statistics, and other disciplines study, game theory analyzes situations in which there are at least two players, who may find themselves in conflict because of different goals or objectives. The outcome depends on the choices of *all* the players. In this sense, decision making is *collective,* but this is not to stay that the players necessarily cooperate when they choose strategies. Indeed, many strategy choices are noncooperative, such as those between combatants in warfare or competitors in sports. In these encounters, the adversaries' objectives may be at cross-purposes: A gain for one means a loss for the other. But in many activities, especially in economics and politics, there may be joint gains that can be realized from cooperation.

## SPOTLIGHT 15.1 The Early History of Game Theory

**John von Neumann**
*(Bettmann/UPI/Corbis.)*

**Oskar Morgenstern**
*(Courtesy of the Institute for Advanced Study, Princeton University Archives.)*

As early as the seventeenth century, such outstanding scientists as Christiaan Huygens (1629–1695) and Gottfried W. Leibniz (1646–1716) proposed the creation of a discipline that would apply the scientific method to the study of human conflict and interactions. Throughout the nineteenth century, several leading economists created simple mathematical models to analyze particular examples of competitive encounters. The first general mathematical theorem on this subject was proved for games of perfect information by the distinguished logician Ernst Zermelo (1871–1953) in 1912. A game is said to have *perfect information* if at each stage of the play, every player is aware of all past moves (by itself and others) as well as all future choices that are possible. The theorem stated that any finite game with perfect information, such as checkers or chess, has an optimal solution in *pure* strategies; that is, no randomization or secrecy is necessary. This theorem is an example of an *existence theorem*: It demonstrates that there must exist a best way to play such a game, but it does not provide a detailed plan for playing a complex game, like chess, to achieve victory.

The famous mathematician F. E. Émile Borel (1871–1956) introduced the notion of a *mixed*, or randomized, strategy when he investigated optimal strategies in duels around 1920. The fact that every two-person *zero-sum* game must have a solution in optimal mixed strategies was proved by the Hungarian-American mathematician John von Neumann (1903–1957) in 1928. Von Neumann's result was extended to the existence of equilibrium outcomes in mixed strategies for multiperson games that are either *constant-sum* or *variable-sum* in 1951 by John F. Nash, Jr. (b. 1928), who was portrayed in the movie, *A Beautiful Mind* (2001).

Modern game theory dates from the publication in 1944 of *Theory of Games and Economic Behavior* by John von Neumann and the Austrian-American economist Oskar Morgenstern (1902–1977). They introduced the first general model and solution concept for multiperson *cooperative games*, which are primarily concerned with coalition formation (by economic cartels, voting blocs, or military alliances) and the resulting distribution of gains or losses. Several other suggestions for a solution to such games have since been proposed. These include the value concept of Lloyd S. Shapley (b. 1923), which relates to fair allocation and serves also as index of voting power (see Chapter 11).

The French artist Georges Mathieu designed a medal for the Musée de la Monnaie in Paris in 1971 to honor game theory. It was the seventeenth medal to "commemorate 18 stages in the development of Western consciousness." Game theory also has a mascot, the tiger, arising from the Princeton University tiger and the Russian abbreviation of the term *game theory* (<u>Т</u>ЕОРИЯ И<u>Г</u><u>Р</u>), where the underlined letters correspond to the sounds of the English *T*, *G*, and *R*, respectively).

Many interactions involve a delicate mix of cooperative and noncooperative behavior. In business, for example, firms in an industry cooperate to gain tax breaks even as they compete for shares in the marketplace.

Game theory has provided important theoretical foundations in economics, starting with microeconomics but now extending to macroeconomics, industrial organization, and international economics. It also has been increasingly applied in political science, especially in the study of voting, elections, and international relations. In addition, game theory has contributed major insights in biology, particularly in understanding the evolution of species and conditions under which animals–humans included–fight each other for territory or act altruistically. It has also illuminated certain fields in philosophy, including ethics, the philosophy of religion, and political philosophy, and inspired many experiments in social psychology.

In the next two sections, we present several simple examples of two-person **total-conflict games**, in which what one player wins the other player loses, so cooperation never benefits the players. We distinguish two different kinds of solutions to such games. Then we analyze two well-known **partial-conflict games**, in which the players can benefit from cooperation but may have strong incentives not to cooperate. We next turn to the analysis of a larger three-person voting game, in which we show how to eliminate undesirable strategies in stages. Finally, we offer some general comments on solving games and discuss different applications of game theory.

## 15.1 Two-Person Total-Conflict Games: Pure Strategies

For some games with two players, determining the best strategies for the players is straightforward. We begin with such a case.

### EXAMPLE 1 ■ A Location Game

Two young entrepreneurs, Henry and Lisa, plan to locate a new restaurant at a busy intersection in the nearby mountains. They agree on all aspects of the restaurant except one. Lisa likes low elevations, whereas Henry wants greater heights–the higher, the better. In this one regard, their preferences are diametrically opposed. What is better for Henry is worse for Lisa, and likewise what is good for Lisa is bad for Henry.

The layout for their location problem is shown in Figure 15.1. Observe that three routes, Avenue A, Boulevard B, and County Road C (blue lines), run in an east–west direction, and that three highways, numbered 1, 2, and 3 (red lines), run in a north–south direction. Table 15.1 shows the altitudes at the nine corresponding intersections. The same information is shown in three dimensions in Figure 15.2.

To maximize the number of customers, Henry and Lisa agree that the restaurant should be at a location where one of the east–west routes intersects one of the three highways. But they cannot agree on which intersection, so they decide to turn their decision into the following competitive game: Henry will select one of the three routes, *A*, *B*, or *C*, and Lisa will simultaneously choose one of the three highways, 1, 2, or 3. Because their choices will be made simultaneously, neither one can predict beforehand what the other will do.

Henry, worried that Lisa will choose a low elevation, tries to determine the highest altitude he can guarantee by picking one of the three routes. For each choice of a route, this means considering the worst-case (lowest) elevation on each route. These

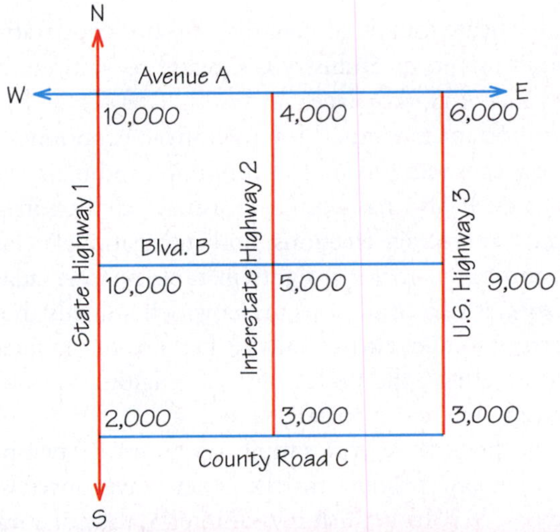


**FIGURE 15.1** The road map for the location of Henry and Lisa's restaurant in Example 1. (The elevations in feet are shown at each intersection.)

**TABLE 15.1** Heights (in thousands of feet) of the Nine Intersections

| | Highways | | |
|---|---|---|---|
| **Routes** | **1** | **2** | **3** |
| ***A*** | 10 | 4 | 6 |
| ***B*** | 6 | 5 | 9 |
| ***C*** | 2 | 3 | 7 |

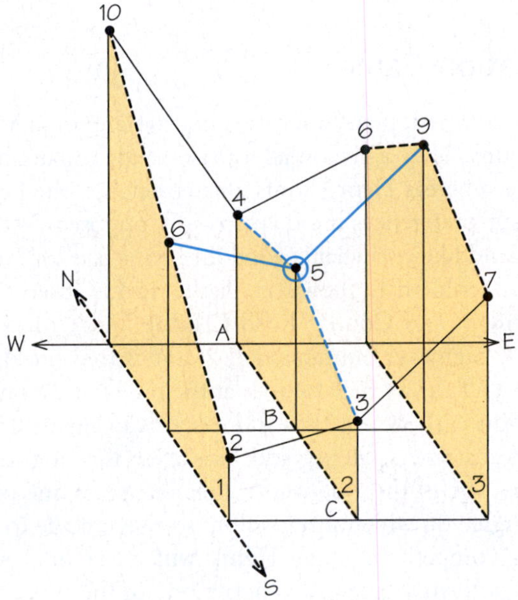


**FIGURE 15.2** Three-dimensional road map showing Henry's and Lisa's possible choices (in thousands of feet).

are the numbers 4, 5, and 2, which are the respective *row minima,* indicated in the right-hand column of Table 15.2. He notes that the highest of these values is 5. By choosing the corresponding route, *B,* Henry can guarantee himself an altitude of at least 5000 feet.

**TABLE 15.2** Heights (in thousands of feet) in Table 15.1, with the Row Minima (maximum circled) and Column Maxima (minimum circled)

| | | *Lisa* Highways | | | |
|---|---|---|---|---|---|
| | Routes | 1 | 2 | 3 | Row Minima |
| | *A* | 10 | 4 | 6 | 4 |
| *Henry* | *B* | 6 | 5 | 9 | ⑤ |
| | *C* | 2 | 3 | 7 | 2 |
| | Column Maxima | 10 | ⑤ | 9 | |

### Maximin — DEFINITION

The **maximin** is the maximum value of the minimum numbers in the rows in a table. The strategy that corresponds to the maximin is called the **maximin strategy**. The number 5 in the right-hand column of Table 5.2, which is circled, is the maximin. Route *B* is Henry's maximin strategy.

Lisa likewise does a worst-case analysis and lists the highest—for her, the worst—elevations for each highway. These numbers, 10, 5, and 9, are the column maxima and are listed in the bottom row of Table 15.2. From Lisa's point of view, the best of these outcomes is 5. If she picks Interstate Highway 2, then she is assured of an elevation of no more than 5000 feet.

### Minimax — DEFINITION

The **minimax** is the minimum value of the maximum numbers in the columns. The strategy that corresponds to the minimax is called the **minimax strategy**. The number 5 at the bottom row of Table 5.2, which is circled, is the **minimax**. Highway 2 is Lisa's minimax strategy.

To summarize, Henry has a strategy that will ensure the height is 5000 or higher, and Lisa has a strategy that will ensure the height is 5000 or lower. The height of 5000 at the intersection of route *B* and Highway 2 is, simultaneously, the lowest value along Boulevard B and the highest along Interstate Highway 2. In other words, the maximin and the minimax are both equal to 5000 for the location game.

### Saddlepoint — DEFINITION

When a row minimum and a column maximum are the same, the resulting outcome is called a **saddlepoint**.

The reason for the term **saddlepoint** should be clear from the saddle-shaped payoff surface shown in Figure 15.2. The middle point on a horse saddle is simultaneously the lowest point along the spine of the horse and the highest point between the rider's legs. (In Figure 15.2, the rider would be facing leftward or rightward.) In our example, one might also think of the saddlepoint as a mountain pass:

As one drives through the pass, the car is at a high point on a highway (in the north–south direction) and at a low point on a route (in the east–west direction).

The resolution of this contest is for Henry to pick *B* and Lisa to pick 2. This puts them at an elevation of 5000, which is simultaneously the maximin and minimax.

**Value** DEFINITION

In total-conflict games, the **value** is the best outcome that both players can guarantee. If a game has a saddlepoint, it gives the value (5 in our example): Players can guarantee this outcome by choosing their maximin and minimax strategies.

Total-conflict games without saddlepoints also have a "value," as we shall see later.

There is no need for secrecy in a game with a saddlepoint. Even if Henry were to reveal his choice of *B* in advance, Lisa would be unable to use this knowledge to exploit him. In fact, both players can use the height information in our example to compute the optimal strategy for their opponent as well as for themselves. In games with saddlepoints, players' worst-case analyses lead to the best *guaranteed* outcome—in the sense that each player can ensure that he or she does not do worse than a certain amount (5 in our example) and may do better (if the opponent deviates from a maximin or minimax strategy).

Another well-known game with a saddlepoint is tic-tac-toe. Two players alternately place an X or an O, respectively, in one of the unoccupied spaces in a $3 \times 3$ grid with 9 cells. The winner is the first player to have three X's, or three O's, in the same row, in the same column, or along a diagonal; if no player does this when all cells are filled in, the game ends in a tie.

An explicit list of all strategies for either the first- or second-moving player in tic-tac-toe is long and complicated, because it specifies a complete plan for all possible contingencies that can arise. For the first-moving player, for example, a strategy might say "put an X in the middle cell, then an X in the corner if your opponent puts an O in a noncorner cell," and so on. While young children initially find this game interesting to play, before long they discover that each player can always prevent the other player from winning by forcing a tie, making the game quite boring. Unlike the game between Henry and Lisa, the list of possible strategies in tic-tac-toe is huge, but only those that force a tie, it turns out, are a saddlepoint.

## EXAMPLE 2 ■ The Restricted-Location Game

Assume in our location game that Henry and Lisa are informed by county officials that it is against the law to locate a restaurant on either Boulevard B or Interstate Highway 2. These two choices, which provided our earlier solution, are now forbidden. The resulting location game without these two strategies is given in Table 15.3 (with payoffs again expressed in thousands of feet).

As before, Henry and Lisa can each do a worst-case analysis. Henry is worried about the minimum number in each row, and Lisa is concerned with the maximum number in each column. These are listed in the right column and bottom row, respectively, in Table 15.4.

Henry sees that his maximin is 6, so he can guarantee a height of 6000 feet or more by choosing route *A*. Likewise, Lisa observes that her minimax is 7, so she can keep the elevation of the restaurant down to 7000 feet or less by selecting Highway 3.

**TABLE 15.3** Heights (in thousands of feet) without Boulevard B and Interstate Highway 2

| | Highways | |
|---|---|---|
| Routes | 1 | 3 |
| *A* | 10 | 6 |
| *C* | 2 | 7 |

**TABLE 15.4** Heights (in thousands of feet) in Table 15.3, with Row Minima (maximum circled) and Column Maxima (minimum circled)

| | | *Lisa* Highways | | |
|---|---|---|---|---|
| | Routes | 1 | 3 | Row Minima |
| *Henry* | *A* | 10 | 6 | ⑥ |
| | *C* | 2 | 7 | 2 |
| | Column Maxima | 10 | ⑦ | |

There is a gap of 7 − 6 = 1 between the minimax and maximin. When the maximin is less than the minimax, as in this case, then a game does *not* have a saddlepoint, but it does have a value (described in the next section).

If Henry plays his maximin strategy, route *A*, and Lisa plays her minimax strategy, Highway 3, then the resulting payoff is 6. However, Henry may be motivated to gamble in this case by playing his other strategy, route *C*. If Lisa sticks to her conservative strategy, Highway 3, then the payoff is 7. Henry will have gained one unit (1000 feet), going from 6 to 7.

This is, however, a risky move. If Lisa suspected it, she might counter by selecting Highway 1. The payoff would then be 2, the best for Lisa and the worst for Henry. So Henry's gamble to gain 1 unit (6 to 7) by moving has the risk that he might lose 4 units (6 to 2) if Lisa also moves.

But then there is no incentive for Lisa to play her nonminimax strategy (that is, to play Highway 1) if she believes Henry, in turn, will move back to his maximin strategy (route *A*), leading to a payoff of 10. This is worse than 6 from her viewpoint. ■

In two-player games that have saddlepoints, like our original 3 × 3 location game and tic-tac-toe, each player can calculate the maximin and minimax strategies for both players before the game is even played. Once the solution has been determined by either mathematical analysis or practical experience (as was probably true of tic-tac-toe), there may be little interest in actually playing the game.

But this is decidedly not the case for much more complex games, like chess, whose solution has not yet been determined–and is unlikely to be in the foreseeable future. Even though computers are able to beat world champions, the computer's winning moves will not necessarily be optimal against those of *all* other opponents. Nevertheless, we know that chess, like tic-tac-toe, has a saddlepoint. (All games of perfect information, in which the players know each other's moves at every step, have a saddlepoint.) What we do not know is whether it yields a win for white, a win for black, or a draw.

Unlike chess, many games, like the $2 \times 2$ restricted-location game, do not have an outcome that can always be guaranteed. These games, which include poker, involve uncertainty and risk. In such games, one does not want to have one's strategy detected in advance, because this information can be exploited by an opponent. It is no surprise, then, that poker players are told to keep a "poker face," revealing nothing about their likely choices. But this advice is not very helpful in telling the players what actually to do in the game, such as how many cards to ask for in draw poker.

We will show that there are optimal ways to play two-person total-conflict games without a saddlepoint so as not to reveal one's choices. But their solution is by no means as straightforward as that of games with a saddlepoint.

## 15.2 Two-Person Total-Conflict Games: Mixed Strategies

Probably most competitive games do not have a saddlepoint like the one we found in our first location-game example. Rather, as is illustrated in our restricted-location game–in which the maximin and minimax are not the same–players must try to keep secret their strategy choices, lest their opponent use this information to his or her advantage.

In particular, players must take care to conceal the strategy they will select until the encounter actually takes place, when it is too late for the opponent to alter his or her choice. If the game is repeated, a player will want to *vary* his or her strategy in order to surprise the opponent.

In parlor games like poker, players often use the tactic of *bluffing*. This tactic involves a player's sometimes raising the stakes when he has a low hand so that opponents cannot guess whether or not his hand is high or low–and may, therefore, miscalculate whether to stay in or drop out of the game (a player would prefer opponents to stay in when he has a high hand and drop out when he has a low hand). In military engagements, too, secrecy and even deception are often crucial to success.

In many sporting events, a team tries to surprise or mislead the opposition. A pitcher in baseball will not signal the type of pitch he or she intends to throw in advance, varying the type throughout the game to try to keep the batter off balance. In fact, we next consider a confrontation between a pitcher and batter in more detail.

### EXAMPLE 3 ■ A Duel Game

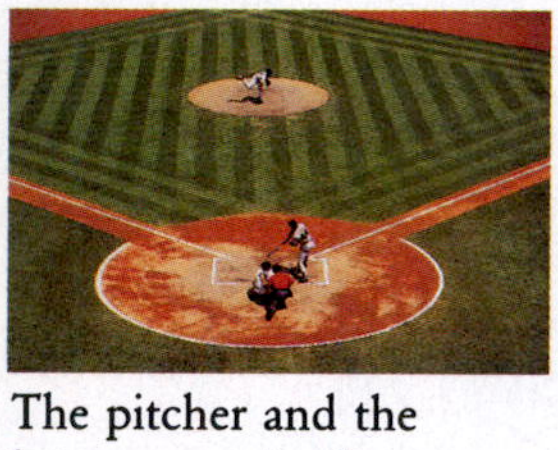
The pitcher and the batter use mixed strategies. (*Alan Schein/ The Stock Market.*)

Assume that a particular baseball pitcher can throw either a blazing fastball or a slow curve into the strike zone and so has two strategies: *fast* (denoted by $F$) and *curve* ($C$). The pitcher faces a batter who attempts to guess, before each pitch is thrown, whether it will be a fastball or a curve, giving the batter two strategies also: guess $F$ and guess $C$. Assume that the batter has the following batting averages, which are known by both players.

- 0.300 if the batter guesses fast ($F$) and the pitcher throws fast ($F$)
- 0.200 if the batter guesses fast ($F$) and the pitcher throws curve ($C$)

- 0.100 if the batter guesses curve ($C$) and the pitcher throws fast ($F$)
- 0.500 if the batter guesses curve ($C$) and the pitcher throws curve ($C$)

A player's batting average is the number of times he hits safely divided by his number of times at bat. If a batter hit safely 3 times out of 10, for example, his average would be 0.300.

This game is summarized in Table 15.5. We see from the right-hand column in the table that the batter's maximin is 0.200, which is realized when he selects his first strategy, $F$. Thus, the batter can "play it safe" by always guessing a fastball, which will result in his batting 0.200, hardly enough for him to remain on the team.

We see from the bottom row of the table that the pitcher's minimax is 0.300, which is obtained when he throws fast ($F$). Note that the batter's maximin of 0.200 is less than the pitcher's minimax of 0.300, so this game does not have a saddle-point. There is a gap of $0.300 - 0.200 = 0.100$ between these two numbers.

Each player would like to play so as to win for himself as much of the 0.100 payoff in the gap as possible. That is, the batter would like to average more than 0.200, whereas the pitcher wants to hold the batter down to less than 0.300.

**TABLE 15.5** Batting Averages in a Baseball Duel

| | | *Pitcher* | | Row Minima |
|---|---|---|---|---|
| | | *F* | *C* | (maximum circled) |
| *Batter* | *F* | 0.300 | 0.200 | (0.200) |
| | *C* | 0.100 | 0.500 | 0.100 |
| Column Maxima (minimum circled) | | (0.300) | 0.500 | |

## A Flawed Approach

If the batter and pitcher in our example consider how they might outguess each other, they may reason along the following lines:

1. *Pitcher* (to himself): If I choose strategy $F$, I hold the batter down to 0.300 (the minimax) or less. However, the batter is likely to guess $F$ because it guarantees him at least 0.200 (his maximin), and it actually provides him with 0.300 against my $F$ pitch. In this case, the batter wins all the 0.100 payoff in the gap.
2. *Batter* (to himself): Because the pitcher will try to surprise me with $C$ by reasoning as in step 1, I should fool him and guess $C$. I would thus average 0.500, which will show him up for trying to gamble and outguess me!
3. *Pitcher* (to himself): But if the batter is thinking as in step 2–that is, guessing $C$–I, on second thought, should really throw $F$. This will lead to an average of only 0.100 for the batter and teach him to not try to outguess me!

This type of cyclical reasoning can go on forever: "I think that he thinks that I think that he thinks. . . ." It provides no resolution to the players' decision problem.

Clearly, there is no pitch, or guess, that is best under all circumstances. Nevertheless, both the pitcher and the batter *can* do better, but not by trying to anticipate each other's choices. The answer to their problem lies in the notion of a **mixed strategy**.

## A Better Idea

The play of many total-conflict games requires an element of surprise, which can be realized in practice by making use of a mixed strategy.

### Pure Strategy — DEFINITION

Each of the definite courses of action that a player can choose is called a **pure strategy**.

All the choices of players–Henry and Lisa, the batter and the pitcher–that we have considered so far are **pure strategies**.

### Mixed Strategy — DEFINITION

A **mixed strategy** is a strategy in which the course of action is randomly chosen from one of the pure strategies in the following way: Each pure strategy is assigned some probability, indicating the relative frequency with which that pure strategy will be played. The specific strategy used in any given play of the game can be selected using some appropriate random device.

Note that a pure strategy is a special case of a mixed strategy, with the probability of 1 assigned to just one pure strategy and 0 to all the rest. When a player resorts to a mixed strategy, the resulting outcome of the game is no longer predictable in advance. (For example, if a pitcher throws a curve ball or a fastball with probability 0.5 each, the batter cannot predict which pitch he or she is about to receive.) Rather, the outcome must be described in terms of the probabilistic notion of an **expected value**.

### Expected Value $E$ — DEFINITION

If each of the $n$ payoffs, $s_1, s_2, \ldots, s_n$, will occur with the probabilities $p_1, p_2, \ldots, p_n$, respectively, then the average, or **expected value** $E$, is given by

$$E = p_1 s_1 + p_2 s_2 + \cdots + p_n s_n$$

We assume that the probabilities sum to 1 and that each probability $p_i$ is never negative. That is, we assume that $p_1 + p_2 + \cdots + p_n = 1$, and $p_i \geq 0$ $(i = 1, 2, \ldots, n)$.

To see how mixed strategies and expected values are used in the analysis of games, we turn to what is perhaps the simplest of all competitive games without a saddlepoint.

## EXAMPLE 4 ■ Matching Pennies

In matching pennies, each of two players simultaneously shows either a head $H$ or a tail $T$. If the two coins match, with either two heads or two tails, then the first player (player I) receives both coins (a win of 1 for player I). If the coins do not match, that is, if one is an $H$ and the other is a $T$, then the second player (player II) receives the two coins (a loss of 1 for player I). These wins and losses for player I are shown in Table 15.6.

**TABLE 15.6** Wins and Losses for Player I in Matching Pennies

| | | Player II | |
|---|---|---|---|
| | | $H$ | $T$ |
| Player I | $H$ | 1 | −1 |
| | $T$ | −1 | 1 |

## Payoff Matrix

DEFINITION

A **payoff matrix** (illustrated by Table 15.6) is a table whose rows and columns correspond to the strategies of the two players. The numerical entries give the payoffs to player I when these strategies are chosen.

Although the entries in our earlier tables for the location game also gave payoffs, they were not monetary, as here. A game represented by a payoff matrix is called *a game in strategic form.*

The two rows in Table 15.6 correspond to player I's two pure strategies, $H$ and $T$, and the two columns to player II's two pure strategies, also $H$ and $T$. The numbers in the table are the corresponding winnings for player I and losses for player II. If two $H$'s or two $T$'s are played, player I wins 1 from player II. When one $H$ and one $T$ are played, player I pays out 1 to player II.

It is fruitless for one player to attempt to outguess the other in this game. They should instead resort to mixed strategies and use expected values to estimate their likely gains or losses.

The best thing for player I to do is randomly to select $H$ half the time and $T$ half the time. This mixed strategy can be expressed as

$$(p_H, p_T) = (p_1, p_2) = (p, 1 - p) = \left(\frac{1}{2}, \frac{1}{2}\right)$$

Note that the probability $p$ of choosing $H$, and $(1 - p)$ of choosing $T$, do indeed sum to 1, as required; in particular, when $p = \frac{1}{2}$, $1 - p = 1 - \frac{1}{2} = \frac{1}{2}$.

This mixture can be realized in practice by the flip of a coin. Player I's resulting expected value is

$$E_H = \frac{1}{2}(1) + \frac{1}{2}(-1) = 0$$

whenever player II plays $H$ (first column of Table 15.6). Whenever player II plays $T$ (second column), player I's resulting expected value is

$$E_T = \frac{1}{2}(-1) + \frac{1}{2}(1) = 0$$

## Mixed-Strategy Value

DEFINITION

A player's expected value is the **mixed-strategy value** of the game. Unlike the use of this notion in games with a saddlepoint, the value here can be realized only by the use of mixed strategies.

The value of 0 in matching pennies is really an expected value and so must be understood in a statistical sense. That is, in a given play of the game, player I will either win 1 or lose 1. However, his or her expectation over many plays of this game is 0. The optimal mixed strategy for player II is likewise a 50-50 mix of $H$ and $T$, which also leads to an expectation of 0, making the game **fair**.

### Fair Game — DEFINITION

A **fair game** has a value of 0 and, consequently, it favors neither player when at least one player uses an *optimal* (mixed) strategy–one that guarantees that the resulting expected payoff is the best that this player can obtain against all possible strategy choices (pure or mixed) by an opponent.

Player II gains nothing by knowing that player I is using the optimal mixed strategy $(\frac{1}{2}, \frac{1}{2})$. However, player I must not reveal to player II whether $H$ or $T$ will be displayed *in any given play* of the game before player II makes his or her own choice of $H$ or $T$. Even without this information, if player II knew that player I was using a particular *nonoptimal* mixed strategy $(p_1, p_2) = (p, 1 - p)$, where $p \neq \frac{1}{2}$ (that is, not choosing a 50-50 mixture between $H$ and $T$), then player II could take advantage of this knowledge and increase his or her average winnings over time to something greater than the value of 0. (See Exercise 14.)

## EXAMPLE 5 ■ Nonsymmetrical Matching

In this game, players I and II can again show either heads $H$ or tails $T$. When two $H$'s appear, player II pays \$5 to player I. When two $T$'s appear, player II pays \$1 to player I. When one $H$ and one $T$ are displayed, then player II collects \$3 from player I. Note that although the sum of player I's gains (\$5 + \$1 = \$6) when there are two $H$'s or two $T$'s, and the sum of player II's gains (\$3 + \$3 = \$6) otherwise, are the same, the game is **nonsymmetrical**.

### Nonsymmetrical Game — DEFINITION

A two-person total-conflict **nonsymmetrical game** is one in which the row player's gains are different from the column player's gains. Note that the row player's gains (\$5 and \$1 in our example) are different from the column player's gains (always \$3). In the original matching pennies, on the other hand, the payoff for winning is the same for each player, so that game is *symmetrical.*

The game just described is given by the payoff matrix in Table 15.7, which shows the payoffs that player I receives from player II. A worst-case analysis, like that which solved our initial location game, is of little help here. Player I may lose \$3 whether he plays $H$ or $T$, making his maximum $-3$. Player II can keep her losses down to \$1 by always playing $T$ (and thus avoiding the loss of \$5 when two $H$'s appear), so player II's minimax is 1. However, if player II chooses $T$ and player I knows this, then player I will also play $T$ and collect \$1 from player II. Can player II do better than lose \$1 in each play of the game?

Consider the situation where player I uses a mixed strategy $(p_H, p_T) = (p, 1 - p)$, which involves playing $H$ with probability $p$ and playing $T$ with

**TABLE 15.7** Payoffs for Player I in a Nonsymmetrical Matching Game

| | | Player II | |
|---|---|---|---|
| | | *H* | *T* |
| **Player I** | *H* | 5 | −3 |
| | *T* | −3 | 1 |

probability $1 - p$, where $0 \le p \le 1$. Against player II's pure strategy *H*, player I's expected value is

$$E_H = (5)(p) + (-3)(1 - p) = 8p - 3$$

Against player II's pure strategy *T*, player I's expected value is

$$E_T = (-3)(p) + (1 - p) = -4p + 1$$

These two linear equations in the variable $p$ are depicted in Figure 15.3. Note that the four points where these two lines intersect the two vertical lines, $p = 0$ and $p = 1$, are the four payoffs appearing in the payoff matrix.

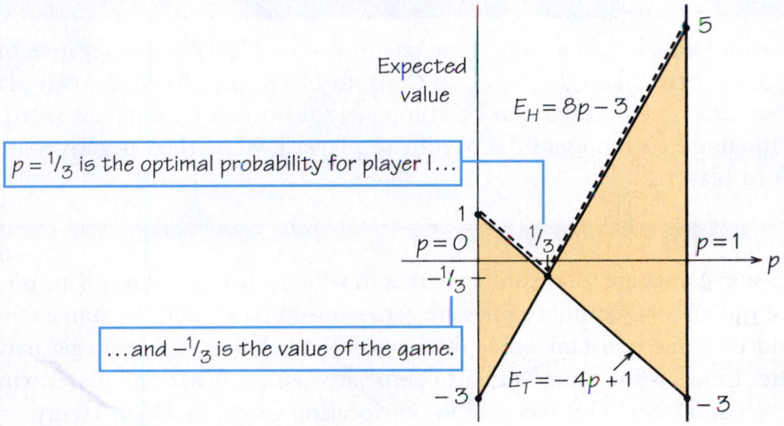


**FIGURE 15.3** Solution to the nonsymmetrical matching pennies.

The point at which the lines given by $E_H$ and $E_T$ intersect can be found by setting $E_H = E_T$, yielding

$$8p - 3 = -4p + 1$$
$$12p = 4$$

so $p = \frac{1}{3}$. To the left of $p = \frac{1}{3}$, $E_T > E_H$, and to the right, $E_H > E_T$; at $p = \frac{1}{3}$, $E_H = E_T$. If player I chooses $(p_H, p_T) = (p, 1 - p) = (\frac{1}{3}, \frac{2}{3})$, he can ensure

$$E_H = 8\left(\frac{1}{3}\right) - 3 = E_T = -4\left(\frac{1}{3}\right) + 1 = -\frac{1}{3}$$

regardless of what player II does.

In other words, player I's optimal mixed strategy is to pick *H* and *T* with probabilities $\frac{1}{3}$ and $\frac{2}{3}$, respectively, which gives player I an expected value of $-\frac{1}{3}$. As can be seen from Figure 15.3, $-\frac{1}{3}$ is the highest expected value that player I can guarantee against *both* strategies *H* and *T* of player II. Although *T* yields player I a higher expected value for $p < \frac{1}{3}$, and *H* yields him a higher expected value for $p > \frac{1}{3}$, player I's choice of $p_H, p_T = (\frac{1}{3}, \frac{2}{3})$ protects him against an expected loss greater than $-\frac{1}{3}$, which neither of his pure strategies does (each may produce a maximum loss of $-3$). Put another way, the intersection of $E_H$ and $E_T$ at $p = \frac{1}{3}$ is the minimum of the function given by $E_T$ to the left and $E_H$ to the right (shown by the dashed line in Figure 15.3). If player II had more than two strategies, this approach to finding a minimum that puts a floor on player I's expected loss can be extended.

A similar calculation for player II results in the same optimal mixed strategy $(\frac{1}{3}, \frac{2}{3})$ and expected value $-\frac{1}{3}$. But because the payoffs for player II are losses, $-\frac{1}{3}$ means that she gains $\frac{1}{3}$ on the average.

Therefore, this game is unfair, even though the sum of the amounts ($6) that player I might have to pay player II when he loses is the same as the sum that player II might have to pay player I when she loses. Interestingly, player II, who will win an average of $33\frac{1}{3}$ cents each time the game is played, is favored, even though she may have to pay more to player I when she loses (a maximum of $5) than player I will ever have to pay her (a maximum of $3).

The symmetrical and nonsymmetrical matching games are examples of what are called **zero-sum games**.

## Zero-Sum Game

DEFINITION

A **zero-sum game** is one in which the payoff to one player is the negative of the corresponding payoff to the other, so the sum of the payoffs to the two players is always zero. These games can be completely described by a payoff matrix, in which the numbers represent the payoffs to player I, while their negatives are the payoffs to player II.

Zero-sum games are total-conflict games in which what one player wins the other loses. But not all total-conflict games are zero-sum–in particular, the sum of the payoffs could be some constant other than zero. Nevertheless, the strategic nature of these latter games is the same as that of zero-sum games: What one player wins, the other player still loses. This was true in our location game, in which Henry's payoff was greater the higher the altitude, and Lisa's greater the lower the altitude.

Scoring in professional chess tournaments usually assigns a payoff of 1 for winning, 0 for losing, and $\frac{1}{2}$ to each player for a tie, making the sum of the payoffs to the two players always 1. Such games, called **constant-sum games**, can readily be converted to zero-sum games. Thus, chess could as well be scored $-1$ for a loss, $+1$ for a win, and 0 for a tie, making the constant 0 in this case. Although constant-sum and zero-sum games have the same strategic nature, constant-sum games are a more general class because the constant need not be zero.

The solution in the symmetrical version of matching pennies illustrated how the mixed strategy of $(\frac{1}{2}, \frac{1}{2})$ guarantees each player the value of 0, but we did not give a *solution technique* for finding optimal mixed strategies. In the nonsymmetrical version of matching pennies, we illustrated a procedure that can be applied to *every* payoff matrix in which each player has only two strategies.

We must use more complex methods, which we will not describe here, to find mixed-strategy solutions when one or both players have more than two strategies. However, one should always check first to see whether a game has a saddlepoint before employing any technique for finding optimal mixed strategies.

In our next example, which is the earlier duel between the pitcher and the batter given by the $2 \times 2$ payoff matrix in Table 15.5, we already showed that there is no saddlepoint. Thus, the solution will necessarily be in mixed strategies. We now proceed to find what mix is optimal.

## EXAMPLE 6 ■ The Duel Game Revisited

In Table 15.8, we add probabilities, which we explain next, to Table 15.5, where $F$ indicates fastball and $C$ indicates curve ball. The pitcher should use a mixed strategy $(p_1, p_2) = (p_F, p_C) = (p, 1 - p)$. The probabilities $p$ and $1 - p$ (where $0 \leq p \leq 1$) are indicated below the game matrix and under the corresponding strategies, $F$ and $C$, for the pitcher. If the pitcher plays a mixed strategy $(p, 1 - p)$ against the two pure strategies, $F$ and $C$, for the batter, he realizes the respective expected values:

$$E_F = (0.3)p + 0.2(1 - p) = 0.1p + 0.2$$
$$E_C = (0.1)p + 0.5(1 - p) = -0.4p + 0.5$$

**TABLE 15.8 A Baseball Duel with Probabilities**

| | | Pitcher | | |
|---|---|---|---|---|
| | | $F$ | $C$ | |
| *Batter* | $F$ | 0.300 | 0.200 | $q$ |
| | $C$ | 0.100 | 0.500 | $1 - q$ |
| | | $p$ | $1 - p$ | |

As in the nonsymmetrical matching-pennies game, the solution to this game occurs at the intersection of the two lines given by $E_F$ and $E_C$. Setting the equations of these lines equal to each other yields $p = 0.6$, giving $E_F = E_C = E = 0.260$.

Thus, the pitcher should use his optimal mixed strategy, which selects $F$ with probability $p = \frac{3}{5}$ and $C$ with probability $1 - p = \frac{2}{5}$. This choice will hold the batter down to a batting average of 0.260, which is the value of the game. We stress that 0.260 is an average and must be interpreted in a statistical manner. It says that about one time in four the batter will get a hit, but not what will happen on any particular time at bat.

Assume that the batter uses a mixed strategy $(q_1, q_2) = (q_F, q_C) = (q, 1 - q)$, as indicated to the right of the game matrix in Table 15.8. This mixed strategy, when played against the pitcher's pure strategies, $F$ and $C$, results in the following expected values:

$$E_F = (0.3)q + 0.1(1 - q) = 0.2q + 0.1$$
$$E_C = (0.2)q + 0.5(1 - q) = -0.3q + 0.5$$

The intersection of these two lines occurs at the point $q = 0.8$, giving $E_F = E_C = E = 0.260$. The batter's optimal mixed strategy is therefore $(q_F, q_C) = (\frac{4}{5}, \frac{1}{5})$, which gives him the same batting average of 0.260.

■

We have seen that the outcome of 0.260, which is the value of the game, occurs when either the pitcher selects his optimal mixed pitching strategy $(\frac{3}{5}, \frac{2}{5})$ or the batter selects his optimal mixed guessing strategy $(\frac{4}{5}, \frac{1}{5})$. This particular result holds true for every two-person zero-sum game; it is the fundamental theorem for such games and is known as the **minimax theorem**.

**Minimax Theorem** DEFINITION

The **minimax theorem** guarantees that there is a unique game value and an optimal strategy for each player, so that either player alone can realize at least this value by playing this strategy, which may be pure or mixed.

The unique value in our example is 0.260.

## 15.3 Partial-Conflict Games

The $2 \times 2$ matrix games (two players, each with two strategies) presented so far have been total-conflict games: One player's gain was equal to the other player's loss. Although most parlor games, like chess or poker, are games of total conflict, and therefore constant-sum, most real-life games are surely not. (Elections, in which there are usually a clear-cut winner and one or more losers, probably come as close to being games of total conflict as we find in the real world.) We will consider two games of partial conflict, in which the players' preferences are not diametrically opposed, that have often been used to model real-world conflicts.

**Variable-Sum Games** DEFINITION

Games of partial conflict are **variable-sum games**, in which the sum of payoffs to the players at the different outcomes varies.

There is some mutual gain to be realized by both players if they can cooperate in partial-conflict games, but this may be difficult to do in the absence of either good communication or trust. When these elements are lacking, players are less likely to comply with any agreement that is made. *Noncooperative games* are games in which a binding agreement cannot be enforced. Even if communication is allowed in such games, there is no assurance that a player can trust an opponent to choose a particular strategy that he or she promises to select.

In fact, the players' self-interests may lead them to make strategy choices that yield both lower payoffs than they could have achieved by cooperating. Two partial-conflict games illustrate this problem.

### EXAMPLE 7 ■ Prisoners' Dilemma

**Prisoners' Dilemma** is a two-person variable-sum game. It provides a simple explanation of the forces at work behind arms races, price wars, and the population problem. In these and other similar situations, the players can do better by cooperating. But there may be no compelling reasons for them to do so unless the players have credible threats of retaliation for not cooperating. The name *Prisoners' Dilemma* was first given to this game by Princeton mathematician Albert W. Tucker (1905–1994) in 1950.

Before defining the formal game, we introduce it through a story.

## Prisoners' Dilemma

STORY

**Prisoners' Dilemma** involves two persons, accused of a crime, who are held incommunicado. Each has two choices: to maintain his or her innocence, or to sign a confession accusing the partner of committing the crime. It is in each suspect's interest to confess and implicate the partner, thereby trying to receive a reduced sentence. Yet if both suspects confess, they ensure a bad outcome–namely, they are both found guilty. What is good for the prisoners as a pair–to deny having committed the crime, leaving the state with insufficient evidence to convict them–is frustrated by their pursuit of their own individual interests.

The game of Prisoners' Dilemma, as we already noted, has many applications, but we will use it here to model a recurrent problem in international relations: arms races between antagonistic countries, which earlier included the superpowers but more recently have included such countries as India and Pakistan and Israel and some of its Arab neighbors. Other countries, such as Iran, may be antagonistic to more than one other country (Israel and the United Staes).

For simplicity, assume there are two nations, Red and Blue. Each can independently select one of two policies:

$A$: Arm in preparation for a possible war (noncooperation).
$D$: Disarm, or at least try to negotiate an arms-control agreement (cooperation).

There are four possible outcomes:

$(D, D)$: Red and Blue disarm, which is *next best* for both because, while advantageous to each, it also entails certain risks.
$(A, A)$: Red and Blue arm, which is *next worst* for both because they spend needlessly on arms and are comparatively no better off than at $(D, D)$.
$(A, D)$: Red arms and Blue disarms, which is *best for Red* and *worst for Blue*, because Red gains a big edge over Blue.
$(D, A)$: Red disarms and Blue arms, which is *worst for Red* and *best for Blue*, because Blue gains a big edge over Red.

This situation can be modeled by means of the matrix in Table 15.9, which gives the possible outcomes that can occur. Here, Red's choice involves picking one of the two rows, whereas Blue's choice involves picking one of the two columns.

We assume that the players can rank the four outcomes from best to worst, where 4 = best, 3 = next best, 2 = next worst, and 1 = worst. Thus, the higher the number, the greater the payoff, making the resulting game an **ordinal game**: It indicates an ordering of outcomes from best to worst but says nothing about the *degree* to which a player prefers one outcome over another. To illustrate, if a player despises the outcome that he or she ranks 1 but sees little difference among the outcomes ranked 4, 3, and 2, the "payoff distance" between 4 and 2 will be less than that between 2 and 1, even though the numerical difference between 4 and 2 is greater.

The ordinal payoffs to the players for choosing their strategies of $A$ and $D$ are shown in Table 15.10, where the first number in the pair indicates the payoff to the row player (Red), and the second number the payoff to the column player (Blue). Thus, for example, the pair (1, 4) in the second row and first column signifies a payoff of 1 (worst outcome) to Red and a payoff of 4 (best outcome) to Blue. This outcome occurs when Red unilaterally disarms while Blue continues to arm, making Blue, in a sense, the winner and Red the loser.

**TABLE 15.9** The Outcomes in an Arms Race, as Modeled by Prisoners' Dilemma

| | | Blue | |
|---|---|---|---|
| | | *A* | *D* |
| ***Red*** | *A* | Arms race | Favors Red |
| | *D* | Favors Blue | Disarmament |

**TABLE 15.10** Ordinal Payoffs in an Arms Race, as Modeled by Prisoners' Dilemma

| | | Blue | |
|---|---|---|---|
| | | *A* | *D* |
| ***Red*** | *A* | (2, 2) | (4, 1) |
| | *D* | (1, 4) | (3, 3) |

Let's examine this strategic situation more closely. Should Red select strategy *A* or *D*? There are two cases to consider, which depend on what Blue does:

- If Blue selects *A*: Red will receive a payoff of 2 for *A* and 1 for *D*, so it will choose *A*.
- If Blue selects *D*: Red will receive a payoff of 4 for *A* and 3 for *D*, so it will choose *A*.

In both cases, Red's first strategy (*A*) gives it a more desirable outcome than its second strategy (*D*). Consequently, we say that *A* is Red's **dominant strategy**, because it is always advantageous for Red to choose *A* over *D*.

In Prisoners' Dilemma, *A dominates* D for Red, so we presume that a rational Red would choose *A*. A similar argument leads Blue to choose *A* as well–that is, to pursue a policy of arming. Thus, when each nation strives to maximize its own payoffs independently, the pair is driven to the outcome (*A*, *A*), with payoffs of (2, 2). The better outcome for both, (*D*, *D*), with payoffs of (3, 3), appears unobtainable when this game is played noncooperatively.

The outcome (*A*, *A*), which is the product of dominant strategy choices by both players in Prisoners' Dilemma, is a **Nash equilibrium**.

### Nash Equilibrium — DEFINITION

When no player can benefit by departing unilaterally (by itself) from its strategy associated with an outcome, the strategies of the players constitute a **Nash equilibrium**. Technically, while it is the set of strategies that define the equilibrium, the choice of these strategies leads to an outcome that we shall also refer to as the equilibrium.

Note that in Prisoners' Dilemma, if either player departs from (*A, A*), the payoff for the departing player who switches to *D* drops from 2 to 1 at (*D, A*) and (*A, D*). Not only is there no benefit from departing, but there is actually a loss, with the *D* player punished with its worst payoff of 1. These losses would presumably deter each nation from moving away from the Nash equilibrium of (*A, A*), assuming the other nation sticks to *A*.

Even if both nations agreed in advance jointly to pursue the socially beneficial outcome, (*D, D*), (3, 3) is unstable. This is because if either nation alone reneges on the agreement and secretly arms (as North Korea did when it developed nuclear weapons), it will benefit, obtaining its best payoff of 4. Consequently, each nation would be tempted to go back on its word and select *A*. Especially if nations have no great confidence in the trustworthiness of their opponents, they would have good reason to try to protect themselves against the other side's defection from an agreement by arming.

### Prisoners' Dilemma — DEFINITION

**Prisoners' Dilemma** is a two-person variable-sum game in which each player has two strategies, cooperate or defect (not cooperate). Defect dominates cooperate for both players, even though the mutual-defection outcome, which is the unique Nash equilibrium in the game, is worse for both players than the mutual-cooperation outcome.

Note that if 4, 3, 2, and 1 in Prisoners' Dilemma were not just ranks but numerical payoffs, their sum would be $2 + 2 = 4$ at the mutual-defection outcome and $3 + 3 = 6$ at the mutual-cooperation outcome. At the other two outcomes, the sum, $1 + 4 = 5$, is still different, illustrating why Prisoners' Dilemma is a variable-sum game.

In real life, of course, people often manage to escape the noncooperative Nash equilibrium in Prisoners' Dilemma. Either the game is played within a larger context, wherein other incentives are at work, such as cultural norms that prescribe cooperation (though this is just another way of saying that defection from (*D, D*) is not rational, rendering the game not Prisoners' Dilemma), or the game is played on a repeated basis–it is not a one-short affair–so players can induce cooperation by setting a pattern of rewards for cooperation and penalties for noncooperation.

In a repeated game, factors like reputation and trust may play a role. Realizing the mutual advantages of cooperation in costly arms races, players may inch toward the cooperative outcome by slowly phasing down their acquisition of weapons over time, or even destroying them (the United States and Russia have been doing exactly this). They may also initiate other productive measures, such as improving their communication channels, making inspection procedures more reliable, writing agreements that are truly enforceable, or imposing penalties for violators when their violations are detected (as has occurred through reconnaissance or spy satellites).

Prisoners' Dilemma illustrates the intractable nature of certain competitive situations that blend conflict and cooperation. The standoff that results at the Nash equilibrium of (2, 2) is obviously not as good for the players as that which they could achieve by cooperating–but they risk a good deal if the other player defects.

While saddlepoints are Nash equilibria in total-conflict games, they can never be worse for *both* players than some other outcome (as in partial-conflict games like

Prisoners' Dilemma). The reason is that if one player does worse in a total-conflict or zero-sum game, the other player must do better.

The fact that the players must forsake their dominant strategies to achieve the (3, 3) cooperative outcome (see Table 15.10) makes this outcome a difficult one to sustain in one-shot play. On the other hand, assume that the players can threaten each other with a policy of tit-for-tat in repeated play: "I'll cooperate on each round unless you defect, in which case I will defect until you start cooperating again." If these threats are credible, the players may well shun their defect strategies and try to establish a pattern of cooperation in early rounds, thereby fostering the choice of (3, 3) in the future. Alternatively, they may look ahead, in a manner that will be described at the end of this chapter, to try to stabilize (3, 3).

## EXAMPLE 8 ■ Chicken

Let's look at one other two-person game of partial conflict, known as **Chicken**, that can also lead to troublesome outcomes. Two drivers approach each other at high speed. Each must decide at the last minute whether to swerve to the right or not swerve. Here are the possible consequences of their actions:

1. Neither driver swerves, and the cars collide head-on, which is the worst outcome for both because they are killed (payoff of 1).
2. Both drivers swerve–and each is mildly disgraced for "chickening out"–but they do survive, which is the next-best outcome for both (payoff of 3).
3. One of the drivers swerves and badly loses face, which is his next-worst outcome (payoff of 2), whereas the other does not swerve and is perceived as the winner, which is her best outcome (payoff of 4).

These outcomes and their associated strategies are summarized in Table 15.11.

**TABLE 15.11** Payoffs in a Driver Confrontation, as Modeled by Chicken

| | | *Driver 2* | |
|---|---|---|---|
| | | **Swerve** | **Not Swerve** |
| *Driver 1* | **Swerve** | (3, 3) | (2, 4) |
| | **Not Swerve** | (4, 2) | (1, 1) |

If both drivers persist in their attempts to "win" with a payoff of 4 by not swerving, the resulting outcome will be mutual disaster, giving each driver his or her worst payoff of 1. Clearly, it is better for both drivers to back down and each obtain 3 by swerving, but neither wants to be in the position of being intimidated into swerving (payoff of 2) when the other does not (payoff of 4).

Notice that neither player in Chicken has a dominant strategy. His or her better strategy depends on what the other player does: Swerve if the other does not, don't swerve if the other player swerves, making this game's choices highly interdependent, which is characteristic of many games. The Nash equilibria in Chicken, moreover, are (4, 2) and (2, 4), suggesting that the compromise of (3, 3) will not be easy to achieve because both players will have an incentive to deviate in order to try to be the winner.

## Chicken

DEFINITION

**Chicken** is a two-person variable-sum game in which each player has two strategies: to swerve to avoid a collision or not to swerve and possibly cause a collision. Neither player has a dominant strategy. The compromise outcome, in which both players swerve, and the disaster outcome, in which both players do not, are not Nash equilibria. The other two outcomes, in which one player swerves and the other does not, are Nash equilibria.

In fact, there is a third Nash equilibrium in Chicken, but it is in mixed strategies, which can be computed only if the payoffs are not ranks, as we have assumed here, but numerical values. Even if the payoffs were numerical, however, it can be shown that this equilibrium is always worse for both players than the cooperative (3, 3) outcome. Moreover, it is implausible that players would sometimes swerve and sometimes not–randomizing according to particular probabilities–in the actual play of this game, compared with either trying to win outright or reaching a compromise.

The two pure-strategy Nash equilibria in Chicken suggest that, insofar as there is a "solution" to this game, it is that one player will succeed when the other caves in to avoid the mutual-disaster outcome. But there are certainly real-life cases in which a major confrontation was defused and a compromise of sorts was achieved in a Chicken-type game. This fact suggests that the one-sided solutions given by the two pure-strategy Nash equilibria may not be the only pure-strategy solutions, especially if the players are farsighted and think about the possible untoward consequences of their actions.

International crises, labor–management disputes, and other conflicts in which escalating demands may end in wars, strikes, and other catastrophic outcomes have been modeled by the game of Chicken (see Spotlight 15.2 for more on game theorists who have analyzed these and other games). But it can be shown that Chicken, like Prisoners' Dilemma, is only one of the 78 essentially different $2 \times 2$ ordinal games in which each player can rank the four possible outcomes from best to worst.

Chicken and Prisoners' Dilemma, however, are especially disturbing, because the cooperative (3, 3) outcome in each is not a Nash equilibrium. Unlike a constant-sum game, in which the losses of one player are offset by the gains of the other, *both* players can end up doing badly–at (2, 2) in Prisoners' Dilemma and (1, 1) in Chicken–in these variable-sum games.

## SPOTLIGHT 15.2 The Nobel Prize in Economics

The Nobel Memorial Prize in Economics was awarded to three game theorists in 1994, marking the 50th anniversary of the publication of von Neumann and Morgenstern's *Theory of Games and Economic Behavior* (see Spotlight 15.1).

- *John C. Harsanyi* (1920–2000) of the University of California, Berkeley, a Hungarian-American who emigrated from Hungary to Australia in 1950 and then to the United States in 1956. He is well known for extending game theory to the study of ethics and showing how societal institutions, each of whose members' satisfaction can be measured against that of others, choose among alternatives. His other major contribution was to give a precise definition to "incomplete information" in games in which players may be thought of as different types, and

*(continued on page 488)*

## The Nobel Prize in Economics *(continued)*

probabilities are assigned to each type.

- *John F. Nash, Jr.* (b. 1928) of Princeton University, an American mathematician who did path-breaking work on both noncooperative game theory (the Nash equilibrium is named after him) and cooperative game theory, especially on bargaining, in which axioms or assumptions are specified and a unique solution that satisfies these axioms is derived. Nash obtained his results in the early 1950s, when he was only in his 20s, after which he became mentally ill and was unable to work. Fortunately, he has made a remarkable recovery and has now resumed research.
- *Reinhard Selten* (b. 1930) of the University of Bonn, a German mathematician who proposed significant refinements in the concept of the Nash equilibrium that help to distinguish those that are most plausible in games (often there are many such equilibria, which creates a selection problem). Selten is also noted for pioneering work on developing game-theoretic models in evolutionary biology.

In 2005 mathematician Robert J. Aumann was awarded the prize for a variety of advances in cooperative and noncooperative game theory, and economist Thomas C. Schelling also was awarded the prize for his contributions to the study of conflicts involving promises, threats, and other kinds of commitments. In 2007 three economists, Leonid Hurwicz, Eric S. Maskin, and Roger B. Myerson received the prize for their work on "mechanism design," which analyzes auction, bargaining, voting and other procedures, especially the incentives of players using them to be truthful in their choices.

John C. Harsanyi

*(Olivier Laude/Gamma-Liaison.)*

John F. Nash, Jr.

*(Reuters/Bettmann.)*

Reinhard Selten

*(Bettmann/Corbis.)*

## 15.4 Larger Games

We have shown how to compute optimal pure and mixed strategies, and the values ensured by using them, in $2 \times 2$ constant-sum games. In $2 \times 2$ variable-sum games, we focused on Nash equilibria as a solution concept in Prisoners' Dilemma and Chicken, but we found that this notion of a stable outcome did not justify the choice of cooperative strategies in either of these games.

We turn next to a somewhat larger game, in which there are three players, each of whom can choose among three strategies, which is technically a $3 \times 3 \times 3$ game. In this game, we eliminate certain undesirable strategies, but in stages, to arrive at a Nash equilibrium that seems quite plausible.

If one of the three players has a dominant strategy in the $3 \times 3 \times 3$ game, we suppose this player will choose it, thereby reducing the game to a $3 \times 3$ game between the other two players. Of course, if no player has a dominant strategy in a three-person game, it cannot be reduced in this manner to a two-person game.

If this game is not one of total conflict, the minimax theorem, which guarantees players the value in a two-person zero-sum game, is not applicable. Even if the game were zero-sum, the fact that we assume the players in the $3 \times 3 \times 3$ game can only rank outcomes, not assign numerical values to them, means that they cannot calculate optimal mixed strategies in it.

The problem in finding a solution to the reduced $3 \times 3$ game is not a lack of Nash equilibria. Rather, there are too many! So the question becomes which, if any, are likely to be selected by the players. Specifically, is one more appealing than the others? The answer is "yes," but it requires extending the idea of dominance, discussed in the previous section, to its successive application in different stages of play.

## EXAMPLE 9 ■ The Status-Quo Paradox

The $3 \times 3 \times 3$ game we analyze involves voting, illustrating the applicability of game theory to politics. We will show that the status quo, or existing state of affairs, may be defeated by another alternative, despite its privileged position as the policy in place at the time.

To illustrate this problem, suppose there is a set of three voters, $V = \{X, Y, Z\}$, and a set of three alternatives, $A = \{x, y, z\}$, from which the voters choose. Assume that voter $X$ prefers $x$ to $y$ to $z$, indicated by $xyz$; voter $Y$'s preference is $yzx$; and voter $Z$'s is $zxy$. These preferences give rise to a *Condorcet voting paradox* (discussed in Chapter 9), because the social ordering, according to majority rule, is *intransitive:* Although a majority (voters $X$ and $Z$) prefer $x$ to $y$, and a majority (voters $X$ and $Y$) prefer $y$ to $z$, a majority (voters $Y$ and $Z$) prefer $z$ to $x$. So there is no **Condorcet winner**–an alternative that would beat all others in separate pairwise contests. Instead, every alternative can be beaten by one other.

Assume that the voting procedure used by the three voters, who choose from among the three alternatives, is the **plurality procedure**, under which the alternative with the most votes wins. If there is a three-way tie (there can never be a two-way tie if there are three voters), we assume that $x$ wins–because it is the status quo–giving $X$ what would appear to be an edge over the other two voters, $Y$ and $Z$.

To begin, assume that voting is **sincere**.

### Sincere Voting — DEFINITION

Under **sincere voting**, every voter votes for his or her most-preferred alternative, based on his or her true preferences, without taking into account what the other voters might do (see Chapter 9).

In this case, $x$ will prevail, because $X$ supports the status quo.

But note that $X$ has a dominant strategy of "vote for $x$": It is never worse and sometimes better than her other two strategies, whatever the other two voters do. Thus, if the other two voters vote for the same alternative, it wins, and $X$ cannot do better than vote sincerely for $x$, so voting sincerely is never worse. On the other hand, if the other two voters disagree, $X$'s tie-breaking vote (along with her regular vote) for $x$ will be decisive in $x$'s selection, which is $X$'s best outcome.

Given the dominant choice of $x$ on the part of $X$, $Y$ and $Z$ face the strategy choices shown in Figure 15.4. $Y$ has one, and $Z$ has two, **dominated strategies**, which are never better and sometimes worse than some other strategy, whatever the other two voters do. For example, observe that "vote for $x$" by $Y$ always leads to his worst alternative, $x$. The dominated strategies are crossed out in the top matrix in Figure 15.4.

This leaves $Y$ with two *undominated* strategies that are neither dominant nor dominated: "vote for $y$" and "vote for $z$." "Vote for $y$" is better than "vote for $z$" if $Z$ chooses $y$ (leading to $y$ rather than $x$), whereas the reverse is the case if $Z$ chooses $z$ (leading to $z$ rather than $x$). By contrast, $Z$ has a dominant strategy of "vote for $z$," which leads to outcomes at least as good and sometimes better than his other two strategies.

If voters have complete information about each other's preferences, then they can perceive the situation represented by the top matrix of Figure 15.4. Reasoning that no player would ever choose a dominated strategy, they would eliminate such strategies from consideration (these have been crossed out in the first reduction).

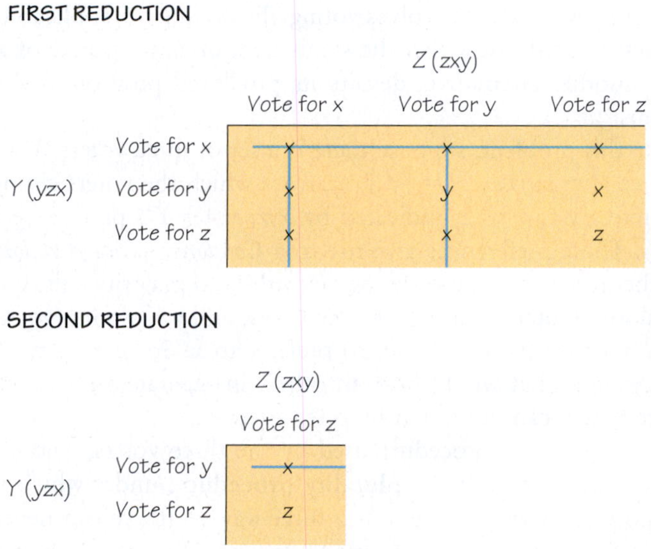


**FIGURE 15.4**
Sophisticated voting, given $X$ chooses "vote for $x$." The dominated strategies of each voter are crossed out in the first reduction, leaving two (undominated) strategies for $Y$ and one (dominant) strategy for $Z$. Given these eliminations, $Y$ would then eliminate "vote for $y$" in the second reduction, making $z$ the sophisticated outcome.

The elimination of these strategies gives the bottom matrix in Figure 15.4. Then $Y$, choosing between "vote for $y$" and "vote for $z$" in this matrix, would cross out "vote for $y$" (second reduction), now dominated because that choice would result in $x$'s winning because it is the status quo. Instead, $Y$ would choose "vote for $z$," which is not $Y$'s sincere strategy. This ensures $z$'s election, which is $Z$'s best outcome but only the next-best outcome for $Y$. In this manner, $z$, which is not the first choice of a majority and could in fact be beaten by $y$ in a pairwise contest, becomes the sophisticated outcome.

## Sophisticated Voting

DEFINITION

The *successive* elimination of dominated strategies by voters (insofar as this is possible), is **sophisticated voting**.

Sophisticated voting results in a Nash equilibrium, because none of the three players can do better by departing from his or her sophisticated strategy when the other two players choose theirs. This is clearly true for *X*, because *x* is her dominant strategy; given *X*'s choice of *x*, *z* is dominant for *Z*; and given these choices by *X* and *Z*, *z* is dominant for *Y*. These "contingent" dominance relations, in general, make sophisticated strategies a Nash equilibrium.

Observe, however, that there are four other Nash equilibria in this game. First, the choice of each of *x*, *y*, or *z* by all three voters are all Nash equilibria, because no single voter's departure can change the outcome to a different one, much less a better one, for that player. In addition, the choice of *x* by *X*, *y* by *Y*, and *x* by *Z*–resulting in *x*–is also a Nash equilibrium, because no voter's departure would lead to a better outcome for him or her.

In game-theoretic terms, sophisticated voting produces a different and smaller game in which some formerly undominated strategies in the larger game become dominated in the smaller game. The removal of such strategies, sometimes in several successive stages, in effect enables sophisticated voters to determine what outcomes eventually *will* be chosen by eliminating those outcomes that definitely *will not* be chosen. Voters can thereby ensure that their worst outcomes will not be chosen by successively eliminating dominated strategies, given the presumption that other voters do likewise.

How does sophisticated voting affect the status quo? If *X* chooses its dominant strategy of voting for *x*, the status quo, and its best outcome, it leads to *X*'s worst outcome (*z*)!

**Status-Quo Paradox** DEFINITION

This situation, in which supporting the apparently favored outcome hurts, is the **status-quo paradox.**

Clearly, the apparent advantage that the status quo enjoys in the absence of a Condorcet winner disappears if the voters are sophisticated. The strategic situation intervenes and causes them to reassess their sincere strategies in light of the favored position of the status quo. In so doing, they may be led to "gang up" against *X*, handing *X* her worst outcome of *z*.

We stress that *Y* and *Z* do not form a coalition against *X* in the sense of coordinating their strategies and agreeing to act together in a way that can be enforced. Rather, they behave as isolated individuals. At most they could be said to form an "implicit coalition." Such a coalition does not imply communication between its members but simply choices based on their common perceived strategic interests.

So far we have used payoff matrices to describe games in strategic form. In these games, the row and column players' choices of strategies led to an outcome from which each player received a payoff. These strategy choices were assumed to be simultaneous, though mentally the players might eliminate some sequentially in the manner illustrated.

In the next example of a larger game, we start by assuming simultaneous choices and show what outcome would occur. Then we assume that the choices of the players need not be simultaneous; it is possible for one player to move first. We will use a **game tree** to analyze the *sequential choices* players can then make, as occurs when first you move, then I move, and so on, which are called *games in extensive form.* As we will see, the outcome in such a game may be wholly different from what it is in a game with simultaneous choices, which raises the question of which kind of game is the most realistic model of a situation.

## EXAMPLE 10 ▪ A Truel

A *truel* is like a duel, except that there are three players. Truels are depicted in several movies, including *The Good, the Bad and the Ugly* (1966), *Reservoir Dogs* (1992), and *Pulp Fiction* (1994).

Each player can either fire, or not fire, his or her gun at either of the other two players. We assume the goal of each player is, *first,* to survive and, *second,* to survive with as few other players as possible. Each player has one bullet and is a perfect shot; no communication (for example, to pick out a common target) leading to a binding agreement with other players is allowed, making the game noncooperative. We will discuss the answers that simultaneous choices, on the one hand, and sequential choices, on the other, give to what is optimal for the players to do in the truel.

If choices are simultaneous, *at the start of play, each player will fire at one of the other two players, killing that player.*

Why will the players all fire at each other? Because their own survival does not depend an iota on what they do. Since they cannot affect what happens to themselves, but can affect how many others survive (the fewer the better, according to the postulated secondary goal), they should all blaze away at each other. Even if the rules of the play permitted shooting oneself, the primary goal of survival would preclude committing suicide. In fact, the players all have dominant strategies to shoot at each other, because whether or not a player survives–we will discuss shortly the probabilities of doing so–he or she does at least as well shooting an opponent.

The game, and optimal strategies in it, would change if the players (1) were allowed more options, such as to fire in the air and thereby disarm themselves, or (2) did not have to choose simultaneously but, instead, a particular order of play were specified. Thus, if the order of play were *A,* followed by *B* and *C* choosing simultaneously, followed by any player with a bullet remaining choosing, then *A* would fire in the air, and *B* and *C* would subsequently shoot each other. (*A* is no threat to *B* or *C,* so neither of the latter will fire at *A* and waste a bullet; on the other hand, if one of *B* or *C* did not fire immediately at the other, that player would not survive to get in the last shot, so both *B* and *C* will fire at each other.) Thus, *A* will be the sole survivor.

David Letterman (*Black Star.*)

Jay Leno (*Black Star.*)

Ted Koppel (*Syracuse Newspapers/Dick Blume/The Image Works.*)

In 1992, a modified version of this scenario was played out in late-night television programming among the three major TV broadcasting networks of the time, with ABC's effectively going first with *Nightline,* its well-established news program, and CBS's and NBC's dueling about which host, David Letterman or Jay Leno, to choose for their entertainment shows. Regardless of their ultimate choices, ABC "won" when CBS and NBC were forced to divide the entertainment audience. In 2002, ABC, presumably to attract a younger audience than that which watched *Nightline,* attempted unsuccessfully to hire Letterman from CBS. Subsequently, Ted Koppel retired, but *Nightline* continues.

To return to the original game (all choose simultaneously), the players' strategies of all firing have two possible consequences: Either one player survives (even if two players fire at the same person, the third must fire at one of them, leaving only one survivor), or no player survives (if each payer fires at a different person). In either event, there is no guarantee of survival. In fact, if each player has an equal probability of firing at one of the two other players, the probability that any particular player will survive is only 0.25.

The reason is that if the three players are *A, B,* and *C, A* will be killed when *B* fires at him or her, *C* does, or both do. The only circumstance in which *A* will survive is if *B* and *C* fire at each other, which gives *A* one chance in four.

*If choices are sequential, no player will fire at any other, so all will survive.*

At the start of the truel, all the players are alive, which satisfies their primary goal of survival, though not their secondary goal of surviving with as few others as possible. Now assume that *A* contemplates shooting *B,* thereby reducing the number of survivors, and cannot fire into the air. Looking ahead, however, *A* knows that by firing first and killing *B,* he or she will be defenseless and be immediately shot by *C,* who will then be the sole survivor.

It is in *A*'s interest, therefore, not to shoot anybody at the start, and the same logic applies to each of the other players. Hence, everybody will survive, which is a happier outcome than when choices are simultaneous, in which case everyone's primary goal of survival is not satisfied–or, quantitatively speaking, satisfied only 25% of the time.

While sequential choices produce a "happier" outcome, do they provide a plausible model of a strategic situation that mimics what people might actually think and do in such a situation? We believe that the players in the truel, artificial as this kind of shoot-out may seem, would be motivated to think ahead, given the dire consequences of their actions. Therefore, they would hold their fire, knowing that if one fired first, he or she would be the next target.

In Figure 15.5, we show this logic somewhat more formally with a *game tree,* in which *A* has three strategies, as indicated by the three branches that sprout from *A:* not shoot ($\bar{S}$), shoot *B,* or shoot *C.* The latter two branches, in turn, give survivors *C* and *B,* respectively, two strategies: not shoot ($\bar{S}$) or shoot *A.*

We assume that the players rank the outcomes as follows, which is consistent with their primary and secondary goals: 4 = best (lone survivor), 3 = next best (survivor with one other), 2 = next worst (survivor with two others), and 1 = worst (nonsurvivor). These payoffs are given for ordered triples (*A, B, C*); thus (3, 3, 1) indicates the next-best payoffs for *A* and *B* and the worst payoff for *C.*

Note that play necessarily terminates when there is only one survivor, as is the case at (1, 1, 4) and (1, 4, 1). To keep the tree simple, we assume that play also terminates when either *A* initially chooses $\bar{S}$, or *B* or *C* subsequently chooses $\bar{S}$, giving outcomes of (2, 2, 2), (3, 3, 1), and (3, 1, 3), respectively. Of course, we could allow the two or three surviving players in the latter cases to make subsequent choices in an extended game tree, but this example is meant only to illustrate the analysis of a game tree, not be the definitive statement on truel possibilities (more will be explored in the exercises).

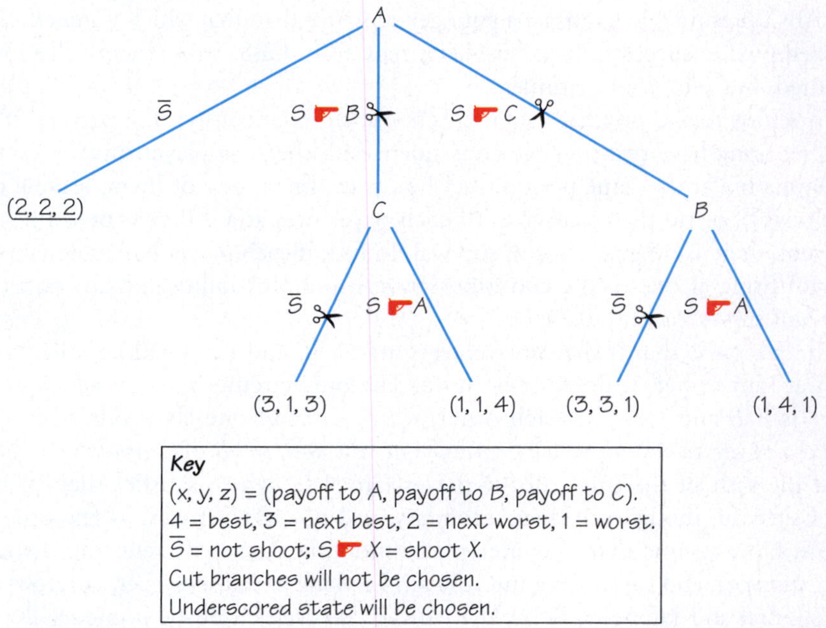


**FIGURE 15.5** A game tree of a truel.

In a game in extensive form represented by a game tree, players work backward, starting the analysis at the bottom of the tree. By "bottom" we mean where play terminates; because this is where the tree branches out, the tree looks upside down in Figure 15.5. The players then work up the tree, using backward induction. **Backward induction** is a reasoning process in which players, working backward from the last possible moves in a game, anticipate each other's rational choices.

To illustrate, because $C$ prefers (1, 1, 4) to (3, 1, 3), we indicate that $C$ would not choose $\bar{S}$ by "cutting" this branch with a scissors. Similarly, $B$ would not choose $\bar{S}$. Thus, if play got to the bottom of the tree, $C$ would shoot $A$ if $C$ were the survivor, and $B$ would shoot $A$ if $B$ were the survivor, following $A$'s shooting $C$ or $B$, respectively.

Moving up to the next level, $A$ would know that if he or she shot $B$, (1, 1, 4) would be the outcome. If he or she shot $C$, (1, 4, 1) would be the outcome, making one or the other the outcome from the bottom level. Choosing between these two outcomes and (2, 2, 2), $A$ would prefer the latter, so $A$ would cut the two branches, $S \rightarrow B$ and $S \rightarrow C$. Hence, $A$ would choose $\bar{S}$, terminating play with nobody shooting anybody else.

This, of course, is the conclusion we reached earlier, based on the reasoning that if $A$ shot either $B$ or $C$, he or she would end up dead, too. Because we could allow each player, like $A$, to choose among his or her three initial strategies in a $3 \times 3 \times 3$ game, and subsequently make moves and countermoves from the initial state (if feasible), the foregoing analysis applies to all players.

Underlying the completely different answers given by the simultaneous and sequential choices in a truel is a change in the rules of play. If play is sequential, the players do not have to fire simultaneously at the start, as assumed in a 3 × 3 × 3 strategic-form game. Rather, a player who moves first (*A* in our example)–and then the later players–would not fire, given that play continues until all bullets are expended or nobody chooses to fire.

In the extensive-form analysis, we ask of each player (it need not be *A*): Given your present situation (all alive), and the situation you anticipate will ensue if you fire first, should you do so? Because each player prefers living to the state he or she would induce by being the first to shoot (certain death), no one shoots. This analysis suggests that truels might be more effective than duels, at least when played sequentially, in preventing the outbreak of conflict.

We will not try to develop this argument into a more general model. The main point is that a game tree allows for a look-ahead approach whereby players compare the present state with possible future states–perhaps several steps ahead–to determine which moves to make. These choices, as we have seen, may lead to radically different outcomes compared with those based on simultaneous choices.

## 15.5 Using Game Theory

### Solving Games

Given any payoff matrix, the first thing we ask is whether it is zero-sum (or constant-sum). If so, we check to see whether it has a saddlepoint by determining the minimum number of each row and the maximum number of each column, as we did in several earlier examples. If the maximum of the row minima (maximin) is equal to the minimum of the column maxima (minimax), then the game has a saddlepoint. The resulting value, and the corresponding pure strategies, provide a solution to the game.

This value will appear in the payoff matrix as the smallest number in its row and the largest in its column. In the 3 × 3 location game in Example 1, this number was 5 (5000 feet).

As in our voting game, dominated strategies can successively be eliminated in the 3 × 3 location game. Thus, route *B* dominates route *C*, and Highway 2 dominates Highway 3. Having made these eliminations, Highway 2 dominates Highway 1. Having made this elimination, route *B* dominates route *A*. Thus, Highway 2 and route *B* survive the successive eliminations, yielding the saddlepoint of 5. The successive-elimination procedure therefore provides an alternative method for finding the saddlepoint in the 3 × 3 location game. Unfortunately, it does not work to find the saddlepoint in *all* two-person zero-sum games larger than 2 × 2.

Recall that instead of eliminating dominated strategies in the 3 × 3 location game, we eliminated route *B* and Highway 2, which dominated other strategies, to obtain the 2 × 2 restricted-location game in Table 15.3. In this game, there were no dominated strategies and, hence, no saddlepoint.

If a two-person zero-sum game does not have a saddlepoint–which was the case not only in the restricted-location game but also for matching pennies, the nonsymmetrical matching game, and the baseball duel–the solution will be in mixed strategies. To find the optimal mix in a 2 × 2 game, we calculate the expected value to a player from choosing its first strategy with probability $p$ and its second with probability $1 - p$, assuming that the other player chooses its first pure strategy (yielding one expected value) and its second pure strategy (yielding another expected value).

Setting these two expected values equal to each other yields a unique value for $p$ that gives the optimal mix, $(p, 1 - p)$, with which the player should choose its first and second strategies. Substituting the numerical solution of $p$ back into either expected-value equation gives the value of the game, which each player can guarantee for itself, whatever strategy its opponent chooses.

Several general algorithms have been developed to find mixed-strategy solutions to large constant-sum games. This work has mostly been done in the field of linear programming, using such algorithms as the simplex method of G. B. Dantzig and the more recent method of N. K. Karmarkar (see Chapter 4).

In variable-sum games, we also begin by successively eliminating dominated strategies, if there are any. The outcomes that remain do not depend on the numerical values we attach to them but only on their ranking from best to worst by the players, as illustrated in the three-person voting example.

Care must be taken in interpreting this solution, however. It began with the choice of a dominant strategy by $X$–and her elimination of her two dominated strategies. Presuming these eliminations, $Y$ and $Z$ were then able to eliminate their own dominated strategies in the first reduction, and $Y$ in turn eliminated a dominated strategy in the second reduction, leading finally to the outcome $z$, supported by $Y$ and $Z$.

This solution is a fairly demanding one, because it assumes considerable calculational abilities on the part of the players. Less demanding, of course, is that players simply choose their dominant strategies, as is possible in Prisoners' Dilemma, but of course games may not have such strategies.

In the game of Chicken, for example, neither player has a dominant (or dominated) strategy, so the game cannot be reduced. In such situations, we ascertain what outcomes are Nash equilibria. There are two (in pure strategies) in Chicken, suggesting that the only stable outcomes in this game occur when one player gives in and the other does not. In Prisoners' Dilemma, by comparison, the choice by the players of their dominant strategies singles out the mutual-defection outcome as the unique Nash equilibrium, which is worse for both players than the cooperative outcome.

In both Chicken and Prisoners' Dilemma, there seems no good reason for the choice of the (3, 3) cooperative outcome, at least if each game is played only once, because this outcome is not a Nash equilibrium. However, there is an alternative theory, called the **theory of moves**, that assumes different rules of play and renders the cooperative outcomes in both Prisoners' Dilemma and Chicken stable, given that the players think ahead.

### Theory of Moves — DEFINITION

**Theory of moves (TOM)** is a dynamic theory that describes optimal strategic choices in strategic-form games in which the players, thinking ahead, can make moves and countermoves.

TOM is *dynamic* in the sense that it allows players, after choosing strategies that lead to an outcome in a payoff matrix, to make subsequent alternating moves and countermoves, with row's being able to move vertically by changing his row strategy, and column's being able to move horizontally by changing her column strategy. The reasoning that the players use in deciding whether to move or not move is backward induction, as illustrated earlier in the truel.

We informally illustrate this reasoning in Prisoners' Dilemma, starting from each of the four possible outcomes:

- If the play starts at the noncooperative (2, 2) outcome, players are stuck, no matter how far ahead they look, because as soon as one player departs, the other player, enjoying its best outcome at (4, 1) or (1, 4), will not move on. *Result:* The players stay at the noncooperative outcome.
- If play starts at the cooperative (3, 3) outcome, then neither player will defect, because if he or she does, the other player will also defect, and both players will end up worse off at (2, 2). Thinking ahead, therefore, neither player will defect. *Result:* The players stay at the cooperative outcome.
- If play starts at one of the (4, 1) or (1, 4) win–lose outcomes, the player doing best (4) will know that if he or she does not move to the cooperative (3, 3) outcome, his or her opponent will move to the noncooperative (2, 2) outcome, inflicting on the best-off player a next-worst (2) outcome. Therefore, it is in this player's interest–as well as the worst-off player's interest–that the best-off player act cooperatively and move first to (3, 3), anticipating that if he or she does not, the (2, 2) rather than the (3, 3) outcome will be chosen. *Result:* The best-off player will move to the cooperative outcome, where play will stop.

Thus, TOM does not predict unconditional cooperation in Prisoners' Dilemma but, instead, makes it a function of the starting point of play. As in the truel, a change in rules from simultaneous choices to sequential choices can induce cooperation.

The calculations we have described for Prisoners' Dilemma, which are grounded in backward induction and could be formalized by a game tree, are not, we believe, beyond the ability of most players. Farsighted players *can* escape the dilemma in Prisoners' Dilemma, provided play begins at a state other than the noncooperative one. But we must be careful in interpreting this result: With the change in the rules, the original game changes, so this "solution" to Prisoners' Dilemma is not for the original dilemma. Similar reasoning in Chicken indicates that if play starts at the cooperative (3, 3) outcome, players will stay at this outcome, but the reasoning in this game is somewhat more complicated than in Prisoners' Dilemma.

## Practical Applications

The element of surprise, as captured by mixed strategies, is essential in many encounters. For example, mixed strategies are used in various inspection procedures and auditing schemes to deter potential violators. By making inspection or auditing choices random, they are rendered unpredictable.

Police and regulatory agencies monitor certain activities to check for illegal actions. Investigators who conduct surveillance include FBI agents, customs agents, bank auditors, insurance investigators, quality-control experts, and drug testers. The National Bureau of Standards is responsible for monitoring the accuracy of measuring instruments and for maintaining reliable standards. The Nuclear Regulatory Agency demands an accounting of dangerous nuclear material as part of its safeguards program. The Internal Revenue Service attempts to identify those cheating on taxes.

Military or intelligence services may wish to intercept a weapon hidden among many decoys, or plant a secret agent disguised to look like a respectable individual. Because it is prohibitively expensive to check the authenticity of each and every possible item or person, efficient methods must be used to check for violations. Both optimal detection and optimal concealment strategies can be modeled as a game between an inspector trying to increase the probability of detection and a violator trying to evade detection. Since the World Trade Center attack on September 11, 2001, we have seen government agencies take much stronger measures to prevent such evasion.

Some inspection games are constant-sum: The violator "wins" when the evasion is successful and "loses" when it is not. On the other hand, cheating on arms-control agreements may well be variable-sum if both the inspector and the cheater would prefer that no cheating occur to there being cheating and public disclosure of it. The latter could be an embarrassment to both sides, especially if it undermines an arms-control agreement both sides wanted and the cheating is minor.

We alluded earlier to the strategy of bluffing in poker, which is used to try to keep the other player or players guessing about the true nature of one's hand. The optimal probability with which one should bluff can be calculated in a particular situation (see Exercise 17). Besides poker, bluffing is common in many bargaining situations, whereby a player raises the stakes (for instance, labor threatens a strike in labor–management negotiations), even if it may ultimately have to back down if its "hand is called."

Perhaps the greatest value of game theory is the framework it provides for understanding the rational underpinnings of conflict in the world today. As a case in point, a confrontation over the budget between the Democratic President Bill Clinton and the Republican Congress resulted in the shutdown of part of the federal government on two occasions between November 1995 and January 1996. Many government workers were frustrated in not being able to do their jobs, even though they knew they would be paid for not working, not to mention the many citizens who were either greatly hurt or substantially inconvenienced by the shutdown. Viewed as a game of Chicken, in which each side wanted not only to get its way for the moment but also to establish a precedent for the future, this conflict was not so foolish as it might seem at first glance.

(*Najlah Feanny/Corbis.*)

The Northern Ireland conflict–settled in principle by a peace agreement in April 1998 after 30 years of fighting and more than 3200 deaths–can be viewed in similar terms. As still another example, the constant price wars among the airlines suggest competitors caught up in a Prisoners' Dilemma, in which they all suffer from lower fares but cannot avoid their dominant strategies of not cooperating, perhaps to try to seize a quick advantage or hurt the competition even more (and possibly even eliminate a competitor). This has led to the bankruptcy of such major airlines as United Airlines and US Airways, though each eventually emerged from bankruptcy.

To be sure, if the airlines cooperate by colluding on fares, which is definitely not advantageous to consumers, the consumers may be thought of as a collective player whose interests are represented by the government. The government can prosecute the airlines for price fixing, or the consumers themselves can file a class-action suit in a "larger" game. The government has frequently been involved in antitrust suits (for example, against Microsoft) and in setting the rules for auctions of airwaves, in which telecommunication companies–advised by game theorists–have paid billions of dollars for the right to construct cellular phone and other networks.

All in all, game theory offers fundamental insights into conflicts at all levels, especially *seemingly* irrational features which, on second look, are often well conceived and effective.

## REVIEW VOCABULARY

**Backward induction** A reasoning process in which players, working backward from the last possible moves in a game, anticipate each other's rational choices. (p. 494)

**Chicken** A two-person variable-sum symmetric game in which each player has two strategies: to swerve to avoid a collision, or not to swerve and cause a collision if the opponent has not swerved. Neither player has a dominant strategy; the compromise outcome, in which both players swerve, is not a Nash equilibrium, but the two outcomes in which one player swerves and the other does not are Nash equilibria. (p. 486)

**Condorcet winner** A candidate that defeats all others in separate pairwise contests. (p. 489)

**Constant-sum game** A game in which the sum of payoffs to the players at each outcome is a constant, which can be converted to a zero-sum game by an appropriate change in the payoffs to the players that does not alter the strategic nature of the game. (p. 480)

**Dominant strategy** A strategy that is sometimes better and never worse for a player than every other strategy, whatever strategies the other players choose. (p. 484)

**Dominated strategy** A strategy that is sometimes worse and never better for a player than some other strategy, whatever strategies the other players choose. (p. 490)

**Expected value** $E$ If each of the $n$ possible payoffs, $s_1, s_2, \ldots, s_n$, occurs with respective probabilities $p_1, p_2, \ldots, p_n$, then the expected value $E$ is

$$E = p_1s_1 + p_2s_2 + \cdots + p_ns_n$$

where $p_1 + p_2 + \cdots + p_n = 1$ and $p_i \geq 0$ $(i = 1, 2, \ldots, n)$. (p. 476)

**Fair game** A zero-sum game is fair when the (expected) value of the game, obtained by using optimal strategies (pure or mixed), is zero. (p. 478)

**Game tree** A symbolic tree, based on the rules of play in a game, in which the vertices, or nodes, of the tree represent choice points, and the branches represent alternative courses of action that the players can select. (p. 491)

**Maximin** In a two-person zero-sum game, the largest of the minimum payoffs in each row of a payoff matrix. (p. 471)

**Maximin strategy** In a two-person zero-sum game, the pure strategy of the row player corresponding to the maximin in a payoff matrix. (p. 471)

**Minimax** In a two-person zero-sum game, the smallest of the maximum payoffs in each column of a payoff matrix. (p. 471)

**Minimax strategy** In a two-person zero-sum game, the pure strategy of the column player corresponding to the minimax in a payoff matrix. (p. 471)

**Minimax theorem** The fundamental theorem for two-person constant-sum games, stating that there always exist optimal pure or mixed strategies that enable the two players to guarantee the value of the game. (p. 482)

**Mixed strategy** A strategy that involves the random choice of pure strategies, according to particular probabilities. A mixed strategy of a player is optimal if it guarantees the value of the game. (p. 475)

**Mixed-strategy value** A player's expected value is the mixed-strategy value of the game. Unlike the use of this notion in games with a saddlepoint, the value here can be realized only by the use of mixed strategies. (p. 477)

**Nash equilibrium** Strategies associated with an outcome such that no player can benefit by choosing a different strategy, given that the other players do not depart from their strategies. (p. 484)

**Nonsymmetrical game** A two-person constant-sum game in which the row player's gains are different from the column player's gains, except when there is a tie. (p. 478)

**Ordinal game** A game in which the players rank the outcomes from best to worst. (p. 483)

**Partial-conflict game** A variable-sum game in which both players can benefit by cooperation but may have strong incentives not to cooperate. (p. 409)

**Payoff matrix** A rectangular array of numbers. In a two-person game, the rows and columns correspond to the strategies of the two players, and the numerical entities give the payoffs to the players when these strategies are selected. (p. 477)

**Plurality procedure** A voting procedure in which the alternative with the most votes wins. (p. 489)

**Prisoners' Dilemma** A two-person variable-sum symmetric game in which each player has two strategies, cooperate or defect. Cooperate dominates defect for both players, even though the mutual-defection outcome, which is the unique Nash equilibrium in the game, is worse for both players than the mutual-cooperation outcome. (p. 482)

**Prisoners' Dilemma (Story)** Prisoners' Dilemma involves two persons, accused of a crime, who are held incommunicado. Each has two choices: to maintain his or her innocence, or to sign a confession accusing the partner of committing the crime. It is in each suspect's interest to confess and implicate the partner, thereby trying to receive a reduced sentence. Yet if both suspects confess, they ensure a bad outcome–namely, they are both found guilty. What is good for the prisoners as a pair–to deny having committed the crime, leaving the state with insufficient evidence to convict them–is frustrated by their pursuit of their own individual interests. (p. 483)

**Pure strategy** A course of action a player can choose in a game that does not involve randomized choices. (p. 476)

**Rational choice** A choice that leads to a preferred outcome. (p. 467)

**Saddlepoint** In a two-person constant-sum game, the payoff that results when a row minimum and a column maximum are the same, which is the value of the game. The saddlepoint has the shape of a saddle-shaped surface and is also a Nash equilibrium. (p. 471)
**Sincere voting** Voting for one's most-preferred alternative in a situation. (p. 489)
**Sophisticated voting** Voting that involves the successive elimination of dominated strategies by voters. (p. 490)
**Status-quo paradox** The status quo is defeated by another alternative, even if there is no Condorcet winner, when voters are sophisticated. (p. 491)
**Strategy** One of the courses of action a player can choose in a game; strategies are mixed or pure, depending on whether they are selected in a randomized fashion (mixed) or not (pure). (p. 467)
**Theory of moves (TOM)** A dynamic theory that describes optimal choices in strategic-form games in which players, thinking ahead, can make moves and countermoves. (p. 496)
**Total-conflict game** A zero-sum or constant-sum game, in which what one player wins the other player loses. (p. 469)
**Value** The best outcome that both players can guarantee in a two-person zero-sum game. If there is a saddlepoint, this is the value. Otherwise, it is the expected payoff resulting when the players choose their optimal mixed strategies. (p. 472)
**Variable-sum game** A game in which the sum of the payoffs to the players at the different outcomes varies. (p. 482)
**Zero-sum game** A constant-sum game in which the payoff to one player is the negative of the payoff to the other player, so the sum of the payoffs to the players at each outcome is zero. (p. 480)

## ✓ SKILLS CHECK

**1.** In the following two-person zero-sum game, the payoffs represent gains to row player I and losses to column player II.

$$\begin{bmatrix} 3 & 7 & 2 \\ 8 & 5 & 1 \\ 6 & 9 & 4 \end{bmatrix}$$

What is the maximin strategy for player I?

**(a)** Play the first row.
**(b)** Play the second row.
**(c)** Play the third row.

**2.** In the following two-person zero-sum game, the payoffs represent gains to row player I and losses to column player II.

$$\begin{bmatrix} 3 & 7 & 2 \\ 8 & 5 & 1 \\ 6 & 9 & 4 \end{bmatrix}$$

The minimax strategy for player II is to play the ______ column.

**3.** In a two-person zero-sum game, suppose the first player chooses the third row as the maximin strategy and the second player chooses the first column as the minimax strategy. Based on this information, which of the following statements is true?

**(a)** This game definitely has no saddlepoint.
**(b)** This game may or may not have a saddlepoint.
**(c)** This game definitely has a saddlepoint.

For Exercises 4–6, consider the following three two-person zero-sum games, wherein the payoffs represent gains to the row player I and losses to the column player II:

$$\begin{bmatrix} 3 & 6 \\ 5 & 4 \end{bmatrix}$$

$$\begin{bmatrix} -1 & 3 \\ 2 & 0 \end{bmatrix}$$

$$\begin{bmatrix} 6 & 5 & 6 & 5 \\ 1 & 4 & 2 & -1 \\ 8 & 5 & 7 & 5 \\ 0 & 2 & 6 & 2 \end{bmatrix}$$

**4.** The ______________ games have a saddlepoint(s)?

**5.** In which two games does neither player have a dominant strategy?

**(a)** The first two games
**(b)** The last two games
**(c)** The first and third games

**6.** ____________ strategies of player I are dominated in the third game.

**7.** In the game of matching pennies, player I wins a penny if the coins match; player II wins a penny if the coins do not match. Given this information, it can be concluded that the $2 \times 2$ matrix that represents this game

**(a)** has two $-1$'s and two 1's.
**(b)** has four 1's.
**(c)** has four $-1$'s.

**8.** If a game has a saddlepoint, then ____________ is the value of the game.

**9.** Which of these games does not have a saddlepoint?

**(a)** Tic-tac-toe
**(b)** Chess
**(c)** Poker

**10.** A mixed strategy uses randomization to ______________________________.

**11.** A game is fair when

**(a)** its value is 0.
**(b)** it requires a mixed strategy.
**(c)** its payoff is unlimited.

**12.** If a game is symmetrical, then _______________ .

**13.** In the following game of batter-versus-pitcher in baseball, the batter's batting averages are given in the game matrix.

| | | *Pitcher* Fastball | Curve |
|---|---|---|---|
| *Batter* | **Fastball** | 0.400 | 0.200 |
| | **Curve** | 0.100 | 0.500 |

What is the pitcher's optimal strategy?

**(a)** Throw more curves than fastballs.
**(b)** Throw more fastballs than curves.
**(c)** Throw about the same number of curves and fastballs.

**14.** In the following game of batter-versus-pitcher in baseball, the batter's batting averages are given in the game matrix.

| | | *Pitcher* Fastball | Curve |
|---|---|---|---|
| *Batter* | **Fastball** | 0.400 | 0.200 |
| | **Curve** | 0.100 | 0.500 |

The batter's exact optimal strategy is _______________ .

**15.** Consider the following partial-conflict game, played in a noncooperative manner.

| | | *Player II* Choice *A* | Choice *B* |
|---|---|---|---|
| *Player I* | Choice *A* | (4, 4) | (1, 3) |
| | Choice *B* | (3, 1) | (2, 2) |

What outcomes constitute a Nash equilibrium?

**(a)** Only when both players select *A*
**(b)** Only when both players select *A* or both select *B*
**(c)** Only when one player selects *A* and the other selects *B*

**16.** Consider the game played between the opposing goalie and a soccer player who, after a penalty, is allowed a free kick. The kicker can elect to kick toward one of the two corners of the net, or else aim for the center of the goal. The goalie can decide to commit in advance (before the kicker's kick) to either one of the sides, or else remain in the center until he sees the direction of the kick. This two-person zero-sum game can be represented as follows, wherein the payoffs are the probability of scoring a goal:

| | | *Goalie* Breaks left | Remains center | Breaks right |
|---|---|---|---|---|
| *Kicker* | Kicks left | 0.5 | 0.9 | 0.9 |
| | Kicks center | 0.1 | 0 | 0.1 |
| | Kicks right | 0.9 | 0.9 | 0.5 |

If we assume that decisions between the left or right side are made symmetrically (with equal probabilities), then this game can be represented by a 2 × 2 matrix, where $0.7 = (\frac{1}{2})(0.5) + (\frac{1}{2})(0.9)$:

| | | *Goalie* Remains center | Breaks side |
|---|---|---|---|
| *Kicker* | Kicks center | 0 | 1 |
| | Kicks side | 0.9 | 0.7 |

The exact optimal strategies of the kicker and goalie are ______________________________ .

**17.** In the following game, player I has the preferences of the row player in Prisoners' Dilemma, and player II has the preferences of the column player in Chicken.

| | | *Player II* Choice *A* | Choice *B* |
|---|---|---|---|
| *Player I* | Choice *A* | (3, 3) | (1, 4) |
| | Choice *B* | (4, 2) | (2, 1) |

Does the player with a dominant strategy benefit from having it?

**(a)** Yes
**(b)** No
**(c)** It doesn't make any difference–both players do equally well from choosing their strategies associated with the Nash-equilibrium outcome.

**18.** In games of partial conflict, cooperation is always rational when _____________________ .

**19.** A game tree is used to

**(a)** determine the possible strategies of a player.
**(b)** anticipate each other's choices through backward induction.
**(c)** plan a deception strategy.

**20.** Theory of moves is dynamic because ___________ .

# CHAPTER 15 EXERCISES

■ Challenge ◆ Discussion

## 15.1 Two-Person Total-Conflict Games: Pure Strategies

Consider the following five two-person zero-sum games, wherein the payoffs represent gains to the row player I and losses to the column player II:

**1.** $\begin{bmatrix} 6 & 5 \\ 4 & 2 \end{bmatrix}$

**2.** $\begin{bmatrix} 0 & 3 \\ -5 & 1 \\ 1 & 6 \end{bmatrix}$

**3.** $\begin{bmatrix} -2 & 3 \\ 1 & -2 \end{bmatrix}$

**4.** $\begin{bmatrix} 13 & 11 \\ 12 & 14 \\ 10 & 11 \end{bmatrix}$

**5.** $\begin{bmatrix} -10 & -17 & -30 \\ -15 & -15 & -25 \\ -20 & -20 & -20 \end{bmatrix}$

**(a)** Which of these games have saddlepoints?
**(b)** Find the maximin strategy of player I, the minimax strategy of player II, and the value for those games given in part (a).
**(c)** List dominated strategies in these games that the players should avoid because the resulting payoffs are worse than those for some alternative strategy.

## 15.2 Two-Person Total-Conflict Games: Mixed Strategies

Solve the following three games of batter-versus-pitcher in baseball, wherein the pitcher can throw one of two pitches and the batter can guess either of these two pitches. The batter's batting averages are given in the game matrix.

**6.**

| | | *Pitcher* | |
|---|---|---|---|
| | | **Fastball** | **Curve** |
| *Batter* | **Fastball** | 0.300 | 0.200 |
| | **Curve** | 0.100 | 0.400 |

**7.**

| | | *Pitcher* | |
|---|---|---|---|
| | | **Fastball** | **Knuckleball** |
| *Batter* | **Fastball** | 0.500 | 0.200 |
| | **Knuckleball** | 0.200 | 0.300 |

**8.**

| | | *Pitcher* | |
|---|---|---|---|
| | | **Blooperball** | **Knuckleball** |
| *Batter* | **Blooperball** | 0.400 | 0.200 |
| | **Knuckleball** | 0.250 | 0.250 |

**9.** A businessperson has the choice of either not cheating on his income tax or cheating and making \$1000 if not audited. If caught cheating, he will pay a fine of \$2000 in addition to the \$1000 he owes. He feels good if he does not cheat and is not audited (worth \$100). If he does not cheat and is audited, he evaluates this outcome as −\$100 (for the lost day). Viewing the game as a two-person zero-sum game between the businessperson and the tax agency, what are the optimal mixed strategies for each player and the value of the game?

**10.** When it is third down and short yardage to go for a first down in American football, the quarterback can decide to run the ball or pass it. Similarly, the other team can commit itself to defend more heavily against a run or a pass. This can be modeled as a 2 × 2 matrix game, wherein the payoffs are the probabilities of obtaining a first down. Find the solution of this game.

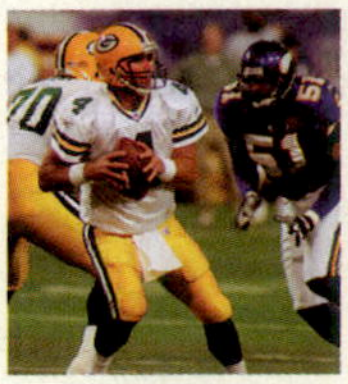
(Layne Kennedy/Corbis.)

| | | *Defense* | |
|---|---|---|---|
| | | **Run** | **Pass** |
| *Offense* | **Run** | 0.5 | 0.8 |
| | **Pass** | 0.7 | 0.2 |

**11.** You have the choice of either parking illegally on the street or parking in the lot and paying \$16. Parking illegally is free if the police officer is not patrolling, but you receive a \$40 parking ticket if she is. However, you are peeved when you pay to park in the lot on days when the officer does not patrol, and you are willing to assess this outcome as costing \$32 (\$16 for parking plus \$16 for your time, inconvenience, and grief). It seems reasonable to assume that the police officer ranks her preferences in the order (1) giving you a ticket, (2) not patrolling with you parked in the lot, (3) patrolling with you in the lot, and (4) not patrolling with you parked illegally.

**(a)** Describe this as a matrix game, assuming that you are playing a zero-sum game with the officer.
**(b)** Solve this matrix game for its optimal mixed strategies and its value.
◆ **(c)** Discuss whether it is reasonable or not to assume that this game is zero-sum.

◆ **(d)** Assuming that you play this parking game each working day of the year, how do you implement an optimal mixed strategy?

**12.** Describe how a pure strategy for a player in a matrix game can be considered as merely a special case of a mixed strategy.

■ **13. (a)** Describe in detail *one* pure strategy for the player who moves first in the game of tic-tac-toe. This strategy must tell how to respond to all possible moves of the other player. (*Hint:* You may wish to make use of the symmetry in the $3 \times 3$ grid in this game; that is, there are one "center" box, four "corner" boxes, and four "side" boxes.)
**(b)** Is your strategy optimal in the sense that it will guarantee the first player a tie (and possibly a win) in the game?

**14.** In the matching-pennies example, consider the case where player I favors heads $H$ over tails $T$. For example, assume that player I plays $H$ three-fourths of the time and $T$ only one-fourth of the time–a nonoptimal mixed strategy. What should player II do if he knows this?

**15.** Assume in the nonsymmetrical matching example that player II is using the nonoptimal mixed strategy $(p, 1 - p) = (\frac{1}{2}, \frac{1}{2})$;; that is, he is playing $H$ and $T$ with the same frequency. What should player I do in this case if she knows this?

**16.** You plan to manufacture a new product for sale next year, and you can decide to make either a small quantity, in anticipation of a poor economy and few sales, or a large quantity, hoping for brisk sales. Your expected profits are indicated in the following table.

| | | *Economy* | |
|---|---|---|---|
| | | **Poor** | **Good** |
| *Quantity* | **Small** | \$500,000 | \$300,000 |
| | **Large** | \$100,000 | \$900,000 |

If you want to avoid risk and believe that the economy is playing an optimal mixed strategy against you in a two-person zero-sum game, then what is your optimal mixed strategy and the resulting expected value? Discuss some alternative ways to go about making your decision.

■ **17.** Consider the following miniature poker game with two players, I and II. Each antes \$1. Each player is dealt either a high card $H$ or a low card $L$, with probability $\frac{1}{2}$. Player I then folds or bets \$1. If player I bets, then player II either folds, calls, or raises \$1. Finally, if II raises, I either folds or calls.

Most choices by the players are rather obvious, at least to anyone who has played poker: If either player holds $H$, that player always bets or raises if he or she gets the choice. The question remains of how often one should bluff–that is, continue to play (by calling or raising) while holding a low card in the hope that one's opponent also holds a low card.

This poker game can be represented by the following matrix game, wherein the payoffs are the expected winnings for player I (depending upon the random deal) and the dominated strategies have been eliminated.

| | | *Player II (when holding L)* | | |
|---|---|---|---|---|
| | | **Folds** | **Calls** | **Raises** |
| *Player I (when holding L)* | **Folds initially** | −0.25 | 0 | 0.25 |
| | **Bets first, folds later** | 0 | 0 | −0.25 |
| | **Bets first, calls later** | −0.25 | −0.25 | 0 |

**(a)** Are there any strategies in this matrix game that a player should avoid playing?
**(b)** Solve this game.
**(c)** Which player is in the more favored position?
**(d)** Should one ever bluff?

■ **18. (a)** Describe in detail *one* pure strategy for the player who moves second in the game of tic-tac-toe.
**(b)** Is your strategy in part (a) optimal in the sense that it will guarantee the second player a tie (and possibly a win) in the game?

**19.** On an overcast morning, deciding whether to carry your umbrella can be viewed as a game between yourself and nature as follows:

| | | *Weather* | |
|---|---|---|---|
| | | **Rain** | **No rain** |
| *You* | **Carry umbrella** | Stay dry | Lug umbrella |
| | **Leave it home** | Get wet | Hands free |

Let's assume that you are willing to assign the following numerical payoffs to these outcomes, and that you are also willing to make decisions on the basis of expected values (that is, average payoffs):

(Carry umbrella, rain) = −2

(Carry umbrella, no rain) = −1

(Leave it home, rain) = −5

(Leave it home, no rain) = 3

**(a)** If the weather forecast says there is a 50% chance of rain, should you carry your umbrella or not? What if you believe there is a 75% chance of rain?
**(b)** If you are conservative and wish to protect against the worst case, what pure strategy should you pick?
**(c)** If you are rather paranoid and believe that nature will pick an optimal strategy in this two-person zero-sum game, then what strategy should you choose?

**(d)** Another approach to this decision problem is to assign payoffs to represent what your *regret* will be after you know nature's decision. In this case, each such payoff is the best payoff you could have received under that state of nature, minus the corresponding payoff in the previous table:

| | | *Weather* | |
|---|---|---|---|
| | | Rain | No rain |
| *You* | Carry umbrella | $0 = (-2) - (-2)$ | $4 = 3 - (-1)$ |
| | Leave it home | $3 = (-2) - (-5)$ | $0 = 3 - 3$ |

What strategy should you select if you wish to minimize your maximum possible regret?

## 15.3 Partial-Conflict Games

Consider the following three two-person variable-sum games. Discuss the players' possible behavior when these games are played in a noncooperative manner (with no prior communication or agreements). The first payoff is to the row player; the second, to the column player. Are the Nash equilibria in these games sensible? Why or why not?

**20.**

| | Player II | |
|---|---|---|
| Player I | (4, 4) | (1, 3) |
| | (3, 1) | (2, 2) |

**21.** Battle of the sexes:

| | | *She buys a ticket for:* | |
|---|---|---|---|
| | | Boxing | Ballet |
| *He buys a ticket for:* | Boxing | (4, 3) | (2, 2) |
| | Ballet | (1, 1) | (3, 4) |

**22.**

| | Player II | |
|---|---|---|
| Player I | (2, 1) | (4, 2) |
| | (1, 4) | (3, 3) |

**23.**

| | Player II | |
|---|---|---|
| Player I | (2, 4) | (4, 3) |
| | (1, 2) | (3, 1) |

**24.**

| | Player II | |
|---|---|---|
| Player I | (3, 4) | (2, 3) |
| | (1, 2) | (4, 1) |

◆ **25.** Assume that two countries in an arms race assign points to all their own weapons so that the total for each is 1000. Each side can then designate weapons of the *other* side, totaling 100 points, that must be eliminated in the next year, thereby effecting a 10% reduction. Would these countries have any reason to lie about how they value their own weapons? Is this procedure practical as an arms-reduction scheme?

## 15.4 Larger Games

**26.** For the preferences of the players given in the text—*xyz* for *X*, *yzx* for *Y*, and *zxy* for *Z*—verify that the strategy choices of *x* by *X*, *y* by *Y*, and *x* by *Z* are a Nash equilibrium. Does this equilibrium seem to you defensible as the social choice by the voters? Under what circumstances might the voters choose these strategies rather than their sophisticated strategies?

■ **27.** Under a voting system called approval voting, a voter can vote for as many alternatives as he or she wishes. (If there are three alternatives, the only undominated strategies of a voter under approval voting are to vote for his or her best, or two best, choices). If voters *X*, *Y*, and *Z* have paradox-of-voting preferences of *xyz*, *yzx*, and *zxy*, and *X* is the status quo, show that *x* is the sophisticated outcome, obtained by all voters voting for their two best choices.

■ **28.** Show by example that approval voting is not immune to the status-quo paradox.

**29.** Odd and Even play Low Person Wins, whose rules are as follows:

**(a)** Odd announces an odd number between 1 and 5 (inclusive).
**(b)** Independently, Even announces an even number between 2 and 6 (inclusive).
**(c)** Whoever announces the lower number gets *twice* this number as its payoff.
**(d)** Whoever announces the higher number gets the *lower* number as its payoff.

What is the Nash equilibrium of this game, based on the successive elimination of dominated strategies? Is there another Nash equilibrium?

■ **30.** Find a two-person zero-sum game with a saddlepoint in which the successive elimination of dominated strategies does *not* lead to the saddlepoint (unlike the 3 × 3 location game; you may restrict yourself to 3 × 3 games).

◆ **31.** Why will the first player to act in a truel shoot in the air (if this option is allowed by the rules)? Is this choice optimal if a second player should succeed in firing in the air at the same time?

**32.** In a sequential truel with no firing in the air allowed, suppose *A*, who hates *B*, goes first; *B*, who hates *C*, goes second; *C*, who hates *A*, goes third. (If a player fires, he will shoot only his *antagonist*–the player he hates.)

**(a)** Which player is in the best position, and why?
**(b)** Does the outcome change if $B$ hates $A$ rather than $C$?

Answer these questions if each player can take only one turn (if alive); and if, after one round, the game continues (if there is more than one player alive) and players can take more than one turn.

**33.** If a fourth player is added to the original truel, show that every player will have an incentive to shoot another player (as in a duel).

**34.** Extend the game tree of the truel in Figure 15.5 to allow the additional possibility that if $A$ does not shoot initially, then $B$ has the choice of shooting or not shooting $C$. Will $A$, in fact, not shoot initially, and will $B$ then shoot $C$?

**35.** Extend the game tree in Exercise 34 to still another level to allow for the possibility that if $A$ does not shoot initially, and $B$ shoots $C$, then $A$ has the choice of shooting or not shooting $B$. What will happen in this case?

**36.** Change Exercise 35 to allow for the possibility that if $A$ does not shoot initially, and $B$ shoots *or does not shoot* $C$, then $A$ has the choice of shooting or not shooting $B$. What will happen in this case?

◆ **37.** What general conclusions would you draw in light of your answers to Exercise 34, 35, and 36?

## APPLET EXERCISES

**To do these exercises, go to www.whfreeman.com/fapp8e.**

What happens to the value of a game if you or your opponent deviates from the optimal strategy? Can you exploit such deviations by your opponent to your advantage? Explore these possibilities in the *Game Theory* applet.

## WRITING PROJECTS

**1.** Consider a conflict that you, personally, had–with a parent, a boss, a girlfriend or boyfriend, or some other acquaintance–in which each of you had to make a choice without being sure of what the other person would do. What strategies did you seriously consider adopting, and what options do you think the other person considered? What plausible outcomes do you think each set of strategy choices would have led to? How would you rank these outcomes from best to worst, and how do you think the other player would have ranked them? In two to three pages, analyze the resulting game, and state whether you think you and the other person made optimal choices. If not, what interfered with your or the other person's rationality?

**2.** It is sometimes argued that game theory does not take account of the (irrational?) emotions of people, such as anger, jealousy, or love. What is your opinion about this question? In one to two pages, give an example, real or hypothetical, that supports your position, paying particular attention to whether the players acted consistently with, or contrarily to, their preferences.

**3.** In tennis, one player often prefers to play from the baseline while her opponent prefers a serve-and-volley game (that is, likes to come to the net). The baseline player attempts to hit passing shots. This player has a choice of hitting "down the line" or "crosscourt." The net player must often correctly guess in which direction the ball will go to cover the shot. In one to two pages, formulate this situation as a duel game and discuss appropriate strategies for the players.

**4.** Quentin Tarantino's films *Reservoir Dogs* (1992) and *Pulp Fiction* (1994) both have truels, but the choices that the characters make in each are completely different. Does the truel analysis offer any insight into why? Discuss in one to two pages.

**5.** In one to two paragraphs, discuss the relationship between the status-quo paradox and the chair's paradox (section 10.4).

## SUGGESTED READINGS

AUMANN, ROBERT J., and SERGIU HART, eds. *Handbook of Game Theory with Economic Applications,* Elsevier, Amsterdam, 1992 (vol. 1), 1994 (vol. 2), 2002 (vol. 3). A comprehensive treatment of game theory and its applications, developed in long chapters written by leading experts.

BINMORE, KEN. *Playing for Real: A Text on Game Theory.* Oxford University Press, Oxford, UK, 2007. A comprehensive and up-to-date intermediate text.

BRAMS, STEVEN J. *Theory of Moves,* Cambridge University Press, New York, 1994. Describes in detail the theory of moves and applies it to a wide variety of conflicts.

DIXIT, AVINASH, and SUSAN SKEATH. *Games of Strategy,* 2nd ed., Norton, New York, 2004. An excellent game-theory text that requires only a minimal mathematical background.

NASAR, SYLVIA. *A Beautiful Mind,* Simon & Shuster, New York, 1998. A biography of John Nash that is also a fascinating account of the early history of game theory. In 2001, a fictionalized version of this biography was made into a movie, which received four Oscars, including Best Picture, in 2002.

OSBORNE, MARTIN J. *An Introduction to Game Theory.* Oxford, New York, 2004. A fine intermediate text with several interesting applications.

## SUGGESTED WEB SITES

**www.economics.utoronto.ca/osborne** Martin Osborne's home page (game theory).

**kuznets.fas.harvard.edu/~aroth/alroth.html** Alvin Roth's Game Theory and Experimental Economics Page.

**www.gametheory.net** is the most comprehensive general Web site.

# The Digital Revolution

# PART V

The advent of computers necessitates mathematical methods to code, store, secure and transmit information accurately and economically. All of these functions were in evidence on July 2, 2005 when Paul McCartney and U2 opened the Live 8 benefit concert in London with the Beatles' classic song "Sgt. Pepper's Lonely Hearts Club Band" before 200,000 people in attendance and a worldwide television audience of millions. Within 45 minutes the song was digitally transmitted by satellite to a London television studio where it was edited, mastered and transmitted to a production center in Germany, then made available for sale by online retailers around the world. It became the fastest-selling digital download and can be currently viewed on YouTube. In this part of the text we examine some of the mathematics that made this event possible. ■

CHAPTER 16



# Identification Numbers

Modern identification numbers serve at least two functions. An identification number should unambiguously identify the person or thing with which it is associated. Less obvious is a "self-checking" aspect of the number.

## 16.1 Check Digits

Look at the 13-digit **International Standard Book Number (ISBN)** printed on the back cover of this book. The number 978-1-4292-0900-7 (978-1-4292-1506-0 for the paperback version) is a **code**. It distinguishes this book from all others. The last digit 7 is there solely to detect errors that may occur when the ISBN is entered into a computer. Grocery items, credit cards, overnight mail, magazines, personal checks, traveler's checks, soft-drink cans, automobiles, and many other items you encounter daily have identification numbers that code data and a digit called a **check digit** for error detection. In this chapter, we examine some of the methods used to assign identification numbers and check digits.

Let's begin by considering the U.S. Postal Service money order shown in Figure 16.1. The first 10 digits of the 11-digit number 63024383845 simply identify the money order. The last digit, 5, serves as an **error-detecting** mechanism. Let's see how this mechanism works. The eleventh (last) digit of a Postal Service money order number is the remainder obtained when the sum of the first 10 digits of the number is divided by 9. In our example, the last digit is 5 because $6 + 3 + 0 + 2 + 4 + 3 + 8 + 3 + 8 + 4 = 41$ and the remainder when 41 is divided by 9 is 5.

Now suppose that instead of the correct number, the number 63054383845 (an error in the fourth position) was entered into a computer programmed for error detection in money orders. The machine would divide the sum of the first 10 digits of the entered number, 44, by 9 and obtain a remainder of 8. Since the last digit of the entered number is 5 rather than 8, the entered number cannot be correct. This crude method of error detection will not detect the mistake of

**FIGURE 16.1** Money order with identification number 6302438384 and appended check digit 5. The check digit is the remainder after dividing the sum of the digits by 9.

replacing a 0 with a 9, or vice versa. Nor will it detect the transposition of digits, such as 63204383845 instead of 63024383845 (the digits in positions three and four have been transposed).

American Express traveler's checks, VISA traveler's checks, and Euro banknotes also use a check digit determined by division by 9. In these cases, the check digit is chosen so that the sum of the digits, including the check digit, is evenly divisible by 9.

## EXAMPLE 1 ■ The American Express Travelers Cheque

The American Express Travelers Cheque with the identification number 387505055 has check digit 7 because $3 + 8 + 7 + 5 + 0 + 5 + 0 + 5 + 5 = 38$ and $38 + 7$ is evenly divisible by 9.

The scheme used on airline tickets, UPS packages, and Avis and National rental cars assigns the remainder after division by 7 of the number itself as the check digit rather than dividing the sum of the digits by 7. For example, the check digit for the number 540047 is 4 because $540047 = 7 \times 77149 + 4$. This method will not detect the substitution of 0 for a 7, 1 for an 8, 2 for a 9, or vice versa. However, unlike the Postal Service method, it will detect transpositions of adjacent digits with the exceptions of the pairs 0, 7; 1, 8; and 2, 9. For example, if 5400474 were entered into a computer as 4500474 (the first two digits are transposed), the machine would determine that the check digit should be 3 since $450047 = 7 \times 64292 + 3$. Because the last digit of the entered number is not 3, the error has been detected.

The scheme used on grocery products, the **Universal Product Code (UPC)**, is more sophisticated. Consider the number 0 38000 00127 7 found on the bottom of a box of corn flakes. The first digit identifies a broad category of goods, the next five digits identify the manufacturer, the next five identify the product, and the last is a check digit. Suppose this number were entered into a computer as 0 58000 00127 7 (a mistake in the second position). How would the computer recognize the mistake?

For any UPC number $a_1a_2a_3a_4a_5a_6a_7a_8a_9a_{10}a_{11}a_{12}$, the computer is programmed to carry out the following computation: $3a_1 + a_2 + 3a_3 + a_4 + 3a_5 + a_6 + 3a_7 + a_8 + 3a_9 + a_{10} + 3a_{11} + a_{12}$. If the result doesn't end with a 0, the computer knows the entered number is incorrect.

For the incorrect corn flakes number, we have $3 \cdot 0 + 5 + 3 \cdot 8 + 0 + 3 \cdot 0 + 0 + 3 \cdot 0 + 0 + 3 \cdot 1 + 2 + 3 \cdot 7 + 7 = 62$. Since 62 doesn't end with 0, the error is detected. Notice that had we used the correct digit 3 in the second position instead of 5, the sum would have ended in a 0 as it should. This simple scheme detects *all* single-position errors and about 89% of all other kinds of errors.

Beginning in January 2005, U.S. retailers were required to have software that could read the 12-digit UPC code used in the United States and the 13-digit European Article Number (EAN) code used in Europe. This change paves the way for the 13-digit EAN to become the worldwide standard. Existing UPC numbers will be converted to EAN numbers by adding an extra 0 at the beginning. The check digit for a 13-digit EAN number $a_1a_2a_3a_4a_5a_6a_7a_8a_9a_{10}a_{11}a_{12}a_{13}$ is selected so that $a_1 + 3a_2 + a_3 + 3a_4 + a_5 + 3a_6 + a_7 + 3a_8 + a_9 + 3a_{10} + a_{11} + 3a_{12} + a_{13}$ ends with 0. Adding an extra 0 in the front of a UPC number does not affect the check digit.

The U.S. banking system uses a variation of the UPC scheme that appends check digits to the numbers assigned to banks. Each bank has an eight-digit routing number $a_1a_2 \cdots a_8$ together with a check digit $a_9$ so that $a_9$ is the last digit of $7a_1 + 3a_2 + 9a_3 + 7a_4 + 3a_5 + 9a_6 + 7a_7 + 3a_8$. The numbers 7, 3, and 9 used in this formula are called the **weights.** (The weights for the UPC scheme are 3 and 1.) The weights were carefully chosen so that all single-digit errors and most transposition errors are detected. The use of different weights in adjacent positions permits the detection of most transposition errors.

## EXAMPLE 2 ■ Bank Identification Number

The First Chicago Bank has the routing number 071000013 on the bottom of all its checks (see Figure 16.2). The check digit 3 is the last digit of $7 \cdot 0 + 3 \cdot 7 + 9 \cdot 1 + 7 \cdot 0 + 3 \cdot 0 + 9 \cdot 0 + 7 \cdot 0 + 3 \cdot 1 = 33$. The first four digits of a nine-digit bank routing number identify the bank's Federal Reserve District, office, state or special collection arrangement; the next four digits are the bank's identification number; the ninth digit is the check digit. The block of numbers 22 63378 shown in Figure 16.2 is the account number. The last block, 0134, is the check number.

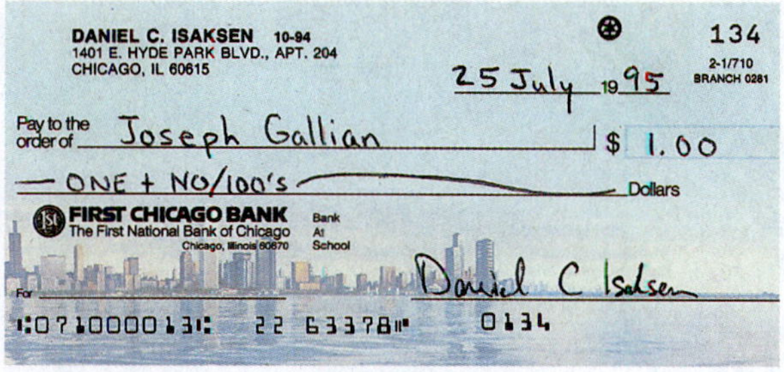


**FIGURE 16.2** A bank check with routing number 071000013. The 3 is the check digit.

You may wonder if there is any advantage to using three weights as opposed to using two, as is the case for the UPC error-detection scheme. The answer is yes. While both the UPC scheme and the bank scheme detect 100% of single-position errors and the same transposition errors involving adjacent digits, the bank scheme will detect most transposition errors of the form $\cdots abc \cdots \to \cdots cba \cdots$, whereas the UPC scheme does not detect such errors.

For example, say we look at a number that begins with 241. In the UPC scheme, these digits contribute $3 \cdot 2 + 4 + 3 \cdot 1 = 13$ toward the total calculation, while the string 142 (the first and third digits are transposed) also contributes $3 \cdot 1 + 4 + 3 \cdot 2 = 13$ toward the total calculation. So the error is not detected. In contrast, using the bank scheme, 241 contributes $7 \cdot 2 + 3 \cdot 4 + 9 \cdot 1 = 35$ toward the total, while 142 contributes $7 \cdot 1 + 3 \cdot 4 + 9 \cdot 2 = 37$ toward the total. Since the total for the correct number ends with 0, the total for the number that had the transposition error would end with the digit 2. Thus, the error is detected.

One of the most efficient error-detection methods is used by all major credit-card companies as well as by many libraries, blood banks, photofinishing companies, and the South Dakota driver's license department. Say a bank intends to issue a credit card with the identification number 312560019643001. It must then add an extra digit for error detection. This is done as follows. Add the digits in positions 1, 3, 5, 7, 9, 11, 13, and 15 and double the result: $(3 + 2 + 6 + 0 + 9 + 4 + 0 + 1) \times 2 = 50$. Next, count the number of digits in positions 1, 3, 5, 7, 9, 11, 13, and 15 that exceed 4 and add this to the total. For our example, only 6 and 9 exceed 4, so the count is 2 and our running total is 52. Finally, take the sum of 52 and the digits in the even-numbered positions: $52 + (1 + 5 + 0 + 1 + 6 + 3 + 0) = 68$.

The check digit is whatever is needed to bring the final tally to a number that ends with 0. Because $68 + 2 = 70$, the check digit for our example is 2. This digit is appended to the end of the number the bank issues for identification purposes. Errors in input data are detected by applying the same algorithm to the input, including the check digit. If the correct number is entered into a computer, the result will end in 0. If the result doesn't end with 0, a mistake has been made. The credit card shown in Figure 16.3 is reproduced from an ad promoting the Citibank VISA card. Notice that the check digit on the card is not valid because the algorithm yields

$$(4 + 2 + 0 + 1 + 3 + 5 + 7 + 9) \times 2 + 3 + (1 + 8 + 0 + 2 + 4 + 6 + 8) + 0 = 94$$

which does not end in 0. This method allows computers to detect 100% of single-position errors and about 98% of other common errors.

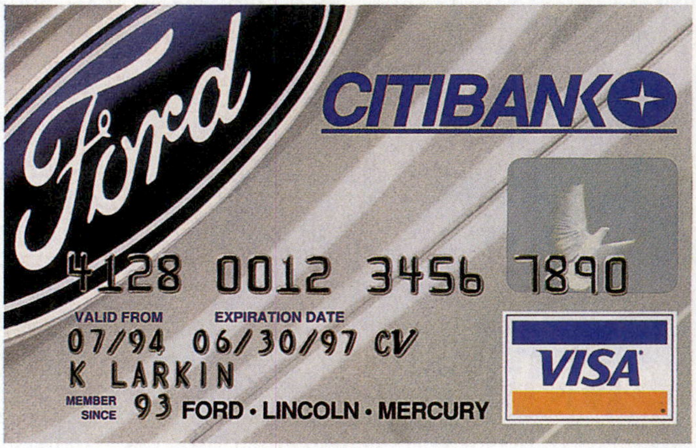


**FIGURE 16.3** VISA card with an invalid number.

Besides detecting errors, the check digit offers partial protection against fraudulent numbers. A person who wanted to create a phony credit card, bank account number, or driver's license number would have to know the appropriate check-digit scheme for the number to go unchallenged by the computer.

Thus far we have not discussed any schemes that detect 100% of single errors and 100% of transposition errors. As seen on the back of this book and most others published since 2007, there are two identifications numbers–a 13-digit number called the *13-digit International Standard Book Number (ISBN-13)* and a 10-digit number, called the *10-digit ISBN (ISBN-10).* The 10-digit ISBN detects 100% of single digit errors and 100% of transposition errors.

A correctly coded 10-digit ISBN $a_1a_2 \cdots a_{10}$ has the property that $10a_1 + 9a_2 + 8a_3 + 7a_4 + 6a_5 + 5a_6 + 4a_7 + 3a_8 + 2a_9 + a_{10}$ is evenly divisible by 11. Consider the 10-digit ISBN of the book you are now reading: 1-4292-0900-3 (1-4292-1506-2 for paperback version). The 1 at the beginning indicates that the book is published in an English-speaking country, the next block of digits, 4292, identifies the publisher, W. H. Freeman and Company. The third block for the hardback edition, 0900, is assigned by the publisher and identifies this particular book. The last digit 3, for the hardback version, is the check digit. Let's verify that this number is a legitimate possibility. We must compute $10 \cdot 1 + 9 \cdot 4 + 8 \cdot 2 + 7 \cdot 9 + 6 \cdot 2 + 5 \cdot 0 + 4 \cdot 9 + 3 \cdot 0 + 2 \cdot 0 + 3 = 176$. Because $176 = 11 \cdot 16$, it is evenly divisible by 11, so no error has been detected.

How can we be sure that this method detects 100% of the single-position errors? Well, let's say that a correct number is $a_1a_2a_3a_4a_5a_6a_7a_8a_9a_{10}$ and that a mistake is made in the second position. (The same argument applies equally well in every position.) We may write this incorrect number as $a_1a_2'a_3a_4a_5a_6a_7a_8a_9a_{10}$, where $a_2' \neq a_2$. For this error to go undetected, it must be the case that $10 \cdot a_1 + 9 \cdot a_2' + 8 \cdot a_3 + 7 \cdot a_4 + 6 \cdot a_5 + 5 \cdot a_6 + 4 \cdot a_7 + 3 \cdot a_8 + 2 \cdot a_9 + a_{10}$ is evenly divisible by 11. Then, since both $10a_1 + 9a_2 + 8a_3 + 7a_4 + 6a_5 + 5a_6 + 4a_7 + 3a_8 + 2a_9 + a_{10}$ and $10a_1 + 9a_2' + 8a_3 + 7a_4 + 6a_5 + 5a_6 + 4a_7 + 3a_8 + 2a_9 + a_{10}$ are divisible by 11, so is their difference:

$$(10 \cdot a_1 + 9 \cdot a_2 + 8 \cdot a_3 + \cdots + a_{10}) - (10 \cdot a_1 + 9 \cdot a_2' + 8 \cdot a_3 + \cdots + a_{10}) = 9 \cdot (a_2 - a_2')$$

Because $a_2$ and $a_2'$ are distinct digits between 0 and 9, their difference must be one of $\pm 1, \ldots, \pm 9$. Thus, the only possibilities for the number $9 \cdot (a_2 - a_2')$ are $\pm 9$, $\pm 18$, $\pm 27$, $\pm 36$, $\pm 45$, $\pm 54$, $\pm 63$, $\pm 72$, $\pm 81$, and none of these is divisible by 11. So a single-position error cannot go undetected.

To verify that the method detects all adjacent transposition errors, let's suppose that the first two digits are transposed (the same argument applies to all positions). Say the correct number is $a_1a_2a_3 \cdots a_{10}$. As before, for the incorrect number $a_2a_1a_3 \cdots a_{10}$ to go undetected, it must be the case that the difference of the correct number and the incorrect number is evenly divisible by 11. That is,

$$(10a_1 + 9a_2 + 8a_3 + \cdots + a_{10}) - (10a_2 + 9a_1 + 8a_3 + \cdots + a_{10})$$

is evenly divisible by 11. This reduces to $a_1 - a_2$ is divisible by 11. But the only possible differences of two numbers between 0 and 9 are plus or minus the numbers between 0 and 9. Thus $a_1 - a_2 = 0$. But then $a_1 = a_2$ and there is no error.

With a bit more work, we could prove that every transposition error is detected, not just the transpositions of adjacent digits. This is possible because 11 is prime.

Since this method, in contrast to the other methods we have described, detects all single-position errors and all transposition errors, why is it not used more? Well, it does have a drawback. Say the next title published by W.H. Freeman is to have 0902 for the third block. (The 10-digit ISBN for all W.H. Freeman books begins with 1-4292.) What check digit should be assigned? Call it $a$. Then the weighted sum is $10 \cdot 1 + 9 \cdot 4 + 8 \cdot 2 + 7 \cdot 9 + 6 \cdot 2 + 5 \cdot 0 + 4 \cdot 9 + 3 \cdot 0 + 2 \cdot 2 + a = 177 + a$. Because the next integer after 177 that is divisible by 11 is 187, we see that $a = 10$. But appending 10 to the existing 9-digit number would result in an 11-digit number instead of a 10-digit one. This is the only flaw in the 10-digit ISBN scheme. To avoid this flaw, publishers use an X to represent the check digit 10. As a result, not all 10-digit ISBNs consist solely of digits–some end with X. Publishers could avoid this inconsistency by simply refraining from using numbers that require an X.

To expand the inventory of ISBNs and make them compatible with the UPC/EAN numbering scheme for other retail items worldwide, publishers began using a 13-digit ISBN in 2007. The 13-digit ISBN is the same as the 10-digit ISBN number except for a prefix of 978 or 979 and the check digit. The check digit for the 13-digit ISBN is calculated so that the weighted sum using the weights 1, 3, 1, 3, … , 1, 3, 1 ends with the 0. Thus the 13-digit ISBN and the 13-digit UPC/EAN numbers used for retail products employ the same check digit method. During a phase-in period publishers will use both the 10-digit and 13-digit numbers.

After single digit errors and adjacent transposition errors, the third most common error is one of the form $\cdots abc \cdots \rightarrow \cdots cba \cdots$. In practice, these kinds of errors commonly occur in phone numbers that have matching digits separated by another digit such as 727 5856. A likely mistake when writing or dialing this number is to switch the 8 and the 6, resulting in the number 727 5658. Such an error is called a *jump transposition*. Remarkably, there is a simple way to encode identification numbers so that the three most common errors are detected 100% of the time without having to introduce an alphabetic character as is done for the 10-digit ISBN numbers. To illustrate the method, suppose that a math instructor wants to publicly post student grades without revealing any information about the students' ID numbers. Assuming the last four digits of each student ID number are different, she could assign each student a six-digit number by multiplying the last four digits of their identification numbers by 13 (adding leading 0s when necessary). For example, a student with an ID number that ends with 8912 is assigned $115856 = 8912 \times 13$. (To preserve confidentiality of the original four digits, students are not informed of the encoding method). Of course, the instructor can recapture the original four-digit numbers by dividing the encoded numbers by 13. Since all encoded numbers are divisible by 13, the jump transposition error 115856 →115658 is detected because 115658 is not divisible by 13.

The arguments for verifying that encoding identification numbers as multiples of 13 detect 100% of all single digit errors, all transposition errors involving adjacent digits, and all jump transposition errors are similar to those used to show that the 10-digit ISBN numbers detect errors. In particular, a single-digit error in the number $a_n a_{n-1} \cdots a_i \cdots a_0$ of the form $a_n a_{n-1} \cdots a'_i \cdots a_0$ where $a'_i \neq a_i$ is not detected if and only if $a_n a_{n-1} \cdots a'_i \cdots a_0$ is a multiple of 13. But if both $a_n a_{n-1} \cdots a_i \cdots a_0$ and $a_n a_{n-1} \cdots a'_i \cdots a_0$ are multiples of 13, then so is their difference $(a_n a_{n-1} \cdots a_i \cdots a_0) - (a_n a_{n-1} \cdots a'_i \cdots a_0) = (a_i - a'_i)10^i$. But 13 does not divide the term on the right when $a_i \neq a'_i$. Similarly, the transposition of adjacent digits $a_i$ and $a_{i-1}$ is undetected if and only if $9(a_i - a_{1-1})10^{i-1}$ is divisible by 13, which happens only when $a_i = a_{i-1}$. And the jump transposition $\cdots a_i a_{i-1} a_{i-2} \cdots \rightarrow \cdots a_{i-2} a_{i-1} a_i \cdots$ is undetected if and only if $99(a_i - a_{i-2})10^{i-2}$ is divisible by 13, which happens only when $a_i = a_{i-2}$. Incidentally, the arguments just given reveal why we used multiplication by 13 rather than some smaller positive integer. For example, if multiplication by 11 were used to transform the identification numbers instead of 13, then all single digit errors and all adjacent transposition errors are detected, but not all jump transpositions are since 11 divides 99.

Many identification numbers use both alphabetic and numerical characters. When a check digit is included, the alphabetic characters are assigned numerical values. The vehicle identification number (VIN) used to identify cars and trucks is one such example, as explained in Spotlight 16.1

## SPOTLIGHT 16.1 The VIN System

Automobiles and trucks are given a vehicle identification number (VIN) by the manufacturer. A typical VIN has 17 alphanumeric characters that code information, such as country where the vehicle was built, manufacturer, make, body style, engine type, plant where the vehicle was built, model year, model, type of restraint, a check digit, and a production sequence number. The check digit is calculated by converting the 26 consecutive letters of the alphabet, respectively, to the numbers 1, 2, 3, 4, 5, 6, 7, 8, 9, 1, 2, 3, 4, 5, 6, 7, 8, 9, 2, 3, 4, 5, 6, 7, 8, 9 (note the skipped digit after the second 9) to obtain a 16-digit number $a_1a_2 \cdots a_{15}a_{16}$ that is weighted with 8, 7, 6, 5, 4, 3, 2, 10, 9, 8, 7, 6, 5, 4, 3, 2. The check digit is the remainder when the weighted sum $8 \cdot a_1 + 7 \cdot a_2 + \cdots + 3 \cdot a_{15} + 2 \cdot a_{16}$ is divided by 11 unless the remainder is 10, in which case an X is used instead. The check digit is inserted in position 9.

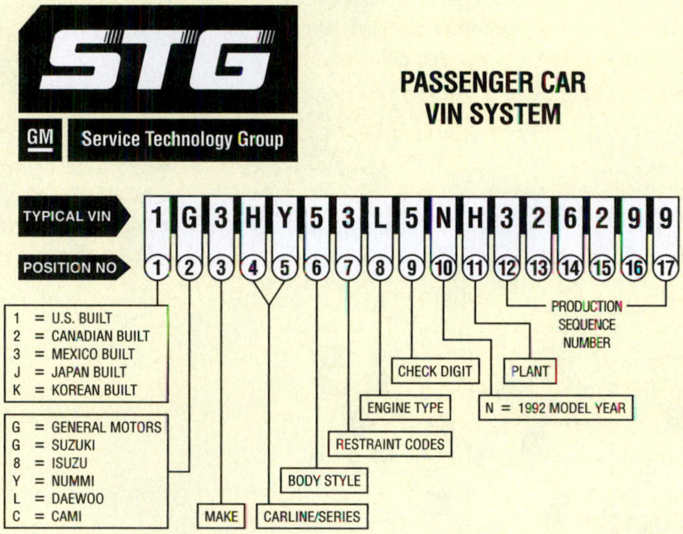


## 16.2 The ZIP Code

Identification numbers sometimes **encode** geographic data. The ZIP code, Social Security numbers, and telephone numbers are the foremost examples. In 1963, the U.S. Postal Service numbered every American post office with a five-digit **ZIP code.** (ZIP is an acronym for Zone Improvement Plan.) The numbers begin with 0's at the points farthest east—00601 for Adjuntas, Puerto Rico—and work up to 9's at the the points farthest west—99950 for Ketchikan, Alaska (see Figure 16.4).

Let's use one of the ZIP codes for Duluth, Minnesota, as an example:

**55812**

**5** The first digit represents one of 10 geographic areas, usually a group of states. The numbers begin at the points farthest east (0) and end at the points farthest west (9).

**58** The second two digits, in combination with the first, identify a central mail-distribution point known as a sectional center. The location of a sectional center is based on geography, transportation facilities, and population density. Although just four centers serve the entire state of Utah, there are six of them to take care of New York City.

12 The last two digits indicate the town or local post office. The order is often alphabetical for towns within a delivery area—for example, towns with names beginning with A usually have low numbers. There are many exceptions to this, such as towns that came into existence after the ZIP code scheme was created. In many cases, the largest city in a region will be given the digits 01 and surrounding towns assigned succeeding digits alphabetically.

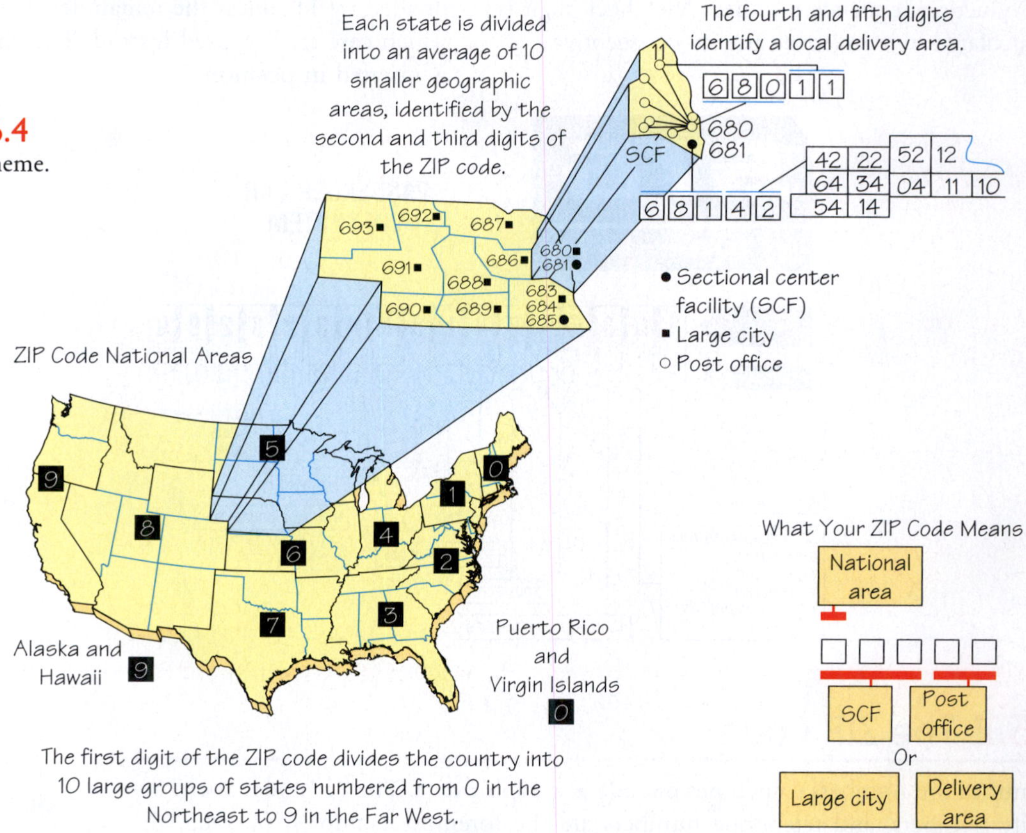


**FIGURE 16.4**
ZIP code scheme.

In 1983, the U.S. Postal Service added four digits to the ZIP code. When four digits are added after a dash—for example, 68588-1234—the number is called the **ZIP + 4 code**. Mail with ZIP + 4 coding is eligible for cheaper bulk rates, being easier to sort with automated equipment. It's also helpful for businesses that wish to sort the recipients of their mailings by geographic location. The first two numbers of the four-digit suffix represent a delivery sector, which may be several blocks, a group of streets, several office buildings, or a small geographic area. The last two numbers narrow the area further. They might denote one floor of a large office building, a department in a large firm, or a group of post office boxes.

For businesses that receive an enormous volume of mail, the ZIP + 4 code permits automation of in-house mailroom sorting. For example, the first seven digits of all mail sent to the University of Minnesota Duluth, are 55812-24. The school has designated nine pairs of digits for the last two positions to direct the mail to the appropriate dormitory or apartment complex.

## 16.3 Bar Codes

In modern applications, bar codes and identification numbers go hand in hand. Bar coding is a method for automated data collection. It is a way to transmit information rapidly, accurately, and efficiently to a computer.

**Bar Code** DEFINITION

A **bar code** is a series of dark bars and light spaces that represent characters.

To **decode** the information in a bar code, a beam of light is passed over the bars and spaces via a scanning device, such as a handheld wand or a fixed-beam device. The dark bars reflect very little back to the scanner, whereas the light spaces reflect much light. The differences in reflection intensities are detected by the scanner and converted to strings of 0's and 1's that represent specific numbers and letters. Such strings are called a **binary coding** of the numbers and letters.

**Binary Code** DEFINITION

Any system for representing data with only two symbols is a **binary code.**

### ZIP Code Bar Code

The simplest bar code is the **Postnet code** used by the U.S. Postal Service and commonly found on business reply forms (see Figure 16.5). For a ZIP + 4 code there are 52 vertical bars of two possible lengths (long and short). The long bars at the beginning and end are called *guard bars* and together provide a frame for the remaining 50 bars. In blocks of five, the 50 bars within the guard bars represent the ZIP + 4 code and a tenth digit for error correction. Each block of five is composed of exactly two long bars and three short bars, according to the pattern shown below:

| Decimal Digit | Bar Code |
|---|---|
| 1 | ııı\|\| |
| 2 | ıı\|ı\| |
| 3 | ıı\|\|ı |
| 4 | ı\|ıı\| |
| 5 | ı\|ı\|ı |
| 6 | ı\|\|ıı |
| 7 | \|ııı\| |
| 8 | \|ıı\|ı |
| 9 | \|ı\|ıı |
| 0 | \|\|ııı |

Handheld scanner reading the shipping bar code on a crate. (*Stewart Cohen/Tony Stone Images.*)

The tenth digit of a Postnet code number is a check digit chosen so that the sum of the nine digits of the ZIP + 4 code and the tenth one is evenly divisible by 10. That is, the check digit $C$ for the ZIP + 4 code $a_1a_2\cdots a_9$ is the digit with the property that the sum $a_1 + a_2 + \cdots + a_9 + C$ ends with 0. For example, the ZIP + 4 code 80321-0421 has the check digit 9, because 8 + 0 + 3 + 2 + 1 + 0 + 4 + 2 + 1 = 21 and 21 + 9 = 30 ends with 0.

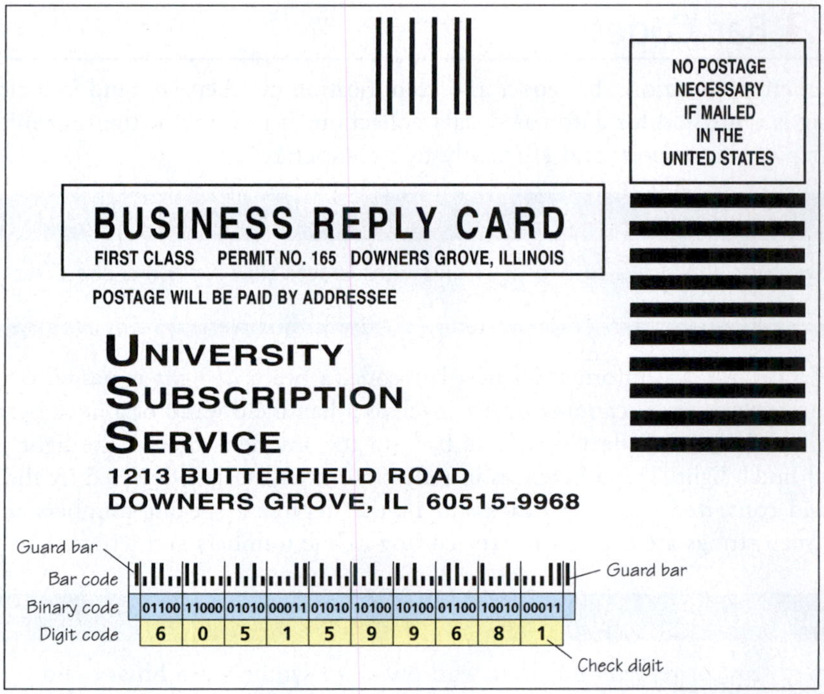


**FIGURE 16.5**
ZIP + 4 bar code.

Because each digit is represented by exactly two long bars and three short ones, any error in reading or printing a single bar would result in a block of five with only one long bar or three long bars. In either case, the error is detected. This is the reason behind the choice of five bars to code each digit rather than four bars. With five bars per digit, there are exactly 10 arrangements composed of two long bars and three short bars. Any misreading of a single bar in such a block is therefore recognizable, because it does not match any other of the blocks for the 10 digits. And because the block location of the error is known, the check digit permits the correction of the error. Let's look at an example of an incorrectly printed bar code and see how the error is correctable.

## EXAMPLE 3 ■ Detecting and Correcting an Error

The scanner ignores the guard bars at the beginning and the end and reads the remaining bars in blocks of five, as shown below. (We have inserted dashed dividing lines for readability.)

3 0 7 2 2 ? 9 0 1 7

The sixth block is an incorrect one because it has only one long bar. To correct the error, the computer linked with the bar-code scanner sums the remaining 9 digits to obtain 31. Because the sum of all 10 digits ends with 0, the correct value for the sixth digit must be 9.

Beginning in 1993, large organizations and businesses that wanted to receive reduced rates for ZIP + 4 bar-coded mail were required to use a 12-digit bar code called the *delivery-point bar code.* This code permits machines to sort a letter into the order in which it will be delivered by the carrier. Mail for the first location on a mail route occurs first, mail for the second location on a route occurs second, and so on.

The 12-digit bar code uses the Postnet bar scheme to code the 12-digit string composed of the 9-digit ZIP + 4 number followed by the last two digits of the street address or box number and a check digit chosen so that the sum of all 12 digits is evenly divisible by 10. For example, a letter addressed to 1738 Maple Street with ZIP + 4 code 55811-2742 would have the Postnet bar code for the digits 558112742384 (38 is from the street address and 4 is the check digit).

Effective January 1, 2009, businesses that desire to receive discounted postage rates from the U.S. postal service must use a new bar code called the *Intelligent Mail Barcode.* The new bar code converts 31 digits of data into 65 vertical bars that encode the type of service, the mail owner, a unique serial number that enables the user to track the letter at every step from arrival at the post office to delivery, and the delivery-point ZIP code. The code uses bars of three lengths and multiple levels to create four states as shown below. For human use, the four states are denoted by the letters T, A, D, and F. The Intelligent Mail Barcode will appear above the printed address.

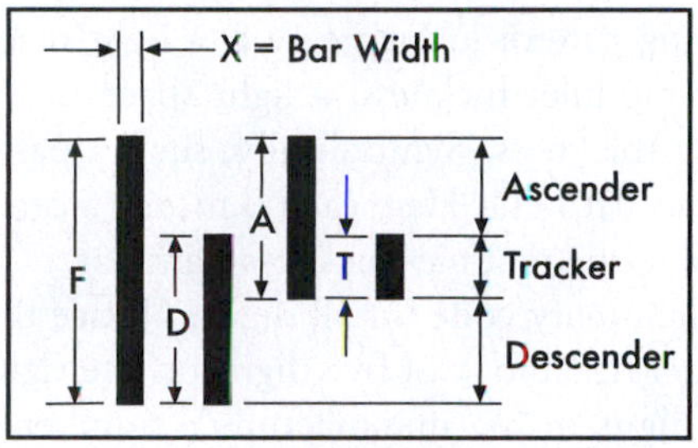

**FIGURE 16.6**
Entomologist Stephen Buchmann developed a reliable, inexpensive way to track bees using the same technology that supermarkets use to speed up the checkout lines and keep track of inventory. He glued bar-code labels onto the backs of 100 bees and placed a laser scanner above the hive. In the past, researchers marked bees with paint or tags, but monitoring activity required the presence of a human observer. (*Scott Camazine/Sue Trainor.*)

## The UPC Bar Code

The bar code that we encounter most often is the UPC, which was first used on grocery items in 1974 and has since spread to most retail products. As Figure 16.6 shows, it has other applications as well. The UPC bar code translates 12-digit UPC identification numbers discussed earlier into bars that can be quickly and accurately read by a laser scanner. The number has four components–two five-digit numbers sandwiched between two single digits–as shown in Figure 16.7.

Here is what the four components represent:

**5** The first digit identifies the kind of product. For example, a 2 signals random-weight items, such as cheese and meat; a 3 means drug and certain other health-related products; a 4 means products marked for price reduction by the retailer; a 5 signals cents-off coupons (see Figure 16.7).

**13000** The next five digits identify the manufacturer.

**22020** The next five digits, assigned by the manufacturer to identify the product, can include size, color, or other important information (but not price).

**5** The final digit is the check digit. This digit is often not printed, but it is always included in the bar code.

MANUFACTURER'S COUPON
EXPIRES 07/29/01
SAVE 20¢
ON (1) 24 oz. OR LARGER BOTTLE OF HEINZ KETCHUP
(Excludes Heinz EZ Squirt™)
HEINZ
TOMATO KETCHUP
CONSUMER: Coupon good only in the USA on purchase of brand/size indicated. Void if copied, transferred, prohibited or restricted. RETAILER: Heinz USA will reimburse you for face value of this coupon plus 8¢ handling if redeemed in compliance with our redemption policy (available upon request). Cash value 1/100¢. Send coupons to: HEINZ USA, P.O. Box 870125, El Paso, TX 88587-0125. LIMIT ONE COUPON PER PURCHASE.
© 2001 H.J. Heinz Company
705528
5 13000 22020 5 (8100)0 70552


**FIGURE 16.7** UPC identification number 5 13000 22020 5. The initial 5 indicates that the number is a manufacturer's coupon. The block 13000 identifies the manufacturer as Heinz. The block 22020 identifies the product. The last digit, 5, is a check digit.

Each digit of the UPC code is represented by a space divided into seven modules of equal width, as illustrated in Figure 16.8. How these seven modules are filled depends on the digit being represented and whether the digit being represented is part of the manufacturer's number or the product number. In every case, there are two light spaces and two dark bars of various thicknesses that alternate. A UPC code has on each end two long bars of one-module thickness separated by a light space of one-module thickness. These two modules are called the *guard bar patterns* (Figure 16.9). The guard bar patterns define the thickness of a single module of each type. They are not part of the identification number. The manufacturer's number and the product number are separated by a center bar pattern consisting of the following five modules: a light space, a (long) dark bar, a light space, a (long) dark bar, and a light space (see Figure 16.9). The center bar pattern is not part of the identification number but merely serves to separate the manufacturer's number and product number. Figure 16.8 shows how the digits 6 and 0 in a manufacturer's number are coded.

Observe the following pattern in Figure 16.8: a light space of one-module thickness, a dark bar of one-module thickness, a light space of one-module thickness, a dark bar of four-module thickness. Symbolically, such a pattern of light spaces and dark bars is represented as 0101111. Here each 0 means a one-module-thickness light space and each 1 means a one-module-thickness dark bar.

Table 16.1 shows the binary code for all digits. Notice that the code for the digits in the product number (the block of five digits on the right side) can be obtained from the code for the digits in the manufacturer's number (the block of digits on the left side), and vice versa, by replacing each 0 by a 1 and each 1 by a 0. Thus, the code 0111011 for 7 in a manufacturer's number becomes 1000100 in the product number. Also notice that each manufacturer's number has an odd number of 1's, whereas each product number has an even number of 1's. This permits a computer linked with an optical scanner to determine whether the bar code was scanned

**TABLE 16.1** Binary UPC Coding

| Digit | Manufacturer's Number | Product Number |
|---|---|---|
| 0 | 0001101 | 1110010 |
| 1 | 0011001 | 1100110 |
| 2 | 0010011 | 1101100 |
| 3 | 0111101 | 1000010 |
| 4 | 0100011 | 1011100 |
| 5 | 0110001 | 1001110 |
| 6 | 0101111 | 1010000 |
| 7 | 0111011 | 1000100 |
| 8 | 0110111 | 1001000 |
| 9 | 0001011 | 1110100 |

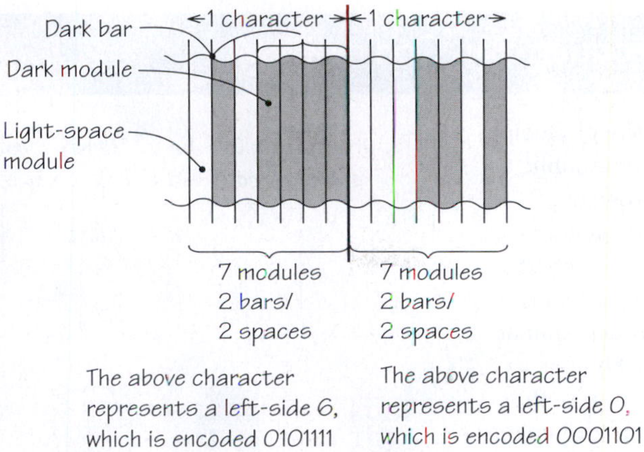


**FIGURE 16.8** UPC bar coding for a left-side 6 and left-side 0, part of the manufacturer's number.

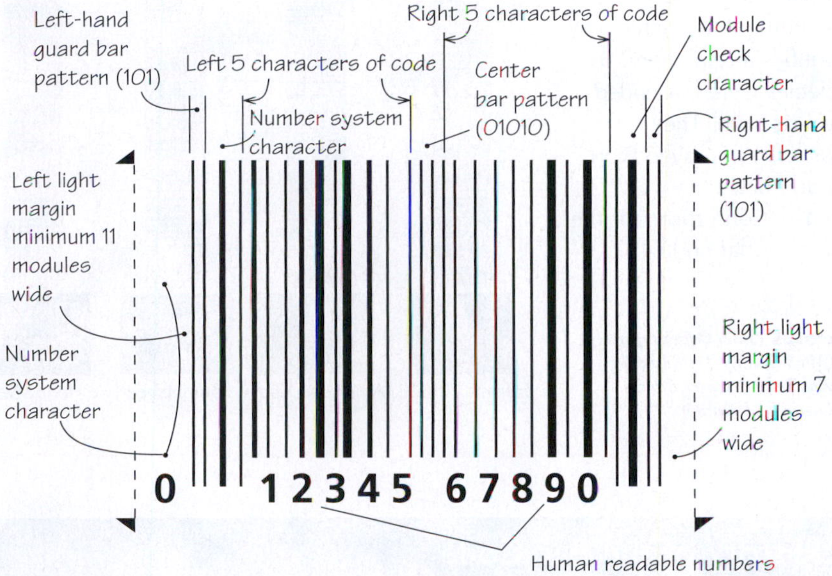


**FIGURE 16.9** UPC bar-code format.

left to right or right to left. (If the first block of digits has an even number of 1's for each digit, the scanning is being done right to left.) Thus, scanning can be done in either direction without ambiguity.

## New Applications of Bar Coding

New applications of bar coding continue to be found. In 2003 a method of bar coding genetic information about animal species that provides a convenient, inexpensive way to identify species was introduced. (See Spotlight 16.2.) In Japan a new generation of bar codes uses mosaics of black and white rectangles that encode much more information than traditional bar codes (see Figure 16.10). These bar codes can be read by specially equipped cell phones to display video, music or text on the screen or link the cell phone to a Web page. Users can point their cell phone at the bar code in a magazine, on a billboard or on the side of a building to receive information about a product or service. A bar code for a movie will allow the viewer to watch a trailer. Scanning the wrapper of a hamburger will provide nutrition information. This new technology is currently in development for use in the United States.

**FIGURE 16.10** Bar code on a building in Japan that can be read by a properly equipped cellphone. (*Ko Sasaki/The New York Times/Redux.*)

## SPOTLIGHT 16.2 New Frontier: Bar Coding DNA

In 2003 Paul Hebert from the University of Guelph in Canada proposed the compilation of a public library of DNA bar codes for animal species. Rather than scanning an animal's entire genome, which is expensive and time consuming, Hebert pinpointed a short piece of a section of a single gene that could be used to distinguish one animal species from another cheaply and quickly. For about $2 per sample the genetic sequence of this tiny gene section can be converted to a four color bar code that corresponds to the four nucleotides that make up the genetic code. The bar code identifies the species of its source in the same way that the UPC bar code identifies a retail item. By 2007 more than 31,000 species were bar coded and a new field of science was born. The technique has already resulted in improved food safety, disease prevention, and better environmental monitoring. The Consortium for the Barcode of Life has set a goal of bar coding 500,000 species by 2012.

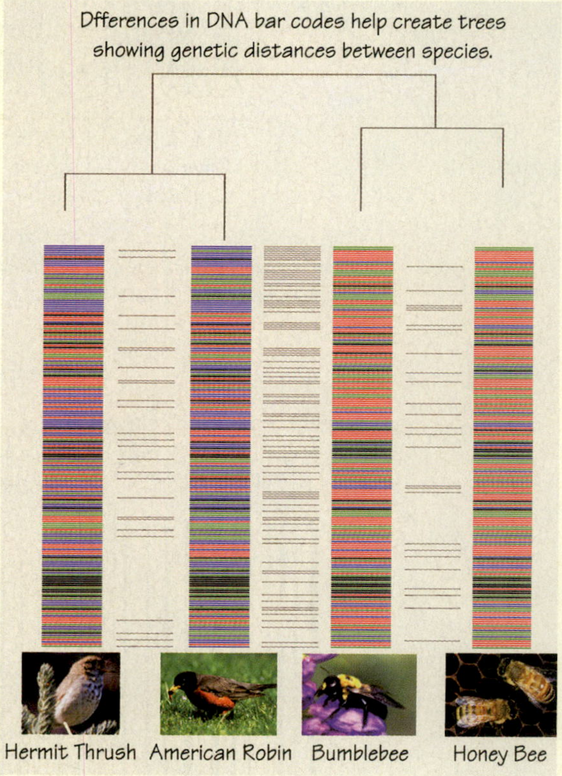
Dfferences in DNA bar codes help create trees showing genetic distances between species.
Hermit Thrush
American Robin
Bumblebee
Honey Bee

Hermit Thrush (*George Jameson*) American Robin (*Jeffrey Lepore/ Photo Researchers, Inc.*) Bumblebee (*Mark Stoeckle/ The Rockefeller University*) Honey Bee (*Scott Camazine/Photo Researchers, Inc.*) DNA bar code (*Mark Stoekle/The Rockefeller University*)

## SPOTLIGHT 16.3 History of the Bar Code

**1948** Graduate students Norman Joseph Woodland and Bernard Silver at Drexel Institute of Technology begin working on a bar code.

**October 7, 1952** Woodland and Silver receive a U.S. patent.

**October 10, 1967** The Association of American Railroad adopted an optical bar code.

**December 1971** The Uniform Code Council, originally called the Uniform Grocery Product Code Council, is formed to administer the UPC.

**1972** U.S. Supermarket Ad Hoc Committee on a Uniform Grocery Product Code recommends the adoption of the 1972 UPC.

**April 3, 1973** An ad-hoc committee composed of grocery executives chooses the linear bar code with 11 digits and a 12th check digit.

**June 26, 1974** A 10-pack of Wrigley's Juicy Fruit chewing gum was the first product with a bar code scanned at a checkout counter in Troy, Ohio. Today, the pack of gum is on display at the Smithsonian Institution's National Museum of American History.

**1974** 95% of the railroad fleet was labeled with a bar code.

**February 1977** European Article Numbering Association formed in Belgium.

**September 1, 1981** United States Department of Defense adopted the use of bar codes for marking all products sold to the United States military.

**1992** Norman Joseph Woodland was awarded the 1992 National Medal of Technology by President George Bush.

## 16.4 Encoding Personal Data

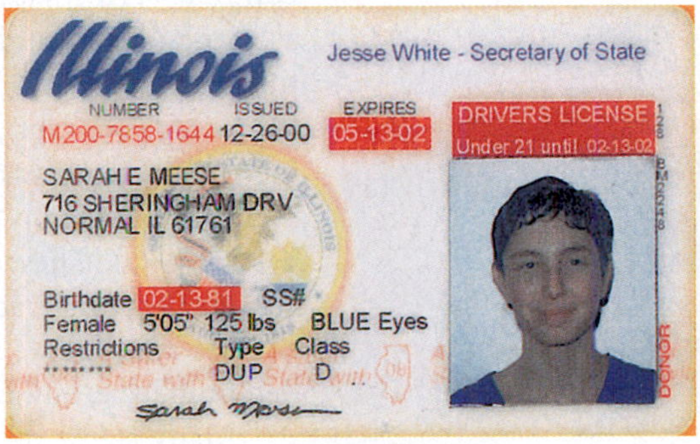


**FIGURE 16.11**
Illinois driver's license.

Consider this Social Security number: 189-31-9431. What information about the holder can be deduced from the number? Only that the holder obtained it in Pennsylvania (see Spotlight 16.4). Figure 16.11 shows an Illinois driver's license number: M200-7858-1644. What information about the holder can be deduced from this number? This time we can determine the date of birth, sex, and much about the person's name.

These two examples illustrate the extremes in coding personal data. The Social Security number has no personal data encoded in the number. It is entirely determined by the place and time it is issued, not the individual to whom it is assigned. In contrast, in some states the driver's license

### SPOTLIGHT 16.4 Social Security Numbers

The first three digits of Social Security numbers show where the number was applied for. Changes in population have forced some numbers to be moved or assigned out of sequence over the years.

| | | | | | |
|---|---|---|---|---|---|
| 001–003 | New Hampshire | 400–407 | Kentucky | 526–527 | Arizona |
| 004–007 | Maine | 408–415 | Tennessee | & 600–601, 764–765 | |
| 008–009 | Vermont | & 756–763 | | 528–529 | Utah |
| 010–034 | Massachusetts | 416–424 | Alabama | & 646–647 | |
| 035–039 | Rhode Island | 425–428 | Mississippi | 530, 680 | Nevada |
| 040–049 | Connecticut | & 587–588, 752–755 | | 531–539 | Washington |
| 050–134 | New York | 429–432 | Arkansas | 540–544 | Oregon |
| 135–158 | New Jersey | & 676–679 | | 545–573 | California |
| 159–211 | Pennsylvania | 433–439 | Louisiana | & 602–626 | |
| 212–220 | Maryland | & 659–665 | | 574 | Alaska |
| 221–222 | Delaware | 440–448 | Oklahoma | 575–576 | Hawaii |
| 223–231 | Virginia | 449–467 | Texas | & 750–751 | |
| & 669–699 | | & 627–645 | | 577–579 | District of Columbia |
| 232–236 | West Virginia | 468–477 | Minnesota | 580 | Virgin Islands |
| 232, 237–246 | North Carolina | 478–485 | Iowa | 580–584 | Puerto Rico |
| & 681–690 | | 486–500 | Missouri | & 596–599 | |
| 247–251 | South Carolina | 501–502 | North Dakota | 586 | Guam |
| & 654–658 | | 503–504 | South Dakota | 586 | American Samoa |
| 252–260 | Georgia | 505–508 | Nebraska | 586 | Philippine Islands |
| & 667–675 | | 509–515 | Kansas | 700–728 | *through July 1, 1963, reserved for railroad employees* |
| 261–267 | Florida | 516–517 | Montana | | |
| & 589–595, 766–772 | | 518–519 | Idaho | | |
| 268–302 | Ohio | 520 | Wyoming | | |
| 303–317 | Indiana | 521–524 | Colorado | | |
| 318–361 | Illinois | & 650–653 | | | |
| 362–386 | Michigan | 525 & 585 | New Mexico | | |
| 387–399 | Wisconsin | & 648–649 | | | |

*Source*: Social Security Administration.

numbers are entirely determined by personal information about the holders. It is no coincidence that the unsophisticated Social Security numbering scheme predates computers. Agencies that have large databases that include personal information such as names, sex, and dates of birth find it convenient to encode these data into identification numbers. Examples of such agencies are the National Archives (where census records are kept), genealogical research centers, the Library of Congress, and state motor vehicle departments.

There are many methods in use to encode personal data such as name, sex, and date of birth. These methods are perhaps most widely used in assigning driver's license numbers in some states. Coding license numbers solely from personal data enables automobile insurers, government entities, and law enforcement agencies to determine the number from the personal data.

Many states encode the surname, first name, middle initial, date of birth, and sex by quite sophisticated schemes.

In one scheme that is based on sound, the first four characters of the license number are obtained by applying the **Soundex Coding System** to the surname as follows.

1. Delete all occurrences of *h* and *w*. (For example, *Schworer* becomes *Scorer* and *Hughgill* becomes *uggill*.)
2. Assign numbers to the remaining letters as follows:

$$a, e, i, o, u, y \to 0 \qquad l \to 4$$
$$b, f, p, v \to 1 \qquad m, n \to 5$$
$$c, g, j, k, q, s, x, z \to 2 \qquad r \to 6$$
$$d, t \to 3$$

3. If two or more letters with the same numeric value are adjacent, omit all but the first. (For example, *Scorer* becomes *Sorer* and *uggill* becomes *ugil*).
4. Delete the first character of the original name if still present. (*Sorer* becomes *orer*).
5. Delete all occurrences of *a, e, i, o, u,* and *y*.
6. Retain only the first three digits corresponding to the remaining letters; append trailing 0's if fewer than three letters remain; precede the three digits obtained in step 6 with the first letter of the surname.

Figure 16.12 shows three examples.

What is the advantage of this method? It is an error-correcting scheme. Indeed, it is designed so that likely misspellings of a name nevertheless result in the correct coding of the name. For example, frequent misspellings of the name *Erickson* are *Ericksen, Eriksen, Ericson,* and *Ericsen*. Observe that all of these yield the same coding as *Erickson*. If a law enforcement official, a genealogical researcher, a librarian, or an airline reservation agent wanted to pull up the file from a data bank for someone whose name was pronounced "Erickson," the correct spelling isn't essential because the computer searches for records that are coded as E-625 for all spelling variations. The search feature of a Web site where many mathematicians post their research papers uses the Soundex Coding System. This system was designed for the U.S. Census Bureau when much census information was obtained orally.

| Step 1 | | Step 2 | |
|---|---|---|---|
| Schworer | → Scorer | → | Scorer 220606 |

| Step 3 | Step 4 | Step 5 | Step 6 | |
|---|---|---|---|---|
| → Sorer 20606 | → orer 0606 | → rr 66 | → | S-660 |

| Step 1 | | Step 2 | |
|---|---|---|---|
| Hughgill | → uggill | → | uggill 022044 |

| Step 3 | Step 4 | Step 5 | Step 6 | |
|---|---|---|---|---|
| → ugil 0204 | → ugil 0204 | → gl 24 | → | H-240 |

| Step 1 | | Step 2 | |
|---|---|---|---|
| Schmidlapper | → Scmidlapper | → | Scmidlapper 22503401106 |

| Step 3 | Step 4 | Step 5 | Step 6 | |
|---|---|---|---|---|
| → Smidlaper 250340106 | → midlaper 50340106 | → mdlpr 53416 | → | S-534 |

**FIGURE 16.12** The Soundex Coding System.

There are many schemes for encoding the date of birth and the sex in driver's license numbers. For example, the last five digits of Illinois and Florida driver's license numbers capture the year and date of birth as well as the sex. In Illinois, each day of the year is assigned a three-digit number in sequence beginning with 001 for January 1. However, each month is assumed to have 31 days. Thus, March 1 is given the number 063 because both January and February are assumed to have 31 days. These numbers are then used to identify the month and day of birth of male drivers. For females, the scheme is identical except that 600 is added to the number. The last two digits of the year of birth, separated by a dash (probably to obscure the fact that they represent the year of birth), are listed in the fifth and fourth positions from the end of the driver's license number. Thus, a male born on October 13, 1940, would have the last five digits 4-0292 ($292 = 9 \cdot 31 + 13$), whereas a female born on the same day would have 4-0892. The scheme to identify birth date and sex in Florida is the same as in Illinois except that each month is assumed to have 40 days and 500 is added for women. Moreover, a dash occurs between the two digits for the year and the three digits for the day. For example, the five digits 49-585 belong to a woman born on March 5, 1949.

In this chapter, we have investigated how mathematics is used to append a check digit to an identification number for error detection. In the next chapter, we will show how codes consisting of 0's and 1's can be devised so that errors can be corrected.

## REVIEW VOCABULARY

**Bar code** A code that employs bars and spaces to represent information. (p. 517)

**Binary code** A coding scheme that uses two symbols, usually 0 and 1. (p. 517)

**Check digit** A digit included in an identification number for the purpose of error detection. (p. 509)

**Code** A group of symbols that represent information together with a set of rules for interpreting the symbols. (p. 509)

**Decoding** Translating code into data. (p. 517)

**Encoding** Translating data into code. (p. 515)

**Error-detecting code** A code in which certain types of errors can be detected. (p. 509)

**International Standard Book Number (ISBN)** An identification number used on books throughout the world that contains a check digit for error detection. (p. 509)
**Postnet code** The bar code used by the U.S. Postal Service for ZIP codes. (p. 517)
**Soundex Coding System** An encoding scheme for surnames based on sound. (p. 524)
**Universal Product Code (UPC)** A bar code and identification number that is used on most retail items. The UPC code detects 100% of all single-digit errors and most other types of errors. (p. 510)
**Weights** Numbers used in the calculation of check digits. (p. 511)
**ZIP code** A five-digit code used by the U.S. Postal Service to divide the country into geographic units to speed sorting of the mail. ZIP stands for Zone Improvement Plan. (p. 515)
**ZIP + 4 code** The nine-digit code used by the U.S. Postal Service to refine ZIP codes into smaller units. (p. 516)

## ✓ SKILLS CHECK

**1.** When a single incorrect digit is entered, an error-detecting code

**(a)** will sometimes detect the error.
**(b)** will always detect the error but may not be able to correct the error.
**(c)** will always detect and correct the error.

**2.** If a U.S. Postal Service money order is numbered 1012065994X, where X indicates that the last digit is obliterated, X is ______________ .

**3.** If the first five digits of a valid U.S. Postal Service money order are rearranged, the resulting number will have the same check digit as the original number

**(a)** Always
**(b)** Sometimes
**(c)** Never

**4.** The sum of the digits of a correctly coded American Express Travelers Cheque identification number is evenly divisible by ______________ .

**5.** Is the number 105408970012 a legitimate airline ticket number?

**(a)** Yes.
**(b)** No, but if the final digit is changed to a 5, the resulting number 105408970015 is legitimate.
**(c)** No, but if the final digit is changed to a 3, the resulting number 105408970013 is legitimate.

**6.** If an American Express Travelers Cheque is numbered X425036790, where X indicates that the first digit is obliterated, X is ______________ .

**7.** If the first two digits of a valid airline ticket identification number are transposed, the resulting number will be valid

**(a)** Always
**(b)** Sometimes
**(c)** Never

**8.** A correctly coded UPC number has a weighted sum that is evenly divisible by ______________ .

**9.** The bank routing number error detection scheme detects

**(a)** all transportation and most single-digit errors.
**(b)** all single-digit errors and most transpositions.
**(c)** all single-digit errors and all transpositions.

**10.** The check digit that should be appended to the UPC code 0-14300-25433 is ______________ .

**11.** The bank routing number error detection scheme

**(a)** detects the same errors as the UPC scheme.
**(b)** detects fewer errors than the UPC scheme.
**(c)** detects more errors than the UPC scheme.

**12.** The check digit that should be appended to the bank routing number 01500085 is ______________ .

**13.** Suppose the 10-digit ISBN 0-1750-3549-0 is incorrectly reported as 0-1750-3540-1. Which of the following statements is true?

**(a)** This error will not be detected by the check digit.
**(b)** While this particular error will be detected, the check digit does not detect all two-digit errors in ISBNs.
**(c)** All two-digit errors in a 10-digit ISBN are detectable by the check digit.

**14.** A correctly coded 10-digit ISBN has a weighted sum that is evenly divisible by ______________ .

**15.** If an error in an identification number is made by transposing the first and third digits, the error is

**(a)** usually detected by the UPC scheme.
**(b)** always detected by the bank scheme.
**(c)** always detected by the ISBN-10 scheme.

**16.** The ISBN-10 error detection scheme detects __________ percent of single digits errors and __________ percent of transposition errors.

**17.** As far as the ability to detect single-position errors and adjacent-digit transposition errors

**(a)** the credit card scheme is superior to the UPC scheme.

**(b)** the credit card scheme is superior to the ISBN-10 scheme.
**(c)** the UPC scheme is superior to the bank scheme.

**18.** If the sixth digit of the Postnet code 20001-5800-7 is incorrect, the correct Postnet code is ____________ .

**19.** If a scanner misreads exactly one bar of a Postnet code the computer will

**(a)** not always detect the error.
**(b)** always detect the error but will not always be able to correct it.
**(c)** always detect the error and correct it.

**20.** If the identification 48945 has been encoded by multiplying the original number by 13, the original number is ______________________________ .

# CHAPTER 16 EXERCISES

■ Challenge ◆ Discussion

## 16.1 Check Digits

**1.** Determine the check digit for a money order with identification number 3953981640.

**2.** Determine the check digit for a money order with identification number 7234541780.

**3.** Determine the check digit for the United Parcel Service (UPS) identification number 873345672.

**4.** Suppose a money order with the identification number and check digit 21720421168 is erroneously copied as 27750421168. Will the check digit detect the error? Explain your reasoning.

**5.** Determine the check digit for the airline ticket number 30860422052.

**6.** Determine the check digit for the Avis rental car with identification number 540047.

**7.** Determine the check digit for the UPC number 38137009213.

**8.** If the packaging of a retail item were damaged in such a way that the first digit of a UPC code was scratched off, but the remaining digits were 88072303584, determine the first digit.

**9.** Determine the check digit for the ISBN 0-669-19493.

**10.** When this edition was in preparation, the publisher sent the author of this section the following ISBNs for the book: ISBN-10: 1-4292-0900-3; ISBN-13: 978-1-4292-0890-0. How did the author know the ISBN-13 was wrong and how did he know how to correct it? (This really happened!)

**11.** When calculating the check digit for a 13-digit ISBN, why can you disregard the first two digits? (Try it for Exercise 10.)

**12.** Determine the check digit for the bank routing number 09100001.

**13.** Determine the check digit for the American Express Travelers Cheque with identification number 461212023.

**14.** Suppose a check digit is assigned to a four-digit number by appending the remainder after division by 7. If the number 96806 has a single-digit error but the check digit is correct, determine the possibilities for the correct number.

**15.** Determine whether the Master Card number 3541 0232 0033 2270 is valid.

**16.** Suppose that the digit indicated by a question mark in the Master Card number 426452002177?337 is unreadable. What is the unreadable number?

**17.** Create a check digit for the UPC number 38137009213 using the weights 7, 1, 7, 1, 7, 1, . . . , 7, 1, instead of 3, 1, 3, 1, 3, 1, . . . , 3, 1. Test to see whether this check digit will detect single-digit errors by trying several examples.

**18.** Create a check digit for the UPC number 38137009213 using the weights 2, 1, 2, 1, 2, 1, . . . , 2, 1, instead of 3, 1, 3, 1, 3, 1, . . . , 3, 1. Is the error caused by replacing the 3 in the first position with an 8 detected? What about the error caused by replacing the 1 in the third position with a 6? Explain why or why not.

**19.** If the weights 5, 1, 5, 1, 5, 1, . . . , 5, 1 were used for the UPC code, which single-digit errors would go undetected?

**20.** Exercises 17, 18, and 19 reveal that using the weights 1, 3, or 7 for a particular position detects all errors in that position, whereas using weights 2 or 5 in a position does not detect all errors. Using this observation, make a guess about error-detection capability using weights 9, 4, 6, or 8.

**21.** Use the credit card scheme to determine the check digit for the number 300125600196431.

**22.** Determine the check digit for the VIN JM1GD222J1581570 (see Spotlight 16.1 for a description of the method to be used).

**23.** For some products, such as soft-drink cans and magazines, an 8-digit UPC number called Version E is used instead of the 12-digit number. The method of calculating the eighth digit, which is the check digit, depends on the value of the seventh digit. The check digit $a_8$ for a UPC Version E identification number $a_1a_2a_3a_4a_5a_6a_7$, where $a_7$ is 0, 1, or 2, is chosen so that $a_1 + a_2 + 3a_3 + 3a_4 + a_5 + 3a_6 + a_7 + a_8$ is divisible by 10. Use this fact to determine the check digit for the following Version E numbers:

(a) 0121690
(b) 0274551
(c) 0760022
(d) 0496580

(georgphotos/Alamy.)

**24.** The check digit $a_8$ for a UPC Version E identification number $a_1a_2a_3a_4a_5a_6a_7$, where $a_7$ is 4, is chosen so that $a_1 + a_2 + 3a_3 + a_4 + 3a_5 + 3a_6 + a_8$ is divisible by 10. Use this fact to determine the check digit for the following Version E numbers:

(a) 0754704
(b) 0774714
(c) 0724444

■ **25.** The 10-digit ISBN 0-669-03925-4 is the result of a transposition of two adjacent digits not involving the first or last digit. Determine the correct ISBN.

**26.** Explain why the bank scheme will detect the error $751 \cdots \to 157 \cdots$ but the UPC scheme will not.

**27.** Suppose the check digit $a_9$ for the bank routing number was chosen to be the last digit of $3a_1 + 7a_2 + a_3 + 3a_4 + 7a_5 + a_6 + 3a_7 + 7a_8$ instead of the way described in this chapter. How would this compare with the actual check digit?

**28.** Explain why an error caused by transposing the first two digits of a Postal Service money order is not detected by the check-digit scheme. Explain why the same is true for the second and third digits. What about the last two digits?

**29.** Suppose a company assigns an extra digit to every employee Social Security number by appending a 0 if the sum of the digits is even and a 1 if the sum is odd. If a 2 were mistakenly read as a 7 would the error be detected? What if a 2 were mistakenly read as an 8? Try a few other experiments with single-digit errors (for experiments you can use three-digit numbers instead of nine-digit numbers). Determine which errors are detected by this method. Explain your reasoning.

**30.** Explain why an error caused by transposing any two digits of an American Express Travelers Cheque is not detected by the check-digit scheme.

**31.** When using the traveler's check, credit card or UPC number algorithms for detecting errors does the computer have to know which digit is the check digit?

**32.** Explain why the Postal Service money order check-digit scheme does not detect the mistake of substituting a 0 for a 9, or vice versa.

**33.** Which digit never appears as a check digit on a Postal Service money order?

**34.** Which digit never appears as a check digit on an American Express Travelers Cheque?

**35.** Which digits never appear as a check digit for an airline identification number?

**36.** Suppose four-digit numbers $a_1a_2a_3a_4$ are assigned a check digit $a_5$ so that $a_1 + 2a_2 + a_3 + 2a_4 + a_5$ is evenly divisible by 10. Test the number 43216 created in this way to see whether the method detects adjacent-digit transposition errors.

**37.** Starting with the 10-digit ISBN 0-7167-4782-0, create three new numbers by transposing any two different digits. (They need not be adjacent.) Are these errors detected by the scheme?

**38.** Suppose in a UPS number an 8 is mistaken for a 5. Is the error detected? What if a 9 is mistaken for a 2?

**39.** Give an argument to show that the 10-digit ISBN error-detection method will detect a transposition error involving the first and third digits. Does the same argument work for the fourth and sixth digits?

■ **40.** Suppose the check digit $a_{10}$ of 10-digit ISBNs were chosen so that $a_1 + 2a_2 + 3a_3 + 4a_4 + 5a_5 + 6a_6 + 7a_7 + 8a_8 + 9a_9 + 10a_{10}$ is divisible by 11 instead of the way described in the chapter. How would this compare with the actual check digit?

■ **41.** Consider a UPC number in which the digits 7 and 2 appear consecutively (that is, the number has the form $\cdots 72 \cdots$). Will the error caused by transposing these digits (that is, the number is taken as $\cdots 27 \cdots$) be detected? What if the digits 6 and 2 were transposed instead? State the general criterion for the detection of an error of the form $\cdots ab \cdots \to \cdots ba \cdots$ by the UPC scheme.

**42.** If the first three digits of a routing number for a checking account are 537 and the 5 and 3 are transposed, will the error be detected? If the first three numbers are 237 and the 2 and 7 are transposed, will the error be detected?

**43.** State a general criterion for the detection of an error of the form $\cdots abc \cdots \to \cdots cba \cdots$ for the routing number of a checking account.

**44.** The state of Utah appends a ninth digit $a_9$ to an eight-digit driver's license number $a_1a_2 \cdots a_8$ so that $9a_1 + 8a_2 + 7a_3 + 6a_4 + 5a_5 + 4a_6 + 3a_7 + 2a_8 + a_9$ is divisible by 10.

**(a)** If the first eight digits of a Utah driver's license number are 14910573, what is the ninth digit?
**(b)** Suppose a legitimate Utah driver's license number 149105767 is miscopied as 149105267. How would you know a mistake was made? Is there any way you could determine the correct number? Suppose you know the error was in the seventh position. Could you correct the mistake?
**(c)** If a legitimate Utah driver's license number 149105767 were miscopied as 199105767, would you be able to tell a mistake was made? Explain.
**(d)** Explain why any transposition error involving adjacent digits of a Utah driver's license number would be detected.

**45.** The Canadian province of Quebec assigns a check digit $a_{12}$ to an 11-digit driver's license number $a_1a_2 \cdots a_{11}$ so that $12a_1 + 11a_2 + 10a_3 + 9a_4 + 8a_5 + 7a_6 + 6a_7 + 5a_8 + 4a_9 + 3a_{10} + 2a_{11} + a_{12}$ is divisible by 10. Criticize this method. Describe all single-digit errors that are undetected by this scheme.

**46.** Speculate on the reason why telephone numbers, Social Security numbers, and serial numbers on most currency do not have check digits.

**47.** Suppose a company uses a check-digit scheme similar to the UPC scheme, except that instead of using the UPC weights 3, 1, 3, 1, . . . it uses $w$, 1, $w$, 1, . . . . If two of the ID numbers used by the company are 73215674 and 73215661, determine $w$.

**48.** If a publishing company has headquarters in both the United States and Germany and publishes the same book in both countries, it is likely that the 10-digit ISBN for the book will be identical except for the first and last digits (because the first digit for U.S. publications is 0 and the first digit for German publications is 3). If the last digit of the U.S. edition is 1, what is the last digit for the German publication?

## 16.2 The Zip Code

**49.** Determine the ZIP + 4 code and check digit for each of the following Postnet bar codes:

(a)

(b)

(c)

**50.** Determine the ZIP + 4 code and check digit for each of the following Postnet bar codes:

(a)

(b)

(c)

**51.** In each of the following Postnet bar codes, exactly one mistake occurs (that is, a long bar appears instead of a short one, or vice versa). Determine the correct ZIP code.

(a)

(b)

(c)

**52.** Below is a 12-digit delivery-point bar code. Determine the ZIP + 4 number, the last two digits of the street address, and the check digit.

**53.** Explain why any two errors in a particular block of five bars in a Postnet code are always detectable. Explain why not all such errors can be corrected.

**54.** Change 173 into a Postnet code.

**55.** Form all possible strings consisting of exactly three $a$'s and two $b$'s and arrange the strings in alphabetical order (for example, the first two possibilities are *aaabb* and *aabab*). Do you see any relationship between your list and the Postnet code?

## 16.3 Bar Codes

**56.** Many recently published books include a bar code on the back cover that has the 10-digit ISBN above the bars and a 13-digit identification number below the bars. Examine several books with a bar code on the back cover. How does the number below the bar code differ from the UPC code? How is the number below the bar code related to the ISBN? Given the fact that the last digit in the number below the bar code is a check digit, determine how it is calculated.

**57.** Suppose the first block of a UPC bar code following the guard bar pattern that a scanner reads is 1000100. Is the scanner reading left to right or right to left?

**58.** The following is an actual identification number and bar code from a roll of wallpaper. What appears to be wrong with them? Speculate on the reason for the apparent violation of the UPC format.

5 011419 194056

Building Regulations: 1985 Class 0
FINE ART WALLCOVERINGS LTD.
HOLMES CHAPEL, CHESHIRE
MADE IN ENGLAND
FABRIQUE EN ANGLETERRE

## 16.4 Encoding Personal Data

**59.** Judging from the information in Spotlight 16.4, which three states were most likely to have had the smallest populations when Social Security numbers were allocated to the states?

**60.** What geographical information was used in allotting the first three digits of Social Security numbers to the states?

**61.** What demographic information was used in allotting the first three digits of Social Security numbers to the states?

**62.** As of 2007 there were seven states with a population under 1 million. Use the data in Spotlight 16.4 to identify those seven states.

◆ **63.** The Canadian postal system has assigned each geographic region a six-character code composed of alternating letters and digits, such as P7B5E1 and K7L3N6. Discuss the advantages this scheme has over the five-digit ZIP code used in the United States.

**64.** Determine the Soundex code for Smith, Schmid, Smyth, and Schmidt.

**65.** Determine the Soundex code for Skow, Sachs, Lennon, Lloyd, Ehrheart, and Ollenburger.

**66.** In Florida, the last three digits of the driver's license number of a female with birth month $m$ and birth date $b$ are $40(m - 1) + b + 500$. For both males and females, the fourth and fifth digits from the end give the year of birth. Determine the last five digits of a Florida driver's license number for a female born on July 18, 1942.

**67.** Explain why an Illinois driver's license number that ends with the last five digits 99817 cannot be valid.

**68.** Determine the last five digits of an Illinois driver's license number for a male born on June 18, 1942.

**69.** In Illinois, one obtains the last three digits of the driver's license number for a female by adding 600 to the number for a male with the same birthday. In Florida 500 is added to the number for a male. Why can't Florida use 600?

**70.** Explain why an Illinois driver's license number that ends with 77061 cannot be valid.

**71.** Determine the birth date of a person whose Illinois driver's license number ends with 58818.

**72.** In Florida, the last three digits of the driver's license number of a male with birth month $m$ and birth date $b$ are $40(m - 1) + b$. For both males and females, the fourth and fifth digits from the end give the year of birth. Determine the birth dates of people with numbers whose last five digits are 42218 and 53953.

**73.** Provide three names that share the same Soundex code as Gallihan.

**74.** Another math book describes the Soundex algorithm for the surname code as follows:

(i) Leave the first letter alone, then cross off all occurrences of the letters a, e, i, o, u, y, h and w.
(ii) Cross off the second of any double letters.
(iii) Leave the first letter alone, and replace each of the other letters with the appropriate number (using the same assignment as given on page 524).
(iv) The code is the first letter of the surname followed by the first three numbers.

Compare the codes for Jackson, Mnack and Shaw using this method and the method given on page 524. (The method on page 524 is the correct one.)

**75.** For driver's license numbers issued in New York before September 1992, the last two digits were the year of birth. The three digits preceding the year encoded the sex and the month and day of birth. For a woman with birth month $m$ and birth date $b$, the three digits were $63m + 2b + 1$ (insert a 0 in front for numbers less than 100). For a man with birth month $m$ and birth date $b$, the three digits were $63m + 2b$. Determine the birth months, birth dates, and sexes of drivers with the three digits 248 and 601 preceding the year.

**76.** The state of Washington encodes the last two digits of the year of birth into driver's license numbers (in positions 8 and 9) by subtracting the two-digit number from 100. For example, a person born in 1942 has 58 in positions 8 and 9, whereas a person born in 1971 has 29 in positions 8 and 9. Speculate on the reason for subtracting the birth year from 100.

**77.** Driver's license number-assignment schemes that use personal data sometimes produce the same number for different people. Speculate about circumstances under which this is more likely to occur.

**78.** Apply the Soundex code to common ways to misspell your name. Do they give the same code as your name does?

**79.** Why would the Soundex system of coding last names be a poor method for encoding names in China?

## WRITING PROJECTS

**1.** Prepare a two-page report on coded information in your location. Possibilities for investigation include driver's license numbers in your state, student ID numbers and bar codes at your school, and bar codes used by your school library and city library. Identify the coding schemes and, when possible, determine whether a check digit is employed. Include samples. The Suggested Readings for this chapter contain information that will assist you.

**2.** Prepare a two-page report on the driver's license coding schemes used by Michigan, Maryland, and Washington (Michigan and Maryland use the same method). J. Gallian's "Assigning Driver's License Numbers" has the information you will need (see the Suggested Readings).

**3.** Use the Web to find material for a two-page report on the history of the bar code.

**4.** Use the Web to find material for a two-page report on the Barcode of Life project.

**5.** Use the Web to find material for a two-page report on Smart Card technology.

## SUGGESTED READINGS

GALLIAN, J. The mathematics of identification numbers, *College Mathematics Journal,* 22 (1991): 194–202. A survey of check-digit schemes associated with identification numbers.

GALLIAN, J. Assigning driver's license numbers, *Mathematics Magazine,* 64 (1992): 13–22. Discusses various methods used by the states to assign driver's license numbers. Several of these methods include check digits for error detection.

GALLIAN, J. Error detection methods, *ACM Computing Surveys,* 28 (1996): 504–517. A detailed description of many error-detection methods.

GALLIAN, J., and S. WINTERS. Modular arithmetic in the marketplace, *American Mathematical Monthly,* 95 (1988): 548–551. A detailed analysis of the check-digit schemes presented in this chapter. In particular, the error-detection rates for the various schemes are given.

KIRTLAND, J. *Identification Numbers and Check Digit Schemes,* Mathematical Association of America, Washington, D.C., 2001. Provides more examples and exercises for the check-digit schemes discussed in this chapter.

## SUGGESTED WEB SITES

**www.d.umn.edu/~jgallian/fapp7** This Web site enables users to calculate check digits using the various methods discussed in this chapter.

# CHAPTER 17



# Information Science

With the enormous volume of email, faxes, Internet traffic, and cellular phone calls, the Digital Revolution has brought about many mathematical challenges. One is how to correct errors in data transmission. Another is how to electronically send and store information economically. A third is how to ensure security of transmitted data. Yet another is how to improve Web search efficiency. In this chapter we illustrate some of the ways in which mathematicians and engineers have responded to these challenges.

## 17.1 Binary Codes

A system for coding data made up of two states (or symbols) is called a **binary code**. Binary codes are the hidden language of computers. The Postnet code (short and long bars) and the Universal Product Code (UPC) bar code (white spaces and dark bars) are two examples of binary codes. Morse code (dots and dashes) and Braille (bumps and flat) are two more. The Ebert and Roeper "thumbs up/thumbs down" rating of films is a binary code with four messages. CDs (compact disks), fax machines, DVDs (digital video disks), high definition television signals, cell phones, and space probes represent data as strings of 0's and 1's rather than the usual digits 0 through 9 and letters *A* through *Z*. In this section we will illustrate one way binary codes can be devised so that errors in the transmission of the code can be corrected.

The idea behind error-correction schemes is simple and one you often use. To illustrate, suppose you are reading the employment section of a newspaper and you see the phrase "must have a minimum of bive years experience." Instantly you detect an error because *bive* is not a word in the English language. Moreover, you are fairly confident that the intended word is *five*. Why so? Because *five* is a word derived from *bive* by changing a single letter and it makes the phrase understandable. In other phrases, words such as *bike* or *give* might be sensible alternatives to *bive*. Using the extra information provided by the context, we are often able to infer the intended meaning when errors occur.

To demonstrate the way error-correcting schemes work, suppose that NASA sends a spacecraft to land at one of 16 possible landing sites on Mars. The spacecraft orbits Mars while surveying the sites for the most favorable landing conditions. NASA officials have coded the 16 landing sites with four-digit strings of 0's and 1's such as 0000, 0001, 0010, 0100 (see Table 17.1 for the complete list). Once the best site has been selected, NASA will inform the spacecraft where to land by sending the code for the site. However, signals sent through space are subject to interference called *noise*. The noise might cause the spacecraft to interpret the signal as 0001 when the signal actually sent was 1001. Fortunately, over the past 60 years mathematicians and engineers have devised highly sophisticated schemes to build extra information into messages composed of 0's and 1's that often permits one to infer the correct message even though the message may have been received incorrectly (see Spotlights 17.1 and 17.2).

As a simple example, let's assume our message is 1001. We will build extra information into this message with the aid of the diagram in Figure 17.1. Begin by placing the four message digits in the four overlapping regions I, II, III, IV, with the digit in the first position (starting at the left of the sequence) in region I, the digit in the second position in region II, and so on. For regions V, VI, and VII, assign 0 or 1 so that the total number of 1's in each circle is even. See Figure 17.2.

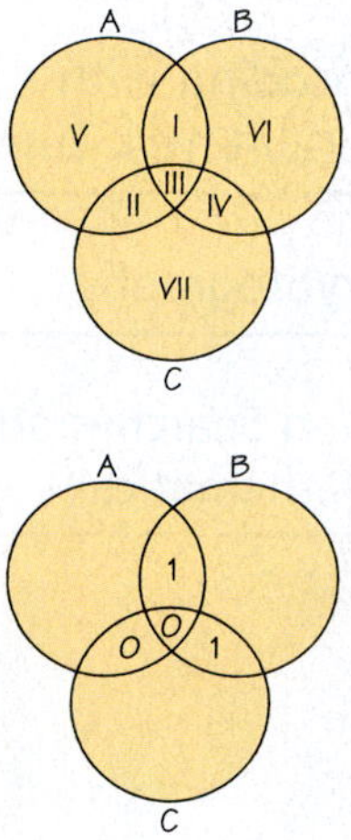


FIGURE 17.1 Diagram for message 1001.

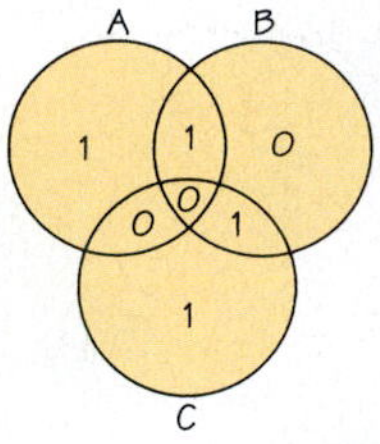


FIGURE 17.2 Diagram for encoded message 1001101.

**TABLE 17.1**

| Message | → | Code Word | Message | → | Code Word |
|---|---|---|---|---|---|
| 0000 | → | 0000000 | 0110 | → | 0110010 |
| 0001 | → | 0001011 | 0101 | → | 0101110 |
| 0010 | → | 0010111 | 0011 | → | 0011100 |
| 0100 | → | 0100101 | 1110 | → | 1110100 |
| 1000 | → | 1000110 | 1101 | → | 1101000 |
| 1100 | → | 1100011 | 1011 | → | 1011010 |
| 1010 | → | 1010001 | 0111 | → | 0111001 |
| 1001 | → | 1001101 | 1111 | → | 1111111 |

We have now encoded our message 1001 using the diagram as 1001101. Now suppose that this encoded message is received as 0001101 (an error in the first position). How would we know an error was made? We place each digit from the received message in its appropriate region, as in Figure 17.3.

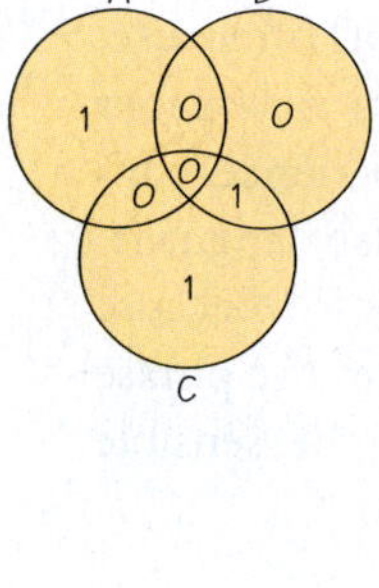


FIGURE 17.3 Diagram for received message 0001101.

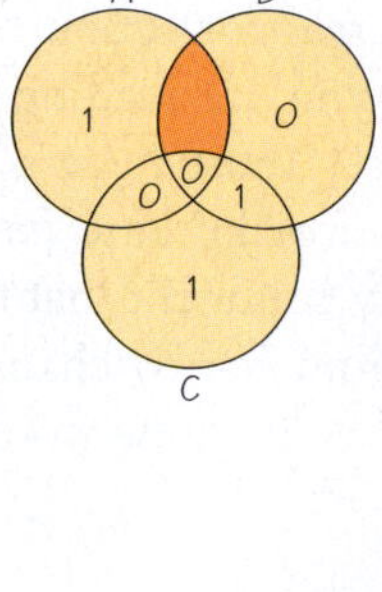


FIGURE 17.4 Circles A and B but not C have wrong parity.

## SPOTLIGHT 17.1 The Ubiquitous Reed-Solomon Codes

One of the mathematical ideas underlying current error-correcting techniques for everything from computer hard-disk drives to CD players was first introduced in 1960 by Irving Reed and Gustave Solomon. Reed–Solomon codes made possible the stunning pictures of the outer planets sent back by the space probes *Voyager 1* and *2*. They make it possible to scratch a compact disc and still enjoy the music.

"When you talk about CD players and digital audio tape and now digital television, and various other digital imaging systems that are coming—all of those Reed–Solomon [codes] are an integral part of the system," says Robert McEliece, a coding theorist at Caltech.

Why? Because digital information consists of 0's and 1's, and a physical device may occasionally confuse the two. *Voyager 2*, for example, was transmitting data at incredibly low power over billions of miles. Error-correcting codes are a kind of safety net, mathematical insurance against the vagaries of an imperfect material world.

In 1960, the theory of error-correcting codes was only about a decade old. Through the 1950s, a number of researchers began experimenting with a variety of error-correcting codes. But the Reed–Solomon paper, McEliece says, "hit the jackpot." "In hindsight it seems obvious," Reed later said. However, he added, "Coding theory was not a subject when we published the paper." The two authors knew they had a nice result; they didn't know what impact the paper would have.

**Irving Reed and Gustave Solomon**

At the Jet Propulsion Laboratory in 1989 to monitor the encounter of *Voyager 2* with Neptune. (*Rex Ridenhouse.*)

Four decades later, the impact is clear. The vast array of applications, both current and pending, has settled the questions of the practicality and significance of Reed–Solomon codes. Billions of dollars in modern technology depend on ideas that stem from Reed and Solomon's original work.

*Source*: Adapted from an article by Barry Cipra, with permission from *SIAM News*, January 1993, p. 1. © by SIAM. All rights reserved.

Noting that in both circles $A$ and $B$ there is an odd number of 1's, we instantly realize that something is wrong, because the intended message had an even number of 1's in each circle. How do we correct the error? Because circles $A$ and $B$ have the wrong parity (parity refers to the oddness or evenness of a number–even integers have **even parity**; odd integers have **odd parity**) and $C$ does not, the error is located in the portion of the diagram in circles $A$ and $B$, but not in circle $C$; that is, region I (see Figure 17.4). Here we also see the advantage of using only 0's and 1's to encode data. If you have only two possibilities and one of them is incorrect, then the other one must be correct. Because the 0 in region I is incorrect, we know 1 is correct.

This technique can be used to encode all 16 possible binary messages of length 4, as shown in Table 17.1. The encoded messages are called *code words*. The three digits appended to each string of length 4 provide the "extra information" that is sufficient to infer the intended four-digit message as long as the received seven-digit message has at most one error. If a received message has two or more errors, this method will never yield the correct message.

## SPOTLIGHT 17.2 Vera Pless

Vera Pless was born on March 5, 1931, to Russian immigrants on the West Side of Chicago. The neighborhood was intellectually stimulating, and there was a tradition of teaching each other things. At age 12, Pless was taught some calculus by a mathematics graduate student. She accepted a scholarship to attend the University of Chicago at age 15. The program at Chicago emphasized great literature but paid little attention to physics and mathematics. At age 18, with no more than one precalculus course in mathematics, she entered the prestigious graduate program in mathematics at Chicago, where, at that time, there were no women on the mathematics faculty nor even women colloquium speakers. After receiving her master's degree, Pless took a job as a research associate at Northwestern University while pursuing a Ph.D. there. In the midst of writing her thesis, she moved to Boston with her husband and continued to work on her thesis at home. She defended her thesis two weeks before her daughter was born.

Over the next several years, Pless stayed at home to raise her children and taught part time at Boston University. When she decided to work full time, she found that women were not welcome at most colleges and universities. Some people told her outright, "I would never hire a woman." Fortunately, there was an Air Force lab in the area that had a group working on error-correcting codes. Although she had never even heard of coding theory, she was hired because of her background in algebra. When the lab discontinued basic research, she took a position as a research associate at MIT. In 1975, she went to the University of Illinois–Chicago, where she remained until her retirement. Having written more than 100 research papers and a widely used book on coding theory, Pless is a leader in the field.

Vera Pless
A leader in the field of coding theory.
*(Courtesy of Vera Pless.)*

## 17.2 Encoding with Parity-Check Sums

Strings of 0's and 1's with extra digits for error correction can be used to send full-text messages. A simple way to do this is to assign a space the string 00000, the letter *a* the string 00001, *b* the string 00010, *c* the string 00100, and so on. Because there are 32 possible binary strings of length 5, the five unassigned strings can be used for special purposes, such as indicating uppercase letters or numerals. For example, we might use the string 11111 to indicate a "shift" from lowercase to uppercase when it precedes the code for a letter (1111100010 represents *B*). This is like the shift key on a keyboard. Similarly, we could use 11110 to indicate we are "shifting" from letters to numerals. Here 11110 followed by the code for *a* represents the numeral 0, 11110 followed by the code for *b* represents the numeral 1, and so on up to 9. Punctuation marks could be handled in the same fashion. However, our diagram method for assigning extra digits does not work for strings with five or more digits. Rather, the messages are encoded by appending extra digits determined by the parity of various sums of certain portions of the messages. We illustrate this method for the 16 messages shown in the left-hand column of Table 17.1. (See also Spotlight 17.3.)

Our goal is to take any binary string $a_1a_2a_3a_4$ and append three check digits $c_1c_2c_3$ so that any single error in any of the seven positions can be corrected. This is done as follows: Choose

$$c_1 = 0 \text{ if } a_1 + a_2 + a_3 \text{ is even}$$
$$c_1 = 1 \text{ if } a_1 + a_2 + a_3 \text{ is odd}$$
$$c_2 = 0 \text{ if } a_1 + a_3 + a_4 \text{ is even}$$
$$c_2 = 1 \text{ if } a_1 + a_3 + a_4 \text{ is odd}$$
$$c_3 = 0 \text{ if } a_2 + a_3 + a_4 \text{ is even}$$
$$c_3 = 1 \text{ if } a_2 + a_3 + a_4 \text{ is odd}$$

The sums $a_1 + a_2 + a_3$, $a_1 + a_3 + a_4$, and $a_2 + a_3 + a_4$ are called **parity-check sums.** They are so named because their function is to guarantee that the sum of various components of the encoded message is even. Indeed, $c_1$ is defined so that $a_1 + a_2 + a_3 + c_1$ is even. (Recall that this is precisely how the value in region V in Figure 17.2 was defined.) Similarly, $c_2$ is defined so that $a_1 + a_3 + a_4 + c_2$ is even, and $c_3$ is defined so that $a_2 + a_3 + a_4 + c_3$ is even.

Let's revisit the message 1001 we considered in Figure 17.1. Then $a_1a_2a_3a_4 = 1001$ and

$$c_1 = 1 \text{ because } 1 + 0 + 0 \text{ is odd}$$
$$c_2 = 0 \text{ because } 1 + 0 + 1 \text{ is even}$$

and

$$c_3 = 1 \text{ because } 0 + 0 + 1 \text{ is odd}$$

So, because $c_1c_2c_3 = 101$, we have $1001 \to 1001101$.

Now how is the intended message determined from a received encoded message? This process is called **decoding.** Say, for instance, that the message 1000, which has been encoded using parity-check sums as $u = 1000110$, is received as $v = 1010110$ (an error in the third position). We simply compare $v$ with each of the 16 code words (that is, the possible correct messages) in Table 17.1 and decode it as the one that differs from $v$ in the fewest positions. (Put another way, we decode $v$ as the code word that agrees with $v$ in the most positions.) This method works even if the error in the message is one of the check digits rather than one of the digits of the original message string. In a situation when there is more than one code word that differs from $v$ in the fewest positions, we do not decode. To carry out this comparison, it is convenient to define the distance between two strings of equal length.

**Distance Between Two Strings** DEFINITION

The **distance between two strings** of equal length is the number of positions in which the strings differ.

For example, the distance between $v = 1010110$ and $u = 1000110$ is 1, because they differ in only one position (the third). In contrast, the distance between 1000110 and 0111001 is 7, because they differ in all seven positions. Thus, our decoding procedure is simply to decode any received message $v$ as the code word $v'$ that is "nearest" to $v$ in the sense that among all distances between $v$ and code words, the distance between $v$ and $v'$ is a minimum. If there is more than one possibility for $v'$,

## Spotlight 17.3 Neil Sloane

In the middle of Neil Sloane's office, which is in the center of AT&T Bell Laboratories, which in turn is at the heart of the Information Age, there sits a tidy little pyramid of shiny steel balls stacked up like oranges at a neighborhood grocery. Sloane has been pondering different ways to pile up balls of one kind or another for most of his professional life. Along the way he has become one of the world's leading researchers in the field of sphere packing, a field that has become indispensable to modern communications. Without it, we might not have modems or compact discs or satellite photos of Neptune. "Computers would still exist," says Sloane. "But they wouldn't be able to talk to one another."

**Neil Sloane**

At work, wearing his famous "Codemart" T-shirt (952 points in a sphere). (*Courtesy of Neil Sloane/AT&T Labs.*)

To exchange information rapidly and correctly, machines must code it. As it turns out, designing a code is a lot like packing spheres: Both involve cramming things together into the tightest possible arrangement. Sloane, fittingly, is also one of the world's leading coding theorists, not least because he has studied the shiny steel balls on his desk so intently.

Here's how a code might work. Imagine, for example, that you want to transmit a child's drawing that uses every one of the 64 colors found in a jumbo box of Crayola crayons. For transmission, you could code each of those colors as a number—say, the integers from 1 to 64. Then you could divide the image into many small units, or pixels, and assign a code to each one based on the color it contains. The transmission would then be a steady stream of those numbers, one for each pixel.

In digital systems, however, all those numbers would have to be represented as strings of 0's and 1's. Because there are 64 possible combinations of 0's and 1's in a six-digit string, you could handle the entire Crayola palette with 64 different six-digit "code words." For example, 000000 could represent the first color, 000001 the next color, 000010 the next, and so on.

But in a noisy signal, two different code words might look practically the same. A bit of noise, for example, might shift a spike of current to the wrong place, so that 001000 looks like 000100. The receiver might then wrongly color someone's eyes. An efficient way to keep the colors straight in spite of noise is to add four extra digits to the six-digit code words. The receiver, programmed to know the 64 permissible combinations, could now spot any other combination as an error introduced by noise and it would automatically correct the error to the "nearest" permissible color.

In fact, says Sloane, "If any of those ten digits were wrong, you could still figure out what the right crayon was."

*Source*: Adapted from an article by David Berreby, *Discover*, October 1990.

we do not decode. Table 17.2 shows the distance between $v = 1010110$ and all 16 code words. From this table, we see that $v$ will be decoded as $u$, because it differs from $u$ in only one position, whereas it differs from all others in the table in at least two positions. This method is called **nearest-neighbor decoding**.

Assuming that errors occur independently, the nearest-neighbor method decodes each received message as the one it most likely represents.

**TABLE 17.2**

| | | | | | | | | |
|---|---|---|---|---|---|---|---|---|
| $v$ | 1010110 | 1010110 | 1010110 | 1010110 | 1010110 | 1010110 | 1010110 | 1010110 |
| Code word | 0000000 | 0001011 | 0010111 | 0100101 | 1000110 | 1100011 | 1010001 | 1001101 |
| Distance | 4 | 5 | 2 | 5 | 1 | 4 | 3 | 4 |
| $v$ | 1010110 | 1010110 | 1010110 | 1010110 | 1010110 | 1010110 | 1010110 | 1010110 |
| Code word | 0110010 | 0101110 | 0011100 | 1110100 | 1101000 | 1011010 | 0111001 | 1111111 |
| Distance | 3 | 4 | 3 | 2 | 5 | 2 | 6 | 3 |

### Nearest-Neighbor Decoding

DEFINITION

The **nearest-neighbor decoding** method decodes a received message as the code word that agrees with the message in the most positions provided that there is only one such message.

The scheme we have just described was first proposed in 1948 by Richard Hamming, a mathematician at Bell Laboratories. It is one of a family of codes that are called the Hamming codes.

Strings obtained from all possible messages of a given length of 0's and 1's by appending extra 0's and 1's using parity-check sums, as illustrated earlier, are called **binary linear codes**. The strings with the appended digits are called **code words**.

### Binary Linear Code

DEFINITION

A set of words composed of 0's and 1's obtained from all possible messages of a given length by using parity-check sums to append check digits to the messages is called a **binary linear code**. The resulting strings are called **code words**.

Think of a binary linear code as a set of *n*-digit strings in which each string is composed of two parts: the message part, consisting of the original messages, and the remaining check-digit part.

The longer the messages are, the more check digits are required to correct errors. For example, binary messages consisting of six digits require four check digits to ensure that all messages with one error can be decoded correctly.

Given a binary linear code, how can we tell whether it will correct errors and how many errors it will detect? It is remarkably easy. We examine all the code words to find one that has the fewest number of 1's, excluding the *zero code word* consisting entirely of 0's. Call this minimum number of 1's in any nonzero code word the *weight* of the code and denote it by $t$.

### Weight of a Binary Code

DEFINITION

The **weight of a binary code** is the minimum number of 1's that occur among all nonzero code words of that code.

If $t$ is odd, the code will correct any $(t - 1)/2$ or fewer errors. If $t$ is even, the code will correct any $(t - 2)/2$ or fewer errors. If we prefer simply to detect errors rather than to correct them (as is often the case in applications), the code will detect any $t - 1$ or fewer errors.

Applying this test to the code in Table 17.1, we see that the weight is 3, so it will correct any $(3 - 1)/2 = 1$ error or it will detect any $3 - 1 = 2$ errors. Be careful here. We must decide *in advance* whether we want our code to correct single errors or detect any two errors. It can do whichever we choose, but not both. If we decide to detect errors, then we will not decode any message that was not among our original list of encoded messages (just as *bive* is not a word in the English language). Instead, we simply note that an error was made and, in most applications, request a retransmission. An example of this occurs when a bar-code reader at the supermarket detects an error and therefore does not emit a sound (in effect, requesting a rescanning). On the other hand, if we decide to correct errors, we will decode any received message as its nearest neighbor.

Here is an example of another binary linear code. Let the set of messages be {000, 001, 010, 100, 110, 101, 011, 111} and append three check digits $c_1$, $c_2$, and $c_3$ using

$$c_1 = 0 \text{ if } a_1 + a_2 + a_3 \text{ is even}$$
$$c_1 = 1 \text{ if } a_1 + a_2 + a_3 \text{ is odd}$$
$$c_2 = 0 \text{ if } a_1 + a_3 \text{ is even}$$
$$c_2 = 1 \text{ if } a_1 + a_3 \text{ is odd}$$
$$c_3 = 0 \text{ if } a_2 + a_3 \text{ is even}$$
$$c_3 = 1 \text{ if } a_2 + a_3 \text{ is odd}$$

For example, if we take $a_1a_2a_3$ as 101, we have

$$c_1 = 0 \text{ because } 1 + 0 + 1 \text{ is even}$$
$$c_2 = 0 \text{ because } 1 + 1 \text{ is even}$$
$$c_3 = 1 \text{ because } 0 + 1 \text{ is odd}$$

So we encode 101 by appending 001, that is, 101 → 101001. The entire code is shown in Table 17.3.

**TABLE 17.3**

| Message | → | Code Word | Message | → | Code Word |
|---|---|---|---|---|---|
| 000 | → | 000000 | 110 | → | 110011 |
| 001 | → | 001111 | 101 | → | 101001 |
| 010 | → | 010101 | 011 | → | 011010 |
| 100 | → | 100110 | 111 | → | 111100 |

Because the minimum number of 1's of any nonzero code word is three, this code will either correct any single error or detect any two errors, whichever we choose.

It is natural for you to ask how the method of appending extra digits with parity-check sums enables us to detect or even correct errors. Error detection is obvious. Think of how a computer spell-checker works. If you type *bive* instead of *five*, the spell-checker detects the error because the string *bive* is not on its list of valid words. On the other hand, if you type *give* instead of *five*, the spell-checker will not detect the error because *give* is on its list of valid words.

Our error-detection scheme works the same way, except that if we add extra digits to ensure that our code words differ in many positions—say, $t$ positions—then even as many as $t - 1$ mistakes will not convert one code word into another code word. And if every pair of code words differ from each other in at least three positions, we can correct any single error because the incorrect received word will differ from the correct code word in one position, but it will differ from all others in two or more positions. Thus, in this case, the correct word is the unique "nearest neighbor." So the role of the parity-check sums is to ensure that code words differ in many positions. For example, consider the code in Table 17.1. The messages 1000 and 1100 differ in only the second position. But the two parity-check sums $a_1 + a_2 + a_3$ and $a_2 + a_3 + a_4$ will guarantee that encoded words for these messages will have different values in positions 5 and 7 as well as in position 2. It is the job of mathematicians to discover the appropriate parity-check sums to correct several errors in long, complicated codes.

## Data Compression

Binary linear codes are fixed-length codes. In a fixed-length code, each code word is represented by the same number of digits (or symbols). In contrast, the Morse code (see Spotlight 17.4), designed for the telegraph, is a **variable-length code**, that is, a code in which the number of symbols for each code word may vary.

Notice that in the Morse code the letters that occur most frequently have the shortest coding, whereas the letters that occur the least frequently have the longest coding. By assigning the code in this manner, telegrams could convey more information per line than would be the case for fixed-length codes or a randomly assigned variable-length coding of the letters. The Morse code is an example of data compression.

### Data Compression — DEFINITION

**Data compression** is the process of encoding data so that the most frequently occurring data are represented by the fewest symbols.

### SPOTLIGHT 17.4 Morse Code

The Morse code is a ternary code consisting of short marks, long marks and spaces (see figure on the right). It was invented in the early 1840s by Samuel Morse as an efficient way to transmit messages using electronic pulses through telegraph wires. The code enabled operators to send strings of short pulses, long pulses and pauses representing characters into indentations on paper tape that could be easily converted back to characters. Although it was widely used up until the mid-twentieth century, it has gradually been supplanted by more machine-friendly codes. Because the Morse code uses data compression, sending messages using Morse code is faster than text messaging. Many Nokia cellphones can convert text messages to Morse code.

| | | | |
|---|---|---|---|
| A | ·— | N | —· |
| B | —··· | O | ——— |
| C | —·—· | P | ·——· |
| D | —·· | Q | ——·— |
| E | · | R | ·—· |
| F | ··—· | S | ··· |
| G | ——· | T | — |
| H | ···· | U | ··— |
| I | ·· | V | ···— |
| J | ·——— | W | ·—— |
| K | —·— | X | —··— |
| L | ·—·· | Y | —·—— |
| M | —— | Z | ——·· |

Morse Code

Figure 17.5 shows a typical frequency distribution for letters in English-language text material.

| | A | B | C | D | E | F | G | H | I | J | K | L | M |
|---|---|---|---|---|---|---|---|---|---|---|---|---|---|
| Percentage: | 8 | 1.5 | 3 | 4 | 13 | 2 | 1.5 | 6 | 6.5 | 0.5 | 0.5 | 3.5 | 3 |
| | N | O | P | Q | R | S | T | U | V | W | X | Y | Z |
| Percentage: | 7 | 8 | 2 | 0.25 | 6.5 | 6 | 9 | 3 | 1 | 1.5 | 0.5 | 2 | 0.25 |

**FIGURE 17.5** A widely used frequency table for letters in normal English usage.

Data compression provides a means to reduce the costs of data storage and transmission. A **compression algorithm** converts data from an easy-to-use format to one optimized for compactness. Conversely, an uncompression algorithm converts the compressed information back to its original form or approximately its original form. Downloaded files in the ZIP format are an example of a particular kind of data compression. When you "unzip" the file, you return the compressed data to its original state. In some applications, such as data sets that represent images, the original data need only be recaptured in approximate form. In these cases, there are algorithms that result in a great saving of space. Graphics Interchange Format (GIF) encoding returns compressed data to its exact original form, while JPEG encoding and MPEG encoding return data only approximately to its original state.

## EXAMPLE 1 ■ Data Compression

Let's illustrate the principles of data compression with a simple example. Biologists are able to describe genes by specifying sequences composed of the four letters A, T, G, and C, which represent the four nucleotides adenine, thymine, guanine, and cytosine, respectively. One way to encode a sequence such as AAACAGTAAC in fixed-length binary form would be to encode the letters as

$$A \rightarrow 00 \qquad C \rightarrow 01 \qquad T \rightarrow 10 \qquad G \rightarrow 11$$

The corresponding binary code for the sequence AAACAGTAAC is then

$$00000001001110000001$$

On the other hand, if we knew from experience that A occurs most frequently, C second most frequently, and so on, and that A occurs much more frequently than T and G together, the most efficient binary encoding would be

$$A \rightarrow 0 \qquad C \rightarrow 10 \qquad T \rightarrow 110 \qquad G \rightarrow 111$$

For this encoding scheme, the sequence AAACAGTAAC is encoded as

$$0001001111100010$$

Notice that this binary sequence has 20% fewer digits than our previous sequence, in which each letter was assigned a fixed length of 2 (16 digits versus 20 digits). However, to realize this savings, we have made decoding more difficult. For the binary sequence using the fixed length of two symbols per character, we decode the sequence by taking the digits two at a time in succession and converting them to the corresponding letters. For the compressed coding, we can decode by examining the digits in groups of three.

## EXAMPLE 2 ■ Decode 0001001111100010

Consider the compressed binary sequence 0001001111100010. Look at the first three digits: 000. Since our code words have one, two, or three digits and neither 00 nor 000 is a code word, the sequence 000 can represent only the *three* code words 0, 0, and 0. Now look at the next three digits: 100. Again, because neither 1 nor 100 is a code word, the sequence 100 represents the *two* code words 10 and 0. The next three digits, 111, can represent only the code word 111 because the other three code words all contain at least one 0. Next consider the sequence 110. Because neither 1 nor 11 is a code word, the sequence 110 can represent only 110 itself. Continuing in this fashion, we can decode the entire sequence to obtain AAACAGTAAC.

The following observation can simplify the decoding process for compressed sequences. Note that 0 occurs only at the end of a code word. Thus, each time you see a 0, it is the end of the code word. Also, because the code words 0, 10, and 110 end in a 0, the only circumstances under which there are three consecutive 1's is when the code word is 111. So, to quickly decode a compressed binary sequence using our coding scheme, insert a comma after every 0 and after every three consecutive 1's. The digits between the commas are code words.

## EXAMPLE 3 ■ Code AGAACTAATTGACA and Decode the Result

Recall: A → 0, C → 10, T → 110, and G → 111. So

AGAACTAATTGACA → 01110010110001101101110100

To decode the encoded sequence, we insert commas after every 0 and after every occurrence of 111 and convert to letters:

| | | | | | | | | | | | | | |
|---|---|---|---|---|---|---|---|---|---|---|---|---|---|
| 0, | 111, | 0, | 0, | 10, | 110, | 0, | 0, | 110, | 110, | 111, | 0, | 10, | 0 |
| A, | G, | A, | A, | C, | T, | A, | A, | T, | T, | G, | A, | C, | A |

## Delta Encoding

For data sets of numbers that fluctuate little from one number to the next, the method of compression called the *delta function* works well. Consider the following closing prices (rounded to the nearest integer) of the Standard & Poor's index of the stock prices of 500 companies in September 2007.

1489 1472 1479 1454 1452 1471 1472 1484 1484 1477 1520 1529
1519 1526 1518 1517 1525 1531 1527 1547 1547 1540 1543 1558

These numbers use 96 characters in all. To compress this data set using the delta method, we start with the first number and continue by listing only the change from each entry to the next. So our list becomes

| | | | | | | | | | | | |
|---|---|---|---|---|---|---|---|---|---|---|---|
| 1489 | −17 | 7 | −25 | −2 | 19 | 1 | 12 | 0 | −7 | 43 | 9 |
| −10 | 7 | −8 | −1 | 8 | 6 | −4 | 20 | 0 | −7 | 3 | 15 |

This time we have used only 44 characters, counting the minus signs, to represent the same data–a savings of 54%.

## SPOTLIGHT 17.5 David Huffman

Large networks of IBM computers use it. So do high-definition televisions, modems, and a popular electronic device that takes the brainwork out of programming a videocassette recorder. All these digital wonders rely on the results of a 58-year-old term paper by an MIT graduate student—a data-compression scheme known as Huffman encoding.

David Huffman

(*Matthew Mulbry.*)

In 1951, David Huffman and his classmates in an electrical engineering graduate course on information theory were given the choice of a term paper or a final exam. For the term paper, Huffman's professor had assigned what at first appeared to be a simple problem. Students were asked to find the most efficient method of representing numbers, letters, or other symbols using binary code. Huffman worked on the problem for months, developing a number of approaches, but none that he could prove to be the most efficient. Finally, he despaired of ever reaching a solution and decided to start studying for the final. Just as he was throwing his notes in the garbage, the solution came to him. "It was the most singular moment of my life," Huffman says. "There was the absolute lightning of sudden realization. It was my luck to be there at the right time and also not have my professor discourage me by telling me that other good people had struggled with the problem," he says. When presented with his student's discovery, Huffman recalls, his professor exclaimed: "Is that all there is to it!"

"The Huffman code is one of the fundamental ideas that people in computer science and data communications are using all the time," says Donald Knuth of Stanford University. Although others have used Huffman's code to help make millions of dollars, Huffman's main compensation was dispensation from the final exam. He never tried to patent an invention from his work and experienced only a twinge of regret at not having used his creation to make himself rich. "If I had the best of both worlds, I would have had recognition as a scientist, and I would have gotten monetary rewards," he says. "I guess I got one and not the other."

But Huffman received other compensation. A few years ago an acquaintance told him that he had noticed that a reference to the code was spelled with a lowercase h. Remarked his friend to Huffman, "David, I guess your name has finally entered the language."

David Huffman died October 7, 1999.

*Source*: Adapted from an article by Gary Stix, *Scientific American*, September 1991, pp. 54, 58.

## Huffman Coding

The methods we have shown previously are too simple for general use, but in 1951 a graduate student named David Huffman (see Spotlight 17.5) devised a scheme for data compression that became widely used. As was the case for the first scheme we discussed, Huffman coding assigns short code words to those characters with high probabilities of occurring and long code words to those with low probabilities of occurring. A Huffman code is made using a so-called *code tree* by arranging the characters from top to bottom according to increasing probability; it proceeds by combining, at each stage, the two least probable combinations and repeating this process until there is only one combination remaining. To illustrate the method, say we have a data set of six letters that occur with the following probabilities.

| | |
|---|---|
| *A* | 0.125 |
| *B* | 0.051 |
| *C* | 0.215 |
| *D* | 0.173 |
| *E* | 0.210 |
| *F* | 0.226 |

Rearranging them in increasing order, we have:

| | |
|---|---|
| *B* | 0.051 |
| *A* | 0.125 |
| *D* | 0.173 |
| *E* | 0.210 |
| *C* | 0.215 |
| *F* | 0.226 |

Because *B* and *A* are the two least likely to occur, we begin our tree by merging them with the one with the smallest probability on the left (that is, *BA* rather than *AB*), adding their probabilities, and rearranging the resulting items in increasing order:

| | |
|---|---|
| *D* | 0.173 |
| *BA* | 0.176 |
| *E* | 0.210 |
| *C* | 0.215 |
| *F* | 0.226 |

This time *D* and *BA* are the two least likely remaining entries so we merge them with *D* on the left since it has smallest probability, add their probabilities, and re-sort from smallest to largest. This gives:

| | |
|---|---|
| *E* | 0.210 |
| *C* | 0.215 |
| *F* | 0.226 |
| *DBA* | 0.349 |

Next we combine *E* and *C* with *E* on the left and re-sort to get:

| | |
|---|---|
| *F* | 0.226 |
| *DBA* | 0.349 |
| *EC* | 0.425 |

Then we combine *F* and *DBA* with *F* on the left and re-sort:

| | |
|---|---|
| *EC* | 0.425 |
| *FDBA* | 0.575 |

And finally,

| | |
|---|---|
| *ECFDBA* | 1.000 |

To assign a binary code word to each letter, we work our way back from the end of the tree to each letter by assigning, at each merging juncture, 0 to the branch with the lower probability, as shown in Figure 17.6. (There is more than one way to draw the tree. Moreover, we can assign the 0's and 1's to the branches in any fashion, but we do it in this specific way for convenience.)

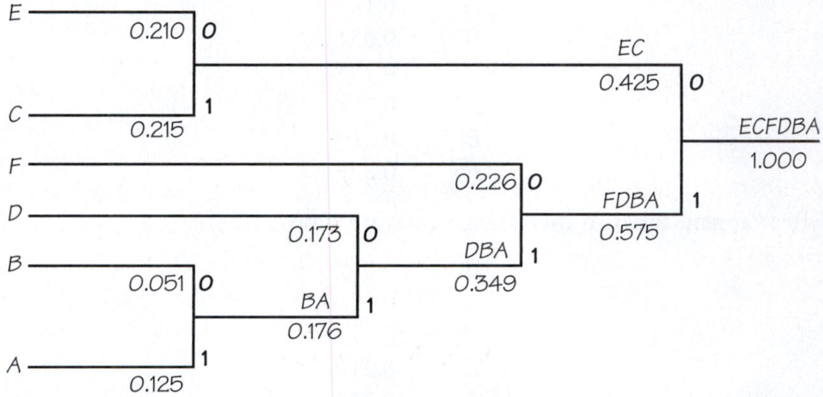


**FIGURE 17.6**
A Huffman tree.

The path to each letter determines the code word for that letter. So we have:

| | |
|---|---|
| $A$ | 1111 |
| $B$ | 1110 |
| $C$ | 01 |
| $D$ | 110 |
| $E$ | 00 |
| $F$ | 10 |

Notice that the letters that occur least often have the longest codes and the letters that occur most often have the shortest codes. Decoding codes created from a Huffman tree is possible because at each stage there is only one way a particular string could have occurred. Here is an example. Consider the Huffman code created using the code words given in the previous display:

$$11101000011010011111010010$$

How can we determine the corresponding string of letters that has this Huffman code? The method is quite simple. Look at the first two digits. If they correspond to a code word, then decode them as that letter. If not, then look at the next digit. If the three digits correspond to a code word, then decode it as the corresponding letter. If not, then these three digits and the next one are a four-digit code word. Looking at our example 11101000011010011111010010 we see that neither 11 nor 111 is a code word but 1110 is the code word for B. So, replacing 1110 with B, we have B1000011010011111010010. The next possibility is 10, which is the code word for F, so we have BF00011010011111010010. Next we have 00, which is the code word for E, giving us BFE011010011111010010. Continuing in this way, we obtain BFECFFCACEF. Of course, in practice, coding and decoding are done by computers.

## 17.3 Cryptography

Thus far, we have discussed ways in which data can be encoded to detect errors or correct errors in transmission. In many situations, there is also a desire for security against unauthorized interpretation of coded data (that is, a desire for secrecy). The process of disguising data is called **encryption**. **Cryptology** is the study of methods to make and break secret codes.

(*Corbis.*)

Historically, encryption was used primarily for military and diplomatic transmissions. Today, encryption is essential for securing electronic transactions of all kinds. Cryptography is what allows you to have a Web site safely receive your credit-card number. Cryptographic schemes prevent hackers from charging calls to your cellular phone. Cryptography is also used for authenticating electronic transactions. In September 1998, history was made when former President Bill Clinton and Ireland's Prime Minister Bertie Ahern used digital signatures to sign an intergovernmental document. Each leader had a unique signing code and a digital certificate that served as a "digital ID," thereby ensuring that the document was approved by them. Although modern encryption schemes are extremely complex, we will illustrate the fundamental concepts involved with a few simple examples.

Among the first known cryptosystems is the so-called **Caesar cipher** used by Julius Caesar to send messages to his troops. To encrypt a message with the method employed by Caesar, we use the following table to replace each letter in the top row with the letter below it.

| | | | | | | | | | | | | | | | | | | | | | | | | | |
|---|---|---|---|---|---|---|---|---|---|---|---|---|---|---|---|---|---|---|---|---|---|---|---|---|---|
| A | B | C | D | E | F | G | H | I | J | K | L | M | N | O | P | Q | R | S | T | U | V | W | X | Y | Z |
| D | E | F | G | H | I | J | K | L | M | N | O | P | Q | R | S | T | U | V | W | X | Y | Z | A | B | C |

For example, the message ATTACK AT DAWN is encrypted as DWWDFN DW GDZQ.

To decrypt the message, replace each letter with the letter above it in the table. Obviously, it would not require much effort for someone to "crack" this code.

To describe more sophisticated schemes for transmitting messages secretly, it is convenient to introduce a special kind of arithmetic used in cryptography. For any positive integers $a$ and $n$, we define $a$ mod $n$ (read: "$a$ modulo $n$" or just "$a$ mod $n$") to be the remainder when $a$ is divided by $n$. Thus,

$$3 \bmod 2 = 1 \text{ because } 3 = 1 \cdot 2 + 1$$
$$6 \bmod 2 = 0 \text{ because } 6 = 3 \cdot 2 + 0$$
$$5 \bmod 3 = 2 \text{ because } 5 = 1 \cdot 3 + 2$$
$$37 \bmod 10 = 7 \text{ because } 37 = 3 \cdot 10 + 7$$
$$38 \bmod 26 = 12 \text{ because } 38 = 1 \cdot 26 + 12$$
$$342 \bmod 85 = 2 \text{ because } 342 = 4 \cdot 85 + 2$$
$$62 \bmod 85 = 62 \text{ because } 62 = 0 \cdot 85 + 62$$

Arithmetic involving mod $n$ is called **modular arithmetic**. Although this arithmetic may appear unfamiliar, you often unconsciously use it. For example, if it is now September, what month will it be 25 months from now? Of course, you answer "October," but the interesting fact is that you didn't arrive at the answer by starting with September and counting off 25 months. Instead, without even thinking about it, you simply observed that $25 = 2 \cdot 12 + 1$ so that 25 mod 12 = 1, and you added one month to September. Similarly, if it is now Wednesday, you know that in 23 days it will be Friday. This time, you arrived at your answer by noting that $23 = 3 \cdot 7 + 2$ (that is, 23 mod 7 = 2), so you added 2 days to Wednesday instead of counting off 23 days. An application of modular arithmetic to genetics is described in Spotlight 17.6.

With modular arithmetic, we can easily describe the Caesar cipher as follows. Begin by saying that the letter A is in position 0, B is in position 1, C is in position 2, and so on. Then the Caesar cipher replaces the letter in position $i$ with the letter in position $(i + 3) \bmod 26$. This formula expresses the fact that the Caesar cipher shifts each letter from A through W three positions to the right while X, Y, and Z are replaced with A, B, and C, respectively. (Think of X, Y, and Z as "wrapping" around to the beginning of the alphabet.) Some authors use the term "Caesar cipher" for any encryption scheme that shifts each letter a fixed number of positions.

## SPOTLIGHT 17.6 Modeling the Genetic Code

The way that genetic material is composed can be conveniently modeled using modulo 4 arithmetic. A DNA molecule is made up of two long strands in the form of a double helix. Each strand is made up of strings of the four nitrogen bases adenine (A), thymine (T), guanine (G), and cytosine (C). Each base on one strand binds to a complementary base on the other strand. Adenine is always bound to thymine, and guanine is always bound to cytosine. To model this situation, we identify A with 0, T with 2, G with 1, and C with 3. Thus, the DNA segment ACGTAACAGGA and its complement segment TGCATTGTCCT are identified by 03120030110 and 21302212332.

Using modulo 4 arithmetic, $0 + 2 = 2$, $2 + 2 = 0$, $1 + 2 = 3$, and $3 + 2 = 1$, and we see that adding 2 to any of the integers 0, 1, 2, or 3 interchanges 0 and 2 and 1 and 3. So, for any DNA segment $a_1a_2 \cdots a_n$ represented by strings of 0's, 1's, 2's, and 3's, we see that its complementary segment is represented by $a_1a_2 \cdots a_n + 22 \cdots 2$, where we add the integers in each component using modulo 4. In particular, $03120030110 + 22222222222 = 21302212332$.

*Source*: Adapted from *Discrete Mathematics* by S. Washburn, T. Marlow, and C. Ryan, Addison-Wesley, 1999.

*(David Mack/Photo Researchers, Inc.)*

## Decimation Cipher

Rather than encrypting a message by adding a fixed number to the position of every letter and using modular arithmetic as is done in the Caesar cipher, another simple way to encrypt a message is to multiply each position by a fixed number and use modular arithmetic. This method of encryption is called the **decimation cipher**. To begin we assign the 26 letters the numbers 0 through 25 in order and select any odd integer $k$ between 3 and 25 except 13. (For the method to work $k$ and 26 must have no prime divisors in common.) Then, in the message, a letter with numerical value $i$ is replaced with the letter with numerical value $ki$ modulo 26. For example, if $k$ is 5, then D is replaced with P since D has value 3 and $5 \times 3 = 15$, which is assigned to P. When $ki$ exceeds 26 we use modulo 26 arithmetic to determine the replacement for the letter. Thus J is replaced by T since J has the value 9 and $5 \times 9 = 45 = 19 \bmod 26$ and T has value 19. The value of $k$ is called the **key**. To decode an encrypted message we use the same method except we multiply the numerical value for each encrypted letter by 21. The number 21 is used to decrypt the message because it has the property that $5 \times 21 = 105 = 1$ modulo 26. As a consequence, for any integer $x$ we have $(x \times 5 \times 21)$ modulo 26 equals $x \times 1 = x$. Thus multiplying an integer $x$ by 5 and then the result by 21 gets us back to $x$ when we use modulo 26. In general, if a number $k$ is used to encrypt a message modulo 26, the value $j$ used to decrypt the message has to be chosen so that $kj = 1 \bmod 26$. Given a particular value for $k$ we can find the corresponding $j$ by trial and error. Table 17.4 shows the values corresponding to each choice of $k$.

**TABLE 17.4**

| Encryption Value | Decryption Value |
|---|---|
| 3 | 9 |
| 5 | 21 |
| 7 | 15 |
| 9 | 3 |
| 11 | 19 |
| 15 | 7 |
| 17 | 23 |
| 19 | 11 |
| 21 | 5 |
| 23 | 17 |
| 25 | 25 |

## EXAMPLE 4 ▪ Decimation Cipher

To illustrate the decimation cipher let's encrypt the message ATTACK AT DAWN using the key 3.

| | | | | | | | | | | | | |
|---|---|---|---|---|---|---|---|---|---|---|---|---|
| Message | A | T | T | A | C | K | A | T | D | A | W | N |
| Position | 0 | 19 | 19 | 0 | 2 | 10 | 0 | 19 | 3 | 0 | 22 | 13 |
| Position × 3 | 0 | 57 | 57 | 0 | 6 | 30 | 0 | 57 | 9 | 0 | 66 | 39 |
| New position | 0 | 5 | 5 | 0 | 6 | 4 | 0 | 5 | 9 | 0 | 14 | 13 |
| Encrypted message | A | F | F | A | G | E | A | F | J | A | O | N |

Modular arithmetic also provides the basis for a more sophisticated cryptosystem called the **Vigenère cipher**. For this method we first select a **key word**, which can be any word. The letters of the key word are then used to determine the amount of shifting for each letter of our message.

## EXAMPLE 5 ▪ Vigenère Cipher

We will use the Vigenère system to encrypt the message ATTACK AT DAWN. Choosing the key word MATH, we shift the first letter of the message by 12 because M is in position 12; the second letter of the message is shifted by 0 (unchanged) because A is in position 0; the third letter of the message is shifted by 19 because T is in position 19, and so on. A shift of $j$ means that the letter in position $i$ is replaced by the letter in position $(i + j)$ mod 26. When we have used all the letters of the key word, we start over at the beginning. To encrypt ATTACK AT DAWN using the key word MATH, we first note that the letters in the key word MATH are in positions 12, 0, 19, and 7, respectively. So, the A in ATTACK is converted to M $(0 + 12 = 12)$, the first T in ATTACK is converted to T $(19 + 0 = 19)$, the second T in ATTACK is converted to M $((19 + 19) \bmod 26 = 12)$, and so on. The first two lines of Table 17.5 show the position numbers for the letters of the message and the key word. The third line of the table is obtained from the first two by adding the values in the columns mod 26 and converting the results back to letters.

**TABLE 17.5**

| | | | | | | | | | | | | |
|---|---|---|---|---|---|---|---|---|---|---|---|---|
| ATTACK AT DAWN | 0 | 19 | 19 | 0 | 2 | 10 | 0 | 19 | 3 | 0 | 22 | 13 |
| MATHMA TH MATH | 12 | 0 | 19 | 7 | 12 | 0 | 19 | 7 | 12 | 0 | 19 | 7 |
| MTMHOK TA PAPU | 12 | 19 | 12 | 7 | 14 | 10 | 19 | 0 | 15 | 0 | 15 | 20 |

## Encrypting Credit-Card Data on the Web

Suppose that you want to purchase a compact disc from Amazon.com. Should you be concerned that a hacker will intercept your credit-card number during the transaction? As you might expect, your credit-card number is sent to Amazon in encrypted form to protect the data.

To describe one way that this encryption can be done, we need to perform addition of binary strings. We add two binary strings $a_1a_2 \cdots a_n$ and $b_1b_2 \cdots b_n$ as follows:

$$\begin{array}{r} a_1a_2 \cdots a_n \\ + \ b_1b_2 \cdots b_n \\ \hline c_1c_2 \cdots c_n \end{array}$$

where $c_i = 0$ if $a_i = b_i$ and $c_i = 1$ if $a_i \neq b_i$. Equivalently, $c_i = (a_i + b_i) \bmod 2$. (Add $a_i$ and $b_i$ in the ordinary way, but replace 2 by 0.)

### EXAMPLE 6 ■ Sum of Binary Strings

$$\begin{array}{r} 11000111 \\ + \ 01110110 \\ \hline 10110001 \end{array} \qquad \begin{array}{r} 00111011 \\ + \ 01100101 \\ \hline 01011110 \end{array} \qquad \begin{array}{r} 10011100 \\ + \ 10011100 \\ \hline 00000000 \end{array}$$

■

We can now explain one way to send credit-card numbers over the Web securely. When you place an order with Amazon, the company sends your computer a randomly generated string of 0's and 1's called a **key**. This key has the same length as the binary string corresponding to your credit-card number, and the two strings are added (think of this process as "locking" the data). The resulting sum is then transmitted to Amazon. Amazon in turn adds the same key to the received string, which then produces the original string corresponding to your credit-card number (adding the key a second time "unlocks" the data).

To illustrate the idea, say you want to send an eight-digit binary string such as $s = 10101100$ to Amazon (actual credit-card numbers have very long strings), and Amazon sends your computer the key $k = 00111101$. Your computer returns the string $s + k = 10101100 + 00111101 = 10010001$ to Amazon, and Amazon adds $k$ to this string to get $10010001 + 00111101 = 10101100$, which is the string representing your credit-card number. If someone intercepts the number $s + k = 10010001$ during transmission, it is of no value without knowing $k$. This method works because of the property of binary addition that $a_1a_2 \cdots a_n + b_1b_2 \cdots b_n = 00 \cdots 0$ if and only if the two strings are identical. Thus, $(s + k) + k = s + (k + k) = s + 00 \cdots 0 = s$. The method is secure because the key sent by Amazon is randomly generated and used only one time.

You can tell when you are using an encryption scheme on a Web transaction by looking to see if the Web address begins with "https" rather than the customary "http." You will also see a small padlock in the status bar at the bottom of the browser window.

## Public Key Cryptography

In the mid-1970s Ronald Rivest, Adi Shamir, and Leonard Adleman devised an ingenious method that permits each person who is to receive a secret message to publicly tell how to scramble messages sent to him or her. Even though the method used to scramble the message is known publicly, only the person for whom it is intended will be able to unscramble the message.

To illustrate their method for transmitting messages secretly, we need the following property of modular arithmetic:

$$(ab) \bmod n = ((a \bmod n)(b \bmod n)) \bmod n$$

This property allows you to replace integers that are greater than or equal to $n$ with integers that are less than $n$ to simplify calculations. You should think of it as saying, "mod before you multiply."

### EXAMPLE 7

### Multiplication Property for Modular Arithmetic

$$\begin{aligned}(17 \cdot 23) \bmod 10 &= ((17 \bmod 10)(23 \bmod 10)) \bmod 10\\ &= (7 \cdot 3) \bmod 10 = 21 \bmod 10 = 1\\ (22 \cdot 19) \bmod 8 &= ((22 \bmod 8)(19 \bmod 8)) \bmod 8\\ &= (6 \cdot 3) \bmod 8 = 18 \bmod 8 = 2\\ (100 \cdot 8) \bmod 85 &= ((100 \bmod 85)(8 \bmod 85)) \bmod 85\\ &= (15 \cdot 8) \bmod 85 = 120 \bmod 85 = 35\end{aligned}$$

We now describe the Rivest, Shamir, and Adleman method by way of a simple example. Say we wish to send the message "IBM." We convert the message to digits by replacing A by 1, B by 2, . . . , and Z by 26. So the message IBM becomes 9213. The person to whom the message is to be sent has picked two primes $p$ and $q$, say, $p = 5$ and $q = 17$. (Recall that a *prime* is an integer greater than 1 whose only divisors are 1 and itself.) The receiver has also picked a number $r$, such as 3, that has no divisors in common with the least common multiple $m$ of $(p - 1) = 4$ and $(q - 1) = 16$ other than 1, and published $n = pq = 85$ and $r = 3$ in a public directory. To decode our message, the receiver must find a number $s$ so that $r \cdot s = 1 \bmod m$ (this is where knowledge of $p$ and $q$ is necessary). That is, $3 \cdot s = 1 \bmod 16$. This number is 11. (The number $s$ can be found by calculating successive powers of $r \bmod m$. When 1 is reached, the previous power of $r$ is $s$. In our example, we have $3 \bmod 16 = 3$, $3^2 \bmod 16 = 9$, $3^3 \bmod 16 = 11$, $3^4 \bmod 16 = 1$, so $s = 3^3 \bmod 16 = 11$.)

To send our message to this person, we consult the public directory to find $n = 85$ and $r = 3$, then send the "scrambled" numbers $9^3 \bmod 85$, $2^3 \bmod 85$, and $13^3 \bmod 85$ rather than 9, 2, and 13, and the receiver will unscramble them. Thus, we send:

$$\begin{aligned}9^3 \bmod 85 &= 49\\ 2^3 \bmod 85 &= 8\\ 13^3 \bmod 85 &= 72\end{aligned}$$

Now the receiver must take the numbers he or she receives–49, 8, and 72–and convert them back to 9, 2, and 13 by calculating $49^{11}$ mod 85, $8^{11}$ mod 85, and $72^{11}$ mod 85.

The calculation of $49^{11}$ mod 85 can be simplified as follows:[1]

$$49 \bmod 85 = 49$$
$$49^2 \bmod 85 = 2401 \bmod 85 = 21$$
$$49^4 \bmod 85 = 49^2 \cdot 49^2 \bmod 85 = 21 \cdot 21 \bmod 85 = 441 \bmod 85 = 16$$
$$49^8 \bmod 85 = 49^4 \cdot 49^4 \bmod 85 = 16 \cdot 16 \bmod 85 = 1$$

So,

$$\begin{aligned} 49^{11} \bmod 85 &= (49^8 \bmod 85)(49^2 \bmod 85)(49 \bmod 85) \\ &= (1 \cdot 21 \cdot 49) \bmod 85 \\ &= 1029 \bmod 85 \\ &= 9 \end{aligned}$$

Thus, the receiver has correctly determined the code for I. The calculations for $8^{11}$ mod 85 and $72^{11}$ mod 85 are left as exercises. Notice that without knowing how $n = pq$ factors, we cannot find the least common multiple of $p - 1$ and $q - 1$ (in our case, 16), and therefore the $s$ that is needed to determine the intended message.

The procedure just described is called the **RSA public key encryption scheme** in honor of Rivest, Shamir, and Adleman, who discovered it. The method is practical and secure because efficient methods exist for finding very large prime numbers (say, about 100 digits long) and for multiplying large numbers, but no one knows an efficient algorithm for factoring large integers (say, about 200 digits long).

The algorithm is summarized below. In practice, the messages are not sent one letter at a time. Rather, the entire message is converted to decimal form, with A represented by 01, B by 02, . . . , and a space by 00. The message is then broken up into blocks of uniform size and the blocks are sent. See step 2 under Sender below.

**Receiver**

1. Pick very large primes $p$ and $q$ and compute $n = pq$.
2. Compute the least common multiple of $p - 1$ and $q - 1$; let's call it $m$.
3. Pick $r$ so that it has no divisors in common with $m$ other than 1 (any such $r$ will do).
4. Find $s$ so that $rs = 1 \bmod m$. (To find $s$, simply compute $r^2 \bmod m$, $r^3 \bmod m$, $r^4 \bmod m$, . . . until you reach $r^t \bmod m = 1$. Then $s = r^{t-1} \bmod m$.)
5. Publicly announce $n$ and $r$, but keep $p$, $q$, and $s$ secret.

**Sender**

1. Convert the message to a string of digits.
2. Break up the message into uniformly sized blocks of digits, appending 0's in the last block if necessary. Call them $M_1, M_2, \ldots, M_k$. For example, for a

[1]To determine $49^2$ mod 85 with a calculator, enter $49 \times 49$ to obtain 2401, then divide 2401 by 85 to obtain 28.247058. Finally, enter $2401 - (28 \times 85)$ to obtain 21. Provided that the numbers are not too large, the search engine Google at www.google.com will do modular arithmetic by simply entering the number and the mod value in the format 49^4 mod 85. Be careful, however, because entering 49^11 mod 85 yields 0, which is incorrect ($m^r$ mod $n$ is never 0 when the greatest common divisor of $m$ and $n$ is 1). Instead, we can use Google to compute smaller powers such as 49^4 mod 85 = 16 and 49^7 mod 85 = 59, then compute $(16 \times 59)$ mod 85 = 9.

string such as 2105092315, we would use $M_1 = 2105$, $M_2 = 0923$, and $M_3 = 1500$.

3. Check to see that the greatest common divisor of each $M_i$ and $n$ is 1. If not, $n$ can be factored and the code is broken. (In practice, the primes $p$ and $q$ are so large that they exceed all $M_i$, so this step may be omitted.)
4. Calculate and send $R_i = M_i^r \bmod n$.

**Receiver**

1. For each received message $R_i$, calculate $R_i^s \bmod n$.
2. Convert the string of digits back to a string of characters.

Let's do another example step by step with $p = 7$, $q = 11$, and the message "HI."

**Receiver**

1. $n = 77$.
2. The least common multiple $m$ of $7 - 1 = 6$ and $11 - 1 = 10$ is 30.
3. We pick $r = 7$.
4. Since $7^4 = 1 \bmod 30$, we have $s = 7^3 \bmod 30 = 13$.
5. Make public $n = 77$ and $r = 7$.

**Sender**

1. HI converts to 89.
2. We will send 8 and 9 individually (that is, our blocks have size 1).
3. The greatest common divisor of 8 and 77 is 1, and the greatest common divisor of 9 and 77 is 1, so we can proceed.
4. Send $8^7 \bmod 77 = 57$ and $9^7 \bmod 77 = 37$.

**Receiver**

1. $57^{13} \bmod 77 = 8$; $37^{13} \bmod 77 = 9$.
2. 89 converts to HI.

This method works because of a basic property of modular arithmetic and the choice of $r$. As a result of choosing the number $m$ as we described, it has the property that for each positive integer $x$ having no common divisors with $n$ except 1, we have $x^m = 1 \bmod n$. So, in the case of our first example with $n = 85$, $m = 16$, and $r = 3$, for the original message 9 and the received message 49, we have mod 85

$$49^{11} = (9^3)^{11} = 9^{33} = 9 \cdot 9^{32} = 9 \cdot (9^{16})^2 = 9 \cdot 1^2 = 9$$

In 2002, Rivest, Shamir, and Adleman received the Association for Computing Machinery A. M. Turing Award, which is considered to be the "Nobel Prize of Computing," for their seminal contribution to public key cryptography.

## 17.4 Web Searches and Mathematical Logic

With the number of Web pages indexed by large Internet search engines such as Google numbering in billions, computer scientists and mathematicians attempt to manage massive data sets by taking advantage of the associated network structure, which represents the interrelations of the data. The algorithm used by the Google search engine, for instance, ranks all pages on the Web using these interrelations to determine their relevance to the user's search. Factors such as the frequency, location near the top of the page of key words, font size, and number of links are taken into account. Spotlight 17.7 discusses the "Kevin Bacon" game that illustrates the so-called small-world phenomenon of the interconnectedness of data. In this section we will show how a branch of mathematics called **Boolean logic**, after the nineteenth-century mathematician George Boole (1815–1864), can be used to make search engine queries more efficient.

## SPOTLIGHT 17.7 Six Degrees of Kevin Bacon

The actor Kevin Bacon is probably better known for a game named after him than for any of the movies he has been in. The game works like this. Every actor who has been in a film with Kevin Bacon has Bacon number 1. Every actor who does not have Bacon number 1, but has been in a film with someone with Bacon number 1, has Bacon number 2. Any actor who does not have Bacon number 1 or 2, but has been in a film with someone who has Bacon number 2, has Bacon number 3, and so on. For example, Nicole Kidman has not been in a film with Bacon but was in *The Interpreter* with Sean Penn, and Sean Penn was in *Mystic River* with Kevin Bacon. So Kidman's Bacon number is 2.

This game was conceived by three college students who first explained it to the public on an MTV show hosted by Jon Stewart. It sometimes goes by the name "Six Degrees of Kevin Bacon" because nearly every actor has a Bacon number that is at most 6. In fact, it is a challenge to think of an actor with a Bacon number exceeding 3. The game is an example of what scientists call the "small-world phenomenon," by which they mean that every person is connected to every other person by a surprisingly short number of links. You can play the Bacon game at www.cs.virginia.edu/oracle/.

The first experiment involving the small-world phenomenon occurred in 1967, when social psychologist Stanley Milgram mailed a series of traceable letters from points in Kansas and Nebraska to "targets" in Boston. The letters could be sent only to someone whom the holders knew on a first-name basis, and who they thought was more likely to know the target than they were themselves. The data revealed a median chain length of about 6.

Mathematicians have their own version of the Bacon game called the "Erdös number," where coauthors of research papers with Paul Erdös have Erdös number 1. The author of this spotlight has Erdös number 3. Ironically, Erdös's Bacon number is 3. In April 2004, a person with Erdös number 4 auctioned on eBay the opportunity for someone to get an Erdös number of 5 by offering to write a joint paper with the highest bidder. The auction was halted when someone bid $1 million as a protest to the idea of selling coauthorships. Before the auction was halted the highest bid was $1031. That bidder also refused to pay as a protest. Interestingly, the actress Danica McKellar, who played Winnie Cooper on the TV series *The Wonder Years*, has a Bacon number 2 and an Erdös number 4.

An *expression* in Boolean logic is simply a statement that is either true or false. For example, the expression, "There are infinitely many prime numbers," is either true or false. For purposes of mathematical reasoning, it is not necessary that we know whether this statement is true or false, but simply that it is one or the other. (It was proved to be true by Euclid more than 2000 years ago.) The expression, "The integer 51 is prime," is an example of an expression that is false, since $51 = 3 \cdot 17$. When combining expressions in logic, we avoid statements that are subject to opinion or various interpretations, such as "math is cool." When we enter a phrase such as "college football" as a query to a search engine, the search engine automatically interprets it as the expression, "This Web page contains the phrase 'college football'." The search engine then returns the list of Web pages for which this expression is true.

In this section we discuss how complex expressions can be constructed by connecting individual expressions with the *connectives* AND, OR, and NOT. For example, to obtain Web pages containing the phrase "football" but not pages that also contain "NFL" or "college," we could formulate the query using the expression "football AND (NOT NFL) AND (NOT college)." The search engine interprets this as, "This Web page contains the phrase 'football' AND it is NOT the case that this Web page contains the word 'NFL' AND it is NOT the case that this Web page contains the word 'college'." The parentheses are not necessary but are sometimes useful, as we will see shortly. Most search engines are not *case sensitive*. That is, no distinction is made between uppercase and lowercase letters.

Each search engine has slightly different conventions for formulating queries. Virtually every search engine has a hyperlink on its Web page that explains how to formulate queries. Although we use traditional terminology and notation from logic for our connectives, most popular search engines employ a more user-friendly format for advanced searches. For Google, "Find results with all of the words" is a substitute for our AND connective; "with at least one of the words" plays the role of our OR connective; and "without the words" is the same as our NOT connective. Some search engines use + for our AND connective and − for our NOT connective. Despite the differences in format, the logic is the same.

Our interest is finding out how we can use Boolean logic to decide whether two different expressions have the same meaning. For example, is the expression, "football AND (NOT NFL) AND (NOT college)," equivalent to "football AND NOT (NFL AND college)"? Or is it equivalent to the expression "football AND NOT (NFL OR college)"? To answer these questions, we will now take a closer look at the connectives AND, OR, and NOT.

The NOT connective allows us to take an expression $P$ and create a new expression NOT $P$, called the *negation* of $P$. If $P$ is true, then NOT $P$ is false. If $P$ is false, then NOT $P$ is true. Rather than writing NOT $P$, we will use the more standard mathematical notation $\neg P$. The negation relationship can be summarized in the following format, known as a **truth table**:

| $P$ | $\neg P$ |
|---|---|
| T | F |
| F | T |

Notice that T and F are used here as shorthand for *true* and *false*, respectively. The left column of the truth table shows the two possible values of $P$: T and F. The right column shows the values of $\neg P$ for each of the corresponding values of $P$.

The AND connective allows us to combine two expressions, $P$ and $Q$, into a new expression $P$ AND $Q$ called the *conjunction* of $P$ and $Q$. The new expression is true when both statements $P$ and $Q$ are true and is otherwise false. The mathematical notation for $P$ AND $Q$ is $P \wedge Q$. (You can remember this by noting that $\wedge$ has a shape like the first letter of AND.) This relationship can also be summarized in a truth table:

| $P$ | $Q$ | $P \wedge Q$ |
|---|---|---|
| T | T | T |
| T | F | F |
| F | T | F |
| F | F | F |

Here, the first two columns are used to show all possible values of $P$ and $Q$, and the right column shows the value of $P \wedge Q$.

Finally, the OR connective allows us to combine two expressions $P$ and $Q$ into a new expression $P$ OR $Q$, called the *disjunction* of $P$ and $Q$, which is true if either $P$ or $Q$, or both, are true and is otherwise false. The mathematical notation for $P$ OR $Q$ is $P \vee Q$. This relationship is summarized by the truth table

| $P$ | $Q$ | $P \vee Q$ |
|---|---|---|
| T | T | T |
| T | F | T |
| F | T | T |
| F | F | F |

In everyday circumstances the word "or" is used in two distinct ways. In some situations "$P$ or $Q$" means either $P$ is valid, or $Q$ is valid, or both are valid, while in other situations "$P$ or $Q$" means exactly one of $P$ or $Q$ is valid. A typical example of the former is the criterion for admission to an entertainment event that states "Must be at least 18 years old or accompanied by an adult." One the other hand, a menu entry that says "Price includes soup or salad" is an example of the latter. To distinguish between these two usages mathematicians call the first the *inclusive or* and the second the *exclusive or*. When you encounter a mathematical statement of the form $P$ OR $Q$, the inclusive or is meant. In nonmathematical, ambiguous situations some people use the term "and/or" to mean the inclusive or.

A statement involving three expressions $P$, $Q$, and $R$ such as $P \wedge Q \wedge R$ appears to be ambiguous. Does this expression mean that $P$ and $Q$ are first combined into a new expression $P \wedge Q$ and this new expression is then combined with $R$? In other words, should we interpret this expression as $(P \wedge Q) \wedge R$? Perhaps the intention was to connect $P$ with the single expression $Q \wedge R$. In this case, the expression is interpreted as $P \wedge (Q \wedge R)$. Just as with the case for arithmetic statements such as $5 + 3 + 6$, which we can interpret to mean $(5 + 3) + 6$ or $5 + (3 + 6)$, it turns out that both interpretations are the same. For example, consider the three requirements for the office of president of the United States. A candidate for president must be at least 35 years old, must be a natural-born U.S. citizen, and must have lived in the United States for at least 14 years. Let $P$ be the statement "a candidate must be at least 35 years old," let $Q$ be the statement "a candidate must be a natural-born U.S. citizen," and let $R$ be the statement "a candidate must have lived in the United States for at least 14 years."

The expression $(P \wedge Q) \wedge R$ can then be interpreted as "a candidate must be at least 35 years old and a natural-born U.S. citizen and also must have lived in the United States for at least 14 years." The expression $P \wedge (Q \wedge R)$ can be interpreted as "a candidate must be at least 35 years old and also a natural-born U.S. citizen who has lived in the United States for at least 14 years." Both of these descriptions are effectively the same. Both expressions are true only in the event that each of $P$, $Q$, and $R$ is true. One way to verify this is to construct the truth table for expression $(P \wedge Q) \wedge R$ and the truth table for $P \wedge (Q \wedge R)$ and show that they give the same values for every possible value of $P$, $Q$, and $R$. Thus, since the order of operations does not matter in this case, we can simply write $P \wedge Q \wedge R$ without worrying about any possible ambiguity. The same thing is true for $P \vee Q \vee R$. The connectives $\wedge$ and $\vee$ have the associative property. Notice this terminology is consistent with the "associative property" of real-number addition and multiplication: $(a + b) + c = a + (b + c)$ and $(ab)c = a(bc)$.

In some cases, however, the ambiguity is not easy to resolve. For example, consider the expression $P \wedge Q \vee R$, which can be interpreted as either $(P \wedge Q) \vee R$ or as $P \wedge (Q \vee R)$. Using the statements $P$, $Q$, and $R$ as before, $(P \wedge Q) \vee R$ can be interpreted as "a candidate must be at least 35 years old and a natural-born U.S. citizen, or must have lived in the United States at least 14 years." On the other hand, $P \wedge (Q \vee R)$ can be interpreted as "a candidate must be at least 35 years old, and be a natural-born citizen or have lived in the United States at least 14 years." These two descriptions are certainly not the same! For example, since Arnold Schwarzenegger is not a natural-born U.S. citizen, the first expression excludes him as a candidate for president, whereas the second one includes him. One way to see exactly how the two statements differ is to compare the truth table for $(P \wedge Q) \vee R$ to the truth table for $P \wedge (Q \vee R)$. The truth table for $(P \wedge Q) \vee R$ is:

| $P$ | $Q$ | $R$ | $(P \wedge Q)$ | $(P \wedge Q) \vee R$ |
|---|---|---|---|---|
| T | T | T | T | T |
| T | T | F | T | T |
| T | F | T | F | T |
| T | F | F | F | F |
| F | T | T | F | T |
| F | T | F | F | F |
| F | F | T | F | T |
| F | F | F | F | F |

The truth table for $P \wedge (Q \vee R)$ is:

| $P$ | $Q$ | $R$ | $(Q \vee R)$ | $P \wedge (Q \vee R)$ |
|---|---|---|---|---|
| T | T | T | T | T |
| T | T | F | T | T |
| T | F | T | T | T |
| T | F | F | F | F |
| F | T | T | T | F |
| F | T | F | T | F |
| F | F | T | T | F |
| F | F | F | F | F |

Because the last columns of these two truth tables differ for some values of $P$, $Q$, and $R$, the two expressions are not equivalent. For example, notice that if $P$ is false and $Q$ and $R$ are both true, then $(P \wedge Q)$ is false. Because $R$ is true, however,

$(P \wedge Q) \vee R$ is true. On the other hand, $P \wedge (Q \vee R)$ is false regardless of whether $(Q \vee R)$ is true or false, because $P$ is false. To avoid ambiguity, it is often best to use parentheses.

A way to avoid ambiguity without using parentheses is to adopt a convention on the order of operations. For example, in arithmetic, the convention is that multiplication takes precedence over addition. Therefore, $3 + 4 \times 5$ is determined by first evaluating $4 \times 5$ and then adding 3. Of course, we could have written $3 + (4 \times 5)$ to avoid the ambiguity altogether. Similarly, in Boolean logic we adopt the convention that $\wedge$ (AND) takes precedence over $\vee$ (OR). Therefore, the expression $P \wedge Q \vee R$, by convention, is to be interpreted as $(P \wedge Q) \vee R$. Furthermore, the convention states that $\neg$ (NOT) takes the highest precedence of all. Thus, $\neg P \wedge \neg Q \wedge R$ is interpreted as $((\neg P) \wedge (\neg Q)) \vee R$.

## EXAMPLE 8 ■ Applying Boolean Logic to a Web Search

Let's revisit the Web queries mentioned before. Let $P$ represent the query "football," which corresponds to the expression, "This Web page contains the phrase 'football'." Let $Q$ represent the expression, "This Web page contains the word 'NFL'." Let $R$ represent the expression, "This Web page contains the word 'college'." We now translate the query "football AND (NOT NFL) AND (NOT college)" as $P \wedge (\neg Q) \wedge (\neg R)$ and write its truth table:

| $P$ | $Q$ | $R$ | $\neg Q$ | $\neg R$ | $P \wedge (\neg Q) \wedge (\neg R)$ |
|---|---|---|---|---|---|
| T | T | T | F | F | F |
| T | T | F | F | T | F |
| T | F | T | T | F | F |
| T | F | F | T | T | T |
| F | T | T | F | F | F |
| F | T | F | F | T | F |
| F | F | T | T | F | F |
| F | F | F | T | T | F |

For every possible value of $P$, $Q$, and $R$, the truth table gives us the value of our expression. The fourth and fifth columns of the table are not strictly necessary, but they are helpful in determining the values in the last column. As expected, this table tells us that the expression $P \wedge (\neg Q) \wedge (\neg R)$ is true precisely when $P$ is true, $Q$ is false, and $R$ is false.

Two expressions are said to be **logically equivalent** if they have the same value, true or false, for each possible assignment of the Boolean variables. To decide whether two expressions are logically equivalent, we construct the truth tables for each one and then check if they have the same values for each of the possible assignments of the Boolean variables. If they do, the expressions are logically equivalent. If they differ for even one case, however, then the expressions are not equivalent.

So, to determine whether the expression, "football AND (NOT NFL) AND (NOT college)," is equivalent to the expression, "football AND NOT (NFL AND college)," we need only compare their corresponding truth tables.

## EXAMPLE 9 ■ Logically Equivalent Expressions

We first determine the truth table for the expression, "football AND NOT (NFL AND college)," which is represented as $P \wedge \neg(Q \wedge R)$. Its truth table is:

| $P$ | $Q$ | $R$ | $Q \wedge R$ | $\neg(Q \wedge R)$ | $P \wedge \neg(Q \wedge R)$ |
|---|---|---|---|---|---|
| T | T | T | T | F | F |
| T | T | F | F | T | T |
| T | F | T | F | T | T |
| T | F | F | F | T | T |
| F | T | T | T | F | F |
| F | T | F | F | T | F |
| F | F | T | F | T | F |
| F | F | F | F | T | F |

This truth table differs in the last column from the truth table for $P \wedge (\neg Q) \wedge (\neg R)$ in the previous example. For instance, when $P$ is true, $Q$ is true, and $R$ is false, we see that $P \wedge (\neg Q) \wedge (\neg R)$ is false but $P \wedge \neg(Q \wedge R)$ is true. Therefore, we must conclude that $P \wedge (\neg Q) \wedge (\neg R)$ is not logically equivalent to $P \wedge \neg(Q \wedge R)$.

On the other hand, the expression, "football AND NOT (NFL OR college)," is represented by $P \wedge \neg(Q \vee R)$. Its truth table is:

| $P$ | $Q$ | $R$ | $Q \vee R$ | $\neg(Q \vee R)$ | $P \wedge \neg(Q \vee R)$ |
|---|---|---|---|---|---|
| T | T | T | T | F | F |
| T | T | F | T | F | F |
| T | F | T | T | F | F |
| T | F | F | F | T | T |
| F | T | T | T | F | F |
| F | T | F | T | F | F |
| F | F | T | T | F | F |
| F | F | F | F | F | F |

Because the last column of this table agrees with the last column for the expression $P \wedge (\neg Q) \wedge (\neg R)$, we know the expression, "football AND (NOT NFL) AND (NOT college)," is equivalent to the expression, "football AND NOT (NFL OR college)."

## Applying Logic to Message Routing

The AND operator in the truth table on page 556 is also used by computers to deliver messages over the Internet with a device called a *router*. Recall that if $P$ and $Q$ are expressions then the statement $P \wedge Q$ is true when both $P$ and $Q$ are true and false otherwise. Computer scientists use an analogous operation on 0 and 1 by allowing $P$ and $Q$ to represent 0 or 1 and defining that $P \wedge Q = 1$ when $P$ and $Q$ are 1 and $P \wedge Q = 0$ otherwise. When $\wedge$ is used in this way it is called the *bitwise AND*. Notice that we can obtain an operation table for the bitwise AND from the table for the logical AND on page 556 by substituting 1 for T and 0 for F. In particular, we have

| $P$ | $Q$ | $P \wedge Q$ |
|---|---|---|
| 1 | 1 | 1 |
| 1 | 0 | 0 |
| 0 | 1 | 0 |
| 0 | 0 | 0 |

The bitwise AND operation can be extended to binary strings of equal length by applying it individually to corresponding entries. Thus $11001001 \wedge 01101101 = 01001001$ since both strings have a 1 only in positions 2, 5 and 8. In general, for any binary string $s$ we can use the bitwise AND to copy whichever entries of $s$ we desire while converting all the other entries of $s$ to 0. For example, if we have a list of binary strings of length 8 and we wish to modify these strings by copying the entries in positions 2, 7, and 8 and changing all other entries to 0, we simply take each string in the list and combine it with 01000011 using the bitwise AND operator. Thus we have, $11101001 \wedge 01000011 = 01000001$; and $00011110 \wedge 01000011 = 00000010$. When doing the bitwise AND operation on binary strings it is convenient to put one string directly above the other. In those positions where both entries are 1 the result is 1; otherwise the result is 0.

$$\begin{array}{rl} s & 11101001 \\ t & \underline{01000011} \\ s \wedge t & 01000001 \end{array} \qquad \begin{array}{rl} s & 00011110 \\ t & \underline{01000011} \\ s \wedge t & 00000010 \end{array}$$

The bitwise AND operation is used by computers to determine when certain entries of two binary strings match. Say, for example, that a computer would take a particular action if two binary strings $s$ and $t$ of equal length match in the first three positions. This happens precisely when $s \wedge 11100000 = t \wedge 11100000$. The reason why this works is that $x \wedge 1 = 1$ only when $x = 1$. Thus, if $s$ and $t$ both begin with 1 then both $s \wedge 11100000$ and $t \wedge 11100000$ will begin with 1; if $s$ and $t$ both begin with 0 then both $s \wedge 11100000$ and $t \wedge 11100000$ will begin with 0; if $s$ and $t$ begin with different digits then $s \wedge 11100000$ and $t \wedge 11100000$ will begin with different digits. The same reasoning applies to positions 2 and 3. The string of five 0's at the end of 11100000 ensures that $s \wedge 11100000$ and $t \wedge 11100000$ will both end with five 0's. So, checking that $s \wedge 11100000 = t \wedge 11100000$ checks whether $s$ and $t$ agree in the first three positions while disregarding the other positions.

In order for devices on a network to communicate with each other each one must be given a unique identifier. This is done with an **Internet Protocol address**.

**Internet Protocol Address** DEFINITION

An **Internet Protocol (IP) address** is a sequence of four numbers between 0 and 255 separated by dots assigned to routers, computers, printers and fax machines that allows those linked electronically to uniquely identify and communicate with each other.

Each computer on the Internet is assigned an IP address. From the IP address, we can determine the **network address** of the computer, which specifies the network or subnet that the computer is a part of. While each computer on the network or subnet will have a different IP address, they will all have the same network address. To demonstrate how computers determine the network address from an IP address, we must first explain how to convert the decimal form of each component of an IP address such as 131.212.66.17, which is convenient for humans, to their binary forms of length 8, which are convenient for computers. To convert an IP address in decimal to binary, we express each decimal number as a sum of distinct powers of 2 ranging from $128 = 2^7$ to $1 = 2^0$. For example, $213 = 128 + 64 + 16 + 4 + 1 = 2^7 + 2^6 + 2^4 + 2^2 + 2^0$. Now, make a row of the powers of 2 from 128 to 1 and beneath each one place a 1 if that power of 2 appears in the sum and a 0 if it does not. For the number 213, we have

| 128 | 64 | 32 | 16 | 8 | 4 | 2 | 1 |
|---|---|---|---|---|---|---|---|
| 1 | 1 | 0 | 1 | 0 | 1 | 0 | 1 |

Reading off the sequence of 0's and 1's, we have that the binary form of 213 is 11010101. Since 00000000 is the binary form of length 8 of 0, and 11111111 is the binary form of $255 = 128 + 64 + 32 + 16 + 8 + 4 + 2 + 1$, all the integers between 0 and 255 can be written as binary strings of length 8 using leading 0's as needed.

Each IP address is assigned a companion number called a **subnet mask** that also consists of four numbers, each of which range from 0 to 255. To determine the network address from the IP address, one performs the bitwise AND operation to the binary forms of the IP address and its companion subnet mask. The resulting number is the network address.

## EXAMPLE 10 ■ Network Address for IP Address 131.212.66.17 With Subnet Mask 255.255.255.0

We determine the network address corresponding to the IP address 131.212.66.17 with subnet mask 255.255.255.0. Since the binary form of 131.212.66.17 is 10000011.11010100.01000010.00010001 and the binary form of the subnet mask is 11111111.11111111.11111111.00000000, combining them using the bitwise AND gives

| | |
|---|---|
| **IP address** | 10000011.11010100.01000010.00010001 |
| **Subnet mask** | 11111111.11111111.11111111.00000000 |
| **Bitwise AND** | 10000011.11010100.01000010.00000000. |

Thus, the network address is 10000011.11010100.01000010.00000000 = 131.212.66.0 (some authors omit one or more 0's when they appear at the end of a network address). ■

From Example 10 it appears that the network address corresponding to a IP address with subnet mask 255.255.255.0 is simply the same as the IP address with the last number changed to 0. This is correct in this case but for other subnet masks, the network address is not readily apparent from the IP address.

## EXAMPLE 11 ■ Network Address for IP Address 131.212.66.56 With Subnet Mask 255.255.255.240

We determine the network address corresponding to the IP address 131.212.66.56 with subnet mask 255.255.255.240. Since the binary form of 131.212.66.56 is 10000011.11010100.01000010.00111000 and the binary form of the subnet mask is 11111111.11111111.11111111.11110000, combining them using the bitwise AND gives

| | |
|---|---|
| **IP address** | 10000011.11010100.01000010.00111000 |
| **Subnet mask** | 11111111.11111111.11111111.11110000 |
| **Bitwise AND** | 10000011.11010100.01000010.00110000. |

In this example, the network address is 10000011.11010100.01000010.00110000 = 131.212.66.48. Since the subnet mask for this IP address ends with four 0's, even if we changed the last four bits in the IP address, we would still get the same network address. Any decimal number whose binary form starts with 0011 is a number between 48 and 63, so any device with a IP address that begins with 131.212.66 and ends with any number between 48 and 63 will be in the network whose address is 131.212.66.48. We note that a device with an IP address that begins with 131.212.66 and ends with a number that is not between 48 and 63 is not on the same network as any device that starts with 131.212.66 and ends with a number between 48 and 63 since the devices will have different network addresses.

The network addresses allow routers to deliver messages to IP addresses in much the same way the postal service delivers mail to home addresses. That is, the postal service first checks to see if the mail is to be sent to a local address and if not, it is relayed to a larger mail center. Likewise, if a message is to be sent from one computer to another, a local router checks to see if both computers are on the same network. If so, the message is sent over that network. Otherwise it is routed to a larger network. In this case the local router searches its memory for the network address that most closely matches that of the destination network address and relays the message to a router in a larger network that has that address in its memory. A postal analogy might be a letter from Duluth, Minnesota addressed to Rochester, Minnesota being routed through Minneapolis.

## REVIEW VOCABULARY

**Binary linear code** A code consisting of words composed of 0's and 1's obtained by using parity-check sums to append check digits to messages. (p. 539)

**Boolean logic** Logic attributed to George Boole that uses operations such as $\wedge$, $\vee$, and $\neg$ to connect statements. (p. 554)

**Caesar cipher** A cryptosystem used by Julius Caesar whereby each letter is shifted the same amount. (p. 547)

**Code word** A string of digits composed of a message and check digits. (p. 539)

**Compression algorithm** A procedure for converting data from one format to another one optimized for compactness. (p. 542)

**Cryptology** The study of how to make and break secret codes. (p. 546)

**Data compression** The process of encoding data so that the most frequently occurring data are represented by the fewest symbols. (p. 541)

**Decimation cipher** A cryptosystem that uses multiplication by a fixed value to shift each letter. (p. 548)

**Decoding** The process of translating received data into code words. (p. 537)

**Distance between two strings** The distance between two strings of equal length is the number of positions in which they differ. (p. 537)

**Encryption** The process of encoding data to protect against unauthorized interpretation. (p. 546)

**Even parity** Even integers are said to have even parity. (p. 535)

**IP address** A sequence of four numbers that uniquely identifies a device on a network. (p. 560)

**Key** A string used to encode and decode data. (p. 550)

**Key word** A word used to determine the amount of shifting for each letter while encoding a message. (p. 549)

**Logically equivalent** Two expressions are said to be logically equivalent if they have the same values for all possible values of their Boolean variables. (p. 558)

**Modular arithmetic** Addition and multiplication involving modulo $n$. (p. 547)

**Nearest-neighbor decoding** A method that decodes a received message as the code word that agrees with the message in the most positions. (p. 538)
**Network address** The portion of an IP address that identifies a local network. (p. 560)
**Odd parity** Odd integers are said to have odd parity. (p. 535)
**Parity-check sums** Sums of digits whose parities determine the check digits. (p. 537)
**RSA public key encryption scheme** A method of encoding that permits each person to announce publicly the means by which secret messages are to be sent to him or her. (p. 552)
**Subnet mask** A companion number to an IP address that allows a router to determine the network portion of an IP address. (p. 561)
**Truth table** A tabular representation of an expression in which the variables and the intermediate expressions appear in columns and the last column contains the expression being evaluated. (p. 555)
**Variable-length code** A code in which the number of symbols for each code word may vary. (p. 541)
**Vigenère cipher** A cryptosystem that utilizes a key word to determine how much each letter is shifted. (p. 549)
**Weight of a binary code** The minimum number of 1's that occur among all nonzero code words of a code. (p. 539)

## ✓ SKILLS CHECK

**1.** Using the circular diagram method to encode the message 1011, the encoded message is

**(a)** 1011001.
**(b)** 1011010.
**(c)** 1010001.

**2.** A four-digit binary message was encoded using Table 17.1 and the message 1010010 was received. Using the nearest neighbor method, the decoded four-digit message is ______________________.

**3.** Using the nearest-neighbor method and the code in Table 17.2, the word 1110011 decodes as

**(a)** 0110010.
**(b)** 1100011.
**(c)** 1010001.

**4.** The distance between received words 1011001 and 1000101 is ______________________.

**5.** The weight of the binary linear code {0000000, 0011111, 0101011, 0110100} is

**(a)** 3.
**(b)** 4.
**(c)** 5.

**6.** If the two messages 0 and 1 are encoded as 000 and 111, respectively, the number of errors the code can correct is ______________________.

**7.** If every pair of code words differs in at least five positions, then nearest-neighbor decoding can accurately decode words that have

**(a)** two mistakes.
**(b)** three mistakes.
**(c)** four mistakes.

**8.** If a binary linear code has weight 4, the maximum number of errors that it will detect is ____________.

**9.** Using the encoding scheme A → 0, B → 10, C → 11, the string 010110 decodes as

**(a)** ABCB.
**(b)** ABCA.
**(c)** ABACA.

**10.** The sum of the binary string 1011001 and 1001101 is ______________________.

**11.** The Caesar cipher encrypts GO HOME NOW as

**(a)** JR KRPH QRZ.
**(b)** DL ELJB KLT.
**(c)** Neither of these.

**12.** Using modular arithmetic, $3^5$ mod 20 is equal to ______________________.

**13.** Using the Vigenère cipher and the key word ADAM to decrypt EIEIO, we obtain

**(a)** ELELR.
**(b)** EFEFL.
**(c)** EFEWO.

**14.** If the message EAPL was encrypted using the decimation cipher with the key 9, the message is _____.

**15.** Using the RSA scheme with $n = 91$ and $s = 5$, the message 4 decodes as

**(a)** 11.
**(b)** 20
**(c)** 23.

**16.** If we use $p = 7$, $q = 17$, and $r = 5$ in the RSA scheme, the value of $s$ is ______________________.

**17.** Which messages are the hardest to break?

**(a)** Messages encrypted with the Vigenère cipher.
**(b)** Messages encrypted with the decimation cipher.
**(c)** Messages encrypted with the RSA scheme.

**18.** The statement "$P$ OR $Q$" is true if and only if ______________________.

**19.** When the statement "$P$ AND NOT $Q$" is true, it must be the case that

(a) $P$ is true.
(b) $P$ is false.
(c) either $P$ or $Q$ is false.

**20.** 00111001 $\wedge$ 11111001 = ______________.

# CHAPTER 17 EXERCISES

■ Challenge ◆ Discussion

## 17.1 Binary Codes

**1.** Use the diagram method shown in Figures 17.1 and 17.2 to verify the code words in Table 17.1 for the messages 0101, 1011, and 1111.

**2.** Use the diagram method to decode the received messages 0111011 and 1000101.

**3.** Find the distance between each of the following pairs of words:

(a) 11011011 and 10100110
(b) 01110100 and 11101100

**4.** Referring to Table 17.1, use the nearest-neighbor method to decode the received words 0000110 and 1110100.

**5.** If the code word 0110010 is received as 1001101, how is it decoded using the diagram method?

**6.** Suppose a received word has the Venn diagram arrangement shown here:

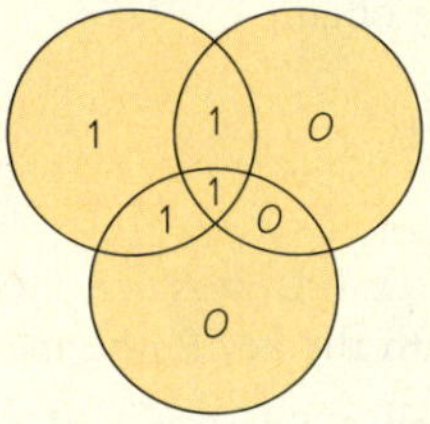


What can we conclude about the received word?

## 17.2 Encoding with Parity-Check Sums

**7.** Determine the binary linear code that consists of all possible three-digit messages with three check digits appended using the parity-check sums $a_2 + a_3$, $a_1 + a_3$, and $a_1 + a_2$. (That is, $c_1 = 0$ if $a_2 + a_3$ is even, $c_1 = 1$ if $a_2 + a_3$ is odd, and similarly for $c_2$ and $c_3$.)

**8.** Let $C$ be the code

{0000000, 1110100, 0111010, 0011101, 1001110, 0100111, 1010011, 1101001}

What is the error-correcting capability of $C$? What is the error-detecting capability of $C$?

**9.** Find all code words for binary messages of length 4 by adding three check digits using the parity-check sums $a_2 + a_3 + a_4$, $a_2 + a_4$, and $a_1 + a_2 + a_3$. Will this code correct any single error?

**10.** Consider the binary linear code

$C$ = {00000, 10011, 01010, 11001, 00101, 10110, 01111, 11100}

Use nearest-neighbor decoding to decode 11101 and 01100. If the received word 11101 has exactly one error, can you determine the intended code word? Explain your reasoning.

**11.** Construct a binary linear code using all eight possible binary messages of length 3 and appending three check digits using the parity-check sums $a_1 + a_2$, $a_2 + a_3$, and $a_1 + a_3$. Decode each of the received words below by the nearest-neighbor method:

001001, 011000, 000110, 100001

**12.** Extend the code words listed in Table 17.1 to eight digits by appending a 0 to words of even weight and a 1 to words of odd weight. What are the error-detecting and error-correcting capabilities of the new code?

**13.** Extend the code words listed in Table 17.2 to eight digits by appending a 0 to words of even weight and a 1 to words of odd weight. What are the error-detecting and error-correcting capabilities of the new code?

**14.** Suppose the weight of a binary linear code is 6. How many errors can the code correct? How many errors can the code detect?

**15.** How many code words are there in a binary linear code that has all possible messages of length 5 with three check digits appended? How many possible received words are there with this code?

■ **16.** Explain why no binary linear code with all possible three-digit messages together with three check digits can correct all possible errors involving two digits.

■ **17.** A *ternary* code is formed by starting with all possible strings of a fixed length composed of 0's, 1's, and 2's and appending extra digits that are also 0's, 1's, or 2's. Form a ternary code by appending to each message $a_1a_2$ the check digits $c_1c_2$ using:

$$c_1 = (a_1 + a_2) \bmod 3$$
$$c_2 = (2a_1 + a_2) \bmod 3$$

■ **18.** Use the ternary code in the preceding exercise and the nearest-neighbor method to decode the received word 1211.

■ **19.** Suppose a ternary code is formed by starting with all possible strings of 0's, 1's, and 2's of length 4 and appending two extra digits that are also 0's, 1's, and 2's. How many code words are there in this code? How many possible received words are there in this code?

## 17.3 Data Compression

**20.** Suppose we code a four-symbol genetic set {A, C, T, G} into binary form as follows:

$$A \to 0 \qquad C \to 10 \qquad T \to 110 \qquad G \to 111$$

Convert the sequence ACAAGTAAC into binary code.

**21.** Use the code in the previous exercise to determine the sequence of symbols represented by the binary code 001100001111000.

**22.** Suppose we code a five-symbol set {*A, B, C, D, E*} into binary form as follows:

$$A \to 0 \qquad B \to 10 \qquad C \to 110$$
$$D \to 1110 \qquad E \to 1111$$

Convert the sequence of *AEAADBAABCB* into binary code. Determine the sequence of symbols represented by the binary code 01000110100011111110.

**23.** Use the code in the previous exercise to convert the sequence *EABAADABB* into binary code. Determine the sequence of letters represented by the binary code 00100011001110111010.

■ **24.** Devise a variable-length binary coding scheme for a six-symbol set {*A, B, C, D, E, F*}. Assume the *A* is the most frequently occurring symbol, *B* is the second most frequently occurring symbol, and so on.

**25.** Judging from the Morse code, what are the three most frequently occurring consonants in English text material? What is the most frequently occurring vowel?

**26.** In English, the letter *H* occurs more often than *D, G, K,* and *W,* but in Morse code, *H* has a longer code than *D, G, K,* and *W.* Speculate on the reason for this apparent violation of data-compression principles.

**27.** Explain why the Morse code must include a space after each letter but fixed-length codes do not.

**28.** Guess the percentage of the occurrences of spaces in typical English text material.

**29.** Following are the closing values (rounded to the nearest integer) of the Dow Jones Industrial Average Index stock market values for the period September 17, 2007–September 28, 2007. Use the delta function method to compress these values. What percentage reduction in characters is there?

13403 13739 13816 13767 13820
13759 13779 13878 13913 13896

**30.** The following numbers were encoded using delta function encoding. Determine the original numbers.

1207 373 −57 −97 −234 −105 178 −73 275
79 −183 −146 −94 129

**31.** Decode the binary string 11100100001001110110011010, which has been encoded using the Huffman code given on page 544.

**32.** Use a Huffman tree code to assign a binary code to the letters that occur with the probabilities

| | |
|---|---|
| *A* | 0.025 |
| *B* | 0.150 |
| *C* | 0.015 |
| *D* | 0.170 |
| *E* | 0.200 |
| *F* | 0.225 |
| *G* | 0.215 |

**33.** Suppose a Huffman tree has been used to create a binary code for the letters *A* through *J*, and the results include *B* = 111110, *J* = 111111, and *G* = 11110. If the code has only two code words of length 6 and one of length 5, what can you say about the probability of the occurrence of the letters *B, J,* and *G*?

## 17.3 Cryptography

**34.** For each part below, explain how modular arithmetic can be used to answer the question.

**(a)** If today is Wednesday, what day of the week will it be in 16 days?

**(b)** If a clock (with hands) indicates that it is now four o'clock, what will it indicate in 37 hours?

**(c)** If a military person says it is now 0400, what time would it be in 37 hours? (Instead of A.M. and P.M., military people use 1300 for 1:00 P.M., 1400 for 2:00 P.M., and so on.)

**(d)** If it is now July 20, what day will it be in 65 days?

**(e)** If the odometer of an automobile reads 97,000 now, what will it read in 12,000 miles?

**35.** Use the Caesar cipher to encrypt the message RETREAT. Determine the intended message corresponding to the message DGYDQFH that was encrypted using the Caesar cipher.

**36.** Suppose you take a message and repeatedly apply the Caesar cipher to it until you return to the original message. How many iterations must be done before this occurs?

**37.** Using 0, 1, 2, . . . , 25 to label the positions of the letters *A*, *B*, *C*, . . . , *Z*, suppose we create a cipher by replacing the letter in position $i$ with the letter in position $(i + 8)$ mod 26. How many iterations of this cipher must be done before a message will return to its original state?

**38.** The message ADDAOS was encrypted using decimation cipher with the key 7. Decrypt it.

**39.** Use the decimation cipher with the key 5 to encrypt RETREAT.

**40.** If you attempted to use the decimation cipher with the key 13, how would the word MESSAGE be encrypted?

**41.** Explain why 2 cannot be used for the key in a decimation cipher.

**42.** Use the Vigenère cipher with the key word HELP to encrypt the message PHONE HOME.

**43.** Given that the BEATLES was used as the key word for the Vigenère cipher to encrypt SSLETRY TXOGPW, decrypt the message.

**44.** Use the Vigenère cipher with the key word CLUE to encrypt the message THE WALRUS WAS PAUL.

**45.** Add the following pairs of binary strings:

**(a)** 10111011 and 01111011
**(b)** 11101000 and 01110001

**46.** All binary linear codes have the property that the sum of two code words is another code word. Use this fact to determine which of the following sets cannot be a binary linear code

**(a)** {0000, 0011, 0111, 0110, 1001, 1010, 1100, 1111}
**(b)** {0000, 0010, 0111, 0001, 1000, 1010, 1101, 1111}
**(c)** {0000, 0110, 1011, 1101}

**47.** Use the RSA scheme with $p = 5$, $q = 17$, and $r = 3$ to determine the numbers sent for the message VIP.

**48.** Use the RSA scheme with $p = 5$, $q = 17$, and $r = 3$ to decode the received numbers 52 and 72.

**49.** In the RSA scheme with $p = 5$, $q = 17$, and $r = 5$, determine the value of $s$.

**50.** Why can't we use the RSA scheme with $p = 7$, $q = 11$, and $r = 3$?

**51.** Explain why we can't employ the RSA scheme to send the message "NO" with $p = 7$ and $q = 11$ using blocks of length 2, but we can send it if we use blocks of length 4.

**52.** Use the search box at www.google.com to compute $13^9$ mod 77, $13^6$ mod 77, and $13^{15}$ mod 77. (To compute $13^9$ mod 77 enter 13^9 mod 77.)

## 17.4 Web Searches and Mathematical Logic

**53.** Show that $P \vee (P \wedge Q)$ is logically equivalent to $P$.

**54.** Show that $\neg(P \vee Q)$ is logically equivalent to $\neg P \wedge \neg Q$.

**55.** Show that $\neg(P \wedge Q)$ is logically equivalent to $\neg P \vee \neg Q$. This relationship and the one in the previous exercise are known collectively as *De Morgan's Laws*.

**56.** Show that $P \vee (Q \wedge R)$ is logically equivalent to $(P \vee Q) \wedge (P \vee R)$.

**57.** Show that $P \wedge (Q \vee R)$ is logically equivalent to $(P \wedge Q) \vee (P \wedge R)$.

**58.** A patron at a restaurant tells the waiter to bring her the chef's recommendation as long as it has "lots of anchovies or is not spicy and in addition the portion must be large." The waiter goes to the kitchen and tells the chef to prepare a dish that has "lots of anchovies and is also large or is spicy and is also large." Did the waiter communicate the patron's wishes correctly to the chef? Use truth tables to support your answer.

**59.** The *implication connective* is defined by the following truth table:

| $P$ | $Q$ | $P \to Q$ |
|---|---|---|
| T | T | T |
| T | F | F |
| F | T | T |
| F | F | T |

Use truth tables to show that $P \to Q$ is logically equivalent to $\neg P \vee Q$.

**60.** The Minnesota Vikings football coach tells his team before the last game of the regular season that if the team wins, they will be in the playoffs. Use the truth table given in the previous exercise to verify that if the Vikings lose and are still in the playoffs, the coach made a truthful statement to the team.

**61.** Using the implication connective and other connectives, variables, and truth tables, determine whether the statement "If it snows, there will be no school" is logically equivalent to the statement "It is not the case that it snows and there is school."

**62.** Suppose $s$ and $t$ are binary strings of length 8. How would you use the bitwise operator $\wedge$ to determine if the last three digits of $s$ and $t$ match? How would you determine if $s$ and $t$ match in positions 2, 4, 6, and 8?

**63.** Using $\wedge$ to denote the bitwise AND operator compute:

(a) $11110001 \wedge 00101110$
(b) $01110001 \wedge 10111110$

**64.** Explain why $01100000 \wedge 1011111$ is undefined.

**65.** In practice, a computer checks to see if $s \wedge 11100000 = t \wedge 11100000$ by checking if $s \wedge 11100000 + t \wedge 11100000 = 00000000$ where addition is done mod 2 in each component. Explain why this works.

**66.** If $s$ is an 8-digit binary string, determine $s \wedge 11111111$ and $s \wedge 00000000$.

**67.** If $s$ is a binary string of length 8 and $s \wedge 01010101 = 00000001$ what is the most that you can say about $s$?

**68.** Given a binary string $s$ of length 8, how could you use the bitwise AND operator to determine if the digits in positions 1, 3, and 5 are 0's?

**69.** Find four binary strings $s$ that satisfy $s \wedge 11100111 = 01100010$.

**70.** How many binary strings $s$ of length 8 are there that satisfy $s \wedge 11100011 = 10000001$?

**71.** Determine the network address for the IP address 8.20.15.1 with subnet mask 255.000.000.000.

**72.** Determine the network address for the IP address 8.20.15.1 with subnet mask 255.255.000.000.

**73.** Determine if the IP address 172.16.17.30 with subnet mask 255.255.255.240 has the same network address as the IP address 172.16.17.15 with the subnet mask 255.255.255.240.

## WRITING PROJECTS

**1.** Prepare a two-page report on applications of modular arithmetic. Explain the calculation of the check digits described in Exercises 7, 9, and 11 with modular arithmetic. Use modular arithmetic to describe the error-detection schemes used in Chapter 16.

**2.** Use the Web to find information for a two-page report on the Braille system of coding.

**3.** Use the Web to find information for a two-page report on the Morse code.

**4.** Use the Web to find information about Smart Card technology and write a two-page report on your findings.

## SUGGESTED READINGS

DENEEN, L. Secret encryption with public keys. *UMAP Journal,* 8 (1987): 9–29. Describes several ways in which modular arithmetic can be used to code secret messages.

PETZOLD, C. *Code,* Microsoft Press, Redmond, Wash., 1999. The first three chapters of this book provide an excellent explanation of the Morse code and the Braille system of coding.

## SUGGESTED WEB SITE

**www.d.umn.edu/~jgallian/fapp7e** This site implements the nearest-neighbor decoding method for seven-digit binary strings using the code given in Table 17.1.

# On Size and Growth PART VI

Mathematics is the study of patterns and relationships. It can explain why there are no King Kongs, analyze designs on ancient pottery, and suggest new and beautiful artistic designs. Mathematicians search for and classify numerical, geometric, and even abstract patterns. In these chapters, we follow some of those searches. We concentrate on geometric patterns but find that those lead to numerical considerations, too.

In Chapter 18, Growth and Form, we look at how the sizes of objects influence their forms. We investigate some big things, such as King Kong, tall trees, mile-high buildings, and mountains. Seeing the underlying principles of scaling will help you to appreciate why objects in the world have the shapes and sizes that they do.

We start with a simple numerical pattern in Chapter 19, Symmetry and Patterns, which leads to questions about esthetically pleasing proportions and the importance of bilateral symmetry. We expand our notion of symmetry and discover surprising limitations that even broader notions of symmetry face. We examine the beauty of fractal patterns, ones that resemble themselves at finer and finer scales, in nature and in traditional art from Africa and elsewhere.

Chapter 20, Tilings, answers the question of how to arrange objects symmetrically on a surface. What shapes can we use? What patterns can arise if the objects themselves are symmetrical, or if we allow irregular shapes but demand that they all face the same way? Most curious of all, you can arrange shapes in a pattern that does not repeat but is nevertheless systematic. ■

CHAPTER 18



# Growth and Form

Fantasy films have made us familiar with giant creatures, including King Kong, Godzilla, and the auliphants in *The Lord of the Rings*. We also find supergiants in literature, such as the giant of "Jack and the Beanstalk," the Big Friendly Giant in Roald Dahl's *The BFG,* and the Brobdingnagians of *Gulliver's Travels.*

Even from an early age, though, we don't really believe in monsters and giants. But could such beings ever exist? What problems would their enormous size cause them? How would they have to adapt to cope? (See Figure 18.1.)

Every species adapts to its environment. In particular, it faces the **problem of scale**: how to adapt and survive at the different sizes from the beginning of life to the final size of a mature adult. For example, the giant panda ranges from barely 1 pound (lb) at birth to 275 lb in adulthood. A baby panda could be crushed by its mother; an adult panda needs to eat a lot.

For contrast, consider the horse. A newborn foal that weighed as little as a newborn panda would be too small to keep up with the herd and could not survive. An adult horse weighs much more than a panda and has to consume far more food, but the horse can move much more quickly and cover greater distances to find sustenance.

There have been large land mammals (mammoths) and huge sea mammals (the blue whale)–not to mention the dinosaurs. But the tallest humans have been only 9 to 10 feet (ft) tall. The largest mammoth was 16 ft at the shoulder (about twice as tall as an elephant). Even the tallest dinosaur, *Supersaurus,* stood only 40 ft high.

What about supergiants and utterly huge monsters? That they have never existed suggests physical limits to size. In fact, with a few simple principles of geometry, we can show that no objects or living beings could exist, unchanged in shape, on a vastly different scale, larger or smaller.

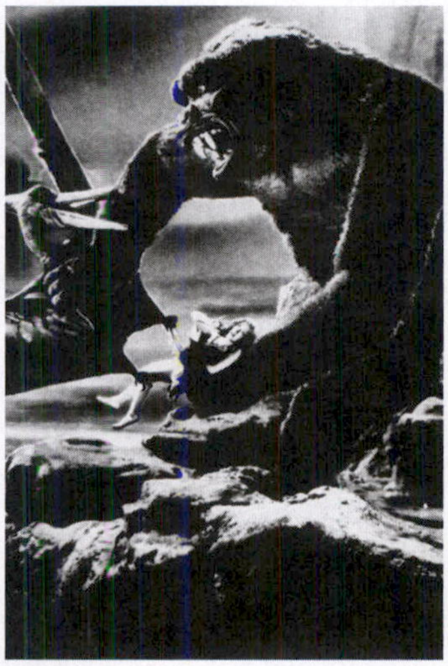

FIGURE 18.1 Could King Kong actually exist? (*Bettmann/Corbis.*)

## 18.1 Geometric Similarity

The powerful mathematical idea that we use is **geometric similarity**.

**Geometric Similarity** DEFINITION

Two objects are **geometrically similar** if they have the same shape, regardless of the materials of which they are made; they do not have to be the same size.

Similar objects need not be the same size, but corresponding distances must be proportional. For example, when a photo is enlarged, it is enlarged by the same factor in both the horizontal and vertical directions–in fact, in any direction (such as a diagonal). We call this factor the **linear scaling factor** or **length scaling factor**.

**Linear (Length) Scaling Factor** DEFINITION

The **linear (length) scaling factor** of two geometrically similar objects is the ratio of a length of any part of the second to the corresponding part of the first.

In Figure 18.2, the linear scaling factor is 3; the enlargement is three times as wide and three times as high as the original. In fact, every pair of points goes to a new pair of points three times as far apart as the original ones.

**FIGURE 18.2** Two geometrically similar photographs. (*David Spurdens/Corbis.*)

Objects can be scaled down as well as up; for example, the smaller photograph in Figure 18.2 is geometrically similar to the larger one, with the linear scaling factor being 1/3.

## How Area and Volume Scale

The enlargement can be divided into $3 \times 3 = 9$ rectangles, each the size of the original. Hence, the enlargement has $3 \times 3 = 3^2 = 9$ times the area of the original. More generally, if the linear scaling factor is some general number $L$ (not necessarily 3), the resulting enlargement has an area $L \times L = L^2$ ("$L$ squared") times the area of the original.

### How Area Scales — RULE

The *area* of a scaled-up object goes up with the *square* of the linear scaling factor.

We symbolize the relationship between the area $A$ and the linear scaling factor $L$ by

$$A \propto L^2$$

where the symbol $\propto$ is read as "is proportional to" or "scales as."

What about enlarging three-dimensional objects? If we take a cube and enlarge it by a linear scaling factor of 3, it becomes three times as long, three times as high, and three times as deep as the original (see Figure 18.3).

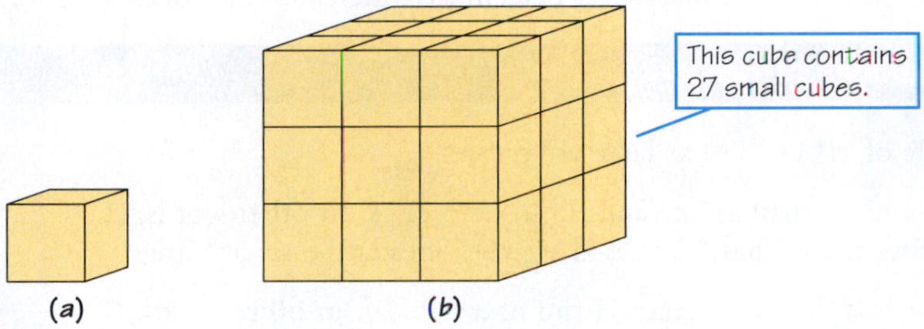
This cube contains 27 small cubes.
(a)
(b)

**FIGURE 18.3** Cube (b) is made by enlarging cube (a) by a factor of 3.

What about volume? How many times as large is the volume of the enlarged cube? The enlarged cube has three layers, each with $3 \times 3 = 9$ little cubes, each the same size as the original. Thus, the total volume is $3 \times 3 \times 3 = 3^3 = 27$ times as much as the original cube.

### How Volume Scales — RULE

The *volume* of a scaled-up object goes up with the *cube* of the linear scaling factor.

Denoting the volume by $V$, we can write

$$V \propto L^3$$

Thus, for an object enlarged by a linear scaling factor of $L$, the enlargement has $L \times L \times L = L^3$ ("$L$ cubed") times the volume of the original. Like the relationship between area and $L^2$, this relationship holds even for irregularly shaped objects, such as science fiction monsters.

You can see, however, that the area of each face (side) of the enlarged cube is $3^2 = 9$ times as large as that of a face of the original cube, just as the area of the photo enlarged by a factor of 3 has nine times the area of the original. The total surface area of the enlarged cube is nine times the total surface area of the original cube.

More generally, for objects of any shape, the total *surface area* of a scaled-up object goes up with the *square* of the linear scaling factor. Thus, the surface area of an object scaled up by a factor of $L$ is $L^2$ times the surface area of the original. This feature holds true even for irregular shapes.

Before we discuss scaling real three-dimensional objects, you should understand the pitfalls of the language for describing increases and decreases.

## The Language of Growth, Enlargement, and Decrease

*[G]eothermal power plants emit* ***35 times less*** *carbon dioxide per kilowatt produced than the average U.S. coal-fueled power plant.*

–Steve Gelsi, "Glitnir goes geothermal," *Wall Street Journal Online* (September 5, 2007), http://blogs.wsj.com/energy/2007/09/05/glitnir-goes-geothermal/trackback/

*The 42,000-seat major league park in Houston was a great deal for the Astros. The total cost of $248.5 million . . . was 38 to* ***108 percent less than*** *the cost of the three other U.S. retractable ballparks.*

–Ken Conrad and Timothy R. Santi, "Sports Engineering Requires Speed, Skill, Endurance," *Design Intelligence,* 10(4) (April 2004), http://www.di.net/articles/archive/2268

Let's explore why the bolded phrases are incorrect and confusing. First, we set out correct ways of discussing percentages, increases, and decreases:

### Meanings of Percentage RULE

"$x$% of $A$" or "$x$% as large as" means $\frac{x}{100} \times A$.

"$x$% more than $A$" means $A$ plus $x$% of $A$, in other words, $(1 + \frac{x}{100}) \times A$. Saying that $A$ has "increased by $x$%" means the same thing.

"$x$% less than $A$" means $A$ minus $x$% of $A$, in other words, $(1 - \frac{x}{100}) \times A$. Saying that $A$ has "decreased by $x$%" means the same thing.

Safeco Field in Seattle cost $517 million to build. To say that the Houston ballpark cost "108% less than" that would mean that it cost less than $0:

$$\begin{aligned}\$517{,}000{,}000 \times \left(1 - \frac{108}{100}\right) &= \$517{,}000{,}000 \times (1 - 1.08)\\ &= \$517{,}000{,}000 \times (-0.08)\\ &\approx -\$41{,}000{,}000\end{aligned}$$

The terms *of, times,* and *as much as* refer to *multiplication* of the original amount, while the terms *more, larger,* and *greater* refer to *adding* to the original amount. For instance, "five times as much" means the same as "four times more than" (the original plus four times as much in addition). Similarly, the relationship of the original amount to the larger amount can be expressed either in multiplicative terms ("one-fifth as much" or "20% as much") or in subtractive terms ("four-fifths less than" or "80% less than")–but don't mix the two.

People often say "five times more than" when they mean "five times as much." About the geothermal plants, the author wrote "35 times less" when he meant "one-thirty-fifth as much." Since one-thirty-fifth is hard to visualize, it would have been more informative to say "about 3% as much." All you can do is be aware of the potential confusion, try to figure out what was meant, and be careful in your own expression. In particular:

**Correct Comparisons** RULE

Don't use *times* with *more* or with *less*.

Finally, we need to distinguish *percent* from percentage *points*: If support for the president decreased from 60% to 30%, it dropped 30 *percentage points* but decreased 50% (because the drop of 30 percentage points is 50% of the original 60 percentage points).

### EXAMPLE 1 ■ What About That Stadium?

Returning to the Houston stadium, how could you state correctly what the author was trying to say?

**SOLUTION** Its cost, \$248.5 million, is about 0.48 (= 248.5/517) times the \$517 million cost of the Seattle stadium, so it cost "48% of the Seattle stadium" or "52% less than the Seattle stadium," or about "half as much." ■

## 18.2 How Much Is That in . . . ?

We are interested in the limits of size and want to compare objects of different sizes, such as a gorilla with King Kong. However, it is not easy to compare two objects if their measurements are given in different units–say, the gorilla in inches and pounds, but King Kong in centimeters and kilograms. Consequently, we explore how to convert units from one measurement system to another.

We introduce units in which physical quantities are measured, and give a table of *conversion factors* and examples of how to convert from one system of units to another.

### U.S. Customary System

Table 18.1 lists units of the *U.S. Customary System* of measurement and their abbreviations. Please note in the table the systematic way of converting from one unit to another and the use of scientific notation. The symbol ≈ means "is approximately equal to."

### Metric System

The world generally uses the metric system in science, industry, and commerce. It was proposed by Gabriel Mouton, a vicar in Lyons, in 1670, and was adopted in France in 1795. The fundamental unit of length, the *meter* (m), was originally defined as one ten-millionth of the distance from the North Pole to the equator, as measured on the meridian through Paris. The length is now defined as the distance that light travels in a vacuum in 1/299,792,458 second (s). The second, in turn, is defined as the time that it takes an atom of the metal cesium to vibrate 9,192,631,770 times.

**TABLE 18.1** Units of the U.S. Customary Systems

*Distance:*

1 mile (mi) = 1760 yards (yd) = 5280 feet (ft) = 63,360 inches (in.)
1 yard (yd) = 3 feet (ft) = 36 inches (in.)
1 foot (ft) = 12 inches (in.)

*Area:*

$$1 \text{ square mile} = 1 \text{ mi} \times 1 \text{ mi} = 5280 \text{ ft} \times 5280 \text{ ft}$$
$$= 27{,}878{,}400 \text{ ft}^2 \approx 2.8 \times 10^7 \text{ ft}^2$$
$$= 63{,}360 \text{ in.} \times 63{,}360 \text{ in.}$$
$$= 4{,}014{,}489{,}600 \text{ in.}^2 \approx 4 \times 10^9 \text{ in.}^2$$
$$= 640 \text{ acres}$$
$$1 \text{ acre} = 43{,}560 \text{ ft}^2$$

*Volume:*

$$1 \text{ cubic mile} = 1 \text{ mi} \times 1 \text{ mi} \times 1 \text{ mi}$$
$$= 5280 \text{ ft} \times 5280 \text{ ft} \times 5280 \text{ ft}$$
$$= 147{,}197{,}952{,}000 \text{ ft}^3$$
$$\approx 147 \times 10^9 \text{ ft}^3$$
$$= 63{,}360 \text{ in.} \times 63{,}360 \text{ in.} \times 63{,}360 \text{ in.}$$
$$\approx 2.5 \times 10^{14} \text{ in.}^3$$

1 U.S. gallon (gal) = 4 U.S. quarts (qt) = 231 in.$^3$, exactly

*Mass:*

1 ton (t) = 2000 pounds (lb)

**TABLE 18.2** Units of the Metric System

*Distance:*

1 meter (m) = 100 centimeters (cm)
1 kilometer (km) = 1000 meters (m)
= 100,000 centimeters (cm) = $1 \times 10^5$ cm

*Area:*

$$1 \text{ square meter } (\text{m}^2) = 1 \text{ m} \times 1 \text{ m}$$
$$= 100 \text{ cm} \times 100 \text{ cm} = 10{,}000 \; (\text{cm}^2) = 1 \times 10^4 \text{ cm}^2$$
$$1 \text{ hectare (ha)} = 10{,}000 \text{ m}^2$$

*Volume:*

$$1 \text{ liter (L)} = 1000 \text{ cm}^3 = 0.001 \text{ m}^3$$
$$1 \text{ cubic meter } (\text{m}^3) = 1 \text{ m} \times 1 \text{ m} \times 1 \text{ m}$$
$$= 100 \text{ cm} \times 100 \text{ cm} \times 100 \text{ cm}$$
$$= 1{,}000{,}000 \text{ cm}^3 = 1 \times 10^6 \text{ cm}^3 \text{ (or cc)}$$

*Mass:*

1 kilogram (kg) = 1000 grams (g) = $1 \times 10^3$ g

All other units of length, area, and volume are *defined* in terms of the meter. For example, a centimeter (cm) is a hundredth of a meter.

Mass is quantity of matter. The metric unit of mass, the *kilogram* (kg), is defined as the mass of a platinum–iridium standard kept in Paris. Since you can't determine the mass of a sack of potatoes by comparing it to that, we measure the mass indirectly by seeing how much force gravity exerts on it–that is, we weigh it, on a scale calibrated in pounds or kilograms. However, a mass of 1 kg would "weigh" (register on the scale) only one-sixth as much on the Moon.

Table 18.2 lists the units of the metric system.

## Converting Between Systems

What are the conversion factors between the U.S. Customary System and the metric system? Since 1959, the fundamental units of the U.S. Customary System, the yard (for length) and the pound (for mass), have been *defined* in terms of metric units, so that we have

$$1 \text{ yd} = 0.9144 \text{ m, exactly}$$

$$1 \text{ lb} = 0.45359237 \text{ kg, exactly}$$

Table 18.3 illustrates the conversion factors. In the following examples we explain how to convert measurements between systems.

**TABLE 18.3 Conversions Between the U.S. Customary System and the Metric System**

*Distance:*

1 in. = 2.54 cm, exactly

1 ft = 12 in. = $12 \times 2.54$ cm = 30.48 cm = 0.3048 m, exactly

1 yd = 0.9144 m, exactly

1 mi = 5280 ft = $5280 \times 30.48$ cm

= 160,934.4 cm, exactly $\approx$ 1.61 km

1 cm $\approx$ 0.393701 in. $\approx$ 0.4 in.

1 m $\approx$ 39.37 in. $\approx$ 3.281 ft

1 km $\approx$ 0.621 mi

*Area:*

1 $\text{ft}^2 \approx 0.09290 \text{ m}^2 = 929.0 \text{ cm}^2$

1 $\text{m}^2 \approx$ 10.76 sq ft

1 hectare (ha) $\approx$ 2.47 acres

*Volume:*

1 $\text{ft}^3 \approx$ 28.32 liters (L)

1 gallon $\approx$ 3.785 liters (L)

1 cubic meter ($\text{m}^3$) = 1000 liters $\approx$ 264.2 U.S. gallons $\approx$ 35.31 $\text{ft}^3$

1 liter (L) = 1000 $\text{cm}^3 \approx$ 1.057 U.S. quarts (qt) $\approx$ 0.2642 U.S. gallons

*Mass:*

1 lb = 0.45359237 kg, exactly

1 kg $\approx$ 2.205 lb

## EXAMPLE 2 ■ What's That in Feet?

An international student tells her American student friends that she is 160 cm tall. They ask how much that is in feet and inches.

We approach this conversion by using the scaling factor 1 cm = $\frac{1}{2.54}$ in.:

$$160 \text{ cm} = 160 \text{ cm} \times 1 \approx 160 \text{ cm} \times \frac{1 \text{ in.}}{2.54 \text{ cm}}$$

$$\approx 63.0 \text{ in.} = 63.0 \text{ in.} \times \frac{1 \text{ ft}}{12 \text{ in.}} \approx \frac{63.0}{12} \text{ ft} \approx 5.25 \text{ ft}$$

However, because we normally give height in feet and a whole number of inches, the height is

$$63.0 \text{ in.} = 5 \times (12 \text{ in.}) + 3.0 \text{ in.} = 5 \text{ ft} + 3.0 \text{ in.} \approx 5 \text{ ft } 3 \text{ in.}$$

Another way to approach the problem is by means of a proportion:

$$\frac{\text{height in in.}}{\text{height in cm}} = \frac{\text{length of 1 in. in inches}}{\text{length of 1 in. in cm}} = \frac{1 \text{ in.}}{2.54 \text{ cm}}$$

so that

$$\text{height in in.} = \text{height in cm} \times \frac{1 \text{ in.}}{2.54 \text{ cm}}$$

$$= 160 \text{ cm} \times \frac{1 \text{ in.}}{2.54 \text{ cm}} \approx 63.0 \text{ in.}$$

## EXAMPLE 3 ■ Got Gas?

In the United States, we measure the efficiency of cars in miles per gallon (mpg); most of the rest of the world measures it in liters per 100 kilometers. The conversion between these two measures is more complicated than other conversions, because the U.S. measure has distance (mi) in the numerator and quantity of fuel (gal) in the denominator, while the other measure has quantity of fuel (L) in the numerator and distance (km) in the denominator. We need to take this difference into account when doing the conversion.

For example, according to the Environmental Protection Administration, the 2008 Toyota Camry hybrid gets 34 mpg on the highway. What is the equivalent in liters per 100 km?

**SOLUTION**

$$34 \text{ mpg} = 34 \times \frac{1 \text{ mi}}{1 \text{ gal}}$$

$$\approx 34 \times \frac{1.609 \text{ km}}{3.785 \text{ L}} = 34 \times \frac{1.609}{3.785} \times \frac{\text{km}}{\text{L}} \approx 14.45 \frac{\text{km}}{\text{L}} = \frac{14.45}{1} \times \frac{100 \text{ km}}{100 \text{ L}}$$

$$= \frac{1}{\frac{1}{14.45}} \times \frac{100 \text{ km}}{100 \text{ L}}$$

$$\approx \frac{100 \text{ km}}{6.9 \text{ L}},$$

or 6.9 L per 100 km. The key steps in the solution are to multiply both units by 100, then divide both numerator and denominator of the fraction by 14.45, so as to get exactly 100 km in the numerator of the result. ■

## 18.3 Scaling a Mountain

Gravity exerts an enormous effect on the size and shape that objects and beings can assume. **Weight** (force under Earth's gravity) is the reading at sea level on a scale (such as your bathroom scale) *calibrated in pounds or kilograms of mass.*

Suppose that the two cubes in Figure 18.3 on p. 573 are made of steel and that the first is 1 ft on a side and the second is 3 ft on a side. A cubic foot of steel weighs about 500 lb; we say that the **density** of steel is 500 lb per cubic foot, or 500 lb/ft$^3$. The cube 1 ft on a side weighs 1 ft$^3$ $\times$ 500 lb/ft$^3$ = 500 lb. The weight $W$ of an object of volume $V$ and uniform density $D$ is

$$W = DV$$

Each cube's bottom face supports the weight of the entire cube. **Pressure** is the force per unit area, so the pressure exerted on the bottom face by the weight of the cube is equal to the weight of the cube divided by the area of the bottom face, or

$$P = \frac{W}{A}$$

The first cube weighs 500 lb and has a bottom face with area 1 ft$^2$, so the pressure exerted on this face is 500 lb/ft$^2$.

The second cube is 3 ft on a side. The area of the bottom face increases with the square of the linear scaling factor, so it is $3^2 \times 1$ ft$^2$ = 9 ft$^2$. As we saw earlier, volume goes up with the cube of the linear scaling factor. So this larger cube has a volume of $3^3 \times 1$ ft$^3$ = 27 ft$^3$. Because both cubes are made of the same steel, the larger cube has 27 times as much steel as the smaller one. Hence, it weighs 27 times as much as the smaller cube, or 27 $\times$ 500 lb = 13,500 lb.

When we divide this weight by the area of the bottom face (9 ft$^2$), we find that the pressure exerted on the bottom face is 1500 lb/ft$^2$, or three times the pressure on the bottom face of the original cube. This makes sense because over each 1 ft$^2$ area stands 3 ft$^3$ of steel. In general, if the linear scaling factor for the cube is $L$, the pressure on the bottom face is $L$ times as much. Using the notation of proportionality, we have $A \propto L^2$ and $W \propto V \propto L^3$, so

$$P = \frac{W}{A} \propto \frac{L^3}{L^2} \propto L$$

### EXAMPLE 4 ■ What About a 10-Foot Cube?

**SOLUTION** If we scale the original cube of steel up to a cube 10 ft on a side, then the dimensions are

$$10 \text{ ft} \times 10 \text{ ft} \times 10 \text{ ft}$$

The total volume is

$$\begin{aligned} V &= \text{length} \times \text{width} \times \text{height} \\ &= 10 \text{ ft} \times 10 \text{ ft} \times 10 \text{ ft} = 1000 \text{ ft}^3 \end{aligned}$$

The weight of the cube is

$$W = D \times V$$
$$= \frac{500 \text{ lb}}{\text{ft}^3} \times 1000 \text{ ft}^3 = 500{,}000 \text{ lb}$$

The area of the bottom face is

$$A = \text{length} \times \text{width}$$
$$= 10 \text{ ft} \times 10 \text{ ft} = 100 \text{ ft}^2$$

The pressure on the bottom face is

$$P = \frac{W}{A} = \frac{500{,}000 \text{ lb}}{100 \text{ ft}^2} = 5000 \text{ lb/ft}^2$$

This is 10 *times*–not "10 times *more* than"–the pressure on the bottom face of the original 1-ft cube.

At some scale factor, the pressure on the bottom face will exceed the steel's ability to withstand that pressure–and the steel will deform under its own weight. That point for steel is reached for a cube about 3 miles (mi) on a side–the pressure exerted by the cube's weight exceeds the **crushing strength** of steel, which under Earth's gravity is about 7.5 million lb/ft$^2$. Because 3 mi = 3 × 5280 ft = 15,840 ft, a 3-mi-long cube of steel would be more than 15,000 times as long as the original 1-ft cube; that is, the linear scaling factor is more than 15,000. The pressure on the bottom face of the cube would therefore be more than 15,000 times as much as for the 1-ft cube, or more than 15,000 × 500 lb/ft$^2$ = 7.5 million lb/ft$^2$.

## EXAMPLE 5 ■ What About Taipei 101?

Taipei 101 in Taiwan, completed in 2004 and named for its 101 floors (see Spotlight 18.1), is the world's tallest skyscraper, at 1671 ft (how much is that in meters?), not counting radio and television antennas. What is the pressure at the bottom of its walls?

**SOLUTION** The building is made of reinforced concrete, which weighs 160 lb/ft$^3$. Although the building tapers a bit toward the top, we are not far off if we model it as straight up and down, that is, as a rectangular solid. Consider one of its supporting walls. The volume of the wall is its height $H$ times the area $A$ of its base, or $V = HA$. The weight of the wall is $W = DV = DHA$. The pressure at the bottom is

$$P = \frac{W}{A} = \frac{DV}{A} = \frac{DHA}{A} = DH$$
$$= \frac{160 \text{ lb}}{\text{ft}^3} \times 1671 \text{ ft} = \frac{267{,}360 \text{ lb}}{\text{ft}^2}$$

or approximately 270,000 lb/ft$^2$.

So the pressure at the bottom of the wall from the wall's weight alone is about 270,000 lb/ft$^2$. That's not counting the contents of the tower, which also must be supported!

Could we have a Super Taipei Tower 10 times as high? The bottom of its walls would have to support about 10 × 270,000 lb/ft$^2$ = 2.7 million lb/ft$^2$. The crushing strength of reinforced concrete under Earth's gravity is 8.5 million lb/ft$^2$, which would leave some safety margin.

If built as planned, the Freedom Tower in New York City, whose cornerstone was laid July 4, 2004, and which was scheduled to be completed in 2009, would be taller still–1776 ft–but contain only 60 floors, and be topped by a 276-ft spire.

Burj Dubai ("Dubai Tower"), a skyscraper in the United Arab Emirates scheduled for occupancy in late 2009, is projected to rise to 818 m (2,684 ft) and have 160 floors, including a spire at the top.

## SPOTLIGHT 18.1 A Mile-High Building?

In 1956, the famous American architect Frank Lloyd Wright (1867–1959) proposed a mile-high tower for the Chicago lakefront. In the text, we focus on the problem of holding up the weight of such a structure.

But there are other limits to the height of a building. For example, the bending of the building in the wind, which can go up dramatically with height, can be controlled by making the building stiffer.

The terrorist destruction of the World Trade Center towers in 2001—resulting not from the aircraft impacts but from the subsequent fires—revealed a vulnerability in the towers' structure.

Even if designed to better resist fires and impacts, however, a mile-high building might not be practical. For example, the enormous number of people (perhaps 100,000) living, working, or visiting in such a building would create enormous traffic problems (pedestrian, parking, deliveries) for blocks around.

Cost per square foot of usable area is an important consideration. Even if the building did not taper, the space in the upper floors might not justify their additional expense. With increasing height, an increasingly larger proportion of the cross-sectional area of all floors must be devoted to services, such as elevators, plumbing, and conduits for heat and air-conditioning. But everyone entering the building and going to any floor needs to start in an elevator on the ground floor, so there must be more elevators and more elevator shafts. In an emergency evacuation, the people must walk down!

Some architects, however, maintain that the main limit on height of a building is human physiology. Differences in air pressure between the top and bottom of a building limit how fast elevators can rise or drop without discomfort to passengers, thereby enforcing long travel times for "vertical commuters." Human psychology might also present some limits.

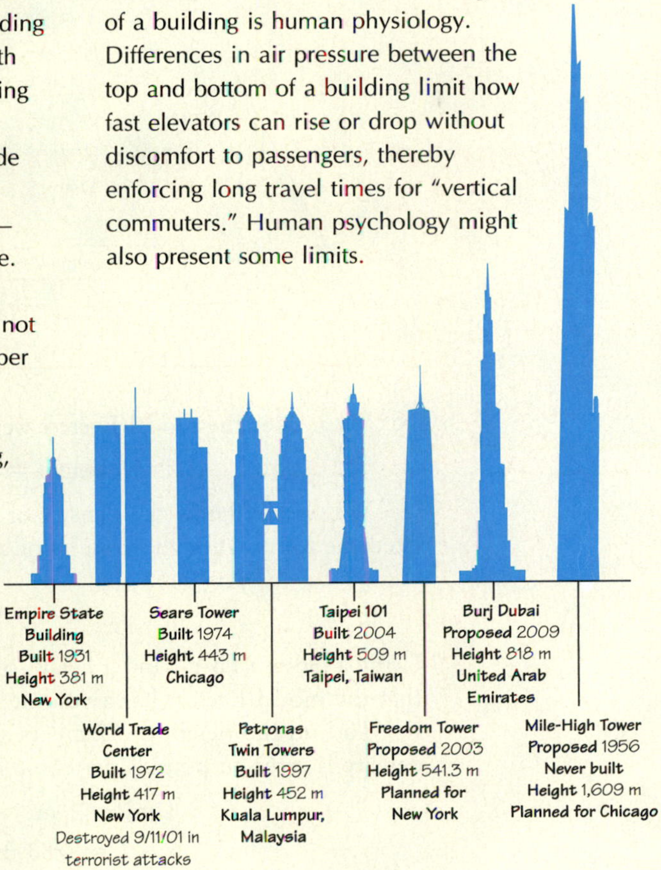


## EXAMPLE 6 ■ How High Can a Mountain Be?

The height of mountains too is limited, by gravity, their composition, and their shape. How tall can a mountain be?

**SOLUTION** We build a simple mathematical model of a mountain. Suppose that it is made of granite, a common material, with uniform density. Granite weighs 165 lb/ft$^3$ and has a crushing strength of about 4 million lb/ft$^2$.

In the interests of both realism and simplicity, we assume that the mountain is a solid cone whose width at the base is the same as its height. Let's model Mount Everest, the tallest mountain on Earth, at about 6 mi high. The base, then, is a circle with diameter (distance across) 6 mi. The radius (half the diameter) is 3 mi (Figure 18.4). Because we took round numbers (6 mi) for the height and width, we record as significant only the first digit or two of the results of the calculations.

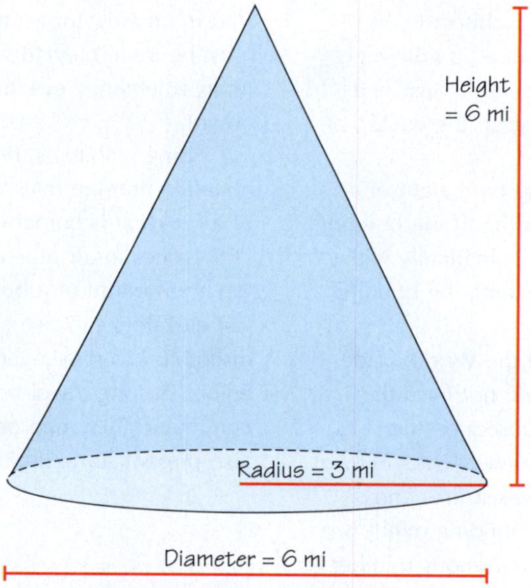


**FIGURE 18.4** Model of Mount Everest as a cone of granite.

What does the model Everest weigh? The relevant formula is $W = DV$, or

$$\text{weight} = \text{density} \times \text{volume}$$

We already know the density of granite (165 lb/ft$^3$), so to find the weight we need the formula for the volume of a cone of radius $r$ and height $h$:

$$V = \text{volume} = \frac{1}{3}\pi r^2 h$$

Using a radius of 3 mi and a height of 6 mi and $\pi$ (pi) approximately 3.14, we find that the model Everest has a volume of about 57 mi$^3$.

To find the weight of 57 mi$^3$ of granite, we need to convert units, because the density is given in pounds per cubic foot (lb/ft$^3$). Let's convert to units of feet:

$$\begin{aligned} 1\ \text{mi}^3 &= 1\ \text{mi} \times 1\ \text{mi} \times 1\ \text{mi} \\ &= 5280\ \text{ft} \times 5280\ \text{ft} \times 5280\ \text{ft} \\ &\approx 1.5 \times 10^{11}\ \text{ft}^3 \end{aligned}$$

Thus

$$57\ \text{mi}^3 \approx 57 \times 1.5 \times 10^{11}\ \text{ft}^3 \approx 8.6 \times 10^{12}\ \text{ft}^3$$

So we have

$$\begin{aligned} W = \text{weight of mountain} &= \frac{165\ \text{lb}}{\text{ft}^3} \times 8.6 \times 10^{12}\ \text{ft}^3 \\ &\approx 1.4 \times 10^{15}\ \text{lb} \\ &\approx 1.4\ \text{quadrillion lb} \end{aligned}$$

Now that we know the weight of the mountain, we want to find the pressure on the base of the cone and compare it with the crushing strength of granite. (Everest is standing, so if our model is any good, that pressure will be below the crushing strength.) Physics tells us that the weight of the mountain is spread evenly over the base of the cone (though we are oversimplifying the geology underlying mountains). Because

$$P = \frac{W}{A} \quad \left(\text{pressure} = \frac{\text{weight}}{\text{area}}\right)$$

we need to calculate the area of the base of the cone. The shape is a circle, and the familiar formula

$$A = \text{area} = \pi r^2$$

gives an area of 28 $\text{mi}^2$ for a radius of 3 mi.

Once again, we need to convert units to express the pressure in pounds per square foot, the units in which the crushing strength is expressed. We get

$$A = \text{area} = 28\ \text{mi}^2 = 28 \times 5280\ \text{ft} \times 5280\ \text{ft} \approx 7.8 \times 10^8\ \text{ft}^2$$

Then

$$\begin{aligned} P &= \frac{W}{A} \\ &= \frac{1.4 \times 10^{15}\ \text{lb}}{7.8 \times 10^8\ \text{ft}^2} \\ &= 1.8 \times 10^6\ \text{lb/ft}^2 \end{aligned}$$

This number is about half the crushing strength of granite under Earth's gravity, $4 \times 10^6$ lb/ft$^2$.

For a mountain to come close to the limitation of the crushing strength of granite, it would have to be only about 10 mi high, not quite twice as high as Everest. Other physical considerations suggest a maximum height of at most 15 mi. That no current mountains are that high may be a consequence of Earth's high amount of volcanic activity and the structural deformation of Earth's crust.

What about mountains made of other materials–glass, ice, wood, old cars? They couldn't be nearly as high. The pressure would cause glass to flow, ice to melt, and old cars to compact. What about mountains on another planet? Or on an asteroid? Their potential height depends on the gravity there.

## 18.4 Sorry, No King Kongs

Unfortunately, the resistance of bone to crushing is not nearly as great as that of steel or granite. This fact helps to explain why there couldn't be any King Kongs (unless they were made of steel or granite!). A King Kong scaled up by a factor of, say, 20 would weigh $20^3 = 8000$ times as much. Though the weight increases with the cube of the linear scaling factor, the ability to support the weight–as measured by the cross-sectional area of the bones, like the area of the bottom face of the cube in Figure 18.3–increases only with the square of the linear scaling factor.

These simple consequences of the geometry of scaling apply not only to supermonsters but also to other objects, such as trees.

FIGURE 18.5 Even giant sequoias can grow no taller than their form and materials allow. (*Michael Rothman.*)

## EXAMPLE 7 ■ How Tall Can a Tree Be?

Galileo suggested that no tree could grow taller than 300 ft (see Spotlight 18.2). The world's tallest trees are giant sequoias (Figure 18.5), which grow only on the West Coast of the United States and hence were unknown to Galileo. The tallest known today is 379 ft.

What limits the height of a tree? If the roots do not adequately anchor it, a tall tree can blow over. (This happened in 1990 to the world's then-tallest tree, the Dyerville Giant, a giant sequoia in Humboldt Redwoods State Park in California.) The tree could buckle or snap under its own weight and the force of a strong wind. The wood at the bottom will crush if there is too much weight above. Finally, there is a limit to how far the tree can lift water and minerals from the roots to the leaves.

**SOLUTION** Could a tree be a mile high? To make a rough estimate of the pressure at the base of the tree due to gravity, let's model the tree as a perfectly vertical cylinder. Over each square foot at the bottom, there is 5280 $ft^3$ of cells of wood, which weighs about half as much as water. A convenient fact of the metric system is that water weighs just about 1 gram (g) per cubic centimeter. So, to calculate the weight, we first translate 1 $ft^3$ into metric measurement:

$$1 \text{ ft}^3 = (12 \text{ in.})^3 = (12 \times 2.54 \text{ cm})^3 \approx 28{,}317 \text{ cm}^3$$

So, 1 $ft^3$ of water weighs about

$$28{,}317 \text{ g} = 28.317 \text{ kg} = 28.317 \times 2.205 \text{ lb} \approx 62.44 \text{ lb}$$

Consequently, 5280 $ft^3$ of water weighs about $5280 \times 62.4$ lb $\approx$ 330,000 lb. The weight of the same volume of wood is about half as much, or about 165,000 lb. Therefore, the pressure at the bottom of the tree would be about 165,000 lb/$ft^2$.

This is an overestimate, because we assumed that the tree does not taper. A tree that tapers steadily looks like an elongated cone; using a more realistic cone model (as we did in the last section for a mountain), you would find that the pressure at the bottom of the tree would be one-third of 164,000 lb/$ft^2$, or about 55,000 lb/$ft^2$.

A biological organism needs a safety factor of at least two to four times the absolute minimum physical limits, so a mile-high tree would need from 110,000 to 220,000 lb/$ft^2$ of upward pressure for water and minerals. Tension in the string of water molecules from root to leaf ranges from 80,000 to 3.2 million lb/$ft^2$, for different kinds and heights of trees, so this consideration does not rule out mile-high trees.

At more than about 500 lb/$in.^2$ = 70,000 lb/$ft^2$, though, the bottom of the tree would begin to crush under the weight above. On this basis, a mile-high tree is barely feasible, with little margin of safety. However, researchers who hauled themselves up to the top of the tallest trees in 2004 found a much lower limit, at least for giant sequoias. With increasing height, leaves are smaller, dryer, and less efficient at photosynthesis. The researchers estimated that trees can't top out higher than 400 to 427 ft. The tallest reliably measured tree was a North American Douglas fir, measured at 413 ft in 1902.

There are other considerations. The taller the tree, the greater the area from which it must draw water and minerals, for which nearby trees also compete. Moreover, for a tree to grow very tall, it would have to live for a very long time. Evolution and time may select against extremely tall trees, or maybe, for no reason at all, they have just never evolved.

### SPOTLIGHT 18.2 Galileo and the Problem of Scale

The Italian physicist Galileo Galilei (1564–1642) was the first to describe the problem of scale, in 1638, in his *Dialogues Concerning Two New Sciences* (in which he also famously discussed the idea of Earth revolving around the Sun):

> You can plainly see the impossibility of increasing the size of structures to vast dimensions either in art or in nature; likewise, the impossibility of building ships, palaces, or temples of enormous size in such a way that their oars, yards, beams, iron-bolts, and, in short, all their other parts will hold together; nor can nature produce trees of extraordinary size because the branches would break down under their own weight, so also would it be impossible to build up the bony structures of men, horses, or other animals so as to hold together and perform their normal functions if these animals were to be increased enormously in height; for this increase in height can be accomplished only by employing material which is harder and stronger than usual, or by enlarging the size of the bones, thus changing their shape until the form and appearance of the animals suggests a monstrosity.
>
> To illustrate briefly, I have sketched a bone whose natural length has been increased three times and whose thickness has been multiplied until, for a correspondingly large animal, it would perform the same function which the small bone performs for its small animal. From the figures shown here you can see how out of proportion the enlarged bone appears. Clearly then if one wishes to maintain in a great giant the same proportion of limb as that found in an ordinary man he must either find a harder and stronger material for making the bones, or he must admit a diminution of strength in comparison with men of medium stature; for if his height be increased inordinately he will fall and be crushed under his own weight. Whereas, if the size of a body be diminished, the strength of that body is not diminished in proportion; indeed the smaller the body the greater its relative strength. Thus a small dog could probably carry on his back two or three dogs of his own size; but I believe that a horse could not carry even one of his own size.

Translated by Henry Crew and Alfonso De Salvo, and published by Macmillan, 1914, and Northwestern University, 1946.

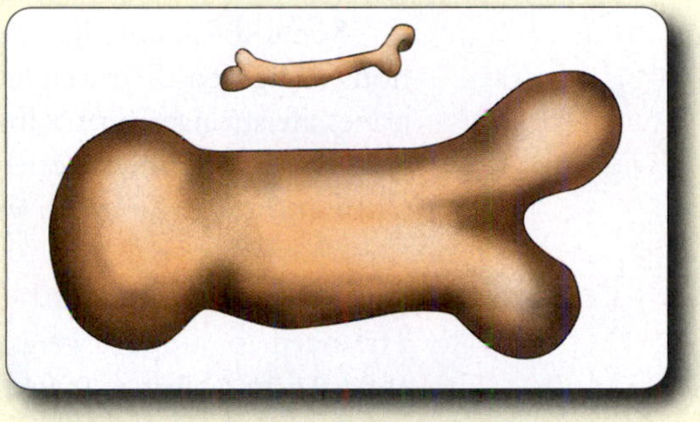

One bone, with another three times as long and thick enough to perform the same function in a scaled-up animal.

## 18.5 Dimension Tension

A large change in scale forces a change in either materials or form. A major manifestation of the scaling problem is the tension between weight and the need to support it. For example, a real building or machine must differ from a scale model: The balsa wood or plastic of the model would never be strong enough for the real thing, which would need aluminum, steel, or reinforced concrete.

Another way to compensate is to redesign the object to distribute its weight better. Let's go back to the original cube. It supports all its weight on its bottom face. In the version scaled up by a factor of 3, each small cube of the bottom layer has a bottom face supporting that cube's weight plus the weight of the two cubes piled on top of it.

Let's redesign the scaled-up cube, concentrating for simplicity only on the front face, with its nine small cubes. We take the three cubes on top and move them to the bottom, alongside the three already there. We take the three cubes on the second level, cut each in half, and put a half-cube over each of the six ground-level cubes (see Figure 18.6). We have the same volume and weight that we started with, but now there is less pressure on the bottom face of each small cube. Of course, the new design is not geometrically similar to the object that we started with–it's no longer a cube. We have solved the scaling problem by changing the proportions.

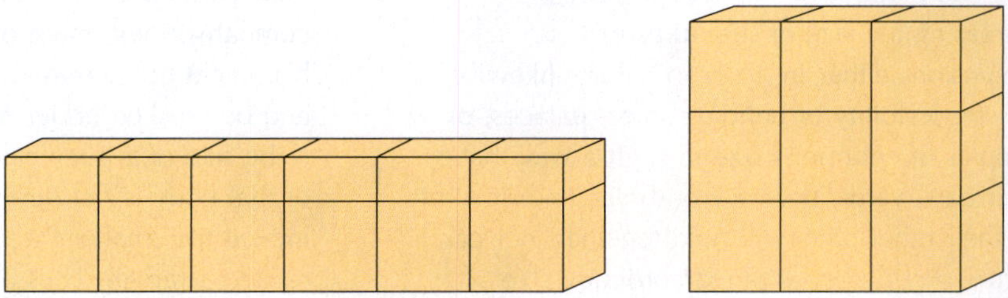

**FIGURE 18.6** Nine small cubes rearranged to support greater weight.

We observe in nature both strategies for scaling: change of materials and change of form. Small animals (such as insects) do not have bony internal skeletons. Larger animals generally do. Animals made of similar materials but differing greatly in size, such as a mouse and an elephant, must differ in shape. If a mouse were scaled up to the size of an elephant, it would need the disproportionately thicker legs of the elephant to support its weight and the elephant's thick hide to contain its tissue.

Some dinosaurs, like *Supersaurus* (which weighed 30 tons), had special adaptations to lighten their weight, such as hollow bones, just as some birds have. Hollow bones are stronger, a paradox that Galileo analyzed. Of two bones of the same weight and length, the hollow one is wider across at its midpoint because of the air it contains; and the greater the width, the greater the resistance to fracture.

## Falls, Jumps, and Flight

The need to support weight can be thought of as a tension between volume and area. As an object is scaled up, its volume and weight go up together, as long as the density remains constant (for example, no air bubbles introduced into the steel to make it into a Swiss cheese!). At the same time, the ability to support the weight goes up with the cross-sectional area, like the bottom face of the steel cube.

### Area-Volume Tension — DEFINITION

**Area–volume tension** is a result of the fact that as an object is scaled up, the volume increases faster than the surface area and faster than areas of cross sections.

Because volume $V$ is proportional to the cube of the linear scaling factor $L$, we have $V \propto L^3$; taking each side to the one-third power, $L \propto V^{1/3}$. The fact that surface area $A$ is proportional to the square of the linear scaling factor becomes

$$A \propto L^2 \propto (V^{1/3})^2 = V^{2/3}$$

so that surface area scales as the two-thirds power of volume.

In any crowded city, you can observe tension between length, area, and volume. Consider an apartment building that spans a city block. The area of parking spaces on the adjacent streets is proportional to the perimeter of (length around) the building. But the number of cars belonging to people in the building is proportional to the number of apartments, which is proportional to the volume of the building. So the higher the building, the greater the parking tension!

In some cities, zoning tries to help the situation by putting shops on the ground floor, which cuts out one floor of apartments. If the residents' cars are away during the day, customers and employees of the shops can park where the apartment dwellers do at night. A more common solution is an underground garage, usually with several levels (with an area for cars proportional to the volume of the building). However, garages that were designed for one car per apartment have proven inadequate now that families tend to have more than one car.

Other examples of dimensional tension solutions include the old-fashioned diner, with its serving counter in the form of S-shapes to expand its effective length, and your small intestine, which coils its 20-ft length to fit into your abdomen.

Area–volume tension has many other practical consequences, some of them related to our childhood fantasies. We can forget about humans "leaping tall buildings in a single bound," "soaring like an eagle," or diving miles below the sea. Consider the following examples.

## EXAMPLE 8 ■ Falls

Area–volume tension affects how animals respond to falling, another of gravity's effects. A mouse may be unharmed by a 10-story fall, and a cat by a two-story fall, but many humans are injured by falling while running, walking, or even just standing.

What is the explanation? The energy acquired in falling is proportional to the weight of the falling object, hence to its volume. This energy must be absorbed either by the object or by what it hits, or must be otherwise dissipated at impact–for example, as sound. The fall is absorbed over part of the surface area of the object, just as the weight of the cube was distributed over its base. With scaling up, volume–hence weight, hence falling energy–goes up much faster than area. As size increases, the hazards of falling from the same height increase.

## EXAMPLE 9 ■ Jumps

A flea can jump about 2 ft vertically, many times its own height. Many people believe that if a flea were as large as a person, it could jump 1000 ft into the air. Imagining–against our earlier arguments–that there could be so large a flea, we know its limits: A scaled-up flea could jump about the same height as a small flea. The strength of a muscle is proportional to its cross-sectional *area* (see Spotlight 18.3). A jump involves suddenly contracting the muscle through its length, so it turns out that the ability to jump is proportional to the *volume* of muscle. But the volume of the flea and the volume of its leg muscles go up in proportion.

Let's say that a real flea's leg muscles account for 1% of its body. If we scale the flea up to the size of a person (without any change in its form), the enlarged flea's leg muscles will still make up 1% of its body. For either flea, each bit of muscle has the same power: In a jump, it propels 100 times its own weight, and it can do so to the same height. Both the weight of the flea and the power of its legs go up proportionately. In fact, the maximum heights that people, fleas, grasshoppers, and kangaroos can jump are all within a factor of 3 of each other.

## SPOTLIGHT 18.3 Scaled to Fit

Big isn't always beautiful when it comes to the U.S. military's physical fitness tests.

Paul Vanderburgh, chair of the University of Dayton Dept. of Health and Sport Science, has spent more than a dozen years researching how a person's body mass affects performance on such tests, which consist of distance runs, push-ups, sit-ups and abdominal crunches. The Arnold Schwarzeneggers of the world actually tend to score *lower*.

Vanderburgh emphasizes that some larger people (like Schwarzenegger) have more muscle, not more fat. Nevertheless, he and fellow researcher Todd Crowder found that scores for larger and heavier (though muscular) men and women are 15–20% lower than for their smaller and lighter counterparts. "A person's strength doesn't increase as fast as their size," explains his student Liz Trouten. "The extra muscle that big people have doesn't make up for their size."

Vanderburgh noticed at the U.S. Military Academy that, even at similar fitness levels, smaller cadets tend to score higher on physical fitness tests than larger cadets. "Fitness testing is a big part of cadets' grade point averages, and the stakes are pretty high. The test results affect class rank and even a cadet's first assignment, so it matters a lot how well a cadet does."

For example, a larger cadet with a fitness test score of 256, which Vanderburgh compares to a grade of C+, may not be eligible for certain awards and assignments. However, a smaller cadet with a perfect score of 300 "would get lots of attention."

*(Cynthia Johnson/Time Life Pictures/ Getty Images.)*

But that doesn't mean the C+ cadet isn't worthy. "In fact, if these two cadets were scale models of each other, these two performances would be biologically the same, and they should receive the same score."

Vanderburgh gives another example using the scale-model approach. Take a woman who is 5 feet 5 inches tall, weighs 130 pounds, and scores a perfect 300 on the fitness test. If she were 5 feet 8 inches tall and 30% heavier, she would score only 250.

To compensate for this body mass "penalty," Vanderburgh and Crowder developed a correction factor, which multiplies the score by a number based on weight, "to place everybody on an even playing field."

This formula is similar to the Flyer Handicap, developed by Vanderburgh and colleague Lloyd Laubach. The handicap adjusts a runner's race time based on age and body weight. "A higher body weight is definitely a handicap for performance, whether it be running a marathon or military physical fitness tests," Laubach said. (A Web calculator for the Flyer Handicap is at http://academic.udayton.edu/PaulVanderburgh/weight_age_grading_calculator.htm.)

*Source*: adapted from an article by Kristen Wicker in the *University of Dayton Quarterly* (Winter 2006–07) 21–22.

## EXAMPLE 10 ■ Flight

Wouldn't it be nice to be able to fly? Well, you have to be able to stay up. The power necessary for sustained flight is proportional to the **wing loading**, which is the weight supported divided by the area of the wings. We know that in scaling up, weight grows with the cube of the length of the bird or plane, and wing area with the square of the length. So the wing loading is proportional to the length of the flying object.

For example, if a bird or plane is scaled up proportionally by a linear scaling factor of 4, it will weigh $4^3 = 64$ times as much but will have only $4^2 = 16$ times as much wing area. So each square foot of wing must support 4 times as much weight.

Once you're up, you have to keep moving. To stay level, an airborne object must fly fast enough to maintain the lift on the wings. The minimum necessary speed is proportional to the square root of the wing loading. Combining this fact with the first consideration, we conclude that the minimum speed goes up with the square root of the length. A bird scaled up by a factor of 4 must fly $\sqrt{4} = 2$ times as fast. (Hovering helicopters, hummingbirds, and insects maintain lift by moving their wings directly rather than through forward motion.)

Take, for instance, a sparrow, whose minimum speed is about 20 miles per hour (mph). An ostrich is 25 times as long as a sparrow, so the minimum speed for an ostrich would be $\sqrt{25} \times 20 = 100$ mph. Have you seen any flying ostriches lately? Heavy birds have to fly fast or not at all!

Of course, ostriches are not just scaled-up sparrows, nor are eagles. Larger flying birds have disproportionately larger wings than a sparrow to keep the wing loading down. The largest animal ever to take to the air was *Quetzalcoatlus northropi,* a flying reptile of 65 million years ago, with a wingspan of 36 ft and a weight of about 100 lb.

You have to stay up, you have to keep moving–and you have to get up there. Here basic aerodynamics imposes further limits. Paleontologists originally thought that *Q. northropi* weighed 200 lb and had a 50-ft wingspan. Even though that works out to about the same wing loading as for 100 lb and a wingspan of 36 ft, other considerations from aerodynamics show that at the larger size, the reptile couldn't have gotten off the ground.

## Keeping Cool (and Warm)

Area–volume tension is also crucial to an animal's thermal equilibrium. Both warm-blooded and cold-blooded animals gain or lose heat from the environment in proportion to body surface area.

### Warm-Blooded Animals

A warm-blooded animal's basal metabolism, or rate of food intake needed to maintain body heat, depends primarily on its surface area, the temperature of its environment, and the insulation provided by its coat or skin. Other factors being equal, a scaled-up mammal scales up its food consumption with *surface area* (proportional to the square of the linear scaling factor), *not with volume* (proportional to its cube). For example, a mouse eats about half of its weight in food every day, while a human consumes only about one-fiftieth of its own weight, because the mouse has more surface area per unit volume.

Thus, the metabolic rate should be proportional to the surface area. Using proportionality notation, we can find how the metabolic rate changes with the mass of the animal. We know that mass is proportional to volume, which in turn is proportional to the cube of length, or

$$M \propto V \propto L^3$$

Taking each side to the one-third power, we have

$$M^{1/3} \propto V^{1/3} \propto L \qquad \text{or} \qquad L \propto M^{1/3}$$

Meanwhile, the metabolic rate (call it $R$) is proportional to surface area, so

$$R \propto A \propto L^2 \propto (M^{1/3})^2 = M^{2/3}$$

So, based on area–volume tension, we would expect metabolic rate to scale as the two-thirds power of body mass. But it doesn't–instead, it scales as the *three-quarters* power of body mass, that is, $R \propto M^{3/4}$. The least-squares line (see section 6.4) through the points in the "mouse-to-elephant" curve of Figure 18.7 has a slope of 0.74, very close to three-quarters. (The logarithmic coordinates used in this graph are explained in section 18.6.)

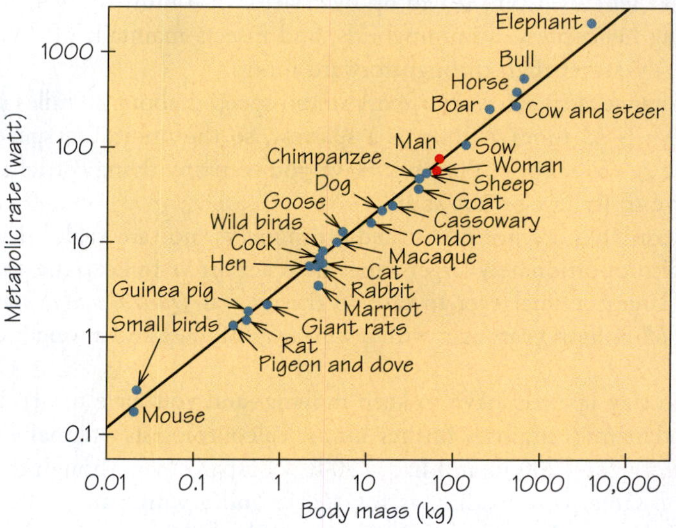


**FIGURE 18.7**
Metabolic rates for mammals and birds, when plotted against body mass on logarithmic coordinates, tend to fall along a single straight line. (*Adapted from Benedict, 1938.*)

Why the difference from the two-thirds that area–volume tension would predict? And does the small difference between two-thirds and three-quarters matter? The answers lie in further considerations from geometry, physiology, and physics. A plant or animal needs a network of vessels (like the blood system) to transport resources to, and wastes away from, every part of the animal's tissues. The terminal branches (in the blood system, capillaries) tend to be just about the same size in all species, for reasons of the physics involved. To minimize the energy involved in transport, the network of vessels needs to be organized as a fractal-like tree, with smaller and smaller vessels branching off (see section 19.5 to learn about fractal patterns). With same-size smallest branches at the ends, minimization of energy demands that the metabolic rate scale as the three-quarters power of body mass. Fractal branching makes it possible for the circulatory system of a whale, with $10^7$ times the mass of a mouse, to have only 70% more branches than the mouse has.

## EXAMPLE 11 ■ Dives

Sperm whales (and some other species) regularly hold their breath and stay under water for an hour. Why can't we? In part, because we aren't as large as whales. A mammal's breath-holding ability depends on how much air it can hold in its lungs, which is proportional to its mass. It also depends on how fast it uses up air–in other words, on its metabolic rate, which is proportional to the three-quarters power of its mass. Hence, the limit of duration of a dive should be proportional to

$$\frac{M}{M^{3/4}} = M^{1/4}$$

For a 90,000-lb sperm whale, this limit is proportional to $90{,}000^{1/4} = 17.3$, while the corresponding figure for a 150-lb human is $150^{1/4} = 3.5$. So the sperm whale should be able to hold its breath for about $17.3/3.5 \approx 5$ times as long. However, humans cannot hold their breath for one-fifth of an hour (12 minutes)! This fact tells us that the whale has special adaptations to make long dives possible. The stars of the 2005 film *The March of the Penguins*, emperor penguins, weigh 80–90 lb, but can dive for as long as 20 minutes. Their special adaptations are more blood per pound of body weight, an abundance of myoglobin (which can store oxygen) in their tissues, and slowing their heart rate during dives.

### Cold-Blooded Animals

Mammals and birds regulate their metabolism and maintain a constant internal body temperature. Cold-blooded animals, such as alligators and lizards, have a somewhat different problem. They absorb heat from the environment for energy, but they must also dissipate any excess heat to keep their temperatures below unsafe levels. The amount of heat that must be gained or lost is proportional to total volume, because the entire animal must be warmed or cooled. But the heat is exchanged through the skin, so the rate is proportional to surface area.

*Dimetrodon* was a large mammal-like reptile that roamed present-day Texas and Oklahoma 280 million years ago (see Figure 18.8). *Dimetrodon* had a great "sail," or fan, on its back. As an individual grew, and as the species evolved, the sail grew. But it did not grow according to *geometric similarity,* the kind of growth we refer to as **proportional growth.**

**Proportional Growth** — DEFINITION

**Proportional growth** is growth according to geometric similarity, where the length of every part of the organism enlarges by the same linear scaling factor.

**FIGURE 18.8** *Dimetrodon* may have evolved a sail to absorb and dissipate heat efficiently. (*Robert F. Walters.*)

Instead, the area of *Dimetrodon*'s sail grew in proportion to the volume of the animal, a fact that strongly suggests to paleontologists that the sail was a temperature-regulating organ. Larger specimens of *Dimetrodon* didn't look like scaled-up smaller ones. We would say that the sail grew disproportionately compared to the rest of the animal. An individual twice as long would have eight ($2^3$) times as much weight and volume and a sail with eight times as much area. If it had grown according to geometric similarity, the sail would have been twice as high and twice as wide, and hence would have had only four times as much area.

*Dimetrodon* was a large animal, but heat regulation is even more important for small animals–like human babies, they can lose heat quickly because of their high ratio of surface area to volume. Paleontologists believe that birds evolved from dinosaurs and that feathers are modified reptilian scales. The wings of birds and insects may have evolved not for flight but as temperature-control devices.

Some scientists have speculated that African Pygmies are small in part because a small body can better lose heat in the hot, humid climate of the Ituri Forest in the Congo, where Pygmies live. The discovery announced in late 2004 of "hobbit-sized" people (1 m tall) who lived on the island of Flores in Indonesia 13,000 years ago, suggests another explanation. Being marooned on the island with a limited food supply (they hunted pygmy elephants) made large size–and a corresponding need for more calories–a disadvantage.

Other scientists have suggested that ancestors of human beings began walking on two legs in part to keep cool in a hot climate. Walking upright exposes less body area to the rays of the sun than walking on all fours and also reduces the amount of water needed by about one-half.

## 18.6 How to Grow

A large change of scale forces adaptive changes in materials or form. However, within narrow limits–in most cases, up to a factor of 2–creatures can grow according to geometric similarity. That is, they can grow proportionally, so that their shape is preserved. A striking example of such growth by a far greater factor is the chambered nautilus (*Nautilus pompilius*). Each new chamber that it adds to its shell is larger than, but geometrically similar to, the previous chamber and also similar to the shape of the shell as a whole–an *equiangular,* or *logarithmic,* spiral (see Figure 18.9).

**FIGURE 18.9** A chambered nautilus shell. (*Photodisc/Punchstock.*)

Most living things grow over the course of their lives by a factor greater than 2. We've seen with *Dimetrodon* that a big specimen was not just a scaled-up small one. Nor is a human adult simply a scaled-up baby: A baby's head is relatively much larger than an adult's, and its arms are disproportionately shorter. In growth from baby to adult, the body does not scale up as a whole. Different parts of the body scale geometrically, each with a different linear scale factor. That is, a baby's eyes grow to perhaps twice their original size, while the arms grow by a factor of about four.

Although the laws for growth can be much more complicated than for proportional growth (or even for the allometric growth that we discuss later), more sophisticated mathematics–for example, differential geometry, the geometry of curves and surfaces–permits analysis of complex and interlocking scalings. For a model of the process in which a baby's head changes shape to grow into an adult head, we can use graph paper: First, we put a picture of the baby's skull on graph paper. Then we determine how to deform the grid until the pattern matches an adult skull (see Figure 18.10 and Spotlight 18.4). The same idea lies at the heart of computerized "morphing," the process in which the face of one person can be changed smoothly into the face of another, with different scalings for different parts of the face.

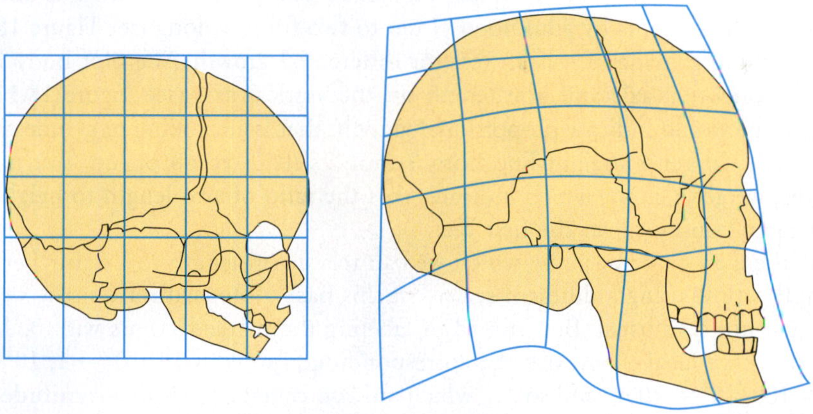

**FIGURE 18.10** Modeling the changes in the shape of a human head from infancy to adulthood.

## SPOTLIGHT 18.4 Helping to Find Missing Children

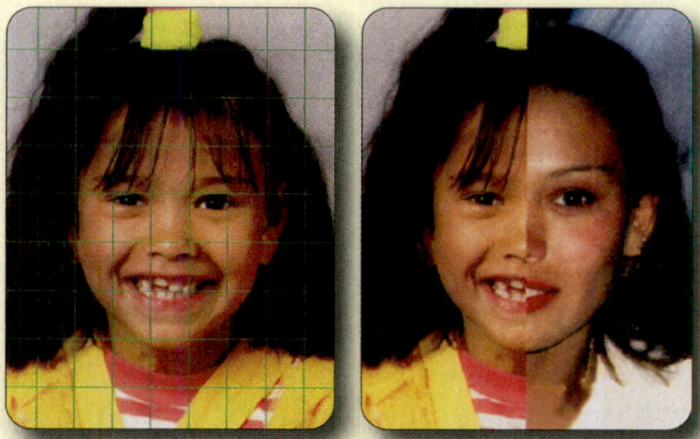

Photograph at age 7; photo showing one side of the child's face and one side of her mother's, to which the child's photo will be stretched.

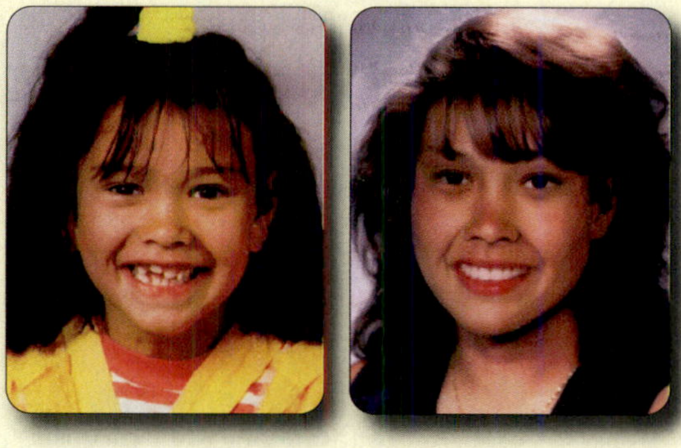

Stretching, merging, and age progression to age 17. *(National Center for Missing and Exploited Children, www.missingkids.com.)*

What does a child who was kidnapped at age 3 look like now, six years later?

At the National Center for Missing and Exploited Children (NCMEC) in Arlington, Virginia, a computer and a more sophisticated version of the graph-paper technique are used to answer such questions. Computer age-progression specialists scan photographs of both the missing child at age 3 and an older sibling or a biological parent at age 9 into a computer. Then the face of the 3-year-old is stretched, depending on age, to reflect craniofacial growth and merged with the image of the sibling or parent at 9 years old. The result is a rough idea of what the missing child may look like. As mathematicians and biologists refine their models of how faces change over time, this technique will improve. It may even become possible to gain an idea of how a child may look at age 40 or 65.

## Allometric Growth

If we measure the arm length or head size for humans of different ages and compare these measurements with body height, we observe that humans do not grow proportionally, that is, in a way that maintains geometric similarity. The head of a newborn baby may be one-third of the baby's length, but an adult's head is usually close to one-seventh of the individual's height. The arm, which at birth is one-third as long as the body, is by adulthood closer to two-fifths as long (see Figure 18.11a).

Graphing provides a way to test for differential growth. We plot body height on the horizontal axis and arm length on the vertical axis (see Figure 18.11b). A straight line would indicate proportional growth, that is, according to geometric similarity. We do get a straight line from 9 months (0.75 years) on up. But up to 9 months, we get a curve, which indicates that the ratio of arm length to height does not remain constant over the first year.

Is there an orderly law by which we can relate arm length to height? Let's plot again, this time using a different scale. For this **base-10 logarithmic scale**, we mark off equal units, as usual. But instead of labeling the marked points with 0, 1, 2, 3, and so on, we label them with the corresponding powers of 10: $10^0 = 1$, $10^1 = 10$, $10^2 = 100$, $10^3 = 1000$, and so on, which are also called **orders of magnitude**. Plotting a point on such a scale is not easy, because the point midway between 1 and 10 is not 5.5, but instead is closer to 3. Special graph paper (available in most college bookstores) marks smaller divisions and makes it easier to plot; paper marked with log scales on both axes is called **log-log paper**, while **semilog paper** has a logarithmic scale on just one axis. Also, many computer plotting packages can produce logarithmic scales.

We could use a logarithmic scale for either height or arm length, or for both. Using logarithmic scales for both, as in Figure 18.11c, the data plot closely to a straight line. Looking carefully, we can discern two different straight lines: a steeper one that fits early development (we will see shortly that it has slope 1.2), and a less steep one (with slope 1.0) that fits development after 9 months of age.

The change from one line to another after 9 months indicates a change in pattern of growth. The pattern after 9 months, characterized by the straight line with slope 1, is indeed proportional growth (sometimes called **isometric growth**). For the pattern before 9 months, the slope is 1.2. The fact that it is greater than 1 means that arm length is increasing relatively faster than height. This early growth also follows a definite pattern, called **allometric growth**.

### Allometric Growth — DEFINITION

**Allometric growth** is growth of the length of one feature at a rate proportional to a power of the length of another.

In geometric scaling, area grows according to the square (second power), and volume according to the cube (third power) of length, so they grow allometrically with length.

If we denote arm length by $y$ and height by $x$, a straight-line fit on log-log paper corresponds to the algebraic relation

$$\log_{10} y = B + a \log_{10} x$$

where $a$ is the slope of the line and $B$ is the point where the graph crosses the vertical axis. If we raise 10 to the power of each side, we get

$$y = bx^a$$

where $b = 10^B$. This equation describes a **power curve**: $y$ is a constant multiple of $x$ raised to a certain power.

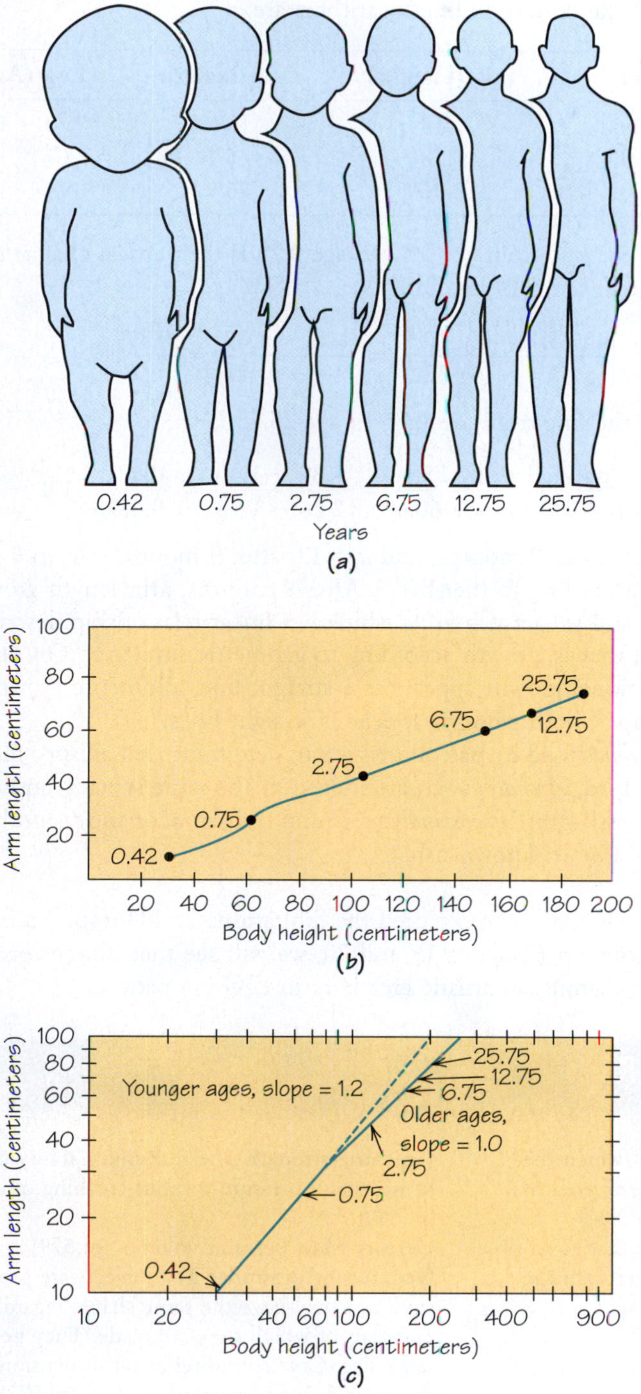


**FIGURE 18.11**
(a) The proportions of the human body change with age. (b) A graph of human body growth on ordinary graph paper. The numbers shown beside the points indicate the age in years; they correspond to the stage of human development shown in part (a). (c) A graph of human body growth on log-log paper.

## EXAMPLE 12 ■ Finding the Power of Growth

How do we arrive at those slopes of 1.0 and 1.2 in Figure 18.11?

**SOLUTION** You could use statistical software or your calculator to find the equation of the least-squares regression line (section 6.4) through the points on the log-log plots. Here we find approximate values for the slope $a$ for each line from the coordinates of the points at the ends of the lines, for ages 0.42, 0.75, and 25.75. The observations and the corresponding logarithms are:

| Age | Height | Log (Height) | Arm Length | Log (Arm Length) |
|---|---|---|---|---|
| 0.42 | 30.0 | 1.48 | 10.7 | 1.03 |
| 0.75 | 60.4 | 1.78 | 25.1 | 1.40 |
| 25.75 | 180.8 | 2.26 | 76.9 | 1.89 |

The slope for the line from age 0.42 to age 0.75 is the vertical change over the horizontal change in terms of log units:

$$\frac{\log 25.1 - \log 10.7}{\log 60.4 - \log 30.0} = \frac{1.40 - 1.03}{1.78 - 1.48} = \frac{0.37}{0.30} \approx 1.2$$

The slope for the line from age 0.75 to age 25.75 is

$$\frac{\log 76.9 - \log 25.1}{\log 180.8 - \log 60.4} = \frac{1.89 - 1.40}{2.26 - 1.78} = \frac{0.49}{0.48} \approx 1.0$$

So $a = 1.2$ up to 9 months, and $a = 1.0$ after 9 months. Up to 9 months, arm length grows according to $(\text{height})^{1.2}$. After 9 months, arm length grows according to $(\text{height})^{1.0}$, and we get $y = bx^{1.0}$, which is a linear relationship describing proportional growth, that is, growth according to geometric similarity. On ordinary graph paper, proportional growth appears as a straight line, allometric growth as a curve. On log-log paper, both patterns appear as straight lines.

Allometry was used by paleontologists to determine that all specimens (just six!) of the earliest bird, *Archaeopteryx,* are indeed of the same species, and that the puzzling minute fossil fish *Palaeospondylus* (found only in Scotland) is probably just the larval stage of a better-known fish. ■

In this chapter we have explored the limitations on life imposed by dwelling in three dimensions. In Chapters 19 and 20, we will see that dimensionality also imposes surprising limits on artistic creativity in devising patterns.

## REVIEW VOCABULARY

**Allometric growth** A pattern of growth in which the length of one feature grows at a rate proportional to a power of the length of another feature. (p. 594)

**Area–volume tension** A result of the fact that as an object is scaled up, the volume increases faster than the surface area and faster than areas of cross sections. (p. 586)

**Base-10 logarithmic scale** A scale on which equal divisions correspond to powers of 10. (p. 594)

**Crushing strength** The maximum ability of a substance to withstand pressure without crushing or deforming. (p. 580)

**Density** Mass per unit volume. (p. 579)

**Geometrically similar** Two objects are geometrically similar if they have the same shape, regardless of the materials of which they are made. They need not be the same size. Corresponding linear dimensions must have the same factor of proportionality. (p. 572)

**Isometric growth** Proportional growth. (p. 594)
**Linear (length) scaling factor** The number by which each linear dimension of an object is multiplied when it is scaled up or down; that is, the ratio of the length of any part of one of two geometrically similar objects to the length of the corresponding part of the second. (p. 572)
**Log-log paper** Graph paper on which both the vertical and the horizontal scales are logarithmic scales; that is, the scales are marked in orders of magnitude 1, 10, 100, 1000, . . . , instead of 1, 2, 3, 4, . . . . (p. 594)
**Orders of magnitude** Powers of 10. (p. 594)
**Power curve** A curve described by an equation $y = bx^a$, so that $y$ is proportional to a power of $x$. (p. 595)
**Pressure** Force per unit area. (p. 579)
**Problem of scale** As an object or being is scaled up, its surface and cross-sectional areas increase at a rate different from its volume, forcing adaptations of materials or shape. (p. 571)
**Proportional growth** Growth according to geometric similarity, where the length of every part of the organism enlarges by the same linear scaling factor. (p. 591)
**Semilog paper** Graph paper on which only one of the scales is a logarithmic scale. (p. 594)
**Weight** Force under gravity. (p. 579)
**Wing loading** Weight supported divided by wing area. (p. 588)

## ✓ SKILLS CHECK

**1.** A penny and a nickel are

**(a)** not geometrically similar, because they are made of different materials.
**(b)** not geometrically similar, because they are of different sizes.
**(c)** geometrically similar, because they have the same shape and proportional dimensions.

**2.** A scale model of a carillon stands 10 in. tall, and the actual carillon stands 100 ft tall. The linear scaling factor of the carillon compared to its model is __________ .

**3.** If a model car is built to a scale of 1 to 40, and the actual car has a turning circle of 37 ft, what should be the turning circle of the model?

**(a)** 11.1 in.
**(b)** 0.925 in.
**(c)** 1480 ft

**4.** You want to enlarge a 3-in. by 5-in. photograph to a 12-in. by 20-in. copy. Assuming that the cost of photographic paper is proportional to its area, and that 3-in. by 5-in. reprints cost 40 cents each, you would expect to pay ________________ for the large copy.

**5.** If a medium 10-in. pizza costs $8 and a similar 14-in. pizza costs $14, which is the better buy?

**(a)** The 10-in. pizza
**(b)** The 14-in. pizza
**(c)** They are about the same price per square inch.

**6.** An artist plans to melt 100 pennies and re-form a larger penny proportional in all dimensions to an ordinary penny. The linear scaling factor of the large penny compared with the ordinary penny is _________ .

**7.** The actor Elijah Wood, who plays Frodo Baggins in the movie of Tolkien's *Lord of the Rings,* is 5 ft 6 in. tall, but his character is barely 4 ft tall. Put correctly, how much shorter is Frodo than Wood?

**(a)** 138% shorter
**(b)** 73% shorter
**(c)** 27% shorter

**8.** The distance for the marathon race was established in 1921 as 42.195 km. Converted to the U.S. Customary System, the distance is __________________________ .

**9.** A kilometer is approximately equal in length to

**(a)** 5 mi.
**(b)** 3 mi.
**(c)** 3/5 mi.

**10.** A weight of 130 lb is approximately the same as ________ kg.

**11.** A 2-liter bottle contains approximately

**(a)** 2 quarts.
**(b)** 1 gallon.
**(c)** 10 pints.

**12.** A common speed limit in European neighborhoods is 30 km/h, which is about __________ mph.

**13.** Coffee costs about $8 per pound in the United States. If a Canadian dollar (Cdn$) exchanges for U.S.$1.02, what is the approximate cost in Canadian dollars of 500 g of coffee?

**(a)** Cdn$2
**(b)** Cdn$4
**(c)** Cdn$8

**14.** A sculpture weighs 140 lb and is supported by three legs, each of which is 0.5 in. by 0.5 in. by 2 in. high. The legs exert a pressure of ________ lb/in.$^2$ on the floor.

**15.** In comparing flight speeds of birds, an analysis of wing loading leads to the conclusion that

**(a)** light birds fly faster than heavy birds.
**(b)** heavy birds fly faster than light birds.
**(c)** heavy and light birds fly at about the same speed.

16. If an object is scaled linearly so that its volume grows to 8 times its original volume, its surface area is scaled to ______ times its original surface area.

17. In comparing the heights that large and small animals jump, analysis of the impact on scaling leads to the conclusion that

(a) smaller jumping animals can jump much higher than larger jumping animals.
(b) larger jumping animals can jump much higher than smaller jumping animals.
(c) all jumping animals jump to about the same height.

18. Assuming that a catfish maintains the same shape and proportions as its grows, and a catfish 8 in. long weighs about 1 lb, a 2-lb catfish is about _____ in. long.

19. The population of Mexico grew by 2.4% in 2002, by another 2.4% (of the 2002 population) in 2003, and by another 2.4% (of the 2003 population) in 2004. The population was growing

(a) proportionally.
(b) allometrically.
(c) by a constant amount each year.

20. The base-10 logarithm of 72 is approximately _____ .

# CHAPTER 18 EXERCISES

■ Challenge ◆ Discussion

Most of these exercises require a calculator; one with square roots will suffice.

## 18.1 Geometric Similarity

1. Your digital camera likely takes pictures with an aspect ratio of 4 to 3, meaning that the longer side is 4/3 times as long (in pixels) as the shorter side. For example, you probably can take a "small" picture with 640 pixels by 480 pixels, or perhaps a "large" picture with 2592 pixels by 1944 pixels (for a total of 2592 × 1944 pixels, or just a little more than 5 megapixels). Photographic prints from your digital camera are available in various sizes of paper, quoted in inches: 4 × 6, 5 × 7, and 8 × 10.

(a) Which of the paper sizes, if any, is geometrically similar to the original digital image?
(b) If a 4 × 6 print is instead made by scaling the shorter side of the digital image to be exactly 4 in., how long should the longer side of the image be on the print?
(c) If a 4 × 6 print is made by scaling the longer side of the digital image to be exactly 6 in., how long should the shorter side of the image be on the print? (*Hint*: The paper isn't wide enough!)

2. The area of a circle of radius $r$ is $\pi r^2$. Expressed in terms of the diameter, $d = 2r$, the area is $\frac{1}{4}\pi d^2$. If we apply a linear scaling factor $L$ to the diameter, then the area of the scaled circle–as in the case of the square that we considered in the text–changes with $L^2$, the square of the linear scaling factor. A natural application of this idea is with pizza. The prices at Domenico's pizza restaurant near Beloit College are \$7, \$8, \$9, and \$10, respectively, for small (10-in.), medium (12-in.), large (14-in.), and extra-large (16-in.) cheese pizzas.

(a) What is the linear scaling factor for an extra-large pizza compared with a small one?
(b) How many times as large in area is the extra-large pizza compared with the small one?
(c) How much pizza does each size give per dollar? What "hidden" assumptions are you making about how the pizzas are scaled up?
(d) The corresponding prices for a pizza with "the works" are \$9.95, \$11.95, \$13.95, and \$15.95. Is there any size of these for which you get more pizza per dollar than some size of the cheese pizzas?
(e) All of the prices are, to the nearest cent, exactly 5% lower than 3 years ago! What kind of scaling is that?

3. The human figures in Lego sets are 4 cm tall (without hats or helmets).

(a) What is the linear scaling factor of a Lego figure if it represents a human who is 160 cm tall?
(b) How does the volume of a real human compare with the volume of a Lego figure?
(c) The car in one Lego set is 10 cm long. Using the linear scaling factor in part (a), how long would a real car be?

4. Dollhouses and their furnishings are usually built to a scale of exactly 1 in. to 1 ft, meaning that an item 1 ft long in a real house is 1 in. long in a dollhouse.

(a) What is the linear scaling factor for a dollhouse?
(b) If a dollhouse were made of the same materials as a real house, how would their weights compare?

5. According to *Time* (March 7, 2005), men's brains on average are 10% larger than women's, even though men on average are only 8% taller. (The article mainly discusses the many differences in brain structure that likely outweigh any size differences.) If the brain scales linearly with height, and men are 8% taller, what percentage larger would you expect their brains to be?

6. At our house, we have some 10-in. frying pans and a 12-in. one; the 12-in. one weighs a lot more, never gets

as hot, and cooks food more slowly. Suppose that a 10-in. frying pan weighs 1 lb, apart from its handle. How much would a geometrically similar 12-in. frying pan weigh? How much would it weigh if it had the same thickness of metal as the 10-in. pan?

**7.** A famous geometry problem of Greek antiquity was *duplication of the cube.* Our knowledge of the history of the problem comes from the third century B.C. from Eratosthenes of Cyrene, famous for his estimate of the circumference of the earth. According to him, the citizens of Delos were suffering from a plague. They consulted the oracle, who told them that to rid themselves of the plague, they must construct an altar to a particular god that would be geometrically similar to the existing one but double the volume.

**(a)** How would the volume of the new altar compare with that of the old if each of its linear dimensions were doubled?

**(b)** What should the linear scaling factor be for the new altar? (The problem intended by the oracle was to construct with straightedge and compasses a line segment equal in length to this particular linear scaling factor. Not until the nineteenth century was the task shown to be impossible. Eratosthenes relates that the Delians interpreted the problem in this sense, were perplexed, and asked Plato about it. Plato told them that the god didn't really want an altar of double the volume but wished to shame them for their "neglect of mathematics and their contempt for geometry.")

**8.** Unfortunately, recent dollar coins (Susan B. Anthony, Sacagawea) have been a failure–the public found them too small and light. It is not yet known if the Presidential series of dollar coins begun in 2007 will be any more successful (they are the same size). Suppose that you are to design a new *$5* coin (whom should it depict?). The sole requirement is that it be made of the same material as the quarter but weigh four times as much. A quarter can be described geometrically as a circular cylinder approximately $\frac{15}{16}$ in. in diameter and $\frac{1}{16}$ in. thick. Because your new dollar should weigh four times as much, it needs to have four times the volume of a quarter. [The formula for the volume of a cylinder is $\pi \times (\text{diameter}/2)^2 \times \text{height}$.]

Susan B. Anthony dollar.

Presidential series dollar.

Coins shown actual size (26.5 mm diameter).
(*United States Mint.*)

**(a)** A member of your public advisory panel suggests just doubling the diameter and doubling the thickness. What do you tell this individual, in the most diplomatic terms?

**(b)** If you double the diameter, how thick does the coin need to be?

**(c)** Another member feels that the result of part (b) would be inconveniently large and proposes instead to scale up the quarter proportionality (she has studied an earlier edition of this book). What would the dimensions be for this coin? (Incidentally, nothing in U.S. law requires a dollar coin to be round. The Canadian $1 coin introduced in 1989 has been very popular; it has the shape of an 11-sided polygon.)

**9.** Criticize the following statement and write a correct version.

"[The Apple Web browser] Safari loads pages up to 2 times faster than Internet Explorer 7 and up to 1.7 times faster than Firefox 2." [http://www.apple.com/safari/, November 2007.]

**10.** Criticize the following statement and write a correct version.

"Wii uses 10 times less power consumption than PS3 & Xbox 360."
[http://www.maxconsole.net/?mode=news&newsid=14580, February 22, 2007.]

**11.** Criticize the following statement and write a correct version.

"The stock of Countrywide Financial Corporation [then the country's largest mortgage lender] has fallen 150% since the start of the year." [Heard on National Public Radio News, September 12, 2007.]

**12.** Criticize the following statement and write a correct version.

"Waiting time for surgery down 500%." [*Evening Standard* (Edinburgh), 5 July 2005.]

**13.** Abuses of the language of comparison aren't hard to find. For example, the phrase "35 times less than" occurs in more than 1900 Internet documents. Search on the Internet and find either an abuse of *times* and *less than* together, or else an abuse of percentages. Figure out what the author meant to say, and write it in correct language.

**14.** Criticize the following statement and write a correct version.

"The average full-time [real estate] agent working in Steubenville [Ohio] sells more than 22 houses per year, whereas the same agent in San Francisco sells fewer than 4–5.7 times less." [Austan Goolsebee, "Bubble-lusions: Why most real-estate agents aren't getting rich," http://www.slate.com/id/2124506, August 26, 2005.]

## 18.2 How Much Is That in...?

**15.** The cost of mailing a lightweight airmail letter from the United States to most of western Europe in 2008 was \$0.90. How much was that in euros (€), the currency of the European Union, when the exchange rate was €1 = \$1.47? (For comparison, the cost then of an airmail letter to the United States varied from country to country in the European Union, ranging from €0.65 to €1.70.)

**16.** The cost of mailing a lightweight letter from the United States to Canada in 2008 was US\$0.69. How much was that in Canadian dollars when the exchange rate was US\$1 = Cdn\$1.02? (The postage cost from Canada to the United States was Cdn\$0.93.)

**17.** In Germany the fuel efficiency of cars is measured in liters of fuel per 100 km (L/100 km). A typical average in a compact station wagon is 7.3 L/100 km. What is that in miles per gallon (mpg)?

**18.** According to Environmental Protection Agency ratings, the highest-mileage 2008 car was the gasoline/electric hybrid Toyota Prius at 48 mpg in the city. How many liters of gasoline does such a Prius use to travel 100 km in the city?

**19.** Consider a real locomotive that weighs 88 tons and an HO-gauge scale model of it, for which the linear scaling factor is 1/87.

**(a)** How much would an exact scale model weigh in tons?
**(b)** What assumptions are involved in your answer to part (a)?
**(c)** How much would an exact scale model weigh in pounds?
**(d)** In kilograms?
**(e)** In metric tonnes (1 metric tonne = 1000 kg)?

**20.** What's wrong in the following quotations?

**(a)** "President Bush visited California, where 12 forest fires have charred more than 700,000 square miles." [Steve Stadelman, WTVO television news, Channel 17, Rockford, Illinois, October 2007.] (Curiously, the same number 700,000 also appeared in news reports for California fires in 2003, 2000, and 1987.)
**(b)** "The population of the USA has topped 300 million.... If current trends continue, it is expected to reach 400 billion by 2043. This makes it an acceleration of growth...." [*Significance* 3 (4) (December 2006) p. 146.]

**21.** Gasoline is sold in the United States by the U.S. gallon and in Europe by the liter (1 U.S. gal = 231 in.$^3$; 1 L = 1000 cm$^3$). What was the equivalent cost, in U.S. dollars per U.S. gallon, for gasoline in Germany priced in euros at €1.42 per liter, when €1 = \$1.25, in September 2005?

**22.** In 1991, Edward N. Lorenz, a meteorologist who was an early researcher into chaos and dynamical systems (discussed in Chapter 23), received the Kyoto Prize in Basic Sciences, consisting of a gold medal and 45 million Japanese yen (¥). If US\$1 = ¥125 at the time, what was the value of the cash award in 1991 U.S. dollars? (In Chapter 21, we show how to convert such an amount to its value in today's dollars.)

**23.** The year 2008 marked the 50th anniversary of the installation of length markers on the sidewalk of the Harvard Bridge across the Charles River between Boston and Cambridge, Massachusetts (where in 1908 Harry Houdini performed one of his "escapes"). The bridge is marked at 10-smoot intervals, where 1 smoot was the height of MIT fraternity pledge, Oliver Smoot. The length of the bridge is 620 m or 364.4 smoots and one ear. How long is a smoot in feet and inches? (Oliver Smoot later became the head of the International Standards Organization.)

## 18.3 Scaling a Mountain

**24.** The two fastest elevators in Taipei 101 go up (or down) at 55 ft/s.

**(a)** With no stops, how long would it take to get to the top of Taipei 101?
**(b)** With no stops, how long would it take to get to the top of the Petronas Towers, which are 1483 ft high and whose elevators run at 41 ft/s?
**(c)** The answers to parts (a) and (b) suggest that building designers find half a minute a reasonable standard for getting to the top of a building (or else elevators don't come any faster). How fast would the elevators in a mile-high building have to be to achieve that standard?

**25.** Calculate the speed, in miles per hour, of the elevators in the three buildings in parts (a), (b), and (c) of Exercise 24.

For Exercises 26–31, refer to the following.

The Canadian dollar for many years fell steadily in value in terms of the U.S. dollar, but that trend has been reversed in the past few years. In January 2002, a U.S. dollar was worth Cdn\$1.60. In January 2008, US\$1 = Cdn\$1.02. How should you measure how much one currency has depreciated (lost value) against another? Let the *home currency* be the one whose change in value you are interested in, and let the *target currency* be the one in terms of which you will measure the change. In our example, we want to track the change in value of the U.S. dollar (home currency) in terms of the Canadian dollar (target currency).

There are two competing practices. Both begin by calculating the change as measured in the target currency, the new value minus the old value–here, Cdn\$1.02 − Cdn\$1.60 = Cdn−\$0.58.

- ▶ Option A, used by the International Monetary Fund and the British periodical *The Economist*, divides this difference by the new trading value (Cdn$1.02) and multiplies by 100 to get a result in percent–here, −57%.
- ▶ Option B, sometimes called the "popular method," divides instead *by the old trading value*, Cdn$1.60, here getting −36% as the percentage change in value of the U.S. dollar in terms of the Canadian dollar.

So, depending on the method used, we can say that the U.S. dollar lost either 57% or 36% in value against the Canadian dollar.

**26.** To get a feeling for how the results of these two options come out, we consider an artificial example where the numbers are simple. Suppose that in January 2008 the imaginary currency of the imaginary country Middle Earth, the Middie ($\mathcal{M}$), traded at US$2 but in January 2009 its value was only $1.

**(a)** Using Option A, calculate how much percentage value the Middie lost against the dollar.
**(b)** Calculate how much percentage value the Middie lost against the dollar using Option B.

**27.** We use the same data as in Exercise 26 but look at matters from the perspective of an American entrepreneur importing rings of power from Middle Earth. In January 2008, $1 = $\mathcal{M}$0.50; in January 2009, $\mathcal{M}$1 = $1.

**(a)** Using Option A, calculate how much percentage value the dollar gained against the Middie.
**(b)** Using Option B, calculate how much percentage value the dollar gained against the Middie.

**28.** In January 2002, when European Union euro (€) currency coins and bills were introduced, the conversion to U.S. dollars was US$1 = €1.160. Six years later, in January 2008, the dollar had declined severely, so that $1 = €0.693.

**(a)** Using Option A, calculate how much percentage value the dollar lost against the euro.
**(b)** Using Option B, calculate how much percentage value the dollar lost against the euro.

**29.** We use the same data as in Exercise 28 but look at matters from the perspective of a European considering a vacation in the United States. In January 2002, €1 = $0.862; in January 2008, €1 = $1.442.

**(a)** Using Option A, calculate how much percentage value the euro gained against the dollar.
**(b)** Using Option B, calculate how much percentage value the euro gained against the dollar.

◆ **30.** Using your results from either Exercises 26–27 or 28–29:

**(a)** Why don't the numbers agree? If the dollar loses a certain percentage against another currency, shouldn't that currency gain the same percentage against the dollar? Why or why not?
**(b)** Which option, A or B, seems to give a better sense of the effect of the change in relative values of currencies?

◆ **31.** For both Options A and B:

**(a)** Can either give a percentage loss that is more than 100%? Would it make sense to speak of a currency declining more than 100%?
**(b)** Is the percentage from Option A–loss or gain–always higher than the Option B percentage? Always lower? Neither, but do you see any pattern?
**(c)** Which option would you expect a person to use who wants to make a decline seem large? To make a gain seem large?

## 18.4 Sorry, No King Kongs

**32.** The weight of a 1-ft cube of steel is 500 lb. What is the pressure on the bottom face in

**(a)** pounds per square inch?
**(b)** atmospheres (1 atm = 14.7 lb/sq in.)?

**33.** In an article on adding organic matter to soil, the magazine *Organic Gardening* (March 1983) said, "Since a 6-inch layer of mineral soil in a 100-square-foot plot weighs about 45,000 pounds, adding 230 pounds of compost will give you an instant 5% organic matter."

**(a)** What is the density of the mineral soil, according to the quotation?
**(b)** How does this density compare with that of steel?
**(c)** How do you think the quotation should be revised to be accurate?

For Exercises 34 and 35, refer to the following.

A mature gorilla weighs 400 lb and stands 5 ft tall; its two feet combined have an area of about 1 ft$^2$.

**34.** **(a)** Give an estimate of the gorilla's weight when it was half as tall.
**(b)** What assumptions are involved in your estimate?
**(c)** When the gorilla is standing, what is the pressure on its feet in pounds per square inch?

**35.** Suppose King Kong is a gorilla scaled up with a linear scaling factor of 10.

**(a)** How much does King weigh?
**(b)** What is the pressure on King's feet in pounds per square inch?

**36.** You may want a waterbed, but waterbeds are not allowed in your building. Apart from the danger of flood if the bed should puncture or leak, the weight is an issue.

**(a)** Suppose that a queen-size waterbed mattress is 80 in. long by 60 in. wide by 12 in. high, and water weighs 1 kg/L. How much does the water in the mattress weigh in pounds?

**(b)** If the weight of the mattress and frame is carried by four legs, each 2 in. by 2 in., what is the pressure, in pounds per square inch, on each leg?
**(c)** How does the pressure on the legs of the waterbed compare with the pressure that a person exerts on his or her feet–for example, a 130-lb person with a total foot area of about one-quarter of a square foot in contact with the ground?

**37.** If you aren't allowed to have a waterbed, how about a spa (hot tub)? Find the weight of the water in a spa that is in the shape of a cylinder 6 ft in diameter and 3.5 ft deep. (*Hint:* The volume of a cylinder is $\pi r^2 h$, where $r$ is the radius and $h$ is the height.)

**38.** What does the largest giant sequoia tree (named "Hyperion") weigh? Model the tree as a (very elongated) cone. Assume that the tree is 379 ft high, has a circumference of 40 ft at the base, and that the density of the wood is 31 lb/ft$^3$. (The volume of a cone of height $h$ and radius $r$ is $\frac{1}{3}\pi r^2 h$.)

**39.** A 6-ft-tall indoor holiday tree needs four strings of lights to decorate it. How many strings of lights are needed for an outdoor tree that is 30 ft high? (Contributed by Charlotte Chell of Carthage College, Kenosha, Wisconsin.)

For Exercises 40 and 41, refer to the following.

An ancient measure of length, the *cubit,* was the distance from the elbow to the tip of the middle finger of a person's outstretched arm. So the length of a cubit depended on the person, though there was some attempt at standardization. Most estimates place the cubit between 17 and 22 in.

**40.** According to classical Greek sources, Pythagoras (sixth century B.C.) used geometric scaling to model the height of Hercules, the heroic figure of classical mythology. Pythagoras compared the lengths of two racecourses, one (according to tradition) paced off by Hercules and the other by a man of average height. Both were 600 "paces" long, but the one by Hercules was longer because of his longer stride. A normal man in the time of Pythagoras would have been about 5 ft tall.

**(a)** If the distance paced off by Hercules was 30% longer than the other racecourse, how tall was Hercules? What does your calculation assume?
**(b)** The sources give two conflicting answers, that Hercules was 4 cubits tall and 4 cubits 1 "foot" tall. What range does this give for his height in feet and inches? In centimeters? (Assume that a Greek "foot" was the length of a modern foot.)

**41.** Goliath [of David and Goliath, as related in the Bible (I Samuel 17:4)] was "six cubits and a span." A "span" was originally the distance from the tip of the thumb to the tip of the little finger when the hand is fully extended, about 9 in. What range of heights would this indicate for Goliath in feet and inches? In centimeters?

For Exercises 42–45, refer to the following.

The *body mass index* (BMI) is the basis for the National Heart, Lung and Blood Institute's weight guidelines. BMI is body weight (in kilograms) divided by the square of height (in meters). A BMI of 25 through 29 is considered "overweight"; a BMI of 30 or over is considered "obese." Some 55% of American adults have a BMI of 25 or above. (*Note:* BMI is not a useful measure for young children, pregnant or breast-feeding women, the frail elderly, or very muscular people.) For practice with this concept, calculate your own BMI.

(*moodboard/Corbis.*)

**42.** Calculate the BMI for a woman 160 cm tall who weighs 65 kg. Is she overweight according to the institute's guidelines?

**43.** How much in kilograms must a man weigh who is 190 cm tall if he is not to be considered overweight according to the institute's guidelines?

**44.** Suppose that weight and height are measured instead in U.S. Customary units of pounds and inches. We can still calculate body weight divided by the square of height using these units. What conversion factor is necessary to convert this number to the BMI?

◆ **45.** Because body weight is average density times body volume, BMI is average density times a quantity that has units of length. Discuss whether BMI makes sense as a measure of being overweight. Would dividing by a different power of height make for a better measure?

## 18.5 Dimension Tension

◆ **46.** Jonathan Swift's Gulliver traveled to Lilliput, where the Lilliputians were human-shaped but only about 6 in. tall. In other words, they were geometrically similar in shape to ordinary human beings but only one-twelfth as tall. What would a Lilliputian weigh?

Are Lilliputians ruled out by the size–shape and area–volume considerations in this chapter? If you think they are, what considerations do you find convincing? If not, why not?

**47. (a)** What would you expect an individual *Quetzalcoatlus northropi* to weigh if it had half the wingspan of an adult?
**(b)** If an individual weighed half as much as an adult, what would you expect its wingspan to be?

**48.** In the children's story *Peter Pan*, Peter and Wendy can fly. We may suppose that they are 4 ft tall, so they are about 12 times as tall as a sparrow is long. What should their minimum flying speed be?

**49.** Icarus of Greek legend escaped from Crete with his father, Daedalus, on wings made by Daedalus and attached with wax. Against his father's advice, Icarus flew too close to the Sun; the wax melted, the wings fell off, and he fell into the sea and drowned. What must have been his minimum cruising speed? What assumptions does your answer involve?

◆ **50.** Recent years have seen the beginnings of human-powered controlled flight, in the *Gossamer Condor* and other superlightweight planes, which have disproportionately large wings compared with geometric scaling up of birds. The *Gossamer Condor* is far longer than an ostrich but it flies at only 12 mph. How can it?

**51.** Justify the claim on p. 589 that a *Q. northropi* weighing 200 lb with a wingspan of 50 ft would have had the same wing loading as one weighing 100 lb with a wingspan of 36 ft.

**52.** The largest and heaviest aircraft in service today is the An-225–and we mean "The" because there is only one! (A second was scheduled to enter service in 2008.) It has been used to bring humanitarian equipment to Iraq, as well as–in a single flight–216,000 meals for American military personnel. The plane has a wing area of 905 $m^2$ and a maximum takeoff weight of 1.3 million lb. What is its wing loading, in $kg/m^2$?

**53.** The cult movie *Them* (1954) features enormous ants (8 m long by 3 m wide). We can investigate the feasibility of such a scaled-up insect by considering its oxygen consumption. A common ant, 1 cm long, needs 24 milliliters (mL) of oxygen per second for each cubic centimeter of its volume. Because an ant has no lungs, it absorbs oxygen through its "skin" at a rate of 6.2 mL per second per square centimeter. Suppose that the tissues of a scaled-up ant would have the same need for oxygen for each cubic centimeter, and that its skin could absorb oxygen at the same rate, as a common ant.

**(a)** Compared with a common ant, how many times as large is an enormous ant's

**(i)** length?
**(ii)** surface area?
**(iii)** volume?

**(b)** What proportion of such an ant's oxygen need could its skin supply?

**(c)** What can you conclude about the existence of such insects? (Adapted from George Knill and George Fawcett, Animal form or keeping your cool, *Mathematics Teacher,* May 1982, 395–397.)

For Exercises 54–57, refer to the following.

Maybe some trees could grow to a mile high, but they just don't live long enough to have the chance. In this problem, we try to determine how fast the height of a tree increases. We can measure indirectly how much mass the tree adds in a year by the area of the annual tree ring added. Here are two relevant facts:

- As you may have noticed from stumps, as a tree grows older, its annual rings get less wide. Although the width of the ring varies somewhat from year to year with the amount of rainfall and other factors, the total *area* of each annual ring is roughly the same over the years, meaning that *the tree adds roughly the same amount of mass each year.* Call that amount $a$; then the mass $M$ of the tree is $M = at$, where $t$ is its age in years.
- Over a large range of tree sizes and tree species, the diameter $d$ of a tree of a species is approximately proportional to the three-halves power of the height $h$ of the tree (different species have different constants of proportionality). Thus, $d \propto h^{3/2}$ (this is shown in Exercise 54).

Now, if we assume that the bulk of the mass of the tree is in the trunk, and if we model the trunk either as a long cylinder or as a thin cone, the mass is proportional to the volume, so $M \propto d^2h$. Then

$$at = M \propto d^2h \propto (h^{3/2})^2h = h^4$$

so $h \propto t^{1/4}$. In other words, *the tree grows in height as the fourth root of its age.*

**54.** Suppose that a tree grows to 20 m in 30 years. How tall will it be (if it lives long enough) when it is 60 years old?

**55.** How long would it take the tree in Exercise 54 to grow to be 40 m tall?

**56.** Giant sequoias can reach 100 m after about 1000 years. If it could keep on growing at the same rate of addition of mass, how long would it take a giant sequoia 100 m tall to grow to 200 m?

■ **57.** The branching of trees is similar to the branching of systems in the bodies of animals. For similar reasons, the area of the cross section of the tree at its base scales as the three-fourths power of the tree's mass, that is, $A \propto M^{3/4}$. Assume that most of the mass is in the trunk and model the tree either as a tall cylinder ($V = \pi r^2h$) or as a cone ($V = \pi r^2h/3$). Show that the diameter $d$ of a tree is approximately proportional to the three-halves power of the height, that is, $d \propto h^{3/2}$.

◆ **58.** Some humans, such as the Bushmen of the Kalahari Desert in Africa, live in desert environments, where it is important to be able to do without water for periods of time. Would you expect such an environment to favor short people or tall ones? (Adapted from A. Zherdev, Horseflies and flying horses, *Quantum,* May–June 1994, 32–37, 59–60.)

◆ **59.** Smaller birds and mammals generally maintain higher body temperatures than do larger ones. Explain why you would expect this to be so. (Adapted from A. Zherdev, Horseflies and flying horses, *Quantum,* May–June 1994, 32–37, 59–60.)

## 18.6 How to Grow

**60.** Listed below are the numbers of species of reptiles and amphibians on some Caribbean islands, together with the approximate areas of the islands. (Suggested by Florence Gordon of the New York Institute of Technology, with contributions from Kevin Mitchell and James Ryan of Hobart and William Smith Colleges, Geneva, N.Y. This table is adapted from Tables 15 and 16 in P. J. Darlington, *Zoogeography: The Geographic Distribution of Animals,* Wiley, New York, 1957, pp. 483–484).

| Island | Area ($mi^2$) | Species |
|---|---|---|
| Redonda | 1 | 3 |
| Saba | 4.9 | 5 |
| Montserrat | 40 | 9 |
| Trinidad | 2,000 | 80 |
| Puerto Rico | 3,400 | 40 |
| Jamaica | 4,500 | 39 |
| Hispaniola | 30,000 | 84 |
| Cuba | 40,000 | 76 |

**(a)** Plot number of species versus area on ordinary graph paper and then on log-log graph paper. If you don't have log-log paper available, use a calculator or spreadsheet to take the logarithms ($\log_{10}$) of all the numbers and graph logarithm of number of species versus logarithm of area on ordinary graph paper. (*Note:* Trinidad is an outlier from the general pattern–see Chapter 6).
**(b)** Is the relationship that you graphed in part (a) proportional? Allometric?
**(c)** What would be the expected number of species on an island of 400 $mi^2$?
**(d)** For each 10-fold increase in the island's size, what happens to the number of species, approximately?

**61.** Listed below are the weights and wingspans of some birds and of some fully loaded airplanes. (Idea and most data contributed by Florence Gordon of the New York Institute of Technology.)

| Bird | Weight (lb) | Wingspan (ft) |
|---|---|---|
| Crow | 1 | 2.9 |
| Harris hawk | 2.6 | 3.2 |
| Blue-footed booby | 4 | 3 |
| Red-tailed hawk | 4 | 4 |
| Horned owl | 5 | 5 |
| Turkey vulture | 6.5 | 6 |
| Eagle | 12 | 7.5 |
| Golden eagle | 13 | 7.3 |
| Whooping crane | 16.1 | 7.5 |
| Vulture | 18.7 | 9.3 |
| Condor | 22 | 9.9 |
| *Quetzalcoatlus northropi* | 100 | 36 |
| **Plane** | | |
| Boeing 737 | 117,000 | 93 |
| DC9 | 121,000 | 93.5 |
| Boeing 727 | 209,500 | 108 |
| Boeing 757 | 300,000 | 156.1 |
| Boeing 707 | 330,000 | 145.7 |
| DC8 | 350,000 | 148.5 |
| DC10 | 572,000 | 165.4 |
| Boeing 747 | 805,000 | 195.7 |
| Boeing 747-400 | 895,000 | 212.6 |
| Anton An-225 | 1,323,000 | 290.2 |

**(a)** Use a calculator or spreadsheet to take the logarithms ($\log_{10}$) of all the numbers and then graph logarithm of weight versus logarithm of wingspan on ordinary graph paper.
**(b)** For the birds, is the relationship that you graphed in part (a) proportional? Allometric? How about for the planes?
**(c)** Does the same relationship of wingspan to weight seem to hold for birds and planes?

## WRITING PROJECTS

**1.** A human infant at birth usually weighs between 5 and 10 lb and has a height (length) between 1 and 2 ft, with the shorter babies having the lesser weight. Considering the weight and height of an adult human, write a paragraph arguing that human growth must not be just proportional growth.

**2.** The principle that area scales with the square of length, and volume with the cube, has important consequences for the depiction and interpretation of data in graphic form. Suppose we wish to indicate in an artistic way that the weekly income of a U.S. carpenter is twice that of a carpenter in (mythical) Rotundia. We draw one moneybag for the Rotundian and another one "twice as large" for the American. (Illustration from Darrell Huff, *How to Lie with Statistics,* Norton, New York, 1954, p. 69.)

What's the problem? Well, first, people tend to respond to graphics by comparing areas. Because the larger moneybag is twice as high and twice as wide as the smaller one, its image has four times the area. Second, we are used to interpreting depth and perspective in drawings in terms of three-dimensional objects. Because the larger bag is also twice as thick as the smaller, it has eight times the volume. The graphic leaves the subconscious impression that the U.S. carpenter earns eight times as much, instead of twice as much. With these ideas in mind, evaluate–in a paragraph each–the following data depictions.

**(a)** Percentages of Ph.D.s earned by women in three fields. (From *Science*, 260, April 16, 1993, 409, as reproduced in Jessica Utts, *Seeing Through Statistics*, Duxbury, Belmont, Calif., 1996, p. 142.)

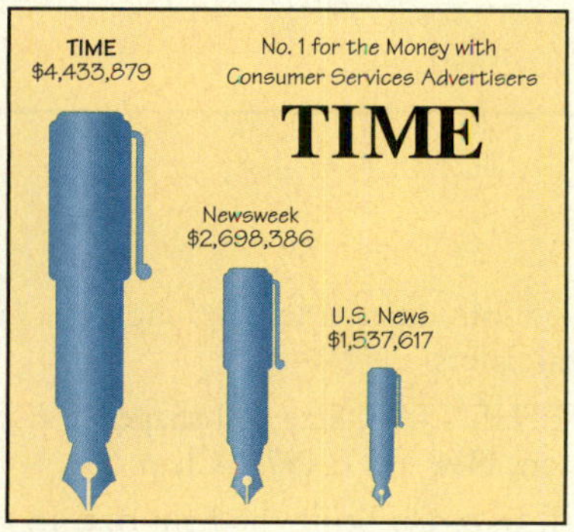


**(b)** Advertising spending in three prominent newsmagazines. (From *Time* magazine, as reproduced in David S. Moore, *Statistics: Concepts and Controversies*, 4th ed., W. H. Freeman, New York, 1997, p. 207.)

**(c)** U.S. colleges as classified by enrollment. (From David S. Moore, *Statistics: Concepts and Controversies*, 4th ed., W. H. Freeman, New York, 1997, p. 217.)

**3.** Evaluate in a paragraph each of the following depictions (a–c). (Illustrations from Edward R. Tufte, *The Visual Display of Quantitative Information*, Graphics Press, 1983, pp. 70, 69, and 57.)

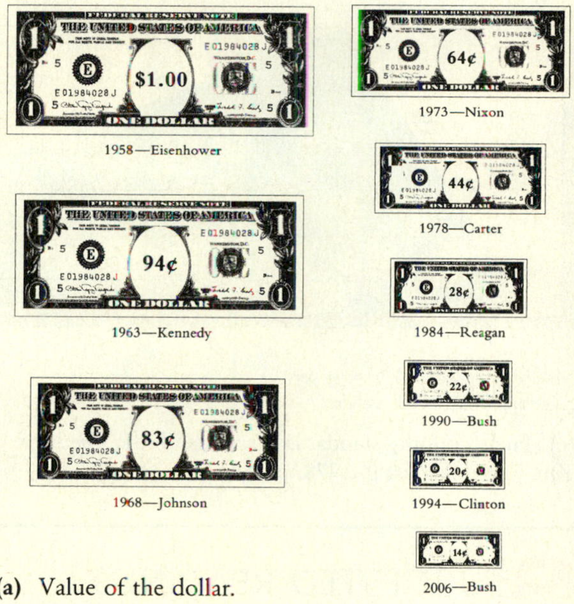


**(a)** Value of the dollar.

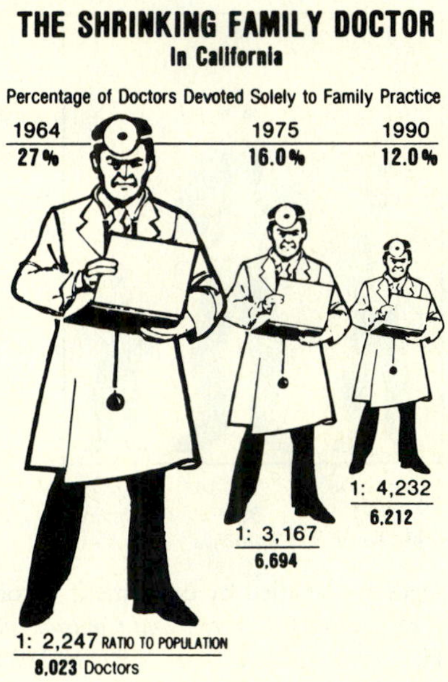


**(b)** The shrinking family doctor.

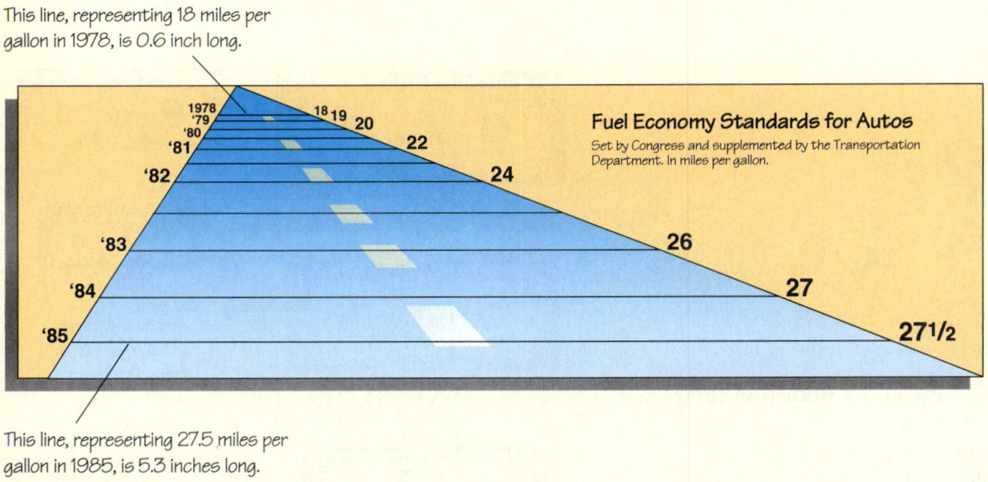


**(c)** Fuel economy standards for autos. (From *The New York Times*, August 9, 1978, p. D2.)

**4.** With the ideas of Writing Projects 2 and 3 in mind, collect and evaluate similar depictions of data from magazines and newspapers.

**5.** Dolls and human figures are usually scaled to be geometrically similar to actual humans. But are dolls designed to represent babies or adult humans? Go to a toy store and measure the height, the vertical height of the head, and the arm length of some dolls and other figures. Scale your measurements to compare them with Figure 18.11; from that comparison, try to estimate the ages of the humans that the figures resemble. Write up your procedure, data, calculations, and conclusions in a page or two.

## SUGGESTED READINGS

ADAM, JOHN A. *Mathematics in Nature: Modeling Patterns in the Natural World*, Princeton University Press, Princeton, N.J., 2003.

BONNER, JOHN TYLER. *Why Size Matters: From Bacteria to Blue Whales*, Princeton University Press, Princeton, N.J., 2006.

DUDLEY, BRIAN A. C. *Mathematical and Biological Interrelations*, Wiley, New York, 1977. Excellent and gentle extended introduction to graphing, scale factors, and logarithmic plots.

GOULD, STEPHEN JAY. Size and shape. In *Ever Since Darwin*, Norton, New York, 1977, Chap. 21.

HALDANE, J. B. S. On being the right size. In *Possible Worlds and Other Papers*, Harper, New York, 1928. Reprinted in James R. Newman (ed.), *The World of Mathematics*, vol. 2, Simon & Schuster, New York, 1956, pp. 952–957. Also reprinted in John Maynard Smith (ed.), *On Being the Right Size and Other Essays by J. B. S. Haldane,* Oxford University Press, Oxford, 1985, pp. 1–8. Succinctly surveys area–volume tension, flying, the size of eyes, and even the best size for human institutions.

McMAHON, T. A., and J. T. BONNER. *On Size and Life,* Scientific American Library, New York, 1983. Astonishingly beautiful and informative book on the effects of size and shape on living things.

SCHMIDT-NIELSEN, KNUT. *Scaling: Why Is Animal Size So Important?* Cambridge University Press, New York, 1984.

WEIBEL, EWALD R. *Symmorphosis: On Form and Function in Shaping Life*, Harvard University Press, Cambridge, Mass., 2000.

## SUGGESTED WEB SITES

**physics.nist.gov/cuu/Units/index.html** In-depth information on SI, the modern metric system.

**www.missingkids.com** National Center for Missing and Exploited Children.

**www.usmint.gov** U.S. Mint.

**www.thusness.com/bmi.t.html** Body mass index calculator and further links.

CHAPTER 19



# Symmetry and Patterns

"The senses delight in things duly proportional." So said the famous philosopher-theologian Thomas Aquinas more than 700 years ago. In this chapter, we examine elements of esthetic appreciation, particularly *symmetry*.

What is symmetry and what does mathematics have to do with it? Symmetry, like beauty, is hard to define. Dictionaries talk about "correspondence of form on opposite sides of a dividing line or plane or about a center or an axis," "correspondence, equivalence, or identity among constituents of an entity," and "beauty as a result of balance or harmonious arrangement" (*American Heritage Dictionary,* 3rd ed.).

In the narrowest sense, symmetry refers to mirror-image correspondence between parts of an object. Crystals, in both their appearance and their atomic structure, provide examples of symmetry in this sense. Taken in a wider sense, though, symmetry includes notions of *balance, similarity,* and *repetition.*

Our wider sense of symmetry leads us to appreciate patterns. *Mathematics is the study of patterns,* and it gives important insights into symmetry.

You are already familiar with mirror-image symmetry, which mathematicians call **reflection symmetry**.

Another kind of symmetry that you know well is **rotation symmetry**, in which rotation of an object about its center leaves it looking the same. A snowflake is a familiar example of both reflection symmetry and rotational symmetry.

Beautiful examples of rotational symmetry arise in nature in the shoots, leaves, and seeds of plants that grow from a central stem. For instance, the scales of a pineapple or a pinecone (Figure 19.1) and the stickers on a cactus follow such a pattern, as do the seeds of a sunflower (Figure 19.2a) and the petals of a daisy. In this pattern, known as **phyllotaxis**, the spirals and their elements are geometrically similar to one another.

The chambered nautilus of Figure 19.2b may stretch your notion of symmetry, as will other examples in this chapter. It has neither reflection symmetry nor rotational symmetry. However, although the successive sections of the nautilus are of different sizes, they are geometrically similar to one another, and the resulting spiral has the same shape at any size: A photographic enlargement superimposed on it would fit exactly. There is balance, similarity, and repetition—the characteristics of symmetry that we identified above.

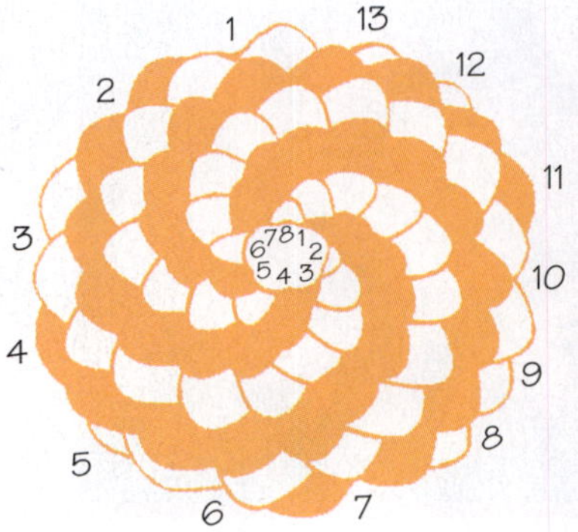

(*Don Hammond/Design Pics/Corbis.*)

**FIGURE 19.1** Spirals of scales on a pinecone: 8 right, 13 left. (*From Verner E. Hoggatt, Jr.,* Fibonacci and Lucas Numbers, *Houghton Mifflin, New York, 1969, p. 81.*)

(a) (b)

**FIGURE 19.2** (a) This sunflower has 55 spirals in one direction and 89 spirals in the other direction. (*Harvey Lloyd/ The Stock Market.*) (b) A chambered nautilus shell. (*James Randkler/Tony Stone Images.*)

This chapter explores and classifies the fundamentally different ways in which a two-dimensional design can be symmetrical. What will be surprising is that there are so few such patterns.

## 19.1 Fibonacci Numbers and the Golden Ratio

### Fibonacci Numbers

Associated with the geometric symmetry of phyllotaxis is a kind of *numeric symmetry,* with a "proportion" in the sense of a ratio of numbers. Strangely, the number of spirals in plants with phyllotaxis is not just any whole number but always comes from a particular sequence of numbers called the **Fibonacci numbers** (see Spotlight 19.1).

**Fibonacci Numbers (Fibonacci Sequence)** DEFINITION

**Fibonacci numbers** occur in the sequence

$$1, 1, 2, 3, 5, 8, 13, 21, 34, 55, 89, 144, 233, 377, \ldots$$

This sequence begins with the numbers 1 and 1 again, and each next number is obtained by adding the two preceding numbers together.

## SPOTLIGHT 19.1 Leonardo of Pisa ("Fibonacci")

Born in Pisa in 1170, Leonardo of Pisa has been known as "Fibonacci" for the past century and a half. This nickname, which refers to his descent from an ancestor named Bonaccio, is modern, and there is no evidence that he was known by it in his own time.

Leonardo was the greatest mathematician of the Middle Ages. His stated purpose in his book *Liber abbaci* (1202) was to introduce calculation with Hindu-Arabic numerals into Italy, to replace the Roman numerals then in use. Other books of his treated topics in geometry, algebra, and number theory.

We know little of Leonardo's life apart from a short autobiographical sketch in the *Liber abbaci*:

> I joined my father after his assignment by his homeland Pisa as an officer in the customhouse located at Bugia [Algeria] for the Pisan merchants who were often there. He had me marvelously instructed in the Arabic-Hindu numerals and calculation. I enjoyed so much the instruction that I later continued to study mathematics while on business trips to Egypt, Syria, Greece, Sicily, and Provence and there enjoyed discussions and disputations with the scholars of those places.

**Leonardo of Pisa ("Fibonacci")**

*A portrait of unlikely authenticity.*

*(From Columbia University, D. E. Smith Collection.)*

(*Source*: L. E. Sigler, *Leonardo Pisano Fibonacci, The Book of Squares: An Annotated Translation into Modern English*, Academic Press, New York, 1987, p. xvi.)

The *Liber abbaci* contains a famous problem about rabbits, whose solution is now called the Fibonacci sequence. Leonardo did not write further about it.

Sometimes a sequence of numbers is specified by stating the value of the first term or first several terms and then giving an equation to calculate succeeding terms from preceding ones. This is called a *recursive rule*, and the sequence is said to be defined by **recursion**. Let's denote the $n$th Fibonacci number by $F_n$; then the Fibonacci sequence can be defined by

### Recursion for the Fibonacci Sequence — PROCEDURE

$$F_1 = 1, F_2 = 1, \quad \text{and} \quad F_{n+1} = F_n + F_{n-1} \quad \text{for } n \geq 2$$

The recursive rule just expresses in algebraic form that the next Fibonacci number is the sum of the previous two.

Look at the sunflower in Figure 19.2a. You see a set of spirals running in the counterclockwise direction and another set in the clockwise direction. It is (just barely) possible to count the number of spirals in both directions. In the sunflower there are 55 in one direction and 89 in the other–two consecutive Fibonacci numbers. In the case of the pineapple, there are three sets of spirals, one each along the three directions through each hexagonally shaped scale. For the common grocery pineapple (*Ananas comosus*), there are always 8 spirals to the right, 13 to the left, and 21 vertically–again, consecutive Fibonacci numbers.

Why are the numbers of spirals in plants the same numbers that appear next to each other in a purely mathematical sequence? There is no easy answer. There are several intricate theories about the dynamics of the plant's growth.

## The Golden Ratio

During the last several centuries, an attractive myth has arisen that the ancient Greeks considered a specific numerical proportion essential to beauty and symmetry. Known in modern times as the **golden ratio**, **golden mean**, or even **divine proportion**, this proportion was investigated by Euclid in Book II of his *Elements*. Recent research reveals little evidence connecting this proportion to Greek esthetics, but let's pursue the golden ratio briefly because of its intimate connection to the Fibonacci sequence and because it does have appeal as a standard for beautiful proportion.

**Golden Ratio** DEFINITION

The value of the **golden ratio**, which is usually denoted by the Greek letter phi ($\phi$), is

$$\phi = \frac{1 + \sqrt{5}}{2} \approx 1.618034\ldots$$

The basic esthetic claim is that a **golden rectangle**—one whose height and width are in the ratio of 1 to $\phi$—is the most pleasing of all rectangles. The Greeks treated lengths geometrically, so for them it was important to construct lengths using straightedge and compass. In Spotlight 19.2 we show how to construct a golden rectangle that is 1 unit by $\phi$ units.

### SPOTLIGHT 19.2 How the Greeks Constructed a Golden Rectangle

In constructing a golden rectangle, the Greeks started from a one-by-one square (shown in black in the figure), which they made by constructing perpendiculars at the two ends of a horizontal segment of unit length. To extend the square to a golden rectangle, they bisected the original segment, getting a new point that divides it into two pieces of length one-half each. Using this new point and a compass opening equal to the distance from it to a far corner of the square (shown by the blue line in the figure), they could add the blue length to the length one-half to get an interval (in red at bottom) with total length $\phi$.

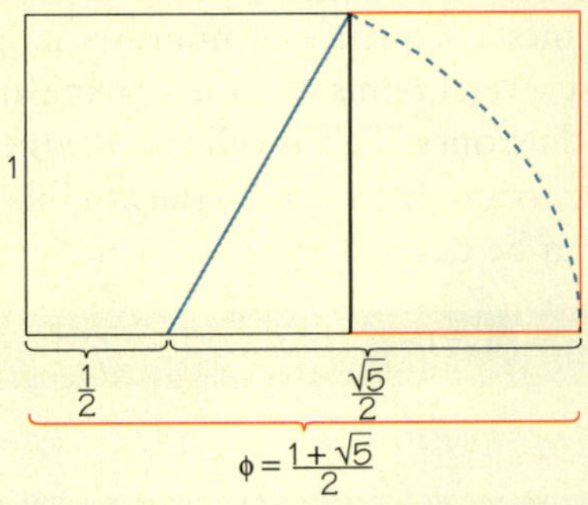


A golden rectangle has the pleasing property that if you cut a square-shaped piece off one end of it, the rectangle that remains is again a golden rectangle.

Why would anyone think that this is an attractive ratio? And where did it come from? The answer lies not in Fibonacci numbers but in the Greeks' pursuit of balance in their study of geometry.

Given two line lengths, one way to find a length that "strikes a balance" between the two is to average them. For lengths $l$ (the larger) and $w$ (the smaller), their average, or *arithmetic mean*, is $m = (l + w)/2$, and it satisfies

$$l - m = m - w$$

The length $m$ strikes a balance between $l$ and $w$, in terms of a common *difference* from the two original lengths. More generally, the arithmetic mean of $n$ numbers or lengths is their sum divided by $n$. (See Chapter 5 for its use in statistics.)

The Greeks, however, preferred a balance in terms of *ratios* rather than differences. They sought a length $s$, the **geometric mean**, that gives a common ratio

$$l \div s = s \div w \quad \text{or} \quad \frac{l}{s} = \frac{s}{w}$$

Hence $lw = s^2$, which expresses the geometric fact that $s$ is the side of a square whose area equals the area of an $l$ by $w$ rectangle (the Greeks thought in terms of geometric objects). In geometry, the geometric mean $s$ is called the *mean proportional* between $l$ and $w$ (see Figure 19.3).

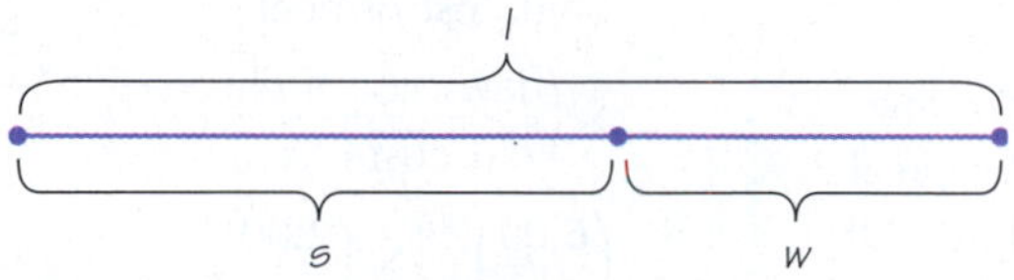


**FIGURE 19.3** The line segment of length $l$ is divided so that the length of $s$ is the geometric mean between $l$ and $w = l - s$. The dividing point divides the length $l$ in the golden ratio.

## Geometric Mean

DEFINITION

The quantity $s = \sqrt{lw}$ is the **geometric mean** of $l$ and $w$. More generally, the geometric mean of $n$ numbers is the $n$th root of the product of all $n$ factors: The geometric mean of $x_1, \ldots, x_n$ is $\sqrt[n]{x_1 \times \cdots \times x_n}$. For example, the geometric mean of 1, 2, 3, and 4 is $\sqrt[4]{1 \times 2 \times 3 \times 4} = \sqrt[4]{24} = 24^{1/4} \approx 2.213$.

The Greeks found symmetry and proportion in the geometric mean, but the geometric mean also has important practical applications (see Spotlight 19.3).

The Greeks were interested in cutting a single line segment of length $l$ into lengths $s$ and $w$, where $l = w + s$, so that $s$ would be the mean proportional between $w$ and $l$. Surprisingly, the ratio $\phi$ arises, as we show. Denote by $x$ the common ratio

$$\frac{l}{s} = \frac{s}{w} = x$$

Substituting $l = s + w$, we get

$$x = \frac{l}{s} = \frac{s + w}{s} = \frac{s}{s} + \frac{w}{s} = 1 + \frac{w}{s}$$

But $w/s$ is just $1/x$, so we have

$$x = 1 + \frac{1}{x}$$

Multiplying through by $x$ gives

$$x^2 = x + 1 \quad \text{or} \quad x^2 - x - 1 = 0$$

This is a quadratic equation of the form

$$ax^2 + bx + c = 0$$

## SPOTLIGHT 19.3 The Consumer Price Index: An Application of the Geometric Mean

The Bureau of Labor Statistics (BLS) uses the geometric mean—not the arithmetic mean—to calculate the Consumer Price Index (CPI), which tracks changes in the cost of the goods and services that people buy.

The geometric mean takes into account substitutions that consumers make when prices change. For example, if the price of beef goes up but the price of chicken doesn't, then consumers may buy less beef and substitute the cheaper chicken for some beef.

Suppose that, overall, U.S. families consume equal dollar values of beef and chicken. A typical family might consume weekly 5 lb of beef at $4/lb and 10 lb of chicken at $2/lb, for $20 each and a total cost of $40. We say that beef and chicken each have a relative *market share* of 0.5 (50% beef, 50% chicken, by dollar value).

What if beef goes up to $6/lb but chicken stays at $2/lb? The *relative price change* in beef is $6/$4 = 1.5 and the relative price change in chicken is $2/$2 = 1.00 (no change). If the average family continues to eat just as much beef and chicken as before, the cost is now $50, an increase of 25%. Because $30 goes for beef and $20 for chicken, the relative market shares (0.6 and 0.4) have changed. The *relative price change* for the family's meat is $50/$40 = 1.25, which is just the arithmetic mean of the two relative price changes (1.50 and 1.00). A more general formulation is:

$$\begin{aligned}
&\text{relative price change}\\
&\quad = (\text{old market share of beef})\ \frac{\text{new cost of beef}}{\text{old cost of beef}}\\
&\qquad + (\text{old market share of chicken}) \times \frac{\text{new cost of chicken}}{\text{old cost of chicken}}\\
&\quad = 0.5 \times \frac{6.00}{4.00} + 0.5 \times \frac{2.00}{2.00}\\
&\quad = \frac{1.50 + 1.00}{2} = 1.25
\end{aligned}$$

A family that eats no beef sees no increase. A family that eats only beef sees an increase of 50%. The CPI is an average over *all* families, weighted by the dollar value that each consumes.

If instead we use the geometric mean, we get a relative price change of $\sqrt{1.50 \times 1.00} = 1.225$. The more general formulation is

$$\begin{aligned}
&\text{relative price change}\\
&\quad = \left(\frac{\text{new cost of beef}}{\text{old cost of beef}}\right)^{(\text{old market share of beef})}\\
&\qquad \times \left(\frac{\text{new cost of chicken}}{\text{old cost of chicken}}\right)^{(\text{old market share of chicken})}\\
&\quad = \left(\frac{6.00}{4.00}\right)^{0.5} \times \left(\frac{2.00}{2.00}\right)^{0.5}\\
&\quad = \sqrt{1.50 \times 1.00} = 1.225
\end{aligned}$$

This relative price change, a 22.5% increase, is less than the 25% using the arithmetic mean.

The intention of the CPI is to measure the change in the cost of goods and services that still yield the same level of satisfaction to consumers. Use of the arithmetic mean presumes that a family buys the same amount of beef and chicken (5 lb beef, 10 lb chicken) as before. Use of the geometric mean presumes that a family buys the same *relative dollar value* of each meat as before, hence $24.50 (12.25 lb) of chicken and $24.50 (4.08 lb) of beef, for a total of $49 = 1.225 × $40. Buying 2.25 lb more chicken and 0.92 lb less beef is supposed to yield the "same satisfaction" as before.

Because the geometric mean is always less than or equal to the arithmetic mean (see Exercise 17), the geometric mean gives a lower figure for inflation than using the arithmetic mean would produce.

Social Security payments, some wage increases, and income tax rates are all automatically geared to the CPI, which we treat in detail in Chapter 21.

with $a = 1$, $b = -1$, and $c = -1$. We apply the famous quadratic formula,

$$x = \frac{-b \pm \sqrt{b^2 - 4ac}}{2a}$$

to get the two solutions

$$x = \frac{1 + \sqrt{5}}{2} \approx 1.618034 \ldots \qquad \text{and} \qquad \frac{1 - \sqrt{5}}{2} \approx -0.618034 \ldots$$

The negative solution does not correspond to a length. The first solution is the golden ratio $\phi$. It occurs often in other contexts in geometry; for example, $\phi$ is the ratio of a diagonal to a side of a regular pentagon (see Figure 19.4).

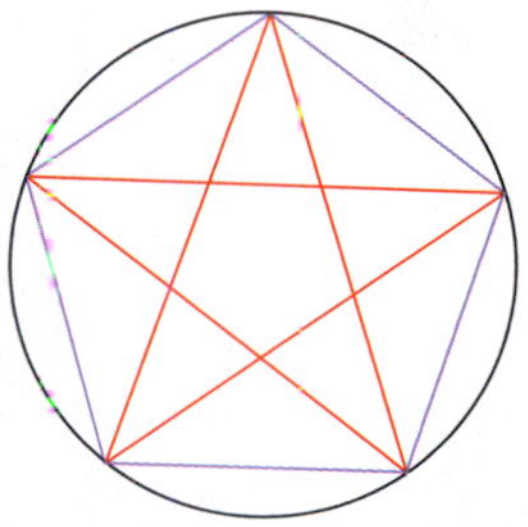

**FIGURE 19.4** In a pentagon with equal sides, $\phi$ is the ratio of a diagonal to a side. The five-pointed star formed by the diagonals was the symbol of the followers of the ancient Greek mathematician Pythagoras.

Thanks to Roger Herz-Fischler (Wilfrid Laurier University) and George Markowsky (University of Maine), we know that the term *golden ratio* was not used in antiquity and that there is no evidence that the Great Pyramid was designed to conform to $\phi$, nor that the Greeks used $\phi$ in the proportions of the Parthenon, nor that Leonardo da Vinci used $\phi$ in proportions for the human figure (Figure 19.5a). The area from the top of the head of the "Mona Lisa" to the top of her bodice may form a golden rectangle, as claimed by Bulent Atalay in his *Math and the Mona Lisa: The Art and Science of Leonardo da Vinci* (2004), but Leonardo left no documents saying that was his intention or design principle. Others have claimed that the impressionists Gustave Caillebotte (1848–1894) and Georges Seurat (1859–1891) used the golden ratio to design some of their paintings, but the painters themselves left no word about it. Wolfgang Amadeus Mozart (1756–1791), who was fascinated by mathematics as a student, may have constructed the lengths of parts of some of his piano sonatas with an eye to the golden ratio; but we do not have evidence that this was his intention.

Moreover, experiments show that people's preferences for dimensions of rectangles cover a wide range, with golden rectangles not holding any special place.

It is true that human bodies exhibit ratios close to the golden ratio, as you can see by comparing your overall height to the height of your navel. The Swiss-born architect Le Corbusier (Charles-Edouard Jeanneret [1887–1965]) used the golden ratio (including a navel-height feature) as the basis for his "Modulor" scale of proportions (Figure 19.5b).

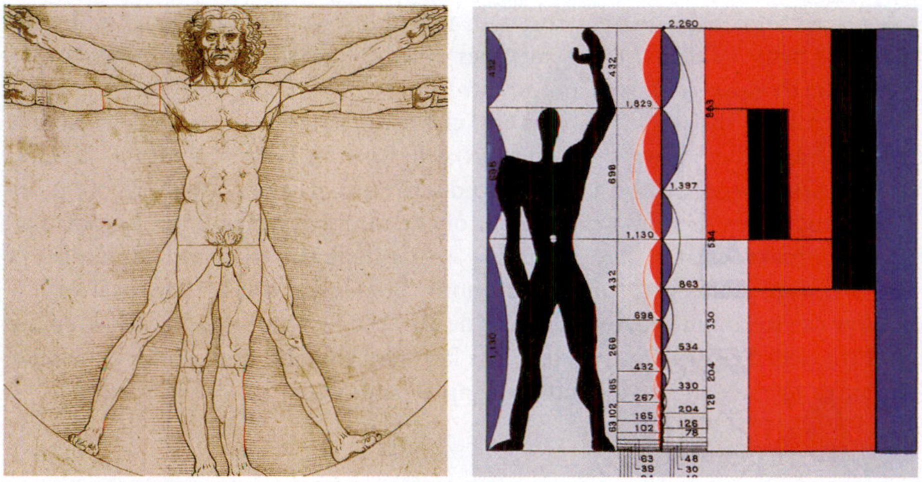

(a) (b)

**FIGURE 19.5** (a) Leonardo da Vinci's "Vitruvian Man" (ca. 1490), based on body proportions by Vitruvius (architect and engineer, first century B.C.). Despite claims on the Web and in the thriller *The Da Vinci Code,* neither Vitruvius nor Leonardo suggested using $\phi$ for human proportions or anything else. (*Accademia, Venice, Italy/Scala/Art Resource, New York.*) (b) Le Corbusier, however, did use $\phi$ in his "Modulor" scale of proportions. (*Le Corbusier, "Le Modulor," 1945. © 2000 Artists Rights Society [ARS], New York/ADAGP, Paris/FLC.*)

The spirals of the sunflower are approximations to an equiangular, or logarithmic spiral (Figure 19.6). The mathematical reason for this connection is that the ratios of consecutive Fibonacci numbers

$$\frac{1}{1} \quad \frac{2}{1} \quad \frac{3}{2} \quad \frac{5}{3} \quad \frac{8}{5} \quad \frac{13}{8} \quad \frac{21}{13} \ldots$$

$$1.0 \quad 2.0 \quad 1.5 \quad 1.666\ldots \quad 1.6 \quad 1.625 \quad 1.615\ldots$$

provide alternately under- and overapproximations to $\phi \approx 1.618034\ldots.$

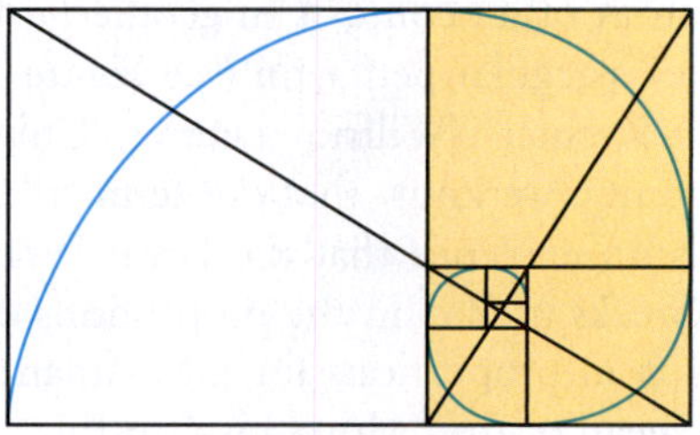

**FIGURE 19.6** A logarithmic spiral determines a sequence of golden rectangles and corresponding squares.

(The spiral of the chambered nautilus shell of Figure 19.2b (p. 610) is also an equiangular spiral, but the rectangles formed have a ratio near 1.3 rather than the golden ratio.)

For reasons that we do not understand, some ratios in the DNA molecule are close to the golden ratio; for example, the length of one full cycle of a strand in the double helix is about 1.62 times its width. Perhaps the most surprising appearance of the golden ratio is in connection with black holes, regions of space in which the gravitational field is so strong that nothing can escape (even light). A rotating black hole loses energy and up to a point heats up as it does so; after that point–when the mass of the hole equals its angular momentum times the square root of $\phi$–the hole starts to cool down instead.

## 19.2 Rosette, Strip, and Wallpaper Patterns

The spiral distribution of the seeds in a sunflower head and the spiraling of leaves around a plant stem are instances of *similarity* and *repetition,* two key aspects of symmetry. They also illustrate *balance,* which refers to regularity in *how* the repetitions are arranged. In considering patterns with repetition, we distinguish the individual element or figure of the design (sometimes called the *motif* ) from the *pattern* of the design–*how the copies of the motif are arranged.*

The problem that we focus on in this chapter is how to explore and classify the fundamentally different ways in which a flat design can be symmetrical. The ideas that we discuss were used by scientists to discover what crystalline forms are possible. Although there is a limitless number of chemical structures, and of motifs that people can make, what is quite surprising is that there is only a limited number of ways to arrange atoms in a structure or motifs in a design in a symmetrical way.

How can we enumerate the ways that designs can be put together without counting all the actual designs themselves? The key mathematical idea is to look at what you can *do* to the pattern without changing its appearance.

### Rigid Motions

Mathematicians describe various kinds of symmetry by using the geometric notion of a **rigid motion**, also known as an **isometry** (which means "same size"). A rigid motion is a specific kind of variation on the original pattern: We pick it up and move it, perhaps rotate it, possibly flip it over–but we *don't change its size or shape.* (The original figure and its image are not just geometrically similar, in the language of section 18.1, but also the same size.)

**Rigid Motion** DEFINITION

A **rigid motion** is one that preserves the size and shape of figures. In particular, any pair of points is the same distance apart after the motion as before.

Figure 19.7 shows the results of various motions applied to the rectangle and its interior of Figure 19.7a. In Figure 19.7b, each side is shrunk by 50%–not a rigid motion, because the size of the rectangle changes. For Figure 19.7c, we shear ("squash") the rectangle–again, this is not a rigid motion because the shape of the rectangle changes. In Figure 19.7d we rotate the rectangle 90° (a quarter-turn) clockwise around the center of the rectangle: This is a rigid motion. Similarly, in Figure 19.7e, rotating by 180° (a half-turn) is a rigid motion.

In Figure 19.7f we reflect the rectangle along a vertical mirror down the middle: Could you tell? The right and left halves exchange places.

Figure 19.7g shows the result of reflecting across a diagonal of the rectangle. All reflections and all rotations are rigid motions. So are all **translations**, which move every point in the plane a certain distance in the same direction.

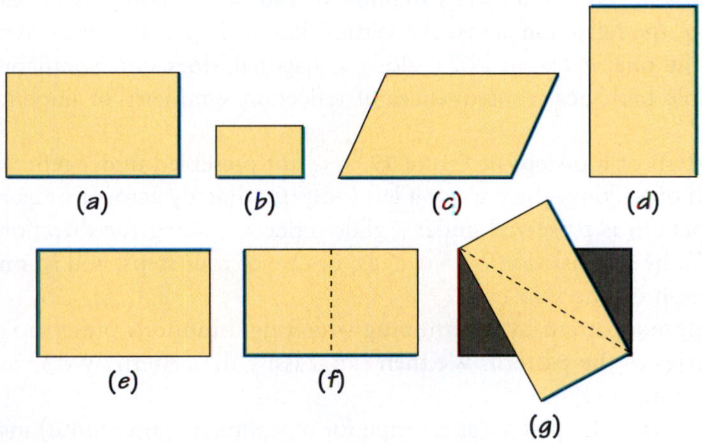


**FIGURE 19.7** Results of various motions applied to a blue-edged rectangle and its interior: (a) the original rectangle and interior; (b) 50% reduction (not a rigid motion); (c) shearing (not a rigid motion); (d) quarter-turn; (e) half-turn; (f) reflection along the vertical line down the middle; (g) reflection along a diagonal line.

The only remaining kind of rigid motion in the plane is a combination of reflection and translation. **Glide reflection** is the kind of pattern that your footprints make as you walk: Each successive element of the design (footprint) is a reflection of the previous one (Figure 19.8). The motion combines, in an integral way, translation ("glide") with a reflection across a line parallel to the direction of the translation.

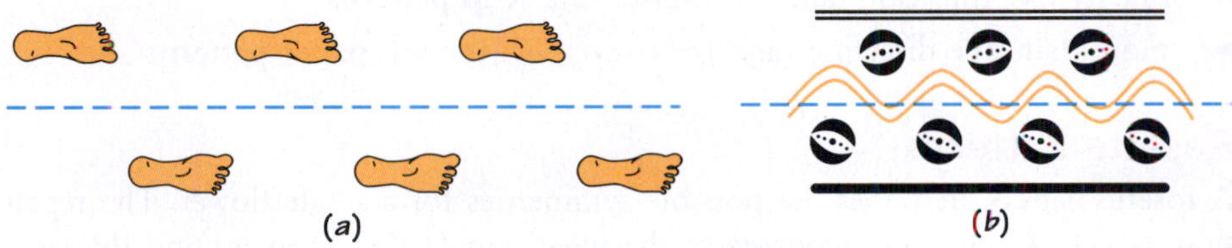


**FIGURE 19.8** Glide reflection of (a) footprints; (b) design elements on a pot from San Ildefonso Pueblo, in New Mexico.

Any rigid motion of the plane must be one of:

- Reflection (across a line)
- Rotation (around a point)
- Translation (in a particular direction)
- Glide reflection (across a line)

Performing one rigid motion after another results in a rigid motion that (surprisingly) must be one of the four types that we have just explored.

## Preserving the Pattern

In terms of symmetry, we are especially interested in rigid motions like those of Figures 19.7e and 19.7f that **preserve the pattern**–that is, ones for which the pattern looks exactly the same, *with all the parts appearing in the same relative places,* after the motion is applied.

You might enjoy thinking of applying these motions as "The Pattern Game": You turn your back, I apply a transformation, then you turn back and see if you can tell whether anything is changed.

The 90° rotation of Figure 19.7a into Figure 19.7d does not preserve the pattern. The moved rectangle doesn't fit exactly over the original rectangle. On the other hand, the 180° rotation in Figure 19.7e does preserve the pattern. It's true that the top of the original rectangle is now on the bottom of the transformed version, but you can't tell. A rotation by any multiple of 180° would also preserve the pattern.

Similarly, the reflection across the vertical line in Figure 19.7f preserves the pattern, while the one in Figure 19.7g, along a diagonal, does not. Spotlight 19.4 discusses possible biological consequences of reflection symmetry or imperfections in it.

The pattern of footsteps in Figure 19.8a is not preserved under reflection along the direction of walking–there is not a left footprint directly across from a right footprint. The pattern is preserved under a glide reflection along the direction of walking, as well as by a translation of two steps, or one of four steps, and so on–but not by a translation of one step.

We analyze a pattern by determining which rigid motions preserve it; they are the **symmetries of the pattern**. We then can classify the pattern by which rigid motions preserve it.

We may think of a pattern as a recipe for repeating a figure (motif) indefinitely. Of course, any pattern in nature or art has only finitely many copies of the figure. If the recipe for repetition is clear, we can imagine that we are looking at just a part of a pattern that extends indefinitely.

Patterns in the plane can be divided into those that have indefinitely many repetitions in

- no direction–the **rosette patterns**
- exactly one direction (and its reverse)–the **strip patterns**
- more than one direction (and their reverses)–the **wallpaper patterns**

## Rosette Patterns

A rosette pattern describes the possible symmetries for a single flower. The repetition aspect of asymmetry consists of the repetition of the petals around the stem. Translations and glide reflections do not come into play. The pattern is preserved under a rotation by certain angles corresponding to the number of petals. There may

## SPOTLIGHT 19.4 "Strive Then to Be Perfect"

Is there no such thing as objective and universal beauty, as claimed by American feminist Naomi Wolf in her book *The Beauty Myth*?

Stand in front of a mirror and look at yourself. Do your left and right sides look exactly symmetrical? What about the part in your hair, freckles on your face, evenness of your shoulders, bending of your ears?

Symmetry may be a proxy for fitness. Symmetrical racehorses tend to run faster; male lions with lopsided facial whisker-spot patterns die younger. The more symmetrical a flower is, the more nectar it produces, making it a better food source for pollinating insects; and correspondingly, insects prefer symmetrical flowers, giving such flowers a better chance of being pollinated.

Perhaps because of association with fitness, symmetry may affect mate selection among animals. Female zebra finches prefer males with symmetrical leg bands. Fruit flies and female barn swallows prefer males with symmetrical tails; a particular parasite can lead to an uneven tail.

What about people? Both male and female Britons, as well as Tanzanian hunter-gatherers, find facial symmetry more attractive than asymmetry. Perfectly symmetrical female faces that are computer-generated from composites of individual photos appear more attractive to men than photos of actual women's faces. Studies also indicate that "symmetrical" men tend to have an earlier first sexual experience, more sexual partners, and more extramarital affairs, while asymmetry of the hands is associated with low sperm count and poor sperm motility. Finally, women with symmetrical breasts tend to be more fertile and less susceptible to breast cancer.

Michalangelo's David

*(Roger Antrobus/Getty Images.)*

or may not be reflections that preserve the pattern, depending on whether the petal itself has reflection symmetry. Most flowers do (Figure 19.9a), but some do not. An everyday example of the rosette pattern—a human-made one—that does not have reflection symmetry is a pinwheel (Figure 19.9b). If there is no reflection symmetry, the motif of the pattern (the element that is repeated) is an entire petal. If there is reflection symmetry, the motif is just half a petal, because the entire pattern can be generated by rotation and reflection of a half petal. The fact that these are the only possibilities is sometimes called *Leonardo's theorem,* after Leonardo da Vinci, who, in the course of planning the design of churches, needed to decide if chapels and niches could be added without destroying the symmetry of the central design.

Leonardo realized that there are two different classes of rosettes, the ones without reflection symmetry (*cyclic rosettes*) and those with it (*dihedral rosettes*) (see Figure 19.9). The respective notations for the patterns are *cn* and *dn*, where *n* is the number of times that the rosette coincides with its original position in one complete turn around the center. A cyclic pattern has no lines of reflection symmetry, while the dihedral pattern *dn* has *n* different lines of reflection symmetry. The flower in Figure 19.9a has dihedral pattern *d34*, because each petal has reflection symmetry, while the pinwheel in Figure 19.9b has pattern *c8*.

**FIGURE 19.9** (a) Flower; each petal has reflection symmetry. *(Gregory G. Dimijian/Photo Researchers, Inc.)* (b) Pinwheel with eight "leaves," each asymmetric symmetrical, hence pattern *c8* (*Bloomimage/ Corbis.*)

## Strip Patterns

We illustrate the different kinds of strip patterns, and their "ingredient" symmetries, with patterns in the art of the Bakuba people of the Democratic Republic of the Congo, who are noted for their fascination with pattern and symmetry (see Spotlight 19.5).

All the strip patterns offer repetition and **translation symmetry** along the direction of the strip. For simplicity, we always position the pattern so that its repetition runs horizontally.

It may be that the pattern has no other rigid motions that preserve it apart from translation, as in Figure 19.10a.

The simplest other rigid motion to check is reflection across a line. For a strip pattern, the center line of the strip may be a reflection line, as in Figure 19.10b; we say that the pattern has symmetry across a horizontal line. There may instead be reflection across a *vertical* axis, such as the vertical lines through or between the V's in Figure 19.10c.

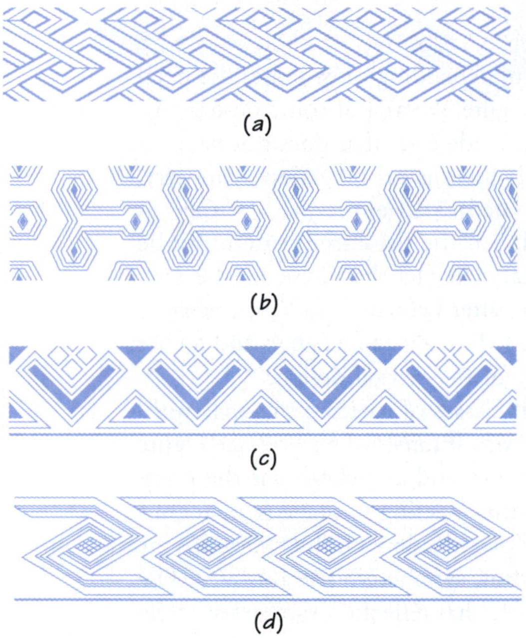


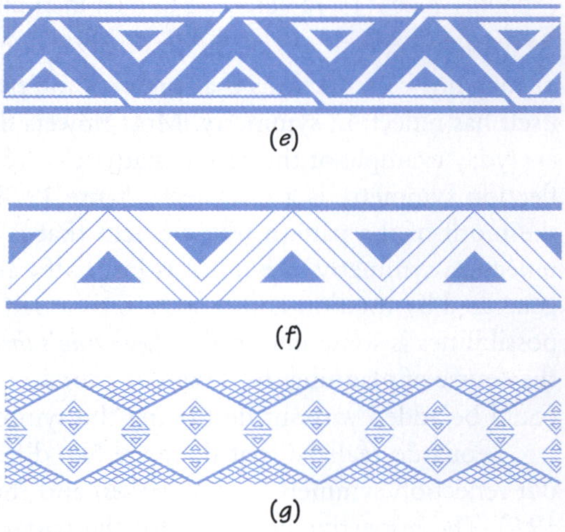


**FIGURE 19.10** Bakuba patterns. (a) Carved stool; (b) pile cloth; (c) pile cloth; (d) embroidered cloth; (e) embroidered cloth; (f) carved back of wooden mask; (g) carved box.

## SPOTLIGHT 19.5 Patterns Created by the Bakuba People

Among the Bakuba people of the Democratic Republic of the Congo (shaded area of map), it is considered an achievement to invent a new pattern, and every Bakuba king had to create a new pattern at the outset of his reign. The pattern was displayed on the king's drum throughout his reign and, for some kings, on his dynastic statue.

When missionaries first showed a motorcycle to a Bakuba king in the 1920s, he showed little interest in it. But the king was so enthralled by the novel pattern the tire tracks made in the sand that he had it copied and gave it his name.

*Source*: Adapted from Jan Vansina, *The Children of Woot*, University of Wisconsin Press, Madison, 1978, p. 221.

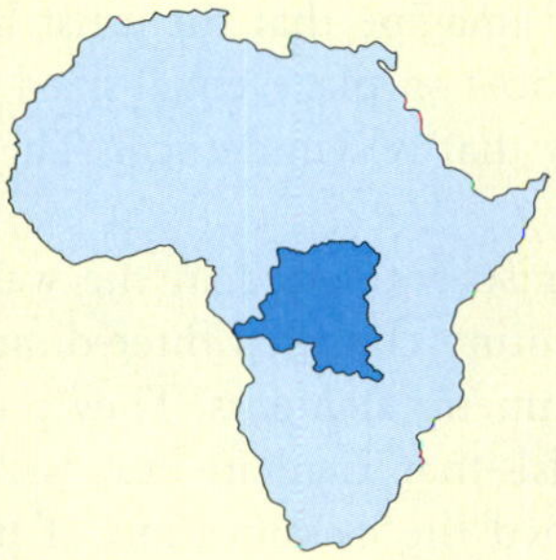

Two women with raffia cloths from the Bakuba village of Mbelo, July 1985. *Left*: Mpidi Muya with embroidered raffia (a kind of fiber) cloth. *Right*: Muema Kenye with plush and embroidered raffia cloth. *(Dorothy K. Washburn.)*

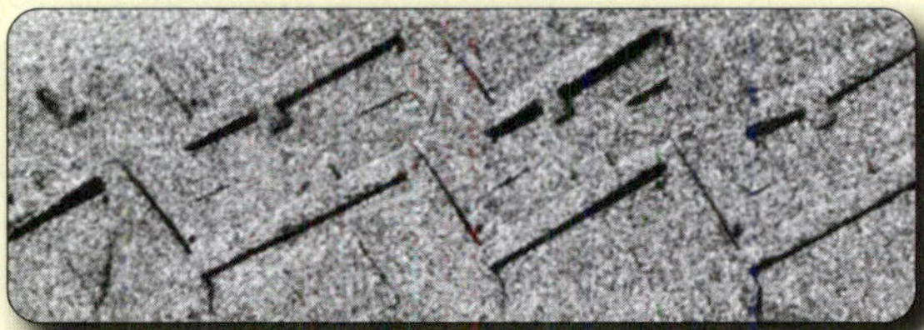

The pattern made by tire tracks fascinated the Bakuba people. *(Travis Amos.)*

What kind of rotational symmetry can a strip pattern have? The only possibility for a strip pattern is a rotation by 180° (a half-turn), because any other angle won't even bring the strip back into itself. (We don't count rotations of 360° or integer multiples [full turns], because any pattern is preserved under these.) Figure 19.10d shows a strip pattern that is unchanged by a 180° rotation about any point at the center of the small crosshatched regions.

What about glide reflections? A row of alternating p's and b's has glide reflection:

| | |
|---|---|
| Glide: | p p p p p p p p p |
| Reflection: | p p p p p p p p p<br>b b b b b b b b b |
| Glide reflection: | p b p b p b p b p |

For glide reflection, a p is translated as far as the next b and is then reflected upside down. Figure 19.10e shows a Bakuba pattern whose only symmetry (except for translation) is glide reflection.

Having examined symmetries on strip patterns, we can ask: What *combinations* of the four are possible? It turns out that apart from the five kinds of patterns we have already seen, there are only two other possibilities: We can have vertical line reflection, half-turns, and glide reflection, either with horizontal line reflection (Figure 19.10g) or without (Figure 19.10f).

Mathematical analysis reveals the following.

**There Are Only Seven Ways to Strip** RULE

There are only seven ways to repeat a pattern along a strip.

That this number is so small is quite surprising, because there are myriad different design elements (motifs). Two designs may look entirely different yet share the same pattern of reproducing their design elements.

## Wallpaper Patterns and Crystal Structures

So far we have classified the patterns that have no translation repetition (the rosette patterns) and those with repetition in one direction (the strip patterns). What about repetitions in more than one direction–say, in two different directions across a plane? It turns out that there are exactly 17 ways to do so, called *wallpaper patterns*. We give illustrations, notation, and a flowchart in Spotlight 19.6.

We emphasize again that "pattern" does not refer to the basic design but to how its repetition is structured across the plane. There is an infinite variety of possible designs that artists can devise. You should imagine that the artist has created one copy of the design and is contemplating how to place equal-sized copies of it in other parts of the (infinite) plane, in a way that is symmetrical. There are very few (17) strategies possible for doing so.

Crystallographers (physicists and chemists interested in the ways that crystals can occur or be built) in the nineteenth century classified three-dimensional crystal structures in terms of combinations of symmetry elements. They proved–after several years of coming up with different totals!–that there are exactly 230 patterns for crystals. Mathematicians have further refined the classification of patterns to take into account colors that are repeated in a symmetrical way.

### SPOTLIGHT 19.6 The 17 Wallpaper Patterns

There are exactly 17 wallpaper patterns. Here we give an example of each, together with a flowchart for identifying them. Crystallographers have standard notations and abbreviations for the patterns. The full notation consists of four symbols:

1. The first symbol is *c* (for "centered") if all rotation centers lie on the reflection lines, or *p* (for "primitive") otherwise.
2. The second symbol indicates rotational symmetry. It is either *1*, *2*, *3*, *4*, or *6*, corresponding to rotational symmetry of, respectively, 360°, 180°, 120°, 90°, or 60°. The symbol is the largest applicable number. For example, if symmetries of 360°, 120°, and 60° are present, the symbol is 6.
3. The third symbol is either *m*, *g*, or *1*, corresponding to the presence of "mirror," "glide," or no reflection symmetry.
4. The fourth symbol (*m*, *g*, or *1*) is for describing symmetry relative to an axis at an angle to the symmetry axis of the third symbol.

(*Note*: The patterns *p31m* and *p3m1* are exceptions to this scheme.)

Below each pattern illustration, we give both the standard abbreviation (on top) and the full notation (below).

(continued on page 623)

## The 17 Wallpaper Patterns *(continued)*

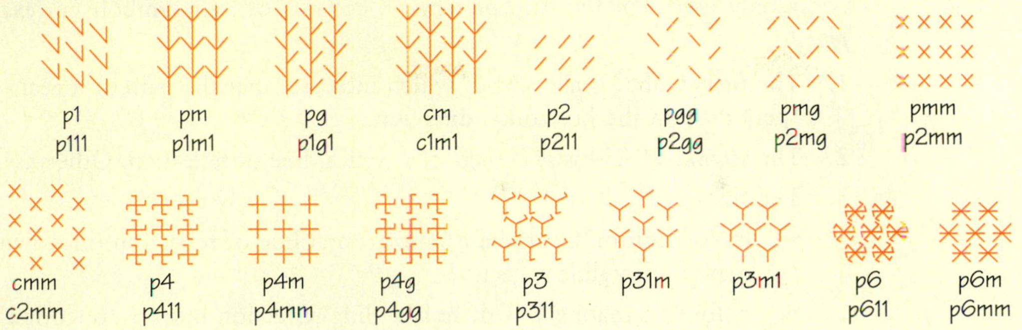


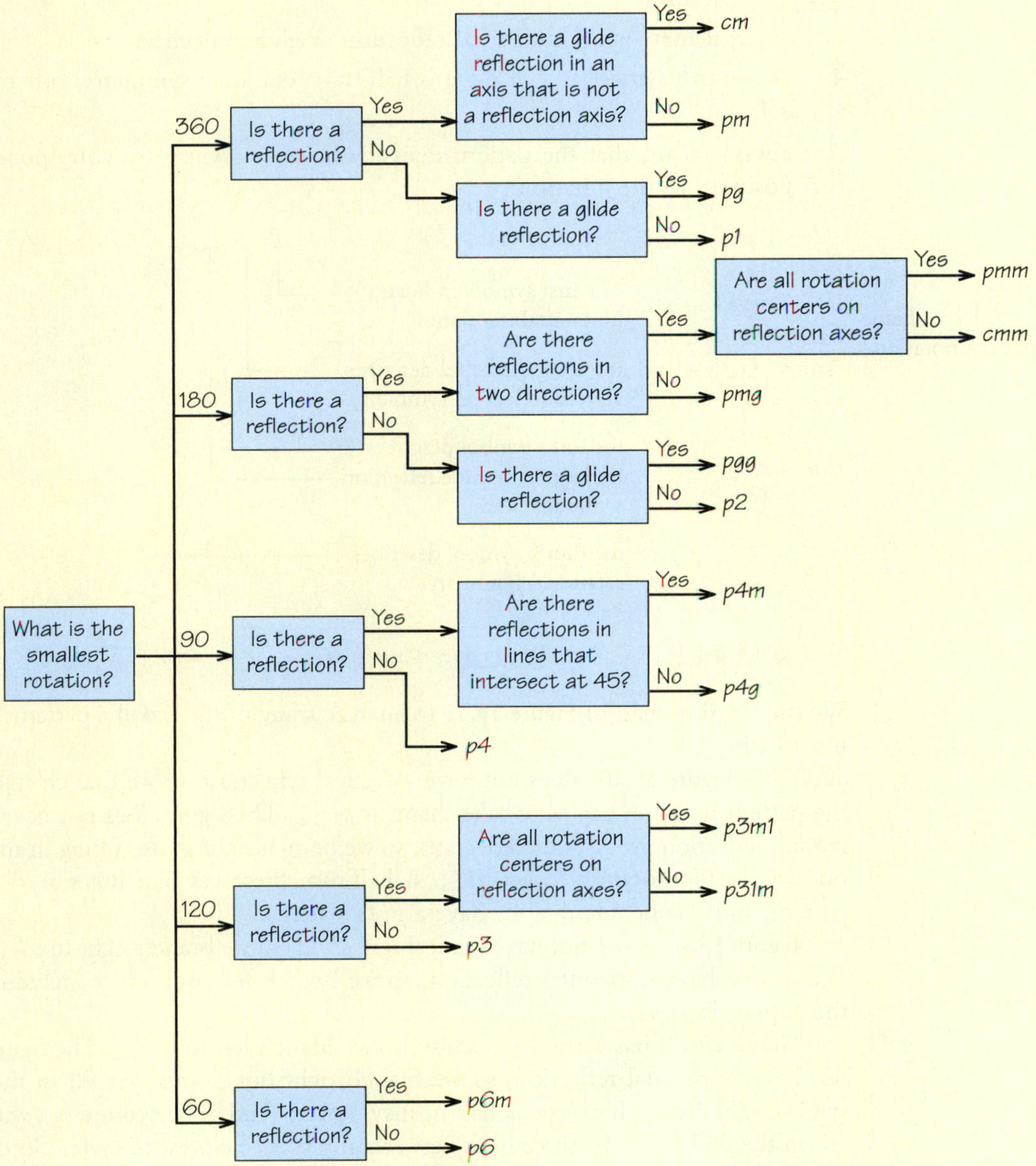

## 19.3 Notation for Patterns

It's useful to have a standard notation for patterns. **Crystallographic notation** is commonly used. For the strip patterns, it consists of four symbols (an example is *pma2*):

1. The first symbol is always a *p*, which indicates that the pattern repeats (is "periodic") in the horizontal direction.
2. The second symbol is *m* if there is a vertical line of reflection. Otherwise, it is *1*.
3. The third symbol is
   - *m* (for "mirror"), if there is a horizontal line of reflection (in which case there is also glide reflection)
   - *a* (for "alternating"), if there is a glide reflection but no horizontal reflection
   - *1*, if there is no horizontal reflection or glide reflection
4. The fourth symbol is *2*, if there is half-turn rotational symmetry; otherwise, it is *1*.

A *1* always means that the pattern does *not* have the symmetry corresponding to that position. In the notation:

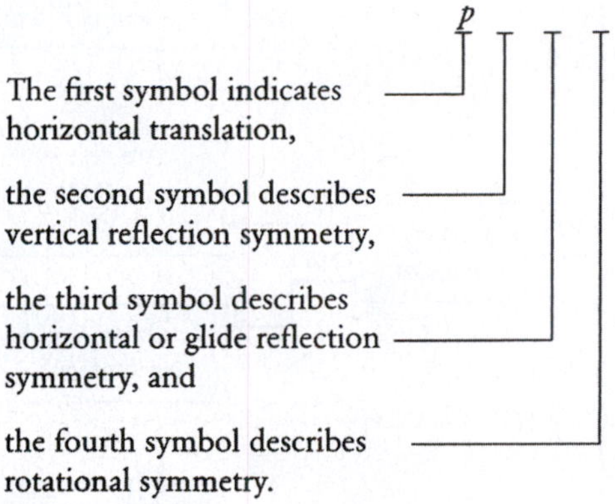

p
The first symbol indicates horizontal translation,
the second symbol describes vertical reflection symmetry,
the third symbol describes horizontal or glide reflection symmetry, and
the fourth symbol describes rotational symmetry.


**FIGURE 19.11** Scheme for strip pattern notation.

### EXAMPLE 1 ■ Bakuba Patterns

We use the flowchart of Figure 19.11 to analyze some of the Bakuba patterns of Figure 19.10.

**SOLUTIONS** Figure 19.10a does not have a vertical reflection, so we branch right, and the pattern notation begins to take shape as *p1_ _*. The figure does not have a horizontal reflection, nor a glide reflection, so we branch right again, filling in the third position in the notation to get *p11_*. A half-turn preserves part but not all of the pattern, so we conclude that we have a *p111* pattern.

Figure 19.10b does not have vertical reflection, so we branch right to *p1_ _*. The figure does have horizontal reflection, so we branch left and left, concluding that the pattern is *p1m1*.

Figure 19.10f has vertical reflection, so we branch left to *pm_ _*. The figure does not have horizontal reflection, so we branch right but cannot yet fill in the third symbol. The figure does have a half-turn symmetry (and glide symmetry), with center on the middle of the three lines between any pair of closest triangles. So the pattern is *pma2*.

Remember the Bakuba people's fascination with tire treadmarks? Apart from esthetic value, certain symmetries are important for practical purposes. The Museum of Transport in Glasgow, Scotland, includes all kinds of vehicle tires. However, only five of the seven strip patterns appear among treads of all the tires there. Examining Figure 19.12, can you guess which two patterns do not appear, what they have in common, and the practical reason why they are not used in tire treads?

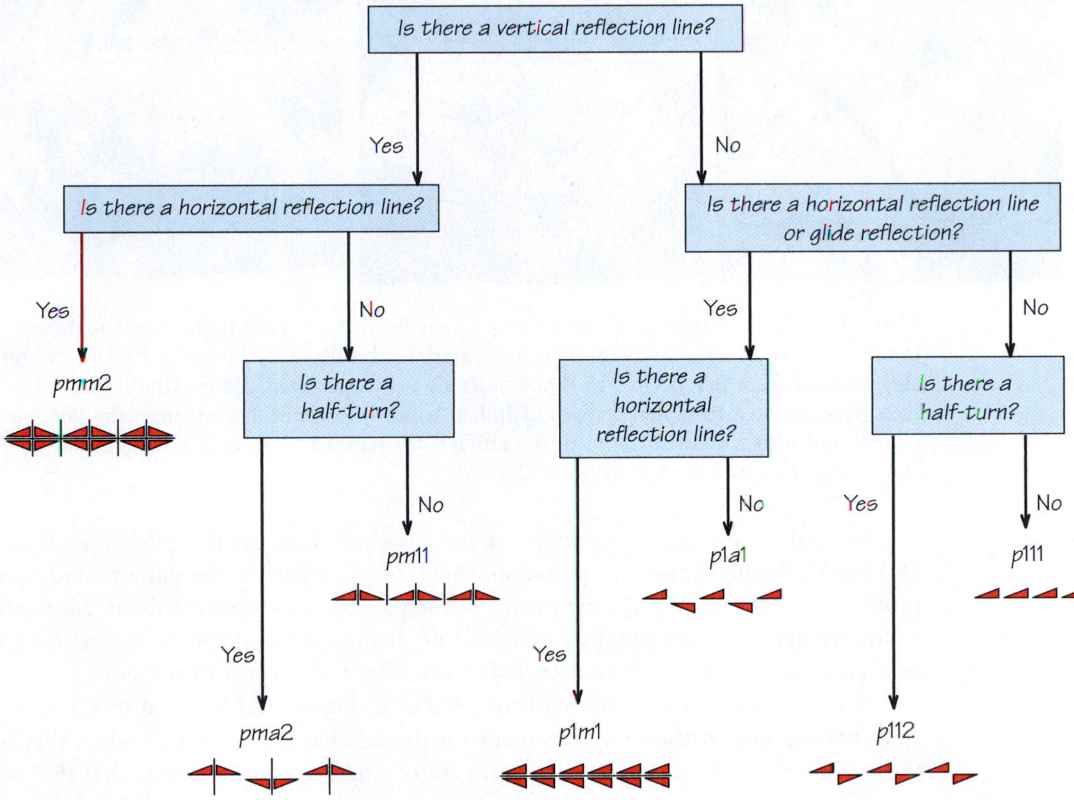


**FIGURE 19.12** A flowchart for identifying the seven strip patterns and classifying them according to crystallographic notation.

## Imperfect Patterns

In applying these classification schemes to patterns on real objects, we need to take into account that the pattern itself may not be perfectly rendered. Also, patterns that are not on flat surfaces—for example, the pattern around the rim of a bowl or around the body of a jar—require some latitude in interpretation.

## EXAMPLE 2 ▪ Patterns on Pueblo Pottery

The pitchers in Figure 19.13 are from a thousand-year-old Pueblo site at Starkweather Ruin near Reserve, New Mexico. Consider the patterns on the bodies of the pitchers, which continue on the back sides. Let's suppose that they could be unwrapped and continued as strip patterns, but we'll disregard the patterns on the spouts and handles.

We immediately come up against the question of the perfectness of the patterns. In Figure 19.13a the "teeth" (represented by the zigzagging of lightning bolts) in the left design element on the main body are "sharper" than those on the right. Is this lack of pattern, or just lack of perfection in executing one? For our analysis, we opt for the latter.

**FIGURE 19.13** Reserve black-on-white pitchers from the Pueblo II "horizon" (culture) (A.D. 900–1100), excavated 1935–1936 from Starkweather Ruin by Professor Paul H. Nesbitt and students from Beloit College. These pots are no longer available for study. In 2005 they were returned to representatives of Indian tribes who claim descent from the Pueblo culture and who requested return of the artifacts for reburial. (*Courtesy of Logan Museum of Anthropology, Beloit College, photos by Paul J. Campbell.*)

Similarly, what are we to make of the diagonal lines on the pitcher in Figure 19.13b? In the narrowest interpretation, these lines are part of the pattern, and rigid motion that is to qualify as a symmetry of the pattern must preserve them. More liberally, we may consider the lines as a kind of shading, a way to make the region appear gray. Indeed, to an observer at a distance, that is the effect of the lines.

For the pattern on the body of the pitcher in Figure 19.13c, we notice that the jagged white line in the design element on the left has three "steps," while that in the one on the right has four. If we were really strict, we would decide that the two are different design elements. But we do detect a similarity of the two that we do not want to deny totally. We attribute the variations in the jagged lines to artistic license and for our purposes consider the two jagged lines to be the same.

**SOLUTIONS** We follow the flowchart in Figure 19.12 and get the following:

- Figure 19.13a: Is there a vertical reflection? *No.* Is there a horizontal reflection or glide reflection? *No.* Is there a half-turn? *No.* Hence the pattern is *p111.*
- Figure 19.13b (narrow interpretation of the diagonal lines): Is there a vertical reflection? *No.* Is there a horizontal reflection or glide reflection? *No.* Is there a half-turn? *Yes* (around the center of each cross). The pattern is *p112.*
- Figure 19.13b (liberal interpretation–diagonal lines as shading, their direction doesn't have to be preserved): Is there a vertical reflection? *Yes* (on a vertical line through the center of a cross). Is there a horizontal reflection? *Yes* (through the center of a cross). The pattern is *pmm2.*
- Figure 19.13c: Is there a vertical reflection? *No.* Is there a horizontal reflection or glide reflection? *No.* Is there a half-turn? *Yes* (around the center of each jagged white line). The pattern is *p112.* (This pitcher has the interesting feature that the patterns on the neck and the body are mirror images of each other.) ■

Women made the pots at Starkweather. They strongly preferred the symmetry of half-turns. Very few of the pots have any reflection symmetry, either reflection or glide. The avoidance of reflection symmetry was a consistent feature of pottery of the indigenous peoples of the Western Hemisphere.

## 19.4 Symmetry Groups

We mentioned earlier that the key mathematical idea about detecting and analyzing symmetry is to look not at the motifs of a pattern but at its symmetries, the transformations that preserve the pattern.

The symmetries of a pattern have some notable properties:

- If we combine two symmetries by applying first one and then the other, we get another symmetry.
- There is an identity, or "null," symmetry that doesn't move anything, but leaves every point of the pattern exactly where it is.
- Each symmetry has an inverse, or "opposite," that undoes it and also preserves the pattern. A rotation is undone by an equal rotation in the opposite direction, a reflection is its own inverse, and a translation or glide reflection is undone by another of the same distance in the opposite direction.
- In applying a number of symmetries one after the other, we may combine consecutive ones without affecting the result ("associativity"). For example, if we have symmetries $A$ followed by $B$ followed by $C$, we can either: first combine $A$ with $B$, apply that symmetry, and then apply $C$; or first apply $A$ and then follow that by applying the combination of $B$ with $C$. That is, we can "associate" adjacent symmetries, but we must observe the overall order ($A$, $B$, $C$) in which they occur.

### EXAMPLE 3 ■ The Symmetries of a Rectangle

Consider the rectangle of Figure 19.14. Its symmetries, the rigid motions that bring it back to coincide with itself (even as they interchange the labeled corners), are as follows:

- The identity symmetry $I$, which leaves every point where it is
- A 180° (half-turn) rotation $R$ around its center
- A reflection $V$ in the vertical line through its center
- A reflection $H$ in the horizontal line through its center

You should convince yourself that the symmetries fulfill the four properties above.

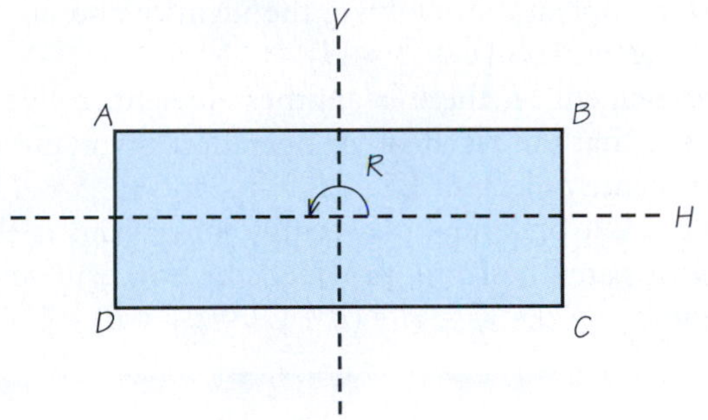


**FIGURE 19.14** A rectangle, with reflection symmetries and 180° rotation symmetry marked.

**SOLUTION**

- Combining any pair by applying first one and then the other is equivalent to one of the others. It's handy to have a notation for this combining; if we apply $V$ first and then $H$, we will write the result as $V \circ H$. You can check that the result is the same as applying $R$, that is, $V \circ H = R$. Check this by following where the corner $A$ goes to under the symmetries. Practice combining symmetries by making yourself a "multiplication table" of them.
- The element $I$ is an identity element.
- Each element is its own inverse.
- Try some examples to verify that associativity holds. For instance, check that $R \circ H \circ V = (R \circ H) \circ V = R \circ (H \circ V)$. In other words, applying $R$ then $H$ then $V$, we get the same result if we combine the first two and then apply the third, or if we apply the first one and then apply the combination of the second two.

The four properties of symmetries of an object are common to many kinds of mathematical objects. The properties characterize what mathematicians call a *group*. Various familiar collections of numbers, together with operations on them, form groups.

## EXAMPLE 4 ■ A Group of Numbers

The positive real numbers form a group under multiplication:

**SOLUTION**

- Multiplying two positive real numbers yields another positive real number.
- The positive real number 1 is an identity element.
- Any positive real number $x$ has an inverse $1/x$ in the collection.
- In multiplying several numbers together, it doesn't matter if we first multiply together some adjacent pairs of numbers; that is, it doesn't matter how we group or parenthesize the multiplication. For instance, $2 \times 3 \times 4 \times 5$ is equal to $2 \times (3 \times 4) \times 5 = 2 \times 12 \times 5$ and also to $(2 \times 3) \times 4 \times 5 = 6 \times 4 \times 5$.

### Group — DEFINITION

A **group** is a collection of elements $\{A, B, \ldots\}$ and an operation $\circ$ between pairs of them such that the following properties hold:

*Closure:* The result of one element operating on another is itself an element of the collection ($A \circ B$ is in the collection).

*Identity element:* There is a special element $I$, called the identity element, such that the result of an operation involving the identity and any element is that same element ($I \circ A = A$ and $A \circ I = A$).

*Inverses:* For any element $A$, there is another element, called its inverse and denoted $A^{-1}$, such that the result of an operation involving an element and its inverse is the identity element ($A \circ A^{-1} = I$ and $A^{-1} \circ A = I$).

*Associativity:* The result of several consecutive operations is the same regardless of grouping or parenthesizing, provided the consecutive order of operations is maintained: $A \circ B \circ C = A \circ (B \circ C) = (A \circ B) \circ C$.

## EXAMPLE 5 ■ A Group of Non-Numbers

With all your experience with arithmetic, numbers are concrete to you, even if thinking of them in terms of a group is not. Here we look at a very simple "abstract" group. The group is a collection of just three elements {A, B, C}, and it is convenient to show how the operation • behaves by giving a table of its results:

| • | A | B | C |
|---|---|---|---|
| A | A | B | C |
| B | B | C | A |
| C | C | A | B |

The table is organized so that, for example, we find the result of A • B by looking in the row for A and the column for B, finding B. So A • B = B. Similarly, as you should check, C • B = A. We confirm that indeed this set is a group under the operation.

**SOLUTION** Since all of the entries in the table are from {A, B, C}, the set is closed under the operation. You should identify which element serves as an identity element. What is the inverse to A? to B? to C? To check associativity would require checking the results of all possible products X • Y • Z, where each of X, Y, and Z can be any of A, B, or C. We won't go to that (tedious) length, but can you see why there are $3^3 = 27$ products to check?

This particular abstract group can be interpreted concretely in several ways. One interpretation is in terms of an equilateral triangle.

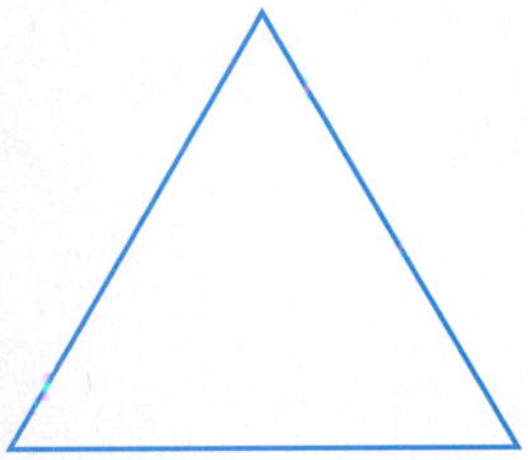

Equilateral triangle.

Each of A, B, and C is a rotation of the triangle about its center. A is a rotation by 0°, B is a rotation clockwise by 120° (one-third of a complete turn), and C is a rotation counterclockwise by 120°. The operation • just gives the result of doing one rotation followed by another. For example, B • C is first to rotate the triangle 120° clockwise, then rotate it 120° counterclockwise–which leaves it as if it had not rotated at all, that is, as rotated by 0°. Hence B • C = A.

We have in fact explored here some of the symmetries of an equilateral triangle. (What other symmetries does an equilateral triangle have? Think of it as a rosette pattern.)

### Symmetry Group of a Pattern — DEFINITION

The symmetries that preserve a pattern form the **symmetry group of the pattern.**

## EXAMPLE 6 ■ Symmetry Groups of Strip Patterns

Each of the strip patterns of Figure 19.10 has a different group of symmetries. What do they have in common, and how do they differ?

**SOLUTION** The pattern of Figure 19.10a is preserved only by translations. If we let $T$ denote the smallest translation to the right that preserves the pattern, then the pattern is also preserved by $T \circ T$ (which we write as $T^2$), by $T \circ T \circ T = T^3$, and so forth. Although the pattern looks the same after each of these translations by different distances, we can tell them apart if we number each copy of the motif and observe which other motif it is carried into under the symmetry. For instance, $T^2$ takes each motif into the motif two to the right. The symmetry $T$ has an inverse $T^{-1}$ among the symmetries of the pattern: the smallest translation to the *left* that preserves the pattern. Moreover, $T^{-1} \circ T^{-1}$ (which we write as $T^{-2}$), $T^{-1} \circ T^{-1} \circ T^{-1} = T^{-3}$, and so forth are also symmetries. The entire collection of symmetries of the pattern is

$$\{\ldots, T^{-3}, T^{-2}, T^{-1}, I, T, T^2, T^3, \ldots\}$$

From this listing, you see that it is natural to think of the identity $I$ as being $T^0$. All the strip patterns are preserved by translations, so the symmetry group of each includes the *subgroup* consisting of all translations in this list. We say that the group is **generated by** $T$, and we write the group as $\langle T \rangle$, where between the angle brackets we list symmetries (**generators**) that, in combination, produce all of the group elements.

The symmetry group of Figure 19.10e includes, in addition, a glide reflection $G$ and all combinations of the glide reflection with the translations. Doing two glide reflections is equivalent to doing a translation, which we express as $G^2 = T$. The glide is only "half as far" as the shortest translation that preserves the pattern. Check that $G \circ T = T \circ G$. The symmetry group of the pattern is

$$\{\ldots, G^{-3}, G^{-2} = T^{-1}, G^{-1}, I, G, G^2 = T, G^3, \ldots\} = \langle G \rangle$$

The pattern of Figure 19.10c is preserved by vertical reflections at regular intervals. If we let $V$ denote reflections at a fixed particular location, the other reflections can be obtained as combinations of $V$ and $T$. To get a handle on what each of the symmetries does, it helps to make a "simplified" copy of the strip (we use V's), number fixed positions on the page, and identify individual copies of the V's with letters, as in the following:

| 1 | 2 | 3 | 4 | 5 |
|---|---|---|---|---|
| $V_a$ | $V_b$ | $V_c$ | $V_d$ | $V_e$ |

The symmetries move the V's among the numbered positions. Let $V$ be the reflection across the vertical line through the middle of position 3, and let $T$ be the translation that moves each V one square to the right. To familiarize yourself with the symmetries, write out the result for each of $V$, $T$, and $VT$ ($V$ followed by $T$). (For convenience, we can omit the operation sign between the two symmetries.)

The symmetry group of the pattern, the list of all of the symmetries, is

$$\{\ldots, T^{-3}, T^{-2}, T^{-1}, I, T, T^2, T^3, \ldots; \quad \ldots, T^{-3}V, T^{-2}V, T^{-1}V, V, TV, T^2V, T^3V, \ldots\}$$

This group is notable because not all of its elements satisfy the *commutative property* that $A \circ B = B \circ A$, which you are used to for numerical operations ($a + b = b + a$; $a \times b = b \times a$). In fact, we do not have $VT = TV$, but instead $VT = T^{-1}V$. Verify this fact by working out the effect of $T^{-1}V$, using your simplified strip from above, and compare with what you got for $VT$ earlier.

We can express this group compactly as $\langle T, V | VT = T^{-1}V \rangle$, where we list the generators and indicate what relations hold among them.

We have made a transition from thinking about patterns in geometrical terms to reasoning about them in algebraic notation–in effect, applying one branch of mathematics to another. This kind of cross-fertilization is characteristic of contemporary mathematics.

The concept of a group is a fundamental one in the mathematical field of abstract algebra. The generality ("abstractness") is exactly why groups and other algebraic structures arise in so many applications, in areas ranging from crystallography, quantum physics, and cryptography, to error-correcting codes (see Chapters 16 and 17) and anthropology (describing kinship systems).

## 19.5 Fractal Patterns and Chaos

We noted earlier that similarity and repetition are key aspects of symmetry, as are balance and proportionality. In most of our examples, the repetitions of a motif have been at the same size. Exceptions were the chambers of the nautilus in Figure 19.2b and the varying sizes of leaves and seeds in plants that feature the spiral pattern of phyllotaxis (Figures 19.1 and 19.2a). These exhibit a kind of "proportion," or numeric symmetry–symmetry with changes of scale. Another example of similarity with changes of scale are the nested dolls ("matrioshka") shown in Figure 19.15. They feature a linear scaling factor (see section 18.1) between one doll and another. Each part of one doll (face, arm, and so forth) has the same proportion (scaling factor) to the corresponding part of a second doll.

**FIGURE 19.15** Nested "matrioshka" dolls from Russia exhibit symmetry at different scales. *(Photodisc Green/Getty Images.)*

### Fractals

Fractals are another example of symmetry in which linear scaling is used. The word **fractal** was invented in 1975 by Benoit Mandelbrot from the Latin word *fractus* meaning "broken into fragments" (of varied sizes), from which we get *fragment* and also *fracture* and *fraction*. Mandelbrot defined a fractal in strict mathematical terms that we formulate more informally as follows:

**Fractal** DEFINITION

A **fractal** is a pattern that exhibits similarity at ever finer scales.

The scaling is usually by a *linear scaling factor*.

We show various fractals in Figure 19.16. Figure 19.16a looks to us like an orchid with pronounced "bee guides" to the pollen. With its vertical mirror line, the overall pattern has *d1* rosette symmetry. However, the basic motif of the lacy wings is repeated at an infinite number of scales. In Figure 19.16b, the "suckers" on the "tentacles" appear in smaller and smaller sizes as the "tentacles" wind their way toward the point at the center. The pattern in Figure 19.16c has overall rosette symmetry of type *c2*, but the "seahorse" motif, with two large "seahorses" foot-to-foot in the center, is repeated in diminishing sizes throughout. Figure 19.16d features (to our imagination) "spikey snowmen," with smaller ones growing out of the sides of larger ones. What do they look like to you? And does the overall pattern as a rosette have symmetry *c1*, *c2*, *d1*, or *d2*?

A famous example of a fractal pattern is Maurits Escher's print "Circle Limit IV" (Figure 19.17). As you examine the angels and devils closer and closer to the boundary of the circle, you notice that they are not necessarily geometrically similar to the ones at the center. However, if you imagine that the print is the image of a hemisphere, then figures farther away from your viewpoint should indeed appear smaller.

(a) "Paradise." (b) "Purgatory." (c) "r-crest." (d) "Scarab 2."

**FIGURE 19.16** Various fractals. (*Courtesy of Noel Giffin and the Spanky Fractal Database.*)

**FIGURE 19.17** M. C. Escher's fractal pattern "Circle Limit IV." (*© 2005 M. C. Escher Company–Holland. All rights reserved. www.mcescher.com.*)

Apart from their beauty and the opportunity that they offer as an art form (there is even "fractal music"!), fractals have two major applications:

- Fractals with very simple rules for replicating the motif mimic very well certain natural phenomena, such as the structure of a leaf, a tree, or a mountain (see Figure 19.18). This fact not only allows us to model leaves, trees, and so on, using fractals but also suggests that such natural phenomena are produced by corresponding simple "rules of nature." Moreover, computer special effects in films can use fractals to mimic nature very closely, as in *Star Trek II: The Wrath of Khan,* for landscapes on the Genesis planet, and in *Return of the Jedi,* for the moons of Endor and the Death Star. In Chapter 23, we investigate in detail one particularly simple replication rule, called an **iterated function system (IFS)**; this complicated-sounding name hides the fact that such a system is just a recursive rule, like the one for forming the Fibonacci numbers. Figure 19.19 shows a simple geometric IFS.

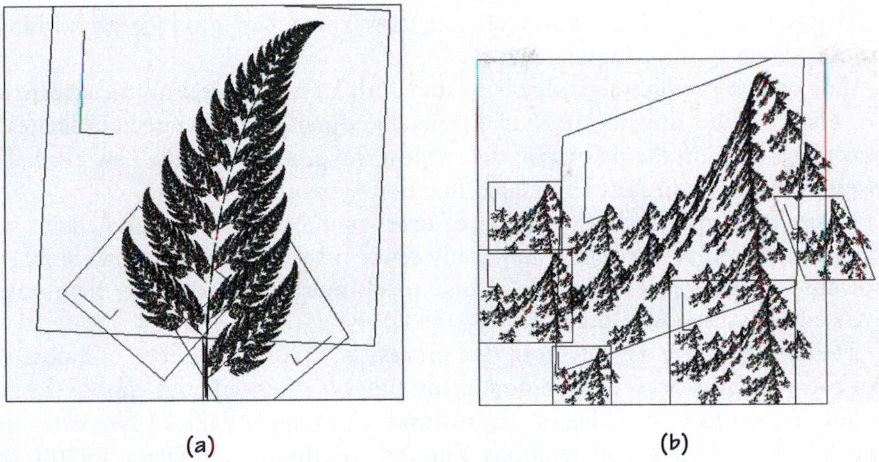


**FIGURE 19.18** (a) Barnsley's fractal fern and (b) a snowy mountain landscape, both with templates showing how they were formed using reflections and linear "distortions" in addition to linear scaling. Each leaf of the fern and each mountain peak is in fact just a smaller copy of the entire image. (*Fred Solomon, Warren Wilson College.*)

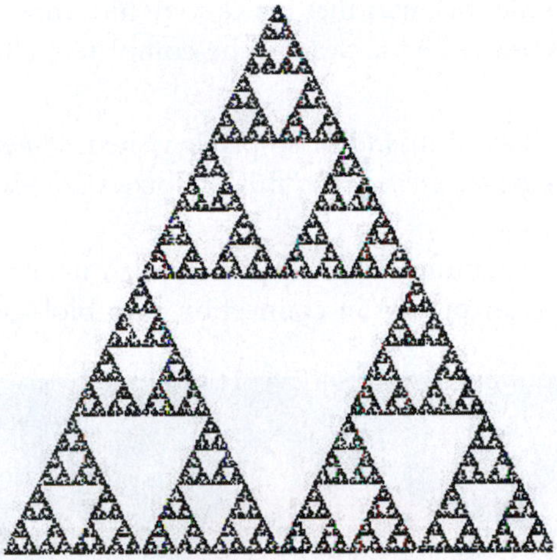

**FIGURE 19.19** A Sierpiński "triangle." Start with the big triangle, and remove its middle triangle. Then do the same for the three remaining smaller triangles. Recursively do the same for each subsequent smaller triangle. Can you guess the area of the resulting figure? (*Hint:* It may be less than you think.) (*Annalisa Crannell, Franklin and Marshall College.*)

- Fractals form the basis for an important method of image compression, simliar in efficiency to the better-known JPEG algorithm. The key idea is to store not the millions of bits that make up an image but instead a much smaller number (maybe thousands) of rules for generating patterns that can be found in the image. A simple example (which doesn't use fractals) is that you can compress a checkerboard of a million pixels that alternate between black and white to just two simple rules: If the current pixel is white, the next one is black, and if the current one is black, the next one is white. A more realistic example is Microsoft's original 1992 Encarta Encyclopedia. It contained thousands of articles and photographs, plus color animations and hundreds of maps–all on one CD-ROM, thanks to fractal data compression.

## Symmetry in Chaos

While the patterns in rosettes, strips, wallpaper patterns, and some fractals can be produced from very simple rules for the symmetries involved, you may be surprised to learn that symmetry can also arise from apparently random behavior.

We think of symmetry as referring to order, and chaos to disorder and randomness. Scientists use the word *chaos* in a technical sense to describe systems whose behavior over time is inherently unpredictable. We explore chaotic systems in Chapter 23 (section 23.5). Here we investigate how chaos can produce astonishingly beautiful designs on a computer screen.

One way to produce a graphic is to start with an initial pixel on the screen, apply a mathematical function (formula) to its coordinates to generate coordinates of a new pixel, light up the new pixel, then repeat the process with the new pixel. The process is recursive, in fact, an iterated function system.

Iterate the process a large number of times–millions or even hundreds of millions of times. Since the screen has many fewer pixels than that, by what mathematicians call the *pigeon-hole principle,* some pixels must be visited more than once–maybe even thousands of times.

The clue to producing art from this process is "color by number": Choose the color for each pixel according to how many times it is visited, and choose the colors with an eye to beauty. Figure 19.20 shows an example with *d5* symmetry that was produced by 30,000,000 iterations. The scale on the right in Figure 19.20 shows the colors for the number of times that pixels were hit; unhit pixels stay black. The order in which pixels are visited appears to be completely chaotic and is irrelevant to the final image.

- If you ignore the first thousand or so pixels visited, *it doesn't matter what pixel you start from–you get the same image!* But the pixels are visited in different orders.
- The formulas are variations on the *logistic map,* an iterated function system that we discuss in Chapter 23 in connection with biological populations.

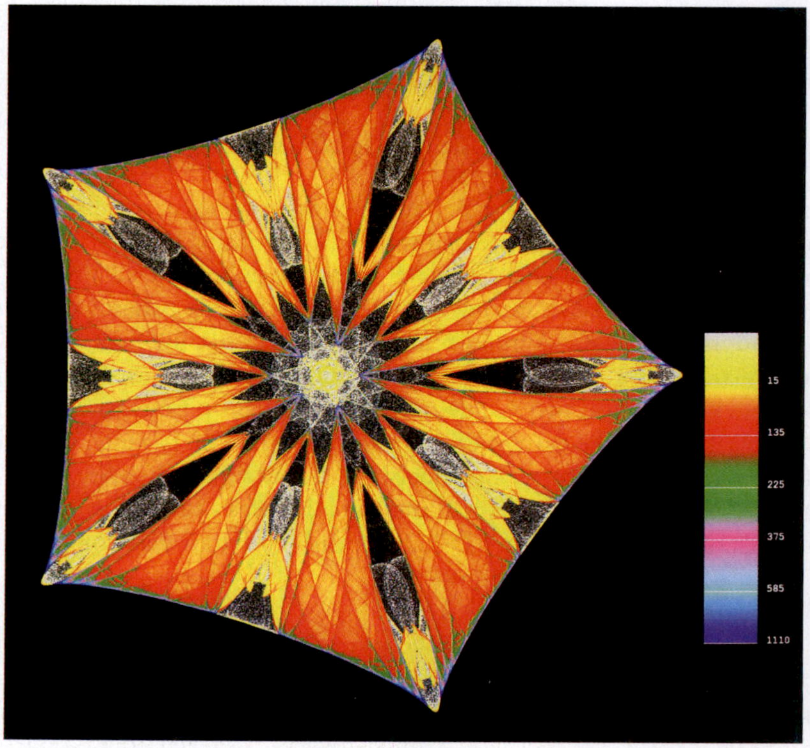


**FIGURE 19.20** "Emperor's Cloak," with *d5* symmetry. This work of art was produced by iterating a chaos-producing function, starting at one point and successively generating new points according to a fixed rule. The color bar shows the coloring of pixels according to how often they are visited by the iterations. (*Figure 1.13, p. 20, of Michael Field and Martin Golubitsky,* Symmetry in Chaos: A Search for Pattern in Mathematics, Art and Nature, *New York, Oxford University Press, 1992.*)

## SPOTLIGHT 19.7 The Father of Fractals

Benoit Mandelbrot

(Roger Ressmeyer/Corbis.)

Benoit Mandelbrot (1924–) was born in Poland, grew up partly in France, and came to the United States to work for IBM. He found that many phenomena feature both repeating patterns and power curves (see section 18.6). The patterns are repeated at a change of scale, as Barnsley's fern in Figure 19.18 and the Sierpiński triangle in Figure 19.19 show, and can be described by very simple rules. In 1975 he coined the term *fractals,* and the systems of rules became known as iterated function systems. Now retired from IBM, he teaches a course in Fractal Geometry at Yale University. The Web site for the course (see Suggested Web Sites) lists 100 or so examples of fractal phenomena in nature and society. Mandelbrot's book, *The (mis)Behavior of Markets: A Fractal View of Risk*, applies fractals to describe how stock market prices vary.

## REVIEW VOCABULARY

**Crystallographic notation** A four-symbol notation used by crystallographers (and mathematicians) to classify strip patterns and wallpaper patterns. (p. 624)

**Divine proportion** Another term for the golden ratio. (p. 612)

**Fibonacci numbers** The numbers in the sequence 1, 1, 2, 3, 5, 8, 13, 21, 34, . . . . Each number after the second is obtained by adding the two preceding numbers. (p. 610)

**Fractal** A pattern that exhibits similarity at ever-finer scales. (p. 631)

**Generated, generators** A group is generated by a particular set of elements (they are the generators) if composing them and their inverse in combinations can produce all elements of the group. (p. 630)

**Geometric mean** The geometric mean of two numbers $a$ and $b$ is $\sqrt{ab}$. (p. 613)

**Glide reflection** A combination of translation (= glide) and reflection in a line parallel to the translation direction. Example: pbpbpb. (p. 617)

**Golden ratio, golden mean** The number

$$\phi = \frac{1+\sqrt{5}}{2} = 1.618034\ldots \quad \text{(p. 612)}$$

**Golden rectangle** A rectangle the lengths of whose sides are in the golden ratio. (p. 612)

**Group** A group is a collection of elements with an operation on pairs of them such that the collection is closed under the operation, there is an identity for the operation, each element has an inverse, and the operation is associative. (p. 628)

**Isometry** Another word for *rigid motion.* Angles and distances, and consequently shape and size, remain unchanged by a rigid motion. For plane figures there are only four possible isometries: reflection, rotation, translation, and glide reflection. (p. 616)

**Iterated function system (IFS)** A sequence of elements (numbers or geometric objects) in which each successive element is determined recursively by applying the same function (rule) to the previous element. (p. 632)

**Phyllotaxis** The spiral pattern of shoots, leaves, or seeds around the stem of a plant. (p. 609)

**Preserves the pattern** A transformation preserves a pattern if all parts of the pattern look exactly the same after the transformation has been performed. (p. 618)

**Recursion** A method of defining a sequence of numbers, in which the next number is given in terms of previous ones. (p. 611)

**Reflection symmetry** Mirror-image symmetry. (p. 609)

**Rigid motion** A motion that preserves the size and shape of figures. In particular, any pair of points is the same distance apart after the motion as before. (Also called *isometry.*) (p. 616)

**Rosette pattern** A pattern whose only symmetries are rotations about a single point and reflections through that point. (p. 618)

**Rotational symmetry** A figure has rotational symmetry if a rotation about its "center" leaves it looking the same, like the letter S. (p. 609)

**Strip pattern** A pattern that has indefinitely many repetitions in one direction. (p. 618)

**Symmetry of the pattern** A transformation of a pattern is a symmetry of the pattern if it preserves the pattern. (p. 618)

**Symmetry group of the pattern** The group of symmetries that preserve the pattern. (p. 629)

**Translation** A rigid motion that moves everything a certain distance in one direction. (p. 617)

**Translation symmetry** An infinite figure has translation symmetry if it can be translated (slid, without turning) along itself without appearing to have changed. Example: AAA (p. 620)

**Wallpaper pattern** A pattern in the plane that has indefinitely many repetitions in more than one direction. (p. 618)

## SKILLS CHECK

**1.** Symmetry includes notions of

**(a)** balance
**(b)** similarity
**(c)** repetition
**(d)** all of the above.

**2.** Many people think that mathematics is just about numbers, but in fact mathematics is the study of _____ .

**3.** Which of the following rectangles is an approximate golden rectangle?

**(a)** 10 by 16
**(b)** 6 by 13
**(c)** 8 by 11

**4.** The geometric mean of 4 and 36 is ____________ .

**5.** Which artist claimed to use the golden ratio in his work?

**(a)** Leonardo da Vinci
**(b)** Wolfgang Amadeus Mozart
**(c)** Neither

**6.** In the Fibonacci sequence ______ follows 13 and 21.

**7.** A rigid motion always moves any pair of points

**(a)** in the same direction.
**(b)** to another pair of points the same distance apart.
**(c)** to their mirror images.

**8.** The capital letters ___, ___, ___, ___, ___, ___, and ___ each have a rotation isometry.

**9.** Assume that the following two patterns continue in both directions. Which of these patterns has a reflection isometry?

ZZZZZZZZZ

UUUUUUUUU

**(a)** ZZZZZZZZZ only
**(b)** UUUUUUUUU only
**(c)** Neither

**10.** This strip pattern

has ____________ and ____________ isometries.

**11.** What isometries does this wallpaper pattern have?

**(a)** Translation and reflection only
**(b)** Translation and rotation only
**(c)** Translation, rotation, and reflection

**12.** This wallpaper pattern

has ____________ and ____________ isometries.

**13.** If a horizontal strip pattern has a glide reflection isometry, then

**(a)** it always has a horizontal reflection isometry.
**(b)** it may also have a horizontal reflection isometry.
**(c)** it cannot have a horizontal reflection isometry.

**14.** If a strip pattern has both vertical and horizontal reflection isometries, then it always has a ____________ isometry.

**15.** Consider the strip pattern in the raffia cloth held by the woman in the photo on the right. What isometries does it have?

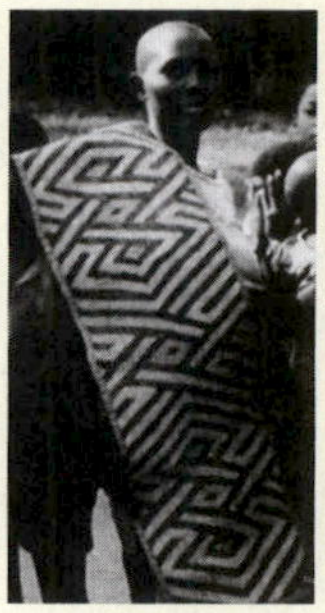

(Dorothy K. Washburn.)

**(a)** Vertical reflection
**(b)** Horizontal reflection
**(c)** Glide reflection

**16.** The symbol *p* indicates that a strip pattern has _____________ symmetry.

**17.** The symbol *2* indicates that a pattern has

**(a)** rotational symmetry.
**(b)** reflection symmetry.
**(c)** too much symmetry.

**18.** The symbol *m* indicates that a wallpaper pattern has ____________ symmetry.

**19.** The symmetry group of a rectangle has how many elements?

**(a)** 4
**(b)** 6
**(c)** 8

**20.** The symmetry group of the strip pattern *pmm2* has ________ elements.

# CHAPTER 19 EXERCISES

■ Challenge ◆ Discussion

## 19.1 Fibonacci Numbers and the Golden Ratio

**1.** Examine the "scales" on the surface of a pineapple, which are arranged in spirals around the fruit. Note that there are spirals in three distinct directions. For each direction, how many spirals are there?

**2.** Repeat Exercise 1, but for a pinecone from your area.

**3.** Repeat Exercise 1, but for a sunflower.

**4.** Here are two primitive models of natural increase of biological populations, similar to those Fibonacci hypothesized around the year 1200. A pair of newborn male and female rabbits is placed in an enclosure to breed.

**(a)** Suppose that the rabbits start to bear young one month after their own birth. This may be unrealistic for rabbits, but we could substitute another species for which it is realistic. Fibonacci used rabbits. At the end of each month, they have another male–female pair, which in turn mature and start to bear young one month later. Assuming that none of the rabbits die, how many pairs of rabbits will there be at the end of six months from the start (just before any births for that month)? (*Hint:* Draw a month-by-month chart of the situation at the end of the month, just before any births.)
**(b)** Repeat part (a), but assume instead that the rabbits start to bear young exactly two months after their own birth.

**5.** New houses are to be built along one side of a street ("Leonardo's Lane"), divided into equal-sized lots. Each house is either a single-family detached house, taking up one lot, or a duplex, taking up two lots. Suppose that there are $n$ lots on the street. How many different arrangements (orderings) of houses are there, for $n = 1$, 2, 3, 4, 5, and in general? (This exercise was inspired by a puzzle by Paul Dixon at the Suggested Web Site by Ron Knott.)

**6.** My wife works in a school district with about 900 faculty and staff. When it snows in the winter, the school district superintendent must decide by 5 A.M. whether to declare a "snow day" and cancel school. The faculty and staff are notified through a binary "telephone tree," in which the superintendent calls two people and each person who receives a call calls two others. Suppose that each call takes exactly one minute.

**(a)** Draw the telephone tree of calls for, and determine how many calls take place in, the first 1 minute, 2 minutes, 3 minutes, 4 minutes, and 5 minutes.
**(b)** How many calls does it take to notify all the faculty and staff? How long does that take?
(This exercise was inspired by a puzzle at the Suggested Web Site by Ron Knott.)

**7.** Here is a trick to "prove" that you can calculate faster than a person with a calculator. Turn your back and ask a friend to write down any two positive integers, then add them to get a third, then add the second and third to get a fourth, and so on, adding each time the last two integers until there are 10 numbers. Have your friend show you the list, whereupon you write down right away the total of all 10, while your friend begins to add them up on the calculator (to prove that you're right). The secret: The total is always 11 times the seventh number, and multiplying by 11 is pretty easy to do in your head–just add each pair of neighboring digits, carrying if necessary. Suppose that your friend writes down $m$ and $n$ as the first two numbers. Show that indeed the total of all 10 numbers is 11 times the seventh number. (Adapted from Martin Gardner, *Mathematical Circus,* Knopf, New York, 1979.)

**8.** The game of Fibonacci Nim begins with $n$ counters. Two players take turns removing at least one counter, but no more than twice as many as the opponent just did. The winner is the player who takes the last counter. One other rule: The first player may not win immediately by taking all the counters on the first turn! (Adapted from Martin Gardner, *Mathematical Circus,* Knopf, New York, 1979.)

**(a)** Play this game taking turns with an opponent and starting with different numbers $n$ of counters and try to come up with a strategy for one player or the other to win. (*Hint:* The key is that any positive integer can be represented uniquely as a sum of Fibonacci numbers.)
**(b)** Proceed as in part (a), but with the rule changes that the player who takes the last counter loses and the first player may not take all but one counter.

**9.** Put the golden ratio $\phi = (1 + \sqrt{5})/2$ into the memory of your calculator.

**(a)** Look at the value of $\phi$. Now square it (either use the $\boxed{x^2}$ button or multiply it by itself). What do you observe?
**(b)** Back to $\phi$. Now take its reciprocal (either use the $\boxed{1/x}$ button or divide it into 1). What do you observe?
**(c)** What formula explains what you saw in part (a)?
**(d)** What formula explains what you saw in part (b)?

**10.** The golden ratio satisfies the equation $x^2 = x + 1$. Show that $(1 - \phi)$ also satisfies the equation, so that $(1 - \phi) = (1 - \sqrt{5})/2$ is the other solution to $x^2 - x - 1 = 0$.

**11.** The geometric mean has interpretations in both arithmetic and geometry.

**(a)** Find the geometric mean of 3 and 27.
**(b)** Find the length of a side of a square that has the same area as a rectangle that is 4 by 64.

**12.** Here's further practice on arithmetic and geometric interpretations of the geometric mean.

**(a)** Find the geometric mean of 4 and 9.
**(b)** You are to make a golden rectangle with 6 inches of string. How wide should it be, and how high?

**13.** What is the geometric mean of 3, 6, and 12?

**14.** What is the geometric mean of 2, 4, 8, 16, and 32? (Such a sequence, in which each successive number is the same constant times the previous one, is called a *geometric sequence.*)

**15.** Another sequence closely related to the Fibonacci sequence is the *Lucas sequence,* which is formed using the same recursive rule but different starting numbers. The $n$th Lucas number $L_n$ is given by

$$L_1 = 1,\ L_2 = 3, \text{ and } L_{n+1} = L_n + L_{n-1} \text{ for } n \geq 2$$

**(a)** Calculate $L_3$ through $L_{10}$.
**(b)** Calculate the ratio of successive terms of the Lucas sequence:

$$\frac{L_2}{L_1}, \frac{L_3}{L_2}, \ldots, \frac{L_{10}}{L_9}$$

What do you notice?

**16.** For a sequence specified by a recursive rule, finding an explicit expression for the $n$th term is not easy, nor is the form necessarily simple. An exact expression for the $n$th term of the Fibonacci sequence is given by the Binet formula:

$$F_n = \frac{1}{\sqrt{5}}\left(\frac{1+\sqrt{5}}{2}\right)^n - \frac{1}{\sqrt{5}}\left(\frac{1-\sqrt{5}}{2}\right)^n$$

**(a)** Verify the formula for $n = 1$ and $n = 2$ by multiplying out, not by using a calculator.
**(b)** Use the Binet formula and your calculator to find $F_5$.
**(c)** In fact, the second term on the right of the equation gets closer and closer to 0 as $n$ gets large. Because we know that the Fibonacci numbers are integers, we can just round off the result of calculating the first term. Find $F_{13}$ by calculating the first term with your calculator and rounding.

**17.** For two positive numbers $x$ and $y$, show that the arithmetic mean $(x + y)/2$ is always greater than or equal to the geometric mean $x^{1/2}y^{1/2} = \sqrt{xy}$. Try some values for $x$ and $y$ and convince yourself, then demonstrate algebraically that it is true in general. When does equality hold? [*Hint:* Suppose that the claim is false, so that $(x + y)/2 < \sqrt{xy}$.) Square both sides of the inequality, bring all terms to one side, factor, and observe a contradiction.]

■ **18.** You may remember having to work problems like, "If Joe can dig a ditch in 3 days, and Sam can dig it in 4, how long will it take the two of them working together?" The answer is related to the *harmonic mean* of 3 and 4. The formula for the harmonic mean of two numbers $x$ and $y$ is

$$\frac{2}{1/x + 1/y}$$

**(a)** Calculate the answer for Joe and Sam, which is *one-half* of the harmonic mean of 3 and 4. Explain why this is the correct answer.
**(b)** Show that the harmonic mean of two positive numbers is always less than or equal to the geometric mean. (Thus, in light of Exercise 17, we have the general conclusion that $H \leq G \leq A$, where $H$ stands for the harmonic mean, $G$ for the geometric mean, and $A$ for the arithmetic mean.) (*Hint:* Suppose that the claim is false. Simplify the fraction that is the harmonic mean, square both sides of the inequality, and proceed as in Exercise 17.)
**(c)** Show once more that the harmonic mean of two positive numbers is always less than the geometric mean, but this time do it with less work: let $A = 1/x$ and $B = 1/y$, and discover one connection (equation) between the harmonic mean of $x$ and $y$ and the arithmetic mean of $A$ and $B$, and a second connection between the geometric mean of $x$ and $y$ and the geometric mean of $A$ and $B$. Then use Exercise 17 on $A$ and $B$.
**(d)** What should be the formula for the geometric mean of three numbers? Of $n$ numbers?
**(e)** Proceed as in part (d), but for the harmonic mean.

**19.** Shari Lynn Levine, a high school student, published an article in *The Fibonacci Quarterly* that investigated the "Beta-nacci" sequence that results if instead of bearing one pair of baby rabbits per month, mature rabbits bear two pairs every month, starting when they reach two months of age. Here we ask you to rediscover some of Shari's results.

**(a)** How many rabbits will there be each month for the first 12 months?

**(b)** What is the recursive rule for the *n*th Beta-nacci number $B_n$?
**(c)** For the terms of the sequence in part (a), calculate the ratios $B_{n+1}/B_n$ of successive terms. (*Motivating hint:* It's not the golden ratio this time.)
**(d)** Suppose that the ratio of successive terms approaches a number $x$. We show how to find $x$ exactly. For very large $n$, we have $B_{n+1} \approx xB_n \approx x^2B_{n-1}$. Substituting these values into the recursive rule for the sequence and dividing by $B_{n-1}$ gives the equation $x^2 = x + 2$. Solve this equation for $x$ (you can use the quadratic formula). Make a table of values of $3B_n$ versus $2^n$. From the evidence, can you suggest a formula for $B_n$?

**20.** Generalize Exercise 19, parts (a) through (d):

**(a)** to the case of each pair of rabbits having three pairs of rabbits (the "Gamma-nacci" sequence).
**(b)** to the case of each pair of rabbits having $q$ pairs of rabbits.

For Exercises 21 and 22, refer to the following.

We have seen that the golden ratio is a positive root of the quadratic polynomial $x^2 - x - 1$. We can generalize this polynomial to $x^2 - mx - 1$ for $m = 1, 2, 3, \ldots$ and consider the positive roots of those polynomials as generalized means–the "metallic means family," as they are sometimes known. In particular, for $m = 2, 3, 4$, and 5, we have respectively the silver, bronze, copper, and nickel means. It is surely surprising that these numbers arise both in connection with quasicrystals (investigated in Chapter 20) and in analyzing the behavior of some dynamical systems (a topic investigated in Chapter 23) as the systems evolve into chaotic behavior.

**21.** Use the quadratic formula to find expressions in terms of square roots for the silver, bronze, copper, and nickel means, and approximate these to three decimal places. Find a general expression in terms of a square root for the *m*th metallic mean.

**22.** Just as the golden mean arises as the limiting ratio of consecutive terms of the Fibonacci sequence, each of the metallic means arises as the limiting ratio of consecutive terms of generalized Fibonacci sequences. A generalized Fibonacci sequence $G$ can be defined by

$$G_1 = 1,\ G_2 = 1, \qquad \text{and} \qquad G_{n+1} = pG_n + qG_{n-1}$$

where $p$ and $q$ are positive integers. The Fibonacci sequence itself is the case $p = q = 1$.

**(a)** Try various small values of $p$ and $q$ and determine which mean they lead to.
**(b)** Divide the equation for $G_{n+1}$ by $G_n$. Assume that $G_{n+1}/G_n$ and $G_n/G_{n-1}$ both tend toward the same number $x$ as $n$ gets large, replace those quantities by $x$, and simplify the resulting equation. What must be the value of $x$?
**(c)** What happens to the sequence and to the mean if we allow one or both of $p$ and $q$ to be negative integers?

## 19.2 Rosette, Strip, and Wallpaper Patterns

**23.** Determine whether each of the following statements is always true or sometimes false. Drawing some sketches may be helpful.

**(a)** A line reflection preserves collinearity of points. That is, if the points $A$, $B$, and $C$ are in a straight line (collinear), then their images reflected in some other line also lie in a straight line.
**(b)** A line reflection preserves betweenness. That is, if the collinear points $A$, $B$, and $C$ (with $B$ between $A$ and $C$) are reflected about a line, then the image of $B$ is between the images of $A$ and $C$.
**(c)** The image of a line segment under a line reflection is a line segment of the same length.
**(d)** The image of an angle under a line reflection is an angle of the same measure.
**(e)** The image of a pair of parallel lines under a line reflection is a pair of parallel lines.

**24.** Determine whether each of the following statements is always true or sometimes false. Drawing some sketches may be helpful.

**(a)** The image of a pair of perpendicular lines under a line reflection is a pair of perpendicular lines.
**(b)** The image of a square under a line reflection is a square.
**(c)** Label the vertices of a square $A$, $B$, $C$, and $D$ in a clockwise direction. Then their images $A'$, $B'$, $C'$, and $D'$ under a line reflection also follow a clockwise direction.
**(d)** The length of the perimeter of a geometric figure is equal to the length of the perimeter of its image under a line reflection.
**(e)** The image of a vertical line under a line reflection is always a vertical line.

**25.** Which of the capital letters of the alphabet, when drawn in the most symmetrical way, has the following symmetries? For example, assume that the upper and lower loops of B are the same size.

**(a)** A horizontal line of reflection symmetry
**(b)** A vertical line of reflection symmetry
**(c)** A rotational symmetry

**26.** Repeat Exercise 25 for the lowercase letters.

**27.** In *The Complete Walker III* (3rd ed., Knopf, New York, 1984, p. 505), Colin Fletcher's answer to "What games should I take on a backpacking trip?" is the game he calls "Colinvert": "You strive to find words with meaningful mirror (or half-turn) images." Some of the words he found are

MOM WOW pod MUd bUM

**(a)** Which of his words reflect into themselves?
**(b)** Which of his words rotate into themselves?
**(c)** Find some more words or phrases of these various types–the longer, the better.

**28.** Repeat Exercise 27, but for words written vertically instead of horizontally.

**29.** For each of the following patterns, identify the rigid motions that preserve the pattern:

**(a)** CCCCCCCCCC
**(b)** GGGGGGGGGG
**(c)** HHHHHHHHHH
**(d)** MMMMMMMMMM

**30.** Repeat Exercise 29, but for

**(a)** SSSSSSSSSS
**(b)** bdbdbdbdbd
**(c)** dbpqdbpqdbpq

## 19.3 Notation for Patterns

**31.** What is the notation (such as *d4* or *c5*) for the symmetry pattern of a regular pentagon (which has all five sides equal)?

**32.** What is the notation for the symmetry pattern of a snowflake?

**33.** Give the notation (such as *d4* or *c5*) for the symmetry patterns of the rosettes in hubcaps (a) through (c) below, disregarding the logos in the centers. (Can you identify the make of car for each hubcap?)

**34.** Repeat Exercise 33 for hubcaps (d) through (f).

**35.** Repeat Exercise 33 for corporate logos (a) through (c) below. (Can you identify the corporations?)

*(a)* *(b)* *(c)*

**36.** Repeat Exercise 33 for automobile logos (d) through (f) below.

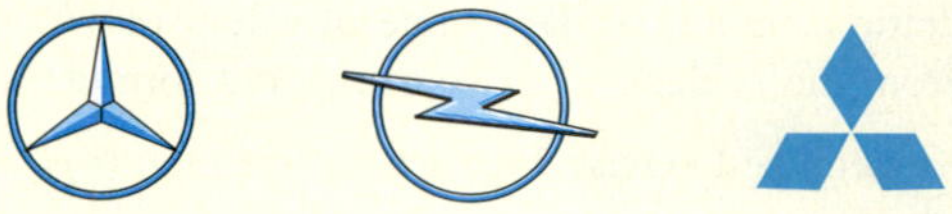

*(d)* *(e)* *(f)*

For Exercises 37–38, refer to the following.

Step patterns are found in Celtic illuminated manuscripts, metal work, and stone crosses. Square ones were constructed by first designing on a square lattice one quarter of the pattern (say, the top right), using horizontal and diagonal lines to produce a prototype such as the following:

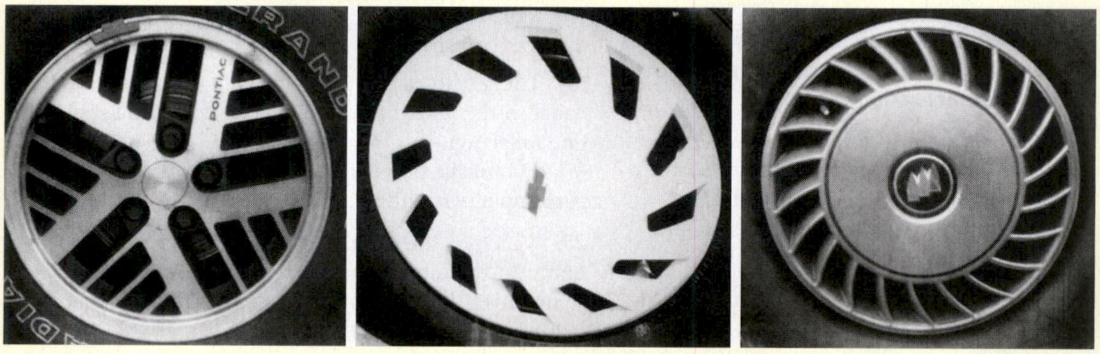


*(a)* *(b)* *(c)*

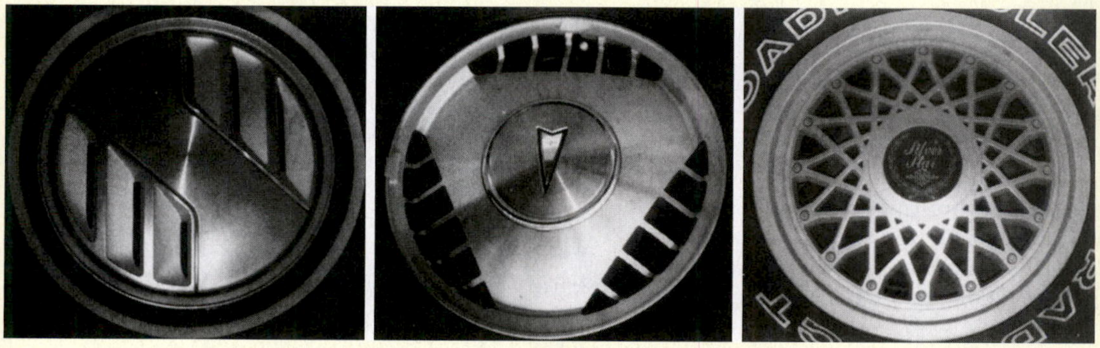

*(d)* *(e)* *(f)*

(*All hubcap photos courtesy of Joe Gallian, University of Minnesota, Duluth.*)

Then three copies were added, either by (1) rotating the original successively by 90° [as in accompanying illustration (a)], or else by (2) reflecting it across its right and bottom edges [as in illustration (b)]. (Based on research by Mark A. M. Lynch of Glasgow Caledonian University, Scotland.)

(a)

(b)

**37.** Identify the rosette pattern for:

**(a)** step pattern (a).
**(b)** step pattern (b).

**38.** Which rosette pattern would result if the prototype, unlike the one above, has reflection symmetry across its diagonal from top left to lower right and

**(a)** strategy (1) is used.
**(b)** strategy (2) is used.

**39.** Use the flowchart in Figure 19.11 to identify the notation for the types of strip patterns from the pottery and basketry, shown in the illustrations below.

**40.** In each of the four accompanying examples, two adjacent triangles of an infinite strip are shown. (Contributed by Margaret A. Owens, California State University, Chico.)

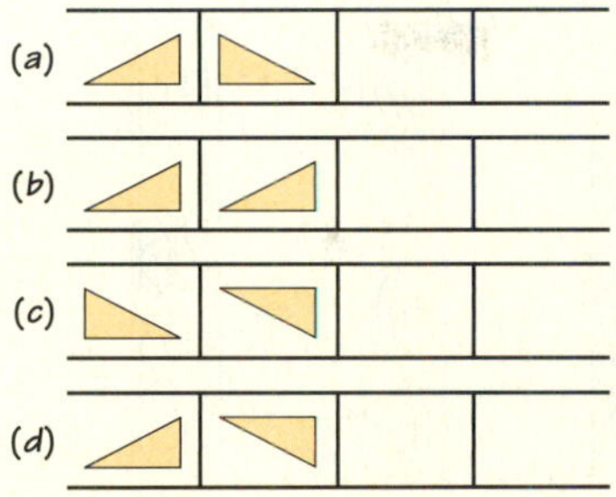


For each example:

**(a)** Determine a motion (translation, reflection, rotation, or glide reflection) that takes the first (= left) triangle to the second (= right) one.
**(b)** Draw the next four triangles of the infinite strip that would result if the second triangle is moved to the next space by another motion of the same kind, and so on.
**(c)** Identify (by notation) the resulting strip as one of the seven possible strip patterns.

(a)

(b)

(c)

(d)

(e)

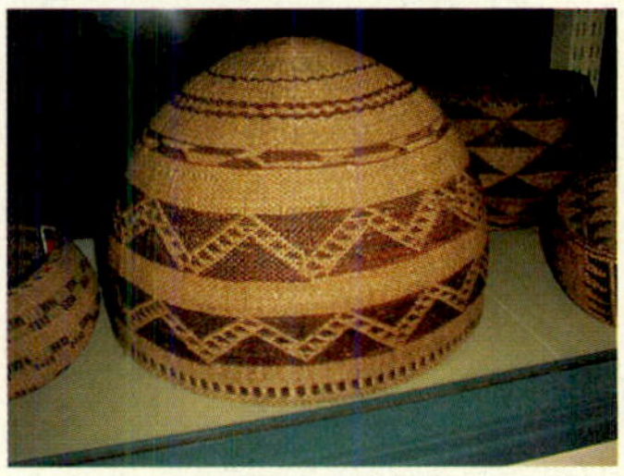
(f)

(g)

Pottery and basketry from the Americas. (a), (b) Mexico, modern. (c) Lower Central America, pre-Columbian. (d) Pomo people, California, early twentieth century. (e) Woodland Indians, central North America, early twentieth century. (f) Pomo people, California, mid-twentieth century; originally from the collection of Dr. Herbert Zim, editor of the "Golden Guides" series of nature books. (g) Woodland Indians, central North America, early twentieth century. (*Courtesy of Logan Museum of Anthropology, Beloit College, photos by Darrah Chavey.*)

**41.** Repeat Exercise 39 for the accompanying eight strip patterns, all of which appear on the brass straps for a single lamp from nineteenth-century Benin in West Africa. (From H. Ling Roth, *In Great Benin.*)

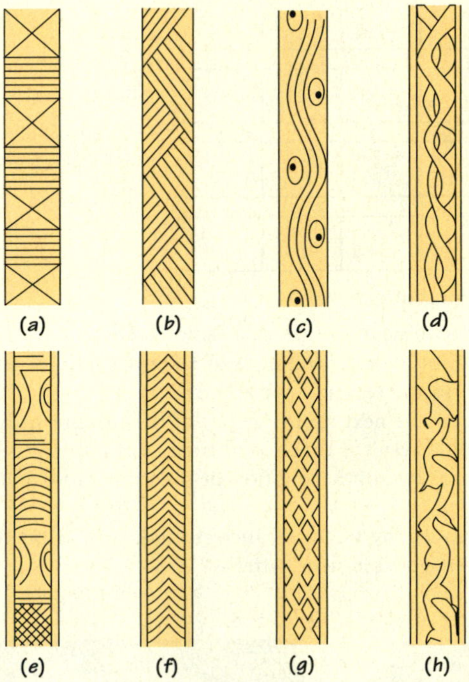


Note that the patterns are roughly carved, so you will need to discern the intent of the artist.

**42.** Repeat Exercise 39 for the accompanying patterns from San Ildefonso Pueblo, New Mexico.

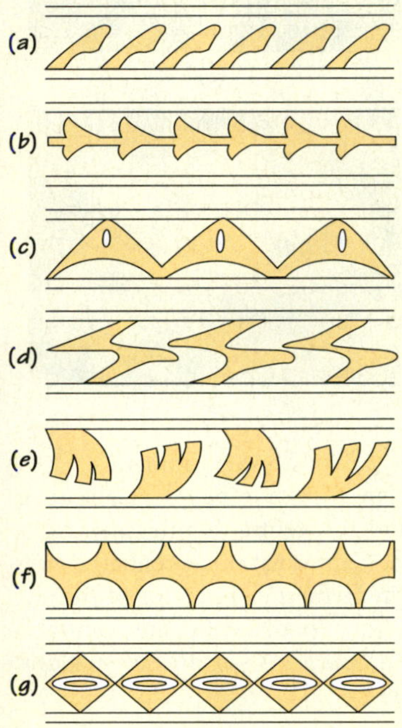


**43.** The following table shows comparative data about the frequency of occurrence of strip designs of various types on pottery (Mesa Verde, Colorado, United States) and smoking pipes (Begho, Ghana, Africa) from two continents.

**Frequency of Strip Designs on Mesa Verde Pottery and Begho Smoking Pipes**

| | Mesa Verde | |
|---|---|---|
| Strip Type | Number of Examples | Percentage of Total |
| *p111* | 7 | 4 |
| *p1m1* | 5 | 3 |
| *pm11* | 12 | 7 |
| *p112* | 93 | 53 |
| *p1a1* | 11 | 6 |
| *pma2* | 27 | 16 |
| *pmm2* | 19 | 11 |
| Total | 174 | |

| | Begho | |
|---|---|---|
| Strip Type | Number of Examples | Percentage of Total |
| *p111* | 4 | 2 |
| *p1m1* | 9 | 4 |
| *pm11* | 22 | 10 |
| *p112* | 19 | 8 |
| *p1a1* | 2 | 1 |
| *pma2* | 9 | 4 |
| *pmm2* | 165 | 72 |
| Total | 230 | |

**(a)** Which types of motions appear to be preferred for designs from each of the two localities?
**(b)** What other conclusions do you draw from the data of this table?
**(c)** On the evidence of the table alone, in which locality is each of the following strip patterns most likely to have been found?

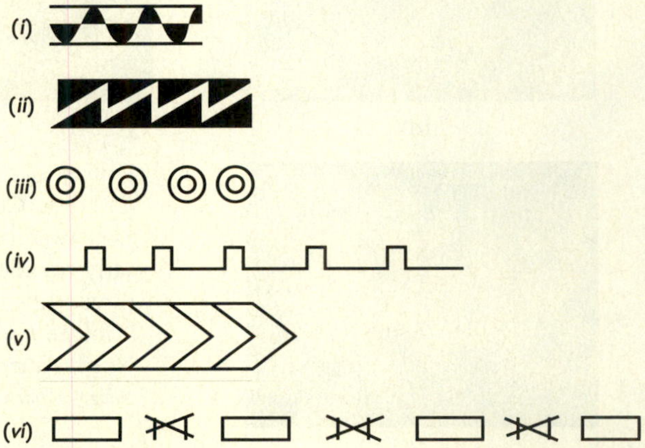

(vii) 

(viii) 

(ix) 

**44.** For the Nigerian Yoruba cloths (a) and (b) in the following illustration, use the flowchart in Spotlight 19.6 to identify (by notation) the type of wallpaper pattern.

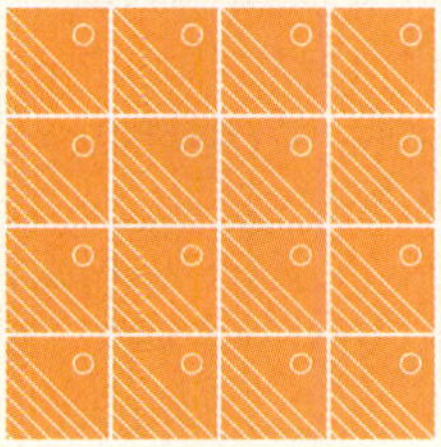

*(a)*

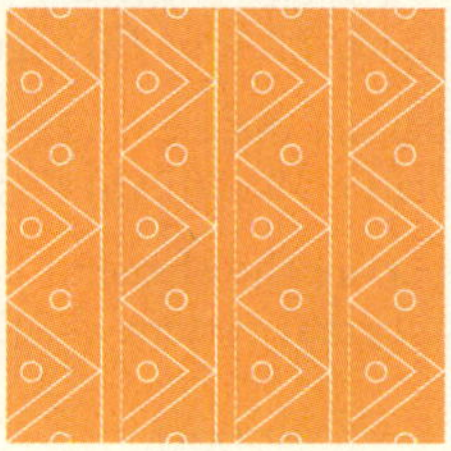

*(b)*

Patterns on Yoruba (West Africa) *adire* cloth, made by starching a pattern onto white cloth, then dyeing the cloth before rinsing out the starch, so that the starched portion remains as a white design against a colored background.

**45.** For the Nigerian Yoruba cloths (c) and (d) in the accompanying illustration, use the flowchart in Spotlight 19.6 to identify (by notation) the type of wallpaper pattern.

*(c)*

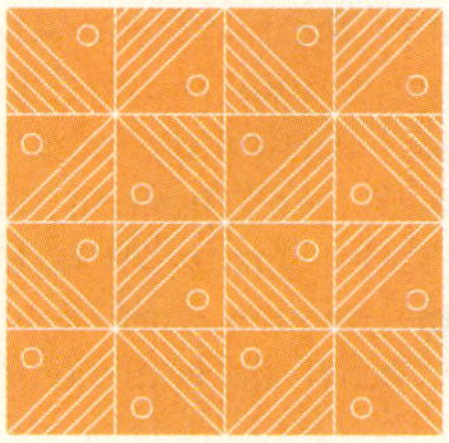

*(d)*

**46.** The triangles in the grid at the top of the following figure show beginning steps in forming instances of several of the wallpaper patterns by putting together a vertical motion and a horizontal motion.

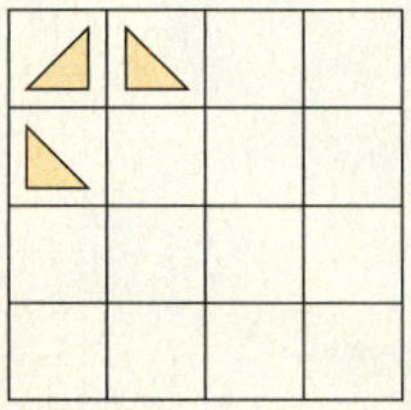

**(a)** Identify the horizontal motion.
**(b)** Identify the vertical motion.
**(c)** Fill in the remaining empty squares.
**(d)** Identify the wallpaper pattern.

■ **47.** Which of the 17 wallpaper patterns can be formed by the technique of Exercise 46?

For Exercises 48–53, refer to the following.

In Chapter 20, we study both repeating and nonrepeating plane patterns, from the point of view of their basic building blocks (tiles). Here we ask you to analyze the repeating patterns from figures in that chapter according to wallpaper type, using the flowchart of Spotlight 19.6. Identify all the symmetries and give the notational type for the wallpaper pattern of:

**48.** Figure 20.10.

**49.** Figure 20.11.

**50.** Figure 20.15.

**51.** The figure in Spotlight 20.2.

**52.** The hexagonal regular tiling at upper right in Figure 20.5.

**53.** The convex hexagon tiling of type 3 in Figure 20.9.

**54.** Visit the Web site escher.epfl.ch/escher/, which features an interactive Java program called Escher Web Sketch. Experiment with choosing wallpaper patterns using crystallographic notation. For each, draw on the screen a colored design for the motif; the program will reproduce the motif using the pattern.

## 19.4 Symmetry Groups

**55.** For positive integers $a$ and $n$, the expression $a$ mod $n$ means remainder when $a$ is divided by $n$. Thus, 23 mod 4 = 3, because $23 = 5 \cdot 4 + 3$, and we say that "23 is equivalent to 3 modulo 4" (see Chapter 17 for further details about this *modular arithmetic*). Every positive integer is equivalent to 0, 1, 2, or 3 modulo 4. Consider the collection of elements {0, 1, 2, 3} and the operation $\oplus$ on them defined by $a \oplus b = (a + b)$ mod 4. Show that under this operation, the collection forms a group.

**56.** Explain, by referring to the properties of a group, whether the collection of all real numbers is a group under the operation of **(a)** addition; **(b)** multiplication.

**57.** Explain why the table for the operation * below shows that the elements indicated do not form a group under *.

| * | A | B | C |
|---|---|---|---|
| A | B | B | B |
| B | B | C | A |
| C | C | A | B |

**58.** Consider the table for the operation # below.

| # | A | B | C | D | E | F |
|---|---|---|---|---|---|---|
| A | A | B | C | D | E | F |
| B | B | A | D | C | F | E |
| C | C | E | A | F | B | D |
| D | D | F | B | E | A | C |
| E | E | C | F | A | D | B |
| F | F | D | E | B | C | A |

**(a)** Explain why the elements form a group under #. (Don't bother to check associativity.)
**(b)** What do you notice about F # E vs. E # F? (This is the smallest example of a group that is noncommutative.)

**59.** For the traditional North American beadwork shown below:

*(Courtesy of Dr. Ron Eglash, RPI. See www.rpi.edu/~Eglash/csdt/na/loom/loom_symm4.html.)*

**(a)** Which rosette pattern does it have?
**(b)** Specify two rigid motions that are generators of the group of the pattern.
**(c)** List the elements of the group.

**60.** Repeat Exercise 59 for the Plains Indian embroidery shown below.

*(Courtesy of Dr. Ron Eglash, RPI. See www.rpi.edu/~Eglash/csdt.html.)*

**61.** What is the group of symmetries of a square?

**62.** What is the group of symmetries of:
**(a)** an equilateral triangle (all three sides equal)?
**(b)** an isosceles triangle (two equal sides) that is not equilateral?
**(c)** a scalene triangle (no pair of sides equal)?

**63.** **(a)** Give a numerical example to show that the operation of subtraction on the integers is not associative.
**(b)** Repeat part (a), but for division on the positive real numbers.

**64.** What are the elements of the group of symmetries of **(a)** Figure 19.10b? **(b)** Figure 19.10f?

**65.** What are the elements of the group of symmetries of **(a)** Figure 19.10d? **(b)** Figure 19.10g?

**66.** What are the elements of the group of symmetries of the dihedral pattern *d8*? (See the flower in Figure 19.9a.)

**67.** What is the group of symmetries of the cyclic pattern *c8*?

■ **68.** What is the group of symmetries of a cube?

■ **69.** What is the group of symmetries of a general rectangular solid (its length, width, and height are all unequal)?

## 19.5 Fractal Patterns and Chaos

◆ **70.** Explore Sprott's Fractal Gallery at sprott.physics.wisc.edu/fractals.htm, which features a "Fractal of the Day" and accompanying fractal music. There are various "rooms" in the gallery–including "Iterated Function Systems," "Natural Fractals" (I particularly like "Broccoli" and "Trees"), and "Publication Quality Attractors" ("SMKBNZQA" is our favorite)–together with PC programs for generating such fractals. What are your favorites, and why?

**71.** Explain how the pattern of the following illustration is fractal.

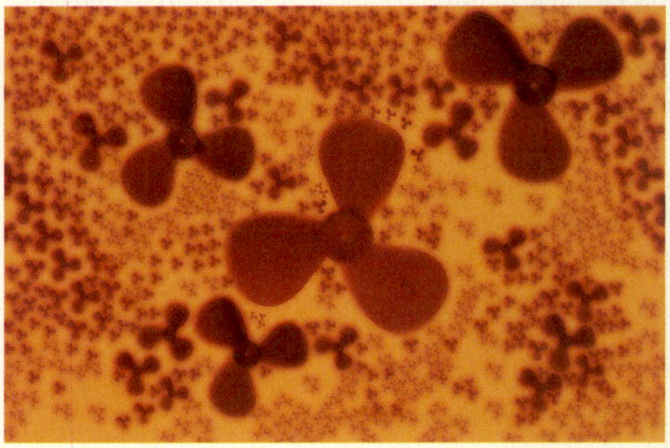

Exercises 72–76 use applets that require a computer with a Web browser equipped with Java and Flash plug-ins. These are available at links from the Web site www.rpi.edu/~eglash/csdt.html.

**72.** The Web site www.ccd.rpi.edu/Eglash/csdt/african/MANG_DESIGN/culture/mang_homepage.html has information about a fractal-patterned ivory hatpin from the Mangbetu culture in Africa. The site includes a tutorial on producing similar designs using reflection, rotation, translation, and scaling. Work your way through the tutorial and then create a Mangbetu-style artifact.

**73.** Cornrow hairstyles are fractal in nature. At the Web site www.ccd.rpi.edu/Eglash/csdt/african/CORNROW_CURVES/, you can see how and why, including a tutorial on designing cornrow hairstyles using reflection, rotation, translation, and scaling. Work your way through the tutorial and then create a hairstyle. The Web site also includes instructions for actual braiding, with a short video.

**74.** The architecture of some African villages follows a fractal pattern–see www.ccd.rpi.edu/Eglash/csdt/african/archi/afractal/afarch.htm. At the Web site www.ccd.rpi.edu/Eglash/csdt/african/archi/mangbetuUpdate2_3_modified_final.swf is an applet, similar to the ones in Exercises 72–73 but without a tutorial. Decide on a plan and use the applet to create a simulated African village.

**75.** The Yupik (Western Eskimo, along the coast of Alaska) make their parka coats with black-and-white patterns. At the Web site www.ccd.rpi.edu/Eglash/csdt/na/yupik/yupik.html is an applet, similar to the ones in Exercises 72–73 but without a tutorial. Yupik designs emphasize reflection symmetry, but the applet allows you to use also rotation, translation, and scaling of a basic motif. Experiment with one of the motifs available and use the applet to create a pattern for a parka.

**76.** Download fractal-creation software and accompanying documentation and use the software to create your own fractal. Recommended software:

For Windows: Fractint, from spanky.triumf.ca/www/fractint/fractint.html

For Macintosh: FractaSketch, from www.info.ucl.ac.be/~pvr/fracta.html with draft of manual (shows how to make fractal trees and leaves).

## WRITING PROJECTS

**1.** Generations of children have enjoyed the popular toy Spirograph®, which allows the user to trace out symmetric patterns. A pencil or pen is placed in a hole in one of several plastic circular disks with teeth on the outside rim. The disk is then meshed in the teeth of another plastic circle and rotated around its inside or outside. Each plastic piece is labeled with the number of teeth that it has on its circumference.

Either obtain a copy of Spirograph® or a closely related toy, or else visit the Web site www.wordsmith.org/~anu/java/spirograph.html, which offers an interactive Java application (which you can download) that mimics what the Spirograph® toy does.

**(a)** Experiment to determine, from the numbers of teeth on the rotating circular disk and the fixed circle, what symmetry pattern the result will have.
**(b)** Choose a rotating circular disk and a fixed circle for which the ratio of the number of teeth reduces to a whole number. For each of several "offsets" (holes to choose for the pencil or pen), trace overlapping designs. What symmetry pattern do you get for the design taken as a whole? Repeat this experiment for other pairs of pieces and try to reach a general conclusion.
**(c)** Write up, in a page or so, a description of your experiments and what conclusions you reached.

**2.** (Project for a team of 2 or 3) Explore your campus looking for symmetrical patterns in decorative elements of walls, floor, carpets, and ceilings. Find one example each of a rosette pattern, a strip pattern, and a wallpaper pattern. Take a digital photo of each and incorporate your photos into a document of three pages or so that explains to the reader where the pattern can be found, what symmetries (translation, rotation, reflection, glide reflection) it has, and how you identified the notation for it.

**3.** (Project for a team of 2 or 3) Visit a store that sells wallpaper and ask for a few old samples. Identify three that have different patterns according to the flowchart in Spotlight 19.6. Write in a page or two your explanations of how you identified the patterns, and attach the wallpaper samples to your report.

## SUGGESTED READINGS

BELCASTRO, SARAH-MARIE, and THOMAS C. HULL. Classifying frieze patterns without using groups, *College Mathematics Journal,* 33 (March 2002): 93–98. Elementary analysis of why there are only seven ways to repeat a pattern along a strip.

CROWE, DONALD W. *Symmetry, Rigid Motions and Patterns,* High School Mathematics and Its Applications (HiMAP) Module 4, COMAP, Lexington, Mass., 1987. Reprinted in smaller format in *The UMAP Journal,* 8(3) (1987): 207–236. Instructional module on rigid motions of the plane, strip patterns, and wallpaper patterns, with worksheets.

LEE, KEVIN D. KaleidoMania!: Interactive Symmetry, Windows/Macintosh program, Key Curriculum Press, 1999. Lets the user construct rosette, strip, and wallpaper patterns.

LIVIO, MARIO. *The Golden Ratio: The Story of Phi, the World's Most Astonishing Number.* Broadway Books, New York, 2002.

POSAMENTIER, ALFRED S., and INGMAR LEHMANN. *The (Fabulous) Fibonacci Numbers,* Prometheus Books, Amherst, N.Y., 2007.

WASHBURN, DOROTHY K., and DONALD W. CROWE. *Symmetries of Culture: Theory and Practice of Plane Pattern Analysis,* University of Washington Press, Seattle, 1988. An introduction to the mathematics of symmetry, splendidly illustrated with photographs of patterns from cultures all over the world. Includes a complete analysis of patterns with two colors, and proofs that there are only four rigid motions in the plane and exactly seven strip patterns.

## SUGGESTED WEB SITES

**www.geom.umn.edu/software/tilings** Tessellation resources. Lists programs for various platforms that allow the user to create designs featuring the rosette, strip, and wallpaper patterns.

**escher.epfl.ch/escher/** Interactive Escher Web Sketch program that allows a user to design repeating patterns. Choose a wallpaper pattern using crystallographic notation and draw on the screen a colored design for the motif; the program then reproduces the motif using the pattern. The software (for Windows, Macintosh, and Unix) can also be downloaded.

**www.geom.uiuc.edu/java/Kali/** Interactive Java Kali Web program that lets the user draw pictures under the action of rosette, strip, or wallpaper groups. Versions for various platforms can be downloaded.

**www.wordsmith.org/~anu/java/spirograph.html** Interactive Spirograph Java application (which you can download) that lets you do electronically what the Spirograph toy does.

**www.rpi.edu/~eglash/eglash.dir/afractal/afractal.htm** African fractals site.

**classes.yale.edu/fractals/index.html** Web site for Mandelbrot's course in fractals at Yale. Features many applets for different kinds of IFS (for example, incorporating randomness), including a fractal music composer.

**maven.smith.edu/~phyllo/index.html** Phyllotaxis: An interactive site for the study of plant pattern formation, by Pau Atela and Christophe Golë.

**www.mcs.surrey.ac.uk/Personal/R.Knott/Fibonacci/fib.html** Fibonacci numbers and the golden section. Splendidly illustrated extensive Web pages by Ron Knott about Fibonacci numbers and the golden ratio: their occurrences in nature, their applications, puzzles, and much more.

# CHAPTER 20



# Tilings

When our ancestors covered the floors and walls of their houses with stones, they selected shapes and colors to form pleasing designs. We can see the artistic impulse at work in mosaics, from Roman dwellings to Muslim religious buildings (see Figure 20.1). The same intricacy and complexity arise in other decorative arts—on carpets, fabrics, baskets, and even linoleum.

Such patterns have one feature in common: They use repeated shapes to cover a flat surface, without gaps or overlaps. If we think of the shapes as tiles, we can call the pattern a **tiling**, or *tessellation*. Even when efficiency is more important than esthetics, designers value clever tiling patterns. In manufacturing, for example, stamping components from a sheet of metal is most economical if the shapes of the components fit together without gaps—in other words, if the shapes form a tiling. We are interested in patterns that could be extended indefinitely far in any direction.

**FIGURE 20.1** Arab mosaic. (*Jose Fuste Raga/Corbis.*)

**Tiling (Tessellation)** DEFINITION

A **tiling (tessellation)** is a covering of the entire infinite plane by nonoverlapping figures.

Some of the tilings in this chapter can be analyzed as wallpaper patterns, as in Spotlight 19.6. But in this chapter, we look at patterns "from the ground up." That is, we start with the basic ingredients of a type or types of tile and ask if or how they can fit together in a pattern. If the result repeats in all directions, it must be one of the wallpaper patterns. We will see, though, that there are other surprising possibilities.

The major mathematical question about tilings is: Given one or more shapes (in specific sizes) of tiles, can they tile the plane? And, if so, how?

The surprising answer to the first question is that it is undecidable. For some particular sets of tiles, we can exhibit tilings; for others, we can prove that there can't be any tiling. In this chapter we will see examples of both situations. But mathematicians have proved that there is no algorithm (mechanical step-by-step process) that can tell for every conceivable set of tile shapes which of the two situations holds. (See Chapter 9, pp. 356–359, for other examples of "unattainable ideals" in regard to voting.)

Given this sobering (and puzzling) limitation, we begin our investigation by considering the simplest kinds of tiles and tilings.

## 20.1 Tilings with Regular Polygons

The simplest tilings use only one size and shape of tile. They are known as **monohedral tilings**.

**Monohedral Tiling** DEFINITION

A **monohedral tiling** is a tiling that uses only one size and shape of tile.

We are particularly interested in tiles that are **regular polygons**, figures all of whose sides are the same length and all of whose angles are equal. A square is a regular polygon with four equal sides and four equal interior angles; a triangle with all sides equal (an **equilateral triangle**) is also a regular polygon. A polygon with five sides is a pentagon, one with six sides is a hexagon, and one with $n$ sides is an ***n*-gon**. Regular polygons are especially interesting because of their high degree of symmetry. Each has the reflection and rotation symmetries of a dihedral rosette pattern (see section 19.2). In three dimensions, the corresponding highly symmetrical figures are called *regular polyhedra* (see Spotlight 20.1).

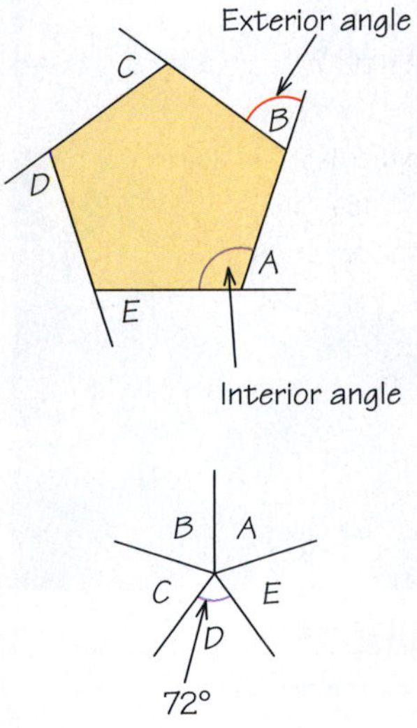


**FIGURE 20.2** The exterior angles of a regular pentagon, like those of any regular polygon, add up to 360°. Each exterior angle measures 72°.

An **exterior angle** of a polygon is an angle formed by one side and the extension of an adjacent side (Figure 20.2). At each vertex of the polygon, there are two exterior angles, depending on which side we extend; but we will consider only one of them. Let us agree to extend the sides consistently in turn as we proceed counterclockwise around the polygon, as in Figure 20.2, producing the set of exterior angles $A$ through $E$, one at each vertex.

By a convention dating back to the ancient Babylonians, angles are measured in degrees, with the total of angles around a point being 360°. If we bring a set of exterior angles together at a point, we can see that they add up to 360° (see Figure 20.2). Hence for a regular polygon with $n$ sides, each exterior angle must measure $360°/n$. For example, a square, with $n = 4$ sides, has 4 exterior angles in a set, each measuring 90°; a regular pentagon, with $n = 5$ sides, has 5, each measuring 72°; a regular hexagon, with 6 sides, has 6, each measuring 60°.

Each exterior angle is paired with a corresponding **interior angle** (the angle inside the polygon formed by the two adjacent sides), and the pair adds up to a straight line, or 180°. For a regular polygon with more than six sides, each interior angle is between 120° and 180°. This last consideration will shortly prove to be crucial to determining how regular polygons can fit together to form tilings.

## SPOTLIGHT 20.1 Regular Polyhedra and Buckyballs

The three-dimensional analogue of a regular polygon is a regular polyhedron, a convex solid whose faces are regular polygons all alike (same number of sides, same size), with each vertex surrounded by the same number of polygons. Although there are infinitely many regular polygons, there are only five regular polyhedra, a fact proved by Theaetetus (414–368 B.C.). They were called the *Platonic solids* by the ancient Greeks.

If the restriction that the same number of polygons meet at each vertex is relaxed, five additional convex polyhedra are obtained, all of whose faces are equilateral triangles. If we allow more than one kind of regular polygon, 13 further convex polyhedra are obtained, known as the *semiregular polyhedra* or *Archimedean solids* (although there is no documented evidence that Archimedes studied them—but Kepler in the early 1600s catalogued them all). The truncated icosahedron, whose faces are pentagons and hexagons, is known throughout the world (once inflated) as a regulation soccer ball. Drawings of it appear in the work of Leonardo da Vinci.

The truncated icosahedron is also the structure of $C_{60}$, a form of carbon known as buckminsterfullerene and, more familiarly, the "buckyball." Sixty carbon atoms lie at the 60 vertices of this molecule, which was discovered in 1985. It is named after R. Buckminster Fuller (1895–1983), inventor and promoter of the geodesic dome. The molecule resembles a dome.

The buckyball is part of a family of carbon molecules, the *fullerenes*, in which each carbon atom is joined to three others. Thirty years before the discovery of fullerenes, mathematicians had shown that a convex polyhedron in which every vertex has three edges must have 12 pentagon faces and may have any number of hexagon faces, from 0 on up, except for 1.

That there must be 12 pentagons follows from a famous equation due to Leonhard Euler (1707–1783). For any convex polyhedron, it must be true that $v - e + f = 2$, where $v$ is the number of vertices, $e$ is the number of edges, and $f$ is the number of faces of the polyhedron.

In 2003 astronomers and mathematicians advanced a remarkable new theory about the shape of the universe, in an effort to explain why it does not show as much historic fluctuation in temperature as other models predict. This lack of fluctuation could be explained by the universe being in the shape of a dodecahedron (the figure shown here with 12 pentagonal sides), with opposite faces coinciding. This theory harks back to Kepler, who had conceived of the universe in terms of the five regular polyhedra nested within one another.

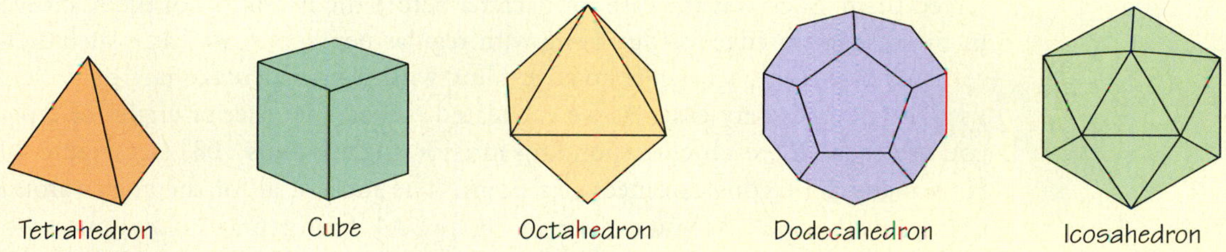


## Regular Tilings

**Regular Tiling** DEFINITION

A monohedral tiling whose tile is a regular polygon is called a **regular tiling**.

A square tile is the simplest case. Apart from varying the size of the square, which would change the scale but not the pattern of the tiling, we can get different tilings by offsetting one row of squares some distance from the next (Figure 20.3a).

However, there is only one tiling using a square that is **edge-to-edge**.

### Edge-to-Edge — DEFINITION

In an **edge-to-edge tiling**, the edge of a tile coincides entirely with the edge of a bordering tile.

Figure 20.3 shows one tiling that is not edge-to-edge and another that is.

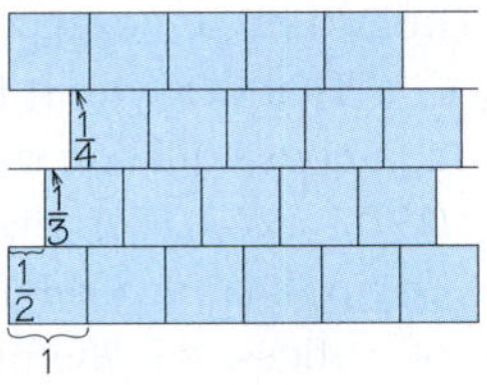


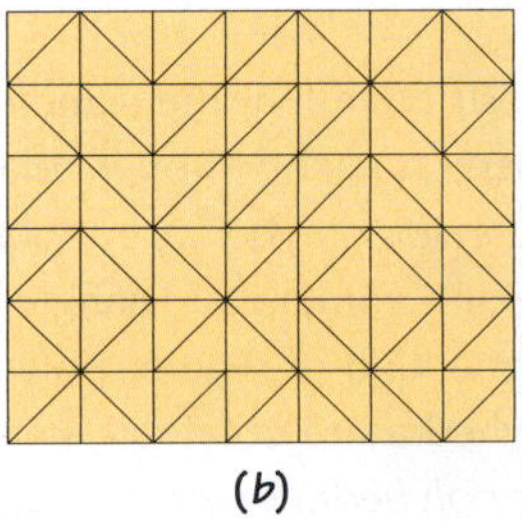


**FIGURE 20.3** (a) A tiling that is not edge-to-edge. The horizontal edges of two adjoining squares do not exactly coincide. (b) A tiling by right triangles that is edge-to-edge.

For simplicity, from now on by *tiling* we mean "edge-to-edge tiling." The edges themselves may be curvy.

Any tiling by squares can be refined to one by triangles by drawing a diagonal of each square, but these triangles are not regular (equilateral). Equilateral triangles can be arranged in rows by alternately inverting triangles. As with squares, there is only one pattern of equilateral triangles that forms an edge-to-edge tiling.

What about tiles with more than four sides? An edge-to-edge tiling with regular hexagons is easy to construct (see the upper right pattern in Figure 20.5).

However, if we look for a tiling with regular pentagons, we won't find one. How do we know whether we're just not being clever enough or there really isn't one to be found? This is the kind of question that mathematics is uniquely equipped to answer. In the other sciences, phenomena may exist even though we have not observed them. Such was the case for bacteria before the invention of the microscope. In the case of an edge-to-edge tiling with regular pentagons, we can conclude with certainty that there is no edge-to-edge tiling with regular pentagons.

The proof is very easy. As we calculated earlier, the exterior angles of a pentagon are each 72°; each corresponding interior angle is thus 108° (see Figure 20.2). How many pentagons can meet at a point? The total of all of the angles around a point must be 360°. As you can see in Figure 20.4, four pentagons at a point would be too many (their angles would add to $4 \times 108° = 432°$, so they'd have to overlap), and three would be too few (their angles would add to $3 \times 108° = 324°$, so some of the area wouldn't be covered). Because 108 does not evenly divide 360, *regular pentagons can't tile the plane.*

With this argument, we can do something that is a favorite with mathematicians–we can *generalize* it to a criterion for when a regular polygon can tile the plane: when the size of its interior angles divides 360 evenly. We can apply this criterion to determine exactly which other regular polygons can tile the plane.

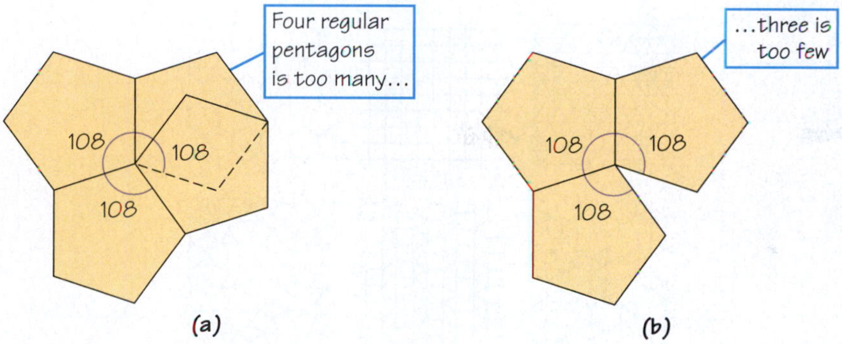


**FIGURE 20.4** Polygons that come together at a vertex in a tiling must have interior angles that add up to 360°–no more, no less.

## EXAMPLE 1 ▪ Identifying the Regular Tilings

**SOLUTION** A regular hexagon has interior angles of 120°; 120 divides 360 evenly, and three regular hexagons fit together exactly around a point. A regular 7-gon (heptagon)–or any regular polygon with more than six sides–has interior angles that are larger than 120° but smaller than 180°. Now 360 divided by 120 gives 3, and 360 divided by 180 gives 2–and there aren't any other possibilities in between. Angles between 180° and 120° divided into 360° will give a result between 2 and 3, and consequently not an integer. So there are no regular tilings of the plane with polygons of more than six sides.

**Only Three Regular Tilings** THEOREM

The only regular tilings are the ones with equilateral triangles, with squares, and with regular hexagons.

The follow-up question, of course, is which *combinations* of regular polygons of different numbers of sides can tile the plane edge-to-edge.

**Vertex Type** DEFINITION

In an edge-to-edge tiling by regular polygons, the **vertex type** of a vertex is the arrangement of the polygons around the vertex.

To describe a vertex type, we list the sizes of polygons, separated by periods, in either clockwise or counterclockwise order starting from the smallest number of sides. For example, 4.4.4.4 (or $4^4$ for short) denotes four squares meeting at a vertex. Similarly, 4.6.12 denotes a square followed by a hexagon then by a dodecagon (12-gon); see the tiling in the middle of the bottom row of Figure 20.5. Two vertices have the same type even if one has the polygons in clockwise order and the other has them in counterclockwise order; both versions of the 4.6.12 type occur in that tiling in Figure 20.5.

**Semiregular Tiling** DEFINITION

A systematic tiling that uses a mix of regular polygons with different numbers of sides but in which all vertex types are alike–the same polygons in the same order–is called a **semiregular tiling** (see Figure 20.5).

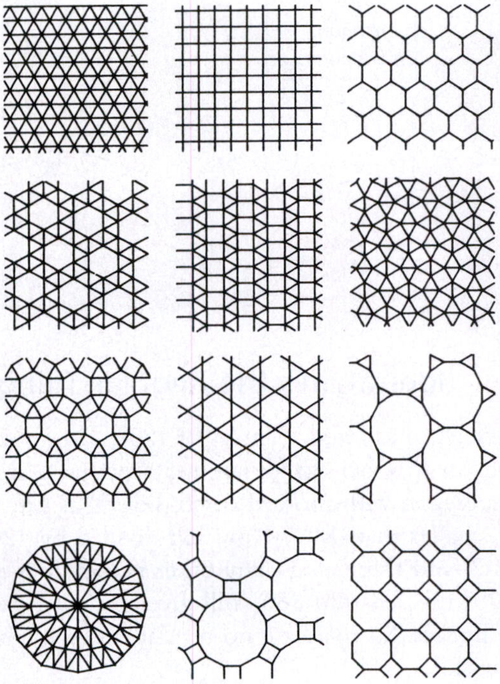

**FIGURE 20.5** The three regular tilings . . . and the eight semiregular tilings, plus a "mystery" tiling that does not belong to either group. Can you identify it? (*Hint:* It uses just one tile, which isn't regular.)

As before, the technique of adding up angles at a vertex (to be 360°) can eliminate some impossible combinations, such as "pentagon, pentagon, pentagon" (Figure 20.4). Once we have found an arrangement that is numerically possible, we must confirm the actual existence of each tiling by constructing it (showing that it is geometrically possible). For example, even though a possible arrangement of regular polygons around a point is "triangle, square, square, hexagon," it is not possible to construct a tiling with that vertex figure at every vertex.

The result of such an investigation is that in a semiregular tiling, no polygon can have more than 12 sides. In fact, polygons with 5, 7, 9, 10, or 11 sides do not occur either. Figure 20.5 exhibits all of the semiregular tilings.

If we abandon the restriction about the vertex types being the same at every vertex, then there are *infinitely many* systematic edge-to-edge tilings with regular polygons, even if we continue to insist that all polygons with the same number of sides have the same size.

## 20.2 Tilings with Irregular Polygons

What about edge-to-edge tilings with irregular polygons, which may have some sides longer than others or some interior angles larger than others? We will look just at monohedral tilings (in which all tiles have the same size and shape) and investigate in turn which triangles, **quadrilaterals** (four-sided polygons), hexagons, and so forth, can tile the plane.

The most general shape of a triangle has all sides of different lengths and all interior angles of different sizes. Such a triangle is called a **scalene triangle**, from the Greek word for "uneven." We can always take two copies of a scalene triangle and fit them together to form a **parallelogram**, a quadrilateral whose opposite sides are parallel (Figure 20.6a). It's easy to see that we can then use such parallelograms to tile the plane by making strips and then fitting layers of strips together edge-to-edge (Figure 20.6b).

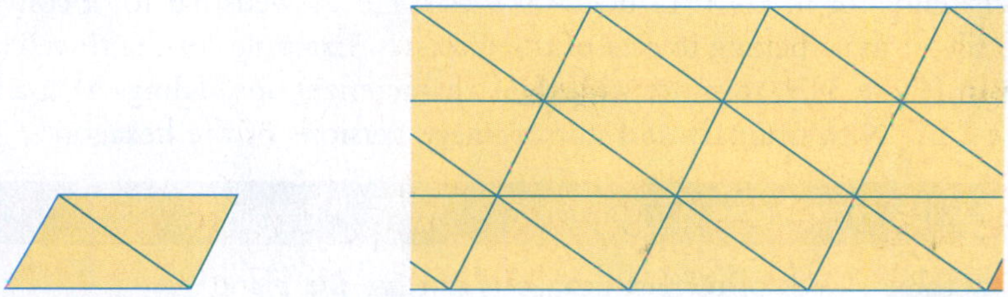
(a) (b)

**FIGURE 20.6** (a) Two scalene triangles form a parallelogram. (b) Every scalene triangle tiles the plane.

## Tiling with Triangles — THEOREM

Any triangle can tile the plane.

What about quadrilaterals? We have seen that squares tile the plane, and rectangles certainly will, too. We have just noted that any parallelogram will tile. What about a quadrilateral (four-sided polygon) with its opposite sides not parallel, as in Figure 20.7a? The same technique as for triangles will work. We fit together two copies of the quadrilateral, forming a hexagon whose opposite sides are parallel. Such hexagons fit next to each other to form a tiling, as in Figure 20.7b.

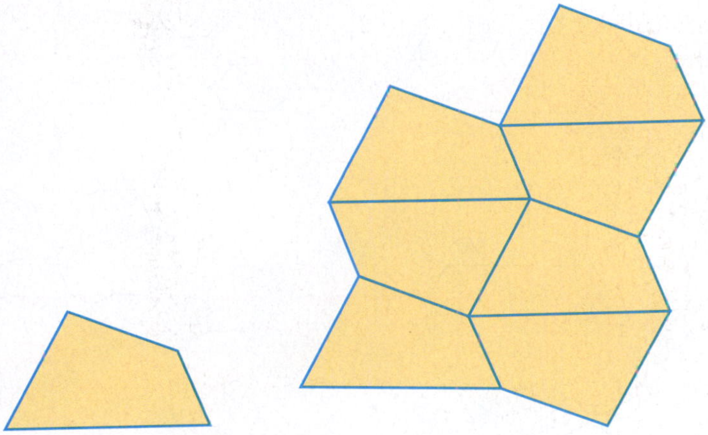
(a) (b)

**FIGURE 20.7** (a) A general quadrilateral. (b) Any quadrilateral tiles the plane.

**FIGURE 20.8** (a) A general nonconvex quadrilateral. (b) Any quadrilateral, convex or not, tiles the plane.

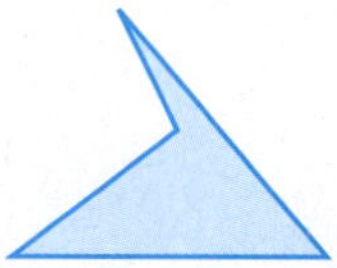
(a)

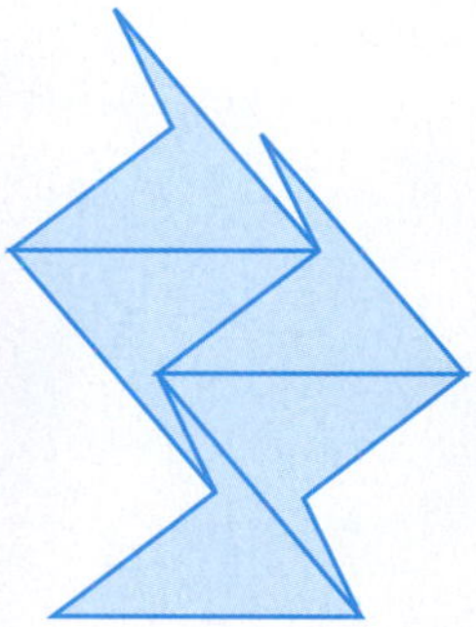
(b)

The quadrilaterals shown in Figure 20.7 are all **convex**. If you take any two points on the tile (including the boundary), the line segment joining them lies entirely within the tile (again, including the boundary). The quadrilateral of Figure 20.8a is not convex, but the same approach works for using it to form a tiling (Figure 20.8b).

## Tiling with Quadrilaterals — THEOREM

Any quadrilateral, even one that is not convex, can tile the plane.

We could hope that such success would extend to irregular polygons with any numbers of sides, but it doesn't. The situation for convex hexagons was determined by Karl Reinhardt in his 1918 doctoral thesis. He showed that for a convex hexagon to tile, it must belong to one of three classes. Examples of the three classes are shown in Figure 20.9, together with their characterizations. Tilings with a hexagon of type 2 use both ordinary and mirror-image versions of the hexagon.

**Tiling with Hexagons** THEOREM

Exactly three classes of convex hexagons can tile the plane.

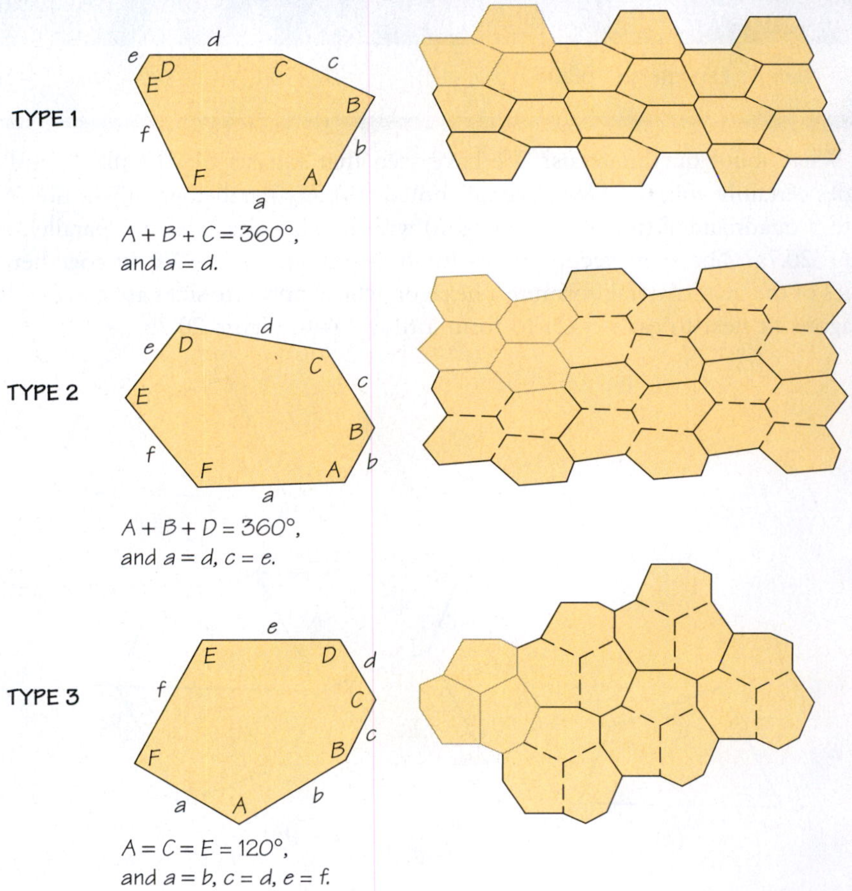


**FIGURE 20.9** The three types of convex hexagon tiles.

Reinhardt also explored convex pentagons and found five classes that tile. For example, any pentagon with two parallel sides will tile. Reinhardt did not complete the solution, as he did for hexagons, by proving conclusively that no other pentagons could tile. He claimed that it would be very tedious to finish the analysis. Still, he felt that he had found them all. In 1968, after 35 years of working on the problem on and off, R. B. Kershner, a physicist at Johns Hopkins University, discovered three more classes of pentagons that will tile. Kershner was sure that he had found all pentagons that tile, but, like Reinhardt, he did not offer a complete proof, which "would require a rather large book."

When an account of the "complete" classification into eight types appeared in *Scientific American* (July 1975), the article provoked an amateur mathematician to discover a ninth type! A second amateur, Marjorie Rice, a housewife with no formal education in mathematics beyond high school "general mathematics" 36 years earlier, devised her own mathematical notation and found four more types (see Spotlight 20.2). A fourteenth type was found by a mathematics graduate student in 1985. Since then, no new types have been discovered, yet no one knows if the classification is complete.

With the situation so intricate for convex pentagons, you might think that it must be still worse for polygons with seven or even more sides. In fact, however, the situation is remarkably simple, as Reinhardt proved in 1927.

**Tiling with Polygons with More Sides** THEOREM

A convex polygon with seven or more sides cannot tile.

### M. C. Escher and Tilings

The Dutch artist M. C. Escher (1898–1972) was inspired by the great variety of decoration in tilings in the Alhambra, a fourteenth-century palace built during the last years of Islamic dominance in Spain. He devoted much of his career of making prints to creating tilings with tiles in the shapes of living beings (a practice forbidden to Muslims). Those prints of interlocking animals and people have inspired awe and wonder among people all over the world. Figures 20.10–20.13 illustrate a few of his drawings and finished works. Like Marjorie Rice, he, too, developed his own mathematical notation for the different kinds of patterns for the tilings.

## 20.3 Using Translations

You may wonder just how much liberty can be taken in shaping a tile and how you might be able to design an Escher-like tiling yourself.

The simplest case is when the tile is just *translated* in two directions; that is, copies are laid edge-to-edge in rows, as in Figure 20.10. Each tile must fit exactly into the ones next to it, including its neighbors above and below. We say that each tile is a **translation** of each other one, because we can move one to coincide with another without doing any rotation or reflection.

When is it possible for a tile to cover the plane in this manner? The boundary of the tile must be divisible into matching pairs of opposing parts that will fit together. Figures 20.10 and 20.11 illustrate two basic ways that this can happen. In the first, two opposite pairs of sides match; in the second, three opposite pairs of sides match.

**Translation Criterion** RULE

A tile can tile the plane by translations if either

1. There are four consecutive points $A$, $B$, $C$, and $D$ on the boundary such that
   **(a)** the boundary part from $A$ to $B$ is congruent by translation to the boundary part from $D$ to $C$, and
   **(b)** the boundary part from $B$ to $C$ is congruent by translation to the boundary part from $A$ to $D$ (see Figure 20.12a) or
2. There are six consecutive points $A$, $B$, $C$, $D$, $E$, and $F$ on the boundary such that the boundary parts $AB$, $BC$, and $CD$ are congruent by translation, respectively, to the boundary parts $ED$, $FE$, and $AF$ (see Figure 20.12b).

## SPOTLIGHT 20.2 In Praise of Amateurs

Marjorie Rice

*(Courtesy Sharon Whittaker.)*

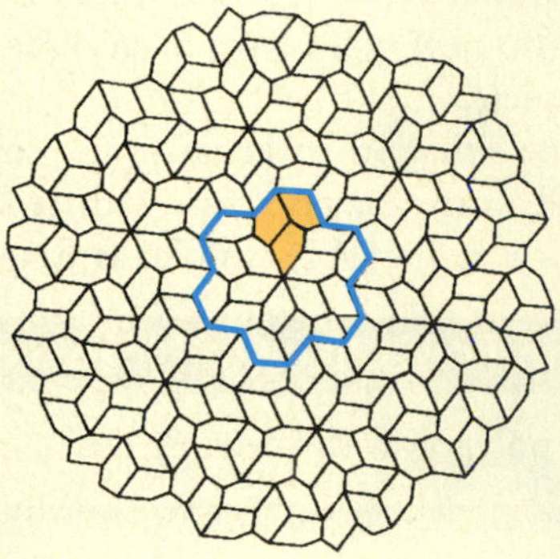

This tiling in the headquarters of the Mathematical Association of America in Washington, D.C., was discovered in 1995 by Marjorie Rice. The angles of each pentagon tile are 60°, 90°, 120°, and 150°—all multiples of 30°. The tiling is periodic, although not every pentagon is surrounded in the same way. Three pentagons form a fundamental block, and the outlined group of 18 pentagons tiles by translation.

R. B. Kershner's claim to have found all convex pentagons that tile was read by many puzzle enthusiasts, including Richard James III and Marjorie Rice. James found a tiling that Kershner had missed.

Rice, a San Diego housewife and mother of five, read about James's new tile. "I thought I would see if I could find still another type. It was a delightful new puzzle to me."

With no formal education in mathematics beyond a high school general mathematics course, she not only worked out her own method of attack but invented her own notation.

"I began drawing little diagrams on my kitchen counter when no one was there, covering them up quickly if someone came by, for I didn't wish to have to explain what I was doing. I was searching for a new type and a few weeks later, I found it." Over the next two years, she found three additional new tilings.

What makes a person pursue a problem so patiently and persistently? She was not trained for it nor paid, but she gained great personal satisfaction.

She was born in 1923 in St. Petersburg, Florida, and went to a one-room country school.

"When I was in the 6th or 7th grade, our teacher pointed out to us the Golden Section in the proportions of a picture frame. This immediately caught my imagination and I never forgot it. I've . . . been especially interested in architecture and the ideas of architects and planners such as Buckminster Fuller. I've come across the Golden Section again in my reading and considered its use in painting and design."

After high school, Rice worked until her marriage in 1945. She was drawn back into mathematics by her children, finding solutions to their homework problems "by unorthodox means, since I did not know the correct procedures." She became especially interested in textile design and the works of M. C. Escher. As she pursued the pentagonal tilings, she produced some imaginative Escher-like patterns (see Figure 20.19 and the figure here).

Intense spirit of inquiry and keen perception are the forte of all such amateurs. No formal education provides these gifts. Lack of a mathematical degree separates these "amateurs" from the "professionals," yet their curiosity and ingenious methods make them true mathematicians.

*Source*: Adapted from Doris Schattschneider, "In Praise of Amateurs," in David A. Klarner (ed.), *The Mathematical Gardner*, pp. 140–166, plus Plates I–III, Wadsworth, Belmont, Calif., 1981.

**FIGURE 20.10** Escher No. 128 (*Bird*), from Escher's 1941–1942 notebook. (*© 1967 M. C. Escher Foundation, Baarn, Holland, all rights reserved.*)

(a)

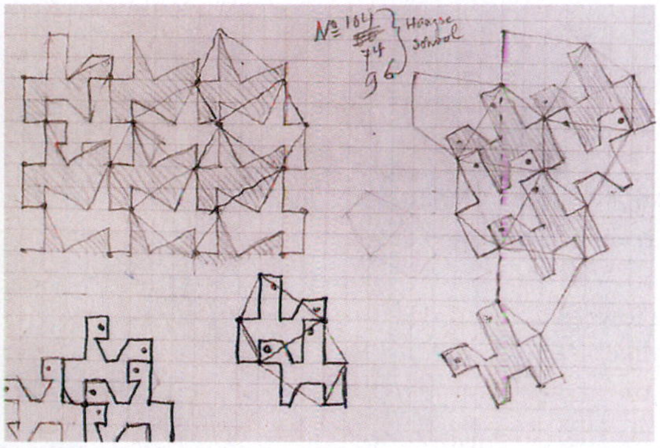

(b)

**FIGURE 20.11** (a) Escher No. 67 (*Horseman*), from Escher's 1941–1942 notebook. (*© 1947 M. C. Escher Foundation, Baarn, Holland, all rights reserved.*) (b) Sketch showing the tile design for the *Horseman* print. (*© 1947 M. C. Escher Foundation, Baarn, Holland, all rights reserved. From the collection of Michael S. Sachs.*)

The tiles for Figures 20.10 and 20.11 are shown in outline form in Figure 20.12, together with points marked to show how the tiles fulfill the criterion.

In fact, alternative 1 of the criterion is a special case of alternative 2 (see Exercises 19–22 and 24). Moreover, alternative 2 *completely characterizes* tiles that can tile by translations. That is, not only is it true that *if alternative 2 is true, then the tile can tile by translations,* but also that the criterion works "in reverse": *If a tile can tile by translations, then alternative 2 must be true* (for some choice of six consecutive points).

A nice feature of the translation criterion is that if you can find points as required for alternative 2, then you can join them in order, as in Figures 20.12a and b, to see how to do the tiling.

To create tilings, though, you can proceed exactly as Escher did. His notebooks show that he designed his patterns in just the way that we now describe.

FIGURE 20.12 Individual tiles traced from the Escher prints of Figures 20.10 and 20.11, with points marked to show they fulfill the criteria for tiling by translations. The two knights form a block that tiles by translation, although a single knight can tile by itself if we allow mirror-image reflections too.

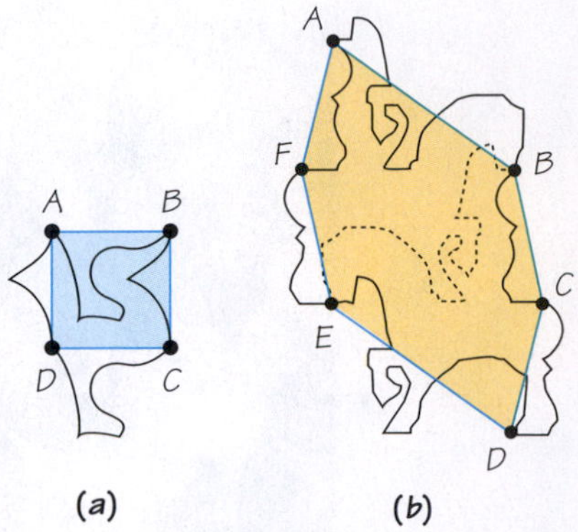


## EXAMPLE 2 ■ Tiling Starting from a Parallelogram

**SOLUTION** For the first alternative of the criterion, start from a parallelogram, make a change to the boundary on one side, then copy that change to the opposite side. Similarly, change one of the other two sides and copy that change on the side opposite it (Figure 20.13). Revise as necessary, always making the same change to opposite sides. You might find it useful (as Escher did) to make your designs on graph paper, or you can work by cutting and taping together pieces of heavy paper.

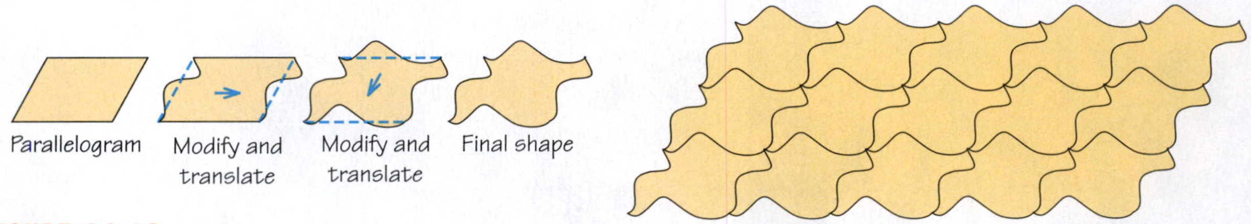


FIGURE 20.13 How to make an Escher-like tiling by translations, from a parallelogram base.

## EXAMPLE 3 ■ Tiling Starting from a Hexagon

**SOLUTION** For the second alternative, start from a **par-hexagon**, a hexagon whose opposite sides are equal and parallel. This is one of the kinds of hexagons that tile the plane. Again, make a change on one boundary and copy the change to the opposite side, and do this for all three pairs of opposite sides (Figure 20.14).

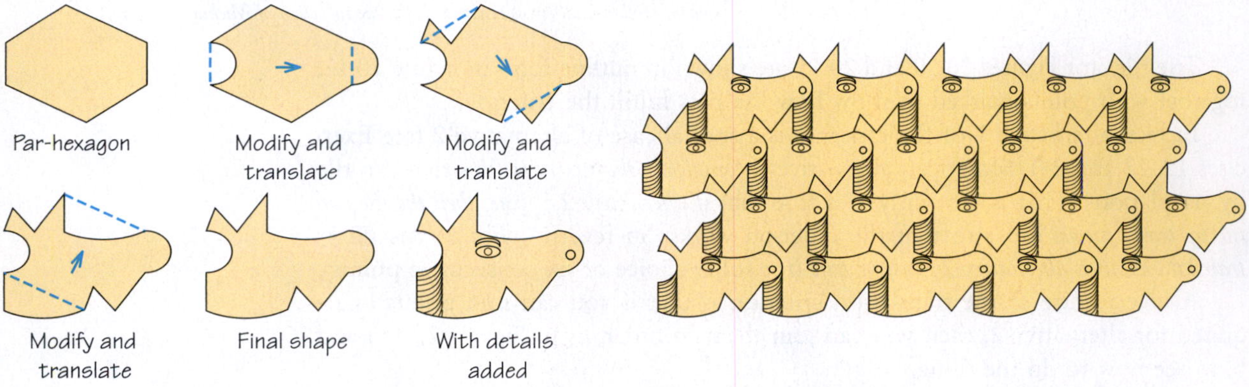


FIGURE 20.14 How to make an Escher-like tiling by translations, from a par-hexagon base.

## 20.4 Using Translations Plus Half-Turns

If the tiling is to allow half-turns, so that some of the figures are "upside down," the part of the boundary of a right-side-up figure has to match the corresponding part of itself in an upside-down position. For that to happen, that part of the boundary must be **centrosymmetric**, that is, symmetric about (unaltered by) a 180° rotation around its midpoint. The key to some of Escher's more sophisticated monohedral designs, and the fundamental principle behind some further easy recipes for making Escher-like tilings, is the **Conway criterion**, formulated by John H. Conway of Princeton University.

### Conway Criterion — RULE

A tile can tile the plane by translations and half-turns if there are six consecutive points on the boundary (some of which may coincide, but at least three of which are distinct)–call them $A$, $B$, $C$, $D$, $E$, and $F$–such that

- the boundary part from $A$ to $B$ is congruent by translation to the boundary part from $E$ to $D$, and
- each of the boundary parts $BC$, $CD$, $EF$, and $FA$ is centrosymmetric.

The first condition means that we can match up the two boundary parts exactly, curve for curve, angle for angle. The second condition means that each of the remaining boundary parts is brought back into itself by a half-turn around its center. Either condition is automatically fulfilled if the boundary part in question is a straight-line segment.

**FIGURE 20.15** Escher No. 6 (*Camel*), from Escher's 1941–1942 notebook. (© *1937–1938 M. C. Escher Foundation, Baarn, Holland, all rights reserved.*)

The tiles for Figures 20.15 and 20.16 are shown in outline form in Figure 20.17, together with points marked to show how the tiles fulfill the Conway criterion. Figure 20.15 shows that Escher sketched little circles exactly where we have red dots in Figure 20.17a.

**FIGURE 20.16** Escher No. 88 (*Sea Horse*). (*© 1947 M. C. Escher Foundation, Baarn, Holland, all rights reserved.*)

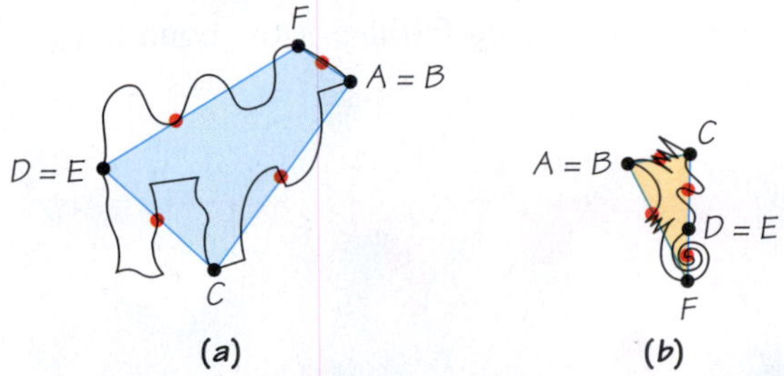


**FIGURE 20.17** Individual tiles traced from the Escher prints of Figures 20.15 and 20.16, with points marked to show they fulfill the Conway criterion for tiling by translations and half-turns (around the red dots).

Mathematicians do not know if the Conway criterion completely characterizes tiles that can tile by translations and half-turns. Tiles that fulfill the Conway criterion can tile by translations and half-turns, but not necessarily vice versa: There could be tiles that can tile that way but do not satisfy the criterion–however, nobody knows of any. (The Conway criterion does, however, completely characterize tiles that produce the wallpaper pattern *p2* of Spotlight 19.6: Any tile that satisfies the criterion can be used to make a *p2* pattern, and any tile that can produce that pattern must satisfy the criterion.)

Once again, you can make Escher-like tilings by starting from simple geometric shapes that tile. This time, the starting geometric tile can be any triangle or any quadrilateral.

## EXAMPLE 4 ■ Tiling Using a Triangle

**SOLUTION** For a triangle, modify half of one side, then rotate that side around its center point to extend the modification to the rest of the side, thereby making the new side centrosymmetric. Then you can do the same to the second and third sides (Figure 20.18).

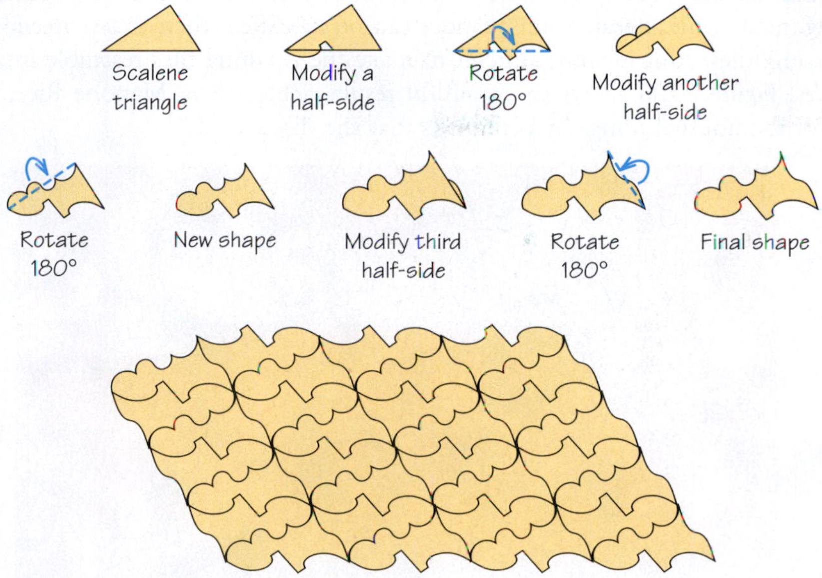


**FIGURE 20.18** How to make an Escher-like tiling by translations and half-turns, from a scalene triangle base.

## EXAMPLE 5 ▪ Tiling Using a Quadrilateral

**SOLUTION** For the quadrilateral, do the same, modifying each of the four sides, or as many as you wish (Figure 20.19). ■

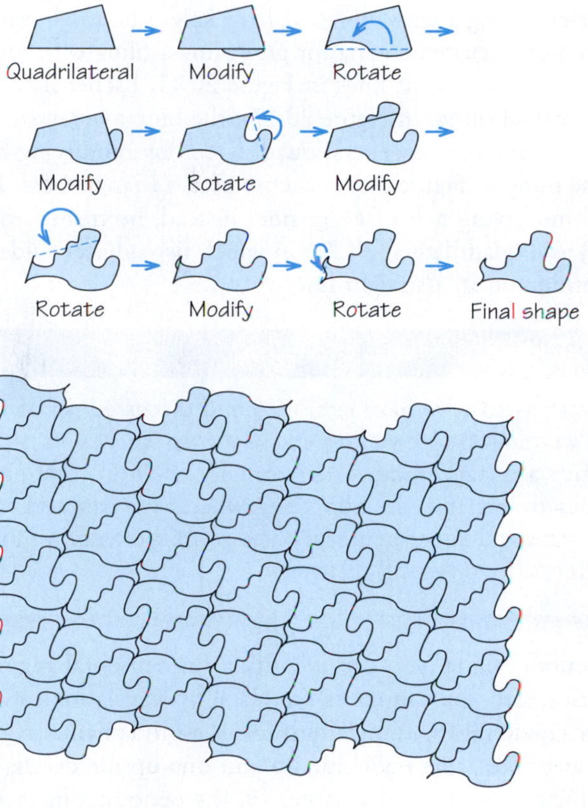


**FIGURE 20.19** How to make an Escher-like tiling by translations and half-turns, from a quadrilateral base.

The same approach will work with some of the sides of some pentagons and hexagons that tile. Because not all sides can be modified, there is less freedom for designing tiles, so it is more difficult to make the resulting tiles resemble intended figures. Figure 20.20 shows the beautiful results achieved by Marjorie Rice, using one of the unusual tilings by pentagons that she discovered.

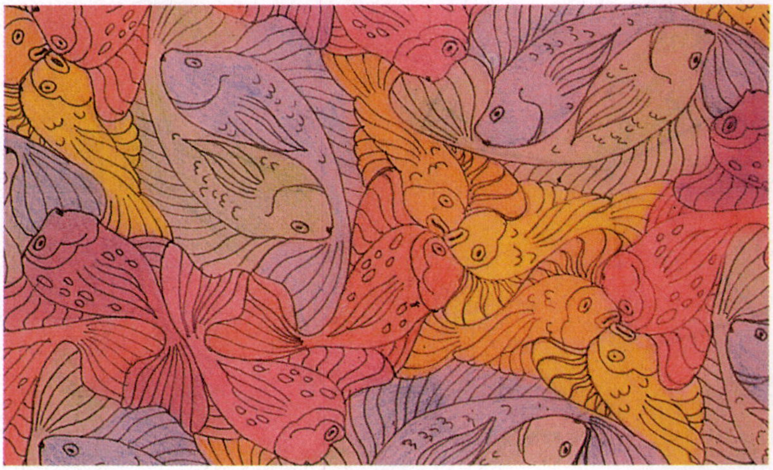

**FIGURE 20.20** *Fish,* by Marjorie Rice, based on one of her unusual tilings by pentagons.

Sketches in Escher's notebook indicate how he designed many of his prints. For the bird tiling of Figure 20.10, the single bird below the tiling shows that he modified the sides of a square. For the knights tiling of Figure 20.11a, the sketches in Figure 20.11b show that he modified the pairs of sides of a par-hexagon. We redraw the two fundamental figures more clearly in Figure 20.12. (The knight tiling also has a reflection symmetry, taking a leftward-facing light knight to a rightward-facing dark knight; but we have not discussed criteria for producing a tiling with such a symmetry.)

As can be seen in faint lines in Figure 20.15, Escher used a parallelogram as a base for the camel tiling. In Figure 20.17a, the blue overlay shows how to make the tiling starting from a more general quadrilateral by modifying half of each side. For the seahorse tiling of Figure 20.16, Escher used a triangle base. However, once more he avoided modifying half of every side; instead, he treated the triangle *ACF* (Figure 20.17b) as a quadrilateral *ACDF* in which two adjacent sides (*CD* and *DF*) happen to continue on in a straight line.

### Periodic Tilings — PROCEDURE

All the patterns that we have exhibited and discussed so far have been **periodic tilings**. If we transfer a periodic tiling to a transparency, it is possible to slide the transparency a certain distance horizontally, without rotating it, until the transparency exactly matches the tiling everywhere. We can also achieve the same result by moving the transparency in some second direction (possibly vertically) by a certain (possibly different) distance.

In a periodic tiling, you can identify a **fundamental region**–a tile, or a block of tiles–with which you can cover the plane by translations at regular intervals. For example, in Figure 20.10, a single bird forms a fundamental region. In Figure 20.15, two adjacent camels, one right side up and one upside down, form a fundamental region. In the terminology of Chapter 19, the periodic tilings are ones that are preserved under translations in more than two directions.

## 20.5 Nonperiodic Tilings

In Figure 20.3a, the second row from the bottom is offset one-half of a unit to the right from the bottom row, the third row from the bottom is offset one-third of a unit further, and so forth. Because the sum $\frac{1}{2} + \frac{1}{3} + \frac{1}{4} + \cdots + \frac{1}{n}$ never adds up to exactly a whole number, there is no direction (horizontal, vertical, or diagonal) in which we can move the entire tiling and have it coincide exactly with itself.

**Nonperiodic Tiling** DEFINITION

A **nonperiodic tiling** is a tiling in which there is no regular repetition of the pattern by translation.

### EXAMPLE 6 ■ A Nonperiodic Tiling through Randomness

**SOLUTION** Consider the usual edge-to-edge square tiling. For each square, flip a coin. Depending on the result, divide the square into two right triangles by adding either a rising or a falling diagonal (see Figure 20.3b). Because what happens in each individual square is unconnected to what happens in the rest of the tiling, this random tiling by right triangles has no chance of being periodic. ■

### Penrose Tiles and Quasicrystals

For all known cases, if a single tile can be used to make a nonperiodic tiling of the plane, then it can also be used to make a periodic tiling. It is still an open question whether this property is true for every possible shape. In 1993, Conway discovered an example in three dimensions of a single convex polyhedron that tiles space nonperiodically but cannot be used to make a periodic tiling.

For a long time, mathematicians also tended to believe the more general assertion that if you can construct a nonperiodic tiling with a set of one *or more* tiles, you can construct a periodic tiling from the same tiles. But in 1964 a set of tiles was found that permits only nonperiodic tiling. It contains 20,000 different shapes! Over the next several years, smaller sets were discovered with the same property, with as few as 100 shapes. But it was still amazing when in 1975 Sir Roger Penrose, a mathematical physicist at Oxford, announced a set that tiles only nonperiodically–consisting of just two tiles! (See Figure 20.21 and Spotlight 20.3.)

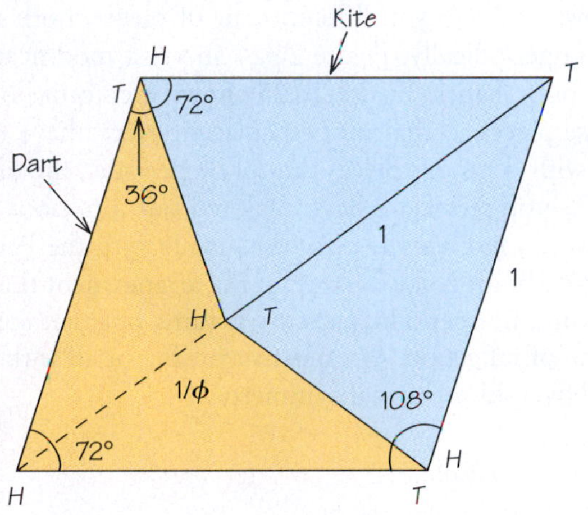


**FIGURE 20.21** Construction of Penrose's "dart" (beige area) and "kite" (blue area). The length $1/\phi \approx 0.618$ is the reciprocal of the golden ratio $\phi$.

## SPOTLIGHT 20.3 Sir Roger Penrose

Sir Roger Penrose
*(Anthony Howarth/Photo Researchers.)*

Sir Roger Penrose, a professor at the University of Oxford, received a doctorate in mathematics but has been seriously interested in physics for many years; he was one of the first to conjecture the existence of black holes. He discovered what are now called the Penrose pieces in 1973. His latest endeavor has been to try to establish that the mind is not a machine, that is, that the ideas and concepts of artificial intelligence cannot explain human consciousness.

A chance meeting with Maurits Escher resulted in Penrose sending him some of his grandfather's art, which helped inspire some of Escher's prints.

Penrose called his tiles "darts" and "kites," and both of these *Penrose tiles* can be obtained from a single rhombus. A **rhombus** is a quadrilateral with four equal sides and equal opposite interior angles. The particular rhombus from which the Penrose tiles are constructed has interior angles of 72° and 108°. If we cut the longer diagonal in two pieces so that the longer piece is the golden ratio, or $(1 + \sqrt{5})/2 \approx 1.618$ times as long as the shorter (see Chapter 19, p. 712), and connect the dividing point to the remaining corners, we split the rhombus into a dart and a kite (Figure 20.21).

Label the front and back vertices of the dart with $H$ (for head) and its two wing tips with $T$ (for tail), and do the reverse for the kite. Then the rule for fitting the pieces together is that only vertices with the same letter may meet: Heads must go to heads, tails must go to tails. Thus the rules don't allow the pieces to fit together as a rhombus (which would allow them to tile periodically).

A prettier method of enforcing the rules, proposed by Conway, is to draw circular arcs of different colors on the pieces and require that adjacent edges must join arcs of the same color. The result is the pretty patterns of Figure 20.22. In fact, Conway thinks of the darts as children, each with two hands. The rule for fitting the pieces together is that children are required to hold hands. Penrose patterns become dancing circles of children.

Figure 20.23 shows a tiling by a different pair of pieces, both rhombuses, that tile the plane only nonperiodically. Figure 20.24 shows a modification of the Penrose pieces into two bird shapes. Figure 20.25 shows a coloring of one particular tiling with the Penrose pieces so that no two adjacent pieces have the same color.

Although tilings with Penrose's pieces cannot be periodic, the tilings possess unexpected symmetry. As you recall, we have explored our intuitions of symmetry in terms of *balance, similarity,* and *repetition.* Patterns made with the Penrose pieces certainly involve repetition, but it is the balance in the arrangement that we seek. What balance can there be in a nonperiodic pattern? It turns out that some Penrose patterns have a single line of reflection. But most surprising of all is that every Penrose pattern has a kind of fivefold rotational symmetry.

**FIGURE 20.22** A Penrose tiling with specially marked tiles, forming what is known as the cartwheel tiling. (*From Sir Roger Penrose.*)

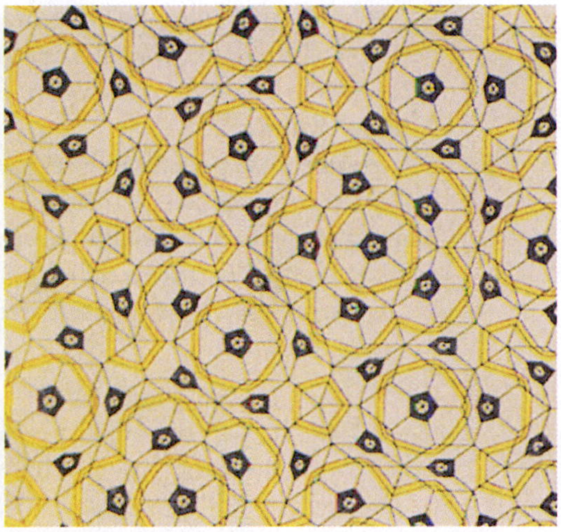

**FIGURE 20.23** A Penrose nonperiodic tiling made with two rhombus shapes, one thin and one fatter. The fatter one has a yellow stripe across one end. (*Tiling by Sir Roger Penrose.*)

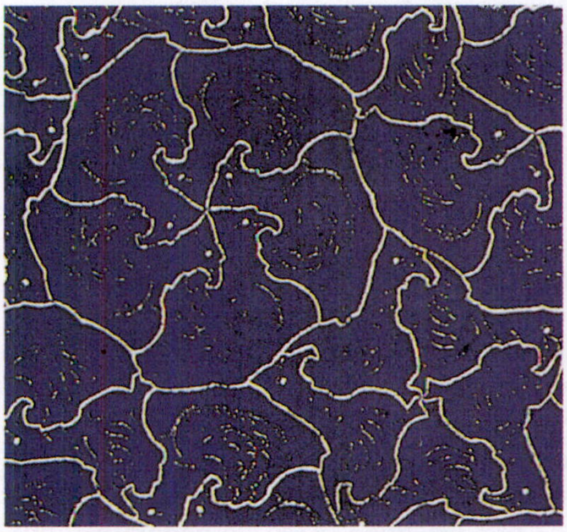

**FIGURE 20.24** A modification of a Penrose tiling by refashioning the kites and darts into bird shapes. (*Tiling by Sir Roger Penrose.*)

**FIGURE 20.25** A Penrose tiling by kites and darts, colored with five colors. A Penrose tiling can always be colored using four colors, in such a way that two tiles that share an edge have different colors. Whether a Penrose tiling can be colored in such a way using only three colors is an unsolved problem. However, we know that if even one Penrose tiling can be colored using three colors, all Penrose tilings can. (*Tiling by Sir Roger Penrose.*)

## EXAMPLE 7

### How Does a Penrose Pattern Have Fivefold Symmetry?

**SOLUTION** Look again at Figure 20.21, which shows how to split a rhombus into the Penrose dart and kite pieces. Except in the recess of the dart and the matching part of the kite, all of the internal angles of the kite and of the dart are either 72° or 36°.

Now, 72° goes into 360° five times, and 36° goes into 360° ten times. If we recall that it is the interior angles that matter in arranging polygons around a point, we see why it might be possible for a Penrose pattern to have fivefold or tenfold rotational symmetry.

A Penrose pattern with tenfold rotational symmetry is impossible, but there are exactly two Penrose patterns that tile the entire plane with fivefold rotational symmetry about one particular point. We show finite parts of these patterns in Figure 20.26. For each pattern, the center of rotational symmetry is at the center of the figure, surrounded by either five darts or five kites.

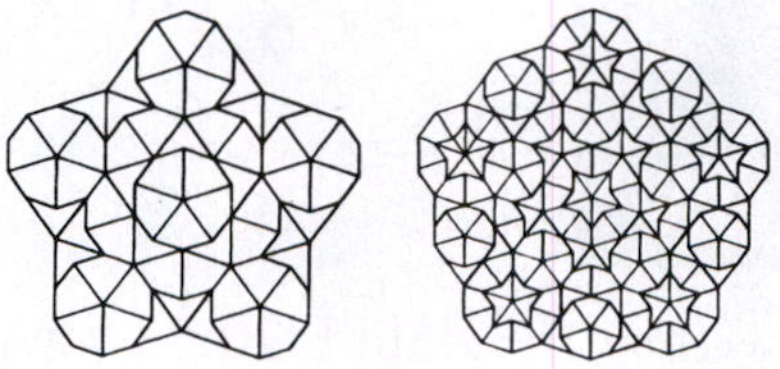

**FIGURE 20.26**
Successful deflation (that is, the systematic cutting up of large tiles into smaller ones) of patches of tiles of a Penrose nonperiodic tiling.

For any other Penrose pattern, the pattern as a whole does not have fivefold rotational symmetry. However, what is surprising is that the pattern must have arbitrarily large finite regions with fivefold rotational symmetry. You can see this feature in the regions of Figure 20.23 that are enclosed by yellow lines. In Conway's metaphor, whenever a chain of children (darts) closes, the region inside has fivefold symmetry.

Conway invented a process called *inflation* that takes any Penrose pattern into a different Penrose pattern with larger darts and kites. The inflation operation (we don't give the details here) systematically cuts up the darts and kites into triangles and regroups the triangles into larger darts and kites.

We can use inflation to show that a Penrose pattern must be nonperiodic. Suppose (contrary to what we want to establish) that some Penrose pattern is periodic; that is, it has translation symmetry. Let $d$ be the distance along the translation direction to the first repetition. Performing inflation does the same thing to each repetition, so the inflated pattern must still have translation symmetry and a distance $d$ along the translation direction to the first repetition. Keep on performing inflation, time after time, until the darts and kites are so large that they are more than $d$ across. The pattern, as we have just argued, must still have translation symmetry at a distance $d$, but it can't, because there's no repetition inside a single tile! We reach a contradiction. So what's wrong? Our initial supposition, that the pattern was periodic in the first place, must have been erroneous. We conclude that all Penrose tilings are nonperiodic.

Despite their being nonperiodic, all Penrose patterns are somewhat alike, in the following remarkable sense.

**Penrose Inside of Penrose** THEOREM

The subpattern of any finite region in one Penrose pattern is contained somewhere inside every other Penrose pattern. In fact, any subpattern occurs infinitely many times in every Penrose pattern.

The nonperiodicity of Penrose filings found a surprising application in 1997–to bathroom tissue. Quilted bathroom tissue is embossed with a pattern to keep the layers together (Figure 20.27). If the pattern is regular, then the multiple layers on the roll can produce lumpy ridges and grooves. Using a nonrepeating Penrose pattern averts the lumpiness. However, the company used Penrose's pattern without his permission, and Penrose sued successfully.

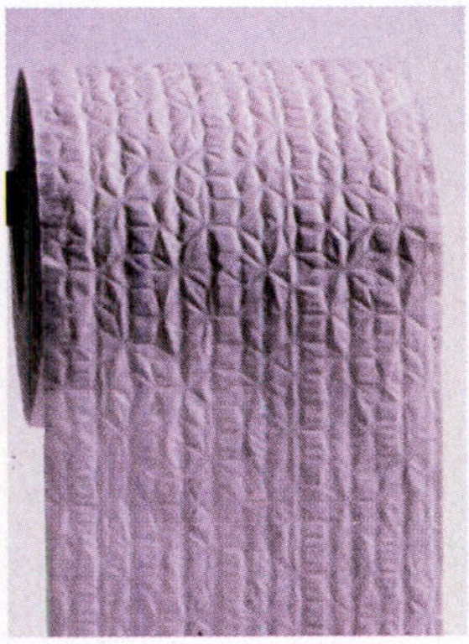

**FIGURE 20.27** Penrose toilet paper. (*Mario Ruiz/Time Magazine.*)

Penrose tilings have another feature that allows us to characterize them as *quasiperiodic,* or somewhere between periodic and random. (Noting the precise definition of random would take us too far afield.) Robert Ammann introduced onto the two rhombic Penrose pieces used in Figure 20.23 lines now known as *Ammann bars.* In any Penrose tiling, these bars line up into five sets of parallel lines, each set rotated 72° from the next, forming a pentagonal grid (Figure 20.28). The distance between two adjacent parallel bars is one of only two values, either $A$ or $B$. Do you want to guess what the ratio of the longer $A$ is to the shorter $B$? You don't think it could possibly be anything but the golden ratio of Chapter 19, do you? And so it is.

## EXAMPLE 8 ■ Musical Sequences

What about the order in which the $A$'s and $B$'s occur, as we move from left to right in Figure 20.28? Is there any pattern to that?

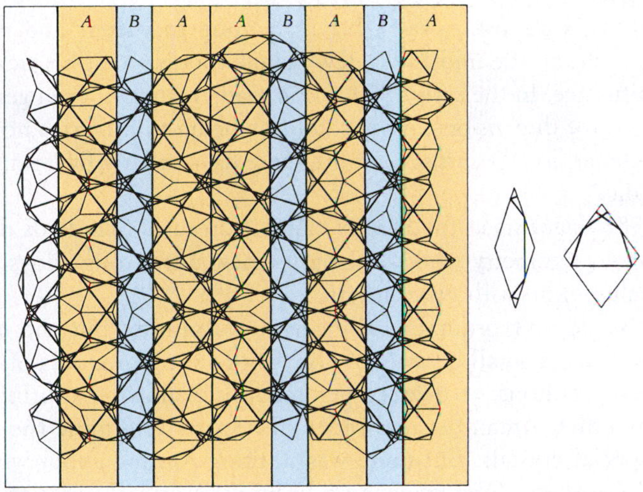


**FIGURE 20.28** Penrose tilings with Ammann bars. Specially placed lines on the tiles produce five sets of parallel bars in different directions.

**SOLUTION** From the limited part of the pattern that we can observe, we see the sequence as

$$ABAABABAABABA$$

You might think from the figure that the pattern continues repeating the group

$$ABAAB$$

indefinitely; after all, there are five symbols in this group. But such is not the case. The sequence of intervals between Ammann bars is nonperiodic–it cannot be produced by repeating any finite group of symbols. We can think of it as a one-dimensional analogue of a Penrose tiling. The notation is reminiscent of the melody pattern of songs: Many popular songs follow the pattern *ABA*, with the first and the last sections having the same melody but the middle section being different. Consequently, a sequence of intervals between Ammann bars is known as a *musical sequence.*

There is some regularity in musical sequences. Two $B$'s can never be next to each other, nor can we have three $A$'s in a row. Just as any finite part of any Penrose tiling occurs infinitely often in any other Penrose tiling, any finite part of any musical sequence appears infinitely often in any other one. The order of the symbols is neither periodic nor random, but between the two–quasiperiodic.

The ratio of darts to kites in an infinite Penrose tiling, or of $A$'s to $B$'s in a musical sequence, is exactly the golden ratio, approximately 1.618. So if you are going to play with sets of Penrose pieces to see what kinds of patterns you can create, you will need about 1.6 times as many darts as kites.

As pointed out by geometers Marjorie Senechal (Smith College) and Jean Taylor (Rutgers University), Penrose tilings have three important properties:

- They are constructed according to rules that force nonperiodicity.
- They can be obtained from a substitution process (inflation and deflation) that features self-similarity at different scales (like the fractals in Chapter 19).
- They are quasiperiodic.

These properties are somewhat independent, meaning that one or two may be true of a tiling without all three being true.

## Quasicrystals and Barlow's Law

Although Penrose's discovery was a big hit among geometers and in recreational mathematics circles in the mid-1970s, few people thought that his work might have practical significance. In the early 1980s some mathematicians even generalized Penrose tilings to three dimensions, using solid polyhedra to fill space nonperiodically. Like the two-dimensional Penrose patterns, these have orderly fivefold symmetry but are nonperiodic.

Yet in 1982 scientists at the U.S. National Bureau of Standards discovered unexpected fivefold symmetry while looking for new ultrastrong alloys of aluminum (mixtures of aluminum with other metals).

Manganese doesn't ordinarily alloy with aluminum, but the experimenters were able to produce small crystals of alloy by cooling mixtures of the two metals at a rate of millions of degrees per second. Following routine procedures, chemist Daniel Shechtman began a series of tests to determine the atomic structure of the special crystals. But there was nothing routine about what he found: The atomic structures of the manganese–aluminum crystals were so startling that it took Shechtman three years to convince his colleagues they were real.

Why did he encounter such resistance? His patterns–and the crystals that produced them–defied one of the fundamental laws of crystallography. Like our discovery that the plane cannot be tiled by regular pentagons, **Barlow's law**, also called the **crystallographic restriction**, says that a crystal must be periodic and hence can have only rotational symmetries that are twofold, threefold, fourfold, or sixfold. If there were a center of fivefold symmetry, there would have to be many such centers. Barlow proved this impossible.

Peter Barlow (1776–1862) argued by contradiction, similar to Conway's proof in which we saw earlier that Penrose patterns are not periodic. Suppose (contrary to what we intend to show) that there is more than one fivefold rotation center. Let $A$ and $B$ be two of these that are closest together (see Figure 20.29). Rotate the pattern of Figure 20.28 by one-fifth of a turn clockwise around $B$, which carries $A$ to some point $A'$. Because the pattern has fivefold symmetry around $B$, the point $A'$, which is the image of the fivefold center $A$, must itself be a fivefold center. Now use $A$ as a center and rotate the pattern by one-fifth of a turn clockwise, which carries $B$ to some point $B'$. As we just argued in the case of $A'$, $B'$ must also be a fivefold center. But $A'$ and $B'$ are closer together than $A$ and $B$, which is a contradiction. Hence our original supposition must be false, and a pattern can have at most one fivefold rotation center (as the patterns in Figure 20.26 in fact do) and so cannot be periodic.

For chemists, crystals are modeled well by periodic three-dimensional tilings; an array of atoms with no symmetry whatever would not be considered a crystal. Since Barlow's law shows that fivefold symmetry is impossible in a *periodic* tiling, no one suspected until Penrose's discovery that there could be symmetric *non*periodic tilings, nor until Schechtman's alloys that real atoms could arrange themselves in such a way.

Schechtman's alloys, since they are not periodic, are not crystals, though in other respects they do resemble crystals. It is scientifically more fruitful to extend the concept of crystals to include them than to rule them out. They are now known as *quasicrystals* (Spotlight 20.4).

Once again, as so often happens in history, pure mathematical research anticipated scientific applications. Penrose's discovery, once just a delightful piece of recreational mathematics, has prompted a major reexamination of the theory of crystals. Barlow's law is not refuted, since it applies only to periodic crystals, not to quasicrystals.

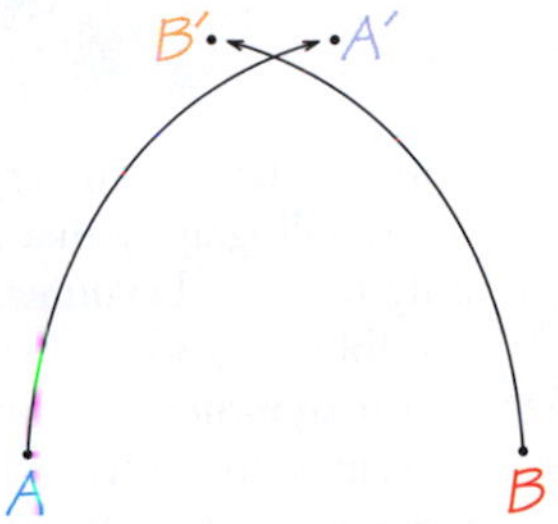


**FIGURE 20.29** Barlow's proof that no pattern can have two centers of fivefold symmetry.

## SPOTLIGHT 20.4 Quasicrystals

In 1984, working at the University of Pennsylvania, Paul Steinhardt and Don Levine did a computer simulation of what a three-dimensional Penrose pattern would be like. They decided to call such structures *quasicrystals*. Later that fall, their chemist colleague Daniel Shechtman showed that quasicrystals really exist. He produced images of an alloy of aluminum and manganese that were amazingly similar to images from the computer simulations. In short order, sevenfold, ninefold, and other symmetries were also shown to occur in real materials.

In 1991, Sergei Burkov showed that quasiperiodic tilings can be made using only a single kind of 10-sided tile, *provided the tiles are allowed to overlap*. With overlaps, the resulting patterns are no longer tilings. They are called *coverages*. In late 1998, scientists presented electron microscope photos that demonstrated that atoms really do form such coverages.

The current theory is that quasicrystals are packings of copies of a single type of atom cluster, with each cluster sharing atoms with its neighbors, that is, overlapping nearby clusters. The clusters form a quasiperiodic pattern that maximizes their density, thereby minimizing the energy of the atoms involved.

In 2007 Steinhardt and Peter J. Lu announced the discovery of decagonal and Penrose tilings in medieval Islamic architecture in Iran.

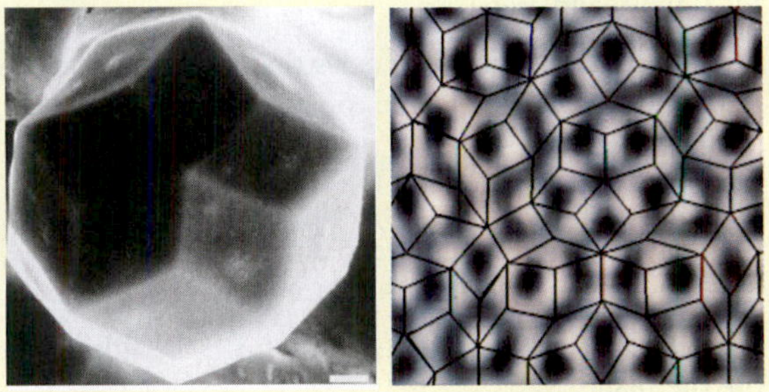

(a) (b)

(a) A scanning electron microscope image of the quasicrystal alloy $Al_{5???}Li_3Cu$ (the question marks indicate uncertainty about how many aluminum atoms are involved). The fivefold symmetry can be seen in the five rhombic faces that meet at a single point in the center of the photograph, forming a starlike shape. (b) This image of the quasicrystal material $Al_{65}Co_{20}Cu_{15}$ was obtained with a scanning tunneling microscope. The resulting image has been overlaid with a nonperiodic tiling to display the local fivefold symmetry. *[Both adapted from Hans C. von Baeyer, "Impossible Crystals,"* Discover *11(2) (February 1990): 69–78, 84.]*

## REVIEW VOCABULARY

**Barlow's law,** or the **crystallographic restriction** A law of crystallography that states that a crystal may have only rotational symmetries that are twofold, threefold, fourfold, or sixfold. (p. 668)

**Centrosymmetric** Symmetric by 180° rotation around its center. (p. 659)

**Convex** A geometric figure is convex if for any two points on the figure (including its boundary), all the points on the line segment joining them also belong to the figure (including its boundary). (p. 653)

**Conway criterion** A criterion for determining whether a shape can tile by means of translations and half-turns. (p. 659)

**Edge-to-edge tiling** A tiling in which adjacent tiles meet only along full edges of each tile. (p. 650)

**Equilateral triangle** A triangle with all three sides equal. (p. 648)

**Exterior angle** The angle outside a polygon formed by one side and the extension of an adjacent side. (p. 648)

**Fundamental region** A tile or group of adjacent tiles that can tile by translation. (p. 662)

**Interior angle** The angle inside a polygon formed by two adjacent sides. (p. 648)

**Monohedral tiling** A tiling with only one size and shape of tile (the tile is allowed to occur also in "turned-over," or mirror-image, form). (p. 648)

***n*-gon** A polygon with $n$ sides. (p. 648)

**Nonperiodic tiling** A tiling in which there is no repetition of the pattern by translation. (p. 663)

**Parallelogram** A convex quadrilateral whose opposite sides are equal and parallel. (p. 652)

**Par-hexagon** A hexagon whose opposite sides are equal and parallel. (p. 658)

**Periodic tiling** A tiling that repeats at fixed intervals in two different directions, possibly horizontal and vertical. (p. 662)

**Quadrilateral** A polygon with four sides. (p. 652)

**Regular polygon** A polygon all of whose sides and angles are equal. (p. 648)

**Regular tiling** A tiling by regular polygons, all of which have the same number of sides and are the same size; also, at each vertex, the same kinds of polygons must meet in the same order. (p. 649)

**Rhombus** A parallelogram all of whose sides are equal–four equal sides and equal opposite interior angles. (p. 664)

**Scalene triangle** A triangle no two sides of which are equal. (p. 652)

**Semiregular tiling** A tiling by regular polygons; all polygons with the same number of sides must be the same size. (p. 651)

**Tiling** A covering of the entire infinite plane by nonoverlapping figures. (p. 647)

**Translation** A rigid motion that moves everything a certain distance in one direction. (p. 655)

**Vertex type** The pattern of polygons surrounding a vertex in a tiling. (p. 651)

## SKILLS CHECK

**1.** In a tiling of the plane, the tiles

**(a)** are allowed to overlap, as long as no area is left uncovered.
**(b)** are not allowed to overlap.
**(c)** must meet edge-to-edge.

**2.** The exterior angle of a regular octagon is __________ .

**3.** In a tiling of the plane, the tiles

**(a)** have to be regular polygons.
**(b)** have to be polygons but need not be regular.
**(c)** can be any shape at all.

**4.** A regular tiling can be constructed using polygons with _____, _____, or _____ sides.

**5.** Regular octagons and squares can form a semiregular tiling of the plane with

**(a)** two octagons and one square at each vertex.
**(b)** two octagons and two squares at each vertex.
**(c)** a varying configuration at the vertices.

**6.** A semiregular tiling has a square, a regular dodecagon (12-gon), and another regular polygon at each vertex. This other polygon has __________ sides.

**7.** A tiling of the plane can be formed using as a tile

**(a)** some but not all convex quadrilaterals.
**(b)** any convex quadrilateral, but no nonconvex quadrilaterals.
**(c)** any quadrilateral.

**8.** The smallest number of sides that a polygon can have and not be able to tile the plane is ____________ .

**9.** A tiling of the plane can be formed using as a tile

**(a)** some but not all pentagons.
**(b)** any pentagon with at least two right angles.
**(c)** any pentagon with at least three right angles.

**10.** A regular polygon with ______ or more sides cannot tile the plane.

11. Can the tile below be used to create a tiling of the plane?

(a) No
(b) Yes, using only translations
(c) Yes, using translations and half-turns

12. The ____________ criterion says that the tile below can be used to create a tiling of the plane using only ________ and ________ .

13. Which of the following is true?

(a) If a polygon fulfills the Conway criterion, it can tile the plane by translations.
(b) If a polygon fulfills the Conway criterion, it can tile the plane by translations and half-turns.
(c) If a polygon doesn't fulfill the Conway criterion, it can't tile the plane at all.

14. Any quadrilateral can tile the plane by __________ .

15. In a nonperiodic tiling of the plane,

(a) the pattern never repeats.
(b) the pattern is not repeated by any translation.
(c) there must be at least three kinds of tiles.

16. Penrose tilings are ___________-periodic.

17. A rhombus always has the property that

(a) opposite sides are unequal in length.
(b) opposite angles are congruent.
(c) it cannot be a square.

18. In a Penrose tiling, the proportion of darts to kites is _______________________ .

19. A Penrose dart always has the trait that

(a) opposite angles are congruent.
(b) it is nonconvex.
(c) the edges are all of different lengths.

20. Barlow's law prohibits the existence of crystals with _______________________ symmetry.

# CHAPTER 20 EXERCISES

■ Challenge ◆ Discussion

*Hint:* For the exercises about determining whether a shape will tile the plane, you should make a number of copies of the shape and experiment with placing them. One easy way to make copies is to trace the shape onto a piece of paper, staple half a dozen other blank sheets behind that sheet, and use scissors to cut through all the sheets along the edges of the traced shape on the top sheet.

## 20.1 Tilings with Regular Polygons

1. Determine the measure of an exterior angle and of an interior angle of a regular octagon (eight sides).

2. Determine the measure of an exterior angle and of an interior angle of a regular decagon (10 sides).

3. Discover a formula for the measure of an interior angle of a regular $n$-gon.

4. Using the formula from Exercise 3 and either your calculator or a short computer program, make a chart of the interior-angle measures of regular polygons with 3, 4, . . . , 12 sides.

■ 5. Use the chart of interior-angle measures from Exercise 4 to determine all of the possible vertex types of regular polygons (with at most 12 sides) surrounding a point.

■ 6. Which of the vertex types of Exercise 5 do not occur in a semiregular tiling?

■ 7. In addition to the vertex types of Exercise 5, exactly five others are possible, each involving one polygon with more than 12 sides. None of these vertex types leads to a semiregular tiling. The five many-sided polygons involved in these five vertex types have 15, 18, 20, 24, and 42 sides. Determine the other polygons in these vertex types.

For Exercises 8 and 9, refer to the lower left corner of Figure 20.5, which shows a tiling by isosceles triangles.

8. Use the center vertex to determine the measures of the angles of the isosceles triangle tile.

9. Every vertex except the center vertex has the same vertex type, in terms of the measures of the angles surrounding the vertex. What is that vertex type?

10. Give a numerical reason why a semiregular tiling could not include both polygons with 12 sides and polygons with 8 sides (with or without any polygons with other numbers of sides).

## 20.2 Tilings with Irregular Polygons

11. You know that a regular pentagon cannot tile the plane. Suppose you cut one in half. Can this new shape

tile the plane? (See Figure 19.4 for a regular pentagon that you can trace.)

**12.** For each of the tiles below, show how it can be used to tile the plane. (Adapted from *Tilings and Patterns,* by Branko Grünbaum and G. C. Shephard, Freeman, New York, 1987, p. 25.)

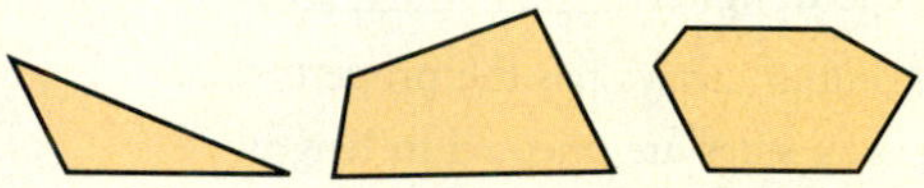

(*a*) (*b*) (*c*)

## 20.3 Using Translations

Refer to tiles (a) through (g) below in doing Exercises 13 and 14.

**13.** For each of the tiles (a) through (c), determine whether it can be used to tile the plane by translations. (From *Tiling the Plane,* by Frederick Barber et al., COMAP, Lexington, Mass., 1989, pp. 1, 8, 9.)

**14.** Repeat Exercise 13, but for tiles (d) through (g).

**15.** Start from a par-hexagon of your choice and modify it to tile the plane by translations. (You will probably find it useful to do your work on graph paper. If you choose a regular hexagon, there is special graph paper, ruled into regular hexagons, that would be particularly useful.) Can you draw a design on the tile so as to make an Escher-like pattern?

**16.** Start from a parallelogram of your choice and modify it to tile the plane by translations. (You will probably find it useful to do your work on graph paper.) Can you draw a design on the tile so as to make an Escher-like pattern?

Refer to the following information in doing Exercises 17–20.

A particularly simple kind of polygon, called a *polyomino,* is one made of squares joined edge-to-edge. The name is a generalization of "domino"; indeed, there is only one kind of domino (two squares joined at an edge to form a rectangle). There are just two "trominos" (short for "triominos"), the straight tromino and the L-tromino.

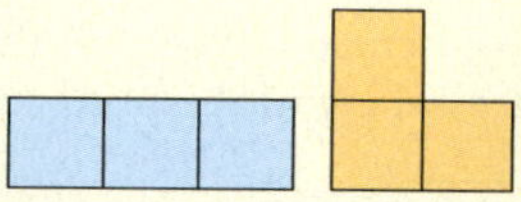

The straight tromino has the shape of a rectangle, so it can tile the plane by translations; and the L-tromino has the shape of a hexagon.

**17.** Is the L-tromino convex? Does the result about what hexagons can tile the plane (p. 654) give any information about whether the L-tromino can tile the plane or not?

**18.** Find a tiling of the plane using just the L-tromino and translations of it. Is there more than one way to do the tiling?

**19.** Show how alternative 2 of the translation criterion can be applied to the L-tromino.

■ **20.** Give an argument why alternative 1 of the translation criterion cannot be applied to the L-tromino. (*Hint:* Label each of the eight corners of the component squares of the tromino with the letters *S*, *T*, . . . , *Z*. Let these be our candidates for the points *A*, *B*, *C*, and *D* of the criterion.) Each of the sides of the tromino that is two units long has nowhere to go under a translation. Any application of the criterion must divide each side

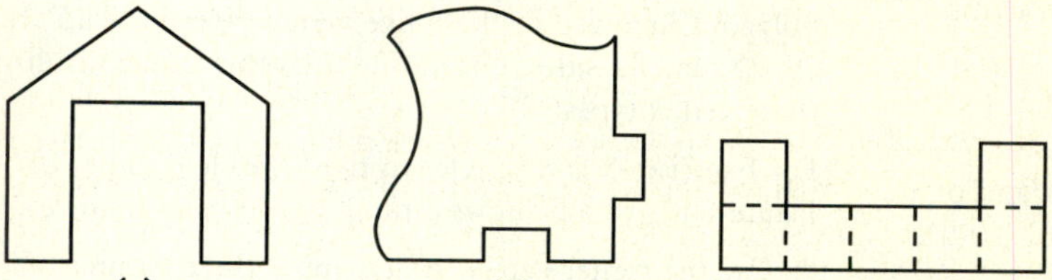

(*a*) (*b*) (*c*)

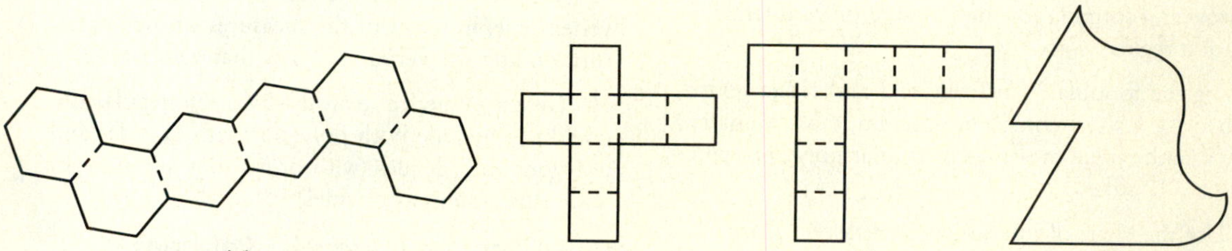

(*d*) (*e*) (*f*) (*g*)

into two pieces, so their midpoints must be two of the points $A$, $B$, $C$, and $D$. Make a similar argument about two corners of the tromino. Thus, we have four points, which can be labeled consecutively $A$, $B$, $C$, and $D$, starting at any one of them. Show that none of the four possibilities "works." (This argument can be generalized to show that trying $A$, $B$, $C$, and $D$ at points other than the corners of the squares won't work either.)

Refer to the following information in doing Exercises 21–26.

Demonstrate to your own satisfaction that there are exactly five shapes of tetrominos (each made of four squares joined at edges)–plus differing mirror images of two of them–as shown below. In the order shown, they are called the square, straight, T, L, and skew tetrominos. The straight and the square tetromino certainly can tile the plane by translations.

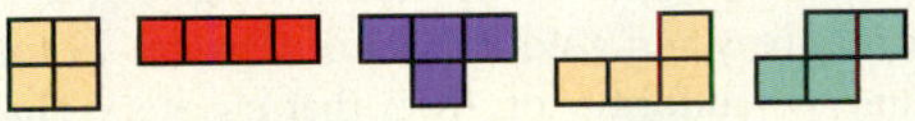

You will definitely find it useful to make yourself several copies of each of the polyominos mentioned below, by cutting them out of graph paper.

**21.** Apply alternative 1 of the translation criterion to the straight-tetromino and show how it can tile.

**22.** Show how alternative 2 of the translation criterion applies to the T-tetromino.

**23.** Apply alternative 1 of the translation criterion to the L-tetromino and show how it can tile.

**24.** Show how alternative 2 of the translation criterion applies to the skew-tetromino.

**25.** Show how alternative 2 of the translation criterion can be applied to the skew-tetromino, and show how it can tile.

**26.** In Exercises 23 and 25, we indulged in what appears to be "overkill," proving the same fact in two different ways. But those exercises should give you the idea that alternative 2 of the translation criterion can reduce to (and hence is more general than) alternative 1 if some points are allowed to coincide. For such a reduction, which pairs of points must coincide? (We are allowed to relabel the remaining four distinct points.)

## 20.4 Using Translations Plus Half-Turns

For Exercises 27 and 28, refer to tiles (a) through (g) on p. 672.

**27.** For each of the tiles (a) through (c), determine whether it can be used to tile the plane by translations and half-turns.

**28.** Repeat Exercise 27, but for tiles (d) through (g).

**29.** Show how an arbitrary pentagon with two parallel sides can tile the plane.

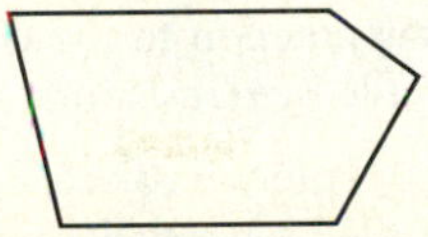

**30.** The following is a pentagonal tile of type 13, discovered by Marjorie Rice. Show how it can tile the plane. (*Hint:* Carefully trace and cut out a dozen or so copies and try fitting them together.)

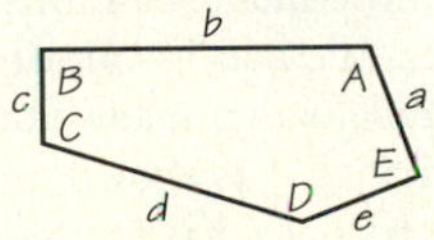


The parts of this pentagon satisfy the following relations: $A = C = D = 120°$, $B = E = 90°$, $a = e$, and $a + e = d$. [Adapted from "In Praise of Amateurs," by Doris Schattschneider, in David A. Klarner (ed.), *The Mathematical Gardner*, Wadsworth, Belmont, Calif., 1981, p. 162.]

**31.** Start from a triangle of your choice and modify it to tile the plane by translations and half-turns. (You will probably find it useful to do your work on graph paper.) Can you draw a design on the tile so as to make an Escher-like pattern?

**32.** Start from a quadrilateral of your choice and modify it to tile the plane by translations and half-turns. (You will probably find it useful to do your work on graph paper.) Can you draw a design on the tile so as to make an Escher-like pattern?

Refer to the information about polyominos preceding Exercise 17, and to the following, in doing Exercises 33–36.

We saw earlier that all the dominos, trominos, and tetrominos tile the plane by translations. Here we investigate the 12 pentominos, shown as follows with a letter notation for each (if you allow mirror images to count as different pentominos, there are 18). It will be useful for you to make several copies of each of the pentominos discussed below.

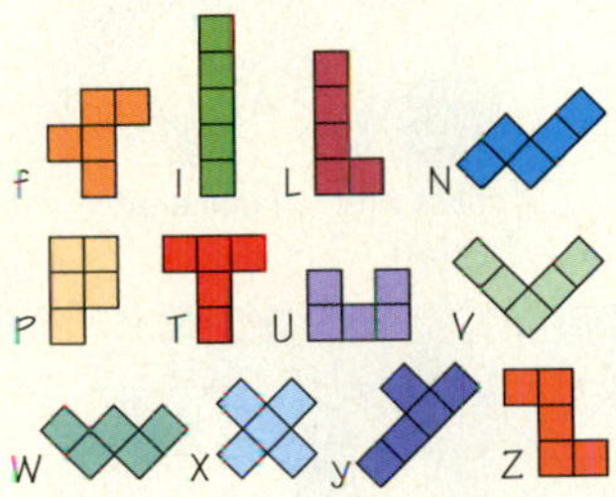


**33.** Just by experimenting, determine which of the pentominos can tile the plane by translations. (*Hint:* There are nine.)

**34.** Apply the Conway criterion to the f-pentomino, and show how it can tile by translations and half-turns.

**35.** Apply the Conway criterion to the U-pentomino, and show how it can tile by translations and half-turns.

**36.** Apply the Conway criterion to the T-pentomino, and show how it can tile by translations and half-turns.

**37.** In the text we discuss criteria and methods for generating Escher-like patterns that involve just translations or translations and half-turns. A slight variation on one of those methods allows construction of tilings that feature a tile and its mirror image.

Begin with a parallelogram made from two congruent isosceles triangles, as shown in the figure below. Each of these triangles has two sides equal. Be sure that the two triangles are arranged so that they have one of the equal sides in common, forming a diagonal of the parallelogram.

Make any modification to half of the third side of one of the triangles. Mirror-reflect that modification across the side, then translate the reflection to become the modification of the other half of the side. Take the complete modification of this side, and translate it to become the modification of the opposite side of the parallelogram.

Modify in any way one of the two remaining sides of the parallelogram, and make the same modification to the opposite side (that is, translate the modification, without rotation or reflection). Then reflect this modification across the diagonal of the parallelogram.

The result is a modified parallelogram that tiles by translation and splits into two pieces that are mirror images of each other. Escher used a similar technique, but starting from a par-hexagon made from two quadrilaterals, in his *Horseman* print, as shown in his sketch in Figure 20.11b.

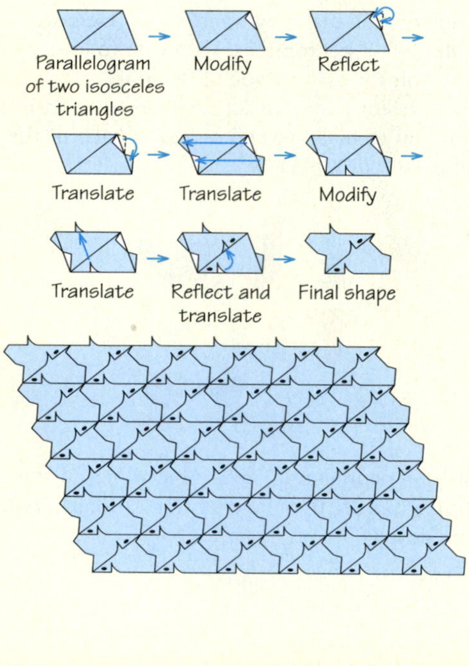


Use this technique to produce a tiling of your own design. Can you draw a design on the tile so as to make an Escher-like pattern?

**38.** Show that the modified parallelogram in Exercise 37 fulfills the Conway criterion, by identifying the six points of the criterion.

## 20.5 Nonperiodic Tilings

**39.** In this chapter we have been concerned mostly with tiling the plane, with some attention to using crystals to fill space. We can also consider a simpler case–tiling the line. For the line, a tile is a line segment of a particular length.

**(a)** What are the monohedral tilings of the line?

**(b)** What are the periodic tilings of the line that use two tiles of different lengths?

Exercises 40–42 connect nonperiodic tilings to the Fibonacci numbers and golden ratio of Chapter 19 but do not require any information from that chapter. (Thanks for this idea to David J. Wright, Oklahoma State University.) Just as the plane can be tiled quasiperiodically with Penrose tiles, the line can be tiled quasiperiodically with a pair of tiles, provided that their lengths are in the right proportion. Let the lengths of the tiles be $a$ and $b$, with $b < a$ and $a$ exactly $c$ times as long as $b$, so that $cb = a$. Thus, scaling up a tile of length $b$ by a factor of $c > 1$ produces a tile of length $a$. Similarly, we determine the scale factor $c$ so that scaling up a tile of length $a$ produces a tile of length $a$ followed by a tile of length $b$, that is, $ca = a + b$.

**40.** **(a)** Using substitution, eliminate $a$ and reduce the two equations to a single equation in just $c$ and $b$ alone.

**(b)** Use the quadratic formula to solve for the two possible values of $c$. Since we want $c > 1$, we choose the larger value.

**41.** We can define an inflation process for a line segment consisting of $a$'s and $b$'s: First, replace each original $b$ by an $a$ and each original $a$ by two adjacent segments $a$ and $b$. For example, we would replace $aba$ by $(ab)(a)(ab)$, where we have inserted parentheses for clarity.

**(a)** Start with just a single $b$ and repeat the inflation process, showing the stages, until you reach a stage with 21 segments.

**(b)** How many segments are there at each stage? (If we continue this process forever, we tile a half-line to the right; we tile the entire line by reflecting this right half-line over to cover the left half-line. The result is called a *Fibonacci tiling* of the line.)

**(c)** If a line segment contains $m$ copies of the $a$ tile and $n$ copies of the $b$ tile, how many tiles will the inflation of the segment contain?

**42.** We can similarly define a deflation process: Replace each original adjacent pair $ab$ by an $a$ and each remaining original $a$ by a $b$.

**(a)** Apply this deflation process repeatedly to the stage in your answer to Exercise 41a that has 21 segments. What do you end up with?
**(b)** Apply this deflation process repeatedly to the periodic sequence of tiles *ababababab*.... What do you end up with?
**(c)** Devise your own periodic sequence of *a* and *b* tiles. Apply the deflation process to it repeatedly. What do you end up with? What do you conjecture?
**(d)** In what sense is the Fibonacci tiling quasiperiodic?

For Exercises 43–46, refer to the following.

The rabbit problem in Chapter 19 (Exercise 4) leads us directly into nonperiodic patterns and musical sequences. Let $A$ denote an adult pair of rabbits and $B$ denote a baby pair. We record the population at the end of each month, just before any births, in a particular systematic way–as a string of $A$'s and $B$'s. At the end of their second month of life, a rabbit pair will be considered to be adult and give birth to a baby pair. At the end of the first month, the sequence is just $A$, and the same is true at the end of the second month. When an adult pair $A$ has a baby pair $B$, we write the new $B$ immediately to the right of the $A$. So at the end of the third month, the sequence is $AB$; at the end of the fourth, it is $ABA$, because the first baby pair is now adult; at the end of the fifth month we have $ABAAB$.

Mathematicians and computer scientists call this manner of generating a sequence a *replacement system.* At each stage we replace each $A$ by $AB$ and each $B$ by $A$.

**43.** What is the sequence at the end of the sixth month?

**44.** Why can't we ever have two $B$'s next to each other?

**45.** Why can't we ever have three $A$'s in a row?

**46.** Show that from the fourth month on, the sequence for the current month consists of the sequence for last month followed by the sequence for two months ago.

For Exercises 47–54, refer to the following.

We can define inflation and deflation of a sequence of $A$'s and $B$'s, and musical sequences themselves, without reference to Penrose patterns, and thereby arrive at an example of a nonperiodic pattern in one dimension. Inflation consists of replacing each $A$ by $AB$ and each $B$ by $A$, and deflation consists of replacing each $AB$ by $A$ and each $A$ by $B$; inflation and deflation undo each other on musical sequences. Call a sequence *musical* if it results from applying inflation to the sequence consisting of a single $B$. Then inflation and deflation preserve musicality: If we inflate or deflate a musical sequence, we get another musical sequence. Another way to think of this relationship is that a musical sequence is self-similar under inflation and deflation.

■ **47.** Let the lone $B$ be considered the first stage of inflation. Show that at the $n$th stage of inflation, for $n \geq 3$, there are $F_n$ (the $n$th Fibonacci number, section 19.1) symbols in the sequence, of which $F_{n-1}$ are $A$'s and $F_{n-2}$ are $B$'s. (*Hint:* Check it for $n$ = 1, 2, 3, and 4.)

**48.** Show that no musical sequence contains $AAA$ or $BB$.

**49.** Show that no musical sequence ends in $AA$ or in $ABAB$.

**50.** Show that apart from the lone sequence $B$, every musical sequence is an initial subsequence of all the succeeding musical sequences.

**51.** Slightly modified, deflation can be used to check whether a finite block of $A$'s and $B$'s can belong to a musical sequence or not. First, if the block has length greater than one, we may suppose that it begins with an $A$ (why?). So at any stage of the deflation with a block beginning with $B$, we may add an initial $A$. Second, we add the additional deflation rule to replace an ending $AA$ with $BA$. If at any stage of this modified deflation we arrive at two or more $B$'s in a row, or three or more $A$'s in a row, then the original block could not be part of a musical sequence. Otherwise, the original block will eventually deflate to a single symbol, at which point we conclude that the original block is a part of a musical sequence. Check the two blocks $ABAABABAAB$ and $ABAABABABA$.

**52.** From Exercise 50 we know that each application of inflation to a musical sequence simply extends it. By successive inflation, then, we build an infinite sequence. Show that as we approach this limiting sequence, the ratio of $B$'s to $A$'s tends toward the golden ratio $\phi$.

**53.** Conclude from Exercise 52 that the sequence cannot be periodic, nor settle into a period after a finite "burn-in" period. Thus, the sequence is nonperiodic. (*Hint:* $\phi$ is not a rational number; that is, it cannot be represented as a ratio $m/n$ of whole numbers $m$ and $n$.)

**54.** Show that any finite block of $A$'s and $B$'s that occurs in the infinite sequence must occur over and over again (just as any patch of tiles in a Penrose pattern occurs infinitely often in the pattern). Thus, the infinite sequence is self-similar.

## WRITING PROJECTS

Get computer software to make some tilings of your own. Check for software at www.geom.uiuc.edu/software/tilings/TilingSoftware.html. Print out your tilings and describe, in a sentence or two each, how you made them.

## SUGGESTED READINGS


LEE, KEVIN. *TesselMania! Deluxe.* Computer program for producing Escher-like tilings for Macintosh or Windows. Available from www.worldofescher.com/store/compaccs.html. Free PC demo downloadable at www.worldofescher.com/down/tessdemo.exe.

LU, PETER J., and PAUL J. STEINHARDT. Decagonal and quasi-crystalline tilings in medieval Islamic architecture, *Science* 315 (23 February 2007) 1106–1110, www.sciencemag.org/cgi/content/full/315/5815/1106.

RANUCCI, ERNEST, and JOSEPH TEETERS. *Creating Escher-Type Patterns,* Creative Publications, Oak Lawn, Ill., 1977.

SCHATTSCHNEIDER, DORIS. *M. C. Escher: Visions of Symmetry*, 2nd ed., Harry N. Abrams, New York, 2004.

SEYMOUR, DALE, and JILL BRITTON. *Introduction to Tessellations,* Dale Seymour Publications, Palo Alto, Calif., 1989. An excellent introduction to tessellations, including how to make Escher-like tessellations.

TEETERS, JOSEPH L. How to draw tessellations of the Escher type, *Mathematics Teacher,* 67 (1974): 307–310.


## SUGGESTED WEB SITES

**www.geom.uiuc.edu/software/tilings** Lists programs for various platforms that allow the user to design tilings.

**www.geocities.com/SiliconValley/Pines/1684/Penrose.html** A Java applet to play with Penrose tiles.

**www.geom.uiuc.edu/apps/quasitiler** Interactive Web program QuasiTiler 3.0 that draws Penrose tilings and their generalizations in higher dimensions.

**www.geometrygames.org/** Interactive Windows and Macintosh program that lets the user design tilings on the plane, the sphere, and the hyperbolic plane. Spherical tilings can be realized as polyhedra.

**www.geometrygames.org/KaleidoTile/index.html** An interactive Windows and Macintosh Classic (PPC) program to create tilings on the plane, a sphere, and the hyperbolic plane.

**demonstrations.wolfram.com/ComplementTiling/** An interactive Web program to make Escher-like tilings, with links to other interactive demonstrations.

**www.eschertiles.com/links.html** Links to the official M.C. Escher site and to sites with Escher-inspired tilings and art.

# Your Money and Resources PART VII

This part concentrates on numerical patterns of growth and decline in the realms of finance, resources, and biology. The unifying concept is a population, whether of dollars, barrels of oil, or tons of fish.

How much interest will your savings account earn in the next year? How much will the payment be on your credit card, your car loan, or a home mortgage? How much would you need to save to pay for a child's college education or for your retirement? What will inflation do to your savings? How much should you pay for a stock? These are problems of daily life for which mathematics provides custom-tailored models. In Chapter 21, Savings Models, and Chapter 22, Borrowing Models, you become familiar with the mathematics and terminology of situations that you will face repeatedly.

Good mathematical models are often versatile and flexible, and the financial models of these chapters apply broadly to important problems in other areas of life. Growth of money at interest is like growth of biological populations. Inflation of a currency or depreciation of an asset is like the decay of a radioactive substance. Finding out how long a retirement "nest egg" will last is similar to determining how long before a nonrenewable resource, such as oil or coal, may be exhausted. Managing a trust fund, such as the endowment of a college, presents problems similar to management of a renewable biological resource, such as a forest or a fishery. In Chapter 23, The Economics of Resources, we explore these similarities, together with the profound effect that economic conditions can have on natural resources.

Finally, you will see the surprisingly large and puzzling consequences that very small changes can produce in a physical system or biological population as a result of behavior that mathematicians call chaos. ■

CHAPTER 21
Savings Models

CHAPTER 22
Borrowing Models

CHAPTER 23
The Economics of Resources

CHAPTER 21



# Savings Models

How much interest will your savings account earn in the next year? How much difference does it make how the interest is calculated? How much should you save for a comfortable retirement in the face of inflation? In this chapter, we consider questions such as these and show how the underlying mathematical models can explain such phenomena as the 2001 crash of "dot-com" stocks and the effect of interest-rate changes on stock prices.

## 21.1 Arithmetic Growth and Simple Interest

When you open a savings account, your primary concerns are the safety and the growth of the "population" of your savings. Suppose that you deposit \$1000 in an account that "pays interest at a rate of 10% annually." (This is an unrealistic rate. We use it because it makes the calculations simple.) Assuming that you make no other deposits or withdrawals, how much is in the account after 1, 2, or 5 years?

The \$1000 is the **principal**, the initial balance of the account. At the end of one year, **interest** is added. The amount of interest is 10% of the principal, or

$$10\% \times \$1000 = 0.10 \times \$1000 = \$100$$

So the balance at the beginning of the second year is \$1100.

We express an interest rate either as a percentage or as a decimal fraction. "Percent" means "per 100," so you can think of the symbol "%" as standing for "1 per 100" or $\frac{1}{100} = 0.01$. So to convert from a percentage to a decimal fraction, divide the percentage by 100 by moving the decimal point two places to the left. An interest rate of 10% is 10/100 or 0.10; an interest rate $r$ (as a decimal number) is $100r\%$.

With **simple interest**, interest is paid only on the original balance, no matter how much interest has accumulated. With simple interest, for a principal of \$1000 and a 10% interest rate, you receive \$100 interest at the end of the first year; so at the beginning of the second year, the account contains \$1100. But at the end of the second year, you again receive only \$100; so at the beginning of the third year, the account contains \$1200. In fact, at the end of each year you receive just \$100 in interest.

The formulas for simple interest are themselves simple.

**Simple Interest** RULE

For a principal $P$ and an annual rate of interest $r$, the interest earned in $t$ years is

$$I = Prt$$

and the total amount $A$ accumulated in the account is

$$A = P + I = P + Prt = P(1 + rt)$$

You may find this method for interest rather strange, since you are accustomed from your savings account to a different system of awarding interest–**compound interest**, which we will consider shortly. However, simple interest is often used for:

- private loans between individuals, because it is easy to calculate;
- commercial loans for less than one year–not just because it is easy to calculate–but also because for low interest rates, simple interest differs negligibly from compound interest; and
- financing of corporations and the government through bonds. A bond is a loan with repayment at the end of a fixed term and simple interest in the mean time, paid usually annually or semiannually.

## EXAMPLE 1 ■ Simple Interest on a Student Loan

Let's suppose that you have exhausted the amount that you can borrow under federal loan programs and need a private direct student loan for \$10,000.

National City Corporation (headquartered in Cleveland, Ohio) quoted a rate in May 2008 of 5.7% for the 2007–08 school year. It offers an interest-only repayment option, under which you make monthly interest payments while you are in school and pay on the principal only after graduation. Under this plan, National City earns simple interest from you while you are in school.

How much monthly interest would you pay for such a \$10,000 loan?

**SOLUTION** The principal is $P = \$10{,}000$, the annual interest rate is $r = 5.7\% = 0.057$ per year, and the number of years is $t = \frac{1}{12}$ year. The interest for one month would be $I = Prt = \$10{,}000 \times 0.057 \times \frac{1}{12} = 47.50$. (Actually, National City would charge an "origination fee" of between 3% and 10.5%, added to the principal, so the payment would be greater; and the initial interest rate might not be 5.7% but could be more than 12%, since it would depend on the creditworthiness of you and your cosigner. Finally, the interest rate could increase each year during the term of the loan.) ■

We frequently observe the kind of growth corresponding to simple interest, called **arithmetic growth** or **linear growth**, in other contexts.

**Arithmetic Growth** DEFINITION

**Arithmetic growth** (also called **linear growth**) is growth by a constant amount in each time period.

For example, the population of medical doctors in the United States grows arithmetically, because the medical schools graduate the same number of doctors each year (and the number of doctors dying is also fairly constant). The concept of linear growth has appeared already in the discussions of linear programming (Chapter 4) and linear regression (Chapter 6).

## 21.2 Geometric Growth and Compound Interest

What you probably expected to happen to the savings account discussed in the last section is that during the second year the account would earn interest of 10% not on the *initial* balance of $1000 (as with simple interest) but on the *new* balance of $1100. Then, at the end of the second year, 10% of $1100, or $110, would be added to the account.

Thus, during the second year you would earn interest on both the principal of $1000 and on the $100 interest that was earned during the first year. You receive more interest during the second year than during the first; that is, the account grows by a greater amount during the second year. At the beginning of the third year the account contains $1210, so at the end of the third year you receive $121 in interest. Again, this is larger than the amount you received at the end of the preceding year. Moreover, the increase during the third year,

$$\text{third-year interest} - \text{second-year interest} = \$121 - \$110 = \$11$$

is larger than the increase during the second year,

$$\text{second-year interest} - \text{first-year interest} = \$110 - \$100 = \$10$$

Thus, not only is the account balance increasing each year, but the amount added also increases each year.

**Compound Interest** DEFINITION

**Compound interest** is interest that is paid on both the original principal and accumulated interest.

Savings institutions usually compound interest and credit it to accounts more often than once a year–for example, quarterly (four times per year). With an interest rate of 10% per year and quarterly compounding, you get one-fourth of the rate, or 2.5%, paid in interest each quarter year. The "quarter" (three months) is the **compounding period**, or the time elapsing before interest is paid.

Consider again a principal of $1000. At the end of the first quarter, you have the original balance plus $25 interest, so the balance at the beginning of the second quarter is $1025. During the second quarter, you receive interest equal to 2.5% of $1025, or $25.625, which in posting to your account is rounded up (since the fraction is half a cent or more) to $25.63. Continuing in this manner, the balance at the end of the first year is $1103.82. (You should "read" all calculations in this chapter by confirming them on your calculator.)

Even though the account was advertised as paying 10% interest, the interest for the year is $103.82, which is 10.382% of the principal of $1000.

Practical note: Without rounding the interest for each quarter, the interest for the year would have been $1103.81, as shown in Table 21.1. This table shows the results of calculation with rounding done only at the end of the year, while savings institutions must round at each compounding and post the amount to your account. A spreadsheet program could duplicate the results of their computer programs; but in the table and in all later calculations, we take the simpler route of rounding only at the final answer. Any differences will be very small; and if your answers differ just a few cents, that will be OK.

**TABLE 21.1** Compound Interest on $1000, at an Interest Rate of 10% Compounded Quarterly

| Date | Beginning Balance | Interest on Principal | Interest on Interest | Total Interest Added | Ending Balance |
|---|---|---|---|---|---|
| January 1 | 1000.00 | | | | |
| March 31 | 1000.00 | 25.00 | 0.00 | 25.00 | 1025.00 |
| June 30 | 1025.00 | 25.00 | 0.63 | 25.63 | 1050.63 |
| September 30 | 1050.63 | 25.00 | 1.27 | 26.27 | 1076.90 |
| December 31 | 1076.90 | 25.00 | 1.92 | 26.92 | 1103.82 |

If interest is compounded monthly (12 times per year) or daily (365 or 366 times per year), the resulting balance is even larger, as shown in Table 21.2, (together with the results of continuous compounding, which we discuss later). We will show you shortly the compound interest formula for these calculations.

## Terminology for Interest Rates

We have seen that an account at a particular annual rate of interest can produce different amounts of interest, depending on how the compounding is done. To help prevent confusion on the part of consumers, the Truth in Savings Act establishes specific terminology and calculation methods for interest.

A **nominal rate** is any stated rate of interest for a specified length of time, such as a 3% annual interest rate on a savings account or a 1.5% monthly rate on a credit-card balance. By itself, a nominal rate *does not indicate or take into account whether or how often interest is compounded.*

Table 21.2 shows that at an annual interest rate of 10% (a nominal rate) compounded daily for one year, $1000 yields $105.16 in interest, which is 10.516% of the principal. Hence the effective annual rate is 10.516%. In other words, $1000 at *simple* interest of 10.516% for one year would earn exactly the same amount of interest.

**Effective Rate and APY** DEFINITION

The **effective rate** is the rate of simple interest that would realize exactly as much interest over the same length of time. For a year, the effective rate is called the **annual percentage yield (APY).**

To keep these different rates straight, we use $i$ for a nominal rate for a specified *compounding period*–a day, a month, or a year–*within which no compounding is done.* Because no compounding is done for shorter intervals than this period, the effective rate and the nominal rate are the same for the compounding period.

**TABLE 21.2** Comparing Compound Interest: The Value of $1000, at 10% Annual Interest, for Different Compounding Periods*

| Years | Compounded Yearly | Compounded Quarterly | Compounded Monthly | Compounded Daily | Compounded Continuously |
|---|---|---|---|---|---|
| 1 | 1100.00 | 1103.81 | 1104.71 | 1105.16 | 1105.17 |
| 5 | 1610.51 | 1638.62 | 1645.31 | 1648.61 | 1648.72 |
| 10 | 2593.74 | 2685.06 | 2707.04 | 2717.91 | 2718.28 |

*Without rounding at posting of interest and neglecting leap years; the difference in most cases is no more than one cent.

**Rate Per Compounding Period** RULE

For a nominal annual rate $r$ compounded $m$ times per year, the rate per compounding period is

$$\text{periodic rate} = \boxed{i = \frac{r}{m}} = \frac{\text{nominal annual interest rate}}{\text{number of compounding periods per year}}$$

For that $1000 in savings at 10% compounded quarterly, we have $r = 10\%$ and $m = 4$, so $i = 2.5\%$ per quarter.

We denote the number of compounding periods per year by $m$. We use $r$ only for an annual rate and $t$ for the number of years.

To avoid confusion, we don't use the terminology *annual percentage rate.* That term has a special meaning just for loans (see Chapter 22, p. 712).

## Geometric Growth

Here we look for the underlying mathematical pattern of compounding. For quarterly compounding, you have at the end of the first quarter

$$\text{initial balance} + \text{interest} = \$1000 + \$1000(0.025) = \$1000(1 + 0.025)$$

and at the end of the second quarter

$$\begin{aligned}\text{initial balance} + \text{interest} &= \$1000(1 + 0.025) \\ &\quad + [\$1000(1 + 0.025)](0.025) \\ &= [\$1000(1 + 0.025)] \times (1 + 0.025) \\ &= \$1000(1 + 0.025)^2\end{aligned}$$

The pattern continues in this way, so that you have $\$1000(1 + 0.025)^4$ at the end of the fourth quarter. You use the calculator button marked $\boxed{y^x}$ to evaluate expressions like $(1.025)^4$; on a spreadsheet, use the caret key ^ (Shift-6), as in 1.025^4.

More generally, with initial principal $P$ and interest rate $i$ (= 100 $i$%) per compounding period, you have at the end of the first compounding period

$$P + Pi = P(1 + i)$$

This amount can be viewed as a new starting balance. Hence, in the next compounding period, the amount $P(1 + i)$ grows to

$$P(1 + i) + P(1 + i)i = P(1 + i)(1 + i) = P(1 + i)^2$$

The pattern continues, and we reach the following conclusion.

### Compound Interest Formula — RULE

An initial principal $P$ in an account that pays interest at a periodic interest rate $i$ per compounding period grows after $n$ compounding periods to

$$A = P(1 + i)^n$$

For convenience, we convert the general interest formula into one specific for years and annual rates. An annual rate of interest $r$ with $m$ compounding periods per year gives a rate $i = r/m$ per compounding period, and $t$ years contain $n = mt$ compounding periods.

### Compound Interest Formula for an Annual Rate — RULE

An initial principal $P$ in an account that pays interest at a nominal annual rate $r$, compounded $m$ times per year, grows after $t$ years to

$$A = P\left(1 + \frac{r}{m}\right)^{mt}$$

### Notation for Savings — DEFINITION

| | |
|---|---|
| $A$ | amount accumulated |
| $P$ | initial principal |
| $r$ | nominal annual rate of interest |
| $t$ | number of years |
| $m$ | number of compounding periods per year |
| $n = mt$ | total number of compounding periods |
| $i = r/m$ | interest rate per compounding period |

The amount added each compounding period is proportional to the amount present. This type of growth is called **geometric growth**.

### Geometric Growth (Exponential Growth) — DEFINITION

**Geometric growth** (also called **exponential growth**) is growth proportional to the amount present.

## EXAMPLE 2 ■ Compound Interest

Suppose that you have a principal of $P = \$1000$ invested at 10% nominal interest per year. Using the compound interest formula $A = P(1 + i)^n$, you can determine the amount in the account after 10 years–once you know the compounding period.

**SOLUTION**

- *Annual compounding.* The annual rate of 10% gives $i = 0.10$, and after 10 years the account has

$$\$1000(1 + 0.10)^{1\times 10} = \$1000(1.10)^{10} = \$2593.74$$

- *Quarterly compounding.* Then $i = r/m = 0.10/4 = 0.025$, and after 10 years ($mt = 4 \times 10 = 40$ quarters) the account contains

$$\$1000\left(1 + \frac{0.10}{4}\right)^{4\times 10} = \$1000(1.025)^{40} = \$2685.06$$

- *Monthly compounding.* Then $i = r/m = 0.10/12 = 0.008333$. The amount in the account after 10 years ($mt = 12 \times 10 = 120$ months) is

$$\$1000\left(1 + \frac{0.10}{12}\right)^{12\times 10} = \$2707.04$$

These entries are found in the last row of Table 21.2.

In doing the calculations, use as many decimal places as your calculator or spreadsheet carries and don't round off until the final result. We show intermediate results with enough decimal places to give the final result to the nearest cent.

## Effective Rate

For an interest rate $i$ per compounding period, a principal of \$1 grows to $(1 + i)^n$ in $n$ periods, so the interest earned on that \$1–which is the effective rate of interest for $n$ periods–is given by:

**Formula for Effective Rate** RULE

$$\text{effective rate} = (1 + i)^n - 1$$

Mostly, we will be interested in the effective rate on an annual basis. For a nominal *annual* interest rate $r$ compounded $m$ times, the interest rate per compounding period is $i = r/m$, and an amount of \$1 grows in one year to

$$\left(1 + \frac{r}{m}\right)^m$$

The effective *annual* rate of interest (the annual percentage yield, or APY) is the amount of interest earned

$$\left(1 + \frac{r}{m}\right)^m - 1$$

divided by the original principal. Since that principal is \$1, we have:

**Formula for Annual Percentage Yield (APY)** RULE

$$\text{APY} = \left(1 + \frac{r}{m}\right)^m - 1$$

### EXAMPLE 3 ■ Finding the Annual Percentage Yield (APY)

With a nominal annual rate of 6% compounded monthly, what is the APY?

**SOLUTION**

$$\left(1 + \frac{0.06}{12}\right)^{12} - 1 = 0.0617 = 6.17\%$$

In some cases you know the principal, the current balance, and the interval of time, and you want to learn the interest rate. For example, money market funds typically report earnings to investors each month, based on interest rates that vary from day to day, but often do not report the average interest rate. We find the equivalent average effective *daily* rate, from which we calculate the APY.

The compound interest formula gives the end-of-month balance as $A = P(1 + i)^n$, where $P$ is the balance at the beginning of the month, $i$ is the average daily interest rate, and $n$ is the number of days that the statement covers. So we have

$$\frac{A}{P} = (1 + i)^n$$

Taking the $n$th root gives

$$1 + i = \left(\frac{A}{P}\right)^{1/n} \qquad i = \left(\frac{A}{P}\right)^{1/n} - 1$$

## EXAMPLE 4 ■ Money Market Account

Suppose that the monthly statement from the fund reports a beginning balance ($P$) of \$7373.93 and a closing balance ($A$) of \$7382.59 for 28 days ($n$). What is the effective daily rate?

**SOLUTION** We thus have

$$i = \left(\frac{7382.59}{7373.93}\right)^{1/28} - 1 = 0.0000419194$$

Thus the average effective daily rate is 0.00419194%. Compounding daily for a year, we would have $(1 + 0.0000419194)^{365} = 1.01542$, for an APY of 1.54%.

## Simple Interest Versus Compound Interest

The amounts in accounts paying interest at 10% per year with compound and simple interest are shown in Table 21.3 and in the graph in Figure 21.1, which dramatically illustrate the exponential growth of money at compound interest compared with the linear growth at simple interest.

In some situations, the contrast is not so immediately dramatic at first glance. The amount of carbon dioxide in the atmosphere, which contributes to global warming, has been growing *superexponentially* since 1750 as a result of constantly increasing burning of fuels. The amount is growing faster than exponentially: The "interest rate," or growth rate, increases every year. The current growth rate is about 0.5% per year–a seemingly low rate of "interest." The international Kyoto Protocol that went into effect in early 2005 (without U.S. participation) aims to lower worldwide emissions. Even at a fixed lower level, though, the "population" of carbon dioxide atoms in the atmosphere would still increase–and global warming would intensify–but just arithmetically, instead of superexponentially. We are in effect "saving" carbon dioxide into the atmosphere, at an unknown future cost.

We noted earlier that the population of U.S. medical doctors grows as if it were at simple interest (arithmetic growth) because the same number of doctors graduate from medical school each year. On the other hand, general human populations tend to grow as if they were at compound interest (geometric growth), because the number of children born–the "interest"–increases as the population–the "balance"–increases.

**TABLE 21.3** The Growth of $1000: Compound Interest Versus Simple Interest

| Years | Amount in Account from Compounded Interest | Amount from Simple Interest |
|---|---|---|
| 1 | 1100.00 | 1100.00 |
| 2 | 1210.00 | 1200.00 |
| 3 | 1331.00 | 1300.00 |
| 4 | 1464.10 | 1400.00 |
| 5 | 1610.51 | 1500.00 |
| 10 | 2593.74 | 2000.00 |
| 20 | 6727.50 | 3000.00 |
| 50 | 117,390.85 | 6000.00 |
| 100 | 13,780,612.34 | 11,000.00 |

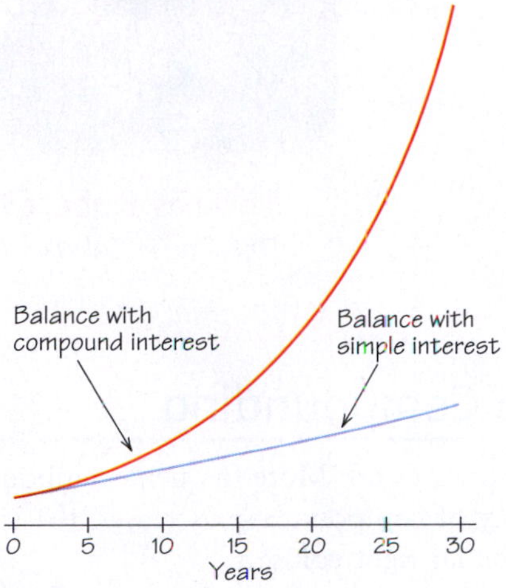


**FIGURE 21.1** The growth of $1000: compound interest and simple interest. The straight line explains why growth at simple interest is also known as linear growth.

The distinction between arithmetic growth and geometric growth is fundamental to the major theory of demographer and economist Thomas Robert Malthus (1766–1834). He claimed that human populations grow geometrically but food supplies grow arithmetically, so that populations tend to outstrip their ability to feed themselves (see Spotlight 21.1).

The situation of nuclear waste generated by a nuclear power plant is more complicated. The absolute volume of waste added each year depends on the size and output of the power plant, not on the growing amount of waste in storage. Hence the volume of waste grows arithmetically. What about the total amount of radioactive material in the storage dump? The waste is a mixture of radioactive and nonradioactive substances. Over time, the radioactive ingredients decay very slowly into nonradioactive ones. While the radioactivity of waste already in storage is decreasing, new amounts of radioactive material are being added each year. The situation requires a hybrid model that incorporates positive arithmetic growth (adding to the dump) accompanied by (much smaller) negative geometric growth (radioactive decay). The situation is like turning on the faucet in the bathtub while leaving the drain hole open a little. The faucet determines how fast water runs in, the height of the water determines how fast it runs out, and those two rates determine what happens to the volume of water in the tub.

**SPOTLIGHT 21.1** Thomas Robert Malthus

Thomas Robert Malthus (1766–1834), a nineteenth-century English demographer and economist, based a well-known prediction on his perception of the different patterns of growth of the human population and growth of the "population" of food supplies.

He believed that human populations increase geometrically but food supplies increase arithmetically—so that increases in food supplies will eventually be unable to match increases in population. He concluded, however, that over the long run there would be restrictions on the natural growth of human populations, too, including war, disease, and starvation—hardly an optimistic forecast and doubtless responsible for the dreary image associated with his views.

Some observers suggest that the genocide in Rwanda in 1994 was indirectly a result of overpopulation compared to available food resources.

Thomas Robert Malthus
*(The Granger Collection, New York.)*

## 21.3 A Limit to Compounding

The rows in Table 21.2 show a trend: More frequent compounding yields more interest. But as the frequency of compounding increases, the interest tends to a limiting amount, shown in the far right column.

Why is this so? Basically, because the extra interest from more frequent compounding is *interest on interest.* For example, in the first row of Table 21.2, the \$3.81 extra interest from compounding quarterly is interest on the \$100 yearly interest. The \$3.81 is less than 10% of the \$100 because the \$100 interest is not on deposit for the whole year. Just part of it is credited to the account (and begins earning interest) at the end of each quarter. As compounding is done more and more often, smaller and smaller amounts of interest on interest are added.

Let's see what happens with the crazy interest rate of 100% per year compounded $m$ times per year. This crazy rate makes the numbers simple (but it is nowhere close to the crazy 66,000 percent rate in Zimbabwe in December 2007, or the incredible 313 million percent rate in Yugoslavia in 1994). Later we examine interest rates closer to those in stable economies. For an initial balance of \$1, the amount at the end of one year–from the compound interest formula, with $P = \$1$ and $i = 100\%$–is

$$A = \$1 \times \left(1 + \frac{100\%}{m}\right)^m = \$\left(1 + \frac{1.00}{m}\right)^m$$

As $m$ increases, this amount, which is just $(1 + 1/m)^m$, gets closer and closer to a special number called $\boldsymbol{e} \approx 2.71828$ (see Spotlight 21.2). This is illustrated in Table 21.4, where the dots (ellipses) indicate that more decimal places follow.

**TABLE 21.4 Yield of $1 at 100% Interest, Compounded $m$ Times per year**

| $m$ | $\left(1 + \frac{1}{m}\right)^m$ |
|---|---|
| 1 | 2.0000000 . . . |
| 5 | 2.4883200 . . . |
| 10 | 2.5937424 . . . |
| 50 | 2.6915880 . . . |
| 100 | 2.7048138 . . . |
| 1,000 | 2.7169239 . . . |
| 10,000 | 2.7181459 . . . |
| 100,000 | 2.7182682 . . . |
| 1,000,000 | 2.7182804 . . . |
| 10,000,000 | 2.7182816 . . . |

**SPOTLIGHT 21.2 The Number e**

The number $e$ is similar to the number $\pi$ in several respects. Both arise naturally, $\pi$ in finding the area and circumference of circles, and $e$ in compounding interest continuously ($e$ is also the base for the system of "natural" logarithms). In addition, neither is rational (expressible as the ratio of two integers, such as 7/2) or even algebraic (the solution of a polynomial equation with integer coefficients, such as $x^2 = 2$); we say that they are *transcendental* numbers. Finally, no pattern has ever been found in the digits of the decimal expansion of either number.

For a general interest rate $r$, as $m$ is made larger and larger, the limiting amount is $e^r$, and the interest method is called **continuous compounding**. The APY, is $(e^r - 1)$. (You can calculate powers of $e$ using the [$e^x$] button on your calculator. On some calculators, this button is the 2nd function of the button marked [LN] or [ln $x$]. For example, to calculate $e^{0.10}$, push [2nd], push [ln $x$], and enter 0.10. You get 1.105170918.)

**Continuous Compounding** DEFINITION

**Continuous compounding** is the method of calculating interest in which the amount of interest is what compound interest tends toward with more and more frequent compounding.

## EXAMPLE 5 ■ Continuous Compounding

For $1000 at an annual rate of 10%, compounded $m$ times in the course of a single year, what is the balance at the end of the year?

**SOLUTION** It is

$$\$1000\left(1 + \frac{0.10}{m}\right)^m$$

This quantity gets closer and closer to $\$1000e^{0.1} = \$1105.17\ldots$ as the number of compoundings $m$ is increased. No matter how frequently interest is compounded–daily, hourly, every second, infinitely often ("continuously")–the original \$1000 at the end of one year cannot grow beyond \$1105.17. The values after 5 and 10 years are shown in the lower rows of Table 21.2.

**Continuous Interest Formula** RULE

For a principal $P$ deposited in an account at a nominal annual rate $r$, compounded continuously, the balance after $t$ years is

$$A = Pe^{rt}$$

We illustrate with \$1000 at 10%. For one year, we have $t = 1$ and

$$A = \$1000e^{0.10} = \$1105.17$$

To find the amount in the account after 5 years, we have $t = 5$:

$$A = \$1000e^{(0.10)(5)} = \$1000e^{0.5} = \$1648.72$$

exactly as shown in the rightmost column of Table 21.2.

It makes virtually no difference whether compounding is done daily or continuously over the course of a year. Most banks apply a daily periodic rate (based on compounding continuously) to the balance in the account each day and post interest daily (rounded to the nearest cent). The daily nominal rate is $r/365$, so each day the balance of the account is multiplied by $e^{r/365}$, the daily effective rate. Except for the rounding in posting interest, the effect is the same as continuous compounding throughout the year, because the compound interest formula gives $A = P(e^{r/365})^{365}$, which is the same as $Pe^{r}$ from the continuous interest formula.

For example, for a principal of \$1000 and an interest rate of 5%, interest compounded daily over a year yields an amount

$$\$1000\left(1 + \frac{0.05}{365}\right)^{365} = \$1051.2675,$$

while continuous compounding yields $\$1000e^{.05} = \$1051.2711$.

## 21.4 A Model for Saving

The compound interest formula tells the fate over time of a single deposited amount, but another common question that arises in finance is: What size deposit do you need to make regularly in an account with a fixed rate of interest, to have a specified amount at a particular time in the future?

This question is important in planning for a major purchase in the future or accumulating a retirement nest egg. Later, in Chapter 22, we apply the same concepts and formula to paying off a mortgage and making installment payments on a car.

(*Sean Justice/Corbis.*)

### EXAMPLE 6 ▪ A Savings Plan

A graduate at her first job saves \$100 per month, deposited directly into her credit union account on payday, the last day of the month. The account earns 1.8% per year, compounded monthly. How much will she have at the end of five years, assuming that the credit union continues to pay the same interest rate?

**SOLUTION** Note that she makes the first deposit at the end of the first month and the last deposit at the end of the sixtieth month. The monthly interest rate is $i = r/12 = 0.018/12 = 0.0015$.

It's easier to look at the deposits in reverse time order. The last deposit is made on the last day of the five years, so it earns no interest and contributes just \$100 to the total.

The second-last deposit earns interest for one month, contributing $\$100(1 + i)$. Similarly, the third-last contribution is on deposit for two months, contributing $\$100(1 + i)^2$.

Continuing in the same way, we find that the first deposit earns interest for 59 months and contributes $\$100(1 + i)^{59}$. The total of all of the contributions is

$$\$100 + \$100(1 + i)^1 + \$100(1 + i)^2 + \cdots + \$100(1 + i)^{59}$$
$$= \$100[1 + (1 + i)^1 + (1 + i)^2 + \cdots + (1 + i)^{59}]$$

This expression is known as a **geometric series**, because the successive terms have geometric growth: Each succeeding term is a constant–in this case, $(1 + i)$–times the preceding term. For the sum of such a series with general ratio $x$, we have:

**Formula for Sum of a Geometric Series** RULE

$$1 + x + x^2 + x^3 + \cdots + x^{n-1} = \frac{x^n - 1}{x - 1}$$

That this formula works for all $x$ (except $x = 1$) can be confirmed by multiplying both sides $(x - 1)$ and watching terms on the left cancel. (You should do this confirmation for $n = 4$.)

In our example, we have $x = 1 + i$, and the formula becomes

$$1 + (1 + i)^1 + (1 + i)^2 + \cdots + (1 + i)^{n-1} = \frac{(1 + i)^n - 1}{i}$$

We have $n - 1 = 59$, or $n = 60$ months, and $i = 0.0015$, the interest rate per month. The total accumulation after five years is

$$A = \$100\left[\frac{(1 + 0.0015)^{60} - 1}{0.0015}\right] = \$6273.37$$

For a uniform deposit of $d$ per compounding period (deposited at the end of the period) and an interest rate $i$ per period, the amount $A$ accumulated after $n$ compounding periods is given by the **savings formula:**

**Savings Formula** RULE

$$A = d\left[\frac{(1 + i)^n - 1}{i}\right] = d\left[\frac{(1 + \frac{r}{m})^{mt} - 1}{\frac{r}{m}}\right].$$

The expression on the right gives the amount accumulated in terms of the nominal annual interest rate $r$, the number $m$ of compounding periods per year, and the number $t$ of years, using the relations $i = r/m$ and $n = mt$.

The savings formula involves four quantities: $A$, $d$, $i$, and $n$. If any three are known, the fourth can be found. A common situation is for $A$, $i$, and $n$ to be known, with $d$ (the regular payment) to be found, because the practical concern for most people is how much their monthly payment will be.

Since we often want to solve for $d$, we solve the savings formula algebraically once and for all for $d$ to get the *payment formula*:

**Payment Formula** RULE

$$d = A\left[\frac{i}{(1+i)^n - 1}\right] = A\left[\frac{\frac{r}{m}}{(1+\frac{r}{m})^{mt} - 1}\right]$$

Sometimes the purpose of saving is to accumulate a fixed sum by a particular date. Such a savings plan is called a **sinking fund**, because you sink money into it. If the same amount is deposited regularly, the sinking fund is called an **annuity**, a term for any series of (usually) equal payments at regular intervals.

**Annuity** DEFINITION

An **annuity** is a specified number of (usually equal) periodic payments.

**Sinking Fund** DEFINITION

A **sinking fund** is a savings plan to accumulate a fixed sum by a particular date, usually through equal periodic deposits.

## EXAMPLE 7 ■ Sinking Fund

Suppose that your parents had started saving for your college education when you were born. How much would they have had to save each month to accumulate \$15,000 over 18 years with an account earning a steady 5% per year, compounded monthly?

**SOLUTION** Applying the payment formula with $A = \$15{,}000$, monthly rate $i = r/m = 0.05/12$, and $n = mt = 12 \times 18 = 216$, we get

$$d = \$15{,}000\left[\frac{\frac{.05}{12}}{(1+\frac{.05}{12})^{216} - 1}\right] = \$42.96$$

This sounds like a manageable amount to contribute, but it doesn't take into account inflation, nor costs beyond the first year, nor the higher cost of a private college. In the next section, we investigate how to take inflation into account.

## Saving for Retirement (It's Never Too Early to Start)

Financial advisers stress the importance of beginning early to save for retirement. Many firms offer a *401(k) plan* (named after a section of law regulating pensions), which allows an employee to make monthly contributions to a retirement account. The plan has the advantage that income tax on the contributions is deferred until the employee withdraws the money during retirement. That means, for example, that an employee making a \$100 monthly contribution may see a reduction in take-home pay of only \$75 or less, since taxes are not withheld on the contribution.

Sometimes a company's pension plan consists of just contributing company stock to the employee's individual 401(k) account. In 2002, the bankruptcy of Enron Corporation resulted in thousands of its employees losing almost their entire retirement savings. Those savings consisted largely of Enron stock contributed by Enron, which fell from \$90 per share to under \$1 per share in just a couple of months. The Enron bankruptcy illustrated how unwise it is for most of an employee's retirement fund to consist of stock in just one company, particularly if–as was the case for Enron–the employee is not free to sell the stock. Even more people lost retirement savings and jobs when the stock of WorldCom declined more than 99% in 2002, after news of financial fraud by its management.

### EXAMPLE 8 ■ Retirement Fund Annuity Savings

Suppose that you start a 401(k) plan when you turn 23 and contribute \$50 at the end of each month until you turn 65 and retire. Suppose (unlike some Enron employees) you put your contributions into a very safe long-term investment that returns a steady 5% annual interest compounded monthly. How much will be in your fund at retirement?

**SOLUTION** Apply the savings formula with $d = \$50$, $i = 0.05/12$, and $n = mt = 12 \times (65 - 23) = 504$. We get

$$A = \$50\left[\frac{(1 + \frac{0.05}{12})^{504} - 1}{\frac{0.05}{12}}\right] = \$85{,}567.43$$

At first glance, that may seem like a lot of money, but it is not so much if that's all you have to live off for the rest of your life (of course, there is also Social Security). In the exercises we explore the effects of saving more each month, getting a higher interest rate, saving on taxes, and–especially–having inflation erode the value of your savings. ■

Annuities are a common way for lotteries to pay out grand prizes and for retirees to receive funds saved up for retirement. We examine an example of each in Chapter 22 (pp. 721–722), where we turn the savings formula around to get a formula (the amortization formula) to relate your savings to a regular payout. In Chapter 22 you find out what monthly income for a fixed period–or for life–\$86,000 could buy.

## 21.5 Present Value and Inflation

Suppose that you want to make a one-time deposit of amount $P$ that will grow to a specific amount $A$ in $n$ compounding periods from now by earning interest at a rate $i$ per period. The quantities $A$, $P$, $i$, and $n$ are related through the compound interest formula, $A = P(1 + i)^n$. The quantity $P$ is called the **present value** of the amount $A$ to be paid $n$ compounding periods in the future.

**Present Value** DEFINITION

The **present value** $P$ of an amount $A$ to be paid $t$ years in the future, earning a nominal annual rate of interest $r$ compounded $m$ times per year–that is, after $n = mt$ compounding periods at a rate $i = r/m$ per compounding period–is

$$P = \frac{A}{(1 + i)^n} = \frac{A}{(1 + \frac{r}{m})^{mt}}$$

## EXAMPLE 9 ■ Certificate of Deposit

A certificate of deposit (CD) pays a fixed rate of interest for a term specified in advance, from a month to 10 years. As I was researching this section, our second son was expecting to enter college in four years. Based on current costs, tuition, room, and board would cost \$12,000 for the first year at the public university in our state capital (and more if he goes to a private college). The best rate that I could find for a four-year CD was 4.80% compounded monthly. How much would we need to set aside now in such a CD to have \$12,000 in four years?

**SOLUTION** We find the present value of \$12,000 four years from now, with $r = 0.048$, $m = 12$, and $t = 4$. The present value formula gives

$$P = \frac{A}{(1 + \frac{r}{m})^{mt}} = \frac{\$12{,}000}{(1 + \frac{0.048}{12})^{12 \times 4}} = \frac{\$12{,}000}{(1 + 0.004)^{48}} = \$9907.48$$

In times of economic inflation, prices increase. When the rate of inflation is constant, the compound interest formula can be used to project prices.

## EXAMPLE 10 ■ Inflation

Suppose that there is constant 3% annual inflation from mid-2009 through mid-2013. What will be the projected price in mid-2013 of an item that costs \$100 in mid-2009?

**SOLUTION** The compound interest formula applies with $P = \$100$, $r = 3\%$, $m = 1$ and $t = 4$. The projected price is $A = P(1 + r)^t = \$100(1 + 0.03)^4 = \$112.55$.

## Inflation and Depreciation as Exponential Decay

During constant-rate inflation, prices grow geometrically (exponentially) and the value of the dollar goes down geometrically: one is growing exponentially; the other is decaying exponentially.

**Exponential Decay** DEFINITION

**Exponential decay** is geometric growth with a negative rate of growth.

Let $a$ (for "additional") represent the annual rate of inflation; what costs \$1 now will cost $\$(1 + a)$ this time next year. For example, if the inflation rate were $a = 25\%$, then what costs \$1 now would cost \$1.25 this time next year. A dollar next year would buy only 0.8 (= 1/1.25) times as much as a dollar buys today. In other words, a dollar next year would be worth only \$0.80 in today's dollars–by next year, a dollar would have lost 20% of its purchasing power. We say that the present value of receiving a dollar next year is \$0.80. Notice that although the inflation rate is 25%, the loss in purchasing power is 20%. For a general inflation rate $a$, a dollar a year from now will buy only a fraction of what a dollar today can buy.

**Present Value of a Dollar a Year from Now with Inflation Rate $a$** RULE

$$\frac{\$1}{1 + a} = \$1 - \frac{\$a}{1 + a}$$

A dollar a year from now is worth $\$[1 - a/(1 + a)]$ today, and the loss in purchasing power is the fraction $a/(1 + a)$. (You should calculate what these expressions become for $a = 25\%$.) The quantity $i = -a/(1 + a)$ behaves like a negative interest rate. We can use the compound interest formula to find the present value of $P$ dollars $t$ years from now as

$$A = P(1 + i)^t = P\left[1 - \frac{a}{1+a}\right]^t.$$

The actual posted price of an item, at any time, is said to be in **current dollars.** That price can be compared with prices at other times by converting all prices to **constant dollars,** dollars of a particular year.

## EXAMPLE 11 ■ Deflated Dollars

Suppose that there is 25% annual inflation from mid-2009 through mid-2013. What will be the value of a dollar in mid-2013 in constant mid-2009 dollars? The inflation figure is unrealistic (we hope!), but it makes the calculations easy, so you can focus on the ideas.

We have $a = 0.25$, so $i = -a/(1 + a) = -0.25/1.25 = -0.20$. This, not 25%, is the negative interest rate, the rate at which the dollar is losing purchasing power. We have $t = 4$ years, so the value of \$1 four years from mid-2009 in 2009 dollars will be

$$\$1(1 + i)^4 = \$(1 - 0.20)^4 = (0.80)^4 = \$0.41$$

In Example 11, we may think of the value of the dollar as "depreciating" 20% per year. Depreciation of the value of equipment is similar.

## EXAMPLE 12 ■ Depreciation

If you bought a car at the beginning of 2009 for \$12,000 and its value in current dollars depreciates steadily at a rate of 15% per year, what will be its value at the beginning of 2012 in current dollars?

**SOLUTION** We have $P = \$12{,}000$, $i = -0.15$, and $n = 3$. The compound interest formula gives

$$A = P(1 + i)^n = \$12{,}000(1 - 0.15)^3 = \$7369.50$$

## The Consumer Price Index

In our preceding model, we supposed that inflation stayed constant over a period of time. That is not generally the case. However, based on regular measures of inflation, we can determine the equivalent today of a price in an earlier year or how much a dollar in that year would be worth today.

The official measure of inflation is the Consumer Price Index (CPI), determined by the Bureau of Labor Statistics. Here we describe and use the CPI-U, the index for all urban consumers, which covers about 80% of the U.S. population and is the index of inflation that is usually referred to in newspaper and magazine articles.

Each month, the Bureau of Labor Statistics determines the average cost of a "market basket" of goods, including food, housing, transportation, clothing, and other items. It compares this cost with the cost of the same (or comparable) goods in a base period. The base period used to construct the CPI-U is 1982–1984. The index for 1982–1984 is set to 100, and the CPI-U for other years is calculated by using the proportion

$$\frac{\text{CPI for other year}}{100} = \frac{\text{cost of market basket in other year}}{\text{cost of market basket in base period}}$$

For example, the cost of the market basket in 1976 (in 1976 dollars) was 0.569 times the cost in 1982–1984 (in 1982–1984 dollars), so the CPI for 1976 is $100 \times 0.569$, or 56.9.

Table 21.5 shows the average CPI for each year from 1913 through 2007, with estimates for 2008 and 2009. This table can be used to convert the cost of an item in dollars for one year to what it would cost in dollars in a different year, using the proportion

$$\frac{\text{cost in year A}}{\text{cost in year B}} = \frac{\text{CPI for year A}}{\text{CPI for year B}}$$

## EXAMPLE 13

### The Price of Our House and the Value of a Dollar

(*Peter Gridley/Getty Images.*)

Where my family and I live, housing is relatively inexpensive. We bought our house in mid-1992 for \$133,000 (close to the median price of U.S. housing at that time). What would be the equivalent cost in mid-2009 dollars?

**SOLUTION** We see from Table 21.5 that the CPI for 1992 is 140.3 and the CPI for 2009 is estimated to be 221.7. The table gives the average value for each year, which is very close to the value at midyear. Month-by-month values are available at the Bureau of Labor Statistics Web site.

Using the proportion, we have

$$\frac{\text{cost in 2009}}{\text{cost in 1992}} = \frac{\text{CPI for 2009}}{\text{CPI for 1992}}$$

or

$$\frac{\text{cost in 2009}}{\$133{,}000} = \frac{227.0}{140.3}$$

so that

$$\text{cost in 2009} = \$133{,}000 \times \frac{227.0}{140.3} \approx \$215{,}000$$

That's what our house would sell for if its price exactly matched inflation.

The ratio $227.0/140.3 = 1.61796$ is the *scaling factor* for converting 1992 dollars to 2009 dollars. What we are observing is a proportion, or *numerical similarity,* between 1992 dollars and 2009 dollars, analogous to the geometric similarity of Chapter 18 (p. 572). To convert from 2009 dollars to 1992 dollars, we would multiply by $1/1.61796 \approx 0.618$.

Spotlight 19.3 (p. 614) describes how the consumer price index is calculated, using the geometric mean (defined and introduced on p. 613).

## Real Growth Under Inflation

It's natural to think that if your investment is growing at 6% per year and inflation is at 3% per year, then the real growth in the value (purchasing power) of your investment is $6\% - 3\% = 3\%$. Such, however, is not the case.

Let's suppose that you invest \$500 for a year at 6%. At the beginning of the year, you have \$500, which at \$5 per pound could buy 100 pounds of steak. At the end of

**TABLE 21.5** U.S. Consumer Price Index (1982–1984 = 100)

| | | | | | | | | | | | |
|---|---|---|---|---|---|---|---|---|---|---|---|
| – | – | 1931 | 15.2 | 1951 | 26.0 | 1971 | 40.5 | 1991 | 136.2 |
| – | – | 1932 | 13.7 | 1952 | 26.6 | 1972 | 41.8 | 1992 | 140.3 |
| 1913 | 9.9 | 1933 | 13.0 | 1953 | 26.7 | 1973 | 44.4 | 1993 | 144.5 |
| 1914 | 10.0 | 1934 | 13.4 | 1954 | 26.9 | 1974 | 49.3 | 1994 | 148.2 |
| 1915 | 10.1 | 1935 | 13.7 | 1955 | 26.8 | 1975 | 53.8 | 1995 | 152.4 |
| 1916 | 10.9 | 1936 | 13.9 | 1956 | 27.2 | 1976 | 56.9 | 1996 | 156.9 |
| 1917 | 12.8 | 1937 | 14.4 | 1957 | 28.1 | 1977 | 60.6 | 1997 | 160.5 |
| 1918 | 15.1 | 1938 | 14.1 | 1958 | 28.9 | 1978 | 65.2 | 1998 | 163.0 |
| 1919 | 17.3 | 1939 | 13.9 | 1959 | 29.1 | 1979 | 72.6 | 1999 | 166.6 |
| 1920 | 20.0 | 1940 | 14.0 | 1960 | 29.6 | 1980 | 82.4 | 2000 | 172.2 |
| 1921 | 17.9 | 1941 | 14.7 | 1961 | 29.9 | 1981 | 90.9 | 2001 | 177.1 |
| 1922 | 16.8 | 1942 | 16.3 | 1962 | 30.9 | 1982 | 96.5 | 2002 | 179.9 |
| 1923 | 17.1 | 1943 | 17.3 | 1963 | 30.6 | 1983 | 99.6 | 2003 | 184.0 |
| 1924 | 17.1 | 1944 | 17.6 | 1964 | 31.0 | 1984 | 103.9 | 2004 | 188.9 |
| 1925 | 17.5 | 1945 | 18.0 | 1965 | 31.5 | 1985 | 107.6 | 2005 | 195.3 |
| 1926 | 17.7 | 1946 | 19.5 | 1966 | 32.4 | 1986 | 109.6 | 2006 | 201.6 |
| 1927 | 17.4 | 1947 | 22.3 | 1967 | 33.4 | 1987 | 113.6 | 2007 | 207.3 |
| 1928 | 17.1 | 1948 | 24.1 | 1968 | 34.8 | 1988 | 118.3 | 2008 (est.) | 219.0 |
| 1929 | 17.1 | 1949 | 23.8 | 1969 | 36.7 | 1989 | 124.0 | 2009 (est.) | 227.0 |
| 1930 | 16.7 | 1950 | 24.1 | 1970 | 38.8 | 1990 | 130.7 | | |

*Note:* This the CPI-U index, which covers all urban consumers, about 80% of the U.S. population. Each index is an average for all cities for the year. The basis for the index is the period 1982–1984, for which the index was set equal to 100. For each year, the figure is the average during the year, hence is usually close to the value at midyear.

*Source:* http://stats.bls.gov/cpi/

the year, you have $500(1 + 0.06) = $530, but steak now costs $5(1 + 0.03) = $5.15 per pound. How much steak would that buy? $530/$5.15 lb = 102.91 lb. In other words, in terms of purchasing power, or real gain, your investment has grown only 2.91%. This is not a great deal different from 3%, but it *is* different, and the difference is greater for higher rates of interest and inflation.

Consider an investment principal $P$ and a market basket of goods of value $m$. Let the annual yield (rate of interest) of the investment be $r$ and the rate of inflation be $a$. We calculate the rate of real growth $g$ of the investment as follows.

At the beginning of the year, the investment would buy quantity $q_{\text{old}} = P/m$ of the market basket. At the end of the year, the investment would buy

$$q_{\text{new}} = \frac{P(1 + r)}{m(1 + a)}$$

market basket. Notice that the gain of $r$ in the investment multiplies the principal by $(1 + r)$, while the erosion due to inflation divides the principal by $(1 + a)$. Here you see directly that the two influences on the investment have directly opposite effects.

The growth of the investment, relative to how many market baskets it could have bought originally, is

$$g = \frac{q_{\text{new}} - q_{\text{old}}}{q_{\text{old}}} = \frac{\frac{P(1+r)}{m(1+a)} - \frac{P}{m}}{\frac{P}{m}} = \frac{1+r}{1+a} - 1 = \frac{r-a}{1+a}$$

In the last expression, the numerator is the difference of the two rates (6% − 3% in our example), which is divided by a quantity greater than 1 if there is inflation. You should confirm that this formula gives 2.91% for $r = 6\%$ and $a = 3\%$.

One way to understand why this is the correct formula is to realize that the gain itself is not in original dollars but in deflated dollars.

The relationship between interest rate, inflation rate, and rate of real growth is called *Fisher's effect,* after the American economist Irving Fisher (1867–1947).

### Real Rate of Growth

RULE

The real (effective) annual rate of growth of an investment at annual interest rate $r$ with annual inflation rate $a$ is

$$g = \frac{r-a}{1+a}$$

### SPOTLIGHT 21.3 Nobel Prize for a Model in Economics

The 1997 Nobel Memorial Prize in Economics was awarded to Robert C. Merton of Harvard University and Myron S. Scholes of Stanford University for their method of valuing financial derivatives. Together with the late Fischer Black in the 1970s, they formulated a mathematical model with appropriate assumptions and solved the resulting equation. At the time, Black was a mathematician with Arthur D. Little Consultants in Boston, Scholes was a professor of finance at MIT, and Merton was an assistant to the economist Paul Samuelson at MIT. All were under age 30 at the time.

The major achievement of Merton, Black, and Scholes was to incorporate variability of market prices into the formula for the value of an option. They realized that the risk involved in the market is already implicitly taken into account in the stock's current price and its volatility (tendency to vary), and they were able to find the right formulation for incorporating the risk into the value of the option. Black and Scholes actually modeled the rate of change of the option value and then used methods from calculus to work backward to calculate the value and the formula itself.

This fairly complicated formula is based on simplifying assumptions (no stock dividends, no transactions costs, fixed price volatility, fixed interest rate, efficient market) plus a major modeling assumption. That modeling assumption is that the change in the price of the stock, as a percentage of the price, has a fixed component proportional to elapsed time (price trends upward over time) plus a random component deriving from volatility (so the price also can jump around). The random component is modeled using the normal distribution of Chapter 5.

## REVIEW VOCABULARY

**Annual percentage yield (APY)** The effective interest rate per year. (p. 682)

**Annuity** A specified number of (usually equal) periodic payments. (p. 692)

**Arithmetic growth** Growth by a constant amount in each time period. (p. 680)

**Compound interest** Interest that is paid on both the original principal and the accumulated interest. (p. 680)

**Compound interest formula** Formula for the amount in an account that pays compound interest periodically. For an initial principal $P$ and an effective rate $i$ per compounding period, the amount after $n$ compounding periods is $A = P(1 + i)^n$. (p. 684)

**Compounding period** The fundamental interval for compounding, within which no compounding is done. Also called simply *period.* (p. 681)
**Constant dollars** Costs are expressed in constant dollars if inflation or deflation has been taken into account by converting the costs to their equivalent in dollars of a particular year. (p. 695)
**Continuous compounding** Payment of interest in an amount toward which compound interest tends with more and more frequent compounding. (p. 689)
**Current dollars** The actual cost of an item at a point in time; inflation or deflation before or since then has not been taken into account. (p. 695)
***e*** The base for continuous compounding, geometric (exponential) growth, and natural logarithms; $e = 2.71828\ldots$ (p. 688)
**Effective rate** The rate of simple interest that would realize exactly as much interest over the same period of time. (p. 682)
**Exponential decay** Geometric growth at a negative rate. (p. 694)
**Exponential growth** Geometric growth. (p. 684)
**Geometric growth** Growth proportional to the amount present. (p. 684)
**Geometric series** A sum of terms, each of which is the same constant times the previous term; that is, the terms undergo geometric growth. (p. 691)
**Interest** Money earned on a savings account or a loan. (p. 679)
**Linear growth** Arithmetic growth. (p. 680)
**Nominal rate** A stated rate of interest for a specified length of time; a nominal rate does not take into account any compounding. (p. 682)
**Present value** The value today of an amount to be paid or received at a specific time in the future, as determined from a given interest rate and compounding period. (p. 693)
**Principal** Initial balance. (p. 679)
**Savings formula** Formula for the amount $A$ accumulated after $n = mt$ periods, with a uniform deposit of $d$ at the end of each compounding period and interest rate $i = r/m$ per period:

$$A = d\left[\frac{(1+i)^n - 1}{i}\right] = d\left[\frac{(1+\frac{r}{m})^{mt} - 1}{\frac{r}{m}}\right]. \quad \text{(p. 691)}$$

**Simple interest** The method of paying interest only on the initial balance in an account, not on any accrued interest. (p. 679)
**Sinking fund** A savings plan to accumulate a fixed sum by a particular date, usually through equal periodic deposits. (p. 692)

## ✓ SKILLS CHECK

**1.** Simple interest is an example of

**(a)** linear growth.
**(b)** variable growth.
**(c)** constant growth.

**2.** If a savings account pays 3% simple annual interest, a deposit of $250 will earn ___________ in 2 years.

**3.** Which of the following pays more interest?

**(a)** 6% compounded annually
**(b)** 6% compounded monthly
**(c)** 6% compounded continuously

**4.** If a bond matures in 3 years and will pay $10,000 at that time, the fair value of it today is ___________, assuming that the bond has an interest rate of 6% compounded annually and there is no inflation.

**5.** If a single deposit is made into a compound interest certificate of deposit, the account

**(a)** earns interest only for the first period.
**(b)** earns the same amount of interest each period.
**(c)** earns more interest in each subsequent period.

**6.** If $800 is invested for one year at 6% compounded quarterly, the amount of interest earned is _________ .

**7.** An 18% annual rate on a credit-card balance is an example of

**(a)** an effective rate.
**(b)** a nominal rate.
**(c)** an adjusted rate.

**8.** If you deposit $1000 at 6.2% simple interest, the balance after three years is _______________ .

**9.** Suppose you invest $250 in an account that pays 4.5% interest compounded quarterly. After 30 months, how much is in your account?

**(a)** $279.08
**(b)** $279.59
**(c)** $279.71

**10.** Suppose you deposit $15 at the end of each month into a savings account that pays 2.5% interest compounded monthly. After a year, ___________ is in the account.

**11.** Which of the following is the most generous interest rate for a one-year CD?

**(a)** 6% simple interest
**(b)** 5.9% compounded annually
**(c)** 5.9% compounded continuously

12. The APY for 5.90% compounded monthly is ____ .

13. The number $e$ is
(a) irrational.
(b) irrelevant.
(c) irrotational.

14. The APY for 5% compounded daily is ________ .

15. The value of $e$ is approximately
(a) 1.414.
(b) 2.718.
(c) 3.14.

16. When $1000 is invested at 8% compounded continuously for 5 years, the balance is ________ .

17. An example of exponential decay is
(a) the depreciation of factory equipment.
(b) a retirement annuity.
(c) the Consumer Price Index.

18. Depositing $100 on a child's annual birth date is an example of ________________ .

19. If your investment is growing at a rate less than the rate of inflation,
(a) you have a positive real growth in your investment.
(b) you do not have a positive real growth in your investment.
(c) you do not have enough information to determine whether real growth is positive or negative.

20. If a new car costs $18,000 and loses value at a rate of 20% per year, its value after 3 years is ________ .

# CHAPTER 21 EXERCISES

■ Challenge ◆ Discussion

The exercises below require a scientific calculator with buttons for powers $\boxed{y^x}$, exponential $\boxed{e^x}$, and natural logarithm $\boxed{\ln x}$.

## 21.1 Arithmetic Growth and Simple Interest

1. Suppose that you need $30,000 for your last year of college. You could go to a private lending institution and apply for a signature student loan; rates range from 7% to 14%. However, your Aunt Sally is willing to loan you the money from her retirement savings, with no repayment until after graduation. All she asks is that in the meantime you pay her each month the amount of interest that she would otherwise get on her savings (since she needs that to live on), which is 6%. What is your monthly payment to her, and how much interest will you pay her over the year (9 months)? (Aunt Sally will be glad to hear from you every month anyway!)

2. On December 28, 2007, you could buy a 10-year U.S. Treasury note ("T-note," a kind of bond) for $10,000 that pays 4.21% simple interest every year through December 28, 2017. How much total interest would it earn by then?

## 21.2 Geometric Growth and Compound Interest

3. An often heard claim is that "the amount of information in the world doubles every three days." Presumably the claim refers to the amount of data, which can be quantified in terms of number of bits. (A bit is the smallest unit of storage in a computer.) Show that the claim is absolutely preposterous by doing a little arithmetic and comparing your result with the estimated number of particles in the universe ($10^{70}$). In particular:

**(a)** Start with one bit of data and double the number of bits every third day. How long does it take to get past $10^{70}$? (*Hint:* Don't just keep multiplying by 2 over and over. Convince yourself that since the amount of data increases by a factor of 2 every 3 days, then it increases by a factor of $2^2 = 4$ every 6 days, a factor of $4^2 = 16$ every 12 days, a factor of $16^2 = 256$ every 24 days, and so forth.)

**(b)** Part (a) involves a lot of multiplying by 2, even if you do it efficiently. Another approach is to use the fact that $2^{10} = 1024 \approx 1000$. Thus, the amount of data increases by a factor of more than 1000 every $3 \times 10 = 30$ days, or every month (except February, but the 31-day months make up for it). By when will the total surely be past $10^{70}$?

4. In a "Foxtrot" cartoon by Bill Amend (9/10/2006) on the next page, the girl Paige confronts a math problem in which "a math teacher assigns one second of homework the first week of school, two seconds the second week, four seconds the third, and so on." She is asked whether she would agree to this weekly homework doubling for the duration of the 36-week school year. How much homework (in hours) would this plan require in week 36?

**5.** You deposit $1000 at 3% per year. What is the balance at the end of one year, and what is the annual yield, if the interest paid is

**(a)** simple interest?
**(b)** compounded annually?
**(c)** compounded quarterly?
**(d)** compounded daily?

**6.** Repeat Exercise 5, but for $1000 at 6% per year.

**7.** I have a CD with National City through 2010 paying 4.69% interest compounded daily. What is the APY for this rate?

**8.** I have an account with First Community Credit Union of Beloit, Wisconsin, which pays dividends on independent retirement accounts (IRAs) at 0.75% per year, compounded monthly, for accounts with balances up to $2000. What is the APY for such a rate?

**9.** U.S. Savings Bonds are a common form of award; for example, in December 2007 the Rodel Exemplary Teacher Initiative, which addresses the shortage of effective teachers in Arizona's neediest schools, chose 12 teachers each to receive a $10,000 U.S. Savings Bond. (The interest is exempt from state and local income taxes and may also be exempt from federal income tax if used to pay for college tuition and fees.) Series EE Savings Bonds issued between November 2007 and April 2008 earn 3.00% interest, compounded semiannually, for the 20 years until their maturity. What is the APY? Will such a bond double in value in 20 years? (If not, the U.S. Treasury will make a one-time adjustment at the end of 20 years to ensure doubling in value.)

**10.** A Paper Series EE Savings Bond is sold at half face value, and the U.S. Treasury Dept. guarantees that it will double in value by 20 years from the issue date. What is the minimum APY for such a bond?

**11.** Suppose that on the statement for a money market account this month, the initial balance was $7744.70, the statement was for 34 days, and the final balance was $7770.84. Calculate the APY.

**12.** Repeat Exercise 11, but for the previous month, which had an initial balance of $7722.54, a period of 27 days, and a final balance of $7744.70.

**13.** *The rule of 72* is a rule of thumb for finding how long it takes money at interest to double: If $r$ is the annual interest rate, then the doubling time is approximately $72/(100r)$ years.

**(a)** Calculate the balance at the end of the predicted doubling time for each $1000, with annual compounding, for the small growth rates of 3%, 4%, and 6%.
**(b)** Repeat part (a) for the intermediate interest rates of 8% and 9%.
**(c)** Repeat part (a) for the larger interest rates of 12%, 24%, and 36%.
**(d)** What do you conclude about the rule of 72?

**14.** More frequent compounding yields greater interest, but with diminishing returns as the frequency of compounding is increased. For small interest rates, there is little difference in yield for compounding annually, quarterly, monthly, daily, or continuously. Investigating doubling times with continuous compounding leads to understanding why the rule of 72 of Exercise 13 works. Recall that for continuous compounding at annual rate $r$, the balance $A$ at the end of $t$ years is $Pe^{rt}$ for an initial principal of $P$. For the initial principal to double, we have $2P = A = Pe^{rt}$, so $e^{rt} = 2$. Taking the natural logarithm of both sides yields $rt = \ln 2$, where ln stands for the natural logarithm, represented on a calculator by a button marked either [ln] or [LN] (not [log] or [$\log_{10}$], which stands for a different kind of logarithm). Using the button gives $\ln 2 = 0.693$. So we have $rt = 0.693$, from which we can determine $t$ if we know $r$.

Calculate the doubling times for continuous compounding at 3%, 6%, and 9%, and compare them with those predicted by the rule of 72. What do you conclude? Why do you think people prefer a rule of 72 over a rule of 69.3?

(FOXTROT 2006 BillAmend. Reprinted with permission of Universal Press Syndicate. All rights reserved.)

## 21.3 A Limit to Compounding

**15.** Use your calculator to evaluate for $n = 1, 10, 100, 1000$, and 1,000,000:

**(a)** $\left(1 + \frac{1}{m}\right)^m$

**(b)** $\left(1 + \frac{2}{m}\right)^m$

**(c)** As $m$ gets larger, what numbers are the expressions in parts (a) and (b) tending toward?

**16.** (Contributed by John Oprea of Cleveland State University.) Use your calculator to evaluate for $m = 1, 10, 100, 1000$, and 1,000,000:

**(a)** $\left(1 - \frac{1}{m}\right)^m$

**(b)** $\left(1 - \frac{2}{m}\right)^m$

**(c)** As $m$ gets larger, what numbers are the expressions in parts (a) and (b) tending toward?

**17.** You have $1000 on deposit at your bank at an annual rate of 3%. How much interest do you receive after one year if the bank compounds

**(a)** continuously?
**(b)** daily, using 365 days in a year?

**18.** Suppose that you have a bank account with a balance of $4532.10 at the beginning of the year and $4632.10 at the end of the year. Your bank advertises "continuous compounding," but in fact compounds continuously over each 24-hour day and posts interest to accounts daily.

**(a)** What effective rate did you receive?
**(b)** What nominal rate is the calculation based on?
**(c)** What difference is there between what the bank is doing and true continuous compounding?

**19.** Suppose that you have an investment that earns 0% in the first year, but 10% in the second year.

**(a)** What rate of interest, compounded annually, would yield the same return after two years?
**(b)** What rate of interest, compounded continuously, would yield the same return after two years?

**20.** Suppose that you have an investment that earns 10% in the first year, 20% in the second year, and 30% in the third year.

**(a)** What rate of interest, compounded annually, would yield the same return after three years? (The answer here is related to the geometric mean of Chapters 14 and 19, but you do not need to know about that to solve the problem.)
**(b)** What rate of interest, compounded continuously, would yield the same return after three years?

(Thanks for the idea to Yi Cheng, Indiana University South Bend.)

**21.** We saw in Example 5 on pp. 689–690 that a nominal rate of 5% compounded continuously yields an effective annual rate of 5.12711%.

**(a)** What effective annual rate does a nominal rate of 4% yield with continuous compounding?
**(b)** The difference between the effective rate under continuous compounding, $e^r - 1$, and the nominal rate $r$ is $e^r - 1 - r$. You can't calculate this formula in your head, but you can approximate it closely with one that you can: $e^r - 1 - r \approx \frac{1}{2}r^2$. Thus, for $r = 4\% = 0.04$, the difference is $\frac{1}{2}(0.04)(0.04) = \frac{1}{2}(0.0016) = 0.0008 = 0.08\%$. So the effective rate is about 4.08%. Apply this formula to approximate the difference for a nominal rate of 5%, and compare the result with the 5.12711%.

**22.** [Suggested by Arthur R. Segal, University of Alabama at Birmingham (retired).] We approximate the smaller difference between regular compounding and continuous compounding. With a nominal annual rate $r$ over $t$ years, continuous compounding yields $e^{rt}$, while compounding $m$ times per year yields $(1 + \frac{r}{m})^{mt}$, for a difference of $D = e^{rt} - (1 + \frac{r}{m})^{mt}$ that can be approximated by

$$D \approx \frac{r^2te^{rt}}{2m + \frac{4r}{3} + \frac{r^2t}{2}}$$

For a $1000 initial investment, calculate both the true difference and the approximate difference between continuous compounding and

**(a)** quarterly compounding at a nominal rate of 4% for 10 years.
**(b)** daily compounding at a nominal rate of 18% for 10 years.

## 21.4 A Model for Saving

**23.** Suppose that you want to save up $2000 for a trip abroad two years from now. How much do you have to put away each month in a savings account that earns 5% interest compounded monthly?

**24.** Repeat Exercise 23, except that you have found a better deal, 7% interest compounded monthly.

**25.** Parents struggle for the first few years after their child is born but are finally able to start saving toward the child's college education when the child goes to school at age 6 (because the parents stop paying for day care). If they save $400 per month in a credit union account paying 5.5% interest compounded monthly, how much will they have for college expenses 12 years later?

**26.** Suppose that you save for retirement by contributing the same amount each month from your

23rd birthday until your 65th birthday, in an account that pays a steady 5% annual interest compounded monthly.

**(a)** How much will be in your fund at age 65 if you save $100 a month?
**(b)** How much will be in your fund if you get a steady return of 7.5% compounded monthly?
**(c)** How much will be in your fund if you get a steady return of 10% compounded monthly? (This is comparable to the average annual return of about 11% for all stocks on the New York Stock Exchange from 1950 to 2000.)

**27.** A colleague feels that he will need $1 million in savings to afford to retire at age 65 and still maintain his current standard of living. A younger colleague, age 30, decides to begin saving for retirement based on that advice. How much does the younger colleague need to save per month to have $1 million at retirement if the fund earns a steady 5% annual interest compounded monthly?

**28.** The younger colleague of Exercise 27 is not satisfied with 5% return, which he could get with long-term certificates of deposit. Instead, he wants to take the riskier route of investing in the stock market, which has over its history returned an APY of about 10% per year (although for 2001–2007, the APY was only 1.6%). Assuming that over the 35 years until his retirement that the stock market behaves just that way (a big assumption!), how much would he need to invest each month to achieve his goal of $2 million by age 65?

**29.** Many young people do not start saving right away for retirement, although by the time that they do, they may be earning more and thus be able to afford to save more each month. How much will be in your fund at age 65 if you don't start saving until age 35 and at that age start saving $100 per month in an account paying a steady 6% annual interest compounded monthly?

**30.** Suppose instead that you have children young, pay for their college expenses, and finally start saving for retirement at age 45. How much do you have to save per month, with a steady return of 7.5% compounded monthly, to accumulate $250,000 by age 65?

■ **31.** Suppose that you are 25, single, and in a 25% bracket for federal income tax and a 7% bracket for state and local income taxes (in 2007 this corresponded to an income, beyond exemptions and deductions, of $32,000–$77,000 for a single person). This means that you pay a total of 32% in income tax on part of your income but a lower rate on the rest (you also pay 7.65% Social Security and Medicare payroll tax). Assume that instead of paying 32% on some income, you put it into a tax-deferred retirement account (TDA) as follows:

**(a)** Suppose that you are willing to commit to $100 a month less take-home pay. You realize that you don't pay income tax on the money that you put into the retirement plan, so you can actually put in more than $100 per month while reducing your take-home pay by only $100. How much can you put into the retirement fund each month?
**(b)** How much will be in your fund at age 65 if you can get a steady return of 7.5% compounded monthly?
**(c)** Suppose that when you turn 65 you withdraw the entire amount in your account and pay the deferred taxes that are owed on it, say a total of 32% (federal, state, and local combined). How much do you net?

■ **32.** We continue the tax-deferral considerations of the previous exercise.

**(a)** Suppose that instead of contributing to a tax-deferred plan, you take the money as income, pay 32% income tax on it, and deposit what remains into a savings account or safe investment that pays a steady 7.5% compounded monthly. (Note that compared with not putting away any money, your paycheck is reduced by just what you contribute, since you still must pay income tax on the $147.06.) How much will be in your account at age 65?
**(b)** Under another alternative, you take $100 per month, pay 32% income tax on it, and deposit what remains into a *Roth IRA* (individual retirement account). For this kind of retirement account, the interest earned is not taxed. Assuming the same savings account or safe investment that pays a steady return of 7.5% compounded monthly, how much will be in your account, tax-free, at age 65?

**33.** Apart from certificates of deposit, returns on investments are rarely the same from year to year, as they vary with prevailing interest rates. How should you calculate an "average" rate of return over several years? Consider a mutual fund that delivers 100% return one year and loses 50% the next year. Calculate just the ordinary average (the arithmetic mean) to get $(100\% + (-50\%))/2 = 25\%$. That sounds good, but check what happens to a $1000 investment: It grows to $2000, then halves back to $1000–for a 0% gain. The customary way used in finance to calculate the "average" return is to use the geometric mean. If the initial value of the portfolio was $P$, and its value after $n$ years is $A$, then the average annual rate of return is the value of $r$ that solves $(1 + r)^n = A/P$, or $r = (A/P)^{1/n} - 1$.

**(a)** Use this formula to determine the average annual rate of return for a portfolio with returns of 15%, 7%, and −20% in three consecutive years.
**(b)** Is the average rate that the formula finds a nominal rate or an effective rate?

■ **34.** We continue the theme of Exercises 25 and 26 by comparing three kinds of investments for retirement: an ordinary after-tax investment, a tax-deferred investment [such as a tax-deferred annuity or an individual retirement account (IRA)], and a Roth IRA. Let an investment earn interest at a steady annual yield $r$ and let your income (in whatever year you receive it) be taxed at rate $\tau$.

**(a)** Ordinary after-tax investment: Explain why if you earn $\$E$, pay taxes on it, let what remains earn interest, and pay tax each year on that year's interest, the $\$E$ grows after $n$ years to $\$E(1-\tau)\times[1+r(1-\tau)]^n$.
**(b)** Ordinary IRA: Explain why if you earn $\$E$, defer taxes on it, let it earn interest, and defer taxes on all the interest, then the $\$E$ grows after $n$ years to $\$E(1+r)^n(1-\tau)$.
**(c)** Roth IRA: Explain why if you earn $\$E$, pay taxes on it, let what remains earn interest, and pay no taxes on all the interest, the $\$E$ grows after $n$ years to $\$E(1-\tau)(1+r)^n$.
**(d)** Which investment gives the best return after $n$ years?
**(e)** The assumptions of constant interest rate and stable tax rate won't necessarily hold, because interest rates fluctuate (though you can lock in a long-term constant interest rate by buying a long-term bond or certificate of deposit) and the tax rate may change (with your income, your state of residence, and changes in tax laws). If your marginal tax rate (the rate you pay on one more dollar of income) is lower in one year than the tax rate you expect to pay in retirement, what kind of retirement investment is better for you that year? If you have a windfall one year and your marginal tax rate is higher that year than the tax rate that you expect to pay in retirement, what kind of retirement investment is better for you that year?

## 21.5 Present Value and Inflation

**35.** Classify the following growth and decay scenarios as linear (arithmetic), exponential (geometric), or neither:

**(a)** The amount of caffeine in the bloodstream decreases by 10% every hour.
**(b)** The amount of trash in a landfill increases by 350 tons per week.
**(c)** The amount of alcohol in the bloodstream decreases by 10 grams (the amount in a standard drink) per hour.
**(d)** Your age increases every day.

(Adapted from Terence Blows, Northern Arizona University.)

**36.** Assume the same situation as in Exercise 35, but for

**(a)** The mean concentration of carbon dioxide in the atmosphere increases by 2 ppm (parts per million) per year.
**(b)** The mean concentration of carbon dioxide in the atmosphere increases 0.5% per year.
**(c)** Your knowledge of mathematics and its applications increases with each section of this book that you study.
**(d)** The number of people in the world increases by 1.3% per year.

(Adapted from Terence Blows, Northern Arizona University.)

**37.** What is the present value of \$10,000, four years from now, at an APY of 5%?

**38.** What is the present value of \$150,000, ten years from now, at an APY of 3%?

**39.** As you will see in Chapter 22, if you have a 30-year \$200,000 mortgage at 8% on a house or apartment, after 22 years of payments you will still owe about \$150,000! What is the present value of \$150,000, 22 years from now, at an interest rate of 8%? (If you put this much more into a down payment, but made the same-size payments as for the 30-year mortgage on \$200,000, you would own the house free and clear after 22 years instead of still owing \$150,000.)

**40.** If you have a 30-year \$200,000 mortgage at 6.48% on a house or apartment, after 10 years of payments you will still owe about \$170,000. What is the present value of \$170,000, 10 years from now, at an interest rate of 6.48%?

**41.** Suppose that inflation proceeds at a constant rate of 3% per year from mid-2009 through mid-2012.

**(a)** Find the cost in mid-2012 of a basket of goods that cost \$1 in mid-2009.
**(b)** What will be the value of a dollar in mid-2012 in constant mid-2009 dollars?

**42.** Suppose that you bought a car in mid-2009 for \$10,000. If its value (in current dollars) depreciates steadily at 12% per year, what will its value (in current dollars) be in mid-2012?

**43.** For the car in Exercise 42, suppose that there is also 3% annual inflation from 2009 through 2015. What will the value of the car be in mid-2015 in inflation-adjusted (mid-2009) dollars?

**44.** I bought my first vinyl record in 1965, at list price, for \$4.98. How much would that be in 2009 dollars? How does that compare with the list price of a CD today?

(mediacolor's/Alamy.)

**45.** My first-semester college mathematics book cost \$10.75 in 1962. What would the equivalent price be in 2009 dollars? How does that compare with what you paid for this book? (My book had black-and-white text and figures, with no photographs, color or otherwise.)

**46.** In 1970, before the OPEC oil embargo, gasoline cost about 25 cents per gallon. In 1974, after the embargo, it cost about 70 cents per gallon. What would the equivalent prices be in 2009 dollars? How do they compare with the price of gasoline today?

Refer to the following in doing Exercises 47 and 48.

From Table 21.5, you can determine the average rate of inflation from one year to another. For example, you find the inflation from 1990 to 2000 by subtracting the two index numbers and dividing by the earlier one: $(172.2 - 130.7)/130.7 = 31.752\%$. However, the average rate of inflation is not this number divided by the number of years (10). We must take into account compounding of the rate of inflation. We set $(1 + a)^{10} = 1.31752$ and find $a = (1.31752)^{1/10} - 1 = 2.80\%$.

**47.** Find the average rate of inflation from 1997 to 2007. Is 3% a good approximation?

**48.** If inflation had been 3% each year from 1997 to 2007, what would the CPI have been in 2007?

**49.** (Suggested by Ed Barbeau's column "Fallacies, Flaws, and Flimflam" in *The College Mathematics Journal*.) Suppose that you get a pay raise of 10%, but in the meantime, there has been inflation of 20%–so in effect you have suffered a pay decrease in terms of what your salary will buy. What is the percentage decrease?

**50.** What is the present value of a \$2000 raise now, which you will enjoy over the course of 40 years more of working, if inflation is a steady 3% per year?

Refer to the following in doing Exercises 51 and 52.

A new assistant professor at a typical American liberal arts college starts at age 30 with a salary of \$45,000, while colleagues retiring at age 65 make about twice that. One college gives annual pay raises of inflation plus one percentage point, plus a promotion raise (to associate professor) of \$1500 after (usually) 6 years and another promotion raise (to full professor) of \$1500 after (usually) another 6 years.

**51.** (Spreadsheet helpful) Can a new assistant professor who starts now expect to be making the equivalent of \$90,000 in today's dollars when she retires 35 years from now if inflation holds steady at

**(a)** 3%?
**(b)** 5%?

**52.** (Spreadsheet helpful) Repeat Exercise 51, but suppose that you are the vice president for academic affairs at the college. Suggest a salary policy that would result in the new assistant professor, when she retires in 35 years, making the equivalent of

**(a)** \$90,000 in today's dollars.
**(b)** \$135,000 in today's dollars. (She would prefer that!–and hence she would be more likely to accept an offer to come work at your college.)

**53.** Surprise! Just for fun, one of your friends wrote your name on an Illinois State Lottery ticket, and you are the sole winner of \$40 million! You discover, however, that you don't get the \$40 million all at once. In fact, it is paid in 20 equal annual installments of \$2 million each. All you get right away is the first installment of \$2 million (minus 20% withheld against federal income tax due and whatever you think your friend deserves for the favor). So, what is the prize really worth to you? That depends on the rate of inflation over the years. Assume a constant rate of 3% inflation over the 19 years until your last payment and calculate the present value of your prize winnings by using the formula for present value combined with the formula for the sum of a geometric series. Do the calculation for a rate of interest of 4%.

Actually, the checks will come not from the state of Illinois but from an insurance company from which Illinois purchases an annuity, whose price depends on current long-term interest rates.

**54.** Repeat Exercise 53, but for an interest rate of 6%.

**55.** (Spreadsheet helpful) Your roommate (a business major) has already planned her retirement and will start funding it in 2009. She plans to retire in 2044 at age 57 on \$100,000 per year in 2044 dollars, living on just the interest on her investments. Assume that she realizes a steady 7.2% and assume a steady 3% annual inflation.

**(a)** What must the size of her nest egg be, and what should her monthly investment be over the 35 years, to achieve this goal?
**(b)** What will be the value in 2009 dollars of her 2044 income of \$100,000?
**(c)** What will be the value in 2009 dollars of her income of \$100,000 in 2072 (when she is 85)?

(Suggested by Terence Blows, Northern Arizona University, Flagstaff, Ariz.)

**56.** (Spreadsheet helpful) You think what your roommate means in Exercise 55 is that she wants to retire in 2044 with a *steady* income of \$100,000 a year in 2009 dollars. You also feel that she should plan to receive that same value of income for 43 years in case she lives to 100 (2% of your classmates will). What is the present value in 2009 of the planned stream of 43 years of retirement income?

In the savings formula, the interest rate $i$ appears twice. The particular ways in which $i$ is involved make it impossible to solve it algebraically to get an explicit formula for $i$. However, with the help of a spreadsheet, you can find $i$ approximately when the other quantities are given. Exercises 57–60 treat such situations.

**57.** (Spreadsheet helpful) Suppose that you decide to lease a car. At the end of the 48-month lease period, you need to make a lump-sum payment of $5000 if you want to keep the car. You decide to save up, just in case you decide to keep the car; if you don't keep this car, you will still have saved a good down payment on a new car. You feel comfortable with saving $70/month (over and above your lease payments). How high an annual nominal interest rate on savings do you need to accumulate $5000 in 48 months, with interest compounded monthly?

**58.** (Spreadsheet helpful). A 1990 advertisement reads, "If you had put $100 per month into this fund starting in 1980, you'd have $37,747 today." Assume that deposits were made on the last day of the month, starting in January 1980, through December 1989, and that interest was paid monthly on the last day of the month (120 months).

**(a)** How much money was deposited during this period?
**(b)** What annual rate of interest, compounded monthly, would lead to the result described in the advertisement? What is the annual yield?

**59.** (Spreadsheet helpful) The Powerball lottery drawing on December 25, 2002, resulted in the largest prize for a single winner anywhere ever, with a jackpot of $314.9 million. The winner had the option of an annuity in 30 equal annual installments of $10.5 million (the first payment being right away) or else an instant lump sum of $170 million (actually, after taxes, about $112 million). Taking the lump sum (most lottery winners do) makes sense, particularly if the winner needs a large sum of money now or else feels that he or she can earn a higher rate of interest than the annuity is based on. The present value of the payment $t$ years from now is given by the compound interest formula $A = P(1 + i)^t$, where $P = 10.5$ million and $i$ is the rate of interest built into the annuity. Hence, we have $P_k = A/(1 + i)^t$ for the $t$th payment. The complete stream of 30 payments has present value (in millions of dollars)

$$170 = 10.5\left[1 + \frac{1}{1+i} + \frac{1}{(1+i)^2} + \cdots + \frac{1}{(1+i)^{29}}\right]$$

We use the geometric series formula with $x = 1/(1 + i)$, finding that the right-hand side is also equal to

$$10.5\left[\frac{1 - \dfrac{1}{(1+i)^{30}}}{1 - \dfrac{1}{1+i}}\right]$$

Enter $i = 0.04$ (annual rate) in the preceding expression; the result is larger than 170. For $i = 0.05$, the result is smaller than 170. Make changes in the value of $i$ until you determine to two decimal places the rate $i$ that gives the closest value to 170. This is the rate on which the annuity is based. Is it a nominal rate or the effective rate?

**60.** (Spreadsheet helpful) Suppose that your parents are willing to lend you $20,000 for part of the cost of your college education and living expenses. They want you to repay them the $20,000, without any interest, in a lump sum 15 years after you graduate, when they will be about to retire and move. Meanwhile, you will be busy repaying federally guaranteed loans for the first 10 years after graduation. But you realize that you can't repay the lump sum without saving up. So you decide that you will put aside money in an interest-bearing account every month for the five years before the payment is due. You feel comfortable with the idea of putting aside $275 a month (the amount of the payment on your government loans). How high an annual nominal interest rate on savings do you need to accumulate the $20,000 in 60 months, if interest is compounded monthly? Enter into a spreadsheet the values $d = 275$, $r = 0.05$ (annual rate), and $n = 60$, and the savings formula with $r$ replaced by $r/12$ (the monthly interest rate). You will find that the amount accumulated is not enough. Change $r$ to 0.09–it's more than enough. Try other values until you determine $r$ to two decimal places.

## APPLET EXERCISES

**To do these exercises, go to www.whfreeman.com/fapp8e.**

How important is it to begin a retirement fund at an early stage of one's career? In the *Saving for Retirement* applet, you will discover that early funding of a retirement plan can make a huge difference in the ultimate amount that will be available when a person retires.

## WRITING PROJECTS

**1.** Exercises 29, 30, and 32 asked you to look at various forms of tax-deferred and ordinary savings and compare them on the basis of amount of tax-free income accumulation at age 65.

Ordinary savings have the important advantage that, at any time, you can do anything you want with the money accumulated so far (buy a car, put down money on a house, and so on). A second advantage is that the money is free and clear, in that taxes have already been paid on it.

A tax-deferred 401(k) retirement fund has the disadvantage that for any funds withdrawn before age $59\frac{1}{2}$, you must pay income tax in the year of withdrawal and, in addition, pay a 10% penalty for "early withdrawal." (These plans were given the advantage of tax deferral to encourage individuals to save for retirement–hence the penalty for withdrawing the money earlier.)

A third option, the Roth IRA, has some of the advantages and disadvantages of each of the other plans.

Look into the details of the rules for 401(k) plans and Roth IRAs, compare your answers in Exercises 29, 30, and 32, and devise and describe your own plan for how you will save for retirement. Your report should run two to three pages.

**2.** Plan your own retirement. Decide on a retirement age and desired income (in today's dollars). Estimate yield on investments, inflation rate, and Social Security benefits. (*A few notes:* By the time of your retirement, a woman retiring in her mid-60s will likely live an average of 20 years more, a man 18 years more. Social Security income goes up with inflation. Annuities are available whose income grows to keep up with inflation. Various other financial products, such as a *life annuity*, can make sure that you don't outlive your retirement income. We treat life annuities in Chapter 22. Also, we neglect here any consideration of any tax deferral of income or earnings, as considered in Exercises 29, 30, and 32.) Write up your assumptions, justifications for them, calculations, and conclusions in three to four pages. Be sure to note any additional factors that you think should be taken into account but which your analysis does not include.

**3.** Should you stock up on "forever" stamps? In 2007 the U.S. Postal Service began selling such stamps, which will suffice for first-class postage at any time in the future, even when rates rise. Until rates rise, these stamps cost $0.41. Would it be a good investment to buy a "lifetime supply" of them? To render your judgment, compare the historic increases postal rates (see www.vaughns-1-pagers.com/economics/postal-rates.htm) with the CPI (p. 816). For example, you could convert postal rates at each of the dates of change to the cost in today's dollars, then see if there appears to be a recent trend that you can project into the future.

**4.** Based on ticket sales and current dollars, the top-grossing domestic film of all time was *Titanic*, which earned $600,788,188 in 1997. But adjusted for inflation of ticket prices, *Gone with the Wind* (1939, plus re-releases) comes out on top at $1,377,944,300 in 2008 dollars. (The comparison, for proceeds rather than audience share, does not adjust for the larger population in 1997.) Use a spreadsheet to analyze how movie ticket prices have risen since 1913 compared to the Consumer Price Index. The CPI data from Table 21.5 can be downloaded from data.bls.gov/cgi-bin/surveymost?cu. (Check the first box, click "Retrieve data," then in the next screen adjust the year "From" to 1913.) Find the data on movie ticket prices at boxofficemojo.com/about/adjuster.htm. Prepare a short report, with a graph. (Thanks to Martin Campbell for the idea.)

**5.** The federal minimum wage is set to rise to $7.25 per hour on July 24, 2009. Has the federal minimum wage kept pace with inflation? Use a spreadsheet to analyze how it has risen since 1938 compared to the Consumer Price Index. The CPI data from Table 21.5 can be downloaded from data.bls.gov/cgi-bin/surveymost?cu. (Check the first box, click "Retrieve data," then in the next screen adjust the year "From" to 1938). Find the data on federal minimum wage rates at www.dol.gov/esa/minwage/chart.htm. Prepare a short report, with a graph.

## SUGGESTED READINGS

KASTING, MARTHA. *Concepts of Math for Business: The Mathematics of Finance.* UMAP Modules in Undergraduate Mathematics and Its Applications: Module 370–372. COMAP, Inc., Arlington, Mass., 1980.

LINDSTROM, PETER A. *Nominal vs. Effective Rates of Interest.* UMAP Modules in Undergraduate Mathematics and Its Applications: Module 474. COMAP, Inc., Arlington, Mass., 1988. Reprinted in Paul J. Campbell (ed.), *UMAP Modules: Tools for Teaching 1988,* COMAP, Inc., Arlington, Mass., 1989, pp. 21–53. A learning module, requiring no more background than this chapter, that teaches about nominal and effective rates of interest and how to calculate them. Gives real examples of banks using different options for calculating interest.

MILLER, CHARLES D., VERN E. HEEREN, and JOHN HORNSBY. Consumer mathematics. In *Mathematical Ideas,* 11th ed., Pearson Education/Addison Wesley Longman, Boston, 2007.

VEST, FLOYD, and REYNOLDS GRIFFITH. The mathematics of bond pricing and interest rate risk. *Consortium (COMAP),* 59 (Fall 1996): HiMAP Pullout Section 1–6.

## SUGGESTED WEB SITES

**www.bls.gov/cpi/** Home page for the inflation tables prepared by the Bureau of Labor Statistics.

**www.bls.gov/data/inflation_calculator.htm** CPI Inflation Calculator. Converts dollar value from any year to its equivalent buying power in any other year.

**www.westegg.com/inflation/** Inflation calculators for the United States (1800–2001), Canada, and Italy, by S. Morgan Friedman, with links to sites about the current purchasing power of amounts of currencies of other countries in the past.

CHAPTER 22



# Borrowing Models

In the previous chapter, we looked at consumer financial models for saving and formulas for calculating the amount accumulated. Savings or investments would not earn interest unless they could be loaned to someone to make productive use of the money.

In this chapter, we examine the other side of consumer finance, borrowing. You may have a student loan, you are likely to borrow to buy a car, you will almost certainly borrow if you buy a house or apartment, and you are borrowing if you use a credit card. For any such loan, you pay "finance charges," which include interest and perhaps other "fees" as well. We investigate and compare some common kinds of loans.

We briefly (re)acquaint you with compound interest and a few formulas from Chapter 21. If you have a grasp of the ideas behind compound interest and can use the compound interest formula (p. 684) and the savings formula (p. 691), which we repeat shortly for your convenience, you can proceed with this chapter without first reading Chapter 21.

## 22.1 Simple Interest

The amount of **interest** charged on a loan is determined by the **principal**, by the amount borrowed, and by the method used to calculate the interest. With **simple interest**, the borrower pays a fixed amount of interest for each period of the loan. The interest rate is usually quoted as an annual rate.

For a principal $P$ and an annual rate of interest $r$, the interest owed after $t$ years is

$$I = Prt$$

and the total amount $A$ due on the loan is

$$A = P(1 + rt)$$

## EXAMPLE 1 ■ Simple Interest on a Federal Student Loan

(*moodboard/Corbis.*)

The federal government offers guaranteed loans through banks to students to pay for tuition, fees, housing, and textbooks (such as this one), with repayment deferred until after graduation. Any eligible student can take out an unsubsidized Stafford loan, in which you are charged interest from the time that you receive the loan until it is repaid in full. The interest rate for new loans after July 1, 2006 is 6.8%. One option is to pay each month the interest due, and defer paying back the principal and interest until six months (the "grace period") after you leave school.

Suppose that you took out a \$3500 Stafford loan on September 1 before your freshman year and begin paying it back on December 1 after graduation (so you will have had the loan for 4 years + 3 months = 51 months).

How much is the monthly interest, how much total interest will you have paid over the 51 months, and how much will you owe when you start to pay back?

**SOLUTION** We have $P = \$3500$ and $r = 6.8\% = 0.068$, and for one month we have $t = \frac{1}{12}$ years. So the interest for one month is $I = Prt = \$3500 \times 0.068 \times \frac{1}{12} = \$19.8333 \approx \$19.83$. Over the 51 months, you will have paid $51 \times \$19.83 = \$1011.33$. You will still owe the original principal of \$3500.

## 22.2 Compound Interest

*Compounding* is the calculation of interest on interest. A common example is the balance on a credit card. As long as there is an outstanding balance owed, the interest owed is calculated on the entire balance, including any part of it that was previously calculated as interest and added to the balance in earlier months.

## EXAMPLE 2 ■ Credit-Card Interest

Suppose that you owe \$1000 on your credit card, the company charges 1.5% interest per month, and you just let the balance ride. How much interest do you pay in the first year?

**SOLUTION** Your interest the first month is 1.5% of \$1000, or $0.015 \times \$1000 = \$15$. The new balance owed is $(1 + 0.015) \times \$1000 = \$1015$. Your interest the second month is not 1.5% of \$1000, or \$15 (as would be the case for simple interest), but 1.5% of \$1015, or $0.015 \times \$1015 = \$15.23$, so the new balance is

$$(1 + 0.015) \times \$1015 = \$(1 + 0.015) \times (1 + 0.015) \times \$1000 = \$1030.23$$

(We neglect the extra charges for your failure to make minimum payments.) After 12 months of letting the balance ride, it has become

$$(1.015)^{12} \times \$1000 = \$1195.62$$

In other words, the actual interest for the year comes to \$195.62, which is 19.562% of \$1000. So, although the quoted rate of interest is 1.5% per month, which seems as if it should amount to $12 \times 1.5 = 18\%$ per year, the interest owed is actually more.

We apply two formulas from Chapter 21: the **compound interest formula** (p. 684) and the **savings formula** (p. 691), phrasing them for loans. Here is the compound interest formula, followed by an example:

**Compound Interest Formula** RULE

If a principal $P$ is loaned at interest rate $i$ per compounding period, then after $n$ compounding periods (with no repayment) the amount owed is

$$A = P(1 + i)^n$$

This formula just generalizes what we saw happen with the credit-card balance. We give the formula in a slightly more elaborate version below, to make the connection to multiple compoundings per year.

**Compound Interest Formula** RULE

For a principal $P$ loaned

- at a nominal annual rate of interest rate $r$
- with $m$ compounding periods per year (so interest rate $i = r/m$ per compounding period), the amount owed
- after $t$ years (hence $n = mt$ compounding periods) with no payment of interest or principal is

$$A = P(1 + i)^n = P\left(1 + \frac{r}{m}\right)^{mt}$$

## EXAMPLE 3 ■ Not Repaying Your Student Loan

As noted above, 6.8% annual simple interest accrues on an unsubsidized Stafford loan from the time that you receive the loan until you begin repayment. When you begin repayment, the interest is (in the terminology of the student loan documents) *capitalized*, meaning that it is added to the principal. The interest rate remains 6.8%, for a monthly rate of 6.8%/12 = 0.566667%. However, if you do not make the payments due, the interest continues to accumulate and this additional interest is usually capitalized every quarter. That means that during a quarter of a year, the interest each month is simple interest on the amount still due at the beginning of each month, and at the end of the quarter the three months of interest is added to the amount due–there is no interest on the quarter's interest until after the quarter is ended. In our terminology, the compounding period is one quarter.

Suppose that you owe \$5000 on your Stafford loan but you fail to make any payments for 9 months. (This would be very foolish, since after 270 days of nonpayment the loan is in default and all kinds of bad things would happen!) How much would you owe then?

**SOLUTION** The principal $P$ is \$5000. The quarterly interest rate is $i = 6.8\%/4 = 1.7\%$ and there are $n = 3$ compounding periods. The compound interest formula gives the amount owed as $A = \$5000\ (1 + 0.017)^3 = \$5259.36$. ■

## Terminology for Loan Rates

The interest on a loan depends on whether or not compounding is done and how the interest is calculated. Just like the Truth in Savings Act mentioned in Chapter 21 (p. 682), the Truth in Lending Act establishes terminology and calculation methods for interest.

A **nominal rate** is any stated rate of interest for a specified length of time. For instance, a nominal rate could be a 1.5% monthly rate on a credit-card balance. By itself, such a rate does not indicate or take into account whether or how often interest is compounded.

The **effective rate** *takes into account compounding.* It is the rate of simple interest that would realize exactly as much interest over the same period of time.

We saw that \$1000 at a yearly interest rate of 18% (a nominal rate), calculated as 1.5% per month compounded monthly, yields \$195.62 in interest owed at the end of the year, which is 19.562% of the original principal. Hence, the effective annual rate is 19.562%. In other words, a \$1000 loan at simple interest of 19.562% for one year would owe exactly the same interest.

Finally, when stated per year ("annualized"), the effective rate is called the **effective annual rate (EAR)**. (The EAR is the same concept as the APY of Chapter 21, p. 682).

To keep the rates straight, we use $i$ for a nominal rate for the specified **compounding period**–such as a day, month, or year–*within which no compounding is done;* this rate is the effective rate for that length of time. For a nominal rate compounded $m$ times per year, we have $i = r/m$. For that \$1000 credit-card balance at 18% compounded monthly, we have $r = 18\%$ and $m = 12$, so $i = 1.5\%$ per month.

The Truth in Lending Act introduced the term **annual percentage rate (APR)**.

### Annual Percentage Rate (APR) — DEFINITION

The **annual percentage rate (APR)** is the number of compounding periods per year times the rate of interest per compounding period.

$$\text{APR} = m \times i$$

In the example of the credit-card balance, the interest is compounded monthly, or $m = 12$ times per year, and the interest rate for the compounding period is $i = 1.5\%$, so the APR is $12 \times 1.5\% = 18\%$. The APR is the rate that the Truth in Lending Act requires the lender to disclose to the borrower. *The APR is not equal to the EAR* (as we have already seen in the credit-card example), and Spotlight 22.1 explains further.

## 22.3 Conventional Loans

A common situation that you are likely to encounter is a loan–for a house, a car, or college expenses–to be paid back in equal periodic installments. Your payments are said to **amortize** (pay back) the loan. In these so-called **conventional loans,** each payment pays the current interest and also repays part of the principal. *As the principal is reduced, there is less interest owed, so less of each payment goes to the interest and more toward paying off the principal.*

We remind you of the savings formula from Chapter 21 (p. 691):

### Savings Formula — RULE

The amount $A$ that is accumulated

- at a nominal annual rate of interest rate $r$
- with $m$ compounding periods per year (so interest rate $i = r/m$ per compounding period)
- after $t$ years (hence $n = mt$ compounding periods)
- by a uniform deposit $d$ at the end of each compounding period

is

$$A = d\left[\frac{(1+i)^n - 1}{i}\right] = d\left[\frac{(1+\frac{r}{m})^{mt} - 1}{\frac{r}{m}}\right]$$

## SPOTLIGHT 22.1 What's the Real Rate?

Financial experts agree that the real, "true" rate of interest for savings or loans is the effective annual rate (EAR).

The 1991 Federal Truth in *Savings* Act requires that savers be told the annual percentage yield (APY), which is the EAR.

The 1968 Federal Truth in *Lending* Act, however, requires that borrowers be told the *annual percentage rate (APR)*, which is *not* the same as the EAR. The APR is the rate of interest per compounding period times the number of compounding periods per year. Thus, a credit-card rate of 1.5% per month translates to an APR of 18%. The APR does not take into account compounding. Hence it is not equivalent to—indeed, it understates—the true cost of borrowing, that is, the EAR. For the credit-card loan, with monthly compounding, the EAR is

$$(1 + 0.015)^{12} - 1 \approx 19.6\%$$

The APR also ignores costs that are sometimes involved in borrowing, such as a flat charge for making the loan in the first place ("loan-processing fee"), charges for late payments, and charges for failing to make a minimum payment.

For home mortgage loans, however, the Truth in Lending Act requires that lenders include in the APR some of the upfront costs referred to as *closing costs*: any "loan origination" fee, "loan-processing" fee, and "points" (additional charges to get a reduced interest rate). The APR does not include title insurance, appraisal, credit-report fees, or transaction taxes.

Closing costs are paid at the closing of the sale, while interest is paid over the life of the loan. However, the APR treats the closing costs included in it as if they were amortized over the term of the mortgage, despite the fact that they are paid beforehand. Here, too, the APR understates the true costs.

However, very few people hold a mortgage to its maturity. The median life of a 30-year mortgage is only about 5 years; that is, half of all mortgage holders pay off their mortgage before 5 years are up, usually because they sell their homes and move elsewhere. Thus, for almost all home loans, the APR also includes interest that will never be paid.

Finally, we must take into account inflation. One advantage of buying a home with a fixed-rate mortgage is that your payment stays the same but your earnings and the value of your home are likely to go up with inflation: You are thus paying back the loan with dollars of lesser value. For any loan in a time of inflation, *Fisher's effect* comes into play: If your loan has an EAR of 7% but inflation is running at 3.5% per year, the true cost to you of the loan is not exactly 7% − 3.5% = 3.5%. Instead, for an EAR of $r$ and an inflation rate of $a$, the cost of the loan at the beginning of the first year is indeed $r - a$ (= 3.5% in our example), but at the end of the first year it is

$$g = \frac{r - a}{1 + a}$$

For $r = 7\%$ and $a = 3.5\%$, we get $g = 3.38\%$. The reason that this is less than the expected 3.5% is that at the end of the first year you are paying back the loan with dollars that have been inflated for a year. As inflation mounts over the term of a mortgage, the cost $g$ goes down steadily each year. For example, at the end of five years of steady inflation at 3.5%, the total inflation has been $a = (1 + 0.035)^5 - 1 = 18.8\%$, and we have $g = 2.95\%$.

A final—and major—consideration is that interest paid on your home mortgage is deductible from taxable income on federal, state, and some local income tax returns. Thus, your home ownership is subsidized by other taxpayers (just as you help subsidize home buyers among them), and the cost to you of the loan is reduced further.

## EXAMPLE 4 ■ Buying a House

Let's suppose that you buy a house with a \$100,000 loan to be paid off over 30 years in equal monthly installments. Suppose that the interest rate for the loan is 6.00%. How much is your monthly payment?

**SOLUTION** Imagine changing the setup slightly so that instead of making monthly payments, you are supposed to pay off the entire principal and interest at the end. Meanwhile, you make payments to a savings fund that you're building up to pay off the loan, and the savings fund earns the same rate of interest as the loan costs. The interest rate of 6.00% on the loan is compounded monthly, so the monthly rate is 0.5%. At the end of 30 years, the principal and interest on the loan will (by the compound interest formula) amount to

$$\$100{,}000 \times (1 + 0.005)^{12\times 30} = \$602{,}257.52$$

On the other hand, saving $d$ each month for 30 years at 6.00% interest compounded monthly, we know from the savings formula (p. 691) that you will accumulate

$$d\left[\frac{(1 + 0.005)^{360} - 1}{0.005}\right]$$

To make $d$ just the right amount to pay off the loan exactly, we need to solve the equation

$$d\left[\frac{(1 + 0.005)^{360} - 1}{0.005}\right] = \$100{,}000 \times (1 + 0.005)^{12\times 30} = \$602{,}257.52$$

for the value of $d$, getting $d = \$599.55$ as your monthly payment. The total of the payments is "only" 360 × \$599.55 = \$215,838.00–on a loan of just \$100,000. (Actually, since the value of $d$ more accurately is \$599.5505, your regular monthly payment would be rounded *up* to \$599.56 but your last payment would be correspondingly slightly less. We will neglect this fine point here and in later calculations.) ■

We put this idea into a more general setting: *Paying off a conventional loan is like saving.* You can think of paying off the loan as making payments to a savings account that earns interest at the same rate as the loan. At the end of the loan term, the savings balance will exactly equal the principal and interest on the loan. Let the loan amount be $P$, the effective interest rate per compounding period be $i$, the number of compounding periods be $n$, and the loan payment be $d$. We equate the principal and interest on the loan (from the compound interest formula) with the savings balance (from the savings formula):

$$P(1 + i)^n = d\left[\frac{(1 + i)^n - 1}{i}\right]$$

The quantity $P$ is sometimes called the *present value of an annuity* of $n$ payments of $d$, each at the end of a compounding period with interest $i$ per period. This terminology is used in the financial mode of some calculators, such as the TI-83.

Solving the above equation for $d$ requires a little algebra. To make things simpler, let $b = (1 + i)^n$, so

$$Pb = d\left[\frac{b - 1}{i}\right]$$

Then

$$d = P\left[\frac{b}{\frac{b-1}{i}}\right] = P\left[\frac{bi}{b-1}\right]$$

Now divide numerator and denominator by $b$, getting

$$d = P\left[\frac{i}{1-b^{-1}}\right]$$

Substituting $(1+i)^n$ back for $b$, we get the usual form of the **amortization payment formula**:

### Amortization Payment Formula RULE

A conventional loan amount $P$

- at a nominal annual rate of interest rate $r$
- with $m$ compounding periods per year (so interest rate $i = r/m$ per compounding period)
- for $t$ years (hence $n = mt$ compounding periods) can be paid off by uniform payments at the end of each compounding period in the amount

$$d = P\left[\frac{i}{1-(1+i)^{-n}}\right] = P\left[\frac{\frac{r}{m}}{1-(1+\frac{r}{m})^{-mt}}\right]$$

## EXAMPLE 5 ■ Repaying Your Student Loan

The standard repayment option for federal student loans is repayment over 10 years with a minimum monthly payment of \$50 at the end of each month, beginning six months after you graduate. For the student loan of Example 1, what will your monthly payments be?

**SOLUTION** With the amortization payment formula, it's easy to figure out your monthly payment. We have $P = \$4511.33$, monthly interest rate $i = r/m = \frac{0.068}{12} = 0.00566667$, and $n = mt = 12 \times 10 = 120$ months for the payback. We find the payment $d$ as

$$d = P\left[\frac{\frac{r}{m}}{1-(1+\frac{r}{m})^{-mt}}\right] = \$4511.33\left[\frac{\frac{0.068}{12}}{1-(1+\frac{0.068}{12})^{-12\times 10}}\right] = \$51.92$$

So your monthly payment will be \$51.92, just a little more than the minimum payment. (That is for this loan; you may owe more for loans for your other years in college.) Hence over the lifetime of the loan, you will pay $120 \times \$51.92 = \$6230.40$, of which $\$6230.40 - \$4511.33 = \$1719.07$ is interest, plus the \$1031.16 interest accrued during the deferment and grace periods, for a total of \$2750.23 in interest. You will pay almost as much in interest as the original principal. If you were to stretch your payments over more years (permissible in some circumstances), an even greater proportion would be interest.

■

## EXAMPLE 6 ■ Buying a Car

You decide to buy a new Wheelmobile car. After a down payment, you need to finance (borrow) \$12,000. Comparing interest rates offered by the car dealership, local banks, and your credit union, the best deal you can find is 4.9% compounded monthly over 48 months. What is your monthly payment?

We have $P = \$12{,}000$, monthly interest rate $i = \frac{0.049}{12}$, and $n = 48$. Using the amortization formula, we have

$$d = \$12{,}000 \left[ \frac{\frac{0.049}{12}}{1 - (1 + \frac{0.049}{12})^{-48}} \right] \approx \$275.81$$

How much interest do you pay? You make payments totaling $48 \times \$275.81 = \$13{,}238.88$, so the interest is $\$13{,}238.88 - \$12{,}000 = \$1{,}238.88$.

What if you could get a 60-month loan at the same rate? Then your monthly payments would be \$225.91; over 60 months, you would pay \$13,554.60, of which \$1554.60 would be interest.

If you had bought a Plushmobile instead, with \$24,000 to finance, you would have borrowed twice as much and your monthly payment would have been twice as much.

A car loan is usually for 48 or 60 months, but when you buy a home, you usually borrow a great deal more money and pay it off over a much longer period. The usual term for a home mortgage is 30 years.

## EXAMPLE 7

### Thirty–Year Mortgage on Median-Priced Home

Let's suppose that you are a family with the U.S. median income of about \$59,000 for a family of four, that you want to buy a median-priced home (\$225,000 in August 2007) with a 30-year fixed-rate mortgage at 6.48%, and that you can make a down payment of only \$10,000. Can you afford such a home?

**SOLUTION** Lenders have "affordability" guidelines that suggest that a family can afford to spend about 28% of its monthly income on housing. Thus, by their guidelines, you can afford $0.28 \times \$59{,}000/12 = \$1376.67$ per month.

What is the monthly payment on the loan? The principal is $P = \$215{,}000$, the monthly interest rate is $i = 0.0648/12 = 0.0054$, and $n = 360$ months. The amortization formula gives a monthly payment of

$$d = \$215{,}000 \left[ \frac{\frac{0.0648}{12}}{1 - (1 + \frac{0.0648}{12})^{-360}} \right] \approx \$1356.12$$

Well, that sounds good. But unfortunately there is more to the mortgage than just the amount needed to amortize the loan. Your payment may also have to cover real estate taxes and homeowner's insurance on the property. On a \$225,000 home, these may add \$450 to the monthly payment, which will then total about \$1800.

So, no, the median family can't afford the median-priced home, at least not without a bigger down payment or a lower-interest loan.

A payment on an amortized loan includes both the current interest and a portion toward repaying the principal. You are "building **equity**" in the house or car that you are paying off.

## Equity — DEFINITION

**Equity** is the amount of principal of a loan that has been repaid.

## EXAMPLE 8 ■ Home Equity

My wife's parents sold their house in rural Minnesota to move to the town where we live. They had bought their house in 1980 for \$100,000 with a 30-year mortgage at an 8% interest rate. After 22 years, how much *equity* did they have in the house—that is, how much of the principal had been repaid? And how much did they still owe on the house?

(Norbert Schwerin/ The Image Works.)

**SOLUTION** What may shock you is that when they sold their house in May 2002—after 269 months of payments, almost exactly three-quarters of the 30 years of the mortgage—they had only \$50,000 in equity (hence still owed \$50,000 on the house) but had already paid \$147,000 in interest. *Three-quarters of their payments had gone to interest.*

We can use the amortization formula to determine just how much equity they had after 269 months of payments, but first we need to determine their monthly payment. We see $P = \$100{,}000$, $n = 360$ months, and $i = \frac{0.08}{12}$ monthly interest, getting $d = \$733.76$.

Now we use the formula again, this time "in reverse." Knowing $i = \frac{0.08}{12}$ and $d = \$733.76$, we find out how much of the loan would have been paid off by the remaining $n = 360 - 269 = 91$ payments:

$$d\left[\frac{1-(1+i)^{-n}}{i}\right] = \$733.76\left[\frac{1-(1+\frac{0.08}{12})^{-91}}{\frac{0.08}{12}}\right] = \$49{,}940.03$$

This is how much my parents-in-law had yet to pay, so their equity was $\$100{,}000 - \$49{,}940.03 = \$50{,}059.97$.

Figure 22.1 and Table 22.1 show that equity builds up very slowly at first but rapidly later. (The values shown do not take into account possible increase or decrease in the value of the house itself or the effect of inflation.) In fact, the amount of principal in a payment grows by a factor of $1 + i$ from one payment to the next, so the equity at any point is the sum of a geometric series (p. 691) whose common ratio is $1 + i$.

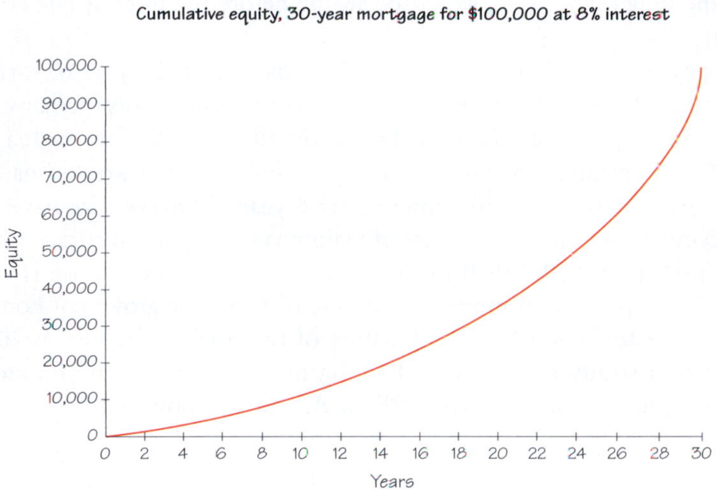


**FIGURE 22.1** Equity grows almost exponentially, especially in the later years of a mortgage.

**TABLE 22.1 Equity in a 30-year Mortgage for $200,000 at 6.48% Interest**

| End of Year | 1 | 2 | 3 | 4 | 5 | 6 | 7 | 8 | 9 | 10 | 15 | 20 | 25 | 30 |
|---|---|---|---|---|---|---|---|---|---|---|---|---|---|---|
| Equity ($\$ \times 10^3$) | 2 | 5 | 7 | 10 | 13 | 16 | 19 | 23 | 26 | 30 | 55 | 89 | 135 | 200 |

When you buy a home, you have several options for the mortgage: a conventional 30-year mortgage, a conventional 15-year mortgage, or a mortgage for either length of time but with an interest rate that can vary.

You might expect the payment on a 15-year mortgage to be double that of a 30-year mortgage. On the contrary, the payment is only 47% more (for a 5% mortgage) to 26% more (for a 9% mortgage). This range includes the prevailing mortgage rates over the past 20 years. Moreover, over the course of a $200,000 mortgage at 6%, you would pay $328,000 in interest over 30 years but only $144,000 over 15 years. At 9%, the interest totals are $380,000 versus $166,000. (Some financial counselors advise taking a 30-year mortgage and making extra payments when you can afford them, rather than incurring the higher payment obligation of a 15-year loan, which, if you encounter tight personal financial circumstances, you might not be able to afford.) In Spotlight 22.2, we discuss what we did in our own circumstances and mention other options.

Since very few mortgages are held for the full term, it is useful to compare the status of mortgages after five years, the median length of time that Americans remain in a home. Table 22.2 shows the equity after five years for a variety of interest rates. For a 30-year mortgage, the equity after five years may be less than the cost of selling the home through a realtor. Of course, the resale value of the home may also be higher after five years.

A mortgage with an interest rate that can vary is called an **adjustable-rate mortgage (ARM)**. Often such mortgages have a substantially lower interest rate (hence a lower payment) than a fixed-rate mortgage. The ARM's interest rate may go up or down with interest rates in the economy. Usually the rate can be raised or lowered only every year or two, and then by a limited percentage. An ARM may be attractive if you plan to pay off the mortgage after only a few years or because it allows lower payments, it facilitates buying a more expensive home, or you do not plan to keep the home long (hence, you would be selling before the interest rate could rise substantially).

Does it pay to buy a house or apartment? Apart from the joys of ownership, you need to take into account the up-front expenses of closing costs (perhaps $3000), plus the back-end expense of selling the house (usually 6%–7% if through a realtor, so say $12,000). Consulting the table, you might think that you would finally be in the black on your house as an investment after 6 years. However, we have not yet taken into account the ongoing expenses of maintenance, repairs, insurance, and real estate taxes (perhaps $4000–$7000 per year). Of course, if your house is rising in value 5% ($10,000) per year or more, it's a different story. The growth of home ownership, which rose to 73% before the bursting of the housing bubble in 2007, depended on such a steady rise in value. Renting may be attractive if you anticipate moving in just a few years; each year, 20% of Americans move.

## SPOTLIGHT 22.2 What We Did with Our House, and What Else You Could Do

We bought our house in 1992. We were offered a choice between an 8.375% fixed-rate 30-year mortgage and an adjustable-rate 30-year mortgage (ARM) at 6.875% whose rate could be raised (or lowered) by up to 2% every year. When we asked, we were also quoted slightly lower rates for corresponding 15-year mortgages.

We were planning to stay in the house much longer than the median of five years, and we were concerned that inflation might force the ARM considerably higher. Also, we did not want the obligation of the higher payments of a 15-year mortgage, in case our circumstances changed (such as through job loss or death). Some loans provide for penalties for paying off the loan early, but in our case (thanks to state law) there was no penalty for making extra payments (if we could afford them).

We chose the 8.375% fixed-rate 30-year mortgage (and made some extra payments). Others in other circumstances, or with a different tolerance for risk, would no doubt have decided otherwise. Had we been sure then that interest rates would not go higher in the 1990s, we would have gone for the ARM. But hindsight is always better than foresight. During the early years of this decade, most homeowners with mortgage interest rates such as ours refinanced at much lower prevailing rates, near 5% for a fixed-rate 30-year mortgage.

Currently, about a third of people take ARMs rather than fixed-rate mortgages as one way to respond to soaring real estate prices in some parts of the country. Newer mortgage "products" include interest-only mortgages and shared appreciation mortgages (SAMs). With an interest-only ARM, payments are (just slightly) lower than for a conventional 30-year mortgage, but you accumulate no equity (still, the market value of the house may rise). After five to seven years, you start also paying off the principal—which means that your payments go up then. In some such loans, the interest rate—and your payments—fluctuates as frequently as every month.

In a SAM, interest payments are lower or absent, but the lender receives a portion of any appreciation (rise in value) when the house is sold. In a nationally reported instance in 2003, a single mother received a no-interest SAM loan to finance the $30,000 down payment on a $223,000 house in Pleasanton, California, through the city's affordable housing program. Four years later, she sold the house for $385,000, and the "affordable housing" lenders got 60% of the $162,000 appreciation, or $97,000. She herself realized $65,000 (minus the cost of the sale) but complained bitterly, saying that she would have been better off to have put the loan on her credit card! Critics have termed SAMs an urban form of sharecropping.

## TABLE 22.2 Equity (in thousands of dollars) on a $200,000 Mortgage After Five Years

| Term (years) | Interest rate 5% | 6% | 7% | 8% | 9% |
|---|---|---|---|---|---|
| 15 | 51 | 48 | 45 | 42 | 40 |
| 30 | 16 | 14 | 12 | 10 | 8 |

## Spotlight 22.3 The Mortgage Crisis

Late 2007 saw the development and widening consequences of what has become known as the "mortgage crisis." To understand what that was, why it took place, and how it will have widespread effects for some time, you need to know what happens when you get a mortgage to buy a house—compared to what used to happen. In the "good" old days, you would go to a local bank (or savings and loan, or credit union). If you proved "credit-worthy,"—meaning that after careful consideration, the personnel felt that you could repay the loan—the bank would loan you *its* money, raised from its depositors. The interest rate depended on your credit rating and down payment. You paid back the loan at a fixed rate of interest, usually over 30 years. If interest rates went down, you could refinance the loan at a lower rate by taking out a new loan to pay off the old one; if rates went up, the bank would still receive your regular payments (but wish that it had charged a higher rate of originally). Meanwhile, the value of your house usually went up 5% to 10% per year, your income went up, and your payments stayed the same. (In fact, if you have only 10% equity in your house and it goes up in value 10% in one year, you have made 100% on your investment! This kind of "leveraging" can make real estate investment very profitable—as long as prices keep rising.)

What changed? Efforts to extend home ownership to a wider proportion of the population resulted in banks making more "subprime" loans (loans to people with poorer credit histories who are less likely to repay), with lower down payments but higher rates of interest (because of the greater risk). Some of those loans were "predatory lending," at high rates of interest to people who could not possibly make the payments. Certainly, real estate speculation played a role, as did greed. House prices rose (the "housing bubble"—house prices doubled from 2000 to 2006) but median income (adjusted for inflation) remained stagnant, which made it increasingly hard to afford houses. Banks countered with adjustable-rate and interest-only mortgages. They also realized that they could maximize current income with up-front charges ("origination fee," and "points" paid by the buyer to lower the interest rate) and at the same time minimize risk by immediately "flipping" the mortgages (selling them to bigger banks or other investors, to whom buyers would then make their payments). All of these factors led to banks making more loans that were riskier.

*(David H. Wells/ Getty Images.)*

What went wrong? Interest rates rose, and people with adjustable-rate mortgages (ARMs) saw their payments rise beyond their ability to pay. At the same time, the "pyramid" of housing prices could not continue with the higher mortgage rates; housing prices fell. When your house becomes worth less than what you would have to pay to keep it, you might be better off just walking away from it (especially since you build up almost no equity in the house in the first few years of the mortgage). Mortgage defaults lowered the value of investments in bundles of mortgages.

Houses are worth perhaps hundreds of billions of dollars less than they were just a year earlier, and big investment banks have mortgages on their hands that are worth hundreds of billions of dollars less than they paid for them. That means that for some time to come, banks have less money to loan and can (should, must) demand more-credit-worthy clients and higher rates of interest.

How does all this affect you? Financial institutions have less money to loan for *any* purpose—buying a home, buying a car, starting a business, etc.—hence the "credit crunch" following the mortgage crisis. We can only hope that the worst will be over before you are in a market for a house, car, or business loan.

Where you won't see direct effect is in the Consumer Price Index (CPI) (see pp. 695–697), which is based on the *rental value* of houses, not their prices. If the doubling of housing prices in 2000–2006 had been taken into account, the CPI would have risen by 5% per year rather than 3%.

## 22.4 Annuities

We recall from Chapter 21 the concept of an **annuity**:

**Annuity** DEFINITION

An **annuity** is a specified number of (usually equal) periodic payments.

We restrict our discussion to *ordinary annuities,* for which payments are made at the end of each interval and the interval is also the compounding period.

An annuity can be interpreted as involving borrowing. For example, winners of lotteries are often offered the choice of receiving either the jackpot amount paid as an annuity over a number of years or else a smaller lump sum to be paid immediately. The cost to the lottery administration is the same. If the winner wants an annuity, the administration buys one from an insurance company for the lump sum. You can think of the insurance company as borrowing the lump sum in exchange for making the payments of the annuity. In effect, the insurance company is amortizing the lump sum over the duration of the annuity.

### EXAMPLE 9 ■ Winning the Lottery

On March 6, 2007, two winning tickets shared a record Mega Millions jackpot of \$390 million. Each ticket's share was one-half of the total, or \$195 million. One option was to receive the \$195 million as an annuity in 26 equal annual installments of \$7.5 million each, the first payment being right away. However, each winner chose instead an instant lump sum of \$116,557,083. What was the interest rate of the annuity?

(*Tim Boyle/Getty Images.*)

**SOLUTION** The insurance company offering the annuity regarded \$116,557,083 as the present value of the annuity. In order to consider it as an ordinary annuity, with payments at the end of each period, we must subtract the first payment, leaving \$195 − \$7.5 million = \$187.5 million to be paid in 25 equal installments at the end of each year, with \$116,557,083 − \$7,500,000 = \$109,057,083 as the present value of the stream of payments.

Solving for $P$ in the amortization payment formula gives

$$P = d\left[\frac{1-(1+i)^{-n}}{i}\right]$$

Converting to millions, we have $P = \$109.057083$, $d = \$7.5$, and $n = 25$. To solve for $i$, we must use either a calculator with financial mode or a spreadsheet. Either way, we find $i = 4.69\%$. ■

As you save for retirement, it is probably wise to save part of your funds in the form of a tax-deferred annuity. If you do not, at retirement you can still sell all of your holdings in other forms and purchase an annuity.

### EXAMPLE 10 ■ How Much Do You Need to Retire?

Suppose that your father wants to retire at age 65 with an annuity that pays \$1000 per month for 25 years and is willing to assume that at retirement the long-range steady interest rate will be 4% per year compounded monthly. What amount should he expect to have to pay for such a stream of income?

**SOLUTION** We apply the amortization formula with $d = \$1000$, $r = 0.04$, $m = 12$, and $t = 25$, to find the amount $P$:

$$P = d\left[\frac{1-(1+\frac{r}{m})^{-mt}}{\frac{r}{m}}\right] = \$1000\left[\frac{1-(1+\frac{0.04}{12})^{-12\times 25}}{\frac{0.04}{12}}\right] = \$189{,}452.48$$

So purchasing such an annuity would cost \$189,452.48.

Such annuities differ in a crucial way from the lottery annuity in Example 9. If you retire at 65 and purchase an ordinary annuity, you would be in trouble if you live longer than the term of the annuity (past 90), because the payments would stop and you would have no further income from the annuity. (About 2% of U.S. children born today can expect to live to age 100.) Similarly, if you die sooner, your estate would still get the payments due after your death, but they wouldn't have helped you meet your living expenses while you were alive.

An approach that avoids these two disadvantages is the *life income annuity*: You receive a fixed amount of income per month for as long as you live. How much you receive per month is based on the life expectancy of people your age, as determined from population data. There are many variations on life annuities, such as payments that increase with anticipated cost-of-living increases, or payments that last until both you and your life partner die (see Spotlight 22.4). But we focus on a simple one-life annuity.

The insurance company that sells you the annuity makes money if you die younger than average and loses money if you die older than average. As in any kind of insurance, over a large number of people, the company can expect gains to balance losses. This is a manifestation of the law of large numbers of Chapter 8. Also, the company's profits vary with the prevailing interest rate during the annuity as compared with the rate built into the annuity.

How much you receive per month depends on your gender. Because women on average live longer than men, the monthly payment to a woman is lower.

## EXAMPLE 11 ■ Life Income Annuity

Suppose that you are a 65-year-old male retiring with \$250,000 in a life income annuity. According to the table from one particular insurance company, you would receive \$6.72 per month for every \$1000, so your monthly income would be \$1680. According to the Social Security Administration, your life expectancy at age 65 is about 16.6 years = 199 months. If you lived exactly that long, you would receive a total of 199 × \$1680 = \$334,320. However, simple algebra cannot be used to find the rate of interest that your annuity would need to earn to last that long. We use the RATE function in Excel; entering =RATE(199,1680,−250000) gives a monthly rate of 0.3065%, for an effective annual rate of 3.74%.

If you are female and retire now at the same age with the same \$250,000 savings in a life income annuity, you would receive \$6.26 per month for every \$1000, or \$1565 per month. Your life expectancy would be about 19.6 years = 235 months. If you lived exactly that long, you would receive a total of 235 × \$1565 = \$367,775. The rate of interest that your annuity would need to earn to last that long can be calculated from the amortization formula; using =RATE(235, 1565, −250000) in

Excel gives a monthly rate of 0.3516%, for an effective annual rate of 4.30%. The difference of this figure from that for a man probably reflects the company's use of different values for life expectancy, which vary with region of the country.

Notice that a man and a woman who save the same amount receive different monthly incomes at retirement: The woman receives less but for longer–about 90% as much for 25% longer. Yet their living expenses are likely to be the same. That consideration has resulted in some companies offering "merged gender" rate schedules for annuity payments, so that the individual receives the same monthly payment regardless of gender.

## SPOTLIGHT 22.4 What Actuaries Do

The Truth in Savings Act and the Truth in Lending Act specify that the APY for savings and the APR for loans must be calculated "according to the actuarial method."

Actuaries are financial experts who assess the costs of risks and investigate the probability of various contingencies—for example, death, default, or cancellation—that might occur. Actuaries are crucially involved in setting premiums. Their calculations take into account historical rates—such as the percentage of female 85-year-olds who live to be 86, or the percentage of unmarried male drivers under age 25 who have auto accidents—and project those rates and the accompanying costs into the future.

Other actuaries concentrate on setting up and evaluating pension and fringe benefit plans. For example, the city of Beloit, Wisconsin, hired a consulting actuary to estimate the current and future costs of free lifetime medical benefits to families of police and firefighters.

Another major activity of actuaries is managing return on investment. Contrary to popular belief, insurance companies (particularly life insurance companies) do not earn all of their money from premiums paid. In fact, a substantial portion of their income comes from return on investment of financial *reserves,* funds that they are required to have to meet current and future insurance obligations.

(David Young-Wolff/PhotoEdit.)

Becoming an actuary requires training in mathematics, statistics, economics, and finance, and includes a sequence of professional exams taken over several years.

## REVIEW VOCABULARY

**Adjustable-rate mortgage (ARM)** A loan whose interest rate can vary during the course of the loan. (p. 718)

**Amortization payment formula** Formula for installment loans that relates the principal $P$, the interest rate $i$ per compounding period, the payment $d$ at the end of each period, and the number of compounding periods $n$ needed to pay off the loan:

$$d = P\left[\frac{i}{1-(1+i)^{-n}}\right], \quad P = d\left[\frac{1-(1+i)^{-n}}{i}\right]. \text{ (p. 715)}$$

**Amortize** To repay in regular installments. (p. 712)

**Annual percentage rate (APR)** The number of compounding periods per year times the rate of interest per compounding period. (p. 712)

**Annuity** A specified number of (usually equal) payments at equal intervals of time. (p. 721)
**Compound interest formula** Formula for the amount in an account that pays compound interest periodically. For an initial principal $A$ and effective rate $i$ per compounding period, the amount after $n$ compounding periods is $A = P(1 + i)^n$. (p. 710)
**Compounding period** The fundamental interval for compounding, within which no compounding is done. Also called simply *period.* (p. 712)
**Conventional loan** A loan in which each payment pays all the current interest and also repays part of the principal. (p. 712)
**Effective annual rate (EAR)** The effective rate per year. (p. 712)
**Effective rate** The actual percentage rate, taking into account compounding. (p. 712)
**Equity** The amount of principal of a loan that has been repaid. (p. 717)
**Interest** Money charged on a loan. (p. 709)
**Nominal rate** A stated rate of interest for a specified length of time; a nominal rate does not take into account any compounding. (p. 711)
**Principal** Initial balance. (p. 709)
**Savings formula** Formula for the amount in an account to which a regular deposit is made (equal for each period) and interest is credited, both at the end of each period. For a regular deposit of $d$ and an interest rate $i$ per compounding period, the amount $A$ accumulated is

$$A = d\left[\frac{(1+i)^n - 1}{i}\right]. \qquad \text{(p. 710)}$$

**Simple interest** The method of paying interest on only the initial balance in an account and not on any accrued interest. For a principal $P$, an interest rate $r$ per year, and $t$ years, the interest $I$ is $I = Prt$. (p. 709)

## ✓ SKILLS CHECK

**1.** A nominal rate of interest
**(a)** takes into account any compounding involved.
**(b)** is always stated as an annual rate.
**(c)** neither of the above.

**2.** (Compound Interest Formula) If you borrow \$1000 at 5% interest per year, compounded quarterly, and pay back the principal and interest after four years, the amount that you pay back is ______________ .

**3.** An effective interest rate
**(a)** always takes inflation into account.
**(b)** is the same as the nominal rate.
**(c)** takes compounding into account.

**4.** (Savings Formula) If you put \$100 at the end of each month for two years in an account that pays 6% annual interest compounded monthly, at the end of the two years you have ______________ .

**5.** Credit-card interest
**(a)** is computed using compound interest.
**(b)** is computed using simple interest.
**(c)** is included in the late fees.

**6.** APR stands for ______________ .

**7.** The nominal rate of interest for a loan is
**(a)** the same as the effective rate.
**(b)** less than the effective rate.
**(c)** never greater than the effective rate.

**8.** If a store credit account charges 1.5% interest each month, the effective annual rate is ______________ .

**9.** In a 30-year mortgage, most of the initial payments
**(a)** go toward reducing the balance.
**(b)** go toward paying the interest.
**(c)** pay insurance costs.

**10.** If a store credit account charges 1.5% interest each month, the APR is ______________ .

**11.** After 15 years of payments on a 30-year mortgage, the balance remaining is
**(a)** about one-third of the original balance.
**(b)** about one-half of the original balance.
**(c)** about two-thirds of the original balance.

**12.** Your credit union offers to finance a \$6000 conventional loan at 4% to be repaid in four years of monthly payments. Your monthly payment is ________ .

**13.** An adjustable-rate mortgage
**(a)** has variable interest rates but maintains a fixed payment amount.
**(b)** has variable payment amounts.
**(c)** is always a better alternative to fixed-rate mortgages.

**14.** If you finance \$15,000 for 3 years at 6% compounded monthly, the monthly payments will be ______________ .

**15.** Equity in a 30-year conventional mortgage grows
**(a)** linearly.
**(b)** logarithmically.
**(c)** exponentially, but slowly.

16. Monthly payments for a 15-year 6% mortgage are about ______ times the payments for a 30-year mortgage of the same amount and the same interest rate.

17. Which of the following arrangements could be an ordinary annuity?

(a) Monthly payments, annual compounding
(b) Annual payments, monthly compounding
(c) Annual payments, annual compounding

18. A convenient rule of thumb is that for a 30-year mortgage at 6%, the monthly payment is about 0.6% of the loan. So, on a $100,000 mortgage, the monthly payment is about $600. About _______ of the first payment goes toward interest.

19. A life income annuity is designed to pay a fixed amount each period until

(a) the annuity runs out of money.
(b) you die.
(c) you reach your life expectancy.

20. If you just won a lottery jackpot paid in 25 equal annual installments of $1 million each at 6% annual effective interest, the present value of the jackpot is __________.

# CHAPTER 22 EXERCISES

■ Challenge ◆ Discussion

## 22.1 Simple Interest

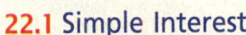

1. Suppose that you take out an unsubsidized Stafford loan on September 1 before your senior year for $5500 and plan to begin paying it back on December 1 after graduation (so you will have had the loan for 15 months, including the six months grace period after leaving school). The interest rate is 6.8%. How much will you owe then, and how much of that will be interest?

2. Assume the same situation as in Exercise 1, but you borrow $5500 on September 1 before your junior year and plan to begin paying it back on December 1 after graduation and grace period 27 months later. How much will you owe then, and how much of that will be interest?

3. Suppose that you borrow $3500 for your first year and $4500 for your second year, as unsubsidized Stafford loans. Suppose that each loan begins on September 1 of its year, that you finish college in four years, and that you begin repayment on December 1 after graduation. What is your total debt then, and how much of that is interest?

4. As in Exercise 3, but you also borrow $5500 for each of your third and fourth years, again on September 1. You finish college in four years, and you begin repayment on December 1 after graduation. What is your total debt then, and how much of that is interest?

## 22.2 Compound Interest

5. If you borrowed $20,000 to buy a new car at 6% interest per year, compounded annually, and paid back the principal and interest at the end of 5 years, how much would you pay back?

6. Assume the same situation as in Exercise 5, but the interest is compounded monthly.

7. If you borrowed $200,000 to buy a house at 6% interest per year, compounded annually, and paid back the principal and interest at the end of 30 years, how much would you pay back?

8. Assume the same situation as in Exercise 7, but the interest is compounded monthly (this is usually the case).

9. A recent credit-card bill of mine showed a daily interest rate of $i = 0.05819\%$.

(a) What is the APR for this rate?
(b) What is the effective annual rate?

10. I received an offer for a credit card with 0% fixed APR for the first 12 months, followed by one of several rates depending on credit history. The highest was a 22.74% APR (and the company reserves the right to change the APR "at any time for any reason").

(a) What is the corresponding daily rate?
(b) What is the effective annual rate?

## 22.3 Conventional Loans

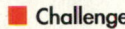

11. A credit-card bill of mine showed $500 due, with a minimum payment of $10 and daily interest rate of $r = 0.04932\%$. If I make no more charges on the card and pay $10 a month as soon as I get each bill, how long would it take to pay off the total? (*Hint:* The amortization payment formula can be changed algebraically into the form

$$(1 + i)^n = \frac{d}{d - iP}$$

Note that making the first payment immediately will reduce the principal $P$ to be amortized to \$500 − \$10 = \$490. If I delay payment, I incur additional daily interest on the amount due. Evaluate the right-hand side. Then, using either a spreadsheet or the power key $\boxed{y^x}$ on your calculator, raise the value of $(1 + i)$ to higher and higher powers $n$ until you find the smallest value for $n$ that makes the left-hand side larger than the right-hand side.)

**12.** (Spreadsheet helpful) Regarding the credit-card bill in Exercise 11: By paying \$10 each month, approximately how much interest would I have paid by the time I pay off the original \$500? (*Hint:* You won't be off by much if you estimate the last payment to be \$10.)

**13.** (Spreadsheet helpful) According to the regulations for the credit card discussed in Exercises 11 and 12, the minimum payment is supposed to be the greater of \$10 or 2% of the balance (rounded to the nearest dollar amount). Suppose that you have such a card and the balance is \$1500. Neglect the rounding down and assume that you pay exactly 2% of the current balance each month until the balance reaches \$500. Notice that if you make a payment of 2% and the bank charges daily interest of 0.04932% in each 30-day month, you in effect reduce the balance to $(1 - 0.02)(1.0004932)^{30} =$ 99.46043% of what it was the previous month. How many 30-day months will it take to reduce the balance of \$1500 to \$500? (Again, assume that you make payments right away.)

**14.** Exercises 9–13 explore how long it might take to pay off a credit-card balance. However, for a high enough interest rate, paying 2% of the balance due will not cover the interest, so the balance actually would increase (this is called *negative amortization*). How high would the APR have to be to make this happen? (Careful: It's not just 12 × 2%.)

**15.** (Spreadsheet helpful.) Because credit-card interest rates are above the rate of Exercise 14, the U.S. Treasury Dept. urged credit-card companies to raise minimum payments for all customers. Many banks now require payment of the "finance charge" (the amount of interest billed) plus 1% of the New Balance (which includes the interest billed). (Although this new policy sounds reasonable, the effect on some consumers was to double their monthly payments.) For a credit card of mine, the minimum payment is the New Balance if less than \$10; otherwise, it is the largest of the three amounts, rounded to the nearest dollar: \$10, 2% of the New Balance, or interest billed plus 1% of New Balance (plus late and over limit fees, which we disregard here). Under this policy, how long would it take to pay off a \$1500 balance with a daily interest rate of 0.04932%, as in Exercises 11 and 13? Include the rounding in your calculations.

**16.** Assume the same situation as in Exercise 15, but for an APR of 22.74%.

**17.** In January 2008 a dealership in Northern Illinois was offering a new 2008 Corolla CE car for \$14,462, about 10.5% lower than the list price. One option for financing was 2.9% over 36 months. What was the monthly payment?

(*Ron Kimball/Kimball Stock.*)

**18.** Assume the same situation as in Exercise 17, but for the second financing option of 3.9% over 48 months.

**19.** Assume the same situation as in Exercise 17, but for the third financing option of 4.9% over 60 months.

**20.** (Spreadsheet helpful.) Put off by even the lowest monthly payment in Exercises 17–19, you might have been attracted to the dealer's roster of "pre-owned" vehicle, including an older but fancier 2007 Toyota Corolla S ("loaded w/ extras, low miles, auto sunroof"). It listed for \$13,995 (after a down payment of \$1000) and a monthly payment of \$245 over 72 months. What was the APR? (Use a spreadsheet and try successive approximations in the amortization payment formula, or a calculator with a financial mode, or a spreadsheet's RATE function.)

**21.** A TV ad in January 2008 offered a 2006 Dodge Stratus car for \$170.75 per month for 84 months and cited a 7.3% APR.

**(a)** What was the price of the car?
**(b)** How much would you pay in interest over the course of the 84 months (7 years, an unusually long period for a car loan)?

**22.** Check the newspapers and pick a car that you would be interested in buying. Give the price, the interest rate, the term, and how much interest you would pay over the course of the loan.

**23.** Suppose that your Stafford loans plus accumulated interest total \$20,000 at the time that you start repayment, and that you elect the standard repayment plan of a fixed amount each month for 10 years, at 6.8% APR.

**(a)** What is your monthly payment?
**(b)** How much will you pay in interest?

**24.** Suppose that your Stafford loans plus accumulated interest total \$40,000 at the time that you start repayment. You could elect the standard repayment plan, as in Exercise 23, and your payments over the 10 years would be double the amount calculated there. Instead, since your accumulated outstanding federal loans total more than $30,000, you can instead elect to repay over 25 years. If you do that:

(a) What is your monthly payment?

(b) How much will you pay in interest?

Refer to the following for Exercises 25 and 26:

Your parents (if their credit rating qualifies) can take out a federal PLUS loan to pay for the total cost of your undergraduate education, less any other financial aid (such as a Stafford loan). The interest rate on a PLUS loan is fixed at 8.5% after July 1, 2006. (There are also fees, which we neglect here.) Unlike the Stafford loan, repayment starts right away (strictly speaking, within 60 days). The standard repayment plan is fixed monthly payments over 10 years.

**25.** Suppose that your parents take out a PLUS loan on your behalf on September 1 before your senior year for $10,000 and begin paying it back a month later. How much is their monthly payment?

**26.** If your parents instead take out a PLUS loan for $10,000 for each of your four years of college, how much is their monthly payment when you graduate?

Refer to the following for Exercises 27 and 28:

An alternative to saddling your parents with a PLUS loan is to augment your Stafford loan with a private federally-guaranteed direct student loan (to you, so you will be repaying). The terms and repayment options of such loans are: The interest is capitalized quarterly and when repayment begins, interest continues during the grace period. (The interest rate is not fixed but can change annually based on interest rate fluctuations.) A bank may charge an origination fee for such a loan, as a percentage of the loan amount; the fee depends on creditworthiness of the borrower and cosigner.

**27.** In December 2007 National City Bank advertised a loan for $10,000 with origination fee of $471.20 (so that the initial principal is $10,471.20), an interest rate of 8.21%, and repayment over 20 years beginning after 45 months in school and 6 months grace. Assuming that the interest rate remains 8.21% throughout the term of the loan:

(a) What is the principal of the loan when repayment begins?

(b) What is the monthly payment?

(c) How much interest would be paid over the course of the loan?

**28.** The bank of Exercise 27 offers a 0.25% reduction in the interest rate at the time repayment begins if the monthly payments are transferred electronically from a bank account, and an additional 0.25% reduction if the first 36 payments are made on time. If the borrower fulfills these conditions:

(a) What is the monthly payment for the first 36 months?

(b) What is the monthly payment after that?

(c) How much interest would be paid over the course of the loan?

**29.** Suppose that you have good credit and can get a 30-year mortgage for $100,000 at 6.5%. What is your monthly payment?

**30.** Assume the same situation as in Exercise 29, except that your credit is not as good and the rate that you are offered is 7.125%.

**31.** Assume the same situation as in Exercise 29, but you inquire about a 15-year loan instead. You are offered 6.125%. What is your monthly payment?

**32.** Assume the same situation as in Exercise 31, but your credit is not as good, and you are offered 6.75%. What is your monthly payment?

**33.** For the mortgage in Exercise 29, how much equity would you have after five years?

**34.** For the mortgage in Exercise 30, how much equity would you have after five years?

**35.** For the mortgage in Exercise 31, how much equity would you have after five years?

**36.** For the mortgage in Exercise 32, how much equity would you have after five years?

**37.** Despite a filter, lots of spam gets into my email. For a while I was getting mortgage offers, such as "$160,000 for less than $735 per month" (for a 30-year mortgage). What would be the corresponding interest rate?

**38.** Suppose that you and two friends decide to live off-campus in your senior year. One of them (who has wealthy parents) suggests that instead of renting an apartment, you could buy a house together, live in it for your senior year, then rent it out or else sell it. Assuming that (with the help of her parents and their good credit rating) you could get a mortgage for $180,000 to buy a house near the campus, what would be the monthly mortgage payment on a 30-year mortgage at 6.75%?

*(amana images/Getty Images.)*

For Exercises 39 and 40, refer to the following.

Payday lenders provide small loans until the borrower's next payday. The borrower receives the desired cash in exchange for a postdated check in the amount of the loan plus a fee, which is usually a percentage of the loan amount. In many states, there are now more payday loan offices than McDonald's fast food outlets. The average loan amount is $300.

**39.** For one payday lender, the fee for a $100 loan for up to two weeks is $15. What is the APR if the loan is for the full two weeks?

**40.** Another payday lender charges $18.62 for a $100 loan for 7 to 14 days. What is the APR if the loan is for 7 days?

For Exercises 41 and 42, refer to the following.

Many income tax preparation services, including the large national chains, offer refund anticipation loans (RALs), or "rapid refunds." These are similar to payday loans in providing an advance on anticipated income–in this case, a tax refund. The loan is repaid when the IRS pays the refund, usually about 7 to 17 days after the loan is made. The cost of the loan is deducted from the proceeds to the client, so this is a discounted loan. The RAL business takes in about $2 billion each year (including tax preparation and check-cashing fees) to arrange payment of the earned-income tax credit to working parents, about 7% of the total of this aid to poor families. A RAL for an anticipated refund usually is issued for a flat fee, often $88, and the average loan is $1500.

**41.** Suppose that the RAL speeds the refund by 7 days. What is the APR for the average RAL?

**42.** Suppose that the RAL speeds the refund by 17 days. What is the APR for the average RAL?

**43.** When interest rates drop, it may become attractive to refinance your home. Refinancing means that you acquire a new mortgage to borrow the current principal due on your home and use the proceeds to pay off your old mortgage. You then begin a new 15- or 30-year mortgage at the new, lower interest rate. A second factor that reduces your monthly payment is that the equity you accumulated under the old mortgage reduces the amount that you have to borrow under the new mortgage. Suppose that you have an existing 30-year $100,000 mortgage at 8.375%, on which you have been paying for five years, and you are considering refinancing at 7.0%.

**(a)** What is your payment under the old mortgage?
**(b)** How much equity do you have in the home?
**(c)** If you use all your equity to reduce the amount of the new mortgage, how much will your monthly payment be?
**(d)** How long is the payback period for the $2000 loan charge–that is, how many months will it take before you have saved $2000 in monthly payments?

**44.** One of the advantages of buying a home with a fixed-rate mortgage is that your payment stays the same but your earnings and the value of your home are likely to go up with inflation. You are paying back the loan with dollars of lesser value.

Consider the following scenario. Suppose that you buy a "starter" two-bedroom home for $105,000 under a special program for first-time home buyers that requires a down payment of only $5000. You have a 30-year fixed-rate mortgage for $100,000 at 7%, on which the monthly payment is $665.30. You also have a $2000 one-time expense in closing costs and annual costs of $200 for insurance and $2000 for property taxes.

You live in the home for five years and spend $10,000 on maintenance, upkeep, and improvements. You then sell the home for $125,000, pay a realtor $9000 to sell it, and pay closing costs of $500 (for title insurance and other costs). Finally, it costs $3000 to move.

**(a)** Make out a balance sheet of revenue and expenses. How did you make out on owning the home?
**(b)** Remember, you also got to live in the home without paying rent! Translate the cost of owning the home into an equivalent monthly rent.

## 22.4 Annuities

**45.** The largest amount won by an individual in a U.S. lottery was $314.9 million, by Jack Whittaker of West Virginia on Christmas Day 2002. (His subsequent life has been far from a fairy tale, as a Google search will reveal.) Instead of receiving $314.9 million in 30 equal annual payments, including one immediately, he chose a lump sum, which came to $170 million. What was the corresponding interest rate of the annuity?

**46.** Today, winners of the Powerball lottery can elect either an immediate lump sum (almost all do) or else an annuity. In the latter case, the advertised jackpot amount is paid in 30 annual payments, including one immediate payment. To keep up with inflation, each payment is 4% more than the previous year's; such an annuity is called a *graduated annuity*. On October 10, 2007, Eugene and Stanislawa Markiewicz took their prize as a $20 million annuity.

**(a)** What was the amount of their first payment, and how much will they receive in their last payment in October 2036?
**(b)** The winners could have chosen instead a lump sum of $9,402,914.90. What was the corresponding interest rate of the annuity?

**47.** Suppose that a man retires at age 65 and in addition to Social Security needs $2000 per month in income. Based on an expected lifetime of 16.6 more years (for men) or 19.6 more years (for women), how much would he have to invest in a life income annuity earning 4% to pay that much per year?

**48.** Assume the same situation as in Exercise 47, but for a woman at age 65, whose expected lifetime is 19.6 more years.

## APPLET EXERCISES

To do these exercises, go to www.whfreeman.com/fapp8e.

There are two ways to buy a car: save up and pay cash or borrow the money. In the *Buying a Car: Cash vs. Loan* applet, you can explore just how much more expensive it is to borrow the money.

## WRITING PROJECTS

**1.** In recent years, incentives from auto manufacturers to potential customers have taken the form of offering either a reduced interest rate on the loan for the car or else a rebate (reduction in price) on the cost itself. In fall 2004 you could buy a 2005 Toyota Camry for $13,570, with one of the following options for payment:

- $750 rebate
- 1.9% APR over 24 months
- 1.9% APR over 36 months
- 2.9% APR over 48 months
- 3.9% APR over 60 months

Suppose that you could afford a $2000 down payment and could get a loan from a credit union at 6.0% over 60 months if you opt for the $750 rebate.

**(a)** What was your monthly payment under each option?
**(b)** Suppose that the rate of inflation over the course of the loan was a steady 3% per year. How do the various options compare in terms of present value of the loan?
**(c)** Locate current advertised incentives for a car that you would like to buy and compare them in an essay of two to three pages.

**2.** A substantial proportion of new cars today are not sold but leased. Contact a local car dealer about a car that you are interested in and find out the details on leasing. Compare the cost of the lease and associated expenses with the cost of purchasing and owning the car. Include estimated maintenance, repair, and insurance costs for each option. Which seems like a better deal, and why? Write two to three pages describing and comparing the two options.

**3.** Banks often offer choices of mortgages with various combinations of interest rates and "points." A point is 1% of the mortgage amount. Points are paid to "buy down" the interest rate for the mortgage; they are paid upfront to the bank at the closing of the house sale. For example, you may have a choice between a mortgage at 6% with 2 points (2%) and a mortgage at 8% and no points. Which would you choose, and why? Does it make a difference how long you are planning to own the home? Or how expensive the home is? Write a page justifying your decision.

**4.** Explore actual costs of homes in your area, mortgages with local banks (including closing costs), and property taxes and insurance. Come up with data on a particular mortgage, and the costs and benefits of refinancing, and make out a corresponding balance sheet for five-year ownership.

## SUGGESTED READINGS

KASTING, MARTHA. *Concepts of Math for Business: The Mathematics of Finance* (UMAP Modules in Undergraduate Mathematics and Its Applications: Module 370–372), COMAP, Inc., Arlington, Mass., 1980.

MILLER, CHARLES D., VERN E. HEEREN, and JOHN HORNSBY. Consumer mathematics. In *Mathematical Ideas,* 11th ed., Pearson Education/Addison Wesley, Reading, Mass. 2008.

YAREMA, CONNIE H., and JOHN H. SAMPSON. Just say "Charge it!" *Mathematics Teacher* 94 (7) (October 2001), 558–564. Shows how to apply the savings formula and the amortization formula and graph the results on the TI-83 calculator. Notes that the 78% of undergraduates in the United States who have credit cards carry an average debt of more than $2700, with 10% owing more than $7000.

## SUGGESTED WEB SITES

**www.lendingtree.com/stmrc/calculators1.asp**
Java applet calculators (for any platform) to calculate payments and amortization schedules for conventional loans, adjustable-rate mortgages, auto loan vs. home-equity loan, and credit-card payoff. (*Note:* Lending Tree, Inc., is a loan broker; mention here of calculators at its Web site does not imply endorsement of its other services by this book's authors, editors, or publisher.)

**www.leaseguide.com/** A guide to how car leasing works, including what "money factor" means, and how leasing cost is determined.

**www.edmunds.com/apps/calc/Calculator Controller?pmtcalAction=apr_cash_calc**
Commercial site offering a calculator to compare rebate vs. interest-rate offers for car purchase. (*Note:* Edmunds is a loan broker; mention here of calculators at its Web site does not imply endorsement of its other services by this book's authors, editors, or publisher.)

# CHAPTER 23



# The Economics of Resources

We use resources all the time: food, money, natural resources, labor, time. Some resources, such as annual flowers, are perishable, while others, such as money and standing timber, can be used now or saved for later use.

Our use of resources involves a complex intermixture of biological, ethical, practical, technical, and economic issues:

- How many people will there be in the world in another 20 or 40 years, and how will those numbers affect resources available to you and to them?
- What should we consume for our own use, give to others more needy, or leave for future generations, in terms of wealth, well-being, and wilderness?
- Will our standard of living keep getting better, or is it not maintainable, even at current levels, in the long run?
- How long will it be, at current patterns of use, until we exhaust a particular resource?
- Can we develop more efficient technology so that we can get more out of the resources that we use?
- How do we balance economics with other important considerations? How much would it cost–how much is it worth–to ensure that we do not let tigers, elephants, or rhinos go extinct?

On a personal level, we face similar but more immediate questions about the balance between consumption and conservation. How much must you save to be able to afford to retire? How do you take into account the fact that \$1000 after you retire won't be worth as much as \$1000 today? Once you retire, how much can you spend without exhausting your nest egg?

In Chapters 21 and 22, we explored mathematical models for saving, accumulating, and borrowing–the building up of resources. From Chapter 21, we use here only two formulas, specialized to an annual interest rate $r$ and number of years $n$.

**Compound interest formula:** If a principal $P$ is deposited into an account that pays interest at rate $r$ per year, then after $n$ years the account contains the amount

$$A = P(1 + r)^n$$

**Savings formula:** For a uniform deposit of $d$ per year (deposited at the end of the year) and an interest rate $r$ per year, the amount $A$ accumulated after $n$ years is

$$A = d\left[\frac{(1 + r)^n - 1}{r}\right]$$

A reader who can use these formulas can proceed in this chapter without first reading Chapters 21 and 22.

In this chapter, we model processes in the other direction–the use, decay, depletion, or spending down of resources, including some resources that replenish themselves regularly. Our models will lead us into the mathematics of dynamical systems and chaos. The models provide important insights into answers to the questions above.

## 23.1 Growth Models for Biological Populations

We encountered geometric growth models for savings accounts in Chapter 21. Growth is proportional to the amount present, and such growth is expressed in terms of compound interest and its formula. We now use a geometric growth model to make rough estimates about sizes of human populations. In addition to the **rate of natural increase**–the annual birth rate minus the annual death rate–we must take into account net migration. The sum of the two, in the terminology of financial models, is the effective rate.

Birth, death, and migration rates rarely remain constant for long, so projections must be made with care. In the short run, however, predictions based on the model may be useful. Let's apply this model to two questions about the population of the United States.

### EXAMPLE 1 ■ Predicting the U.S. Population

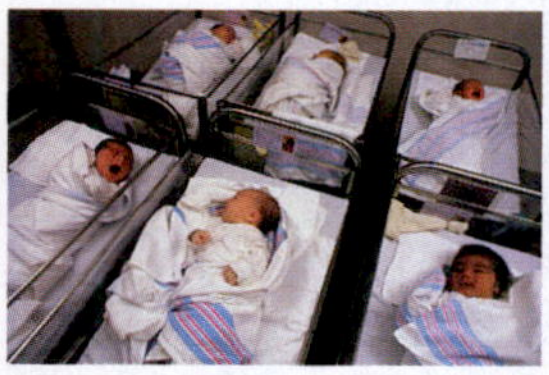

*(Blaine Harrington III/ Corbis.)*

The U.S. population increased at an average effective growth rate of 0.95% per year (including immigration) to 303.1 million at the beginning of 2008. What is the anticipated population at the beginning of 2012? What is it if the effective rate of growth changes to 1.2% per year or to 0.7% per year?

**SOLUTION** We apply the compound interest formula with initial population size ("principal") 303.1 million. Using a year as the compounding period and the formula $A = P(1 + r)^n$, where $n = 4$, the projected population size in 2012 for a rate $r = 0.0095$ is

$$\begin{aligned}\text{population in 2012} &= (\text{population in 2008}) \times (1 + \text{growth rate})^4 \\ &= 303{,}100{,}000\ (1 + 0.0095)^4 \\ &= 303{,}100{,}000\ (1.0385) \\ &\cong 315{,}000{,}000\end{aligned}$$

Because of the limited accuracy of the estimates of population and growth rate, we round off the final answer. The result of a calculation can't be more precise than the ingredients.

In the same way, with a growth rate of 1.2% per year, we predict a population of 318 million, while a growth rate of 0.7% per year yields 312 million.

So an uncertainty of one-fourth of one percentage point in the growth rate has major implications, even over fairly short time horizons. The presence or absence of 6 million people would have a significant impact on our social and economic systems! Indeed, much of the concern over long-range funding of Social Security programs results from uncertainties over birth and immigration rates. Figure 23.1a gives a graph of the U.S. population in 2007, structured by age and sex. Figure 23.1b shows possible futures for India.

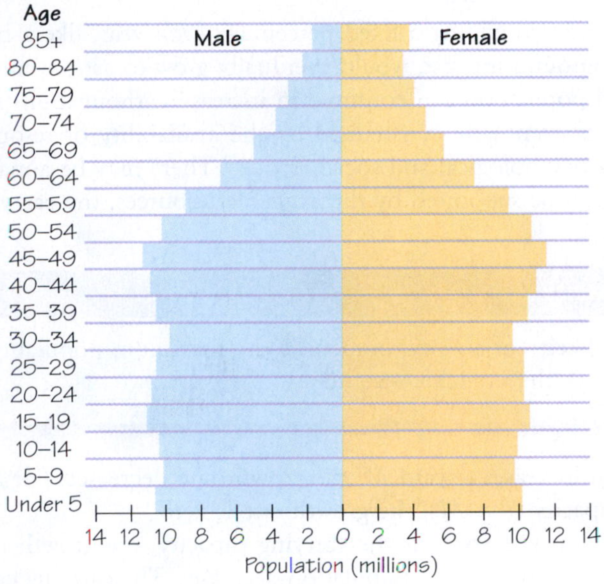

(a)

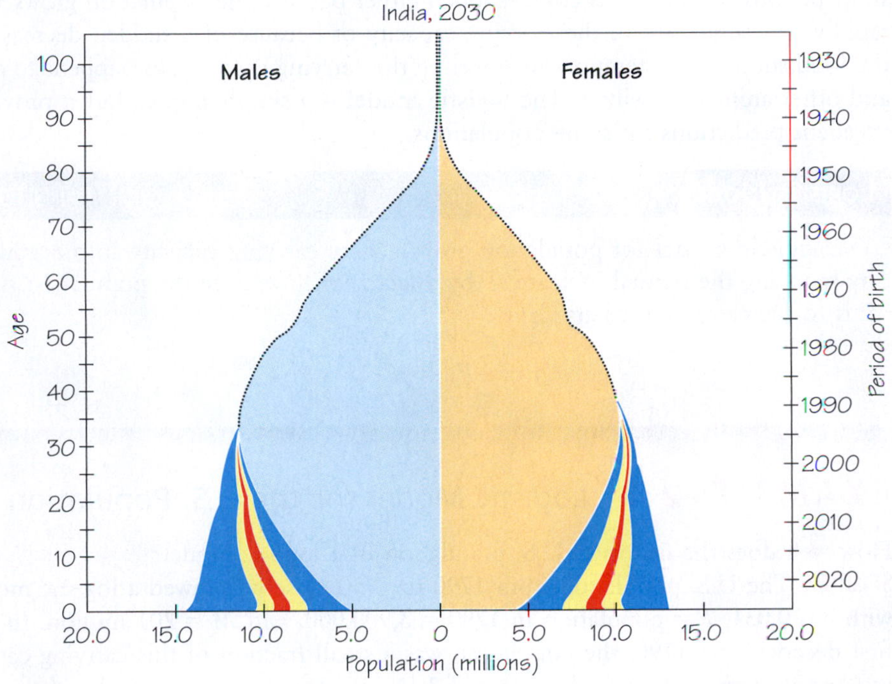

(b)

**FIGURE 23.1** Graphs of the population of the United States in 2007 grouped by age and gender (a), and of scenarios for the population of India in 2030 (b).

Rates of increase in most developing nations are much higher than in industrialized nations. With a growth rate of 2.5%, Nigeria, Africa's most populous country, will grow from 144 million in mid-2007 to 254 million in mid-2030, an increase of three quarters. Such projections raise concern over providing sufficient food and resources for all people.

It is not just the number of people that is crucial, but also the **population structure**. In poorer countries, the proportion of the population over 60 years of age will be 20% by 2050, compared with 8% now; in Japan, it will be 40%.

## Limitations on Growth

A population that keeps adding a fixed percentage each year, like a bank account accumulating compound interest, would eventually grow to astronomical numbers. But no biological population can continue to increase without limit (see Spotlight 23.1). Its growth is eventually constrained by the availability of resources such as food, shelter, and psychological and social "space." There may be a maximum population size that can be supported by the available resources, the **carrying capacity** of the environment.

### Carrying Capacity — DEFINITION

The **carrying capacity** of an environment is the maximum population size that it can support with the available resources.

As the population increases toward $M$, the growth rate decreases. For a population at the carrying capacity ($P = M$), the growth rate is zero.

If the population ever exceeds the carrying capacity, the growth rate becomes negative (because $P > M$) and the population decreases. The carrying capacity is the long-range capacity to support the population, so the population could exceed it for short periods of time. This could happen either because the population grows very rapidly and surges above the carrying capacity or because of a sudden decrease in the food supply, thus temporarily lowering the carrying capacity, as happens to deer and other animals in winter. The **logistic model** is a simple model, but it provides excellent predictions for some populations.

### Logistic Model — DEFINITION

The **logistic model** for population growth takes carrying capacity into account by reducing the natural increase $rP$ by a factor of how close the population size $P$ is to the carrying capacity $M$:

$$\text{growth rate } P' = rP\left(1 - \frac{P}{M}\right)$$

## EXAMPLE 2 ■ Logistic Model for the U.S. Population

How well does the historical U.S. population fit a logistic model?

**SOLUTION** The U.S. population from 1790 to 1950 closely followed a logistic model with $r = 0.031$, $P =$ population in 1790 = 3,900,000, and $M = 201$ million. In the first decades after 1790, the population was a small fraction of this carrying capacity, and it grew at close to the rate $r$ of 3.1% per year (a rate higher than in many

## SPOTLIGHT 23.1 12 Billion by 2050–or Only 9 Billion?

How many people can the world hold? Are developing countries heading for a population disaster? Will falling fertility play havoc with Social Security in the United States? Will aging result in 50% of Japanese being over 60 in 2100?

Answers come from mathematical modeling of the future from predicted trends. The best analyses suggest a probability distribution over a range of estimates. They project separately by age, gender, education, and other characteristics. They try to factor in improvements in agriculture, spread of diseases (especially HIV), changes in urbanization, increases in economic aspirations, and the potential for climate change (for example, from global warming).

A basic concept is *total fertility rate* (TFR), the average number of births per woman. Absent catastrophes (such as war or disease), a rate of 2.1 continues a population at the same size. A model that assumes a value above 2.1 will predict an ever-increasing population; one that assumes a value below 2.1 will predict an ever-dwindling one. Most of the world's population growth will occur in the lesser-developed countries, whose TFR values are well above 2.1 (for example, it is 4.6 for Africa). Many countries in Europe have TFR values near 1; without immigration, they will lose population and struggle with fewer workers to provide social benefits to the elderly.

China's situation illustrates *demographic momentum*. Even though its fertility rate has been below replacement level for 20 years, the number of women in the childbearing years was (and still is) so large that China's population will continue to grow until 2040.

The most sophisticated models try to assess how the TFR will vary with changing social circumstances. The most important single factor impacting fertility is education of women. There is a strong negative association between level of education and TFR, and the effect can be very large: In India and China, women with some college education have on average only half as many children as women with no education. Hence, a country's policies about education, and their success, may directly impact future population levels.

(Will & Deni McIntyre/Corbis.)

As we have revised this book for successive editions, we have seen population projections change. The estimates have decreased, because fertilities have declined. The key questions are how to model such declines, whether they will continue, and how they will adapt to other world changes. In Spotlight 21.1 on p. 688, we saw Malthus's over-simplified prediction that mere arithmetic growth in food supplies would limit the geometric increase of human populations. Some demographers now think that population growth will remain a serious problem in some parts of the world but that global population may stabilize or even decline after 2050.

How many people the world will have depends on how well we as a world conserve the environment, distribute food, provide jobs, produce and consume energy, and make other critical decisions about our money and resources. The key concern is the quality of life of *all* people. Political and economic events in far corners of the world, and even natural disasters such as the tsunami at the end of 2004, impact us all. Neglecting problems faced by increasing numbers of poor people provides no security, peace, or moral refuge for anyone.

developing nations today). By 1920, the U.S. population had reached 106 million, and the growth rate had slowed by about one-half, to 1.5% per year (see Figure 23.2).

The 2008 U.S. population of 303.1 million far exceeds the hypothesized carrying capacity of 201 million. The structure of the U.S. population changed, from a large proportion of people making their livings on family farms to a highly urbanized society. The average number of children per family shrank. As the structure changed, the model based on the prior structure gradually became invalid.

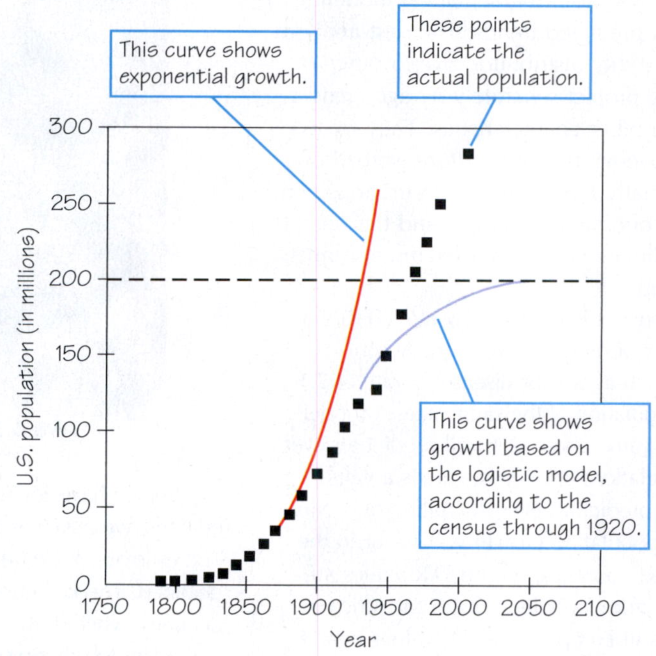


**FIGURE 23.2** U.S. population by year, showing actual growth, exponential (geometric) growth, and logistic growth.

The logistic model applies not only to population growth limited by carrying capacity but also to modeling the spread of a technology or product, such as DVD players or flat-screen plasma TVs. Initially, sales are slow. Then they begin to climb rapidly. Finally, as the market gets saturated, new sales slow. We will see in the next section that the logistic model also can apply to exhaustion of nonrenewable resources.

## 23.2 How Long Can a Nonrenewable Resource Last?

People use resources, some of which are renewable, but others are not. In this section, we model depletion of **nonrenewable resources**. In the next, we treat renewable resources.

**Nonrenewable Resource** DEFINITION

A **nonrenewable resource** is one that does not tend to replenish itself.

Gasoline, coal, and natural gas are important examples, while lottery winnings and inheritances could be examples from personal affairs. There is no practical way to recover or reconstitute these resources after use. Some substances, such as aluminum

(*Joseph Baylor Roberts/ Getty Images.*)

or the sand used to make glass, are potentially recyclable, but to the extent that we do not recycle them, they, too, are nonrenewable.

For a nonrenewable resource, only a fixed supply $S$ is available. Even without human population increases, we face dwindling nonrenewable resources. We are interested in the question: How long will the supply of a resource last?

As long as the rate of use of the resource remains constant, the answer is easy. If we are using $U$ units per year and continue using $U$ units per year, then the supply will last $S/U$ years. This kind of calculation is the basis for statements such as those claiming that at the current rate of consumption, U.S. recoverable coal reserves will last 250 years or that the U.S. strategic reserve of gasoline (stored in underground salt domes in the South) will last 60 days.

However, the rate of use of resources tends to increase with population and with a higher standard of living. For example, projections for use of electric power often assume that use will increase by a fixed percentage each year. This is the simplest situation (apart from constant usage) and one that we can easily model.

Suppose that $U_1 = U$ is the rate of use of the resource in the first year (this year), and that usage increases $r = 0.05 = 5\%$ each year. Then the usage in the second year is

$$U_2 = U_1 + 0.05U_1 = 1.05U$$

and usage in the third year is

$$U_3 = U_2 + 0.05U_2 = 1.05U_2 = 1.05(1.05U) = (1.05)^2U$$

Generalizing, we see that usage in year $i$ will be $(1.05)^{i-1}U$. Total usage over the next five years, for example, will be

$$U + (1.05)^1U + (1.05)^2U + (1.05)^3U + (1.05)^4U$$

This situation should remind you of the accumulation of regular deposits plus interest (see Chapter 21). Here the usage $U$ corresponds to a deposit, and the increasing rate of use $r$ corresponds to the annual interest rate. We may think of the situation as making regular withdrawals (with interest) from a fixed supply of the nonrenewable resource. The savings formula gives

$$A = d\left[\frac{(1 + r)^n - 1}{r}\right]$$

In the resource situation, $A$ is the accumulated amount of the resource that has been used up at the end of $n$ years, and $U$ is the initial rate of use. We have

$$A = U\left[\frac{(1+r)^n - 1}{r}\right]$$

To find out how long the supply $S$ will last, we set the supply $S$ equal to the cumulative use $A$ over $n$ years and then determine what $n$ has to be. We have

$$S = U\left[\frac{(1+r)^n - 1}{r}\right]$$

We perform some algebra to isolate the term involving $n$, to get

$$(1+r)^n = 1 + \frac{S}{U}r$$

At this point, to isolate $n$, we have to take the natural logarithm of both sides.

$$\ln[(1+r)^n] = n\ln(1+r) = \ln\left(1 + \frac{S}{U}r\right)$$

thus gives the final expression

$$n = \frac{\ln[1 + (S/U)r]}{\ln(1+r)}$$

which may look complicated but is quite easy to evaluate on a calculator. The expression $S/U$ is called the **static reserve**, and $n$ is called the **exponential reserve**.

### Static Reserve and Exponential Reserve — DEFINITION

The **static reserve** is how long the supply $S$ will last at a particular constant annual rate of use $U$, namely, $S/U$ years. The **exponential reserve** is how long the supply $S$ will last at an initial rate of use $U$ that is increasing by a proportion $r$ each year, namely

$$\frac{\ln[1 + (S/U)r]}{\ln(1+r)} \text{ years.}$$

## EXAMPLE 3 ▪ U.S. Coal Reserves

Coal accounts for 30% of U.S. energy use, including 50% of U.S. electricity. Recoverable reserves of U.S. coal would last about 250 years at the current rate of use, so the static reserve is 250 years. How long would the supply last if the rate of use increases 1% per year, as it did from 2001 to 2007 (that is, at about the same rate that the U.S. population grew)?

**SOLUTION** The corresponding exponential reserve is

$$n = \frac{\ln[1 + (250)(0.01)]}{\ln 1.01} = \frac{\ln 3.5}{\ln 1.01} \approx 126 \text{ years}$$

That's quite a difference!

We must not take such projections as exact predictions. Estimates of supplies of a resource may underestimate how much is available, and previously unknown sources may be discovered or the technology improved to extract previously unavailable supplies. In addition, as supplies dwindle, the economic considerations of supply, demand, and price come into play. We will never completely run out of oil. It will always be available "at a price."

We must not take such projections lightly, either, because we are discussing resources that, once used, are gone forever. In any projection, it is very important to examine the assumptions, because small differences in the rate of increase of use can make big differences in the exponential reserve.

## EXAMPLE 4 ■ Using Up Retirement Savings

Suppose that you begin retirement with \$1 million in savings, and you don't trust banks or the stock market, so you keep it all under your mattress. Suppose that it costs you \$50,000 per year to live at your accustomed standard of living and there is no inflation. How long will your retirement nest egg last? How long will it last if inflation is constant at 3%?

**SOLUTION** The static reserve is \$1,000,000/\$50,000 per year = 20 years. If, however, there is constant 3% per year inflation (as during 2001–2007), then it will cost you increasingly more per year to live, so you should realize that your savings will last only for the length of the exponential reserve, which is

$$n = \frac{\ln(1 + 20(0.03))}{\ln 1.03} = 15.9 \text{ years}$$

You have a fine strategy if you expect to live just 16 more years and want to die broke!

In our examples so far, we have assumed that the resource is just sitting there, waiting to be used up. For many natural resources, however, we have to find and develop new sources. As doing that becomes more difficult and more costly, at some point the exponentially increasing demand outstrips the ability to meet that demand.

Such a situation is modeled well by the logistic model famously applied to oil by M. King Hubbert, director of Shell Oil Company's research laboratory. Figure 23.3 shows data for cumulative U.S. oil production through 2001 compared with the logistic curve for the ultimate production of $M = 240$ gigabarrels (240 billion barrels). In 1956, Hubbert predicted that U.S. production would peak in the early 1970s (it did) and decline steadily thereafter (it did that, too, except for a blip from Alaska in the 1980s) (see Figure 23.4, whose curve is similar to but "heavier in the tails" than the normal distribution curve of Chapter 5).

If we rearrange the logistic equation on p. 734 by dividing both sides by $P$ and doing a little algebra, we get

$$\frac{P'}{P} = r - \frac{r}{M}P$$

You can recognize this as the equation of a straight line $y = a - bx$, where $P'/P$ takes the role of $y$ and $P$ takes the role of $x$. In other words, for a logistic model, if we graph $P'/P$ against $P$, we get a straight line. Figure 23.5 shows that the data fit the Hubbert model well.

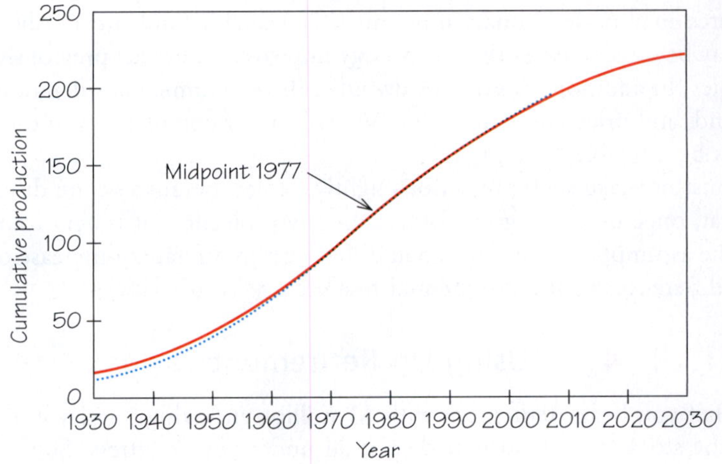


**FIGURE 23.3** Logistic model (solid curve), assuming ultimate production of 240 billion barrels, and actual data (points) for cumulative U.S. crude oil production, in billions of barrels, through 2007.
*Source*: Adapted from Seppo A. Korpela, Oil depletion in the United States and the world, www.mecheng.osu.edu/files/u57/opmatalk.pdf, with revised and further data from the Energy Information Administration through 2007.

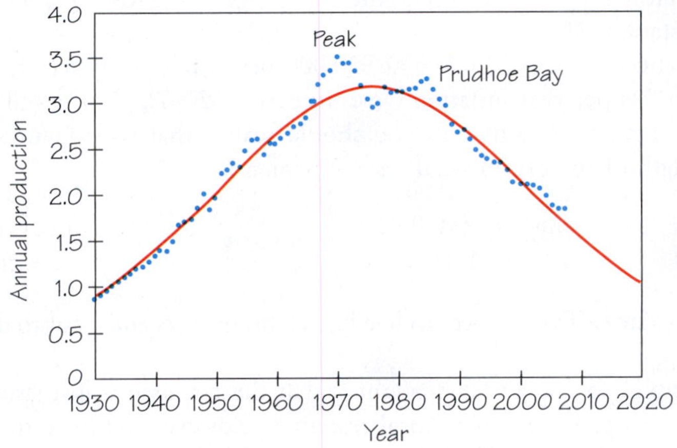


**FIGURE 23.4** U.S. crude oil production, in billions of barrels versus year through 2007, and production as predicted by the logistic model, assuming ultimate production of 240 billion barrels.
*Source*: Adapted from Seppo A. Korpela, Oil depletion in the United States and the world, www.mecheng.osu.edu/files/u57/opmatalk.pdf, with revised and further data from the Energy Information Administration through 2007.

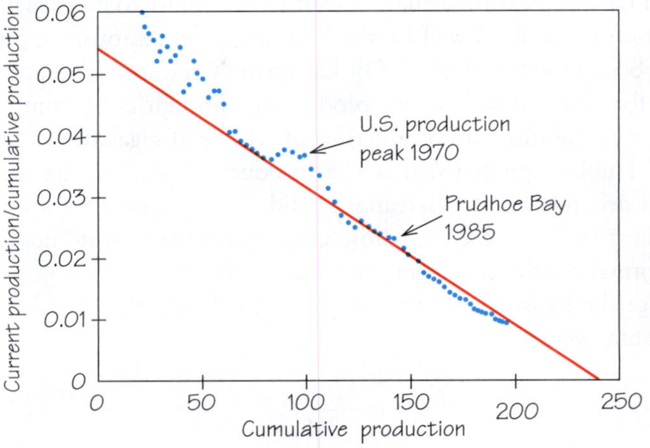


**FIGURE 23.5** Display of fit of data to the logistic model (solid line) for U.S. oil production. The line follows the equation $y = 0.054 - 0.000225x$.
*Source*: Adapted from Seppo A. Korpela, Oil depletion in the United States and the world, www.mecheng.osu.edu/files/u57/opmatalk.pdf, with revised and further data from the Energy Information Administration through 2007.

The world's oil and gas are running out far faster than most people realize or than their governments are willing to acknowledge (see Exercises 11 and 12). The need for "affordable" fuels will likely soon dominate the political agendas of the entire world, particularly since experts have long predicted that world oil output would peak around 2006. In fact, production was level for 2005, 2006, and 2007.

## 23.3 Sustaining Renewable Resources

A **renewable natural resource** is a resource that tends to replenish itself, such as fish, wildlife, and forests. How much can we harvest and still allow for the resource to replenish itself?

**Renewable Resource** DEFINITION

A **renewable resource** is one that tends to replenish itself.

Other renewable resources are biological populations. We concentrate on the subpopulation with commercial value. For a forest, this might be trees of a commercially useful species and appropriate size. We measure the population size as its **biomass**, the physical mass of the population. For example, we measure a fish population in pounds rather than in number of fish, and a forest not by counting the trees but by estimating the number of board feet of usable timber.

### Reproduction Curves

Our models for growth include many simplifications. Real populations may behave according to one of our models or another known model. Complicated factors that can affect populations, such as climatic or economic change, may mean that the only way to understand a population is to plot a graph of its size over time. Either from such a graph or from a model, we can construct a **reproduction curve**, which predicts next year's population size (biomass) based on this year's size. Although the precise shape of the curve varies from one population to another, the shape in Figure 23.6 is typical. For all possible sizes, it shows for the size this year (on the horizontal axis), the size next year (on the vertical axis), taking into account growth in size of continuing members and addition of new members, minus losses due to death and other factors.

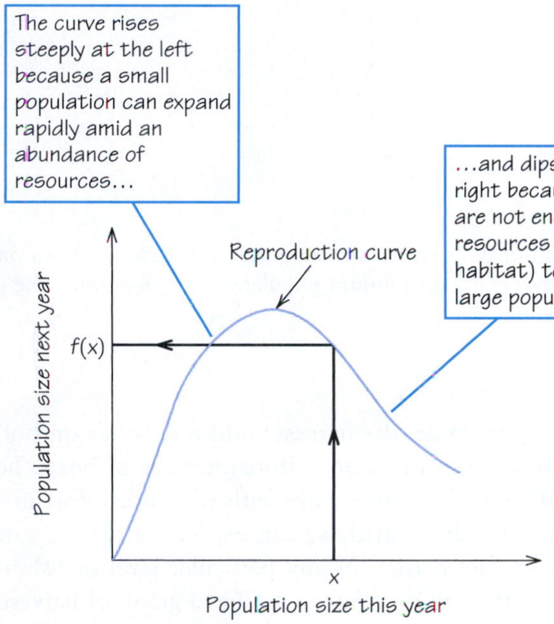
The curve rises steeply at the left because a small population can expand rapidly amid an abundance of resources...
...and dips on the right because there are not enough resources (food, good habitat) to support a large population.
Reproduction curve
Population size next year
f(x)
x
Population size this year

**FIGURE 23.6** A typical reproduction curve.

Let $x$ on the horizontal axis be a typical size of the population in the current year. The size *next* year is given by the height of the curve above the point marked $x$. This value is denoted by $f(x)$. (You can think of $f$ as standing for "function of," or even as "forthcoming.")

Figure 23.7a shows the same reproduction curve, plus the broken line $y = x$ (which makes a 45° angle with the horizontal axis). You can trace what happens for various choices for $x$. For an $x$ for which the curve is above the broken line, next year's size, $f(x)$, is larger than this year's, $x$. In Figure 23.7b, the **natural increase**, or gain in population size, is shown as the length of the green vertical line from the broken line to the curve, which in algebraic terms is $f(x) - x$. For an $x$ for which the curve is below the broken line, next year's size is smaller than this year's and $f(x) - x$ is negative. For the size labeled $x_e$, for which the curve crosses the broken line, the size is the same next year as it was this year. This is the **equilibrium population size**.

**Equilibrium Population Size** DEFINITION

An **equilibrium population size** does not change from year to year.

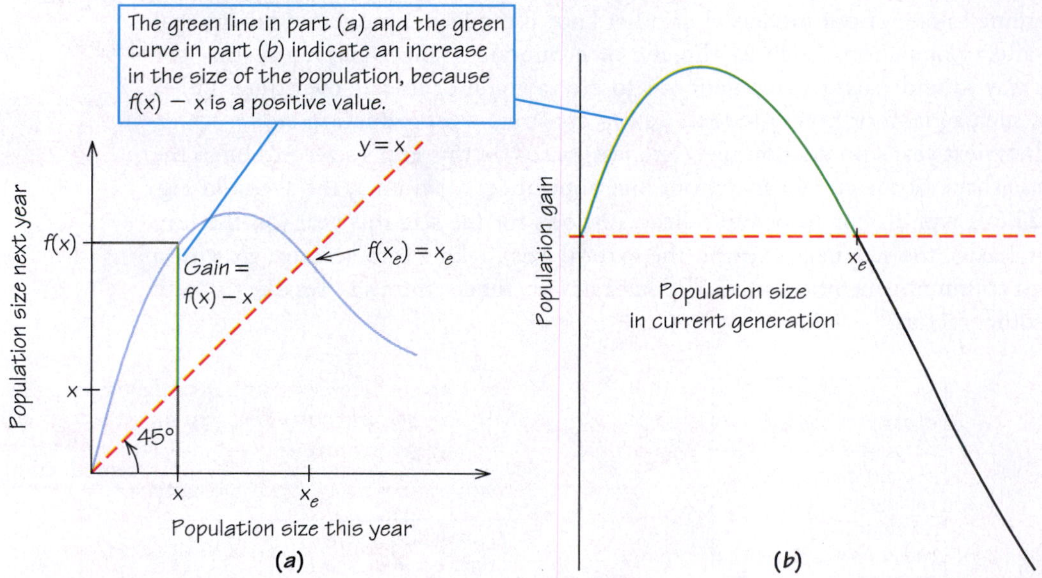


**FIGURE 23.7** Depiction of the natural increase (gain) in population from one year to the next. The population size $x_e$ is the equilibrium population size, for which the population one year later is the same, or $f(x_e) = x_e$.

## Sustained-Yield Harvesting

In the case of fishing, for example, the harvest **yield** $h$ depends on both the population $x$ of fish and the amount of fishing effort (number of boats, hours of fishing). If the fishing fleet fishes the same banks with the same effort in a year when there are only half as many fish as usual, we can expect the fleet to catch only half as many fish. A simple model is that for any particular level of fishing effort, the harvest is proportional to the fish population, so that a graph of harvest versus population would be a straight line with a steeper slope corresponding to greater effort.

For each level of fishing effort, there is a level of **sustainable yield**, one that could be sustained year after year because the fish population would recover to the same level after each harvest. Figure 23.8 shows the gain curve and two possible harvest lines, with gain and harvest on the vertical axis and current population along the horizontal axis. We focus first on comparing the gain curve with the solid black harvest line. Where the curve lies above the harvest line, the gain exceeds the number harvested. Where the curve lies below the line, the harvest exceeds the annual growth. Where the curve and the line intersect, harvest equals growth; *next year the population will return to the same initial size and we can harvest the same amount.*

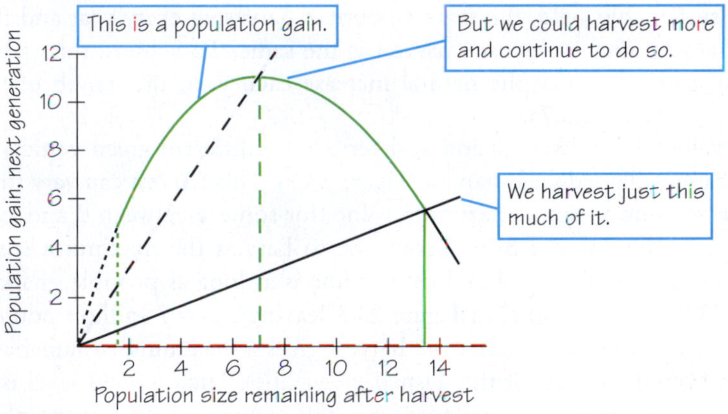


**FIGURE 23.8** This population curve shows the effects of harvesting on a fish population. The scales for population and gain sizes are in millions of pounds.

The reason is that we harvest all of the gain and bring the population back to the same level as before the current season's growth, from which the population should generate the same gain next year. In the figure, the curve and the black harvest line intersect at 5 million pounds, where the harvest equals the population gain (indicated by height of the green line). The post-harvest population is 13.5 million pounds (on horizontal axis at foot of green line); the pre-harvest population was 13.5 + 5 = 18.5 million pounds. The 13.5 million pounds remaining after harvest will again grow by 5 million pounds (height of green line) to 18.5 million pounds by next fishing season, when we can again harvest the gain, 5 million pounds.

However, at a higher level of fishing effort, shown by the steeper dashed black harvest line, the fleet could harvest 11 million pounds, leaving a post-harvest population of 7 million pounds (at foot of dashed green line). The pre-harvest population was 11 + 7 = 18 million pounds, about the same as in the previous scenario. The blue population gain curve indicates that the 7 million pounds left after the harvest can be expected to grow by 11 million pounds (height of dashed green line) by the start of next fishing season–the same total pre-harvest level of 18 million pounds as the previous year. As before, the cycle can repeat, but this time with more than double the harvest each year.

Finally, we consider the dotted black harvest line at far left. It corresponds to harvesting 5 million pounds (height of the dotted green line) and leaving just 1.5 million pounds (at foot of dotted green line). The main difference between this and the first scenario is that here *if the fleet takes 6.5 million pounds instead of 5 million, it wipes out the fish altogether*. Even if it takes only 5.5 million pounds, the fish grow back the next year by only 3 million, to a total of 4 million pre-harvest; and trying to harvest 5 million then extinguishes the fish population.

You can see that how much can be harvested "safely" and on a sustainable basis depends on knowing where we are on the population gain curve. But that curve is difficult to determine, as is estimating how many fish are out there. Moreover, the curve is an ideal that varies with changes in the weather, the fishes' environment, and other fish populations.

**Sustained-Yield Harvesting Policy** DEFINITION

A **sustained-yield harvesting policy** is a policy that if continued indefinitely will maintain the same yield.

For a sustainable yield, the same amount is harvested every year and the population remaining after each year's harvest is the same. To achieve this stability, the harvest must exactly equal the natural increase each year, the length of the green vertical line in Figure 23.7a.

Each value of $x$ between 0 and $x_e$ determines a different green vertical line and corresponding sustained-yield harvest (Figure 23.9). This harvest can vary from 0 (for $x = 0$ or $x = x_e$) up to some maximum value (for some $x$ between 0 and $x_e$). A goal for a timber company or a fishery could be to harvest the **maximum sustainable yield**: to select an $x$ whose colored vertical line is as long as possible, marked as $x_M$ in Figure 23.9b. For example, in Figure 23.8 leaving $x_M = 7$ million pounds (horizontal axis) of fish in the sea after the harvest gives a maximum sustainable yield of 11 million pounds (height of the dashed green line). Such a yield level is difficult to determine, and we will see that economic and environmental factors–plus the inevitable involvement of mathematical chaos–also come into play.

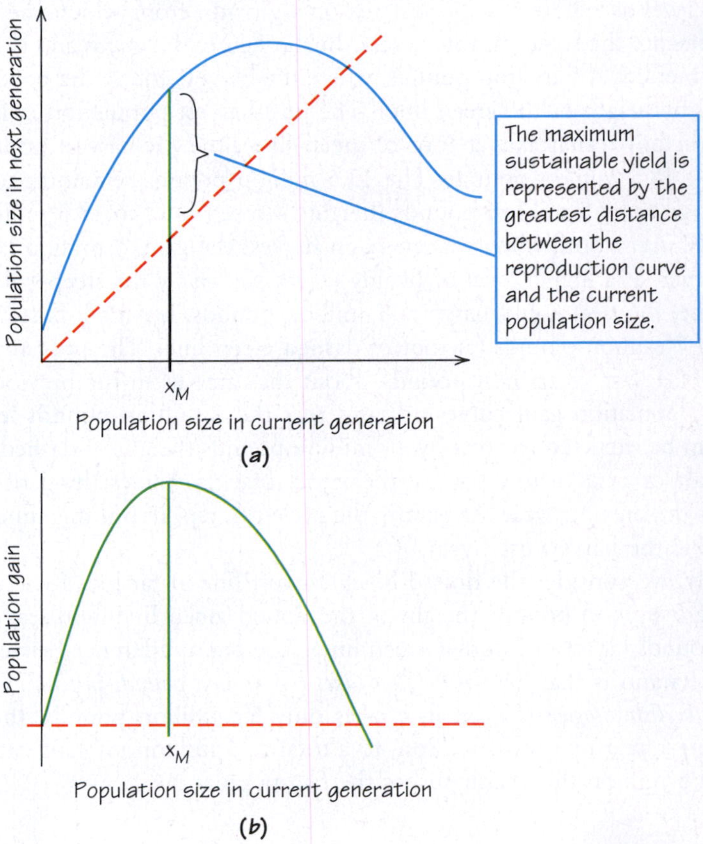


**FIGURE 23.9** The reproduction curve, with the population size $x_M$ corresponding to the maximum sustainable yield.

## EXAMPLE 5 ■ Decline of a Fishery

A successful fishing effort attracts more fishers and more boats. A lower price for fish causes fishers to fish longer hours to maintain their incomes. Whichever is the case, the long-run effect can be catastrophic. How so?

**SOLUTION** Figure 23.10 shows the sudden decline of a fishery with only a 20% increase beyond the original fishing effort. Extinction of the resource could result from additional fishing effort. The yield, shown as a vertical green line, is much greater for fishing effort geared to maximize sustainable yield (Figure 23.10a) than for effort 20% greater (Figure 23.10b). The fishing effort can be measured by the slope of the dashed red harvest line; the slope is steeper in Figure 23.10b. With the greater fishing effort, the yield declines from the maximum sustainable yield of 2.4 million pounds to less than 1 million pounds, and the population of fish pre-harvest shrinks from 9 million pounds to 7.5 million pounds.

(*Stuart Westmorland/Corbis.*)

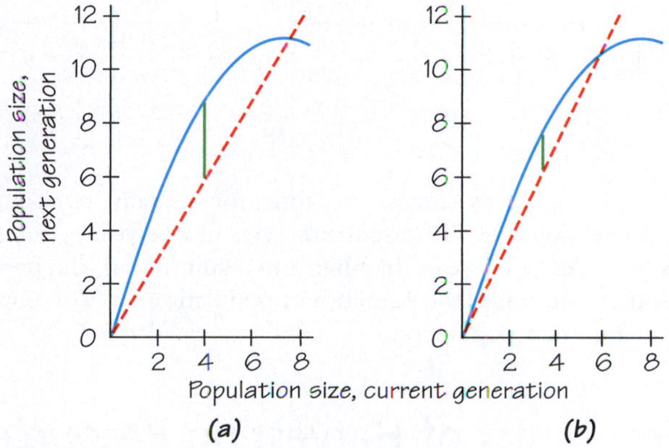


**FIGURE 23.10** Deterioration of a fishery with increasing fishing pressure. The yield, shown as a vertical green line, is much greater in (a), which shows fishing effort geared to maximum sustainable yield, while (b) shows a 20% greater effort (as measured by the slope of the dashed red line).

## Dynamics of a Population over Time

The line $y = x$ provides a convenient way to trace the evolution of the population over several years (see Figure 23.11a), by alternating steps vertically to the curve and horizontally to the line $y = x$. Begin with the first year's population on the horizontal axis and go up vertically to the curve. The height is the population in the second year. Proceed horizontally from the curve over to the line $y = x$. Proceeding vertically from there to the curve yields a height that is the population in the third year. The result is a **cobweb diagram**, so-called because it resembles a spider web. A cobweb diagram gives a convenient visual representation of the evolution of the population over time.

Figure 23.11 shows several traces for the same reproduction curve, each starting from a different initial population on the horizontal axis. The resulting variation is surprising and can be "chaotic" in a very specific mathematical sense, showing how apparently random behavior can result from strict rules.

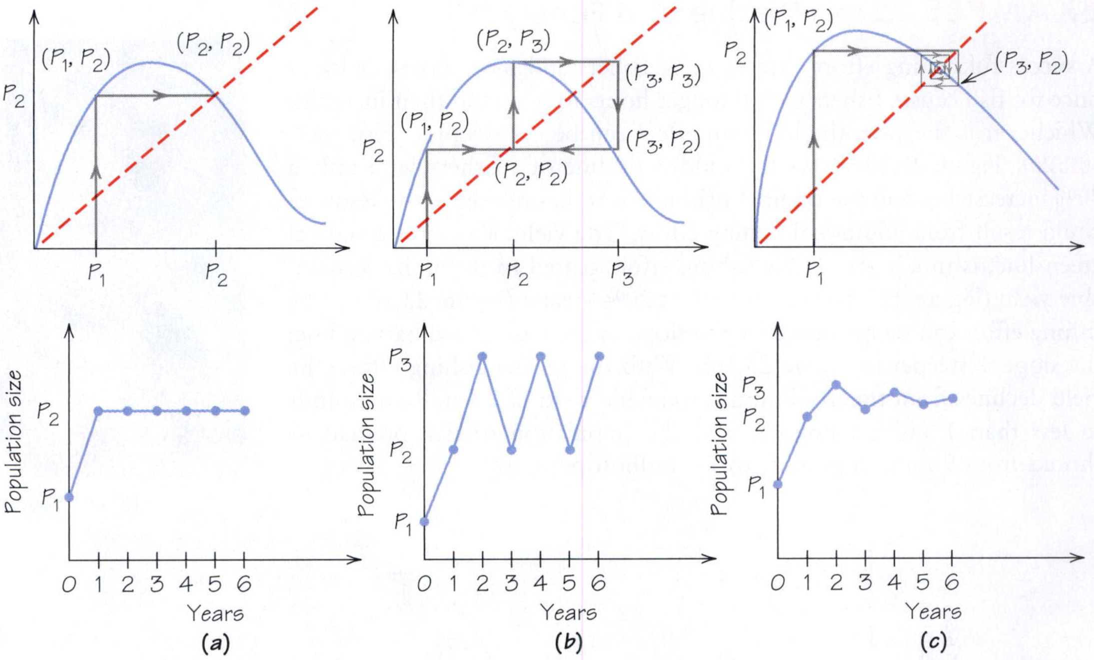


**FIGURE 23.11** Examples of the dynamics, over time, for the same reproduction curve but different starting populations. (a) The population goes in one year to the equilibrium population and stays there year after year. (b) After initial adjustment, the population cycles between values over and under the equilibrium population. (c) The population spirals in toward the equilibrium population.

## 23.4 The Economics of Harvesting Resources

We consider two models: one for a cattle ranch and one that applies to both a fishing boat and a tree farm.

We assume that the price $p$ received is the same for each harvested unit and does not depend on the size of our harvest. In effect, we assume that our operation is a small part of the total market, not substantially affecting overall supply and hence price.

We want to stay in business, so we do not extinguish the resource for quick profits. For any given population size, we harvest just the natural increase.

### EXAMPLE 6 ■ Cattle Ranching

We assume that the cost of raising and bringing a steer to market is the same for every steer and does not depend on how many steers we bring to market. What should our sustainable-yield harvesting policy be?

**SOLUTION** Because the cost does not depend on the population size, the cost curve is a horizontal line (Figure 23.12).

As long as the selling price per unit is higher than the harvest cost per unit, we make a profit. The points of view of economics and biology agree, because the maximum profit occurs for the maximum sustainable yield.

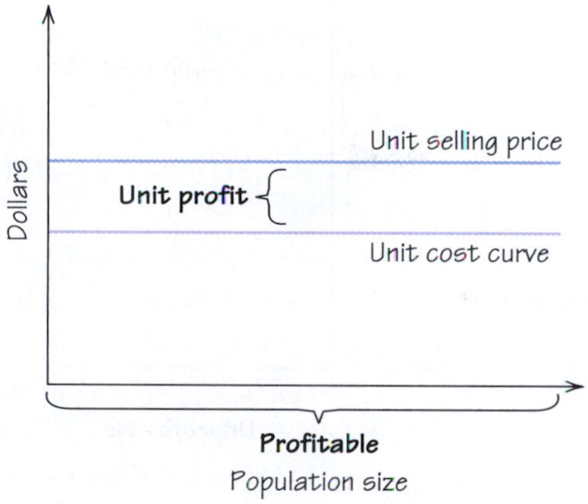


**FIGURE 23.12** The unit cost, unit revenue, and unit profit of harvesting one unit, as a function of population size, for the cattle ranch.

## EXAMPLE 7 ■ Fishing and Logging

In this model, we assume that the cost of harvesting a unit of the population decreases as the size of the population increases. This is the familiar principle of **economy of scale**. For example, the same fishing effort yields more fish when fish are more abundant. Similarly, a logger's harvest costs per tree are less when the trees are clumped together. This is the logger's motivation to clear-cut large stands. What should our sustainable-yield harvesting policy be?

**SOLUTION** The cost curve slopes downward and to the right, as in Figure 23.13. The size of the population from which one unit is harvested is shown on the horizontal axis. The cost of harvesting a single unit is measured on the vertical axis.

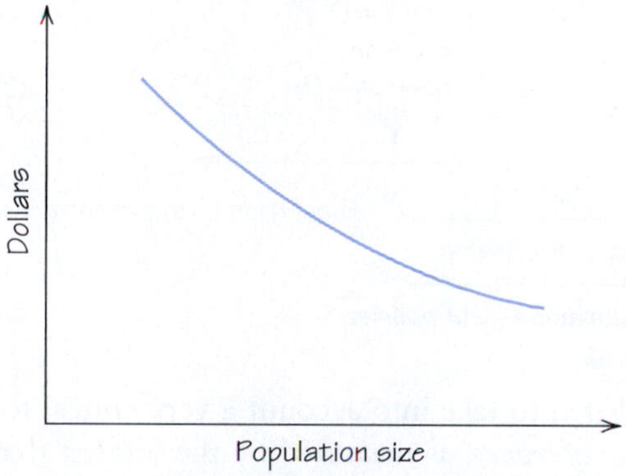


**FIGURE 23.13** The unit cost, as a function of population size, for fishing or logging.

An optimal sustainable-yield harvesting policy depends on the relation between price and costs. There are two cases, as shown in Figure 23.14.

- *The unit cost curve lies entirely above the unit price line.* The price for a harvested unit is less than the cost of harvesting it, no matter how large the population. It is impossible to make a profit.
- *The unit cost curve intersects the unit price line.* Above a certain population size, the price for a harvested unit is more than the cost of harvesting it, so profit is possible. Some population size, call it $x_Q$, gives a maximum net profit. Using calculus, it can be shown that $x_Q$ is actually larger than $x_M$, the population that gives the maximum sustainable yield (see Figure 23.15). Compared to harvesting the maximum sustainable yield, economic considerations result in harvesting less, but also in maintaining a larger stock of the population.

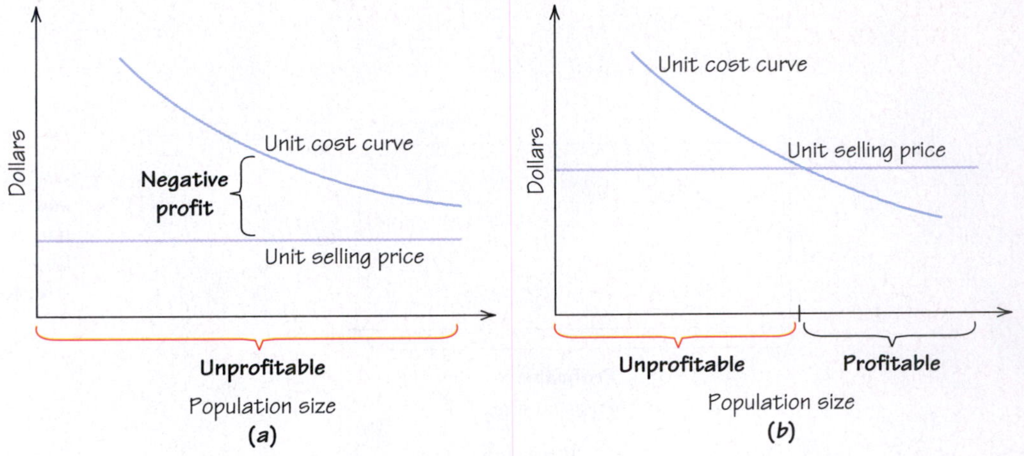


**FIGURE 23.14** The unit cost, unit revenue, and unit profit of harvesting one unit, as a function of population size, for fishing or logging. (a) The market price is below the harvesting cost for all population sizes. (b) The operation is profitable for populations above a certain minimum size.

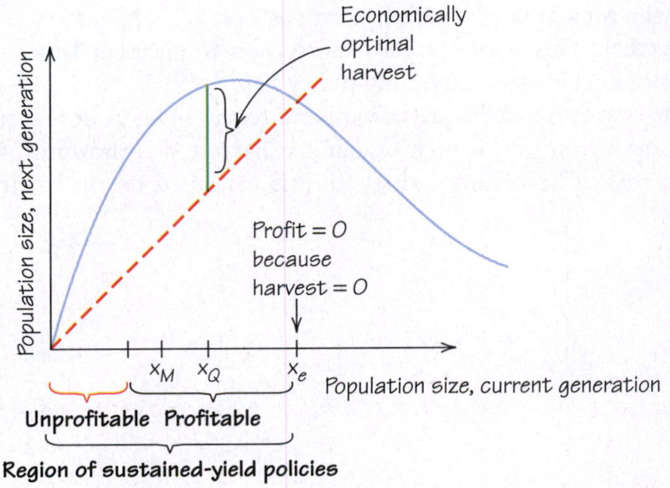


**FIGURE 23.15** Reproduction curve showing regions of profitability for sustained-yield policy, with the economically optimal population size $x_Q$ marked.

Our simple models fail to take into account a very critical feature of a modern economy: the *time value of money,* as measured by the interest that capital can earn. We now investigate why biological populations are susceptible to overexploitation and even extinction.

## Why Eliminate a Renewable Resource?

Some species, such as the passenger pigeon, were harvested to extinction. In other cases, an entire ecosystem has been destroyed (see Spotlight 23.2). Why would anyone eliminate a renewable resource? Our approach helps explain why.

Sustained-yield policies involve revenues that will be received, year after year, in the future. The value of these revenues should be discounted to reflect the lost investment income that we could earn if instead we had the revenues today. For funds invested at a return of $100r\%$ per year, compounded annually, the present value $P$ of an amount $A$ to be received in $n$ years in the future is related to $A$ by the compound interest formula $A = P(1 + r)^n$.

## SPOTLIGHT 23.2 The Tragedy of Easter Island

**Easter Island**
*(Tom Till/Tony Stone Images.)*

Easter Island is famous for its isolation—1400 miles to the nearest island—and for its hundreds of huge stone statues. For 30,000 years before the arrival of people in about 400, Easter Island maintained a lush forest, with several species of land birds. By the time of the first visit by Europeans in 1722, the island was barren, denuded of all trees and bushes over 10 feet high, and with no native animals larger than an insect. The 2000 or so islanders had only three or four leaky canoes made of small pieces of wood.

What happened? Careful analysis of pollen in soil samples tells the sad story. The settlers and their descendants cut wood to plant gardens, build canoes, make sledges and rollers to move the huge statues, and burn for cooking and warmth in the winter. In addition to crops they raised and chickens they had brought to the island and cultivated, they ate palm fruit, fish, shellfish, the meat and eggs of birds, and the meat of porpoises that they hunted from seagoing canoes. The population of the island grew to 7000 (or perhaps even 20,000).

By 1500, the forest was gone. Most tree species, all land birds, half of the seabirds, and all large and medium-sized shellfish had been extinguished. There was no firewood, no wood for sledges and rollers to transport hundreds of statues at various stages of completion, and no wood for seaworthy canoes. Without canoes, fishing declined and porpoises could not be taken. Stripping the trees exposed the soil, which eroded, so crop yields fell. The people continued raising chickens, but warfare and cannibalism ensued. By 1700, the population had crashed to 10–25% of its former size.

Why didn't the people notice earlier what was happening, imagine the consequences of keeping on as they had been, and act to avert catastrophe? After all, the trees did not disappear overnight.

From one year to the next, changes may not have been very noticeable. The forests may have been regarded as communal property, with no one charged with limiting exploitation or ensuring new growth. There was no quantitative assessment of the resources available and need for conservation versus the long-term needs of the "public works" program of erecting statues. Moreover, the religion of the people, the prestige of the chiefs, and the livelihood of hundreds depended on the statue industry. There was no perceived need to limit the population and no technology for birth control. Once the large trees were gone, there was no means for excess population to emigrate.

Adapted from Jared Diamond, "Easter's end," *Discover*, 16(8) (August 1995): 63–69.

The economic goal is to maximize the sum of the present values of all future receipts from harvesting. The optimal harvesting policy thus must depend on the expected rate of return $r$. We don't delve into the details of the calculations here but instead just give the results of the analysis.

Again, there are several cases to consider:

1. The unit cost of harvesting exceeds the unit price received, for all population sizes. Then it is impossible to make a profit.

2. For small $r$: for some population size $x$, the unit cost of harvesting equals the unit price received. Then there is a size between $x$ and $x_e$ (the equilibrium population size) for which the present value of the total return is maximized and the population and its yield are sustained.
3. For larger $r$, *the economically optimal policy may be to harvest the entire population, immediately extinguish the resource, and invest the proceeds.* The unit price exceeds the unit cost for *all* population sizes.

Let's put this in the simplest and starkest terms. Suppose that you own a resource, such as a forest, whose cost of harvesting is small relative to its value. If the rate at which the forest population grows is greater than what you can earn on other investments, it pays to let the forest keep on growing.

On the other hand, if the forest grows more slowly than the rate of return on other investments, the economically optimal harvesting policy is to cut down all the trees now and invest the money. You could then start raising cattle on the land–and right there you have the scenario that is resulting in deforestation all over the world.

The sobering fact is that *very few economically significant renewable resources can sustain annual growth rates over 10%.* Many, like whales and most forests, have growth rates in the 4% to 5% range. These values–even a growth rate of 10%–are far below the return that many investors expect on their investment. For example, over the long run, the U.S. stock market has yielded an average 11% return, but venture capital firms expect to exceed a 25% profit.

The concept of maximum sustainable yield is an attractive ideal if the expectations of investors are low enough. However, there are still difficult problems:

- One problem is "the tragedy of the commons," discussed by ecologist Garrett Hardin. Several hundred years ago, English shepherds would graze their flocks together on common land. The grass of the commons could support only a fixed number of sheep. Each shepherd could reasonably think that adding just one or two more sheep to his flock would not overtax the commons. Yet if each did so, there could be disaster, with all the sheep starving. Many natural-products industries, such as fisheries, are a form of commons. Small overexploitation by each harvester can produce disastrous results for all. Global warming may be a tragedy of a worldwide commons.
- How, in the presence of human needs or greed, can we anticipate and prevent overexploitation and possible extinction of a resource? By and large, it has been politically impossible to force a harvesting industry to reduce current harvests to ensure stability in the future.
- In some industries, such as a fishery, growth of the population may be abundant one year but meager another, so that a steady yield cannot be sustained without damaging the resource. A few good years in a row may provoke increased investment in fishing capacity. Then attempting to harvest at the same levels in succeeding normal or below-normal years results in overfishing. This exact scenario destroyed the California sardine fishery in the 1930s, the Peruvian anchovy fishery in 1972, and much of the North Atlantic fishery in the 1980s. The ocean off northwest Africa was nearly picked clean in the past decade.

Were the fishers and regulators mentioned above at fault for extinguishing these fisheries by overexploiting a dependable resource? Or were the extinctions due to chance variations of the fish stocks? In the next section, we examine a third possibility—that the fish stocks followed simple rules that nevertheless produced "chaotic" behavior of stocks, that is, wide variation from one year to the next. When we do not see the pattern, we interpret such behavior as randomness, much as the moves in a chess game may appear random and inexplicable to someone who does not know the rules of the game.

## 23.5 Dynamical Systems and Chaos

In this and the two previous chapters, we have considered systems that change over time: bank accounts, the amount due on a loan, and the size of a population. Other examples are a dripping faucet, a playground swing, a pinball play, the solar system, the business cycle, epidemics, the passage of a drug through the human body, and the weather. Some of these are very predictable (interest on a bank account), while others are notoriously unpredictable (the weather). Some involve no outside influences (the amount due on a loan, assuming that you don't get behind on payments!), while others are the result of many contributing factors (the business cycle).

In some systems (such as the population of a country), the state of the system may depend largely on its states at previous times (e.g., last year's population), while in other systems (such as an epidemic) chance may play a large role (e.g., in who and how many become infected).

We are interested in modeling systems, such as a fishery, as they operate without influence from outside or from chance. The applicable mathematical tool is a **dynamical system**.

**Dynamical System** DEFINITION

A **dynamical system** is a mathematical model for a system whose state evolves with time and whose future states depend deterministically on its present and past states.

To make this definition meaningful, we need to be explicit about what we mean by **deterministically**.

**Deterministic** DEFINITION

A system is **deterministic** if its changes through time depend only on natural and mathematical laws and are not substantially affected by what we consider to be chance or free will.

An example of a deterministic system is the path of a golf putt, which is governed by gravity, terrain, wind, and the force imparted by the golfer. A non-example is the outcome of a vigorous toss of a coin or a random number generator; although the result, like the golf putt, is determined by physical laws, we consider the result to be random. Another non-example is the outcome of an election, which involves choices by humans.

### Mathematical Chaos

We think of **chaos** as referring to general confusion, unpredictability, and apparent randomness. Mathematicians and other scientists use the word to describe systems whose behavior over time is inherently unpredictable.

## Chaos — DEFINITION

A dynamical system exhibits **chaos** if it is:

1. *Near-periodic*–any state is near one that eventually will repeat;
2. *Transitive*–from any state you can eventually get close to any other; and
3. *Sensitive*–a small change in the initial state can produce widely diverging results later.

## EXAMPLE 8 ■ Chaos in Manhattan

You may already know from experience that getting around Manhattan can be a chaotic experience, in the ordinary sense of the word. We show here that Manhattan's transit system is also chaotic in the mathematical sense.

Consider an urban area, such as Manhattan, where several modes of public transport are available. Subway trains and buses leave from set stops, and taxis wait at taxi stands; call these locations *transit points*. Trains and buses retrace their routes, hence are periodic; taxis are not. The system is closed: Vehicles do not leave the city nor do they travel through the fourth dimension.

This system fits the definition of chaos:

1. *Near-periodic*: Wherever you live, there is a transit point nearby.
2. *Transitive*: Wherever you live, and wherever you want to go, there is a transit point near you whose vehicle will take you near to where you want to go.
3. *Sensitive*: Vehicles at two different transit points near you can eventually take you to vastly different places.

If the system covers the city, then (1) is a consequence of (2). Also, anyone who has gotten on the wrong subway train or bus realizes that (3) is an inevitable consequence of (1) and (2). So in fact (1) and (3) both follow from (2)–a conclusion that is true not just of this Manhattan example but of a large class of dynamical systems. ■

The most noticeable property of a chaotic system is sensitivity–that a small change now can make a big difference later.

This feature is sometimes known as the **butterfly effect**, from the title of a 1979 talk by meterologist E. N. Lorenz: "Predictability: Does the Flap of a Butterfly's Wings in Brazil Set Off a Tornado in Texas?" (The phrase probably traces to a 1953 science fiction story by Ray Bradbury, "A Sound of Thunder," in which history is changed by a time-traveler who steps on and kills a prehistoric butterfly.)

We can get a feel for chaotic systems by playing with some **iterated function systems (IFS)**. The fancy name just means that we take an initial value, apply a function to it, then repeat over and over. This is exactly what we did earlier, geometrically, with reproduction curves for populations. (See section 19.5, pp. 631–635, for more about IFS and their connection to fractals.)

## EXAMPLE 9 ■ Doubling on a "Stone Age" Calculator

Imagine that you have a calculator that keeps only the last two digits of a number. It has a special key marked [DBL] that doubles the number in the display and keeps *only* the last two digits. For example, [DBL] applied to 52 gives 04 (*not* 104).

Let's start with two numbers that are as close together as can be on this calculator, such as 37 and 38. As we push the DBL key over and over again, will the result stay close?

**SOLUTION**

37, 74, 48, 96, 92, 84, 68, 36, 72, 44, 88, 76, 52, 04, . . .
38, 76, 52, 04, 08, 16, 32, 64, 28, 56, 12, 24, 48, 96, . . .

Already, by the fourth iteration, the two sequences are far apart.

## EXAMPLE 10 ■ The Solar System

The American moon landings in 1969 and later, as well as all other space missions, were possible because of the predictability, or *determinism,* of the solar system. The moon and planets follow their orbits like clockwork. So how could the solar system be chaotic?

**SOLUTION** Over tens of millions of years, the orbit of each planet is chaotic, meaning that the slightest change in its position or velocity–due to, say, a comet passing nearby–could produce a huge difference later.

More down-to-earth examples of physical systems that can exhibit chaos include the fluttering of a falling autumn leaf, heart arrhythmias, and the Tilt-A-Whirl amusement park ride.

### Chaos in Biological Populations

If we measure this year's population as a fraction $x$ of the carrying capacity, and do the same for next year's population as a fraction $f(x)$, the logistic model takes the form

$$f(x) = \lambda x(1 - x)$$

where the Greek letter lambda $\lambda = 1 + r$ is the amount by which the population is multiplied each year. When expanded, the equation has the familiar form of a quadratic in $x$:

$$f(x) = -\lambda x^2 + \lambda x$$

## EXAMPLE 11 ■ The Logistic Population Model

What behaviors can occur in the logistic model?

**SOLUTION** For different values of the parameter $\lambda$ and different starting values for the population fraction, each of the behaviors of Figure 23.11 on p. 746 can occur:

- $\lambda = 2.8$ and starting population fraction $x = 0.36$ produces Figure 23.11a.
- $\lambda = 3.1$ and starting population fraction $x = 0.235$ produces Figure 23.11b.
- $\lambda = 3.0$ and starting population fraction $x = 0.4$ produces Figure 23.11c.

In other words, for population growth rates (values of $\lambda$) that are fairly close together (2.8, 3.1, 3.0), the population evolves very differently. This is a surprising and nonintuitive conclusion.

But there is more. For $\lambda = 4$ and any starting population fraction, the population does not settle down into any of the patterns of Figure 23.11; year after year it wanders "unpredictably" all over the place (Figure 23.16). This is *chaotic behavior:* It is deterministic, complex, and–in the long run–unpredictable. In the short run, the behavior is completely predictable. For example, from this year's population fraction, the equation tells us exactly what next year's will be. Repeating the use of the equation, we can determine what it will be the following year. But as the years pass, any sense of pattern gets lost in the complexity.

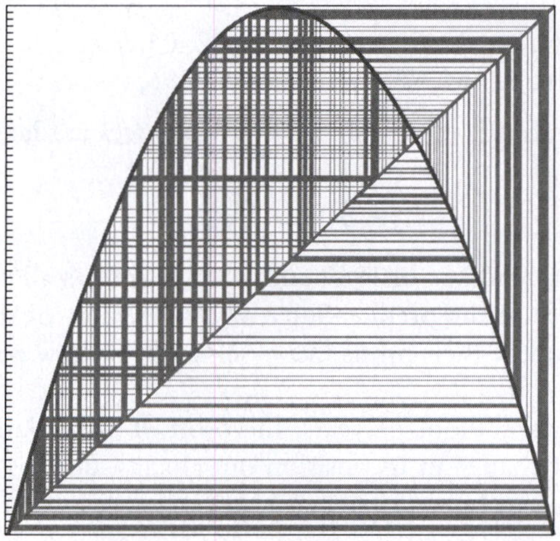

**FIGURE 23.16** Chaotic behavior of a population.

This potentially chaotic behavior of a biological population is bad news for those who manage a ranch or any biological population in the wild or in captivity. In recent years, the lobster catch in Maine has been much higher than in previous years, reaching record levels, for no discernible reason. On the other hand, in the late 1950s the annual harvest of Dungeness crabs off the central California coast declined from 12 million pounds to less than 1 million pounds without any evidence of disease, heightened predation, or increased crabbing effort. Researchers who modeled that population in 1994 found that booms and busts are the rule. The population can remain nearly level for generations before suddenly exploding or crashing without warning.

Searching for an environmental cause for these fluctuations could be futile, because there may not be one. Moreover, observing the population over a few generations provides no help in predicting future behavior.

## EXAMPLE 12 ▪ Childhood Disease Epidemics

The incidence of childhood diseases such as chickenpox and measles varies greatly from year to year. Why?

**SOLUTION** There are three plausible explanations for the fluctuations:

- There is an underlying regular cycle that is perturbed and occasionally overwhelmed by random events ("noise").
- There is no pattern, because the fluctuations are due solely to chance.
- There is no discernible pattern, because such fluctuations are inherent in the epidemiology of the disease, a chaotic system.

The first explanation, a perturbed cycle, fits chickenpox, with a cycle of one year. For measles, either the second or the third explanation may be correct, depending on the size of the community. For small communities, chance is an adequate explanation. For large communities, historical data from before the era of mass immunization suggest that measles cases were chaotic. That doesn't mean that they occurred at random but rather that they were unpredictable. Research also shows that there is a critical community size above which a disease will not die out solely by chance. For measles, this size is about 250,000.

What you need to understand about chaos is that behavior that appears to be random can be produced even with very simple systems that are completely governed by deterministic rules. Just because the behavior appears chaotic does not mean a lack of underlying order and structure, though discovering that structure may be difficult.

Even if we discover the structure, prediction may elude us because of chaos. If we had an absolutely correct model of how weather behaves and measurements at every location on and above the Earth, we might still not be able to forecast the weather accurately a week ahead.

What about the fishery extinctions? Perhaps the fishers and the fish were victims not of greed or chance but of the chaotic nature of the reproduction curve for the fish.

## REVIEW VOCABULARY

**Biomass** A measure of a population in common units of equal value. (p. 741)

**Butterfly effect** A small change in initial conditions of a system can make an enormous difference later on. (p. 752)

**Carrying capacity** The maximum population size that can be supported by the available resources. (p. 734)

**Chaos** Complex but deterministic behavior that is unpredictable in the long run. (p. 751)

**Cobweb diagram** A kind of graphical portrayal of the evolution of a dynamical system, such as a population. (p. 745)

**Compound interest formula** Formula for the amount in an account that pays compound interest periodically. For an initial principal $P$ and effective rate $r$ per year, the amount after $n$ years is $A = P(1 + r)^n$. (p. 732)

**Deterministic** A system is deterministic if its future behavior is completely determined by its present state, past history, and known laws. (p. 751)

**Dynamical system** A system whose state depends only on its states at previous times. (p. 751)

**Economy of scale** Costs per unit decrease with increasing volume. (p. 747)

**Equilibrium population size** A population size that does not change from year to year. (p. 742)

**Exponential reserve** How long a fixed amount of a resource will last at a constantly increasing rate of use. A supply $S$, at an initial rate of use $U$ that is increasing by a proportion $r$ each year, will last

$$\frac{\ln\left(1 + \frac{S}{U} r\right)}{\ln(1 + r)} \text{ years. (p. 738)}$$

**Iterated function system (IFS)** A sequence of elements (numbers or geometric objects) in which the next element is produced from the previous one according to a function (rule). (p. 752)

**Logistic model** A particular population model that begins with near-geometric growth but then tapers off toward a limiting population (the carrying capacity). (p. 734)

**Maximum sustainable yield** The largest harvest that can be repeated indefinitely. (p. 744)

**Natural increase** The growth of a population that is not harvested. (p. 742)

**Nonrenewable resource** A resource that does not tend to replenish itself. (p. 736)

**Population structure** The division of a population into subgroups. (p. 734)

**Rate of natural increase** Birth rate minus death rate; the annual rate of population growth without taking into account net migration. (p. 732)

**Renewable natural resource** A resource that tends to replenish itself; examples are fish, forests, wildlife. (p. 741)

**Reproduction curve** A curve that shows population size in the next year plotted against population size in the current year. (p. 741)

**Savings formula** Formula for the amount in an account to which a regular deposit is made (equal for each period) and interest is credited, both at the end of each period. For a regular deposit of $d$ and an effective interest rate $r$ per year, the amount $A$ accumulated after $n$ years is

$$A = d\left[\frac{(1 + r)^n - 1}{r}\right]. \text{ (p. 732)}$$

**Static reserve** How long a fixed amount of a resource will last at a constant rate of use; a supply $S$ used at an annual rate $U$ will last $S/U$ years. (p. 738)
**Sustainable yield** A harvest that can be continued at the same level indefinitely. (p. 743)
**Sustained-yield harvesting policy** A harvesting policy that can be continued indefinitely while maintaining the same yield. (p. 744)
**Yield** The amount harvested at each harvest. (p. 742)

## ✓ SKILLS CHECK

**1.** The carrying capacity of a population is

**(a)** the largest recorded population.
**(b)** the largest supportable population.
**(c)** the change in population.

**2.** The shape of the reproduction curve reflects that a small population has abundant resources and can grow quickly by ________________ steeply at the left.

**3.** The logistic curve is a model for a population that is growing

**(a)** linearly.
**(b)** exponentially.
**(c)** with a ceiling.

**4.** U.S. oil use has grown according to a(n) ________ model.

**5.** Management of a nonrenewable resource can be modeled by

**(a)** an annuity.
**(b)** a savings account.
**(c)** an add-on loan.

**6.** If we have enough reserves of a product to last 200 years at the current rate of use, but the rate of use increases by 10% per year, the supply will last ________ years.

**7.** The equilibrium population size is the same as

**(a)** the carrying capacity.
**(b)** the intersection point of the reproduction curve and the diagonal.
**(c)** the natural increase of the population.

**8.** If we have enough reserves of a product to last 1000 years at the current rate of use, but the rate of use increases by 1% per year, the supply will last ________ years.

**9.** If the starting population for a reproduction curve is changed, the subsequent population pattern

**(a)** will still drift to the same pattern.
**(b)** will change to a different pattern.
**(c)** will sometimes change to a different pattern.

**10.** For the logistic model $f(x) = 3x(1 - x)$, if the starting population fraction is 0.5, what is the next population fraction is __________ .

**11.** Economic considerations

**(a)** always work against the conservation of a resource.
**(b)** never interfere with the conservation of a resource.
**(c)** always affect in some way the conservation of a resource.

**12.** For the logistic model $f(x) = 3x(1 - x)$, if the starting population fraction is 0.4, the next population fraction is __________ .

**13.** The population size that leads to the maximum net profit under sustainable-yield harvesting is always

**(a)** the same as the maximum sustainable yield.
**(b)** smaller than the maximum sustainable yield.
**(c)** larger than the maximum sustainable yield.

**14.** For the logistic model $f(x) = 4x(1 - x)$, if the starting population fraction is 0.1, the next population fraction is ________ .

**15.** A system whose current state depends solely on its previous states is called

**(a)** a dynamical system.
**(b)** a stable system.
**(c)** an optimal system.

**16.** For the logistic model $f(x) = 4x(1 - x)$, a starting population fraction that will immediately lead to 0 population is __________ .

**17.** Chaotic behavior appears to be random

**(a)** but is actually not random.
**(b)** and is random.
**(c)** and is sometimes random.

**18.** The "butterfly effect" refers to a feature of chaos called __________________ .

**19.** Pressing a digit and then repeatedly pressing the SIN key is a model of

**(a)** an iterated function system.
**(b)** chaos.
**(c)** randomness.

**20.** Elimination of a natural resource can be caused by ______________________________ .

# CHAPTER 23 EXERCISES

■ Challenge ◆ Discussion

## 23.1 Growth Models for Biological Populations

**1.** (Spreadsheet helpful) For many years, China has been the world's most populous country. However, India has been catching up, with 1132 million in mid-2007 and growing at 1.6% per year, versus China then with 1318 million and growing at only 0.6% per year. If these rates continue, when will India have more people than China?

**2.** The population of the less-developed countries (excluding China) in mid-2007 of 4.086 billion is expected to grow at 1.8% per year (this is an annual yield, so you may think of it as compounded annually). If this growth rate continues until mid-2025, what will be the size of the population then?

**3.** If the growth rate of the less-developed countries of Exercise 2 had changed suddenly to 1.7% in mid-2007, what would be the size of the population in mid-2025?

**4.** (Spreadsheet helpful) If the growth rate of the less-developed countries of Exercise 2 decreased by $\frac{1}{25}$ of a percentage point (0.04%) per year from 2007 through 2011, beginning in mid-2007, what would be the size of the population in mid-2012?

**5.** An advertisement for Paul Kennedy's book *Preparing for the Twenty-First Century* (Random House, 1993) asked: "By 2025, Africa's population will be: 50%, 150% or 300% greater than Europe's?" The population of Europe in mid-2007 of 729 million is expected to stay constant through 2025. The population of Africa in mid-2007 of 925 million is expected to increase at about 2.4% per year. What answer would you give to the question?

**6.** In its estimates for doubling times for populations in the world, the Population Reference Bureau uses a rule of 70, similar to (but slightly more accurate than) the rule of 72 used in banking and explained in Exercises 13 and 14 in Chapter 22. The rule of 70 says that if a country's population continues to grow at a constant rate of $r$% per year, then it will double in size every $70/r$ years. (As noted in Chapter 22, a rule of 69.3 would be even more accurate, but the difference between that and the rule of 70 is only 1%.) Apply the rule of 70 to estimate the doubling times for the following populations (figures are for mid-2007):

**(a)** Africa, 944 million, 2.5%
**(b)** United States, 302 million, 0.6% (not including immigration)

**7.** Do the calculations as in Exercise 6, but for:

**(a)** China, 1.318 billion, 0.5%
**(b)** The world as a whole, 6.625 billion, 1.2%

**8.** Wisconsin's electricity demand increased 3.03% per year for the 35 years from 1970 to 2005. If that trend continues, when will Wisconsin need to have twice as much generating capacity as it did in 2005?

(James Schnept/Getty Images.)

For Exercises 9 and 10, refer to the following.

Is Warren Sanderson right about world population growth slowing down (see Spotlight 23.1)? How much difference does it make in projections if we look at the world as a whole or break it down by countries or regions? In Exercises 9 and 10, we investigate this question, first projecting as a whole and then projecting by regions and adding.

**9.** The population of the world in mid-2007 of 6.625 billion is expected to increase 1.2% per year.

**(a)** Project the population to mid-2025 (by then you may have finished having children, if you have any) and to mid-2050 (by then you may be thinking about retiring).
**(b)** What are the assumptions involved in your projections?

**10.** Divide the countries of the world into three groups with differing rates of increase (see the table). (Why is this useful?)

| Group | Population Mid-2007 (billions) | Rate of Growth (%) |
|---|---|---|
| More-developed countries | 1.221 | 0.1 |
| Less-developed countries (excluding China) | 4.086 | 1.8 |
| China | 1.318 | 0.5 |

**(a)** Redo the projections in Exercise 9a for the years 2025 and 2050 by projecting each group separately and adding the totals. Is there a major difference from the results in Exercise 9a?
**(b)** Will the world be able to support the numbers of people that you project? What problems will these greater numbers of people cause? What could be done to avert those problems? Do you think that anything

will be done before there is some kind of worldwide crisis?

(In mid-1995, the world population was 5.7 billion and growing at 1.5% per year. Those figures led to projections as in Exercise 9 of 8.3 billion for 2020 and 11.1 billion for 2040. The corresponding growth rates for the groups of Exercise 10 were 0.2%, 2.2%, and 1.1%, which led to projections of 8.9 billion for 2020 and 12.7 billion for 2040.)

## 23.2 How Long Can a Nonrenewable Resource Last?

**11.** In 2005, world oil reserves totaled at most 2900 billion barrels, while daily consumption was 84.7 million barrels in 2006. The U.S. Geological Survey (USGS) projected then that world consumption would increase 1.9% per year through 2025.

**(a)** What was the static reserve for oil in 2005?
**(b)** What was the exponential reserve for oil in 2005?
**(c)** What considerations may affect the answers to parts (a) and (b) over time?

**12.** In 2008, world natural gas proven reserves totaled 6185 trillion cubic feet, while annual consumption was 105.5 trillion cubic feet in 2006. The USGS projected then that world consumption would increase 2.2% per year through 2025.

**(a)** What was the static reserve for natural gas in 2008?
**(b)** What was the exponential reserve for natural gas in 2008?
**(c)** What considerations may affect the answers to parts (a) and (b) over time?

**13.** Can our energy problems be solved by increasing the supply? [Thanks for the idea to Evar D. Nering of Arizona State University, in "The mirage of a growing fuel supply," *The New York Times* (June 4, 2001) Op-Ed page.]

**(a)** Suppose that we have a 100-year supply of a resource (such as oil, for which known world reserves will last less than 100 years at the current world rate of use). That is, the resource would last 100 years at the current rate of consumption. Suppose that the resource is consumed at a rate that increases 2.5% per year (this is the average increase in consumption for oil in the United States since 1973). How long will the resource last?
**(b)** Suppose that we underestimated the supply and actually have a 1000-year supply at the current rate of use. How long will that last if consumption increases 2.5% per year?
**(c)** Let's think big and suppose that there is 100 times as much of the resource as we thought–a 10,000-year supply. How long will that last if consumption increases 2.5% per year?

**14.** In this problem we explore the consequences of reducing the rate of growth of oil use. Suppose that we halve the growth rate from the 2.5% per year given in Exercise 13 to 1.25% per year. [Thanks for the idea to Evar D. Nering of Arizona State University, in "The mirage of a growing fuel supply," *The New York Times* (June 4, 2001) Op-Ed page.]

**(a)** How long will the 100-year supply last?
**(b)** How long will the 1000-year supply last?
**(c)** How long will the 10,000-year supply last?

**15.** We continue the ideas of Exercises 13 and 14, but with a more radical hypothesis.

**(a)** How long would the 100-year supply last if we reduced our consumption by just $\frac{1}{2}$% per year–that is, if we used $\frac{1}{2}$% less each year instead of 2% more?
**(b)** If we used 1% less each year?

**16.** By the time there is concern about using up a nonrenewable resource, it may be too late. Suppose that a resource has a static reserve of 10,000 years, but consumption is growing at 3.5% per year.

**(a)** How long will the resource last?
**(b)** How long before half the resource is gone?
**(c)** How much longer will the resource last if after half of it is gone, consumption is stabilized at the then-current level?
**(d)** What implications do you see to your answers?

**17.** Do a calculation to criticize the claim in the following quotation: "The United States holds 437 billion tons of known (coal) reserves, enough energy to keep 100 million large electric generating plants going for the next 800 years or so." [*Forbes* (December 15, 1975), p. 28; thanks to Albert A. Bartlett.]

For Exercises 18–20, refer to the following.

The formula for the average growth rate over a period of time is even simpler than the one for the exponential reserve of a resource. If usage at the beginning is $N_0$ and at the end of an interval of $t$ years it is $N$, then the average annual rate of growth is

$$\frac{1}{t}\ln\left(\frac{N}{N_0}\right)$$

**18.** The U.S. population was 62.95 million in 1890 and 302 million in 2007 (at roughly the same times of the year). What was the annual rate of growth, to two decimal places?

**19.** The U.S. population was 3.93 million in 1790 and 62.95 million in 1890. What was the average annual rate of growth, to two decimal places?

**20.** **(a)** The average increase in oil consumption in the United States during 1993–2000 was nearly 2%. In fact, consumption in 1993 was 6.291 billion barrels and consumption in 2000 was 7.211 billion barrels. What is a more accurate (two-decimal-place) estimate of the average annual percentage increase in oil consumption? **(b)** Oil consumption in the United States was 19.701 million barrels per day in 2000 and 20.731 million barrels per day in 2004 (roughly one-fourth of world consumption). What was the average annual rate of growth? (From 2004 through 2007, growth was zero, no doubt due to much higher prices for oil.)

## 23.3 Sustaining Renewable Resources

## 23.4 The Economics of Harvesting Resources

For Exercises 21–25, refer to the following.

We suppose that a population has the reproduction curve shown in the following figure, with units of thousands of tons of biomass. The mathematical description is that the population in the following year, $x_{n+1}$, depends on the population $x_n$ in the current year (after any harvest) according to

$$x_{n+1} = f(x_n) = \tfrac{1}{5}x_n(20 - x_n) = 4x_n(1 - 0.05x_n)$$

for $x_n$ between 0 and 10, in units of millions of pounds. We start with a population this year of $x_1$ whose value we vary. (For these exercises, a spreadsheet or a programmable calculator is useful.)

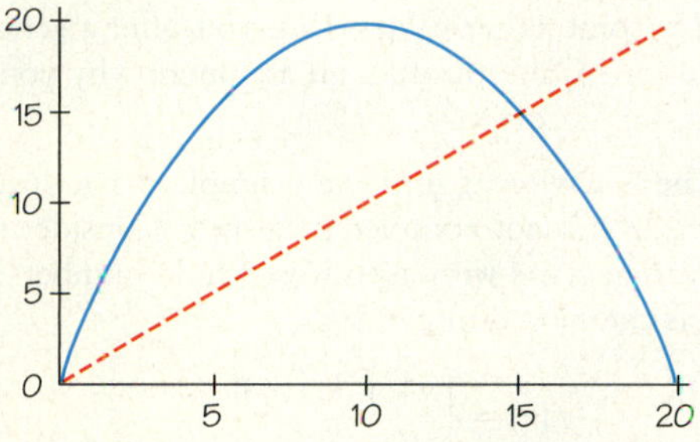


**21.** Start with $x_1 = 5$. Calculate numerically the population in the first few years, draw a cobweb diagram, and briefly describe the qualitative behavior of the population.

**22.** Repeat Exercise 21 with the starting value $x_1 = 10$.

**23.** Repeat Exercise 21 with the starting value $x_1 = 7$, going at least as far as $x_{10}$.

**24.** Try to find a starting value (besides 0, 5, 10, and 15) that leads to a stable population over time.

**25.** What is the equilibrium population size?

For Exercises 26–36, refer to the following.

We suppose that the population has the reproduction curve shown in the following figure, with units of thousands of tons of biomass, whose mathematical description is

$$x_{n+1} = f(x_n) = 3x_n(1 - 0.05x_n)$$

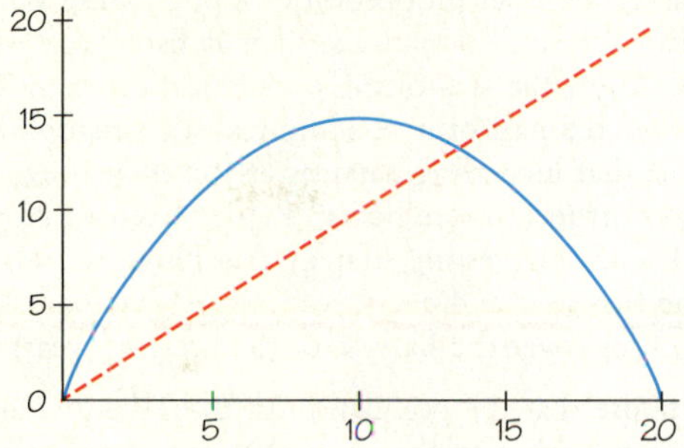


**26.** Start with $x_1 = 5$. Calculate numerically a population in the first 10 years, draw a cobweb diagram, and briefly describe the qualitative behavior of the population.

**27.** Repeat Exercise 26 with the starting value $x_1 = 10$.

**28.** Try to find a starting value (besides 0 and 20) that leads to extinction of the population.

**29.** What is the equilibrium population size?

◆ **30.** Which of the reproduction curves–the one for Exercises 21–25 or the one for Exercises 26–29–seems to you more realistic as a model of a biological population, and why?

**31.** What is the significance of the red dashed line in the preceding figures and its intersection with the blue curve?

**32.** The following figure shows the annual population gain in the absence of any harvesting. Determine the maximum sustainable yield to one decimal point.

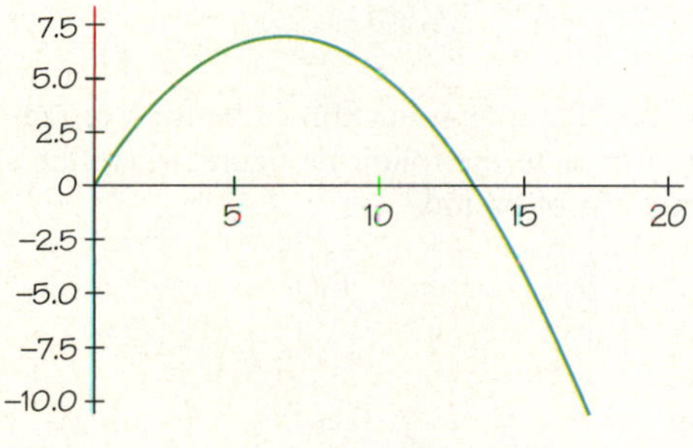


**33.** Suppose that the population of Exercises 26–32 starts with $x_1 = 11$ and each year we harvest half of the population. For example, in year 1, we harvest 5.5 million pounds, leaving the remaining 5.5 million pounds to reproduce (according to the reproduction curve) for the next year. Calculate numerically the population in the first 10 years, draw a cobweb diagram, and briefly describe the qualitative behavior of the population.

**34.** Repeat Exercise 33 but for a starting population $x_1 = 5$.

**35.** Harvesting a set proportion of a population is unrealistic for some situations, such as fishing, in which we can't know the size of the population or when we have harvested half of it. A more realistic situation for fishing is that increasing harvests attract increasing fishing effort (e.g., more boats). Repeat Exercise 33 with $x_1 = 11$ and a harvesting strategy that harvests 1 million tons the first year and every year harvests an extra 1 million tons (over the harvest of the previous year).

**36.** Suppose that the population of Exercises 26–35 has been overharvested to the point that only 1 million tons remain at the end of a particular year. If there is no harvesting at all until a year after a year with a population of 11 million tons, when can harvesting resume?

**37.** A reproduction curve for a population is shown in the following figure. Estimate the equilibrium population size and the maximum sustainable yield. (The units are in millions of pounds.)

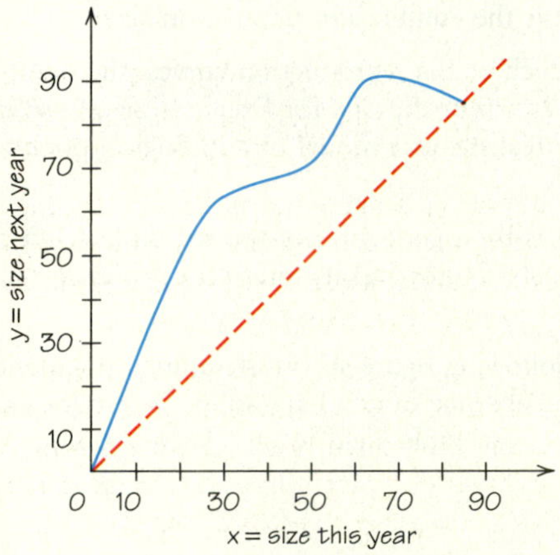


**38.** Suppose that a reproduction curve for a certain population is as in the following figure, where the units are in millions of pounds.

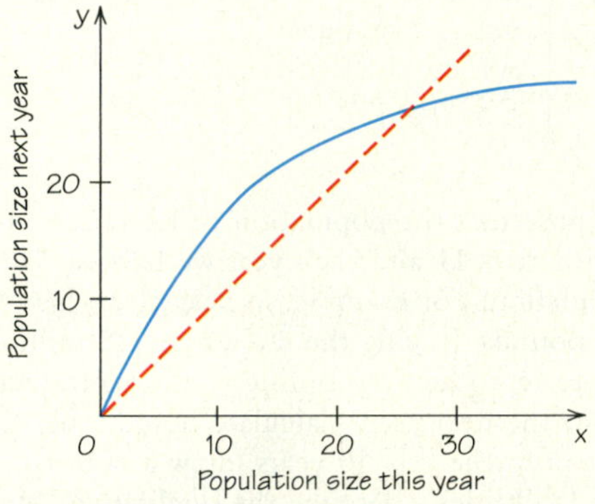


**(a)** Estimate the sustainable yields corresponding to a population of size 10 remaining after the harvest.
**(b)** Estimate the maximum sustainable yield.

## 23.5 Dynamical Systems and Chaos

**39.** In doubling on a "Stone Age" calculator (see Example 9):

**(a)** What do you notice about the two sequences that were produced?
**(b)** Suppose that we had started with a different seed, say 39. What would happen as we iterate the doubling?
**(c)** Explain what you observe in parts (a) and (b) and give a general argument about why it is true.

**40.** Explain why the logistic model on p. 734 is the same as the one on p. 753, where we have $f(x) = \lambda x(1 - x)$.

**41.** You saw in Figure 23.11 that a logistic model can result in a stable value, produce cycling between several values, or result in chaos. Other dynamical systems can exhibit similar behavior. Here we examine the system in which we start with a positive whole number $n$ and iterate the following function:

$$f(n) = \text{sum of the squares of the digits of } n$$

For example, we have $f(133) = 1^2 + 3^2 + 3^2 = 19$.

**(a)** Calculate what happens as $f$ is applied repeatedly, starting with 133. What do you observe?
**(b)** Pick a number different from 133 and different from 1, and iterate $f$ repeatedly. What do you observe?
**(c)** Why did we exclude 1 in part (b)?
**(d)** Try some other values. Can you offer a general conclusion? Can you offer an argument why your conclusion is correct?

**42.** The behavior of some very simple dynamical systems is still not completely known. Consider the system that starts with a positive whole number $n$ and gives as the next number

$$f(n) = \begin{cases} 3n + 1 & \text{if } n \text{ is odd} \\ n/2 & \text{if } n \text{ is even} \end{cases}$$

[This iterative function system was devised by Lothar O. Collatz, later of the University of Hamburg, during his student days before World War II. It is sometimes called the "$3n + 1$ problem" or the "Syracuse problem" (because it became popular at the Mathematics Department of Syracuse University), and the sequences generated are sometimes called "hailstone numbers."]

**(a)** Start with $n = 1$. What happens?
**(b)** Start with $n = 13$. What happens?
**(c)** Start with $n = 12$. What happens?

What you observe is known to happen for all $n < 10^{40}$, but after more than 60 years mathematicians have been unable to show that it happens for every $n$ whatsoever.

**43.** (Requires programmable calculator, spreadsheet, or BASIC programming) A population model slightly different from the logistic model is given by the iterative function system

$$g(x) = x + rx(1 - x)$$

where $x$ is a fraction of the limiting population and $r$ is a growth rate.

**(a)** Set $r = 3$, start with $x = x_1 = 0.01$, and calculate the first 20 values $x_1, \ldots, x_{20}$.
**(b)** In part (a), you should have found $x_{10} = 0.722914$. Replace this value with the rounded-up value $x_{10} = 0.723$ and continue on to calculate $x_{20}$.
**(c)** Now replace $x_{10}$ with the rounded-down value $x_{10} = 0.722$ and continue on to calculate $x_{20}$.

**44.** A dynamical system expressed as an iterated functional system $f(x)$ has an *equilibrium point* at a value $x_0$ if, once the system reaches $x_0$, it always stays at that value. In terms of an equation, an equilibrium point exists at $x_0$ if $f(x_0) = x_0$.

**(a)** For the dynamical system of Exercise 43, find all equilibrium points.
**(b)** For the logistic population model of Exercise 40, find all equilibrium points.

■ **45.** (Requires programmable calculator, spreadsheet, BASIC programming, or preferably use of software available under Suggested Web Sites) The behavior of the logistic population model $f(x) = \lambda x(1 - x)$ depends on the value of the positive parameter $\lambda$. As $\lambda$ increases from 0 to 4, the system changes from one behavior to another, through the following possible states:

- The population simply dies out.
- The population tends toward a nonzero equilibrium point.
- The population oscillates between 2 points.
- The population oscillates between 4 points, then 8 points, then 16 points, and so on.
- The population oscillates between numbers of points that are not powers of 2, until at last . . . the population oscillates between 3 values.
- The population behaves chaotically.

Explore what happens for various values of $\lambda$ between 0 and 4, trying to identify where the shifts in the system's behavior take place.

## WRITING PROJECTS

**1.** Based on the calculations you did in Exercises 9 and 10, write a one- to two-page guest editorial for a newspaper. Describe your projections and how you arrived at them, how serious a problem you think population growth is, what problems it is likely to cause, what you think needs to be done, and what the implications are for your own life.

**2.** Identify a particular regional, national, or world nonrenewable primary resource (such as coal) or a secondary resource (one, such as electric power, that is produced from primary resources). Research how much of it is available now and what the current rate of consumption is. Determine the static reserve. Estimate the growth rate in consumption, taking into account human population increase, and determine the exponential reserve. What social and technological factors contribute to the increasing rate of consumption? Brainstorm how those factors could be changed. Write an essay of three to five pages.

**3.** Identify a particular regional, national, or world renewable resources (such as timber or clean drinking water). Research how much of it is produced now, how much is harvested now, and what the current rate of consumption is. Estimate the growth rate in consumption, taking into account human population increase. For how long can this resource continue to meet the demand? What social and technological factors contribute to the increasing rate of consumption? Brainstorm how those factors could be changed. Write an essay of three to five pages.

## SUGGESTED READINGS

BARTLETT, ALBERT A. *The Essential Exponential! For the Future of Our Planet.* Center for Science, Mathematics, & Computer Education, University of Nebraska–Lincoln (126 Morrill Hall, Lincoln, NE 68588-0350), 2004.

CLOVER, CHARLES. *The End of the Line: How Overfishing Is Changing the World and What We Eat*, The New Press, New York, 2006.

COHEN, JOEL. *How Many People Can the Earth Support?* Norton, New York, 1995.

GLEICK, JAMES. *Chaos: Making a New Science*, Viking, New York, 1987.

PETERSON, IVARS. *Newton's Clock: Chaos in the Solar System*, W. H. Freeman, New York, 1993.

SCHWARTZ, RICHARD H. *Mathematics and Global Survival*, 4th ed., Ginn Press, Needham Heights, Mass., 1998.

## SUGGESTED WEB SITES

**www.popin.org/** UN Population Division population statistics and estimates for all countries.

**www.prb.org/** Population Reference Bureau population statistics and rates of growth by regions.

**www.maths.anu.edu.au/~briand/chaos** Downloadable Java applets to illustrate chaos.

**www.ac.wwu.edu/~stephan/Animation/pyramid.html** Animated display of population structures of individual countries from 1950 to 2050.

**math.bu.edu/DYSYS/applets/** Downloadable Java applets designed to accompany Robert L. Devaney's *A Toolkit of Dynamics Activities,* but can be used independently.

**staff.science.uva.nl/~alejan/dynamicstour.html** Java applet for the logistic population model of Figures 23.11 and 23.16.

**www.census.gov/ipc/www/idb** International data base (IDB) of demographic data for 227 countries, with projections and capability to produce population pyramids for various years.

# ANSWERS TO SKILLS CHECK EXERCISES

## Chapter 1

1. a
2. 7; 8
3. a
4. 3
5. c
6. *B*; *E*
7. b
8. 6
9. c
10. 6
11. a
12. 4
13. b
14. 12
15. a
16. 3
17. b
18. digraph; graph; digraph
19. b
20. 8; 13

## Chapter 2

1. c
2. 27
3. b
4. 26
5. b
6. 33
7. c
8. *V*
9. a
10. 54
11. b
12. 2600
13. c
14. 6; 8; 10
15. b
16. 9
17. a
18. 18
19. c
20. 16

## Chapter 3

1. c
2. 14 min
3. b
4. 2 min
5. a
6. 4; 2; 3
7. c
8. 4; 2; 3
9. c
10. 10 min
11. b
12. 14
13. c
14. 3
15. b
16. 1
17. c
18. 2
19. a
20. 3

## Chapter 4

1. a
2. 6; 2
3. a
4. 2; 5
5. b
6. 25; 6
7. c
8. $12
9. b
10. convex
11. c
12. 3; 2
13. c
14. 3
15. c
16. 3; 1
17. c
18. 57
19. c
20. −6

## Chapter 5

1. a
2. 3
3. a
4. right
5. c
6. 14
7. c
8. 124.9
9. b
10. left
11. b
12. 50
13. c
14. 98, 120, 125.5, 132, 147
15. b
16. grams
17. a
18. standard deviation
19. c
20. 95

## Chapter 6

1. a
2. positive
3. c
4. 70
5. c
6. −5
7. b
8. $500 + 100x$
9. b
10. 702.26
11. b
12. 0.86
13. c
14. $0.02x + 3$
15. b
16. 0.96
17. a
18. $-0.88 + 0.05x$
19. b
20. (8,3)

## Chapter 7

1. b
2. 52
3. c
4. more
5. c
6. 25
7. a
8. 6694
9. b
10. 40
11. a
12. observational
13. b
14. placebo
15. c
16. 0.35
17. b
18. 0.065 (or 6.5%)
19. a
20. 0.06 (or 6%)

## Chapter 8

1. a
2. 24
3. b
4. 0.85
5. a
6. $\frac{1}{18}$
7. b
8. $\frac{3}{10}$
9. b
10. 24
11. b
12. 0.3038
13. c
14. 0.4
15. b
16. less
17. a
18. 1511
19. c
20. 19.4

## Chapter 9

1. b
2. if any two voters exchange ballots, the election outcome is unchanged
3. b
4. a switch in a ballot from being a vote for the loser to being a vote for the winner doesn't change the election outcome
5. c
6. defeats every other candidate in a one-on-one contest
7. d
8. sometimes produces no winner at all
9. c
10. receives the most first-place votes
11. a
12. has the highest Borda score
13. a
14. reverses the order in which this non winner and the winner were ranked
15. c
16. one-on-one contests take place according to an ordering of the candidates called an "agenda"
17. b
18. they are not monotone
19. c

20. satisfies the Condorcet winner criterion and independence of irrelevant alternatives, and always produces at least one winner in every election

## Chapter 10

1. c
2. Borda
3. c
4. either an insincere ballot or a disingenuous ballot
5. a
6. monotonicity
7. b
8. treats both candidates equally and all voters equally
9. b
10. there are only three candidates
11. c
12. placing the additional *j* candidates at the bottom of each ballot (in any order whatsoever)
13. c
14. agenda manipulation
15. a
16. group manipulation
17. d
18. manipulable
19. a
20. the chair has the most power, but fares the worst

## Chapter 11

1. c
2. 11
3. c
4. *C*
5. b
6. $\frac{1}{4}$
7. b
8. 720
9. c
10. the voters with weights 3 and 4
11. b
12. 256; The motion is defeated
13. a
14. 20
15. c
16. 8
17. a
18. $2^n$
19. b
20. $\{A,B,C\}$, $\{A,B,D\}$, $\{A,C,D\}$

## Chapter 12

1. b
2. *B*
3. c
4. just to the left or just to the right of *M*
5. c
6. *C*
7. a
8. only one candidate can win, and a median choice is not too far away from anybody, and two department stores at ends of a main street are closer to most consumers than one in the center
9. c
10. when he or she is one of the top two candidates identified by the poll
11. b
12. *B*
13. a
14. less power than voters in large toss-up states
15. b
16. about three times
17. c
18. it favors citizens who live in the largest states
19. b
20. the law mandates that the popular-vote winner wins if states with a majority of electoral votes pass it

## Chapter 13

1. a
2. reflects the relative worth of each issue to that party
3. a
4. the transfer of items (or parts thereof) from one party to the other until points are equalized
5. d
6. the boat, car, and part of the land
7. b
8. cash only
9. b
10. never willingly choose his or her least-preferred item, and avoid wasting a choice on an item that he or she knows will remain available and can be chosen later
11. b
12. no other player received more than he or she did
13. a
14. each nondivider receives a portion that he or she has approved
15. b
16. leaves the game
17. a
18. one player separates the remainders into two portions and the other chooses
19. b
20. the first player

## Chapter 14

1. a
2. 352; 44; 4.545; 2.273; 1.182
3. c
4. 14; 13; 17; 17; 19
5. b
6. 300
7. c
8. 0; 0; 100
9. b
10. 1; 1; 98
11. a
12. 0.5
13. a
14. 2; 3
15. a
16. Hill–Huntington
17. c
18. Jefferson
19. c
20. Webster

## Chapter 15

1. c
2. third
3. b
4. third
5. a
6. Three
7. a
8. the value of the saddlepoint
9. c
10. prevent a player from being exploited by always choosing a pure strategy
11. a
12. each player's strategic choices are the same
13. c
14. to expect more fastballs than curves
15. b
16. more often kick side and break side
17. b
18. it yields greater payoffs
19. b
20. players must think ahead about what moves are optimal in the future in order to make optimal choices in the present

## Chapter 16

1. a
2. 1
3. a
4. 9
5. b
6. 0
7. b
8. 10
9. b
10. 9
11. c
12. 9
13. b
14. 11
15. c
16. 100; 100
17. a
18. 20001-2800-7

19. c
20. 3765

## Chapter 17

1. b
2. 1011
3. b
4. 3
5. a
6. one
7. a
8. 3
9. b
10. 0010100
11. a
12. 3
13. c
14. MATH
15. c
16. 29
17. c
18. either $P$ or $Q$ is true
19. a
20. 00111001

## Chapter 18

1. c
2. 120
3. a
4. $6.40
5. b
6. $100^{1/3} \approx 4.6$
7. c
8. approximately 26.22 mi
9. c
10. 60
11. a
12. 19
13. c
14. 187
15. b
16. 4
17. c
18. 10
19. a (or b)
20. 1.86

## Chapter 19

1. d
2. patterns
3. a
4. 12
5. c
6. 34
7. b
8. H; I; N; O; S; X; Z
9. b
10. translation; rotation
11. c
12. translation; reflection
13. b
14. half-turn rotation
15. c
16. translation
17. a
18. reflection
19. a
20. infinitely many

## Chapter 20

1. b (or c)
2. 45°
3. c
4. 3; 4; 6
5. a
6. six
7. c
8. five
9. a
10. seven
11. b
12. Conway; translations; half-turns
13. b
14. translations and half-turns
15. b
16. quasi
17. b
18. the golden ratio
19. b
20. fivefold

## Chapter 21

1. a (or c)
2. $15
3. c
4. $8396.19
5. c
6. $49.09
7. b
8. $1186.00
9. b
10. $182.08
11. c
12. 6.06%
13. a
14. 5.13%
15. b
16. $1491.82
17. c
18. a sinking fund
19. b
20. $9216

## Chapter 22

1. c
2. $1219.89
3. c
4. $2543.20
5. a
6. annual percentage rate
7. c
8. 19.56%
9. b
10. 18%
11. c
12. $135.47
13. b
14. $456.33
15. c
16. $1\frac{1}{4}$ to $1\frac{1}{2}$
17. c
18. $500
19. b
20. $13.55 million if you get the first payment right away or $12.78 million if you have to wait a year for the first payment

## Chapter 23

1. b
2. rising
3. c
4. exponential
5. a
6. 32
7. b
8. 241
9. c
10. 0.75
11. c
12. 0.72
13. b
14. 0.36
15. a
16. 0 or 1
17. a
18. sensitivity to initial conditions
19. a
20. greed, chance, or chaotic variation

# ANSWERS TO ODD-NUMBERED EXERCISES

## Chapter 1

**1. (a)** 8
**(b)** 12
**(c)** *A*: 3; *B*: 2; *C*: 3; *D*: 2; *E*: 4; *F*: 4; *G*: 3; *H*: 3
**(d)** *A*, *D*, and *F*
**(e)** *E*, *G*, and *H*

**3. (a)** No
**(b)** *EC*, *AD*, *BD*, and *AC*
**(c)** 5; 6

**5.** *E*: 0; *A*: 1; *H*, *D*, and *G*: 2; *B* and *F*: 3; *C*: 5

**7. (a)** *BCGDFB*
**(b)** **(i)** *BD*; *BFD*; **(ii)** *CBF*; *CGDF*; *CGDBF*
**(c)** *GDBCG*

**9. (a)** 4; 4
**(b)** 7; 6
**(c)** 10; 14

**11.** 2

**13.** Drawings can vary. Possible renderings for **(a)** and **(b)** include the following:
**(a)**
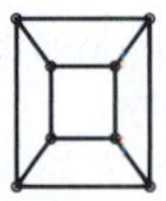
**(b)**
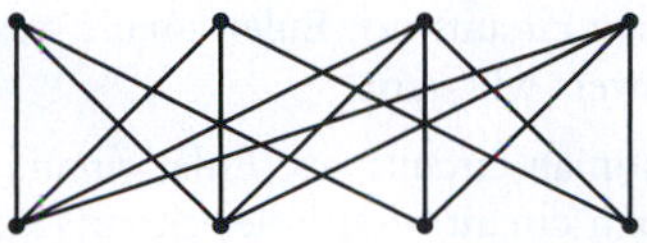
**(c)** Yes

**15.** Drawings can vary. Possible renderings include the following:
**(a)**

**(b)**
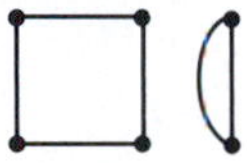

**17.** Drawings can vary. Possible renderings include the following:
**(a)**

**(b)**
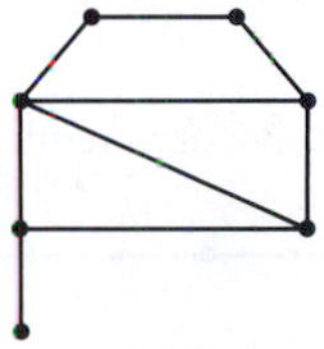

**19.** (a) Not all edges are traveled by worker; (b) end of route not the same as beginning of route; not realistic; no Euler circuit in graph

**21.** Since this graph is connected and even-valent, it has an Euler circuit.

**23.**
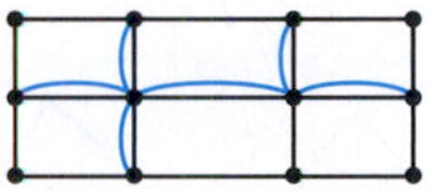

**25. (a)** 3; **(b)** and **(c)** Answers will vary.

**27.** Do not choose edge 2, but edges 1 or 10 could be chosen.

**29.** Answers will vary.

**31.** 2

**33.** Answers will vary; no

**35.** Answers will vary. Possible answers include *AECDABDCBEA*.

**37.** Drawings can vary. Possible renderings for **(a)**, **(b)**, and **(c)** include the following:
**(a)**
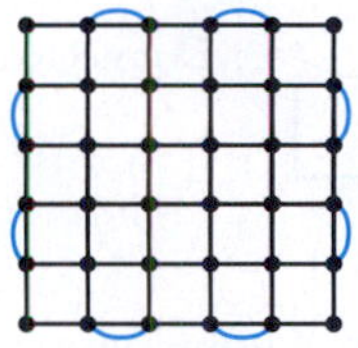
**(b)**
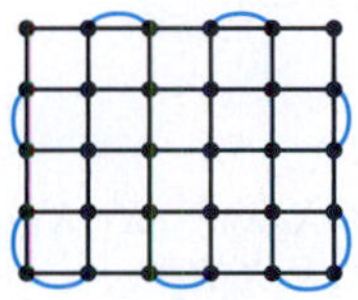
**(c)**
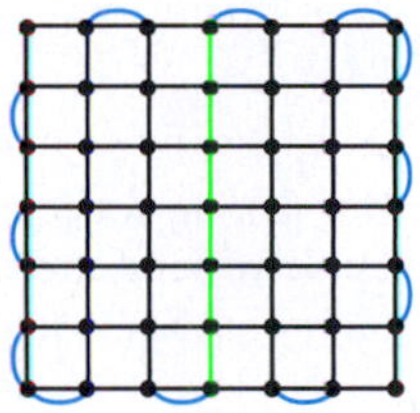
**(d)** Yes; no

**39. (a)** 2
**(b)** Yes

(c) 2
(d) No

41. Answers will vary.

43. Drawings can vary. Possible renderings include the following:

45. There are many circuits that achieve a minimum length of 44,000 feet.

47. (b) and (c); Additional answers will vary.

49. Answers will vary.

51. (a) Drawings can vary. Possible renderings include the following:

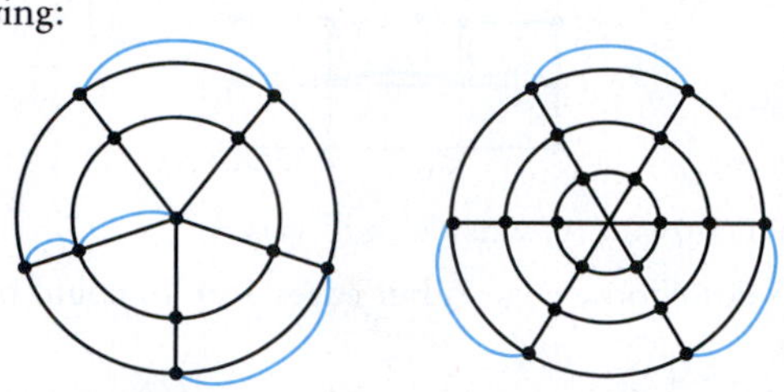

(b) The best eulerization for the four-circle, four-ray case adds two edges.
(c) Answers will vary.

53. Yes

55. Answers will vary.

57. 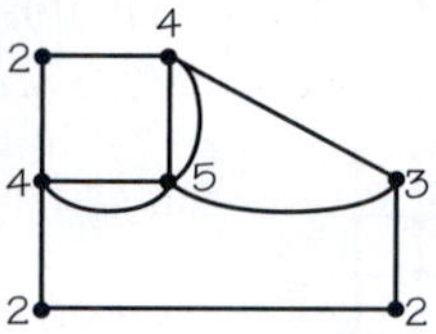

; Connected

59. Answers will vary.

61. Answers will vary. Possible answers include *ABDEFBEBFEDBACDCA*.

## Chapter 2

1. (a) $X_5X_6X_1X_3X_4X_2X_5$
(b) $X_5X_4X_3X_2X_1X_6X_7X_8X_9X_{10}X_{11}X_{12}X_5$
(c) $X_5X_4X_3X_1X_2X_7X_6X_9X_8X_5$

3. (a) Yes
(b) Yes
(c) Yes

5. (a) No for (a); yes for (b); no for (c)
(b) No longer be possible to send messages between these two sites

7. Answers will vary.

9. (a) Yes for both.
(b) Answers will vary.
(c) Add edges $X_2X_8$, $X_8X_6$, $X_6X_4$, and $X_4X_2$.

11. Drawings can vary. Possible renderings include the following:

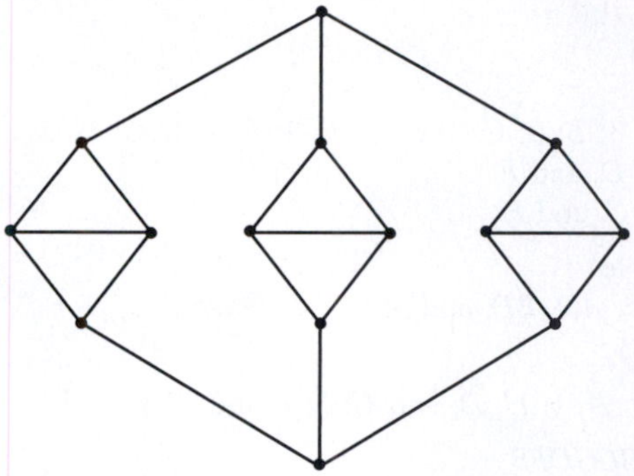

13. (a) No
(b) No

15. (a) Yes
(b) No
(c) No

17. (a) Yes
(b) No
(c) Answers will vary.

19. (a) Hamiltonian circuit: yes; Euler circuit: yes; Additional answers will vary.
(b) Hamiltonian circuit: yes; Euler circuit: yes
(c) Hamiltonian circuit: yes; Euler circuit: no
(d) Hamiltonian circuit: no; Euler circuit: yes; Additional answers will vary.

21. (a) Hamiltonian circuit: yes; Euler circuit: no
(b) Hamiltonian circuit: yes; Euler circuit: no
(c) Hamiltonian circuit: yes; Euler circuit: no
(d) Hamiltonian circuit: no; Euler circuit: no

23. (a) Drawings will vary.
(b) Drawings will vary.
(c) Answers will vary.

25. (a) 2520
(b) 16,807
(c) 16,800

27. (a) 15,120
(b) 17,576

29. Yes; 172

31. (a) 17,558,424
(b) Answers will vary.

33. 10,000,000; 900

35. Drawings will vary.; 6, 10, and 15 edges, respectively; $\frac{n(n-1)}{2}$ edges; 3, 12, and 60, respectively

**37.** **(a)** Possible drawings include the following:

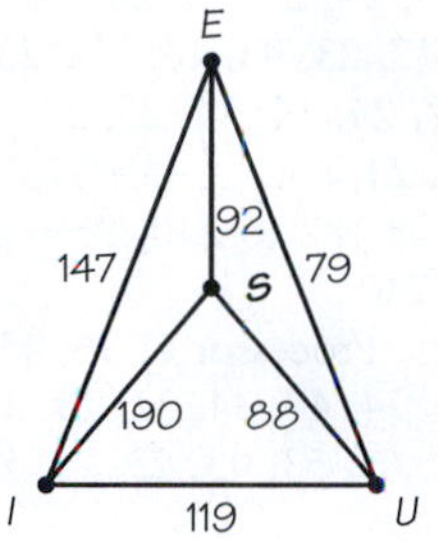


**(b)** Tour (1): *UISEU*: 480
Tour (2): *USIEU*: 504
Tour (3): *UIESU*: 446
**(c)** Tour (3)
**(d)** No
**(e)** Tour (1); yes; yes; no
**(f)** Tour (2); no

**39.** *FMCRF*

**41.** *MACBM*

**43.** A traveling salesman problem

**45.** Yes; Hamiltonian circuit; Chinese postman problem; Answers will vary and requires at least 9 reuses of edges.

**47.** Answers will vary.

**49.** The optimal tour is the same but its cost is now 4700.

**51.** Diagram (a): **(a)** There is a circuit and wiggled edges do not include all vertices. **(b)** The circuit does not include all the vertices of the graph.
Diagram (b): **(a)** The tree does not include all vertices of the graph. **(b)** Not a circuit
Diagram (c): **(a)** Not a tree **(b)** Not a circuit
Diagram (d): **(a)** Not a tree **(b)** Not a circuit

**53.** **(a)** 1, 2, 3, 4, 5, 8; Cost is 23.
**(b)** 1, 1, 1, 2, 2, 3, 3, 4, 5, 6, 6; Cost is 34.
**(c)** 1, 1, 1, 2, 2, 2, 2, 2, 3, 3, 3, 3, 4, 4, 4, 5, 5, 6, 7; Cost is 60.
**(d)** 1, 2, 2, 3, 3, 3, 4, 5, 5, 5, 6, 6; Cost is 45.

**55.** 27; 27; at least 26

**57.** Yes

**59.** Yes; Additional answers will vary.

**61.** Yes; yes

**63.** There are three different trees with the same cost.

**65.** **(a)** True
**(b)** False (unless all the edges of the graph have the same weight)
**(c)** True
**(d)** False
**(e)** False

**67.** **(a)** Answers will vary for each edge.
**(b)** 5; one less than the number of vertices in the graph
**(c)** No (*CD* must be included.)

**69.**

| | A | B | C | D |
|---|---|---|---|---|
| A | 0 | 16 | 13 | 5 |
| B | 16 | 0 | 19 | 11 |
| C | 13 | 19 | 0 | 8 |
| D | 5 | 11 | 8 | 0 |

**71.** **(a)** 22; $T_3T_2T_5$
**(b)** 30; $T_3T_5T_7$

**73.** $T_1$, $T_5$, and $T_7$; 28; $T_1$, $T_4$, and $T_7$

**75.** Answers will vary.

**77.** Drawings can vary. Possible renderings include the following:

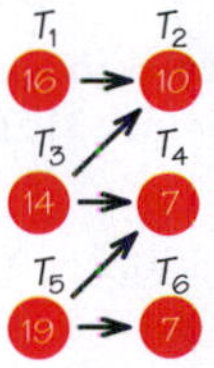


## Chapter 3

**1.** Answers will vary.

**3.** Answers will vary.

**5.** **(a)** Processor 1: $T_1$, $T_2$, $T_3$, $T_5$, $T_7$; Processor 2: Idle 0 to 2, $T_4$, $T_6$, idle 4 to 5
**(b)** Processor 1: $T_1$, $T_2$, $T_3$, $T_6$, $T_7$; Processor 2: Idle 0 to 2, $T_4$, $T_5$, idle 4 to 5
**(c)** Yes
**(d)** No
**(e)** $T_3$ and $T_5$

**7.** **(a)** **(i)** Processor 1: $T_1$ from 0 to 13, $T_3$ from 13 to 25, $T_6$ from 25 to 45; Processor 2: $T_2$ from 0 to 18, $T_4$ from 18 to 27, $T_5$ from 27 to 35, idle from 35 to 45
**(ii)** Processor 1: $T_1$ from 0 to 13, $T_3$ from 13 to 25, $T_4$ from 25 to 34, $T_5$ from 34 to 42; Processor 2: $T_2$ from 0 to 18, $T_6$ from 18 to 38, idle from 38 to 42
**(b)** Yes
**(c)** $T_2$, $T_6$, and 38; Sum of the task times divided by 2 is 40.

**9.** **(a)** Yes
**(b)** No

**11.** **(a)** Processor 1: $T_1$, $T_6$, idle 15 to 21, $T_7$, idle 27 to 31; Processor 2: $T_2$, $T_5$, $T_8$; Processor 3: $T_3$, $T_4$, idle from 13 to 31
**(b)** Processor 1: $T_1$, $T_6$, idle 15 to 21, $T_7$, idle 27 to 31; Processor 2: $T_3$, $T_4$, idle from 13 to 21, $T_8$; Processor 3: $T_2$, $T_5$, idle from 21 to 31
**(c)** Processor 1: $T_4$, idle 10 to 11, $T_6$, idle 18 to 21, $T_8$; Processor 2: $T_2$, $T_5$, $T_7$, idle 27 to 31; Processor 3: $T_1$, $T_3$, idle 11 to 31

**13.** Answers will vary.

**15.** Yes

**17.** **(a)** No
**(b)** $T_2$ should have been scheduled at time 0.
**(c)** Use the digraph with no edges and the list: $T_2$, $T_1$, $T_3$, $T_4$, $T_5$.

**19.** **(a)** $T_1$, $T_2$, $T_3$, and $T_6$
**(b)** No tasks require that $T_1$ and $T_6$ be done before these other tasks can begin.
**(c)** $T_6$
**(d)** Processor 1: $T_1$, $T_6$; Processor 2: $T_2$, $T_4$, idle from 18 to 30; Processor 3: $T_3$, $T_5$, idle from 12 to 30
**(e)** No
**(f)** Processor 1: $T_6$, idle from 20 to 22; Processor 2: $T_3$, $T_5$, $T_1$; Processor 3: $T_2$, $T_4$, idle from 18 to 22
**(g)** Yes
**(h)** Yes

**21.** **(a)** 120
**(b)** No; $T_1$ must be assigned to the first machine at time 0.
**(c)** No; Since when 2 divides 31, there is a remainder of 1.
**(d)** No

**23.** Yes

**25.** Answers will vary.

**27.** No

**29.** **(a)** Task times: $T_1 = 3$, $T_2 = 3$, $T_3 = 2$, $T_4 = 3$, $T_5 = 3$, $T_6 = 4$, $T_7 = 5$, $T_8 = 3$, $T_9 = 2$, $T_{10} = 1$, $T_{11} = 1$, and $T_{12} = 3$. This schedule would be produced from the list: $T_1$, $T_3$, $T_2$, $T_5$, $T_4$, $T_6$, $T_7$, $T_8$, $T_{11}$, $T_{12}$, $T_9$, $T_{10}$.
**(b)** Task times: $T_1 = 3$, $T_2 = 3$, $T_3 = 3$, $T_4 = 2$, $T_5 = 2$, $T_6 = 4$, $T_7 = 3$, $T_8 = 5$, $T_9 = 8$, $T_{10} = 4$, $T_{11} = 7$, $T_{12} = 9$, and $T_{13} = 3$. This schedule would be produced from the list: $T_1$, $T_5$, $T_7$, $T_4$, $T_3$, $T_6$, $T_{11}$, $T_8$, $T_{12}$, $T_9$, $T_2$, $T_{10}$, $T_{13}$.

**31.** **(a)** **(i)** Processor 1: $T_1$, $T_3$, $T_5$, $T_7$, idle from 16 to 20; Processor 2: $T_2$, $T_4$, $T_6$, $T_8$
**(ii)** Processor 1: $T_8$, $T_5$, $T_4$, $T_1$; Processor 2: $T_7$, $T_6$, $T_3$, $T_2$
**(b)** Yes

**33.** Answers will vary.

**35.** In part (a), 33 is not exactly divisible by 4; In part (b) 56 is not exactly divisible by 5.

**37.** **(a)** **(i)** Machine 1: 12, 9, 15, idle from 36 to 50; Machine 2: 7, 10, 13, 20
**(ii)** Machine 1: 12, 13, 20; Machine 2: 7, 9, 15, 10, idle from 41 to 45
**(iii)** Machine 1: 20, 12, 9, idle from 41 to 45; Machine 2: 15, 13, 10, 7
**(b)** An optimal schedule is possible.
**(c)** The critical path list is $T_6$, $T_5$, $T_4$, $T_1$, $T_7$, $T_2$, $T_3$, using the first processor.

**39.** **(a)** Machine 1: 129; Machine 2: 129
**(b)** Machine 1: 123; Machine 2: 123
**(c)** Yes

**41.** **(a)** Processor 1: 12, 13, 45, 34, 63, 43, 16, idle 226 to 298; Processor 2: 23, 24, 23, 53, 25, 74, 76; Processor 3: 32, 23, 14, 21, 18, 47, 23, 43, 16, idle 237 to 298
**(b)** Processor 1: 12, 24, 14, 34, 25, 23, 16, 16, 76; Processor 2: 23, 23, 21, 63, 43, idle 173 to 240; Processor 3: 32, 23, 53, 74, idle 182 to 240; Processor 4: 13, 45, 18, 47, 43, idle 166 to 240
**(c)** Three machines: Processor 1: 76, 45, 43, 24, 23, 18, 16, 13; Processor 2: 74, 47, 34, 32, 23, 21, 14, 12, idle 257 to 258; Processor 3: 63, 53, 43, 25, 23, 23, 16, idle 246 to 248
Four machines: Processor 1: 76, 43, 24, 23, 16, idle 182 to 194; Processor 2: 74, 43, 25, 23, 16, 13; Processor 3: 63, 45, 32, 23, 18, 12, idle 193 to 194; Processor 4: 53, 47, 34, 23, 21, 14, idle 192 to 194
**(d)** Processor 1: 84, 45, 43, 25, 23, 23, 16, 12; Processor 2: 82, 55, 34, 32, 23, 18, 14, 13; Processor 3: 71, 61, 43, 24, 23, 21, 16, idle 259 to 271

**43.** Answers will vary.

**45.** Each task heads a path of length equal to the time to do that task.

**47.** 9; Number of bins would not change, but the placement of the items in the bins would differ.

**49.** **(a)** 17
**(b)** 16
**(c)** 16
**(d)** 13

**51.** Yes, both are acceptable.

**53.** **(a)** Answers will vary.
**(b)** It is possible.

**55.** No; yes

**57.** Answers will vary.

**59.** Answers will vary.

**61.** **(a)** 152 min; 124 min
**(b)** 155 min; 120 min
**(c)** Yes; five-processor decreasing-time schedule
**(d)** 11
**(e)** NFD: 13; WFD: 11
**(f)** An optimal packing with 10 bins exists.

**63.** **(a)** Answers will vary.
**(b)** Answers will vary.
**(c)** Packing rectangles of width 1 in an $m \times 1$ rectangle is a special case of the two-dimensional problem.
**(d)** Answers will vary.

**65.** Answers will vary.

**67.** **(a)** Graph (a) no; (b) yes; (c) no; (d) yes; (e) yes; (f) yes
**(b)** Graph (a) yes; (b) yes; (c) no; (d) yes; (e) yes; (f) yes
**(c)** Graph (a) 4; (b) 3; (c) 5; (d) 3; (e) 2; (f) 2

**69.** **(a)** Drawings can vary. Possible renderings include the following:

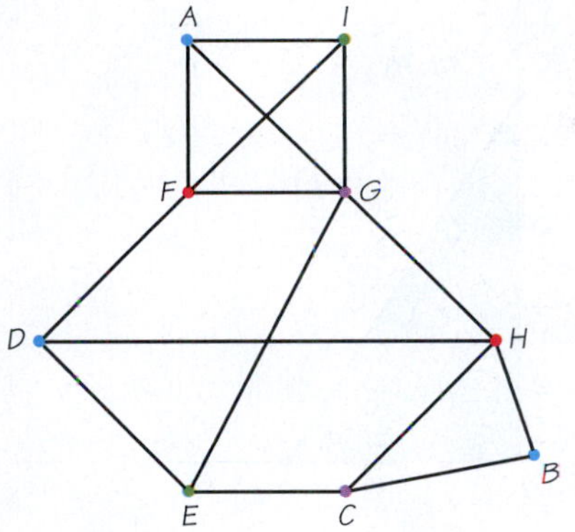


**(b)** 4

**(c)** The coloring in **(a)** indicates one possible arrangement.

**71.** **(a)** Drawings can vary. Possible renderings include the following:

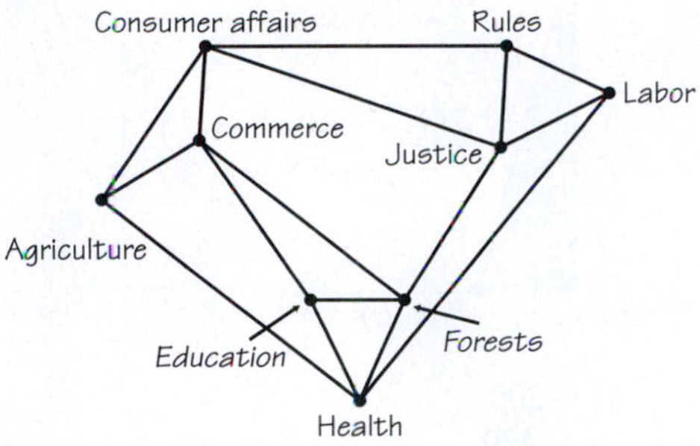


**(b)** 3

**(c)** 3; Additional answers will vary.

**73.** **(a)** Drawings can vary. Possible renderings include the following.

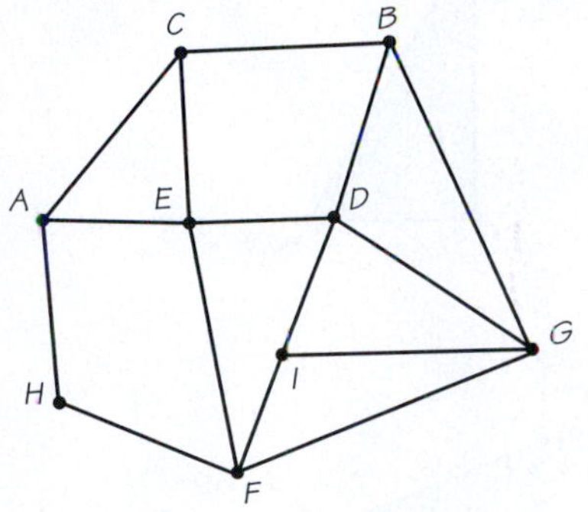


**(b)** 3

**(c)** 3

**75.** Answers will vary.

**77.** Graph (a) 6; (b) 8; (c) 6; (d) 3; (e) 3; (f) 4; Minimum is either the maximal valence of any vertex or one more than the maximal valence.

**79.** **(a)** Graph (a) 4; (b) 2; (c) 4; (d) 4; (e) 2; (f) 3

**(b)** Answers will vary.

**81.** 3

**83.** 3; only if each child formed his/her own play group

## Chapter 4

**1.** **(a)**

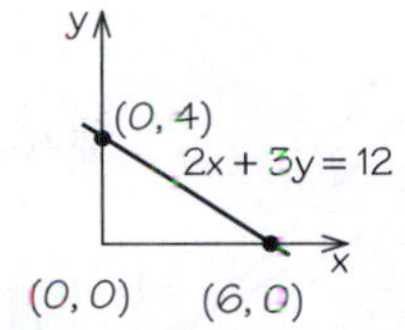


**(b)**

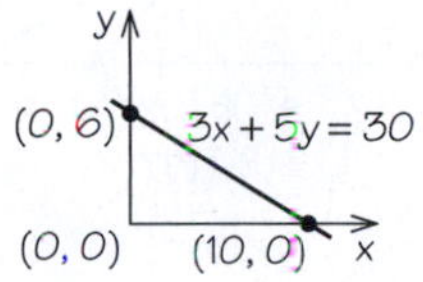


**(c)**

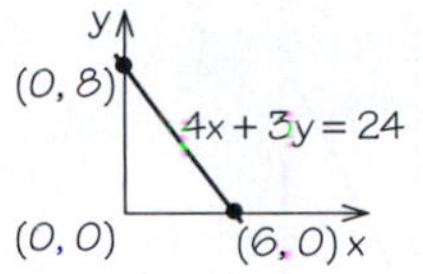


**3.** **(a)**

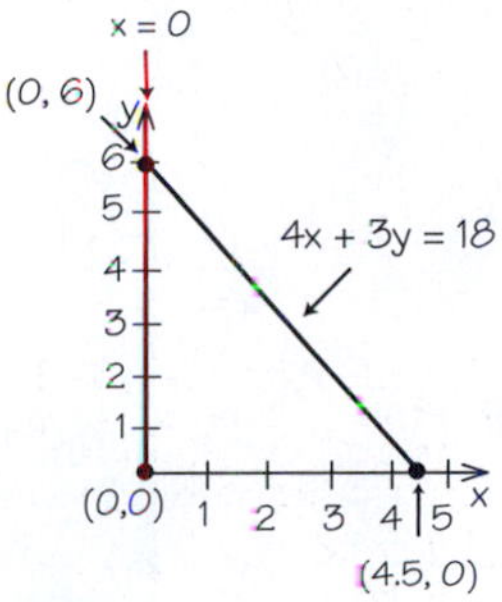


**(b)**

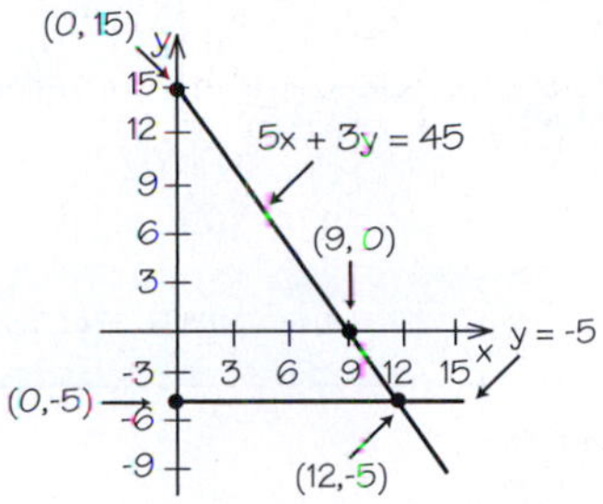


**(c)**

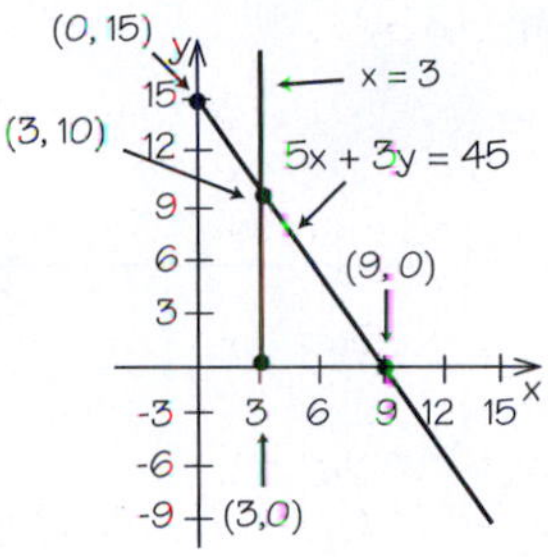

Note: For Exercises 5 and 7, first quadrant only is shown. Point of intersection is labeled.

**5. (a)**

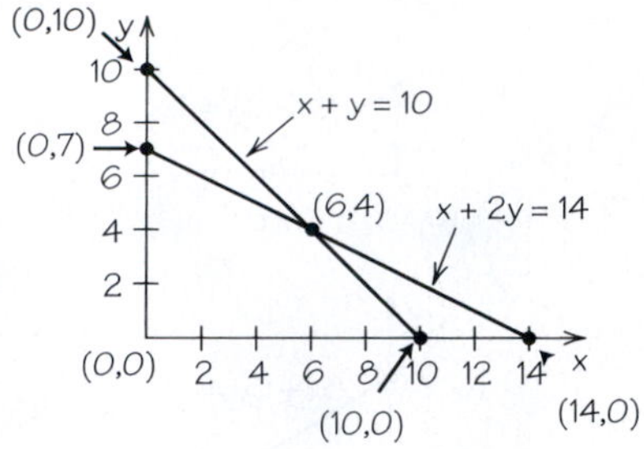


**(b)**

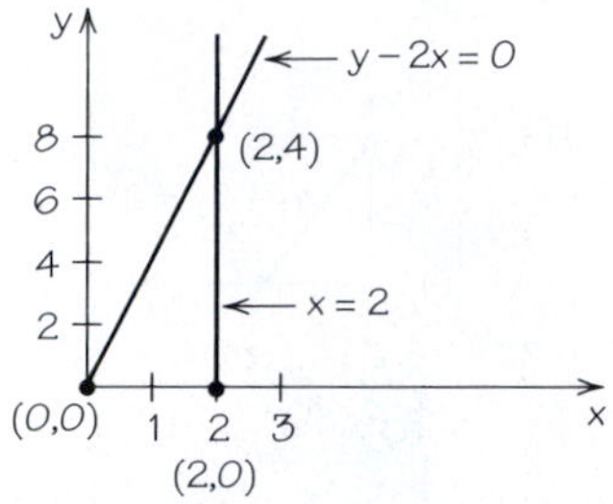


**7. (a)**

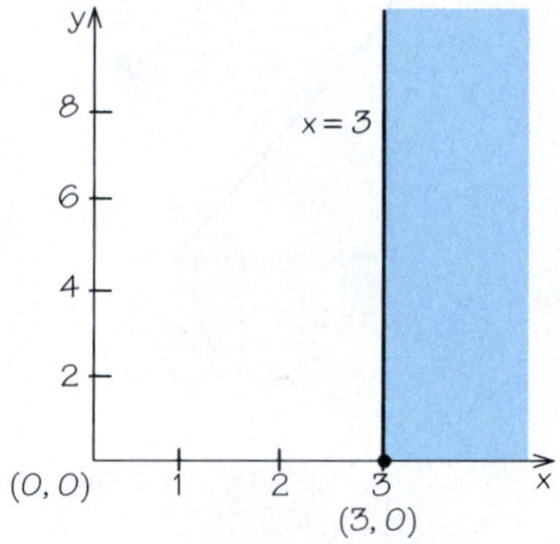


**(b)**

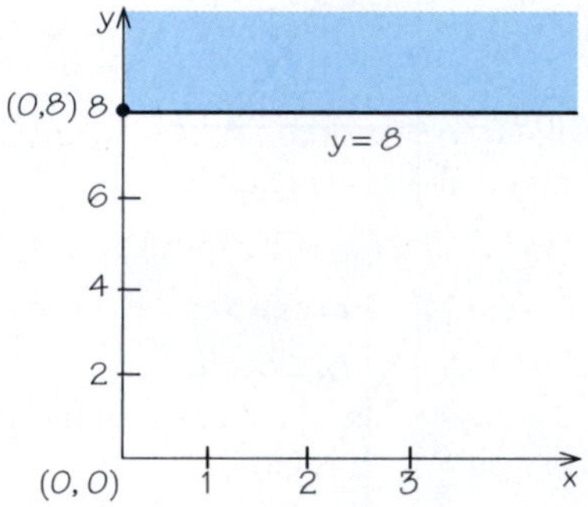


**(c)**

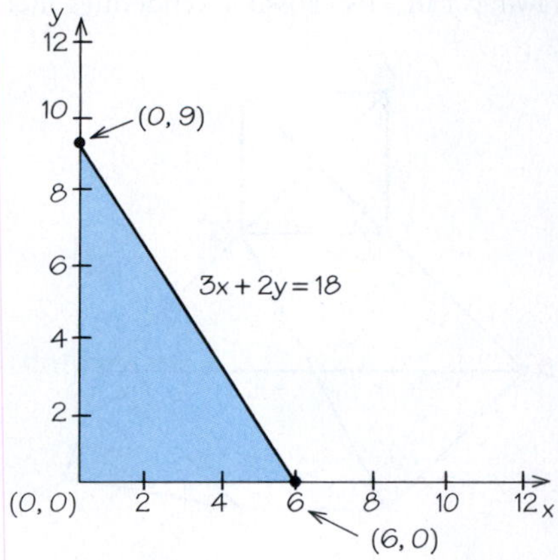


**(d)**

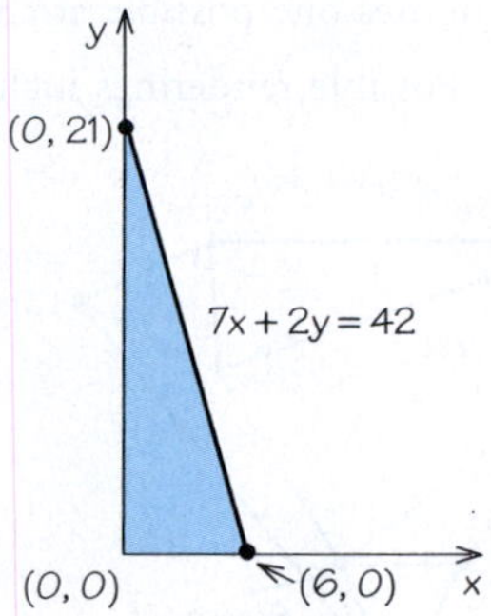


**9. (a)** $6x + 4y \le 300$
**(b)** $30x + 72y \le 420$

**11.**

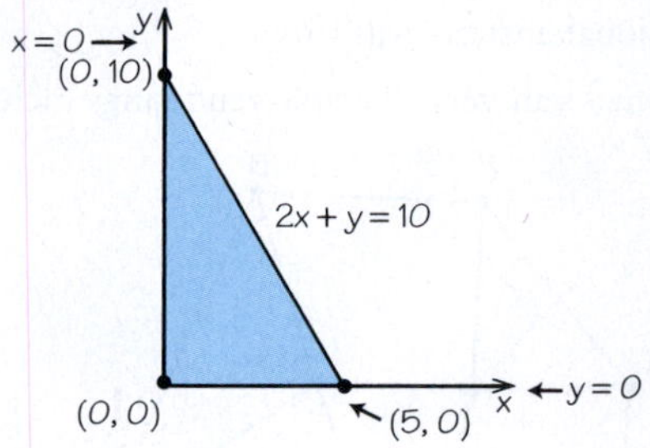


**13.**

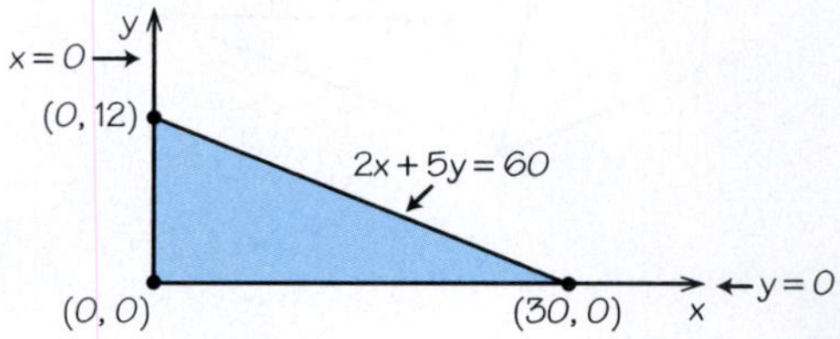

**15.**

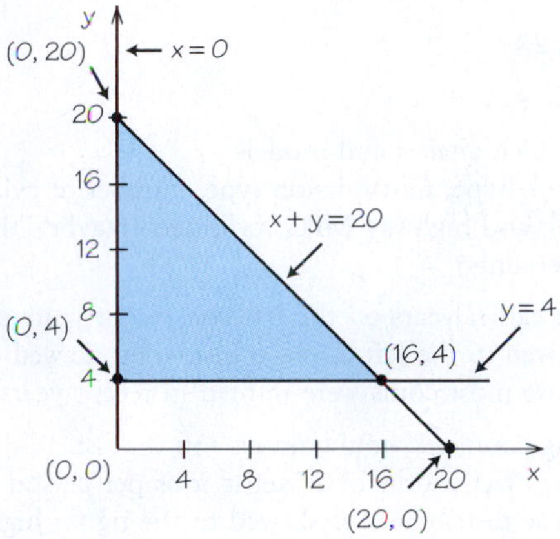


**17.** Exercise 11: (2, 4): yes; (10, 6): no; Exercise 13: (2, 4): yes; (10, 6): yes; Exercise 15: (2, 4): yes; (10, 6): yes

**19.** Make 0 skateboards and 30 dolls for a profit of $111.

**21.** Note: These situations are shown only for the first quadrant.

**(a)**

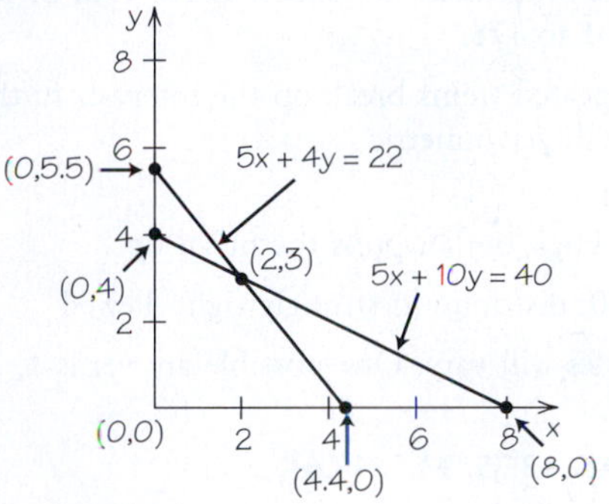


**(b)**

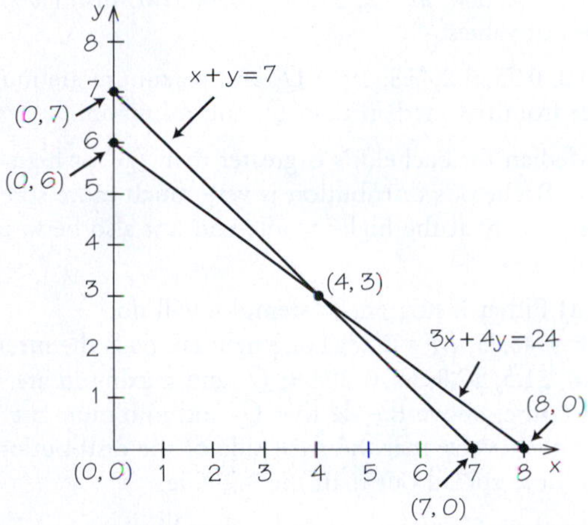


**23.**

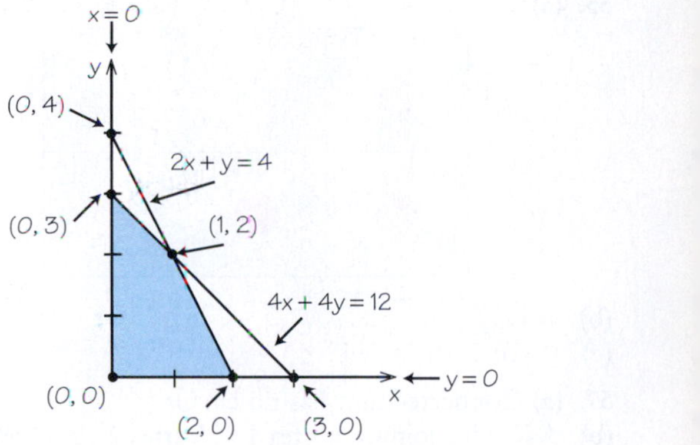


**25.**

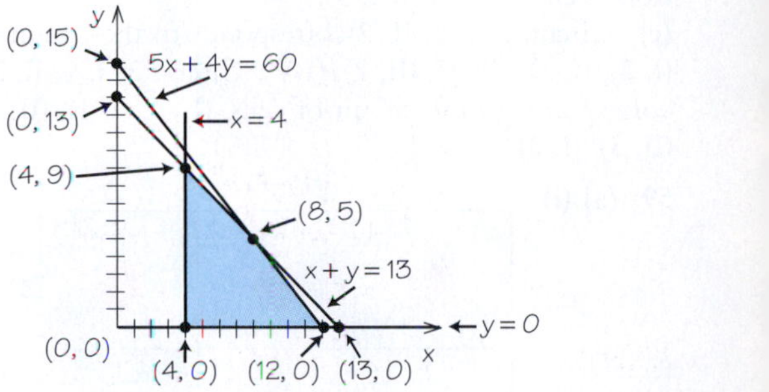


**27.** 28

**29.** 38

**31.** **(a)** (2, 0)
**(b)** It is not.
**(c)** Profit at (2, 0) greater than the profit at $\left(\frac{13}{7}, 0\right)$
**(d)** Yes; Profit at $R$ is less than the profit at $Q$.
**(e)** Answers will vary.

**33.** Schedule 400 oil changes and no tune-ups.; Schedule 300 oil changes and 20 tune-ups.

**35.** Schedule 360 routine visits and no comprehensive visits.; Schedule 210 routine visits and 30 comprehensive visits.

**37.** Take four math courses and no other courses.; Take two math courses and 2 other courses.

**39.** Make 2 grade A and 5 grade B batches in both cases.

**41.** Make 3000 cartons of regular and 2000 cartons of diet in both cases.

**43.** Make no desk lamps and 1200 floor lamps.; Make 150 desk lamps and 1080 floor lamps.

**45.** Make 50 chairs, 10 tables, and no beds each month.

**47.** Make 470 pounds of Excellent, none of Southern, 2400 pounds of World, and 320 pounds of Special.

**49.** 43

**51.** Make 60 business and no charity calls.; Make 45 business and 10 charity calls.

**53.** Make 3 bikes and 2 wagons in both cases.

**55. (a)**

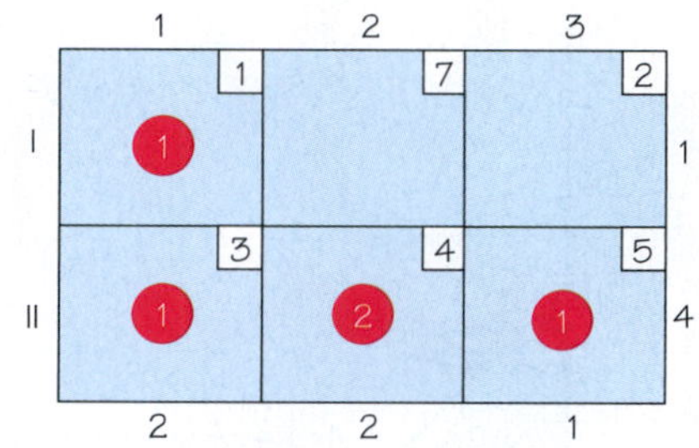


**(b)** 17
**(c)** (I, 2): 5; (I, 3): −1

**57. (a)** Connected and has no circuit
**(b)** Add edge joining Vertex I to Vertex 2.; Add edge from Vertex I to Vertex 3.
**(c)** Circuit 2, I, 1, II, 2 corresponds to the circuit of cells, (I, 2), (I, 1), (II, 1), (II, 2), (I, 2). Circuit 3, I, 1, II, 3 corresponds to the circuit of cells, (I, 3), (I, 1), (II, 1), (II, 3), (I, 3).

**59. (a) (i)**

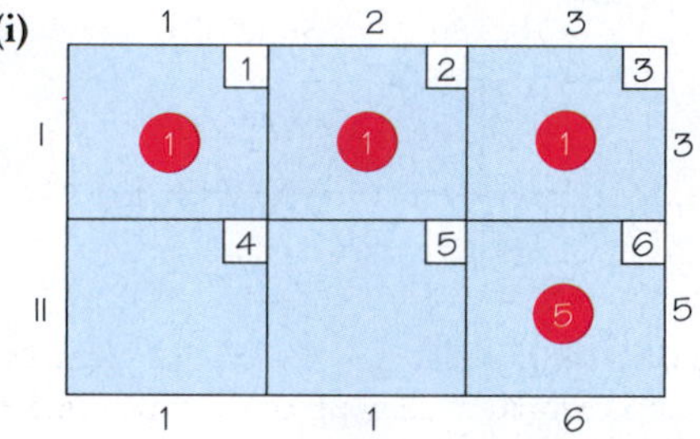


**(ii)**

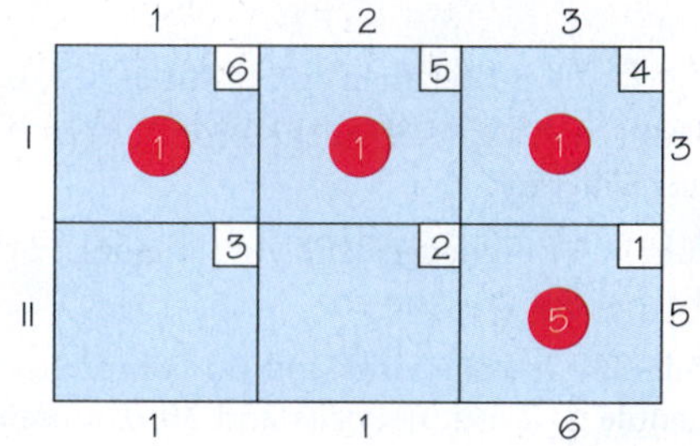


**(iii)**

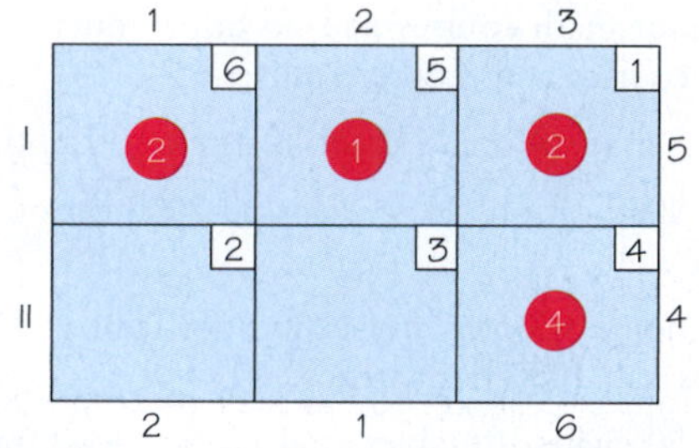


**(b)** For both **(i)** and **(ii)** the tableaux shown are optimal. However, there are also other optimal tableaux. For **(iii)** the tableau shown is not optimal. Using the stepping stone algorithm, the cost can be reduced to 16.

**61. (i) (a)** 33
**(b)** No
**(c)** Cost: 24
**(ii) (a)** 37
**(b)** No
**(c)** Cost: 28

## Chapter 5

**1. (a)** Vehicle makes and models
**(b)** Vehicle type, transmission type, number of cylinders, city MPG, and highway MPG; cylinders (maybe), the two MPGs (certainly)

**3.** Given earlier years on the left and recent years on the right, skewed to the left; Draw a histogram skewed to the left because most coins were minted in recent years.

**5. (a)** Big countries would always top the list.
**(b)** Using class widths of 2 metric tons per person starting 0.0–1.9, the distribution is skewed to the right.; high outliers: Canada, Australia, and United States

**7. (a)** Alaska: 5.7%, Florida: 17.6%.
**(b)** Single-peaked and roughly symmetric, center near 12.7%, spread from 8.5% to 15.6%

**9.** There is one high outlier, 200. The center of the 17 observations other than the outlier is 137 (9th of 17). The spread is 101 to 178.

**11. (a)** Repeated stems break up the intervals further.
**(b)** Reasonably symmetric

**13. (a)** 141.1
**(b)** 137.6; High outlier pulls the mean up.

**15.** $66,570; distribution strongly right-skewed

**17.** Examples will vary. One possible answer is 1, 2, 2, 2, 3, 3, 4, 17.

**19.** 5.7, 11.7, 12.75, 13.5, 17.6

**21.** 5799, 20000, 25942, 34986, 36700; two distinct clusters of values

**23.** 0.0, 0.75, 3.2, 7.8, 19.9; $Q_3$ and maximum are much farther from the median than $Q_1$ and minimum.

**25.** Median for bachelor's is greater than $Q_3$ for high school. Bachelor's distribution is very much more spread out, especially at the high-income end but also between the quartiles.

**27. (a)** Either histogram or stemplot will do.
**(b)** $\bar{x} = 48.25$, $M = 37.8$; Long right tail pulls the mean up.
**(c)** 2.0, 21.5, 37.8, 60.1, 204.9; $Q_3$ and maximum are much farther above the $M$ that $Q_1$ and minimum are below it, showing that the right side of the distribution is much more spread out than the left side.

**29.** Arizona, California, Nevada, New Mexico, and Texas

**31. (a)** $\bar{x} = 5.448$, $s = 0.221$
**(b)** $M = 5.46$; yes

**33.** For both datasets, $\bar{x} = 7.50$ and $s = 2.03$. Data A have two low outliers and Data B have one high outlier.

**35. (a)** $s = 0$ is smallest possible. Examples will vary. One possible answer is 1, 1, 1, 1.

**(b)** 0, 0, 10, 10
**(c)** Yes
**(d)** No

**37.** About 27

**39.** Mean: $A$, median: $B$

**41. (a)** 327 to 345 days
**(b)** 16%

**43.** $Q_1 = 1381$, $Q_3 = 1641$

**45. (a)** −20.51% to 41.53%
**(b)** 20.51% or greater

**47. (a)** Median = 10% by symmetry
**(b)** 9.6% to 10.4%
**(c)** 9.866% to 10.134%

**49. (a)** 50%, 2.5%
**(b)** 0.37 to 0.43

**51.** Red: somewhat right-skewed (with no outliers), yellow: quite symmetrical (with no outliers)

**53.** Red: $\overline{x} = 39.71$, $s = 1.799$; yellow: $\overline{x} = 36.18$, $s = 0.975$; yellow distribution

**55.** Between 2.5% and 16%

**57.** Mode, mean, and median

## Chapter 6

**1. (a)** Latter case; study time
**(b)** Relationship only
**(c)** Latter case; rainfall
**(d)** Relationship only

**3. (a)** Life expectancy increases with GDP in a curved pattern. The increase is very rapid at first, but levels off for GDP above roughly $5000 per person.
**(b)** Answers will vary.

**5.** Strong positive straight-line relationship

**7. (a)** Speed
**(b)** The regression line for Exercise 31 is included.

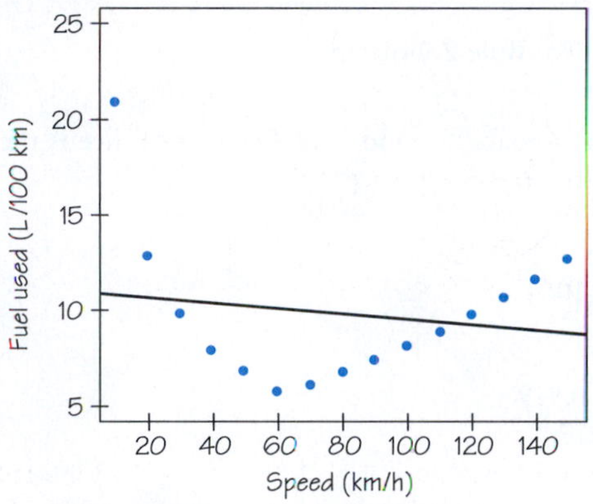


Relationship is curved; Additional answers will vary.
**(c)** None overall
**(d)** Quite strong; little scatter

**9.** 2.31

**11. (a)** *Hint*: Choose two values of weeks, preferably near 1 and 150.
**(b)** 5.42 and 4.64
**(c)** −0.0053; On average pH declined by 0.0053 per week during the study period.

**13.** 1

**15.** 0.9353; strong straight-line pattern

**17.** 0.9674; Point extends (strengthens) the straight-line pattern.

**19.** −0.1700; relationship strong but curved

**21.** 1

**23. (a)** Negative
**(b)** Negative
**(c)** Positive
**(d)** Small

**25. (a)** Dividend Growth: 0.98; Small Cap Stock: 0.81; Emerging Markets: 0.35
**(b)** No, just moved in the same direction

**27. (a)** Predicted highway mpg = 10.48 + 0.89 × (city mpg)
**(b)** 25.61 mpg
**(c)** Yes; Points follow a straight line.

**29.** Predicted highway mileage of a car that gets 18 mpg in the city is approximately 26 mpg or 27 mpg (26.5 mph).

**31.** See Exercise 7 answer for plot; Predicted values are approximately 10.91, 10.03, and 8.85, respectively.

**33.** Predicted height of husband = 33.67 + 0.54 × (height of woman); 69.85 in.

**35.** In $\hat{y} = mx + b$, substitute $b = \overline{y} - \left(r\frac{s_y}{s_x}\right)\overline{x}$, $m = r\frac{s_y}{s_x}$, and $x = \overline{x}$.

**37.** Answers will vary.

**39. (a)** All four have $r = 0.816$ and $y = 3.0 + 0.5x$.
**(c)** Data Set A; Additional answers will vary.

**41.** Answers will vary.

**43.** Answers will vary.

**45.** Lead level and reading score, respectively; negative; Answers will vary.; yes

**47. (a)** Positive; not
**(b)** $r = 0.3602$

**49. (a)** 0.42; For each additional inch of women's height, the height of the next person dated goes up by 0.42 inch on average.
**(b)** 69.22 in.

## Chapter 7

**1. (a)** U.S. residents aged 18 and older
**(b)** 1027

**3.** Answers will vary.

**5.** **(a)** Answers will vary.
**(b)** Larger; bias due to voluntary response

**7.** Taylor, Brianna, Alexis

**9.** **(a)** 001 to 371
**(b)** 214, 235, 119

**11.** Repeated samples of the same size from the same population will always be the same.

**13.** **(a)** 35, 75, 115, 155, 195
**(b)** Answers will vary.

**15.** **(a)** All people aged 18 and over living in the United States
**(b)** 30%
**(c)** Answers will vary.

**17.** Answers will vary.

**19.** No imposed treatment; observational

**21.** Answers will vary.

**23.** Classes 1, 2, 7, 8, 9, 13, 14, 15, 17, 18, 19, 22, 25, 27, and 29 receive the treatment.

**25.** This is a randomized comparative experiment with four branches, similar to Figure 7.3 with two more branches. The "flow chart" outline must show random assignment of subjects to groups, the four treatments, and the response variable (healthcare spending). We can't show the group sizes because we don't know how many people or households are available to participate.

**27.** **(a)** There are six treatments:

| | Discount Level | | |
|---|---|---|---|
| | **20%** | **40%** | **60%** |
| 50% on sale | 1 | 2 | 3 |
| 100% on sale | 4 | 5 | 6 |

**(b)** Label the subjects 01 to 60. The first group contains subjects labeled 7, 8, 10, 15, 25, 27, 54, 55, 58, 60.

**29.** **(a)** This is a randomized comparative experiment similar to Figure 7.3.
**(b)** Tea group contains rats 4, 6, 7, 8, 9, 11, and 12.

**31.** **(a)** This is a randomized comparative experiment with four branches. Best to use groups of size 216.
**(b)** 253, 296, 304, 470, 731
**(c)** Neither the subjects nor those working with the subjects know the contents of the pill each subject took daily.
**(d)** Differences could be due to the chance assignment of subjects to groups.
**(e)** Answers will vary.

**33.** Answers will vary.

**35.** Observational

**37.** Both are statistics.

**39.** **(a)** Approximately normal with mean 0.14 and standard deviation 0.0155.
**(b)** 0.109 to 0.171

**41.** **(a)** 60% of the digits 0–10 are 0–5.
**(b)** 29; 72.5%; 1; 6

**43.** 0.397 to 0.443

**45.** **(a)** 0.167 to 0.221
**(b)** It is likely that more than 171 ran a red light.

**47.** **(a)** 0.5
**(b)** $\frac{1}{\sqrt{n}}$

**49.** **(a)** 11.3%
**(b)** Answers will vary.

**51.** **(a)** No
**(b)** Yes

**53.** Smaller

**55.** 2% to 12.4%

**57.** About 0.16

## Chapter 8

**1.** **(a)** Results will vary.
**(b)** Results will vary.

**3.** 0.105

**5.** **(a)** {0, 1, 2, 3, 4, 5, 6, 7, 8, 9, 10}
**(b)** {0, 10, 20, 30, 40, 50, 60, 70, 80, 90, 100}
**(c)** {Yes, No}

**7.** **(a)** {HHHH, HHHM, HHMH, HMHH, MHHH, HHMM, HMMH, MMHH, HMHM, MHHM, MHMH, HMMM, MMMH, MHMM, MMHM, MMMM}
**(b)** {0, 1, 2, 3, 4}

**9.** **(a)** 0.19
**(b)** 0.287

**11.** $\frac{1}{216}$

**13.** **(a)** Histograms should include grades 0–4, inclusive, on the horizontal axis and the associated probabilities on the vertical axis.
**(b)** 0.64

**15.** **(a)** No; Rule 2 violated
**(b)** Yes

**17.** The probability model for this pair of dice is the same as the one for regular dice.

**19.** $\frac{8}{15}$

**21.** **(a)** 1024
**(b)** $\frac{1}{512}$

**23.** **(a)** 0.919
**(b)** 0.377

**25.** **(a)** 6
**(b)** *asp, pas, sap, spa*
**(c)** 66.7%

**27.** **(a)** 4

(b) 2,598,960
(c) 0.00000154

29. (a) $\frac{1}{2} \times \text{base} \times \text{height} = \frac{1}{2}(2)(1) = 1$
(b) $\frac{1}{2}$
(c) 0.125

31. 0.25

33. 2.78; 0.8669

35. Owner-occupied units: 6.248; rented units: 4.321

37. Both models have mean 1, because both density curves are symmetric about 1.

39. \$30

41. \$0.45

43. (a) Loss of a $\frac{1}{4}$ point
(b) Yes

45. Between 0.11 and 0.19

47. (a) 5.77 mg
(b) 4; Answers will vary.

49. (a) Sketch not included in answers.
(b) 4580 to 4620
(c) 4588.46 to 4611.54

51. (a) About 0.16
(b) Mean: 20.8; standard deviation: 1.6
(c) About 0.0015

53. (a) 45,697,600
(b) 2600
(c) 0.0000569

55. (a) 0.18
(b) 0.39

57. (a) 1.156 days
(b) Between 0.54 day and 1.00 day

59. (a) 100%
(b) 80%
(c) 12%

## Chapter 9

1. Answers will vary.

3. Answers will vary.

5. Each one-on-one score will have a winner because there cannot be a tie.

7.

| | Number of voters (4) | | | |
|---|---|---|---|---|
| **Rank** | 1 | 1 | 1 | 1 |
| First | *A* | *B* | *C* | *D* |
| Second | *B* | *C* | *D* | *A* |
| Third | *C* | *D* | *A* | *B* |
| Fourth | *D* | *A* | *B* | *C* |

9. (a) Yes
(b) Alfonse D'Amato (*D*)

11. (a) *C*
(b) *A*
(c) *D*
(d) *D*

13. (a) Five-way tie
(b) *C*
(c) Five-way tie
(d) *E*

15. (a) *C*
(b) *E*
(c) *E*
(d) *E*

17. (a) *E*
(b) Answers will vary. One possible answer is the following:

| | Number of Voters (2) | |
|---|---|---|
| **Rank** | 1 | 1 |
| First | *A* | *C* |
| Second | *B* | *B* |
| Third | *C* | *A* |

19. (a) If everyone prefers *B* to *D*, for example, then *D* has no-first place votes at all.
(b) Moving a winning candidate up one spot on some list neither decreases the number of first-place votes for the winning candidate nor increases the number of first-place votes for any other candidate.

21. (a) Condorcet winner always wins this kind of one-on-one contest.
(b) Moving a candidate up on some list only improves that candidate's chances in one-on-one contests.

23. In order to have one candidate's ranking be consistently higher than another candidate's would imply that only one candidate would be considered.

25. Answers will vary.

27. Answers will vary. One possible answer is the following:

| | Number of Voters (5) | |
|---|---|---|
| **Rank** | 3 | 2 |
| First | *A* | *B* |
| Second | *B* | *C* |
| Third | *C* | *A* |

29. (a) Since *D* has the least number of first-place votes, *D* is eliminated. Since *B* has the least number of first-place votes, *B* is eliminated. Thus, *A* is the unique winner.
(b) *B*

31. Answers will vary.

**33.** **(a)** A three-way tie with both methods
**(b)** That alternative is the sole winner with both methods.
**(c)** No; Either the situation in part (a) or the situation in part (b) must occur.

**35.** **(a)** $D$
**(b)** $A$, $B$, $D$, and $F$
**(c)** $B$, $D$ and $F$
**(d)** $A$, $B$, $D$, and $F$

## Chapter 10

**1.** Answers will vary. One example of two such elections is the following:

| Rank | Election 1<br>Number of Voters (3) | | |
|---|---|---|---|
| First | *A* | *A* | *B* |
| Second | *B* | *B* | *A* |

| Rank | Election 2<br>Number of Voters (3) | | |
|---|---|---|---|
| First | *B* | *A* | *B* |
| Second | *A* | *B* | *A* |

**3.** Answers will vary. One example of two such elections is the following:

| Rank | Election 1<br>Number of Voters (3) | | |
|---|---|---|---|
| First | *A* | *B* | *B* |
| Second | *B* | *A* | *A* |

| Rank | Election 2<br>Number of Voters (3) | | |
|---|---|---|---|
| First | *B* | *B* | *B* |
| Second | *A* | *A* | *A* |

**5.** **(a)** Doesn't treat all voters the same
**(b)** A dictatorship in which Voter #1 is the dictator
**(c)** A dictatorship in which Voter #2 is the dictator and a dictatorship in which Voter #3 is the dictator

**7.** Consider the leftmost voter changes his or her preference ballot to the following:

*C*
*B*
*D*
*A*

**9.** Consider the leftmost voter changes his or her preference ballot to the following:

*A*
*C*
*D*
*B*

**11.** Answers will vary.

**13.** Consider the leftmost voter changes his or her preference ballot to the following:

*B*
*A*
*D*
*C*

**15.** Consider the leftmost voter changes his or her preference ballot to the following:

*A*
*C*
*B*

**17.** Consider the leftmost voter changes his or her preference ballot to the following:

*B*
*A*
*C*

**19.** **(a)** Consider the agenda $D$, $A$, $C$, $B$.
**(b)** Consider the agenda $B$, $D$, $A$, $C$.
**(c)** Consider the agenda $B$, $A$, $C$, $D$.

**21.** Consider the voters in the 7% group to change their ballots to the following:

*H*
*J*
*D*

**23.** **(a)** To go from having a unique winner to a different unique winner occurs if the winning alternative in the first election has exactly two first-place votes, and one of these two voters changed his or her ballot by moving some other alternative into first place (yielding a worse outcome for this voter).
**(b)** Tie in second election
**(c)** Answers will vary.

**25.** **(a)** Answers will vary.
**(b)** Answers will vary.

**27.** 1 and 4

**29.** Answers will vary.

## Chapter 11

**1.** **(a)** A winning or blocking coalition would be 50 senators plus the vice president, or more than 50 senators.
**(b)** At least 41 senators

**3.** Weight-5 and weight-4 voters; weight-3 voter

**5.** **(a)** 1958: $B$, $G$, $L$; 1964: $N$, $G$, $L$; later years: none
**(b)** 1958 and 1964: $N$, $B$, $G$, $L$; 1970 and 1976: none; 1982: $G$

**7.** The last juror in the permutation is the pivotal voter.

**9.** **(a)** $\left(\frac{5}{12}, \frac{1}{4}, \frac{1}{4}, \frac{1}{12}\right)$
**(b)** $\left(\frac{1}{3}, \frac{1}{3}, \frac{1}{6}, \frac{1}{6}\right)$
**(c)** $\left(\frac{1}{4}, \frac{1}{4}, \frac{1}{4}, \frac{1}{4}\right)$

**11.** Nevada

**13.** **(a)** NNNN, NNNY, NNYN, NNYY, NYNN, NYNY, NYYN, NYYY, YNNN, YNNY, YNYN, YNYY, YYNN, YYNY, YYYN, YYYY
**(b)** { }, {*D*}, {*C*}, {*C, D*}, {*B*}, {*B, D*}, {*B, C*}, {*B, C, D*}, {*A*}, {*A, D*}, {*A, C*}, {*A, C, D*}, {*A, B*}, {*A, B, D*}, {*A, B, C*}, {*A, B, C, D*}
**(c)** Each subset corresponds to a set of "yes" voters.
**(d)** **(i)** 1; **(ii)** 4; **(iii)** 6

**15.** *A* has a critical vote in 6 coalitions; *B, C,* and *D* each have critical votes in 2. The Banzhaf power index is (12, 4, 4, 4).

**17.** **(a)** 35
**(b)** 0
**(c)** 105
**(d)** 105

**19.**

| Year | Banzhaf Index |
|---|---|
| 1958 | (32, 0, 0, 0, 0) |
| 1964 | (32, 0, 0, 0, 0) |
| 1970 | (32, 2, 2, 2, 2) |
| 1976 | (32, 2, 2, 2, 2) |
| 1982 | (28, 4, 4, 0, 4) |

**21.** $\frac{5}{32}$

**23.** **(a)** {*A, C, D*} and {*A, B*}
**(b)** Each winning coalition includes *A*, so *A* has veto power. That makes {*A*} a minimal blocking coalition. There are two other minimal blocking coalitions: {*B, C*} and {*B, D*}.
**(c)** (10, 6, 2, 2)
**(d)** Answers will vary. Possible answers include [5 : 3, 2, 1, 1]
**(e)** $\left(\frac{7}{12}, \frac{1}{4}, \frac{1}{12}, \frac{1}{12}\right)$

**25.** Answers will vary.

**27.** **(a)** [4 : 2, 1, 1, 1]
**(b)** [6 : 2, 2, 1, 1, 1]

**29.** (24, 24, 24, 24, 20, 20, 20); faculty

**31.** If there were only two minimal winning coalitions, those two coalitions must overlap; there is a voter we'll call *V* who is in both. This *V* is therefore in every winning coalition, and hence has veto power.

**33.** **(a)** We will omit *E* from the list: One could include him in any winning or losing coalition without altering the vote total.

| Winning Coalition | Extra Votes | Losing Coalition | Votes Needed |
|---|---|---|---|
| {*A, B, C, D*} | 49 | {*A*} | 3 |
| {*A, B, C*} | 42 | {*B, C*} | 6 |
| {*A, B, D*} | 27 | {*B, D*} | 21 |
| {*A, C, D*} | 26 | {*C, D*} | 22 |
| {*A, B*} | 20 | {*B*} | 28 |
| {*A, C*} | 19 | {*C*} | 29 |
| {*A, D*} | 4 | {*D*} | 44 |
| {*B, C, D*} | 1 | { } | 51 |

**(b)** 4
**(c)** 19
**(d)** 4

**35.** **(a)** [8 : 6, 1, 1, 1, 1, 1, 1, 1, 1]
**(b)** (492, 16, 16, 16, 16, 16, 16, 16, 16)
**(c)** $\left(\frac{2}{3}, \frac{1}{24}, \frac{1}{24}, \frac{1}{24}, \frac{1}{24}, \frac{1}{24}, \frac{1}{24}, \frac{1}{24}, \frac{1}{24}\right)$
**(d)** No

**37.** [9 : 4, 4, 4, 1, 1, 1, 1, 1]

**39.** The three weight-3 voters or two weight-3 voters and one weight-1 voter form minimal winning coalitions. The Banzhaf power index is (30, 30, 30, 6, 6, 6).

**41.** Case 1 (three weight-1 voters): $\left(\frac{17}{60}, \frac{17}{60}, \frac{17}{60}, \frac{1}{20}, \frac{1}{20}, \frac{1}{20}\right)$ and Case 2 (four weight-1 voters): $\left(\frac{9}{35}, \frac{9}{35}, \frac{9}{35}, \frac{2}{35}, \frac{2}{35}, \frac{2}{35}, \frac{2}{35}\right)$; In Case 1, the weight-3 voter is $5\frac{2}{3}$ times as powerful as the weight-1 voter and in Case 2 the weight-3 voter is $4\frac{1}{2}$ times as powerful as the weight-1 voter.

## Chapter 12

**1.** Assume a distribution is skewed to the left. The heavier concentration of voters on the right means that fewer voters are farther from the median. Because there are fewer voters "pulling" the mean rightward, it will be to the left of the median. Likewise, a distribution skewed to the right will have a mean to the right of the median.

**3.** While there is no median position such that half the voters lie to the left and half to the right, there is still a position where the middle voter (if the number of voters is odd) or the two middle voters (if the number of voters is even) are located, starting either from the left or right. In the absence of a median, less than half the voters lie to the left and less than half to the right of this middle voter's (voters') position (positions).

Hence, any departure by a candidate from a position of a middle voter to the position of a non-middle voter on the left or right will result in that candidate's getting less than half the votes—and the opponent's getting more than half. Thus, the middle position (positions) is (are) in equilibrium, making it (them) the extended median.

**5.** When the four voters on the left refuse to vote for a candidate at 0.6, his opponent can do better by moving to 0.7, which is worse for the dropouts.

**7.** Yes

**9.** When it is the median or the extended median; yes; Possible examples include the following:

| **Position $i$** | 1 | 2 | 3 | 4 | 5 | 6 | 7 |
|---|---|---|---|---|---|---|---|
| **Location ($l_i$) of position $i$** | 0.1 | 0.2 | 0.3 | 0.5 | 0.6 | 0.8 | 0.9 |
| **Number of voters ($n_i$) at position $i$** | 7 | 8 | 1 | 2 | 1 | 2 | 1 |

**11.** Not necessarily

**13.** Answers will vary.

**15.** Since the districts are of equal size, the mayor's median or extended median must be between the leftmost and rightmost medians or extended medians.; If, say, the left-district positions are much farther away from the mayor's median or extended median than the right-district positions then the mayor's mean would be in the interval of the left-district positions.

**17.** If, say, $A$ takes a position at $M$ and $B$ takes a position to the right of $M$, $C$ should take a position just to the left of $M$ that is closer to $M$ than $B$'s position, giving $C$ essentially half the votes and enabling him or her to win the election. If neither $A$ nor $B$ takes a position at $M$, $C$ should take a position next to the player closer to $M$; the position that $C$ takes to maximize his or her vote may be either closer to $M$ (if the candidates are far apart) or farther from $M$ (if the players are closer together), but this position may not be winning.; no

**19.** Following the hint, $C$, will obtain $\frac{1}{3}$ of the vote by taking a position at $M$, as will $A$ and $B$, so there will be a three-way tie among the candidates. Because a non-unimodal distribution can be bimodal, with the two modes close to $M$, $C$ can win if he or she picks up most of the vote near the two modes, enabling $C$ to win with more than $\frac{1}{3}$ of vote.

**21.** $B$ should enter just to the right of $\frac{3}{4}$ making it advantageous for $C$ to enter just to the left of $A$, giving $C$ essentially $\frac{1}{4}$ of the vote.

**23.** $A$: $\frac{1}{6}$; $B$: $\frac{5}{6}$; $C$: $\frac{1}{2}$; $D$: indifferent between entering just to the left of $A$, just to the right of $B$, or in between $A$ and $C$ at $\frac{1}{3}$, or between $C$ and $B$ at $\frac{2}{3}$

**25.** No

**27.** Answers will vary.

**29.** By definition, more voters prefer the Condorcet winner to any other candidate. Thus, if the poll identifies the Condorcet winner as one of the top two candidates, he or she will receive more votes when voters respond to the poll by voting for one or the other of these candidates. The possibility that the Condorcet winner might not be first in the poll, but win after the poll is announced, shows that the plurality winner may not be the Condorcet winner. Some argue that the Condorcet winner is always the "proper" winner, but others counter that a non-Condorcet winner who is, say, everybody's second-most-preferred candidate is a better social choice than a 51%-Condorcet winner who is ranked last by the other 49%.

**31.** $D$; Yes, due to a poll that identifies either the top two or the top three candidates would not include $D$.

**33.** $A$; Answers will vary.

**35.** Assume a voter votes for just a second choice. It is evident that voting for a first choice, too, can never result in a worse outcome and may sometimes result in a better outcome (if the voter's vote for a first choice causes that candidate to be elected).

**37.** Following the hint, the voter's vote for a first and third choice would elect either $A$ or $C$. If the voter also voted for $B$, then it is possible that if $A$ and $B$ are tied for first place, then $B$ might be elected when the tie is broken, whereas voting for just $A$ and $C$ in this situation would elect $A$.

**39.** No

**41.** No; $D$ and $C$; Class I; These voters cannot bring about a preferred outcome by voting for candidates different from $A$ and $B$.

**43.** Without polling: (i): $A$, (ii): $D$, (iii): $B$ and $D$; with polling: (i): $B$, (ii): $D$, (iii): $D$

**45.** 9.5

**47.** Substitute into the formula for $r_i$ in Exercise 46, $d_i = (n_i/N)D$ and $D = R$ The proportional rule is "strategy-proof" in the sense that if one player follows it, the other player can do no better than to follow it. Hence, knowing that an opponent is following the proportional rule does not help a player optimize against it by doing anything except also following it.

**49.** No; yes

**51.** The Democrat can win the election by winning in any two states or in all three. The first three expressions in the formula for $PWE_D$ give the probabilities of winning in the three possible pairs of states, whereas the final expression gives the probability of winning in all three states.

**53.** Yes

## Chapter 13

**1.** Donald receives the Trump Tower triplex and about 87% ownership of the Palm Beach mansion. Ivana gets the rest.

**3.** Phil gets his way on the stereo level issue, the smoking rights issue, the phone time issue, the visitor policy issue, and about 87% of his way on the alcohol issue. Mike gets his way on the rest.

**5.** Answers will vary.

**7.** *Allocation 1*:
**(a)** Not proportional
**(b)** Not envy-free
**(c)** Not equitable
**(d)** Example: Give Bob *X*, Carol *Y*, and Ted *Z*.
*Allocation 2*:
**(a)** Not proportional
**(b)** Not envy-free
**(c)** Not equitable
**(d)** Example: Give Bob *Y*, Carol *X*, and Ted *Z*.
*Allocation 3*:
**(a)** Not proportional
**(b)** Not envy-free
**(c)** Not equitable
**(d)** It is Pareto-optimal.
*Allocation 4*:
**(a)** Not proportional
**(b)** Not envy-free
**(c)** Not equitable
*Allocation 5*:
**(a)** It is proportional.
**(b)** Not envy-free
**(c)** It is equitable.

**9.** Mary gets both the car and the house and pays John $43,831.25.

**11.** *A* receives the farm plus $7333. *B* receives $132,334. *C* receives both the house and the sculpture, while paying $139,667.

**13.** The allocation is as follows:

| Potential Recipient | Total Points |
|---|---|
| *A* | 16.0 |
| *B* | 13.5 |
| *C* | 17.0 |
| *D* | 20.5 |

**15.** Answers will vary. Possible answers include: One could use an absolute measure, awarding; for example, one point for each month a potential recipient has been waiting.

**17.** Carol first chooses the investments, and the final allocation has her also receiving the boat and the washer-dryer.

**19.** Fred first chooses the boat, and the final allocation has him also receiving the car and the motorcycle.

**21.** Ivana first chooses the Connecticut estate, and the final allocation has her also receiving the Trump Plaza apartment and the cash and jewelry.

**23.** Answers will vary. Possible ways include having Bob divide the cake into four pieces and letting Carol choose any three.

**25.** **(a)** Divider
**(b)** Bob knows the preferences of the other party. In Exercise 22, we assumed that the divider didn't know the preferences of the other party.

**27.** **(a)** See figures below:

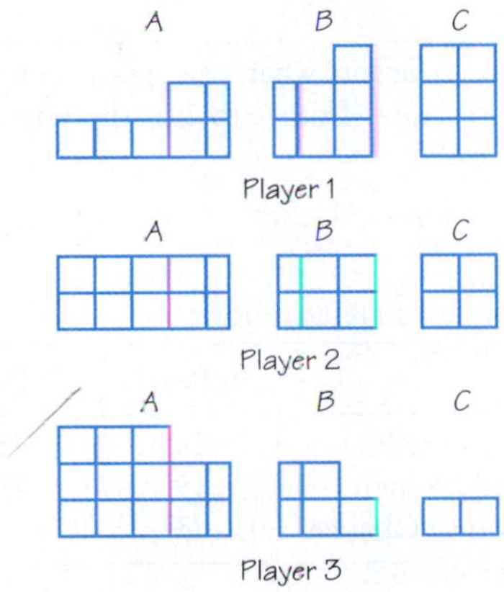


**(b)** Player 2 finds *A* acceptable, but not *B* or *C*. Player 3 finds *A* acceptable, but not *B* or *C*.
**(c)** *C*
**(d)** (i) Player 1: 6 square units; Player 2: 7 square units; Player 3: 10 square units; (ii) Player 1: 6 square units; Player 2: $8\frac{2}{3}$ square units; Player 3: 8 square units.

**29.** **(a)** Ted thinks he is getting at least one-third of part of the cake (Bob's piece) plus one-third of the rest of the cake (Carol's piece).
**(b)** Bob gets to keep exactly two-thirds (in his own view) of the piece that he initially received and thought was at least of size one-half. Two-thirds times one-half equals one-third.
**(c)** Answers will vary.

**31.** Answers will vary.

**33.** **(a)** At the point where the other knife started
**(b)** Answers will vary.

## Chapter 14

**1.** Rent: 43.66%, Food: 22.54%; Transportation: 9.86%; Gym: 16.90%; Miscellaneous: 7.04%; no

**3.** The simplest solution is to use the original apportionment.

**5.** $0 + 2 + 1 + 2 + 3 + 2 = 10$

**7.** Answers will vary.

**9.** Geometry: 3 sections, Algebra: 1 section, Calculus: 1 section

**11.** **(a)** Abe: 36, Beth: 19, Charles: 23, David: 20, Esther: 2
**(b)** Charles gives the coin to Esther.
**(c)** Population paradox

**13.** The apportionments are as follows:

| State | Apportionments | | |
|---|---|---|---|
| $A$ | 27 | 28 | 28 |
| $B$ | 7 | 7 | 7 |
| $C$ | 16 | 17 | 17 |
| $D$ | 37 | 37 | 38 |
| $E$ | 2 | 1 | 1 |
| **Totals** | 89 | 90 | 91 |

The Alabama paradox occurs when the apportionment for the smallest state decreases from 2 to 1 as the house size increases from 89 to 90.

**15.** Geometry: 4 sections, Algebra: no sections, Calculus: 1 section

**17.** The apportionments are as follows:

| | | 36 Pearls | 37 Pearls |
|---|---|---|---|
| **Jefferson method:** | Abe | 14 | 15 |
| | Beth | 19 | 19 |
| | Charles | 3 | 3 |

| | | 36 Pearls | 37 Pearls |
|---|---|---|---|
| **Webster method:** | Abe | 14 | 15 |
| | Beth | 19 | 19 |
| | Charles | 3 | 3 |

Additional answers will vary.

**19.** Hamilton method: 88 + 2 + 2 + 1 + 1 + 1 + 1 + 1 + 1 + 1 + 1 = 100%;
Jefferson method: 90 + 1 + 1 + 1 + 1 + 1 + 1 + 1 + 1 + 1 + 1 = 100%;
Webster method: The apportionment is the same as the Jefferson apportionment, so it, too, violates the quota condition.

**21.** Before excise tax: Abe: 36, Beth: 19, Charles: 22, David: 20, Esther: 3; After excise tax: Abe: 36, Beth: 19, Charles: 23, David: 20, Esther: 2; Esther gives the coin to Charles.

**23.** The apportionments are as follows:

| State | Apportionments | | |
|---|---|---|---|
| $A$ | 27 | 28 | 28 |
| $B$ | 7 | 7 | 7 |
| $C$ | 16 | 16 | 17 |
| $D$ | 37 | 37 | 37 |
| $E$ | 2 | 2 | 2 |
| **Totals** | 89 | 90 | 91 |

There are no paradoxes.

**25. (a)** One quota will be rounded up, and the other down to obtain the Webster apportionment. The quota that is rounded up will have fractional part greater than 0.5, and will be greater than the fractional part of the quota that is rounded down. The Hamilton method will give the party whose quota has the larger fractional part an additional seat. Thus the apportionments will be identical.
**(b)** These paradoxes never occur with the Webster method, which gives the same apportionment in this case.
**(c)** The Hamilton method, which always satisfies the quota condition, gives the same apportionment.
**(d)** Not always

**27.** 10.77%

**29. (a)** 0.5068 seats per million and 45.88%
**(b)** 0.7218 seats per million and 48.53%
**(c)** Yes

**31.** 0; 1.4142; 2.4495; 3.4641

**33.** The apportionments are as follows:

| | Webster | Hill–Huntington |
|---|---|---|
| Algebra | 2 | 2 |
| Geometry | 3 | 2 |
| Calculus | Cancelled! | 1 |

It's likely that the principal would prefer the Webster method.

**35.** 2240

**37.** The tentative apportionments: 2, 9, and 10. As the house is overfilled, one apportionment must be reduced. The critical divisors are $100{,}000/\sqrt{2}$, $600{,}000/\sqrt{72}$, and $700{,}000/\sqrt{90}$. The first two critical divisors are equal, with an approximate value of 70,711; the third is approximately 73,786. Therefore, the two smaller districts are tied, because the district with the least critical divisor must have a reduced apportionment.

**39. (a)** States with small populations
**(b)** Yes
**(c)** Yes
**(d)** The apportionment is as follows:

| State | Apportionments |
|---|---|
| Virginia | 18 |
| Massachusetts | 14 |
| Pennsylvania | 12 |
| North Carolina | 10 |
| New York | 10 |
| Maryland | 8 |
| Connecticut | 7 |
| South Carolina | 6 |
| New Jersey | 5 |
| New Hampshire | 4 |
| Vermont | 3 |
| Georgia | 2 |
| Kentucky | 2 |
| Rhode Island | 2 |
| Delaware | 2 |
| **Total** | 105 |

**41. (a)** Not unless a state's quota is a whole number
**(b)** $d_i = \frac{p_i}{n_i - 1}$
**(c)** Small states
**(d)** No

**43. (a)** Let $n = \lfloor q \rfloor$. If $q$ is between $n$ and $n + 0.4$, then the Condorcet rounding of $q$ is equal to $n$. Since $q + 0.6 < n + 1$ in this case, it is also true that $\lfloor q + 0.6 \rfloor = n$. On the other hand, if $n + 0.4 \leq q < n + 1$, then the Condorcet rounding of $q$ is $n + 1$, and also $n + 1 \leq q + 0.6 < n + 1.6$, so $\lfloor q + 0.6 \rfloor = \mathrm{n} + 1$.
**(b)** Small states
**(c)** If the sum of the tentative apportionments is less than the house size, $d_i = \frac{p_i}{n_i + 0.4}$. If the sum of the tentative apportionments is more than the house size, $d_i = \frac{p_i}{n_i - 0.6}$.

**45.** Answers will vary.

## Chapter 15

**1. (a), (b)** Saddlepoint at row 1 (maximin strategy), column 2 (minimax strategy), giving value 5
**(c)** Row 2 and column 1

**3. (a)** No saddlepoint
**(b)** Rows 1 and 2 are both maximin strategies.; Column 1 is the minimax strategy.
**(c)** None

**5. (a), (b)** Saddlepoint at row 3 (maximin strategy), column 3 (minimax strategy), giving value −20
**(c)** Column 3 dominates columns 1 and 2, so column player should avoid strategies from columns 1 and 2.

**7.** Batter's optimal mixed strategy: $\left(\frac{1}{4}, \frac{3}{4}\right)$; pitcher's optimal mixed strategy: $\left(\frac{1}{4}, \frac{3}{4}\right)$; value 0.275

**9.** Saddlepoint is "not cheat" and "audit," giving value −$100.

**11. (a)**

| | Officer Does Not Patrol | Officer Patrols |
|---|---|---|
| You park in street | 0 | −$40 |
| You park in lot | −$32 | −$16 |

**(b)** Your optimal mixed strategy: $\left(\frac{2}{7}, \frac{5}{7}\right)$; officer's optimal mixed strategy: $\left(\frac{3}{7}, \frac{4}{7}\right)$; value: −$22.86
**(c)** It is unlikely that the officer's payoffs are the opposite of yours.
**(d)** Answers will vary.

**13. (a)** Answers will vary.
**(b)** Showing that your strategy is optimal involves showing that it guarantees at least a tie.

**15.** Always play *H*.

**17. (a)** "Bet, then call" should be avoided by player I.
**(b)** Player I: $\left(\frac{1}{3}, \frac{2}{3}, 0\right)$; player II: $\left(\frac{2}{3}, 0, \frac{1}{3}\right)$; value $-\frac{1}{12}$.
**(c)** Player II
**(d)** Yes

**19. (a)** 50% chance of rain: leave umbrella; 75% chance of rain: carry umbrella
**(b)** Carry umbrella
**(c)** Saddlepoint at "carry umbrella" and "rain," giving value −2
**(d)** Leave umbrella

**21.** The Nash equilibrium outcomes are (4, 3) and (3, 4).

**23.** The Nash equilibrium outcome is (2, 4), which is the product of dominant strategies by both players.

**25.** Players have no incentive to lie about the value of their own weapons unless they were sure about the preferences of their opponents and could manipulate them to their advantage. But if they do not have such information, lying could cause them to lose more than 10% of their weapons, as they value them, in any year.

**27.** Answers will vary.

**29.** The payoff matrix is as follows:

| | | Even | | |
|---|---|---|---|---|
| | | 2 | 4 | 6 |
| Odd | 1 | (2, 1) | (2, 1) | (2, 1) |
| | 3 | (2, 4) | (6, 3) | (6, 3) |
| | 5 | (2, 4) | (4, 8) | (10, 5) |

Odd will eliminate strategy 1, and Even will eliminate strategy 6, because they are dominated. In the reduced 2 × 2 game, Odd will eliminate strategy 5. In the reduced 1 × 2 game, Even will eliminate strategy 4. The resulting outcome will be (2, 4), in which Odd chooses strategy 3 and Even chooses strategy 2. The outcome (2, 1) in which Odd chooses strategy 1 and Even chooses strategy 2, is also in equilibrium.

**31.** Answers will vary. Consider that the first player shoots in the air will be no threat to the two other players, who will then be in a duel and shoot each other.

**33.** Answers will vary.

**35.** Nobody will shoot.

**37.** The possibility of retaliation deters earlier shooting.

## Chapter 16

**1.** 3

**3.** 3

**5.** 5

**7.** 1

**9.** X

**11.** They contribute 30 to the weighted sum.

**13.** 6

**15.** Yes

**17.** 7; will detect single-digit errors

**19.** Odd-numbered positions, odd digit replaced with odd digit or even digit replaced with even digit

**21.** 2

**23.** **(a)** 7
**(b)** 4
**(c)** 7
**(d)** 2

**25.** 0-669-09325-4

**27.** The actual check digit and the check digit calculated with the weighted sum are both 0 or their sum is 10.

**29.** Yes; no; An error is detected when an odd digit is misread as an even one or vice versa.

**31.** No

**33.** 9

**35.** 7, 8, 9

**37.** Yes

**39.** Answers will vary.

**41.** Not detected; will be detected

**43.** The error will be detectable unless $a - c = \pm 5$.

**45.** Substitution of $b$ for $a$ where $b - a = \pm 5$ in positions 1, 5, 7, 9 and 11 is undetected.; All errors in position 3 are undetected.; Substitution of $b$ for $a$ where $b - a$ is even in position 8 is undetected.

**47.** 7

**49.** **(a)** 51593-2067; 2
**(b)** 50347-0055; 1
**(c)** 44138-9901; 1

**51.** **(a)** 20782-9960
**(b)** 55435-9982
**(c)** 52735-2101

**53.** Answers will vary.

**55.** If you replace each short bar in the bar code table by an $a$ and replace each long bar in the bar code table by a $b$ the resulting strings are listed in alphabetical order.

**57.** Right to left

**59.** Wyoming, Nevada, Alaska

**61.** Population size

**63.** The Canadian scheme detects any transposition error involving adjacent characters. Also, there are more possible Canadian codes.

**65.** S-000; S-200; L-550; L-300; E-663; O-451

**67.** A person born in 1999 is too young for a driver's license.

**69.** Certain instances will yield a number requiring four digits.

**71.** August 1, 1958

**73.** Answers will vary. Possible answers include Gallian, Galliam, Gilliam, and Galin.

**75.** 248: female born on March 29; 601: male born on September 17

**77.** Answers will vary.

**79.** Consider short names and a large population.

## Chapter 17

**1.** No answer provided.

**3.** **(a)** 6
**(b)** 3

**5.** 1001101

**7.** 000000, 100011, 010101, 001110, 110110, 101101, 011011, 111000

**9.** 0000000, 1000001, 0100111, 0010101, 0001110, 1100110, 1010100, 1001111, 0110010, 0101001, 0011011, 1110011, 1101000, 1011010, 0111100, 1111101; no

**11.** 000000, 100101, 010110, 001011, 110011, 101110, 011101, 111000; 001011; 111000; 010110; 100101

**13.** 00000000, 00010111, 00101110, 01001011, 10001101, 11000110, 10100011, 10011010, 01100101, 01011100, 00111001, 11101000, 11010001, 10110100, 01110010, 11111111; The code will detect any three errors or correct any single error.

**15.** 32; 256

**17.** 0000, 1012, 2021, 0111, 0222, 1120, 2210, 2102, 1201

**19.** 81; 729

**21.** AATAAAGCAA

**23.** 111101000111001010; *AABAACAEADB*

**25.** $t$, $n$, $r$; $e$

**27.** In the Morse code, a space is needed to determine where each code word ends. In a fixed-length code of length $k$, a word ends after each $k$ digits.

**29.** 13403 336 77 −49 53 −61 20 99 35 −17; 46%

**31.** *BCEEFCDDCFF*

**33.** $B$ is the least likely letter, $J$ is the second least likely, and $G$ is the third least likely letter.

**35.** UHWUHDW; ADVANCE

**37.** 13

**39.** HURHUAR

**41.** There is no integer $j$ such that $2j = 1 \bmod 26$.

**43.** ROLLING STONES

**45.** **(a)** 11000000
**(b)** 10011001

**47.** 23, 49, 16

**49.** 13

**51.** N converts to 14 and O converts to 15, but 14 and 77 have a greatest common divisor of 7. Using blocks of length 4, NO converts to 1415 and the greatest common divisor of 77 and 1415 is 1.

**53.** Entries in the column for the variable $P$ and $P \vee (P \wedge Q)$ are both TTFF.

**55.** Entries in the column for $\neg(P \wedge Q)$ and $\neg P \vee \neg Q$ are both FTTT.

**57.** Entries in the column for $P \wedge (Q \vee R)$ and $(P \wedge Q) \vee (P \wedge R)$ are both TTTFFFFF.

**59.** Entries in the column for $\neg P \vee Q$ and $P \rightarrow Q$ are both TFTT.

**61.** They are logically equivalent.

**63.** **(a)** 00100000
**(b)** 00110000

**65.** $s \wedge 11100000 = t \wedge 11100000$ is the same as $s \wedge 11100000 - t \wedge 11100000 = 00000000$, but in mod 2, subtraction is the same as addition since $1 + 1 = 0$.

**67.** $s$ has a 0 in position 2, 4, and 6 and a 1 in position 8.

**69.** 01100010, 01101010, 01110010, 01111010

**71.** 8.0.0.0

**73.** It does not.

## Chapter 18

**1.** **(a)** None
**(b)** $5\frac{1}{3}$ in.
**(c)** $4\frac{1}{2}$ in.

**3.** **(a)** 0.025
**(b)** Volume of the real person is 64,000 times that of the Lego.
**(c)** 13.1 ft

**5.** 26%

**7.** **(a)** The new altar would have a volume 8 times as large.
**(b)** About 1.26

**9.** Answers will vary.

**11.** Nothing can decrease 150% without becoming negative.

**13.** Answers will vary.

**15.** €0.61

**17.** 32 mpg

**19.** **(a)** 0.00013364 tons
**(b)** We assume that all parts of the scale model are made of the same materials as the real locomotive.
**(c)** 0.267 lb
**(d)** 0.121 kg
**(e)** 0.000121 metric tonnes

**21.** $6.72/gal

**23.** About 5 ft 7 in.

**25.** **(a)** 37.5 mph
**(b)** 28.0 mph
**(c)** 120 mph

**27.** **(a)** 50%
**(b)** 100%

**29.** **(a)** 40%
**(b)** 67%

**31.** **(a)** For Option A, yes; for Option B, no
**(b)** For either a loss or a gain, the absolute value of the percentage is higher for Option B.
**(c)** Either way, use Option B.

**33.** **(a)** 900 lb/ft$^3$
**(b)** About 2 times as dense as steel
**(c)** The mineral soil weighs about 4500 to 4600 lb.

**35.** **(a)** 400,000 lb
**(b)** 28 lb/in.$^2$

**37.** 6100 lb or 2800 kg

**39.** Answers will vary.

**41.** 9 ft 3 in. to 11 ft 9 in. (In modern times, there have been men over 9 ft tall.); 282 cm to 358 cm

**43.** Less than 90.25 kg

**45.** Answers will vary.

**47.** Answers will vary.

**49.** Answers will vary.

**51.** The square of wingspan is proportional to wing area, so the wing loading is proportional to weight divided by square of wingspan. For the 200-pounder, that ratio is 0.080, while for the 100-pounder it is 0.077. They are close.

**53.** **(a)** **(i)** 800; **(ii)** 640,000; **(iii)** 512,000,000
**(b)** One eight-hundredth
**(c)** There couldn't be any such giant ants.

**55.** 480 years

**57.** $A \propto d^2$ and $A \propto M^{3/4} \propto (d^2h)^{3/4} = d^{3/2}h^{3/4}$, so $d^2 \propto d^{3/2}h^{3/4}$, hence $d^{1/2} \propto h^{3/4}$, and $d \propto h^{3/2}$.

**59.** A small warm-blooded animal has a large surface-area-to-volume ratio. Pound for pound, it loses heat more rapidly than a larger animal, hence must produce more heat per pound, resulting in a higher body temperature.

**61.** **(a)** On log-log paper:

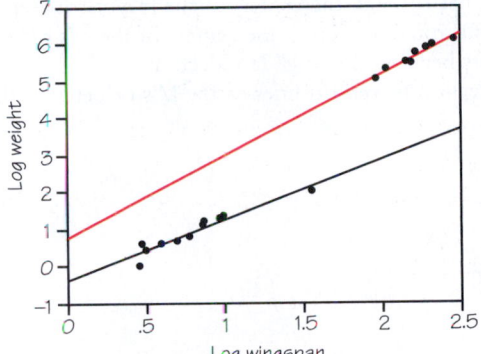

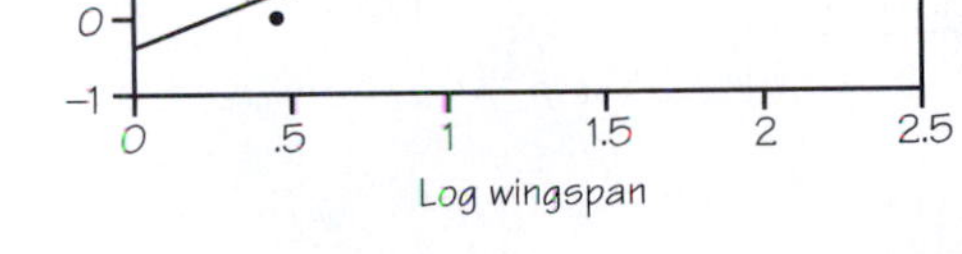

**(b)** Both allometric
**(c)** Slope for birds less steep than for planes

## Chapter 19

**1.** 5, 8, and 13

**3.** Answers will vary but will be Fibonacci numbers.

**5.** 1, 2, 3, 5, 8, ..., $F_{n+1}$

**7.** Seventh number: $5m + 8n$; total: $55m + 88n$

**9. (a)** The digits after the decimal point do not change.
**(b)** The digits after the decimal point do not change.
**(c)** $\phi^2 = \phi + 1$
**(d)** $\frac{1}{\phi} = \phi - 1$

**11. (a)** 9
**(b)** 16

**13.** 6

**15. (a)** 4, 7, 11, 18, 29, 47, 76, 123
**(b)** 3, 1.333, 1.75, 1.571, 1.636, 1.611, 1.621, 1.617, 1.618; The ratios approach $\phi$.

**17.** Answers will vary.

**19. (a)** 1, 1, 3, 5, 11, 21, 43, 85, 171, 341, 683, 1365
**(b)** $B_n = B_{n-1} + 2B_{n-2}$
**(c)** 1, 3, 1.667, 2.2, 1.909, 2.048, 1.977, 2.012, 1.994, 2.003, 1.999
**(d)** $B_n = \frac{2^n - (-1)^n}{3}$

**21.** Silver: 2.414, bronze: 3.303, copper: 4.236, nickel: $\frac{1}{2}(5 \pm \sqrt{29}) \approx 5.193$; general: $\frac{1}{2}\left(m + \sqrt{m^2 + 4}\right)$

**23.** All are true.

**25. (a)** B, C, D, E, H, I, K, O, X
**(b)** A, H, I, M, O, T, U, V, W, X, Y
**(c)** H, I, N, O, S, X, Z

**27. (a)** MOM, WOW; MUd and bUM reflect into each other, as do MOM and WOW.
**(b)** pod rotates into itself.; MOM and WOW rotate into each other.
**(c)** Answers will vary. Possible answers include: CHECK BOOK BOX.

**29.** For all parts, translations
**(a)** Reflection in the horizontal midline of the C's
**(b)** None other than translations
**(c)** Reflection in the horizontal midline; reflections in vertical lines through the centers of the H's or between them; 180° rotation around the centers of the H's or the midpoints between them; glide reflections.
**(d)** Reflection in vertical lines of the M's or between them

**31.** *d5*

**33. (a)** *c5*
**(b)** *c12*
**(c)** *c22*

**35. (a)** *c6*
**(b)** *d2* (CBS)
**(c)** *d1* (Dodge Ram)

**37. (a)** *c4*
**(b)** *d2*

**39. (a)** *p1a1*
**(b)** *p1m1*
**(c)** *p111*
**(d)** *p112*
**(e)** *pm11*
**(f)** *pma2*
**(g)** *pmm2*

**41. (a)** *pmm2*
**(b)** *p1a1*
**(c)** *pma2*
**(d)** *p112*
**(e)** *pmm2* (perhaps)
**(f)** *p1m1*
**(g)** *pma2*
**(h)** *p111*

**43. (a)** Half-turns: Mesa Verde pottery; reflections: Begho smoking pipes
**(b)** Neither culture completely excludes any strip type. Begho designs are heavily concentrated in *p1m1*, *p112*, or *pmm2*, while Mesa Verde designs are more evenly distributed over the seven patterns.
**(c)** (i) *pm11* or *pma2*: Mesa Verde; (ii) *p112*: Mesa Verde; (iii) *pmm2*: Begho; (iv) *pm11*: Begho; (v) *p1m1*: Difficult to say; (vi) *pmm2*: Begho; (vii) *pmm2*: Begho; (viii) *pma2*: Mesa Verde; (ix) *p1a1*: Mesa Verde

**45. (c)** Smallest rotation is 90°.; There are reflections.; There are reflections in lines that intersect at 45°: *p4m*.
**(d)** Smallest rotation is 90°.; There are no reflections: *p4*.

**47.** None of the five patterns with hexagonal symmetry can be realized, nor *p4g*, *p4*, *cm*, and *cmm*. The remaining eight can all be formed by the technique.

**49.** *pg*, if color is disregarded; if not, then *p1*

**51.** *p6*

**53.** *p3*

**55.** Answers will vary.

**57.** There is no identity element.

**59. (a)** *d2*
**(b)** Any two of: $R$ (180° rotation around the center), $V$ (reflection in vertical line through its center), H (reflection in horizontal line through its center)
**(c)** $\{I, R, V, H\}$

**61.** There are four rotation symmetries (including the identity), two reflection symmetries, and two reflections across diagonal lines.;
$\{I, R, R^2, R^3, H, V, RH = VR, RV = HR\}$

**63.** Answers will vary.

**65.** Label with letters copies of the pattern elements in the positions, and pick a fixed position about which to make a half-turn $R$.
**(a)** $\langle T, R \mid R^2 = I, T \circ R = R \circ T^{-1} \rangle$
**(b)** $\langle T, R, H \mid R^2 = H^2 = I, T \circ H = H \circ T, R \circ H = H \circ R, (R \circ T)^2 = I \rangle$

**67.** $\langle R \mid R^8 = I \rangle$, where R is a rotation by 45°

**69.** There are four rotation symmetries (including the identity), three reflection symmetries, and an inversion through the center.

**71.** The carved head is reproduced in the same shape at different scales.

**73.** Answers will vary.

**75.** Answers will vary.

## Chapter 20

**1.** Exterior: 45°; interior: 135°

**3.** $180° - \frac{360°}{n}$

**5.** Notation is to denote a regular $n$-gon by $n$, separate the sizes of polygons by periods, list in clockwise order starting from the smallest $n$.; possible vertex types: 3.3.3.3.3.3, 3.3.3.3.6, 3.3.3.4.4, 3.3.4.3.4, 3.3.4.12, 3.4.3.12, 3.3.6.6, 3.6.3.6, 3.4.4.6, 3.4.6.4, 3.12.12, 4.4.4.4, 4.6.12, 4.8.8, 5.5.10, 6.6.6

**7.** 3.7.42, 3.9.18, 3.8.24, 3.10.15, 4.5.20

**9.** At each of the vertices except the center one, six triangles meet, with angles (in clockwise order) of 75°, 75°, 30°, 30°, 75°, and 75°.

**11.** Yes

**13.** **(a)** No
**(b)** No
**(c)** No

**15.** Answers will vary.

**17.** No; no

**19.** Answers will vary.

**21.** Label the four corners consecutively $A$, $B$, $C$, and $D$.

**23.** Place the skew-tetromino on a coordinate system with unit length for the side of a square and with the lower left corner at (0, 0). Then $A = (1, 2)$, $B = (3, 2)$, $C = (2, 0)$, and $D = (0, 0)$ works.

**25.** Place the skew-tetromino on a coordinate system with unit length for the side of a square and with the lower left corner at (0, 0). Then $A = (0, 1)$, $B = (1, 2)$, $C = (3, 2)$, $D = (3, 1)$, $E = (2, 0)$, and $F = (0, 0)$ works.

**27.** **(a)** Yes
**(b)** No
**(c)** No

**29.** See figure below:

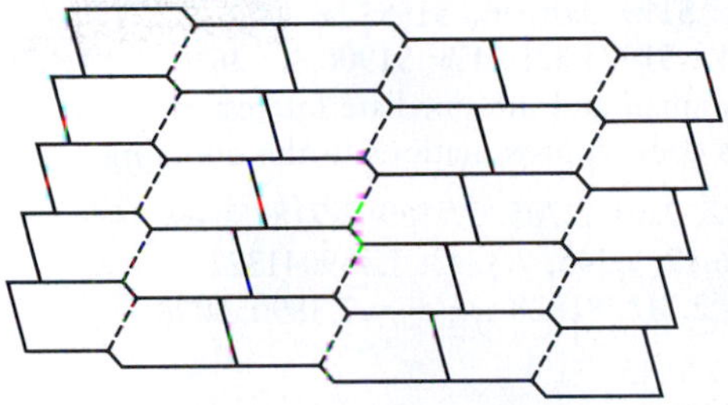

**31.** Answers will vary.

**33.** N, Z, W, P, y, I, L, V, X

**35.** Place the U on a coordinate system with unit length for the side of a square and with the lower left corner at (0, 0). Then $A = (2, 2)$, $B = (3, 2)$, $C = (3, 0)$, $D = (1, 0)$, $E = (0, 0)$, and $F = (0, 1)$ works.

**37.** Answers will vary.

**39.** **(a)** Consecutive segments, all of the same length; Each length gives a different tiling.
**(b)** Consecutive repetition of any finite sequence of $a$'s and $b$'s

**41.** **(a)** *b*; *a*; *ab*; *aba*; *abaab*; *abaababa*; *abaababaabaab*; *abaababaabaababaababa*
**(b)** $F_n$ segments at stage $n$
**(c)** $2m + n$

**43.** *ABAABABA*

**45.** The two leftmost $A$'s would have had to come from two $B$'s in a row in the preceding month.

**47.** Answers will vary.

**49.** If a sequence ends in $AA$, its deflation ends in $BB$, which is impossible for a musical sequence. Similarly, if a sequence ends in $ABAB$, its deflation ends in $AA$, which we just showed to be impossible.

**51.** The first is, the second is not.

**53.** If the sequence were periodic, the limiting ratio of $A$'s to $B$'s would be the same as the ratio in the repeating part, which would be a rational number, contrary to the result of Exercise 52.

## Chapter 21

**1.** \$150; \$1350

**3.** **(a)** 698 days
**(b)** After 24 months

**5.** **(a)** \$1030.00; 3%
**(b)** \$1030.00; 3%
**(c)** \$1030.34; 3.034%
**(d)** \$1030.45; 3.045%

**7.** 4.801%

**9.** 3.023%; no

**11.** 3.684%

**13.** **(a)** 3%: $2032.79; 4%: $2025.82; 6%: $2012.20
**(b)** 8%: $1999.00; 9%: $1992.56
**(c)** 12%: $1973.82; 24%: $1906.62; 36%: $1849.60
**(d)** For small and intermediate interest rates, the rule of 72 gives good approximations to the doubling time.

**15.** **(a)** 2, 2.59, 2.705, 2.7169, 2.718280469
**(b)** 3, 6.19, 7.245, 7.3743, 7.389041321
**(c)** $e = 2.718281828\ldots$; $e^2 = 7.389056098\ldots$

**17.** **(a)** $30.45
**(b)** $30.45

**19.** **(a)** 4.881%
**(b)** 4.766%

**21.** **(a)** 4.081%
**(b)** 5.125%; very slightly less than the true effective rate

**23.** $79.41

**25.** $81,327.45

**27.** $880.21

**29.** $100,451.50

**31.** **(a)** $147.06
**(b)** $444,683.29
**(c)** $302,384.64

**33.** **(a)** −0.523%
**(b)** Effective rate

**35.** **(a)** Exponential (decay)
**(b)** Linear
**(c)** Exponential (decay)
**(d)** Linear

**37.** $8227.02

**39.** $27,591.08

**41.** **(a)** $1.09
**(b)** $0.92

**43.** Approximately $3900

**45.** $78.97; Answers will vary.

**47.** 2.592%; (rounding 2.592% to the nearest whole percent) yes

**49.** −8.333%.

**51.** Nowhere close
**(a)** $63,541.55
**(b)** $61,524.30

**53.** $29.8 million; taking into account 4% on funds received, $33.4 million through last payment

**55.** **(a)** $1,388,888.89; $735.16
**(b)** $35,538.34
**(c)** $15,532.98

**57.** 1.60% per month, or an annual rate of 19.2%

**59.** 4.97%; effective rate

## Chapter 22

**1.** $5967.50; $467.50

**3.** $10,006; $2006

**5.** $26,764.51

**7.** $1,148,698.24

**9.** **(a)** About 21.24%
**(b)** About 23.66%

**11.** All but a tiny amount after 91 months

**13.** About 203 months

**15.** 193 months (more than 16 years)

**17.** $419.94

**19.** $272.25

**21.** **(a)** $11,204.25
**(b)** $3138.75

**23.** **(a)** $230.16
**(b)** $7619.20

**25.** $123.99

**27.** **(a)** $14,971.03
**(b)** $127.66
**(c)** $19,687.20

**29.** $632.07

**31.** $850.62

**33.** $6388.73

**35.** $23,813.20

**37.** 3.68%

**39.** 391% if calculated as 14 days out of 365 (390% if calculated as 2 weeks out of 52)

**41.** 325% if calculated as 7 days out of 365 (324% if calculated as 1 week out of 52)

**43.** **(a)** $760.07
**(b)** $4612.20
**(c)** $634.62
**(d)** 16 months

**45.** Approximately 4.97% (Answers can vary due to rounding.)

**47.** $290,580.05

## Chapter 23

**1.** Beginning of 2023

**3.** 5.534 billion

**5.** Almost 100% greater than Europe's population

**7.** **(a)** 140 years
**(b)** 58 years

**9.** **(a)** 8.211 billion; 11.065 billion
**(b)** No change in growth rate, no change in death rates, no global catastrophes, etc.

**11.** **(a)** 34 years
**(b)** 27 years
**(c)** Answers will vary.

**13.** **(a)** 51 years
**(b)** 132 years
**(c)** 224 years

**15.** **(a)** 138 years
**(b)** Forever

**17.** 30 lb/plant/day is unreasonable.

**19.** 2.77%

**21.** After the first year, the population stays at 15.

**23.** 7, 18.2, 6.6, 17.6, 8.4, 19.5, 2.0, 7.3, 18.6, 5.3

**25.** 0, 15

**27.** The population is oscillating but slowly converging to about 13.3.

**29.** 0, $\frac{40}{3}$

**31.** The red dashed line indicates the same size population next year as this year; Where it intersects the blue curve is the equilibrium population size.

**33.** 11, 12.0, 12.6, 12.9, 13.1, 13.2, 13.3, 13.3, 13.3, 13.3 (rounded)

**35.** The population sizes are 11, 15.0, 13.7, 14.9, 14.9, 15.0, 14.8, 14.3, 13.0, 9.5–and the following year the population is wiped out.

**37.** 15 million pounds; 35 million pounds for an initial population of 25 million pounds

**39.** **(a)** The last entry shown for the first sequence is the fourth entry of the second sequence.
**(b)** 39, 78, 56, and we have "joined" the second sequence.
**(c)** Regardless of the original number, after the second push of the key we have a number divisible by 4, and all subsequent numbers are divisible by 4. There are 25 such numbers between 00 and 99.

**41.** **(a)** Sequence stabilizes at 1.
**(b)** Answers will vary.
**(c)** That would trivialize the exercise!
**(d)** Answers will vary.

**43.** **(a)** 0.0397, 0.15407173, 0.545072626, 1.288978, 0.171519142, 0.59782012, 1.31911379, 0.0562715776, 0.215586839, **0.722914301**, 1.32384194, 0.0376952973, 0.146518383, 0.521670621, 1.27026177, 0.240352173, 0.78810119, 1.2890943, 0.171084847, **0.596529312**
**(b)** **0.723**, 1.323813, 0.0378094231, 0.146949035, 0.523014083, 1.27142514, 0.236134903, 0.777260536, 1.29664032, 0.142732915, **0.509813606**
**(c)** **0.722**, 1.324148, 0.0364882223, 0.141958718, 0.507378039, 1.25721473, 0.287092278, 0.901103183, 1.16845189, **0.577968093**

**45.** Period 2 begins at $\lambda = 3$, period 4 at about 3.449, period 8 at 3.544, period 3 at about 3.828, and chaotic behavior onsets at about 3.57.

# INDEX

## D



## E

## F



## G

## T

## U



## V

D0216918

The Samuel Gompers Papers

# THE
# Samuel Gompers
# PAPERS

## VOLUME
## 1
## The Making of a Union Leader, 1850-86

*Editor*
Stuart B. Kaufman

*Associate Editors*
Peter J. Albert
Elizabeth A. Fones-Wolf
Dolores E. Janiewski
David E. Carl

*Contributing Editors*
Dorothee Schneider
Grace Palladino

HD
6508
.S218
1986

UNIVERSITY OF ILLINOIS PRESS
*Urbana and Chicago*

© 1986 by the Board of Trustees of the University of Illinois
Manufactured in the United States of America

*This book is printed on acid-free paper.*

Library of Congress Cataloging in Publication Data

Main entry under title:
The Samuel Gompers papers.
Bibliography: v. 1, p.
Includes index.
Contents: v. 1. The making of a union leader, 1850-1886.
1. Trade-unions—United States—History—Sources.
2. Labor and laboring classes—United States—History—Sources.
3. Gompers, Samuel, 1850-1924—Archives. I. Gompers, Samuel, 1850-1924. II. Kaufman, Stuart Bruce. HD6508.S218 1986 331.88'32'0924 84-2469
ISBN 0-252-01137-6 (vol. 1)
ISBN 0-252-01138-4 (set)

8

*To*
*Louis R. Harlan*

# CONTENTS

# INTRODUCTION

"For years," journalist Benjamin Stolberg wrote in 1925, "Samuel Gompers was American Labor." Calling him the "clarion consciousness" of craft unionism, Stolberg described Gompers as the Moses of labor's "forty years in the wilderness, its daily struggle for manna, its defense against inner rebellion and outer attack."[1] And yet today, some sixty years after his death, Gompers' name is as likely to evoke the image of a labor bureaucrat as it is that of a working-class hero. If his oratory and organizational abilities alarmed opponents during his lifetime, his strategy of "pure and simple" trade unionism and his demand for "more" remain the most readily identified components of his complex legacy.

Conventional wisdom to the contrary, however, Gompers' place in history cannot be easily assessed. If some historians see him as an exemplary mediator and conciliator, others argue that he proved to be no more than a typical "corporate liberal" and still others go so far as to label him a "class collaborator" and opportunist. Recent examinations of Gompers' radical roots and of the forward-looking elements of his strategy have served to reemphasize his complexity. He was, David Brody writes, "a man of extraordinary parts—a Jewish cigar-maker schooled in the Marxist debates of the 1870s, a vigorous thinker capable of turning those radical ideas to the purposes of American trade unionism, a crafty and ambitious leader with a sixth sense for the thinking of his followers."[2]

The dispersal of Gompers' documentary record has compounded the problem of understanding this complex personality and the formative period of the American Federation of Labor. Certainly Gompers was well aware of the difficulties his life would pose for future historians and he lamented the fact that "so much which exists in the archives of the labor movement and elsewhere . . . baffles gathering and research."[3] Those who have searched for scattered sources, often in vain, will agree with his appraisal. This edition of *The Samuel Gompers Papers* attempts to ease the problem of access by offering a collection

of documents by and about Samuel Gompers culled from numerous manuscript repositories, libraries, and institutional collections.

In this first volume the editors of the Samuel Gompers Papers project have assembled the documentary history of Gompers' rise from an obscure cigarmaker to the presidency of the American Federation of Labor and have sought to illuminate the connection between Gompers' experience in the New York City labor movement and the distinctive philosophy and strategy developed by the AFL. Although the files of the Samuel Gompers Papers project now contain more than half a million pages of documents, the material on Gompers' early years is very sparse. To compensate for this the editors have included many documents in this volume that were not written to or by Gompers himself; in this sense, *The Making of a Union Leader* is uncharacteristic of the volumes to follow. The inclusion of such a relatively large proportion of "non-Gompers" documents reflects the editors' attempt to create a structure within which one can easily follow the path of Gompers' development. In the process of collecting and annotating these documents for publication, the editors have also identified and put into context the many individuals and groups who played a role in this very fluid period of American labor history.

Because his work in New York City's cigar shops proved central to Gompers' formative years, the cigar industry figures prominently in this book. During the period covered by the documents in volume 1, the industry was undergoing fundamental changes—in its markets, the composition of its workforce, and in its production methods—that led to deteriorating working conditions and declining wage rates. The depression of the 1870s accentuated the impact of these changes and culminated in the "Great Strike" of the New York City cigarmakers in 1877-78. Newspaper articles from the daily, socialist, and trade union press, accounts of cigarmakers' meetings, and the testimony of Gompers and his colleagues detail these developments here. They describe the workers of the cigar industry engaged in factory and home production, the conditions of their employment, the course of their protests through mass meetings, agitation, and the strike of 1877-78, and their attempts to create a strong trade union organization.

Gompers' association with a group of predominantly German-American socialists in New York cigar shops is also documented here, since this relationship nourished his developing philosophy and honed his skills as a political strategist. These men proved particularly influential in shaping Gompers' convictions that trade unions could play a major role in transforming society and emancipating the wage earner, and that they must remain independent of political parties.

Gompers' involvement in the local cigarmakers' unions in New York City—the United Cigarmakers and local 144 of the CMIU—and his attempts to introduce fundamental changes into the CMIU as a whole provide another focus of this volume. They serve to illustrate the evolution of his trade union ideology and reflect the cultural, intellectual, and political turmoil that shaped the cigarmakers' movement in New York City.

The editors have followed the activities of Gompers and his colleague, Adolph Strasser, through the socialist press and the *Cigar Makers' Official Journal* (*CMOJ*) from their first efforts to unite New York City cigar workers outside the existing national union to their reorganization efforts within the CMIU. They introduced, at both the local and national levels, a system of benefits supported by high dues and accompanied by centralized bureaucratic controls. At the same time they worked to shape a political strategy for the union that emphasized lobbying among existing political institutions, particularly the prohibition of the production of cigars in tenements. These tactics met with opposition from many workers and labor leaders from the late 1870s on. Some socialists favored independent partisan politics over lobbying within the political establishment. Some resented the centralized organizational structure that Gompers and Strasser had created in the CMIU. Gompers' struggle with these dissenters, especially between 1882 and 1886, receives extensive coverage here.

Gompers' rise to prominence in the CMIU and particularly in local 144 paralleled his emergence as a major labor figure on the city, state, and national levels. This development, made clear by his participation in national trade union conventions between 1881 and 1886 and his testimony before the Senate Education and Labor Committee in 1883, is well documented in this volume. So, too, is the growing antagonism that Gompers and a number of other trade unionists felt toward the KOL, which, during the 1880s, was becoming the largest labor organization in the United States. The editors have tried to follow the course of Gompers' relationship with the Knights, particularly as it served as a foil against which he was able to further define and articulate his beliefs in trade unionism.

Volume 1 culminates with the founding convention of the American Federation of Labor. The documents in this book illuminate the fact that in many ways the foundations for the rise of Gompers and the emergence of the AFL had been laid during the preceding decade. In the fall of 1886, however, few observers realized that they were witnessing the establishment of what would become North America's most powerful and most enduring labor organization.

This volume, then, although organized around the career of an

individual, at the same time strives to capture the political, intellectual, social, and industrial climate that informed and affected the American working classes in the late nineteenth century. In this sense the editors have attempted to combine the methods of the "old" trade union history with the insights of the "new" labor history. By presenting Gompers and his trade union strategies in the context of a changing social and industrial milieu, *The Making of a Union Leader* seeks to offer the basis for a reexamination of Samuel Gompers as he grew from labor activist into labor statesman.

## THE COLLECTION

In a sense the history of the Samuel Gompers Papers project began with the attention Gompers himself gave to preserving papers relating to the labor federation. From the beginning of his presidency of the AFL in 1886, Gompers later wrote, he retained "correspondence, documents, publications, historical material upon labor in the United States and abroad, letterbooks, and complete records" in the Federation's files along with "personal material which I had been saving since the early 'seventies. . . . No one," he claimed, "ventured to throw away a piece of paper which had printing or writing on it without first consulting me."[4] In 1894, however, John McBride defeated Gompers for the AFL presidency. When the Federation prepared to move its headquarters from New York to Indianapolis, its new secretary, August McCraith, discarded many of the records Gompers had collected. Consequently, few documents now remain in the AFL-CIO's files from the years between 1886 and 1894 and still fewer date from before 1886. Those from 1886-94 consist primarily of the president's letterbooks (volumes containing letterpress copies of outgoing correspondence), some Executive Council records such as minutes of meetings and vote books, a run of the AFL's first journal, the *Union Advocate,* an account book, and other miscellaneous correspondence.

Gompers was reelected president of the AFL in 1895 and, during the next twenty-nine years, generated a vast amount of correspondence and other documentary material. Some of these records he retained in a personal collection of papers and other memorabilia. After his death, the AFL donated part of this to the New York Public Library,[5] exhibited a small portion of it in a memorial room in the AFL headquarters building,[6] and stored the rest of it, known as "President Gompers' Files," with the Federation's other noncurrent records.[7] The bulk of Gompers' papers, including most of his voluminous outgoing and incoming correspondence as AFL president, were integrated with the Federation's other records and their disposition

must be considered as part of the story of the AFL's records as a whole.

The AFL's central file room was the principal repository of the organization's noncurrent records, particularly the papers of its executive officers; other such material remained in the president's and secretary-treasurer's offices. The Federation's departments apparently maintained their files independently of the central file. With the passage of time, the shortage of space in the central file room and in departmental offices necessitated storing an increasing quantity of noncurrent records in the basement of the headquarters building.[8] Vaughn Bornet, a scholar who did research at the AFL in the early 1950s, vividly depicted the contents of the basement storage area: "Several hundred standard four-drawer metal filing cabinets crowd every spare inch of space. They are stacked eight drawers high in orderly array. Aisles provide the only open space except for a rough work bench by the door. Although the floor is swept regularly, the coal dust of several decades covers cabinet interiors, file folders, and before long the researcher in an impartial blanket of black."[9]

In 1952 the Federation initiated a program of microfilming its noncurrent records because of the need for space. E. Logan Kimmel, the official in charge of the central file, initially received authorization to rent a Recordak rotary microfilm camera and to film and then discard some of the records in the central file room.[10] In 1953 the program assumed more comprehensive proportions. Kimmel purchased the rotary microfilm camera, completed filming in the relatively recent and well-organized files in the central file room, and moved on to the older records stored in the basement that were in his custody. The physical condition of these documents made their filming more difficult.[11] Members of his staff screened the files, culling documents they considered to have administrative importance or historical significance, disposing of the rest, and arranging the selected documents for filming. After microfilming them, they discarded these as well.[12]

When Bornet learned of the AFL's microfilming program in March 1955, he contacted the director of the State Historical Society of Wisconsin, Clifford Lord, and urged the Society to attempt to acquire the documents that the Federation was discarding. Edwin Witte, a labor scholar at the University of Wisconsin who had worked with George Meany as a member of the National War Labor Board, acted as an intermediary and in April assured Lord that "all facilities of the Federation would be open" to him.[13] Lord visited AFL headquarters in May, surveyed the Federation's records program, and then wrote Meany to ask for the bulk of the original records that the

Federation was processing.[14] The AFL Executive Council approved the transfer in August on the condition that "the Society will agree to return to us upon request any files we may consider necessary to the conduct of our operations."[15]

By the time the Society began acquiring AFL records, Kimmel's staff had already screened, filmed, and disposed of a large number of documents under his control. In addition, some departments—the Department of Organization among them—had discarded their noncurrent records without filming them.[16] The Gompers-era records that the AFL microfilm preserved included material relating to resolutions at conventions beginning with the 1909 meeting; correspondence with national and international unions, the earliest dating from 1890; files involving jurisdictional disputes between affiliates, the earliest dated 1896; circulars and mimeographed materials beginning in 1906; and a small number of records of directly affiliated locals.[17] In addition the AFL filmed "President Gompers' Files" and the strike files; the Federation sent the bulk of these documents to the Society.[18]

Most of the unfilmed material the AFL sent to Wisconsin dated from after 1924. Gompers-era records included some files from the legislative department, early national and international union files, and mining department correspondence.[19] During the 1960s the AFL-CIO made two further donations of records from the Gompers era. In 1966 and 1969 it gave the president's copybooks (1883-1925) to the Library of Congress, and it donated the secretary's copybooks (1904-25) to Duke University in 1969.[20] The AFL-CIO retained still other important records from the Gompers era at its headquarters in Washington, D.C. These range from long runs of the Executive Council minutes and vote books to scattered items in storage files and in the executive offices and the Federation's departments.[21]

While the body of documents that can be referred to as the Samuel Gompers Papers begins with the records that the AFL retained or deposited with other institutions, the project is drawing upon many other sources for documents written by or to Gompers. For example, more than a dozen unions affiliated with the AFL during Gompers' presidency retained correspondence with him in their files and made these papers available to us.[22] The large cache of Gompers' letters in the National Civic Federation (NCF) papers at the New York Public Library reflects his instrumental role in the NCF. As chairman of the Labor Committee of the Advisory Commission of the Council of National Defense, he was involved in most of the correspondence in the committee's files, which are preserved at the National Archives. This and other record groups at the Archives reflect his national importance, as do the collections in other archives and repositories—

such as the Library of Congress, the Catholic University of America, the State Historical Society of Wisconsin, the Tamiment Collection at New York University, the Archives of Labor and Urban Affairs at Wayne State University, the Labor-Management Documentation Center at Cornell University, and the Chicago Historical Society—and in private hands. These have yielded further material illustrating his wide range of contacts with workers and labor officials, radicals and reformers, industrialists, civic leaders, government officials and politicians, students, teachers, and intellectuals. Finally, his career as a labor journalist and writer yielded a lifetime of editorials, articles, and books, and the thousands of reports on his activities in the daily and labor press and the journals of his day further document his stature.

## THE PROJECT

The Samuel Gompers Papers project began in 1973, setting as its goals to gain as full an understanding as possible of the total corpus of Gompers documents, to collect copies of them, and to disseminate them in a comprehensive microfilm and a selective printed edition. The search for documents involved surveying finding aids and labor history literature, corresponding with virtually every repository that might have acquired Gompers material as well as with many key individuals and organizations, advertising in various publications, and undertaking research trips to examine promising collections and photocopy or microfilm relevant material. The Gompers Papers staff has collected hundreds of thousands of documents in this manner, and the process of search and collection will continue for the duration of the project.

In 1975, with special authorization from Lane Kirkland, then AFL-CIO secretary-treasurer, the project staff began locating and filming the extant records from the Gompers era housed at the AFL-CIO headquarters in Washington, D.C. These comprise a large portion of the documents published in 1979 in the project's first series of microfilm, *American Federation of Labor Records: The Samuel Gompers Era.* A joint undertaking with the State Historical Society of Wisconsin, the *AFL Records* brought together approximately 300,000 pages of documents held by the AFL-CIO and the Society. Subsequently, the project conducted a search for Gompers correspondence in the records of unions affiliated with the AFL in Gompers' time, collecting some 15,000 pages of documents for a second microfilm publication, *The American Federation of Labor and the Unions.* Internal evidence in records already located, together with the opening of new collections,

continually suggests additional sources of Gompers material. Records collected from other repositories will be included in subsequent microfilm publications.

The project's selective, annotated, printed edition, of which this volume is the first, complements our microfilm publications. In the process of searching for Gompers materials, evaluating and organizing them, and preparing them for filming, the editors are in an excellent position to identify the documents that illuminate the rich themes, important insights, and defining contours of this vast collection. Gompers' immense energy, his longevity, and his volubility served to create an enormous body of manuscripts in which middle- and lower-level labor and ethnic leaders, a wide range of working-class people, and those interested in or associated with them play the supporting roles. The printed edition should serve as one key for scholars interested in probing the larger body of Gompers documents, give students, teachers, and writers ready access to a distilled and annotated research collection of the first importance, and provide the large number of serious readers within the labor movement and among the general public with a more intimate sense of the history of organized labor and the working-class experience.

The first volume in the project's multivolume series has been a joint effort. Working part-time for the project in New York City from 1975 on, David E. Carl conducted a search of the repositories in that city as well as helping with documentary collection, annotation, and research in other ways; he left the project in 1981. The project's three associate editors at the University of Maryland (Peter J. Albert, Elizabeth A. Fones-Wolf, and Dolores E. Janiewski, who joined the project, respectively, in 1974, 1977, and 1979) have collaborated with the editor in all aspects of the work of the project. Searching discussion and penetrating criticism of the volume among the editors have alternated with meticulous and taxing editorial work. These exchanges have sometimes involved questioning and refining the project's editorial method and at other times prompted a search for new information or documents to fill a gap in the story, supply missing details in developing events, and provide a better focus on the background against which the events developed. In brief, the editors have shaped this volume together.

## Editorial Method

The project's editorial method begins with the assumption that these volumes are a form of literature rather than simply a compendium of historical documents. In selecting material for this edition the

editors have tried to assemble a collection that conveys a sense of the unfolding of events over time. The act of selection is inevitably a subjective process. No historian could achieve a mathematical or scientific balance mirroring the components of a total collection of almost a million pages of documents. In any case, considering the accidents and vagaries responsible for the survival of any particular collection of letters and the large amount of routine and trivial material in twentieth-century bureaucratic correspondence, a sampling of the less significant materials seems sufficient to preserve a sense of the texture of the collection. The overall objective of the selection is to recover the rich multilevel perspective that Gompers' papers provide, one that cuts through fifty formative years in the development of the modern labor movement.

The editors have organized the documents chronologically. Each document is preceded by a headnote that identifies it and is followed by an endnote describing the physical form of the original and specifying its location. The editors are using an editorial apparatus that intervenes as little as possible between the documents and the reader. We have silently standardized the format of correspondence to the extent of positioning place and date line at the top right, the internal address above the salutation at the top left, and the complimentary close on the same line as the signature. Other silent adjustments include converting underlined passages to italics and omitting most material that the writer has stricken through. In the few cases where the document contains a word or passage of significance that the author appears to have stricken out, we reproduce it with strikeovers.

Other aspects of our policy of silent standardization depend upon the nature of the document. In holograph documents—those written in an individual's hand—texts appear with all misspellings, errant punctuation, stray dashes, and other distinguishing characteristics intact. This respects the direct personal relationship between the writer and a document shaped by his or her own hand. Mechanically reproduced documents, such as articles printed in newspapers or typed letters, require a different treatment. Typographical and spelling errors seem to have been endemic to the labor bureaucracy and press of the late nineteenth and early twentieth centuries. For the most part it is impossible to determine which of these errors were the work of the documents' authors and which were anonymously introduced by the typographers and secretaries of the day. Since their sheer numbers make them distracting, however, the editors have silently corrected many of them, with the following exceptions. There are no changes in the original syntax of mechanically produced documents and no

changes that would substantively alter their meaning. Errors remain if they seem essential to the particular mood or style of a document. Certain features of the documents generally appear as found, including capitalization, the spelling of contemporary personal names, errors of usage, such as the use of a noun form for a verb, and common spelling conventions of the time, for instance, the use of "segar" for "cigar," the spelling of Pittsburgh, Pennsylvania, without the final "h," or lowercasing the word "street" as part of a street name. The original punctuation usually appears as found, except where it is an apparent typographical error and tends to garble the document's meaning.

The policy on silent corrections enables the editors to avoid an excessive use of brackets and other editorial cosmetics. This edition uses brackets to present a reading of portions of a document that are difficult to decipher or to insert material that is not in the original document. The latter may include letters or whole words obliterated in the original text, words or punctuation marks necessary to make the text sensible, and alternate words that clarify the meaning of the text, such as "respectively" for "respectably." Ellipses within brackets indicate indecipherable text; a question mark within brackets signifies uncertainty or speculation on the editors' part.

Supplementing the documents, the editors have prepared notes, a glossary, and a number of editorial essays. The notes briefly identify individuals, organizations, events, and the like or explain developments that are not clear from the text. The glossary, located at the end of the volume, provides the reader with information on important individuals (such as members of Gompers' family, leaders of national or state labor bodies, and editors of labor papers), and on national and international labor organizations. Names of individuals and organizations for whom there are glossary entries are printed in large and small capital letters at their first mention in the notes. The editorial essays supply important background information or give an overview of complex developments that evolved over a fairly lengthy period of time. Though the primary purpose of the essays is to facilitate the reading of the documents, they also tend to alert the reader to the interpretative judgments that informed the editors' selection of particular documents for inclusion in this edition. The notes, glossary entries, and editorial essays do not generally indicate sources of information unless the material is quoted or has a unique aspect that appears to require attribution. In general the project's annotation research has involved extensive use of city directories, manuscript censuses, local histories, trade union journals, and newspaper accounts as well as standard biographical and reference works and labor history literature.

ACKNOWLEDGMENTS

The Samuel Gompers Papers project is indebted to Florence Gompers MacKay for her permission to publish the papers of her grandfather and for locating documents for the project in the possession of members of her family. We would also like to single out the contribution of Louis R. Harlan, editor of the Booker T. Washington Papers, under whom Stuart Kaufman served an apprenticeship as a Fellow-in-Editing of the National Historical Publications Commission (later the National Historical Publications and Records Commission; NHPRC) and as an assistant editor. In editorial philosophy and methodology, the point of departure for the Samuel Gompers Papers has been the approach Harlan imparted a decade and a half ago. For this reason we have dedicated this volume to him.

The editors are also grateful to the AFL-CIO for its help and support in this undertaking. Lane Kirkland, formerly secretary-treasurer and now president of the Federation, granted permission to photocopy and publish the voluminous records from the Gompers era that remain in the Federation's files. Over a period of many months, AFL-CIO officials helped us locate pertinent records and cooperated in every conceivable way with our program of copying the documents—from providing us with access to the Federation's own photocopying facilities and permitting us to bring in our own microfilming equipment to allowing us to borrow the Federation's microfilms of Gompers-era records for duplication in a professional film laboratory. Finally, the Executive Council of the AFL-CIO has contributed financially to the project, and on several occasions the Federation has used its good offices in support of our search for documents and our fund-raising efforts elsewhere.

Officers and staff members at AFL-CIO headquarters have lent their sympathy and cooperation to the project's efforts. Especially helpful, in addition to President Kirkland, were President George Meany, Secretary-Treasurer Thomas R. Donahue, and Assistants to the Secretary-Treasurer Wesley Reedy and James J. Kennedy, Jr.

Many labor unions have provided access to their records and have aided the project in searching their files and copying documents. These include the Bakery, Confectionery and Tobacco Workers International Union, the United Brotherhood of Carpenters and Joiners of America, the Amalgamated Clothing and Textile Workers Union, the American Flint Glass Workers Union, the United Food and Commercial Workers International Union, the United Hatters, Cap and Millinery Workers International Union, the Laborers' International Union of North America, the International Ladies' Garment Workers'

Union, the United Mine Workers of America, the International Brotherhood of Painters and Allied Trades, the United Paperworkers International Union, the Brotherhood of Railway, Airline and Steamship Clerks, Freight Handlers, Express and Station Employes, the Retail, Wholesale and Department Store Union, the Seafarers International Union of North America, and the International Brotherhood of Teamsters, Chauffeurs, Warehousemen, and Helpers of America.

The staffs of many archives and repositories have facilitated our research and cooperated with us in locating and photocopying Gompers documents. In particular we acknowledge our heavy debt to the archives of the Catholic University of America, the U.S. Department of Labor Library, the Library of Congress, the National Archives, the New York Public Library, the Tamiment Institute of New York University, the Labor-Management Documentation Center at Cornell University, and the State Historical Society of Wisconsin.

We are proud to have earned the financial support of a significant group of labor organizations despite difficult economic times. The project has received financial contributions from the AFL-CIO Executive Council, the Joseph Anthony Beirne Memorial Foundation of the Communications Workers of America, and the following unions: the Associated Actors and Artistes of America, the Bakery, Confectionery and Tobacco Workers International Union, the International Brotherhood of Boilermakers, Iron Ship Builders, Blacksmiths, Forgers and Helpers, the International Union of Bricklayers and Allied Craftsmen, the International Brotherhood of Electrical Workers, the International Union of Electronic, Electrical, Technical, Salaried and Machine Workers, the United Food and Commercial Workers International Union, the Glass, Pottery, Plastics and Allied Workers International Union, the American Federation of Government Employees, the Laborers' International Union of North America, the National Association of Letter Carriers, the International Longshoremen's Association, the Newspaper Guild, the International Union of Operating Engineers, the International Brotherhood of Painters and Allied Trades, the United Paperworkers International Union, the United Association of Journeymen and Apprentices of the Plumbing and Pipe Fitting Industry of the United States and Canada, the Brotherhood of Railway Carmen of the United States and Canada, the Service Employees International Union, the United Steelworkers of America, the International Brotherhood of Teamsters, Chauffeurs, Warehousemen and Helpers of America, and the International Alliance of Theatrical Stage Employes and Moving Picture Machine Operators of the United States and Canada.

Our project was generously supported by our two sponsoring uni-

versities. Officials at the University of Maryland stretched their resources and ingenuity on the project's behalf and in general provided a climate of support for an enormous undertaking. Among those who have aided the project have been President John S. Toll, Chancellors Robert L. Gluckstern and John Slaughter, Acting Chancellor William E. Kirwan, Provost Shirley Strum Kenny of the Division of Arts and Humanities, History Department Chairmen Emory G. Evans and Walter Rundell, Jr., and Acting Director Paul A. Weinstein of the Industrial Relations and Labor Studies Center. We would also like to thank the director of the University of Maryland Libraries, H. Joanne Harrar, as well as the staff of the McKeldin Library, where the project has its offices. We have drawn daily upon the resources of McKeldin Library and have found its facilities and staff indispensable to the project. The University of Maryland General Research Board provided a small grant to help us with translation work. In addition, Pace University enabled us to utilize the services of David Carl from 1975 to 1981 to accomplish an extensive search of New York City sources. President Edward J. Mortola and William M. Welty, former associate dean of the Pace Graduate School, were instrumental in establishing this arrangement.

We could not have undertaken a project of this magnitude without the financial support and the backing of the NHPRC and the National Endowment for the Humanities (NEH). Oliver Wendell Holmes, as executive director of the Commission, encouraged the initiation of this project, and the NHPRC's present executive director, Frank G. Burke, along with Roger A. Bruns and George L. Vogt, provided a sounding board over the years as we worked out our philosophy of editing and defined the concept and scope of the project. Commission staff member Sara Dunlap Jackson guided us through the military records in the National Archives, and Mary A. Guinta helped orient our researchers to the Archives civil records. The Commission also arranged for an independent committee composed of Edward C. Carter II, Jonathan Grossman, and David F. Trask to review the work of the project; the committee made valuable suggestions on many aspects of our work. Helen Aguera, George F. Farr, Jr., and Kathy J. Fuller, members of the NEH staff, have visited the project and helped the editors to describe more precisely the methodology and identify more clearly the audience for the printed volumes. A timely grant from the Ford Foundation provided clerical, secretarial, and translation support and photographic services.

The project is indebted to David Brody, Philip S. Foner, David Montgomery, Maurice F. Neufeld, and Irwin Yellowitz, members of our board of editorial advisors who critically read the manuscript of

this volume and provided detailed written critiques. Other members of our board, who also read the volume and made helpful suggestions for revision, and who have been of assistance to the project in a variety of ways, include Emory G. Evans, Herbert G. Gutman, Louis R. Harlan, Walter Rundell, Jr., and Philip Taft.

The editors have also benefited from the work of scholars who have either directly shared their own research with us or worked with our staff for a time. Leslie Rowland loaned her valuable collection of material on New York cigarmakers gleaned from newspaper, magazine, and secondary sources. Judith Goldberg provided information on the KOL in Philadelphia and George Pearlman sent us documents and information on P. J. McGuire. Martha Woodward took time out from her year in England to help us authenticate Gompers' birth certificate and his parents' marriage certificate, and Kenneth Fones-Wolf has helped out repeatedly as a source of information. Dorothee Schneider shared her own research and insights on the German labor movement of New York City and helped to refine the selection and notes for this volume. She, Hannelore Jarausch, and Patrick McGrath produced the volume's careful translations of the German-language documents; they received advice on specialized aspects of their translations from Vernon Lidtke of the Johns Hopkins University and Karl Stowasser of the University of Maryland and utilized the previous work of David Carl and Karl Moehlmann at Pace University. Grace Palladino joined our staff in time to undertake a rigorous critique and rewriting of our introduction and editorial essays as well as to help with the last stages of the volume's production. Debra Herman, an NHPRC Fellow-in-Editing, expanded the project's holdings of photographic material, from which we selected the photographs and drawings for this volume. Raymond W. Smock joined the staff briefly to critique the volume and to assist with the collection of photographs. Several other members of the University of Maryland faculty provided timely advice, James Harris advising us on the annotation of individuals in the German socialist movement, John McCusker suggesting British sources of documents, and Ronald Weissman and Sibylle Sampson consulting with us in our selection of word-processing equipment. The map for the volume was prepared by Jean Locke of the Cartographic Services Laboratory at the University of Maryland, Joel R. Miller, director.

It would be difficult to do justice to the quantity of careful and tedious work involved in searching, collecting, logging-in, selecting, transcribing, proofreading, annotating, and publishing the documents in a volume such as this. An acknowledgment can only barely begin to recognize the long hours of dedicated work many staff workers

have contributed in support of the editors' efforts. Celia Ramos Gray, the project secretary, began work in 1980 enabling the staff to utilize word-processor technology to its maximum potential. Many graduate assistants aided the project, particularly in the area of document search and annotation. These include Frederick Augustyn, Patricia A. Cooper, David Corbin, Joseph Dees, Anna Leon, Lizette LeSavage, Harold Eugene Mahon, Katherine Kidd Morin, Lawrence Myers, Timothy Newell, Nora Oakes, Judy Riordan, Elizabeth Robertson, William Ross, Kevin Swanson, and Richard Wilkoff. Graduate students in the Historical Editing Seminar at the University of Maryland over the last seven years spent a portion of their semester interning with the project and contributed importantly to the annotation research of this volume. Undergraduate student assistants logged in and transcribed documents, helped with proofreading, handled a variety of clerical tasks, and undertook routine searches. They include: Margaret Rauner Brodnick, Susan Compton, Deona Dichoso, Tonya Little, Richard F. Measell, Joan Bellistri Simison, and Marie Thomas.

## *Notes*

1. Benjamin Stolberg, "What Manner of Man Was Gompers?" *Atlantic Monthly* 135 (1925):404.
2. See, for example, Gerald N. Grob, *Workers and Utopia: A Study of Ideological Conflict in the American Labor Movement, 1865-1900* (Evanston, Ill., 1961), p. 145; Harold C. Livesay, *Samuel Gompers and Organized Labor in America* (Boston, 1978), pp. 7-8; James Weinstein, "Gompers and the New Liberalism, 1900-1909," and Ronald Radosh, "The Corporate Ideology of American Labor Leaders from Gompers to Hillman," in James Weinstein and David W. Eakins, eds., *For a New America: Essays in History and Politics from* Studies on the Left, *1959-1967* (New York, 1970), pp. 101-14, 125-51; Philip S. Foner, "Class Collaborator," in Gerald E. Stearn, ed., *Gompers* (Englewood Cliffs, N.J., 1971), p. 143. David Brody, *Workers in Industrial America: Essays on the Twentieth Century Struggle* (New York, 1980), p. 23.
3. SG, *Seventy Years,* 2:524.
4. Ibid., 1:361, 329.
5. A partial list of the material the AFL sent the New York Public Library is to be found in "The Samuel Gompers Collection," Mar. 25, 1926, in "File of Correspondence in Connection with the Life and Death of Samuel Gompers, President, American Federation of Labor, 1924-1937," reel 8, track 2, President Gompers' Files, George Meany Memorial Archives, AFL-CIO, Washington, D.C. Also see the New York Public Library *Bulletin* 30 (1926):155-56. The Library has since microfilmed some of this material on four reels of film. The filmed records include biographical sketches; addresses before meetings of the National Civic Federation, labor conventions and meetings, university students, and general audiences; press statements and interviews; and articles by SG.
6. "The Samuel Gompers Collection," Mar. 25, 1926.
7. "American Federation of Labor Archives," Mar. 2, 1937, AFL File,

Manuscripts Division, Library of Congress, Washington, D.C. This includes an inventory made by Thomas Powderly Martin, acting chief of the Manuscripts Division, that lists fifty-four file drawers of correspondence and printed material in this collection.

8. Stuart B. Kaufman and Peter J. Albert interviews with E. Logan Kimmel, Apr. 28, 1976, and July 27, Aug. 22, 1977.

9. ["The New Labor History," draft], Vaughn Davis Bornet Papers, State Historical Society of Wisconsin, Madison. In 1952 Bornet prepared an inventory of the collection entitled "The American Federation of Labor Archives" and a list of recommendations as to its disposition, "The American Federation of Labor to Posterity," which he presented to Florence Thorne; both the inventory and the recommendations are in his papers at Madison. Also see Bornet, "The New Labor History: A Challenge for American Historians," *Historian* 18 (1955):4.

10. Kaufman and Albert interview with Kimmel, July 27, 1977; Kimmel memo to George Meany, Feb. 20, 1953, Meany Archives; Robert Land memo of conversation with Kimmel, Mar. 17, 1953, Manuscripts Division, Library of Congress.

11. Kimmel to Meany, memo, May 19, 1953, Meany Archives; Land memo, Aug. 21, 1953, Manuscripts Division, Library of Congress.

12. Kaufman and Albert interview with Kimmel, July 27, 1977; Kimmel to William Schnitzler, memo, Sept. 26, 1958, Meany Archives.

13. Bornet to Clifford Lord, Mar. 28, 1955, Bornet Papers; Edwin Witte to Lord, Apr. 11, 1955, AFL-CIO File, State Historical Society of Wisconsin.

14. Lord field report, May 12-15, 1955, Lord to Meany, July 9, 1955, ibid.

15. AFL Executive Council Minutes, Aug. 9, 1955; Schnitzler to Lord, Nov. 9, 1955, AFL-CIO File, State Historical Society of Wisconsin.

16. Lord field report, May 12-15, 1955; Kimmel to Meany, memo, July 12, 1955, Meany Archives; Richard Erney field report, June 5-30, 1961, AFL-CIO File, State Historical Society of Wisconsin.

17. Except for the last, the Gompers Papers project published these records as reels 25 through 57 of *American Federation of Labor Records: The Samuel Gompers Era* (Sanford, N.C., 1979).

18. Kimmel to Wesley Reedy, memo, Mar. 12, 1964, Meany Archives. The Society reorganized and refilmed the Gompers files; they are published as reels 59 through 137 of the *AFL Records.*

19. Kimmel to Wesley Reedy, memo, Mar. 12, 1964, Meany Archives. The Society filmed and published the national and international union correspondence (1885, 1890-1911) and the mining department records (1911-15) as reels 138 through 144 of the *AFL Records.*

20. John Lorenz to Schnitzler, Oct. 7, 1966, Meany Archives; Paul Heffron to Jean Webber, Nov. 23, 1973, AFL-CIO Library.

21. The Gompers Papers project filmed and published the Executive Council material, as well as much of the Gompers correspondence, letterbooks, scrapbooks, and some miscellaneous items retained by the Federation, as reels 1 through 24 of the *AFL Records.*

22. The project published these as its second microfilm publication, *The American Federation of Labor and the Unions: National and International Union Records from the Samuel Gompers Era* (Sanford, N.C., 1982).

## SYMBOLS AND ABBREVIATIONS

| | |
|---|---|
| A and HLS | Autograph and handwritten letter, signed |
| ADS | Autograph document, signed |
| AFL | The American Federation of Labor |
| ALtS | Transcript of autograph letter, signed |
| ALS | Autograph letter, signed |
| CMIU | The Cigar Makers' International Union |
| DCU | The Catholic University of America, Washington, D.C. |
| DNA | The National Archives of the United States, Washington, D.C. |
| FOTLU | The Federation of Organized Trades and Labor Unions of the United States and Canada |
| HLS | Handwritten letter, in a hand other than the author's, signed |
| HLtSr | Transcript of handwritten letter, signature representation |
| KOL | The Knights of Labor |
| MiDW | Wayne State University, Detroit, Mich. |
| PD | Printed document, not signed |
| PLSr | Printed letter, signature representation |
| PRO | The Public Record Office, London, England |
| SG | Samuel Gompers |
| SLP | The Socialistic or Socialist Labor party |
| TLp | Typed letter, letterpress copy, not signed |
| TLpS | Typed letter, letterpress copy, signed |
| WHi | The State Historical Society of Wisconsin, Madison |

## SHORT TITLES

| | |
|---|---|
| *AFL Records* | Peter J. Albert and Harold L. Miller, eds., *American Federation of Labor Records: The Samuel Gompers Era,* microfilm (Sanford, N.C., 1979) |
| AFL, *Proceedings,* 1886 | AFL, *Proceedings of the First Annual Convention of the American Federation of Labor* (1886?; reprint ed., Bloomington, Ill., 1905) |
| FOTLU, Proceedings, 1881 | FOTLU, *Report of the First Annual Session of the Federation of Organized Trades and Labor Unions of the United States and Canada, . . . December 15, 16, 17 and 18, 1881* (1881?; reprint ed., Bloomington, Ill., 1905) |
| FOTLU, Proceedings, 1884 | FOTLU, *Report of the Fourth Annual Session of the Federation of Organized Trades and Labor Unions of the United States and Canada, . . . October 7, 8, 9 and 10, 1884* (1884?; reprint ed., Bloomington, Ill., 1905) |
| FOTLU, Proceedings, 1885 | FOTLU, *Report of the Fifth Annual Session of the Federation of Organized Trades and Labor Unions of the United States and Canada . . . December 8, 9, 10 and 11, 1885* (1885?; reprint ed., Bloomington, Ill., 1905) |
| FOTLU, Proceedings, 1886 | FOTLU, *Report of the Sixth Annual Session of the Federation of Organized Trades and Labor Unions of the United States and Canada* (1886?; reprint ed., Bloomington, Ill., 1905) |
| SG, *Seventy Years* | Samuel Gompers, *Seventy Years of Life and Labor: An Autobiography,* 2 vols. (New York, 1925) |

## CHRONOLOGY

| | | |
|---|---|---|
| 1850 | Jan. 26 | Birth of SG in London |
| 1856-60 | | SG attends Jews' Free School |
| 1860 | | SG enters cigarmaking trade |
| 1863 | June 10–July 29 | Gompers family immigrates to New York City |
| 1867 | Jan. 28 | Marriage of SG and Sophia Julian |
| 1869 | Dec. | KOL founded |
| 1872 | Oct. 4 | SG becomes a U.S. citizen |
| 1873 | ca. Mar. | Founding of the United Cigarmakers |
| | May | Publication of Carl Hillmann's *Praktische Emanzipationswinke,* Leipzig |
| 1874 | Jan. 13 | Tompkins Square riot |
| | Nov. 7 | Publication of New York City Board of Health report on tenement-house cigarmaking |
| 1875 | Nov. 24 | The CMIU charters local 144 with SG as president |
| 1876 | July | Founding of the Workingmen's Party of the United States, in Philadelphia |
| 1877 | Spring | Founding of the Amalgamated Trades and Labor Union of New York and Vicinity |
| | July | National railroad strikes |
| | Aug. 30–Sept. 2 | CMIU convention, held in Rochester, N.Y., elects Adolph Strasser president of the union |
| | ca. Sept.–ca. Jan. 1878 | New York City cigarmakers' strike |
| | Dec. | Founding of the SLP |
| 1878 | | First General Assembly of KOL |
| 1879 | Sept. | KOL General Assembly elects Terence Powderly grand master workman |
| 1881 | Aug. 2-3 | National labor convention held in Terre Haute, Ind. |
| | Oct. 31–Nov. 14 | SG's tenement-house articles published in the *New Yorker Volkszeitung* |
| | Nov. 15-18 | Founding convention of the FOTLU, held in Pittsburgh |
| 1882 | Feb. | Founding of the New York City Central Labor Union |
| | Apr. 1 | Disputed election of CMIU 144 officers |
| | July 1 | KOL District Assembly 49 organized in New York City |
| | July 23 | Founding of local 1 of the Cigarmakers' Progressive Union of America |

| | | |
|---|---|---|
| | Nov. 21-24 | Second FOTLU convention, held in Cleveland |
| 1883 | Jan. | SG and Strasser found KOL Local Assembly 2458 (the "Defiance Assembly") |
| | Mar. 12 | New York legislature passes bill prohibiting tenement-house cigar manufacture |
| | Aug. 16, 18, 27 | SG testifies before the Education and Labor Committee of the U.S. Senate |
| | Aug. 21-24 | Third FOTLU convention, held in New York City |
| 1884 | May 12 | New York legislature passes second bill prohibiting tenement-house cigar manufacture |
| | Oct. 7-10 | Fourth FOTLU convention, held in Chicago, proposes eight-hour campaign to culminate on May 1, 1886 |
| 1885 | Dec. 8-11 | Fifth FOTLU convention, held in Washington, D.C. |
| 1886 | Jan. 1 | Seceding faction of Cigarmakers' Progressive Union joins CMIU |
| | Jan. 20 | Lockout of New York City cigarmakers by United Cigar Manufacturers' Association firms |
| | Feb. | Cigarmakers' Progressive Union, New York City Central Labor Union, and KOL District Assembly 49 reach agreement with cigar manufacturers, ending lockout. CMIU continues its strike. |
| | May 4 | Bombing in Chicago's Haymarket Square |
| | May 18 | Trade union conference, Donaldson Hall, Philadelphia, meets to consider differences with KOL |
| | May 25–June 3 | Special session of the KOL General Assembly, held in Cleveland, receives proposed treaty from trade union committee |
| | Aug. 7 | CMIU and Cigarmakers' Progressive Union mass meeting at Cooper Union denounces KOL District Assembly 49 |
| | Aug. 11 | Cigarmakers' Progressive Union convention votes to amalgamate with CMIU |
| | Sept. 28 | Trade union committee meets with KOL General Executive Board, in Philadelphia |
| | Oct. 4-20 | KOL General Assembly, held in Richmond, Va., requires member cigarmakers to leave CMIU or suffer expulsion from the Knights |
| | Oct. 5 | Henry George accepts nomination for mayor of New York City |
| | Dec. 7-11 | Sixth convention of the FOTLU and founding convention of the AFL, held in Columbus, Ohio |

*Documents*

## *Samuel Gompers' Childhood in London and His Early Years in New York City*

Samuel Gompers' paternal grandfather, Samuel Moses Gompers, worked for a time as a calico printer in Amsterdam before becoming an import-export merchant. In 1845 he settled with his wife and their five children in the Spitalfields district of East London, not far from what was known as "the London Ghetto." Gompers remembered the area, one of the poorest in the city, for its "endless rows of shabby houses bordered by pavements—nothing else, no trees, no green grass, no flowers."[1]

As a merchant, Samuel Moses traveled regularly between London and the Continent, and in Amsterdam he became acquainted with the Roods, a Dutch family of "well-to-do tradesmen."[2] Their daughter Sarah eventually accompanied him back to England and moved in with the Gompers family; some two years later, on November 27, 1849, she married Solomon Gompers, Samuel Moses's eldest son who was a cigarmaker by trade. Shortly thereafter Samuel Gompers was born in the house in Tenter Street where his parents and his grandparents lived.

The two families subsequently moved to Fort Street, where they lived directly across from a silk factory. Their house, "worn gray with the passing years,"[3] according to Gompers, provided shelter during the next decade for Solomon's growing family on the ground floor, Samuel Moses's family on the second, and still a third family on the top floor. Solomon's apartment consisted of only two rooms—a large front room that served as a sitting room, bedroom, dining room, and kitchen; and a smaller back room that was used for storage in the winter and as the children's bedroom in the summer.

From 1856 to 1860, Gompers was enrolled at the Jews' Free School on Bell Lane, one of the largest elementary schools in Great Britain. At the age of ten, however, although he ranked third in his class, the financial condition of his family required him to leave school. Gompers' father apprenticed him to a shoemaker, but this arrangement lasted only eight weeks because the noise of the shop repelled the boy. Gompers then started working for a cigarmaker in Bishopsgate Street. According to Gompers, he preferred that trade because "there

was a society among the cigarmakers but none among the shoemakers."[4] Although he attended classes at the Jews' Free School at night, studying the Talmud, Hebrew, French, and music, Gompers also continued his education in the cigar factory, where he listened to fellow workers debate the issues surrounding the American Civil War.

As Solomon's family grew, its expenses increased and its economic situation became more difficult. Deteriorating conditions in the cigarmaking trade further complicated family finances. Import duties that the United States levied against European tobaccos and cigars in 1862 made North America a less attractive market, and the trade suffered a sudden decline. To relieve the depression in the industry, the Cigarmakers' Society of England created a fund to help cigarmakers emigrate. After much deliberation, Solomon Gompers decided to take advantage of this financial assistance and immigrate to the United States where he had friends and relatives and where the cigar industry was expanding. The Gompers family left England on the ship *London* on June 10, 1863, and arrived in New York City on July 29.

They settled on the Lower East Side of Manhattan, an area well known for its immigrant population. Although the ethnic mix of the neighborhood varied over the years, by the late 1860s and 1870s German immigrants predominated there to such a degree that English-speaking residents called the district around First and Second avenues "Dutchtown" and Avenue A "Dutch Broadway."[5] If the Gompers' first home in New York, a four-room apartment on Attorney Street, was an improvement over their housing in London, the family still lived in an overcrowded, squalid, tenement district that also housed manufacturing and commercial concerns. Gompers lived on the Lower East Side for most of the next two decades, although he and his family moved frequently within the area, "seeking some little improvement in comfort or rent."[6] Their living quarters—on Houston, Sheriff, Lewis, Columbia, Stanton, and Rivington streets—were small two- or three-room apartments with only one room open to the outside air. The unventilated rooms were stifling in the summer and cold in the winter, and indoor sanitary facilities were nonexistent.

In the front room of their first apartment in New York Solomon Gompers set up a workshop where he and Samuel made cigars. Samuel worked with his father for some eighteen months; then, after gaining what he called the "self-confidence that goes with mastery of a trade,"[7] he began working in cigarmaking shops in the city. His first position was at Stachelberg's, where, he later wrote, he emerged as a shop

leader, defending fellow workers and presenting their grievances to the employer.

Changing market conditions, lockouts, and strikes forced Samuel to move from job to job. Although working conditions were generally poor, he later regarded this life in the factory to be one of his "most pleasant memories." "I loved the touch of soft velvety tobacco," he remembered, "and gloried in the deft sureness with which I could make cigars grow in my fingers, never wasting a scrap of material." Gompers appreciated the "mind-freedom that accompanied skill as a craftsman," which left accomplished cigarmakers "free to think, talk, listen, or sing" because they could perform their tasks "more or less mechanically." Occasionally they chose one of their shopmates to read to them—each donating some of their finished cigars to reimburse him for his lost time—and such reading was generally followed by discussion. This atmosphere, Gompers maintained, encouraged a free exchange of ideas and an enduring fellowship among shopmates.[8]

Samuel Gompers and his father joined local 15 of the Cigar Makers' National Union of America—New York City's English-speaking local—in 1864, but Samuel attended meetings only occasionally in the 1860s. Instead, he became active in fraternal organizations that, he found, provided a better outlet for his "idealism and sentiment." At the age of fourteen he helped form and became president of the Arion Base Ball and Social Club, which organized athletics, debated public issues, and held mock court. He participated as well in the Ancient Order of Foresters, helped found the Independent Order of Foresters and the Rising Sun Lodge of the Independent Order of Odd Fellows, and took an active role in the Hand-in-Hand Society, a Jewish mutual benefit organization. During these years Gompers supplemented his brief formal education by taking free classes at Cooper Union in "history, biography, music, mechanics, measurement of speed, elocution, economics, electric power, geography, astronomy, and travels," attending Saturday evening lectures there, and participating in the debating club.[9]

In the summer of 1866 Gompers met Sophia Julian, a London-born tobacco stripper who worked in the same cigarmaking shop. A few months later, mutual friends suggested that Samuel and Sophia should marry to celebrate his seventeenth birthday. Thus, "without consultation or announcement of plans," as Gompers put it, the two married on January 28, 1867, but then returned to their own homes at the evening's end. After the marriage license was published, Gompers recalled, "there was a sort of a hullabaloo about it and then it was simply all right and my wife came over to our house and lived with us."[10] Subsequently they went to Hackensack, N.J., where Gom-

pers found work at the shop of George Edmonson. Sophia moved back to Gompers' parents' home in New York City just prior to the birth of her first child, Samuel Julian Gompers, on September 4, 1868. Gompers continued to work in Hackensack and later in Lambertsville, N.J.; then, probably in the fall of 1868, he returned to work in the New York City cigar shops that would become the center of his activity during the next decade.

### Notes

1. SG, *Seventy Years,* 1:2, 16.
2. Ibid., p. 3.
3. Ibid., p. 1.
4. Ibid., p. 18.
5. Ibid., p. 72.
6. Ibid., p. 493.
7. Ibid., p. 34.
8. Ibid., pp. 66, 56-57, 45.
9. Ibid., pp. 42, 30.
10. Ibid., p. 36.

# The Marriage Certificate of Solomon Gompers[1] and Sarah Rood[2]

[November 27, 1849]

1849. Marriage Solemnized In the Great Synagogue chambers in the Parish of St. James Dukesplace in the City of London

| No. | When Married. | Name and Surname. | Age. | Condition. | Rank or Profession. | Residence at the Time of Marriage. | Father's Name and Surname. | Rank or Profession of Father. |
|---|---|---|---|---|---|---|---|---|
| 134 | Twenty Seventh day of November 1849 | Solomon Gompertz | Full | Bachelor | Cigar Maker | 19 Tenter Street Spitalfields | Samuel Gompertz[3] | Dealer |
| | | Sarah Root | Minor | Spinster | ——— | 19. Tenter Street Spitalfields | Moses Root[4] | Deceased |

Married at the Great Synagogue chambers according to the Rites and Ceremonies of the Jewish Religion by me,
L E Pyke[5]

| | | | |
|---|---|---|---|
| This Marriage was solemnized between us, | O The mark of Solomon Gompertz<br>O The mark of Sarah Root) | in the Presence of us, | Mark Marks<br>Hyam Jonas |

Simeon Oppenheim[6]
Secy of the Great Synagogue Dukesplace

Printed form, with autograph entries, signed, book 5, p. 67, Register Office, District of Islington, Finsbury Town Hall, London.

1. Solomon GOMPERS, a cigarmaker, was SG's father.
2. Sarah Rood GOMPERS, SG's mother.
3. Samuel Moses (Salomon Mozes) GOMPERS (1803?-81), an import-export merchant, was SG's paternal grandfather.
4. Moses Isayes ROOD, a porter, was SG's maternal grandfather.
5. Lewis Eleazar Pyke (1789-1851) for many years served as shomer or beadle of the Great Synagogue.
6. Simeon Oppenheim (1808-74), a London merchant, served as secretary of the Great Synagogue between 1843 and 1866.

# The Birth Certificate of Samuel Gompers

[March 5, 1850]

1850. BIRTHS in the Spitalfields District of Whitechapel in the county of Middlesex

| No. | When Born. | Name, if any. | Sex. | Name and Surname of Father. | Name and Maiden Surname of Mother. | Rank or Profession of Father. | Signature, Description, and Residence of Informant. | When Registered. | Signature of Registrar. | Baptismal Name, if added after Registration of Birth. |
|---|---|---|---|---|---|---|---|---|---|---|
| 159 | Twenty sixth[1] January 1850 19 Tenter Street Christ Church | Samuel | Boy | Solomon Gumpertz | Sarah Gumpertz formerly Root | Cigar Maker | O The Mark of Sarah Gumpertz Mother 19 Tenter Street Christ Church | Fifth March 1850 | George Deboos Registrar | |

Printed form, with autograph entries, signed, Spitalfields book 13, p. 32, Register Office, District of Stepney, London.

1. In his autobiography SG states that he was born on Jan. 27 at "No. 11 Tentor Street" (*Seventy Years,* 1:1, 36).

# Entries for the Gompers Household in the 1851 Census Return for London

[1851]

| Liberty Old Artillery Ground | | Ecclesiastical District of St Mary | | Borough of Tower Hamlets | | | | |
|---|---|---|---|---|---|---|---|---|
| No. of House-holder's Schedule | Name of Street, Place, or Road, and Name or No. of House | Name and Surname of each Person who abode in the house, on the Night of the 30th March, 1851 | Relation to Head of Family | Condition | Age of males | Age of females | Rank, Profession, or Occupation | Where Born |
| 113 | [2 Fort St.] | Catherine Gompers[1] | Wife | M | | 44 | General Dealer | Holland |
| | | Solomon | Son | M | 22 | | Cigar Maker | do |
| | | Sarah | Sons Wife | M | | 24 | | do |
| | | Fanny[2] | Dau | UM | | 14 | Cap Maker | do |
| | | Mary[3] | Dau | UM | | 12 | do | do |
| | | Catherine[4] | Dau | | | 10 | do | do |
| | | Sarah[5] | Dau | | | 8 | | do |
| | | Simeon[6] | Son | | 2 | | | Middlesex Spitalfields |
| | | Samuel[7] | Gd Son | | 14m | | | do do |
| | | May Duggan | Servant | UM | | 20 | General Servant | Ireland |
| | | Aaron [Boseen?] | Lodger | M | 65 | | | Holland |
| | | Isaac Goldsmith | Lodger | UM | 17 | | General Dealer | ″ |
| | | Joel [Driazeloor?] | Lodger | UM | 22 | | Hawker | ″ |
| | | Abraham Martin | Lodger | UM | 35 | | Hawker | Holland |

Printed form, with handwritten entries, Census of 1851, H.O. 107/1543, folio 46, pp. 24-25, PRO. The following column was left blank by the census taker: "Whether Blind, or Deaf-and-Dumb."

1. Actually Henrietta (Jette) Salomon Haring GOMPERS (1807?-79), SG's paternal grandmother.
2. Fanny (Femmetje) Gompers COHEN, SG's aunt.
3. Clara (Mietje) Gompers LE BOSSE, SG's aunt.
4. Catherine (Grietje) GOMPERS, SG's aunt.
5. Sarah Gompers LEVY, SG's aunt.
6. Simon GOMPERS, SG's uncle and childhood companion, became a shoemaker.
7. SG.

# An Excerpt from the Forty-Third Report of the Jews' Free School[1]

May, 5620—1860.

. . .

The Committee[2] venture to assert, that, whether as regards magnitude, completeness, or success, the Jews' Free School is not second to any School in the British dominions. In respect to its size, it may be stated that there is accommodation for 2,000 children, giving to each child a healthful amount of space. In respect to completeness, it may be mentioned that every department is specially adapted for its peculiar purpose. Galleries are provided for simultaneous teaching; separate groups of parallel desks (of the most approved construction) are set apart for individual instruction; spacious and even elegant class-rooms are appropriated for those higher branches of study which demand isolation and fixity of purpose. A good library for reference and circulation (with a reading room attached) forms a prominent feature in the Establishment. A museum supplied with almost everything that can be required to illustrate lessons on art, science, and manufacture, and which is continually receiving additions, has also been formed. Text-books and class-books, selected with the greatest care, drawing-copies and models, maps, diagrams; every article, in short, that was formerly regarded as a luxury of school life, has here been considered a necessity, and has been copiously supplied. Nor has attention been wanting in other details. Every improvement in heating, ventilation, and drainage, has been carried out, and a constant and adequate supply of water secures the comfort and cleanliness of the pupils. To the Girls' School is attached a laundry replete with every convenience for its particular purpose; and Girls are trained to habits of industry and order, which must exercise a beneficial influence on their future career. In both playgrounds there is a covered way for protection, during wet weather; in the Boys' play-ground this covered way is also used as an open class space, during the heat of summer.

In respect to the educational advantages afforded, it is proper to say, that every department of Hebrew and English is sedulously cultivated. In the Boys' School the *Talmud Torah* is incorporated with a number of other pupils, and these together study *Tanakh*[3] Commentary, Mishna,[4] Hebrew Grammar, and Composition. Other classes are made familiar with the translation of the Pentateuch and Prayer-book, and with Hebrew writing; and throughout the school, the sacred language engrosses its proper share of attention. In English, beside the popular reading, writing, and arithmetic, Euclid, Mensuration,

Algebra, Natural Science, History, Geography, and Grammar, enter into the course of study. During the past year, William Ellis, Esq.,[5] in the true spirit of philanthropy, has conducted a class for the study of social economy, which important branch of Education will in future form an integral portion of the system of the School. Physiology, as applied to health, is taught in the highest class, who are also instructed in vocal music from notes, and in drawing. Object and gallery lessons form a prominent feature in the daily manual of instruction; these latter are made subservient to a knowledge of Biblical history.

In the Girls' School, less attention is, of necessity, devoted to the higher branches of intellectual cultivation, in order to afford time for instruction in duties more especially useful to females. Needle-work in all its applications to domestic or ornamental purposes, washing, ironing and other household economies, are sedulously inculcated, while care is taken that every Girl shall acquire at least a knowledge of her prayers, not only in the original, but also in the vernacular. History, Geography, and Grammar also form parts of the course of study in the Girls' School.

But the exertions of the Committee have not been confined to supplying the mental wants of the children only; the Free School has acquired all the characteristics of a normal School, and is not only training young men and women to the important duties of teaching within its walls, but is also fitting them for similar and more important functions elsewhere. The Boys' School being more regularly under Government Inspection,[6] the pupil teachers, now thirteen in number, are, of necessity, conducted through a course of study to fit them for their annual examination by H. M. Inspector; and it is gratifying to the Committee to state, that hitherto all the apprentices have acquitted themselves satisfactorily.[7] But besides this course, the pupil teachers receive regular and systematic instruction in Latin, French, and Hebrew, and the higher branches of literature; the same advantages are also offered to all the adult teachers, who desire to avail themselves of the opportunity for improvement. In the Girls' School two classes of teachers are being trained by masters specially adapted for that purpose.

The results of the system are already apparent; seven pupil teachers have successfully completed their apprenticeship under Minutes of Council. Of these, one, Mr. G. J. Emanuel, has been appointed to the English and classical mastership of the Jews' College School; one, Mr. I. Hart has been elected Hebrew Master to the Manchester Jews' School; one, Mr. J. H. Cohen, has undertaken a private School at Ramsgate; one has become a teacher in the United States; the other three are engaged as Assistant Teachers; and, while undergoing in

the School the necessary probation of three years' service before becoming eligible to receive Certificates, they have entered at the University of London, and are contemporaneously striving for degrees in arts. Miss Lipman, educated under the direction of the Committee, has also been found capable of discharging the duties of Assistant Governess in the Girls' School.

The Boys' School is under the special supervision of Mr. M. Angel,[8] whose invaluable services as Head Master, and whose energetic and wide-spread exertions to serve the interests of the School in every way, have long secured the completest confidence and respect of the Committee. Mr. M. H. Myers[9] continues to occupy the position of senior Hebrew Master which he has so long filled with honor to himself, and advantage to the Institution. In addition to these gentlemen, there are two certificated, and seven other assistant teachers, and thirteen pupil teachers. The Boys' School now contains 1,060 pupils. Of these, 210 occupy the new wing or class-rooms. They are divided into four sections; and are taught, besides the ordinary branches of an elementary education, translation of Bible and prayers, geography, grammar, history, and social economy; the highest class learning in addition, Euclid, Algebra, and Physics. Four classes of 200 boys occupy one compartment of the great school, and are being qualified by sound elementary instruction to fill the vacancies constantly occurring in the class-rooms. The central division of the School consists of six sections of 350 boys, who learn reading, writing, arithmetic, and prayers, so as to permit of their being drafted into higher classes. The lowest division consists of six sections of 300 boys, who are taught on the simultaneous system. It will thus be seen, that while there is a well-defined gradation in the education given in the school, the instruction is as widely diffused as possible, and great care is taken that every boy shall receive a share proportioned to his capabilities. It may not be superfluous here to add, that by a simple architectural arrangement, these three compartments are convertible into one vast area, in which nearly 3,000 persons are accommodated during the *Yamim Nora'im*[10] (New Year and Day of Atonement), when the School is fitted up as a temporary place of worship, subsidiary to the Metropolitan Synagogues.

The Girls' School is under the direction of Mrs. Phillips,[11] whose management completely justifies the confidence reposed by the Committee in her judgment, perseverance, and capabilities. Under her there are, one assistant governess, two teachers for needlework, and fifteen assistant teachers. There are also ten young ladies under training, who assist in the duties of the School. The number of the children in attendance is 675. They are divided as follows:—117 occupy the

lowest class-room, and are taught simultaneously. 400 are in the great school-room; these are separated into twelve groups by means of curtains, and receive individual instruction in the ordinary branches of female education, so as to fit them gradually to advance into the class-rooms. The highest three sections, of 158 girls, are taught in separate rooms; their course of instruction carries forward the education imparted in the great school-room. A systematic plan thus ensures that every child shall receive a due proportion of information, and diffuses the blessings of education to the widest possible extent.

. . .

Jews' Free School, *Forty-third Report of the H[avurat] K[odesh] Talmud Torah ve-Hinukh Yeladim*[12] *or Jews' Free School, Bell Lane, Spitalfields* (London, 1860), pp. 11-16.

1. Established as a boys' school in 1817, the Jews' Free School in Bell Lane had grown out of a small institution for religious education founded by the Great Synagogue in 1732. During the nineteenth century, with substantial patronage from the Rothschild family, it became one of the largest elementary schools in Great Britain and was opened to both boys and girls.

2. The Committee of the Jews' Free School, the governing body of the institution.

3. A Hebrew term for the Old Testament.

4. The collection of Jewish traditions forming the basic part of the Talmud.

5. William Ellis (1800-1881), economist, educator, and philanthropist, introduced the teaching of political economy in schools.

6. The school was inspected annually as a condition of receiving Parliamentary educational grants. At this time, the inspections of the Jews' Free School were conducted by Matthew Arnold (1822-88), a government inspector of schools from 1851 to 1886.

7. In addition to serving as an elementary school, the Jews' Free School was a normal school, operating under "Minutes of Council," regulations of the Committee of Council on Education, by which pupil-teachers received stipends while completing five-year apprenticeships. Renewal each year was subject to the trainees' successfully passing examinations by a government inspector.

8. Moses Angel (1819-98) was the headmaster of the Jews' Free School from 1840 to 1897.

9. Morris H. Myers retired in October 1860 after thirty-two years of service at the Jews' Free School.

10. "Days of Awe," the term applied to the period from the first day of Rosh Hashanah through Yom Kippur.

11. Hannah B. L. Phillips (d. 1889) was headmistress at the Jews' Free School for twenty-five years.

12. "Holy Society for the Study of the Torah and the Education of Children."

# An Excerpt from a News Account of an Award Ceremony at the Jews' Free School

[August 3, 1860]

. . .

. . . Mr. Sampson Samuel,[1] at the invitation of the President,[2] addressed a few words to the children, representing to the prizeholders that they should not estimate the prizes by their monetary value, but consider them as tokens of the approbation of the committee of their moral and intellectual progress, and urging upon the unsuccessful competitors to try again, as true merit sooner or later was sure to be recognised. He concluded by forcibly pointing out that intellectual progress, after all, must be subordinate to moral and religious advancement, upon which individual and collective happiness much more depended than upon the most splendid talent.

. . .

We now subjoin a list of the names of the pupils that received prizes:—

. . .

3rd Class.—Abraham Lyons, Barnett Myers, Morris Duparque, Joseph Eagleman, Phillip Camper, David Magus, Moss Cohen, Tobias Cohen, Jacob Morris, Benjamin Solomon, Wolfe Isaacs, Samuel Gompertz, Henry Ford, Jacob Barnett, Asher Bernstein, Eliazer Bernstein.

. . .

*Jewish Chronicle and Hebrew Observer,* Aug. 3, 1860.

1. Sampson Samuel was a member of the Committee of the Jews' Free School.
2. Sir Anthony Rothschild (1810-76) was president of the Committee of the Jews' Free School and first president of the United Synagogue, London.

# Entries for the Households of Samuel Gompers' Parents and Grandparents in the 1861 Census Return for London

| The undermentioned Houses are situate within the Boundaries of the | | | | |
|---|---|---|---|---|
| Parish [or Township][1] of Old Artillery Ground | | | Parliamentary Borough of Tower Hamlets | |
| No. of Schedule | Road, Street, &c., and No. or Name of House | Name and Surname of each Person | Relation to Head of Family | Condition |
| 3 | [2 Fort Street] | Solomon Gomperts[2] | Head | Mar |
| | | Sarah do | Wife | Mar |
| | | Samuel do | Son | |
| | | Henry do[3] | Son | |
| | | Alexander do[4] | Son | |
| | | Levy do[5] | Son | |
| | | Harriett do[6] | Daur | |
| 4 | ″ | Samuel Gompers | Head | Mar |
| | | Harriet do[7] | Wife | Mar |
| | | Solomon do | Son | Mar |
| | | Fanny do | Daur | Mar |
| | | Clara do | Daur | Mar |
| | | Kate do | Daur | un |
| | | Sarah do | Daur | un |
| | | Simon do | Son | |
| | | Amelia Garaty | | un |

1. These brackets are on the original form.
2. Solomon Gompers is listed twice in this return—as the head of his own household (schedule 3) and as the son of Samuel Moses Gompers (schedule 4).
3. Henry GOMPERS (b. 1853), SG's brother, became a cigarmaker.
4. Alexander GOMPERS (1857-1926), SG's brother, became a cigarmaker.
5. Louis GOMPERS, SG's brother, became a cigar manufacturer.
6. Harriett GOMPERS, SG's sister.
7. Henrietta Gompers.

[1861]

| | | | |
|---|---|---|---|
| | | Ecclesiastical District of St Marys | |
| Age of | | Rank, Profession, or Occupation | Where Born |
| Males | Females | | |
| 32 | | Cigar Maker | Amsterdam Holland |
| | 33 | | Do |
| 11 | | | Spitalfields |
| 8 | | | Whitechapel |
| 4 | | | Spitalfields |
| 2 | | | Do |
| | 1 mo | | Do |
| 57 | | General Dealer | Amsterdam Holland |
| | 53 | | Do |
| 32 | | Cigar Maker | Do |
| | 25 | | Do |
| | 23 | | Do |
| | 20 | Cap Maker | Do |
| | 17 | do | Do |
| 12 | | | Tenter Ground Spitalfields |
| | 12 | | Amsterdam Holland |

Printed form, with handwritten entries, Census of 1861, R.G. 9/264, folio 110, p. 1, PRO. The following columns were left blank by the census taker: "Houses, Inhabited, Uninhabited . . . ," "Whether Blind, or Deaf-and-Dumb."

# Entries for the Gompers Family on the Passenger List of the Ship *London*[1]

[July 30, 1863]

DISTRICT OF NEW YORK—PORT OF NEW YORK.

I, W B Moore[2] Master of the Ship London do solemnly, sincerely and truly Swear that the following List or Manifest, subscribed by me, and now delivered by me to the Collector of the Customs of the Collection District of New York, is a full and perfect list of all the passengers taken on board of the said Ship at London from which port said Ship has now arrived; and that on said list is truly designated the age, the sex, and the occupation of each of said passengers, the part of the vessel occupied by each during the passage, the country to which each belongs, and also the country of which it is intended by each to become an inhabitant; and that said List or Manifest truly sets forth the number of said passengers who have died on said voyage, and the names and ages of those who died.

Sworn to this July 30 1863, [A?] P. W B Moore So help me God.
before me [Illegible]
List or Manifest of all the Passengers taken on board the Ship London whereof
Moore is Master, from London burthen 11 tons.

. . .

| NAMES. | Age. Years. | Age. Months. | SEX. | OCCUPATION. | The country to which they severally belong. | The country in which they intend to become inhabitants. |
|---|---|---|---|---|---|---|
| Solomon Gomperts | 35 | | M | Segar Maker | Holland | do [U.S. America] |
| Sarah Gomperts | 36 | | F | None | do | do |
| Samuel Gomperts | 11 | | M | " | England | do |
| Henry Gomperts | 9 | | M | " | do | do |
| Alexander Gomperts | 5 | | M | " | do | do |
| Louis Gomperts | 3 | | M | " | do | do |
| Harriett Gomperts | 2 | | F | " | do | do |
| Jacob Gomperts[3] | Inft | | M | " | do | do |

Printed form, with autograph and handwritten entries, signed, Passenger Lists of the Port of New York, RG 36, Records of the Bureau of Customs, DNA. The following columns were left blank: "Died on the voyage" and "Part of the vessel occupied by each passenger during the voyage."

1. The American packet *London.* According to the manifest, the ship carried 251 passengers—108 men, 65 women, and 78 children. They were predominantly from England, Holland, and Germany; twenty were cigarmakers.

2. William B. Moore of New York City owned one-eighth share of the ship *London.*

3. Jacob GOMPERS, SG's brother, became a diamond polisher.

# The Marriage Certificate of Samuel Gompers and Sophia Julian[1]

Brooklyn, January 28th, 1867

CERTIFICATE OF MARRIAGE.

STATE OF NEW YORK. 101

I hereby Certify, that Samuel Gompers and Sophia Julian were joined in Marriage by me, in accordance with the Laws of the State of New York, in the City of Brooklyn this 28th day of January 1867

Witnesses to the Marriage,
Jacob Davis[3]
Mary D. Ancona[4]

Attest, James Buckley[2]
Official Station,
Justice of the Peace
Residence, Brooklyn

STATE OF NEW YORK.

1. Full Name of Groom, Samuel Gompers
2. Place of Residence, 198 Second St—New York
3. Age, 19 Years[5]
4.
5. Occupation, Segar Maker
6. Place of Birth, London
7. Father's Name, Solomon Gompers
8. Mother's Maiden Name, Sarah Rood
9. No. of Groom's Marriage,
10. Full Name of Bride, Sophia Julian
    Maiden Name, if a Widow,—
11. Place of Residence, No 9 Carroll St—Brooklyn
12. Age, 18 Years—[6]
13.
14. Place of Birth, London
15. Father's Name, David Julia[7]
16. Mother's Maiden Name, Catharine Solomons
17. No. of Bride's Marriage,—

N.B.—At Nos. 4 and 13 state if Colored; if other races, specify what. At Nos. 9 and 17 state whether 1st, 2d, 3d, &c. Marriage of each

We, the Groom and Bride named in the above Certificate, hereby Certify that the information given is correct, to the best of our knowledge and belief.

Samuel Gompers Groom.
her
Sophia x Julian Bride.
mark

Signed in presence of Jacob Davis
and Mary D. Ancona

Printed form, with autograph entries, signed, Office of City Clerk, Brooklyn, N.Y.

1. Sophia Julian GOMPERS, SG's first wife.

2. James Buckley (1837-72) was a justice of the peace in Brooklyn's Second District from about 1865 to 1872.

3. Jacob Davis (b. 1849), a Dutch-born tailor, immigrated to the United States in 1862.

4. Mary D. Ancona was born in England in 1850 and immigrated to the United States in 1857.

5. SG was actually seventeen years old.

6. Born in August 1850, Sophia Julian was actually sixteen years old.

7. David H. JULIAN, SG's father-in-law, was in the furniture business.

# Samuel Gompers' Application to Become a Citizen of the United States

[October 4, 1872]

Superior Court of the City of New York.

IN THE MATTER OF
Samuel Gompers
On his application to become a Citizen of the United States. } Minor.

State of New York,
City and County of New York, } ss. Mark Levy[1]
of 91 Columbia St. N.Y. Cigar Maker
being duly sworn, doth depose and say, that he is well acquainted with the above-named applicant; that the said applicant has resided in the United States for three years next preceding his arrival at the age of twenty-one years; that he has continued to reside therein to the present time; that he has resided five years within the United States, including three years of his minority, and that he has resided in the State of New York one year at least, immediately preceding this application; and that during that time he has behaved as a man of good moral character, attached to the principles of the Constitution of the United States, and well disposed to the good order and happiness of the same; and deponent verily believes, that for three years

next preceding this application, it has been the real and honest intention of the said applicant to become a citizen of the United States.

Sworn in open Court, this 4" day
of Oct. 1872 } x Mark Levy
Thomas Boese[2] Clerk.

State of New York,
City and County of New York } ss. Saml. Gompers
of 350 Third St. N.Y. Cigar Maker

the above-named applicant, being duly sworn, says that he has arrived at the age of twenty-one years; that he has resided in the United States three years next preceding his arrival at that age, and has continued to reside therein to the present time; that he has resided five years within the United States, including the three years of his minority, and that he has resided one year at least, immediately preceding this application, within the State of New York, and that for three years next preceding this application, it has been his real and honest intention to become a citizen of the United States.

Sworn in open Court, this 4" day
of Oct. 1872 } x Samuel Gompers
Thomas Boese Clerk.

I do declare on oath, that it is my bona fide intention, and has been for the three years next preceding this application, to become a Citizen of the United States; and to renounce forever all allegiance and fidelity to any foreign Prince, Potentate, State or Sovereignty whatever; and particularly to the Queen of the United Kingdom of Great Britain and Ireland, of whom I was before a subject.

Sworn in open Court, this 4" day
of Oct. 1872 } x Samuel Gompers
Thomas Boese Clerk.

I Samuel Gompers
solemnly swear, that I will support the Constitution of the United States; and that I do absolutely and entirely renounce and abjure all allegiance and fidelity to any foreign Prince, Potentate, State or Sovereignty whatever; and particularly to the Queen of the United Kingdom of Great Britain and Ireland, of whom I was before a subject.

Sworn in open Court, this 4" day
of Oct 1872 } x Samuel Gompers
Thomas Boese Clerk.

Printed form, with autograph entries, signed, Superior Court Records, New

York County Court House. Initialed twice in margin, "W.E.C." (?); also a third notation, "M.S." (?).

1. Possibly Marcus LEVY, a cigarmaker who was SG's uncle through marriage to SG's aunt, Sarah Gompers.

2. Thomas Boese (1827-1904), a lawyer, was clerk of the court from 1872 to 1896.

## *Samuel Gompers' Introduction to Socialist Thought*

Gompers maintained that his serious initiation into the labor movement began in 1873, when he started working in a cigar factory owned by David Hirsch, a socialist and a political immigrant from Hamburg. Hirsch had opened the shop on the Lower East Side in the early 1870s, and there he gave work to some of his fellow socialist exiles. While working in Hirsch's shop Gompers was "constantly in the company of German-speaking shopmates," most of them "familiar with the writings of the leading German labor thinkers of the age" and many connected with the Arbeiter Union, the central organization of New York City's German workers.[1]

At Hirsch's Gompers met Ferdinand Laurrell, a man who profoundly influenced his thinking on trade unionism. Six years older than Gompers, Laurrell was born in Sweden and had served as secretary of the Scandinavian section of the International Workingmen's Association (IWA) until forced into exile in the early 1870s. Coming to New York after working for a time in Hamburg, he remained an important figure in the IWA and was elected a member of the International's General Council.[2] Gompers later wrote that Laurrell's "kindly talks and warnings did more to shape my mind upon the labor movement than any other single influence," and that the trade union principles he learned from his mentor "remained the basis upon which my policies and methods were determined in the years to come."[3] As a testimony to this influence, Gompers dedicated his autobiography to Laurrell.

It was Laurrell who introduced Gompers to the *Communist Manifesto,* translating it for him from the German and interpreting it "paragraph by paragraph." The pamphlet, Gompers felt, offered "an interpretation of much that before had been only inarticulate feeling," since Marx "grasped the principle that the trade union was the immediate and practical agency which could bring wage-earners a better life."[4] Spurred on by this encounter with Marx, Gompers taught himself to read German and then began studying, as he put it, "all the German economic literature that I could lay hands on—Marx, Engels, Lassalle, and the others."[5]

It was during this period that Gompers encountered the document

that follows, the pamphlet *Praktische Emanzipationswinke* by Carl Hillmann, a journalist and an active member of the Sozialdemokratische Arbeiterpartei (Social Democratic Workingmen's party) in Saxony. It greatly impressed Gompers because it demonstrated what he called "the fundamental possibilities of the trade union."[6] Hillmann argued forcefully that trade unions functioned as the catalysts of fundamental social and political change and that they were the "natural, historically-based" instruments for achieving "the emancipation of the working class." He emphasized that their pursuit of such practical objectives as shorter hours, higher wages, and better working conditions, far from being limited or self-serving goals, provided "the elementary schooling and drill exercises of the proletariat" and had a significant impact on politics and legislation. Hillmann warned that it would be "a fatal error to subordinate the trade union movement directly to the purely political party movement" and, instead, urged workers to develop strong local unions, organized democratically and providing health, death, and travel benefits. National and international trade union organizations could then be established "on this firm foundation," unifying locals through a general constitution and the publication of a trade union journal.[7] These ideas became the pillars of Gompers' trade union philosophy.

*Notes*

1. SG, *Seventy Years,* 1:72.
2. The International Workingmen's Association (IWA), later known as the First International, was founded in September 1864 in London by a group of French and English trade unionists. Karl Marx, a major figure in the German socialist exile community in London, wrote the party platform and emerged as the leading theoretician of the organization. The first American delegate to attend one of the IWA congresses was Andrew C. Cameron, who represented the National Labor Union at the 1869 Basel congress; the first American affiliate, a group composed largely of German-speaking trade unionists in New York, joined the IWA the same year as Section 1. By 1872 there were several American sections, and in July these held the first national convention of the North American Federation of the IWA. Later that year the IWA transferred its executive body, the General Council, from London to New York in an attempt by Marx to remove the IWA from the influence of and potential control by European anarchists led by Bakunin. Friedrich Adolf Sorge, the North American Federation's corresponding secretary, became general secretary of the Council; Ferdinand LAURRELL was among the three cigarmakers elected to it.
3. SG, *Seventy Years,* 1:70.
4. Ibid., pp. 74, 83.
5. Ibid., p. 75.
6. Ibid.
7. See "A Translation of a Pamphlet by Carl Hillmann," May 1873, below.

# A Translation of a Pamphlet by Carl Hillmann[1]

Hamburg, late May 1873.

## PRACTICAL SUGGESTIONS FOR EMANCIPATION.

### A Word toward the Advancement of Workingmen's Associations

### FOREWORD.

The members of the General German Workingmen's Association[2] are not the only ones who have been justly accused of not knowing how to deal with historical facts and human diversity; the Social Democratic Workingmen's party[3] has met with similar criticism and, to some extent, rightly so. The political movement of the German working class is still making one-sided and doctrinaire assertions. These must be completely abandoned if one wants to construct a new governmental edifice. Theory must be transformed into practice and the new basis and structure of society so organized that the rotten pillars of the old order in their collapse will not bury the young sprouts of the new society.

The following articles were composed with this viewpoint in mind. Their popular tone explains their purpose since they are intended for working people who are striving, with unprecedented energy, to found a state with "equal rights and equal duties." I have expressed myself as briefly and clearly as possible, always keeping in mind the historical facts, the diversity of conditions, and the extensive evidence, which has been supplemented by my practical experience and my work with various workingmen's and labor publications. Those in particular who have as their goal the healing of individual parts of the diseased body politic may learn from the following articles that in such an effort the entire constitution of society must be considered and an individual member cannot be healed without attending to the body as a whole.

One comment for those men involved in the emancipation of the working class: I wrote the following articles in my spare time. May the whole be read and criticized with this in mind. The author has attempted to improve on his poor primary-school education in a small Saxon town by continuing self-education.

Carl Hillmann.

### I.

### THE HISTORICAL AND NATURAL NECESSITY OF WORKINGMEN'S ASSOCIATIONS.

In his *Misère de la philosophie,* written in response to Proudhon[4] in 1847, Karl Marx[5] had already demonstrated conclusively that English

workingmen's associations have played the same role in the organization of the working class as the formation of towns during the Middle Ages did for the middle classes of bourgeois society. Since then, other economists, such as Lujo Brentano[6] in his *Arbeitergilden der Gegenwart,* have shown in great detail that contemporary workers' organizations, the workingmen's associations or workingmen's assemblies, have the same importance for the solution of the social question as the craft guilds of the Middle Ages had for the rise of bourgeois society. Even though Brentano vehemently denies the existence of social-democratic tendencies in his arguments, and even though this same economist blames all of Social Democracy and the "Internationals" for failing to make either an intellectual, internal, or external distinction between the practical efforts of the English working class and the social democratic movement, it is still worth noting that the most significant scholar of Social Democracy is in agreement on workingmen's associations with an economist who teaches at the University of Breslau. Even Brentano states that although the workingmen's associations in their early stages excluded purely political goals, as did the craft guilds, they can nonetheless exert strong pressure on the policies of governments and on the rule of the strong and powerful. It was in this way that the craft guilds of the Middle Ages were the unwitting instrument for the emancipation of bourgeois society. Similarly, a great number of English workingmen's associations are today the instruments for the emancipation of the working class. We must hold to both propositions all the more: not only because the English workingmen's associations or workingmen's assemblies, just like the old craft guilds, have survived every period of political reaction, all countermovements of the propertied and ruling classes, all periods of inflation and trade crises, but also because the above propositions prove to us once and for all that despite persecution, reaction, police, and militarism it is possible to promote the organization of the working class and bring about its emancipation.

From the day of its inception, the International Workingmen's Association has fully understood this, and even if the foaming waves of defamation, lies, and delusion crash over it today, sunny days of greater conviction and recognition will follow upon the storm. The workingmen's associations convened in Sheffield in 1866[7] not only granted full recognition to the endeavors of the International Workingmen's Association to unite the workers of every nation in a common bond of brotherhood, they urged every organization represented there to enter into fraternal association with this body in the belief that this was of the utmost importance for the progress and prosperity of the entire working class. The International Workingmen's Association

became the instrument through which was conveyed to the English workingmen's associations the historical thesis that they were the natural, historically based organizations by means of which it would be possible to achieve political and social demands, and at the same time it brought this thesis to a conscious realization. In fact, the most advanced English workingmen's associations have become fully aware that even in pursuit of their immediate goals, they must not forget the political and social emancipation of the working class as a whole.

At the Geneva congress of 1866,[8] the International Workingmen's Association adopted the following resolution: "The formation and advancement of workingmen's associations must and should remain the primary task of the working class in the near and distant future; apart from their efforts to counteract the encroachments of capitalism, they must learn to act consciously as the focus for the organization of the working class, in the interest of its total emancipation;—they must support every social and political movement which strives for this goal and consider themselves the advocates and representatives of the entire class, concerning themselves attentively with the interests of the most poorly paid trades, for example, the agricultural workers who as a result of their exceptionally unfortunate circumstances, through their dispersion and low level of education, are not able to offer the slightest organized resistance.—This cannot fail to attract those outside the workingmen's organizations and impress upon the mass of workers the conviction that their goal, far from being a limited, self-seeking one, is the universal emancipation of the oppressed millions."

The Basel congress of 1869[9] resolved that the formation of workingmen's assemblies (workingmen's associations) should be actively promoted, that the different workingmen's groups should unite in national organizations and confer jointly on the measures to be taken in order to eliminate the present wage system through cooperative labor, and that the General Council[10] should work to bring about international ties.

At the conference of delegates in London, September 17 to 23, 1871,[11] the importance of the trade union movement was once again emphasized, just as the last congress at The Hague had pointed out to the workingmen's associations of all countries that the General Council is the mediator of international ties. In this connection, we must call attention to the Official Proclamation of the General Council in New York on January 26, 1873.*

*The Official Proclamation of the New York General Council states:

In consideration: (1) That the conflict between labor and capital is neither local nor national, but rather a problem that encompasses all nations in which modern

Given the preceding historical data, resolutions, and compilations, it should be sufficiently clear both what importance workingmen's associations or workingmen's assemblies must have for the labor movement as a whole, and how false and worthless is the opinion of those socialist and non-socialist workers who believe that the task of the trade unions can be accomplished merely through activity in resistance, protection, and other supportive measures, and that these bodies could ultimately be considered executive organizations that exist for negotiations with employer coalitions.

The full significance of the labor congress in Erfurt in June of last year, which was convened in order to found an "association of trade unions" independent of a purely political party movement, was not understood by most of our party members, nor did it achieve any practical results. Let us examine the reasons. Once we have found the cause for this phenomenon, we can try to do better in the future.

In his writings, Lassalle[12] tries to make us understand that a lively, energetically conducted, purely political agitation would quickly help the working class achieve its rights. This opinion is widespread even

---

society exists; (2) That there is an international agreement among capitalists for the exploitation and oppression of the working class, and therefore resistance efforts on the part of the workers fail mostly because of a lack of solidarity among the different sectors of labor within each country and because of the absence of fraternal bonds of union among the working classes of different nations; (3) That the principle of solidarity imposes upon workers the duty of assisting each other at home and abroad; (4) That emigration, and/or the export of the labor force from one country to another, fosters competition with the workers of the latter country.

For these reasons the General Council of the I.W.A. submits the following plan to the various workingmen's associations of all nations for an organization to promote the effectiveness and prosperity of workingmen's associations everywhere: (1) All associations of a trade in each country should unite in order to elect an executive committee for their country; (2) These executive committees should be in constant communication with other nations through the mediation of a General Executive Council so that they remain informed at all times about the condition of trades and labor in every country; (3) Treasuries should be established and placed under the control of the executive committees in order to support needy union members in any country and to meet the expenses of the General Executive Committee; (4) All executive committees of the different trades in each country should unite in order to provide assistance in those cases where, because of a lack of means, a given trade is unable to continue the struggle against the exploiter; (5) In the case of emigration, every member of such an international workingmen's association should have the same rights in his new country as the older members in that country; (6) If a union member must leave his country because of political persecution, he should receive the same support to which he was entitled in his former country; (7) The international workingmen's associations, through their executive committees, should consider it their obligation to prevent, to the best of their ability, the import and export of the labor force under any type of contract system.

among our party members, not only because our party has continuously recruited from among the followers of Lassalle, but also because our party program has a prominent political character. In it, social demands are emphasized sharply enough; only the individual points for practical agitation are not sufficiently detailed. One must remember that the abuse to which workers have been subjected has frequently made them politically indifferent and has driven many of them to define their demands from the opposite direction, that is, purely socially. The causes for the misunderstanding of the trade union movement among many people are, on the one hand, Lassalle's provocative phrase "purely political agitation," and on the other, the workers' suspicion of political parties. The former are rushing like a storm 10 years ahead of the movement which is supposed to include and unite all elements of the working class; the latter do not understand that workingmen's organizations with purely social programs have a tremendous impact on legislation, and thus on politics. These are two paths to the same goal. And now that we have discovered this, now that we have first of all learned that the trade union organization is the natural and historical instrument that will enable the workers to eliminate class rule, why then still argue about the proper form, name, and appearance? We must hold fast to the heart, the soul, the essence of the thing. That is why it is exceedingly difficult to argue with people who cannot see the forest for the trees. While they smilingly and condescendingly look down upon a strike or a lockout, and they are pleased with any conflict that annihilates the phrases of the opposition and the rhetoricians, they refuse to admit that the trade union organization is destined to eliminate all party quarrels among the workers and is preparing them for social and political emancipation. On the contrary. Such purely theoretical skirmishers consider the practical and healthy trade union movement to be something flawed, reactionary, and nostalgic, only retarding and dividing the whole labor movement and costing a great deal of money and energy that could be better spent in political agitation. And unfortunately it is often the case that these same party members work 12 to 13 hours a day in their shops, or submit to offensive shop regulations and even allow wife and child to work. A carpenter I know who called himself a good socialist told me one day he could not join the trade union because that was not allowed under the iron law of wages.[13] He said there would first have to be government aid (credit from the state to the associations), then the situation would change. It is sad indeed to hear such words from a man who attends all mass meetings and who—according to him—can recite Lassalle's responses from memory. This type of worker is the exact opposite of

those who spit in Lassalle's face at his first appearance. The phraseology of the Progressive party[14] has surrounded the workers' brains with a thick slime that allows them to think independently only in rare cases. The purely theoretical elaboration of social, political, and economic propositions by the dogmatic followers of Lassalle has driven them from one corner to another and unfortunately the devil has exorcised Beelzebub.

There are still a large number who prefer to think of the union movement as the tail of the political movement, but under the weight of overwhelming and unalterable facts they too will have to strike their sails.

It is obvious that anyone who wants to attain practical results must deal with all actual conditions and circumstances that impede the implementation of practical attempts to organize the working class.

It has been commented upon before that by far the greatest majority of workers has no political sense, that is, they are not interested in the Reichstag or in legislation, in matters of tariffs, taxes, commerce, provinces, or princes, in republic or monarchy. It is difficult to rouse them to action. They are most easily reached through pay increases, shorter working hours, and travel and health benefits. This purely practical sense of the working class must be used by those who have experienced and realized that the trade union organization is the natural and historically determined means for gradually bringing labor to power. It is a fact that certain trade unions such as the printers, hatmakers, and goldsmiths are making such rapid strides because they do not belong to any political movement and as a matter of principle refrain from all political disputes while not denying any member the right to join a purely political party outside the trade union. Matured through bitter experience in social struggles, these workers are confirmed pioneers in independent political organizations and representation, and these elements not only protect us from revolts and putsches, they encourage more serious work, a sense of duty, and an acceleration of the solid organization of the working class. They supply firm economic and social underpinnings for the political life in today's state, without the development of which bloody dramas would become historical necessities. The party of the Fourth Estate[15] must attempt to prevent the latter, despite the provocations of the ruling and propertied.

These remarks should demonstrate that it is a fatal error to subordinate the trade union movement directly to the purely political party movement. I will attempt to prove this while at the same time trying to explain the intellectual and inner nature of the workingmen's associations with respect to the program of the Social Democratic

Workingmen's party, concluding with suggestions for the development and propagation of workingmen's organizations.

## II.
## The Purely Political Party Movement and Workingmen's Associations.

It need not be demonstrated here that today and always it is political power that the working class must use in order to obtain full equality and the elimination of class rule. Whoever has only half observed the labor movement and has been in touch with the working class directly must and will admit (in whatever form or shape he has encountered the labor movement) that it is primarily or at least secondarily a political movement. Furthermore, we do not need to prove that neither the immediate elimination of the present economic system (capitalist production) nor the rapid removal of the present governmental structure (the monarchical state) will allow the rise and liberation of the working class to become a reality. We know that the economic, social, political, and spiritual liberation of the working class and the overcoming of race and class hatred, as well as the complete establishment of a free people's state, cannot be accomplished in a decade. Furthermore, it is useless to put the masses off with the idea of social revolution, since it unfolds by itself through the superior power of capital and would be impossible without the total dominance of the propertied. We must above all rouse the consciousness of the people for the emancipation of the Fourth Estate. Everything depends on this.

The inner drive of the working class toward emancipation is no longer a dream; it is neither an idea, nor an invention, nor anything forced: It is a fact which cannot be debated away. Whether this aspiration expresses itself only in purely political agitation, or in the desire for the elimination of oppressive factory regulations, for the shortening of the working day, for increasing demands, for regulating the labor market, etc., its goal is once and for all the achievement of independence. Herein lies the crux, the focal point, and the total essence of the social question.

Throughout history the external characteristic of the oppressed has been that they first roused themselves to action and began to fight at the point where the shoe pinched most. Thus, for example, the rebellious country folk fought on the side of Dr. Luther's Reformation movement in the Peasants' Revolt because they were under the illusion that the priests and papal officials were responsible for their serfdom, when in reality it was the feudal nobility with its privileges and absolute

autocracy over the peasants who used the priests and religion to keep the peasants in bondage and stupidity. The petite bourgeoisie thinks competition, free trade, and the efforts of the working class are responsible for its decline, when it is an established fact that it is concentrated capital with its privileges and power over the labor force that expropriates the petite bourgeoisie and forces the working class to defend and seize its rights. It is no wonder then that the working class first turns its attention to the elimination of oppressive factory regulations, obtaining shorter working hours, regulating wages, and a higher valuation of labor. Such means of defense are nothing but the elementary schooling and drill exercises of the proletariat which not only enrich its experience and keep it from straying onto false paths, but also establish the solid positions from which the workers will finally recognize and strive to eliminate the true source of their bondage. It is a matter and only a matter of encouraging this awareness in the masses so that they can consciously carry on the battle and learn to eliminate the source of social strife. By eliminating this source, the working class can achieve that independence which is the central issue in the solution of the social question.

The question arises: By what means can one most rapidly achieve, advance, and hasten the *conscious* struggle for independence?

The answer to this question is implicit in the foregoing hints. We have seen that:

1) in the same way as the craft guilds of the Middle Ages were the unwitting means for the emancipation of bourgeois society, so the workingmen's associations of today are the means of emancipating the working class;
2) the great mass of workers mistrusts all purely political parties because, on the one hand, they have often been abused and deceived by them, and on the other, their lack of information about the social movement prevents them from seeing the importance of the political side. Moreover, the workers show more understanding and practical sense for issues of immediate interest, such as shorter working hours, higher valuation of labor, elimination of repugnant factory regulations, etc.;
3) a purely trade oriented organization exerts continuous pressure on legislation and the government; consequently, this form of the labor movement is also political, even if only secondarily so;
4) the establishment of the free people's state, that is, the economic, social, political, and spiritual liberation of the working class and the establishment of the freedom of the worker, requires preparatory development and training;
5) the activities of the workingmen's associations ripen ideas leading

to the emancipation of the working class, and therefore these natural organizations must be treated as equal to purely political agitation and must not be considered either a reactionary development or a tail of the political movement.

The preceding statements suggest the tactical position and approach which the Social Democratic Workingmen's party must adopt vis-à-vis the trade union movement. They must, as a matter of logical consistency, be the same as those acknowledged by the International Workingmen's Association many years ago as the most appropriate, and which result in the promotion of the independent trade union movement and in bringing the conscious effort for emancipation to precise expression in and through these organizations.

This effort has already begun. The older associations of cigar and tobacco workers, printers, hatmakers, and gold- and silversmiths must be designated as organizations which, independent of political parties, have already tested their powers in vigorous struggles and have wrung respect from their opponents. In the case of the workingmen's associations it is not a matter of deceitful phrases, but rather of a strong bastion and bulwark of defense against even further deterioration and degradation of the working class. They must fulfill not only this initial obligation but can also push wages at least to a level where it becomes possible to expand and increase demands; and since wages, according to the iron law of wages, adjust themselves according to the demands of a people, nothing can be more obvious than the necessity of increasing demands. Through an increase in demands one not only combats the plague of hunger, but also the worker learns to value shorter working hours. He not only gives the power of labor a higher value, he protects himself from overproduction and trade crises, thereby expanding his social, political, and economic education without estranging himself from family life, but rather coming closer to it. Finally, the trade unions nurture those most awesome weapons in the hands of the proletariat—statistics and mass discipline—which, supported by political agitation and organization, will eventually shatter the bourgeois world and inaugurate the new society.

The sense for cooperative work awakened by the trade unions is of no less importance, and the very fact that excellent producer cooperatives have been founded in England and Germany in the spirit of the total liberation of the working class demonstrates the impact that workingmen's associations have in the elimination of capitalist production.

Finally, one should consider how tremendously influential the workingmen's associations have been on the policies of England's government. One might recall the great results they achieved under the

banner of the "Anti-Corn League"[16] against the landed gentry of England. Even though the workers were shamefully deceived by the Liberal party to gain their cooperation—and cleverly used by the Conservative party for its own efforts and goals, the English workers alone, through their own energy and politics, are responsible for the ten-hour day and the restrictions placed on female and child labor. When, during the American war, it appeared that the war-minded London cabinet would intervene in that murderous affair, the workers forced the cabinet to remain neutral by threatening agitation. Since it is still fresh in our memories, we scarcely need to remind anyone of the great demonstrations of our own times, in Newcastle and London in April and May of this year, drawing a crowd of more than 200,000 people in support of universal suffrage.

In contrast to such successes how insignificant appear our popular assemblies in which we still argue about concocted socialist systems of earlier epochs, authoritarian dogmas, articles of faith, and infallible means and cures for a solution to the social question. How naive and narrow-minded sounds the argument that the workingmen's associations have no political influence and are to be considered only a necessary evil! And yet the Newcastle workers carry out a demonstration which would be impossible in a German context because German workers lack an organization essential for winning the respect of governments, legislators, and employers. Also consider this: English workers began their struggle in the face of the opposition of the entire European bourgeoisie, in the face of the raging cry of a one-sided class government from which every concession had to be wrested step by step over the last 50 years.

Therefore it was a sound idea for the trade union congress in Erfurt in June 1872 to support the independence of the trade union movement from political party machinery. Whoever builds a house must build it on solid ground. The workingmen's associations are the bedrock and the strong foundations upon which and with which alone it is possible to give solid support, steadiness, and weight to political agitation. Therefore the trade union movement cannot harm but only benefit the political consciousness of the working class and it would be a disastrous and almost unforgivable mistake to give the workingmen's associations the veneer of political agitation from the outset. It is a crime to want to tear down in the name of "universal, equal, and direct suffrage" these purely natural organizations that have arisen from actual conditions, and, like the last congress of the General German Workingmen's Association in Berlin, to pass resolutions whose aim it is to transform the trade unions into purely political associations as quickly as possible. Let the workers keep their eyes open!—Un-

fortunately even the Social Democratic Workingmen's party is not completely devoid of elements that negate and destroy the workingmen's associations, and therefore I saw it as my duty to demonstrate clearly the full significance of these historically determined and naturally developing organizations in order to shield them from evil hands and fanatical dogmatists.

Let us finally rid ourselves of the unfortunate error which suggests that the bourgeoisie and the legislators do not recognize, observe, and study the importance of the trade union movement for the liberation of the working class. Today it is a tooth and nail fight and every effort is made to annihilate and destroy the movement, while tomorrow a pact is made with it in hopes of discovering its weaknesses and lulling the workers into a sense of security in order to test the dagger of treachery and deceit anew. Building on the ignorance and forgetfulness of the masses, they play the young organizations false in order to stab them in the back. The form and organization of the unions will grow such a tough and solid hide in the process of the struggle that the unnatural tyrants in uniforms and white ties will become as hollow as reeds that break in the wind!

## III.
## The Nature of Workingmen's Associations and the Program of the Social Democratic Workingmen's Party.

First of all a laurel wreath for the annointed head of Max Hirsch,[17] the wonder doctor, who, because of his devotion to the organization of the working class and the establishment of workingmen's associations, is a friend of the emancipation of the working class, and therefore our friend and my friend. Because of our friendship for this great wonder doctor we have not infrequently been thrown into the same pot with professional "social demagogues" on the grounds that Mr. Wonder Doctor has not been able to make plausible that lovely phrase about the harmony between labor and capital. Even though every edition of *Dr.* Max Hirsch's *Gewerkverein* brings new proof of the ridiculousness of this theory—no matter—they continue to harmonize. Together with other learned gentlemen, *Dr.* Max Hirsch, the lawyer of the Federation of German Workingmen's Associations, strictly denies the political tendency of the workingmen's associations. Now of course the trade unions are not political clubs in the sense that they argue over republic or monarchy, imperial glory and the advantages or disadvantages of particularism, or over heroic military deeds and beautiful cavalry charges. But by denying the political tendency

of the workingmen's associations *Dr.* Max Hirsch has deceived himself and this was demonstrated by the second meeting of the Federation of German Workingmen's Associations on April 17. The 25 delegates of the harmonious German Workingmen's Associations passed a resolution in favor of participating in the Reichstag and Landtag elections. The resolution calls for the naming of a slate of their own candidates and rejects any compromise with parties hostile to the workingmen's associations. At the same time they held fast to the declaration that workingmen's associations have no inherently political character.—First harmony between labor and capital, then rejection of any compromise with elements hostile to workingmen's associations, the latter, an equivalent to a declaration of war! So the workingmen's associations have no political character, yet their representatives pass *ex officio* a resolution in favor of participation in the Reichstag and Landtag elections. But in order to avoid the suspicion of "social demagoguery" a final passage explains that workingmen's associations are not inherently political! We can see that those learned men who consciously or unconsciously refuse to draw logical consequences and practical applications from a certain thing will always find themselves in conflict with their own actions and historical facts.

The program of the Social Democratic Workingmen's party can be summarized in one sentence: Equal rights and equal duties! This is why our program stresses in point II, paragraph 2: "The struggle for the liberation of the working classes is not a struggle for class privileges and prerogatives, but for equal rights and equal duties and for the elimination of all class rule."—Let us compare this point with the ideal goal of the workingmen's associations or workingmen's assemblies that intend to put organization in place of the disorganization of society and—as C. Marx and L. Brentano have explained—let us further remember that in the same way as the craft guilds in the Middle Ages were the unwitting means for the emancipation of bourgeois society, so today workingmen's associations are the means for the emancipation of the working class. Consequently, just as the feudal state had to agree to recognize the organization of the guilds and extend their laws and provisions to communal, state, and police governments, so the state will sooner or later have to recognize the workingmen's associations or workingmen's assemblies: not only recognize them but eventually have to extend the organizational form of the trade unions to all of political and communal life. Now any trade union member knows that the unions maintain, even in the smallest detail, the principle "Equal rights, equal duties," which means the same dues, no special privileges for individuals, as equal an administration as possible, the same enjoyment of rights! If we imagine this

system developed, matured, and extended to the state and community in almost the same way in which inequality between master, journeyman, and apprentice established the pattern for the privileges of the master and the triumph of the grande bourgeoisie, we have before us the ideal of social democracy: The state of equal rights and equal duties, a free people's state.

"The worker's economic dependence on the capitalist is the basis for servitude in every form and therefore the Social Democratic party seeks the full labor value for each worker through an elimination of the present mode of production (wage system) and its replacement with cooperative labor."—The workingmen's associations or trade unions pursue the material betterment and spiritual elevation of their members and since, in their specific and general goals they advocate the independence of the entire working class, they struggle against the capitalists with all available means; and almost all statutes which provide rich material and extensive subject matter for the preparation of these articles, even those of the Hirsch-Duncker Workingmen's Associations, emphatically endorse "cooperative labor"—even if the more radical workingmen's associations call for this to be achieved by "any means whatever" including state subvention.—Simple logic tells us that from the day on which the organizational regulations of the trade unions are accepted as legislation, the trade union movement has become a political one. This demonstrates the validity of that further proposition of our party program which states that the social question is inseparable from the political and that the trade unions, since they are social organizations, also have political goals. Here again we can see how the nature of the trade unions accords with the political party program.

"The solution of the social question is only possible in the democratic state," continues our party program. When we once more examine the nature of the workingmen's organizations, we will find that they are the most profoundly popularly ruled (democratic) organizations that can be conceived. Their administrative officials are entrusted only with executive and not with legislative power, their only authority resides in the general will. The legislative authorities are the general assemblies, the congresses, in special cases a committee or control commission, and the strike ballot of the members. These are at the same time the basic features of direct legislation through the people, the necessary elementary schooling through which the people can exercise and develop their right of initiative and veto.

All of those workingmen's associations or workingmen's assemblies that have passed beyond the embryonic stage, however, have given their organizations a strongly unified character that grew in part out

of the struggle against the bourgeoisie, in part from economizing on administrative costs, etc., and, finally, on its own for practical reasons. This further demonstrates the correctness of the statement in the Social Democratic Workingmen's party program, according to which "the political and economic liberation of the working class is possible only if it conducts its struggle in a united and uniform manner and provides itself with a unified organization." The unified organization of the individual trades is the prerequisite and foundation for the achievement of such a unified general organization as exists already in England and for which we are striving in Germany where many believe it can be achieved by storm, which is contrary to the natural development of the trade union movement. The international organization of trade unions is inconceivable without a unified national organization. The former depends on the latter and the workers of different nationalities will not be restrained by anything in the world from forming international ties as soon as unified national organizations have been formed. In fact the most advanced workingmen's associations in Germany, which have no external political party affiliation, are far closer to an international trade union organization that is based particularly on reciprocity of association statutes concerning rights and duties of members, than those who have adopted the innocent little word "international" for their titles, a little word that regularly provides politicians and policemen with the opportunity to test the vitality of the trade union movement. The nature of the trade unions and the solidarity of interests of the workers of all nations drives them, as well as the coalitions of capitalists, toward international association. This is further proof of the vitality of the International Workingmen's Association whose sole task is to show the workers how to strive consciously and independently toward this goal.

At this point it is hardly necessary to compare the next 10 demands of Social Democracy with the character of the workingmen's associations. The principle of the workingmen's associations, "Equal rights and equal duties," has not only abolished the privileges of master and journeyman estates but also transformed the capital and property of the trade unions into the common property of all members; Catholics, Lutherans, Jews, and religious as well as social sectarians are united in them, without exceptions and without preferences.—Universal, equal, direct, and secret suffrage is practiced, as is the remuneration of representatives and delegates, as well as direct legislation (initiative and veto rights) through the people. The creation of a people's army instead of a standing army is, to be sure, not mentioned in any trade union statutes, nor is the separation of church and state or of school and church,—but all organized trade union members, regardless of

where they work, are so unenthusiastic about standing armies and Bible study in the schools that there is a general effort to demonstrate to the workers the necessity of these old customs. Compulsory primary education and free instruction in all public schools are contained in the trade unions' goal of materially and spiritually elevating their members.—Furthermore, since the trade unions fight for the independence of the working class in general, they struggle in particular for the independence (democratization) of the courts, the introduction of juries and industrial arbitration courts, public and oral court proceedings, and free administration of justice; here English models have offered many positive examples, especially since the arbitration boards can be viewed as forerunners of industrial arbitration courts and the simplification of judicial procedures.—Police and government persecution of the independent labor press awakens the desire to abolish all press laws. The disciplining and suppression of associations and coalitions heightens the desire to eliminate association laws. The normal working day, the restriction of female and child labor, and the elimination of competition from prison and workhouse labor are self-evident goals of the trade unions. The trade unions' social demands are far more specialized and developed; they include, for example, the regulation of wages, working hours, night and Sunday labor, apprentice and benefit systems, and statistical inquiries about wages, working hours, food prices, illnesses, and fatalities, etc.—Concerning taxes, trade unions reject the indirect system and some have even imposed progressive voluntary income taxes on their members (for instance, the organization of printers' helpers in Berlin and Hamburg-Altona during their fight with the master printers).—Finally, as far as the last point in the party program is concerned, "governmental promotion of the cooperative system and state credits for free producer cooperatives with democratic guarantees,"—nothing more need be said since this has already been discussed.

Three essential points of the trade union movement have been brought out: first, the natural and historical necessity of trade unions; then, the demonstration of their independence from and equality with political agitation; finally, the identification of the political program of the Social Democratic Workingmen's party with the trade union movement.

There is nothing new in what has been said up to now. It is not only the bourgeoisie that hates and persecutes the trade union movement, and seeks to suppress it because it knows its political character and significance in both practical and theoretical terms. Even scholars and high government officials keep it under close observation and attempt to divide the workers, true to the principles of Jesuitry, ex-

ploiting the weaknesses and passions of the individuals and the masses for the advantage of the ruling classes and the disadvantage of the oppressed. We can no longer turn back. The workers' freedom of association and assembly can be abridged but no longer taken away. Were a class government to attempt this, a permanent state of war and siege would be declared, since pressure produces counterpressure, and the working class would be drawn to the barricades, ready for a revolution. However, the workers will learn to resist such temptations through organization, and since no government can exist in a state of siege with its people for long, only two things remain: Either the governments themselves must help the people to organize in order to avoid catastrophes and put social-democratic principles into practice,—or: The people will fight for their rights through organizations such as the world has never before seen! Both paths, in their final objectives, lead to the solution of the social question and thereby to the elimination of class rule, the victory of the social-democratic or labor movement, and the collapse of old traditions and the cult of personal authority, as well as to an end to the exploitation of the masses.

These are not utopian dreams!

## IV.
## The Organization of the Trade Unions.

After the preceding observations there need be no more argument about the significance or insignificance of the trade union organization. It is not only determined by natural and historical necessity rather than being the creation of individual instigators or agitators, but it is also of a political nature, if only secondarily, and the essence of the trade unions is in agreement with the program of the Social Democratic Workingmen's party. It has been mentioned previously that for all of these reasons the equality of the trade union movement with the purely political party movement is not only useful but even essential.

There remains for me only to make some observations about workingmen's organizations themselves, but these should not be considered exhaustive or definitive.

Everyone who has observed the social movement is aware of the fact that it is especially the educated workers and those in better positions who took the initiative in the founding of workingmen's organizations. Despite the worse laws of an earlier day and the restrictive influence of the guilds that held sway in Germany longer than anywhere else, it was the hatmakers, shipbuilders, printers and

typesetters, masons and carpenters, machinists and iron workers who had, in part, freed themselves from their masters in the old guilds and formed journeymen's and fraternal associations among themselves, whose purpose it was to resist the oppression of the masters who had attained comfort and wealth. Most of these were local organizations that only rarely extended beyond city boundaries. They usually communicated through passwords and codes, in quarterly assemblies, financial subscriptions, and meetings in journeymen's hostels. When it was essential to stand up to a brutal master, a few words in the hostel room were enough to make this clear to everyone. The journeymen sought to avoid the house of such a tyrant. Forcing such masters to use inferior workers and depriving them of new ones brought even them to a better understanding of the demands of living, thinking, feeling, and acting "goods."—Following the internal demoralization and final dissolution of the old guilds through freedom of trade and movement, the local journeymen's associations took the opportunity to further expand their scope by means of the freedom to form associations. A great number of masters, who had withdrawn from guild life and resentfully and peevishly shut the door on the spirit of the times, and yet were beginning to revel in the eternal realm of free competition, were divested of control of the benefit system and of health and death benefit funds by those journeymen who understood the importance of freedom of association; these institutions were then administered independently. Unfortunately, most of the trades let the opportunity slip by to put the control of these funds into the hands of the workers who had earlier contributed by far the greatest amount to them. Only a very few cities and a few trades shook off the tutelage of the masters.—These journeymen's associations, with the support of health benefits, are the predecessors and foundations for a national workingmen's organization. Those trades that have until now failed to build further on these foundations must above all take care to strengthen local organizations in each city and, where they do not exist, to form new associations. Here, however, it is the duty of educated and enlightened workers not to inquire into the political or religious creeds of people. They should above all try to attract members through popular lectures, pamphlets, and informative leaflets. They must attempt to establish contact with their colleagues in other cities and arouse those sunk into lethargy. If possible, their goal should be to attract every fellow worker in a trade through suitable means such as certificates of employment, travel support, etc.—Bad experience has taught us how foolish it is to give such organizations names that have a political-sounding appearance for the uninitiated worker. We have earlier explained how suspicious

the workers are of such appearances. Call them simply shoemakers', carpenters', coopers', tinsmiths', etc., associations or trade unions. Since these organizations will always be secondarily political, and since the social cannot be separated from the political, a paragraph of the statutes may very well state: religion and politics are to be kept out. Common interests unite the workers. Those who jointly pursue and protect their interests already practice politics. No ironclad paragraphs can suppress politics and its consequences, or keep it at a distance. As soon as these local organizations unite in a national federation, the political tendency will come more to the fore. And we will already be dealing with laws concerning shorter working hours, female, child, and prison labor, etc. At that moment the workingmen's associations or workingmen's assemblies transcend the embryonic phase and make themselves felt in all sorts of movements that find their expression in convulsive strikes and lockouts. The persecutions that the workers experience at the hands of the police and government officials, the abuses that the daily press heaps upon them, finally the "wisdom" they receive at the hands of the wine- and beer-drinking philistines, makes them feel more than ever the subservience in which the present government, in league with the bourgeoisie, seeks to hold them. Through independent newspapers, people's assemblies, and elections, trade union members will realize that their endeavors are identical with those of the Social Democrats. The ice of suspicion will melt with better understanding and loyalty to the conviction that people's organizations need only be developed and cultivated in order to replace the dying organism of the contemporary state with a new one. This new organism contains all the prerequisites for prosperous growth and the solution of the social question. The more the worker accustoms himself to this, the more are revealed to him the treasures of an economy based on the social interdependence of all men, and laws based on the governmental unity of all public institutions. He will abandon the well-meaning but unrealistic proposals of Fourier[18] and Cabet[19] in favor of the real business of the day; rendered wise by his own experiences and by the visible, and complete or developing achievements of the labor movement in all civilized nations, he will be convinced that socialization or socialism is an historical necessity which needs only to be systematically promoted. Firm in this conviction, he will carefully nurture and protect this young growth in order to avoid a premature birth, yet helping to hasten the delivery of the new society at the correct time as painlessly as possible.

With these happy thoughts the worker will hasten to greet his brothers who have been degraded to pariahs through the division of labor by capitalists and exploiters, and whose children and wives have

been brutalized at the loom, in the mines, in the fields, and snatched by death in their early years by plagues, hunger, and war. He will ask them to join the great cultural movement and point to the English agricultural, mine, and manual laborers who rush ahead of their brothers on the Continent toward the day of decision as in a great storm cloud.

The rancor which has been spread among the combatants and the bitterness evoked by senseless slogans and polemics will yield to the recognition of the common goal of all, which can be reached all the more quickly as historical facts and embryonic organizations are used as the means best suited to awaken and further in the workers the class consciousness that is longing for fulfillment. Another proven means of agitation is the indication that modern society more and more hinders the journeyman from gaining his independence as a master and allows only few exceptions to the rule: "Whoever is a wage earner (proletarian) shall remain a proletarian."

The following points must be especially emphasized in the organization of trade unions:

1) Local journeymen's associations must be respected and the associations of an earlier day that are now almost completely ruled in a dictatorial and absolute fashion by senior journeymen must be organized democratically. The principle "equal rights, equal duties" must be maintained under all circumstances. Even the chairman must not have any privileges. He must be considered an executive only and not a policymaker. The executive committee must assist the chairman and must, to a certain degree, be granted authority in subordinate and administrative matters. Health, death, and travel benefits must, if not already in existence, be connected with the workingmen's associations. In all these matters, no importance should be given to names, appearances, and forms; one should adhere to the spirit and content of a phrase in deliberations about statutes.—In local associations a control or revision committee must be provided for, representing the members independently of the executive committee, accepting complaints concerning the administration, and auditing account statements, etc.
2) Before one proceeds to national unification of a trade, one must be concerned with establishing the greatest possible uniformity of regulations in the local associations; therefore the local associations must exchange statutes, issue receipts to arriving and departing members for contributions made, and at the same time indicate the place, time, and name of the local association in which the concerned party was a member. If there are in-

sufficient fellow workers in one place to form an association, the different localities of a county or district must join together. One should pay close attention to the integrity of members, and tolerate no arrogance, pretensions, or coarse acts of violence. Arbitration courts must be entrusted with the task of settling disputes among members. Changes in the local organization may only be undertaken with a ⅔ majority and all the members of a local organization must be responsible for every action.

3) When these spiritual prerequisites for a national organization have been achieved in the local associations, one can continue to build on this firm foundation. A congress or an assembly of representatives from the various local assemblies can unite the individual member groups through a central statute to which all associations must be subject. In order to confirm and maintain equality with and independence from political parties, the larger bodies must attempt to create their own journals and publications (as, for example, the cigarmakers, printers, hatmakers, and goldsmiths have already done). Smaller and related trades can work hand in hand. When every trade has a publication, then, just as the national organization is formed on the basis of the local, the international on the national, a central trade publication will rise over the publications of the individual trades. The resolution passed by the trade union congress in Erfurt did the proper thing in this respect. Unfortunately there are still too few good local and national associations to support such an undertaking. In this case as well the independent workingmen's organizations have to do the preliminary work.

When all these preconditions are fulfilled, the trade union organization will get into full swing and international agreements and cooperative institutions will materialize. When a house is built one must start from the bottom and make every effort to dig out the foundation carefully. Whoever fails to do this, and instead considers the trade union movement a game, should not be surprised when despite all efforts everything collapses at the first opportunity.

The author of these articles is a trade union member of long standing and has lived through the developmental stages of the workingmen's organization without ignoring political movements. He has acquired this conviction: Without social groundwork there can be no lasting political organization or agitation, and therefore there should be no political tinge to young organizations that will, according to their nature and essence once they have matured, make their mark on politics as noteworthy members of society. It is a fruitless effort, a Sisyphean task, to want to skip developmental stages in the history

of civilization. Whoever attempts to do so in spite of this fact should not be surprised when he is disillusioned. Detours will bring even him to the insight that the young sprouts of society must be carefully tended and nurtured so that with their growth and development they can eventually fully replace rotten and outmoded conditions.

May this insight finally come to life among German workers!

*Praktische Emanzipationswinke. Ein Wort zur Förderung der Gewerksgenossenschaften* (Leipzig, 1873). Translated from the German by Hannelore Jarausch.

1. Carl HILLMANN, a journalist, was active in trade union and socialist affairs.

2. Allgemeiner Deutscher Arbeiterverein (ADAV). The socialist General German Workingmen's Association was organized in May 1863 under the leadership of Ferdinand Lassalle who served as its first president. Johann Baptist von Schweitzer became its leader after Lassalle's death in 1864. The ADAV advocated suffrage for all workers and the formation of an autonomous working-class party. It favored state-supported cooperatives but otherwise had no clear-cut economic program.

3. Sozialdemokratische Arbeiterpartei (SDAP). The Social Democratic Workingmen's party was founded in 1869 by Wilhelm Liebknecht and August Bebel. The group's political and social aims were closer to Marx's than Lassalle's. It favored the abolition of wage labor and sought to build up a political party as well as trade unions to further that goal. The SDAP's main difference with the ADAV, however, was its hostility to German unification under Prussian hegemony and its adherence to a radical republican federalism. After these divisions were rendered irrelevant by German unification in 1871, the two parties moved closer to each other, finally uniting at the Gotha congress in 1875 to form the Sozialistische Arbeiterpartei Deutschlands (Socialist Labor Party of Germany).

4. Pierre Joseph Proudhon (1809-65), a French political theorist, proposed the reinvigoration of the institution of private property, particularly through a new credit system, in order to transform the peasants and workers into small-scale owners. He envisioned the workers organizing cooperative societies for production, consumption, mutual aid, and insurance. Marx wrote *Misère de la philosophie* (1847) in rebuttal of Proudhon's *Système des contradictions économiques, ou philosophie de la misère* (1846). Marx argued that Proudhon's theories left no room for the development of modern industrial society.

5. Karl Marx (1818-83), a revolutionary political economist, sociologist, philosopher, and journalist, played a critical role in the International Workingmen's Association (IWA). He was the author of the *Manifest der kommunistischen Partei* (*The Communist Manifesto*; 1848) with Friedrich Engels, and of *Das Kapital* (3 vols., 1867-94). His belief that working-class politics should result in autonomous organizations independent of bourgeois parties and movements influenced Gompers' conception of the labor movement.

6. Lujo Brentano (1844-1931), a German academician, liberal social reformer, and pacifist. His two-volume *Arbeitergilden der Gegenwart* (*Workers' Guilds of the Present Day*), published in 1871-72, is a study of English trade unions.

7. The 1866 trade union conference held at Sheffield, England, adopted a resolution urging the unions represented to join the IWA.

8. The first general congress of the IWA convened in Geneva in September 1866.

9. The IWA congress met in Basel in September 1869.

10. The General Council was the executive body of the IWA, elected by the International's annual congresses. Its duties included carrying out the decisions of

the congresses, linking the member organizations by correspondence, collecting information concerning the condition of workers, and issuing regular reports to the members. The 1872 Hague congress of the IWA transferred the General Council from London to New York to prevent an anarchist takeover.

11. The beginning of the Franco-Prussian War prevented the IWA from holding a congress in 1870, but delegates attended a private conference in London in September 1871.

12. Ferdinand Lassalle (1823-64), a German socialist labor leader and a founder of the ADAV, sought to achieve working-class control of the state through universal suffrage and to have the state finance producer cooperatives in which workers could secure the full value of their labor.

13. The "iron law" or subsistence theory of wages holds that if wages rise above the subsistence level, competition resulting from the ensuing increase in the number of workers will drive wages down to the subsistence level again. Conversely, a wage rate below the subsistence level has the effect of reducing the work force, thereby leading to an increase in wages. David Ricardo was one of the principal exponents of this theory, which was taken up by Lassalle to support his claim that trade unions could not ameliorate the workers' condition and that only a political change and a radical restructuring of society could better their lot. While it was no longer a central belief for most members of the ADAV after the mid-1860s, it nevertheless proved to be an impediment to the development of trade union support within the Lassallean movement.

14. Fortschrittspartei. The liberal Progressive party, founded in 1861, represented large sections of the middle class. Its influence was strongest in Berlin and the towns of East Prussia.

15. The workingmen's estate (Arbeiterstand) was analogous to the preindustrial craftsmen's estate (Handwerkerstand), which, together with the other three estates (nobility, bourgeoisie, and peasantry), made up medieval society. In many cases German socialists used the term workingmen's estate instead of the Marxist term proletariat.

16. The Anti–Corn Law League, founded under middle-class leadership in Manchester, England, in 1839, pressed for repeal of the Corn Laws passed by Parliament in 1815 to protect British grain prices from foreign competition. Prompted by the Irish famine, Parliament repealed the laws in 1846.

17. Max Hirsch (1832-1905), with Progressive party leaders Franz Duncker and Hermann Schulze-Delitzsch, founded the Verband der deutschen Gewerkvereine (Federation of German Workingmen's Associations) in 1869 in Berlin and served as its counsel. The federation, also referred to as the Hirsch-Duncker Workingmen's Associations, was established to counteract socialism, and its program emphasized sick and disability benefits and freedom from politics. Its journal was *Der Gewerkverein.* Hirsch later served in the Reichstag as a representative of the Progressive party.

18. François Marie Charles Fourier (1772-1837), a French social scientist and reformer, wrote a series of volumes on the rebuilding of society around communities, called phalansteries, to be established by voluntary action.

19. Étienne Cabet (1788-1856), a prominent utopian socialist and organizer of urban workers in France, published his novel *Voyage en Icarie* in 1840 describing an ideal communist society achieved through the elimination of private property and the establishment of absolute equality. Cabet hoped to achieve his utopia through a political movement cutting across class lines.

## *Samuel Gompers and Early Cigarmakers' Unions in New York City*

During the 1870s, when Gompers was becoming an active participant in the labor movement, the cigar industry in New York City was undergoing fundamental changes—changes to which the documents that follow make repeated reference. One major development was the shift from small-shop production to cigarmaking in factories and tenements, a result of the American tariff on imported cigars and growing domestic markets, tax laws that favored larger units of production, and the introduction of a device known as the mold by German cigarmakers in the late 1860s. The mold was a mechanical press whose use broke down the skilled craft of cigarmaking into two distinct stages, bunch-making and rolling, thereby allowing manufacturers to subdivide the craft. Bunch-making involved placing filler tobacco in a binder leaf, laying these "bunches" in the grooves of a mold, and then pressing them into shape. Rollers wrapped these bunches and finished them into cigars. The growing size of the manufacturing unit in the trade reflected this change in production. Prior to the 1870s most cigars were manufactured in shops having from one to three workers; by 1877, however, half the production in the city came out of seventy large units, a fraction of the approximately 1,400 cigar firms in New York.

The tenement-house system of cigar manufacture—under which families made cigars in tenement-house apartments that they rented from their employers—also emerged in New York City during the 1870s. It enabled cigar manufacturers to operate with less capital because they did not need to provide a work place for their employees, and it allowed them to benefit from the extensive use of female and child labor, since families put as many of their members to work as possible. Some manufacturers relied solely upon the tenement-house system; others employed workers in shops and factories as well as in tenement houses, where they generally produced their lowest-priced cigars. These tenement cigar "factories" could be found throughout Manhattan's Lower East Side as well as in midtown, sections where the manufacturers could find recent immigrants desperate for work.

The development of the tenement-house system coincided with the

entrance of a new ethnic group, the Bohemians, into the trade. Bohemian immigration into the New York cigar trade followed the Austrian government's monopolization of the Bohemian tobacco industry and the recruitment of Bohemian cigarmakers by some New York firms in the 1860s. Bohemians in other lines of work who followed their countrymen to New York found that cultural and language barriers inhibited their pursuit of work in their own trades. According to Thomas Čapek they "drifted into the tobacco shops, and soon butchers, blacksmiths, students, tailors, musicians, men, women, and children toiled at tobacco."[1] Some Bohemians worked in large cigar factories that predominated in the Bohemian district just north of Houston Street on Manhattan's East Side, but the majority worked in the tenement houses clustered there and in adjacent neighborhoods. The *New-York Tribune* reported that the tenement workers were "nearly, or quite all natives of Bohemia."[2]

Initially the skilled craftsmen who comprised the membership of the CMIU attempted to resist some of these changes. The 1870 CMIU convention, for example, prohibited union members from working with bunch-makers, and many cigarmakers, including Gompers' local, struck unsuccessfully against the introduction of the mold into their shops. Furthermore, although some Bohemian cigarmakers organized into local 90 of the CMIU, most were ineligible to join the International because of their system of work.

In the summer of 1872, however, Adolph Strasser, a Hungarian bunch-maker and a member of the International Workingmen's Association, organized an independent union for cigarmakers in New York City that was open to all "regardless of sex, method, or place of work, or nationality,"[3] as Gompers put it. The union, which was organized into German, Bohemian, and, for a time, English-language sections coordinated by a central council, became known as the United Cigarmakers. By March 1873 the New York Cigarmakers' State Union had endorsed the new organization.

The leaders of the United Cigarmakers did not completely sever their connection with the CMIU. Strasser, who became a member of CMIU 15, joined Gompers and Laurrell and other members of the local in working toward the liberalization of the parent body. While local 15 claimed less than fifty members in 1873, the United Cigarmakers reached a membership of 1,700 in the fall of 1873, before declining significantly during the depression of the mid-1870s. In September 1875 the CMIU convention voted to open its membership to workers without regard to sex or system of work and also authorized the dissolution of local 15 and the chartering of the United Cigarmakers as local 144.

Just prior to its chartering the United Cigarmakers reorganized, according to a plan that Gompers later claimed to have developed. The new constitution, printed below, provided for the collection of dues by shop stewards and the representation of shops by delegates in a central board of administration. It deferred establishing ethnic sections until the organization's membership exceeded 500. The members' dues supported a centrally administered system of unemployment and strike benefits.

Gompers served as president of CMIU 144 from its founding in November 1875 until June 1876. He was elected corresponding secretary for a term and then resumed the presidency at the beginning of 1877, serving until the middle of 1878. At first, he recalled, he presided over an almost "peripatetic organization"[4] that availed itself of meeting rooms maintained by saloonkeepers, first in the Bowery and then at 10 and 80 Stanton Street. When the union became financially stable, it was able to rent a permanent office. Initially, however, its influence remained limited, and its membership was less than 200 in 1876. Separate German and Bohemian cigarmakers' groups in New York City, not affiliated with the CMIU, continued to exist throughout the decade.

Although union cigarmakers in New York City accommodated themselves to the new technology in their trade and to the ethnic diversity of those who worked in it, they remained highly critical of the production of cigars in tenement houses. Spokesmen at agitation meetings held during the 1870s generally advocated the abolition of the tenement-house system altogether. One of the most important of these meetings, held on September 27, 1874, appointed a committee to ask the New York City Board of Health to take measures against tenement-house cigarmaking. The board conducted an investigation but its report, published in the *City Record* of November 7, 1874, claimed that the health and working conditions of the tenement-house cigarmakers compared favorably with those working in factories. The report outraged cigarmakers and fueled the next decade's campaign against the tenement-house system.

*Notes*

1. Thomas Čapek, *The Čechs (Bohemians) in America: A Study of Their National, Cultural, Political, Social, Economic, and Religious Life* (Boston and New York, 1920), p. 72.
2. Sept. 23, 1874.
3. SG, *Seventy Years,* 1:110.
4. Ibid., p. 116.

# A Translation of an Article in the *Arbeiter-Zeitung*

[July 5, 1873]

THE CIGARMAKERS' MOVEMENT.

(Sent in.)

For quite some time now great efforts have been made to organize the cigarmakers on a basis that could put them in a position to stop the further depression of their already seriously depressed industry. But every attempt has failed so far, on the one hand because of the poor participation of the employed workers who do not understand their position, and, on the other, because of the great power of the employers.

Proof of this is provided by the International Cigar Makers' Union,[1] which has been in existence for quite some time and which was a fine organization in the days of its founding considering the conditions prevailing then. It had strict regulations that were strictly carried out. But the power of conditions crushed it because it did not take them into account; consequently, it now finds itself in a vegetating rather than an active state. For while the federal tax collector shows us that the number of cigar manufacturers in the United States totals 12,291 and that these employ 71,491 persons, the C.M.I.U. has only something over 2,000 members. This number decreases from day to day because members cannot find work under the regulations of this union. The C.M.I.U. must take these facts into account at its next convention, otherwise it is doomed. A development in this direction is already underway, begun by a significant number of locals in the corporation itself. This group will forge ahead because it is abreast of the times; if it does not, there is already a union[2] in this city that, hand in hand with the opposition, is taking every possible measure to force the old union to realize its mistakes or to surpass that union, that is, to create a new organization of cigarmakers in the United States that will cast off everything old-fashioned and will know how to deal with the different conditions of the times.

The conditions in our industry are truly as diverse as they are bad. In the southern states the industry is almost totally in the hands of Negroes, who were excluded from the organization up until last year because they did not happen to be white. Thus they were powerless against their employers because they were unorganized. In California the Chinese have taken over the entire industry. They have no organization either, because they are still so uneducated, ignorant of

the language, and inaccessible to the movement. They are flooding the western market with cheap goods and labor. In the cities of the West the cigarmakers' union is being forced to capitulate and withdraw in one city after another because it cannot maintain the union system under the pressure of conditions. The situation is no better in the North or the East. The importation and immigration of new workers, the introduction of the roller system and molds, and the resulting overproduction have created an almost unbearable state of affairs that must make every thinking person afraid for the future or, if he is of an heroic nature, must waken the spirit of energy in him.—1,500 cigarmakers is the average number who have been unemployed every week since the end of November last year and no improvement in sight! On the contrary, shop closings, a sort of lockout, are expected next week, and this is bound to increase the above mentioned figure. Where do we see the necessity here for the 12- or 16-hour workday when the employers are not even capable of employing the existing laborers a full 6 hours daily? True, in other trades the situation is not much better, and this is partly the cause of the calamity in this trade. In spite of all this, some speculators in Syracuse have managed to obtain permission from the state legislature to allow a large number of convicts in Auburn and Sing Sing to learn the cigar industry—a sort of state aid to oppress the free workers of America.

What is going to come of all this? we must ask. An organization used to exist, they say, and a strong one at that, and what was it able to accomplish? It provided the proof of its strength in the great strike of 1870![3]

It is true that the organization existed, but it was not a strong one because a great mass of workers was excluded from participation as a result of difficulties in membership requirements and the organization's inner structure.—At that time the unions had around 1,500 members, while here in this city and its surrounding areas 13,000 non-members had work; thus union members not only had to struggle against the 2,500 employers but against these workers as well. Plenty of money was there, but solidarity, perseverence, true organization were not. This is why the result of all the efforts was a pitiful fiasco.

In order to create a strong union, all of the locals here have called a meeting for the 6th of July.[4] If the organization can there increase its membership to 4,000, then the cigarmakers will be in a position to assert their interests and make them count at the proper moment, just as the bosses can do now.

Therefore, forward cigarmakers, toward unification! Long live work and the unity of all workers.

*Arbeiter-Zeitung,* July 5, 1873. Translated from the German by Patrick McGrath.

1. The CIGAR Makers' International Union of America.

2. The United Cigarmakers.

3. In late December 1869 and early January 1870 several New York City cigar manufacturers cut wages. CMIU locals 15, 87, 90, and 97 struck to resist these reductions, and during March of 1870 there were 350 cigarmakers out on strike. The unions' effort failed, however, and in July they were forced to call off their strike and resume work at the reduced wage.

4. On July 6 Adolph Strasser presided over a mass meeting of about 500 workers from the English, German, and Bohemian sections of the United Cigarmakers at Germania Hall.

# An Article in the *New York Sun*

[September 26, 1874]

## PESTILENCE IN THE CIGAR.

The following notice in English, German, and Bohemian is being extensively circulated among the 25,000 and more persons who in this city are dependent upon cigar making for their daily bread:

The cigar makers of the city are invited to attend a meeting of citizens to be held on Sunday, Sept. 27, 1874, at 2 o'clock P.M., in the Germania Assembly Room, 291 and 293 Bowery, to give a helping hand by their presence to abolish the breeding places for contagious cases as created chiefly by the manufacturing of cigars in tenement houses. The means for a remedy shall be a memorial to the Board of Health,[1] with an illustration of the evil. Able speakers will attend.

The Committee of Arrangements.

Under the Internal Revenue laws there are registered as cigar makers (i.e., technically, the bunch makers), employed at licensed and registered manufactories, about 14,000 adult persons. With the exception of a nationality to be hereafter mentioned, women and children, though employed in several branches of the business, are not registered as operatives by the United States officials. That exception, and it may be worthwhile for the women's rights agitators to take note of it, is in the case of the Bohemian women, who from the skill acquired in the factories at home, where their sex is almost wholly employed at this business, have been the chief workers, the Sclavic men being only helpers. There are about 4,000 Bohemian women registered as operative cigar makers and between 2,000 and 3,000 men of the same stock. About 2,000 German women are also employed. Of the remaining half of those registered over 6,000 are

Germans or German Poles. There are about 200 Americans and perhaps 500 Cubans and Spaniards employed. There is scarcely an Irishman in the trade. Besides the 14,000 registered operatives there are at least 10,000 others, mainly women and children, engaged in this occupation.

The hours of labor in the factories are nine a day, or fifty-four a week. The prices paid range from $6 to $21 per thousand, according to the quality and care required. In the factory, as distinguished from the tenement, the employer pays for stripping the leaf. The price of this ranges from $1 to $3 per thousand. The average may be fairly set down at $1.50. He also employs a "caser"—an average of one for every fifty of the cigar makers proper. Such a man will get about $12 a week. The rental of a building fit for manufacturing purposes ranges from $3,000 to $5,000 a year. The large majority are in the neighborhood of Chatham square and the lower Bowery.

### The Bohemian Quarter

of New York is on the east side, mainly between Third avenue and Avenue D, and from First to Sixth street, as well as including parts of Pitt, Chrystie, Ridge, Lewis, and Sheriff streets. The cigarmakers' tenements are mostly within those limits, though there are five in Elizabeth street, on the west side of the Bowery, and one up town as far as Seventy-fifth street, between First avenue and Avenue A. In all there are about forty tenement factories registered on the internal revenue books.

The tenement factories are all on the same plan, four stories above the basements, each containing four tenements, invariably of two rooms, opening into one another, and a dark bedroom. These houses are usually twenty-five feet by fifty, and so the lighted rooms are as a rule ten by ten feet in area. The dark room, ventilated only from a small window in the passage, is barely six feet square at the utmost. This gives sixteen tenements, or thirty-two lighted and sixty-six dark rooms, which with two back basement tenements, generally with two rooms each, make a total of fifty-two rooms, excluding the front basements, which are commonly used for a lager beer saloon or a grocery. In these houses, when full, as they usually are, will be found at least 100 workers. Reference is had of course to the manufactories of cigars.

### Where Disease Is Bred.

Entering the narrow hall, into which the steep stairs project close to the street door, the olfactories are at once startled by a pungent

odor, so strong in some instances as to make a sensitive person sneeze "on sight," or rather "on smell." This is, of course, from the tobacco. Ascending flight after flight the guide stops at different doors and inquires for persons who live or are employed there. He leaves the door open so that the inquisitive observer who accompanies him can take in the general aspect of the rooms without arousing suspicions or intruding unduly on the courtesy due the humblest. In all cases there were young children and women at work; in most of these were also one or more male adults. It was said by the veteran operative who guided *The Sun* reporter's tour, that in cold weather the odor was so overpowering and pungent, doors and windows being closed, that persons unaccustomed thereto were compelled to shut their eyes in pain. Yet about four-thousand persons eat, cook, and sleep, as well as work, in these places. Young children fall asleep from the narcotic effects of the pervading odor. Women suffer greatly from it, especially in diseases peculiar to the sex. It is also a prolific source of eye diseases.

A number of the principal cigar manufacturers have taken to hiring these tenements and subletting them to their employees, who are therefore compelled to live in the same place and atmosphere as that in which they work.

The occupation of tenements for this purpose began about three years ago, but until this year there were not over half a dozen so occupied. The system is growing rapidly. With employers merely governed by avarice, this is no wonder, when the profits are considered.

### The Employers' Profits.

Take them in order. In the first place, the operative is, under this atrocious system, compelled to pay the employer's factory rent and furnish a profit thereon. The shrewd manufacturer leases his tenement, rents it to his own employees and their families, charging for the tenements described an average rent of $12 per month, and of $8 for the back basement. One tenement is usually occupied for office and packing purposes, storing boxes, &c., leaving seventeen for rent. When busy the rental income will be per month at least $196, or $2,352 per annum. Add to this the office rent, $140, and the total will be about $2,500. The average rent for similar tenements in the neighborhood was stated, in reply to inquiries, at $11 per month. Very few of the factory houses will compare favorably for cleanliness or convenience with others in their neighborhood, and some of them are reported to be very bad in a sanitary sense, and that apart from the occupation pursued.

The first profit to the manufacturer then is in dispensing with the

outlay necessary for the rent of a factory. The second is in that derived from the percentage gained by the subletting to tenants by whom, it must be remembered, the employing landlord runs no risk, as the rent is invariably collected in advance from the amounts due for labor.

But large as are the profits thus derived from the necessities of the people, it by no means follows that this is all or most of that ground out by the system. The prices paid—from $6 to $21—in the factories must be borne in mind while considering this part of the investigation. Most of those employed in the tenement factories are

BOHEMIANS,

the women as a rule being the "bunch makers," which is the branch that constitutes legally the "cigarmaker" proper. The work done is of the poorer kind. The rates paid are usually from $3.50 to $8 per thousand, and this includes both "stripping" and "casing." The tenement employer thus saves factory rental; derives profit from letting the tenements; saves the average $1.50 per thousand for "stripping," which on a week's work of 5,000 for one of these Bohemian families, is at least $7.50 per week, as also the $12 paid as "casers' " wages. The weekly profit on the stripping will be about $7.50 for each tenement, or nearly $650 per annum for each house. The pay of one "caser" for each house at least is saved, thus making an additional gain of about $450 per annum. But the profit is enormous on the prices paid for labor. The reporter's guide said that at the same rate and working the same number of hours, a man who could earn in a factory a little over $2 a day, would net only about $5 a week.

Five persons make up the average Bohemian family—usually the husband and wife, a half-grown child or two, and probably an old person, father or mother. This family, working seven days in the week, for about seventeen hours a day, at the indicated prices, can earn collectively about $20 a week. In many cases there are more than five in the tenement, for often boarders are taken or two families share it together. The room in which the work is chiefly performed is also kitchen and dining room.

THE CHILDREN.

That family is most fortunate that possesses the largest number of children, as they are all made useful. Under the laws of New York, children under ten are not allowed to be employed in factory labor. Though there are no means of enforcing this law, without complaint of a citizen, yet it is generally respected by cigar manufacturers who

are not engaged in the tenement system of "sweating." The needs of the parents, as well as their ignorance, combine in this matter with the cupidity of the manufacturers, largely to recruit the "dangerous classes," if an early journey to "God's acre" is escaped.

The operative cigar makers, as well as their employers, who are not parties to this tenement system, are of course bitterly hostile. The latter see the destruction of what they regard as the only legitimate mode of doing business. But the operatives see in unmistakable form in its continuance the utter demoralization and degradation of their craft. They realize that it is rapidly reducing their wages.

THE DECLINE

already felt through it being estimated at not less than forty per cent., and they believe this must go on in a more rapid ratio hereafter. The more intelligent among them are most painfully affected by the effects of the system on the health of those engaged, and of the neighborhood, as well as the danger that arises from the spreading of infectious diseases through the articles manufactured. Instances of small pox occurring in these tenements, the patients remaining some time before their removal was effected, the operations of cigar making going on as usual, were mentioned and authenticated.

This danger can be readily seen and comprehended, and the cigar makers wisely propose to call for the community's interference as a matter of sanitary precaution and self protection. There is little doubt that when the facts are all laid before the public there will be some proper action taken by the health authorities. The "sweating" manufacturers are alarmed at the agitation, and it is said that the tenant employees have received instructions not to allow visitors or to give information, probably on pain of loss of employment. It was evident that the presence of *The Sun*'s representative accompanying the leader of this agitation aroused suspicious fears.

In answer to questions, the Trades Union President[2] acknowledged that his trade was demoralized. Formerly the Union had a membership of about 1,400, or one tenth of the registered cigarmakers. Now its membership is 48.

*New York Sun,* Sept. 26, 1874.

1. The New York City Board of Health, established in 1866.
2. Probably Martin Diederich PLATE, a cigarmaker and president of CMIU 90 of New York City.

# A Translation of an Article in the *Social-Demokrat*

[December 27, 1874]

ABOUT THE CIGARMAKERS.

(Sent in.)

During the summer of 1872 an association of cigar workers was formed here[1] in response both to the reactionary movement of the "Cigarmakers' International Union" and to mass production, which had arisen by that time. This new association aimed at uniting, without regard to nationality, all men and women working in this trade into the common struggle against the increasingly more powerful forces of Big Capital in order to help them assert their rights and interests. Immediately a German and a Bohemian section were formed, with a central body; an English section was founded, but it was disbanded shortly thereafter. International Union No. 90[2] was in sympathy with this association. In addition, a death benefit fund was instituted that has already been a blessing for many members. All contributions were used (with the exception of a few strike support payments) exclusively for organizing; the officers did not receive a single cent for their work but rather, in addition to their time, had to sacrifice money as well for the cause.

Despite the greatest efforts, mass meetings, and the reorganization that was carried out, the German section did not accomplish anything noteworthy in the autumn of last year. Those of our colleagues who were opponents at the founding of this organization and who disgracefully deserted us after reorganization might be considered the main reason for this partial failure. We do not want to attack anyone personally here, so as not to harm the good cause which we serve, but our call goes out to all of you: "Fight with us shoulder to shoulder so that we can put up a dam to our opponents which, in spite of all their capitalist machinations, they will not be able to break through.["] The Bohemian section has worked much better than the German section, and has grown visibly: proof enough that there are not so many equivocating elements among them, who instead of improving the situation only make matters worse.—At its last meeting the German section decided in future, after settling the routine business of the meeting, to discuss the socio-political questions of the day, so that everyone understands them. The section also agreed with the decision of the central body that a protest meeting should be held on the 3d day of Christmas at 1 o'clock in the afternoon in Concordia Hall,

Ave. A., in order to give the Board of Health a blunt reply to its one-sided capitalist resolution concerning the manufacture of cigars in tenement houses.[3] Every worker true to his principles is obliged to urge his colleagues to attend this meeting, so that this board can be shown by a mass demonstration that the working class is not willing to accept every lie as the truth. Finally, we want to mention that the German section meets every 2d and 4th Wednesday of the month at No. 80 Stanton st., the Bohemian section every 1st and 3d Monday of the month at 220 2d Street, and the central body every 2d and 4th Friday of the month at 220 3d St.

The Central Body of the United Cigarmakers.

*Social-Demokrat*, Dec. 27, 1874. Translated from the German by Patrick McGrath.

1. New York City.
2. CMIU 90, a New York City Bohemian local, was organized in 1869 and consolidated with CMIU 144 in April 1876.
3. The meeting, held Dec. 27, 1874, protested against the recent report of the New York City Board of Health which maintained that the working conditions of cigarmakers in tenement houses compared favorably with those found in cigar factories.

## A Translation of an Article in the *Social-Demokrat*

[January 24, 1875]

REPORT ON THE TRADE UNION ACTIVITIES OF THE UNITED CIGARMAKERS OF N[ORTH] A[MERICA] IN 1874.

During the first meeting of the Central Committee, on January 3, 1874, it was reported that a large-scale strike was to be expected since the workers at Kerbs & Spies,[1] Bondy's,[2] and at other factories wanted to regain their pre-depression wage scales. The Central Committee advised against this since the time did not seem ripe for the desired outcome. The strike broke out nevertheless, and was partially successful in the beginning, with the firm of Kerbs & Spies giving in to the demands. Consequently, the same demands were voiced by workers at other large factories, in hopes of pushing them through with or without a strike. Thus, the strike took on greater dimensions. One enthusiastic mass meeting followed another; money was collected among factory and tenement-house workers in order to help the needy. But since the strike occurred in just those factories where the workers

were not organized and even fought against the existing organizations, the necessary funds and proper leadership were lacking at the very outset. Dubious elements (in English, "suckers") became involved, conspired against the strikers, and created confusion until the striking workers saw that the unions and the organized workmen would have to become involved in order to reestablish order and achieve success. The organization—that is, the Central Committee of the United Cigarmakers—intervened, purged the ranks of undesirable elements, took over the relief system, and reestablished order. All those who applied and were recognized as truly needy received 6 dollars a week in support payments whether they had been members of the organization before the outbreak of the strike or had joined recently. Within a few weeks 580 dollars had been paid in relief payments and every cent had been accounted for.

Yet the strike was lost. The main reasons were: 1. the bad economic situation of the trade at the time and the high unemployment this caused since all the factories were very soon filled with workers who had not participated in the strike; 2. the lack of funds to support everyone without exception; 3. plots and conspiracies by unscrupulous individuals; and 4. the dishonest conduct of the biggest manufacturer involved in the strike, Kerbs & Spies, both by their dismissal of the alleged ringleaders of the strike and by breach of promise (they had made concessions to a workers' committee, which they simply ignored or denied after work had been resumed).

Apart from exhausting efforts and much time, the strike cost the organization more than 80 dollars, which had to be paid through an extra tax levied on the members.

But along with its negative consequences the strike also had some positive points; chief among them was that many workers realized the position of their class. The membership of the United Cigarmakers increased fourfold, an English section was formed, and a health benefit society, which continues to flourish, was founded by the Bohemian members of the organization. The dismissal of the ringleaders of the strike led to the establishment of a cooperative shop by the Bohemian cigarmakers that still exists and is doing quite well.

Among the negative results of the lost strike are, above all, the drop of membership from about 1,700 to 1,200, the dissolution of the English section, and consequently the complete defenselessness against the wage cuts below the subsistence level that soon followed. The huge increase of tenement-house work and the reduction in shop work contributed to the situation. Shop work, which used to employ a large number of cigarmakers at good pay in the past, has sunk to a minimum.

Another factor was the mass unemployment and the consequent inability of the members to pay their dues. During the year 1874 over 1,000 members had to be dropped because they had not paid their dues, and it seems this will continue even now. The spirit of lassitude and numbness contributes to the situation.

Even the budget for death benefits has increased exceptionally this year. While only one member died in 1872, five died in 1874, for whom 500 dollars in death benefits were paid. Four of them died of consumption.

This year great efforts and many sacrifices were made in campaigning, mainly against the spread of tenement-house work, which threatens everybody's existence. While there were about 25 houses occupied by tenement workers in 1873, their number had increased to 48 at the beginning of the campaign and today, after the inspectors submitted their famous report, their number has increased to 78. Whole shops have closed, others are partly empty. The tenement factories are filled to the brim and work without pause. The campaign was begun after smallpox and other diseases broke out in the tenement factories. In 7 houses, smallpox had broken out, most of the cases ending in death. After an inspection by the health commissioner, 12 more cigarmakers were taken to the hospital with smallpox and two houses were closed to the public. Nevertheless the system was expanded to include 50 more houses.

The failure of the campaign was caused by the terrible lack of participation on the part of the cigarmakers in general, but also by the counterpropaganda of the cigar manufacturers involved—mainly Liechtenstein Brothers and Co.[3]—who, supported by a cigarmaker named D. Rise,[4] even organized a counter-demonstration with pitiable result. But after the Board of Health's report became public the above mentioned firm announced its victory by renting 14 additional houses and filling them with tenement workers.

All these facts have been communicated to the health commissioner, the mayor, and the governor.

The membership of the organization (that is, those in good standing) is at 735. The balance was $45.60 at the end of 1874.

J. Schäfer,[5]
Recording Secretary.

*Social-Demokrat,* Jan. 24, 1875. Translated from the German by Dorothee Schneider.

1. Kerbs and Spiess became the largest manufacturer of cigars in New York City during the 1870s.

2. By 1877 Charles Bondy and Co. had 600 workers, 350 in tenements.

3. Lichtenstein Brothers became the second largest cigar manufacturer in New York City during the 1870s.

4. Possibly David Rice, who worked as a cigarmaker and packer in New York City from 1874 to 1880.

5. John SCHÄFER, a New York City cigarmaker, was active in the Social Democratic Workingmen's Party of North America and, in the late 1870s, in the SLP.

## Hugh McGregor[1] to the Editor of the *National Labor Tribune*

New York, Sept. 17, 1875.

Editor National Labor Tribune:

Throughout the United States there exist numerous organized bodies of workingmen who declare the present political and social systems are false, and require to be changed from their very foundation; that the present degrading dependence of the workingman upon the capitalist for the means of livelihood is the cause of the greater part of the intellectual, moral and economic degradation that afflicts society; that every political movement must be subordinate to the first great social end, viz., the economic emancipation of the working classes. Roughly stated, this is the creed of an immense number of workingmen, and they are the principles held in common by the organized bodies above referred to, but there are minor points on which we differ. To clear up these points of difference, and unite all the various bodies in one grand, invincible organization, has for some time past been the earnest desire of us all, and now it seems that this desire is about to be realised. From all quarters comes a cry for unity. Suggestions have been made to hold a congress in some convenient place, for the purpose of deciding upon a common plan of action, but the time and place of meeting is not yet fixed. Many persons hostile to the cause of labor—and notably the editors of the stock-jobbing press—have sought to bring this radical labor movement into disrepute by persistently asserting that the movement is French, German or Russian, but nothing can be further from the truth. Just as well might these venal writers assert that the "law of gravitation" because it was discovered by Newton, is therefore an English movement. For all men who know anything about social science know that the emancipation of labor is neither a local nor a national, but a social problem embracing all countries in which modern society exists. We, are workingmen, born or naturalized in the United States, therefore such attempts to make aliens of us, only show to what depth of villainy our opponents will proceed. "Divide and conquer!" has ever been

the cry of the oppressor, and it has been successful through all the ages; but now the working classes are rallying to the voice of common sense. They appeal to the Tribune of Labor, and we are confident that the appeal will not be made in vain. We know that the Tribune will speak in thunder tones for unity.

Finally, we shall be happy to enter into correspondence with any body of *bona fide* workingmen, so as to effect that unity without which all attempts to achieve social progress by political means must be unsuccessful. Hoping that all intelligent workingmen and organized bodies will coalesce of their own free will from an instinctive sense of a real genuine community of interest. I remain,

Yours &c., H. McGregor,
Gen. Sec., N.Y. Com. on Organization.[2]

*National Labor Tribune,* Sept. 25, 1875.

1. Hugh McGregor, a jeweler, was a member of the Social Democratic Workingmen's Party of North America and the Economic and Sociological Club. He later served as SG's secretary in the office of the AFL.

2. In his autobiography SG wrote that he and other members of the Economic and Sociological Club wrote this letter and authorized McGregor to publish it as secretary for their group (*Seventy Years,* 1:210).

# A Translation of a News Account of a Meeting of the United Cigarmakers and the Constitution of the United Cigarmakers

New York. 18 Oct. [1875]

CONCERNING THE CIGARMAKERS.

A delegation of weavers from Fall River appeared at the last meeting of the United Cigarmakers, held on October 11 at No. 642 5th Street; they briefly described the reprisals on the part of the manufacturers[1] and stressed the solidarity of all workers. Since the financial circumstances of the union are very weak, only a $10 donation from the treasury could be approved for the present. However, it was decided to arrange collections in all factories, and lists were printed for this purpose. The new constitution was then read and accepted with few revisions. It was decided to make it effective immediately, and to have it published in the German and English workers' press. A letter from the president[2] of the International Cigarmakers' Union was read, in which the United Cigarmakers were urged to join the International

Union. Since it was already rather late the matter was postponed until the next session. . . .[3]

## CONSTITUTION.

### INTRODUCTION.

The continual attacks and acts of repression by the capitalists, as well as the recognition by the working class of its position in society, make it the duty of every worker to unite with his fellow workers in order to improve his situation and to achieve an existence commensurate with his dignity as a human being. One of the main tasks of this organization shall be to raise the worst-paid workers to the level of the best-paid. To this end, we the members of this organization pledge to support each other, one for all and all for one.

Experience has proven to the working class that individual workingmen's associations are not strong enough to oppose the unified capitalists; therefore, the goal of this organization shall always be to unite all workers into one, unified, central association.

The fact that the worker's existence depends on the capitalist forms the basis for every form of servitude; this is why the cigarmakers of New York are striving to institute free, cooperative work to replace wage labor, in order to secure for every worker the full fruits of his labor.

Toward the achievement of these goals, the following constitution shall provide our guidelines:

### ARTICLE 1.

The name of the union shall be: "United Cigarmakers of New York."

### ARTICLE 2.

Duties and Rights of the Members.

1. Every person employed in the cigar trade without regard to nationality, sex, or mode of production who has reached the age of 17 and works for wages can become a member of the organization.

Members who become foremen or employers shall receive a retiring card and, should they return to their former occupation, shall be reinstated to their former rights without further payment.

2. Every person must pay an entrance fee of 50 cents, payable in bimonthly installments.

3. A weekly fee of 10 cents is to be paid to the shop steward.

4. Members who are unemployed or become unable to work because of illness pay no fees, but a stamp saying "out of work" shall be put in the membership book in place of a receipt for the weekly fees. This is only applicable in cases where the member was employed less than three days per week. The executive board has the right to require sufficient evidence in any such matter.

5. Members who are more than 5 weeks behind in their payment of fees lose all rights to financial support; those who owe payment for a period of 3 months and 3 days shall be struck from the records without notice and must pay the entrance fee should they want to return.

6. Every member is obliged to use all of his influence in finding work for unemployed members.

ARTICLE 3.

Financial Support during Unemployment.

1. To be qualified for financial support, the member must have been in the union one year. This support shall be provided for 5 weeks and any further payment may be claimed only after a period of 6 weeks, even if the worker is unemployed during this period.

2. Support consists of 1 dollar a day and 50 cents for half a workday. This payment may only be collected in person at the weekly meeting of the executive board by signing a receipt.

3. Members who receive support must accept work wherever it is found if union wages are paid. Should they refuse this employment, they are not entitled to support for 8 weeks—inability to work is here an exception.

4. If a member receives work before the 5-week support period has ended and becomes unemployed within 6 weeks after that, then he is entitled only to that part of the 5-week support that he was not paid.

5. A worker who is forced to leave his employment must have a slip made out by the director in which the date and hour of his leaving are given. As soon as he returns to work, the shop director must be notified immediately.

6. Members who willfully give up their employment are entitled to no support for the following 4 weeks.

7. A member who works against the interests of the association or breaks the statutes can be put on trial and, after a committee has thoroughly reviewed the case, can be penalized in a manner acceptable to the association.

ARTICLE 4.

Officers.

1. The union elects by ballot in the months of December and June the following officers for a period of 6 months to form the executive board: a president, a vice-president, a recording secretary, a corresponding secretary, a financial secretary, a treasurer, 3 trustees, 3 members of the auditors' committee, and one sergeant at arms.

The duties of the officers are:

1. The executive board meets every week in order to collect monies, take in new members, and receive the reports of the shop stewards. The day, time, and place of meetings are determined by the union itself.

2. It is the duty of the financial secretary to take in all monies and to keep a correct account of them, as well as to deliver them to the treasurer in exchange for a receipt at the end of each meeting; to read off the names of new members and members who are to be struck from the records at the first business meeting of every month; and to give a financial report every 3 months containing the number of members, revenues, and expenditures.

3. The treasurer shall not keep more than 50 dollars in cash and is to give security for this amount. All other monies shall be deposited by the trustees in the bank decided upon by the union for this purpose. He must also pay all expenses authorized by the union, and at the end of each quarter must submit all vouchers and bills to the auditors' committee.

4. The auditors' committee must examine all reports, books, and vouchers of the financial secretary and of the treasurer, and must certify their correctness.

5. The entire executive board must make sure that the treasurer and the trustees are doing their duty.

ARTICLE 5.

Duties and Rights of the Shop Steward.

1. In each factory where there are members one shop steward (or, if necessary, two) shall be elected for a period of 3 months. If there is only one member, he is to act as shop steward.

The duties of the shop steward are:

1. To collect membership dues every week and deliver them to the executive board.

2. If a member leaves a factory, the shop steward must make out a slip for him giving the exact reason for his leaving and the date;

he must also do this when members who have received financial support arrive to work.

3. To inform the executive board about everything happening at the factory that is of interest to the association.

4. To appear at the meetings called by the executive board.

5. To make sure that all monies delivered are entered in the books while he is present.

6. In necessary cases, or at the request of one-sixth of the members, to call shop meetings.

ARTICLE 6.

Strikes.

[1.] Justified grounds for a work stoppage are the following:

1. Wages that are lower than those set by the union.
2. Extension of the normal work day.
3. Unhealthy working conditions.
4. Oppressive measures, like dismissals or oppressive factory regulations.

[2.] As soon as the executive board is notified of a strike it must, if possible, make inquiries at the place in question; if the number of strikers is more than 15, it must call a special general meeting of the union within 5 days.

3. The executive board may give its approval of the strike only after two-thirds of the workers from the factory have decided by secret ballot to strike.

4. The executive board has the right to reject every strike application as long as other union members have a lower average wage. Workers may, however, appeal to the union, and a special general meeting must be called for this purpose within 5 days. The vote shall be taken by ballot and two-thirds of those present must vote in favor of the strike in order for it to be valid.

5. The executive board shall have the right to stop members from working for employers who conspire against the existence of the union or violate the rights of workers repeatedly, provided that such a course of action is feasible, has been presented to the union, and approved by a two-thirds majority.

6. An exact statistic of average wages should be kept, to serve as a guideline for the executive board.

7. Only those members who have been in the union 3 months and are no more than 5 weeks behind in the payment of their dues are entitled to financial support during a strike. The payment is 7 dollars per week.

ARTICLE 7.

Meetings.

Meetings are scheduled twice a month, and members must show their membership card to be admitted.

Every 3 months there is a general meeting in which officers must give complete reports on their activities.

In extraordinary cases, the executive board may call a general or mass meeting.

ARTICLE 8.

General Provisions.

1. As soon as the organization has 500 members, language sections shall be formed. Every section shall have the same statutes.

2. The task of administration shall be given over to a central committee, in which each union shall be represented.

3. If the number of unemployed exceeds one-tenth of the total membership during a time of crisis or stagnation in the trade, the association may decide in a special general meeting to reduce the amount of support payments.

4. Members who work for the association and lose time during the day, receive a reimbursement of 40 cents per hour.

5. This organization cannot be dissolved as long as it still has 50 active members.

ARTICLE 9.

Revisions or Additions to the Statutes.

Revisions or additions to the statutes may only be made during the quarterly general meetings in January, April, July, and October with a two-thirds majority; such requests, however, must have been presented to the executive board 4 weeks prior to a quarterly meeting, and must have been read in a previous business meeting.

ARTICLE 10.

The business of the union is to be conducted as follows.

1. Reading of the minutes.
2. Registration of new members.
3. Reading of correspondence.
4. Reports from the executive board and committees.
5. Unfinished business.
6. New business.
7. General union business.
8. Close of meeting.

These statutes were passed in the general meeting of 11 October and became effective immediately.

*Social-Demokrat,* Oct. 24, 1875. Translated from the German by Patrick McGrath.

1. In what was known as the "Great Vacation," 15,000 workers struck the Fall River, Mass., mills in August 1875 protesting a wage reduction. When the workers returned to work after four weeks, the owners instituted a lockout. They subsequently opened the mills only to nonunion workers willing to sign a contract pledging not to join a union. The strike resulted in the blacklisting of union leaders and the breaking up of the Weavers' Protective Association, along with the carders' and loom fixers' unions. Only the mule spinners' organization survived the struggle.

2. George Hurst was president of the CMIU from 1875 to 1877.

3. The balance of the account reported on the donations accepted by the United Cigarmakers' Executive Board at its Oct. 18 meeting.

## A Translation of an Article in the *Social-Demokrat*

New York, 24 October [1875].

It will be of interest to many readers of the "Social-Demokrat" who work in the cigar industry to learn something about the New York cigarmakers' movement of the past few years. When, in the summer of 1872, the 8-hour movement[1] was reviving the unions to new life, it began to dawn on the cigarmakers that something would have to happen to improve their situation which had been significantly worsened by the development of industry, the division of labor, and the mass hiring of women and children. Mass meetings were called for the purpose of creating an organization that was to embrace all persons employed in the cigar industry without regard to nationality, sex, or method of production. The New York cigarmakers were forced to create an independent organization because the statutes of the Cigar Makers' International Union did not allow rollers and wrapper makers to become members, and in New York this method of production was already dominant. In the beginning several hundred Bohemian and German cigarmakers, both male and female, were successfully organized and their number grew to 1,700 in the winter of 1873. During this time, a strike took place at Kerps and Spies,[2] the largest manufacturer of this city, and was lost as a result of breach of promise on the part of these exploiters and the joint action of the manufacturers' association. This reduced the number of members significantly. The

modern home-industry,[3] which was introduced the year before on a small scale, spread at a frightening rate and has grown in such dimensions that more than half of all the cigars are produced during 14- to 18-hour-long workdays in these houses of misery. Haggard, pale, and dull faces, and smallpox and other diseases are the results of such excessive and inhuman working hours. During the autumn of 1874, uninterrupted agitation was begun in order to put a stop—through the power of the law—to this unbounded exploitation. Workers appealed to public opinion, and the Board of Health was called upon to eliminate these ills. But since the Board of Health stands in the service of capitalism, it referred to these ills as a triumph of progress and made its opinion known in the bourgeois press. This caused a further reduction in membership in the union. In the spring of 1875 conferences were held at which it was decided that a reorganization would have to take place in order to protect ourselves against the insatiable greed of the manufacturers, and after six months of debate and parliamentary discussion with the Bohemian union, the new constitution was passed. In the beginning, plans to form an international union were made. These were temporarily set aside because it was expected that at the congress of the international union in September[4] regulations would be adopted which would make joint cooperation possible. At the last meeting of the United Cigarmakers this matter was thoroughly considered and it was decided to make inquiries about the present state of the international union in order to be able to come to a final decision in the next meeting. It is to be hoped that the cigarmakers in this country, where no truly strong unions exist, will organize on a similar basis since the new organization is continually gaining ground in New York and will in the course of time enjoy sweeping success.

*Social-Demokrat,* Oct. 31, 1875. Translated from the German by Patrick McGrath.

1. The eight-hour movement of 1872 consisted of a series of spontaneous strikes for the eight-hour day from early April to late June in New York City. Some strikers also demanded the enforcement of the federal eight-hour law, which applied to federal contracts. The strikes began with the Brooklyn painters and quickly spread throughout the building trades. As the painters, carpenters, and some furniture trades achieved the eight-hour day without a reduction in wages, other trades initiated their own strikes. A variety of factors defeated their efforts, however, including large surpluses of stock held by some manufacturers, threats by others to relocate their plants, and the workers' lack of a central organization to coordinate the strikes. Most workers returned to their jobs on the old terms by the end of June.
2. Kerbs and Spiess.
3. That is, the production of cigars in tenements.
4. The CMIU held its 1875 convention in Paterson, N.J., Sept. 6-10.

# The Charter of CMIU 144

[November 24, 1875]

THE CIGAR MAKERS'
INTERNATIONAL UNION. OF AMERICA,
DOTH GRANT THIS CHARTER TO

Saml. Gompers President.
Levie. Bossie[1] Vice President. Ad. Strasser[2] Fin. Secretary.
Henry Baer[3] Rec. Secretary. K. Nickelsberg[4] Treasurer.
David Straus[5] Cor. Secretary. H. Fronhafer.[6] Door Keeper.

And to their successors legally elected, to constitute a Union, to be known as Union No. 144 of the City of New. York. N.Y. for the purpose of effecting a thorough organization of the trade. And the said Union being duly formed, is hereby authorized and empowered to initiate into the Union, any person or persons, duly proposed and approved According to the Constitution adopted by the International Union, and to enact By-Laws for the government of their Union.

Provided Always, That the said Union do conform to the Constitution, Laws, Rules and Regulations of the International Union; and provided also that said Union be held in the City of New York and not removed therefrom without the consent of the International Union, and in default thereof, or any part, this Charter may be suspended or taken away by the decision of the International Union.[7] And further in consideration of the due performance of the above, the International Union do bind themselves to support the said Union No. 144 in the exercise of their rights and privileges as a Subordinate Union.

In Witness Whereof, we have subscribed our Names and affixed the Seal of the International Union, this 24th day of November A.D. One Thousand Eight Hundred and Seventy-five

George. Hurst President, C.M.I.U.
W. J. Miller[8] 1st. Vice President, C.M.I.U.
L. P. Rohner[9] 2d. Vice President, C.M.I.U.
Jno. J. Junio[10] Pres. Cigar Makers' State Union of N.Y.

Printed form, with handwritten entries, signature representations, George Meany Memorial Archives, Washington, D.C.

1. Levie (variously Levi, Louis) Bossie (b. 1837), a cigarmaker born in Holland, served as vice-president of CMIU 144 until August 1876.

2. Adolph STRASSER, a cigarmaker, played a leading role in the United Cigarmakers and was president of the CMIU from 1877 to 1891.

3. Henry Baer, a cigarmaker from Hamburg, was active in the International Workingmen's Association and a member of the Economic and Sociological Club. During the later 1870s and early 1880s he served as auditor and treasurer of CMIU 144.

4. Kaufman Nickelsberg (1845-1919), a Dutch-born cigarmaker, immigrated to the United States in 1863. He served as treasurer of CMIU 144 from 1875 to 1877, when, on charges brought by SG, a New York court sentenced him to eighteen months imprisonment for embezzling union funds.

5. David Straus (variously Strauss; b. 1843), a cigarmaker born in New York City, served as corresponding secretary of CMIU 144 until August 1876.

6. Possibly Henry Frohnhoefer (b. 1853), a Brooklyn cigarmaker.

7. Two lines are blocked out here on the original document.

8. William J. MILLER was first vice-president of the CMIU from 1875 to 1877.

9. Louis P. ROHNER was second vice-president of the CMIU from 1875 to 1876.

10. John Joseph JUNIO, active in reform politics and, later, in the KOL, served as CMIU president in 1867 and president of the CMIU's New York State body from 1875 to 1877.

## *Samuel Gompers, Adolph Strasser, and the Reform of the CMIU*

The structure and function of CMIU 144 reflected Gompers' and Strasser's belief that building a financially secure, well-organized trade union should take precedence over political party activity. "Those who . . . fight primarily for the abolition of wage labor" without meeting the immediate needs of the working class, Strasser argued, "have not fully grasped the idea of modern socialism."[1] Following the British trade union model, Gompers and Strasser worked to incorporate local 144's system of centralized administration, high dues, and benefits into the CMIU as a whole, and to limit the union's political involvement to non-partisan, issue-oriented campaigns.

Gompers proposed sick and out-of-work benefits at the CMIU's 1877 convention, where he represented local 144. While the convention rejected these proposals, it did follow his lead in two other important respects. First, it voted in favor of his recommendation to table a motion by Frank Hirth, a socialist cigarmaker from Michigan, that urged CMIU locals to "form themselves into labor bodies upon the basis and platform of the W.P.U.S." (Workingmen's Party of the United States).[2] Second, after parliamentary maneuvering by Gompers, it elected his nominee, Adolph Strasser, to the presidency of the CMIU. On becoming president, Strasser carefully reviewed the benefit features of such English unions as the Amalgamated Carpenters and Joiners and noted that their "Protective and benevolent . . . character" was "*the secret* of the growth and power of trade unions in England."[3] Together with Gompers, Ferdinand Laurrell, and Fred Bloete, he planned the reorganization of the CMIU along these lines.

At the next CMIU convention, held in 1879, Gompers and Strasser won approval for a number of their proposals. Two of these, introduced by Gompers, exemplified the non-partisan political activity they envisioned for the CMIU: the first called on the International to work "to obtain legislation in the interest of the craft" while the second declared that no local would be permitted to "aid, cooperate or identify itself with any political party whatsoever."[4] The convention also ratified resolutions, introduced by Gompers but initially presented in Strasser's presidential address, amending the CMIU constitution to

provide for uniform dues, a standard initiation fee, and the quarterly equalization of funds held by local unions. Further, it endorsed a travel benefit for members seeking employment and established a strike fund that could be disbursed only after approval by the Executive Board or the general membership. "Heretofore," Strasser noted after the convention, "the International Union has been a combination of local unions, very loosely connected, either for defensive or aggressive purposes."[5] The new constitution transformed them into interdependent units of a centralized national union. While the convention did not approve two of the most important proposals—those for sick and death benefits—it did refer them to the locals for consideration. They were adopted at the CMIU's 1880 meeting and subsequently approved by referendum. Thus, by 1881, the CMIU had developed a system of benefits and controls that, in its comprehensiveness and financial soundness, was a model of what came to be known as "business unionism."

*Notes*

1. *Social-Demokrat,* July 9, 1876, translated in H. M. Gitelman, "Adolph Strasser and the Origins of Pure and Simple Unionism," *Labor History* 6 (1965):75-76.
2. See "Excerpts from the Proceedings of the Eleventh Convention of the CMIU," Sept. 1, 1877, below.
3. *Cigar Makers' Official Journal,* Apr. 10, 1879.
4. See "Excerpts from the Proceedings of the Twelfth Convention of the CMIU," Sept. 4, 1879, below.
5. *Cigar Makers' Official Journal,* Sept. 15, 1879.

# Adolph Strasser to the Editor of the *Cigar Makers' Official Journal*[1]

New York, Dec. 4. 1875.

A Mass Meeting was called by the United Cigar Makers of New York, for the purpose of discussing the present condition of the trade, which took place Sunday Nov. 28. at Concordia Assembly Rooms. The hall was well filled and great enthusiasm prevailed. The Officers of the meeting were L. Berliner,[2] chairman, Paul Tesse,[3] 2. chairman, G. Walier[4] and L. Baer,[5] Secretaries.

After a brief speech by the chairman, the first speaker of the day, Mr. S. Gompers, President of the organization, was introduced, who addressed the meeting as follows:

Fellow Cigar Makers. All who are present to day, should be convinced that our condition is growing worse every day, and the future is threatened with danger. Who can deny, that reductions are almost of daily occurrence, because the Capitalists' only ambition is profits, and the time has come when we must assert our rights as workingmen. Every one present has the sad experience, that we are powerless in an isolated condition, while the capitalists are united; therefore it is the duty of every Cigar Maker to join the organization.

Some weeks ago the Union was reorganized on a basis, where the experience of past years has been felt in operation and grievous mistakes avoided.

One of the main objects of the organization is, the elevation of the lowest paid worker to the standard of the highest, and in time we may secure for every person in the trade an existence worthy of human beings.

In conclusion let me tell you, we have resolved in our last general meeting, to join the International Union and [are] expecting the Charter every day.

The second speaker, A. Strasser, addressed the meeting in German as follows. Fellow Workingmen; it is to be supposed, the Cigar Makers assembled here are determined to act, considering the fact, that since the panic our condition has grown steadily worse. The history of each factory will prove, that reductions from three to eight Dollars per M[6] have taken place and will continue so. First the Cigar Makers will recognize, that a strong organization only can save them from becoming abject slaves to their master.

The tenement house system is increasing and threatens through the evil of free competition to destroy the existing factories. Day and night, Sundays and holidays, cigars are made in these houses of misery and slavery, for wages which merely keep body and soul together.

Some, influenced by the public press, entertain the opinion, that wages are regulated by the laws of supply and demand, but they seem to forget that one individual may offer his labor for $12, another for $10 and still another for $7, while in a Union a uniformity of action and will prevails, which destroys the disastrous competition between the members.

In our new Constitution, adopted Oct. 11. 1875, we recognize the solidity of the whole working class to act harmoniously against their common enemy, the Capitalist.

We also pledge ourselves to support the unemployed, because hunger will force the best workmen to work for low wages, therefore rally round the Union. United we are a power to be respected, divided we shall be the slaves of capital.

R. Nicelsberg,[7] the next speaker, made the following remarks; the only way to improve our condition, is to join the Union, let the Cigar Makers take an example of their brethren in England, they commenced to organize in 1857,[8] but without success, until 1863, and at present only 11 persons can be found not belonging to the Union. Therefore the past sad experience may not deter us from becoming good Union men.

The chairman called upon those wishing to join to step forward and pay their initiation fee. 39 members were enrolled and a great many promised to call in the next meeting of the Union, not having the necessary funds at hand.

By order of the Board of Officers of Union No. 144 United Cigar Makers of New York.

A. Strasser
Fin. Secretary

Head Quarters 642 Fifth St.

Regular meetings every second and fourth Monday in the month at 8 P.M. Board of Officers meets every Monday.

*Cigar Makers' Official Journal,* Jan. 1876.

1. The *Cigar Makers' Official Journal,* the monthly journal of the CMIU, began publication in November 1875.

2. Louis BERLINER was a member of the United Cigarmakers and, later, of the New York American Section of the Workingmen's Party of the United States.

3. Paul C. Jesse was born in Germany in 1846 and immigrated to the United States in 1869. He was a trustee of CMIU 144 in 1877 and later that year moved to Providence, R.I.

4. G. Walder was elected auditor of CMIU 144 in early 1877.

5. Louis Baer, from Hamburg, Germany, was an early shopmate of SG's and a member of the International Workingmen's Association. The brother of Henry Baer, he helped organize CMIU 144 and in 1877 was elected trustee.

6. That is, per thousand.

7. Probably Kaufman Nickelsberg.
8. The Cigar Makers' Mutual Association of London was founded in 1835.

# To George Hurst

New York, Dec. 21. 1875.

Mr. Geo. Hurst,
Pres. C.M.I.U. of America.
Sir.

In conformity with your request for information of the condition of the trade in our respective localities, I herewith submit a brief report, suggesting that you publish as much thereof as you deem fit in the Official Journal.

Two thirds of the shops in this city have either closed entirely, or from Dec. 7. until the middle of January; and the remainder have discharged more than half their hands thus throwing out of employment thousands of our fellow craftsmen. Considering the large number previously idle, the number will now reach about five or six thousand.

I will here give you the names of a few of the factories, that have closed or nearly so. [as they occur to me][1] Orgler[2] discharged 10 out of 16. Jacoby all hand workmen, the form workmen on half time, Stahl 8 out of 24, Doring[3] entirely, Greenhall[4] all, Hirsch[5] 14, Stratton & Storm[6] nearly all and have reduced from one to two Dollars per M in their tenement houses; and others might be mentioned that would take up the whole space of our valuable journal.

Now, these being facts, we as workingmen, should throw off this lethargy and become thinking people who are able and willing to work for our own interests.

When we see on the one hand the factories decreasing, and hand work with it, and many out of employment, on the other the tenement houses with its thousands crowded together, like hogs in a pen, [only the hogs have the advantage they need not work] working from 16 to 20 hours a day, producing treble the number of Cigars, that they properly should, it compels us to ask ourselves the question, what shall we do to alter or prevent its further growth? I will answer, join your Trade Union, be earnest and energetic, persevering to make your Trade Union a place where the Working-people will properly learn the condition of their trade, and what is most advantageous to it, impress upon their minds the necessity of shortening the hours of labor, and you will have done your duty for present, but in order to

make this principle more explicit, I copy the following Article from a labor Paper, it being so logical, that I will give a verbatim extract.

Suppose there is one third more men than there is work for, [in many cases there are one half too many] now this one third compels the other two thirds to work for less in order to have employment, the two thirds work say nine hours a day, if the other third were allowed to do their share of the work, all would only work six hours a day, suppose this was done! you say in reply, each man would earn only two thirds as much in the six hours as he did when he worked nine hours. Granted for the present. But look again, this six hour rule employs every man and stops the competition, which heretofore existed, what then? Why simply this, that the reduction in Wages caused and maintained by the competition of that unemployed one third is cut off.

The two thirds finding the danger of competition removed, at once find themselves able to ask and demand more wages, and when they do want more, the employer cannot turn upon his heel and say, "get out if you don't like your wages, there are plenty to take your place." There are none to take their places; the other one third can make a similar demand with like results, the point thus secured is, to cut off the present surplus labor in the market and prevent that reduction of labor caused by its competition in the labor market, this is the principle observed. And so I claim, for whatever effects one trade effects another, and we find in ours people working too many hours and creating an over production, earning very little in consequence of our one third or more out of work, driving a competition with those who are employed and the result is that we all eke out a miserable existence, either through our small wages or our non-employment.

I would also like to inform the trade, that our Union, No. 144, is under a fair way of progression, having received 46 new members since Dec. 1.

More anon from,

Yours Fraternally S. G.

*Cigar Makers' Official Journal,* Jan. 1876.

1. All brackets in this letter are in the original.
2. Solomon Orgler.
3. Sebastian Döring.
4. Abraham Greenhall.
5. Probably David Hirsch's factory, where SG was employed from 1873 to 1877. At the time he first began working there, SG later recalled, the shop employed between fifty and sixty men.
6. Straiton and Storm manufactured cigars in a factory and in a tenement. By 1877, SG later recalled, the firm operated at least ten tenement houses.

# To George Hurst

New York, Jan. 22. 1876

Mr. George Hurst,
Pres. C.M.I.U. of America.
Sir.

Having promised in my last note to the journal that I would give a report at some future day of the condition of trade in this locality, I now fulfil the same.

A few hundred [of the thousands of unemployed Cigar Makers][1] have been given work since the commencement of the New Year, but in the majority of these cases and elsewhere reduction of wages has taken place, so that notwithstanding those that are out of work are starving, and those that are fortunate to be employed are leading a life of misery in consequence of their meagre earnings.

There are in this city and vicinity Cigar Makers who have not been in work ranging from three months to one year, and instances are known where even this long period of non-employment has been exceeded. Fellow Craftsmen you can imagine the keen suffering, misery, degradation and utter demoralization of our unfortunate fellows.

These are plain and unvarnished facts, [not exaggerated in the least] and in view of them and for the purpose of preventing the further spread of these accursed wrongs and the amelioration of our fellow working people, a Mass Meeting was called by the Bohemian Cigar Makers of the city on Monday evening, Jan. 17 at which meeting the Organizers of this District in company with the Board of Officers of Union No. 144 were present. Although not many strangers [were] present, good feeling prevailed, before dispersing the meeting was addressed by Mr. A. Strasser and others, it was then resolved to call another Mass Meeting on Feb. 7. they anticipating a good attendance then.

A Mass Meeting was also resolved upon by Union No. 144 to take place on Sunday March 5. at 2. P.M. at Concordia Assembly Rooms, [Large Hall] No. 28 & 30 Ave. A.

A committee of five were appointed to make all necessary arrangements for the same, 5,000 circulars have been ordered printed in the Bohemian, German and English Languages and steps have been taken for their proper distribution, there are to be speakers in the three languages, all means have been taken will have a tendency to induce the Cigar Makers to join the Union.

The above fact of calling upon all who are employed at Cigar Making to attend the Mass Meeting must be recognized by all Trade Unions as the only means by which we can ever attain our desired object, Unity of purpose that all have a right to live and enjoy the

comforts appertaining thereto for those who are willing and able to work, for we must remember that while we are divided against ourselves upon the most vital questions of our existence, we remain a prey to the avarice of our employers.

And I therefore assert, to effect any good by our Unions, we must bring all elements working in our trade into one Organization, for the wrongs heaped upon one element today are merely the precursor for another to morrow. Therefore let us up to the work, the questions of the day are.

1st. Union of all working people;
2nd. The reduction of the hours of labor.
3rd. Bring the wages of the lowest paid toiler to the standard of the highest.

The fourth will be given at some future time by others if not by me.

More anon from,

Yours in the cause S. G.

*Cigar Makers' Official Journal,* Feb. 1876.

1. All brackets in this letter are in the original, except [were].

## A News Account of a Mass Meeting of CMIU 144 at Concordia Hall

[March? 1876]

MASS MEETING OF CIGAR MAKERS UNION NO. 144.

The above Union held a Mass Meeting at Concordia Halle on the afternoon of Sunday, March 5, 1876, which meeting was very largely attended.

There were several speakers present, who addressed the meeting in the Bohemian, German and English languages, they urgent all employed at Cigar Making to unite with the Union, their safeguard.

The attention given to the speakers by those present showed that they felt that something was wanted to improve the condition in the city of New York, and that something was a lack of unity among the Cigar Makers of that city, feeling thus many from a sense of duty united themselves with Union 144 before the meeting adjourned. There were 68 new members added to the Union and a great many who did not have the required initiation fee with them, promised to join at the regular meeting of the Union on Monday Evening, March 6. The Mass Meeting, considering the general depression in our trade,

was a decided success, before the meeting adjourned the chairman invited anyone who might wish to say a word to join in a general discussion relative to Unionism.

The arrangements for this meeting, [thanks to the Committee][1] were almost perfection, to secure a large attendance of the Craft.

The Organizers for this district, A. Strasser and L. Gomperts, were among the speakers and did a noble work for the Union as did many others.

Union No. 144 has among its members many earnest workers, men who are well fitted for this arduous yet noble work of re-organization the trade in the city of New York, men who think but to act, and with such men success is certain, would that we had more like them in our organization; then our success would be more speedy and we could sooner show that Cigar Makers can unite and become a power.

*Cigar Makers' Official Journal,* Apr. 1876.

1. All brackets in this letter are in the original.

## A News Account of a Meeting of CMIU 144

[April 29, 1876]

International Cigar Makers Union No. 144, met last Monday evening at 642 Fifth Street. Mr. Gompertz, President in the chair. A. Strasser reported that he had received a communication from a boss cigar-packer to the effect that he would be willing to give $100 to inaugurate a movement against the tenement house system of cigar-making. It was stated that this system was very demoralising to the persons engaged in this branch of trade. A suggestion was made to unite with the bosses in order to expose this abominable system of slavery and agitate for state intervention. The negligence of the Board of Health was strongly condemned.

*Socialist* (New York), Apr. 29, 1876.

## To the Editor of the *Cigar Makers' Official Journal*

[September 1876]

To the Editor of the Cigar Makers Official Journal.

It will be observed that the Shop Directors duty is to collect the

dues of members weekly (which accrue weekly) having a settled conviction that when the same is collected in this manner the members are more willing to pay, and at the same time it affords them the opportunity to pay their dues without any trouble (which in the main they are not fond of incurring) members thereby retaining good standing; while on the other hand there is organization in the shops; the Shop Directors acting in the capacity as Delegates, being able to report proceedings of the Board on the following morning to the members working in their shops. It is the desire of Union No. 144 that these amendments[1] be read at the meetings of local unions under the jurisdiction of the International Union.

Yours Fraternally Samuel Gompers,
Cor. Sec'y, C.M.U. No. 144.

ARTICLE II.

*Sec. 5.* Members out of employment or unable to work through sickness shall be exempt from payment of dues, and their cards shall be receipted with a seal (out of work) but this shall not apply to members receiving out of work assistance. This shall not be applicable if the member is employed less than three (3) days per week, in all cases the Board of Officers may demand satisfactory proof.

*Sec. 6.* Members more than five (5) weeks in arrears of dues shall not be entitled to out of work assistance for a term of four (4) weeks thereafter, those in arrears more than three (3) months and three (3) days shall be stricken from the roll without notice, and upon rejoining shall be required to pay the regular initiation fee.

ARTICLE III.

*Sec. 1.* Every member in good standing one (1) year after joining this Union (provided forty weeks dues have been contributed thereto by said member) shall be entitled to out of work assistance for a term of three weeks, to commence with the second week of non-employment.

*Sec. 6.* Any member quitting a job without proper cause shall not be entitled to assistance during the following four (4) weeks.

*Sec. 8.* Any member discovered making cigars between the hours of eight o'clock A.M. and five o'clock P.M. shall forfeit all claims for assistance for a period of not less than eight weeks.

ARTICLE IV.

*Sec. 1.* The officers of this organization shall be a President, Vice President, one English and one German Recording Secretary, Corresponding Secretary, Financial Secretary, Treasurer, three Trustees,

three Auditors and a Doorkeeper. The Officers in connection with the Shop Directors shall constitute the Board.

*Sect. 1.* A. The Officers of this Union shall be voted for separately by ballot and must receive a majority of all votes cast to entitle them to election. At every unsuccessful ballot the name of the candidate receiving the lowest number of votes shall be withdrawn.

*Sec. 1.* B. The election for officers of this Union shall take place semi-annually at the last meetings in June and December.

*Sec. 1.* C. Nominations for officers of this union shall take place at the meetings before election.

ARTICLE V.

Paragraph 1. The duties of Directors shall be to collect the dues weekly and deliver them to the Board and participate in the proceeding thereof.

Paragraph 2. To give passes to members quitting the factory, stating the cause and time of leaving, also to new-comers receiving assistance.

Paragraph 3. To report to the board all subjects of interest occurring in their factories.

Paragraph 4. To attend all meetings called by the board.

Paragraph 5. To ascertain that all monies paid by them are entered in the book of the Financial Secretary.

Paragraph 6. To call shop meetings upon notice from one-sixth of the members or whenever necessary.

Paragraph 7. To make regular statements of the number of workers and the average rate of wages.

ARTICLE VIII.

*Sec. 5.* Any member removing his residence shall immediately notify the Financial Secretary.

*Cigar Makers' Official Journal,* Sept. 1876.

1. That is, amendments to CMIU 144's constitution.

## *Samuel Gompers' Participation in Workingmen's Organizations in the 1870s*

Gompers began attending meetings of the International Workingmen's Association (IWA) in 1873 and soon became a participant in an inner circle of the Association that called itself "Die Zehn Philosophen" or the ten philosophers.[1] By the mid-1870s "Die Zehn Philosophen" had evolved into the loosely organized Economic and Sociological Club, whose members included Gompers and many of his early associates, including David Kronburg, Ferdinand Laurrell, Fred Bloete, Louis Berliner, Henry Baer, Hugh McGregor, and J. P. McDonnell. George E. McNeill and Ira Steward, leaders of the Boston eight-hour movement, were also associated with the club.

The Economic and Sociological Club rejected what Gompers, in his memoirs, called "Socialist partyism," and favored "trade unions, amalgamated trades unions, and national or international amalgamation of all labor unions" as the instruments of social change. Members of the group, Gompers remembered, considered the shorter workday to be "the first step in bettering industrial conditions" and joined in efforts to unify the fragmented IWA around this and other trade union demands.[2]

In July 1876 a unity conference of disparate American socialist groups met in Philadelphia and established the Workingmen's Party of the United States (WPUS). Gompers became active in the New York American (that is, English-speaking) Section of the WPUS, which emphasized trade union activities. The New York German Section, on the other hand, leaned toward political activism, and the short history of the WPUS was marked by a continuing dispute between political activists and trade union–oriented members. Illustrative of this conflict is the series of letters below, written to the editor of the party organ the *Labor Standard* by Martin Doyle, Louis Berliner, John Schäfer, Gompers, and Berthold Kaufmann. In December 1877 the political activists took control of the party at its national convention and renamed it the Socialistic Labor party (later, the Socialist Labor party).

Gompers' associates in the Economic and Sociological Club subsequently played a role in establishing two organizations that reflected

their views. In 1877 they participated in the founding of the Amalgamated Trades and Labor Union of New York and Vicinity, which, unlike the existing English- and German-speaking central bodies in the city, admitted only representatives of trade unions. "Our membership fluctuated," Gompers later related, "but uniformly contained the strong trade unions of the city. . . . the leaders in these unions were almost invariably trained in the school of the International Workingmen's Association, which taught the primary importance of economic organization." In 1878 members of the club helped organize the International Labor Union (ILU). Gompers recalled that its organizers intended to make the ILU a national institution to strengthen trade unions but that it "failed of its national purpose and became an organization of textile workers." Undaunted, the club members corresponded and met with like-minded individuals and began laying the groundwork for another body to bind the national unions together in mutual support.[3]

*Notes*

1. SG, *Seventy Years,* 1:87.

2. Ibid., p. 210. The depression of the 1870s intensified a debate within the International Workingmen's Association on the issue of political versus trade union activism, creating an open split in the group. In the parlance of the debate, the political activists were often called "Lassalleans," a term that in America loosely embraced any of a broad spectrum of socialists who supported independent labor party activity. These political activists formed the Social Democratic Workingmen's Party of North America (SDWP) in 1874. Adolph Strasser and P. J. McGuire were among the SDWP's members who later became Gompers' close associates. Another group of Internationalists, with whom Gompers was closer at the time and including such men as Laurrell, Kronburg, and McDonnell, founded the Association of United Workers of America (AUWA) in 1874 to promote trade unions. In September 1876 the AUWA joined with elements of the SDWP in New York City to form the New York American Section of the Workingmen's Party of the United States.

3. SG, *Seventy Years,* 1:127, 212.

# Martin Doyle[1] to the Editor of the *Labor Standard*

New York, October 7th, 1876.

Editor *Labor Standard.*

The members of the American Section[2] met on last saturday night in Jefferson Hall, 253 Avenue A to discuss the relations between the working class and the political parties.[3] The discussion was participated in by Messrs. J. P. McDonnell,[4] L. Berliner, John McNally,[5] D. Kronberg,[6] E. P. Rice, S. Gompers, L. Huck[7] and J. H. Monckton.[8] At the conclusion of the debate Mr. McDonnell moved and L. Berliner Seconded the following which was unanimously passed:—

Whereas—All issues raised by the political parties of the propertied class are in the interest of that class and of no advantage to the working class.

Be it resolved, That it is the duty of the working class to ignore not only all the existing political parties but their issues and to support only those questions raised by and coming from recognized Trade and Labor bodies, made in the interest of the working class.

On next saturday night Mr. McNally will speak on "Trades Unions and their relation to the Workingmens Party of the U.S." All members are urged to be present.

On Tuesday, the 18th inst. Mr. McDonnell will lecture on "A workers review of the past century." Admission free to all.

Donations amounting to $9.10 were subscribed for the *Labor Standard* after which the meeting adjourned.

M. Doyle,
organizer.

*Labor Standard* (New York), Oct. 14, 1876.

1. During the 1870s Martin F. DOYLE, a New York City carpenter, was active in the Association of United Workers of America (AUWA), the Workingmen's Party of the United States (WPUS), and the International Labor Union (ILU).

2. The New York American Section of the WPUS.

3. During the fall of 1876, J. P. McDonnell, editor of the *Labor Standard,* conducted an editorial campaign against immediate political action by the WPUS. The paper experienced financial difficulty when the party's politically oriented faction withdrew support.

4. Joseph Patrick MCDONNELL was a leading figure in the AUWA and a member of the Economic and Sociological Club. As the longtime editor of the *Labor Standard,* he was a major contributor to the defense of trade unionism within the Marxist tradition. He was a leader in the New Jersey Federation of Trades and Labor Unions in the 1880s and 1890s.

5. John MCNALLY, a leader of the Paterson, N.J., Silk Weavers Association, was active in the New York American Section of the WPUS and, later, in the ILU.

6. David KRONBURG, a member of the Economic and Sociological Club, was a founder of the AUWA and the WPUS.

7. Louis HUCK served as an officer of CMIU 144 in the 1870s and was active in the WPUS.

8. James H. MONCKTON, a stairbuilder, was a member of the Economic and Sociological Club, helped organize the AUWA, and was active in the WPUS.

## Louis Berliner to the Editor of the *Labor Standard*

New York, December 5th, 1876.

Editor *Labor Standard.*

At a public meeting of the German Section W.P.U.S. on last Saturday eve, one of the members, John Schäfer, gave a lecture upon "The Solution of the Social Question." He started by commenting at length upon the theory set forth by a certain Dr. Schütz,[1] who spoke upon the same question a few days before, and ended by saying that he, the Dr., practiced evasion, because he did not tell the truth. Mr. Schäfer, and all the other speakers who coincided with him used the same subterfuge when they endeavored to impress upon the minds [of] their hearers that a political class organization and the Ballot box could solve the question. Mr. Schäfer and his adherents committed a double wrong when they recommended the above remedy, for the simple reason that they do not believe it themselves. Not being able to restrain myself any longer I obtained the floor and said that it was a sophistry on the part of the speakers to belie the workpeople by telling them they could better their condition, much more emancipate themselves from the present system of wrong, by offering the above remedy as a solution, that they are committing a wrong which they never can answer for. I continued by saying that our first step must be to organize, strengthen and centralize our trade and labor unions. By making them strong and solid we have not only accomplished more and better bread, but a power felt by the ruling classes. We will leave politics and political organizations to politicians, and if ever any one of them attempts to enter our trade or labor organization for the mere purpose of advancing their own personal interest, I should be one of the first to have him kicked out. One speaker then arose and said: "We as socialists are not opposed to trade unions but we know that trade unions only deal with individuals while we deal with the whole human race. What we want is the abolition of the wages system, and that can only be accomplished by the Ballot box." What a beautiful

dream to imagine oneself the redeemer of mankind and worlds regenerator. I am none of those for I got bravely over it. I take the labor movement for what it is, a question of facts and interest. It is my firm conviction Mr. Editor, that if the masses of the workpeople would look upon the Labor movement in the above light they would have no need of being the dependent slave to King Dollar for one hour longer.

How much more practicable, earnest and sensible it would have sounded had this so-called socialist said: We workingmen heartily favor trade unions, for they are our only hope, our only salvation. In them and with them well organized we are free and independent men, unorganized and without them we are ready and willing tools to capital where or whenever our masters demand it, immaterial what political party we belong to or what we may know about the iron law of wages. These so-called socialists seem to forget everything that is transpiring around them. It does not depend upon what we want but mainly what we can get. There are a very small minority of the workpeople who thoroughly comprehend what the abolition of the wages system means.

But they all understand what eight hours and more pay means. Therefore I maintain that there is and only can be one class organization and that is the economical or industrial organization of the whole wages class into one solid phalanx. Class rule and class privileges will then be abolished and the whole present system, *without the use of one Ballot.*

L. Berliner.

*Labor Standard* (New York), Dec. 9, 1876.

1. Fritz Schütz of Milwaukee spoke at a meeting of the club Vorwärts on Nov. 25, 1876.

# John Schäfer to the Editor of the *Labor Standard*

[December 16, 1876]

Editor *Labor Standard.*

Under this title[1] Mr. L. Berliner published in No. 35 of the Labor Standard an article offensive to me and other members of our party, who took part in the meeting. I protest against his letter, because Mr. Berliner was not authorized to give a report of this meeting. He got the floor three or four times in the meeting, and had time enough to speak against all points which he did not consider right.

Mr. Berliner's means is Trades Unionism alone. He says Trades Unions only can settle the labor question, and he has a right to say so. I and other members of our party just have the same right to say a *social-political class organization,* what our party is and shall be, with our principles and platform, is the best to help us. I say so and I believe what I say.

John Schaefer.

*Labor Standard* (New York), Dec. 16, 1876.

1. "Labor Organization."

## To the Editor of the *Labor Standard*

[January 6, 1877]

Editor *Labor Standard.*

In Number 36 of the *Labor Standard* under the above caption[1] you published a letter from Mr. John Schäfer which is intended to be an answer to one published the week previous signed by L. Berliner and upon the same topic; but whether the object sought to be attained by Mr. S. in his answer was successfully carried out or not, I trust you will permit me to show.

L.B. in his first paragraph says in substance, that at a meeting of the German Section, W.P.U.S., Mr. John Schäfer delivered a lecture upon "the solution of the social question," a subject, lectured upon by one Dr. Schütz some time previous, and, that Mr. S. commented on the theory of Dr. Schütz, saying the Dr. did not tell the *truth* and practises *evasion.* L. Berliner adds to this that Mr. Schäfer and all who coincided with him used the same *subterfuge* when they told their hearers that a political class organization and the ballot box *could solve* the question; he further says that Mr. S. and his adherents committed a double wrong when they recommended the above remedy for the reason that they *do not believe it* themselves (the italics are mine). Now what is the answer given to these assertions by Mr. S.? He says the article published by Mr. B. was offensive to him and other members of the Party, and protests against the same for the reason that L.B. was not authorized to give a report of said meeting. I now ask in the name of common sense is this an answer to what Mr. B. asserts; for my own part I say No, aye a thousands times No, for it will not do for Mr. Schäfer to hide himself behind a *protest,* without another word of explanation, or contradiction of assertions, which, if true, must place him in such position for which he must forever hold his peace, and for which he must some time be held responsible; the lines quoted by Pruning Knife[2] are so appropriate that I will copy them:

"Thou must be true thyself,
If thou the truth wouldst teach;
It needs the overflow of heart,
To give the lips full speech."

And as for the offensive part, if it be the truth (and Mr. S. does not contradict it) he should not take umbrage at it for as a Socialist, he should hail the truth with genuine delight, but now Mr. S. also takes exception to the unauthorized report. Does J. S. forget that it was a *public* meeting and was not a regular business meeting in which secrecy was enjoined and which any man had a right to criticise as he best saw fit and in this instance I will remind him of the story of the man who dwelt in the Glass house.

L. B. says further in substance that by organizing and centralizing our Trade and Labor Unions we have not only accomplished more and better bread, but a power felt by the ruling classes and continues further by saying that it would have been more practical, earnest, and sensible, had Mr. J. S. and his supporters spoken in favor of organizing the masses into purely industrial or economical class organizations with *less hours* and *more wages* for their motto.

What does Mr. S. say to this? He says these are Mr. Berliner's means, and he has a right to say so, and claims for himself the same right, which, permit me, Mr. Editor, to deny him, for I maintain, that so long as Mr. S. does not refute the arguments set forth by Mr. B. he cannot claim for himself the right stated above.

Yours, &c, Samuel Gompers.

*Labor Standard* (New York), Jan. 6, 1877.

1. "The Economical Question."
2. In January 1878 the *Labor Standard* stated that Pruning Knife was the editor of the St. Louis *Voice of Labor.*

## Berthold Kaufmann[1] to the Editor of the *Labor Standard*

[January 6, 1877]

Editor *Labor Standard.*

A communication by Mr. L. Berliner to the Labor Standard of Dec. 9th, about a meeting of the German Section, does not contain the views of our speakers on our whole question. Therefore, Mr. Editor, I feel it my duty, and in the interest of truth necessary to correct

him. In said meeting it was expressed that the wrong system of wages slavery could only be abolished by an organization of all workers. Two ways are open for us: to organize as Trade Unions, or as Political Party.—Now the question: Can all people be organized in trades unions under the present existing system? The increase of machinery and the division of labor throw many out of employment, who will work for every price. In a good business time trades unions may arise, and become strong, but in times like the present they will become weak. They have too much to do to keep full ranks and have no time in addition to the transaction of business or the agitation of new ideas. They will have higher wages and short work time, but not abolition of the inhuman system of capitalistic rules.

The German socialists were never against trades unions, but I ask you, Mr. Editor: can the workingmen fight in an economical way against their bosses with any success, if they leave the political power in the hands of their enemies? Conspiracy laws and others will be the result of such conduct. The German members and all good socialists, who are long in the labor movement with open eyes and ears, have learned that the working people must be organized as a class, with feelings of this class, and fight in every way against their rulers! not only as workingmen against their bosses—no, as citizens of each country against the political parties, in which our bosses are organized to seek only their own interests. In this country in the last election thousands of workingmen voted more for a change of system than years ago. Can we, the organized minority of workingmen, sweep away the feelings of our unknown brothers? No; we have to keep these up, and use them for our principle; we have to give them [a] chance to vote for better members of Assembly or Congress than they have now. Such men must be responsible to their voters, and by this way only [we] can get what the trades unions want: a normal work-day, protection of labor, and better conditions; and in this way may be settled the whole social question; if not, we are organized on a good principle; and powerful as we are, we can do how it pleases us, or we are compelled to do.

Fraternally greeting,

yours sincerely B. Kaufmann,
Organizer of the German Section, New York.

*Labor Standard* (New York), Jan. 6, 1877.

1. In the 1870s Berthold W. KAUFMANN, a cabinetmaker, was an organizer and vice-president of the New York German Section of the Workingmen's Party of the United States and a leader of the Amalgamated Union of Furniture Workers of North America.

# To John Walker[1]

New York, March 1st, 1877.

Sir:

In conformity with a resolution adopted by Union No. 144 of the C.M.I.U. of America, the undersigned was induced to write a letter to the *Journal,* for the purpose of clearing up the false impressions that might be engendered through the reference made by you in your letters to the Journal,[2] in the matter of the differences that exist between Union No. 144 of this city and the organization Mr. Saqui[3] says he represents.

I do not, nor does our Union, wish to be understood to cast any reflection upon you; on the contrary: our convictions and sympathies are in unison with yours, upon all matters as far contributed by you to the Journal, and for the reason that you may not be misled, as much as for any other, this letter is written.

In the first instance, I will state that we are not at a decided disadvantage,[4] for notwithstanding we are aware, that Mr. Saqui has complained against our Union, we are left in a state of utter ignorance of the cause thereof. For I am firmly of the opinion that if we were put in possession of the information, we could successfully refute the arguments or cause (so called) complaint against Union No. 144 by either Mr. Saqui or any other cigar maker.

2nd. We will confess though that if Mr. Saqui complains that any member of our Union who is employed and does not pay his dues for a term of three months, he will be stricken from the roll of membership, there he is perfectly right, and Mr. Saqui knows that this is so by experience, for he will remember (for I hope he will see this to refresh his memory) that he became a member of Union No. 144, was shop-collector a considerable time, never uttered a word against our Union, then left the city of New York, returned about two months after and wanted his card of membership receipted (No Work) for whole time of his absence, and when this was refused, we never saw Mr. Saqui again in our midst. So that if this is one of the causes of complaint, he is also in the right.

3d. If Mr. Saqui complains that at one of our meetings a forty pound parcel of political circulars that was sent by one of the political parties, and the same were consigned to the waste basket against his wish, then he is unquestionably right. But if Mr. Saqui says aught other than what I have just stated, then, sir, he is not right nor has he cause for complaint.

To verify my word I will state that at some near day (of which due

notice will be given) Union No. 144 will hold a Mass Meeting[5] for the purpose of organizing the cigar makers in our Union.

I do hereby ask Mr. Saqui to be present, and there give his grounds for organizing a separate Union from ours, or anything he may have to say against our Union. I hope, Mr. Saqui will not absent himself, thinking that he will not have a fair opportunity to be heard.

You, Sir and Mr. Saqui knows as well as myself, that the unorganized cigar makers (of which I hope and expect a large number will be present) are only too glad to hear anything that may be said against a Union, and that in itself will guarantee him a good opportunity. I will add that this last proposition is not made by authority of the Union, but upon my own responsibility, and I feel confident, in fact, am assured of being sustained in it by the Union. And now, dear sir, in conclusion I will re-echo your words; knowing as I do by experience, that, when in any trade societies are so established (separate Unions), with their birth are born the elements of internal discord and discontent, which will be a work in years to come of great magnitude to remove.

Hoping for a return of your health and vigor, so that you may be enabled to prosecute your good work in the cause of oppressed labor, I am

Fraternally yours Samuel Gompers,
Pres't C.M. Union No. 144.

*Cigar Makers' Official Journal,* Apr. 1877.

1. John P. WALKER was general secretary of the Cigar Makers' Mutual Association of London.

2. In December 1876 and February 1877 the *Cigar Makers' Official Journal* published letters from Walker relating that New York City cigarmaker A. J. Saqui, representing a union "distinct" from the CMIU, had written seeking affiliation with the British union. Walker said he told Saqui that he supported "centralization" rather than "localization" and deplored the splintering of unions.

3. Probably Alexander Saqui (1853-1932), an English-born cigarmaker who immigrated to the United States in 1863, was president of CMIU 15 of New York City in early 1876 before the local surrendered its charter, and later became a cigar manufacturer and insurance broker.

4. The May 1877 *Cigar Makers' Official Journal* indicated that this "Should read 'That we *are* at a decided disadvantage.'"

5. Possibly the meeting described in "A News Account of a Mass Meeting of Cigarmakers at Turner Hall," May 15, 1877, below.

# A News Account of a Hearing at the Essex Market Police Court

[March 17, 1877]

A DEFAULTING TREASURER.

Samuel Gompels,[1] of No. 264 Stanton street, President of Cigarmakers' Union, No. 144, of the United Cigar Makers of New York, preferred a charge of forgery yesterday, at the Essex Market Police Court, against Kauffmann Nicholsbey,[2] a former treasurer of the society. On September 23, 1876, it is charged that Gompels gave $100 to Nichelsbey to deposit in the Dry Dock Savings Bank, which deposit was never made, and a false entry of the deposit was made in the passbook. Mr. Gompels produced the book and showed where the teller of the bank had marked the entry supposed to have been made on that date, "fraud."

Mr. Howe,[3] counsel for defendant, admitted that no money had been deposited in the Dry Dock Savings Bank by his client on September 25 nor on any day subsequent, but asked for a dismissal of the complaint on the ground that the passbook was the property of the depositor and not of the bank, and that he had a right to make any entry in it he pleased. The section of the Revised Statutes in regard to the question of forgery was read to the judge, but he decided to hold the prisoner in $1,000 bail to answer.[4]

*New York Herald,* Mar. 17, 1877.

1. SG.
2. Kaufman Nickelsberg.
3. Possibly Walter Howe (b. 1849?), a New York City lawyer.
4. Nickelsberg was subsequently found guilty of embezzlement and sentenced to eighteen months' imprisonment.

# A News Account of a Mass Meeting of Cigarmakers at Turner Hall

[May 15, 1877]

SEGAR MAKERS MASS MEETING.

A large and enthusiastic meeting of segar makers of New York city and vicinity was held on Sunday last at 2 P.M., at the large Turner hall, East 4th street, for the purpose of devising means to better the condition of the segar makers, and if not influence an increase of wages, at least resist any reduction, which it is claimed may take place shortly.

Mr. Berliner was elected chairman of the meeting.

A Mr. Strasser addressed the assemblage in German. He very emphatically denounced the actions of segar manufacturers in their treatment of segar makers in their employ. He claimed that it is impossible for segar makers to even eke out a pitiable existence at the present schedule of wages. In a great many shops not over $3.25 per thousand is paid, and even at prices such as six, seven, eight, ten and 12 dollars, no more than eight to ten dollars can be earned, as too many demands are made by the employers. The continual introduction of segar machinery threatens to drive segar makers out of the trade, and famine was staring them in the face. A united action of all segar makers would therefore become an imperative necessity. An appeal to join the existing Trade Unions was made.

A Mr. Gamperts[1] then ventured some very forcible remarks, also describing the unfortunate condition of segar makers, in terms which elicited tremendous cheers from the assemblage. He read a copy of the following rules and regulations for segar makers in the employ of a Chicago firm.

1. Every segar maker employed in this shop, must leave one weeks wages in the possession of the firm.

2. Any segar maker or bunch maker not having entered the shop at 7:30 A.M., cannot commence to work before 1 P.M.

3. Any segar maker, bunch maker or stripper who takes a segar from the premises, shall forfeit a week's wages.

4. Every person employed by the firm must submit to having his body searched.

5. Any segar maker or bunch maker who shall make a segar so tight as to prevent it from smoking, shall forfeit a week's wages.

6. Any segar maker or roller who shall make a segar not properly filled, shall pay a fine of from one to two and a half cents, according to the grade of the segar, and make another one instead.

The reading of these rules produced intense excitement, and if a member of that Chicago firm had been present, he would undoubtedly never have discovered the cause of his departure from this world of misery.

Several other speakers addressed the meeting in the same strain as the preceding ones, and several hundred new names were enrolled upon the list of members of the Segar Makers' Union No. 144. The initiation fee is fifty cents, and the weekly contribution ten cents. The meeting then adjourned.

*United States Tobacco Journal*, May 15, 1877.

1. SG.

## *The New York City Cigarmakers' Strike of 1877-78 and Its Aftermath*

During the last two weeks of July 1877 a series of railroad strikes, called in opposition to wage reductions, already low pay, and irregular pay practices, swept through seventeen states. In Pittsburgh, Martinsburg, West Virginia, Baltimore, and other cities, local police, state militia, and federal troops clashed with strikers and strike sympathizers, evoking widespread mass meetings in support of the railroad workers. "The railroad strike of 1877," Gompers later wrote, "was the tocsin that sounded a ringing message of hope to us all."[1]

It was in this climate that a series of strikes broke out in New York City cigar shops against pay cuts and factory rules restricting conversation and smoking and regulating other aspects of shop floor life. The first strike, against Frederick De Bary and Co., began on August 16 and was concluded successfully several weeks later. A month-long strike against M. Stachelberg and Co. that began during the last week in August also proved successful. Other cigarmakers, including tenement-house workers, were encouraged by these results and, Gompers recalled, "went out on strike without organization or discipline." By mid-October over 10,000 cigarmakers had joined in what Gompers termed "The Great Strike."[2]

Fearing that this mass walkout, which he subsequently described as "reckless" and "precipitate,"[3] would jeopardize their recent gains, Gompers and other union cigarmakers took a leading role in the strike. Adolph Strasser drafted a code of bylaws that established the Central Organization of the Cigarmakers of New York, comprised of delegates from the factories and tenement houses, and served as president on its first Executive Committee. The bylaws empowered the Central Organization to fix minimum wage rates for shops and tenements and to prohibit striking workers from returning to work without its authorization. The Central Organization also coordinated strike activities.

Various groups provided financial support for the strike. CMIU 144 pledged its treasury, about $4,000, and cigarmakers throughout the country sent donations. German workers in New York City contributed funds and German merchants extended credit to striking

cigarmakers. Bohemian mutual aid societies were especially strong financial supporters. By the end of the strike the Executive Committee of the Central Organization had disbursed nearly $40,000 to provide food for the strikers, lodging for some of those who had been evicted from tenements, and medical assistance. Finally, the organization arranged to operate a factory to provide both work and funds for the strikers. Gompers left his job at Hirsch's shop to serve as superintendent and foreman of this factory. Despite the efforts of the Central Organization, however, the strike began to dissipate by mid-November, and within a month the manufacturers considered it to be effectively over. Although the Amalgamated Trades and Labor Union attempted to raise additional funds to continue the struggle, members of the Central Organization voted on January 24 to end the strike and return to work.

The end of the strike left Gompers in difficult personal straits. Blacklisted by the manufacturers as a strike leader, he was unable to find a job for four months. His family now included four children, and a baby was on the way. He later vividly recalled the hardship of this period: family dinners of soup prepared from water, salt, pepper, and flour; contributions to his table coming from the none-too-full larders of his relatives in New York City; his threatening the life of a physician who was reluctant to attend his wife in February 1878 at the birth of their "strike baby," Alexander Julian; pawning everything of value except his wife's wedding ring; and, finally, finding cheaper accommodations for the family in Brooklyn. Unable to separate himself from what he called "the active work" in the labor movement, he soon brought his family back to the city and again immersed himself in union work. This, he later maintained, was the "turning point" of his life.[4]

Ethnic divisions among workers in the trade persisted in the aftermath of the strike, frustrating the efforts of Strasser and Gompers to translate the cooperation achieved through the strike's Central Organization into a permanent unification of New York City's cigarmakers. In an article in the *Cigar Makers' Official Journal,* Strasser argued that dividing cigarmakers into separate language organizations perpetuated discord. Groups based on nationality "get into fights sooner or later," playing into the hands of employers who had always tried to "fan national hatreds."[5] Local 144 was able to arrange a merger in June 1879 with Bohemian CMIU local 20 and a German cigarmakers' organization, and the CMIU convention that year included the concept of one local per city in its constitution. Many non–English-speaking cigarmakers, however, continued to belong to ethnic local trade organizations unaffiliated with the CMIU.

Political differences reinforced these ethnic divisions and led to tensions within local 144 itself. Gompers and Strasser's emphasis on lobbying and electoral support of mainstream politicians aroused the antagonism of those who were members of both the CMIU and the political wing of the Workingmen's Party of the United States and its 1877 successor, the SLP. Their criticism of the Strasser leadership often appeared in the German-language newspapers associated with their party. In this spirit, for instance, the *Arbeiter-Stimme* condemned the "philosophical, practical, American socialists" among the cigarmakers who "copied all kinds of hair-splitting ideas from the ruling groups of robbers,"[6] and the *New Yorker Volkszeitung* characterized the political activity of the cigarmakers' leaders as "money-bag politics."[7] The strong undercurrent of dissatisfaction in local 144 led to a small secession of members in 1880 and a much deeper division in the union in 1882.

*Notes*

1. SG, *Seventy Years,* 1:140.
2. Ibid., p. 147.
3. Ibid.
4. Ibid., pp. 157, 163.
5. Feb. 10, 1878.
6. May 12, 1878.
7. Jan. 2, 1880.

# Two News Accounts of a Mass Meeting at Cooper Union

[July 27, 1877]

## MEETING OF WORKING MEN.

The labor meeting at the Cooper Institute last evening was well attended, and passed off quietly. At 8:15, when the delegates of the Amalgamated Labor Union[1] filed upon the platform, every seat was occupied and a number of persons were standing, principally around the doors. It was a decent assemblage, evidently composed of genuine working men. Germans preponderated. Mr. John Fortune,[2] President of the Amalgamated Union, was chosen Chairman, Mr. Hubert Vice-Chairman, and Messrs. Straussen[3] and Ward Secretaries. Mr. Fortune, in his opening speech, enlarged upon the sufferings of the working men, and said that such oppression should no longer be tolerated in a free country. [Applause.][4] Now was the time to give expression to the real sentiments of the laboring classes. No longer should they display the cowardice of the past. "When force is brought against us," said the speaker, "I contend that we have the right to meet it in the same spirit." [Cries of "That's so."] In conclusion he introduced Mr. J. P. McDonald,[5] who read in English a series of resolutions, reciting the inability of the working men to exist at the present rate of compensation; demanding increased pay and shorter hours of labor; expressing sympathy with the strikers,[6] and pledging them financial aid; protesting against the use of Militia, and greeting those soldiers who "Fraternized with their fellow-workmen," and urging the necessity of organization. Mr. Weil,[7] of the Typographical Union, read the same resolutions in German. They were loudly applauded, especially the clause relating to the Militia. They were adopted unanimously. Mr. McDonald then came forward, amid cries of "Beecher"[8] and hisses and delivered a speech that evinced capabilities sufficient to make the fortune of a political stump orator. He began by describing how for three years past working men have been giving up the marrow of their bones to fatten persons who never earned an honest dollar in their lives. [Applause.] The silk weavers of Paterson, he said, have no longer anything but rags to wear, although they make silk dresses. [Great applause.] The spectacle had been presented of 70,000 men marching with torchlights to exhibit the rags on their backs and the hunger in their cheeks, at the beck of politicians who were only passing laws to oppress them. [Applause.] The present war is the outgrowth of those three years. No matter how swiftly it may subside, it will leave marks behind that will never be forgotten. [Tremendous applause.] The strike is the result of desperation. There was no concert

of action at the start. It spread because the workmen of Pittsburg felt the same oppression that was felt by the workmen of West Virginia. [Great applause, and three cheers for the men of Pittsburg.] The barrier of creed and nationality has been broken. Hereafter there shall be no North, no South, no East, no West, only one land of labor, and the working men must own and possess it. [Tremendous applause.] There was a time when one trade would stand idly by and see another ground in the dust, but this has past. When one is oppressed all are oppressed, and this is why thousands of other workmen flocked to the [aid?] of the firemen and brakemen. [Applause.] Suppose, as has been charged, the tramps and thieves did join the workmen, who are the tramps and who are the thieves—the poor thieves, not the rich ones. [Loud and long continued applause and cheers repeated over and over.] The strikers have a right to ask their fellow workmen not to take their places and thus play into the hands of the common enemy. The papers say the strike was right but the men were wrong. [Shouts "That's what they always say," and hisses for the *Herald.*] Mr. Bergh[9] di[v]es to the rescue of a horse that is underfed and overworked, and he does well, but society has no care for a man. [Applause.] Vanderbilt[10] and Jay Gould[11]—[Hisses and shouts "Hang them up."] No, they are only doing in a big way what every employer is doing in a small way. They are merely the result of a bad system. [That's it.] But when they array themselves against the people it is our duty to array ourselves against them. [Tremendous cheers.]

The speaker then attacked Henry Ward Beecher, whom he characterized as the theologian of nest-hiding notoriety.[12] If, he said, the working man had had the chance to treat Mr. Beecher several years ago as he wanted them to be treated, he would never have got into court. A bread and water diet would have curbed his passions effectually. [Cheers, laughter, hisses, groans, and shouts of "put him in a coalmine."] He only does the bidding of the manufacturers and employers who attend his church and pay him $25,000 a year. [A voice—He ought to strike for another $75,000.] Heretofore, he said, there had been too much grunting and growling. "[If?]," he shouted, "we have any manhood in us we will grunt and growl no more; we will act. [Wild cheers, and cries of "That's it."] We can do nothing, however, while we remain disorganized. While we are disorganized we are only a mob and a rabble; when organized we become a power to be respected. [Great cheers.] If the working men had been organized in every city the strike would now be more successful than it now is. The working people are in a maze. They don't know what to do. They wait, hesitating for the next telegraph dispatch. This condition of things is their own fault." [Applause.]

Taking this as a basis the speaker made a strong appeal for the

reorganization of the trades unions and their federation under a common international head; also for funds in aid of the "brave men injured in Pittsburg, West Virginia, &c." The "shedding of blood and the destruction of property" he said "were only the desperation of maddened men, and are acts which must be justified because human nature cannot lie down to be trampled upon.["] In conclusion Mr. McDonald said: "I ask you to be very prudent under the present circumstances. Do not commit any rash act in response to the voice of any one who may urge you to do so. [Applause.] You have no power, because you have no organization, and it would be simply suicidal to do what would be sure to bring more misery upon yourselves and your families." [Great applause.]

Mr. Straussen, of the Cigar Makers' Union, followed with an effective speech in German, which was mainly a paraphrase of what Mr. McDonald had said. He too was cheered to the echo. Robert Blissert,[13] of the Tailors' Union, was the next speaker. Blissert, though evidently illiterate, was full of points, which, added to a natural gift of oratory, quite carried away his audience. He was very severe on the politicians, especially on Mayor Ely[14] and John Kelly.[15] He next attacked the Governor[16] roundly. Justice, he claimed, would have made the strike the subject of arbitration to see whether the companies or the men were in the wrong, and the ones found guilty would be compelled to succumb. Beecher came in for another dose. His salary of $25,000 was, the speaker said, enough to support 50 families. Was he so big that he could eat as much as 50 families? Continuing, he said that the working men must show the capitalists that they have rights—peaceably if they can, but anyhow. "I tell any legislator, Governor, militiaman, or policeman," he said, "that if he assails my household he falls or I do. If he tries to deprive me of my bit I will give him a wrestle for it." [Great applause.] The soldiers were praised for refusing to shoot workmen, and Beecher was compared to Judas on the ground that he wrote the life of Christ, and belied Him for a few pieces of gold and silver. [Cries of "Give it (to) him."] "Let him tell us next Sunday," yelled Blissert, "What he does with all the dollars of his salary above $1 per day; also whether there are any more Elizabeths."[17] It is a mistake to suppose, said the speaker, that the workmen are not organized. They are organized. What the papers think is dead is alive. The unions have taken a lesson from the political spies that followed them in years past. They meet in secret, (the most dangerous thing capital ever had to contend with,) and they are planning and plotting now. [Great cheers.]

There was a great deal more of the same sort. A drunken old man who had caused frequent interruptions by maudlin shouts, at length aroused the ire of the assemblage and there was a general movement

to have him put out. Everybody rose to their feet, cries of vengeance filled the air on all sides, and several rushed forward to execute them. The greatest excitement and disorder existed for a few moments. Many persons made a break for the doors in alarm. The Chairman earnestly appealed for order, and the trouble finally subsided. The speaker, before resuming, said that if any man interrupted him again until he finished he would take him by the collar and put him out himself. The old fellow raised a subsequent row, however, and was only saved from rough treatment by the interference of the officers of the meeting, who insisted that no violence should be offered. The last speaker was Mr. Gompers, of the Cigarmakers' Union, who made a miserable fist of it. The meeting broke up peaceably.

A Police force of 300 men was kept at the Seventeenth Precinct Station house, in anticipation of any trouble which might have arisen from the meeting. A reserve of 150 men was also kept in readiness at the Police Central Office.

*New York Times,* July 27, 1877.

1. The Amalgamated Trades and Labor Union of New York and Vicinity (ATLU), founded in the summer of 1877 by members of the Economic and Sociological Club, promoted labor organizing, lobbying, and the defeat of legislators who voted against labor measures. It opposed the nomination of labor candidates, however, believing this would be premature and divisive to a labor movement composed of young and financially weak unions. During 1884 it unsuccessfully sought to consolidate with the New York City Central Labor Union (CLU), a rival organization that supported political action. By 1886 the CLU had completely displaced the ATLU.

2. John FORTUNE, a tailor, was president of the ATLU.

3. Adolph Strasser.

4. All brackets are in the original, except for [aid?], [If?], and ["].

5. J. P. McDonnell.

6. A reference to the railroad strikes of July 1877.

7. Jean WEIL, a printer, was secretary of the German-American Typographia from 1876 to 1883.

8. Henry Ward Beecher (1813-87) was an abolitionist reformer, editor, and pastor of the Plymouth Congregational Church, Brooklyn, N.Y. (1847-87). An exponent of the gospel of wealth, he had attacked trade unions and strikes in a sermon delivered on July 22, 1877. Responding to the complaint that wages were too low, he had asserted that a workingman and his family could live on a dollar a day, an income he said was sufficient to provide them with bread and water. "The man who cannot live on bread and water," he said, "is not fit to live" (*New York Times,* July 23, 1877).

9. Henry Bergh (1811-88) was the founder and first president of the American Society for the Prevention of Cruelty to Animals and a founder of the Society for the Prevention of Cruelty to Children.

10. William Henry Vanderbilt (1821-85) was a New York financier, railroad magnate, and philanthropist.

11. Jay Gould (1836-92) was a speculator whose manipulation of the gold market helped precipitate the panic of 1873. His holdings in the railroad and telegraph industries involved him in some of the most notable strikes of the 1880s, particularly the telegraphers' strike of 1883 and the Southwest railroad strike of 1886.

12. Beecher's trial and acquittal on charges of adultery in 1875 (*Tilton vs. Beecher*) had received national attention.

13. Robert BLISSERT, a tailor, was active in the International Workingmen's Association, participated in the ATLU, and became a leader in the New York City CLU.

14. Smith Ely, Jr. (1825-1911), a lawyer and merchant, was elected as a Democrat to the U.S. Congress in 1871 and 1875 and was mayor of New York City in 1877 and 1878.

15. John Kelly (1821-86), a former Democratic congressman (1855-58), was the most powerful figure in Tammany Hall from the early 1870s until his death.

16. Lucius Robinson (1810-91), formerly a Republican New York assemblyman and comptroller, was elected governor on the Democratic ticket in 1876; he was defeated when he ran for reelection.

17. A reference to Elizabeth Tilton (Mrs. Theodore Tilton).

[July 27, 1877]

SYMPATHY WITH STRIKERS.

. . .

Samuel Gompers, who made the last address, spoke in the same tone as the previous speakers, ridiculing the advice to the engineers and firemen to go West. They were not farmers, but men who stuck to their engines and trains. All that they demanded was that they should have a voice in fixing the terms for which they should work, so as to obtain such wages as would enable them to support their families in that decency which belonged to American workmen. He appealed to all present to enroll themselves, and become agitators for their respective Trades Unions.

The meeting was then adjourned.

*New-York Tribune*, July 27, 1877.

## A News Account of a Mass Meeting of Cigarmakers at Concordia Hall

[August 6, 1877]

DISSATISFIED CIGAR-MAKERS.

The Bohemian and German journeymen cigar-makers, including a number of female operatives, held what they called a mass-meeting at Concordia Hall yesterday afternoon, to effect a joint organization on socialistic principles. A German and a Bohemian President were chosen to conduct the meeting, in the persons of Messrs. Pollmann and Pratske; and addresses were delivered by several speakers, from which it appeared that one of the chief causes of dissatisfaction among this class of workmen is the system of tenement-house labor introduced

in this city by the manufacturers. It was shown that in many instances from ten to twenty families working at cigar-making, live in one house, each family occupying one room, which at the same time is used as a workshop, the result of which is, it is alleged, disease and frequent deaths. A resolution was passed, protesting against the inaction in this matter of the Board of Health, whose attention, it was asserted, has been repeatedly called to it. It was demanded that the system of tenement labor in the manufacture of cigars be entirely abolished, and it was intimated that if this complaint was not attended to by the proper authorities, like means would be adopted as in the railroad strikes. Many of those present were enrolled as members of the organization, and have to contribute each five cents a week to defray its expenses. It was noticed that a number of Socialist leaders were present, representing other organizations; and several of the members of Cigar-makers Union, No. 144, who however keep aloof from this Socialistic movement. The last named Union is a strong organization, numbering about two thousand members, with ample funds, which are used exclusively for the benefit of members in cases of disease and inability to work.

*New-York Tribune,* Aug. 6, 1877.

## Excerpts from the Proceedings of the Eleventh Convention of the CMIU[1]

Rochester, N.Y., Aug 31st, 1877.

SECOND DAY.

. . .

AFTERNOON SESSION.

. . .

The committee on constitution then read their report. On motion of Rep. Burke[2] it was resolved to discuss the constitution by sections.

. . .

Rep. Burke moved that the preamble be adopted as read.

Rep. Gompers[3] offered an amendment to insert after ["]hours["] a clause for the support of the unemployed, sick and travelling members. In the affirmative voted: Goldsmith,[4] Gompers. In the negative: Walsh,[5] Phillips,[6] Burke, Bichler,[7] Hirth.[8]

The preamble was then adopted as read.

. . .

[September 1, 1877]

THIRD DAY.

. . .

Article VIII, section I; adopted.

Repr. Gompers moved the following amendment: That the revenue of the Int. Union shall be derived from an annual capitation tax of 60 cents per member, payable monthly.

The money to be forwarded to the Int. Pres. with the monthly report by the respective officers.

The yeas and nays were called and resulted as follows:

Yeas—Phillips, Gompers.

Nays—Walsh, Burke, Bichler, Hirth, Goldsmith.

. . .

Article XI, section I; adopted.

Repr. Gompers moved the following substitute to Section I. Every member in good standing six months after joining shall be entitled to the sick benefit. It shall be $4 for the first 3 weeks, or 65 cents per day, and $3 for the next 3 weeks, or 50 cents per day.

Section II. Any member being able to return to his employment before receiving six weeks assistance, and being reattacked before six weeks have elapsed shall be entitled to the balance of six weeks only.

Section III. Any member receiving assistance shall upon evidence that his conduct is irregular, or being found in a state of intoxication forfeit all claims for a term of eight weeks. Any fraudulent attempt to obtain sick benefit shall be subject to the same rule or expulsion as the union may determine.

Repr. Hirth moved that the substitute be laid on the table; carried.

Repr. Gompers moved the following substitute:

Sect. I. Every member in good standing one year after joining shall be entitled to the out-of-work benefit, to commence with the second week of non-employment. It shall be $4 for the first three weeks or 65 cents per day, and $3 for the next three weeks or 50 cents per day.

Repr. Goldsmith moved that the substitute be acted upon by sections. The yeas and nays were called and resulted as follows:

Yeas—Goldsmith, Gompers.

Nays—Walsh, Phillips, Burke, Bichler, Hirth.

Repr. Hirth moved to lay section I. on the table; carried.

. . .

Repr. Gompers offered the following as a substitute for article XI., section 1.

Every union shall establish a Labor Bureau for the purpose of designating work to unemployed members; carried.

. . .

AFTERNOON SESSION.

. . .

Repr. Hirth offered the following resolution:

That a lyceum be established for the purpose of educating the workingmen, and to give them practical knowledge as to their social relations; such knowledge to be disseminated by the reading of essays and by discussions of labor questions. We feel confident that such actions will prevent the misguidance of the workingmen, which leads them to failures and disastrous defeats.

Whereas a most reckless competition of aggregated capital, the introduction of labor saving machinery and consequently the employment of children to an unnatural extent has made manual labor superfluous, which forces the competition against each other, and

Whereas the trade unions are utterly incompetent? to remove the pressure resting upon them, caused by the above mentioned social infirmities; the delegates in convention do hereby recommend to and urge upon all local unions to form themselves into labor bodies upon the basis and platform of the W.P.U.S.

Repr. Gompers moved that the foregoing resolutions be laid on the table.

Yeas—Walsh, Phillipps, Goldsmith, Gompers.

Nays—Burke, Bichler, Hirth.

(TO BE CONTINUED.)

*Cigar Makers' Official Journal*, Oct. 15, 1877.

[September 1, 1877]

. . .

THIRD DAY—AFTERNOON SESSION.

Repr. Hirth moved that the delegates of this Convention do further recommend to all local Unions the establishment of sick relief societies, the same to be of a local character only.

Rep. Philips moved that the foregoing resolution be laid on the table. Carried.

Rep. Goldsmith offered the following resolution:

Resolved, That we the C.M.I.U. of America declare ourselves to be in harmony with the platform of the W.P. of the U.S.

. . .

After a lengthy discussion of the resolution offered by Rep. Gold-

smith, the previous question was asked and the yeas and nays ordered.

Yeas—Walsh, Burke, Bichler, Goldsmith, Hirth.

Nays—Phillips, Gompers.

. . .

EVENING SESSION.

Rep. Gompers submitted to the consideration of the Convention a telegram from the Cigarmakers Union 144, advising them of a strike in Stachelberg & Co.'s shop.[9]

. . .

*Cigar Makers' Official Journal,* Mar. 10, 1878.

1. The CMIU held its 1877 convention in Rochester, N.Y., Aug. 30–Sept. 2.
2. James BURKE represented CMIU 5 of Rochester, N.Y. He was second vice-president of the CMIU from 1877 to 1879.
3. CMIU 144 elected SG as its delegate to the 1877 convention. During the convention he secured Adolph Strasser's election as CMIU president.
4. Marcus K. GOLDSMITH represented CMIU 39 of New Haven, Conn. He was active in the Workingmen's Party of the United States.
5. Francis N. Walsh (1852-1910?), an English-born cigarmaker who immigrated to the United States in 1873, represented Philadelphia CMIU 3.
6. Hyman Phillips, a Cincinnati cigarmaker, represented CMIU 4 of that city. Born in England in 1843, he immigrated to the United States in 1857. He was corresponding secretary of CMIU 4 during 1877 and of CMIU 10 of Cincinnati from 1878 to 1879. Later he became a sanitary officer in the Cincinnati Department of Health.
7. William Bichler (1850-1914?), a German-born cigarmaker who immigrated to the United States in 1864, represented CMIU 11 of Chicago. In the mid-1880s he became a cigar manufacturer.
8. Frank HIRTH represented CMIU 22 of Detroit and CMIU 147 of East Saginaw, Mich. He edited SLP papers in Detroit and Chicago in the late 1870s.
9. M. Stachelberg and Co.

# A News Account of a Mass Meeting of Cigarmakers at Germania Hall

[October 15, 1877]

## THE CIGARMAKERS' STRIKE.

The strike of the cigarmakers of this city and vicinity began about seven weeks ago. It was for some days confined to one or two manufactories, but it now reaches large proportions. It is estimated that there are from twenty thousand to twenty-two thousand cigarmakers in New York, Brooklyn and Jersey City and that 10,000 of these are now on strike. Yesterday a large mass meeting of the strikers and those who are in sympathy with them was held at Germania Hall, Bowery. The meeting lasted from two o'clock to half-past five, and

during that time it was thought that nearly five thousand persons visited the hall. Many will be surprised to learn that more than one-half of the cigarmakers of this city are women and girls, and probably about five hundred of these were present at the mass meeting yesterday. There were speeches in English, German and Bohemian. At one time two meetings were in progress, a temporary platform having been erected in the east wing of the great hall. The meeting was full of enthusiasm, and some of the speaking in regard to the relations of capital and labor was very spirited, and a part rose to the point of genuine eloquence. When the meeting was opened probably three-quarters of the men present were smoking cigars. A request from the platform that smoking be not indulged in was looked upon as a curious thing for a convention of cigarmakers, and the request was not generally complied with. Lager flowed freely, but there was good order.

### The Addresses.

The meeting was called to order by Mr. Vandepoorten,[1] and Mr. Daniel Harris[2] was chosen chairman. In a brief speech the chairman stated the objects of the meeting, and then introduced Mr. J. Palda,[3] who spoke in Bohemian, advising the cigarmakers to put aside all feeling born of nationality and to organize under one head with their fellow craftsmen of whatever clime. He stated the measures proposed by the striking cigarmakers. These are briefly as follows:—Each shop which employs 300 hands and more to elect three delegates, and from the shops where a less number are employed one delegate to be elected; some of the smaller shops to combine in the election of delegates; these to form the Executive Board of the Cigarmakers' Union.

### A Revelation for Smokers.

Mr. Samuel Gompers, President of Union No. 144, of the Cigarmakers' International Union of America, then made an address in English. He spoke of the reductions which have been made in the wages of cigarmakers, whereby the pay is now almost at a starvation point, and then commented upon the evils of the tenement house system of cigar manufacture by which the people are compelled to work from sixteen to eighteen hours a day, Sundays included, sleeping and eating in the small rooms where the cigars are made, this system not only breeding disease among the people thus employed, but liable to be the means of carrying contagion to those who consume the cigars. Cases, he said, have been known in this city where persons barely recovered from smallpox have been engaged in making cigars in tenement houses. He advocated the union of all the cigarmakers,

claiming that among its benefits would be the education of the working classes in regard to their rights and interests. He bitterly criticised the statement said to have been made by one of the largest manufacturers that for the last two years he had not made a dollar profit.

Mr. Strasser, President of the Cigarmakers' International Union of America and editor of the *Cigarmakers' Official Journal*, made a speech, in German, referring to the trade unions as the practical factors in the all-absorbing question of labor and capital, as being the "great arms of defense against the encroachments of employers." Mrs. Hallsler[4] made an address, in Bohemian, treating principally of the evils of the tenement house system of cigar manufacture, in which everybody, from the grandfather to the babe, was a maker of cigars, and which tended so much to the degradation of women. She spoke eloquently and was heartily applauded. Another lady spoke in Bohemian. Addresses were made in English by Messrs. Vanderpoorten, J. P. McDonnell, H. Hexter[5] and John McNally, lately the President of the Silk Weavers' Union of Paterson; and by Mr. M. Jonas[6] and John Moranete in Bohemian. Mr. Louis Berliner, financial secretary of Union 144, presided at the smaller meeting.

THE STRIKE SPREADING.

A committee appeared upon the platform, representing the employes of Messrs. Sutro & Newmark,[7] who reported that the workmen had in a meeting that day concluded to demand $1 more on each thousand cigars made for that firm. This announcement was received with great applause. The chairman called attention to the report in the press of the settling of the strike of Martinez & Co.,[8] corner of Rivington and Attorney streets, and said that the report was untrue and advised all cigarmakers to stay away from the factory. It was reported that after a strike of seven weeks the firm of M. Stackelberg & Co., Nos. 92 and 94 Liberty street, had yielded and the men go to work to-day. The speakers seemed to be confident of the final complete success of the strike, and that it would not only sweep the city, but stir up the cigarmakers all over the country to demand "living" wages.

*New York Herald*, Oct. 15, 1877.

1. Probably Abraham Vanderpoorten (b. 1830?), a Dutch-born cigarmaker who in 1878 was elected auditor of CMIU 144.

2. Daniel HARRIS, a leader in the Central Organization of the Cigarmakers of New York, became president of CMIU 144 in the late 1880s and subsequently held major offices in New York state labor organizations.

3. Lev J. PALDA cofounded *Dělnické Listy*, a Bohemian-language workingmen's paper, with Frank Skarda in Cleveland in 1875. They moved the paper to New York in 1877.

4. Mary HAUSLER, a Bohemian, was vice-president of the Central Organization of the Cigarmakers of New York during the strike.

5. Also called H. Heckster. In 1882 an H. Heckscher was nominated as vice-president of the National Cigar Packers' Union.

6. A cigar packer named Morris Jonas lived in New York City in 1877.

7. Sutro and Newmark was one of the largest tenement manufacturers in Manhattan, employing 600 workers. At one point during the strike the firm ejected fifty-two striking cigarmakers' families from its tenements.

8. The firm of Vincente Ybor y Martinez closed its Key West, Fla., factory in 1874 and two years later opened a large establishment in Manhattan. It returned to Key West in 1878.

# An Excerpt from a News Account of a Meeting of Cigarmakers at Concordia Hall

[October 25, 1877]

THE CIGAR-MAKERS' STRIKE.

. . .

At the afternoon meeting at Concordia Assembly Rooms of the striking shop delegates the high spirits and gaining confidence of the strikers in their success was more marked than ever before. The roll-call showed all present. Mr. Gompers, President of Cigar-Makers' Union No. 144, was announced as present, and in a neat speech said that although the strikers had already been informed that the private fund of the union would be at their disposal next Monday, he now presented them with the $200 extra he then held in his hand, which had been raised by subscription, and that he desired to say that the same amount would be raised, if necessary, every day during the strike, if it lasted all winter. Long and loud applause followed this announcement.

. . .

*New York World*, Oct. 25, 1877.

# A News Account of a Mass Meeting of Cigarmakers at Cooper Union

[October 31, 1877]

THE CIGAR-MAKERS' STRIKE.

General Roger A. Pryor[1] has been engaged as counsel[2] by the striking cigar-makers. He charges them nothing for his services. The

Executive Committee has fixed on $7 as the lowest price to be received for the lowest grade of cigars made at Seeger's,[3] of Chatham street. The delegates from that shop considered this excessive, and said such a rate would keep them out of work forever. Still the rate was fixed. At the request of Union No. 144 the name of George Winter[4] was struck off the list of speakers at the meeting last evening, on the ground that he was a Communist and did not believe in trades-unions.

The meeting at Cooper Institute in the evening [was] a large one, the whole place being filled and crowds gathering outside. Captain McCullough,[5] of the Seventeenth Precinct, was on hand with a force of fifty men stationed in and around the building, but long before the adjournment he reduced the number by one-half, saying that he had never seen a meeting of the kind conducted so quietly. The only display made was by the Cigar Packers' unions, who came in a body headed by a band of music and carrying the American and German flags. At least one-half of the assembly, both in the building and outside, were German and Bohemian women who seemed to take as much, if not more, interest in the proceedings than the men. Those occupying positions upon the stage were the speakers, to the number of ten, the delegates composing the Central Organization[6] of strikers and a few members of other trade-unions. Mr. Strasser presided, and Mrs. Mary Hausler, the Bohemian Vice-President of the Central Organization, was Vice-President. President Strasser stated that the object of the meeting was to lay before the American people and the working classes particularly the position of the cigar-makers. They had been doing two days' work for one-half day's pay. Five dollars a week would not supply a family with food, clothing and education for the children. They had better drown themselves than leave the union. He also desired the assembled people to show how well they could behave, if they were a mass of strikers.

Mrs. Hausler, the Vice-President, then explained in the Bohemian language the object of the meeting. Her voice was clear and ringing, while her gestures were given in quite a dramatic manner. Loud applause greeted her at the close of her remarks.

Samuel Gompers, President of Cigar-Makers' Union No. 144, delivered the most telling speech of the evening. He said the strike was not of their seeking, but had been forced upon them. The tenement-houses were perfect pest-holes. One mother he knew had wished that her child might die, as she had no time to care for it. In the shop of Kerbs & Speiss[7] one year ago they made a rule that the strippers should not even whisper to each other while at work. The first and second times a fine would be imposed, and the third offense would be punished by dismissal. A twelve-year-old lad having been caught

the third [time] was sent home, where, upon being chided, he drowned himself in the East River. They would rather die by starvation now than to do so in one year, and in the mean time work like dogs.

A delegate from the Custom Tailors' union, Mr. John Fortune, made a quiet little speech, complimenting the method of the strikers and abusing the city government for first introducing a decrease of wages.

Mr. Smith,[8] of the Brooklyn Cigar-Makers' Union No. 87,[9] next spoke, making a ridiculous picture of the "liberality of the manufacturer." "But if we are wrong," said he, "let us go to work. (Cries of No, no, never.) Then what shall we do? (Voices—Stick to the strike; Hold the fort.) Yes, I say, Hold the fort!"

Outside the hall a stand was raised, and the American and German colors were displayed. A calcium light over in Fourth avenue threw a glare over an acre of upturned faces. At the opening of the meeting there were about 3,000 persons gathered thickly around the stand, a large share of the auditors being women. Daniel Harris presided, and William Kunart[10] and Joseph Colopovert were elected vice-presidents. Most of the speakers, as soon as they had addressed the crowd, left the stand and joined the indoor meeting, and those who had spoken inside came out and stood in turn to address the crowd in the square. All the speeches were peaceable but firm. Said one of the speakers: "I don't believe in strikes. I think they are bad in every way. Strikes always do harm. They should only be resorted to after everything else has failed. The present is a time when a strike by the cigar-makers is justifiable. Nothing else would secure our rights. What else could we do? We have learned that we can't depend on the politicians to do anything for us. I advise all of you not to have anything to do with the politicians. They are no good, and they are all alike. Even the Board of Health and the Sanitary Commission[11] are composed of politicians."

The speeches were in English, German and Bohemian.

The manufacturers have reaffirmed their determination not to yield to the demands of their late workmen, or to reinstate them in employment while they are members of the Cigar-Makers' Union. They also invited their former workmen to meet them at any time for conference. The statement was also made at the meeting that President Strasser, of the central body of strikers, was not himself a working cigar-maker, but an avowed German Communist.

*New York World,* Oct. 31, 1877.

1. Roger Atkinson PRYOR was associated with prolabor and antitrust causes during his career as a New York City attorney.

2. The Central Organization of the Cigarmakers of New York hired Pryor to defend tenement-house cigarmakers against ejection suits filed in the city courts by manufacturers.

3. Siegel and Co. (variously M. Siegel or M. Seegleis) operated a small factory at 135 Chatham St.

4. Ernst George WINTER, a German-born cigarmaker, was editor of the *Arbeiter-Stimme.*

5. John H. McCullough (variously McCullagh), born in Ireland about 1842, was a New York City police precinct commander.

6. The Central Organization of the Cigarmakers of New York.

7. Kerbs and Spiess, the largest New York cigar manufacturer, had about 1,100 employees in its factory and tenements. The firm's workers were among the most active strikers.

8. Walter T. Smith served several terms as president of CMIU 87 of Brooklyn, N.Y., during the 1870s.

9. The CMIU chartered local 87 of Brooklyn in 1867.

10. William Kunart (b. 1848), a Bohemian cigarmaker, immigrated to the United States in 1872. In 1900 he lived in the Bronx and worked as a clerk in an insurance office.

11. Possibly a reference to the four sanitary commissioners who, along with the health officer of the port of New York and the city's four police commissioners, composed the New York City Board of Health.

## An Article in the *New York World*

[November 9, 1877]

THE CIGAR-MAKERS' STRIKE.

The interest and excitement created by the troubles of Wednesday[1] led the delegates of the striking cigar-makers forming the Central Organisation to meet in full force yesterday afternoon at the Concordia Assembly Rooms. During the reading of communications the applause was long and loud, caused by the unusually large amount of money received as contributions. The cigar-makers of San Francisco sent the sum of $2,000, Boston (Mass.) Union, $90, and from other sources at least $100 more. The sixty hands employed by the large firm of Buchanan & Lyle,[2] No. 44 Tiffany street, Brooklyn, struck yesterday for higher wages. More ejectment warrants were served during the day and the families turned out upon the streets. In the District Court the firms of Holzman & Deutschberger,[3] No. 154 Attorney street; Adler & Landauer[4] and L. Hirschhorn & Co.,[5] No. 89 Water street, had their ejectment cases postponed until next Monday. In the Fifty-seventh Street Court the cases of Simon Brothers[6] were adjourned on account of the Judge's[7] sickness. About 10 o'clock in

the morning Kerbs & Spiess served warrants upon six families occupying rooms in their tenement-house, and shortly after a sergeant with fifteen policemen made their appearance and ordered tenants to move out at once. The Relief Committee of the strikers were soon on hand, accompanied by a number of wagons furnished gratuitously by a livery stable in the neighborhood. The goods were loaded and moved off as quickly as possible to the quarters already furnished for them by the committee, and in two hours' time all of those ejected were comfortably cared for. No disturbance of any kind was made, but the police on duty were loudly jeered at. Some of the strikers brought to the scene a band of music, but they were prevailed upon by the calmer portion to restrain their enthusiastic indignation. Mr. Spiess,[8] it is said, sent a notice to all the house agents located in the Nineteenth Ward not to rent rooms to any of Spiess's hands, as the strikers were not able to pay for them. For this the Executive Committee, through their lawyers, threaten to call Mr. Spiess to account. In the case of the sick woman in the tenement-house of Levy & Ullman,[9] on Broome street, Marshal Slimp yesterday denied that he had acted unreasonably, only doing his duty as directed, and that no harsh measures were used.[10] During the day Judge Dinkel,[11] hearing of the case, sent an order to the marshal not to further interfere and to replace the furniture in the rooms, which was done. A citizen living at No. 254 Broome street has taken it upon himself to supply her with all groceries and provisions, and, if required, to pay three weeks' rent for her in advance. J. Cohn,[12] of Pitt street, has offered the cigar-makers the use of two dwellings owned by him on that street free of rent until such a time as they may be able to pay. Thirteen rooms in as many houses have been offered gratuitously. The Sick Committee notified the delegates that they were in good working condition, and had collected $418 in money. Two cigar-makers were arrested on Tuesday for resisting the arrest of two women pickets stationed in front of Kerbs & Spiess's shop. It seems that the firm had complained of them, and in driving them away the police, it is said, used their clubs, accompanied by kicks, without cause. These two men interfered, were arrested and fined $5 each. This was refunded yesterday by the Central Organization. Mr. Park,[13] of Belcher, Park & Co.,[14] also caused the arrest of a picket, who was discharged yesterday by Judge Duffy[15] with a warning.

A mass-meeting of the cigar-makers of Brooklyn, New York, Williamsburg and Newark was held last evening, under the auspices of Brooklyn Union No. 87, at Froelich's Hall, corner of Court and Carroll streets, in that city. There were some four or five hundred present. Walter Smith, President of the Brooklyn Union, occupied

the chair, assisted by Aaron Samuels[16] as Vice-President. Mr. Smith stated the objects of the meeting to be the formation of a large local union and to assist to the best of their ability the cigar-makers now on a strike. He also gave a clear explanation of the present situation of the New York workmen. Resolutions were then passed bidding their fellow-workmen of New York to be of good cheer; that their cause was just, and that they pledged themselves to stand by and aid them all in their power.

Mr. Samuel Gompers, President of New York Union No. 144, was then introduced, and spoke to the meeting. He said: "The time has now come when speaking is of little avail. We are now called upon as men and fathers of families to make one grand effort to better our miserable condition, which has been thrust upon us for the past five years, and, if required, we must force a different condition of affairs." Speaking of the duration of the strike he said that the manufacturers had stated that their hands had not the grit to hold out one week, but they would soon see their ability to keep it up for a year; that there were other considerations besides that of a higher rate of wages—their manhood and rights as fellow-men were at stake, and that the only hope and expectation of success was in the union as an united body, which must be held intact throughout the Union. They would succeed this time or die in the attempt. (Loud and prolonged applause.) The Chinese story mentioned in *The World* of yesterday[17] was referred to, and Mr. Gompers said he desired that a warning should be given that if the manufacturers imported any large number to New York they would be responsible for any violent action that might be taken to protect their wives and children, and provide them with bread. (Applause.) Mr. Gunsburg, of the Williamsburg Union,[18] then delivered an address in German, covering the same ground as Mr. Gompers. A. Strasser, President of the International Union of Cigarmakers, sent an excuse for his non-appearance, and Mr. Smith made a few remarks in his place, giving an amusing and sarcastic description of the "justness" of the manufacturers, also stating that the Chinese question of cheap labor was merely a scare; that no cigar-makers would leave San Francisco to come to New York, as the wages were so much greater there, and that the only hands possible to obtain would be coolies thrown out of employment for incompetency. Quite a squad of police were stationed in the hall and outside, but not the slightest disturbance was made, and in the course of an hour the police were withdrawn.

*New York World,* Nov. 9, 1877.

1. The first court-approved evictions of tenement-house cigarmakers took place on Wednesday, Nov. 7.

2. Buchanan and Lyall, Brooklyn's largest cigar manufacturer, had about sixty employees at this time, all of whom worked in one factory. The firm went bankrupt in the early 1880s.

3. Holzmann and Deutschberger was one of the largest tenement manufacturers at the time of the strike, employing eighty families with about 200 cigarmakers in two tenement houses.

4. Adler and Landauer owned two tenement factories where forty workers made cigars.

5. L. Hirschhorn and Co. housed 300 workers in eight tenement houses.

6. Simon Brothers employed some thirteen tenement families.

7. Walter S. Pinckney (1831?-88), a lawyer and Republican leader who ran unsuccessfully for the New York Senate in 1873, served as an elected justice of the Seventh District Court of New York City from 1875 to 1879.

8. Louis Spiess (1843-78), born in Germany, was the junior partner of Kerbs and Spiess and was elected treasurer of the National Cigar Manufacturers' Association during the 1877 strike.

9. Levy and Ullmann had at least one tenement house, where about thirty families were employed. The firm also appears to have operated a factory.

10. On Nov. 7, City Marshal Slimp, on a warrant from the Fourth District Court of New York City, ejected striker Joseph Fabrosky and his wife, who was in labor, from Levy and Ullmann's tenement house. After a neighbor guaranteed the rent and the strikers' physician stated the woman was too ill to be moved, the marshal allowed her to return but refused to readmit the furniture. The following day Fabrosky obtained a stay of proceedings, but the daily papers reported that the experience had left his wife in critical condition.

11. John A. Dinkel (1841-93), an immigrant from Württemberg and a lawyer, ran unsuccessfully for the New York Assembly as a Democrat in 1874. He was elected justice of the Fourth District Court of New York City from 1875 to 1881, and later filled an unexpired term as city alderman.

12. Possibly Joseph Kohn, a cigar manufacturer.

13. Noel R. Park was a partner in Belcher, Park and Co., a New York City cigar factory. In the 1880s he was in the grocery and coffee business.

14. Belcher, Park and Co. operated a small cigar factory that went out of business in the early 1880s.

15. Patrick Gavan Duffy (1834?-95), an Irish immigrant, was a teacher and a Tammany Hall politician who served as a New York City police justice between 1875 and 1893.

16. Aaron Samuels (variously Samuel; 1823?-87), a Brooklyn, N.Y., cigarmaker, was president of CMIU 87 of that city in the early 1880s.

17. By the mid-1860s the Chinese apparently constituted a majority of the cigarmakers in California. Non-Chinese West Coast cigarmakers, organized independently of the CMIU until the mid-1880s, adopted a label in 1875 to certify that their cigars were made by Caucasian labor. East Coast cigarmakers do not appear to have become generally interested in Chinese labor competition until the strike of 1877; their concern at that time was possibly prompted by a report in the *New York World* of Nov. 8 that Straiton and Storm had requested a San Francisco firm to send 300 Chinese cigarmakers to take the places of striking New York City workers. While the threat did not materialize, the *Cigar Makers' Official Journal* thereafter regularly devoted attention to the problem of competition from Chinese cigarmakers. The journal

subscribed to the view generally expressed in the American press of the time that Chinese workers, unlike other immigrant groups, could not be assimilated into American culture.

18. The CMIU granted charters to two Williamsburg, N.Y., locals in late 1877, locals 43 and 97. Both appeared on the CMIU's list of locals until late 1878.

## An Article in the *New-York Tribune*

[December 7, 1877]

### THE CIGAR-MAKERS' STREET PARADE.

The faces of the striking cigar-makers beamed with smiles of triumph as they assembled, yesterday, in front of Concordia Hall, to take part in the procession of the organization. Men, women and children came singly and in groups from shops, each shop bearing a flag and banners with inscriptions. The flags were those of the United States, Germany, Bohemia and Austria, to represent the nationalities of the strikers. One banner bore, in large letters, the words: "No more tenement-house work." Nearly 1,000 persons were in line at 1:30 P.M., when the procession started. At the head marched President Strasser, followed by the Central Organization. Then came, in succession, workmen in the new cooperative shop,[1] members of the Cigar-makers' Union No. 144, workmen in various shops in the city, and the Cigar-makers' Unions of Brooklyn and Williamsburg. As the procession passed by shops where union men were employed, it was greeted with loud cheers; at others in which non-union men worked it was as loudly hissed. The procession finally halted in front of the new cooperative shop at No. 42 Vesey-st. On the way the strikers were joined by new recruits, while a number of women and children wearied with the long and rapid march, dropped out of line near Worth and Chambers-sts. In front of the shop, Messrs. Strasser and Gompertz, congratulated the strikers on the victory they had won, and urged them to stand together so as to insure permanent success. Mr. M. Smith,[2] the manager of the new shop, made a brief address. The strikers then returned to their headquarters.

Mr. Smith, in conversation with a *Tribune* reporter, yesterday, said that he had received large orders which he had been unable to fill on account of the strike. He waited for nine weeks, and as he saw no indications of yielding on the part of the strikers, he concluded to consult his own interest and make a compromise with them. He knew that the Cigar Manufacturers' Association[3] would be indignant at the steps he had taken. The shop will be managed by Mr. Smith, and all

business with customers will be conducted by him at present. Implements and labor will be supplied by the men, who will receive union wages. After some time the shop will become cooperative, or rather a joint stock company. Mr. Smith said that he had already heavy orders, and expected that there would be work enough to keep the men busy day and night.

After the procession, the Central Committee held its usual meeting. Sixteen families were reported to have been ejected from their homes. A ball, to be given by several trades-unions for the benefit of the strikers, was announced for December 17, at Tammany Hall. The day's receipts were $200.

*New-York Tribune,* Dec. 7, 1877.

1. See "An Interview with Samuel Gompers in the *New York World,*" Dec. 9, 1877, below.

2. Matthias M. Smith was expelled from the National Cigar Manufacturers' Association (NCMA) for allowing the striking cigarmakers to operate his factory. He served as wholesale agent for the factory during the strike and subsequently repossessed it.

3. In October representatives of sixteen firms met to revive the NCMA, an organization that had been inactive for at least the previous year.

## An Interview with Samuel Gompers in the *New York World*

[December 9, 1877]

THE STRIKING CIGAR-MAKERS.

The central organization of the Cigar-Makers' Union is again holding secret sessions each afternoon, and liberal promises of important disclosures within the next forty-eight hours about the pending strike are freely made. A *World* reporter yesterday visited the new union shop at No. 42 Vesey street, and Mr. Gomperts, the foreman, explained the objects of and the system pursued in the establishment. "It is not, as many suppose," he said, "a cooperative shop, dividing the profits of the work with the hands employed. It is owned by the Central Organization, who select the workmen from the shops on the strike, using only the best. At present, owing to limited room, we are able to employ only about four hundred and fifty hands altogether. The organization has succeeded in renting the two buildings above, however, Nos. 44 and 46, and workmen will be employed during the coming week to open large arches or doors through the intervening

walls, thus throwing the three buildings into one. When this is completed at least 1,200 hands can be set to work. It is the intention of the manager, Mr. Smith, to work both night and day for some time to come, in order, if possible, to get some stock on hand, so that 2,400 hands will be employed altogether. The shop is, of course, working on union principles, both as to the price of wages and in shop rules. Although the employees are allowed to talk and sing, the most perfect order prevailed. The union, not wishing to cripple themselves by paying the rent for the building out of their present funds, the hands now employed and those to be employed have all agreed to allow 25 per cent. of each week's wages to remain with the manager, to go towards the payment of the rent. This amount is to be considered a personal loan to Union No. 144, to be repaid whenever circumstances warrant it. Over 4,000 circulars have already been sent out to cigar-dealers throughout the country, announcing the readiness of the factory to supply all of the most popular brands of cigars not patented. A large number of orders have been received in the past two days and there is every prospect of success." Many of the manufacturers are said to be much troubled by this last move of their late employees, and one of the largest said yesterday to a *World* reporter that he seriously contemplated manufacturing none but the most common brands of cigars—such as retailed at 5 cents or less. "If the present high tax on imported cigars could be reduced," said the speaker, "the majority of the manufacturers would undoubtedly become importers, for the reason that a cigar which could in Havana be bought for $15 per 1,000, with an added tax of $25—making the cost $40 per 1,000—could be sold to the dealer here for $50, who could retail it easily at 10 cents, making his usual profit of 100 per cent."

Contributions to the relief fund were received yesterday as follows: Cincinnati Union No. 1, $70; Cincinnati Packers, $14.80; Chicago, $129; Chicago Upholsterers, $23; Boston, $30; Rockland, Me., $20; Warren, Pa., $18; Guelph, Ontario, $10; St. Joseph, Mo., $5.50; and Marlborough, Mass., $8.50. George Bence[1] ejected two families, and Straiton & Storm one during the day.

*New York World,* Dec. 9, 1877.

1. George Bence (1829-89) was a New York City tenement-house cigar manufacturer. A wealthy businessman who had lived on the West Coast and later Nevada, he began manufacturing cigars in New York City about 1864.

# A News Account of a Mass Meeting of Cigarmakers at Cooper Union

[December 29, 1877]

LABOR'S VOICE HEARD.

The large hall of the Cooper Union was densely packed last evening with men and women who had assembled in response to the call for a mass meeting of the striking cigarmakers. On the platform were seated members of the strikers' Central Committee and officers of other trades unions. The hall was gayly decorated with American, German, and Bohemian flags, and the mottoes: "Union forever," "United we stand, divided we fall," "No one shall break the Union," and "Heaven helps those who help themselves," were conspicuously displayed.

Mr. A. Strasser, who presided, said:

"It is now twelve weeks since the cigarmakers quit work, and we intend to hold out until we are victorious. It is a battle between twenty thousand half-starving people and a few monopolists. One thousand families have been turned out of their homes by the landlords on account of the strike. The police treatment of the pickets who have been placed in front of the shops to prevent non-union men from going to work is scandalous. The police have no right to club our pickets."

Mr. S. Gompert said: "The manufacturers have bribed the police to club our pickets. There are many policemen who are smoking cigars given them by the manufacturers, and all this in spite of the rule prohibiting policemen from receiving bribes. Chief Justice Daly[1] decided that all persons have the right to walk in the street and speak to whom they please. We insist that our pickets have this right."

Mr. L. J. Palda urged the cigarmakers to maintain the strike, and Mrs. Mary Heisler[2] said that there must be no faltering among the strikers. She was loudly applauded.

Resolutions were passed declaring that the unanimous uprising of 10,000 men and women is the natural result of systematic oppression on the part of the manufacturers; that the struggle is to be continued for an indefinite time, till the recognition of the Union has been complied with; and that the strikers are strongly opposed to any compromise whatever. The resolutions also condemn the tenement house system of working in all its forms as an invention to renew chattel slavery, opposed to civilization, and to the spirit of the age.

*New York Sun,* Dec. 29, 1877.

1. Charles Patrick Daly (1816-99), a former Democratic New York assemblyman, served as judge of the Court of Common Pleas of the City of New York from 1844 to 1886, the last twenty-seven years as chief justice. In July 1877, during a tailors' strike, Daly ruled that strikers had the right to picket.

2. Mary Hausler.

## To the Editor of the *Cigar Makers' Official Journal*

New York, Jan. 22, 1878.

Editor *Cigar Makers Official Journal*:

Whereas it has been openly stated by many of our co-workers (erroneously, in my opinion) that the amalgamation of the three Unions[1] will prove detrimental to the interests of the cigarmakers, I do hereby challenge any member or members of either of our three Unions to a friendly discussion of the above question, at any time or place he or they may appoint for that purpose.

Samuel Gompers,
President Union 144.

*Cigar Makers' Official Journal*, Feb. 10, 1878.

1. CMIU 144 amalgamated with local organizations of Bohemian (CMIU 20) and German cigarmakers in June 1879.

## A Translation of an Article in the *Arbeiter-Stimme*

[January 27, 1878]

THE ECONOMIC STRUGGLE.

—New-York. Cigarmakers' strike.—It is all over. In its causes one of the most justified, in its organization one of the most perfectly conducted, and, as far as the number of participants is concerned, one of the greatest strikes of this or any country—has been lost for the workers after a hard 15-week struggle.

Even as recently as last week it looked as though the fight would erupt with new strength but this was merely a last flaring up of dying embers. A general panic followed this short burst of enthusiasm and then everything was lost. The fact that of the original 11,000 strikers

only 1,146 took part in the ballot for or against continuing the strike shows how profound this panic was. The vote was 534 for continuation and 612 against. A 78-vote majority decided what could no longer be averted. It was a regular "Plevna."[1] Hunger remained the victor. Last Thursday the central body decided to take a vote on whether or not the strike should be continued; a secret vote was also to be taken in all shop meetings and the results were to be made known on Saturday, the 19th of January. The results were as above.

The central body, realizing that the strike was now out of its control, declared it at an end and allowed all members to go back to work. Those who can find no work or who have been subjected to disciplinary action will receive financial support from the central body as before.

In view of the fact that these represent a considerable number, and further that the Central Organization has a moral obligation to pay its enormous debts consisting of cash loans, grocery deliveries, and rent for those who were fired, and so on, it is the duty of all who took part in or supported the strike to make every effort to ward off further suffering and not to compromise the reputation of the Central Organization.

Union 144 and the German and Bohemian union held extra meetings last week in order to bring about amalgamation, but this was rejected by the German and Bohemian union. However, there seems to be an inclination toward introducing identical statutes and then uniting the separate organizations by means of a central body, thereby insuring common action.

Concurrent with the discontinuation of the strike was the failure of the so-called cooperative shop on 42 Vesey St. Mr. M. Smith declared the contract broken on the grounds that the strike was ended before payment of the agreed sum of money. Since that time he has given receipts to those working in the shop for the money they had already paid in. Last Saturday he refused to give out any receipts, declared the contract null and void, and announced at the same time a significant reduction in wages. The workers did not agree to this; they have already paid off 1,100 dollars and are now forced to go to court if they want to get any of their money back. So much for today. We will have more to say about this matter.

*Arbeiter-Stimme,* Jan. 27, 1878. Translated from the German by Patrick McGrath.

1. Plevna, a town in northern Bulgaria, was the site of a Russian siege during the Russo-Turkish War of 1877-78, where the Turks were defeated after holding out against hopeless odds for five months.

## An Item in the *Cigar Makers' Official Journal*

[April 10, 1878]

—Union 144 is reorganizing on a benevolent and protective basis. It is proposed to introduce out of work, sick, burial and strike benefits.

*Cigar Makers' Official Journal,* Apr. 10, 1878.

## An Article in the *New York Sun*

[April 20, 1878]

A CO-OPERATIVE FAILURE.

During the recent great and long strike of the cigarmakers of this city it was announced that the Central Union had organized an association for the manufacture of cigars on cooperative principles, and had bought the factory of M. M. Smith at 42 Vesey street. The operatives were all to be paid the twenty-five per cent. advance demanded by the strikers, and Mr. Smith was to be employed as the agent of the association for selling the cigars manufactured. The cooperators numbered two hundred and fifty and were chiefly women. They were pleased with the principle on which they were working, but in six weeks the enterprise was relinquished through want of capital to purchase the tobacco with which to work and the stamps for the boxes. They had calculated on a longer continuance of the strike, and on advances of funds from the strikers and from sympathizers in other trades for the necessary purchases. As the strike closed, the money was no longer obtainable, and no chance existed of competition with the capitalists in the business. Mr. Smith may, in one sense, have been employed by the association to sell the goods, but was in reality the employer. The strike inconvenienced him, and he desired that his business should continue. He therefore agreed to pay the old wages with the understanding that the extra twenty-five per cent. demanded should be the profit of the operatives. It was, however, to remain in his hands, and to be paid to the earners only in case the association continued in operation for a specified distant period. Otherwise, he was to retain the money. As he anticipated, it did not so continue, and the twenty-five per cent. remained in his hands. Mr. Smith thinks that cooperative enterprises are commendable, but that they need a despotic head, like any private business. He has little faith in the democratic principle as applied to them, especially in the case

of working people, who understand only manual labor, and have no experience in management. The association under his auspices had no voice in the conduct of affairs, and no chance to wrangle in regard to them, but their disagreements and jealousies found frequent scope in smaller matters. Discontent arose whenever an individual was paid, through his skill, a trifle more than others, and carping was freely indulged in when any favoritism appeared to be shown by Mr. Smith or Mr. S. Gomez,[1] the superintendent, in awarding seats in coveted positions or in any other way.

*New York Sun,* Apr. 20, 1878.

1. SG.

## To the Editor of the *New York Sun*[1]

N.Y., April 23d 1878.

To the Editor of the Sun.
Sir:

In your issue of last Sunday a statement appeared under the heading of "A Co-operative Failure," containing many misstatements calculated to give the public an erroneous impression concerning the shop known as the "Union Shop," 42 Vesey str. I confess my surprise that the same has not already been contradicted by Mr. Smith, scarcely believing he would utter the words attributed to him. Having waited until the present time and the contradiction not forthcoming, I feel constrained to place the matter in its true and just light before a public, who so generously aided us in our struggle, and respectfully request you to publish the same.

In the first place the Central Organization never intended the shops to be carried on on co-operative principles, nor was it relinquished for the want of capital to purchase tobacco, stamps, or boxes, for the very good reason that not one fraction had ever been expended by us for these purposes, even during the time the Organization was carrying on the scheme. Neither did we calculate upon the longer continuation of the strike, nor for funds from friends to make these purchases, which I will show further on.

Mr. Smith says that "discontent arose whenever an individual was paid through his skill a trifle more than others and that carping was freely indulged in, when favoritism appeared to be shown by Mr. Smith or Mr. Gompers, the superintendent, in awarding seats in coveted positions or in any other way."

I assert and can prove that there was no discontent by the larger earnings of one over another, nor of the seats referred to, nor, in fact, about anything else, and as for the appearance of favoritism, there never was any, nor did there any appear, as every cigar maker employed in that shop can testify.

What Mr. Smith has to say about workingmen only understanding manual labor, and not management, can only have reference to me, and for that reason I do not desire to enter into the matter.

And now I will briefly and frankly state the objects the Central Organization had in view when this scheme was gone into.

1st. To decrease the number requiring relief by finding employment for them.

2nd. To increase the income for the relief of the ever increasing needy. (which the workmen very liberally did)

3d. To create consternation in the employers association.

You will observe that when the strike was declared at an end these objects had ceased to exist, and with them the necessity for the continuance of the so-called Union Shops, it having outlived its usefulness, which the workmen in the shop and the Organization as a whole, voluntarily relinquished.

Respectfully Yours Samuel Gompers.

*Cigar Makers' Official Journal,* May 10, 1878.

1. The *New York Sun* did not publish SG's letter.

## An Article in the *New-York Tribune*

[November 5, 1878]

CIGAR MAKERS AGAIN IN TROUBLE.

The recent decision[1] in the United States Court that cigars must not be retailed in the same room where they are manufactured, but that the boxes must be stamped in the room where the cigars are made, has induced the cigar makers to strive for the abolition of the tenement-house work system. In this movement, they claim they will be aided by a number of large shop cigar manufacturers and the cigar-makers generally throughout the country. On Sunday a mass-meeting[2] was held at the Germania Assembly Rooms, and a series of resolutions was passed, declaring that: Since, in consequence of the tenement-house system many of the workers are obliged to barter cigars for groceries and other necessary things, and since this causes a loss of

revenue to the Government of $250,000 per annum; since the manufacturers extort large rents from the workers, and shops are thus converted into dwelling apartments which are injurious to the health of the inmates, the cigar makers request the Commissioner of Internal Revenue,[3] and Congress, to abolish the tenement-house work system. A committee of five, consisting of A. Strasser, S. Gompertz, C. Daltow,[4] T. Curtis and A. Perina,[5] were appointed to present the resolutions to the Commissioner.[6]

S. Gompertz, president of one of the unions, said that these tenement-house shops were dens of filth, exhaling miasmatic gases that poisoned the air and bred disease; that children of tender years were compelled to work there the entire day, and thus were kept away from school; that the system was one that led to gross immorality and crime. In sixty-three tenement-houses there lived 950 families who worked, cooked, washed and slept in the same rooms. A. Strasser, president of the International Union, and who conducted the cigar makers' strike last year, also made a speech in which he said that when the Chinese cigar makers in California began to crowd the whites out of the work, the latter succeeded in passing a State law requiring every cigar factory to affix on each box a union stamp,[7] by which cigars made by white men could be distinguished from those made by Chinese; and the public showed their condemnation of Chinese work by refusing to smoke cigars made by Chinamen.

The German Tobacco Inquiry Commissioners,[8] who are now in this city, have visited several bonded warehouses and tobacco manufactories. The commission came to this city from Chicago, having visited Washington and other of the large cities. The names of the members of the commission were published when they first arrived in New York.

*New-York Tribune,* Nov. 5, 1878.

1. In order to make it difficult for cigar manufacturers to sell their cigars without paying the tax on them, Commissioner of Internal Revenue Green B. Raum on Mar. 21, 1878, ordered that by May 1, 1878, all establishments that manufactured and sold cigars on the same premises must erect partitions between their factories and store rooms. The U.S. Circuit Court upheld Raum's decision in October. The cigarmakers seized upon Raum's definition of a factory as an entire room separated from other parts of the building by actual and permanent partition as an instrument for challenging tenement-house manufacture of cigars. Tenement manufacturers commonly bonded a whole block as a single factory covered by a single licensing fee. If the Bureau of Internal Revenue had charged an annual licensing fee for each manufacturing room, as the cigarmakers believed Raum's definition required, the increased cost would have added prohibitively to the manufacturing cost.

2. This meeting was held Oct. 27, 1878.

3. Green Berry Raum (1829-1909), a former Republican congressman (1867-69),

served as U.S. Commissioner of Internal Revenue from 1876 to 1883 and later served as U.S. Commissioner of Pensions (1889-93).

4. Probably Conrad Detlef (1848?-97?), a New York City cigarmaker who emigrated from Hesse.

5. Anton? Perina, secretary for the Bohemians in the Executive Committee of the Central Organization of the Cigarmakers of New York during the 1877 strike, served as president of CMIU 20 of New York City from 1878 to 1879.

6. Strasser sent the resolutions to Raum on Nov. 17, 1878.

7. In 1874 the Cigar Makers' Association of the Pacific Coast began using a white label that said in part, "The Cigars herein contained are made by White Men" (Ira Cross, *History of the Labor Movement in California* [Berkeley, 1935], p. 136). In the spring of 1878 the California State Senate considered but did not pass a bill (No. 568) requiring manufacturers to stamp all cigar boxes with the name of the race of the workers who made the contents.

8. In September 1878 the German government sent an Imperial Commission to study the system used by the U.S. Bureau of Internal Revenue for taxing cigars and tobacco. During a six-week tour the commission visited most of the prominent tobacco and cigar factories in the United States.

# Adolph Strasser to Officers and Members of CMIU Locals

New York, December 21, 1878.

To the Officers and Members of Local Unions.

The subject of tenement house labor has been brought to the attention of the Commissioner of Internal Revenue, but it seems doubtful to expect a material change of any account from that source. It will probably silence his hostility to our interests in the present and future. The proper place to look for a remedy of our grievances on that most important subject is the National Congress in Washington.[1]

Mr. Hewitt,[2] the Chairman of the Investigation Committee on Labor distinctly stated that Congress has the power to enact laws in that direction, owing that the cigar industry being controlled by the government.

In the past the sending of petitions to Congress was deemed effective by most of the trade and labor unions of the country, but experience has shown that almost without exception they were thrown into the waste basket. The past mode of redressing our grievances being ineffective should teach us a new road which may lead to a better result.

Both houses of Congress have adjourned until the 6th of January, thus giving us a chance to visit them personally in their respective districts. Every Congressman and Senator, without distinction of party politics should be visited and the passage of the enclosed law de-

manded. If your Congressman or Senator should not reside in the same town, then send him a letter with a copy of the law to be proposed. Be energetic in your demands.

Public meetings with the help of the daily press will do a great deal to bring the subject before the attention of Congress, and at the same time will in a measure counteract the opposition of the tenement house manufacturers.

If your union should not meet within the above mentioned time, then call a special meeting at once for the purpose. But if this should be impossible, then the officers of the union should act as a Committee in the matter. Do not delay, every minute is precious.

Yours fraternally, By order of the Executive Board.
A. Strasser,
Int. Pres.

*Cigar Makers' Official Journal,* May 10, 1879.

1. The CMIU called for the passage of an amendment to the internal revenue bill (H.R. 4414, 45th Cong., 2d sess., 1878) outlawing tenement-house cigar manufacturing. The Senate failed to pass the amendment, and beginning in the spring of 1880 the CMIU looked to the New York legislature for relief.

2. Abram S. HEWITT, an iron manufacturer and philanthropist, served five terms as a Democrat in the U.S. Congress beginning in 1874.

# A Proposal for Amending the Constitution of the CMIU

[February 10, 1879]

SUBSTITUTE FOR
ARTICLE XII[1] OF THE INT. CONSTITUTION.

(Proposed by Union 144.)

*Sect. I.* The International Union guarantees its moral and pecuniary support to all its members, in difficulties which may arise between them and their employers. The amount of assistance shall be six dollars per week, and shall commence on the day when the union mailed the application to the Int. Pres.; provided the difficulty be approved by the proper authority.

*Sect. II.* When any difficulty arises between the member[s] of any union and their employers, three officers of the union shall furnish a full and official statement of the same to the Int. Pres., who shall submit a copy thereof to the other officers composing the Ex. Board

and Board of Appeal,[2] and if after a full and sufficient investigation of all facts in the case they approve of the same, the Int. Pres. shall issue a circular setting forth the facts to all local unions, and the number of members who are idle through such difficulty, and ordering them to their assistance he shall state the person or persons appointed to receive the same.

*Sect. III.* After any difficulty being approved the Int. Pres. shall levy a weekly assessment upon every local union in proportion to their membership, that will be sufficient to secure to every one on strike a proper relief in accordance with the constitution.

*Sect. IV.* In case the Ex. Board and Board of Appeal fail to approve of any difficulty, the local unions shall have the right to appeal, within 15 days after the decision being rendered to a general vote of all the local unions. The appeal, with three dollars for incurring expenses, shall be forwarded to the Int. Pres., who shall submit the same to a vote of all local unions, and if approved by a majority he shall proceed as in Sect. III.

*Sect. V.* Every difficulty involving more than 10 members shall be submitted at once by the Int. Pres. to a vote of all local unions, and if a majority of those having voted approve the same he shall proceed as in Sect. III. Any difficulty for an increase of wages shall not be considered legal unless approved by a two-thirds majority of all votes cast. Unions failing to vote within 8 days, commencing on the day when the circular being mailed shall be fined 3.00 dollars. They shall have the right to return their vote by telegram at the expense of the Int. Union, provided their location is over 200 miles away from where the office of the Int. Pres. is seated.

*Sect. VI.* The vote of local unions on difficulties shall be in proportion to their membership; one vote from 7 to 50 members, two votes from 50 to 100 members, or a fraction of not less than 75, three votes from 100 to 200 or a fraction of not less than 160, and one additional vote for every one hundred more.

*Sect. VII.* Every local union shall retain for every member on the books the amount of 15 cents per month as a striking fund, and 25 cents shall be added to said fund for every new member. This fund shall not be used or appropriated for any other purpose, but shall remain in the custody of the local union subject to the order of the officers authorised by the Int. Union. Any local union failing to comply with this section shall after a notice of thirty days was given be suspended by the Ex. Board.

*Sect. VIII.* Every local union shall send a monthly report to the Int. Pres. stating the amount in the strike fund, also the number of members initiated and suspended. Failing to send this report the union

shall be fined 50 cts. The Int. Pres. shall issue a quarterly statement of the amount every union has retained in the striking fund.

*Sect. IX.* Any local union being over two weeks in arrears with the assessment in aid of a difficulty, shall be notified by the Int. Pres.; failing to pay the arrears within 30 days the union shall be suspended by the Ex. Board.

*Sect. X.* Any local union applying for assistance shall be at least three months under the jurisdiction of the Int. Union, and no taxation in aid of difficulties shall be levied on a new union for a similar term.

*Sect. XI.* No member of any local union shall be entitled to any strike benefit, unless he is a member in good standing for at least three months.

*Sect. XII.* Unions sending money to each other must remit the same by express or post office order. The Secretary transmitting such money shall notify the Pres. of the said union at their separate address. The Secretary receiving such money shall immediately send a receipt to the Secretary and a copy of such receipt to the Pres. of the union from whence the money came.

*Cigar Makers' Official Journal,* Feb. 10, 1879.

1. The existing article consisted of only sections I, II, and XII of this proposal. The amended Article XII was approved by the locals and went into force in May 1879. It was incorporated, with some modifications, as Article XIII of the new constitution adopted by the CMIU later that year. (See "The Constitution Proposed by the Twelfth Convention of the CMIU," Sept. 15, 1879, below.)

2. Under the 1877 constitution the Executive Board of the CMIU consisted of the president and the first and second vice-presidents. A Board of Appeals, composed of five members living in different cities from the members of the Executive Board, was responsible for hearing appeals from Executive Board decisions.

# An Article in the *New York Herald*

[February 22, 1879]

## TENEMENT HOUSE CIGAR MAKING.

About five hundred cigar makers, many of them accompanied by their wives and daughters, assembled last evening in Cooper Union Hall to protest against the rejection by the United States Senate, on Tuesday last, of the proposed amendment to the internal revenue laws providing that no cigar manufactory should be used as a dwelling or for any other household or domestic purpose. The manufacture of cheap cigars in tenement houses, where whole families are often

engaged at the trade, eating, sleeping and working in the same apartment, has long been a source of trouble to others of the same calling, who are employed in factories and are members of the Cigarmakers' International Union of America. These tenement house operatives, it is claimed, are compelled to work from sixteen to eighteen hours a day for starvation wages, and live under the rod of an overseer as unrelenting as any who ever stood over a plantation of slaves. The International Union has for years been fighting the tenement house system and at length brought the matter to the attention of the government. The meeting last night was an enthusiastic one in its denunciation of the action of the Senate in refusing to amend the revenue laws so as to give the protestants the relief they seek.

Mr. L. Berliner was chosen chairman, and in his opening address declared that the amendment had been rejected by a trick; that every Senator who voted against it had committed a crime. It was the continuance and perpetuation of the meanest kind of slavery the world ever knew, and were the honorable Senators on trial before him for the act he would ask no greater sentence upon them than that they be condemned to the life of a tenement house cigar-maker. It was their firm purpose to denounce and expose that system until Americans would blush to own its existence in their country.

Mr. Berliner then introduced a Mr. Palda, a Bohemian, who spoke at length in his native tongue. A Mr. Taylor followed with an address in German, following up the position of Mr. Berliner.

### EVILS OF THE SYSTEM.

Mr. Gompers spoke in English at some length. He believed it was only necessary that the public should know the evils of the tenement house system to secure its abolition. He was informed by an agent of the union in Washington that at ten o'clock on Monday night last the proposed amendment was adopted without a dissenting vote. The next day it was reconsidered and defeated by a vote of 34 to 25 on the yeas and nays. His informant added in his letter that the lobby was active in the interest of the tenement house manufacturers, and that money was freely used. After presenting the worst phases of the obnoxious tenement house system of cigar manufacturing, Mr. Gompus accused the Board of Health of complicity with the capitalists who maintain them, and asserted that the inspection of those places by the Sanitary Department was only a farce, the owners having been notified in advance to prepare for the ordeal. This proceeding he denounced as infamous, and wanted to know what the public would

think of a police captain who would send word to any den of infamy of his intention to raid upon it.

There were several other addresses, and the speakers were frequently and loudly applauded.

RESOLUTIONS.

The following resolutions were unanimously adopted by a vote which made the walls of the hall resound:—

Whereas the Senate Committee on Finance have approved and reported an amendment to section 3,399 of the Revised Statutes for an improved description of what should constitute a factory and a better means for collecting the revenues; and whereas the Senate of the United States first unanimously agreed to said report, and upon the day subsequent reconsidered and then defeated the said amendment; whereas this said amendment would not only secure a larger revenue to the United States but also materially improve the condition of thousands who are employed at the trade and give the honest manufacturer a fair chance in the market; therefore, be it

Resolved, That we, the persons employed in the trade of cigar making, in mass meeting, at Cooper Institute, in the city of New York, assembled, demand, in the interest of the government and the trade (except the few traffickers in this system) that the House of Representatives do adopt the above named amendment and insist upon its agreement by the Senate.

Resolved, That we will never rest until this nefarious tenement house system of cigar manufacturing is forever abolished.

Resolved, That we will hold in grateful remembrance all those who have done or may do anything to promote the passage of the above amendment.

Resolved, That we hold up to public contumely, contempt and scorn those who have opposed the passage of the above amendment.

Resolved, That a copy of these resolutions be immediately forwarded to the President of the United States, to the President of the Senate, to the Speaker of the House of Representatives, to the United States Senators from New York State, to members of Congress from this city and to the Commissioner of Internal Revenue.

After further routine proceedings the meeting adjourned.

*New York Herald,* Feb. 22, 1879.

# A Committee of CMIU 144 to the Editor of the *Cigar Makers' Official Journal*

New York, March 31, 1879.

Editor Cigarmakers Official Journal.

At a meeting of Union 144 held this evening a resolution was adopted, by which the undersigned were elected a Committee for the purpose of assigning and making public the reasons and objects this union had in view in proposing the substitute for Article 12 of the constitution of the Int. Union. They are in brief as follows;

1. The section providing for telegraphic communication has this advantage. It will give a Union that has a strike first, the opportunity of making it known first, notwithstanding the distance it may be from the seat of the body possessing the approving power.

2. It is proposed to give the power of approving a strike, to a larger body of men, i.e. the Ex Board and *Board of Appeals,* whose official position shall enable them to decide whether a strike is practicable or not, and also to give this body such power to enforce the assistance of a strike if approved, not as now, where the Ex Board is merely an approving body without the power of enforcing its mandates, or the laws as they even now stand.

3. When a strike occurs involving a large number of men, and probably a drain upon the funds of the Unions, the right of approving a large strike in that case is reserved to the members by a general vote.

4. The section providing for a time, within which the vote must be taken, and assistance rendered is based upon the standpoint that where a blow is struck it must be a quick and an effective one.

5. The advantages of the proposed "Strike-fund" over the assessments now in vogue are quite apparent. It is in time of peace preparing for war. The members will not feel the sudden strain upon their almost empty pockets for assessments to support a strike as now, and the publication of the statement at regular and stated intervals of the condition of this fund, will, we believe, have the following effects: 1. That members will not so readily go on strike when the funds do not warrant such action. 2. The members will not be so ready to sanction a strike where the fund is low, and they themselves liable to be assessed! We know too well by recent experience, that men on strike were forced to return to their shops, not so much on account of others taking their places, as from lack of assistance in their struggle—a natural result of the assessment system just now in vogue.

6. Experience has demonstrated that heretofore some of the unions

have been backward in forwarding their assessment in aid of a strike, whether in consequence of a theory or a lack of discipline is not for our consideration, but the fact remains, that a union, failing to act in accord with the laws of an organization it is allied with, must be compelled either to obey the laws or quit the organization. It is not sufficient, (our union holds) for any Union to shelter itself behind a theory that they are opposed to strikes and will not support them. We are aware that notwithstanding opposition from whatever quarter, strikes will occur; that they can not be argued out of existence; we know that strikes are but preliminary skirmishes to the great battle of labor, and will occur wherever low wages, long hours, and oppressive rules are the conditions under which workingmen toil. While we would advise the toilers to avoid strikes otherwise than as a last resort, yet in view of the foregoing facts we hold ourselves in duty bound to array ourselves on the side of labor, and rendering them our hearty and effectual support. Believing that these reasons may guide some of the Unions in the questions submitted for their adoption and help them to a sound conclusion in the premises, we are yours in the cause of labor.

Samuel Gompers
B. Rosenberg[1]
John Haggerty[2]
Committee.

*Cigar Makers' Official Journal,* Apr. 10, 1879.

1. B. (possibly Bernard) Rosenberg served as president of CMIU 144 during 1879 and 1882.

2. Possibly John Haggerty (b. 1836?), a New York cigarmaker and packer who in 1877 was elected trustee of CMIU 144.

## Two Items in the *Cigar Makers' Official Journal*

[June 10, 1879]

MONTHLY REPORT OF THE STATE OF TRADE.

. . .

New York—Trade very fair on hand work. The amalgamation of the three unions took effect during this month, viz. Union 144, Union 20[1] (Bohemian) and the German Union. We are now working under

one charter. By the united action of all nationalities a rapid increase of numbers is expected.

. . .

*Cigar Makers' Official Journal,* June 10, 1879.

1. Local 20 of New York was affiliated with the CMIU from 1878 to 1879.

[August 10, 1879]

. . .

New York—Trade dull on hand-made cigars, plenty work in filthy tenement houses. Mr. S. Gompers was elected delegate to the Buffalo Convention.[1] Geo. Wolder[2] was elected president in place of B. Rosenberg resigned.

. . .

*Cigar Makers' Official Journal,* Aug. 10, 1879.

1. The CMIU held its 1879 convention in Buffalo, N.Y., Sept. 1-4. (See "Excerpts from the Proceedings of the Twelfth Convention of the CMIU," Sept. 2-4, 1879, below.)

2. Probably George Wolders (1850?-1925), who served as a trustee of the CMIU and as an officer of CMIU 144.

## Excerpts from the Proceedings of the Twelfth Convention of the CMIU

Buffalo, September 2nd, 1879.

Afternoon session of the committee of the whole. Walter Kelby[1] of No. 14[2] in the chair.

Rep. No. 55,[3] Callaghan[4]—*Moved.*—Any member quitting a job without proper cause shall not be entitled to assistance from the union under whose jurisdiction he was employed. Carried.

Rep. No. 144 Gompers—*Moved.*—That the committee of the whole instruct the committee on constitution, that a uniform initiation fee be adopted by them. Carried.

Rep. No. 144, Gompers—*Moved*—Also that a uniform amount of dues be charged in all local unions. Carried.

Rep. No. 144 Gompers.—*Moved*—That we recommend an out of work benefit[5] to unemployed members; to be presented separately to local unions for its adoption. Carried.

Rep. No. 144, Gompers.—*Resolved.*—That the delegates be, and hereby are instructed to submit for the consideration of the Int. Union, at the forthcoming convention the following suggestion, that the convention take into serious consideration the question of making better provision for the necessary expenditures of the Int. Union. Carried.

Rep. No. 6,[6] Fitzgerald[7]—*Moved*—Every union shall establish a labor bureau for the purpose of designating work for the unemployed. Carried.

Rep. No. 144, Gompers.—*Moved.*—That the committee on constitution be instructed to adopt some sick benefit law[8] for sick members, and that the same be duly considered. Carried.

. . .

Rep. No. 144, Gompers.—*Moved*—That the committee on constitution be instructed to adopt some equalization of monies of all local unions. Carried.

Rep. No. 144, Gompers.—*Moved*—That the International Convention meet annually. Carried.

Rep. No. 144, Gompers—*Moved.*—We recommend that the Executive Board of the International Union be empowered to communicate with the officers of the various National Organizations of Cigarmakers with a view to bringing about a closer unity and identity of purpose, and for the mutual gratuitous acceptance of members holding valid traveling cards. Carried.

Rep. No. 144, Gompers.—*Moved*—That the International Executive Board be authorized to grant charters to local unions, upon application of at least 7 members, all such applications to be accompanied by a fee of $5.00; but not more than one charter shall be granted in any one city. Carried.

. . .

Buffalo, September 3d, 1879.

AFTERNOON SESSION.

. . .

COMMUNICATIONS.

To the Officers and Members of the Cigar Makers' International Union of America.

Gentlemen.—Although I have not the honor to be a delegate to your convention, I desire, with your kind permission, to make, in my judgment, a few timely and well meant remarks.

I doubt not, Mr. Chairman and Gentlemen, that some of the leading topics to be discussed and acted upon at this convention will involve the question of devising some plan for the assistance of *Traveling Cigarmakers,* a provision for the sick and needy, and for members out of employment &c. In order to obtain this much desired end, Union 144 of New York respectfully submits among many other amendments to the Int. Constitution, the following article:

*Article I, Section 3:*

Every member shall pay into the funds of the union to which he belongs, the sum of 25 cents per week.

Now, Gentlemen, while I honor the intention of Union 144 to improve our condition, I am, nevertheless, of the opinion that the passage of such a law or a similar one in the present state of trade, and in view of the general hard times from which we have not yet entirely emerged, would drive scores of cigarmakers, who are now members, out of the Union. As far as I am concerned, and I am only *one* out of *many,* I cannot afford to pay $13.00 per annum. I believe in *being* what our name implies; I believe in maintaining the Union, but I do *not* believe in converting it into a benevolent association.

I did not join a benevolent association when I became a member. I would ask some of the members of Union 144, why they do not join one of the many benevolent associations with which the country is flooded, and which would hardly cost them any more.

With these few brief remarks I close, hoping that these few lines will be received in the same spirit in which they are written. They are the fruits of hard thinking and of mature reflection; *it is after a great deal of reflection that I sign them.*

I remain, Gentlemen, very respectfully yours, &c.

Jacob Dewes,[9]

Ex-President, Union No. 2[10] of Buffalo.

. . .

Buffalo, September 4th, 1879.

AFTERNOON SESSION.

. . .

To the Convention.

Gentlemen—We intended to propose to this Body for its adoption a system of out of work and sick benefits; but since the Convention has rejected the clause for dues of 25 cents per week, the proposing

of the above benefits would be fruitless. We respectfully request that this statement be incorporated in the record of proceedings.

Yours respectfully
Samuel Gompers, Rep. No. 144.
A. F. Plate,[11] Rep. No. 122.[12]
William V. Todd,[13] Rep. No. 27[14]

To the Convention.

Gentlemen—Your Committee on Constitution to which was referred the matter of an out of work or sick benefit, would respectfully report the following resolutions:

*Resolved*—That while this Convention is in favor of both an out of work and sick benefit, we believe that the members are not yet sufficiently prepared to pay into the Union such a sum as will be necessary for the maintenance of the same.

*Resolved*—That we recommend the discussion of the above questions by the Local Unions, so that the delegates to the next convention may be exactly informed upon the subject, and we further recommend the same to the earnest consideration of the next convention of the International Union.

*Resolved*—That these resolutions together with the plan of these laws be published in the Official Journal.

*Sick Benefit.*

*Section 1.* Every member in good standing six months after joining the International Union, shall be entitled to benefit for a term of six weeks—to commence with the second week of such member's sickness—and at the rate of three dollars a week or fifty cents per day. A period of three months must have elapsed before he shall be entitled to any further benefit.

*Section 2.* Any member having brought sickness upon himself through intoxication or venereal diseases, shall not be entitled to any benefit whatever.

*Section 3.* Any member going into the streets or into a place of public resort during his sickness, without the permission of his physician, shall not be entitled to any benefit for one week.

*Section 4.* Any member doing any work or business while sick, and receiving benefit without the written permission of his physician, shall be deprived of all sick benefits for the next three months.

*Section 5.* Every Local Union shall appoint in rotation according to the list of members a Visiting Committee consisting of seven, to visit the sick members daily. Any member refusing or neglecting to visit shall be fined 25 cents.

*Section 6.* Any member becoming sick shall report to the President of the Union under whose jurisdiction he is, who shall report on the same day to the Financial Secretary of the Union, and failing to do so shall be fined 50 cents.

*Section 7.* Any member refusing to admit the Visiting Committee shall be deprived of his benefit for one week.

*Section 8.* Members appointed to visit the sick shall attend to their duty at such hours as may be designated or the physician permit, but shall be excused visiting persons having *contagious* diseases.

*Section 9.* Females shall not be entitled to any sick benefits two weeks before and four weeks after confinement.

*Out Of Work Benefit.*

*Section 1.* Every member in good standing six months after joining this Union shall be entitled to the out of work benefit for a term of four weeks, to commence with the first week of non-employment.

*Section 2.* The out of work benefit shall be $4 per week but no fractional part of a week shall be paid. The amount shall be paid at the weekly meeting of the Board of Administration. Members claiming assistance must personally apply, show certificate from the proper officer, and sign receipt for the amount received.

*Section 3.* Every member claiming assistance shall report daily to the Labor Bureau, and must take employment in the shop designated. Members refusing to work in such shop, shall not be entitled to any assistance for the following four weeks.

*Section 4.* Any member entitled to the out of work assistance, neglecting to report daily to the Labor Bureau, commencing from the day of his discharge, shall have 60 cents deducted from his benefit for every such omission.

*Section 5.* Every member obtaining employment before having received four weeks' assistance, and being again thrown out of work before six weeks shall have elapsed, shall be entitled to the balance of the four weeks assistance.

*Section 6.* Any member being compelled to quit work, shall receive a pass from the Collector of the shop, stating date and cause of quitting; every member having received benefit, when commencing work in another shop shall report to the Shop Collector. The pass shall read as follows.

This is to certify that ______ has quitted the shop of ______
Cause ______ Date ______

______ Collector.

*Section 7.* Any member quitting a job without proper cause, shall not be entitled to assistance during the following four weeks.

*Section 8.* Any member discovered making cigars or doing any other business between 8 A.M. and 5 P.M. shall forfeit all claims to out of work assistance for a period of not less than eight weeks.

. . .

EXTRA SESSION.

. . .

Rep. No. 144, Gompers—*Resolved.* That the International Union desires and will endeavour to obtain legislation in the interest of the craft.

*Resolved.* That we deem it both injudicious and detrimental to the best interests and welfare of the organization for any union as such to contract political affiliations. And for these reasons no union under the jurisdiction of the International Union shall be permitted to aid, cooperate or identify itself with any political party whatsoever. Carried.

*Resolved*—That the Executive Board be empowered to appoint an organizer for the purpose of organizing the trade in general, whenever they deem it expedient. And shall also be empowered to levy an assessment upon each member of the International Union to defray such expense. Carried.

. . .

*Cigar Makers' Official Journal,* Oct. 10, 1879.

1. Walter Kelby, a Chicago cigar packer, served as president of CMIU 11 in the early 1870s and as secretary of CMIU 14 from 1879 to 1880.

2. The CMIU chartered local 14 of Chicago in 1879.

3. The CMIU chartered local 55 of Hamilton, Ontario, in the late 1860s.

4. John O'CALLAGHAN, an officer of CMIU 55, Hamilton, Ontario, served as first vice-president of the CMIU from 1879 to 1880.

5. The CMIU adopted an out-of-work benefit in 1889.

6. The Syracuse body that participated in the founding of the Cigar Makers' National Union in 1864 was originally local 5; it became local 6 of the CMIU in 1867.

7. Theodore Fitzgerald, a Syracuse cigarmaker, served as president of CMIU 6 from 1878 to 1879.

8. The sick benefit drafted by the 1879 CMIU Committee on Constitution was endorsed, with some modification, by the 1880 convention and adopted by referendum. It appeared as Article IX in the 1881 constitution.

9. Jacob Dewes (d. 1888) served as vice-president of local 2 of Buffalo, N.Y., during the 1870s and president during part of 1879 and 1880.

10. Local 2 of Buffalo was one of the founding locals of the Cigar Makers' National Union in 1864.

11. Augustus F. Plate (b. 1855), president of CMIU 122 of Warren, Pa., was born in New York. From 1876 to 1886 he held offices in CMIU locals at Buffalo, N.Y., and Warren and Corry, Pa.

12. The CMIU chartered local 122 of Warren, Pa., in 1877.

13. William V. TODD was an officer of CMIU 27, Toronto, and third vice-president of the CMIU in 1880-81 and 1890-93.

14. CMIU 27 of Toronto received its charter in 1865.

# The Constitution Proposed[1] by the Twelfth Convention of the CMIU

[September 15, 1879]

## CONSTITUTION OF THE CIGAR MAKERS' INTERNATIONAL UNION OF AMERICA.

### PREAMBLE.

"Labor has no protection—the weak are devoured by the strong. All wealth and all power centre in the hands of the few, and the many are their victims and their bondsmen."

So says an able writer in a treatise on association, and in studying the history of the past, the impartial thinker must be impressed with the truth of the above quotation. In all countries and at all times capital has been used by those possessing it to monopolize particular branches of business until the vast and various industrial pursuits of the world have been under the immediate control of a comparatively small portion of mankind. Although an unequal distribution of the world's wealth, it is perhaps necessary that it should be so.

To attain to the highest degree of success, in any undertaking, it is necessary to have the most perfect and systematic arrangement possible; to acquire such a system it requires the management of a business to be placed as nearly as possible under the control of one mind; thus concentration of wealth and business tact conduce to the most perfect working of the vast business machinery of the world. And there is, perhaps, no other organization of society so well calculated to benefit the laborer and advance the moral and social condition of the mechanic of the country, if those possessed of wealth were all actuated by those pure and philanthropic principles so necessary to the happiness of all. But, alas! for the poor of humanity, such is not the case. "Wealth is power," and practical experience teaches us that it is power too often used to oppress and degrade the daily laborer.

Year after year the capital of the country becomes more and more concentrated in the hands of the few, and in proportion as the wealth of the country becomes centralized, its power increases, and the la-

boring classes are impoverished. It therefore becomes us, as men who have to battle with the stern realities of life, to look this matter fair in the face. There is no dodging the question. Let every man give it a fair, full and candid consideration and then act according to his honest convictions. What position are we, the cigarmakers, to hold in society? Are we to receive an equivalent for our labor sufficient to maintain us in comparative independence and respectability, to procure the means with which to educate our children and qualify them to play their part in the world's drama; or must we be forced to bow the suppliant's knee to wealth, and earn by unprofitable toil a life too void of solace to confirm the very chains that bind us to our doom?

"In union there is strength," and in the formation of a Cigarmakers International Union, embracing every cigarmaker in the country, a union founded upon a basis broad as the land in which we live, lies our only hope. Single-handed we can accomplish nothing, but united there is no power of wrong we may not openly defy.

Let the cigarmakers of such places as have not already moved in this matter organize as quickly as possible and connect themselves with the International Union. Do not be humbugged with the idea that this thing cannot succeed. We are no theorists: this is no visionary plan, but one eminently practicable. Nor can injustice be done to any one; no undue advantage can be taken of any of our employers. There is not, there cannot be any good reason why they should not pay us a fair price for our labor. If the profits of their business are not sufficient to remunerate them for their trouble of doing business, let the consumer make the balance. The Stereotype argument of our employers, in every attempt to reduce wages, is that their large expenses and small profits will not warrant the present prices for labor; therefore, those just able to live now must be content with less hereafter.

In answer, we maintain the expenses are not unreasonable, and the profits are large, and the aggregate great. There is no good reason why we should not receive a fair equivalent for our labor. A small reduction seriously diminishes the already scanty means of the operative and puts a large sum in the employer's pocket; and yet some of the manufacturers would appear charitable before the world.

We ask, is it charitable? Is it humane? Is it honest? To take from the laborer, who is already fed, clothed and lodged too poorly, a portion of his food and raiment, and deprive his family of the necessaries of life—by the common resort—a reduction of his wages? It must not be so.

To rescue our trade from the condition into which it has fallen, and raise ourselves to that condition in society to which we, as me-

chanics, are justly entitled, and to place ourselves on a foundation sufficiently strong to secure us from further encroachments, and to elevate the moral, social and intellectual condition of every cigarmaker in the country, is the object of our International Union, and to the consummation of so desirable an object, we, the delegates in convention assembled, do pledge ourselves to unceasing effort.

ARTICLE I.

*Sec. 1.*—This organization shall be known as the "Cigarmakers' International Union of America," and shall consist of local unions who acknowledge the jurisdiction of the International Union, such unions to be represented by delegates who shall establish their claims to membership by certificates of election, signed by the President and Secretary, and bearing the seal of the Union attached.

*Sec. 2.*—No local unions shall permit the rejection of an applicant for membership on account of sex, color, or system of working.

*Sec. 3.*—No employer or foreman shall be eligible to become a member of any local union, but this section in no wise shall debar members of cooperative factories from membership.

ARTICLE II.

*Sec. 1.*—The session of the International Union shall be held annually, on the third Tuesday of the month of September of each year, at such places as the delegates may have provided for at the last preceding session.

*Sec. 2.* *Forms of Credentials.*

Cigarmakers' Union No. ______ of ______

This is to certify that Mr. ______ of ______ Union was duly elected at a stated meeting as a representative thereof to your body.

(Seal.) President.

Secretary.

*Sec. 3.*—No representative of a local union shall be admitted to a seat in the International Union until the capitation tax of the union he represents is paid in full.

*Sec. 4.*—Each delegate shall be entitled to one (1) vote per every one hundred (100) members that the union he represents has paid per capita tax for. Delegates representing less than one hundred (100) members shall be entitled to one (1) vote.

*Sec. 5.*—The basis of representation in the International Union shall be one delegate for each local union, provided that said union shall not contain less than 25 members. Unions having less than 25

members shall combine with their nearest sister unions for that purpose.

*Sec. 6.*—The International Union shall allow to local unions the mileage of delegates to and from the convention, provided that no portion of the general fund, except for mileage, shall be called upon to defray the expenses of delegates.

*Sec. 7.*—That the mileage shall be three cents per mile.

*Sec. 8.*—All delegates to the International Convention shall be elected at the first meeting in August of their respective unions preceding the convention.

ARTICLE III.

*Sec. 1.*—The officers of the International Union consist of a President, First and Second Vice President, a Treasurer, and two Trustees.

*Sec. 2.*—The President, First and Second Vice President shall comprise the Executive Board.

*Sec. 3.*—The International Union shall elect a Board of Appeals, to consist of five, and locate them at five different places where no officer of the Executive Board is stationed.

*Sec. 4.*—The election of President, First and Second Vice Presidents and Board of Appeals shall be held immediately after the adoption of the reports of committee on officers' reports. It shall require a majority of all the votes polled to constitute an election; at every unsuccessful balloting the name of the candidate receiving the lowest number of votes shall be withdrawn until an election takes place.

*Sec. 5.*—Every member of the International Union shall be eligible to any office in the International Union.

ARTICLE IV.

*Sec. 1.*—It shall be the duty of the President to preside at all meetings of the International Union, conduct the same according to parliamentary rules, examine all documents, audit all bills, preserve all documents and receipts, and present the same at the next session of the International Union, and see that all officers perform their duty. He shall, at the opening of the session, with the consent of the Union, appoint a clerk for the session, whose compensation shall be fixed by the Union. He shall conduct all correspondence between the International Union and local unions, as shall hereinafter be provided for, keep a just and true account between the several local unions and International Union, designate from what sources the finances have been derived, and shall announce before the adjournment of the Union the amount received, expended, and for what purpose, and pay over to the Treasurer all monies over and above $75.00. He shall

publish an official journal at least once a month, stating the condition of trade of all unions connected with the International Union and all other business appertaining to local unions. He shall furnish traveling, membership and retiring cards upon application from the local unions, subject to the laws hereinafter provided, and shall perform all other duties enjoined by the constitution. He shall receive for his services the sum of two hundred and fifty dollars per annum.

ARTICLE V.

*Sec. 1.*—In case of any vacancy, the Executive Board shall forthwith elect said officers, but in all cases the vacancy of President shall be filled by the Vice Presidents in their respective order.

ARTICLE VI.

*Sec. 1.*—The Executive Board shall decide all points of law and differences that may arise between the local unions under the jurisdiction of the International Union; in all cases the parties interested may appeal from the decision of the Executive Board to the Board of Appeals within thirty days. The decision of the Board of Appeals shall be valid unless appealed from until the next session of the International Union.

ARTICLE VII.

*Sec. 1.*—It shall be the duty of the Board of Appeals to decide all differences that may arise between local unions and the Executive Board, when their decision has been appealed from and decide as to legality of strikes as hereinafter provided in all cases. The Board of Appeals shall make known their decision within one week.

ARTICLE VIII.

*Sec. 1.*—The Treasurer shall make no disbursement without an order countersigned by the President and Trustees. He shall deposit, in connection with the Trustees, all sums over seventy-five dollars in the appointed bank in his and their names for the International Union. He shall publish a monthly report of receipts and expenditures in the official journal.

*Sec. 2.*—No monies shall be drawn from the bank without an order of the International President with his seal and the signatures of the Trustees attached, a copy of this clause signed by the International President and Trustees shall be placed in the hands of the banker, and another pasted conspicuously in bank book.

### ARTICLE IX.

*Sec. 1.*—The revenue of the International Union shall be derived from a capitation tax of sixty cents per member annually, payable monthly, such money to be forwarded to the International President with the monthly report by the respective officers.

*Sec. 2.*—When a deficiency exists in the revenue of the International Union, the Executive Board, in conjunction with the Board of Appeals, shall be empowered to levy an assessment not to exceed ten cents per member, to meet such deficiency, said assessment not to occur more than once in a period of six months.

*Sec. 3.*—Any member of a union which has been suspended for non-payment of per capita tax, desiring a traveling card, shall upon proper application to the International President, receive the same upon payment of the amount of per capita tax against him.

*Sec. 4.*—All unions neglecting to comply with sec. 1 of this article shall after due notice be suspended from the International Union, and shall not be reinstated until all arrears are paid in full, provided the reinstatement shall take place within one year from the date of its suspension.

### ARTICLE X.

*Sec. 1.*—The International Union shall issue traveling, retiring, and membership cards, which shall be in possession of the President thereof. He shall distribute them per order of local unions, for the use of any member in good standing, and no local union shall have authority to grant or receive any other card but those provided by the International Union.

*Sec. 2.*—Any member of the International Union going from one locality to another shall provide himself with a traveling card, said card shall bear a certificate of membership of his local union, and any member obtaining such card must present the same before gaining admittance or membership in any other union. No member shall be entitled to receive a traveling card, unless he is in good standing and clear on the Secretary's book.

*Sec. 3.*—Any member moving from one locality to another and obtaining employment, shall immediately deposit his card in the union nearest where he may be employed, and all cards not deposited within one week thereafter shall be annulled.

### ARTICLE XI.

*Sec. 1.*—The International Executive Board is authorized to grant charters to local unions upon application of at least seven persons

employed in the trade, all applications for a charter must be accompanied by a fee of five dollars, but no more than one charter shall be granted in one city.

ARTICLE XII.

*Sec. 1.*—Every union shall establish a labor bureau for the purpose of designating work for the unemployed.

ARTICLE XIII.

*Sec. 1.*—The International Union guarantees its moral and pecuniary support to all its members in difficulties, which may arise between them and their employers. The amount of assistance shall be four dollars per week, and shall commence on the day when the union made the application to the International President, provided the difficulty be approved by the proper authority.

*Sec. 2.*—When any difficulty arises between the members of any union and their employers, three officers of the union shall furnish a full and official statement of the same to the International President, who shall submit a copy thereof to the other officers comprising the Executive Board and Board of Appeals, and if after a full and sufficient investigation of all the facts in the case, they approve of the same, the International President shall issue a circular setting forth the facts to all local unions and the number of members who are idle, through such difficulty, and ordering them to their assistance, he shall state the person or persons to receive the same.

*Sec. 3.*—In case the Executive Board and Board of Appeals fail to approve of any difficulty, the local unions can appeal within fifteen days after the decision being rendered to a general vote of all the local unions. The appeal with three dollars for expenses incurred shall be forwarded to the International President, who shall submit the same to a vote of all local unions, and if approved by a majority, he shall proceed as in this constitution provided.

*Sec. 4.*—Every difficulty involving more than twenty-five members shall be submitted at once by the Int. Pres. to a vote of all Local Unions, and if a majority of those having voted approve the same, he shall proceed as the constitution directs. Any difficulty for an increase of wages shall not be considered legal unless approved by a two-thirds majority of all votes cast; Unions failing to vote within one week, commencing on the day of the circular being mailed, shall be subject to such action as the Executive Board may deem fit; they shall have the right to return their vote by telegram at the expense of the Int. Union, provided their location is over two hundred miles away from where the office of the Int. President is seated.

*Sec. 5.*—The vote of Local Unions on difficulties shall be in proportion to their membership:—One vote from 7 to 50 members; two votes from 50 to 100 members or fraction of not less than 75; three votes from 100 to 200 or fraction of not less than 160; and an additional vote for every 100 more.

*Sec. 6.*—Every Local Union shall retain for every member on the books the amount of 15 cts. per month as a striking fund, and 25 cts. shall be added to said fund for every new member. This fund shall not be used or appropriated for any other purpose, but shall remain in the custody of the Local Union, subject to the order of the officers authorized by the International Union. Any Local Union failing to comply with this section, shall after a notice of thirty days be suspended by the Executive Board.

*Sec. 7.*—Any Local Union being directed by the Executive Board to forward money to another Local Union, and failing to comply within five days from date of said notice, shall be suspended.

*Sec. 8.*—No member of the International Union shall be entitled to any strike benefit, unless he is a member in good standing for at least three months.

*Sec. 9.*—Unions sending money to each other must remit the same by express or post office order. The Secretary transmitting such money shall notify the President of the said Union at their separate address. The Secretary receiving such money shall immediately send a receipt to the Secretary and a copy of such receipt to the President of the Union from whence the money came.

## ARTICLE XIV.

### Traveling Benefit.

*Sec. 1.*—Any cigarmaker being a member in good standing for six months in the International Union, and not being able to obtain employment, wishing to leave the jurisdiction of the union under which he has been working, to seek employment, shall be entitled to the loan of 3 cents per mile to carry him to the nearest union; such loan or loans not to exceed in the aggregate $20.00. But in no case shall any member, working under the jurisdiction of any union one week, be entitled to such benefits from said union.

*Sec. 2.*—Any member quitting a job without a sufficient cause, shall not be entitled to the traveling benefit for a term of three months.

*Sec. 3.*—Any member receiving loans on card shall after getting employment pay to the collector of the shop in which he is employed, 10 per cent of his earnings weekly.

*Sec. 4.*—Every shop shall elect a collector, and in every shop in

which there is but one union man employed, he shall be constituted shop collector.

*Sec. 5.*—It shall be the duty of the collector to collect and pay to the secretary of the union all amounts that he may have received within 48 hours after receiving the same.

*Sec. 6.*—Any member refusing to comply with the provisions contained in this Article shall not be entitled to receive further loans for the period of three months after he has paid his arrears in full.

### ARTICLE XV.

### Monthly Report.

*Sec. 1.*—The Financial Secretary of each Local Union shall within five days after the last day of each month, forward to the Int. President a financial statement of the receipts (from what sources received, the amounts expended for what purposes) all of which must be itemized, the number of members at the commencement of the month, the names of the members initiated or admitted by card, the names of members retired, withdrawn or expelled.

*Sec. 2.*—The Corresponding Secretary of each Local Union shall within five days after the last day of each month, forward to the International President a report of the state of trade, prices paid, hours of labor, system of working, cost of living, and all other general conditions of the craft in his locality.

*Sec. 3.*—Any Local Union that shall suffer any of the provisions of this Article to be neglected or violated after due notification shall be suspended until they are complied with.

### ARTICLE XVI.

*Sec. 1.*—Every Local Union shall elect an Executive Committee at the same time that they elect the remainder of the officers of their union, whose duty it shall be to meet at least once in every week, and the Executive Committee shall be empowered to comply with the directions of the Executive Board of the International Union as provided by the laws.

### ARTICLE XVII.

*Sec. 1.*—Any person wishing to become a member of any Local Union shall pay the sum of $1.00 as an initiation fee which may be paid by four weeks instalments.

*Sec. 2.*—Every member shall pay in to the funds of the union to which he belongs the sum of 10 cents per week.

*Sec. 3.*—Any member drawing a traveling card shall pay dues for the week in which his card is issued to the union from which he receives it, and no other union shall charge dues for the same week.

*Sec. 4.*—Any member [union?] receiving dues from members for a longer period of time than they may remain members thereof shall return the excess when they draw their traveling or retiring card.

*Sec. 5.*—Any member of any Local Union who shall fail to pay the dues as provided by law for a term of eight weeks shall be suspended from the union.

*Sec. 6.*—Any member suspended from any Local Union cannot become a member of any union under the jurisdiction of the International Union unless he pays an initiation fee of $2.00 which may be paid in four weekly instalments.

### ARTICLE XVIII.

### Equalization.

[*Sec. 1.*—]The International President shall compile quarterly the monthly reports, equalize the several amounts, and direct such unions as may have expended less than their *pro ratio* amount for the benefits provided by the laws, to forward to those unions that may have expended in excess thereof such amounts as they in virtue of such equalization may be entitled to, and any union failing to comply with any of the provisions contained in this Article within ten days within date of said notice, shall be suspended from the International Union.

*Sec. 2.*—No union shall be permitted to expend for general expenses including hall rent etc. more than the following percentages of its gross receipts, excluding the striking fund; Unions numbering 30 members or less 25 per cent, from 30 to 50 members 15 per cent, from 50 members and upwards 10 per cent.

*Sec. 3.*—In the event of the funds of any Local Union becoming exhausted, the Executive Board of the International Union shall immediately upon notice thereof direct any other Local Union to forward to the aforesaid union such amounts as may be deemed necessary.

*Sec. 4.*—Any Union failing to comply with the provisions of this Article within six days from the receipt of the notice from the Executive Board, shall be suspended.

### ARTICLE XIX.

*Sec. 1.*—It shall be the duty of the Corresponding Secretary of each Local Union immediately after the election of President and

Secretary to forward to the International President the addresses of such officers for publication in the Official Journal.

ARTICLE XX.

*Sec. 1.*—The Official Journal shall be edited in conformity with the principles and resolutions of the International Union. Every member of every Local Union shall receive a free copy of every issue.

*Sec. 2.*—If deemed necessary by at least five Unions to suspend, amend, or add parts of this constitution, the Executive Board shall submit the same to Local Unions to be voted upon within 30 days. In case the vote should be equally divided, those Unions not having voted must be counted in the affirmation, but when all Unions shall have voted and a tie occurs, the question shall be considered lost.

ARTICLE XXI.

*Sec. 1.*—All propositions for the alterations of the laws of the International Union to be voted at the annual session shall be published in the Official Journal at least four weeks prior thereto.

ARTICLE XXII.

Sinking Fund.

*Sec. 1.*—The International Union shall raise a sinking fund which shall consist of the funds of Local Unions and shall amount to the sum of two (2) dollars per member.

*Sec. 2.*—Whenever the sinking fund of the International Union shall fall below the sum as provided in Section 1 of this Article, the Executive Board shall be empowered to levy an assessment on each member to replenish the same.

*Sec. 3.*—This Article shall not take effect until six months after its ratification by the Local Unions.

ARTICLE XXIII.

*Sec. 1.*—This Union shall not be dissolved while there are three dissenting Unions, nor shall this section be subjected to any alteration whatever.

*Sec. 2.*—This constitution shall take effect and remain in full force from November 1st 1879, provided the same has been duly ratified by that time by a majority of Local Unions under the jurisdiction of the International Union, and all Local Unions shall vote separately upon each article of the constitution and forward the result thereof to the International President before the 25th day of October 1879.

All of which is respectively [respectfully?] submitted by Committee on Constitution.

J. O'Collaghan.
Walter Kelby.
William V. Todd.
William C. Schaefer.[2]
P. H. Rowe.[3]
C. W. Smith.[4]
Samuel Gompers.

*Cigar Makers' Official Journal,* Sept. 15, 1879.

1. The Nov. 10, 1879, issue of the *Cigar Makers' Official Journal* announced the ratification of the constitution by CMIU locals. The major new features of the constitution of 1879 included uniform fees, the loaning system for traveling cigarmakers, quarterly equalization of funds among union locals, the sinking fund, separate voting on each article of proposed constitutions, and centralized control of strikes and strike benefits.

2. William C. SCHAEFER was an officer of CMIU 25, Milwaukee.

3. John H. Rowe, a Richmond cigarmaker, acted as a vice-president of local 133 of that city in 1873 and as treasurer after its reorganization in 1879. During the 1880s Rowe became a cigar manufacturer.

4. Christian W. Smith (1847-1919?), a Prussian-born cigarmaker, served as president, treasurer, and trustee of CMIU 2 of Buffalo, N.Y., in the late 1870s and early 1880s. Later he became a cigar manufacturer in Buffalo, where he resided until his death.

# A News Account of a Mass Meeting at Cooper Union

[January 27, 1880]

THE DEMANDS OF WORKING MEN.

A mass-meeting of working men, under the auspices of the Amalgamated Labor Union, was held last evening in the large hall of the Cooper Institute. About 2,000 persons were present. The purpose of the meeting was to indorse bills before the Legislature for the establishment of a Bureau of Labor Statistics, the abolition of prison labor, the enforcement of the Eight-hour law, the abolition of tobacco factories in tenement-houses,[1] and the Mechanics' Lien law. These bills were indorsed in a series of resolutions. The first speaker was John Blissard,[2] who, reciting the oft-repeated grievances of the working men as rehearsed at their public meetings, urged them to do all in their power to assist in the movement of the union to secure the passage of the bills enumerated in the resolutions. John Gambos,[3] of

the Cigar-makers' Union, drew a graphic picture of the evils of the tenement-house system of cigar-making, showing how families were huddled together, in poverty, filth, and wretchedness, in loathsome tenements, and were forced to sleep and eat in the same apartments where they worked in the preparation of the weed for smokers. He passed some severe censure on Prof. Chandler[4] for not taking action in regard to the matter suggested by his union, and also on Assemblyman Hayes,[5] for refusing, as he said, to support the passage of the bill. J. P. McDonnell and others made addresses, after which the meeting adjourned.

*New York Times,* Jan. 27, 1880.

1. A bill prohibiting the manufacture of cigars in New York City tenement houses was introduced unsuccessfully as a sanitary measure in the New York legislature in 1880, 1881, and 1882.

2. Robert Blissert.

3. SG.

4. Charles Frederick Chandler (1836-1925), a longtime professor of chemistry at Columbia University, was president of the New York City Board of Health from 1873 to 1883 and a member of the New York State Board of Health in the 1880s.

5. Isaac Israel Hayes (1832-81), a physician and Arctic explorer, served as a Republican assemblyman in the New York legislature from 1876 to 1881.

# Entries for the Households of Samuel Gompers and Various Relatives in the 1880 Census Return for New York City

City of New York, . . . 1st day of June, 1880.

| | In Cities. | | Dwelling houses numbered in order of visitation. | Families numbered in order of visitation. | The name of each Person whose place of abode, on 1st day of June, 1880, was in this family. | Personal Description. | | |
|---|---|---|---|---|---|---|---|---|
| | Name of Street. | House Number. | | | | Color—White, W; Black, B; Mulatto, Mu; Chinese, C; Indian, I. | Sex—Male, M; Female, F. | Age at last birthday prior to June 1, 1880. If under 1 year, give months in fractions, thus: 2/12. |
| | | | 1 | 2 | 3 | 4 | 5 | 6 |
| 50 | Columbia Street | [Front 85] | [10] | 62 | Gompert Emanuel[1] | [W] | M | 34 |
| 1 | Columbia Street | Front 85 | 10 | 62 | Gompert Kate | W | F | 39 |
| 2 | ″ | ″ | ″ | ″ | — Louis | — | M | 3 |
| 3 | ″ | ″ | ″ | ″ | — Sarah | — | F | 1 |
| 4 | ″ | ″ | ″ | 63 | Gompert Solomon | — | M | 53 |
| 5 | ″ | ″ | ″ | ″ | — Sarah | — | F | 53 |
| 6 | ″ | ″ | ″ | ″ | — Louis | — | M | 22 |
| 7 | ″ | ″ | ″ | ″ | — Jacob | — | M | 18 |
| 8 | ″ | ″ | ″ | ″ | — Simon[2] | — | M | 15 |
| 9 | ″ | ″ | ″ | ″ | ″ Bella[3] | — | F | 12 |
| 10 | ″ | ″ | ″ | ″ | — Harriet[4] | — | F | 11 |
| 11 | ″ | ″ | ″ | 64 | Gompert, Sam | — | M | 29 |
| 12 | ″ | ″ | ″ | ″ | — Sophia | — | F | 29 |
| 13 | ″ | ″ | ″ | ″ | ″ Solomon[5] | — | M | 11 |
| 14 | ″ | ″ | ″ | ″ | — Rosetta[6] | — | F | 8 |
| 15 | ″ | ″ | ″ | ″ | — Henry.[7] | — | M | 6 |
| 16 | ″ | ″ | ″ | ″ | — Abraham[8] | — | M | 4 |
| 17 | ″ | ″ | ″ | ″ | — Alexander[9] | — | M | 2 |
| 18 | ″ | ″ | ″ | 65 | Gompert Simon | — | M | 30 |
| 19 | ″ | ″ | ″ | ″ | — Lizzie[10] | ″ | F | 30 |
| 20 | ″ | ″ | ″ | ″ | — Samuel[11] | — | M | 10 |
| 21 | ″ | ″ | ″ | ″ | — Minnie | — | F | 8 |
| 22 | ″ | ″ | ″ | ″ | — Solomon | — | M | 7 |
| 23 | ″ | ″ | ″ | ″ | — Abraham[12] | — | M | 2 |
| 24 | ″ | ″ | ″ | ″ | — Fannie[13] | — | F | 1 |

1. Emanuel GOMPERS, a cigarmaker, was SG's cousin and also his uncle by virtue of his marriage to SG's aunt, Catherine Gompers.

2. Simon GOMPERS (1865-1953), SG's brother.

3. Bella GOMPERS, SG's sister.

4. Henrietta GOMPERS (1869-1954), SG's sister.

5. Samuel Julian GOMPERS (1868-1946), son of SG and Sophia Gompers, became a printer and later chief clerk of the U.S. Department of Labor.

6. Rose GOMPERS, daughter of SG and Sophia Gompers.

| Relationship of each person to the head of this family—whether wife, son, daughter, servant, boarder, or other. | Civil Condition. Single, /. | Married, /. | Occupation. Profession, Occupation or Trade of each person, male or female. | Number of months this person has been unemployed during the Census year. | Education. Attended school within the Census year, /. |
|---|---|---|---|---|---|
| 8 | 9 | 10 | 13 | 14 | 21 |
| H | | /. | Cigar Maker | 0 | |
| W | | /. | Keep House | | |
| son | / | | at Home | | |
| Daugh | / | | — | | |
| H | | /. | Cigar Maker | 0 | |
| W | | /. | Keep House | | |
| Son | /. | | Cigar Maker | 0 | |
| son | /. | | — | 0 | |
| son | /. | | School | | /. |
| Daugh | /. | | at Home | | |
| Daugh | /. | | School | | /. |
| H | | /. | Cigar Maker | 0 | |
| W | | /. | Keep House | | |
| son | /. | | School | | /. |
| Daugh | / | | — | | /. |
| son | / | | — | | /. |
| son | / | | at Home | | |
| son | / | | — | | |
| H | | /. | Slipper Maker | 0 | |
| W | | /. | Keep House | | |
| son | /. | | School | | /. |
| Daugh | / | | — | | /. |
| son | / | | — | | /. |
| son | / | | at Home | | |
| Daugh | | | — | | |

7. Henry Julian GOMPERS (1874-1938), son of SG and Sophia Gompers, became a granite cutter.

8. Abraham Julian GOMPERS (1876-1903), son of SG and Sophia Gompers, became a cutter in the clothing industry.

9. Alexander Julian GOMPERS (1878-1947), son of SG and Sophia Gompers, became a cigarmaker.

10. Elizabeth Tate GOMPERS, SG's aunt through her marriage to SG's uncle, Simon Gompers.

11. Samuel GOMPERS (b. 1869), son of Simon and Elizabeth Tate Gompers.

12. Abraham GOMPERS (b. 1877), son of Simon and Elizabeth Tate Gompers.

13. Fannie GOMPERS, daughter of Simon and Elizabeth Tate Gompers.

City of New York, . . . 1st day of June, 1880.

| Nativity. Place of Birth of this person, naming State or Territory of United States, or the Country, if of foreign birth. | Place of Birth of the Father of this person, naming the State or Territory of United States, or the Country, if of foreign birth. | Place of Birth of the Mother of this person, naming the State or Territory of United States, or the Country, if of foreign birth. | |
|---|---|---|---|
| 24 | 25 | 26 | |
| Holland | Holland | Holland | 50 |
| Holland | Holland | Holland | 1 |
| New York | — | — | 2 |
| — | — | — | 3 |
| Holland | — | — | 4 |
| — | — | — | 5 |
| England | — | — | 6 |
| — | — | — | 7 |
| New York | — | — | 8 |
| — | — | — | 9 |
| — | — | — | 10 |
| England | — | — | 11 |
| — | England | England | 12 |
| New York | — | | 13 |
| — | — | | 14 |
| — | — | — | 15 |
| — | — | — | 16 |
| — | — | — | 17 |
| England | Holland | Holland | 18 |
| — | England | England | 19 |
| New York | — | — | 20 |
| — | — | — | 21 |
| — | — | — | 22 |
| — | — | — | 23 |
| — | — | — | 24 |

Printed form, with autograph entries, signed, Census of 1880, supervisor's dist. 1, enumeration dist. 154, 50:5-6, RG 29, Records of the Bureau of the Census, DNA. Columns left blank by the census taker were: born within census year (7); widowed or divorced (11); married during census year (12); sick or temporarily disabled (15); blind (16); deaf and dumb (17); idiotic (18); insane (19); maimed, crippled, bedridden, or otherwise disabled (20); cannot read (22); cannot write (23).

# To the Editor of the *Cigar Makers' Official Journal*

New York, December 1, 1880.

Simultaneously with the news of the ratification of the new Constitution, including the Sick and Death Benefits,[1] we shall, of course, enter upon an era of dissension, discouragement and perhaps open revolt. But if we may form a judgment from what was stated by some of the Delegates to the Chicago Convention[2] in the course of private conversation, there is more to be feared from acts of *omission* than acts of *commission.* In other words we have cause to apprehend that some will fail or avoid doing anything which tends to promote the success of these measures, that everything will be left undone which ought to be done, and the proper measures left to take care of themselves. How far this policy if carried out, will succeed, I am of course not able to tell. Nevertheless I desire to venture an opinion in the premises, and candidly state what I deem to be the duty of the hour. Every one, who is acquainted with my views, knows that I am the advocate of beneficial and benevolent features being incorporated in Trades unions; it being my mature conviction that upon such features mainly depends the permanent success of Trades Organizations. At the Convention, and subsequently there has been much said upon these questions, and it is but just and proper that all members having and taking an active interest in the welfare of our trade, should pause to reflect upon what has thus far been done, and what still remains to be done. We have an undoubted right to oppose any measure to the last extremity; and also exert all our efforts for the purpose of having engrafted on our Constitution such views as we entertain. But when we have availed ourselves of all these rights and privileges, and been defeated, our opposition to certain questions having failed to meet with the response of our constituents, there is in such case an opportunity presented to us, of showing, how far we have attained that degree of intelligence, that honesty and sincerity of purpose which is clearly manifested in the act of "bowing to the will of the majority." We should thus do all that in our power lies to make the laws and principles of our organization a living fact, and an undoubted success. At the Convention the only line of argument pursued by the opponents of the features was that, our constituents were opposed to the measures suggested. All that the advocates thereof asked was, that our members in *their entirety* should have an opportunity of passing their opinion upon the matter. This request was reluctantly granted. The result must naturally surprise those members who predicted its positive

defeat by the referendum. Of course much may be done to thwart the efficacy and intention of these laws, by inaction or non-attention to the enforcement of the restrictive rules, which are meant to operate as safeguards against imposition. Such omission may be so used as to prove the futility of these benefits. The authors of such a policy would then set to work a-bellowing, that fraud can not be prevented, and thus appeal to the worst prejudices of the unthinking. But I believe that the sound judgment of the large majority of our members will prevail, and counteract the malicious intentions of pseudo friends and avowed enemies. In conclusion I would ask, nay I would earnestly appeal to our entire membership, to rise to the occasion, do all they possibly can to render our laws successful in their operation; to restore our organization to the proud position it once held, to crystalize it into a solid and permanent shape, and by so doing earn the lasting gratitude of a downtrodden people.

Samuel Gompers,
Pres. Union 144.

*Cigar Makers' Official Journal,* Dec. 10, 1880.

1. The 1880 CMIU convention endorsed constitutional amendments establishing sick and death benefits. The sick benefit provided members who had been in good standing for at least six months with $4 a week for up to eight weeks of their illness. The death benefit provided $25 toward funeral costs for members who had been in good standing for at least a year. The membership adopted the sick and death benefits in referendum, and they appeared as Article X of the 1881 constitution.
2. The 1880 CMIU convention was held in Chicago, Ill., Sept. 21-25.

## To the Editor of the *Cigar Makers' Official Journal*

New York, Jan. 1881.

A discussion has arisen in regard to the proper construction to be put upon that portion of the "Sick and Death Benefit" laws, having specific reference to the time when members become entitled to them.[1] Our friend, Mr. Blend,[2] in the last issue of the Journal,[3] suggests that I should deliver my opinion in reference thereto. Being desirous to avoid misunderstanding, I embrace this opportunity of plainly expressing my opinion, that either Mr. Blend or myself is laboring under a complete misapprehension on this point. To begin with, if my recollection serves me faithfully, no reference was made in the Committee on Constitution, either to my views on the subject, or to those en-

tertained thereon by Mr. Blend. It appears to me, that by fair implication, no member is entitled to those benefits, until they shall have been in operation *six* and *twelve* months respectively. Most assuredly, had I known that it was intended otherwise, I would have been compelled, however reluctantly, to strenuously oppose the proposed measure. Now I would like to recall an incident of the Convention to Mr. Blend's recollection, viz., that the opponents of these measures hurled at us the following taunt: They said, "We would be unable to fulfil these obligations, with an increase of dues of only five cents." I also distinctly recollect my reply thereto, which was to the following effect: "The funds we would accumulate within the first six months, would in all probability more than cover the losses that would ensue during the six months following until next Convention, when such measures could be enacted, as would remedy the defect." I also challenge Mr. Blend's attention to another point. Let him compare the phraseology of the "Sick and Death Benefit Laws," I mean such part as bears specifically upon the question at issue, with the corresponding portion of the "Traveling Benefit Law"; he will at once observe the similarity. And yet no one has claimed, in reference to this latter, that members become entitled to it, until six months shall have elapsed, after its ratification, by the Local Unions, subsequent to its adoption by the Buffalo Convention. I will further state that I do not share Friend Blend's apprehension, as to the entire laws being endangered at the next Convention. These measures are accomplished facts. They have made deeper inroads into the minds of the members, than the most sanguine advocates could have anticipated. They are progressive. They are of a more decidedly protective character, more amply surrounded with efficient safeguards, than any other laws of the same nature, that have hitherto been incorporated in the International Constitution. They bind the members more firmly together. They are necessary to the well-being of our craft, and as such cannot be obliterated. The next convention, can, aye and must improve upon these laws; but can and will and dare not annul them! Forward, and ever forward, be our watchword.

Samuel Gompers.

*Cigar Makers' Official Journal*, Mar. 10, 1881.

1. Adolph Strasser interpreted the new constitutional provisions for sick and death benefits to mean that no sick benefits should be paid before June 1, 1881, and that no death benefits should be paid before Dec. 1, 1881.

2. Fred BLEND, an Ohio-born cigarmaker, was president (1869-71) and later first vice-president (1880-85) of the CMIU.

3. Fred Blend to the Editor, Dec. 27, 1880, *Cigar Makers' Official Journal*, Jan. 10, 1881.

## *The KOL, the Trade Unions, and the Founding of the FOTLU*

"The want of unity among the various labor organizations," Adolph Strasser observed, "is the cause of all past defeats, the enslavement and degradation of the working classes."[1] Echoing these sentiments during the last half of the 1870s several national trade unions, including the iron molders, iron and steel workers, printers, and cigarmakers, called for the establishment of a national federation of trade unions. By the early 1880s two competing organizations had emerged, the Knights of Labor and the Federation of Organized Trades and Labor Unions of the United States and Canada. The debate over which of the two—the KOL or the FOTLU and its successor, the American Federation of Labor—would lead the labor movement continued into the 1890s.

The Order of the Knights of Labor, founded in 1869, proved to be the leading general organization of workingmen during the early 1880s. Originally a secret society open to all members of the producing classes, the Knights excluded only stockbrokers, gamblers, lawyers, bankers, liquor dealers, and other such "parasites." The Order's local assemblies were either single-trade organizations or mixed bodies; during the 1870s local trade assemblies of coal miners predominated. District assemblies that grouped local assemblies geographically were the main administrative centers during the KOL's first decade.

It was not until 1878, when the KOL's General Assembly first met, and 1879, when Terence Powderly was elected grand master workman, that the Order achieved national scope or prominence. For Powderly, the "pressing demand of the present" was to "consolidate, unify, all production and distributive labor into one harmonious and homogeneous organization,"[2] and he campaigned constantly to draw all workers into the ranks of the KOL. To broaden the Knights' appeal, for example, Powderly urged the organization to abandon its secret ritual, oath of allegiance, and the rule prohibiting members from mentioning the Order's name in public, and under his leadership it gradually stripped itself of these trappings.

By 1881 the KOL had chartered over 1,500 local assemblies and was becoming a powerful force in hundreds of communities. At-

tempting to unite workers across craft, ethnic, religious, racial, and gender lines, the Knights offered a vision of a cooperative moral industrial order. Inspired by this vision, members supported a wide variety of reform movements ranging from radical greenbackism and land nationalization to temperance, socialism, and cooperation. As members of the Order grew in number, they moved into the local political arena where they waged impressive and often successful campaigns.

Members of the KOL were also ardent defenders of shop floor autonomy. The skilled workers who led the organization perceived that the growing power of capital threatened their independence. The flexible structure of the KOL allowed these workers not only to pursue traditional craft goals but also to forge links with semiskilled and unskilled workers. Because many local assemblies concerned themselves primarily with issues like work rules and jurisdiction, it was not unusual for them to enter into alliances with trade unions, or for trade unionists to be members of the KOL as well as of their trade organizations.

Not all Knights approved of these alliances, however. Many district and national leaders opposed strikes, for example, and urged workers to change conditions primarily through education, cooperation, and political action. Nevertheless the autonomous local assemblies were prone to strike when threatened by intransigent employers, and although KOL leaders felt compelled to sustain their members, they endorsed such activity only reluctantly. When the 1880 General Assembly voted to reserve 30 percent of its resistance fund for strike activity and a full 60 percent for producer and distributive cooperatives, trade unionist members protested vigorously. In January 1881 St. Louis District Assembly (DA) 17 called for a special assembly to air these growing criticisms, but without success.

In June 1881 an organization known as the Amalgamated Labor Union issued a call for a labor convention to be held in Terre Haute that August. P. J. McGuire, active in DA 17 after his move to Missouri in 1878, later explained that despite the call's stated object of establishing a national labor union, dissatisfied Knights intended to use the convention to create a new secret order that would supplant the KOL. According to McGuire, two organizations, the Amalgamated Labor Union and the Knights of Industry, sponsored the conference. He described the Amalgamated as "an offshoot" of the KOL, "composed of disaffected members of that order,"[3] organized in 1878 and concentrated in Ohio and Indiana. The Knights of Industry appears to have been an auxiliary benevolent association that McGuire's St. Louis

district assembly created about six months prior to the convention call.

The response to the Terre Haute convention was limited. During the summer of 1881 Powderly and other KOL national leaders considered attending, but in the end they did not do so. Gompers was elected to represent the CMIU, but the death of a newborn child prevented his attendance. Only three national trade unions—the lake seamen, the tanners and curriers, and the printers—eventually sent delegates to Terre Haute.

Lyman Brant, the delegate of the International Typographical Union (ITU), reported afterward that because so few trade unionists participated, the convention "was not entitled to be considered a representative meeting of trade-unionists of the United States and Canada." On its second day, Brant related, a local delegate introduced "a crude plan of organization," whose preamble "declared opposition to the existing system of trade-union organization." It proposed forcing trade unions "out of existence, and out of the chaos thus created to form *another* secret society of workingmen."[4] Believing that the only way to prevent the adoption of this plan lay in deferring action on it, Brant, his ITU colleague Mark Crawford, McGuire, and other trade unionists decided to propose convening a second convention, to be held in Pittsburgh later in the year, where they hoped to form a national organization more representative of trade union interests. According to Crawford, McGuire drafted the call for the Pittsburgh convention and all the trade union delegates then signed it. After the trade unionists threatened to quit the convention if that assembly did not support their proposal, the delegates adopted it nearly unanimously.

The leaders of the KOL responded more aggressively to the call for the Pittsburgh convention than they had to the Terre Haute call. Recognizing that Pittsburgh and surrounding Allegheny County were the heart of the Knights' strength in Pennsylvania, newly elected Knights Grand Secretary Robert Layton worked to assure a large KOL attendance. At the Pittsburgh convention, forty-seven of the sixty-nine delegates from Pennsylvania represented Knights' assemblies and, altogether, Knights accounted for fifty of the 107 delegates. Gompers was among those representing trade unions. Crawford described the atmosphere in Pittsburgh as "surcharged with all kinds of rumors as to schemes to capture the Trades Congress."[5]

Layton played a prominent role in the convention's debate over the title of the new organization. When the Committee on Plan of Organization, which Gompers chaired, proposed the name "Federation of Organized Trades Unions of the United States of America

and Canada," several delegates objected that the title was overly restrictive. Layton, representing KOL DA 3, declared: "I am opposed to excluding from the Federation all organizations except those of skilled mechanics. The Knights of Labor wish to be with you, and they embrace all laborers."[6] The *Pittsburgh Evening Chronicle*'s coverage of the same debate reported Layton as having said, "There seems to be something singular about the manner in which we are changing base. This Congress was widely advertised as a Labor Congress and now we are talking about trades. Why not make the Knights of Labor the basis for the Federation?"[7] In a compromise, the delegates resolved the issue by choosing the name "Federation of Organized Trades and Labor Unions of the United States and Canada."

Despite the large number of delegates from KOL bodies who attended the founding convention of the FOTLU, the constitution of the new body provided for representation at future meetings only by national and international trade unions—on a graduated basis proportionate to membership—and by local trades assemblies or councils—with a single delegate each. In 1882 the FOTLU modified this formula to allow representation by KOL district assemblies and by local trade unions, including KOL local assemblies, on the basis of one delegate apiece. Only one KOL body, however, sent a representative to any subsequent FOTLU meetings.

Soon after the FOTLU's founding convention, Myles McPadden, a member of the KOL Executive Board, assessed the KOL's performance at Pittsburgh. Writing to Powderly he explained: "I tell you these men dreaded the K of L. and nearly in every instance these men were K. of Labor." On the other hand, he assured Powderly that the KOL had little to fear from the new organization. "The congress," he contended, "accomplished nothing, as the delegates had no instructions to do any thing posative hence it was merely advisory they have reccomended for adoption several things that would be of benafit to the Labouring men of the country, but, I have no faith the congress will accomplish any thing."[8]

*Notes*

1. *Cigar Makers' Official Journal,* June 10, 1878.

2. *Trades,* Apr. 5, 1879.

3. *Carpenter,* June 15, 1888.

4. International Typographical Union, *Report of Proceedings of the Thirtieth Annual Session of the International Typographical Union Held in St. Louis, Missouri, June, 1882* (Chicago, 1882), p. 47.

5. "Large Oaks from Little Acorns Grow," *American Federationist* 27 (1920):145.

6. FOTLU, Proceedings, 1881, p. 16.
7. *Pittsburgh Evening Chronicle,* Nov. 17, 1881.
8. Nov. 23, 1881, Terence Vincent Powderly Papers, DCU.

# The Call for an International Trade Union Congress at Pittsburgh

Terre Haute, Ind., September 15th, 1881.

To all Trades and Labor Unions of the United States and Canadas:

Fellow Workingmen:

The time has now arrived, for a more perfect combination of Labor—one that will concentrate our forces, so as to more successfully cope with concentrated capital.

We have numberless trades unions, trades assemblies or councils, and various other local, national and international labor unions, all engaged in the noble task of elevating and improving the condition of the working classes. But, great as has been the work done by these bodies, there is vastly more that can be done, by a combination of all these organizations, in a federation of trades.

In Great Britain and Ireland, annual trades union congresses are held; France and other countries have, also similar gatherings. The work done by these assemblages of workmen, speaks more in their favor than a volume of other arguments.

Only in such a body, can proper action be taken to promote the general welfare of the industrial classes. There we can discuss and examine all questions, affecting the national interests of each and every trade, and by a combination of forces secure that justice, which isolated and separated trade and labor unions, can never fully command.

A National Trades Union Congress, can prepare labor measures, and agree upon laws they desire passed by the Congress of the United States; and a Congressional Labor Committee, after the manner of the Parliamentary Committee of Trades Unions in England, could be elected, to urge and advance legislation, at Washington on all such measures, and report to the various trades.

In addition to this, an annual congress of trades unions, could organize a systematic agitation to propagate trades union principles, and to impress the necessity of protective trade and labor organizations, and to encourage the formation of such unions, and their amalgamation in trades assemblies. Thus we could elevate trades unionism, and obtain for the working classes, that respect for their rights, and that reward for their services, to which they are justly entitled.

A federation of this character can be organized with a few simple rules, and no salaried officers. The expenses of its management will be trivial, and can be provided for by the Trades Union Congress.

Impressed with the necessity of such a federation, and the impor-

tance of an International Trades Union Congress, to perfect the organization, we the undersigned, delegates, in a preliminary national convention, assembled at Terre Haute, Indiana, held August 2nd, 1881, do hereby resolve to issue the following call:

That all international and national unions, trades assemblies or councils, and local trades or labor unions, are hereby invited to send delegates to an international union congress, to be held in Pittsburgh, Pa., on Tuesday, November 15, 1881. Each local union will be entitled to one delegate, for one hundred members or less, and one additional delegate for each additional 500 members, or major part thereof; also, one delegate for each international or national union, and one delegate for each trades assembly or council.

J. E. Coughlin,[1] Pres't National Tann' and Curr. Unions.[2]
R. Powers,[3] Gen. Pres't Lake Seamen's Union.[4]
Lyman A. Brant,[5] International Typographical Union.[6]
P. J. McGuire,[7] St. Louis Trades and Labor Assembly.[8]
T. Thompson,[9] I.M.U. Dayton, O.[10]
George W. Osborne,[11] I.M.U. of Springfield, O.[12]
W. C. Pollner,[13] Cleveland Trades Assembly.[14]
Sam'l L. Leffingwell,[15] Indianapolis Trades Assembly.[16]
J. R. Backus,[17] Terre Haute Amalgamated Labor Union.

The following gentlemen have also sent in their names as endorsers of the Call:

Geo. Clark,[18] Pres't International Typographical Union.
P. F. Fitzpatrick,[19] Pres't Iron Molders' Un. of America.[20]
John Kinnear,[21] Pres't Cent. Tr. & L. Assembly, Boston.[22]
Geo. Rodgers,[23] Pres't Chicago Trades Assembly.[24]

Note—Several large organized Labor Unions, such as the Amalgamated Association of Iron and Steel Workers,[25] International Cigar Makers, National Granite Cutters,[26] and others assure us that they will be represented, but some of them hold their annual conventions, and others wish to consult their local unions, and their responses to the invitation to sign the Call will be too late for use; but I think each and all will heartily co-operate with any movement, looking to the solidifying of our Labor associations.

Mark W. Moore,[27]
Cor. Sec'y and Treas. 230 N. 12th Street.

*Cigar Makers' Official Journal,* Oct. 10, 1881.

1. John E. COUGHLIN was a currier in Chicago.
2. The National Tanners' and Curriers' Union was headquartered in Chicago.
3. Richard POWERS was president of the Lake Seamen's Benevolent Association of Chicago from 1878 to 1887 and a member of the FOTLU Legislative Committee from 1881 to 1885.

4. The Lake SEAMEN'S Benevolent Association of Chicago.

5. Lyman A. BRANT was corresponding secretary of the International Typographical Union (ITU) between 1880 and 1881, and president of the Detroit Trade and Labor Council in 1881.

6. The International TYPOGRAPHICAL Union.

7. Peter James MCGUIRE, secretary of the St. Louis Trades Assembly, was a founder of the Brotherhood of Carpenters and Joiners of America in 1881 and, as its secretary, was its leading official until 1901. Between 1886 and 1900 he served as a member of the AFL Executive Council.

8. The St. Louis Trades Assembly was organized in 1878 and participated in the formation of the FOTLU in 1881 and the AFL in 1886. In 1887 the Trades Assembly merged with the Central Labor Union to form the St. Louis Central Trades and Labor Union.

9. Thomas Thompson (b. 1849) was born in Canada and immigrated to the United States in 1862. He worked as a molder in Dayton until 1903.

10. Iron Molders' Union of North America (IMUNA) 181 of Dayton, Ohio, was founded in 1877.

11. George W. Osborne was probably a member of IMUNA 72 of Springfield, Ohio. In 1883 he helped found KOL Mad River Assembly 2582, serving as its treasurer.

12. IMUNA 72 was organized in 1864, surrendered its charter in 1872, and was reorganized in 1878.

13. William C. POLLNER was president of CMIU 17 of Cleveland in 1881.

14. The Cleveland Trades and Labor Assembly, the successor to the Cleveland Industrial Council, was active until the late 1880s. In 1887 a division in the assembly between KOL bodies and trade unions led the trade unionists to withdraw and form the Central Labor Union, affiliated with the AFL.

15. Samuel Langdale LEFFINGWELL was president of the Indianapolis Trades Assembly in the early 1880s.

16. The Indianapolis Trades Assembly (variously Trades and Labor Assembly of Indianapolis and Vicinity), founded in 1873, apparently declined during the depression of the 1870s. Reorganized in July 1880, it changed its name to the Indianapolis Central Labor Union in 1883.

17. Possibly Josephus R. Backus, a Terre Haute shoemaker, born in Ohio in 1833.

18. George CLARK served as president of the ITU from 1881 to 1883.

19. Patrick Francis FITZPATRICK was president of the IMUNA from 1879 to 1890.

20. The IRON Molders' Union of North America.

21. John Kinnear (1836-1915?), a Scottish-born printer who immigrated to the United States in 1840, represented the Boston Central Trades and Labor Union at the November trade union congress. In the late 1880s he became a doorkeeper at the Massachusetts state house.

22. In 1878 local unions founded the Boston Workingmen's Central Union. It later changed its name to the Boston Central Trades and Labor Union and participated in the forming of the FOTLU in 1881.

23. George RODGERS was president of the Trade and Labor Assembly of Chicago from 1881 to 1884 and was later president of the Illinois State Federation of Labor.

24. The Chicago Trade and Labor Council, formed in 1877, changed its name to the Trade and Labor Assembly of Chicago in 1880. The Chicago Federation of Labor superseded it in 1896.

25. The National Amalgamated Association of IRON and Steel Workers.

26. The GRANITE Cutters' National Union of the United States of America.

27. Mark W. Moore (d. 1914?), a Terre Haute printer in 1881, became a manager and vice-president of the Law Reporter Co. of Washington, D.C., in the 1890s. The company published the *American Federationist* after the AFL's headquarters was moved to Washington.

## Robert Layton[1] to Terence Powderly[2]

Pittsburg Oct 26th 1881.

Brother Powderly

I recd the Commissions and forwarded one to Pierce[3] and another to Mac.[4] Will send Shinn[5] one to day. Mac. told me in a letter in which I recd a note for $1,500.00 that you would receive said note.

*All* the Locals in Dist No *3*[6] will or promise to send Delegates.[7] Also the Amalgamated Association,[8] and if that don't capture it, we will order D.A. No 9[9] to forward 15 or 20 delegates, and if the thing is still uncaptured, we will call on Delegates from the Iron Moulders Union of North America, and if we can't get control of it then, I think we will send for the US. Government to send us a detatchment of armed men to assist. That Convention cannot take very many wild flights with such a stronghold of organization as Pittsburg to resist and clip its wings, if necessary. Do you know what it means, when I say D.A. 3. will be there? It means that nearly 70 Locals will each have a Delegate there to represent them if wanted. I don't know just what the exact No of Locals in this District is, but its *some Local* Won't Scranton make the effort necessary to equel this grand array of organized strength?

As ever Robt D. Layton
G S

ALS, Terence Vincent Powderly Papers, DCU.

1. Robert D. LAYTON was grand secretary of the KOL from 1881 to 1883.

2. Terence Vincent POWDERLY, a machinist, was the leader of the KOL as grand master workman (1879-83) and general master workman (1883-93).

3. George P. Pierce, a glass bottle blower and organizer for the KOL in Glassboro, N.J., represented District Assembly 2 at the KOL General Assemblies in September 1879 and in 1882.

4. In 1881 and 1882 Myles MCPADDEN, an iron molder, represented District Assembly 3 at the KOL General Assemblies and served on the Knights' General Executive Board.

5. Dexter L. Shinn (1837-1918?), a bricklayer, was an organizer for the KOL in Clarksburg, W.Va. Later he was a bookkeeper and a Mormon minister.

6. District Assembly 3 of the KOL was founded as a mixed assembly in Pittsburgh in 1875 and was reorganized in 1876.

7. That is, to the trade union congress at Pittsburgh.

8. The National Amalgamated Association of Iron and Steel Workers.

9. District Assembly 9 of the KOL, founded in 1877 in West Elizabeth, Pa., was almost exclusively a miners' assembly.

2 Fort Street, the address of Samuel Gompers' early childhood home in London, as sketched in 1918 (George Meany Memorial Archives, AFL-CIO).

Sarah Rood Gompers, circa 1889 (George Meany Memorial Archives, AFL-CIO).

Solomon Gompers, circa 1888 (George Meany Memorial Archives, AFL-CIO).

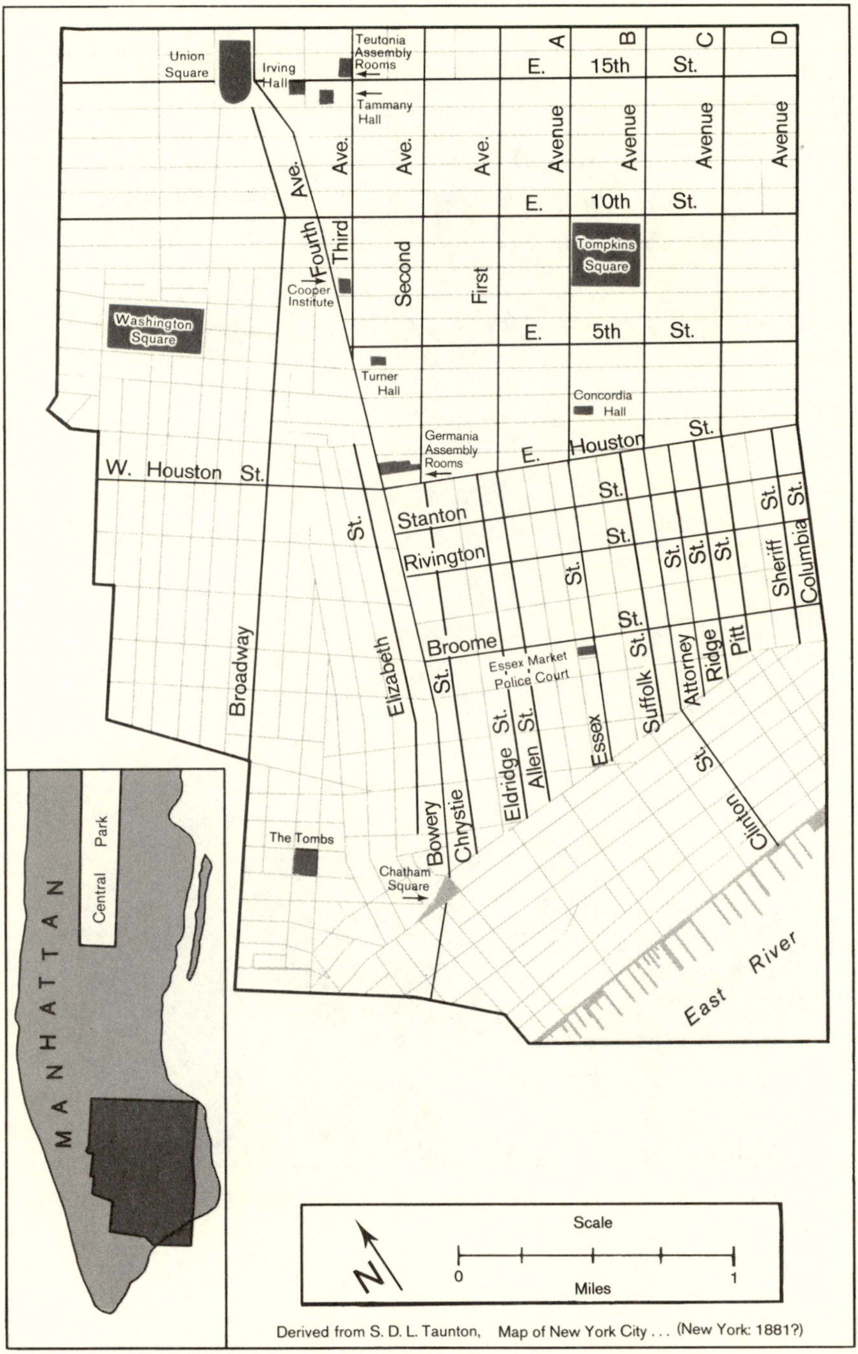


Locations in lower Manhattan pertinent to Samuel Gompers' early career.

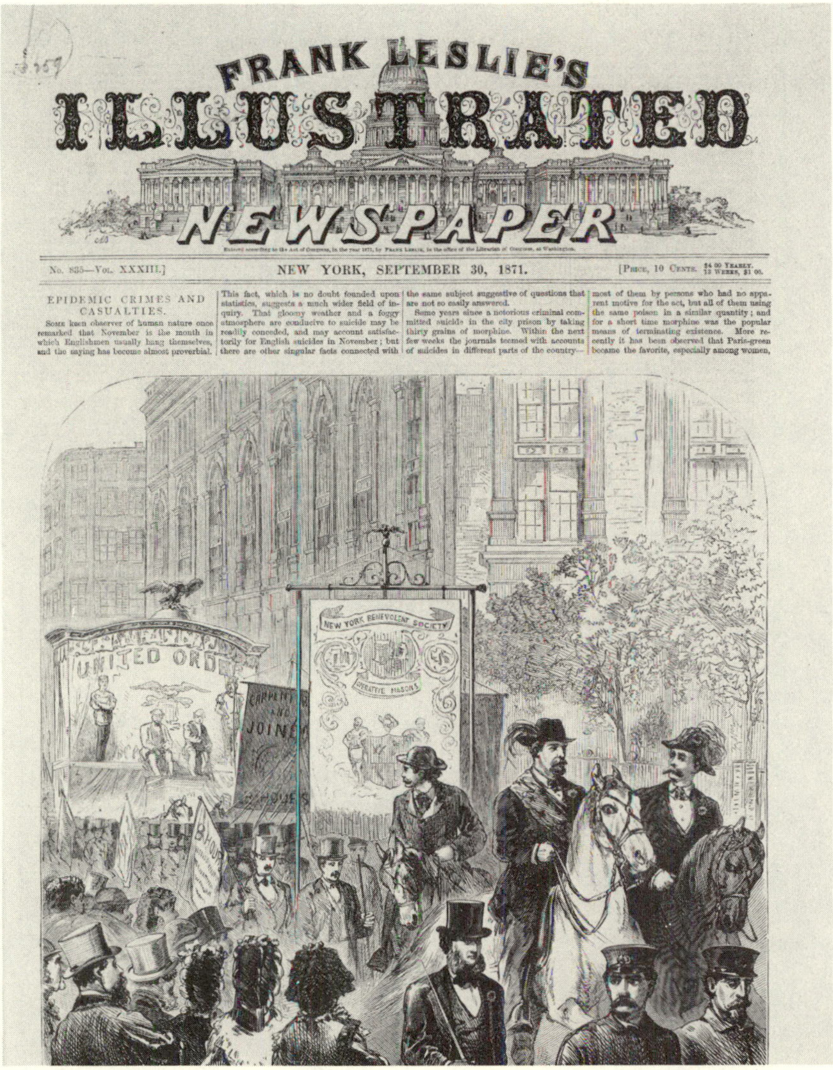

FRANK LESLIE'S ILLUSTRATED NEWSPAPER

No. 835—Vol. XXXIII.] NEW YORK, SEPTEMBER 30, 1871. [Price, 10 Cents. $4 00 Yearly. 13 Weeks, $1 00.

EPIDEMIC CRIMES AND CASUALTIES.

Some keen observer of human nature once remarked that November is the month in which Englishmen usually hang themselves, and the saying has become almost proverbial. This fact, which is no doubt founded upon statistics, suggests a much wider field of inquiry. That gloomy weather and a foggy atmosphere are conducive to suicide may be readily conceded, and may account satisfactorily for English suicides in November; but there are other singular facts connected with the same subject suggestive of questions that are not so easily answered.

Some years since a notorious criminal committed suicide in the city prison by taking thirty grains of morphine. Within the next few weeks the journals teemed with accounts of suicides in different parts of the country—most of them by persons who had no apparent motive for the act, but all of them using the same poison in a similar quantity; and for a short time morphine was the popular means of terminating existence. More recently it has been observed that Paris-green became the favorite, especially among women,



Eight-hour demonstration, New York, 1871 (*Frank Leslie's Illustrated Newspaper*).

Cigarmaking in a New York tenement, 1877 (*New York Daily Graphic*).

Cigarmaking in a New York tenement, 1879 (*New York Daily Graphic*).

Cigarmaking in a New York factory, 1877 (*Frank Leslie's Illustrated Newspaper*).

Cigarmaking in a shop, 1873 (*Practical Magazine*).

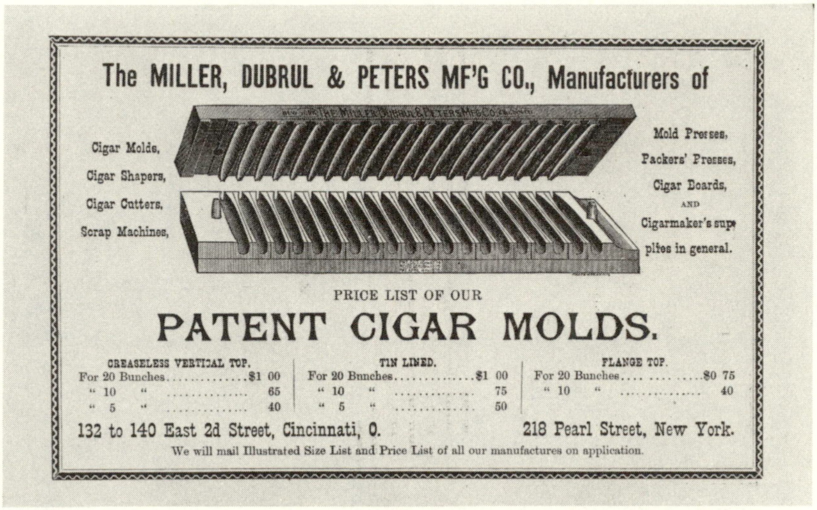


A cigar mold (*Cigar and Tobacco Manufacturers' Directory, 1882-83*).

L. Hirschhorn and Co., 1878 (*United States Tobacco Journal*).

Lichtenstein Brothers and Co., 1878 (*United States Tobacco Journal*).

Adolph Strasser (Photoworld/FPG).

THE
Cigar Makers
INTERNATIONAL UNION,
OF AMERICA,
DOTH GRANT
THIS
CHARTER

TO Saml. Gompers President

Levie Bossie Vice President

Henry Baer Rec Secretary

David Straus Cor Secretary

Fin. Secretary

Treasurer.

Door Keeper.

And to their successors legally elected, to constitute a Union, to be known as Union No. 144 of the City of New York N. Y. for the purpose of effecting a thorough organization of the trade. And the said Union being duly formed is hereby authorized and empowered to initiate into the Union, any person or persons, duly proposed and approved ACCORDING TO THE CONSTITUTION adopted by the INTERNATIONAL UNION, and to enact By-Laws for the government of their Union.

Provided Always, That the said Union [illegible] to the Constitution, Laws, Rules and Regulations of the INTERNATIONAL UNION; and provided also that said Union be held in New York and not removed therefrom without the consent of the INTERNATIONAL UNION, and in default thereof, or any part, [illegible] may be suspended or taken away [illegible] of the INTER-[illegible]

[illegible] themselves to [illegible] Union No. 144 in the exercise of their rights and privileges as a Subordinate Union

In Witness Whereof, we have subscribed our Names and affixed the Seal of the International Union, this 24th day of November A.D. One Thousand Eight Hundred and Seventy-five

George Hurst President, C. M. I. U.

W. Miller [illegible] Vice President, C. M. I. U.

L. P. Rehmer [illegible] Vice President, C. M. I. U.

Charter of CMIU 144 (George Meany Memorial Archives, AFL-CIO).

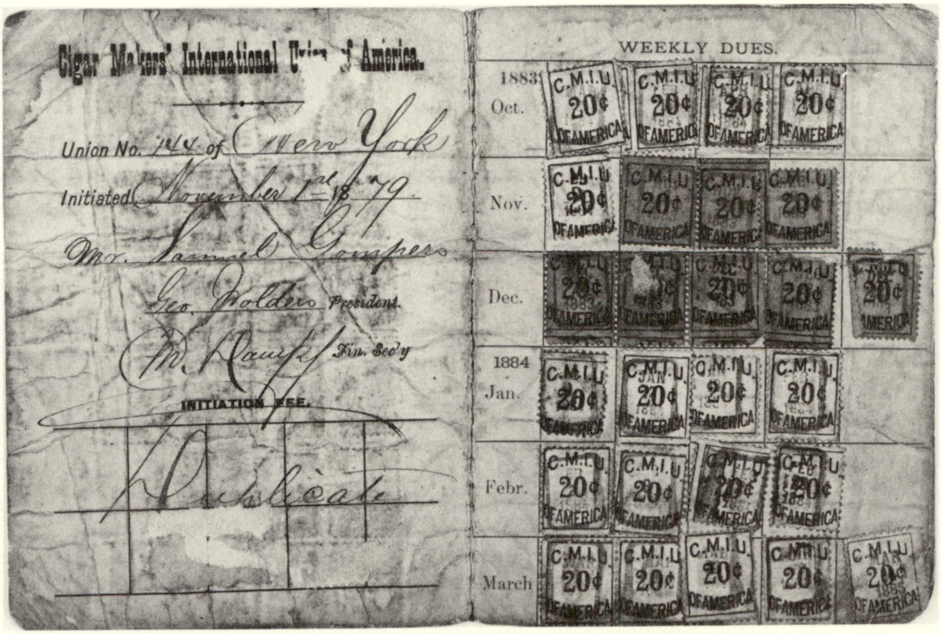


Samuel Gompers' dues book, 1883-84 (Samuel Gompers Collections, Rare Books and Manuscript Division, New York Public Library, Astor, Lenox and Tilden Foundations).

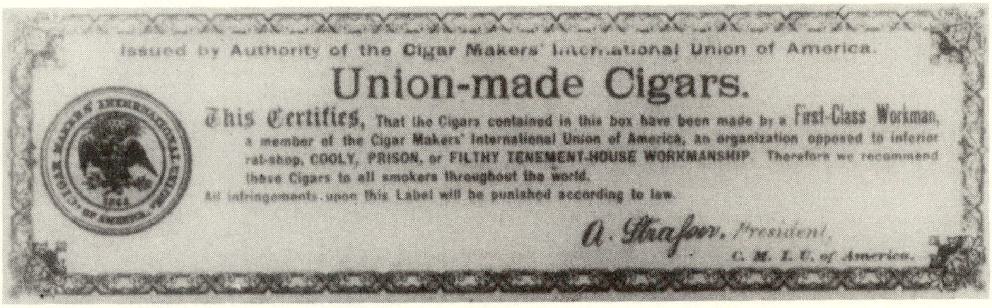

Union-made Cigars.
This Certifies, That the Cigars contained in this box have been made by a First-Class Workman, a member of the Cigar Makers' International Union of America, an organization opposed to inferior rat-shop, COOLY, PRISON, or FILTHY TENEMENT-HOUSE WORKMANSHIP. Therefore we recommend these Cigars to all smokers throughout the world.
All infringements upon this Label will be punished according to law.
President,
C. M. I. U. of America.


The CMIU label, 1881 (*Cigar Makers' Official Journal*).

P. J. McGuire (George Meany Memorial Archives, AFL-CIO).

J. P. McDonnell, 1878 (State Historical Society of Wisconsin).

FOTLU Legislative Committee, 1881. Front row, left to right: Charles F. Burgman, Richard Powers, Alexander C. Rankin. Back row: Samuel Gompers, William H. Foster (George Meany Memorial Archives, AFL-CIO).

Terence V. Powderly, 1886 (*Frank Leslie's Illustrated Newspaper*).

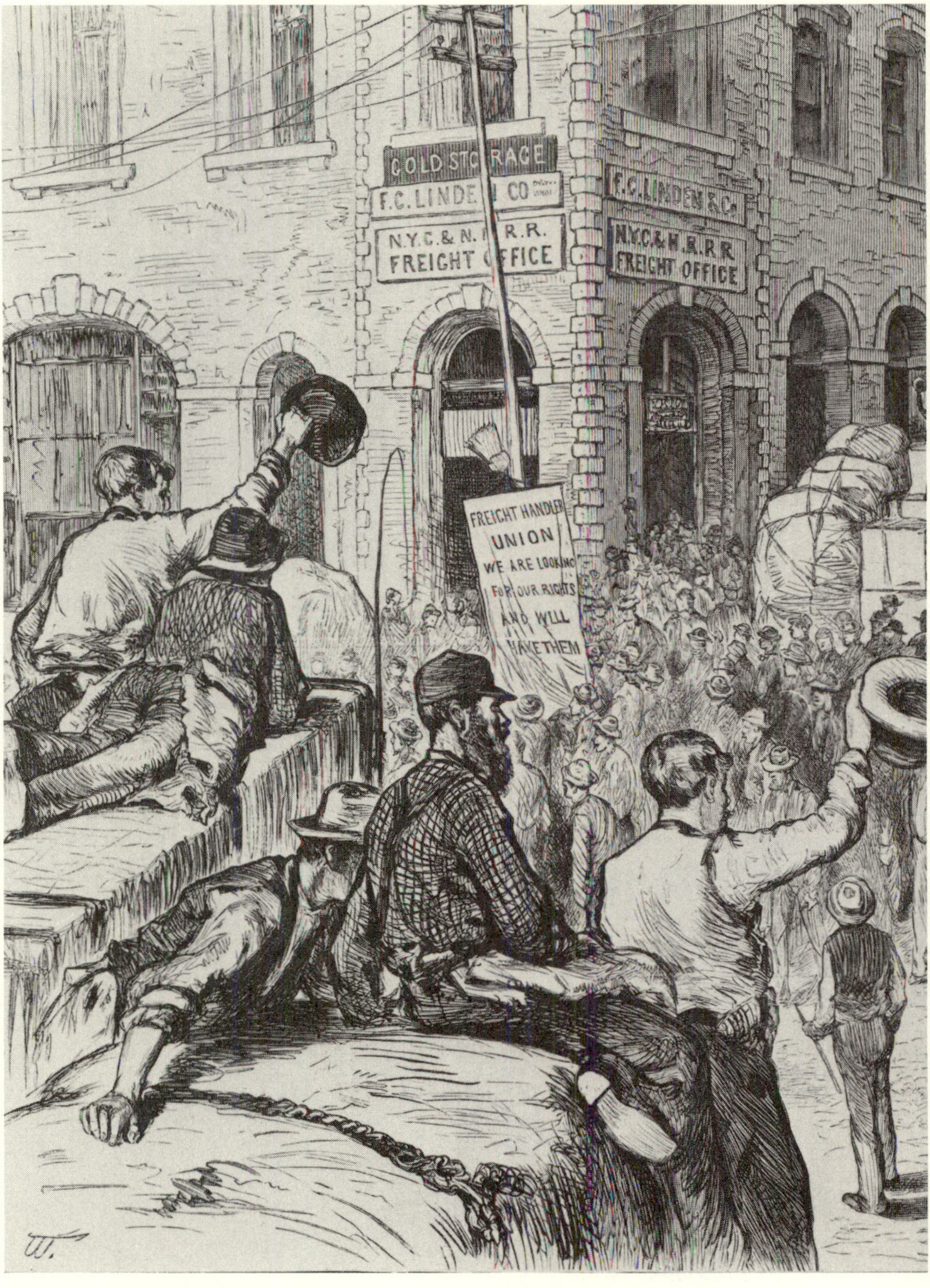


The New York freight handlers' parade, 1882 (*Harper's Weekly*).

STORAGE

Samuel Gompers, 1886 (*Harper's Weekly*).

## *The CMIU Campaign against Tenement-House Cigar Manufacture in New York City*

Although the CMIU worked throughout the 1870s to abolish tenement-house cigar production, local 144 stepped up the campaign after the New York City cigarmakers' strike of 1877-78. The local sponsored mass meetings to urge the New York City Health Commissioner to shut down tenement workshops and tried to induce the U.S. Bureau of Internal Revenue to deny licenses to tenement-house manufacturers. These efforts failed, however.

Late in 1879 Strasser and Gompers began a lobbying campaign in Albany, and, subsequently, a bill prohibiting tenement cigar manufacture did reach the floor of the assembly. It was defeated in April 1880, however, as was a similar measure in the senate early the next year. During the fall of 1881 the CMIU campaigned to elect legislators sympathetic to tenement-house legislation and to defeat those opposed to it.

In an attempt to awaken the public to the unsavory working conditions of tenement cigarmakers, Gompers made a survey of tenement workshops and published his findings, in German, in the *New Yorker Volkszeitung* in late 1881. Much of the report, in an English version, appeared in the *Cigar Makers' Official Journal* between February and May 1882. His first attempt at serious journalism, these articles reveal Gompers' indignation and dogged determination as well as his sense of irony and powers of expression.

In 1882 the anti-tenement campaign met with greater but not unqualified success. Although the senate passed a tenement-house bill in the spring, the document disappeared mysteriously before final consideration in the assembly. According to the *New York Sun,* this measure and a bill prohibiting hat manufacture in prisons were stolen from the assembly clerk's desk just prior to the final reading, possibly at the suggestion of Tammany Hall leader John Kelly. After a year of further CMIU lobbying and a state-wide campaign to defeat legislators opposed to regulating tenement-house manufacture, the New York legislature on March 12, 1883, passed a bill prohibiting tenement-house cigar manufacturing in New York City.

The cigarmakers' victory was short-lived, however. Manufacturers

circumvented the law by incorporating their tenement-house workers into their factory work forces and, in some cases, by moving their workshops to Long Island or Pennsylvania. In addition, they successfully challenged the law in court. The legislature passed a second anti-tenement law in May 1884, but the courts again ruled against it, deciding that the law abrogated private property rights.

This second failure convinced Gompers, he later wrote, that legislation would not bring about better working conditions because the "average judge" could not understand that "industrial justice" was not an "abstract matter," but in fact "must be shaped to meet working relations and the needs of workers for a better life." The CMIU abandoned its legislative campaign and turned instead to what Gompers called "organization work," that is, an attempt to harass manufacturers into abolishing the tenement-house system.[1] The industry was slow to abandon tenement-house manufacture, however, and did so only after technological innovation—especially the introduction of the suction table—and new managerial strategies rendered it obsolete.

*Note*

1. SG, *Seventy Years,* 1:197.

# The Committee on Tenement-House Agitation of CMIU 144 to the Executive Boards of Local Unions of New York State

New York, October 28, 1881.

To the Executive Boards of Local Unions of New York.
Gentlemen:

We the undersigned Committee, have been instructed to take steps with a view to abolish the infamous system of manufacturing cigars in tenement-houses. The prices paid for the cigars made in these pest-houses are from $2 to $6 less per thousand than those made in factories. The tenement-house system would have been abolished long ago, were it not for the European immigration which arrives daily, and supplies these manufacturers with cheap labor. There is a constant endless stream, which, unless checked, will inevitably destroy us. We appeal to you to help us in the struggle, as by helping us you will help yourselves.

The Committee, elected by Union No. 144, New York, have decided to pledge every candidate for the Assembly and Senate upon this issue. It is a cigar makers' issue, pure and simple, in behalf of which we intend to use our political influence. We suggest to you to call at once a special meeting of the Executive Board, then resolve to visit all candidates without distinction of party, or at least those who are likely to be elected, and pledge them to vote for the abolishment of tenement-house cigar factories during the next session of the Legislature. We also request you to secure the co-operation of other trade and labor unions, thus creating a greater influence and pressure upon our so-called representatives.

Hoping that you will act at once, and notify us of the result, we remain

Yours fraternally, Samuel Gompers, Secretary,
Frederick Reibetanz,
C. G. Bloete,[1]
Committee.

*Cigar Makers' Official Journal,* May 15, 1882.

1. Charles George BLOETE held various offices in CMIU 144 between 1877 and 1881 and represented the local in the Amalgamated Trades and Labor Union of New York and Vicinity.

# A Translation of a Series of Articles by Samuel Gompers on Tenement-House Cigar Manufacture in New York City

[October 31, 1881]

## TENEMENT-HOUSE CIGAR MANUFACTURE.

For many years the system of tenement-house cigar manufacture has formed one of the most dreadful, cancerous sores in our city: In every way, whether in regard to the wage conditions of the workers—and not only that of the tenement-house workers—or their existence as human beings or family members, or the influence it has upon the immediate surroundings of the tenement factories and indirectly upon all the working population of the city, every year this system proves itself more of a veritable plague spot in the already quite corrupt economic and social life of New York. The truth of this assertion has been recognized often enough in so-called "decisive" places, and most recently Mayor Grace[1] expressed his intention of attacking this pernicious institution.

Unfortunately, the impressions produced in people by general assertions of grievances, no matter how well-grounded they may be, are usually not very deep or lasting. Only when precise and authentic details allow the public to gain insight into the actual character of the evils we reproach are people set into motion and one can count on finding the appropriate support for an agitation to abolish the grievances. One of the most striking examples of all times of this tendency is the fall of the Tweed Ring.[2] Long before its collapse there was scarcely anyone in the city of New York who was not convinced that Tweed and his companions were daily and hourly plundering the city in the most shameful way. Despite all this, no movement to oust the scoundrels could be launched successfully. Only when the "Times" produced that famous exposé[3] which gave very precise details about the specific fraudulent transactions of city officials did a storm of indignation arise in the public which then swept the whole gang of political crooks out of public life.

Proceeding from this standpoint we have made it our business, through an exact examination of the facts and through publication of the results obtained—we begin the publication elsewhere in today's paper—to provide the necessary factual foundation for those oft-repeated general assertions about the dreadful conditions produced by the tenement-factory system. This is no longer a matter of phrases

but of facts and figures that cannot be argued away and that are well suited to horrify the reader who has had no previous idea of the depth of this abyss.

We hope the results of these publications will provide a lever that will at least contribute to preparing as quick an end as possible to that institution which is a burning humiliation to the so highly praised culture of our day and of our country.

*New Yorker Volkszeitung,* Oct. 31, 1881. Translated from the German by Hannelore Jarausch.

1. William Russell Grace (1832-1904), an Irish businessman who immigrated to New York City in 1865, was mayor of New York City for two terms (1881-83, 1885-87). He later endowed Grace Institute, an educational institution for young women.

2. William Marcy Tweed (1823-78), born in New York City, was a U.S. congressman in the mid-1850s and the leader of the New York City Tammany Hall Democratic organization in the 1860s. He spent most of the 1870s in prison on graft convictions.

3. The *New York Times* exposé began Sept. 20, 1870, and continued through most of 1871.

[October 31, 1881]

## THE CURSE OF TENEMENT-HOUSE CIGAR MANUFACTURE.

In presenting to the readers of the "N.Y. Volkszeitung" the results of a careful examination of tenement-house cigar manufacture, of its system, the circumstances under which it takes place, and the dreadful consequences that inevitably result from it, we ask that attention be paid to the fact that the information depends in part on the degree of willingness of those most immediately involved, that is, the workers themselves, to inform us about their conditions. Since these workers are in constant fear not only of losing their jobs but also of being evicted from their apartments, it is natural for them to regard with suspicion every stranger who attempts to gain precise information about their situation. Although it is fairly safe to assume that everything which the reporter learned about the conditions of the workers, not through his own observations but through what the workers told him themselves, still makes the situation appear far more favorable than it really is, there were nevertheless no attempts made in the following reports to express speculations or assumptions that would alter the actual facts. On the contrary, the following remarks, without any coloration or exaggeration, are a true mirror of that which our thoroughly objective reporter saw and heard. Since we will refrain from making the required commentary on the results of the investigation, without further introduction we begin with the presentation of our reporter which in its almost photographic way speaks an eloquent and terrible language.

HERMANN BLASKOPF'S TENEMENT-HOUSE FACTORY,

No. 90 Cannon St., was the first of these buildings visited by this writer. It is a five-story double tenement house. Already from a distance it gave the impression of not having been repaired for a generation; we did not wish to rely on external impressions, however, but wanted to see with our own eyes what it looks like on the inside, see the rooms in which people live and work, are born and die. Fifteen families live in the house, an average of four on each floor. Each family has a room and a bedroom; the size of the room is 11 by 13 feet, the bedroom 5½ by 7½ feet. In the wall adjoining the dark corridor is a hole measuring 18 inches square, a so-called window; the ceiling height is 7½ feet. 52 people live in this house; moreover the whole bottom floor and half of one of the upper stories are used as an office and a place for packing cigars and as storage space for tobacco and tobacco stems. In two families, four people work in the apartment rooms described above; in three families, three. The remaining families have two workers each but the strippers are not included in this figure since this task is usually carried out by an old person who is useless for other occupations or by a child. Tobacco in every stage of preparation is found in all the rooms; mostly it lies spread out over the floor to dry. In the bedroom we find casks, chests, and rusty milk cans that contain tobacco and tobacco stalks, called "stems" by the workers. Working hours are from 6 or 6:30 in the morning until 10, 11, or even 12 o'clock at night. Wages vary from $4.25 to $6.00 per thousand and a family in which two people work can produce 2,800 cigars a week on the average, but the families with more working members do not produce proportionately more, but significantly less for their number. Rent is from $7.00 to $9.00 per month. What one finds as furniture in these apartments usually consists first and foremost of a worktable, kitchen table, and cook stove, two or three wooden chairs, a bedstead, and a few cheap pictures of saints. We go through the rooms, the hallway, down the stairs, and into the yard, and everywhere we come across tobacco, tobacco scraps, tobacco stems, and other filth. Even in the yard where the children who are still too young to be able to work—and they have to be *very* young not to—are playing, great piles of drying tobacco are lying about. One structure in the yard arouses a curious impression; it looks like a small model of a dilapidated palace or castle, but when we come nearer our sense of smell quickly tells us what the purpose of this "little palace" is. We go to the door but the stench drives us back; this breeding-ground of disease has no drain to the sewer and consists only of a pit which is emptied out when it is filled to the brim. Great

piles of tobacco stems, some 60 to 70 pounds, lie, rotting and moldy, in the entry way next to the stairs. We could not ascertain how often these piles are removed to make space for new ones, but the atrocious smell emitted by these deposits indicates that it does not happen often. And this odor of tobacco hovers over everything, the infant's cradle, the marriage bed, and the food set before the children. The cellars are dank, damp, and filthy, and the store on the ground floor, which serves as office and packing area, is encrusted with filth. The condition of the upper floors is made even worse by the fact that no water rises to them—and this is the case not only during a period of drought—which, of course, in no way encourages cleanliness. We asked each family that did not regard us with too much suspicion what they thought about efforts to eliminate the tenement factory system and the answer was always the same: "We wish it could be abolished and the sooner the better."

Hermann Blaskopf has a small house with three families at No. 80 Cannon St. In one family four people were working and two children were lying in the tobacco, all of them in a room of 7 by 9 feet with a 7½ foot ceiling.

### George Bence's Tenement Factories[1]

are the next ones that the writer visited; the above-named has several establishments of this type, six houses on 2nd St. and one on Avenue B.

No. 222 East 2nd St. is a five-story tenement house that is occupied by 19 families, consisting of 83 people, not counting the two families in the store on the ground floor. Each family has a room with a kitchen and bedroom. The dimensions are as follows: room 9 by 12 feet, kitchen 8 by 9 feet, bedroom 6 by 6½ feet, ceiling height 8 feet. A family of eight persons, almost all of them adults, lives in one of these apartments; their parents were not cigarmakers but originally farmers and thus had such a large family. Rent varies between $9 and $10.50. Four people work in some of the rooms, three in others, but in most of them only two; children who usually strip tobacco are not included in these figures. The question of working hours was answered most reluctantly and the answer that they work from 6 in the morning until 9 or 10 at night came only after we assured them that we didn't mean them specifically. Wages vary between $5.25 and $6.25 per thousand cigars; a family consisting of two workers produces an average of 2,500 cigars per week. 12 people work on the second floor; two children have died of smallpox within the last two months in this house.

Nos. 228-230 2nd St. each house the same number of families, with the total of 156 people, not including 9 strippers, 15 cigarmakers who work in the house but do not live there, and three families in the store on the ground floor.[2] Several months ago two children in No. 230 came down with smallpox, one of whom died. Only 6 families who make cigars at home reside at No. 234, and they refused to divulge the number of their members; the other 13 families work as cigarmakers in regular shops or at other trades. The condition of the houses is most miserable; wherever one steps one trips over tobacco in every possible form. In the rooms the odor of tobacco is stupefying; tobacco usually lies spread out around the stove to dry; piles of it lie in the bedrooms or packed in crates, cans, or barrels. Tobacco lies on the steps, in the hallways, even in the yard. In the yard it is amusing to find a large sign attached to a shed on which shiny letters proclaim a refreshing inscription: "Lager Beer." We step closer, thinking that the yard is a strange place for the sale of beer but find to our disgust that the shed is nothing other than an exceedingly filthy water closet. On the side of it there is a gutter but its drain is clogged and the stench coming from its contents of reeking urine mixes with the equally foul smell emanating from numerous piles of filth consisting of rubbish, garbage, musty tobacco scraps, and moldy tobacco strips. Nauseated, we stride through all this, hoping that the next house will offer us something more cheerful, drive away our depression, and restore our lost appetite.

We therefore went to the house owned by Mr. Bence across the street, No. 200 E. 2nd Street, occupied by twelve families, three of whom do not make cigars. 56 people live there; two families, consisting of eight people, live downstairs in the stores. Each family has a room and a bedroom with the following dimensions: room 10 by 13 feet, bedroom 6 by 7½ feet, ceiling height 8 feet. In the bedroom there is a window opening on the hallway which is filled with an Egyptian darkness, and this is the only ventilation for a room in which people must sleep. The rent for these "residences" is between $6.50 and $7.50. Our hope of finding something pleasant here was not fulfilled; as always, in every room, tobacco, filth, and human beings, thrown together, among it all children, pathetic little creatures who should romp a few years in the fresh air before being sent to school and who here are already tortured by the yoke of labor, stripping, making wrappers, or completing cigars. The halls and stairways are as dark and gloomy as if it were night and not broad daylight; we can smell and touch tobacco all around us but if we want to see it we must light a match. In the yard we found an old table around which five young lads between 14 and 18 were sitting, on wobbly chairs and a milk

can, playing cards, smoking bad cigars, and drinking even worse beer, sold to them by the landlord of the house. Should they have a few pennies to spare they can satisfy their gambling urges in the Policy Shop that is in full swing in the same house. The exterior of the house is an incredible picture in itself and would remind us of the crumbling old buildings found in novels except for the filth. The rear wall is also worth a glance; it was once white but has long since abandoned any such pretensions. In front of each window there is a human trap, given the name of fire escape, piled high with crates and baskets that are, in their turn, filled with rubbish and tobacco stems; in the yard puddles of stagnant water strike the eye and nose.

(To be continued.)

*New Yorker Volkszeitung,* Oct. 31, 1881. Translated from the German by Hannelore Jarausch.

1. George Bence (variously Bentz) produced exclusively in tenements. In 1881 he employed about 400 workers in several locations in Manhattan and the Bronx. By the late 1880s the firm was no longer operating.

2. Almost all tenement houses had stores on the ground floor. Some served as offices for the manufacturers, others for sorting and packing. Still others were rented out separately to saloonkeepers, grocers, bakers, and the like. The families of the storekeepers also lived in the tenements.

[November 1, 1881]

## BARRACKS OF MISERY.

We continue our rounds through the tenement houses used for cigar manufacture that are such an unedifying spectacle and proceed first to some additional houses belonging to

### George Bence

some of whose houses we described yesterday. Our route takes us to No. 223 East 2nd St., a house with 16 families, of which 3 do not make cigars. The two families living in the shops are not included in this figure. The number of people is 60. The families who live on the street side each have a room, bedroom, and kitchen; those in the rear apartments only a room with a kitchen. The dimensions are: room 10 by 14 feet, kitchen 10 by 8 feet, bedroom 6 by 7 feet, ceiling height 8 feet. Four people of one family work in some of these back rooms; wages, so we were told, vary from $5.25 to $6 per thousand. If there are people in the family who are too young or too old to make cigars, they strip tobacco and cut cigars; if a family lacks such members they hire a stripper and pay him 53 to 60 cents per thousand cigars for which tobacco is stripped. Working hours are usually from

6 in the morning until 11 at night. Tobacco and filth are as in other houses, and as far as light is concerned, we can only say that there is as good as none; darkness, impenetrable gloom everywhere, and it is really a wonder that not more people break their neck going up and down the stairs. We had to use a match again and saw the tobacco dust, rubbish, and stems around us on all sides. We went into the yard and, as small as it is, it was considered superfluous and so filled up with old timbers that one cannot take two steps straight ahead. The condition of the house is not improved by the fact that a junk dealer occupies one of the stores downstairs and a beer hall is located in the other. Visitors to the latter do their best not to let the reeking puddles in the yard dry up.

No. 26 Avenue B houses 14 families, including 58 people; not counted in that number are four people who work in the house but do not live there and two families who live in the shop on the first floor. Each of the families living on the street side has a room, kitchen, and bedroom; those who live in the back only a room and bedroom. The dimensions are: room 13 by 10 feet, kitchen 10 by 9 feet, ceiling height 8 feet. In the bedroom there is a 15-inch-square window that looks out onto the dark hallway. In the apartments in the back below the third floor there is no natural light at all since it is totally blocked out by the wall of another house rising directly in front of the windows; people are forced to use kerosene lamps all day long in order to be able to work. There have been four cases of smallpox in this house, three of which were fatal; two were in the same family. This house has no yard because of the thick wall that rises close to the windows. When we looked through the cellar grating we saw large piles of filth in the gloom and we found a puddle of stagnant water along the board fence that separates the narrow path from the neighboring house. Working hours are from 6 in the morning until 10 in the evening. Wages are different from the other houses and vary from $5 to $5.75 per thousand; a family of two workers produces 2,800 cigars per week on the average. Payday comes only once a month which makes the residents of this house dependent on the store owners who give them credit from one payday to the next.

The next visit on our rounds was to the tenement factory of

### ROSENTHAL BROTHERS & CO.[1]

They own three houses on 15th Street and four on 16th St. No. 623 East 15th St. shelters 20 families numbering 98 people, plus ten people who work there but live elsewhere. Each family has a room of 8 by 10 feet, kitchen of 8 by 6 feet, bedroom of 7 by 5½ feet,

ceiling height 7¾ feet. There is only one window in each room in this house, a window 2 feet high and 9 inches wide in the kitchen and an even smaller one in the bedroom, which understandably lets in almost no light or air at all. There is no hearth in the whole house, no mantelpiece, nothing but a round hole in the chimney through which a stove pipe could be put. In several families three or four people work in one room, surrounded on all sides by huge quantities of tobacco, with the fire blazing with all its might to dry the tobacco sufficiently; obviously under these conditions the air in the rooms is thick and steamy. In one of the rooms the father, mother, and small girl had an eye infection; the mother and a small boy also had sores on their lips.[2] Last week a child died in this house. Rent varies from $7.50 to $8.50; wages are $3.75 per thousand; working hours go from 5 in the morning until 10 or 11 at night. We were told: "We begin around 5 o'clock in the morning and work as long as we can." When calculating the working hours, one must not forget that Sunday is not a day of rest in these tenement factories; most families, despite their religious scruples, work until 2 or 3 in the afternoon on Sunday, some the whole day through as on any other day. A family with two workers produces an average of 2,800 cigars per week, provided a third person does the necessary stripping.

No. 621 East 15th St. has 18 families, totaling 92 people; 10 others work in the house but do not live there. Seven people were working in an 8-by-10-foot room; two small children were lying in the tobacco. A cookstove was in the room, spreading unbearable heat, next to it a small kitchen table where the family takes its meals. The bedrooms have no windows and neither air nor light can reach them. The room in which this large family works has only one window; the haze and stench are unbearable, the quantity of tobacco enormous, rubbish piled up everywhere. In every way this house resembles the first one we described; a short while ago two children died there of diarrheal illnesses.

No. 619 houses 19 families, all together 90 people; four of these families do not make cigars. Eight people work in a room of 8 by 10 feet; next to them are huge piles of tobacco, stems, and refuse. Several weeks ago two children died of bronchitis in this house; at the time of our visit five other children were ill with bronchitis and whooping cough. In one family, father and son had a lung disease; these people do not make cigars but attribute their illness to the tobacco fumes and filth in the house. All the families in the house suffer because there is no running water; the water closets are in frightful condition. In this row of houses owned by Rosenthal Brothers & Co., rent, wages,

working hours, size of rooms, etc., resemble those in the buildings described above.

When we went around the corner we came upon the houses owned by the same company on 16th Street; No. 634 is the first with which we wish to acquaint our readers. 16 families, consisting of 73 people, live in it; 20 others who do not live in the house work there as well. The families on the street side have a room, kitchen, and bedroom; those who live in the back, only a room and bedroom. The dimensions are: room 10 by 13 feet, kitchen 10 by 9 feet, bedroom 6 by 6½ feet, ceiling height 8 feet. Here as usual we found the only bedroom window 15 inches square looking out on the dark corridor. Six people work in one family and a seventh is hired by it; since last summer four children have died of measles in this house. Great heaps of tobacco lie about the rooms; one is constantly stumbling over tobacco rubbish and stems in the halls and on the steps. Rent varies from $7 to $9. The water closet is a drainless cesspool full of filth, as are the seats and floors in it.

No. 636 East 16th St. houses 16 families with 75 people; six people work in the house but do not live there. The ground floor is used as an office; the apartments are the same as at No. 634. Five people work in some families, four in others, two or three in the remainder. Three children have died of diarrheal ailments in this house since last summer. The condition of this house is the same as the earlier one, but it is even dirtier and more dilapidated. The walls, partitions, and stairways are defective and unsafe; the staircase steps are covered with dirt and tobacco refuse. Eight- or nine-year-old children work in several rooms making wrappers; despite its youth, the oldest of these pitiful creatures looks as if it will soon say farewell forever to all work. The water supply is deficient here as well, and the water closets are very filthy.

No. 638 East 16th St. is occupied by 16 families with 67 people. One family consisting of eight people lives in three rooms, the size of which is the same as that given above; the apartments are no different from those in the neighboring houses. Some families work with four people, most with three, several with two here as well. In this house several people were ill with diseases of the eyes and glands, and rashes on the cheeks and lips, and had wrapped cloths around their faces and heads. The filth and tobacco haze were dreadful.

No. 640 East 16th St. has 17 families with 70 people under its roof, plus 17 people who do not reside in the house but work there. In one of the small apartments, in no way different from those of the neighboring houses, lives a family of nine people. The filth, stench of tobacco, etc., are nauseating; two children recently died of measles

in this house and we must say that the houses of Rosenthal Brothers & Co. do the greatest honor to the "business sense" of the company.

(To be continued.)

*New Yorker Volkszeitung,* Nov. 1, 1881. Translated from the German by Hannelore Jarausch.

1. Rosenthal Brothers was a relatively new firm that appeared after the strike of 1877-78. It expanded very rapidly in the early 1880s and was among the twenty largest firms in the city during the late 1880s.

2. Anti-tenement reformers consistently emphasized the danger of having cigarmakers with respiratory and mouth diseases make cigars, since part of the cigarmaking routine involved the worker biting off the overlapping end of the wrapper leaf after finishing the cigar.

[November 2, 1881]

## WORK DENS AS LIVING QUARTERS.

What we have seen, heard, and smelled so far on our rounds through the atrociously unnatural cigar factories does indeed not encourage us to continue the investigative tour we have begun, and this writer would certainly prefer reporting more pleasant and appetizing things to the reader. But he who has once made it his duty to drag out of its dark hiding place into the light of day the total horror of the system which poisons men, demeans women, and murders children, who has undertaken to show his colleagues, the workers struggling for their daily bread, what a devouring, poisonous cancer the pursuit of the almighty dollar, through exploitation, oppression, and sacrifice of our fellow men, has created in our midst, he must be willing to get his hands dirty in this duty and can say to anyone who turns away, disgusted by the unfolding picture: It is up to us to change it! Therefore we continue on our chosen path. Much remains to be revealed, much to be described, unfortunately nothing comforting, nothing inviting. Having finished with the "model establishments" of Rosenthal & Co., today's path leads us to the tenement-house factories of

### LICHTENSTEIN BROS. & CO.[1]

The firm has seven tenement factories in a row, Nos. 305-317 East 44th Street. No. 313, the first one we entered, houses 16 families with a total of 55 people. Each family has a room, kitchen, and bedroom; the room measures 11 by 14, kitchen 10 by 8, and bedroom 6½ by 8 feet, ceiling height of 8½ feet. A 15-inch-square window looks out on the dark hall. In these rooms we also found the reeking piles of tobacco as well as crates, casks, milk cans, and vats filled with tobacco, clippings, and stems, along with presses, molds, and work-

tables so that the "shop" has completely taken over the living quarters. In the staircases and corridors there is the same dust and dirt, thick as a finger, in which tobacco refuse and stems lie scattered. When we visited one of the rooms at quarter past 8 in the evening, the father was sitting at the table and working, dressed in an undershirt, drawers, and apron; two small children were lying on an old lounge not far from the worktable, waiting for sleep to close their weary eyes and perhaps bring them dreams of green meadows and gardens where there is fresh air and no tobacco. But it is probably impossible to dream of anything but tobacco in this atmosphere. Next to the sofa there stood a bed, or at least something that might be called such, which was waiting for a 12-year-old boy who was also still working. He was to go to bed, but when?—when "working hours" were over and he was allowed to lie down, to dream, like his sisters—of tobacco! Working hours here are from 6 or 6:30 in the morning until 10 or 11 in the evening. A family with two workers produces an average of 2,800 cigars per week; wages go from $5 to $5.75.

There are only a few families in No. 317 since the rocks that lie along the eastern wall of the building make several stories uninhabitable as water literally pours into the rooms.

No. 315 has 14 families with 49 people; some of the apartments in this house are vacant.

No. 311 has 16 families with 59 people.

No. 309 has 14 families with 42 people.

No. 307 has 16 families with 60 people.

No. 305 has 15 families with 46 people.

Conditions in these houses are almost everywhere the same; in every room there lie the inevitable piles of tobacco, and the dirt cannot be driven out despite the efforts of many a housewife to keep her apartment somewhat clean. Tobacco and the lengthy, tiring working hours make cleanliness impossible, and eventually "people get used to it." Huge piles of tobacco stems in the yard, which are kept in a decaying process of fermentation by the rain and water that trickles through the rocks, make the surrounding apartments very hazardous to health; while looking at them we instinctively thought of the outcries of press and public last spring when sanitation workers were attacked because snow and ashes had accumulated on the streets. Here, under the windows of rooms inhabited by hundreds of people, lie great heaps of rubbish, exhaling a pestilential stench, which public health officials neither see nor want to see and to which public opinion is indifferent. And in the apartments round about, husbands, wives, and children sit unceasingly at their work, day and night, except for a few hours

of sleep, and their careworn, joyless existence is harder and more hopeless than that of a convicted criminal. As if the stench spread by the rotting stems was not enough to poison the atmosphere, this is joined by the offensive smells that stream out of the nearby slaughter houses and out of the brewery in the next street. In No. 305 we found several small children busy stripping tobacco; in the remaining houses others, a bit older, at the same task.

C. HEINE[2]

owns a tenement factory on East 15th Street. No. 561 in the above street is inhabited by 16 families with 61 people. The families living in the front have a room and two bedrooms; those in the back only one bedroom as well as a living room. Cooking and eating, actually all of family life, take place in this one room, the "shop," of 9 by 11½ feet, where the work is done as well. The dimensions of the bedrooms are 5½ by 7 feet, ceiling height 8 feet; rent between $8 and $10.50. In every room there is the usual 15-inch-square window. The "living rooms" differ in no way from those described above. Tobacco rules everything—it lies spread about the stove and layered in piles next to the beds, the children, dog, and cat roll about in it, and it is hard to find a corner free from tobacco. Of course we came across the same thing on the stairs and in the hallways—tobacco dust, tobacco scraps, and tobacco stems wherever we tread. Here we were also told that the inhabitants of the house suffer greatly because of the lack of water; water is almost never to be had. No water in such filth!! When we wanted to go out into the yard we discovered that there was none; there had been one once but it had been replaced by a structure used as a cigar factory employing many workers. Because of the lack of a yard, the water closets are located in the cellar; these are very dirty and spread a stench through the house which wins out over that of the tobacco. In one room we saw a mother who had just begun to nurse her child but had not interrupted her work of making wrappers. That is how people work in these factories; not a moment must be lost—the mother with the babe at her breast, the father with his evening meal next to the worktable, the weary child, drunk with sleep. All work without rest or leisure, and if they could make cigars in their sleep the boss would lower wages and force them to do that as well. Working hours here are from 5 or 5:30 in the morning until 10, 11, or 12 at night, and we were told that those who start the earliest are also the latest to stop. Wages in this house are between $4.25 and $6.00 per thousand; a family of two workers

produces an average of 2,500 cigars per week. Just a few weeks ago a two-year-old child died in this house.

(To be continued.)

*New Yorker Volkszeitung*, Nov. 2, 1881. Translated from the German by Hannelore Jarausch.

1. Lichtenstein Brothers, the city's second largest cigar manufacturing firm and largest tenement manufacturer, employed around a thousand workers in tenements and in a factory on the Bowery in the early 1880s. Later in the decade the firm lost ground; it closed in 1897.

2. C. Heine was a very small tenement firm that operated only in the early 1880s. The factory in the backyard of the tenement described by SG did not belong to this firm.

[November 3, 1881]

## SLAVES OF THE TOBACCO INDUSTRY.

We continue our description of the tenement-house cigar factories and since we have found the same filth and miserable conditions everywhere, we entered the first house that we came across with no hope of finding anything more pleasant. It is a house owned by

### BONDY & SCHWARTZKOPF.[1]

This firm has two tenement factories at Nos. 222-224 East 2nd Street. Adjoining those houses of George Bence which we first described, No. 222 contains 19 families consisting of 63 people, along with two families living in the stores on the ground floor and six cigar packers who work in the house but do not live there. Each family has a room, kitchen, and bedroom. The dimensions of the rooms are: living room 9 x 13 feet, kitchen 9 x 9 feet, bedroom 6 x 6½ feet, ceiling height 7½ feet; rent between $9 and $10.50. Here there are five-, six-, and seven-year-old children who do go to school but, as soon as it is over, go to work and strip until 9 or 10 at night when, exhausted, they collapse in the tobacco and fall asleep, at which time they are carried to their bed; a small ten-year-old fellow daily makes cigars. When we visited the house, the corpse of a little child who had died a day or two before was lying in one of the rooms. First we will report on the next house and thereby describe the conditions in both. No. 224 East 2nd Street is inhabited by 17 families, 73 people; not included in these figures are six packers and one stripper who work there but live elsewhere. Each family has a room, bedroom, and kitchen; the room measures 9 x 13 feet, the kitchen 9 x 9 feet, and the bedroom 6 x 6½ feet, ceiling height 7½ feet. In one family two children under 7 were stripping; the parents told us that they went

to school, yet our visit took place during school hours and at that time the children were working. In one room a picture greeted us that seemed created to demonstrate the poison of tenement-house work: a young woman was sitting at the worktable, rolling cigars with her hands, her feet rocking a cradle in which lay a baby; the poor child was sick, terribly emaciated, and its shrunken features seemed to say: "How can I stay alive in such a place?!" So the babe in the cradle is sacrificed to the Moloch of wage slavery! Working hours in both houses are from 6 in the morning until 10 or 11 at night; a man told us there were even families who start already at five and do not stop before midnight. Wages vary from $4.50 to $5.50 per thousand; a family with two workers produces 2,500 cigars per week. The rooms, hallways, and steps are pitch dark. Tobacco and stems lie about everywhere, especially in the bedrooms, which have no windows except for the one measuring 15 inches square and opening onto the dark corridor. Everything that could make a home comfortable or livable is absent in these apartments. There is not even any water and this last condition alone, in an "apartment factory" and especially a cigar factory, allows one to judge what it must look like in the house. In the last 14 days two children have died in this house.

### WM. FELTMAN

has two tenement house factories at East 15th Street. No. 615, the first of the houses, shelters 23 families with 72 people, as well as seven people employed there but living elsewhere. Each family has a room 9 by 11 feet and two bedrooms, 6 by 7½ feet, ceiling height 7½ feet. The rooms have only one window each. In the bedrooms are the usual 15-inch-square openings in the wall, looking out on the dark and dirty hallway. We found neither fireplace nor mantelpiece in any of the rooms, but like all the others we have seen they are flooded with tobacco and the filth it creates. In two of the apartments we did not find even the scanty "furniture" that we had seen in other places. Although the people work day and night, signs of the bitterest poverty and misery greeted us everywhere. Since it is pitch black in the corridors and stairways even in the daytime, it is not thought worth the trouble to provide light at night; there is a total lack of illumination. We found several small children at work in this house. Working hours are from 8:30 in the morning until 11 or 12 at night. Wages vary from $5.00 to $6.00 per thousand; a family of two workers produces an average of 2,600 cigars per week.

No. 617 has 17 families with 62 persons; seven people not resident

in the house work there. The office is on the second floor. In no way are conditions in this house different from the neighboring houses just described; we find the same heaps of tobacco, the same filth, the same poverty, and the same relentless, profitless labor.

### KAUFMAN BROS. & BONDY[2]

have three tenement factories on East 45th Street. No. 316 is inhabited by 14 families with 46 people (two apartments are vacant); 16 other people, not residing in the house, work there. Each family has a room 10 by 12 feet, kitchen 8 by 8 feet, and bedroom 6 by 7 feet, ceiling height 8 feet; in the bedrooms the usual hole looks out on the corridor. By now we are so accustomed to the tobacco, filth, etc., that we accept these as obvious and will not describe them further. In one room we found a sick, emaciated child lying in a cradle; the mother told us this pitiable creature was one year old, but it looked scarcely older than one month. The child was suffering from a dysentery-like disease; its large, weary eyes seemed to tell us that death would soon come as his deliverer. The mother had to work, making wrappers next to the dying child, and could only rock the cradle or nurse the child for a few moments. In one family we saw a six-year-old child stripping, in another family an eight-year-old, in a third a ten-year-old at the same task, or making wrappers and cigars. The eastern wall abuts rocks and allows water to seep into the apartment. One family has already been prostrated by illness because of this nuisance. Working hours are from 5 in the morning until 10, 11, or 12 at night, sometimes longer, especially at the end of the week. Wages vary from $5.25 to $5.75 per thousand; a family of two workers produces an average of 2,700 cigars per week. Rent ranges from $8 to $9.50. The same apartments which now cost $9 cost only $6 five months ago before the house was turned into a cigar factory. The water closet is only a drainless pit and is in horrible condition here as well. Between this house and the other houses of the firm lies a hospital, the "Metropolitan Throat Dispensary"; admirable juxtaposition!

No. 312 houses 16 families with 54 people; several very small children work here from early morning until late at night.

In No. 310 there are 16 families with 49 people; in the rooms there are the usual heaps of tobacco, and tobacco is spread out around the stoves. In one family the father lay ill in bed and the wife worked with all her might to protect herself, the sick one, and their five children from hunger. As we entered another apartment we surprised a family of six people at what they call their dinner, and how were these people eating?! There were only two chairs in the whole apart-

ment; the father and mother were sitting on them at the table. The mother held the youngest child on her lap, a second child was standing, and the remaining two knelt at the table and ate in this position. The "dinner" consisted, as we can point out right away, mainly of cooked or smoked sausage or something similar which does not need to be prepared first. The housewife would lose too much valuable time cooking a meal; she has other work to do: making wrappers and rolling cigars.

In No. 316 there were complaints that the house could not be heated; whenever it was tried the whole house, from top to bottom, filled with suffocating smoke. Another example of despicable meanness must be stated, namely that the firm compels every family to add ten cigars, for which they are not paid, to each thousand.

### HIRSCHORN AND BENDHEIM.[3]

The above firm has three houses on East 45th St., adjoining the houses of Kaufman Bros. & Bondy. No. 304 has 16 families, counting 58 people. Each family has a room 11 x 12 feet, kitchen 11 x 6 feet, bedroom 8 x 5½ feet, ceiling height 7¾ feet. The kitchen has no window, the bedroom a 15-inch-square opening onto the dark corridor. Tobacco filth is everywhere. In one room a working mother was rocking, with her feet, a cradle in which lay a crying child. Working hours are from 5:30 until 10 or 11 at night. Wages are from $5 to $6.25 per thousand; two people produce about 2,700 cigars per week.

No. 306 has 15 families, 56 people; one apartment is vacant. Dimensions of the rooms are like those of No. 304. We saw several small children working, one not more than seven years old. We found the same conditions in No. 308, which houses 16 families with 62 people.

### WM. KUTSCHER & CO.

The above firm owns two tenement factories on 1st Ave.; No. 1885 has 16 families, 60 people; moreover, 20 people who do not live in the house work there. The front apartments consist of a room 9 by 14 feet, kitchen 9 by 9 feet, bedroom 6 by 7 feet—the apartments in the back have only one room and a bedroom of the same size—ceiling height 8 feet; rent $7.25 to $9; working hours from 5 or 6 in the morning until perhaps midnight. Wages are $5 per thousand; two workers produce an average of 2,600 cigars per week. One of the men told us: "You may believe me that some families sleep only three hours and work the rest of the time." No. 1887 houses 14 families, 53 people, and 14 other persons who do not live in the house

but work there. The conditions resemble those in the earlier houses, filth, etc., as in other tenement factories.

### LANDAUER, STRAUSS & CO.[4]

are the owners of two tenement factories on 1st Ave., located hardly 100 feet from the East River. The stores in No. 726 are also used by the firm as cigar factories and 8 to 10 people are employed there; besides them 8 families with 38 people live in the house. The apartments mostly consist of a room 9 by 13 feet and bedroom 7 by 7 feet, ceiling height 8 feet. Wages are from $4 to $5 per thousand. When we asked one family about working hours, the woman said: "Oh, there are some who scarcely sleep three or four hours, but we don't want to work ourselves to death, we stop already at 10 (!!) o'clock every evening." The cellar of this house is used for tobacco storage.

In No. 728 there are 6 families, 17 people, but 19 others work in the house. One of the stores on the ground floor is used as an office, the other as a fermenting or "sweating" room for tobacco. The fumes from the latter are something dreadful, permeate the whole house, and take the breath away from anyone entering the house. The "water closets," with no sewer connection, are also very dirty.

(To be continued.)

*New Yorker Volkszeitung*, Nov. 3, 1881. Translated from the German by Hannelore Jarausch.

1. Bondy and Schwartzkopf was apparently a firm recently formed by two manufacturers who had operated separate tenements in the late 1870s. This large company, which also had a factory on the Bowery, was no longer in business under that name in the late 1880s.

2. Kaufmann Brothers and Bondy was a large tenement firm with eight or nine tenements throughout the city. The partnership between the Kaufmann brothers and Charles Bondy dissolved in the mid-1880s, and both tenement firms prospered separately during the latter half of the decade.

3. Hirschhorn and Bendheim succeeded Adolph Hirschhorn, a declining firm that in 1877 had been one of the largest tenement-house manufacturers. With Bendheim as a financial partner, the company employed some 200 workers in tenements and in a factory in the early 1880s.

4. Landauer, Strauss and Co. was a recently formed and relatively small firm. In 1888 it was no longer operating in New York.

[November 4, 1881]

## IN THE WORK DUNGEONS.

We have already described a whole series of tenement-house cigar factories and now continue with more of the same.

### ADOLPH MOONELIS[1]

owns two tenement-house factories on 2nd Ave. In No. 647 we found 16 families with 87 people, not counting the families in the stores on the ground floor. Each apartment consists of a room 8 x 13 feet, and two bedrooms 7 x 9 and 6 x 8 feet, respectively, ceiling height 8½ feet. The back rooms have only one window each; one of the bedrooms is windowless, the second has the usual opening onto the hall. Rent ranges from $10.50 to $11.50, working hours from 5:30 until 11 or 12 at night. Wages are from $4.50 to $5.50 per thousand; two workers produce an average of 2,600 cigars per week. We were told: "There are families here who never look at the clock, they work as long as they can, sleep a couple of hours, and then start over again." No. 349 has 13 families with 69 people. The house has some good characteristics but none that could further cleanliness; the filth is as bad as in every other tenement factory. Several small children died here recently of a diarrheal sickness. We were not pleased to discover that Peter Cooper[2] is the owner of these two houses and rents them to Mr. Moonelis.[3] We do not believe that the famous elderly philanthropist has, since the time of the rental, so much as glanced into these buildings; otherwise he would know how little honor his property brings him now.

### SAMUEL JOSEPHS & CO.[4]

As we continue on 1st Avenue we see a huge sign between 39th and 40th Streets telling passersby in huge letters that the above firm owns three tenement factories here. No. 693 1st Ave. has 18 families of 64 people; 12 persons besides work there but do not live in the house. Two families living in the stores are not included in the above figures. The front apartments consist of a room, kitchen, and bedroom, measuring 10 x 13, 9 x 8, and 7 x 6 feet, respectively, ceiling height 8 feet; in the rear apartments there is no kitchen. The kitchen has no windows; the bedrooms have the usual hole onto the hall. Rent varies from $7 to $10, wages from $4 to $5 per thousand; two workers produce an average of 2,900 cigars per week. As everywhere else, working hours are from daybreak until almost midnight. A man told us that in the room next to his, where a childless couple lived, someone was always working; while the husband slept, the wife worked, and vice versa. No. 691 has 18 families with 67 people; No. 689, 17 families with 74 people. The apartments resemble those in the first house. The filth and the gigantic accumulations of tobacco in these three houses surpass everything we have seen so far; dust lies so thick in the pitch-black halls and stairways that it rises in dense clouds when one walks through them.

The same firm has another tenement factory at No. 313 East 39th St.; 14 families with 52 people live there, 6 others work in the house. The apartments in front consist of a room, kitchen, and bedroom, 10 x 14, 8 x 8, and 7 x 7 feet, respectively, ceiling height 8 feet. Those apartments in the back have only a room and a bedroom, the latter with the 15-inch-square opening onto the dark hall. In many of the rear apartments the bedstead stands in the workroom, which means that a 10 x 14 foot room serves as living room, shop, kitchen, dining room, and bedroom. When you also take into account that it is uncommonly difficult to obtain water in these apartments, you can imagine what conditions reign in these work-apartments overflowing with filth and tobacco. Rent for these pens is from $7.50 to $10.00 per month; working hours are the usual and pushed to the furthest limit. We were told wages went from $4-5 per thousand but the firm is very arbitrary since some receive only $3.75 for a $4 task, and when people object to this they are told that they will be paid the whole wage only when they deliver "good work."

### Silverthau & Co.[5]

The above firm has three tenement factories at Nos. 340-344 East 36th Street. No. 340 is vacant at the moment. In No. 342 there live 15 families with 65 people; 29 others work but do not live in the house. The front apartments have two bedrooms—those in the back only one—7 by 6 feet, besides a room of 9 by 13 feet, with a ceiling of 8 feet. The bedrooms have the usual opening onto the dark hall. Rent is from $7.50 to $9.50. One stumbles over stinking piles of filth as well as the inevitable tobacco in all parts of the house. Wages are from $4.50 to $5.50 per thousand but a worker told us that these prices are not firm since the company has already put out feelers to see if people would not be satisfied (!) with less. In one of the rooms seven persons of one family were working; in two families, four each; in three, three each; in the remainder, two persons not including the children who strip. A short time ago, four people—a man and three children—died of typhoid fever in this house. While wandering through the rooms we caught a glimpse of a scene that would permanently destroy the appetite of many a cigar smoker. On the floor sat a 6-year-old girl stripping tobacco; directly in front of her, in the middle of the tobacco, stood a chamber pot on which sat a small child busy with a job that need not be described further. Seven people live and work in one apartment; there is scarcely any furniture, but to make up for it, cheap pictures of saints hang on the walls. No. 344 houses only 7 families with 30 people at the moment; the remaining apartments are still vacant. The cellar of the completely vacant house at

No. 340 is used for tobacco storage and for collecting the stems; on these sat five children stripping tobacco for the "shop." Stairways and halls in all of these houses are uncommonly filthy and dark. When we came into the yard a command painted on the fence caught our eye: "Please keep off the grass." To appreciate the irony of these words one must stand in the midst of this desolate, filthy wasteland. We peeked into every corner but there was not one blade of grass to be seen. The fresh green of nature is banished from these caverns, grass does not grow—only human beings vegetate here under the ban of a blighted and sorrowful existence. The firm that owns these houses had the impudence to advertise for 40 families in the "Delnicky Listy"[6] and to use such words as "good wages, good work, good tobacco, clean houses, and cheap rents."

SIMON BROS.[7]

have six tenement factories at Nos. 355-365 East 76th St. No. 363 houses 14 families, 65 people, besides 10 others who work in the house. The front apartments consist of a room 11 by 14 feet, and two bedrooms 8 by 7 feet; the rear ones, of only a room and a sleeping room. Ceiling height is 8 feet. Rent ranges from $7.50 to $9.50, wages from $5.25 to $6.00; working hours are the same as everywhere else. Stems and refuse from all the other houses are brought to the cellar of this one; when the cellar is full, it is emptied and the collection begins again. As a result the stench of rotting stems permeates the house. The apartments are the same in all the houses; the five remaining ones together lodge 58 families with 235 people. The cellar of No. 355 is used to "sweat" tobacco, i.e., it is fermented in crates; only the initiated know what a "fragrant" neighborhood results from such a sweating room. In addition, the "scraps" are sifted and chopped up here. Tobacco is moistened in the cellar of No. 357, usually with stagnant water, which is preferred to clean, and this process further creates a penetrating stench. Along with this the water closets have no sewer connection and are in horrible condition. The halls and stairways are as dark and dirty as all the others we have seen until now in the tenement factories. There is no lack of working children either. Two circumstances must still be mentioned: several of these unfortunate workers become "exploiters" by employing others and paying them even less than they earn themselves; the other circumstance typical of the ruthlessness and despotism of the employers is that about nine months ago the firm reduced wages by 50 cents per thousand cigars while raising the rent by $1 and informed the "renter workers" of this only at the end of the week in which this edict went

into effect. Recently three people—two men and a woman—died of consumption in these houses.

### Foster & Hilson[8]

own several tenement factories on East 10th Street, near Ave. D.; No. 444 houses 14 families with 65 people. The front apartments have a room 10 by 13, kitchen 10 by 8, bedroom 6 by 7 feet, ceiling height 8 feet. Rent is from $8 to $10.50. No. 440 is a simple tenement house and shelters 8 families, 35 people; 13 more work in the house. Rent is from $8.50 to $11.50. No. 442 is a small two-story house, inhabited by 2 families, a total of 10 people. The floor of the first story is only 9 inches above ground; the house has no cellar and is therefore very damp. As a result of the damp, dusty, and pestilential atmosphere the children in one family were sick. No. 446 has 13 families, 41 people; seven others work in the house. The apartments consist only of a room and a bedroom, 13 by 9 and 8½ by 6, respectively, ceiling height 8½ feet. The bedrooms have no windows except for the usual opening onto the hall. Rent is from $6 to $8. Three children died here recently of diphtheria and one man of dysentery. Wages are from $4.25 to $5.35. In all four houses the filth is as bad as elsewhere.

(To be continued.)

*New Yorker Volkszeitung*, Nov. 4, 1881. Translated from the German by Hannelore Jarausch.

1. Adolph Moonelis (b. 1851) was born in Austria and immigrated to the United States in 1870. His firm, a small one in the 1870s, had recently expanded into tenement manufacture. It employed about 200 workers in the tenements and in a factory in the early 1880s and later in the decade became one of the largest manufacturers in the city.

2. Peter Cooper (1791-1883) of New York City was an inventor, manufacturer, politician, and philanthropist. In 1876 he ran for president of the United States on the National Independent (Greenback) ticket. He founded the Cooper Union for the Advancement of Science and Art in New York City in 1859 to provide tuition-free scientific education and free weekly lectures for the poor.

3. Tenement manufacturers did not usually own houses but rented them from the owners and sublet the apartments to the cigarmakers. Many of the tenements in the 11th and 17th wards belonged to old New York landholding families, among them the Astors, the Coopers, and the Stuyvesants.

4. Samuel Josephs and Co. employed about 250 workers in 1881. In the late 1880s it was among the smaller cigar manufacturers in the city.

5. Silverthau and Co., a recently established tenement firm, was expanding rapidly and by 1886 was among the fifteen largest cigar manufacturers in the city.

6. *Dělnické Listy*, a Czech newspaper edited in New York, was at this time the most important socialist newspaper for the Bohemian immigrant community in the United States.

7. Simon Brothers began as a small tenement firm in the 1870s. By 1881 it employed about 350 workers, all in tenements, and was among the largest tenement manufacturers in the city. By the mid-1880s, however, the firm was out of business.

8. Foster and Hilson was one of the larger tenement firms in the city, employing about 250 workers in 1881. It remained one of the largest tobacco manufacturers in the city throughout the 1880s.

[November 5, 1881]

## THE CURSE OF TENEMENT-HOUSE CIGAR MANUFACTURE.

Today's article concludes the descriptions of the apartment factories of our city; those not yet depicted follow:

### SUTRO & NEWMARK.[1]

The above firm has five tenement factories at Nos. 315-323 East 74th Street. No. 315 contains 17 families with 72 people. The apartments consist of a room 10 by 13½, kitchen 7 by 10½, and bedroom 6 by 7 feet, ceiling height 8 feet. No. 316 has 16 families with 75 people; No. 321, 17 families with 78 people; No. 323, 17 families with 88 people; No. 317, 18 families of 76 people. In the five houses together there are 85 families with 389 people. There are also 24 people who work in the houses but do not live in them. Rent varies from $7 to $9, wages from $5.50 to $6.00 per thousand. When we asked a man about working hours, he replied precisely as follows: "Write that I often got up at 5 in the morning and saw some people already at work, and those same people worked until midnight of that day. I said to myself, my God, how long can they stand it? Now I am not surprised by it any more, it has become ordinary." Another one told us: "Some work from 5 in the morning until 1 or 2 at night; then they throw themselves, fully clothed, on their bed, sleep a few hours, then go back to work without wasting a minute, and thus they continue until they become ill and must stop." We are tired of describing the filth; it is enough to say that we found it the same everywhere. Several small children work here from morning until night; in the last months eight children died in these houses, most of them of diphtheria, and one man of consumption. Another man contracted a disease of the lungs and kidneys; the doctor suggested he stop working in the house. The man followed this advice as soon as the opportunity offered itself, and he is now on the road to recovery.

### LEVY BROTHERS.[2]

This firm has three tenement factories on East 13th Street. No. 704 contains 15 families with 59 people; No. 706, 14 families with 68 people; No. 708, 15 families with 67 people; the office is located in the stores of No. 704. There is a beer hall and a restaurant on the ground floor of No. 708; those who live on this floor are not

included in the above figures. Ten persons who do not reside in the houses work there. The apartments in the three houses are all the same; those on the street side consist of a room 9 by 14, kitchen 9 by 9, and bedroom 6 by 6 feet, ceiling height 8 feet. Those apartments in the rear have no kitchen. The bedroom has the familiar opening onto the dark hall. Five people in one family in No. 704 work in one room; several families work with four people, most with three. Two children in this house died of diarrheal ailments since last summer; also one man and one 16-year-old boy—we could not ascertain of which diseases they were the victims. When we inspected the houses several children were ill with a throat disease; in one family, and not one of the worst off, we found three sick children whose heads and throats were wrapped in cloths. There were also several sick children in No. 706 but the parents would not tell us what ailed the children. Rent in these houses is from $6.75 to $9.50. Wages are given as $5 to $6 by some people, as $5.50 to $6 by others; we could not determine if this is due to an error or if the firm pays varying wages. Complaints were often made that it is very difficult to obtain water.

### FREY BROTHERS.[3]

The three tenement factories of this firm are located at Nos. 1329-1333 1st Avenue. No. 1329, on the ground floor of which the office is located, has 14 families with 51 people; No. 1331 houses 12 families with 47 people, besides two families in the stores; and 16 families with 58 people live in No. 1333, on the ground floor of which there is also a beer hall and a cigar factory with 15 workers. Each family has a room 10 by 13½, kitchen 10 by 8, and bedroom 7 by 6 feet, ceiling height 8 feet. Rent goes from $8 to $9.50; wages are $5.75 per thousand. Two workers produce an average of 2,400 cigars per week. Several small children work in these houses; in No. 1333 three of them were at work, among them a seven-year-old child who was stripping tobacco.

### KERBS & SPIESS.[4]

The above firm has several tenement factories on East 54th St. and 2nd Ave. No. 316 East 54th St. houses 8 families with 42 people, in apartments consisting of a room 14 by 12, kitchen 9 by 6, and two bedrooms of 7 by 6 feet each, ceiling height 8½ feet. One bedroom has no window at all, the second has one opening onto the hall. Rent varies from $9.50 to $10.50. In No. 314 work 20 families with 95 people. Here each apartment has a room 13 by 10, kitchen 10 by 10, and bedroom 6 by 5½ feet, ceiling height 8 feet; rent is $10. In this house a child died recently of diphtheria and almost all of the

children in the house came down with the measles. In No. 312, with apartments similar to those in No. 314, there are 20 families with 90 persons; two men, one woman, and six children died in one year in this house. No. 1014 2nd Ave. contains, besides two families in the stores, 16 families with 77 persons; No. 1016 has 16 families with 76 persons. The apartments are like those on 54th St. Wages vary from $5.45 to $6.00. Quite a number of small children work in these homes.

C. DOMCHSKY[5]

has a tenement factory in the rear building of No. 135 Allen St. Four of its apartments are vacant; 6 families with 21 people work in the remaining apartments, consisting of a room 12 by 10 feet and bedroom 7 by 6 feet, ceiling height 7½ feet. The bedroom has a window two feet high and just as wide, but the rear wall of a house on Eldridge St. rises directly in front of it, not six inches away, and cuts off all light and air. Rent is $7.00, wages from $5.00 to $5.50. A family of two workers produces an average of 2,600 cigars per week. The apartments here are unusually cramped with almost no light or ventilation.

LOUIS ASCH[6]

owns some tenement factories on Lewis St. near 5th St. No. 185 Lewis St. has 6 families with 36 people. Each family has a room 13 by 10, kitchen 9 by 6, and two bedrooms 6 by 6 feet each, ceiling height 7 feet; rent is from $9.50 to $10.50. No. 187 Lewis St. contains 8 families with 40 people; the "office" is located on the ground floor. Each apartment consists of two rooms 12 by 7 each, and a bedroom 9 by 7, ceiling height 7 feet. No. 821 5th St. is an old rattletrap which has only three working families. Wages range from $5.00 to $6.00; working hours are the usual ones. Filth is the same as everywhere else—a pile of at least 200 pounds of rotting stems was lying in one corner of the yard. Everything—rooms, stairs, yard, and cellar—is in a horribly dilapidated condition and the inhabitants suffer greatly because of lack of water.

MENDEL BROTHERS.[7]

This firm has some tenement factories on East 3rd St., but there the families who make cigars are mixed in with regular renters. In No. 241, however, there are 18 working families with 67 people. Wages are from $5 to $6 per thousand; two workers produce an average of 2,600 cigars per week. Two small children work in this house.

BIER & CO.

began manufacturing cigars in tenement houses about one month ago; their houses are located on East 39th St. but are not yet fully occupied. No. 317 East 39th St. is designed for 16 families, but so far contains only ten families with 41 people. The apartments in front contain a room 10 by 13, and two bedrooms 7 by 6½ each, ceiling height of 8 feet; the rear apartments have one bedroom less. Instead of a window the bedrooms have the usual opening onto the dark hall. No. 319 has 15 families with 52 people; ten others work in the house. The apartments are the same as in the house described above. Three small children work here all day, stripping and making wrappers. Wages are from $5 to $6; a family of two workers produces an average of 2,500 cigars per week. Rent is from $8 to $10.50; filth in these houses defies all description.

BONDY & LEDERER.[8]

The tenement factories of the above firm are found at Nos. 94-100 Attorney St. In the apartments in front there are a room 13½ by 9½, kitchen 9 by 8, and bedroom 7 by 6½, ceiling height 7½ feet. The rear apartments have no kitchen, the bedrooms have only the 15-inch-square opening onto the dark hall. No. 94 is inhabited by 16 families with 64 people; No. 96, by 16 families with 61 people; No. 98, by 16 families with 59 people; and No. 100, by 16 families with 67 people. Rent is from $7.50 to $9.75, wages $5.50 per thousand. Filth is as pervasive as in the houses hitherto visited. Eight or ten children are condemned here to perpetual labor. The hallways and stairs are pitch black; in the yard we found puddles of stagnant water. In No. 100 three children died lately, two from diarrheal illnesses; two deaths also occurred recently in No. 98 and No. 94.

M. JACOBY & CO.[9]

have two tenement factories on Broome St. No. 127 contains 14 families with 53 people, and No. 125, the same number of families with 60 people; four others live in the stores. Two apartments in No. 127 were vacant at the time of our visit. The apartments in front consist of a room 13½ by 9, kitchen 9 by 9, bedroom 6½ by 6, ceiling height 7¾ feet; the rear apartments have no kitchen. The bedrooms above the third floor on the east side of No. 125 have a window to the outside since there is no tall building nearby; all the others have only the opening onto the very dark hallway. The filth in these houses is nauseating. Rent is from $8 to $10, wages from $5 to $5.50. In connection with these houses we must mention an unexpected incident

that is not without interest. In order to collect the data that we have presented to our readers we naturally had to direct many questions to the inhabitants of the houses and could not allow ourselves to be put off by evasive or even rude replies. At No. 127, when we wanted to question a couple sitting together at the worktable, the wife seemed to guess the purpose of our visit and suspiciously evaded all our questions. "I don't know!" was almost invariably her answer and as we proceeded with our questions the characteristic response slipped out: "I'm for my boss." Circumstances, misery, and need together with unceasing toil have turned this woman into a timid, "boss"-fearing slave and we were forced to seek our information elsewhere.

### HOLZMANN & DEUTSCHBERGER.[10]

This firm has several tenement-house cigar factories on Attorney St., some allegedly also on Ridge St., but we were not able to find the latter. No. 156 Attorney St. has 16 families with 65 people; No. 154, 16 families with 50 people, 7 others work in the house; No. 166 has only 8, and No. 168 only 4 families who make cigars. The apartments in all the houses are the same: those located in front have a room 13 by 10, kitchen 10 by 8, bedroom 6½ by 6, ceiling height 7½ feet; the apartments in the back have no kitchen. Rent varies from $8 to $10.50, wages from $4.75 to $5.25 per thousand; two workers produce an average of 2,500 cigars per week. Working children were not missing here either. We wish to say that the filth in these houses surpasses the worst that we have seen until now and, moreover, not one drop of water is to be had all day long.

### KAUFMANN BROTHERS

have only recently begun tenement manufacture in their new houses, Nos. 330-334 East 63rd St., and the houses are not more than half occupied. They have apartments for 60 families; each apartment consists of a room 12⅓ by 9½, and two bedrooms 9½ by 7 and 6 by 6½, respectively, ceiling height 7½ feet. One bedroom has the 15-inch-square opening, the second lacks even that. The firm's "new" houses are very old and dilapidated, filth and tobacco are rampant; even in the vacant apartments, huge piles of tobacco were stacked up to dry. The water closets are pits with no drains and are in horrible condition; one of them was swarming with vermin, and one look spoiled our appetite for dinner. What the condition of these houses will be when the apartments are fully occupied is more easily imagined than described. Rent ranges from $8.75 to $9.25, wages $5.75 per thousand; a family of two workers produces an average of 2,400 cigars per week.

We hereby conclude our sad rounds through filth and misery. We found the same conditions in all of these houses: the greatest deprivation and an existence beneath human dignity despite the most diligent, industrious efforts, unceasing toil with no rest or leisure, taxing the total strength of the body. The people we encountered in these work dens, especially the women, are emaciated, pale, and their faces have a fixed, anxious expression. What we have designated as "families" throughout our report does not always mean a couple and its children; more often than not, relatives, brothers and sisters, in-laws, etc., sometimes a few friends, work in one and the same apartment. Something that we noticed time and again was that meals were often taken at the worktable and during work; pieces of food therefore end up in the tobacco and are carelessly incorporated into the "tenement house cigar." Most to be pitied are the small children whom we so often found enslaved to work. They spend a childhood full of misery and care; there is no hope for them and since their sad fate brings them so early on to the thorny path of earning their own bread, the sad certainty is that their young life will not last long. We hereby close our investigative report; we will discuss the results of our work shortly.

*New Yorker Volkszeitung*, Nov. 5, 1881. Translated from the German by Hannelore Jarausch.

1. In 1881 Sutro and Newmark was one of the largest tenement manufacturers in New York, with 600 workers. Throughout the decade it maintained its position as one of the city's largest manufacturers.

2. Levy Brothers had a factory and several tenement houses with a combined work force of about 600 in 1881. It was among the largest cigar firms in the city throughout the 1880s.

3. Frey Brothers, a small firm in 1881, expanded its tenement production rapidly to become a major manufacturer by the mid-1880s.

4. Kerbs and Spiess was the largest cigar manufacturer in the city through the mid-1880s, its work force having increased from some 500 employees in the late 1870s to 1,500 in 1881. It operated a large factory on the Upper East Side and many tenements.

5. C. Domschke (alternately Domshky) employed about 150 workers in several tenement houses in the Lower East Side. The firm went out of business in the mid-1880s.

6. Louis Ash had grown from a shop of fifty workers in the late 1870s to become a large tenement manufacturer with about 250 workers in 1881.

7. Mendel Brothers was among the largest tenement manufacturers in New York during the 1880s. It employed approximately 450 workers in several houses in 1881. In 1886 it was still one of the ten largest firms in the city.

8. Bondy and Lederer had about 300 employees in tenements in 1881 and by 1887 ranked third among the city's cigar manufacturers.

9. M. Jacoby and Co.'s work force expanded from 100 in 1877 to 400 in 1881, but in the mid-1880s its relative position in the industry had begun to decline.

10. Holzmann and Deutschberger, one of the largest tenement manufacturers in the city, with about 500 workers in 1881, gradually declined in the late 1880s.

# Total Results of the Investigation of Tenement-House Cigar Manufacture.

[November 7, 1881]

We present below a tabular summary of the results of our investigations. This table is not only of the highest interest for the present but can serve as a source of information for the future. The figures are as precise as possible. A potential difference of some million cigars produced annually (in tenement houses) is of little importance.

| Name of Firm: | Number of Families. | Number of Rooms Occupied by Each Family. | Rents. | Wages per 1,000 Cigars. | Number of Cigars Produced, on the Average of 3,800 from a Family of 3 Persons. | Hours of Work per Day. |
|---|---|---|---|---|---|---|
| H. Blaskopf | 18 | 2—3 | $7.00— 9.00 | $4.25—6.00 | 68,400 | 16—18 |
| Geo. Bence | 109 | 2—3 | 7.00— 8.00 | 5.00—5.75 | 414,200 | 16—18 |
| Rosenthal Bros. & Co. | 122 | 2— | 7.50— 8.50 | 3.75—4.50 | 463,600 | 17—20 |
| Lichtenstein Bros. & Co. | 154 | —3 | 7.00— 9.50 | 5.38—5.70 | 585,200 | 15—19 |
| C. Heine. | 16 | 2—3 | 8.00—10.50 | 4.25—6.00 | 60,800 | 17—19 |
| Bondy & Schwartzkopf | 36 | —3 | 9.00—10.50 | 4.50—5.50 | 136,800 | 16—17 |
| Wm. Heltmann | 40 | —3* | 6.50— 9.50 | 5.00—6.00 | 152,000 | 15—17½ |
| Kaufmann Bro. & Bondy | 46 | —3 | 8.00— 9.50 | 5.25—5.75 | 174,800 | 17—19 |
| Hirschkorn & Bendheim | 47 | —3 | 8.00— 9.50 | 5.00—6.25 | 178,600 | 16½—18 |
| Wm. Kutcher & Co | 30 | —3 | 7.25— 9.00 | —5.00 | 114,000 | 18—19 |
| Landauer, Strauss & Co. | 15 | 2— | 6.50— 8.00 | 4.00—5.00 | 57,000 | 16—20 |
| A. Moonelis. | 29 | —3* | 10.50—11.50 | 4.50—5.50 | 110,200 | 17½—18½ |
| Samuel Joseph & Co. | 67 | 2—3 | 7.00—10.00 | 3.75—5.00 | 254,600 | 17—20 |
| Silverthau & Co. | 24 | 2—3* | 7.50— 9.50 | 4.50—5.50 | 91,200 | 16—18 |
| Simon Bros. | 72 | 2—3 | 7.50— 9.50 | 5.25—6.00 | 273,600 | 16—17 |
| Foster & Hilson. | 40 | 2—3 | 8.00—11.50 | 4.25—5.35 | 152,000 | 16—18 |
| Sutro & Newmark | 117 | —3 | 7.50—10.50 | 5.50—6.00 | 444,600 | 16—18 |
| Levy Bros. | 44 | 2—3 | 6.75— 9.50 | 5.00—6.00 | 167,200 | 16—17 |
| Frey Bros. | 42 | —3 | 8.00— 9.50 | —5.75 | 159,600 | 15—18 |
| Kerbs & Spiess. | 110 | —3 | 9.00—10.50 | 5.45—6.00 | 418,000 | 15—19 |
| C. Domschky | 6 | 2— | — 7.00 | 5.00—5.50 | 22,800 | 16—17 |
| Louis Asch. | 18 | —3 | 9.50—10.50 | 5.00—6.00 | 68,400 | 17—18 |
| Bier & Co. | 26 | 2—3* | 3.00—10.50 | 5.00—6.00 | 98,800 | 16—18 |
| Mendel Bros. | 33 | —3 | 8.00—10.00 | 5.00—6.00 | 125,400 | 16—17 |
| Bondy & Lederer | 64 | 2—3 | 7.50— 9.75 | —5.50 | 250,800 | 15½—17 |
| M. Jacoby & Co. | 28 | 2—3 | 8.00—10.00 | 5.00—5.50 | 106,400 | 16—18 |
| Holzmann & Deutschberger | 44 | 2—3 | 8.00—10.50 | 4.75—5.25 | 167,200 | 16—19 |
| Kaufmann Bros. | 30 | —3* | 8.75— 9.25 | —5.75 | 114,000 | 16—17½ |
| Totals | 1,427 | | | | 5,422,600 | |

Notes: Rooms marked with 2 means room and bedroom.
" " " 3 means room, bedroom, and kitchen.
" " " 3* means room, two bedrooms, and kitchen.[1]

| | |
|---|---|
| Total production in tenement houses per year, approximately | 281,977,000 |
| " " " the city of New York generally | 670,000,000 |
| Total production in factories and shops | 389,000,000 |
| In the year 1877 1250 families in tenement houses had an annual production of approximately | 247,000,000 |
| Out of a total production of | 350,000,000 |
| Which gives a production in factories and shops of | 103,000,000 |

From this it can be seen that while total production in this city increased by about 320 million pieces, the increase in tenement houses was only 44 million.

It is undeniable that this beneficial effect is purely and simply the result of the stubborn agitation of this question by the Cigar Makers' Union. If the organization and agitation had not been continuously sustained, tenement-house manufacture would certainly have increased to the same extent, and only few workers would be employed outside of tenement factories. May the workers always bear that in mind.

*New Yorker Volkszeitung,* Nov. 7, 1881. Translated from the German by Hannelore Jarausch.

1. This should not read "room, two bedrooms, and kitchen" but simply "room and two bedrooms" (see installment of Nov. 14, 1881, below).

[November 11, 1881]

## REPORTS ON TENEMENT-HOUSE CIGAR MANUFACTURE.

### Some Implications.

Anyone who has carefully read our reports about conditions in tenement-house cigar factories must admit these conditions raise questions that demand serious consideration and cannot be thoroughly enough discussed. For years these questions, in all their significance, have arisen time and again, and to remain silent in face of them would be an inexcusable crime. We have seen that the capitalist who manufactures cigars in tenement houses leases a number of such tenement houses and rents the apartments in them to his workers at enormously high prices, thereby not only sparing himself rent for his factory but also making a considerable profit. As we have seen and described them the apartments are filthy beyond belief; the air is foul because of all sorts of unclean and unhealthy exhalations. In most cases the buildings themselves are in such a terrible condition that they are hardly suitable for human habitation. We have seen that in these apartments sick people are at work, completing cigars which find their way to market and into the mouths of consumers; that children stricken with contagious diseases—smallpox, diphtheria, measles, scarlet fever, etc.—lie in the "fabrication room"; that tobacco and cigars must inevitably be impregnated with disease germs and when they reach consumers they must continue to spread such plagues. When we are asked, in light of all these facts, if something has been done to put an end to these situations and if public officials are aware of the evils of this system, we can answer both questions in the affirmative. But if we are asked if public officials have done anything to put an end to the abuses or even to mitigate them, we must unfortunately answer: "No; public officials have not only failed to do their duty, they have intentionally contributed to the perpetuation of the system.["] We have in our hands the documentation necessary to support this allegation. It is of the utmost significance and importance, but so extensive that we must limit ourselves here to excerpts which will nevertheless be so positive and convincing that any impartial reader will admit that our observations are correct, our grievances justified. In the fall of 1874 steps were already taken to abolish the tenement-house manufacture of cigars and a large mass meeting[1] was called to protest against the system. Mr. John Swinton[2] presided over that meeting and subsequently appeared before the New York Board of Health along with two well-known doctors,[3] and in a clear and truthful manner he portrayed the abuses of the system. The Board of Health promised

to undertake an investigation of the conditions in the houses in question, but in the following week a mass meeting[4] of the tenement-house workers was called in protest against implementing the announced investigation. At the meeting several of those types who were already notorious as traitors to the cause of their fellow workers were seen in the forefront, and their intentions, as well as their employers', were immediately obvious. The worker, vegetating in his apartment factory and considered ignorant, was once again to appear in a role prepared for him, with no will of his own, so that the general population would believe he liked his position, was content with his fate, and opposed any agitation in the matter. But the cleverly conceived plan of the employers was thwarted, the true purpose of the agitation revealed and enthusiastically greeted; the bootlickers saw themselves forced to leave the hall, naturally not until they had called the police, and the true workers remained in control of the situation. Now the employers had to find a new strategy, and the compliant Board of Health gave them all the necessary time before starting the investigation. Quickly the tenement-house factories were freshly painted, whitewashed, plastered, and cleaned; through the cooperation of the officials, the employers learned the precise day on which their visits would take place. On that day not one shred of tobacco could lie about; the residents had to take a holiday, not work, and receive the visitors in their Sunday best. When everything was in order, the Board of Health made its rounds, the results of which appeared as a report in the "City Record" of November 6, 1874, signed by Doctors Emerson[5] and Tracy,[6] which said:

"In this inspection we visited all rooms in which cigars were being manufactured as well as all other rooms, saw people at their work and at their meals and spoke with them and their employees (should probably read 'employers,' editor's note) whenever we had the opportunity to see the latter."

In their discussion of conditions in the apartment factories the doctors used four of the most miserable factories in the city as points of comparison and then said that each resident of the apartment factories had more cubic feet of air than workers in the workshops; also that apartment factories can be well ventilated by "opening windows and doors," though to be sure during working hours doors and windows were closed to conserve heat, and the air in the rooms was bad. Then they said, quite jesuitically: "Workers in the shops begin work at 7 in the morning and stop at 6 in the evening; those who work at home can, of course, begin as early and stop as late as they like. Through the tenement-house system many members of a family, the aged and infirm, housewives, nursing mothers, daughters whose

parents do not want to expose them to the temptations and depraved atmosphere of the shops, and even school children in their free time can lend a hand and increase the income of the family. . . . Also the time that others use to get to and from work is saved. . . . In this respect, the results of home work are beneficial; as regards the feasibility of increasing working hours, we have nothing to say (!), for that is more the province of economists than health officials.["]

Then it says in the report: "There are in these tenement factories as many round- and rosy-cheeked infants and no greater number of emaciated and sickly ones than in tenement houses of equal size in which cigars are not made. We mean, of course, in proportion to the total number of children in such houses, since it cannot be denied that the number of children of tenement-house workers is astonishingly small: in 66 families we found only 70 children, and in 148 others, only 234. . . . Several cases of smallpox have occurred here in families that were only reported to health officials when they resulted in death. The fact that smallpox is prevalent in one of these families would probably have seriously damaged their business." (So, according to these guardians of health, business, not that of the worker but that of the employers, is of greater interest than public health.) The gentlemen continue in their report: "We found that home workers have more air space and ventilation and live under more hygienic (healthful) conditions than other workers. . . , that their sleep takes place under more favorable circumstances, that as a rule they offer no 'nuisance' to their neighbors, but there is nonetheless enough evidence that the manufacture of cigars is not a healthy occupation." The report concludes with the following words: "We found a family with seven children and since this large number deviates so much from the norm, we looked into the background of the family and learned that the parents had been making cigars for only four years and had previously been farm workers; the children were born during the time when their parents still lived in the country.["]

We now ask the reader to examine carefully the above excerpts from the report of the doctors and then ask himself which of the following conclusions is correct. It is indisputable that the doctors failed completely to carry out their duty appropriately; either they were induced, through some means or considerations, to submit a false, distorted, and contradictory report, or both gentlemen were painfully ignorant and did not themselves understand what they were reporting to the public. The latter supposition would be plausible enough had not one of the two, Dr. Tracy, later written a certain book—certainly not for those workers who wanted to struggle against the abuses of tenement-house factories—in which he develops ideas

totally opposed to those in his report, and in which he gives lie to and accuses himself. Thereby it would be proven that only one of the above conclusions is the correct one. Excerpts from the text intended only for doctors will introduce our next article and we will not fail to present testimony from other doctors.

(To be concluded.)

*New Yorker Volkszeitung,* Nov. 11, 1881. Translated from the German by Hannelore Jarausch.

1. Held on Sept. 27, 1874.

2. John SWINTON was active in New York City politics and in labor affairs. He was a member of the editorial board of the *New York Sun* from 1875 to 1883 and published *John Swinton's Paper,* an influential labor and reform newspaper in New York City during the mid-1880s.

3. Both the *New York Sun* and the *New York Herald* of Oct. 1, 1874, identify only one doctor among the members of the committee appearing before the Board of Health, Dr. George C. Stiebeling (d. 1896?), a physician who had a large medical practice on the East Side. A German immigrant, Stiebeling was a leader of the International Workingmen's Association in New York City in the early 1870s. He was subsequently a contributor to the *Social-Demokrat,* the organ of the Social Democratic Workingmen's Party of North America, and in 1877 was appointed coeditor with J. P. McDonnell of the newly founded *Labor Standard,* the organ of the Workingmen's Party of the United States. The party nominated him as a candidate for the New York state senate in 1877.

4. Cigarmakers supporting the tenement-house system held a mass meeting in Concordia Hall on Oct. 4, 1874.

5. Nathaniel Bright Emerson (1839-1915), an 1869 graduate of New York City's College of Physicians and Surgeons, was a city health official in the 1870s. During the 1880s and 1890s he was a leading member of the Board of Health of Oahu, Hawaii.

6. Roger S. Tracy (1841-1926), an 1868 graduate of New York City's College of Physicians and Surgeons, was a leading city health official for three decades and author of several books in the field of public health. His "On the Hygiene of Occupation" appeared in volume 2 of Albert Henry Buck, ed., *A Treatise on Hygiene and Public Health,* 2 vols. (New York, 1879), which comprised volumes 18 and 19 of Hugo von Ziemssen, ed., *Cyclopaedia of the Practice of Medicine* (New York, 1874-81).

[November 12, 1881]

## TENEMENT-HOUSE CIGAR MANUFACTURE.

(Continuation instead of conclusion.)

At the end of yesterday's article about the agitation against the tenement-house factory system we stated that Dr. Tracy, one of the doctors from the Board of Health who submitted the infamous report about the tenement factories, contradicted his report in a text he wrote himself and gave lie to his own statement. The article[1] is included in Volume 2 of Buck's "Hygiene and Public Health," p. 42,

and in it Dr. Tracy says: "It has been generally observed that this occupation is very injurious to the female sex; according to Kostial,[2] out of one hundred female cigar workers between the ages of 12-16, 72 became ill within six months of starting work. They suffer from anxiety, heart palpitations, anemia, exhaustion, insomnia, and fever spells. . . . Layett[3] does not hesitate to attribute these incidents to the premature working age and the unhealthy conditions of the apartments. Kostial indicates that miscarriages, due to the death of the fetus, are frequent among women who work in tobacco factories, and Rulf[4] has found nicotine in the amniotic fluid. Heurteux,[5] Bondel, and Schneider have demonstrated the presence of the same alkaloid in the urine. Kostial showed that the milk given by the women has a peculiar odor of tobacco."

We would like to pursue the words of Dr. Tracy and hear from him the opposite of that which he said in his report. It says in the text: "Through observations in cigar factories and through my experience as a hospital doctor, I could see that the sexual development of young girls who work in cigar factories is decidedly delayed, and during an investigation that Dr. R. B. Emerson and I undertook concerning the situation of those cigarmakers who work at home in overcrowded tenement houses, we were most surprised at the small size of the families. . . . When we consider what swarms of children usually grow up in tenement houses and in the families of workers and craftsmen, then the small number of offspring in a certain class becomes significant. . . . The reason for this low rate of reproduction among cigarmakers has not yet been investigated; Hertaux's and Kostial's observations of nicotine in the body fluids lead one to assume that the scarcity of children is the result of miscarriages." This passage of the text concludes with the words: "One should forbid child labor under the age of puberty and perhaps the labor of all women."

Here now we have his conviction as a falsifier and perverter of the truth, an enemy of public health, and an official who has made himself guilty of criminal contempt and negligence of his duties, not in the words of an opponent or an enemy of Dr. Tracy, but in his own. How can one reconcile his behavior with the words of a joint letter concerning the same affair, written by the doctors Willard Parker,[6] Ludwig Borry,[7] Joseph Wiener,[8] and Francis Serr,[9] in which they say: "While it is the doctor's responsibility always to use his full powers to cure illness, it is his highest duty to teach mankind in such a way that illness is prevented." We ask every impartial person to measure the behavior of Doctors Tracy and Emerson by the criterion that the words of these doctors offer us; he will on his own come to a conclusion that we need not force upon him. But we need not be satisfied with that;

the words of other experts should be brought forth as testimony in order to convict these betrayers of their duty so that sooner or later public opinion can hold them responsible. In the "Encyclopedia of State Hygiene," volume IV, page 261, it says: "A shriveling or withering of the breasts is a common ailment of women who work in tobacco factories." It goes on to say, "The bodily construction of cigarmakers is by no means a robust one and the newborn children of working women are often sick; the stooped and sitting position that this work requires hampers circulation during pregnancy through constant abdominal pressure on the uterus and leads to a premature separation of the placenta during the last months of pregnancy. Premature births and miscarriages are the result, as well as an engorging of the brain and the spinal cord of the fetus with blood, to which must be added the constant poisoning of the latter by nicotine which the mother constantly inhales. . . . Because of the 'intoxication' of the milk, which itself accumulates excessively during the long working hours, even it becomes especially damaging to the delicate constitution of the newborn, which is why most of the deaths among children occur between the ages of two and four months. . . . The child has very little desire for the breast—even refuses to take it. In its sleep it is restless and often moans; and suddenly, after the child has been nursed with the milk which has collected for hours and is saturated with nicotine, illness sets in with such speed and force that it is almost always fatal."

The previously mentioned Doctors Borry, Wiener, and Serr state, concerning this business: "The air in such rooms must necessarily be poisoned completely by tobacco and thereby becomes extremely damaging to the health of those people who not only work in the rooms but also live and sleep there."

Dr. Granget[10] says: "After having treated more than 1,400 children in different institutions, I certify that the constant inhalation of tobacco, especially in the young—the children living in tenement factories—at first leads to an inability to digest food, emaciation, and general weakness of the body; then the mucous membranes of the nose, mouth, and windpipe become irritated and affected, which produces conditions that are difficult to treat as long as the harmful influence lasts, and forms the breeding ground for future consumption."

Dr. F. H. Goodwin[11] says: "Where a number of people are crowded together in one or two rooms, the tobacco dust combined with poor ventilation is highly injurious and inevitably results in chronic illness."

Dr. Wm. H. F. Fleming writes: "The manufacture of tobacco and

cigars should take place only in specially built factories and should be banished from all residential apartments."

Dr. E. Landers says, concerning the same matter, "Every doctor and layman must admit that the overcrowding of tenement houses is extremely unfavorable to health; if you add to this misery the inhalation of dusty air, as is the case in our tenement houses used both as apartments and work places, the harmful influences must be significantly increased."

Were it not for lack of space we could still cite hundreds of the best doctors who have spoken about the abuses of tenement factories in the same way as the doctors already referred to; mentioning some of their names will suffice: Prof. Peters of Prague, the doctors Roberts, Hoffmann, Stiebeling, Lang, Messemer,[12] Barry, Balser, Schönemann,[13] Kiernan,[14] Finnell,[15] and others.

Our next article, with which we shall close the reflections on our inspection of the tenement factories, will deal with the history of the agitation against the system and will give a description of the work and efforts which have been made in this direction, unfortunately with little success so far.

(To be concluded.)

*New Yorker Volkszeitung*, Nov. 12, 1881. Translated from the German by Hannelore Jarausch.

1. Roger S. Tracy, "On the Hygiene of Occupation," in volume 2 of Albert Henry Buck, ed., *A Treatise on Hygiene and Public Health*, 2 vols. (New York, 1879).

2. In 1868 T. Kostial's article on the health of female cigar workers in the city of Iglau, Czechoslovakia, appeared in the *Wochenblatt der kaiserlich königlich Gesellschaft der Aerzte zu Wien*, a weekly publication of Viennese doctors.

3. Possibly Alexandre Layet, a professor at the School of Naval Medicine in Rochefort, France, who was the author of *Hygiène des professions et des industries* (Paris, 1875).

4. Maurice Ruef, a member of the Medical Society of Strasbourg, published an article on the health of tobacco workers in the *Gazette médical de Strasbourg* in 1845.

5. Alfred Heurtaux was a Parisian physician and author of *Du cancroïde en général* (Paris, 1860).

6. Probably Willard Parker (1800-1884), a professor at the New York College of Physicians and Surgeons (1839-69) and one of the first commissioners of the New York Board of Health.

7. Possibly Robert A. Barry (1824-82), an 1851 graduate of New York City's College of Physicians and Surgeons, who was one of the founders of the New York Academy of Medicine and a school officer.

8. Joseph Wiener (1828-1904), who immigrated to New York City from Bohemia in 1849, was one of the founders of the department of pathology at the New York College of Physicians and Surgeons.

9. Franz Serr (b. 1848) received his M.D. from the University of Munich and published a book on children's diseases in 1875. The following year he immigrated to New York City.

10. John P. Granget (1851?-90), born in New York, received his M.D. from the

University of the City of New York in 1873, and in 1886 was a physician to the New York Children's Aid Society.

11. Possibly Francis Goodwin (1829-82), a Brooklyn, N.Y., physician.

12. Possibly Michael Jean Baptist Messemer (1848-94), who received his M.D. from New York City's Bellevue Hospital Medical College in 1875 and was deputy coroner (1881-84) and coroner (1884-94) of New York City.

13. Probably Charles P. R. Schonemann, who received his M.D. from the University of the City of New York in 1875 and practiced in New York City until 1907.

14. Possibly James George Kiernan (1852-1923), who was assistant physician at Ward's Island, N.Y. (1874-78), and subsequently head of the Cook County (Illinois) Hospital for the Insane.

15. Possibly Thomas Constantine Finnell (1827-90), a New York City physician and former president of the New York State Medical Society (1868-69).

[November 14, 1881]

## TENEMENT-HOUSE CIGAR MANUFACTURE.

(Conclusion.)

We rejoin the agitation against the tenement-house system begun in the year 1874. Since nothing could be accomplished through the "investigation" carried out by public health officials, a protest was sent to Governor Dix;[1] the latter promised an investigation but never ordered one.

The movement was then quiet until the great cigarmakers' strike of 1877-78, which began with no more than 600 members of the union, involved about 9,000 cigarmakers and lasted 102 days. During that time the owners of the tenement-house factories evicted nearly 1,000 families; this fact, together with a description of the abuses connected with the system of tenement-house manufacture, was made public. The population of the entire country became aware of it and the result was that after the strike the number of families engaged in such manufacture decreased by about 400. The next step was that Mr. Strasser, the president of the International Cigar Makers' Union, appeared before the congressional committee chaired by Mr. Abram Hewitt that had, some time earlier, investigated the reasons behind the stagnation of business.[2] Mr. Strasser's arguments were so masterfully presented that Mr. Hewitt felt compelled to say: "Mr. Strasser, you have made out a good case," as was reported the following day in the "Volkszeitung." Subsequently a bill was introduced in the U.S. Senate that would amend tax laws in such a way as to eliminate tenement factories. The bill was approved by the Finance Committee of the Senate and met with equal approval in the Committee of the Whole but when the plenary session was concluded and the bill reported to the House, Senator P. Whyte[3] of Maryland rose and stated that if this law were approved, it would ruin all small manufacturers

in the country. Subsequently the bill was rejected by a vote of 27 to 24. A misapprehension determined this action by Senator Whyte. There existed in Baltimore an organization of small manufacturers[4] whose purpose it was to monitor the positions of tax commissioner Green B. Raum, whom they considered an enemy of small manufacturers. They believed that the bill in question was the work of Raum's and selected a committee, to which even a former union member[5] belonged, who has since seen his error and admitted it. This committee called on Senator Whyte, depicted its anxiety about Raum and its fears about the bill, and this was the reason behind the bill's defeat.

In the year 1879 a committee of the union appeared once again before the Board of Health and received the promise that a new investigation would take place, but this promise has not been kept.

During the summer of 1879 Commissioner Raum, at the behest of the union, appointed a man to undertake an investigation of the tenement-house factories; but since this man was a New Yorker and it was known that the Tax Commission of New York was hostile to the cause of the workers there was a protest against his appointment and finally Col. Whittfield[6] of Cincinnati was appointed in his place. Accompanied by an equally authorized union member, he undertook an inspection of the houses and reported that the system should be abolished on moral, social, and economic grounds; but since the tax department was responsible only for the collection of taxes and could not involve itself in moral and other issues, it had no jurisdiction in the matter.

In the winter of 1879-80 a bill was introduced in the Assembly of the State of New York, urging the abolition of tenement factories on sanitary grounds; the bill was referred to committee. During this time the union was bitterly cursed and reviled, and frightful vengeance sworn against the sponsors of the bill.The manufacturers appeared before the committee, hired an outstanding attorney as counsel, published a memorandum, quoted the "Report" of the Board of Health, and spent huge sums of money, but despite all this, initially remained unsuccessful since the bill was reported favorably out of committee. At that point we visited every assemblyman of this city and all but four promised to vote for the bill. The same was done in other parts of the state where there were cigarmaker unions, and all assemblymen from such places, with the exception of Alvord[7] of Syracuse and the four from New York City, voted for the bill. Nevertheless, after a long struggle and a debate which lasted all day, it was rejected by a vote of 49 to 40. At that time the organized cigarmakers in the state numbered not more than 1,500; if they had such an effect then, how much more could they accomplish now that their power has grown so significantly!

But in last year's legislature a bill was introduced again (without the efforts of the organization) and during its debate Alvord admitted having lost 700 votes because the "unionists and communists" had worked against him since he voted against the first bill the year before.

The union has striven to agitate continuously about this question, to make every effort to have it discussed unceasingly and spoken about, favorably or not. It deserves praise for its continuous efforts, and the opposition has served only to direct its strength and make it even more determined to reach its goal. This question, of recent date and familiar at first only here in our city, today finds cigarmakers everywhere on the front lines of the struggle against this system and often better organized than in New York. As an example we will cite the cities of Buffalo, Binghamton, and Rochester where every candidate for office who has voted against the measure or is unwilling to commit himself to vote for it has no chance of success.

The matter will again be brought up and agitated about in the newly elected legislature, possibly to be defeated once more. But whatever may happen, despite all disappointments and defeats, the Cigar Makers' Union has pledged itself to destroy this system of exploitation, a system that even in our age of predatory industry is considered unjust and dishonorable, a system that almost irretrievably demoralizes and degrades thousands of human beings, that lives and feeds on the lifeblood of innocent children, old and infirm men, and innocent women, that drives hundreds to despair and into a premature grave, a system that is a curse, an abscess, and a cancer on the social body of our day.

As conclusion to our description of this system and our struggles against it, I appeal to the working class in general and the cigarmakers in particular: "Let the word pass from mouth to mouth: Go to it! Down with the tenement pestholes. Work for the salvation of one class of mankind and thereby earn the gratitude of all humanity.["]

Yours in the cause of labor
Samuel Gompers.

P.S. I would like to advise my readers, in case one or another wants to undertake an investigation in order to see for himself the truth of my description, that he should not do it right now but only after a month or two, since now, after the publication of these articles, the "landlord-bosses" will strive to do something for appearances. For instance, Samuel Joseph & Co. is now having its houses whitewashed.*

S.G.

* Corrections. (1) Concerning the table on the total results of data gathered on tenement-house cigar manufacture, it is to be noted the

number 3* (see table) means not room, two bedrooms, and kitchen but only room and 2 bedrooms.

(2) We have received a letter from the firm Kerbs & Spiess stating that their rent is $8.50 per month (rather than $9.50 as we had reported) and their wages for home work vary from $5.40 to $7.00 (rather than $5.45 to $6.00 as we reported). After careful inquiries it has been shown that in respect to the rent, the company was correct. Concerning the wage of $7.00 we note that this wage is paid only for an unusually large cigar (5½ inch) and only rarely, when orders are placed for them and only to two families.

*New Yorker Volkszeitung*, Nov. 14, 1881. Translated from the German by Hannelore Jarausch.

1. John Adams Dix (1798-1879) was a Democratic U.S. senator from New York (1845-49), secretary of the U.S. Treasury (1861), and Republican governor of New York (1873-75).

2. Strasser testified before the committee on Aug. 5, 1878 (U.S. Congress, House Select Committee, *Investigation . . . Relative to the Causes of the General Depression in Labor and Business, etc.*, 45th Cong., 3d sess., 1879, Misc. Doc. 29 [Washington, D.C., 1879], pp. 99-105).

3. William Pinkney Whyte (1824-1908) was a Democratic U.S. senator (1868-69, 1875-81, 1906-8) and governor of Maryland (1872-74).

4. On Feb. 18, 1879, Edwin Johnson and E. J. Oppeit, representing the Cigar Manufacturers' Association of Baltimore, appealed to Maryland legislators to oppose the anti–tenement-house bill.

5. Edwin Johnson was president of Baltimore CMIU 1 during the early 1870s and president of the CMIU between 1871 and 1872. He opened a tobacco shop and cigar factory about 1875 and then, during the 1880s, became a salesman.

6. Smith Adams Whitfield (1844-95) worked for the U.S. Bureau of Internal Revenue between 1866 and 1879 and as a postmaster in Cincinnati during the 1880s and early 1890s.

7. Thomas Gold Alvord (1810-97) was elected to fifteen terms in the New York Assembly between 1843 and 1881, first as a Democrat and then a Republican, and was lieutenant governor of New York between 1864 and 1866.

## A Series of News Accounts of the Founding Convention of the FOTLU in Pittsburgh[1]

[November 15, 1881]

### CHAMPIONS OF LABOR.

Between fifty and sixty delegates to the National Labor Congress, which meets at Turner Hall this afternoon, had arrived in the city

last evening. A number of additional delegates from a distance are expected to arrive to-day, and at least 120 are expected to be present when the roll is called. Every branch of organized labor from Boston to San Francisco will be represented, and the delegates have been selected with especial reference to their fitness for the duties they are expected to perform. Among those from a distance are H. S. Streat,[2] of Chicago Typographical Union;[3] R. Powers, President of the Chicago Lake Seamen's Union; M. L. Crawford,[4] Chicago Trades Assembly; L. A. Brant, of Detroit, International Typographical Union; Samuel Gompers, of New York, International Cigar Makers' Union; H. Foster,[5] Cincinnati Trades Assembly,[6] and Samuel L. Leffingwell, Indianapolis Trades Assembly. The local organizations, the glass-workers associations, of which there are three, the Amalgamated Association, Knights of Labor and coal miners, will be represented by their best men, and the cause of labor is expected to derive lasting benefit from the deliberations of this and future congresses.

THE OBJECTS OF THE CONGRESS.

"The object of the gathering," said one of the delegates to a *Commercial Gazette* reporter, "is to concentrate the forces of labor, in order that needed reforms may be more easily obtained. In the federation which it is proposed to form, all questions affecting the national interests of the various trades will be discussed, a Congressional Labor Committee will be appointed to secure the passage by the United States Congress of such laws as are needed by the various trades to better their condition. In addition to this the yearly meetings of the association will bring the principles of trade unionism before the public in their proper light, and new organizations will be formed in localities which are now apparently hostile to organized labor. In Great Britain an organization[7] similar to the one proposed has long existed; yearly meetings are held, and the result has been to greatly ameliorate the condition of the working people. Committees are appointed each year which exercise a general supervision over the affairs of all trades, direct agitation when it is needed, and pay particular attention to securing the passage of beneficial and the repeal of obnoxious laws. The result has been satisfactory in the highest degree, and promises to continue so, as the Secretary of the Parliamentary Committee, Mr. Henry Broadhurst,[8] has recently been elected to Parliament, and within the House of Commons can accomplish much more than he could outside.

"The advisability of organizing a federation of all the trades in this country has been under discussion for a long time, but it was not until last August, when a national convention of trade unionists was

held in Terre Haute, Indiana, that any decisive steps were taken. The subject was fully discussed at that convention, and as a result the Congress which meets to-day was called. The federation will be organized with a few simple rules and no salaried officers. The expenses of its management will be trivial, and can easily be provided for."

THE WORK TO BE DONE.

The convention will assemble at two o'clock this afternoon and will likely remain in session several days. All meetings will be open to the public. Mr. M. L. Crawford, who represents the Chicago Trades Assembly, will call the Congress to order, and President Jarrett,[9] of the Amalgamated Association, will probably be chosen temporary chairman. Steps will be taken to secure the repeal of obnoxious conspiracy laws in the several States; the tariff, and the most effective means to make it protective in the highest degree, will be discussed; and some action will be taken looking to the curtailment of the power of monopolies of all descriptions. The advisability of advocating a law making eight hours a day's work will also be considered, and will likely excite a protracted discussion, as in many of the trades represented it will be an impossibility to conform to any such limitation. The great work of the congress, however, will be to perfect a federal organization, and all other considerations will be made secondary at this time.

*Pittsburgh Commercial Gazette,* Nov. 15, 1881.

1. See "The KOL, the Trade Unions, and the Founding of the FOTLU," above.
2. Harrison S. Streat (1841-1906), a printer for the *Chicago Tribune,* was born in Virginia and became a member of International Typographical Union (ITU) 16 of Chicago, serving as vice-president (1886) and as president (1887-88).
3. ITU 16 of Chicago was organized in 1852.
4. Mark L. CRAWFORD, a prominent member of ITU 16, was a leader during the 1880s in the Trade and Labor Assembly of Chicago and ITU president in 1883-84.
5. William Henry FOSTER, an Irish-born printer, was the founder of the *Cincinnati Exponent* and an officer of ITU 3 of Cincinnati.
6. The Cincinnati Trades and Labor Assembly was organized in 1878; it merged with the Central Labor Union in 1892 to form the Central Labor Council.
7. The TRADES Union Congress of Great Britain.
8. Henry BROADHURST.
9. John JARRETT was president of the National Amalgamated Association of Iron and Steel Workers from 1880 to 1883.

[November 15, 1881]

## THE LABOR CONGRESS.

The Allegheny county delegates to the Labor Congress, which met this afternoon at Turner Hall, were in caucus last night and this

morning. The object was to consult as to rules of the Congress and also to select subjects which the delegation consider as most desirable for the Congress to take action upon. Committees were appointed last night and these reported this morning. It is understood two of the resolutions will include a deliverance in relation to convict labor and the systematic appointment of Congressional committees. As far as regards the first named the system will be condemned in emphatic terms. There is no doubt this will be adopted unanimously by the Congress. As to Congressional committees, the idea is to follow the example set in Great Britain, where Parliamentary committees have long existed. The duty of each committee will be to observe all laws relating remotely or directly to labor, emanating from its district, and take steps to prevent the passage of inimical acts and advocate the adoption of favorable laws. It is not expected that this proposition will meet with any opposition.

The fact that the Allegheny county delegates, who are some fifty in number, have been in caucus, has created the impression among some of the foreign[1] delegates that the object was to devise means to capture the organization of the Congress. Nothing was further from their thoughts. It is said that there will be developed an inharmonious spirit, however, by the Socialistic element. There are a number of radical Socialists from abroad who will attend the Congress, and as is the wont of that peculiar class, unless they shall be permitted to shape the proceedings to suit their views, they will protest. If they do not bolt it will be the turning of a new leaf in the history of labor conventions in which they have taken part. The practical labor men are supposed to be by far the stronger, so that the actions of the Convention will most likely be according to their way of thinking. Pittsburgh is an eminently practical place full of workingmen who earn their living by plying their skill and muscle, and who have not any sympathies with individuals who run on wind and sentiment.

Since yesterday quite a number of delegates arrived and it is thought the Congress will go to work with not less than three hundred men on the floor. Morning, night and afternoon sessions will be held. Among other suggestions of subjects mentioned unofficially by delegates there will be: The repeal of the obnoxious conspiracy laws in the several States; the tariff, and the most effective means to make it protective in the highest degree; the curtailment of the power of monopolies of all descriptions; and the advisability of advocating a law making eight hours a day's work. The main work of the convention, however, will be, as stated in the *Chronicle* yesterday, to perfect a national federation of trades.

*Pittsburgh Evening Chronicle,* Nov. 15, 1881.

1. That is, from outside the Pittsburgh region.

[November 16, 1881]

## THE LABOR CONGRESS.

At two o'clock yesterday afternoon the delegates to the National Labor Congress assembled in Turner Hall, on Sixth avenue. Mr. L. A. Brant, of Detroit, called the Congress to order, and, after the object of the gathering had been stated, a prayer for guidance was offered by A. M. McKenzie,[1] a delegate from New York. President Jarrett, of the Amalgamated Association, was chosen Temporary Chairman, and M. L. Crawford, of Chicago, and H. H. Bengough,[2] of Pittsburgh, were chosen Temporary Secretaries. A committee on credentials, consisting of one delegate from each State, was then appointed, and this developed the fact that delegates were present from Massachusetts, New York, Pennsylvania, Ohio, Michigan, Illinois, Missouri, California, Maryland, West Virginia, Indiana and Wisconsin.

The committee retired to prepare their report, and during their absence short speeches, advancing ideas to be discussed at some future time, were made by a number of gentlemen. All were conservative in tone, and, while the difference between labor and capital was referred to as a conflict that was irrepressible, none of the speeches were in the slightest degree communistic. On the contrary, the intelligence and moderation displayed was remarkable. All the speakers expressed themselves as being in favor of the greatest moderation. Mr. Gompers, the representative of the International Cigar Makers' Union, said he had come to Pittsburgh, not to air his opinions, but to work, not to build a bubble, but to lay the foundation for a superstructure that would be solid, and that would be a true federation of trade unions. He was in favor of progressing slowly, and wanted the organization to be emphatically a workingmen's organization; one that is not defiled by money, but which will in itself contain the elements of strength. His plan for the organization was similar to that outlined in the *Commercial Gazette* yesterday morning, and this, he said, could not be good unless the founders were good, could not be honest unless they were honest; therefore, the elements essential to success were goodness, honesty, industry and practicability.

The Committee on Credentials returned about half-past four o'clock and reported that ninety delegates were entitled to seats. As their names were called the delegates reported the number of constituents they represented. The aggregate was 215,634. A number of additional delegates have arrived since the committee reported, and it is thought that at least half a million workingmen will be represented.

After the adoption of the report of the Committee on Credentials,

a Committee on Permanent Organization was appointed, the time of meeting was agreed upon, and a committee consisting of one delegate from each State represented was appointed to prepare a declaration of principles.

J. S. Shattuck,[3] of Beaver Falls, Pa., presented a resolution endorsing the action of the local trade organizations in reference to the daily papers of this city which do not employ union printers, and prohibiting any representatives of those papers from attending the meetings of the Congress. The resolution was adopted, and the meeting adjourned.

In the evening a meeting was held for the interchange of views on questions of interest to the trades represented. President Cline,[4] of the Window Glass Workers' Association,[5] was made Chairman for the evening. Mr. John Flanigan[6] was the first speaker. He took for his subject, "Convict Labor," and spoke at length of the injustice of the services of criminals being sold to contractors at prices which enabled them to undersell manufacturers of all classes who employ free labor. The extent of the evil was shown by statistics, and the Congress earnestly exhorted to give the subject their closest attention.

Mr. T. W. Taylor,[7] of Homestead, Pa., who has acquired a wide reputation as an advocate of principles calculated to advance the best interests of the workingmen, and who is familiarly known as "Old Beeswax," was the next speaker. He contended that if workingmen did not enjoy all the rights belonging to them they had no person to blame but themselves, as they were not united as a whole, and, in their divided efforts to ameliorate their condition, had contented themselves with feebly attempting to lop off a few branches of the forest which stood in their way instead of striking at the root of the evil and pulling it up. As workingmen they did not take sufficient interest in politics, and thereby showed that they did not respect their own power. He did not advocate the principles of any party, but predicted that as long as workingmen voted for men whose sympathies were not with them, they could never expect to advance. Another of the follies of working men was intemperance, and as long as they continue to spend $7,000,000 annually for drink, all their efforts at reform would prove futile. In his opinion, the most effective means of bettering their condition was co-operation. Until working men joined together in co-operative societies and established mills and manufactories there would continue to be a conflict between capital and labor. It was by this means alone that their interests would become identical. He spoke at length on this subject, reciting the success of co-operation in England, and backed up his assertions with numerous instances. In conclusion he urged the delegates and visitors to consider

what he had said, and in future take an active interest in politics, avoid intemperance, and establish co-operative societies as rapidly as possible. When Mr. Taylor had finished, the chairman made a few remarks and the meeting adjourned.

The meeting to-day, at which a permanent organization will be effected, is expected to be very interesting. The business to be transacted is of great importance and will likely determine the usefulness of the federation. It is conceded that the success of the Congress depends greatly upon the character of the organization effected.

Among the delegates mentioned for permanent Chairman are R. Powers, of Chicago, President of the Lake Seamen's Union; M. L. Crawford, representing the Chicago Trades Assembly; M. Walsh,[8] ex-President of the International Typographical Union, representing the Milwaukee Trades Assembly;[9] L. A. Brant, of Detroit, representing the International Typographical Union; President Jarrett, of the Amalgamated Association; President Cline, of the Window Glass Workers' Association; A. W. Rankin,[10] of the Moulders' Union of this city;[11] and M. Gompers, of the International Cigar Makers' Union. The latter is the leader of the Socialistic element, which is pretty well represented in the Congress, and one of the smartest men present. It is thought that an attempt will be made to capture the organization for Mr. Gompers, as the representative of the Socialists, and if such an attempt is made, whether it succeeds or not, there will likely be some lively work, as the delegates opposed to Socialism are determined not to be controlled by it. If the Socialists do not have their own way, they may bolt, as they have always done in the past. If they do bolt, the power of the proposed organization will be so seriously crippled as to almost destroy its usefulness.

The majority of the delegates realize the importance of effecting an organization that will harmonize all differences likely to arise, and last evening seemed hopeful that this could be accomplished. They think that the Committee on Organization will present the name of Mr. Rankin of this city, or some western man, for permanent Chairman, and that the Socialistic element will be prevented not only from capturing the organization, but from introducing any of their peculiar ideas into the declaration of principles to be prepared.

*Pittsburgh Commercial Gazette,* Nov. 16, 1881.

1. Kenneth Mackenzie (1828-1909), vice-president of the Amalgamated Trades and Labor Union of New York and Vicinity (ATLU), was a New York bookbinder and itinerant Methodist minister. He was president of the ATLU from July 1882 to 1885.

2. Herbert H. Bengough (1845-1926?), a Pittsburgh printer, represented KOL Local Assembly 1630 at this convention. He was later a member of the International

Typographical Union (ITU). In the 1890s Bengough became a U.S. pension agent and a court clerk.

3. John S. Shattuck (b. 1828), a machinist born in Massachusetts, represented KOL Local Assembly 1569 at this convention. The following year he represented District Assembly 3 at the KOL General Assembly.

4. Isaac CLINE, a Pittsburgh window-glass worker, was president of KOL Window Glass Workers' Local Assembly 300 from 1881 to 1887.

5. WINDOW Glass Workers' Local Assembly 300.

6. Possibly John Flanigan (variously Flannigan; b. 1820?), a Pittsburgh iron molder born in Ireland. He was a member of the Allegheny County committee of the Greenback-Labor party and recording secretary of KOL District Assembly 3 in the early 1880s.

7. Thomas W. Taylor (b. 1817?) was born in England. A member of the cooperative movement led by Robert Owen, he later became a close associate of Owen's son, Robert Dale Owen, in the New Harmony cooperative community in Indiana. Settling in Homestead, Pa., he was active in the Greenback-Labor party in the late 1870s and early 1880s, running for the Pennsylvania legislature in 1884.

8. Michael P. Walsh (1838-1919), a printer, was born in Ireland and came to the United States in 1840. Walsh was president of ITU 23 of Milwaukee and the Milwaukee Trades Assembly in the early 1880s. He was elected to the Wisconsin state legislature in 1883 and 1885 and in 1885 began working as a collector for the U.S. Bureau of Internal Revenue. He served as sheriff of Milwaukee County from about 1890 to 1893 when he returned to printing and the presidency of ITU 23.

9. The Milwaukee Trades Assembly was organized in 1880; it was succeeded by the Milwaukee Central Labor Union in 1886.

10. Alexander C. RANKIN, a Pittsburgh iron molder, represented Iron Molders' local 14 at this convention.

11. At the founding of the National Union of Iron Molders in 1859, the Pittsburgh molders' union became local 14.

[November 17, 1881]

## LABOR REFORMERS.

The second day's business of the International Labor Congress was commenced yesterday morning at eight o'clock. The credentials of a number of additional delegates were presented, and then Mr. Howard,[1] of Massachusetts, presented the report of the Committee on Organization. It recommended that the Congress should be officered by a President, two Vice Presidents, a Secretary and two Assistants, and on behalf of the majority of the committee the following names were presented: For President—Samuel Gompers of New York, representing the Cigarmakers' International Union; for Secretary—William Wilson,[2] of St. Louis Typographical Union No. 16;[3] for Assistant Secretary—H. H. Bengough, of Local Assembly No. 1,630.[4]

Mr. Gompers took occasion to deny the statement that he was a leader of the Socialistic element, and that the committee had been captured for him, saying that he had attended the Congress only for the purpose of assisting in the federation of labor organizations. A

number of delegates differed with him, however, as Mr. Street, on behalf of the minority of the Committee on Organization, offered the following persons to be voted for as officers: For President—Richard Powers, of the Lake Seamen's Union, Chicago; First Vice President—Robert Howard, of Spinners' Union, Fall River, Mass.; Second Vice President—William Wilson, of Printers' Union, St. Louis, Mo.; Secretary—R. H. Weber,[5] of Printers' Assembly, No. 1630, Pittsburgh; First Assistant Secretary—Robert Burns,[6] of Buffalo, N.Y.; Second Assistant Secretary—W. C. Polliner, of the Cleveland Trades Assembly.

Mr. James Maloy[7] thought the Amalgamated Association should be represented on the ticket, and one of the delegates nominated Mr. John Jarrett for Permanent Chairman. Mr. Walsh, of Milwaukee, spoke in favor of the election of Mr. Powers, stating that the labor of the West needed to be organized, and that his candidate was the man to do the work successfully. For a time it looked as if the chairmanship would be hotly contested, but Mr. Gompers poured oil on the troubled waters by stating that he was thoroughly devoted to trade unionism, and in order to facilitate the work of completing the organization, would withdraw his name. Mr. Powers gracefully followed suit, and Mr. Jarrett was unanimously chosen Permanent Chairman. Messrs. Powers and Gompers were chosen vice presidents, and the temporary secretaries, Messrs. Crawford, of Chicago, Bengough, of Pittsburgh, and Pollener, of Cleveland, were retained.

The organization having been effected, the names of the committee to prepare a declaration of principles was announced, and then the Congress got down to work. The roll of States was called, and resolutions were introduced denouncing Chinese labor and favoring the protection of trade organizations by a law providing for their incorporation; demanding the repeal of conspiracy laws so far as they related to labor organizations regulating their own wages, and denouncing convict contract labor, demanding its abolition and pledging themselves not to support candidates for the Legislature who did not pledge themselves to attempt its abolishment.

These resolutions were received with many demonstrations of applause, and after being referred to the committee, a recess of an hour was taken for dinner.

Upon reassembling it was decided [to appoint a?] committee [consisting] of one delegate from each State to prepare a plan for organization.

The call of States was resumed, and a resolution was presented expressing the sympathy of the workingmen for the oppressed people in Ireland, and enunciating the idea that land was property which

should be owned only by actual tillers of the soil, and that no person should be allowed to control more land than he could till. This resolution, on account of the peculiar idea expressed, excited considerable discussion. A majority of the speakers held that the idea was not correct, and that it would not be proper for the Congress to endorse such a proposition. The resolution was finally referred to a special committee.

Other resolutions were presented, as follows: By Daniel Rodgers,[8] of Pennsylvania, providing for the reduction of the hours of labor, and the enforcement of all laws relating to miners and mining; by John Axtel,[9] of Pennsylvania, on behalf of the boiler makers, condemning button-set rivets on boilers; by James Michaels,[10] of Pennsylvania, denouncing the wholesale importation of cheap foreign labor; by Mr. Powers, asking State Legislatures to enact laws requiring stationary engineers to be licensed; by Mr. Gompers, asking for the establishment of a bureau of statistics, and by Mr. Cline, asking that Congress levy a tax on all labor imported for contract purposes. These resolutions were all referred to the committee, after which Mr. David Rogers, representing the miners of Western Pennsylvania, presented a resolution of respect to the memory of Alexander McDonald,[11] M.P., recently deceased, and directing that a message of condolence be sent to the miners of Great Britain, which was adopted unanimously.

The report of the Committee on Rules was presented and adopted, after which the report of Mr. Moore, Secretary and Treasurer of the Terre Haute convention, giving a statement of moneys contributed and dispersed, was presented. Its consideration was deferred until today.

The Committee on Resolutions were, after some discussion, authorized to have their report printed. During the discussion of the motion the character of the organization to be formed was alluded to. Mr. Jarrett was of the opinion that the work of this Congress could only be preliminary, as he, for one, could not act until he had obtained the consent of his organization. Other delegates differed in opinion, but the majority thought with Mr. Jarrett that the only thing to be done was to agree upon a plan of organization, make a declaration of principles, and submit the same to the various trades. Then, if it were deemed advisable, a permanent federation of trades could be formed.

After this discussion the Congress adjourned to meet again this morning at Schiller Hall. The principal business to be considered is the report of the committee to which was referred the Irish resolution; the plan of organization, and the declaration of principles. The discussion of the first and last subjects will likely be animated, and the

reference to the rights of landowners will be expunged if the resolution is adopted, as the majority of the delegates, while sympathizing with the Irish people, do not think it comes within their province to dictate how many acres of land shall be owned by any person. Unless some new questions are sprung upon the Congress it will likely finish its work to-day.

*Pittsburgh Commercial Gazette,* Nov. 17, 1881.

1. Robert HOWARD, a spinner, was secretary of the Fall River Mule Spinners' Association from 1878 to 1897 and principal officer of the Amalgamated Mule Spinners' Association from 1878 to 1887.

2. William Wilson was a member of International Typographical Union (ITU) 8 of St. Louis and was its delegate to the St. Louis Trades Assembly in 1881.

3. Actually ITU 8 of St. Louis, derived from one of the locals that founded the National Typographical Union in 1852.

4. KOL Printers' Local Assembly 1630 of Pittsburgh, which was organized in 1881.

5. Robert E. Weber (variously Webber; d. 1885?), recording secretary of ITU 7 of Pittsburgh in 1881, represented KOL Printers' Local Assembly 1630 at this convention, and District Assembly 3 at the KOL General Assembly the following year. In 1883 he moved to Wheeling, W.Va., joining ITU 79.

6. Michael J. Byrne (1851-1927?), born in Ireland, represented the Buffalo Operative Plasterers' Union at this convention. He later was a contractor and also operated a saloon for a short time.

7. James Maloy (b. 1838), a Pittsburgh iron worker, represented National Amalgamated Association of Iron and Steel Workers 70 of Pittsburgh at this convention. He emigrated from Ireland in 1850 and lived in Tennessee and Kentucky before moving to Pennsylvania in the 1870s.

8. Daniel Rogers represented the Amalgamated Association of Pittsburgh Miners and Drivers.

9. John W. Exler (1858-1930?), a boilermaker born in Pennsylvania, was a delegate at this convention from KOL Boiler Makers' Local Assembly 1595 of Pittsburgh. In 1888 he was foreman of the Iron City Boiler Works. He became president of the James Lappan Manufacturing Co. around 1916.

10. James L. Michels (variously Michaels), a Pittsburgh glassblower, represented KOL Window Glass Workers' Local Assembly 300 at this convention.

11. Alexander MACDONALD, a Scottish miners' leader, was a Member of Parliament from 1874 until his death in 1881.

[November 18, 1881]

## THE LABOR CONGRESS

The delegates to the International Labor Congress assembled at Schiller Hall yesterday morning, and as all the preliminaries had been arranged, proceeded to transact the real business for which they were called together—the adoption of a plan of organization and a platform or declaration of principles. The minutes of the last session having been read and approved, Mr. Gompers submitted the first article of the "plan of organization," which was as follows:

"This association shall be known as 'The Federation of Organized Trades Unions of the United States of America and Canada,' and shall consist of such Trades Unions as shall, after being duly admitted, conform to its rules and regulations, and pay all contributions required to carry out the objects of this Federation."

A number of delegates thought that labor unions, as well as trade organizations, should be recognized in the name, and after some discussion the article was amended to read "Trade and Labor organizations."

A resolution of sympathy for D. R. Jones,[1] President-elect of the Miners' Association,[2] who is dangerously ill of smallpox, was unanimously adopted, and then the second and third sections of the "Plan" were offered and unanimously adopted. They were as follows:

"The objects of this Federation shall be the encouragement and formation of Trade and Labor Unions; the encouragement and formation of Trade and Labor Assemblies and Councils; the encouragement and formation of National and International Trade and Labor Unions; to secure legislation favorable to the interests of the industrial classes.

["]The sessions of this Federation shall be held annually on the third Tuesday in November, at such places as the delegates have selected at the preceding congress."

Mr. Somers[3] of Joliet Ill., offered a resolution asking that a tariff plank be embodied in the declaration of principles, after which a recess was taken until one o'clock.

THE FIRST HITCH.

Vice President Gompers called the delegation to order at the appointed time, and the fourth section of [the] "plan" which read as follows, was [offered:?]

["The] basis of representation in the sessions of [the] Federation shall be, for the national or international Trades and Labor unions, one delegate for 5,000 members or less and one delegate for every 5,000 or major portion thereof above the first 5,000. For local trades assemblies or councils, one delegate. No local trade or labor union shall be entitled to a representation in the sessions of this Federation where international or national unions of said craft exist, or where there are trades assemblies or councils in the locality."

This caused a lively discussion. Mr. Lynch,[4] of Wisconsin, thought the basis of representation too high, especially as the purpose of the organization was to bring isolated unions together, and to join organizations of craftsmen in different parts of the country. Mr. Rankin

favored the article, stating that the power of the Federation should belong to the highest executive body of the different organizations, and they should therefore select the delegates to the congress. Mr. Powers was of the opinion that local unions, if they number only twenty-five members, should have the right to send delegates to the Congress. Mr. Brennan,[5] of Pittsburgh, favored the classification of the trades, and making the representation according to States. He did not think that a delegate representing twenty-five workingmen should have the same power as the representative of 10,000 workingmen. The debate continued until after three o'clock. When a vote was taken the section was rejected by fifty nays to twenty-eight ayes.

Mr. Gompers called attention to the necessity of incorporating some basis of representation into the plan, and Mr. Brant, of Detroit, offered the following as a substitute:

"That all international and national unions, trades assemblies or councils and local trades or labor unions shall be entitled to one delegate for 100 members or less, and one additional delegate for each additional 500 members, or major part thereof; also, one delegate for each international or national union, and one delegate for each trades assembly or council."

Mr. Brant spoke at length in favor of the adoption of the substitute. The representation was the same as that allowed by the English Federation, and it had been successful. He thought if there was any danger of a congress being captured by the enemies of labor, it would be greatly increased if that congress was composed of a small number of delegates; but there was absolutely no danger if the body was large.

Mr. Gompers did not think Mr. Brant's substitute a fair one, as, if it were adopted, one class of men would always be more numerously represented than the other. In fact, the substitute gave the next city in which the congress was held absolute control of the body, as every insignificant local union, if it had not more than half a dozen members, would send delegates, while unions at a distance could not afford to do so. The congress, he thought, should be representative in its character, and the entire body of workingmen should be benefited, no matter where they were located.

At this point the substitute was referred to the committee, and a letter from the Washington, D.C., Confederation of Labor,[6] expressing sympathy with the Congress and regret at their inability to be represented, was read. A number of the delegates were in favor of tabling the communication because it was not gotten up in official style, but it was finally received.

APPLYING THE GAG LAW.

By this time it was four o'clock, and the whole day had been consumed in the discussion of four or five subjects. President Jarrett took the chair, however, and by a little ruse succeeded in expediting business wonderfully. At a previous meeting a rule had been adopted making it imperative for a vote to be taken on any questions whenever seven members called for the "question." This rule had not been enforced by Mr. Gompers, but when President Jarrett took the chair he enforced it in a manner that made it resemble a self-inflicted gag law. As soon as a motion had been stated, he would ask, "Are you ready for the question?" Immediately the "question" would be called for by a number of delegates, who thought that by so doing they would place the motion in proper shape for debate. But Mr. Jarrett was not of the same mind, and the last three sections of the "plan" were railroaded through with a speed that was highly creditable to Mr. Jarrett's conception of the rule, but not entirely satisfactory to those delegates who thought they should be permitted to air their opinions on every question that came before the house. The last sections were as follows:

"At the annual sessions of this Federation the delegates shall elect a legislative committee[7] consisting of five delegates, one of which shall be the Federation Secretary, who shall be elected separately.

"The duties of the legislative committee shall be to exercise a supervision over the organization and the execution of the laws, and such instruction as may from time to time be given them at the session of the Federation.

"The legislative committee shall choose from among themselves a chairman, first and second vice chairman and treasurer, who shall serve for one year."

This done the special committee to which the Irish resolutions were referred, reported the following as a substitute for the original, explaining their action by stating that while the original resolution was doubtless written by a person actuated by the best motives, the committee did not deem it wise to have it go out with the endorsement of the congress. The substitute was as follows:

*Whereas,* We greatly deplore the unjust land laws that have been enforced against the Irish p[eople in the] past; and

*Whereas,* Hundreds of Ireland's bravest spirits are unjustly imprisoned in consequence of their heroic attempts to ameliorate the condition of her oppressed people; therefore be it

*Resolved,* That we extend to these champions, battling in the cause of human liberty, our hearty sympathy, and that we also extend to

the oppressed of all nations struggling for liberty and right the same encouraging words of sympathy.

The substitute was railroaded through by President Jarrett as the last sections of the plan of organization had been, although several delegates wished to speak on the subject, and one proposed an amendment. This action on the part of the chairman doubtless saved a hot discussion, but it did not meet with the approval of some of the members.

THE DECLARATION OF PRINCIPLES.

The chairman of the committee appointed to prepare a declaration of principles, then read their report, which, as adopted, is as follows:

"Gentlemen of the Labor Union Congress: Your committee, appointed to prepare a declaration of principles, and to whom was referred the various resolutions presented to this congress, having carefully considered all matters pertaining to the subject in question, have the honor to submit the following as the result of their labors:

["]*Whereas,* A struggle is going on in the nations of the civilized world, between the oppressors and the oppressed of all countries, a struggle between capital and labor which must grow in intensity from year to year and work disastrous results to the toiling millions of all nations, if not combined for mutual protection and benefits. The history of the wage workers of all countries is but the history of constant struggle and misery, engendered by ignorance and disunion, whereas the history of the non-producers of all countries proves that a minority thoroughly organized may work wonders for good or evil. It behooves the representatives of the workers of North America in congress assembled, to adopt such measures and disseminate such principles among the people of our country as will unite them for all time to come, to secure the recognition of the rights to which they are justly entitled. Conforming to the old adage, 'In union there is strength,' a formation embracing every trade and labor organization in North America, a union founded upon the basis as broad as the land we live in, is our only hope. The past history of trade unions proves that small organizations, well conducted, have accomplished great good, but their efforts have not been of that lasting character which a thorough unification of all the different branches of industrial workers is bound to secure.

["]Conforming to the spirit of the times, and the necessities of the industrial classes, we declare the following:

["]*Resolved,* That all organizations of workingmen into what is known as a Trade or Labor Union should have the right to the protection

of their property in like manner as the property of all other persons and societies is protected, and to accomplish this purpose we insist upon the passage of laws in the State Legislatures and in Congress for the incorporation of trade unions and similar labor organizations.

["]*Resolved,* That we are in favor of the passage of such legislative enactments as will enforce by compulsion the education of children; that if the State has the right to exact certain compliance [with] its demands, then it is also the right of the [State to e]ducate its people to the proper under[standing of such] demands.

["*Resolved,* That we] are in favor of the passage of [laws in the several States] forbidding the employ[ment of children under the age] of fourteen in any capacity under the penalty of fine and imprisonment.

["]*Resolved,* That necessity demands the enactment of uniform apprentice laws throughout the country; that the apprentice to a mechanical trade may be made to serve a sufficient term of apprenticeship, from three to five years, and that he be provided by his employer in his progress to maturity with proper and sufficient facility to finish him as a competent workman.["]

EIGHT HOURS A DAY'S WORK.

["]*Resolved,* That the 'national eight-hour law'[8] is one intended to benefit labor and to relieve it partly of its heavy burdens; that the evasion of its true spirit and intent is contrary to the best interests of the nation. We therefore demand the enforcement of said law in the spirit of its design.

["]*Resolved,* That it is hereby declared the sense of this congress that convict or prison labor as applied to the contract system in several of the States is a species of slavery in its worst form; that it pauperizes labor, demoralizes the honest manufacturer and degrades the very criminal whom it employs; that as many articles of use and consumption made in our prisons under the contract system come directly and detrimentally in competition with the products of honest labor, we demand that the laws providing for labor under the contract system herein complained [of], be repealed, so as to discontinue the manufacture of all articles which will compete with those of the honest workingman or mechanic.

["]*Resolved,* That what is known as the 'truck' system of payment, instead of lawful currency as a value for labor performed, is not only a gross imposition, but a downright swindle to the honest laborer and mechanic, and calls for entire abolition; and we recommend that active measures shall be enforced to eradicate the evil by the passage of laws

imposing fine and imprisonment upon all individual firms or corporations who continue to practice the same.

[“]*Resolved,* That we favor the passage of such laws as will secure to the mechanic and workingman the first lien upon property, the product of his labor, sufficient in all cases to justify his legal and just claims, and that proper provision be made for legally recording the same.

[“]*Resolved,* That we demand the repeal and erasure from the statute books of all acts known as conspiracy laws, as applied to organizations of labor in the regulation of wages which shall constitute a day's work.

[“]*Resolved,* That we recognize the wholesome effects of a Bureau of Labor Statistics as created in several States, and urge upon our friends in Congress the passage of an act establishing a National Bureau of Statistics, and recommend for its management the appointment of a proper person, identified with the laboring classes of the country.

[“]*Resolved,* That railroad land grants forfeited by reason of non-fulfillment of contract should be immediately reclaimed by the government, and henceforth the public domain reserved exclusively as homes for actual settlers.[”]

A Protective Tariff Wanted.

[“]*Resolved,* That we recommend to the Congress of the United States the adoption of such laws as shall give to every American industry full protection from the cheap labor of foreign countries.

[“]*Resolved,* That we demand the passage of a law by the United States Congress to prevent the importation of foreign laborers under contract.

[“]*Resolved,* That we recommend to all trades and labor organizations to secure proper representation in all law-making bodies by means of the ballot, and to use all honorable measures by which this result can be accomplished.[”]

The preamble and first and second resolutions were adopted without dispute, but the third, which related to the employment of children under fourteen years of age, excited a protracted discussion. It was opened by Mr. Brennan, of Pittsburgh, who thought it almost too wide in its bearings. He thought it should be qualified by amending it to read “children under fourteen years of age shall not be employed at any mechanical labor.” The majority of the delegates differed with Mr. Brennan, and spoke at length reciting the results of their own observation. None of them urged the prohibition of child labor, because it might decrease their own wages, but all said they wanted the

influence of the congress to go out against it, because of its cruelty and demoralizing effects. Mr. Bergman[9] painted a graphic picture of the misery of the child laborers of the Pacific slope, and how they were growing up in ignorance and without enjoying the brightest days of their lives. Mr. Gompers spoke of his investigations among the tenement cigar makers of the metropolis; how he found little children who were too young to understand any of the questions he asked of them, but yet were compelled to work from before daylight until after dark, and how he often found the little ones fast asleep beside their work, being unable to hold out. Other delegates recounted their experiences, and if these stories, coming from men who knew what they were talking about, and which were pathetic enough to bring tears to most eyes, could be published in full, they would form a powerful argument in favor of keeping the little ones out of the work shops and sending them to school where they belong. The resolution as it appears above was adopted unanimously.

The other resolutions were adopted with very little debate, until one was read which declared in favor of all the railroads and telegraph lines being purchased and controlled by the government. This was [dis?]approved by a number of delegates, on the ground that if the government obtained the control favored, it would make the power of the ascendant political party perpetual, by reason of the vast number of employes which would be placed at its mercy. President Jarrett ruled the resolution out of order, as having no relation to the objects of the congress. An appeal was taken from his decision, but the Chair was sustained and the resolution left out.

PITTSBURGH PRIDE VS. WESTERN PRINCIPLE.

The next discussion was on that plank declaring in favor of the protection of American industries. It was a fight of Pittsburgh pride against Western principle, and the debate was warm. Mr. Brant, of Detroit, opened the ball by stating that he did not think the Congress [sh]ould commit itself to the tariff, as by many the protective policy was not held to be necessary to the success of the nation.

President Jarrett replied at length, enumerating the benefits which American workingmen had derived from a high tariff, and contrasting the wages they received with those paid in Great Britain.

Mr. Somers, of Joliet, Illinois, said that while he was a Western man, he appreciated the force of Mr. Jarrett's remarks, and was radically in favor of having the Congress declare for a high tariff. They were all workingmen, and all would be benefited.

Mr. Cummins,[10] of Boston, thought the discussion was not in order,

or at least not necessary. The protectionists were in power, and he thought it would be time to take the bull by the horns when it approached. He did not favor the resolutions, anyhow, and attempted to prove that the workingmen of Great Britain were as well off as those of this country.

Mr. Bergman, of San Francisco, was not a protectionist, but was willing to vote for the resolution as an experiment and out of courtesy to his Pittsburgh friends. He was of the opinion, however, that Capital was more benefited by a protective tariff than Labor.

Mr. Crawford, of Chicago, thought it was a mistake to force the resolution through, as it would only cause dissension. As long as the east and west were situated as they are at present they would not agree on the subject. Therefore he was in favor of not making any reference to the tariff, promising at the same time that if the East offered no tariff resolutions, none advocating free trade would come from the West.

Mr. Michaels, of Pittsburgh, was sorry the question had been broached, but now that it had been mentioned it must go through.

An attempt was then made to lay the resolution on the table, but it was voted down, and after another strong argument by Mr. Jarrett in favor of the resolution it was adopted.

The other resolutions were adopted without debate, and at six o'clock the Congress adjourned to meet at the same place this morning.

### The Work Yet to Be Done.

The delegates expect to get through with all business to-day, and adjourn *sine die*. If they succeed, President Jarrett will have to enforce the rule which worked so admirably yesterday, as considerable business yet remains to be transacted. The Committee on Resolutions have a number of resolutions which were not incorporated in the platform, and which will have to be disposed of. Then the basis of representation in future congresses must be settled, and this is likely to excite prolonged discussion. In addition to this the officers provided for in the plan of organization will have to be chosen. For the Legislative Committee the most prominent names mentioned are Jarrett, Cline, Rankin and Layton of Pittsburgh; Gompers and Lynch, of New York; Cummins and Howard, of Massachusetts; Crawford and Powers, of Chicago; Brant, of Detroit; Walsh, of Wisconsin, and Bergman, of San Francisco. They will be elected off the floor, as will also be the Secretary, who will be the only salaried officer, and it is not certain that he will be salaried. As the committee elects its own officers, and

one will be President of the Federation, it is impossible to make any reasonable prediction as to who will draw the prizes. The delegates seem determined to sink all personal considerations, however, and put the best men at the head of the Federation. It is likely that those who are considered the best organizers will be chosen, as the work of the next year will be principally confined to solidifying the ranks of trade unionism, healing local differences, and organizing trades which at present have no unions.

*Pittsburgh Commercial Gazette,* Nov. 18, 1881.

1. David R. JONES was chief executive officer of the Amalgamated Association of Pittsburgh Miners and Drivers from 1879 to 1882.

2. The Amalgamated Association of Pittsburgh Miners and Drivers was organized in late 1879 and was apparently succeeded in 1883 by the Miners' Amalgamated Association of Pennsylvania.

3. Perry G. Somers of Joliet, Ill., was vice-president of the Fourth District of the National Amalgamated Association of Iron and Steel Workers from 1881 to 1882.

4. James LYNCH, a New York City carpenter, was president of the Amalgamated Trades and Labor Union of New York and Vicinity.

5. William J. Brennen (1850-1924?), a machinist born in Pittsburgh, represented Pittsburgh KOL Local Assembly 791. Brennen was involved in the Greenback-Labor movement in the 1870s and was active for most of his life in the Democratic party, serving as a city alderman in the early 1880s and running unsuccessfully for Congress in 1890. He became a lawyer in 1883 and during his career provided legal services for organized labor in Pittsburgh.

6. The Washington (D.C.) Federation of Labor was organized in the spring of 1881; it was supplanted by the Washington Central Labor Union in 1896.

7. The Legislative Committee, elected at the annual sessions of the FOTLU, served as the executive body of the organization. It initially consisted of five officers: a secretary, designated by the annual session, who was the predominant official; a chairman; first and second vice-chairmen; and a treasurer. In 1883 the FOTLU changed the title "chairman" to "president" in its constitution and enlarged the committee to nine by including six vice-presidents in place of the vice-chairmen. These officers were to be elected by the annual session of the FOTLU, with the election of the secretary taking precedence.

8. An act of June 25, 1868 (U.S. *Statutes at Large,* 15:77), established an eight-hour workday for all laborers and mechanics employed by or on behalf of the U.S. government.

9. Charles F. BURGMAN, a tailor, was a delegate to the 1881 FOTLU convention from the Representative Assembly of Trades and Labor Unions of the Pacific Coast.

10. Sherman Cummin (b.1849), a printer born in New Brunswick, Canada, immigrated to the United States in 1867. He represented International Typographical Union (ITU) 13 of Boston at this convention. About 1884 he moved to New York City, joining ITU 6.

[November 19, 1881]

## ADJOURNED SINE DIE.

The closing sessions of the Labor Congress were held at Schiller

Hall yesterday. At the morning session Mr. Samuel L. Leffingwell, of Indiana, Chairman of the committee which prepared the declaration of principles, reported the resolutions which had been presented, but not incorporated in the report. One of the resolutions pledged the delegates to use their best efforts to get rid of the Chinese labor evil, and others asked for the prohibition of tenement house cigar making. Both resolutions were adopted, and then the committee on the "plan of organization" finished their report. By it the basis of representation in future Congresses will be as follows: From national or international unions, for one thousand members or less, one delegate; for four thousand, two delegates; for eight thousand, three delegates; for sixteen thousand, four delegates; for thirty-two thousand, five delegates, and so on. From local trades assemblies or councils, one delegate. The other sections provided for the manner of conducting the business of future congresses, defined the duties of the Legislative Committee, and authorized the committee to solicit subscriptions to defray their expenses.

After the report had been adopted President Jarrett stated that he was compelled to attend to other duties, and turned the convention over to Vice President Powers. A vote of thanks was tendered Mr. Jarrett, and then a number of resolutions of minor importance were disposed of. This consumed the balance of the morning.

After dinner the Congress proceeded to the election of officers. A Secretary was first chosen. Messrs. Gompers of New York; Foster, of Cincinnati; Crawford, of Chicago; Leffingwell, of Indiana; Pollner, of Cleveland, and Weber, of Pittsburgh, were nominated. The last two gentlemen declined, and after three ballots had been taken Mr. Foster was declared elected. His election was made unanimous on motion of Mr. Gompers.

In order that all sections of the country might be represented in the Legislative Committee, it was decided to appoint a committee consisting of one delegate from each State to present names to be voted upon.

During the absence of this committee, Mr. Brant, of Detroit, offered a series of resolutions declaring that the bill introduced in Congress in 1880, as part of the report of the Public Land Commission[1] would have the effect, if passed, to place the bulk of the public lands at the disposal of Western cattle kings and other capitalists at a nominal figure; that those lands in a few years would be found very valuable for farming purposes, and that persons wishing to cultivate them would have to do so in the capacity of tenant farmers or hirelings in competition with Chinese labor. In view of these facts the resolutions

urged that all labor assemblies pass resolutions, giving their Congressional representatives to understand that if they voted for the measure, they would be punished by the political opposition of the workingmen. After a short discussion the resolutions were adopted, as were also the following, which were presented by Mr. Rodgers of Pennsylvania:

*Resolved,* That we demand strict laws for the inspection and ventilation of mines, factories and workshops, and sanitary supervision of all food and dwellings.

*Resolved,* That strict laws be enacted making employers liable for all accidents resulting from their negligence or incompetence to the injury of their employes.

A series of resolutions from the Brass Finishers' and Brass Moulders' Union,[2] appealing to the Congress to support the efforts of the Detroit Union in organizing the trade throughout the country, were presented by Mr. Lynch.[3] A number of similar resolutions were presented by other delegates, and the whole were referred to a special committee with instructions to condense them into one resolution.

The committee appointed to select names for the Legislative Committee by this time had returned, and reported that they had agreed upon Samuel Gompers, of New York; Richard Powers of Illinois; C. F. Bergman, of California, and A. C. Rankin, of Pennsylvania. Their choice was unanimously confirmed, and the officers-elect were loudly called upon for speeches. Each responded briefly, pledging themselves to perform the duties devolving upon them faithfully, and predicting glorious results.

It was then agreed that the next Congress should be held in Cleveland, on the third Tuesday of November, 1882,[4] and after a resolution offered by Mr. Gompers, thanking the *Commercial Gazette,* and other Union papers of Pittsburgh, for the fair reports of the proceedings, had been passed a fervent prayer for success was offered by Mr. McKenzie, of New York, and the Congress adjourned *sine die.*

The Legislative Committee will meet for organization to-day. It is hard to predict who will be chairman, as all are good men and equally popular. The Secretary, Mr. Foster, is employed as a compositor on the Cincinnati *Enquirer,* and although a young man, is President of the Trades Assembly of Cincinnati. Mr. Gompers is organizer of the International Cigar Makers Union, of New York. Mr. Powers is General President of the Lake Seamen's Union, and is considered one of the best organizers in the West. Mr. Bergman is Treasurer of the Trades Assembly of San Francisco[5] and President of the Tailors' Union of that city. Mr. A. W. Rankin is a member of the Iron Moulders' Union of this city, and is well known. All the officers are men of more

than ordinary intelligence, conservative in their disposition, and their choice gives general satisfaction to the delegates.

*Pittsburgh Commercial Gazette,* Nov. 19, 1881.

1. H.R. 4805 (46th Cong., 2d sess.), introduced Mar. 1, 1880, proposed to sell certain public lands as pasturage. The bill provided that the initial price would be $1.25 per acre and that it would decline to 12½¢ per acre by 1890. The bill never became law.

2. Detroit Brass Finishers' and Molders' Protective Union 1.

3. Charles D. Lynch (1842-1922?) was born in Ireland and immigrated to the United States in 1848. He came to Detroit around 1869, where he was a brass finisher, saloon keeper, and traveling agent for brass and iron work companies.

4. The 1882 FOTLU convention met in Cleveland, Nov. 21-24.

5. The Representative Assembly of Trades and Labor Unions of the Pacific Coast was organized in San Francisco in 1878; it disbanded about 1884.

## Excerpts from the Minutes of the Legislative Committee of the FOTLU

Pittsburg, Nov. 19, 1881.

An adjourned meeting[1] of the Legislative Committee of the Federation of Organized Trades and Labor Unions of the United States and Canada was held in Room 22, St. Clair Hotel, at 1 P.M. of the above date, Chairman Richard Powers presiding and all members present.

The Committee on Seal reported having selected a design of hands crossed over a globe and clasped, which had been approved by the Legislative Committee, and they had ordered a seal accordingly. Report received.

The Committee on Address to Unions submitted the following form as a report:

Cincinnati, Ohio,——— 1881.

To the Trades and Labor Unions of the United States and Canada:
Fellow-workingmen:

At the Labor Congress assembled at the City of Pittsburg a Federation of the Trades and Labor Unions was resolved upon, a Legislative Committee elected, a Plan of Organization and a Declaration of Principles adopted.

The necessity of such a Federation has long been recognized by the organized workingmen of our land, the practical benefits that

would result therefrom almost incalculable. We would respectfully call your attention and earnest consideration to the following Declaration of Principles and Plan of Organization, and solicit your hearty co-operation in the endeavor to bring to perfection and successful consummation this Federation.

[Insert Declaration of Principles and Plan of Organization.][2]

You will observe that for the ensuing year the existence and efficiency of the Legislative Committee will greatly depend upon the liberality displayed by your respective organizations. The following resolution was unanimously adopted by the Congress, bearing upon this subject:

"Resolved, That the Legislative Committee are empowered and hereby authorized to solicit subscriptions from the Trade and Labor Unions and labor sympathizers for the purpose of assisting in defraying the necessary expenses of the Committee for the ensuing year."

We thus, in making this, our first address to the labor organizations, do at the same time make an appeal for assistance, so that we may be placed in a position to perform at least a part of the arduous duties the exigencies of the times and the instructions of the Congress demand.

The only thing remaining to insure success is available funds to maintain our propaganda upon those measures deemed of the most immediate importance, so that at the close of our official term we may be in a position to lay before the representation at the next Congress a favorable report of our stewardship; but without both your co-operation and financial aid nothing can be accomplished. We, therefore, repeat in the most emphatic but respectful manner the appeal for aid.

Contrary to the prevailing charge that workingmen are incompetent to transact their own affairs in a becoming manner, we would take occasion to express our deep sense of gratification at the very able, intelligent and harmonious manner in which the business before the Congress was conducted.

In conclusion we would add that it now becomes the duty of every labor organization to ratify the action of the Pittsburg Trades and Labor Union Congress. Then we shall have, not only in name, but in truth, a grand Federation of Labor.

We have the honor to subscribe ourselves:

Legislative Committee: Richard Powers
Samuel Gompers
C. F. Burgman
A. C. Rankin
W. H. Foster, Secretary

P.S.—Forward all moneys to A. C. Rankin, 61 Robinson Street, Allegheny City, Pa., and please notify W. H. Foster, 14 Eastbourne Terrace, Cincinnati, Ohio, thereof.

Copies of the full proceedings of the Congress, price ——— each, can be had by applying to the address of any member of the Legislative Committee.

The report was, on motion, adopted.

On motion of Mr. Burgman it was ordered that the Treasurer be required to furnish bond in the sum of $1,000, and the Secretary in $500, both to be secured by real estate.

Mr. Gompers moved that the Secretary be empowered to have on hand for temporary use a sum not exceeding $50 at any one time. Carried.

The same gentleman moved, and it was seconded by Mr. Burgman, that when questions of urgent importance require quick action members of the Legislative Committee may answer questions to be voted on by half-rate telegram.

Mr. Gompers also moved that the Secretary's compensation for time devoted to the performance of his official [duties?] be at the rate of $3 per day and traveling and hotel expenses in addition when such are necessarily incurred, this to be also the rate for the other members of the Legislative Committee when officially employed. The motion was agreed to.

On motion of Messrs. Burgman and Foster the Treasurer and Secretary were instructed to forward their bonds to the Chairman of the Legislative Committee within one month from date.

A motion by Mr. Gompers was carried, authorizing the Secretary to purchase a lithogram.

Adjourned to meet next day.

Attest: W. H. Foster,
Secretary

ADS, Minutes of Meetings, Legislative Committee, FOTLU, reel 1, *AFL Records*.

1. The morning session of the meeting included appointment of committees, authorization of the secretary to print the proceedings of the convention, and election of the following officers: Richard Powers, chairman; SG, first vice-chairman; Charles F. Burgman, second vice-chairman; and Alexander C. Rankin, treasurer. William H. Foster presided, the FOTLU convention having elected him secretary.

2. These brackets are in the original document.

Pittsburg, Nov. 20, 1881.

On the morning of the above date the Legislative Committee met

in Room 22, St. Clair Hotel, Chairman Richard Powers presiding and all members present.

The minutes of the previous day were read and approved.

The Committee on Seal reported having secured one according to design at a cost of $5.50.

On motion an order on the treasury was drawn for that amount, the report received and the Committee discharged.

The following, offered by by Mr. Rankin, was adopted:

Resolved, That the Secretary be empowered to procure 1,000 letter-heads each for the Treasurer and Secretary, and 250 each for the other members of the Legislative Committee.

Mr. Gompers moved that each member of the Committee be empowered to procure, at a cost not exceeding 50 cents, a book in which he shall keep a record of all official acts and time devoted to the work of Committee. Carried.

The following resolution was offered and unanimously adopted:

Resolved, That it is the sense of this Committee that no member thereof should publicly advocate the claims of any of the political parties; but this shall not preclude the advocacy to office of a man who is pledged purely and directly to labor measures.

On motion of Mr. Gompers the Secretary was authorized to procure electrotype plates of the Plan of Organization and Platform of Principles.

A similar course was taken with another motion of Mr. Gompers authorizing the Secretary to insert in the address to Unions the number of delegates and an approximate estimate of the total representation.

On motion the Secretary was authorized to remain in Pittsburg until, with the assistance of Mr. Bengough (Mr. Pollner having been relieved), he had the minutes in shape to complete them at his home, and the Treasurer was authorized to pay Messrs. Pollner and Bengough and Secretary Foster, for their services at the prescribed rate.

The Secretary stated that in his capacity as Secretary of the Federation it was his duty to send greeting to the Trades Congress of Great Britain, and through that body to the workingmen of the United Kingdom, and also to forward a letter of condolence on the death of Mr. McDonald, M.P.

On motion he was instructed to give in the former communication the greetings of this Committee, and in the latter its sense of sorrow.

A motion prevailed requiring each member of the Legislative Committee to correspond with the Secretary at least once a month.

After giving some instructions of a general nature to the Secretary, the Committee adjourned.

Attest: W. H. Foster,
Secretary

ADS, Minutes of Meetings, Legislative Committee, FOTLU, reel 1, *AFL Records.*

# A Report of a Committee of Cigarmakers' Unions of Greater New York City

New York, Nov. 1881.

The committee appointed by the Unions of New York, Brooklyn, Hoboken, Williamsburgh, and Jersey City, in accordance with a resolution of the last convention,[1] met and approved the following Article; and submit the same to the vote of local Unions, to be incorporated and classified in the Int. Constitution.

ARTICLE.

Sec. I. The Unions of New York City, Brooklyn, Hoboken, Williamsburgh, and Jersey City, shall be constituted the District of New York, and shall consist of four delegates of the New York Union,[2] and one delegate from each of the other Unions and shall be known as the "Committee on Grievances."

Sec. II. The committee shall elect as its chairman, a member in good standing, of any bona fide Trades Union (other than the Cigar Makers Union) whose duty it shall be to preside at all meetings, and have the casting vote when a tie occurs. His term of office shall be three months.

Sec. III. The Grievance committee shall meet at least once a month, in the City of New York, at such time and place as they may determine.

Sec. IV. Any Union of this district, feeling aggrieved, may call a meeting of this committee, provided the other Unions have three days notice thereof.

Sec. V. It shall be the duty of this committee to decide on all questions that may be brought before them, such decision to be subject to an appeal to the Int. Ex. Board, which shall be final.

Sec. VI. Members of the Unions comprising this district, shall be subject to the laws, rules, local assessments and fines of the Union under whose jurisdiction they work, they shall report to the Secretary

of the Union under whose jurisdiction they have obtained employment within three days; failing to perform which duty, they shall be fined ten (10) cents per day, for every day thereafter, payable to said Union.

Sec. VII. Members of this district must deposit their cards in the Union under whose jurisdiction they either work or reside.

Sec. VIII. Difficulties arising in any shop, in either of the cities comprising this district, shall be subject to and controlled by the Union of said City only.

Sec. IX. All laws or parts of laws, conflicting with this Article shall not apply to this district.

| | | | | |
|---|---|---|---|---|
| H. Gutstadt[3] | of | Brooklyn | Union | 87 |
| R. Kuenstler[4] | of | Hoboken | " | 8[5] |
| A. J. Walters | of | Jersey City | " | 131[6] |
| A. M. Lewis[7] | of | Williamsburgh | " | 132[8] |
| R. Pinner[9] | of | New York City | " | 144 |
| C. G. Bloete | of | " " " | " | " |
| M. Levy | of | " " " | " | " |
| S. Gompers | of | " " " | " | " |

*Cigar Makers' Official Journal,* Nov. 10, 1881.

1. The September 1880 CMIU convention established a board composed of four delegates from CMIU 144 and one from each of the other New York City area unions to settle difficulties arising among them.

2. Prior to adopting this proposal in 1883 as Article XXV of its constitution, the CMIU reduced the number of local 144's delegates from four to two and provided that in case more locals organized in the New York City area, local 144 was to have only one delegate.

3. Herman GUTSTADT, a German-born cigarmaker, was president of Brooklyn CMIU 87 in 1880 and one of the few prominent German members in the KOL Spread the Light Club—Local Assembly 1562—and in District Assembly 49 in New York City. In 1885 he moved to San Francisco where he became a leader in the anti-Chinese movement.

4. Rudolph Kuenstler (1846?-97), a German-born cigarmaker, was the first president of CMIU 8 of Hoboken, N.J. Elected in 1877, he served as an officer of the local until 1882. He later opened his own small cigarmaking shop.

5. The CMIU chartered local 8 of Hoboken, N.J., in 1877.

6. The CMIU chartered local 131 of Jersey City, N.J., in 1881.

7. Ansel M. Lewis (1855-1940), born in England, served as president of CMIU 144 and corresponding secretary of CMIU 132 before moving to Indiana about 1882. In 1890 he came to Syracuse, N.Y., where he was a pawnbroker and later a jeweler.

8. The CMIU chartered local 132 of Williamsburg, N.Y., in 1881.

9. Reuben E. PINNER, a New York–born cigarmaker.

# To the Officers and Members of the CMIU

[December 15, 1881]

To the Officers and Members of Local Unions of the Cigar Makers International Union of America:

Fellow Workers:

It becomes my pleasant duty as your delegate to the Trade and Labor Union Congress held at the city of Pittsburgh, Pennsylvania, on Nov. 15, 16, 17, 18, 19, to make this my official report of the proceedings (and a few comments thereon) of said body. You will understand that the means to give you a report in detail is precluded by the fact, that it would be unwise to devote more space of our Journal to this matter than is absolutely necessary; you will then follow with me the points of greatest importance.

The Congress was called to order, Mr. John Jarret of the Amalgamated Association of Iron and Steel Workers, made Temporary Chairman. The committee on credentials reported (together with those afterwards admitted) 114 delegates present, each delegate as his name was called, stated the number of members he represented, which footed up made a total of nearly 300,000. Some of the members thus mentioned were represented in two, and in a few instances, three different ways, such as for instance, the Trade Assembly[1] of New York City, represented Union 144 as a part of that body, while I at the same time represented the C.M.I.U., of course, Union 144 included. With Chicago, Union 14, it was the same. Therefore, deducting this double representation, it is my opinion that there were represented at this Congress between 200,000 and 250,000 working people. Another fact must be noted in connection with the delegates.

By the call issued, each local union could be represented by a delegate, in consequence of which the entire unions of Pittsburgh were represented, and which gave a majority to this locality. Whether this worked for good or evil, is not the question; that it was wrong in principle, no one will deny. I can however say that this delegation were honorably inclined, and that the danger of one locality overbalancing the entire organization of the whole Federation at future congresses is obviated by the plan of organization adopted, and to which I will hereafter refer. It is but fair to state, and I have a great pleasure in recording the fact, that the Congress was composed of, taken as a whole, the most intelligent men in our Trades Unions. Every man seemed to put forth his best efforts to accomplish that much-desired object, a Federation of our organizations. Of course, it must be borne in mind, that this was the first Congress, and that it

was considerably handicapped by the fact that we were compelled to place the organization upon such a basis, that it would be pleasing to all. The delegates came there with no power to ratify, but hoping and depending upon their respective organizations ratifying their actions and making their declarations and plan of organizations, a living fact.

The congress lasted four full days, and was marked for its cool deliberations and strict attention to the business before it. The following committees were appointed, consisting of one delegate from each of the States represented: Declaration of Principles, Permanent Organization, Plan of Organization, etc. etc. While the committees were engaged in their work, short addresses were made, the purport of which was the means of furthering the object of the Congress. The first of the committees to report were the committee on permanent organization, which recommended your delegate, as permanent chairman of the convention, but owing to the fact that one of the daily papers scurrilously attacked me, which was undoubtedly inspired by some evil-disposed person, stating that I desired to capture the congress for a faction.[2] To prove that this was entirely malicious and untrue, I withdrew my candidature, and the temporary officers were made permanent with the addition that Mr. Powers of the Lake Seamen's Union and myself were added as Vice-Presidents. The delegates next presented their various instructions in the shape of proposed laws, repeal of laws, resolutions and petitions. Those confided to my care, by the convention at Cleveland, were faithfully presented, and we have the satisfaction of knowing that they are all included in the declaration of principles.[3]

The following is a synopsis of the plan of organization:

1st. The name of the Association is: The Federation of organized Trade and Labor Unions of the United States and Canada.

2nd. The objects are, to encourage and form trade and labor unions, and secure legislation in the interest of labor.

3rd. The Sessions to be held annually, on the third Tuesday in November.

4th. The representation to be from *National* and *International* Trades and Labor Unions, and *Trades Councils* or *Assemblies,* in proportion to their membership.

5th. The Officers to consist of a Legislative Committee.

6th. The duty of the Legislative Committee is to exercise a faithful supervision over the organization and secure the passage of laws in the interest of labor.

7th. The revenue of the Federation is to be derived from a capitation tax of three cents per member, per annum, payable quarterly.

8th. The financial matters must be closed two weeks prior to the convening of the Congress, and a balance sheet presented.

9th. The remuneration for loss of time by the Legislative Committee to be three dollars per day; traveling and incidental expenses to be defrayed.

Article 4 was the one which created quite a discussion; it being the article to which I have already called attention as the one designed to prevent any one locality from an undue preponderance of delegates, and placing the representation where it properly belongs, to wit, National and International Unions and Trades Assemblies, or Councils of Trades which have no National or International Unions.

The committee reported a large number of standing orders or rules; but while they are important for the transaction and the proper fulfilment of the business of the Federation, they are too numerous to be touched upon in this report.

Resolutions were then adopted with a view to bringing about a recognition of the identity of interests and fraternal feeling with the Trades Unions of Great Britain.

The declaration of principles was then read and adopted, of which I will also only give a synopsis. The preamble, however, being so deftly set forth, that it would almost amount to neglect, not to give it in full. The following is therefore a verbatim copy of the same:

"*Whereas,* A struggle is going on in the nations of the civilized world, between the oppressors and the oppressed of all countries, a struggle between capital and labor which must grow in intensity from year to year and work disastrous results to the toiling millions of all nations, if not combined for mutual protection and benefits. The history of the wage workers of all countries is but the history of constant struggle and misery, engendered by ignorance and disunion, whereas the history of the non-producers of all countries proves that a minority thoroughly organized may work wonders for good or evil. It behooves the representatives of the workers of North America in congress assembled, to adopt such measures and disseminate such principles among the people of our country as will unite them for all time to come, to secure the recognition of the rights to which they are justly entitled. Conforming to the old adage, 'In union there is strength,' a formation embracing every trade and labor organization in North America, a union founded upon a basis as broad as the land we live in, is our only hope. The past history of trade unions proves that small organizations, well conducted, have accomplished great good, but their efforts have not been of that lasting character which a thorough unification of all the different branches of industrial workers is bound to secure."

Conforming to the spirit of the times, and the necessities of the industrial classes, we declare the following

DECLARATION OF PRINCIPLES.

1st. Demands the passage of laws permitting trades unions to become chartered and recognized as legal bodies.

2nd. Demands compulsory education of all children.

3rd. Demands the passage of laws prohibiting the employment of children under the age of 14 years in any capacity, violators to be punished.

4th. Demands the enactment of uniform apprentice laws.

5th. Demands the enforcement of the National Eight Hour Law in the spirit of its designers.

6th. Demands the abolition of all contract convict labor.

7th. Protests against the "order" or "truck" system, and demands the adoption of laws, and the enforcement of those that exist for the abolition of this nefarious system.

8th. Demands the adoption of laws that will give to mechanics and workingmen the first lien upon property, the product of their labor.

9th. Demands the repeal of all laws known as conspiracy laws, as applying to labor organizations, in the endeavor to regulate wages and the hours of labor.

10th. Favors the passage of laws creating a Bureau of Labor Statistics, the manager to be a person identified with the laboring classes of the country.

11th. Demands a protective tariff to American industries.

12th. Demands the passage of laws preventing the importation of foreign labor under contract.

13th. Recommends to all labor organizations to secure proper representation in all law-making bodies.

The first question that brought forth a discussion, was the declaration against the employment of children under the age of 14 years, but which was finally adopted.

The next was the one declaring in favor of a protective tariff. How far this question affects the industrial classes is, of course, of some moment. If it performed what its advocates claim for it, the protection of labor, it is of the greatest importance and should be adopted. But, when we find that while the industries are protected by preventing the importation of foreign manufactured articles, *it does not prevent the importation of the cheapest and most servile labor* even under contract, and competing with us right in our shops, which savors so much of slavery, that the difference can scarcely be discovered. I am therefore

of the opinion that this declaration, which was however adopted, to say the least, was superfluous.

Resolutions were adopted declaring the presence and competition of Chinese with free white labor as extremely dangerous and demanding the passage of laws entirely prohibiting their importation.

The following resolutions in reference to the manufacture of cigars in tenement houses were adopted:

Resolved, That the manufacture of cigars in "tenement houses" is detrimental both to the interest of the manufacturers (except about 30 who unfairly profit by this nefarious system) and working people, and upon this ground should be prohibited.

Resolved, That inasmuch as the manufacture of cigars in tenement houses in the city of New York is highly injurious to the health of the tenant workers and on sanitary, economical, moral and social grounds, should be prohibited.

Resolutions demanding the passage of laws for inspection and ventilation of mines, factories and workshops, and sanitary supervision of all food and dwellings, also the passage of laws making employers responsible for injury to their employes, resulting from negligence or incompetency of employers or their agents.

Believing that the organized working people of America should sympathize with the oppressed of all nations, a resolution extending our sympathies to the people of Ireland and other countries in their struggle for emancipation, was adopted.

Several other resolutions were introduced, some of which were adopted; they in most cases refer to matters connected with special trades or callings, except one, of which the following is the substance: That a bill was introduced in Congress by the Public Land Commissioners, which purposes to sell public domain in large tracts of land at the nominal figure of $1.25 per acre; that by this means large land monopolists will be fostered, and those who desire to live, will be the serfs of the land kings. Congress is therefore called upon to denounce and defeat the wholesale robbery of the public lands.

The next business in order was the election of the Legislative Committee; the Secretary was to be elected by the Congress separately. The following is a copy of the official proceedings in reference thereto:

The nomination and election of a secretary for the Legislative Committee was then taken up.

C. F. Burgman, of California, nominated Samuel Gompers of New York.

Mr. Brant, of Michigan, nominated W. H. Foster, of Ohio.

H. S. Street, of Chicago, nominated M. L. Crawford, of Illinois.

Mr. Byrne, of New York, nominated Samuel Leffingwell, of Indiana.

A committee of tellers, consisting of Messrs. Street, Hennebury,[4] Rankin, Burgman and Byrne were appointed.

The first ballot resulted as follows:

| | | | |
|---|---|---|---|
| Gompers, N.Y. | 32 | Crawford, Ill. | 17 |
| Foster, Ohio | 16 | Leffingwell, Ind. | 15 |

The name of Mr. Leffingwell was dropped, he having received the lowest number of votes. The second ballot was then taken, as follows:

| | | | |
|---|---|---|---|
| Gompers | 34 | Foster | 25 |
| Crawford | 21 | | |

The name of Mr. Crawford was dropped and the third ballot proceeded, resulting in the election of Mr. Foster, of Cincinnati, Ohio, by the following vote:

| | | | |
|---|---|---|---|
| Foster | 44 | Gompers | 31 |

The next business was the election of the Legislative Committee. A number of names were placed in nomination, but were withdrawn, and a motion passed to submit the matter to a committee consisting of one from each State.

The Committee on Suggestion of a Legislative Committee having returned, presented the following report and names:

Samuel Gompers, of New York.

A. C. Rankin, of Pennsylvania.

Richard Powers, of Illinois.

C. F. Burgman, of California.

The congress, on motion, made the nominations unanimous, and afterward elected the above committee."

The place decided upon to hold the next session of the Congress, was Cleveland, Ohio.

The delicacy of the subjects before the congress, more especially with these heterogeneous delegations, had to be proceeded with, with great difficulty and care. That more was not accomplished could be expected, it was but the initiative, the beginning of the end.

Since the congress adjourned, I have read an adverse *criticism* by a so-called *advanced* labor paper, in which the characters of men have not even escaped the venom of opposition to the trade unions. The organization of trades unions is the thorn in the side of these would-be "world regenerators," with a federation of trades they would cry out with Shylock, "*Why not take my life, as take the means by which I live.*" We should always be open to criticism, whether favorable or otherwise, but when this degenerates to exaggeration, misconstruction and downright falsehood and actual libel, it deserves the scathing rebuke and contempt of all fair-minded men. I will add that this is not prompted by a spirit of revenge, for I was rather gently handled

by this *Critic,* but I can not maintain a silence when the characters of men, good and true, are unjustly attacked.

It is perhaps necessary to mention that in the performance of my duty as your delegate, upon the questions that came before the congress in which I took part, in my advocacy or opposition to measures I was prompted not so much by my own opinions or convictions, as by the wish to reflect the expressed or implied opinions and what I deemed the standard of intelligence of my constituents. I did not nor do I now, believe in the advocacy of a project, which before hand I believe would be rejected, when the same would be referred to a general vote for ratification. I was further prompted, as I stated on the first day of the session, that I came there not to air my opinions, but to work; not to build a bubble, but to lay the foundation for a superstructure that will be solid and that would be a true federation of the trades and labor unions. I was in favor of progressing slowly, if working too fast was dangerous, and was strongly in favor of making the movement emphatically a working-class organization, one that would not be defiled by money improperly obtained; that would contain within itself all the elements of strength, through the combined efforts of the Federated Trades Unions of our common country. To say that I took an active part in the proceedings is perhaps superfluous. That I have endeavored to perform my duty to you honestly and faithfully, I can assure you.

In commencing this report I mentioned that the same of necessity could only be brief. You will observe that this is true, for it does but little justice to the congress. In a week or two an appeal will be issued by the Legislative Committee which will contain the full declaration of principles and plan of organization, and asking for aid, which, I hope, you will render.

The proceedings of the congress will be printed and published in book form, and sold at the nominal price of 10 cents per copy, just sufficient to cover the cost. I cordially recommend that the members feeling sufficient interest in the matter, do supply themselves with copies which they can procure by writing and enclosing cost to W. H. Foster, 14 Eastbourne Terrace, Cincinnati, Ohio.

In conclusion, fellow-workers, remember that this Federation can only be what we the constituent parts make of it. If it is to be good, thorough and efficient, we must possess these qualities, and I therefore urgently appeal to you, to give this young movement the support it now needs to make it successful, *ratification* by all labor unions. Ratify it then and earn the lasting gratitude of a downtrodden people.[5]

Fraternally yours, Samuel Gompers,
Delegate to the Labor Congress.

N.B. Having been elected as a member of the Legislative Committee, it was necessary to stop in Pittsburgh for a few days after the congress adjourned, to settle up matters of importance, and which were referred to this committee.

S.G.

*Cigar Makers' Official Journal,* Dec. 15, 1881.

1. The Amalgamated Trades and Labor Union of New York and Vicinity.

2. This is a reference to the rumor conveyed in the *Pittsburgh Commercial Gazette* concerning a socialist attempt to capture the organization for SG, whom the article described as "the leader of the Socialistic element" (see "The Labor Congress," Nov. 16, 1881, above).

3. The 1881 CMIU convention met in Cleveland, Sept. 20-23. The convention proceedings report only one proposal that the delegates conveyed to SG. It dealt with convict labor.

4. Thomas T. HENNEBERRY, a Cleveland cooper, was treasurer and then president of the Coopers' International Union during the 1870s and early 1880s.

5. The Mar. 15, 1882, issue of the *Cigar Makers' Official Journal* reported that the CMIU had ratified the proposal to join the FOTLU.

## *The Cigarmakers and the Labor Movement in New York City*

Although CMIU 144 grew rapidly in the late 1870s and early 1880s, the local was plagued by internal discord. Struggles between socialists and other elements in 144, and criticism of Strasser's autocratic leadership of the CMIU, reflected deep ethnic and political divisions among New York's cigarmakers. Socialists who espoused independent political activity were critical of Gompers' and Strasser's extensive anti-tenement lobbying campaign in Albany, their willingness to work with main line politicians, and their apparently growing opposition to socialism and the SLP. These socialists were usually proponents of community-based ethnic locals and they resented both the CMIU's policy of limiting the number of locals to one per city and the high dues upon which the union's system of centralized benefits and controls was based.

The extent of dissatisfaction within 144 became evident in July 1880 when a small group of German socialists seceded from the union and formed the United Cigarmakers of North America. Other socialists, who remained in the local, continued to criticize its leadership and that of the CMIU. Their dissatisfaction came to a head on April 1, 1882, when they elected socialist leader Samuel Schimkowitz as president of local 144.

Schimkowitz's opponents contested the election, claiming that he was an employer and not a workman and was therefore ineligible to serve as an officer. Strasser temporarily suspended the results of the vote, pending an investigation by 144. Schimkowitz rejected this ruling and, instead, brought charges against Strasser before the CMIU's Executive Board. As he awaited the board's response, Schimkowitz attempted to assert his authority as the local's president at a meeting in mid-April, but was rebuffed when 144's vice-president, B. Rosenberg, Gompers, and their adherents walked out. The Rosenberg faction retained control of the local's $17,000 treasury and its membership rolls, and in a special meeting of its own suspended Schimkowitz and other socialist members of the local for "conduct unbecoming of union men."[1] It also elected new officers for 144. From this point on,

both factions claimed that they constituted local 144 and each proceeded to hold separate weekly meetings.

On May 2 the local's investigating committee, chaired by Gompers, submitted its report to Strasser. The committee concluded that Schimkowitz was a manufacturer, despite evidence that he had apparently given up his manufacturer's license before the election. On the basis of these findings, Strasser made permanent Schimkowitz's suspension from the local's presidency. The Schimkowitz group ignored Strasser's ruling, however, and continued to meet as local 144. On June 2 the CMIU executive board acted on Schimkowitz's appeal and, faulting Strasser for his hasty action in the case, ordered 144 to conduct another investigation. Subsequently, on June 15, the Rosenberg faction's board of administration completed this second review and again suspended Schimkowitz from office. In July Schimkowitz and his adherents seceded from 144 and joined with the United Cigarmakers of North America to form local 1 of the Cigarmakers' Progressive Union of America (CMPU). The Progressives began to issue their own newspaper, *Progress,* and affiliated with the New York City Central Labor Union (CLU), a recently formed central body.

The CLU grew out of a mass demonstration by forty-five New York unions held on January 20, 1882, in support of land reform in Ireland. The organization, established in March, drew together socialists in the Irish and German communities who favored independent labor party activity. They were opposed by Gompers, Strasser, and others who advocated closer cooperation with the existing Amalgamated Trades and Labor Union of New York and Vicinity (ATLU) rather than the establishment of a new body. According to the *New Yorker Volkszeitung,* the ATLU group "used all conceivable means to warn the delegates to keep away from independent political action, but without success."[2] In July the CLU launched the United Labor party to participate in the fall elections; the party nominated Schimkowitz for state assemblyman.

Although CMIU 144 was affiliated with the ATLU, and CMPU 1 with the CLU, both cigarmakers' unions developed connections with the KOL, whose growing strength in New York was reflected in the organization of District Assembly 49 in July 1882. Gompers, Strasser, and other CMIU members founded Local Assembly (LA) 2458, known as the Defiance Assembly, in January 1883. Members of the CMPU established LA 2814, the Progressive Labor Club, the same year.

For the time, the Knights' role in the New York cigar trade was limited. During 1884, when discussions aimed at merging the CMIU and the CMPU faltered on what the Progressives called a "difference of vital principles," the Progressive Labor Club turned to the KOL

General Assembly for help to "harmonize the elements of these two organizations, with a view to their amalgamation."[3] Whatever steps the KOL may have taken, the talks ended inconclusively. During 1885 and 1886, however, the Knights' activities in the New York cigar trade would become much more extensive and would arouse bitter resentment among cigarmakers.

*Notes*

1. *New Yorker Volkszeitung,* Apr. 15, 1882.
2. Feb. 27, 1882.
3. KOL, *Record of the Proceedings of the Eighth Regular Session of the General Assembly, . . . Sept. 1-10, 1884* (N.p., n.d.), p. 727.

# A Translation of Excerpts from a News Account of a Meeting of CMIU 144

[January 16, 1882]

. . .

Yesterday afternoon a meeting of Cigarmakers' Union 144 was held in the Germania Assembly Rooms. After Mr. Brinkmann[1] had stated that workers would have to represent their own interests and that they would have to become involved in politics, Mr. Walther[2] asked permission to pose a question. But the chairman, Mr. Pfrommer,[3] denied his request, saying it was out of order. The members who had come—about 1,200 strong—wanted to hear Mr. Walther. Since the chairman did not comply with their wishes, a great commotion ensued and the chairman adjourned the meeting. He then left the hall with a number of members. The remaining members were urged to keep calm and to stay in their places, which they did. The meeting was then opened anew by Mr. Brückmann,[4] this time as a *public meeting of cigarmakers.* Mr. Brückmann asked for unity and for adherence to the union, and he said one should not be misguided by such events. Mr. Walther was then allowed to speak: As to the membership meeting, he said it was unfortunately true that the constitution of the union gave the president the right to act as he did, but the members themselves could change this if each of them participated in the election of officers and delegates. On the question, Which political direction should the cigarmakers choose? he said many still held the opinion that if one wanted to achieve gains for the workers this would have to be accomplished by a politician. But he said that this was not the right way, and that the workers would have to recognize their interests and represent these themselves. He added that the capitalists recognized *their* interests only too well. . . .

Mr. Ecks[5] admonished them to hold together, and added that the workers had to gain all political power for themselves. After Mr. Woyticzeck[6] had spoken in English, the following resolution was voted upon and accepted unanimously:

Whereas: The system of manufacturing cigars in tenement houses is harmful

(1) because cigarmakers employed in tenements work for much less than those in the shops and are therefore forced to work 14 to 18 hours daily in order to acquire the basic necessities of life;

(2) because they therefore ruin themselves and their families not only physically and mentally but also morally;

(3) because workers become dependent on the manufacturers in a way which is worse than the dependency of the former slaves to their masters and

(4) because we are convinced that the manufacturers are unwilling to abolish the system voluntarily; be it

Resolved that: it is our duty as organized workers to bring about the legal abolition of this accursed system. But since Democrats, Republicans, etc., have up to now dominated all legislative bodies as representatives of capitalist interests, and since not much can be expected from them in favor of the workers, we therefore consider it our duty to attempt to involve all workers in politics independent of other political parties. They can then nominate and vote for their own candidates for congressional and assembly elections, so that laws furthering the interests and advantages of the workers can be passed.

After this Mr. Leib[7] said: Even if the general staff had made itself invisible, the army had remained. He said that this was a good sign which proved that workers understood their position. The speaker then gave a survey of the labor movement during the last ten years. After this the chair was ordered to call another cigarmakers' meeting at some other time. It should be noted that Messrs. Pfrommer, Strasser, etc., tacked two slips of paper to the entrance when they left, which stated: "The cigarmakers' meeting has ended; this is a meeting of socialists." It has therefore been established that the vast majority of Union 144's members are socialists.

*New Yorker Volkszeitung*, Jan. 16, 1882. Translated from the German by Dorothee Schneider.

1. Probably William Brinkmann (b. 1854), a German-born cigarmaker and later cigar manufacturer.

2. Herman Walther served as a member of the Cigarmakers' Progressive Union (CMPU) National Board of Supervision at New Haven during 1883. In 1885 he moved to New York City and two years later was a delegate to the SLP convention.

3. Julius Pfrommer, president of CMIU 144 in early 1882, worked as a cigarmaker in New York from 1881 to 1883.

4. Englebert Brückmann (d. 1898?) was a cigarmaker in New York during the 1880s and 1890s.

5. Henry Ecks (1837?-1917?), a New York City cigarmaker, served as an officer of the CMPU's May 1885 convention, which he attended as a representative of CMPU 1.

6. Vincent William Woytisek was a Bohemian socialist who served as CMIU 144's president in late 1881.

7. Francis Leib (b. 1832?), a Prussian-born cigarmaker, was a cigar manufacturer in New York from the mid-1880s until 1898 or later.

# To the Editor of the *Cigar Makers' Official Journal*

New York, January, 1882.

Will you permit me to submit the following matter to the consideration of the Cigarmakers of the entire country, and give an opportunity, which is denied me, in other quarters. That no misunderstanding may arise, it is best to briefly state the circumstances, which led to the necessity of this statement.

Within the last few months a number of cigarmakers, arriving from Europe, have joined our Union (No. 144), without waiting to become informed as to the practical operations of our Unions, both International and Local. They have declared, that our efforts for the abolition of the Tenement House Cigar Factories were useless; "that we ought to wait until we have elected our own men to Congress, and then it will be easily accomplished." These men forget, that if even they had what they wanted in this direction, there is still a President of the United States, with the vetoing power; and a Supreme Court to decide as to its constitutionality, to overcome. To wait for the achievement of all this would be worse than criminal. We are organized, and must be active; we must endeavour to rid our trade of this pest, and that, before another panic sets in, or when that event comes, it will find us all in these houses, rendering us unfit to take up the gauntlet with concentrated capital, and there will be one more trade almost beyond redemption.

To endeavor to show these members their error at the quarterly meeting of our Union, a discussion was held, upon the following subject. "What is the most practical political action, our Union can take for abolition of the Tenement House Cigar Factories?"[1]

To the astonishment of our members, there appeared in the following issue of the "New Yorker Volks-Zeitung," a paper which has for its heading, "Dedicated to the interests of the working classes," a report of our meeting, winding up with a remark of which the following is a translation. "The next meeting will show upon which side the majority will be, and whether the organized Cigarmakers will allow themselves *longer* to be utilized by the shrewd politicians."[2] This was the statement, after throwing odium upon the defenders of our Union, who desired to carry out the resolution repeatedly adopted at our conventions.

The Board Members of our Union (composed of delegates of all the shops where Union-men are employed) seeing in this a miserable slander upon the good name of our Union, passed a resolution de-

claring the attack as "contemptible and scurrilous," which it was; for it is no use hiding the fact, that "if our Union is *no longer* to be utilized by the politicians, &c." then the only logical deduction is, that it has been so in the past; and for that to have taken place, the older members must have been aware of it, and naturally parties to it.

We forwarded a copy of the resolution, to the Editors, and what did this paper, "dedicated to the interest of the working classes" do. It repeated, and added to, this vile attack, the following question. "Have the organized Cigarmakers never been utilized by the corrupt politicians of both parties?["] This was not all; an entire harangue assailing anyone who dared to differ with them.[3]

Feeling it incumbent upon me, as a member of our Union, not to allow these slanders to go unanswered, I wrote a copy of the following letter, to that paper:

New York, Jan. 11th, 1882.

To the Editor of the "N.Y. Volks-Zeitung."

The repetition of your attack upon Union 144, Cigarmakers' International Union, is only increased by the addition to the first report of Monday. You ask, "Have the organized Cigarmakers never been utilized by the corrupt politicians of both parties." I answer distinctly and emphatically no! and any member of our Union can if necessary, prove the contrary.

Is it not rather this that prompts the attack, that knowing the Bill for the abolition of the Tenement House Cigar Factories is now pending in the Legislature, a point is stretched in order to assist in defeating this Bill, and prove, that, as some please to call it "pressions politic" (a kind of force or trading politics) but which I term the Trades Union Politics, "can accomplish nothing." I wish to add that I once remarked, and see no reason to change my opinion, that any politics that is inconsistent with the politics of Trades Unions, is capitalistic.

Samuel Gompers

P.S.—This letter was short and to the point, and moreover true. (I might have added to the last word, "middle class"), but did this *Labor Paper* publish it? decidedly not. In their "answers to correspondents" they falsely say, that the reason they refuse to publish my letter, is because "I am throwing a monstrous suspicion upon our members, and stirring up hatred; and for the benefit of our Union they will not publish it."[4] Just observe the audacity! after first attacking us in the most shameful manner possible, (besides publishing us as knaves or asses before) refuse to publish a denial and contradiction, under the plea, that they want to save the Union this dissension. It

would make me laugh, were it not so serious. I ask, and challenge any man of that paper, to show in which way they can construe any word, line or sentence of that letter, as casting a suspicion upon our members. It is not that the objects are more far-reaching. Our Union is one which believes, it is quite a grown body able to take care of itself, and not needing any-one to teach it how to walk, and which fact is very unpalatable to them, hence their opposition. In this their latest effort, they lied, they know they lied, but stick to it.

Samuel Gompers.

*Cigar Makers' Official Journal,* Feb. 15, 1882.

1. SG, Adolph Strasser, Samuel Schimkowitz, and Herman Walther spoke at a Jan. 8, 1882, meeting called by CMIU 144.
2. *New Yorker Volkszeitung,* Jan. 9, 1882.
3. Ibid., Jan. 11, 1882.
4. Ibid., Jan. 12, 1882.

## Adolph Strasser to the Officers and Members of CMIU 144

[April 12?, 1882]

To the Officers and Members of Union 144, New York.
Fellow workmen!

I have received a complaint from several members of your Union, stating that at the last election of officers, a manufacturer[1] was elected to fill the position of President, which is clearly a violation of the Constitution, and of all decisions rendered previously on the subject. The men or women who work for wages have no interest in common with manufacturers, be they small or large.

From all information received, the complaint seems to be reliable, and you are therefore requested to defend your action as soon as possible, in writing, signed by five officers or members of the Union.

Knowing the dangers which beset Unions when manufacturers or foremen are members, I decide, that Samuel Schimkowitz shall cease to act as President, while the investigation is pending. No member has a right to recognize him in that capacity until a final decision shall have been rendered.

Yours fraternally, A. Strasser,
Int. Pres.

P.S. The complaint was signed by ten members of the Union.

*Cigar Makers' Official Journal,* June 15, 1882.

1. Samuel SCHIMKOWITZ.

# A Translation of a News Account of a Meeting of CMIU 144

[April 14, 1882]

REPORT OF THE ADMINISTRATIVE BOARD OF CIGAR MAKERS' UNION 144.

A special meeting was called by President S. Schimkowitz, and Vice-President Rosenberg as well, for 7:30 yesterday evening at 89 First Ave. Schimkowitz opened the meeting. Rosenberg questioned Schimkowitz concerning Schimkowitz's suspension, which was decreed by the Intern. President, Strasser. Schimkowitz responded that he would remain president of union 144 until the question had been finally decided. At this, Rosenberg, Gompers, and comrades started a terrible ruckus, and put the finishing touches to it when Rosenberg opened a special meeting in the same tavern.[1] Not long after this, 12 uniformed policemen and a few detectives appeared; the proprietor then asked Rosenberg, Gompers, and comrades to leave the tavern because they had provoked the ruckus. They did not leave, however, whereupon we suggested that we would leave the tavern for the time being. We returned shortly afterward, however, and met as the administrative board of union 144; 58 delegates and officers were present. The following resolutions were passed: (1) A complaint is to be filed with the International Executive Board against Strasser for gross violation of duty. (2) A committee was elected to examine the results of the election. (3) The steps which President Schimkowitz has taken in order to obtain his rights were approved. After some routine business was taken care of, the meeting was adjourned.

N.B. Members in the shops and districts are herewith requested to send their delegates only to the meeting of the administrative board called by President Schimkowitz. The collectors will soon receive instructions as to how they are to proceed in this matter.

Delnicke Listy is requested to copy.

*New Yorker Volkszeitung,* Apr. 14, 1882. Translated from the German by Patrick McGrath.

1. At this meeting the Rosenberg group elected him president, Louis Berliner vice-president, Meyer Dampf financial secretary, and H. F. Hallahan corresponding secretary.

# Samuel Schimkowitz et al. to the Executive Board of the CMIU

New York, April 14, 1882.

To the Executive Board of the Cigar Makers' International Union of America:
Fred. Blend, Esq., First International Vice-President—

Dear Sir,—

The undersigned, officers of Union 144, respectfully inform you that Union 144, of New York, resolved, at the regular Board meeting on Thursday, April 13, 1882, to prefer charges against Mr. Strasser, International President, for violating the Constitution and overstepping his power as International President, and in accordance with Sec. III, of Art. 2.[1] of the International Constitution, we submit the following specification of charges to you:

I. That the International President has, without authority, or without consulting the Board of Administration of Union 144, directed a letter to this Union, in which he states that he has suspended Mr. Samuel Schimkowitz, President of Union 144, acting only from private information received, that said S. Schimkowitz is a cigar manufacturer, and the fact being that the said S. Schimkowitz was no cigar manufacturer at the time of his acting President of Union 144, and denying him a right of trial, shows that the International President is acting in conjunction with some men whose sole purpose is to have said Samuel Schimkowitz removed from office or suspended from the organization.

II. That the International President has, by suspending Mr. S. Schimkowitz, the legally-elected and installed President of Union 144, created a split in this Union—the great majority acknowledging Mr. S. Schimkowitz as President—not taking notice of the communication of Mr. Strasser and a small faction favored by Mr. Strasser recognizing the authority of Mr. Rosenburg, the Vice-President.

III. That the International President, in the past five months, has tried to disharmonize and create a split in Union 144, by continuously slandering the members of the Union, and has driven thousands out of the Organization.

IV. We further charge Mr. Strasser with using the *Cigar Makers' Official Journal* for a partisan purpose, allowing only his own favored class to abuse, attack and slander through its columns, while denying the abused the right of publishing the defense.

Therefore, we most respectfully ask the Executive Board of the

International Union to elect a committee to investigate the charges against Mr. Strasser, and that he may remain suspended from office while the investigation is pending.

We also ask for a committee to be sent to New York, to investigate the late election held in this Union, in which, under the auspices of Mr. Strasser, a Vice-President, a Financial Secretary, a Trustee and an Auditor were fraudulently counted into office. It is creating dissensions in our ranks and threatens to break up our Union, and in order to pacify the disturbing element, the committee may do some good work here.

We will also call your attention to the circular enclosed, in which you will find questions asked by Union 144 of the International President in regard to the vote on Art. V.[2] of the Constitution, as there is great doubt existing here whether Art. V. was carried, if allowing Unions to vote according to the financial report of October, 1881; but by some reason of his own despotic will, refused to answer some of the questions put to him.

Hoping that you will take immediate action on the matter, and see that justice will be done to Union 144, and that you will press the charges against Mr. Strasser, we remain,

Fraternally, S. Schimkowitz, President.
John Winkler,[3] Trustee.
A. Forschner,[4] Auditor.

Charles Solkey,[5]
Cor. Sec. Union 144, 103 Suffolk street, New York.

*Cigar Makers' Official Journal,* Supplement, Sept. 1883.

1. Article II, section 3, of the 1881 CMIU constitution provided for appeals from decisions of the president to the Executive Board and the annual sessions of the CMIU. In such cases the CMIU first vice-president, in this case Fred Blend, was to receive all communications.

2. The CMIU convention of September 1881 in Cleveland amended Article V of the constitution, increasing the dues from fifteen to twenty cents.

3. John WINKLER served as a trustee of the Cigarmakers' Progressive Union (CMPU) in 1883-84.

4. August J. FORSCHNER was treasurer of the CMPU from 1884 to 1886.

5. Possibly the Charles Solkey (b. 1853?), who was corresponding secretary of CMIU 9 of Troy, N.Y., from 1878 to 1880.

# An Article in the *Cigar Makers' Official Journal*

[April 15, 1882]

A LOCAL CONFLICT.

After years of struggle and hard work, the cigarmakers of New York joined Union 144, with the hope of bettering their condition materially and socially. Experience had taught them the lesson, that they were powerless outside of the union and unable to resist the encroachments of the manufacturers. They joined the union in large numbers, as many as 300 per week. Of course these large forces could not be disciplined in a week or a month, nor educated up to trades union principles in a very short time. For some months it was a chaos, especially on questions relating to strikers, until gradually everything seemed to run in the regular channel. A great victory was accomplished; a powerful organization effected; the experience of years utilized in a systematic form, enabling us to handle with dispatch an army of men and women. Harmony, so essential to the success of trades organizations, prevailed; everything appeared to be lovely, until a few designing and evil disposed persons commenced to undermine the organization by creating factions. The increase of dues was the first weapon in the hands of those who intended to create mischief. They appealed to the weak elements, to the men who were compelled to join the union or quit the shops, and insisted upon it that those who advocated 20 cents per week, intended to break up the union. All arguments about the probabilities of lock-outs and strikes, where large amounts of money would be needed, were in vain. They intended to gain a point by creating two parties, one which was termed the 15c. and the other 20c. party.

The second cause which helped these evil disposed persons to further their schemes, was the method by which the tenement-house committee of union 144 intended to secure legislation against filthy tenement-house factories. It was before the November election when Mr. Edward Grosse,[1] a former member of Typographical Union No. 7,[2] and delegate to the Trades Assembly,[3] was nominated on a so-called independent ticket,[4] as a candidate for the legislature. Mr. Grosse was a staunch advocate of the bill in the legislature of 1880, the debates on the subject having been published in the Journal. He also advocated a bureau of labor statistics, a mechanics' lien law, the reduction of fares on the Elevated Road, and was the only assemblyman who dared to oppose the infamous tramp law[5] now upon the

statutes of the State. Before the election he appealed to the cigar-makers and workingmen in general for a public endorsement of services rendered. Mr. James Lynch, President of the Trades Assembly and John Fortune, Ex-president of the Trades Assembly, volunteered to address a public meeting for the purpose of securing his return to the legislature. It was the interest of the cigar-makers to have somebody in the legislature upon whom they could rely, and who has proved faithful in the past, in promoting the desired legislation.

The committee of union 144 called a public meeting,[6] and invited Mr. Grosse to explain his programme in relation to labor questions, but what did happen? a number of cigar makers (being in a rather intoxicated condition) who had but lately arrived from Germany, claimed that this was not in accordance with socialistic principles, as established by the Socialistic Party. Their argument was that nothing should be done, before we elect our own Assemblymen, Senators, Governors and Judges, as every concession wrested from the representatives of capital is a worthless gift. If this theory be true, then every increase of wages is a worthless gift. From that day the conflict in the union was a permanent one; hardly any business having been transacted since the month of November, and the time having been wasted in calling the roll, the ayes and noes on every question, and in appealing from the decision of the chair. We publish in another column a fable by George McNeill,[7] which is a good picture of the situation.

It must be borne in mind, that within the last two years, over 3,000 cigarmakers arrived in the city, (nearly all from Germany), who claim to be socialists and followers of Ferdinand Lassalle. They started the battle-cry, that all officers of union 144, N.Y., who did not favor the socialistic method of agitation, had to be bounced, even boasting that they would ultimately control the International Union. Their aim was to capture the funds of the union, then either dictate to or control the International Union, and if not successful to secede, and have a local union according to their own fashion. Not particular about the means, fair or foul, they commenced to spread rumors, that $400 were missing in the treasury; then stretched it to $4,000. Every week they invented a new slander and falsehood, and well informed men stated that they were educated in the school of scandal. The election was to be held on the first Saturday of April, for which occasion they had prepared a circular in three languages for distribution in various shops, containing a tissue of falsehoods from beginning to end. They also utilized the newspapers for the same purposes, reminding us of a political election, with the only difference, that the candidates received a larger share of abuse. They succeed[ed] in electing some

officers, but failed to capture the treasury. They elected the president, who unfortunately for them turned out to be a manufacturer. Now they insist upon it that a manufacturer has a right to be in the union.

*Cigar Makers' Official Journal,* Apr. 15, 1882.

1. Edward GROSSE, a printer, was an independent New York assemblyman, elected in 1880. He introduced the first tenement-house bill in the New York state legislature.

2. The New York German Printers' Union, organized in 1869, participated in the formation of the German-American Typographia in 1873 and became local 7.

3. The Amalgamated Trades and Labor Union of New York and Vicinity.

4. The German-American Independent Citizens' Association of the Tenth District of New York City nominated Grosse for the New York Assembly. He was also nominated by the Irving Hall Democrats.

5. On May 5, 1880, the New York legislature passed "An act concerning tramps" (Laws of 1880, ch. 176), providing six months imprisonment at hard labor for "all persons who rove from place to place begging, and all vagrants living without labor or visible means of support."

6. The meeting took place on the evening of Nov. 5, 1881. SG, Strasser, and Meyer Dampf were among those signing an announcement endorsing Grosse that appeared in the *New Yorker Volkszeitung* of that date.

7. George Edwin MCNEILL, a Boston printer, was an eight-hour advocate, labor leader, and labor editor. His fable, "The Dam Builders," was serialized in the *Cigar Makers' Official Journal* beginning with the Apr. 15, 1882, issue.

## An Article in the *Cigar Makers' Official Journal*

[April 15, 1882]

ON THE WARPATH.

The tenement-house manufacturers have resolved to oppose and defeat, by all means (bribery not excepted) the bill introduced in the legislature, with a view to abolishing tenement-house cigar factories. They engaged a lawyer,[1] and appointed a committee consisting of Messrs. Haas[2] of the firm of Kerbs, Spiess & Co., B. Lichtenstein,[3] Geo. Bence and A. Moonelies to proceed to Albany. They arrived there on March 21, and succeeded in having the Assembly bill, introduced by Mr. John C. Niglutsch,[4] recommitted. When we reached Albany the Thursday following, we recognized the danger which threatened the bill, and proceeded immediately in conjunction with M. Makenzie[5] of the Bookbinders Union, to the Senate Chamber, where we urged Senator Grady[6] to push the bill through, if possible the same day. Mr. Grady readily complied with our request; the clerk read the following bill:

*An Act* to improve the public health in the city of New York, by regulating the manufacturing of cigars in the tenement-houses of said city.

The People of the State of New York, represented in Senate and Assembly, do enact as follows:

*Section* 1. The manufacturing of cigars or preparation of tobacco in any form in any rooms or apartments which in the city of New York are used as dwellings, that is for the purpose of living, sleeping or doing any household work therein, is hereby prohibited.

§2. No part of any section of any floor in any tenement-house in the city of New York, in which the manufacturing of cigars or the preparation of tobacco is carried on, shall hereafter be used for dwelling purposes.

§3. The term "any section of any floor" shall be construed to comprehend any number of rooms on any floor of a tenement-house that adjoin each other and extend in a contiguous line from the windows opening into the street, to the windows opening into the yard of such tenement-houses.

§4. The first floor of said tenement-houses on which there is a store for the sale of cigars and tobacco, shall be exempt from the prohibition provided in sections one and two of this act.

§5. It shall be the duty of every sanitary inspector of said city to report any violation of this act, coming to his knowledge, forthwith to a police magistrate and to procure the punishment of the person or persons having committed such violation. But this provision shall not be construed to preclude any other citizen to perform the duty herein assigned to said sanitary inspectors.

§6. Every person who shall be found guilty of a violation of this act, or of having caused another to commit such violation, shall be punished for every offence by a fine of not less than ten dollars and not more than one hundred dollars, or by imprisonment for not less than six months, or both such fine and imprisonment.

§7. All acts and parts of acts inconsistent with this act are hereby repealed.

§8. This act shall take effect on the first day of May, eighteen hundred and eighty-two.

22 Senators voted in the affirmative: the names will appear in the next issue of the Journal.

Negative 3—their names are:

Norman M. Allen,[7] R., Dayton, N.Y.
Addison P. Jones,[8] D., Catskill, "
Dennis McCarthy,[9] R., Syracuse, "

This was a bombshell in the enemies' camp. Tuesday following the

manufacturers and their lawyer appeared before the Committee on Cities. The lawyer stated that the bill was unconstitutional, and that the people working in tenement-houses are perfectly happy and contented.

Mr. George Bence claimed that the trade would go to Baltimore, thus throwing 10,000 out of employment, and that socialism and communism were at the bottom of it. Two years ago, Mr. Bence insisted upon it that the trade would go to Jersey, Binghamton and Westfield.

Mr. Haas said, the cigarmakers of New York did not oppose tenement-house factories; it was the Int. Union who intended to scatter the trade all over the country, and drive it out of the city. The cigarmakers employed in the tenement-houses are earning over $20 per week; nobody could expect that they should live in houses like Vanderbilt.

Mr. Joseph Vanderburgh[10] of union 68,[11] Albany, made a telling and impressive speech against the sophistries of the manufacturers, citing some comments of the newspapers on this public nuisance. Messrs. Kenneth Makenzie, S. Gompers and the Int. President also refuted the statements of the manufacturers' lawyer.

### An Investigating Committee.

The result of the hearing was the appointment of a committee, consisting of five Assemblymen, who were instructed to investigate the tenement-house cigar factories in New York.[12] We immediately volunteered to act as the guide of the Committee, which was reluctantly granted. As soon as the news reached New York, that the Investigating Committee would arrive there, the tenement-house manufacturers commenced to clean, calsomine and fumigate the tenement-houses. Out of the tenement-houses of Blasskopf & Co. not less than 15 cases of garbage were removed in one day. They instructed the people employed in the houses to put on their Sunday dress, and ordered the children out in the street, refusing to give them any tobacco.

Mr. David Gideon,[13] Chairman of the Legislative Committee, insisted upon going first to Kerps & Spiess, to which we objected, as it looked like a whitewashing affair. When the committee arrived there, the cigarmakers, with rags in their hands, [were] cleaning the windows and trying to create the impression that they were always clean. The committee also visited the houses of Hirschhorn & Bendheim, and Samuel Josephs & Co., which were in a filthy and miserable condition, not expecting a visit of the Legislative Committee.

*Cigar Makers' Official Journal,* Apr. 15, 1882.

1. Morris S. Wise (1850-1905?) served as general counsel of the National Cigar Manufacturers' Association in 1881. He was a founder in 1881 and was later secretary of the Legal Protective Association of Segar Manufacturers.

2. Louis Haas (1847-1905), a Prussian immigrant, worked as a superintendent at Kerbs and Spiess and later operated a New York–based importing business.

3. Benjamin Lichtenstein, a leading New York City cigar manufacturer, was born in Bavaria about 1840. He served as president of the Legal Protective Association of Segar Manufacturers between 1882 and 1884.

4. John C. Niglutsch (1850-87), a musician and surgeon, was a Republican assemblyman in 1882. The CMIU successfully opposed his reelection that year, because of his role in defeating efforts to enact a tenement-house bill. He was later a clerk at Castle Garden.

5. Kenneth Mackenzie.

6. Thomas Francis Grady (1853-1913), a prominent Tammany Hall Democrat and former New York assemblyman, was a longtime state senator (1882-83, 1889, 1896-1912).

7. Norman M. Allen (1828-1909), a Republican and longtime member of the Cattaraugus County and Dayton, N.Y., boards of supervisors, served three terms as state senator (1864-65, 1872-73, 1882-83).

8. Addison Porter Jones (1822-1910), a Catskill, N.Y., businessman and Democrat, was a member of the Greene County Board of Supervisors and served twice in the state senate (1878-79, 1882-83).

9. Dennis McCarthy (1814-86) was a Republican member of the New York Assembly (1846), mayor of Syracuse (1853), U.S. congressman (1867-71), state senator (1876-85), and lieutenant governor (1885-86).

10. Joseph Vanderburgh (b. 1840?), an Albany, N.Y., cigarmaker, was a delegate to the 1882 session of the New York State Workingmen's Assembly.

11. The CMIU chartered local 68 in 1879.

12. The *New Yorker Volkszeitung* reported on Apr. 4, 1882, that three committee members visited the tenement houses but that they rejected CMIU 144's offer to guide them around the tenements. The three were New York City assemblymen David Gideon, John McManus, and Theodore Roosevelt. In his autobiography Roosevelt recalled visiting the tenement houses several times—with other committee members, separately with some labor union representatives, and by himself. SG wrote in his autobiography that he guided Roosevelt through the tenements.

13. David Gideon (1847-1929) was a Tammany Democrat in the New York Assembly during 1882.

## Fred Blend to Charles Solkey

Evansville, Ind., May 2d, 1882.

Charles Solkey, Esq.
Dear Sir:

Your registered letter was received to-day. On account of my absence from the city, your former letter[1] did not reach me until yesterday, and I immediately answered it in a manner that I thought was right

and proper. I very much regret the situation in New York, and when I first heard of the difference of opinions existing there, I was in hopes that with mature reflection the trouble might blow over, but instead it seems the breach has widened.

I am of the opinion that you are in the wrong course when you prefer charges against President Strasser and request the Executive Board to suspend him from office pending investigation. I question the right of the Executive Board to assume this authority. I think the proper course for you would have been to appeal from the decision of the President, lay all the facts before the Executive Board, and unquestionably all your rights under the Constitution would have been sustained. Always willing to concede justice to all alike, and not willing to assume the responsibility of this very important question alone, I have at once concluded to make copies of your communication for each of the Executive Board, and advise with them as to future proceedings in this matter, and be governed by their decision. Of course you understand it will take some little time to do this, but will advise you as to the result as soon as received. In both of your communications you request me to come to New York to investigate the situation. This would be out of the question, as the Int. Union makes no provisions for expense of that nature, and I could not afford the expense myself.

Very respectfully, Fred. Blend,
First Vice-President.

*Cigar Makers' Official Journal,* Supplement, Sept. 1883.

1. Solkey sent two copies of the Apr. 14 appeal, one by registered mail.

## An Item in the *Paterson Labor Standard*

[May 13, 1882]

The Cohoes labor troubles[1] continue. The Harmony Company not only refused to pay living wages themselves, but have stooped to the meanness of asking the knitting mills not to employ any of the locked out operatives. On next Monday night a grand picnic will be held at Lansingburgh[2] for the benefit of the strikers. The editor[3] of this paper has been requested by the Cohoes operatives to be the speaker on the occasion, but it is impossible for him to go there owing to poor health and increased responsibilities. His place will however, be taken by Mr. Samuel Gompers of New York, Vice-President of the National Trades Federation, a gentleman in every way qualified as a labor man,

an able speaker, and a wise adviser. We ask our Cohoes friends to give him a hearty reception and to hearken to his advice.

*Paterson Labor Standard,* May 13, 1882.

1. On Apr. 6, 1882, Harmony Mills, a cotton manufacturing company in Cohoes, N.Y., posted a notice reducing wages 10 percent. In response over 10,000 workers struck by the middle of the month and continued the struggle for four months with support from labor unions throughout the state. On Aug. 26, 1882, however, discouraged by insufficient strike relief and evictions from company-owned housing, the workers voted to return to the mills under the reduced rate.

2. Lansingburgh, N.Y., was across the Hudson River from Cohoes.

3. J. P. McDonnell.

## A Decision of the Executive Board of the CMIU

New York, June 2d, 1882.

The Executive Board, in conference in New York City, upon an appeal taken from the decision of the President, by a certain portion of the members of Union 144, relative to the suspension of Samuel Schimkowitz as President of Union 144, and relative to the falsification charged, of the vote taken on Art. 5 of the Constitution. After a thorough and careful examination we arrived at the following conclusions. 1. That after fully investigating and counting the vote of local unions, upon the adoption or rejection of Art. 5 referred to, we hereby sustain the President's decision. 2. In reference to the suspension of Samuel Schimkowitz, upon thorough investigation of all the facts appertaining to his election and suspension, we have unanimously arrived at these conclusions, to wit., that said Samuel Schimkowitz was duly elected President of Union 144, and that the Int. President did, without warrant of law, suspend Samuel Schimkowitz from office. We find upon investigation that the International President, upon representation made to him by five members of Union 144, that said Samuel Schimkowitz was an employer, did suspend him without first giving Union 144 a reasonable time to investigate and act upon the information filed. The Executive Board is unanimously of the opinion, that local Unions should have the authority to first investigate the eligibility of its members, subject to an appeal to the Int. President, Executive Board and Convention. In consideration of the facts above enumerated, we hereby reverse the decision, and restore to membership with Union 144 all persons affected by said

decision, upon the following conditions: 1st. That we hereby direct that Union 144, within ten days after the promulgation of this decision, thoroughly investigate and act upon the information filed, that said Sam. Schimkowitz was an employer at the time of his election and installation to office; if it is found that Samuel Schimkowitz was an employer beyond a reasonable doubt,—then Union 144 shall be required to enforce the provision of the Constitution. 2d. The Organization represented by Chas. Solkey, as Co. Sec., is hereby directed to convey into the treasury of Union 144 all monies, books, papers and vouchers of expenditures made, for which the members of that organization shall receive complete credit, the same as though paid to Union 144; and that all members of said organization shall be furnished Stamps of the International Union, to cover the amount paid, as dues, fines, and assessments. And it is further directed that all members affected by the decision of Pres. Strasser, who became delinquent in the payment of their dues and assessments, upon payment of all arrears, without fines, within 30 days after the promulgation of this decision, shall be restored to full membership and their benefits unimpaired. 3d. It is the unanimous opinion of the Executive-Board, in consideration of the many prevailing opinions and the complicated condition of affairs existing in the City of New York, and for the purpose of restoring harmony, we recommend that Union 144 go into an election of officers for the unexpired term within thirty days after the promulgation of the above decision. 4th. It is the unanimous opinion of the Executive Board, in consideration of the fact of the difficulties in the management of a large local organization, we recommend that Union 144 devise and agree upon a plan as soon as practicable, for more than one charter in New York City. And it is further directed that the Trustees of Union 144, holding the charter of said Union at the present time, shall be the custodians of the bank books and other property of Union 144, until such time that the Union may elect a new set of regular officers, or until the election of officers, in accordance with the recommendation herein provided. 5th. We hereby recommend that Union 144 cause these decisions and recommendations to be printed in at least three languages, in circular form, for general distribution among the cigar makers of New York City.

(Copy as rendered by the Executive Board.)

J. L. Aubry,[1]
Sec.

*Cigar Makers' Official Journal*, June 15, 1882.

1. James L. AUBREY, a leader in the Trades and Labor Assembly of St. Joseph, Mo., was second vice-president of the CMIU in 1881-82.

# An Excerpt from an Article in the *New York Unionist*[1]

[June 3, 1882]

## THE COHOES STRIKE

. . .

. . . At the recent festival of the operatives at Lansingburg, Mr. Samuel Gompers of New York City, in the course of a very able address referred to the panic of 1873, and the reduction in wages dating from that period, when thousands were out of employment, and spoke of the legislature of 1880 making tramps and criminals of men looking for work, and making poverty a crime. He then referred to the return of prosperous times and told of his visit to Cohoes, where the rich Harmony Mills company seemed to own half of the city.

He paraphrased Tennyson's "Charge of the Light Brigade," as he looked on the magnificent building:

Harmony to the right of them,
Harmony to the left of them;
Oh! how Harmony blustered!
Laborers not to ask the reason why,
Theirs but to work and die—
Among the dead they'll be mustered.

He thought the reduction demanded by the Harmony company was but a flank movement. It was feared that the "hands" wanted an increase, and were going to ask for it. He told of a committee of female weavers who waited on the plethoric, well fed, and

### IMPORTANT ROBERT JOHNSON,[2]

and were told in a gentlemanly way that they could "go to work for the reduction, or pack up their duds and git." He rehearsed a tale of eviction as cruel and startling as any that comes from Ireland. Mrs. McMarrow had been an employe of this grasping monopoly for 28 years. Her sister, Miss Walter, lay dying in the house. She went to Robert Johnson and asked him to let her remain for a while, and not disturb the family, as a member was on her death bed. She was ordered to get out. Her sister died three days afterward.

### How They Grow Rich.

"This company," said Mr. Gompers, ["h]ave grown rich on the blood of children who should be at school, instead of losing life and strength in arduous toil in the fetid atmosphere of these mills. The families of the managers and overseers are said to be christians. They teach these little children in the Sunday school, because it looks well, and they think they will get a pass for heaven for so doing, but they do not know the little ones on the street, and hold their silken robes as they pass the emaciated forms and ragged calico gowns of the mill workers.["] He compared the Harmony authorities to the old fable of the frog who wished to appear as big as a bull, and said if they kept on inflating themselves up they will soon burst. If this strike goes on for six months, said he, the operatives will raise the question, what right have they to let their mills stand idle when we are ready to work for reasonable wages? In the words of Will Fern, in Dickens' "Chimes," the very words of the Bible seemed changed with these people until the reading is: "Where thou goest I cannot go, where thou lodgest I cannot lodge. Thy people are not my people, nor thy God my God."

### Union the Remedy.

He urged organization. First—a local organization; second a national organization, and, third—a federation with other national labor unions in the United States. Had this been done at the time of the great strike two years ago[3] there would be no trouble now. He urged the friends of the operatives to Boycott the productions of the Harmony mills. "Let them be isolated and alone."

*New York Unionist,* June 3, 1882.

1. The *New York Unionist* was the organ of the Amalgamated Trades and Labor Union of New York and Vicinity and was edited by J. P. McDonnell.

2. Robert Johnston (1807-90) was born in England and immigrated to New York in 1833. He worked as a mill supervisor and was subsequently superintendent and general manager of Harmony Mills.

3. Workers at Harmony Mills in Cohoes, N.Y., struck twice during the spring of 1880. Female weavers, supported by some 4,000 other workers, struck from Feb. 25 to Mar. 8 and won a 10 percent wage increase, limitations on the docking of workers for imperfectly woven material, and a fifty-minute lunch break. Spinners initiated what became a second general strike on Mar. 26 over the discharge of two of their number for union activities. The company retaliated by refusing meals to strikers in company boardinghouses, giving workers in company tenements notice to vacate, withholding back pay for failure to give proper notice before leaving work, and threatening to hire Canadian workers to fill the strikers' positions. In a settlement on Apr. 21 the spinners agreed to restrict their organization to benevolent activities, while the company agreed to take back all but two strikers, refrain from interfering with the spinners' society, and release withheld pay.

# An Excerpt from a News Account of a Mass Meeting of Cigarmakers in Jersey City, N.J.[1]

[June 17, 1882]

MASS MEETING OF CIGAR MAKERS IN JERSEY CITY.

. . .

Mr. Gompers was the next speaker. He said: This is one of a series of meetings arranged by the C.M.I.U. I would like to call your attention to facts occurring right around us. The cost of living has become so enhanced, which has been at the expense of the workingmen, that now they are commencing to move in order to increase their wages, so as to compel the employers to pay for the increase. Most all trades have organized except the cigarmakers, who have permitted the tide to almost pass away without taking advantage of the opportunity. In reference to the ballot he said the corporations who control the stomachs of the workingmen control their ballot. The party which compels men to accept starvation wages, to live with a large family in a room and bedroom, that party can control the vote of workingmen. Mr. Gompers cited the case of Butler[2] and Talbot,[3] in Massachusetts, where the operatives were compelled to vote for Talbot or get discharged. Only organized workingmen can resist such impositions of employers. Mr. Gompers showed that high wages did not mean dear products, as that increased consumption made the articles proportionately cheaper. In an eloquent appeal the speaker urged his hearers to organize for their own benefit and that of their families. He said that the dangerous elements in society were not the well-paid men but the ill-paid and overworked. Intelligence was the [. . . .]

. . .

Mr. Gompers again took the floor and said that if the workingmen of New Jersey had Trades Councils in every city the Child Labor bill would not have been defeated[4] nor would the Cigar bill have been *stolen*[5] if all the Trades Assemblies in New York State were like that of Buffalo. After some further remarks in favor of benevolent features in Trades Unions, Mr. Gompers concluded amidst applause. . . .

. . .

*New York Unionist,* June 17, 1882.

1. The New York District of the CMIU held a mass meeting on June 11, 1882, in Jersey City, N.J. It was one of a series of mass meetings the District held during the spring and summer of 1882.

2. Benjamin Franklin Butler (1818-93), a Civil War general, served as a Republican

in the U.S. Congress (1867-75). He was an unsuccessful candidate for governor of Massachusetts in four elections between 1871 and 1879, including 1878 when he was nominated by the Greenback-Labor party and the Democratic party but lost to Thomas Talbot in a vigorously contested campaign. He was elected governor in 1882 and in 1884 ran unsuccessfully for the presidency of the United States on the Greenback and Anti-Monopolist tickets.

3. Thomas Talbot (1818-86) served as a Democrat in the Massachusetts legislature during the 1860s and, after joining the Republican party, was elected lieutenant governor in 1872 and 1873. In 1874 he became governor, succeeding William B. Washburn, who had been elected to the U.S. Senate. He served until 1875 and was then reelected governor in 1878.

4. Probably House Bill 184, introduced in the 1882 New Jersey assembly, which prohibited the employment of children under twelve, limited the hours of minors under twenty-one and of adult women, and provided for an inspector of factories. The New Jersey assembly passed the bill; the senate passed it with an amendment with which the House refused to concur.

5. A reference to an incident in May 1882, when the New York senate passed a tenement-house bill only to have it disappear prior to the final vote in the assembly.

# An Excerpt from a News Account of a Mass Meeting of Cigarmakers in Williamsburg, N.Y.[1]

[June 24, 1882]

## CIGARMAKERS' MASS MEETING.

. . .

. . . He[2] then introduced Mr. S. Gompers, of Union 144, who spoke in his usual able and reasoning manner, and riveted the attention of his hearers. He was frequently interrupted by hearty applause. Comparing the condition of the cigarmakers with other trades, he concluded that of all trades the cigarmakers, who claim to be at least amongst the most intelligent, were not paid sufficient wages to enable them to replace the waste of energy and physical strength caused by their labor. It was not to be wondered at that cigarmakers are quoted in the statistics of the Board of Health as giving the largest percentage to the death rates. How much longer is this to last? Something must be done, said Mr. Gompers, or the next panic, which may possibly come within a year, will find us unprepared and leave us worse than the mill operatives in the New England States—too low for redemption. What have we gained by the so-called prosperity of the last two years? Our wages are no higher to-day than they were during the worst years of the last panic; we may be more steadily employed, but

no more. Therefore, if the wages are no higher to day than they were four years ago, the much talked of prosperity has not only not served us, but it has, on the contrary, done us injury. The cost of living, rent, etc., has increased at least twenty-five per cent., and in consequence thereof we are working for lower wages than during the days of the panic. Just think of it. In spite of the *great prosperity*, you are working at a reduction in wages. What a pretty state of affairs! What do you propose to do? Do you not desire a better state of things? Then organize, for in Trades Union organization is the only safeguard for wage workers.

. . .

*New York Unionist*, June 24, 1882.

1. The New York District of the CMIU held an organizing meeting on June 18, 1882, in Fitten's Union Hall, Williamsburg, N.Y.
2. The chairman of the meeting was Herman Davidson, president of CMIU 132 of Williamsburg, N.Y.

# An Excerpt from a News Account of a Mass Meeting of Cigarmakers in Hoboken, N.J.[1]

[July 8, 1882]

## A MASS MEETING OF CIGARMAKERS.

. . .

Mr. Gompers, of Union 144, the next speaker, said: It has been asserted that the working people of this country are not so thoroughly alive to the labor question nor as active in the movement as some of their European comrades, but I deny this charge, and claim, without fear of contradiction, that the American stands second to but one other nationality in the labor movement, and that is the English. The Americans may be moving somewhat slow, but it is the gradual sure development, the progression from step to step, the continual going ahead, which is the only natural movement of labor. As to not enough, let me say, the working class can never do enough; there is and will always be an open field, for just as soon as we shall have enough we shall cease to exist, or go backward. It is too true that the labor movement is divided, and I will show who has divided it. From my own experience in the [union] of which I am a member, I can quote that those parties who claim the monopoly of sincerity and honesty,

in the labor movement, are a class of men who are Union men only in proportion to the amount of dues they pay; for instance, at ten cents a week they are fair, at fifteen cents per week they are middling, but at twenty cents per week the Union gets too expensive, and must either come down or be broken up. That is about as far as their Union principle goes. As the dues increase, so their Unionism decreases, and vice versa, so that if the Union were to give them 50 cents per week for belonging to it they would be the best Union men out. But they also have other objects in view, such as to, if possible, subordinate the Union to a so-called political party, and if they cannot succeed in that, they are willing to break it up. They fully understand that New York organized, means the whole country organized, and for that reason they are doing all in their power to keep New York city disorganized, and that of course is only consistent with their great friendship for Trades Unions.

The power of Trades Unions can be seen when we look at that great strike of iron workers[2] going on at present. It is said by some of the so-called labor men that the organization mentioned is conservative; why, they even have as a declaration of demands the protective tariff. Well they may be conservative in *their* sense, but let me tell you that this four weeks' strike has spread more intelligence amongst them than could all the works of the so-called, would-be scientific, economic, and social thinkers. As to the amount of dues to be paid, I would say that you may judge the intelligence of a Union by the amount of dues the members pay for its support. High dues means high state of intelligence, greater facilities to fight, and better chances to win.

. . .

*New York Unionist,* July 8, 1882.

1. The New York District of the CMIU held a mass meeting on July 2, 1882, in Hoboken, N.J.

2. The National Amalgamated Association of Iron and Steel Workers struck unsuccessfully for a wage increase, June 1–Sept. 20, 1882. The strike was centered in Pittsburgh.

## A Translation of a News Account of the Founding Meeting of Local 1 of the Cigarmakers' Progressive Union of America

[July 24, 1882]

The former members of Union 144, C.I.U. of A., who have constituted an independent organization since July 17, as well as cigar-

makers interested in joining this organization, held a general meeting yesterday in the Concordia Assembly Rooms, 28-30 Ave. A, under the leadership of S. Schimkowitz. First a letter from the cigarmakers of New Haven, Conn., was read in which they expressed their sympathy with the decision of the cigarmakers here to declare themselves independent, as well as their wishes concerning the new organization. The letter was received with great enthusiasm, and it was decided to have a reply sent by the secretary along with all the resolutions made at the meeting. Then came the acceptance of the following declaration of principles as the basis for the new constitution: In all the civilized nations of the world, workers are now trying to organize in order to resist together the pressure of capital, which grows stronger and stronger every year. If this battle is conducted solely on the economic front, it can have no lasting success. As a class, workers must take control of the legislative process, for in the legislative bodies each and every political freedom of the working class is destroyed by the now-ruling class. The union founded today by the cigarmakers of New York will therefore try above all to mitigate the strikes that are of course unavoidable in the struggle against capital, or, as much as possible, try to direct them by means of enlightened action to the advantage of the workers. Secondly, this organization will make it its duty to elevate its members in an intellectual way by means of informative speeches and to enlighten them about their situation as a class. Thirdly, this organization will work very closely with the organizations of other trades for the purpose of mutual support in the struggle against capital and for the purpose of joint action on the political front. Our motto shall be: "Only workers in the legislature." They know the suffering of the people and know best where to begin in order to assure the working populace the full return on its work. After the acceptance of this declaration of principles it was decided: (1) The union shall carry the name "Cigarmakers Progressive Union of America No. 1." (2) The membership fee is $1, which can be paid in 4 weekly installments. An exception is made for all who join the union by October 1. For these, the membership fee is 50 cents, to be paid in 2 weekly installments. (3) Cigarmakers belonging to the Int. Union who are members in good standing of their organization—that is, are no more than 7 weeks behind—and want to join our union, should be able to join without charge by September 1 and retain any rights to benefits they might have. (4) The health plan is obligatory, that is, all who belong to the union must also belong to the health plan. (5) Dues are a total of 15 cents per week. (6) The union pays benefits of $5 for strikes, $4 sick pay, and $30 payment in case of death. (7) The payment of the old dues continues up to

and including July 29. After this date the new tax comes into effect. (8) The assessments written out by the International Executive Board are to be paid to the fund of our organization up to the 4th assessment. (9) The above is to be printed and circulated, along with a call to organize, in 6,000 German, 3,000 English, and 2,000 Bohemian copies.

A committee that was sent to the "German-Bohemian Union"[1] to invite them to work together with the "Cigarmakers Progressive Union of America" reports to have found willing cooperation from this group. It immediately put its newspaper, which appears in German and Bohemian, at our disposal and elected a conference committee that is to prepare the way for a common organization. Steps in this connection were given over to the Administrative Board, meeting today in Lincoln Hall.

*New Yorker Volkszeitung*, July 24, 1882. Translated from the German by Patrick McGrath.

1. German socialists who seceded from CMIU 144 in July 1880 formed the independent American Cigar and Tobacco Workers' Association, which they subsequently renamed the United Cigarmakers of North America. By 1882 this union contained five sections, at least one of which was Bohemian, and was issuing its own newspaper, *Die Wahrheit* (*The Truth*).

## *The FOTLU, the KOL, and the Eight-Hour Campaign of 1886*

In his memoirs Gompers described the FOTLU as an organization "committed to relief by legislation" that was modeled in part on the Trades Union Congress of Great Britain. Its platform reflected the founding delegates' view that the Federation should function primarily as a labor lobby, but in fact the FOTLU lacked the financial resources to do so. Although the typographers', cigarmakers', granite cutters', and carpenters' unions, as well as several city central bodies, consistently contributed to the FOTLU and attended its conventions, it generally received only meager support from trade unions. The Federation had no paid executive officer to carry out its programs—its executive body was the Legislative Committee—and, as Gompers later explained, its legislative activities depended "upon the interest and opportunities of individuals." Despite its weakness, however, the FOTLU, in its emphasis on organization along trade union lines, offered workers an alternative to the KOL.[1]

Competition between the KOL and the FOTLU had reached serious proportions by the time the Federation's second convention met in Cleveland in November 1882. P. J. McGuire, chief executive officer of the Brotherhood of Carpenters and Joiners of America and a Knight, had already made known his belief that the two organizations should carve out distinct spheres of activity. "It is a mistake to think the K of L should absorb everything," he wrote Carpenters' president Gabriel Edmonston in February 1882. "The K of L has not accomplished what some expect. It can only succeed by bringing in the select few of each trade. To bring in whole trades is a mistake." McGuire now wrote to the FOTLU convention convinced that the "friction" between the rival organizations had "already become serious and irritating." He urged that the FOTLU seek industrial rather than political unity, to be achieved through the federation of existing trade unions rather than their absorption into one general organization. "While industry prevails, trades unions will exist, and this necessitates organization by trades," he argued, and this "principle of unionism" must be extended from the local to the national level, culminating in an overarching federation of trade unions. In Gompers' opinion, this document sounded a "new note" for the FOTLU.[2]

McGuire's letter reflected some of the internal quarrels that plagued the KOL at this time. McGuire was a Knight who sought the Order's return to its earlier traditions. He was a member of Local Assembly (LA) 1562—the Spread the Light Club—a Brooklyn-based local assembly whose leaders strongly opposed the Knights' decision to abandon secrecy. "The very publicity of the K of L is its destruction," McGuire wrote Edmonston in July 1882. "It is filling up now with politicians who think it some political machine for their use. It will be swallowed up by them unless it becomes as secret as formerly." Later that summer members of LA 1562 formed the "Committee of Safety" and issued a secret circular that criticized KOL leader Terence Powderly for his lack of "backbone" in abandoning secrecy. In his letter to the FOTLU McGuire raised the issue again when he suggested that the KOL had an important role to play as a secret labor organization alongside an open federation of trade unions.[3]

The official Federation proceedings do not reveal what action the convention took in response to McGuire's letter. McGuire's brief history of the AFL, written in 1888, indicates that the FOTLU endorsed the document and issued it—whether in part or *in toto* is not clear—as a "manifesto" to its affiliates. Gompers later remembered McGuire's communication as an "able letter of counsel pointing out the need and the possibilities of federated action in the industrial field, national and international." In particular he cited McGuire's proposal for unified action on the part of workers to establish the eight-hour day, a call that became the central focus of FOTLU activity during its final years.[4]

At the 1884 FOTLU convention the Legislative Committee recommended that all labor organizations should vote on whether to participate in a general strike for the eight- or nine-hour day, to be held no later than May 1, 1886. Gompers recalled that he and Edmonston drafted the convention's formal motion that declared "eight hours shall constitute a legal day's labor from and after May 1, 1886," and urged that "labor organizations . . . so direct their laws as to conform to this resolution by the time named."[5] The 1885 convention urged nonstriking unions to support those that did strike.

The FOTLU Legislative Committee repeatedly appealed to the KOL leadership to support the campaign. Powderly, however, who advocated achieving the eight-hour day through legislative means, opposed the FOTLU's proposal. On March 13, 1886, he issued a secret circular advising KOL assemblies that the leadership of the Order did not endorse the strikes called for May 1. Nevertheless, numerous rank-and-file Knights supported the movement. In the course of the eight-hour agitation in the months preceding May 1, 1886,

thousands of workers in cities across the United States joined existing local unions and KOL assemblies or formed new ones. On May 1, workers participated in parades and demonstrations, and over 350,000 struck.

Although many workers achieved shorter hours as a result of this agitation, their gains often proved short-lived. The antilabor sentiment generated by the Haymarket Square bombing in Chicago on May 4, 1886, brought the movement to an abrupt halt. In that city, on May 3, police had killed two people and wounded several more during a melee between strikers and nonunion workers at the McCormick Harvesting Machine Company. In protest, anarchist August Spies, a strike sympathizer, held a meeting at Haymarket Square on May 4. When the police began forcibly to disperse the small crowd, an unidentified assailant threw a bomb that killed one policeman and fatally wounded six others. During the summer eight anarchists, including Spies, were tried and convicted of murder in connection with the deaths of the police officers. The coincidence of the Haymarket incident with the demonstrations sponsored by the FOTLU to promote the eight-hour day encouraged employers, the press, and the public to link trade unions with anarchists as threats to public order. Gompers believed that the association of Haymarket with trade unions in the public mind "killed the eight-hour movement."[6] It was three years before the FOTLU's successor, the AFL, renewed the campaign.

The demise of the FOTLU followed the collapse of the eight-hour movement. By the fall of 1886, according to Gompers, "All were convinced that the old Federation could not do the effective work required." What was needed, he added, was a "consolidated organization for the promotion of trade unionism" with "a central office and officers who could give all their time to the Federation work."[7] Growing tensions between trade unions and the KOL during 1886 provided the final impetus for the replacement of the FOTLU with a new and stronger organization, the AFL.

*Notes*

1. SG, *Seventy Years,* 1:264, 236. SG served on the Legislative Committee for four terms: as first vice-chairman in 1881-82, chairman in 1882-83, first vice-president in 1883-84, and president in 1885-86. He never held the organization's leading position of secretary, however.

2. McGuire to Edmonston, Feb. 21, 1882, reel 1, *AFL Records*; *Carpenter,* July 15, 1888; "P. J. McGuire to the Officers and Delegates of the 1882 FOTLU Convention," Nov. 20, 1882, below; SG, *Seventy Years,* 1:233.

3. July 8, 1882, reel 1, *AFL Records*; Committee of Safety to Dear Brother, Aug. 12, 1882, Terence Vincent Powderly Papers, DCU.

4. *Carpenter,* July 15, 1888; SG, *Seventy Years,* 1:233.
5. SG, *Seventy Years,* 1:237; FOTLU, Proceedings, 1884, p. 14.
6. *New York Sun,* Nov. 10, 1887.
7. SG, *Seventy Years,* 1:264.

# P. J. McGuire to the Officers and Delegates of the 1882 FOTLU Convention

Philadelphia, Pa., Nov. 20, 1882.

To the Officers and Delegates of the Federation of Trades.
Fellow Workers:

In behalf of the Brotherhood of Carpenters, I am instructed to extend to you our most hearty greetings, and we assure you of our warmest sympathy and most cordial co-operation.

We favor a Federation of Trades and Labor Unions, organized as an industrial body, and not as a political one. Its work should be to bring all trades together in closer unity for the better protection of our interests as workmen, and for the wider extension of the principles of unionism, so that all organized bodies of Labor may make common cause, and that none may suffer for want of that moral and pecuniary assistance, which, isolated and detached, we can not secure, and which, united and consolidated, we are bound to obtain.

We favor this Federation, because it is the most natural and assimilative form of bringing the trades and labor unions together. It preserves the industrial autonomy and distinctive character of each trade and labor union, and without doing violence to their feelings or traditions, blends them all in one harmonious whole—a Federation of Trades.

Such a body looks to the organization of the working classes as workers, and not as soldiers or politicians. It organizes them in their respective trades unions, and makes the qualities of the man as a worker—the only test of fitness, and sets up no political or religious test of membership. It strives for the unification of all Labor not by straining at a forced unity of diverse thought and widely separated methods—not by prescribing a uniform plan of organization, regardless of their experience or necessities—not by antagonizing or destroying existing organizations, but by preserving all that is integral and good in them, and by widening their scope that each, without submerging its individual character, may act with the other in all that concerns them.

While industry prevails, trades unions will exist, and this necessitates organization by trades, for the men of one craft will more readily unite for their collective interests as they are brought closer in contact with each other, and become impressed with the necessity of organization. None can deny that trades unions are the product of our civilization, and are not only necessary but beneficial and profitable to their [members.]

If this then be true, it should logically follow that the principle of unionism, good in its local application, should be extended to effect the national and international organization of each trade. And it equally follows that these trades and labor unions should be combined in a Federation of Trades and Labor Unions.

That such a form of organization is necessary is amply proven by the experience of many trades during the past twelve months. Isolated Labor in conflict with consolidated Capital in many cases has been driven to the wall, defeated for a time, but destined to emerge victorious only as it learns from its defeats, the lesson that our Trades Councils and our National and International Unions must be affiliated practically, so as to form one unbroken chain of union for each other's defense and welfare. To accomplish this task, your Congress has assembled. Workmen, everywhere look with expectation to your deliberations, that they may be marked with a higher degree of statesmanship than marks the acts of our State Legislature, or of the United States Congress. It is for you to settle questions that these politicians can never comprehend, nor do they care to understand them. But these questions will be forced onward in spite of their Penal Codes and Conspiracy Laws.

The first thing to understand is the dual character of labor organization—its public and its secret side. We desire to be organized publicly wherever possible, so that the existence of our unions may be known and that they may be legalized. But where by acts of the capitalists this has become impossible then we favor secret organization. Both forms are necessary and instead of being hostile, they should be harmonized so as to cooperate with each other. There is no need for one trying to swallow the other, nor is there need for any conflict. The open trade unions, local, national and international, can and ought to work side by side with the Knights of Labor, and this would be the case, were it not for men—either overzealous or ambitious who busy themselves in attempting the destruction of existing unions to serve their own whims and mad iconoclasm.

This should cease and each should understand its proper place and work in that sphere. And if they desire to come under one head or to affiliate their forces, then let all trade and labor societies—secret and public—be represented in the Federation of Trades and Labor Unions.

The benefit of this Federation is not only to render pecuniary and moral assistance in case of strikes, or lock outs, but its very existence can lessen strikes, by playing with the fears of employers who in many cases would then hesitate to provoke strikes, that they knew would be well supported. In addition to this, new unions could be organized,

local trade councils could be formed, and national and international unions could be established and a systematic plan of propaganda inaugurated that would strengthen and enliven our unions.

By proper support of the Federation, a weekly official journal might be published and a labor literature of tracts and pamphlets could be printed at small cost for public distribution.

We find the workmen of Europe, in Great Britain, France, Spain, Italy, Switzerland, Belgium, Holland, Austria and Denmark, have held largely attended Trades Congresses the past year. And just as these National Labor Congresses develop, they will tend to correspondence with each other and open up friendly relations between the workmen of all countries, that will lead undoubtedly to an "era of peace and good will among nations." And hence before many years we may expect an International Trades Union Congress, to determine among other questions the importation of labor from one country to another in case of strikes, and to discuss the evil immigration of labor to flood a country at the behests of cheap labor capitalists.

While we consider these subjects there are other matters we must deal with. We desire a reduction of the hours of labor, not so much by enactment of laws, which go on our statute books never to be enforced. But we wish to make it an enactment of the workmen themselves, that on a given day they will agree to work no longer than eight hours per day and to enforce that rule themselves. But where the government is an employer—on public works—we demand the enforcement of the Eight Hour law. And in the next session of Congress the Legislative Committee of the Federation should demand the enforcement of that law, and also the creation of a National Bureau of Labor Statistics and the Legalization of Trades Unions. Let the Congressional work of the Federation be centred *primarily* on these three points; and *secondarily* on such others as in your wisdom you may determine.

In asking these measures from Congress let us by no means sink to the level of party politicians. Let us not sacrifice our unions for the sake of any political party. But as union men let us work in harmony for those issues we do comprehend and upon which we are fully agreed, and let us ignore all those questions likely to disrupt and divide us.

Yours Fraternally, P. J. McGuire.

*Carpenter,* Dec. 1882, Jan. 1883.

# Excerpts from Two News Accounts of the Second Annual Convention of the FOTLU

[November 24, 1882]

## THE WORKINGMEN.

. . .

### Private Ownership of Land.

The following was offered by Mr. Grinnell,[1] of Detroit:

*Resolved,* That we believe that the present system of land tenure in America is detrimental to our interests as wage workers, and that some system should be adopted that will guarantee to all an equitable share in the National resources of the land.

A long paper on the subject was ready by the gentleman who introduced the resolution in favor of its adoption.

Mr. Powers spoke adversely upon the adoption of the above, claiming that the subject was too gigantic for the members to grapple, but should be left to the body politic. Mr. Foster,[2] of Massachusetts, spoke likewise on the matter, stating that as trades unionists they had no right to take up such questions that do not properly come within their scope. Mr. Gompers made extended remarks on the subject. He denied that land controlled wages, and declared that wages controlled land and everything else. High wages, he said, bring about labor-saving machinery, and every introduction of that kind meant that articles can be purchased cheaper by the working class. For instance, to have a box of goods taken to California from Cleveland it would cost say five dollars with wages of men at three dollars per day. To have it taken an equal distance in China, in the absence of the necessary machinery, at wages at seven cents a day, it would cost as much as ten dollars. If the private ownership of land is wrong, the private ownership of the products of that land is wrong. "It is not the ownership of land that should be fought, but the doings of the capitalists that we are organized to oppose," declared the speaker.

Mr. McIntyre[3] thought that too much time had been taken in an argument which trades unionists have no right to consider. The resolution was voted down. It was resolved to grant admission to the federation of delegates at its next meeting of delegates from the Women's National Labor League[4] of Washington. District assemblies of the Knights of Labor were granted representation on the same basis as local trades unions, although considerable argument was made

before the point was carried. The congress will reassemble at 9 o'clock this morning.

*Cleveland Herald,* Nov. 24, 1882.

1. Judson GRENELL, a labor editor, was an officer of International Typographical Union (ITU) 18 of Detroit and of the Detroit Trade and Labor Council in the early 1880s.

2. Frank Keyes FOSTER, a Boston printer and labor editor in the 1880s and 1890s, was active in the ITU, the KOL, and the FOTLU.

3. David McIntyre, a New York City patternmaker, represented the Amalgamated Society of Engineers, Machinists, and Millwrights, New York Central District.

4. The Women's National Labor League (WNLL) was organized in August 1882 in Washington, D.C., in response to the exclusion of women from appointments in the departments of Interior and War. Its constitution condemned political corruption, the exploitation of labor by capital, and sexual discrimination in civil service appointments; its members elected Charlotte Smith as the WNLL's first president.

[November 24, 1882]

## LABOR LEADERS.

. . .

A resolution that the Knights of Labor be entitled to a representation in the assembly on the basis of one representative from each lodge was adopted by a vote of fourteen for, two against, and two not voting. This was done after a long and heated discussion, Mr. Gompers, of New York, being bitterly opposed to it.

### In the Evening

about three hundred people, among whom were many ladies, assembled in the City Armory to hear Hon. T. V. Powderly, mayor of Scranton, Pennsylvania, speak upon the subject of "The Labor Question in Europe and America." Although the lecture was given under the auspices of the McNevin Club, a Cleveland organization, many delegates to the Labor Congress were in attendance and occupied seats on the platform. Hon. M. A. Foran[1] presided and introduced the speaker in his usually happy manner. Mr. Powderly is a fine looking man, is neatly dressed, and is a pleasant speaker. He said he had come to Cleveland scarcely prepared to speak upon any subject, but the labor question was one upon which something could always be said. He would not follow closely the text laid down for him. He asked the audience to go back with him fifteen years to the time when three men, Allen, Larkin, and O'Brien,[2] walked upon a scaffold in England, innocent of any crime, and were swung into eternity. Why had those men died like murderers? because they loved the land of their birth and had struggled, with no thought of malice, to do the best they could for that land and make it free. He then pointed to

THE POVERTY-STRICKEN PEOPLE

of that land who had struggled for seven centuries for a free flag on the soil of Ireland. He said they had thus far accomplished nothing, but he hoped to see the day when the Irish would throw off the English yoke and make for themselves a republic. There were men as noble and good as any in America lying in graves in Ireland who had struggled through life in poverty. They did not go down to their graves in that condition, because they did not want to work, but because they were oppressed by men who knew they were crushing the manhood out of them. In Ireland he said the people did not have the means in their hands that American laborers have. If they did they would teach the world a lesson. They have risen and asked that the land which belongs to them be given back to them.[3] That was all they could do. He said he did not believe one man on the face of the earth had a right to a handful of land; that all the land belonged to God and to his subjects in common. He favored the "no rent" manifesto with all his heart, and thought the Irish people should have stood by it. Some said the Irish were prosperous, and only grumbled because it was natural for them to do so. That was false. They were not prosperous, and they grumbled because they were asked to give up their heritage to the

HERDS OF ENGLISH CATTLE

which were fattened to feed and fill the pockets of the landlords. Notwithstanding all this, if Ireland could make her own laws—and she has the ability—she would prosper.

The speaker then referred to trades' unions, saying they were first formed for the purpose of bettering the condition of laboring men. All organizations had failed, however, in the grand object—the unification of labor. All trades should be banded together, so that every man should know the condition of his fellow. The Knights of Labor, of which he is an officer, was organized several years since at Philadelphia, as an organization into which all workingmen could come. It had grown rapidly until it had branches in every part of this country, and in some portions of Canada. It had a great power for good and for evil. Judicious management was needed that its influence should be exerted in the right direction. The work was bound to go on, and nothing could stop it. He then related his experience among the miners of Maryland, and told how men had struck against an unjust demand, and their places had been filled by other workmen.[4] He said such things were wrong, as it made slaves of men. There must be a remedy for the evil, and he hoped it would come out of the Knights of Labor

organization. He asked why it was that laborers built buildings for rich people when they had no roofs to cover their own heads, and answered it by saying that

THE GIANT HAND OF MONOPOLY

has been allowed to grasp every enterprise in the land until it has gained control of Congress, and wrested from the people their rightful heritage, and placed it in the hands of corporations. He explained his answer by showing that Congress had granted to various railroad companies 296,000,000,000 acres of land.[5] Those grants he characterized as rank steals. He hoped to see the day come when men shall not have the power to vote away the people's heritage. That was one of the things asked by the Knights of Labor. Another thing asked by that organization was that children should not be allowed to work in factories and mines, but be sent to school and fitted to carry on the work now being done. It also asked that eight hours constitute a day's work; he would be in favor of five hours. As a reason he said that machines were now in use, run by girls, which would do as much work as one hundred men could do one hundred years ago. If the production can be so greatly increased by using machines, the people who run the machines should not be compelled to work so many hours.

The Knights of Labor also sought to obliterate race lines and religious prejudices in questions of labor, which was one of the grandest things the organization ever accomplished. The speaker concluded by saying that the condition of affairs in America today was similar to that in France before the revolution; that the warning note had been sounded and it behooved everybody to heed it.

. . .

*Cleveland Leader*, Nov. 24, 1882.

1. Martin Ambrose FORAN, a cooper, was an organizer in 1870 of the Coopers' International Union and its president for three years. He served as a Democratic congressman from Ohio from 1883 to 1889.

2. William O'Meara Allen, Michael Larkin, and Michael O'Brien (also known as William Gould), the "Manchester Martyrs," were executed in Manchester, England, on Nov. 23, 1867, for their participation in an unsuccessful attempt to free two Fenian comrades from police custody.

3. In the late 1870s an agricultural depression in Ireland and a succession of poor potato harvests resulted in widespread suffering and numerous evictions of tenant farmers. Michael Davitt founded the Irish National Land League in October 1879, and during the period of popular unrest known as the "Land War" of 1879-82 it spearheaded resistance to excessively high rents and evictions. The unrest subsided after the British government conceded to the League's principal demands in the Land

Act of 1881 and in a subsequent agreement (the "Kilmainham Treaty") with League president Charles Stewart Parnell in the spring of 1882. Davitt and others founded an American branch of the League, the American Land League, in the spring of 1880, linking existing local organizations in the United States that provided financial support for the Irish movement. Terence Powderly, a founder and president of the Scranton branch of the League, was elected to the Central Council of the American Land League in 1880 and became a vice-president in 1881.

4. Maryland miners, many of them members of KOL District Assembly 25, struck Mar. 14, 1882, over western Maryland coal companies' proposals to reduce wages and increase hours. The miners lost the strike and returned to work on Aug. 24 at the companies' original terms.

5. Federal land grants to railroads consisted of a net total of about 131 million acres.

## An Excerpt[1] from the Minutes of the Legislative Committee of the FOTLU

Cleveland, Ohio, Nov. 24, 1882.

. . .

It was carried as the sense of the Committee that the Federation of Trades Unions of the District of Columbia[2] might be empowered to present to Congress statistics, but not measures.

The Secretary was, on motion, directed to draft and issue an address to Trades Unions, Trades Assemblies, &c.[3]

The Committee, on motion, adjourned.

Attest: W. H. Foster,
Secretary.

ADS, Minutes of Meetings, Legislative Committee, FOTLU, reel 1, *AFL Records.*

1. During the first portion of this meeting, the Legislative Committee elected the following officers: SG, chairman; Richard Powers, first vice-chairman; Gabriel Edmonston, second vice-chairman; and Robert Howard, treasurer. William H. Foster presided, having been elected secretary by the 1882 FOTLU convention.

2. The Washington Federation of Labor.

3. Possibly incorporating P. J. McGuire's letter of Nov. 20 to the 1882 FOTLU convention (see above).

## *The 1883 Hearings on the Relations between Labor and Capital*

In 1882, in response to a recent outbreak of strikes, the U.S. Senate authorized its Committee on Education and Labor to undertake a broad investigation of wages, hours, working conditions, the causes of strikes, and possible means of promoting harmony between labor and capital. Chaired by Senator Henry Blair of New Hampshire, the committee held brief hearings in Washington, D.C., in February 1883, more extensive hearings in New York City in the late summer and fall of that year, and took testimony in still other cities in the fall. Witnesses included employers, union officials, workers, and reformers.

Because the FOTLU convention was meeting in New York City in August, trade unionists from all over the country were available to testify there before the committee. In their testimony, Gompers, Gabriel Edmonston, Adolph Strasser, W. H. Foster, and Frank Foster all supported the establishment of a federal bureau of labor statistics, the abolition of the contract labor system, the enforcement of the federal eight-hour law, and the incorporation of trade unions and criticized such companies as Western Union Telegraph, whose employees were required to pledge that they would refrain from union activities before they could be rehired after a strike in 1883. KOL witnesses supported trade unionists on these issues.

Labor witnesses did not agree on all matters, however. KOL Grand Secretary Robert Layton exemplified the faith that some leaders of the Order had in arbitration and cooperatives, while P. J. McGuire described strikes as "a revolt against the class rule of the capitalists" and declared that "no strike is a loss or a failure to the workers."[1] Strasser maintained that well-organized and experienced trade unions kept strikes from occurring and claimed that his union had prevented some 200 of them in the previous three years.

When the testimony touched on the broader goals and philosophy of the labor movement, differences among the labor witnesses became more apparent. Strasser, for example, contended that "socialistic feeling" had no place in the CMIU. "We have no ultimate ends," he insisted. "We are going on from day to day. We are fighting only for immediate objects—objects that can be realized in a few years." John

McClelland, secretary of the KOL General Executive Board, however, suggested that the government take over the telegraph system as "a starting point" on the way to more general public control over industry, and Henry George and John Swinton called for government-controlled railroads and telegraphs accompanied by land reform. Swinton also advocated currency reform, an income tax, and the creation of national departments of industry, health, education, and public works. Edward King, representing the New York City Central Labor Union, declared that workingmen had come to understand that "there is no relief for the workingman except by political action" and spoke in favor of legislation that would free labor to fight with capital "according to any just conception of competition." McGuire, on the other hand, expressed contempt for contemporary politics and hoped that an "industrial" government would replace the present "political" one and would be run by statesmen who would "deal with industrial questions in the halls of legislation, and will not waste time in strife about sectionalism or patronage."[2]

Despite the varied and sometimes contradictory positions taken by labor witnesses, the hearings did encourage Congress to implement some of the proposals not requiring extensive federal intervention. For example, Congress created a federal Bureau of Labor Statistics in the Department of the Interior in 1884 and, in 1885, prohibited the importation of foreign contract labor. Three years later it strengthened the federal eight-hour law and created an independent Department of Labor; the department was not accorded cabinet status until 1913, however.

*Notes*

1. U.S. Congress, Senate Committee on Education and Labor, *Report of the Committee of the Senate upon the Relations between Labor and Capital. . .* , 48th Cong., 4 vols. (Washington, D.C., 1885), 1:322.
2. Ibid., pp. 466, 460, 215, 559, 562, 347.

# Excerpts from Samuel Gompers' Testimony before the Education and Labor Committee of the U.S. Senate[1]

New York, August 16, 1883.

Samuel Gompers sworn and examined.

By the Chairman:[2]

Question. What is your full name and where do you reside?—Answer. Samuel Gompers; I reside in the city of New York.

Q. What is your age?—A. I shall be 34 years of age next January.

Q. How long have you resided in the city of New York?—A. With an intermission of about ten months,[3] over twenty years.

Q. What is your employment?—A. I am a cigar-maker.

Q. That has been your occupation during your life?—A. Yes, sir.

Q. Have you been connected with the labor movements of the country to any extent; and, if so, how closely and for how long a time?—A. I have; I joined the Cigar-makers' Union[4] about eighteen years ago or a little over; remained a member about four years; left the city; retained my membership in a branch organization; returned and continued my membership until about fifteen years ago. My membership ceased on account of the general decline of the organization at that time. I am not sure of the time, but about three months or six months later I joined a new organization of cigar-makers, and remained in membership there some years. Thereafter I joined Union 144 of the Cigar-makers' International Union; that is, I was a member of an organization[5] from which Union 144 grew or was established. I was one of its charter members, and have been a member of that organization ever since. Some three years ago the trades unions and labor organizations of the country called a convention to meet in Terre Haute, Ind., at which a confederation of trade and labor unions was proposed, and they held a session there, and adjourned to meet in November, 1881, in Pittsburgh, Pa., where the Federation of Organized Trades and Labor Unions was instituted. I was a delegate from the Cigar-makers' International Union to that convention, and was elected the first vice-chairman of the legislative committee. At the second session, held in the city of Cleveland, Ohio, I was elected the chairman of the legislative committee. I am a member of the order of the Knights of Labor,[6] and I belong also to one or two workingmen's clubs, in which lectures, discussions, &c., are indulged in for the improvement of the minds of our members.

Q. Have you now described fully your various connections with labor organizations?—A. Yes, I think I have described the connection that I have held with labor organizations; that is, so far as membership

is concerned. I have attended, of course, many labor meetings and associations.

Q. It would seem that your opportunities of learning the objects and scope of these labor organizations has been as good as any one's; won't you give us, therefore, a general idea of what labor organizations there are in this country, what their purposes are, and what they seem to be accomplishing as a whole. Give us an idea of the general labor movement and of those labor organizations which are leading it. Go on and open up the matter as it is in your own mind.—A. Then I will not start with the organizations. I would rather speak first of the general condition of labor as I find it, as I know it and believe it to be.

### Condition of Working People.

Q. Well, take up the subject in your own way, but before you get through I would like you to answer the question I have put with regard to the extent and the actual objects and results of these organizations.—A. Oh, certainly; I shall endeavor to give you that to the best of my ability. The condition of the working people appears to be coming to what may rightly be termed a focus. On the one hand it would be well to note the underlying motives that frequently break out in what are generally termed strikes. Strikes are the result of a condition, and are not, as is generally or frequently understood, the cause. For instance, in the State of Massachusetts they have a ten-hour law, intended to benefit the female and child operatives there,[7] yet the employers (and the same is true in Cohoes, in this State, and other places where the hours of labor are recognized as settled) or their agents start up the mills several minutes, sometimes seven, eight, nine, or ten minutes, before the time commencing to work according to rule and law. In other instances they close them at "noon" several minutes after 12 o'clock and open them again several minutes before the hour, or half hour rather, has elapsed, closing again for the day several minutes after the rule requires. These employers are pretty well described by some of the English economists and labor advocates—not labor advocates, but men who have made economic questions a study; they call them "minute thieves." In Massachusetts an instance is given in the report of the Massachusetts Bureau of Labor Statistics[8] for 1882,[9] page 302, showing that twenty-two minutes a day have been stolen from the operatives. The average wages of the operatives in the Lowell mills in Massachusetts, taking it from the Massachusetts report of the Bureau of Labor Statistics, is from 85½ to 90⁹⁄₁₀ cents per day. The highest wages per day is $1.05, or $6.30 per week, the lowest wages being those of the females employed in

that department of the trade known as drawers, whose weekly wages are $2.05. This can be found in the Massachusetts Report of 1882, page 331. The employes of those corporations have time and again requested the remedying of such grievances as they may have, and Mr. Carroll D. Wright[10] reports in that same volume that the answer these operatives received was, "There is no redress here and no appeal; if you do not like it, get out." In the branch of industry in which I work we have a bane to contend with, a curse, known as the manufacturing of cigars in tenement houses, in which the employer hires a row of tenements four or five stories high, with two, three, or four families living on each floor, occupying a room and bed-room, or a room, bed-room, and an apology for a kitchen. The tobacco for the work is given out by the manufacturer or his superintendent to the operatives who work there, the husband and wife, and they seldom work without one or more of their children, if they have any. Even their parents, if they have any, work also in the room, and any indigent relative that may live with them also helps along. I myself made an investigation of these houses about two years ago; went through them and made measurements of them, and found that however clean the people might desire to be they could not be so. The bed-room is generally dark, and contains all the wet tobacco that is not intended for immediate use, but perhaps for use on the following day; while in the front room (or back room, as the case may be) the husband and wife and child, or any friend or relative that works with them, three or four or five persons, are to be found. Each has a table at which to work. The tobacco which they work and the clippings or cuttings, as they are termed, are lying around the floor, while the scrap or clip that is intended to be used immediately for the making of cigars is lying about to dry. Children are playing about, as well as their puny health will permit them, in the tobacco. I have found, I believe, the most miserable conditions prevailing in those houses that I have seen at any time in my life.

Q. How many families are thus engaged in the manufacture of cigars in this city?—A. Between 1,900 and 2,000. The lowest ascertained number was 1,920 families. That was about five or six months ago.

Q. About 10,000 people, taking the average to a family of five?—A. Probably. . . .

By Mr. George:[11]. . . .

Q. Do you mean to say that about 1,900 families, engaged in the manufacture of cigars, live in the manner which you have just described?—A. Four-fifths of them, I think. Within this last year one of the manufacturers has endeavored to build a row of houses that are an improvement upon the old ones; but notwithstanding all at-

tempts to keep these places clean, that is impossible, in consequence of the long hours of toil and the fact that all of the family are employed right at the work of cigar-making.

By the Chairman:

Q. I do not know but you have covered the point, but I would like to know what proportion of the cigar manufacturing here located is carried on in these tenement houses which you have described.—A. About one-seventh, I think.

By Mr. George:

Q. The rest is done in shops or factories?—A. Yes, sir; I can give you the exact figures hereafter.

By the Chairman:

Q. Can you give us some idea of the cigar manufacturing operatives—men, women, and children—what you would call the cigar-manufacturing population of this city and vicinity?—A. I cannot, exactly.

Q. You think that those 8,000 or 10,000 people represent about one-seventh of the manufacture?—A. Yes; and there are a great many children in those families.

Q. The other six-sevenths of the work is done by workmen, boys, and girls in the shops?—A. Yes, sir. There are many who work in shops in New York who live in Brooklyn, Jersey City, and Newark, and some even in Paterson and adjacent cities, twelve or fifteen miles away. They come over here every morning. The fares, through commutation tickets, are cheap, taking into consideration also the lower rents that they pay in these places than they would have to pay here. Of course, they are compelled to rise from bed about an hour earlier and to travel an hour longer in the evening than a man living in the city of New York would have to do.

The Cigar-makers' International Union adopted a system of agitation against the tenement-house cigar manufacture some years ago, believing that it was a public nuisance, and the press of the city of New York, together with that of the entire country, took this matter in hand, discussed it ably, exposed the iniquity of the system and the greed and avarice to which many men will resort in unfair competition, even with their fairer rivals in the trade. The opinions of the press, several of them, were extracted and printed by us and spread broadcast. . . .

For three successive years the legislature of the State of New York has been appealed to to abolish this system. More than two years ago the sympathies and the co-operation of the public were enlisted in this cause, and upon sanitary grounds the abolition of the system was advocated. At its last session the legislature of the State of New York abolished the system of making cigars in tenement houses, and

on the 1st of October next it will be unlawful to make cigars here in that way.

By Mr. Call:[12]

Q. Can you state briefly the provisions of the law?—A. The law is known, I think, by the title, "An act to improve the public health of the city of New York by prohibiting the manufacture of cigars or the preparation of tobacco in the tenement houses of said city in any form."

Q. It only relates to tenement houses, then?—A. Yes, sir; that is, houses where people live. It provides that no room or suite of rooms used for dwelling purposes can be used for the purpose of the manufacture of cigars or the preparation of tobacco.

Q. What is the penalty for violating the law? Is it upon the owner of the house, or upon the parties engaged in the manufacture?—A. Upon the manufacturer, and the persons in his employ. It is believed that the public good is decidedly of more importance than the individual's interest, no matter who he may be. For years the Cigar-makers' International Union, an organization comprising about 185 branches throughout the United States, with many branches in Canada, took every means of fighting against this production of cigars in tenement houses, and adopted a label,[13] which has been patented, and through the general making known of this label the business of making cigars in tenements was considerably reduced. . . .

By Mr. George: . . .

Q. You say that was invented by whom?—A. By the Cigar-makers' International Union, and issued by authority of the union. The label reads:

UNION-MADE CIGARS.

This certifies that the cigars contained in this box have been made by first-class workmen and members of the Cigar-makers' International Union of America, an organization opposed to inferior, rat-shop, coolie, prison, or filthy tenement-house workmanship. Therefore we recommend these cigars to all smokers throughout the world. All infringements upon this label will be punished according to law.

A. STRASSER,
Pres't Cigar-makers' International Union of America.

And it is attested by the seal of the organization.

Q. That label is to be attached to every box of cigars, to let the purchaser know that they were properly made?—A. Yes, sir; that was the intention.

## Cohoes, N.Y.

I visited Cohoes, N.Y., during the strike there, about a year ago. That strike was organized against a proposed reduction of 10 per

cent. in the operatives' wages. There were certain conditions surrounding the people in Cohoes that struck me very forcibly. On meeting the committee who received me (as I had been invited to attend), I made inquiries as to an immense building which I saw in the town, that being the first time I had visited Cohoes, and upon all hands was I informed, "That belongs to the Harmony Mills." Inquiring further as to another building, I was told, "That belongs to the Harmony Company." Everything belonged to the Harmony Company. The hotel was the Harmony Hotel. The boarding-houses were Harmony boarding-houses; the tenements in which the people lived belonged to the Harmony Company. The water is controlled by the Harmony Company. The water-power by which the mills are run, the water which the people drink, the water which the other manufacturers are compelled to use, all is under the control of the Harmony Company.

By Mr. Pugh:[14]

Q. How many persons are there in the employ of that company?—A. Over 5,000.

Q. Where is Cohoes?—A. It is within an hour's travel from Albany, on the Mohawk River. As to the church there, I am informed that the minister in that church is a brother-in-law of the superintendent of the Harmony Mills. When the Harmony Company are in want of water to run their mills, and the people want water to drink, they have to go thirsty and the mills are run.

By the Chairman:

Q. Is the water supply of the town taken from the river?—A. From the river; supplied through works first constructed by the Harmony Company.

Q. Are the city and the Harmony Company substantially identical? Does the company own the city pretty much?—A. Pretty much.

Q. Has not the city, the municipality, any reasonable opportunity of freeing itself of this dependence for water upon the Harmony Company? Can they not get a supply of water elsewhere?—A. Not very easily. I think it would require a great outlay, more than the people of Cohoes would be able to bear, outside of the interest of the Harmony Company. I was informed while there that several attempts had been made to start competitive mills in Cohoes, but that in consequence of the ownership by the Harmony Company, and their control of the water supply of Cohoes, competition was strangled at once; and while I have not traveled very extensively, I have seen some mills, and I am of opinion that no greater water facilities exist in this country than in Cohoes for the running of mills.

Q. I interrupted your statement to draw closer attention to your assertion that when water was scarce the people went thirsty in order

that the mills might run. You, perhaps, were never thirsty in that city yourself, but you may know of the complaints of people who reside there. I would like to know what your information is on that point.—A. The complaints were general. Of course scarcity of water in a place of so few inhabitants is not apt to occur very frequently, but when it does occur, and it has occurred several times, then complaint is general.

Q. Then the dearth is of water for purposes of cleanliness and ablution, rather than for drinking?—A. Sometimes it is.

Q. But still you do understand that the corporation restricts the people in the necessary amount of water for sanitary purposes?—A. No, sir, I do not; but I say that when there is a natural drought or scarcity of water they do. I do not wish to be understood as saying that the Harmony Company are willfully depriving the people of water, but that when there is a natural scarcity of water they first run the mills, even though the people have to go dirty and thirsty.

Q. That you understand from common conversation and from complaints that you have heard yourself?—A. Yes, sir.

Q. Complaints that you have heard on the ground?—A. Yes, sir; during my visit there.

Q. Was that a time of scarcity of water or not?—A. I could not answer that question.

Q. Do you believe that statement?—A. If I did not believe it, if I did not place some credit in it, I would not mention it.

Q. You think it is a fact?—A. Yes, sir.

Q. It satisfies your own judgment as a true statement?—A. Yes, sir; I believe it to be a truth; I have no reason to doubt it; I made inquiries after the persons told me that, and the statements were verified. I will say, by the way, that so much was I impressed with the information continually given that this and that and the other thing belonged to Harmony Mills, that although I am not of a poetical turn of mind I paraphrased Tennyson's Charge of the Light Brigade, so that instead of "cannon to right of them, cannon to left of them," it was "Harmony to right of them, Harmony to left of them." The operatives there were striking against a reduction of 10 per cent. in wages which was proposed, notwithstanding the fact that during that period we had had the greatest era of prosperity that this country had known.

Q. Do you know what the wages were from which the deduction of 10 per cent. was proposed; how were they compared with the wages of similar operatives at other places?—A. The wages of the operatives at Cohoes, N.Y., were less than those in the mills in Massachusetts.

Q. You mean that they were less after the reduction of 10 per cent. at that time?—A. Yes.

Q. Did you investigate the question and become satisfied of that fact?—A. I could not state that as a positive fact, but I am under the impression that it is true.

Q. You are not sure about that?—A. I am not; in the Lawrence, Massachusetts, mills the operatives were subjected to a reduction of 10 per cent. about that time, and the employes were generally under the impression that it was designed to counteract the intended efforts of the operatives to strike for an increase of wages; the operatives were starved into submission at Cohoes, and in Lawrence there was a compromise.[15]

Q. You say, then, that the result was that the Cohoes operatives were starved into accepting the reduction of 10 per cent. wages?—A. Yes, sir; during this era of prosperity.

Q. And the Lawrence operatives secured a compromise?—A. Yes, sir; the reason that the Lawrence operatives succeeded in forcing a compromise was—I would rather withdraw the remark, if I can, with reference to the Lawrence mills making a compromise; I have a full note at home, but my note that I have here merely makes mention that Lawrence also made a reduction of 10 per cent., and my memory fails me whether the operatives secured a compromise or were defeated.

### New York Freight-Handlers.

I will proceed now to another branch of inquiry, in reference to one of the most hardworked class of people under the sun, the freight-handlers of the city of New York. They are a body of men, very sinewy, working for 17 cents an hour for the railroad corporations. Last year they had the hardihood to ask for three cents more an hour, making 20 cents an hour, when the railroads informed them that they would not pay it. The freight-handlers were, after a struggle,[16] starved into submission, and are working now for 17 cents an hour.

Q. Now, you are here and see these people: what sort of life does a freight-handler have on 17 cents an hour?—A. He generally lives in very poor quarters; his home is but scantily furnished; he can eat only of the coarsest food; his children, like too many others, are frequently brought into the factories at a very tender age; in some instances his wife takes in sewing and does chores for other people, while in other instances that I know of they work in a few of the remaining laundries where women are still engaged, the work not having been absorbed by the Chinese. By this means the home, of course, is broken up; indeed there is hardly the semblance of a home, and in these instances where the wife goes out to work no meal is cooked. Many of the stores have for sale dried meats or herrings,

cheese, or some other article which does not require any cooking. Of course, when the wife is at home although the living is very poor, it is cooked; she cooks what can be purchased with the portion of the 17 cents per hour remaining after the payment of rent, and the cost of light, fuel, &c.

Q. You speak of these men working by the hour?—A. Yes, sir.

Q. Why are they not employed by the day or the month? For how many hours daily do they usually work?—A. Ten hours a day usually. Did I understand you to ask me why are they hired by the hour?

Q. Yes.—A. Employers generally know how best to take advantage of their help; and the motive to obtain labor at the cheapest rate possible and to get as much work for that cheapest rate is always a sufficient incentive to find out which is the cheapest way, and they have found that to be the way in that case.

Q. Well, I suppose employment in handling freight is less continuous during the day, or it extends over the whole twenty-four hours, perhaps, in such a way that the actual time during which the ten hours' work is done extends through the twenty-four, or through a longer period, is it so?—A. No, sir.

Q. Then I do not understand exactly why it is unjust to hire these men by the hour rather than by the day, if they make ten hours a day.—A. That, I say, is the average, but there are some days that they work eleven and twelve hours; and in the event of a train being late they will have little work for an hour, and then have to work after six or seven o'clock.

Q. Do you understand that if a train is belated and the freight-handler awaits the arrival of the train he loses the time of waiting?—A. Yes, sir, he loses it.

Q. And then, between trains, he loses that time also?—A. Yes, sir; but when the trains are regular, and there is a large number of freight cars always on hand, the men work at loading them.

Q. So that you think their work averages about ten hours a day?—A. Yes; so that a freight-handler earns about $1.70 a day.

## Importation of Labor.

About a year and half ago there was a strike of the cigar-makers of the city of Milwaukee.[17] They were striking for an increase. At that time the manufacturers there acceded to the demands of the cigar-makers, except Ashmun & Co.,[18] who, being the largest manufacturers, declined to accede. In consequence of Mr. Ashmun taking that position he induced the other manufacturers, who had already acceded to the wishes of the strikers, to lock them out and not to re-

employ them while they would continue to belong to the Union or would insist upon the increase of wages. He accomplished this result, and advertising was going on throughout the country for cigar-makers to go to Milwaukee. He at the same time sent agents to Germany and had advertisements inserted in the public papers there drawing a very rose-colored picture of Milwaukee and of the surroundings and of the advantages of employment by Ashmun & Co. He not only did this, but he had circulars sent out, and the consuls of the United States stationed in different parts of Germany endorsed his rose-colored pictures and descriptions with their signatures and with their official seals. Mr. Ashmun made very good use of the representives of this country in Germany to assist him in defeating and undermining the workingmen of Milwaukee.

Q. You mean to say that he secured his help abroad?—A. Yes, sir; with the aid of United States consuls.

By Mr. George:

Q. How do you know that fact?—A. The circulars have been sent here.

Q. Have you seen any of them?—A. I have.

Q. Have you got one?—A. I have not one with me, but I could obtain one.

Q. I wish you would do it.—A. If I can, I will. It is now a year and a half ago, and matters sometimes begin to heal.

Q. But you have seen and read the circulars?—A. Yes, sir.

By the Chairman:

Q. You say matters sometimes begin to heal; what do you mean?—A. I mean that after a strike, and after either side is defeated, for several weeks or months, perhaps, the feeling between the employer and the employes is not one of very marked friendship, except in very isolated instances; but after perhaps a year (although they are continually preparing for trouble) the old sore has healed in a measure; but when a further attempt is being made to reduce wages or an increase of the hours of labor, or to adopt or enforce some certain obnoxious shop rule upon the employes, then it is revived with double force. What I meant was, that the Milwaukee strike has in a measure healed.

Q. Do you mean by that that it may be difficult for you to get this desired testimony?—A. I do not know; it may be.

Q. You spoke of it in connection with the request that you should get that testimony.—A. I spoke of it in this strain: that in consequence of that fact I might not have attached so much importance to that document and might not have retained it, but I shall endeavor to obtain the document or a copy of it, if that is possible.

By Mr. Call:

Q. What did you understand the certificate of the consuls to be?—A. Indorsing the trustworthiness of the statements made by Ashmun & Co.

Q. What were those statements?—A. That Milwaukee was a very beautiful place; that the firm of Ashmun & Co. were a responsible firm; that there was no difficulty existing there; and that Mr. Ashmun was a very good man to his employes.

Q. Was there anything said about the strike?—A. Oh, no; that was omitted—of course, unintentionally. I think we have a complaint to make against our Government for a system permitting consuls of the United States to be made use of when there is a struggle between capital and labor.

By Mr. Call:

Q. Have you any evidence that those consuls intended to in any way give aid or comfort to the employers?—A. I have no evidence that they knew the facts, but I believe it is the duty of any person, no matter who he may be, especially when he is in a responsible official position, to inquire what use it is to be put to before he affixes his signature to any document.

By the Chairman:

Q. Do you believe that any private individual has a right to avail himself of the official character of a United States officer, at home or abroad, to assist him in the prosecution of his private enterprise or business in that way?—A. No, sir.

Q. Irrespective of whether the statements indorsed by such officer are true or false?—A. No, sir.

Q. To allow that would be to give one citizen an advantage over another, would it not?—A. Yes, sir.

By Mr. Pugh:

Q. Have you any information as to the number of people that they secured to come to this country by that circular?—A. A sufficient number, together with the small number that they obtained from the surrounding region and from New York (where they got very few by the way)—a sufficient number to defeat the strike.

By Mr. George:

Q. They won the victory by it?—A. Yes, sir.

By Mr. Call:

Q. Do you not understand our consuls to be commercial agents, and that it is their business to assist individual citizens in their commercial relations with foreign countries?—A. Yes, sir.

By the Chairman:

Q. Do you understand that this firm was engaged in foreign com-

merce? I thought it was a cigar strike out in Milwaukee. This cigar manufacturer in Milwaukee engaged in a struggle with his help and went abroad to get other help; is not that it?—A. Yes, sir.

Q. Do you understand that he was then engaged in foreign commerce, or could legitimately call for aid from the commercial agents of the United States?—A. I think not.

By Mr. Call:

Q. Is not bringing people over here as much commerce as bringing goods is?—A. I think not.

By Mr. Pugh:

Q. What is your idea of the wrong that the manufacturers committed in supplying the places of those who struck in that way?—A. I am not complaining of the manufacturers. I am complaining that the persons who were consuls abroad of this country should have permitted their personal and official signatures to be used in furtherance of private interests in the struggle between capital and labor, and should have thrown their influence into the scale in favor of capital.

By Mr. Call:

Q. Any person might have gone abroad for the purpose of obtaining immigrants. Now, might not the consul's sentiments and sympathies have been entirely with the strikers and yet might not he have felt called upon officially to do what he did?—A. I believe that if his sympathies were with the strikers of Milwaukee he should have said so and should have subscribed his name to the statement that there was a strike there, a struggle between Mr. Ashmun and his employes, so as to let the people know the facts. I do not believe it is proper to advise people in Germany to come to America and then put them in the position that they are unable to return, and are compelled by circumstances to undermine their fellow-workmen.

By the Chairman:

Q. Had there been no strike you would not have complained?—A. No, sir; I have no objection to the people of any country coming to America, Chinese excepted (I am not so sentimental as all that), provided they come here of their own free will, and not influenced by deception.

Mr. Call. I do not understand the witness as charging these officials with suppressing the truth, but as saying that they did not inform themselves properly of the facts.

By the Chairman:

Q. What was the fact in this case; did the employer get help from abroad?—A. A large number of them. The people were taken to Milwaukee and laid down there penniless, and they could not return,

but were either compelled to work and be dubbed with a name which workmen generally do not like to bear, or else starve. To avoid starvation they had to undermine their fellow-workmen.

Q. On that point, do you think there is any feeling among American laborers or workmen adverse to free and open competition with foreign laborers from European countries when they come here?—A. No, sir; I believe that they have no objection. They do wish, however, to put a stop to the introduction of the Chinese into this country, at least for a period, so as to give the American workmen a breathing spell. Our people had hardly recovered from the panic, and they were not going to be trodden down by the Chinese undermining them.

Q. But the European laborer will work more cheaply than the American laborer, will he not?—A. He becomes easily acclimatized and soon harmonizes with the American people.

Q. And he soon wants as much wages as anybody?—A. Yes, sir; and, as a certain Senator said, "it is a question whether the workingmen of America shall eat rats, rice, or beefsteak." I choose beefsteak. I will vote for that every time. I do not want it understood that my vote can be purchased for a beefsteak, but that I will vote always for measures that will improve the condition of the workingmen.

Q. You speak of this opposition to the Chinese being designed to give the American workingmen a breathing spell after the panic. Do I understand you to mean that the opposition to Chinese emigration is temporary?—A. No, sir.

Q. Then, there is a permanent opposition, you think, to that immigration?—A. Decidedly.

I desire to say a word or two further on in reference to this case of Ashmun & Co. and its relevance to this thread-bare theory of "supply and demand," but just now I wish to pass on to another branch, and will try to go along rapidly.

### Grievances of Car-Drivers in New York.

The car-drivers of the city of New York are working from fourteen to sixteen hours a day in all weathers, and receive $1.75 a day.

Q. Now, why is not that enough?—A. Because it will not purchase the commonest necessaries of life.

Q. You understand, of course, that my question is designed to draw you out fully in regard to that class of workmen, their condition, &c. I understand your assertion to be that it is not enough; it does not seem to me, either, that it is enough; but I want to know from you what chance a man has to live on $1.75 a day?—A. He has this chance: his meals are served to him by his wife or friend or child, as

the case may be, in a kettle, while he is driving his team, and at the end of the route he may possibly have two or three minutes to swallow his food. It is nothing more than swallowing it, and when he comes home he is probably too tired or perhaps too hungry to eat.

Q. There is no cessation in his work during the day of any consequence, then?—A. If there is, that which is termed relays or switches, he has still the same number of hours to work.

Q. Do you mean that that is deducted from his fourteen or sixteen hours?—A. Yes, sir.

Q. Then, if the relays amounted to an hour, he would be absent from his home seventeen hours?—A. Yes, sir.

Q. And if two hours, eighteen?—A. Yes, sir. And in the matter of these relays, in some instances men who do not and cannot live, on account of the meagerness of their wages, on the route of the railroad, are compelled to live at some distance, and when they have these relays or switches it takes them sometimes twenty or thirty minutes to reach their homes, and to return again takes another half or three-quarters of an hour.

Q. Then, do I understand you that these relays and the time occupied morning and evening going to and returning from their work are to be added to the fourteen or sixteen hours of actual service required?—A. The actual service is from fourteen to fifteen hours. Then there is the looking after their horses and cleaning the car besides.

Q. From the time that a car-driver leaves home in the morning until he returns for the night how much of the twenty-four hours will ordinarily be consumed?—A. I cannot tell you exactly as to how long a time they have at home, for the reason that it depends to some extent upon how far they live from the route of travel.

Q. State it approximately as near as you can.—A. Well, I do not believe that they have more than seven and a half hours out of the twenty-four.

By Mr. Call:

Q. At what hour in the morning do they commence ordinarily, and what time do they quit?—A. Several of the street railroads of this city run all day and night; and on those, of course, the men commence at various hours. During the day the traffic on some routes is not so much as on others, and then they will be relayed; and, although they may go on to work at 5 o'clock in the morning, they probably would not get off before 11 or 12 o'clock at night or probably later still. I would not say later still positively, but I think in some instances later.

By Mr. Pugh:

Q. Have they ever been paid higher wages?—A. Yes, sir. About

two years ago they were on a strike[19] to obtain, I think, $2 a day, but were starved into submission.

Q. What do they get now?—A. One dollar and seventy-five cents.

By Mr. George:

Q. Does the conductor get the same wages, or more?—A. I think he gets 25 cents more, by reason of his position of trust.

By the Chairman:

Q. Have you any knowledge with regard to those who operate the elevated railways?—A. The men who work at ticket collecting or at the boxes where the tickets are deposited receive $1.25 a day, I think. I would rather wait until I can give you information definitely. I think I can do so now, but I prefer to wait.

By Mr. George:

Q. Are the car-drivers allowed to have seats?—A. They are not. They have to stand all the time.

By Mr. Call:

Q. How many hours do they stand?—A. Fourteen or fifteen.

Q. Do you mean fourteen hours' standing without intermission?—A. Very little intermission. They sometimes rest back against the door of the car for a while. They also, in some instances, have to act as conductors; that is, give change, count the passengers, and register the number of passengers on an indicator. And then they are sometimes held responsible when somebody is run over on account, perhaps, of their having to perform two men's work. The greed of the horse-railroad companies has been such that they have introduced on several lines what is known as the bobtailed car, and have dispensed with the services of a conductor.

By the Chairman:

Q. Don't you think that is because they cannot afford to pay any more?—A. I hardly believe that. Judging from the traffic, they are capable of paying it, and judging from what is currently reported as their dividends, they are more than capable of paying it. I must acknowledge, though, that so far as their dividends are concerned, I am personally uninformed. I take merely current rumor and the appearance of the traffic, the number of passengers I see on the cars.

## Wages of Tailoresses.

Among some of the tailoresses in the city I have made a personal investigation. They make a regular heavy pantaloon, working pants, for seven cents a pair. They are capable of making ten pairs per day of twelve hours. Boys' pantaloons they make for five to six cents per pair, making fourteen to sixteen pairs per day of twelve hours. They

work mostly seven full days in the week; sometimes they will stop on Sunday afternoon, but all work on Sunday, and their average weekly wages is about $3.81, providing no time is lost.

They are compelled to provide their own cotton out of this, and their own needles and thimbles, and other small things that are necessary in the work. Overalls and jumpers (a kind of calico jacket used by laborers in warm weather sometimes, to prevent the dirt getting to the shirt or underclothing) they make for thirty to thirty-five cents per dozen. They generally work in "teams" of two, and they make about three dozen per day, or in a working day of thirteen to fifteen hours they earn from forty-five to fifty-two and a half cents each. They work generally in the shop, but usually finish some work at home on Sunday.

### Tobacco Strippers' Work and Wages.

In the manufacture of cigars in shops there is a branch termed "stripping." I am not sure as to these statistics that I am going to give you, but I believe them to be correct. Nine-tenths of these strippers, or about that proportion, are females. Their average hours of labor are ten per day. Their wages range between three and seven dollars a week when at work. About one-half of these girls are employed at the former wage, two-thirds at $5 a week, and the remaining third at a higher wage.

They lose days and weeks' work frequently, or have lost them in the past more than at present, and in very rare instances are they paid for loss of time, even when it is caused by national or other holidays. In the shops, more especially the larger ones, they are prohibited from holding any conversation under pain of fine or dismissal. Even if they were disposed to converse they could not. The very positions in which they work, or are placed to work (which are not necessary to the work), in long rows, in which each faces the back of the girl in front of her, precludes them from holding conversation. They suffer in every way the disciplinary measures of imprisonment at hard labor. They cannot hold conversation. One sits with her face to the back of the other, and that is the rule in almost all the factories. Where there are only a few of them of course it makes very little difference. It is believed that this plan of placing them gets more work out of them.

### Shop Girls.

In this city we have a number of establishments, retail stores generally, large ones, which employ a number of small girls usually known

as cash-girls. They labor ten hours daily, on Saturday about sixteen, except at holiday times, such as Christmas and New Year's, when they have to work about fourteen hours a day. The earnings of these girls amount to from two to three dollars a week. The girls range in age generally from about ten or eleven to thirteen or fourteen.

Q. What is their duty?—A. To carry cash from the customer to the cashier.

Q. Or to receive it?—A. They sit in rows generally, and when a sales-lady wishes to make known that a sale is made, she does it by calling out "Cash!"—she does not call for a girl, she calls "cash," that being more important than the girl. The girl is called in that way and must go just as fast as she can. From sitting quietly—because they are compelled to sit quietly, without conversation—she must start up from her seat immediately, or from the place where they are kept standing, having no seats. I see that in Wanamaker's great store in Philadelphia he provides seats for his cash-boys (he has no cash-girls), but in many instances these places have no seats for their cash-girls. Some months ago while investigating this condition of affairs I saw the subject-matter about that time made public through one of the New York papers, and it spoke of the endurance of these girls, or, rather, how puny these poor children looked. In one of our largest establishments the children all dropped off after a while. They could not stand it. It is such a strain to be almost motionless, and then to make a sudden start.

Q. That, you understand, gives the system a nervous shock?—A. Yes, sir.

Q. And you seem to think that that sudden spring is more apt to be a source of injury to the system than the work itself?—A. Probably more than the work itself, although I think it is decidedly too exacting to have children work in stores ten hours a day, even if those were the longest hours; but they are not.

Q. What ages do you say they are?—A. Ten to thirteen or fourteen. When they become that age, fourteen, or sometimes older, they try to obtain situations in other places as sales-ladies.

Q. Do you understand that this work, of which you have been speaking, breaks down their health?—A. Yes, sir.

Q. Then how happens it that they change from that to this other employment?—A. It is not so exacting upon their system to be a sales-lady, more especially within the last few years, through the agitation of many of the ladies who go shopping and who have taken up the matter and urged that the sales-ladies ought to be provided with seats when not actually waiting on customers.

DISTRICT MESSENGER BOYS.

We have an institution in this city (I suppose it is a mere branch of the large institution throughout the country generally) known as the District Messenger Company. They employ boys. They advertise generally for boys between the ages of fourteen and sixteen. The boys, though, are very frequently and more frequently below the age of fourteen than above it. Their work is taking messages and carrying small parcels from one place to another. Their hours of work are ten; at least they are supposed to be ten per day. The companies, or some of them, have their boys to work over time and they pay them first $4 a week, I believe, and for over time six or seven cents an hour. If any of these boys in carrying a message or a parcel is delayed or is a little beyond the tariff time he is fined, not in money but, in time, and if he should be ten minutes beyond the time he must work one hour overwork to make up that fine. I speak more especially of the boys employed by the American District Telegraph Company. The other companies are by no means charitable institutions, but they are not so exacting. They have not yet attained the first stage of old Scrooge in Dickens's story.

Q. You think those boys are worse off than the telegraph boys of the Western Union?—A. Do you mean the operators by the word "boys"?

Q. No; I have reference to those who deliver messages.—A. There is some kind of an agreement or understanding between the District Telegraph Company and the Western Union Company. What the agreement is I do not know, except that within a very short time, within a year, I believe, the headquarters of the American District Telegraph Company have been removed to the Western Union building, and their boys deliver the telegraph company's messages. It is one concern so far as the telegraph service of the Western Union is concerned.

Q. Is this the only class of messenger or errand boys about the city?—A. No; I have mentioned that there are others, for instance, the Mutual District messenger boys.

Q. Is the service rendered substantially the same?—A. Yes; but while they have a system of tariffs and fines, they are not quite so exacting as the American District Telegraph Company. The boys when once entered upon the "force" as it is called (they are drilled with "sergeants," &c.), are compelled to be provided with a suit of clothes to be obtained from the company. Having seen these clothes and having some acquaintance with tailors and cap makers, I am informed

and believe that these suits could be purchased in any store made to order for one-half the price charged by the company to these boys.

Q. Do you mean to be understood that the companies make money upon these uniforms?—A. I mean to be understood that they use this means to defraud the boys out of certain moneys; that they overcharge them, and use the employment as a means to force these boys to purchase a suit of clothes from them, made by somebody with whom they have a contract undoubtedly at a much less price.

### Newsboys and Bootblacks.

Q. Now, about the newsboys and the other little fellows that we see around the streets, the bootblacks. Those little waifs seem to be pretty busy doing something all the time. What pay do they get out of their labors—how do they live?—A. Well, the newsboys earn very small sums. I do not believe more than one-half of them live at home with their parents. The others, out of the papers they sell or earn, try to purchase a ticket for some variety show, and buy cigarettes, of course, and keep just sufficient to get a meal in a five-cent restaurant and to pay their lodging in a newsboys' lodging-house, which costs about half a dollar a week.

Q. What chance is there of their attending school?—A. Without answering that question I would like to make a statement that I read in one of the papers (and the paper said that the superintendent of the Newsboys' Home acknowledged it to be true) that the newsboys were required to pay for one week's lodging in advance; that one boy was taken sick while in the lodging-house, and sent to the hospital after the second night of the week for which he had paid, and when he came out of the hospital he thought that he had five nights good yet to sleep in the lodging-house, but when he came there he was informed that he had forfeited that money by not sleeping in the lodging-house during the week.

Q. Is this newsboys' lodging-house a private institution?—A. I think it is.

Q. Then the boy has no special right there only as he pays for it, and it would not be kept up unless it was a source of money-making to the owner, would it?—A. I cannot say as to that.

Q. Are the newsboys employed by the newspapers, or do they just get so much for every paper they sell?—A. They get so much for every paper they sell, and sometimes a man can buy two-cent papers for a penny. Some will offer you two papers for a cent.

Q. When they have a supply left which they do not sell what becomes of it?—A. It is their own loss.

By Mr. George:

Q. They buy the papers themselves and make what they can?—A. Yes, sir; and it is quite a sight to see some of the boys running after the wagons that contain the papers, the evening papers more especially; to see one hundred or two hundred of them, and as one drops off that has been served with his papers another one takes his place, the others coming up continually and keeping up the crowd. If the poor boys were on the point of starvation and their only hope of life was in that wagon I do not believe they could run much faster or risk their lives much more than they do sometimes.

Q. How about the bootblacks?—A. How the bootblacks do I cannot say, any more than their position in life is very hard.

Q. Does the newsboy get a chance for school at all?—A. I do not see where that comes in, except that possibly one here and there may have an opportunity of going to a night school, and that, I think, is not generally taken advantage of by them. The boy fails to see the importance of an education himself, and there are very few who are willing to lend a hand to guide him.

(Recess.)

STEAMBOAT HANDS—TRUCKMEN.

After recess the witness proceeded as follows:

A short time ago I had occasion to make two trips to New Haven, or by way of New Haven, and I took the New Haven steamboat route. While on the night trip I was considerably interested by the work of the deck-hands. I made inquiries of those employed in transferring the freight from the docks to the boat on trucks, two wheeled trucks, and I was impressed with the peculiar nature of their work and the severity of the toil, in consequence of the boat being higher, when unloaded, than the pier, and lower as the load became heavier on board and bore down the gang-plank as the freight was put in. I made inquiry as to the wages paid and the hours of labor, and I learned that those who were employed as truckmen there were working, on an average, fourteen hours a day. I made two trips each way last week, and I did not see one of those truckmen idle for one half-minute. On the contrary, when their trucks were emptied they did not walk from the boat to the dock, or *vice versa*, but ran, and ran also when they possibly could with their loads. They earn $38 per month. The hours of labor are divided through the twenty-four hours thus: The boat starts from New York at eleven o'clock at night. From that time they rest until they arrive in New Haven. Immediately upon the landing of the boat, as soon as the gang-plank is placed, they

proceed to unload, and they load again immediately after the unloading has been accomplished. They then come back to New York at ten o'clock, and reach here at three; from which time they work until eleven incessantly, when the boat starts again. That is on one boat; the other I was on board of coming from New Haven to this city. I arrived at 10.5 P.M. The men were working, and I made inquiry, and they had been working the whole afternoon. I was upon the deck and viewed them working. I do not know the position of the official in charge, whether it is baggage-master or package superintendent, or whatever his office may be termed, but he bullied every man there that appeared to make even one step a little slower than another. There was a small engine, but rather weighty, that one man carried on his truck, and the man was just putting his truck below it, and another of the deck-hands, that is, a man who assists in putting freight on the trucks, did not move fast enough, and hallooing did not suffice for this boss or superintendent, but he had a rattan cane in his hand and he whacked the man with it across the back.

By Mr. George:

Q. What did the man do?—A. He ran very fast with that truck.

By the Chairman:

Q. He did not resent it?—A. No, sir; he did not even look up. I am a very small man, but I think that if that man felt the way I felt about it, on the impulse of the moment he would have knocked that boss down if he could. I was struck with the remarkable appearance of one man there, a man not very stout looking, but very sinewy, and very intelligent, and I might almost say of refined appearance. Whatever rest these men have is on board the boat. They cannot go ashore except with their trucks to haul goods on board or from the boat.

Q. Does their pay include their food on board the boat?—A. Yes; I think it does. The deck-hands, those who are assisting in the placing of the freight on board, get only $38. When I asked the question what they were getting, I was told it was $38 a month, but one of them said: "By golly, we ought to get $60, if anything." But I suppose they have no organization.

Q. Do you think it was wrong for him to swear?—A. Well, I think sometimes a man untutored emphasizes a statement by an oath which a man of culture will do without, and they have probably no means of emphasizing their statements other than that.

Q. You think it was justified under the circumstances, I suppose?—A. Yes, sir.

I desire to say that the statement I am about to make, or to attribute to a certain person, I only have from having read it in the papers, and the statement was criticized considerably at the time. That is the

only authority I have for it, but judging from the manner in which the statement was criticized it is my belief that it was made, and if it was made, as I believe it was, then this man was decidedly more frank in his statements than employers generally are. A Massachusetts manufacturer some two or three years ago, a member of the legislature, speaking upon the question of the hours of labor, said in the course of his argument: "I regard my employes as I do a machine, to be used to my advantage, and when they are old and of no further use I cast them in the street." That, of course, is the general mode of dealing with working people when they are incapable of further performing work; the employers being not charitable institutions, but men of business, they really do cast their employes off into the streets. But this remark was very frank, and it exhibits a considerable degree of the existing feeling.

Q. Do you think that the obligation to take care of a man who has worked for another when he ceases to work, if he has been paid proper wages while at work, devolves upon the employer any more than upon the community at large?—A. I will not say that it does; but if the Government will take men as soldiers to fight their battles, and, in consequence of injuries received, give them pensions, I think that a man who works from the earliest time that he can work is entitled to some consideration in his old age. How that is to be arranged is probably more than I should attempt to say. To whom the responsibility belongs I cannot say, but I think that it is right that such a man should be cared for, and that there is that responsiblity somewhere.

By Mr. George:

Q. Name the party to whom that statement is attributed.—A. If I knew the name I would mention it. At the time when it was published I saw it in the papers, in a conservative paper, not a paper devoted to the interests of labor or following out any doctrine or any theory, the New York Staats Zeitung.

By the Chairman:

Q. The statement is, I understand, that some member of the Massachusetts legislature gave utterance to that sentiment in the legislature itself?—A. Yes, sir.

Q. When was it?—A. I cannot say the exact time.

Q. You would not convey the impression that you think that such an utterance is illustrative of the *general* feeling on the part of employers toward the employed, would you?—A. I think I would, as a general rule.

Q. Then you used that case to illustrate your idea of the feeling that exists on the part of those who employ labor towards their la-

borers?—A. Yes, sir; as a rule. Of course there are honorable exceptions, men who believe that a man who has spent his life in their employ and trying to do his duty is entitled to some consideration, that when he becomes old and unable to perform work as well as he did in his former years is entitled to something else than to be cast into the streets.

Q. I do not think that such a feeling is prevalent.—A. I hope it is not. I am afraid it is.

By Mr. Call:

Q. I suppose you mean to convey the idea that that is the result of present conditions rather than an intentional thing as a matter of feeling?—A. Oh, decidedly. I do not believe that the question of feeling enters into the matter at all. An employer may make his calculation of his income and his expenditures and find that if he employs this person he will make somewhat less than if he employs that person; and he is governed generally by that consideration.

By the Chairman:

Q. Can that absence of feeling that you speak of be well avoided under the competitive method of doing business which prevails in our day?—A. I doubt it, sir.

Q. So that the employers are victims of the system under which the business is done as well as the employes, are they not; and the real fault you find is with the system?—A. Well, that would be driving me into the theoretical sphere, which for the present at least I would like to avoid.

By Mr. Pugh:

Q. What is your opinion as to whether that idea of regarding the laborer as a machine exists more now than it has existed in the past?—A. I think it exists now in a greater degree than it did formerly. Not only do I think that, but I am forced to the opinion that it is increasing and intensifying even as we go along.

Q. Anyhow, that, you say, is the view that the employes take of the sentiments entertained towards them by their employers?—A. Yes, sir. They find that employers are no longer—when I speak of employers I speak of them generally—that they are no longer upon the same footing with them that they were on formerly. They find that where a man who may have worked at the bench with them employs one or two hands they and he may have full social intercourse together, but as that man increases his business and employs a larger number of hands they find that his position has been removed so far above that of his old friends that they meet no more socially. Probably they may meet occasionally in the factory, when there will be a passing remark of "Good-morning" or "Good-day"; and then, after a while,

the employer fails to see the employes at all; the superintendent does all the business and the employer does not bother himself any more about the men. That is how the position of the two has been changed since both were workingmen at the bench. The difference is considerably greater when the employer and the employe did not know each other before, and when the employer's resources are already large. In such cases he and the men do not know each other at all, and in most such instances the employes are not known as men at all, but are known by numbers—"1," "2," "3," or "4," and so on.

By the Chairman:

Q. You mean in kinds of business where the employes are numerous?—A. Yes, sir. Anybody coming to the shop asking for a certain man by name will find that he is not known unless he has been working there for months or perhaps years, and when anybody asks for "Mr. Johnson" one of the boys will go up to the shop and say, "Who is Mr. Johnson? Oh, that's you, 24," or whatever the number may be.

By Mr. George:

Q. You say that statement applies to all shops where there are a large number of employes?—A. Yes, sir; wherever there is a large number employed that system prevails. The men lose in a great measure their individuality and become parts of the great machine.

By the Chairman:

Q. That is the practice in the police department, is it not? You do not take it that there is any special inhumanity in that manner of numbering the employes; it is only a matter of convenience in doing business, is it not?—A. I should not wonder. It is only a matter of convenience, I suppose, but the tendency of it is like that of everything else in modern society which makes man, the worker, a part of the machine.

Q. So that his individuality is lost?—A. Yes.

By Mr. George:

Q. I have noticed that they use paper checks in the cigar-making business; are the men numbered on that paper, or named?—A. That check has three or four separate headings, showing the amount of stock received of the different kinds, then the amount of cigars made and delivered.

Q. Is the man numbered on that check or named?—A. He is numbered, except when the shop is not an extraordinarily large one. In such cases they probably have not the checks, but instead they have the names of the hands on the pay-roll and they are known by their names. They are known by numbers too, however.

### FEMALE LABOR—FRAUDS UPON WOMEN.

In speaking of female labor, I should say that we have many schemes set afloat by persons who have no honest intentions in the matter, some of them employers of a class who are equally an incubus on honest employers and upon credulous working people.

A case of note took place here. A gentleman in Division street in this city advertised for good sewing girls. A large number were waiting in the morning at the door of the store, and a good number of them were accepted for "a trial," as it was said; but about one-half of them were discharged after three days' work, without any payment whatever, upon the pretext that the work did not suit. Saturday arrived and the remainder of those engaged waited for their week's wages, but about 6 o'clock they were told that the firm had missed the time to draw money from the bank, but that on Saturday next everything would be attended to. Another week passed by, and on that next Saturday, when the poor sewing girls expected a fortnight's pay, they came as usual in the morning at 7 o'clock, and found the doors of the shop closed, the firm having "removed" to no one knew where. This is one case that attention was called to at that time; but such cases frequently occur. There are some employers in this city who advertise to teach girls trades, advertise for hands, and do this same thing time after time, defrauding poor girls out of their few hard-earned dollars. The Labor Report of Massachusetts for 1872[20] tells us of the case of a sewing girl who applied for work at a large establishment in Boston. She was told that she could have it for so many weeks anyway, at such a price per week. "But what shall I do," she said, "for the rest of the year?" "Oh," was the answer, "you can do as many others do; some gentleman will pay your rent in a private room, and pay your board in full or in part, for the privilege of occasionally visiting you in a friendly way!"

### WAGES AND PAUPERISM.

I had these statistics that I am about to read before the compendium of the last census came out. I have selected five of the greatest industrial States, the States in which the manufacturing industries have developed more than in any of the others. These are Illinois, Massachusetts, New York, Pennsylvania, and Ohio. The inhabitants of Illinois in 1870 were 2,539,891. The products of that State were in value $205,620,672. Of Massachusetts the inhabitants were 1,457,357, and the products were in value $785,194,651. Of Ohio the inhabitants were 2,655,260 and the value of the products was $269,713,610. Of Pennsylvania the inhabitants were 3,557,212, and the value of the

products $711,894,344; giving a total population for these five States of 14,557,212, and a total value of products, $2,526,325,845. Those five States, with a little more than one-third of all the inhabitants of the country, produced nearly six-eighths of all the manufactured products made in the United States.

By Mr. George:

Q. Are the products there included confined exclusively to manufactures?—A. Exclusively to manufacturers.

My next statement shows the numbers and the classes employed in these five States, with their average wages:

Illinois employed 73,045 males, 6,717 females, and 3,217 youths, at an average wage of $1.25 per day.

Massachusetts employed 179,032 males, 86,229 females, and 14,119 youths, at an average of $1.40 per day.

New York employed 267,378 males, 63,795 females, and 20,627 youths, at an average of $1.34 per day.

Ohio employed 119,686 males, 11,575 females, and 5,941 youths, at an average of $1.14 per day.

Pennsylvania employed 256,543 males, 43,712 females, and 19,232 youths, at an average of $1.35.

Totals, 895,684 males, 212,020 females, 63,136 youths.

These figures are based on 300 working-days to the year; and this statement shows that not quite one-third but more than one-fourth of all those employed are females and children.

Now, as to the pauperism of these five States:

Illinois had 6,054 paupers, and the cost of their support was $556,061.

Massachusetts had 8,056, and the expenditure for their support was $1,121,604.

New York had 26,152, and supported them by an expenditure of $2,661,385.

Ohio had 6,383, who were supported at a cost of $566,280.

Pennsylvania had 15,872, and supported them at a cost of $1,256,024.

The total number of paupers in these five States, as officially ascertained, was 62,497, and the cost of their support was $6,161,354. Thus, these five States, with their immense yearly product of wealth, have 4,000 more than one-half of all the paupers in all the United States, and spend three-fifths of all the money expended in the United States to support paupers.

By the Chairman:

Q. What does the population of those five States aggregate?—A. As I have said, these figures were taken from the census of 1870,

when the population of those States was some fourteen millions and odd.

Q. At that time our total population was 38,000,000; so that those States had a little more than one-third of the population and a little more than one-half the paupers, including the foreign population?—A. Yes. Now, the average wages of the laborers in the manufacturing industries of the five greatest manufacturing States, as per census of 1870, was \$1.29$\frac{6}{10}$ per day, or \$405.64$\frac{8}{10}$ per working year. Averaging five to the family, the amount to support each individual was \$81.14$\frac{9}{10}$. At the same time, there were in those States 62,497 paupers, maintained by the States at a cost of \$6,161,354, or a fraction over \$95 per individual. Thus it appears that the workingman was compelled to support himself above the degree of "pauperism" on \$14 less per annum than the State spent to support paupers as paupers. These figures are of course old, but they can be depended upon, except that the wages of the workingmen is less now than it was in 1870.

By Mr. George:

Q. You arrive at that result by assuming the families to consist of five each and dividing the wages among them on that basis?—A. Yes, sir.

Q. And the cost of the "plant" of the poor-house or alms-house is not included in your figures there?—A. No, sir; only food and management, so that the laborer working and receiving this average pay, with five to support, including himself, was receiving less for the support of each individual member of his family than the State was paying for the support of each pauper.

By the Chairman:

Q. It seems, then, that the laborer takes care of his family cheaper than the State takes care of its paupers?—A. Yes, sir; but he works for it.

Q. That is, he maintains his family with more prudence and economy than even the State does its paupers?—A. Yes, sir; apparently.

### REDUCTION OF THE HOURS OF LABOR.

Mr. Ira Stuwart[21] has remarked that improvements in machinery and the fact of its invention discharges labor faster than new industries are founded, and tends to increase the hours of labor; and that among the reasons for this may be named the following:

["]Capital must reproduce itself and bear interest" is the first and main maxim of modern industry. What does this mean? It means that capital is the accumulated value of past labor employed in production for the purpose of further accumulation, and when invested in machinery or other enterprises must reappear in its original form as money, not alone to an amount

equal to the amount invested, but in an increased amount. The individual possessor of capital invested in any industry wants his interest, and the corporations have a more artificial term for it; they term it dividends. The question is, how is this result to be obtained by machine work? It is a well-settled fact that the only producer of value is labor; there is a well-established value in labor; hence if machinery displaces manual labor and performs the labor itself it must create some value; but what is the amount of it, especially taking into consideration that human labor and that of machinery are always combined?

Q. You adopt that view as your own, I suppose?—A. I do. Otherwise I would not read it. I intend to speak of the hours of labor, and that is why I use this as a preliminary statement. The hours of labor have been discussed by many thinkers on the labor question, and by many from different stand-points. During my attendance upon this committee I have heard a good many questions asked and answered, and in my humble opinion some of the answers were not what they ought to have been. I maintain that the hours of labor ought to be reduced. From every stand point the hours are too long in modern industries, more especially where the individual, the worker, is but a part of the machine and is compelled to keep in motion in accordance with the velocity with which the machine turns. The production of goods is not, as many have been led to believe, lessened by a reduction of the hours of labor; but, on the contrary, the productivity of labor increases. In all countries, in all States in this country, in all factories where in certain branches of trade the experiment has been made, wherever the hours of labor have been reduced, there the productivity of labor has become greater.

By the Chairman:

Q. Absolutely, or in proportion to the time occupied?—A. Absolutely.

Q. One day with another, more goods produced?—A. More produced, one day with another. I am saying that the productivity of labor has increased; not from the desire, or probably not from the ability, of the laborer to produce more in the shorter number of hours, but as a consequence of the fact that, owing to the reduction of the hours of labor, machinery has been improved, new tools have been made, and the different industries have been divided and subdivided, so that as a consequence of the reduction of the hours there has come increased production.

Q. Then the increased productivity is the result of the improved machinery and not of the shorter hours of labor?—A. But the improved machinery is the result of the reduction of the hours of labor.

Q. But we are speaking of the direct cause; and the reduction of the hours of labor is not the direct cause of the increased production,

you say, but the reduction of the hours of labor has led to invention and improved machinery, and by the machinery, combined with shorter hours of labor, more is produced?—A. Decidedly.

Q. Then you attribute the invention not to genius alone, but to genius and *opportunity*?—A. Yes, sir; I hold that the necessity for inventions brings them forth, and that they do not come forth without that. A man might go to China and live there a hundred years and probably never think of a Morse telegraph machine.

Q. But do you think that the inventors of this country are stimulated to invent by reason of the reduction of the hours of labor?—A. I think that the necessity created by reduction of the hours of labor for other means of supplying wants that need to be satisfied is the cause of inventive genius becoming active.

Q. Then your ground is this, that the *immediate* result of the reduction of hours of labor is decreased production, and that that creates the necessity of supplying that deficiency of production by improved machinery?—A. Let me answer by saying, as one of our greatest economists says, that there is but one sure and permanent way by which the customs and habits of the people can be improved. Or rather that, if you wish to improve the condition of the people, you must improve their habits and customs. The reduction of the hours of labor reaches the very root of society. It gives the workingman better conditions and better opportunities, and makes of him what has been too long neglected—a consumer instead of a mere producer.

Q. You think, then, that he will consume more in consequence of the reduction of the hours of labor?—A. I do, positively.

Q. In a certain way he will have more time to consume in?—A. Yes, sir. And another thing: A man who goes to his work before the dawn of day requires no clean shirt to go to work in, but is content to go in an old overall or anything that will cover his members; but a man who goes to work at 8 o'clock in the morning wants a clean shirt; he is afraid his friends will see him, so he does not want to be dirty. He also requires a newspaper; while a man who goes to work early in the morning and stays at it late at night does not need a newspaper, for he has no time to read, requiring all the time he has to recuperate his strength sufficiently to get ready for his next day's work. I agree with Mr. Ira Stuart in his view. I have regretted very much that his work has not been so widely circulated as it deserved to be. I say this because I think it contains the fundamental truths of the labor question. The reduction in the hours of labor reaches the lowest stratum of society. I say I agree entirely with Mr. Stuart, and I think he is (or rather has been, for he is dead now) the ablest thinker on the economic question, more especially in regard to the application

of the reduction of the hours of labor and to the movement and its effects upon society in general. That labor deserves a reduction of the hours of toil I believe hardly any one will dispute, unless when he is on "the other side of the house" and labor is seeking to enforce such a reduction against his interest, as he thinks. The general reduction of the hours of labor to eight per day would reach further than any other reformatory measure; it would be of more lasting benefit; it would create a greater spirit in the working man; it would make him a better citizen, a better father, a better husband, a better man in general. The "voting cattle," so called, those whose votes are purchased on election day, are drawn from that class of our people whose life is one continuous round of toil. They cannot be drawn from workingmen who work only eight hours. A man who works but eight hours a day possesses more independence both economically and politically. It is the man who works like his machine and never knows when to stop, until in his case perpetual motion is almost arrived at—*he* is the man whose vote you can buy. The man who works longest is the first to be thrown out on the side-walk, because his recreation is generally drink.

Q. Do you know of anybody who works only eight hours?—A. Yes, sir.

Q. Where is that system in operation? I know there are laws to that effect, but where does the eight-hour system prevail actually?—A. About twenty-six years ago, in Australia, the workingmen obtained, against the protests (because the arguments for the reduction of the hours of labor were not then understood as they are to-day)—despite the protests of the employing class, I say, the workingmen obtained the adoption of the eight-hour law,[22] and that law is in force up to the present day; and such general satisfaction does that law give now, and the operation of it, that the 21st of April each year is observed as a national holiday almost equal to our Fourth of July here. Not only do the different classes of workingmen enjoy the festivities of the day, but the officers of banks, and of the chamber of commerce, join in these festivities. We maintain that the reduction of the hours of labor to eight per day in the United States or in any industrial country means the improvement of the condition of every man, woman, and child in that country.

Q. Do I understand you to say that in Australia, as a matter of fact, the working classes engaged in manufacturing and those engaged in agricultural occupations, and also the shop-keepers and others in commercial occupations, actually work only eight hours per day?—A. Eight hours per day.

Q. And they are really dismissed at the end of the eight hours;

there are no special agreements by which they shall work longer?—A. There are no special agreements. Just the same, for instance, as to-day, while there are a large number of employes that work much more than ten hours, yet the general standard of a day's work here is ten hours.

Q. I understood you to speak of Australia as an illustration of the benefit of reducing the hours of labor to eight?—A. Yes, sir; I do.

Q. And I was desirous of knowing whether that is an *actual* reduction or only a theoretical one. In our own country I do not think that the hours of labor are generally reduced to-day even to ten. I mean *as a fact.* I know that the farmer works his twelve to fourteen hours; I know that the professional man often works from fourteen to sixteen hours; I know that people get a good deal more time out of me than they could claim under an eight-hour law, and I do not claim to be an extra industrious man. All through this country I think people work from ten to fourteen hours, although we have laws prescribing fewer hours; so I did not know but human nature got the better of the law in Australia too.—A. No; the law is pretty generally in force there, or rather a system of eight hours' work.

Q. How about wages there? But perhaps you are coming to that.—A. No; I have not made that a study. I am not informed as to the rate of wages there.

By Mr. Call:

Q. State how you know this information that you have given us about Australia to be authentic.—A. I am in receipt of papers at times from Melbourne making mention of it. I have heard of the matter for the last five or six years, and then I have received papers of that country which have published accounts of the festivities that have taken place upon that particular day in April.

Q. The anniversary of the day on which the law was passed?—A. Yes, sir.

Q. Then it is a law?—A. Oh, it is a law for Government employes. But as to labor done for individuals or firms, the law does not apply to that, but merely the fact. Now, there are some industries here in which the workmen work only eight hours in this country.

By the Chairman:

Q. That practice has existed in Australia, according to your statement, for nearly a generation; now, are the Australians superior to the English and the Americans?—A. In what particular?

Q. Well, you have spoken of what would be the effects upon society generally of the reduction of the hours of labor. The Australians are of the same race as ourselves and the Englishman. Now, can you state whether, on the whole, Australia is better off under the operation of

that law or practice than the Englishman, the Scotchman, the Irishman, or the American?—A. I spoke of the question of the reduction of the hours of labor and its application to industrial countries, and merely mentioned that case to show that, although Australia may not be properly classified as an industrial country, yet despite the opposition of those who opposed that change it has been made there and is now regarded with satisfaction by all classes. I merely mentioned that as a fact, that the opponents of the measure at first became afterwards friends to it, believing that it had worked so much good in comparison with the condition that existed prior to the adoption of the eight-hour system.

Q. Do you believe, in such a manner that you would act upon it if you were called upon to act or not to act, that a law of the United States or of the States reducing the hours of labor to eight per day would be or could be actually enforced in this country?—A. For private employers, do you mean?

Q. Yes; for corporations and factories—private employers?—A. I think, first, that the general Government, under the Constitution, possesses no such power. I am speaking upon this question because I believe that it is a wrong that, with all our modern inventions, the working people should be called upon to work the long hours that they do; and while in this instance I am not seeking redress from the general Government, or asking it to reduce the hours of labor generally, I do say that it is more than negligence, more than wrong, on the part of the Government to permit its own statute on that subject to remain unenforced. The Government has adopted a national eight-hour law for all Government employes. If that law is wrong it should be obliterated from the statute books; but as it is good, it ought to remain; and so long as it is a law it is worse than neglect on the part of any officer of the Government to set the example of ignoring or violating it.

Q. The real difficulty comes in here: Government employment and other employments must have more or less relation to each other, and the Government employe does not like to take eight hours' pay when another man outside may work a little longer and get ten hours' pay.—A. That may be true; but the fact is that the operatives in the Government employ were paid ten hours' pay for eight hours' of labor until the panic[23] arose, and that then, for what reason I cannot say, the eight-hour law was construed to mean a reduction of wages, in spite of the fact that two proclamations[24] of President Grant[25] had been issued setting forth that no reduction in pay should result from the reduction of the hours of labor from ten to eight. That was during

the panic, and when the labor organizations were considerably crippled.

Q. I admit your point as against the apparent evasion of what would seem to be the intent of the law; but the other difficulty, which is a substantial one, still remains. Can any such law be properly enforced so as to reach the great mass of laborers or wage workers of the country in private employ?—A. I do not think it can under our present competitive system.

Q. Then the suggestion of an eight-hour law would seem to be of no use?—A. No, sir; I do not take that view. I hold that our representatives, possessing more than the average intelligence, or at least the average intelligence, should be in the van, should in a measure teach the people, and should adopt the eight-hour law for the Government employes, not so much to benefit those employes, as to set an example to be imitated by private employers, to be requested by the employed, to be agitated for, to be organized for, to be attained. That law ought to be enforced to enable the workingmen to look to the Government as an example and say, "Here, our Government has adopted the eight-hour law, and *it* is worthy of imitation."

Q. Do you think that private employers would be much influenced by sentimental considerations of that kind?—A. Probably not all; but I know that a good many employes would.

Q. A portion of the employes would refuse to work more than eight hours probably; but would not they find as a result that they would have to take the eight hours' pay?—A. They would not refuse to work more than eight hours until the movement became somewhat general, at least in their own trades.

By Mr. Pugh:

Q. A law of Congress is an expression of public opinion. Now, would not that expression of public opinion in favor of the national eight-hour law have necessarily a great influence upon the same question between individuals?—A. Yes, sir; I think it would.

Q. That would be its moral effect, and the force of public opinon would compel conformity to it in practice in the course of time?—A. That is what I have been attempting to get at.

By the Chairman:

Q. Here is this law, more than a dozen years old, still on the statute-book, but entirely disregarded; so that that theory is absolutely disproved by the fact.—A. And the National Legislature is continually banged at to have that law enforced.

Q. Yes; the mass of public opinion would seem to be the other way, or else Congressional action is not an expression of public opinion. But here is a law on the statute-book, and yet not only is its principle

not adopted among the people at large, but even the Government itself fails to enforce its own edict.—A. It is a very peculiar Government in that respect.

Q. Now, I have suggested the question whether an eight-hour law, or any law regulating the rates of wages, in order to be of benefit must not reach not only the employer but also the employed, with penalties attached, so that no man should be allowed to work more than eight hours, under penalty of fine and imprisonment; and if an employer permitted him to work for more than eight hours he also should be subjected to punishment. With such a law enforced you would have something which would make room for this surplus labor in regard to which there has been so much testimony here. But just so long as compensation depends upon the amount of labor performed, and just so long as the individual is capable of doing his ten hours' work, he will work his ten hours, and, by private agreements between himself and his employer, they will increase wages and production by the evasion of any general law which applies only to the employer.—A. If the eight-hour law were enforced by the Government, and at least it ought to enforce its own edict—

Q. [Interposing.][26] Do you mean that the Government should go into the labor market just as any other employer does, and see what it can obtain service for; or that, finding that competent labor for ten hours a day can be obtained at a certain price, it should hire that labor, work it only eight hours, and pay it the same price as for ten hours' work? Would that be fair towards private employers?—A. I do say that the Government of the United States ought to be in advance of its people. It is the duty of a legislator, as I understand it, to frame and adopt measures for the welfare of the people. I believe that the duty of the legislature is to propose laws for the benefit of the people. The Constitution of the country, I believe, does not give our National Government the right to adopt a law which would be applicable to private employments; yet for its own employes it ought to be in advance; it ought not to enter the labor market, as you have suggested, Mr. Chairman, in competition with all other employers, but ought to be in advance. The selfish, mercenary, or other such motives which govern individuals in their struggle to accumulate wealth ought not to exist in our Government, although they do exist to a morbid degree in too many of our employers. The Government having adopted the eight-hour law, it seems to me to be hardly a debatable question, the efficiency or the good of the reduction of the hours of labor; and the Government having adopted that law, it should be faithfully executed, not as some one during the panic did construe it, or as others do now, but in accordance with its language and spirit.

Of course I will bow in submission to a decision upon the law, but I cannot for the life of me see how that eight-hour law was construed to mean anything but what it plainly says on the face of it. It seems to cover and imply everything that is needed, and not only was that the first construction that was placed upon the law, but several of the executive officers were desirous of reducing the wages of the Government employes when the law came into force, and President Grant, in a proclamation which he issued, quoted the law, and stated that from its plain language and meaning it was clear that no reduction of wages could or should ensue from the reduction of the hours of labor. There was a subsequent proclamation to that same effect. Yet, in violation of (1) the law and (2) the two proclamations of the President—and it is to be presumed that the President upon a question of construction of law is not going to issue a proclamation without having his legal adviser advise him as to whether the construction is correct or not—that law was disregarded! As representatives of the organized trades and labor unions of this country we have met in council within a few days, and on next Tuesday we shall meet in our third annual session[27] in this city. From time to time we have formulated our demands and requests for things that we thought ought to be done by legislation or otherwise. We know very well that the Government of the United States and the legislative power cannot be more probably than a step in advance of the people. If they go much further they are apt to have the platform pulled away from under them, leaving them floating in the air without any support. That is not the intention of organized labor; but we do say that our legislators ought to be at least that one step in advance of the general public. Now, if they will have the national eight-hour law enforced, we do not ask anything further of them in reference to the reduction of the hours of labor. Let the question of endeavoring to enforce a reduction of the hours of labor among private employers be a question to be settled amicably, if possible, between ourselves and our employers, and I think it will not be many years before it will be generally settled. This seems to me to be the question of questions, the reduction of the hours of labor.

By Mr. George:

Q. If that eight-hour law were enforced, would it not be mere favoritism to certain men who might be favorites of officers of the Government, enabling them to get soft places?—A. That may probably be true so far as regards sinecures, if any such things exist.

Q. Well, they do.—A. But, that is not the question. We are speaking of the men who toil, not of those men who draw their salaries for doing nothing.

Q. You see, the Government does not employ more than a thousandth part of the labor of the country. Now, would not the payment of ten hours' wages for eight hours of labor in the Government employ be a sort of political favoritism which the party in power would secure for their friends, and would not the eight-hour law so operate?—A. No; I think not. I think that when we want anything done on our dry-docks, or in our navy-yards, or our machine shops, or our foundries, the men are engaged to do that work not because of favoritism but because of their competency. I am not speaking now as to whether men who may be employed—I do not know how careful to be in this matter, because I don't care to speak with reference to the generally prevailing cry about political patronage.

Mr. George. I do not speak of it as a party matter.

The Chairman. It was an apprehension of the evil which Senator George speaks of which prevented the Republicans from enforcing the eight-hour law.

By Mr. George:

Q. The Government employs clerks at good salaries, who work short hours; but has that had the effect of reducing the hours of labor of clerks in private employments, or of increasing their salaries?—A. Well, the clerks in private employments do not work anything like the number of hours that other employes do; but the employes that I have reference to are those engaged in manufacturing and in manual labor.

By the Chairman:

Q. There is one difficulty in the practical application of the law. The labor market changes from day to day and from year to year. Your idea of that law is that the Government shall pay higher wages than are paid by outside employers, as an example and for its formative effect upon public opinion. You start, then, with a law which prescribes the compensation at so much a day, or so much a week, or so much a month, or so much a year, being, say, 20 per cent. higher than the then existing outside rate; but it may be that in the course of a year the outside market rate will have risen to be as high as the Government is paying. How would you adjust that difficulty? Would you provide that the Government should as a rule pay a certain percentage higher than was paid in outside employment, and which outside employment would you make the basis?—A. No; I would not make that a prescribed rule, nor would I borrow trouble about what may occur some years hence. I believe that, while being as prudent as possible, it is better to leave what may arise in a given event to be adjusted by the legislature or the people. Whenever it does arise, I believe they will find a way out of the difficulty, and I prefer to leave it to them rather

than to advance a theory as to what could or should be done, and which might be mistaken; as I believe rather in the combined sense of the mass of the people than in the theories of the philosophers.

Q. Do you think the combined sense of the mass of the people would say, "Give a Government employe more than a private employe receives who does the same work"?—A. I believe that were the eight-hour law submitted to the combined sense of the mass of the people they would vote decidedly in favor of it.

Q. Yes, I can see that; but my question is whether the member of a trades union would say, "Pay the fellow who works for the Government 20 per cent. more than I receive"?—A. Yes. I am confident that if you leave this matter to the trades unions they will pronounce in favor of it. I am positive of that; and when I speak of myself I do not refer to my individuality, but I am confident that the workingmen would be willing to bear their share of the burden of the enforcement of the eight-hour law.

Q. The retroactive effect upon themselves would be so great?—A. Yes, sir. I am confident that the discussion and demonstration of the benefits that would result generally from the reduction of the hours of labor have had a great influence on the members of our organizations who might at one time have been probably opposed to it, but who are now positively in favor of the reduction of the hours of labor.

Q. You seem to go upon the principle in your reasonings that the Government cannot regulate the hours of labor as between private employers and employes. If it is for the general good to do so, why not? making the law efficient by applying it to both parties by affixing penalties and enforcing them for violation?—A. I am not a very good expounder of the Constitution, so I will not contend upon that point.

By Mr. Call:

Q. Let me see if I understand your idea about this eight-hour law. I understand that, in the first place, you propose a rule of action by which shall be prohibited the exaction of more than eight hours of labor daily of any one man. Now, suppose the man wants to work more than eight hours, would you favor a law prohibiting it?—A. No; I would not favor such a law. I believe that the regulation of that would easily evolve out of the organized efforts of labor and the means that would be taken to agitate the question and educate the workers to understand that it would be to their benefit, to their lasting benefit, to abstain from more than eight hours' work.

Q. You think that would become the universal custom and rule?—A. Yes, sir.

By Mr. George:

Q. We might all be in favor of the eight-hour law, and in favor of

its enforcement by Congress, provided we could be made to believe that it would have the effect which you expect it would have. But there may be some doubt in our minds as to whether the enforcement of the eight-hour law would have the slightest effect upon the general business of the country. There is a class of Government employes now who have largely the advantage in salaries and in everything else, and I think that advantage, instead of operating for the good of the people, operates to the injury of the whole country. Government employs very few mechanics, and, supposing that law was enforced, they would employ as laborers how many men do you think; five hundred?—A. Probably that many.

Q. Now, would not those be places that everybody would want to get, because they would be easier than outside places?—A. If that was so it would have a tendency to make the other working people try to imitate the example, and, as they could not all become Government employes, they would endeavor to have the eight-hour law enforced in their private employments.

Q. Is there not a very great desire on the part of workingmen to get Government employment, and would not that be regarded as a very great advantage? I ask you to think about that. It would become a mere plum, it seems to me, to be given to the party in power.—A. I do not think it would have that effect only to a very slight degree; but if it is the opinion that upon the enforcement of the national eight-hour law immediately as a result the private employers would try to enforce it in their business of course whoever expected that would be disappointed; but still I am confident that it would tend to bring that about.

[Examination suspended.][28]

Adjourned.

U.S. Congress, Senate Committee on Education and Labor, *Report of the Committee of the Senate upon the Relations between Labor and Capital. . .* , 48th Cong., 4 vols. (Washington, D.C., 1885), 1:270-301.

1. Omitted from SG's testimony are his remarks on the size and number of rooms in tenement houses, the rents charged for them, and the sanitary facilities available to the residents—information similar to that found in "A Translation of a Series of Articles by Samuel Gompers on Tenement-House Cigar Manufacture in New York City," Oct. 31–Nov. 14, 1881, above. Also omitted are his presentation of newspaper reports on tenement-house cigar manufacture, and his discussion of labor legislation in Great Britain and Switzerland and of the French conseil prud'homme.

2. Henry William Blair (1834-1920), a New Hampshire lawyer and Civil War officer, served two terms in the state legislature and two in the U.S. House of Representatives (1875-79) prior to his election to the U.S. Senate in 1879. He chaired the Senate Committee on Education and Labor from 1881 to 1891. President Benjamin Harrison appointed him envoy to China after Blair failed to receive his party's nomination for

a third term in 1891. He returned to the U.S. House of Representatives for a single term in 1893.

3. After his marriage in January 1867, SG lived briefly in Hackensack, N.J., and then in Lambertsville, N.J.

4. In 1864 SG joined local 15 of the Cigar Makers' National Union.

5. The United Cigarmakers.

6. SG was a member of Local Assembly 2458, the Defiance Assembly, founded in January 1883 and consisting of members of the CMIU.

7. In 1874 Massachusetts enacted a measure limiting the working hours of women and children under eighteen to ten per day and sixty per week (Laws of 1874, ch. 221).

8. The Massachusetts Bureau of Statistics of Labor, created in 1869 by a resolution of the Massachusetts legislature, was the first institution of its kind in the United States.

9. Massachusetts Bureau of Statistics of Labor, *Thirteenth Annual Report* (Boston, 1882).

10. Carroll Davidson WRIGHT served as chief of the Massachusetts Bureau of Statistics of Labor from 1873 to 1888.

11. James Zachariah George (1826-97) was a Democratic U.S. senator from Mississippi from 1879 to 1897.

12. Wilkinson Call (1834-1910), a Florida Democrat and former Confederate officer, served in the U.S. Senate from 1879 to 1897.

13. Article IX of the constitution adopted by the 1880 CMIU convention authorized issuance of a label, free of charge, to employers of CMIU members. The label was light blue.

14. James Lawrence Pugh (1820-1907) represented Alabama in the Confederate Congress during the Civil War and served in the U.S. Senate as a Democrat from 1881 to 1897.

15. In mid-March 1882 over 5,000 textile workers struck against the Pacific Mills in Lawrence, Mass., protesting a wage reduction of 20 percent and the change from day- to piecework. The company replaced the strikers, and by August the mills were fully operational and the strike broken.

16. New York City railroad freight handlers struck for a wage increase in June 1882; the strike ended unsuccessfully in August.

17. Members of Milwaukee CMIU 25 struck for higher wages in October 1881. The cigar manufacturers initially agreed to the union's demands but then organized a manufacturers' association and locked out the workers. The strike continued through 1882, ending in failure and leaving the local decimated.

18. Edward Aschermann (1834-1904) emigrated from Germany in 1850. He and a partner operated a leading cigar manufacturing firm in Milwaukee under the name of Ed. Aschermann and Co.

19. On June 27, 1880, New York City horse-railroad car drivers on the Second Avenue line struck, demanding a wage increase. The employers were able to run the line with new workers under police protection, and by July 7 the drivers had returned to work at the old wages.

20. Massachusetts Bureau of Statistics of Labor, *Third Annual Report* (Boston, 1872).

21. Ira STEWARD, leader of the movement for the eight-hour workday.

22. On Apr. 21, 1856, building workers in Melbourne, capital of the state of Victoria, Australia, conducted a demonstration on behalf of the eight-hour day. That state apparently did not require an eight-hour day for work on government contract until 1870, however.

23. The financial panic of 1873.

24. Dated May 19, 1869, and May 11, 1872.

25. Ulysses Simpson Grant (1822-85) was president of the United States from 1869 to 1877.

26. These brackets are in the original document.

27. The 1883 FOTLU convention was held in New York City, Aug. 21-24.

28. These brackets are in the original document.

New York, August 18, 1883.

Samuel Gompers recalled and further examined.

REDUCTION OF THE HOURS OF LABOR.

The Chairman. You may proceed with any additional statement which you desire to make.

The Witness. In consequence of the reduction of the hours of labor of which I have spoken, inventions and improvements in machinery are made, labor is divided and subdivided, new tools are devised and made available in consequence of the greater simplicity of the labor that is then assigned to each man. The laborers work in what are known as "teams," and the productivity of their labor generally increases much beyond the proportion of the reduction in the hours of labor by reason of that reduction, and further, by reason of the improved machinery and the division and subdivision of labor. Further, the operative or workingman is better able to perform the work assigned to him, for he can do more work in proportion in eight hours than he could have done in ten. The Department of State of the United States, in the report for the year 1881, and the report of the bureau of labor statistics in New Jersey,[1] page 81,[2] speaking of the cotton industries, show the average run of the spindle in several countries, and I have copied a number of their statements. The American spindles are run at 64½ passages a minute; in Germany the spindles run at the rate of 39; in France, 24; in Russia, 19; in England, 83. America, in the running of her spindles, is in advance of all other countries except England.

By Mr. George:

Question. What does that statement mean?—Answer. It means the rate of motion of the spindle.

Mr. George. I have always heard the instrument that passes through the fabric called a shuttle.

The Witness. Yes, sir; that is a shuttle. I am not acquainted myself with the technicalities of the cotton industry, but I merely use these figures to show that a greater number of motions is made per minute in America than anywhere else, except in England.

By the Chairman:

Q. Those figures are a test of the relative efficiency of the machinery in the different countries, are they?—A. Decidedly; and an illustration of the greater productivity, resulting from the reduction of the hours of labor; improved machinery, as a natural consequence, being brought into operation when the hours of labor are reduced.

Q. Do you mean that from those figures it would appear that an operative in England can perform more work in a given time than one in this country?—A. Yes, sir; and the operatives of this country more than the operatives of Germany, France, or Russia. Mr. Robert Howard, the secretary of the Mule-spinners' National Union,[3] a man who has worked in the mills of England as well as in this country for a number of years, wrote me a letter some few months ago, when I asked him for information as to the hours of labor and the condition of the operatives engaged in factory labor in Massachusetts other than that reported by the Massachusetts bureau of labor statistics. The information was intended to be used before this committee, which was then sitting in Washington. With the permission of the committee, I will submit that letter in connection with my testimony here. It is as follows:

February 3, 1883.

Dear Friend: I transmit a lengthy investigation on the condition of factory labor, furnished by Carroll D. Wright, chief of the "Bureau of Statistics on Labor," in Massachusetts. You will find it instructive and entertaining, no doubt, and gather some good points from it. I am of opinion that a law should be framed by the National Government making the hours of labor uniform in all textile factories in the United States, because I am satisfied from past experience that a State working short hours has to labor at great disadvantages in comparison with a State where no restrictions in the hours exist. Many hardships are endured by its people, also, which never come to light. For instance, in Massachusetts we have a ten-hour law, governing our textile industries; crossing the boundary line into Rhode Island, Connecticut, New Hampshire, and thence to Maine, and New York, employers can work their employes any number of hours without any interference by law; the result is, in order with 60 hours per week in Massachusetts, against 66 and, in some instances, 70 in the other states, machinery has been speeded up to such an extent, as to make the operatives work almost beyond human endurance. You might show, in favor of less hours, that Massachusetts, with its "ten-hour law," from Carroll D. Wright's report of 1881, is producing more yarn to the spindle, and more cloth to the loom, than other states working long hours. Many employers are of opinion that ten hours will produce as much as twelve. A few mills in the surrounding States have voluntarily reduced their hours to ten per day, and yet, paying the same wages as in neighboring eleven-hour mills, have found their product and their profit satisfactory, and not reduced by the change. The Willimantic Linen Co., Connecticut, one of the finest mills in the country, reduced the

hours from eleven to ten, and the wages earned are the same and the production shows no loss. In another mill, a superintendent says, that he asked permission of the directors of the firm where he was employed to be allowed to reduce the hours of labor from thirteen to eleven (11) per day. With the thirteen hours per day, the production was 90,000 pieces of print cloths annually, and with the eleven (11) per day, the production swelled to 120,900 pieces annually. In Switzerland, Europe, the inspectors in their factory laws report similar results. I believe shortening the hours of labor would be a boon to the worker without cost to anybody else. More time for the cultivation of his moral, mental, and physical capabilities, would make him better, stronger, and wiser. I hold the better the workman the better the work. Improvements in the man should not be thought useless; every step in bettering his condition, lightening the slavery of his toil, teaching him thrift and hope, will return something in the quality, economy, and effectiveness of his work. A general reduction in the hours of labor may be made worth more to the peace of society than a hundred prisons. Our school or educational system in Massachusetts is grand, and I must say, is rigidly enforced. In all other New England States children may be put in the factories at any age and worked any number of hours, while in Massachusetts no child is permitted under ten years of age to work in any textile factory under penalty of fine both to its parents and employers, and for twenty weeks in the year it is compelled to leave the mill to attend school until it attains the age of 14 years, when it is allowed to work not more than 60 hours in any one week, according to the law of the State.

I hope these few points may be of service.

Remaining yours, in the cause of labor,

Robt. Howard.

Samuel Gompers.

The Chairman. Does it appear whether there was any improved machinery which might account, in part, for the increased production stated there to have resulted from reduced hours of labor?

The Witness. I cannot say; I am merely reading from Mr. Howard's letter what he says on that point. And I may say right here, that I believe the National Government has not the power to legislate in regard to the operatives in any factory except its own workshops; but, wherever the legislation comes from, the children ought to be protected in their infancy from the avarice alike of their parents and of their employers. In no time, in the wildest and most barbarous countries, not even in this country two hundred years or more ago, did the Indian ever send his child to work or to maintain itself at so tender an age as children are sent to these factories. It is only in our civilized society that parents are compelled or impelled through avarice on their own part, or on the part of the employing class, to set children of tender age to work, frequently to the supplanting of man's labor.

The Chairman. Even the wild beasts are tender of their offspring until maturity, I believe?

STATISTICS OF MANUFACTURES.

The Witness. Yes. Before proceeding to another branch of the general subject that you are investigating, I desire to say that there are some statistics that I have collected from the last census in reference to the cost of the materials used in our manufactures, the amount of wages paid, the value of the products, the number of laborers employed, and the profits and the number of establishments. There were 2,738,950 workingmen or operatives employed in industrial pursuits in this country. This is taken from the compendium of the last census.

By Mr. George:

Q. That means only manufacturing industries, excluding agricultural?—A. Yes, sir. There were 253,840 establishments or employers. The cost of the raw material was $3,304,340,029. The amount paid in wages was $947,919,674. The value of the product was $5,369,677,706, showing a profit of $1,027,408,003. You will observe the contrast between the wages paid to nearly 2¾ million of laborers and the profits paid to a little more than one-fourth of a million of employers, including both large and small employers, those who may employ one man or two men being included in this number of establishments as well as those who employ a thousand.

By the Chairman:

Q. Suppose that you add the ordinary interest on the fixed and the floating capital involved. There ought to be a fair interest, computed on the floating capital, and then should there not be a fair rent allowed upon the fixed capital, and should not that be deducted from the billion which goes to the capitalist there, in order to show his real profit?—A. I believe it is not necessary that after the profits are stated a man should have interest besides upon his fixed or floating capital.

Q. But it is very often the case that the man who manufactures is not the man who owns the floating capital. The man who builds up the establishment often hires the capital, and in order that he may be able to get his establishment equipped with all the appliances required for production, he is obliged to borrow money and pay interest upon it. You cannot deal with him, therefore, as if he were the owner of the capital as well as the controller of the establishment, can you?—A. It is evident that if labor has produced an article, and the employer gets $1 for the production of that article, and, by whatsoever means, that article is sold and costs the consumer $5, without any additional labor, there is clearly a profit of $4 on that article which the laborer does not get, $4 over and above his wages.

Q. Now where is that amount distributed?—A. It is apparent, I

think, that in the first place it goes to the employer of labor; afterwards, in many instances, to middle men, to go-betweens between the producer and consumer.

Q. Your main point may be true, that labor does not get a fair share.—A. That is what I wish to demonstrate.

By Mr. George:

Q. In getting at the profit, you add the cost of the wages to the cost of the raw material, and deduct that sum from the value of the manufactured article?—A. Yes, sir; I take the amount of the raw material and the wages (adding them together) and then subtract that from the aggregate value, to arrive at the result as profit.

By the Chairman:

Q. Have you anything from the census showing the amount of capital invested in those 253,000 establishments?—A. I have not; but it can be easily obtained.

Q. Now, to get at the real value of that statement, should or should not something be allowed in the way of rent or interest upon the capital invested in those establishments, and also upon the capital invested in the raw material?—A. Interest, of course, is allowed, and may probably be right. I will not defend it; but it is usually allowed.

By Mr. George:

Q. But is not the interest itself, the ordinary legal interest, the profit upon capital?—A. That is what I was going to mention; that there is over $1,274,000 of profit; and if that does include interest and profit, if it does not cover more than it should cover in the way of interest and profit of every description, then we have no reason to complain. But I maintain that the profits here shown on labor are more than capital should receive for its outlay.

By the Chairman:

Q. I do not controvert your main point at all; but I call your attention to those elements which must be considered in order to make a fair statement which would strike the public mind in such a way as to give the testimony value. I do not say the worker should not receive ample wages before any profit goes to capital, because the impulse given to machinery by the effort of the living worker is the fundamental condition of production. But should not these other elements be considered?—A. I misunderstood the drift of your observation. As our lamented martyr President, Lincoln, said in his second message, "Labor is prior to and independent of capital. Capital is only the fruit of labor, and could never have existed if labor had not first existed. Labor is the superior of wealth, and deserves much higher consideration."

By Mr. George:

Q. Suppose a man has a million dollars invested in manufactures and gets 6 per cent. interest on that, ought or ought not that interest to stand in the place of profit on his capital? Has he a right to get an additional profit besides that?—A. He has a legal right. The law provides the rate of interest that a man shall get in return for money loaned to another, but when that money is invested by himself in productive, or perhaps unproductive, business, he can take as much interest as he pleases, even if it be 100 per cent.

Q. Well, suppose he has only half a million and he borrows half a million, then if he charges interest on the half million that he borrows and wants an additional profit on that, does he not make the laborer pay for his want of capital—pay twice—pay to the man from whom he borrowed it, and also pay to him?—A. Yes, sir; I think so. In fact, I know it.

Mr. Call. How much does he make the machinery pay?

Mr. George. The machinery and the real estate are capital.

By the Chairman:

Q. That "if" is a misleading "if" so far as your data there are concerned. Here is the original plant: The laborer can get no employment in that factory unless the factory is built. That has first to be built and paid for. Suppose yourself to be a gentleman designing to establish a factory and that you are without the necessary means. You are a manufacturer simply—nothing else. Another man has capital and he is no manufacturer. Now, you first go to him and ask him for money to build your factory. Of course he loans it to you, if you get it at all, at some rate per cent. of interest. If all things were owned in common it would be different, but I am speaking of the existing conditions of society. First, you must hire the money, and you have got to pay something for it. You do so, and proceed to erect your factory, put in your machinery, and fix up the whole establishment ready to produce. Next, you must have something to work upon, your raw material, and that must be bought. From the same man, or from somebody else, you hire the money to purchase the raw material, and for that you are obliged to pay a rate of interest, or else you cannot get it. Thus you have obtained the first conditions, the plant, the permanent capital, and the raw material, the floating capital, and upon the cost of the whole you are paying interest and will continue to pay interest so long as you continue to produce, year by year. The next thing is the labor; and the laborer has a necessity which is not upon capital, for he must live from day to day; he must be clothed and fed. Those are the necessities of a living body and soul and they are primary necessities. The laborer therefore is worthy in a peculiar sense of his hire. He should be properly rewarded, and if he is not

to receive enough of the necessaries of life your establishment must break down and he must go to work in some other place where he can be better provided for. But suppose that these three elements, the plant, the raw material, and the laborer combine; the result is production. You have got something to sell. That is taken to market, and you, the manufacturer, handle the article and get the most that you can for it. But now in this statement which you have brought up as an illustration, the cost of the original plant is not given, and therefore that element is not represented upon which interest is to be paid. Your raw material is given. Then comes labor. Your figures show $3,500,000,000 of capital hired to buy the raw material with. Wages gets another billion; that makes $4,500,000,000. Now, there should be added to that, as it seems to me, interest upon that $3,500,000,000 which bought the raw material, and interest upon the amount invested in the plant, that would make, it may be, $5,000,000,000. That should be subtracted from the $5,500,000,000, which is what you receive for the entire product in the market, and the remainder, whatever it may be, is the amount that you, as the manufacturer, put in your pocket as your profit. Now, is not that a fair statement of all the conditions that should be embraced in your data in order to present the case fairly?—A. In the first instance, I would say that your hypothetical question exposes a condition of affairs that is really deplorable. As you have suggested that I should be the person selected as the manufacturer, if I have no money, instead of borrowing money to go on with a factory, instead of hiring a number of hands and borrowing money to buy raw material, I ought to go into a workshop and earn an honest living instead of undertaking to make money from nothing, absolutely nothing. I believe, however, that the fact is as you have supposed it in many cases. Let the man who has the money hire the man who has the skill to carry on the factory; but do not let the man who has no money hire capital to build the factory and to buy the raw material, and pay interest on that money, and then hire money to pay his labor, and make a profit above all that, and hire a superintendent at a good salary to manage the business, and be himself one of the lords of the land—so-called.

Q. Then when you had changed the manufacturer into the capitalist, you would really feel as though that capitalist should be deprived of all interest upon his capital?—A. No; but I mean to say that this man who is the manufacturer in your hypothetical question, is a man who is also an incubus upon labor.

Q. You would not think that a man who was capable of taking the management of a great manufacturing establishment was an incubus, would you?—A. No; not as a manager; but on your supposition he

would receive not the return of a manager, but the return of an employer, which is generally considerably higher.

Q. But sometimes the employer's returns are much lower, because sometimes employers fail, do they not?—A. Yes; and when they do it is evident that some of the conditions in the labor field require alteration. But I believe that in modern society and so long as the competitive system lasts (and I do not know when it will end, and do not desire to prophesy on the subject either), I believe that under this system the employer is entitled to a return. That is, if he is willing to pay living wages. And if he does not do that, then I say that between the upper and the nether millstone, between the consumer and the workingman, he ought to be crushed out as a manufacturer and forced to take to the field as a laborer.

Q. Undoubtedly labor ought to be first paid out of the whole result. That is your main point, I believe?—A. Yes, sir; and then if there is a profit I do not deny that the manufacturer may be entitled to it.

STRIKES.

I desire to say a few words now in reference to strikes. Before doing so, I wish to say (as it will save time by avoiding the necessity for repetition) that I indorse fully the statements made here in relation to strikes by Mr. McGuire in his testimony[4] which has intervened between that which I gave here the day before yesterday and my present testimony. While fully indorsing what he said, I desire to add a few considerations on that subject. The strikes that I have mentioned as having taken place, such as in the Lawrence mills and the Harmony mills at Cohoes, in this State, the freight-handlers strikes here, the car-drivers' strike, and others, those were in trades and callings in which the operatives or workingmen were either not at all organized, very poorly organized, or organized just previous to or during the strike, and those cases have no bearing whatsoever, as showing the value of strikes in general. While I am in the labor movement and take a stand opposed to strikes whenever they can be avoided, I have no sympathy with, nor can I indorse or echo, the statements of many men who are too ready to condemn strikes. Strikes have their evils, but they have their good points also, and with proper management, with proper organization, strikes do generally result to the advantage of labor, and in very few instances do they result in injury to the workingmen, whether organized or unorganized.

By Mr. George:

Q. You mean ultimate injury?—A. Ultimate injury. Strikes ought to be, and in well-organized trades unions they are, the last means

which workingmen resort to to protect themselves against the almost never satisfied greed of the employers. Besides this, the strike is, in many instances, the only remedy within our reach as long as legislation is entirely indifferent to the interests of labor. The bureau of statistics of Ohio[5] in 1876 reports eleven strikes in that State in that year, strikes of a general character. Two of those strikes were gained by the operatives (coal miners); two were lost; five were compromised, and as to the two remaining, the result was unknown. Two won and five compromised, make seven out of eleven that were either won by the strikers or compromised. This is the report for the year closing January, 1878, including the year 1877, when we had not yet fully emerged from the panic. A few years ago a labor society of England offered a prize for the best essay on labor strikes. The prize was awarded to a member of Parliament, Mr. McDonnell,[6] a representative of the coal-miners of England, and in that essay (of which I have tried to obtain a copy lately, but have failed) he reports that in the ten years previous to that time there were about ninety-seven strikes of a general character, not mere local or circumscribed strikes, but important ones, of which seventy were gained, twenty were lost, and seven were either compromised or the result was unknown. The results of organization in my own trade, cigar-making, are instructive. From the year 1873 to 1878 the cigar-makers of this country were reduced in wages systematically *every spring and every fall.* The reductions in wages were sometimes large and sometimes not quite so large, but a reduction was the order of the day at those periods. At that time the cigar-makers' organization was in a very weak and puerile condition. Further, the manufacturers of cigars throughout that period managed to introduce a system of truck or "pluck-me" payments, by which the workingmen were paid in kind, cigars, and were required to go out and sell them to any grogshop or other place of any description where they could sell them; or they would receive store orders, or, in the case of single men, they would be required to board at certain hotels or boarding houses. In the city of Elmira, in this State, a manufacturer paid his workingmen $6 per thousand if they were taking their wages out in truck or kind, while he paid only $5 a thousand to those single men who were in boarding houses, and but $4 a thousand to those cigar-makers who wanted cash, legal tender. In one instance a man who expected at the end of the week to receive a dollar or two besides his board, was informed by his employer that he could not give it to him because he had just made an investment and he was sorry for it. The investment was that he had just bought a horse, but "to tell you the truth," said he, "I had to take it in trade. That being so, of course it is not a very high price that I want for

it, and you can have that horse for the balance of your wages," which was a little over $3. The cigar-maker believing that he might possibly be able to sell the animal, and finding it impossible to obtain any money, took the horse. He went around with it all that Saturday evening, and not having any money, he could not purchase the horse a meal, and if he had done so subsequent events showed that the horse might not have been in a condition to eat it, for about 11 o'clock that night it died and he was required to furnish a wagon and to see that the carcass was removed and taken to the proper place of interment, or he would have been fined by the city authorities. The consequence to him was a night's effort to get rid of the horse and the payment of about $4 to have it carted away after it had died. This was told me by the man to whom it occurred. Last year the same man was the candidate of the workingmen for member of assembly.

THE CIGAR-MAKERS' UNION.

What I wish to show is the condition of the cigar-makers at that period when there was no organization. When our organizations commenced to emerge and reorganize throughout the country, the first year there were seventeen strikes in our trade, of which twelve or thirteen were successful. The rest were either lost or compromised. In the year following we had forty-six strikes, of which thirty-seven, I think, were successful, three lost, and six compromised. In these last two years, since which we have held no convention (we will hold one next year and we will hear the result), I am convinced that we have had over one hundred and sixty or one hundred and seventy strikes, and the strikes have been successful except in, perhaps, twenty instances, where they may have been lost or compromised. The truck system of which I spoke exists no longer in our trade. We have adopted a course of action which our experience has taught us, and that is, in certain periods of the year, when it is generally dull, not to strike for an advance of wages. Formerly, before the organization, men would probably strike for an advance of wages in the dull season, and be content that they were not reduced in the busy season. Our experience has taught us to adopt a different mode of action.

Q. You strike now when business is active?—A. Yes, sir; and then, when we obtain an increase of wages when times are fair, our object is to endeavor to obtain fair wages during the dull season also, and, while we have made provision not to strike for an increase of wages during those periods, we are always in a position to strike against a reduction of wages or the introduction of the truck system, or other obnoxious rules. We have found that, for the purpose of accomplishing

this object, it is entirely valueless to organize a union during a strike, and that it is little better than valueless to organize just immediately before a strike. We have found that if we are desirous of gaining anything in a strike, we must prepare in peace for the turbulent time which may come. And the Cigar-makers' International Union, of which I now speak especially, is an organization that has in its treasury between $130,000 and $150,000 ready to be concentrated within five days at any time at any given point. I hold in my hand a copy of the constitution of that organization. Of course I am not desirous of making a propaganda for it, but to illustrate what I have been saying I will read from it this provision: "Any union being directed by the executive board to forward money to another local union, and failing to comply within five days from date of said notice, shall be suspended." That is, in the event of a strike at a given point, the international president of the organization is directed to direct or request the nearest union to immediately send on its whole treasury if that is necessary, and the unions throughout the entire country and Canada to forward their entire treasury if necessary, to be placed at the disposal of the organization that is in trouble.

By the Chairman:

Q. The funds are already deposited or under the control of these various subunions all over the country?—A. Yes, sir; the funds are not concentrated at any given point, but are subdivided out among the local branches, of which there are 185.

Q. So that the custody of the money is, of course, very safe?—A. Yes, sir. And there is a special feature in the organization, the equalization of its funds. That is, at the end of a certain period, every six months, the president of the general international union is directed to make an aggregate account, from the reports that are made by the local organizations, of the moneys in their hands and their membership; the aggregate funds are calculated and a certain per capita is arrived at, and at the end of this period the union or the branch that has expended less for local expenses under the laws than its neighbor or any other union, is directed to forward to that union which may have expended more a certain sum of money. So that if the organization has, say (as it has), 18,000 members in round numbers, $160,000 would be about $9 a head. Now, at the end of a certain period an organization which would have 500 members would have $4,500 in its treasury; an organization with ten members would have $90 in its treasury; an organization with 2,000 members would have $18,000 in its treasury. Besides this, the organization pays sick benefits, death benefits, and strike benefits.

Q. It is really a life and health assurance company?—A. It has those features.

Q. It is like an ordinary assurance company with the strike benefit added, and a health benefit if the need for it arises?—A. Yes, sir. I should like to state the reasons for the adoption of those features, after I have gone through my general statement on the matter of strikes.

OTHER TRADES UNIONS.

During the last panic the granite-cutters of this country were able to maintain their rates of wages against reduction, and their hours of labor at eight per day. And it was the force of their organization which exempted them from being compelled to work under a reduction of wages or to submit to the lengthening of their hours of labor. The locomotive engineers[7] were compelled but once during the panic to submit to a reduction of wages. The mule-spinners' national organization have accomplished, through their organization, a result which was denied them by legislation. They had endeavored for years to compel the mills to pay them their wages weekly. They had been paid fortnightly or monthly. The legislature, however, repeatedly told them that it could not be done, that it would be too inconvenient, and should not be insisted upon. Through the efforts of the mule-spinners' union they are now paid weekly. So that what they were unable to accomplish by legislation they have accomplished by their organization.

Q. What are considered by the laboring people to be the advantages of these short payments?—A. The fact that on Saturday, or least once a week, the laborer can go home with the few dollars that he has earned, give them to his wife, and be placed upon a footing with all other persons in the purchase of his articles of consumption; while if he is paid fortnightly or monthly he is compelled to purchase from a certain butcher, grocer, or clothier (if he can buy clothes), even though that dealer may take advantage of him, and thus the laborer be compelled to pay higher prices than his neighbor, and at the same time to accept articles of inferior quality. The purchasing power of his wages is reduced by the fact of his not receiving his wages at least once a week.

Q. Do you think that if he is paid the larger sum at the end of the month some of it is more likely to be expended injudiciously, so that he will not have the money by him during the succeeding month and will therefore buy on credit many of the necessaries of life?—A. Yes, sir; also that. The Iron and Steel Workers' Association in 1881 en-

deavored to obtain an increase of wages, but failed in their attempt.[8] If they had been loosely organized or if they had become disorganized the manufacturers would have accomplished on the first of last June a reduction of wages of ten or fifteen per cent. At or about that time the press of the entire country, with very few exceptions, either advised the iron and steel workers to accept the reduction, on the ground that they were unable to cope with the iron masters, or wrote such discouraging reports and articles that it required men of nerve to withstand the pressure from the general opinion prevailing that they would not be able to hold out, in consequence of the employers insisting on the reduction. The iron and steel workers, however, adopted a resolution that they would under no consideration submit to the reduction, and the fact that they had demonstrated to their employers the year previous that they were fully competent to inflict great damage and injury upon them satisfied the employers that they were going to make a hard fight, and they preferred to recede from the enforcement of the proposed reduction of ten per cent. rather than to encounter the serious opposition of the amalgamated iron and steel workers' organization. This victory was obtained without a strike; by a mere resolution that, if the reduction was insisted upon they would strike. And again, the fact that they had held together after they were defeated before prevented a reduction of wages last June.

In the report of the bureau of labor statistics of New Jersey[9] a quotation is made, and while I have not the quotation with me, I will give the substance of it. It mentions the fact of England being one of the richest countries on this earth; the fact that during the panic of 1878 few if any of the members of trades unions in England suffered the evils from which so large a number suffered who were not members of the unions, and the fact that the unions had, by means of strikes and other measures, improved the condition of the working people in England materially and to a very marked extent in the instance of the farm laborers of England. Formerly, it says, the poor-houses, the work-houses, and the jails were filled after the harvest was reaped, and many of the laborers were compelled to either beg, steal, or starve, but after the organization of the Farm Laborers' Union[10] their condition was materially improved, so that more than one-half of the jails, alms-houses, and prisons formerly required were unnecessary and vacant. Mr. Joseph Arch,[11] one of the organizers and the head of that organization, was called upon some few years ago by the British ministry to tell them what could be done for the farm laborers. He answered that the time when they could have done them

any material good had passed, and that when they could have done them any good they had never offered their aid.

There is nothing in the labor movement that employers who have had unorganized laborers dread so much as organization; but organization alone will not do much unless the organization provides itself with a good fund, so that the operatives may be in a position, in the event of a struggle with their employers, to hold out.

THE "BENEFIT" FEATURE IN TRADES UNIONS.

In answer to your inquiry, Mr. Chairman, a while ago, I wish to say that these benefit features of which I have spoken have been introduced into these organizations for a dual purpose. The first object is benevolence and humanity, trying to help each other in the event of any of our number falling out of work, or becoming sick, or dying. It is an old maxim with us that no trades unionist can be buried in potter's field. We wish to have a trades unionist cared for even after death, and given a decent burial. We are now even considering the question of introducing a benefit in the event of the death of a member's wife. The advantage of a loan or traveling benefit is also great. The laws of the State of New York, passed during the panic, made poverty a crime, and the man who went from one place to another in quest of work was often punished and sent to prison as a "tramp." Our union at that time introduced a traveling benefit, so that the members of our craft were saved from the operation of that law, and could go in search of work with sufficient means to carry them from place to place.

Q. I would like to ask you for such knowledge as you have in regard to so-called "tramps" in the period of panic when the law to which you refer was passed. You may remember that at that time there were a great many persons who were called tramps, moving from one place to another, in quest of work sometimes and sometimes for other purposes. Now, of those migratory persons, about what proportion, in your opinion, were honest men reduced to poverty, anxious and willing to work if they could get work? They were denounced generally, we all know, as "tramps," and laws were enacted against tramps, and I would like to get some idea of how many of them were, in your opinion, the people.—A. I will answer that question later, to the best of my ability; but I prefer now to continue my statement of the purposes and advantages of these benefits. I have stated the first purpose. The second purpose is that men who become members of these organizations shall find that there are other advantages connected with trades unions than the mere matter of protection against

reduction of wages. In the event of a strike or a struggle either for an increase of wages or against a reduction of wages, or an effort for the reduction of the hours of labor, and in the event of the workingman being defeated in that struggle or effort, the fact is in many of these organizations where these benevolent features are not incorporated, that the members fall apart, thus throwing themselves upon the tender mercies of the employer, who can thus take advantage of each one; while in the organizations having these benevolent and beneficial features the members are held together by mutual interests other than those connected directly with the question of strikes; and this fact that they are held thus together prevents the reduction of wages and the taking advantage of them individually by their employers. These are the dual purposes of those beneficial features.

"TRAMPS."

In reference to "tramps" so called, or, as I should call them, workingmen who for a time at least have become superfluous in society, men rather, whom the employing class have made superfluous—the best evidence that these, or a majority of them, were honest-intentioned workingmen is the fact that since the era of our so-called prosperity set in we find that the complaint against this immense number of our people has been reduced almost to a minimum. The fact is that, with the return of prosperity, most of those men have come to the cities and have returned to work, while those who were longer out of employment may have become demoralized, through want of employment, and, as they got a dollar or ten cents on their travels, may have spent it in drink or something else; but that they were brought to that condition, or that the larger portion of them were brought to that condition, by circumstances over which they had no control, I am certainly convinced. The number of such people could not be easily estimated. I have seen it variously stated, by the workingmen, who probably stated the largest number, and by Mr. Carroll D. Wright, the chief of the Massachusetts bureau of labor statistics, who stated probably the lowest number. I have seen it stated from 60,000 or 70,000 up to 2,000,000; but of course neither of those numbers can be anything like accurate.

Q. They probably counted the same men in a good many different localities?—A. Very probably. On this question of strikes I have nothing to add, except, that if you desire to question me, I shall be pleased to give any further information that I may possess.

By Mr. George:

Q. You understand that this recent strike of the telegraphers[12] has failed.—A. Yes, sir. I understand that that strike is finally at an end.

Q. Do you know the cause of its failure, or have you a well-considered opinion as to the cause?—A. I have an opinion as to the cause of the failure of that strike.

WHY SOME STRIKES FAIL.

Q. What is you opinion?—A. My opinion is that the first few strikes that workingmen generally indulge in are lost, from the fact that their employers are unable to comprehend the idea that labor has certain rights which they ought to respect; second, because they are really unaware that the laborers who are on a strike are capable of inflicting an injury upon them; and third, that when they are once in a strike and hold out for a considerable period they do not like to weaken and accede to the terms of their employes, but prefer to make large sacrifices from their wealth or capital rather than to accede to those demands. This being the second telegraph strike[13] (though it may really be termed the first thorough strike of the telegraphers), the company, that is, Jay Gould, was unable to comprehend at first, what he was forced to comprehend in the case of the locomotive engineers and firemen last year,[14] that the workingmen were in a position to inflict a considerable damage upon the company. This strike has another instructive feature. It will teach the telegraphers this, that if they are desirous of holding out for a long period and fighting a concern of the magnitude of the Western Union Telegraph Company they will have in time of peace to prepare for war.

Q. They will have to have a treasury, you mean?—A. They will have to have a treasury. Further, the accumulation of such a treasury, the payment of benefits to members, and the demonstration to members that the organizations are fully capable of keeping their promises, not making promises and violating them to their members, but making promises and keeping them, will insure the confidence of the members in their organization, and they will find that when they next indulge in or threaten a strike Mr. Jay Gould will be more willing to lend an ear to their complaints and grievances. As to this general outcry among employers (and sometimes the press will echo the cry), about the men on strike being turbulent, about their being destroyers of property and violators of the public peace, the truth is that where these offenses are committed they are very seldom the work of the men on strike, nor are they countenanced by the men on strike, but that they are not committed by overzealous friends I will not say.

ADVANTAGES OF TRADES UNIONS.

Even in such instances, however, the organizations of labor are the conservators of the public peace; for when strikes occur among men who are unorganized, often acting upon ill-considered plans, hastily adopted, acting upon passion, and sometimes not knowing what they have gone on strike for, except possibly some fancied grievance, and hardly knowing by what means they can or may remedy their grievances, each acts upon his own account without the restraint of organization, and feels that he serves the cause of the strike best when he does something that just occurs to him; while the man who belongs to a trades union that is of some years' standing is, by the very fact of his membership of the organization and his experience there, taught to abide by the decision of the majority. Therefore when anything of that kind I have mentioned occurs or is heard of in the organizations that are of long standing, it is condemned in the most strenuous terms and action is taken to prevent the accomplishment of any such purpose, or if it is accomplished to prevent the recurrence of it. The members of our organization are made to well understand that such a mode of warfare in strikes is not tolerated in any well-regulated or well-organized trades union. So high an authority as the Duke of Argyle,[15] in his work, The Reign of Law, states that "combinations of workingmen for the protection of their labor are recommended alike by reason and experience." When we strike as organized workingmen, we generally win, and that is the reason of the trouble that our employers go to when they try to show that strikes are failures, but you will notice that they generally or always point to unorganized workers. That is one reason also why when the employers know that the workingmen are organized and have got a good treasury strikes are very frequently avoided. There are fewer strikes among organized workingmen, but when they do strike they are able to hold out much longer than the others, and they generally win. The trades unions are not what too many men have been led to believe they are, importations from Europe, if they are imported, then, as has been said, they were landed at Plymouth Rock from the Mayflower. Modern industry evolves these organizations out of the existing conditions where there are two classes in society, one incessantly striving to obtain the labor of the other class for as little as possible, and to obtain the largest amount or number of hours of labor; and the members of the other class being, as individuals, utterly helpless in a contest with their employers, naturally resort to combinations to improve their condition, and, in fact, they are forced by the conditions which surround them to organize for self-protection. Hence trades unions. Trades

unions are not barbarous, nor are they the outgrowth of barbarism. On the contrary they are only possible where civilization exists. Trades unions cannot exist in China; they cannot exist in Russia; and in all those semi-barbarous countries they can hardly exist, if indeed they can exist at all. But they have been formed successfully in this country, in Germany, in England, and they are gradually gaining strength in France. In Great Britain they are very strong; they have been forming there for fifty years, and they are still forming, and I think there is a great future for them yet in America. Wherever trades unions have organized and are most firmly organized, there are the rights of the people most respected. A people may be educated, but to me it appears that the greatest amount of intelligence exists in that country or that State where the people are best able to defend their rights, and their liberties as against those who are desirous of undermining them. Trades unions are organizations that instill into men a higher motive-power and give them a higher goal to look to. The hope that is too frequently deadened in their breasts when unorganized is awakened by the trades unions as it can be by nothing else. A man is sometimes reached by influences such as the church may hold out to him, but the conditions that will make him a better citizen and a more independent one are those that are evolved out of the trades union movement. That makes him a better citizen and a better man in every particular. There are only a few who can be reached by the church so as to affect their daily walk in life compared with the numbers reached by these organizations.

### Trades Unions Not Communistic.

By the Chairman:

Q. The outside public, I think, very largely confound the conditions out of which the trades union grows or is formed, with the, to the general public mind, somewhat revolutionary ideas that are embraced under the names of socialism and communism. Before you get through, won't you let us understand to what extent the trades union is an outgrowth or an evolution of those ideas, and to what extent it stands apart from them and is based on different principles?—A. The trades unions are by no means an outgrowth of socialistic or communistic ideas or principles, but the socialistic and communistic notions are evolved from some of the trades unions' movements. As to the question of the principles of communism or socialism prevailing in trades unions, there are a number of men who connect themselves as workingmen with the trades unions who may have socialistic convictions, yet who never gave them currency; who say, "Whatever ideas we may

have as to the future state of society, regardless of what the end of the labor movement as a movement between classes may be, they must remain in the background, and we must subordinate our convictions, and our views and our acts to the general good that the trades-union movement brings to the laborer." A large number of them think and act in that way. On the other hand, there are men—not so numerous now as they have been in the past—who are endeavoring to conquer the trades-union movement and subordinate it to those doctrines, and in a measure, in a few such organizations that condition of things exists, but by no means does it exist in the largest, most powerful, and best organized trades unions. There the view of which I spoke just now, the desire to improve the condition of the workingmen by and through the efforts of the trades union, is fully lived up to. I do not know whether I have covered the entire ground of the question.

By Mr. George:

Q. You state, then, that the trades unions generally are not propagandists of socialistic views?—A. They are not. On the contrary, the endeavors of which I have spoken, made by certain persons to conquer the trades unions in certain cases, are resisted by the trades unionists; in the first place for the trades unions' sake, and even persons who have these convictions perhaps equally as strong as the others will yet subordinate them entirely to the good to be received directly through the trades unions. These last help those who have not such convictions to resist those who seek to use the trades unions to propagate their socialistic ideas.

Q. Do you think the trades unions have impeded or advanced the spread of socialistic views?—A. I believe that the existence of the trades-union movement, more especially where the unionists are better organized, has evoked a spirit and a demand for reform, but has held in check the more radical elements in society.

By the Chairman:

Q. You may remember the statement of Mr. Lenz,[16] that they originated in conservatism, but that they are tending to radicalism or to socialism, with the chances very decidedly in the direction that socialism will come to control the trades union instead of the former conservative spirit. What is the truth about that, from your knowledge of the labor organizations?—A. As to the views Mr. Lenz holds, I think very little heed ought to be paid to them, for the reason that the views he now expresses are not the views he expressed a year or two ago, and what he expressed a year ago he expressed views contrary to and acted in conflict with a few months ago, and he has turned

another somersault here. I would rather answer an independent question that you may propound than answer or endeavor to correct any statement that that person has made.

Q. Well, I did ask you a question, whether, from your stand-point, socialism was gaining control of the trades unions and becoming their actuating spirit?—A. I think not. Some few years ago, when the trades-union movement was down, or rather when, from the effects of the panic, the workingmen had not organized upon the present plan—had not adopted the benevolent and beneficial features, and consequently had become dispersed to a great extent—in that condition of things this element thought now was the time for them to capture the trades-union movement, and they did capture the unions in a few instances.

Q. But the general drift of the movement is conservative, merely remedial or protective to the laborer, and recognizes the existing order of society?—A. As I said, Mr. Chairman, the views of some individuals must not be confounded with those of whole organizations. Some of the men, of course, may not have high aspirations as to the future state of society, but, as I said, a large number of our able men, good men I believe, have convictions that the state of society under which we live, the competitive system, is not one that ought to last as the highest system of civilization that we can arrive at; yet they subordinate their theories or convictions to the general good, and many of them are regarded as very conservative, and so act.

COMPETITION TO BE DISPLACED BY COOPERATION.

Q. What would they call that condition of society, or condition of industrial trade, which they anticipate as the legitimate or desirable successor of the competitive system? What would be the term or phrase they would apply to it?—A. A social state.

Q. Well, "competitive" relates to the method of production. What change do they propose in that respect?—A. They propose a universal co-operative system to supplant and be a substitute for the competitive system.

By Mr. George:

Q. What is the feeling on the part of the laborers of the country towards their employers; is it one of confidence, unity, and good will, or one of distrust, suspicion, and enmity? Explain, as well as you understand it, what the feelings and sentiments of the wage-receivers are towards the wage-payers.—A. I think I have answered that question in my previous testimony. I have quoted a statement made by a Massachusetts manufacturer, a member of the legislature, when he

said in substance, "I regard my employes the same as I would an old machine, which, when it becomes rusty, I thrust into the street."

Q. That was an expression of the feeling of the employers toward the employed. Now, I ask you about the feelings of the employed towards the employer.—A. Well, I think I said then, in substance, that the employed, the workingmen, believe that that is about the view generally entertained by employers as to their help, and they resent it by about the same feeling—not quite the same, but they resent it and feel it strongly.

Q. Is the present tendency of affairs to increase or intensify that feeling, or to remove it?—A. To intensify it. The views are gaining upon every side that the classes in society are becoming decidedly more distinct, and, as the lines are drawn, so does this feeling become intensified; in fact, I believe I can best describe that by reading from the preamble adopted by the Confederation of the Labor Unions of the United States and Canada at Pittsburg, Pa. (of which I am at present chairman), in November, 1881:

> Whereas a struggle is going on in the nations of the civilized world between the oppressors and the oppressed of all countries, a struggle between capital and labor, which must grow in intensity year by year, and work disastrous results to the toiling millions of all nations, if not combined for mutual protection and benefit, for the history of the wage-workers of all countries is but the history of constant struggle and misery engendered by ignorance and disunion; and, whereas the history of the non-producers of all ages proves that a minority thoroughly organized may work wonders for good or evil, and so on.

That extract, I believe, sets forth in as few words as possible the feeling that prevails among the working classes, that there is an ever-recurring conflict between the two classes, and that the employers are ever on the watch to see whether they cannot take advantage of their employes, the same as the stock speculator looks at the ticking of the indicator to see whether he cannot take advantage of those with whom he is dealing; except that in this instance the fellow stock-broker is generally as alert to take advantage as the first, and is at the other end looking at the wire with the same object, while the workingman is not so vigilant in looking out for himself.

Q. Explain fully, if you can, the nature and extent of the social intercourse between the wage-receivers and the wage-payers.—A. Where the wage-payers are small manufacturers, employing one or two hands, a spirit of cordiality and friendship may exist; but as the employer engages a larger number of hands, just in proportion does he become removed from the social status of his employe. His own status is raised, and the intercourse which may have formerly existed

is cut off. The ties of friendship gradually become lessened, and there is no cordiality existing between the two after that. The workingman works as well as he possibly can to retain his employment, and when he looks after the interests of his employer it is because he wishes to retain his employment, and not because he loves him.

By the Chairman:

Q. I would like to draw out your idea a little more definitely as to the exact thing that is to be accomplished before you undertake to state your remedies. All concede, I suppose, that whatever is to go to the capitalist or to the laborer must be derived from the sale of the article produced. That is so, is it not?—A. Decidedly.

Q. Now, is it or is it not your claim that, before any portion of what is produced goes to the capitalist, a full and fair remuneration to the laborer should be deducted, upon the ground that his reasonable and just claim is primary to all others. That is the main point, is it not?—A. Yes, sir.

Q. That is the first thing to be secured—that the laborer's remuneration shall take precedence, in the order of time, of any compensation to the capitalist?—A. Yes, sir.

Q. And the next thing you wish to secure is that the amount of his compensation shall be just and reasonable?—A. Yes, sir.

Q. Those are two objects that are to be attained?—A. Yes, sir. And further, the treatment of the men in the factories—not as slaves, but as men. The machinery is guarded against rust, and, when passing, if one of the arms or wheels or belts is rather lower than the other, the employer will take off his hat and pass beneath it. We do not ask the employer to take off his hat to his employes, but we do say that "good morning" will not hurt him, more especially when he is spoken to.

### ARBITRATION.

Q. Proceed with the statement of the remedies.—A. I would like to say a few words first with reference to arbitration. I am in favor of arbitration when that can be accomplished; but, as Mr. McGuire said here yesterday, arbitration is only possible when the workingmen have, by the power of their organization, demonstrated to the employers that they are the employers' equal. Arbitration may then be successful. But there are certain conditions in arbitration to which I wish to call attention, because, though there may appear to be a fair arbitration, yet the conditions may make it unfair to the laborer. For instance, the employer bids in the market for a contract for the production of goods. He bids lower than any other manufacturer and

obtains the contract; not necessarily for the public service, but for the production of something for private consumption or use in the regular channels of commerce; for instance, the production of certain articles of clothing. Or he bids in the market, whether by contract or otherwise, and accepts an order for the production of a hundred thousand suits of clothes at a certain price, or a hundred thousand tons of pig iron at a certain price. He has calculated, "My machinery will cost so much; so much for my plant; so much for my raw material, and so much for my labor. This price will leave me a margin." He knows that he cannot get his plant cheaper; he knows he cannot get his raw material cheaper. What will he do in such a case? He will say: "There is no margin for me there, and how can I sell unless I sell somewhat cheaper than these other men, I being new in the market." So, instead of paying his laborers as much wages as they may have received in other factories, or as he may have paid them before, he says: "In consequence of my having taken this contract at this low figure, my employes will have to submit to a reduction." Now, if the employes have demonstrated their equality in power with the employer, this matter is arbitrated upon, and, if it is really a question between the employes of this man and this individual man, arbitration, even though it be a fair arbitration as between these two particular parties, will really be unfair to the workingman. The arbitrators, looking only at the particular case, will probably allow the reduction of wages; yet that will be really unfair, because if the workingmen in that case submit to the decision it will mean a general reduction of wages throughout that entire industry. Now, I hold that under such circumstances arbitration, to accomplish its true purpose, should require that this manufacturer, if he is desirous of procuring contracts or the production of goods at a lower rate than other manufacturers would produce them for, and desires to draw his profit out of the laborer, that such a man, out of regard to both the honest manufacturer and the laborer, ought to be crushed out of business. I merely wished to add that consideration to what Mr. McGuire has already stated in reference to arbitration.

By Mr. George:

Q. Suppose he was crushed out of business, what would become of his employes?—A. They would find work from another employer who would get the same order, and the business would not be undermined by this new fledgling.

By the Chairman:

Q. The trades unions, then, will have a tendency to eliminate the men from the class of employers who are not fitted to be employers either by natural qualifications or by the possession of capital or

credit?—A. Yes; provided such men are desirous of having their work done cheaper than others.

Q. You do not get the point of my question. I say that the natural tendency of these trades organizations, as you understand it, is to eliminate from the class of employers those who, by lack of capacity or of credit, are really unfit to act as employers, and who, if they become employers, naturally are a burden to the market?—A. Yes, sir.

Q. Do you think that if society chose to make regulations as to who should be permitted to work at any trade or manual employment it would be equally within its power to establish rules as to who should act as employers? If a man without capacity and without credit assumes to act as an employer, is he not likely to create quite as great a demoralization in the community as a man undertaking to work as a workman or a professional man who is unfit for his work?—A. Yes, sir. But I have my doubts whether the General Government could regulate who should and who [should] not be employers, except in dealing with lunatics, and then a sheriff's jury would send the man to a lunatic asylum and assign the care of his property to an administrator.

Q. Which do you think does the most harm in society, the man who is unfit to set up in business and who consequently fails and produces panic and bankruptcy, or the individual laborer or professional man who undertakes to act in any given capacity without a proper preparation?—A. I think if you were to eliminate from the class of employers the incompetent men, you would not even then prevent the panics which periodically come. I think that panics come, not through those men's idiocy or incompetency, but are attributable to causes not generally understood.

REMEDIES.

The remedies that I suggest, and which I think the Government can and ought to adopt, are the following:

(1.) Strict enforcement of the national eight-hour law. The workingmen of this country, in all their organizations where they have come together, either in private or in public, either as local, State, National, or confederated unions, have set forth that demand for the enforcement of the national eight-hour law.

(2.) The passage of a law[17] by Congress giving the trades and labor unions the right to become chartered under the general laws of our Government. The laws written and now in operation to protect the property of the capitalist and the moneyed class generally are almost

innumerable, yet nothing has been done to protect the property of the workingmen, the only property that they possess, their working power, their savings bank, their school, and trades union; and we ask that our existence as organizations may be legalized, not for the purposes of strikes, as has been said, but for such reasons and objects as have been recognized in England and France. . . .

. . .

A measure which we think would give sufficient authority and assurance for the legalization and incorporation of trades unions would be covered perhaps by the bill introduced by the chairman of this committee at the last session in the United States Senate.[18] We think that bill would cover the ground, and we ask the adoption of that by Congress. We say that if men who are organized for the purposes of dealing in "corners" in grain and in the other farm products of the country, those who are speculating upon the hungry stomachs of the world, if such men can be incorporated, how much more are the laborers of the country entitled to be considered in the general legislation of the country.

(3.) We ask also, for the purpose of procuring information for the legislators of our country (who frequently find a very good excuse for non-action by saying that they are ignorant as to the true condition of the working people), the establishment of a national bureau of labor statistics. Such a bureau would give our legislators an opportunity to know, not from mere conjecture, but actually, the condition of our industries, our production and consumption, and what could be done by law to improve both. Our State governments would undoubtedly follow the lead of the national Congress, and legislate in the interest of labor; but we see that so long as our national legislators have an excuse for saying that they do not know the condition of labor, there is very little chance of obtaining legislation.

There is a bill of whose provisions I know but little, but I have been requested by Mr. Richard Powers, the president of the Lake Seamen's Union, to speak of it. It is a bill which was introduced in the House of Representatives by Mr. McLean,[19] of Missouri, and upon which that gentleman delivered a speech in the House on the revival of American shipping. He moved an amendment to House bill 7061 which covered the legislation desired by the Lake Seamen's Union.

I know that they have considerable grievances which they have brought forward, and I should regret very much if Mr. Powers should not have an opportunity of appearing before this committee. I believe he will be in this city next week to attend our confederation as a delegate.

There are several other measures to which I might call attention and which I might suggest as remedies, but the best organized trades unions of the world are eminently practical. They are composed of men who are desirous of obtaining reforms by gradual means, and in that spirit we ask the adoption of these measures which I have set forth here, because we believe and know that they will redound to our benefit as workingmen and to the benefit of society. If the legislators of this country are desirous of acting in this matter and alleviating the distress that is too prevalent, and if they desire to assist those who are working in this cause to mount a step higher, let them adopt these measures and they will receive the thanks of the working people of the world and of all posterity. But in any event, they ought not to continue to be so indifferent to the condition of labor as they have been in the past.

Adjourned.

U.S. Congress, Senate Committee on Education and Labor, *Report of the Committee of the Senate upon the Relations between Labor and Capital. . . ,* 48th Cong., 4 vols. (Washington, D.C., 1885), 1:361-82.

1. The New Jersey Bureau of Statistics of Labor and Industries was created on Mar. 27, 1878.

2. The *Fourth Annual Report* of the New Jersey Bureau of Statistics of Labor and Industries (Somerville, N.J., 1881), pp. 80-81, contains an abstract of the U.S. Department of State report *Commercial Relations of the United States,* no. 12 (Washington, D.C., 1881).

3. The Amalgamated MULE Spinners' Association.

4. P. J. McGuire testified on Aug. 17.

5. The Ohio State Bureau of Labor Statistics was created on May 5, 1877, and issued its first annual report on Dec. 20, 1877.

6. Alexander MacDonald.

7. The Brotherhood of LOCOMOTIVE Engineers.

8. Actually 1882. On June 1, 1882, the National Amalgamated Association of Iron and Steel Workers struck for a fifty cent raise. The manufacturers successfully resisted the strike, which was concentrated in the Pittsburgh district, and on Sept. 20 the workers returned to work at the old scale.

9. Probably the *Fourth Annual Report,* pp. 118-21.

10. The National AGRICULTURAL Labourers' Union.

11. Joseph ARCH was a founder of the National Agricultural Labourers' Union.

12. The Brotherhood of TELEGRAPHERS, KOL District Assembly 45, unsuccessfully struck the Western Union Telegraph Co. from July 19 to Aug. 11, 1883. In the aftermath, the telegraphers were critical of the KOL's poor support of the strike and withdrew their organization from the Order.

13. The Telegraphers' Protective League struck unsuccessfully against the Western Union Telegraph Co. in January 1870 to protest wage reductions and the discharge of a union member.

14. In September 1882 the Brotherhood of Locomotive Engineers and the Gould-owned Southwestern railroad system averted a strike by arriving at a settlement that

included a wage increase and the establishment of 100 miles a day or less as a standard day's work, with overtime pay for more than 100 miles.

15. George Douglas Campbell (1823-1900), duke of Argyll. *The Reign of Law* was published in 1867.

16. Charles Lenz, a German-born journalist living in New York City, edited *Capital and Labor*, a national manufacturers' journal. In his testimony before the committee on Aug. 15, he depicted the labor movement as divided among trade unionists, socialists, communists, and those in favor of establishing cooperatives; he contended that the major question facing the movement was whether the socialists or the trade unionists would control it.

17. It was not until 1886 that Congress provided for the voluntary incorporation of national trade unions (U.S. *Statutes at Large,* 24:86).

18. S. 2474 (47th Cong., 2d sess.), introduced in 1883, did not become law.

19. James Henry McLean (1829-86), a Republican congressman (1882-83), sought unsuccessfully to amend H.R. 7061 (47th Cong., 2d sess.) in January 1883 to require all ships in the merchant marine to maintain proper sleeping accommodations and comfortable quarters for seamen.

New York, August 27, 1883.

Samuel Gompers again appeared and made the following statement:

Mr. Chairman and gentlemen of the committee: At the recent Congress of the Federation of Organized Trades and Labor Unions, held in this city, a resolution was adopted directing me, as president of the association, to appear before you and make a statement in the name of organized labor.

### Conditions Imposed on Employes by the Western Union Telegraph Company.

The subject-matter of the resolution and of the statement I am to submit is the fact that the Western Union Telegraph Company has made it a condition precedent to the employment of operators and line-men that they shall be oath-bound to the effect that they do not belong and will not ever belong to any labor organization, the Telegraphers' Brotherhood in particular; not merely the brotherhood, however, but any organization that may have for its object the regulation of wages or hours of labor, or other conditions of employment. This oath, we are informed, has been made a condition precedent to the employment of these men by this corporation, and I have been instructed to bring the matter to the attention of this committee.

By Mr. George:

Question. Do you understand it to be an oath that is required of the men, or simply a written obligation?—Answer. A written obligation to which they swear. They may not be called upon to make a positive affidavit and to go through that regular form, but they are

compelled to sign their names to this obligation with all the solemnity necessary to such a transaction, before employment will be given them by the Western Union Company.

Q. You do not understand, however, that an actual *oath* is administered to them, a religious oath, do you?—A. I don't know whether that is positively enforced or not. The delegates assembled there in our meeting were of the opinion that if a corporation could prescribe the industrial associations or organizations that its employes should or should not belong to, as a condition precedent to their employment, the same right might equally well be extended to political organizations, and these corporations might go on to prescribe what political party or association a man might belong to as a condition precedent to his employment. If the laws of the United States inhibit, as they do, any attempt to intimidate a citizen in his political rights or opinions, we claim that it is equally proper that an employe should be protected in and should be permitted to exercise his right of association with any organization that seeks to accomplish an amelioration of his condition or of the condition of workingmen generally. We believe that if it is punishable by our laws for an employer or any other person to attempt to intimidate anybody in his employ, or to interfere with an employe's political rights as a citizen, it should be equally punishable for an employer to attempt to prescribe the organizations of any kind to which an employe should or should not belong. This being the view of our body, I was delegated to lay the matter before this committee, and to express our hope that you will agree upon some measure by which such things may be prohibited and prevented in the future. We believe that so long as a worker, man or woman, performs the appointed labor to the satisfaction of the employer and exhibits good behavior during the employment, that is all that should be required. These, we believe, are the essential conditions upon which workingmen or women should be employed; and the conduct of a workingman or workingwoman outside of the shop, so far as belonging or failing to belong to any organization, is not and should not be made a condition of employment. That is all that I wish to say upon this point.

When I testified here before there were one or two things which I promised that I would ascertain and state to the committee, and I am now ready to give the committee that information.

### Strike in the Lawrence Mills.

The strike in the Lawrence mill (in Massachusetts), of which I spoke in my former testimony, was against a proposed 10 per cent. reduction

of wages and it was lost. I had forgotten that when I testified in regard to it the other day.

WAGES ON THE ELEVATED ROADS IN NEW YORK.

With respect to the employes on the Metropolitan Elevated Railroad, I am informed that they are receiving an average wage of from $1.25 to $2.00 per day.

PROFITS OF RAILROAD COMPANIES.

I was asked also in reference to the profits of the companies owning and operating railroads in the State and in the city of New York. Some two weeks ago, either the railroad commissioners or the superintendent of insurance, I am not sure which, published the annual statement of the earnings of the several railroads, and that statement is easily obtainable from the officer whom I have designated, or I may be able to obtain a copy myself and furnish it to the committee.

The Chairman. We shall be glad to receive the information.

The Witness. If I can obtain it I will send it to the committee. I have nothing more to say at present.

U.S. Congress, Senate Committee on Education and Labor, *Report of the Committee of the Senate upon the Relations between Labor and Capital. . .* , 48th Cong., 4 vols. (Washington, D.C., 1885), 1:685-87.

## An Excerpt from an Article in *John Swinton's Paper*

[October 21, 1883]

THE GREAT TRADE UNIONS.

. . .

PROGRESSIVE CIGARMAKERS' UNION NO. 1.

If success in carrying out the objects for which men band together in Trades Unions entitles an organization to the first place among its fellows, then Progressive Cigarmakers' Union No. 1 of this city should take that place; for the work done by that organization in bettering the condition of its members has been simply marvelous. Though not quite eighteen months old, the Union has succeeded in increasing the pay received by cigarmakers of all grades; it was a potent factor in bringing about the abolishment of the tenement-house cigar manu-

facturing system, and it has forced unscrupulous manufacturers to acknowledge that workingmen have rights which bosses must respect.

The Progressive Union was organized because it was a necessity. Originally, the leading spirits of this new organization were members of the International Union No. 144, and the cause of their withdrawal from that body, primarily, was politics.

The more radical members of the old Union had become dissatisfied with the poor policy of begging favors from the politicians of the old parties, and proposed, in order to secure labor legislation in the future, that independent candidates be put in the field. The conservative element opposed this. But the climax was reached when last year the President of the International organization, in violation of the laws of the Union, as was afterward shown, suspended a newly-chosen President of No. 144, who had been elected by a large majority of the members.

Thereupon the Progressive Union on July 17, 1882, was formed. Its growth was very rapid, drawing for its members upon the old organization, until last April, when it numbered about 2,600 members. About that time the Progressives resolved to ask for an increase of $1 per thousand for making cigars of all grades, and by energetic agitation they induced the cigarmakers in other cities of the United States to make a similar demand. It was a hard struggle in this city, lasting over six weeks, but the Progressive Union finally triumphed. Two remarkable things about this strike were that, instead of losing members and exhausting the treasury—which has generally been the experience of all Unions during a strike—the Progressives came out of the fight with nearly 2,000 more members than they had when they went into it and with $1,000 added to their funds.

Their success in this matter placed the Progressives at the head of the labor unions of this city.

One of the results of this successful strike was the organization by fifteen cigar firms of the Employers' United Manufacturing Association,[1] its object being to check, if possible, the growing power of the Progressive Union. Soon after its organization a member of the Employers' Association made certain changes in his method of paying off his hands which were so manifestly unjust that a strike ensued, which, by a vote of the Employers' Association, shortly afterward was changed to a general lockout, thus throwing out of work over 2,500 cigarmakers.

This was a critical time for the Progressives. In the fight they had not only the bosses to contend with, but also the old Cigarmakers' Union. It is not necessary to tell the story again here, for all who take an interest in Trade Union matters remember the circumstances.

However, the Progressives again came out victorious, after a contest lasting nearly a month, and this time with a membership of nearly 8,000! And an increased treasury also!

Since the tenement house law[2] has gone into effect, though the Progressive Union is constantly taking in large numbers of new members, the membership has decreased to 7,500. But this can be accounted for by the fact that a great many of the men are going to other cities to work. In the Secretary's office is a notice that members of the Progressive Union desiring to leave the city will be told where they can receive plenty of work at good wages. There are over 12,000 cigarmakers in this city. As stated above, the Progressive Union has 7,500 of them, the Internationals about 600, and about 4,000 are non-union workers. The Progressives are actively and continuously working to draw these latter into the fold, and if we are to judge by the past, not more than a year will go by before the cigarmakers of New York City will be organized into one compact body.

The manufacturers are also organizing, having now three distinct protective associations[3] of their own.

Last Spring delegates from a number of Progressive Unions throughout the country met in this city, and organized the Progressive Cigarmakers' Union of America, and now twenty Unions are represented in it, hailing from New York City, New Haven, Brooklyn, Hoboken, Philadelphia, Union Hill, Wellsboro, Pa.; Boston, Springfield, Cigarville, N.Y.; Packers' Branch, New York City; Wilkesbarre, Chicago, Milwaukee, and elsewhere.

The executive officer and Secretary of Progressive No. 1 is Mr. Vincent W. Woytisek. The office is at 156 East Fourth street, where he or his assistants are always on hand to give information to people who earn their living by cigarmaking.

*John Swinton's Paper*, Oct. 21, 1883.

1. The United Cigar Manufacturers' Association (UCMA) was organized in New York in late 1882. Between 1882 and 1883 it had about fourteen members, most of them large manufacturers with more than 200 employees. In late 1886 it amalgamated with the American Cigar Manufacturers' Association.

2. On Mar. 12, 1883, the New York legislature passed a bill (Laws of 1883, ch. 93) prohibiting the manufacture of cigars in tenement houses. The law took effect on Oct. 1, 1883, but was declared unconstitutional in February 1884.

3. Probably the UCMA, the New York Cigar Manufacturers' Association, and the Tenement House Manufacturers' Association.

## A News Account of a Mass Meeting of Cigarmakers in Philadelphia

Philadelphia, April 8. [1884]

PHILADELPHIA CIGARMAKERS IN SESSION.

A mass meeting of cigarmakers of this city was held last night at the Assembly Building, Tenth and Chestnut streets, under the auspices of the Cigarmakers' International Union, No. 100,[1] of this city. I. W. Bisbing[2] presided. He said that the meeting had been called for the purpose of strengthening the hands of the International Union in this city. Addresses were made by W. M. Eggleston,[3] of Baltimore, the General Organizer of the International Union, and S. Gompers, of New York, both of whom declared that the best way to prevent misunderstandings between employers and employes was by the organization of the latter in one body that could speak with authority. This was the purpose of the International Union.

*United States Tobacco Journal,* Apr. 12, 1884.

1. CMIU 100 of Philadelphia was organized and chartered in 1881; it consolidated with CMIU 165 of that city in 1902.

2. Ingham Wood Bisbing (1838-1921), president of CMIU 100 of Philadelphia and active in the Philadelphia Central Labor Union (CLU), was born in the city and played an important role in the Greenback-Labor party there. In April 1885 he was elected treasurer of the CLU, resigning in July to become a CMIU organizer. For the next three decades he was active in CMIU organizing.

3. William M. Eggleston (d. 1885), a member of CMIU 11 of Baltimore, was appointed as the CMIU's first full-time organizer in 1883.

## A Translation of Excerpts from a News Account of a Meeting of Cigarmakers in New York City

[July 25, 1884]

UNIFICATION AND COOPERATION.

On Sunday, the 20th of July, a large joint meeting of the "Progressives" and the "Internationals" was held in New York in order to discuss orally and eye-to-eye, within a larger circle of members from both camps, the question—"unification or cooperation?"—that has been so fervently debated by both sides on the letter page of the "New Yorker Volkszeitung." The meeting had been called by the

Progressive Union in accordance with a resolution it had passed a week earlier. Consequently, more than 2,000 cigarmakers from both unions came to the meeting in the Germania Assembly Rooms.

. . .

SPEAKERS OF THE INTERNATIONAL UNION:

Gompers: "The members of the Progressive Union seem to have some mistaken ideas about the Internationals. Whoever joins the International Union can have any political views he wants; if the Progressives are really as progressive as they say, they have an immense field for their ideas within the International." He said his union could not espouse mere cooperation because it absolutely could not recognize another union. If it did so, he said, then any CMIU local in the country could dissociate itself—partly or entirely—from the International when it felt like it and rely on cooperative action when the need arose. The group whose finances stretched the farthest would have to support the rest in order not to lose a strike. "You believe in independent politics; we, as you like to say, in pressure politics. Let us give up our politics for a while and fight together against the evils within our trade, if you are honest and willing to do so. It is time for peace with honor and a unity under one banner. In the quarrels so far, only our bosses have won, only they have been pleased. Do your grand principles give you any comfort in this matter? I call upon you again: let us unite under the banner of the International Union!"

. . .

*Progress* (New York), July 25, 1884. Translated from the German by Dorothee Schneider.

## A News Account of a Meeting of the New York State Workingmen's Assembly[1] in Albany

[January 25, 1885]

On Tuesday morning last the delegates to the Assembly met in Albany and listened to the address of President Thayer.[2] New York City was well represented, especially the Printers. Phil. Scannell,[3] delegated from the Concord;[4] "Doc" Kean,[5] of the Amalgamated Trades and Labor Union; "Dave" Kells,[6] Donovan[7] and Finn[8] of No. 6,[9] and Berry[10] of the Faustis. John Devitt,[11] McMackin[12] and Daly[13]

were there from the Operative Painters.[14] Sam Gompers and Gutstadt, of the Cigarmakers' International Union; F. N. Eppenetter[15] of the United Varnishers;[16] Dan Clancy,[17] of the Granite Cutters, and Geo. Blair.[18] From Brooklyn, John P. Eagan[19] and John J. Hopkins,[20] of the Hat Makers and Finishers;[21] William Barry[22] and John Smith[23] of the K. of L. A long list of resolutions and bills, covering almost every subject discussed by Trades Unionists, was introduced and referred to their respective committees. Among them were factory and child labor, convict contract labor, incorporating Trades Unions, providing for a State printing office, mechanics' lien law, and life and limb bill, against the "truck system," in favor of paying wages not less than monthly, and to place the Labor Bureau at Castle Garden[24] under the charge of Trades Unions.

The Assembly resolved to call on Congress to speedily resume operations on public works.

Several committees reported during the week, and their reports were adopted.

The convention adjourned *sine die* on Thursday evening, after electing Walter N. Thayer, President; Samuel Gompers, New York, First Vice-President; Charles E. Renuck,[25] Oneida, Second Vice-President; John Phillips,[26] Brooklyn, Secretary; W. H. Coughlin,[27] Albany, Treasurer.

*John Swinton's Paper,* Jan. 25, 1885.

1. The New York State Workingmen's Assembly (NYSWA) was organized in 1865 to lobby for legislation beneficial to the state's working people. The Assembly accepted KOL and trade union affiliates. In 1897 it united with the New York State Branch of the AFL under the name Workingmen's Federation of the State of New York. The Federation changed its name to the New York State Federation of Labor in 1910.

2. Walter Nelson THAYER, a Troy printer, was president of the NYSWA from 1884 to 1886.

3. Philip J. Scannel, a compositor born in New York in 1845, was president of the Concord Cooperative Printing Co., a member of International Typographical Union (ITU) 6, and master workman of KOL District Assembly 64.

4. The Concord Labor Club, founded in 1882, was composed of members of ITU 6 of New York City.

5. Edward J. KEAN, a New York City printer, was president of the Amalgamated Trades and Labor Union of New York and Vicinity (ATLU).

6. David Kells (1838?-92) was an Irish-born Brooklyn printer and leader of ITU 6, serving as its president in 1879-80.

7. Florence F. Donovan (1850?-1910), a printer born in New York, was active in ITU 6 and in state organizing for the ITU. He was master workman of KOL District Assembly 64 in 1887 and, from 1886 until 1893, he served as commissioner of the New York State Board of Mediation and Arbitration. Later he became a hotel proprietor.

8. John J. Finn, a New York City printer, was elected vice-president of the ATLU in 1882 and worked as a clerk after 1886.

9. The New York Printers Union was organized in 1850. Two years later it participated in the formation of the National Typographical Union and became local 6.

10. Thomas Berry, a Brooklyn printer, represented the Faustus Labor Club, a Brooklyn printers' society.

11. John Devitt, walking delegate of the New York Mutual Benevolent and Protective Society of Operative Painters, was born in Ireland in 1847. He represented the painters in the New York City Central Labor Union (CLU) and during 1886 was a member of the executive committee directing the Henry George mayoralty campaign.

12. John McMACKIN, a painter, was active in the New York City CLU.

13. Francis Daly represented the New York Mutual Benevolent and Protective Society of Operative Painters.

14. The New York Mutual Benevolent and Protective Society of Operative Painters was organized in 1860.

15. Frank N. Eppenetter, a varnisher and later an upholsterer, was born in New York City about 1858. During 1885 he served as walking delegate for the United Varnishers of New York and Vicinity and as secretary of the New York City Amalgamated Building Trades Council.

16. In December 1884 Varnishers Union 18 withdrew from the International Furniture Workers' Union of America, complaining of a high per capita tax, and reorganized independently under the name United Varnishers of New York and Vicinity.

17. Daniel Clancey (variously Clancy), a member of the Albany branch of the Granite Cutters' National Union, represented its New York City branch at the NYSWA from 1885 to 1887.

18. George BLAIR, a boxmaker and president of the NYSWA from 1874 to 1883, was active in politics as a Greenbacker and as a Democrat for several decades.

19. John P. Egan (1816-99) was president of the Brooklyn Hat Finishers Association. In 1878 he ran unsuccessfully for the New York legislature as an independent candidate. He became a clerk in 1886.

20. John J. Hopkins (1850-1918), an iron molder in Albany, N.Y., had emigrated from Ireland. He represented KOL Local Assembly 2791 of Albany at the 1885 NYSWA, not the Brooklyn Hat Finishers' Association. Beginning in 1900, Hopkins worked as a fireman in the New York State Capitol's boiler house.

21. The Brooklyn Hat Finishers' Association was organized in 1868.

22. William J. Barry, a Brooklyn painter and later a real estate agent, represented KOL Local Assembly 3170.

23. John Smith represented the Brooklyn Mutual Union.

24. Castle Garden, located at the southern-most point of Manhattan, was an immigrant-receiving station from 1855 to 1892, when it was replaced by the facility at Ellis Island. From the 1850s the German Society of New York, in cooperation with the Irish Immigrant Society, operated an employment bureau at Castle Garden for immigrants of all nationalities. The service was free. In the 1880s over a thousand workers a month, men and women, found employment through the bureau.

25. Charles E. Remick (b. 1840), a dry goods merchant in Oneida, N.Y., represented KOL Local Assembly 1791. In 1889 he served as president of the village of Oneida and from 1890 to 1893 was sheriff of Madison County, N.Y.

26. John PHILLIPS was secretary of the National Trade Association of Hat Finishers (1880-96) and the United Hatters of North America (1896-1904).

27. William H. Coughlin (d. 1893) represented IMU 8 of Albany. He was later a mail clerk.

# A Call for Establishing the Eight-Hour Day

Washington, July 7, 1885.

The Federation of the Organized Trades and Labor Unions of the United States and Canada to All Trades and Labor Unions throughout the United States and Canada, and to All Workmen Who Hope for Future Betterment:

We deem it important to call your attention to the resolution adopted by the 4th annual session of the "International Federation of Trades,"[1] fixing May 1, 1886, for the general adoption of 8 hours as a day's work. The multiplication and use of labor-saving machines makes it our first duty to shorten the hours of labor, if we would share in the benefits of their introduction, otherwise they must result in driving out the existence of free labor, the boast of an enlightened age and civilization. We hold that this is the only practical solution of a much needed reform, on the importance of which we all agree. The objections of waiting for a law effecting all workmen alike are serious, involving an invaluable right, tedious delay and a loss of faith in our own powers.

Your right to put into practical operation such law is superior and undisputed, and it only remains for you to say if the Eight Hour law passed by your representatives shall become a fixed rule of action from May 1, 1886.

The ratification of this measure by every local trade and labor union or assembly of Knights of Labor on this continent, publicly announced, would add to the impetus now gained and give it a prestige of success.

No fair minded intelligent employer will oppose it, because it is an undisputed fact that every measure which improves the condition of employees carries with it equal advantages to employers and the community as well. In support of this statement we refer to the fact that it was the high price and scarcity of labor in this country, due to the independence of American labor that stimulated the inventive genius of America to supply an obedient substitute, and made us a nation of inventors and manufacturers.

It is not difficult to follow this first cause to its logical effect, the

accumulate[d] wealth of our country, and to it we confidently look for still further development in this direction.

If this wealth is now unequally distributed, a large part of the blame must rest on our own shoulders, because of our failure to shorten the hours of labor.

In conclusion we ask you to remember this is our Eight Hour law, and upon us depends its failure or success.

Fraternally yours, G. Edmonston,[2]
Secretary.

*Progress* (New York), Aug. 21, 1885.

1. The FOTLU held its fourth annual session in Chicago, Oct. 7-10, 1884.

2. Gabriel EDMONSTON was one of the founders of the Brotherhood of Carpenters and Joiners of America and was a member of the FOTLU Legislative Committee from 1882 to 1886.

## To the Editor of *John Swinton's Paper*

Bureau of Statistics of Labor, New York, July 13, 1885.

To the Editor:

The Bureau of Statistics of Labor[1] has opened a branch office at 744 Broadway (Room 12), this city, where blank forms may be obtained by any person feeling interested in and desiring the furtherance of the objects for which the bureau was created.

The branches of inquiry for this year upon which the bureau has determined are "female labor," in all its phases, and "strikes," their causes, cost, duration and result.

Will you kindly publish the above in your valued paper and help the bureau in reaching a large body of people who are deeply interested in these questions?

Samuel Gompers,
Special Agent.

*John Swinton's Paper*, July 19, 1885.

1. In May 1883 the New York legislature created a bureau of labor statistics. SG helped establish the bureau's New York City office in 1885.

## *The Culmination of the Rivalry between the CMIU and the Cigarmakers' Progressive Union of America*

During the mid-1880s the rivalry between New York cigarmakers' unions embroiled the KOL in a struggle that eventually contributed to the broader rupture between trade unions and the Order. In 1885 the CMIU and the Cigarmakers' Progressive Union (CMPU) attempted unsuccessfully to reconcile their differences. A joint meeting in Rochester in October developed a merger plan that included amendments to the CMIU's constitution that would have curtailed the power of the union's president, Adolph Strasser.[1] Strasser, however, would allow CMIU locals to vote only on aspects of the proposal that required no constitutional changes, a condition that had been specified by the CMIU's 1885 convention, and the plan in this form could not gain approval from the leaders of the Progressive Union. Nevertheless, some Progressives seceded from the CMPU and joined the CMIU on January 1, 1886, under the terms of the Rochester agreement as modified by Strasser.

In the meantime the divided cigarmakers continued to pay the price of intense interunion rivalry. In November, in the midst of the two unions' consideration of the Rochester proposal, the CMIU broke a KOL boycott instituted on behalf of the CMPU against the firm of Straiton and Storm, and CMIU workers replaced Progressives who had been locked out by the firm. On January 2, 1886, the United Cigar Manufacturers' Association (UCMA) took advantage of the internecine struggle to promulgate a new price list. Intended to "equalize" the pay scale throughout the New York City cigar industry, it reduced the wages of higher-paid craftsmen while raising those of other cigarmakers, including tenement-house workers. While the two unions at first cooperated in opposing the new scale—on January 14 each struck a selected firm that was a member of the association—the UCMA countered with a general lockout, beginning January 20, that affected 15,000 workers. Moreover, cooperation between the CMIU and the CMPU quickly evaporated amid mutual recrimination. The CMIU's separate agreement with the firm of Kerbs and Spiess on January 18 and the Progressive's settlement with the manufac-

turers' association on February 11 placed the two unions at odds once again.

The continuation of the CMIU strike left the Progressives unable to supply enough cigar packers for the manufacturers' association shops, a provision stipulated by their settlement with the UCMA. Consequently, the manufacturers concluded an agreement at the end of February with KOL District Assembly (DA) 49. The district assembly agreed to furnish them with packers and the KOL label, while the manufacturers promised to cease tenement-house production by May 1 and to require their employees to join the Order. The CMIU reacted by protesting unsuccessfully to Terence Powderly and the KOL General Executive Board and by boycotting the KOL label. In addition, it sent Gompers, Fred Haller, and John Kirchner on a speaking campaign to inform local unions and KOL assemblies of DA 49's actions. Finally, it began publication of the *Picket*, a four-page weekly paper that Gompers edited for about six months beginning in the spring of 1886.

By the summer of 1886 the district assembly's inability to provide a sufficient number of workers for UCMA factories was jeopardizing its agreement with the manufacturers. DA 49's relationship with the CMPU was deteriorating as well. The assembly's leaders alienated the Progressives by first threatening to deprive any manufacturer of the KOL label for introducing machinery that reduced wages, but then allowing the use of machinery upon payment of a royalty to the assembly. More important in exacerbating tensions, however, was the district assembly's insistence that the Progressives give up their union and join the KOL.

In July the district assembly put increased pressure upon the manufacturers to have all their employees enroll in the Order. As a result, over 8,000 cigarmakers in UCMA shops struck that month in resistance to demands that they join the KOL. During August the CMIU and CMPU moved swiftly toward a merger. On August 7 representatives of the two unions jointly denounced DA 49, and, four days later, the CMPU's national convention voted to amalgamate with the CMIU. DA 49 carried the dispute to the KOL General Assembly that met in Richmond in October, and there secured the passage of a resolution that required cigarmakers to choose between membership in the Order or in the CMIU. This action brought to a head the growing tension between the CMIU and the KOL and played a significant part in precipitating the final rupture between the trade unions and the Order.

*Note*

1. It provided that union members at large elect the president, that he have no jurisdiction over issues involving his own local, and that a special committee control the union's journal.

# Excerpts from the Proceedings of the Sixteenth Convention[1] of the CMIU

[September 22, 1885]

. . .

Reps. Gompers and Kirchner[2] offered the following resolution:

*Whereas,* A disposition has been shown by the members of the Cigar Makers' Progressive Union to again become part of, and to consolidate with, the International Union, as evidenced by the adoption of resolutions at their recent Convention,[3] and the ratification of the same by their members, electing a committee for the purpose of meeting a like committee from this body; and,

*Whereas,* We deem the consolidation and organization of the cigar makers into one grand and powerful union of paramount importance to the entire craft and the cause of labor in general; therefore be it

*Resolved,* That a committee of five be elected by this body, to be provided with proper credentials to meet the committee above referred to, with power to enter into a conference with said committee with a view of agreeing upon questions at variance and the consolidation of both organizations.

*Resolved,* That the question upon which said committee may agree shall be referred to a general vote of the members, and if agreed to by a majority of the members voting, to be binding on all concerned.

. . .

Rep. Gompers moved to refer the resolution to a special committee of five, which was agreed to.

. . .

[September 23, 1885]

The Special Committee on Consolidation reported as follows:

The undersigned committee, specially appointed to act upon the resolution offered by Reps. Gompers and Kirchner, referring to the request of the Cigar Makers' Progressive Union of America, for consolidation with the Cigar Makers' International Union, beg leave to offer the following report with recommendation relative thereto:

1st. The committee is unanimously of the opinion that a consolidation of all members of the craft under one compact and harmonious bond of unity and fraternity is vitally essential for a further progress of the cause of labor reform, and we therefore recommend the resolution referred to this committee for adoption.

2nd. We further recommend that the Committee of Conference

(requested by the resolution referred to us), and which is to confer with a similar committee appointed by the Cigar Makers' Progressive Union of America, with a view of effecting a basis upon which to consolidate, shall be elected immediately after the election of the regular officers of the International Union.

3rd. We recommend that after the election of the said committee, the International President be directed to immediately send a communication to the Secretary of the Executive Board of the Progressive Union, notifying him of the action of this Convention.

Signed, Respectfully, Chas. Erb,[4] Chairman.
Geo. Birkenhauer,[5]
Samuel Gompers,
Alex. Roswog,[6]
John S. Kirchner, Secretary.

Report of the committee was adopted.

. . .

*Cigar Makers' Official Journal,* special issue, Oct. 1885.

1. The convention met in Cincinnati, Sept. 21-30, 1885.
2. John S. KIRCHNER, a founder and leader of CMIU 100 of Philadelphia, was CMIU fourth vice-president (1885-87) and secretary of the FOTLU (1886).
3. The Cigarmakers' Progressive Union convention in Philadelphia, May 10-15, 1885, adopted resolutions urging unity in the cigar industry and appointed a committee to meet with the CMIU.
4. Charles M. Erb (b. 1850), a cigarmaker born in Michigan, was president of CMIU 22 of Detroit during 1879 and its delegate to the 1885 CMIU convention. He served as treasurer of the Cigar Makers' Benevolent Association of Detroit and as an organizer of the Detroit section of the SLP. About 1900 he became a cigar manufacturer.
5. George Birkenhauer was corresponding secretary of CMIU 56 in Leavenworth, Kan., from 1885 to 1886.
6. Alexander Roswog (variously Roszwog; b. 1856), an Illinois-born cigarmaker, served as corresponding and financial secretary of CMIU 47 of Quincy, Ill., in the mid-1880s.

## To the Members of the CMIU

New York, November 13, 1885.

You are about to perform one of the most important functions, as members of our organization i.e., voting upon the amendments to the constitution of our organization, as proposed by the last Conven-

tion. In doing this responsible work you ought to exercise your deliberate and best judgment.

There is one amendment upon which you are called to vote, to which I desire to call your especial attention and calm consideration. It is the amendment to Sec. 2 of Art. XI.

The present section provides and has provided ever since the Union Label was introduced into our International Union, that the printed matter on it shall be as follows:

(COPY OF UNION LABEL.)

Sec. 2.—Issued by authority of the Cigar Makers' International Union of America.

UNION MADE CIGARS.

(LOCAL STAMP.)

All infringements upon this label will be punished according to law.

Prest. C.M.I.U., of America.

The proposed amendment makes a radical change in this, merely stating that the cigars in the box are "Union made cigars."

This change, I am convinced, is dangerous in the extreme to the future efficiency of the label, and neutralizes the good it has already accomplished. I will try briefly to point out my reasons for asking you to defeat this amendment.

1. The label is established, known and recognized, at a glance, as the International Union Label.

2. We would have to go over the same work of making known the new label that we did in establishing the popularity of our present one; during which time you will readily see how much ground we would lose.

3. The proposed change in the reading matter on the label would only appeal to organized labor and its sympathizers; while the present label not only secures this support, but appeals to those who care not a straw for organized labor, but who are naturally opposed to the filthy tenement-house and Chinese coolie-made cigars.

4. Our present union label is an agitator and an argument in itself. It is an appeal to all classes of society, and creates the feeling most required to make it a success. The proposed label would narrow down the field of our operations.

For the reasons set forth above, I appeal to the members to vote against the amendment to Sec. 2, Art. XI.; and thus avoid the loss for a long time to come of the good that has resulted from that true-blue union label.

I believe the Convention made a mistake in adopting this change, and many others with whom I have conversed upon the subject think

likewise, and I trust that the members will remedy the matter by voting the change down.[1]

Samuel Gompers.

*Cigar Makers' Official Journal,* Nov. 1885.

1. The CMIU members approved the amendment by a vote of 2,192 to 1,959.

## An Excerpt from the Minutes of the Legislative Committee of the FOTLU

Washington, D.C., Dec. 11, 1885.

The Legislative Committee chosen by the Fifth Annual Congress[1] of the Federation of Trades and Labor Unions of the United States and Canada met in Grand Army Hall, Washington, D.C., on the evening of the above date. Present: Samuel Gompers, W. E. Tomson,[2] P. F. McAuliffe,[3] Hugo A. Miller,[4] H. Emrich,[5] Gabriel Edmonston, W. H. Foster—7. Absent: Samuel S. Green,[6] George G. King[7]—2.

The Chair referred to the appeal for aid in bringing about the 8-hour rule, and asked if the committee desired to give the Secretary any instructions in reference to it.

Mr. Edmonston moved that an appeal be issued about the middle of January, accompanied by subscription lists bearing the seal of the Federation, and asking Secretaries to receive same for the Federation. Carried.

On motion of Mr. Emrich the Secretary was directed to issue first the 8-hour call, asking Secretaries to report the result of a canvas in their trades. Carried.

The same gentleman moved that with this call should be issued a form of agreement to be governed by the 8-hour rule, to be submitted to employers for signature.

. . .

Attest: W. H. Foster,
Secretary.

ADS, Minutes of Meetings, Legislative Committee, FOTLU, reel 1, *AFL Records.*

1. The FOTLU held its fifth annual session Dec. 8-11, 1885, in Washington, D.C.
2. William E. TOMSON, a Chicago machinist, was elected second vice-president of the FOTLU at its 1885 convention.
3. Patrick F. MCAULIFFE, a stonecutter from Washington, D.C., was elected third vice-president of the FOTLU in 1885.

4. Hugo A. MILLER served the FOTLU from 1882 until 1886 as its German secretary; he was secretary of the German-American Typographia from 1886 to 1894.

5. Henry EMRICH, elected sixth vice-president of the FOTLU in 1885, was secretary of the International Furniture Workers' Union from 1882 to 1891 and treasurer of the AFL in 1888-89.

6. Samuel S. GREEN, a Louisville printer, was elected first vice-president of the FOTLU at its 1885 convention.

7. George G. KING, a printer from Baltimore, was elected fifth vice-president at the 1885 FOTLU convention.

# An Excerpt from a News Account of a Mass Meeting in Washington, D.C.

[December 11, 1885]

## A LABOR DEMONSTRATION

. . .

Mr. Samuel Gompers of the International Cigarmakers' Union of New York, who was next introduced, said that his reason for entering the cause of trades unions was to assist in ameliorating and bettering the condition of the people of his own class—the workingmen of the world. The best refutation of the flings of such men as Judge Snell[1] was the grand meeting of honest toilers before him. Those who sneer at labor do a great injustice and evince a degree of ignorance that is in itself lamentable. If the troubles between capital and labor of to-day be compared with the troubles of twenty years ago the deduction will be made that there is now less violence and less bloodshed. Unionism has instilled into the minds of their members that more can be accomplished by deliberation than by wild outbreaks of disorganized labor. While the unions have a due regard for the safety of the property of others, they will not permit themselves or their rights to be trampled upon. The working man organized can compel the most arrogant employer to come to terms. [Applause.][2] There are too many among the working men who do not deem it well to reduce the hours of labor, lest the wages should be reduced accordingly. That was not necessarily the case; in fact the contrary was the truth. The rule is that the people who work the greatest number of hours receive the least pay. Take China and other countries where long hours are given to labor, and what is the state of things? Invariably low wages. Look at the people who get the large pay for their services; are they not the ones who work the least number of hours? The laborer on short hours has time to improve his mind and possess himself of a weapon

with which he can assert himself with justice. He is by the eight-hour rule enabled to devote necessary time to his family and bring up his children as they should be raised.

Referring to the Chinese, the speaker said that if there is a law on the statute books of this country against Chinese importation, in view of the recent developments that law ought to be rigidly enforced. If the working people fail to get just legislation from the Congress of the United States they must turn and seek it at the hands of the Labor Congress, which was in session this week in the shadow of the dome of the great United States Capitol. [Applause.]

. . .

*Washington Post,* Dec. 11, 1885.

1. William B. Snell (1822-90) was a Washington, D.C., police court judge from 1870 until his retirement in 1888. The 1885 FOTLU convention demanded Snell's removal from office because he "expressed himself in violent language towards labor unions when a suit was being tried before him involving the right of unions to employ pickets during strikes" (FOTLU, Proceedings, 1885, p. 15).
2. The brackets in this document are in the original.

## An Agreement between Kerbs and Spiess and CMIU Locals 10[1] and 144

New York, Jan. 18, 1886.

The following conditions have been agreed upon between the firm of Kerbs & Spiess, cigar manufacturers, and local unions under the jurisdiction of the International Union:

1st. The firm of Kerbs & Spiess agrees to employ only members of the International Union—cigarmakers, rollers, and bunch-breakers.

2d. The firm agrees to pay the annexed scale of prices for the term of one (1) year, seven dollars[2] to be the lowest price for any cigar to be made in the shop.

3d. The firm agrees to abolish their tenement-house cigar factories and to employ their workpeople in regular factories only.

4th. The firm agrees to make no general discharges or lay-offs on the upper floor where the better class of cigars are made without at the same time proportionate discharges on the lower floor where the cheaper cigars are made.

5th. The firm agrees to reduce the hours of labor during the slack season in case they should conclude not to discharge any workmen on both floors.

6th. The firm agrees to arbitrate with a committee appointed by the union in order to fix the prices on new jobs to be introduced in the factory.

7th. The representatives of the union agree not to object to any workman joining the International Union who belongs to another association of cigarmakers.

8th. The representatives of the union agree in case the employes in the factory violate the agreement to supply the firm of K. & S. with sufficient workmen for the management of business.

9th. The representatives of the union agree not to furnish the union label to any factory paying less than the firm of K. & S.

10th. The representatives of the union agree to allow the firm 120 rollers and the requisite number of bunch-breakers to be known as the inferior work, and to be employed on a separate floor, at the scale of prices hereto annexed, no cigar to be made for less than $7 a 1,000.

(Signed) Kerbs & Spiess.
Fred Haller,[3]
Union No. 10.
A. Strasser,
Samuel Gompers,
For Union 144.

*United States Tobacco Journal*, July 31, 1886.

1. The CMIU chartered Cigar Makers' Progressive International Union 10 on Jan. 1, 1886. It consisted of former Progressive cigarmakers who joined the International under those portions of the Rochester agreement that were recognized by the CMIU.

2. Per thousand.

3. Frederick HALLER was a leader of the Cigarmakers' Progressive Union and the New York City Central Labor Union.

# A Translation of an Excerpt from an Article in the *New Yorker Volkszeitung*

[February 19, 1886]

. . .

## PACKER MEETINGS

Yesterday morning another meeting of the former packers from Levy Bros.' shop was held in No. 80, Ave. C. After much discussion

about whether they should resume work or not, they decided to go back to work and to inform the company to this effect by means of a committee. Shortly thereafter, T. Gompers[1] appeared on instructions of the Int. Union and impressed upon the remaining packers in a 1½-hour long speech, that they should reconsider their decision and not go back to work until the International came to an agreement with Levy Bros. He also said, if they were dissatisfied with the support they had received so far, he would see to it that the support was increased to $25, because it was important, he argued, that the company respect the just demands of the Internationals. Afterwards it was decided not to go back to work.

. . .

*New Yorker Volkszeitung*, Feb. 19, 1886. Translated from the German by Patrick McGrath.

1. SG.

# A Circular Letter Issued by the CMIU Locals of New York City and Vicinity

[February 20, 1886]

Fellow Workmen:

The events of the past few weeks in this city in reference to the lockout of the cigarmakers, the accounts of which no doubt [you have] either read or heard, but we ask, we appeal to you, not to lay this circular aside, as many others are, but read it yourself, then read it to the organization with [. . . con]tents known to all those with whom you come in contact in the labor movement. Give it the widest publicity you can. It contains the truth, it exposes the knavery of people and a supposed labor organization who have presumed upon the credulity of the workingmen of the country.

On the 2d of January the Cigar Manufacturers' Association of New York posted a "Price List" in their fourteen shops. When the cigarmakers went to work on the 4th the reductions in wages were so flagrant that they all seemed bent upon leaving the shops in a body; in other words, what the manufacturers evidently designed, was about to be accomplished, that we were to be overwhelmed with a general strike. The leaders of the almost defunct Progressive Union seemed bent upon doing all they could to foment the flame by urging all to go out with a rush, while the best defined policy proved to be to

protest against the reduction in one shop, for according to the constitution of the Manufacturers' Association, this in itself would be sufficient cause for them to lockout all the cigarmakers in the other shops a few days later, which they did. You will readily see the good policy of this move. It is briefly this: If there would be a general strike we would be compelled to try and keep "scabs" out of the shops. By the manufacturers declaring a lockout, they involuntarily did this work for us.

The so-called Progressive Union, knowing their own weakness, asked the Central Labor Union[1] of this city to elect a committee to act with them in the lockout. That committee came to our meeting, and stated the object of their acting was to bring about harmony (?) of action among the cigarmakers' unions during this trouble.

We gave the committee all the information in our possession, revealed to them our plan of operations, and invited them to attend the meetings of our strike committee every evening to take part in our deliberations, adding: "Gentlemen, but by all means keep your hands off our difficulty, as we believe we understand and can better arrange the technical matters in our trade than any man working at another trade." This the committee of the Central Labor Union agreed to and expressed themselves as gratified at the information given and the treatment received at our hands. We told them also that we would defend the action of Levy Bros. cigarmakers in striking against the reduction, before a committee of the Manufacturers' Association on the following Friday (at which they were represented and took part), and that no important step would be taken by us without informing them thereof. To prevent any conflict we sent an invitation to the Progressives (?), requesting them to elect a committee to act and cooperate with our committee to fight the manufacturers. A committee put in an appearance, and instead of acting with us, put questions to our committee whether we would change the constitution of the International Union in several instances, when they were answered that that was beyond our power; but let us lay all differences aside at least during this trouble and battle against the manufacturers shoulder to shoulder. They persisted in demanding unconstitutional changes before they would do so.

You will readily see that they never intended to act with us, but merely used this as a pretence to betray the interests of the cigarmakers. But now mark the treacherous course of these people. Within ten days from the time the Central Labor Union committee made us the promise they did, they, with a committee of the Progressives (?), without so much as informing the cigarmakers involved in this gigantic strike and lockout, let alone not informing the Cigarmakers' Inter-

national Union, without mentioning a word, secretly held a conference with the Manufacturers' Association and agreed upon a price list which would reduce the wages of cigarmakers about $2.00 per week, as a compromise, instead as the manufacturers at first wanted, to reduce to the amount of about $2.25 per week (glorious result after a two weeks' battle). Our Union, knowing the treachery of the Progressives (?), (having had previous experience of their actions) had pickets stationed at the factories and at the manufacturers' headquarters (Grand Union Hotel). We were thus continually kept informed of the movements of the Progressives (?), Manufacturers' and Central Labor Union committees, and within an hour after this shameful compromise was effected we were in possession of the new "Price List." Energetic action was then necessary to defeat this bold, bad and disgraceful scheme to reduce our wages. We adopted the following resolutions, which were printed in all the morning papers:

*Whereas,* A committee of the Cigarmakers' Progressive Union and of the Central Labor Union are holding secret conferences with the Cigar Manufacturers' Association at Grand Union Hotel, without either being authorized or directed so to do by the striking employees of Levy Bros.;

*Whereas,* Said committees were assured by the Strike Committee of the Cigarmakers' International Union that no conference would be held with the manufacturers unless said committees were notified.

*Whereas,* The committee of the Central Labor Union gave positive assurance not to interfere in the management of the strike and lockout and to leave it in the hands of the cigarmakers' unions; therefore be it

*Resolved,* That we condemn the action of the Progressive and Central Labor Union committees as treacherous to the cause of labor, and to the cigarmakers especially.

*Resolved,* That we emphatically protest against any arrangement or agreement entered into by any parties in which the committee authorized by the striking employees took no part; and be it further

*Resolved,* That we insist upon the old scale of prices, and positively refuse to recognize any compromise involving a reduction of wages.

The publication of these resolutions had the desired effect. It was a bombshell in the ranks of the conspirators (Progressive and Central Labor Union and Manufacturers' committees) who determined to rob the cigarmakers of thousands of dollars every week in reduced wages. The conspirators had no time to recover from their astonishment and counteract the effect of the publication of these resolutions on the Progressive members who voted this scheme down. When this result was made known, one of the largest manufacturers said: "If I thought these Progressive leaders did not know more tricks I would have given them a pointer"; and subsequent events proved that he did.

When the scheme had been defeated we again presented a bold front to the manufacturers, and it seemed that we were nearer than

ever to successful resistance to the reduction in wages, and that victory would crown our efforts. Never in the history of the labor movement has success for the cigarmakers appeared so certain as in this strike and lockout, never was the prospects better, not a cigarmaker was in the shops. The cigar packers voluntarily left their places to help the cigarmakers in their struggle, thus rendering even the stock on hand unavailable. The manufacturers could not fill an order, but—scarcely had three days passed over our heads when these self-same Progressive and Central Labor Union committees again re-entered into a secret conference with the manufacturers, and agreed upon a basis for a bill of prices which [we?] desire you will note. They have agreed to equalize prices prevailing in the factories of the Manufacturers' Association. Now bear in mind, the highest price paid by any of them is not as high as several of the largest manufacturers in the trade, as for instance Straiton & Storm, Kerbs & Spiess, and many others. Yet there are only four manufacturers in the association who pay the highest among them, six who pay what will be the average or equalized prices, while there are four who pay the lowest. Now consider further that those four who pay the highest, employ over 1,400 cigarmakers in shops, while the four who pay the lowest prices employ only about 250 cigarmakers in shops, then according to this shameful compact, the 250 will be raised for the purpose of reducing the wages of 1,400 direct[ly,] while the great indirect loss will be, that all other manufacturers outside of the Manufacturers' Association will claim the right to have their work done at prices according to the equalized, or more properly styled, reduced prices.

We have done all we possibly could to prevent the Progressive (?) and Central Labor Union from committing this outrage on our trade. These Progressive (?) and Central Labor Unions have resolved that the cigarmakers shall work at a reduction of wages. The cigarmakers on the other hand have declared that they will not.

So dishonorably and treacherously have the Progressive and Central Labor Unions acted in this entire trouble, that the Executive Board of Cigar Packers have refused to jointly confer with them and the manufacturers.

In consequence of this shameful compact the manufacturers have declared the lockout at an end, they have grasped at this opportunity to crawl out of the most distressing position any number of employers ever got themselves into; they would have been compelled to make an unconditional surrender, to pay old prices was merely a question of a few days, and then as opportunity arose compelled the lowest to pay higher wages, to bring about a price list, but upon the basis of the highest not the lowest, but as one of the employers said: "The

only chance we have is with the Progressives, I know they will accept conditions the Internationals will never consent to."

We appeal to you to form your opinion of a body of men calling themselves a union to enter into a conspiracy with an association of manufacturers to reduce the wages of workingmen and women.

But true to the principles and record of the Cigarmakers' International Union, we will not submit to a reduction of our wages! We will not surrender!

For the purpose of concentrating our forces, we have declared three shops on strike where the largest reduction of wages will ensue from this so-called price list, namely, Levy Bros., McCoy & Co., and Browne & Earle.

We have resolved to support every cigarmaker, bunch maker and cigar packer of those shops whether they are members of the International Union, non-union, or even Progressive members, as long as the fight against the reduction lasts.

Our employed members have assessed themselves 10 per cent. of their weekly earnings.

We will strain every nerve and expend every cent at our command to bring victory to the cigarmakers even out of the jaws of defeat. Notwithstanding the conspiracy of the Central Labor Union, cigar manufacturers and Progressives (?) Union.

Hoping that you will fully conceive the nefarious scheme of these bold, bad men, relegate them into oblivion where they rightfully belong and take some action to give us your moral support to battle against wrong, more especially when combined with the oppressor.

We are respectfully yours, in the cause of labor.

Cigarmakers' International Union of America—Unions Nos. 144, 10, 141,[2] 87, 132, 131, 13[3] and 213,[4] district of New York and vicinity.

Boycott all cigars that do not bear the blue label of the Cigarmakers' International Union of America.

*United States Tobacco Journal,* Feb. 20, 1886.

1. The New York City Central Labor Union (CLU) was formed in March 1882 by workers who advocated independent political action in contrast to the Amalgamated Trades and Labor Assembly that SG had helped to form. After defeat in the fall 1882 election it concentrated until 1886 on economic action and the use of the boycott. In 1886 it managed Henry George's mayoralty campaign. In early 1889 socialist trade unionists, claiming the central body was under the influence of corrupt politicians, seceded to form the Central Labor Federation (CLF), which the AFL chartered; before the end of the year the two groups reunited. In mid-1890, however, the socialist faction again broke off and reorganized the CLF. The CLU refrained from political action after 1892 and in January 1899 merged with the CLF to form the Central Federated Union; the new body affiliated with the AFL and was represented at the 1901 convention.

2. The CMIU granted a charter to local 141 of New York City in 1882. Its members were Bohemian.

3. The CMIU chartered local 13, composed of German-speaking New York City cigarmakers, in 1882.

4. CMIU 213, an organization of New York City cigar packers, was chartered in 1883.

# Adolph Strasser to Terence Powderly

Buffalo, N.Y., February 25 1886.

T. V. Powderly Esq.
General M.W. of the K. of L.
Dear sir,

At the last convention of the C.M. Int. Union a resolution was adopted, instructing the officers to protest against the issuance of the K of L label,[1] on cigars made by scabs, and persons receiving a lower compensation than that recognised by the branches under the jurisdiction of the Int. Union.

The General Assembly was also requested to adopt the blue label of the Int. Union. That request was embodied in a series of resolutions which I have mailed to you when at Hamilton.[2]

I am still waiting for a reply, believing that as a matter of courtesy of one organisation to another, you should have notified me of the action of the General Assembly.

I am again receiving complaints from New York City, that the K. of L. label is issued to tenement house manufacturers, and to manufacturers paying from $1.00 to $2.00 less per 1000 than the scale adopted by the Int. Union.

Will you tolerate such action?

Hoping for an early reply, I remain,

Yours fraternally A. Strasser
Int Pres.

ALS, Terence Vincent Powderly Papers, DCU.

1. As early as 1880 the KOL in the Pittsburgh area had furnished a blue label for cigars. The Knights adopted a white label for all KOL-produced goods in 1884.

2. The 1885 KOL General Assembly met in Hamilton, Ontario, Oct. 5-13. The resolutions were not presented there and on Mar. 16 Strasser accused Powderly of suppressing them. Powderly told the General Assembly in May that he had never received the resolutions (KOL, *Record of the Proceedings of the Special Session of the General Assembly, . . . May 25 to June 3, 1886* [N.p., n.d.], p. 31).

# An Excerpt from an Article in the *United States Tobacco Journal*

[February 27, 1886]

QUARRELING UNIONS.

The Cigarmakers' International Union is still doing their utmost to beat the Cigarmakers' Progressive Union. They are rapidly gaining all the shops outside of the United Cigar Manufacturers' Association and bulldozing the packers in the latter. There are a number of packers, however, who sympathize with the Progressive Union, and who will not be guided by the Internationals. The Internationals are endeavoring to call out all the packers of McCoy & Co., Brown & Earle and Levy Bros. So far they have not met with much success.

The most important step of the United Cigar Manufacturers' Association is the abolition of their tenement-houses. The association wants the Knights of Labor label, and, to get it, the tenement houses must go.

The Knights of Labor comes into the fight as the mediator between the two quarreling unions. They don't take sides, and their label is proof that the [. . . .] for the benefit of our readers we give a *fac simile* copy of the label:

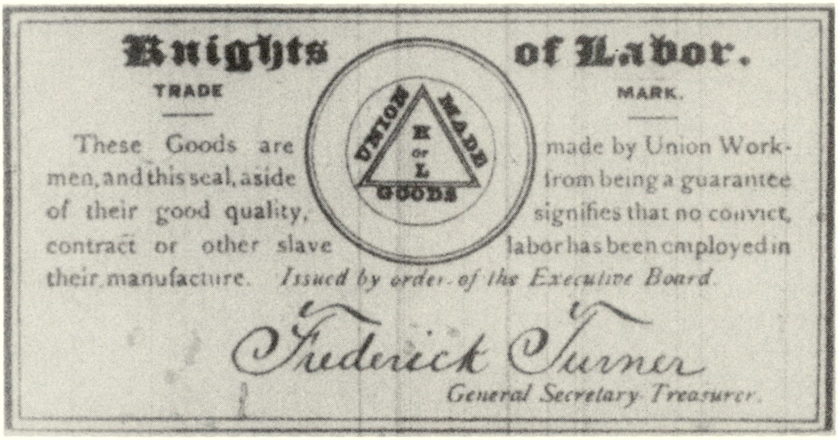


. . .

*United States Tobacco Journal*, Feb. 27, 1886

## Adolph Strasser to Terence Powderly

Buffalo, N.Y., March 6, 1886.

T. V. Powderly, Esq.,
G.M.W. of the Knights of Labor:
Dear Sir:—

The recent action of your Organizers and of District 49[1] of the Knights of Labor of New York City in interfering with the management of our strike compels me to submit my protest against the action of such persons.

The Cigar-makers' International Union has, by an almost unanimous vote, approved a strike against a reduction of wages in several shops, which is still in progress. From personal investigation and information received, the fact appears that the scabs in the shops of Levy Bros., Frank McCoy, and Brown & Earle have been organized as members of the Knights of Labor. The employers of these scabs have also been promised the white label of the Knights of Labor on the production in their factories, evidently as a reward for employing scab labor and paying low wages.

I consider the action of your Organizers in New York City and of District 49 of the Knights of Labor a bold and unscrupulous attack upon recognized trades' union principles, and as hostile to the Cigar-makers' International Union in particular. The President and Secretary of the Manufacturers Association admitted in my presence to you that they would be compelled to surrender, in case you should refuse to help them out of the trap in which they have caught themselves.

Should you fail to denounce the action of your Organizers in New York City, you will merit the condemnation of the Cigar-makers' International Union and of every National trades' union in the country.

At the same time I call your attention to a letter which you have mailed, on December 26, 1885, to W. H. Foster,[2] Secretary of the Federation of Trades. In that letter you disapprove of such crimes as have been committed against the Cigar-makers' International Union in New York City.

I now demand action on your part. Should you fail to listen to the warning contained in my letter, the Cigar-makers' International Union will be compelled to protect itself against unscrupulous employers and so-called labor reforms.

Yours, respectfully, A. Strasser,
International President.

KOL, *Record of the Proceedings of the Special Session of the General Assembly, . . . May 25 to June 3, 1886* [N.p., n.d.], p. 30.

1. KOL District Assembly 49 was organized July 1, 1882.
2. In his letter to Foster, Powderly asserted that he had "always held that the man who proved untrue to his trade union was unfit for membership in the K of L" and that he had never consented to Knights' organizers interfering in trade union affairs (Terence Vincent Powderly Papers, DCU).

## An Item in the *Laborer*

[April 3, 1886]

"The Picket" is a new labor paper just at hand, printed in the interest of the blue label. Sam Gompers is at the helm, which is a guarantee that The Picket will be "on guard."

*Laborer* (Haverhill), Apr. 3, 1886.

## The Employees of Sutro and Newmark to Terence Powderly

New York April 19th 1886.

T. V. Powderly,
G.M.W. K. of L.

After having exhausted all honorable means, & after having respectfully applied time and again, to the officers of our local for such information, as would let us know where & how we stood; & after each and every such effort, compelled to confess, that information which we had a right to possess, was witheld for some purpose, which we could not devine, We, members of L.A. 2458 do hereby, earnestly appeal to you, to give, or have given to us, such information & advice, as lies in your power, & as one brother is in duty bound to give to another.

Our position is this: We are members of Cigar Makers Int. U. of A. Nos. 144 & 10. After the lockout & settlement, and after an ineffectual attempt to force us into the Union having made the settlement,[1] our employer[2] appeared & informed us, (in the presence of a committee of K. of L.) that it was immaterial to him, as to what union we adhered to, but commanded that each & every one *must* join the K. of L. a refusal to be punished with discharge.

Acting upon, what in our hearts we thought to be the proper course, we consulted our leaders & officers, the result being our initiation in Defiance L.A. 2458. For the last two weeks, committees have appeared, with permission of our employer, have taken our members without our consent, & have informed us, that we must proceed to take a certain step, in order that our firm, Messrs. Sutro & Newmark be enabled to receive the "White Label." We are to go to headquarters of D.A. 49, there deliver our books or cards, pay sixty cents for dues, sign a printed card, in which we acknowledge that we have taken a false step, & that we are anxious to retract &c.; & then we will receive what is called a "Working Card" which will allow us to work with the other & more favored men. If this step be refused on our part, Mr. Newmark has told us, he would be compelled to discharge us. Now, sir, imagine our quandary; we are anxious & willing, & always have been, to do what is right. We are not of that class, who fight their fellow workman to satisfy their pride or spleen, but have in our desire to do what is proper, followed the advice given us by our leaders; if they have misled us, would you hold 3000 or 4000 honest people responsible for the actions of a few?

We place our faith in you: we assure you, that if you inform us & convince us that we have been misled, that all our efforts will be concentrated to effect a reconciliation in the ranks of the cigarmakers, regardless of the wishes of our former leaders. We have pleaded for information from the proper parties, & have been put off with ambiguous replies. Pray help us, in this the winter of our need & distress, & you may rest assured that your advice & counsel shall be as our guiding star.

Yours Fraternally The employees of Sutro & Newmark
Address. L. Steinbach (Pres't of the shop).
c/o J. Niederstein.[3]
1393 Second Ave.
N.Y City.

ALS, Terence Vincent Powderly Papers, DCU.

1. The Cigarmakers' Progressive Union.

2. Bernhard Newmark (1836?-98) was a Russian immigrant.

3. John Niederstein (b. 1837), who was a saloonkeeper on Manhattan's East Side, was born in Prussia and emigrated in 1867. By 1900 he had become a hotelkeeper in the borough of Queens.

## *The KOL, the Trade Unionists, and the Founding of the AFL*

Relations between the KOL and national and international trade unions deteriorated rapidly during the spring of 1886, and the growing conflict was exacerbated by the struggle between the cigarmakers and the Knights in New York City. A central source of tension was the overlapping structure of the KOL and the unions. The Order's local and district assemblies, representing single trades or mixed bodies, often functioned as trade unions, and this led to jurisdictional conflicts with trades already having national organizations. The KOL's dramatic increase in membership—from 100,000 to over 700,000 between July 1885 and June 1886—made a clash more likely. The International Typographical Union (ITU), CMIU, Bricklayers, Iron Molders, Granite Cutters, Carpenters, and other unions complained bitterly that the Knights did not respect their strikes and boycotts, undercut union wages and hours, and accepted as members workers who had been suspended or expelled by their trade unions. Terence Powderly dismissed the charges as "purely imaginary"[1] and deflected these criticisms by accusing trade unions of similar transgressions. Such charges and countercharges set the tone for a struggle that ultimately spurred trade unionists to found the AFL.

In an attempt to resolve differences between the Order and the trade unions, P. J. McGuire, Adolph Strasser, Josiah B. Dyer, P. F. Fitzpatrick, and W. H. Foster called a conference of union leaders that met in Philadelphia on May 18. Many of the twenty-two official delegates who attended the meeting were conciliatory toward the Knights. William Weihe, for example, told the gathering that the Iron and Steel Workers had "no grievous complaint" and saw "no present indications of any serious trouble from the Knights of Labor." ITU leader M. R. H. Witter, whose organization presented a series of charges against the KOL, nevertheless wrote Powderly afterward of his convictions that "mutual understanding—not antagonism—should be the common purpose" and that the continued growth of the KOL was important to the future of the printers' union. Similarly Dyer wrote Powderly that "we who are both Trades Unionists and Knights of Labor from principle and not from policy, . . . feel that there should

be some means adopted to bring about a closer affiliation" between the two parties since, as he noted, they shared common objectives and common enemies in "the Capitalists and Anarchists."[2]

Despite the moderate positions of some of the participants, the Philadelphia meeting signaled a turning point in the relations between the trade unions and the KOL. Although several unions, including the Iron and Steel Workers and the ITU, had been weighing the advantages of joining the KOL en masse without loss of trade autonomy, the delegates passed a resolution declaring that "we do not deem it advisable for any Trade Union to be controlled by or to join the Knights of Labor in a body." Further, they condemned the actions of "a certain element of the Knights of Labor" that had "the avowed purpose" of destroying the trade unions. They resolved to hold an annual conference of the trade unions' chief officers to "protect and promote" their interests and "establish fraternal relations between the different branches of industry." Finally, they appointed a committee of five—Weihe, Chris Evans, Strasser, McGuire, and Fitzpatrick—and one alternate, David Boyer, to draw up a "treaty" with the KOL that would confine the Knights to working with unorganized workers and engaging in educational activities.[3]

When the committee presented its treaty to the KOL General Assembly meeting in Cleveland the following week, the delegates took no direct action on it but voted that a KOL committee should be chosen to negotiate directly with the unions. During the next few months there was little improvement in the strained relations. Throughout the summer McGuire urged Powderly to name the KOL committee, but Powderly proposed the General Executive Board, a body most trade unionists considered hostile to their interests. At the same time the KOL sought the affiliation of national unions in an attempt to unify the labor movement. Several small unions joined the Order, but the ITU, Iron and Steel Workers, Bricklayers, Carpenters, and others rejected these overtures. As the summer wore on Powderly began to view the matter personally and claimed that warfare against the Knights was made "not by the trade unions but by a few men at the head of the Int. Cigar makers union."[4] On several occasions he publicly denounced Gompers as a drunkard with whom no business could be transacted.

The trade union committee finally met with the KOL General Executive Board in Philadelphia on September 28. The trade unionists proposed that the treaty be referred to the forthcoming KOL General Assembly, which was to meet in Richmond in October, that a special KOL committee investigate the unions' past grievances, and that the KOL and the trade unions establish a mechanism for adjusting any

future difficulties. Although Powderly promised to recommend these proposals, the Richmond General Assembly widened the breach between the KOL and the trade unionists. Not only did Powderly fail to present the committee's proposals, but the assembly adopted a resolution introduced by District Assembly 49 that required KOL cigarmakers to leave the CMIU or suffer expulsion from the Knights.

Meanwhile the trade union committee issued a call for a national labor congress to be held at Columbus, Ohio, on December 8 "for the purpose of forming a closer alliance of Trades Unions and of establishing an American Federation of Labor."[5] Seeking to reinvigorate the failing FOTLU, Gompers, McGuire, and John Kirchner arranged for its sixth annual meeting to be held in Columbus at the same time. Initially the two conventions remained technically separate; after successful negotiations between delegations from the two bodies, however, the FOTLU merged with the new American Federation of Labor and the FOTLU delegates joined in the sessions of the trade union conference. The combined convention included forty-two delegates representing twenty-five organizations with a membership of over 300,000.

As a member of the Committee on Constitution, Gompers played a key role in shaping the new organization. The AFL was to be solely an organization of trade union bodies; unlike the FOTLU, there was no provision for affiliation by KOL assemblies. The constitution provided for the formal chartering of Federation affiliates and created a stronger financial system than that of the FOTLU, replacing the FOTLU's annual assessments with a monthly per capita tax. Affiliated bodies that fell three months in arrears were to be suspended. Finally, the AFL was to have a full-time president with an initial salary of $1,000 per year.

A variety of issues commanded the delegates' attention. Aroused by the recent political elections, they debated a resolution advocating the formation of an independent labor party. After considerable discussion they amended it to urge "a most generous support to the independent political movement of the workingmen."[6] They also demanded the further restriction of Chinese immigration and endorsed steps to achieve the eight-hour day.

A key issue at the conference was the relationship of the Federation with the KOL. Powderly had appointed a committee, headed by John W. Howes, to meet with the trade unionists in Columbus. The trade union committee and the Knights met on December 9 and 10 but reached an impasse. The former insisted on offering the treaty drawn up at the Philadelphia conference in May, while the Knights contended that the Cleveland General Assembly had rejected the proposal. Howes

could only promise to report the unions' grievances to the KOL General Executive Board, and the trade unionists concluded that they "might have terminated the conference the first half hour of the meeting, for it was evident the K. of L. Committee had no power to act." The Knights' representatives later reported to the KOL General Executive Board that there was an "atmosphere of insincerity" during the proceedings and that some of the trade union leaders used their grievances "to accomplish selfish purposes."[7] Hostility toward the KOL surfaced repeatedly at Columbus. The FOTLU Legislative Committee blamed the failure of the eight-hour movement on "the expressions of hostility coming from leading members of the Knights of Labor." Similarly, the new Federation condemned the Knights for persistently attempting to undermine and disrupt the unions. The convention refused to seat the delegate representing the Window Glass Workers because that organization was a KOL assembly, and it endorsed the cigarmakers' struggle with the KOL, declaring that the CMIU blue label was "the only union label in that trade."[8] Finally, Gompers rose to defend himself against the charge published in the Knights' circular *The Order and the Cigar-Makers* that he was habitually intoxicated.

The convention elected Gompers as president. The *United States Tobacco Journal* reported his election as the convention's attempt "to vindicate the reputation of one of its most prominent members and most persistent fighters against the order."[9] The feud between the KOL and the trade unions thus culminated in the founding of a rival organization that would, under the leadership of Gompers, finally supplant the Order.

*Notes*

1. Terence Powderly to P. J. McGuire, May 11, 1886, Terence Vincent Powderly Papers, DCU.
2. *Boycotter,* May 22, 1886; M. R. H. Witter to Powderly, May 6, 1886, Josiah B. Dyer to Powderly, May 5, 1886, Powderly Papers.
3. *Boycotter,* May 22, 1886.
4. Powderly to Henry Dettman, Aug. 11, 1886, ibid.
5. *Carpenter,* Nov. 1886.
6. AFL, *Proceedings,* 1886, p. 16.
7. Ibid., p. 18; Dec. 13, 1886, in KOL, *Proceedings of the General Assembly of the Knights of Labor of America Eleventh Regular Session, . . . October 4 to 19, 1887* (N.p., n.d.), p. 1447.
8. FOTLU, Proceedings, 1886, p. 6; AFL, *Proceedings,* 1886, p. 19.
9. Dec. 18, 1886.

# A Call for a Conference of National and International Trade Union Officers

April 26, 1886.

*Strictly confidential*
*to all whom it may concern.*
Dear Sir.

we the undersigned officers of the organizations named, deem it highly important to hold a conference, at an early date, of the chief Executive Officers of each and every National and International Trades Union in America.

It is suggested that said *conference* be an informal one, and be held in *Phila. Pa on Tuesday May 18, 1886.* (meeting place to be announced later)

The object of the conference is to devise ways and means, to protect our respective organizations from the malicious work of an element, who openly boast that Trades Unions must be destroyed. This element urges our local unions to disband, and it is doing incalculable mischief, by arousing antagonisms and dissensions in the labor movement. Under cover of the Knights of Labor, and as far as we can learn, without authority from that body, this element pursues its evil work. "Rats," "Scabs" and unfair employes are backed up by this element, suspended and expelled members of Trades Unions are welcomed into their ranks, and these elements use the K of L. as an instrument through which to vent their spite against Trades Unions. That this has been the case, can be amply demonstrated by the Cigar-Makers and Typographical Int Unions. Other trades have been more or less affected.

These and other subjects concerning the relations of the Trades Unions and the K. of L. require that the above conference be held as speedily as possible that we may agree upon some plan to submit to the General Officers of the K of L. to cease this hostility and antagonism towards Trades Unions.

In view of the urgency of the case, we call upon you, to personally attend the conference or appoint a substitute. If you will attend sign this circular, and return in envelope. Then you will be informed as to place of meeting. *Consider this circular strictly confidential.* It is mailed to all nat and Int Unions—

*signed* Fraternally Yours P. J. McGuire
Gen Sect Brotherhood of Carpenters[1]
A. Strasser. Pres Cigar makers Int Union
Josiah Dyer.[2] Gen Sect Granate Cutters. Nat. Union
F. J. Fitzpatrick. Pres. Iron Moulders union of N.A.
W. H. Foster Secty of Federation of Trades of N.A.

P.S. The following is list of parties invited

| | | |
|---|---|---|
| A. Strasser | Pres. Cigar Makers Int Union | |
| F. Fitzpatrick | " Iron Moulders " | " |
| Josiah Dyer | Secty Granat Cutters Nat | " |
| P. J. McGuire | " Brotherhood of Carpenters | |
| W. H. Foster | Secty Fed of Trades | |
| E. S. McIntosh[3] | " Typo Int Union | |
| R. Howard | " Spinners Nat " | |
| P. M. Arthur[4] | Pres. Locom. Engineers | |
| Wm. Werhe[5] | " Amal Iron and Steel workers. | |
| J. Phillips | Sect. Hatters Int Union[6] | |
| Thos O Dea[7] | " Bricklayers Int Union[8] | |
| G. Gessner[9] | " Glass Workers | |
| J. Ehmann[10] | " Ohio Valley Trade & Labo[r] Assembly[11] | |
| F. Melke[12] | Sect of German printers Int Union[13] | |
| Tom O Reilly[14] | Pres. Telegraphers Nat | " |
| H. Emrich | Secty Cabnet Makers Int | "[15] |
| A. H. Sharp[16] | " Tailors Nat | "[17] |
| Chris Evans[18] | " Miners[19] | |
| Geo Harris[20] | Pres. " Amal | "[21] |
| W. W. McClelland | " Amalg Engineers | " |
| W. M. Shultz | Sect Metal workers Int. | "[22] |
| | Pres[23] Locomotive Firemen[24] | |

I would be pleased to have you suggest additional names.

HLtSr, Terence Vincent Powderly Papers, DCU.

1. The Brotherhood of CARPENTERS and Joiners of America.

2. Josiah Bennett DYER was national secretary of the Granite Cutters' International (later National) Union from 1878 to 1895.

3. Edward S. MCINTOSH was secretary-treasurer of the International Typographical Union from 1885 to 1886.

4. Peter M. ARTHUR served as grand chief engineer of the Brotherhood of Locomotive Engineers from 1874 to 1903.

5. William WEIHE was president of the National Amalgamated Association of Iron and Steel Workers from 1884 to 1892.

6. The National Trade Association of HAT Finishers of the United States of America.

7. Thomas O'DEA served as secretary of the Bricklayers' and Masons' International Union from 1884 to 1887 and from 1888 to 1900.

8. The BRICKLAYERS' and Masons' International Union of America.

9. Frank M. GESSNER was secretary of Window Glass Workers' Local Assembly 300 of the KOL between 1883 and 1886.

10. John Ehmann (variously Ehman) was a member of the Prosperity Assembly of the KOL, located in Wheeling, W.Va., and editor of the *Ohio Valley Budget*.

11. The Ohio Valley Trades and Labor Assembly was formed by local unions in the area around Wheeling, W.Va., in March 1882 and affiliated with the AFL in 1915.

12. Frederick K. MILKE was secretary of the German-American Typographia in 1886.

13. The German-American TYPOGRAPHIA.

14. Thomas O'REILLY was elected president of the Brotherhood of Telegraphers in 1884 and in 1886 became master workman of KOL National District Assembly 45, the telegraphers' trade assembly. He was an editor of the *Journal of the Knights of Labor* from 1889 to 1893.

15. The International FURNITURE Workers' Union of America.

16. Charles H. SHARP was elected president of the Journeymen Tailors' National Union in August 1885.

17. The Journeymen TAILORS' National Union of the United States.

18. Christopher EVANS was secretary of the National Federation of Miners and Mine Laborers (NFMML) from 1885 to 1888 and AFL secretary from 1889 to 1894.

19. The National Federation of MINERS and Mine Laborers.

20. George HARRIS served as president of the Miners' Amalgamated Association of Pennsylvania (subsequently the Miners' and Mine Laborers' Amalgamated Association of Pennsylvania) from 1883 until 1887 and helped organize the NFMML in 1885. He was an organizer and district leader of the United Mine Workers of America in the 1890s.

21. The Miners' Amalgamated Association of Pennsylvania, a state organization affiliated with the Amalgamated Association of Miners of the United States, was organized in 1883. In 1886 it changed its name to Miners' and Mine Laborers' Amalgamated Association of Pennsylvania, and in late 1888 it merged with a faction of KOL National Trade Assembly 135 and with the NFMML to form the National Progressive Union of Miners and Mine Laborers.

22. The METAL Workers' National Union of North America.

23. Frank Pierce SARGENT was grand master of the Brotherhood of Locomotive Firemen from 1885 to 1902.

24. The Brotherhood of LOCOMOTIVE Firemen.

## Terence Powderly to L. Steinbach

Scranton Pa. May 1 1886.

L. Steinbach Esq.
New York N.Y.
Dear sir and brother:—

Your letter sickens me, I am sick and tired of hearing complaints of how workmen wrong each other. I have been made to feel the lash also within the last few months for actions of which I knew nothing whatever.

The General Assembly will meet at Cleveland on the 25 of this month[1] and I am in hopes that all of these troubles in the cigar trade will be ended at that session. In the meantime make an appeal to the Gen Ex. Bd. for a stay of proceedings in your case. I do not think it is right to deprive a man of his right to toil for a trifle and will never consent to it so far as I am concerned.

Lay the case before the Assembly that demands the card and ask of them to set you right. I am so busy with the affairs of the order that I cannot attend to details and if I had the time one half of the trouble which our order is now in would never have occurred. With kind wishes I am

Fraternally yours, T. V. Powderly

TLpS, Terence Vincent Powderly Papers, DCU.

1. The special session of the KOL General Assembly met in Cleveland, May 25–June 3, 1886.

## An Item in the *Blue Label*

[May 1, 1886]

Our brothers of Union 144, New York, sends a sample copy of their boycotter, called the *Picket.* It is spicy and fearless, and knifes the scab manufacturers in good shape, as the following will show: "Scab shops, McCoy & Co., factory No. 6, 3d district, New York; Levy Bros., factory No. 401, 3d district, New York; Brown & Earle, factory No. 1307, 3d district, New York. They are tyrants and enslavers of labor and do not pay the international cigar-makers' price. Boycott them until the Blue Label appears on their box."

*Blue Label* (Chicago), May 1, 1886.

## Lichtenstein Brothers and Co. to Frederick Turner[1]

New York, May 12th, 1886

Fred R Turner Esq
Genl Secretary K of L
Box 834, Philadelphia Pa—
Dear Sir!

For some time past we have been considerably annoyed by the receipt of numerous letters from our customers, almost daily stating that the Cigar Makers International Union has been persistently boycotting all Cigars not bearing their blue label. Special complaints reached us from Louisville Ky where an International travelling orator

had addressed meetings for the apparent purpose of discrediting the Knights of Labor label, stating that the blue (International) label was the only one entitled to the patronage of honest workingmen &c &c. The effect of all these efforts has been to create confusion amongst the dealers and even induce some of them in different parts of the country to discriminate against the Knights of Labor label by refusing to handle goods bearing that label. We have refrained from troubling you in this matter until now, hoping that proper remedies would be applied by your order to stop these outrageous, malicious and unprincipled attacks by the organs of the International Cigar Makers, especially as many of the Internationals professed to be Knights of Labor, whom we have employed in our own shops as such, but so far instead of a cessation of these persistent onslaughts, they seem to be continued by them with increased venom. A few days ago one of our principal customers, the firm of R. W. Tansill of Chicago wrote to us that the sale of their "Punch" Brand of Cigars (made by us for them during years past, which they have run for upwards of ten years past, introducing it and keeping it before the public at an enormous outlay of money) has been considerably interfered with by the efforts of the International Cigar Makers Union.

The tricky and deceitful wording used by them that the Punch brand had been extensively advertised as an "*International*" Union made Cigar (which to our knowledge is not the case), making it appear as if the K of L label on them is a counterfeit and fraud, has undoubtedly deterred many a conscientious smoker from using the Cigar. We enclose herewith *marked* copies of the Blue Label Pacific Coast Boycotter, issued by the Cigar Makers International Union, which will show conclusively the despicable falsehoods resorted to for the mere purpose of venting their spite against parties who prefer using the K. of L. label instead of the "Blue" Label. These parties can not help knowing that we, who are using the white label of the K of L, have no work done in Tenement houses, and if we did, that we could not get the K of L label, still they unblushingly in every issue of their papers make the statement that our goods, bearing the white or K of L. label,[2] are Tenement house work, thus trying to prevent the smokers from using goods made by Union hands. The above papers are sent to you by request of Messrs R. W. Tansill of Chicago, as also the enclosed inquirial[?] letter of E. Brennest Denver Col, one of their large customers and will conclusively show, that the pernicious efforts of the Cigar Makers International Union are not confined to one locality alone, but are scattered all over the country. Unless the most decisive steps are taken by your order at the earliest possible moment, this uncalled for persecution of the goods made by Knights of Labor

and bearing their label will result in depriving thousands of honest working men, now earning their living in Knights of Labor shops of their means of existence. We are already working on three fourths time with reduced force, mainly if not wholly owing to the uncertainty in the minds of parties, handling our product as to which label is the correct one and if the nuisance is not abated soon, less working time and more discharges will be made necessary, if our goods can not be marketed without this interference and discrimination of parties pretending to be friends of the workingmen against workingmen of their own trade. We trust you will kindly use your best efforts towards the abatement of these detrimental actions on the part of the International Cigar Makers Unions, which although mainly directed against us as Manufacturers will undoubtedly be of greater injury to the members of the Knights of Labor, now in our employ. We remain awaiting an early reply

very respectfully yours Lichtenstein Bros & Co

ALS, Knights of Labor Papers, WHi.

1. Frederick TURNER, a goldbeater and later a grocer, was general secretary-treasurer of the KOL from 1883 to 1886 and general treasurer from 1886 to 1888.

2. In April 1886 the KOL General Executive Board discontinued use of the white label for Knights' cigars, instead adopting a blue label.

## An Item in the *Laborer*

[May 15, 1886]

The Pickett uses some pretty strong language in regard to the circumstances under which the white label has been granted to non-union cigarmakers in New York City. If the facts are as stated there is surely a great grievance. You must remember this, Sam, that your fight should not be against the entire Order, but solely with those who are false to the first principles of labor organization, and are acting in either malice or ignorance in thus prostituting the label of the K. of L.

*Laborer* (Haverhill), May 15, 1886.

# A Committee of the Philadelphia Trade Union Conference to the Cleveland General Assembly of the KOL

Philadelphia Pa., May 18, 1886.

To the Officers and Members of the General Assembly of the Knights of Labor, in special session convened at Cleveland O. May 25. 1886: Fellow Workmen and Brothers: —

In our capacity as a committee of six, selected by the conference of Chief Officers of National and International Trade Unions, held in Philadelphia, Pa, May 18, 1886, beg leave to submit for your consideration, and with hope of approval, the following terms with a view to secure complete harmony of action and fraternity of purpose among all the various branches of Organized Labor:

TREATY

1st That in any branch of Labor, having a National or international Trade Union the Knights of Labor, shall not initiate any person or form any Assembly of persons following a trade or calling organized under such National or International Union; without the consent of the nearest Local Union of the National or International Union affected

2nd No person shall be admitted to membership in the Knights of Labor, who works for less than the regular scale of wages fixed by the trade Union of his craft or calling and none shall be admitted to membership in the Knights of Labor, who have ever been convicted of ["]Scabbing," "ratting," Embezzlement or any other offence against the Union of his trade or calling, Until Exonerated By said Union

3rd That the charter of any Knights of Labor Assembly of any trade having a National or International Union, shall be revoked And the members of the Same be requested to Join a mixed Assembly or form a Local Union, Under the Jurisdiction of their National or International Trade Union.

4th That any organizer of the Knights of Labor who endeavors to induce Trade Unions to disband; or tampers with their growth or privileges Shall have his Commission forthwith revoked.

5th That wherever a strike of any Trade Union is in progress No Assembly or District Assembly of the Knights of Labor, shall Interfere until settled to the satisfaction of the Trade Union affected

6th That the Knights of Labor shall not establish or issue any trade mark or label in Competition with any trade mark or label Now

issued, or that may Be hereafter issued by any National or International Trade Union

Wm Weihe
Christopher Evans
A. Strasser
P. J McGuire
P. F. Fitzpatrick
David P. Boyer,[1] Alternate

A and HLS, Edward A. Wieck Papers, MiDW.

1. David P. BOYER served as chief organizer for the International Typographical Union from 1885 to 1888.

## Martin Witter[1] to Terence Powderly

St. Louis, Mo., May 21, 1886.

My Dear Sir:

I have not yet been advised of action of Philadelphia Conference. I hope nothing was done to increase complications. I advised the adoption of this programme: A request to the Knights to abstain from dictation in affairs pertaining distinctly to trades. 2d. a request that men who do not belong to unions, where unions exist, be not admitted to K. of L. I fear the greatest obstacle to the adoption of this plan for the future lies in the relations of the Knights and the Cigar-makers. It is clear to me that their relations have become so complicated that—putting out of sight the origin of dispute, and to some extent its merits—an understanding can only be reached through compromise. My anxiety for a friendly understanding, prompts me to propose, if necessary, arbitration—for if the dispute goes on both parties to it must inevitably suffer. I will write Mr. Boyer, who informs me he will be in Cleveland, to use his influence for peace and harmony.

Apologizing for this intrusion

I remain with best wishes M. R. H. Witter

ALS, Terence Vincent Powderly Papers, DCU.

1. Martin R. H. WITTER was president of the International Typographical Union between 1884 and 1886.

# A News Account of a Meeting of Cigarmakers in Detroit

[May 24, 1886]

DON'T LIKE THE LABEL.

Germania Hall, corner of Russell and Mullett streets, was about two-thirds filled yesterday afternoon at a meeting advertised to discuss the subject of "Scabs, Knights of Labor and Trades Unions." The gathering was held under the auspices of the International Cigarmakers' Union, which has a grievance against the Knights of Labor for placing labels on cigar boxes. They ask that the label be withdrawn and that of their particular union be indorsed. Frank Miller[1] acted as Chairman and A. Stuermer[2] as Secretary.

The first speaker was Fred. Haller, of New York. He is a slender young man and a recognized leader in labor movements, of good voice and is a forcible speaker. He detailed the action of the Knights of Labor in labeling their cigar boxes, denouncing it as a detriment to the union cigarmakers. He favored the trades unions, which he described as the pioneers in the labor movements. Every day they are growing stronger in their battle for principle and right. "We desire to work in peace and harmony with the Knights. They have accomplished much, and have a great duty to perform; their future is a great one, and they have a large field to work in. Most of the trades union men are Knights of Labor. They constitute the back bone of the order, and they should therefore be respected. Never yet has there been such a commotion as that made by the action of the General Executive Board and District Assembly No. 49 in the matter of the cigar makers. We cannot afford to fight, but if necessary we will be compelled to defend ourselves. Trades unions are the outgrowth of long experience on the part of organizers who have the cause of labor at heart. The existing difficulty should be amicably settled by the conference of the Knights in Cleveland. We desire to co-operate with them and to avoid these conflicts. We want to avoid a breach with the Knights, but will defend ourselves and preserve our organization at all hazards. [Great applause.][3] We will not see our organization destroyed by men who are against us. It is the duty of every trades unionist and of every member of the International Cigarmakers' Union to stand by the cigar organization and to use his best endeavors to push it. The Trades Union has fulfilled everything which it has promised and it cannot be conquered."

S. Gompers, a thick-set man with dark hair and mustache, was the

next speaker. He has the faculty of holding his listeners. Mr. Gompers spoke in the same strain as Mr. Haller, and was frequently applauded. He said he stood on the same platform as Mr. Powderly. In a letter written by the latter about six months ago he said that a man who was not true to his trades union is not true to the Knights of Labor. "We demand," said Mr. Gompers, "that Mr. Powderly live up to this as well as the humblest member of the order." The speaker exhibited contempt for workmen not connected with the union, called "scabs," miserable fellows, scalawags, and urged that they be crushed out and that honest labor be encouraged. He was very bitter against the Knights of Labor on the label question and cited several instances in which he claimed that "scabs" were admitted into assemblies in order to give manufacturers a chance to put a Knight of Labor label on their goods.

A voice—"The same is done in Detroit."

Mr. Gompers paused a moment and said, "It's wrong; that is the long and short of it." He said that in Albany a firm called in an organizer to form an assembly among their men, but they were not admitted. The next night the employes took a boat for New York, where they were made Knights of Labor, and when they returned to work wages were cut down and the Knights of Labor label placed on the goods made by them. "The label is of blue and in the shape of a seal, and I want you to crush it when you see it." [Applause.][4] ["]The Knights of Labor first had a white label, and when they noticed the rare success of the International label they changed the color to correspond with the latter. No man ought to be forced to join the Knights of Labor. Mr. Powderly is always absolutely right in defending the trades unions, but others of the Executive Council are singularly always against him. The trades unions are the natural outgrowth of the system of production under which we live. The unions are attacked on many sides as being conservative. This is not so. To give up the organization is a folly, and to attempt to disrupt it is a crime."

John Kirchner, of Philadelphia, also spoke, and was followed by others.

*Detroit Free Press,* May 24, 1886.

1. Frank C. Miller (1859-1932?) was a cigarmaker and a cigar manufacturer in Detroit.

2. J. Adam Stuermer (1852-1902?), a German-born cigarmaker, served as president of CMIU 22 of Detroit in 1880 and 1881.

3. These brackets are in the original.

4. These brackets are in the original.

# A Report of the General Executive Board[1] to the General Assembly of the KOL

[May 28, 1886]

To the General Assembly, Knights of Labor:

Your Executive Board respectfully submit for your consideration the evidence concerning the grievances that have arisen between the International Cigar-makers' Union and the Knights of Labor.

March 3, 1886, while the Board was in session at Philadelphia, Pa., Messrs. A. Strasser, President, and John S. Kirchner, Fourth Vice-President, of the I.C.M.U., appeared before the Board regarding a trouble existing in New York City between the said I.C.M.U. and members of the Order in that city.

After listening to a statement from the gentlemen named, it was decided that three members of the Board should go to New York and investigate the matter at the earliest possible moment.[2] The pressure of duty upon the Board made it impossible to go until March 13, when Brothers Hayes,[3] Bailey[4] and Barry[5] left Philadelphia, arriving in New York the following morning.

Upon their arrival notice was sent to Brother Samuel Gompers and others of L.A. 2458, who were also officers of the I.C.M.U., in New York City, and interested in the difficulty, as well as to officers of D.A. 49 and Local Assembly 2814.[6]

Representing the I.C.M.U. and L.A. 2458, there appeared Brothers Samuel Gompers, Frederick Haller and a large number of others.

To avoid wrangling between the contending parties and to utilize to the fullest extent the limited time which, owing to press of other grave duties, the Board could bestow upon the examination, it was deemed advisable to examine each side separately, and that each side should select its best witnesses, whose testimony in full should be taken as given, and submitted to the full Board for a decision upon the points involved.

Before the examination was completed the Board was compelled to suspend it in order to go to Troy, N.Y., to consider an exigency that had arisen there, and to prevent a serious complication that threatened to arise in relation to Fuller, Warren & Co.[7]

After a few days' absence they returned to New York, resumed the taking of testimony, and continued each day for about ten days, working from twelve to fifteen hours a day, until called to St. Louis, Mo., by reason of the great strike on the Southwestern system of railroads.[8]

From their arrival in St. Louis until the meeting of the General

Assembly in this special session, the attention of a majority of the Board, and especially of those members who conducted the investigation, has been entirely occupied in connection with that great strike.

We present herewith the evidence taken both on behalf of the International Cigar-makers' Union and of those representing the opposite side of the controversy. We also give papers and documents showing the attacks made upon the general officers and members of the Order.

Let it be understood that these attacks were made before the examination had been concluded, and while of necessity the attention of the general officers was engaged in a struggle that required the wisest judgment and closest attention to manage.

Your Board could not help feeling that these bitter and uncalled for attacks at such a time were, to say the least, unkind, and that they showed very little of the fraternal and mutual sympathy which ought to be manifested between representatives of different branches of the army of organized labor.

It has been continuously asserted that the K. of L. label was given to employers that pay less than the prices of the International Union.

The Board has received numbers of boxes of cigars from manufacturers upon which the label of the International Union was used, and for which the prices paid the cigar-makers in the employ of these manufacturers were from three to four dollars per thousand. In view of this fact, we, the Board, in issuing a new label to be used entirely on cigars, limited the price to be paid for the making of cigars [to] not less than six dollars for scrap work.

Box No. 1 is an instance of the use of the K. of L. label on cigars made in a factory where no Knights of Labor were employed. The label having been obtained by L.A. 2458. It also has the label of the International Cigar-makers' Union.

Box No. 2 represents a grade of cigars that sold for nineteen dollars per thousand, and on the box will be found the K. of L. label and the International Union label.

Box No. 3 is a cigar manufactured in the Ninth District of Pennsylvania, wrapped with a Sumatra wrapper, and for which the price paid the cigar-makers is four dollars per thousand. The International Union label is upon the box.

Box No. 4 is a box with the stamp K. of L. on the cover, on which is the International Union label, presumably stamped to mislead members of the Order, and without objection from the International Union.

Boxes 5, 6, 7 and 8 represent grades of cigars on which the In-

ternational Union label is used, and the price for making is only three dollars per thousand; coming from different manufacturers.

L.A. 2458 was suspended for continued violation of the law, taking packers and strippers after notification repeatedly given, while there were Assemblies of those branches in existence; initiating and making members without even presenting them to the Assembly for initiation; giving a receipt book indicating membership in the Order of the Knights of Labor upon the payment of one dollar, one of which is in our possession, given to the Board by the party so introduced to membership.

As regards the conference with the committee appointed by the representatives of the Trades' Unions at their Philadelphia session, the Board submits the proposed treaty to the General Assembly without any recommendation, inasmuch as the same was not presented to the Board until just previous to the opening of the session, and was not known until the same was read in the session by the General Master Workman, on Wednesday. Consequently, there was no time to give it the consideration which the paper deserves.

Respectfully submitted, T. V. Powderly, Chairman.
Frederick Turner, Secretary.
John W. Hayes.
William H. Bailey.
Thomas B. Barry.

KOL, *Record of the Proceedings of the Special Session of the General Assembly, . . . May 25 to June 3, 1886* [N.p., n.d.], pp. 28-29.

1. The 1885 constitution of the KOL provided for a General Executive Board of five members, including the general master workman and the general secretary-treasurer, to exercise general supervision and control over the Order and consider appeals from the general master workman's decisions. The May 1886 special session of the General Assembly elected six additional members who served as a temporary auxiliary to the General Executive Board until the Richmond General Assembly in October. At that time the General Assembly amended the constitution, removing the general secretary-treasurer from the board and increasing its membership to six. In 1888 the General Assembly gave the general master workman the power to nominate members of the board and reduced its size to four.

2. Members of the KOL General Executive Board conducted hearings on the dispute in the cigar industry in New York City, Mar. 16-25, 1886. Their findings were published in a report entitled *The Order and the Cigar-Makers* (Philadelphia, 1886).

3. John William HAYES, a grocer, served on the KOL General Executive Board from 1884 to 1888 and subsequently was KOL general secretary-treasurer (1888-1902) and general master workman (1902-16).

4. William H. BAILEY, a miner, was a member of the KOL General Executive Board from 1884 to 1887.

5. Thomas B. BARRY, an axemaker, was a leader in an 1885 strike against Saginaw

Valley lumber mill operators. In 1886 he was sent by the KOL General Executive Board, on which he served from 1885 to 1888, to help settle the Chicago stockyards strike.

6. KOL Local Assembly 2814, composed of New York City cigarmakers, organized in 1883. Many of its members were also members of the Cigarmakers' Progressive Union.

7. In April 1884 Fuller, Warren and Co., a stove manufacturing firm in Troy, N.Y., attempted to impose a system of contract work and a 20 percent reduction of wages on the stove mounters in its Clinton foundry and refused to recognize their union, KOL Local Assembly 3275. The workers walked out. After three months the company, with the assistance of the Stove Manufacturers' Association, broke the strike by blacklisting the strikers and assembling a work force composed of new and former workers. KOL District Assembly 68 expelled members who continued to work for the company, and the KOL and the New York State Workingmen's Assembly (NYSWA) instituted a boycott against the firm. On Mar. 9, 1886, Fuller and Warren agreed not to discriminate against members of the Knights, to suspend workers that the Knights expelled, to hire no new employees until all the striking Knights were reinstated, and to settle all future differences by arbitration. In July 1886, however, the Knights reinstated the boycott, claiming the company had broken its agreement, and the NYSWA followed suit. The boycott continued into the next decade.

8. In 1885 the KOL successfully resisted wage reductions and the discharge of labor leaders through widely publicized strike victories across Jay Gould's Southwestern railroad system. The victories contributed to the rapid growth of the Knights in 1885-86. Another strike of the Gould railroad system by the KOL began in early March 1886, when about 9,000 workers walked out to protest the firing of a KOL member and to demand union recognition along with a daily wage of $1.50 for the unskilled. The dispute affected more than 5,000 miles of track. Powderly participated in the negotiations, but on May 3, acknowledging defeat, the KOL General Executive Board called off the strike.

# An Excerpt from a News Account of the Cleveland General Assembly of the KOL

Cleveland, June 3. [1886]

## GENERAL ASSEMBLY.

The air this morning is full of prophecies of "blue ruin" for the Knights of Labor, on account of the triumph of the extreme radical and anti-Trades-Union wing of the Order, in yesterday's election of new members[1] of the General Executive Board. The four members elected yesterday—Quinn[2] of New York, Mullen[3] of Richmond, Cavanaugh[4] of Cincinnati, and Gibson[5] of Hamilton (Ontario), were the candidates of the "oath-bound Home Club"[6] of New York, about which the papers are telling most terrible stories; and they defeated

such veterans as Carlton[7] of Massachusetts, Buchanan[8] of Denver, and Murray[9] of New York. The excitement over the matter in the Order itself is hardly surpassed by that in the newspapers; and the scenes in the General Assembly, during the progress of the election, are said to have been as feverish as such things are in the bitterest political campaign. The "Home Club" radicals are overjoyed, and the more conservative element are proportionately dumbfounded over an incident that was wholly unexpected to the latter. There is also dissatisfaction among the delegates from the far West, because the Executive Board, with the new additions, is almost wholly Eastern in its membership. But the two other members of the six who were provided for, are to be chosen to-day, and I presume that the offices will be given to the West.

[*Later.*—This forenoon the two additional new members of the Board were elected—Buchanan of Denver and Aylsworth[10] of Baltimore.][11]

But Mr. Powderly and the other older and more experienced members of the General Assembly do not seem to think that the radical victory of the "Home Club" means any such thing as the destruction or splitting of the Order. Mr. Powderly said to-night that he had heard for years that the "Home Club" was opposed to him and his system of administration; but that all this noise about the club looked to him like the holding of a post-mortem examination of a four-year-old corpse, which had been reduced to dust. It will probably be found, after the organization of the new and enlarged Board, that it will work with better judgment than its enemies anticipate.

The Trades Unionists of course feel that the result of the election is a knock-down for them, and both last night and to-day their leaders have been predicting a disruption and bitter war between the Trades Unions and the Knights of Labor.

One thing must be said. The election shows that there is an abundance of personal independence in the General Assembly; and that, no matter how popular any man may be, the delegates intend that the government and the business of the Order shall remain in their hands.

. . .

*John Swinton's Paper,* June 6, 1886.

1. The Cleveland General Assembly of the KOL elected a six-member auxiliary board to assist the General Executive Board in the face of the rapid increase in KOL membership and the growing number of strikes, lockouts, and boycotts.

2. James E. QUINN, a bookbinder, was a leader of the inner circle of KOL District Assembly 49 known as the Home Club.

3. William H. MULLEN, a printer, helped organize KOL District Assembly 84 in 1885 and served as its master workman.

4. Hugh CAVANAUGH, a shoemaker, served as secretary (1884-86) and district master workman (1886-88) of KOL District Assembly 48. From 1890 to 1893 he was general worthy foreman of the KOL.

5. David R. GIBSON was vice-president of the Bricklayers' and Masons' International Union in 1883 and a frequent delegate to the Bricklayers' conventions through 1896.

6. The Home Club emerged among socialist elements of District Assembly (DA) 49 in the early 1880s who opposed the elimination of secrecy from the KOL and sought to remove Powderly from office. Its inner circle consisted of nine members who included the leaders of the District Assembly. Despite some Knights' claims that the Home Club controlled the KOL national leadership and the fact that it clearly symbolized anti–trade union sentiment within the KOL, its influence beyond New York City was questionable. DA 49, however, played an important role in the KOL General Assembly, particularly in 1886, and it actively pursued a policy of opposing autonomous trade unions in New York City in the name of organizing all workers under its leadership. Members of the CMIU believed that the machinations of the Home Club were responsible for their difficulties with the KOL during 1886 that culminated in a DA 49–sponsored resolution at the 1886 General Assembly demanding the expulsion of CMIU members from the Order (see "A Resolution Introduced by District Assembly 49 at the General Assembly of the KOL," Oct. 19, 1886, below).

7. Albert A. CARLTON, a shoemaker, was master workman of KOL District Assembly 30 from 1879 to 1886 and a member of the Knights' General Executive Board from the fall of 1886 to 1888.

8. Joseph Ray BUCHANAN, a printer, organized western railroad workers between 1884 and 1886 and led several successful railroad strikes. He served on the KOL General Executive Board from 1884 to 1885 and served on the auxiliary board in 1886.

9. George F. Murray (1855-1936), a New York City printer and editor for sixty years, was active in International Typographical Union 6 and a delegate to the KOL's General Assembly in 1885, 1886, and 1888.

10. Ira B. AYLSWORTH, a stairbuilder, was a member of the KOL General Executive Board from 1886 to 1888.

11. Brackets are in the original document.

## An Interview with Samuel Gompers in the *United States Tobacco Journal*

[June 5, 1886]

### WHAT ARE THEY GOING TO DO ABOUT IT?

Mr. Samuel Gompers, editor of *Picket* and President of the Boycott Committee of the Cigarmakers' International Union, was interviewed late yesterday afternoon by a representative of the *U.S. Tobacco Journal* regarding the rather disadvantageous position in which the Inter-

nationals are placed by the result of the Cleveland conference. Mr. Gompers was very frank in expressing himself and his remarks will be found interesting. When asked what he thought would be the effect on the cigarmakers of the measures taken at Cleveland Mr. Gompers said:

"In regard to the cigarmakers I can say that by the election of James E. Quinn, alias Monroe, the Assembly practically endorsed the action of District Assembly 49 in its fight against the Cigarmakers' International Union. What shall we do about it? We may await developments and see what is done with our petition[1] (since the General Assembly has failed to act upon it) by the committee which has the matter in hand. We shall expect justice, and will accept nothing else. We shall take just such action as the circumstances will warrant us in taking to defend the interests of our organization. We deem the election of the auxiliary members of the Executive Board as the elevation of the cheap labor advocates of the Knights of Labor as opposed to the Cigarmakers' International Union, which seeks to obtain as high wages as possible. We certainly regard the course of the Knights of Labor in permitting their members to work ten hours a day, when our members only work eight, as an attack upon our organization. It is rather premature to say what we will do next, but come what may the members of the International Union will defend themselves in all ways possible."

Mr. Gompers thought it was a great mistake that a satisfactory adjustment of the troubles of the cigarmakers had not been effected.

"What do you think is the significance of the result of the Cleveland conference?" Mr. Gompers was asked.

"I believe it fails to satisfy the wants of the cigarmakers of the country. It shows them that they have nothing to expect from the Knights of Labor. I think in the end they will have to rely on the Cigarmakers' International Union to protect them in the fight with employe[r]s."

Mr. Gompers said that the Cigarmakers' International Union would go right on in the course it had taken in the boycott of employe[r]s. It was confident of a victory in the end. When asked whether he thought the fight between the Internationals and the Knights of Labor would result in a reduction of prices, he said it would not so far as those of the former was concerned. They would not be changed.

*United States Tobacco Journal,* June 5, 1886.

1. Possibly a reference to SG and Adolph Strasser's presentation of the CMIU's complaints against the Knights' interference in the cigarmaking trade, at the Cleveland session of the General Assembly. SG asserted that in New York City the KOL label appeared on tenement-house cigars.

# An Article in the *United States Tobacco Journal*

[June 5, 1886]

WAR TO THE TEETH.

*Picket,* the organ of the Cigarmakers' International Union to-day publishes a lengthy manifesto, in consequence of the failure to settle the differences between the union and the Cigarmakers' Knights of Labor at the Cleveland conference during the past week. The tone of the document which we republish below is extremely doleful. It concedes that the victory obtained by Knights of Labor was because the leaders of the latter were too sharp for the Internationals, and hence, as *Picket* admirably expresses it sprung a trap on the latter. The manifesto alluded to also recites the history of the troubles between the Internationals and Knights of Labor in a manner that is rather refreshing not to say amusing. Notwithstanding the doleful spirit of the document, it shows a determination on the part of the Internationals to suffer no further interference from the Knights of Labor. The document is as follows:

It is with feelings of genuine regret that we pen these lines, being written in the light of the latest information received from Cleveland. We say regret, because we have come to look to the meeting of the Delegates in Special Sessions as a blessing, and an end to the war that has systematically been waged against the Trades Unions, a settlement of the differences that have existed between them and the K. of L., the prevention of the encroachments of over-zealous organizers, and a check being placed upon the acceptance as members of the Order of persons who have proved themselves unfaithful to their Union and fellow men.

Our latest advices inform us that little if anything can be expected to settle these difficulties, and further that the representatives of the worst element of the order have been elected to the highest office within the gift of the delegates; that able, honest and true men who represented not only the intelligence of the order, but those who live in peace with the unions, have been defeated. It is the victory of ignorance over intelligence, and there is not an anti-unionist or "scab" in the country who will not feel encouraged.

We say we regret this state of affairs, but accept the situation as it presents itself. We know what we can expect, and we are prepared for it. If anybody thinks that trades unions are going to lie down and be swallowed up, he is mistaken. If the thought prevails that unions

will turn tail if attacked, we assure them they will be mistaken. We want not war, we want to live in peace, more especially with the men in the labor movement.

We have always held that there was room for both the trades unions and the K. of L., but we say now that we will not tolerate no interference with our trade matters; that we shall insist that the scabs and rats in the trades shall not be covered by the shield of Knighthood.

That we shall resent a wrong or a blow intended to be aimed at the unions, and shall call upon organized labor of the country to bear witness to it, and appeal for that aid innate in man and which requires but "one touch of nature" to make "the whole world akin."

To the workingmen we say, stand by your unions. They are your defense, your guard.

The delegates of District Assembly 49 have sprung a new trap. They are determined by fair means or foul to make the General Assembly whitewash the dirty deeds done against the New York cigarmakers.

The delegates of District Assembly 49 came here with a printed circular, in which they try to defend the action of the tenement-house ring in reducing wages.

The circular begins its attack upon the International Union by saying that, "While the bosses were threatening their employees with a lockout, the International Union, through some manipulation only known to themselves, made the shop of Kerbs & Spies an International Union shop." It omits to say that Kerbs & Spies, by these manipulations, were made to abolish their tenement-house cigar factory, and that they raised the wages of their employees in some instances as [much as] $1.60 a thousand.

It also omits to state that Kerbs & Spies is to-day working under the union rules, that it is one of the best shops in the country, and that it runs under the eight-hour regulation.

The writer of the circular has the effrontery to say that the International Union "placed their locked-out people in Kerbs & Spies', and then had to take members from other shops yet to fill the shop of Kerbs & Spies." No mention is made of the fact that up to the very last the International men in the strike-shops were so greatly in the majority that every time a vote was taken on a party question the opposition, calling itself Progressive, struck tail and left the room. This trick to hide the weakness of the opposition to the International Union was beautifully left unmentioned.

The International Union Strike Committee is accused with having broken faith with a committee of the Progressives and of the Central Labor Union, two days after making an agreement to work hand in

hand with them. This charge is clinched fast by the assertion that "abundant proof exists." But none of this proof is vouchsafed. It may be that this "abundant proof" exists in that so-called affidavit of Jos. Oppenheim,[1] of Levy Bros., the head of the ring of New York tenement-house cigar manufacturers. The writer of the circular evidently realized the fact that if he based his assertions upon the evidence given by an employer of a scab and tenement-house labor, and in that way attempted to vilify one of the greatest labor organizations of the country, he would place himself in an unenviable light before the working people of the United States.

The circular then makes the broad assertion that the International Union tried to settle the lockout in such a manner that every employee of the firms involved would have to join the Cigarmakers' International Union. The writer, it seems, did not intend his circular to reach New York, for in that city it is known that the Cigarmakers' International Union was fighting for the restoration of the old rates of wages only, whilst the Progressives and District 49 were fighting for nothing else but to destroy the Cigarmakers' International Union.

John Love[2] is mentioned in the circular as being the first manufacturer who applied for the K. of L. (white) label, and his shop is mentioned as being fair, his employees having been organized in the K. of L. Prices paid for labor have nothing to do with the use of the K. of L. label if the employees are only organized in the K. of L.

It is known to every cigarmaker in New York that John Love pays less even than the scale called for by the sell-out "equalization." Of course, if John Love was entitled to the K. of L. label, there is no reason why they should be withheld from the dirtiest tenement-house cigars.

After reciting a number of charges against L.A. 2,458, of which the officers of that L.A. have even never been informed, and upon which L.A. 2,458 was suspended without even the semblance of a trial, the circular goes on to say that "the entire matter had by mutual consent been referred to the General Executive Board for investigation, and D.A. 49 was ready to stand by the decision whatever it may have been."

One would expect that the circular would then state what the General Executive Board did. But nothing is said of what the General Executive Board did, for the simple reason that it did absolutely nothing.

The grievances of the Cigarmakers' International Union have been investigated by the working people of the country, and they have decided to do something, and they are doing something. They are

calling for the blue label of the Cigarmakers' International Union.

*United States Tobacco Journal,* June 5, 1886.

1. Joseph Oppenheim (1852?-88), a member of the Levy Brothers cigar firm, was secretary of the United Cigar Manufacturers Association in 1886.

2. John W. Love.

## The General Executive Board to Members of the KOL

Philadelphia, Pa., July 2, 1886.

To the Order everywhere, Greeting:

The General Executive Board has decided to issue the following circular[1] that our members may fully understand the exact position of the Knights of Labor and its relation to the International Cigar-makers' Union.

On March 3d, Messrs. Strasser, Kirchner and other officers of the International Cigar-makers' Union met with the General Executive Board, in Philadelphia, and made complaint that unfair people were organized into the Order. After an all-day session on the case, we promised, as soon as time and opportunity would permit, to go to New York, make an investigation, and, if proven that such charges were well founded, to revoke their charter, for this organization will not be made a refuge for unprincipled and unfair people.

The first act of Mr. Strasser after leaving us, and before his complaint could get any consideration, was to issue a letter and mail it to Brother Powderly in New York City, when he was positive Powderly was still in Philadelphia; all of which proved conclusively that he had no confidence in the proposed investigation vindicating his complaints. The letter was dated March 6th,[2] and may be found in the minutes of the special session of the General Assembly, page 30.

Since the date of the above letter, Mr. Strasser and his colleagues have been constantly sending circulars and men through the Order to boycott all goods except those bearing their International blue label, and have charged the General Master Workman and the balance of the General Executive Board with co-operating in the organization of scabs into the Order.

The officers of the Cigar-makers' International Union knew when they were flinging their charges broadcast that there were upwards of 60,000 members of the Knights of Labor and trade unions engaged in a life and death struggle for living wages. The Southwest strike

was in progress. And at that time, when it was the plain duty of every man (with a spark of union feeling in his breast) to stretch forth the hand of sympathy, the chief officers of the Cigar-makers' International Union (to their eternal shame be it said) refused to exercise a moment's patience, and violated every principle of unionism by charging on our rear, while the militia of Illinois, in obedience to the order from corporate wealth, were drowning the cry of the oppressed in the roar of musketry, and feeding the hungry with cold lead and steel. Such conduct cannot be explained. Nowhere in the history of the labor movement does its parallel exist. But one reason can be assigned for the unaccountable actions of the officers of the International Union, and that is that men who indulge to excess in the use of intoxicants cannot transact business with cool heads. On two occasions the men who came to Philadelphia to confer with the General Executive Board were too full for utterance. The General Executive Board has never had the pleasure of meeting with Mr. Gompers when he was sober.[3]

What these men expect to accomplish by their acts and efforts to destroy the Knights of Labor is hard to understand, the Order being the power that has made it possible for their Union to prosper.

All offers made in the interest of peace and harmony were rejected and scorned by the officers of the International Union, saying their Constitution will not allow them to work with Knights of Labor who are not members of the International Union. We then requested them to change their Constitution so as to enable them to receive our cards on equality.

We have never discriminated in the past in favor of our label as against theirs—only asked that our members see that cigars bear a union label, assuring them that the goods were made by honest labor. The position we have always occupied, and still adhere to, is that our cards be received on an equality with theirs; that our members be allowed to work in shops under the control of the International Union, and *vice versa.* In other words, we are willing to place our organization, with its hundreds of thousands of members, on the same footing with their organization, containing but 18,000.

Up to the present we have not said or done anything to the injury of this Cigar-makers' Union; neither do we intend to lay a straw in the way of the success of that or any other organization. But, in defence of the Knights of Labor, and our actions as General Officers, it is necessary that the Order should hear a full statement of the facts bearing on the case. With that purpose in view, we have compiled a full and complete review of all the facts elicited through the investigation made by the General Executive Board as to the cause of the trouble and complaints made by the Executive Officers of the Inter-

national Union, a copy of which is mailed with this. And we trust that every person in the Order will read this statement, so that if the International Union charge that we are organizing scabs, rats or black sheep, the Order and its officers can be defended.

The pressure of duty upon the Board made it impossible to go to New York until March 13th, when Brothers Hayes, Bailey and Barry left Philadelphia. The investigation was begun on March 16th, after due notice had been given to the parties interested. It was continued with intervals until near the meeting of the General Assembly on May 24th, when we were compelled to close the case, notwithstanding that the Representatives of D.A. 49 had not been fully heard—a fact which, nevertheless, is of little moment, as nearly all the testimony directed against that District Assembly was merely based on hearsay evidence.

Knights of Labor must not boycott goods bearing either the blue seal or white label of the Order. This order is imperative, and must be obeyed.

T. V. Powderly,
Frederick Turner,
John W. Hayes,
W. H. Bailey,
T. B. Barry,
General Executive Board.

KOL, *The Order and the Cigar-Makers* (Philadelphia, 1886), pp. 1-3.

1. In addition to this letter, the forty-eight-page circular entitled *The Order and the Cigar-Makers* contained a review of the testimony taken by the KOL General Executive Board on the cigarmakers' difficulties and the text of the testimony taken Mar. 16-25, 1886, in New York City.

2. Printed above.

3. SG responded to the General Executive Board's accusation at the Columbus convention of the AFL (see "War among Labor Leaders," *New-York Tribune*, Dec. 10, 1886, below).

# To the Officers and Delegates of the New York City Central Labor Union

New York, July 18, 1886.

To the Officers and Delegates of the Central Labor Union—
Fellow Workmen:—

The cigarmakers of Buffalo are on strike[1] for the past six weeks for an increase of wages. In that entire time the employers have failed

to secure any others to take the places of the men on strike. Everything indicates success if men who pretend to act in the name of labor do not prove treacherous to the interests of labor.

It has been publicly expressed by the Progressive Union that they will send fifty "scabs" to Buffalo to take the places of the men on strike. They have gone further than any other "scab" agency now extant. They have advertised in the daily papers for cigarmakers to go to that city.

As an officer of the Cigarmakers' International, I deem it my duty to call your attention to this state of affairs of a body represented in your organization; also to ask you to take such action as will prevent the consummation of so treacherous an act, or at least place your seal of condemnation upon it.

Ought we to wait until such a schism has been made as to render a thorough unity impossible; or to once for all place the brand traitor on the brow of those who would betray our interests—Judas like—to unscrupulous employers?

Samuel Gompers,
Second Vice-President.[2]

*Tobacco Leaf,* July 24, 1886.

1. The CMIU Buffalo strike began May 29. In mid-July the Cigarmakers' Progressive Union (CMPU), asserting that CMIU members were scabbing on Progressives in a lockout in Hornellsville, N.Y., retaliated by encouraging members of the CMPU to break the Buffalo strike. A week later, however, the CMPU rescinded this action in anticipation of a reconciliation with the CMIU, and before the end of the month the striking Buffalo cigarmakers won a wage increase.

2. Gompers was elected CMIU second vice-president in the spring of 1886 to replace P. J. Powers, who had resigned. He served in this office until 1896, when he became first vice-president, and he held that position until his death in 1924.

## An Article in the *United States Tobacco Journal*

[July 24, 1886]

### CIGARMAKERS LET LOOSE.

The streets of this city were full of idle cigarmakers on Saturday last, who, pursuant to prearrangement, were bent upon what the jolly individual from Limerick would call a "spree." The cigarmakers put a glamor of dignity over the occasion by calling it a parade. According to *Picket,* the cigarmakers' mouthpiece, all had a grand time. Below

is an extract of *Picket*'s report. The enthusiasm of the little leaflet will cause our readers to smile. Says *Picket*:

On last Saturday the cigarmakers outdid themselves, and all around where they worked, lived or went, put on a holiday appearance. For weeks preparations were going on under an energetic and competent committee to make the day's parade and picnic an immense success, many of the shops vying with each other in making a grand display.

Glorious weather cheered the hearts of those who participated in the largest single-trade parade that has taken place in this city for years. Nearly 6,000 members of the International Cigarmakers' Union marched through the streets on the east side preliminary to enjoying a grand afternoon and evening picnic at Karl's Park, 148th street and 3d avenue.

The procession represented the unions comprised in the District of New York of the International Union and visiting unions from Williamsburgh, Brooklyn, Greenville, N.J., Hoboken, Newark, Jersey City, and delegates from Philadelphia, South Norwalk, Conn., Elizabeth, N.J., Poughkeepsie, N.Y., and other places. There were fourteen bands of music, two drum corps, and any number of banners, flags, streamers and transparencies, the latter bearing matters pertaining to the blue label and the Labor problem. There were also twenty-two stages for the accommodation of the female operatives and those who did not care to walk, thirty-five open barouches and many wagons laden with flowers and pretty little girls attired in gay costumes.

As the different divisions began to form in line at 10 A.M., thousands of people from the surrounding neighborhood crowded the side-walks, and flags and bunting were displayed on the house fronts along Avenue B and in the streets thereabouts. As the different detachments arrived, they were placed in line according to the programme, which had been carefully arranged by Grand Marshal Charles Alt[1] and his aids.

The procession formed at Avenue B, between Fourth and Fifth streets. The first division was composed of visiting unions and delegations and the employes of Carl Upman, Kerbs & Spiess and Glacum & Condit, formed on 5th street, with the right resting on Avenue B. The second division, which consisted of the employes of Straiton & Storm, Howard Ives, Kahner & Merkel, Union No. 237,[2] of Morrisania, and all union men working in open shops, formed in lines on East 4th street, between Avenues A and B, with the right resting on Avenue B. The third division was made up of the employes of Foster & Hilson, Bogert & Heyden, Louis Ash, Mendel & Mohr, Wertheim & Schiffer, and Unions Nos. 132, 149[3] and 87, of Brooklyn, which

were drawn up on East 4th street, between Avenues A and B, with the right resting on Avenue B.

The aids to the grand marshal were Michael Ledderman,[4] marshal of the first division, and aids L. Wagner,[5] N. Reveira, and Louis Lelyveld;[6] marshal of the second division Ernest J. Arundel,[7] and aids William Wismar,[8] David Pronk[9] and Henry Lozier; marshal of the third division A. Rosenbaum,[10] and aids George Pape,[11] P. Stoltz[12] and Christopher Branch.

In the procession there were fourteen bands of music, two Grand Army drum corps, twenty-three stages, thirty-five open carriages for the female members, flags, banners, mottoes and painted advice to all smokers to "Buy Blue Label Cigars." The girls looked "real sweet."

At 11 A.M. the order to march was given. Every man in the parade wore a boutonniere in the lapel of his coat, below which floated a satin faced *fac simile* of the blue label of the International Union. They all carried canes, and wore black, blue, or gray polo caps.

The appearance of thousands of men in marching order all uniformed alike was a sight worth seeing and one not easily forgotten.

*United States Tobacco Journal,* July 24, 1886.

1. Charles Alt (b. 1859) was a New York City cigarmaker.

2. Local 237 of Morrisania, N.Y., was affiliated with the CMIU from 1885 to 1889.

3. The *Cigar Makers' Official Journal* announced the chartering of local 149 of Brooklyn, N.Y., in January 1886.

4. Michael C. Lederman was a secretary of CMIU 131 in Jersey City, N.J., from 1883 to 1887. In 1885 he was a member of the Hudson County Central Labor Union.

5. Leopold Wagner, a cigarmaker born in Prussia about 1857, represented CMIU 144 at the 1886 convention of the Political Branch of the NYSWA.

6. Louis Lelyfeld (d. 1887) was a member of CMIU 144.

7. Ernest J. Arundel (1859-1906), a cigarmaker born in England, became a member of CMIU 334, Saratoga Springs, N.Y., in 1883 and later served as its secretary (1904) and president (1905).

8. William Wismar, a cigarmaker born in Prussia about 1856, was assistant financial secretary of CMIU 144 in 1878. In 1881 and 1882 Wismar was president of CMIU 132 of Williamsburgh, N.Y.

9. David Pronk, born in Holland about 1850, was a cigarmaker who lived in Brooklyn, N.Y.

10. Adolph Rosenbaum, born about 1825 in Germany, was elected vice-president of the newly consolidated New York City Cigar Packers' Union in 1877.

11. George Pape (1851-1903) was born in Germany and immigrated to New York in 1870. Pape was a founding member of CMIU 144 and its president in 1879 and 1882.

12. Peter Stoltz (1855-1908) was a German-born cigarmaker. He immigrated to New York City in 1865 and became a member of CMIU 144.

# An Excerpt from an Article in the *Tobacco Leaf*

[July 31, 1886]

CIGARMAKERS FIGHTING EACH OTHER.

For two or three months past the great cigar manufacturing industry of this city has been pursuing the even tenor of its way. This means that it has been paying taxes on a production of cigars ranging between 55,000,000 and 60,000,000 per month. Until last week it looked as if a season of quietude would continue long enough to enable employers and employees alike to make up for their respective losses arising from the memorable lockout inaugurated at the commencement of the present year. Appearances were deceptive, as they often are; and now war, internecine war, rages all along the line. Through the smoke of the battle it is difficult to discern who is right and who wrong, or who will be victorious in the conflict. All that is positively clear to the vision of the ordinary looker-on is that—

1st—Wages are not in dispute.

2d—Knights of Labor demand abolition of the Progressive Union.

3d—Fourteen of the large cigar manufacturers of New York favor the Knights in their demand.

4th—The Progressive Union prefers coalition with the International Union to absorption by the Knights of Labor.

It seems to be a war of labor upon labor—one of the most unfortunate events that could occur to the labor interest. As gleaned from contemporary reports the particulars relating to the affair are as annexed:—

THE COMMENCEMENT.

Last week District Assembly 49 of the Knights of Labor informed the Cigarmakers' Progressive Union that the open union must be given up or the charter returned. On Sunday last about 2,000 men of the Progressive Union met in the Germania Assembly Rooms to decide whether they would abandon their union or cease to be Knights of Labor. A vote was taken, and it was unanimously decided to stick to the union. These resolutions, with others, were adopted:—

*Resolved,* That we, as cigarmakers, condemn the action of District Assembly 49, and pledge ourselves to stick by the union, irrespective of the consequences.

*Resolved,* That we consider it our duty and a matter of honor to take this position against the despots in the Knights of Labor.

*Resolved,* That we, as members of the Cigarmakers' Progressive Union, pledge ourselves to oppose the demand of the Knights of Labor to give up to them our books, whereby we would lose our liberty.

A committee was appointed to obtain for the union a charter from the International Union; in other words, secure a consolidation with the Internationals.

. . .

*Tobacco Leaf,* July 31, 1886.

## An Excerpt from an Article in the *United States Tobacco Journal*

[July 31, 1886]

IN AN UPROAR.

. . .

Mr. Samuel Gompers, who is the editor of *Picket* and second vice-president of the Cigarmakers' International Union, when asked by a reporter of the *U.S. Tobacco Journal* whether in case the Progressives consolidated with the Internationals it would mean raising the prices hitherto paid in the Knights of Labor shops, said that it would not for the present, but, of course, it would have a tendency to bring about that result eventually. Mr. Gompers when reminded of the difference in New York prices paid Internationals as compared with those in other sections of the country, said that he was aware that a uniform price-list was desirable. An effort had been made to make prices uniform, but as yet it had not succeeded.

Questioned regarding the conditions upon which the Progressives would be admitted into the International Union Mr. Gompers said:

"Oh, there won't be any trouble about amalgamating the Progressive and International Unions, if the former are really in earnest. Each individual union can have its own 'principles,' provided it adheres to the general constitution and rules. The Progressives can come in as the others did in January. The only difficulty, perhaps, will be with the funds. Our initiation fee is $3, but we have on hand at the present time a per capita fund of $12. You can figure what that amounts to for 26,000 members. Over $300,000, isn't it? So you see, if the Progressives come in at $3 a head they would be entitled to a share in that large sum. If they didn't have that amount on hand—the $3 per head—I would be willing to hold a monster picnic now and a

ball in the winter to raise funds for them. I will go to any length to have them admitted if they want to come in. The question of wages would not necessarily have anything to do with their admission. They would not have to comply with our scale immediately, unless the 'blue label' was desired by the employer. They could not work in tenement houses—that, of course, is one of our strictest conditions. An employer can use the bunch machines, but he could not get the label. Yes, I am decidedly in favor of admitting the Progressives. If we coalesce the days of the tenement house factories are numbered. If the Knights of Labor are victorious it means an indefinite perpetuation of the tenement house system."

. . .

*United States Tobacco Journal*, July 31, 1886.

## An Article in *John Swinton's Paper*

[August 1, 1886]

THE CIGARMAKERS.

Since this paper went to press last week strange events have happened in this city among the cigarmakers. It was first rumored that the Progressive Cigarmakers were informed that they would have either to leave the Union and cleave to the Knights of Labor, or *vice versa*. Later on the rumor became fact. On Sunday last the Union met in Germania Assembly rooms and resolved not to give up their autonomy—at least to the K. of L. On Monday, it again met in the same place and resolved to make an effort to join the Internationals. On Tuesday morning a committee of the Knights of Labor waited on Levy Bros. and informed the cigarmakers—rollers and bunchers, that if they refused to continue in the Order they must leave, and 450 (all of the men and women) left. They met and resolved that they would retain their independence, would not be dictated to by the Order, and would not return to work until the notice was withdrawn. The firm closed its shop, and a meeting of the United Cigar Manufacturers' Association was held in the Grand Union Hotel the same afternoon. The K. of L. had a committee to consult with the bosses, and the latter resolved to stand by the Order. The National Executive Board of the Progressive Union met on Tuesday night and issued a call for a convention of the twenty unions, preparatory to disbanding and joining their old antagonists, the Internationals.

If the present differences between the K. of L. and the Progressive Union will lead to a consolidation of the two cigarmakers' unions, it will be for the best interests of the labor movement of this city and wherever the two unions exist. Not alone have the cigar-makers suffered, but every trades union likewise, by the quarrel. It is to be hoped that the International, exercising its usual wisdom, will see fit to accept the overtures of the Progressives.

On Wednesday the excitement among the Progressives continued. At 10 o'clock the K. of L. committee visited Lichtenstein Bros. & Co., in Second avenue, and gave out the same notice as at Levy Bros. the day before. At 2 P.M. the 400 employes held a meeting, and adopted resolutions similar to those of Levy Bros.' workmen—to stand by the union. Sutro & Newmarck were next visited by Messrs. Daley & Wolf,[1] the committee, with like results. Other firms in the association were also waited upon, all the workmen expressing a unanimous purpose to abide by the trade union. About 6,000 persons will be on the street if all persist in standing by the union.

The International officers are willing to aid the Progressives if they desire to amalgamate, and that is just what they most sincerely wish. The manufacturers don't want them to do that. "Do anything you like," said Secretary Oppenheim to the Progressive pickets on Wednesday, "but don't unite with the Internationals."

President Hammerstein,[2] of the American Association of Cigar Manufacturers,[3] expressed about the same views. "If they unite," said he, "the price for cigar-making will go up from $1 to $3 a thousand. We shall help the United Association."

*John Swinton's Paper,* Aug. 1, 1886.

1. Probably Jacob Wolf, a cigarmaker and master workman of KOL Local Assembly 2814.

2. Oscar Hammerstein (1847-1919), a New York City composer, theatrical manager, and inventor of cigarmaking labor-saving devices, was founder and editor (1876-89) of the *United States Tobacco Journal* and president of the American Cigar Manufacturers' Association (ACMA) in 1886.

3. The ACMA was organized in the spring of 1886. Later that year the United Cigar Manufacturers' Association merged with the ACMA.

## Three Items in *Tobacco*

[August 6, 1886]

### NOTES ON THE STRIKE.

. . .

The officers of the various Cigarmakers' International and Progressive Unions held a very friendly conference last Saturday evening

at No. 101 Avenue A. There were represented International Unions Nos. 10, 13, 141 and 144, of New York; Nos. 87, 132, and 149, of Brooklyn; No. 131, of Jersey City; No. 8, of Hoboken; also, Cigar Packers' Unions Nos. 213 and 251,[1] and the Progressive Cigarmakers' Unions. A resolution was passed stating as the sense of the conference that the International and Progressive cigarmakers throughout the country should be united, under the jurisdiction of the general Cigarmakers' International Union; that the International District of New York be advised to consent to the amalgamation, and that the Internationals will do their utmost to help their Progressive brethren in their present conflict with District Assembly No. 49 of the Knights of Labor.

In accordance with these resolutions the New York District of the International Cigarmakers' Union held a meeting on Sunday at No. 52 Stanton Street, to consider the proposal of its own unions and of the Progressives to amalgamate. The recommendation was regarded with great favor, and a resolution was passed, and will be sent to the International local unions throughout the country, asking them to approve of an assessment of ten per cent. to be levied upon the wages of all Internationals for the relief of the Progressive strikers as soon as the amalgamation shall become an accomplished fact. Meanwhile subscriptions have been started in International shops for the same object, and it is believed that the Internationals will cheerfully give ten per cent. for their striking brethren, so enthusiastic are they over the prospect of amalgamation. Frederick Haller, president of International Union No. 13, says that the Knights could not possibly bring cigarmakers from any part of the country to fill the shops that are on strike, as the Internationals have a network of unions all over the country and control nearly all the men.

. . .

Part of the money in the treasury of Assembly No. 49, Knights of Labor, was derived from the tax levied on the newly patented bunching machines. Some of the manufacturers were using Herman Schultze's bunching machine,[2] under contracts with Lewyn & Martin, paying that firm $200 for each machine. The Board of Arbitration and Strikes of District Assembly 49 growled over this, and said if bunching machines were brought into general use cigarmakers would have to learn another trade. They said that, with a girl to guide the machine, 4,000 cigars could be turned out in a day. It was finally agreed that if the Knights would permit the use of a certain number of machines for three years, Lewyn & Martin would turn over $25 of each $200 to

the treasury of the Progressive Local Assembly to assist its benefit fund.

. . .

The large hall at Cooper Union was packed to its utmost capacity Wednesday evening with cigarmakers of both the International and Progressive Unions, to denounce the action of District Assembly No. 49 of the Knights of Labor to force the Progressive Union to disband. Among the vice-presidents for the evening were representatives of various trades unions, to show their approval of the stand made by the cigarmakers. Samuel Gompers, one of the speakers, drew a contrast between the declaration of Powderly, that a good Knight was a good trades union man, and the action of the rulers of the Home Club and of District Assembly No. 49, showing that a good unionist could not be a good Knight. This Home Club consisted of truck driver L. B. Maguire,[3] who never drives trucks, and who in the southwest traveled under the alias of Brown; of a bookbinder who does not dare to join a bookbinders' union; of a bricklayer without a trowel; of a waiter who takes tips not from customers, but from manufacturers. Such were the men who undertook to dictate to the cigarmakers what they were to do, and negotiated with the manufacturers in the name of their workmen without any authority from those men. These men obtained the Knights of Labor label for cigars at the rate of $1 per thousand, and sold them to the manufacturers at $1.50 per thousand.

Robert Haller,[4] another of the speakers, said that he had been asked to keep the quarrel between the cigarmakers and Knights from the public press. He would answer that the cigarmakers had done all in their power in that direction, but District Assembly No. 49 insisted on interfering with their affairs, and when they protested the Home Club replied that it had wiped the Internationals out of existence and that the Progressives must go too. An appeal was made to the General Executive Board, and the cigarmakers were only sneered at. Then the matter was taken before the General Assembly, but the Home Club's influence was so great that the issue was dodged. John Morrison,[5] Master Workman of District Assembly No. 26,[6] also denounced the action of the Home Club in vigorous language. Resolutions were unanimously adopted denouncing District Assembly No. 49 and the Home Club for their unwarranted attack upon the cigarmakers' unions, and approving of the proposed amalgamation of the unions.

*Tobacco,* Aug. 6, 1886.

1. In March 1886 the *Cigar Makers' Official Journal* announced the chartering of local 251, composed of New York City cigar packers.

2. In 1886 a new power-driven bunching machine was developed that mechanically shaped and rolled the cigar bunch and allowed manufacturers to substitute unskilled for semiskilled workers. Its use was limited because it produced only a short-filler cigar while the predominant product at this time was the long-filler cigar.

3. Thomas B. McGuire, a marble polisher and truck driver, was a member of the KOL Executive Board and a leader in KOL District Assembly 49.

4. Frederick Haller.

5. John Morrison, a machinist and a founding member of the Home Club, later had charge of advertising for the *American Federationist.*

6. District Assembly 126, a trade assembly of carpet workers, was headquartered in New York City. The KOL chartered the assembly in 1886 and expelled it in 1887.

## An Excerpt from an Article in the *United States Tobacco Journal*

[August 7, 1886]

OUT IN THE WARM.

. . .

Mr. Samuel Gompers, Second Vice President of the Cigarmakers' International Union, said that the cigarmakers expected to have a prolonged fight with the Knights of Labor and the United Cigar Manufacturers' Association. The attitude of the latter indicated that they were going to starve the Progressives into submission to the Knights of Labor, if possible; but the cigarmakers were prepared for them, and he thought the manufacturers would very soon come to their senses when they got tired of the embarrassment resulting from not having sufficient skilled workmen to keep their factories in operation.

. . .

*United States Tobacco Journal,* Aug. 7, 1886.

## An Item in the *Workmen's Advocate*

[August 15, 1886]

SHAKING HANDS.

*The Picket,* a little weekly journal printed in the interests of the cigarmakers, with a leaning toward the old Union, tenders its horny hand to the Progressives in the following straight-forward manner:

"Of the questions that divided the Unions we think it best to say nothing; we would both maintain our relative positions and accomplish no good results. What we wish to do is of paramount importance to aid as far as we can, to assist by all means in our power, the manifest desire on the part of the members of the Progressive Union to join forces, to amalgamate with the International, to wipe out the blot that has existed in our trade the last few years—again become one, in fact and in deed—to bleed when they bleed, to enjoy peace when peace is at hand, to be brothers not only in name but fact, to prevent those who, ghoul-like, would goad us on to a sanguinary end, that they might feed on the decaying carcasses of both."

That's well, boys; and may the Union be permanent and lead the united cigarmakers forward to great achievements in the field of labor reform.

*Workmen's Advocate* (New Haven, Conn.), Aug. 15, 1886.

## An Article in the *United States Tobacco Journal*

[September 4, 1886]

A BIG STRIKE COMING.

A general strike of the cigarmakers in this city is regarded as only a question of time by all the larger cigar manufacturers. In order to get something definite as to the cigarmakers' intentions, a reporter sought Mr. Samuel Gompers, the Vice-President of the Cigarmakers' International Union.

"Mr. Gompers," he said, "what truth is there in the report that the cigarmakers in this city are on the eve of a general strike for higher wages?"

An impressive shrug was the answer.

"The manufacturers expect a strike. The United and the American Cigar Manufacturers' Associations have amalgamated and a new association will be formed, embracing nearly all of the manufacturers in the city."

"A strike by the cigarmakers is not improbable. Polonius' advice to Laertes is applicable to the cigarmakers just now: 'Beware of entrance in a quarrel; but, being in, bear't that the opposed may beware of thee.' I don't know what the union will do. No demand for higher

wages has yet been made, and the union has no official knowledge that a strike is contemplated."

"But aren't the cigarmakers getting ready for a strike?"

Another impressive shrug was the only answer.

"Will a demand for more wages in International shops be made?"

"We have a contract with Kerbs & Spiess, Foster, Hilson & Co., and others, which is for the term of one year. In the shops where this contract exists there will be no demand during its duration. We shall live up to the letter of our contract—on that you can depend. Of the shops that are not strict International shops, but where Internationals and Progressives are working, I am unable to say."

"Has anything been done about exacting royalty for labor-saving machines?"

"Nothing; but we have been struck with the absurdity of accepting, say $25 a year in lieu of the wages of two or three men, which the machine displaces. I am not opposed to bunch machines, or, in fact, any kind of cigar labor-saving machinery; only the cigarmaker ought to share justly in the benefits with the manufacturer."

"Is the consolidation of the Progressive Union with the International Union a fact?"

"Yes; it is a fact. The final returns of the vote will be in on September 8.[1] The consolidation has to be voted on as a whole. In some districts where there are already International unions, and where the district is not large, there has been some objection to the granting of a charter to another union, but here in New York, where all the trouble of the Progressives has been almost entirely borne, we can afford to be magnanimous, and I expect Internationals all over the country will see the matter in the same light."

The reporter afterward saw a member of the executive committee of the Progressive Union, who was not as diplomatic as Mr. Gompers, and who needed little incentive to talk right out in meeting. Said he: "We cigarmakers are only biding our time. We are going to strike, but not before our plans are fully matured, so that the strike will be general. We ain't going to jeopardize the success of the strike by striking only in a few shops. The cigarmakers are now working very quietly, and are only waiting for the consolidation of the Progressive and International Unions. The ratification will be completed on September 8, and after that look out for squalls."

"Will there be strikes in strictly International shops?"

"There won't in shops where we have a contract; but in the shops of the Big 14[2] there is bound to be trouble, and in many more factories besides. The wages received in all of the shops are too low. The cigarmakers feel that they are not getting enough pay, and our aim

will be to get a raise of from $1 to $3 a thousand to make the prices uniform."

The cigarmaker chuckled over the prospects, and went away with every appearance of satisfaction at the news which had just been imparted.

*United States Tobacco Journal,* Sept. 4, 1886.

1. That is, the vote of the CMIU membership. The Progressives approved the consolidation in August.

2. The fourteen members of the United Cigar Manufacturers' Association.

## The National Executive Board of the Cigarmakers' Progressive Union of America to Its Local Unions

New York, September 10, 1886.

To all Local Unions of the Cigarmakers' Progressive Union of America:

Members—

The articles of agreement proposed by the last Special Convention of the Cigarmakers' Progressive Union of America, held on August 10, 1886, is also accepted by the members of the International Union of America, and the consolidation is therefore effected. We, the undersigned members of the National Executive Board, do to-day declare the Cigarmakers' Progressive Union of America disbanded.

Our duties as officers are ended with date, and are herewith laid into the hands of the Executive Board of the Cigarmakers' International Union of America.

All local unions having applied for a charter from the Cigarmakers' International Union will from date be under the jurisdiction of the Cigarmakers' International Union of America.

At such places where no charter is applied for, our members will at once merge with the Local International Union of their respective city. Books of membership will legitimize.

Members: there is only one Cigarmakers' Organization now, and we therefore appeal to you to fulfill your duty as Union men and maintain the principles of organizing the working people of the world.

National Executive Board
Cigarmakers' Progressive Union of America.
Anton Frank,[1]
Secretary.

*United States Tobacco Journal,* Sept. 11, 1886.

1. Anton FRANK, a New York cigarmaker, became treasurer of Cigar Makers' Progressive International Union 90 after the Progressives merged with the CMIU.

# A Committee of the Philadelphia Trade Union Conference to the Officers and Members of All Trade Unions

[September 28, 1886]

To the Officers and Members of all Trades Unions:

The standing committee of the national and international trades union conference held in this city May 18 last, have performed their work, and presented to the General Assembly of the Knights of labor, in special session at Cleveland, O., the treaty prepared by the trades union[s]. To-day the said trades union committee have waited upon the general executive board of the Knights of Labor to secure some definite action at their hands at the forthcoming general assembly at Richmond, Va.[1] We desire to report that we have the most positive assurance from General Master Workman T. V. Powderly and some members of the general executive board of the Knights of Labor that they will use every endeavor at Richmond to establish proper and satisfactory relations with the trades unions. It is with the utmost satisfaction that we note that all the trades unions of America have had an unparalleled growth since last May. The trades union movement is by no means a failure, but the past short-lived attempt at organizing for strike purposes and then to disband, whether successful or defeated, have been failures.

The life, strength and perpetuity of a labor organization is in high dues, sufficient benefit[s] and strict discipline. This is the basis upon which trades unions are founded and they have stood the test of time and are destined to have a more glorious future in America. They have braved the fierce opposition of capitalists, and though in their day they have had to contend with conspiracy laws and blacklist, they have raised the level of wages and reduced the hours of labor and removed many evils affecting the working classes. To accomplish this work has cost unmeasured sacrifices on the part of trades union men, and for years the active men in the movement have labored, not alone for their own trade, but for all classes of labor as well, so that after all there is a fraternity of feeling among trades union men in behalf of all branches of honest toil. This was amply proven when the trades

unions rushed to the aid of the telegraphers during their memorable strike and in the lockout in Hocking Valley.[2]

At all times trades unions have been ready to render moral and financial assistance to all branches of struggling labor. We hold it to be the duty of every workman to organize for his own protection, and we believe there will be of necessity various forms of labor organizations, but this requires by no means that there should be any antagonism between any of these organizations. They should work together in harmony and avoid clashing. This is the desire of all true workmen. We look to the Richmond General Assembly of Knights of Labor to outline a policy whereby all phases of the labor movement may work together side by side and move on with majestic and powerful strides and uplift the working people.

William Weihe, Chairman.
P. J. McGuire, Secretary.
P. F. Fitzpatrick.
Christopher Evans.
Adolph Strasser.

*Granite-Cutters' Journal,* Oct. 1886.

1. The KOL General Assembly met in Richmond, Va., Oct. 4-20, 1886.
2. A strike of 3,000 miners, caused by a reduction of wages, began in June 1884 in the Hocking Valley, Ohio, coal fields. It became a lockout when the mine operators would not allow the miners to return without signing a contract eliminating their union. In March 1885 the lockout ended with the miners accepting the reduction and the operators withdrawing their demand for a nonunion contract.

## An Article in the *Philadelphia Public Ledger*

[September 30, 1886]

TRADES UNIONS AND THE KNIGHTS OF LABOR.

A committee consisting of William Weihe, of the Amalgamated Association of Iron and Steel Workers; P. J. McGuire, of the Brotherhood of Carpenters and Joiners; A. Strasser, of the Cigar Makers' International Union; D. M. Pascoe,[1] of the International Typographical Union; Christopher Evans, of the National Federation of Miners and Mine Laborers; P. F. Fitzpatrick, of the Iron Moulders' National Union, and J. B. Dyer, of the National Granite Cutters' Union, on Tuesday held a conference at the Bingham House, with General Master Workman Powderly and John W. Hayes and Thomas B. Barry, of the General Executive Board of the Knights of Labor. Their object

was to secure some decisive action at the General Assembly of the Knights of Labor at Richmond towards securing friendly relations between trades unions and the Knights.

In an address to trades unions, the committee, of which Mr. Weihe is chairman, says: "We have the most positive assurance from T. V. Powderly and some members of the General Executive Board of the Knights of Labor that they will use every endeavor to establish proper and satisfactory relations with trades unions."

Before leaving the city yesterday a member of the Trades Unionists' Committee said: "In my opinion the two great bodies of organized labor are no nearer a settlement of existing differences than they were before, and it is improbable that the desired end will be accomplished at Richmond. The Knights desire [us?] to recognize the working cards of a number of expelled members of trades unions who have since been organized into local assemblies of the Knights of Labor, and this we will never agree to do."

*Philadelphia Public Ledger,* Sept. 30, 1886.

1. David M. PASCOE was secretary-treasurer of the International Typographical Union (1886-87) and editor of the *Tocsin* (1884-87).

## *Labor Politics, the Henry George Campaign, and the United Labor Party*

In 1886 a coalition of trade unionists, socialists, Knights of Labor, and reformers, coordinated by the Central Labor Union (CLU), conducted a campaign to elect Henry George as mayor of New York City.[1] This resurgence of independent labor politics in New York, which was triggered by a significant increase in labor organization membership, had its counterpart in many other cities that year. George, a well-known reformer and author of the popular *Progress and Poverty,* ran on a program that denounced monopoly and called for the reform of land ownership, the sanitary inspection of buildings, equal pay for equal work, an end to government use of contract labor, and the public ownership of utilities. His platform demanded the democratization of the judicial system and the abolition of special privileges, and affirmed the right to private property in everything except land.

The CLU's political campaign overcame deeply rooted ethnic and political divisions within the ranks of New York City workers. As Gregory Weinstein, a delegate to the CLU, pointed out, that organization itself was "the battleground of all the mixed, antagonistic, warring elements" that comprised the city's labor movement. Nevertheless, men who previously had been locked in bitter controversy came together in the George campaign, and conflicts between socialists, greenbackers, land reformers, trade unionists, and members of the KOL subsided, if only for the moment. Together these diverse forces established a campaign newspaper, the *Leader,* which was printed at the *New Yorker Volkszeitung*'s facilities, edited and written by New York City journalists who donated their labor, and supported by contributions from local trade unions. They held scores of mass meetings, established local organizations in every precinct and every trade in the city, and climaxed the campaign with a procession in a driving rainstorm on the last weekend in October. The campaign, in Gompers' later view, "united people of unusual abilities from many walks of life" and "proved a sort of vestibule school for many who later undertook practical work for human betterment."[2]

Democrat Abram Hewitt eventually won the election for mayor with 90,552 votes to George's 68,110 and Republican Theodore

Roosevelt's 60,435. George's strong showing, despite opposition from the business community, the Catholic Church, and the Tammany Hall organization, encouraged the CLU, the SLP, and members of the KOL to organize the United Labor party (ULP) in January 1887. While the *Leader,* edited by the socialist Sergius Shevitch, initially supported the new party, George founded the *Standard* in January to promulgate his views. The paper was dedicated to the single tax, that is, the proposal that the government utilize its power of taxation to absorb 100 percent of all rents.

The alliance weakened, however, after the SLP drew up a platform attacking private ownership of the means of production. George and his allies, fearing that the socialist plank would alienate farmers and small businessmen, continued to emphasize land reform and the single tax. SLP supporters were excluded from the ULP's state convention in August that nominated George for secretary of state, the highest office open during that election. The socialists subsequently founded the Progressive Labor party and ran their own slate in the fall. Both parties disintegrated, however, after poor showings in the election.

*Notes*

1. The conviction and imprisonment of five New York City trade unionists on charges of extortion arising from a boycott of the Theiss Music Hall sparked labor participation in the 1886 mayoralty campaign. The firm's owner had agreed to pay $1,000 toward the costs of the boycott in exchange for its termination, but subsequently sought prosecution of the boycott's leaders. Their sentencing outraged the trade union community.

2. Remarks at the Edward King Memorial Meeting, Jan. 28, 1923, Edward King Papers, Labor-Management Documentation Center, Cornell University, Ithaca, N.Y.; SG, *Seventy Years,* 1:313.

# An Excerpt from a News Account of a Mass Meeting at Cooper Union

[October 6, 1886]

HURRAHS FOR GEORGE.[1]

. . .

A Grand Rally of Workingmen at Cooper Union.

. . .

The next speaker was Samuel Gompers, President of the State Workingmen's Union, who said: "At election time the politicians stand near the polls loaded to the muzzle with crisp new two-dollar bills. On election day men who would not give you a cent to save you from starving are only too anxious to hand you a two-dollar bill. Boycott them and their two-dollar bills. Vote for the candidate of progress and of poverty, Henry George. (Cheers.) It will be necessary to have a Labor representative at the polls. Let the tailor, the carpenter, the bricklayer and all of our trades have a representative at the polls to see that every friend of labor votes for Henry George and for him only. To those who have misconstrued the law, imprisoned our brothers, indicted our fellows and held the menace of the penitentiary over our heads, to those let us show that we cannot be clubbed into submission."

. . .

*New York World*, Oct. 6, 1886.

1. The well-known labor reformer Henry GEORGE accepted the nomination for the mayoralty on Oct. 5. In November he ran second as the liberal and labor candidate in a three-way contest against Theodore Roosevelt and the victorious Abram S. Hewitt.

# Excerpts from Two News Accounts of a Meeting of the Tobacco Section of the New York City Central Labor Union[1] at Nilsson Hall

[October 11, 1886]

FORMING A GEORGE LEGION.

About 500 cigar makers and tobacco workers met yesterday morning at Nilsson Hall, East Fifteenth street, and formed a Henry George Legion like that of the printers. Henry Zoatzer had charge of the

bookstand in the hallway outside the meeting room door, and was offering for sale the works of Henry George, Karl Marx and others. As each person went into the hall he was handed a card to fill out, with blank spaces for his name and residence, that he might readily indicate his desire to join the legion.

On the platform were Samuel Gompers, president of the State workingmen's assembly; Frederick Haller, Ludwig Jablinousky,[2] Adolph Lustig,[3] Raymond De Armay, R. Casuss, Joseph E. Nejedlz,[4] Joseph Vince[5] and Max Pavelk. Jablinousky called the meeting to order, and named Joseph Vince of the Cigar Makers' International Union, No. 144, for president. Mr. Vince in taking the chair outlined the object of the meeting and urged unity and industry in the work for their candidate, Henry George. He introduced Samuel Gompers as the first speaker. Mr. Gompers spoke for about thirty minutes. He spoke on the land question as treated by Henry George, and then addressed himself to the task of rousing the enthusiasm of his audience. He attacked both the old political parties, referred humorously to the boodle Aldermen,[6] and said that Jay Gould was even more independent than the late William H. Vanderbilt, because Gould said: "The public be ——— and the law be ——— too!" The telegraph wires were still over ground.[7] While he was talking a good-looking man with a full beard entered the hall. The audience thought he was Henry George, and they began to cheer and applaud him. The good-looking man blushed and sat down, and then the speaker thought it was time to read the following letter from Henry George:

Samuel Gompers, Esq.:

My Dear Sir—I much regret my inability to be present at the meeting of the tobacco section of the Central Labor Union on the 10th, the more so that my name has been announced. Please express to the members of the section my high appreciation of this invitation, and my cordial thanks for their support. I hope some other opportunity for meeting them will present itself before election.

Fraternally yours, Henry George.

After Mr. Gompers' speech the following preamble and resolutions were adopted:

*Whereas,* The workingmen of this city, in accord with the spirit of the times, regard the corruption of our officials, the despoilers of the people's moneys, the peculation in office, the perversion, over-riding and trampling under foot of constitutionally guaranteed rights as a series of directed blows aimed to overawe and coerce us into a submission not to law but to illegal force.

*Whereas,* The workingmen of this city, in their organized capacity, have nominated Henry George, that able, fearless and honest expo-

nent of reform in politics as well as economics, for the Mayoralty of this city, as a protest against the corruption and the usurpation of our liberties; therefore be it

*Resolved,* That we, the organized cigar makers and tobacco workers of this city, in mass meeting assembled, heartily indorse the nomination of Henry George, and pledge ourselves, individually and collectively, to do all that honest citizens can do to secure his election.

*Resolved,* That we call upon all workingmen and other honest, well-meaning, liberty-loving men of our city to enlist in the great cause for the perpetuation of our republic, remembering that "eternal vigilance is the price of liberty," and that those who would be free themselves must strike the blow.

The resolutions were adopted unanimously.

. . .

*New York Star,* Oct. 11, 1886.

1. During the summer of 1886, in order to provide more time for political discussion, the New York City Central Labor Union (CLU) reorganized itself and formed ten trade sections that met separately to consider all trade difficulties.

2. Ludwig JABLINOWSKI, who became secretary of Cigarmakers' Progressive Union 1 in 1885, was financial secretary of the New York City CLU from 1884 to 1886.

3. Adolph Lustig (b. 1857), a Bohemian-born cigarmaker who immigrated to the United States in 1881, held positions as the recording, financial, and corresponding secretary of New York City CMIU 141 during the mid-1880s.

4. Joseph E. Nejedly (1862-1918), a Bohemian-born New York City cigarmaker, represented CMIU 141 at the New York State Workingmen's Assembly in 1887. In 1900 he was a deputy collector for New York City.

5. Joseph Vince (1842-1904), born in England, became a member of CMIU 144 and later of CMIU 97 in Boston.

6. A report of the New York Senate Committee on Railroads in 1886 revealed that the Broadway and Seventh Avenue Railroad Co. had illegally created a paper organization known as the Broadway Surface Railroad Co. that, through bribery, won a franchise from the city to build a Broadway surface railroad. The committee found that there was no legal authority to construct the railroad, and the twenty aldermen implicated in the scheme were either prosecuted or left the city to avoid prosecution.

7. An 1884 New York City statute required the removal of all telegraph, telephone, electric light, or other wires and cables installed above city streets. The Western Union Co., controlled by Jay Gould, was able to continue stringing lines, first by defying the regulation and subsequently by arranging a compromise with the enforcing agency, the Board of Commissioners of Electrical Subways.

[October 11, 1886]

## LABOR'S POLITICAL WAVE.

. . .

The day began with a meeting of the tobacco workers of the city at Nilsson Hall, East Fifteenth street, in the morning. About 600

orderly, neatly dressed cigarmakers assembled in the hall. Chairman Julian Vince[1] opened the meeting with a brief address, urging his listeners to stand by the candidate of organized labor and to make good the pledge it made to Henry George.

The first speaker, Samuel Gompers, president of the Workingmen's State Trades Assembly, said he had always opposed independent political action on the part of labor unions, but he thought the time had come when workingmen should select as well as vote for candidates. He said that if workingmen were true to themselves Mr. George's election was certain.

. . .

*New York Morning Journal,* Oct. 11, 1886.

1. Joseph Vince.

# An Excerpt from an Article in the *New York Star*

[October 17, 1886]

## TALKING TO THE WORKMEN.

Last night more than thirty thousand citizens of New York shouted and threw up their hats for Henry George for Mayor. There were sixteen meetings. Marion square, the home of Hon. Patrick Henry Doody,[1] again resounded with the loud huzzas of the third district Georgites. The crude and fiery eloquence of Sam Gompers and the calm, sensible remarks of Thomas J. Ford,[2] the brass worker, commanded the attention of the audience. Three cheers for George and all dispersed. . . .

*New York Star,* Oct. 17, 1886.

1. Patrick Henry Doody, a bookkeeper and clerk, lived and worked in Manhattan into the 1920s. He was active in the New York City Central Labor Union and the Henry George campaign, participated in the Greenbacker political movement, and served as secretary of KOL District Assembly 49.

2. Thomas J. Ford, a metalworker, became a walking delegate for the brass and silver workers' union in New York City.

# A Resolution Introduced by District Assembly 49 at the General Assembly of the KOL

[October 19, 1886]

DOCUMENT 160.

The Local Assemblies of the Cigar Trade Industry combined in an Executive Board, known as the "Executive Board of the Cigar Trade," attached to D.A. 49, Knights of Labor.

At a meeting of the above mentioned Board it was resolved, to present the following for the consideration of the General Assembly.

Through the continuous and unremitting attacks and inimical action on the part of the International Cigar-maker's Union, it is plain to see that the above Union is doing all in its power to injure the Order of the Knights of Labor, and their attitude being such that it is impossible to arrive at an agreeable understanding with said Union—the attempts at boycotting the goods bearing the Knights of Labor label, and also the shops that are loyal to the Order are progressing without intermission on their part; therefore, be it

*Resolved,* That the General Assembly be petitioned to indorse the order issued by D.A. 49, to the effect that all cigar-makers, packers or whosoever are employed in the cigar trade who are members of the Knights of Labor and also members of the International Union, to withdraw from said Union or leave the Order.

The Executive Board is actuated in presenting this petition by an earnest desire to promote the interest, welfare and harmony of the institution, and in asking the protection of the Order for all true Knights wherever dispersed in the United States and Canada, feeling assured that such action cannot but conserve to the best interests of the Order, whereby its dignity and usefulness may not be impaired. The present status of the matter is such that there are members of the International Union, who are also brothers in the Knights of Labor, who are doing all in their power to the prejudice of the Order.

We, therefore, hope and pray that the General Assembly will, in its best wisdom, devise and enact such laws bearing on this important matter as will redound to its dignity and future usefulness, and desire that the petitioners may receive that which the petition sets forth.[1]

KOL, *Record of Proceedings of the General Assembly of the Knights of Labor of America, . . . Richmond, Virginia, October 4 to 20, 1886* (N.p., 1886), p. 200.

1. The General Assembly passed this resolution. The 1887 General Assembly revoked it and ordered local assemblies to reinstate members of the CMIU.

# An Excerpt from an Article in the *United States Tobacco Journal*

Richmond, Va., Oct. 19. [1886]

WAR ON THE CIGARMAKERS.

The trades unions might as well begin to fortify, for war was declared against them to-day by the General Assembly of the Knights of Labor. The attack on the unions was made by District Assembly No. 49, of New York. With commendable foresight Maguire[1] has kept not only his own men, but every delegate who would vote as they did, on the ground up to date. No. 49 mustn't be financially comfortable, for this convention will cost the district a stiff sum. Many of the unionists have been compelled to quit the ground for lack of funds, and to-day Maguire saw that any measure proposed by him was likely to be adopted without difficulty. Powderly had just finished a statement that at the proper time he intended to introduce a resolution containing a plan for the settlement of any difficulty that might arise between labor organizations. He thought such difficulties should be referred to the heads of the organizations and by them settled. Right on the heels of this statement Maguire introduced a resolution providing that members of the Cigarmakers' International Union who were also Knights should sever their connection with one or the other organization. In spite of the protests of union delegates the resolution was adopted, but its adoption has generated a spirit of insubordination that ventures on open mutiny, for the unionists openly declare that though the resolution refers only to cigarmakers, it is really directed against all labor organizations outside of the Knights of Labor.

. . .

*United States Tobacco Journal,* Oct. 23, 1886.

1. Thomas B. McGuire.

# From P. J. McGuire

Cleveland, Ohio, Oct 21/86

Saml. Gompers
Pres. Fed of Trades. &c
Dear Sir & Bro—

I have just issued a call for a Trades Union Convention to be held

at Columbus O. Wednesday Dec 8, 1886 at 10 A.M. This is under instructions of Standing Trades Union Committee which met Sept 28, at Phila.

I have written J. S. Kirchner, Sec of Federation—as I was instructed—to invite and urge the Fed of Trades & Labor Unions to change its meeting place from St. Louis to Columbus for Dec 7, 1886, and then on Dec 8 to go into joint session with the convention.

Yours Truly P. J. McGuire.
Sec'y Committee

ALtS, The Papers of Gabriel Edmonston, reel 1, *AFL Records.*

## An Article in the *New York World*

[October 22, 1886]

THE KNIGHTS AND THE UNIONS.

The announcement in recent despatches from Richmond that the General Assembly of the Knights of Labor had adopted a resolution that members of the Cigar-Makers' International Union who were also Knights must sever their connection with one or the other organization, created much talk among members of trades-unions. More would have been said and the excitement more developed had the men not been so deeply engrossed in the political struggle. This was voiced by a prominent house-painter who belongs to a trades-union and the order. He said: "I have nothing to say just now, but I shall have in about two weeks." The cigarmakers were not so reticent and expressed themselves in forcible language.

George E. McNiell, who is regarded on all sides as one of the most conservative labor leaders in the United States, arrived in this city yesterday from Richmond on his way to Boston. He said the resolution was not exactly worded as given in the despatch. The action was decidedly contrary to the principles of the order. If the resolution was to apply to all trades it would take out the ablest men in the order, as there would be no question that the trades-unionists would prefer their unions to the Knights of Labor, because the former regulated the wages of the trades. In the Veteran Labor Corps more than half of the fifty members in Richmond had been in the labor movement for over twenty years, long before the order was established. In the corps were printers, shoemakers, cigarmakers, ironworkers and miners.

Samuel Gompers, First Vice-President of the Cigarmakers' International Union, said: "We are not going to give up either our union or the order. We shall permit them to expel us if they want to do so, but we shall maintain our union at all hazards. If the resolution is enforced it will take out of the Knights about 10,000 members. We shall of course prepare to retaliate."

M. Dampf,[1] Secretary-Treasurer of Cigarmakers' Union No. 144, said: "I have been a member of the Knights of Labor for over four years in the Defiance Association, which was formed when District 49 was expelled[2] from the order, and we have our charter from the General Assembly. The result of the resolution adopted yesterday will be bad for the labor movement in general and the order in particular. Outside of New York our members are among the most active Knights of Labor. The resolution really applies to all trades-unionists, because there is no reason why they should exclude one trade more than another. In my opinion District 49 will begin the war on our union as soon as the election is over. The manufacturers will take advantage of the situation and force wages down."

James W. Sullivan,[3] of Typographical Union No. 6, said: "If the resolution adopted in Richmond was applied to members of Typographical Union No. 6 who are also members of the Knights of Labor, the result would simply be to cause them to retain their membership in the union. The fight is between Forty-nine and the International Union, and this is only one of the phases. At present the printers have no trouble with Forty-nine. The International Union is the model trades-union of the country, and I should like to see the undoubted strength of District Assembly 49 exerted in other directions than in that of quarreling with fellow-workmen."

*New York World*, Oct. 22, 1886.

1. Meyer Dampf (1842?-95), a German-born cigarmaker, served as financial secretary of CMIU 144 from 1880 to 1895. His daughter, Sophia, married SG's eldest son, Samuel Julian Gompers.

2. In November 1882 Terence Powderly revoked the charter of Brooklyn Local Assembly 1562 for issuing circulars abusing the KOL general officers and for maintaining an illegal boycott against the Duryea Starch Company. Against Powderly's protests, District Assembly (DA) 49 admitted delegates from the expelled local, and in February 1883 Powderly suspended DA 49 and attached locals directly to the General Assembly. The General Executive Board investigated the suspensions and overruled Powderly, and DA 49 was reinstated in August.

3. James W. SULLIVAN, a printer, worked for the *Leader* during the Henry George campaign and edited the *Standard* with George from 1887 to 1889. He was a longtime close associate of SG's, serving with him in the National Civic Federation and helping edit the *American Federationist.*

## An Item in the *Picket*

[October 23, 1886]

A STRANGE EPIDEMIC.

From the Picket.

A strange epidemic, known as nervous excitement, is creating havoc among the ring and hall politicians of New York City. Having paid close attention to the symptoms of the disease and the behavior of the patients, we have arrived at the firmly fixed opinion that the disease is in its first stage only. We will venture to predict that it will be nervous and political prostration. The cause of this disease is the Coun(ty) Tammin(y)ation[1] of the atmosphere in the vicinity of the City Hall, brought about by defective plumbing of the drains heading from the city treasury. The people of New York are going to send a plumber down there, who is a union man, to do a union job and stop the leaks. That plumber is Henry George. When he gets in his work, we shall have a healthier class of statesmen.

*Boycotter* (New York), Oct. 23, 1886.

1. A reference to the alliance between two wings of the New York City Democratic party, the County Democracy and the Tammany Hall political machine. This alliance dominated the city government.

## An Item in the *United States Tobacco Journal*

[October 23, 1886]

THE CIGARMAKERS WILL STICK.

Samuel Gompers, one of the Vice Presidents of the Cigarmakers' International Union, was asked by a *U.S. Tobacco Journal* reporter what the union would do after the success of District Master Workman T. B. Maguire[1] in passing a resolution through the Richmond Convention that the union must either disband or leave the Knights of Labor. Mr. Gompers replied that the International Union would neither disband nor leave the order, but would fight Maguire and District Assembly No. 49. Of the 22,000 members of the International Union 10,000 left the order after the last fight of District Assembly No. 49 against trades unions. The rest would remain in the order until they were expelled, but they would never abandon their union.

*United States Tobacco Journal,* Oct. 23, 1886.

1. Thomas B. McGuire.

## To the Legislative Committee of the FOTLU

New York Oct. 25th 1886

the Legislative Committee of the Fed. &c.
Dear Sirs.

A crisis in the labor movement has arrived by reason of the action of certain persons who would crush out the rock-bed of labor reform[,] the Trades Unions.

At the last session of the Federation our late Secretary W H Foster was directed to ascertain from the authorities of the K of L whether they would discountenance the evident and systematic warfare that was being waged against Trades-Unions. The officers of that Order were most prolific in their disavowels of knowledge and condemnation of such actions.

In spite of these statements however not only the officers but the Order through its General Assembly has resolved to wage that war fiercer and with greater energy than ever.

You will remember that in Ex session the Officers of this Federation were directed to expose the action of these men and their organization if they failed to satisfy us that their warfare would cease. As Officers we have refrained from so doing but have notwithstanding kept a close watch upon their doings and shall report at the proper place, the Congress.

The Trades Unions of the country in the past year were forced to take decisive action against the actions above recited and a conference of the Ex Officers of 23 National and International Unions was held in Philadelphia several months ago (which I attended) and such action taken by them to more closely draw the Unions together. It is possible to make our Federation a success if we adopt the suggestion and recommendation of that Committee, a copy of which I enclose herewith.

I therefore propose, That we—by the power in us vested as the Ex Committee of the Fed. to do all we can to further the interests of and bring about a greater Unity of the forces of labor—change the locality of meeting of the session of the Federation from St Louis to Columbus O. on December 7th 1886.

Please forward your vote as soon as possible (as time is pressing) to J S Kirchner 543 Dillwyn St Philadelphia Pa, upon [the] proposition here submitted.

In the hope that we may arise to the necessities of this great occasion. I am,

Fraternally Yours Samuel Gompers.
Pres.

ALS, The Papers of Gabriel Edmonston, reel 1, *AFL Records.*

# An Excerpt from a News Account of a Mass Meeting at Cooper Union

[October 27, 1886]

GEORGE'S GREATEST RALLY.

. . .

Samuel Gompers said that the movement they were engaged in was a constitutional one; they had a perfect right to engage in it. But it was unwise for a man in Mr. Hewitt's position to say that the success of their movement would be dangerous to the commonwealth. It might be the other way. It might be that that would come to pass which Dickens had put into the mouth of one of his characters: "Gentle folks, be careful, for it may come to pass that even the Bible will change in our altered minds." The passage that we have often read may read that "whither thou goest I cannot go; where thou lodgest I cannot lodge; thy people are not my people, nor thy God my God." Do not attempt to spread the schism too far, Mr. Hewitt, the speaker went on, "for fear that you may bring forth, instead of this constitutional and legal method, men who will say:

Oh, angels shut thine eyes!
Let conflagration light the outraged skies
And a red Nemesis burn the hellish clan,
And chaos and the slavery of man.[''] (Cheers.)

. . .

*New York World*, Oct. 27, 1886.

# An Article in the *United States Tobacco Journal*

[October 30, 1886]

THE ELECTION.

The Mayoralty campaign, which will be fought out next Tuesday, will be memorable from the interest taken in it by the cigarmakers. Their candidate, as is well known, is Henry George, and for once the cigarmakers, with other labor organizations, are united in his interest. Every cigar and tobacco factory in this city is organized into a Henry George Club, who not only announce their purpose of voting for him

for mayor, but cheerfully contribute from their earnings toward accomplishing this result.

In a canvass made by a reporter of the *U.S. Tobacco Journal* among the cigar and tobacco factories, every one was found to be enthusiastic and unanimous in his favor.

There was not a single exception. There are totally about 15,000 cigarmakers in the city. Between nine and ten thousand belong to the Cigarmakers' International Union. The officers of the union aver that two thirds of these are voters. But a large proportion are women and a great many foreigners are known to be in the ranks of the cigarmakers. There are about 6,000 voters among the cigarmakers, all told. This is a very liberal estimate of their voting strength, although not as large as what they themselves claim. The cigarmakers will parade to-night, and muster fully 12,000 men; but this is not to be taken as an indication of their voting strength. The parade will comprise, besides cigarmakers, packers, strippers and tobacco factory employees. It will be under the marshalship of L. Wagner, with Assistant Marshals Jacobs and Benke.

Heretofore an election has passed without any excitement, except now and then one of their number has run on the Socialistic ticket, and has created a little flurry among the cigarmakers. That has been confined, hitherto, to wards and assembly districts. Never before has there been a candidate upon whom the cigarmakers could with enthusiasm unite. You ask why this is so? The answer is plain. Henry George is a powerful writer and subtle talker, whose ideas are peculiarly in harmony with the professed doctrines of the workingman, particularly the cigarmaker. The cigarmaker is a philosophical student or he is nothing. As he sits at his bench making bunches or rolling up cigars, his employment does not prevent his mind having full scope and his tongue great activity. He has by dint of argument solved, to his complete satisfaction, all the vexed social questions, and now he is jubilant at the chance of putting his theories in force. Any cigarmaker will tell you, in all seriousness, that with Henry George elected, he expects forever after to toil not and live a life of ease. The millennium, with Henry George, has come, and hereafter the cigarmaker, if he chooses to work, will expect to get a higher reward than a fine house, a rich dinner, good clothes, with the countless anxieties of the present bosses. They have looked at the manufacturer's lot in life, and have taken it for granted that all is gold that glitters.

Meantime, what are the manufacturers doing? The movement for Henry George has been so strong, that any effort of theirs has been powerless to stem the tide. They let the men in their factories have their way, confident that the common sense of workingmen, not ci-

garmakers, will result in a man being elected for Mayor of this first city of the Union, who will not be Henry George.

The cigar manufacturers of this city, with one accord, don't wish their political views made known, but with not a solitary exception, they are not for George. Abram S. Hewitt is their choice. He is regarded as an eminently practical man. He is known as an honest man, who is not a doubtful experiment. He knows perfectly the requirements of the city, and is above the dealings of politicians, and withal strictly honorable. Roosevelt[1] is generally thought to be too young, although he will get some supporters from cigar manufacturers. George is dismissed as a mere theorist.

A cigar manufacturer thinks he sees troublous times in the near future. He predicts, in the event of Henry George's election, the early closing of his factory. "Before thirty days," he says, "there will be no resisting the demands of the cigarmakers. They will be so elated at the idea of displacing the office-holders that they think they will have the right to go for the manufacturers. Thirty days more," says this pessimist, "and riots will be the order of the day."

Even should the calamity of Henry George's election occur, things won't be as bad as this alarmist predicts. There will be, no doubt, trouble with the cigarmakers. Before they find out that there is not going to be the Elysium that the cigarmaker thinks, there will doubtless be many strikes; but it won't be long before order resumes her sway. Henry George can't override the laws and can't create a law unto himself. His functions as Mayor will be distinctly prescribed. His peculiar ideas about land and taxation are in no danger of being fulfilled. The Mayor is not, at the same time, the Governor, the Legislature and the courts. They can be depended on to resist any encroachments of the Mayor; but it is not expected that any such duel will take place, and on next Tuesday, instead, Henry George can harmlessly give forth his theories, and the cigarmakers can as harmlessly talk over their work in the factories, and wonder how they were defeated.

*United States Tobacco Journal*, Oct. 30, 1886.

1. Theodore Roosevelt (1858-1919), president of the United States from 1901 to 1909, was a Republican New York assemblyman (1882-84), president of the New York Police Board (1895-97), and governor of New York from 1899 to 1900. He was elected vice-president of the United States in 1900, assuming the presidency in 1901 after William McKinley's assassination.

# An Excerpt from a News Account of a Mass Meeting at Irving Hall

[October 30, 1886]

. . .

Samuel Gompers, President of the Federated Trades and Labor Unions of America and Canada, said that by the appearance of the hall "it did not look very much as if the George boom was dying out, as certain papers had stated." If the dying continued he said Abram S. Hewitt would not be heard of after next Tuesday. In accepting the nomination of the United Democracy Mr. Hewitt had stigmatized the workingmen as Anarchists. "If we are always stigmatized as Anarchists," he said, "when we do anything to better our condition; if we continue to be so stigmatized, by and by we shall consider that there is no harm in anarchy. If, when we resist employers reducing our wages below a living basis, we are called Anarchists, we shall know that anarchy is not wrong. If they tell us to appeal to the ballot-box and when we do so they call us Anarchists, then anarchy is not wrong. I am not a friend of Anarchism but its enemy, because it is not a friend of labor organizations. What we seek is to improve our conditions. We have yet left to us certain inalienable rights, one of which is the expression of our opinion through the ballot-box, and if we, as workingmen, choose to exercise that right, it is in the interest of the country that we should do so.

"The politicians are accusing us of not being respectable. If the party that has produced the Jaennes[1] and Squires[2] and the boodle Aldermen is respectable I don't want to be respectable. The spoilsmen are reduced to great straits when they have to hire professional liars to issue bogus reports of so-called workingmen's meetings, and a renegade like Carl Schurz,[3] who has never been true to any cause he ever espoused. Would you like to know why the Democrats haven't had a parade this year? I'll tell you. Because they are afraid to show their weakness. They have the politicians, but we have the men, and we will parade on our own account to-morrow night."

. . .

*New York World,* Oct. 30, 1886.

1. Henry Jaehne (b. 1849) was a New York City alderman between 1883 and 1886. In May 1886 he was convicted of accepting a bribe to vote for granting the Broadway Surface Railroad Co. a franchise and served six years in prison.

2. John Rollin M. Squire (1836-99) was commissioner of public works for New York City between 1885 and 1886. In August 1886 he was indicted for conspiring with a contractor but was later acquitted.

3. Carl Schurz (1829-1906), a Republican U.S. senator from Missouri (1869-75)

and secretary of the interior (1877-81), was denounced as a renegade by regular Republicans for his leadership of the Liberal Republican movement in 1872, which supported Horace Greeley over the Republican nominee, Ulysses S. Grant, for the presidency of the United States.

# An Announcement of a Demonstration for Henry George

[October 31, 1886]

THE PARADE.

Proclamation to the People of New York City:

On Saturday evening, Oct. 30, a grand torchlight demonstration of the supporters of Henry George for the Mayoralty will take place in the streets of the city.

All bodies of organized labor and all political organizations attached to the new party of equal rights, social reform, true Republicanism and universal Democracy are invited to participate.

It is intended to make this demonstration historic. In it will be represented the tens of thousands who, abreast the thought of the age and animated by the same spirit which gave birth to the Declaration of Independence, believe that the principles and institutions inherited by the citizens of America from the sages and heroes of the Revolution will, when applied to the vexed problems of the day, promote the general welfare and secure life, liberty and happiness to the individual. In it will be seen the majority of the workers of the city, who sustain society, who respect the laws of society, and who have been unjustly deprived of the blessings which should be secured by society. In it will be seen thousands who are deprived of their birthright of a footing on the earth, who must work one hundred days a year for a landlord, while the common methods of transportation could carry them conveniently to suburban homes if the city should justly resume its right to its thoroughfares and possess itself of the land nearby now held by speculators; who pay thrice the value of fuel because avaricious men have possessed themselves by means of unrepublican laws of the free gifts of nature; who pay enhanced prices for articles of general consumption by reason of tax laws bearing unequally upon classes, whose opportunities, owing to a monopoly by capital of the advantages of modern inventions, of obtaining work are curtailed by the employment of machinery in the various industries, and who know that the disparity in the condition of man is

created rather by injustice than by differences in men. As Thomas Jefferson believed that the earth in its fruits belonged to the living; as DeWitt Clinton[1] demonstrated in his day through the canal system that the means of transportation should be the property of the State; as Horace Greeley[2] asserted that every child born in the State of New York by right was entitled to a part of the soil of New York—so the men of this new party declare that certain functions of society should of right be administered by the State and certain privileges belong to all men by right of their existence.

All members of labor organizations, trades unions and Knights of Labor, and all citizens of the city of New York in favor of social and political reform are hereby invited to take part in the procession, and to assemble at the place indicated on Saturday, Oct. 30, at 7 P.M.

Sam Gompers, Secretary.
Wm. McCabe,[3] Chairman.

*John Swinton's Paper*, Oct. 31, 1886.

1. DeWitt Clinton (1769-1828), as New York canal commissioner and governor of New York, played a major role in the completion of the Erie Canal.

2. Horace Greeley (1811-72), founder and editor of the *New-York Tribune* and Liberal Republican candidate for president in 1872, was a leader of the land reform movement centered on providing citizens with free homesteads.

3. William McCabe, a printer, was a founder of the New York City Central Labor Union.

## A Statement by the New York City Members of the Executive Committee of the Political Branch[1] of the New York State Workingmen's Assembly

New York, Oct. 31, 1886.

The New York City members of the Executive Committee of the Workingmen's State Assembly met this morning, at No. 52 Stanton street, to examine the pledges made by all candidates for the Assembly, and find that candidates of all parties, with few exceptions, have signed the pledges in the interest of labor legislation. We recommend the election of Daniel E. Finn,[2] First Assembly District; Charles A. Binder,[3] Tenth District; Edward P. Hagen,[4] Sixteenth District; William Dalton,[5] Seventeenth District; John F. Kenny,[6] Eighteenth District; J. A. Cantor,[7] Twenty-third District, and John B. Shea,[8] Twenty-fourth District, as all these candidates were especially mentioned by the State Convention

as friends. We also, after investigation, recommend the re-election of James E. Powers,[9] of the Third, and Charles Smith,[10] of the Eighth District, as they have voted for our measures, except upon the two occasions when they satisfactorily explained their absence. We urge the defeat of Robert Ray Hamilton,[11] of the Eleventh Assembly District, as a pronounced enemy; one who has systematically opposed all legislation in the interest of labor.

We also recommend, in accordance with the resolutions adopted in our State convention that in the local contest in New York City, the candidate of the workingmen for Mayor, Henry George, shall be supported. We urge every member of organized labor, as well as unorganized labor, to lay aside all other work on the day of election and work for the election of Henry George.

George Blair,
Thomas Bernard,
Samuel Gompers,
John H. Dougherty,[12]
John J. Doyle,
Leopold Wagner,
W. W. Stone,
Oliver Smith,
Richard F. Aull.[13]

*New York World,* Nov. 1, 1886.

1. The New York State Workingmen's Assembly (NYSWA) formed its political branch in 1882 and replaced it with a legislative committee in 1888.

2. Daniel E. Finn (1845-1910) was an Irish-born saloon owner and later a lawyer. A Tammany Hall Democrat, he was elected to four terms in the New York Assembly (1885-87, 1895) and was appointed city magistrate in 1905.

3. Charles Archimedes Binder (1857?-91), a New York City lawyer, was a Republican assemblyman in 1883 and 1885.

4. Edward P. Hagan (1846-93), a Tammany Hall Democrat who was in the real estate and liquor business, was elected seven times to the New York Assembly (1879-80, 1885-89) and in 1891 to the state senate.

5. William Dalton (1852-1923), a butcher shop owner and a Tammany Hall Democrat, represented the Seventeenth Assembly District between 1886 and 1888.

6. John F. Kenny (b. 1853), a house painter born in South Carolina, represented the Excelsior Labor Club and Painters' Union 1 during 1884 in the New York City Central Labor Union and was elected to the New York Assembly as a Democrat in 1884 and 1886.

7. Jacob Aaron Cantor (1854-1921), a Democrat from New York City, served in the state assembly from 1885 to 1887 and in the state senate from 1887 to 1898. In 1913 he was elected to the U.S. House of Representatives for one term.

8. John B. Shea (1855?-1906), a civil engineer, was a Democratic assemblyman between 1885 and 1888 and was elected to the New York City Board of Aldermen in 1890.

9. James Edward Power (b. 1855?), a saloon owner and former patternmaker, was elected as a Democrat to the New York Assembly in 1885 and 1886.

10. Charles J. Smith (1851-99), a German-born saloon keeper, served five terms as New York assemblyman, four as a Republican (1884-87) and a fifth as a Tammany Hall Democrat (1889). He was also a New York City alderman.

11. Robert Ray Hamilton (1851-90), a Republican lawyer, served in the New York Assembly in 1881 and between 1886 and 1889.

12. John H. Dougherty represented KOL District Assembly 64 at the 1886 session of the Political Branch of the NYSWA.

13. Richard F. Aull, a printer born in Germany in 1843, immigrated to the United States in 1870. He represented International Typographical Union 6 at the 1886 convention of the Political Branch of the NYSWA.

# An Excerpt from a News Account of a Mass Meeting at Cooper Union

[November 7, 1886]

. . .

### DEFEAT CANNOT DISHEARTEN.

Mr. Samuel Gompers, President of the State Trades Assembly, said they had often been reminded that the George boom had died out, and that all that was left were a few men who were fond of agitating strife. Well, the agitators in all movements for progress in the world were the men who sounded the gong and called upon people to arouse. "They say to us," he continued, "that we have not elected our man. Well, we haven't, but we have done this[:] we have elected a leader. They talk to us of defeat—defeat to the workingmen, who have grown strong by defeat. Why, it was our defeat that made it possible to concentrate on Henry George. We will rise up stronger than ever, taking advantage of faults that we have committed. They talk of the reasons for our existence! I will venture to say that if called before a jury of intelligent citizens Tammany Hall and the County Democracy can't give a good reason for being alive." (Great applause.) Continuing, he said he had received hundreds of labor papers and each one of them gloried in the work of the workingmen last Tuesday. "The press of this city had been brought to acknowledge that something must be done for the workingmen; that the large vote showed that there were evils that must be remedied. Among the reforms that will take place as a result of the work of Tuesday last is the tenement-house reform; another is that labor shall not continue for more than eight

hours a day; there must be immediate reform in this, that our young children shall not be compelled to enter the factories before a certain age; the trades unions, too, must be legalized, the same as the Stock Exchange or the Cotton Exchange, and if we decide to boycott a fellow it shall not be construed to be an illegal act." In conclusion the speaker said: "We have accomplished a wonderful result. I did not at first believe that we should get over 30,000 votes, and I say now if there had not been the amount of bribery at the polls that there has been[,] Henry George would have been elected. (Great applause.) I am confident that if we stand by our unions in the future victory is sure to be ours.["] (Applause.)

. . .

*New York World,* Nov. 7, 1886.

## An Item in the *New York Sun*

[November 10, 1886]

MR. GOMPERS'S EYES WERE BLACKENED.

Samuel Gompers of the Central Labor Union, accompanied by Mr. Keene,[1] chief clerk of the Labor Bureau at Albany, went into the Kenwood House, corner of Bayard street and the Bowery, on Monday night and asked for cigars. John Howard,[2] the proprietor, handed them a box to select from. Gompers said the cigars did not have the union label on, and would not take any. Howard said they were union-made cigars, and took Gompers and Keene across the street to the factory of Thomas Plunket,[3] who made the cigars, and who says he employs 150 union men. The discussion grew very warm, a crowd gathered, and blows were struck. Yesterday Gompers appeared before Justice Duffy, at the Tombs, with both eyes in deep mourning, and charged Henry Hersch,[4] aged 56, with being his assailant. Hersch's lawyer proved that Hersch was innocent, and he was discharged.

*New York Sun,* Nov. 10, 1886.

1. Edward J. Kean.
2. John Howard (1844-99), a saloonkeeper and former cabinetmaker, was born in Ireland and came to the United States about 1862. In 1886 he purchased the Kenwood House, a hotel in the Bowery district.
3. Thomas J. Plunket (1843?-1901), born in Ireland in 1843, worked as a printer until 1886 when he began operating a cigar factory at 22 Bowery.
4. Possibly Henry Hirsch, a cigarmaker, tobacco merchant, and real estate agent in New York City.

# P. J. McGuire to the Officers and Members of All Trade Unions

Cleveland, O., November 10, 1886.

To The Officers and Members of All Trades Unions of America:
Fellow Workers:—

On May 18, 1886, a conference of the chief officers of various National and International Trades Unions was held in Philadelphia, Pa., at which twenty National and International Unions were represented, and twelve more sent letters of sympathy tendering their support to the conference. This made at that time thirty-two National and International Trades Unions, with 367,736 members in good standing.

Since then quite a number of Trade Union Conventions have been held, at all of which the action of the Trades Union conference has been emphatically and fully endorsed, and a desire for a closer federation or alliance of all Trades Unions has been very generally expressed. Not only that, but a great impetus has been given to the formation of National Trades Unions, and several new National Unions have recently been formed, while all the trades societies, with national or international heads, have increased in membership and grown stronger in every respect.

The time has now arrived to draw the bonds of unity much closer together between all the Trades Unions of America! We need an annual Trades Congress that shall have for its object:

1—The formation of Trades Unions and the encouragement of the Trades Union movement in America.

2—The organization of Trades Assemblies, Trades Councils, or Central Labor Unions in every city in America, and the further encouragement of such bodies.

3—The founding of State Trades Assemblies, or State Labor Congresses to influence State legislation in the interest of the working masses.

4—The establishment of National and International Trades Unions, based upon the strict recognition of the autonomy of each trade, and the promotion and advancement of such bodies.

5—An American Federation or Alliance of all National and International Trades Unions, to aid and assist each other, and, furthermore, to secure national legislation in the interest of the working people, and influence public opinion by peaceful and legal methods in favor of Organized Labor.

6—To aid and encourage the Labor Press of America, and to disseminate tracts and literature on the labor movement.

With these objects in view a convention of all Trades Unions in the United States and Canada will be held at Druid Hall, 146 South Fourth street, Columbus, O., to begin on Wednesday, December 8th, 1886, at 10 A.M.

The basis of representation will be: From National or International Unions, less than 4,000 members, one delegate; 4,000 or more, two delegates; 8,000 or more, three delegates; 16,000 or more, four delegates; 32,000 or more, five delegates, and so on. From each local Trades Union, not having a National or International Union, one delegate. But no Trades Union shall be entitled to representation which has not been organized three months prior to the session of this convention.

We respectfully urge upon your organization to elect its proper number of delegates, and be represented at this convention. It is highly important, so don't fail to have this matter brought before your Union and promptly attended to. No tax will be charged for the admission of delegates to this Convention.

By order of the Trades Union Committee.

Yours Fraternally, P. J. McGuire
Secretary.

W. Weihe, Iron and Steel Workers,
P. F. Fitzpatrick, Iron Molders,
A. Strasser, Cigar Makers,
Chris. Evans, Coal Miners,
P. J. McGuire, Carpenters,
Committee.

PLSr, The Papers of Gabriel Edmonston, reel 1, *AFL Records.*

## The Legislative Committee of the FOTLU to the Trade and Labor Unions of America

New York, Nov. 16, 1886.

To the Trades and Labor Unions of America:
Greeting—

The Annual session of the Federation will be held in Druid Hall, 146 South Fourth street, Columbus, O., Tuesday, December 7, 1886, at 10 A.M., instead of St. Louis.

The change in place of meeting was brought about by the fact that the Trades Union Conference Committee have called a Convention of all National and International Trades Unions to meet at Columbus, O., on December 8, which promises to be largely attended.

In view of this fact the Legislative Committee deemed it inadvisable, and that they would be recreant to the trust reposed in them if the two bodies having the same purposes in view were allowed to meet in different cities, and allow the opportunity thus presented to pass, of possibly bringing them together and forming a thorough and permanent Federation, a unity of all the forces of labor into a solid phalanx, yet each Trade Union to preserve its own identity and autonomy. The alarming concentration of capital into the hands of fewer persons, the tendency of the employing class to not only refuse the just demands of labor, but to use the power of their possessions to coerce labor into more degrading conditions and wrest from the toilers the rights which were considered sacred and achieved, makes it all the more necessary for the Trades Unions to make this effort of unification.

It will be observed therefore how important this session of the Federation will be. It already promises to be the most successful and eventful in the history of American Trades Unionism.

We respectfully urge your organization to elect its proper number of delegates, and be represented at this convention. It is highly important, so don't fail to have this matter brought before your Union and promptly attended to.

Unions having already elected delegates will kindly notify them of the change of meeting place from St. Louis to Columbus, O.

Fraternally yours, Samuel Gompers, President.
S. S. Green,
W. E. Tomson,
P. F. McAuliffe,
Hugo A. Miller,
Geo. S. King,
Henry Emerich,
Gabr'l Edmonston,
J. S. Kirchner, Sec.
Legislative Committee.

*Tocsin* (Philadelphia), Nov. 27, 1886.

# A Series of News Accounts of the 1886 FOTLU Convention and the Founding Convention of the AFL in Columbus

Columbus, Ohio, Dec. 7 [1886]

## TRADES UNION CONVENTION.

The Federation of Trades and Labor Unions of the United States and Canada met here to-day. It is the sixth session of the federation and the recent attack of the Knights of Labor on some of the trades unions, lends a peculiar interest to the proceedings of the body. The delegates began to arrive early and by this morning there were representatives of nearly 500,000 skilled mechanics gathered at the United States Hotel which is the headquarters of the federation.

The place of meeting was at Druid's Hall on Fourth-st., opposite the Central Market. President Samuel Gompers, of the Cigarmakers' International Union, called the meeting to order. Grafton Pearce,[1] of the Ohio Trades Assembly,[2] delivered the address of welcome. He said in part:

[“]It is your purpose to devise some means by which the organized mechanics of the country may be so cemented and bound together that an attack on any one trade shall be resisted by the united power of all the trades; while at the same time, each shall be left perfectly free to manage its own affairs in such a manner as its members, the only persons qualified by knowledge and experience to act, shall from time to time, deem best for their interests. It is greatly to be regretted that there exists any feeling of antagonism between different organizations. There is room enough for all. If the bettering of the condition of the masses is all that is desired, no conflict of authority need occur, no clashing of theories or methods need disturb that unity of action that should be the distinguishing characteristics of all labor organizations. The members of trades unions are not lawless. Their intelligence has long taught them that a resort to violence—to arson, murder and an indiscriminate destruction of property—is not only opposed to the principles of good government and personal rights, but is the surest way of forfeiting the respect and support of the community. There seems to be a growing tendency of late on the part of labor organizations to mix politics with the more essential features of their organizations. The two will not mix. Once promulgate a political platform and all harmony becomes discord, all union becomes disunion. The two cannot exist together. Mechanics are like all free, intelligent, thinking citizens—they think for themselves, and

vote as they think. What kind of a political platform would it be on which Henry George and T. V. Powderly could stand together—the one a radical free trader, the other a radical protectionist, and what sort of a political platform would it be that omitted all mention of the tariff question. So it would be as to other questions equally important. We hold the balance of power, and by proper use of that power we can accomplish vastly more than any one can possibly hope to accomplish by independent political action, or by assuming in any way the attitude of a political party, or by becoming the advocates or opponents of any particular political policy.["]

President Gompers in reply said that trades unions were conservators of the public peace. By them the vicious and ignorant are held in check, and society, the people and property are kept in safety. Trades unions believe in independent manhood, the right of free speech and free assemblage. At the afternoon session the committee on legislation handed in its report. It was in part as follows:

["]A number of National unions have been organized the past year and others are in course of formation. The new unions will be represented next year. The membership of the old has increased 100 per cent. In reviewing the work done in effecting a reduction of hours we can record a large measure of success; never before has there been such a general movement as has taken place during the past year for shorter hours. We are not able to record a complete success, but the eight hour agitation was the means of reducing the hours of work of over 200,000 workers. The trades unions as a rule responded most zealously to the appeal of this federation, and had their efforts been met with that co-operation by the organization of the Knights of Labor that the identity of interests required, there would have been still greater results attained.["]

The report favors international laws about labor and a labor holiday. A supplementary act is suggested for the strictest enforcement of the present Anti-Chinese law.[3] The Pinkerton system of mercenaries[4] was denounced. In regard to strikes the report says:

["]We do not as a federation or as individuals wish to be understood as advocating strikes; on the contrary it is known that the best regulated trades unions have fewest strikes. Yet, while we deprecate this method of warfare, we cannot and will not join in the general hue and cry of condemnation of them. Strikes are bad, but only are they so when they are failures.["]

The following officers were elected: J. N. Smith,[5] of the National Tailors' Union, chairman; J. L. Wright,[6] of International Typographical Union, vice-chairman; John S. Keerdener,[7] of Cigarmakers' International Union, secretary; Henry Emerich, of the Furniture Mak-

ers' Union, assistant secretary. The legislation committee was instructed to confer with a committee of the conference of trades unions, which meets to-morrow. There is little doubt but that the two organizations will amalgamate and that action inimical to the Knights of Labor will be taken.

Last night, Mr. Hall,[8] of the Knights of Labor, accompanied by Messrs Arrington[9] and McFeely,[10] arrived in town. They handed P. J. Maguire, the secretary of the conference which meets to-morrow, the following letter from T. V. Powderly, under date of Philadelphia, December 5:

[“]I have selected a committee of five, consisting of Brothers Howes,[11] Hall, McFeely, Arrington and Gant,[12] to discuss past grievances between the trades unions and the Knights of Labor and pave the way for the avoidance of future ones.[”]

Messrs. Howes and Gant, the remaining members of the committee of the Knights of Labor, are expected to arrive in town to-morrow and there is a possibility that some sort of treaty will be patched up.

*New-York Tribune,* Dec. 8, 1886.

1. Grafton Pearce (d. 1891), a printer in Columbus from the early 1850s, was a long-standing member of International Typographical Union (ITU) 5.

2. The Ohio State Trades and Labor Assembly was founded in June 1883 primarily by members of the Columbus Trades Assembly. Its membership included trade unions as well as KOL assemblies. In 1896 it voted to affiliate with the AFL; it was chartered by the Federation in 1897, at which time it changed its name to the Ohio Federation of Labor.

3. The Chinese Exclusion Act of 1882 (22 U.S. Stat. 58) made the immigration of Chinese laborers to the United States unlawful for a period of ten years.

4. The Pinkerton detective agency was founded in 1855 by Allan Pinkerton. From the late 1870s it and a growing number of other agencies were regularly involved in strikes, infiltrating union organizations, guarding property and strikebreakers, and supplying workers for struck firms. Congressional hearings on the use of private armed guards during the Homestead steel strike of 1892 provided an important impetus in the passage of state legislation restricting this activity. By 1899 twenty-four states and the District of Columbia had passed laws against the use of armed guards, and in the late 1890s the Pinkerton agency sharply curtailed its involvement in strikes.

5. James W. SMITH was a vice-president of the Journeymen Tailors' National Union and first vice-president of the FOTLU Legislative Committee. He served as second vice-president of the AFL from 1886 to 1887.

6. Julian L. Wright (1849?-96) was a member of ITU 101, the Columbia Typographical Union of Washington, D.C.

7. John S. Kirchner.

8. Goldsmith P. Hall (b. 1846), a blacksmith from Bridgeton, N.J., was active in KOL District Assembly 2. He served as a deputy inspector of factories and workshops for New Jersey in 1886-87.

9. Louis (variously Lewis) Arrington was a Milwaukee glassblower and a member of KOL District Assembly 143.

10. James M. McFeely (b. 1840?) was a Philadelphia shoemaker during the 1870s and 1880s and a member of KOL District Assembly 1.

11. John Howes, born about 1833 in Massachusetts, listed his occupations as carpenter, inventor, agent, and water filter manufacturer. He was district worthy foreman of KOL District Assembly 30 during the early 1880s and master workman in 1886, and represented Worcester, Mass., at the KOL General Assemblies between 1884 and 1886.

12. Thompson Fulton Gantt, a clerk in the U.S. Surgeon General's office in Washington, D.C., from 1883 to 1886, was appointed a member of the KOL committee to the AFL convention but did not sign the final report of that committee to the KOL General Executive Board.

Columbus, Ohio, Dec. 8. [1886]

## FEDERATION OF TRADES.

The interest among the trades unionists and the Knights of Labor in this city depends on the action which will be taken by the Conference of Trades and Labor Unions, which met this afternoon at Druids' Hall, in regard to the visiting committee from the General Executive Board of the Knights of Labor. Among the credentials which have been presented to the conference are those of A. G. Denny,[1] of Pittsburg, the representative of the Glass Workers' Assembly of the Knights of Labor. Mr. Denny is a prominent Knight and there was much discussion of the question of admitting him as a delegate. It was finally decided that his organization was not eligible. The visiting committee of the Knights has asked to be admitted to the meetings, but that privilege has not as yet been granted and it is doubtful if it will be. The feeling among a number of delegates is that the movement has shown such strength that it is the desire of the General Executive board to head off any action that might be taken as inimical to the Knights here and the committee was appointed with that end in view. One of the delegates said yesterday: "There is no use in our fooling away our time on this committee. It has no power to act and is simply sent here to prevent us from acting. We are too old birds to be caught with such chaff. The feeling against the Knights is bitter and it will be a victory for the conservatives if they are able to prevent an open declaration of war."

The committee of the Knights will not talk much now. One of them said: "Until the conference decides what it is going to do in our case, it would be ill-advised for us to talk, but I am at liberty to say that if they decline to confer with us in a spirit of fairness now, I don't think that they will have another chance." The composition of the committee is not considered favorable to trades unions or to a conciliatory policy.

Seldom has a more intelligent lot of men gathered at a labor meet-

ing. There are only thirty-eight delegates present, but they represent nearly all of the big trades unions of the country, and have a constituency of 400,000 skilled mechanics. The leaders in the movement are P. J. McGuire, secretary of the Brotherhood of American Carpenters and Joiners, one of the coolest and shrewdest men in the labor movement; Adolf Strasser, of the Cigarmakers' Union who has behind him one of the best organized unions in the country and is himself regarded as the best organizer and fighter in the movement; "Sam" Gompers, of the same union; Christopher Evans and John McBride,[2] of the Federation of Miners and Mine Laborers, and P. J. Fitzpatrick of the iron moulders. Among the New-Yorkers present are H. Casserly,[3] of the United Order of Carpenters;[4] George Block,[5] of the National Bakers' Union;[6] Edward Finkleston[7] of the Barbers' Union;[8] and Messrs. Weiner,[9] of the Waiters';[10] Hanlon,[11] of the Operative Painters; H. Emerich, of the Furniture Workers; and M. Davis,[12] of the German Trades Unions.[13]

The meeting of the federation was called to order by Chairman Smith, at 10 A.M. Communications were received from the Federation of Cuban Cigarmakers,[14] of New-York; the Amalgamated Society of Engineers of Great Britain and America, and from the Brewers' Union, of St. Louis, expressing regret at being unable to send delegates at present, but stating that they were in sympathy with the movement. Committees were appointed on standing orders, on resolutions, finance, and the Legislative Committee's report. A resolution was adopted that the federation attend the Trades Union Conference in a body and an adjournment was taken until the conference has finished its work.

The first session of the Conference of Trades Unions was called to order at noon in Druids' Hall by P. J. McGuire. He read the call for the conference. John McBride, of the Miners, was unanimously elected chairman and P. J. McGuire, secretary. A Committee on Credentials, consisting of Messrs. Scott,[15] Edmondson, Miller and McDermot,[16] was appointed and an adjournment was taken until 2 P.M. in order to permit the committee to do its work. The delegates to the Federation of Trades were admitted as a body and given full privileges as delegates. The following committees were appointed:

On Constitution—Gompers, Mulrany,[17] Emerich, Fitzpatrick, McGuire, McLaughlin,[18] Winders.[19]

On Rules—Smith, Gompers, Kirshner, Kane[20] and Evans.

On Permanent Organization—Strasser, Apple[21] and Fitzpatrick.

On Resolutions—Casserly, Harris, Finkleston, Mahony[22] and Kane.

President McBride made a short address in which he said that he hoped members would be temperate in the discussions. In considering

the questions to come up, they should abstain from personalities and treat each other with courtesy and respect. The secretary read a letter from William Weihe, of the Amalgamated Iron and Steel Workers, who wrote that he was unable to attend, but that his organization would be represented at the next conference provided no action favoring free trade was taken. Thomas Odea, of Cohoes, of the National Bricklayers and Masons' Union; H. H. Lane,[23] of the New-Haven Trade Council;[24] the president[25] of the Amalgamated Flint Glass Workers;[26] J. P. Donnelly,[27] of the Milwaukee Plumbers and Gasfitters;[28] Frank Rodney,[29] of the Federation of Trade and Labor Unions of the Pacific Coast,[30] and the Federation of Cuban Cigar-makers, of New-York, all sent communications expressing good will and sympathy with the objects of the convention.

A resolution was introduced that all of the sessions of the convention be made executive and after a long and rather heated debate it was carried by a majority of 8, and the newspaper men were asked to retire. After the exclusion of the press, circulars on the memorial of the last secretary of the Federation, W. H. Foster, who died recently were read. An informal discussion was had on the formation of a National organization of trade and labor unions. A number of speeches were made, all of which were favorable to the idea. New credentials were received from J. J. Black, of the New-York Stereotypers;[31] E. L. Daily[32] and W. H. Marden,[33] Lynn, Mass., of the New-England Shoe Lasters.[34]

*New-York Tribune*, Dec. 9, 1886.

1. Albert G. DENNY, a glassworker, was secretary of KOL Local Assembly 300 and a KOL general organizer in the mid-1880s.

2. John McBRIDE was president of the Ohio Miners' Amalgamated Association from 1882 to 1889 and a founder and first president of the National Federation of Miners and Mine Laborers (NFMML). He helped found the United Mine Workers of America (1890) and served as its president (1892-95). In 1894 he defeated SG for the AFL presidency, serving one term.

3. James A. Casserly, a New York City carpenter, was a leader in Branch 1 of the United Order of American Carpenters and Joiners in the mid-1880s. Casserly also served as treasurer of the New York City Central Labor Union (CLU) from September 1886 until August 1887.

4. The United Order of American Carpenters and Joiners, an independent New York area union, was founded in 1872; it merged with the Brotherhood of Carpenters and Joiners of America in May 1888.

5. George G. BLOCK, secretary of the Journeymen Bakers' National Union from 1886 to 1888, was a founder of the New York City CLU and a leader in the Henry George mayoralty campaign.

6. The Journeymen BAKERS' National Union of the United States.

7. Edward FINKELSTONE was president of the Journeymen Barbers Protective Union of New York.

8. The Journeymen Barbers Protective Union of New York was organized in early 1886. It was a charter local of the Journeymen Barbers' National Union that was organized the following year.

9. Julius WIENER of Waiters' Union 1 of New York became first secretary of the Waiters' and Bartenders' National Union in 1891.

10. New York City waiters were organized in January 1885 and, as Waiters' Union 1, were represented in AFL conventions in the late 1880s. The union received a charter from the AFL in 1887.

11. John F. Hanlon, a painter, was born in New York City in 1855, where he lived until 1907. From 1896 to 1906 he owned a small painting and decorating business on Manhattan's West Side.

12. Bernhard Davis (b. 1860) was a cigarmaker, born in Germany, who immigrated to the United States in 1881. He was a member of CMIU 10 of New York City and served as that union's delegate to the 1893 CMIU convention.

13. Twelve unions organized the United German Trades of the City of New York in 1885 to provide support for the labor press. During its four-year existence it worked to increase the circulation and advertising of the *New Yorker Volkszeitung,* initiated the movement to establish the *Leader*—the organ of the 1886 Henry George mayoralty campaign—and supported the SLP.

14. This union was active during 1886 and 1887 in both New York City and Tampa.

15. John Scott (1833-1915) was president of International Typographical Union (ITU) 91 of Toronto in 1885 and 1886. He apparently moved to New York City in the early 1890s and became a member of ITU 6. He resided at the Union Printers' Home in Colorado Springs, Colo., from 1895 until his death.

16. James McDermott (1842?-91) was an active member of Bricklayers' Union 1 of Cincinnati, Ohio, serving as local corresponding secretary in 1890 and 1891.

17. Edward Mulraney was active through the 1890s in both the Chicago Trade and Labor Assembly and the Bricklayers' and Stone Masons' Union of Chicago and Vicinity (after 1897 local 21 of the Bricklayers' and Masons' International Union of America).

18. Daniel MCLAUGHLIN, an organizer and first treasurer (1885-88) of the NFMML, served as first vice-president of the AFL for one term (1887-88).

19. John R. Winders, a San Francisco printer born in California in 1850, was an officer of ITU 21 in the 1880s, and from 1888 to 1891 he was a district organizer of the ITU.

20. John Kane was a member of United Brotherhood of Carpenters and Joiners 27 of Toronto.

21. George W. APPEL was general secretary of the Metal Workers' National Union of North America.

22. Jeremiah F. Mahoney (1849-1908) was a member of CMIU 49 of Springfield, Mass., and an AFL organizer from the 1890s until 1902.

23. Horatio H. Lane (1842-1909?), first president of the New Haven Trade Council, was a woodcarver and an active member of the New Haven Branch of the International Woodcarvers' Association of North America. He attended the first annual convention of the Connecticut State Branch of the AFL in 1887.

24. The New Haven Trade Council was formed in 1881, composed of delegates from New Haven trade unions. It changed its name in the late 1930s to the Central Labor Council of New Haven.

25. William J. SMITH served as president of the American Flint Glass Workers' Union from 1885 to 1900.

26. The American FLINT Glass Workers' Union of North America.

27. James P. Donnelly (1839-1917?) of Cincinnati was president of the International Association of Journeymen Plumbers, Steam Fitters, and Gas Fitters from 1886 to 1887. He moved to Covington, Ky., about 1900.

28. That is, the International Association of Journeymen PLUMBERS, Steam Fitters, and Gas Fitters.

29. Frank RONEY was vice-president of the Iron Molders' Union of North America (1886-88) and a founder and president of the Representative Council of the Federated Trades and Labor Organizations of the Pacific Coast (1885-87).

30. The Representative Council of the Federated Trades and Labor Organizations of the PACIFIC Coast.

31. The New York Stereotypers Association was organized in 1863.

32. Edward L. DALEY was the general secretary of the New England Lasters' Protective Union (NELPU) from 1885 to 1895.

33. William Henry MARDEN was treasurer of the NELPU in the late 1880s and fourth vice-president of the AFL in 1893-94.

34. The New England LASTERS' Protective Union.

Columbus, Ohio, Dec. 9. [1886]

## KNIGHTS AT A DISCOUNT.

The trades-unionists are working slowly, most of the time up to the present having been taken up with efforts to effect an amalgamation between the Federation of Trades and the Trades Union Conference. A committee consisting of Adolf Strasser, Larkin McHugh,[1] Daniel McLoughlin, J. Hanlon and George Block has been appointed to confer with the Legislative Committee of the Federation as to a plan of amalgamation, and they will in all likelihood be ready to report to-morrow. The Knights of Labor Committee has succeeded in accomplishing nothing as yet. Messrs. Howes and Grant of the committee arrived here to-day, and as Messrs. Arrington, Hall and McFeeley were already here this makes the committee complete. They sent word to the conference to-day that they would like to hold a consultation. The original Trades Union Committee, consisting of P. F. Fitzpatrick, Christopher Evans, P. J. McGuire, David P. Boyer and Adolf Strasser, called on them at the Exchange Hotel to-night and made arrangements for a conference to-morrow. There is little hope that the conference will amount to anything, and the Knights have little hopes of preventing the trades unions from taking a decided stand against the order. A. C. Denny, the general organizer of the Knights, whose credentials as a delegate from Local Assembly No. 300 of glass workers were refused, has not left the city, but will use his powers as a diplomat to bring about an amicable arrangement between the two organizations.

An early session of the Conference was held this morning, and it was just 8 A.M. when Chairman McBride's gavel fell. The report of the Committee on Credentials was favorable to all except Denny, and

was adopted unanimously. The first resolution reported and adopted was the following:

*Resolved,* That the delegates of the various trade and labor unions here represented do hereby form a federation of all trades and labor unions of America, to insure the attainment of the objects for which this convention was called.

A committee of five was appointed under this resolution to confer with the Legislative Committee of the Federation of Trades in regard to the matter. Communications were read from the United Shoe Salesmen's Protective Union of New-York[2] favoring the objects of the Conference, and from the National Executive Board of the Tailors' Progressive Union[3] favoring independent political action. An adjournment was then taken to allow the Federation a chance to meet.[4]

A resolution offered by Mr. Mulraney, of Chicago, in the latter's meeting, favoring a general movement for eight hours all over the country was adopted by the Federation, as was a resolution making some arrangements for the printers who work in the Government Printing Office at Washington. A communication was received from the Chicago Central Labor Union[5] asking the Federation to pass resolutions of sympathy with the Anarchists[6] who are under sentence and asking for a new trial for them. The matter was referred to the Committee on Resolutions. It is hardly likely to be heard of again, as the general feeling among the delegates is to let the Anarchists alone and to express no opinion in regard to them. In case the matter is pushed it has been decided to offer a resolution stating that trades unions are founded on law and order and have no sympathy for riot and bloodshed or people who incite them.

At the afternoon session of the Conference an invitation was received from State Factory Inspector Henry Dorn[7] for the delegates to visit the State institutions. It was accepted. Most of the afternoon session was taken up in the reports from the various organizations represented. The reports give the birth of the organizations, their growth and present strength and the complaints which exist against the Knights of Labor. Some of the reports on the latter topic are said to be breezy. It was decided, however, not to make them public until to-morrow, when they will all have been received and put in shape. The Committee on Constitution reported that they would be prepared to submit a constitution on Saturday.

In the evening such of the delegates as were not busy on committee work attended a trades-union meeting at Gumble's Hall, in High-st. Speeches approving the objects of the Conference were made, and Knights of Labor were at a discount. Considerable interest is felt in the forthcoming report of the committee which was appointed at the

Philadelphia Conference in May last to see the Knights of Labor. The report will be submitted either to-morrow or Saturday and is said to be exceedingly interesting, handling the action of the Knights in plain, vigorous English. This report will form the basis of whatever action against the Knights may be taken. The convention hopes to finish its work by Saturday.

---

The Knights of Labor in New-York were not surprised yesterday to learn that their committee at Columbus was refused admission to the meetings of the trade unions' delegates. "It is the old row between the two factions of the labor party," said one of the Knights yesterday. "You see, the Knights are accused of trying to get control of the labor element in this country, and the local unions don't want to lose the grip that they have already on the laboring people."

*New-York Tribune,* Dec. 10, 1886.

1. Larkin C. McHugh (1836?-94?) was a member of Iron Molders' National Union 4 of Cincinnati.

2. The United Shoe Salesmen's Protective Union of New York, founded with the help of the New York City Central Labor Union (CLU) in early 1886, may have been affiliated with the KOL from 1886 to 1887 as Local Assembly 5371.

3. The Progressive TAILORS' National Union.

4. At this FOTLU session, five members of the Legislative Committee (SG, John Kirchner, Henry Emrich, Gabriel Edmonston, and Hugo Miller) reported on their meeting with the delegation from the trade union conference. They proposed that the FOTLU merge with the AFL and turn over its money, records, and effects to the new organization. The FOTLU delegates unanimously approved the recommendation.

5. In 1883 radical members of Chicago CMIU 14 withdrew from the local to form Progressive Cigarmakers' Union 15. The following year the Progressives and four other local unions that had seceded from the Trade and Labor Assembly of Chicago formed the Chicago CLU. Led by Albert Parsons and August Spies, it had ties to the anarchist International Working People's Association. A strong organization in the 1880s, it gradually declined after the formation of the Chicago Federation of Labor in 1896.

6. A reference to the eight men tried and convicted of murder in connection with the Haymarket incident, namely, George ENGEL, Samuel FIELDEN, Adolph FISCHER, Louis LINGG, Oscar W. NEEBE, Albert R. PARSONS, Michael SCHWAB, and August SPIES.

7. Henry Dorn (1843-1912) was born in Prussia. He immigrated to the United States and worked in Cleveland as a machinist until being appointed first Ohio inspector of shops and factories in 1884. In 1885 the Ohio legislature created the position of chief state inspector; Dorn served in that capacity from 1885 to 1889. He later worked in Columbus as a mechanical engineer.

Columbus, O., Dec. 10 [1886]

FEDERATION OF TRADES.

The Federation of Trades-unions are still proceeding very slowly with the work of perfecting a more effective organization. They meet with closed doors and the filtering process is kept up, so far as pertains to the actual business of the session. The Powderly Committee met with a committee from the congress this afternoon, but nothing was accomplished beyond establishing the point that the Knights had no power to act on anything. Mr. Gompers, a leading spirit in the congress, was asked by *The Press* correspondent to-night concerning the conference between the committees, and said:

"From members of the committee who heard the Knights of Labor it is learned that the latter have no power to act. The Trades-union Committee insisted upon the ratification of the treaty prepared by Messrs. Weihe, Strasser, McGuire, Fitzgerald[1] and Boyer, and presented to the Knights at their convention in Cleveland, or such answer or modification of that document as the Knights may be prepared to make. We do not propose to allow this matter to hang fire any longer. The Knights of Labor have held two conventions since the proposition of that treaty, but we heard nothing from that body until the committee came to this city, and then without previous notification to us. It is the belief of delegates to the congress that the circumstances under which this committee was appointed and came here was one of the strategic moves of Mr. Powderly to prevent this convention from taking any action tending towards the formation of a permanent federation and to prevent any expression of condemnation of the course pursued by the Knights to destroy the trades-union movement."

"What did the Knights do in regard to approving the treaty proposed?" "They stated that they were without authority to act or make any proposition to modify the treaty tending to harmony between the trades-unions and Knights of Labor."

"What did they come here for?" "They wanted to discuss past grievances that are so well-known as to need no discussion. What we want is action, which they are not prepared to take. We believe the time for parleying, except that which will tend to a strict understanding of the relations between the Knights and trades-unions is past."

One of the Knights of Labor stated that the meeting between the committees was conciliatory, though no final result was reached. The main point of the Cleveland treaty, he said, was that the Knights should give up their organizations in the ranks of tradesmen where unions are organized. This the Knights refused to discuss, on the ground that the treaty had been taken out of their hands. They

consented, however, to refrain from attempting to disrupt the unions, reserving the right to organize assemblies in response to requests. It is not probable that the congress will close its work to-night.

*Philadelphia Press,* Dec. 11, 1886.

1. P. F. Fitzpatrick.

Columbus, O., Dec. 10. [1886]

WAR AMONG LABOR LEADERS.

General Master Workman Powderly's fondness for writing secret circulars has made serious trouble for the order at last. It has come to light that much of the bitterness which has been exhibited by the delegates of the Cigarmakers' International Union toward the Knights of Labor is due to an attack on the officers of that organization which is contained in a circular in regard to the cigarmakers' business which was sent out some time ago by the General Executive Board of the Knights of Labor. It was hinted that copies of this circular are in Columbus, but it was not intended that these should be made public. A *Tribune* correspondent succeeded in getting hold of a copy, however. It is addressed to "the Order and the Cigarmakers," and is dated at Philadelphia on July 2, 1886. . . .[1]

. . .

This was signed by Mr. Powderly and the other members of the General Executive Board. When it became known to President Strasser to-day that the circular was to be made public, he wrote the following open letter to the General Executive Board from Columbus:

["]During the month of July, 1886, a forty-eight page pamphlet entitled 'The Order and the Cigarmakers,' was circulated to the entire order of the Knights of Labor, containing the following sentences: 'But one reason can be assigned for the unaccountable actions of the officers of the International Union and that is that men who indulge to excess in the use of intoxicants cannot transact business with cool heads. On two occasions the men who came to Philadelphia to confer with the General Executive Board were too full for utterance.' It is needless for me to reiterate my sentiments so frequently expressed, relating to the necessity of abstaining from the use of intoxicants by officers of labor organizations, because my views on that question are so well known by the delegates attending the various conventions of the Cigarmakers' International Union, over which I have had the honor to preside. One week previous to the Richmond Convention, I called on you, in company with Messrs. Fitzpatrick, McGuire, Evans, Weihe, Pasco and Dyer, on which occasion you promised me to print

the names of the officers of the International Union referred to in the pamphlet. I have patiently waited until now for such an explanation, and probably would not have mentioned it now, had it not come to my knowledge that since the adjournment of the Richmond Convention a number of individuals made it their business to circulate those slanders broadcast throughout the United States and Canada. After you failed to abide by your promise, given in the presence of witnesses, I have come to the conclusion that the entire matter is a base fabrication and infamous slander, calculated to destroy the well-established and honorable reputation of the organization I represent. It is needless to say that the attempt will prove a complete failure and eventually recoil on its originators.["]

Mr. Gompers was exceedingly indignant at the direct attack made on him, and when the conference met in the afternoon asked leave to make a personal explanation. He said:

["]I wish to read part of a secret pamphlet, published some months ago and spread broadcast throughout the land, in which the Executive Board of the Knights of Labor seem to sink to the lowest level of vituperation and abuse because I dare be a union man. Before proceeding to read, I desire to say that it is well known that I have been an officer of the Federation for four terms, an officer of the union with which I am connected, president of the Workingmen's Assembly of the State of New-York, and have held many other positions of honor and trust in the labor movement. Many questions have arisen upon which I have been opposed, but until recently, no matter how bitter the opposition to my views, I have enjoyed the confidence of my friends and the personal respect of my opponents. During the month of January, 1886, the cigar manufacturers of New-York reduced the wages of their employes. Our union refused to submit to the reduction, and as a result the manufacturers locked out 10,000 people. Because I was active in my opposition to outside interference with the matters of our trade, I have been the butt for the abuse and slander of the General Executive Board of the Knights of Labor. In my hand I hold the pamphlet to which I refer. It says: 'The General Executive Board has never had the pleasure of meeting Mr. Gompers sober.' As you will see, this is a forty-eight-page circular. I deny the correctness of the testimony published, but admitting that it is correct, it still shows that I appeared before that board as the representative of the cigarmakers, acted as their counsel and was also a witness. I desire to add that I made a verbatim copy of the testimony as given. How a man drunk could do this I will leave to your own judgment. But these people, understanding that life is short, that it is tiresome to wade through forty pages of testimony, publish their scurrilous

slanders, which are ordered to be read in the assemblies of the order first, while that which would give them the lie, their own version of the testimony, will scarcely be touched. It seems to me, that abuse from that quarter comes with bad grace. No man has ever charged me with selling to either private or corporate wealth the interests of workingmen with which I have been intrusted. The strikes of the collar and cuff laundry people of Cohoes,[2] the clothing cutters of New-York,[3] the railroad people in the Southwest, the beef and pork butchers of Chicago,[4] whom the General Executive Board of the Knights of Labor have had in charge, have all had one fate, that where the employers refused to accede to the work-people, the Board has acceded to the demands of the employers. It is a matter of public notoriety that the order from Mr. Powderly that the Chicago butchers should return to work was made public some time before the officers of the order in Chicago received it. I do not charge that Mr. Powderly was bribed to betray the interests of the working people; but a man in his position who was bribed could do the work of the employers no better.

[“]These men are not, I claim, the representatives of the bonafide labor movement of the country. Two of them are grocery store keepers, Turner and Hayes; another, Bailey, is ex-chief of the Shawnee police; while the other two[5] have been floating like scum on the top of a portion of the labor movement, continually seeking to divert it to their own personal ends. Friends have inquired of me why I have not prosecuted these people for criminal libel. I am not willing to lend additional opportunities for them to pose as martyrs for the cause of labor. A large number of witnesses, men widely known throughout the country, are willing to asseverate, on their word of honor, that the reflections made upon me by the General Executive Board of the Knights of Labor are the grossest slanders, false and malicious.[”]

Mr. Gompers’s speech and the publicity given the circular have destroyed any slight chance that there may have been for a settlement of the troubles between the Knights and the unions.

The Knights of Labor committee has left the city. A meeting of the committees representing the two organizations was held this afternoon at the United States Hotel. Mr. Hall, of the Knights, said that his committee could make no arrangements that would be binding on the order. They were simply here to discuss the situation with the trades unionists and to report. Mr. Fitzpatrick, for the trades unionists, said that the only basis of settlement that would be satisfactory to his body would be that of the proposed Cleveland treaty. A general discussion followed, but after a three hours’ session the meeting adjourned without result. The Conference subsequently adopted a res-

olution continuing the Committee on Knights of Labor, adding Mr. Daly, of Massachusetts, to it, and instructing them to treat with the Knights only on a basis of the trades union treaty. At the afternoon session a resolution was passed stating that "the time had now arrived when the working people should decide upon the necessity of action as citizens at the ballot-box, independent of existing political parties, and this convention urges a most generous support to the independent movement of the working people." A National Apprenticeship law was also demanded. There were present at the conference forty-two delegates, representing twenty-five unions with a membership of 320,000.

A meeting of the Federation of Trades was held in the afternoon. There was a heated discussion on a motion favoring a new trial for the Chicago Anarchists. It was finally referred to the Conference for action.

*New-York Tribune,* Dec. 11, 1886.

1. See "The General Executive Board to Members of the KOL," July 2, 1886, above.

2. On May 13, 1886, women members of the Joan of Arc Assembly of the KOL in Troy, N.Y., working in the collar and cuff department of George P. Ide and Co., struck for higher wages. The thirty-three shirt and collar establishments of the Collar and Shirt Manufacturers' Association locked out some 4,000 KOL members, who were joined by the remainder of the approximately 10,000 to 15,000 workers in the industry. They returned to work on June 23, 1886, after KOL General Secretary John Hayes accepted the manufacturers' price list.

3. During the last week of August 1886 KOL clothing cutters in two New York City shops struck to protest the employment of nonunion men. In response eighty firms in the Clothing Manufacturers' Association locked out 950 cutters on Aug. 30. District Assembly 49 decided that the strike violated the cutters' contract, which required that a conference committee of the workers and manufacturers first consider the difficulty. The assembly brought the dispute to conciliation and the workers returned to work on Sept. 8 without the employers conceding the union shop.

4. Chicago packinghouse workers struck successfully for the eight-hour day in May 1886. When their employers reinstated the ten-hour day in October, 20,000 to 25,000 workers, three-quarters of them KOL members, walked out. After fruitless discussions with the packers, KOL negotiators ordered the men back to work on a ten-hour basis. In November they struck again and initiated a boycott of Armour & Co. Terence Powderly settled the strike, however, by ordering KOL workers back to work at the old hours. He justified his action on the basis of his Mar. 13, 1886, circular that Knights should refrain from striking for the eight-hour day.

5. Presumably Powderly and Thomas B. Barry, the other members of the KOL General Executive Board who signed the pamphlet.

Columbus, Dec. 11. [1886]

## THE AMERICAN FEDERATION.

The Trades Union Convention wound up its work this afternoon by giving a direct slap at the General Executive Board of the Knights

of Labor in the election to the presidency of the New American Federation of Trades of Samuel Gompers, of the Cigar Makers' International Union. Mr. Gompers has been one of the most persistent fighters against the order in the ranks of the trades unions. The convention has done much work. An amalgamation has been formed which will result, it is hoped, in the establishment of an organization fully as powerful, better disciplined and more conservative than the Knights of Labor. Action has been taken favorable to independent political action, but the trouble with the Knights of Labor has not been settled, and the feeling against the order is, if anything, more bitter than ever. The matter was settled by the adoption of a report in which the committee appointed to confer with the Knights of Labor recommend that the Executive Council of the newly organized federation shall issue an address to the public on the subject at an early date.

The work of to-day was principally confined to the discussion of the constitution of the new American Federation and the election of officers. The principal points in the new constitution are as follows:

The association shall be known as the American Federation of Labor, and shall consist of such trades unions as shall conform to its rules and regulations. The objects of the Federation shall be the encouragement and formation of local trades and labor unions and the closer federation of such societies through the organization of central trades and labor unions in every city, and the further combination of such bodies into State, Territorial, or Provincial organizations to secure State legislation in the interests of the working masses; the establishment of national and international trades unions based upon a strict recognition of the autonomy of each trade; to secure national legislation in the interests of the working people and to influence public opinion by peaceful and legal methods in favor of organized labor. The basis of representation in the convention shall be: from a national, or international union less than 4,000 members, one delegate; 4,000 or more, two delegates; 8,000 or more, three delegates; 16,000 or more, four delegates; 32,000 or more, five delegates, and so on and from each local district or trades union not connected with, or having a national or international union affiliated with this federation, one delegate. No organization which has seceded from any local, national or international organization shall be allowed a representation or recognition in this federation. The officers shall be an executive council with power to watch legislative measures directly affecting the interests of working people, and to initiate whenever necessary such legislative action as the convention may direct. The Executive Council shall use every possible means to organize new national or interna-

tional trades unions and to organize local trades unions and connect them with the federation. While we recognize the right of each trade to manage its own affairs, it shall be the duty of the Executive Council to secure the unification of all labor organizations so far as to assist each other in any justifiable boycott and with voluntary financial help in the event of a strike, or lockout, duly approved by the Executive Council. When a strike has been approved by the Executive Council, the particulars of the difficulty, even if it be a lockout, shall be explained in a circular issued by the president of the federation to the unions affiliated therewith. It shall then be the duty of all affiliated societies to urge their local unions and members to make liberal financial donations in aid of the working people involved. The revenue of the federation shall be derived from a per capita tax of one-half cent per month for each member in good standing. Whenever the revenue of the Federation shall warrant such action, the Executive Council shall authorize the sending out of trades union speakers from place to place in the interest of the federation. The remuneration for the loss of time by the Executive Council shall be at the rate of $3 per diem, travelling and incidental expenses to be also defrayed. Any seven wage workers of good character and favorable to trades unions and not members of any body affiliated with the federation, who will subscribe to this constitution shall have the power to form a local body to be known as a "federal labor union," and shall hold regular meetings for the purpose of strengthening and advancing the trades union movement and shall have the power to make their own rules in conformity with the constitution and shall be granted a local charter by the president of the federation, provided the request for a charter be indorsed by the nearest local or National trades union officers connected with this federation. This constitution shall go into effect March 1, 1887.

There are certain portions of this document which will hurt the Knights of Labor. The section which relates to strikes and boycotts will give to individual trades the support of a strong national body which they find now only among the Knights, and will take away one of the strongest inducements to the formation of trade districts, and the provision for the formation of Federal trades unions, will obviate the difficulty of forming unions in small towns where there are not enough of one trade to form a union and will take the place of the mixed local assemblies of the Knights of Labor.

After the constitution had been adopted, Samuel Gompers, of New-York, was elected president; George Harris, of Pennsylvania, and James W. Smith, of Springfield, Ill., vice-presidents; P. J. McGuire, of Philadelphia, secretary, and Gabriel Edmunds,[1] of Washington, treas-

urer. Mr. Gompers was born in London thirty-seven years ago, and has been a cigarmaker all his life. He has held many important offices in his own union, has been president of the Workingmen's State Assembly of New-York, and was the president of the old Federation of Trades. Mr. Harris is the president of the Federation of Miners and Mine Laborers, of Pennsylvania. Mr. Smith represents the National Progressive Tailors' Union,[2] and was the chairman of the Federation of Trades. Mr. McGuire is the general secretary of the Brotherhood of Carpenters and Joiners, and was the secretary of the Trades Union Committee and of the conference. He was born in New-York City about forty years ago and for many years has been one of the most prominent men in the labor movement. He is regarded as one of the ablest men in the movement. Mr. Edmondston is the financial secretary of the local union of carpenters in Washington, and was born in Baltimore.

The Federation adjourned to meet in Baltimore on the second Tuesday in December, 1887. The delegates have nearly all started for home. They are well pleased with the result of the session and are confident that the American Federation of Labor will quickly grow to be the most powerful labor organization of the country.

*New-York Tribune*, Dec. 12, 1886.

1. Gabriel Edmonston.
2. According to the proceedings, Smith represented the Journeymen Tailors' National Union.

Columbus, Ohio, Dec. 12 [1886]

## TRADES UNION LEADERS CONTENTED.

Adolph Strasser, president of the Cigarmakers' International Union, in commenting on the work of the Federation of Trades, said: "I am perfectly satisfied with the result of the convention. We have perfected an organization on a good financial system and the result will be to bring all the national and international trades unions into the American Federation. The Knights of Labor are dead and it's useless to talk about a corpse. They have lost seventy five per cent of their membership in Massachusetts since October 1, through the failure of the Peabody strike."[1]

John McBride, of the Federation of Miners, said: "Our action will not necessarily bring us into direct conflict with other organizations, but we have preserved the trades unions."

P. F. Fitzpatrick, of the International Iron Moulders, said: "The trades have done what they should have done long ago—united in one solid body."

H. Emrich, of the National Furniture Workers' Union, said: "We have laid the foundation of a great labor movement."

J. R. Wilders,[2] of the International Typographical Union, said: "I think the difficulties between the unions and the Knights of Labor could have been satisfactorily settled if District Assembly 49, of New-York, and some of our own radicals could have been thrown into the ocean."

George Harris, of the Pennsylvania Miners' Union, thought that the whole trouble with the Knights of Labor arose from the efforts of that organization to organize in trades and districts already covered by trades unions.

*New-York Tribune,* Dec. 13, 1886.

1. In early July 1886 tanners and curriers in Salem, Peabody, and Stoneham, Mass., struck for shorter hours. On July 12 the leather manufacturers declared a lockout. Nonunion workers took most of the strikers' places and on Nov. 28 a mass meeting of the striking KOL assemblies voted to give up the strike.

2. J. R. Winders.

# GLOSSARY

## Individuals

APPEL, George W. (b. 1860?), served as the general secretary of the Metal Workers' National Union of North America from 1886 to 1889. Born in Maryland, he worked as a silver plater and brass finisher in Baltimore.

ARCH, Joseph (1826-1919), was a founder (1872) and leader of the National Agricultural Labourers' Union and a Member of Parliament (1885-86, 1892-1900).

ARTHUR, Peter M. (1831-1903), a Scottish immigrant, was grand chief engineer of the Brotherhood of Locomotive Engineers (BLE) from 1874 until his death. He was a charter member of BLE Division 46, the Brotherhood of the Footboard, in Albany, N.Y., was its chief engineer in 1868, and represented it in BLE conventions from 1866 to 1874. Arthur served as second grand assistant engineer of the BLE from 1869 to 1874. As grand chief engineer he maintained the BLE's independence from the AFL and other labor organizations.

AUBREY, James L., served in the early 1880s as an officer of CMIU 95 of St. Joseph, Mo., president of the newly organized Trades and Labor Assembly of St. Joseph, and second vice-president (1881-82) of the CMIU, resigning during the summer of 1882 when he stopped working in the trade. He was probably the same individual who had been an officer of CMIU 56 of Leavenworth, Kans., and CMIU 98 of St. Paul, Minn., in the early 1870s.

AYLSWORTH, Ira B. (b. 1854), a Baltimore stairbuilder, was born in Canada and immigrated to the United States in 1868. He was a founding member of local 29 of the Brotherhood of Carpenters and Joiners of America and its secretary from 1883 to 1886. He also served as a member of the KOL auxiliary board in 1886 and of the KOL General Executive Board from 1886 to 1888. During 1887 and 1888 Aylsworth was a leading figure in the movement to organize a KOL national trades assembly of carpenters.

BAILEY, William H., a miner, was born in Hamilton, Ontario, and

lived in Shawnee, Ohio. He served as a member of the KOL General Executive Board (1884-87) and was master workman of the miners' National Trade Assembly 135 (1886-87).

BARRY, Thomas B. (1852-1909), was a member of the KOL General Executive Board from 1885 to 1888. He was born in Cohoes, N.Y., and moved to East Saginaw, Mich., in 1880 to work as an axemaker. He was elected to the Michigan state legislature in 1885 as a candidate of the Labor, Democratic, and Greenback parties, and sponsored a ten-hour bill that passed in July 1885. Barry led a protracted strike in 1885 against Saginaw Valley lumber mill operators, and in 1886 the KOL General Executive Board sent him to Chicago to help settle the Chicago stockyards strike. He grew increasingly dissatisfied with the Knights' leadership, however, and particularly with Terence Powderly, whom he criticized as having been too willing to compromise the workers' demands in the Saginaw Valley and Chicago strikes. He resigned from the General Executive Board in September 1888 and publicly accused the Order's leaders of misuse of funds, maladministration, and autocratic rule. The General Assembly expelled Barry from the Order in November. After a brief attempt at organizing dissident Knights into a rival body, the Brotherhood of United Labor, Barry left the labor movement and worked as a lecturer and traveling agent.

BERLINER, Louis (b. 1831), was a member of the United Cigarmakers, and served CMIU 144 as financial secretary during 1877 and 1879 and as vice-president in 1882. A native of New York City, he was a member of the city's American Section of the Workingmen's Party of the United States and one of the founders of the Amalgamated Trades and Labor Union of New York and Vicinity. He moved to Brooklyn in the mid-1880s.

BLAIR, George (1845-1920), a German-born boxmaker and a Union sailor during the Civil War, was secretary and president of the Box Makers' Union, president of the New York City Workingmen's Central Council, and, from 1874 to 1883, president of the New York State Workingmen's Assembly (NYSWA). Blair joined Local Assembly 28 of the KOL in early 1875 and participated in calling the first KOL General Assembly in 1878. Between 1877 and 1879 he ran for several offices as a Greenbacker, including mayor of New York, comptroller of New York state, and U.S. congressman. He organized and chaired the Political Branch of the NYSWA in the early 1880s. In 1886 Governor David B. Hill appointed him to a commission to study prison labor problems. In the 1890s Blair emerged as a leader in Tammany Hall, and beginning in 1898 he served several years as superintendent

of the outdoor poor in the New York City Department of Public Charities.

BLEND, Fred (1845-1913), an Ohio-born cigarmaker, was president (1869-71) and first vice-president (1880-85) of the CMIU. He also served as an officer of cigarmakers' locals in Columbus and Toledo, Ohio, and Evansville, Ind., during the 1870s and 1880s. Blend became a cigar manufacturer in the early 1890s; he later moved to Louisville, Ky., where he continued in cigarmaking.

BLISSERT, Robert (1843-99), a socialist and Irish nationalist, was a tailor by trade. Born in England of Irish parents, he served in India with the British army and was later blacklisted for labor agitation during the London tailors' strike of 1867. The following year he immigrated to New York City, where he was active in the Tenth Ward Council of the International Workingmen's Association and participated in the KOL Excelsior Labor Club and the Amalgamated Trades and Labor Union of New York and Vicinity. He helped found the New York City Central Labor Union in 1882 and participated in its leadership throughout the 1880s.

BLOCK, George G. (1848-1925), secretary of the Journeymen Bakers' National Union (JBNU) from 1886 to 1888, was born in Bohemia and immigrated to New York City in 1870. Moving to Philadelphia in the 1870s, he worked as a pocketbook maker and a journalist, joined the Social Democratic Workingmen's Party of North America, and, during 1877, was organizer for the Philadelphia American-speaking section of the Workingmen's Party of the United States. Block returned to New York City in the early 1880s where he joined the staff of the *New Yorker Volkszeitung*. He helped found the New York City Central Labor Union and was secretary of the Executive Committee of the Henry George campaign. In 1885 he established the *Deutsch-Amerikanische Bäcker-Zeitung* through which he helped generate interest in organizing the JBNU in 1886. He served as secretary of the union until 1888 and editor of the journal until 1889. Around 1889 he went into the liquor business.

BLOETE, Charles George (1839-1908), a Prussian-born cigarmaker, came to New York City in 1866, joining CMIU 90. He became a member of the United Cigarmakers of New York soon after its organization in 1873 and was subsequently a prominent member of CMIU 144, serving successively as German recording secretary, auditor, and treasurer for the union between 1877 and 1881. He represented his local in the Amalgamated Trades and Labor Union of New York and Vicinity, where he served as financial secretary during

1882, and in the New York City Central Labor Union during 1886 and 1887. In 1887 he was appointed special agent of the New York Bureau of Labor Statistics and held that office until 1904.

BOYER, David P. (1842-1931), a Columbus printer, was born in Ohio. During the 1870s and 1880s he held offices in International Typographical Union (ITU) 5, the Columbus Trades Assembly, and the Ohio State Trades and Labor Assembly. Boyer served as chief organizer for the ITU from 1885 to 1888.

BRANT, Lyman A. (1848-95), was an officer of International Typographical Union (ITU) 18 of Detroit between 1871 and 1880 and was president of the Detroit Trade and Labor Council in 1881. While serving as corresponding secretary of the ITU between 1880 and 1881, he played a major role in organizing the FOTLU. He was elected to the Michigan state legislature as a Democrat in 1883 and later worked as a collector in the U.S. Customs House and clerk of the Michigan House of Representatives.

BROADHURST, Henry (1840-1911), a leader of the London stonemasons, was secretary of the British Trades Union Congress's parliamentary committee (1875-90) and a Liberal member of Parliament (1880-92, 1894-1906).

BUCHANAN, Joseph Ray (1851-1924), was born in Missouri and moved to Denver in 1878, where he became an editor and an organizer of the International Typographical Union. In 1882 he helped organize KOL Local Assembly 2327 and began publishing the *Labor Enquirer.* The following year Buchanan helped form the Rocky Mountain division of the International Workingmen's Association. Between 1884 and 1886 he organized western railroad workers and led several successful railroad strikes. Buchanan served as a member of the KOL General Executive Board from 1884 to 1885 and served on the KOL auxiliary board in 1886. In 1887 the KOL expelled him as a result of his disagreement with the decision to expel CMIU cigarmakers. Buchanan moved to New Jersey in 1888 and twice ran unsuccessfully for Congress. In 1892 he helped organize the People's party, serving on its national committee during the 1892, 1896, and 1900 elections. He was labor editor of the *New York Evening Journal* (1904-15) and a member of the conciliation council of the U.S. Department of Labor (1918-21).

BURGMAN, Charles F., a tailor and a socialist, was a delegate to the 1881 FOTLU convention from the Representative Assembly of Trades and Labor Unions of the Pacific Coast and second vice-president of the FOTLU Legislative Committee during 1881-82. At the 1881

convention he successfully advocated the passage of a resolution prohibiting further Chinese immigration. In 1882 Burgman worked with Burnette G. Haskell in founding the Pacific Coast Branch of the International Workingmen's Association, and between 1883 and 1885 he served as business manager of Haskell's paper, *Truth.* Burgman operated a tailoring establishment in San Francisco into the late 1890s.

BURKE, James (1848-1906), a New York-born cigarmaker, was second vice-president of the CMIU from 1877 to 1879 and an officer of Rochester local 5 in the late 1870s and early 1880s.

CARLTON, Albert A. (1847-1915), a Massachusetts-born shoemaker and Civil War veteran, was a member of the Knights of St. Crispin and the Lynn, Mass., Workingmen's Association. In 1877 he joined KOL Local Assembly 1715 and two years later was elected the first master workman of District Assembly 30, serving until 1886. Carlton was appointed a KOL lecturer in 1886 and served as a member of the Knights' General Executive Board from 1886 to 1888.

CAVANAUGH, Hugh (b. 1850), was an Irish-born shoemaker who immigrated to the United States in 1852. He became a member of the Knights of St. Crispin in Boston when he was nineteen, settled in Cincinnati in 1875, and helped organize KOL Local Assembly 280 there in 1877. Cavanaugh served as secretary (1884-86) and district master workman (1886-88) of KOL District Assembly 48 and was a member of the KOL auxiliary board in 1886. In 1887 he ran unsuccessfully on the Union Labor party ticket for Ohio state senator. About 1888 Cavanaugh became a shoe salesman; he was later an insurance agent. He remained active in the Knights, serving as general worthy foreman from 1890 to 1893.

CLARK, George (1836-88), president of the International Typographical Union (ITU) from 1881 to 1883, was born in Scotland. He edited the *Central Baptist* in St. Louis for eight years, beginning in the mid-1860s. Clark served as secretary of ITU 8 of St. Louis in 1880 and as its president from 1886 to 1888.

CLINE, Isaac (1835-1906), was a Pittsburgh window-glass worker. Born in New Jersey, and a Civil War veteran, Cline was connected with the Window Glass Blowers' Union in the 1860s, representing it at the National Labor Union convention in 1866. He was president of KOL Window Glass Workers' Local Assembly 300 from 1881 to 1887. In 1884 he went to Europe to help organize glass workers, and the following year he became the first president of the Universal Federation of Window Glass Workers.

COHEN, Fanny (Femmetje) Gompers (b. 1836), SG's aunt, married

Jacob Cohen, a union cigarmaker. They immigrated to the United States and had three sons and five daughters.

COUGHLIN, John E. (1853?-99), was born in Massachusetts. He was a currier in Chicago during the 1880s and was president of the National Tanners' and Curriers' Union in 1881. Living in Milwaukee during the 1890s, he served as general president of the United Brotherhood of Tanners and Curriers of America.

CRAWFORD, Mark L. (1848-1932), born in Indiana, served as secretary-treasurer (1882-83), president (1883-84), and chief organizer (1884-85) of the International Typographical Union (ITU). A prominent member of ITU 16 of Chicago, he was an officer of the Trade and Labor Assembly of Chicago and participated in the organization of the Illinois State Federation of Labor. Crawford became superintendent of the Chicago House of Corrections (1890-97) and was later an inspector for the U.S. Bureau of Immigration and Naturalization.

DALEY, Edward L. (1855-1904), was born in Danvers, Mass., and apprenticed in the shoemakers' trade at age thirteen. A member of KOL Local Assembly 715, Daley organized the Lynn, Mass., Lasters' Protective Union in 1878 and was its first secretary. He was the general secretary of the New England Lasters' Protective Union from 1885 to 1895. In 1891 he was elected to a single term in Congress as a Democrat.

DENNY, Albert G. (1845-1904?), a Pittsburgh glassworker, was secretary of KOL Local Assembly 300. He served as a KOL general organizer in the mid-1880s and organized workers for the KOL in Europe in 1884. During 1886 Denny was a member of the executive committee of the Trades Assembly of Western Pennsylvania. In 1887 he served as agent for the *Commoner and Glassworker* and from 1889 to his death was an insurance and real estate agent.

DOYLE, Martin F. (1826?-86), a New York City carpenter, was a member of the Central Committee of the Association of United Workers of America, served as an organizer for the Workingmen's Party of the United States in New York City, and was secretary of New York City Branch 1 of the International Labor Union.

DYER, Josiah Bennett (1843-1900), granite cutters' leader, was born in England and came to the United States in 1871. He was an early member of the KOL in Boston and helped organize a branch of the Granite Cutters' International Union in Graniteville, Mass., in 1877. The following year he was elected secretary of the Granite Cutters' International (later National) Union, serving in that office until 1895.

EDMONSTON, Gabriel (1839-1918), a founder and the first president of the Brotherhood of Carpenters and Joiners of America (1881-82), was born in Washington, D.C., and served in the Confederate army. He helped organize Washington carpenters in 1881 and was carpenter of the House of Representatives in the 1880s. A member of the FOTLU Legislative Committee from 1882 to 1886, he was elected its secretary in 1884. He introduced a series of resolutions at the FOTLU 1884 convention calling for the inauguration of the eight-hour movement. From 1886 to 1888 Edmonston served as treasurer of the AFL.

EMRICH, Henry (b. 1846?), a cabinetmaker born in Prussia, immigrated to New York in 1866 and joined the Cabinet Makers' Union two years later. Emrich was active in the political organization of the New York City Central Labor Union in the 1880s. He served as secretary of the International Furniture Workers' Union between 1882 and 1891 and was its delegate to the FOTLU and AFL conventions between 1885 and 1889. He was elected sixth vice-president of the FOTLU in 1885 and treasurer of the AFL in 1888 and 1889.

ENGEL, George (1836-87), a printer born in Kassel, Germany, immigrated to the United States in 1873, moved to Chicago in 1874, and was a member of the International Workingmen's Association, the SLP, and, after 1883, the anarchist International Working People's Association, or Black International. He wrote for the *Anarchist.* Engel was convicted of murder in connection with the Haymarket incident and was hanged on Nov. 11, 1887.

EVANS, Christopher (1841-1924), an Ohio miners' leader and AFL secretary from 1889 to 1894, was born in England and immigrated to the United States in 1869. He helped found the Ohio Miners' Amalgamated Association, serving as its president in 1889, and the National Federation of Miners and Mine Laborers (NFMML). As secretary of the NFMML from 1885 to 1888, he participated in joint conferences between miners and operators to establish annual scales of prices and wages in the Midwest coal region. After 1895 he became an organizer for the United Mine Workers of America (UMWA) and the AFL, and in 1901 he was appointed UMWA statistician.

FIELDEN, Samuel (b. 1847), born in Lancashire, was a cotton mill worker in England. After coming to the United States in 1868, he was an itinerant laborer hauling stone and working the canals, railroads, and levees of the Midwest and South. He settled in Chicago in 1871, was involved in labor organization among teamsters, and by the early 1880s was a prominent labor agitator. He joined the anarchist

International Working People's Association in 1884. Fielden was convicted of murder in connection with the Haymarket incident. His sentence was commuted to life imprisonment by Governor Richard J. Oglesby, and he was pardoned by Governor John Peter Altgeld in 1893.

FINKELSTONE, Edward (1863-1927), president of the Journeymen Barbers Protective Union of New York from 1886 to 1887, was born in Germany and immigrated to the United States in 1873. He led the movement that founded the Journeymen Barbers' National Union in December 1887 and served for one year as its first president.

FISCHER, Adolph (1858?-87), a compositor born in Bremen, Germany, immigrated to the United States about 1871. He worked in Little Rock, Ark., and St. Louis, Mo., before coming to Chicago in 1883 where he was employed by the *Chicagoer Arbeiter-Zeitung*. Fischer was convicted of murder in connection with the Haymarket incident and was hanged on Nov. 11, 1887.

FITZPATRICK, Patrick Francis (1835-99), president of the Iron Molders' Union of North America (IMUNA), was born in County Caven, Ireland, and immigrated to the United States at the age of sixteen. Apprenticed to the trade of iron molding in Troy, N.Y., he eventually settled in Cincinnati, where he held offices in IMUNA 4 until 1898 and was one of the founders of the Building Trades Council. From 1879 to 1890 Fitzpatrick was president of the IMUNA and helped restore confidence in a union that had been shaken by charges of corruption and by the depression of the 1870s.

FORAN, Martin Ambrose (1844-1921), Democratic congressman from Ohio (1883-89), was a Cleveland cooper, an organizer in 1870 of the Coopers' International Union, and its president for three years. During the early 1870s he was active in the movement to form a federation of national trade unions. He became a lawyer in 1874 and served as prosecuting attorney of Cleveland between 1875 and 1877. As a congressman, he supported labor legislation, and in 1884 he introduced a bill to prevent the importation of contract labor.

FORD, Thomas J. (b. 1850?), an Irish-born metal worker, was a walking delegate for the brass and silver workers' union in New York City, and in 1887 he served as secretary of the national union of brass workers that, the following year, became KOL National Trades Assembly 242.

FORSCHNER, August J. (b. 1838), a German-born cigarmaker who immigrated to the United States in 1880, was a Cigarmakers' Pro-

gressive Union National Executive Board trustee (1883-84) and treasurer (1884-86).

FORTUNE, John (b. 1840?), president of the Amalgamated Trades and Labor Union of New York and Vicinity, was born in Ireland. He was an early leader of the New York Custom Tailors' Protective and Benevolent Union.

FOSTER, Frank Keyes (1855-1909), was born in Massachusetts and worked as a printer in Connecticut before settling in Boston in 1880. He was active in the International Typographical Union and represented the Boston Central Trades and Labor Union at the 1883 FOTLU convention, where he was elected secretary of the Legislative Committee. A member of KOL Local Assembly 2006, he was elected secretary of District Assembly 30 and a member of the Knights' General Executive Board in 1883. In 1884 he began editing the KOL organ in Massachusetts, the *Laborer* (Haverhill). Foster ran unsuccessfully for lieutenant governor of Massachusetts in 1886 on the Democratic ticket. In 1887 he helped found the Massachusetts State Federation of Labor and served as its secretary. During the same year he founded the *Labor Leader* in Boston, editing it until 1897.

FOSTER, William Henry (1848?-86), an Irish-born printer who immigrated to the United States in 1873, was the founder of the *Cincinnati Exponent,* an officer of International Typographical Union (ITU) 3 of Cincinnati, and delegate from the Cincinnati Trades and Labor Assembly to the 1881 FOTLU convention. He was elected FOTLU Legislative Committee secretary in 1881, 1882, and 1885. In 1883 he joined the staff of the *Philadelphia Evening Call* and in 1884 became president of ITU 2 of Philadelphia. He helped found the city's Central Labor Union and served as its secretary. He joined KOL Local Assembly 3879 in 1886 and was its delegate to District Assembly 1.

FRANK, Anton (1856-1925), a New York cigarmaker born in Germany, immigrated to the United States in 1879. He was secretary of the Cigarmakers' Progressive Union of America (CMPU) in 1886 and treasurer of Cigar Makers' Progressive International Union 90 after the merger of the CMPU and the CMIU. Moving to New Haven, Conn., about 1892, Frank served as an officer in CMIU 39 during the 1890s.

GEORGE, Henry (1839-97), a Philadelphia-born journalist, labor reformer, and anti-monopolist, began his newspaper career in 1860 as a printer and then worked as an editor for several San Francisco papers. George published *Progress and Poverty,* his most influential work, in 1879 and *The Irish Land Question* in 1881, and subsequently

served in the British Isles as correspondent for the *Irish World.* In 1886 he ran second as the liberal and labor candidate in a three-way contest for mayor of New York City against Theodore Roosevelt and the victorious Abram S. Hewitt. His supporters gave serious consideration to a presidential race in 1888, but their hopes were dashed by his disappointing showing in the 1887 campaign for secretary of state of New York. In the campaign's aftermath supporters launched the single-tax movement. George meanwhile continued his writing, edited the *Standard* from 1887 to 1890, undertook several speaking tours, and traveled extensively. In 1897, against medical advice, he again ran for mayor of New York City; he died four days before the election.

GESSNER, Frank M. (1850-1906), secretary of Window Glass Workers' Local Assembly 300 of the KOL between 1883 and 1886, was born in Pittsburgh where he worked as a glassblower and brickmaker. He founded the *American Glass Worker* in 1886 and later was a glass trade journal editor. During the late 1890s Gessner was secretary of the Pennsylvania state committee of the SLP.

GIBSON, David R., was vice-president of the Bricklayers' and Masons' International Union in 1883 and a frequent delegate to the Bricklayers' conventions through 1896. He represented District Assembly 61 of Hamilton, Ontario, at the 1885 KOL General Assembly and served on the KOL auxiliary board in 1886.

GOLDSMITH, Marcus K., a leader in CMIU 39 of New Haven, Conn., was active in the Social Democratic Workingmen's Party of North America and in 1877 was corresponding secretary of the Board of Supervisors of the Workingmen's Party of the United States.

GOMPERS, Abraham (b. 1877), was the son of Simon and Elizabeth Tate Gompers. He was working as a longshoreman in 1900.

GOMPERS, Abraham Julian (1876-1903), was the son of SG and Sophia Gompers; he worked in New York City in the clothing industry as a cutter. In 1901, after he contracted tuberculosis, his parents sent him to convalesce at the Denver, Colo., home of Max Morris, secretary and treasurer of the Retail Clerks' International Protective Association. Abraham worked briefly for the association before his death.

GOMPERS, Alexander (1857-1926), SG's brother, was born in London and immigrated with the family to the United States in 1863. He was a cigarmaker and an early member of CMIU 144. He married Rachel Bickstein, who was born in London, and they had six children.

GOMPERS, Alexander Julian (1878-1947), was the son of SG and

Sophia Gompers. He was a cigarmaker and cigar manufacturer in New York City and in Washington, D.C. From 1914 to 1947 he served as an official of the New York State Department of Labor. He and his wife, Ella Appelbaum Gompers, had three children: Esther, Sophia, and May.

GOMPERS, Bella (b. 1867), SG's sister, was born in New York, and about 1886 married Samuel Isaacs, a Boston cigarmaker.

GOMPERS, Catherine (Grietje; b. 1841), SG's aunt, married Samuel Pennamacour, and they immigrated to the United States in 1864. They returned to England, where her husband died. In 1869 Catherine came back to New York City, where she married SG's cousin Emanuel Gompers.

GOMPERS, Elizabeth Tate (b. 1848), was born in England and immigrated to New York in 1869. She married Simon Gompers, SG's uncle, and they had eight children.

GOMPERS, Emanuel (b. 1846), SG's cousin and uncle by virtue of his marriage to Catherine Gompers, SG's aunt, was born in Holland and immigrated to the United States in 1872. He was a cigarmaker and beginning in the late 1890s operated a cigar shop.

GOMPERS, Fannie (b. 1879), was the daughter of Simon and Elizabeth Tate Gompers. She was working in a tin foil factory in 1900.

GOMPERS, Harriett (b. 1861), SG's sister, was possibly the daughter that Solomon Gompers mentioned in the family history as having died when very young (SG, *Seventy Years,* 1:504).

GOMPERS, Henrietta (1869-1954), SG's sister, was born in New York, and about 1887 married Harry Isaacs, a tailor. They settled in Boston.

GOMPERS, Henrietta (Jette) Salomon Haring (1807?-79), SG's paternal grandmother, was born in Holland where she married SG's grandfather, Samuel Gompers.

GOMPERS, Henry (b. 1853), SG's brother, was born in London and immigrated with the family to America in 1863. A cigarmaker and member of the CMIU, he married Sarah Wennick, a Dutch immigrant, and they had two sons.

GOMPERS, Henry Julian (1874-1938), was the son of SG and Sophia Gompers. In 1887 he was the AFL's first office boy, and he later became a granite cutter. About 1914 he moved from New York City to Washington, D.C., where he ran Gompers' Monumental Works. He and his wife, Bessie Phillips Gompers, had four children: Sophia, Samuel, Alexander, and Louis.

GOMPERS, Jacob (1863-1906), SG's brother, was born in London and immigrated with the family to America in 1863. A diamond polisher and first president of the Diamond Workers' Union of America, he married Sophia Spero, and they had three sons and two daughters.

GOMPERS, Louis (1859-1920), SG's brother, was born in London and immigrated with the family to the United States in 1863. He became a Brooklyn, N.Y., cigar manufacturer and president of the Retail Tobacco Dealers' Association. He married Sophia Bickstein, and they had seven children.

GOMPERS, Rose (1872-99), was the daughter of SG and Sophia Gompers. She married Samuel Mitchell, a U.S. postal employee and a member of the Letter Carriers' Union, about 1890. They lived in New York City and had two children, Henrietta and Ethel.

GOMPERS, Samuel (b. 1869), was the son of Simon and Elizabeth Tate Gompers. He became a cigarmaker.

GOMPERS, Samuel Julian (1868-1946), was the son of SG and Sophia Gompers. Born in New York City, he left school at the age of fourteen to work in a New York City print shop. He moved to Washington, D.C., about 1887 and worked as a printer in the Government Printing Office, a compositor in the U.S. Department of Commerce and Labor, and a clerk in the U.S. Census Office. He was a member of the Association of Union Printers and the Columbia Typographical Union. In 1913 he became chief of the Division of Publication and Supplies of the U.S. Department of Labor, and in 1918 he became chief clerk of the Department of Labor, a position he held until 1941. Gompers and his wife, Sophia Dampf Gompers, had one child, Florence.

GOMPERS, Samuel Moses (Salomon Mozes; 1803?-81), SG's paternal grandfather, was born in Amsterdam. Originally a calico printer, he later became an import-export merchant. He married Henrietta Haring, and they had six children: Solomon, Fannie, Clara, Kate, Sarah, and Simon. In 1845 he immigrated to London, and about 1869 came to the United States, returning to London in 1876.

GOMPERS, Sarah Rood (1827-98), SG's mother, was born in Amsterdam. About 1847 she came to London to live in the Gompers household and in 1849 married Solomon Gompers. She immigrated to the United States with her family in 1863; in 1872 she gave birth to her fourteenth and last child.

GOMPERS, Simon (1865-1953), SG's brother, was born in New York and worked there as a sheet metal worker. He married Leah Lopez,

and they had five children. The family moved to Norwalk, Conn., in 1924.

GOMPERS, Simon (1849-96), SG's uncle and close childhood companion, was born in London and was a union shoemaker. Immigrating to the United States in 1868, he married Elizabeth Tate; they had six sons and two daughters.

GOMPERS, Solomon (1827-1919), SG's father, was a cigarmaker who was born in Amsterdam and immigrated to England with his family in 1845. He became a member of the cigarmakers' union there in 1848. In 1863 he immigrated to the United States with his wife Sarah and their six children.

GOMPERS, Sophia Julian (1850-1920), SG's first wife, was born in London and immigrated to the United States about 1858. She was living with her father and stepmother in Brooklyn and working as a tobacco stripper in a cigar factory when she married SG in 1867. Between 1868 and 1885 she and SG had at least nine children, six of whom lived past infancy: Samuel, Rose, Henry, Abraham, Alexander, and Sadie.

GREEN, Samuel S., a Louisville printer and member of International Typographical Union 10, was elected first vice-president of the FOTLU at its 1885 convention.

GRENELL, Judson (1847-1919?), a labor editor born in New York, served as an officer of International Typographical Union (ITU) 47 of New Haven during the early 1870s and as an officer of ITU 18 of Detroit and the Detroit Trade and Labor Council during the early 1880s. Joining the KOL, he became master workman of Henry George Local Assembly 2697 and delegate to District Assembly 50. Grenell edited a variety of labor papers in the 1870s and 1880s, including such Detroit papers as the *Socialist*, an SLP organ, and, with Joseph Labadie, the *Advance and Labor Leaf*. He was active in the single-tax movement and in 1886 was elected as an Independent Labor party candidate to the Michigan state legislature. Later he worked as an editor and reporter with several city dailies.

GROSSE, Edward (1845-97), born in Germany and trained as a printer, served as secretary to Johann Baptist von Schweitzer, president of the Allgemeiner Deutscher Arbeiterverein, before immigrating to the United States in 1869. In New York City he joined German-American Typographia 7, became active in the International Workingmen's Association, and was a member of the Economic and Sociological Club. He became a lawyer in 1878 and two years later won a seat as an independent in the state legislature, where he introduced

the first tenement-house bill. He served as an assistant district attorney between 1888 and 1891; and in 1894 he was appointed as a collector of internal revenue in New York City.

GUTSTADT, Herman (1853-1918?), a German-born cigarmaker, immigrated to the United States in 1867 where he became active in the Brooklyn labor movement. A member of the Brooklyn Spread the Light Club—KOL Local Assembly 1562—and of District Assembly 49, he was one of the few prominent German-speaking members of the New York City KOL in the early 1880s. He served as an officer of CMIU 87 of Brooklyn and in late December 1885 led a group of over 100 New York City cigarmakers to San Francisco under CMIU auspices to replace Chinese cigarmakers. There he became an officer of CMIU 228, vice-president of the Federated Trades Council (FTC), and a prominent leader in the anti-Chinese movement, promoting a blue union label in early 1886 to differentiate union- from Chinese-made cigars. During 1887, dissatisfied with the weakness of the FTC's anti-Chinese campaign, Gutstadt helped organize another central body, the Trades Union Mutual Alliance, to which he was elected president. When the two organizations reconciled later in the year, the Alliance dissolved and Gutstadt returned to the vice-presidency of the FTC. In the early 1890s he became a cigar manufacturer; he was later a storekeeper, a clerk, and an insurance agent. With SG he coauthored *Meat vs. Rice: American Manhood against Asiatic Coolieism* (Washington, D.C., 1902).

HALLER, Frederick (b. 1853), was born in Augusta, Ga., and was later apprenticed to a Savannah cigar manufacturer. He moved to New York City in 1880, becoming a leader of the Cigarmakers' Progressive Union and the Central Labor Union. In 1886 he led a large group of Progressives into the CMIU, joined SG in the cigarmakers' struggle with the KOL, and was active in the Henry George mayoralty campaign. In 1888 Haller moved to Buffalo, N.Y., where he began studying law and was elected president of the New York State Branch of the AFL. He resigned at the end of 1890 to practice law, became assistant district attorney of Erie County in 1896, and five years later prosecuted Leon Czolgosz, the assassin of President William McKinley.

HARRIS, Daniel (1846-1915), a Civil War veteran and cigarmaker, was born in England and immigrated to the United States in the early 1860s. During the 1877-78 cigarmakers' strike the Central Organization of the Cigarmakers of New York appointed Harris to its Committee on Organization for Pennsylvania. In the late 1880s Harris was president of CMIU 144. He served as president of the New York

State Workingmen's Assembly from 1892 to 1897 and the New York Federation of Labor in 1899 and from 1906 until his death.

HARRIS, George (1853-1935), a Pennsylvania miners' leader, was born in England and immigrated to the United States in 1880, settling in Reynoldsville, Pa. He served as president of the Miners' Amalgamated Association of Pennsylvania (after 1886 the Miners' and Mine Laborers' Amalgamated Association of Pennsylvania) from 1883 until 1887, helped organize the National Federation of Miners and Mine Laborers (NFMML) in 1885, and was elected first vice-president of the AFL in 1886. The following year he resigned his position with the miners' federation to become an organizer for KOL National Trade Assembly 135. In 1888 he participated in its merger with the NFMML to form the National Progressive Union of Miners and Mine Laborers, reorganized as the United Mine Workers of America (UMWA) in 1890. Harris was an organizer for the UMWA in the 1890s and in 1897 was elected president of UMWA Pennsylvania District 2 and vice-president of the newly organized UMWA Pennsylvania State Association, serving until 1899. He was an active campaigner for the Republican party in the 1880s and 1890s. In 1922 he moved to Wilmington, Del.

HAUSLER, Mary (variously Marie; 1847?-85), a Bohemian cigarmaker who immigrated to the United States about 1872, served as vice-president of the Central Organization of the Cigarmakers of New York during the 1877-78 strike and later became a member of CMIU 141, a Bohemian local. Hausler was particularly active in recruiting women cigarmakers for the CMIU in New York.

HAYES, John William (1854-1942), was initiated into the KOL while employed as a brakeman with the Pennsylvania Railroad in 1874. Soon after joining the Order he was commissioned as an organizer and, after the loss of his right arm in an 1878 railroad accident, became a telegrapher. Losing his position in 1883 because of union activities, he entered the grocery business. He was a member of the KOL General Executive Board from 1884 to 1916, serving as general secretary-treasurer from 1888 until 1902 and as the Knights' last general master workman from 1902 until the Order closed its central office in 1916. Hayes took an active part in the People's party in the early 1890s. He was manager of the Atlantic Gas Construction Co. in Philadelphia and, later, president of the North Chesapeake Beach Land and Improvement Co.

HENNEBERRY, Thomas T. (b. 1844?), a Cleveland cooper, was born in Ireland. He was treasurer and then president of the Coopers'

International Union during the 1870s and early 1880s. Between 1880 and 1882 he was also a wholesale liquor dealer.

HEWITT, Abram S. (1822-1903), an iron manufacturer, philanthropist, and reform politician, was elected as a Democrat to five terms in the U.S. Congress, beginning in 1874. In 1886 he defeated Henry George and Theodore Roosevelt for the mayoralty of New York City.

HILLMANN, Carl (1841-97), was an active member of the International Workingmen's Association (IWA) and an ardent trade unionist in Germany. Born in Saxony, Hillmann was working as a typesetter and was active in trade union and socialist affairs in Hamburg at the time of Ferdinand Laurrell's sojourn there. A zealous follower of Karl Marx, Hillmann supported the Marxist wing of the German socialist movement led by Wilhelm Liebknecht and August Bebel, co-founders in 1869 of the Sozialdemokratische Arbeiterpartei (Social Democratic Workingmen's party; SDAP).

His first work was a brief history of the IWA (*Die Internationale Arbeiterassociation, 1864-1871* [1871]). His second publication, *Praktische Emanzipationswinke* (1873), first appeared in four installments in May 1873 in *Der Volksstaat,* organ of the SDAP and the IWA, while Hillmann was serving as a party organizer. In 1874 Hillmann became editor of the *Süddeutsche Volkszeitung,* published in Stuttgart, and in September 1874 he was imprisoned on charges of violating imperial press laws. Briefly released, he was again imprisoned in January 1875 and was still in jail in Rottenburg when he completed *Die Organisation der Massen* (1875). Upon his release he resigned his editorship and returned to Hamburg, where he found employment as one of the editors of the *Gerichtszeitung,* a local party organ. The government proscribed this newspaper in 1881 and exiled seventy-five people, including Hillmann, from the city. After an unsuccessful effort to continue the publication of the paper, he renounced socialism in 1882, though continuing to work as a journalist.

HIRTH, Frank (b. 1837?), was born in Baden, Germany, and lived in Canada before settling in Detroit in the early 1870s. There he edited the *Socialist,* an SLP paper, and was corresponding secretary of CMIU 22. In 1878 he moved to Chicago where, with Albert Parsons, he edited another SLP organ, also called the *Socialist,* until it expired the following year. He was vice-president of CMIU 25 of Milwaukee in 1882 and became superintendent of a cigar manufacturing cooperative set up after local 25's strike that year. After the cooperative closed, Hirth remained in Milwaukee as a cigarmaker.

Howard, Robert (1845-1902), a spinner and union organizer in Lancashire, England, immigrated to the United States in 1873. In 1878 he became secretary of the Fall River Mule Spinners' Association, serving until 1897. At the same time he played a leading role in national organizations of the trade—the Amalgamated Mule Spinners' Association, of which he was principal officer from 1878 to 1887, and the National Cotton Mule Spinners' Association. Howard was a campaigner for shorter-hour legislation and other labor measures. He was elected to a term in the Massachusetts House of Representatives in 1880 and to the Massachusetts Senate in 1886. He was treasurer of the FOTLU Legislative Committee in the early 1880s and was elected master workman of KOL District Assembly 30 in 1886. In the mid-1890s he conducted an organizing campaign among southern textile workers for the AFL.

Huck, Louis, served as an officer of CMIU 144 in the 1870s and of locals 39 (New Haven) and 71 (Pontiac, Ill.) in the 1880s. He was active in the Workingmen's Party of the United States, serving as treasurer for the party organ the *Labor Standard* and running on the party ticket for New York City alderman in 1877. In 1882 he undertook a major organizing tour for the CMIU through the Northeast and Midwest.

Hurst, George (1845-1923), born in New York, served as secretary of CMIU 23 of Suffield, Conn., and as president (1875-77) of the CMIU. In the early 1880s he moved to Hartford, Conn., where he was financial secretary of local 42 in 1901 and 1902.

Jablinowski, Ludwig (b. 1856), a cigarmaker born in Germany, immigrated to the United States in 1880 and in 1885 became secretary of Cigarmakers' Progressive Union 1. Following the Progressives' amalgamation with the CMIU in 1886, he remained secretary of the local, renamed Cigar Makers' Progressive International Union 90, until 1889. Jablinowski was a financial secretary of the New York City Central Labor Union (CLU) between 1884 and 1886 and was active in the Henry George mayoralty campaign. He was one of the founders in 1889 of the New York City Central Labor Federation, which opposed the CLU during the following decade. In the 1890s he was a reporter for the SLP's organ the *People* and for the *New Yorker Volkszeitung;* he was later an editor of the *People.*

Jarrett, John (1843-1918), president of the National Amalgamated Association of Iron and Steel Workers (AAISW), was born in England where he became an iron puddler. He immigrated to Pennsylvania in 1862 but returned to England six years later and became

active in the Amalgamated Ironworkers Association. Immigrating to Pennsylvania again in 1872, he became a leader in the Sons of Vulcan and, as vice-president, helped form the AAISW, serving as that union's president from 1880 to 1883. Jarrett led his union out of the FOTLU in 1882 when the Federation refused to endorse a high tariff. After leaving the presidency of the Amalgamated, he was secretary of the American Tin Plate Co. and a lobbyist for the Tin Plate Association, American consul at Birmingham, England (1889-92), and an executive of the tin plate and sheet steel trade association (1892-1900).

JONES, David R. (b. 1853), was born in West Virginia and worked as a miner before studying law. In 1879 he helped organize the Amalgamated Association of Pittsburgh Miners and Drivers and served as its chief executive officer (1879-82) before turning to the practice of law.

JULIAN, David H. (b. 1816?), SG's father-in-law, emigrated from England and later established himself in the furniture business in Brooklyn, N.Y.

JUNIO, John Joseph (b. 1842), born in Boston, was president of the CMIU (1867), president of the New York Cigar Makers' State Union (1875-77), and an officer of local 6 of Syracuse, N.Y. He was active in labor reform, representing the Mechanical Order of the Sun at the National Labor Union Congress in 1868, running for New York secretary of state on the Workingmen's Party of the United States ticket in 1877, serving as state chairman of the Greenback-Labor party, and representing District Assembly 152 at the KOL General Assembly in 1887. He was an organizer for the CMIU in the mid-1880s.

KAUFMANN, Berthold W. (b. 1845), a German immigrant, was a cabinetmaker. He was a member of the Social Democratic Workingmen's Party of North America and subsequently organizer and vice-president of the New York German Section of the Workingmen's Party of the United States. He served as secretary of the Central Committee of the Amalgamated Union of Furniture Workers of North America during 1875 and 1876, and was president of the Socialistic Cooperative Publishing Association, the organization that published the *New Yorker Volkszeitung,* from 1881 to 1884.

KEAN, Edward J. (b. 1851), a New York City printer, was born in Maryland. He served as president of the Amalgamated Trades and Labor Union of New York and Vicinity and as editor of International Typographical Union 6's organ, the *Boycotter.* Kean was chief clerk of the New York Bureau of Labor Statistics between 1885 and 1891.

KING, George G. (1848?-88?), a Maryland-born printer and member of International Typographical Union 12 of Baltimore, represented the Baltimore Federation of Labor at the 1885 FOTLU convention and was elected fifth vice-president.

KIRCHNER, John S. (1857-1912), a cigarmaker, was born in Maryland and became active in the labor movement in 1877 when he joined a Baltimore local assembly of the KOL. After moving to Philadelphia he helped organize CMIU 100, serving as its secretary (1881, 1884-90), and was CMIU fourth vice-president (1885-87) and organizer for Pennsylvania. In 1886 he was appointed secretary of the FOTLU upon the death of William H. Foster.

KRONBURG, David (variously Kronberg), a German immigrant, was one of the inner circle of the North American Federation of the International Workingmen's Association, a member of the Economic and Sociological Club, and a founder of the Association of United Workers of America and the Workingmen's Party of the United States.

LAURRELL, Carl Malcolm Ferdinand (1844-1921), a cigarmaker, was born in Sweden where he served as secretary of the Scandinavian section of the International Workingmen's Association until forced into exile. After working in Hamburg, Germany, for several years, he immigrated to the United States in the early 1870s and became active in the New York City labor movement. He became SG's mentor during the 1870s and subsequently remained close to SG as an active member of the United Cigarmakers and the CMIU.

LAYTON, Robert D. (1847?-1909), grand secretary of the KOL from 1881 to 1883, was born in Pennsylvania. A Civil War veteran, toolmaker, and railroad worker, Layton joined the Knights in 1873 and served as financial secretary and district master workman of Pittsburgh District Assembly 3. In the early 1880s he became an insurance salesman. He served on the U.S. Immigration Commission during the administration of Benjamin Harrison, for whom he campaigned in the 1888 presidential election.

LE BOSSE, Clara (Mietje) Gompers (b. 1838), SG's aunt, married François Le Bosse, a Frenchman who worked in the London mint. They had two sons and one daughter.

LEFFINGWELL, Samuel Langdale (1830-1903), a Mexican War and Civil War veteran, was a leader of the Cincinnati and Columbus typographical unions in Ohio during the 1850s. In 1875 he joined the KOL in Indianapolis, working as an organizer and attending several sessions of the General Assembly. He also was a leader of the Indianapolis Trades Assembly and attended the 1881 FOTLU convention

as a representative of the Indianapolis trade unions. Leffingwell edited several labor papers, including an organ of the International Typographical Union. In 1885 he helped organize the Indiana State Federation of Trade and Labor Unions, twice serving as its president.

LEVY, Marcus (b. 1840), was an Amsterdam-born cigarmaker. He married Sarah Gompers, SG's aunt, in 1863, immigrated with her to New York City in 1865, and later became a cigar manufacturer.

LEVY, Sarah Gompers (b. 1842), SG's aunt, married Marcus Levy in 1863 and they immigrated to New York City in 1865.

LINGG, Louis (1864-87), a carpenter born in Baden, Germany, was active in the German labor movement before immigrating to Chicago in 1885. He was a member and organizer for the Brotherhood of Carpenters and Joiners and a delegate to the Chicago Central Labor Union. Lingg was convicted of murder in connection with the Haymarket incident and was sentenced to death. He committed suicide in jail.

LYNCH, James, one of the founders of the Amalgamated Trades and Labor Union of New York and Vicinity (ATLU), its president and later its treasurer, was a New York City carpenter, and a member of the Economic and Sociological Club. He headed the ATLU's special committee appointed to work with the CMIU in its campaign for a tenement-house bill. During the early 1880s Lynch served as the New York City walking delegate for the United Order of American Carpenters and Joiners and was a delegate to its Grand Council.

MCAULIFFE, Patrick F. (b. 1850), a Washington, D.C., stonecutter, was born in Ireland and immigrated to the United States in 1865. A delegate from the Granite Cutters' National Union to the 1885 FOTLU convention, he was elected third vice-president of the Federation that year.

MCBRIDE, John (1854-1917), the only person to defeat SG for the presidency of the AFL, presided over the formation of the United Mine Workers of America (UMWA) and was its second president. The son of an Ohio miner, he was elected president of the Ohio Miners' Protective Union in 1877 and master workman of KOL District Assembly 38 in 1880, and served as president of the Ohio Miners' Amalgamated Association from 1882 to 1889. In 1885 he was a founder and first president of the National Federation of Miners and Mine Laborers and in 1886 presided over the founding convention of the AFL, declining the Federation's nomination for president. He served as president of the National Progressive Union of Miners and Mine Laborers in 1889 and was a leader in merging that union with

KOL National Trade Assembly 135 in 1890 to form the UMWA. He became president of the UMWA in 1892 and served until 1895.

McBride served as a Democrat in the Ohio legislature from 1884 to 1888, was commissioner of the Ohio Bureau of Labor Statistics from 1890 to 1891, and later became active in the populist movement. He was elected president of the AFL over SG in 1894, and narrowly lost to SG the next year. McBride purchased the *Columbus Record* in 1896, which he edited until 1917. He left that position in 1917 due to illness and moved to Arizona, where he served as a federal labor conciliator.

McCABE, William (1844-1924), a leader in the single-tax movement, emigrated from New Zealand as a child and later served in the Union army. He became a printer and associate of Henry George in San Francisco and served as president of the Sacramento Typographical Union (International Typographical Union 46). He moved to New York in 1880, where he was a founder of the Central Labor Union in 1882 and was active in Henry George's 1886 mayoralty campaign. For the last thirty years of his career he worked as a compositor for the *New York Herald,* and in 1924 he was nominated for governor of New York by the Commonwealth Labor (formerly Single Tax) party.

MACDONALD, Alexander (1821-81), was a Scottish miners' leader and Member of Parliament from 1874 until his death. A leading advocate of mine reform, he visited the United States frequently between 1867 and 1873, touring mines and speaking on the importance of building well-financed unions. He regularly corresponded with American labor leaders.

McDONNELL, Joseph Patrick (1847-1906), born in Ireland and active in the Fenian movement, joined the International Workingmen's Association (IWA) after moving to London in 1868 and served as Irish secretary of the IWA's General Council. Immigrating to New York City in 1872, he became a leading figure in the Association of United Workers of America and a member of the Economic and Sociological Club. He began editing the *Labor Standard,* an organ of the Workingmen's Party of the United States (WPUS), in 1876. Associated with the trade union faction of the WPUS, he moved the paper to Boston in 1877 in the midst of the WPUS's dissolution over the dispute between trade unionist and political tactics. In 1878 he moved the paper to Paterson, N.J., where he helped organize the International Labor Union. McDonnell was imprisoned twice between 1878 and 1880 for publishing libelous material against strikebreakers and manufacturers. In 1883 he helped organize the New Jersey Federation of Trades and Labor Unions; he served as its chairman until

1897. In 1884 he was a founder of the Paterson Trades Assembly. He became New Jersey's first factory inspector in 1884 and, in 1892, was appointed to the New Jersey Board of Arbitration. McDonnell continued to publish the *Labor Standard* until his death.

McGREGOR, Hugh (1840-1911), was an English-born jeweler. He served as a volunteer with Garibaldi's army and immigrated to the United States in 1865. During the 1870s he was a member of the International Workingmen's Association and a founder and active organizer for the Social Democratic Workingmen's Party of North America. He served as secretary of its New York branch in 1875 and its Philadelphia branch in 1876, returning to New York in the spring of that year to edit the new English-language organ of the party, the *Socialist.* A participant in the Economic and Sociological Club, he apparently left the socialist movement and became active in a small circle of New York City positivists. During the late 1880s he served as SG's secretary, directing the AFL office during the president's absence. He helped organize seamen on the Atlantic coast and between 1890 and 1892 served as secretary of the short-lived International Amalgamated Association of Sailors and Firemen. He later worked briefly in the AFL's Washington office.

McGUIRE, Peter James (1852-1906), chief executive officer of the United Brotherhood of Carpenters and Joiners of America (UBCJA), was born in New York City. He joined a local carpenters union there in 1872 and became a member of the International Workingmen's Association. He was involved in relief efforts in New York City during the depression of the 1870s and played a major role in organizing the Tompkins Square demonstration of January 1874. In 1874 he helped organize the Social Democratic Workingmen's Party of North America and was elected to its Executive Board; that same year he joined the KOL. During the late 1870s McGuire traveled widely, organizing and campaigning first on behalf of the Workingmen's Party of the United States and then for the SLP. After living for a time in New Haven, Conn., he moved to St. Louis in 1878 and the following year was instrumental in the establishment of the Missouri Bureau of Labor Statistics, to which he was appointed deputy commissioner. He resigned in 1880 to campaign for the SLP and for the Greenback-Labor party. In 1881 he was elected secretary of the St. Louis Trades Assembly and, as a member of the provisional committee to organize a carpenters' national union, began editing the *Carpenter.* Through that journal he generated interest in organizing the Brotherhood of Carpenters and Joiners (later UBCJA); he was elected secretary of the union at its founding convention later that year and held the

position until 1901. McGuire moved to New York City in 1882, was a founder of the New York City Central Labor Union, and became a member of Spread the Light KOL Local Assembly 1562. He served as secretary of the AFL from 1886 to 1889, second vice-president from 1889 to 1890, and first vice-president from 1890 to 1900.

MCGUIRE, Thomas B. (b. 1849), a leader of KOL District Assembly 49 and a member of the Home Club, was a marble polisher and truck driver. Born in New York City, he served in the Union army during the Civil War. He became active in the labor movement in the early 1870s and subsequently served as an officer in KOL local assemblies 2234 and 1974. McGuire was master workman of KOL District Assembly 49 in 1886. He was a member of the Knights' General Executive Board from 1886 to 1888 and from 1892 to 1897, and lectured for the Order in the interim. In 1893 he was a member of the Advisory Committee of the People's party. During the 1890s he resided in Amsterdam, N.Y.

MCINTOSH, Edward S. (1843-1910), secretary-treasurer of the International Typographical Union (ITU) from 1885 to 1886, was a Civil War veteran. Born in Doylestown, Pa., where he edited the *Beacon,* he became active in the Republican state campaign committee. In 1881 he joined ITU 2 of Philadelphia and served two terms as its president beginning in 1897.

MCLAUGHLIN, Daniel (1831-1901), was first vice-president of the AFL for one term (1887-88). He was born in Lanarkshire, Scotland, and was involved with Alexander MacDonald in the Scottish miners' movement before immigrating to Braidwood, Ill., in 1869. There he emerged as a leader of the KOL, served as a member of the KOL General Executive Board (1880-81), and successfully ran for town mayor on the Greenback ticket in 1877 and 1881.

McLaughlin was active in the Illinois State Federation of Labor and in several miners' unions. He served as president of the Coal Miners Benevolent and Protective Association of Illinois (1885-88) and helped organize the National Federation of Miners and Mine Laborers in 1885. As its treasurer (1885-88) he proposed joint conferences between miners and operators to establish annual scales of prices and wages in the Midwest coal region. McLaughlin was elected to the state legislature on the Republican ticket in 1886 and 1888. Involved in the contest with the KOL for control of the coal fields, he helped organize the National Progressive Union of Miners and Mine Laborers in 1888. About 1890 he left Illinois to become a mine superintendent in Starkville, Colo.

McMackin, John (1852-1906), a painter and leader in the New York City Central Labor Union, was born in Ireland and immigrated to the United States in 1865. He was active in the Henry George mayoralty campaign and later, as a Republican, served as special inspector of customs for New York City (1889-94), deputy commissioner (1897-99) and commissioner (1899-1901) of the New York Bureau of Labor Statistics, and New York labor commissioner (1901-5).

McNally, John (b. 1830), a weaver and leader of the Paterson, N.J., Silk Weavers' Association, was born in England and immigrated to the United States in 1873. He was active in the New York American Section of the Workingmen's Party of the United States and in 1878 participated in the formation of the International Labor Union, serving as vice-president of its Executive Committee.

McNeill, George Edwin (1837-1906), a Boston printer born in Massachusetts, was secretary of the Grand Eight-Hour League and president of the Boston Eight-Hour League. He helped lobby for the establishment of the Massachusetts Bureau of Statistics of Labor and served as its deputy director from 1869 to 1873. McNeill was an officer of the Sovereigns of Industry, president of the International Labor Union in 1878, and secretary-treasurer of KOL District Assembly 30 from 1884 to 1886. He was editor or associate editor of several papers including the *Labor Standard* in Boston, Fall River, Mass., and Paterson, N.J., and in 1887 published *The Labor Movement: The Problem of To-Day.* He helped organize the Massachusetts Mutual Accident Association in 1883 and was elected its secretary and manager in 1892. In 1897 he served as the AFL's fraternal delegate to the Trades Union Congress of Great Britain.

McPadden, Myles (1852-1906?), an Irish-born iron molder, was an active member of the Iron Molders' Union of North America and the KOL in St. Louis, Pittsburgh, and Chicago. In 1881 and 1882 he represented District Assembly 3 at the KOL General Assemblies and served on the Knight's General Executive Board; in 1882, as a KOL organizer, he assisted miners in western Maryland and Clearfield, Pa., in their unsuccessful strike. McPadden went to work for the McCormick Harvesting Machine Co. in Chicago the same year where, as a leader of the Molders' Union, he became involved in a series of labor disputes with the company. After the defeat of the 1886 strike at McCormick, which culminated in the Haymarket incident, McPadden kept a store and then worked for the U.S. Bureau of Internal Revenue in Chicago until his death.

Marden, William Henry (1843-1903), a leader of the Massachusetts

shoemakers, was born in that state and served in the Union army. He was secretary of the Knights of St. Crispin Lodge in Stoneham, Mass., became a member of KOL Local Assembly 2340, and served as treasurer of the New England Lasters' Protective Union in the late 1880s. In 1893 he was elected fourth vice-president of the AFL. He was a member of the Massachusetts House of Representatives from 1895 to 1899.

MILKE, Frederick K. (1845-1902), a German-born Brooklyn printer who immigrated to the United States in 1879, was secretary of the German-American Typographia in 1886.

MILLER, Hugo A. (1856-1926), secretary of the German-American Typographia, was born in Freiberg, Saxony, and entered the printing trade at the age of fifteen. In 1873 he came to New York City, joined the Typographia, and became active in union affairs. He represented the union at the 1882 FOTLU convention and served the Federation from 1882 until 1886 as German secretary. Miller was secretary of his union from 1886 until 1894, when it amalgamated with the International Typographical Union (ITU) and moved its offices to Indianapolis. Thereafter he served as a vice-president of the ITU and secretary-treasurer of its German branch. From 1886 to 1926 he edited the Typographia's organ, the *Deutsch-Amerikanische Buchdrucker-Zeitung* (later, the *Buchdrucker-Zeitung*).

MILLER, William J., president of CMIU 4 of Cincinnati, was first vice-president of the CMIU between 1875 and 1877.

MONCKTON, James H. (1825?-98), a stairbuilder, was born in New York. He was a member of the Economic and Sociological Club, helped organize the Association of United Workers of America, and in 1877 ran on the Workingmen's Party of the United States ticket for local New York City office. In the early 1890s he became a teacher.

MORRISON, John (1864-1938), a New York–born machinist, was a founding member of KOL District Assembly (DA) 49's inner circle, the Home Club. In 1886 he revealed its history and leveled charges against the group at the Cleveland KOL General Assembly. The same year he helped organize carpet workers' DA 126 and was elected its district master workman. A close friend of SG's, he later was in charge of advertising for the *American Federationist* and various publications of the building trades unions.

MULLEN, William H. (b. 1850), a Richmond KOL leader and editor of the *Labor Herald,* was a Virginia-born printer. He was an officer of International Typographical Union (ITU) 90 of Richmond during the late 1870s and early 1880s. In 1885 he helped organize KOL

District Assembly 84 and served as its master workman; he was a member of the auxiliary board of the KOL in 1886. Richmond Knights nominated him for Congress in the fall of 1886. In the mid-1890s Mullen was again an officer of ITU 90.

NEEBE, Oscar W. (b. 1850), a tinsmith born in New York City, became involved in the labor movement in Chicago after moving there in 1875. He was a member of the anarchist International Working People's Association and was operating a yeast business in 1886. He was convicted of murder in connection with the Haymarket incident and was sentenced to fifteen years in prison; he was pardoned by Governor John Peter Altgeld in 1893.

O'CALLAGHAN, John (d. 1889), filled several official positions for CMIU 55 of Hamilton, Ontario, including the presidency. He also served as the first vice-president of the CMIU from 1879 to 1880 and was elected financial secretary and member of the Executive Board in 1883.

O'DEA, Thomas (1846-1926), an Irish immigrant and Civil War veteran, served as secretary of the Bricklayers' and Masons' International Union (BMIU; 1884-87 and 1888-1900). After retirement from the BMIU, O'Dea worked in Cohoes, N.Y., as a contractor.

O'REILLY, Thomas (b. 1854), a onetime candidate for the priesthood and a telegrapher from Edinburgh, Scotland, came to New York in 1882 and became a leader of the 1883 telegraphers' strike in New York City. In 1884 he was elected president of the Brotherhood of Telegraphers and in 1886 became master workman of KOL National District Assembly 45, the telegraphers' trade assembly. O'Reilly acted as Terence Powderly's representative in discussions with Catholic prelates who feared that the KOL posed a threat to the Catholic Church. He was an editor of the *Journal of the Knights of Labor* from 1889 to 1893; in 1901 he became Powderly's confidential clerk at the Bureau of Immigration.

PALDA, Lev J. (1847-1912), a weaver, journalist, and leader of the Czech-American socialist movement, immigrated to the United States in 1867. From 1868 to 1871 he edited the *Národní Noviny* in St. Louis and Chicago, and in 1875 he and Frank Skarda founded *Dělnické Listy* in Cleveland. In 1877 they moved the paper to New York City where Palda helped organize the Czech-Slavic International Workingmen's Association of New York. In 1892 he helped found and was president of the Czech-American National Committee. He later moved to Cedar Rapids, Mich., where he operated a small cigar factory.

PARSONS, Albert R. (1848-87), a Confederate army veteran from

Montgomery, Ala., moved to Chicago after the war, working as a printer and joining International Typographical Union 16 and the KOL. He was active in the Workingmen's Party of the United States and the SLP. The SLP nominated him for the presidency of the United States in 1879, but he declined because he was underage. Parsons was a leader of the eight-hour movement and a founder and leader of the American affiliate of the anarchist International Working People's Association and edited its organ, the *Alarm.* He was convicted of murder in connection with the Haymarket incident and was hanged on Nov. 11, 1887.

PASCOE, David M. (1859-1923), secretary-treasurer of the International Typographical Union (ITU) from 1886 to 1887, served as secretary of ITU 2 of Philadelphia (1884-86), helped organize KOL printers' Local Assembly 3879, and edited the labor paper the *Tocsin* (1884-87).

PHILLIPS, John (1837?-1904), longtime secretary of the hatters union, was born in Ireland and immigrated to the United States in 1851. A leader of the Brooklyn hat finishers' union in the early 1870s, he was secretary of the National Trade Association of Hat Finishers from 1880 to 1896. After its merger with the National Hat Makers' Association in 1896, he served as secretary of the new organization, the United Hatters of North America, until 1904. Phillips also served variously as secretary, treasurer, and president of the New York State Workingmen's Assembly in the late 1880s and early 1890s and as the president of the Brooklyn Central Labor Union (1899-1904).

PINNER, Reuben E. (1846-1916), a New York–born cigarmaker, served as financial secretary of local 144 in 1878 and 1879 and as trustee of the CMIU during 1880.

PLATE, Martin Diederich (1829-99?), president of CMIU 90, was born in Hanover and immigrated to the United States in 1852. He was a longtime board member of the Cigar Makers' Benevolent Society and served as CMIU 90's first secretary from 1867 to 1870 before becoming its president. Plate was a member of the Executive Board of the New York Cigar Makers' State Union in 1873. In the 1890s he became a cigar manufacturer.

POLLNER, William C. (1854-1922?), born in Germany, immigrated to the United States in 1872. He was employed as a marble cutter and polisher in Pennsylvania before settling in Cleveland in 1878. There he worked as a cigarmaker and, in 1881, became president of CMIU 17. Pollner represented the Cleveland Trades and Labor Assembly at the 1881 FOTLU convention. Shortly afterward he became

a cigar manufacturer and during the late 1880s held a political appointment with the U.S. Bureau of Internal Revenue. He then served as manager of a brewing company, director of the Cleveland police, member of the Cleveland city council, and about 1900 became a lawyer.

POWDERLY, Terence Vincent (1849-1924), general master workman of the KOL, was born in Carbondale, Pa. Apprenticed as a machinist, he moved to Scranton and joined the International Machinists and Blacksmiths of North America in 1871, becoming president of his local and an organizer in western Pennsylvania. After being dismissed and blacklisted for his labor activities, Powderly joined the KOL in Philadelphia in 1876 and shortly afterward founded a local assembly of machinists and was elected its master workman. In 1877 he helped organize District Assembly 5 (number changed to 16 in 1878) and was elected corresponding secretary. He was elected mayor of Scranton on the Greenback-Labor ticket in 1878 and served three consecutive two-year terms. At the same time he played an important role in calling the first General Assembly of the KOL in 1878, where he was chosen grand worthy foreman, the KOL's second highest office. The September 1879 General Assembly elected him grand master workman, and he continued to hold the Order's leading position (title changed to general master workman in 1883) until 1893. Active in the secret Irish nationalist society, *Clan na Gael,* Powderly was elected to the Central Council of the American Land League in 1880 and was its vice-president in 1881. He became an ardent advocate of land reform and temperance and, as master workman, favored the organization of workers into mixed locals rather than craft unions, recommended that they avoid strikes, encouraged producers' cooperatives, and espoused political reform.

In 1894 Powderly was admitted to the Pennsylvania bar and in 1897 President William McKinley, for whom he had campaigned, appointed him commissioner general of immigration. President Theodore Roosevelt removed him from his position in 1902 but in 1906 appointed him special representative of the Department of Commerce and Labor to study European immigration problems. Powderly was chief of the Division of Information in the Bureau of Immigration and Naturalization from 1907 until his death.

POWERS, Richard (1850-1929), an Irish sailor who immigrated to the United States in 1870, served as president of the Lake Seamen's Benevolent Association of Chicago from 1878 to 1887. He was a member of the Legislative Committee of the FOTLU from 1881 to 1885, and that body sent him to Washington, D.C., to lobby for

seamen's measures. He was also active in the KOL, representing District Assembly 136 at the 1886 General Assembly, and in the *Clan na Gael,* a secret Irish nationalist society. Later he served as a sewer inspector and revenue collector in Chicago.

PRYOR, Roger Atkinson (1828-1919), a Virginia Democratic congressman from 1858 to 1860, served as a Confederate general during the Civil War. He then moved to New York City where, as a lawyer, he advocated the rights of labor unions and challenged the validity of trusts. In 1886 he defended New York City eight-hour strikers and in 1887 represented the Haymarket anarchists before the U.S. Supreme Court. In the 1890s he was a judge and became a justice of the New York Supreme Court.

QUINN, James E. (1836-1901), a bookbinder born in Boston, was elected master workman of KOL District Assembly (DA) 49 in New York City in 1883 and was active in the Central Labor Union. He was a leader of the inner circle of DA 49 known as the Home Club and an activist in the United Labor party. He served on the auxiliary General Executive Board of the KOL in 1886. In 1887 he defied the KOL leadership by publicly urging clemency for the condemned Haymarket anarchists.

RANKIN, Alexander C. (b. 1849?), a Pittsburgh iron molder associated with the Greenback-Labor movement, was born in Pennsylvania. He represented Iron Molders' local 14 at the 1881 FOTLU convention and was elected treasurer of the FOTLU Legislative Committee. He was also a leader of KOL District Assembly 3 until 1889.

RODGERS, George (b. 1844), a Chicago iron molder and president of the Trade and Labor Assembly of Chicago between 1881 and 1884, was born in Wales and immigrated to the United States in 1850. He joined KOL Local Assembly 400 in the late 1870s and, in the early 1880s, served District Assembly 24 as worthy foreman, organizer, and General Assembly delegate; he was also president of Iron Molders Union 23. In 1884 he helped found and served as first president of the Illinois State Federation of Labor.

ROHNER, Louis P. (b. 1843), a cigarmaker born in Maryland, was secretary of CMIU 1 of Baltimore during the 1870s and second vice-president of the CMIU from 1875 to 1876.

RONEY, Frank (1841-1925), vice-president of the Iron Molders' Union of North America (1886-88) and a founder and president of the Representative Council of the Federated Trades and Labor Organizations of the Pacific Coast (1885-87), was an immigrant from Ireland who came to the United States to avoid imprisonment for his

participation in the Fenian movement. He settled in Omaha, Neb., where he worked as an iron molder, served as an officer of his local union, and became active in reform politics. In 1875 he moved to San Francisco and was a leader of the Workingmen's Party of California in 1878. He organized the Seamen's Protective Union in 1880, served as president of the San Francisco Representative Assembly of Trades and Labor Unions in 1881 and 1882, and was one of the promoters of the anti-Chinese label on cigars and shoes. He was a strong proponent of a national federation of trades. Blacklisted in his trade, Roney found work as a foundry worker in Vallejo, Calif., helped organize the town's Trades and Labor Council in 1899, and served one term as its president. In 1909 he moved to Los Angeles and in 1915 served as secretary-treasurer of the city's Iron Trades Council.

ROOD, Moses Isayes (b. 1790?), SG's maternal grandfather, was a porter in Amsterdam in the late 1820s.

SARGENT, Frank Pierce (1854-1908), grand master of the Brotherhood of Locomotive Firemen (BLF) from 1885 to 1902, was born in Vermont and was a textile operative, farm laborer, and a United States cavalryman before becoming a railroad worker. He joined Lodge 94 of the BLF in Tucson, Ariz., in 1881 and in 1883 was elected vice grand master of the Brotherhood. As grand master he urged moderation and negotiation with employers and refused to support sympathetic strikes. He was appointed to the U.S. Industrial Commission in 1898, and Theodore Roosevelt appointed him U.S. commissioner general of immigration in 1902. Sargent was active in the National Civic Federation and was one of the labor representatives to its Division of Conciliation and Mediation.

SCHÄFER, John, a New York City cigarmaker, was a leader of the United Cigarmakers and a member of the International Workingmen's Association in the early 1870s. In 1875 he became secretary of the New York City branch of the Social Democratic Workingmen's Party of North America, and the following year, administrator of the *Arbeiter-Stimme,* an organ of the Workingmen's Party of the United States. In the late 1870s he was active in promoting the SLP in the New York City area.

SCHAEFER, William C. (1830?-1900), a Milwaukee cigarmaker, participated in the reorganization of CMIU 25 in 1878 and served during the next few years as a local officer. He was a member of the CMIU Board of Appeals from 1879 to 1880.

SCHIMKOWITZ, Samuel, a member of the SLP, helped organize the Cigarmakers' Progressive Union of America (CMPU) in 1882. He served as president of CMPU 1 and as a delegate to the New York

City Central Labor Union. In the fall of 1882 he ran unsuccessfully for the New York Assembly on the United Labor party ticket.

SCHWAB, Michael (b. 1853), a bookbinder born in Kitzingen, Germany, and a member of the Sozialdemokratische Arbeiterpartei, immigrated to the United States in 1879 where he became a member of the SLP. In the early 1880s he was a reporter in Chicago for the *Chicagoer Arbeiter-Zeitung* and organized the socialist clubs that participated in the founding of an American branch of the anarchist International Working People's Association in 1883. He was convicted of murder in connection with the Haymarket incident. His sentence was commuted to life imprisonment by Governor Richard J. Oglesby, and he was pardoned by Governor John Peter Altgeld in 1893.

SHARP, Charles H. (b. 1844), a Philadelphia tailor, was born in England and immigrated to the United States in 1877. He was elected president of the Journeymen Tailors' National Union (JTNU) in August 1885 but resigned from office when his local withdrew from the JTNU.

SMITH, James W. (1837-1903), a vice-president of the Journeymen Tailors' National Union, was born in Ireland and immigrated to the United States in 1850, settling finally in Springfield, Ill. He served as first vice-president of the FOTLU Legislative Committee (1885-86), second vice-president of the AFL (1886-87), and president of the Illinois State Federation of Labor (1885-86, 1888-89).

SMITH, William J. (1853-1906), of Pittsburgh, served as president of the American Flint Glass Workers' Union (AFGWU) from 1885 to 1900 and held concurrent positions as secretary in 1885 and as treasurer from 1887 to 1900. As a lamp chimney blower in Pittsburgh in the 1870s, Smith joined KOL Local Assembly 320. He was a delegate to the convention establishing the AFGWU in 1878. After leaving the union, he served as an agent and later as an auditor for the Macbeth-Evans Glass Co. of Pittsburgh.

SPIES, August Vincent Theodore (1855-87), born in Landsberg, Germany, came to the United States in 1872 and moved to Chicago in 1873, joining the SLP and the KOL. He was a founder of an American affiliate of the International Working People's Association in 1883. Spies, who became editor of the *Chicagoer Arbeiter-Zeitung* in 1880, was convicted of murder in connection with the Haymarket incident and was hanged on Nov. 11, 1887.

STEWARD, Ira (1831-83), leader of the movement for the eight-hour workday, was born in Connecticut and apprenticed as a machinist, becoming a leading figure in the International Machinists and

Blacksmiths of North America. During the mid-1860s in Boston he helped organize several organizations devoted to the establishment of the eight-hour workday, and he successfully promoted the passage of a Massachusetts ten-hour law for women and children in 1874. During the 1870s he developed close ties with New York City trade unionists in the International Workingmen's Association and he helped found and served as an organizer for the International Labor Union. Steward believed that freeing workers from long hours of labor would stimulate their desire for a better life and facilitate their education and organization, leading eventually to the abolition of the wage system.

STRASSER, Adolph (1843-1939), was born in Hungary and immigrated to the United States about 1872. He became a cigarmaker, helped organize New York City cigar workers excluded from membership in the CMIU, and played a leading role in the United Cigarmakers. Strasser was a member of the International Workingmen's Association and, in 1874, helped organize the Social Democratic Workingmen's Party of North America, serving as its executive secretary. He was also a founder of the Economic and Sociological Club. In 1876 he was a delegate to the unity congress that organized the Workingmen's Party of the United States, and he aligned with the trade unionist faction of the party. During 1876 and 1877 he worked to establish a central organization of New York City trade unions, and his efforts culminated in the founding of the Amalgamated Trades and Labor Union of New York and Vicinity in the summer of 1877. Strasser was elected vice-president of the CMIU in 1876 and president in 1877 and successfully promoted the reorganization of the union in the late 1870s and early 1880s. After retiring as president in 1891, he continued to work for the CMIU as an organizer, auditor, and troubleshooter. In addition he served as an AFL lecturer, member of the Federation's legislative committee, and AFL arbitrator of jurisdictional disputes. He ended his labor career in 1914, becoming a real estate agent in Buffalo, and in 1919 he moved to Florida.

SULLIVAN, James W. (1848-1938), a printer from Carlisle, Pa., moved to New York City in 1882 after serving his apprenticeship in Philadelphia. He worked for the *New York Times* and the *New York World* and joined International Typographical Union (ITU) 6, becoming a leading figure in the ITU. He was a strong supporter of land reform and edited the *Standard* with Henry George from 1887 to 1889. A close associate of SG, he participated in the Social Reform Club and the People's Institute of the Ethical Culture Society. He served with SG in the National Civic Federation, accompanied him

to Europe in 1909, helped to edit the *American Federationist,* and served with SG on the Advisory Commission of the Council of National Defense during World War I. His European travels encouraged him to support the movement for direct legislation through the initiative and the recall. He opposed labor movement involvement in socialist political activities, publishing *Socialism as an Incubus on the American Labor Movement* (New York, 1909) and a report critical of English socialism (in Commission on Foreign Inquiry, National Civic Federation, *The Labor Situation in Great Britain and France* [New York, 1919]).

SWINTON, John (1830-1901), a Scottish-born journalist, emigrated in 1843 and apprenticed as a printer in Montreal. He was on the editorial board of the *New York Times* throughout the 1860s, and the *New York Sun* from 1875 to 1883 and again from 1892 to 1897. He published the influential New York City labor reform newspaper *John Swinton's Paper* between October 1883 and August 1887. Active in New York City politics, he ran for mayor in 1874 on the Industrial Political party ticket and worked in the mayoralty campaign of Henry George in 1886.

THAYER, Walter Nelson (1849-1929), a leader of International Typographical Union 52 of Troy, N.Y., and an official in the Troy city government, served as president of the New York State Workingmen's Assembly in 1884-86. In 1892 he became warden of New York's Dannemora Prison.

TODD, William V. (b. 1852), a Quebec-born cigarmaker, was active in the Toronto local of the CMIU, serving as president and later secretary of local 27 from 1879 to 1881. He was a member of the CMIU Board of Appeals in 1880 and third vice-president of the CMIU from 1880 to 1881 and again from 1890 to 1893. Todd frequently represented his local at the conventions of the Canadian Trades and Labour Congress and at meetings of the Toronto Trade and Labour Council.

TOMSON, William E., a machinist and financial secretary of the Chicago Trade and Labor Assembly, was elected second vice-president of the FOTLU at its 1885 convention. In the 1890s he became a salesman.

TURNER, Frederick (b. 1846), general secretary-treasurer of the KOL from 1883 to 1886, was born in England and immigrated to the United States in 1856, settling in Philadelphia. A Civil War veteran, he became a goldbeater and helped organize local assemblies of the KOL in that trade in Philadelphia, New York, and Boston in 1873 and 1874. As a result of his organizing activities, he was black-

listed in 1877 and became a grocer. He served as financial secretary and recording secretary of Pennsylvania District Assembly 1 and from 1880 was a member of the Knights' General Executive Board. After the Order divided the secretary-treasurer's position in 1886, Turner served as general treasurer until 1888.

WALKER, John P. (d. 1886?), was general secretary of the Cigar Makers' Mutual Association of London for over twenty-two years. An advocate of international worker cooperation, he helped organize and served as president of the International Conference of the Representatives of the Tobacco Trades Associations of Europe in 1871. During 1876 and 1877 the *Cigar Makers' Official Journal* published a series of letters from him describing the conditions of British and European cigarmakers.

WEIHE, William (1845-1908), president of the National Amalgamated Association of Iron and Steel Workers from 1884 to 1892, was born in Pennsylvania. He served as a local officer of the Sons of Vulcan before becoming a member of the executive committee of the Amalgamated in 1876. In 1882 he was elected as a Democrat to the Pennsylvania state legislature; he later became a Republican. Weihe was employed as a deputy immigration inspector for New York from 1896 until his death.

WEIL, Jean (1850-1915), a German-born printer who immigrated to New York City in 1870, helped organize the German-American Typographia and served as its secretary from 1876 to 1883. He later became editor and manager of the *Brewers' Journal,* a trade publication.

WIENER (variously Weiner), Julius, was an officer in Waiters' Union 1 of New York. About 1890 he moved to Brooklyn and became a secretary of Waiters' Union 2. Wiener was first secretary of the Waiters' and Bartenders' National Union, chartered by the AFL in 1891. He and the other members of the Brooklyn local were suspended that year, however, in an internal dispute with local 1.

WINKLER, John (1845?-84), was elected by the Cigarmakers' Progressive Union's 1883 convention to serve as a National Executive Board trustee.

WINTER, Ernst George (b. 1842), a German-born cigarmaker, immigrated to the United States in 1874. He was a member of the Social Democratic Workingmen's Party of North America and subsequently editor of the *Arbeiter-Stimme,* the New York–based journal of the Workingmen's Party of the United States. He moved to St. Louis, Mo., in 1880, and the following year served as one of the official representatives of the SLP to the St. Louis Trades Assembly. Winter was

elected president of CMIU 44 of St. Louis in 1883. In the 1890s he became a cigar manufacturer and, later, a postmaster.

WITTER, Martin R. H. (1847-1917), was a member of International Typographical Union 8 of St. Louis for fifty years and president of the international union between 1884 and 1886. He held the position of St. Louis city register for three terms beginning in 1908.

WOYTISEK, Vincent William (b. 1855), a Bohemian cigarmaker who immigrated to the United States in 1875, served as president of CMIU 144 during late 1881 and as secretary of local 1 of the Cigarmakers' Progressive Union until about 1885. He ran on the SLP ticket for the New York Assembly in 1882, and in 1884 was elected treasurer of the New York City Central Labor Union. During the late 1880s and 1890s he operated a saloon and sold real estate. About 1898 he became a lawyer.

WRIGHT, Carroll Davidson (1840-1909), born in New Hampshire, studied law in Vermont and moved to Boston in 1871 after service in the Civil War. He was elected to the Massachusetts Senate as a Republican in 1871 and served as chief of the Massachusetts Bureau of Statistics of Labor from 1873 to 1888 and as commissioner of the U.S. Bureau of Labor from 1885 to 1905. During the last four years of his life he was president of Clark College.

## ORGANIZATIONS

The National AGRICULTURAL Labourers' Union was founded in May 1872; it apparently dissolved in the mid-1890s.

The Journeymen BAKERS' National Union of the United States was organized in 1886, participated in the formation of the AFL that year, and was chartered by the AFL in 1887. In 1890 it adopted the name Journeymen Bakers' and Confectioners' International Union of America and, in 1903, Bakery and Confectionery Workers' International Union of America.

The BRICKLAYERS' and Masons' International Union of America was organized in 1865. It did not affiliate with the AFL until 1916.

The Brotherhood of CARPENTERS and Joiners of America was organized in 1881 and was chartered by the AFL in 1887. In 1888 the Brotherhood and the United Order of American Carpenters and Joiners merged, forming the United Brotherhood of Carpenters and Joiners of America.

The Cigar Makers' National Union of America was organized in 1864 and changed its name to the CIGAR Makers' International Union

of America in 1867. It participated in the formation of the FOTLU in 1881. The following year, seceding New York City locals formed the Cigarmakers' Progressive Union of America; the Progressives rejoined the International in 1886. The AFL chartered the CMIU in March 1887.

The American FLINT Glass Workers' Union of North America was organized in 1878 by locals formerly affiliated with the KOL. It joined the AFL in 1897.

The Furniture Workers' Association of North America was organized in 1873, and in 1882 it changed its name to the International FURNITURE Workers' Union of America. It affiliated with the AFL in 1887 and in 1896 merged with the Machine Wood Workers' International Union of America to form the Amalgamated Wood Workers' International Union of America.

The Granite Cutters' International Union of the United States and the British Provinces of America was formed in 1877. In 1880 it changed its name to the GRANITE Cutters' National Union of the United States of America, and in the following year participated in the formation of the FOTLU. It joined the AFL in 1888 and in 1905 adopted the name Granite Cutters' International Association of America.

The National Trade Association of HAT Finishers of the United States of America was organized in 1854. In 1896 the union merged with the National Hat Makers' Association of the United States, forming the United Hatters of North America, which affiliated that year with the AFL.

The National Amalgamated Association of IRON and Steel Workers was organized in 1876 and in 1887 was chartered by the AFL. In 1897 it changed its name to the National Amalgamated Association of Iron, Steel, and Tin Workers and in 1908 dropped "National" from its name.

The National Union of Iron Molders (after 1874 the IRON Molders' Union of North America) was organized in 1859 and in 1881 participated in the formation of the FOTLU. It was chartered by the AFL in 1887. In 1907 it changed its name to the International Molders' Union of North America.

The New England LASTERS' Protective Union was organized in 1878 and affiliated with the AFL in 1887. It changed its name to the Lasters' Protective Union of America in 1890, and in 1895 merged with KOL National Trade Assembly 216 and the Boot and Shoe

Workers International Union to form the Boot and Shoe Workers' Union.

The locomotive engineers organized the Brotherhood of the Footboard in 1863. In 1864 the organization became the Brotherhood of LOCOMOTIVE Engineers.

The Brotherhood of LOCOMOTIVE Firemen was organized in 1873 and in 1878 merged under its name with the International Firemen's Union. In 1906 it adopted the name Brotherhood of Locomotive Firemen and Enginemen.

The METAL Workers' National Union of North America was organized in 1882 and was chartered by the AFL in 1887. It disbanded in 1889.

The National Federation of MINERS and Mine Laborers was organized in 1885. Beginning in 1886, it was involved in bitter jurisdictional disputes with KOL miners' National Trade Assembly (NTA) 135. In 1888 it merged with a faction of NTA 135 and the Miners' and Mine Laborers' Amalgamated Association of Pennsylvania to form the National Progressive Union of Miners and Mine Laborers.

The Amalgamated MULE Spinners' Association, formed in 1861, participated in the formation of the FOTLU. It reorganized in 1887 as the National Cotton Mule Spinners' Association of America and affiliated with the AFL in 1889.

The Representative Council of the Federated Trades and Labor Organizations of the PACIFIC Coast was organized in 1885. It originated at a Pacific Coast anti-Chinese convention called by the San Francisco District Assembly of the KOL. As originally conceived, it was to be a federation of all western unions with an authority over them similar to that of a national trade union over its locals. It failed to achieve significant influence outside of San Francisco and in 1891 was superseded by the Council of Federated Trades of the Pacific Coast.

In 1885 the year-old National Association of Plumbers, Steam Fitters, and Gas Fitters split between local trades assemblies in New York and Brooklyn that wished to remain affiliated with the KOL, and those outside of New York that organized the International Association of Journeymen PLUMBERS, Steam Fitters, and Gas Fitters. The new union established its headquarters in Milwaukee in order to manage the cooperative shops established by striking Milwaukee plumbers. By 1887 the cooperatives' financial demands had severely weakened the union, and in 1889 twenty-three locals withdrew and

formed the United Association of Journeymen Plumbers, Gas Fitters, Steam Fitters, and Steam Fitters' Helpers of the United States and Canada.

The Lake SEAMEN'S Benevolent Association of Chicago was organized in 1878 and in 1886 briefly affiliated with the KOL as Seamen's District Assembly 136. In 1892 it joined with seamen's unions from the Atlantic, Pacific, and Gulf coasts to form the National Seamen's Union, which changed its name to the International Seamen's Union of America in 1895.

The Journeymen TAILORS' National Union of the United States, composed of custom tailors, was organized in 1883 and was chartered by the AFL in 1887. It changed its name in 1889 to the Journeymen Tailors' Union of America and in 1913 to the Tailors' Industrial Union. The following year it merged with the Amalgamated Clothing Workers of America (ACWA) but in 1915 seceded from the ACWA and again assumed the name of Journeymen Tailors' Union of America. The union's membership declined thereafter, and in 1938 the AFL revoked its charter.

In 1886 socialist members of the United Tailors' Union, founded the previous year, took over the leadership of the organization and renamed it Progressive Tailors' Union 1. They participated in the August 1886 convention that created the Progressive TAILORS' National Union (PTNU), a body in which ready-made or shop tailors predominated but which also included custom tailors. The PTNU affiliated with the AFL and later attempted unsuccessfully to merge with the Journeymen Tailors' National Union. It was apparently absorbed by the United Garment Workers, which members of PTNU 11 helped found in 1891.

The Brotherhood of TELEGRAPHERS, District Assembly 45 of the KOL, was organized in 1882 in New York.

The TRADES Union Congress of Great Britain, the central organization of that country's trade union movement, was founded in 1868.

The German-American TYPOGRAPHIA was organized in 1873 and was chartered by the AFL in 1887. It merged with the International Typographical Union as an autonomous unit in 1894.

The National Typographical Union was organized in 1852 by a group of locals that had held national conventions in 1850 and 1851 under the name Journeymen Printers of the United States. In 1869 it adopted the name International TYPOGRAPHICAL Union (ITU).

Although ITU members participated in the formation of the FOTLU in 1881 and in the organizing of the AFL in 1886, the union did not affiliate with the Federation until 1888.

In 1880 the KOL merged several local assemblies of glass blowers, gatherers, cutters, and flatteners to form WINDOW Glass Workers' Local Assembly 300, an organization with national jurisdiction over these trades. This assembly disbanded in 1910, and the National Window Glass Workers absorbed its membership.

# INDEX

Names of persons or organizations for whom there are glossary entries are followed by an asterisk.

ST. JOHN FISHER COLLEGE LIBRARY
HD6508 .S218 1986
010101 000
The Samuel Gompers papers / ed
0 1219 0075550 9

DATE DUE

NOV 2 3 2002

GAYLORD
PRINTED IN U.S.A.